AF822122

LAND ROVER DEFENDER 2007 - 2016 Workshop Manual

TDCi Puma Engine

2.2L & 2.4L Diesel

This Workshop Manual draws together the information necessary to maintain and repair all Defender models fitted with the Puma engine from 2007 to 2016.

Brooklands Books Ltd., PO Box 904, Amersham, Bucks, HP6 9JA, England.

ISBN: 9781783182046 2812 Ref: LRD6WH

Contents

General Information

100-00 - General Information 7

100-01 - Identification Codes 79

100-02 - Jacking and Lifting 82

Chassis

204-00 - General Information - Suspension 85

204-01 - Front Suspension 93

204-02 - Rear Suspension 99

204-04 - Wheels and Tires 105

205-01 - Driveshaft 111

205-02 - Rear Drive Axle and Differential 115

205-03 - Front Drive Axle/Differential 131

206-00 - Brake System 143

206-03 - Front Disc Brake 155

206-04 - Rear Disc Brake 157

205-05 - Parking Brake and Actuation 162

206-06 - Hydraulic Brake Actuation 165

207-07 - Power Brake Actuation 170

206-09A - Anti-Lock Control - Traction Control 172

206-09B - Anti-Lock Control - Stability Assist 178

211-00 - Steering System 179

211-02 - Power Steering 189

211-03 - Steering Linkage 199

211-04 - Steering Column 202

211-05 - Steering Column Switches 206

Powertrain

303-00 - Engine System: 2.2L 2007-2011 and 2.4L 2012-2016 Diesel 209

303-01 - Engine 2.2L 2007-2011 and 2.4L 2012-2016 Diesel 217

303-03 - Engine Cooling - ID4 2.2L Diesel 270

303-04A - Fuel Charging and Controls 286

303-04B - Fuel Charging and Controls - Turbocharger - ID4 2.2L Diesel 302

303-05 - Accessory Drive - ID4 2.2L Diesel 311

303-06 - Starting System - 2.2L Diesel & 2.2L Diesel 316

303-07 - Glow Plug System - 2.2L Diesel & 2.2L Diesel 319

303-08 - Engine Emission Control - 2.2L Diesel & 2.4L Diesel 323

303-12 - Intake Air Distribution and Filtering 328

303-14 - Electronic Engine Controls - 2.2L Diesel & 2.4 Diesel 337

308-00 - Manual Transmission/Transaxle and Clutch MT82 6-Speed Manual Transmission 355

308-01 - Clutch: MT82 6-Speed Manual Transmission 356

308-02 - Clutch Controls - Vehicles With: MT82 6-Speed Manual Transmission 361

308-03 - Manual Transmission/Transaxle MT82 6-Speed Manual Transmission 377

308-06 - Manual Transmission/Transaxle External Controls - Vehicles MT82 6-Speed Manual Transmission 429

308-07A - Transfer Case - Vehicles MT82 6-Speed Manual Transmission 431

308-07B - Four-Wheel Drive Systems - Vehicles With MT82 6-Speed Manual Transmission 459

309-00 - Exhaust System - ID4 2.2L Diesel 461

310-00 - Fuel System 483

310-01 - Fuel Tank and Lines - ID4 2.2L Diesel 486

Contents continued

Electrical

412-00 - Climate Control System 496

412-01 - Air Distribution and Filtering 511

412-02 - Heating and Ventilation 513

412-03 - Air Conditioning - ID4 2.2L Diesel 519

412-04 - Control Components 526

413-01 - Instrument Cluster 528

413-06 - Horn 535

413-07 - Clock 536

413-09 - Warning Devices 536

414-00 - Battery and Charging System 537

414-01 - Battery, Mounting and Cables 545

414-02 - Generator and Regulator ID4 2.2L Diesel 548

415-01A - Audio Unit 550

415-01B - Information and Entertainment System 557

415-02 - Antenna 559

415-03 - Speakers 561

417-01 - Exterior Lighting 568

417-02 - Interior Lighting 578

418-00 - Module Communications Network 581

418-02 - Wiring Harnesses 582

419-01A - Anti-Theft - Active 596

419-01B - Anti-Theft - Passive 602

419-12 - Winch 606

Body and Paint

501-02 - Front End Body Panels 624

501-03 - Body Closures 628

501-05 - Interior Trim and Ornamentation 633

501-08 - Exterior Trim and Ornamentation 644

501-09 - Rear View Mirrors 647

501-10 - Seating 647

501-11 - Glass, Frames and Mechanisms 664

501-12 - Instrument Panel and Console 689

501-14 - Handles, Locks, Latches and Entry Systems 697

501-16 - Wipers and Washers 709

501-17 - Roof Opening Panel 721

501-18 - Convertible Top 722

501-19 - Bumpers 724

501-20 - Seatbelt System 725

501-25A - Body Repairs - General Information 743

501-25B - Body Repairs - Corrosion Protection 766

501-25C - Body Repairs - Water Leaks 780

501-26 - Body Repairs - Vehicle Specific Information and Tolerance Checks 798

501-27 - Front End Sheet Metal Repairs 801

501-28 - Roof Sheet Metal Repairs 805

501-29 - Side Panel Sheet Metal Repairs 805

501-30 - Rear End Sheet Metal Repairs 807

Detailed Contents List 810

LAND ROVER OFFICIAL FACTORY PUBLICATIONS

Land Rover Series 1 Workshop Manual	4291
Land Rover Series 1 1948-53 Parts Catalogue	4051
Land Rover Series 1 1954-58 Parts Catalogue	4107
Land Rover Series 1 Instruction Manual	4277
Land Rover Series 1 & II Diesel Instruction Manual	4343
Land Rover Series II & IIA Workshop Manual	AKM8159
Land Rover Series II & Early IIA Bonneted Control Parts Catalogue	605957
Land Rover Series IIA Bonneted Control Parts Catalogue	RTC9840CC
Land Rover Series IIA, III & 109 V8 Optional Equipment Parts Catalogue	RTC9842CE
Land Rover Series IIA/IIB Instruction Manual	LSM64IM
Land Rover Series 2A and 3 88 Parts Catalogue Supplement (USA Spec)	606494
Land Rover Series III Workshop Manual	AKM3648
Land Rover Series III Workshop Manual V8 Supplement (edn. 2)	AKM8022
Land Rover Series III 88, 109 & 109 V8 Parts Catalogue	RTC9841CE
Land Rover Series III Owners Manual 1971-1978	607324B
Land Rover Series III Owners Manual 1979-1985	AKM8155
Military Land Rover (Lightweight) Series III Parts Catalogue	61278
Military Land Rover Series III (L.W.B.) User Handbook	608179
Military Land Rover (Lightweight) Series III User Manual	608180
Land Rover 90/110 & Defender Workshop Manual 1983-1992	SLR621ENWM
Land Rover Defender Workshop Manual 1993-1995	LDAWMEN93
Land Rover Defender 300 Tdi & Supplements Workshop Manual 1996-1998	LRL0097ENGBB
Land Rover Defender Td5 Workshop Manual & Supplements 1999-2006	LRL0410BB
Land Rover Defender Electrical Manual Td5 1999-06 & 300Tdi 2002-2006	LRD5EHBB
Land Rover 110 Parts Catalogue 1983-1986	RTC9863CE
Land Rover Defender Parts Catalogue 1987-2006	STC9021CC
Land Rover 90 • 110 Handbook 1983-1990 MY	LSM0054
Land Rover Defender 90 • 110 • 130 Handbook 1991 MY - Feb. 1994	LHAHBEN93
Land Rover Defender 90 • 110 • 130 Handbook Mar. 1994 - 1998 MY	LRL0087ENG/2
Military Land Rover 90/110 All Variants (Excluding APV & SAS) User Manual	2320-D-122-201
Military Land Rover 90 & 110 2.5 Diesel Engine Versions User Handbook	SLR989WDHB
Military Land Rover Defender XD - Wolf Workshop Manual - 2320D128 -	302 522 523 524
Military Land Rover Defender XD - Wolf Parts Catalogue	2320D128711
Discovery Workshop Manual 1990-1994 (petrol 3.5, 3.9, Mpi & diesel 200 Tdi)	SJR900ENWM
Discovery Workshop Manual 1995-1998 (petrol 2.0 Mpi, 3.9, 4.0 V8 & diesel 300 Tdi)	LRL0079BB
Discovery Series II Workshop Manual 1999-2003 (petrol 4.0 V8 & diesel Td5 2.5)	VDR100090/6
Discovery Parts Catalogue 1989-1998 (2.0 Mpi, 3.5, 3.9 V8 & 200 Tdi & 300 Tdi)	RTC9947CF
Discovery Parts Catalogue 1999-2003 (petrol 4.0 V8 & diesel Td5 2.5)	STC9049CA
Discovery Owners Handbook 1990-1991 (petrol 3.5 V8 & diesel 200 Tdi)	SJR820ENHB90
Discovery Series II Handbook 1999-2004 MY (petrol 4.0 V8 & Td5 diesel)	LRL0459BB
Freelander Workshop Manual 1998-2000 (petrol 1.8 and diesel 2.0)	LRL0144
Freelander Workshop Manual 2001-2003 ON (petrol 1.8L, 2.5L & diesel Td4 2.0)	LRL0350ENG/4
Land Rover 101 1 Tonne Forward Control Workshop Manual	RTC9120
Land Rover 101 1 Tonne Forward Control Parts Catalogue	608294B
Land Rover 101 1 Tonne Forward Control User Manual	608239
Range Rover Workshop Manual 1970-1985 (petrol 3.5)	AKM3630
Range Rover Workshop Manual 1986-1989 (petrol 3.5 & diesel 2.4 Turbo VM)	SRR660ENWM & LSM180WS4/2
Range Rover Workshop Manual 1990-1994 (petrol 3.9, 4.2 & diesel 2.5 Turbo VM, 200 Tdi)	LHAWMENA02
Range Rover Workshop Manual 1995-2001 (petrol 4.0, 4.6 & BMW 2.5 diesel)	LRL0326ENGBB
Range Rover Workshop Manual 2002-2005 (BMW petrol 4.4 & BMW 3.0 diesel)	LRL0477
Range Rover Electrical Manual 2002-2005 UK version (petrol 4.4 & 3.0 diesel)	RR02KEMBB
Range Rover Electrical Manual 2002-2005 USA version (BMW petrol 4.4)	RR02AEMBB
Range Rover Parts Catalogue 1970-1985 (petrol 3.5)	RTC9846CH
Range Rover Parts Catalogue 1986-1991 (petrol 3.5, 3.9 & diesel 2.4 & 2.5 Turbo VM)	RTC9908CB
Range Rover Parts Catalogue 1992-1994 MY & 95 MY Classic (petrol 3.9, 4.2 & diesel 2.5 Turbo VM, 200 Tdi & 300 Tdi)	RTC9961CB
Range Rover Parts Catalogue 1995-2001 MY (petrol 4.0, 4.6 & BMW 2.5 diesel)	RTC9970CE
Range Rover Owners Handbook 1970-1980 (petrol 3.5)	606917
Range Rover Owners Handbook 1981-1982 (petrol 3.5)	AKM8139
Range Rover Owners Handbook 1983-1985 (petrol 3.5)	LSM0001HB
Range Rover Owners Handbook 1986-1987 (petrol 3.5 & diesel 2.4 Turbo VM)	LSM129HB

Engine Overhaul Manuals for Land Rover & Range Rover

300 Tdi Engine, R380 Manual Gearbox & LT230T Transfer Gearbox Overhaul Manuals	LRL003, 070 & 081
Petrol Engine V8 3.5, 3.9, 4.0, 4.2 & 4.6 Overhaul Manuals	LRL004 & 164
Land Rover/Range Rover Driving Techniques	LR369
Working in the Wild - Manual for Africa	SMR684MI
Winching in Safety - Complete guide to winching Land Rovers & Range Rovers	SMR699MI

Workshop Manual Owners Edition

Land Rover 2 / 2A / 3 Owners Workshop Manual 1959-1983
Land Rover 90, 110 & Defender Workshop Manual Owners Edition 1983-1995
Land Rover Discovery Workshop Manual Owners Edition 1990-1998

Description and Operation

How to use This Manual

This manual covers all aspects necessary in order to service the vehicle effectively.

The manual is structured into five main sections, General Information, Chassis, Powertrain, Electrical and Body and Paint with each section dealing with a specific part of a vehicle system.

Each of the five main sections contain sub-sections dealing with items which form a part of that specific system.

Pages at the start of the manual list all sections available. Each section has a contents list detailing, where applicable, Specifications, Description and Operation, Diagnosis and Testing, General Procedures and Repair Procedures.

Where components need to be removed or disassembled in sequence, each operation in the sequence will be identified numerically and also graphically in an accompanying illustration.

△ NOTE: Dimensions quoted are to design engineering specifications with service limits quoted, where applicable.

Workshop Manual Organization

The five main sections, together with the areas which they cover are given below:

- **Section 1** - General Information.
- **Section 2** - Chassis.
- **Section 3** - Powertrain.
- **Section 4** - Electrical.
- **Section 5** - Body and Paint.

Sub-section numbers appear after the initial section number, for example, **Section 412-03** covers air conditioning, which is part of the electrical section.

In the number given above, the first digit of the number **'4'** indicates the section **i.e. Electrical**.

The second and third digits **'12'** of the number indicate the vehicle system **i.e. Air Conditioning**.

The last two digits of the number **'03'** indicate the part of the system covered by the sub-section **i.e. Air Conditioning Compressor**.

General Service Information

Description and Operation

Introduction

This manual has been written in a format that is designed to meet the needs of Land Rover technicians worldwide and to assist them in the efficient repair and maintenance of Land Rover vehicles.

This manual provides descriptions and methods for accomplishing adjustment, service and repair work using tested and effective procedures. Following these procedures will help ensure product reliability.

Special Tools

The Special Tool(s) Table provided at the beginning of each procedure lists the special tool(s) required to carry out repair operations within that specific procedure. Wherever possible, illustrations are provided which will assist technicians in identifying the special tool(s) required and also showing such tool(s) in use.

Special tools may be obtained from the manufacturer, SPX Tools, the addresses of their branches will be found in the Special Tools Glossary contained within this Section.

Important Safety Instructions

Appropriate service methods and correct repair procedures are essential for the safe and reliable operation of all motor vehicles as well as ensuring the personal safety of the individual carrying out the work.

This manual cannot possibly anticipate all such variations and provide advice or cautions as to each. Any person who departs from the instructions provided in this manual must first establish that they compromise neither their personal safety nor the vehicle integrity by their choice of methods, tools or parts.

Individuals who undertake their own repairs should have some skill or training and limit repairs to components which could not affect the safety of the vehicle or its passengers. Any repairs required to safety critical items such as steering, brakes, suspension or supplemental restraint system should be carried out by a Land Rover Dealer. Repairs to such items should NEVER be attempted by untrained individuals.

Warnings, Cautions and Notes which appear in this manual

As you read through this manual, you will come across Warnings, Cautions and Notes. A Warning, Caution or Note is placed at the beginning of a series of steps. If the warning, caution or note only applies to one step, it is placed at the beginning of the specific step after the step number.

Warnings, Cautions and Notes have the following meanings:

WARNING Procedures which must be followed to avoid the possibility of personal injury.

Caution: Calls attention to procedures which must be followed to avoid damage to components.

Note: Gives helpful information.

References

References to the Left Hand (LH) or Right Hand (RH) side given in this manual are made when viewing the vehicle or unit from the rear.

Fault Diagnostic Equipment

The vehicle is equipped with a number of electronic control systems to provide optimum performance of the vehicle's systems.

Diagnostic Equipment (T4) is available and must be used where specified. The use of this equipment will assist with the fault diagnostic abilities of the Dealer workshop. In particular, the equipment can be used to interrogate the electronic systems for diagnosis of faults which may become evident during the life of the vehicle.

This manual is produced as a reference source to supplement T4.

Features of the equipment include:

a. Fully upgradeable support for the technician

b. Structured diagnostics to accommodate all skill levels

c. Direct print-out of screen information and test results

Testing the vehicle

Operations covered in this manual do not include reference to testing the vehicle after repair. It is essential that work is inspected and tested after completion and if necessary, a road test of the vehicle is carried out, particularly where safety related items are concerned.

Repairs and Replacement Parts

Land Rover parts are manufactured to the same exacting standards as the original factory fitted components. For this reason, it is essential that only genuine Land Rover parts are used during maintenance or repair.

Attention is particularly drawn to the following points concerning repairs and the fitting of replacement parts and accessories.

Safety features and corrosion prevention treatments embodied in the vehicle may be impaired if other than Land Rover recommended parts are fitted. In certain territories, legislation prohibits the fitting of parts not to manufacturer's specification. Torque wrench setting figures, where given, must be adhered to and locking devices, where specified must be used. If the efficiency of a locking device is impaired during removal it must be replaced.

Owners purchasing accessories whilst travelling abroad must ensure that the accessory and its fitted location on the vehicle conform to legal requirements.

The terms of the vehicle warranty may be invalidated by the fitting of parts other than those recommended by Land Rover.

△ NOTE: The fitting of non-approved Land Rover parts and accessories or the carrying out of non-approved alterations or conversions may be dangerous. Any of the foregoing could affect the safety of the vehicle and occupants; also, the terms and conditions of the vehicle warranty may also be invalidated.

All Land Rover recommended parts have the full backing of the vehicle warranty.

Land Rover Dealers are obliged to supply only Land Rover recommended parts.

Specifications

Land Rover are constantly seeking to improve the specification, design and production of their vehicles and alterations take place accordingly. Whilst every effort is made to ensure the accuracy of this Manual, it should not be regarded as an infallible guide to current specifications of any particular vehicle.

This Manual does not constitute an offer for sale of any particular vehicle. Land Rover dealers are not agents of Land Rover and have no authority to bind the manufacturer by any expressed or implied undertaking or representation.

Standard Workshop Practices

Description and Operation

Vehicle in Workshop

When working on a vehicle in the workshop always make sure that:

1. Where practicable, the parking brake is applied and the wheels are securely chocked to prevent the vehicle moving forwards or backwards.
2. Whenever possible, the ignition key is removed before any work is carried out on the vehicle.
3. If the engine is to be run, there is adequate ventilation, or an extraction hose is used to remove exhaust fumes.
4. There is adequate room to raise the vehicle and remove the wheels, if necessary.
5. Fender covers are always installed if any work is to be carried out in the engine compartment.
6. Where practicable, the battery is disconnected if working on the engine, underneath the vehicle, or if the vehicle is raised.

⚠ CAUTION: When electric arc welding on a vehicle, always disconnect the generator wiring to prevent the possibility of a surge of current causing damage to the internal components of the generator.

1. If using welding equipment on the vehicle, a suitable fire extinguisher is readily available.

Battery - General

⚠ WARNING: **It is essential that a period of 10 minutes elapses after the battery is disconnected before any work is undertaken on any part of the SRS system.**

CAUTIONS:

⚠ Prior to carrying out any procedures which involve disconnecting/ or connecting the battery, refer to the Electrical Section of this manual - Battery disconnection/connection.

Refer to: Battery (414-01 Battery, Mounting and Cables, Removal and Installation)

⚠ A discharged battery condition may have been caused by an electrical short circuit. If this condition exists there will be an apparently live circuit on the vehicle even when all normal circuits are switched off. This can cause arcing when the jumper cables are connected.

Jump Starting a Vehicle

CAUTIONS:

⚠ While it is not recommended that a vehicle is jump started, it is recognized that this may occasionally be the only practical way to mobilize a vehicle. Reference should be made to the following.

⚠ It is advisable not to use starter/charger sets for jump starting but if this is unavoidable, make sure that the sets are not used in the START mode.

1. Always make sure that the jumper cables are adequate for the task.
2. Always make sure that the slave battery is of the same voltage (12 volts) as the vehicle battery. The batteries must be connected in parallel.
3. Make sure that the battery terminals of both batteries are fully tightened.
4. Where another vehicle is used to jump start a disabled vehicle, make sure that the two vehicles are not touching.
5. It is advisable that the engine of the donor vehicle is switched off during jump starting; take care to make sure that the battery of the donor vehicle does not also become discharged.
6. Always make sure that switchable electric circuits are OFF before connecting jump cables. This reduces the risk of arcing occurring when the final connection is made.

Following jump starting of a disabled vehicle, the discharged battery must be checked for serviceability and recharged as soon as possible to avoid permanent damage.

Do not rely on the generator to restore a discharged battery. For a generator to recharge a battery, it would take in excess of eight hours continuous driving with no additional loads placed on the battery.

Trickle charging (defined as voltages <16 volts) may be carried out with the battery connected. Make sure that the battery terminals are fully tightened prior to trickle charging.

⚠ CAUTION: Boost charging may only be carried out with the battery disconnected from the vehicle.

Towing the Vehicle

CAUTIONS:

⚠ The vehicle has permanent four-wheel drive. The following towing instructions must be adhered to:

⚠ The brake servo and power assisted steering system will not be functional without the engine running. Greater pedal pressure will be required to apply the brakes and the steering system will require greater effort to turn the front road wheels. The vehicle tow connection should be used only in normal conditions, snatch recovery should be avoided.

Towing the vehicle on all four wheels with driver operating steering and brakes.

Turn ignition key to position 1 to release steering lock.

Select neutral in main gearbox and transfer box.

Secure tow rope, chain or cable to front towing eyes.

Release the parking brake.

Rear suspended tow by breakdown vehicle

⚠ CAUTION: The steering wheel and/or linkage must be secured in the straight ahead position. Do not use the steering lock mechanism for this purpose.

If the front axle is to be trailed turn ignition key to position **1** to release steering lock.

Select neutral in main gearbox and transfer box.

Transporting the vehicle by trailer

⚠ CAUTION: Underbody components must not be used as lashing points.

Lashing/towing eyes are provided on front and rear of the chassis side members to facilitate the securing of the vehicle to a trailer or other means of transportation.

Position vehicle on trailer and apply the parking brake. Select neutral in main gearbox.

General installation Instructions

Component removal

Whenever possible, clean components and the surrounding area before removal.

- Blank off openings exposed by component removal.
- Following disconnection, seal fuel, oil or hydraulic lines immediately using suitable blanking plugs or caps.
- Seal open ends of exposed oilways using suitable tapered hardwood plugs or conspicuous plastic plugs.
- Immediately a component is removed, place it in a suitable container; use a separate container for each component and its associated parts.
- Clean bench and provide marking materials, labels and containers before disassembling components.

Disassembling

Observe scrupulous cleanliness when disassembling components, particularly when brake, fuel, air suspension or hydraulic system parts are disassembled. A particle of dirt or cloth fragment could cause a serious malfunction if trapped in these systems.

- Blow out all tapped holes, crevices, oilways and fluid passages with dry, compressed air.

⚠ WARNING: Suitable eye protection must be worn.

- Use suitable marker ink to identify mating parts, do not use a scriber or centre punch as they could initiate cracks or distortion.
- Wire or tape mating parts together where necessary to prevent accidental interchange.
- Suitably identify parts which are to be renewed and to those parts requiring further inspection. Keep these parts separate.
- To make sure that the correct replacement part has been obtained, do not discard a part due for renewal until after comparing it with the new part.

Cleaning components

Always use cleaning agents which are suitable for the work being undertaken and the components being cleaned. NEVER use gasoline (petrol) as a cleaning agent (degreaser). Always make sure that the component being cleaned is compatible with the cleaning agent.

Always follow the manufacturer's instructions regarding the use of cleaning agents and make sure that the environment in which the work is being undertaken is suitable. See Health and Safety Precautions for further information regarding cleaning.

General inspection of components

All components should be inspected for wear or damage before reassembling.

- Always make sure that component to be inspected is clean and free from oil or grease.
- When a component is to be checked dimensionally against design specified values, use the appropriate measuring equipment i.e. micrometers, verniers, surface plates, dial test indicators (DTI).
- Always make sure that all measuring equipment is correctly calibrated before use.
- Reject a component which is not within specified values/limits or if it appears to be damaged.
- A component may be re-installed if dimensions obtained during checking are at the maximum tolerance limit and it is in an undamaged condition.
- Bearing journal clearances should be checked where necessary using Plastigage.
- Gaskets, seals and O-ring seals are to be re-used unless damaged.

Joints and Joint Faces

All gaskets should be installed dry unless stated otherwise. Always apply the specified lubricant to O-rings and install O-rings using the fingers only.

Use gasket removal spray and/or plastic scrapers to remove traces of old gasket.

⚠ CAUTION: DO NOT use metal scrapers or emery cloth as these may damage the sealing surfaces.

Many joints use sealants instead of gaskets as the sealing medium. Where this is the case, the sealant together with its part number will be found listed in the relevant repair operation and also in the sealants table.

⚠ CAUTION: Always remove all traces of the old sealant prior to reassembly. Use plastic scrapers, specified solvents where available or dry, lint free cloth. DO NOT use metal scrapers or emery cloth as these may damage the sealing surfaces. Make sure that sealing surfaces are free from oil or grease as sealants will not adhere properly to contaminated surfaces.

Do not allow sealant to enter tapped holes or oilways.

Locking Devices

Always replace locking devices with one of the same design and of the correct size.

Tab washers

Always release locking tabs before loosening fixings, do not re-use tab washers.

Locknuts

Always use a backing spanner when loosening and tightening locknuts, brake and fuel pipe unions.

Roll pins

Always install new roll pins of the correct size.

Circlips

Always install new circlips ensuring that they are of the correct size for the groove.

Woodruff keys

Woodruff keys may be re-used provided there is no indication of wear or distortion.

Remove any burrs from edges of keyways using a fine file.

Split pins

Never attempt to straighten and re-use a split pin, always make sure that replacement pins are of the correct size for the hole in which they are to be installed.

Screw Threads

- Damaged nuts, bolts and screws must always be discarded. Attempting to recut or repair damaged threads with a tap or die impairs the strength and install of the threads and is not recommended.

△ NOTE: During certain repair operations, it may be necessary to remove traces of thread locking agents using a tap. Where this is necessary, the instruction to do so will appear in the relevant operation and it is essential that a tap of the correct size and thread is used.

- Some bolts are coated with a thread locking agent and unless stated otherwise, they must not be re-used. New bolts having the same part number as the original must always be installed. When nuts or bolts are to be discarded, the repair operation and relevant torque chart will include an instruction to that effect. Do not use proprietary thread locking agents as they may not meet the specification required. See also Encapsulated ('Patched') Bolts and Screws.
- Always make sure that replacement nuts and bolts are at least equal in strength to those that they are replacing. Castellated nuts must not be loosened to accept a split pin except in recommended cases when this forms part of an adjustment.
- Do not allow oil or grease to enter blind holes, the hydraulic action resulting from tightening the bolt or stud can split the housing and also give a false torque reading.
- Always tighten a nut, bolt or screw to the specified torque figure, damaged or corroded threads can give a false torque reading.
- Nut and bolt loosening and tightening sequences, where given, must ALWAYS be followed. Distortion of components or faulty sealing of joints will result if the sequences are not followed. Where an instruction is given to tighten in stages, these stages must be adhered to; do not attempt to combine stages particularly where certain stages involve tightening by degrees.
- To check or re-tighten a fixing to a specified torque, first loosen a quarter of a turn, then retighten to the specified torque figure.
- Unless instructed otherwise, do not lubricate bolt or nut threads prior to installing.

Where it is stated that bolts and screws may be re-used, the following procedures must be carried out:

- Check that threads are undamaged.
- Remove all traces of locking agent from the threads.

CAUTION: DO NOT use a wire brush; take care that threads are not damaged.

- Make sure that threads are clean and free from oil or grease.
- Apply the specified locking agent to the bolt threads.

Bolt and Nut Identification

An ISO metric bolt or screw made of steel and larger than 6 mm in diameter can be identified by either of the symbols ISO M or M embossed or indented on top of the bolt head.

In addition to marks identifying the manufacturer, the top of the bolt head is also marked with symbols indicating the strength grade e.g. 8.8, 10.9, 12.9, 14.9.

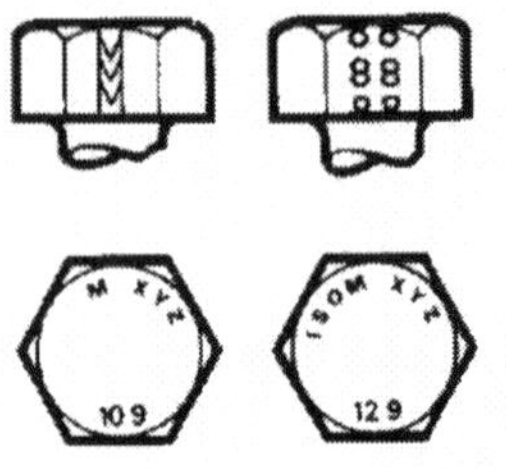

Alternatively, some bolts and screws have the M and strength grade symbol stamped on the flats of the hexagon.

In addition to marks identifying the manufacturer, the top of the bolt head is also marked with symbols indicating the strength grade e.g. 8.8, 10.9, 12.9, 14.9. Alternatively, some bolts and screws have the M and strength grade symbol stamped on the flats of the hexagon.

Encapsulated ('Patched') bolts and screws

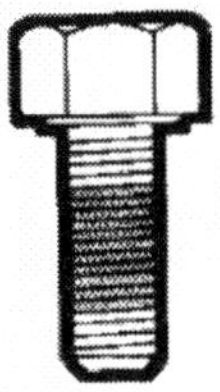

Encapsulated ('patched') bolts and screws have a thread locking agent applied to the threads during manufacture. Most thread locking agents are coloured, the band of colour extending for 360° around the thread. Some locking agents however, are neutral in colour and may not be so easily identified apart from a slightly darker area of thread where the locking agent has been applied. The locking agent is released and activated by the tightening process and is then chemically cured to provide the locking action.

Self-locking bolts and screws

Unless stated in a specific repair procedure, self-locking bolts and screws i.e. nylon patched or trilobular thread can be re-used provided that resistance is felt when the locking portion enters the female thread.

Nylon patched bolts and screws have a locking agent either applied to, or inserted in the threaded portion. They are identified by the presence of a coloured section of thread extending approximately 180° around the thread or by a coloured plug inserted into the bolt.

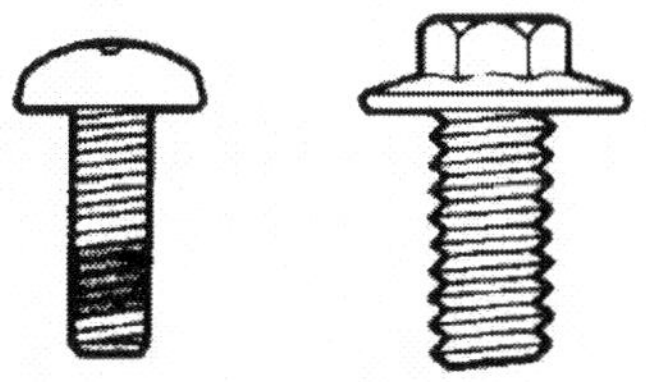

Trilobular bolts have a special thread form which creates a slight interference with the thread of the hole or nut into which it is screwed.

CAUTION: Do Not re-use self-locking fasteners in critical locations e.g. drive plates/flywheel or engine bearings. Do not install non self-locking fasteners where a self-locking fastener is specified.

Trilobular bolts should not be used as a substitute for patched bolts.

Nut identification

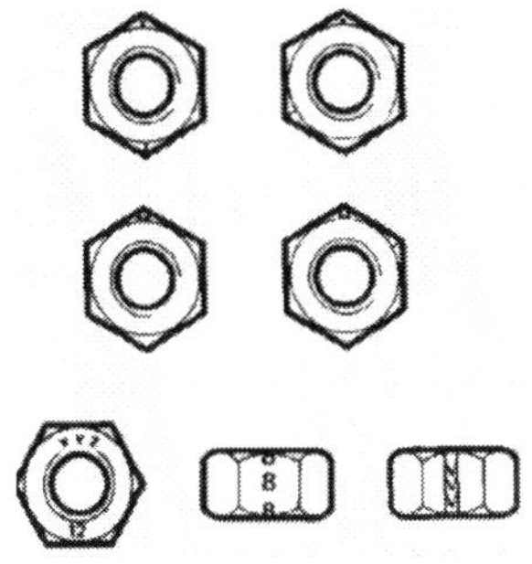

E48630

A nut with an ISO metric thread is marked on one face or one of the hexagonal flats with the strength grade symbol 8, 12, 14. Some nuts with the strength grade 4, 5 or 6 are also marked and some have the metric symbol M on the hexagonal flat opposite the strength grade marking.

A clock face system is sometimes used as an alternative method of indicating the strength grade. The external chamfers or a face of the nut is marked in a position relative to the appropriate hour mark on a clock face to indicate the strength grade.

A dot is used to locate the 12 o'clock position and a dash to indicate the strength grade. If the grade is above 12, two dots identify the 12 o'clock position.

When tightening a slotted or castellated nut, never loosen it to insert a split pin except where specified as part of an adjustment procedure. If difficulty is experienced in correctly positioning the slot, alternative washers or nuts should be selected.

Where a nut is tightened to adjust or maintain bearing pre-load, the tightening procedure must be adhered to.

Self-locking nuts

Unless stated otherwise, self-locking nuts once removed must be discarded and new nuts of the same type and strength grade installed.

Ball and Roller Bearings

When removing and installing bearings, make sure that the following practices are observed to make sure component serviceability:

CAUTION: Service tools have been developed for removing the majority of bearings; these must always be used where specified.

- Remove all traces from bearing under inspection by cleaning with a suitable degreasant; maintain absolute cleanliness throughout operations.
- Conduct a visual inspection for markings on rolling elements, raceways, outer surfaces of outer or inner surfaces of inner rings. Reject any bearings found to be marked since marking in these areas indicates onset of wear.
- Hold inner race of bearing between finger and thumb of one hand and rotate outer race to check that it revolves absolutely smoothly. Repeat holding outer race and rotating inner race. DO NOT spin the bearing.
- Rotate outer ring gently using a reciprocating movement whilst holding inner ring; feel for any check or obstruction to rotation. Reject bearing if movement is not absolutely smooth.
- Check bearing for blueing or signs of overheating.
- Lubricate bearing with the specified lubricant.
- Inspect bearing surface of shaft and bearing housing for discolouration or other markings which indicate overheating of bearing or movement between bearing and seating.
- Before installing bearing, make sure that shaft and bearing housing are clean and free from burrs.
- If one bearing of a pair shows signs of wear, overheating etc., it is advisable to replace bearings as a pair unless it is suspected that one bearing may have been faulty when installed, was installed incorrectly or the fault arose due to oil seal failure.
- Never reinstall a bearing unless it is in a fully serviceable condition.

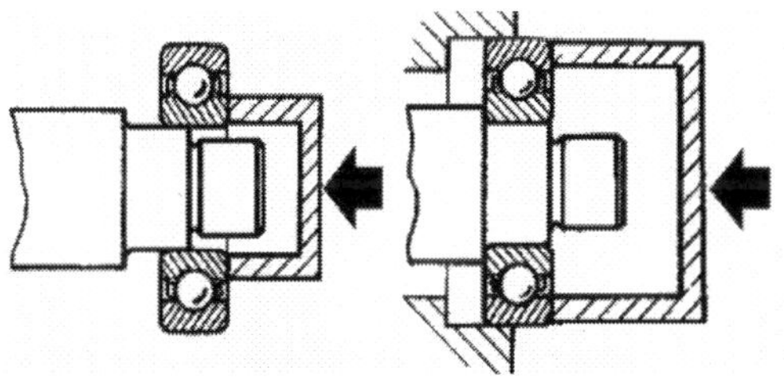

- When installing a bearing to a shaft, only apply force to the inner ring of the bearing. When installing a bearing into a housing, only apply force to the outer ring of the bearing.

CAUTION: Service tools have been developed for installing the majority of bearings; these must always be used where specified.

- In the case of grease lubricated bearings, fill the space between the bearing and outer seal with the recommended grade of grease before installing the seal.

CAUTION: When a waxed oil seal (installed dry) type of oil seal is to be installed, take great care that grease does not contaminate the running surface of the seal.

- Always make suitable reference marks between the components of separable bearings e.g. taper roller bearings when disassembling to make sure correct location of components when assembling. Never install new rollers in an outer ring, always install a new bearing assembly.

Brake Pads and Linings

Always install the correct grade and specification of brake pads and linings. When replacing these items, always replace as complete axle sets.

Brake Hydraulics

Always observe the following recommendations when working on the braking system:

WARNING: Do not intermix brake fluid of different specifications.

- Always use two spanners when loosening or tightening brake pipes or hose connections.
- Make sure that hoses run in a natural curve and are not kinked or twisted.
- Install brake pipes and hoses securely in their retaining clips and make sure that they cannot contact a potential chafing point.
- Containers used for brake fluid must be kept absolutely clean.
- Do not store brake fluid in unsealed containers, the fluid will absorb water which will lower the boiling point of the fluid.
- Do not allow brake fluid to be contaminated with other fluids such as mineral oil and do not put brake fluid in a container which has previously been used for storing other fluids.
- Do not re-use brake fluid which has been bled from the system.
- Always use brake fluid or a suitable brake cleaning fluid to clean hydraulic components.
- Unless stated otherwise, use only clean brake fluid to lubricate hydraulic seals and components.
- Always install blanking plugs to hoses, pipes or components immediately after disconnection.
- Check thread compatibility of original equipment with replacement components.
- Observe absolute cleanliness when working with hydraulic components.

Pipes and Hoses

When removing or installing flexible hydraulic pipes and hoses, make sure that the following procedures are observed to make sure component serviceability:

- Prior to removal, clean area around hose or pipe end which is to be disconnected.
- Obtain appropriate blanking plugs or caps before disconnecting hose or pipe end fittings in order that connections can be plugged immediately following disconnection.
- Always install blanking plugs or caps to pipes and unions immediately following disconnection.
- Clean hose or pipe and blow through with an air line.

WARNING: Suitable eye protection must be worn.

- Check hoses externally for cracks, separation of plies, security of end fittings and external damage; replace faulty hoses.
- Check pipes for signs of corrosion and chafing, replace as necessary.

CAUTION: If pipes are found to be chafed, rectify clips, mounting points etc., to prevent further problems in service.

- When installing hoses, make sure that no unnecessary bends are introduced and that hoses are not kinked, twisted or positioned close to potential chafing points.
- When installing pipes, make sure that pipes are positioned and clipped clear of potential chafing points.
- Always replace sealing washers installed to banjo bolts, sealing plugs etc.
- Always use a backing spanner when tightening unions and do not overtighten union nuts or banjo bolts.
- After engagement of 'quick-fit' connection hoses, perform a 'tug' test to make sure connection is securely installed.
- After any work on hydraulic systems, always check for fluid leaks whilst a second operator applies working pressure to the brake pedal or operates the system that has been worked on.

Fuel system hoses

Some fuel hoses are made up of two laminations, an armoured rubber outer sleeve and an inner viton core. Whenever a hose is removed, make sure that the inner bore is inspected to check that the viton lining has not become separated from the outer sleeve.

WARNING: Never attempt to repair fuel hoses or rectify leaking quick-fit connectors. The fuel hose and connectors must be replaced as an assembly.

Fuel system hose clips

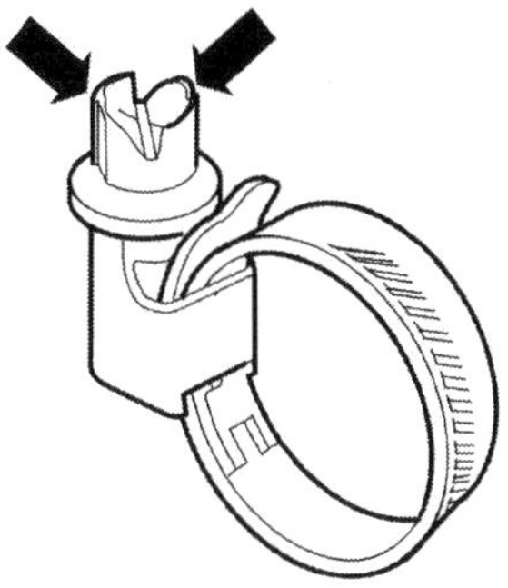

Certain fuel system hose clips are of the 'break-off head' type where a slot in the screw head shears off when the clip is tightened to a specific torque. These clips may be removed using a screwdriver and must be replaced with new clips on reassembly. Clips must be tightened until the portion of the slot shears off. Do not attempt to tighten clips by any other method, do not install any other type of clip.

'Quick-fit' connections are also installed to certain fuel hoses. After engagement of 'quick-fit' connections, perform a 'tug' test to make sure connection is securely installed.

Other fuel system hose clips are of the 'Jubilee' type and there may be a tamper proof cover installed over the screw head. These cover must be carefully removed before slackening the clip and should be replaced after final tightening, ensuring that the internal hexagon on the cover is correctly located on the clip screw.

Cooling system hoses

CAUTION: The following precautions must be observed to make sure that the integrity of the cooling system hoses and their connection to the system is maintained.

Hose orientation and connection

Correct orientation of cooling system hoses is important to make sure that hoses do not become fatigued or damaged through contact with adjacent components.

Where orientation marks are provided on the hose and corresponding component, the marks must be aligned when the hose is installed. Hoses must be installed fully on to their connection points, usually a moulded form on a pipe provides a positive indicator.

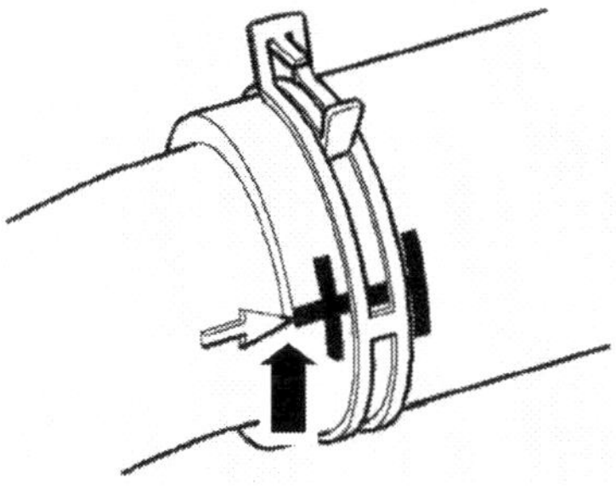

Hose clips

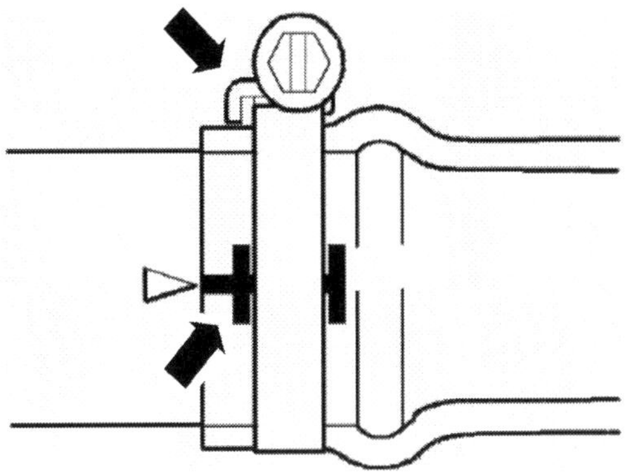

Markings are usually provided on the hose to indicate the correct clip position. If no markings are provided, position the clip directly behind the retaining lip at the end of the stub pipe. Worm drive clips should be orientated with the crimped side of the drive housing facing towards the end of the hose or the hose may become pinched between the clip and the stub pipe retaining lip. Unless otherwise stated, worm drive clips should be tightened to 3 Nm (2 lb-ft). Make sure that hose clips do not foul adjacent components.

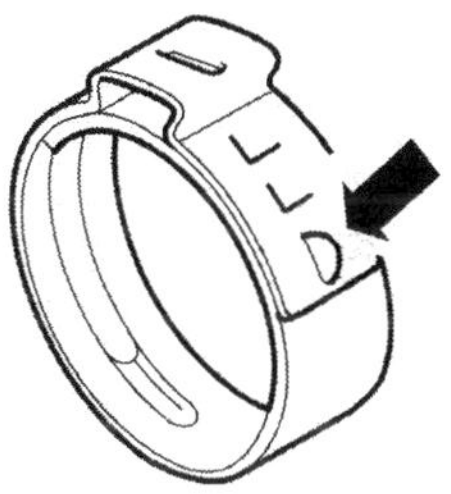

Oetiker clips may be removed by bending the tag (arrowed) and releasing the free end of the clip. Clips must not be re-used. When installing new clips, make sure clip is positioned on hose before tightening and make sure that when clip is tightened, the tag is located in the longitudinal slot in the free end of the clip (arrowed in illustration).

'Quick-fit' connections are also installed to certain hoses/pipes. Inspect 'quick-fit' connections for damage, prior to connection. Replace if damaged. After engagement of 'quick-fit' connections, perform a 'tug' test to make sure connection is securely installed.

Heat protection

Always make sure that heat shields and protective sheathing are in good condition; replace if damage is evident. Particular care must be taken when routing hoses close to hot engine components such as the exhaust manifolds and exhaust gas recirculation (EGR) pipes. Hoses will relax and deflect slightly when hot, make sure this movement is taken into account when routing and securing hoses.

Electrical Precautions

General

The following guidelines are intended to make sure the safety of the operator whilst preventing damage to the electrical and electronic components of this vehicle.

Equipment

Prior to commencing any test procedure on the vehicle, make sure that the relevant test equipment is working correctly and that any harness or connectors are in good condition. It is particularly important to check the condition of all plugs and leads of mains operated equipment.

Polarity

Never reverse connect the vehicle battery and always make sure the correct polarity when connecting test equipment.

High voltage circuits

Whenever disconnecting live ht circuits, always use insulated pliers and never allow the open end of the ht lead to contact other components, particularly ECU's.

Connectors and harnesses

The engine compartment of a vehicle is a particularly hostile environment for electrical components and connectors. Always observe the following:

- Make sure electrically related items are dry and oil free before disconnecting/connecting test equipment.
- Make sure that disconnected multiplugs and sensors are protected from any possible oil, coolant or other liquid contamination. Any such contamination could impair performance or lead to component failure.
- Never force connectors apart or pull on the wiring harness.
- Always make sure locking tabs are disengaged before disconnecting multiplugs etc. and make sure that correct orientation is achieved before connection.
- Make sure that any protection covers, insulation etc. are replaced if disturbed.

Having confirmed that a component is faulty, carry out the following:

- Switch off the ignition and disconnect the battery.
- Remove the component and support the disconnected harness.
- When replacing electrical components, keep oily hands away from electrical connections and make sure that locking tabs on connectors are fully engaged.

Battery Disconnection/Connection

Always refer to the Electrical Section of this manual - Battery Connection/Disconnection prior to attempting to connect or disconnect the battery.

Refer to: Battery (414-01 Battery, Mounting and Cables, Removal and Installation)

Fuel Handling Precautions

The following information lists basic precautions which must be observed if fuel is to be handled safely. It also outlines other areas of risk which must not be ignored. As this information is issued for basic guidance only, consult your local Fire Department where any doubt as to personal and environmental safety exists - See also Health and Safety Precautions.

General precautions

Always have the correct type of fire extinguisher containing Foam, CO2, Gas or powder accessible when handling or draining fuel or dismantling fuel systems. Fire extinguishers must also be located in areas where fuel is stored.

Make sure that suitable warning signs are exhibited.

Keep all sources of ignition well away from areas where fuel is being handled.

Make sure that any leadlamps are flameproof and kept clear of spillage.

WARNINGS:

Do not disassemble or reassemble fuel system components whilst vehicle is over a pit.

No one should be permitted to repair components associated with fuel without first having specialist training.

Always disconnect the vehicle battery before carrying out disassembly, reassembly or draining work on a fuel system.

Fuel tank and system draining

Draining must be carried out in accordance with the procedures given in the relevant Fuel System section of this manual.

WARNINGS:

Never drain fuel or work on a fuel system while the vehicle is over a pit. Extraction or draining of fuel must be carried out in a well ventilated area.

Never switch on or operate mobile (cellular) phones in the vicinity of vehicles when operations are being carried out on the fuel system.

Always attach fuel vapour warning labels to fuel tanks immediately after draining.

Containers used for storing fuel must be clearly marked with the contents and placed in a safe storage area which meets the requirements of the local authority.

CAUTION: Some fuel lines are now installed with quick release connectors. If a connector is damaged, no attempt must be made to repair the connector, a new fuel line and connector(s) assembly must be installed.

Always release pipe clips fully before attempting to disconnect fuel pipes.

Fuel tank repairs

CAUTION: No attempt should be made to repair a fuel tank. If the structure of the tank is damaged, a new tank must be installed.

Oil seals

Never use a seal which has been improperly stored or handled.

- Take great care when removing old seals that the sealing surfaces and seal housing are not damaged.
- Carefully examine seal before installing to make sure that it is clean and undamaged.
- Make sure that the surface on which the seal is to run and also the seal housing is clean and free from burrs or scratches. Renew the component if the sealing surface cannot be restored.
- Special tools and protection sleeves are provided for installing the majority of seals and must be used when specified.
- Many seals are now coated with a protective wax and DO NOT need to be lubricated prior to installing. Always check the relevant repair procedure which will state if a seal must be installed dry. Never touch these seals with oily hands as the oil will contaminate the protective coating and affect the sealing properties of the seal; also, make sure that installing tools and protection sleeves are free from oil and grease. Seals which must be lubricated prior to installing should have the recommended lubricant applied to the areas specified in the repair procedure.
- Make sure that a seal is installed the correct way round. For example, the lip of the seal must face towards the lubricant which it is sealing.
- When installing an oil seal, make sure that it is positioned square to shaft and housing. Where the seal is to be installed to a housing prior to installing over a shaft, take care not to allow the weight of an unsupported shaft to rest on the seal.

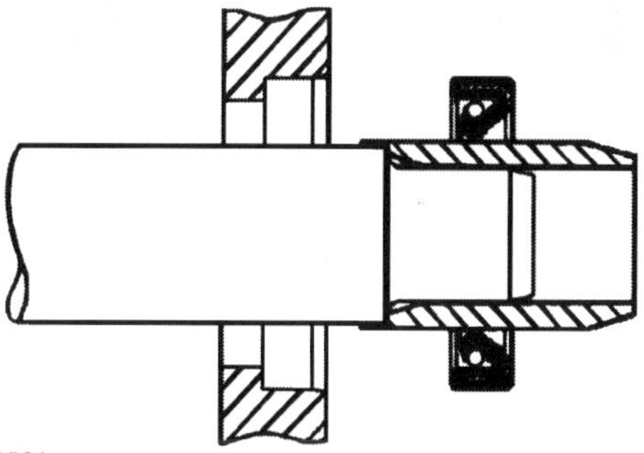

E48561

- Always use the recommended special tool and protection sleeve to install an oil seal. If no tool is specified, use a suitable mandrel approximately 0.4 mm (0.015 in) smaller than the outside diameter of the seal. Use adhesive tape on the shaft to protect the sealing lip of the seal.
- Press or drift the seal in to the depth of its housing if the housing is shouldered or flush with the face of the housing where no shoulder is provided. Make sure that the seal is not tilted in the housing when it is installed.

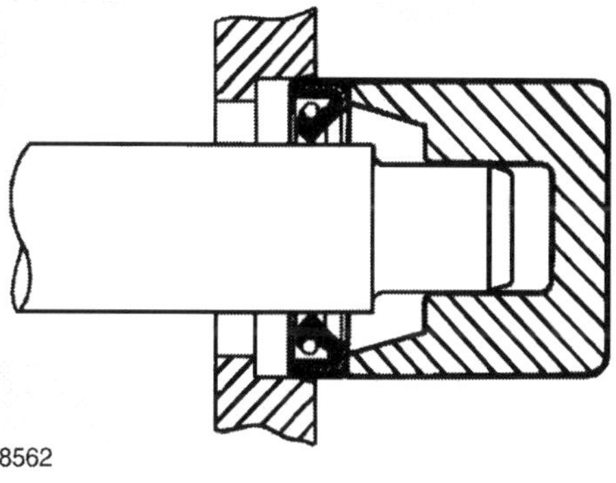

E48562

(A/C) System Precautions

The A/C system contains fluids and components which could be potentially hazardous to the service engineer or the environment if not serviced and handled correctly. The following guidelines are intended to alert the service engineer to potential sources of danger and emphasise the importance of ensuring the integrity of the A/C operating conditions and components installed to the vehicle.

Where necessary, additional specific precautions are detailed in the relevant sections of this Manual and also in the Health and Safety Section. These precautions must be referred to prior to commencing repair operations.

The refrigerant used in the A/C system is HC-134a (Hydro fluorocarbon) R134a.

WARNINGS:

⚠ Servicing must only be carried out by personnel familiar with both the vehicle system and the charging and testing equipment. All operations must be carried out in a well ventilated area away from open flame and heat sources.

⚠ R134a is a hazardous liquid and when handled incorrectly can cause serious injury. Suitable protective clothing, consisting of face protection, heat proof gloves, rubber boots and rubber apron or waterproof overalls, must be worn when carrying out operations on the A/C system.

Remedial actions

⚠ WARNING: Due to its low evaporating temperature, R134a must be handled with care. R134a splashed on any part of the body will cause immediate freezing of that area. Also, refrigerant cylinders and replenishment trolleys when discharging will freeze skin to them if contact is made.

If an accident involving R134a should occur, conduct the following remedial actions:

- If liquid R134a enters the eye, do not rub it. Gently run large quantities of eye wash over affected eye to raise the temperature. If an eye wash is not available, cool, clean water may be used to flush the eye. After rinsing, cover the eye with a clean pad and seek immediate medical attention.
- If liquid R134a is splashed onto the skin, run large quantities of water over the affected area to raise the temperature. Implement the same action if the skin comes in contact with discharging cylinders. Wrap the contaminated body parts in blankets (or similar materials) and seek immediate medical attention.
- If the debilitating effects of inhalation of R134a vapour are suspected, seek fresh air. If the affected person is unconscious, move them away from the contaminated area to fresh air and apply artificial respiration and/or oxygen and seek immediate medical attention.

Service precautions

Observe the following precautions when handling components used in the system:

- A/C units must not be lifted by their hoses, pipes or capillary lines.
- Hoses and lines must not be subjected to any twist or stress; the efficiency of the system will be impaired by kinks or restrictions. Make sure that hoses are correctly positioned before tightening couplings, and make sure that all clips and supports are utilised.
- Flexible hoses should not be positioned closer than 100 mm (4.0 in) to the exhaust manifold unless protected by heat shielding.
- Completed assemblies must be checked for refrigeration lines touching metal panels. Any direct contact of components and panels may transmit noise and so must be eliminated.
- The appropriate torque wrench must be used when tightening refrigerant connections to the stipulated value. An additional spanner must be used to hold the union to prevent twisting of the pipe when tightening connections.
- Before connecting any hose or pipe, make sure that refrigerant oil is applied to the seat of the new O-rings, **BUT NOT** to the threads of the connection.
- All protective plugs or caps must remain in place in the component until immediately prior to connection.
- Make sure components are at room temperature before uncapping/unplugging, to prevent condensation of moisture from the air that enters it.
- When disconnecting, immediately plug or cap all pipes to prevent ingress of dirt and moisture into the system.
- Components must not remain uncapped/ unplugged, if a system has been left uncapped/ unplugged for 24 hours or longer, a new receiver/drier must be installed.
- The receiver/drier contains desiccant which absorbs moisture. It must be positively sealed at all times. A receiver/drier that has been left uncapped for longer than 24 hours must not be used; install a new unit.
- The receiver/drier should be the last component connected to the system to make sure optimum dehydration and maximum moisture protection of the system.
- Whenever a component of the refrigeration system is replaced, it will also be necessary to install a new receiver/drier unit.

- Use alcohol and a clean lint-free cloth to clean dirty connections.
- Make sure that all new parts installed are marked for use with R134a.
- When a major repair has been completed, a leak test should be conducted; refer to the Repairs Section of this manual for the correct procedure.

Refrigerant oil

⚠ CAUTION: Refrigerant oil (ND-8 PAG) easily absorbs water and must not be stored for long periods. Do not pour unused refrigerant oil back into the container. Always use an approved refrigerant oil.

When replacing components in the system, drain the refrigerant oil from the component being replaced into a graduated container. On assembly, add the quantity of refrigerant oil drained to the new component - See Compressor Replacement in this Section.

A/C Compressor

A new compressor is sealed and pressurised with Nitrogen gas. When installing a new compressor, slowly release the sealing cap; gas pressure should be heard to vent as the seal is broken.

⚠ CAUTION: A new compressor should always be sealed and could be pressurised with nitrogen gas. To avoid possible oil loss, release the sealing cap(s) slowly. Do not remove the cap(s) until immediately prior to connecting the pipes to the compressor.

Rapid refrigerant discharge

If the A/C system is damaged as a result of an accident and the system is punctured, the refrigerant will discharge rapidly. The rapid discharge of refrigerant will also result in the loss of most of the oil from the system. The compressor must be removed and all the remaining oil in the compressor drained and refilled as instructed in the air conditioning section of this manual.

Precautions for refrigerant recovery, recycling and recharging

When the A/C system is recharged, any existing refrigerant is first recovered from the system and recycled. The system is then charged with the required weight of refrigerant and volume of refrigerant oil.

⚠ WARNING: Refrigerant must always be recycled before re-use to make sure that the purity of the refrigerant is high enough for safe use in the system. Recycling should always be carried out with equipment which is design certified by Underwriter Laboratory Inc. for compliance with SAE J1991. Other equipment may not recycle refrigerant to the required level of purity.

CAUTIONS:

⚠ A R134a Refrigerant Recovery Recycling Recharging Station must not be used with any other type of refrigerant. Refrigerant R134a from domestic and commercial sources must not be used in motor vehicle systems.

⚠ The system must be evacuated immediately before recharging commences. Delay between evacuation and recharging is not permitted.

A/C Compressor Replacement

A new compressor is supplied filled with a full charge (X cm^3) of refrigerant oil.

A calculated quantity of oil must be drained from the new compressor before installing. To calculate the quantity of oil to be drained:

- Remove the drain plug from the old compressor.
- Invert the compressor and gravity drain the oil into a calibrated measuring cylinder. Rotate the compressor clutch to make sure the compressor is completely drained.
- Note the quantity of oil drained (Y cm^3).
- Calculate the quantity of oil to be drained from the new compressor using the following formula: $\mathbf{X\ cm^3 - (Y\ cm^3 + 20\ cm^3) = Q\ cm^3}$
- Remove the drain plug from the new compressor and drain Q cm^3 of oil. Install and tighten the compressor drain plug.

Vehicle Weights from 07/2001

When loading a vehicle to its maximum (Gross Vehicle Weight) consideration must be taken of the unladen vehicle weight and distribution of the payload to make sure that axle loadings do not exceed the permitted maximum values.

It is the customer's responsibility to limit the vehicle's payload in an appropriate manner such that neither maximum axle loads nor Gross Vehicle Weight are exceeded.

NOTES:

△ Axle weights are not accumulative. The individual maximum axle weights and gross vehicle weights must not be exceeded.

△ EEC Kerb weight = Unladen weight + full fuel tank + 75 Kg (165 lb).

90 Models

Vehicle axle weights - 90 models	Station wagon	Station wagon	Utility	Utility
Front axle	1200 Kg	2645 lb	1200 Kg	2645 lb
Rear axle	1500 Kg	3307 lb	1500 Kg	3307 lb
Maximum Gross Vehicle Weight (GVW) -	2550 Kg	5622 lb	2400 Kg	5291 lb
EEC Vehicle kerb weights				
Soft top	1770 Kg	3402 lb	1993 Kg	4393 lb
Pick-up	1770 Kg	3402 lb	1993 Kg	4393 lb
Hard top	1815 Kg	4001 lb	1987 Kg	4380 lb
Station wagon	1870 - 1885 Kg	4122 - 4155 lb	1989 - 1998 Kg	4385 - 4404 lb

110 Models - Non Japanese specification

Vehicle axle weights -	Station wagon	Station wagon	Utility	Utility
Front axle	1200 Kg	2645 lb	1200 Kg	2645 lb
Rear axle	1750 Kg	3858 lb	1850 Kg	4078 lb
Maximum Gross Vehicle Weight (GVW) -	2950 Kg	6503 lb	3050 Kg	6724 lb
EEC Vehicle kerb weights	**Standard**	**Standard**	**Heavy duty**	**Heavy duty**
Soft top	1885 Kg	4155 lb	2080 Kg	4585 lb
High capacity Pick-up	1920 Kg	4232 lb	2122 Kg	4678 lb
Hard top	1920 Kg	4232 lb	2110 Kg	4651 lb
Station wagon	2055 Kg	4530 lb	2229 Kg	4914 lb

110 Models - Japanese specification

Vehicle axle weights -	Metric	Imperial
Front axle	1115 Kg	2457 lb
Rear axle	2180 Kg	4805 lb
Maximum Gross Vehicle Weight (GVW) -	3295 Kg	7262 lb

130 Models

Vehicle axle weights	Metric	Imperial
Front axle	1580 Kg	3483 lb
Rear axle	2200 Kg	4850 lb
Maximum Gross Vehicle Weight (GVW) -	3500 Kg	7716 lb
EEC Vehicle kerb weights	**Heavy duty**	**Heavy duty**
Crew cab and high capacity pick-up	2177 - 2286 Kg	4667 - 5039 lb

110 Models - Non Japanese specification

Towing weights

NOTES:

⚠ * Only applies to vehicles modified to accept coupled brakes.

⚠ All weight figures are subject to local restrictions

Towing weights	On-road	On-road	Off-road	Off-road
Unbraked trailers	750 Kg	1653 lb	500 Kg	1102 lb
Trailers with overrun brakes	3500 Kg	7716 lb	1000 Kg	2204 lb
4 wheel Trailers with coupled brakes *	4000 Kg	8818 lb	1000 Kg	2204 lb

Off Road Performance

90 Models

⚠ NOTE: Departure angles do not account for the addition of a tow hitch.

Item	Metric	Imperial
Max. gradient (EEC Kerb weight)	45°	
Approach angle		
Soft top and Pick-up (EEC Kerb weight)	48°	
Hard top and Station wagon (EEC Kerb weight)	51.5°	
Departure angle		
Soft top and Pick-up (EEC Kerb weight)	49°	
Hard top and Station wagon (EEC Kerb weight)	53°	
Wading depth	500 mm	20 in

Min. ground clearance (unladen)		
Soft top and Pick-up (EEC Kerb weight)	191 mm	7.5 in
Hard top and Station wagon (EEC Kerb weight)	229 mm	9.0 in

110 and 130 Models

NOTE: Departure angles do not account for the addition of a tow hitch.

Item	Metric	Imperial
Max. gradient (EEC Kerb weight)	45°	
Approach angle	50°	
Departure angle		
110 Models	35°	
130 Models	34°	
Wading depth	500 mm	20 in
Min. ground clearance (unladen)	215 mm	8.5 in

Vehicle dimensions

90 Models

Item	Metric	Imperial
Overall length		
Soft top and Pick-up	3722 mm	146.5 in
Hard top and Station wagon	3883 mm	152.9 in
Overall width	1790 mm	70.5
Overall height		
Soft top and Pick-up	1965 mm	77.4 in
Pick-up and Station wagon	1963 mm	77.3 in
Hard top	1972 mm	77.6 in
Wheelbase	2360 mm	92.9 in
Track front/rear	1486 mm	58.5 in
Width between wheel boxes	925 mm	36.4 in

110 Models

Item	Metric	Imperial
Overall length		
Soft top and Pick-up	4438 mm	175 in
Hard capacity Pick-up	4631 mm	182 in
Hard top/station and county	4599 mm	181 in
Overall width	1790 mm	70.5 in
Overall height		
Non Japanese specification	2035 mm	80.1 in
Japanese specification	2060 mm	81.1 in
Wheelbase	2794 mm	110 in
Track front/rear	1486 mm	58.5 in
Width between wheel boxes		
High capacity Pick-up	1090 mm	43 in
All other models	925 mm	36.4 in

130 Models

Item	Metric	Imperial
Overall length	5132 mm	202 in
Overall width	1790 mm	70.5 in
Overall height	2035 mm	80.1 in
Wheelbase	3266 mm	127 in
Track front/rear	1486 mm	58.5 in
Width between wheel boxes	1090 mm	43 in

Health and Safety Precautions

Description and Operation

Introduction

Modern vehicles contain many materials and liquids which if not handled with care can be hazardous to both personal health and the environment. Also, many of the procedures associated with vehicle maintenance and repair involve physical hazards or other risks to health.

This subsection lists some of these hazardous operations and the materials and equipment associated with them. Precautions necessary to avoid these hazards are identified.

The list is not exhaustive and all operations and procedures and the handling of materials, should be carried out with health and safety in mind.

Before using any product the Materials Safety Data Sheet supplied by the manufacturer or supplier should be consulted.

WARNINGS:

⚠ Many liquids and other substances used in motor vehicles are poisonous and should under no circumstances be consumed and should, as far as possible, be kept from contact with the skin. These liquids and substances include acid, anti-freeze, brake fluid, fuel, windscreen washer additives, lubricants, refrigerants and various adhesives.

⚠ Observe due care when working on or near hot components. This may include, but is not limited to:

- Exhaust systems
- Emissions control systems
- Cooling systems
- Climate control systems
- Braking systems
- Lubrication systems
- Vehicle powertrain systems
- Electric vehicle systems.

Acids and Alkalis

For example - alkalis such as caustic soda used in cleaning materials; acids such as sulphuric acid used in batteries.

Both alkalis and acids are irritant and corrosive to the skin, eyes, nose and throat. They cause burns and can destroy ordinary protective clothing.

Avoid splashes to the skin, eyes and clothing. Wear suitable protective impervious apron, gloves and goggles. Do not breath mists.

Make sure access to eye wash bottles, shower and soap are readily available for splashing accidents.

Display Eye Hazard sign.

Air Conditioning Refrigerant

Highly flammable, combustible – observe No Smoking policy.

Skin contact may result in frostbite.

Instructions given by the manufacturer must be followed. Avoid naked lights, wear suitable protective gloves and goggles.

If refrigerant comes into contact with the skin or eyes, rinse the affected areas with water immediately. Eyes should also be rinsed with an appropriate irrigation solution such as a solution of 9% Sodium Chloride and Purified Water. **DO NOT RUB THE EYES AND SEEK IMMEDIATE MEDICAL ATTENTION.**

Air Conditioning Refrigerant

Do Not

1. Do Not expose refrigerant bottles to sunlight or heat.
2. Do Not expose refrigerant bottles to frost.
3. Do Not drop refrigerant bottles.
4. Do Not vent refrigerant to atmosphere under any circumstance.
5. Do Not mix refrigerants.

Adhesives and Sealants

Many adhesives and sealants are highly flammable – OBSERVE NO SMOKING POLICY. These items, should be stored in flameproof cabinets in No Smoking areas. Cleanliness and tidiness in use should be observed, for example disposable paper covering benches. All adhesives and sealants should be dispensed from applicators where possible; containers, including secondary containers, should be labelled appropriately.

Anaerobic, Cyanoacrylate (super-glues) and other Acrylic Adhesives

Many are irritant, sensitizing or harmful to the skin and respiratory tract. Some are eye irritants.

Skin and eye contact should be avoided and the manufacturer's instructions followed.

Cyanoacrylate adhesives (super-glues) MUST NOT contact the skin or eyes. If skin or eye tissue is bonded, cover with a clean moist pad and **SEEK IMMEDIATE MEDICAL ATTENTION.** Do not attempt to pull skin tissue apart. Use in well ventilated areas as vapors can cause irritation to the nose and eyes.

For two-pack systems see Resin-based and Isocyanate Adhesives/Sealers.

Solvent-based Adhesives/Sealers - See Solvents

Follow manufacturers instructions.

Water-based Adhesives/Sealers

Those based on polymer emulsions and rubber/latex may contain small amounts of volatile, toxic and harmful chemicals. Skin and eye contact should be avoided and adequate ventilation provided during use.

Hot Melt Adhesives

In the solid state, they are safe. In the molten state they may cause burns and health hazards may arise from the inhalation of toxic fumes.

Use appropriate protective clothing and a thermostatically controlled heater with a thermal cut-out and adequate extraction.

Resin-based Adhesives/Sealers, for example Epoxide and Formaldehyde Resin-based

Mixing should be carried out in well ventilated areas as harmful or toxic volatile chemicals may be released.

Skin contact with uncured resins and hardeners can result in irritation, dermatitis, and absorption of toxic or harmful chemicals through the skin. Splashes can damage the eyes.

Provide adequate ventilation and avoid skin and eye contact.

Isocyanate (Polyurethane) Adhesives/ Sealers

See also Resin-based Adhesives

Individuals suffering from asthma or respiratory allergies should not work with or near these materials as sensitivity reactions can occur.

Over exposure is irritating to the eyes and respiratory system. Excessive concentrations may produce effects on the nervous system including drowsiness. In extreme cases, loss of consciousness may result. Long term exposure to vapour concentrations may result in adverse health effects.

Prolonged contact with the skin may lead to skin irritation and in some cases, dermatitis.

Splashes entering the eye will cause discomfort and possible damage.

Any spraying should preferably be carried out in ventilated booths which incorporate facilities for removing vapors and spray droplets from the breathing zone. Wear appropriate gloves, eye and respiratory protection.

Antifreeze

May be flammable when undiluted.

Vapors may be given off from coolant antifreeze when heated. Avoid breathing these vapors.

Antifreeze may be absorbed through the skin in toxic or harmful quantities. Antifreeze, if swallowed, can be fatal; **SEEK IMMEDIATE MEDICAL ATTENTION.**

Battery Acids

See also Alkalis and Acids.

Gases released during battery charging are explosive. Always remove the battery from the vehicle prior to charging. Never use naked flames or allow sparks near charging or recently charged batteries. NEVER add acid to a battery, the chemical reaction produced will be violent and explosive. In cases of eye contact, wash affected area with copious amounts of water and **SEEK IMMEDIATE MEDICAL ATTENTION.**

Make sure there is adequate ventilation during battery charging, observe NO SMOKING POLICY.

Brake Pads and Linings

Always fit the correct grade and specification of brake pads and linings. When renewing pads and linings, always replace as complete axle sets.

Brake and Clutch Fluid

Splashes to the skin and eyes are irritating and in the long term can be damaging, avoid prolonged skin contact. In cases of eye contact, wash affected area with copious amounts of water and SEEK IMMEDIATE MEDICAL ATTENTION.

Chemical Materials

All chemical materials should always be used with caution and stored and handled with care. They may be toxic, harmful, corrosive, irritant or highly flammable and give rise to hazardous fumes and dusts.

The effects of excessive exposure to chemicals may be immediate or delayed; briefly experienced or permanent; cumulative; superficial; life threatening; or may reduce life expectancy.

Chemical Materials - Do's

1. Do carefully read and observe hazard and precaution warnings given on material containers (labels) and in any accompanying leaflets, posters or other instructions. Material health and safety data sheets can be obtained from manufacturers.
2. Do remove chemical materials from the skin and clothing as soon as practicable after soiling. Change heavily soiled clothing and have it cleaned.
3. Do organize work practices and protective clothing to avoid soiling of the skin and eyes.
4. Do avoid breathing vapors, aerosols, dusts or fumes; inadequate container labelling; fire and explosion hazards.
5. Do wash before job breaks, before eating, smoking, drinking or using toilet facilities when handling chemical materials.
6. Do keep work areas clean, uncluttered and free of spills.
7. Do store chemical materials according to national and local regulations.
8. Do keep chemical materials out of the reach of children.

Chemical Materials - Do Not

1. Do Not mix chemical materials except under the manufacturers instructions; some chemicals can form other toxic or harmful chemicals, give off toxic or harmful fumes or become explosive when mixed together.
2. Do Not spray chemical materials, particularly those based on solvents, in confined spaces, for example when people are inside a vehicle.
3. Do Not apply heat or flame to chemical materials except under the manufacturers instructions. Some are highly flammable and some may release toxic or harmful fumes.
4. Do Not leave containers open. Fumes given off can build up to toxic, harmful or explosive concentrations. Some fumes are heavier than air and will accumulate in confined areas such as pits.
5. Do Not transfer chemical materials to unlabelled containers.
6. Do Not clean hands or clothing with chemicals. Chemicals, particularly solvents and fuels, will dry skin and may cause irritation leading to dermatitis or be absorbed through the skin in toxic or harmful quantities.
7. Do Not use emptied containers for other materials except when they have been cleaned under supervised conditions.
8. Do Not sniff or smell chemical materials, even brief exposure to high concentrations of fumes can be toxic or harmful.

Corrosion Protection Materials

Some corrosion protection materials are highly flammable – observe NO SMOKING POLICY.

These materials are varied and the manufacturers instructions must always be followed. The materials may contain solvents, resins or petroleum products. Skin and eye contact should be avoided. They should only be sprayed in conditions of adequate ventilation and not in confined spaces.

Dust

Dust or powder produced during repair operations may be irritant, harmful or toxic. Avoid breathing dusts from powdery chemical materials or those arising from dry abrasion operations. Wear respiratory protection if ventilation is inadequate.

Fine dusts of combustible material can present an explosion hazard. Avoid explosive limits and sources of ignition.

Electrical Equipment

Electric shock can result from the use of faulty electrical equipment or from the misuse of equipment in good condition.

Make sure that electrical equipment is maintained in good condition and frequently tested. Faulty equipment should be labelled and preferably removed from the work station.

Make sure that flexes, cables, plugs and sockets are not frayed, kinked, cut, cracked or otherwise damaged. If using cable reel extension equipment, ALWAYS ensure that the cable is fully unwound from the reel.

Make sure that electrical equipment and flexes do not come into contact with water.

Make sure that electrical equipment is protected by the correct rated fuse.

Never misuse electrical equipment and never use equipment which is in any way faulty. The results could be fatal.

Make sure that the cables of mobile electrical equipment cannot get trapped and damaged, such as in a vehicle hoist.

Make sure that the designated electrical workers are trained in basic First Aid.

In cases of electrocution:

1. Switch off the power supply before approaching the victim.
2. If this is not possible, **DO NOT TOUCH THE VICTIM** but push or drag the person from the source of electricity using dry, non-conductive material.
3. Commence resuscitation if trained to do so.
4. **SEEK IMMEDIATE MEDICAL ATTENTION.**

Exhaust Fumes

These contain asphyxiating, harmful and toxic chemicals and particles such as carbon oxides, nitrogen oxides, aldehydes, lead and aromatic hydrocarbons. Engines should be run only under conditions of adequate exhaust extraction or general ventilation and not in confined spaces.

Gasoline (Petrol) engine

There may not be adequate warning of odour or of irritation before toxic or harmful effects arise. These may be immediate or delayed.

Gas Oil (Diesel engine)

Soot, discomfort and irritation usually give adequate warning of hazardous fume concentrations.

Fibre Insulation

The fibrous nature of surfaces and cut edges can cause skin irritation. This is usually a physical and not a chemical effect.

Precautions should be taken to avoid excessive skin contact through careful organization of work practices and the use of gloves.

Fire

Many of the materials found on or associated with the repair of vehicles are highly flammable. Some give off toxic or harmful fumes if burnt; others such as fluoroelastomers when burnt or damaged by excessive heat can break down and produce highly corrosive hydrofluoric acid - See Fluoroelastomers.

Should any material be in a burnt or overheated condition, handle with extreme caution and wear protective clothing when handling such items. Dispose of such material in accordance with local regulations.

Decontaminate and dispose of protective clothing immediately after use.

Observe strict fire safety when storing and handling flammable materials or solvents, particularly near electrical equipment or welding processes.

Make sure, before using electrical or welding equipment, that there is no fire hazard present.

Have a suitable fire extinguisher available when using welding or heating equipment.

First Aid

Apart from meeting any legal requirements it is desirable for someone in the workshop to be trained in First Aid procedures.

Splashes in the eye should be flushed carefully with clean water for at least ten minutes.

Soiled skin should be washed with soap and water.

In case of cold burns, from alternative fuels, place affected area in cool to cold water.

Individuals affected by inhalation of gases and fumes should be removed to fresh air immediately. If effects persist, consult a doctor.

If liquids are swallowed inadvertently, consult a doctor giving him the information on the container or label. Do not induce vomiting unless this action is indicated on the label.

Fluoroelastomers (Synthetic Rubber)

Many 'O' rings, seals, hoses, flexible pipes and other similar which appear to be manufactured from natural rubber are, in fact, made of synthetic materials called Fluoroelastomers.

Under normal operating conditions, these materials are safe and do not constitute a health hazard. However, if the materials are damaged by burning or exposure to excessive heat, they can break down and produce highly corrosive hydrofluoric acid.

WARNING: Contact with hydrofluoric acid can cause serious burns on contact with the skin. If skin contact does occur, carry out the following steps immediately:

Remove any contaminated clothing.

SEEK IMMEDIATE MEDICAL ATTENTION

Irrigate affected area of skin with copious amounts of cold water or limewater for 15 to 60 minutes.

Foams - Polyurethane

Used in sound and noise insulation. Cured foams used in seat and trim cushioning.

Unreacted components are irritating and may be harmful to the skin and eyes. Wear gloves and goggles.

Individuals with chronic respiratory diseases, asthma, bronchial medical problems, or histories of allergic diseases should not work in or near uncured materials.

The components, vapors or spray mists can cause direct irritation, sensitivity reactions and may be toxic or harmful.

Vapors and spray mists must not be inhaled. These materials must be applied with adequate ventilation and respiratory protection. Do not remove the respirator immediately after spraying, wait until the vapour/mists have cleared.

Burning of the uncured components and the cured foams can generate toxic and harmful fumes. Smoking, naked flames or the use of electrical equipment during foaming operations and until vapors/mists have cleared should not be allowed. Any heat cutting of cured foams or partially cured foams should be carried out in areas having suitable fume extraction equipment.

Fuels

Avoid skin contact with fuel where possible. Should contact occur, wash the affected skin with soap and water.

Gasoline (Petrol)

Highly flammable - OBSERVE NO SMOKING POLICY.

Swallowing gasoline (petrol) can result in mouth and throat irritation and absorption from the stomach can result in drowsiness and unconsciousness. Small amounts can be fatal to children. Inhalation into the lungs, through vomiting, is a very serious hazard.

Gasoline (petrol) dries the skin and can cause irritation and prolonged or repeated contact may cause dermatitis; if it is allowed to enter the eyes, it will cause severe smarting. Wash affected area with copious amounts of water and **SEEK IMMEDIATE MEDICAL ATTENTION.**

Gasoline (petrol) may contain appreciable quantities of benzene, which is toxic upon inhalation and the concentration of vapors must be kept very low. High concentrations will cause eye, nose and throat irritation, nausea, headache, depression and symptoms of drunkenness. Very high concentrations will result in rapid loss of consciousness.

Make sure there is adequate ventilation when handling and using gasoline (petrol). Great care must be taken to avoid the serious consequences of inhalation in the event of vapour build up arising from spillages in confined spaces.

Special precautions apply to cleaning and maintenance operations on gasoline (petrol) storage tanks.

Gasoline (petrol) should not be used as a cleaning agent. It must not be siphoned by mouth.

Gas-oil (Diesel Fuel)

Combustible.

Prolonged skin contact with high boiling point gas oils (diesel fuel) may cause serious skin disorders including skin cancer.

Inhalation into the lungs will cause internal bleeding - **SEEK IMMEDIATE MEDICAL ATTENTION.**

If swallowed, DO NOT induce vomiting - SEEK IMMEDIATE MEDICAL ATTENTION.

Kerosene (Paraffin)

Used also as heating fuel, solvent and cleaning agent.

Flammable - OBSERVE NO SMOKING POLICY.

Irritation of the mouth and throat may result from swallowing. The main hazard from swallowing arises if liquid aspiration into the lungs occurs.

Liquid contact dries the skin and can cause irritation or dermatitis. Splashes in the eye may be slightly irritating.

In normal circumstances the low volatility does not give rise to harmful vapors. Exposure to mists and vapors from kerosene at elevated temperature should be avoided (mists may arise in dewaxing). Avoid skin and eye contact and make sure there is adequate ventilation.

If swallowed, DO NOT induce vomiting - SEEK IMMEDIATE MEDICAL ATTENTION.

Gas Cylinders

Gases such as oxygen, acetylene, argon and propane are normally stored in cylinders at pressures of up to 138 bar (13800 kPa) (2000 lbf/in^2) and great care should be taken in handling these cylinders to avoid mechanical damage to them or to the valve gear attached. The contents of each cylinder should be clearly identified by appropriate markings.

Cylinders should be stored in well ventilated enclosures, and protected from ice and snow or direct sunlight. Fuel gases, for example acetylene and propane should not be stored in close proximity to oxygen cylinders.

Care should be exercised to prevent leaks from gas cylinders and lines and also to avoid sources of ignition.

Only trained personnel should undertake work involving gas cylinders.

General Workshop Tools and Equipment

It is essential that all tools and equipment are maintained in good condition and the correct safety equipment is used where required.

Never use tools or equipment for any purpose other than that for which they were designed. Never overload equipment such as hoists, jacks, axle and chassis stands or lifting slings. Damage caused by overloading is not always immediately apparent and may result in a fatal failure the next time that the equipment is used.

Do not use damaged or defective tools or equipment, particularly high speed equipment such as grinding wheels. A damaged grinding wheel can disintegrate without warning and cause serious injury.

Wear suitable eye protection when using grinding, chiselling or sand blasting equipment.

Wear a suitable breathing mask when using abrasive blasting equipment or using spraying equipment.

Make sure there is adequate ventilation to control dusts, mists and fumes.

High Pressure Air, Lubrication and Oil Test Equipment

Always keep high pressure equipment in good condition, and regularly maintained, particularly at joints and unions.

Never direct a high pressure nozzle, for example diesel injector, at the skin as the fluid may penetrate to the underlying tissue and cause serious injury.

Jacking

Always refer to the Jacking and Lifting section of this manual prior to raising the vehicle off the ground.

When vehicle is to be raised by means of a jack, ensure that it is standing on level ground, that parking brake is applied and wheels are chocked. ALWAYS use the recommended jacking points and ensure that vehicle jack has sufficient load capacity for the weight of the vehicle.

WARNING: Do not work on or under a vehicle supported only by a jack. Always support the vehicle on safety stands.

Ensure that hoists have sufficient load capacity for the weight of the vehicle.

Legal Aspects

There are many laws and regulations relating to health and safety in the use and disposal of materials and equipment in a workshop.

For a safe working environment and to avoid environmental pollution, workshops should be familiar, in detail, with the many health and safety laws and regulations within their country, published by both national and local authorities.

Lubricants and Greases

Avoid all prolonged and repeated contact with mineral oils. All lubricants and greases may be irritating to the eyes and skin.

Used Engine Oil

Prolonged and repeated contact with engine oil will result in the removal of natural fats from the skin, leading to dryness, irritation and dermatitis. In addition, used engine oil contains potentially harmful contaminants which may cause skin cancer. Adequate means of skin protection and washing facilities must be provided.

Do not employ used engine oils as lubricants or for any application where appreciable skin contact is likely to occur.

Health Protection Precautions

1. Avoid prolonged and repeated contact with oils, particularly used engine oils.
2. Wear protective clothing, including impervious gloves where practicable.
3. Do not put oily rags into pockets.
4. Avoid contaminating clothes, particularly underpants, with oil.
5. Heavily soiled clothing and oil-impregnated footwear should not be worn. Overalls must be cleaned regularly.
6. First Aid treatment should be obtained immediately for open cuts and wounds.
7. Use barrier creams, applying them before each work period, to help the removal of oil from the skin.
8. Wash with soap and water to make sure all oil is removed (skin cleansers and nail brushes will help). Preparations containing lanoline replace the natural skin oils which have been removed.
9. Do not use gasoline (petrol), kerosene (paraffin), diesel fuel (gas oil), thinners or solvents for cleaning skin.
10. If skin disorders develop, obtain medical advice without delay.
11. Where practicable, degrease components prior to handling.
12. Where there is a risk of eye contact, eye protection should be worn, for example chemical goggles or face shields; in addition an eye wash facility should be provided.

Environmental Precautions

This section provides general information which can help to reduce the environmental impacts from the activities carried out in workshops.

Emissions to air

Many of the activities that are carried out in workshops emit gases and fumes which can contribute to global warming, depletion of the ozone layer and/or the formation of photochemical smog at ground level. By considering how the workshop activities are carried out, these gases and fumes can be minimised, thus reducing the impact on the environment.

Exhaust fumes

Running car engines is an essential part of workshop activities and exhaust fumes need to be ventilated to atmosphere. However, the amount of time engines are running and the position of the vehicle should be carefully considered at all times, to reduce the release of poisonous gases and minimise the inconvenience to people living nearby.

Solvents

Some of the cleaning agents used are solvent based and will evaporate rapidly to atmosphere if used carelessly, or if containers are left unsealed. All containers must be firmly closed when not required and solvent should be used sparingly. Wherever possible, solvents having a low toxicity and flammability should be selected. Always follow the instructions supplied by the solvent manufacturer. Similarly, many paints are solvent based and the spray should be used in such a way as to reduce emissions to a minimum.

Refrigerant

It is illegal to release any refrigerant into the atmosphere. Discharge and replacement of these materials from air conditioning units should only be carried out using the appropriate equipment.

Discharges to water

Most workshops will have two systems for discharging waste water - storm drains and foul drains. Storm drains should only receive clean water i.e. rainwater. Foul drains will accept many of the normal waste water i.e. washing water, detergents and domestic type waste BUT NOT oil, petrol, solvent, acids, hydraulic fluid, antifreeze and similar fluids. If in doubt, always consult the local authority or water company.

Spillages

Every precaution must be taken to prevent spillage of oil, fuel, solvents etc., reaching the drains. All handling of such materials must take place well away from drains and preferably in an area with a suitable containing wall to prevent discharge into drains or watercourses. If a spillage occurs, it must be soaked up immediately using a spill kit where provided.

Checklist

Spillage prevention:

- Store liquids in a secure area.
- Make sure that taps on liquid containers are secure and cannot be accidentally turned on.
- Protect bulk storage tanks from vandalism by locking the valves.
- Transfer liquids from one container to another in an area away from open drains.
- Ensure lids are replaced securely on containers.
- Have spill kits available near to points of storage and liquid handling areas.

Spill Kits

Special materials are available to absorb a number of different substances. They can be in granular form, ready to use and are supplied in suitable containers. Disposal of used spill absorbing material is dealt with in Waste management.

Land contamination

Oils, fuels and solvents etc. can contaminate any soil with which they come into contact. Such materials MUST never be disposed of by pouring on to soil and every precaution must be taken to avoid spillage reaching soil. Waste materials stored on open ground could either leak or have contaminating substances washed off them that would contaminate the land. Always store these materials in suitable skips or similarly robust containers.

Legal compliance

Some sites may have a discharge consent for effluent discharge to the foul drain for a car wash etc. It is essential to know the types of effluent which are allowed to be discharged into the drain and to check the results of any monitoring carried out by the Water Company.

Where paint spraying operations are carried out it may be necessary to apply to the Local Authority for an air emissions licence to operate the plant. If such a licence is necessary, additional precautions will be necessary to comply with the requirements and the results of any air quality monitoring must be checked regularly.

Checklist

Always adhere to the following:

- Know what legal consents and licences apply to the operations.
- Check that the emissions and discharges comply with legal requirements.

Waste Management

Pollution can be reduced by careful handling, storage and disposal of all waste materials that occur on sites. Legislation makes it illegal to dispose of waste materials other than to licensed waste carriers and disposal sites.

This means that it is necessary to not only know what the waste materials are but also to have the necessary documentation and licences.

Handling and storage of waste

Ensure that waste materials are not poured down the drain or on to soil and are stored in such a way that they do not escape on to land or soil.

All waste must be segregated into individual types e.g. oils, metals, batteries, scrap components etc. This will prevent any reaction between different materials and assist in disposal.

Disposal of waste

Dispose of waste in accordance with the following guidelines:

- **Fuel, hydraulic fluid, anti-freeze and oil:** Keep separate and dispose of to specialist contractors.
- **Refrigerant:** Collect in specialist equipment and reuse.
- **Detergents:** Safe to pour down the foul drain if diluted.
- **Paint, thinners:** Keep separate and dispose of to specialist contractor.
- **Components:** Return to supplier for refurbishment or disassemble and reuse any suitable parts. Dispose of remainder in ordinary waste.
- **Small parts:** Reuse any suitable parts, dispose of the remainder in ordinary waste.
- **Metals:** Can be sold if separate from general waste.
- **Tyres:** Keep separate and dispose of to specialist contractor. DO NOT attempt to dispose of tyres by burning.
- **Components/materials containing asbestos:** Keep separate and dispose of to specialist contractor.
- **Oil and fuel wastes (e.g. rags, used spill kit material):** Keep separate and dispose of to specialist contractors.
- **Air filters:** Keep separate and dispose of to specialist contractors.
- **Rubber/plastics:** Dispose of in ordinary waste.
- **Hoses:** Dispose of in ordinary waste.
- **Batteries:** Keep separate and dispose of to specialist contractors.
- **Air bags - DANGER EXPLOSIVES:** Keep separate and dispose of to specialist contractors.
- **Electrical components:** Return to supplier for refurbishment or disassemble and reuse any suitable components. Dispose of remainder in ordinary waste.
- **Catalytic converters:** May be sold if kept separate from general waste.
- **Packaging:** Compact/recycle as much as possible and dispose of in ordinary waste.
- **Office/paper waste:** Recycle paper and toner and ink cartridges, dispose of remainder in ordinary waste.

Noise

Car alarm testing, panel beating, running engines, using air tools etc. are operations which invariably produce a large amount of noise. The location of such activities and also the time of day must be carefully considered having regard to the proximity of houses schools etc.

Some operations may produce high noise levels which could, in time, damage hearing. In these cases, suitable ear protection must be worn.

Solder

Solders are mixtures of metals such that the melting point of the mixture is below that of the constituent metals (normally lead and tin). Solder application does not normally give rise to toxic lead fumes, provided a gas/air flame is used. Oxy-acetylene flames should not be used, as they are much hotter and will cause lead fumes to be produced.

Some fumes may be produced by the application of any flame to surfaces coated with grease, and inhalation of these should be avoided.

Removal of excess solder should be undertaken with care, to make sure that fine lead dust is not produced, which can give toxic effects if inhaled. Respiratory protection may be necessary.

Solder spillage and filings should be collected and removed promptly to prevent general air contamination by lead.

High standards of personal hygiene are necessary in order to avoid ingestion of lead or inhalation of solder dust from clothing.

Solvents

For example acetone, white spirit, toluene, xylene, trichloroethane.

Used in cleaning and dewaxing materials, paints, plastics, resins and thinners.

Some may be highly flammable or flammable.

Skin contact will degrease the skin and may result in irritation and dermatitis following repeated or prolonged contact. Some can be absorbed through the skin in toxic or harmful quantities.

Splashes in the eye may cause severe irritation and could lead to loss of vision.

Brief exposure of high concentrations of vapors or mists will cause eye and throat irritation, drowsiness, dizziness, headaches and, in the worst circumstances, unconsciousness.

Repeated or prolonged exposure to excessive but lower concentrations of vapors or mists, for which there might not be adequate warning indications, can cause more serious toxic or harmful effects.

Aspiration into the lungs, for example through vomiting, is the most serious consequence of swallowing.

Avoid splashes to the skin, eyes and clothing. Wear protective gloves, goggles and clothing if necessary.

Make sure there is good ventilation when in use, avoid breathing fumes, vapors and spray mists and keep containers tightly sealed. Do not use in confined spaces.

When spraying materials containing solvents, for example paints, adhesives, and metal coatings, use extraction ventilation or personal respiratory protection in the absence of adequate general ventilation.

Do not apply heat or flame except under specific and detailed manufacturers instructions.

Suspended Loads

CAUTION: Never improvise lifting tackle.

There is always a danger when loads are lifted or suspended. Never work under an unsupported, suspended or raised load, for example a suspended engine.

Always make sure that lifting equipment such as jacks, hoists, axle stands and slings are adequate and suitable for the job, in good condition and regularly maintained.

Viton

In common with many other manufacturers vehicles, some components installed to Land Rover vehicles have seals, 'O' rings or gaskets which contain a material known as 'Viton'.

Viton is a fluoroelastomer, that is a synthetic rubber type which contains Fluorine. Although Viton is the most well known fluoroelastomer, there are others, including Fluorel and Tecmoflon.

When used under design conditions fluoroelastomers are perfectly safe. If, however, they are exposed to temperatures in excess of 400°C, the material will not burn, but will decompose, and one of the products formed is hydrofluoric acid.

This acid is extremely corrosive and may be absorbed directly, through contact, into the general body system. **WHERE CASES OF SKIN CONTACT OCCUR, SEEK IMMEDIATE MEDICAL HELP.**

O-rings, seals or gaskets which have been exposed to very high temperatures will appear charred or as a black sticky substance.

DO NOT, under any circumstances touch them or the attached components.

Enquiries should be made to determine whether Viton or any other fluoroelastomer has been used in the affected O-ring, seal or gasket. If they are of natural rubber or nitrile there is no hazard. If in doubt, be cautious as the material may be Viton or any fluoroelastomer.

If Viton or any other fluoroelastomers have been used, the affected area should be decontaminated before the commencement of work.

Disposable heavy duty plastic gloves should be worn at all times, and the affected area washed down using wire wool and a limewater (calcium hydroxide) solution to neutralise the acid before disposing of the decomposed Viton residue and final cleaning of the area. After use, the plastic gloves should be discarded carefully and safely.

Welding

Welding processes include Resistance Welding (Spot Welding), Arc Welding and Metal Active Gas (MAG) Welding.

Resistance Welding

This process may cause particles of molten metal to be emitted at a high velocity, and the eyes and skin must be protected.

Arc Welding

This process emits a high level of ultra-violet radiation which may cause arc-eye and skin burns to the operator and to other persons nearby. Gas-shielded welding processes are particularly hazardous in this respect. Personal protection must be worn, and screens used to shield other people.

CONTACT LENS WEARERS ARE ADVISED TO REVERT TO ORDINARY SPECTACLES WHEN ARC WELDING as the arc spectrum is believed to emit microwaves which dry out the fluid between the lens and the eye. This may result in blindness when the lens is removed from the eye.

Metal spatter will also occur, and appropriate eye and skin protection is necessary.

The heat of the welding arc will produce fumes and gases from the metals being welded, the rods and from any applied coatings or contamination on the surfaces being worked on. These gases and fumes may be toxic and inhalation of these should be avoided. The use of extraction ventilation to remove the fumes from the working area may be necessary particularly in cases where the general ventilation is poor, or where considerable welding work is anticipated. In extreme cases or confined spaces where adequate ventilation cannot be provided, air-fed respirators may be necessary.

⚠ CAUTION: Some of the components installed to the vehicle e.g. the interior cross beam and under bonnet cross member are manufactured from magnesium alloy. On no account should any welding operations be attempted on these components.

Gas Welding (and Cutting)

SPECIAL PRECAUTIONS MUST BE TAKEN BEFORE ANY WELDING OR CUTTING TAKES PLACE ON VESSELS WHICH HAVE CONTAINED COMBUSTIBLE MATERIALS, FOR EXAMPLE BOILING OR STEAMING OUT OF FUEL TANKS.

Warning Symbols on Vehicles

Decals showing warning symbols will be found on various vehicle components.

These decals must not be removed. The warnings are for the attention of owners/operators and persons carrying out service or repair operations on the vehicle.

Special Tool Glossary

Description and Operation

Service Tools

Special service tools have been developed to facilitate removal, dismantling and assembly of mechanical components in a cost effective and time efficient manner. The use of such special tools also helps prevent the potential for damage to components.

Some operations described in this manual cannot be carried out properly without the aid of the relevant service tools.

All orders and enquiries should be sent direct to:

https://jlrequipment.service-solutions.com.

About This Manual

Description and Operation

Introduction

This manual has been written in a format that is designed to meet the needs of technicians worldwide. The objective is to use common formats and include similar content in each manual.

This manual provides general descriptions for accomplishing diagnosis and testing, service and repair work with tested and effective techniques. Following them will help to ensure reliability.

Important Safety Instructions

Appropriate service methods and correct repair procedures are essential for the safe, reliable operation of all motor vehicles as well as the personal safety of the individual carrying out the work.

Anyone who departs from the instructions provided in this manual must first establish that personal safety or vehicle integrity is not compromised by the choice of method, tools or components.

Warnings, Cautions and Notes in This Manual

⚠ WARNING: Warnings are used to indicate that failure to follow a procedure correctly may result in personal injury.

⚠ CAUTION: Cautions are used to indicate that failure to follow a procedure correctly may result in damage to the vehicle or equipment being used.

△ NOTE: Notes are used to provide additional essential information required to carry out a complete and satisfactory repair.

Generic warnings or cautions are in their relevant description and operation procedure within section 100-00. If the generic warnings or cautions are required for a procedure, there will be a referral to the appropriate description and operation procedure.

If a warning, caution or note only applies to one step, it is placed at the beginning of the specific step.

Trustmark Authoring Standards (TAS) Removal and Installation Procedures

△ NOTE: TAS style procedures can be identified by steps that have no accompanying step text and the magenta color of the electrical connectors and fasteners such as nuts, bolts, clamps or clips.

A TAS removal and installation procedure uses a sequence of color illustrations to indicate the order to be followed when removing/disassembling or installing/assembling a component.

Many of the TAS procedures will have the installation information within the removal steps. These procedures will have the following note at the beginning of the procedure:

△ NOTE: Removal steps in this procedure may contain installation details.

Items such as O-ring seals, gaskets, seals, self-locking nuts and bolts are to be discarded and new components installed unless otherwise stated within the procedure. Coated nuts or bolts are to be reused, unless damaged or otherwise stated within the procedure.

Specification procedures will contain all technical data that are not part of a repair procedure.

TAS Graphics*

Due to the nature of this publication we can only show these graphics in Grey - but generally they are as are as follows:

- Blue - Indicates the target item, item to be removed/installed or disassembled/assembled
- Green and Brown - Indicates a secondary item that needs to be detached, removed/installed or disassembled/assembled prior to the target item
- Yellow - Component that is touched or affected in a way but remains in the vehicle. It may be detached, attached, moved, modified, checked, adjusted etc.
- Magenta - Indicates electrical connectors and fasteners such as nuts, bolts, clamps or clips
- Pale Blue - is for the special tool(s) and general equipment.

There may be multiple steps assigned to one illustration.

Numbered pointers are used to indicate the number of electrical connectors and fasteners such as nuts, bolts, clamps or clips.

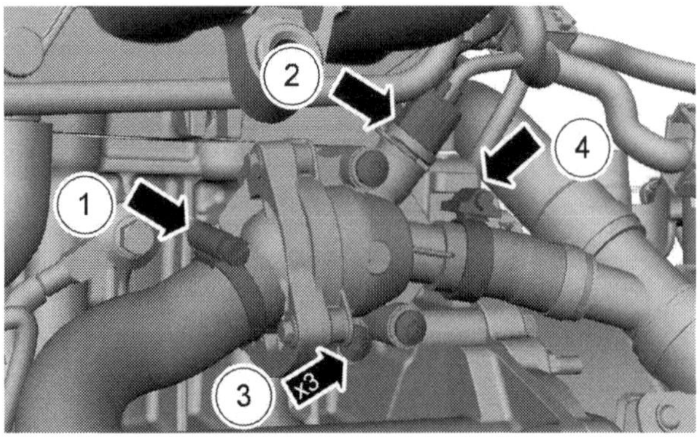

Items in the illustration can be transparent or use cutouts to show hidden detail(s).

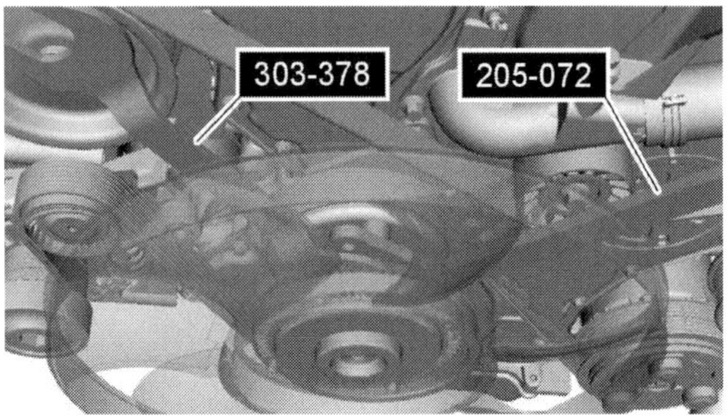

Warning symbols are used to indicate potential risks resulting from a certain component or area

Instruction symbols are used to apply sealer, lubricant, weight, tape or cleaning detergent to a component

Location symbols are used to show the location of a component or system within the vehicle.

TAS Symbols

Symbols are used inside the graphics and in the text area to enhance the information display. The following paragraphs describe the various types and categories of symbols.

Prohibition symbols advise on prohibited actions to either avoid damage or health and safety related risks.

Health and Safety symbols recommend the use of particular protection equipment to avoid or at least reduce the risk or severity of possible injuries.

Warning symbols are used to indicate potential risks resulting from a certain component or area

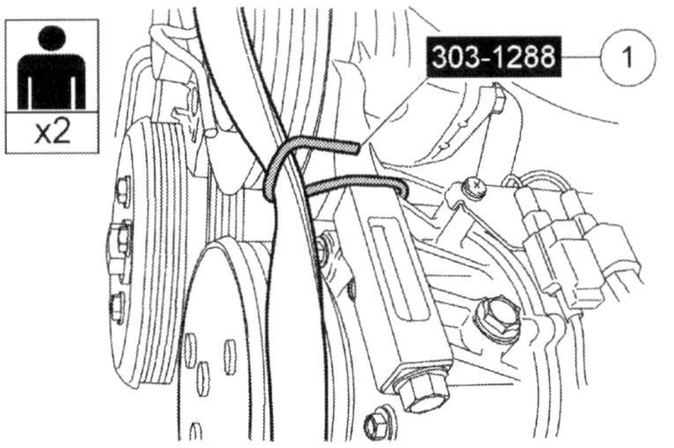

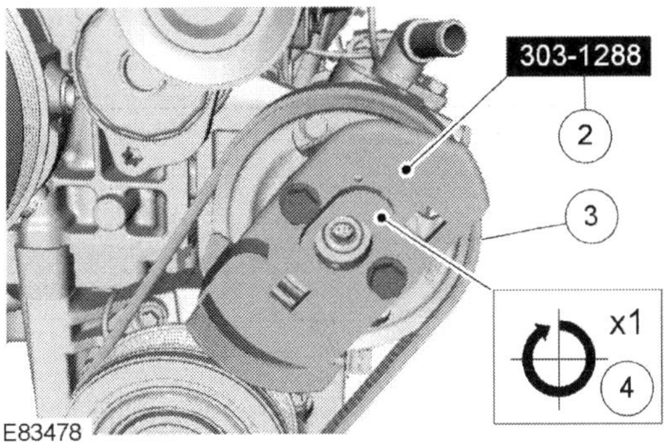

2.

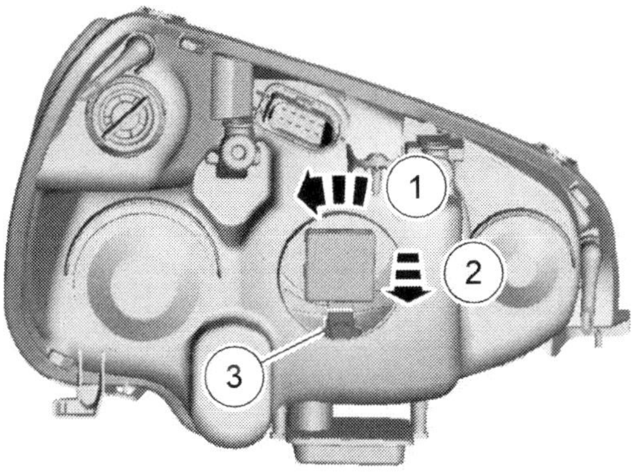

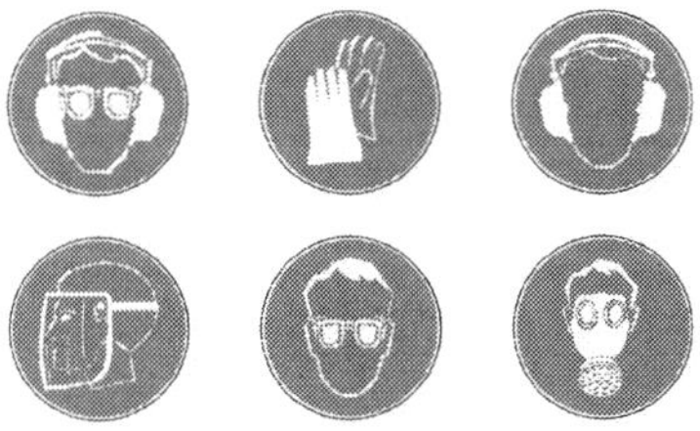

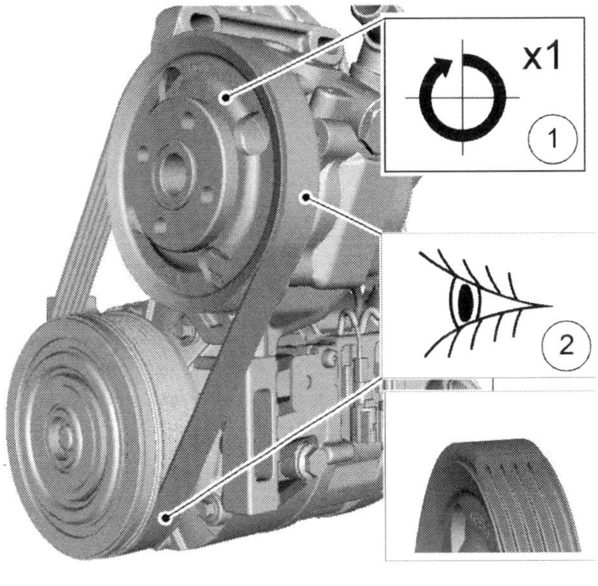

3.

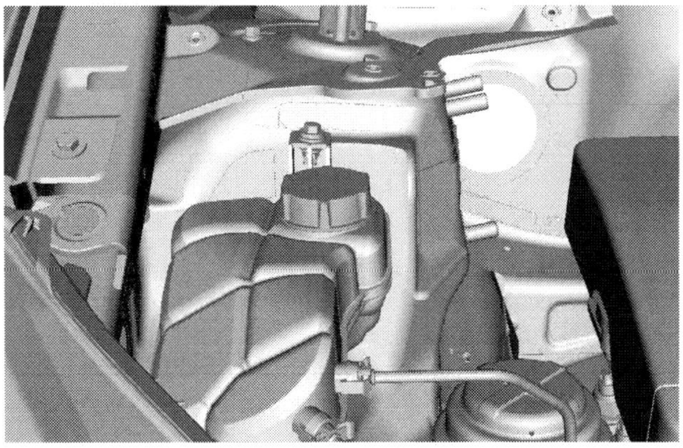

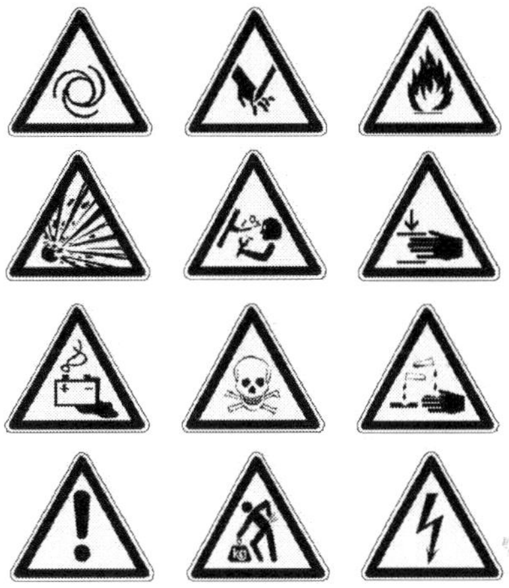

4.

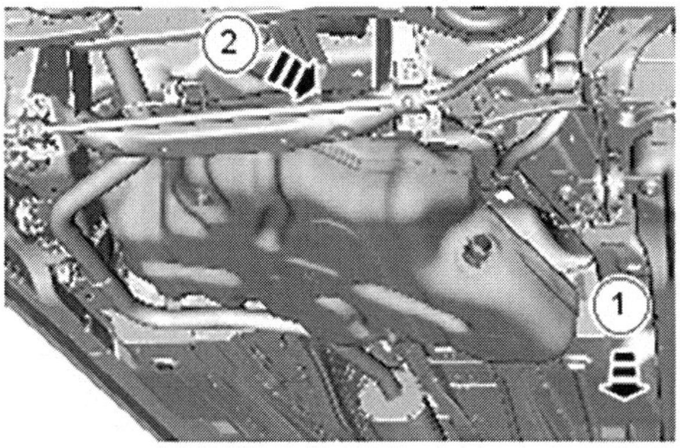

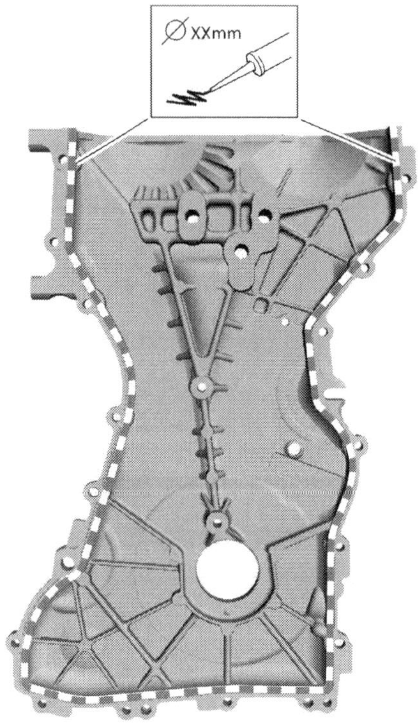

Gearshift lever or selector lever position symbols are used to show which gearshift lever or selector lever position is to be set.

Pointer symbols are used to draw the attention to components and give special instructions such as a required sequence or number of components. The number of components is reflected by the value inside the luty arrow. A sequence number is located inside the circle. Numbers inside circles are also used to allocate special information such as tightening torques or chemicals to a particular component.

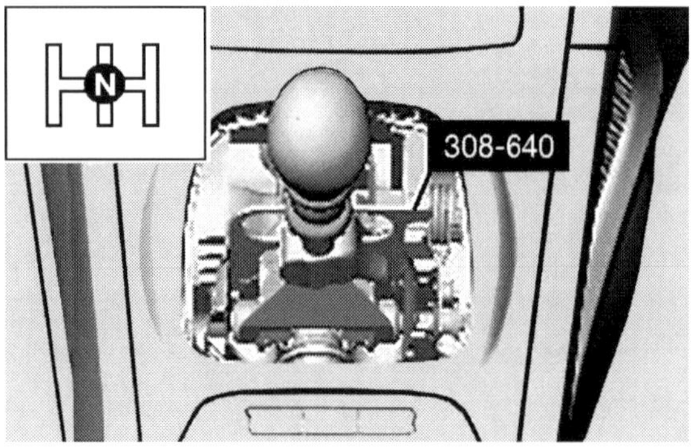

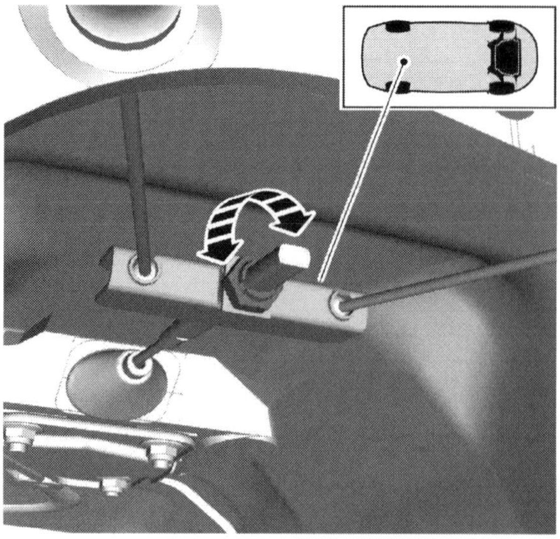

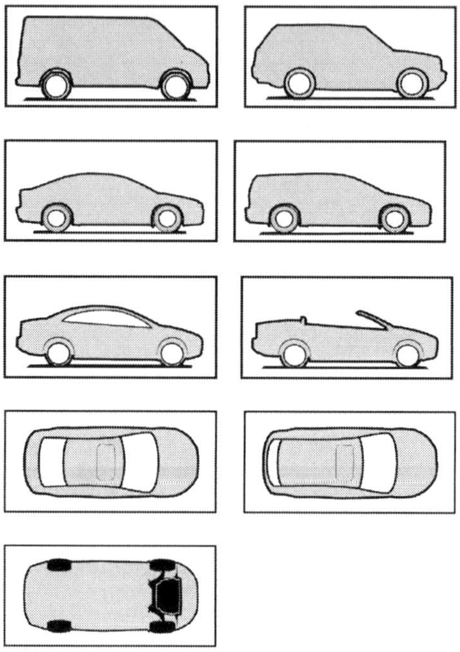

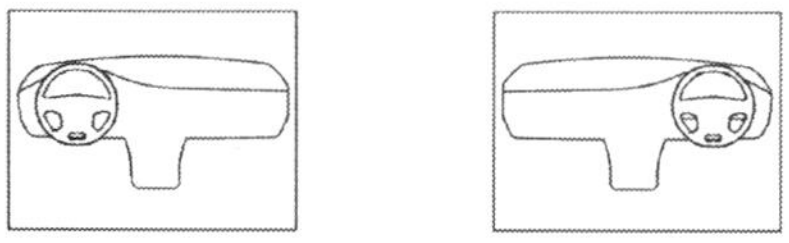

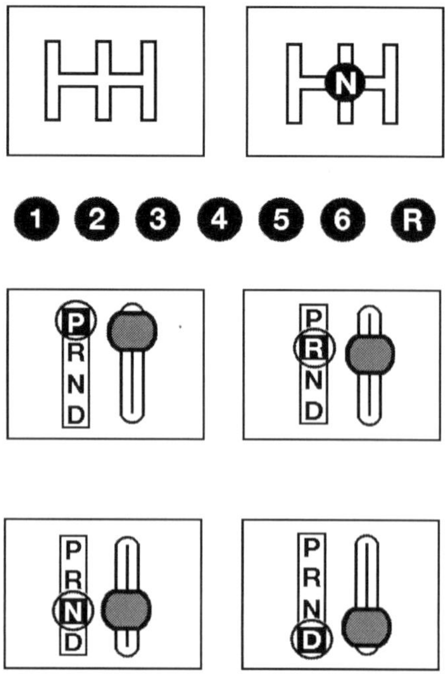

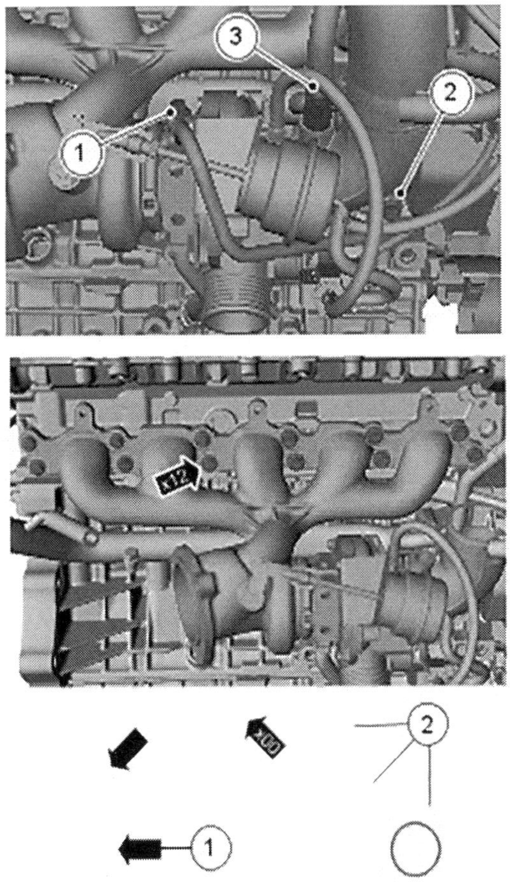

Movement arrows are used to show three dimensional or rotational movements. These movements can include specific values inside the symbol if required.

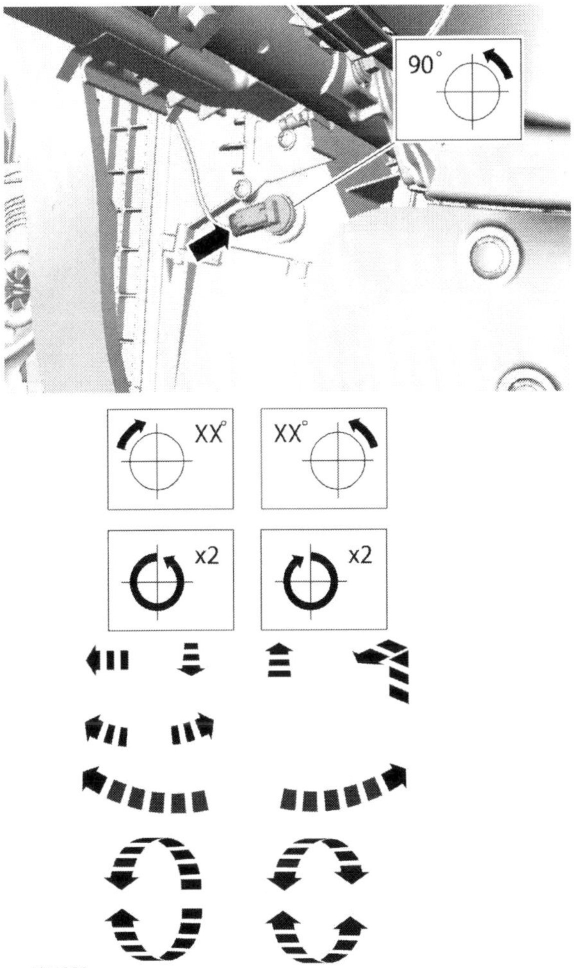

E84838

Standard tool symbols recommend the use of certain standard tools. These tools can include dimension values if required.

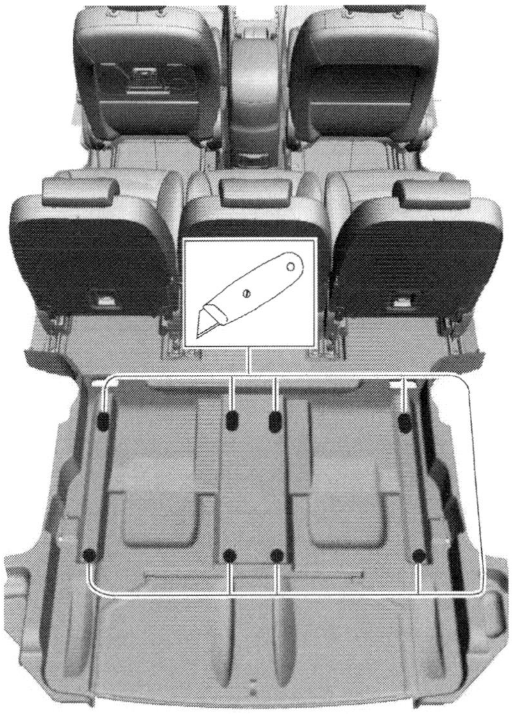

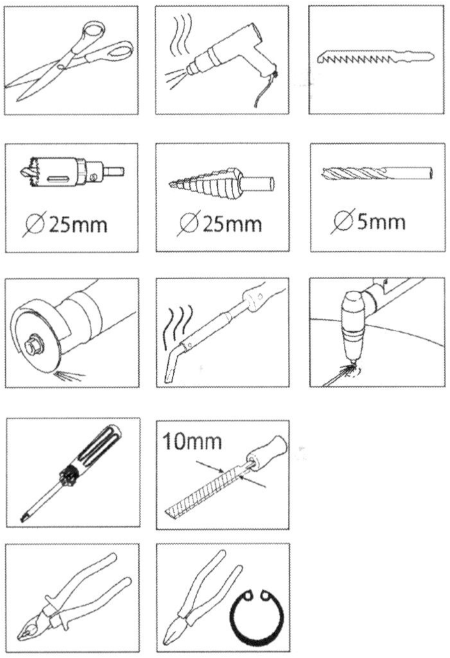

The following graphic illustrates a set of symbols that are used to provide detailed information on where to apply a material.

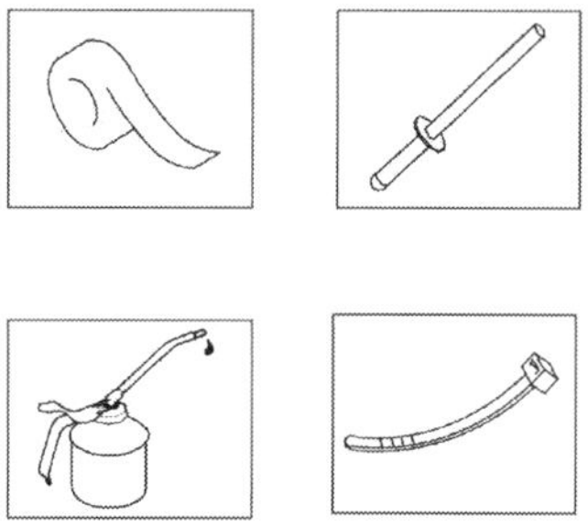

Measurement symbols provide detailed information on where to carry out a specific measurement. These symbols can include specific values if required.

Diagnostic Trouble Code Index - DTC: Engine Control Module (ECM)

Description and Operation

Engine Control Module (ECM)

⚠ CAUTION: Diagnosis by substitution from a donor vehicle is **NOT** acceptable. Substitution of control modules does not guarantee confirmation of a fault, and may also cause additional faults in the vehicle being tested and/or the donor vehicle.

NOTES:

△ Guided Diagnostics must be followed and the repair advised by the process completed. Failure to do so may result in the rejection of any warranty claim made.

△ If a control module or a component is suspect and the vehicle remains under manufacturer warranty, refer to the Warranty Policy and Procedures manual, or determine if any prior approval programme is in operation, prior to the installation of a new module/component.

△ Generic scan tools may not read the codes listed, or may read only 5-digit codes. Match the 5 digits from the scan tool to the first 5 digits of the 7-digit

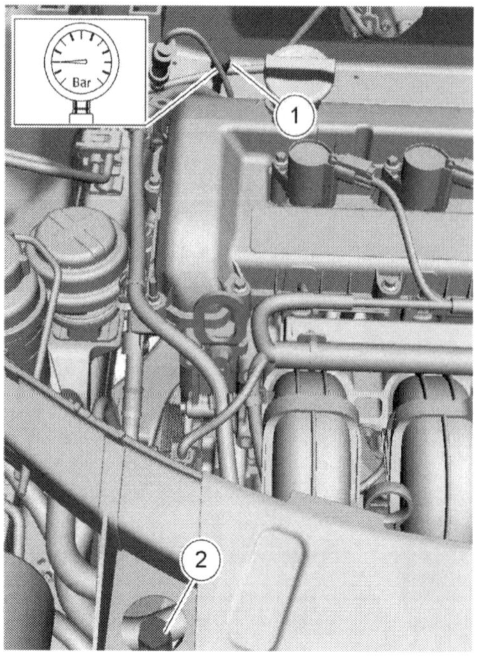

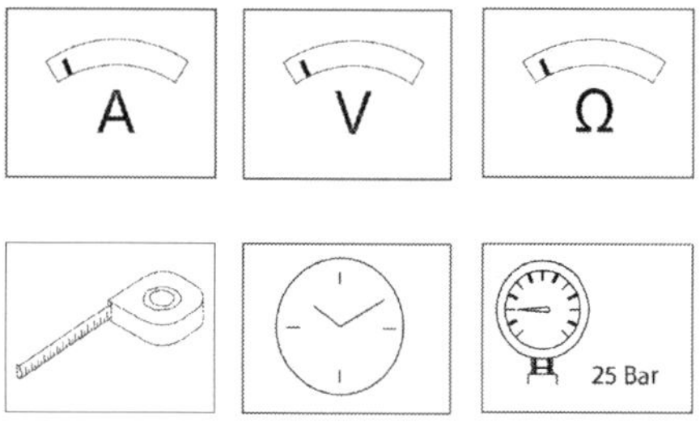

Special Tools and Torque Figure(s)

Special tools will be shown with the tool number in the illustration. The special tool number(s), general equipment, material(s) and torque figure(s) used for the procedure step will be shown in the text column.

code listed to identify the fault (the last 2 digits give extra information read by the manufacturer-approved diagnostic system).

△ When performing voltage or resistance tests, always use a digital multimeter accurate to three decimal places, and with an up-to-date calibration certificate. When testing resistance always take the resistance of the digital multimeter leads into account.

△ Check and rectify basic faults before beginning diagnostic routines involving pinpoint tests.

△ Inspect connectors for signs of water ingress, and pins for damage and/or corrosion.

△ If DTCs are recorded and, after performing the pinpoint tests, a fault is not present, an intermittent concern may be the cause. Always check for loose connections and corroded terminals.

The table following lists all Diagnostic Trouble Codes (DTCs) that could be logged in the Engine Control Module (ECM). For additional diagnosis and testing information, refer to the relevant Diagnosis and Testing section in the workshop manual.

Refer to: Electronic Engine Controls (303-14 Electronic Engine Controls, ID4 2.2L Diesel Diagnosis and Testing)

DTC	Description	Possible Causes	Action
B10A2-12	Crash Input - Circuit short to battery	NOTE: Circuit reference - INSW - • Inertia switch circuit short circuit to power	• Refer to the electrical circuit diagrams and check the inertia switch circuit for short circuit to power
B10A2-14	Crash Input - Circuit short to ground or open	NOTE: Circuit reference - INSW - • Inertia switch circuit short circuit to ground, open circuit, high resistance • Inertia switch triggered	• Refer to the electrical circuit diagrams and check the inertia switch circuit for short circuit to ground, open circuit, high resistance • Reset the inertia switch
B1206-00	Crash Occurred - No sub type information	NOTE: Circuit reference - INSW - • Inertia switch circuit short circuit to ground, short circuit to power, open circuit, high resistance • Inertia switch triggered	• Refer to the electrical circuit diagrams and check the inertia switch circuit for short circuit to ground, short circuit to power, open circuit, high resistance • Reset the inertia switch
P0030-11	HO2S Heater Control Circuit (Bank 1, Sensor 1) - Circuit short to ground	NOTE: Circuit reference - U HEATER - • Heated oxygen sensor heater circuit short circuit to ground	• Refer to the electrical circuit diagrams and check the heated oxygen sensor heater circuit for short circuit to ground
P0030-12	HO2S Heater Control Circuit (Bank 1, Sensor 1) - Circuit short to battery	NOTE: Circuit reference - U HEATER - • Heated oxygen sensor heater circuit short circuit to power	• Refer to the electrical circuit diagrams and check the heated oxygen sensor heater circuit for short circuit to power
P0030-13	HO2S Heater Control Circuit (Bank 1, Sensor 1) - Circuit open	NOTE: Circuit reference - U HEATER - • Heated oxygen sensor heater circuit open circuit, high resistance	• Refer to the electrical circuit diagrams and check the heated oxygen sensor heater circuit for open circuit, high resistance
P0045-13	Turbocharger/Supercharger Boost Control A Circuit/Open - Circuit open	NOTE: Circuit reference - VNT+ - • Variable geometry turbine vane actuator circuit open circuit, high resistance	• Refer to the electrical circuit diagrams and check the variable geometry turbine vane actuator circuit for open circuit, high resistance. Repair the circuit as necessary. Using the manufacturer approved diagnostic system, clear the DTCs and perform routine - Turbo, EGR and Air Path Dynamic Test
P0046-00	Turbocharger/Supercharger Boost Control A Circuit Range/Performance - No sub type information	NOTE: Circuit reference - VNT+ - • Variable geometry turbine vane actuator circuit short circuit to ground, open circuit, high resistance	• Refer to the electrical circuit diagrams and check the variable geometry turbine vane actuator circuit for short circuit to ground, open circuit, high resistance. Repair the circuit as necessary. Using the manufacturer approved diagnostic system, clear the DTCs and perform routine - Turbo, EGR and Air Path Dynamic Test
P0046-19	Turbocharger/Supercharger Boost Control A Circuit Range/Performance - Circuit current above threshold	NOTE: Circuit reference - VNT+ - • Variable geometry turbine vane actuator circuit short circuit to ground, short circuit to power	• Refer to the electrical circuit diagrams and check the variable geometry turbine vane actuator circuit for short circuit to ground, short circuit to power. Repair the circuit as necessary. Using the manufacturer approved diagnostic system, clear the DTCs and perform routine - Turbo, EGR and Air Path Dynamic Test
P0046-4B	Turbocharger/Supercharger Boost Control A Circuit Range/Performance - Over temperature	NOTE: Circuit reference - VNT+ - • Variable geometry turbine vane actuator circuit short circuit to ground, short circuit to power, open circuit, high resistance	• Refer to the electrical circuit diagrams and check the variable geometry turbine vane actuator circuit for short circuit to ground, short circuit to power, open circuit, high resistance. Repair the circuit as necessary. Using the manufacturer approved diagnostic system, clear the DTCs and perform routine - Turbo, EGR and Air Path Dynamic Test
P0047-00	Turbocharger/Supercharger Boost Control A Circuit Low - No sub type information	NOTE: Circuit reference - VNT+ - • Variable geometry turbine vane actuator circuit short circuit to ground, short circuit to power, open circuit, high resistance	• Refer to the electrical circuit diagrams and check the variable geometry turbine vane actuator circuit for short circuit to ground, short circuit to power, open circuit, high resistance. Repair the circuit as necessary. Using the manufacturer approved diagnostic system, clear the DTCs and perform routine - Turbo, EGR and Air Path Dynamic Test

P0047-11	Turbocharger/Supercharger Boost Control A Circuit Low - Circuit short to ground	NOTE: Circuit reference - VNT+ - • Variable geometry turbine vane actuator circuit short circuit to ground	• Refer to the electrical circuit diagrams and check the variable geometry turbine vane actuator circuit for short circuit to ground. Repair the circuit as necessary. Using the manufacturer approved diagnostic system, clear the DTCs and perform routine - Turbo, EGR and Air Path Dynamic Test
P0048-00	Turbocharger/Supercharger Boost Control A Circuit High - No sub type information	NOTE: Circuit reference - VNT+ - • Variable geometry turbine vane actuator circuit short circuit to ground, short circuit to power, open circuit, high resistance	• Refer to the electrical circuit diagrams and check the variable geometry turbine vane actuator circuit for short circuit to ground, short circuit to power, open circuit, high resistance. Repair the circuit as necessary. Using the manufacturer approved diagnostic system, clear the DTCs and perform routine - Turbo, EGR and Air Path Dynamic Test
P0048-12	Turbocharger/Supercharger Boost Control A Circuit High - Circuit short to battery	NOTE: Circuit reference - VNT+ - • Variable geometry turbine vane actuator circuit short circuit to power	• Refer to the electrical circuit diagrams and check the variable geometry turbine vane actuator circuit for short circuit to power. Repair the circuit as necessary. Using the manufacturer approved diagnostic system, clear the DTCs and perform routine - Turbo, EGR and Air Path Dynamic Test
P0069-92	MAP - Barometric Pressure Correlation - Performance or incorrect operation	• Manifold absolute pressure and temperature sensor circuit short circuit to ground, short circuit to power, open circuit, high resistance	• Refer to the electrical circuit diagrams and check the manifold absolute pressure and temperature sensor circuit for short circuit to ground, short circuit to power, open circuit, high resistance. Repair the circuit as necessary. Using the manufacturer approved diagnostic system, clear the DTCs and perform routine - Turbo, EGR and Air Path Dynamic Test
P007C-11	Charge Air Cooler Temperature Sensor Circuit Low (Bank 1) - Circuit short to ground	NOTE: Circuit reference - ACT - • Charge air temperature sensor circuit short circuit to ground	• Refer to the electrical circuit diagrams and check the charge air temperature sensor circuit for short circuit to ground
P007D-15	Charge Air Cooler Temperature Sensor Circuit High (Bank 1) - Circuit short to battery or open	NOTE: Circuit reference - ACT - • Charge air temperature sensor circuit short circuit to power, open circuit, high resistance	• Refer to the electrical circuit diagrams and check the charge air temperature sensor circuit for short circuit to power, open circuit, high resistance
P007E-27	Charge Air Cooler Temperature Sensor Circuit Intermittent/Erratic (Bank 1) - Signal rate of change above threshold	NOTE: Circuit reference - ACT - • Charge air temperature sensor circuit short circuit to ground, short circuit to power, open circuit, high resistance	• Refer to the electrical circuit diagrams and check the charge air temperature sensor circuit for short circuit to ground, short circuit to power, open circuit, high resistance
P0088-00	Fuel Rail/System Pressure - Too High - No sub type information	• Fuel injector fault • Fuel pipes restricted	• Using the manufacturer approved diagnostic system, check for fuel injector related DTCs and perform the relevant corrective actions • Check the fuel pipes for any signs of damage, crimping or restrictions
P0088-94	Fuel Rail/System Pressure - Too High - Unexpected operation	• Fuel injector fault • Fuel pipes restricted	• Using the manufacturer approved diagnostic system, check for fuel injector related DTCs and perform the relevant corrective actions • Check the fuel pipes for any signs of damage, crimping or restrictions
P0088-9A	Fuel Rail/System Pressure - Too High - Component or system operating conditions	• Fuel injector fault • Fuel pipes restricted	• Using the manufacturer approved diagnostic system, check for fuel injector related DTCs and perform the relevant corrective actions • Check the fuel pipes for any signs of damage, crimping or restrictions
P00BC-00	Mass or Volume Air Flow A Circuit Range/Performance - Air Flow Too Low - No sub type information	NOTE: Circuit reference - FMAF - • Intake system blocked/leaking • Mass air flow and temperature sensor internal failure	• Check the intake system for blockages/leaks. Using the manufacturer approved diagnostic system, clear the DTCs and perform routine - Turbo, EGR and Air Path Dynamic Test • Install a new mass air flow and temperature sensor as necessary
P00BD-00	Mass or Volume Air Flow A Circuit Range/Performance - Air Flow Too High - No sub type information	NOTE: Circuit reference - FMAF - • Intake or exhaust gas recirculation system fault • Mass air flow and temperature sensor internal failure	• Using the manufacturer approved diagnostic system, clear the DTCs and perform routine - Turbo, EGR and Air Path Dynamic Test • Install a new mass air flow and temperature sensor as necessary

P0100-36	Mass or Volume Air Flow Sensor "A" Circuit - Signal frequency too low	⚠ NOTE: Circuit reference - FMAF - • Intake system blocked/leaking • Mass air flow and temperature sensor circuit short circuit to ground, short circuit to power, open circuit, high resistance • Mass air flow and temperature sensor internal failure	• Check the intake system for blockages/leaks. Using the manufacturer approved diagnostic system, clear the DTCs and perform routine - Turbo, EGR and Air Path Dynamic Test • Refer to the electrical circuit diagrams and check the mass air flow and temperature sensor circuit for short circuit to ground, short circuit to power, open circuit, high resistance • Install a new mass air flow and temperature sensor as necessary
P0101-16	Mass or Volume Air Flow Sensor "A" Circuit Range/Performance - Circuit voltage below threshold	⚠ NOTE: Circuit reference - FMAF - • Intake system blocked/leaking • Mass air flow and temperature sensor circuit short circuit to ground, open circuit, high resistance • Mass air flow and temperature sensor internal failure	• Check the intake system for blockages/leaks. Using the manufacturer approved diagnostic system, clear the DTCs and perform routine - Turbo, EGR and Air Path Dynamic Test • Refer to the electrical circuit diagrams and check the mass air flow and temperature sensor circuit for short circuit to ground, open circuit, high resistance • Install a new mass air flow and temperature sensor as necessary
P0101-17	Mass or Volume Air Flow Sensor "A" Circuit Range/Performance - Circuit voltage above threshold	⚠ NOTE: Circuit reference - FMAF - • Intake system blocked/leaking • Mass air flow and temperature sensor circuit short circuit to power • Mass air flow and temperature sensor internal failure	• Check the intake system for blockages/leaks. Using the manufacturer approved diagnostic system, clear the DTCs and perform routine - Turbo, EGR and Air Path Dynamic Test • Refer to the electrical circuit diagrams and check the mass air flow and temperature sensor circuit for short circuit to power • Install a new mass air flow and temperature sensor as necessary
P0104-38	Mass or Volume Air Flow Sensor "A" Circuit Intermittent - Signal frequency incorrect	⚠ NOTE: Circuit reference - FMAF - • Intake system blocked/leaking • Mass air flow and temperature sensor circuit short circuit to ground, short circuit to power, open circuit, high resistance • Mass air flow and temperature sensor internal failure	• Check the intake system for blockages/leaks. Using the manufacturer approved diagnostic system, clear the DTCs and perform routine - Turbo, EGR and Air Path Dynamic Test • Refer to the electrical circuit diagrams and check the mass air flow and temperature sensor circuit for short circuit to ground, short circuit to power, open circuit, high resistance • Install a new mass air flow and temperature sensor as necessary
P0107-16	Manifold Absolute Pressure/Barometric Pressure Sensor Circuit Low - Circuit voltage below threshold	⚠ NOTE: Circuit reference - MAP - • Manifold absolute pressure and temperature sensor circuit short circuit to ground, open circuit, high resistance	• Refer to the electrical circuit diagrams and check the manifold absolute pressure and temperature sensor circuit for short circuit to ground, open circuit, high resistance. Repair the circuit as necessary. Using the manufacturer approved diagnostic system, clear the DTCs and perform routine - Turbo, EGR and Air Path Dynamic Test
P0108-17	Manifold Absolute Pressure/Barometric Pressure Sensor Circuit High - Circuit voltage above threshold	⚠ NOTE: Circuit reference - MAP - • Mass air flow and temperature sensor aperture blocked • Manifold absolute pressure and temperature sensor circuit short circuit to power	• Check the integrity of the mass air flow and temperature sensor • Refer to the electrical circuit diagrams and check the manifold absolute pressure and temperature sensor circuit for short circuit to power. Repair the circuit as necessary. Using the manufacturer approved diagnostic system, clear the DTCs and perform routine - Turbo, EGR and Air Path Dynamic Test
P0112-11	Intake Air Temperature Sensor 1 Circuit Low (Bank 1) - Circuit short to ground	⚠ NOTE: Circuit reference - IAT - • Mass air flow and temperature sensor temperature signal circuit short circuit to ground	• Refer to the electrical circuit diagrams and check the mass air flow and temperature sensor temperature signal circuit for short circuit to ground
P0113-15	Intake Air Temperature Sensor 1 Circuit High (Bank 1) - Circuit short to battery or open	⚠ NOTE: Circuit reference - IAT - • Mass air flow and temperature sensor temperature signal circuit short circuit to power, open circuit, high resistance	• Refer to the electrical circuit diagrams and check the mass air flow and temperature sensor temperature signal circuit for short circuit to power, open circuit, high resistance
P0113-17	Intake Air Temperature Sensor 1 Circuit High (Bank 1) - Circuit voltage above threshold	⚠ NOTE: Circuit reference - IAT - • Mass air flow and temperature sensor temperature signal circuit short circuit to power	• Refer to the electrical circuit diagrams and check the mass air flow and temperature sensor temperature signal circuit for short circuit to power
P0114-27	Intake Air Temperature Sensor 1 Intermittent/Erratic (Bank 1) - Signal rate of change above threshold	⚠ NOTE: Circuit reference - IAT - • Mass air flow and temperature sensor temperature signal circuit short circuit to power, open circuit, high resistance	• Refer to the electrical circuit diagrams and check the mass air flow and temperature sensor temperature signal circuit for short circuit to power, open circuit, high resistance

P0116-26	Engine Coolant Temperature Sensor 1 Circuit Range/Performance - Signal rate of change below threshold	NOTE: Circuit reference - ECT - • Cooling system fault • Engine coolant temperature sensor signal circuit short circuit to ground, short circuit to power, open circuit, high resistance	• Refer to the relevant section of the workshop manual and test the cooling system • Refer to the electrical circuit diagrams and check the engine coolant temperature sensor signal circuit for short circuit to ground, short circuit to power, open circuit, high resistance
P0117-11	Engine Coolant Temperature Sensor 1 Circuit Low - Circuit short to ground	NOTE: Circuit reference - ECT - • Engine coolant temperature sensor signal circuit short circuit to ground	• Refer to the electrical circuit diagrams and check the engine coolant temperature sensor signal circuit for short circuit to ground
P0118-15	Engine Coolant Temperature Sensor 1 Circuit High - Circuit short to battery or open	NOTE: Circuit reference - ECT - • Engine coolant temperature sensor signal circuit short circuit to power, open circuit, high resistance	• Refer to the electrical circuit diagrams and check the engine coolant temperature sensor signal circuit for short circuit to power, open circuit, high resistance
P0130-00	O2 Sensor Circuit (Bank 1 Sensor 1) - No sub type information	NOTE: Circuit reference - UEGO RCD - • Heated oxygen sensor signal circuit short circuit to ground, short circuit to power, open circuit, high resistance	• Refer to the electrical circuit diagrams and check the heated oxygen sensor signal circuit for short circuit to ground, short circuit to power, open circuit, high resistance
P0130-11	O2 Sensor Circuit (Bank 1 Sensor 1) - Circuit short to ground	NOTE: Circuit reference - UEGO VIP - • Heated oxygen sensor signal circuit short circuit to ground	• Refer to the electrical circuit diagrams and check the heated oxygen sensor signal circuit for short circuit to ground
P0130-12	O2 Sensor Circuit (Bank 1 Sensor 1) - Circuit short to battery	NOTE: Circuit reference - UEGO VIP - • Heated oxygen sensor signal circuit short circuit to power	• Refer to the electrical circuit diagrams and check the heated oxygen sensor signal circuit for short circuit to power
P0130-28	O2 Sensor Circuit (Bank 1 Sensor 1) - Signal bias level out of range / zero adjustment failure	• Heated oxygen sensor signal circuit short circuit to ground, short circuit to power, open circuit, high resistance	NOTE: If additional intake/exhaust system related DTCs are also set, perform the relevant corrective action(s) first. • Refer to the electrical circuit diagrams and check the heated oxygen sensor signal circuit for short circuit to ground, short circuit to power, open circuit, high resistance

P014C-00	O2 Sensor Slow Response - Rich to Lean (Bank 1 Sensor 1) - No sub type information	• Exhaust leakage between the heated oxygen sensor and the catalytic converter • Heated oxygen sensor tip damaged • Heated oxygen sensor tip blocked • Heated oxygen sensor tip contaminated by excessive oil consumption • Heated oxygen sensor tip contaminated by poor or incorrect fuel	NOTE: If additional intake/exhaust system related DTCs are also set, perform the relevant corrective action(s) first. • Check the installation of the heated oxygen sensor • Check the heated oxygen sensor tip for damage • Check the heated oxygen sensor tip for blockage • Check the heated oxygen sensor tip for oil contamination. Carry out an oil consumption check • Check the heated oxygen sensor tip for contamination by poor or incorrect fuel. Check that the customer is using the correct grade of fuel
P0181-26	Fuel Temperature Sensor A Circuit Range/Performance - Signal rate of change above threshold	NOTE: Circuit reference - FTS - • Fuel temperature sensor signal circuit short circuit to ground, short circuit to power, open circuit, high resistance	NOTE: If additional fuel system related DTCs are also set, perform the relevant corrective action(s) first. • Refer to the electrical circuit diagrams and check the fuel temperature sensor signal circuit for short circuit to ground, short circuit to power, open circuit, high resistance
P0182-11	Fuel Temperature Sensor A Circuit Low - Circuit short to ground	NOTE: Circuit reference - FTS - • Fuel temperature sensor signal circuit short circuit to ground	NOTE: If additional fuel system related DTCs are also set, perform the relevant corrective action(s) first. • Refer to the electrical circuit diagrams and check the fuel temperature sensor signal circuit for short circuit to ground
P0183-15	Fuel Temperature Sensor A Circuit High - Circuit short to battery or open	NOTE: Circuit reference - FTS - • Fuel temperature sensor signal circuit short circuit to power, open circuit, high resistance	NOTE: If additional fuel system related DTCs are also set, perform the relevant corrective action(s) first. • Refer to the electrical circuit diagrams and check the fuel temperature sensor signal circuit for short circuit to power, open circuit, high resistance

P0184-27	Fuel Temperature Sensor A Circuit Intermittent - Signal rate of change above threshold	NOTE: Circuit reference - FTS - • Fuel temperature sensor signal circuit short circuit to ground, short circuit to power, open circuit, high resistance	NOTE: If additional fuel system related DTCs are also set, perform the relevant corrective action(s) first. • Refer to the electrical circuit diagrams and check the fuel temperature sensor signal circuit for short circuit to ground, short circuit to power, open circuit, high resistance
P0191-23	Fuel Rail Pressure Sensor A Circuit Range/Performance - Signal stuck low	• Fuel system fault • Fuel rail pressure sensor circuit short circuit to ground • Fuel system restricted/leaking • Fuel rail pressure senor internal failure	• Using the manufacturer approved diagnostic system, check for fuel system related DTCs and perform the relevant corrective actions • Refer to the electrical circuit diagrams and check the fuel rail pressure sensor circuit for short circuit to ground • Check the fuel system for restrictions and leaks • Install a new fuel rail pressure sensor as necessary
P0191-24	Fuel Rail Pressure Sensor A Circuit Range/Performance - Signal stuck high	• Fuel system fault • Fuel rail pressure sensor circuit short circuit to power • Fuel rail pressure senor internal failure	• Using the manufacturer approved diagnostic system, check for fuel system related DTCs and perform the relevant corrective actions • Refer to the electrical circuit diagrams and check the fuel rail pressure sensor circuit for short circuit to power • Install a new fuel rail pressure sensor as necessary
P0192-16	Fuel Rail Pressure Sensor A Circuit Low - Circuit voltage below threshold	• Fuel system fault • Fuel rail pressure sensor circuit short circuit to ground, open circuit, high resistance • Fuel rail pressure senor internal failure	• Using the manufacturer approved diagnostic system, check for fuel system related DTCs and perform the relevant corrective actions • Refer to the electrical circuit diagrams and check the fuel rail pressure sensor circuit for short circuit to ground, open circuit, high resistance • Install a new fuel rail pressure sensor as necessary
P0193-17	Fuel Rail Pressure Sensor A Circuit High - Circuit voltage above threshold	• Fuel system fault • Fuel rail pressure sensor circuit short circuit to power, open circuit, high resistance • Fuel rail pressure senor internal failure	• Using the manufacturer approved diagnostic system, check for fuel system related DTCs and perform the relevant corrective actions • Refer to the electrical circuit diagrams and check the fuel rail pressure sensor circuit for short circuit to power, open circuit, high resistance • Install a new fuel rail pressure sensor as necessary
P0194-27	Fuel Rail Pressure Sensor A Circuit Intermittent/Erratic - Signal rate of change above threshold	• Fuel system fault • Fuel rail pressure sensor circuit short circuit to ground, short circuit to power, open circuit, high resistance • Fuel rail pressure senor internal failure	• Using the manufacturer approved diagnostic system, check for fuel system related DTCs and perform the relevant corrective actions • Refer to the electrical circuit diagrams and check the fuel rail pressure sensor circuit for short circuit to ground, short circuit to power, open circuit, high resistance • Install a new fuel rail pressure sensor as necessary
P0196-26	Engine Oil Temperature Sensor Range/Performance - Signal rate of change below threshold	NOTE: Circuit reference - OTS - • Engine oil temperature sensor signal circuit short circuit to ground, short circuit to power, open circuit, high resistance	• Refer to the electrical circuit diagrams and check the engine oil temperature sensor signal circuit for short circuit to ground, short circuit to power, open circuit, high resistance
P0197-11	Engine Oil Temperature Sensor Circuit Low - Circuit short to ground	NOTE: Circuit reference - OTS - • Engine oil temperature sensor signal circuit short circuit to ground	• Refer to the electrical circuit diagrams and check the engine oil temperature sensor signal circuit for short circuit to ground
P0198-15	Engine Oil Temperature Sensor Circuit High - Circuit short to battery or open	NOTE: Circuit reference - OTS - • Engine oil temperature sensor signal circuit short circuit to power, open circuit, high resistance	• Refer to the electrical circuit diagrams and check the engine oil temperature sensor signal circuit for short circuit to power, open circuit, high resistance
P0199-27	Engine Oil Temperature Sensor Circuit Intermittent/Erratic - Signal rate of change above threshold	NOTE: Circuit reference - OTS - • Engine oil temperature sensor signal circuit short circuit to ground, short circuit to power, open circuit, high resistance	• Refer to the electrical circuit diagrams and check the engine oil temperature sensor signal circuit for short circuit to ground, short circuit to power, open circuit, high resistance
P0200-00	Injector Circuit - No sub type information	• Fuel injector circuit short circuit to ground, short circuit to power, open circuit, high resistance • Engine control module internal failure	• Refer to the electrical circuit diagrams and check the fuel injector circuits for short circuit to ground, short circuit to power, open circuit, high resistance • Using the manufacturer approved diagnostic system, clear the DTCs and retest. If the fault persists, install a new engine control module

P0200-01	Injector Circuit - General electrical failure	• Fuel injector circuit short circuit to ground, short circuit to power, open circuit, high resistance • Engine control module internal failure	• Refer to the electrical circuit diagrams and check the fuel injector circuits for short circuit to ground, short circuit to power, open circuit, high resistance • Using the manufacturer approved diagnostic system, clear the DTCs and retest. If the fault persists, install a new engine control module
P0200-11	Injector Circuit - Circuit short to ground	• Fuel injector circuit short circuit to ground, open circuit, high resistance • Engine control module internal failure	• Refer to the electrical circuit diagrams and check the fuel injector circuits for short circuit to ground, open circuit, high resistance • Using the manufacturer approved diagnostic system, clear the DTCs and retest. If the fault persists, install a new engine control module
P0201-01	Cylinder 1 Injector Circuit / Open - General electrical failure	⚠ NOTE: Circuit reference - INJ A L / INJ A H - • Fuel injector 1 circuit short circuit to ground, short circuit to power, open circuit, high resistance • Fuel injector 1 internal failure	• Refer to the electrical circuit diagrams and check the fuel injector 1 circuit for short circuit to ground, short circuit to power, open circuit, high resistance • Using the manufacturer approved diagnostic system, clear the DTCs and retest. If the fault persists, install a new fuel injector 1
P0201-13	Cylinder 1 Injector Circuit / Open - Circuit open	⚠ NOTE: Circuit reference - INJ A L / INJ A H - • Fuel injector 1 circuit open circuit, high resistance • Fuel injector 1 internal failure	• Refer to the electrical circuit diagrams and check the fuel injector 1 circuit for open circuit, high resistance • Using the manufacturer approved diagnostic system, clear the DTCs and retest. If the fault persists, install a new fuel injector 1
P0202-01	Cylinder 2 Injector Circuit / Open - General electrical failure	⚠ NOTE: Circuit reference - INJ D L / INJ D H - • Fuel injector 2 circuit short circuit to ground, short circuit to power, open circuit, high resistance • Fuel injector 2 internal failure	• Refer to the electrical circuit diagrams and check the fuel injector 2 circuit for short circuit to ground, short circuit to power, open circuit, high resistance • Using the manufacturer approved diagnostic system, clear the DTCs and retest. If the fault persists, install a new fuel injector 2
P0202-13	Cylinder 2 Injector Circuit / Open - Circuit open	⚠ NOTE: Circuit reference - INJ D L / INJ D H - • Fuel injector 2 circuit open circuit, high resistance • Fuel injector 2 internal failure	• Refer to the electrical circuit diagrams and check the fuel injector 2 circuit for open circuit, high resistance • Using the manufacturer approved diagnostic system, clear the DTCs and retest. If the fault persists, install a new fuel injector 2

P0203-01	Cylinder 3 Injector Circuit / Open - General electrical failure	⚠ NOTE: Circuit reference - INJ B L / INJ B H- • Fuel injector 3 circuit short circuit to ground, short circuit to power, open circuit, high resistance • Fuel injector 3 internal failure	• Refer to the electrical circuit diagrams and check the fuel injector 3 circuit for short circuit to ground, short circuit to power, open circuit, high resistance • Using the manufacturer approved diagnostic system, clear the DTCs and retest. If the fault persists, install a new fuel injector 3
P0203-13	Cylinder 3 Injector Circuit / Open - Circuit open	⚠ NOTE: Circuit reference - INJ B L / INJ B H- • Fuel injector 3 circuit open circuit, high resistance • Fuel injector 3 internal failure	• Refer to the electrical circuit diagrams and check the fuel injector 3 circuit for open circuit, high resistance • Using the manufacturer approved diagnostic system, clear the DTCs and retest. If the fault persists, install a new fuel injector 3
P0204-01	Cylinder 4 Injector Circuit / Open - General electrical failure	⚠ NOTE: Circuit reference - INJ C L / INJ C H- • Fuel injector 4 circuit short circuit to ground, short circuit to power, open circuit, high resistance • Fuel injector 4 internal failure	• Refer to the electrical circuit diagrams and check the fuel injector 4 circuit for short circuit to ground, short circuit to power, open circuit, high resistance • Using the manufacturer approved diagnostic system, clear the DTCs and retest. If the fault persists, install a new fuel injector 4
P0204-13	Cylinder 4 Injector Circuit / Open - Circuit open	⚠ NOTE: Circuit reference - INJ C L / INJ C H- • Fuel injector 4 circuit open circuit, high resistance • Fuel injector 4 internal failure	• Refer to the electrical circuit diagrams and check the fuel injector 4 circuit for open circuit, high resistance • Using the manufacturer approved diagnostic system, clear the DTCs and retest. If the fault persists, install a new fuel injector 4
P0219-00	Engine Overspeed Condition - No sub type information	• Incorrect gear selection (low gear at high road speed) • Oil ingestion • Fuel injector fault	• Using the manufacturer approved diagnostic system, clear the DTCs and retest • Check for sources of oil leaking into the intake system (turbocharger, positive crankcase ventilation system, etc) • Using the manufacturer approved diagnostic system, check for fuel injector related DTCs and perform the relevant corrective actions
P0251-11	Injection Pump Fuel Metering Control A - Circuit short to ground	⚠ NOTE: Circuit reference - VCV - • Fuel volume control valve circuit short circuit to ground	• Refer to the electrical circuit diagrams and check the fuel volume control valve circuit for short circuit to ground

P0251-12	Injection Pump Fuel Metering Control A - Circuit short to battery	⚠ NOTE: Circuit reference - VCV - • Fuel volume control valve circuit short circuit to power	• Refer to the electrical circuit diagrams and check the fuel volume control valve circuit for short circuit to power
P0251-13	Injection Pump Fuel Metering Control A - Circuit open	⚠ NOTE: Circuit reference - VCV - • Fuel volume control valve circuit open circuit, high resistance	• Refer to the electrical circuit diagrams and check the fuel volume control valve circuit for open circuit, high resistance
P0251-71	Injection Pump Fuel Metering Control A - Actuator stuck	⚠ NOTE: Circuit reference - VCV - • Fuel volume control valve internal failure	• Using the manufacturer approved diagnostic system, clear the DTCs and retest. If the fault persists, install a new fuel volume control valve
P0263-00	Cylinder 1 Contribution/Balance - No sub type information	• Fuel injector 1 circuit short circuit to ground, short circuit to power, open circuit, high resistance • Fuel injector 1 fault • Cylinder 1 mechanical fault	• Refer to the electrical circuit diagrams and check the fuel injector 1 circuit for short circuit to ground, short circuit to power, open circuit, high resistance • Using the manufacturer approved diagnostic system, check for fuel injector related DTCs and perform the relevant corrective actions • Perform a compression test or cylinder leakage test to assess the mechanical condition of the cylinder
P0266-00	Cylinder 2 Contribution/Balance - No sub type information	• Fuel injector 2 circuit short circuit to ground, short circuit to power, open circuit, high resistance • Fuel injector 2 fault • Cylinder 2 mechanical fault	• Refer to the electrical circuit diagrams and check the fuel injector 2 circuit for short circuit to ground, short circuit to power, open circuit, high resistance • Using the manufacturer approved diagnostic system, check for fuel injector related DTCs and perform the relevant corrective actions • Perform a compression test or cylinder leakage test to assess the mechanical condition of the cylinder
P0269-00	Cylinder 3 Contribution/Balance - No sub type information	• Fuel injector 3 circuit short circuit to ground, short circuit to power, open circuit, high resistance • Fuel injector 3 fault • Cylinder 3 mechanical fault	• Refer to the electrical circuit diagrams and check the fuel injector 3 circuit for short circuit to ground, short circuit to power, open circuit, high resistance • Using the manufacturer approved diagnostic system, check for fuel injector related DTCs and perform the relevant corrective actions • Perform a compression test or cylinder leakage test to assess the mechanical condition of the cylinder
P0272-00	Cylinder 4 Contribution/Balance - No sub type information	• Fuel injector 4 circuit short circuit to ground, short circuit to power, open circuit, high resistance • Fuel injector 4 fault • Cylinder 4 mechanical fault	• Refer to the electrical circuit diagrams and check the fuel injector 4 circuit for short circuit to ground, short circuit to power, open circuit, high resistance • Using the manufacturer approved diagnostic system, check for fuel injector related DTCs and perform the relevant corrective actions • Perform a compression test or cylinder leakage test to assess the mechanical condition of the cylinder
P029A-00	Cylinder 1 - Fuel Trim at Max Limit - No sub type information	• Fuel system fault • Fuel injector 1 fuel pipe damaged/blocked/leaking • Fuel injector 1 internal failure	• Using the manufacturer approved diagnostic system, check for fuel system related DTCs and perform the relevant corrective actions • Check fuel injector 1 fuel pipe for damage/blockage/leaks • Install a new fuel injector 1 as necessary
P029B-00	Cylinder 1 - Fuel Trim at Min Limit - No sub type information	• Fuel system fault • Fuel injector 1 fuel pipe damaged/blocked/leaking • Fuel injector 1 internal failure	• Using the manufacturer approved diagnostic system, check for fuel system related DTCs and perform the relevant corrective actions • Check fuel injector 1 fuel pipe for damage/blockage/leaks • Install a new fuel injector 1 as necessary
P029D-00	Cylinder 1 - Injector Leaking - No sub type information	• Fuel system fault	⚠ NOTE: Ignore this DTC if it is not set with other fuel system or misfire DTCs. • Using the manufacturer approved diagnostic system, check for fuel system and misfire related DTCs and perform the relevant corrective actions
P029D-96	Cylinder 1 - Injector Leaking - Component internal failure	• Fuel injector 1 leaking	• Install a new fuel injector 1 as necessary
P029E-00	Cylinder 2 - Fuel Trim at Max Limit - No sub type information	• Fuel system fault • Fuel injector 2 fuel pipe damaged/blocked/leaking • Fuel injector 2 internal failure	• Using the manufacturer approved diagnostic system, check for fuel system related DTCs and perform the relevant corrective actions • Check fuel injector 2 fuel pipe for damage/blockage/leaks • Install a new fuel injector 2 as necessary

P029F-00	Cylinder 2 - Fuel Trim at Min Limit - No sub type information	• Fuel system fault • Fuel injector 2 fuel pipe damaged/blocked/leaking • Fuel injector 2 internal failure	• Using the manufacturer approved diagnostic system, check for fuel system related DTCs and perform the relevant corrective actions • Check fuel injector 2 fuel pipe for damage/blockage/leaks • Install a new fuel injector 2 as necessary
P02A1-00	Cylinder 2 - Injector Leaking - No sub type information	• Fuel system fault	⚠ NOTE: Ignore this DTC if it is not set with other fuel system or misfire DTCs. • Using the manufacturer approved diagnostic system, check for fuel system and misfire related DTCs and perform the relevant corrective actions
P02A1-96	Cylinder 2 - Injector Leaking - Component internal failure	• Fuel injector 2 leaking	• Install a new fuel injector 2 as necessary
P02A2-00	Cylinder 3 - Fuel Trim at Max Limit - No sub type information	• Fuel system fault • Fuel injector 3 fuel pipe damaged/blocked/leaking • Fuel injector 3 internal failure	• Using the manufacturer approved diagnostic system, check for fuel system related DTCs and perform the relevant corrective actions • Check fuel injector 3 fuel pipe for damage/blockage/leaks • Install a new fuel injector 3 as necessary
P02A3-00	Cylinder 3 - Fuel Trim at Min Limit - No sub type information	• Fuel system fault • Fuel injector 3 fuel pipe damaged/blocked/leaking • Fuel injector 3 internal failure	• Using the manufacturer approved diagnostic system, check for fuel system related DTCs and perform the relevant corrective actions • Check fuel injector 3 fuel pipe for damage/blockage/leaks • Install a new fuel injector 3 as necessary
P02A5-00	Cylinder 3 - Injector Leaking - No sub type information	• Fuel system fault	⚠ NOTE: Ignore this DTC if it is not set with other fuel system or misfire DTCs. • Using the manufacturer approved diagnostic system, check for fuel system and misfire related DTCs and perform the relevant corrective actions
P02A5-96	Cylinder 3 - Injector Leaking - Component internal failure	• Fuel injector 3 leaking	• Install a new fuel injector 3 as necessary
P02A6-00	Cylinder 4 - Fuel Trim at Max Limit - No sub type information	• Fuel system fault • Fuel injector 4 fuel pipe damaged/blocked/leaking • Fuel injector 4 internal failure	• Using the manufacturer approved diagnostic system, check for fuel system related DTCs and perform the relevant corrective actions • Check fuel injector 4 fuel pipe for damage/blockage/leaks • Install a new fuel injector 4 as necessary
P02A7-00	Cylinder 4 - Fuel Trim at Min Limit - No sub type information	• Fuel system fault • Fuel injector 4 fuel pipe damaged/blocked/leaking • Fuel injector 4 internal failure	• Using the manufacturer approved diagnostic system, check for fuel system related DTCs and perform the relevant corrective actions • Check fuel injector 4 fuel pipe for damage/blockage/leaks • Install a new fuel injector 4 as necessary
P02A9-00	Cylinder 4 - Injector Leaking - No sub type information	• Fuel system fault	⚠ NOTE: Ignore this DTC if it is not set with other fuel system or misfire DTCs. • Using the manufacturer approved diagnostic system, check for fuel system and misfire related DTCs and perform the relevant corrective actions
P02A9-96	Cylinder 4 - Injector Leaking - Component internal failure	• Fuel injector 4 leaking	• Install a new fuel injector 4 as necessary
P02CC-00	Cylinder 1 Fuel Injector Offset Learning at Min Limit - No sub type information	• Fuel injector 1 fuel pipe damaged/blocked/leaking • Fuel injector 1 internal failure	• Check fuel injector 1 fuel pipe for damage/blockage/leaks • Install a new fuel injector 1 as necessary
P02CD-00	Cylinder 1 Fuel Injector Offset Learning at Max Limit - No sub type information	• Fuel injector 1 fuel pipe damaged/blocked/leaking • Fuel injector 1 internal failure	• Check fuel injector 1 fuel pipe for damage/blockage/leaks • Install a new fuel injector 1 as necessary
P02CE-00	Cylinder 2 Fuel Injector Offset Learning at Min Limit - No sub type information	• Fuel injector 2 fuel pipe damaged/blocked/leaking • Fuel injector 2 internal failure	• Check fuel injector 2 fuel pipe for damage/blockage/leaks • Install a new fuel injector 2 as necessary
P02CF-00	Cylinder 2 Fuel Injector Offset Learning at Max Limit - No sub type information	• Fuel injector 2 fuel pipe damaged/blocked/leaking • Fuel injector 2 internal failure	• Check fuel injector 2 fuel pipe for damage/blockage/leaks • Install a new fuel injector 2 as necessary
P02D0-00	Cylinder 3 Fuel Injector Offset Learning at Min Limit - No sub type information	• Fuel injector 3 fuel pipe damaged/blocked/leaking • Fuel injector 3 internal failure	• Check fuel injector 3 fuel pipe for damage/blockage/leaks • Install a new fuel injector 3 as necessary

P02D1-00	Cylinder 3 Fuel Injector Offset Learning at Max Limit - No sub type information	• Fuel injector 3 fuel pipe damaged/blocked/leaking • Fuel injector 3 internal failure	• Check fuel injector 3 fuel pipe for damage/blockage/leaks • Install a new fuel injector 3 as necessary
P02D2-00	Cylinder 4 Fuel Injector Offset Learning at Min Limit - No sub type information	• Fuel injector 4 fuel pipe damaged/blocked/leaking • Fuel injector 4 internal failure	• Check fuel injector 4 fuel pipe for damage/blockage/leaks • Install a new fuel injector 4 as necessary
P02D3-00	Cylinder 4 Fuel Injector Offset Learning at Max Limit - No sub type information	• Fuel injector 4 fuel pipe damaged/blocked/leaking • Fuel injector 4 internal failure	• Check fuel injector 4 fuel pipe for damage/blockage/leaks • Install a new fuel injector 4 as necessary
P02EE-00	Cylinder 1 Injector Circuit Range/Performance - No sub type information	• Fuel injector 1 circuit short circuit to ground, short circuit to power, open circuit, high resistance • Fuel injector 1 internal failure	• Refer to the electrical circuit diagrams and check the fuel injector 1 circuit for short circuit to ground, short circuit to power, open circuit, high resistance • Install a new fuel injector 1 as necessary
P02EE-01	Cylinder 1 Injector Circuit Range/Performance - General electrical failure	• Fuel injector 1 circuit short circuit to ground, short circuit to power, open circuit, high resistance • Fuel injector 1 internal failure	• Refer to the electrical circuit diagrams and check the fuel injector 1 circuit for short circuit to ground, short circuit to power, open circuit, high resistance • Install a new fuel injector 1 as necessary
P02EE-21	Cylinder 1 Injector Circuit Range/Performance - Signal amplitude < minimum	• Fuel injector 1 fault	• Using the manufacturer approved diagnostic system, check for fuel injector 1 related DTCs and perform the relevant corrective actions
P02EE-22	Cylinder 1 Injector Circuit Range/Performance - Signal amplitude > maximum	• Fuel injector 1 fault	• Using the manufacturer approved diagnostic system, check for fuel injector 1 related DTCs and perform the relevant corrective actions
P02EE-29	Cylinder 1 Injector Circuit Range/Performance - Signal invalid	• Fuel injector 1 fault	• Using the manufacturer approved diagnostic system, check for fuel injector 1 related DTCs and perform the relevant corrective actions
P02EF-00	Cylinder 2 Injector Circuit Range/Performance - No sub type information	• Fuel injector 2 circuit short circuit to ground, short circuit to power, open circuit, high resistance • Fuel injector 2 internal failure	• Refer to the electrical circuit diagrams and check the fuel injector 2 circuit for short circuit to ground, short circuit to power, open circuit, high resistance • Install a new fuel injector 2 as necessary
P02EF-01	Cylinder 2 Injector Circuit Range/Performance - General electrical failure	• Fuel injector 2 circuit short circuit to ground, short circuit to power, open circuit, high resistance • Fuel injector 2 internal failure	• Refer to the electrical circuit diagrams and check the fuel injector 2 circuit for short circuit to ground, short circuit to power, open circuit, high resistance • Install a new fuel injector 2 as necessary
P02EF-21	Cylinder 2 Injector Circuit Range/Performance - Signal amplitude < minimum	• Fuel injector 2 fault	• Using the manufacturer approved diagnostic system, check for fuel injector 2 related DTCs and perform the relevant corrective actions
P02EF-22	Cylinder 2 Injector Circuit Range/Performance - Signal amplitude > maximum	• Fuel injector 2 fault	• Using the manufacturer approved diagnostic system, check for fuel injector 2 related DTCs and perform the relevant corrective actions
P02EF-29	Cylinder 2 Injector Circuit Range/Performance - Signal invalid	• Fuel injector 2 fault	• Using the manufacturer approved diagnostic system, check for fuel injector 2 related DTCs and perform the relevant corrective actions
P02F0-00	Cylinder 3 Injector Circuit Range/Performance - No sub type information	• Fuel injector 3 circuit short circuit to ground, short circuit to power, open circuit, high resistance • Fuel injector 3 internal failure	• Refer to the electrical circuit diagrams and check the fuel injector 3 circuit for short circuit to ground, short circuit to power, open circuit, high resistance • Install a new fuel injector 3 as necessary
P02F0-01	Cylinder 3 Injector Circuit Range/Performance - General electrical failure	• Fuel injector 3 circuit short circuit to ground, short circuit to power, open circuit, high resistance • Fuel injector 3 internal failure	• Refer to the electrical circuit diagrams and check the fuel injector 3 circuit for short circuit to ground, short circuit to power, open circuit, high resistance • Install a new fuel injector 3 as necessary
P02F0-21	Cylinder 3 Injector Circuit Range/Performance - Signal amplitude < minimum	• Fuel injector 3 fault	• Using the manufacturer approved diagnostic system, check for fuel injector 3 related DTCs and perform the relevant corrective actions
P02F0-22	Cylinder 3 Injector Circuit Range/Performance - Signal amplitude > maximum	• Fuel injector 3 fault	• Using the manufacturer approved diagnostic system, check for fuel injector 3 related DTCs and perform the relevant corrective actions
P02F0-29	Cylinder 3 Injector Circuit Range/Performance - Signal invalid	• Fuel injector 3 fault	• Using the manufacturer approved diagnostic system, check for fuel injector 3 related DTCs and perform the relevant corrective actions

P02F1-00	Cylinder 4 Injector Circuit Range/Performance - No sub type information	• Fuel injector 4 circuit short circuit to ground, short circuit to power, open circuit, high resistance • Fuel injector 4 internal failure	• Refer to the electrical circuit diagrams and check the fuel injector 4 circuit for short circuit to ground, short circuit to power, open circuit, high resistance • Install a new fuel injector 4 as necessary
P02F1-01	Cylinder 4 Injector Circuit Range/Performance - General electrical failure	• Fuel injector 4 circuit short circuit to ground, short circuit to power, open circuit, high resistance • Fuel injector 4 internal failure	• Refer to the electrical circuit diagrams and check the fuel injector 4 circuit for short circuit to ground, short circuit to power, open circuit, high resistance • Install a new fuel injector 4 as necessary
P02F1-21	Cylinder 4 Injector Circuit Range/Performance - Signal amplitude < minimum	• Fuel injector 4 fault	• Using the manufacturer approved diagnostic system, check for fuel injector 4 related DTCs and perform the relevant corrective actions
P02F1-22	Cylinder 4 Injector Circuit Range/Performance - Signal amplitude > maximum	• Fuel injector 4 fault	• Using the manufacturer approved diagnostic system, check for fuel injector 4 related DTCs and perform the relevant corrective actions
P02F1-29	Cylinder 4 Injector Circuit Range/Performance - Signal invalid	• Fuel injector 4 fault	• Using the manufacturer approved diagnostic system, check for fuel injector 4 related DTCs and perform the relevant corrective actions
P0300-00	Random Misfire Detected - No sub type information	• Catalytic converter / exhaust system restricted • Fuel injector circuit short circuit to ground, short circuit to power, open circuit, high resistance • Engine system fault	• Check the catalytic converter / exhaust system for restrictions • Refer to the electrical circuit diagrams and check the fuel injector circuit for short circuit to ground, short circuit to power, open circuit, high resistance • Using the manufacturer approved diagnostic system, check for engine system related DTCs and perform the relevant corrective actions
P0301-00	Cylinder 1 Misfire Detected - No sub type information	• Catalytic converter / exhaust system restricted • Fuel injector 1 circuit short circuit to ground, short circuit to power, open circuit, high resistance • Engine system fault	• Check the catalytic converter / exhaust system for restrictions • Refer to the electrical circuit diagrams and check the fuel injector 1 circuit for short circuit to ground, short circuit to power, open circuit, high resistance • Using the manufacturer approved diagnostic system, check for engine system related DTCs and perform the relevant corrective actions
P0302-00	Cylinder 2 Misfire Detected - No sub type information	• Catalytic converter / exhaust system restricted • Fuel injector 2 circuit short circuit to ground, short circuit to power, open circuit, high resistance • Engine system fault	• Check the catalytic converter / exhaust system for restrictions • Refer to the electrical circuit diagrams and check the fuel injector 2 circuit for short circuit to ground, short circuit to power, open circuit, high resistance • Using the manufacturer approved diagnostic system, check for engine system related DTCs and perform the relevant corrective actions
P0303-00	Cylinder 3 Misfire Detected - No sub type information	• Catalytic converter / exhaust system restricted • Fuel injector 3 circuit short circuit to ground, short circuit to power, open circuit, high resistance • Engine system fault	• Check the catalytic converter / exhaust system for restrictions • Refer to the electrical circuit diagrams and check the fuel injector 3 circuit for short circuit to ground, short circuit to power, open circuit, high resistance • Using the manufacturer approved diagnostic system, check for engine system related DTCs and perform the relevant corrective actions
P0304-00	Cylinder 4 Misfire Detected - No sub type information	• Catalytic converter / exhaust system restricted • Fuel injector 4 circuit short circuit to ground, short circuit to power, open circuit, high resistance • Engine system fault	• Check the catalytic converter / exhaust system for restrictions • Refer to the electrical circuit diagrams and check the fuel injector 4 circuit for short circuit to ground, short circuit to power, open circuit, high resistance • Using the manufacturer approved diagnostic system, check for engine system related DTCs and perform the relevant corrective actions
P0335-78	Crankshaft Position Sensor A Circuit - Alignment or adjustment incorrect	• Crankshaft position sensor alignment incorrect • Crankshaft position sensor reluctor ring fault • Foreign matter on crankshaft position sensor face	• Check the crankshaft position sensor alignment • Check the integrity of the crankshaft position sensor reluctor ring • Check for foreign matter on the crankshaft position sensor face
P0336-31	Crankshaft Position Sensor A Circuit Range/Performance - No signal	• Crankshaft position sensor circuit short circuit to ground, short circuit to power, open circuit, high resistance • Crankshaft position sensor alignment incorrect • Crankshaft position sensor reluctor ring fault • Crankshaft position sensor internal failure	• Refer to the electrical circuit diagrams and check the crankshaft position sensor circuit for short circuit to ground, short circuit to power, open circuit, high resistance • Check the crankshaft position sensor alignment • Check the integrity of the crankshaft position sensor reluctor ring • Install a new crankshaft position sensor as necessary

P0336-38	Crankshaft Position Sensor A Circuit Range/Performance - Signal frequency incorrect	• Crankshaft position sensor circuit short circuit to ground, short circuit to power, open circuit, high resistance • Crankshaft position sensor alignment incorrect • Crankshaft position sensor reluctor ring fault • Crankshaft position sensor internal failure	• Refer to the electrical circuit diagrams and check the crankshaft position sensor circuit for short circuit to ground, short circuit to power, open circuit, high resistance • Check the crankshaft position sensor alignment • Check the integrity of the crankshaft position sensor reluctor ring • Install a new crankshaft position sensor as necessary
P0336-64	Crankshaft Position Sensor A Circuit Range/Performance - Signal plausibility failure	• Crankshaft position sensor circuit short circuit to ground, short circuit to power, open circuit, high resistance • Crankshaft position sensor air gap incorrect • Foreign matter on crankshaft position sensor face • Crankshaft position sensor alignment incorrect • Crankshaft position sensor reluctor ring fault • Crankshaft position sensor internal failure	• Refer to the electrical circuit diagrams and check the crankshaft position sensor circuit for short circuit to ground, short circuit to power, open circuit, high resistance • Check the crankshaft position sensor air gap at 90° intervals (maximum air gap 4.5mm) • Check for foreign matter on the crankshaft position sensor face • Check the crankshaft position sensor alignment • Check the integrity of the crankshaft position sensor reluctor ring • Install a new crankshaft position sensor as necessary
P0336-66	Crankshaft Position Sensor A Circuit Range/Performance - Signal has too many transitions/events	• Crankshaft position sensor circuit short circuit to ground, short circuit to power, open circuit, high resistance • Crankshaft position sensor alignment incorrect • Crankshaft position sensor reluctor ring fault • Crankshaft position sensor internal failure	• Refer to the electrical circuit diagrams and check the crankshaft position sensor circuit for short circuit to ground, short circuit to power, open circuit, high resistance • Check the crankshaft position sensor alignment • Check the integrity of the crankshaft position sensor reluctor ring • Install a new crankshaft position sensor as necessary
P0336-67	Crankshaft Position Sensor A Circuit Range/Performance - Signal incorrect after event	• Crankshaft position sensor circuit short circuit to ground, short circuit to power, open circuit, high resistance • Crankshaft position sensor alignment incorrect • Crankshaft position sensor reluctor ring fault • Crankshaft position sensor internal failure	• Refer to the electrical circuit diagrams and check the crankshaft position sensor circuit for short circuit to ground, short circuit to power, open circuit, high resistance • Check the crankshaft position sensor alignment • Check the integrity of the crankshaft position sensor reluctor ring • Install a new crankshaft position sensor as necessary
P0341-29	Camshaft Position Sensor A Circuit Range/Performance (Bank 1 or single sensor) - Signal invalid	• Camshaft position sensor circuit short circuit to ground, short circuit to power, open circuit, high resistance • Camshaft position sensor alignment incorrect • Camshaft position sensor internal failure	• Refer to the electrical circuit diagrams and check the camshaft position sensor circuit for short circuit to ground, short circuit to power, open circuit, high resistance • Check the camshaft position sensor alignment • Install a new camshaft position sensor as necessary
P0341-3A	Camshaft Position Sensor A Circuit Range/Performance (Bank 1 or single sensor) - Incorrect has too many pulses	• Camshaft position sensor circuit short circuit to ground, short circuit to power • Camshaft position sensor alignment incorrect • Camshaft position sensor internal failure	• Refer to the electrical circuit diagrams and check the camshaft position sensor circuit for short circuit to ground, short circuit to power • Check the camshaft position sensor alignment • Install a new camshaft position sensor as necessary
P0342-31	Camshaft Position Sensor A Circuit Low (Bank 1 or single sensor) - No signal	• Camshaft position sensor circuit open circuit, high resistance • Camshaft position sensor alignment incorrect • Camshaft position sensor internal failure	• Refer to the electrical circuit diagrams and check the camshaft position sensor circuit for open circuit, high resistance • Check the camshaft position sensor alignment • Install a new camshaft position sensor as necessary
P0380-81	Glow Plug/Heater Circuit A - Invalid serial data received	⚠ NOTE: Circuit reference - GPC / GPFB - • Glow plug control module power or ground circuit open circuit, high resistance • Glow plug control module control circuit short circuit to ground, short circuit to power, open circuit, high resistance • Glow plug control module feedback circuit short circuit to ground, short circuit to power, open circuit, high resistance • Glow plug control module internal failure	• Refer to the electrical circuit diagrams and check the glow plug control module power and ground circuits for open circuit, high resistance • Refer to the electrical circuit diagrams and check the glow plug control module control circuit for short circuit to ground, short circuit to power, open circuit, high resistance • Refer to the electrical circuit diagrams and check the glow plug control module feedback circuit for short circuit to ground, short circuit to power, open circuit, high resistance • Using the manufacturer approved diagnostic system, clear the DTCs and retest. If the fault persists, install a new glow plug control module
P0381-12	Glow Plug/Heater Indicator Circuit - Circuit short to battery	⚠ NOTE: Circuit reference - GPFB - • Glow plug control module feedback circuit short circuit to power	• Refer to the electrical circuit diagrams and check the glow plug control module feedback circuit for short circuit to power

P0381-13	Glow Plug/Heater Indicator Circuit - Circuit open	NOTE: Circuit reference - GPFB - • Glow plug control module feedback circuit open circuit, high resistance	• Refer to the electrical circuit diagrams and check the glow plug control module feedback circuit for open circuit, high resistance
P0383-11	Glow Plug Control Module 1 Control Circuit Low - Circuit short to ground	NOTE: Circuit reference - GPC - • Glow plug control module control circuit short circuit to ground	• Refer to the electrical circuit diagrams and check the glow plug control module control circuit for short circuit to ground
P0384-12	Glow Plug Control Module 1 Control Circuit High - Circuit short to battery	NOTE: Circuit reference - GPC - • Glow plug control module control circuit short circuit to power	• Refer to the electrical circuit diagrams and check the glow plug control module control circuit for short circuit to power
P0401-21	Exhaust Gas Recirculation A Flow Insufficient Detected - Signal amplitude < minimum	• Exhaust gas recirculation gas transfer system, mixer pipe or cooler leakage • Exhaust gas recirculation valve mechanical failure - Not opened to commanded position	• Check the exhaust gas recirculation gas transfer system for leakage. Using the manufacturer approved diagnostic system, clear the DTCs and retest • Using the manufacturer approved diagnostic system, clear the DTCs and perform routine - Turbo, EGR and Air Path Dynamic Test. Install a new exhaust gas recirculation valve as necessary
P0402-22	Exhaust Gas Recirculation A Flow Excessive Detected - Signal amplitude > maximum	• Exhaust gas recirculation valve mechanical failure - Not closed to commanded position	• Using the manufacturer approved diagnostic system, clear the DTCs and perform routine - Turbo, EGR and Air Path Dynamic Test. Install a new exhaust gas recirculation valve as necessary
P0403-00	Exhaust Gas Recirculation "A" Control Circuit / Open - No sub type information	NOTE: Circuit reference - EVA+ / EVA- - • Exhaust gas recirculation valve circuit short circuit to ground, short circuit to power, open circuit, high resistance	• Refer to the electrical circuit diagrams and check the exhaust gas recirculation valve circuit for short circuit to ground, short circuit to power, open circuit, high resistance
P0403-11	Exhaust Gas Recirculation "A" Control Circuit / Open - Circuit short to ground	NOTE: Circuit reference - EVA+ / EVA- - • Exhaust gas recirculation valve circuit short circuit to ground, open circuit, high resistance	• Refer to the electrical circuit diagrams and check the exhaust gas recirculation valve circuit for short circuit to ground, open circuit, high resistance
P0403-12	Exhaust Gas Recirculation "A" Control Circuit / Open - Circuit short to battery	NOTE: Circuit reference - EVA+ / EVA- - • Exhaust gas recirculation valve circuit short circuit to power, open circuit, high resistance	• Refer to the electrical circuit diagrams and check the exhaust gas recirculation valve circuit for short circuit to power, open circuit, high resistance
P0403-13	Exhaust Gas Recirculation "A" Control Circuit / Open - Open circuit	NOTE: Circuit reference - EVA+ / EVA- - • Exhaust gas recirculation valve circuit open circuit, high resistance	• Refer to the electrical circuit diagrams and check the exhaust gas recirculation valve circuit for open circuit, high resistance
P0404-21	Exhaust Gas Recirculation A Control Circuit Range/Performance - Signal amplitude < minimum	NOTE: Circuit reference - EVA+ / EVA- - • Exhaust gas recirculation gas transfer system, mixer pipe or cooler leakage • Exhaust gas recirculation valve circuit short circuit to ground, short circuit to power, open circuit, high resistance • Exhaust gas recirculation valve mechanical failure - Not closed to commanded position	• Check the exhaust gas recirculation gas transfer system for leakage. Using the manufacturer approved diagnostic system, clear the DTCs and retest • Refer to the electrical circuit diagrams and check the exhaust gas recirculation valve circuit for short circuit to ground, short circuit to power, open circuit, high resistance • Using the manufacturer approved diagnostic system, clear the DTCs and perform routine - Turbo, EGR and Air Path Dynamic Test. Install a new exhaust gas recirculation valve as necessary
P0404-22	Exhaust Gas Recirculation A Control Circuit Range/Performance - Signal amplitude > maximum	NOTE: Circuit reference - EVA+ / EVA- - • Exhaust gas recirculation valve circuit short circuit to ground, short circuit to power, open circuit, high resistance • Exhaust gas recirculation valve mechanical failure - Not closed to commanded position	• Refer to the electrical circuit diagrams and check the exhaust gas recirculation valve circuit for short circuit to ground, short circuit to power, open circuit, high resistance • Using the manufacturer approved diagnostic system, clear the DTCs and perform routine - Turbo, EGR and Air Path Dynamic Test. Install a new exhaust gas recirculation valve as necessary

P0405-14	Exhaust Gas Recirculation Sensor A Circuit Low - Circuit short to ground or open	NOTE: Circuit reference - EVP - • Exhaust gas recirculation valve position sensor circuit short circuit to ground, open circuit, high resistance	• Refer to the electrical circuit diagrams and check the exhaust gas recirculation valve position sensor circuit for short circuit to ground, open circuit, high resistance
P0406-12	Exhaust Gas Recirculation Sensor A Circuit High - Circuit short to battery	NOTE: Circuit reference - EVP - • Exhaust gas recirculation valve position sensor circuit short circuit to power	• Refer to the electrical circuit diagrams and check the exhaust gas recirculation valve position sensor circuit for short circuit to power
P0409-77	Exhaust Gas Recirculation Sensor A Circuit - Commanded position not reachable	• Exhaust gas recirculation valve mechanical failure	• Using the manufacturer approved diagnostic system, clear the DTCs and perform routine - Turbo, EGR and Air Path Dynamic Test. Install a new exhaust gas recirculation valve as necessary
P0426-27	Catalyst Temperature Sensor Circuit Range/Performance (Bank 1, Sensor Circuit 1) - Signal rate of change above threshold	• Engine system fault • Exhaust gas temperature sensor 1 internal failure	• Using the manufacturer approved diagnostic tool check for associated DTCs/events at the same timeline which would influence fuelling or rapid changes in exhaust temperature • Install a new exhaust gas temperature sensor 1 as necessary
P0427-11	Catalyst Temperature Sensor Circuit Low (Bank 1, Sensor Circuit 1) - Circuit short to ground	• Exhaust gas temperature sensor 1 signal circuit short circuit to ground	• Refer to the electrical circuit diagrams and check the exhaust gas temperature sensor 1 signal circuit for short circuit to ground
P0428-15	Catalyst Temperature Sensor Circuit High (Bank 1, Sensor Circuit 1) - Circuit short to battery or open	• Exhaust gas temperature sensor 1 signal circuit short circuit to power, open circuit, high resistance	• Refer to the electrical circuit diagrams and check the exhaust gas temperature sensor 1 signal circuit for short circuit to power, open circuit, high resistance
P042A-11	Catalyst Temperature Sensor Circuit (Bank 1, Sensor Circuit 2) - Circuit short to ground	• Exhaust gas temperature sensor 2 signal circuit short circuit to ground	• Refer to the electrical circuit diagrams and check the exhaust gas temperature sensor 2 signal circuit for short circuit to ground
P042B-27	Catalyst Temperature Sensor Circuit Range/Performance (Bank 1, Sensor Circuit 2) - Signal rate of change above threshold	• Engine system fault • Exhaust gas temperature sensor 2 internal failure	• Using the manufacturer approved diagnostic tool check for associated DTCs/events at the same timeline which would influence fuelling or rapid changes in exhaust temperature • Install a new exhaust gas temperature sensor 2 as necessary
P042D-15	Catalyst Temperature Sensor Circuit High (Bank 1, Sensor Circuit 2) - Circuit short to battery or open	• Exhaust gas temperature sensor 2 signal circuit short circuit to power, open circuit, high resistance	• Refer to the electrical circuit diagrams and check the exhaust gas temperature sensor 2 signal circuit for short circuit to power, open circuit, high resistance
P0486-11	Exhaust Gas Recirculation Sensor B Circuit - Circuit short to ground	NOTE: Circuit reference - EVP - • Exhaust gas recirculation valve position sensor circuit short circuit to ground • Exhaust gas recirculation valve internal failure	• Refer to the electrical circuit diagrams and check the exhaust gas recirculation valve position sensor circuit for short circuit to ground • Using the manufacturer approved diagnostic system, clear the DTCs and perform routine - Turbo, EGR and Air Path Dynamic Test. Install a new exhaust gas recirculation valve as necessary
P0486-15	Exhaust Gas Recirculation Sensor B Circuit - Circuit short to battery or open	NOTE: Circuit reference - EVP - • Exhaust gas recirculation valve position sensor circuit short circuit to power, open circuit, high resistance • Exhaust gas recirculation valve internal failure	• Refer to the electrical circuit diagrams and check the exhaust gas recirculation valve position sensor circuit for short circuit to power, open circuit, high resistance • Using the manufacturer approved diagnostic system, clear the DTCs and perform routine - Turbo, EGR and Air Path Dynamic Test. Install a new exhaust gas recirculation valve as necessary
P0487-13	Exhaust Gas Recirculation Throttle Control Circuit A / Open - Open circuit	• Throttle actuator circuit open circuit, high resistance	• Refer to the electrical circuit diagrams and check the throttle actuator circuit for open circuit, high resistance
P0488-19	Exhaust Gas Recirculation Throttle Control Circuit A Range/Performance - Circuit current above threshold	NOTE: Circuit reference - EVA+ / EVA- - • Throttle actuator circuit short circuit to ground, short circuit to power	• Refer to the electrical circuit diagrams and check the throttle actuator circuit for short circuit to ground, short circuit to power
P0488-21	Exhaust Gas Recirculation Throttle Control Circuit A Range/Performance - Signal amplitude < minimum	NOTE: Circuit reference - EVA+ / EVA- - • Throttle actuator circuit short circuit to ground, short circuit to power, open circuit, high resistance • Throttle actuator mechanical failure	• Refer to the electrical circuit diagrams and check the throttle actuator circuit for short circuit to ground, short circuit to power, open circuit, high resistance • Using the manufacturer approved diagnostic system, clear the DTCs and perform routine - Turbo, EGR and Air Path Dynamic Test. Install a new throttle actuator as necessary

P0488-22	Exhaust Gas Recirculation Throttle Control Circuit A Range/Performance - Signal amplitude > maximum	NOTE: Circuit reference - EVA+ / EVA- - • Throttle actuator circuit short circuit to ground, short circuit to power, open circuit, high resistance • Throttle actuator mechanical failure	• Refer to the electrical circuit diagrams and check the throttle actuator circuit for short circuit to ground, short circuit to power, open circuit, high resistance • Using the manufacturer approved diagnostic system, clear the DTCs and perform routine - Turbo, EGR and Air Path Dynamic Test. Install a new throttle actuator as necessary
P0488-4B	Exhaust Gas Recirculation Throttle Control Circuit A Range/Performance - Over temperature	NOTE: Circuit reference - EVA+ / EVA- - • Throttle actuator circuit short circuit to ground, short circuit to power • Throttle actuator mechanical failure	• Refer to the electrical circuit diagrams and check the throttle actuator circuit for short circuit to ground, short circuit to power • Using the manufacturer approved diagnostic system, clear the DTCs and perform routine - Turbo, EGR and Air Path Dynamic Test. Install a new throttle actuator as necessary
P0489-00	Exhaust Gas Recirculation Control Circuit Low - No sub type information	NOTE: Circuit reference - EVA+ / EVA- - • Exhaust gas recirculation valve actuator circuit short circuit to ground	• Refer to the electrical circuit diagrams and check the exhaust gas recirculation valve actuator circuit for short circuit to ground
P049D-00	Exhaust Gas Recirculation A Control Position Exceeded Learning Limit - No sub type information	• Engine system fault • Exhaust gas recirculation valve internal failure	• Using the manufacturer approved diagnostic tool check for associated DTCs/events at the same timeline which would influence the adaptation values of the exhaust gas recirculation valve • Using the manufacturer approved diagnostic system, clear the DTCs and perform routine - Turbo, EGR and Air Path Dynamic Test. Install a new exhaust gas recirculation valve as necessary

P0504-62	Brake Switch A / B Correlation - Signal compare failure	• Brake pedal switch incorrectly installed • Stop lamp switch circuit short circuit to ground, short circuit to power, open circuit, high resistance • Brake test switch circuit short circuit to ground, short circuit to power, open circuit, high resistance	NOTES: This DTC is set when there is a mismatch between the stop lamp switch signal and the brake test switch signal. This DTC may be induced by the driver depressing the accelerator and brake pedals simultaneously for 120 seconds. • Check the installation of the brake pedal switch • Refer to the electrical circuit diagrams and check the stop lamp switch circuit for short circuit to ground, short circuit to power, open circuit, high resistance • Refer to the electrical circuit diagrams and check the brake test switch circuit for short circuit to ground, short circuit to power, open circuit, high resistance
P0512-11	Starter Request Circuit - Circuit short to ground	• Starter motor relay circuit short circuit to ground	• Refer to the electrical circuit diagrams and check the starter motor relay circuit for short circuit to ground
P0512-12	Starter Request Circuit - Circuit short to battery	• Starter motor relay circuit short circuit to power	• Refer to the electrical circuit diagrams and check the starter motor relay circuit for short circuit to power
P0512-13	Starter Request Circuit - Circuit open	• Starter motor relay circuit open circuit, high resistance	• Refer to the electrical circuit diagrams and check the starter motor relay circuit for open circuit, high resistance
P0531-12	A/C Refrigerant Pressure Sensor A Circuit Range/Performance - Circuit short to battery	NOTE: Circuit reference - ACPT - • Refrigerant pressure sensor circuit short circuit to power	• Refer to the electrical circuit diagrams and check the refrigerant pressure sensor circuit for short circuit to power
P0531-14	A/C Refrigerant Pressure Sensor A Circuit Range/Performance - Circuit short to ground or open	NOTE: Circuit reference - ACPT - • Refrigerant pressure sensor circuit short circuit to ground, open circuit, high resistance	• Refer to the electrical circuit diagrams and check the refrigerant pressure sensor circuit for short circuit to ground, open circuit, high resistance

P0531-27	A/C Refrigerant Pressure Sensor A Circuit Range/Performance - Signal rate of change above threshold	NOTE: Circuit reference - ACPT - • Refrigerant pressure sensor circuit short circuit to ground, short circuit to power, open circuit, high resistance	• Refer to the electrical circuit diagrams and check the refrigerant pressure sensor circuit for short circuit to ground, short circuit to power, open circuit, high resistance
P0606-00	Control Module Processor - No sub type information	• Engine control module is not configured correctly	• Using the manufacturer approved diagnostic system, re-configure the engine control module with the latest level software. If the fault persists, install a new engine control module
P0606-48	Control Module Processor - Supervision software failure	• Engine control module is not configured correctly	• Using the manufacturer approved diagnostic system, re-configure the engine control module with the latest level software. If the fault persists, install a new engine control module
P0606-49	Control Module Processor - Internal electronic failure	• Engine control module is not configured correctly	• Using the manufacturer approved diagnostic system, re-configure the engine control module with the latest level software. If the fault persists, install a new engine control module
P0607-00	Control Module Performance - No sub type information	• Engine control module is not configured correctly	• Using the manufacturer approved diagnostic system, re-configure the engine control module with the latest level software. If the fault persists, install a new engine control module
P0607-48	Control Module Performance - Supervision software failure	• Engine control module is not configured correctly	• Using the manufacturer approved diagnostic system, re-configure the engine control module with the latest level software. If the fault persists, install a new engine control module
P0607-68	Control Module Performance - Event information	• Missing/invalid data from the instrument cluster • Missing/invalid data from the anti-lock brake system control module	• Using the manufacturer approved diagnostic system, check the instrument cluster for related DTCs and refer to the relevant DTC index • Using the manufacturer approved diagnostic system, check the anti-lock brake system control module for related DTCs and refer to the relevant DTC index
P060A-00	Internal Control Module Monitoring Processor Performance - No sub type information	• Engine control module is not configured correctly	• Using the manufacturer approved diagnostic system, re-configure the engine control module with the latest level software. If the fault persists, install a new engine control module
P060A-48	Internal Control Module Monitoring Processor Performance - Supervision software failure	• Engine control module is not configured correctly	• Using the manufacturer approved diagnostic system, re-configure the engine control module with the latest level software. If the fault persists, install a new engine control module
P060B-48	Internal Control Module A/D Processing Performance - Supervision software failure	• Engine control module is not configured correctly	• Using the manufacturer approved diagnostic system, re-configure the engine control module with the latest level software. If the fault persists, install a new engine control module
P060D-46	Internal Control Module Accelerator Pedal Position Performance - Calibration/parameter memory failure	• Engine control module is not configured correctly	• Using the manufacturer approved diagnostic system, re-configure the engine control module with the latest level software. If the fault persists, install a new engine control module
P0610-00	Control Module Vehicle Options Error - No sub type information	• Engine control module is not configured correctly • Car configuration file mismatch with vehicle specification	• Using the manufacturer approved diagnostic system, re-configure the engine control module with the latest level software. If the fault persists, install a new engine control module • Using the manufacturer approved diagnostic system, check and up-date the car configuration file as necessary. Clear the DTCs and retest
P061A-48	Internal Control Module Torque Performance - Supervision software failure	• Engine control module is not configured correctly	• Using the manufacturer approved diagnostic system, re-configure the engine control module with the latest level software. If the fault persists, install a new engine control module
P061B-00	Internal Control Module Torque Calculation Performance - No sub type information	• Engine control module is not configured correctly	• Using the manufacturer approved diagnostic system, re-configure the engine control module with the latest level software. If the fault persists, install a new engine control module

P061B-46	Internal Control Module Torque Calculation Performance - Calibration/parameter memory failure	• Engine control module is not configured correctly	• Using the manufacturer approved diagnostic system, re-configure the engine control module with the latest level software. If the fault persists, install a new engine control module
P061B-48	Internal Control Module Torque Calculation Performance - Supervision software failure	• Engine control module is not configured correctly	• Using the manufacturer approved diagnostic system, re-configure the engine control module with the latest level software. If the fault persists, install a new engine control module
P0620-01	Generator Control Circuit - General electrical failure	NOTE: Circuit reference - LIN - • Battery/charging system fault • LIN bus circuit short circuit to ground, short circuit to power, open circuit, high resistance	• Refer to the relevant section of the workshop manual and test the battery and charging system • Refer to the electrical circuit diagrams and check the LIN bus circuit for short circuit to ground, short circuit to power, open circuit, high resistance
P0627-12	Fuel Pump A Control Circuit Open - Circuit short to battery	NOTE: Circuit reference - LPRL - • Fuel pump relay circuit short circuit to power	• Refer to the electrical circuit diagrams and check the fuel pump relay circuit for short circuit to power
P0628-11	Fuel Pump A Control Circuit Low - Circuit short to ground	NOTE: Circuit reference - LPRL - • Fuel pump relay circuit short circuit to ground	• Refer to the electrical circuit diagrams and check the fuel pump relay circuit for short circuit to ground
P0629-13	Fuel Pump A Control Circuit High - Circuit open	NOTE: Circuit reference - LPRL - • Fuel pump relay circuit open circuit, high resistance	• Refer to the electrical circuit diagrams and check the fuel pump relay circuit for open circuit, high resistance
P062B-00	Internal Control Module Fuel Injector Control Performance - No sub type information	• Engine control module is not configured correctly	• Using the manufacturer approved diagnostic system, re-configure the engine control module with the latest level software. If the fault persists, install a new engine control module
P062B-46	Internal Control Module Fuel Injector Control Performance - Calibration/parameter memory failure	• Engine control module is not configured correctly	• Using the manufacturer approved diagnostic system, re-configure the engine control module with the latest level software. If the fault persists, install a new engine control module
P062B-48	Internal Control Module Fuel Injector Control Performance - Supervision software failure	• Engine control module is not configured correctly	• Using the manufacturer approved diagnostic system, re-configure the engine control module with the latest level software. If the fault persists, install a new engine control module
P0630-41	VIN Not Programmed or Incompatible - ECM/PCM - General checksum failure	• Engine control module previously installed on another vehicle • New engine control module installed and VIN not yet programmed	• Install the original or a new engine control module as necessary • Using the manufacturer approved diagnostic system, re-configure the engine control module with the latest level software
P0630-81	VIN Not Programmed or Incompatible - ECM/PCM - Invalid serial data received	• Engine control module is not configured correctly	• Using the manufacturer approved diagnostic system, re-configure the engine control module with the latest level software
P0642-16	Sensor Reference Voltage A Circuit Low - Circuit voltage below threshold	NOTE: Circuit reference - VSENS EVP - • Exhaust gas recirculation valve position sensor supply circuit short circuit to ground	• Refer to the electrical circuit diagrams and check the exhaust gas recirculation valve position sensor supply circuit for short circuit to ground
P0643-17	Sensor Reference Voltage A Circuit High - Circuit voltage above threshold	NOTE: Circuit reference - VSENS TPS- • Throttle position sensor supply circuit short circuit to power	• Refer to the electrical circuit diagrams and check the throttle position sensor supply circuit for short circuit to power
P0645-11	A/C Clutch Relay Control Circuit - Circuit short to ground	NOTE: Circuit reference - ACCR - • Air conditioning relay circuit short circuit to ground	• Refer to the electrical circuit diagrams and check the air conditioning relay circuit for short circuit to ground
P0645-12	A/C Clutch Relay Control Circuit - Circuit short to battery	NOTE: Circuit reference - ACCR - • Air conditioning relay circuit short circuit to power	• Refer to the electrical circuit diagrams and check the air conditioning relay circuit for short circuit to power
P0645-13	A/C Clutch Relay Control Circuit - Circuit open	NOTE: Circuit reference - ACCR - • Air conditioning relay circuit open circuit, high resistance	• Refer to the electrical circuit diagrams and check the air conditioning relay circuit for open circuit, high resistance

P0652-16	Sensor Reference Voltage B Circuit Low - Circuit voltage below threshold	NOTE: Circuit reference - VSENS CID - • Camshaft position sensor circuit short circuit to ground	• Refer to the electrical circuit diagrams and check the camshaft position sensor circuit for short circuit to ground
P0653-17	Sensor Reference Voltage B Circuit High - Circuit voltage above threshold	NOTE: Circuit reference - VSENS CID - • Camshaft position sensor circuit short circuit to power	• Refer to the electrical circuit diagrams and check the camshaft position sensor circuit for short circuit to power
P065B-16	Generator Control Circuit Range/Performance - Circuit voltage below threshold	NOTE: Circuit reference - LIN - • Generator LIN circuit short circuit to ground • Generator internal failure	• Refer to the electrical circuit diagrams and check the generator LIN circuit for short circuit to ground • Refer to the relevant section of the workshop manual and test the generator
P065B-17	Generator Control Circuit Range/Performance - Circuit voltage above threshold	NOTE: Circuit reference - LIN - • Generator LIN circuit short circuit to power • Generator internal failure	• Refer to the electrical circuit diagrams and check the generator LIN circuit for short circuit to power • Refer to the relevant section of the workshop manual and test the generator
P065B-87	Generator Control Circuit Range/Performance - Missing message	NOTE: Circuit reference - LIN - • Generator LIN circuit short circuit to ground, short circuit to power, open circuit, high resistance • Generator internal failure	• Refer to the electrical circuit diagrams and check the generator LIN circuit for short circuit to ground, short circuit to power, open circuit, high resistance • Refer to the relevant section of the workshop manual and test the generator
P065C-00	Generator Mechanical Performance - No sub type information	• Primary drive belt fault • Generator pulley slipping on shaft • Generator internal failure	NOTE: If additional generator related DTCs are also set, perform the relevant corrective action(s) first. • Check the primary drive belt condition and tension • Check the installation of the generator pulley • Refer to the relevant section of the workshop manual and test the generator
P0667-00	Control Module Internal Temperature Sensor 'A' Range/Performance - No sub type information	• Engine control module internal temperature higher than expected	• Check if the vehicle has been used in extreme conditions which could have resulted in excessively high under bonnet temperatures. Using the manufacturer approved diagnostic system, clear the DTCs and retest. If the fault persists, install a new engine control module
P0667-27	Control Module Internal Temperature Sensor 'A' Range/Performance - Signal rate of change above threshold	• Engine control module internal temperature higher than expected	• Check if the vehicle has been used in extreme conditions which could have resulted in excessively high under bonnet temperatures. Using the manufacturer approved diagnostic system, clear the DTCs and retest. If the fault persists, install a new engine control module
P0668-11	Control Module Internal Temperature Sensor 'A' Circuit Low - Circuit short to ground	• Engine control module internal failure	• Using the manufacturer approved diagnostic system, clear the DTCs and retest. If the fault persists, install a new engine control module
P0669-15	Control Module Internal Temperature Sensor 'A' Circuit High - Circuit short to battery or open	• Engine control module internal failure	• Using the manufacturer approved diagnostic system, clear the DTCs and retest. If the fault persists, install a new engine control module
P0670-13	Glow Plug Control Module 1 Control Circuit / Open - Circuit open	• Glow plug control circuit open circuit, high resistance	• Refer to the electrical circuit diagrams and check the glow plug control circuit for open circuit, high resistance
P0693-11	Fan 2 Control Circuit Low - Circuit short to ground	NOTE: Circuit reference - EDF 2 - • Air conditioning condensor fan relay circuit short circuit to ground	• Refer to the electrical circuit diagrams and check the air conditioning condensor fan relay circuit for short circuit to ground
P0693-13	Fan 2 Control Circuit Low - Circuit open	NOTE: Circuit reference - EDF 2 - • Air conditioning condensor fan relay circuit open circuit, high resistance	• Refer to the electrical circuit diagrams and check the air conditioning condensor fan relay circuit for open circuit, high resistance
P0694-12	Fan 2 Control Circuit High - Circuit short to battery	NOTE: Circuit reference - EDF 2 - • Air conditioning condensor fan relay circuit short circuit to power	• Refer to the electrical circuit diagrams and check the air conditioning condensor fan relay circuit for short circuit to power

P06B9-00	Cylinder 1 Glow Plug Circuit Range/Performance - No sub type information	• Cylinder 1 glow plug circuit short circuit to ground, short circuit to power, open circuit, high resistance • Glow plug control module internal failure	• Refer to the electrical circuit diagrams and check the cylinder 1 glow plug circuit for short circuit to ground, short circuit to power, open circuit, high resistance • Using the manufacturer approved diagnostic system, clear the DTCs and retest. If the fault persists, install a new glow plug control module
P06BA-00	Cylinder 2 Glow Plug Circuit Range/Performance - No sub type information	• Cylinder 2 glow plug circuit short circuit to ground, short circuit to power, open circuit, high resistance • Glow plug control module internal failure	• Refer to the electrical circuit diagrams and check the cylinder 2 glow plug circuit for short circuit to ground, short circuit to power, open circuit, high resistance • Using the manufacturer approved diagnostic system, clear the DTCs and retest. If the fault persists, install a new glow plug control module
P06BB-00	Cylinder 3 Glow Plug Circuit Range/Performance - No sub type information	• Cylinder 3 glow plug circuit short circuit to ground, short circuit to power, open circuit, high resistance • Glow plug control module internal failure	• Refer to the electrical circuit diagrams and check the cylinder 3 glow plug circuit for short circuit to ground, short circuit to power, open circuit, high resistance • Using the manufacturer approved diagnostic system, clear the DTCs and retest. If the fault persists, install a new glow plug control module
P06BC-00	Cylinder 4 Glow Plug Circuit Range/Performance - No sub type information	• Cylinder 4 glow plug circuit short circuit to ground, short circuit to power, open circuit, high resistance • Glow plug control module internal failure	• Refer to the electrical circuit diagrams and check the cylinder 4 glow plug circuit for short circuit to ground, short circuit to power, open circuit, high resistance • Using the manufacturer approved diagnostic system, clear the DTCs and retest. If the fault persists, install a new glow plug control module
P06DF-00	Glow Plug Control Module 1 Memory Checksum Error - No sub type information	• Engine control module is not configured correctly • Glow plug control module internal failure	• Using the manufacturer approved diagnostic system, re-configure the engine control module with the latest level software • Using the manufacturer approved diagnostic system, clear the DTCs and retest. If the fault persists, install a new glow plug control module
P0830-00	Clutch Pedal Switch A Circuit - No sub type information	NOTE: Circuit reference - CPP-TT - • Clutch pedal switch circuit short circuit to ground, short circuit to power, open circuit, high resistance	• Refer to the electrical circuit diagrams and check the clutch pedal switch circuit for short circuit to ground, short circuit to power, open circuit, high resistance
P0A09-16	DC/DC Converter Status Circuit Low - Circuit voltage below threshold	• Engine control module internal failure	• Using the manufacturer approved diagnostic system, clear the DTCs and retest. If the fault persists, install a new engine control module
P0A10-17	DC/DC Converter Status Circuit High - Circuit voltage above threshold	• Engine control module internal failure	• Using the manufacturer approved diagnostic system, clear the DTCs and retest. If the fault persists, install a new engine control module
P0A3B-68	Generator Over Temperature - Event information	• Generator over temperature	NOTES: If additional generator related DTCs are also set, perform the relevant corrective action(s) first and retest. Consider environmental conditions before condemning the generator. • Using the manufacturer approved diagnostic system, clear the DTCs and retest. If the fault persists, install a new generator
P115A-00	Low Fuel Level - Forced Limited Power - No sub type information	• Low fuel level condition • Fuel level sensor fault	NOTE: This DTC is set when the fuel level falls below 6 litres. • Check the fuel level and add fuel if necessary • Using the manufacturer approved diagnostic system, check the instrument cluster for related DTCs and refer to the relevant DTC index
P115B-68	Low Fuel Level - Forced Engine Shutdown - Event information	• Low fuel level condition • Fuel level sensor fault	NOTE: This DTC is set when the fuel level falls below 6 litres. • Check the fuel level and add fuel if necessary • Using the manufacturer approved diagnostic system, check the instrument cluster for related DTCs and refer to the relevant DTC index

P117D-21	Fuel Volume Regulator Control Exceeded Maximum Control Limit - Signal amplitude > minimum	• Fuel system fault • Fuel volume control valve circuit short circuit to ground, short circuit to power, open circuit, high resistance • Fuel volume control valve internal failure	• Using the manufacturer approved diagnostic system, check for fuel system related DTCs and perform the relevant corrective actions • Refer to the electrical circuit diagrams and check the fuel volume control valve circuit for short circuit to ground, short circuit to power, open circuit, high resistance • Using the manufacturer approved diagnostic system, clear the DTCs and retest. If the fault persists, install a new fuel volume control valve
P117E-22	Fuel Volume Regulator Control Exceeded Minimum Control Limit - Signal amplitude > maximum	• Fuel system fault • Fuel volume control valve circuit short circuit to ground, short circuit to power, open circuit, high resistance • Fuel volume control valve internal failure	• Using the manufacturer approved diagnostic system, check for fuel system related DTCs and perform the relevant corrective actions • Refer to the electrical circuit diagrams and check the fuel volume control valve circuit for short circuit to ground, short circuit to power, open circuit, high resistance • Using the manufacturer approved diagnostic system, clear the DTCs and retest. If the fault persists, install a new fuel volume control valve
P1259-00	Immobilizer to PCM Signal Error - No sub type information	• Immobilizer control module power or ground circuit open circuit, high resistance • Immobilizer signal circuit short circuit to ground, short circuit to power, open circuit, high resistance • Immobilizer control module internal failure • Engine control module internal failure	• Refer to the electrical circuit diagrams and check the immobilizer control module power and ground circuits for open circuit, high resistance • Refer to the electrical circuit diagrams and check the immobilizer signal circuit for short circuit to ground, short circuit to power, open circuit, high resistance • Install a new immobilizer control module as necessary • Using the manufacturer approved diagnostic system, clear the DTCs and retest. If the fault persists, install a new engine control module
P1259-64	Immobilizer to PCM Signal Error - Signal plausibility failure	• Immobilizer control module power or ground circuit open circuit, high resistance • Immobilizer signal circuit short circuit to ground, short circuit to power, open circuit, high resistance • Immobilizer control module internal failure • Engine control module internal failure	• Refer to the electrical circuit diagrams and check the immobilizer control module power and ground circuits for open circuit, high resistance • Refer to the electrical circuit diagrams and check the immobilizer signal circuit for short circuit to ground, short circuit to power, open circuit, high resistance • Install a new immobilizer control module as necessary • Using the manufacturer approved diagnostic system, clear the DTCs and retest. If the fault persists, install a new engine control module
P132B-22	Turbocharger/Supercharger Boost Control A Performance - Signal amplitude > maximum	• Turbocharger fault • Intake system blocked/leaking	• Check the variable geometry turbine vane actuator linkage. Using the manufacturer approved diagnostic system, check for turbocharger related DTCs and perform the relevant corrective actions • Check the intake system for blockages/leaks. Using the manufacturer approved diagnostic system, clear the DTCs and perform routine - Turbo, EGR and Air Path Dynamic Test
P1336-76	Crankshaft/Camshaft Sensor Range/Performance - Wrong mounting position	• Crankshaft position sensor air gap incorrect • Crankshaft position sensor alignment incorrect • Crankshaft position sensor reluctor ring fault • Crankshaft position sensor internal failure • Camshaft position sensor alignment incorrect • Camshaft position sensor internal failure	• Check the crankshaft position sensor air gap at 90° intervals (maximum air gap 4.5mm) • Check the crankshaft position sensor alignment • Check the integrity of the crankshaft position sensor reluctor ring • Install a new crankshaft position sensor as necessary • Check the camshaft position sensor alignment • Install a new camshaft position sensor as necessary
P1336-78	Crankshaft/Camshaft Sensor Range/Performance - Alignment or adjustment incorrect	• Crankshaft position sensor air gap incorrect • Crankshaft position sensor alignment incorrect • Crankshaft position sensor reluctor ring fault • Crankshaft position sensor internal failure • Camshaft position sensor alignment incorrect • Camshaft position sensor internal failure	• Check the crankshaft position sensor air gap at 90° intervals (maximum air gap 4.5mm) • Check the crankshaft position sensor alignment • Check the integrity of the crankshaft position sensor reluctor ring • Install a new crankshaft position sensor as necessary • Check the camshaft position sensor alignment • Install a new camshaft position sensor as necessary

P138A-11	Glow Plug Control Module Control Circuit Range/Performance - Circuit short to ground	NOTE: Circuit reference - GPFB - • Glow plug control module feedback circuit short circuit to ground • Glow plug control module internal failure	• Refer to the electrical circuit diagrams and check the glow plug control module feedback circuit for short circuit to ground • Using the manufacturer approved diagnostic system, clear the DTCs and retest. If the fault persists, install a new glow plug control module
P138A-12	Glow Plug Control Module Control Circuit Range/Performance - Circuit short to battery	NOTE: Circuit reference - GPFB - • Glow plug control module feedback circuit short circuit to power • Glow plug control module internal failure	• Refer to the electrical circuit diagrams and check the glow plug control module feedback circuit for short circuit to power • Using the manufacturer approved diagnostic system, clear the DTCs and retest. If the fault persists, install a new glow plug control module
P138A-15	Glow Plug Control Module Control Circuit Range/Performance - Circuit short to battery or open	NOTE: Circuit reference - GPC - • Glow plug control module control circuit short circuit to power, open circuit, high resistance • Glow plug control module internal failure	• Refer to the electrical circuit diagrams and check the glow plug control module control circuit for short circuit to power, open circuit, high resistance • Using the manufacturer approved diagnostic system, clear the DTCs and retest. If the fault persists, install a new glow plug control module
P1500-00	Vehicle Speed Sensor - No sub type information	• Vehicle speed sensor circuit short circuit to ground, short circuit to power, open circuit, high resistance • High speed CAN bus circuit short circuit to ground, short circuit to power, open circuit, high resistance	• Refer to the electrical circuit diagrams and check the vehicle speed sensor circuit for short circuit to ground, short circuit to power, open circuit, high resistance • Using the manufacturer approved diagnostic system, perform a CAN network integrity test. Refer to the electrical circuit diagrams and check the high speed CAN bus circuit for short circuit to ground, short circuit to power, open circuit, high resistance
P150A-01	Cylinder 1 Injector Circuit Range/Performance - General electrical failure	• Cylinder 1 fuel injector circuit short circuit to ground, short circuit to power, open circuit, high resistance • Cylinder 1 fuel injector internal failure	• Refer to the electrical circuit diagrams and check the cylinder 1 fuel injector circuit for short circuit to ground, short circuit to power, open circuit, high resistance • Using the manufacturer approved diagnostic system, clear the DTCs and retest. If the fault persists, install a new cylinder 1 fuel injector
P150B-01	Cylinder 2 Injector Circuit Range/Performance - General electrical failure	• Cylinder 2 fuel injector circuit short circuit to ground, short circuit to power, open circuit, high resistance • Cylinder 2 fuel injector internal failure	• Refer to the electrical circuit diagrams and check the cylinder 2 fuel injector circuit for short circuit to ground, short circuit to power, open circuit, high resistance • Using the manufacturer approved diagnostic system, clear the DTCs and retest. If the fault persists, install a new cylinder 2 fuel injector
P150C-01	Cylinder 3 Injector Circuit Range/Performance - General electrical failure	• Cylinder 3 fuel injector circuit short circuit to ground, short circuit to power, open circuit, high resistance • Cylinder 3 fuel injector internal failure	• Refer to the electrical circuit diagrams and check the cylinder 3 fuel injector circuit for short circuit to ground, short circuit to power, open circuit, high resistance • Using the manufacturer approved diagnostic system, clear the DTCs and retest. If the fault persists, install a new cylinder 3 fuel injector
P150D-01	Cylinder 4 Injector Circuit Range/Performance - General electrical failure	• Cylinder 4 fuel injector circuit short circuit to ground, short circuit to power, open circuit, high resistance • Cylinder 4 fuel injector internal failure	• Refer to the electrical circuit diagrams and check the cylinder 4 fuel injector circuit for short circuit to ground, short circuit to power, open circuit, high resistance • Using the manufacturer approved diagnostic system, clear the DTCs and retest. If the fault persists, install a new cylinder 4 fuel injector
P1551-01	Cylinder 1 Injector Circuit Range/Performance - General electrical failure	• Cylinder 1 fuel injector circuit short circuit to ground, short circuit to power, open circuit, high resistance • Cylinder 1 fuel injector internal failure	• Refer to the electrical circuit diagrams and check the cylinder 1 fuel injector circuit for short circuit to ground, short circuit to power, open circuit, high resistance • Using the manufacturer approved diagnostic system, clear the DTCs and retest. If the fault persists, install a new cylinder 1 fuel injector
P1552-01	Cylinder 2 Injector Circuit Range/Performance - General electrical failure	• Cylinder 2 fuel injector circuit short circuit to ground, short circuit to power, open circuit, high resistance • Cylinder 2 fuel injector internal failure	• Refer to the electrical circuit diagrams and check the cylinder 2 fuel injector circuit for short circuit to ground, short circuit to power, open circuit, high resistance • Using the manufacturer approved diagnostic system, clear the DTCs and retest. If the fault persists, install a new cylinder 2 fuel injector

P1553-01	Cylinder 3 Injector Circuit Range/Performance - General electrical failure	• Cylinder 3 fuel injector circuit short circuit to ground, short circuit to power, open circuit, high resistance • Cylinder 3 fuel injector internal failure	• Refer to the electrical circuit diagrams and check the cylinder 3 fuel injector circuit for short circuit to ground, short circuit to power, open circuit, high resistance • Using the manufacturer approved diagnostic system, clear the DTCs and retest. If the fault persists, install a new cylinder 3 fuel injector
P1554-01	Cylinder 4 Injector Circuit Range/Performance - General electrical failure	• Cylinder 4 fuel injector circuit short circuit to ground, short circuit to power, open circuit, high resistance • Cylinder 4 fuel injector internal failure	• Refer to the electrical circuit diagrams and check the cylinder 4 fuel injector circuit for short circuit to ground, short circuit to power, open circuit, high resistance • Using the manufacturer approved diagnostic system, clear the DTCs and retest. If the fault persists, install a new cylinder 4 fuel injector
P1598-68	Engine Driven Backwards - Forced Engine Shutdown - Event information	• The engine has been stalled and driven backwards	NOTE: This DTC is for information only. • Using the manufacturer approved diagnostic system, clear the DTCs and retest
P1631-72	Main Relay (power hold) - Actuator stuck open	• Main relay low current circuit open circuit, high resistance • Main relay high current circuit short circuit to ground, open circuit, high resistance	• Refer to the electrical circuit diagrams and check the main relay low current circuit for open circuit, high resistance • Refer to the electrical circuit diagrams and check the main relay high current circuit for short circuit to ground, open circuit, high resistance
P1631-73	Main Relay (power hold) - Actuator stuck closed	• Main relay high current circuit short circuit to power	• Refer to the electrical circuit diagrams and check the main relay high current circuit for short circuit to power
P1631-92	Main Relay (power hold) - Performance or incorrect operation	• Main relay low current circuit open circuit, high resistance • Main relay high current circuit short circuit to ground, short circuit to power, open circuit, high resistance	NOTE: This DTC may be set if the battery is disconnected before the engine management system has powered down. • Refer to the electrical circuit diagrams and check the main relay low current circuit for open circuit, high resistance • Refer to the electrical circuit diagrams and check the main relay high current circuit for short circuit to ground, short circuit to power, open circuit, high resistance
P164C-00	Internal Control Module Start-Stop Performance - No sub type information	• Starter has been disabled by the engine control module following a high number of failed crank attempts	NOTE: This DTC is set when the starter has been disabled by the engine control module to protect the starting system following a high number of failed crank attempts • Using the manufacturer approved diagnostic system, check for associated DTCs at the same timeline which could indicate a reason for the failed crank attempts and refer to the relevant DTC index. Clear the DTC and retest the system
P1798-00	CAN TCM/INST Circuit Malfunction - No sub type information	• High speed CAN bus circuit short circuit to ground, short circuit to power, open circuit, high resistance	• Using the manufacturer approved diagnostic system, perform a CAN network integrity test. Refer to the electrical circuit diagrams and check the high speed CAN bus circuit for short circuit to ground, short circuit to power, open circuit, high resistance
P2122-16	Throttle/Pedal Position Sensor/Switch D Circuit Low - Circuit voltage below threshold	NOTE: Circuit reference - VSENS APP1 / APP1 / GND APP1 - • Accelerator pedal position sensor 1 circuit short circuit to ground, open circuit, high resistance	• Refer to the electrical circuit diagrams and check the accelerator pedal position sensor 1 circuit for short circuit to ground, open circuit, high resistance
P2123-17	Throttle/Pedal Position Sensor/Switch D Circuit High - Circuit voltage above threshold	NOTE: Circuit reference - VSENS APP1 / APP1 / GND APP1 - • Accelerator pedal position sensor 1 circuit short circuit to power, open circuit, high resistance	• Refer to the electrical circuit diagrams and check the accelerator pedal position sensor 1 circuit for short circuit to power, open circuit, high resistance
P2127-16	Throttle/Pedal Position Sensor/Switch E Circuit Low - Circuit voltage below threshold	NOTE: Circuit reference - VSENS APP2 / APP2 / GND APP2 - • Accelerator pedal position sensor 2 circuit short circuit to ground, open circuit, high resistance	• Refer to the electrical circuit diagrams and check the accelerator pedal position sensor 2 circuit for short circuit to ground, open circuit, high resistance
P2128-17	Throttle/Pedal Position Sensor/Switch E Circuit High - Circuit voltage above threshold	NOTE: Circuit reference - VSENS APP2 / APP2 / GND APP2 - • Accelerator pedal position sensor 2 circuit short circuit to power, open circuit, high resistance	• Refer to the electrical circuit diagrams and check the accelerator pedal position sensor 2 circuit for short circuit to power, open circuit, high resistance

P2138-62	Throttle/Pedal Position Sensor/Switch D / E Voltage Correlation - Signal compare failure	⚠ NOTE: Circuit reference - VSENS APP1 / APP1 / GND APP1 - VSENS APP2 / APP2 / GND APP2 - • Accelerator pedal position sensor 1 circuit short circuit to ground, short circuit to power, open circuit, high resistance • Accelerator pedal position sensor 2 circuit short circuit to ground, short circuit to power, open circuit, high resistance	• Refer to the electrical circuit diagrams and check the accelerator pedal position sensor 1 circuit for short circuit to ground, short circuit to power, open circuit, high resistance • Refer to the electrical circuit diagrams and check the accelerator pedal position sensor 2 circuit for short circuit to ground, short circuit to power, open circuit, high resistance
P213E-77	Fuel Injection System Fault - Forced Engine Shutdown - Commanded position not reachable	• Engine system fault	• Using the manufacturer approved diagnostic system, check for engine system related DTCs and perform the relevant corrective actions
P213E-92	Fuel Injection System Fault - Forced Engine Shutdown - Performance or incorrect operation	• Engine system fault	• Using the manufacturer approved diagnostic system, check for engine system related DTCs and perform the relevant corrective actions
P2141-00	Exhaust Gas Recirculation Throttle Control Circuit A Low - No sub type information	• Throttle actuator circuit short circuit to ground	• Refer to the electrical circuit diagrams and check the throttle actuator circuit for short circuit to ground
P2141-12	Exhaust Gas Recirculation Throttle Control Circuit A Low - Circuit short to battery	• Throttle actuator circuit short circuit to power	• Refer to the electrical circuit diagrams and check the throttle actuator circuit for short circuit to power
P2142-11	Exhaust Gas Recirculation Throttle Control Circuit A High - Circuit short to ground	• Throttle actuator circuit short circuit to ground	• Refer to the electrical circuit diagrams and check the throttle actuator circuit for short circuit to ground
P2142-12	Exhaust Gas Recirculation Throttle Control Circuit A High - Circuit short to battery	• Throttle actuator circuit short circuit to power	• Refer to the electrical circuit diagrams and check the throttle actuator circuit for short circuit to power
P2148-00	Fuel Injector Group A Supply Voltage Circuit High - No sub type information	• Fuel injector circuit fault	• Using the manufacturer approved diagnostic system, check for fuel injector related DTCs and perform the relevant corrective actions
P2228-16	Barometric Pressure Sensor A Circuit Low - Circuit voltage below threshold	• Engine control module internal failure	⚠ NOTE: The barometric pressure sensor is integral to the engine control module. • Using the manufacturer approved diagnostic system, clear the DTCs and retest. If the fault persists, install a new engine control module
P2229-17	Barometric Pressure Sensor A Circuit High - Circuit voltage above threshold	• Engine control module internal failure	⚠ NOTE: The barometric pressure sensor is integral to the engine control module. • Using the manufacturer approved diagnostic system, clear the DTCs and retest. If the fault persists, install a new engine control module
P2230-27	Barometric Pressure Sensor A Circuit Intermittent/Erratic - Signal rate of change above threshold	• Engine control module internal failure	⚠ NOTE: The barometric pressure sensor is integral to the engine control module. • Using the manufacturer approved diagnostic system, clear the DTCs and retest. If the fault persists, install a new engine control module
P2263-21	Turbocharger/Supercharger Boost System Performance - Signal amplitude < minimum	• Intake system blocked/leaking • Manifold absolute pressure and temperature sensor fault • Variable geometry turbine vane actuator fault • Turbocharger failure	• Check the intake system for blockages/leaks. Using the manufacturer approved diagnostic system, clear the DTCs and perform routine - Turbo, EGR and Air Path Dynamic Test • Using the manufacturer approved diagnostic system, check for manifold absolute pressure and temperature sensor related DTCs and perform the relevant corrective actions • Check the variable geometry turbine vane actuator linkage. Using the manufacturer approved diagnostic system, check for turbocharger related DTCs and perform the relevant corrective actions • Check the turbocharger bearing and oil seal. Install a new turbocharger as necessary

P2263-22	Turbocharger/Supercharger Boost System Performance - Signal amplitude > maximum	• Intake system blocked/leaking • Manifold absolute pressure and temperature sensor fault • Variable geometry turbine vane actuator fault • Turbocharger failure	• Check the intake system for blockages/leaks. Using the manufacturer approved diagnostic system, clear the DTCs and perform routine - Turbo, EGR and Air Path Dynamic Test • Using the manufacturer approved diagnostic system, check for manifold absolute pressure and temperature sensor related DTCs and perform the relevant corrective actions • Check the variable geometry turbine vane actuator linkage. Using the manufacturer approved diagnostic system, check for turbocharger related DTCs and perform the relevant corrective actions • Check the turbocharger bearing and oil seal. Install a new turbocharger as necessary
P226B-00	Turbocharger/Supercharger Boost Pressure Too High - Mechanical - No sub type information	• Variable geometry turbine vane actuator fault	• Check the variable geometry turbine vane actuator linkage. Using the manufacturer approved diagnostic system, check for turbocharger related DTCs and perform the relevant corrective actions
P228C-77	Fuel Pressure Regulator 1 Exceeded Control Limits - Pressure Too Low - Commanded position not reachable	• Fuel system fault • Fuel injector stuck open	• Using the manufacturer approved diagnostic system, check for fuel system related DTCs and perform the relevant corrective actions • Check for fuel injector stuck open
P2297-00	O2 Sensor Out of Range During Deceleration Bank 1, Sensor 1 - No sub type information	• Fuel system fault • Heated oxygen sensor circuit short circuit to ground, short circuit to power, open circuit, high resistance • Heated oxygen sensor internal failure	• Using the manufacturer approved diagnostic system, check for fuel system related DTCs and perform the relevant corrective actions • Refer to the electrical circuit diagrams and check the heated oxygen sensor circuit for short circuit to ground, short circuit to power, open circuit, high resistance • Using the manufacturer approved diagnostic system, clear the DTCs and retest. If the fault persists, install a new heated oxygen sensor
P242B-21	Exhaust Gas Temperature Sensor Circuit Range/Performance Bank 1 Sensor 3 - Signal amplitude < minimum	• Engine system fault • Exhaust gas temperature sensor 3 contaminated • Exhaust gas temperature sensor 3 signal circuit short circuit to ground, open circuit, high resistance • Exhaust gas temperature sensor 1 internal failure	• Using the manufacturer approved diagnostic tool check for associated DTCs/events at the same timeline which would influence fuelling or rapid changes in exhaust temperature • Check exhaust gas temperature sensor 3 for contamination • Refer to the electrical circuit diagrams and check the exhaust gas temperature sensor 3 signal circuit for short circuit to ground, open circuit, high resistance • Install a new exhaust gas temperature sensor 3 as necessary
P242B-22	Exhaust Gas Temperature Sensor Circuit Range/Performance Bank 1 Sensor 3 - Signal amplitude > maximum	• Engine system fault • Exhaust gas temperature sensor 3 contaminated • Exhaust gas temperature sensor 3 signal circuit short circuit to power, open circuit, high resistance • Exhaust gas temperature sensor 1 internal failure	• Using the manufacturer approved diagnostic tool check for associated DTCs/events at the same timeline which would influence fuelling or rapid changes in exhaust temperature • Check exhaust gas temperature sensor 3 for contamination • Refer to the electrical circuit diagrams and check the exhaust gas temperature sensor 3 signal circuit for short circuit to power, open circuit, high resistance • Install a new exhaust gas temperature sensor 3 as necessary
P242B-27	Exhaust Gas Temperature Sensor Circuit Range/Performance Bank 1 Sensor 3 - Signal rate of change above threshold	• Engine system fault • Exhaust gas temperature sensor 3 contaminated • Exhaust gas temperature sensor 3 signal circuit short circuit to ground, short circuit to power, open circuit, high resistance • Exhaust gas temperature sensor 1 internal failure	• Using the manufacturer approved diagnostic tool check for associated DTCs/events at the same timeline which would influence fuelling or rapid changes in exhaust temperature • Check exhaust gas temperature sensor 3 for contamination • Refer to the electrical circuit diagrams and check the exhaust gas temperature sensor 3 signal circuit for short circuit to ground, short circuit to power, open circuit, high resistance • Install a new exhaust gas temperature sensor 3 as necessary
P242C-11	Exhaust Gas Temperature Sensor Circuit Low Bank 1 Sensor 3 - Circuit short to ground	⚠ NOTE: Circuit reference - EGT 3 - • Exhaust gas temperature sensor 3 contaminated • Exhaust gas temperature sensor 3 signal circuit short circuit to ground	• Check exhaust gas temperature sensor 3 for contamination • Refer to the electrical circuit diagrams and check the exhaust gas temperature sensor 3 signal circuit for short circuit to ground

P242D-15	Exhaust Gas Temperature Sensor Circuit High Bank 1 Sensor 3 - Circuit short to battery or open	NOTE: Circuit reference - EGT 3 - • Exhaust gas temperature sensor 3 contaminated • Exhaust gas temperature sensor 3 signal circuit short circuit to power, open circuit, high resistance	• Check exhaust gas temperature sensor 3 for contamination • Refer to the electrical circuit diagrams and check the exhaust gas temperature sensor 3 signal circuit for short circuit to power, open circuit, high resistance
P242D-17	Exhaust Gas Temperature Sensor Circuit High Bank 1 Sensor 3 - Circuit voltage above threshold	NOTE: Circuit reference - EGT 3 - • Exhaust gas temperature sensor 3 contaminated • Exhaust gas temperature sensor 3 signal circuit short circuit to power	• Check exhaust gas temperature sensor 3 for contamination • Refer to the electrical circuit diagrams and check the exhaust gas temperature sensor 3 signal circuit for short circuit to power
P243F-76	Diesel Particulate Filter Restriction - Soot Accumulation Too High (Bank 2) - Wrong mounting position	• Diesel particulate filter incorrectly installed • Diesel particulate filter requires regeneration • Customer driving routine does not allow the system to clean the particulate filter	• Check the diesel particulate filter is correctly installed. Clear the DTC and retest the system • Perform a diesel particulate filter regeneration • Advise customer of driving routine required to regenerate diesel particulate filter as stated in the vehicle handbook
P244A-64	Diesel Particulate Filter Differential Pressure Too Low (Bank1) - Signal plausibility failure	• Diesel particulate filter differential pressure sensor circuit short circuit to ground, short circuit to power, open circuit, high resistance • Diesel particulate filter differential pressure sensor internal failure	NOTE: If a new diesel particulate filter pressure sensor or hose lines have been installed or incorrectly routed, or any pressure sensor circuit repairs carried out, the engine control module must learn and store the new diesel particulate filter pressure sensor offset value. The following conditions must be met to allow the diesel particulate filter pressure sensor offset value to be learnt and stored: Using the manufacturer approved diagnostic system, clear DTCs from engine control module, then monitor the datalogger signal 'sump oil temperature - measured' ensuring a minimum of 50 degrees C is achieved. Start engine, run above 500rpm for 2 minutes, then a further 30 seconds at idle. Ensure the engine cooling fan is not running. Set vehicle in park and set ignition status to off. Wait 30 seconds for the engine control module to power down, learn and store diesel particulate filter pressure sensor offset value. This process must be carried out six times, to allow a large negative offset value to adapt back to 0 Hpa • Refer to the electrical circuit diagrams and check the diesel particulate filter differential pressure sensor circuit for short circuit to ground, short circuit to power, open circuit, high resistance • Install a new diesel particulate filter differential pressure sensor as necessary

P2453-11	Diesel Particulate Filter Pressure Sensor A Circuit Range/Performance - Circuit short to ground	• Diesel particulate filter differential pressure sensor circuit short circuit to ground • Diesel particulate filter differential pressure sensor internal failure	NOTE: If a new diesel particulate filter pressure sensor or hose lines have been installed or incorrectly routed, or any pressure sensor circuit repairs carried out, the engine control module must learn and store the new diesel particulate filter pressure sensor offset value. The following conditions must be met to allow the diesel particulate filter pressure sensor offset value to be learnt and stored: Using the manufacturer approved diagnostic system, clear DTCs from engine control module, then monitor the datalogger signal 'sump oil temperature - measured' ensuring a minimum of 50 degrees C is achieved. Start engine, run above 500rpm for 2 minutes, then a further 30 seconds at idle. Ensure the engine cooling fan is not running. Set vehicle in park and set ignition status to off. Wait 30 seconds for the engine control module to power down, learn and store diesel particulate filter pressure sensor offset value. This process must be carried out six times, to allow a large negative offset value to adapt back to 0 Hpa • Refer to the electrical circuit diagrams and check the diesel particulate filter differential pressure sensor circuit for short circuit to ground • Install a new diesel particulate filter differential pressure sensor as necessary
P2453-15	Diesel Particulate Filter Pressure Sensor A Circuit Range/Performance - Circuit short to battery or open	• Diesel particulate filter differential pressure sensor circuit short circuit to power, open circuit, high resistance • Diesel particulate filter differential pressure sensor internal failure	NOTE: If a new diesel particulate filter pressure sensor or hose lines have been installed or incorrectly routed, or any pressure sensor circuit repairs carried out, the engine control module must learn and store the new diesel particulate filter pressure sensor offset value. The following conditions must be met to allow the diesel particulate filter pressure sensor offset value to be learnt and stored: Using the manufacturer approved diagnostic system, clear DTCs from engine control module, then monitor the datalogger signal 'sump oil temperature - measured' ensuring a minimum of 50 degrees C is achieved. Start engine, run above 500rpm for 2 minutes, then a further 30 seconds at idle. Ensure the engine cooling fan is not running. Set vehicle in park and set ignition status to off. Wait 30 seconds for the engine control module to power down, learn and store diesel particulate filter pressure sensor offset value. This process must be carried out six times, to allow a large negative offset value to adapt back to 0 Hpa • Refer to the electrical circuit diagrams and check the diesel particulate filter differential pressure sensor circuit for short circuit to power, open circuit, high resistance • Install a new diesel particulate filter differential pressure sensor as necessary

P2453-21	Diesel Particulate Filter Pressure Sensor A Circuit Range/Performance - Signal amplitude < minimum	• Diesel particulate filter differential pressure sensor circuit short circuit to ground • Diesel particulate filter differential pressure sensor internal failure	NOTE: If a new diesel particulate filter pressure sensor or hose lines have been installed or incorrectly routed, or any pressure sensor circuit repairs carried out, the engine control module must learn and store the new diesel particulate filter pressure sensor offset value. The following conditions must be met to allow the diesel particulate filter pressure sensor offset value to be learnt and stored: Using the manufacturer approved diagnostic system, clear DTCs from engine control module, then monitor the datalogger signal 'sump oil temperature - measured' ensuring a minimum of 50 degrees C is achieved. Start engine, run above 500rpm for 2 minutes, then a further 30 seconds at idle. Ensure the engine cooling fan is not running. Set vehicle in park and set ignition status to off. Wait 30 seconds for the engine control module to power down, learn and store diesel particulate filter pressure sensor offset value. This process must be carried out six times, to allow a large negative offset value to adapt back to 0 Hpa • Refer to the electrical circuit diagrams and check the diesel particulate filter differential pressure sensor circuit for short circuit to ground • Install a new diesel particulate filter differential pressure sensor as necessary
P2453-22	Diesel Particulate Filter Pressure Sensor A Circuit Range/Performance - Signal amplitude > maximum	• Diesel particulate filter differential pressure sensor circuit short circuit to power • Diesel particulate filter differential pressure sensor internal failure	NOTE: If a new diesel particulate filter pressure sensor or hose lines have been installed or incorrectly routed, or any pressure sensor circuit repairs carried out, the engine control module must learn and store the new diesel particulate filter pressure sensor offset value. The following conditions must be met to allow the diesel particulate filter pressure sensor offset value to be learnt and stored: Using the manufacturer approved diagnostic system, clear DTCs from engine control module, then monitor the datalogger signal 'sump oil temperature - measured' ensuring a minimum of 50 degrees C is achieved. Start engine, run above 500rpm for 2 minutes, then a further 30 seconds at idle. Ensure the engine cooling fan is not running. Set vehicle in park and set ignition status to off. Wait 30 seconds for the engine control module to power down, learn and store diesel particulate filter pressure sensor offset value. This process must be carried out six times, to allow a large negative offset value to adapt back to 0 Hpa • Refer to the electrical circuit diagrams and check the diesel particulate filter differential pressure sensor circuit for short circuit to power • Install a new diesel particulate filter differential pressure sensor as necessary

P2453-27	Diesel Particulate Filter Pressure Sensor A Circuit Range/Performance - Signal rate of change above threshold	• Diesel particulate filter differential pressure sensor circuit short circuit to ground, short circuit to power, open circuit, high resistance • Diesel particulate filter differential pressure sensor internal failure	NOTE: If a new diesel particulate filter pressure sensor or hose lines have been installed or incorrectly routed, or any pressure sensor circuit repairs carried out, the engine control module must learn and store the new diesel particulate filter pressure sensor offset value. The following conditions must be met to allow the diesel particulate filter pressure sensor offset value to be learnt and stored: Using the manufacturer approved diagnostic system, clear DTCs from engine control module, then monitor the datalogger signal 'sump oil temperature - measured' ensuring a minimum of 50 degrees C is achieved. Start engine, run above 500rpm for 2 minutes, then a further 30 seconds at idle. Ensure the engine cooling fan is not running. Set vehicle in park and set ignition status to off. Wait 30 seconds for the engine control module to power down, learn and store diesel particulate filter pressure sensor offset value. This process must be carried out six times, to allow a large negative offset value to adapt back to 0 Hpa • Refer to the electrical circuit diagrams and check the diesel particulate filter differential pressure sensor circuit for short circuit to ground, short circuit to power, open circuit, high resistance • Install a new diesel particulate filter differential pressure sensor as necessary
P2453-29	Diesel Particulate Filter Pressure Sensor A Circuit Range/Performance - Signal invalid	• Diesel particulate filter differential pressure sensor circuit short circuit to ground, short circuit to power, open circuit, high resistance • Diesel particulate filter differential pressure sensor internal failure	NOTE: If a new diesel particulate filter pressure sensor or hose lines have been installed or incorrectly routed, or any pressure sensor circuit repairs carried out, the engine control module must learn and store the new diesel particulate filter pressure sensor offset value. The following conditions must be met to allow the diesel particulate filter pressure sensor offset value to be learnt and stored: Using the manufacturer approved diagnostic system, clear DTCs from engine control module, then monitor the datalogger signal 'sump oil temperature - measured' ensuring a minimum of 50 degrees C is achieved. Start engine, run above 500rpm for 2 minutes, then a further 30 seconds at idle. Ensure the engine cooling fan is not running. Set vehicle in park and set ignition status to off. Wait 30 seconds for the engine control module to power down, learn and store diesel particulate filter pressure sensor offset value. This process must be carried out six times, to allow a large negative offset value to adapt back to 0 Hpa • Refer to the electrical circuit diagrams and check the diesel particulate filter differential pressure sensor circuit for short circuit to ground, short circuit to power, open circuit, high resistance • Install a new diesel particulate filter differential pressure sensor as necessary

P2454-00	Diesel Particulate Filter (DPF) Pressure Sensor A Circuit Low - No sub type information	• Diesel particulate filter differential pressure sensor circuit short circuit to ground, short circuit to power, open circuit, high resistance • Diesel particulate filter differential pressure sensor internal failure	NOTES: ⚠ If a new diesel particulate filter pressure sensor or hose lines have been installed or incorrectly routed, or any pressure sensor circuit repairs carried out, the engine control module must learn and store the new diesel particulate filter pressure sensor offset value. The following conditions must be met to allow the diesel particulate filter pressure sensor offset value to be learnt and stored: Using the manufacturer approved diagnostic system, clear DTCs from engine control module, then monitor the datalogger signal 'sump oil temperature - measured' ensuring a minimum of 50 degrees C is achieved. Start engine, run above 500rpm for 2 minutes, then a further 30 seconds at idle. Ensure the engine cooling fan is not running. Set vehicle in park and set ignition status to off. Wait 30 seconds for the engine control module to power down, learn and store diesel particulate filter pressure sensor offset value. This process must be carried out six times, to allow a large negative offset value to adapt back to 0 Hpa ⚠ This DTC is set when the particulate pressure sensor fails a plausibility check. • Using the manufacturer approved diagnostic system, check datalogger signal - Particulate Filter Differential Pressure Sensor Voltage Bank 1 (0x03DB). Refer to the electrical circuit diagrams and check the diesel particulate filter differential pressure sensor circuit for short circuit to ground, short circuit to power, open circuit, high resistance • Install a new diesel particulate filter differential pressure sensor as necessary
P2454-21	Diesel Particulate Filter (DPF) Pressure Sensor A Circuit Low - Signal amplitude < minimum	• Diesel particulate filter differential pressure sensor circuit short circuit to ground, short circuit to power, open circuit, high resistance • Diesel particulate filter differential pressure sensor internal failure	NOTES: ⚠ If a new diesel particulate filter pressure sensor or hose lines have been installed or incorrectly routed, or any pressure sensor circuit repairs carried out, the engine control module must learn and store the new diesel particulate filter pressure sensor offset value. The following conditions must be met to allow the diesel particulate filter pressure sensor offset value to be learnt and stored: Using the manufacturer approved diagnostic system, clear DTCs from engine control module, then monitor the datalogger signal 'sump oil temperature - measured' ensuring a minimum of 50 degrees C is achieved. Start engine, run above 500rpm for 2 minutes, then a further 30 seconds at idle. Ensure the engine cooling fan is not running. Set vehicle in park and set ignition status to off. Wait 30 seconds for the engine control module to power down, learn and store diesel particulate filter pressure sensor offset value. This process must be carried out six times, to allow a large negative offset value to adapt back to 0 Hpa ⚠ This DTC is set when the particulate pressure sensor fails a plausibility check. • Using the manufacturer approved diagnostic system, check datalogger signal - Particulate Filter Differential Pressure Sensor Voltage Bank 1 (0x03DB). Refer to the electrical circuit diagrams and check the diesel particulate filter differential pressure sensor circuit for short circuit to ground, short circuit to power, open circuit, high resistance • Install a new diesel particulate filter differential pressure sensor as necessary

P2455-00	Diesel Particulate Filter (DPF) Pressure Sensor A Circuit High - No sub type information	• Diesel particulate filter differential pressure sensor circuit short circuit to ground, short circuit to power, open circuit, high resistance • Diesel particulate filter differential pressure sensor internal failure	NOTES: ⚠ If a new diesel particulate filter pressure sensor or hose lines have been installed or incorrectly routed, or any pressure sensor circuit repairs carried out, the engine control module must learn and store the new diesel particulate filter pressure sensor offset value. The following conditions must be met to allow the diesel particulate filter pressure sensor offset value to be learnt and stored: Using the manufacturer approved diagnostic system, clear DTCs from engine control module, then monitor the datalogger signal 'sump oil temperature - measured' ensuring a minimum of 50 degrees C is achieved. Start engine, run above 500rpm for 2 minutes, then a further 30 seconds at idle. Ensure the engine cooling fan is not running. Set vehicle in park and set ignition status to off. Wait 30 seconds for the engine control module to power down, learn and store diesel particulate filter pressure sensor offset value. This process must be carried out six times, to allow a large negative offset value to adapt back to 0 Hpa ⚠ This DTC is set when the particulate pressure sensor fails a plausibility check. • Using the manufacturer approved diagnostic system, check datalogger signal - Particulate Filter Differential Pressure Sensor Voltage Bank 1 (0x03DB). Refer to the electrical circuit diagrams and check the diesel particulate filter differential pressure sensor circuit for short circuit to ground, short circuit to power, open circuit, high resistance • Install a new diesel particulate filter differential pressure sensor as necessary
P2455-22	Diesel Particulate Filter (DPF) Pressure Sensor A Circuit High - Signal amplitude > maximum	• Diesel particulate filter differential pressure sensor circuit short circuit to ground, short circuit to power, open circuit, high resistance • Diesel particulate filter differential pressure sensor internal failure	NOTES: ⚠ If a new diesel particulate filter pressure sensor or hose lines have been installed or incorrectly routed, or any pressure sensor circuit repairs carried out, the engine control module must learn and store the new diesel particulate filter pressure sensor offset value. The following conditions must be met to allow the diesel particulate filter pressure sensor offset value to be learnt and stored: Using the manufacturer approved diagnostic system, clear DTCs from engine control module, then monitor the datalogger signal 'sump oil temperature - measured' ensuring a minimum of 50 degrees C is achieved. Start engine, run above 500rpm for 2 minutes, then a further 30 seconds at idle. Ensure the engine cooling fan is not running. Set vehicle in park and set ignition status to off. Wait 30 seconds for the engine control module to power down, learn and store diesel particulate filter pressure sensor offset value. This process must be carried out six times, to allow a large negative offset value to adapt back to 0 Hpa ⚠ This DTC is set when the particulate pressure sensor fails a plausibility check. • Using the manufacturer approved diagnostic system, check datalogger signal - Particulate Filter Differential Pressure Sensor Voltage Bank 1 (0x03DB). Refer to the electrical circuit diagrams and check the diesel particulate filter differential pressure sensor circuit for short circuit to ground, short circuit to power, open circuit, high resistance • Install a new diesel particulate filter differential pressure sensor as necessary

P2455-22	Diesel Particulate Filter (DPF) Pressure Sensor A Circuit High - Signal amplitude > maximum	• Diesel particulate filter differential pressure sensor circuit short circuit to ground, short circuit to power, open circuit, high resistance • Diesel particulate filter differential pressure sensor internal failure	NOTES: If a new diesel particulate filter pressure sensor or hose lines have been installed or incorrectly routed, or any pressure sensor circuit repairs carried out, the engine control module must learn and store the new diesel particulate filter pressure sensor offset value. The following conditions must be met to allow the diesel particulate filter pressure sensor offset value to be learnt and stored: Using the manufacturer approved diagnostic system, clear DTCs from engine control module, then monitor the datalogger signal 'sump oil temperature - measured' ensuring a minimum of 50 degrees C is achieved. Start engine, run above 500rpm for 2 minutes, then a further 30 seconds at idle. Ensure the engine cooling fan is not running. Set vehicle in park and set ignition status to off. Wait 30 seconds for the engine control module to power down, learn and store diesel particulate filter pressure sensor offset value. This process must be carried out six times, to allow a large negative offset value to adapt back to 0 Hpa This DTC is set when the particulate pressure sensor fails a plausibility check. • Using the manufacturer approved diagnostic system, check datalogger signal - Particulate Filter Differential Pressure Sensor Voltage Bank 1 (0x03DB). Refer to the electrical circuit diagrams and check the diesel particulate filter differential pressure sensor circuit for short circuit to ground, short circuit to power, open circuit, high resistance • Install a new diesel particulate filter differential pressure sensor as necessary
P2456-00	Diesel Particulate Filter (DPF) Pressure Sensor A Circuit Intermittent/Erratic - No sub type information	• Diesel particulate filter differential pressure sensor circuit short circuit to ground, short circuit to power, open circuit, high resistance • Diesel particulate filter differential pressure sensor hoses restricted/split • Diesel particulate filter differential pressure sensor internal failure	NOTE: If a new diesel particulate filter pressure sensor or hose lines have been installed or incorrectly routed, or any pressure sensor circuit repairs carried out, the engine control module must learn and store the new diesel particulate filter pressure sensor offset value. The following conditions must be met to allow the diesel particulate filter pressure sensor offset value to be learnt and stored: Using the manufacturer approved diagnostic system, clear DTCs from engine control module, then monitor the datalogger signal 'sump oil temperature - measured' ensuring a minimum of 50 degrees C is achieved. Start engine, run above 500rpm for 2 minutes, then a further 30 seconds at idle. Ensure the engine cooling fan is not running. Set vehicle in park and set ignition status to off. Wait 30 seconds for the engine control module to power down, learn and store diesel particulate filter pressure sensor offset value. This process must be carried out six times, to allow a large negative offset value to adapt back to 0 Hpa • Refer to the electrical circuit diagrams and check the diesel particulate filter differential pressure sensor circuit for short circuit to ground, short circuit to power, open circuit, high resistance • Check the diesel particulate filter differential pressure sensor hoses for restrictions/splits • Install a new diesel particulate filter differential pressure sensor as necessary

P2463-00	Diesel Particulate Filter Restriction - Soot Accumulation (Bank 1) - No sub type information	• Diesel particulate filter requires regeneration • Customer driving routine does not allow the system to clean the particulate filter	• Perform a diesel particulate filter regeneration • Advise customer of driving routine required to regenerate diesel particulate filter as stated in the vehicle handbook
P246C-00	Diesel Particulate Filter Restriction - Forced Limited Power (Bank 1) - No sub type information	• Diesel particulate filter requires replacement	• Check and install a new diesel particulate filter as required
P2505-13	ECM / PCM Power Input Signal - Circuit open	• Main relay low current circuit short circuit to power, open circuit, high resistance • Main relay high current circuit open circuit, high resistance	• Refer to the electrical circuit diagrams and check the main relay low current circuit for short circuit to power, open circuit, high resistance • Refer to the electrical circuit diagrams and check the main relay high current circuit for open circuit, high resistance
P2507-11	ECM / PCM Power Input Signal Low - Circuit short to ground	• Main relay high current circuit short circuit to ground	• Refer to the electrical circuit diagrams and check the main relay high current circuit for short circuit to ground
P2508-12	ECM / PCM Power Input Signal High - Circuit short to battery	• Main relay high current circuit short circuit to power	• Refer to the electrical circuit diagrams and check the main relay high current circuit for short circuit to power
P253F-00	Engine Oil Deteriorated - No sub type information	• The engine control module has detected deterioration of the engine oil	• Refer to the relevant section of the workshop manual and drain and refill the engine oil. Using the manufacturer approved diagnostic system, clear the DTCs and retest
P2564-23	Turbocharger Boost Control Position Sensor A Circuit Low - Signal stuck low	• Variable geometry turbine vane position sensor circuit short circuit to ground • Variable geometry turbine vane actuator circuit short circuit to ground, short circuit to power, open circuit, high resistance	• Refer to the electrical circuit diagrams and check the variable geometry turbine vane position sensor circuit for short circuit to ground. Repair the circuit as necessary. Using the manufacturer approved diagnostic system, clear the DTCs and perform routine - Turbo, EGR and Air Path Dynamic Test • Refer to the electrical circuit diagrams and check the variable geometry turbine vane actuator circuit for short circuit to ground, short circuit to power, open circuit, high resistance. Repair the circuit as necessary. Using the manufacturer approved diagnostic system, clear the DTCs and perform routine - Turbo, EGR and Air Path Dynamic Test
P2456-27	Diesel Particulate Filter (DPF) Pressure Sensor A Circuit Intermittent/Erratic - Signal rate of change above threshold	• Diesel particulate filter differential pressure sensor circuit short circuit to ground, short circuit to power, open circuit, high resistance • Diesel particulate filter differential pressure sensor hoses restricted/split • Diesel particulate filter differential pressure sensor internal failure	⚠ NOTE: If a new diesel particulate filter pressure sensor or hose lines have been installed or incorrectly routed, or any pressure sensor circuit repairs carried out, the engine control module must learn and store the new diesel particulate filter pressure sensor offset value. The following conditions must be met to allow the diesel particulate filter pressure sensor offset value to be learnt and stored: Using the manufacturer approved diagnostic system, clear DTCs from engine control module, then monitor the datalogger signal 'sump oil temperature - measured' ensuring a minimum of 50 degrees C is achieved. Start engine, run above 500rpm for 2 minutes, then a further 30 seconds at idle. Ensure the engine cooling fan is not running. Set vehicle in park and set ignition status to off. Wait 30 seconds for the engine control module to power down, learn and store diesel particulate filter pressure sensor offset value. This process must be carried out six times, to allow a large negative offset value to adapt back to 0 Hpa • Refer to the electrical circuit diagrams and check the diesel particulate filter differential pressure sensor circuit for short circuit to ground, short circuit to power, open circuit, high resistance • Check the diesel particulate filter differential pressure sensor hoses for restrictions/splits • Install a new diesel particulate filter differential pressure sensor as necessary
P2459-65	Diesel Particulate Filter Regeneration Frequency (Bank 1) - Signal has too few transitions/events	• Diesel particulate filter requires regeneration • Customer driving routine does not allow the system to clean the particulate filter	⚠ NOTE: If DTC P2459-65 or AMBER warning light is displayed with no other reported messages, no repair is required. If the vehicle is driven on a highway AS DIRECTED IN THE HANDBOOK then the light will be extinguished and the system self healed, nothing more than this is required • Perform a diesel particulate filter regeneration • Advise customer of driving routine required to regenerate diesel particulate filter as stated in the vehicle handbook

P2565-24	Turbocharger Boost Control Position Sensor A Circuit High - Signal stuck high	• Variable geometry turbine vane position sensor circuit short circuit to power • Variable geometry turbine vane actuator circuit short circuit to ground, short circuit to power, open circuit, high resistance	• Refer to the electrical circuit diagrams and check the variable geometry turbine vane position sensor circuit for short circuit to power. Repair the circuit as necessary. Using the manufacturer approved diagnostic system, clear the DTCs and perform routine - Turbo, EGR and Air Path Dynamic Test • Refer to the electrical circuit diagrams and check the variable geometry turbine vane actuator circuit for short circuit to ground, short circuit to power, open circuit, high resistance. Repair the circuit as necessary. Using the manufacturer approved diagnostic system, clear the DTCs and perform routine - Turbo, EGR and Air Path Dynamic Test
P268C-41	Cylinder 1 Injector Data Incompatible - General checksum failure	• Injector calibration data held in the engine control module is different to that read from the injector • Injector calibration data being entered does not match the expected values	• Using the manufacturer approved diagnostic system reprogram the injector correction codes
P268C-55	Cylinder 1 Injector Data Incompatible - Not configured	• Injector calibration data held in the engine control module is different to that read from the injector • Injector calibration data not stored/programmed	• Using the manufacturer approved diagnostic system reprogram the injector correction codes
P268D-41	Cylinder 2 Injector Data Incompatible - General checksum failure	• Injector calibration data held in the engine control module is different to that read from the injector • Injector calibration data being entered does not match the expected values	• Using the manufacturer approved diagnostic system reprogram the injector correction codes
P268D-55	Cylinder 2 Injector Data Incompatible - Not configured	• Injector calibration data held in the engine control module is different to that read from the injector • Injector calibration data not stored/programmed	• Using the manufacturer approved diagnostic system reprogram the injector correction codes
P268E-41	Cylinder 3 Injector Data Incompatible - General checksum failure	• Injector calibration data held in the engine control module is different to that read from the injector • Injector calibration data being entered does not match the expected values	• Using the manufacturer approved diagnostic system reprogram the injector correction codes
P268E-55	Cylinder 3 Injector Data Incompatible - Not configured	• Injector calibration data held in the engine control module is different to that read from the injector • Injector calibration data not stored/programmed	• Using the manufacturer approved diagnostic system reprogram the injector correction codes
P268F-41	Cylinder 4 Injector Data Incompatible - General checksum failure	• Injector calibration data held in the engine control module is different to that read from the injector • Injector calibration data being entered does not match the expected values	• Using the manufacturer approved diagnostic system reprogram the injector correction codes
P268F-55	Cylinder 4 Injector Data Incompatible - Not configured	• Injector calibration data held in the engine control module is different to that read from the injector • Injector calibration data not stored/programmed	• Using the manufacturer approved diagnostic system reprogram the injector correction codes
U0001-88	High Speed CAN Communication Bus - Bus off	• High speed CAN bus circuit short circuit to ground, short circuit to power, open circuit, high resistance	• Using the manufacturer approved diagnostic system, perform a CAN network integrity test. Refer to the electrical circuit diagrams and check the high speed CAN bus circuit for short circuit to ground, short circuit to power, open circuit, high resistance
U0300-00	Internal Control Module Software Incompatibility - No sub type information	• Car configuration file mismatch with vehicle specification	• Using the manufacturer approved diagnostic system, check and up-date the car configuration file as necessary. Clear the DTCs and retest

Diagnostic Trouble Code Index - DTC: Instrument Cluster (IPC)

Description and Operation

⚠ CAUTION: Diagnosis by substitution from a donor vehicle is **NOT** acceptable. Substitution of control modules does not guarantee confirmation of a fault, and may also cause additional faults in the vehicle being tested and/or the donor vehicle.

NOTES:

△ Guided Diagnostics must be followed and the repair advised by the process completed. Failure to do so may result in the rejection of any warranty claim made.

△ If a control module or a component is at fault and the vehicle remains under manufacturer warranty, refer to the Warranty Policy and Procedures manual, or determine if any prior approval program is in operation, prior to the installation of a new module/component.

△ Generic scan tools may not read the codes listed, or may read only 5-digit codes. Match the 5 digits from the scan tool to the first 5 digits of the 7-digit code listed to identify the fault (the last 2 digits give extra information read by the manufacturer-approved diagnostic system).

△ When performing voltage or resistance tests, always use a digital multimeter accurate to three decimal places, and with an up-to-date calibration certificate. When testing resistance always take the resistance of the digital multimeter leads into account.

△ Check and rectify basic faults before beginning diagnostic routines involving pinpoint tests.

△ Inspect connectors for signs of water ingress, and pins for damage and/or corrosion.

△ If DTCs are recorded and, after performing the pinpoint tests, a fault is not present, an intermittent concern may be the cause. Always check for loose connections and corroded terminals.

The table below lists all Diagnostic Trouble Codes (DTCs) that could be set. For additional diagnosis and testing information, refer to the relevant Diagnosis and Testing section in the workshop manual.

DTC	Description	Possible Causes	Action
B1A75-01	Fuel Sender 1 - General electrical failure	• Fuel level sensor circuit short circuit to ground, short circuit to power, open circuit, high resistance	• Refer to the electrical circuit diagrams and check the fuel level sensor circuit for short circuit to ground, short circuit to power, open circuit, high resistance
B1A81-64	Internal Trip Switch - Signal plausibility failure	• Trip switch stuck active • Instrument cluster internal failure	• Test the operation of the trip switch • Using the manufacturer approved diagnostic system, clear the DTCs and retest. If the fault persists, install a new instrument cluster
P1602-31	Immobilizer/ECM Communication Error - No signal	• Immobilizer signal circuit short circuit to ground, short circuit to power, open circuit, high resistance • Immobilizer control module power or ground circuit open circuit, high resistance • Engine system fault • Immobilizer control module internal failure	• Refer to the electrical circuit diagrams and check the immobilizer signal circuit for short circuit to ground, short circuit to power, open circuit, high resistance • Refer to the electrical circuit diagrams and check the immobilizer control module power and ground circuits for open circuit, high resistance • Using the manufacturer approved diagnostic system, check the engine control module for related DTCs and refer to the relevant DTC index • Using the manufacturer approved diagnostic system, clear the DTCs and retest. If the fault persists, install a new immobilizer control module
U0001-88	High Speed CAN Communication Bus - Bus off	• High speed CAN bus circuit short circuit to ground, short circuit to power, open circuit, high resistance	• Using the manufacturer approved diagnostic system, perform a CAN network integrity test. Refer to the electrical circuit diagrams and check the high speed CAN bus circuit for short circuit to ground, short circuit to power, open circuit, high resistance

U0100-87	Lost Communication With ECM/PCM A - Missing message	• Engine control module power or ground circuit open circuit, high resistance • High speed CAN bus circuit short circuit to ground, short circuit to power, open circuit, high resistance • Engine system fault	• Refer to the electrical circuit diagrams and check the engine control module power and ground circuits for open circuit, high resistance • Using the manufacturer approved diagnostic system, perform a CAN network integrity test. Refer to the electrical circuit diagrams and check the high speed CAN bus circuit for short circuit to ground, short circuit to power, open circuit, high resistance • Using the manufacturer approved diagnostic system, check the engine control module for related DTCs and refer to the relevant DTC index
U0401-68	Invalid Data Received from ECM/PCM A - Event information	• Missing/invalid data from the engine control module	• Using the manufacturer approved diagnostic system, check the engine control module for related DTCs and refer to the relevant DTC index
U2101-00	Control Module Configuration Incompatible - No sub type information	• Instrument cluster is not configured correctly	• Using the manufacturer approved diagnostic system, re-configure the instrument cluster with the latest level software
U3000-17	Control Module - Circuit voltage above threshold	• Battery/charging system fault	NOTE: This DTC may be set if the vehicle has been jump started. • Refer to the relevant section of the workshop manual and test the battery and charging system
U3000-46	Control Module - Calibration/parameter memory failure	• Instrument cluster power or ground circuit open circuit, high resistance • Instrument cluster internal failure	• Refer to the electrical circuit diagrams and check the instrument cluster power and ground circuits for open circuit, high resistance • Using the manufacturer approved diagnostic system, clear the DTCs and retest. If the fault persists, install a new instrument cluster

Diagnostic Trouble Code Index: Anti-lock Brake System (ABS)

Description and Operation

CAUTION: Diagnosis by substitution from a donor vehicle is **NOT** acceptable. Substitution of control modules does not guarantee confirmation of a fault, and may also cause additional faults in the vehicle being tested and/or the donor vehicle.

NOTES:

If the control module or a component is suspect and the vehicle remains under manufacturer warranty, refer to the Warranty Policy and Procedures manual (section B1.2), or determine if any prior approval programme is in operation, prior to the installation of a new module/ component.

Generic scan tools may not read the codes listed, or may read only 5-digit codes. Match the 5 digits from the scan tool to the first 5 digits of the 7-digit code listed to identify the fault (the last 2 digits give extra information read by the manufacturer-approved diagnostic system).

When performing voltage or resistance tests, always use a digital multimeter accurate to three decimal places, and with an up-to-date calibration certificate. When testing resistance always take the resistance of the digital multimeter leads into account.

Check and rectify basic faults before beginning diagnostic routines involving pinpoint tests.

Inspect connectors for signs of water ingress, and pins for damage and/or corrosion.

If DTCs are recorded and, after performing the pinpoint tests, a fault is not present, an intermittent concern may be the cause. Always check for loose connections and corroded terminals.

The table below lists all Diagnostic Trouble Codes (DTCs) that could be logged in the Anti-lock Brake System Control Module (ABS). For additional diagnosis and testing information, refer to the relevant Diagnosis and Testing section in the workshop manual.

Refer to: Anti-Lock Control - Traction Control (206-09A Anti-Lock Control - Traction Control, Diagnosis and Testing)

DTC	Description	Possible Causes	Action
B1A84-56	Car Configuration Data - Invalid/incomplete configuration	• Anti-lock brake system control module is not configured correctly	• Using the manufacturer approved diagnostic system, re-configure the anti-lock brake system control module with the latest level software - Configure New Module
C0030-38	Left Front Tone Wheel - Signal frequency incorrect	• Front left wheel speed sensor reluctor ring damaged • Front left wheel bearing failure • Front left wheel/tire size incorrect	• Check the integrity of the front left wheel speed sensor reluctor ring • Check the front left wheel bearing • Check that the wheels and tires comply with the manufacturer's specification for the vehicle
C0031-14	Left Front Wheel Speed Sensor - Circuit short to ground or open	• Front left wheel speed sensor signal circuit short circuit to ground, open circuit, high resistance	• Refer to the electrical circuit diagrams and check the front left wheel speed sensor signal circuit for short circuit to ground, open circuit, high resistance
C0031-25	Left Front Wheel Speed Sensor - Signal shape/waveform failure	• Front left wheel speed sensor signal circuit short circuit to ground	• Using the manufacturer approved diagnostic system, check datalogger signal - Left Front Wheel Speed Sensor Input (0x2B06). Refer to the electrical circuit diagrams and check the front left wheel speed sensor signal circuit for short circuit to ground
C0031-2F	Left Front Wheel Speed Sensor - Signal erratic	• Front left wheel speed sensor signal circuit short circuit to ground, short circuit to power, open circuit, high resistance	• Using the manufacturer approved diagnostic system, check datalogger signal - Left Front Wheel Speed Sensor Input (0x2B06). Refer to the electrical circuit diagrams and check the front left wheel speed sensor signal circuit for short circuit to ground, short circuit to power, open circuit, high resistance
C0031-31	Left Front Wheel Speed Sensor - No signal	• Front left wheel speed sensor signal circuit short circuit to ground, short circuit to power, open circuit, high resistance • Front left wheel speed sensor reluctor ring damaged	• Refer to the electrical circuit diagrams and check the front left wheel speed sensor signal circuit for short circuit to ground, short circuit to power, open circuit, high resistance • Check the integrity of the front left wheel speed sensor reluctor ring
C0031-62	Left Front Wheel Speed Sensor - Signal compare failure	• Front left wheel speed sensor power circuit short circuit to ground, short circuit to power, open circuit, high resistance • Front left wheel speed sensor signal circuit short circuit to ground, short circuit to power, open circuit, high resistance	• Refer to the electrical circuit diagrams and check the front left wheel speed sensor power circuit for short circuit to ground, short circuit to power, open circuit, high resistance • Refer to the electrical circuit diagrams and check the front left wheel speed sensor signal circuit for short circuit to ground, short circuit to power, open circuit, high resistance
C0031-64	Left Front Wheel Speed Sensor - Signal plausibility failure	• Front left wheel speed sensor power circuit short circuit to ground, short circuit to power, open circuit, high resistance • Front left wheel speed sensor signal circuit short circuit to ground, short circuit to power, open circuit, high resistance • Front left wheel/tire size incorrect	• Refer to the electrical circuit diagrams and check the front left wheel speed sensor power circuit for short circuit to ground, short circuit to power, open circuit, high resistance • Refer to the electrical circuit diagrams and check the front left wheel speed sensor signal circuit for short circuit to ground, short circuit to power, open circuit, high resistance • Check that the wheels and tires comply with the manufacturer's specification for the vehicle
C0033-38	Right Front Tone Wheel - Signal frequency incorrect	• Front right wheel speed sensor reluctor ring damaged • Front right wheel bearing failure • Front right wheel/tire size incorrect	• Check the integrity of the front right wheel speed sensor reluctor ring • Check the front right wheel bearing • Check that the wheels and tires comply with the manufacturer's specification for the vehicle
C0034-14	Right Front Wheel Speed Sensor - Circuit short to ground or open	• Front right wheel speed sensor signal circuit short circuit to ground, open circuit, high resistance	• Refer to the electrical circuit diagrams and check the front right wheel speed sensor signal circuit for short circuit to ground, open circuit, high resistance
C0034-25	Right Front Wheel Speed Sensor - Signal shape/waveform failure	• Front right wheel speed sensor signal circuit short circuit to ground	• Using the manufacturer approved diagnostic system, check datalogger signal - Right Front Wheel Speed Sensor Input (0x2B07). Refer to the electrical circuit diagrams and check the front right wheel speed sensor signal circuit for short circuit to ground
C0034-2F	Right Front Wheel Speed Sensor - Signal erratic	• Front right wheel speed sensor signal circuit short circuit to ground, short circuit to power, open circuit, high resistance	• Using the manufacturer approved diagnostic system, check datalogger signal - Right Front Wheel Speed Sensor Input (0x2B07). Refer to the electrical circuit diagrams and check the front right wheel speed sensor signal circuit for short circuit to ground, short circuit to power, open circuit, high resistance
C0034-31	Right Front Wheel Speed Sensor - No signal	• Front right wheel speed sensor signal circuit short circuit to ground, short circuit to power, open circuit, high resistance • Front right wheel speed sensor reluctor ring damaged	• Refer to the electrical circuit diagrams and check the front right wheel speed sensor signal circuit for short circuit to ground, short circuit to power, open circuit, high resistance • Check the integrity of the front right wheel speed sensor reluctor ring
C0034-62	Right Front Wheel Speed Sensor - Signal compare failure	• Front right wheel speed sensor power circuit short circuit to ground, short circuit to power, open circuit, high resistance • Front right wheel speed sensor signal circuit short circuit to ground, short circuit to power, open circuit, high resistance	• Refer to the electrical circuit diagrams and check the front right wheel speed sensor power circuit for short circuit to ground, short circuit to power, open circuit, high resistance • Refer to the electrical circuit diagrams and check the front right wheel speed sensor signal circuit for short circuit to ground, short circuit to power, open circuit, high resistance

C0034-64	Right Front Wheel Speed Sensor - Signal plausibility failure	• Front right wheel speed sensor power circuit short circuit to ground, short circuit to power, open circuit, high resistance • Front right wheel speed sensor signal circuit short circuit to ground, short circuit to power, open circuit, high resistance • Front right wheel/tire size incorrect	• Refer to the electrical circuit diagrams and check the front right wheel speed sensor power circuit for short circuit to ground, short circuit to power, open circuit, high resistance • Refer to the electrical circuit diagrams and check the front right wheel speed sensor signal circuit for short circuit to ground, short circuit to power, open circuit, high resistance • Check that the wheels and tires comply with the manufacturer's specification for the vehicle
C0036-38	Left Rear Tone Wheel - Signal frequency incorrect	• Rear left wheel speed sensor reluctor ring damaged • Rear left wheel bearing failure • Rear left wheel/tire size incorrect	• Check the integrity of the rear left wheel speed sensor reluctor ring • Check the rear left wheel bearing • Check that the wheels and tires comply with the manufacturer's specification for the vehicle
C0037-14	Left Rear Wheel Speed Sensor - Circuit short to ground or open	• Rear left wheel speed sensor signal circuit short circuit to ground, open circuit, high resistance	• Refer to the electrical circuit diagrams and check the rear left wheel speed sensor signal circuit for short circuit to ground, open circuit, high resistance
C0037-25	Left Rear Wheel Speed Sensor - Signal shape/waveform failure	• Rear left wheel speed sensor signal circuit short circuit to ground	• Using the manufacturer approved diagnostic system, check datalogger signal - Left Rear Wheel Speed Sensor Input (0x2B08). Refer to the electrical circuit diagrams and check the rear left wheel speed sensor signal circuit for short circuit to ground
C0037-2F	Left Rear Wheel Speed Sensor - Signal erratic	• Rear left wheel speed sensor signal circuit short circuit to ground, short circuit to power, open circuit, high resistance	• Using the manufacturer approved diagnostic system, check datalogger signal - Left Rear Wheel Speed Sensor Input (0x2B08). Refer to the electrical circuit diagrams and check the rear left wheel speed sensor signal circuit for short circuit to ground, short circuit to power, open circuit, high resistance
C0037-31	Left Rear Wheel Speed Sensor - No signal	• Rear left wheel speed sensor signal circuit short circuit to ground, short circuit to power, open circuit, high resistance • Rear left wheel speed sensor reluctor ring damaged	• Refer to the electrical circuit diagrams and check the rear left wheel speed sensor signal circuit for short circuit to ground, short circuit to power, open circuit, high resistance • Check the integrity of the rear left wheel speed sensor reluctor ring
C0037-62	Left Rear Wheel Speed Sensor - Signal compare failure	• Rear left wheel speed sensor power circuit short circuit to ground, short circuit to power, open circuit, high resistance • Rear left wheel speed sensor signal circuit short circuit to ground, short circuit to power, open circuit, high resistance	• Refer to the electrical circuit diagrams and check the rear left wheel speed sensor power circuit for short circuit to ground, short circuit to power, open circuit, high resistance • Refer to the electrical circuit diagrams and check the rear left wheel speed sensor signal circuit for short circuit to ground, short circuit to power, open circuit, high resistance
C0037-64	Left Rear Wheel Speed Sensor - Signal plausibility failure	• Rear left wheel speed sensor power circuit short circuit to ground, short circuit to power, open circuit, high resistance • Rear left wheel speed sensor signal circuit short circuit to ground, short circuit to power, open circuit, high resistance • Rear left wheel/tire size incorrect	• Refer to the electrical circuit diagrams and check the rear left wheel speed sensor power circuit for short circuit to ground, short circuit to power, open circuit, high resistance • Refer to the electrical circuit diagrams and check the rear left wheel speed sensor signal circuit for short circuit to ground, short circuit to power, open circuit, high resistance • Check that the wheels and tires comply with the manufacturer's specification for the vehicle
C0039-38	Right Rear Tone Wheel - Signal frequency incorrect	• Rear right wheel speed sensor reluctor ring damaged • Rear right wheel bearing failure • Rear right wheel/tire size incorrect	• Check the integrity of the rear right wheel speed sensor reluctor ring • Check the rear right wheel bearing • Check that the wheels and tires comply with the manufacturer's specification for the vehicle
C003A-14	Right Rear Wheel Speed Sensor - Circuit short to ground or open	• Rear right wheel speed sensor signal circuit short circuit to ground, open circuit, high resistance	• Refer to the electrical circuit diagrams and check the rear right wheel speed sensor signal circuit for short circuit to ground, open circuit, high resistance
C003A-25	Right Rear Wheel Speed Sensor - Signal shape/waveform failure	• Rear right wheel speed sensor signal circuit short circuit to ground	• Using the manufacturer approved diagnostic system, check datalogger signal - Right Rear Wheel Speed Sensor Input (0x2B09). Refer to the electrical circuit diagrams and check the rear right wheel speed sensor signal circuit for short circuit to ground
C003A-2F	Right Rear Wheel Speed Sensor - Signal erratic	• Rear right wheel speed sensor signal circuit short circuit to ground, short circuit to power, open circuit, high resistance	• Using the manufacturer approved diagnostic system, check datalogger signal - Right Rear Wheel Speed Sensor Input (0x2B09). Refer to the electrical circuit diagrams and check the rear right wheel speed sensor signal circuit for short circuit to ground, short circuit to power, open circuit, high resistance
C003A-31	Right Rear Wheel Speed Sensor - No signal	• Rear right wheel speed sensor signal circuit short circuit to ground, short circuit to power, open circuit, high resistance • Rear right wheel speed sensor reluctor ring damaged	• Refer to the electrical circuit diagrams and check the rear right wheel speed sensor signal circuit for short circuit to ground, short circuit to power, open circuit, high resistance • Check the integrity of the rear right wheel speed sensor reluctor ring
C003A-62	Right Rear Wheel Speed Sensor - Signal compare failure	• Rear right wheel speed sensor power circuit short circuit to ground, short circuit to power, open circuit, high resistance • Rear right wheel speed sensor signal circuit short circuit to ground, short circuit to power, open circuit, high resistance	• Refer to the electrical circuit diagrams and check the rear right wheel speed sensor power circuit for short circuit to ground, short circuit to power, open circuit, high resistance • Refer to the electrical circuit diagrams and check the rear right wheel speed sensor signal circuit for short circuit to ground, short circuit to power, open circuit, high resistance

C003A-64	Right Rear Wheel Speed Sensor - Signal plausibility failure	• Rear right wheel speed sensor power circuit short circuit to ground, short circuit to power, open circuit, high resistance • Rear right wheel speed sensor signal circuit short circuit to ground, short circuit to power, open circuit, high resistance • Rear right wheel/tire size incorrect	• Refer to the electrical circuit diagrams and check the rear right wheel speed sensor power circuit for short circuit to ground, short circuit to power, open circuit, high resistance • Refer to the electrical circuit diagrams and check the rear right wheel speed sensor signal circuit for short circuit to ground, short circuit to power, open circuit, high resistance • Check that the wheels and tires comply with the manufacturer's specification for the vehicle
C0051-28	Steering Wheel Position Sensor - Signal bias level out of range / zero adjustment failure	• Front or rear left/right wheel speed sensor circuits transposed • Incorrect steering angle sensor installed	• Check the wheel speed sensor circuits for correct installation • Using the manufacturer approved diagnostic system, clear the DTCs and retest. If the fault persists, install a new steering angle sensor
C0051-67	Steering Wheel Position Sensor - Signal incorrect after event	• Steering angle sensor internal failure • Anti-lock brake system control module internal failure	• Using the manufacturer approved diagnostic system, clear the DTCs and retest. If the fault persists, install a new steering angle sensor • Using the manufacturer approved diagnostic system, clear the DTCs and retest. If the fault persists, install a new anti-lock brake system control module. Perform routine - Calibrate Acceleration Sensor(s)
C0062-28	Longitudinal Acceleration Sensor - Signal bias level out of range / zero adjustment failure	• Yaw rate lateral acceleration sensor incorrectly installed • Yaw rate lateral acceleration sensor internal failure	• Check the installation of the yaw rate lateral acceleration sensor • Using the manufacturer approved diagnostic system, clear the DTCs and retest. If the fault persists, install a new yaw rate lateral acceleration sensor. Perform routine - Calibrate Acceleration Sensor(s)
C0062-54	Longitudinal Acceleration Sensor - Missing calibration	• Yaw rate lateral acceleration sensor incorrectly installed prior to performing the calibration routine • Vehicle unevenly laden or parked on an uneven surface prior to performing the calibration routine	NOTE: This DTC will set when the calibration routine is performed and fails. • Check the installation of the yaw rate lateral acceleration sensor. Using the manufacturer approved diagnostic system, perform routine - Calibrate Acceleration Sensor(s) • Ensure that the vehicle is unladen and parked on a level surface. Using the manufacturer approved diagnostic system, perform routine - Calibrate Acceleration Sensor(s)
C0062-64	Longitudinal Acceleration Sensor - Signal plausibility failure	• Yaw rate lateral acceleration sensor incorrectly installed • Yaw rate lateral acceleration sensor not calibrated • Yaw rate lateral acceleration sensor power or ground circuit open circuit, high resistance • Private CAN bus circuit short circuit to ground, short circuit to power, open circuit, high resistance • Yaw rate lateral acceleration sensor internal failure	• Check the installation of the yaw rate lateral acceleration sensor • Using the manufacturer approved diagnostic system, perform routine - Calibrate Acceleration Sensor(s) • Refer to the electrical circuit diagrams and check the yaw rate lateral acceleration sensor power and ground circuits for open circuit, high resistance • Refer to the electrical circuit diagrams and check the private CAN bus circuit for short circuit to ground, short circuit to power, open circuit, high resistance • Using the manufacturer approved diagnostic system, clear the DTCs and retest. If the fault persists, install a new yaw rate lateral acceleration sensor
C0063-14	Yaw Rate Sensor - Circuit short to ground or open	• Yaw rate lateral acceleration sensor power or ground circuit open circuit, high resistance • Private CAN bus circuit short circuit to ground, short circuit to power, open circuit, high resistance • Yaw rate lateral acceleration sensor internal failure	• Refer to the electrical circuit diagrams and check the yaw rate lateral acceleration sensor power and ground circuits for open circuit, high resistance • Refer to the electrical circuit diagrams and check the private CAN bus circuit for short circuit to ground, short circuit to power, open circuit, high resistance • Using the manufacturer approved diagnostic system, clear the DTCs and retest. If the fault persists, install a new yaw rate lateral acceleration sensor
C0063-1C	Yaw Rate Sensor - Circuit voltage out of range	• Yaw rate lateral acceleration sensor power or ground circuit open circuit, high resistance	• Refer to the electrical circuit diagrams and check the yaw rate lateral acceleration sensor power and ground circuits for open circuit, high resistance
C0063-27	Yaw Rate Sensor - Signal rate of change above threshold	• Yaw rate lateral acceleration sensor power or ground circuit open circuit, high resistance • Private CAN bus circuit short circuit to ground, open circuit, high resistance • Yaw rate lateral acceleration sensor incorrectly installed • Yaw rate lateral acceleration sensor internal failure • Anti-lock brake system control module internal failure	• Refer to the electrical circuit diagrams and check the yaw rate lateral acceleration sensor power and ground circuits for open circuit, high resistance • Refer to the electrical circuit diagrams and check the private CAN bus circuit for short circuit to ground, short circuit to power, open circuit, high resistance • Check the installation of the yaw rate lateral acceleration sensor • Using the manufacturer approved diagnostic system, clear the DTCs and retest. If the fault persists, install a new yaw rate lateral acceleration sensor. Perform routine - Calibrate Acceleration Sensor(s) • Using the manufacturer approved diagnostic system, clear the DTCs and retest. If the fault persists, install a new anti-lock brake system control module. Perform routine - Calibrate Acceleration Sensor(s)

C0063-28	Yaw Rate Sensor - Signal bias level out of range / zero adjustment failure	• Yaw rate lateral acceleration sensor incorrectly installed • Yaw rate lateral acceleration sensor internal failure	• Check the installation of the yaw rate lateral acceleration sensor • Using the manufacturer approved diagnostic system, clear the DTCs and retest. If the fault persists, install a new yaw rate lateral acceleration sensor. Perform routine - Calibrate Acceleration Sensor(s)
C0063-41	Yaw Rate Sensor - General checksum failure	• Yaw rate lateral acceleration sensor internal failure • Anti-lock brake system control module internal failure	• Using the manufacturer approved diagnostic system, clear the DTCs and retest. If the fault persists, install a new yaw rate lateral acceleration sensor. Perform routine - Calibrate Acceleration Sensor(s) • Using the manufacturer approved diagnostic system, clear the DTCs and retest. If the fault persists, install a new anti-lock brake system control module. Perform routine - Calibrate Acceleration Sensor(s)
C0063-49	Yaw Rate Sensor - Internal electronic failure	• Yaw rate lateral acceleration sensor power or ground circuit open circuit, high resistance • Yaw rate lateral acceleration sensor internal failure	• Refer to the electrical circuit diagrams and check the yaw rate lateral acceleration sensor power and ground circuits for open circuit, high resistance • Using the manufacturer approved diagnostic system, clear the DTCs and retest. If the fault persists, install a new yaw rate lateral acceleration sensor. Perform routine - Calibrate Acceleration Sensor(s)
C0063-4A	Yaw Rate Sensor - Incorrect component installed	• Yaw rate lateral acceleration sensor internal failure	• Using the manufacturer approved diagnostic system, clear the DTCs and retest. If the fault persists, install a new yaw rate lateral acceleration sensor. Perform routine - Calibrate Acceleration Sensor(s). If the fault persists, install a new anti-lock brake system control module. Perform routine - Calibrate Acceleration Sensor(s)
C0063-64	Yaw Rate Sensor - Signal plausibility failure	• Yaw rate lateral acceleration sensor incorrectly installed • Mismatch between yaw rate lateral acceleration sensor and steering angle sensor signals • Yaw rate lateral acceleration sensor internal failure	⚠ NOTE: This DTC can be set if the vehicle is being tested on a dynamometer/rolling road. • Check the installation of the yaw rate lateral acceleration sensor • Using the manufacturer approved diagnostic system, check the anti-lock brake system control module for steering angle sensor related DTCs and perform the relevant corrective actions • Using the manufacturer approved diagnostic system, clear the DTCs and retest. If the fault persists, install a new yaw rate lateral acceleration sensor. Perform routine - Calibrate Acceleration Sensor(s)
C0063-81	Yaw Rate Sensor - Invalid serial data received	• Yaw rate lateral acceleration sensor incorrectly installed • Yaw rate lateral acceleration sensor internal failure	⚠ NOTE: This DTC can be set if the vehicle is being tested on a dynamometer/rolling road. • Check the installation of the yaw rate lateral acceleration sensor • Using the manufacturer approved diagnostic system, clear the DTCs and retest. If the fault persists, install a new yaw rate lateral acceleration sensor. Perform routine - Calibrate Acceleration Sensor(s)
C0063-86	Yaw Rate Sensor - Signal invalid	• Yaw rate lateral acceleration sensor internal failure	• Using the manufacturer approved diagnostic system, clear the DTCs and retest. If the fault persists, install a new yaw rate lateral acceleration sensor. Perform routine - Calibrate Acceleration Sensor(s). If the fault persists, install a new anti-lock brake system control module. Perform routine - Calibrate Acceleration Sensor(s)
C0063-95	Yaw Rate Sensor - Incorrect assembly	• Yaw rate lateral acceleration sensor internal failure	• Using the manufacturer approved diagnostic system, clear the DTCs and retest. If the fault persists, install a new yaw rate lateral acceleration sensor. Perform routine - Calibrate Acceleration Sensor(s). If the fault persists, install a new anti-lock brake system control module. Perform routine - Calibrate Acceleration Sensor(s)
C0063-96	Yaw Rate Sensor - Component internal failure	• Yaw rate lateral acceleration sensor power or ground circuit open circuit, high resistance • Yaw rate lateral acceleration sensor internal failure	• Refer to the electrical circuit diagrams and check the yaw rate lateral acceleration sensor power and ground circuits for open circuit, high resistance • Using the manufacturer approved diagnostic system, clear the DTCs and retest. If the fault persists, install a new yaw rate lateral acceleration sensor. Perform routine - Calibrate Acceleration Sensor(s). If the fault persists, install a new anti-lock brake system control module. Perform routine - Calibrate Acceleration Sensor(s)
C0064-64	Roll Rate Sensor - Signal plausibility failure	• Yaw rate lateral acceleration sensor incorrectly installed • Yaw rate lateral acceleration sensor internal failure	• Check the installation of the yaw rate lateral acceleration sensor • Using the manufacturer approved diagnostic system, clear the DTCs and retest. If the fault persists, install a new yaw rate lateral acceleration sensor. Perform routine - Calibrate Acceleration Sensor(s)
C006A-54	Multi-axis Acceleration Sensor - Missing calibration	• Yaw rate lateral acceleration sensor not calibrated	• Using the manufacturer approved diagnostic system, perform routine - Calibrate Acceleration Sensor(s)

C101F-49	Generic Valve Failure - Internal electronic failure	• Anti-lock brake system hydraulic control unit internal failure	• Using the manufacturer approved diagnostic system, clear the DTCs and retest. If the fault persists, install a new anti-lock brake system hydraulic control unit
C1109-24	Vehicle Dynamics Control Switch - Signal stuck high	• Dynamic stability control switch stuck active • Dynamic stability control switch circuit short circuit to ground	• Test the operation of the dynamic stability control switch • Refer to the electrical circuit diagrams and check the dynamic stability control switch circuit for short circuit to ground
C1A77-16	Valve Relay Supply Circuit - Circuit voltage below threshold	• Anti-lock brake system hydraulic control unit valves power or ground circuit open circuit, high resistance	• Refer to the electrical circuit diagrams and check the anti-lock brake system hydraulic control unit valves power and ground circuits for open circuit, high resistance
C1A95-2F	Wheel Speed Sensor Supply - Signal erratic	• One or more wheel/tire sizes incorrect • One or more wheel speed sensor circuits subject to electromagnetic interference • One or more wheel speed sensor circuits short circuit to ground, short circuit to power, open circuit, high resistance • One or more wheel speed sensor reluctor rings damaged/de-magnetised	• Check that the wheels and tires comply with the manufacturer's specification for the vehicle • Check the installation of the wheel speed sensor wiring harnesses for correct routing • Refer to the electrical circuit diagrams and check the wheel speed sensor circuits for short circuit to ground, short circuit to power, open circuit, high resistance • Check the wheel speed sensor reluctor rings for damage or de-magnetisation
C1A95-64	Wheel Speed Sensor - Signal plausibility failure	• One or more wheel/tire sizes incorrect • One or more wheel speed sensor circuits subject to electromagnetic interference • One or more wheel speed sensor circuits short circuit to ground, short circuit to power, open circuit, high resistance • One or more wheel speed sensor reluctor rings damaged/de-magnetised	• Check that the wheels and tires comply with the manufacturer's specification for the vehicle • Check the installation of the wheel speed sensor wiring harnesses for correct routing • Refer to the electrical circuit diagrams and check the wheel speed sensor circuits for short circuit to ground, short circuit to power, open circuit, high resistance • Check the wheel speed sensor reluctor rings for damage or de-magnetisation
C1A96-14	Brake Light Switch - Circuit short to ground or open	• Brake pedal switch circuit short circuit to ground, open circuit, high resistance	• Refer to the electrical circuit diagrams and check the brake pedal switch circuit for short circuit to ground, open circuit, high resistance
C1A96-64	Brake Light Switch - Signal plausibility failure	• Brake pedal switch incorrectly installed • Brake pedal switch circuit short circuit to ground, short circuit to power, open circuit, high resistance	• Check the brake pedal switch installation • Refer to the electrical circuit diagrams and check the brake pedal switch circuit for short circuit to ground, short circuit to power, open circuit, high resistance
C1A97-24	Lateral Accelerometer - Signal stuck high	• Yaw rate lateral acceleration sensor internal failure	• Using the manufacturer approved diagnostic system, clear the DTCs and retest. If the fault persists, install a new yaw rate lateral acceleration sensor and perform routine - Calibrate Acceleration Sensor(s)
C1A98-2F	Yaw Rate Sensor - Signal erratic	• Yaw rate lateral acceleration sensor signal erratic (for > 2 minutes)	NOTE: This DTC is for information only. • Using the manufacturer approved diagnostic system, check for yaw rate lateral acceleration sensor related DTCs and perform the relevant corrective actions
C1A98-96	Yaw Rate Sensor - Component internal failure	• Yaw rate lateral acceleration sensor power or ground circuit open circuit, high resistance • Yaw rate lateral acceleration sensor incorrectly installed • Yaw rate lateral acceleration sensor internal failure	• Refer to the electrical circuit diagrams and check the yaw rate lateral acceleration sensor power and ground circuits for open circuit, high resistance • Check the installation of the yaw rate lateral acceleration sensor • Using the manufacturer approved diagnostic system, clear the DTCs and retest. If the fault persists, install a new yaw rate lateral acceleration sensor and perform routine - Calibrate Acceleration Sensor(s)
C1B00-29	Steering Angle Sensor - Signal invalid	• Steering angle sensor incorrectly installed • Steering angle sensor internal failure	• Check the installation of the steering angle sensor • Using the manufacturer approved diagnostic system, clear the DTCs and retest. If the fault persists, install a new steering angle sensor and perform routine - Calibrate Steering Angle Sensor
C1B00-49	Steering Angle Sensor - Internal electronic failure	• Steering angle sensor incorrectly installed • Steering angle sensor internal failure	• Check the installation of the steering angle sensor • Using the manufacturer approved diagnostic system, clear the DTCs and retest. If the fault persists, install a new steering angle sensor and perform routine - Calibrate Steering Angle Sensor
C1B00-54	Steering Angle Sensor - Missing calibration	• Steering angle sensor not calibrated	• Using the manufacturer approved diagnostic system, perform routine - Calibrate Steering Angle Sensor
C1B00-64	Steering Angle Sensor - Signal plausibility failure	• Steering angle sensor incorrectly installed • Steering angle sensor internal failure • Yaw rate lateral acceleration sensor incorrectly installed	• Check the installation of the steering angle sensor • Using the manufacturer approved diagnostic system, clear the DTCs and retest. If the fault persists, install a new steering angle sensor and perform routine - Calibrate Steering Angle Sensor • Check the installation of the yaw rate lateral acceleration sensor

C1B00-81	Steering Angle Sensor - Invalid serial data received	• Steering angle sensor power or ground circuit open circuit, high resistance • Steering angle sensor incorrectly installed • Steering angle sensor internal failure	• Refer to the electrical circuit diagrams and check the steering angle sensor power and ground circuits for open circuit, high resistance • Check the installation of the steering angle sensor • Using the manufacturer approved diagnostic system, clear the DTCs and retest. If the fault persists, install a new steering angle sensor and perform routine - Calibrate Steering Angle Sensor
C1B00-92	Steering Angle Sensor - Performance or incorrect operation	• Steering angle sensor power or ground circuit open circuit, high resistance • Steering angle sensor incorrectly installed • Steering angle sensor internal failure	• Refer to the electrical circuit diagrams and check the steering angle sensor power and ground circuits for open circuit, high resistance • Check the installation of the steering angle sensor • Using the manufacturer approved diagnostic system, clear the DTCs and retest. If the fault persists, install a new steering angle sensor and perform routine - Calibrate Steering Angle Sensor
C1B02-16	Return Pump - Circuit voltage below threshold	• Anti-lock brake system hydraulic control unit pump power or ground circuit open circuit, high resistance	• Refer to the electrical circuit diagrams and check the anti-lock brake system hydraulic control unit pump power and ground circuits for open circuit, high resistance
C1B02-49	Return Pump - Internal electronic failure	• Anti-lock brake system hydraulic control unit internal failure	• Using the manufacturer approved diagnostic system, clear the DTCs and retest. If the fault persists, install a new anti-lock brake system control module. Perform routine - Calibrate Acceleration Sensor(s)
C2009-64	Front Axle Wheel Speed Sensors Swapped - Signal plausibility failure	• Front wheel speed sensor circuits transposed	• Refer to the electrical circuit diagrams and check the front wheel speed sensor circuits for correct connection to the anti-lock brake system control module
C200A-64	Rear Axle Wheel Speed Sensors Swapped - Signal plausibility failure	• Rear wheel speed sensor circuits transposed	• Refer to the electrical circuit diagrams and check the rear wheel speed sensor circuits for correct connection to the anti-lock brake system control module
U0001-88	High Speed CAN Communication bus - Bus off	• High speed CAN bus circuit short circuit to ground, short circuit to power, open circuit, high resistance	• Using the manufacturer approved diagnostic system, perform a CAN network integrity test. Refer to the electrical circuit diagrams and check the high speed CAN bus circuit for short circuit to ground, short circuit to power, open circuit, high resistance
U0074-88	Control Module Communication Bus B off - Bus off	• Private CAN bus circuit short circuit to ground, short circuit to power, open circuit, high resistance	• Refer to the electrical circuit diagrams and check the private CAN bus circuit for short circuit to ground, short circuit to power, open circuit, high resistance
U0100-86	Lost Communication With ECM/PCM A - Signal invalid	• Engine control module power or ground circuit open circuit, high resistance • High speed CAN bus circuit short circuit to ground, short circuit to power, open circuit, high resistance • Engine system fault	• Refer to the electrical circuit diagrams and check the engine control module power and ground circuits for open circuit, high resistance • Using the manufacturer approved diagnostic system, perform a CAN network integrity test. Refer to the electrical circuit diagrams and check the high speed CAN bus circuit for short circuit to ground, short circuit to power, open circuit, high resistance • Using the manufacturer approved diagnostic system, check the engine control module for related DTCs and refer to the relevant DTC index
U0100-87	Lost Communication With ECM/PCM A - Missing message	• Missing/invalid data from the engine control module	• Using the manufacturer approved diagnostic system, check the engine control module for related DTCs and refer to the relevant DTC index
U0123-87	Lost Communication With Yaw Rate Sensor Module - Missing message	• Yaw rate lateral acceleration sensor power or ground circuit open circuit, high resistance • Private CAN bus circuit short circuit to ground, short circuit to power, open circuit, high resistance • Yaw rate lateral acceleration sensor internal failure	• Refer to the electrical circuit diagrams and check the yaw rate lateral acceleration sensor power and ground circuits for open circuit, high resistance • Refer to the electrical circuit diagrams and check the private CAN bus circuit for short circuit to ground, short circuit to power, open circuit, high resistance • Using the manufacturer approved diagnostic system, clear the DTCs and retest. If the fault persists, install a new yaw rate lateral acceleration sensor and perform routine - Calibrate Acceleration Sensor(s)
U0126-87	Lost Communication With Steering Angle Sensor Module - Missing message	• Steering angle sensor power or ground circuit open circuit, high resistance • Private CAN bus circuit short circuit to ground, short circuit to power, open circuit, high resistance • Steering angle sensor internal failure	• Refer to the electrical circuit diagrams and check the steering angle sensor power and ground circuits for open circuit, high resistance • Refer to the electrical circuit diagrams and check the private CAN bus circuit for short circuit to ground, short circuit to power, open circuit, high resistance • Using the manufacturer approved diagnostic system, clear the DTCs and retest. If the fault persists, install a new steering angle sensor and perform routine - Calibrate Steering Angle Sensor

U0155-81	Lost Communication With Instrument Panel Cluster (IPC) Control Module - Invalid serial data received	• Instrument cluster power or ground circuit open circuit, high resistance • High speed CAN bus circuit short circuit to ground, short circuit to power, open circuit, high resistance • Instrument cluster fault	• Refer to the electrical circuit diagrams and check the instrument cluster power and ground circuits for open circuit, high resistance • Using the manufacturer approved diagnostic system, perform a CAN network integrity test. Refer to the electrical circuit diagrams and check the high speed CAN bus circuit for short circuit to ground, short circuit to power, open circuit, high resistance • Using the manufacturer approved diagnostic system, check the instrument cluster for related DTCs and refer to the relevant DTC index
U0155-87	Lost Communication With Instrument Panel Cluster (IPC) Control Module - Missing message	• Missing/invalid data from the nstrument cluster	• Using the manufacturer approved diagnostic system, check the instrument cluster for related DTCs and refer to the relevant DTC index
U0300-81	Internal Control Module Software Incompatibility - Invalid serial data received	• Incorrect anti-lock brake system control module installed	• Using the manufacturer approved diagnostic system, clear the DTCs and retest. If the fault persists, install a new anti-lock brake system control module. Perform routine - Calibrate Acceleration Sensor(s)
U0401-86	Invalid Data Received From ECM/PCM - Signal invalid	• Missing/invalid data from the engine control module	• Using the manufacturer approved diagnostic system, check the engine control module for related DTCs and refer to the relevant DTC index
U0428-83	Invalid Data Received From Steering Angle Sensor Module - Value of signal protection calculation incorrect	• Steering angle sensor power or ground circuit open circuit, high resistance • Private CAN bus circuit short circuit to ground, short circuit to power, open circuit, high resistance • Steering angle sensor internal failure	• Refer to the electrical circuit diagrams and check the steering angle sensor power and ground circuits for open circuit, high resistance • Refer to the electrical circuit diagrams and check the private CAN bus circuit for short circuit to ground, short circuit to power, open circuit, high resistance • Using the manufacturer approved diagnostic system, clear the DTCs and retest. If the fault persists, install a new steering angle sensor and perform routine - Calibrate Steering Angle Sensor
U0428-86	Invalid Data Received From Steering Angle Sensor Module - Signal invalid	• Steering angle sensor power or ground circuit open circuit, high resistance • Private CAN bus circuit short circuit to ground, short circuit to power, open circuit, high resistance • Steering angle sensor internal failure	• Refer to the electrical circuit diagrams and check the steering angle sensor power and ground circuits for open circuit, high resistance • Refer to the electrical circuit diagrams and check the private CAN bus circuit for short circuit to ground, short circuit to power, open circuit, high resistance • Using the manufacturer approved diagnostic system, clear the DTCs and retest. If the fault persists, install a new steering angle sensor and perform routine - Calibrate Steering Angle Sensor
U1A14-00	CAN Initialisation Failure - No sub type information	• Incorrect anti-lock brake system control module installed	• Using the manufacturer approved diagnostic system, clear the DTCs and retest. If the fault persists, install a new anti-lock brake system control module. Perform routine - Calibrate Acceleration Sensor(s)
U2015-08	Control Module Boot Software - Bus signal/message failures	• Event information - The ignition must be cycled off and on after re-configuring the anti-lock brake system control module	• Using the manufacturer approved diagnostic system, clear the DTCs and retest. If the fault persists, contact dealer technical support
U2101-56	Control Module configuration Incompatible - Invalid/incomplete configuration	• Anti-lock brake system control module is not configured correctly	• Using the manufacturer approved diagnostic system, re-configure the anti-lock brake system control module with the latest level software
U2101-68	Control Module configuration Incompatible - Event information	• Anti-lock brake system control module is not configured correctly • Incorrect anti-lock brake system control module installed	• Using the manufacturer approved diagnostic system, re-configure the anti-lock brake system control module with the latest level software • Install a new anti-lock brake system control module as necessary
U3000-00	Control Module - No sub type information	• Incorrect anti-lock brake system control module installed	• Using the manufacturer approved diagnostic system, clear the DTCs and retest. If the fault persists, install a new anti-lock brake system control module. Perform routine - Calibrate Acceleration Sensor(s)
U3000-45	Control Module - Program memory failure	• Anti-lock brake system control module internal failure	• Using the manufacturer approved diagnostic system, clear the DTCs and retest. If the fault persists, install a new anti-lock brake system control module. Perform routine - Calibrate Acceleration Sensor(s)
U3000-49	Control Module - Internal electronic failure	• Anti-lock brake system control module internal failure	• Using the manufacturer approved diagnostic system, clear the DTCs and retest. If the fault persists, install a new anti-lock brake system control module. Perform routine - Calibrate Acceleration Sensor(s)
U3000-4B	Control Module - Over temperature	• Excessive use of brakes and/or traction control system	NOTE: This DTC may be set if the manufacturer approved diagnostic system has been operating the valves for a prolonged period. • Allow the vehicle to cool. Using the manufacturer approved diagnostic system, clear the DTCs and retest

U3000-68	Control Module - Event information	• Tire pressure(s) incorrect • Wheel speed sensor fault • Yaw rate lateral acceleration sensor fault • Steering angle sensor fault	NOTE: This DTC is set when the dynamic stability control has been active for an unfeasibly long period of time. • Check and adjust the tire pressures as necessary • Using the manufacturer approved diagnostic system, check for wheel speed sensor related DTCs and perform the relevant corrective actions • Using the manufacturer approved diagnostic system, check for yaw rate lateral acceleration sensor related DTCs and perform the relevant corrective actions • Using the manufacturer approved diagnostic system, check for steering angle sensor related DTCs and perform the relevant corrective actions
U3002-81	Vehicle Identification Number - Invalid serial data received	• Instrument cluster power or ground circuit open circuit, high resistance • High speed CAN bus circuit short circuit to ground, short circuit to power, open circuit, high resistance • Instrument cluster fault • Anti-lock brake system control module previously installed on another vehicle	• Refer to the electrical circuit diagrams and check the instrument cluster power and ground circuits for open circuit, high resistance • Using the manufacturer approved diagnostic system, perform a CAN network integrity test. Refer to the electrical circuit diagrams and check the high speed CAN bus circuit for short circuit to ground, short circuit to power, open circuit, high resistance • Using the manufacturer approved diagnostic system, check the instrument cluster for related DTCs and refer to the relevant DTC index • Install the original anti-lock brake system control module
U3002-87	Vehicle Identification Number - Missing message	• Instrument cluster power or ground circuit open circuit, high resistance • High speed CAN bus circuit short circuit to ground, short circuit to power, open circuit, high resistance • Instrument cluster system fault	• Refer to the electrical circuit diagrams and check the instrument cluster power and ground circuits for open circuit, high resistance • Using the manufacturer approved diagnostic system, perform a CAN network integrity test. Refer to the electrical circuit diagrams and check the high speed CAN bus circuit for short circuit to ground, short circuit to power, open circuit, high resistance • Using the manufacturer approved diagnostic system, check the instrument cluster for related DTCs and refer to the relevant DTC index
U3006-16	Control Module Input Power A - Circuit voltage below threshold	• Anti-lock brake system control module power or ground circuit open circuit, high resistance • Battery/charging system fault	• Using the manufacturer approved diagnostic system, check datalogger signal - ECU Power Supply Voltage (0xD112). Refer to the electrical circuit diagrams and check the anti-lock brake system control module power and ground circuits for open circuit, high resistance • Refer to the relevant section of the workshop manual and test the battery and charging system
U3006-17	Control Module Input Power A - Circuit voltage above threshold	• Battery/charging system fault	• Using the manufacturer approved diagnostic system, check datalogger signal - ECU Power Supply Voltage (0xD112). Refer to the relevant section of the workshop manual and test the battery and charging system
U3006-1C	Control Module Input Power A - Circuit voltage out of range	• Anti-lock brake system control module power or ground circuit open circuit, high resistance • Battery/charging system fault	• Using the manufacturer approved diagnostic system, check datalogger signal - ECU Power Supply Voltage (0xD112). Refer to the electrical circuit diagrams and check the anti-lock brake system control module power and ground circuits for open circuit, high resistance • Refer to the relevant section of the workshop manual and test the battery and charging system

Description and Operation

VIN Number

The VIN number is found in three locations:

1. Stamped on the RH side of the chassis towards the rear of the front lashing eye.
2. At the bottom of the windshield glass on the RH side of the vehicle and is visible from the outside.
3. On a plate attached to the brake pedal box in the engine compartment.

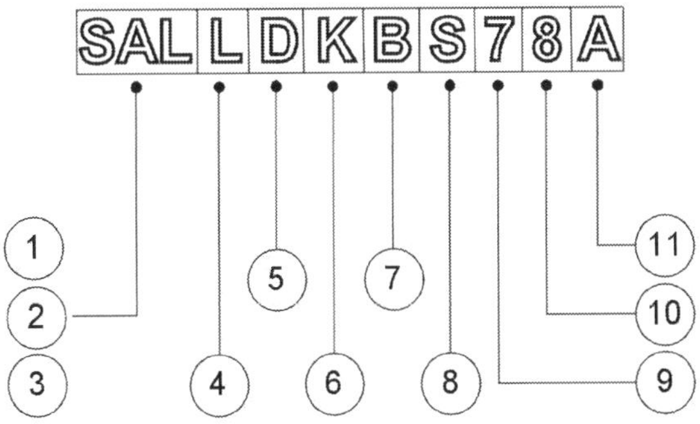

VIN number - UK, EU and ROW

VIN Position	Character	Identifies
1 - 3 - World identifier	SAL	Jaguar Land Rover Limited
4,5 - Vehicle type	LD	Defender
6 - Class	H	110 inch
	K	130 inch
	V	90 inch
	B	110 inch Extra heavy duty
	W	90 inch Extra heavy duty

7 - Body style	A	Regular
	B	3 Door station wagon
	F	4 Door Crew Cab non h/cap
	H	H/cap with or without 4 Door Crew Cab
	L	Soft Top
	M	5 Door Station Wagon
	N	Hard Top
	R	Single Cab Pick Up
	S	Double Cab Pick Up
	T	2 Door High Capacity Pick Up
	V	Crew Cab High Capacity Pick Up
	W	2 Door Chassis Cab
	X	4 Door Chassis Cab
	Y	5 Door Utility Station Wagon
8 - Engine	P	2.2L Euro spec (EU5)
	R	2.2L ROW spec (non EU5)
	S	2.4L Euro spec (EU4)
	T	2.4L ROW spec (EU2)
	N	2.4L 24V Military Euro spec (EU4)
	W	2.4L 24V Military ROW spec (EU2)

9 - Transmission and steering	7	RHD 6 speed manual transmission
	8	LHD 6 speed manual transmission
10 - Model year	7	2007 Model year
	8	2008 Model year
	9	2009 Model year
	A	2010 Model year
	B	2011 Model year
	C	2012 Model year
	D	2013 Model year
	E	2014 Model year
	F	2015 Model year
	G	2016 Model year
11 - Plant	A	Solihull

	C	Zimbabwe
	D	Thailand
	F	KD
	J	Malaysia
	K	Kenya
	P	Pakistan
	W	Turkey

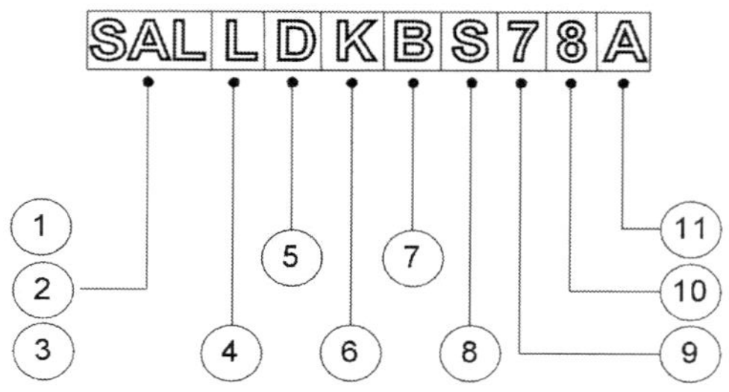

E91381

VIN number - NAS

VIN Position	Character	Identifies
1 - 3 - World identifier	SAL	Jaguar Land Rover Limited
4 - Vehicle type	D	Defender
5 - Class	H	110 inch
	W	90 inch Extra heavy duty
6 - Body style	B	3 Door station wagon
	M	5 Door Station Wagon
7 - Engine	P	2.2L Euro spec (EU5)
	R	2.2L ROW spec (non EU5)
	S	2.4L Euro spec (EU4)
8- Transmission and steering	8	LHD 6 speed manual transmission
9- Check digit	*	Derived by calculation
10 - Model year	7	2007 Model year
	8	2008 Model year
	9	2009 Model year

10 - Model year	7	2007 Model year
	8	2008 Model year
	9	2009 Model year
	A	2010 Model year
	B	2011 Model year
	C	2012 Model year
	D	2013 Model year
	E	2014 Model year
	F	2015 Model year
	G	2016 Model year
11 - Plant	A	Solihull

VIN number - UK, EU and ROW

Vehicle Identification Plate

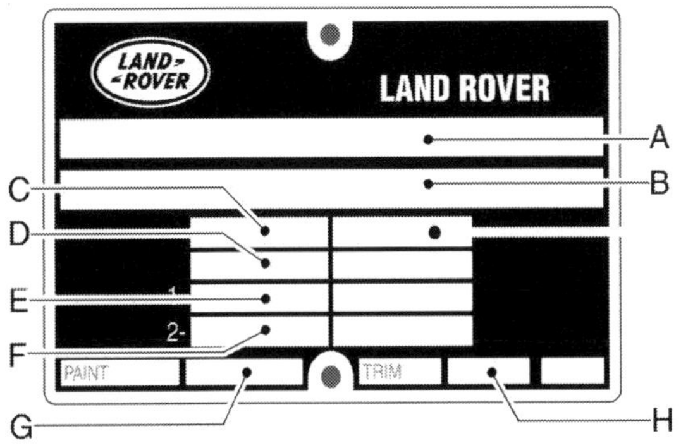

M010312

The VIN plate contains the following information:

- **A** - Type approval number
- **B** - Vehicle identification number
- **C** - Maximum permitted laden weight for vehicle
- **D** - Maximum vehicle and trailer weight
- **E** - Maximum road weight - front axle
- **F** - Maximum road weight - rear axle
- **G** - Paint code
- **H** - Trim level

Unit/Assembly Serial Number Locations

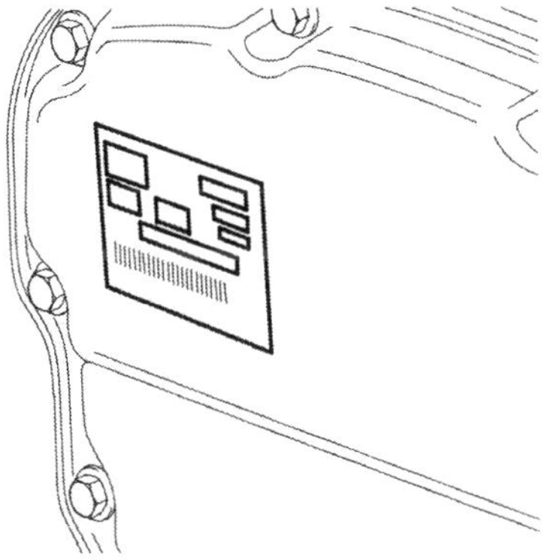

Engine Identification Label

- Date and time of manufacture
- Engine part number

Manual Transmission Serial Number

The Manual Transmission Serial Number is stamped on a cast pad on the bottom RH side of the transmission casing.

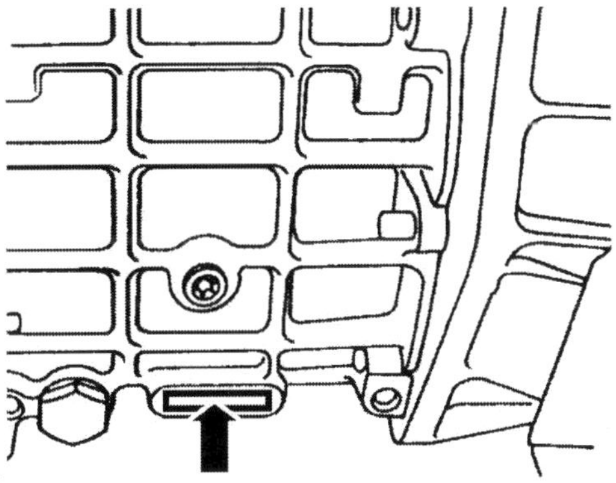

J6091

Front Axle Serial Number

The front axle serial number is stamped on the LH side of the front axle tube, inboard of radius arm mounting bracket.

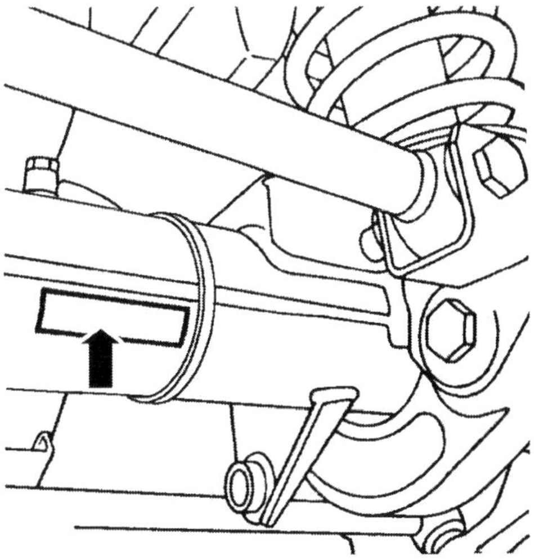

Rear Axle Serial Number

The rear axle serial number is stamped on the rear axle tube on the LH side, inboard of the spring mounting.

Transfer Case Serial Number

The transfer case serial number is stamped on the LH side of the transfer case below the mainshaft rear bearing housing adjacent to the bottom cover.

Jacking

Description and Operation

General

WARNING: The following instructions must be adhered to before raising the vehicle off the ground:

- Position vehicle on a solid, level surface.
- Apply the parking brake.
- Select 'P' - PARK on automatic transmission selector or 1st gear on manual transmission and 'H' High on transfer case.
- Select low range in the transfer box.

CAUTION: To avoid damage occurring to the under body components of the vehicle the following jacking procedures must be adhered to.

DO NOT POSITION JACKS OR AXLE STANDS UNDER THE FOLLOWING COMPONENTS.

- **Body structure**
- **Bumpers**
- **Fuel lines**
- **Brake lines**
- **Front radius arms**
- **Panhard rod**
- **Steering linkage**
- **Rear trailing links**
- **Fuel tank**
- **Engine sump**
- **Gearbox bell housing**

Vehicle jack

The jack provided with the vehicle is only intended for use in an emergency such as changing a tire. DO NOT use the jack for any other purpose. Refer to the Owner's Handbook for the vehicle jack location points and jacking procedures.

WARNING: Never work under a vehicle supported solely by the vehicle jack.

Hydraulic jack

A hydraulic jack with a minimum lifting capacity of 1500 kg, (3,300 lbs) must be used.

WARNINGS:

Do not commence work on the underside of the vehicle until suitable axle stands have been placed in the correct position.

Always chock the wheels when jacking. The parking brake may be ineffective when the wheel(s) are off the ground.

Raising the Front of the Vehicle

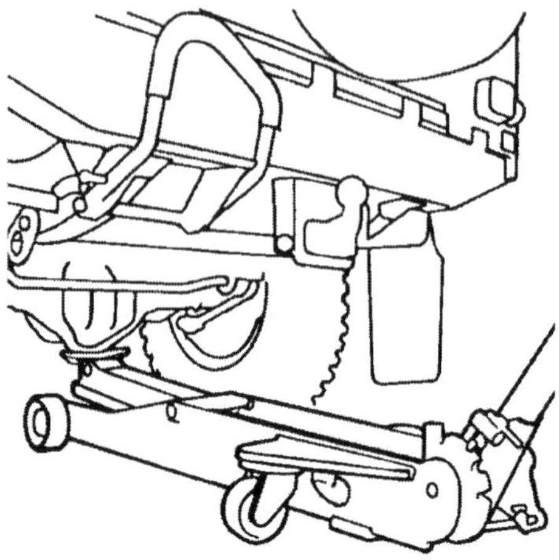

- Position cup of hydraulic arm under differential casing.

NOTE: The differential casing is not central to the axle. Care should be taken when raising the front road wheels off the ground as the rear axle has less sway stiffness.

- Raise front road wheels to enable an axle stand to be installed under left hand axle tube.
- Position an axle stand under right hand axle tube, carefully lower jack until axle sits securely on both axle stands, remove trolley jack.
- Before commencing work on underside of vehicle re-check security of vehicle on stands.
- Reverse procedure when removing vehicle from stands.

Raising the Rear of the Vehicle

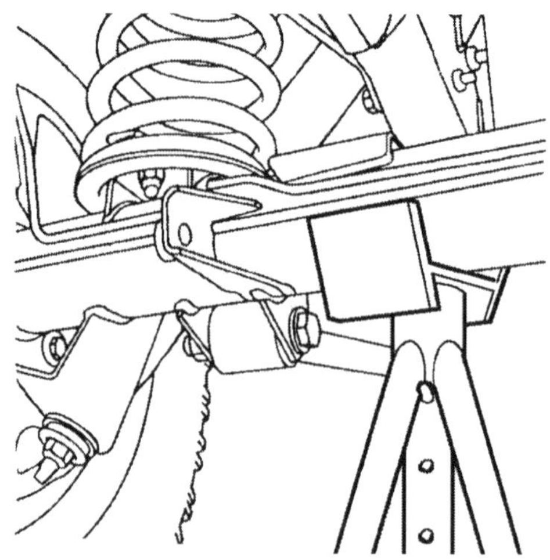

- Position cup of hydraulic arm under differential casing.

- Raise vehicle to enable axle stands to be installed under left and right hand axle tubes.
- Lower jack until axle sits securely on axle stands, remove trolley jack.
- Before commencing work on underside of vehicle re-check security of vehicle on stands.
- Reverse procedure when removing vehicle from stands.

Vehicle Recovery

Description and Operation

⚠ CAUTION: The vehicle has permanent four-wheel drive. The following towing instructions must be adhered to:

Towing the vehicle on all four wheels with driver operating steering and brakes.

- Turn ignition key to position '1' to release steering lock.
- Select neutral in main gearbox and transfer gearbox.

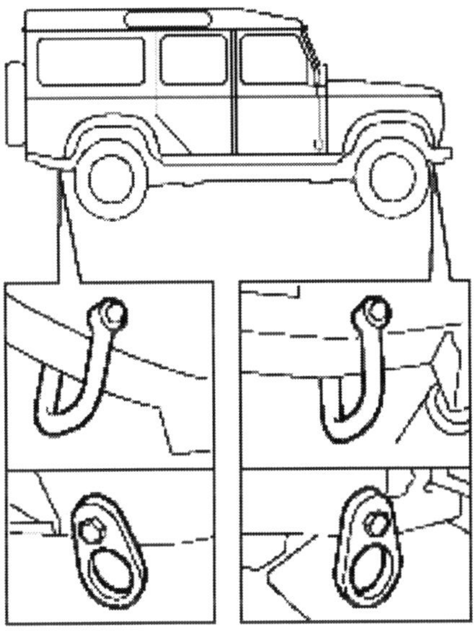

- Secure tow rope, chain or cable to front towing eyes (alternative types shown).
- Release the parking brake.

⚠ CAUTION: The brake booster and power assisted steering system will not be functional without the engine running. Greater pedal pressure will be required to apply the brakes, the steering system will require greater effort to turn the front road wheels. The vehicle tow connection should be used only in normal road conditions, snatch recovery should be avoided.

Transporting the vehicle by trailer

Lashing/towing eyes are provided on front and rear of the chassis side members, see J6085, to facilitate the securing of the vehicle to a trailer or other means of transportation.

⚠ CAUTION: Underbody components must not be used as lashing points.

Install vehicle on trailer and apply park brake. Select neutral in main gearbox.

Rear suspended tow by breakdown vehicle

- If the front axle is to be trailed turn ignition key to position '1' to release steering lock.
- Select neutral in main gearbox and transfer box.
- The steering wheel and/or linkage must be secured in a straight ahead position. DO NOT use the steering lock mechanism for this purpose.

Jump starting

⚠ WARNING: Hydrogen and oxygen gases are produced during normal battery operation. This gas mixture can explode if flames, sparks or lighted tobacco are brought near battery. When charging or using a battery in an enclosed space, always provide ventilation and shield your eyes.

Keep out of reach of children. Batteries contain sulphuric acid. Avoid contact with skin, eyes, or clothing. Also, shield eyes when working near battery to protect against possible splashing of acid solution. In case of acid contact with skin, eyes, or clothing, flush immediately with water for a minimum of fifteen minutes. If acid is swallowed, drink large quantities of milk or water, followed by milk of magnesia, a beaten egg, or vegetable oil.

SEEK MEDICAL AID IMMEDIATELY.

To Jump Start - Negative Ground Battery

⚠ WARNING: To avoid any possibility of injury use particular care when connecting a booster battery to a discharged battery.

- Position vehicles so that jump leads will reach, ensuring that vehicles DO NOT TOUCH, alternatively a fully charged slave battery may be positioned on floor adjacent to vehicle.
- Ensuring that ignition and all electrical accessories are switched off, that parking brake is applied and neutral is selected, connect the jump leads as follows;

A. Connect one end of first jumper cable to positive (+) terminal of booster battery.

B. Connect other end of first jumper cable to positive (+) terminal of discharged battery.

C. Connect one end of second jumper cable to negative terminal of booster battery.

D. Connect other end of second jumper cable to a good earth point on the disabled vehicle (e.g. engine front lifting eye, as shown in J6086), NOT TO NEGATIVE TERMINAL OF DISCHARGED BATTERY.

Keep jumper lead away from moving parts, pulleys, drive belts and fan blade assembly.

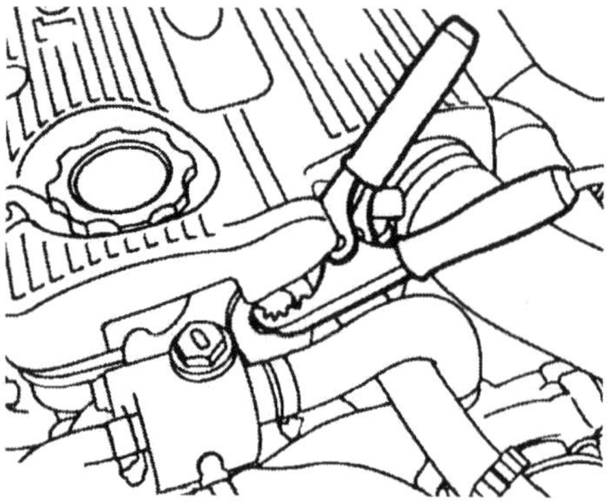

J6086

WARNING: Making final cable connection could cause an electrical arc which if made near battery could cause an explosion.

- If booster battery is installed in another vehicle, start engine and allow to idle.
- Start engine of vehicle with discharged battery, following starting procedure in Owners' Manual.

CAUTION: If vehicle fails to start within a maximum time of 12 seconds, switch ignition off and investigate cause. Failing to follow this instruction could result in irreparable damage to catalyst, if fitted.

- Remove negative (-) jumper cable from the engine and then terminal of booster battery.
- Remove positive (+) jumper cable from positive terminals of booster battery and discharged battery.

Lifting

Description and Operation

Four post vehicle ramps

Position the vehicle on the ramp with the front and the rear of the vehicle equidistant from the ends of the ramp. Chock the wheels, select neutral on the transmission and where practicable apply the parking brake.

If a wheel free condition is required, use a wheel-free system which supports beneath the axle casings.

Two post vehicle ramps

WARNING: When raised to the desired height, make sure the vehicle is stable before starting work.

CAUTION: If the drive shafts are to be disconnected, release the parking brake and select neutral on the transmission in order that the shaft(s) can be rotated when the vehicle is raised to the desired height.

1. Position the vehicle with the centre of the lifting pillars aligned with the front of the driver/ passenger seat cushions.
2. Extend the lifting arms and position the pads of each lifting arm beneath the body frame longitudinal lifting points, arrowed in the illustration.
3. Raise the vehicle until the wheels are just clear of the ground and check that the pads of each lifting arm are still correctly positioned and are in full contact with the body frame longitudinals.
4. Raise the vehicle to the desired height.
5. Ensure that the vehicle is correctly supported on all four lifting pads, that the pads are still correctly positioned and are in full contact with the body frame longitudinals.

Road roller testing

Four wheel Rolling Road

WARNING: DO NOT attempt to drive individual wheels with vehicle supported on floor jacks or stands.

Provided that front and rear rollers are rotating at identical speeds and that normal workshop safety standards are applied, there is no speed restriction during testing except any that may apply to the tyres.

Two wheel rolling road

IMPORTANT: Use a four wheel rolling road for brake testing if possible.

If brake testing on a two wheel rolling road is necessary, it must be carried out with propeller shaft to the stationary axle removed and neutral selected in BOTH main gearbox and transfer gearbox. When checking brakes, run engine at idle speed to maintain brake servo vacuum.

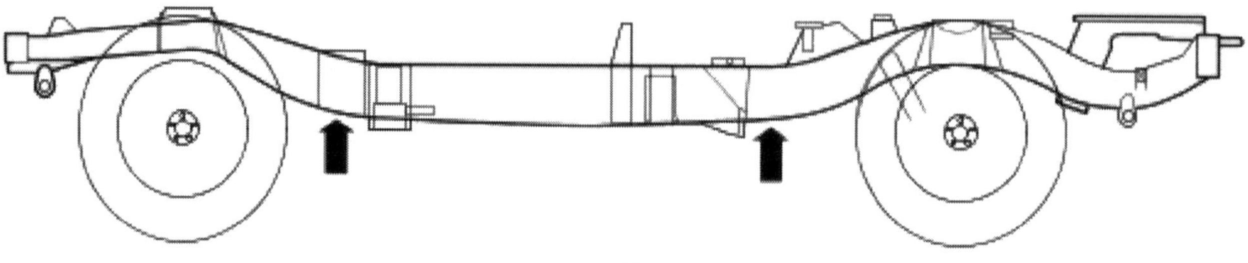

Specifications

Suspension	
90	
Front	Live beam axle, linear rate coil springs, coaxial telescopic nitrogen-charged hydraulic shock absorbers. Lateral axle location by fabricated steel Panhard rod and longitudinal axle location by forged steel radius arms. Stabilizer bar if alloy wheels fitted.
Rear	Live beam axle, linear rate coil springs (dual rate for heavy duty derivative), telescopic nitrogen-charged hydraulic shock absorbers. Lateral axle location by forged steel 'A' frame arms and longitudinal axle location by fabricated steel trailing links. Stabilizer bar if alloy wheels fitted.
110	
Front	Live beam axle, dual rate coil springs, coaxial telescopic nitrogen-charged hydraulic shock absorbers. Lateral axle location by fabricated steel Panhard rod and longitudinal axle location by forged steel radius arms. Stabilizer bar if station wagon or crew cab derivative.
Rear	Live beam axle, triple rate coil springs, telescopic nitrogen-charged hydraulic shock absorbers. Lateral axle location by forged steel 'A' frame arms and longitudinal axle location by fabricated steel trailing links. Stabilizer bar if station wagon or crew cab derivative.
110 heavy duty and 130	
Front	Live beam axle, linear rate coil springs, coaxial telescopic nitrogen-charged hydraulic shock absorbers. Lateral axle location by fabricated steel Panhard rod and longitudinal axle location by forged steel radius arms. Stabilizer bar.
Rear	Live beam axle, triple rate coil springs, coaxial linear rate helper springs, telescopic nitrogen-charged hydraulic shock absorbers. Lateral axle location by forged steel 'A' frame arms and longitudinal axle location by fabricated steel trailing links. Stabilizer bar.

Road spring data

Road spring	Part number	Colour code
90 2 seat utility (2400 Kg)		
Front - Driver's side	REB500200	Red/green/white
Front - Passenger side	REB500220	Red/green/orange
Rear - Driver's side	NRC 9448	Blue/red
Rear - Passenger side	NRC 9449	Yellow/white
90 2 seat utility (2550 Kg)		
Front - Driver's side	REB500200	Red/green/white
Front - Passenger side	REB500220	Red/green/orange
Rear - Driver's side	RKB101230	Green/yellow/red
Rear - Passenger side	RKB101240	Green/yellow/white
90 4 seat station wagon (2400 Kg)		
Front - Driver's side	REB500200	Red/green/white
Front - Passenger side	REB500220	Red/green/orange
Rear - Driver's side	RKB500290	Red/green/green
Rear - Passenger side	RKB500280	Red/green/red
90 4 seat station wagon (2550 Kg)		
Front - Driver's side	REB500200	Red/green/white
Front - Passenger side	REB500220	Red/green/orange
Rear - Driver's side	RKB500270	White/green/pink
Rear - Passenger side	RKB500310	White/green/purple
LHD 110 2 seat utility, 5 seat station wagon and 5 seat crew cab (3050 Kg)		
Front - Driver's side	NRC 8044	White/white
Front - Passenger side	NRC 8045	Yellow/yellow
Rear - Both sides	RKB101111	Purple/brown
RHD 110 2 seat utility, 5 seat station wagon and 5 seat crew cab (3050 Kg)		
Front - Both sides	NRC 8045	Yellow/yellow
Rear - Both sides	RKB101111	Purple/brown
LHD 110 7 seat station wagon (3050 Kg)		
Front - Driver's side	NRC 8044	White/white
Front - Passenger side	NRC 8045	Yellow/yellow
Rear - Both sides	RKB500300	White/green/green
RHD 110 7 seat station wagon (3050 Kg)		
Front - Both sides	NRC 8045	Yellow/yellow
Rear - Both sides	RKB500300	White/green/green
110 heavy duty and all 130 (3500 Kg)		
Front - Driver's side	NRC 9448	Blue/red
Front - Passenger side	NRC 9449	Yellow/white
Rear - Both sides	RKB101111	Purple/brown
Rear helper springs - Both sides	RRC 3266	No colour code

Wheel Alignment Specification - Front

Item		Left-hand		Right-hand		Total/Balance	
Camber		**Nominal**	**Tolerance**	**Nominal**	**Tolerance**	**Nominal**	**Tolerance**
	Decimal degrees	-0.17°	± 0.75°	-0.17°	± 0.75°	0°	± 1°
	Degrees/minutes	-10’	± 45’	-10’	± 45’	0’	± 1°
		Minimum	**Maximum**	**Minimum**	**Maximum**	**Minimum**	**Maximum**
	Decimal degrees	-0.92°	0.58°	-0.92°	0.58°	-1°	1°
	Degrees/minutes	-55’	35’	-55’	35’	-1°	1°
Castor		**Nominal**	**Tolerance**	**Nominal**	**Tolerance**	**Nominal**	**Tolerance**
	Decimal degrees	3.25°	± 0.75°	3.25°	± 0.75°	0°	± 1°
	Degrees/minutes	3°15’	± 45’	3°15’	± 45’	0’	± 1°
		Minimum	**Maximum**	**Minimum**	**Maximum**	**Minimum**	**Maximum**
	Decimal degrees	2.50°	4°	2.50°	4°	-1°	1°
	Degrees/minutes	2°30’	4°	2°30’	4°	-1°	1°
Toe		**Nominal**	**Tolerance**	**Nominal**	**Tolerance**	**Nominal**	**Tolerance**
	Decimal degrees	-0.08°	±0.08°	-0.08°	±0.08°	-0.17°	±0.17°
	Degrees/minutes	-5’	±5’	-5’	±5’	-10’	±10’
		Minimum	**Maximum**	**Minimum**	**Maximum**	**Minimum**	**Maximum**
	Decimal degrees	-0.17°	0°	-0.17°	0°	-0.33°	0°
	Degrees/minutes	-10’	0’	-10’	0’	-20’	0’

Wheel Alignment Specification - Rear

Item		Left-hand		Right-hand		Total/Balance		Thrust Angle	
Camber		**Nominal**	**Tolerance**	**Nominal**	**Tolerance**				
	Decimal degrees	-0.17°	± 0.75°	-0.17°	± 0.75°				
	Degrees/minutes	-10’	± 45’	-10’	± 45’				
		Minimum	**Maximum**	**Minimum**	**Maximum**				
	Decimal degrees	-0.92°	0.58°	-0.92°	0.58°				
	Degrees/minutes	-55’	35’	-55’	35’				
Toe		**Nominal**	**Tolerance**	**Nominal**	**Tolerance**	**Nominal**	**Tolerance**	**Nominal**	**Tolerance**
	Decimal degrees					0°	± 0.20°	0°	± 0.40°
	Degrees/minutes					0’	± 12’	0’	± 24’
		Minimum	**Maximum**	**Minimum**	**Maximum**	**Minimum**	**Maximum**	**Minimum**	**Maximum**
	Decimal degrees					-0.20°	0.20°	-0.40°	0.40°
	Degrees/minutes					-12’	12’	-24’	24’

Torque values - front suspension

	Nm
Drag link ball joint nut	40 + tighten to next castle slot on nut
Tie rod ball joint nut	40 + tighten to next castle slot on nut
Securing ring for mounting turret	14
Panhard rod to axle	230
Panhard rod to chassis bracket	230
Panhard rod chassis bracket to chassis	123
Radius arm to axle	210
Radius arm to chassis	176
Shock absorber (upper and lower mountings)	38
Bump stop to chassis frame	22
Stabilizer bar to stabilizer link	75
Stabilizer link to axle	40 + tighten to next castle slot on nut
Stabilizer bar to chassis bracket	45
Tie bar to Panhard rod mounting bracket	85
Tie bar to steering box	85

Torque values - Rear suspension

	Nm
Trailing link to chassis bush	176
Trailing link to chassis	62
Trailing link to axle	176
‘A’ frame ball joint to axle	176
‘A’ frame to ball joint	115
‘A’ frame to chassis bracket	
Vehicles fitted with ¾ inch UNF bolts	176
Vehicles fitted with M16 bolts	150
‘A’ frame to chassis bracket to chassis (if applicable)	63
Shock absorber to axle	38
Shock absorber to shock absorb-er upper mounting bracket	75
Shock absorber upper mounting bracket to chassis	44
Stabilizer bar to stabilizer link	75
Stabilizer link to axle	40 + tighten to next castle slot on nut
Stabilizer bar bush bracket to chassis	22
Bump stop to chassis frame	22

Front hub assembly end float spacer selection.

End float (mm)	Spacer size (mm)	Colour code
0.00	15.5	Purple
0.025	15.4	Yellow
0.050	15.4	Yellow
0.075	15.4	Yellow
0.10	15.3	Red
0.125	15.3	Red
0.150	15.3	Red
0.175	15.2	Blue
0.200	15.2	Blue
0.225	15.2	Blue
0.250	15.2	Blue
0.275	15.1	Green
0.300	15.1	Green
0.325	15.1	Green
0.350	15.1	Green
0.375	15.0	Black
0.400	15.0	Black
0.425	15.0	Black
0.450	15.0	Black
0.475	14.9	White
0.500	14.9	White
0.525	14.9	White
0.550	14.9	White

Suspension System

Diagnosis and Testing

Principles of Operation

For a detailed description of the suspension system, refer to the relevant Description and Operation section of the workshop manual.

Refer to: Front Suspension (204-01 Front Suspension, Description and Operation)

Refer to: Rear Suspension (204-02 Rear Suspension, Description and Operation)

Inspection and Verification

WARNING: Before carrying out a road test, make sure the vehicle is safe to do so. Failure to follow this instruction may result in personal injury.

CAUTION: Diagnosis by substitution from a donor vehicle is **NOT** acceptable. Substitution of control modules does not guarantee confirmation of a fault and may also cause additional faults in the vehicle being checked and/or the donor vehicle

NOTE: Check and rectify basic faults before beginning diagnostic routines involving pinpoint tests

1. Visually inspect for obvious signs of mechanical or electrical damage.

Visual Inspection

Mechanical
• Tire pressures • Wheels or tires • Suspension components • Steering system components • Wheel bearing(s)

2. If an obvious cause for an observed or reported concern is found, correct the cause (if possible) before proceeding to the next step.
3. If the cause is not visually evident, verify the symptom and refer to the Symptom Chart.

Symptom Chart

Symptom	Possible Cause	Action
Evidence of fluid on shock absorber	• Fluid on shock absorber from an external source • Fluid leaking from shock absorber	• Shock absorber not faulty, do not renew • GO to Pinpoint Test A
Snapping, clicking or popping noise	• Damaged or worn front/rear suspension components • Suspension fasteners loose • Front or rear wheel bearing(s) worn • Constant velocity (CV) joint worn	• GO to Pinpoint Test C
Grinding noise when the vehicle is in motion, intensifies when turning or when there is a shift in load	• Foreign object(s) in brakes • Front/rear, left/right brake pads/discs excessively worn • Damaged or worn front/rear suspension components • Suspension fasteners loose • Front or rear wheel bearing(s) worn	• GO to Pinpoint Test C

Symptom	Possible Cause	Action
Humming, rumbling or growling noise when the vehicle is in motion, intensifies when turning the steering slightly to the left or right	• Suspension fasteners loose • Damaged or worn front/rear suspension components • Uneven/irregular tire wear • Front or rear wheel bearing(s) worn	• GO to Pinpoint Test C
△ NOTE: All Wheel Drive (AWD) specific Humming, rumbling, or growling noise from the front of the vehicle at all speeds	• Suspension fasteners loose • Transmission internal failure • Front half-shaft support bearing • Damaged or worn front suspension components • Worn or damaged driveshaft joint • Front wheel bearing worn	• GO to Pinpoint Test C
Humming, rumbling or growling noise from the rear of the vehicle varying at different vehicle speed and load	• Suspension fasteners loose • Damaged or worn rear suspension components • Rear differential internal failure • Worn or damaged driveshaft joint • Rear wheel bearing worn	• GO to Pinpoint Test C

Pinpoint Tests

PINPOINT TEST A : SHOCK ABSORBER FLUID LEAK DIAGNOSIS	
TEST CONDITIONS	**DETAILS/RESULTS/ACTIONS**
A1: ASSESS LEAK	
NOTES: △ Residual oil left over from the shock absorber assembly process may create oil staining on the shock absorber tube. This will not affect the function of the shock absorber.	
△ Slight seepage is considered normal.	
	1. Assess the extent of the oil leakage
	Is the leakage serious enough to indicate that the shock absorber seal has failed? **Yes** - GO to Pinpoint Test B **No** - Shock absorber not faulty, do not renew

PINPOINT TEST B : CONFIRM LEAK	
TEST CONDITIONS	**DETAILS/RESULTS/ACTIONS**
B1: ROAD TEST	
	1. Clean all traces of oil from the shock absorber
	2. Drive the vehicle over a speed bump or similar ten times
	Is any fluid visible on the outside of the shock absorber? **Yes** - Install a new shock absorber **No** - Shock absorber not faulty, do not renew

PINPOINT TEST A : NOISY WHEEL BEARING TESTS	
TEST CONDITIONS	**DETAILS/RESULTS/ACTIONS**
A1: NOISY WHEEL BEARING INITIAL CHECKS	
	1. Confirm customer noise
	Has the customers concern been confirmed? **Yes** - GO to 2 **No** - No further action required
A2: CHECK FOR CAMPAIGNS, BULLETINS AND SSMS	

	1. Check the Jaguar Land Rover claims submission system to clarify if the vehicle has any outstanding campaigns. Refer to the corresponding bulletins and SSMs which may be valid for the specific customer complaint
	Has a corresponding campaign, bulletin or SSM been located? **Yes** - Carry out recommendations as required **No** - GO to 3
A3: ORIGIN/CORNER CHECK	
	1. Confirm the origin of the noise symptom, if available using the Jaguar Land Rover approved diagnostic equipment (Steelman ChassisEAR)
	Has the noise origin been located to a wheel/suspension corner assembly? **Yes** - GO to 4 **No** - No further action required
A4: SUSPENSION SYSTEM CHECK	
	1. Check suspension joints and links for wear/damage and check suspension system fixings/ fasteners are torqued to the correct specification
	Were there any worn/damaged or loose fixings/fasteners found? **Yes** - Rectify vehicle as necessary **No** - GO to 5
A5: TIRE CHECK	
	1. Check the tire for uneven/irregular tire wear or being out of shape
	Was there any uneven/irregular tire wear found or is the tire out of shape? **Yes** - Swap the position of the suspect tire to the opposite side of the vehicle and carry out a road test. If the noise has reduced/gone or swapped origin, reposition the wheels to their original position and advise the customer to fit a new tire **No** - GO to 6
A6: WHEEL BEARING CHECK	
	1. Check that the noise is coming from within the knuckle/hub assembly, if available using a stethoscope
	Can you confirm that the noise is coming from the wheel bearing? **Yes** - Rectify the vehicle as necessary **No** - Refer to the relevant section of the workshop manual for Drive Axle/Differential, Transfer Case, Steering System **Refer to:** Steering System (211-00 Steering System - General Information, Diagnosis and Testing) or Brake system diagnosis and testing procedures

DTC Index

For a list of diagnostic trouble codes (DTCs) that could be logged on this vehicle, please refer to Section 100-00.

Front Toe Adjustment

General Procedures

⚠ NOTE: Recognised front wheel alignment and tracking equipment should be used for this operation. Only the use of basic equipment is described below. No adjustment is provided for castor, camber or swivel pin inclinations.

1. Set vehicle on level ground with road wheels positioned straight ahead.
2. Push vehicle back and forwards to settle linkage.
3. Set up the equipment to manufacturers instructions and check alignment as advised by equipment supplier.
4. Position trammel probes on inner face of wheel, not the rims, if the latter are damaged.
5. Measure toe-out at horizontal centre-line of wheels.
6. Check tightness of clamp bolt fixings. Tighten to 14 Nm (10 lbf/ft).
7. Slacken clamps at both ends of track rod.
8. Rotate track rod to increase or decrease its effective length until correct toe-out is obtained.

Refer to: (211-00 Steering System - General Information, Specifications)

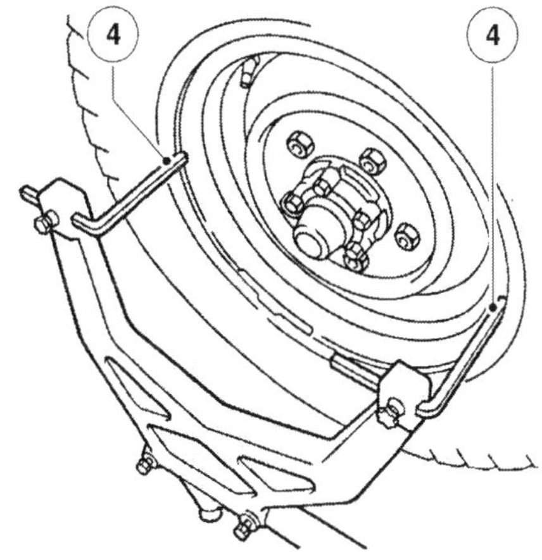

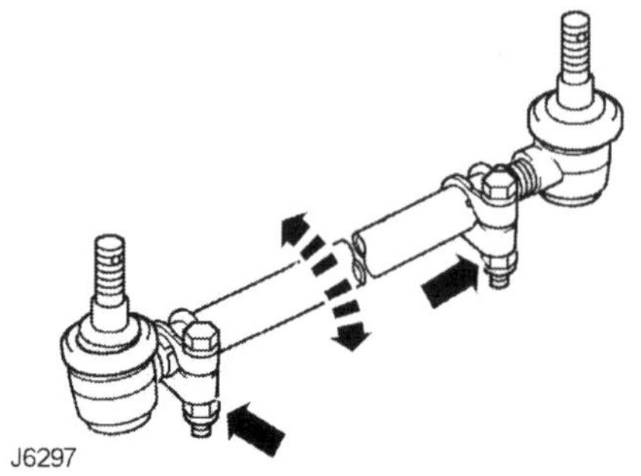

9. Push vehicle rearwards turning steering wheel from side to side to settle ball joints. With road wheels set in straight ahead position, push vehicle forward a short distance.
10. Recheck track and adjust if necessary.
11. When alignment is correct, tap ball joints in direction of arrows to maximum of travel, to ensure full unrestricted movement of track rod.
12. Tighten clamp bolts to 14 Nm (10 lbf/ft).

Steering Wheel Vibration/Shimmy While Driving

General Procedures

Check

NOTES:

Factory wheel alignment and wheel balancing are covered for 12 months/20,000 Km (12.500 mile), whichever occurs first, for OEM equipment only. Refer to Warranty Policies and Procedures manual for further information.

Until flat-spots are removed, significant vibration/shimmy may be present. A road test is required to make sure temporary tire flat-spots are removed. For longer term flat-spots, a longer road test may be required. The road test should be completed on normal open roads (highway is optimal) to allow the highest speed that speed limits and road/traffic conditions safely allow.

All road tests must be performed with tire pressures at 29 psi / 2.0 bar / 200 kPa.

Make sure that the Inline diagnostic unit (IDU)2 noise, vibration and harshness (NVH) is fully charged before carrying out the diagnostic procedure.

Make sure that the Inline diagnostic sensor (IDU)2 noise, vibration and harshness (NVH) sensor is paired through Bluetooth® to the JLR approved diagnostic equipment.

1. NOTE: **MENA and China markets only:** If the retailer is equipped with tire heaters, follow the manufacturers instructions to warm the tires prior to balancing the wheels and tires, if a customer has a steering vibration/shimmy concern:

    1. Raise the vehicle using a suitable 2-Post ramp.

Refer to: Lifting - [+] Without Armour (100-02, Description and Operation).

Refer to: Lifting - [+] With Armour (100-02, Description and Operation).

2. Install the tire heater to the tire.
3. Pull the tire heater together and secure with the velcro fastener.
4. Pull the toggles to tighten around the tire to form a good fit.
5. Repeat Steps above to the other tires.
6. Connect all the electrical connectors to the control box and run the tire heating process.
7. When the heating process is complete, remove the tire heaters and lower the vehicle.
8. Adjust tire pressures to the recommended (cold) settings.

Refer to: Specifications - [+] Without Armour (204-04, Specifications).

Refer to: Specifications - [+] With Armour (204-04, Specifications).

9. Road test the vehicle for approximately 8 Km (5 miles), if possible road test the vehicle at speeds up to approximately 80 Km/h (50 mph).
10. If the amount of steering vibration/shimmy is considered unacceptable following the initial road test completed in the diagnostic procedure, continue with this procedure.

2. If steering wheel vibration/shimmy is still present following the normal length PDI road test, continue to the diagnostic procedure below.
3. CAUTION: The Jaguar Land Rover-approved diagnostic tool must be loaded with SDD152.05 Software Management Pack v290 (or later).

NOTE: The Jaguar Land Rover (JLR) approved diagnostic equipment will read the Vehicle Identification Number (VIN) for the vehicle and automatically take the vehicle out of 'Transportation mode' if required.

Connect the JLR approved battery support unit.

General Equipment: Jaguar Land Rover approved battery support unit

4. Connect the JLR approved diagnostic equipment to the vehicle and begin a new session.

    General Equipment: Jaguar Land Rover approved diagnostic equipment

5.
   1. Follow the JLR approved diagnostic equipment prompts.
   2. Click here to run: Special applications - Inline diagnostic unit 2 noise vibration and harshness diagnostic test - Wheel imbalance.

6.
   1. If the hyperlink is not available, the application can be found as follows:
   2. Select the 'Diagnosis' Session Type.
   3. Select the following symptom:
   4. Chassis - Steering system – Steering system symptoms.
   5. From recommendations, run: Special applications - Inline diagnostic unit 2 noise vibration and harshness diagnostic test - Wheel imbalance.

7. Follow the on-screen information and select 'Accept'.
8. Connect the IDU2 NVH sensor to the JLR approved diagnostic equipment and complete the IDU2 NVH sensor self test/reset and select 'Continue'.
9. Wait for the IDU2 NVH sensor self test/reset to complete.
10. Follow the on-screen information and select 'Continue'.
11. Disconnect the JLR approved battery support unit.
12. Follow the on-screen information and select 'Continue'.
13. ⚠ WARNING: Make sure that it is legal and safe to perform the wheel imbalance test at the speed required prior to carrying out the road test.

Select the speed that the customer has highlighted creates the vibration and select Continue.

14. Follow the on-screen information and select 'Continue'.
15. Check the IDU2 NVH sensor number is seen on the display and select the correct number, then select 'Continue'.
16. The IDU2 NVH sensor will connect to the Jaguar Land Rover approved diagnostic equipment through Bluetooth®. Wait for the IDU2 NVH sensor to pair with the Jaguar Land Rover approved diagnostic equipment.
17. Install the IDU2 NVH sensor to the holder.
18. Install the install the IDU2 NVH sensor and holder assembly to the steering wheel in the orientation shown in the illustration.
19. NOTES:

△ This step will require the aid of another technician.

△ The IDU2 application will show the percentage complete for each test.

△ The application will perform 3 tests to calculate

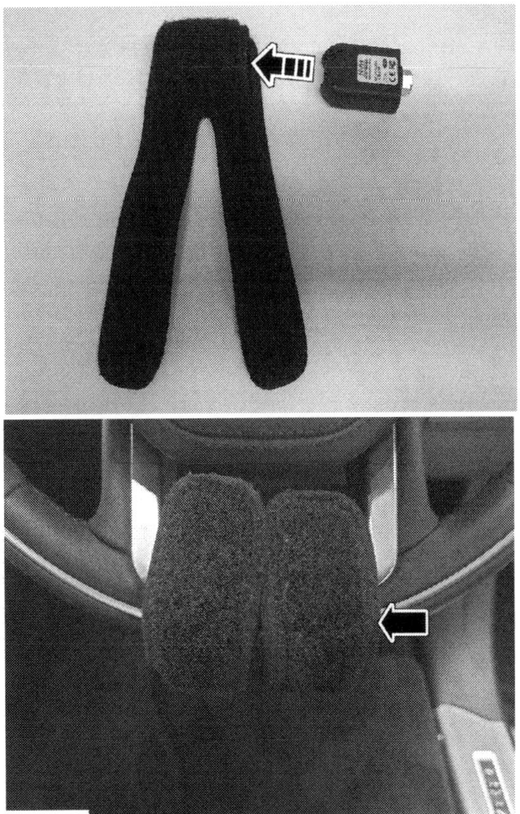

E252048

the average value, but must remain/driven 'Within limits' to register the values. The display will indicate the status of the vehicle.

   1. Drive the vehicle until all 3 tests have been completed and the screen has changed.

20. △ NOTE: This information must be added to the retailer verbatim of the claim, otherwise the claim will be rejected.

Make a record of the 'Test reference' number produced at the end of the diagnostic test in the warranty claim submission.

21.
   1. If the amount of steering vibration/shimmy is considered acceptable following the initial road test completed in the diagnostic procedure. Return the vehicle to the customer with a full explanation of the situation/flat-spot scenario and how it can/does affect the drive of the vehicle until the wheels and tires reach normal operating temperature.
   2. If the amount of steering vibration/shimmy is considered unacceptable following the initial road test completed in the diagnostic procedure, balance the wheel and tire(s) as necessary.

Refer to: Wheel and Tire Assembly Balancing

Wheel and Tire Assembly Balancing

General Procedures

Check

CAUTION: **Only Jaguar Land Rover approved wheel and tire balancing equipment must be used during this procedure.**

NOTES:

If a retailer sub-contracts tire replacement and/ or balancing the contractor must be suitably equipped with the JLR approved wheel and tire balancing equipment.

Until flat-spots are removed, significant vibration/ shimmy may be present. A road test is required to make sure temporary tire flat-spots are removed. For longer term flat-spots, a longer road test may be required. The road test should be completed on normal open roads (highway is optimal) to allow the highest speed that speed limits and road/traffic conditions safely allow.

All road tests must be performed with tire pressures at 29 psi / 2.0 bar / 200 kPa.

1. Remove the wheel and tire(s).

Refer to: Wheel and Tire (204-04, Removal and Installation).

2.
    1. Adjust the tire pressure(s) to 36 psi / 2.5 bar / 250 kPa.
    2. Using the JLR approved tire balance equipment, balance the wheel and tire assembly until the residual imbalance dynamic is minimized.
    3. Target is 0g (zero grams) on each plane; maximum is inner 5g, outer 5g.
    4. Use the 'bulls-eye' balancing mode to minimize residual imbalance.

3.
    1. The maximum Radial Force Variation (RFV) values should be as follows:
    2. 60N (6.1 kg, 13.5 lbf) First Harmonic.
    3. 100N (10.2 kg, 22.5 lbf) Peak to Peak.

4.
    1. If these levels cannot be achieved, perform the following:
    2. Remove the tire.
    3. Follow the match mounting procedure as detailed by the wheel balance equipment.
    4. Mark the high point of the 1st harmonic RFV on the outer (and inner for future reference) sidewall of the tire.
    5. On vehicles installed with electric power steering: Install the lowest 1st harmonic RFV wheel/tire assemblies to the rear axle. On vehicles installed with hydraulic power steering: Install the lowest 1st harmonic RFV wheel/tire assemblies to the front axle.
    6. Print out the results of the balance and force variation (before and after), attach to repair order.

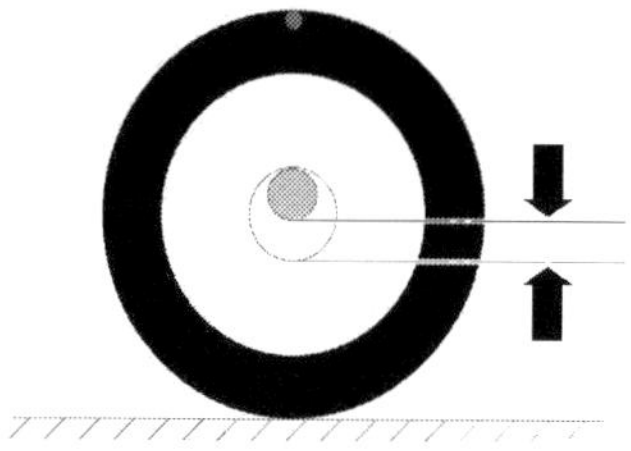

5. Prior to tightening the wheel nuts, the high-point of RFV should be marked and the road wheel and tire assembly installed to the vehicle with the RFV high point at the top.

6. NOTE: Nothing should be used to brace the wheel while tightening the nuts as this can disturb the match mounting.

   Install the wheel and tire(s).

Refer to: Wheel and Tire (204-04, Removal and Installation).

7. Lower the vehicle.

8.
    1. Tighten the wheel nuts to the specified torque.

Refer to: Specifications (204-04, Specifications).

    2. Adjust tire pressures to recommended (cold) settings.

9. Road test the vehicle for approximately 8 Km (5 miles), if possible road test the vehicle at speeds up to approximately 80 Km/h (50 mph).

10. Return the vehicle to the customer with a full explanation of the/flat-spot scenario, and how it can/does affect driving the vehicle until the wheel and tire reaches normal operating temperature.

Front Suspension

Description and Operation

Front suspension

Description

The front suspension design allows maximum wheel travel and axle articulation providing good ground clearance without loss of traction or directional stability.

Long radius arms are fitted to the front axle and provide maximum axle articulation which is vital for off road performance. The radius arms are secured to fabricated mounting brackets welded to the front axle. Flexible rubber bushes are used on a stem end joint to secure the rear of the radius arms to mountings on the chassis cross member.

A Panhard rod, which ensures that the front axle remains centrally located, is fitted transversely and also uses ferrule rubber bush mountings at both axle and chassis locations.

Two rubber bearing bushes, with retaining straps, secure the rear of the stabilizer bar to the chassis mountings, while bushed links support the front of the bar to the front axle.

Conventional long travel coil springs and hydraulic shock absorbers are used to control body movement in all conditions. The shock absorbers are secured to fabricated towers which are bolted to the chassis. The upper and lower fixings use a single location stud with flexible rubber bushes, support washers and securing nuts. Retaining plates are used to secure the coil springs to the fabricated towers and axle mountings.

Rubber bump stops are fitted underneath the chassis, adjacent to the front road springs, and prevent possible damage that could occur should there be excessive axle to chassis movement.

Changes have been incorporated to improve the durability of the suspension. The current Panhard rods are forged. New Panhard rods are introduced which are fabricated from tubular steel and are handed.

The rods are fitted with larger bushes which use M16 bolts in place of the M14 bolts. The torque for the new bolts is raised to 230 Nm (170 lb.ft).

The Panhard rod support brackets are now fabricated from upgraded steel which allows the higher torque figure to be applied to the fixing bolts.

Item	Description
1	Panhard rod
2	Radius arms
3	Coil springs
4	Shock absorbers
5	Stabilizer bar

Front suspension 02MY and Later

Front Stabilizer Bar

Removal and Installation

Removal

1. Mark for reassembly position of rubber bushes on stabilizer bar.
2. Remove 4 nuts, bolts and washers securing both stabilizer bar bush straps to chassis mounting brackets.

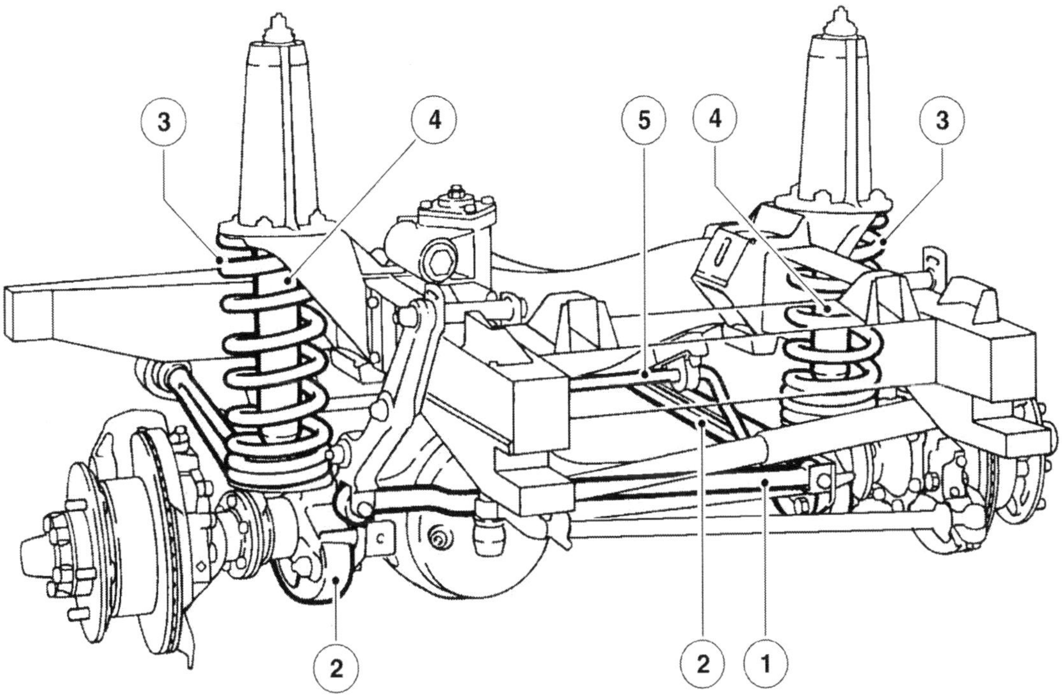

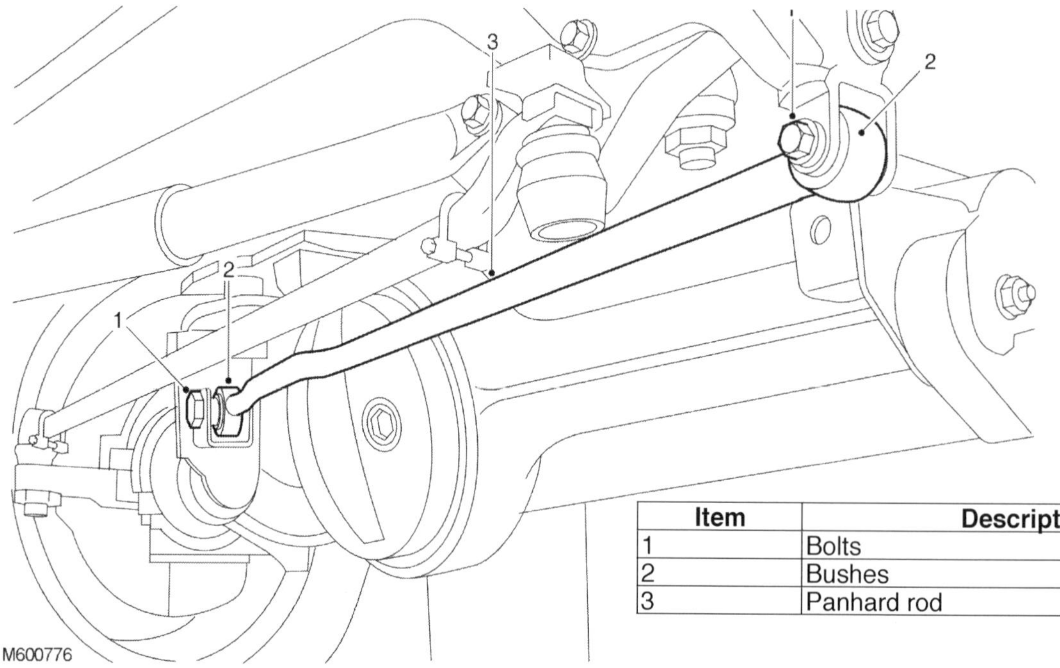

Item	Description
1	Bolts
2	Bushes
3	Panhard rod

3. Remove nuts, bolts, washers and rubber bushes securing stabilizer bar to both links
4. Remove stabilizer bar.

Installation

1. Position bushes on stabilizer bar. Ensure split points towards axle on RH bush and away from axle on LH bush.
2. Install stabilizer bar with two straps. To ensure correct fit angled sides of bar should point down. Loosely Install the bolts, washers and nyloc nuts.
3. Install bolt, washers and rubber bushes. Using new nuts fit stabilizer bar to links and tighten to 68 Nm (50 lbf.ft).
4. Tighten nuts securing straps to 30 Nm (22 lbf.ft).

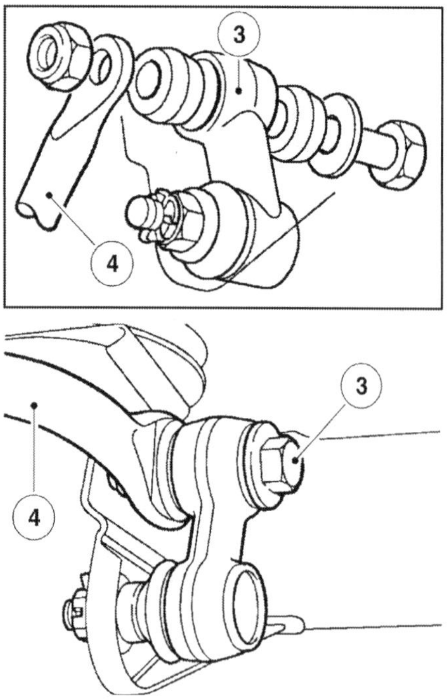

Front Stabilizer Bar Link

Removal and Installation

Removal

1. Raise vehicle on ramp.
2. Remove 2 nuts, bolts, washers and rubber bushes from ball joint links.
3. Remove cotter pin and loosen castellated nut a few turns.
4. Release link joint using tool LRT-57-018 as shown.
5. Remove castellated nut and link.

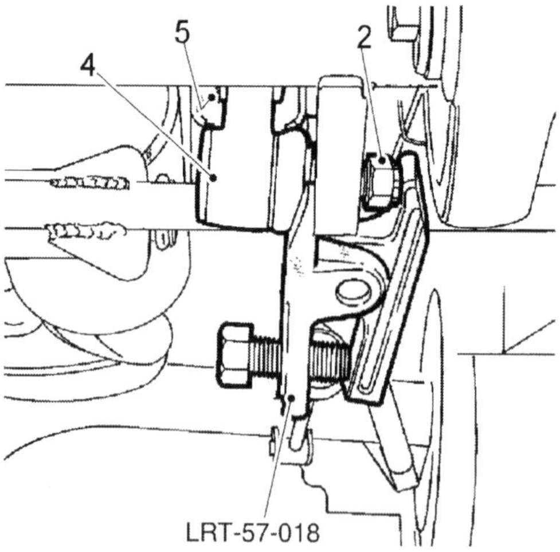

Installation

1. Fit link and castellated nut. Ensure ball joint link arm points up. Tighten nut to 40 Nm (30 lbf/ft) and fit new cotter pin.
2. Align stabilizer bar to links.
3. Fit bolts, washers and rubber bushes using new self locking nuts and secure stabilizer bar to links. Tighten fixings to 68 Nm (50 lbf/ft).
4. Lower vehicle.

Front Suspension - Panhard Rod

Removal and Installation

Removal

1. Remove fixings at mounting arm.

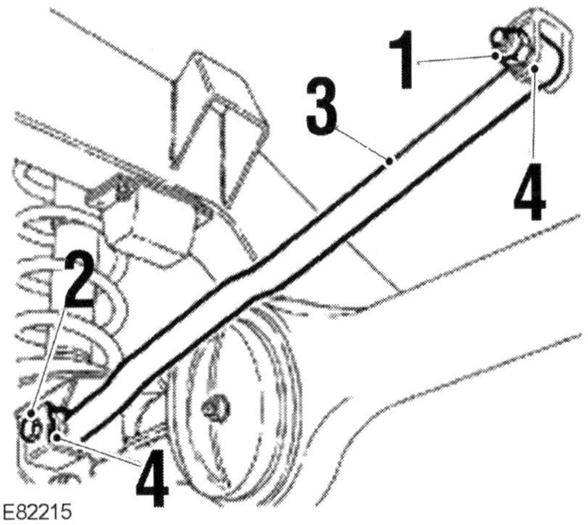

E82215

2. Remove fixings at axle bracket.
3. Remove panhard rod.
4. Using a suitable length of steel tubing, press out flexible bushes. Ensure tubing locates on outer edge of bush and not on rubber inner.

Installation

1. CAUTION: Apply pressure to outer edge of bush, and not rubber inner.

Install replacement bushes.

2. Install panhard rod to axle bracket and mounting arm. Tighten chassis fixing (**1**) to 200 Nm (148 lbf. ft). Tighten axle fixing (**2**) to 250 Nm (184 lbf.ft).
    - **Note**: If you are re-using fixings on a vehicle built prior to VIN 735937, then tighten the axle fitting to 250 Nm (184 lbf.ft) and the chassis fixing to 230 Nm (170 lbf.ft). If a new fixing is used on any vehicle, then use the torque settings of 200 Nm and 250 Nm, respectively.

Front Suspension - Spring

Removal and Installation

Removal

1. Remove front shock absorber.

   Refer to: Front Shock Absorber (204-01 Front Suspension, Removal and Installation)

CAUTION: Avoid over stretching brake hoses. If necessary, loosen hose connector locknuts to allow hoses to follow axle.

2. Lower axle sufficient to free road spring.

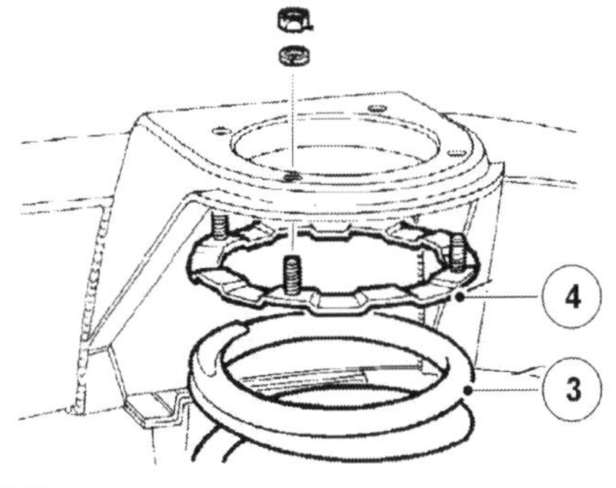

RR1593M

3. Withdraw road spring.
4. Withdraw shock absorber bracket securing ring.

Installation

1. Install shock absorber bracket retaining ring. Retain in position with a nut.
2. Position road spring and raise axle.
3. Remove nut retaining securing ring.
4. Install front shock absorber.

Refer to: Front Shock Absorber (204-01 Front Suspension, Removal and Installation)

Front Wheel Bearing and Wheel Hub

Removal and Installation

Part(s)

STEP	PART NAME	QUANTITY
Installation Step 4	Renew Part Wheel hub seal(s)	1
Installation Step 9	Renew Part Half shaft locknut(s)	1
Installation Step 18	Renew Part Wheel hub gasket	1
Installation Step 19	Renew Part Drive-shaft flange bolts	5

Removal

1. Raise front of vehicle.
2. WARNING: Support on safety stands.

   Remove front road wheel.

3. Pull back front brake caliper jump hose shield and clamp brake hose.
4. Position container collect brake fluid.
5. Loosen brake pipe to jump hose union and disconnect.

CAUTION: Use 2 spanners when loosening or tightening unions.

6. Remove 2 bolts securing brake caliper to hub.
7. Remove brake caliper.
8. Remove dust cap.
9. Remove circlip and shim(s) from drive shaft.
10. Remove and discard 5 bolts securing driving member to hub.
11. Remove driving member and discard gasket.

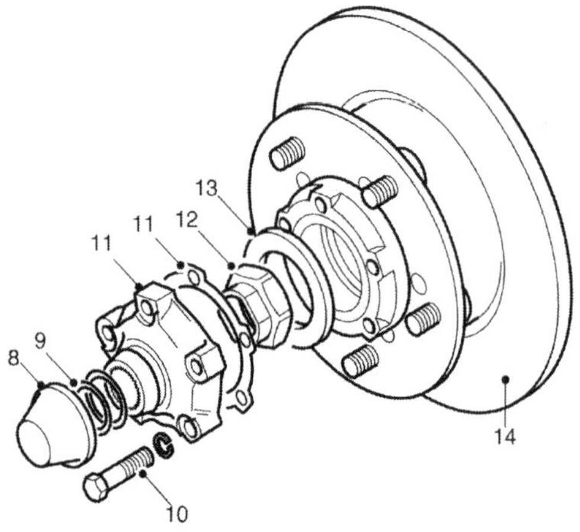

12. Knock back staking and using a suitable socket, remove and discard hub nut.
13. Remove washer from hub.
14. Remove hub and brake disc assembly complete with bearings.
15. Remove outer bearing and spacer from hub.
16. Remove grease seal and inner bearing from hub.
17. Remove inner and outer bearing tracks from hub.

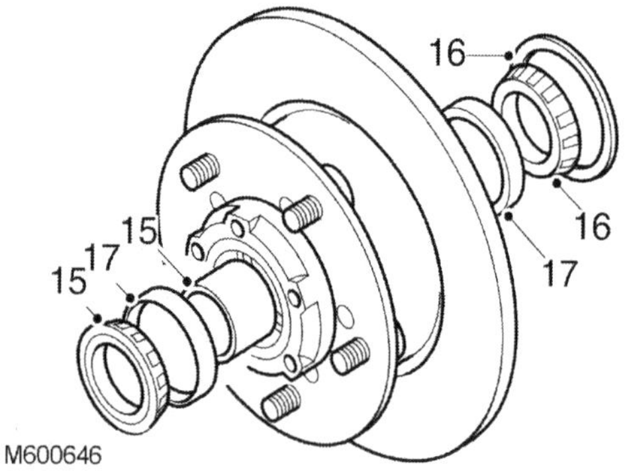

Installation

1. Clean hub and bearing locations.
2. Install inner and outer bearing tracks to hub.
3. Pack inner bearing with grease and fit to hub.
4. Install new seal flush with rear face of hub using LRT-54-003 and LRT-99-003.

   Renew Part: Wheel hub seal(s)
5. Clean stub axle.
6. Pack outer bearing with grease, fit spacer and bearing to hub.
7. Position LRT-54-019 over hub nut threads on axle casing.
8. Install hub assembly to stub axle, remove LRT-54-019.
9. Install washer and new hub nut.

   Renew Part: Half shaft locknut(s)

   Torque: 30 Nm
10. Rotate and push/pull hub to settle bearings. Tighten hub nut.

    Torque: 210 Nm
11. To check hub assembly end float, mount a dial gauge using bracket LRT-99-503 to driving member bolt hole.
12. Ensure dial gauge is contacting hub nut face.
13. Move hub assembly in and out noting dial gauge reading.
14. If end float is present refer to specifications table for correct spacer and change spacer as necessary.

Refer to: (204-00 Suspension System - General Information, Specifications)

15. When no end float is evident, remove the dial gauge and mounting bracket.
16. Stake the hub nut.
17. Clean hub and axle shaft faces.
18. Install new driving member gasket.

    Renew Part: Wheel hub gasket
19. Position driving member to hub and tighten new bolts.

    Renew Part: Driveshaft flange bolts

    Torque: 65 Nm
20. Install original shim(s) to drive shaft and secure with circlip.
21. Position brake caliper to hub, align fixings, Install bolts.

    Torque: 82 Nm
22. Remove plugs from brake pipe connections.
23. Connect brake pipe union to jump hose and tighten union.

⚠ CAUTION: Use 2 spanners when tightening or loosening unions.

24. Remove brake hose clamp from jump hose.
25. Bleed brake system.

Refer to: Brake System Bleeding (206-00 Brake System - General Information, General Procedures)

26. Ensure that retaining studs and nuts are clean.

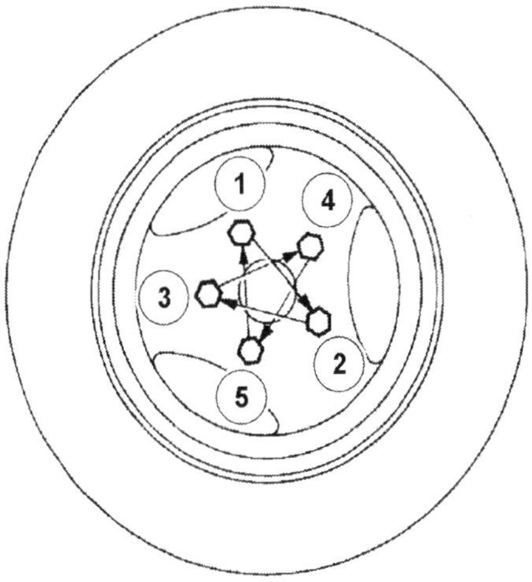

RR2142E

27. **Alloy wheels:** Lightly coat wheel mounting spigot face with a suitable anti-seize compound to minimise possibility of adhesion between wheel and spigot face.
28. Refit wheel taking care not to damage stud threads. (Do not apply oil).
29. Fit wheel nuts and turn by hand for at least three full threads before using any form of wheel wrench.
30. Tighten nuts as much as possible using a suitable wrench.
31. Lower vehicle and finally tighten nuts to correct torque sequence shown.

    Alloy wheels 130 Nm

    Steel wheels 105 Nm

    Heavy duty wheels 170 Nm
32. Operate foot brake to locate brake pads before taking vehicle on road.

Wheel Studs

Removal and Installation

Removal

1. Remove wheel.

   Refer to: Wheel and Tire (204-04 Wheels and Tires, Removal and Installation)
2. Drive stud out of driveshaft flange.

Installation

1. Position stud in flange.
2. Install a suitable spacer over stud.
3. Using a M16 x 1.5 nut, a slave wheel nut is suitable, pull stud into flange until shoulder of stud abuts flange.
4. Refit wheel.

   Refer to: Wheel and Tire (204-04 Wheels and Tires, Removal and Installation)

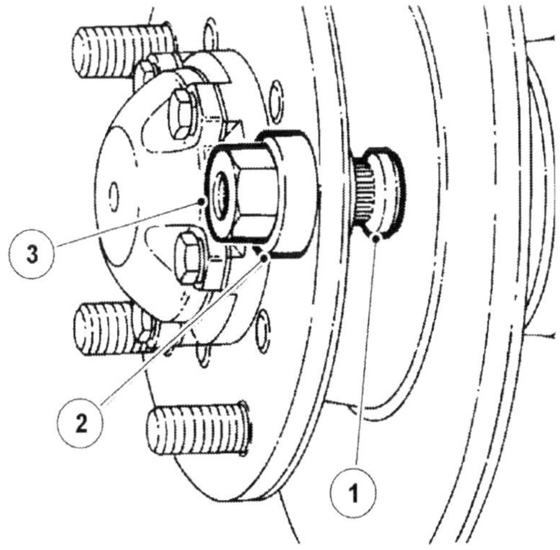

Front Shock Absorber

Removal and Installation

Removal

1. Loosen road wheel retaining nuts.
2. Support chassis on stands and remove road wheel.
3. Support axle weight with jack.
4. Remove shock absorber lower fixing and withdraw cupwasher, rubber bush and seating washer.

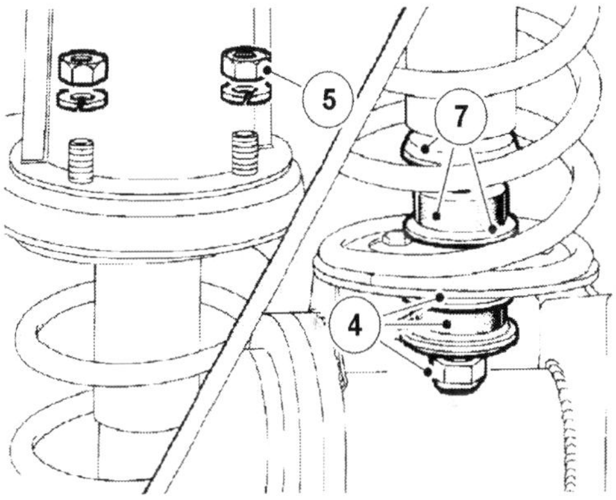

5. Remove four shock absorber bracket fixings.
6. Withdraw shock absorber and bracket assembly.
7. Withdraw lower seating washer, rubber bush and cupwasher.
8. Remove fixings, shock absorber to mounting bracket.
9. Withdraw mounting bracket.
10. Lift off top seating washer, rubber bush and cupwasher.

Installation

1. Assemble shock absorber components.
2. Position shock absorber, complete with bracket and secure with 4 fixings.

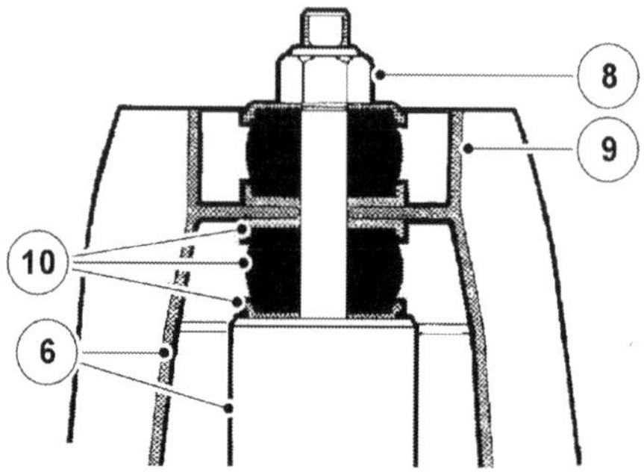

CAUTION: When tightening the nuts on both the upper and lower fixings, ensure that at least two threads are visible on the section of the bolt that extends beyond the nut.

3. Secure shock absorber lower fixing.
4. Install road wheel, remove chassis stands and jack. Tighten wheel nuts to correct torque:
    1. Alloy wheels - 130 Nm (96 lbf.ft)
    2. Steel wheels - 100 Nm (80 lbf.ft)
    3. Heavy duty wheels - 170 Nm (125 lbf.ft)

Bump Stop

Removal and Installation

Removal

1. Remove the bump stop.

Installation

1. Position bolts in slots in chassis brackets.
2. Install the bump stop, secure with washers and nuts.

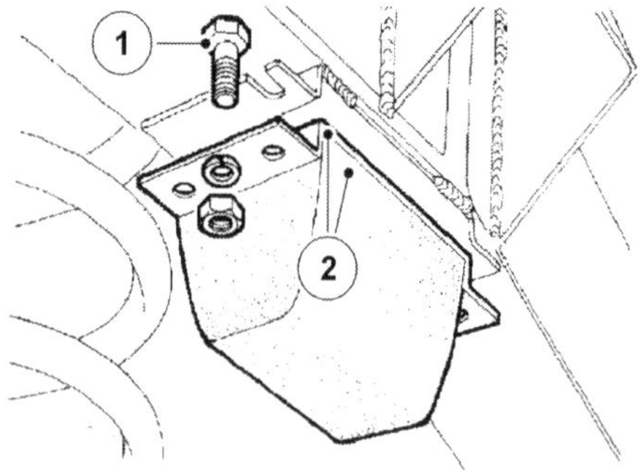

Radius Arm

Removal and Installation

Removal

1. Loosen road wheel retaining nuts.
2. Raise front of vehicle. Support chassis on stands and remove wheel.
3. Support front axle weight with jack.
4. Remove radius arm to chassis side member fixings.

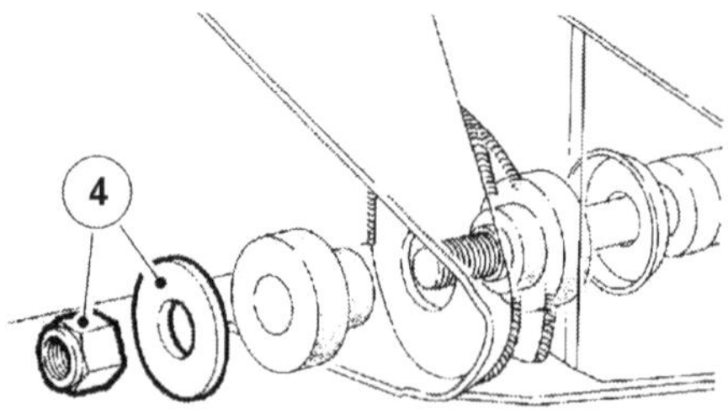

5. Disconnect track rod at ball joint using tool LRT-57-018.

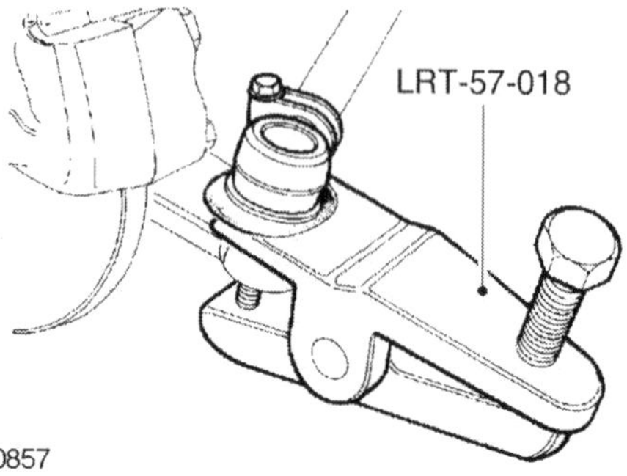

M600857

6. Remove fixings, radius arm to axle.
7. Lower radius arm front end to clear axle and remove from vehicle.
8. Using suitable length of steel tubing, press out flexible bushes.

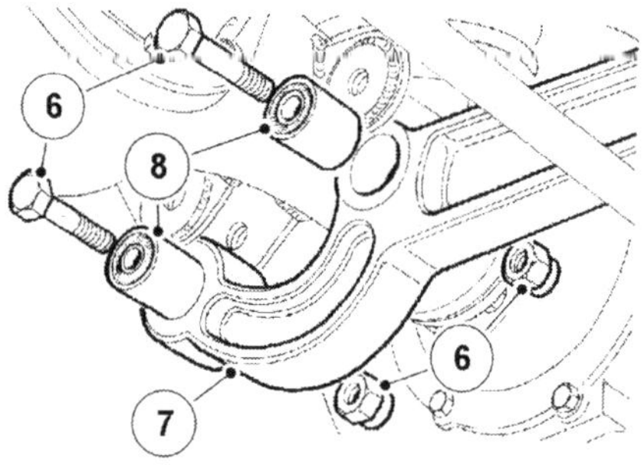

RR1592

Installation

1. Press in replacement bushes.

CAUTION: When pressing in new bushes press on outer edge of bush and not rubber inner.

2. Install radius arm to axle mounting.
3. Install track rod at ball joint.
4. Install radius arm to chassis. Tighten bolts to 176 Nm (130 lbf.ft).
5. Fully tighten radius arm to axle fixings to 197 Nm (145 lbf.ft).
6. Install road wheel, remove chassis stands and jack. Tighten wheel nuts to correct torque:
    1. Alloy wheels - 130 Nm (96 lbf.ft)
    2. Steel wheels - 100 Nm (80 lbf.ft)
    3. Heavy duty wheels - 170 Nm (125 lbf.ft)

Description and Operation

Rear axle suspension

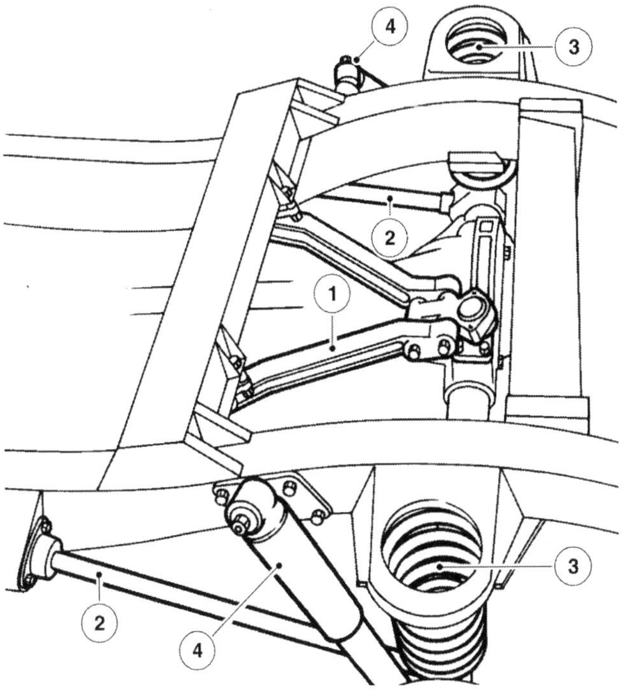

Item	Description
1	'A ' frame, upper link assembly
2	Lower link
3	Coil springs
4	Shock absorbers

DESCRIPTION

The rear suspension design locates the rear axle with two round section steel lower link arms and a forged 'A' frame, upper link assembly. This system allows maximum axle articulation and wheel travel while maintaining roll stiffness and directional stability.

The link arm is secured by a single retaining nut to the chassis mounting, comprising a rubber bushed bracket, which is retained by three fixings. A ferrule rubber bush with a single retaining bolt is used to secure the link arm to its axle mounting.

The upper link assembly is located on the rear differential housing by a pivot ball-pin assembly. Two brackets bolted to the chassis crossmember support both sides of the 'A' frame of the link assembly, secured by single retaining bolts.

A Boge Hydromat self levelling unit can be fitted, as an option, on 110/130 models to give additional support when the vehicle is used to carry heavier loads.

Two rubber bearing bushes, with retaining straps, secure the rear of the stabilizer bar, if fitted, to the chassis mountings, while bushed links support the front of the anti-roll bar to the axle.

Conventional long travel coil springs and hydraulic shock absorbers are used to control body movement.

The shock absorbers are secured to chassis mounting brackets and fabricated lower mountings welded to the rear axle. Retaining plates are used to secure the coil springs to the axle mounting while fabricated brackets, welded to the chassis, are used for the upper spring location.

Wheel Bearing and Wheel Hub

Removal and Installation

Part(s)

STEP	PART NAME	QUANTITY
Installation Step 8	Renew Part Half shaft locknut(s)	1
Installation Step 17	Renew Part Wheel hub gasket	1
Installation Step 18	Renew Part Wheel hub bolt(s)	5

Removal

1. ⚠ WARNING: Support on safety stands.

Raise rear of vehicle.

2. Remove rear road wheel.
3. Remove 2 bolts securing brake caliper to hub.
4. Remove bolt from rear brake pipe valve to differential casing. Release both rear brake pipes from the rear axle retaining clips. Release caliper from disc and tie aside.
5. Remove 5 bolts securing axle shaft.
6. Remove axle shaft and discard gasket.
7. Knock back staking, and using a suitable socket, remove and discard stake nut.
8. Remove hub nut washer.

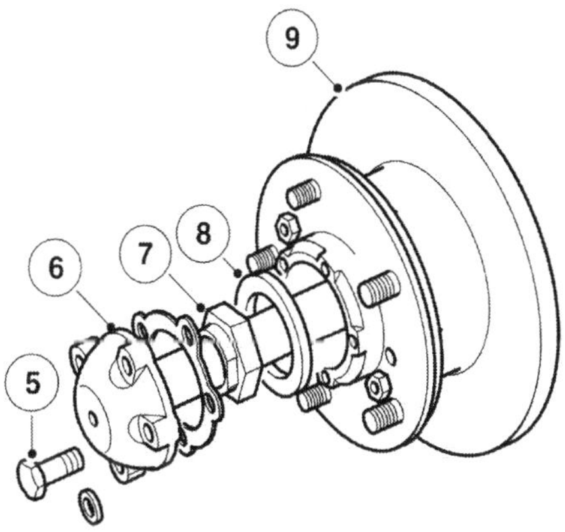

9. Remove hub and brake disc assembly complete with bearings.
10. Remove outer bearing and spacer from hub.

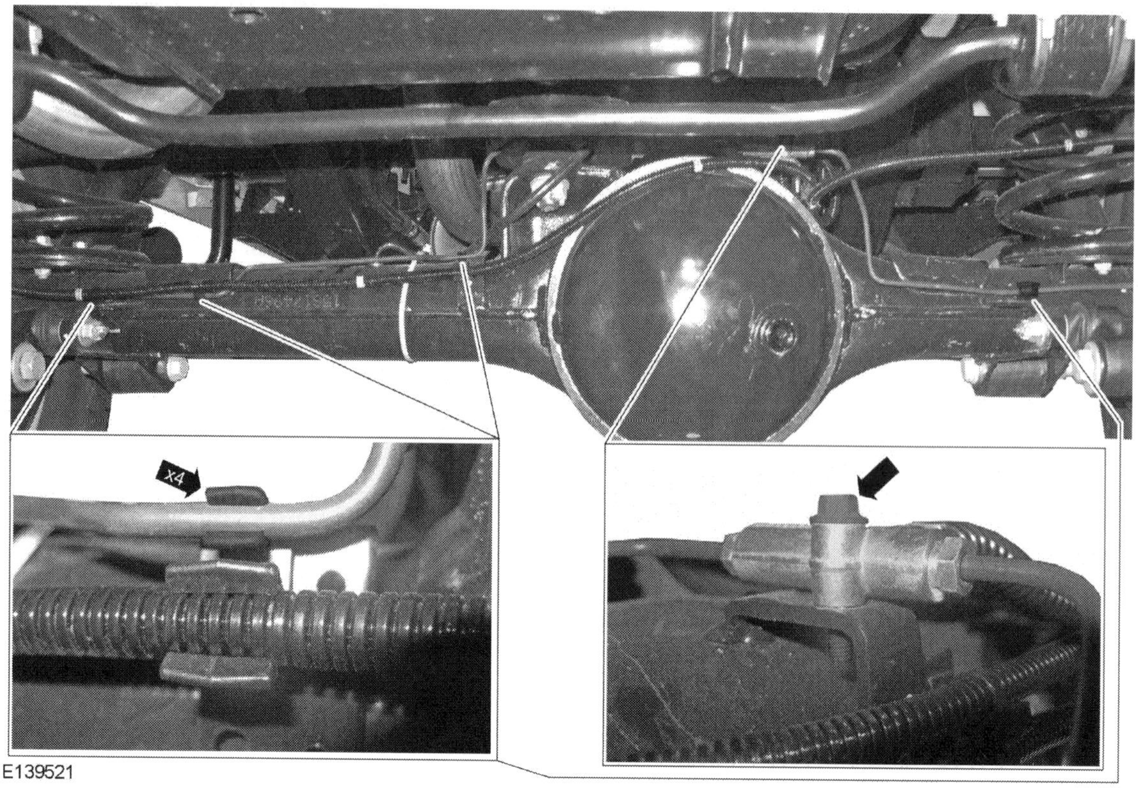

E139521

11. Remove grease seal and inner bearing from hub.
12. Remove inner and outer bearing tracks from hub.

E75817

Installation

1. Clean hub and bearing locations.
2. Fit inner and outer bearing tracks to hub.
3. Pack inner bearing with grease and fit to hub.
4. Fit new grease seal flush with rear face of hub using LRT-54-003 and LRT-99-003.

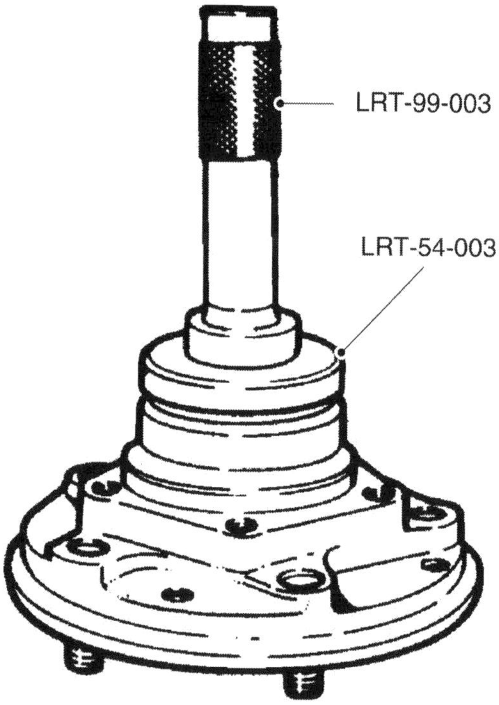

5. Clean stub axle.
6. Pack outer bearing with grease, fit spacer and bearing to hub.
7. Fit hub assembly to stub axle, remove LRT-54-019.
8. Fit washer and new hub nut and tighten the nut.

Renew Part: Half shaft locknut(s)

Torque: 30 Nm

9. Rotate and push/pull hub to settle bearings. Tighten hub nut.

   Torque: 210 Nm

10. To check hub end float, mount a dial gauge using bracket LRT-99-503 to driving member bolt hole.
11. Ensure dial gauge is contacting hub nut face.
12. Move hub in and out noting dial gauge reading.
13. If end float is present refer to the specification section for the correct spacer and change spacer as necessary.

Refer to: (204-02 Rear Suspension, Specifications)

14. When no end float is evident, remove the dial gauge and mounting bracket.
15. Stake the hub nut.
16. Clean hub and axle shaft faces.
17. Fit new driving member gasket.

    Renew Part: Wheel hub gasket
18. Position driving member to hub and tighten new bolts.

    Renew Part: Wheel hub bolt(s)

    Torque: 65 Nm
19. Fit original shim(s) to drive shaft and secure with circlip.
20. Position caliper to hub, fit bolts and tighten the bolts.

    Torque: 82 Nm
21. Secure the rear brake pipes to the rear axle. Install rear brake pipe valve bolt to differential casing and tighten the bolt to correct torque. Secure both rear brake pipes to the rear axle retaining clips.

    Torque: 15 Nm
22. Fit road wheel, remove axle stands and tighten road wheel nuts.

    Torque: 130 Nm
23. Operate brake pedal to locate brake pads before driving vehicle.
24. Check and top up brake fluid.

Rear Suspension - Specifications

End float (mm)	Spacer size (mm)	Colour code
0.00	15.5	Purple
0.025	15.4	Yellow
0.050	15.4	Yellow
0.075	15.4	Yellow
0.10	15.3	Red
0.125	15.3	Red
0.150	15.3	Red
0.175	15.2	Blue
0.200	15.2	Blue
0.225	15.2	Blue
0.250	15.2	Blue
0.275	15.1	Green
0.300	15.1	Green
0.325	15.1	Green
0.350	15.1	Green
0.375	15.0	Black
0.400	15.0	Black
0.425	15.0	Black
0.450	15.0	Black
0.475	14.9	White
0.500	14.9	White
0.525	14.9	White
0.550	14.9	White

Rear Suspension - Spring

Removal and Installation

Part(s)

STEP	PART NAME	QUANTITY
Installation Step 4	Renew Part Shock absorber mounting retaining nut(s)	2

Removal

1. Loosen rear road wheel retaining nuts.
2. Support chassis on stands and remove wheels.
3. Support rear axle weight with jack.
4. Disconnect shock absorbers at one end.
5. Position coil spring compressor correctly on road spring.
6. Compress spring evenly to facilitate removal.
7. Lower axle to free road spring from upper seat.
8. Remove spring retainer plate.

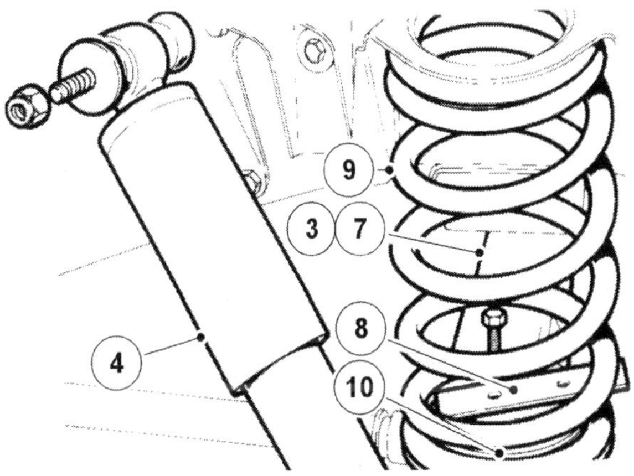

⚠ CAUTION: Avoid lowering axle further than rear brake flexible hose will allow.

9. Withdraw road spring.
10. Collect spring seat.

Installation

1. Position spring seat on axle location.
2. Install road spring into chassis location and, using a turning motion, fit to spring seat.
3. Install spring retainer plate. Tighten bolts.

    Torque: 45 Nm
4. △ NOTE: For vehicles built prior to VIN 737011, discard the old fixings and tighten new fixing to the torque specified

    Secure shock absorber. Tighten fixings to correct torque.

    Renew Part: Shock absorber mounting retaining nut(s)

    Upper fixing 85 Nm

    Lower fixing 28 Nm

5. Install road wheels, remove chassis stands and jack. Tighten wheel nuts to correct torque:

   Alloy wheels 130 Nm

   Steel wheels 100 Nm

   Heavy duty wheels 170 Nm

Wheel Studs

Removal and Installation

Removal

1. Remove wheel.

   Refer to: Wheel and Tire (204-04 Wheels and Tires, Removal and Installation)

2. Drive stud out of driveshaft flange.

Installation

1. Position stud in flange.
2. Install a suitable spacer over stud.
3. Using a M16 x 1.5 nut, a slave wheel nut is suitable, pull stud into flange until shoulder of stud abuts flange
4. Refit wheel.

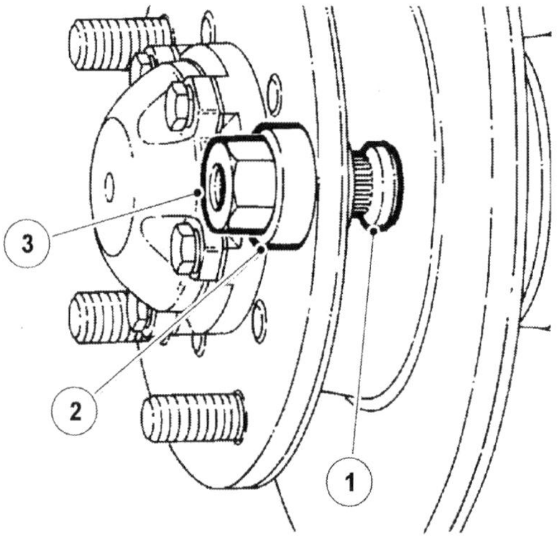

Refer to: Wheel and Tire (204-04 Wheels and Tires, Removal and Installation)

Rear Stabilizer Bar

Removal and Installation

Removal

1. Note for reassembly, position of rubber bushes on stabilizer bar.
2. Remove 4 nuts, bolts and washers securing both stabilizer bar bush straps to chassis mounting brackets.
3. Remove nuts, bolts, washers and rubber bushes securing stabilizer bar to links.
4. Remove stabilizer bar.

Installation

1. Position rubber bushes on stabilizer bar. Install joint towards axle.
2. Install stabilizer bar with two straps. Ensure link arms point down as shown. Loosely instal, bolts, washers and new nyloc nuts.
3. Install bolt, washers and rubber bushes. Install anti-roll bar to links and tighten to 68 Nm (50 lbf.ft).
4. Tighten nuts securing straps to 30 Nm (22 lbf.ft).

Rear Shock Absorber

Removal and Installation

Part(s)

STEP	PART NAME	QUANTITY
Installation Step 3	Renew Part Shock absorber upper mounting nut	2

Removal

1. Loosen road wheel retaining nuts.
2. Support chassis on stands. Remove road wheel and support rear axle weight with jack.
3. Remove fixings and withdraw shock absorber from axle bracket.

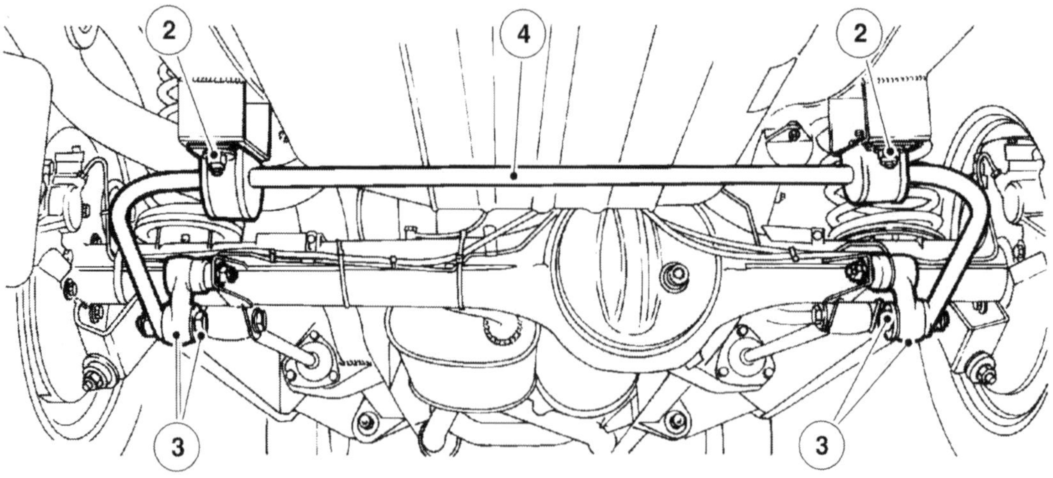

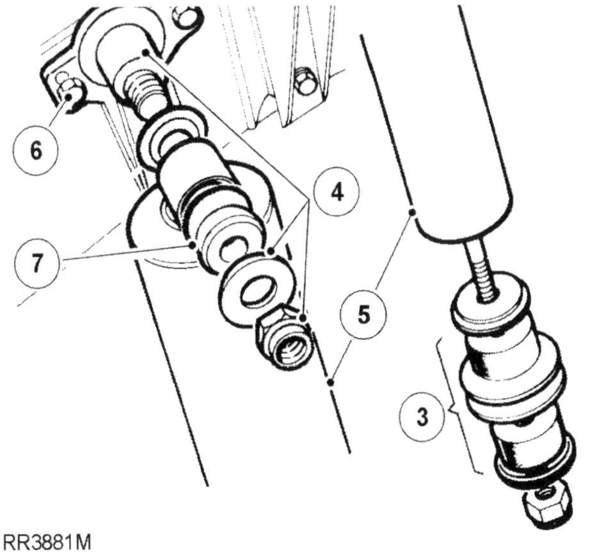

RR3881M

4. Remove upper fixings.
5. Withdraw shock absorber.
6. If required, remove mounting bracket
7. If required, remove mounting rubbers.

Installation

1. Position shock absorber and install upper fixings.
2. Secure shock absorber with lower fixings to axle bracket.
3. △ NOTE: For vehicles built prior to VIN 737011, discard the old fixings and tighten new fixing to the torque specified.

   Tighten upper and lower fixings.

Renew Part: Shock absorber upper mounting nut

Upper fixing 85 Nm

Lower fixing 28 Nm

4. Install road wheels, remove chassis stands and jack. Tighten wheel nuts to correct torque:

   Alloy wheels 130 Nm

   Steel wheels 100 Nm

   Heavy duty wheels 170 Nm

Rear Stabilizer Bar Link

Removal and Installation

Removal

1. Remove 2 nuts, bolts, washers and rubber bushes from links and lower stabilizer bar to clear links.
2. Remove cotter pin and loosen castellated nut a few turns.
3. Release link using tool LRT-57-018 as shown.
4. Remove castellated nut and remove link.

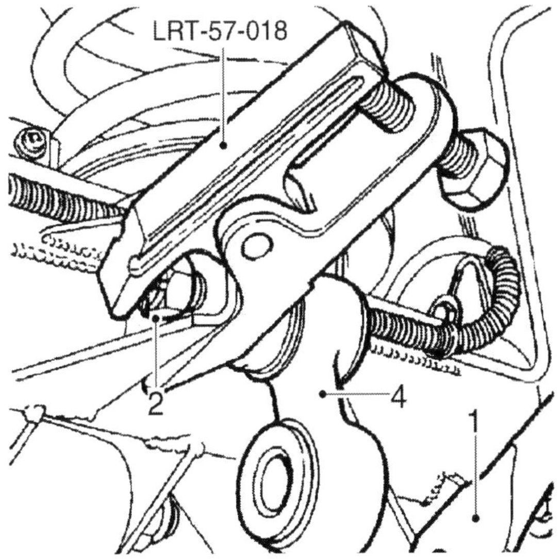

Installation

1. Install stabilizer bar link arm and castellated nut. Point link arm down as shown. Tighten fixing to 40 Nm (30 lbf.ft) and fit new cotter pin.
2. Align stabilizer bar to links.
3. Install bolts, washers and rubber bushes using new self locking nuts and secure stabilizer bar to links. Tighten to 68 Nm (50 lbf.ft).

Upper Arm

Removal and Installation

Removal

1. Support rear of chassis on stands, allow axle to hang freely.
2. Remove fixings securing upper arm bracket to frame.

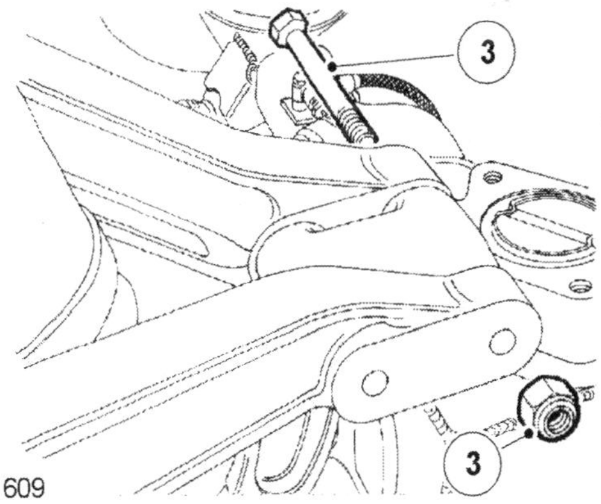

RR1609

3. Remove fixings securing upper arms to pivot bracket.
4. Remove upper arm, complete with frame bracket.
5. Remove bolt.
6. Separate upper arm from bracket.
7. Renew bush
8. Press out rubber bushes.

9. CAUTION: Apply pressure to outer edge of bush, and not rubber inner.

Fit bush centrally in housing.

Installation

1. NOTE: Do not fully tighten fixings until all components are in position.

2. Fit upper arm to pivot bracket and tighten fixings to 115 Nm (84 lbf/ft).

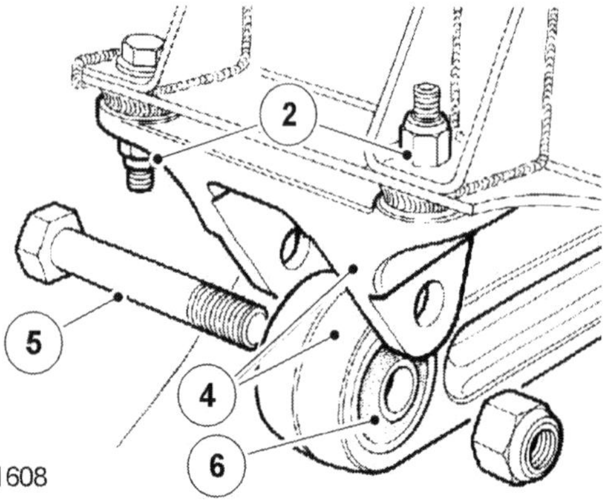

RR1608

3. Fit frame bracket to chassis mounting.

4. If you are fitting a new nut to the upper arm bolt, fully tighten to 115 Nm (84 lbf/ft). If you are refitting an existing nut, then tighten to 176 Nm (130 lbf/ft).

Lower Arm

Removal and Installation

Removal

1. Site vehicle on a ramp.

2. Alternatively, support vehicle on stands under rear axle.

3. Remove lower arm rear fixings.

4. Remove mounting bracket fixings at side member bracket.

5. Remove lower arm complete.

6. Remove locknut.

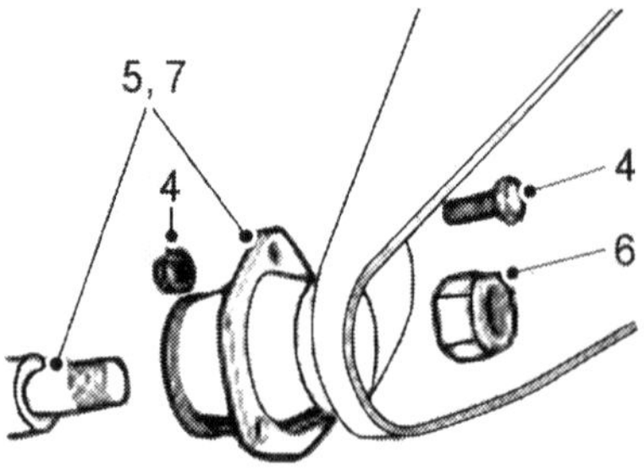

7. Remove mounting bracket from lower arm.

8. Renew bush

9. Press out rubber bushes.

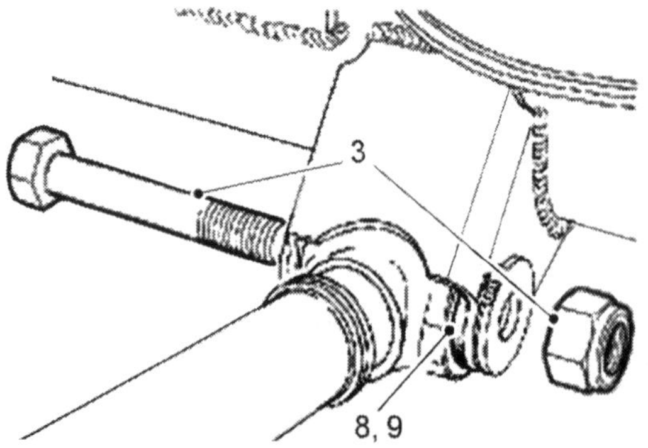

Installation

1. CAUTION: Apply pressure to outer edge of bush, and not rubber inner.

Fit bush centrally in housing.

2. Fit mounting rubber to lower arm.

3. Secure mounting rubber to chassis bracket, but do not fully tighten locknut.

4. Fit lower arm to axle mounting and secure fixing to 176 Nm (130 lbf/ft).

5. Lower vehicle, allow axle to take up static laden position, and fully tighten lower arm to chassis fixing to 176 Nm (130 lbf/ft).

Axle Bump Stop

Removal and Installation

Removal

1. Remove fixings.

2. Remove bump stop.

Installation

1. Position bolts in slots in bracket.

2. Install bump stop, secure with washers and nuts.

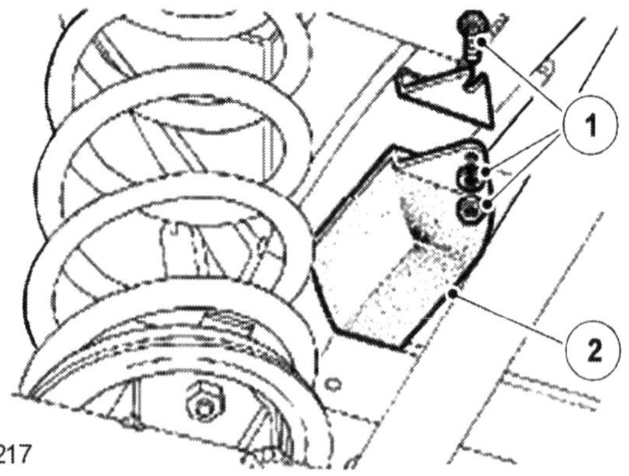

E82217

Specifications

Wheels - 90 Models

WARNING: Always use the same make and type of radial-ply tyres, front and rear. DO NOT use cross-ply tyres, or interchange tyres from front to rear.

Wheel type	Wheel size
Steel wheel - UK and Western Europe	6F x 16
Steel wheel - Other markets	5.5F x 16
Alloy wheel	7J x 16

Wheels

WARNING: Always use the same make and type of radial-ply tyres, front and rear. DO NOT use cross-ply tyres, or interchange tyres from front to rear.

If the wheel is marked TUBED, an inner tube MUST be fitted, even with a tubeless tire. If the wheel is marked TUBELESS, an inner tube must NOT be fitted.

Wheel type	Wheel size
Steel wheel - UK and Western Europe	6F x 16
Steel wheel - Other markets except Japan	5.5F x 16
Steel wheel - Japan	6.5J x 16
Alloy wheel	7J x 16

Tire sizes

Model	Tire size
90	205/80 R16 Radial
	265/75 R16 Radial (Multi terrain)
	7.50 R16 Radial
110 - except Japan	7.50 R16 Radial
110 Japan	7.50 R16C
130	7.50 R16 radial

Tire pressures

Model - Tire size	Front	Rear
90 - 205/80 R16	1,9 bar	2,6 bar
	28 lbf/in^2	38 lbf/in^2
	2,0 kgf/cm^2	2,7 kgf/cm^2
90 - 265/75 R16	1,9 bar	2,4 bar
	28 lbf/in^2	35 lbf/in^2
	2,0 kgf/cm^2	2,7 kgf/cm^2
90 - 7.50 R16	1,9 bar	2,6 bar
	28 lbf/in^2	38 lbf/in^2
	2,0 kgf/cm^2	2,7 kgf/cm^2
110 - 7.50 R16 (except Japan)	1,9 bar	3,3 bar
	28 lbf/in^2	48 lbf/in^2
	2,0 kgf/cm^2	3,4 kgf/cm^2
110 - 7.50 R16C (Japan)	2,2 bar	4,1 bar
	32 lbf/in^2	60 lbf/in^2
	2,3 kgf/cm^2	4,3 kgf/cm^2
130 - 7.50 R16	3,0 bar	4,5 bar
	44 lbf/in^2	65 lbf/in^2
	3,1 kgf/cm^2	4,6 kgf/cm^2

Road Wheel Nut Torque Specifications

Wheel type	Nm	lb-ft
Steel wheels	108	80
Alloy wheels	130	96
Heavy duty wheels	170	125

Wheel nuts must be tightened by diagonal selection

Wheels and Tires

Description and Operation

GENERAL

Dependent on specification and model type, the vehicle is equipped with pressed steel or alloy wheel rims, both using tubeless radial ply tires.

Tire codes

The text, codes and numbers moulded into the tire wall vary between tire manufacturers, however most tires are marked with the information shown in the illustrated example.

NOTE: The illustration is an example of the type of markings moulded into tires and is for guidance only. For specific tire specifications.

Refer to: (204-04 Wheels and Tires, Specifications)

Item	Description
1.	Type of tire construction - **Radial Ply**
2.	Load index - **104**
3.	Speed symbol - **S or T**
4.	USA Tyre quality grading - **Tread wear 160 Traction A temperature B**
5.	Tread wear indicators moulded into tread pattern are located at intervals around the tire and marked by a code - **E66 103S6**
6.	Tyres with 'Mud Snow' type tread pattern are marked - **M and S**
7.	Tyre reinforcing mark - **reinforced**
8.	USA Load and pressure specification - **(900 Kg (1984 LBS) at 340KA (50PSI) MACS PRESS**
9.	Tyre size - **205 16 or 235/70 R16**
10.	Type of tire - **TUBELESS**
11.	Country of manufacture - **MADE IN GREAT BRITAIN**
12.	USA Compliance symbol and identification - **DOT AB7C DOFF 267**
13.	European type approval identification - **E11 01234**
14.	Tire construction - **SIDE WALL 2 PLIES RAYON. TREAD 2 RAYON 2 STEEL**
15.	Manufacturers brand name/type - **TRACTION PLUS mzx M**

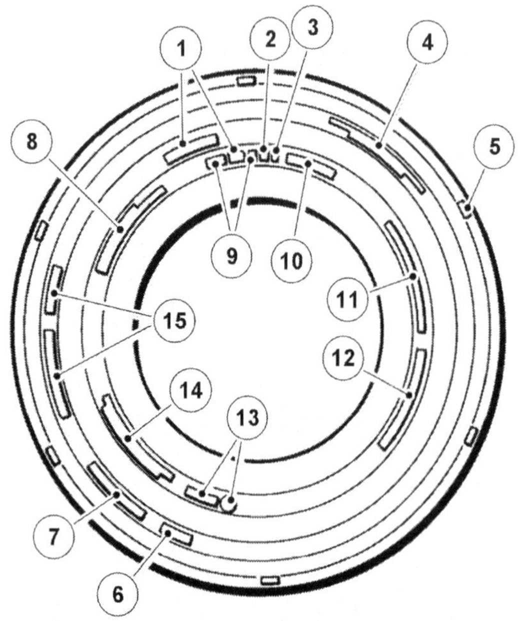

GENERAL INFORMATION

⚠ WARNING: This is a multi-purpose vehicle with wheels and tires designed for both on and off road usage. Only use wheels and tires specified for use on the vehicle.

The vehicle is equipped with tubeless S, T or H rated radial ply tires as standard equipment. The tires are of European metric size and must not be confused with the P size metric tires available in N. America.

Vehicle wheel sets, including spare wheel, must be fitted with the same make and type of tire to the correct specification and tread pattern. Under no circumstances must cross-ply or bias-belted tires be used.

⚠ WARNING: DO NOT fit an inner tube to an alloy wheel.

For tire specification and pressures.

Refer to: (204-04 Wheels and Tires, Specifications)

Steel wheels

Tubeless tires are mounted on 5.5 or 6.5 inch wide by 16 inch diameter steel wheels.

Alloy Wheels

Tubeless tires are mounted on 7.0 inch wide by 16 inch diameter cast aluminium alloy wheels. The surface has a paint finish covered with a clear polyurethane lacquer. Care must be taken when handling the wheel to avoid scratching or chipping the finish. The alloy wheel rim is of the asymmetric hump type incorporating a safety hump to improve location of the tire bead in its seat. If difficulty is experienced in fitting tires to this type of rim.

Refer to: Wheel and Tire (204-04 Wheels and Tires, Removal and Installation)

TIRE INSPECTION

Inspect tires at weekly intervals to obtain maximum tire life and performance and to ensure compliance with legal requirements. Check for signs of incorrect inflation and uneven wear, which may indicate a need for balancing or front wheel alignment, if the tires have abnormal or uneven wear patterns. See Tire Wear Chart in this section.

Check tires at least weekly for cuts, abrasions, bulges and for objects embedded in the tread. More frequent inspections are recommended when the vehicle is regularly used in off road conditions.

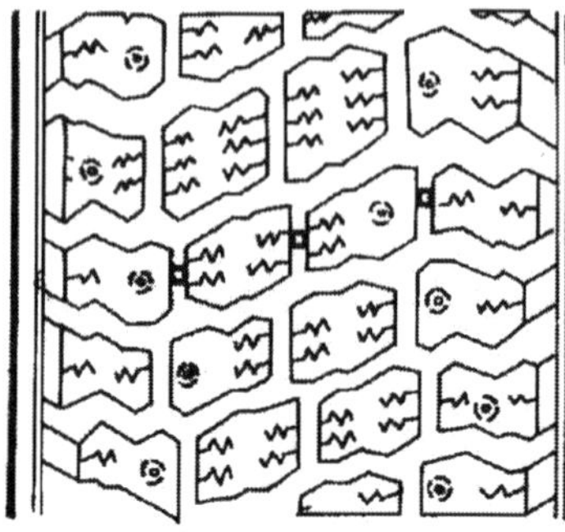

RR2145E

To assist tire inspection, tread wear indicators are moulded into the bottom of the tread grooves, as shown in the illustration above.

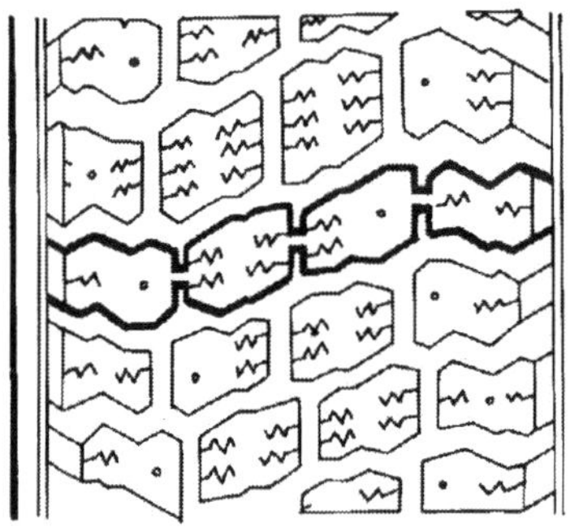

RR2146E

When the tread has worn to a depth of 1.6 mm the indicators appear at the surface as bars which connect the tread pattern across the width of the tread as shown in the illustration above.

NOTES:

⚠ DO NOT attempt to interchange tires, e.g. from front to rear, as tire wear produces characteristic patterns depending on their position. If tire position is changed after wear has occurred, the performance of the tire will be adversely affected.

⚠ Territorial vehicle regulations governing tire wear MUST be adhered to.

When the indicators appear in two or more adjacent grooves, at three locations around the tire, a new tire must be fitted.

Wheel inspection

Regularly check the condition of the wheels. Replace any wheel that is bent, cracked, dented or has excessive runout.

Valve inspection

Check condition of inflation valve. Replace any valve that is worn, cracked, loose, or leaking air.

Tire pressures

Maximum tire life and performance will be obtained only if tires are maintained at the correct pressures.

Tyre pressures must be checked at least once a week and preferably daily, if the vehicle is used off road.

The tire inflation pressure is calculated to give the vehicle satisfactory ride and steering characteristics without compromising tire tread life. For recommended tire pressures in all conditions.

Refer to: (204-04 Wheels and Tires, Specifications)

Always check tire inflation pressures using an accurate gauge and inflate tires to the recommended pressures only.

Check and adjust tire pressures ONLY when the tires are cold, vehicle parked for three hours or more, or driven for less than 3.2 km (2 miles) at speeds below 64 km/h (40 mph). Do not reduce inflation pressures if the tires are hot or the vehicle has been driven for more than 3.2 km (2 miles) at speeds over 64 km/h (40 mph), as pressures can increase by 0.41 bars (6 lb/in^2) over cold inflation pressures.

Check ALL tire pressures including the spare. Refit the valve caps as they form a positive seal and keep dust out of the valve.

WHEEL BALANCING

CAUTION: It is essential that all wheel balancing is carried out off the vehicle. The use of on the vehicle balancing could cause component damage or personal injury and MUST NOT be attempted.

NOTE: Before attempting to balance a wheel and tire assembly clean all mud and dirt deposits from both inside and outside rims and remove existing balance weights.

Remove stones from the tire tread in order to avoid operator injury during dynamic balancing and to obtain the correct balance.

Inspect tires for damage and correct tire pressures and balance according to the equipment manufacturers instructions.

Steel wheels

Clean area of wheel rim and attach balance weights in position shown.

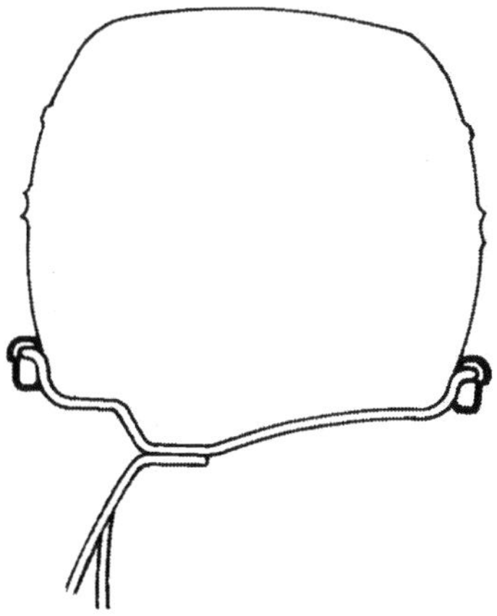

Alloy wheels

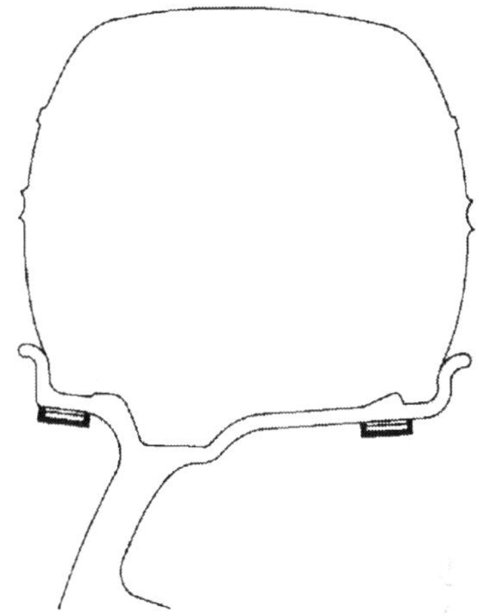

Clean area of wheel rim and attach adhesive balance weights in position shown. Cut through rear face of weight strip to detach required weights.

CAUTION: Use only correct adhesive balance weights to avoid damage to aluminium wheel rim. DO NOT attempt to use a steel wheel weight on an aluminium wheel.

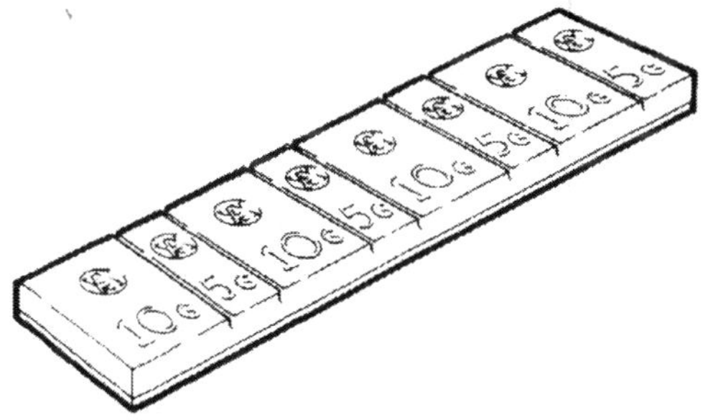

Static Balance

Wheel tramp

Static balance is the equal distribution of weight around the wheel. A statically unbalanced wheel will cause a bouncing action called wheel tramp. This condition will eventually cause uneven tire wear.

Item	Description
A	Heavy spot.
B	Add balance weights here.
C	Centre line of spindle.

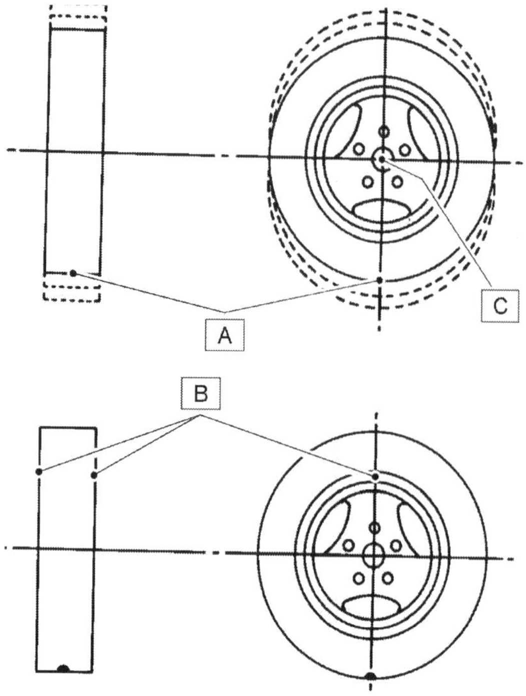

Dynamic Balance

Wheel shimmy

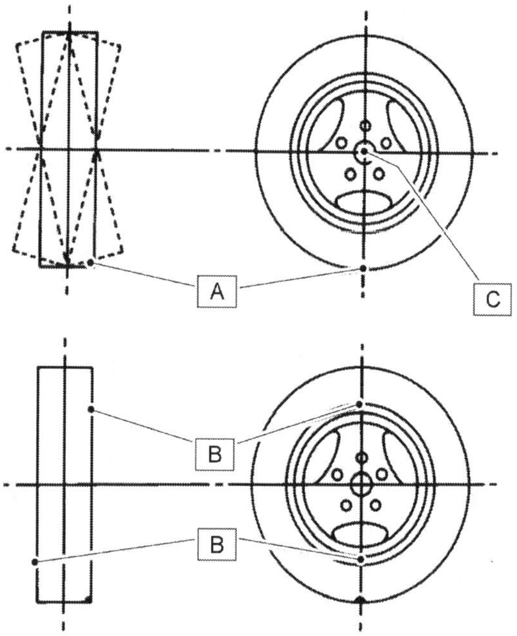

Item	Description
A	Heavy spot.
B	Add balance weights here.
C	Centre line of spindle.

Dynamic balance is the equal distribution of weight on each side of the centre line so that when the wheel spins there is no tendency for side to side movement. A dynamically unbalanced wheel will cause wheel shimmy.

Off Vehicle Balancing

Balance wheel assembly referring to equipment manufacturer's instructions.

It is essential that the wheel is located by the centre hole NOT the stud holes. To ensure positive wheel location the diameter of the locating collar on the machine shaft must be 112,80 to 112,85 mm (4.441 to 4.443 in). This diameter will ensure that the collar fits correctly within the centre hole of the wheel.

Where possible, always use the vehicle wheel retaining nuts to locate the wheel on the balancer, to avoid damaging the wheel. If this is not possible, the locating nuts must be of a similar pattern to the original wheel nuts. The use of conical type wheel nuts for this purpose may damage the surface on alloy wheels.

Cleaning

Wash the aluminium wheels using a suitable wash and wax concentrate, correctly diluted and rinse with cold clear water. DO NOT use abrasives or aluminium wheel cleaners containing acid, as they will destroy the lacquer finish.

Tire changing

Use only tire changing equipment to mount or demount tires, following the equipment manufacturer's instructions. DO NOT use hand tools or tire levers, as they may damage tire beads or the wheel rim.

Puncture repair

Remove punctured tire from wheel and repair using a combination service plug and vulcanising patch. Always follow manufacturer's instructions when using a puncture repair kit.

Only punctures in tread area are reparable, DO NOT attempt to repair punctures in tire shoulders or sidewalls.

Do not attempt to repair a tire that has sustained the following: bulges or blisters, ply separation, broken or cracked beads, wear indicators visible and punctures larger than 6 mm diameter.

⚠ CAUTION: Do not use tire sealants that are injected through valve stem to repair punctured tires, they may produce wheel corrosion and tire imbalance.

Aluminium wheel rim bead seats should be cleaned using a non-abrasive cleaner to remove the mounting lubricants and old rubber. Before mounting or demounting a tire, bead area should be well lubricated with a suitable tire lubricant.

FAULT- SYMPTOMS

- Check tyre pressures.

Refer to: (204-04 Wheels and Tires, Specifications)

- Check condition of tyres.
- Check front wheel alignment.

Refer to: Front Toe Adjustment (204-00 Suspension System - General Information, General Procedures)

- Check wheel balance.

NOTES:

⚠ In the event that any apparent vibration is not eliminated at this stage.

⚠ In the event that any apparent vibration is not eliminated at this stage, go to steering Fault Diagnosis, Fault - Symptom (Steering vibration, road wheel shimmy/wobble).

Refer to: Steering System (211-00 Steering System - General Information, Diagnosis and Testing)

⚠ Radial ply tyres have a flexible sidewall, which produces a sidewall bulge making the tyre appear under-inflated. This is a normal condition for radial ply tyres. Do not attempt to reduce this bulge by over-inflating the tyre.

Tire wear chart

⚠ CAUTION: This chart is for general guidance only and does not necessarily include every cause of abnormal tire wear.

Item	Description
1.	Correct inflation.
2.	Under-inflation.
3.	Over-inflation.
4	Tread contact with road.

Fault	Cause	Remedy
Rapid wear at shoulders	Tires under inflated	Inflate to correct pressure
	Worn suspension components; i.e. ball joints, Panhard rod bushes, steering damper	Replace worn components
Rapid wear at centre of tread	Tires over inflated	Inflate to correct pressure
Wear at one shoulder	Track out of adjustment	Adjust track to correct figure
	Bent Panhard rod	Check and replace worn or damaged components
Bald spots or tire cupping	Wheel out of balance	Balance wheel and tire assembly
	Excessive radial run out	Check run out and replace tire if necessary
	Shock absorber worn	Replace shock absorber
	Excessive braking	
Tire scalloped	Track out of adjustment	Adjust track to correct figure
	Worn suspension components	Check and replace worn or damaged components, and replace tire
	Excessive cornering speeds	

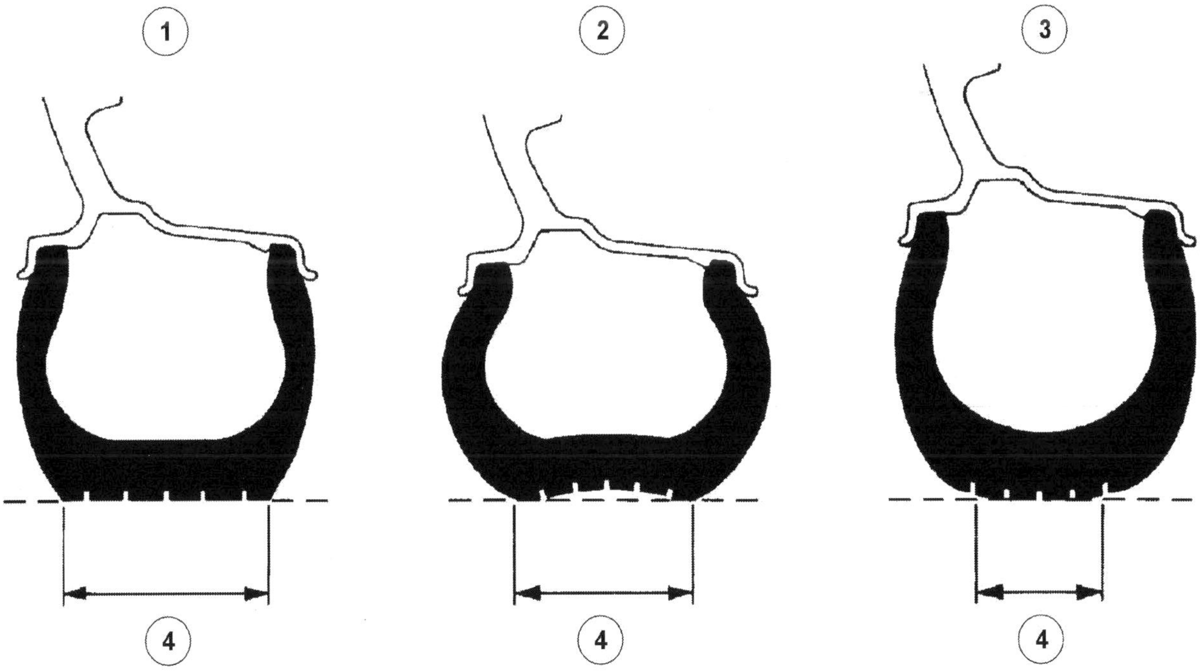

Wheel and Tires

Removal and Installation

Removal

WARNING: The parking brake acts on transmission, not rear wheels, and may not hold vehicle when jacking unless following procedure is used. If one front wheel and one rear wheel is raised no vehicle holding or braking effect is possible. Wheels MUST be chocked in all circumstances.

1. Apply parking brake, select a gear in main gearbox and engage low gear in transfer box.
2. CAUTION: Using power tools may result in damage to the wheel nuts. Make sure the wheel nuts are released using hand tools only.

Loosen 5 wheel nuts.

3. Using a suitable trolley jack, raise vehicle and place on axle stands.

Refer to: Jacking (100-02 Jacking and Lifting, Description and Operation)

4. CAUTION: Using power tools may result in damage to the wheel nuts. Make sure the wheel nuts are released using hand tools only.

Remove wheel nuts and carefully withdraw wheel over studs.

Installation

1. Ensure that retaining studs and nuts are clean.
2. **Alloy wheels:** Lightly coat wheel mounting spigot face with a suitable anti-seize compound to minimize possibility of adhesion between wheel and spigot face.
3. Refit wheel taking care not to damage stud threads. (Do not apply oil).
4. Fit wheel nuts and turn by hand for at least three full threads before using any form of wheel wrench.
5. CAUTION: Using power tools may result in damage to the wheel nuts. Make sure the wheel nuts are installed using hand tools only.

Tighten nuts as much as possible using a suitable wrench.

6. CAUTION: Using power tools may result in damage to the wheel nuts. Make sure the wheel nuts are installed using hand tools only.

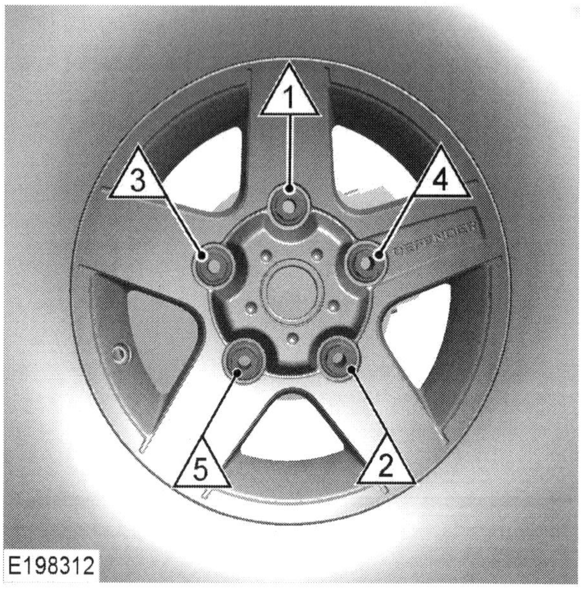

Lower vehicle and finally tighten nuts to correct torque sequence shown.

1. Steel wheels - 108 Nm
2. Alloy wheels - 130 Nm
3. Heavy duty wheels - 170 Nm

Description and Operation

Description

The front and rear drive shafts have non-constant velocity type universal joints, with needle roller bearings. The bearing cups are pre-packed with lubricant on assembly and a grease nipple is fitted for servicing as specified, in maintenance section.

Both shafts have ball splines to accommodate the variation in distance between the axles and transmission. The splines are pre-packed with lubricant and protected by a rubber gaiter.

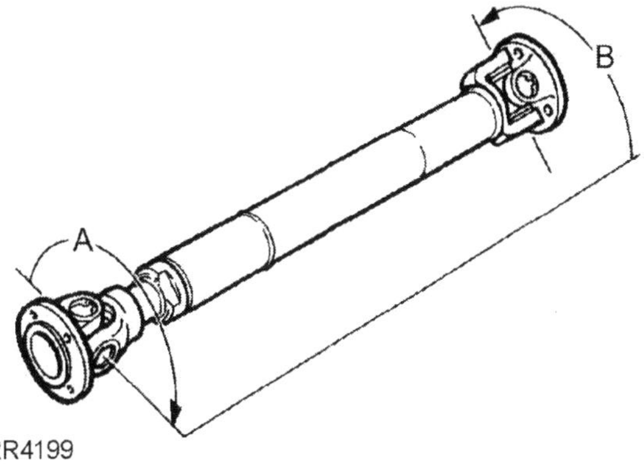

RR4199

NOTE: This joint does not require lubrication.

The front and rear driveshafts are 'phased', with the joints at each end, A and B mis-aligned as shown. The phasing is necessary to allow for greater variation in angular changes. On reassembly it is essential the driveshafts are realigned correctly.

Item	Description
1.	Flanged yoke
2.	Grease nipple
3.	Journal spider
4.	Needle roller bearing
5.	Circlip
6.	Splined shaft
7.	Rubber gaiter (dust cap)
8.	Ball spline shaft

Vibration Harshness

Check the propeller shaft universal joints are not seized or worn by checking for excessive radial and axial movement. This can be done by inserting a suitable lever into the joint and checking for movement. In the event that both shafts are satisfactory, but the vibration/ harshness is still present, the transfer box operation and balance of the road wheels should be checked.

Diagnosis and Testing

Principles of Operation

For a detailed description of the Driveshaft, refer to the relevant Description and Operation section in the workshop manual.

Refer to: Driveshaft (205-01 Driveshaft, Description and Operation)

Inspection and Verification

NOTE: If a control module or a component is suspect and the vehicle remains under manufacturer warranty, refer to the Warranty Policy and Procedures manual, or determine if any prior approval programme is in operation, prior to the installation of a new module/ component.

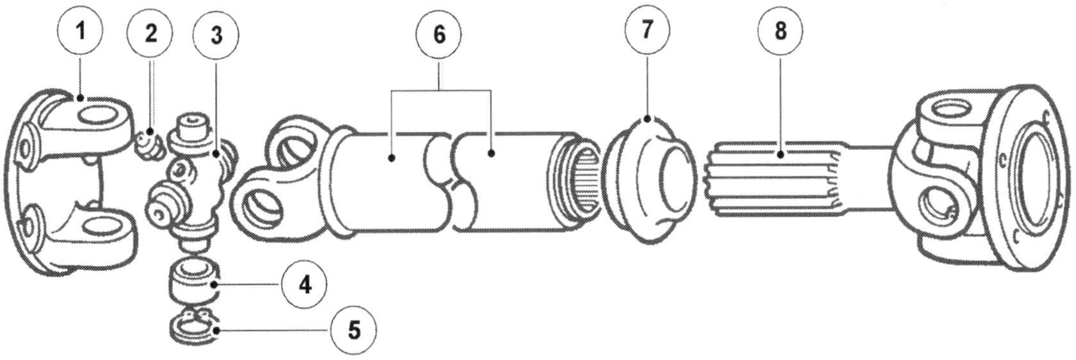

Symptom Chart

Symptom	Possible Causes	Action
Excessive freeplay in front/ rear driveshaft universal joint	• Universal joint bearing(s) worn/damaged	• Remove the universal joint from the driveshaft. Check the integrity of the yokes. If neither yoke is damaged or deformed, install a new universal joint. If either yoke is damaged or deformed, install a new driveshaft
Knocking noise from front/ rear of the vehicle	• Universal joint bearing(s) worn/damaged	• Remove the universal joint from the driveshaft. Check the integrity of the yokes. If neither yoke is damaged or deformed, install a new universal joint. If either yoke is damaged or deformed, install a new driveshaft

1. Visually inspect for obvious signs of damage and system integrity.
2. If an obvious cause for an observed or reported concern is found, correct the cause (if possible) before proceeding to the next step.
3. If the cause is not visually evident, verify the symptom and refer to the Symptom Chart, alternatively check for Diagnostic Trouble Codes (DTCs) and refer to the DTC Index.

Visual Inspection

Mechanical
• Driveshafts
• Universal joints
• Sliding joints
• Grease nipples
• Fixings
• Sliding joint boots

Front Driveshaft

Removal and Installation

Part(s)

STEP	PART NAME	QUANTITY
Installation Step 2	Renew Part Front Half shaft nut	4
Installation Step 3	Renew Part Front Half shaft nut	4

Removal

1. WARNING: Do not work on or under a vehicle supported only by a jack. Always support the vehicle on safety stands.

   Raise and support the vehicle.

2. NOTE: Mark the front driveshaft to front differential drive flange.

   Release the driveshaft from the front differential. Remove and discard the 4 nuts.

3. NOTE: Mark the front driveshaft to transfer case drive flange.

   Remove the driveshaft. Remove and discard the 4 nuts

Installation

1. NOTE: Clean the driveshaft drive flanges and mating faces.

   To install, reverse the removal procedure.

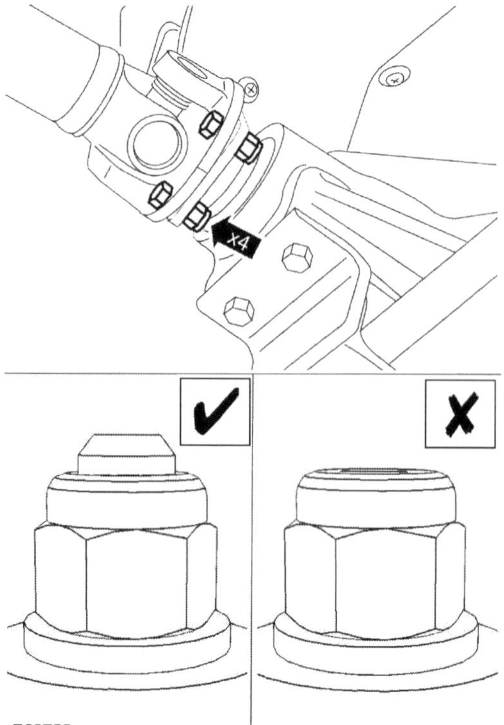

E90729

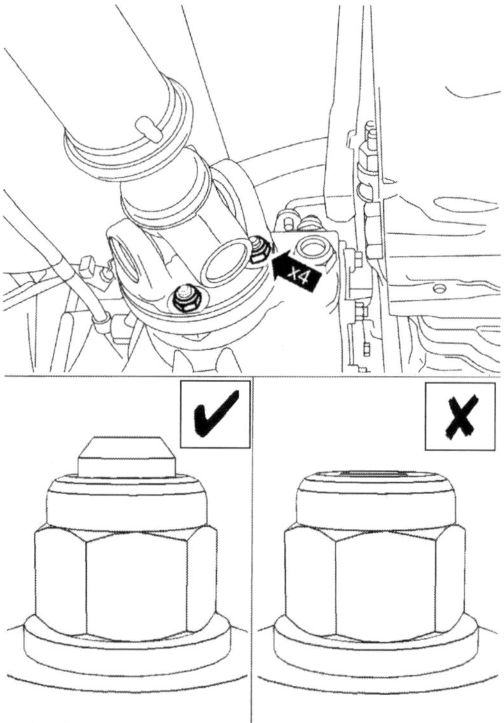

E90730

2. NOTE: Install new nuts.

   Align the position of the driveshaft in relation to the drive pinion flange.

   Renew Part: Front Half shaft nut

   Torque: 47 Nm

3. NOTE: Install new nuts.

   Align the position of the driveshaft in relation to the drive pinion flange.

   Renew Part: Front Half shaft nut

   Torque: 47 Nm

Rear Driveshaft

Removal and Installation

Part(s)

STEP	PART NAME	QUANTITY
Installation Step 3	Renew Part Driveshaft flange nut(s)	4
Installation Step 4	Renew Part Driveshaft flange nut(s)	4

Removal

1. ⚠ WARNING: Support on safety stands and chock front wheels.

   Raise rear of vehicle.

2. △ NOTE: Mark the rear driveshaft to parking brake flange.

   Remove and discard 4 nuts securing driveshaft to parking brake flange.

△ NOTE: Rotation of driveshaft may be required during the above procedure.

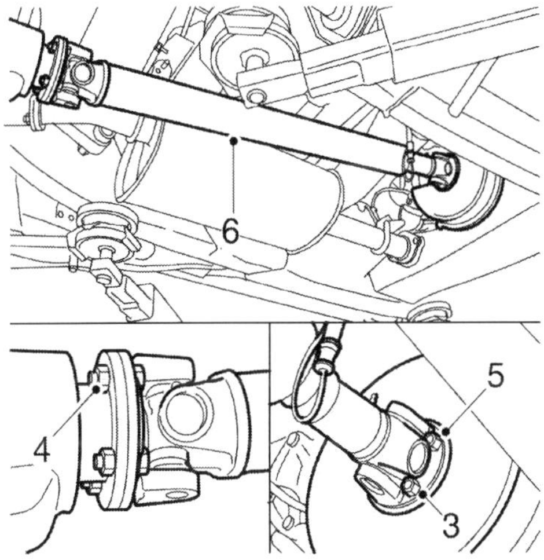

3. NOTES:

△ Mark the rear driveshaft to rear differential flange.

△ Rotation of driveshaft may be required during the procedure.

   Remove and discard 4 nuts securing driveshaft to rear axle flange.

△ NOTE: Rotation of driveshaft may be required during the above procedure.

4. Release driveshaft from parking brake flange.
5. Remove driveshaft.

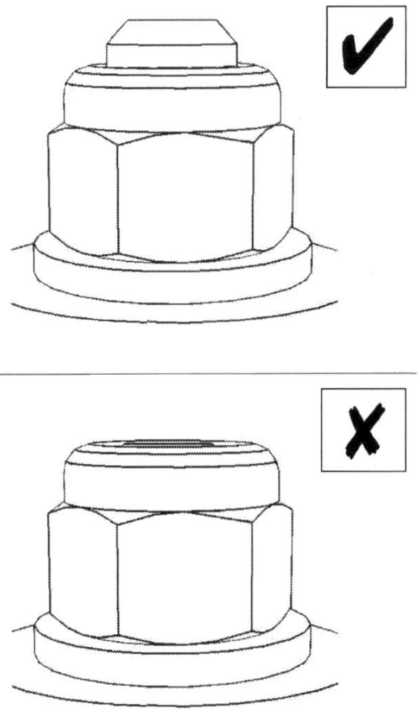

Installation

1. Clean driveshaft flange mating faces.
2. Fit driveshaft to park brake flange bolts.

△ NOTE: Ensure relationship marks align.

3. Position driveshaft to rear axle flange and fit bolts.

   Renew Part: Driveshaft flange nut(s)

4. △ NOTE: Rotation of driveshaft may be required during the procedure.

Fit nuts and bolts securing driveshaft to rear axle flange and tighten to correct torque.

   Renew Part: Driveshaft flange nut(s)

   Torque: 47 Nm

5. Fit nuts securing driveshaft to park brake and tighten to correct torque.

   Torque: 47 Nm

△ NOTE: Rotation of driveshaft may be required during the procedure.

6. Remove stands and lower vehicle.
7. Apply parking brake.

Universal Joint

Disassembly and Assembly

Disassembly

1. Remove driveshaft.

Refer to: Rear Driveshaft (205-01 Driveshaft, Removal and Installation)

Refer to: Rear Driveshaft (205-01 Driveshaft, Removal and Installation)

2. Thoroughly examine universal joints for signs of damage or wear.

3. Clean universal joint bearing cups and circlips.

⚠ CAUTION: To ensure correct assembly and reduce possibility of imbalance, before removing driveshaft joint mark position of spider pin relative to journal yoke ears.

4. Remove circlips.

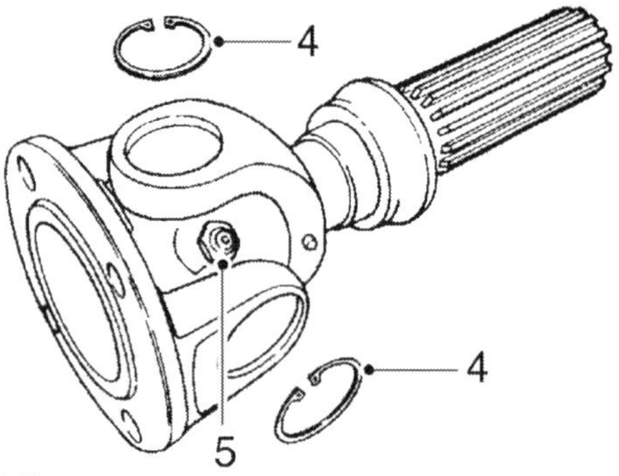

M470249

5. Note position and remove grease nipple.

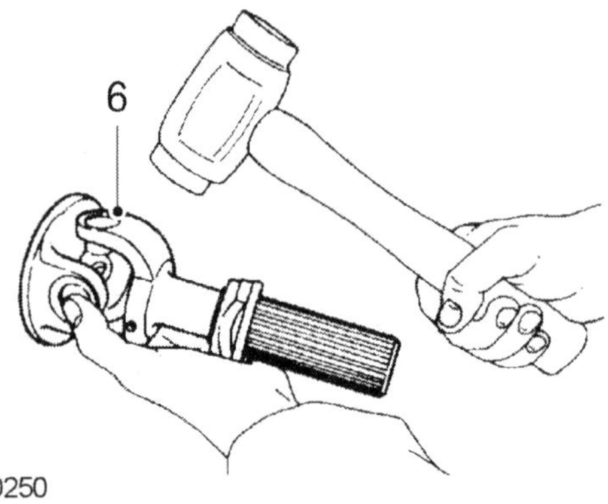

M470250

6. Tap yokes to eject bearing cups.

7. Remove bearing cups.

8. Remove spider.

9. Clean yokes and bearing cup locations.

Assembly

1. Remove bearing cups from new spider.

2. Check all needle rollers are present and positioned in bearing cups.

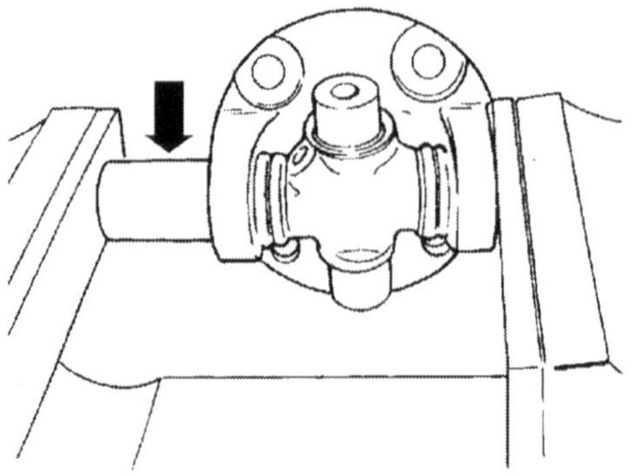

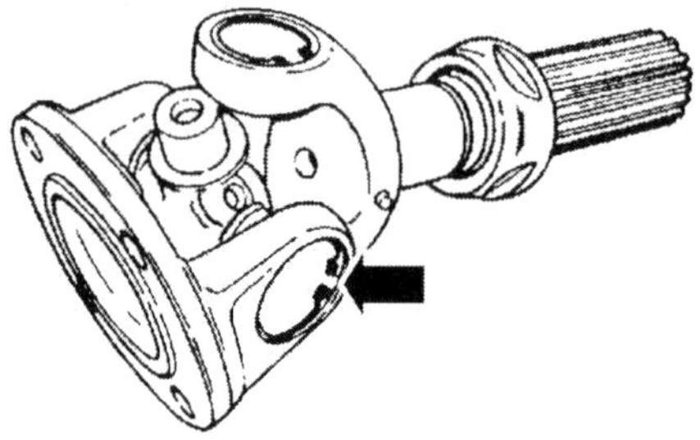

3. Enter new spider with seals into yokes of propeller shaft flange.

4. Partially insert one bearing cup into flange yoke and enter spider trunnion into bearing cup.

5. Insert opposite bearing cup into flange yoke.

6. Press both cups into place.

7. Press each cup into its respective yoke up to lower land of circlip groove. Damage may be caused to cups and seals if cups pass this point.

8. Fit circlips and check no end float exists.

9. Fit grease nipple and lubricate

10. Repeat instructions for opposite end of driveshaft as described in 3 to 9.

11. Fit driveshaft.

Refer to: Rear Driveshaft (205-01 Driveshaft, Removal and Installation)

Refer to: Rear Driveshaft (205-01 Driveshaft, Removal and Installation)

Specifications

Rear axle	
Type	Spiral bevel, fully floating shafts
Ratio	3.54:1

Sealers

Item	Land Rover Part No.
Rear axle differential case	STC 3811
Crown wheel to differential case bolts	STC 50552
Differential case bolts	STC 50552
Hub drive member bolts	STC 50552

Lubricants

Item	Specification
Final drive *	Castrol EPX 80W-90 or Texaco Geartex EP85W-90

*** Do not use any lubricant other than that specified**

Capacities

Unit	Capacity
Differential - 90	1.70 litres (2.9 pints) (1.80 US quarts)
Differential - 110/130	1.85 litres (3.3 pints) (1.95 US quarts)

Torque Specifications

Description	Nm	lb-ft
Pinion housing to axle case	41	30
Crown wheel to differential case	58*	43*
Differential bearing cap to pinion housing	90	65
Differential drive flange to drive shaft	47	34
Bevel pinion nut	129	96
Lower link to axle	176	130
Pivot bracket ball joint to axle	176	130
Brake disc to hub	73*	53*

*** Apply sealant, Part No. STC 50550 to threads**

Rear Drive Axle and Differential

Description and Operation

Description

The welded steel rear axle casing houses a separate spiral bevel type differential unit, which is off set to the right of the vehicle centre line. The differential unit drives the rear wheels via the axle shafts and fully floating hubs which are mounted on tapered roller bearings.

Lubrication

The differential is lubricated with oil and the hub bearings with grease.

The hub bearings are fitted with inner and outer seals. The outer seals prevent the differential oil mixing with the hub grease and the inner seals prevent dirt ingress into the hub.

Ventilation

Ventilation of the hub bearings is through the outer oil seals and the differential ventilation pipe, which terminates at a high level.

Rear Drive Axle and Differential

Diagnosis and Testing

Principles of Operation

For a detailed description of the Rear Drive Axle and Differential, refer to the relevant Description and Operation section in the workshop manual.

Refer to: Rear Drive Axle and Differential (205-02 Rear Drive Axle/Differential, Description and Operation)

Inspection and Verification

⚠ CAUTION: Diagnosis by substitution from a donor vehicle is **NOT** acceptable. Substitution of control modules does not guarantee confirmation of a fault, and may also cause additional faults in the vehicle being tested and/or the donor vehicle.

NOTES:

△ If the control module or a component is suspect and the vehicle remains under manufacturer warranty, refer to the Warranty Policy and Procedures manual (section B1.2), or determine if any prior approval programme is in operation, prior to the installation of a new module/component.

△ When performing voltage or resistance tests, always use a digital multimeter accurate to three decimal places, and with an up-to-date calibration certificate. When testing resistance always take the resistance of the digital multimeter leads into account.

△ Check and rectify basic faults before beginning diagnostic routines involving pinpoint tests.

1. Visually inspect for obvious signs of damage and system integrity

Rear axle hub - 90

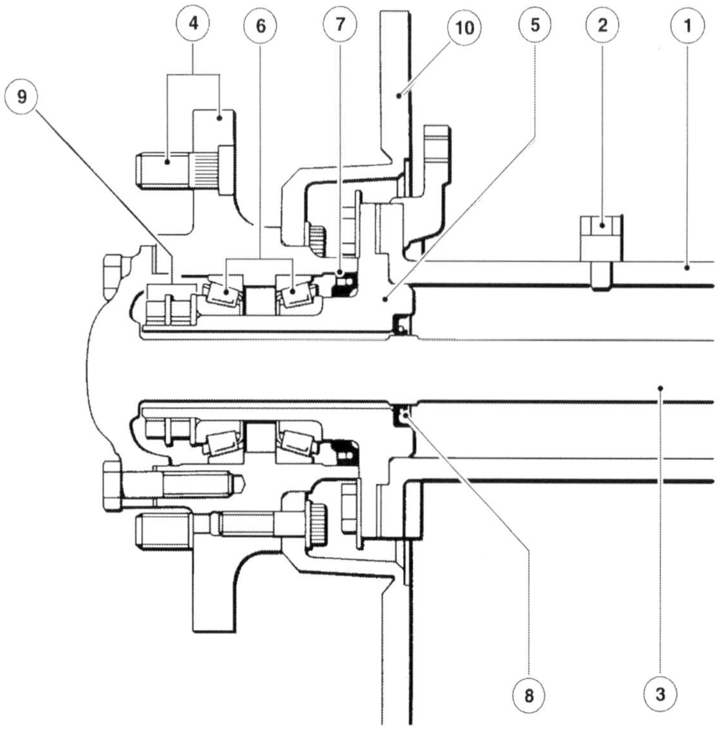

Item	Description
1	Axle casing
2	Ventilation pipe
3	Axle shaft
4	Wheel studs and hub
5	Wheel bearing stub axle
6	Wheel bearings
7	Inner hub seal
8	Outer hub/axle shaft seal
9	Hub lock plate, thrust washer and nuts
10	Brake disc

Rear axle hub - 110/130

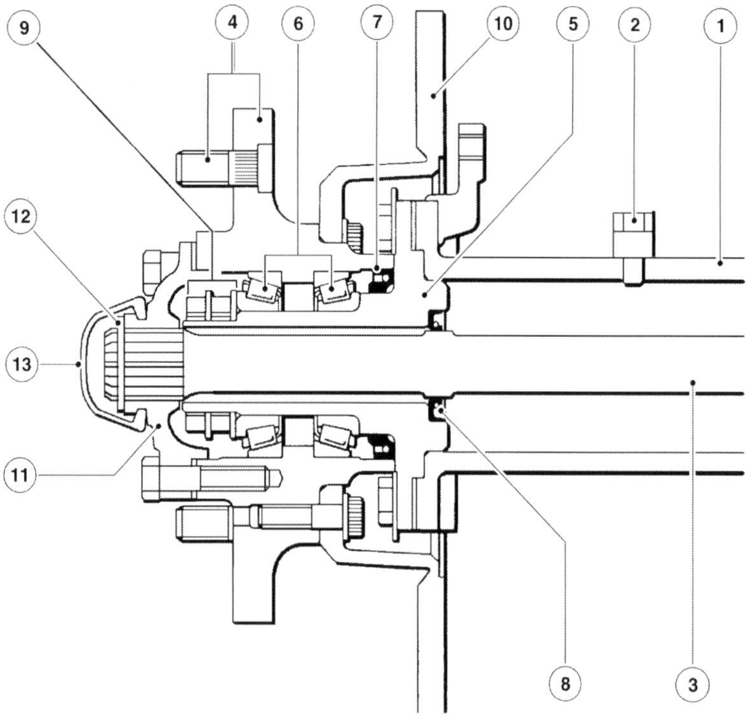

Item	Description
1	Axle casing
2	Ventilation pipe
3	Axle shaft
4	Wheel studs and hub
5	Wheel bearing stub axle
6	Wheel bearings
7	Inner hub seal
8	Outer hub/axle shaft seal
9	Hub lock plate, thrust washer and nuts
10	Brake disc
11	Drive flange
12	Drive shaft circlip
13	Dust cap

Visual Inspection

Mechanical
• Rear axle
• Rear differential
• Rear wheel hubs

2. If an obvious cause for an observed or reported concern is found, correct the cause (if possible) before proceeding to the next step

3. If the cause is not visually evident, verify the symptom and refer to the Symptom Chart, alternatively check for Diagnostic Trouble Codes (DTCs) and refer to the DTC Index

Symptom Chart

Symptom	Possible Causes	Action
Grease leak from the rear left/right wheel hub	• Rear left/right hub seal failure	• Install a new rear left/right hub seal
Oil leak from the rear left/right wheel hub	• Rear left/right Half shaft seal failure • Rear axle breather blocked	NOTE: After rectifying a rear axle oil leak, refer to the relevant section of the workshop manual and check the rear differential oil level. Top up as necessary. • Install a new rear left/right Half shaft seal • Unblock the rear axle breather
Oil leak from the rear differential pinion seal	• Rear differential pinion seal failure • Rear axle breather blocked	NOTES: After rectifying a rear axle oil leak, refer to the relevant section of the workshop manual and check the rear differential oil level. Top up as necessary. When installing a new differential pinion seal, a new differential pinion flange should also be installed. • Install a new rear differential pinion seal and flange • Unblock the rear axle breather
Rear differential oil level too high	• Rear differential oil overfilled • Water ingress to the rear differential via a defective rear left/ right Half shaft seal during wading	• Refer to the relevant section of the workshop manual and check the rear differential oil level. Correct as necessary • Install a new rear left/right Half shaft seal. Refer to the relevant section of the workshop manual and drain and refill the rear differential oil
Loss of drive - 90	• Rear left/right Half shaft failure • Rear differential failure	• Install a new rear left/right Half shaft as necessary • Install a new rear differential as necessary
Loss of drive - 110/130	• Rear left/right driving member splines worn • Rear left/right Half shaft failure • Rear differential failure	• Install a new rear left/right driving member as necessary • Install a new rear left/right Half shaft as necessary • Install a new rear differential as necessary

Differential Case - 90/110/130

In-vehicle Repair

Part(s)

STEP	PART NAME	QUANTITY
Step 11	Renew Part Differential case nut(s)	1

1. Using suitable container, drain axle oil.
2. Mark differential and propeller shaft flanges to facilitate reassembly.
3. NOTE: Discard the 4 nuts.

Remove 4 nuts and bolts. Release the propeller shaft from differential and tie aside.

4. Remove 5 hub drive member bolts and withdraw axle half shafts sufficiently to disengage from differential unit.
5. Remove 10 nuts securing differential to axle case.
6. Withdraw differential unit.
7. NOTE: The differential unit can only be serviced as a complete assembly with matching drive pinion. For advice ring Land Rover Service Department.

Ensure mating faces are clean and apply a bead of sealant, Part No. STC 3811 to axle case.

8. Support differential unit and position on axle casing.
9. Secure with self locking nuts and tighten nuts to correct torque.

   Torque: 40 Nm

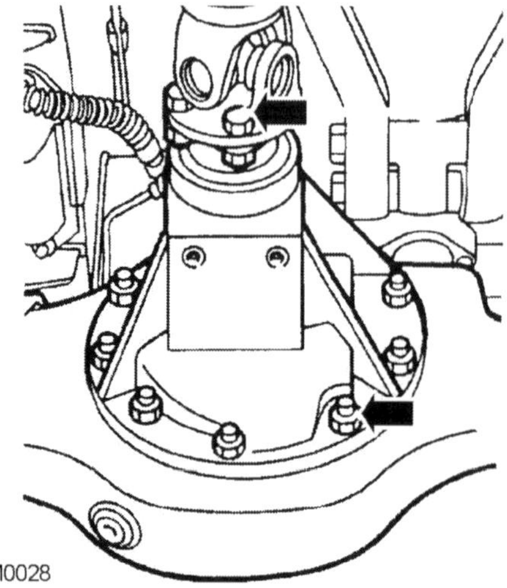
51M0028

10. Position propeller shaft to differential housing and align reference marks.
11. NOTE: New nuts must be installed.

    Install and tighten flange nuts.

    Renew Part: Differential case nut(s)

    Torque: 47 Nm
12. Refit half shafts, using new drive member gaskets. Tighten bolts.

    Torque: 65 Nm
13. Refill axle oil with approved lubricant.

Refer to: (205-02 Rear Drive Axle/Differential, Specifications)

Drive Pinion Seal - 110/130

In-vehicle Repair

Part(s)

STEP	PART NAME	QUANTITY
Installation Step 6	Renew Part Rear differential pinion flange kit	1
Installation Step 7	Renew Part Differential pinion shaft bolt	1
Installation Step 9	Renew Part Drive-shaft flange nut(s)	4

Removal

NOTE: *This procedure is applicable to 110 and 130 models from the following VIN's:

- 110 Non ABS - 638164
- 110 With ABS - 638249
- 110 Heavy Duty and 130, Non ABS - 638224
- 110 Heavy Duty and 130, With ABS - 638134

1. Raise rear of vehicle.

WARNING: Do not work under a vehicle supported only by a jack. Always support the vehicle on safety stands.

2. Reference mark the propeller shaft flanges for reassembly.
3. NOTE: Discard the 4 nuts.

Remove 4 nuts and bolts securing propeller shaft to differential housing. Release propeller shaft and tie aside.

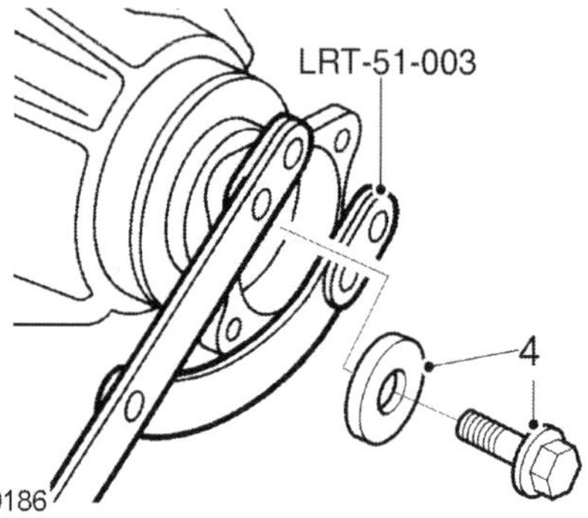

M510186

4. NOTE: Discard the bolt.

Using LRT-51-003 to restrain the pinion flange, remove and discard the bolt securing the pinion flange.

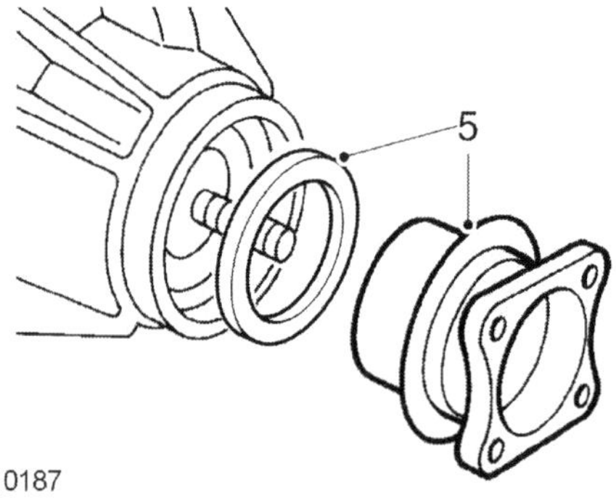

M510187

5. CAUTION: Take care to avoid damage to oil seal recess.

Remove and discard the pinion flange. Position container to catch oil spillage and using a suitable lever, remove and discard the pinion oil seal.

Installation

1. Clean pinion oil seal recess.
2. Lubricate NEW oil seal lip with clean axle oil.
3. Using LRT-51-009 install pinion oil seal.
4. Clean the mating splines of the differential pinion shaft.

5. Apply high strength retainer STC50554 to the pinion flange splines.
6. Install new pinion flange.

Renew Part: Rear differential pinion flange kit

7. NOTE: A new bolt must be installed.

Restrain flange using LRT-51-003 and fit bolt. Tighten bolt.

Renew Part: Differential pinion shaft bolt

Torque: 100 Nm

8. Position propeller shaft to differential housing and align reference marks.
9. NOTE: New nuts must be installed.

Install flange bolts and tighten.

Renew Part: Driveshaft flange nut(s)

Torque: 47 Nm

10. Top-up differential oil level.
11. Remove stands and lower vehicle.

Drive Pinion Seal - 90

In-vehicle Repair

Part(s)

STEP	PART NAME	QUANTITY
Installation Step 3	Renew Part Differential pinion seal	1
Installation Step 6	Renew Part Rear differential flange hub	1
Installation Step 7	Renew Part Differential pinion shaft nut	1
Installation Step 9	Renew Part Driveshaft flange nut(s)	4

Removal

1. Raise rear of vehicle.

WARNING: Do not work under a vehicle supported only by a jack. Always support the vehicle on safety stands.

2. Reference mark the propeller shaft flanges for reassembly.
3. NOTE: Discard the 4 nuts.

Remove 4 nuts and bolts securing propeller shaft to differential housing.

4. Release propeller shaft and tie aside.
5. NOTE: Discard the bolt.

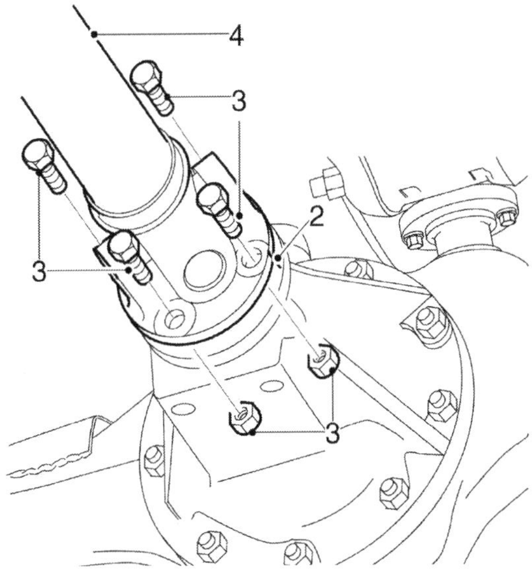

Using LRT-51-003 to restrain the pinion flange, remove and discard the bolt securing the pinion flange.

6. Remove and discard the pinion flange.
7. Position container to catch oil spillage.
8. CAUTION: Take care to avoid damage to oil seal recess.

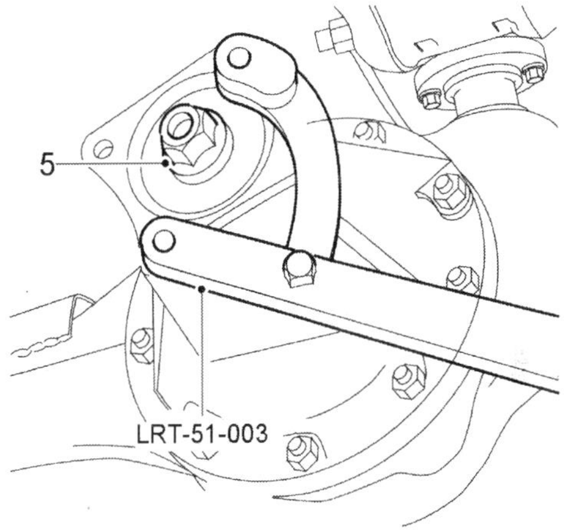

Using a suitable lever, remove and discard pinion oil seal.

Installation

1. Clean pinion oil seal recess.
2. Lubricate NEW oil seal lip with clean oil.
3. Using LRT-51-010 fit pinion oil seal.

Renew Part: Differential pinion seal

4. Clean the mating splines of the differential pinion shaft.
5. Apply high strength retainer STC50554 to the pinion flange splines.

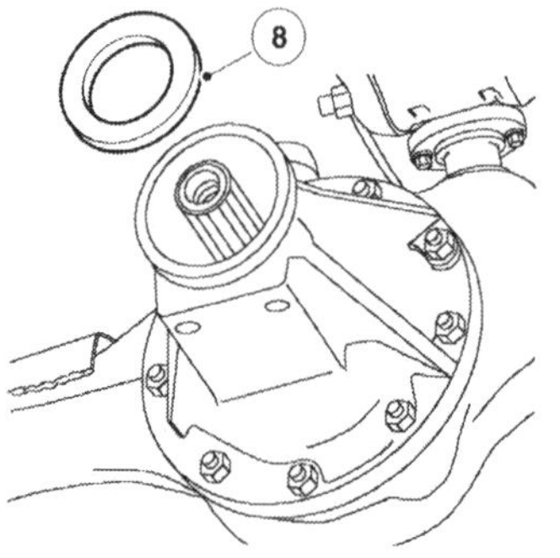

51M0050A

6. Install new pinion flange

   Renew Part: Rear differential flange hub

7. △ NOTE: A new nut must be installed.

Restrain flange using LRT-51-003 and Install bolt and tighten.

   Renew Part: Differential pinion shaft nut

   Torque: 100 Nm

8. Position propeller shaft to differential housing and align reference marks.

9. △ NOTE: New nuts must be installed.

Install flange bolts and tighten.

   Renew Part: Driveshaft flange nut(s)

   Torque: 47 Nm

10. Remove stands and lower vehicle.

11. Top-up differential oil level. In-vehicle Repair

Stub Shaft Pilot Bearing and Seal - 90 and 110/130

In-vehicle Repair

Removal

△ NOTE: This procedure covers removal and installation of the stub shaft and oil seal.

1. 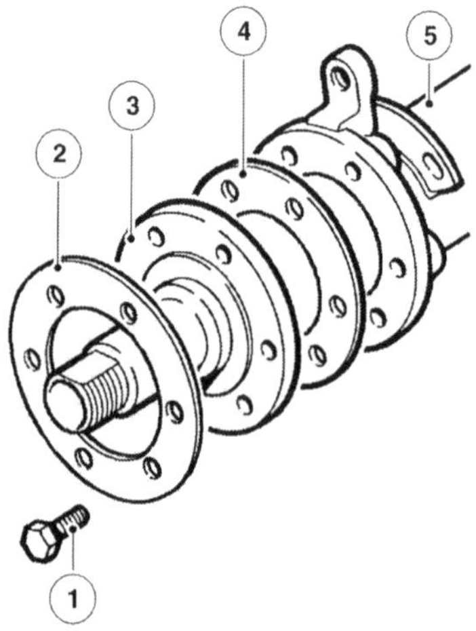

   1. Stub shaft to axle casing bolt
   2. Mud shield
   3. Stub shaft
   4. Stub shaft gasket
   5. Axle case

2. Remove hub assembly.

Refer to: Wheel Bearing and Wheel Hub (204-02 Rear Suspension, Removal and Installation)

3. Remove 6 bolts from stub shaft to axle casing; remove mud shield.
4. Remove stub shaft and gasket.
5. Remove and discard oil seal.

Installation

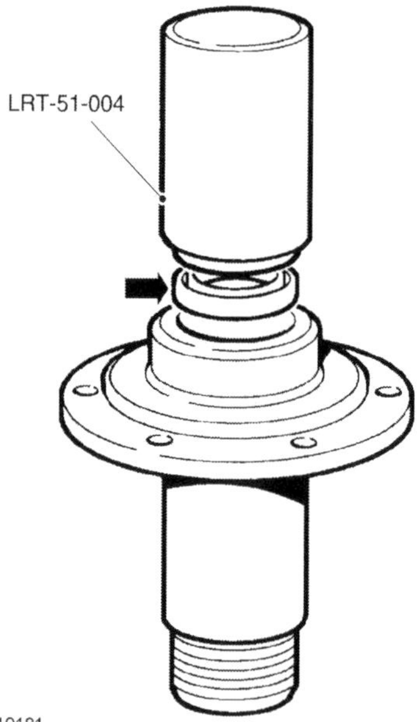

M510181

1. Lubricate replacement seal with EP90 oil. Fit seal, lip side trailing using tool LRT-51-004. Drive seal into housing until it is flush with rear face of stub shaft.
2. Fit new gasket, stub shaft and mud shield. Fit and tighten bolts.

   Torque: 65 Nm

3. Refit hub assembly.

Refer to: Wheel Bearing and Wheel Hub (204-02 Rear Suspension, Removal and Installation)

Axle Assembly

Removal and Installation

Removal

⚠ WARNING: Remove and refit of axle requires a further two persons to steady the axle when lowering or repositioning axle.

1. Drain brake system.
2. Support chassis rear.
3. Remove road wheels.
4. Support axle weight with hydraulic jack.
5. Disconnect shock absorbers.
6. Disconnect flexible brake hose at RH chassis side member and breather hose at banjo connection on axle casing.
7. Disconnect lower links at axle.
8. Mark differential and propeller shaft flanges with identification marks for assembly.
9. Remove 4 nuts and bolts, lower propeller shaft and tie to one side.
10. Disconnect pivot bracket ball joint at axle bracket.
11. Release bolts and remove coil spring retaining plates.
12. Lower axle and remove road springs.
13. If applicable, remove stabilizer bar links at axle.

Refer to: Rear Stabilizer Bar (204-02 Rear Suspension,Removal and Installation)

Refer to: Rear Stabilizer Bar (204-02 Rear Suspension,Removal and Installation)

14. Remove axle assembly.

Installation

1. Position axle and fit lower links. Tighten fixings to 176 Nm (130 lbf/ft).
2. If applicable, fit stabilizer bar links to axle.

Refer to: Rear Stabilizer Bar (204-02 Rear Suspension, Removal and Installation)

3. Raise axle and locate road springs.
4. Fit coil spring retaining plates and secure with fixing bolts.
5. Secure pivot bracket ball joint to axle bracket. Tighten fixing to 176 Nm (130 lbf/ft).
6. Align propeller shaft to differential drive flange and tighten fixings to 47 Nm (35 lbf/ft).
7. Reconnect flexible brake hose and axle breather hose.
8. Refit shock absorbers.
9. Fit road wheels and tighten to correct torque: Alloy wheels - 130 Nm (96 lbf/ft) Steel wheels - 100 Nm (80 lbf/ft) Heavy Duty wheels - 170 Nm (125 lbf/ft)
10. Remove rear chassis support.
11. Bleed brake system.

Refer to: Brake System Bleeding (206-00 Brake System - General Information, General Procedures)

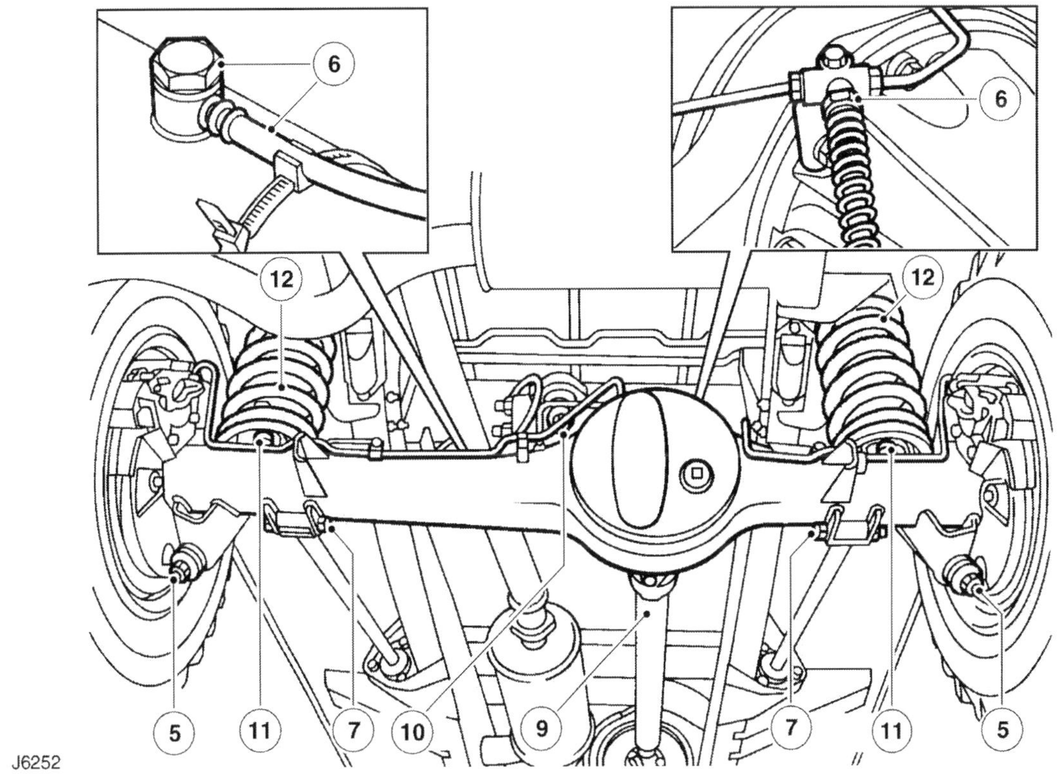

Disassembly and Assembly

Part(s)

STEP	PART NAME	QUANTITY
Disassembly Step 20	Renew Part Differential pinion shaft roll pin	1
Step 5	Renew Part Differential crown wheel retaining bolt(s)	10
Step 25	Renew Part Rear differential pinion shim	1
Step 33	Renew Part Rear differential pinion flange bolt	1
Step 37	Renew Part Driveshaft flange bolts	1

Disassembly

1. Remove differential assembly.
2. Secure differential assembly in a vice or stand.

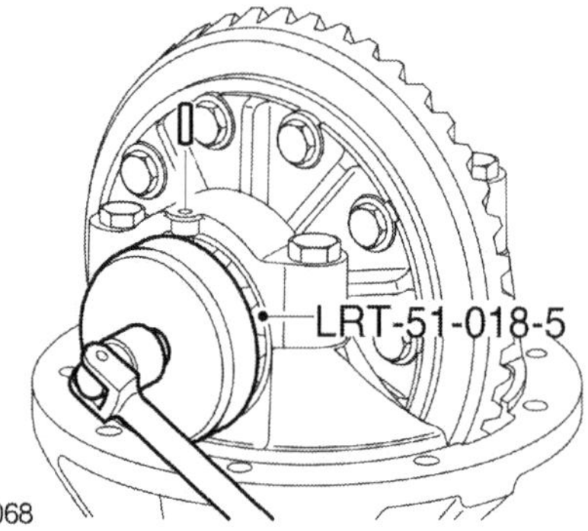

M510068

3. Remove roll pins securing adjusting nuts, using LRT 51-018/5 loosen adjusting nuts.

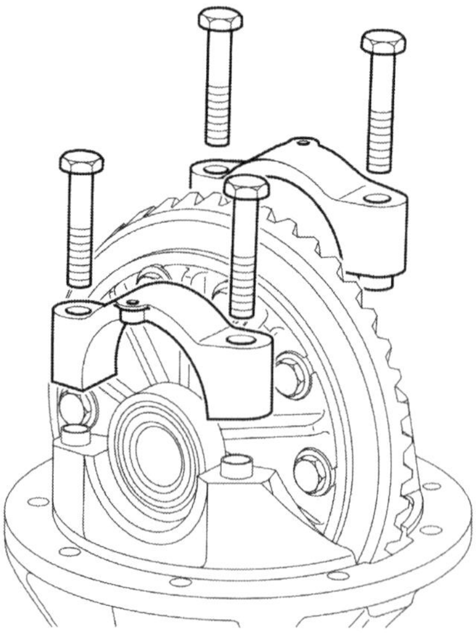

4. Reference mark bearing caps to aid assembly.
5. Remove bolts securing bearing caps and remove caps.
6. Remove crown wheel assembly and collect bearing outer tracks.

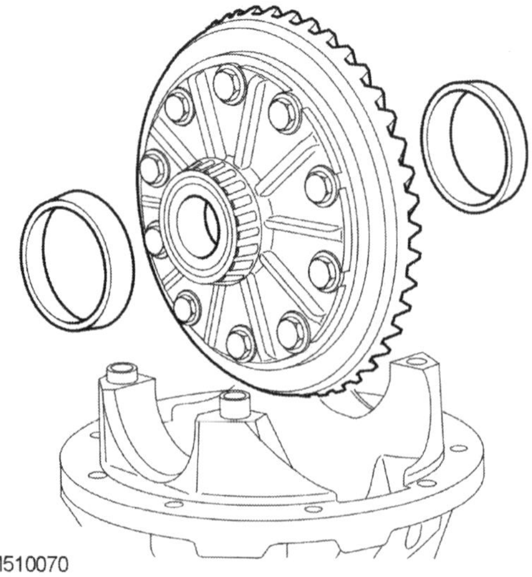

M510070

7. Mark outer tracks to bearings to aid assembly, if bearings are to be re-used.

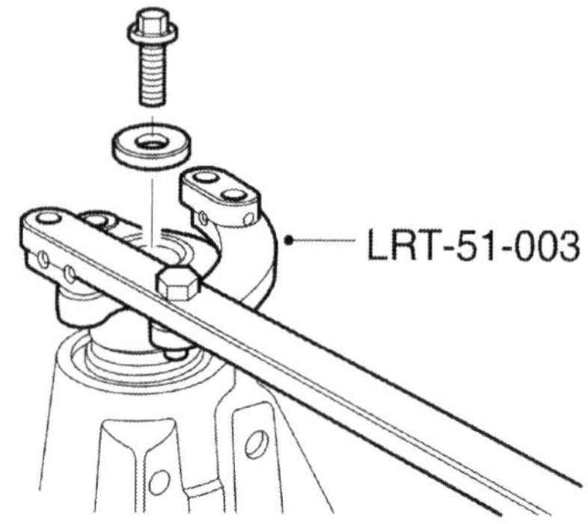

M510071

8. Using LRT-51-003 to restrain pinion flange remove bolt and collect washer.

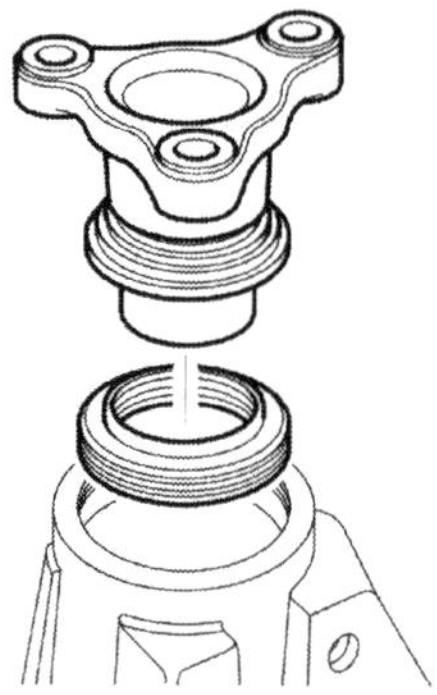

M510072

9. NOTE: Older front differentials have a square flange and an extra spacer fitted, this spacer must be removed. Later front differentials have a round flange but no spacer fitted.

   Remove pinion flange

10. CAUTION: Take care to avoid damage to oil seal recess.

    Using a lever, remove pinion oil seal.

11. Carefully tap pinion from housing, collect pinion and tail bearing.
12. Remove pinion tail bearing shim and record shim size.

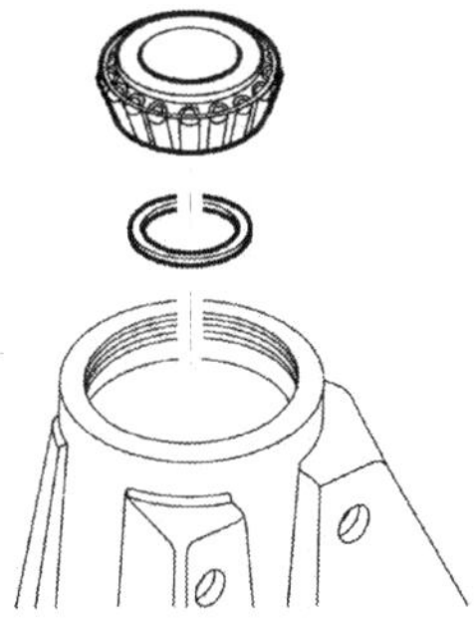

M510073

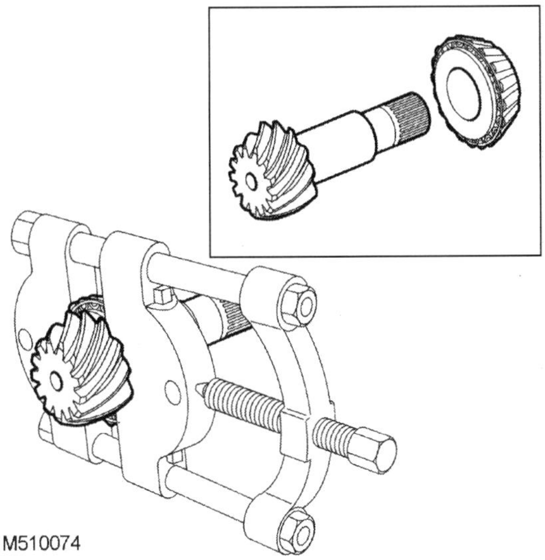

M510074

13. Using a puller, remove pinion head bearing.

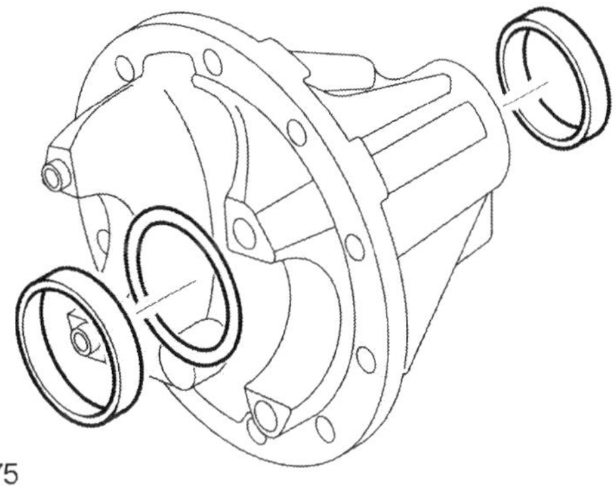

M510075

14. Remove pinion bearing races.
15. Remove pinion head bearing shim and record shim size.
16. Using a two legged puller and LRT 51-018/6, remove the differential bearings.
17. Secure the crown wheel assembly in a vice.
18. Remove and discard 10 bolts securing the crown wheel to carrier.
19. Carefully remove the crown wheel from the carrier.
20. ⚠ NOTE: For vehicles with a 4mm roll pin.

    Remove and discard roll pin securing carrier cross shaft and remove cross shaft.

    Renew Part: Differential pinion shaft roll pin

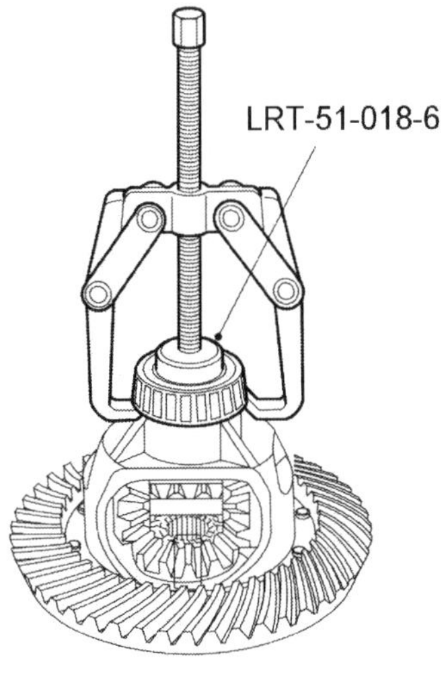

M510076

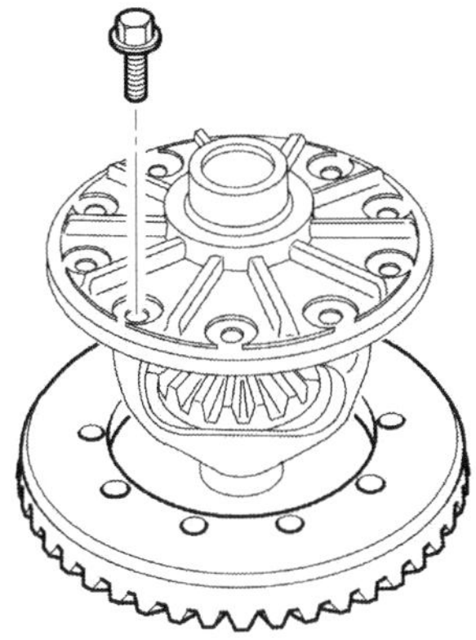

M510077

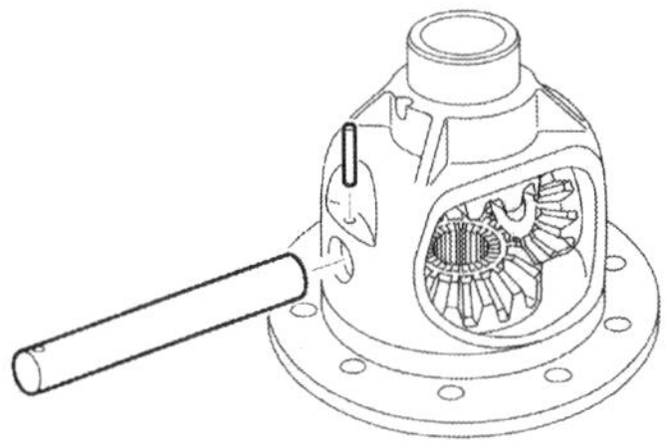

M510078

21. ⚠ NOTE: For vehicles with 6mm roll pin.

    Secure the planetary gear housing in the press using suitable V blocks and press out the cross shaft. Press out the cross shaft and shear the pin.

22. Rotate gears to the open part of carrier and remove planet gears.

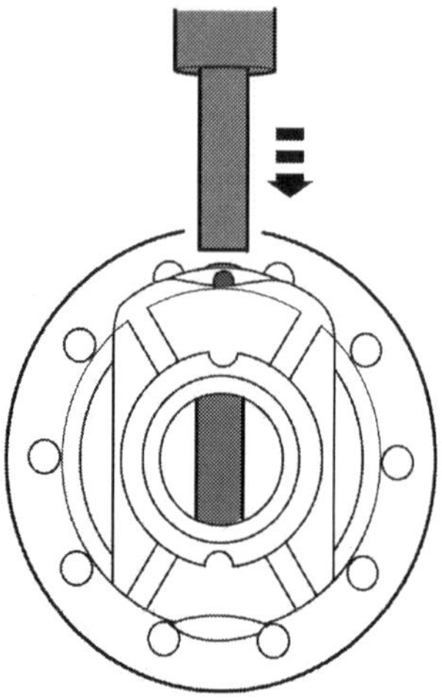

23. NOTE: If the planetary gear housing has been fitted with a 4mm roll pin hole for the cross shaft, discard the planetary gear housing and replace it with one with a 6mm roll pin hole.

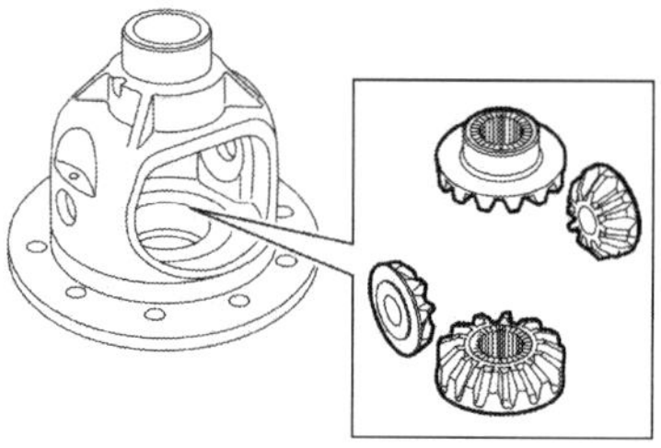

M510079

Remove sun gears.

24. Clean and inspect all components for wear and damage.

Assembly

1. Install planet gears and rotate to align cross shaft holes.

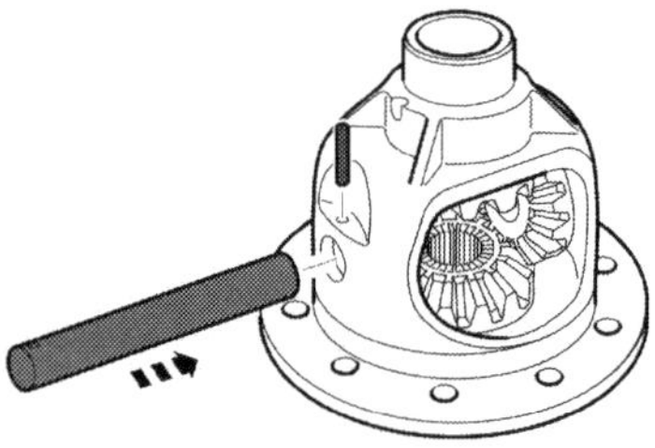

E131981

2. Install cross shaft, ensure roll pin hole is aligned.
3. NOTE: Press the roll pin fully into the planetary gear housing (approximately 8-10mm will protrude from the housing when correctly installed).

   Press the pin into the planetary gear housing.
4. Secure cross shaft with new roll pin.
5. Install crown wheel to carrier, Install and tighten new bolts.

   Renew Part: Differential crown wheel retaining bolt(s)

   Torque: 60 Nm

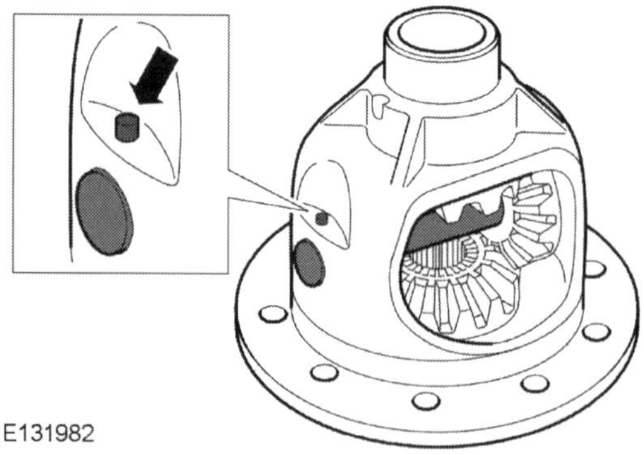

E131982

6. Ensure original head bearing shim is clean and free from burrs and Install under bearing race.

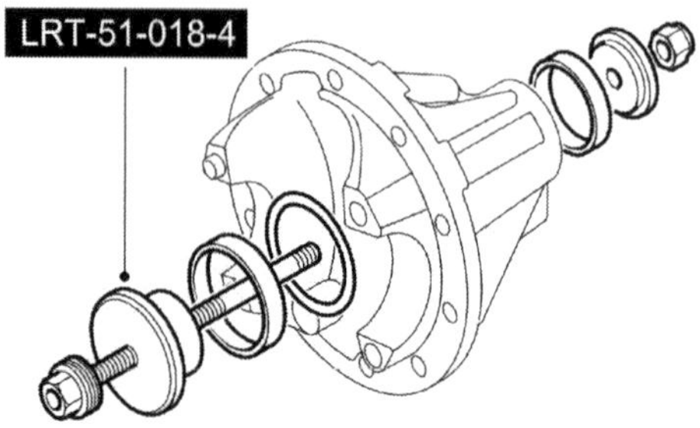

M510105A

7. Ensure pinion bearing cup recesses are clean and free of burrs and using LRT 51-018-4 Install pinion head and tail bearing races.
8. Install pinion head bearing to pinion.
9. Lubricate bearings with thin oil.
10. Ensure original tail bearing shim is clean and free from burrs and Install under bearing race.
11. Install pinion and pinion tail bearing.
12. Install pinion flange, washer and bolt.
13. Use LRT-51-003 to restrain pinion flange.
14. Tighten pinion flange bolt.

    Torque: 100 Nm
15. Check pinion for end float. Should read zero.
16. Rotate pinion several times to settle bearings, check pinion torque to turn. Torque to turn should be recorded during pinion rotation. Pinion torque to turn should be 4 to 6 Nm (3 to 4.5 lbf.ft).
17. NOTE: To increase torque to turn, Install narrower spacer; to decrease torque to turn Install wider spacer.

Adjust size of tail bearing shim to obtain correct pinion torque to turn, (0.025 mm = 1 Nm (0.001 = 0.7 lbf.ft) approx).

18. Position LRT-51-018/7 on surface plate, establish zero and reference DTI.
19. Ensure pinion height setting block, setting gauge and mating faces are clean and free from burrs.
20. Locate setting block LRT 51-018/1 over pinion head, ensure it is fully seated in position.

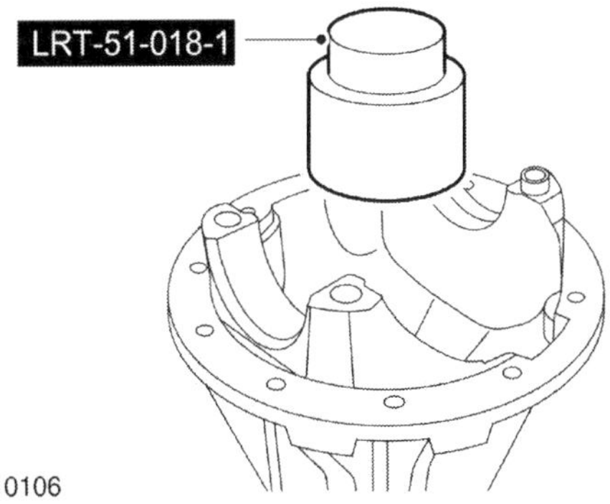

M510106

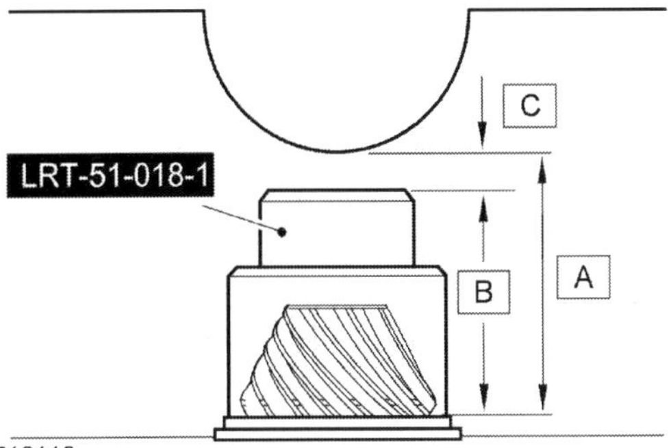

M510116

21. ⚠ CAUTION: Setting block height must be checked using figures on side of block.

A = Nominal pinion height setting, 74.390

B = Setting block height.

C = Head height setting. Pinion height setting procedure:

C = A - B. Subtract nominal pinion height A from setting block height B (on side of setting block). Example: 74.390 - 73.130 = 1.26 mm

(2.929 - 2.88 = 0.049)

Therefore pinion head height reading is 1.260 mm ± 0.025 mm (0.049 ± 0.001).

22. Align setting gauge LRT-51-018/7 to setting block, rock gauge to obtain minimum reading. If reading is lower than required reading, decrease shim size. If reading is higher than required reading, increase shim size.

23. Using LRT-51-003 to restrain pinion flange, remove bolt and washer. Remove pinion flange.

24. Remove pinion, collect tail bearing and tail bearing shim.

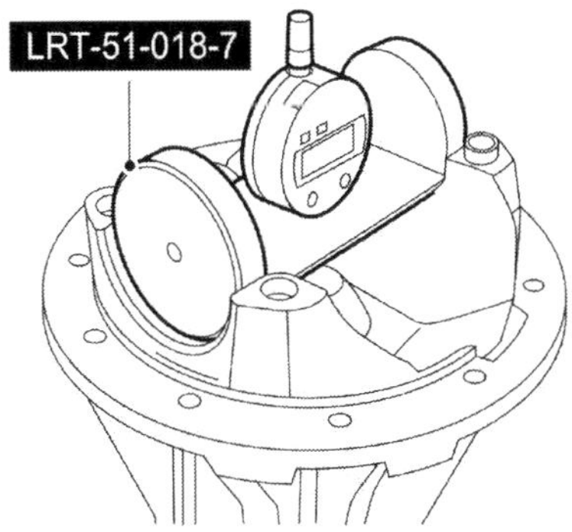

M510082

25. Remove pinion head bearing outer race and shim. Discard shim. Ensure bearing race recess is clean and free from burrs.

Renew Part: Rear differential pinion shim

26. Install calculated shim, and using LRT-51-018-4 Install head bearing outer race.

27. Install pinion, pinion tail bearing and tail bearing shim.

28. Install pinion flange and bolt and washer. Using LRT-51-003 to restrain pinion flange, tighten bolt.

Torque: 100 Nm

29. Rotate pinion in both directions to settle bearings.

30. Recheck pinion torque to turn, adjust if necessary.

31. Recheck pinion head height.

32. Using LRT-51-003 to restrain pinion flange, remove bolt and washer. Remove pinion flange.

33. Discard bolt.

Renew Part: Rear differential pinion flange bolt

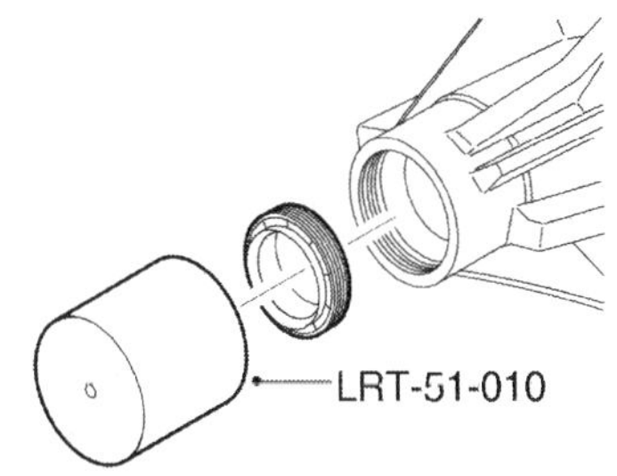

M510086

34. Using LRT-51-010 Install pinion seal.

35. Ensure spacer and tail bearing are correctly located.

36. Install pinion, pinion flange and washer.

37. Install and tighten new pinion flange bolt.

Renew Part: Driveshaft flange bolts

Torque: 100 Nm

38. Lightly oil differential bearings.

39. Ensure spring dowels are fitted in bearing caps.

40. Install differential bearing outer races and locate differential assembly into housing.

41. Install bearing caps and tighten bolts.

Torque: 10 Nm

42. Install adjusting nuts, tighten crown wheel side nut. Ensure opposing nut is loose.

Torque: 22 Nm

43. Position DTI to check crown wheel backlash. Adjust opposing nut to obtain correct crown wheel backlash.

44. Rotate pinion in both directions to settle bearings.

45. NOTE: Crown wheel backlash should be within 0.076mm to 0.177mm (0.003 in to 0.007in).

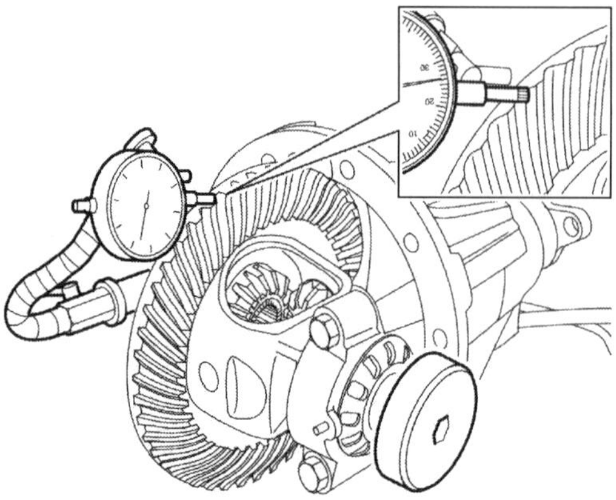

Measure in 3 places to obtain correct crown wheel backlash.

46. Align adjusting nuts to next roll pin slot, do not loosen nuts to align slots.

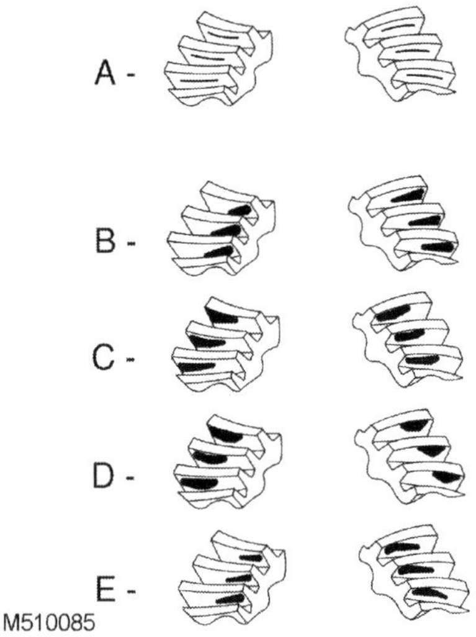

47. Tighten bearing cap bolts.

Torque: 90 Nm

48. Secure adjusting nuts with new roll pins.

49. Apply Prussian Blue to crown wheel teeth to check tooth contact.

50. NOTE: Note assembly torque to turn when checking tooth contact. Total torque to turn should not exceed 10.85 Nm (8 lbf.ft).

A = Normal pattern: The drive pattern should be centered on the gear teeth. The coast pattern should be centered on the gear teeth but may be towards the toe. There should be some clearance between the pattern and the top of the gear teeth.

B = Backlash incorrect: Thinner pinion shim required.

C = Backlash incorrect: Thicker pinion shim required. D = Pinion shim incorrect: Decrease backlash.

E = Pinion shim incorrect: Increase backlash. Rotate pinion several times to obtain full tooth contact.

51. Install differential assembly.

Axle - 110/130

Disassembly and Assembly

Part(s)

STEP	PART NAME	QUANTITY
Step 6	Renew Part Differential crown wheel retaining bolt(s)	10
Step 27	Renew Part Rear differential pinion shim	1
Step 39	Renew Part Rear differential pinion flange bolt	1

Disassembly

1. NOTE: *This procedure is applicable to 110 and 130 models from the following VINs:

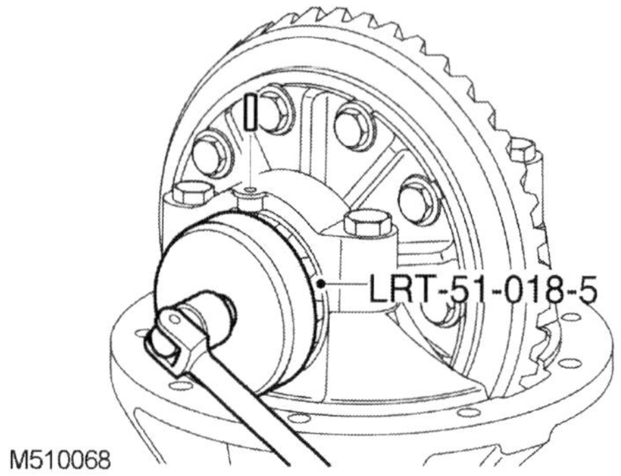

2. 110 Non ABS - 638164
3. 110 Heavy Duty and 130, Non ABS - 638224
4. 110 Heavy Duty and 130, With ABS - 638134

Remove differential assembly.

2. Secure differential assembly in a vice or stand.
3. Remove roll pins securing adjusting nuts, using LRT 51-018/5 loosen adjusting nuts.
4. Reference mark bearing caps to aid assembly.
5. Remove 4 bolts securing bearing caps and remove caps.
6. Remove crown wheel assembly and collect bearing outer tracks.
7. Mark outer bearing tracks if bearings are to be re-used.
8. Using LRT-51-003 to restrain pinion flange remove bolt and collect washer.
9. NOTE: Older front differentials have a square flange and an extra spacer fitted, this spacer must be removed. Later front differentials have a round flange but no spacer fitted.

Remove pinion flange

10. Using a lever, remove pinion oil seal.

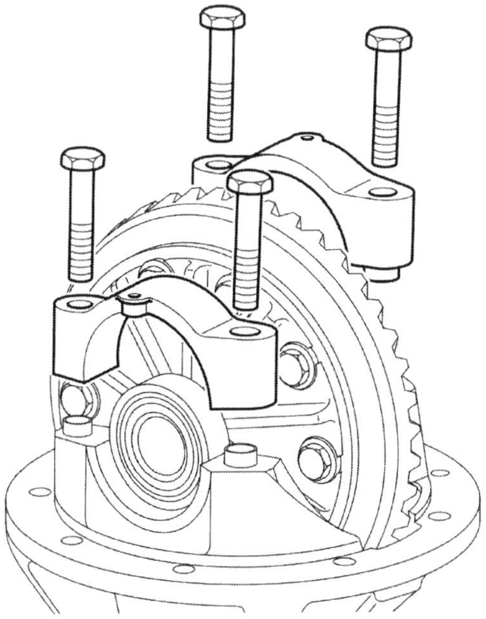

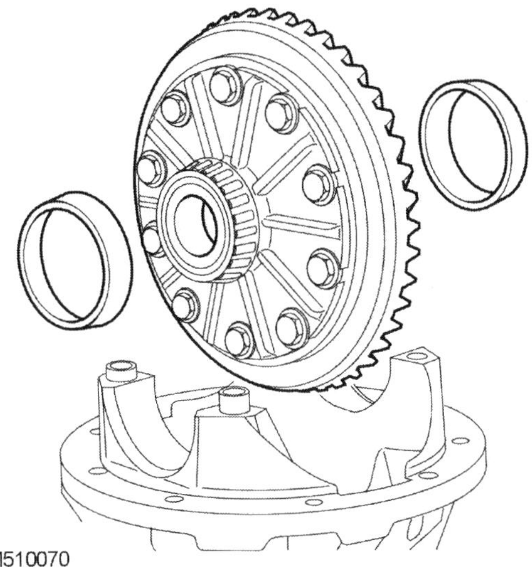

M510070

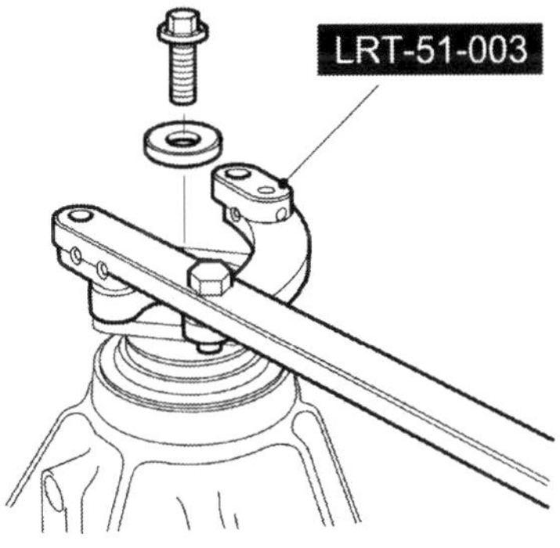

M510095

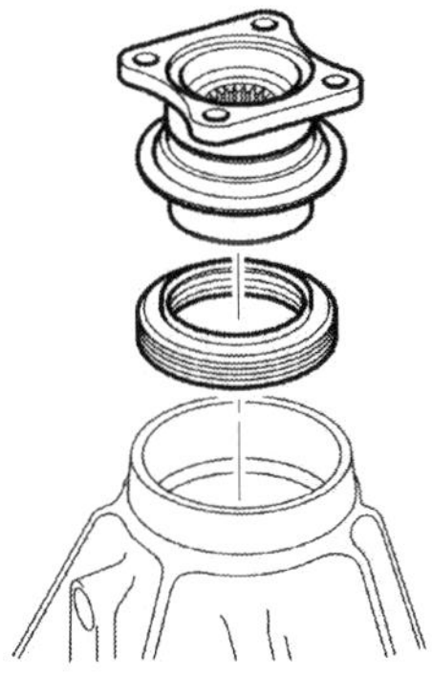

M510096

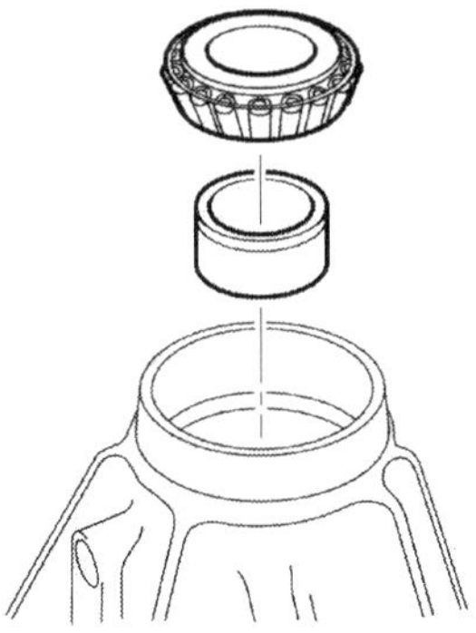

M510097

CAUTION: Take care to avoid damage to oil seal recess.

11. Carefully tap pinion from housing, collect pinion and tail bearing.
12. Remove pinion tail bearing spacer and record spacer size.

CAUTION: Do not discard spacer at this stage.

13. Using a bearing puller, remove pinion head bearing, if bearing is to be replaced.

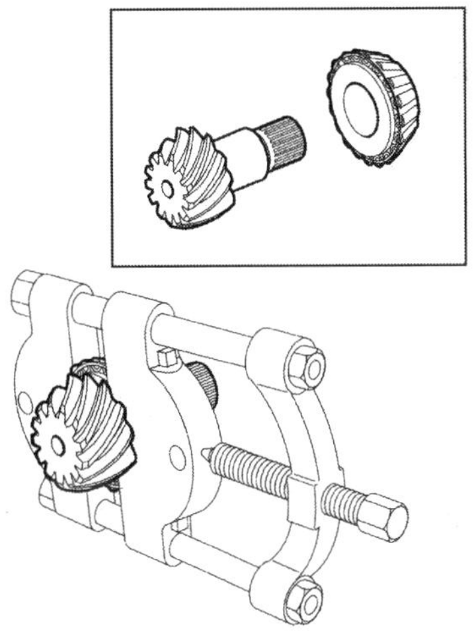

14. Remove pinion head and tail bearing tracks.

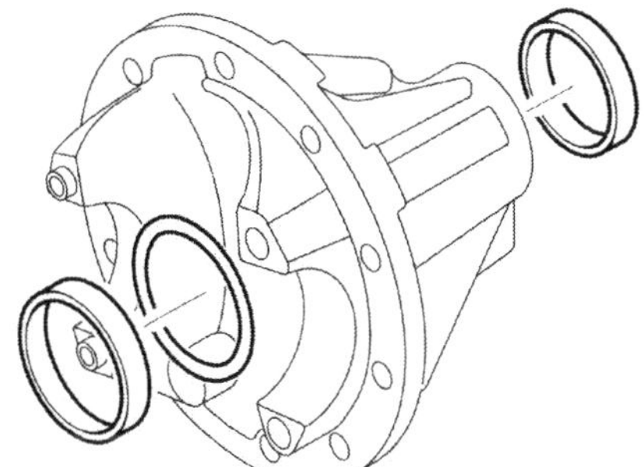

M510075

15. Remove pinion head bearing shim and record shim size.
16. Using a two legged puller and LRT 51-018/6, remove the differential bearings.
17. Secure the crown wheel assembly in a vice.

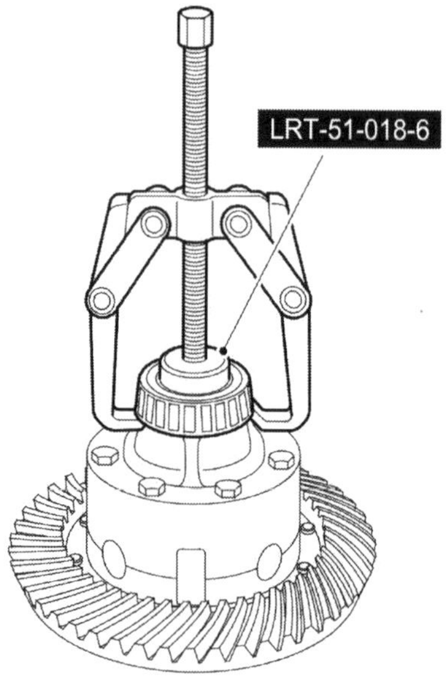

18. Reference mark position of crown wheel to carrier to aid assembly.

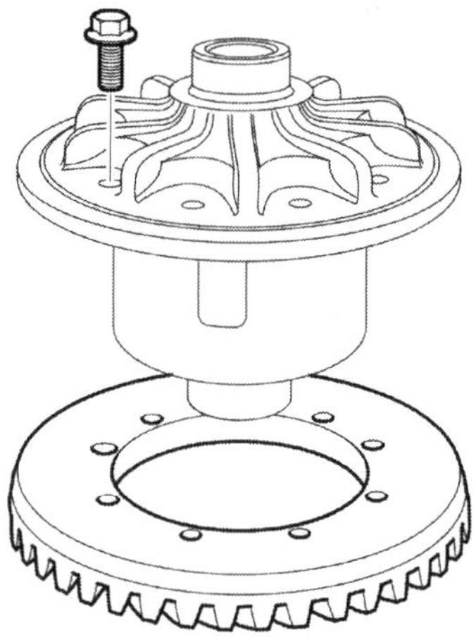

19. Remove and discard 10 bolts securing the crown wheel to the carrier.
20. Carefully remove the crown wheel from the carrier.
21. Reference mark position of differential casing to aid assembly.
22. Remove 8 bolts securing differential gear upper casing and remove casing.
23. Remove upper sun gear, recover shim.
24. Noting their fitted position, remove planet gears, cross pins; recover spherical washer from each planet gear.
25. Remove lower sun gear from differential gear lower casing, recover shim.
26. Clean and inspect all components for wear and damage.

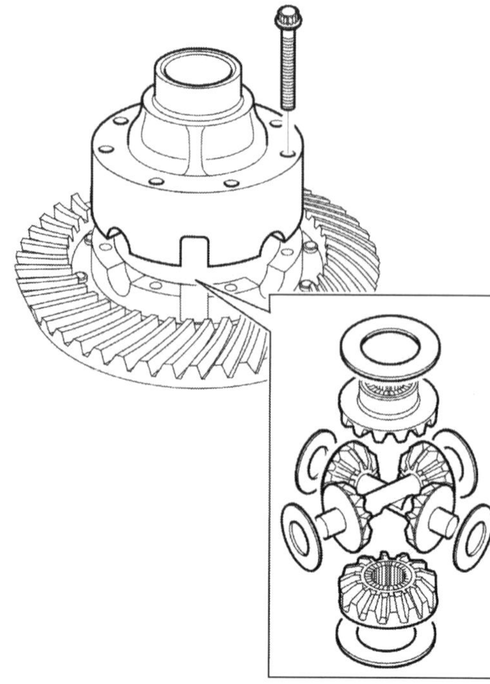

Assembly

1. Lubricate sun and planet gears, spherical washers, shims and cross pins.
2. Fit shim to lower sun gear and fit sun gear to differential gear lower casing.
3. Fit all planet gears, cross shafts and spherical washers into lower casing. Ensure gears and cross shafts are fully engaged.
4. Fit shim to upper sun gear, position upper sun gear to planet gears ensuring full teeth engagement.
5. Ensuring differential gear casing reference marks are aligned, position differential gear upper casing. Fit bolts and tighten progressively by diagonal selection.

   Torque: 32 Nm

6. Fit crown wheel to carrier ensuring reference marks are aligned, fit and tighten new bolts progressively.

Renew Part: Differential crown wheel retaining bolt(s)

   Torque: 60 Nm

7. Ensure original pinion head bearing shim is clean and free from burrs, fit shim in bearing outer track recess.

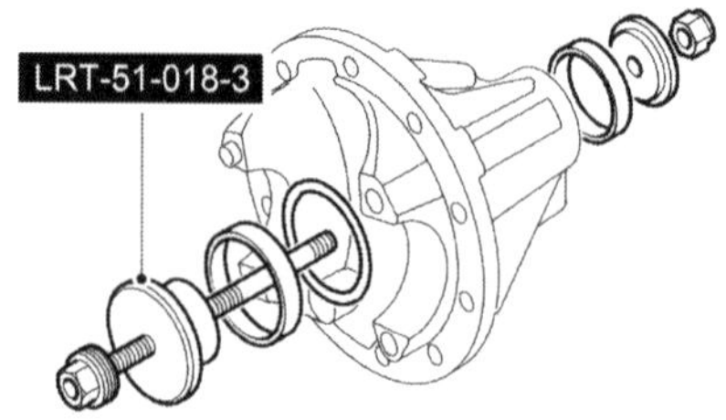

8. Using LRT-51-018/3 fit pinion head and tail bearing outer tracks.

9. Lubricate bearings with thin oil.
10. Fit pinion head bearing to pinion.
11. Fit pinion into pinion housing and hold in place.
12. Ensure original spacer is clean and free from burrs, fit to pinion shaft ensuring that groove in spacer is facing towards drive flange. Push spacer hard against pinion head bearing.
13. Fit pinion tail bearing.
14. Fit pinion flange, washer and bolt.
15. Use LRT-51-003 to restrain pinion flange.
16. Tighten pinion flange bolt.

    Torque: 100 Nm
17. Check pinion for end float. Should read zero.

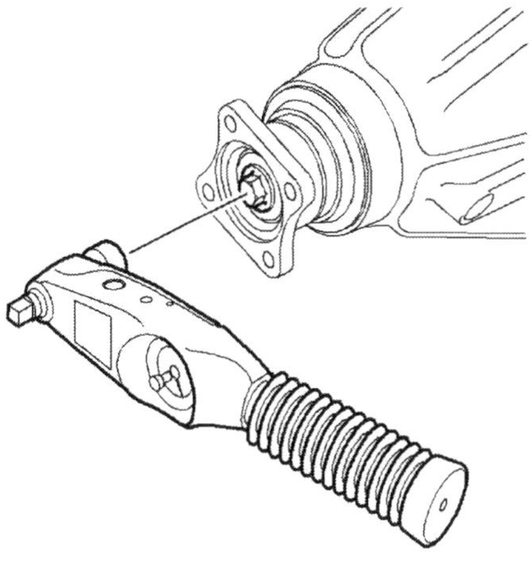

18. Rotate pinion several times to settle bearings, check pinion torque to turn. Torque to turn should be recorded during pinion rotation. Pinion torque to turn should be 4 to 6 Nm (3 to 4.5 lbf.ft).
19. Adjust size of tail bearing spacer to obtain correct pinion torque to turn (0.025 mm = 1 Nm (0.001' = 0.7 lbf.ft) approx).

NOTE: To increase torque to turn, fit narrower spacer; to decrease torque to turn fit wider spacer.

20. Position LRT-51-018/7 on surface plate, secure DTI with grub screw, establish zero and reference DTI.
21. Ensure pinion height setting block, setting gauge and mating faces are clean and free from burrs.
22. Locate setting block LRT- 51-018/2 over pinion head, ensure it is fully seated in position.
23. 'A' = Nominal pinion height setting, 76.04.

    'B' = Setting block height.

    'C' = Head height setting. Pinion height setting procedure:

    'C' = 'A' - 'B'. Subtract nominal pinion height 'A' from setting block height 'B' (on side of setting block) Example: 76.04 - 74.7 = 1.34 mm (2.993' - 2.941' = 0.053'). Therefore pinion head height reading is 1.34 mm ± 0.025 mm (0.053' ± 0.001').

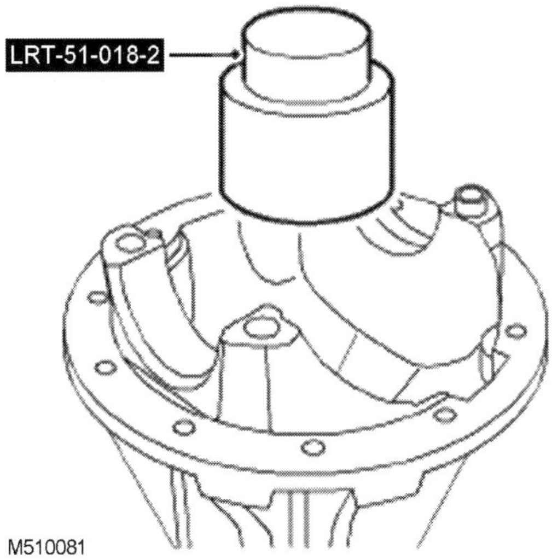

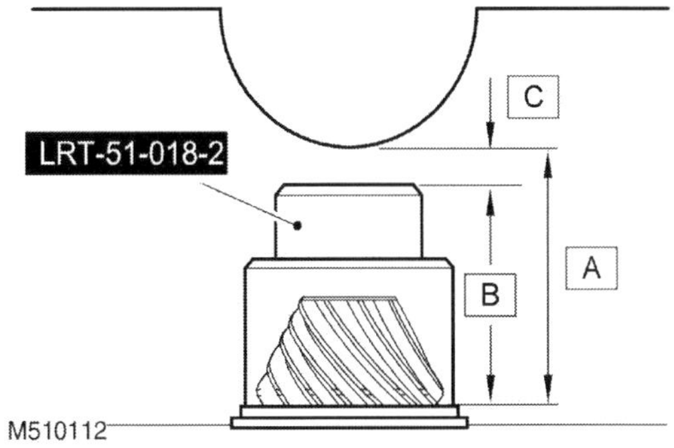

CAUTION: Setting block height must be checked using figures on side of block.

24. Align setting gauge LRT-51-018-7 to setting block, rock gauge to obtain minimum reading. If reading is lower than required reading, decrease shim size. If reading is higher than required reading, increase shim size.
25. Using LRT-51-003 to restrain pinion flange, remove bolt and washer. Remove pinion flange.
26. Remove pinion, collect tail bearing and tail bearing spacer.
27. Remove pinion head bearing outer track and shim, discard shim. Ensure bearing track recess is clean and free from burrs.

    Renew Part: Rear differential pinion shim
28. Fit calculated shim and using LRT-51-018/3, fit pinion head bearing outer track.
29. Fit pinion, spacer and pinion tail bearing.
30. Fit pinion flange, bolt and washer. Using LRT-51-003 to restrain pinion flange, tighten bolt to correct torque.

    Torque: 100 Nm
31. Rotate pinion in both directions to settle bearings.
32. Recheck pinion torque to turn, adjust if necessary.
33. Recheck pinion head height.
34. Using LRT-51-003 to restrain pinion flange, remove bolt and washer. Remove pinion

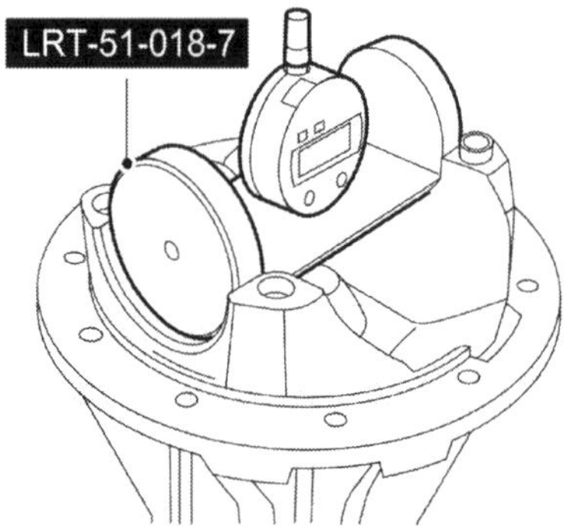

M510082

35. Discard bolt.

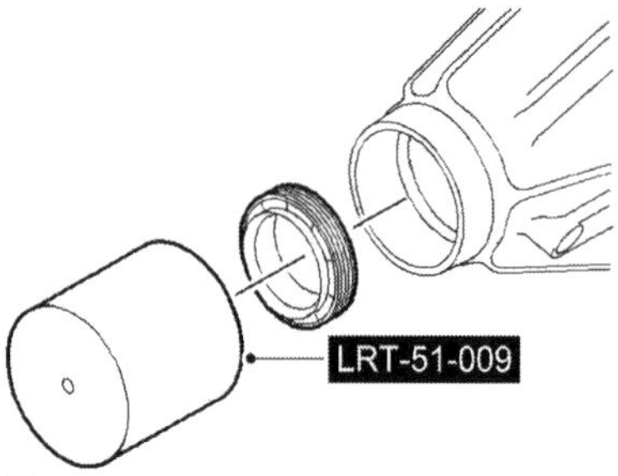

M510103

36. Using LRT-51-009, fit pinion seal.
37. Ensure spacer and tail bearing are correctly located.
38. Fit pinion, flange and washer.
39. Fit and tighten new pinion flange bolt.

    Renew Part: Rear differential pinion flange bolt

    Torque: 100 Nm

40. Lightly oil differential bearings.
41. Ensure dowels are fitted in bearing caps.
42. Fit differential bearing outer races and locate differential assembly into housing.
43. Fit bearing caps and tighten bolts.

    Torque: 10 Nm

44. Fit adjusting nuts, tighten crown wheel side nut. Ensure opposing nut is loose.

    Torque: 22 Nm

45. Position DTI to check crown wheel backlash. Adjust opposing nut to obtain correct crown wheel backlash.

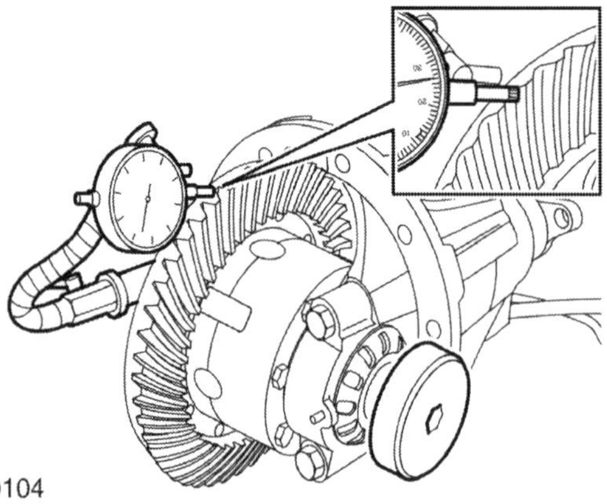

M510104

46. Rotate crown wheel in both directions to settle bearings.
47. Measure crown wheel backlash in 3 positions, adjust as required.

    △ NOTE: Crown wheel backlash should be within 0.076 mm to 0.177 mm (0.003 in to 0.007 in).

48. Align adjusting nuts to next roll pin slot, do not loosen nuts to align slots.
49. Tighten bearing cap bolts.

    Torque: 180 Nm

50. Secure adjusting nuts with new roll pins.
51. Apply Prussian Blue to crown wheel teeth to check tooth contact.

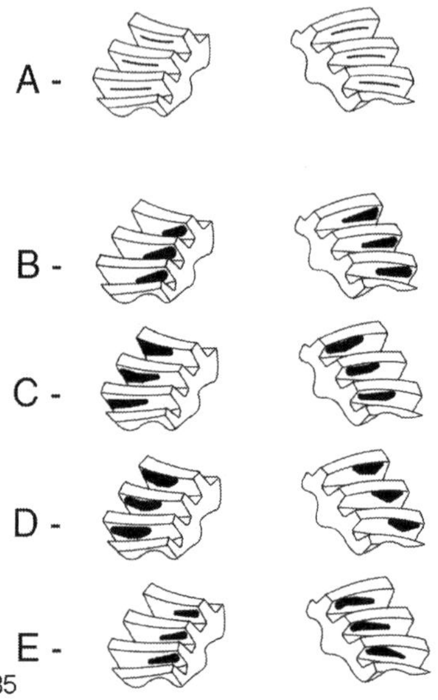

M510085

52. △ NOTE: Note assembly torque to turn when checking tooth contact. Total torque to turn should not exceed 10.85 Nm (8 lbf.ft).

    A = Normal pattern: The drive pattern should be centered on the gear teeth. The coast pattern should be centered on the gear teeth but may be towards the toe. There should be some clearance between the pattern and the top of the gear teeth.
    B = Backlash correct: Thinner pinion shim required.
    C = Backlash correct: Thicker pinion shim required.
    D = Pinion shim correct: Decrease backlash.
    E = Pinion shim correct: Increase backlash. Rotate pinion several times to obtain full tooth contact.

53. Fit differential assembly.

Specifications

Front axle	
Type	Spiral bevel, enclosed constant velocity joints, fully floating shafts
Ratio	3.54:1
Angularity of universal joint on full lock	32°

Sealers

Front axle and final drive	Land Rover Part No.
Hub retaining bolts	STC 50552
Swivel bearing housing to axle casing bolts	STC 50552
Stub axle bolts	STC 50552
Swivel pin housing grease	STC 3435*

*Swivel pin grease used on later vehicles with filler plug only in swivel pin housing

Lubricants

Item	Specification / Land Rover part number
Final drive	Castrol EPX 80W-90 or Texaco Geartex EP85W-90
Swivel pin	Molytex EP90 OR 80

Do not use any lubricant other than that specified

Capacities

Item	Capacity
Front differential	1.70 litres (2.9 pints) (1.80 US quarts)
Swivel pin housing oil/grease (each)	0.35 litres (0.60 pints) (0.37 US quarts)

Torque Specifications

Description	Nm	lb-ft
Hub driving member to hub	65*	48
Brake disc to hub	73	53
Stub axle to swivel pin housing	65*	48
Brake caliper to swivel pin housing	82	61
Upper swivel pin to swivel pin housing	65*	48
Lower swivel pin to swivel pin housing	25*	18
Oil seal retainer to swivel pin housing	11	8
Swivel bearing housing to axle case	73*	53
Pinion housing to axle case	41	30
Crown wheel to differential housing	58	43
Differential bearing cap to pinion housing	90	65
Differential drive flange to drive shaft	47	34
Mudshield to bracket lower swivel pin	11	8
Bevel pinion nut	130	
Draglink to hub arm	40	30
Panhard rod to axle bracket	88	65
Radius arm to axle	190	140
Radius arm to chassis side member	190	140

*** Apply sealant, Part No. STC 50552 To bolt threads**

Front Drive Axle and Differential

Description and Operation

Operation

The welded steel front axle casing houses a separate spiral bevel type differential unit, which is off-set to the right of the vehicle centre line. The differential unit drives the front wheels via the axle shafts and constant velocity joints which are totally enclosed in the spherical and swivel housings.

The front wheels are pivoted on tape roller bearings at the top and bottom of the swivel housing. The wheel hubs on all axles are supported by two taper bearings and driven by drive flanges which are splined to the one piece, stub shaft/constant velocity joint.

Lubrication

The differential, swivel pin housing and wheel hubs are individually lubricated and separated by oil seals (7) and (8), see J6267A, to prevent oil transfer across the axle when the vehicle is traversing steep inclines. The wheel bearings are lubricated with grease and the swivel housing and differential with oil. On later vehicles, identified by having only a filler plug in the swivel housing, grease is used to lubricate the housing assembly,

Ventilation

Ventilation of the differential is through a plastic pipe (2) which terminates at a high level in the vehicle axle. The swivel housings ventilate through axle shaft oil seals (8) into the differential and the hub bearings vent via the oil seals into the swivel housing.

GENERAL

Front axle hub and swivel housing

Item	Description
1	Axle casing
2	Ventilation pipe
3	Axle shaft
4	Wheel studs and hub
5	Stub axle
6	Wheel bearings
7	Inner and outer hub seals
8	Axle shaft seal
9	Hub lock plate, thrust washer and nuts
10	Brake disc
11	Drive flange
12	Shim washer and circlip
13	Dust cap
14	Constant velocity joint/shaft
15	Thrust collar for CV joint
16	Roller bearing
17	Spacer
18	Circlip
19	Top and bottom swivel taper bearing
20	Top and bottom swivel pins
21	Spherical housing, seal and retainer
22	Swivel housing
23	Constant velocity shaft seal

The front differential is mounted on the LH side of the chassis

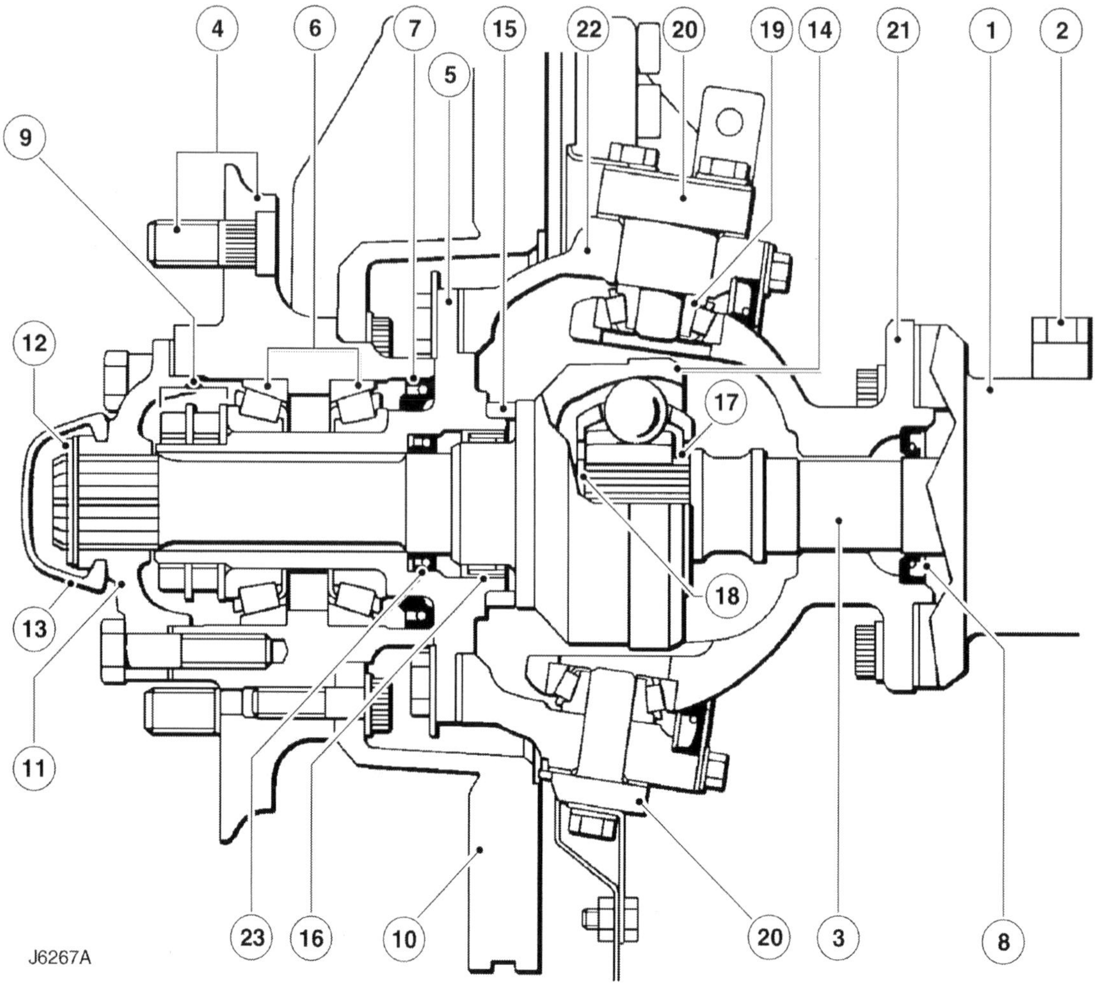

Front Drive Axle

Diagnosis and Testing

Principles of Operation

For a detailed description of the Front Drive Axle and Differential, refer to the relevant Description and Operation section in the workshop manual.

Refer to: Front Drive Axle and Differential (205-03 Front Drive Axle/Differential, Description and Operation)

Inspection and Verification

CAUTION: Diagnosis by substitution from a donor vehicle is **NOT** acceptable. Substitution of control modules does not guarantee confirmation of a fault, and may also cause additional faults in the vehicle being tested and/or the donor vehicle.

NOTES:

If the control module or a component is suspect and the vehicle remains under manufacturer warranty, refer to the Warranty Policy and Procedures manual (section B1.2), or determine if any prior approval programme is in operation, prior to the installation of a new module/component.

When performing voltage or resistance tests, always use a digital multimeter accurate to three decimal places, and with an up-to-date calibration certificate. When testing resistance always take the resistance of the digital multimeter leads into account.

Check and rectify basic faults before beginning diagnostic routines involving pinpoint tests.

1. Visually inspect for obvious signs of damage and system integrity

Visual Inspection

Mechanical
• Front axle
• Front differential
• Front wheel hubs
• Front swivel housings

2. If an obvious cause for an observed or reported concern is found, correct the cause (if possible) before proceeding to the next step
3. If the cause is not visually evident, verify the symptom and refer to the Symptom Chart, alternatively check for Diagnostic Trouble Codes (DTCs) and refer to the DTC Index

Symptom Chart

Symptom	Possible Causes	Action
Grease leak from the front left/ right wheel hub	• Front left/right hub seal failure • Front left/right constant velocity shaft seal failure	• Install a new front left/right hub seal • Install a new front left/right constant velocity shaft seal
Grease leak from the front left/ right swivel housing	• Front left/right swivel housing seal failure • Spherical housing surface damaged	• Install a new front left/right swivel housing seal • Install a new front left/right spherical housing and swivel housing seal
Oil leak from the front left/right swivel housing	• Front left/right Half shaft seal failure • Front axle breather blocked	⚠ NOTE: After rectifying a front axle oil leak, refer to the relevant section of the workshop manual and check the front differential oil level. Top up as necessary. • Install a new front left/right Half shaft seal • Unblock the front axle breather
Oil leak from the front differential pinion seal	• Front differential pinion seal failure • Front axle breather blocked	NOTES: ⚠ After rectifying a front axle oil leak, refer to the relevant section of the workshop manual and check the front differential oil level. Top up as necessary. ⚠ When installing a new differential pinion seal, a new differential pinion flange should also be installed. • Install a new front differential pinion seal and flange • Unblock the front axle breather
Front differential oil level too high	• Front differential oil overfilled • Water ingress to the front differential via a defective front left/right Half shaft seal during wading	• Refer to the relevant section of the workshop manual and check the front differential oil level. Correct as necessary • Install a new front left/right Half shaft seal. Refer to the relevant section of the workshop manual and drain and refill the front differential oil
Loss of drive	• Front left/right driving member splines worn • Front left/right Half shaft failure • Front left/right constant velocity joint failure • Front differential failure	• Install a new front left/right driving member as necessary • Install a new front left/right Half shaft as necessary • Install a new front left/right constant velocity joint as necessary • Install a new front differential as necessary

Drive Pinion Seal - 90

In-vehicle Repair

Part(s)

STEP	PART NAME	QUANTITY
Installation Step 4	Renew Part Rear differential pinion flange kit	1
Installation Step 5	Renew Part Differential pinion shaft bolt	1
Installation Step 7	Renew Part Driveshaft flange nut(s)	4

Removal

1. WARNING: Do not work under a vehicle supported only by a jack. Always support the vehicle on safety stands.

Raise front of vehicle.

2. Reference mark the drive shaft flanges for reassembly.
3. NOTE: Discard the 4 nuts.

Remove 4 nuts and bolts securing drive shaft to differential housing.

4. Release drive shaft and tie aside.
5. NOTE: Discard the nut.

Using LRT-51-003 to restrain the pinion flange, remove and discard the bolt securing the pinion flange.

6. Remove and discard the pinion flange.

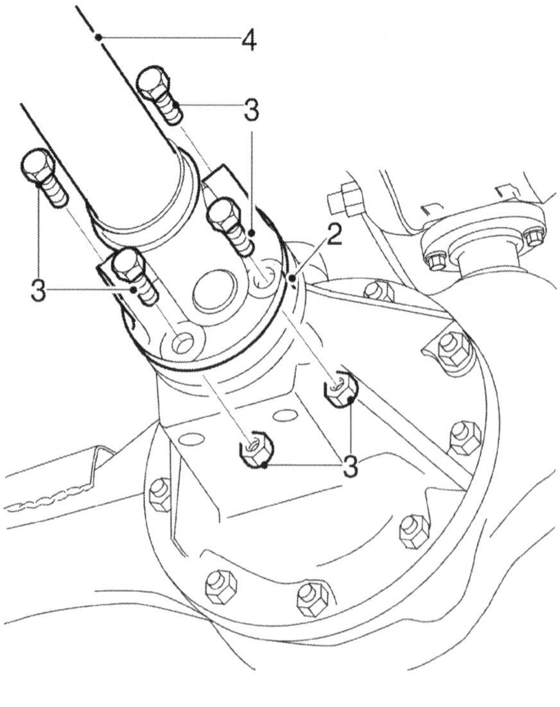

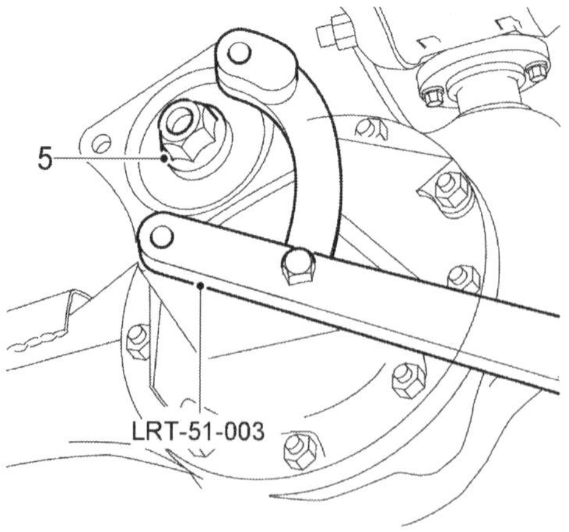

7. Position container to catch oil spillage.
8. CAUTION: Take care to avoid damage to oil seal recess.

Using a suitable lever, remove and discard pinion oil seal.

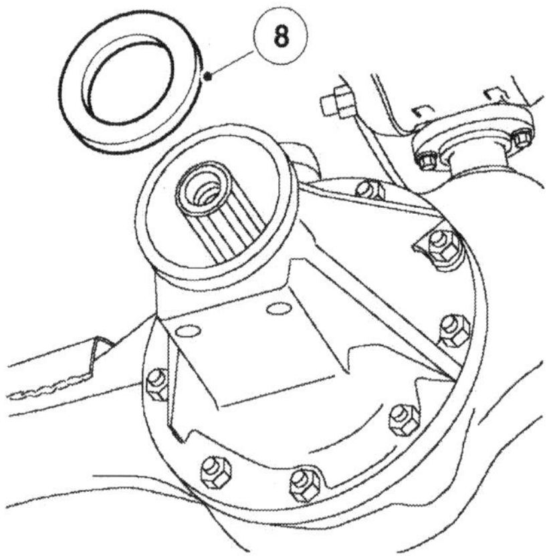

Installation

1. Clean pinion oil seal recess.
2. Lubricate oil seal lip with clean oil.
3. Using LRT-51-010 fit pinion oil seal.
4. Fit new pinion flange.

   Renew Part: Rear differential pinion flange kit

5. NOTE: A new bolt must be installed.

Restrain flange using LRT-51-003 and fit bolt. Tighten bolt.

Renew Part: Differential pinion shaft bolt

Torque: 100 Nm

6. Position drive shaft to differential housing and align reference marks.
7. NOTE: New nuts must be installed.

Fit flange bolts and tighten.

Renew Part: Driveshaft flange nut(s)

Torque: 47 Nm

8. Remove stands and lower vehicle.
9. Top-up differential oil level.

Axle Assembly

Removal and Installation

Removal

⚠ WARNING: Remove and refit of axle requires a further two persons to steady axle when lowering or repositioning axle.

1. Support chassis front.
2. Remove road wheels.
3. Support axle weight with hydraulic jack.

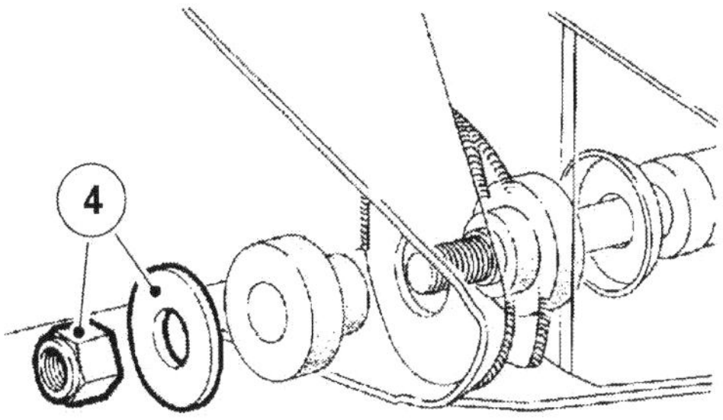

4. Remove radius arms to chassis frame nuts.
5. Disconnect steering damper from track rod. Using a extractor remove track rod links from swivel pin arms.
6. Remove four nuts and bolts securing radius arms to axle bracket.
7. Remove radius arms.
8. Remove bolts securing brake hose brackets. Refit bolts to prevent oil leakage.
9. Remove bolts from brake calipers and tie to one side.
10. Remove nuts and washers securing shock absorbers to axle.
11. Disconnect drag link from swivel pin housing arm.
12. Remove two nuts and bolts securing panhard rod to axle bracket. Lift rod clear of axle.
13. Mark for reassembly drive shaft flanges. Remove four nuts and bolts, tie propeller shaft to one side.
14. Release axle ventilation pipe banjo and lower axle assembly. Remove road springs.
15. Disconnect stabilizer bar link.

Refer to: Front Stabilizer Bar Link (204-01 Front Suspension, Removal and Installation)

16. Remove axle assembly.
17. Transfer components to new axle, if appropriate.

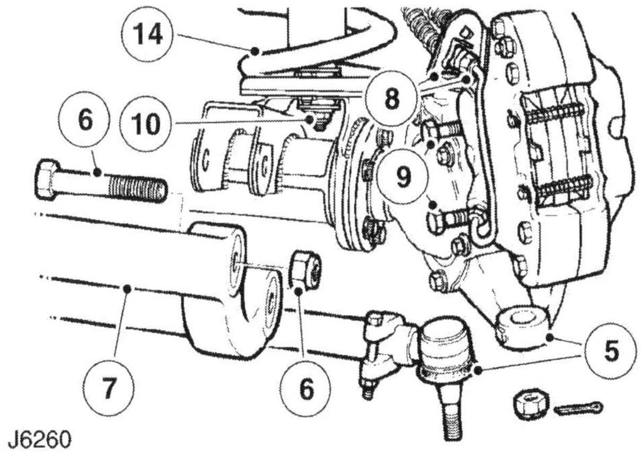

Installation

1. Position axle under vehicle, supporting left side of axle, and Install anti-roll bar links.

Refer to: Front Stabilizer Bar Link (204-01 Front Suspension, Removal and Installation)

2. Install propeller shaft. Tighten bolts to 47 Nm (35 lbf.ft).
3. Install panhard rod to axle bracket. Tighten bolts to 88 Nm (65 lbf.ft).
4. Install drag link to swivel pin arm. Tighten fixings to 40 Nm (30 lbf.ft).
5. Install shock absorbers to axle.
6. Install brake calipers. Tighten bolts to 82 Nm (60 lbf. ft).
7. Tighten upper swivel pin bolts to 78 Nm (58 lbf.ft).
8. Install radius arms to axle brackets. Tighten bolts to 197 Nm (145 lbf.ft).
9. Install steering damper to track rod.
10. Install radius arms to chassis side member. Tighten fixings to 197 Nm (145 lbf.ft).
11. Tighten track rod end to 40 Nm (30 lbf.ft) and Install new split pin.
12. Remove chassis supports, Install road wheels and tighten to correct torque:
    1. Alloy wheels - 130 Nm (96 lbf.ft)
    2. Steel wheels - 100 Nm (80 lbf.ft)
    3. Heavy duty wheels - 170 Nm (125 lbf.ft)

Differential Carrier

Removal and Installation

Special Tool(s)

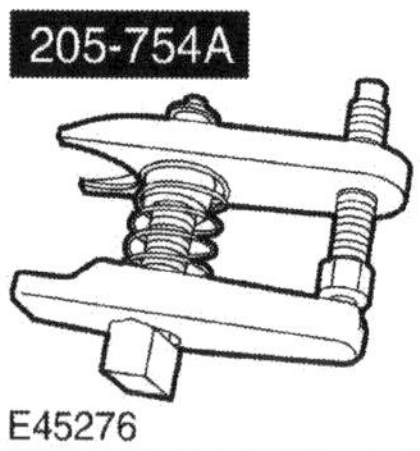 E45276	205-754A Splitter, Ball Joints

Part(s)

STEP	PART NAME	QUANTITY
Installation Step 2	Renew Part Drive-shaft flange nut(s)	4
Installation Step 3	Renew Part Front stabilizer bar to sub-frame nuts	4
Installation Step 4	Renew Part Half shaft gasket	1
Installation Step 5	Renew Part Split pin	1
Installation Step 9	Renew Part Half shaft gasket	1
Installation Step 10	Renew Part Split pin	1
Installation Step 11	Renew Part Split pin	1

Removal

1. CAUTION: Do not work on or under a vehicle supported only by a jack. Always support the vehicle on safety stands.

Raise and support the vehicle.

2. NOTE: Clean the area around the front axle assembly filler plug and the front axle assembly drain plug.

Drain the front axle assembly. Position a container to collect the fluid. Remove the filler plug. Remove the drain plug.

3. Remove the front wheels and tires.

4. Remove the tie rod protector. Remove the 2 bolts.

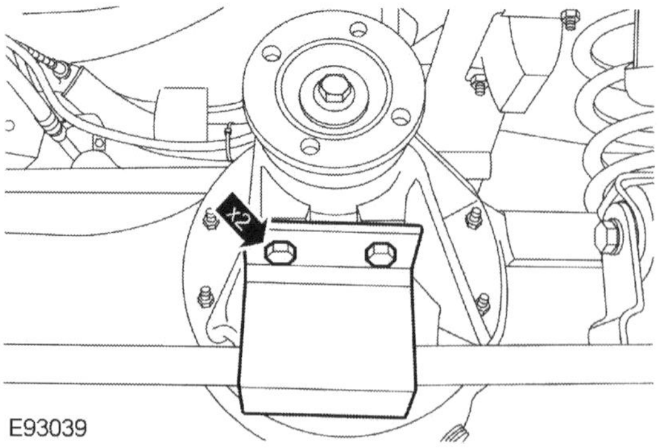

E93039

5. CAUTION: Before disconnecting or removing components, make sure the immediate area around joint faces and connections are clean. Plug open connections to prevent contamination.

Release the LH front wheel speed sensor. Release the harness from the clip.

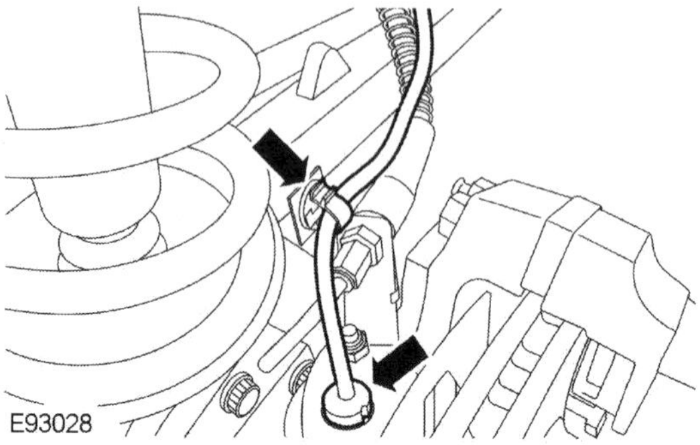
E93028

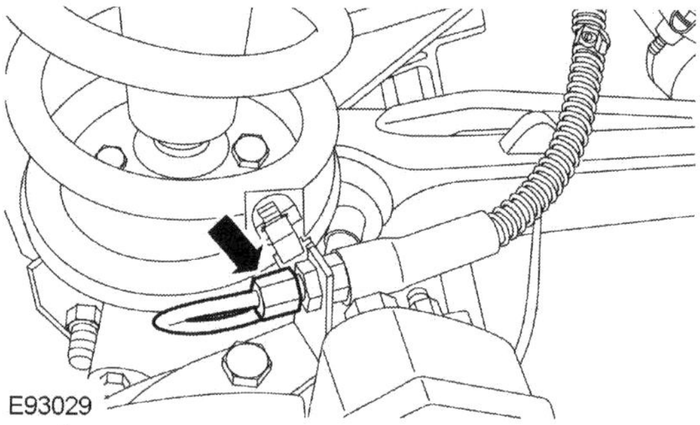
E93029

6. CAUTION: Before disconnecting or removing components, make sure the immediate area around joint faces and connections are clean. Plug open connections to prevent contamination.

Disconnect the LH front brake pipe. Clamp the brake hose to prevent fluid loss.

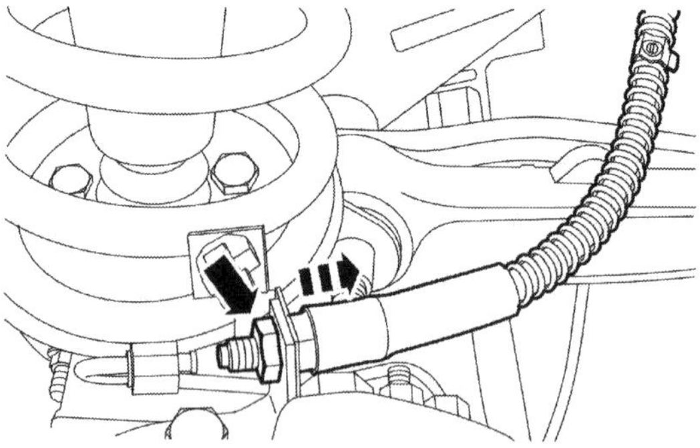

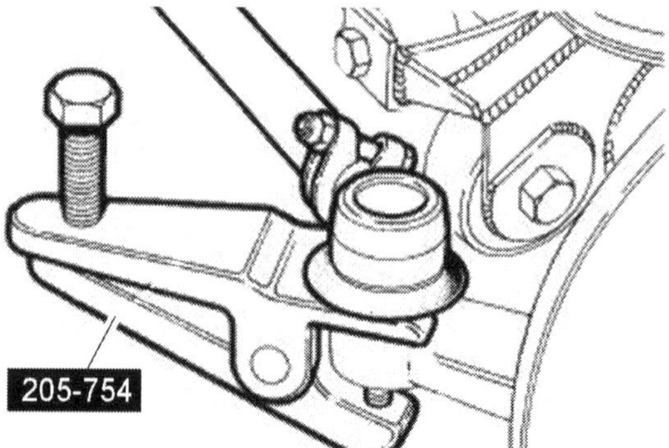

E93031

7. Release the LH front brake hose. Remove the nut.
8. Using the special tool, release the drag link end. Remove and discard the split pin. Remove the nut. Collect the washer.

   Special Tool(s): 205-754A

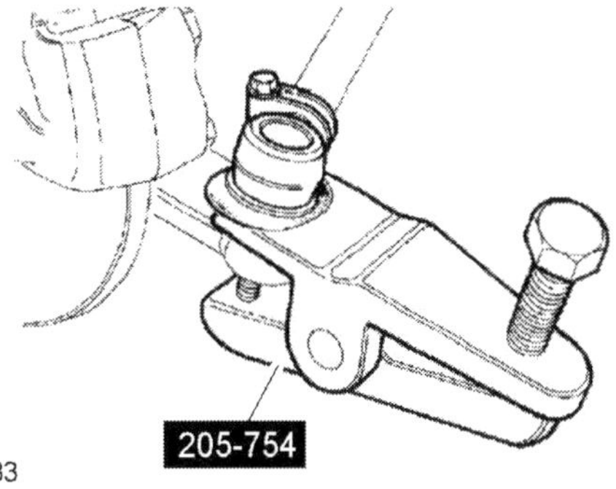

E93033

9. Using the special tool, release the LH tie rod end. Remove and discard the split pin. Remove the nut. Collect the washer.

   Special Tool(s): 205-754A

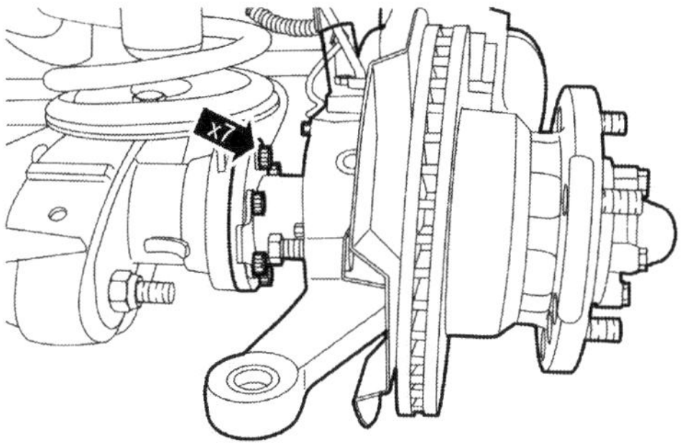

10. CAUTION: Do not allow the swivel pin housing to hang on the Half shaft, failure to follow this instruction may result in damage to the vehicle.

With assistance, remove the LH swivel pin housing. Remove the 7 bolts. Remove and discard the gasket.

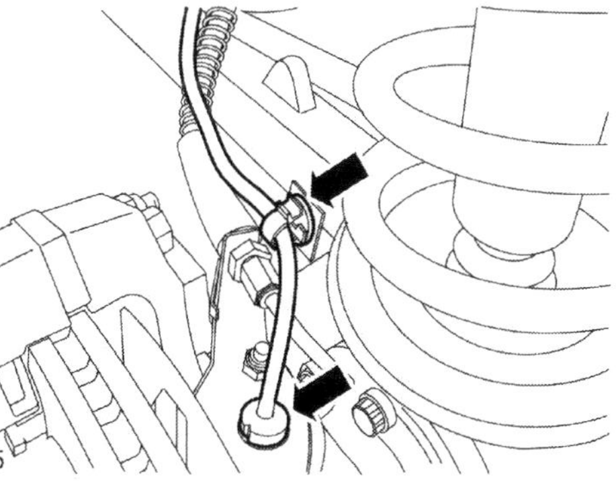

E93035

11. CAUTION: Before disconnecting or removing components, make sure the immediate area around joint faces and connections are clean. Plug open connections to prevent contamination.

Release the RH front wheel speed sensor. Release the harness from the clip.

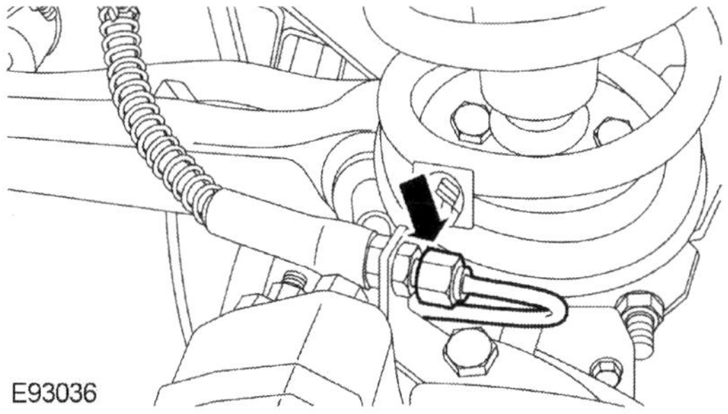

E93036

12. CAUTION: Before disconnecting or removing components, make sure the immediate area around joint faces and connections are clean. Plug open connections to prevent contamination.

Disconnect the RH front brake pipe. Clamp the brake hose to prevent fluid loss.

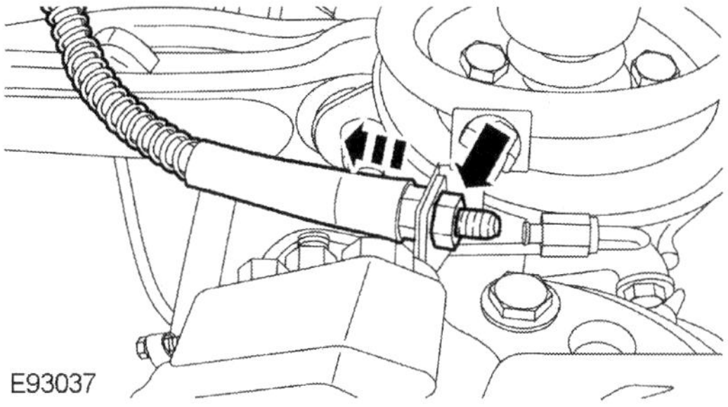

E93037

13. Release the RH front brake hose. Remove the nut.
14. Using the special tool, release the RH tie rod end. Remove and discard the split pin. Remove the nut. Collect the washer.

    Special Tool(s): 205-754A

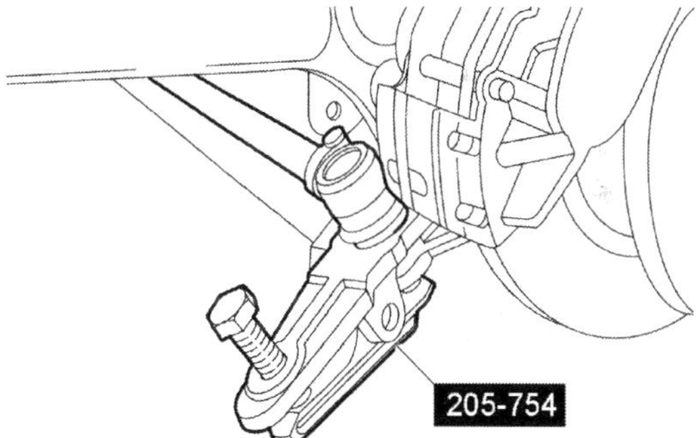

E93032

15. CAUTION: Do not allow the swivel pin housing to hang on the Half shaft, failure to follow this instruction may result in damage to the vehicle.

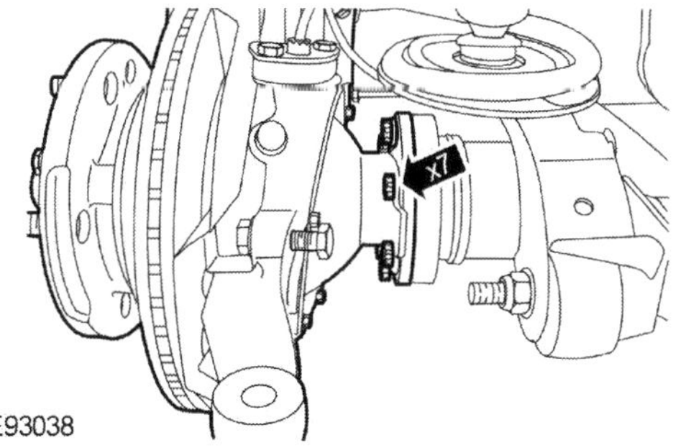

E93038

With assistance, remove the RH swivel pin housing. Remove the 7 bolts. Remove and discard the gasket.

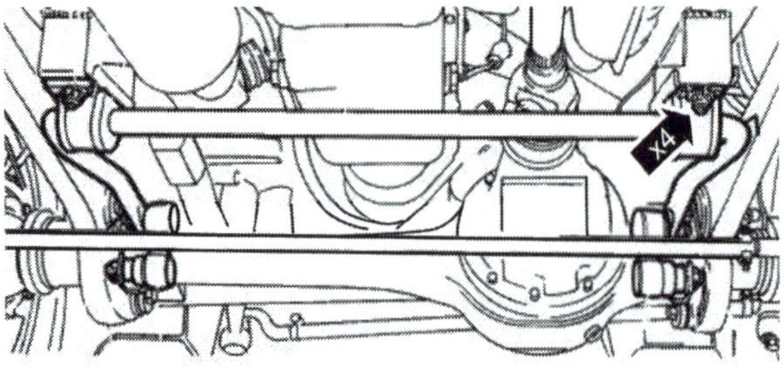

E93040

16. △ NOTE: Discard the nuts.

Reposition the front stabilizer bar. Remove the 4 nuts and bolts.

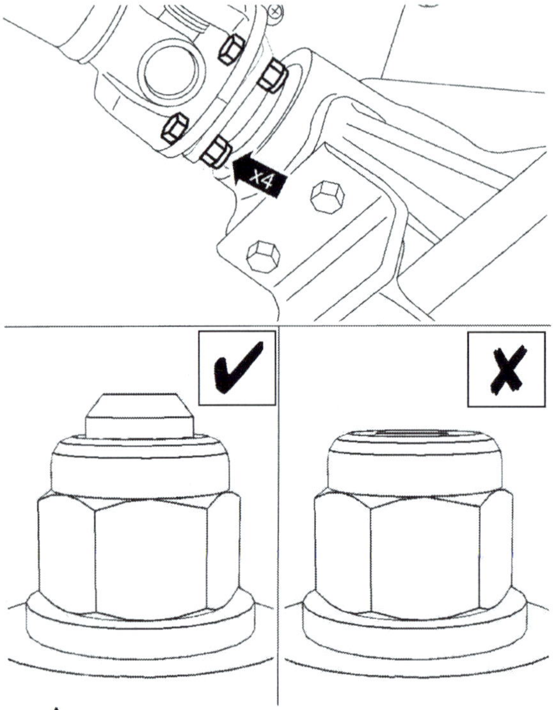

17. △ NOTE: Discard the nuts.

Release the front driveshaft. Remove the 4 nuts and bolts. Support the driveshaft using a suitable tie strap.

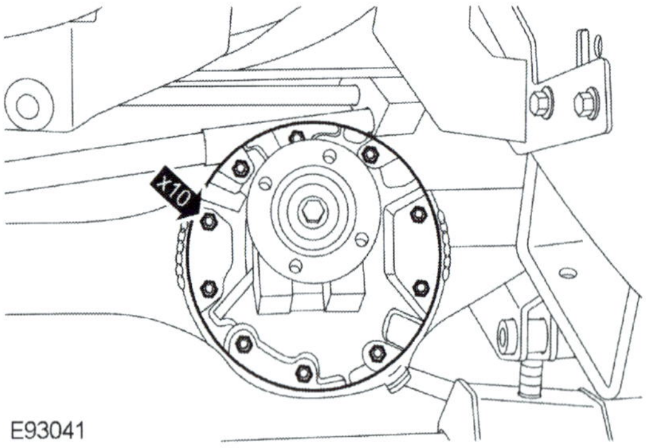

E93041

18. Remove the front differential carrier. Remove the 10 nuts.

Installation

1. △ NOTE: Clean the component mating faces.

   Install the front differential carrier. Apply a bead of the specified sealant to the differential carrier mating face. Tighten the nuts.

   Torque: 41 Nm

2. △ NOTE: Install new nuts.

   Secure the front driveshaft. Tighten the nuts and bolts.

   Renew Part: Driveshaft flange nut(s)

   Torque: 47 Nm

3. △ NOTE: Install new nuts.

Secure the front stabilizer bar. Tighten the nuts and bolts.

   Renew Part: Front stabilizer bar to subframe nuts

   Torque: 68 Nm

4. ⚠ CAUTION: Support the weight of the Half shaft when installing the swivel pin housing, failure to follow this instruction may result in damage to the vehicle.

   △ NOTE: Clean the component mating faces.

With assistance, install the RH swivel pin housing. Install a new gasket. Apply the specified sealant to the bolts. Tighten the bolts.

   Renew Part: Half shaft gasket

   Torque: 73 Nm

5. Secure the RH tie rod end. Install the washer. Tighten the nut. Install a new split pin.

   Renew Part: Split pin

   Torque: 40 Nm

6. Secure the RH front brake hose. Tighten the nut.

   Torque: 10 Nm

7. △ NOTE: Remove and discard the blanking caps.

Connect the RH front brake pipe. Tighten the union. Remove the hose clamp.

   Torque: 15 Nm

8. △ NOTE: Remove and discard the blanking caps.

Secure the RH front wheel speed sensor. Secure the harness in the clip.

9. ⚠ CAUTION: Support the weight of the Half shaft when installing the swivel pin housing, failure to follow this instruction may result in damage to the vehicle.

NOTE: Clean the component mating faces.

With assistance, install the LH swivel pin housing. Install a new gasket. Apply the specified sealant to the bolts. Tighten the bolts.

Renew Part: Half shaft gasket

Torque: 73 Nm

10. Secure the LH tie rod end. Install the washer. Tighten the nut. Install a new split pin.

Renew Part: Split pin

Torque: 40 Nm

11. Secure the drag link end. Install the washer. Tighten the nut. Install a new split pin.

Renew Part: Split pin

Torque: 40 Nm

12. Secure the LH front brake hose. Tighten the nut.

Torque: 10 Nm

13. NOTE: Remove and discard the blanking caps.

Connect the LH front brake pipe. Tighten the union. Remove the hose clamp.

Torque: 15 Nm

14. NOTE: Remove and discard the blanking caps.

Secure the LH front wheel speed sensor. Secure the harness in the clip.

15. Install the tie rod protector. Tighten the bolts.

Torque: 30 Nm

16. Fill the front axle assembly with the correct amount of the specified oil. Install and tighten the drain plug and the filler plug. Remove the container.

Torque: 60 Nm

17. Bleed the front brakes.

Refer to: Brake System Bleeding (206-00 Brake System - General Information, General Procedures)

18. Install the front wheels and tires. Tighten the wheel nuts.

Torque: 140 Nm

Front Stub Axle, Constant Velocity Joint and Swivel Pin Housing

Item	Description
1.	Swivel pin housing
2.	Top swivel pin and brake hose bracket
3.	Upper and lower swivel pin bearings
4.	Shim
5.	Retaining plate and washer
6.	Oil seal
7.	Joint washer
8.	Swivel bearing housing
9.	Joint washer
10.	Lower swivel pin
11.	Mudshield bracket
12.	Swivel housing inner oil seal

Disassembly and Assembly

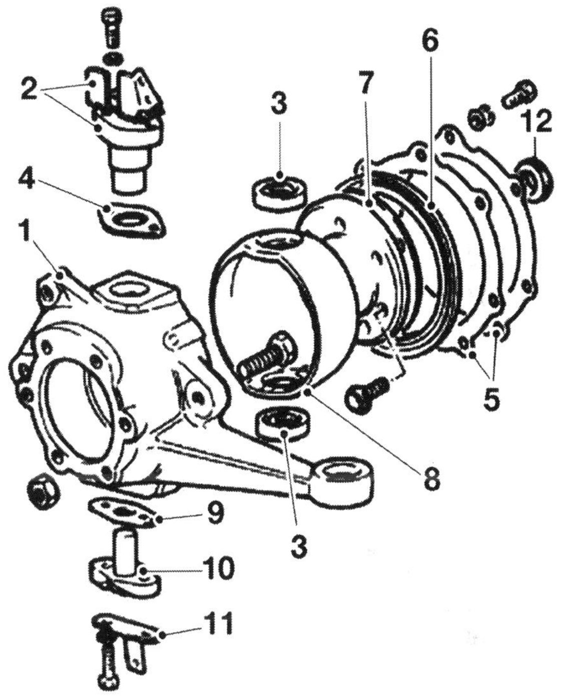

Disassembly

1. Remove front hub assembly.

Refer to: Wheel Bearing and Wheel Hub - Vehicles Built Up To: 01/1999 (204-01, Removal and Installation).

Refer to: Front Wheel Bearing and Wheel Hub (204-01 Front Suspension, Removal and Installation).

2. NOTE: On later vehicles the swivel pin housing is filled with grease for life, the level and drain plugs being deleted.

Drain swivel pin housing and refit plug.

3. Remove 6 bolts retaining stub axle to swivel housing.
4. Remove mud shield.
5. Remove stub axle and joint washer.

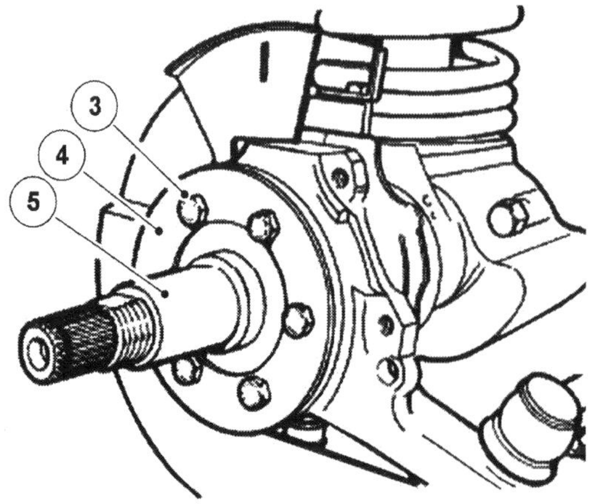

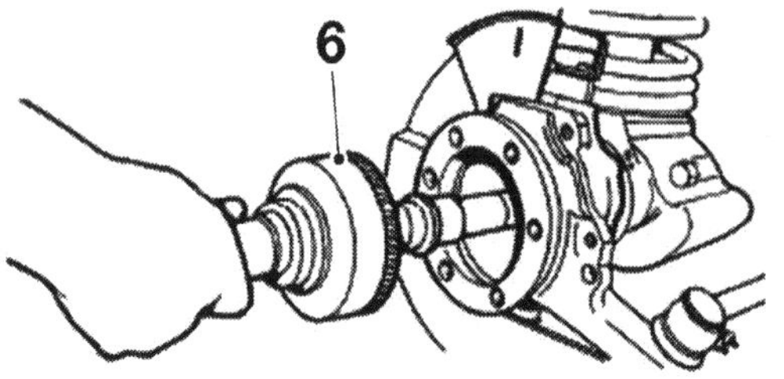

6. Withdraw axle shaft and constant velocity joint from axle casing.

7. Hold axle shaft firmly in a soft jawed vice.

8. Using a soft mallet drive constant velocity joint from shaft.

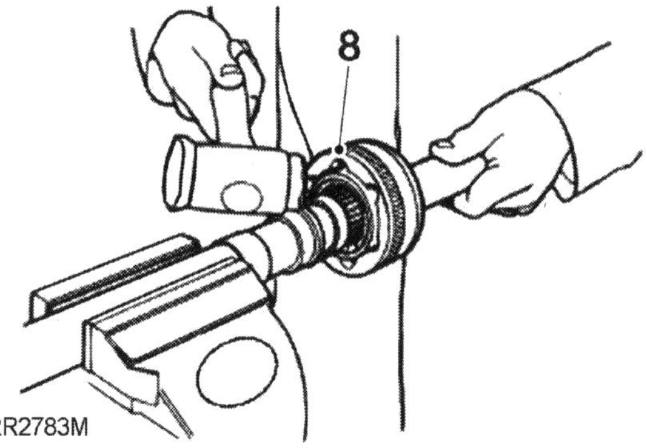

RR2783M

9. Remove circlip and collar from axle shaft.

10. Mark positions of constant velocity joint, inner and outer race and cage for reassembly.

11. Swivel cage and inner race to remove balls.

12. Examine all components, in particular, inner and outer track, cage balls and bearing surfaces for damage and excessive wear.

13. Maximum acceptable end-float on assembled joint 0.64mm. Renew if worn or damaged. Lubricate with a recommended oil during assembly.

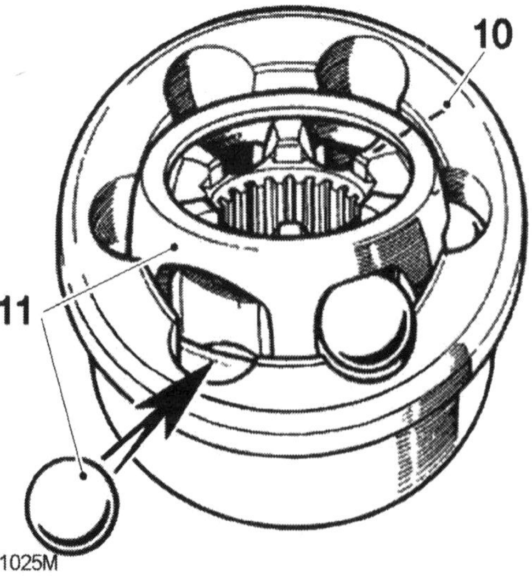

ST1025M

Assembly

1. Install collar and a new circlip.

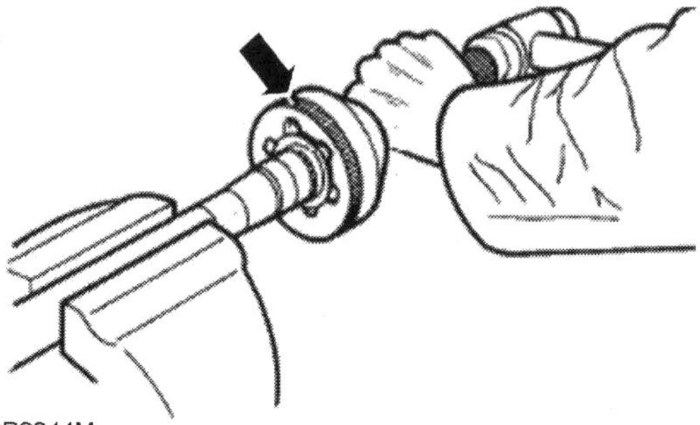

RR3844M

2. Engage constant velocity joint on axle shaft splines and using a soft mallet, drive joint in fully.

3. Drill and chisel off thrust ring taking care to avoid damaging stub axle.

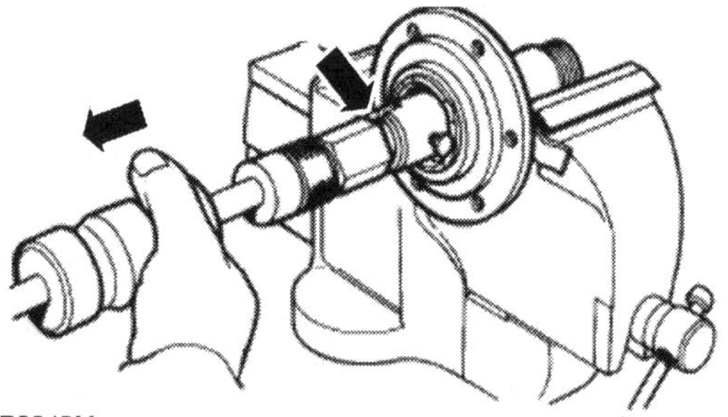

RR3843M

4. Remove bearing and oil seal using special tool LRT-37-004 and slide hammer LRT-99-004. Ensure lip of tool locates behind bearing to drive it out.

5. Repeat instruction for removal of oil seal.

6. Lubricate seal and lip with EP90 oil and with cavity side leading press in a new oil seal using special tool LRT-54-004.

7. Using special tool LRT-54-005, fit bearing with its part number visible when fitted, and flush with end face of stub axle.

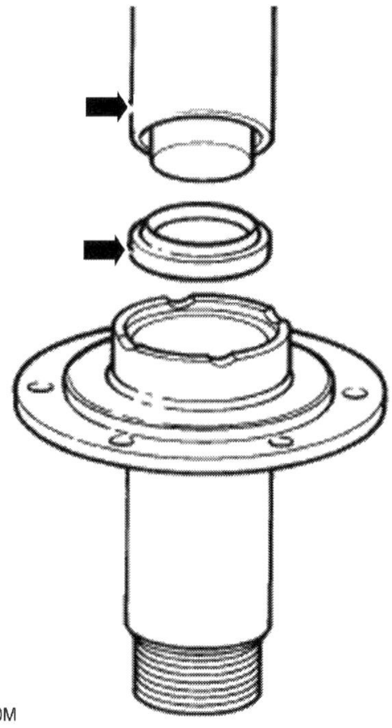
RR3840M

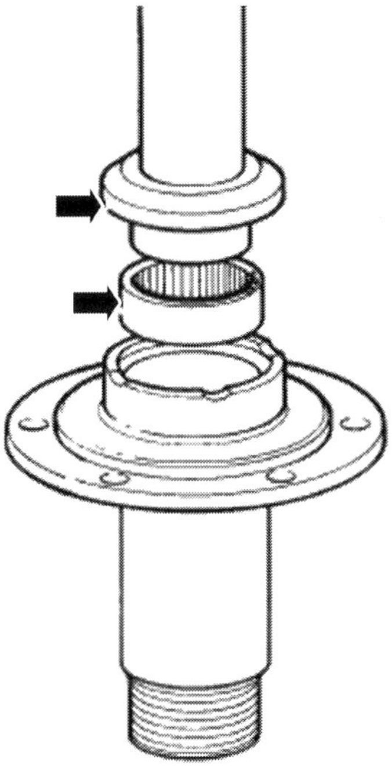
RR3839M

8. Press fit a new thrust ring onto stub axle.

9. NOTE: Removal of oil seal and retaining plate is achieved when swivel bearing housing is removed.

Remove bolts securing oil seal retaining plate and joint washer. Release assembly from swivel pin housing.

10. Remove 2 bolts, retaining lower swivel pin to housing.

11. Remove brake disc shield bracket.

12. Tap lug to remove lower swivel pin and joint washer.

13. Remove two bolts retaining brake hose bracket and top swivel pin.

14. Remove bracket, top swivel pin and shims.

15. Remove swivel pin housing while retrieving lower and upper bearings.

16. NOTE: Use upper bearing opening to gain access to lower bearing track.

    Remove lower bearing track from swivel bearing housing.

17. Remove 7 bolts retaining swivel bearing housing to axle case.

18. Remove inner oil seal from back of housing.

19. NOTE: Use lower bearing opening to gain access to upper bearing track.

    Remove top bearing track from swivel bearing housing.

20. If worn, pitted or damaged, renew housing.

21. CAUTION: Ensure bearing tracks are fitted square or damage could occur.

    Install upper and lower bearing tracks into swivel bearing housing.

22. With seal lips trailing, install swivel housing inner oil seal into rear of housing. Grease seal lips.

23. Coat swivel bearing housing to axle casing bolts with sealant, Part No. STC 50552.

24. Coat both sides of joint washer with a sealing compound. Position swivel bearing housing to axle mating face.

25. Place retaining plate, joint washer and oil seal over axle flange ready for assembly.

26. Install swivel bearing housing to axle flange with 7 bolts. Tighten to 73 Nm (54 lbf.ft).

27. Grease and install upper and lower swivel pin taper roller bearings.

28. Position swivel pin housing over swivel bearing housing.

29. Coat joint washer both sides with sealing compound and position on lower swivel pin.

30. Loosely install brake shield bracket plus lower swivel pin with lug outboard to swivel pin housing.

31. Loosely install top swivel pin plus existing shims and brake hose bracket to swivel pin housing.

32. Apply sealant, Part No. STC 50552 to threads of lower swivel pin bolts; tighten bolts to 25 Nm (18 lbf.ft), bend over lock tabs.

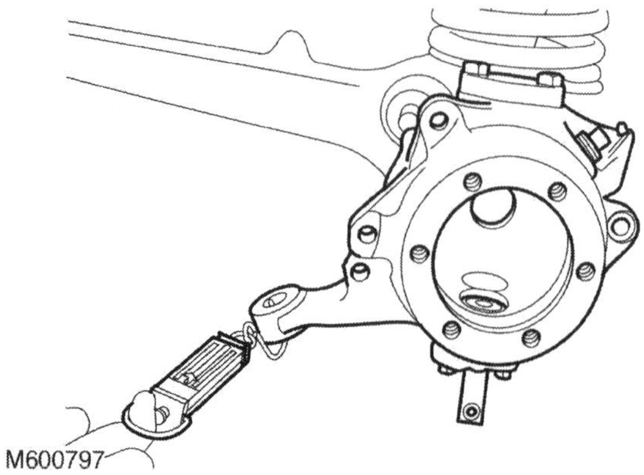

33. Apply sealant Part No. STC 50552 to threads of top swivel pin bolts, install bolts and tighten to 65 Nm (48 lbf.ft).

34. NOTE: Swivel housing oil seal and axle should not be fitted.

    Attach a spring balance to track rod ball joint bore and pull balance to determine effort required to turn swivel pin housing. Resistance, once initial inertia has been overcome, should be 1.16 to 1.46 kg. Adjust by removing or adding shims to top swivel pin.

35. When setting is correct remove top swivel bolts, apply sealant, Part No. STC 50552 to threads of bolts. Refit bolts and tighten to 65 Nm (48 lbf.ft), and bend over lock tabs.

36. Apply recommended grease between lips of swivel oil seal.

37. Install oil seal, joint washer and retaining plate with 7 bolts and spring washers. Tighten to 10 Nm (7 lbf.ft).

38. Install tie rod and drag link and secure with new cotter pins. Tighten fixing to 40 Nm (30 lbf.ft).

39. Install brake disc shield.

40. Loosely install lock stop bolt and nut.

41. Apply a recommended grease between lips of swivel housing oil seal.

42. Secure oil seal with retaining plate and securing bolts. Tighten to 10 Nm (7 lbf.ft).

43. Install track-rod and drag link and secure with new cotter pins.

44. Loosely install lock stop bolt for later adjustment.

45. Install brake disc shield.

46. CAUTION: Take care not to damage axle shaft oil seals.

    Insert axle shaft, and when differential splines are engaged, push assembly in fully.

47. Place a new joint washer in position on swivel pin housing to stub axle mating face. Coat threads of stub axle bolts with sealant, Part No. STC 50552.

48. CAUTION: Ensure that constant velocity joint bearing journal is butted against thrust ring on stub axle before stub axle is secured.

    Install stub axle with flat at 12 O'clock position.

49. Place mud shield in position and secure stub axle to swivel pin housing with 6 bolts and tighten evenly to 65 Nm (48 lbf.ft).

50. Install brake jump hoses to brake jump hose bracket.

51. Install complete front hub assembly.

Refer to: Wheel Bearing and Wheel Hub - Vehicles Built Up To: 01/1999 (204-01, Removal and Installation).

Refer to: Front Wheel Bearing and Wheel Hub (204-01 Front Suspension, Removal and Installation).

52. Check swivel pin housing oil drain plug is fitted.

53. Remove swivel pin level and filler plugs.

54. NOTE: On later vehicles fill swivel pin housing with 0.33 Litres of Molytex EP 00 grease.

    Check and top up with new oil until oil runs out from level hole. Allow excess oil to drain and wipe clean.

Refer to: Specifications (205-03 Front Drive Axle/ Differential, Specifications).

55. Install swivel pin level and filler plugs.

56. Set steering lock stop bolts.

Refer to: Steering Lock Stop Adjustment (211-00 Steering System - General Information, General Procedures).

General Information

Specifications

Brake Hydraulic Fluid

Item	Specification
* Recommended hydraulic fluid	SHELL DONAX YB DOT4 ESL FLUID

CAUTION: * If the above fluid is not available, use a low viscosity DOT 4 brake fluid meeting ISO 4925 Class 6 and Land Rover LRES22BF03 requirements.

General Specification

NOTE: Pressure reducing valves are not fitted to all 110 specifications.

Item	Specification
Front calliper	AP Lockheed, four opposed pistons
Operation	Hydraulic, self adjusting
Front disc:	
- 90 Models	Solid, outboard
- 110/130 Models	Ventilated, outboard
Disc diameter	298 mm (11.73 in)
Front disc thickness:	
- 90 Models	14.1 mm (0.56 in)
- 110/130 Models	24 mm (0.95 in)
Wear limit	1 mm (0.04 in) per side of disc
Disc run-out maximum	0.15 mm (0.006 in)
Pad area	58 cm² (9.0 in²)
Total swept area	801.3 cm² (124.2 in²)
Pad material	Ferodo 3440 non asbestos
Pad minimum thickness	3 mm (0.12 in)
Rear calliper	AP Lockheed, four opposed pistons
Operation	Hydraulic, self adjusting
Rear disc	Solid, outboard
Rear disc diameter:	
- 90 Models	290 mm (11.42 in)
- 110/130 Models	298 mm (11.73 in)
Rear disc thickness:	
- 90 Models	12.5 mm (0.49 in)
- 110/130 Models	14.1 mm (0.56 in)
Wear limit:	
- 90 Models	0.38 mm (0.015 in) per side of disc
- 110/130 Models	1 mm (0.04 in) per side of disc
Disc run-out maximum	0.15 mm (0.006 in)
Pad area:	
- 90 Models	30.5 cm² (4.37 in²)
- 110/130 Models	36.2 cm² (5.61 in²)
Total swept area	694 cm² (106.98 in²)
Pad material	Ferodo 3440 non asbestos
Pad minimum thickness	3 mm (0.12 in)
Parking brake	
Parking brake type	Mechanical, cable operated drum brake on the rear of the transfer gearbox output shaft
Parking brake drum internal diameter	254 mm (10.0 in)
Width	70 mm (2.75 in)
Pad material	Ferodo 3611 non asbestos
Master cylinder/Booster	
Master cylinder manufacturer	Lucas
Master cylinder type	Tandem, 25.4 mm (1.0 in) diameter
Booster type	LSC 80
Pressure reducing valve, failure conscious:	
- 90 Models	Cut-in pressure 24 bar (360 lbf/in²), ratio 4.0:1
- 110/130 Models	Cut-in pressure 43 bar (645 lbf/in²), ratio 2.9:1

Torque specifications

Description	Nm	lbf-ft
Brake pipe to pressure reducing valve	15	11
Brake pipe to brake calipers	15	11
Brake pipe to master cylinder	15	11
Flexible hoses to callipers	15	11
Bleed screws	15	11
Brake caliper to swivel pin housing	82	60
Brake disc bolts*	73	54
Brake drum screw	8**	6
Brake drum adjuster bolt	25	18
Booster assembly to pedal box	26	19
Pedal box to bulkhead	25	18
Master cylinder to booster	26	19

* Apply Loctite 270 before assembly

** If you are refitting old screws, tighten to 25 Nm (18 lbf-ft).

COMPONENT LOCATION - ANTI-LOCK CONTROL

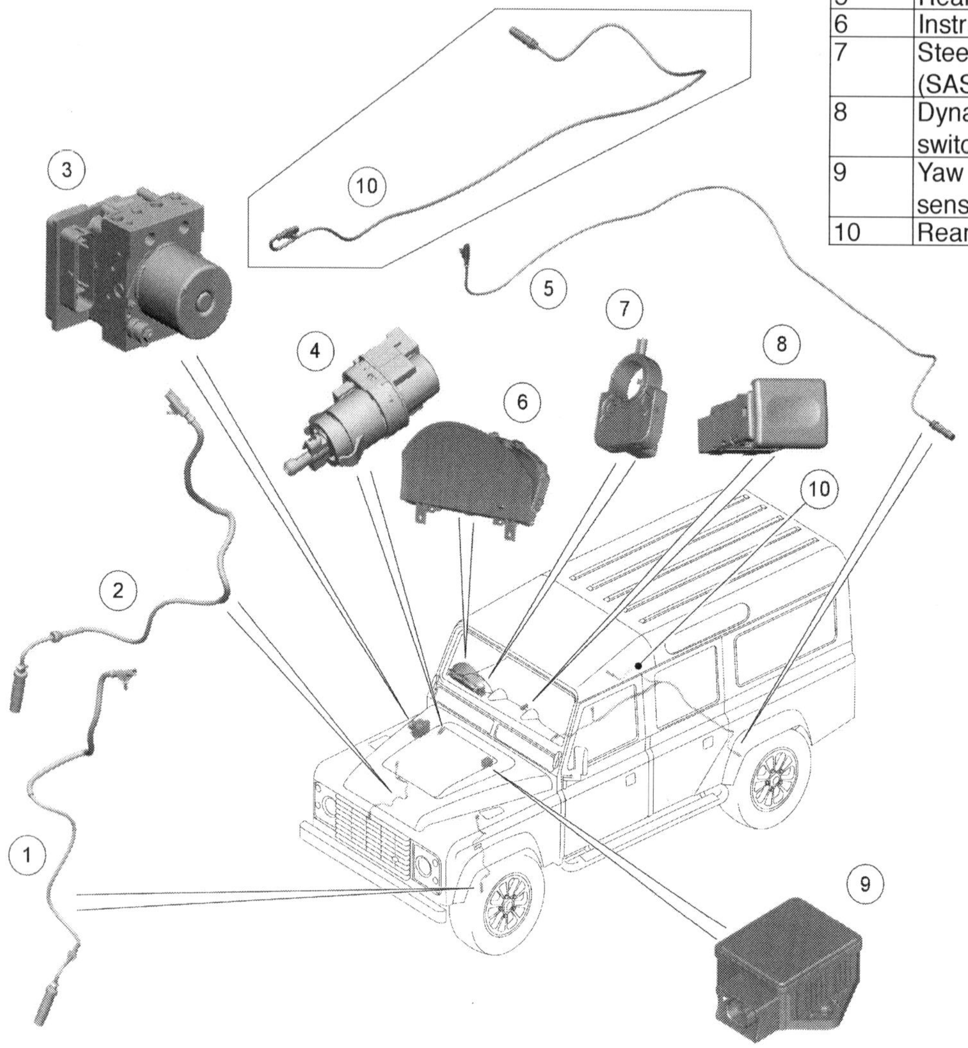

Item	Description
1	Front left wheel speed sensor
2	Front right wheel-speed sensor
3	Anti-lock Brake System (ABS) control module
4	Brake pedal switch
5	Rear left wheel speed sensor
6	Instrument Cluster (IC)
7	Steering Angle Sensor Module (SASM)
8	Dynamic Stability Control (DSC) switch
9	Yaw rate Lateral acceleration sensor
10	Rear right wheel speed sensor

OVERVIEW

The anti-lock control and traction control system is based on the 4 channel Bosch 8.0 system and provides the following brake functions:

- ABS (Anti-lock Brake System)
- EBD (Electronic Brake Force Distribution)
- ETC (Electronic Traction Control)
- DSC (Dynamic Stability Control).

The system consists of the following components:

- An integrated HCU (Hydraulic Control Unit) and ABS (Anti-lock Brake System) control module
- A Yaw rate Lateral acceleration sensor
- A SASM (Steering Angle Sensor Module)
- Four wheel speed sensors
- A brake pedal switch
- Four warning indicators in the IC (Instrument Cluster).

COMPONENT DESCRIPTION

BRAKE PEDAL SWITCH

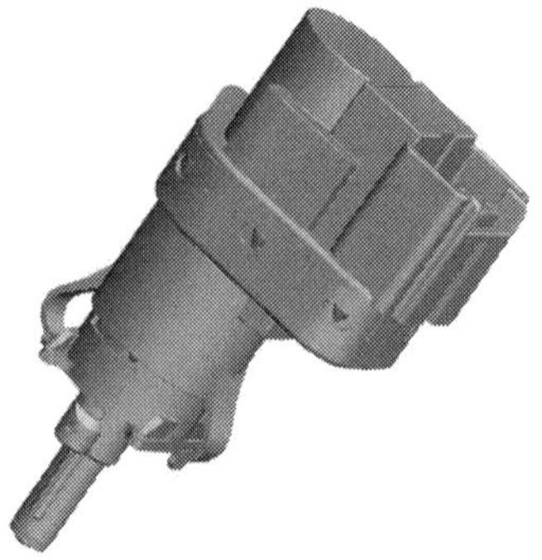

E125620

The brake pedal switch is a two-pole switch, mounted in the brake pedal bracket and operated by the brake pedal:

- The 'brake test switch' pole supplies a brake pedal status signal to the ECM (Engine Control Module).
- The 'stop lamp switch' pole operates the stop lamps and supplies a brake pedal status signal to the ABS control module and the ECM.

When the brake pedal is released:

- The 'brake test switch' contacts are closed, connecting an ignition power feed from the CJB (Central Junction Box) to the ECM only.
- The 'stop lamp switch' contacts are open.

When the brake pedal is pressed:

- The 'brake test switch' contacts open.
- The 'stop lamp switch' contacts close, and connects an ignition power feed from the CJB to the ABS and ECM.

DYNAMIC STABILITY CONTROL (DSC) SWITCH

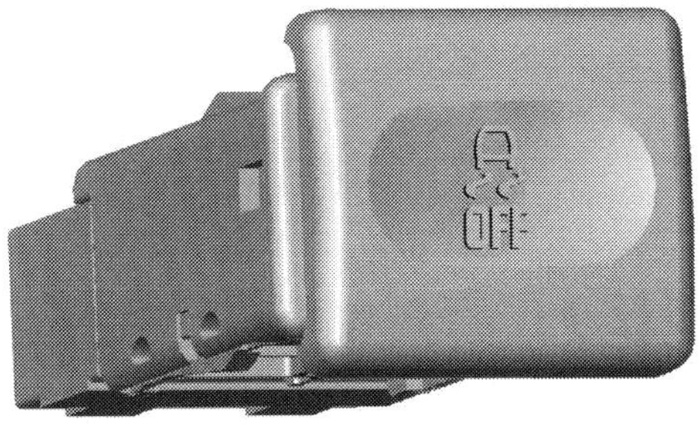

The DSC switch is a momentary switch installed in the center console. Operating the DSC switch connects a ground to the ABS control module. With the first press of the DSC switch, the ABS control module disables the DSC functions. When the DSC switch is operated again, the ABS control module re-enables the DSC functions. The DSC switch must be operated for a minimum of 0.3 second for the ABS control module to react. The DSC function is re-enabled at the beginning of each ignition cycle.

To confirm that the DSC function is selected off, the ABS control module send a message to the IC via the HS (High Speed) CAN (Controller Area Network) bus, and the amber colored 'DSC OFF' warning indicator is continuously illuminated in the IC. When the DSC function is selected on again, the 'DSC OFF' warning indicator will extinguish.

If a fault is detected with the DSC switch, the ABS control module enables the DSC function and any switch requests are ignored.

WHEEL SPEED SENSORS

A passive wheel speed sensor is installed in each wheel hub to provide the ABS module with a rotational speed signal from each road wheel. A lead connects each sensor to the vehicle wiring.

Each wheel speed sensor has a signal connection with the ABS module. The ABS module monitors the signals at vehicle speeds of 8 km/h (5 mph) and above, the rotation of the Half shafts induces the signals which are converted into individual wheel-speeds and the overall vehicle speed by the ABS module.

The ABS module monitors the wheel-speed sensor for faults. If a fault is detected the ABS module stores a related fault code and illuminates the appropriate warning indicators, depending on the system functions affected to either: ETC, ABS, EBD, DSC.

Refer to: Brake System (206-00 Brake System - General Information, Description and Operation)

YAW RATE AND LATERAL ACCELERATION SENSOR

The yaw rate and lateral acceleration sensor is located below the climate control assembly and secured to the floor tunnel with two bolts.

The yaw rate and lateral acceleration sensor receives an ignition power supply from the CJB. The sensor measures the yaw rate and lateral acceleration of the vehicle and providing values to the ABS control module via a Private CAN (Controller Area Network) bus. If a sensor fault is detected, the ABS control module stores a related fault code and illuminates the ABS warning indicator on the IC. The fault codes can be retrieved using Land Rover approved diagnostic equipment.

STEERING ANGLE SENSOR MODULE (SASM)

SASM (Steering Angle Sensor Module) is located under the steering column cowl and secured to the turn signal indicator switch with two clips.

A fly lead connects the SASM to the passenger compartment wiring harness via a four pin multiplug.

The SASM measures the steering wheel angle and the rate of change of the steering wheel angle. These measurements are received by the ABS control module via a Private CAN bus.

The center (straight ahead) position of the steering wheel has to be learned by the ABS control module every time the ignition is switched ON.

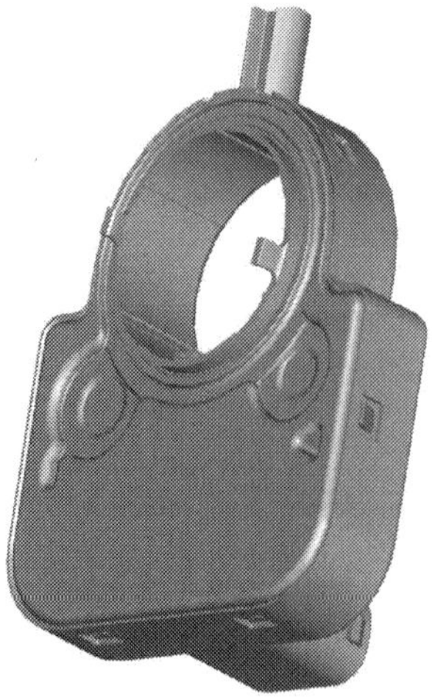

This process begins when the vehicle starts to move. The SASM is unable to determine the center position so inputs from the yaw rate and lateral acceleration sensor and wheel speed signals are also used by the ABS control module to help it perform this process. If extreme conditions are present, for example ice causing extreme wheel spin or understeer/oversteer, the ABS control module may not be able to determine the center position of the steering wheel. In this situation the amber DSC warning indicator will illuminate. The ABS control module will continue to try to learn the straight ahead position and will extinguish the warning once it has done so.

DIFFERENTIAL LOCK UNIT

The ABS control module monitors the state of the differential-lock unit (engaged or disengaged) and changes the strategy of calculating the vehicle reference speed. When the differential-lock is engaged the front and rear axles are driven at the same speed and therefore when cornering the front axle will be traveling faster than if the differential-lock was disengaged. The vehicle reference speed is calculated using the yaw rate and lateral acceleration values in conjunction with the actual wheel speed. The IC receives the differential-lock unit state signal via a hardwired connection and sends a message to the ABS via the HS (High Speed) CAN bus.

If the ABS did not change the strategy of calculating the reference speed it would assume that the reference speed calculated with the differential-lock engaged, as incorrect and consequently a fault code would be stored.

INSTRUMENT CLUSTER (IC) WARNING INDICATOR LAMPS

The IC (Instrument Cluster) contains warning indicators to display the operating status of the Anti-lock control and Stability assist functions. The warning indicators provide a visual notification of either a system warning or information indication to the driver.

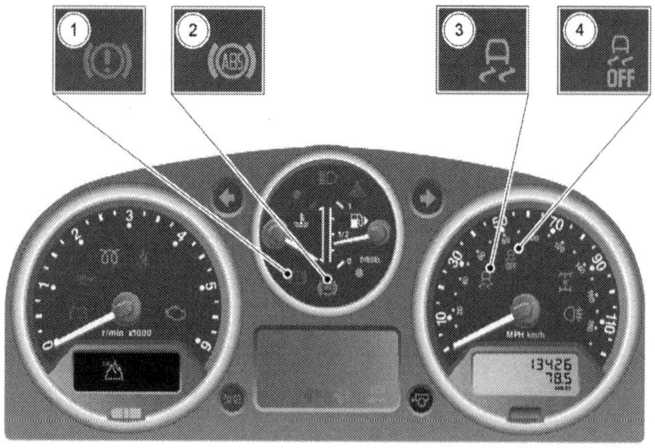

Item	Description
1	Brake warning indicator
2	Anti-lock Brake System (ABS) warning indicator
3	Dynamic Stability Control (DSC) warning indicator
4	Dynamic Stability Control (DSC) OFF indicator

Brake Warning Indicator (Red)

When the ignition switch is first turned to 'ON' position, the brake warning indicator illuminates for approximately 3 seconds for a bulb check.

The brake warning indicator is shared by three brake systems:

- **Parking brake:** illuminates when the parking brake is applied and while the ignition is ON, it will extinguish when the parking brake is fully released.
- **Brake fluid level:** illuminates when the brake-fluid level is low.
- **EBD fault:** illuminates when a fault is detected with EBD.

Operation of the brake warning indicator is controlled by a hardwired connection from the parking brake switch and the brake fluid reservoir cap. The ABS control module is connected to the IC via the HS CAN bus.

ABS Warning Indicator (Amber)

The ABS indicator is amber colored and is continuously illuminated if there is a fault that affects ABS performance or causes the ABS function to be disabled.

Operation of the ABS warning indicator is controlled by the ABS control module via the HS CAN bus.

When the ignition switch is first turned to 'ON' position, the ABS warning indicator illuminates for approximately 3 seconds for a bulb check.

Dynamic Stability Control (DSC) Warning Indicator (Amber)

When the ignition switch is first turned to 'ON' position, the DSC indicator illuminates for approximately 3 seconds for a bulb check.

Operation of the DSC warning indicator is controlled by the ABS control module via the HS CAN bus.

Illuminates for minimal periods when the DSC function is active; continuous illumination indicates a fault.

Dynamic Stability Control (DSC) OFF Indicator (Amber)

When the ignition switch is first turned to 'ON' position, the DSC OFF indicator illuminates for approximately 3 seconds for a bulb check.

Operation of the DSC OFF indicator is controlled by the ABS control module via the HS CAN bus.

Illuminates continuously when the DSC function is disabled.

HYDRAULIC CONTROL UNIT (HCU)

Item	Description
1	Anti-lock Brake System (ABS) control module electrical connector
2	Outlet port to Left front brake caliper
3	Outlet port to Right rear brake caliper
4	Outlet port to Left rear brake caliper
5	Outlet port to Right front brake caliper
6	Primary circuit inlet port
7	Secondary circuit inlet port

The HCU (Hydraulic Control Unit) is a 4 channel unit that modulates the supply of hydraulic pressure to the front and rear brakes under the control of the ABS control module.

The HCU is attached by three mounting bushes to a bracket on the driver-side of the engine compartment. Hydraulic pipes connect the HCU to the master cylinder and the brake calipers.

The primary and secondary outlets of the master cylinder are connected to primary and secondary circuits within the HCU. The primary circuit in the HCU has separate outlet ports to the front brakes. The secondary circuit in the HCU has separate outlet ports to the rear brakes. Each of the circuits in the HCU contains the following components to control the supply of hydraulic pressure to the brakes:

- A normally open, solenoid operated pilot valve, to enable passive braking.
- A normally closed, solenoid operated priming valve, to connect the brake fluid reservoir to the return pump during active braking.
- A return pump, to generate hydraulic pressure for active braking and return brake fluid to the reservoir.
- Normally open, solenoid operated inlet valves and normally closed, solenoid operated outlet valves to modulate the hydraulic pressure in the individual brakes.
- Filters, to protect the components from contamination.

The primary circuit also incorporates a pressure sensor to provide the ABS control module with a hydraulic pressure signal.

Contact pins on the HCU mate with contacts on the ABS control module to provide the electrical connections from the ABS control module to the return pump motor and the pressure sensor. The solenoids that operate the valves are installed in the ABS control module.

The replacement HCU is supplied pre-filled. After installation on the vehicle, Land Rover approved diagnostic equipment must be used to operate the solenoid valves and the return pump to ensure correct bleeding of the HCU and brake circuits.

The HCU has three operating modes:

- Normal braking / EBD mode
- ABS braking mode
- Active braking mode

Normal Braking / EBD Mode

Initially, all of the solenoid operated valves are de-energized. Operating the brake pedal produces a corresponding increase or decrease of pressure in the brakes, through the open pilot valves and inlet valves. If the ABS control module determines that EBD is necessary, it energizes the inlet valves for the brakes of the trailing axle, to isolate the rear brakes from any further increase in hydraulic pressure.

ABS Braking Mode

If the ABS module determines that ABS braking is necessary, it energizes the inlet and outlet valves of the related brake and starts the return pump. The inlet valve closes to isolate the brake from pressurized fluid; the outlet valve opens to release pressure from the brake into the accumulator and the return pump circuit; the reduced pressure allows the wheel to accelerate. The ABS module then operates the inlet and outlet valves to modulate the pressure in the brake to apply the maximum braking effort without locking the wheel. Control of the valves for each wheel takes place individually.

Schematic of the Hydraulic Control Unit

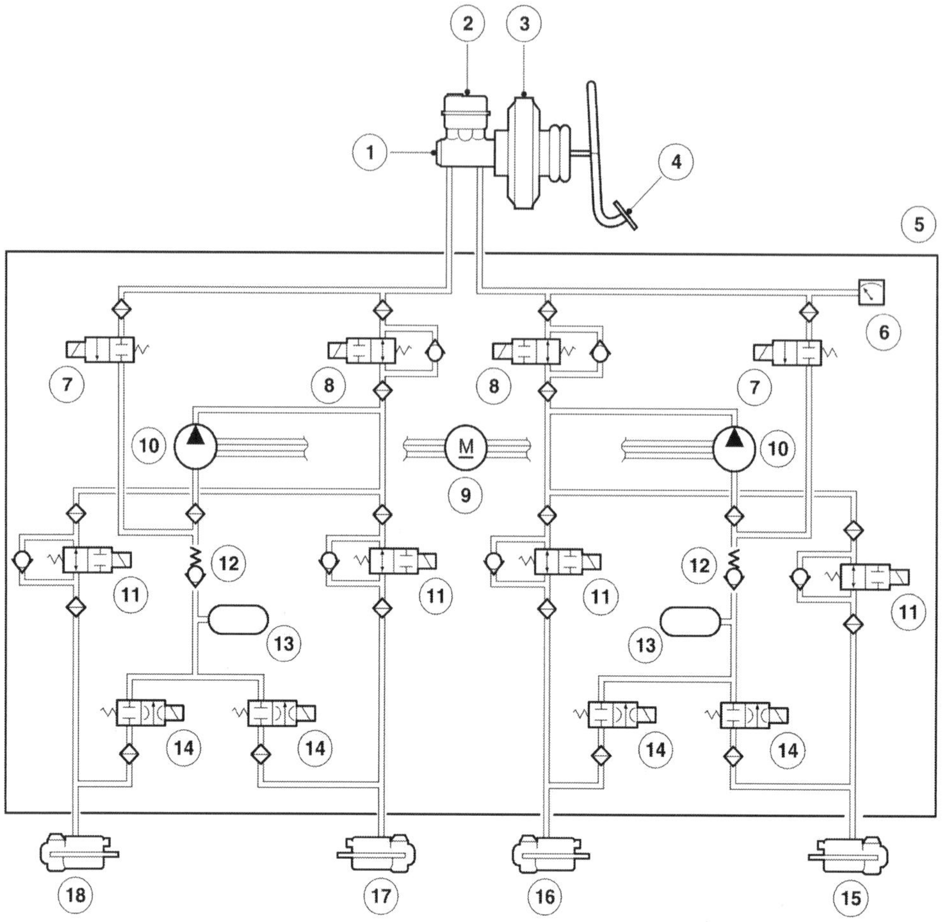

Item	Description
1	Brake master cylinder
2	Brake fluid reservoir
3	Brake booster
4	Brake pedal
5	Hydraulic Control Unit (HCU)
6	Pressure sensor
7	Priming valve
8	Pilot valve
9	Return pump motor
10	Return pump
11	Inlet valve
12	Relief valve
13	Accumulator
14	Outlet valve
15	Front left brake caliper
16	Front right brake caliper
17	Rear right brake caliper
18	Rear left brake caliper

Active Braking Mode

The active braking mode is used to generate and control hydraulic pressure to the brakes for ETC.

For active braking, the ABS module energizes the pilot valves and priming valves, starts the return pump and energizes all of the inlet valves. Brake fluid, drawn from the reservoir through the master cylinder and priming valve, is pressurized by the return pump and supplied to the inlet valves. The ABS module then operates the inlet valves and outlet valves, as required, to modulate the pressure in the individual brakes. Some noise may be generated during active braking.

ABS CONTROL MODULE

The ABS control module controls the brake functions using the HCU to modulate hydraulic pressure to the individual wheel brakes.

The ABS control module is attached to the HCU on the driver-side of the engine compartment. A 38 pin connector provides the electrical interface between the ABS control module and the vehicle wiring.

SYSTEM OPERATION

Anti-lock Brake System (ABS)

ABS controls the speed of all road wheels to ensure optimum wheel-slip when braking at the adhesion limit. This prevents the wheels from locking, which helps to retain effective steering control of the vehicle.

Electronic Brake Force Distribution (EBD)

EBD limits the brake pressure applied to the rear wheels. When the brakes are applied, the weight of the vehicle transfers forwards, which reduces the ability of the rear wheels to transfer braking effort to the road surface. This can cause the rear wheels to slip and make the vehicle unstable.

EBD uses the anti-lock braking hardware to automatically optimize the pressure of the rear brakes, below the point where anti-lock braking would be invoked. Only the rear axle is under EBD control.

Electronic Traction Control (ETC)

ETC attempts to optimize forward traction by braking a spinning wheel until it regains grip.

ETC is activated if an individual wheel speed is above that of the vehicle reference speed (positive slip) and the brake pedal is not pressed. The brake is applied to the spinning wheel, allowing the excess torque to be transmitted to the non-spinning wheels through the drive line.

Dynamic Stability Control (DSC)

DSC uses brakes and powertrain torque control to assist in maintaining the lateral stability of the vehicle. While the ignition is energized the DSC function is permanently enabled, unless selected off using the DSC switch.

DSC enhances driving safety in abrupt maneuvers and in under-steer or over-steer situations that may occur in a bend. The ABS control module monitors the yaw rate and lateral acceleration of the vehicle and the steering input, then selectively applies individual brakes and signals for powertrain torque adjustments to reduce under-steer or over-steer. In general:

- In an under-steer situation; the inner wheels are braked to counteract the yaw movement towards the outer edge of the bend.
- In an over-steer situation; the outer wheels are braked to prevent the rear end of the vehicle from pushing towards the outer edge of the bend.

The ABS control module monitors the tracking stability of the vehicle using inputs from the wheel speed sensors, the SASM (Steering Angle Sensor Module), and the yaw rate and lateral acceleration sensor. The tracking stability is compared with stored target data. Whenever the tracking stability deviates from the target data, the ABS control module intervenes by applying the appropriate brakes. The following interactions occur in an intervention situation:

- HS CAN signal to the ECM, to reduce engine torque.
- Application of braking to the appropriate corner of the vehicle.

CONTROL DIAGRAM - ANTI-LOCK CONTROL

⚠ NOTE: **A** = Hardwired; **D** = HS (High Speed) CAN (Controller Area Network) bus; **U** = Private CAN (Controller Area Network) bus.

Item	Description
1	Battery
2	Battery Junction Box (BJB)
3	Ignition switch
4	Central Junction Box (CJB)
5	Front left wheel-speed sensor
6	Front right wheel-speed sensor
7	Rear right wheel-speed sensor
8	Rear left wheel-speed sensor
9	Dynamic Stability Control (DSC) switch
10	Brake pedal switch
11	Engine Control Module (ECM)
12	Differential-lock switch
13	Handbrake switch
14	Brake fluid level switch
15	Anti-lock Brake System (ABS) control module
16	Steering Angle Sensor Module (SASM)
17	Yaw rate and lateral acceleration sensor
18	Instrument Cluster (IC)

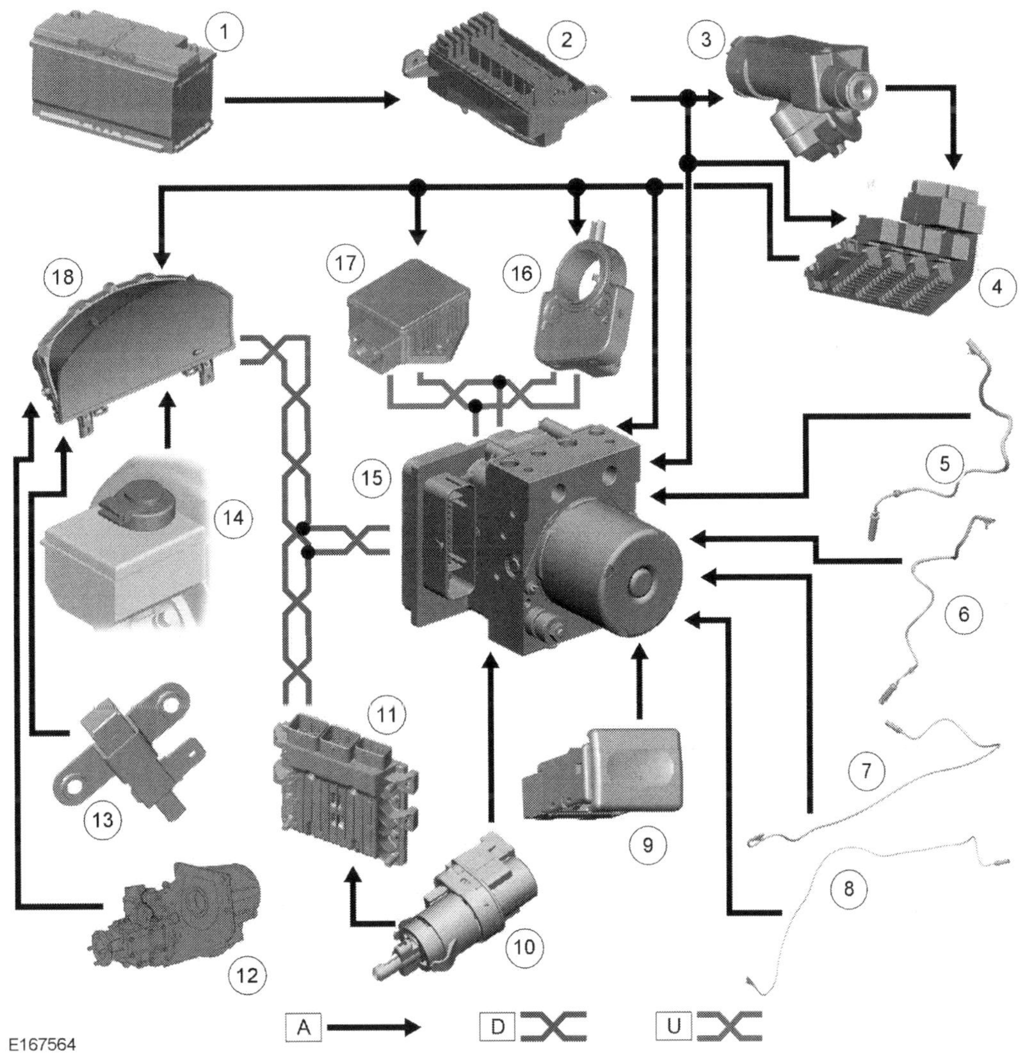

Master cylinder component

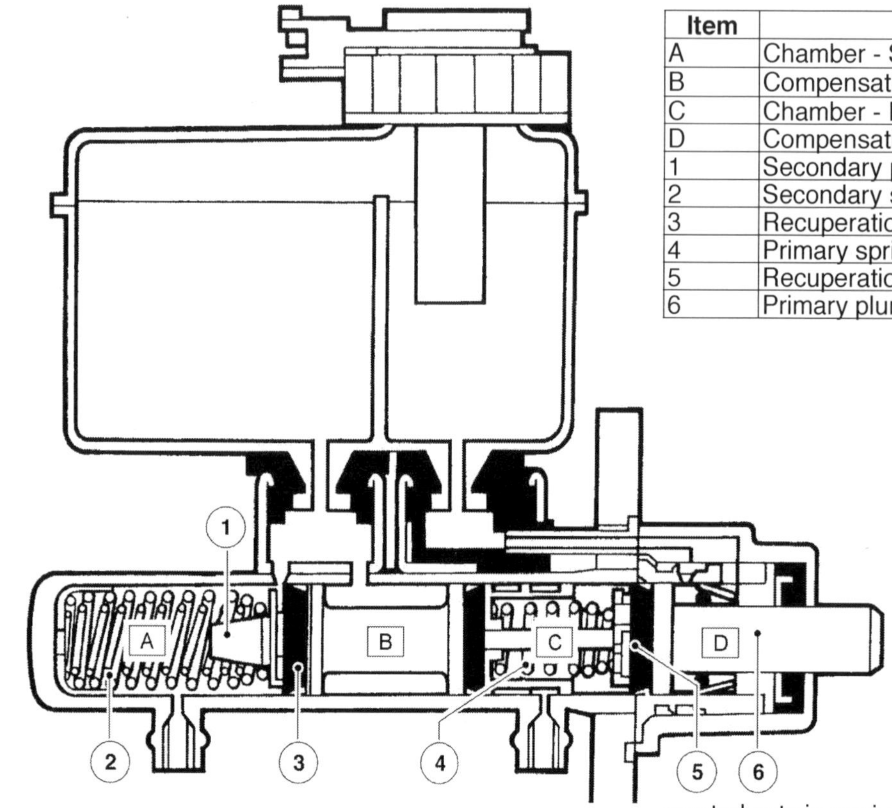

Item	Description
A	Chamber - Secondary circuit
B	Compensation chamber
C	Chamber - Primary circuit
D	Compensation chamber
1	Secondary plunger
2	Secondary spring
3	Recuperation seal
4	Primary spring
5	Recuperation seal
6	Primary plunger

BASE BRAKING SYSTEM

Master cylinder description

The mechanical components of the hydraulic braking system consists of four piston caliper disc brakes at the front and two piston caliper disc brakes at the rear.

Vented front brake discs are fitted as standard on 110/130 models, while 90 models have solid discs. However, on 90 models with a heavy duty chassis, vented front discs may also be fitted.

A cable controlled parking brake operates a single drum brake mounted on the output shaft of the transfer gearbox and is completely independent of the main braking system.

The basic hydraulic system involves 2 separate and independent primary and secondary circuits which permits a degree of braking should a fault occur in one of the circuits. The primary circuit operates the rear brake calipers and the secondary circuit the front brake calipers.

Master cylinder operation

A tandem master cylinder, which is assisted by a light weight, short, compact brake booster, is fed by an internally divided brake fluid reservoir. The rear section supplies the brake fluid for the primary circuit and the front section supplies the brake fluid for the secondary circuit. When the brake pedal is not pressed the fluid can move unrestricted between the primary- and the secondary circuit.

When the brake pedal is applied, the primary plunger assembly moves up the cylinder bore and the pressure created acts in conjunction with the primary spring to overcome the secondary springs, thus moving the secondary plunger assembly up the bore. At the same time initial movement of both plungers takes the recuperating seals past the cut-off holes in the Primary- and Secondary chambers and applies pressure to the fluid in those chambers, which is directed to the respective circuits.

The fluid in the compensation chambers is unaffected by movement of the plungers and can move unrestricted between the separate chambers and respective reservoirs in the brake fluid reservoir, both before and during brake application. When the brakes are released, the plunger assemblies, aided by the return springs are retracted faster than the brake fluid; this creates a depression between the brake fluid in the Primary- and Secondary chambers and the recuperation seals.

The recuperation seals momentarily collapse allowing fluid in the compensation chambers to flow through the holes in the plungers, over the collapsed seals and into the Primary- and Secondary chambers respectively. The movement of fluid from one set of chambers to the other is compensated for by fluid from the separate the other is compensated for by fluid from the separate reservoirs in the brake fluid reservoir moving through the feed holes in the cylinder. Conversely, the final return movement of the plunger assemblies causes the extra fluid in the Primary- and Secondary chambers to move through the cut off holes into the brake fluid reservoir.

The brake booster provides controlled power assistance to the brake pedal when pressure is applied. Power is obtained from the brake vacuum pump, located on the right side of the engine cylinder block. The vacuum is applied to both sides of a flexing diaphragm, and by admitting atmospheric pressure to the rear diaphragm, assistance is obtained. The brake booster is mounted between the brake pedal and master cylinder and is linked to these by push rods. If a failure occurs in the brake vacuum system, the two push rods will act as a single rod allowing the brakes to function in the normal way, although more effort will be required to operate the brake pedal.

The brake system should be drained and flushed at the recommended service intervals.

On 90 models a pressure reducing valve fitted to the right-hand bulkhead in the engine compartment, maintains the braking balance. Pressure to the rear calipers is regulated by the pressure reducing valve. This valve is of the failure by-pass type, allowing full system pressure to the rear brake calipers in the event of a front (secondary) circuit failure.

Master cylinder overhaul components

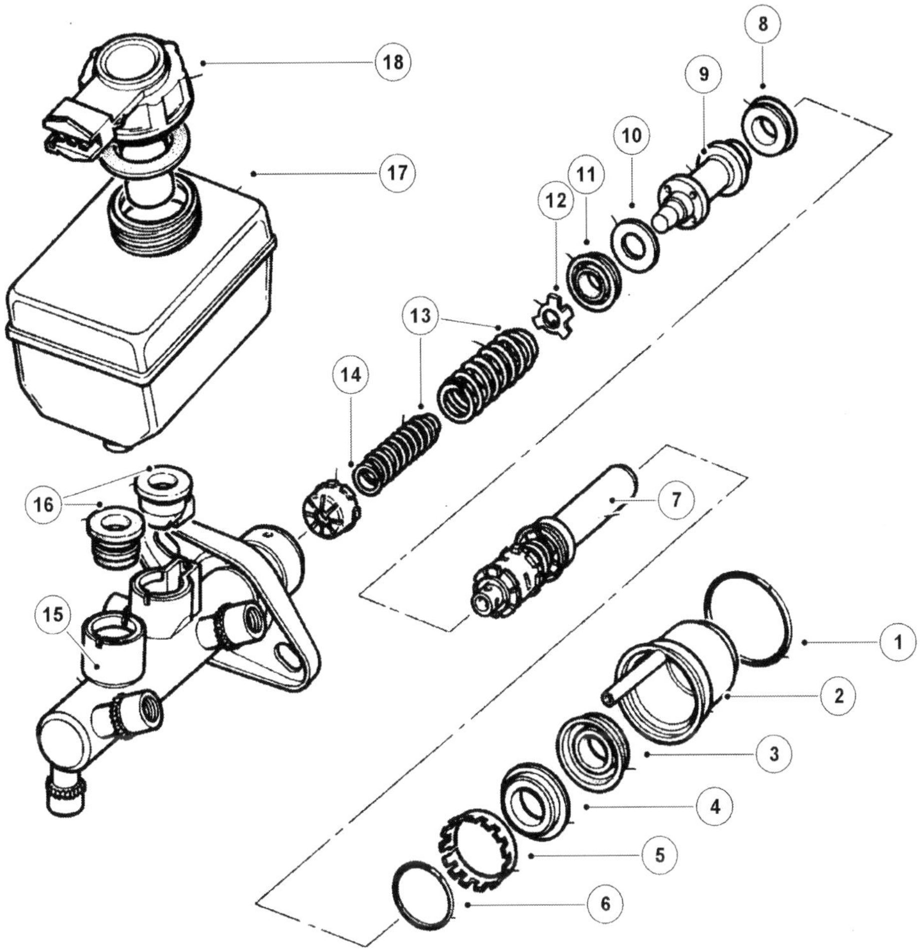

Item	Description
1	Water ingress seal
2	Transfer housing
3	Vacuum seal
4	Guide ring
5	Retaining ring
6	O-ring seal
7	Primary plunger assembly
8	'L' seal
9	Secondary plunger
10	Washer
11	Recuperating seal (primary cup)
12	Seal retainer
13	Springs
14	Swirl tube
15	Master cylinder body
16	Reservoir seals
17	Reservoir
18	Low fluid level switch and cap

Hydraulic system

Hydraulic system schematic - 90 Models

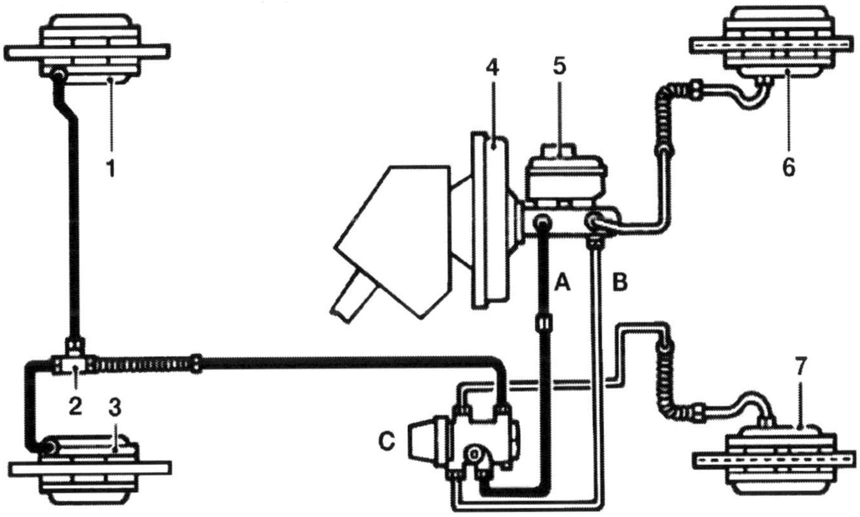

Item	Description
A	Primary circuit
B	Secondary circuit
C	Pressure reducing valve
1	Left rear brake caliper
2	'T' connector
3	Right rear brake caliper
4	Brake booster
5	Brake master cylinder and reservoir
6	Left front brake caliper
7	Right front brake caliper

Hydraulic system schematic - 110/130 Models

△ NOTE: In some countries, a pressure reducing valve may be fitted to 110 models to conform to legal requirements.

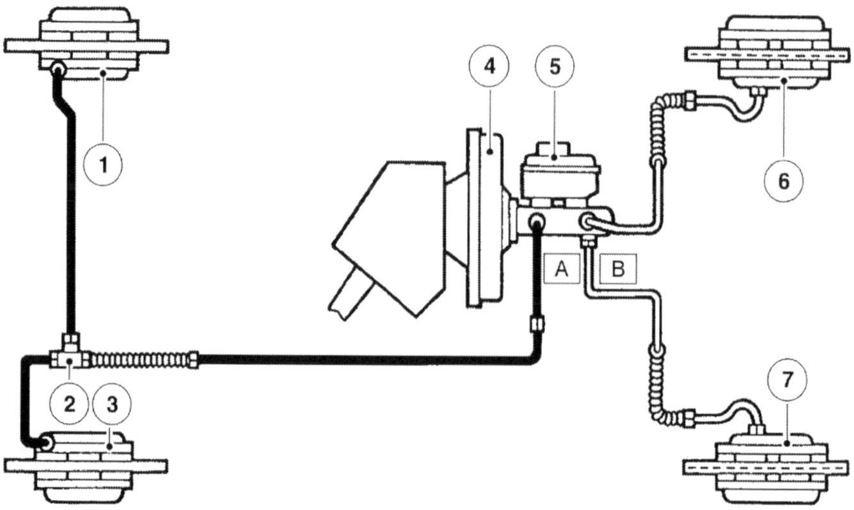

Item	Description
A	Primary circuit
B	Secondary circuit
1	Left rear brake caliper
2	'T' connector
3	Right rear brake caliper
4	Brake booster
5	Brake master cylinder and reservoir
6	Left front brake caliper
7	Right front brake caliper

Front brake caliper assembly

Front brake caliper components

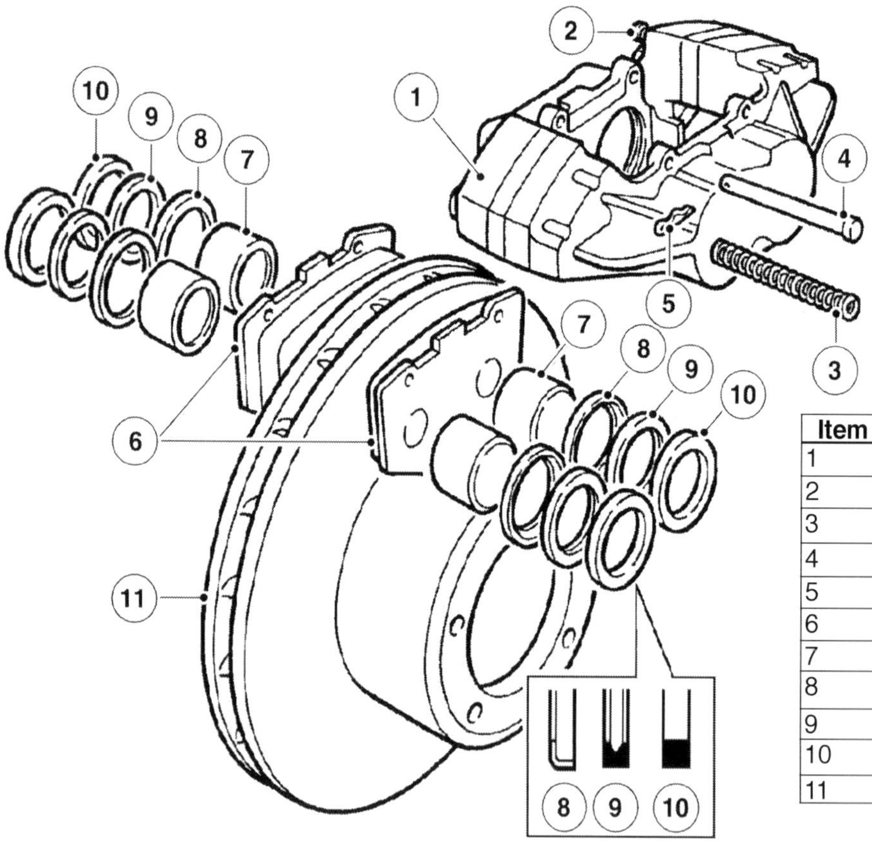

Item	Description
1	Caliper
2	Bleed screw
3	Anti-rattle spring (2 off)
4	Pad retaining pin (2 off)
5	Split pin (2 off)
6	Friction pads
7	Piston
8	Wiper seal retainer
9	Wiper seal
10	Fluid seal
11	Brake disc

Modifications to the rear brakes have been introduced on 110 and 130 models. This is to increase the service life of the rear brake pads in harsh environments.

New rear brake calipers are introduced which allow the fitment of brake pads with a thicker friction material. On the 90 model rear brake pads has a chamfer on leading edge only. On 110 and 130 models the brake pads have increased leading and trailing edge chamfers which improve efficiency.

Aerodynamic deflectors are located forward of the rear wheels to protect the rear calipers from the ingress of dirt, which can cause excessive pad wear. The deflectors are fitted to 90, 10 and 130 models in markets where environmental conditions can cause excessive brake pad wear.

Rear brake caliper assembly

Rear brake caliper components

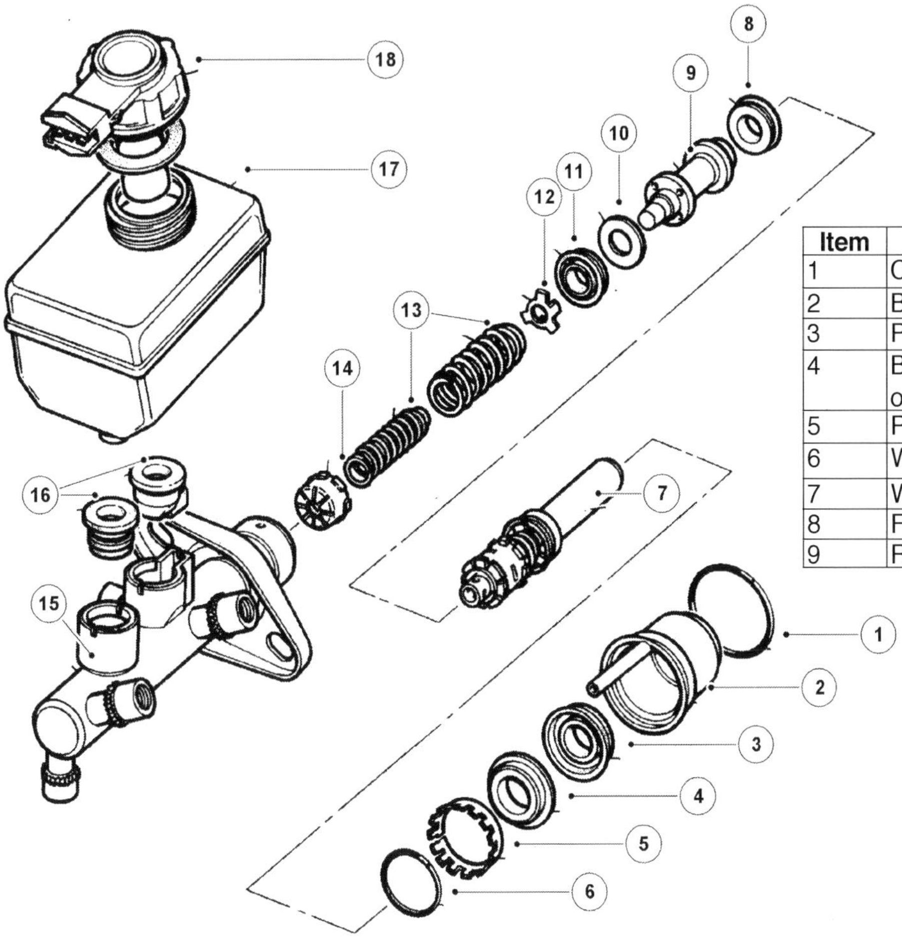

Item	Description
1	Caliper
2	Bleed screw
3	Pad retaining springs
4	Brake pads (for reference only)
5	Piston
6	Wiper seal retainer
7	Wiper seal
8	Fluid seal
9	Retraining pins

Rear brake caliper installation

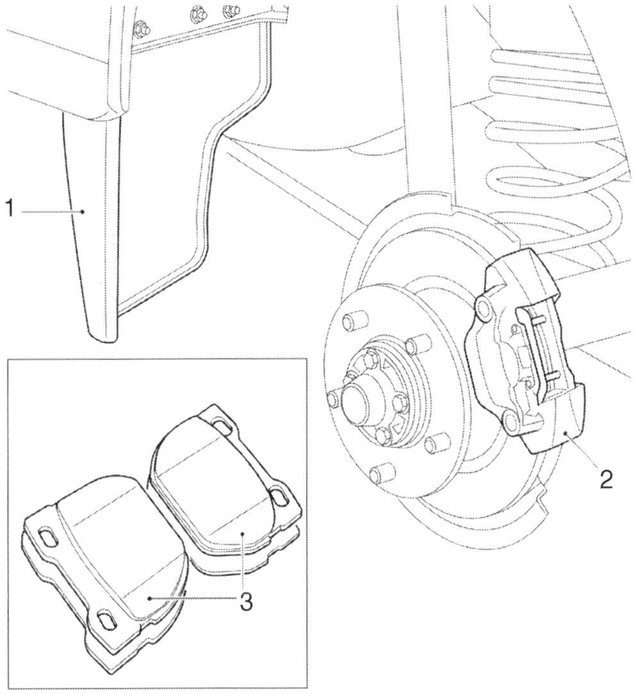

Item	Description
1	Deflector
2	Rear brake caliper
3	Rear brake pads (110 and 130 models)

Brake System Bleeding

General Procedures

WARNING: Do not allow brake fluid to come into contact with eyes or skin.

CAUTIONS:

Brake fluid can damage paintwork, if spilled wash off immediately with plenty of clean water.

Use only correct grade of brake fluid. If an assembly fluid is required use ONLY brake fluid. Do NOT use mineral oil, i.e. engine oil etc.

Thoroughly clean all brake calipers, pipes and fittings before commencing work on any part of the brake system. Failure to do so could cause foreign matter to enter the system and cause damage to seals and pistons which will seriously impair the efficiency of the brake system. To ensure the brake system efficiency is not impaired the following warnings must be adhered to :-

Parking brake

Parking brake components

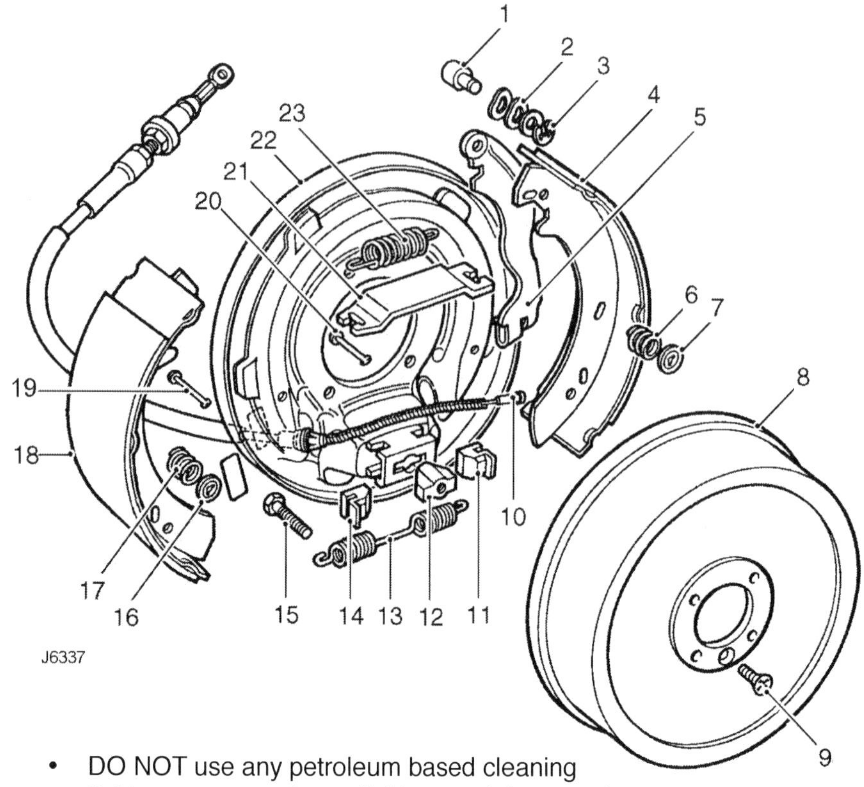

Item	Description
1	Pin
2	Washer
3	'C' clip
4	Brake shoe
5	Cable lever
6	Hold down spring
7	Dished washer
8	Brake drum
9	Screw
10	Brake cable
11	Adjuster slide
12	Adjuster nut
13	Spring
14	Adjuster slide
15	Adjuster bolt
16	Dished washer
17	Hold down spring
18	Brake shoe
19	Hold down pin
20	Hold down pin
21	Abutment plate
22	Back plate
23	Spring

- DO NOT use any petroleum based cleaning fluids or any proprietary fluids containing petrol.
- DO NOT use brake fluid previously bled from the system.
- DO NOT flush the brake system with any fluid other than the recommended brake fluid. The brake system should be drained and flushed at the recommended service intervals. Cover all electrical terminals carefully to make absolutely certain that no fluid enters the terminals and plugs.
- Cover all electrical terminals carefully to make absolutely certain that no fluid enters the terminals and plugs.
- During bleed procedure, brake fluid level must not be allowed to fall below the MIN mark. Keep reservoir topped up to the MAX mark.
- To bleed the hydraulic circuits, four bleed nipples are provided, one at each caliper.

1. There are two methods by which air can be removed from the braking system:
    1. MANUAL BLEED PROCEDURE
    2. PRESSURE BLEED PROCEDURE
2. **Pressure bleed procedure**
    1. Purpose designed equipment for pressure filling and bleeding of hydraulic systems may be used on Land Rover vehicles. The equipment manufacturers instructions must be followed and the pressure must not exceed 4.5 bar, 65 lb/in.
3. **Manual bleed procedure**

    Equipment required
    1. Clean glass receptacle
    2. Bleed hose
    3. Wrench
    4. Approx. 2 litres (3 pts) brake fluid
4. **Master cylinder bleed**
    1. Disconnect battery negative lead
    2. Depress brake pedal fully and slowly 5 times.
    3. Release pedal and wait 10 seconds.
    4. Repeat until firm resistance is felt at the pedal.
5. **Complete circuit bleed**
    1. Disconnect battery negative lead
    2. Fit bleed hose to caliper bleed screw.
    3. Dip free end of bleed hose into brake fluid in bleed bottle.
    4. Open bleed nipple.
    5. Depress brake pedal fully several times until fluid is clear of air bubbles.
    6. Keeping pedal fully depresses, tighten bleed nipple then release pedal.
    7. Repeat procedure for remaining calipers.
    8. Fit bleed screw protection caps.
    9. Check/top-up fluid level when bleeding is complete.

Description and Operation

Front Disc Brake

Refer to: Brake System (206-00 Brake System - General Information, Description and Operation)

Removal and Installation

Removal

1. Remove hub assembly.

Refer to: Front Wheel Bearing and Wheel Hub (204-01 Front Suspension, Removal and Installation)

2. Remove 5 hub to disc bolts.

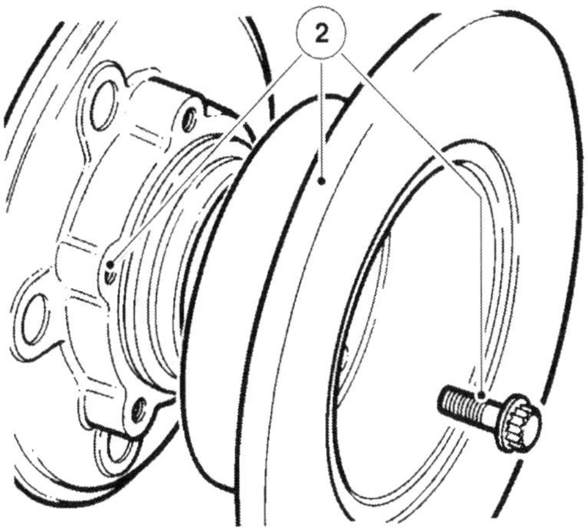

3. NOTE: On 110/130 vehicles ventilated discs are fitted as standard.

Tap disc to separate from hub.

Installation

1. Locate disc to hub.
2. Apply Loctite 270 to disc bolts and tighten to 73 Nm (54 lbf.ft).
3. Check total disc runout with a dial indicator, this must not exceed 0,15 mm, 0.006 in. If necessary reposition disc.
4. Install hub assembly.

Refer to: Front Wheel Bearing and Wheel Hub (204-01 Front Suspension, Removal and Installation)

5. Disc reclamation
6. NOTE: The disc MUST BE renewed if the minimum running thickness stamped on the disc is recorded.

Check disc thickness. This dimension may be machined to minimum thickness of 12 mm (0.47 in.) - solid discs, 22 mm (0.90 in) - ventilated discs. Machine equal amount off each face.

Front Disc Brake - Brake Pads

Removal and Installation

Removal

1. Remove front road wheels.
2. Clean exterior of calipers.

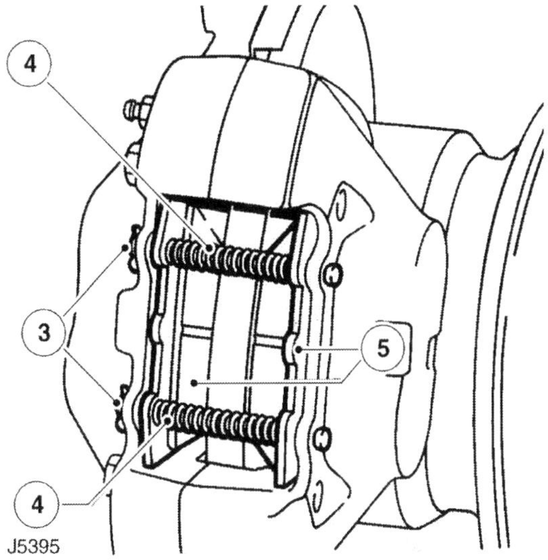

3. Remove split pin from retaining pins.
4. Remove pad retaining pins and anti-rattle springs.
5. Remove brake pads.
6. Clean exposed parts of pistons, using new brake fluid. Wipe away excess with a lint free cloth.

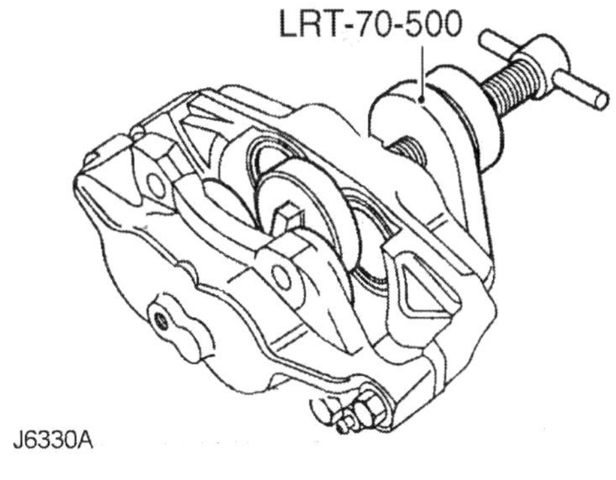

7. Using piston clamp LRT-70-500 press each piston back into its bore. Ensure that displaced brake fluid does not overflow from reservoir.

Installation

1. Install brake pads.

NOTE: Ensure pads are correctly fitted, with leading edge chamfer fitted as shown.

2. Install pad retaining pins and anti-rattle springs. Secure with new split pins.
3. Apply brake pedal several times to locate pads.
4. Install road wheels, remove chassis stands and jack. Tighten wheel nuts to 130 Nm (96 lbf.ft).

5. Check fluid reservoir. Top up if necessary, using correct grade of fluid.

Refer to: (206-00 Brake System - General Information, Specifications)

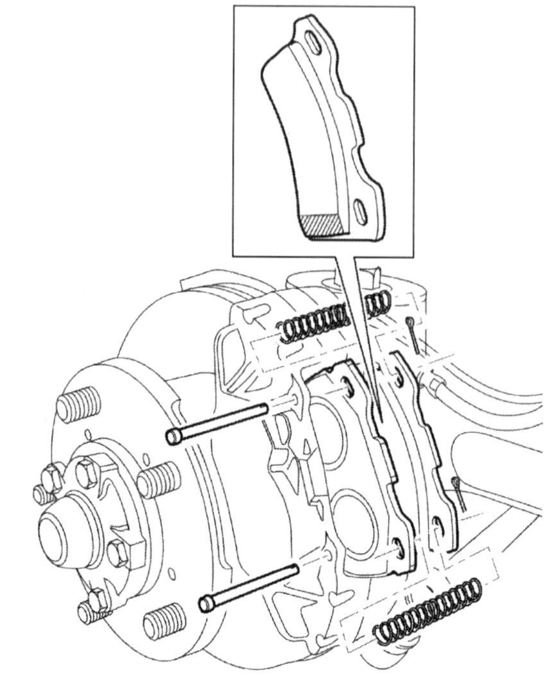

Front Disc Brake - Brake Caliper

Removal and Installation

Removal

1. Loosen front road wheel nuts.
2. ⚠ WARNING: Support on safety stands.

   Raise front of vehicle.
3. Remove road wheel.
4. Remove split pins from pad retaining pins.
5. Remove brake pad retainers and springs.
6. Remove brake pads and shim if fitted.
7. Pull back brake hose shield.
8. Clamp brake hose using a suitable clamp.
9. Position container beneath brake caliper to collect spillage.
10. Loosen brake pipe union and disconnect from caliper.
11. Plug brake pipe and caliper connections.
12. Remove 2 bolts securing caliper.
13. Remove brake caliper.

⚠ WARNING: Do not separate caliper halves

14. Clean outer surfaces of caliper using aerosol brake cleaner.
15. Using special tool LRT-70-500, clamp pistons in inboard half of caliper.

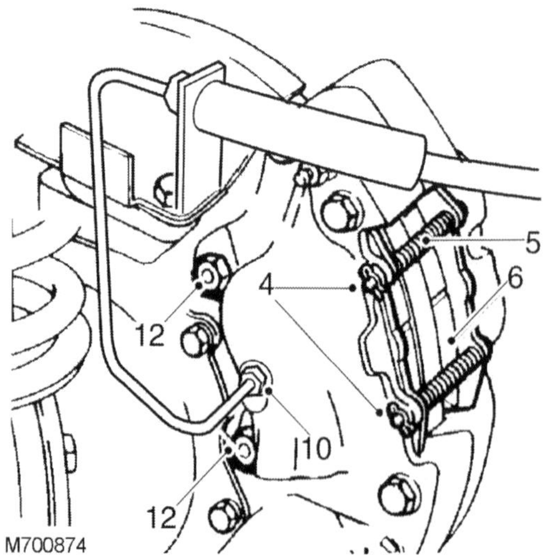

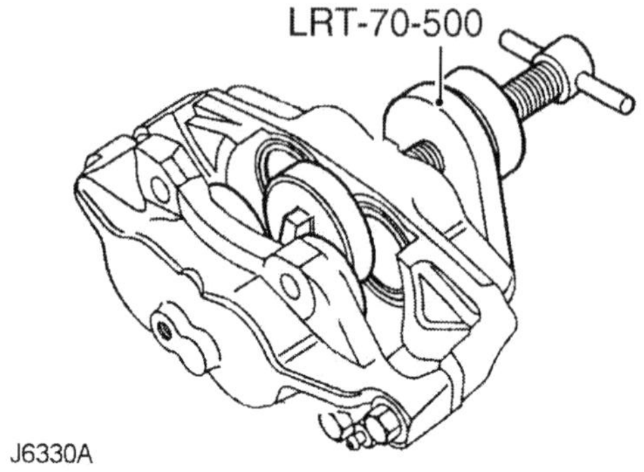

Gently, clear, and with CAUTION, apply air pressure to fluid inlet port to expel pistons. It is unlikely that pistons will expel at same time, regulate rate with a suitable piece of wood between appropriate piston and caliper.

16. Finally remove pistons, identifying them with their respective bores.
17. Remove wiper seal retainer by inserting a blunt screwdriver between retainer and seal. Pry retainer carefully from mouth of bore.
18. Taking care not to damage seal grooves, extract wiper seal and fluid seal.
19. Clean bores, pistons and seal grooves using clean brake fluid only. If caliper or pistons are corroded, or their condition is not perfect, new parts must be fitted.
20. Coat new fluid seal with brake fluid. Ease seal into groove in bore using only fingers, ensuring it is properly seated. Fluid seal and groove are not same in section, so when seal is seated it feels raised to touch at edge furthest away from mouth of bore.
21. Coat appropriate piston with brake fluid. Insert it squarely into bore by hand only. Do not tilt piston during insertion, leave approximately 8mm projecting from bore.
22. Coat new wiper seal with brake fluid and fit to new seal retainer. Slide assembly, seal first, over protruding piston and into bore recess. Use piston clamp to press home seal retainer and piston.

23. Clamp outboard pistons and carry out same procedure for removing and fitting outboard pistons and seals.

Installation

1. Clean caliper to hub mating faces.
2. Position caliper to hub, install bolts and tighten to 82 Nm (60 lbf.ft).
3. Remove plugs from pipe to caliper connections and wipe.
4. Connect pipe to caliper, tighten union to 15 Nm (11 lbf.ft).
5. Apply a light film of grease to rear of brake pads.
6. Install pads to calipers. From 02MY, ensure pads are correctly fitted, with leading edge chamfer fitted as shown.
7. Install pad retainers and springs.
8. Install split pins to retainers.
9. Remove brake hose clamp.
10. Bleed brakes.

Refer to: Brake System Bleeding (206-00 Brake System - General Information, General Procedures)

11. Depress brake pedal to seat pads onto discs.
12. Install road wheels, remove axle stands and tighten nuts to 130 Nm (96 lbf.ft)

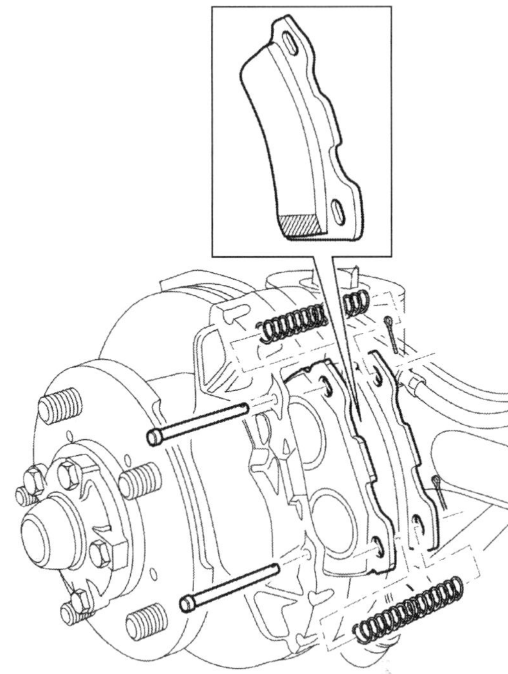

Rear Disc Brake

206-04

Description and Operation

Rear Disc Brake

Refer to: Brake System (206-00 Brake System - General Information, Description and Operation) Removal and Installation

Removal

1. Remove rear wheel bearing and hub assembly.

Refer to: Wheel Bearing and Wheel Hub (204-02 Rear Suspension, Removal and Installation)

2. Remove disc bolts.
3. Remove disc from rear hub.

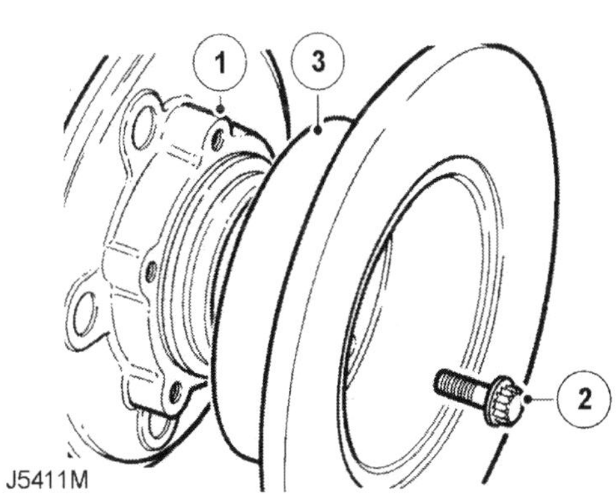

Installation

1. Install disc to rear hub.
2. Install disc bolts. Tighten to 73 Nm (54 lbf.ft).
3. Check total disc run out, this must not exceed 0,15 mm (0.006 in). If necessary reposition disc.
4. Install rear wheel bearing and hub assembly.

Refer to: Wheel Bearing and Wheel Hub (204-02 Rear Suspension, Removal and Installation)

5. Disc reclamation.
6. △ NOTE: The brake disc MUST BE renewed if the minimum running thickness stamped on the disc is recorded.

Check disc thickness. This dimension may be machined to minimum thickness of 12 mm. Machine equal amounts off each face.

Rear Disc Brake - Brake Pads - 90

Removal and Installation

Removal

WARNING: Brake pads must be renewed in axle sets only. Failure to follow this instruction may result in braking efficiency being impaired.

1. WARNING: Make sure to support the vehicle with axle stands.

Raise and support the vehicle.

2. Remove both rear wheels and tires.

Refer to: Wheel and Tire (204-04 Wheels and Tires, Removal and Installation)

3. WARNING: Do not use compressed air to clean brake components. Dust from friction materials can be harmful if inhaled.

Clean the brake caliper using brake cleaning fluid.

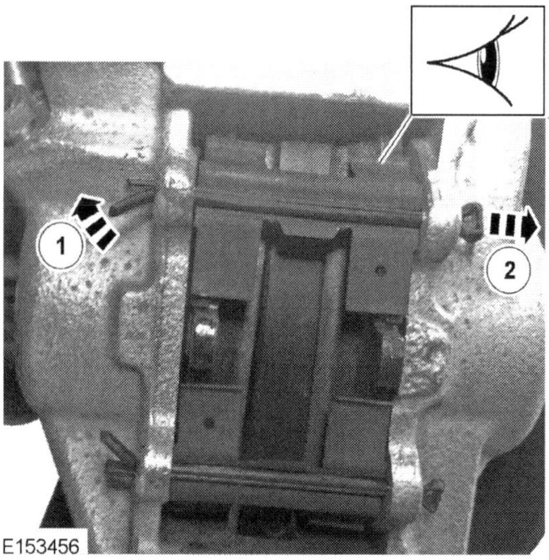

4. NOTE: Note the position of the components prior to removal..

Remove the anti-rattle springs.

1. Release the pin.
2. Remove the pin.
3. Repeat to the other anti-rattle spring.

5. CAUTIONS:

As the piston is pushed back into the caliper housing, the brake fluid level in the reservoir will rise. Do not allow the reservoir to overflow.

Make sure that the surface of the brake disc is not damaged during this step.

Using suitable tools, fully retract the brake caliper pistons.

- Remove the brake pads.

Installation

1. CAUTION: Make sure that the brake pads are installed with the chamfers in the lower position.

Install brake pads.

2. NOTE: Make sure that these components are installed to the noted removal position.

Install the anti-rattle springs.

1. Install the pin.
2. Secure the pin.
3. Repeat to the other anti-rattle spring.

3. Apply brake pedal several times to locate pads.
4. Repeat the procedure for the other side.
5. Install both rear wheels and tires.

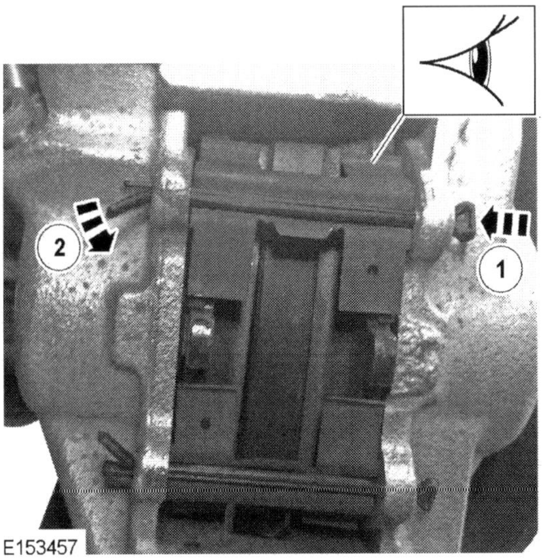

Refer to: Wheel and Tire (204-04 Wheels and Tires, Removal and Installation)

6. Check fluid reservoir, top up if necessary, using correct grade of fluid.

Refer to: (206-00 Brake System - General Information, Specifications)

Rear Disc Brake - Brake Caliper - 90

Removal and Installation

Removal

1. Remove rear road wheels.
2. Using a recognised hose clamp, clamp flexible brake hose above rear axle.
3. Remove brake pipe from rear brake caliper.
4. Seal pipe ends to prevent ingress of dirt.
5. Remove retaining pins and springs and withdraw pads. If same pads are to be refitted, identify them for assembly in original positions.
6. Remove 2 bolts and withdraw caliper from rear axle.

WARNING: Do not separate caliper halves.

7. Clean outer surfaces of caliper with aerosol brake cleaner.
8. WITH CAUTION expel pistons from their bores by applying air pressure to fluid inlet port. It is unlikely both pistons will expel at same time, regulate rate with a suitable piece of wood inserted between two pistons.
9. Finally, remove pistons keeping them identified with their respective bores.
10. Remove wiper seal retainer by inserting a blunt screwdriver between retainer and seal and pry retainer carefully from mouth of bore.

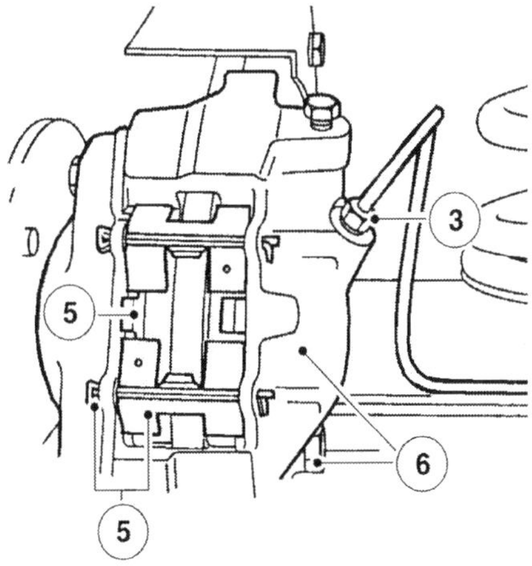

11. Taking care not to damage seal grooves, extract wiper seal and fluid seal.

Installation

1. Clean bores, pistons and particularly seal grooves using clean brake fluid only. If caliper or pistons are corroded or their condition is not perfect new parts must be fitted.
2. Apply brake fluid to new seals. Fit seals into groove in bores. When seals are seated they feel raised to touch at edge furthest away from mouth of bores.
3. Coat pistons with brake fluid. Insert them squarely into bores. Do not tilt pistons during insertion and leave 8mm projecting from bores.
4. Coat new wiper seals with brake fluid and fit to new seal retainers. Slide assemblies, seal first, over protruding piston into bore recesses.
5. Using special tool LRT-70-500 - piston clamp, press home seal retainer and piston. Carry out same procedure for fitting outboard pistons and seals.
6. Install calipers and pads to vehicle.
7. Install caliper to axle, tighten 2 bolts evenly to 82 Nm (60 lbf.ft).
8. Connect brake pipe to caliper. Tighten to 15 Nm (11 lbf.ft).
9. Remove clamp from flexible brake hose.

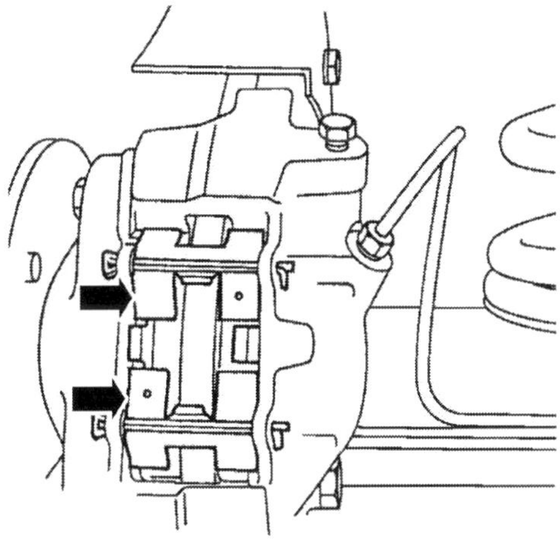

10. **Up to 2002 MY,** insert pads and retaining springs, secure in position with new retaining pins and spread ends or fit new split pins, depending on vehicle model.

11. ⚠ NOTE: Vehicles from 02 MY have leading and trailing edge chamfers on rear pads.

2002MY on, ensure shim are correctly fitted, see illustration.

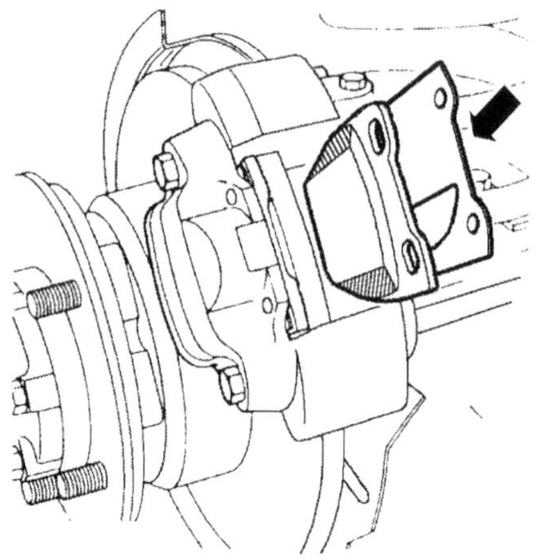

12. Bleed brake system.

Refer to: Brake System Bleeding (206-00 Brake System - General Information, General Procedures)

13. Press brake pedal firmly several times to locate pads.

14. Fit road wheels, remove axle stands. Finally tighten road wheel nuts to correct torque:

    1. Alloy wheels - 130 Nm (96 lbf.ft)

    2. Steel wheels - 100 Nm (80 lbf.ft)

    3. Heavy duty wheels - 170 Nm (125 lbf.ft)

15. Road test vehicle. Note new brake pads require 'bedding-in', for several hundred miles before brakes are at maximum efficiency.

Rear Disc Brake - Brake Caliper - 110/130

Removal and Installation

Removal

1. Remove rear road wheels.

2. Using a recognised hose clamp, clamp flexible brake hose above rear axle.

3. Remove brake pipe from rear brake caliper.

4. Seal pipe ends to prevent ingress of dirt.

5. Remove retaining pins and springs and withdraw pads. If same pads are to be refitted, identify them for assembly in original positions.

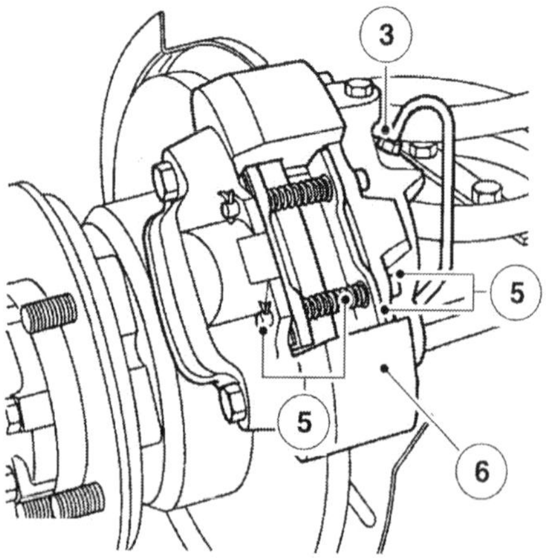

6. Remove 2 bolts and withdraw caliper from rear axle.

⚠ WARNING: Do not separate caliper halves.

7. Clean outer surfaces of caliper with aerosol brake cleaner.

8. WITH CAUTION expel pistons from their bores by applying air pressure to fluid inlet port. It is unlikely both pistons will expel at same time, regulate rate with a suitable piece of wood inserted between two pistons.

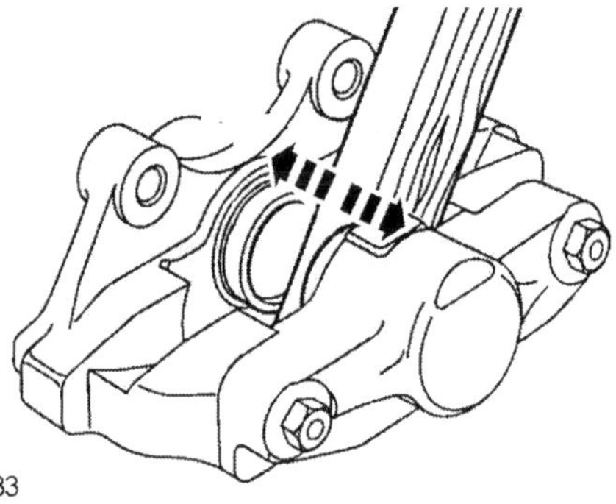

J6333

9. Finally, remove pistons keeping them identified with their respective bores.

10. Remove wiper seal retainer by inserting a blunt screwdriver between retainer and seal and pry retainer carefully from mouth of bore.

11. Taking care not to damage seal grooves, extract wiper seal and fluid seal.

Installation

1. Clean bores, pistons and particularly seal grooves using clean brake fluid only. If caliper or pistons are corroded or their condition is not perfect new parts must be fitted.

2. Apply brake fluid to new seals. Fit seals into groove in bores. When seals are seated they feel raised to touch at edge furthest away from mouth of bores.

3. Coat pistons with brake fluid. Insert them squarely into bores. Do not tilt pistons during insertion and leave 8mm projecting from bores.

4. Coat new wiper seals with brake fluid and fit to new seal retainers. Slide assemblies, seal first, over protruding piston into bore recesses.
5. Using special tool LRT-70-500 - piston clamp, press home seal retainer and piston. Carry out same procedure for fitting outboard pistons and seals.
6. Install calipers and pads to vehicle
7. Install caliper to axle, tighten 2 bolts evenly to 82 Nm (60 lbf.ft).
8. Connect brake pipe to caliper. Tighten to 15 Nm (11 lbf.ft).
9. Remove clamp from flexible brake hose.

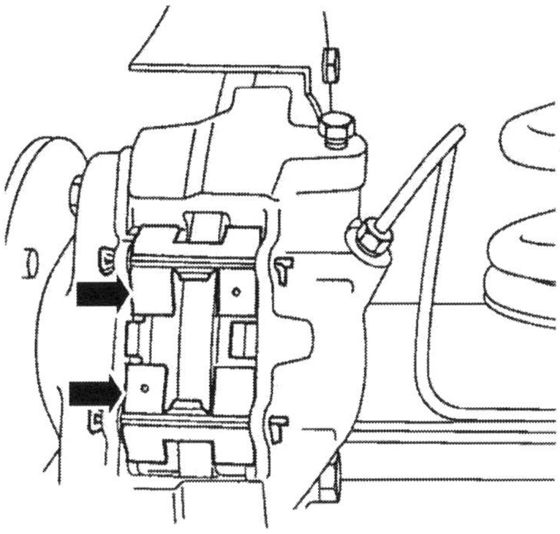

10. **Up to 2002 MY,** insert pads and retaining springs, secure in position with new retaining pins and spread ends or fit new split pins, depending on vehicle model.
11. NOTE: Vehicles from 02 MY have leading and trailing edge chamfers on rear pads.

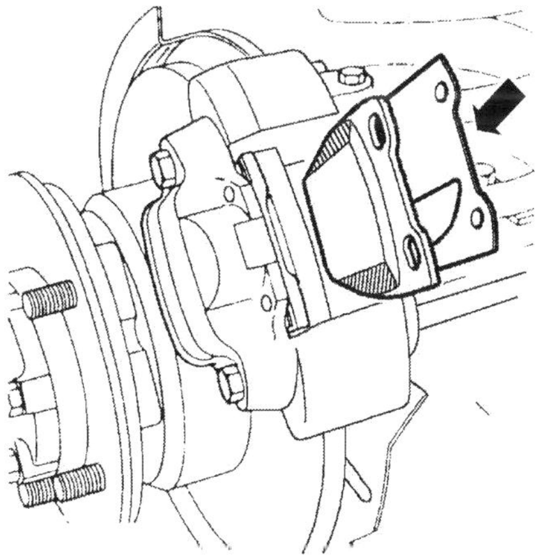

2002MY on, ensure shim are correctly fitted, see illustration.

12. Bleed brake system.

Refer to: Brake System Bleeding (206-00 Brake System - General Information, General Procedures)

13. Press brake pedal firmly several times to locate pads.
14. Install road wheels, remove axle stands. Finally tighten road wheel nuts to correct torque:
    1. Alloy wheels - 130 Nm (96 lbf.ft)
    2. Steel wheels - 100 Nm (80 lbf.ft)
    3. Heavy duty wheels - 170 Nm (125 lbf.ft)
15. Road test vehicle. Note new brake pads require 'bedding-in', for several hundred miles before brakes are at maximum efficiency.

Rear Disc Brake - Brake Pads - 110/130

Removal and Installation

Removal

1. Remove rear road wheels.
2. Clean exterior of calipers.

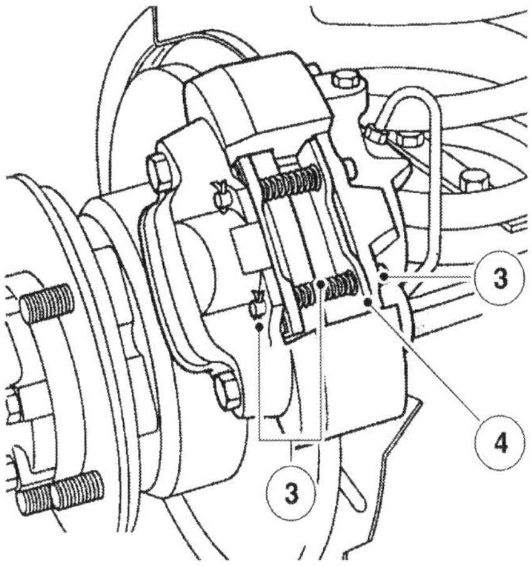

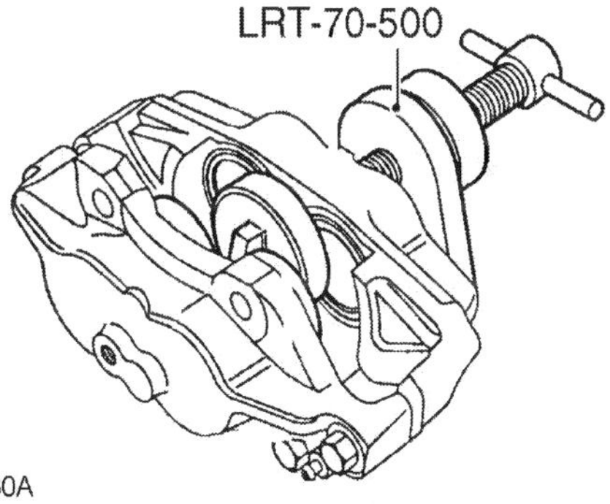

J6330A

3. Remove pad retaining pins and anti-rattle springs.
4. Remove brake pads and shims.
5. Clean exposed parts of pistons, using new brake fluid.
6. Wipe away excess fluid with lint free cloth.
7. Using piston clamp LRT-70-500 press each piston back into its bore. Ensure that displaced brake fluid does not overflow from reservoir.

Installation

1. Install brake pads and shims. Ensure shims are correctly installed, see illustration.

NOTE: Vehicles from 02MY have leading and trailing edge chamfers on rear pads. Shims are also fitted.

2. Install anti-rattle springs and secure with retaining pins.
3. Apply brake pedal several times to locate pads.
4. Install road wheels, remove chassis stands and jack. Tighten wheel nuts to 130 Nm (96 lbf.ft).
5. Check fluid reservoir, top up if necessary, using correct grade of fluid.

Refer to: (206-00 Brake System - General Information, Specifications)

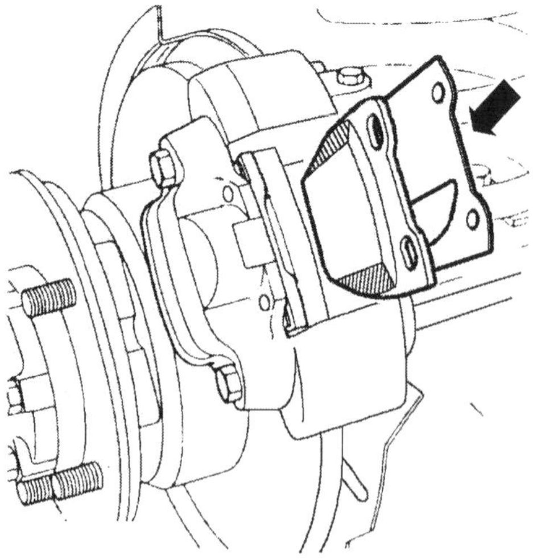

206-05

Parking Brake and Actuation

Description and Operation

Parking Brake

Refer to: Brake System (206-00 Brake System - General Information, Description and Operation)

Parking Brake Shoe and Lining Adjustment

General Procedures

NOTE: The parking brake should be fully operational on third notch of ratchet.

1. Raise one rear wheel clear of ground and support on axle stand.

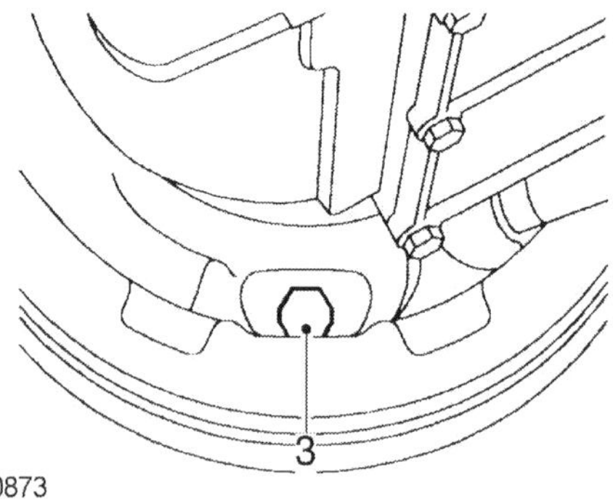

2. Release parking brake lever.
3. Tighten brake adjuster to 25 Nm (18 lbf.ft) to fully expand shoes to drum.
4. Back off adjuster 1½ turns, check that drum is free to rotate.
5. CAUTION: Cable adjustment must ONLY be used for initial setting and to compensate for cable stretch. It MUST NOT be used to take up brake shoe wear, which MUST be adjusted at brake drum.

   Check operation of parking brake lever to give pawl 2 notches free movement on ratchet before being fully operational on third notch of ratchet. Adjust parking brake accordingly if lever travel exceeds the above tolerance.

6. Remove axle stand and wheel chock.

Parking Brake Cable

Removal and Installation

Removal

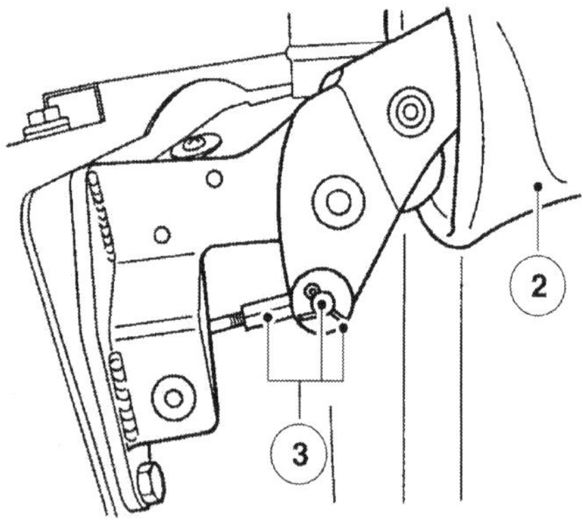

1. Park vehicle on level ground, chock road wheels and release parking brake. Alternatively, raise vehicle on ramp.
2. Remove 3 trim studs and lift up parking brake gaiter.
3. Remove split pin, clevis pin, washer and disconnect cable from parking brake lever.
4. Slacken off transmission brake drum adjusting screw.
5. Disconnect drive shaft from output flange.
6. Remove retaining screw and withdraw brake drum.

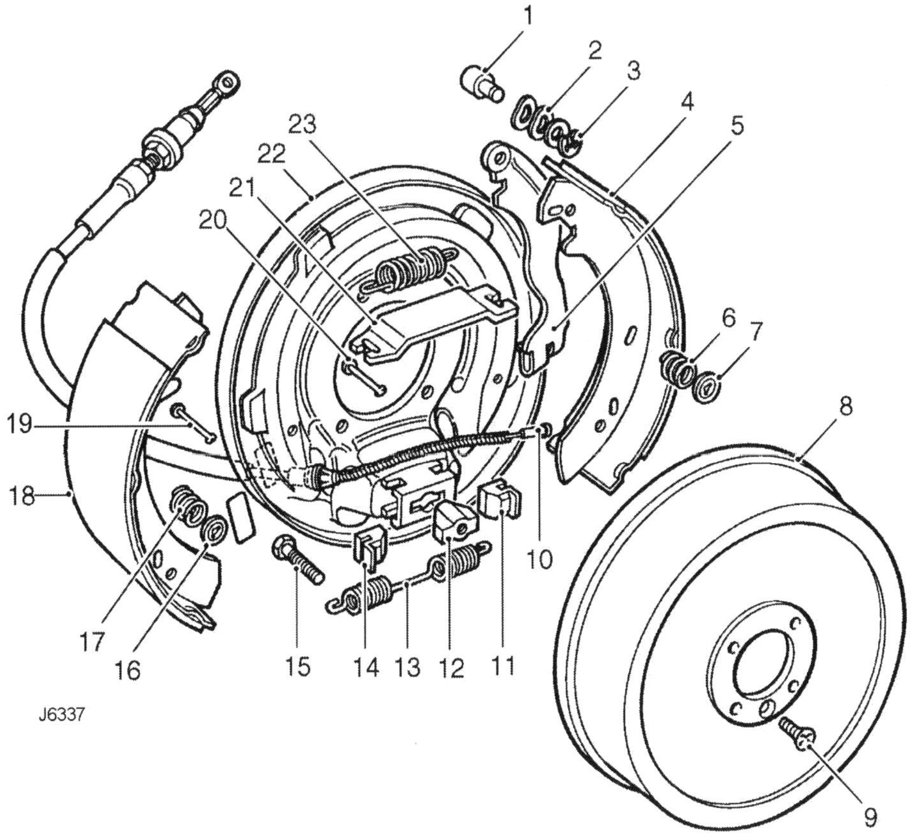

Item	Description
1	Pin
2	Washer
3	'C' Circlip
4	Brake shoe
5	Cable lever
6	Hold down spring
7	Dished washer
8	Brake drum
9	Screw
10	Brake cable
11	Adjuster slide
12	Adjuster nut
13	Spring
14	Adjuster slide
15	Adjuster bolt
16	Dished washer
17	Hold down spring
18	Brake shoe
19	Hold down pin
20	Hold down pin
21	Abutment plate
22	Back plate
23	Spring

7. Release parking brake cable clevis from abutment on cable lever, see J6337, and pull through aperture in back plate.

8. Pull cable from heelboard and remove from vehicle.

Installation

1. Feed new cable through heelboard ensuring rubber grommet is correctly located.

2. Position cable over guide plate, insert through backplate and connect to cable lever.

3. Install cable to parking brake lever and secure with clevis pin and split pin.

4. Install parking brake gaiter.

5. Install brake drum. Tighten screw to 25 Nm (18 lbf. ft).

6. Screw in and tighten adjuster bolt until brake drum will not rotate by hand.

7. Tighten adjuster bolt further to 25 Nm (18 lbf.ft) to ensure brake drum is locked.

8. Slacken off adjuster bolt by 1.5 turns to give brake shoes running clearance. Check that the drum is free to rotate.

9. Slacken locknut and adjust cable to give the parking brake pawl two notches free movement on the ratchet before being fully operational on third notch (brake shoes are fully expanded against drum).

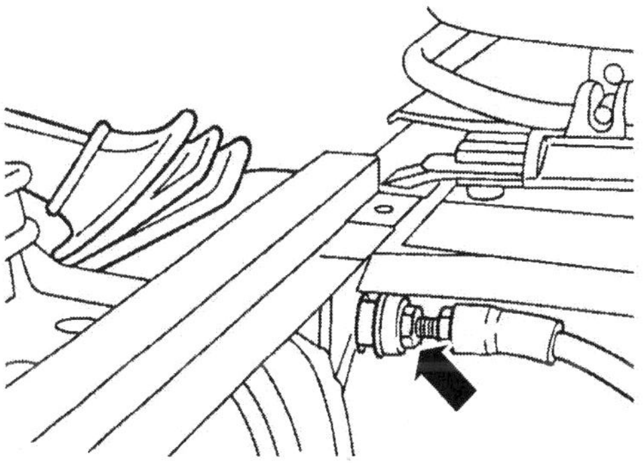

⚠ NOTE: Cable adjustment is for a new cable or to compensate for cable stretch. Cable adjustment must not be used to take up brake shoe wear.

10. Install drive shaft to output flange. Tighten fixings to 46 Nm (34 lbf.ft).

11. Remove wheel chocks and check operation of parking brake.

Parking Brake and Actuation - Shoes

Removal and Installation

Removal

1. Park vehicle on level ground, chock road wheels and release parking brake. Alternatively, raise vehicle on a ramp.

2. Disconnect rear propeller shaft from transmission output flange at brake drum.

3. Slacken off parking brake drum adjustment bolt.

4. Remove single screw securing brake drum to output flange.

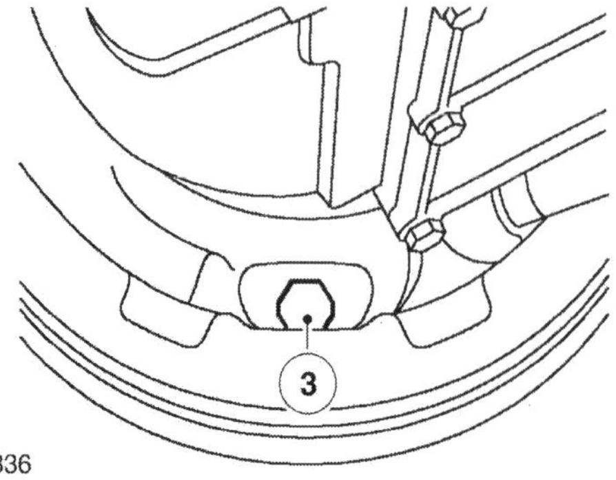

J6336

5. Withdraw drum to expose brake assembly.
6. Release top and bottom springs from brake shoes, see J6337.

Refer to: Parking Brake Cable (206-05 Parking Brake and Actuation, Removal and Installation)

7. Grip dished washer with a pair of pliers, depress washer and turn through 90°.
8. Remove dished washer, complete with hold down spring and pin from both shoes.
9. Move brake shoes out from adjuster slides, release from abutment plate and remove from backplate.
10. Check that springs are satisfactory for continued use. If new brake shoes are to be fitted, the springs should also be renewed.

Installation

1. Locate RH brake shoe in slide and secure brake shoe and lever assembly to backplate with hold down pin, spring and dished washer.
2. Locate LH brake shoe in slide and fit abutment plate between both brake shoes. Secure LH shoe with hold down pin, spring and dished washer.
3. Install pull-off springs to brake shoes.
4. Install brake drum. Tighten screw to 25 Nm (18 lbf.ft).
5. Check that hand brake lever is released.
6. Screw in and tighten adjuster bolt until brake drum will not rotate by hand.
7. Tighten adjuster bolt to 25 Nm (18 lbf.ft) to ensure brake drum is locked.
8. Slacken off adjuster bolt by 1.5 turns to give shoes a running clearance. Check that the drum is free to rotate.
9. Install propeller shaft to output flange. Tighten fixings to 46 Nm (34 lbf.ft).
10. Remove wheel chocks and check operation of parking brake.

Parking Brake Switch

Removal and Installation

Removal

1. Release cover from parking brake lever.

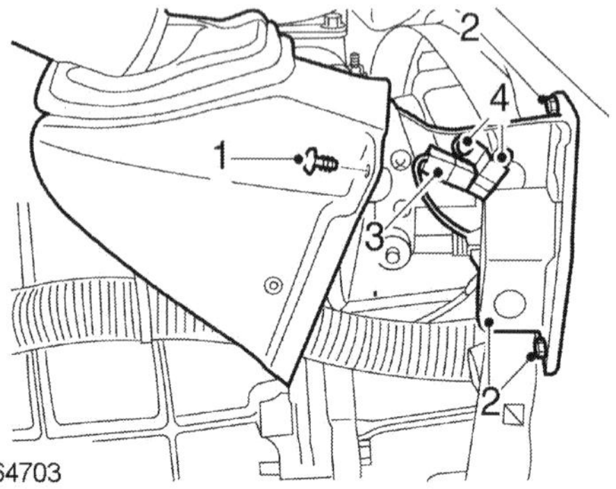

M864703

2. Remove 2 bolts and release parking brake lever from body.
3. Release connector from parking brake switch.
4. Remove 2 screws securing parking brake switch to parking brake lever and remove switch.

Installation

1. Install parking brake switch to parking brake lever and tighten screws.
2. Install connector to parking brake switch.
3. Position parking brake lever and tighten bolts to 22 Nm (16 lbf.ft).
4. Position cover to parking brake lever.

Brake Pedal and Bracket

Removal and Installation

Removal

1. Disconnect the battery ground cable.
2. Remove the brake booster.
3. Remove the stoplamp switches.

Refer to: Brake Booster (206-07 Power Brake Actuation, Removal and Installation)

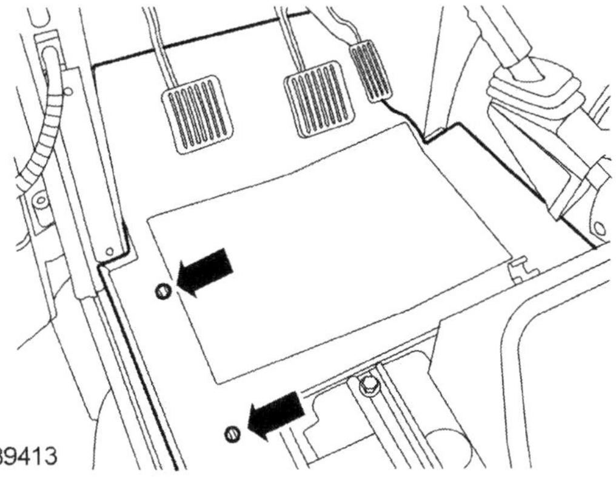

E89413

4. Remove the floor covering.
   - Remove the 2 clips.

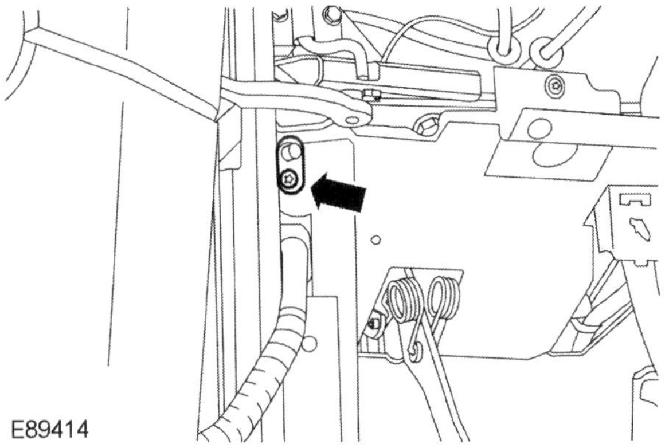

E89414

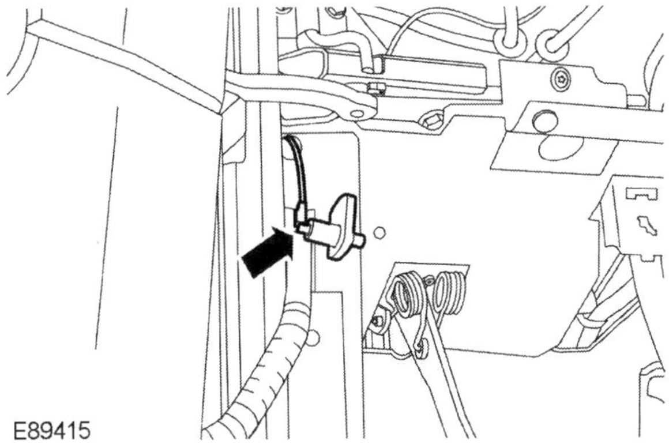

E89415

5. Release the interior lamp switch.
   - Remove the screw.
6. Remove the interior lamp switch.
   - Disconnect the electrical connector.

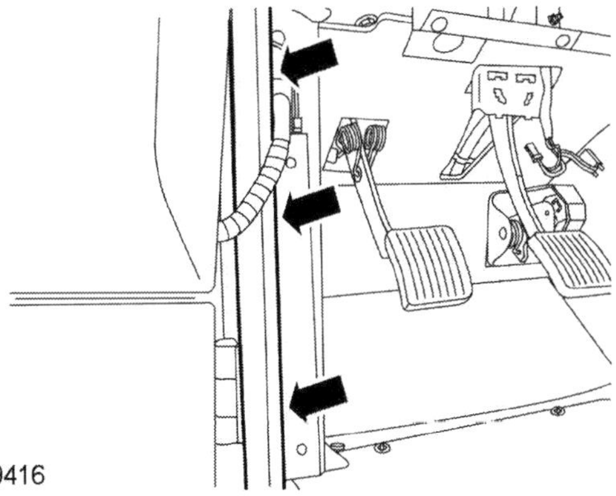

E89416

7. Release the front door aperture seal.

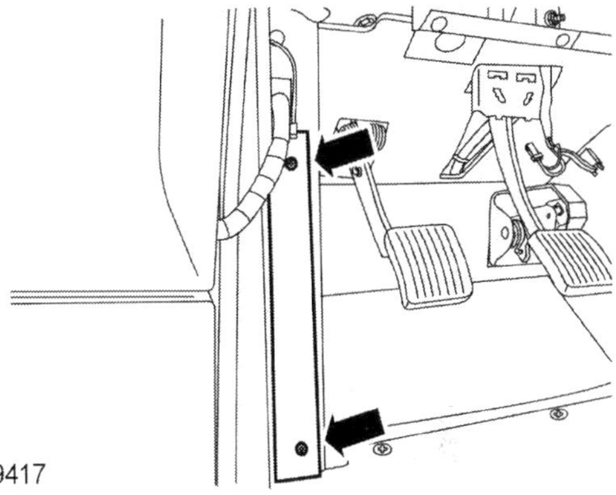

E89417

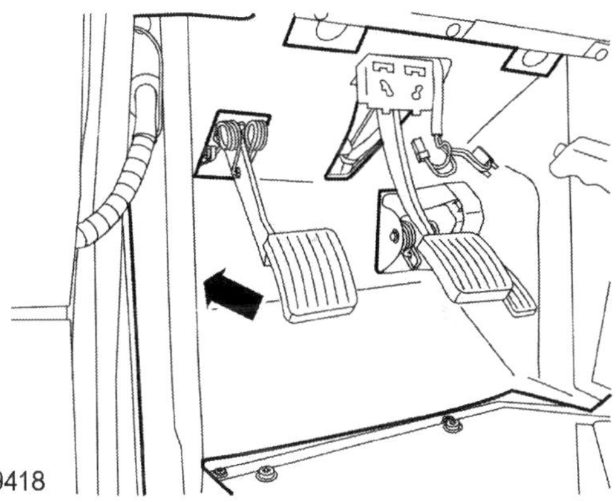

E89418

8. Remove the A-pillar trim panel.
   - Remove the 2 screws.
9. Release the floor covering away from the bulkhead.

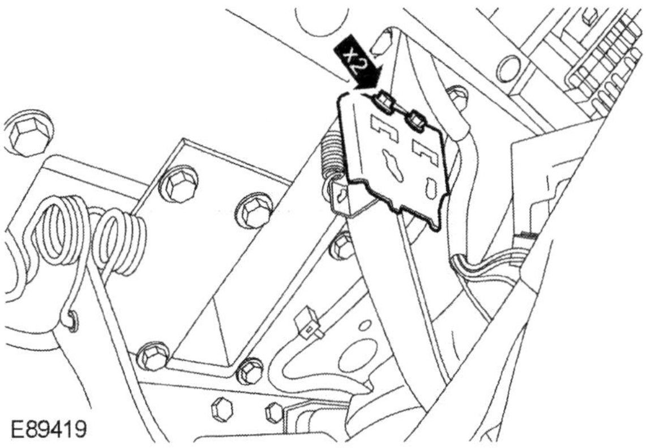

E89419

10. Remove the stoplamp switch support bracket.
    - Remove the 2 bolts.

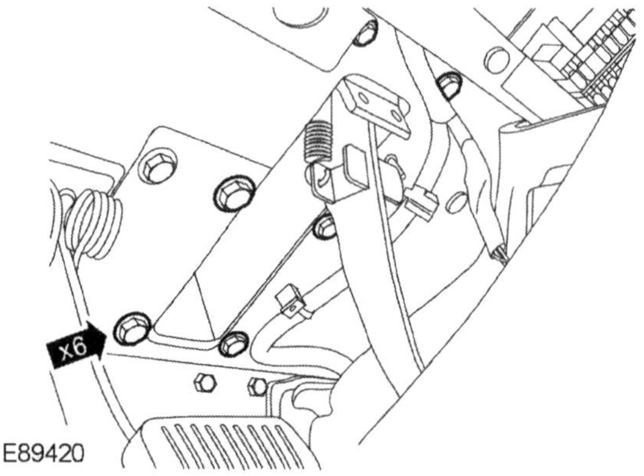

E89420

11. Release the brake pedal and bracket.
    - Remove the 6 bolts.

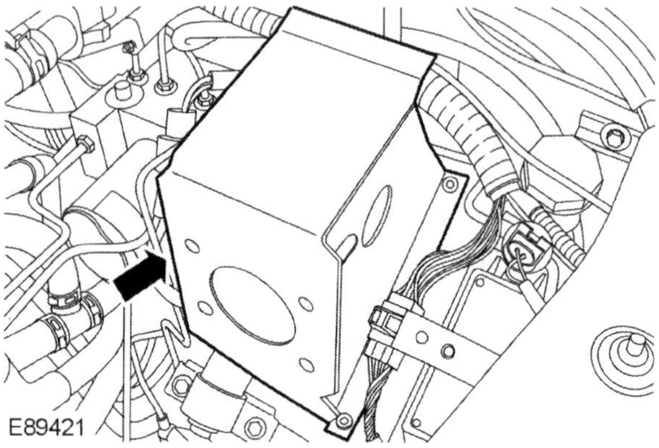
E89421

12. Remove the brake pedal and bracket.

Installation

1. To install, reverse the removal procedure.
    - Tighten to 25 Nm (18 lb.ft).
2. Tighten to 9 Nm (7 lb.ft).
3. Connect the battery ground cable.

Refer to: Brake Booster (206-07 Power Brake Actuation, Removal and Installation)

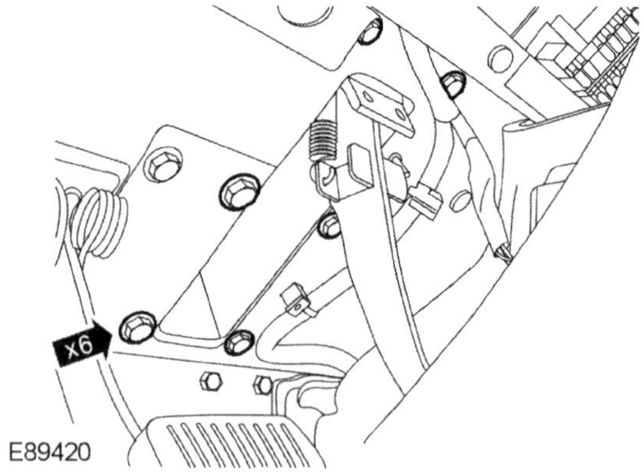

E89420

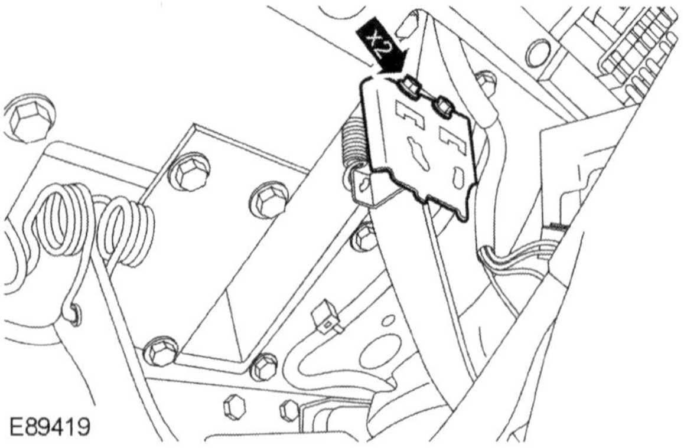

E89419

Brake Pressure Control Valve

Removal and Installation

Removal

1. Disconnect battery.
2. Clean area around brake pressure control valve ports.
3. Place a container under valve to catch escaping brake fluid.
4. Disconnect primary circuit pipe unions 'A' from brake pressure control valve.
5. Disconnect secondary circuit pipe unions 'B' from brake pressure control valve.
6. Cover pipes to prevent ingress of dirt.
7. Remove single retaining nut and bolt securing brake pressure control valve to engine bulkhead.
8. Remove brake pressure control valve.

Installation

1. Fit brake pressure control valve to engine bulkhead. Tighten bolt to 15 Nm (11 lbf/ft).
2. Connect primary and secondary circuit pipes to brake pressure control valve. Tighten unions to 16 Nm (12 lbf/ft).
3. Fill brake reservoir with recommended brake fluid.

Refer to: (206-00 Brake System - General Information, Specifications)

4. Bleed the brake system.

Refer to: Brake System Bleeding (206-00 Brake System - General Information, General Procedures)

5. Reconnect battery and road test vehicle.

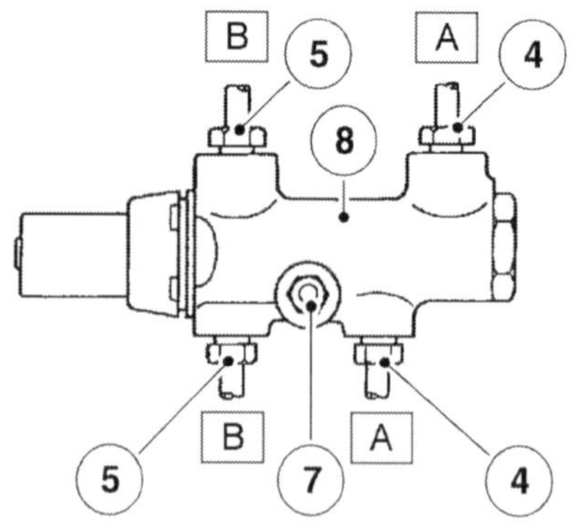

J6340

Brake Master Cylinder

Disassembly and Assembly

Part(s)

STEP	PART NAME	QUANTITY
Step 12	Renew Part Brake master cylinder	1

Disassembly

1. Disconnect battery and remove master cylinder from booster.

Refer to: Startup Battery Disconnect and Connect (414-01 Battery, Mounting and Cables, General Procedures)

2. Before commencing overhaul procedure thoroughly clean master cylinder and inspect outer surfaces for damage and condition, renew complete assembly if necessary.
3. The reservoir is a push fit in master cylinder and secured by seals. Carefully ease reservoir from master cylinder by rolling it from seals as illustrated.

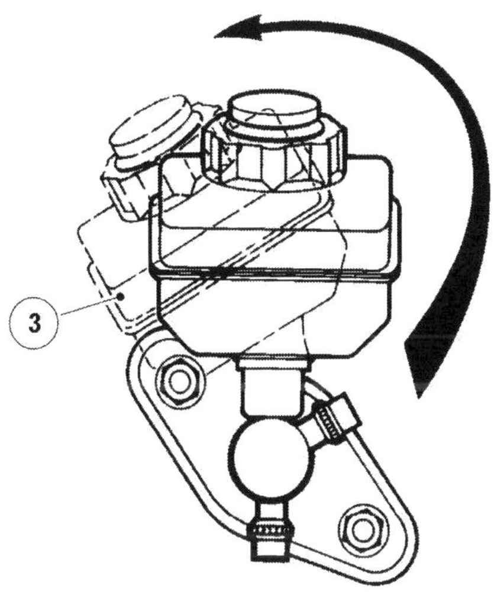

4. Using soft jaws, one either side of master cylinder flange and clamp flange in a suitable vice. Remove water ingress O-ring seal from master cylinder to booster flange and discard.
5. Hold outside of transfer housing with a suitable pair of grips, carefully pull, while working pliers in a backwards and forwards rocking motion to ease housing off master cylinder, discard housing and vacuum seal.

ST3316M

6. Withdraw 2 reservoir seals from master cylinder and note their positions in inlet ports for reassembly. Discard both seals.

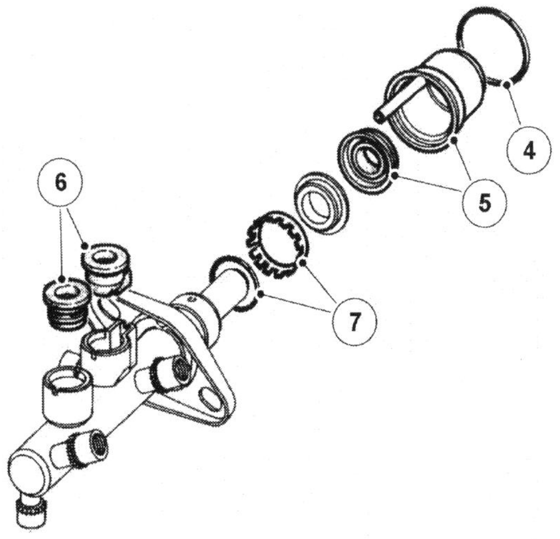

ST3318M

7. Remove retaining ring and O-ring seal from machined outer surface of master cylinder, discard both seal and retaining ring.
8. Remove guide ring from mouth of master cylinder which supports primary plunger assembly and place to one side, this component is not part of master cylinder service kit and is to be refitted on assembly of unit.
9. Pull primary plunger assembly out of master cylinder.

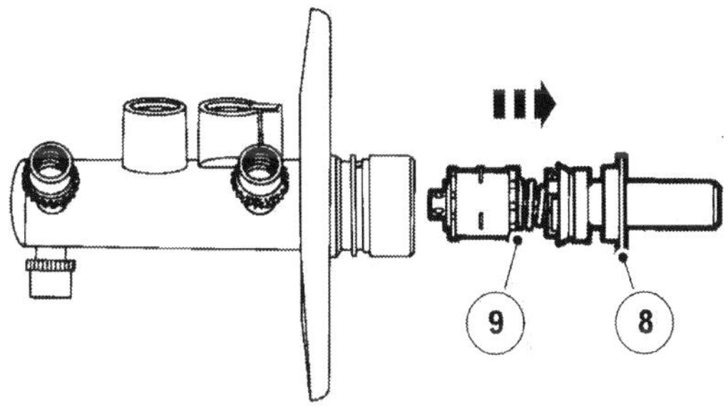

ST3319M

△ NOTE: The primary plunger assembly cannot be broken down any further and is serviced as a complete unit. Discard assembly.

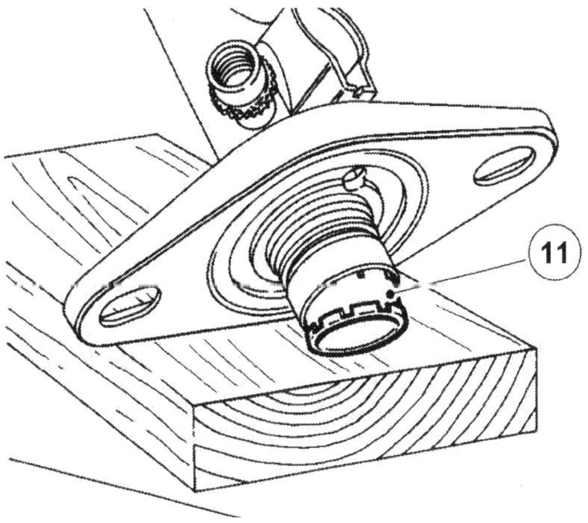

10. The secondary plunger assembly will remain at bottom of master cylinder bore, plunger can be easily expelled by tapping assembly on a piece

of timber until plunger appears at cylinder mouth, carefully pull plunger from master cylinder.

11. If swirl tube was not expelled at same time as secondary plunger, repeat above operation to expel it from bottom of master cylinder bore and discard.
12. Clean all parts with Girling cleaning fluid or unused brake fluid and place cleaned parts on to a clean sheet of paper. Inspect cylinder bore and plungers for signs of corrosion, ridges and score marks. Provided working surfaces are in perfect condition, new seals from a Girling Service repair kit may be used.

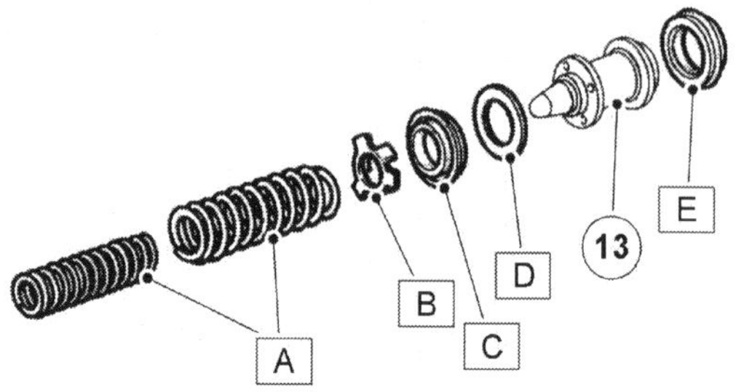

13. Remove components above from secondary plunger and discard: A Springs B Seal retainer C Recuperating seal (primary cup) D Washer E ‘L’ seal.

NOTE: A small screwdriver with end rounded and polished is required to remove ‘L’ seal. DO NOT damage secondary plunger.

14. Coat new seals in unused brake fluid and firstly install ‘L’ seal to plunger.
15. Install washer followed by recuperating seal. Install seal retainer and springs, ensure springs are correctly seated.

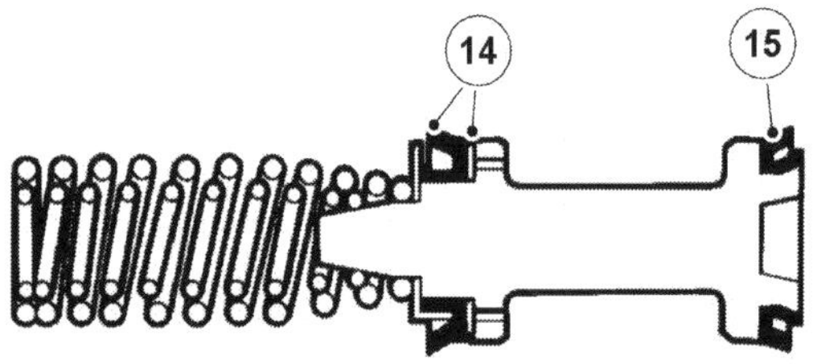

Assembly

CAUTION: It is important that the following instructions are carried out precisely, otherwise damage could be caused to new seals when inserting plungers into cylinder bore. Generous amounts of new brake fluid should be used to lubricate parts during assembly.

NOTE: Thoroughly check that no debris is lodged in fluid passageways and drillings. If debris is found, carefully remove, re-clean cylinder and re-check.

1. Install new swirl tube to bottom of cylinder bore.
2. Lubricate secondary plunger and cylinder bore. Offer plunger assembly to cylinder until

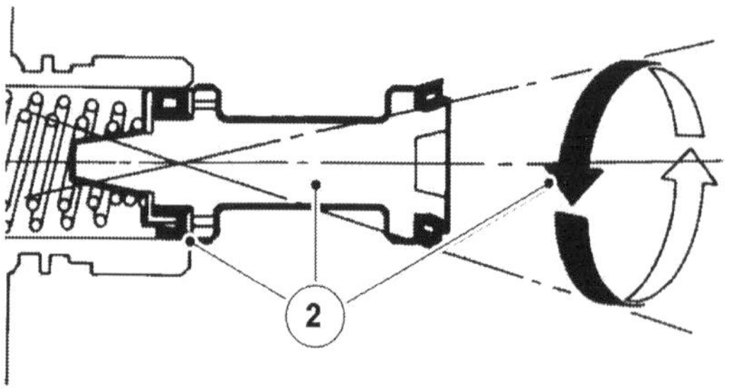

recuperation seal is resting centrally in mouth of bore. Gently introduce plunger with a circular rocking motion, as illustrated. Ensuring that seal does not become trapped, ease seal into bore and slowly push plunger down bore in one continuous movement.

3. Install primary plunger assembly using same method as for secondary plunger, push plunger down bore.
4. Install original guide ring to support primary plunger.
5. Coat a new O-ring with brake fluid and install to its respective groove on outer location surface of master cylinder.

CAUTION: O-ring should not be rolled down outer location surface of master cylinder but should be slightly stretched and eased down cylinder and into its groove. Do not over stretch seal.

6. Install a new retaining ring on outer surface of master cylinder ensuring that serrations of ring are facing mounting flange.
7. Install two new reservoir seals in their respective ports.
8. Install a new vacuum seal to either primary plunger or to bottom of transfer housing bore, open face of seal towards primary plunger guide ring.
9. Lubricate vacuum seal with brake fluid, install transfer housing to master cylinder, push housing fully up to cylinder mounting flange. Do not adjust transfer housing after fitting.
10. Lubricate a new water ingress seal with brake fluid, slightly stretch seal and ease it down housing until seal is in correct position between housing and flange.
11. Roll reservoir into top of master cylinder, reversing procedure described in instruction 3.
12. Install master cylinder to booster.

Refer to: Brake Master Cylinder (206-06 Hydraulic Brake Actuation, Removal and Installation)

Renew Part: Brake master cylinder

13. Reconnect battery, and road test vehicle.

Brake Master Cylinder

Removal and Installation

Removal

1. Disconnect battery negative lead.
2. ⚠ CAUTION: Do not allow brake fluid to contact paint finished surfaces as paint may be damaged. If spilled, remove fluid and clean are with clean warm water.

   Place a container under the master cylinder to collect any brake fluid spillage.
3. Clean area around master cylinder ports.
4. Loosen 2 unions securing brake pipes to master cylinder ports.
5. Disconnect both brake pipes from master cylinder. Cover, not plug, pipe ends to prevent entry of dirt.
6. Release 2 connectors from reservoir cap.
7. Remove 2 nuts securing master cylinder to brake booster.
8. Withdraw master cylinder from booster and remove.
9. Carefully ease reservoir from master cylinder by rolling it from seals.
10. △ NOTE: Master cylinder to reservoir seals are different sizes.

Remove seals from master cylinder.

Installation

1. Install NEW seals to master cylinder, ensuring seals are fitted to correct ports.
2. Install reservoir to master cylinder.

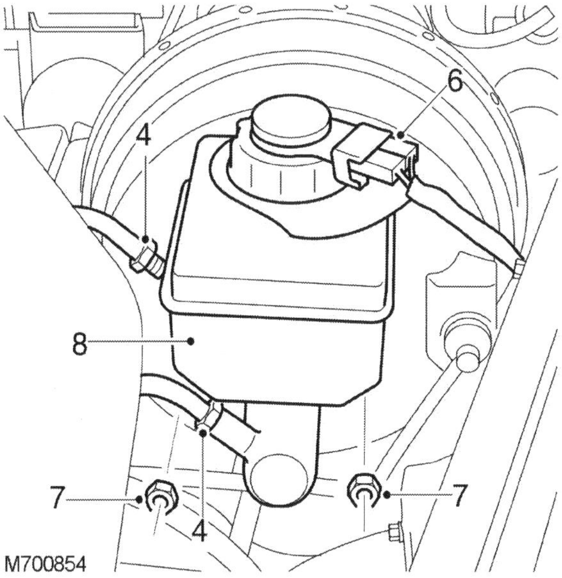

3. Ensuring that water ingress seal is in position, Install master cylinder to booster.
4. Install nuts securing master cylinder to booster and tighten to 26 Nm (19 lbf.ft).
5. Connect brake pipes to master cylinder and tighten unions to 15 Nm (11 lbf.ft).
6. Install connectors to reservoir cap.
7. Fill reservoir with recommended brake fluid.

Refer to: (206-00 Brake System - General Information, Specifications)

8. Bleed the brake system.

Refer to: Brake System Bleeding (206-00 Brake System - General Information, General Procedures)

Specifications

Description	Nm	lb-ft
Brake vacuum pump bolts	23	17
Brake booster nuts	14	10
Brake master cylinder nuts	26	19

Brake Booster

Removal and Installation

Part(s)

STEP	PART NAME	QUANTITY
Removal Step 7	Renew Part Brake booster to brake pedal pin	1

Removal

1. Disconnect the battery ground cable.

Refer to: Startup Battery Disconnect and Connect (414-01 Battery, Mounting and Cables, General Procedures)

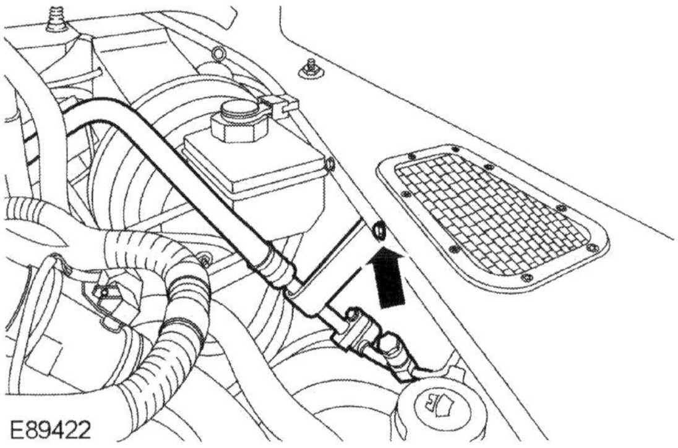
E89422

2. Remove the bolt. Release the air conditioning (A/C) high-pressure pipe.

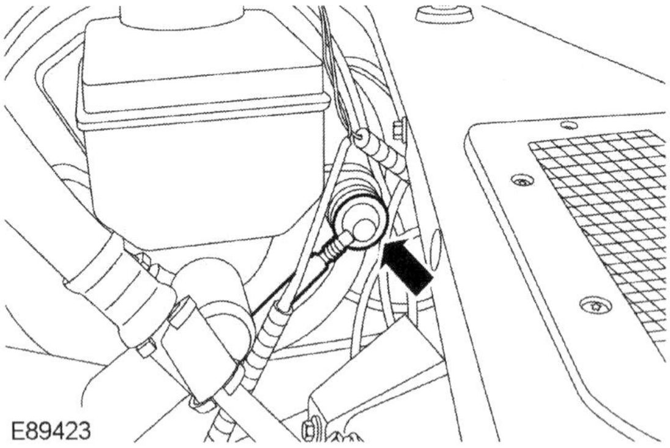
E89423

3. Release the brake booster vacuum hose.

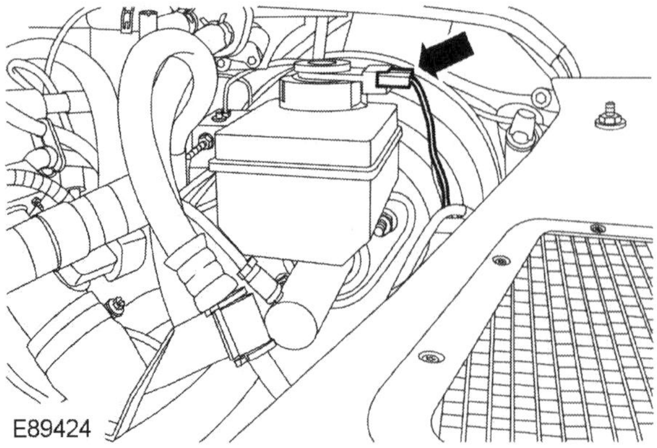
E89424

4. Disconnect the 2 electrical connectors from the brake fluid low level warning indicator.

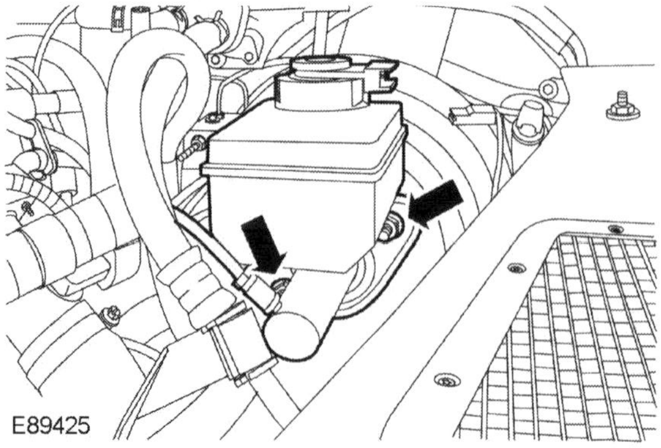
E89425

5. Remove the 2 nuts. Release the brake master cylinder.

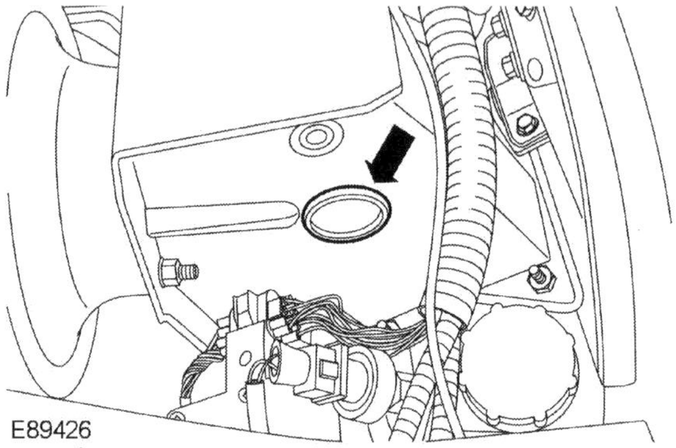
E89426

6. △ NOTE: RH shown, LH similar.

Remove the 2 grommets from the brake pedal box.

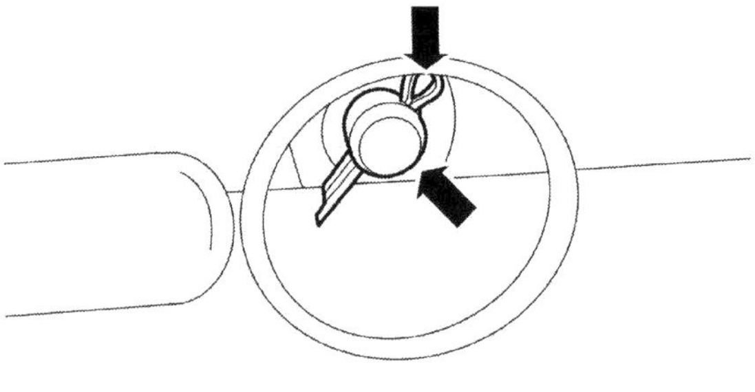
E89427

7. Remove the brake booster clevis pin. Remove and discard the split pin.

Renew Part: Brake booster to brake pedal pin

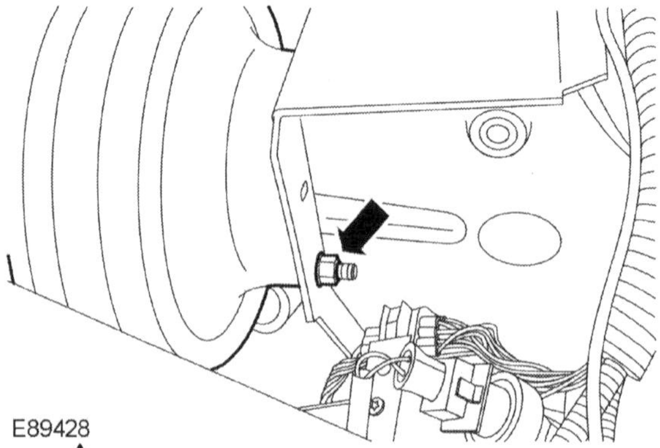
E89428

8. △ NOTE: RH shown, LH similar.

Remove the 2 nuts. Remove the brake booster.

Installation

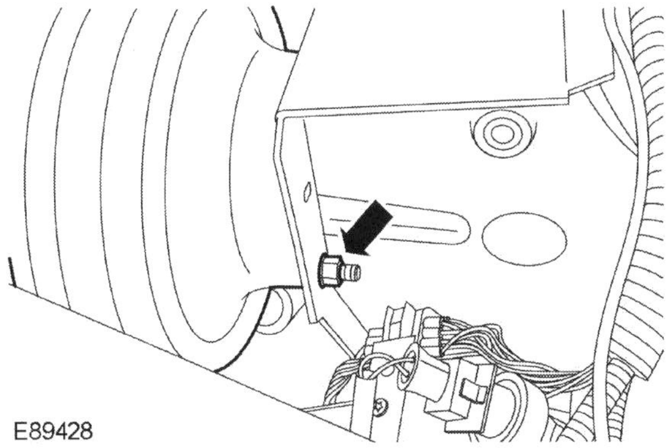

1. To install, reverse the removal procedure and tighten.

Torque: 14 Nm

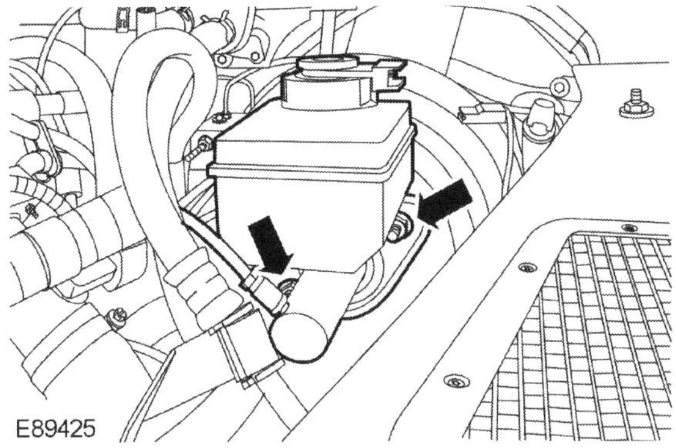

2. *Torque:* 26 Nm
3. Connect the battery ground cable.

Refer to: Startup Battery Disconnect and Connect (414-01 Battery, Mounting and Cables, General Procedures)

Brake Booster Non-Return Valve

Removal and Installation

Removal

1. Disconnect brake vacuum hose from booster non return valve.
2. Carefully prise valve out with a screwdriver blade between valve and grommet. Take care not to exert too much pressure on the vacuum chamber.
3. Remove rubber grommet but be careful not to allow it to fall into the vacuum chamber.
4. Check the valve for correct operation; it should not be possible to pass air through into the booster in direction of arrow. Do not use compressed air.

Installation

1. Install rubber grommet.
2. Smear ribs of the valve with Lucas Girling rubber grease to assist assembly, and push valve fully home.
3. Connect vacuum hose to the valve.
4. Road test vehicle.

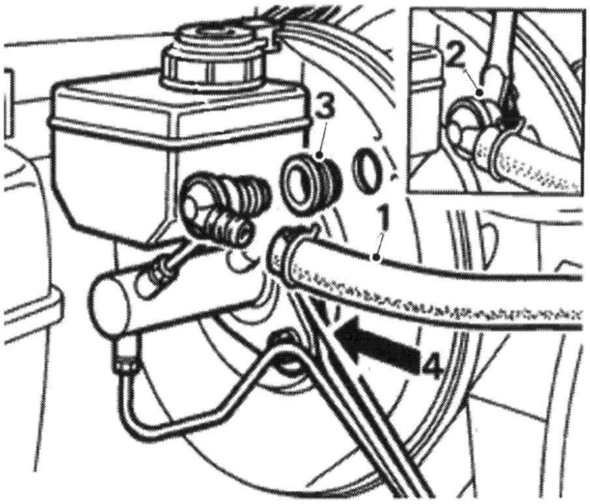

Brake Vacuum Pump

Removal and Installation

Removal

NOTE: Removal steps in this procedure may contain installation details.

1. CAUTION: Before disconnecting any components, make sure the area is clean and free from foreign material. When disconnected all openings must be sealed.

Torque: 23 Nm

Installation

1. NOTES:

Remove and discard the blanking caps.

Clean the component mating faces.

To install, reverse the removal procedure.

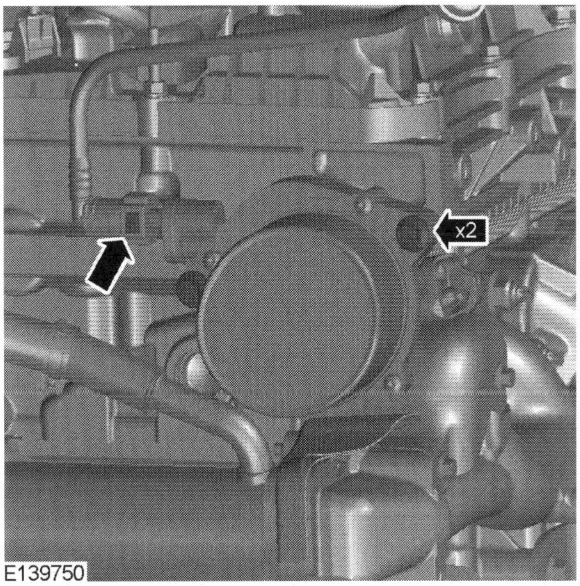

Description and Operation

Anti-Lock Control - Traction Control

Refer to: Brake System (206-00 Brake System - General Information, Description and Operation)

Diagnosis and Testing

Principles of Operation

For a detailed description of the Anti-Lock Control - Traction Control, refer to the relevant Description and Operation section in the workshop manual.

Refer to: Anti-Lock Control - Traction Control (206-09 Anti-Lock Control - Traction Control, Description and Operation)

Inspection and Verification

CAUTION: Diagnosis by substitution from a donor vehicle is **NOT** acceptable. Substitution of control modules does not guarantee confirmation of a fault, and may also cause additional faults in the vehicle being tested and/or the donor vehicle.

NOTES:

If the control module or a component is suspect and the vehicle remains under manufacturer warranty, refer to the Warranty Policy and Procedures manual (section B1.2), or determine if any prior approval programme is in operation, prior to the installation of a new module/component.

When performing voltage or resistance tests, always use a digital multimeter accurate to three decimal places, and with an up-to-date calibration certificate. When testing resistance always take the resistance of the digital multimeter leads into account.

Check and rectify basic faults before beginning diagnostic routines involving pinpoint tests.

1. Visually inspect for obvious signs of damage and system integrity
2. If an obvious cause for an observed or reported concern is found, correct the cause (if possible) before proceeding to the next step
3. If the cause is not visually evident, verify the symptom and refer to the Symptom Chart, alternatively check for Diagnostic Trouble Codes (DTCs) and refer to the DTC Index

DTC Index

For a list of Diagnostic Trouble Codes (DTCs) that could be logged on this vehicle, please refer to Section 100-00.

Refer to: Diagnostic Trouble Code Index: Anti-lock Brake System (ABS) (100-00 General Information, Description and Operation)

Visual Inspection

Mechanical	Electrical
• Tire size, condition and installation • Wheel speed sensor condition and installation • Wheel speed sensor target ring • Wheel bearings • Yaw rate lateral acceleration sensor • Steering angle sensor • Hydraulic Control Unit (HCU) condition and installation	• Fuses • Wiring harnesses and connectors • Warning lamp operation • Wheel speed sensors • Battery junction box • Stop lamp switch • Differential lock switch • Dynamic stability control switch • Yaw rate lateral acceleration sensor • Steering angle sensor • Anti-lock Brake System (ABS) control module • Controller Area Network (CAN)

Hydraulic Control Unit - LHD

Removal and Installation

Removal

NOTE: Removal steps in this procedure may contain installation details.

1. CAUTIONS:

   Be prepared to collect escaping fluids.

   If brake fluid is spilt on the paintwork, the affected area must be immediately washed down with cold water.

   Before disconnecting or removing the components, make sure the area around the joint faces and connections are clean.

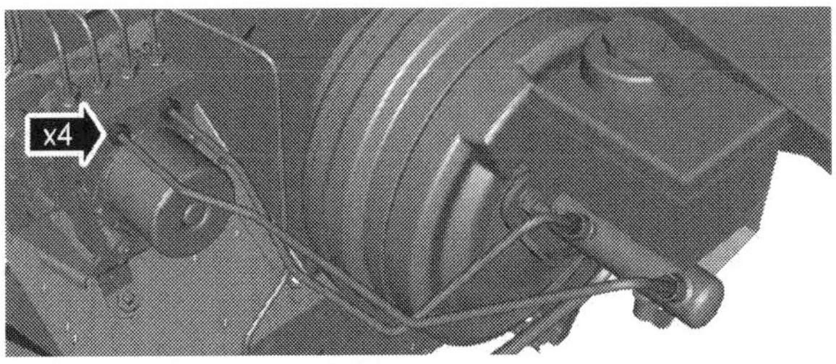

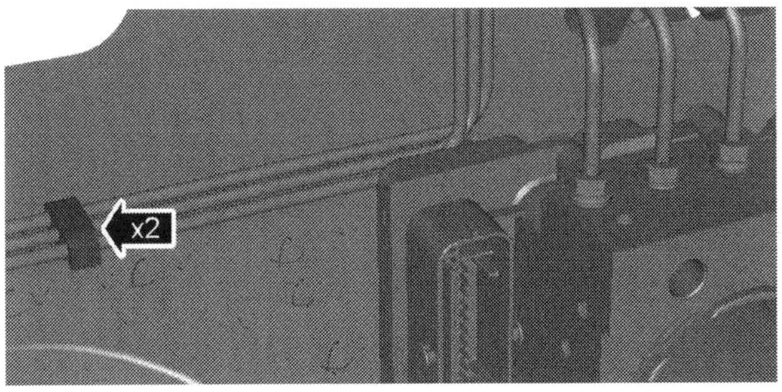

2. NOTE: Note the orientation of the securing clips and foam padding.

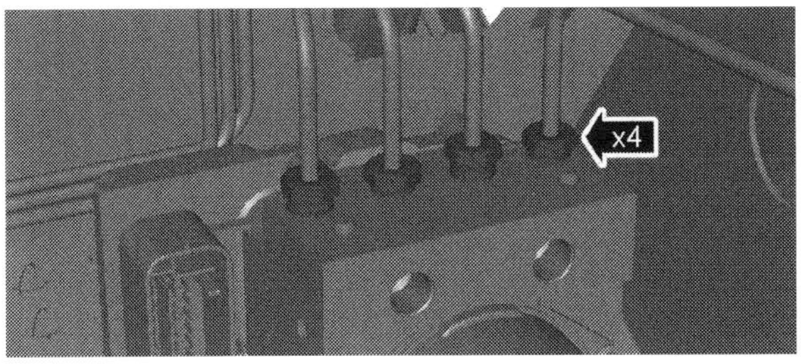

3. CAUTIONS:

Before disconnecting any components, make sure the area is clean and free from foreign material. When disconnected all openings must be sealed.

Be prepared to collect escaping fluids.

Before disconnecting or removing the components, make sure the area around the joint faces and connections are clean.

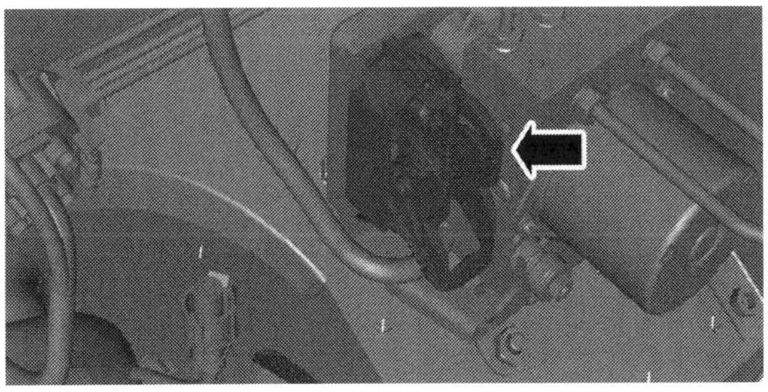

4.

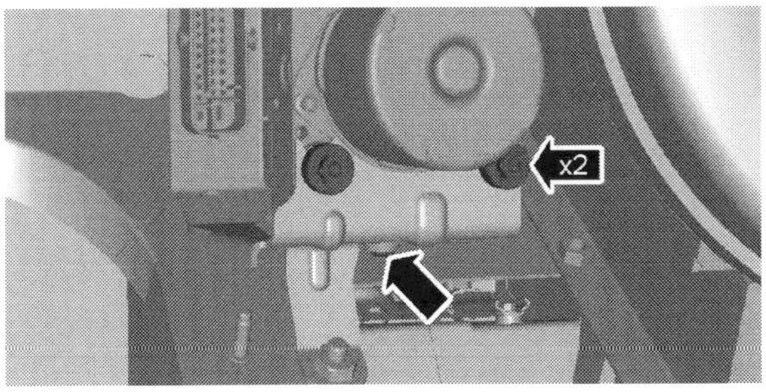

5.

Installation

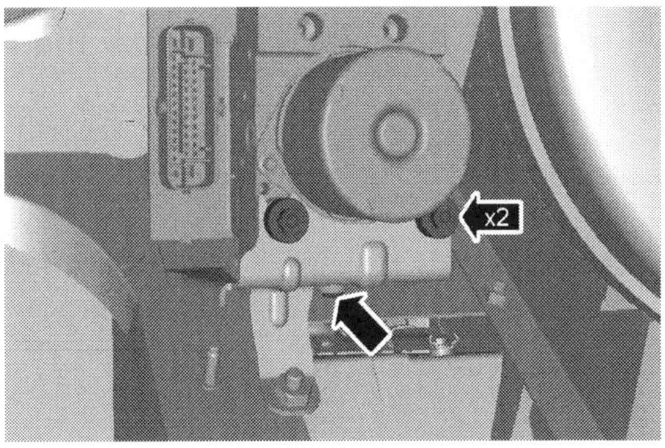

1. *Torque:* 8 Nm

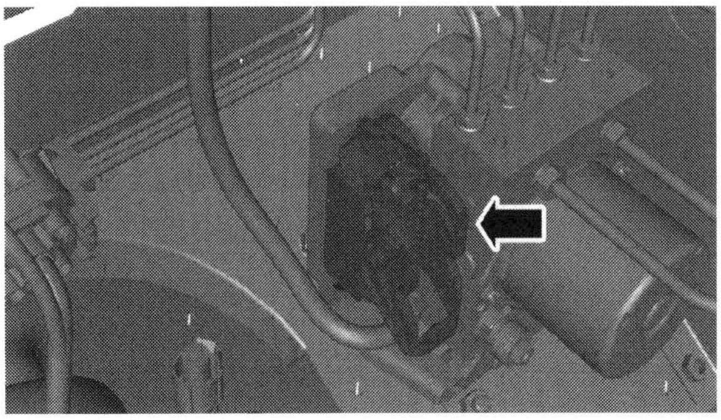

2.

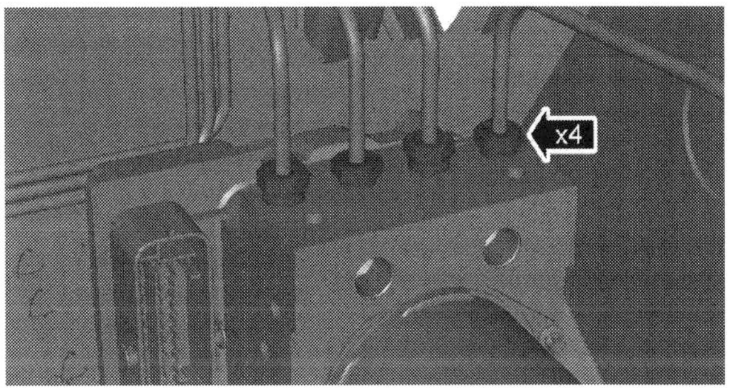

3. CAUTION: If brake fluid is spilt on the paintwork, the affected area must be immediately washed down with cold water.

NOTES:

Remove and discard the blanking caps.

Install the brake pipe unions finger tight.

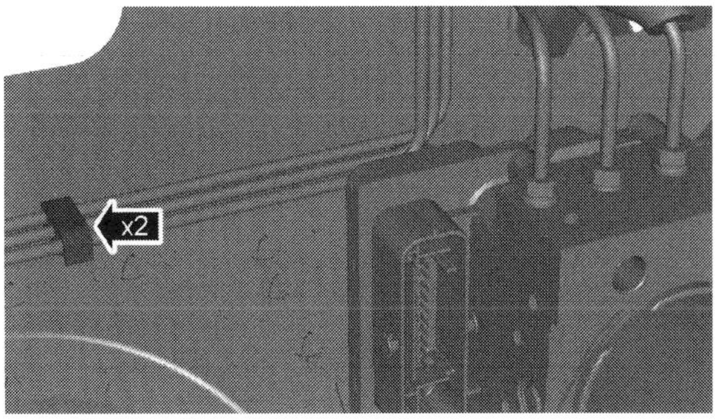

4. NOTE: Make sure that the securing clips and foam padding are installed in the same orientation as noted in the removal procedure.

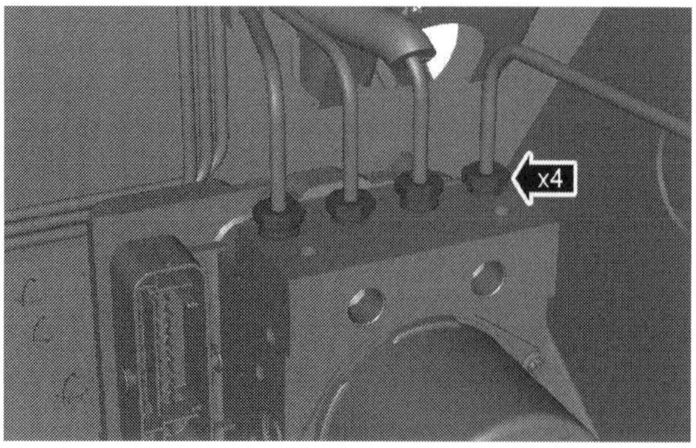

5. CAUTION: If brake fluid is spilt on the paintwork, the affected area must be immediately washed down with cold water.

   Torque: 15 Nm

6. CAUTION: If brake fluid is spilt on the paintwork, the affected area must be immediately washed down with cold water.

   Torque: 15 Nm

7. Configure the new module using the JLR approved diagnostic equipment.
8. **Refer to:** Brake System Bleeding (206-00 Brake System - General Information, General Procedures)Removal and Installation

Front Wheel Speed Sensor

Removal

CAUTION: If the sensor is removed for any reason, a new sensor bush must be installed.

NOTES:

Left-hand shown, right-hand similar.

Removal steps in this procedure may contain installation details.

Some variation in the illustrations may occur, but the essential information is always correct.

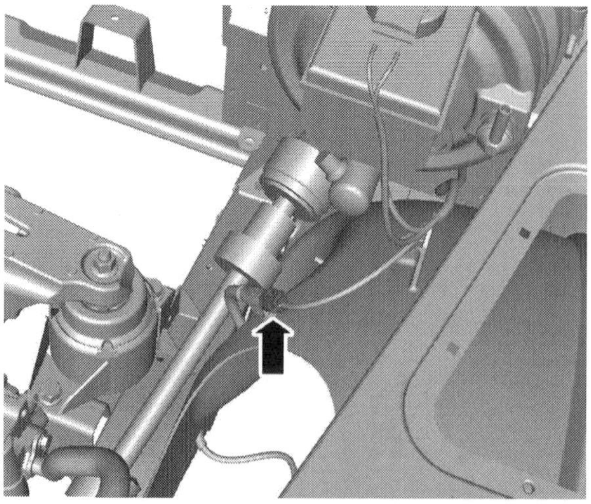

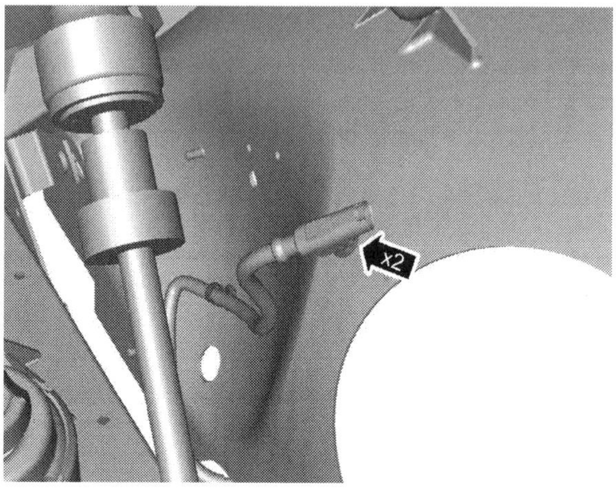

1. NOTE: LHD illustration shown, RHD is similar.
2. NOTE: LHD illustration shown, RHD is similar.
3. WARNING: Make sure to support the vehicle with axle stands.

Raise and support the vehicle.

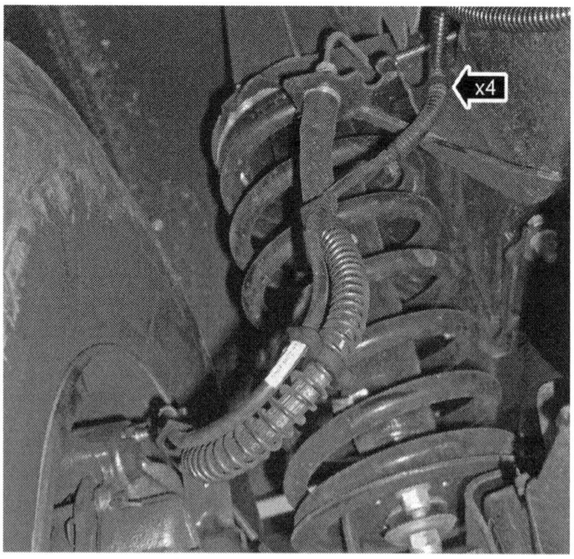

4.

5. CAUTION: Carefully prise ABS sensor from the front hub.

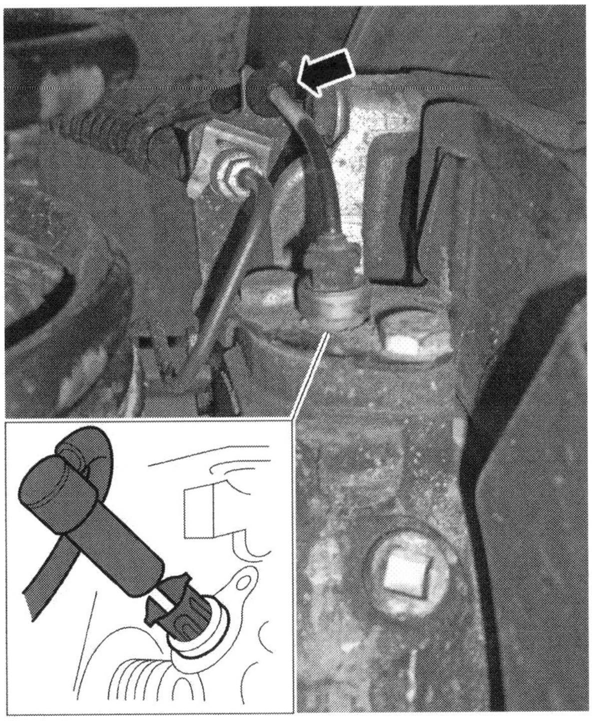

Installation

1. CAUTION: Clean the ABS sensor, smear with anti-seize grease.

To install, reverse the removal procedure.

Rear Left Wheel Speed Sensor

Removal and Installation

Removal

CAUTION: If the sensor is removed for any reason, a new sensor bush must be installed.

NOTES:

Some variation in the illustrations may occur, but the essential information is always correct.

Removal steps in this procedure may contain installation details.

1. WARNING: Make sure to support the vehicle with axle stands.

Raise and support the vehicle.

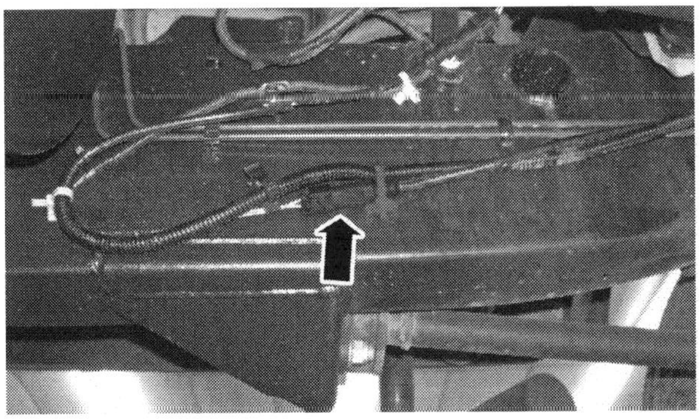

2.

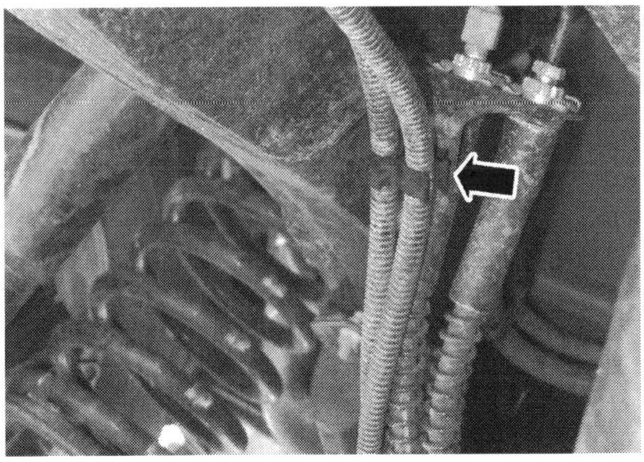

3.

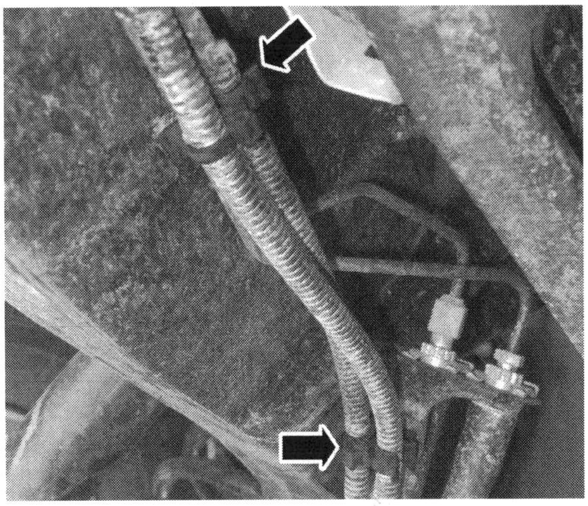

4.

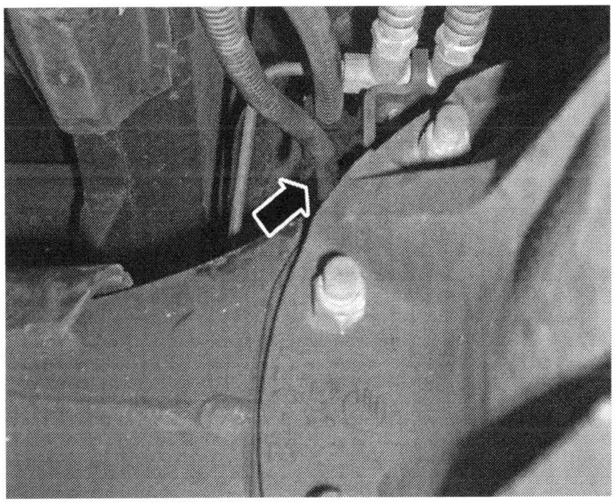

5.

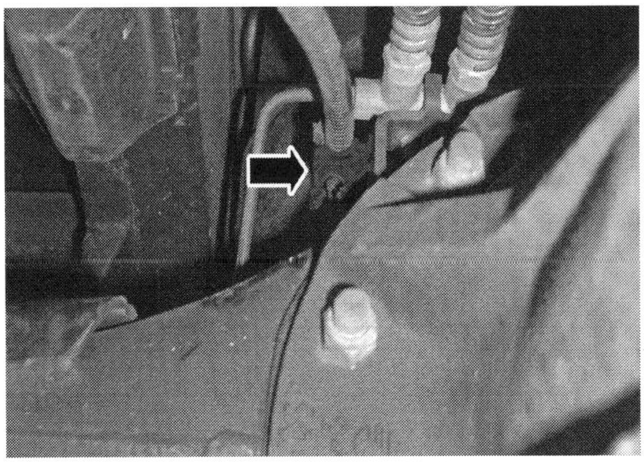

6.

7.

8. CAUTIONS:

⚠ Carefully pries ABS sensor from the front hub.

⚠ If the sensor is removed for any reason, a new sensor bush must be installed.

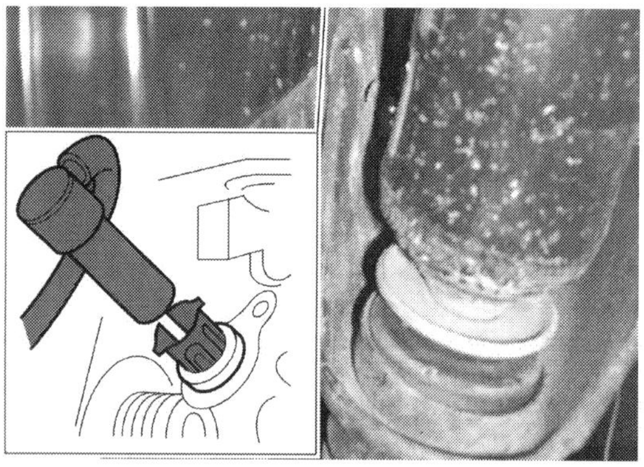

Installation

1. ⚠ CAUTION: Clean the ABS sensor, smear with anti-seize grease.

To install, reverse the removal procedure.

Rear Right Wheel Speed Sensor

Removal and Installation

Removal

⚠ CAUTION: If the sensor is removed for any reason, a new sensor bush must be installed.

NOTES:

△ Removal steps in this procedure may contain installation details.

△ Some variation in the illustrations may occur, but the essential information is always correct.

1. ⚠ WARNING: Make sure to support the vehicle with axle stands.

Raise and support the vehicle.

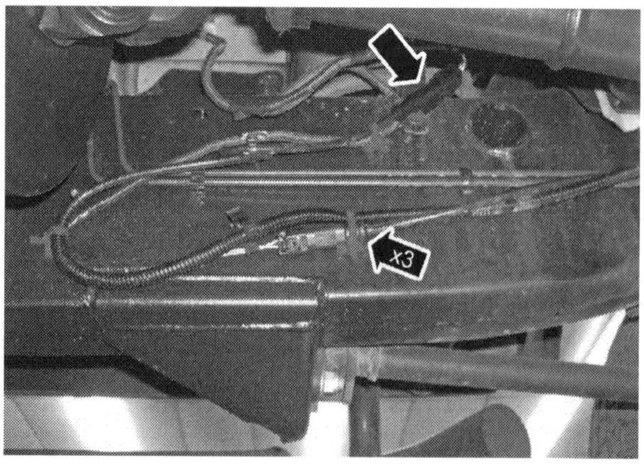

2.

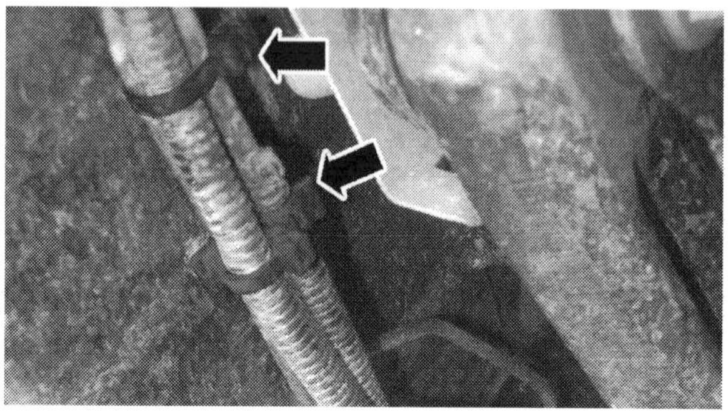

3.

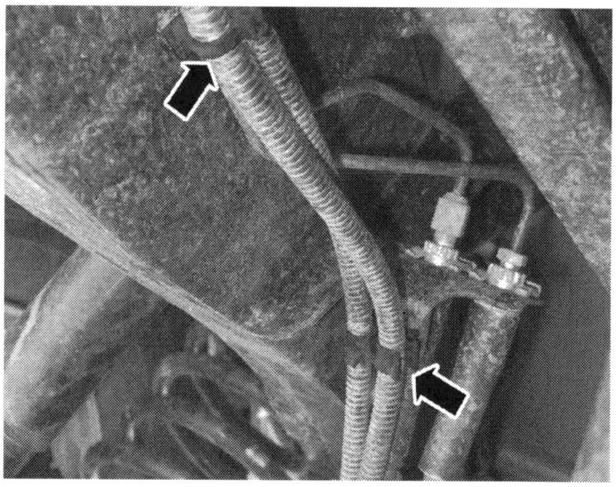

4.

⚠ Carefully prise ABS sensor from the front hub.

⚠ If the sensor is removed for any reason, a new sensor bush must be installed.

5.

6.

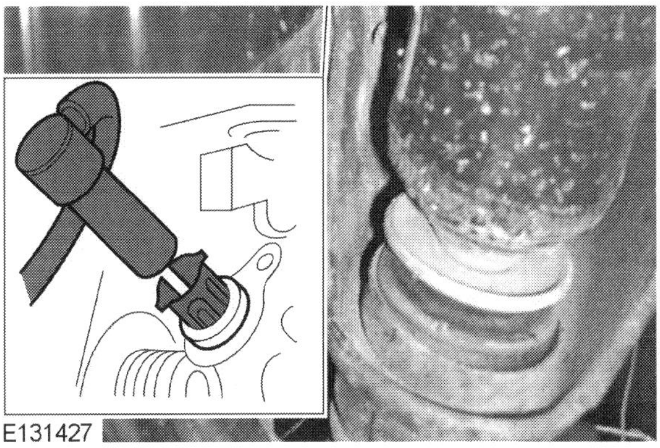

7. CAUTIONS:

Installation

1. CAUTION: Clean the ABS sensor, smear with anti-seize grease.

To install, reverse the removal procedure.

Hydraulic Control Unit - RHD

Removal and Installation

Removal

NOTE: Removal steps in this procedure may contain installation details.

1. CAUTIONS:

Be prepared to collect escaping fluids.

If brake fluid is spilt on the paintwork, the affected area must be immediately washed down with cold water.

Before disconnecting or removing the components, make sure the area around the joint faces and connections are clean.

Torque: 15 Nm

2. CAUTIONS:

Before disconnecting any components, make sure the area is clean and free from foreign material. When disconnected all openings must be sealed.

Be prepared to collect escaping fluids.

Before disconnecting or removing the components, make sure the area around the joint faces and connections are clean.

Torque: 15 Nm

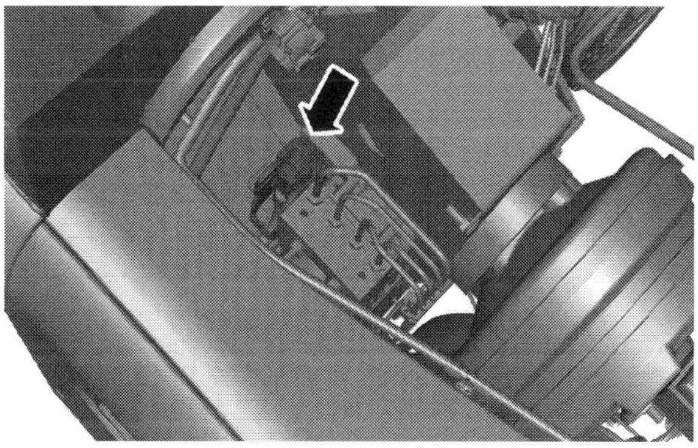

3.

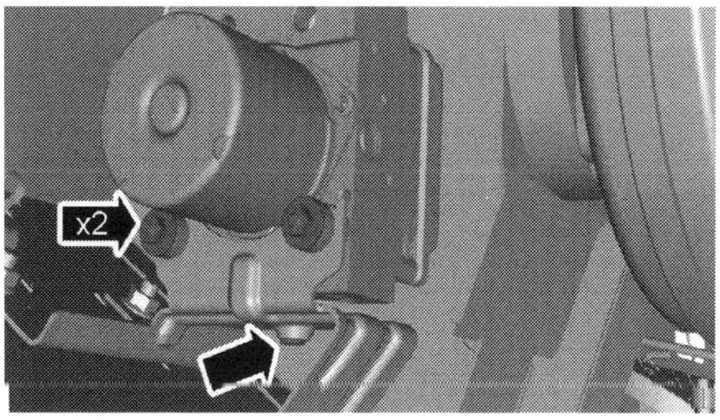

4. *Torque:* 8 Nm

Installation

1. To install, reverse the removal procedure.
2. Configure the new module using the JLR approved diagnostic equipment.

Description and Operation

Anti-Lock Control - Stability Assist

Refer to: Brake System (206-00 Brake System - General Information, Description and Operation)

Yaw Rate Sensor

Removal and Installation

Removal

⚠ CAUTION: If accidentally dropped or knocked install a new sensor.

NOTES:

△ Some variation in the illustrations may occur, but the essential information is always correct.

△ Removal steps in this procedure may contain installation details.

1. **Refer to:** Heater Core and Evaporator Core Housing (412-02, Removal and Installation).
2. *Torque:* 7 Nm

Installation

1. ⚠ CAUTION: The Yaw rate sensor is extremely sensitive to vibrations. Make sure that no components are in contact with the Yaw rate sensor.

To install, reverse the removal procedure.

Yaw Rate Sensor

Removal and Installation

Removal

⚠ CAUTION: If accidentally dropped or knocked install a new sensor.

NOTES:

△ Some variation in the illustrations may occur, but the essential information is always correct.

△ Removal steps in this procedure may contain installation details.

1. **Refer to**: Heater Core and Evaporator Core Housing (412-02, Removal and Installation).

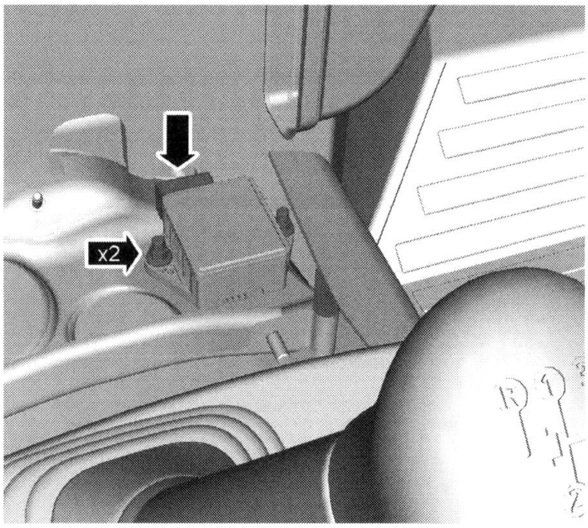

2. *Torque:* 7 Nm

Installation

1. ⚠ CAUTION: The Yaw rate sensor is extremely sensitive to vibrations. Make sure that no components are in contact with the Yaw rate sensor.

To install, reverse the removal procedure.

Specifications

Steering

Steering geometry	
Steering wheel diameter	412 mm (16.22 in)
Front wheel alignment	-10' ± 10' toe-out
Rear wheel alignment	+5' ± 15' toe-in
Camber angle - Front	-10' ± 45'
Camber angle - Rear	-10' ± 45'
Castor angle	3° 15' ± 45'
Cross castor	1° Maximum
Cross camber - Front	1° Maximum
Cross camber - Rear	1° Maximum
Swivel pin inclination - static	7° Check with vehicle on level ground at EEC kerb weight. Rock the front of the vehicle up and down to allow it to take up normal static position.

Turning circle between kerbs	
90 Models	
265/75 x 16 tires	12.65 m (41.5 ft)
All other tires	11.70 m (38.4 ft)
110 Models	
750 x 16 tires	13.41 m (44 ft)
130 Models	
750 x 16 tires	15.24 m (50 ft)

Power steering system

Fluids and capacities	
Power steering fluid	Texaco Cold Climate 33270
Fluid capacity - steering gear and reservoir	1.6 litres (3.4 pints)

Torque specifications

Description	Nm	lbf-ft
Drag link to drop arm	40	30
Ball joint nuts	40	30
Clamp bolt (M6)	9	7
Steering column bracket nuts	22	16
Steering wheel nut	43	32
Tie bar to mounting nut initial torque	80	60
Tie bar to mounting nut final torque	110	80
Universal joint pinch bolt	30	22
Steering gear adjuster locknut	60	45
Steering gear drop arm nut	175	130
Steering gear sector shaft cover to steering box	75	55
Steering gear to chassis	80	60
Steering gear fluid pipes 14mm thread	15	11
Steering gear fluid pipes 16mm thread	20	15
Steering gear tie bar	80	60
Power steering pump high pressure fluid pipe	20	15
Power steering pump mounting	35	25
Power steering pump pulley bolts	10	7
Power steering pump hose clamp	3	2
Power steering pump front mounting plate bolts	9	7
Power steering reservoir hose clamp	3	2

Fixing size	Nm	lbf-ft
M5	6	4.5
M6	9	7
M8	25	18
M10	45	33
M12	90	65
M14	105	77
M16	180	133
1/4 in	9	7
5/16 in	24	18
3/8 in	39	29
7/16 in	78	58
½ in	90	65
5/8 in	135	100

Torque specifications in the table above are for all screws and bolts used, except for those specifically referred to in the Torque specifications table.

Steering System

Diagnosis and Testing

Principles of Operation

For a detailed description of the Steering System, refer to the relevant Description and Operation section in the workshop manual.

Refer to: Steering System (211-00 Steering System - General Information, Description and Operation)

Inspection and Verification

CAUTIONS:

⚠ Diagnosis by substitution from a donor vehicle is NOT acceptable. Substitution of control modules does not guarantee confirmation of a fault, and may also cause additional faults in the vehicle being tested and/or the donor vehicle.

⚠ Do not hold steering wheel on full lock for more than 30 seconds in any one minute to avoid overheating fluid and possibly damaging seals.

Excessive pressure in the system is almost always caused by a faulty relief valve in the PAS pump.

Insufficient pressure in the system is usually caused by low fluid level or PAS pump drive belt slip, or one of the following: PAS system leaks, faulty PAS pump relief valve, fault in steering box valve and worm assembly, leak at piston in steering box, worn components in PAS pump or box.

NOTES:

△ If the control module or a component is suspect and the vehicle remains under manufacturer warranty, refer to the Warranty Policy and Procedures manual (section B1.2), or determine if any prior approval programme is in operation, prior to the installation of a new module/ component.

⚠ When performing voltage or resistance tests, always use a digital multimeter accurate to three decimal places, and with an up-to-date calibration certificate. When testing resistance always take the resistance of the digital multimeter leads into account.

⚠ Check and rectify basic faults before beginning diagnostic routines involving pinpoint tests.

1. Visually inspect for obvious signs of damage and system integrity

Visual Inspection

2. If an obvious cause for an observed or reported concern is found, correct the cause (if possible) before proceeding to the next step

3. If the cause is not visually evident, verify the symptom and refer to the Symptom Chart, alternatively check for Diagnostic Trouble Codes (DTCs) and refer to the DTC Index

Mechanical
1. Tire pressures 2. Panhard rod 3. Steering column 4. Steering box 5. Drop arm 6. Drag link 7. Tie rod 8. Ball joints 9. Steering damper 10. Power steering fluid level 11. Power steering pump 12. Power steering pipes and hoses 13. Primary drive belt

Symptom Chart

Symptom	Possible Causes	⚠ Action
Excessive freeplay in steering	• Panhard rod insecure • Ball joint worn • Steering box incorrectly adjusted	⚠ NOTE: Adjustments to the steering box should not be required during the warranty period. • Check the installation of the panhard rod • Check the ball joints • Adjust the steering box
Steering stiff/binding	• Tire pressure(s) incorrect • Steering column lower shaft universal joint seized • Tie rod damaged and fouling on axle • Steering damper failure • Swivel pin bearing pre-load incorrect	• Check and adjust the tire pressures • Install a new steering column lower shaft • Install a new tie rod • GO to Pinpoint Test B • Check and adjust the swivel pin bearing pre-load
Excessive kickback through steering wheel	• Steering damper failure • Ball joint worn • Suspension bushes worn • Swivel pin bearing pre-load incorrect	• GO to Pinpoint Test B • Check the ball joints • Install new suspension bushes • Check and adjust the swivel pin bearing pre-load

180

Insufficient power assistance	• Tire pressure(s) incorrect • Steering column lower shaft universal joint seized • Power steering fluid leak • Primary drive belt slipping/loose • Insufficient fluid pressure	⚠ CAUTION: The steering wheel must not be held on full lock for more than 30 seconds in one minute, as this may overheat the fluid and cause damage to the oil seals. • Check and adjust the tire pressures • Install a new steering column lower shaft • Rectify the leak as necessary. Top up and bleed the power steering fluid • Check the condition and tension of the primary drive belt • GO to Pinpoint Test A
Vibration through steering wheel	• Wheel/tire sizes incorrect • Wheel balance incorrect • Steering damper failure • Steering box incorrectly adjusted • Wheel bearing(s) excessive freeplay • Swivel pin bearing pre-load incorrect • Wheel alignment adjusted incorrectly	⚠ NOTE: Adjustments to the steering box should not be required during the warranty period. • Check that the wheels and tires comply with the manufacturer's specification for the vehicle • Balance the wheels • GO to Pinpoint Test B • Adjust the steering box • Check the wheel bearings end-float • Check and adjust the swivel pin bearing pre-load • Check and adjust the wheel alignment
Vehicle pulls to left/right - When not braking	• Wheel/tire sizes incorrect • Tyres worn irregularly • Steering box not centralised	• Check that the wheels and tires comply with the manufacturer's specification for the vehicle • Check the tires • Centralise the steering box

Vehicle pulls to left/right - When braking	• Air in brake system • Brake pads seized in caliper • Swivel pin bearing pre-load incorrect	• Refer to the relevant section of the workshop manual and bleed the brake hydraulic system • Check the installation of the brake pads • Check and adjust the swivel pin bearing pre-load
Poor directional stability	• Vehicle overloaded or unevenly loaded • Wheel balance incorrect • Rear suspension bushes worn • Worn shock absorbers • Steering damper failure	• Refer to the Owner Handbook for vehicle loading information • Balance the wheels • Install new rear suspension bushes • Install new shock absorbers • GO to Pinpoint Test B
Power steering noisy	• Power steering fluid level low • Primary drive belt slipping/loose • Power steering pipes/hoses fouling the body/chassis	⚠ CAUTION: The steering wheel must not be held on full lock for more than 30 seconds in one minute, as this may overheat the fluid and cause damage to the oil seals. • Check the power steering system for leaks. Rectify the leak as necessary. Top up and bleed the power steering fluid • Check the condition and tension of the primary drive belt • Check the power steering pipes/hoses for correct installation

181

Power Steering System Pressure Test

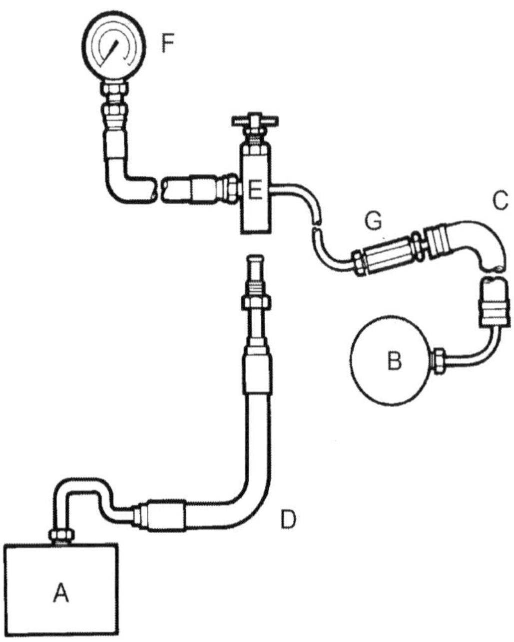

Item	Description
A.	Steering box
B.	Power steering pump
C.	Hose (power steering pump to steering box)
D.	Hose LRT-57-030
E.	Test adaptor LRT-57-001
F.	Pressure gauge LRT-57-005
G.	Thread adaptor LRT-57-004
H.	Thread adaptor LRT-57-022

Pinpoint Tests

PINPOINT TEST A : POWER STEERING PRESSURE TESTS	
TEST CONDITIONS	**DETAILS/RESULTS/ACTIONS**
A1: POWER STEERING PRESSURE TEST 1	
CAUTIONS: ⚠ The gauge is able to measure a pressure of up to 140kgf/cm². The steering wheel should be turned slowly during testing to avoid exceeding this pressure and damaging the gauge.	
⚠ The steering wheel must not be held on full lock for more than 30 seconds in one minute, as this may overheat the fluid and cause damage to the oil seals.	
NOTE: Rectify any power steering fluid leaks before commencing the pressure test.	
	1. Connect the pressure gauge to the vehicle, as detailed above
	2. Set the test adapter valve to open
	3. Start the engine
	4. Top up and bleed the power steering system

	5. Allow the engine to idle
	6. Centralise the steering
	Does the gauge read less than 7kgf/cm²? **Yes** GO to 2 **No** GO to 4

A2: POWER STEERING PRESSURE TEST 2

CAUTIONS:

⚠ The gauge is able to measure a pressure of up to 140kgf/cm². The steering wheel should be turned slowly during testing to avoid exceeding this pressure and damaging the gauge.

⚠ The steering wheel must not be held on full lock for more than 30 seconds in one minute, as this may overheat the fluid and cause damage to the oil seals.

	1. Allow the engine to idle
	2. Turn the steering wheel fully to the left and note the pressure value on the gauge
	3. Turn the steering wheel fully to the right and note the pressure value on the gauge
	4. Centralise the steering
	In both instances, did the gauge read approximately 28kgf/cm²? **Yes** GO to 3 **No** GO to 4

A3: POWER STEERING PRESSURE TEST 3

CAUTIONS:

⚠ The gauge is able to measure a pressure of up to 140kgf/cm². The steering wheel should be turned slowly during testing to avoid exceeding this pressure and damaging the gauge.

⚠ The steering wheel must not be held on full lock for more than 30 seconds in one minute, as this may overheat the fluid and cause damage to the oil seals.

	1. Raise the engine speed to 1000rpm
	2. Turn the steering wheel fully to the left and note the pressure value on the gauge
	3. Turn the steering wheel fully to the right and note the pressure value on the gauge
	4. Centralise the steering
	5. Allow the engine to idle

	In both instances, did the gauge read 70kgf/cm² to 77kgf/cm²? **Yes** Power steering pressure test passed **No** GO to 4
A4: POWER STEERING PRESSURE TEST 4	
	1. Allow the engine to idle
	2. Set the test adapter valve to closed (for a maximum of 5 seconds) and note the pressure value on the gauge
	Did the gauge reading increase? **Yes** Install a new steering box and repeat the power steering pressure test. GO to 1 **No** Install a new power steering pump and repeat the power steering pressure test. GO to 1

PINPOINT TEST B : STEERING DAMPER TESTS	
TEST CONDITIONS	**DETAILS/RESULTS/ACTIONS**
B1: STEERING DAMPER TEST 1	
	1. Inspect steering damper for damage and fluid leaks
	Is the steering damper damaged or leaking? **Yes** Install a new steering damper **No** GO to 2
B2: STEERING DAMPER TEST 2	
	1. Remove the steering damper
	2. Secure one end of the steering damper in a vice (using soft jaws)
	3. Extend and retract the steering damper by hand
	Is the resistance equal in both directions? **Yes** Install the original steering damper to the vehicle **No** Install a new steering damper

Power Steering System Filling and Bleeding

General Procedures

1. Check that fluid level is at maximum level.
2. Run engine to normal operating temperature.
3. Recheck reservoir fluid level, top up if necessary.
4. NOTE: Maintain maximum fluid level in reservoir. Do not increase engine speed or move steering wheel.

With engine at idle speed, slacken bleed screw. When fluid seeps past bleed screw re-tighten screw.

5. Check fluid level.
6. Clean fluid from around bleed screw.
7. Check hose connections, pump and steering box for fluid leaks by holding steering on full lock in both directions.

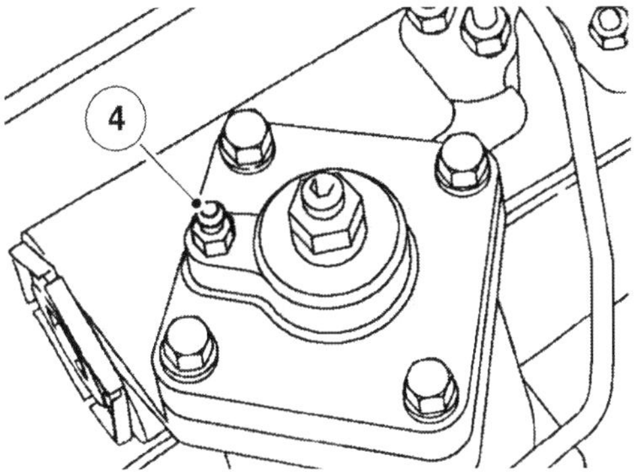

CAUTION: Do not maintain this pressure for more than 30 seconds in any one minute to avoid overheating fluid and possibly damaging seals.

8. Carry out road test.
9. Check and top-up fluid level.

Power Steering System Flushing

General Procedures

NOTES:

If heavy steering or contamination within the power steering system is found, it is necessary to carry out the system flush procedure as detailed below. If any components have been replaced in the power steering system the procedure below must be carried out in full.

Some variation in the illustrations may occur, but the essential information is always correct.

1. Remove the power steering fluid reservoir cap.
2. Using a suitable syringe, remove the power steering fluid from the power steering fluid reservoir.

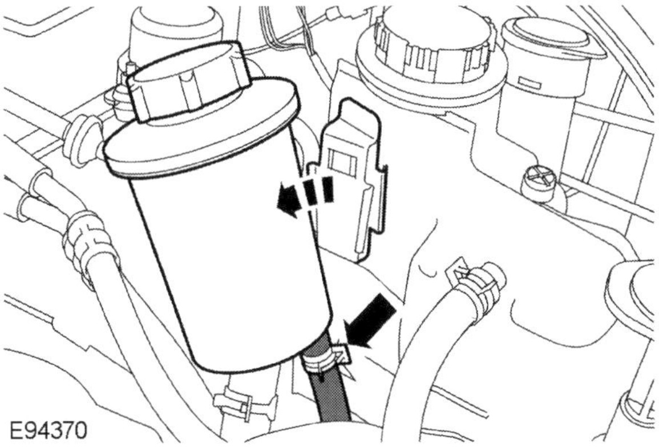

E94370

3. CAUTION: Be prepared to collect escaping fluids.

NOTE: Note the orientation of the clip.

Detach the power steering fluid reservoir.

- Detach but do not remove the power steering fluid reservoir.
- Release the power steering fluid return hose from the power steering fluid reservoir.
- If a quick release coupling is fitted to the power steering return hose, release the power steering fluid return hose from the coupling by removing the clip.

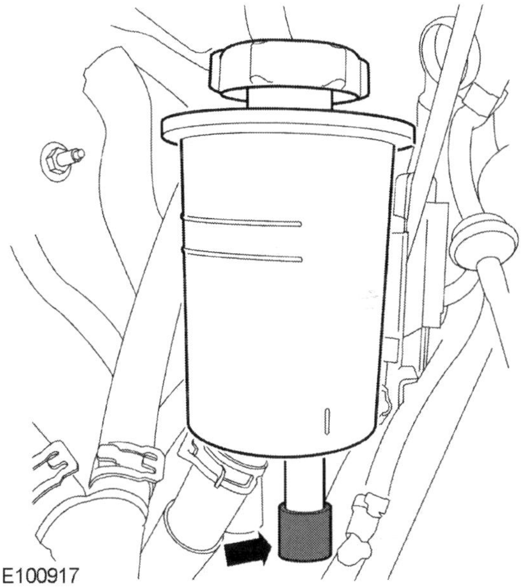

E100917

4. CAUTION: Be prepared to collect escaping fluids.

NOTE: Before disconnecting any components, make sure the area is clean and free from foreign material. When disconnected all openings must be sealed.

Using a suitable blanking cap, cap the power steering reservoir return pipe.

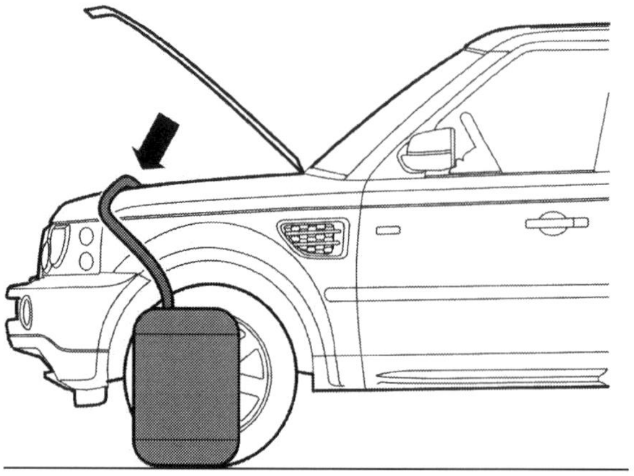

5. CAUTION: Be prepared to collect escaping fluids.

NOTE: Make sure the extended pipe is not kinked or twisted and is correctly secured with hose clips.

Attach a suitable pipe to the power steering return hose to allow the fluid to drain.

6. NOTES: The suitable funnel should have the a capacity of 4 liters and an O-ring seal

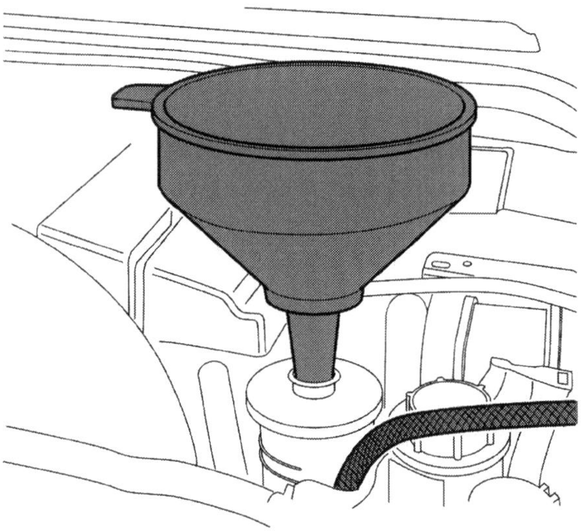

E94372

The suitable funnel must be tightly sealed to the power steering fluid reservoir to avoid fluid leakage.

Install a suitable funnel onto the power steering fluid reservoir.

7. WARNING: Do not work on or under a vehicle supported only by a jack. Always support the vehicle on safety stands.

Raise and support the vehicle with the wheels just clear of the ground.

8. CAUTIONS:

Steps 8 and 9 must be carried out within 2 - 3 seconds of each other. Failure to follow this instruction may result in damage to the power steering system.

Be prepared to collect escaping fluids.

Using the suitable funnel, top up the power steering system with the specified fluid. Make sure the fluid level is maintained at two thirds full in the funnel.

9. CAUTIONS:

Be prepared to collect escaping fluids.

Do not allow the power steering fluid level in the power steering fluid reservoir to fall below the minimum power steering fluid level. Failure to follow this instruction may result in damage to the power steering system.

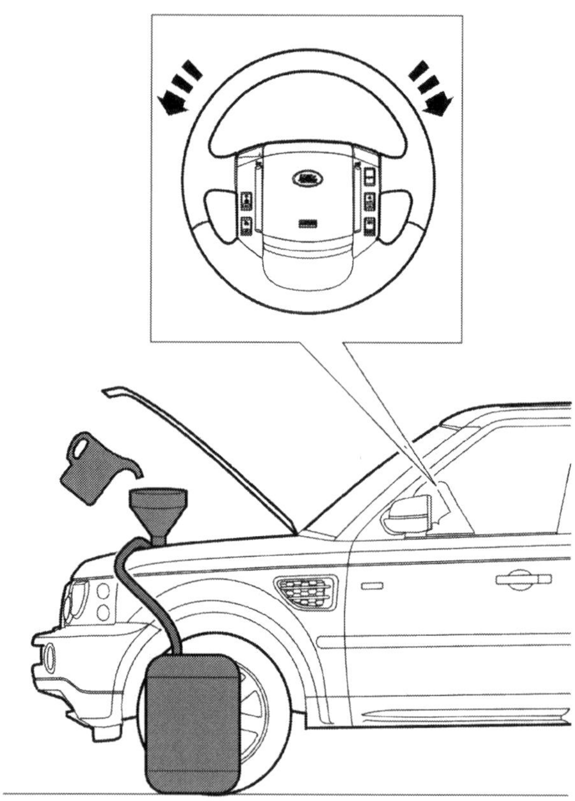

E94373

Make sure the engine is switched off as soon as the full 4 liters of power steering fluid has entered the power steering fluid reservoir.

Flush the power steering system.

- Start the engine
- With assistance turn the steering slowly lock to lock 3 times at approximately 1 revolution every 5 seconds.
- Continue to flush the power steering system until 4 liters of power steering fluid has been added to the power steering reservoir. This should take approximately 30 seconds.

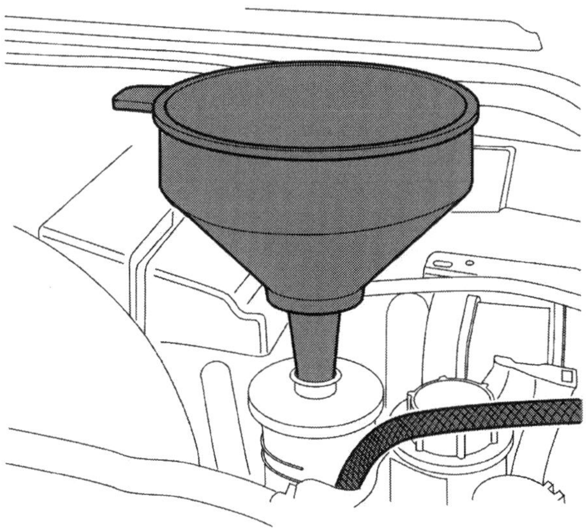

E94372

10. CAUTION: Be prepared to collect escaping fluids.

Remove the suitable funnel.

11. CAUTION: Be prepared to collect escaping fluids.

Remove the suitable pipe to the power steering return hose.

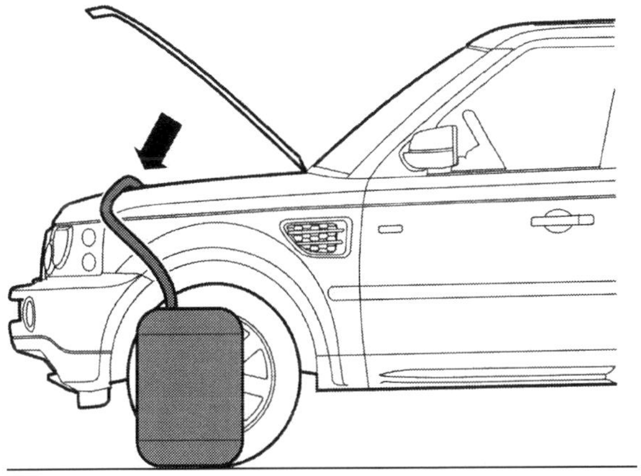

E100918

12. CAUTION: Be prepared to collect escaping fluids.

NOTE: Note the orientation of the clip.

If a quick release coupling is fitted to the power steering return hose, connect the power steering fluid return hose to the coupling by installing the clip.

13. Install a new power steering fluid reservoir.

Refer to: Power Steering Fluid Reservoir (211-02 Power Steering, Removal and Installation)

Steering Gear Adjustment

General Procedures

1. WARNING: Adjustments of steering box should not be required while in warranty period. If box is stiff or tight and within warranty, it must be returned to manufacturer. No attempt must be made to introduce backlash.

Apply park brake, chock wheels and jack up front of vehicle until wheels are clear of ground.

2. Support chassis front on axle stands.
3. Disconnect drag link from steering drop arm.
4. Check torque to turn.

Refer to: Steering System (211-00 Steering System - General Information, Diagnosis and Testing)

5. NOTES:

Only check for no backlash when steering box is in central position.

If steering wheel is not straight, it should be repositioned

Refer to: Steering Wheel (211-04 Steering Column, Removal and Installation)

Centralise steering box.

Refer to: Steering Gear Centralization (211-00 Steering System - General Information, General Procedures)

6. Adjustment is obtained by rocking the drop arm about centre whilst an assistant slowly tightens the steering box adjuster screw.

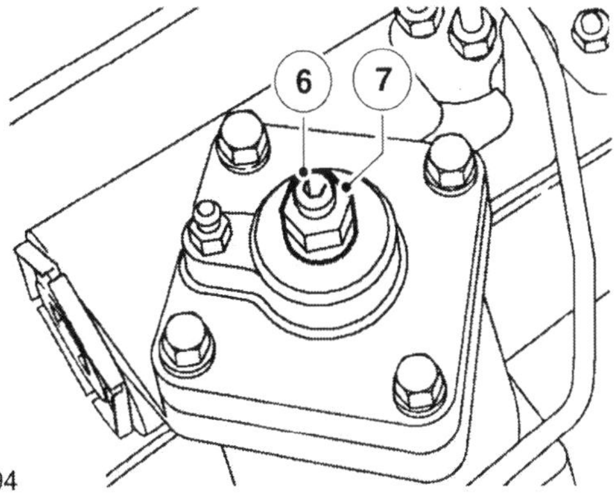

J6294

7. Tighten locknut when all backlash has been removed.
8. Repeat the check for backlash. If backlash exists loosen locknut and repeat adjustment procedure.
9. Turn steering wheel lock to lock and check no tightness exists.

10. Ensure front wheels are aligned and in straight ahead position.
11. Adjust drag link 924 mm between ball joint centres.
12. Connect drag link and tighten to 40 Nm (30 lbf.ft).
13. Lower vehicle to ground level and remove chocks.
14. Road test vehicle

Refer to: Steering System (211-00 Steering System - General Information, Diagnosis and Testing)

15. RH drive vehicles - if steering wheel is to right, drag link is too long. If steering wheel is to left drag link is too short. LH drive vehicles - if steering wheel is to right, drag link is too short. If steering wheel is to left drag link is too long.
16. Adjust drag link until steering wheel points straight ahead when vehicle is travelling in a straight line.

Steering Gear Centralization

General Procedures

All vehicles

1. Disconnect sector shaft arm drag link from drop arm.

Refer to: Sector Shaft Arm Drag Link (211-03 Steering Linkage, Removal and Installation)

Right-hand drive vehicles

1. Turn steering wheel on full LH lock.
2. Turn steering wheel back exactly two turns.

Left-hand drive vehicles

1. Turn steering wheel on full RH lock.
2. Turn steering wheel back exactly two turns.

All vehicles

1. Fit sector shaft arm drag link.

Refer to: Sector Shaft Arm Drag Link (211-03 Steering Linkage, Removal and Installation)

Steering Lock Stop Adjustment

General Procedures

1. Measure clearance between tire wall and radius arm at full lock. This must be not less than 20 mm.

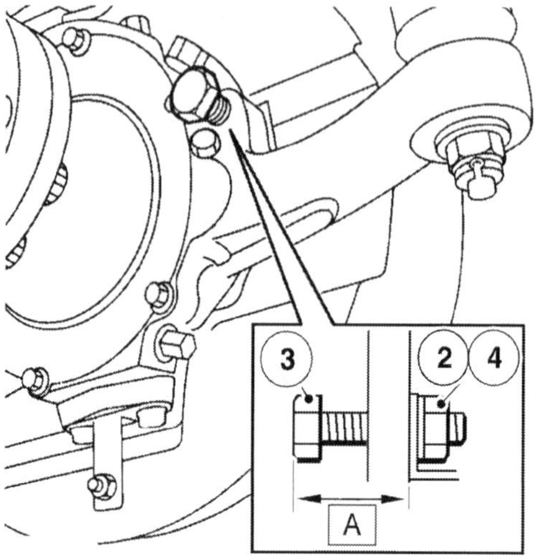

2. Loosen stop bolt locknut.
3. Turn stop bolt as required.
4. Tighten locknut. 5. Check clearance between tire wall and radius arm on each lock.
6. △ NOTE: Alternatively lock stop adjustment may be carried out using following procedure.

Measure stop bolt protrusion A. Refer to table for correct setting.

7. Loosen stop bolt locknut.
8. Turn stop bolt as required.
9. Tighten locknut.
10. Check wheel position at full lock.

Alloys	**Size**	**Setting**
BF Goodrich Mud Terrain	265	59.7 mm
Goodyear GT+4	235	55.7 mm
Michelin M+S 4X4	235	52.2 mm
Steels	**Size**	**Setting**
Goodyear	205	55.2 mm
Michelin	205	55.2 mm
Avon	7.50	56 mm
Michelin	7.50	56 mm
Goodyear	7.50	56 mm

Specifications

Description	Nm	lb-ft
Power steering pump bolts	23	17
Power steering fluid reservoir nuts	4	3
Power steering high-pressure pipe union	20	15

Removal and Installation

Part(s)

STEP	PART NAME	QUANTITY
Installation Step 2	Renew Part Steering gear drop arm nut	1
Installation Step 2	Renew Part Steering gear drop arm washer	1
Installation Step 7	Renew Part Steering gear bolts	4
Installation Step 7	Renew Part Steering gear tab washer	2
Installation Step 14	Renew Part Panhard rod nut(s)	2

Removal

1. Remove the steering column lower shaft.

Refer to: Steering Column Lock and Ignition Switch Housing (211-05 Steering Column Switches, Removal and Installation)

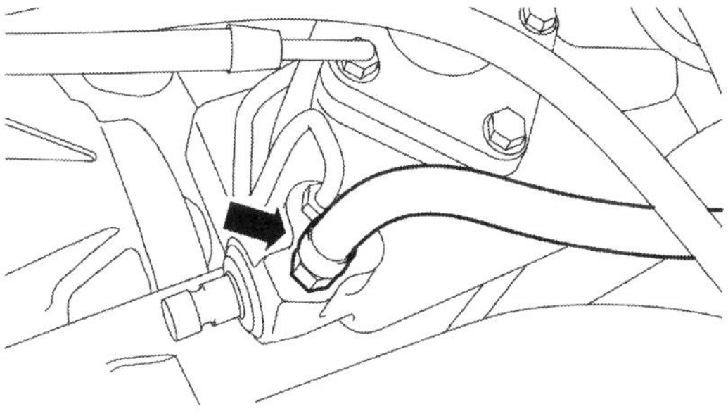

2. CAUTIONS:

If power steering fluid comes into contact with the paintwork, the affected area must be immediately washed down with cold water.

Before disconnecting or removing the components, make sure the area around the joint faces and connections are clean. Plug open connections to prevent contamination.

Position a suitable container to collect any power steering fluid spillage. Disconnect the power assisted steering (PAS) return hose from the steering gear.

3. CAUTIONS:

If power steering fluid comes into contact with the paintwork, the affected area must be immediately washed down with cold water.

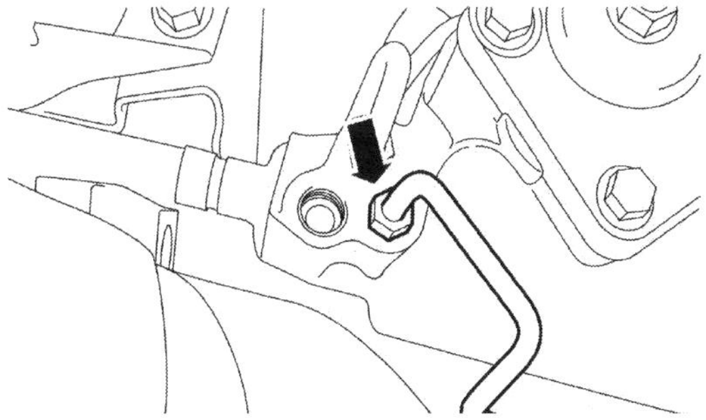

Before disconnecting or removing the components, make sure the area around the joint faces and connections are clean. Plug open connections to prevent contamination.

Disconnect the PAS high pressure feed pipe from the steering gear.

4. WARNING: Do not work on or under a vehicle supported only by a jack. Always support the vehicle on safety stands.

Raise and support the vehicle.

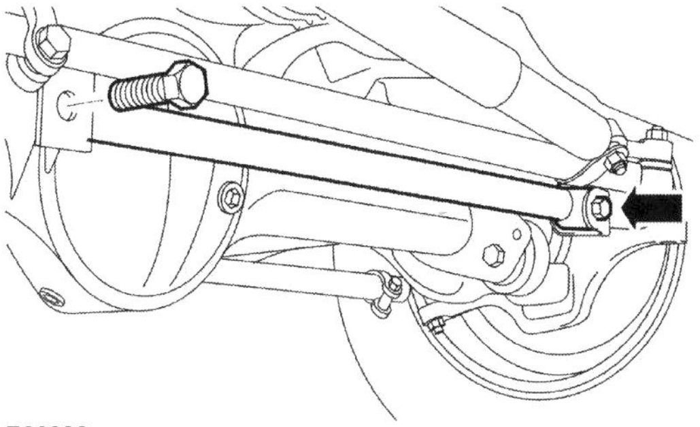

E90339

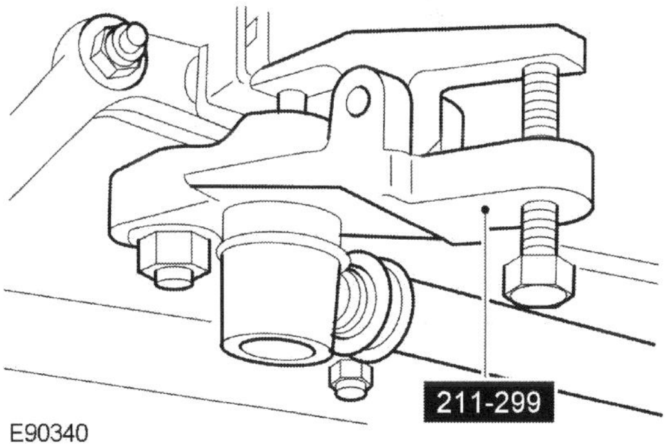

E90340

5. Remove the 2 panhard rod nuts and bolts and remove the panhard rod. Discard the 2 panhard rod nuts.

6. CAUTION: Make sure the ball joint seal is not damaged. A damaged seal will lead to the premature failure of the joint.

Using the special tool, remove the split pin and remove the drag link to drop arm nut, release the drag link from the drop arm.

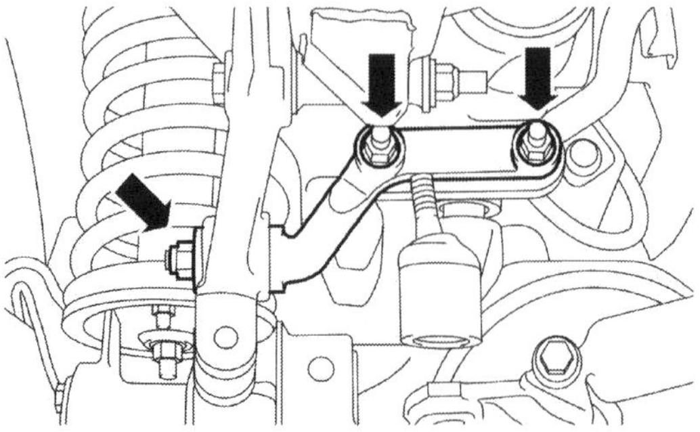

7. Remove the 2 nuts and bolts and remove 1 nut. Remove the tie bar.
8. Release the tab washers from the 4 bolts. Remove and discard the 4 steering gear bolts. Remove and discard the 2 tab washers. With assistance, remove the steering gear.

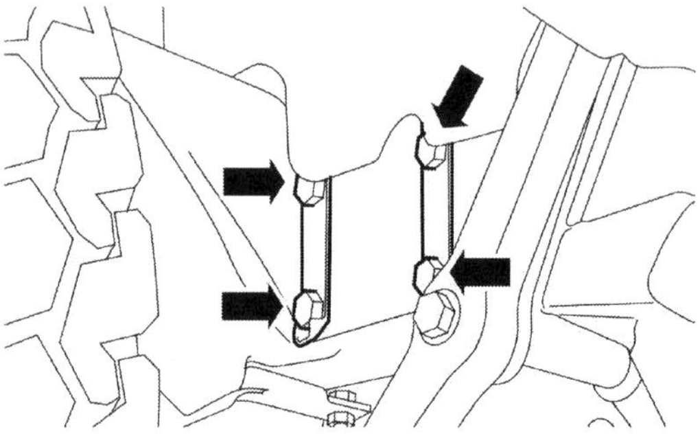

9. NOTE: When installing a reconditioned steering gear, complete steps 12, 13 and 14.

Release the drop arm nut lock washer. Secure the steering gear assembly in a suitable vice.

10. Discard the drop arm nut lock washer. Remove and discard the drop arm nut.
11. NOTE: Some variation in the illustrations may occur, but the essential information is always correct.

Using the special tool, remove the drop arm from the steering gear.

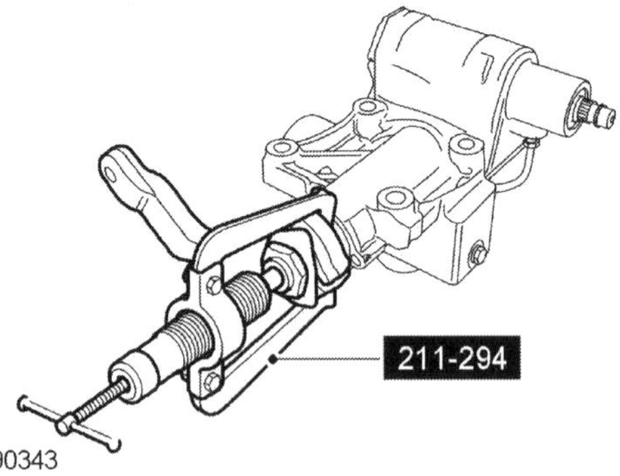

E90343

Installation

1. Clean the drop arm and steering gear mating faces.
2. NOTE: Install a new nut and lock washer.

Install the drop arm to the steering gear. Install the lock washer. Install but do not fully tighten the drop arm nut.

Renew Part: Steering gear drop arm nut

Renew Part: Steering gear drop arm washer

3. Secure the drop arm in a suitable vice.
4. Tighten the drop arm nut.

   Torque: 176 Nm

5. Secure the drop arm nut using the lock washer.
6. Remove the steering gear assembly from the vice.
7. NOTE: Install new bolts and tab washers.

With assistance, install the steering gear. Install the tab washers. Install and tighten the steering gear bolts.

Renew Part: Steering gear bolts

Renew Part: Steering gear tab washer

Torque: 65 Nm

8. Secure the steering gear bolts using the tab washers.
9. Install the tie bar to the steering gear. Loosely install the nuts and bolts.
10. Tighten the tie bar to panhard rod mount bracket nut.

    Torque: 65 Nm

11. Tighten the tie bar to steering gear nuts and bolts.

    Torque: 85 Nm

12. Install and tighten the nut. Install the split pin. Secure the drag link to the drop arm.

    Torque: 40 Nm

13. NOTE: Steering gear shown removed for clarity.

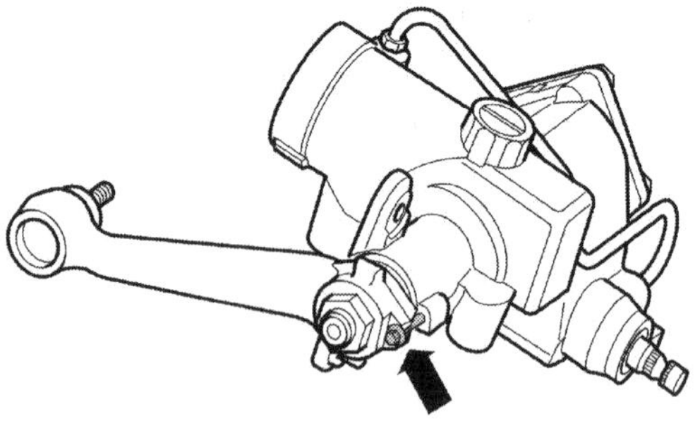

Install a centralizing bolt to the steering gear.

14. NOTE: Install new nuts.

Install the panhard rod. Tighten the nuts and bolts.

Renew Part: Panhard rod nut(s)

Torque: 230 Nm

15. CAUTION: If power steering fluid comes into contact with the paintwork, the affected area must be immediately washed down with cold water.

NOTE: Inspect the PAS high pressure feed pipe O-ring seal, if there is any indication of damage install a new seal.

Connect the PAS high pressure feed pipe to the steering gear. Tighten as required.

Torque: 18 Nm

16. CAUTION: If power steering fluid comes into contact with the paintwork, the affected area must be immediately washed down with cold water.

NOTE: Inspect the PAS return hose O-ring seal, if there is any indication of damage install a new seal.

Connect the PAS return hose to the steering gear. Tighten as required.

Torque: 22 Nm

17. Install the steering column lower shaft.

Refer to: Steering Column Lock and Ignition Switch Housing (211-05 Steering Column Switches, Removal and Installation)

18. Refill and bleed the PAS system.

Refer to: Power Steering System Filling and Bleeding (211-00 Steering System - General Information, General Procedures)

Power Steering Fluid Reservoir

Removal and Installation

Removal

1. Position drain tin beneath reservoir.
2. Slacken mounting bracket clamp bolt and raise reservoir to gain access to feed and return hose retaining clips.
3. Slacken clips, disconnect hoses from reservoir, allow fluid to drain.

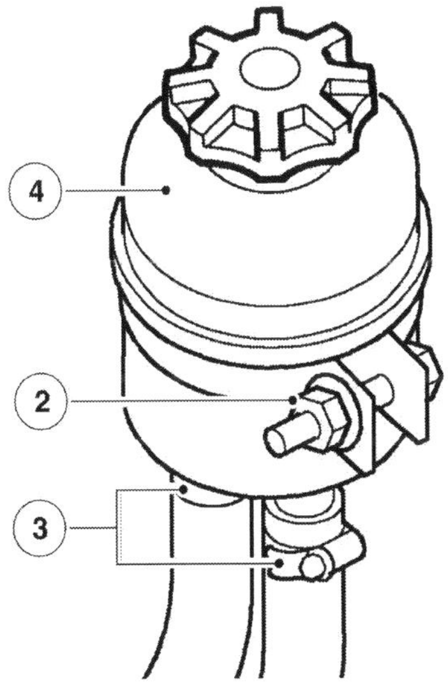

4. CAUTION: Plug connections to prevent ingress of dirt.

Discard fluid drained from system. Do not allow fluid to contact paintwork, if spilled, remove fluid and clean area with warm water.

Remove reservoir.

Installation

1. Position reservoir, connect feed and return hoses and tighten clips.

   Torque: 3 Nm

2. Fit reservoir in clamp, tighten clamp bolt.
3. Fill reservoir between upper mark and end of dipstick with power steering fluid.

Refer to: (211-00 Steering System - General Information, Specifications)

4. Bleed power steering system.

Refer to: Power Steering System Filling and Bleeding (211-00 Steering System - General Information, General Procedures)

Power Steering Pump

Removal and Installation

Part(s)

STEP	PART NAME	QUANTITY
Installation Step 5	Renew Part Power steering hose O-ring seal(s)	1

Removal

1. Disconnect the battery ground cable.

Refer to: Startup Battery Disconnect and Connect (414-01 Battery, Mounting and Cables, General Procedures)

2. Remove the cooling fan.

Refer to: Cooling Fan (303-03 Engine Cooling, ID4 2.2L Diesel Removal and Installation)

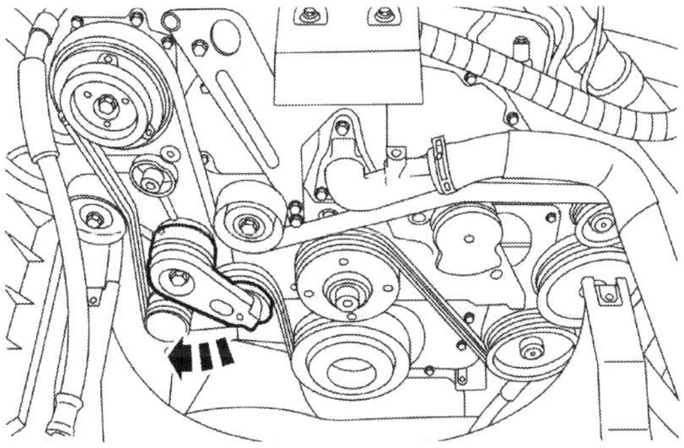

3. Rotate the accessory drive belt tensioner clockwise. Release the tension from the accessory drive belt.

4. WARNING: Do not work on or under a vehicle supported only by a jack. Always support the vehicle on safety stands.

Raise and support the vehicle.

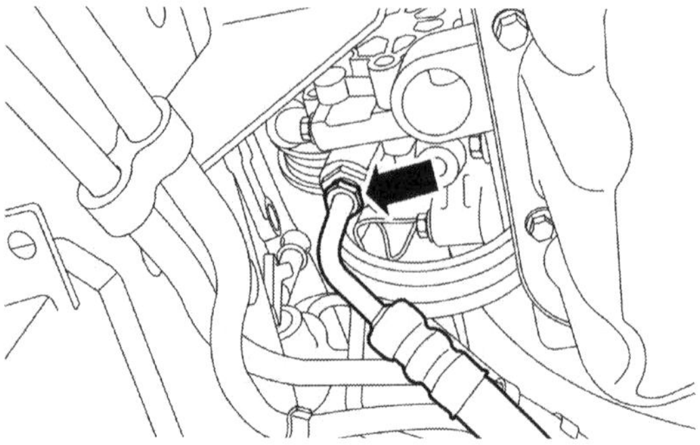

5. CAUTIONS:

Make sure that the area around the component is clean and free of foreign material.

Before disconnecting any components, make sure the area is clean and free from foreign material. When disconnected all openings must be sealed.

If power steering fluid comes into contact with the paintwork, the affected area must be immediately washed down with cold water.

Disconnect the power steering high-pressure pipe union. Remove and discard the O-ring seal.

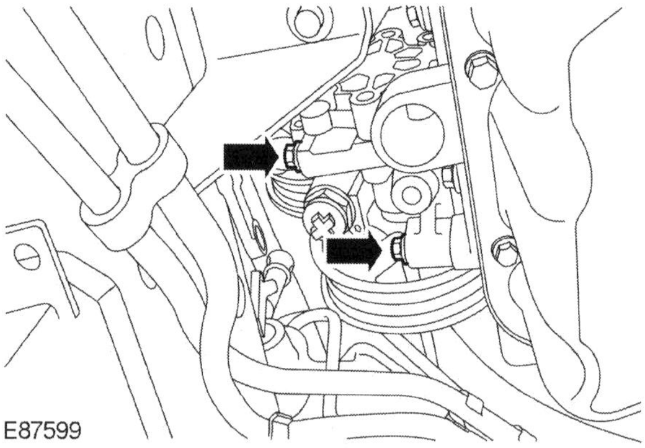

E87599

6. Remove the 2 bolts from the power steering pump.

7. Remove the 2 nuts. Release the power steering fluid reservoir.

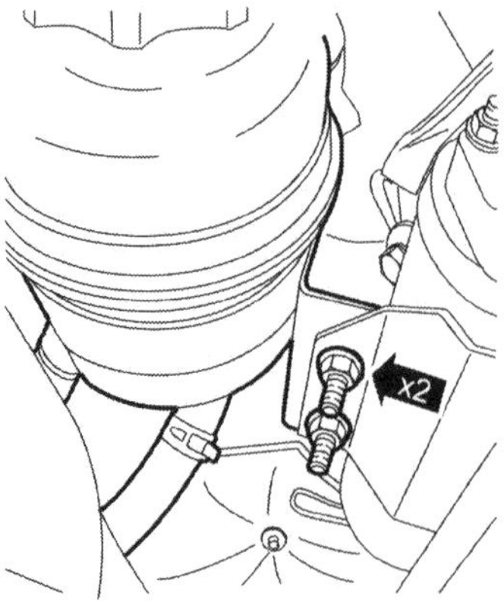

E87600

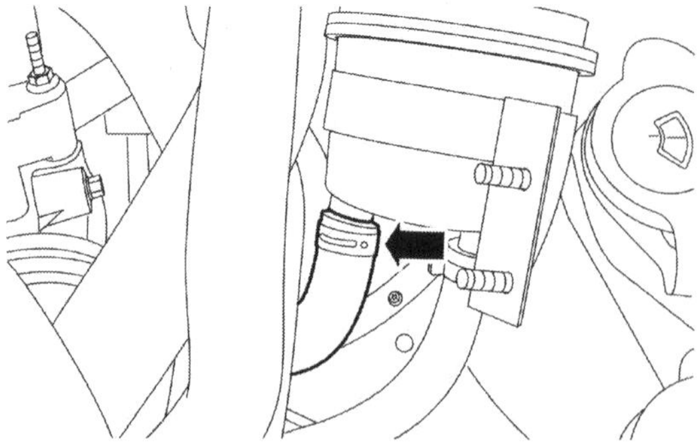

8. Release the clip. Disconnect the power steering fluid reservoir supply hose.

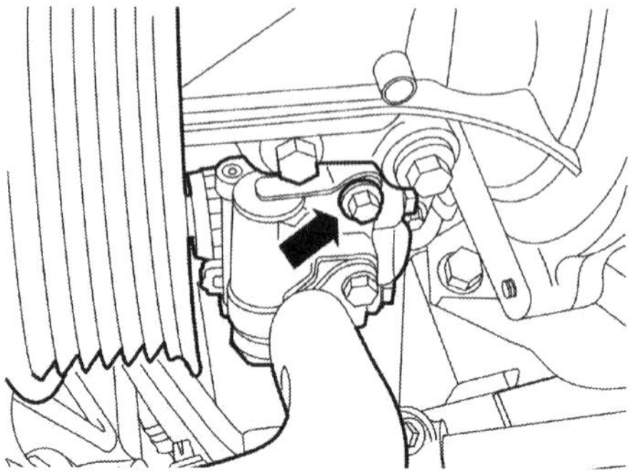

9. Remove the bolt. Remove the power steering pump.

Installation

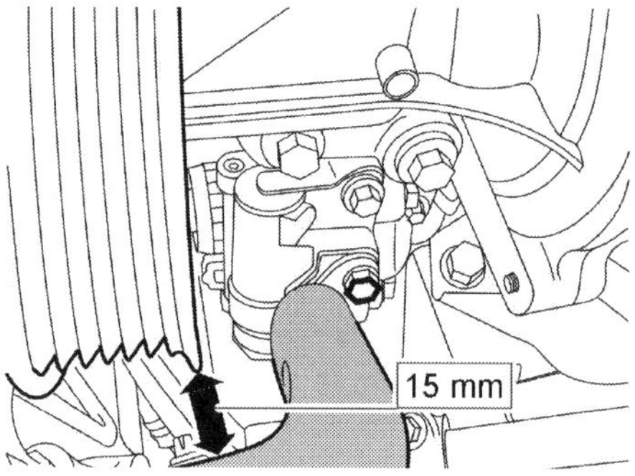

1. Install the power steering pump and tighten.

   Torque: 23 Nm

2. CAUTION: A 15 mm gap, or greater, must exist between the power steering pump pulley and the hose, at the rearmost edge of the pulley. Adjust the hose if required.

Connect the power steering fluid reservoir supply hose. Secure the clip. Check that the correct gap exists between the power steering pump pulley and the hose.

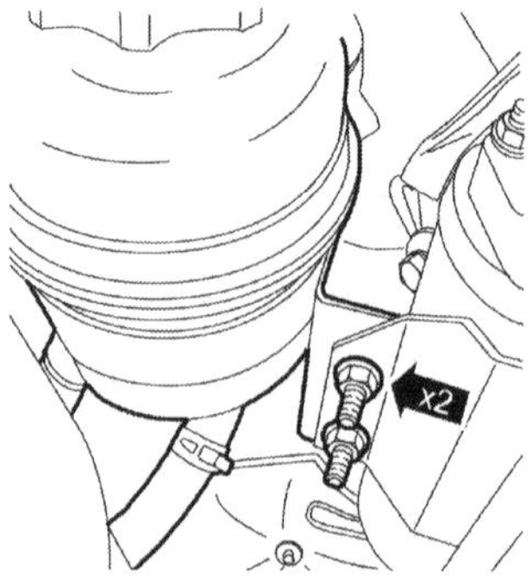

3. Secure the power steering fluid reservoir and tighten.

   Torque: 4 Nm

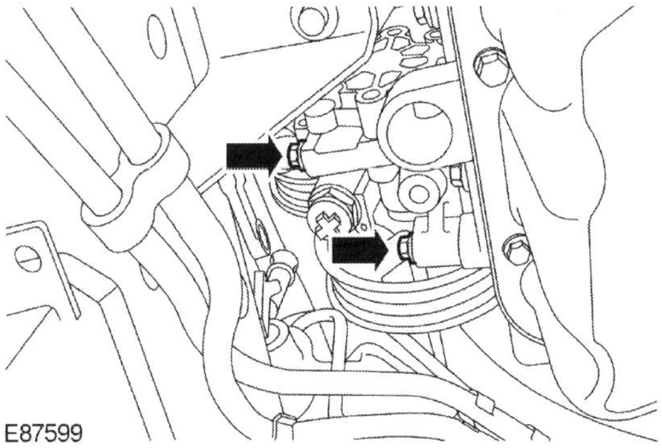

4. *Torque:* 23 Nm

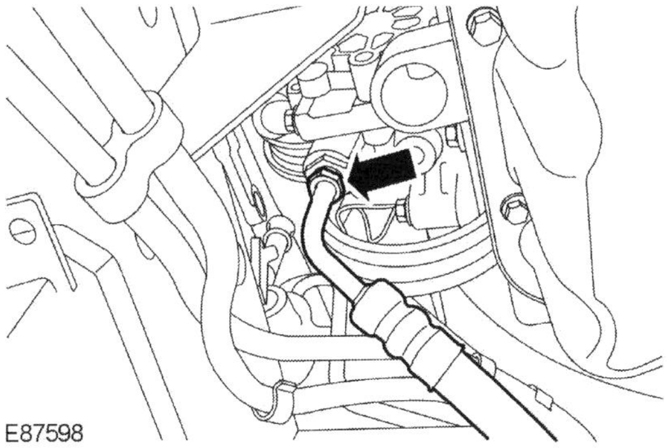

5. NOTES:

△ Remove and discard the blanking caps.

△ Install a new O-ring seal.

Connect the power steering high-pressure pipe union and tighten.

Renew Part: Power steering hose O-ring seal(s)

Torque: 25 Nm

6. Rotate the accessory drive belt tensioner clockwise. Secure the accessory drive belt.

7. Install the cooling fan.

Refer to: Cooling Fan (303-03 Engine Cooling, ID4 2.2L Diesel Removal and Installation)

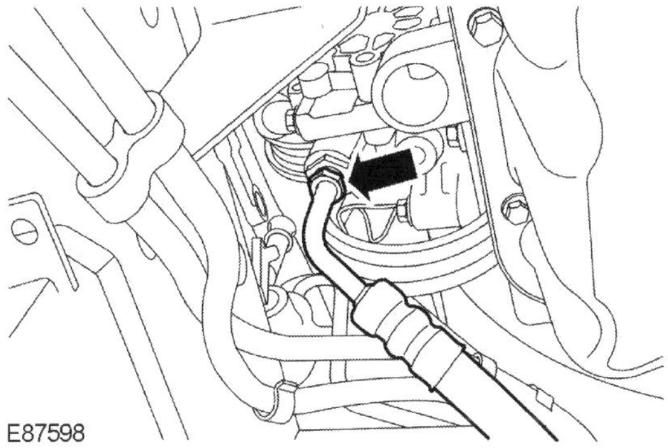

8. Connect the battery ground cable.

Refer to: Startup Battery Disconnect and Connect (414-01 Battery, Mounting and Cables, General Procedures)

9. Fill and bleed the power steering system.

Refer to: Power Steering System Filling and Bleeding (211-00 Steering System - General Information, General Procedures)

Steering Angle Sensor

Removal and Installation

Removal

NOTES:

△ Removal steps in this procedure may contain installation details.

△ Some variation in the illustrations may occur, but the essential information is always correct.

△ Make sure that the road wheels are in the straight ahead position.

1. **Refer to:** Startup Battery Disconnect and Connect (414-01 Battery, Mounting and Cables, General Procedures)
2. **Refer to:** Steering Wheel (211-04 Steering Column, Removal and Installation)

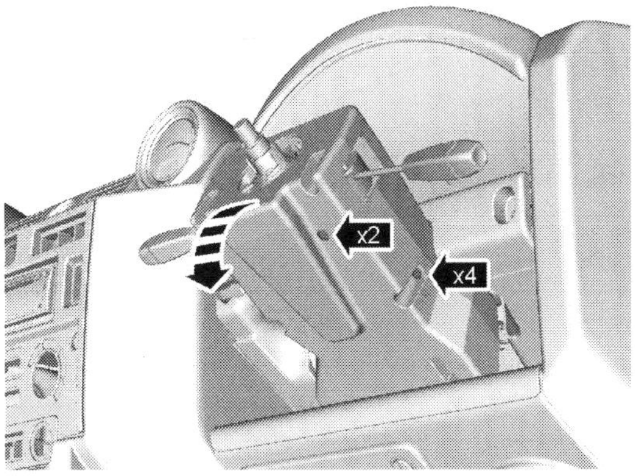

3.

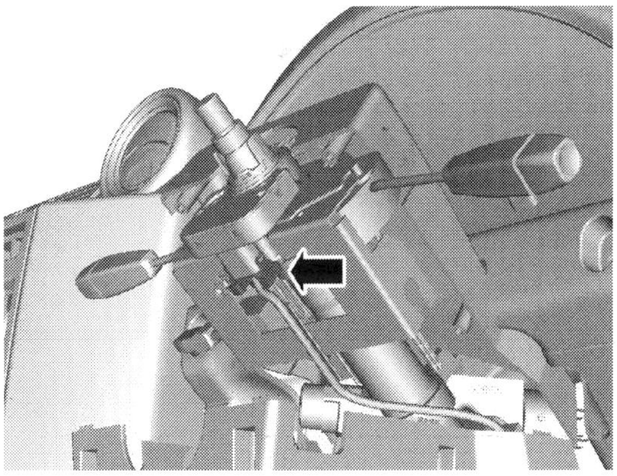

4.

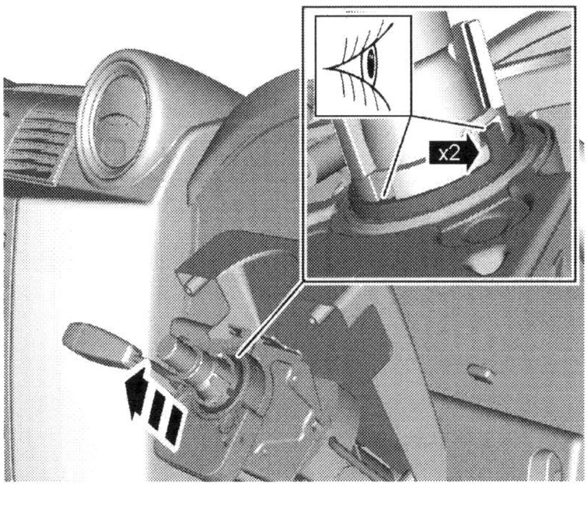

5.

Installation

1. NOTE: If a new component has been installed, configure using Land Rover approved diagnostic equipment.

   To install, reverse the removal procedure.

Steering Gear

Disassembly and Assembly

Part(s)

STEP	PART NAME	QUANTITY
Step 1	Renew Part Power steering control valve actuator kit	1
Step 12	Renew Part Power steering control valve actuator kit	1
Step 23	Renew Part Steering Box assembly	1
Step 48	Renew Part Steering gear drop arm washer	1

Disassembly

WARNING: Wear safety glasses while removing and refitting circlips and retaining ring.

CAUTION: Absolute cleanliness is essential when overhauling steering gear.

NOTES:

Overhaul of steering gear should not be carried out during the warranty period.

This procedure is for Adwest steering gear only.

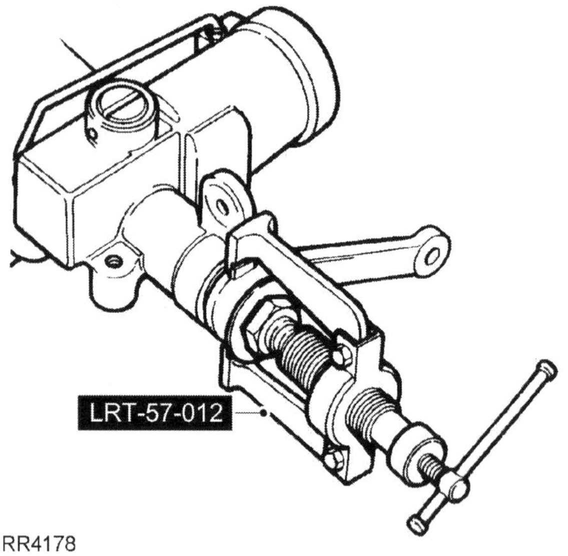

RR4178

1. Remove steering gear from vehicle. Mark drop arm and steering gear for realignment on assembly. Remove drop arm using extractor LRT-57-012. Loosen drop arm securing nut, but do not remove before using extractor. Remove dirt excluder from output shaft.

Refer to: Steering Gear (211-02 Power Steering, Removal and Installation)

2. Drain oil, remove blanking plugs and bleed screw. Hold steering gear over suitable container, turn input shaft from lock to lock, until oil is drained. Refit bleed screw.
3. Rotate retainer ring until one end is 12 mm from extractor hole. Using a drift through hole in cylinder, lift retaining ring from groove in cylinder bore.
4. Remove retainer ring, using a screwdriver.

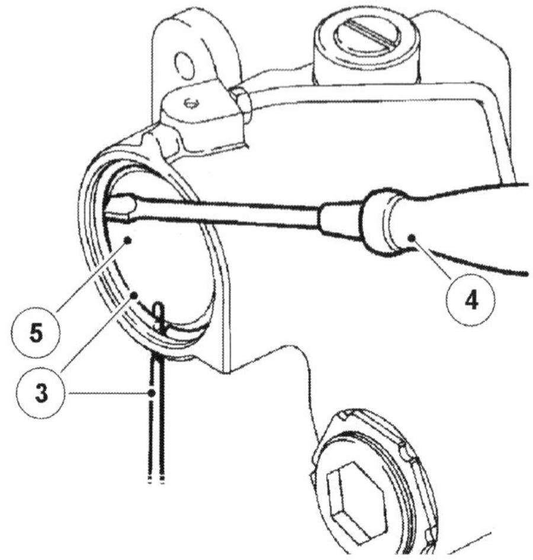

5. Turn input shaft (left lock on left hand drive vehicles, right lock on right hand drive vehicles) until piston pushes out cover. Turn input shaft fully in opposite direction, applying pressure to piston.
6. Remove set screw retaining rack pad adjuster.
7. Remove rack adjuster and pad.
8. Remove sector shaft adjuster locknut.
9. Remove four bolts from sector shaft cover.
10. Screw in sector shaft adjuster until cover is removed.

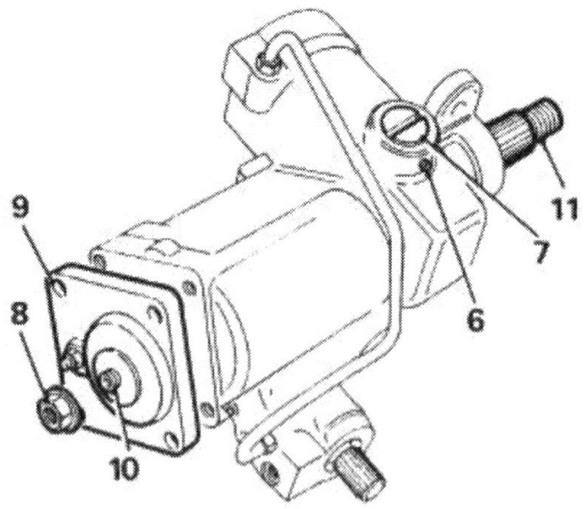

11. Slide out sector shaft.
12. Remove piston, a bolt screwed into piston will assist removal.
13. Remove input shaft dirt excluder.
14. Remove worm adjuster locknut using ‘C’ wrench LRT-57-007.

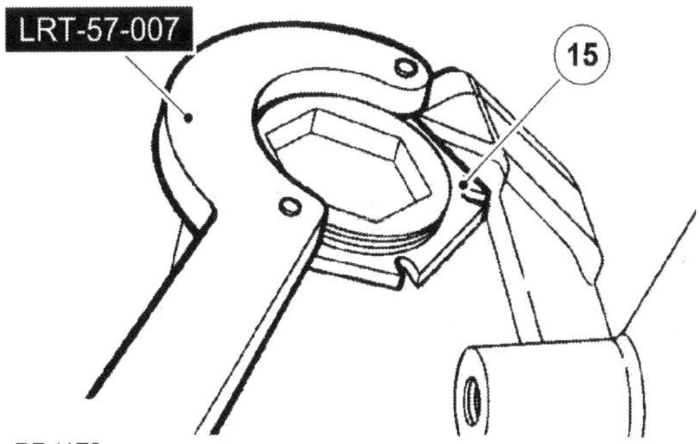

RR4179

15. Remove worm adjuster using wrench LRT-57-006.
16. Tap splined end of shaft to free bearing.
17. Remove bearing cup and caged ball bearing assembly.
18. Remove valve and worm assembly.
19. NOTE: If difficulty is experienced warm the casing and bearing assembly. Cool bearing cup using a mandrel and tap steering gear on a bench.

Remove inner bearing cage, cup and shim washers. Retain shims for reassembly.

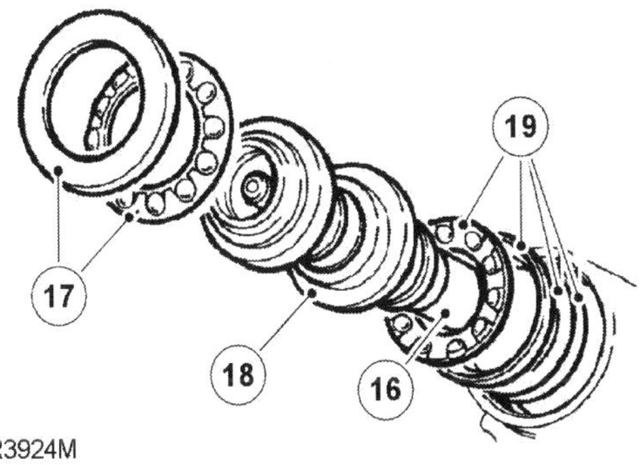

RR3924M

20. CAUTION: Do not remove sector shaft bearings from casing. Replacement parts are not available. If sector shaft bearings are worn fit a new steering gear.

Remove circlip and seal from sector shaft housing bore.

21. CAUTION: The use of a seal puller is recommended to prevent damage to casing, resulting in possible oil leaks.

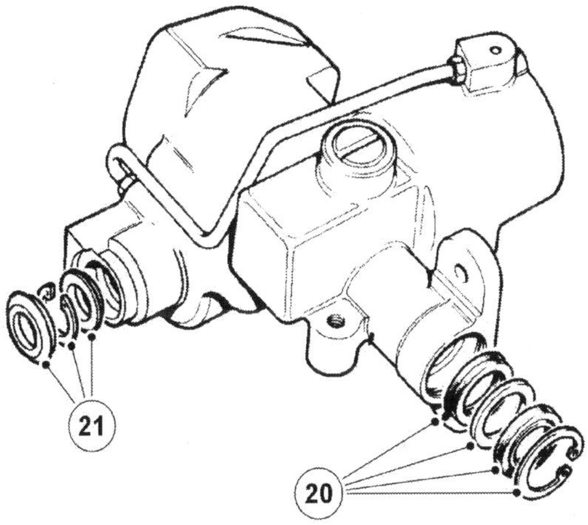

Remove dirt excluder, circlip and seal from input shaft housing bore.

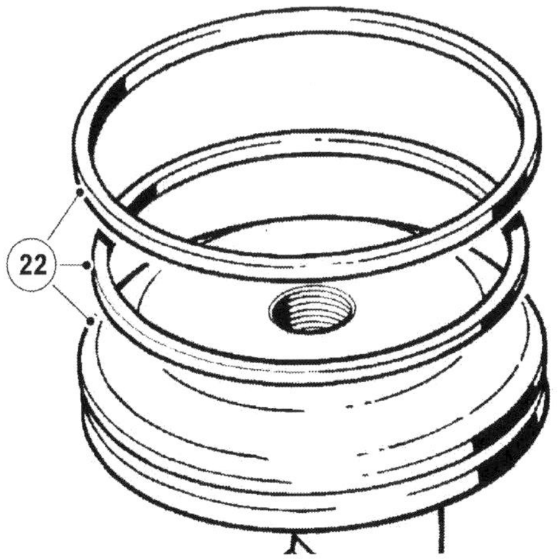

22. Discard all rubber seals and obtain replacements.

NOTE: A rubber seal is fitted behind plastic ring on rack piston. Discard seal and plastic ring.

23. Examine piston bore for scoring and wear.
24. Examine feed tube.

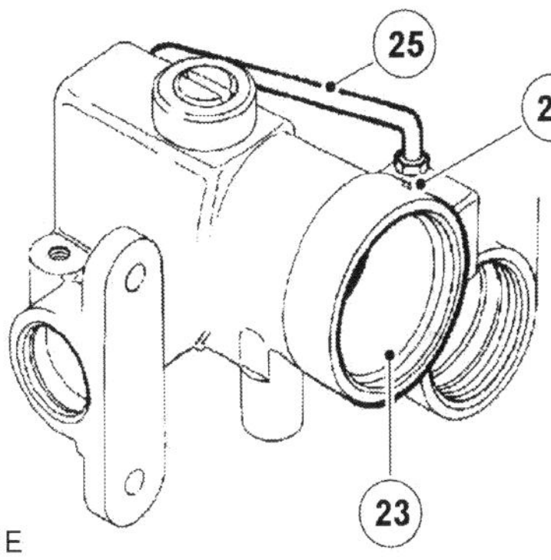

RR2351E

25. Fit a new feed tube if damaged. Tighten union.

   Torque: 22 Nm

26. Check there is no side play on roller.
27. If side play on roller exists fit a new sector shaft.
28. Check condition of adjuster screw threads. Check adjuster end float. Fit new adjuster if end float exceeds 0.15 mm.
29. Examine bearing areas on shaft for excessive wear.
30. Examine gear teeth for uneven or excessive wear.
31. Inspect cover and bearing. If worn or damaged, fit a new steering gear.

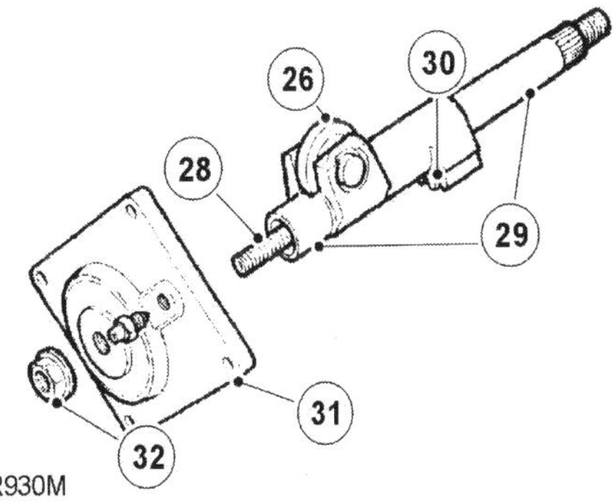

RR930M

32. The locknut is also a fluid seal. Fit new nut during assembly.

33. Examine bearing areas for wear. The areas must be smooth and not indented.

34. Examine worm track which must be smooth and not indented.

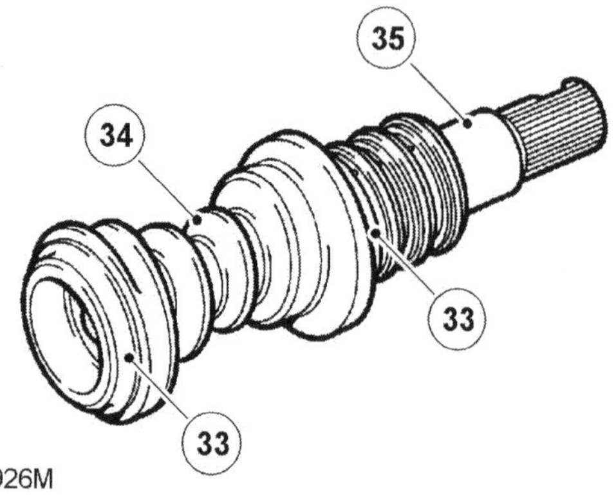

RR3926M

35. △ NOTE: Any sign of wear makes it essential to fit new valve and worm assembly.

Check for wear on torsion bar assembly pin. No free movement should exist between input shaft and torsion bar or between torsion bar and worm.

36. Examine valve rings for cuts, scratches and grooves. The valve rings should be free to rotate in grooves. Renew the valve and worm assembly if any faults are found.

37. Examine ball races and cups for wear and general condition.

38. If ball cage has worn against bearing cup, fit replacements.

39. Bearing balls must be retained by cage.

40. Bearing and cage repair is carried out by complete replacement of assembly.

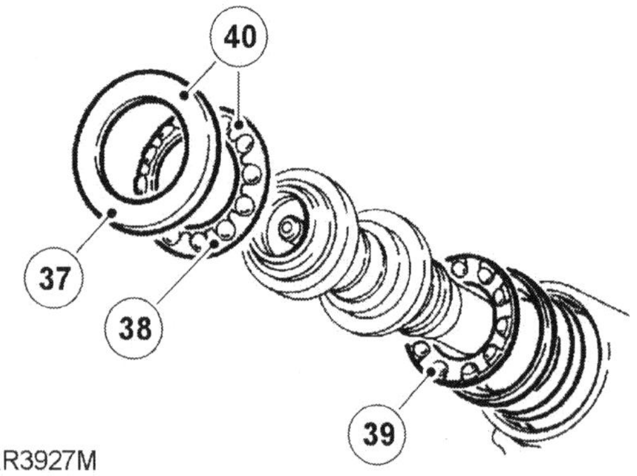

RR3927M

41. Examine thrust pad for scores.

42. Examine adjuster for wear in pad seat.

43. Fit new sealing ring to rack adjuster.

44. Examine for excessive wear on rack teeth.

45. Ensure thrust pad bearing surface is free from scores and wear.

46. Ensure piston outer diameters are free from burrs and damage.

47. Examine seal and ring groove for scores and damage.

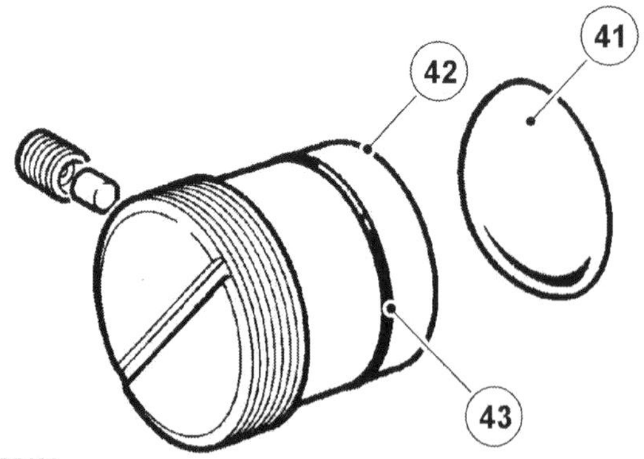

RR3928M

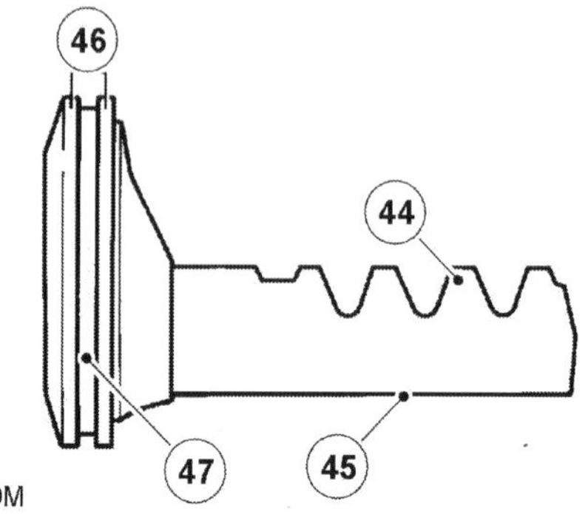

RR3929M

Assembly

1. Fit new ring to piston. Warm nylon seal and fit to piston.

Renew Part: Power steering control valve actuator kit

2. Slide piston assembly into cylinder with rack tube outwards.

△ NOTE: When fitting replacement oil seals lubricate with recommended fluid and ensure absolute cleanliness.

3. Fit oil seal, lip side first.

4. Fit extrusion washer.

5. Fit dust seal, lipped side last.

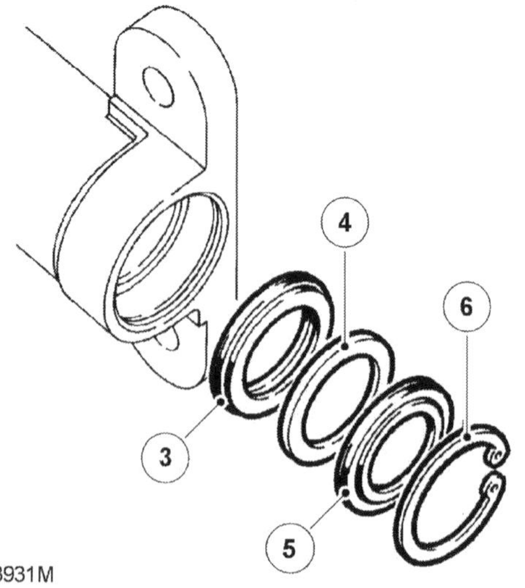

RR3931M

6. Fit circlip.

7. Refit original shims and inner bearing cup. Use Petroleum Jelly to aid assembly.

NOTE: If original shims are not used, fit shims of 0.76 mm thickness.

8. Fit inner cage and bearings assembly.
9. Fit valve and worm assembly.
10. Fit outer cage and bearings assembly.

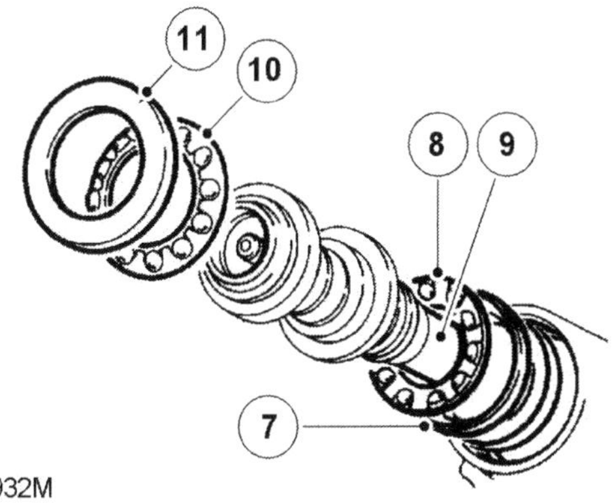

RR3932M

11. Fit outer bearing cup.
12. Fit new worm adjuster sealing ring.

Renew Part: Power steering control valve actuator kit

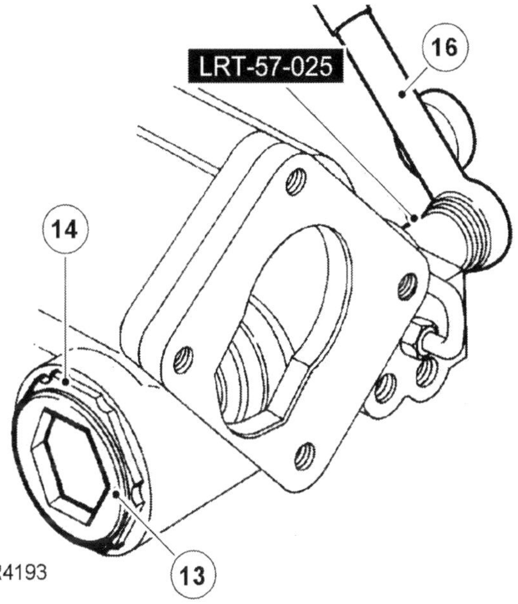

RR4193

13. Loosely screw adjuster into casing.
14. Fit locknut, do not tighten.
15. Turn in worm adjuster until end float is almost eliminated. Ensure bearing cages are seated correctly.
16. Measure maximum rolling torque of valve and worm assembly, using a torque wrench and spline socket LRT-57-025.
17. Turn in worm adjuster while rotating shaft to increase figure measured.

    Torque: 0.56 Nm

18. Back off worm adjuster ¼ turn. Turn in worm adjuster to increase reading by 0.21 - 0.34 Nm with locknut tight. Use worm adjusting wrench LRT-57-006 and locknut wrench LRT-57-028.

    Torque: 100 Nm

19. Screw slave bolt into piston to aid assembly.
20. Fit piston and rack so piston is 70 mm (2.75 in) from outer end of bore.

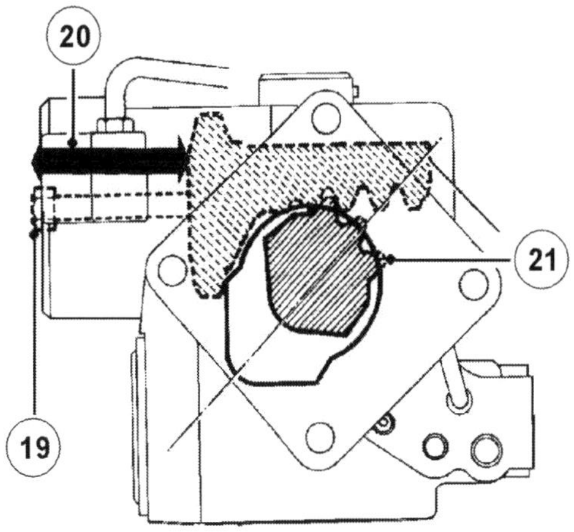

21. Fit sector shaft using seal saver LRT-57-021. Align roller with cut out in casing as shown. Push in sector shaft while rotating input shaft to allow sector roller to engage worm.
22. Fit rack adjuster and thrust pad to engage rack. Back off half turn on adjuster.

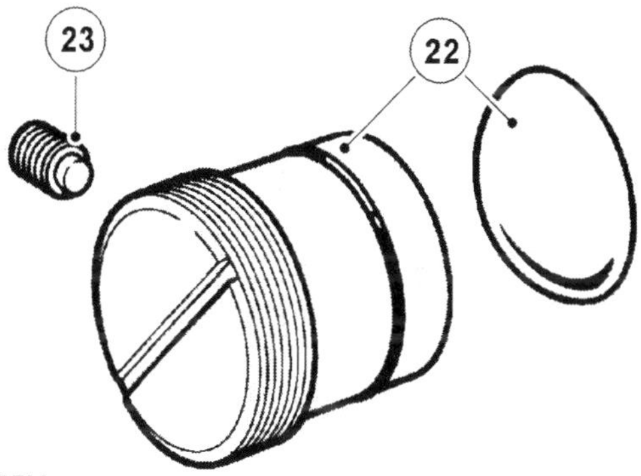

RR3935M

23. Loosely fit new nylon pad and adjuster set screw assembly.

Renew Part: Steering Box assembly

24. Fit new sealing ring to cover.
25. Align cover with casing.
26. Screw cover assembly fully on to sector shaft adjuster screw.
27. If necessary back off sector shaft adjuster screw. Tap cover in place to allow cover to joint fully with casing.

NOTE: Before tightening fixings, rotate input shaft to ensure sector shaft roller is free to move in valve worm. If initial resistance is left, turn adjuster screw approximately two turns in a clockwise direction.

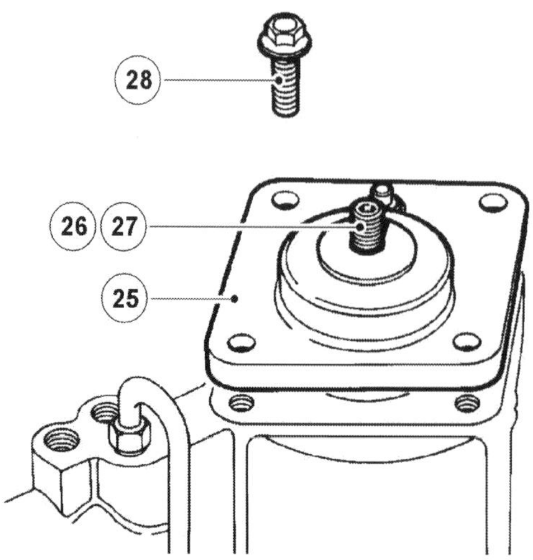

28. Fit cover bolts.

Torque: 75 Nm

29. Fit new square section seal to cover.
30. Remove slave bolt from piston. Press cover into cylinder just to clear retainer ring groove.

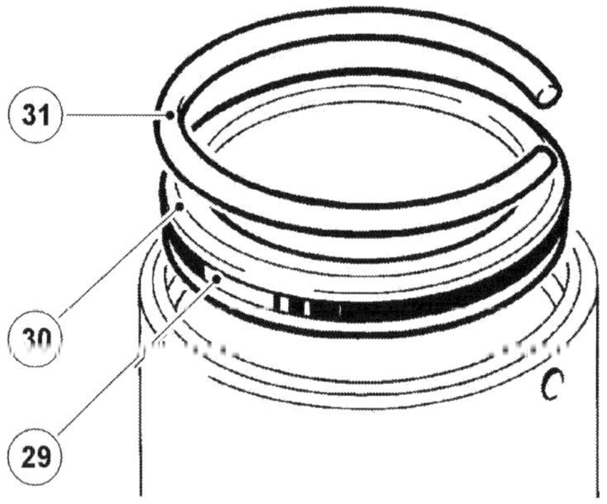

31. Fit retaining ring to groove with one end of ring positioned 12 mm from extractor hole.
32. Refit drop arm and tighten nut sufficiently to ensure that no backlash exists between drop arm and sector shaft.
33. To set worm on centre, rotate input shaft to full inner-lock (full right lock for a left hand drive vehicles, full left lock for a right hand drive vehicles). Rotate input shaft back towards centre two full turns.
34. The box is now on centre and can be adjusted.
35. Hold input shaft and rock the drop arm to establish backlash is present. Continue rocking and slowly turn sector shaft adjusting screw clockwise. Continue turning adjuster screw until backlash has almost been eliminated.
36. Refit locknut and tighten.

⚠ NOTE: It is important steering gear is centralised before any adjustments are made.

37. Check maximum rolling torque one and a quarter turns either side of centre position, using a torque wrench and spline socket LRT-57-025. Rotate adjuster screw to obtain across centre torque of 0.34 Nm plus torque figure at one and a quarter turns. Tighten adjuster locknut.

Adjuster screw 0.34 Nm

Adjuster locknut 60 Nm

38. Turn in rack adjuster to increase figure measured in previous instruction by 0.23 - 0.34 Nm. **The final figure may be less, but must not exceed 1.35 Nm.**
39. Lock rack adjuster in position with grub screw and tighten. With input shaft rotated from lock to lock, rolling torque figures should be greatest across centre position and equally disposed about centre position. The condition depends on the value of the shimming fitted between the valve and worm assembly inner bearing cup and casing. The original shim value will give correct torque peak position unless major components have been replaced.

Torque: 5 Nm

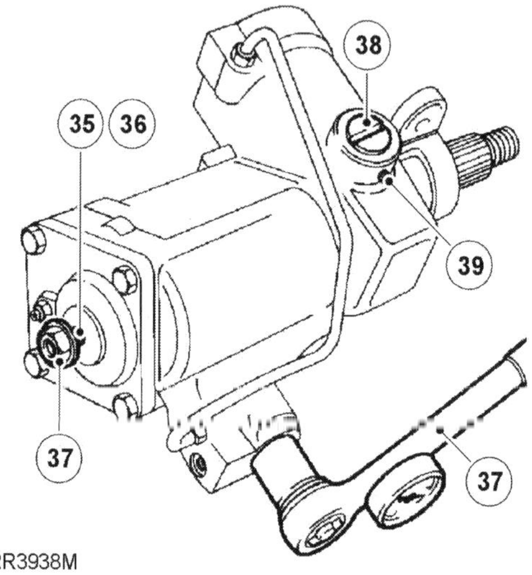

RR3938M

40. With input coupling shaft toward the operator, turn shaft fully anti-clockwise.

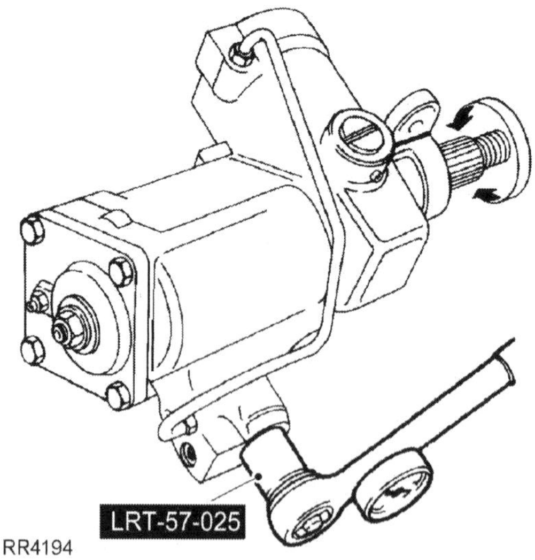

RR4194

41. Check torque figures obtained from lock to lock using torque wrench and spline socket LRT-57-025.
42. Check also for equal engagement either side of centre.
43. Note where greatest figures are recorded relative to steering position. If greatest figures are not

recorded across centre of travel (steering straight-ahead), adjust as follows: If torque peak occurs before centre position, add to shim washer value. If torque peak occurs after centre position, subtract from shim washer value. Shims are available in the following thicknesses; 0.03 mm, 0.07 mm, 0.12 mm and 0.24 mm.

CAUTION: When reshimming valve and worm, extreme caution must be exercised to prevent seal damage during reassembly.

NOTE: Adjustment of 0.07mm to shim value will move torque peak area by ¼ turn on the shaft.

44. Fit seal, lip side first, into housing. Use seal saver LRT-57-016 and seal installer LRT-57-026. Note that seal is fitted to a depth of 4.75 - 5.00 mm from face of box.
45. Secure seal with circlip.
46. Smear inner lip of dirt excluder with PTFE grease. Fit dirt excluder using LRT-57-027. When fitted correctly outer shoulder of excluder is 4.00 - 4.50 mm from face of box, dimension X.
47. Remove drop arm. Smear inner lip of dirt excluder with PTFE grease and refit, ensuring outer lip is flush with casing.
48. With input shaft on centre, align assembly marks on drop arm and steering gear. Fit drop arm to steering gear using a new tab washer. Bend over tab and tighten.

Renew Part: Steering gear drop arm washer

Torque: 176 Nm

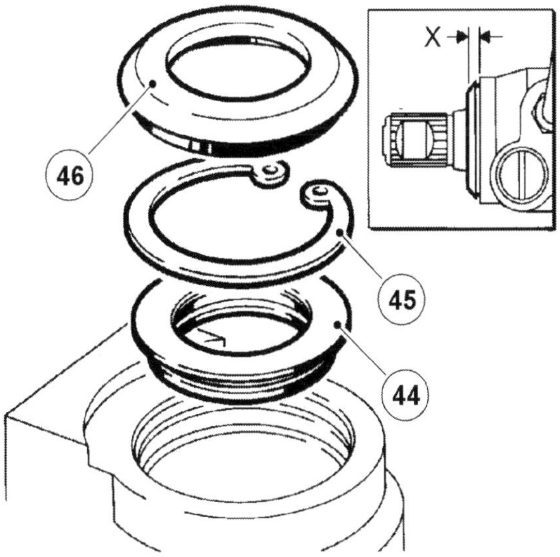

49. Fit steering gear.

Refer to: Steering Gear (211-02 Power Steering, Removal and Installation)

Steering Linkage

211-03

Tie Rod End

Removal and Installation

Removal

1. Park vehicle on level ground and chock rear wheels.
2. Raise vehicle and locate axle stands or use a ramp.
3. Centralise steering.

Refer to: Steering Gear Centralization (211-00 Steering System - General Information, General Procedures)

4. Unscrew 2 bolts and remove tie rod protection bracket from axle differential housing.
5. Remove split pin and castellated nut securing tie rod to swivel housing arms.
6. Disconnect tie rod ends using LRT-57-018
7. Remove tie rod.
8. Slacken clamp bolts.
9. Unscrew tie rod ends.

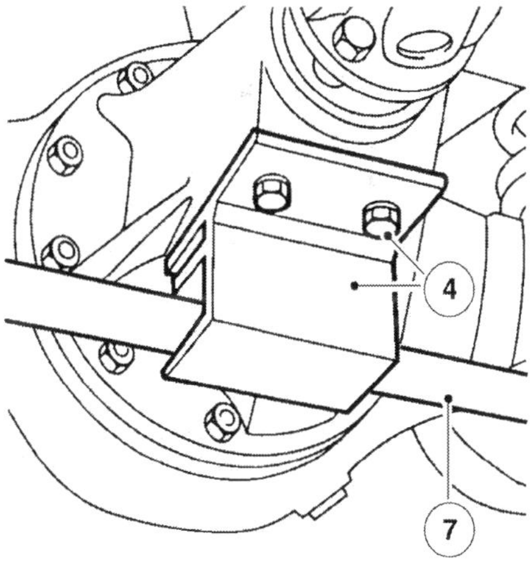

Installation

1. Fit new tie rod ends to tie rod and loosely tighten clamp bolts.
2. Screw in tie rod ends to full extent of threads and fully tighten clamp bolts.

CAUTION: A tie rod that is damaged or bent must be renewed. DO NOT attempt to repair or straighten

3. Fit tie rod to swivel housing arms and tighten tie rod end nuts to 44 Nm (32 lbf/ft). Fit new split pin.

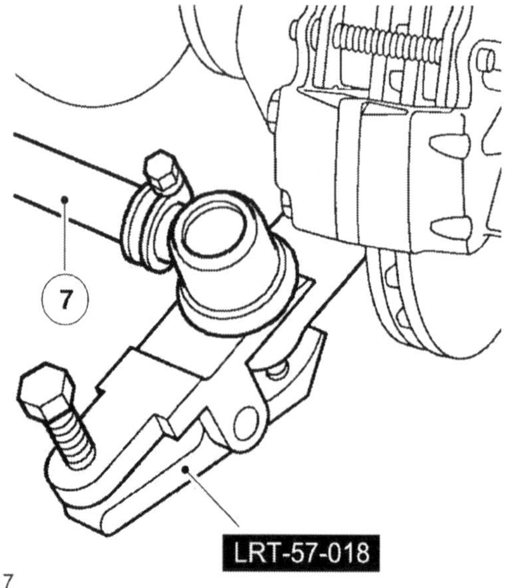

J6317

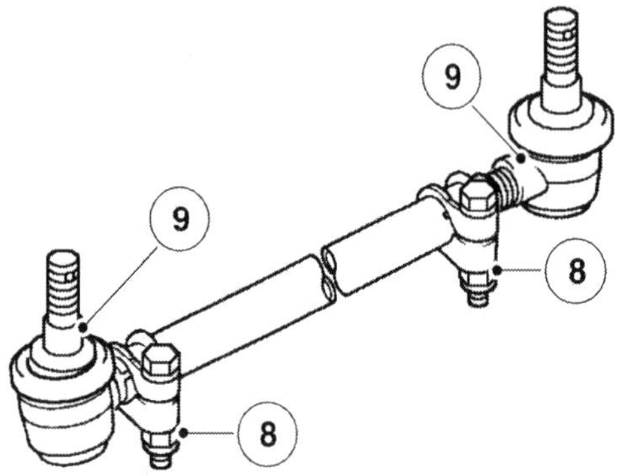

J6318

4. Fit tie rod protection bracket to axle differential housing and tighten to 30 Nm (22 lbf/ft).

5. Check wheel alignment.

Refer to: Front Toe Adjustment (204-00 Suspension System - General Information, General Procedures)

6. Remove axle stands or vehicle from ramp.

Steering Linkage Damper

Removal and Installation

Removal

1. Park vehicle on level ground and chock rear wheels.

2. Raise vehicle and locate axle stands or use a ramp.

3. Remove bolt securing steering linkage damper to drag link bracket.

4. Remove retaining nuts, washers and rubber bush securing steering linkage damper to chassis mounting.

5. Withdraw steering linkage damper.

6. Check rubber bushes and washers, renew if necessary

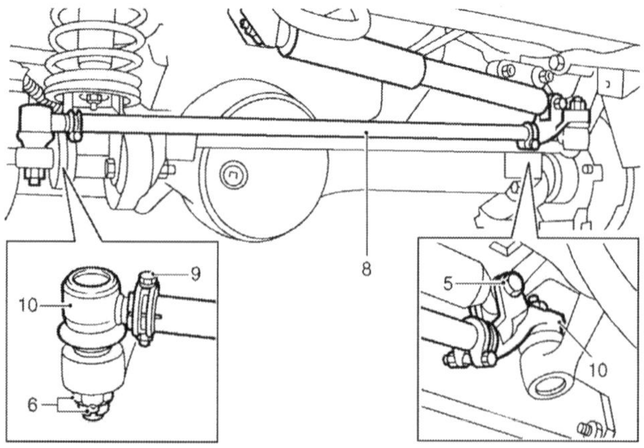

Installation

1. Ensuring rubber bushes and washers are correctly positioned, fit steering linkage damper to chassis mounting.

2. Install steering linkage damper to drag link bracket.

3. Remove axle stands or vehicle from ramp.

Sector Shaft Arm Drag Link

Removal and Installation

Removal

1. Park vehicle on level ground and chock rear wheels.

2. Set road wheels in straight ahead position.

3. Raise vehicle and fit axle stands or use a ramp.

4. Remove front road wheel.

5. Disconnect steering damper at sector shaft arm drag link bracket.

6. Remove split pin and castellated nut securing sector shaft arm drag link ball joint to swivel housing arm.

7. Disconnect sector shaft arm drag link ball joints using LRT-57-018.

8. Remove sector shaft arm drag link.

9. Slacken clamp bolts.

10. Unscrew ball joints and remove from sector shaft arm drag link.

Installation

1. Clean internal threads of sector shaft arm drag link.

2. Fit new ends to sector shaft arm drag link and loosely fit clamp bolts.

3. Set ball joints to sector shaft arm drag link measurement to 28,5 mm.

4. Adjust ball pin centres to nominal length of 924 mm, this length is adjusted during refit.

5. Centralise steering box.

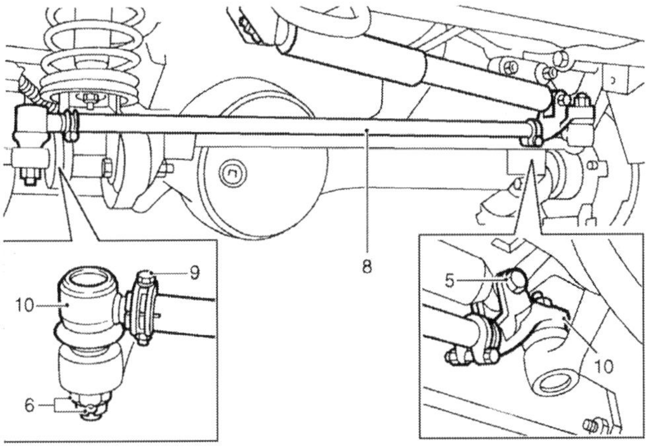

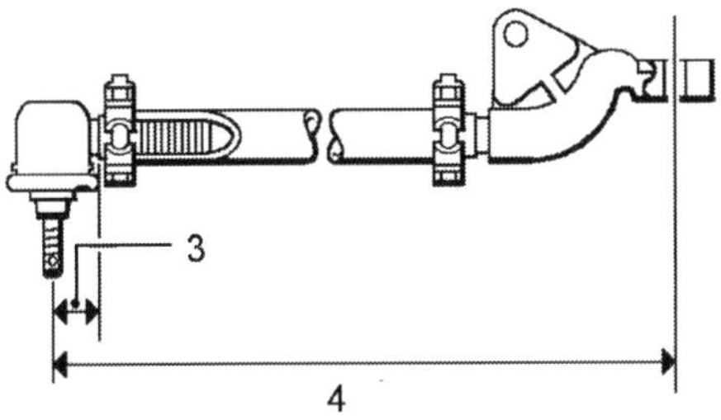

Refer to: Steering Gear Centralization (211-00 Steering System - General Information, General Procedures)

6. Align steering wheel, if necessary.

7. CAUTION: A sector shaft arm drag link that is damaged or bent must be renewed. DO NOT attempt repair.

Fit sector shaft arm drag link to swivel housing arms and tighten nuts to 40 Nm (30 lbf/ft). Fit new split pins.

8. Ensure full steering travel is obtained between lock stops.

Refer to: Steering Lock Stop Adjustment (211-00 Steering System - General Information, General Procedures)

Adjust sector shaft arm drag link length to suit.

9. Tap ball joints in direction shown so both pins are in same angular plane.

10. Tighten clamp bolts to 14 Nm (10 lbf/ft).

11. Refit road wheel and remove axle stands or vehicle from ramp.

12. Road test vehicle.

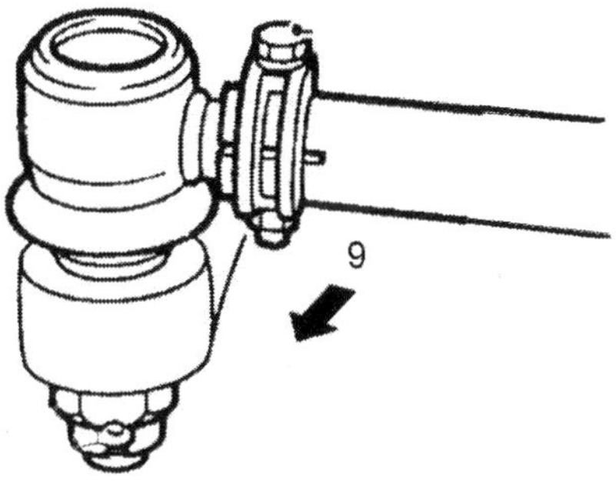

13. WARNING: To correct steering wheel deviations greater than ± 5° remove and reposition steering wheel.

If driving straight ahead and steering wheel is offset by 0° ± 5° in either direction, correct by adjusting sector shaft arm drag link length.

Steering Gear Drop Arm

Removal and Installation

Part(s)

STEP	PART NAME	QUANTITY
Installation Step 3	Renew Part Steering gear tab washer	1

Removal

1. Park vehicle on level surface and chock rear wheels.

2. Raise vehicle and locate axle stands or use a ramp.

3. Disconnect steering linkage damper from drag link.

Refer to: Steering Linkage Damper (211-03 Steering Linkage, Removal and Installation)

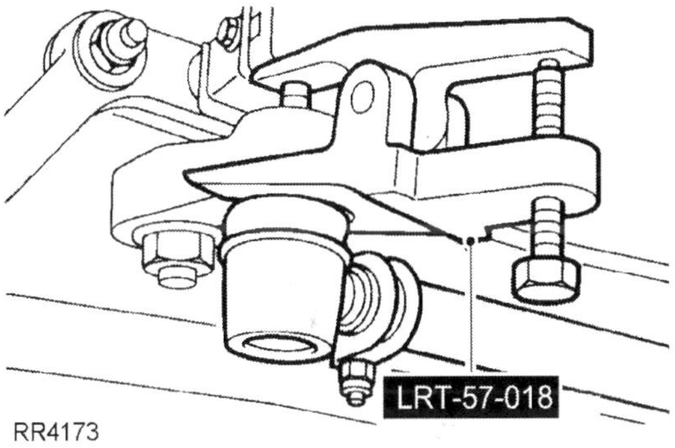

4. Disconnect drag link ball joint from drop arm using extractor LRT-57-018.

5. Mark drop arm and steering box for reassembly.

6. Bend back tabs on locking washer, slacken retaining nut, but do not remove.

7. Fit extractor LRT-57-012 and release drop arm from steering box spline.

8. Remove nut and discard locking washer.

9. Remove drop arm.

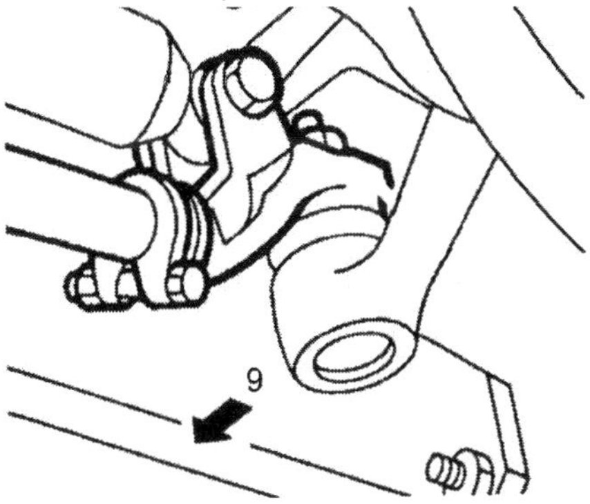

Installation

1. Centralise steering box.

Refer to: Steering Gear Centralization (211-00 Steering System - General Information, General Procedures)

2. Align reassembly marks and fit drop arm onto steering box splines.
3. Install new tab washer and retaining nut. Tighten and bend over tab washer.

   Renew Part: Steering gear tab washer

   Torque: 176 Nm
4. Install drag link to drop arm. Tighten ball joint nut

   Torque: 40 Nm
5. Remove axle stands or vehicle from ramp.

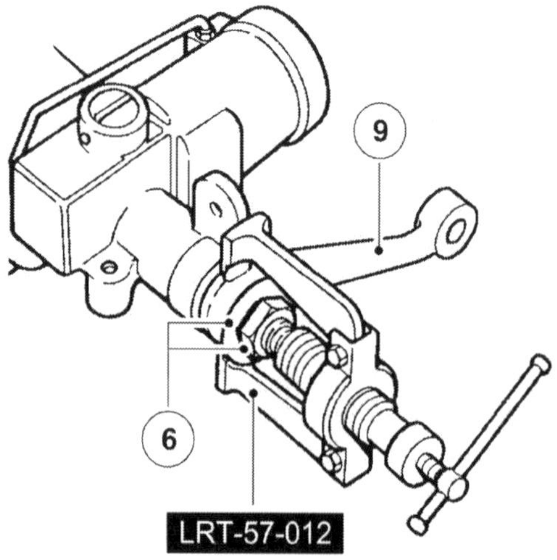

Steering Column

211-04

Removal and Installation

△ NOTE: Make sure that the wheels and tires are in the straight-ahead position.

1. Disconnect the battery ground cable.

For additional information, **Refer to:** Battery Disconnect and Connect (414-01 Battery, Mounting and Cables, General Procedures).

2. Remove the instrument panel.

For additional information, **Refer to:** Instrument Panel (501-12)

Instrument Panel and Console, Removal and Installation).

3. Remove the steering column lock and ignition switch housing.

For additional information, **Refer to:** Steering Column Lock and Ignition Switch Housing (211-05 Steering Column Switches, Removal and Installation).

4. Release the central junction box (CJB) from the bulkhead.

   Remove the 4 nuts.

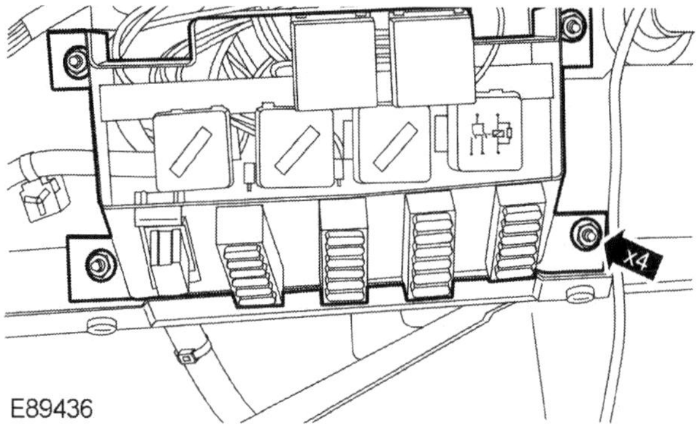

E89436

5. Release the steering column.

   Remove the bolt.
6. Remove the brake pedal and bracket.

For additional information, **Refer to:** Brake Pedal and Bracket

(206-06 Hydraulic Brake Actuation, Removal and Installation).

7. Release the steering column from the lower shaft.

   Mark the relationship between both shafts.

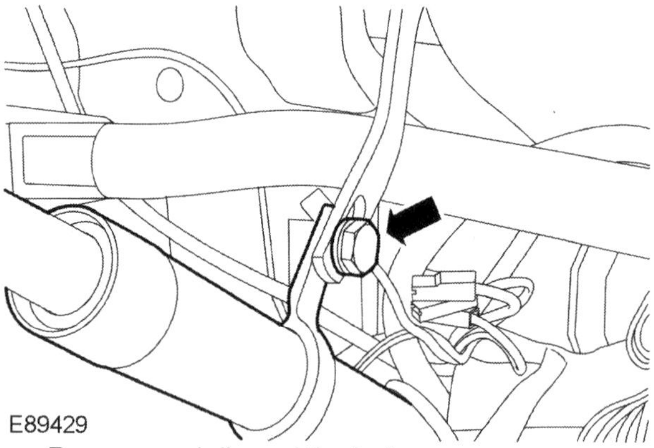

E89429

   Remove and discard the bolt.
8. Release the steering column from the vehicle body.

   Remove the 2 bolts.

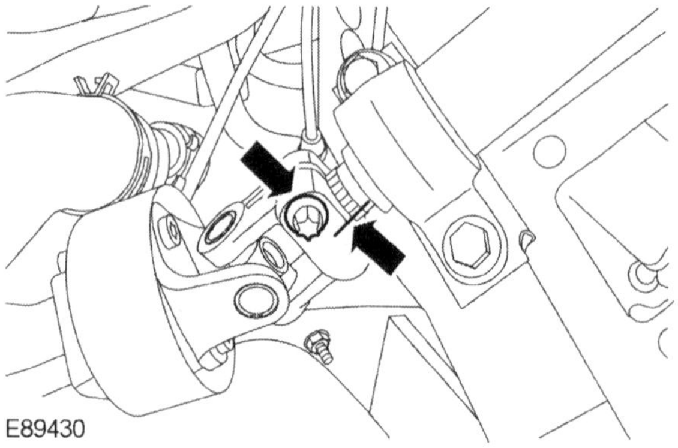

E89430

9. Release the heater control valve and bracket.

Remove the 2 bolts.

10. Remove the steering column upper support bracket.

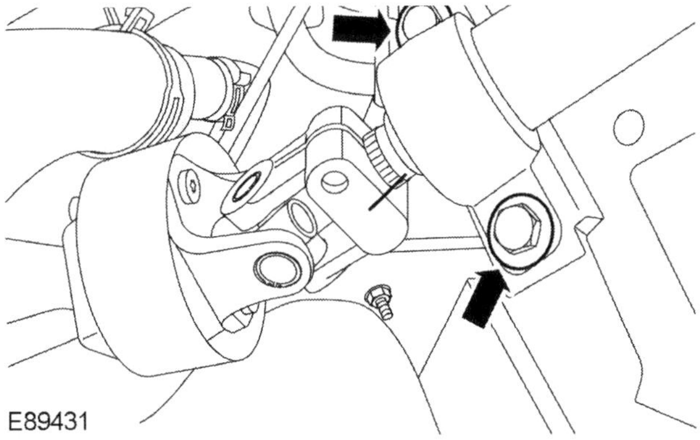
E89431

E89432

Remove the 4 bolts.

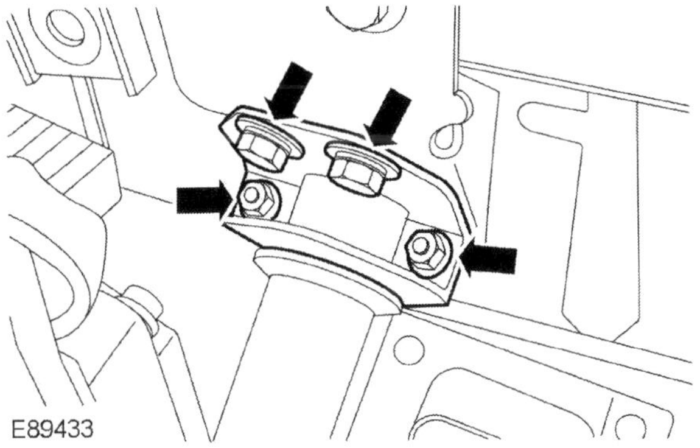
E89433

Collect the rubber strip from around the steering column shaft.

11. Remove the steering column from the steering column lower shaft.

12. Release the steering column main support bracket.

Remove the bolt.

13. NOTE: The steering column must be positioned correctly to pass through the bulkhead.

Remove the steering column.

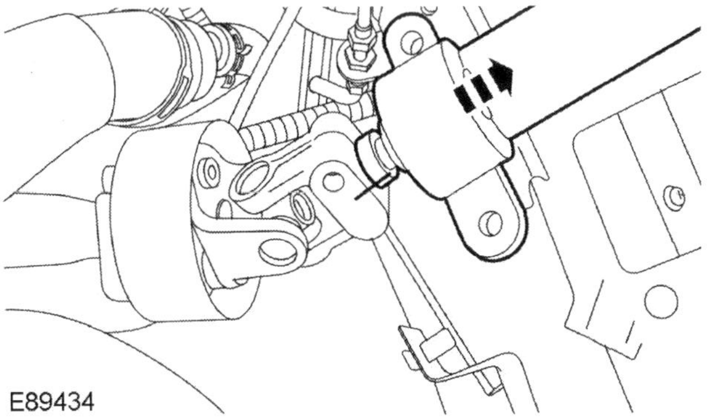
E89434

Installation

CAUTION: Care must be taken when installing the steering column to the steering column lower shaft, failure to follow this instruction may result in damage to the steering column lower shaft.

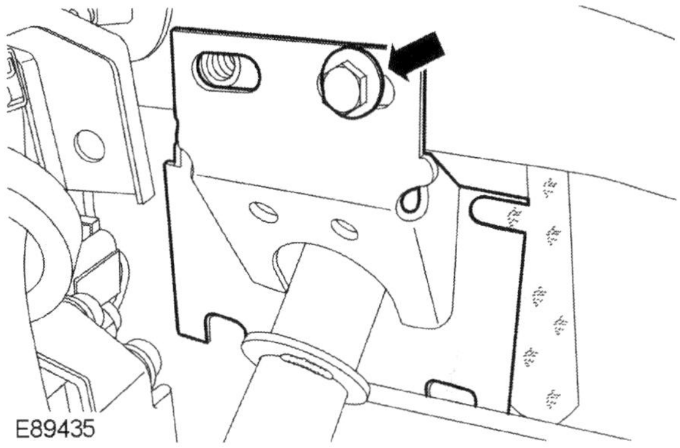
E89435

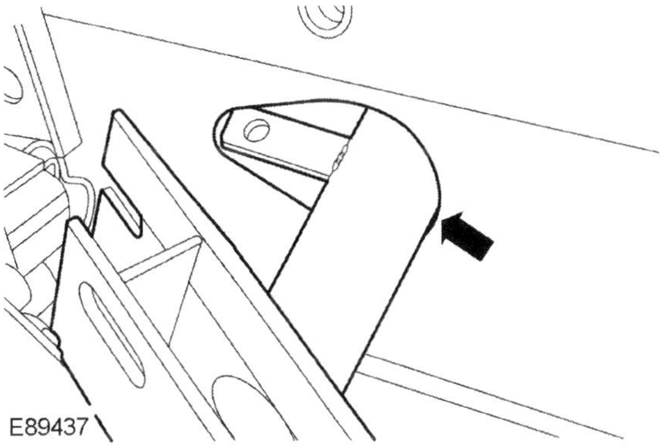
E89437

1. To install, reverse the removal procedure.

   Tighten to 22 Nm (16 lb.ft).

2. Tighten to 22 Nm (16 lb.ft).

3. Install the steering column upper support bracket.

   Tighten the M6 bolts to 9 Nm (7 lb.ft).

   Tighten the M8 bolts to 22 Nm (16 lb.ft).

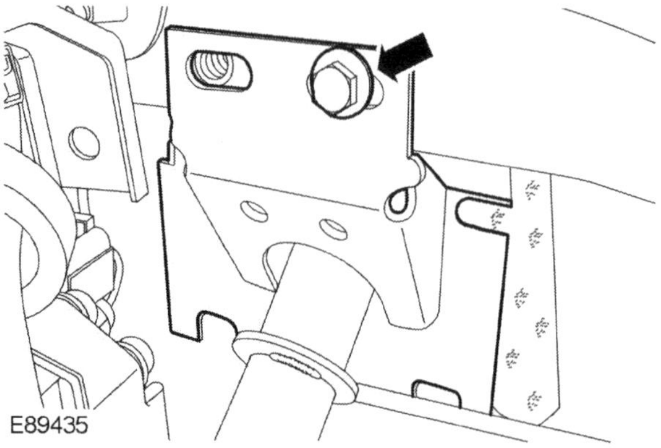

4. Tighten to 22 Nm (16 lb.ft).
5. NOTE: Install a new bolt. Tighten to 30 Nm (22 lb.ft).
6. Tighten to 48 Nm (35 lb.ft).
7. Connect the battery ground cable.

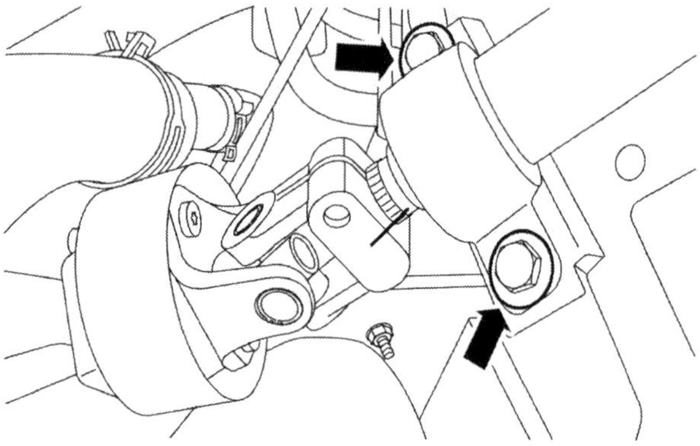

For additional information, refer to: Battery Connect (414-01 Battery, Mounting and Cables, General Procedures).

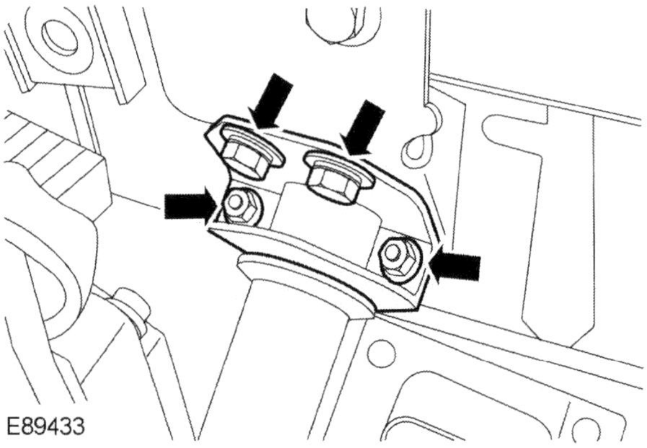

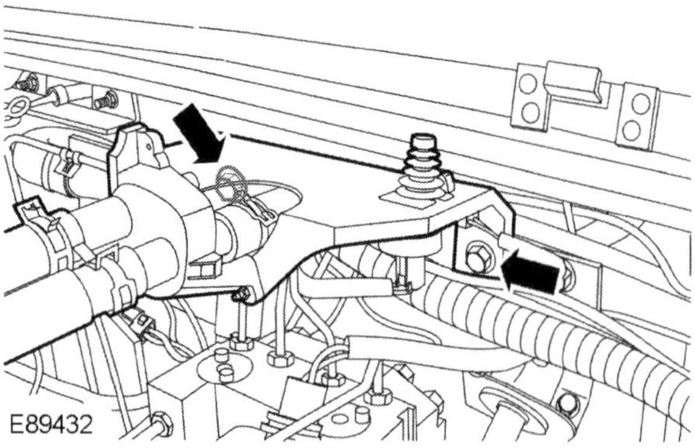

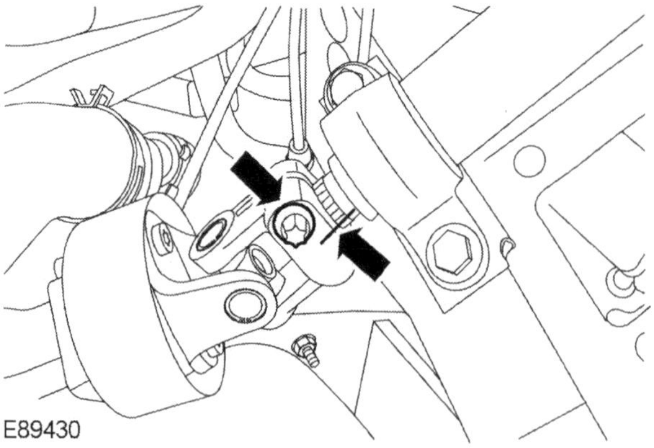

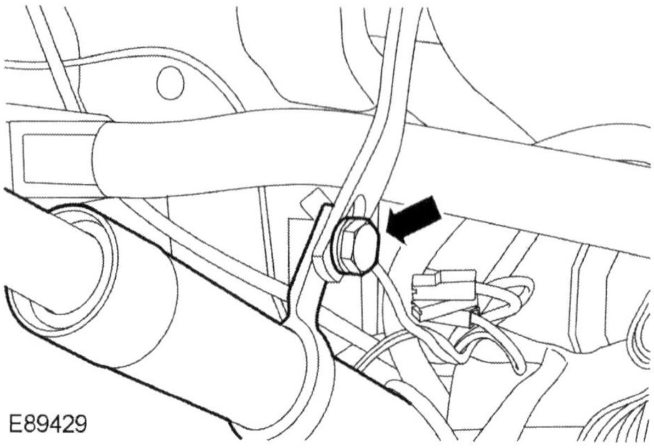

Steering Column Lower Shaft

Removal and Installation

Removal

⚠ NOTE: Make sure that the wheels and tires are in the straight-ahead position.

1. Release the steering column lower shaft.

Remove and discard the 2 bolts.

2. Remove the steering column lower shaft.

Mark the relationship between both shafts.

Remove and discard the bolt.

Installation

1. To install, reverse the removal procedure.

Tighten to 30 Nm (22 lb.ft).

2. Tighten to 30 Nm (22 lb.ft).

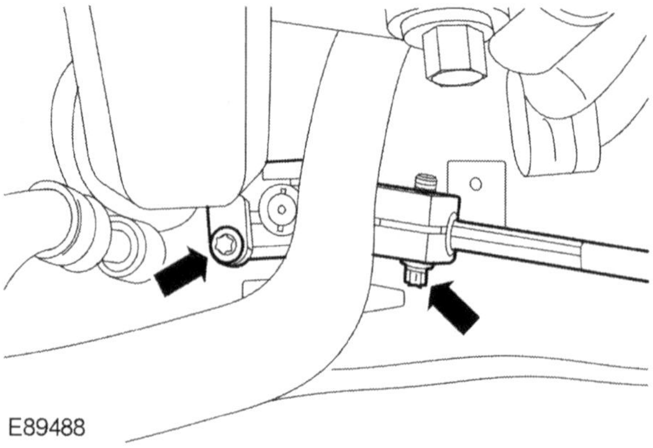

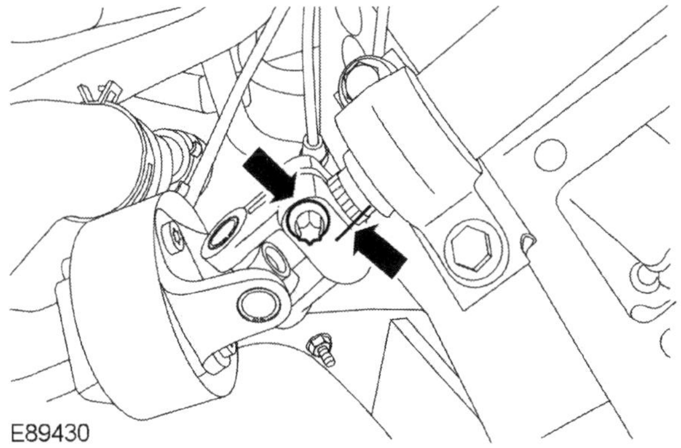

E89430

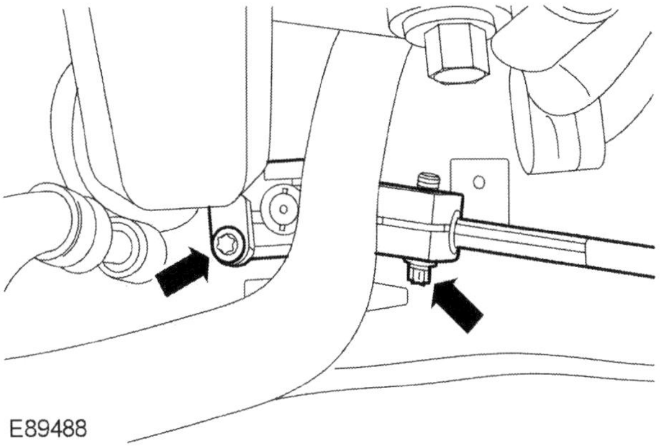

E89488

Steering Wheel

Removal and Installation

Special Tool(s)

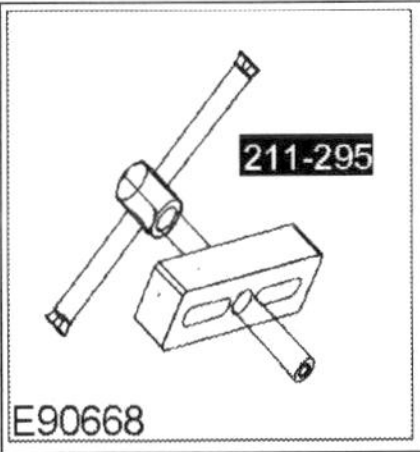 E90668	Remover, Steering Wheel 211-295

Removal

⚠ NOTE: Make sure that the wheels and tires are in the straight-ahead position.

E88430

1. Remove the steering wheel center pad.
2. Remove the steering wheel nut and locking washer.

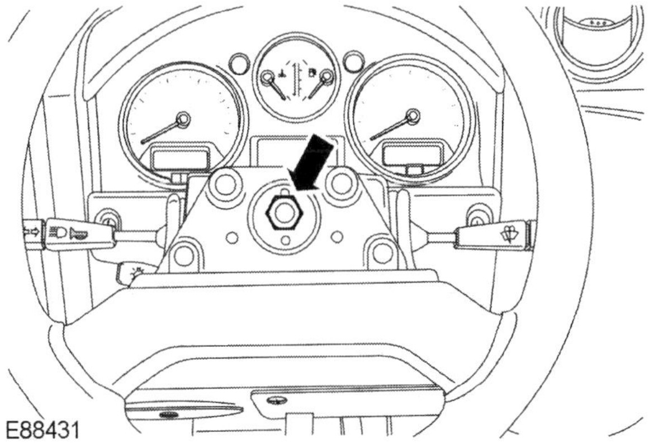

E88431

3. Mark the relationship between the steering wheel and steering column.
4. Using the special tool, remove the steering wheel.

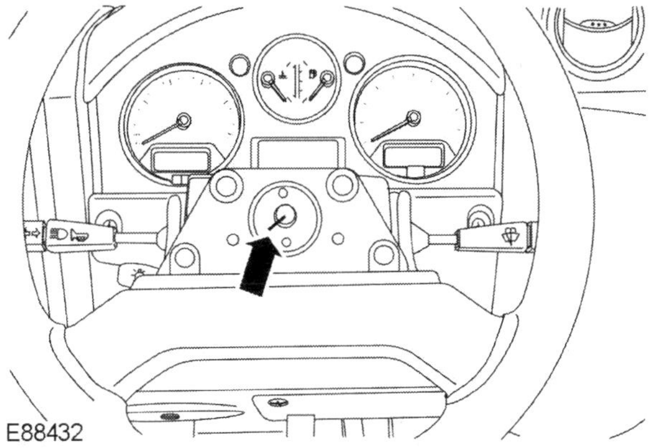

E88432

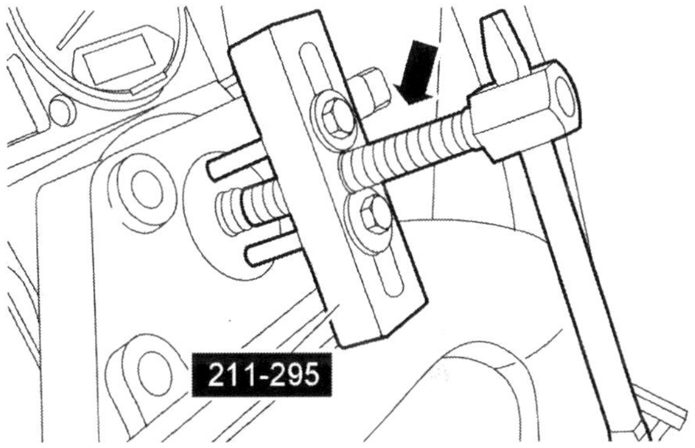

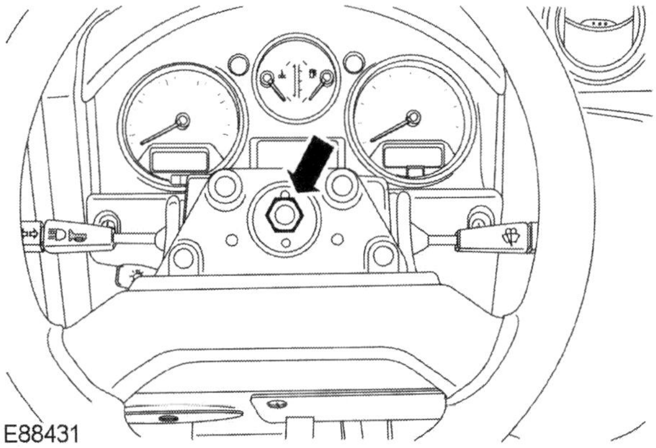

E88431

Installation

1. ⚠ NOTE: Make sure the alignment marks are correctly aligned.

Install the steering wheel.

- Tighten to 43 Nm (32 lb.ft).

2. Install the steering wheel center pad.

OVERVIEW

The steering column switches comprise the steering column multifunction switch and the ignition switch. Both switches are located on the steering column assembly.

The steering column multifunction switch comprises a case, which houses a turn signal indicator switch assembly and a windshield wiper switch assembly. The multifunction switch is located behind the steering wheel and is secured at the top of the steering column assembly and at the bottom of the column lock housing.

LEFT HAND COLUMN SWITCH

Turn Signal Indicator Switch

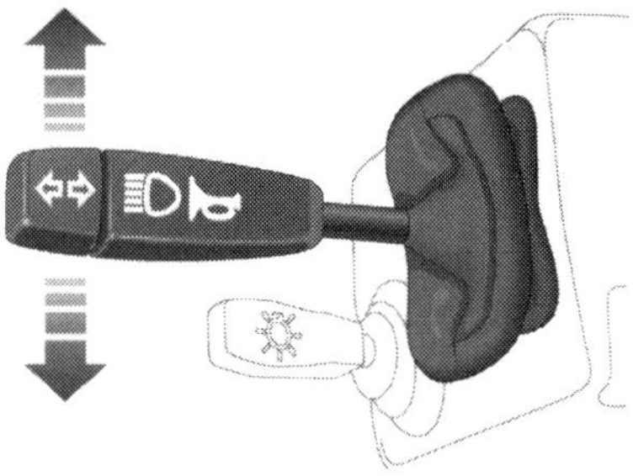

The turn signal indicator switch assembly is located in the Left Hand (LH) side of the case. The switch is connected to the main harness via a connector on the back of the switch. The switch controls the following functions:

- Left/right turn signal operation
- High/low beam operation
- Headlamp flash
- Horn operation

The turn signal indicators are operated by pushing the switch up for Right Hand (RH) indicators and down for LH indicators. The switch has a detent position, which locks the switch in the selected position until it is moved to the central off position. The LH and RH turn signal indicator switch positions are connected on separate wires to the hazard warning relay, via the hazard warning switch. When a switch position selection is made, a circuit is completed from the hazard warning relay to ground, via the selected switch position. The hazard warning relay detects the completed circuit and operates the selected turn signal indicator until the switch is moved to the central off position. The turn signal indicators can be cancelled either manually by the driver or automatically when the steering wheel is rotated to the straight ahead position.

High/Low Beam and Headlamp Flash

High beam is operated by pushing the switch forwards. The switch is latched in this position and the high beam is active until the switch is manually pulled rearwards.

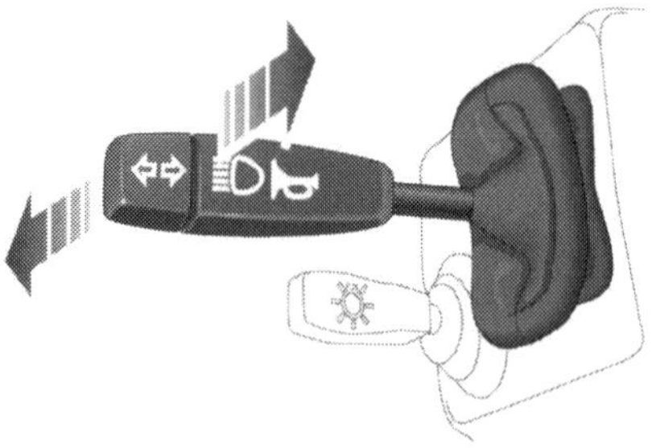

The headlamp flash function is operated by pulling the switch rearwards. The switch contacts complete a circuit and the headlamps are activated for as long as the switch is operated. The switch is non-latching in this position and the headlamp flash is switched off when the switch is released and it returns to its off position. The high beam and headlamp flash positions are connected on separate wires to the high/low beam relay and ground. When a switch selection is made, a circuit is completed from the high/low beam relay to ground via the switch contacts.

RIGHT HAND COLUMN SWITCH

Windshield Wiper and Washer Switch Assembly

Wiper

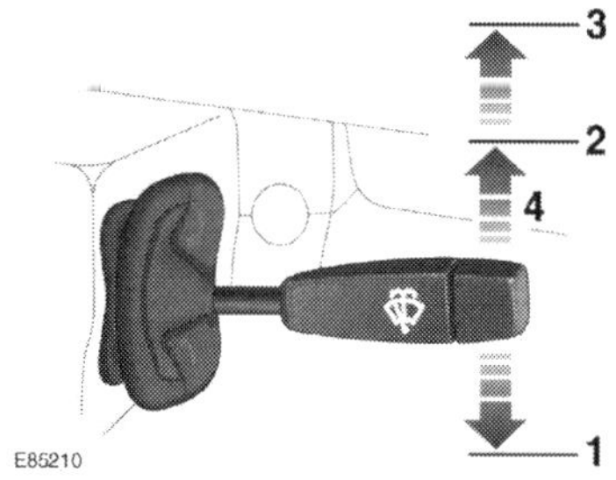

Item	Description
1	Intermittent wipe
2	Low speed wipe
3	High speed wipe
4	Single wipe

Washer

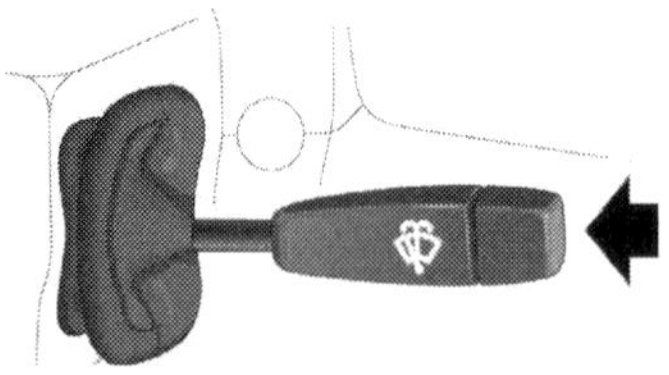

The windshield wiper functions are operated by moving the switch up or down. Flick wipe is selected by pushing the switch up against the spring pressure. The flick wipe switch contact is connected on a single wire to the windshield wiper delay Electronic Control Unit (ECU) and ground. When the switch is operated the circuit is completed between the delay ECU and ground. With the circuit completed the wipers operate for as long as the switch contact is made.

Intermittent is selected by pushing the wiper switch down, to the detent position, the wipers operate at a delay period, which is controlled by the windshield wiper delay ECU. The wipers remain in the intermittent mode until the wiper switch is moved to the off or slow/fast speed positions. The intermittent switch contact is connected between the delay ECU and ground. When the switch is moved to the intermittent position the circuit is completed. With the circuit completed, the wipers operate in intermittent for as long as the switch contact is made.

Slow speed operation is selected by pushing the wiper switch up, to the second detent position. The wipers operate at slow speed until the wiper switch is moved to the off, intermittent or fast speed positions. The slow speed switch contact is connected between the delay ECU and ground. When the switch is moved to the slow speed position the circuit is completed. With the circuit completed, the wipers operate at slow speed for as long as the switch contact is made.

Fast speed operation is selected by pushing the wiper switch up, to the second detent position. The wipers operate at fast speed until the wiper switch is moved to the off, intermittent or slow speed positions. The fast speed switch contact is connected between the delay ECU and ground. When the switch is moved to the fast speed position the circuit is completed. With the circuit completed, the wipers operate at fast speed for as long as the switch contact is made.

The windshield washers are operated by pressing the button on the end of the wash/wipe stalk located on the RH side of the steering column. When the button is pressed, the washers operate immediately and stop immediately the button is released. The washer button contact is connected between the delay ECU and ground. When the button is pressed the circuit is completed. With the circuit completed, the washers operate for as long as the contact is made.

IGNITION SWITCH

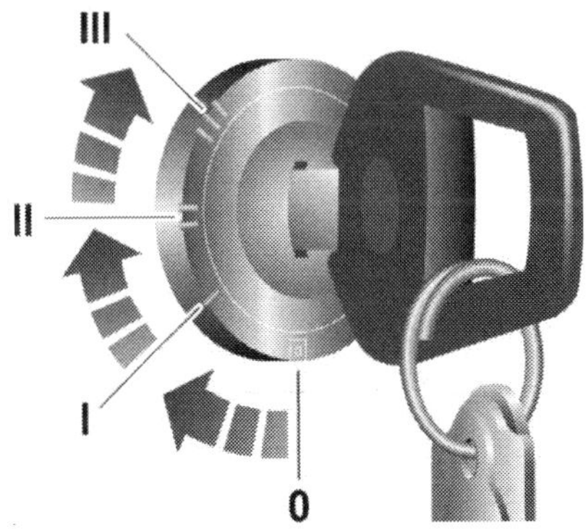

The ignition switch is located in the LH end of the steering column assembly. The switch is held in the column lock casting with 2 locking tabs, which engage in slots in the column lock casting.

The switch has a slot, which provides for the location of the drive shaft. The drive shaft passes through the column lock. This shaft is rotated by the driver when the ignition key is turned in the key barrel. This rotation turns a drum inside the ignition switch, which moves 2 electrical contacts to select the required ignition position. A spring loaded ball locates in a seat for each of the 3 ignition switch positions, allowing the driver to feel when the required position is reached.

Right Steering Column Multifunction Switch

Removal and Installation

Removal

1. Remove steering column shroud.

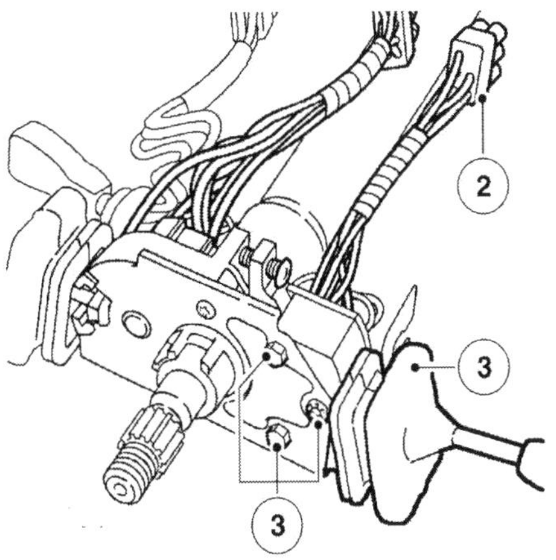

Refer to: Steering Column Shrouds (501-05 Interior Trim and Ornamentation, Removal and Installation)

2. Disconnect switch multiplug from main harness.
3. Remove 2 screws and shake proof washer securing switch to mounting bracket.
4. Remove switch complete with harness.

Installation

1. Fit new switch to mounting bracket.
2. Reconnect switch multiplug.
3. Fit steering column shroud.

Refer to: Steering Column Shrouds (501-05 Interior Trim and Ornamentation, Removal and Installation)

Left Steering Column Multifunction Switch

Removal and Installation

Removal

1. Remove the steering wheel.

Refer to: Steering Wheel (211-04 Steering Column, Removal and Installation)

2. Remove the steering column shrouds.

Refer to: Steering Column Shrouds (501-05 Interior Trim and Ornamentation, Removal and Installation).

3. Disconnect the 2 electrical connectors.
4. Disconnect the electrical connector.

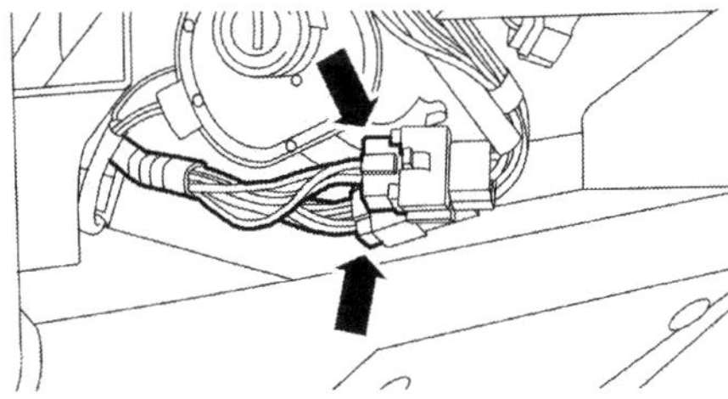

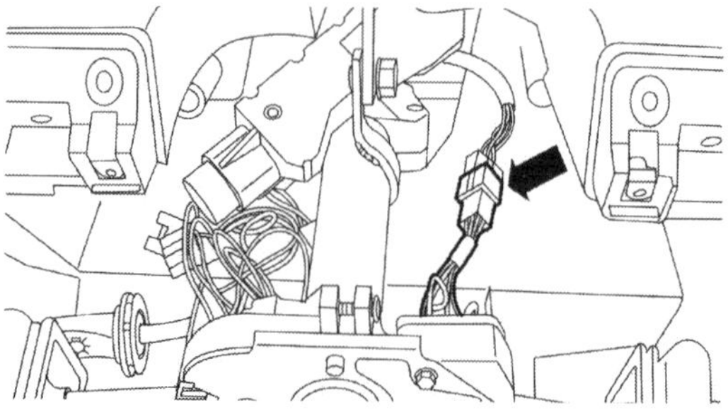

5. Remove the steering column multifunction switch assembly.
   - Undo but do not remove the multifunction switch clamp screw.

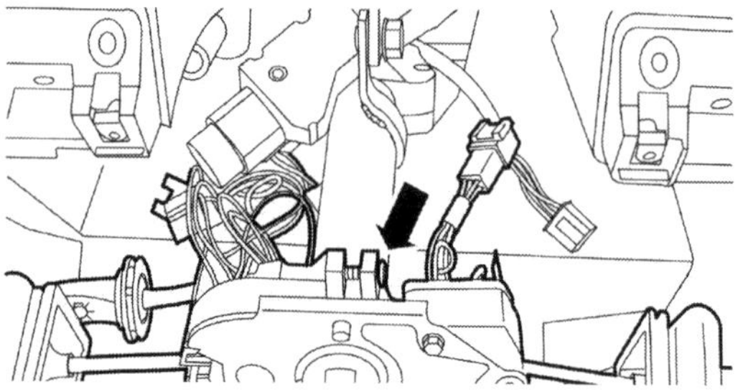

Installation

1. To install, reverse the removal procedure.

Ignition Switch

Removal and Installation

Removal

1. Disconnect the battery ground cable.

Refer to: Battery Disconnect and Connect (414-01, General Procedures).

2. Remove steering column shrouds.

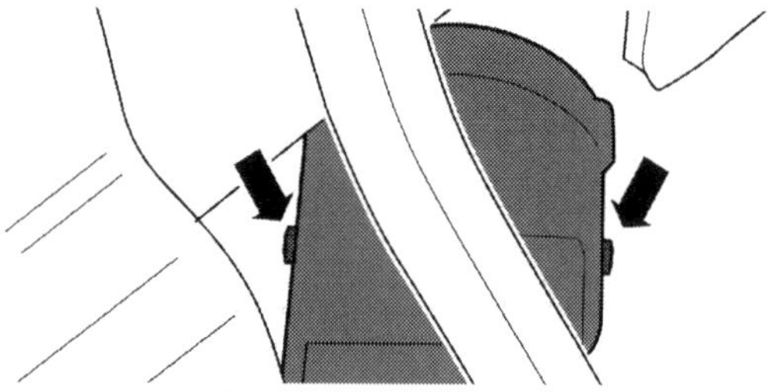

Refer to: Steering Column Shrouds (501-05, Removal and Installation).

3. Remove the instrument cluster.

Refer to: Instrument Cluster (413-01, Removal and Installation).

4. Remove the passive anti theft system (PATS) transceiver from the ignition barrel.
5. Remove the 2 screws.

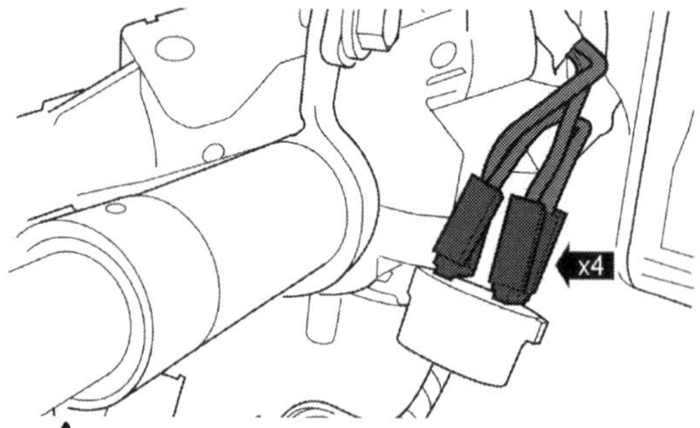

6. ⚠ CAUTION: Note the position of the electrical connectors prior to removal, failure to carry out this instruction may cause damage to the vehicle.

Disconnect the 4 ignition switch electrical connectors.

Installation

1. Connect harness leads to ignition switch as noted on removal.

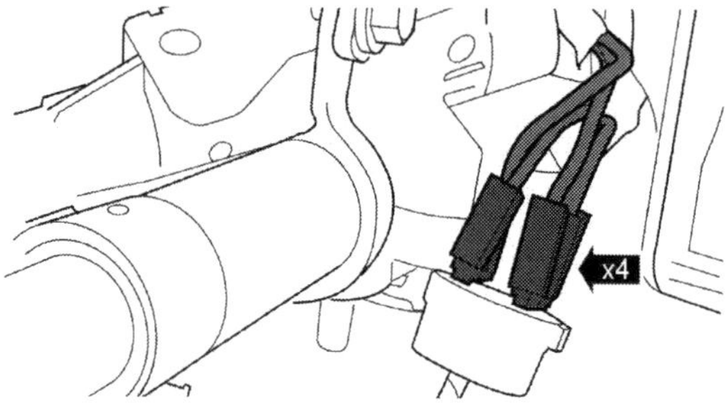

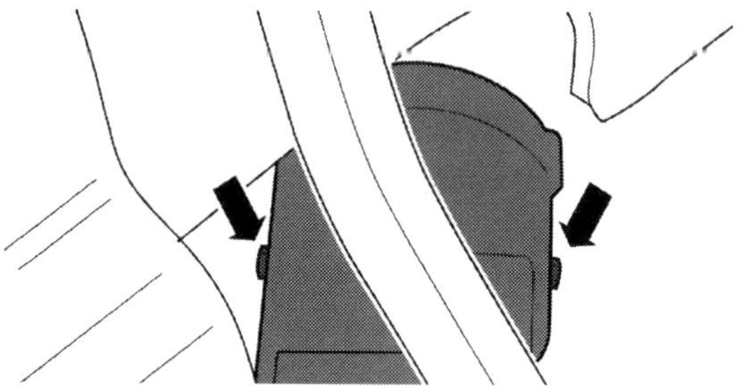

2. Install the 2 screws.

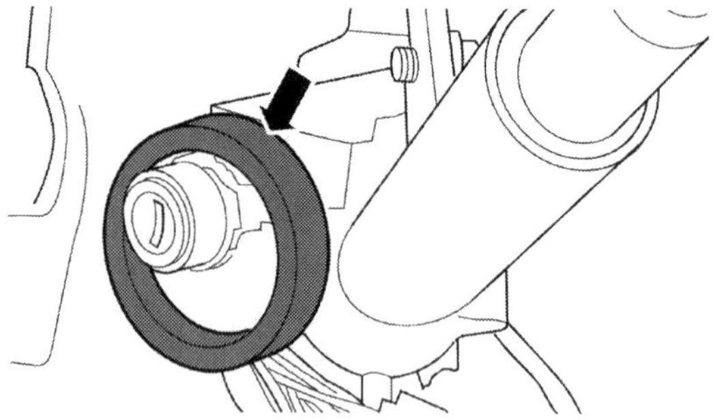

3. Install the PATS transceiver.
4. Install the instrument cluster.

Refer to: Instrument Cluster (413-01, Removal and Installation).

5. Install steering column shrouds.

Refer to: Steering Column Shrouds (501-05, Removal and Installation).

6. Connect the battery ground cable.

Refer to: Battery Disconnect and Connect (414-01, General Procedures).

Powertrain

Important Powertrain Information

Two different versions of the Ford Puma or TDCi engine were used in these models:

The 2007-2011 models had a 2.4-litre size (2401cc) and the 2012-2016 models had a 2.2-litre size (2198cc). The main physical difference was that the 2.4-litre had an 88.9mm bore and the 2.2-litre had an 86mm bore. They both had the same stroke of 94.6mm, and both delivered the same power and torque.

The other key difference between the two engines is that the later (2.2) engine had a Diesel Particulates Filter (DPF) for most countries and the 2.4 engine did not. There were also invisible differences in the Engine Control Unit, which on the 2.4-litre engine only incorporates a speed limiter (to 82mph).

In all other major respects, the two engines were the same. This Workshop Manual uses the pages for the later 2.2-litre type, with the addition of notes where there are differences from the 2.4-litre engine. Neither Jaguar Land Rover nor Brooklands Books accepts any liability for workshop errors arising from additional differences not described here.

Engine: 2.2L 2007-2011 and 2.4L 2012-2016 Diesel

Procedures for 2.4-litre 2007-2011 and the 2.2-litre 2012-2016 are the same unless otherwise indicated

Bearing Inspection

General Procedures

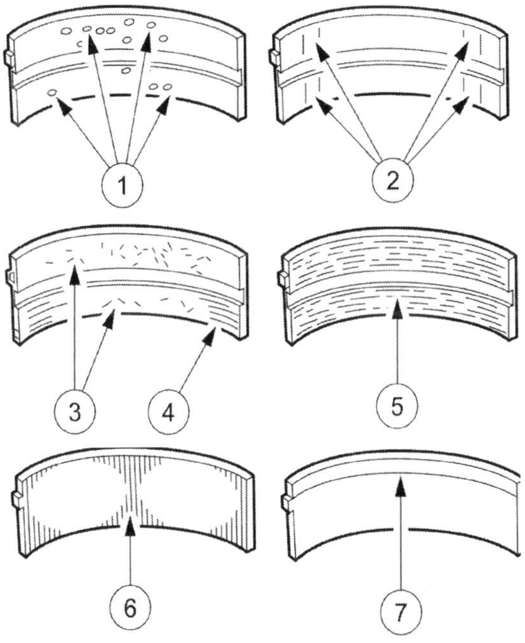

1. Inspect bearings for the following defects.
    1. Cratering - fatigue failure
    2. Spot polishing - incorrect seating.
    3. Imbedded dirt engine oil.
    4. Scratching - dirty engine oil.
    5. Base exposed - poor lubrication.
    6. Both edges worn - journal damaged.
    7. One edge worn - journal tapered or bearing not seated.

Camshaft Bearing Journal Clearance

General Procedures

1. NOTES:

⚠ Make sure that the following stages are followed exactly. The tappets or followers must be removed to carry out this measurement.

⚠ Make sure that the camshaft is to specification.

⚠ The bearing caps and journals should be free from engine oil and dirt.

Position on a length of plastigage on the bearing cap.

- Insert the camshaft, without lubrication, into the cylinder head.
- Position a plastigage strip, which should be equal to the width of the bearing cap, on the bearing journal.

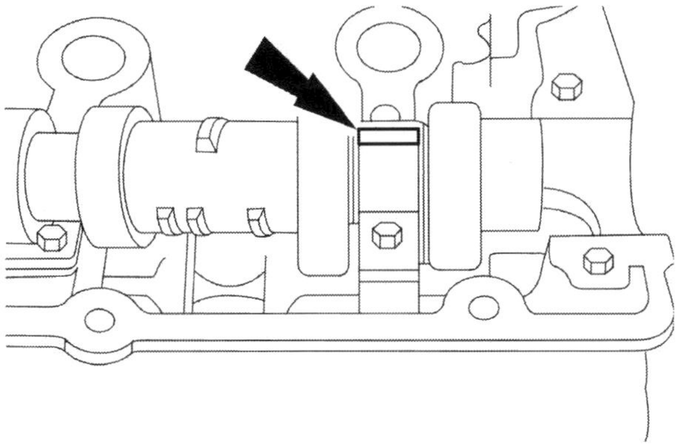

2. Install the camshaft bearing caps.
    - Follow the relevant tightening sequence.

3. ⚠ NOTE: Do not strike the bearing caps.

Remove the camshaft bearing caps.

- Follow the relevant loosening sequence.

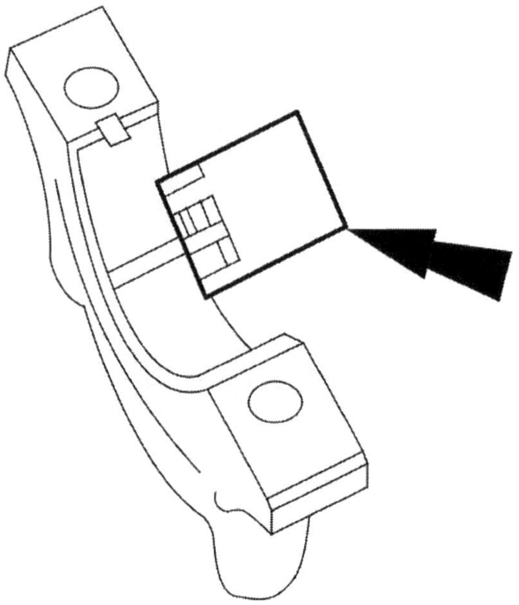

4. Using the special tool, read off the measurement.
   - Compare the width of plastigage with the plastigage scale.
   - The value that is read off is the bearing clearance.
   - If the values are not to specification install a new camshaft.

Camshaft Bearing Journal Diameter

General Procedures

1. Determine the diameter of the camshaft journals.
   - Using a micrometer measure the diameter at 90 degrees intervals to determine if the journals are out-of-round.
   - Measure at two different points on the journal to determine if there is any tapering.
   - If the measurements are out of the specified range, install a new camshaft.

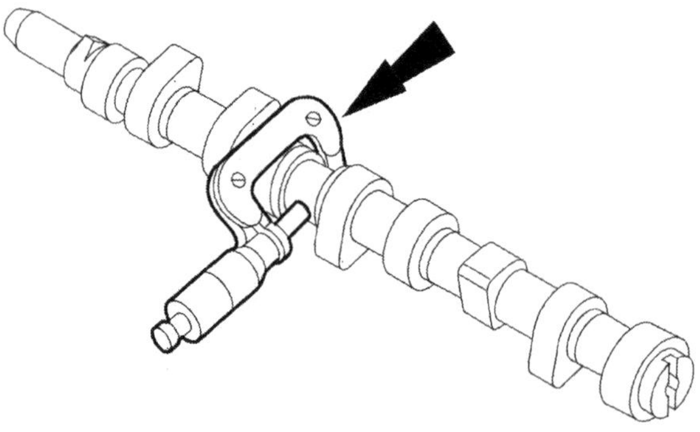

VUJ0001695

Camshaft End Play

General Procedures

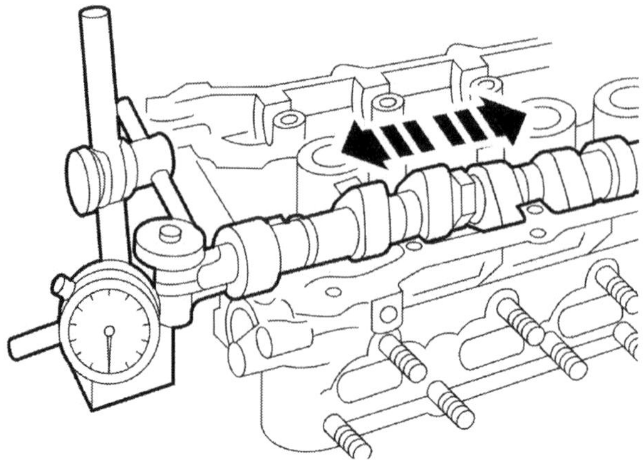

1. ⚠ NOTE: Make sure that the camshaft is to specification.

Using the special tool, measure the end play.

- Slide the camshaft in both directions. Read and note the maximum and minimum values on the dial indicator gauge.

1. End play = maximum value minus minimum value.
   - If the measurement is out of specification, install new components.

Camshaft Lobe Lift

General Procedures

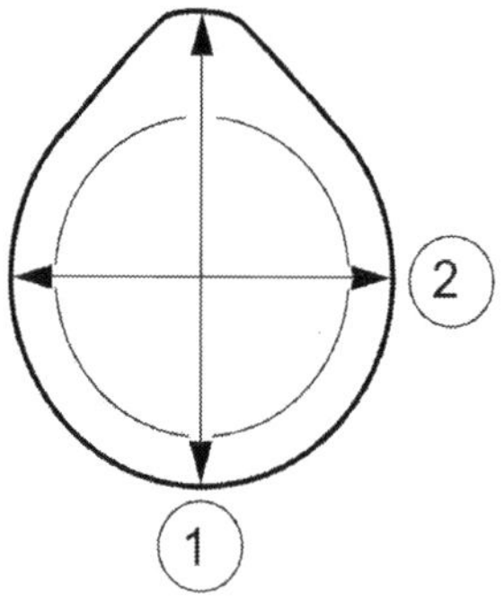

1. Measure the diameter (1) and diameter (2) with a vernier caliper. The difference in measurements is the lobe lift.

Camshaft Surface Inspection

General Procedures

1. Inspect camshaft lobes for pitting or damage in the active area. Minor pitting is acceptable outside the active area.

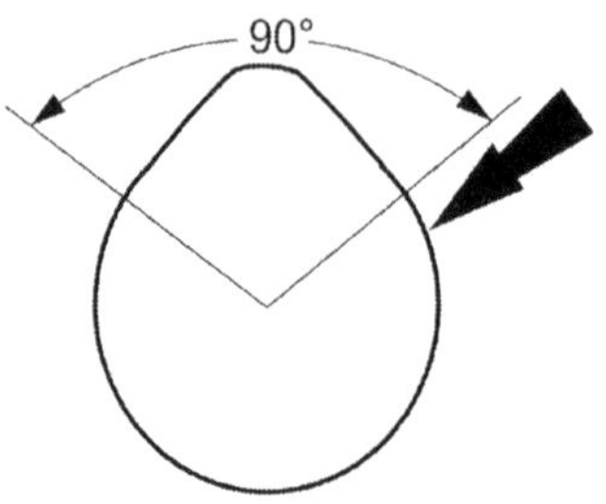

Connecting Rod Cleaning

General Procedures

1. CAUTION: Do not use a caustic cleaning solution or damage to connecting rods may occur.

Mark and separate the parts and clean with solvent. Clean the oil passages.

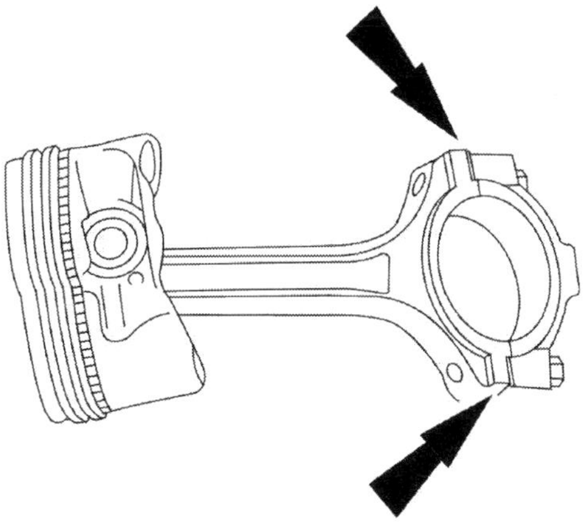

Connecting Rod Large End Bore

General Procedures

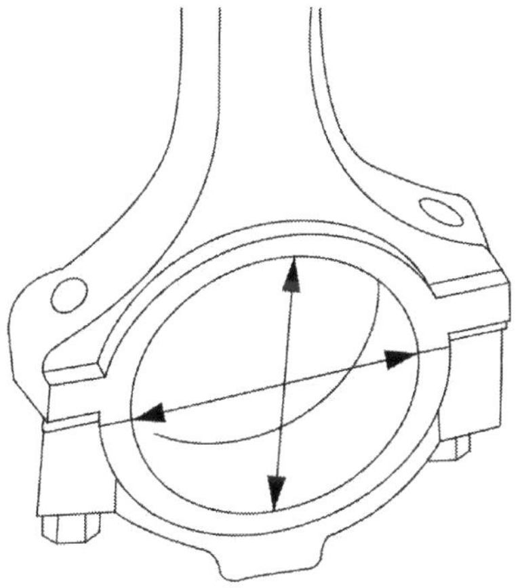

1. Measure the bearing bore in two directions. The difference is the connecting rod bore out-of-round. Verify the out-of-round is within specification.

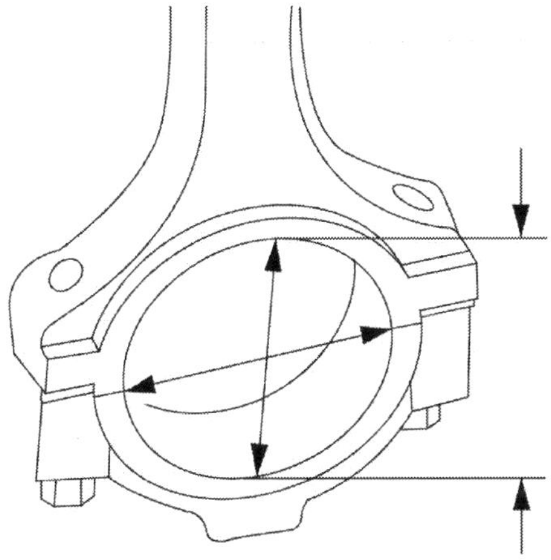

2. Measure the bearing bore diameter in two directions. Verify the bearing bore is within specification.

Cylinder Bore Out-of-Round

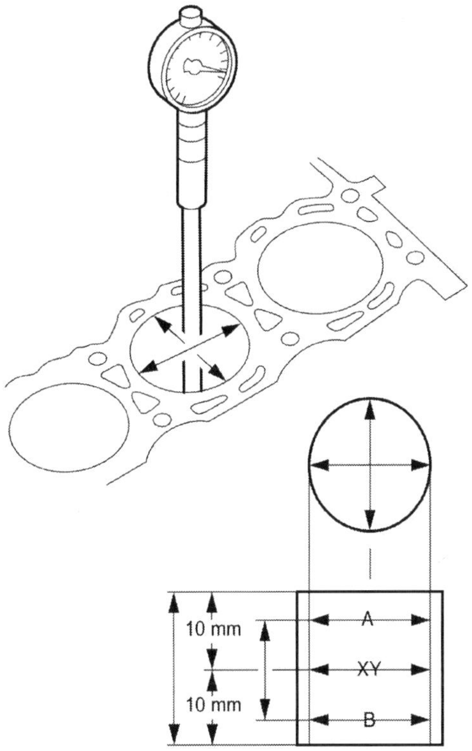

1. NOTE: The main bearing caps or lower crankcase must be in place and tightened to the specified torque; however, the bearing shells should not be installed.

Measure the cylinder bore with an internal micrometer.

- Carry out the measurements in different directions and at different heights to determine if there is any out-of-roundness or tapering.
- If the measurement is out of the specified range, hone out the cylinder block or install a new block.

Cylinder Block Distortion

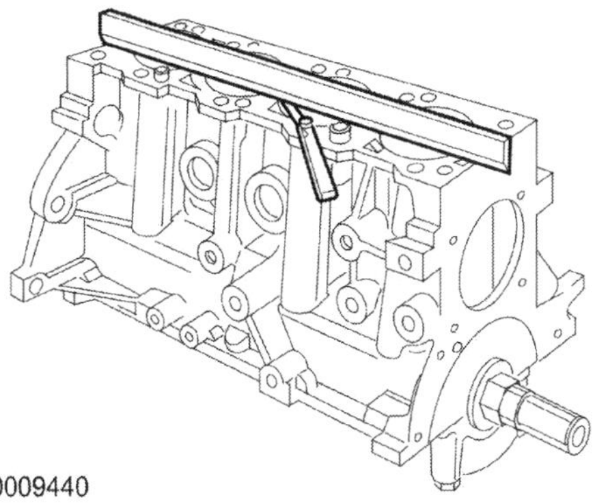

ELE0009440

General Procedures

1. Using a Straight Edge and a Feeler Gauge, measure the cylinder block/cylinder head distortion.

- Measure the mating face distortion.
- If the value is not to specification rework the mating face (if allowed).

Exhaust Manifold Cleaning and Inspection

General Procedures

1. Inspect the cylinder head joining flanges of the exhaust manifold for evidence of exhaust gas leaks.
2. Inspect the exhaust manifold for cracks, damaged gasket surfaces, or other damage that would make it unfit for further use.

Piston Inspection

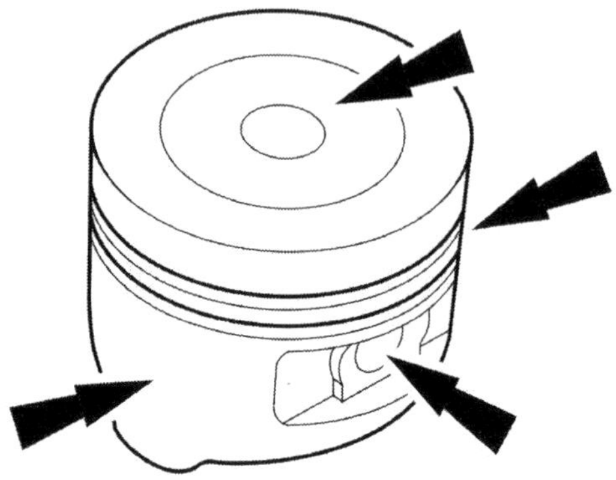

General Procedures

1. CAUTION: Do not use any aggressive cleaning fluid or a wire brush to clean the piston.

Carry out a visual inspection.

- Clean the piston skirt, pin bush, ring grooves and crown and check for wear or cracks.
- If there are signs of wear on the piston skirt, check whether the connecting rod is twisted or bent.

Piston Pin Diameter

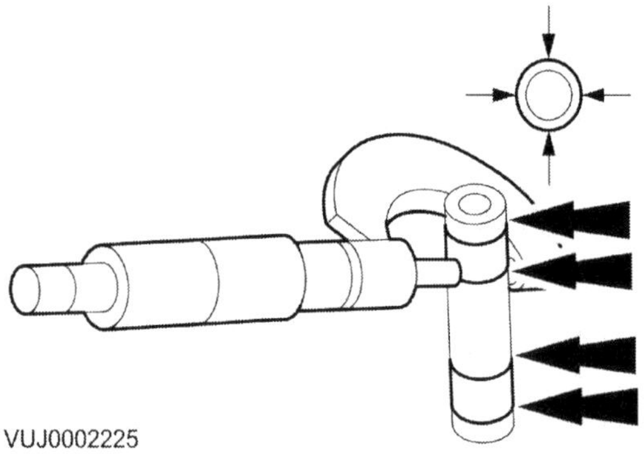

General Procedures

1. NOTE: The piston and piston pin are a matched pair. Do not mix up the components.

Measure the piston pin diameter.

- Measure the diameter in two directions.
- If the values are not to specification, install a new piston and a new piston pin.

Piston Pin to Bore Diameter

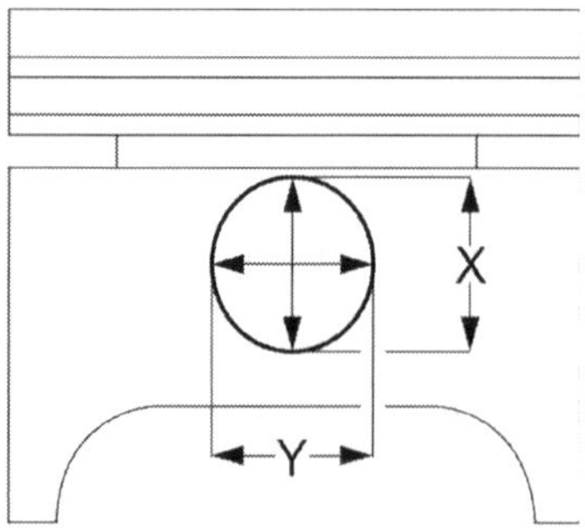

General Procedures

1. NOTE: The piston and piston pin form a matched pair. Do not mix up the components.

Measure the diameter of the piston pin bore.

- Measure the diameter in two directions.
- If the values are not to specification, install both a new piston and a new piston pin.

Piston Ring End Gap

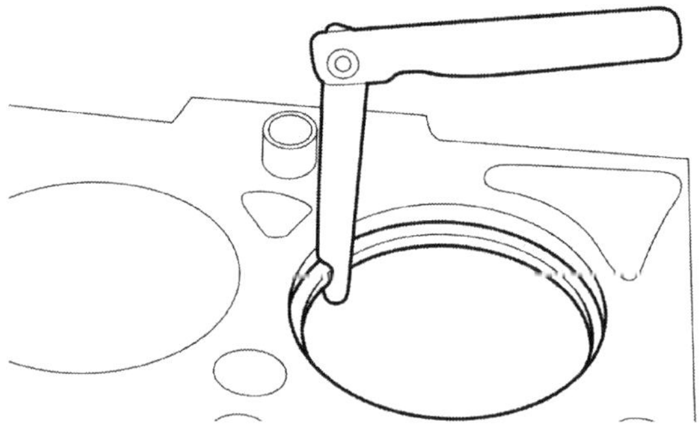

General Procedures

1. CAUTION: Do not mix up the piston rings. Install the piston rings in the same position and location.

Using the Feeler Gauge, measure the piston ring gap.

- The values given in the specification refer to a gauge ring used during production.

Piston Ring-to-Groove Clearance

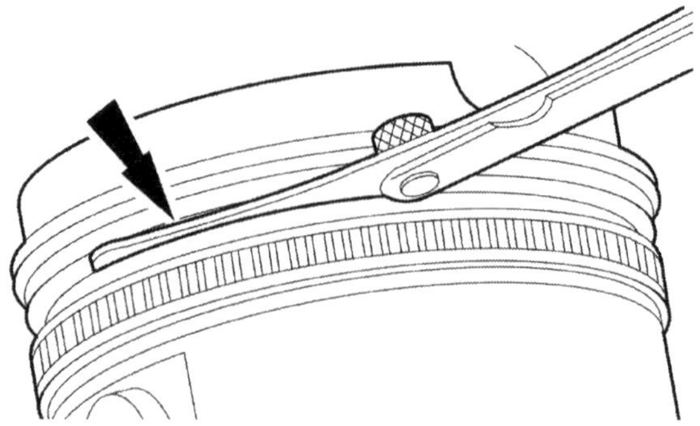

General Procedures

1. NOTE: The piston ring must protrude from the piston groove. To determine the piston ring clearance, insert the Feeler Gauge right to the back of the groove, behind the wear ridge.

Using the Feeler Gauge, measure the piston ring clearance.

Valve Spring Free Length

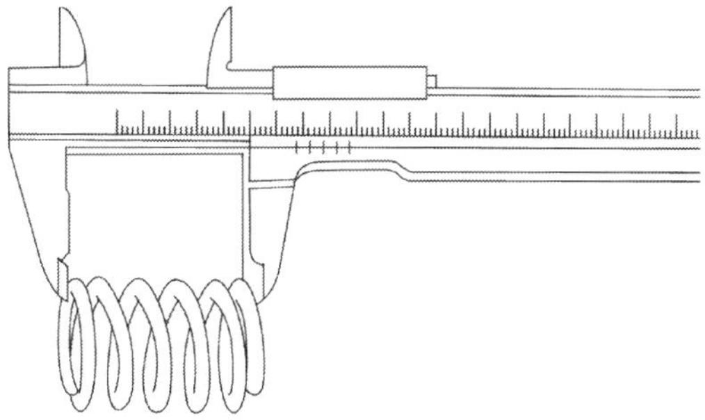

General Procedures

1. Using a vernier gauge, measure the free length of each valve spring. Verify the length is within specification.

Valve Stem Diameter

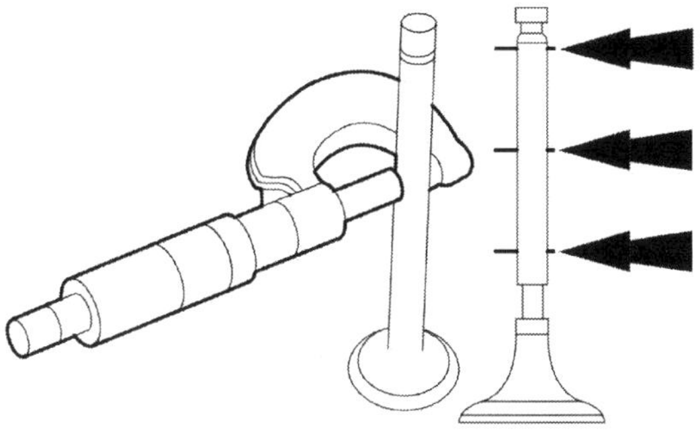

General Procedures

1. Using a micrometer measure the diameter of the valve stems.
    - If the measurements are not to specification, install a new valve.

Leakage Test Using Smoke Test Equipment

General Procedures

CAUTION: Do not exceed 5 psi test pressure, when testing the engine or cooling systems.

NOTES:

The vehicle battery must be in good condition and fully charged before carrying out this procedure.

On vehicles with more than one air cleaner, it will be necessary to insert smoke at both air cleaner outlet pipes independently if both sides of the engine require testing.

In some cases it may be necessary to remove undertrays, trim or engine covers to obtain access to all potential leak locations.

For further information regarding operation of the test equipment refer to the manufacturers operators manual supplied with the kit.

This procedure contains some variation in the illustrations depending on the vehicle specification, but the essential information is always correct.

This procedure contains illustrations showing certain components removed to provide extra clarity.

1. WARNING: Make sure the smoke test equipment is secure. Failure to follow this instruction may result in personal injury.

Install the smoke test equipment to a suitable location.

2. NOTES:

The illustration shows a 2.0L Diesel engine, other engines are similar.

It may be necessary to remove the air cleaner(s) to allow access to the air cleaner outlet pipes.

It will be necessary to cap one of the air cleaner outlet pipes. Use the blanking caps supplied in the kit to cap the open orifice.

Release the air cleaner outlet pipe(s) from the air cleaner.

3. CAUTION: Make sure that the adaptor is correctly fitted.

Install the appropriate adaptor to the air cleaner outlet pipe.

E177551

4. CAUTION: Make sure that the adaptor is correctly fitted.

If required, install the appropriate blanking adaptors to the air cleaner outlet pipes.

5. Connect a suitable compressed air line to the smoke test equipment.

6. Connect the smoke test equipment positive power cable to the battery positive terminal.

7. WARNING: Do not connect the smoke test equipment negative cable to the battery negative terminal.

Connect the smoke test equipment negative cable to a suitable body ground point.

8. NOTE: Make sure the smoke test equipment adapter is a good fit to the air cleaner outlet pipe. This must be an air tight seal.

For further information regarding operation of the test equipment refer to the manufacturers operators manual supplied with the kit

Cylinder Head Distortion

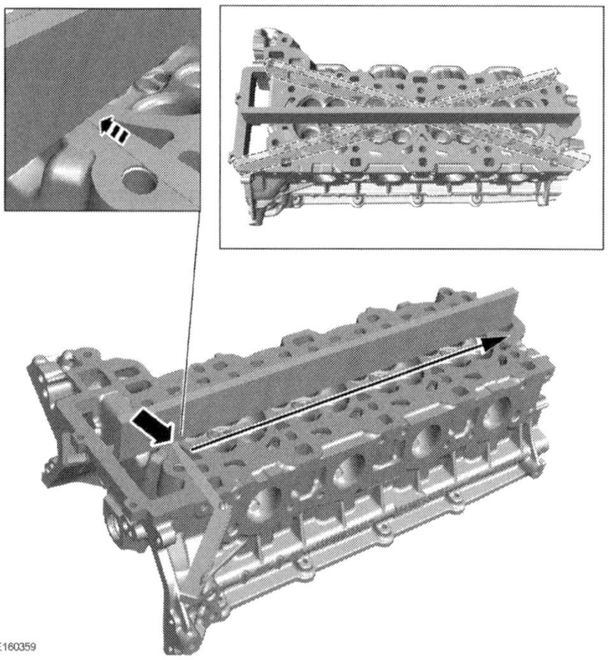

General Procedures

Check

1. Using a suitable metallic straight edge and feeler gauge, measure the cylinder head face in the areas illustrated. **Note the maximum value**.

2. CAUTION: Machine the **minimum** thickness of material from the cylinder head to meet specification. If a selection of cylinder head gaskets are available, increase the thickness of the cylinder head gasket by one size.

NOTES:

Prior to having the cylinder head machined, prior approval is required by Jaguar or Land Rover engineering.

⚠ If the cylinder head requires machining, this must be carried out by a local engineering company.

If the cylinder head exceeds the maximum value, the cylinder head must be machined.

Refer to: (303-01 Engine, Specifications)

Cylinder Compression Test

General Procedures

Check

⚠ CAUTION: Before disconnecting or removing components, make sure the area around the joint faces and connections are clean. Plug open connections to prevent contamination.

NOTES:

⚠ Removal steps in this procedure may contain installation details.

⚠ Some variation in the illustrations may occur, but the essential information is always correct.

⚠ The vehicle battery must be in good condition and fully charged before carrying out this procedure.

1. **Refer to:** Glow Plug (303-07 Glow Plug System, ID4 2.2L Diesel Removal and Installation)
2. ⚠ CAUTION: Make sure the fuel injection system is disabled before carrying out a cylinder compression test. Failure to follow this step may result in damage to the vehicle.
3. ⚠ CAUTION: Make sure the fuel injection system is disabled before carrying out a cylinder compression test. Failure to follow this step may result in damage to the vehicle.

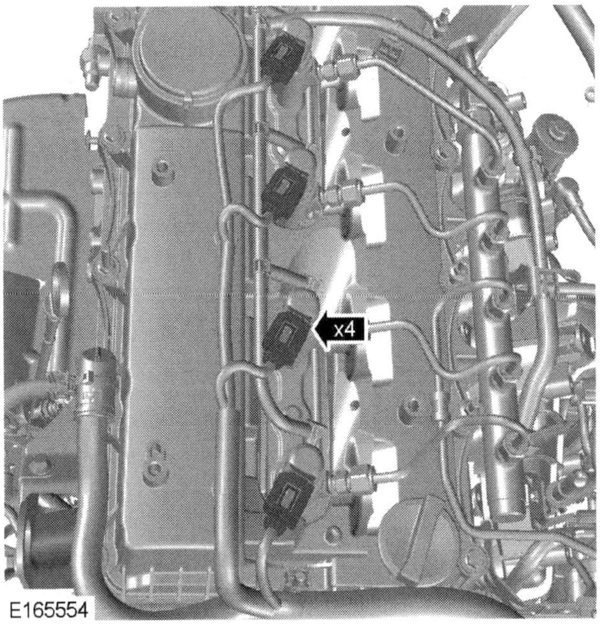

E165554

Crank engine for approximately five seconds to remove any remaining fuel in the cylinders.

4 Install the compression test gauge.

5 NOTES:

⚠ This step requires the aid of another technician.

⚠ Make a rough copy of graphic number E165557 in (over page), and use it to record each cylinder compression figure.

⚠ The vehicle battery must be in good condition and fully charged before carrying out this procedure.

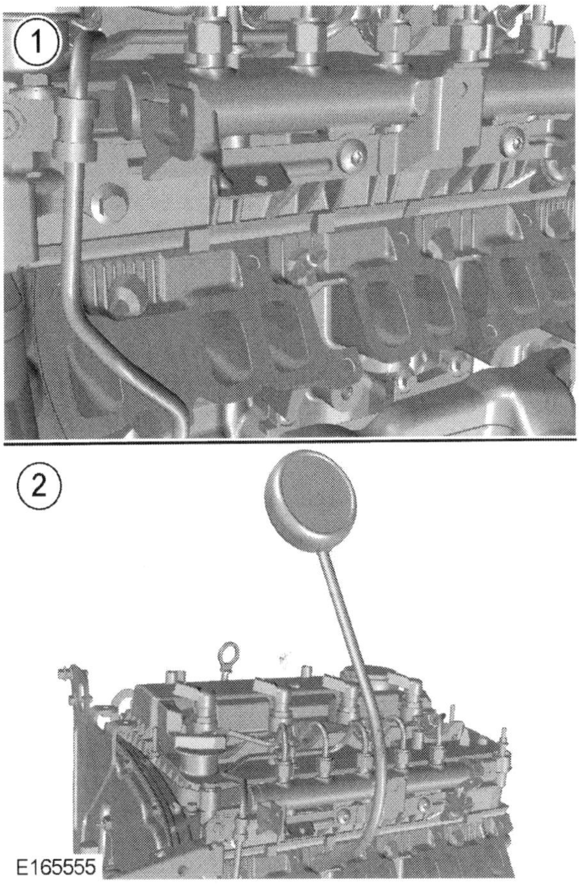

E165555

⚠ The vehicle battery must be in good condition and fully charged before carrying out this procedure.

Crank the engine for 10 seconds and record the figure displayed on the compression test gauge. Make sure the pressure is released from the compression test gauge after each cylinder recording. Repeat the process for all cylinders.

6. NOTE: Remove and discard all blanking caps.

To install, reverse the removal procedure.

7. The minimum cylinder compression reading recorded must be within 10% of the maximum cylinder compression reading recorded. If the difference across the cylinders is higher than 10% please contact dealer technical support (DTS) for further assistance.

8. Using Land Rover approved diagnostic equipment, read and clear any diagnostic trouble codes (DTCs).

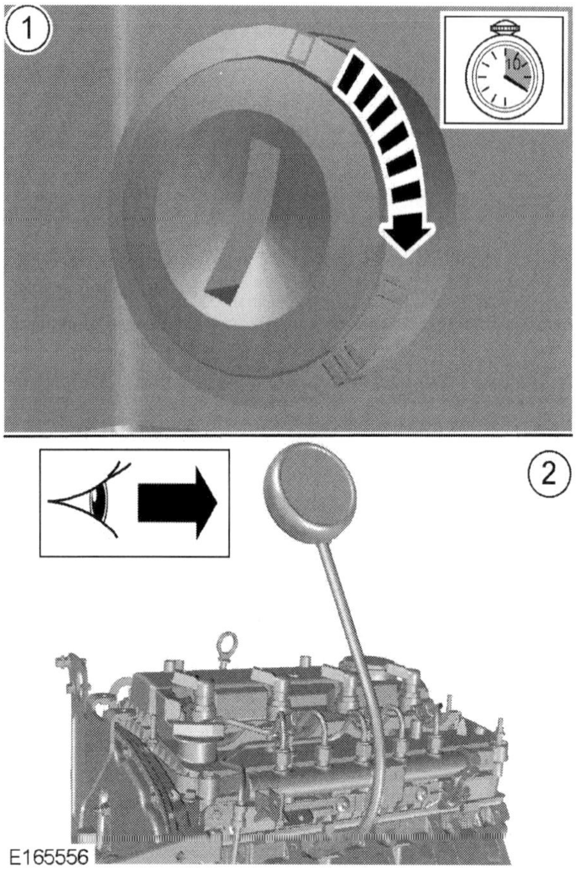

E165556

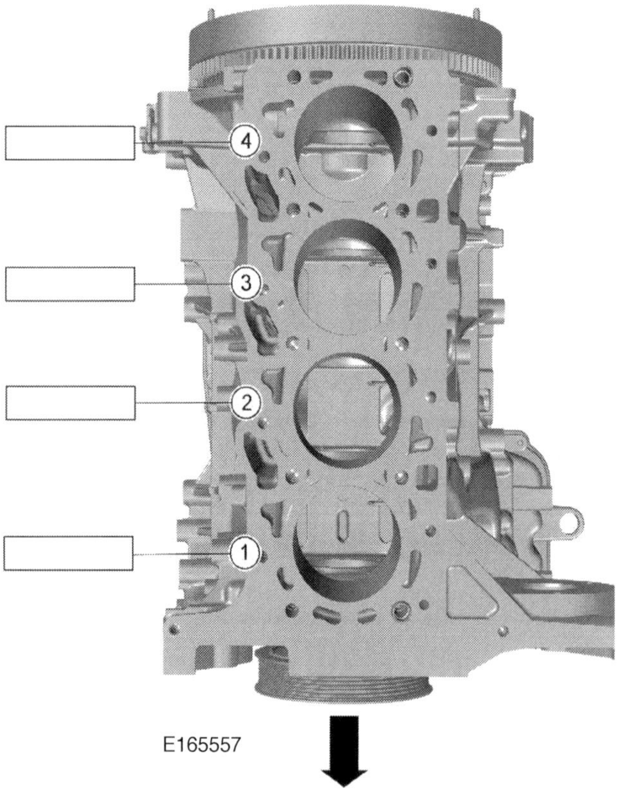

E165557

Thread repair

General Procedures

Repair

CAUTION: Jaguar Land Rover (JLR) approve the use of thread inserts on all engines except V6 Cologne, on all areas of the engine except power conversion elements, for example, pistons, crankshafts, con-rods etc.

NOTE: Where applicable please use this method of repair. Do not replace the affected component, for example, the complete engine or cylinder head.

1. Thread repair kits are to be sourced locally, Heli-Coil™, Time-Sert™ or a similar brand.
2. CAUTION: Extra care should be taken, especially during the tapping phase of the repair. This is critical to make sure the repair is correct.

Please follow any instructions that are supplied with the thread repair kit.

Specifications

Lubricants, Fluids and Sealers

Item	Specification
Vehicles with diesel particulate filter	SAE 5W30 STJLR.03.5005
Vehicles without diesel particulate filter	SAE 5W30 STJLR.03.5005
Sealant — oil pan, camshaft carrier and engine front cover	WSE-M4G323-A4

Capacities

Description	Litres
Engine oil, service fill with oil filter change	7.0
Engine oil MIN to MAX on dipstick	2.0

Cylinder Head Dimensions

Description	mm
***2.2* only Cylinder head maximum permitted warp** (flatness specification)	**0.2**
***2.4* only Cylinder head maximum permitted warp** (flatness specification)	**0.1**
Peak to valley height of mating surface	0.02
Cylinder head gasket thickness	
Piston protrusion of 0.310 - 0.400 mm	1.1 (one tooth)
Piston protrusion of 0.401 - 0.450 mm	1.15 (two teeth)
Piston protrusion of 0.451 - 0.500 mm	1.2 (three teeth)

Oil pressure specifications

Description	bar
Minimum oil pressure at idle speed	1.25
Minimum oil pressure at 2000 rpm	2.0

Bolt Torque Values

Description	Nm	lb-ft
Transmission to engine bolts	40	30
Engine mount nuts	80	59
Engine mount bolts	80	59
Engine mount bracket bolts	63	46
Valve cover bolts	10	7
+ Rocker shaft bolts		
Stage 1	10	7
Stage 2	Further 30 degrees	
Timing chain guide bolts	15	11
Timing chain tensioner nut	15	11
Timing chain tensioner bolts	15	11
Oil pump chain tensioner bolts	22	16
Camshaft sprocket bolts	35	24
Fuel injection pump sprocket bolts	33	26
Generator bolts	48	35
Generator mounting bracket bolts	25	18
Exhaust gas recirculation (EGR) valve to EGR valve outlet tube bolts	10	7
EGR valve to EGR cooler bolts	23	17
Exhaust manifold to EGR cooler bolts	23	17
Intake manifold bolts	15	11
Intake manifold to EGR valve outlet tube bolts	10	7

217

Coolant pump bolts	23	17
Oil level indicator tube bolt	10	7
Oil pump pickup tube bolts	10	7
Oil filter housing assembly bolts	23	17
Oil pump bolts	10	7
Oil pan drain plug	23	17
Oil pressure switch	15	11
Oil pan bolts		
Stage 1	7	5
Stage 2	14	10
Coolant manifold bolts	10	7
EGR cooler to cylinder head bolts	23	17
+ Cylinder head bolts	**A**	-
Engine front cover bolts	14	10
Engine front cover nuts	10	7
Glow plugs	10	7
Power steering pump bolts	23	17
Power steering pump bracket bolts	23	17
Exhaust manifold bolts	40	30
+ Exhaust manifold nuts	40	30
+ Exhaust manifold studs	20	15
Turbocharger oil return tube bolts	10	7
Turbocharger oil return tube clamp bolt	22	16
Turbocharger oil supply line banjo bolt	35	26
Engine lifting eye bolts	22	16
Crankshaft position sensor (CKP) bolt	7	5
+ Crankshaft pulley bolts	**A**	-
Crankshaft rear oil seal carrier bolts	10	7
+ Flywheel bolts		
Stage 1	25	18
Stage 2	40	30
Stage 3	Further 48 degrees	
Crankshaft main bearing cap bolts		
Stage 1	45	33
Stage 2	80	59
Stage 3	Further 105 degrees	
Connecting rod bearing cap bolts		
Stage 1	30	22
Stage 2	Further 100 degrees	
Ladder frame bolts	23	17
Camshaft carrier bolts	A	-
Fuel injector clamp bolts		
Stage 1	6	4
Stage 2	Further 360 degrees *Further 180 degrees (2.4-litre)	

+ New nuts/bolts/studs must be installed

A - Refer to the procedure in this section

Engine - 2.2L Diesel & 2.4L Diesel

Description and Operation

EXTERNAL VIEW

Engine Structure

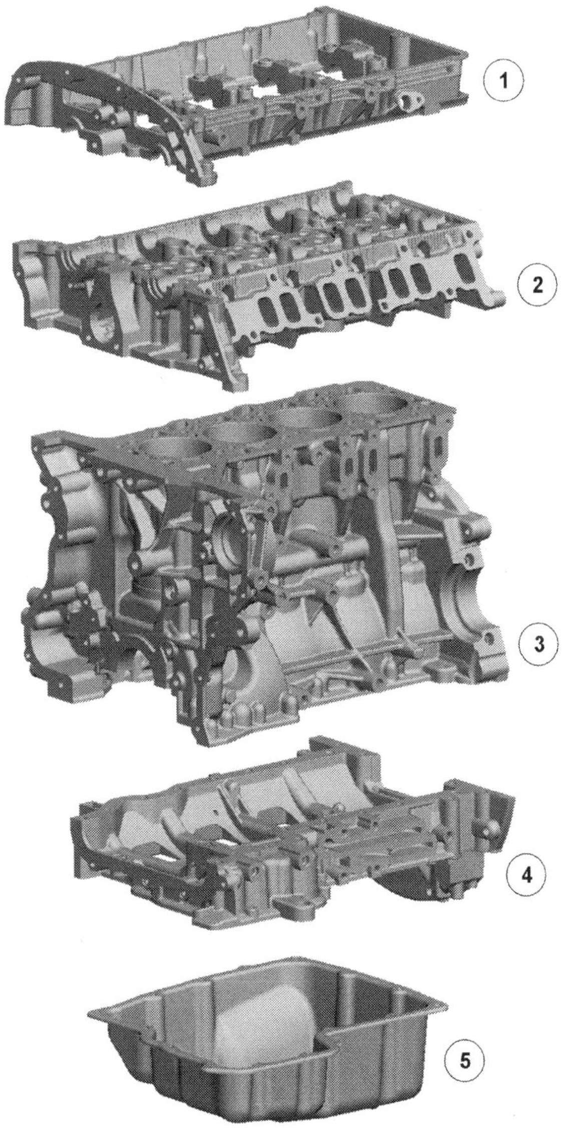

Item	Description
1	Camshaft carrier
2	Cylinder head
3	Cylinder block
4	Skirt stiffener
5	Oil pan

OVERVIEW

The 2.2 liter diesel engine is a 4 cylinder unit with common rail fuel injection and with four valves per cylinder operated by dual overhead camshafts. The unit is liquid cooled and turbocharged.

The cylinder block is manufactured in Compacted Graphite Iron (CGI) and is coupled with a separate aluminum 'ladder-frame' skirt stiffener to provide a lightweight, compact and very stiff bottom end of the engine. The cylinder head and camshaft carrier are cast aluminum. The single-piece oil sump is formed from pressed steel. The exhaust manifold is cast iron and includes connections for the turbocharger and EGR (exhaust gas recirculation) cooler. A plastic acoustic cover is fitted over the upper engine to absorb engine-generated noise.

Engine Identification Stamp

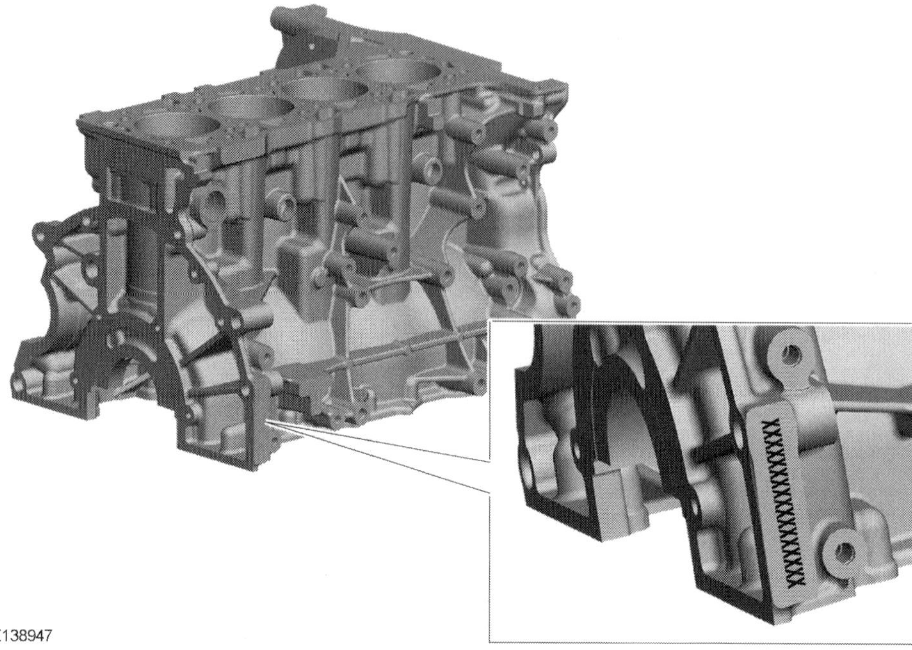

E138947

The alpha-numeric engine type number (5 digits) and numeric identification number (12 digits) are stamped on the rear right side of the cylinder block. These numbers are repeated on a tamper-proof engine identification label on the timing cover.

Engine Identification Label

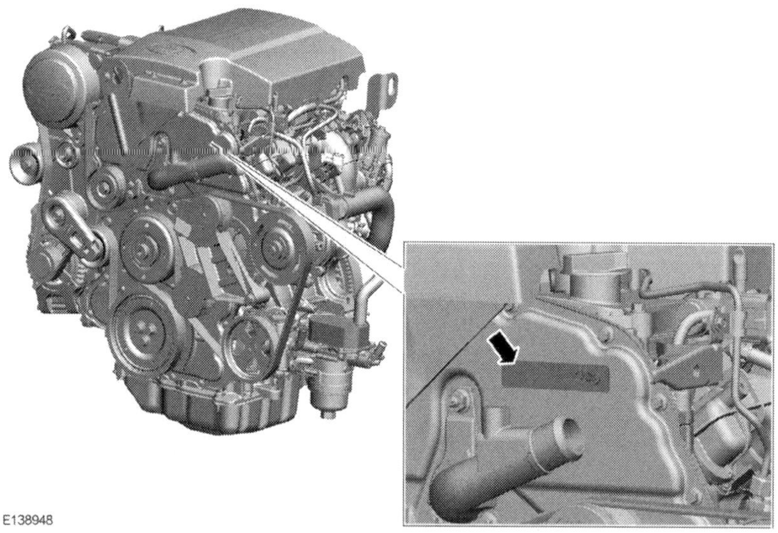

E138948

The engine type number, DT224, is derived from **D**iesel **T**urbo, **2**. **2** liter, **4** cylinder. The engine identification number is derived from the date and timo (24 hour clock) the block was manufactured, **DD/ MM/ YY/ HH/ MM/ SS**.

CYLINDER BLOCK

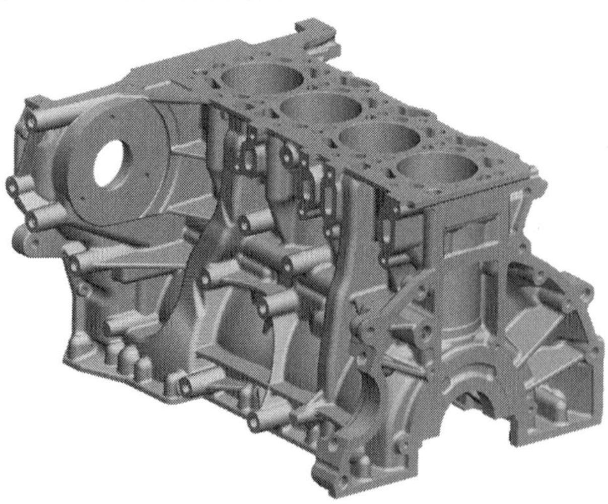

The cylinders and crankcase are contained in the cylinder block, which is of single cast CGI construction with a hollow beam structure. With this type of construction less material is required than for a conventional cast iron block, therefore reducing engine weight and length.

The cylinders are numbered consecutively from 1 to 4, with cylinder 1 at the front of the engine.

The cylinder bores are machined directly into the block. Cylinder cooling is achieved by coolant circulating through chambers in the cylinder block casting.

Passages in the cylinder block distribute engine oil to the main bearings, piston cooling jets, cylinder head and turbocharger. The passage that supplies oil to number 1 main bearing and the turbocharger contains a filter, to prevent debris entering the turbocharger and damaging the bearings.

Piston Cooling Jets

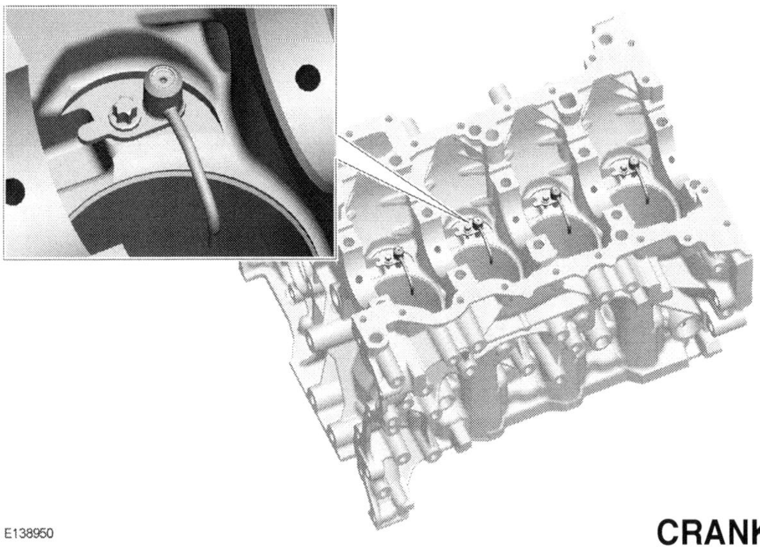

Jets located in the cylinder block provide piston and piston pin lubrication and cooling. These jets spray oil on to the inside of the piston. The oil then flows through an internal wave shaped oil cooling channel to help cool the piston crown.

Piston Oil Cooling

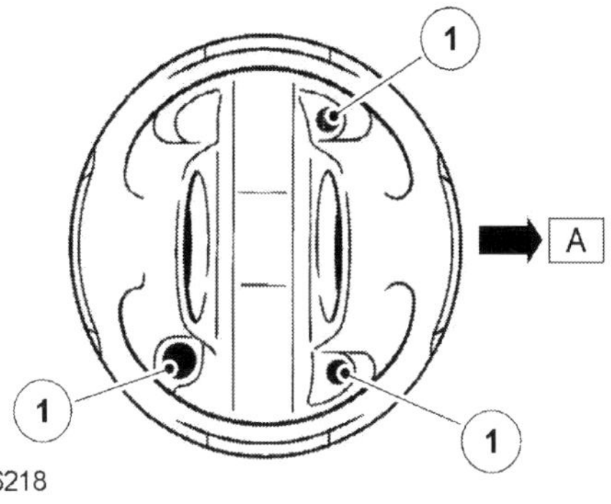

Item	Description
A	Front of the engine
1	Oil cooling channels

CRANKSHAFT

The crankshaft is forged steel and fillet rolled with induction hardened journals, which run in 5 bearings with clamped 2 layer bearing shells.

The main bearing caps are double and cross-bolted, this adds to the strength and rigidity of the engine block.

The main bearings are aluminum/tin split plain selective bearings. An oil groove in the top half of each bearing transfers oil into the crankshaft for lubrication of the connecting rod bearings. The upper and lower shells of bearing number 3 contain integral thrust washers, which limit the end float of the crankshaft.

The arrow on the bearing caps must point towards the front of the engine and are identified as follows:

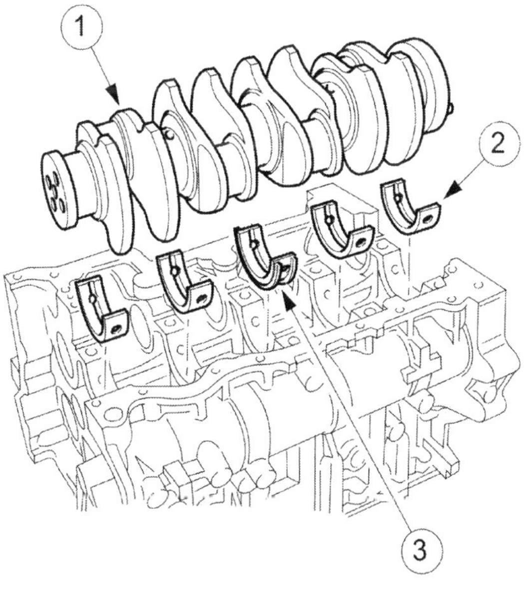

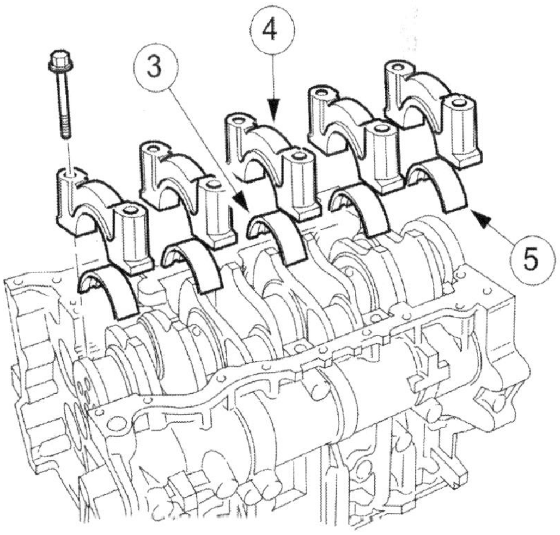

Item	Description
1	Crankshaft
2	Upper main bearing shell
3	Upper main bearing shell (No. 3 is a thrust bearing)
4	Bearing cap
5	Lower main bearing shell

Bearing Cap	Identification
1	F (front)
2	2
3	3
4	4
5	R (rear)

Crankshaft Rear Oil Seal

The crankshaft rear oil seal consists of a rubber sealing element installed in a metal case attached to the rear of the cylinder block and skirt stiffener. A rubber gasket seals the metal case to the cylinder block and skirt stiffener.

A new crankshaft rear oil seal is supplied with an installation sleeve that must not be removed until the seal is fully installed.

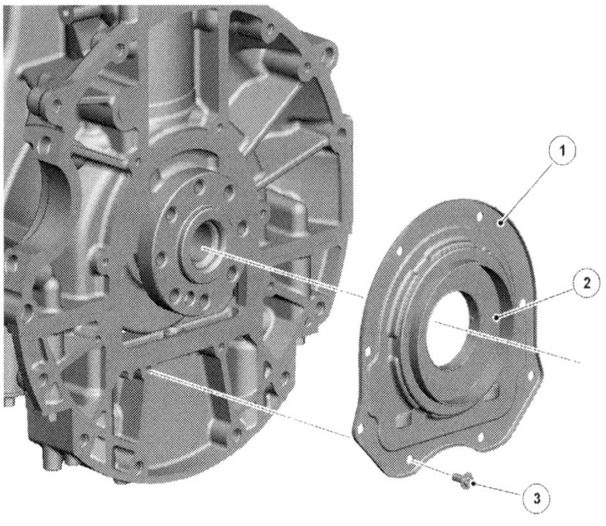

Item	Description
1	Rear oil seal
2	Installation sleeve
3	Bolt (8 off)

Crankshaft Front Oil Seal

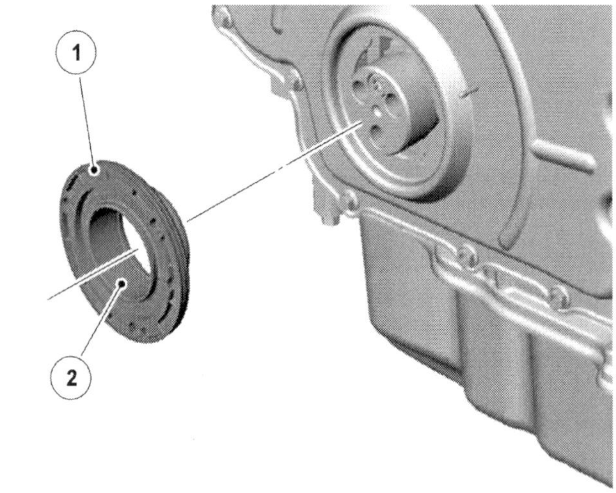

Item	Description
1	Front oil seal
2	Installation sleeve

The crankshaft front oil seal consists of a PTFE (polytetrafluoroethylene) seal assembly installed in the front cover. Three lugs on the outer case of the seal assembly engage with the front cover to lock the seal assembly in position.

A new crankshaft front oil seal is supplied with an installation sleeve that is pushed out when the seal is installed.

CONNECTING RODS AND PISTONS

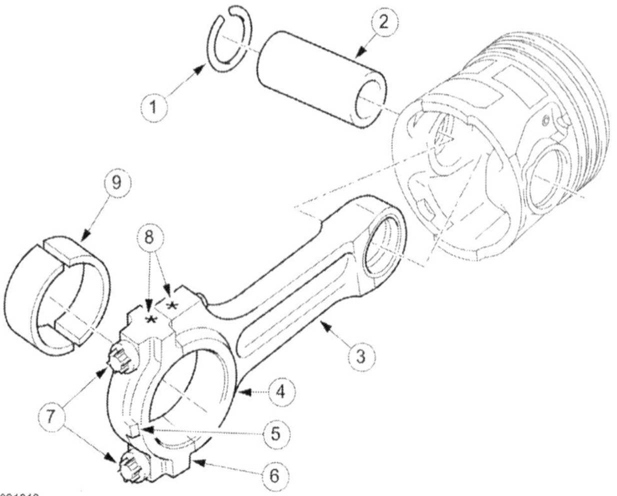

031819

Item	Description
1	Circlip
2	Piston pin
3	Connecting rod
4	Connecting rod bearing
5	Projection
6	Bearing cap
7	Retaining bolts
8	Identification
9	Bearing shell

The connecting rods are manufactured from sinter-forged steel. The selective connecting rod bearings are aluminum/tin split plain bearings. The connecting rod bearing is 'sputter coated', which is a manufacturing process that layers the bearing material to produce a higher load capacity for improved durability.

Engine oil for lubrication of the connecting rod bearings is supplied from the main bearings through passageways in the crankshaft.

The connecting rods are available in 3 different lengths. For identification, the connecting rod and bearing cap are marked with the corresponding identification code.

Piston Rings

The pistons are made from aluminum alloy and are fitted with three rings. The piston crown incorporates a pronounced bowl, which forms the combustion chamber and promotes the swirl and turbulence necessary for good combustion and improved emissions. In addition, the piston also incorporates oil cooling galleries within the piston crown to enhance piston cooling.

The arrow on the piston crowns must point towards the front of the engine.

Each piston is installed on a wrist pin located in an aluminum/tin bushing in the connecting rod.

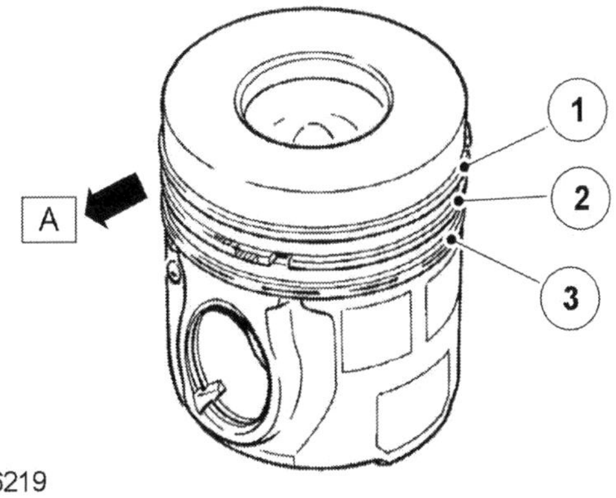

E86219

Item	Description
A	Front of the engine
1	Upper compression ring
2	Lower compression ring
3	Oil control ring

CYLINDER HEAD

Cylinder Head and Valves The cylinder head is made of gravity die cast aluminum. Four valves per cylinder provides improved cylinder charging. A compact combustion chamber and vertical fuel injectors guarantee optimum distribution of the fuel in the combustion chamber.

Eighteen deep-seated bolts help reduce distortion and secure the cylinder head to the cylinder block. The 8 cylinder head outer bolts are shorter than the inner bolts and are located beneath the camshafts, 4 under the inlet camshaft and 4 under the exhaust camshaft. Two hollow dowels align each cylinder head with the cylinder block.

The cylinder head has four ports machined at each cylinder location, 2 exhaust ports and 2 inlet ports. One of the inlet ports is helical and functions as a swirl port, the other is arranged laterally as a tangential port and functions as a charge port.

NOTE: The cylinder head cannot be reworked.

All valves are supported in sintered metal seats and guide inserts. Collets, valve collars and spring seats locate single valve springs on both intake and exhaust valves. Valve stem seals are integrated into the spring seats.

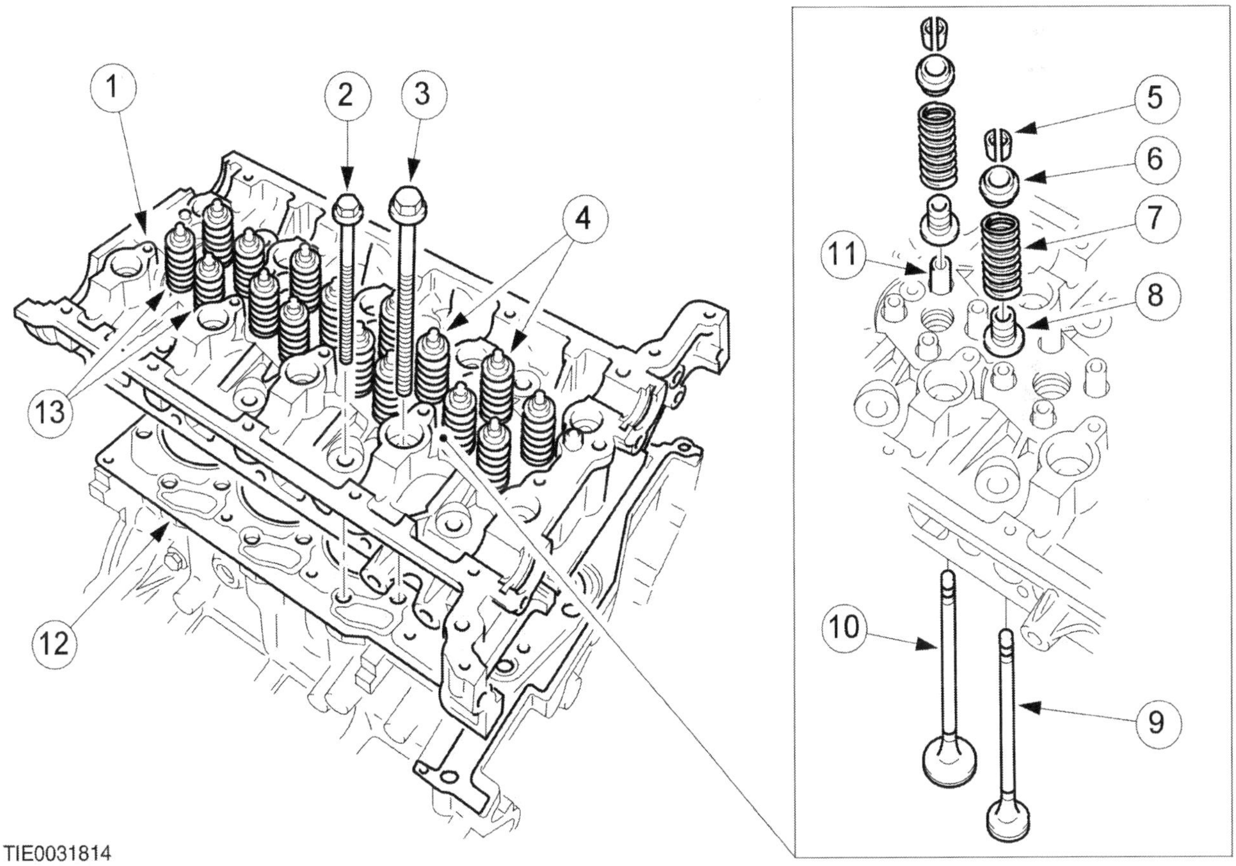

TIE0031814

Item	Description
1	Cylinder head
2	Outer cylinder head bolt (8 off)
3	Inner cylinder head bolt (10 off)
4	Intake valves (16 off)
5	Valve collets (32 off)
6	Spring retainer (16 off)
7	Valve spring (16 off)
8	Valve stem oil seal (16 off)
9	Exhaust valve (8 off)
10	Intake valve (8 off)
11	Valve guide (16 off)
12	Cylinder head gasket
13	Valve assembly

The camshafts are of a hollow steel tube construction, with pressed on sintered lobes and have a machined face at the front to accept the camshaft gear sprocket. The camshafts are retained in the cylinder head by the camshaft carrier. The reluctor, machined towards the rear of the inlet camshaft, enables the CMP (camshaft position) sensor to provide a signal, which enables the ECM (engine control module) to determine the position of the camshaft relative to the crankshaft.

Refer to: Electronic Engine Controls (303-14 Electronic Engine Controls, ID4 2.2L Diesel Description and Operation)

Camshafts

The camshafts are of a hollow steel tube construction, with pressed on sintered lobes and have a machined face at the front to accept the camshaft gear sprocket. The camshafts are retained in the cylinder head by the camshaft carrier. The reluctor, machined towards the rear of the inlet camshaft, enables the CMP (camshaft position) sensor to provide a signal, which enables the ECM (engine control module) to determine the position of the camshaft relative to the crankshaft.

Refer to: Electronic Engine Controls (303-14 Electronic Engine Controls, ID4 2.2L Diesel Description and Operation)

The aluminum camshaft carrier is located on top of the cylinder head and is retained by 24 bolts and a stud. Two alignment pins are used to accurately locate the carrier to the cylinder head.

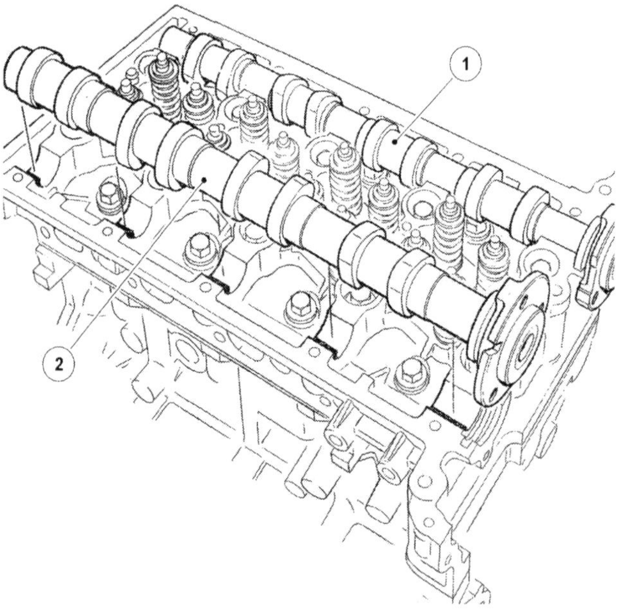

Item	Description
1	Inlet camshaft
2	Exhaust camshaft

Camshaft Carrier

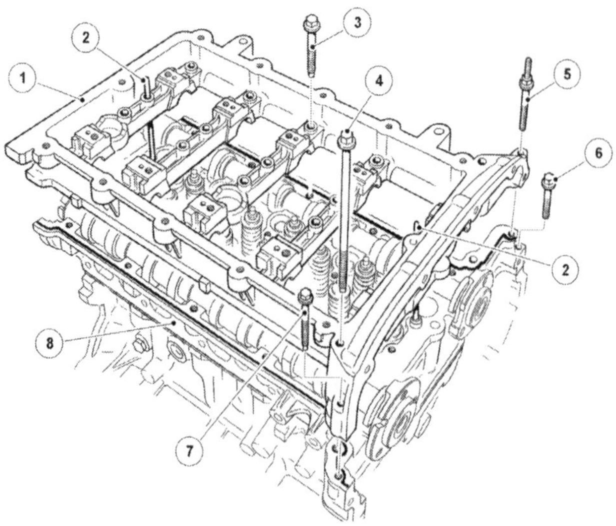

Item	Description
1	Camshaft carrier
2	Alignment pin
3	Bolt (20 off)
4	Bolt (2 off)
5	Stud
6	Bolt
7	Bolt
8	Cylinder head

Rocker Arms

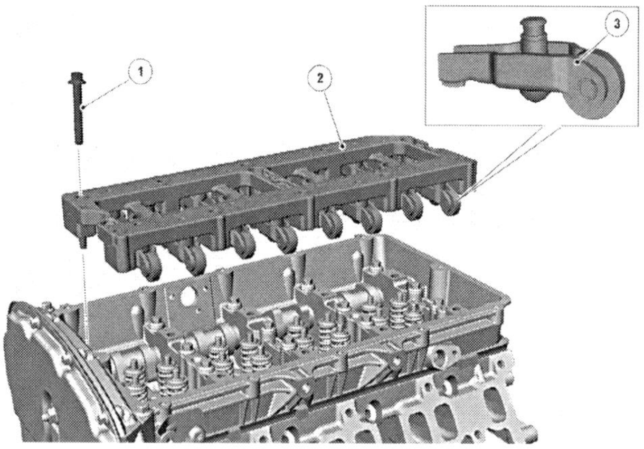

Item	Description
1	Bolt (10 off)
2	Frame
3	Rocker assembly (16 off)

The valves are operated through roller-type rocker arms with integrated hydraulic lash adjusters, actuated by the camshaft lobes. The use of this type of actuation method helps reduce friction in the valve operating mechanism. The rocker arms are installed in a frame located in the camshaft carrier and secured with 10 bolts.

Lash Adjuster

The body of the hydraulic lash adjuster contains a plunger and 2 chambers for oil feed and pressurized oil

The pressurized oil is supplied to the adjusters via the main oil galleries in the cylinder head and through a hole in the side of the adjuster body.

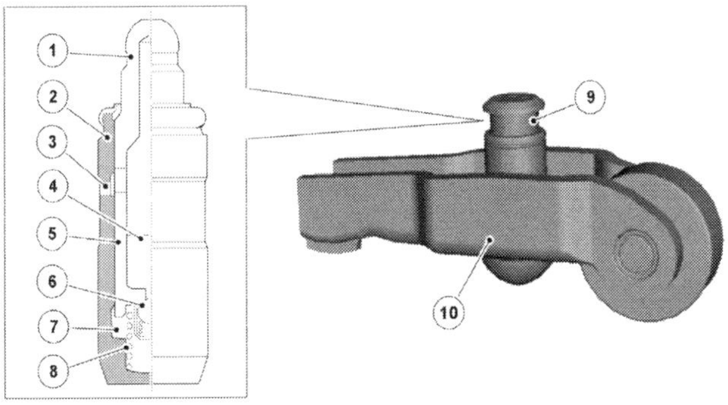

Item	Description
1	Plunger cap
2	Hydraulic lash adjuster body
3	Oil hole
4	Reservoir chamber
5	Plunger
6	Check ball
7	High-pressure chamber
8	Plunger spring
9	Hydraulic lash adjuster
10	Rocker arm

The body of the hydraulic lash adjuster contains a plunger and 2 chambers for oil feed and pressurized oil. The oil passes into a feed chamber in the adjuster and then through to a separate pressure chamber via a one-way ball valve.

Oil flow from the pressure chamber is determined by the amount of clearance between the adjuster outer body and the center plunger. Oil escapes up the side of the plunger every time the adjuster is operated, the downward pressure on the plunger forcing a corresponding amount of oil in the adjuster body to be displaced. When the downward pressure from the camshaft and rocker arm is removed (i.e. after the trailing flank of the camshaft lobe has passed), oil pressure forces the adjuster's plunger up again. This pressure is not sufficient to effect the valve operation, but eliminates the clearance between the finger rocker and top of the valve stem.

Cylinder Head Gasket

The cylinder head gasket is a 4-layer, laminated steel type that is available in 3 thickness grades. The choice of gasket grade is dependent on the maximum piston protrusion in the cylinder block. The grade of a gasket is shown by the number of identification holes and teeth incorporated into the gasket.

Two dowels ensure correct alignment of the cylinder head and gasket with the cylinder block.

⚠ CAUTION: If a cylinder head gasket is being changed, only use a new gasket with the same identification markings. If new pistons or connecting rods are fitted, measure the maximum piston protrusion height and select the appropriate gasket.

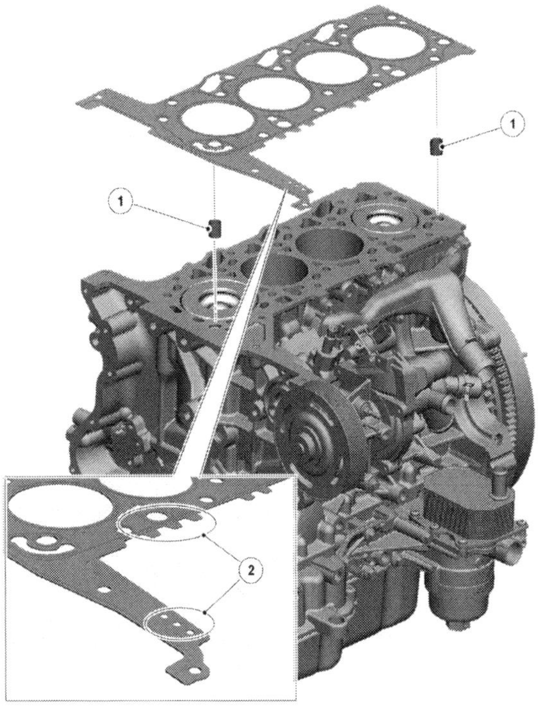

Item	Description
1	Dowel
2	Grade identification holes/teeth

SKIRT STIFFENER

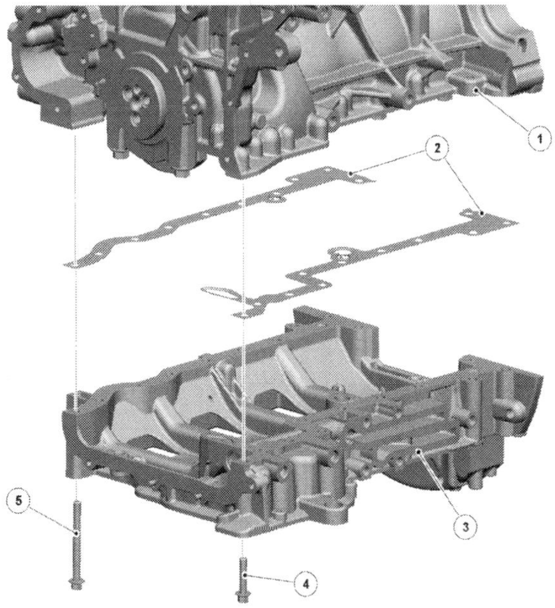

Item	Description
1	Cylinder block
2	Gaskets
3	Skirt stiffener
4	Bolt (3 off)
5	Bolt (19 off)

The aluminum alloy skirt stiffener is fitted to the lower cylinder block to stiffen the base structure of the engine thus helping to reduce NVH (noise, vibration and harshness) . The frame is made of high-pressure die cast aluminum and also incorporates an oil baffle plate to reduce oil foaming and splash.

The alignment of the sides of the skirt stiffener with the cylinder block is set using a special tool. The rear face alignment is set using a suitable straight edge.

Refer to: Engine (303-01, Assembly).

The skirt stiffener is secured to the cylinder block with 22 bolts.

Gaskets seal the joints between the skirt stiffener and the cylinder block.

OIL PAN

The pressed steel oil pan is attached to the skirt stiffener with 11 bolts and 5 nuts. The engine oil drain plug is located in the rear face of the oil pan.

A bead of RTV (room temperature vulcanizing) sealant seals the joint between the oil pan and the skirt stiffener.

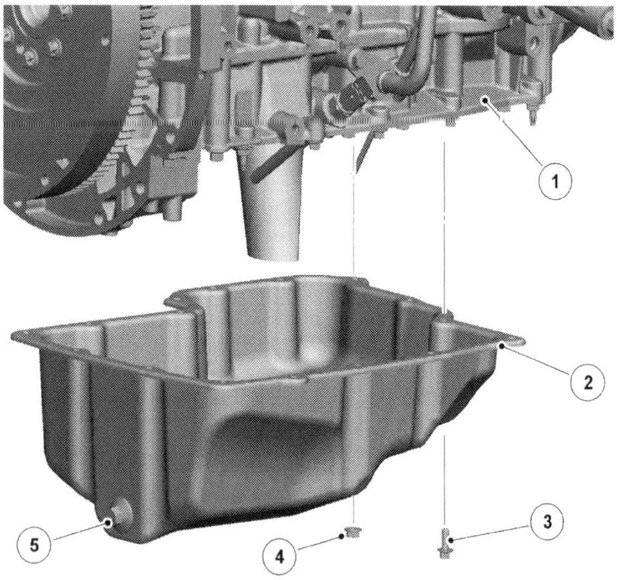

Item	Description
1	Skirt stiffener
2	Oil pan
3	Bolt (11 off)
4	Nut (5 off)
5	Drain plug

CAMSHAFT DRIVE

Drive Gear

The primary drive chain transmits the drive from the crankshaft sprocket to the exhaust camshaft, the inlet camshaft and the high-pressure fuel pump. The secondary drive chain transmits drive from the crankshaft sprocket to the oil pump located on the underside of the skirt stiffener.

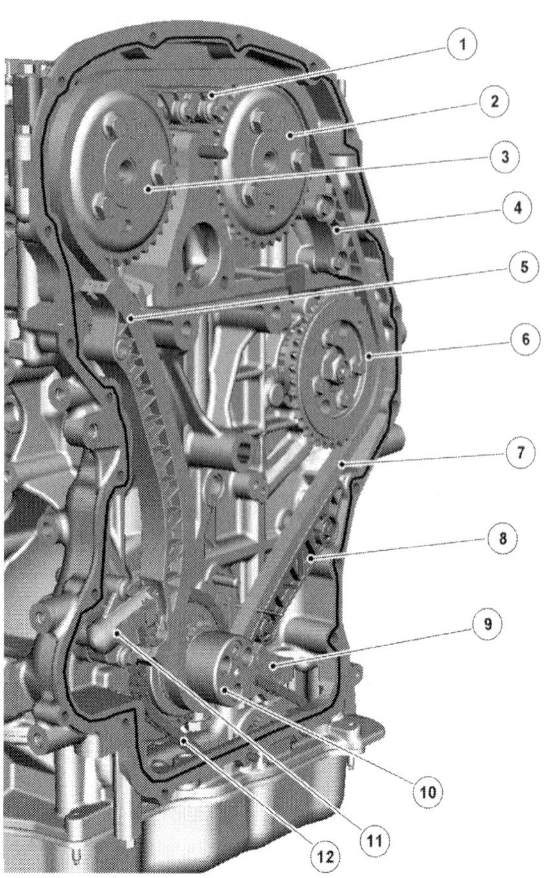

Item	Description
1	Chain guide
2	Inlet camshaft sprocket
3	Exhaust camshaft sprocket
4	Chain guide
5	Tensioner blade
6	High-pressure fuel pump sprocket
7	Primary drive chain
8	Lower chain guide
9	Secondary drive chain tensioner
10	Crankshaft sprocket
11	Hydraulic tensioner
12	Secondary drive chain

One bolt locates the crankshaft sprocket on the crankshaft, and three bolts are used to attach the engine crankshaft pulley/ torsional vibration damper to the crankshaft via holes in the crankshaft sprocket. The inlet and exhaust camshaft sprockets are aligned on their respective camshafts using a spigot tool and a timing peg. Once they are in their correct position, three bolts are used to secure each camshaft sprocket to their camshaft.

The primary drive chain has an hydraulic tensioner, operated by engine oil, which acts on a pivoting flexible tensioner blade. Chain guides are installed on the drive side of the primary chain. The primary chain is splash lubricated with oil from the skirt stiffener and oil pan assembly.

The secondary chain tensioner is spring operated and acts directly on the chain. The secondary chain is splash lubricated with oil from the skirt stiffener and oil pan assembly.

Front Cover

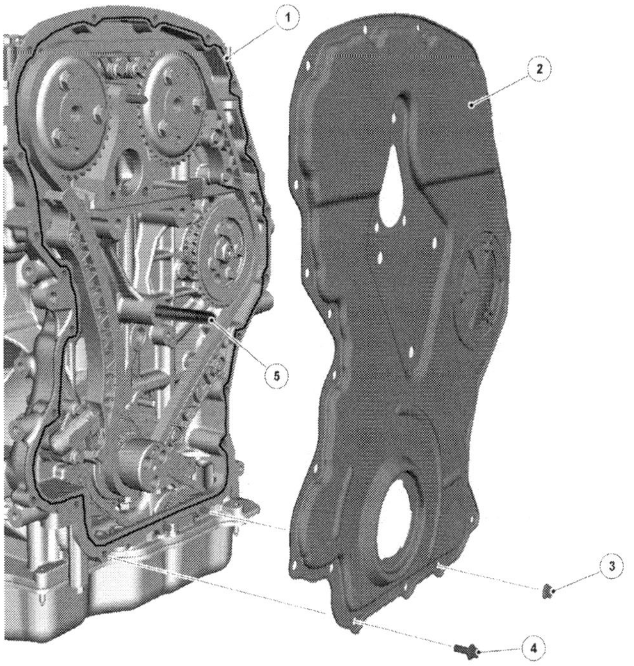

Item	Description
1	Sealant bead
2	Front cover
3	Nut (2 off)
4	Bolt (20 off)
5	Guide pin

The pressed steel front cover is attached to the front of the cylinder block, the cylinder head and the skirt stiffener with 20 bolts and 2 nuts, and sealed with a bead of RTV silicon sealant.

When installing the front cover, a special locating tool is required, which fits in the crankshaft sprocket aperture, to aid the alignment of the front cover.

INTAKE MANIFOLD

The plastic intake manifold is mounted on the LH (left-hand) side of the cylinder head using 9 bolts and isolator bushes. Rubber gaskets, integrated into the manifold, seal the joints between the manifold and the cylinder head.

Mounting flanges on the manifold provide for the attachment of the electronic throttle and the MAP sensor.

Refer to: Electronic Engine Controls (303-14 Electronic Engine Controls, ID4 2.2L Diesel Description and Operation)

A mounting boss on the manifold provides for the attachment of the EGR valve outlet tube.

Refer to: Engine Emission Control (303-08 Engine Emission Control, ID4 2.2L Diesel Description and Operation)

Captive nuts are installed in the manifold for the securing bolts of the EGR valve outlet tube and a fuel pipe bracket.

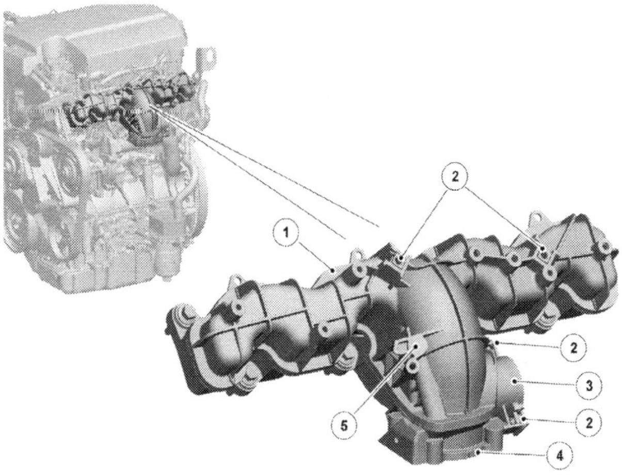

Item	Description
1	Intake manifold
2	Captive nut
3	EGR valve outlet tube mounting boss
4	Electronic throttle mounting flange
5	MAP (manifold absolute pressure) sensor mounting flange

EXHAUST MANIFOLD

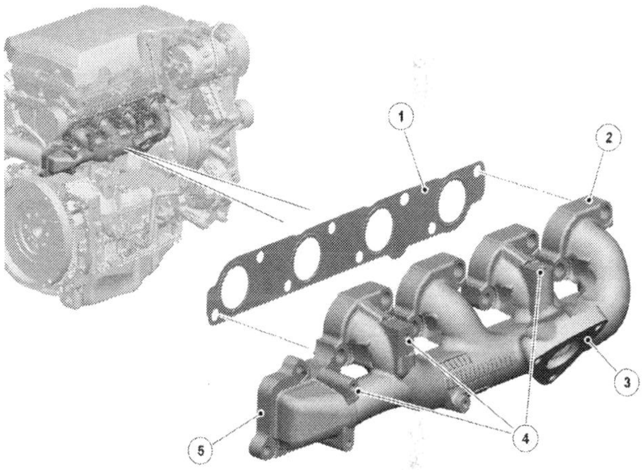

Item	Description
1	Gasket
2	Exhaust manifold
3	Turbocharger mounting flange
4	Heat shield attachment points
5	EGR cooler mounting flange

The cast iron exhaust manifold is secured to the RH (right-hand) side of the cylinder head using 2 nuts and 6 bolts. A laminated metal gasket seals the manifold to the cylinder head. A mounting flange towards the front of the manifold is provided to attach the turbocharger.

Refer to: Turbocharger (303-04 Fuel Charging and Controls - Turbocharger, ID4 2.2L Diesel Description and Operation)

A second mounting flange, located on the rear of the manifold, is provided for attachment of the EGR cooler. Tapped holes incorporated into bosses on the manifold provide attachment points for the exhaust heat shields.

Refer to: Engine Emission Control (303-08 Engine Emission Control, ID4 2.2L Diesel Description and Operation)

CAMSHAFT COVER & ENGINE COVER

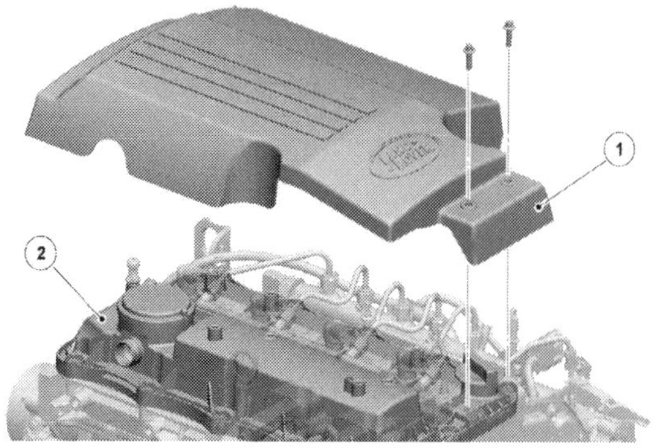

Item	Description
1	Engine cover
2	Camshaft cover

LUBRICATION SYSTEM

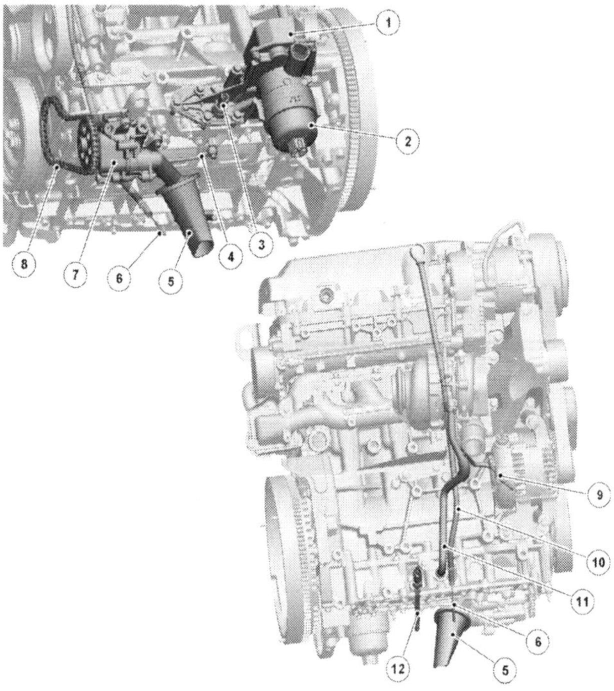

Item	Description
1	Oil cooler
2	Oil filter
3	Oil pressure switch
4	Feedback tube
5	Oil pick-up tube
6	Dipstick
7	Oil pump
8	Secondary drive chain
9	Turbocharger oil feed tube
10	Dipstick outer tube
11	Turbocharger oil drain tube
12	Oil temperature/low level sensor

The engine is lubricated by a forced feed oil circulation system. The main components of the system are:

- Oil pick-up
- Oil pump
- Oil filter and cooler assembly.

Oil Pick-up

The plastic oil pick-up is immersed in the oil reservoir to provide a supply to the oil pump at all normal vehicle attitudes. A mesh screen in the inlet prevents debris from entering the oil system.

Oil Pump

The oil pump is located on the LH front underside of the skirt stiffener and is secured by 4 bolts. The unit is a spur gear pump driven by the crankshaft sprocket via the secondary drive chain. An integral pressure relief valve diverts oil back to the inlet side of the pump to limit the pump outlet pressure.

Oil Filter and Cooler Assembly

The oil filter and cooler assembly is attached to the LH side of the skirt stiffener and consists a full-flow, disposable canister-type filter and an oil cooler attached to a casting. The casting aligns with the oil galleries in the skirt stiffener and is sealed by a gasket.

The engine cooling system cools the oil in the oil cooler and is regulated by means of a separate thermostat, which prevents the flow of coolant through the oil cooler when the engine is cold, ensuring the engine oil warms up quickly. The thermostat opens at 75±2 °C(167±35 °F).

Oil to and from the oil cooler passes through galleries in the skirt stiffener. Hoses from the engine cooling system are connected to 2 pipes on the oil cooler for the supply and return of coolant. An oil pressure switch is located in the casting to sense the pressure of the oil as it leaves the oil filter and cooler. A warning lamp in the instrument cluster illuminates if low oil pressure is detected.

Refer to: Instrument Cluster (413-01 Instrument Cluster, Description and Operation)

Lubrication System Operation

Oil is drawn through the oil pick-up into the oil pump, then supplied to the oil filter and cooler assembly through the oil galleries in the skirt stiffener. After passing through the filter, a proportion of the oil (controlled by a restrictor in the oil filter housing) passes through the oil cooler. The return flow from the oil cooler combines with the remainder of the oil from the filter, then passes through the skirt stiffener into the cylinder block main oil gallery.

The main oil gallery has drillings that direct the oil to the cylinder head and the main bearings. Cross drillings in the crankshaft main bearings carry the oil to the connecting rod big-end bearings. Oil galleries in the cylinder head carry the oil to the camshafts and the hydraulic lash adjusters.

Oil at reduced pressure is directed towards the cylinder head via a restrictor in the cylinder block/cylinder head locating dowel. Oil then passes through a drilling in the cylinder head to the camshaft carrier, where it is directed via separate galleries to the camshaft bearings and hydraulic tappet housings. Return oil from the cylinder head drains into the sump via the cylinder head bolt passages.

Engine - Diagnosis and Testing

Principles of Operation

For a detailed description of the Engine, refer to the relevant Description and Operation section in the workshop manual.

Refer to: Engine (303-01 Engine, ID4 2.2L Diesel Description and Operation)

Inspection and Verification

⚠ CAUTION: Diagnosis by substitution from a donor vehicle is **NOT** acceptable. Substitution of control modules does not guarantee confirmation of a fault, and may also cause additional faults in the vehicle being tested and/or the donor vehicle.

NOTES:

△ If the control module or a component is suspect and the vehicle remains under manufacturer warranty, refer to the Warranty Policy and Procedures manual (section B1.2), or determine if any prior approval programme is in operation, prior to the installation of a new module/component.

△ When performing voltage or resistance tests, always use a digital multimeter accurate to three decimal places, and with an up-to-date calibration certificate. When testing resistance always take the resistance of the digital multimeter leads into account.

△ Check and rectify basic faults before beginning diagnostic routines involving pinpoint tests.

1. Visually inspect for obvious signs of damage and system integrity.
2. If an obvious cause for an observed or reported concern is found, correct the cause (if possible) before proceeding to the next step
3. If the cause is not visually evident, verify the symptom and refer to the Symptom Chart, alternatively check for Diagnostic Trouble Codes (DTCs) and refer to the DTC Index

Visual Inspection Chart

Mechanical	Electrical
• Engine oil level • Coolant level • Transmission fluid level • Fuel level • Coolant leaks • Oil leaks • Fuel leaks • Visibly damaged or worn parts • Loose or missing nuts or bolts • Fuel contamination/ grade/quality • Sensor fitment/ condition • Viscous fan • Routing of pipework etc. to avoid vibrations • Contact anywhere between the driveline and chassis	• Fuses • Wiring harnesses and connectors • Injectors • Glow plugs • Sensor(s) • Engine control module (ECM)

Oil Pressure Check

△ NOTE: Prior to checking the engine oil pressure, a road test of 6 miles (10 kilometres), must be carried out. Do not attempt to attain engine normal operating temperature by allowing the engine to idle.

1. WARNINGS:

⚠ The spilling of hot engine oil is unavoidable during this procedure, care must be taken to prevent scalding. Failure to follow this instruction may result in personal injury.

⚠ Wear protective gloves.

Remove the oil pressure sensor.

Refer to: Engine Oil Pressure Sensor (303-14 Electronic Engine Controls, ID4 2.2L Diesel Removal and Installation)

2. Install the oil pressure gauge and the oil pressure gauge connector.

3. Check and top-up the engine oil, if required.
4. Start and run the engine.
5. Note the oil pressure readings with the engine running at idle and at 2,000 rpm.

Refer to: (303-01 Engine, ID4 2.2L Diesel Specifications)

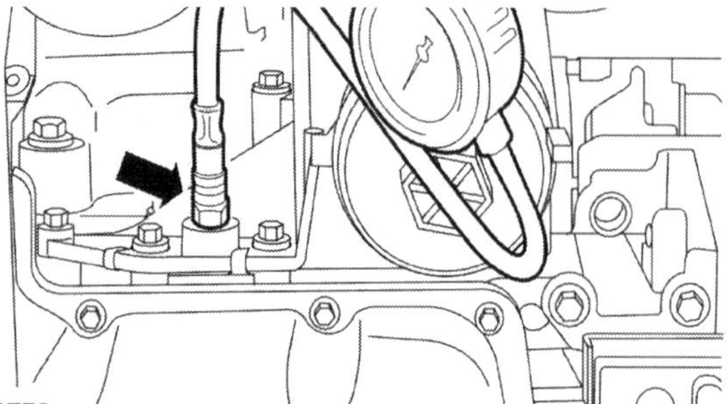
E92759

6. Turn off the engine.
7. Remove the special tools.
8. Install the oil pressure sensor.

Refer to: Engine Oil Pressure Sensor (303-14 Electronic Engine Controls, ID4 2.2L Diesel Removal and Installation)

9. Check and top-up the engine oil, if required.

DTC Index

For a list of Diagnostic Trouble Codes (DTCs) that could be logged on this vehicle, please refer to Section 100-00.

Engine Oil Draining and Filling

General Procedures

Part(s)

STEP	PART NAME	QUANTITY
Step 6	Renew Part Oil pan drain plug	1
Step 7	Renew Part Oil filter and O-ring seal kit	1

Activation

1. ⚠ WARNING: Do not work on or under a vehicle supported only by a jack. Always support the vehicle on safety stands.

Raise and support the vehicle.

Refer to: Lifting (100-02 Jacking and Lifting, Description and Operation)

2. Loosen the oil filter cover 3 turns and allow the engine oil to drain.
3. ⚠ NOTE: Position cloth to collect fluid spillage - remove the oil filter cover and oil filter element.

4. Remove and discard the oil filter element. Remove and discard the oil filter cover O-ring seal.
5. Drain the engine oil. Position a container to collect the engine oil. Remove and discard the drain plug. Allow the engine oil to drain.
6. Install a new drain plug. Clean the component mating faces. Tighten the drain plug. Remove the container.

E90316

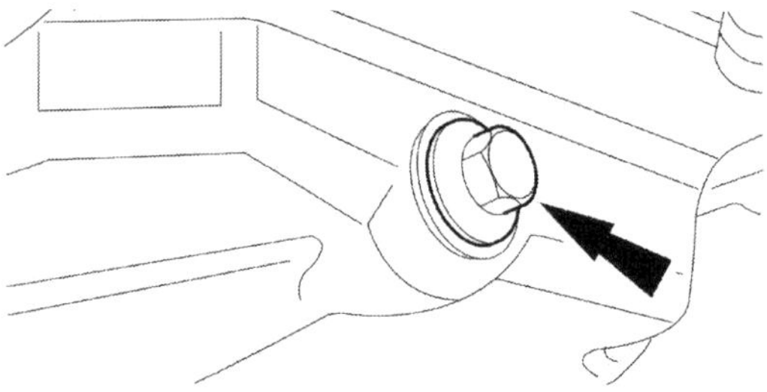

Renew Part: Oil pan drain plug

Torque: 23 Nm

7. Install a new oil filter cover O-ring seal. Clean the components. Lubricate the O-ring seal with clean engine oil.

Renew Part: Oil filter and O-ring seal kit

8. Install a new oil filter element to the oil filter cover.
9. Install the oil filter cover and oil filter element. Tighten the oil filter cover.

Torque: 35 Nm

10. Fill the engine with the recommended oil to the correct level.
11. Connect the battery ground cable.

Refer to: Startup Battery Disconnect and Connect (414-01 Battery, Mounting and Cables, General Procedures)

12. Check and top-up the engine oil.

Engine Removal

1. **Refer to:** Air Conditioning System Recovery, Evacuation and Charging (412-00 Climate Control System - General Information, General Procedures)
2. ⚠ WARNING: Do not work on or under a vehicle supported only by a jack. Always support the vehicle on safety stands.

Raise the vehicle on a twin-post lift.

3. **Refer to:** Startup Battery Disconnect and Connect (414-01 Battery, Mounting and Cables, General Procedures)

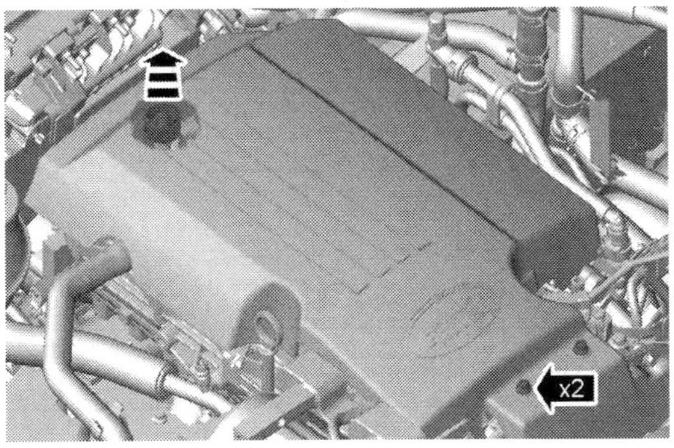

4.

5. **Refer to:** Cooling System Draining, Filling and Bleeding (303-03 Engine Cooling, ID4 2.2L Diesel General Procedures)
6. **Refer to:** Cooling Fan (303-03 Engine Cooling, ID4 2.2L Diesel Removal and Installation)

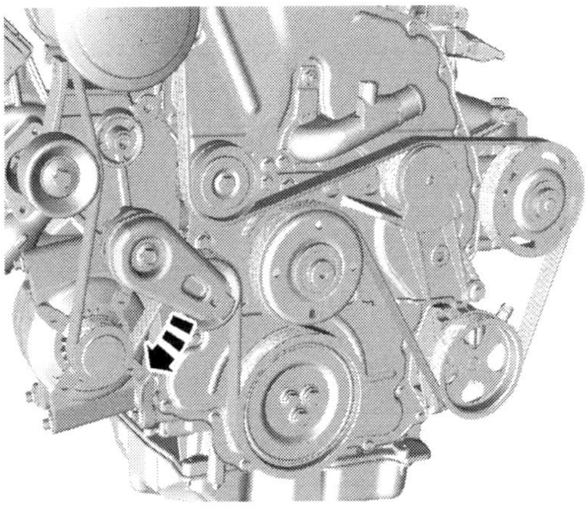

7.

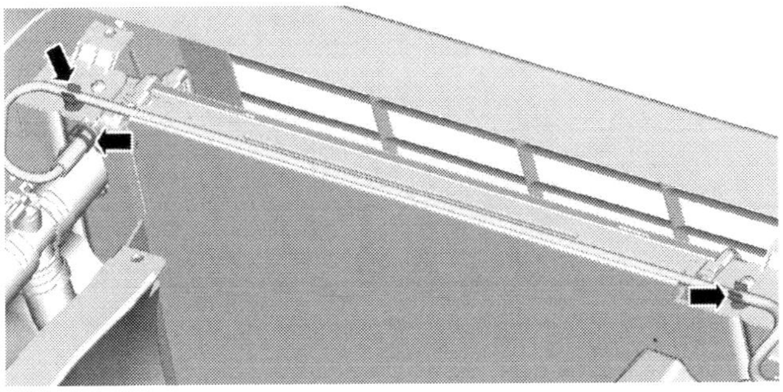

8.

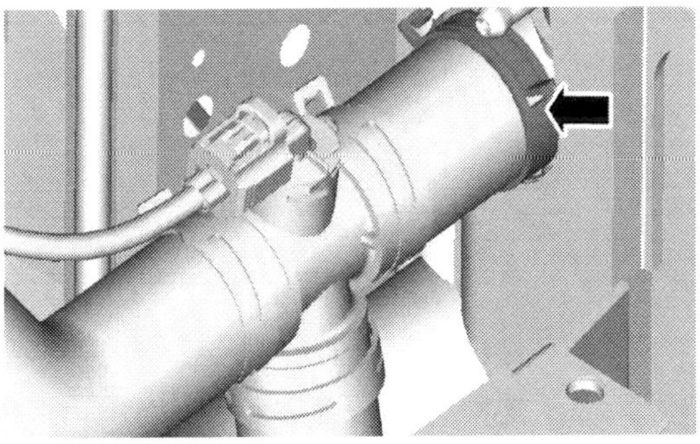

9.

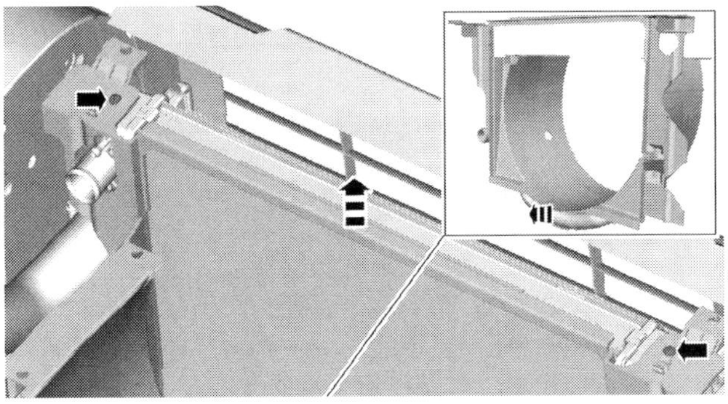

10.

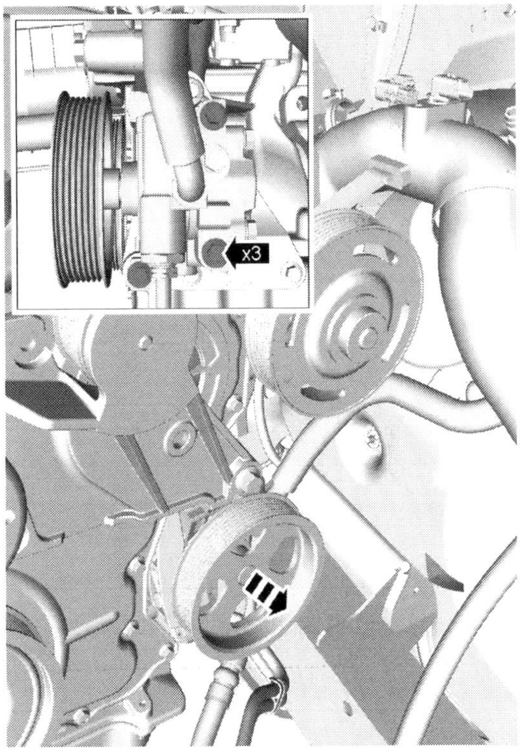

11.

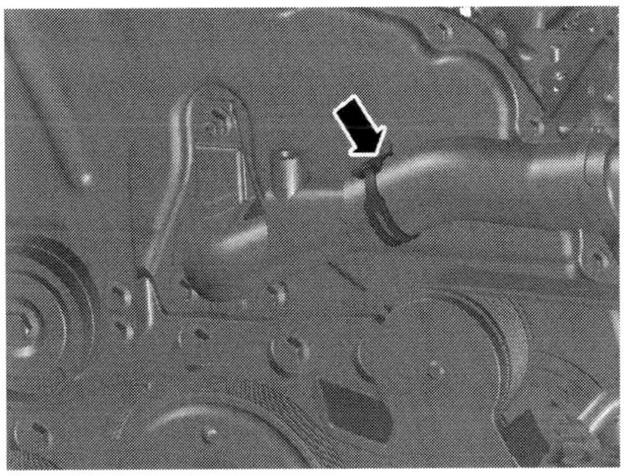

12.

13.

14.

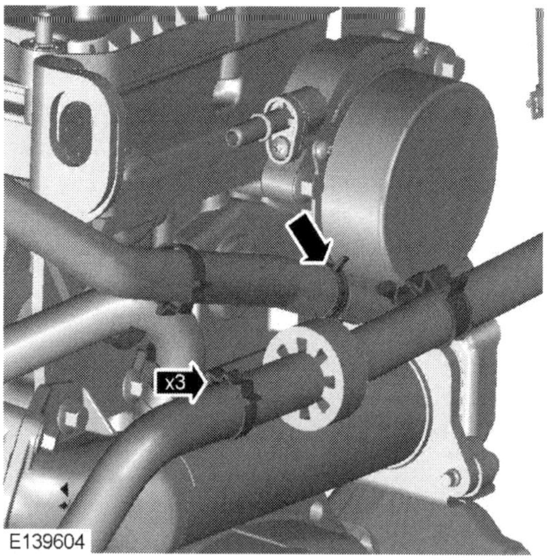

15.

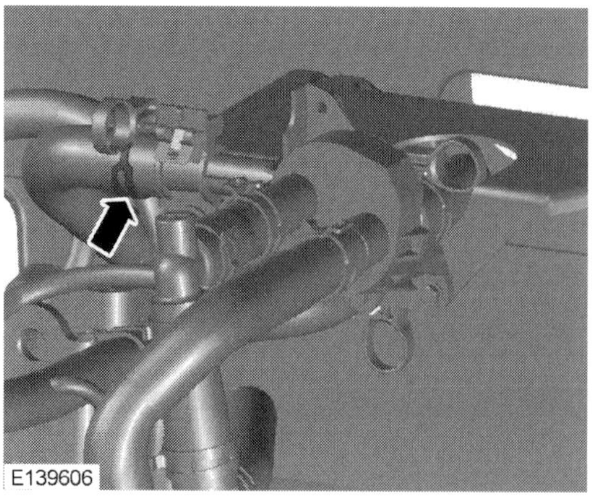

16.

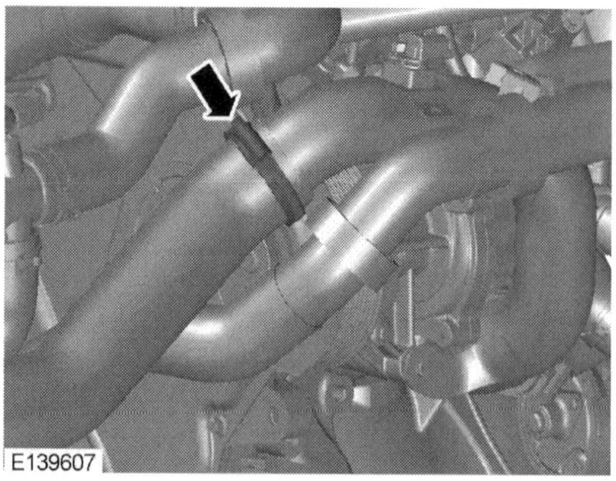

17.

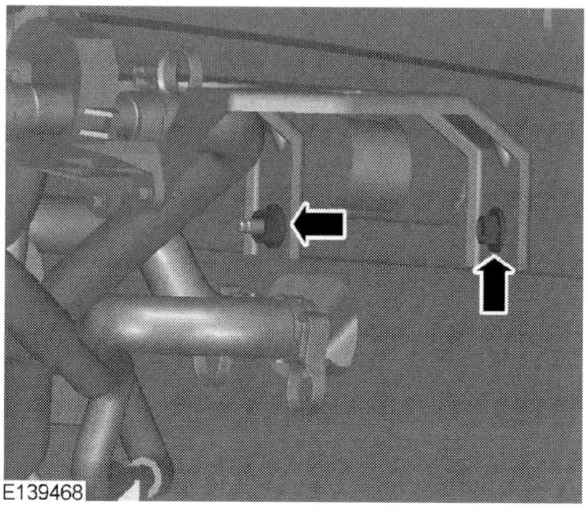

18.

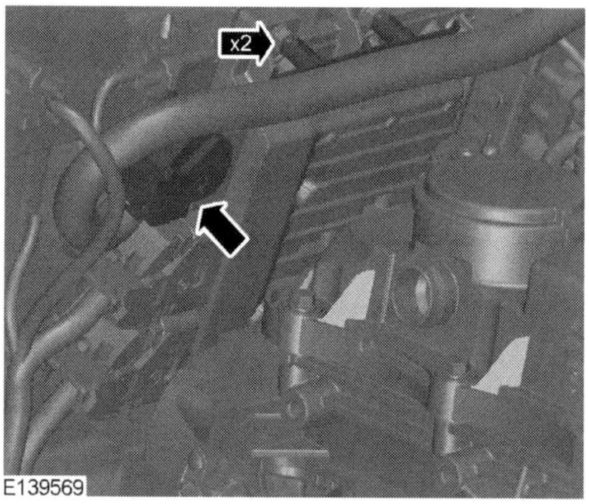

19.

20.

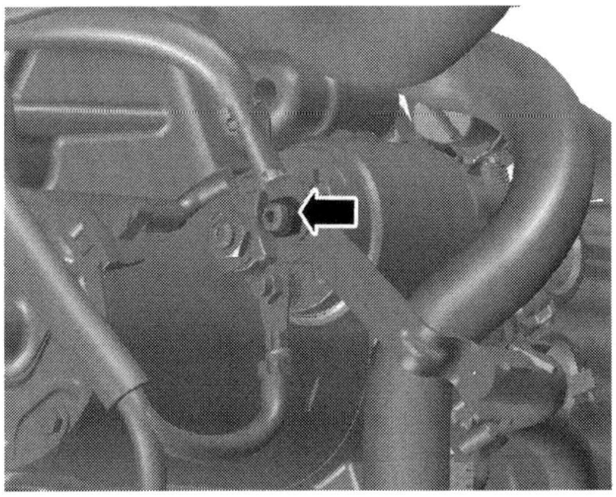

21.

22.

23.

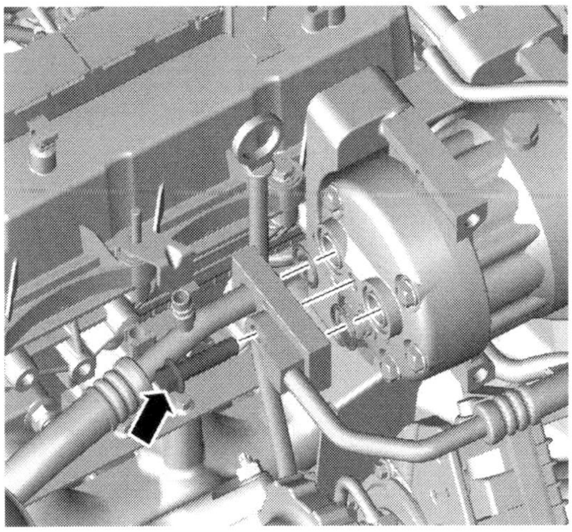

24. CAUTIONS:

CAUTION: Remove and discard the O ring seals.

Before disconnecting any components, make sure the area is clean and free from foreign material. When disconnected all openings must be sealed.

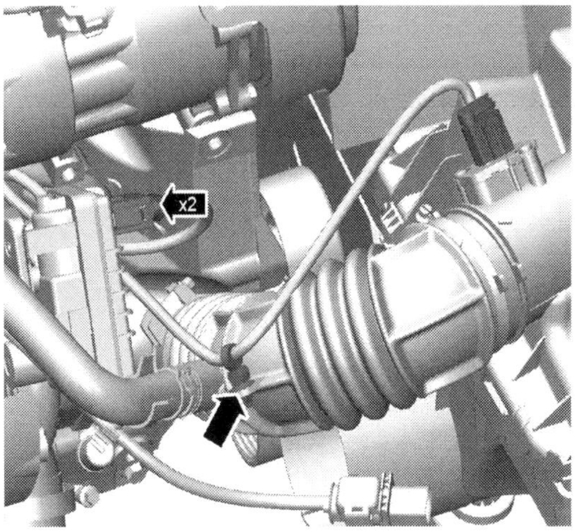

25.

26.

27.

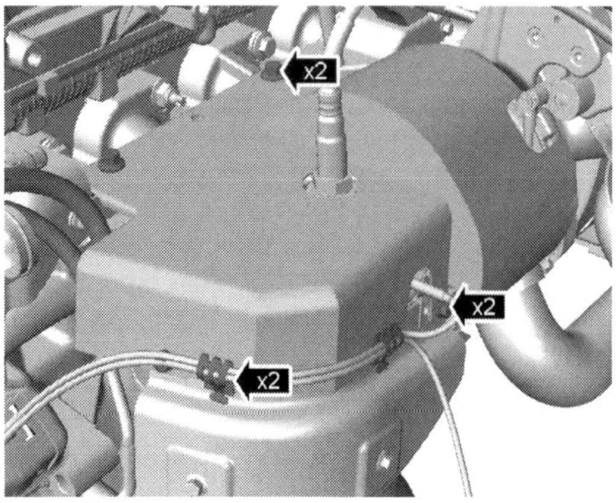

28. NOTE: Vehicles with diesel particulate filter (DPF).

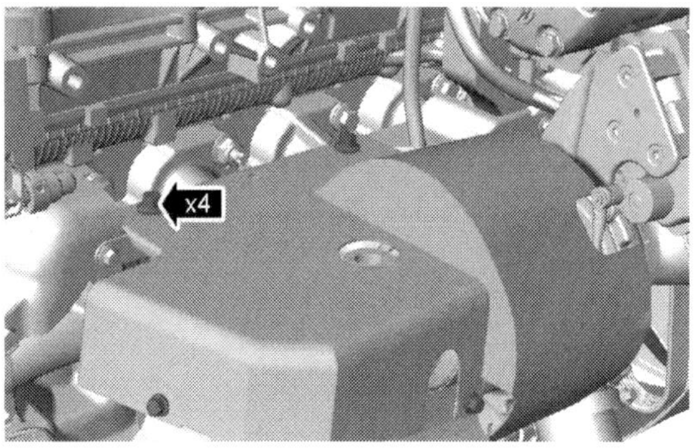

29.

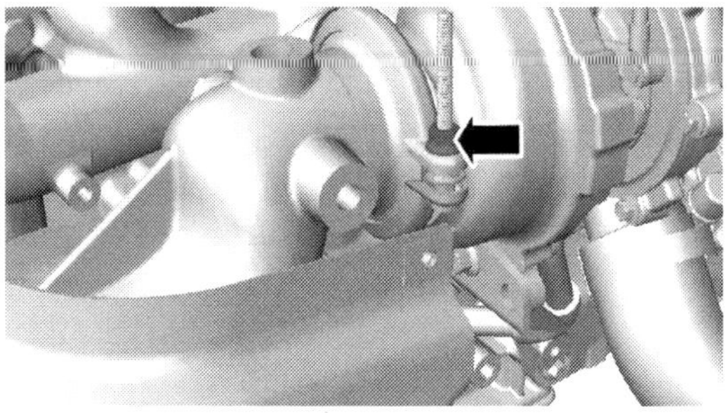

30.

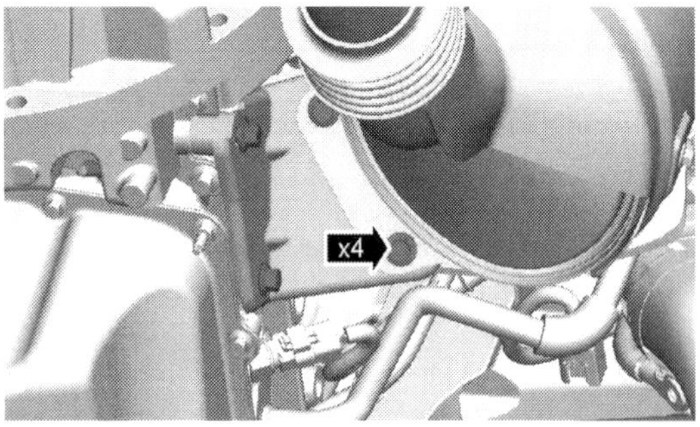

31.

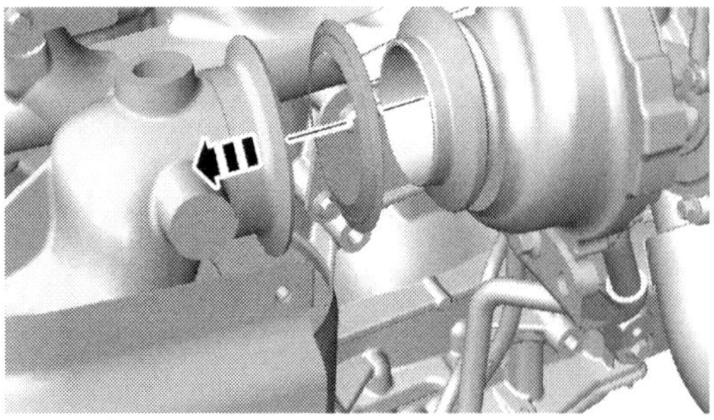

32. NOTE: Remove and discard the gasket.

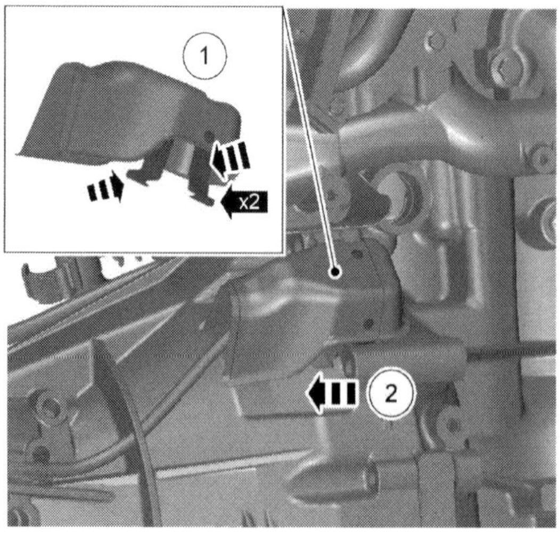

33.

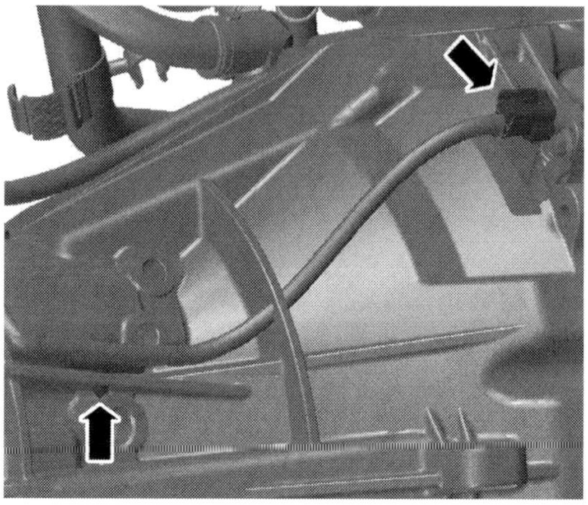

34.

35.

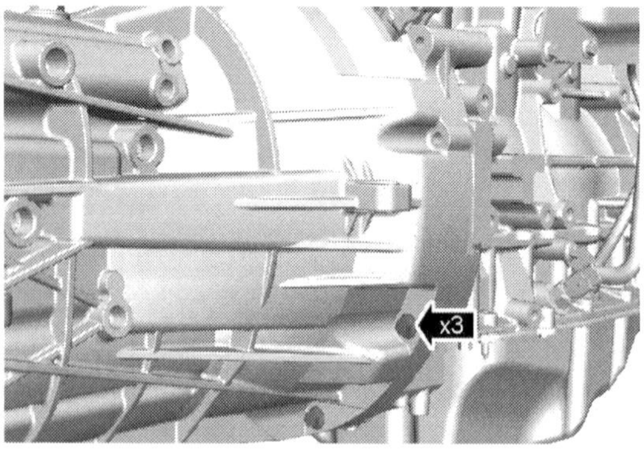

36.

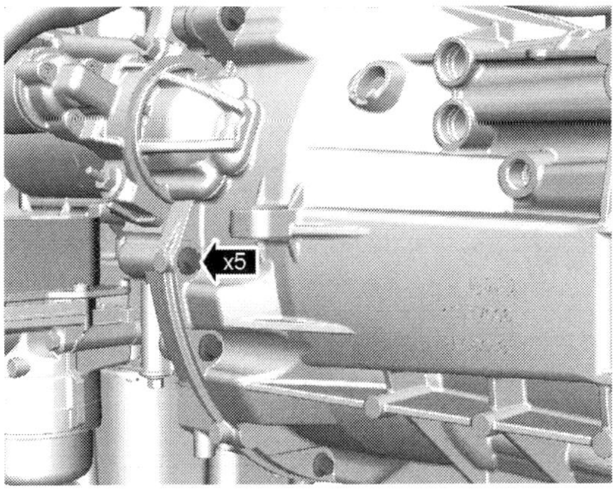

37.

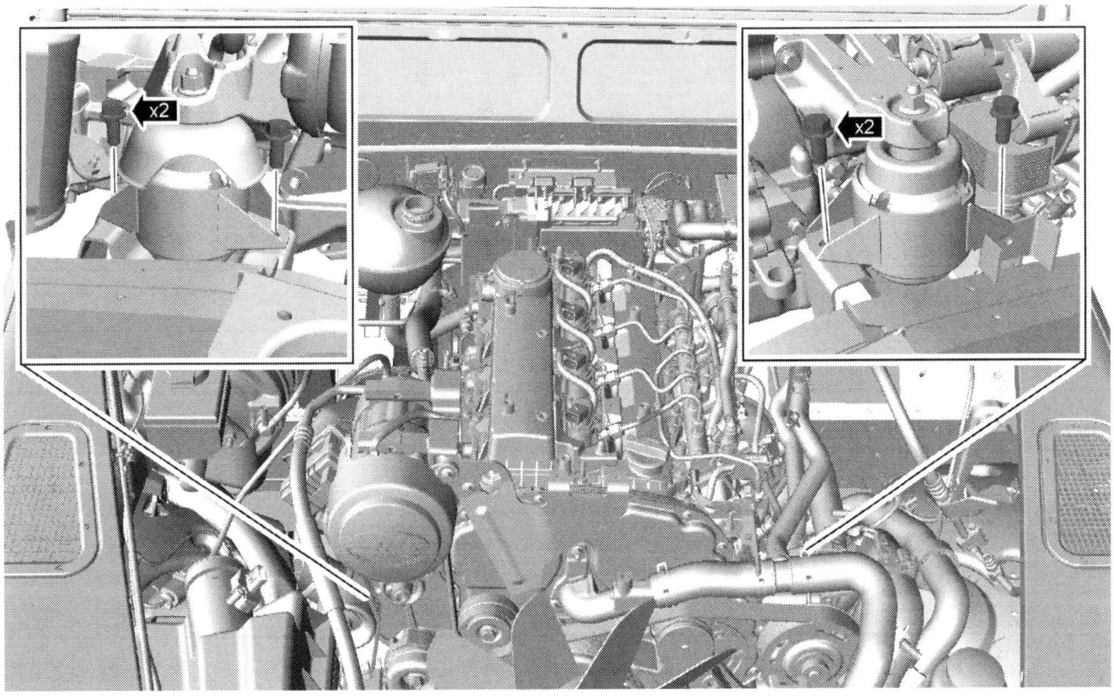

38.

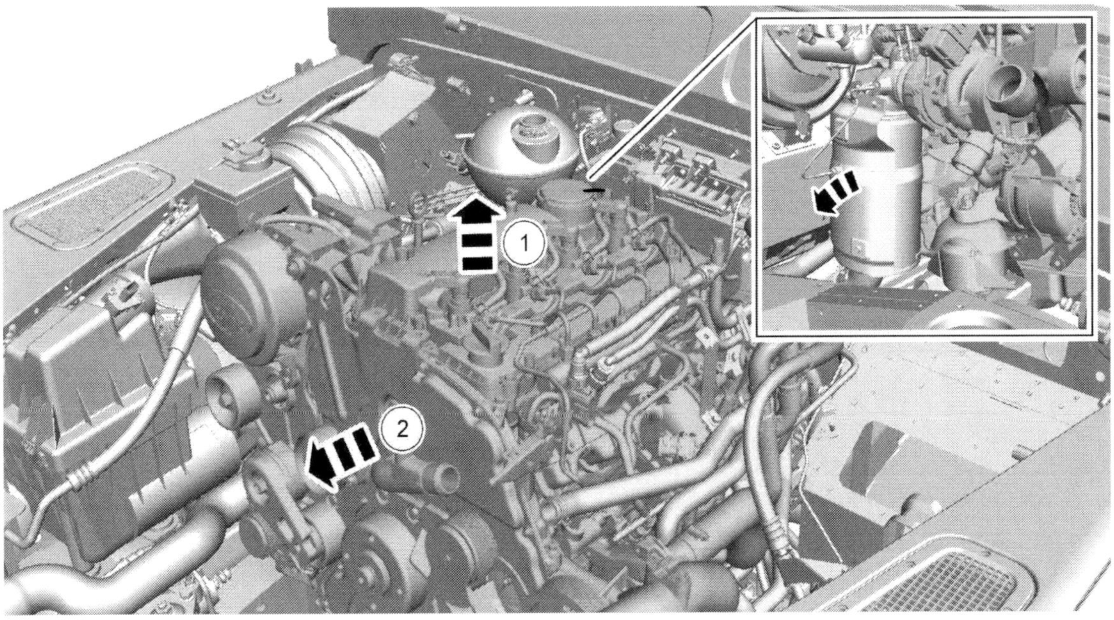

39. CAUTION: Care must be taken to avoid damage to surrounding components.

Engine Installation

Part(s)

STEP	PART NAME	QUANTITY
Step 4	Renew Part Exhaust pipe to turbocharger gasket	1
Step 19	Renew Part Air conditioning pipe to compressor O-ring seal(s)	2

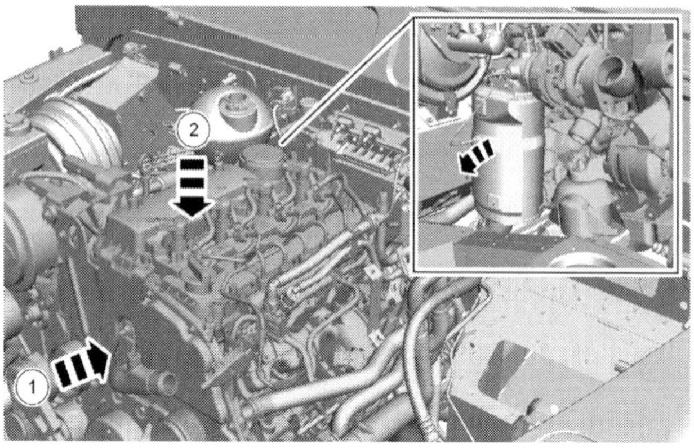

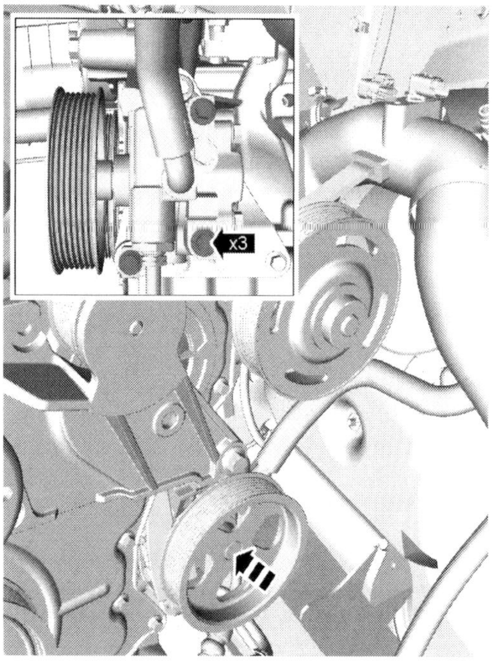

1. CAUTION: Care must be taken to avoid damage to surrounding components.

2. Tighten the bolts.

   Torque: 23 Nm

3. Tighten the bolts.

   Torque: 80 Nm

4. NOTE: Make sure a new gasket is installed.

Renew Part: Exhaust pipe to turbocharger gasket

5. Tighten the nut.

   Torque: 10 Nm

6. WARNING: Do not work on or under a vehicle supported only by a jack. Always support the vehicle on safety stands.

   Raise and support the vehicle.

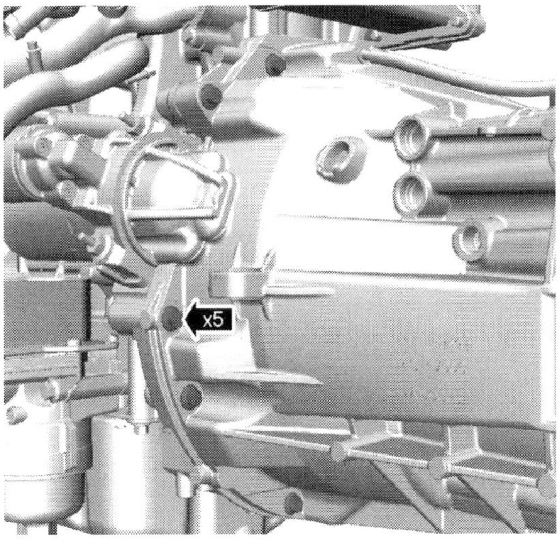

7. Tighten the bolts.

   Torque: 40 Nm

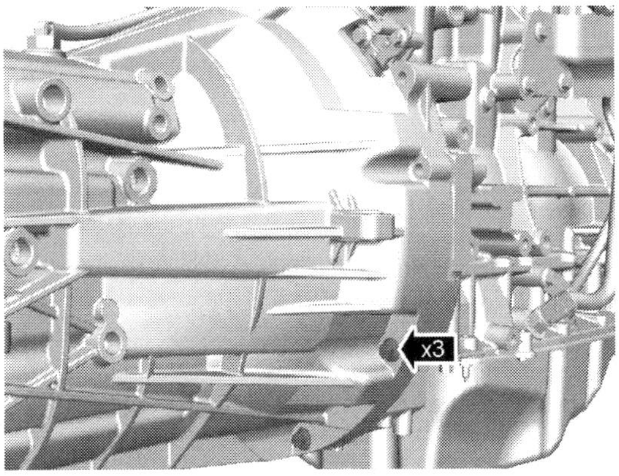

8. Tighten the bolts.

   Torque: 40 Nm

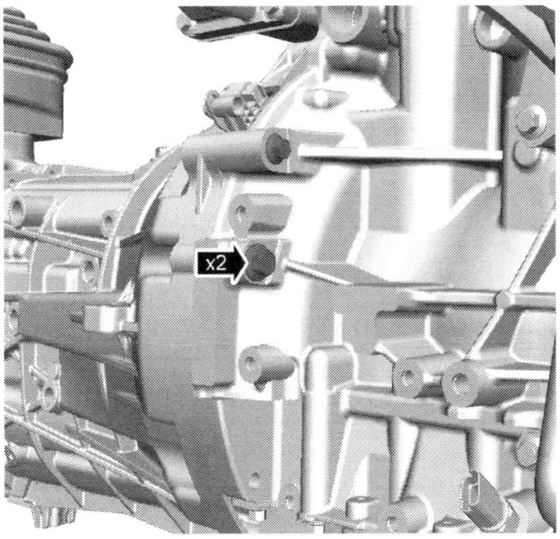

9. Tighten the bolts.

   Torque: 40 Nm

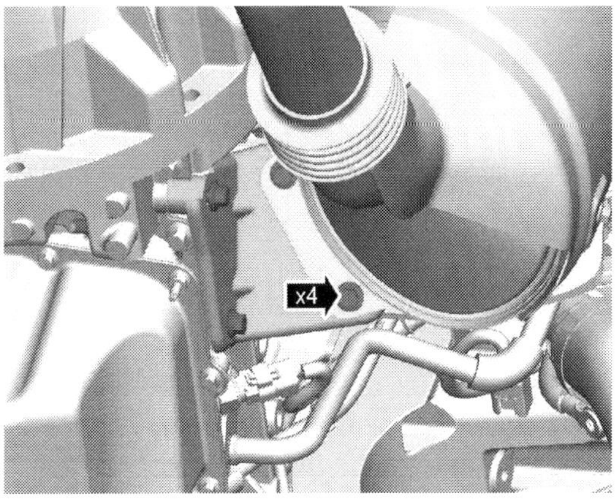

10. Tighten the bolts.

    Torque: 25 Nm

11. Lower the vehicle.

12.

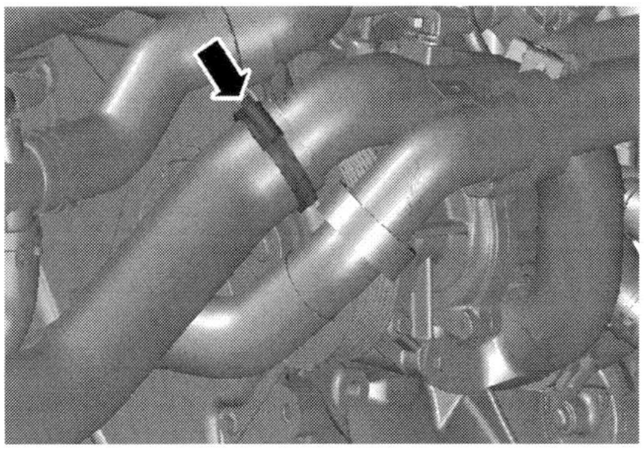

13.

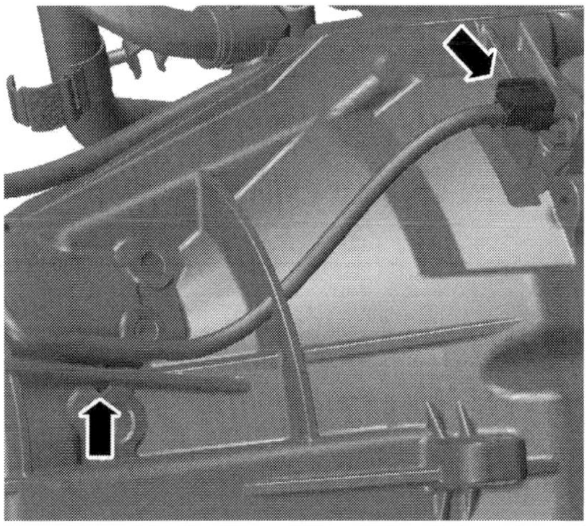

14.

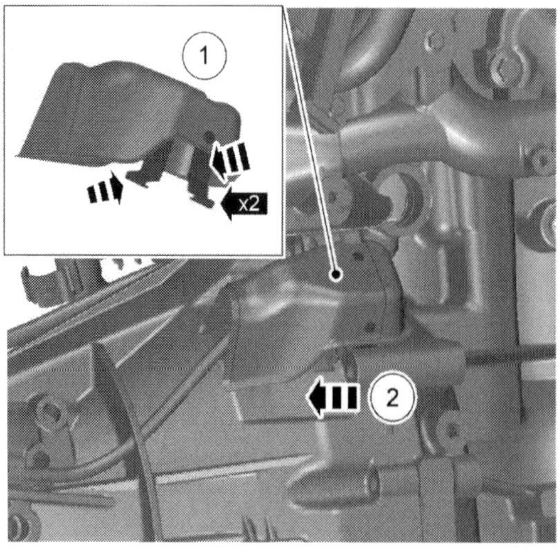

15.

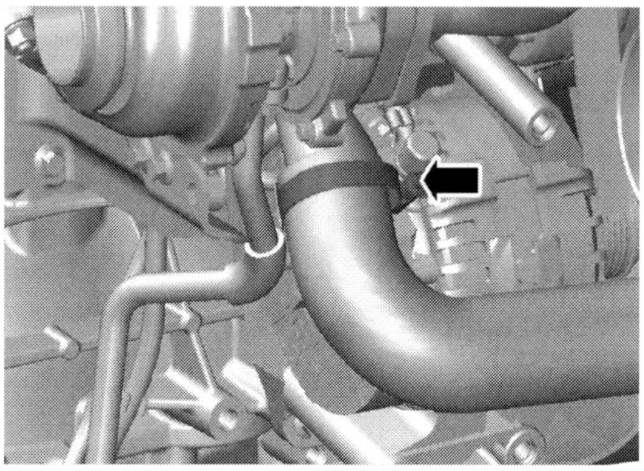

16. Tighten the bolt.

    Torque: 3 Nm

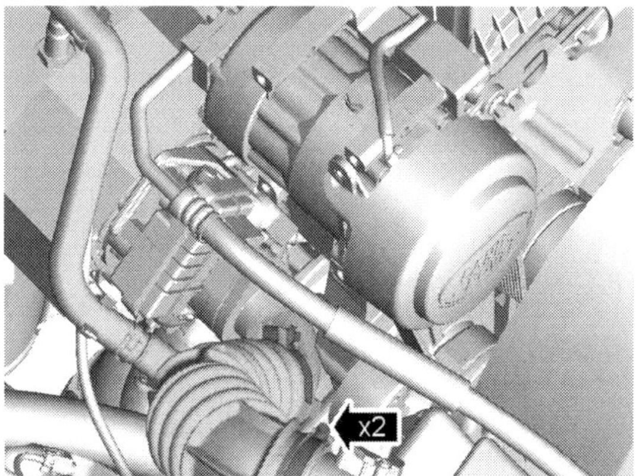

17. Tighten the bolts.

    Torque: 3 Nm

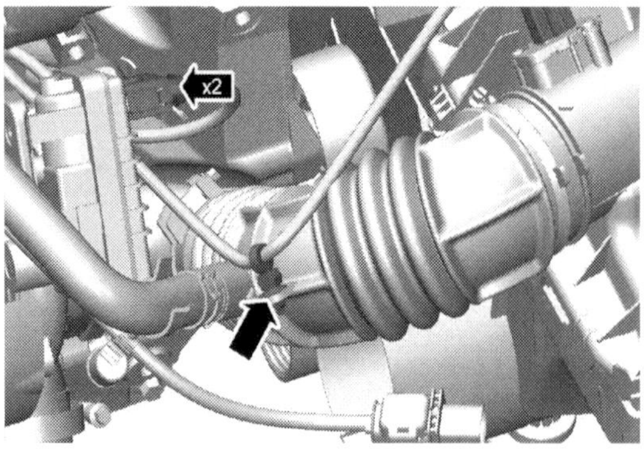

18.

19. NOTES:

△ Remove and discard the blanking caps.

△ Make sure that a new seals are installed.

Tighten the bolt.

Renew Part: Air conditioning pipe to compressor O-ring seal(s)

Torque: 30 Nm

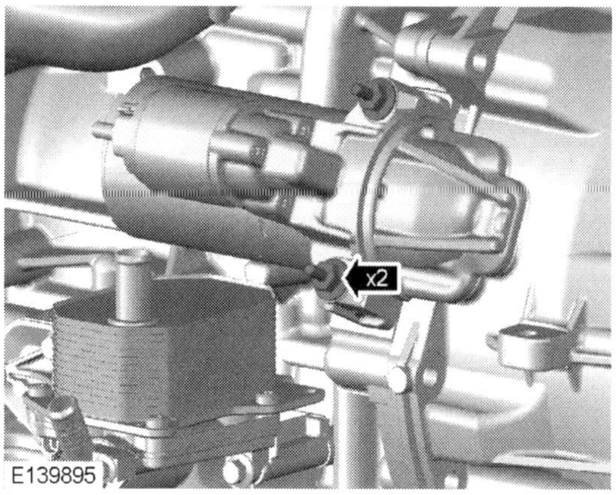

20. Tighten the nuts.

    Torque: 35 Nm

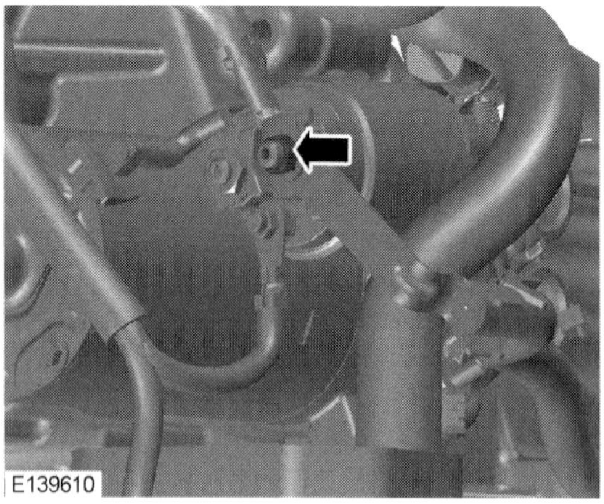

21. Tighten the nut.

    Torque: 12 Nm

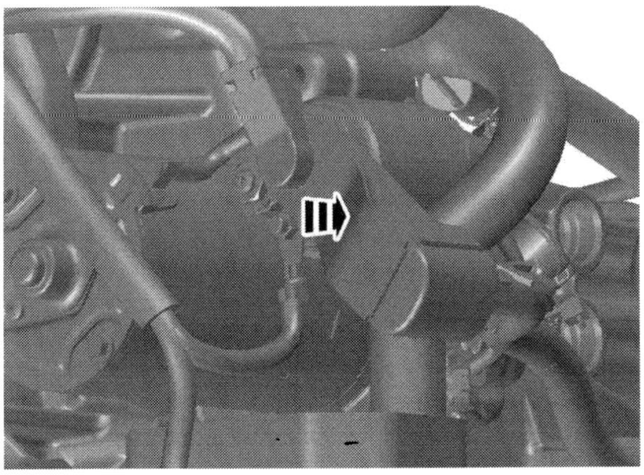

22.

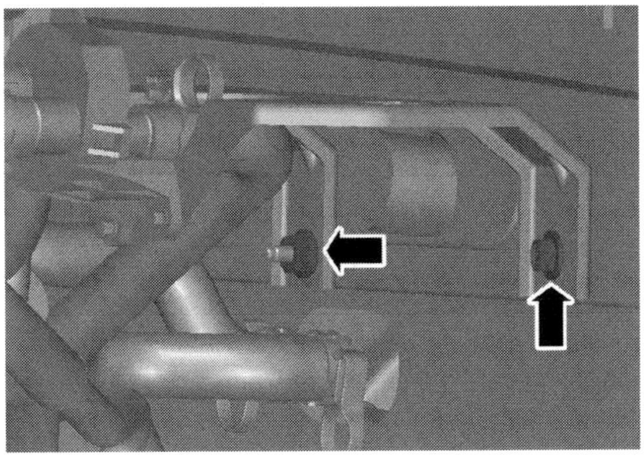

23. Tighten the nuts.

Torque: 10 Nm

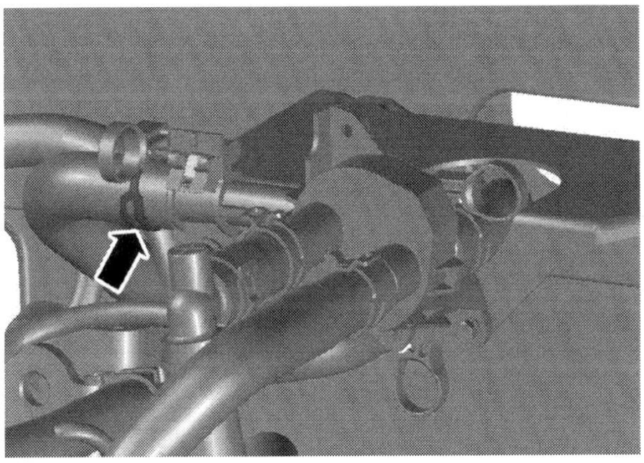

24.

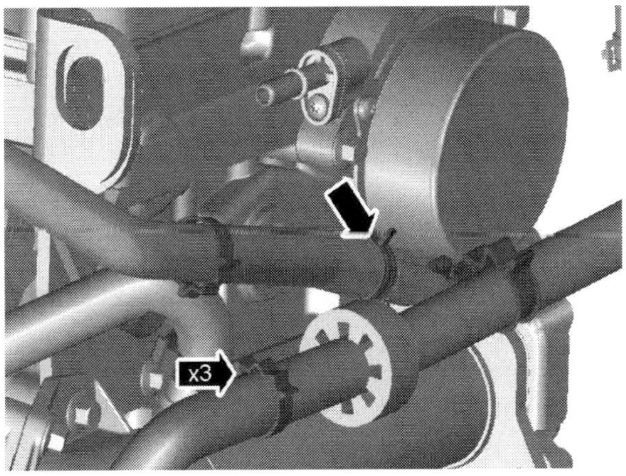

25.

26.

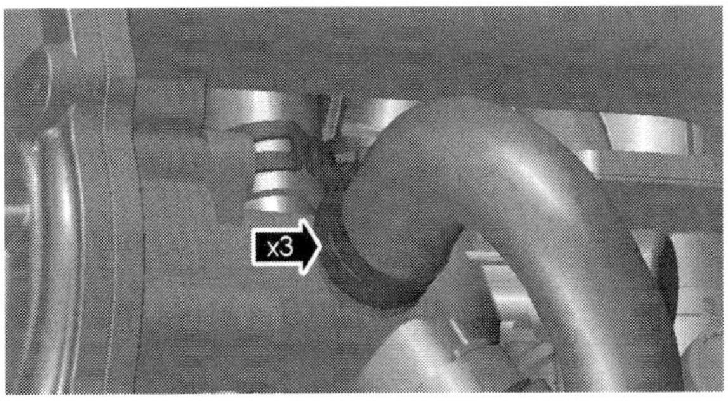

27.

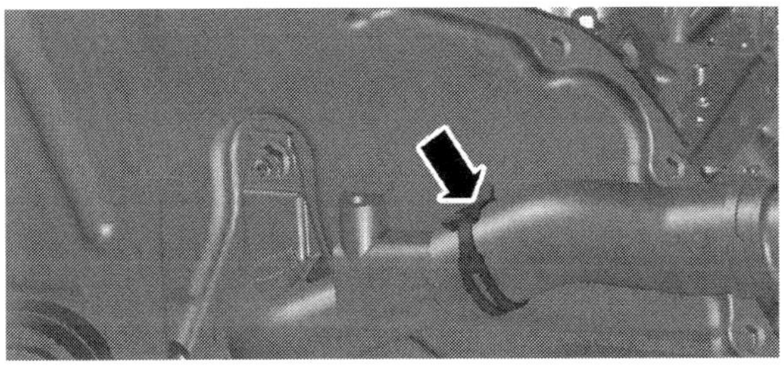

28.

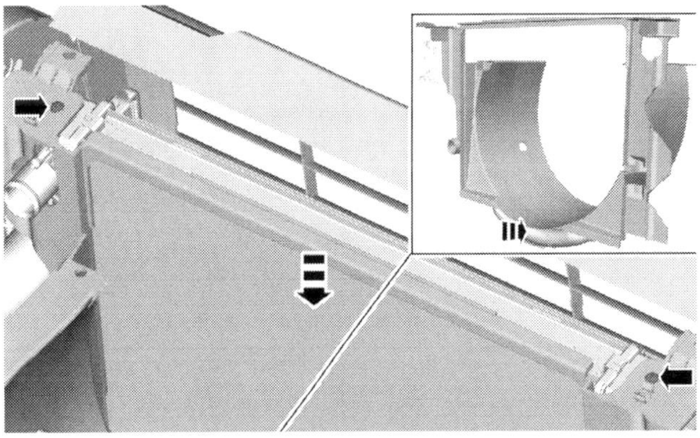

29.

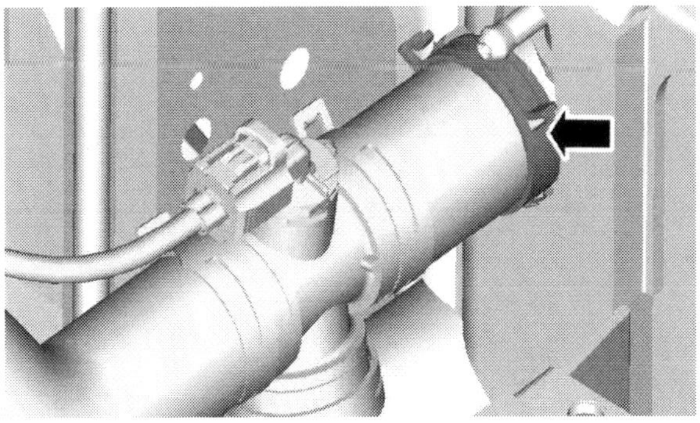

30.

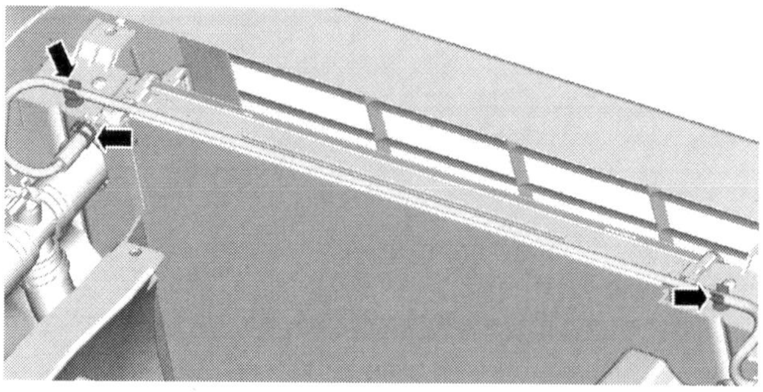

31.

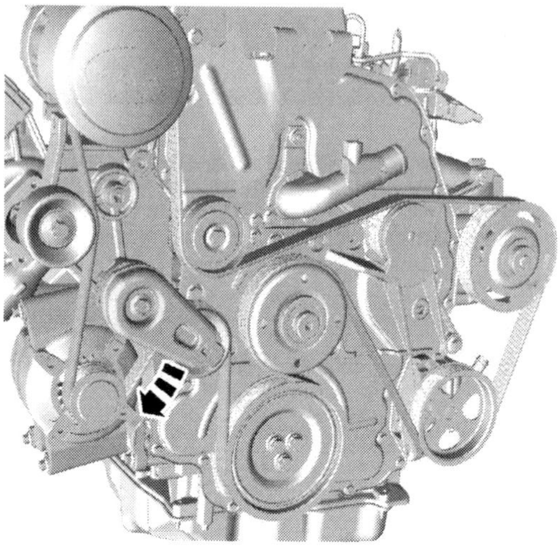

32.

33. **Refer to:** Cooling Fan (303-03 Engine Cooling, ID4 2.2L Diesel Removal and Installation)

34.

35. **Refer to:** Startup Battery Disconnect and Connect (414-01 Battery, Mounting and Cables, General Procedures)

36. **Refer to:** Cooling System Draining, Filling and Bleeding (303-03 Engine Cooling, ID4 2.2L Diesel General Procedures)

37. **Refer to:** Air Conditioning System Recovery, Evacuation and Charging (412-00 Climate Control System - General Information, General Procedures

Camshafts

Removal and Installation

Removal

⚠ NOTE: Removal steps in this procedure may contain installation details.

1. **Refer to:** Brake Vacuum Pump (206-07 Power Brake Actuation, Removal and Installation)
2. **Refer to:** Generator (414-02, Removal and Installation).
3. **Refer to:** Air Conditioning (A/C) Compressor (412-03, Removal and Installation).
   - Vehicles with Air Conditioning.

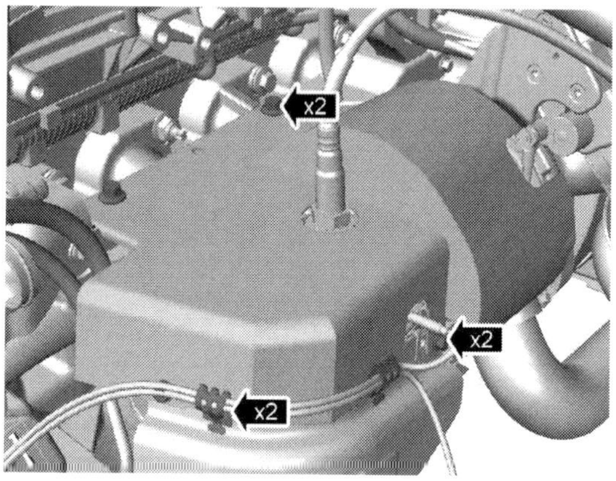

4. Torque: 7 Nm
   - Vehicles with DPF.

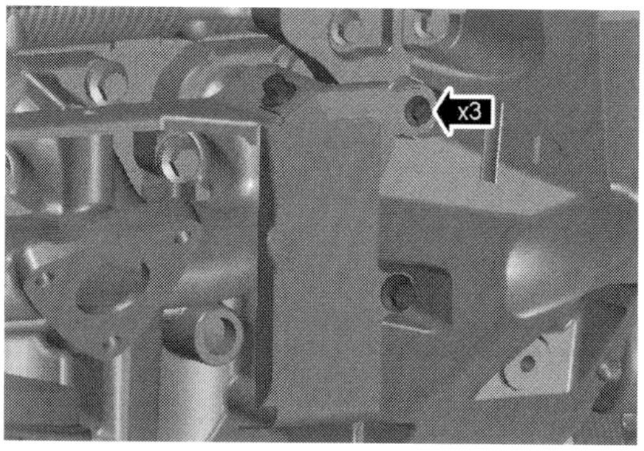

5. Torque M8: 20 Nm, M6: 10 Nm

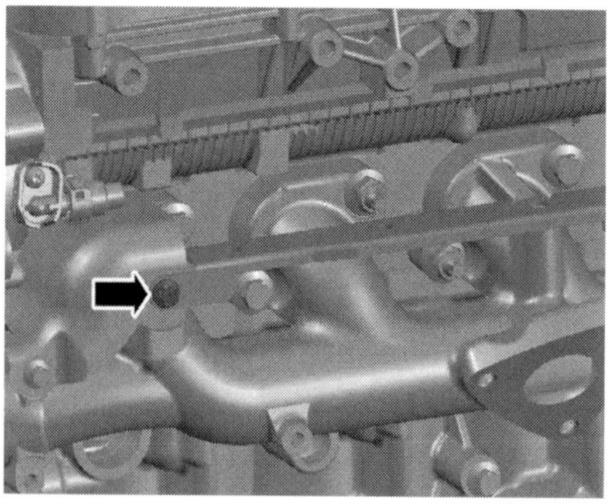

6. Torque: 20 Nm

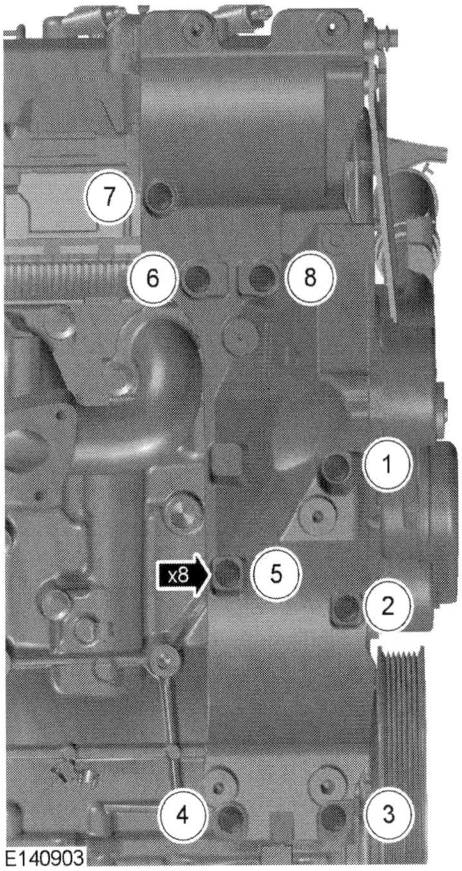

E140903

7. CAUTION: Tighten the bolts in the sequence shown, failure to follow this instruction may result in damage to the vehicle.

   Torque Stage 1: 4 Nm, Stage 2: 25 Nm

8. **Refer to:** Valve Cover (303-01 Engine, ID4 2.2L Diesel Removal and Installation)

9. **Refer to:** Timing Chain (303-01 Engine, ID4 2.2L Diesel Removal and Installation)

10.

11. Torque stage 1: 8 Nm, Stage 2: 23 Nm

12. **Refer to:** Camshaft Position Sensor (303-14 Electronic Engine Controls, ID4 2.2L Diesel Removal and Installation)

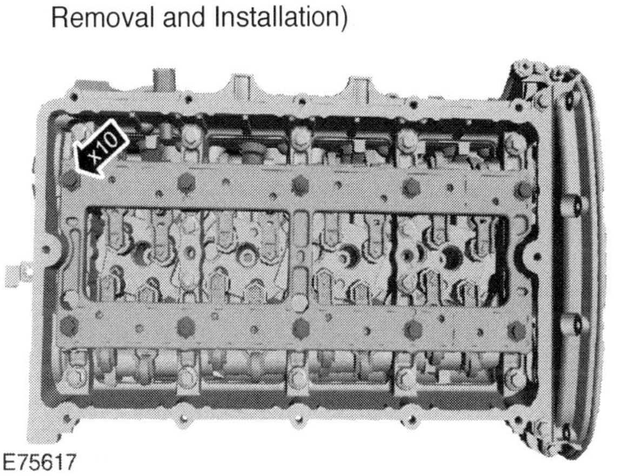

E75617

13. CAUTIONS:

   Make sure new bolts are installed.

   Remove and tighten the bolts in the sequence shown, failure to follow this instruction may result in damage to the vehicle.

   Visually inspect rocker arms to ensure they are seated correctly.

   Torque stage 1: 10 Nm, Stage 2: 45°

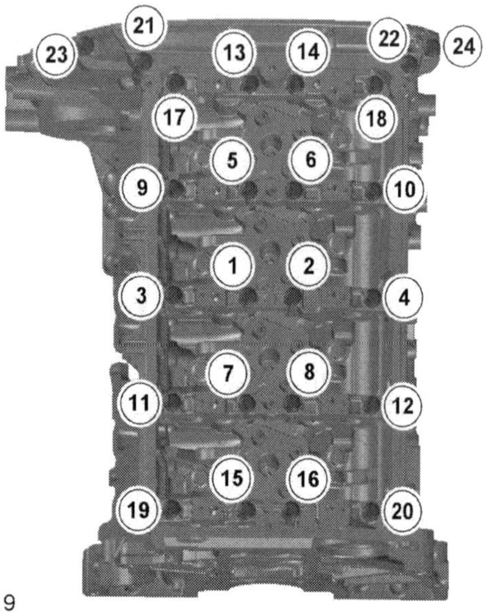

E140019

14. Torque: 10 Nm

15. ⚠ CAUTION: Remove and tighten the bolts in the sequence shown, failure to follow this instruction may result in damage to the vehicle.

 Torque Stage 1: 4 Nm, Stage 2: Bolts 1 to 22 15 Nm, bolts 23 to 24 8 Nm, Stage 3: Bolts 1 to 22 23 Nm, bolts 23 to 24 10 Nm.

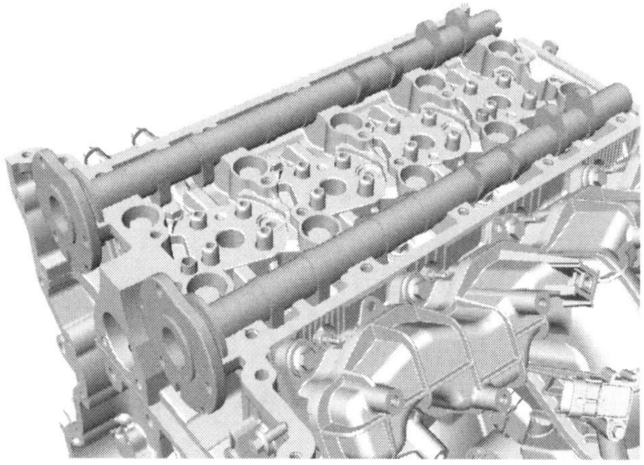

16.

Installation

1. ⚠ NOTE: Lubricate the camshaft journals and the camshaft lobes with clean engine oil.

2. NOTES:

⚠ Clean the component mating face.

⚠ Apply sealant meeting Land Rover specification. WSE – M4G323 – A4 RTV Sealant.

Apply a 2.5 mm (0.1 inches) diameter bead of the specified sealant to the cylinder head.

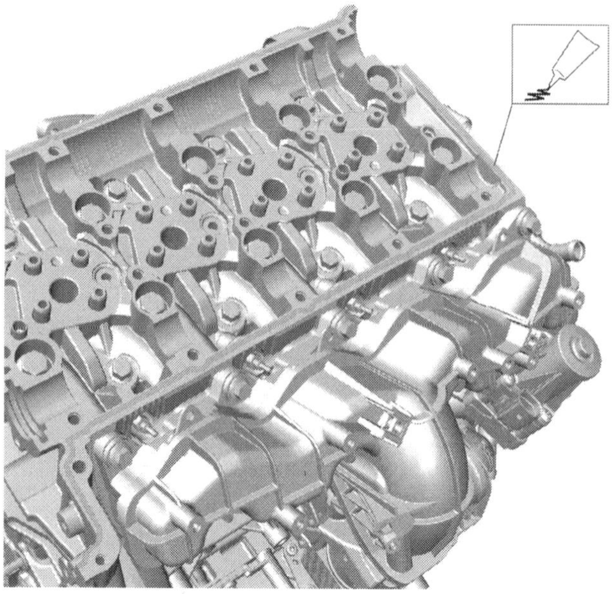

3. To install, reverse the installation procedure.

Crankshaft Pulley

Removal and Installation

Special Tool(s)

Aligner, Engine Front Cover

303-1310

E87614

Removal

1. **Refer to:** Battery Disconnect and Connect (414-01, General Procedures).

2. **Refer to:** Cooling Fan (303-03 Engine Cooling, ID4 2.2L Diesel Removal and Installation)

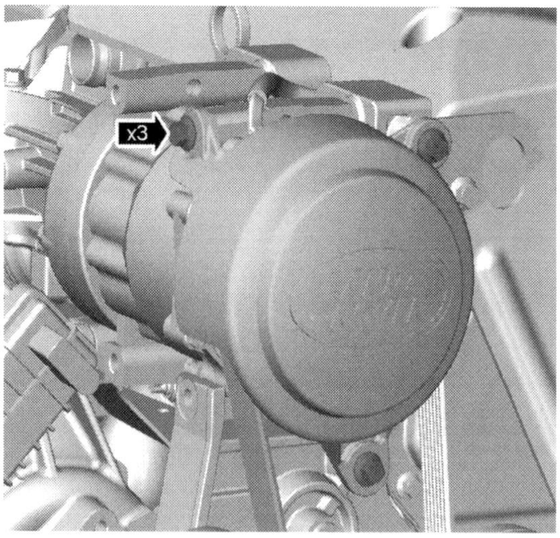

3. *Torque:* 10 Nm

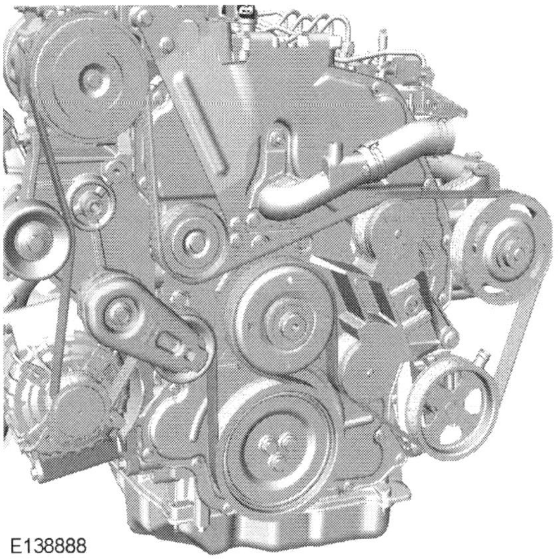

4.

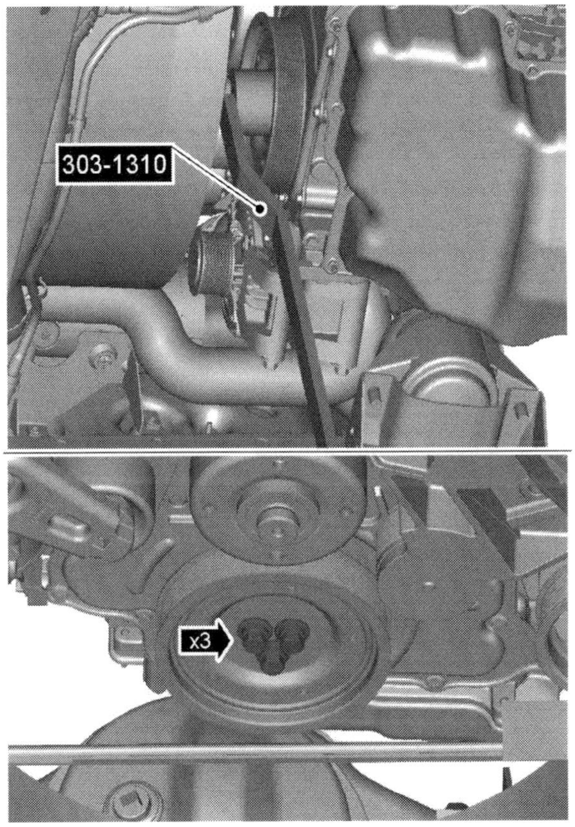

5. NOTE: Apply the foot and hand brake, select 1st gear to prevent crankshaft rotation.

Torque:

1. Stage : Run down 3 bolts to 20 Nm
2. Stage : Run down 3 bolts to 45 Nm
3. Stage : Run down 3 bolts to 90 Nm
4. Stage : Turn on all 45 degrees
5. Stage : Turn on all 45 degrees

Installation

1. To install, reverse to removal procedure.

Crankshaft Front Seal

Removal and Installation

Special Tool(s)

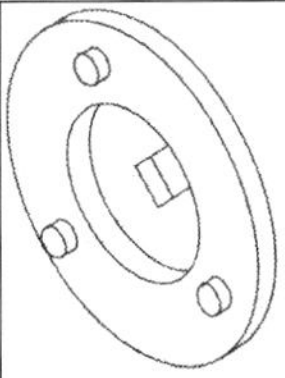 303679	Remover/Installer, Front Oil Seal JLR-303-1619

Removal

1. WARNING: Do not work on or under a vehicle supported only by a jack. Always support the vehicle on safety stands.

   Raise and support the vehicle.

2. **Refer to:** Crankshaft Pulley (303-01 Engine, ID4 2.2L Diesel Removal and Installation)

3. Tool: JLR-303-1619 (Remover/Installer, Front Oil Seal)

Installation

1. CAUTIONS:

Make sure that the mating faces are clean and free of foreign material.

Do not over tighten the crankshaft front seal. Failure to follow this instruction may result in damage to the vehicle.

Take extra care not to damage the seal.

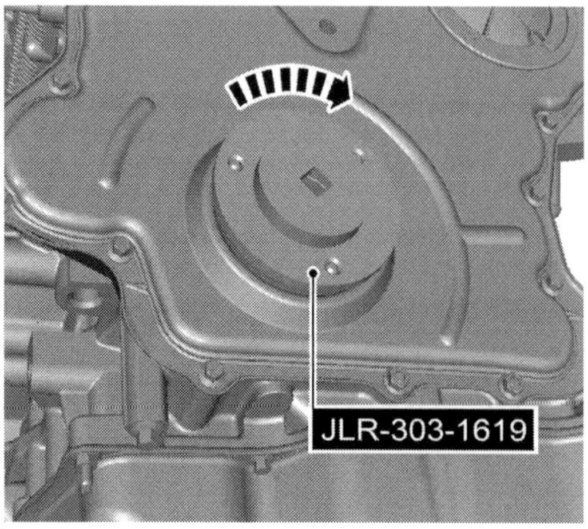

NOTE: A new crankshaft front oil seal is supplied with an alignment sleeve that will be pushed out during installation.

Tool: JLR-303-1619 (Remover/Installer, Front Oil Seal)

- Torque: 19 Nm

2. To install, reverse to removal procedure.

Crankshaft Rear Seal

Removal and Installation

Part(s)

STEP	PART NAME	QUANTITY
Installation Step 2	Renew Part Crankshaft rear seal	1

Removal

1. **Refer to:** Startup Battery Disconnect and Connect (414-01 Battery, Mounting and Cables, General Procedures)
2. **Refer to:** Flywheel (303-01 Engine, ID4 2.2L Diesel Removal and Installation)
3. WARNING: Do not work on or under a vehicle supported only by a jack. Always support the vehicle on safety stands.

Raise and support the vehicle.

4.

Installation

1. NOTE: Clean the component mating faces.
   1. Check the 2 foam pads are located on the ladder frame gasket.
   2. If the ladder frame gasket tabs are present, trim using a suitable tool flush to the engine block and ladder frame.
   3. Apply a thin layer of sealant to the areas shown.
2. CAUTIONS:

A new crankshaft rear seal is supplied with an alignment sleeve that must not be removed until the crankshaft rear seal is fully installed. Failure to follow this instruction may result in damage to the vehicle.

Install the new crankshaft rear seal within five minutes of applying the recommended sealant.

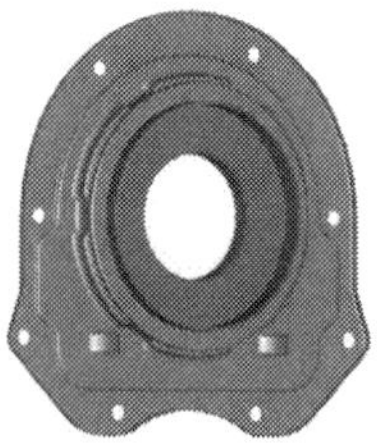

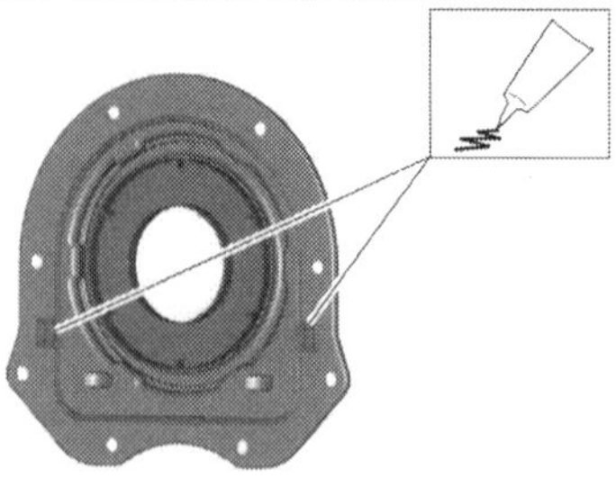

Install the new crankshaft rear seal within five minutes of applying the recommended sealant.

Do not add the sealant anywhere other than the area shown.

Using the recommended sealant completely fill the square areas shown.

Refer to: (303-01 Engine, ID4 2.2L Diesel Specifications)

Renew Part: Crankshaft rear seal

3. Hand start all bolts and tighten the bolts in sequence.

   Torque:

   Stage 1: 5 Nm

   Stage 2: 9.5 Nm

   Stage 3: 9.5 Nm

4. Remove and discard the seal alignment sleeve.
5. To install, reverse the removal procedure.

Cylinder Head

Removal and Installation

Removal

1. WARNING: Do not work on or under a vehicle supported only by a jack. Always support the vehicle on safety stands.

   Raise and support the vehicle.

2. **Refer to:** Battery Disconnect and Connect (414-01 Battery, Mounting and Cables, General Procedures).

3. **Refer to:** Camshafts (303-01 Engine, ID4 2.2L Diesel Removal and Installation)
4. **Refer to:** Intake Manifold (303-01 Engine, ID4 2.2L Diesel Removal and Installation)
5. Torque: 30 Nm

6. Torque: 25 Nm

7.
   1. Stage 1: Tighten the M10 bolts and nuts to 15 Nm (11 lb.ft).
   2. Stage 2: Tighten the M10 bolts and nuts to 35 Nm (26 lb.ft).

8. Torque: 3 Nm

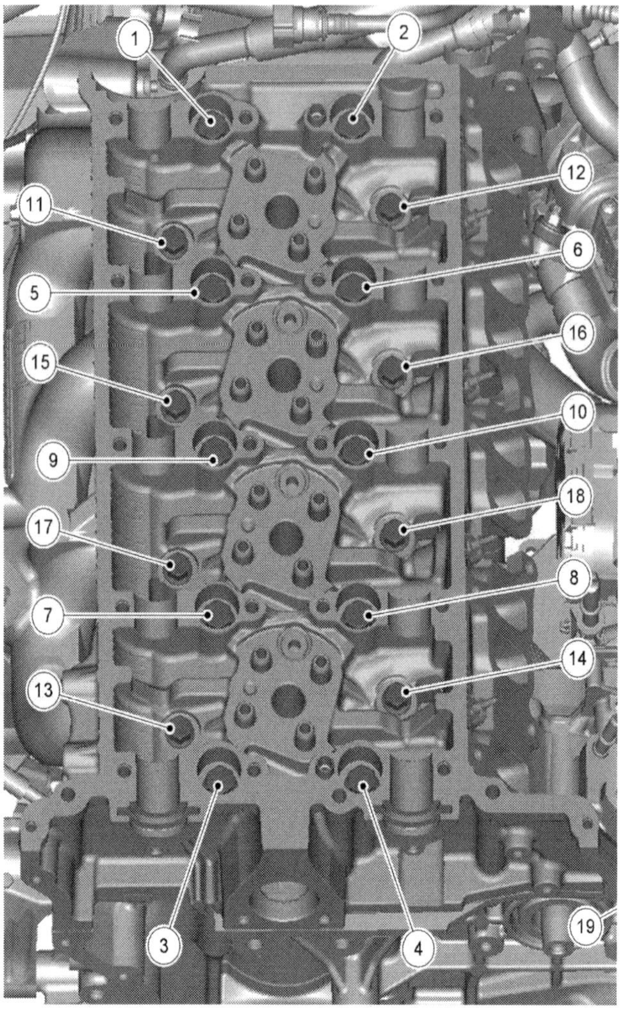

9.

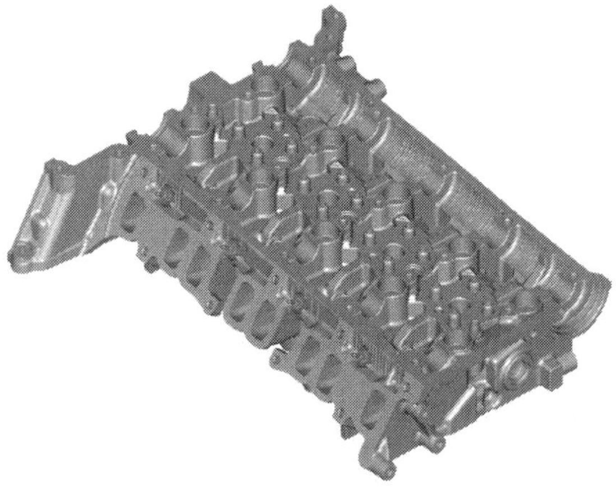

10. NOTE: Do not disassemble further if the component is removed for access only.

Torque: 13 Nm

Installation

1. CAUTION: Make sure that the mating faces are clean and free from foreign material.

NOTE: Measure the piston protrusion of each cylinder at top dead center (TDC).

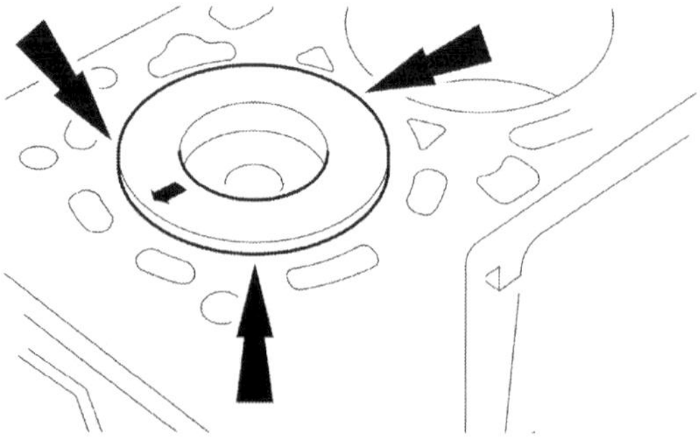

Measure the distance between the piston crown and the cylinder block at the points indicated.

2. Determine the cylinder head gasket thickness.

Refer to: (303-01 Engine, ID4 2.2L Diesel Specifications)

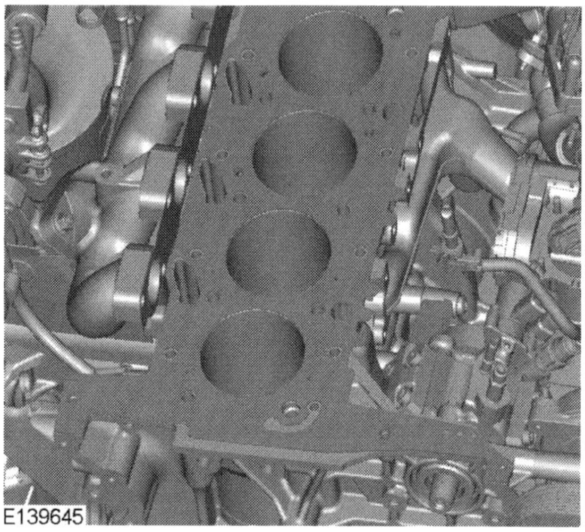

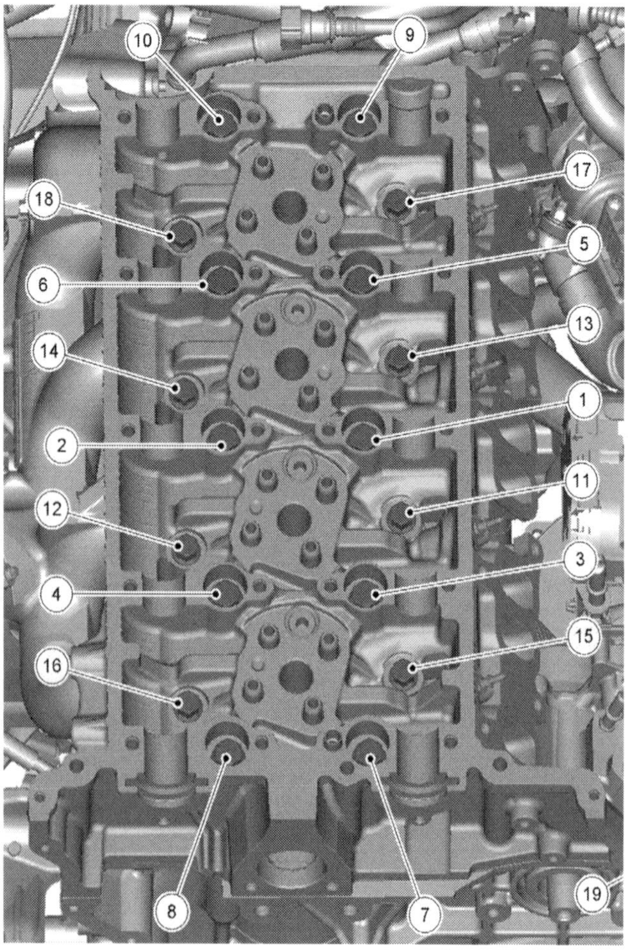

3. CAUTIONS:

The thickness of the new cylinder head gasket depends on the piston protrusion (hole/tooth marked).

Make sure that the mating faces are clean and free from foreign material.

Fit new cylinder head gasket, dry, to cylinder block.

4. CAUTION: Make sure that the mating faces are clean and free from foreign material.

NOTE: New bolts must be installed.

With assistance, install the cylinder head.

1. Stage 1: Tighten bolts 1 through 10 to 10 Nm.
2. Stage 2: Tighten bolts 11 through 18 to 5 Nm.
3. Stage 3: Tighten bolts 1 through 10 to 20 Nm.
4. Stage 4: Tighten bolts 11 through 18 to 10 Nm.
5. Stage 5: Tighten bolts 1 through 10 to 35 Nm.
6. Stage 6: Tighten bolts 11 through 18 to 20 Nm.
7. Stage 7: Tighten bolts 1 through 10 to 45 Nm.
8. Stage 8: Tighten bolts 11 through 18 to 26 Nm.
9. Stage 9: Tighten bolts 1 through 10° to 90°.
10. Stage 10: Tighten bolts 11 through 18° to 90°.
11. Stage 11: Tighten bolts 1 through 10° to 90°.
12. Stage 12: Tighten bolts 11 through 18° to 90°.
13. Stage 13: Tighten the M6 bolt to 10 Nm.

5. To install, reverse to removal procedure.

Left Engine Mount

Removal and Installation

Removal

NOTE: Some variation in the illustrations may occur, but the essential information is always correct.

1. Disconnect the battery ground cable. For additional information, refer to:

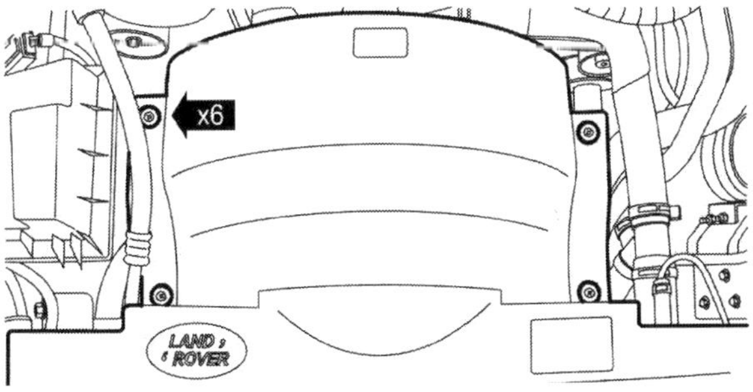

Refer to: Battery Disconnect and Connect (414-01, General Procedures).

2. 3. Remove the cooling fan upper shroud.
3. Remove the air cleaner outlet pipe. For additional information, refer to:

Refer to: Air Cleaner Outlet Pipe (303-12 Intake Air Distribution and Filtering, ID4 2.2L Diesel Removal and Installation)

4. Remove the RH engine mount nut.
5. Disconnect the charher air cooler outlet hose.
6. Remove the RH engine mount nut.

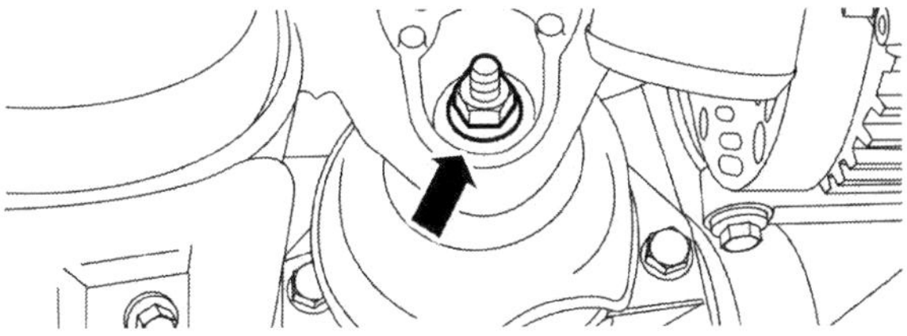

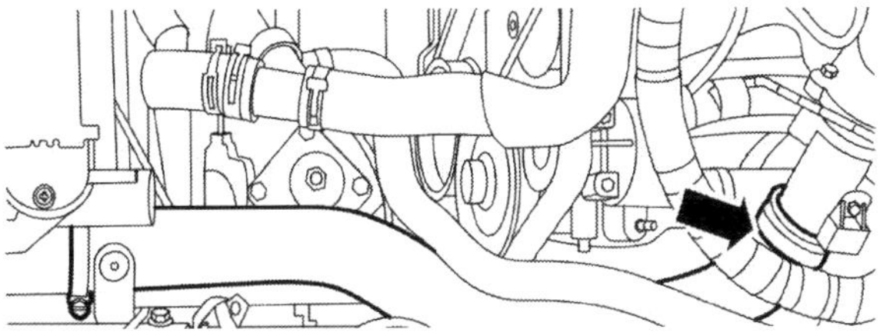

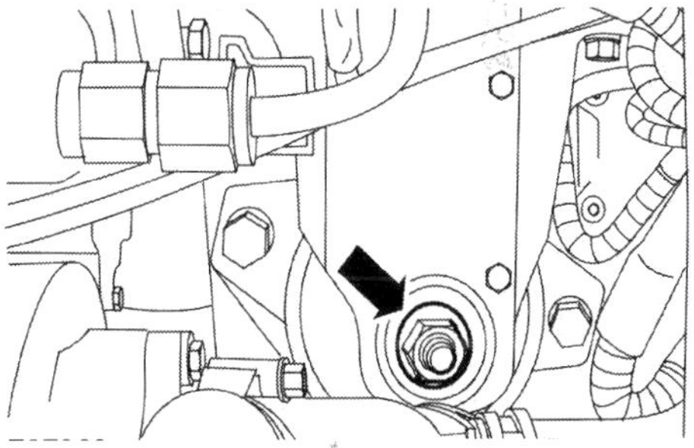

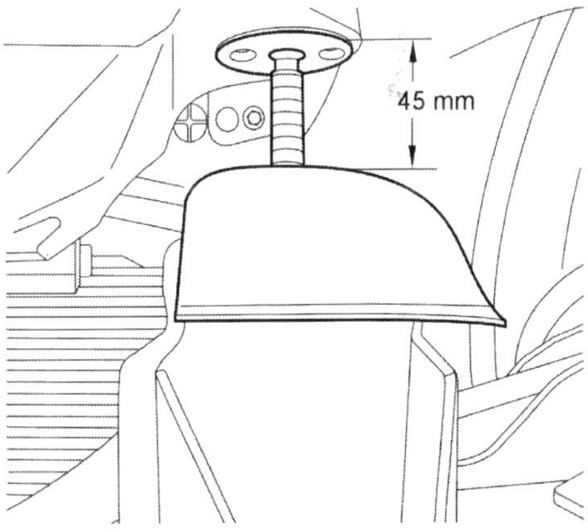

7. Using a suitable hydraulic jack, raise the engine to a maximum of 45 mm (1.77 inches).

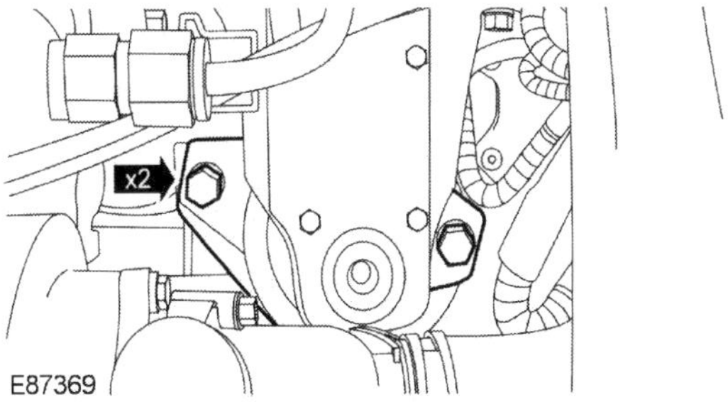

8. Remove the LH engine mount.

Installation

1. To install, reverse the removal procedure.

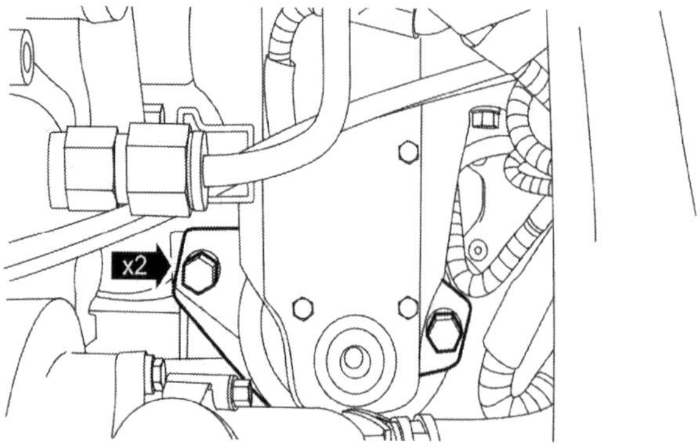

2. Tighten to 80 Nm (59 lb.ft).

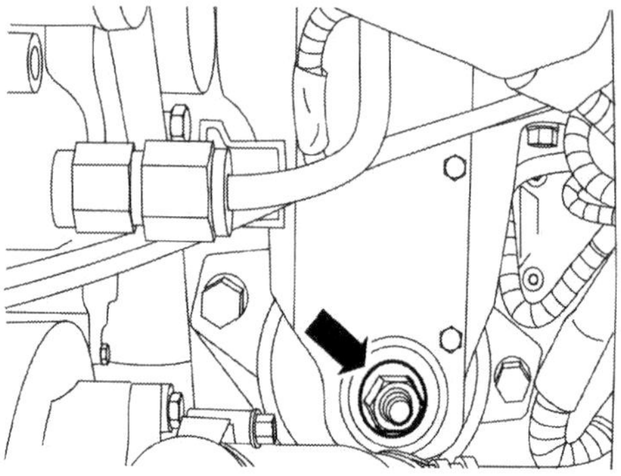

3. Tighten to 80 Nm (59 lb.ft).

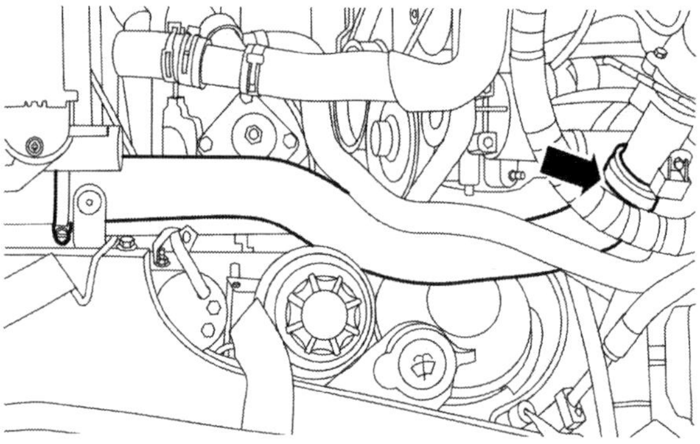

4. Tighten to 3 Nm (2 lb.ft).

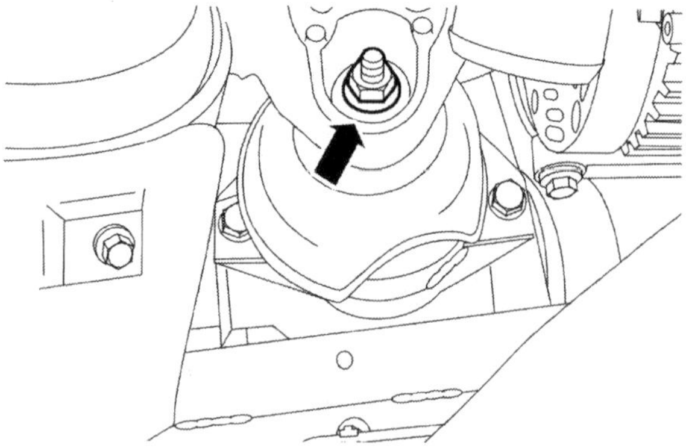

5. Tighten to 80 Nm (59 lb.ft).

Right Engine Mount

Removal and Installation

Removal

NOTE: Some variation in the illustrations may occur, but the essential information is always correct.

1. Disconnect the battery ground cable. For additional information, refer to:

Refer to: Battery Disconnect and Connect (414-01, General Procedures).

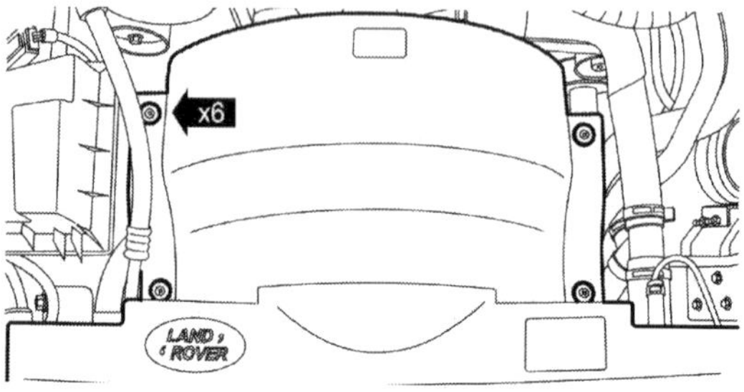

2. 3. Remove the cooling fan upper shroud.

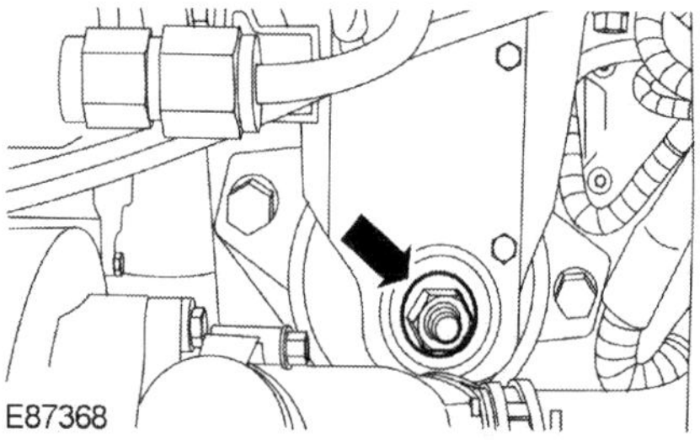

3. Remove the LH engine mount nut.

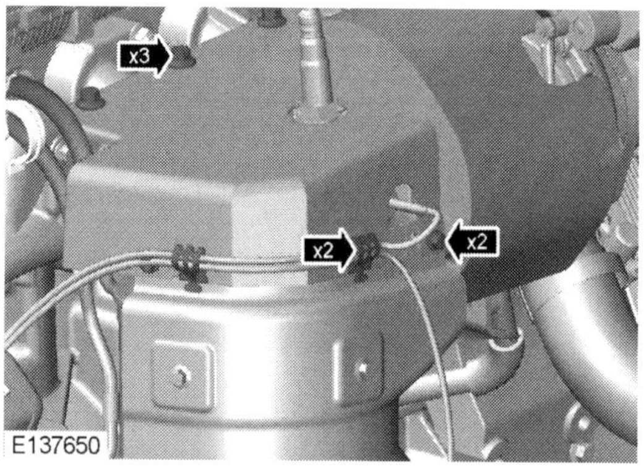

4. Remove the turbocharger heat shield.

5. Remove the air cleaner outlet pipe. For additional information, refer to:

Refer to: Air Cleaner Outlet Pipe (303-12 Intake Air Distribution and Filtering, ID4 2.2L Diesel Removal and Installation)

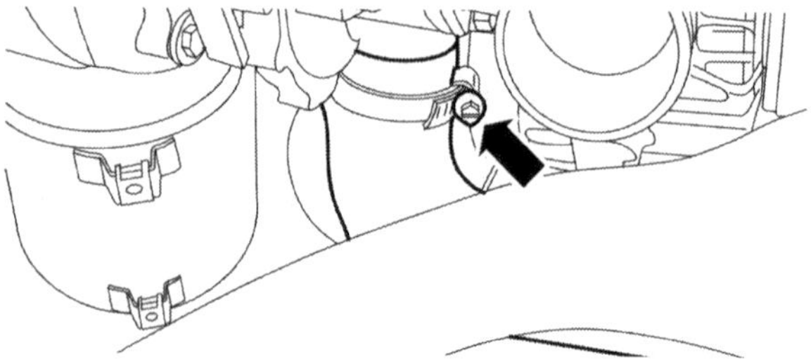

6. Disconnect the charger air cooler inlet hose.

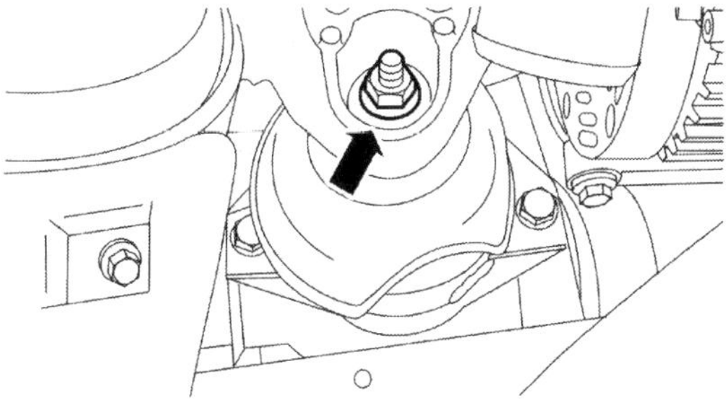

7. Remove the RH engine mount nut.

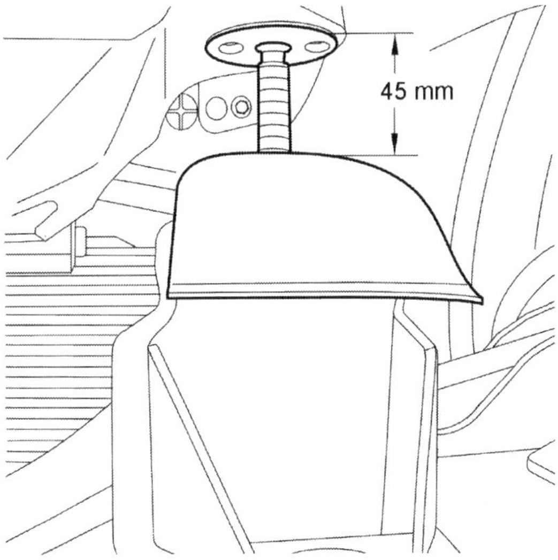

8. Using a suitable hydraulic jack, raise the engine to a maximum of 45 mm (1.77 inches).

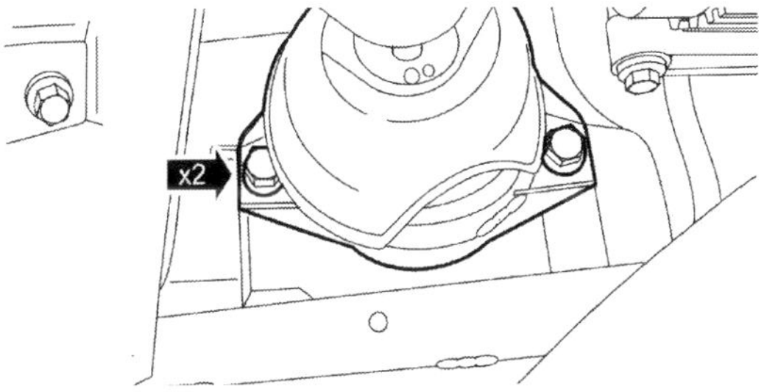

9. Remove the RH engine mount.

Installation

1. To install, reverse the removal procedure.

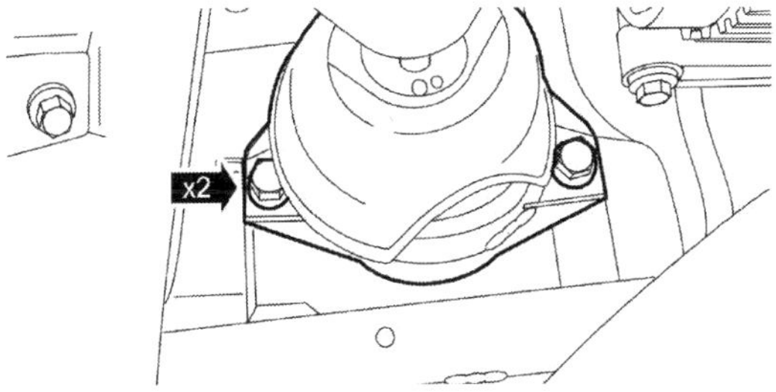

2. Tighten to 80 Nm (59 lb.ft).

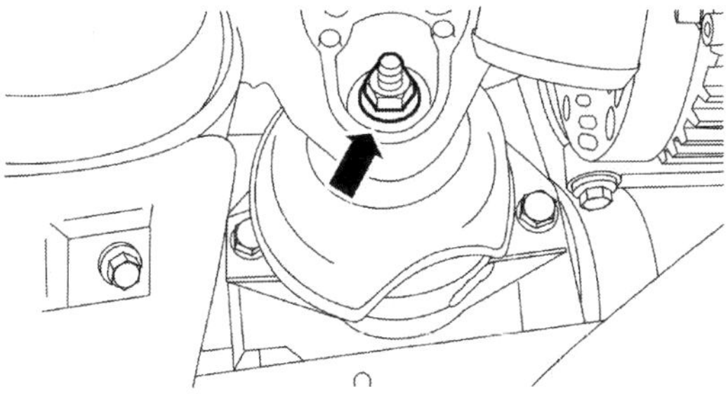

3. Tighten to 80 Nm (59 lb.ft).

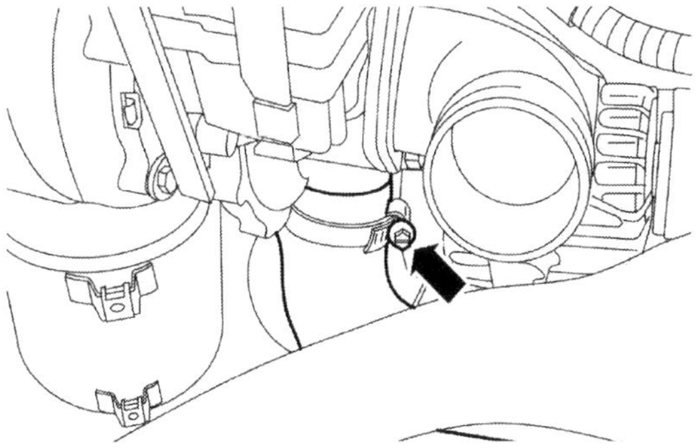

4. Tighten to 3 Nm (2 lb.ft).

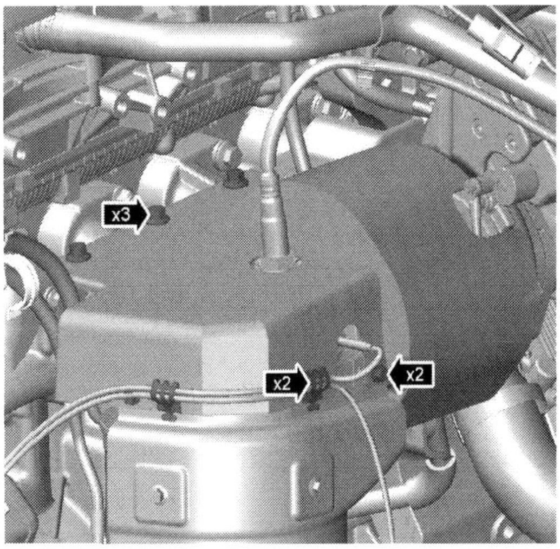

5. Tighten to 10 Nm (7 lb.ft).

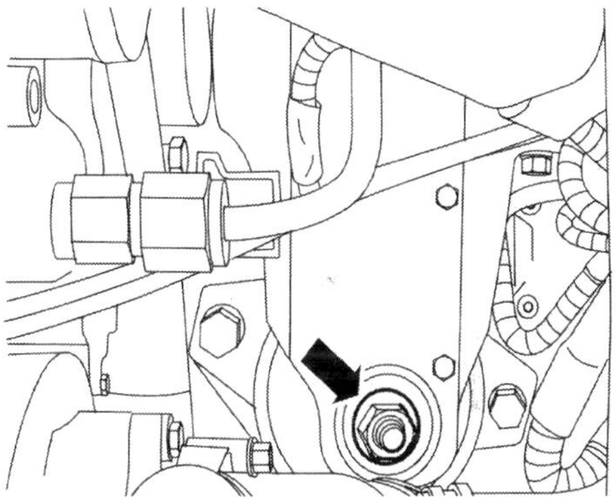

6. Tighten to 80 Nm (59 lb.ft).

Exhaust Manifold

Removal and Installation

Removal

1. **Refer to:** Battery Disconnect and Connect (414-01, General Procedures).

2. ⚠ WARNING: Do not work on or under a vehicle supported only by a jack. Always support the vehicle on safety stands.

   Raise and support the vehicle.

3. **Refer to:** Turbocharger (303-04B, Removal and Installation).

4. Torque: 10 Nm

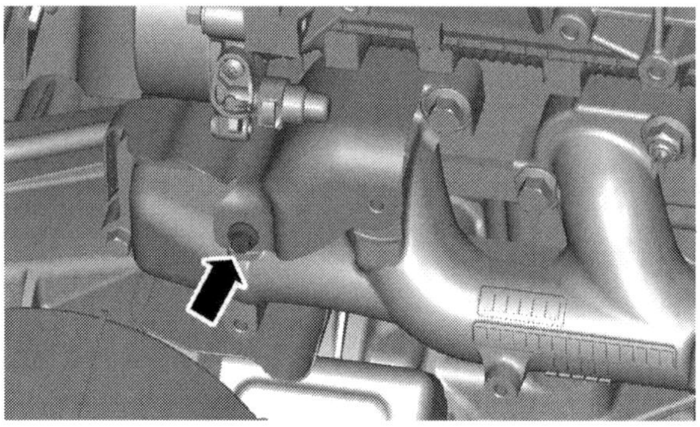

5. Torque: 9 Nm

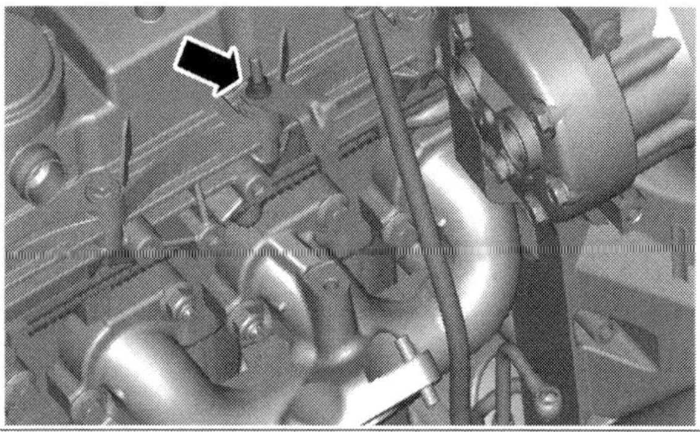

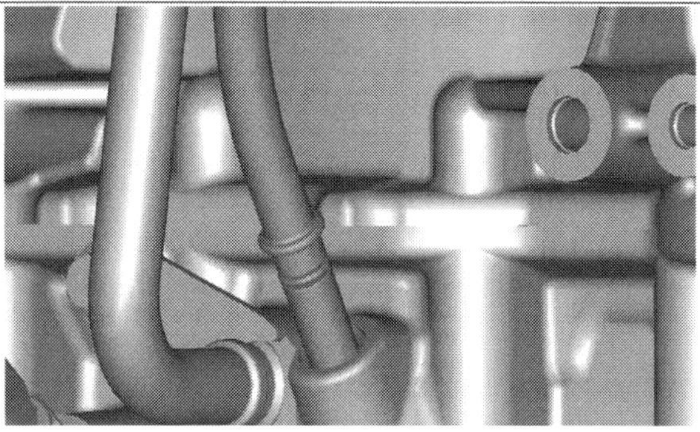

6. Torque: 10 Nm

7. CAUTION: Make sure that the mating faces are clean and free from foreign material.

Installation

1. NOTE: Install a new gasket.
    1. Stage 1: Tighten the M10 bolts and nuts to 15 Nm (11 lb.ft).
    2. Stage 3: Tighten the M10 bolts and nuts to 35 Nm (26 lb.ft).
    3. Stage 4: Repeat the M10 bolts and nuts to 35 Nm (26 lb.ft).

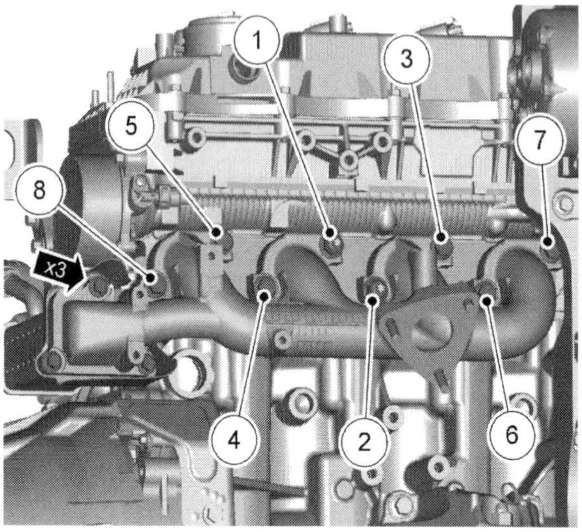

1. NOTE: Install a new gasket.
    1. Stage 1: Tighten the M10 bolts and nuts to 15 Nm (11 lb.ft).
    2. Stage 3: Tighten the M10 bolts and nuts to 35 Nm (26 lb.ft).
    3. Stage 4: Repeat the M10 bolts and nuts to 35 Nm (26 lb.ft).
    4. Tighten the M8 bolts to 25 Nm (18 lb.ft).
2. To install, reverse to removal procedure.

Intake Manifold

Removal and Installation

Removal

NOTE: Removal steps in this procedure may contain installation details.

1. **Refer to:** Exhaust Gas Recirculation Valve (303-08 Engine Emission Control, ID4 2.2L Diesel Removal and Installation).
2. **Refer to:** Throttle Body (303-04 Fuel Charging and Controls, ID4 2.2L Diesel Removal and Installation).

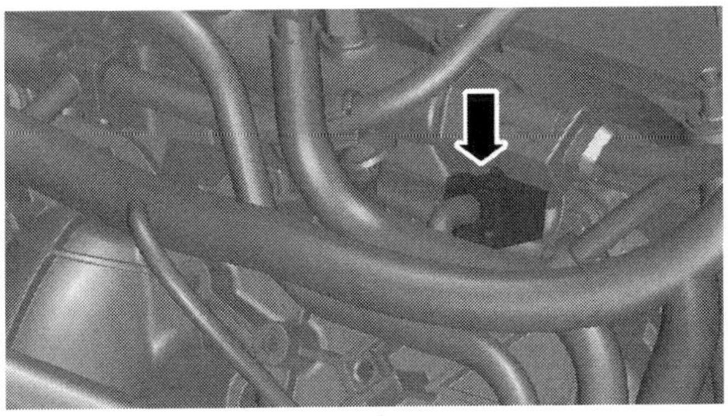

3.

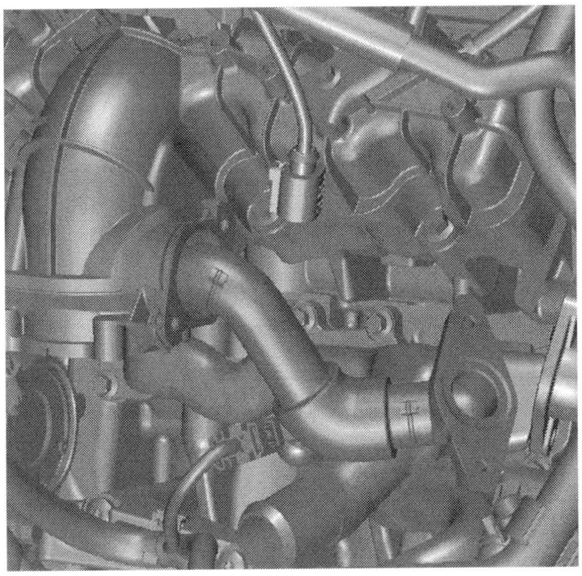

4.

5.

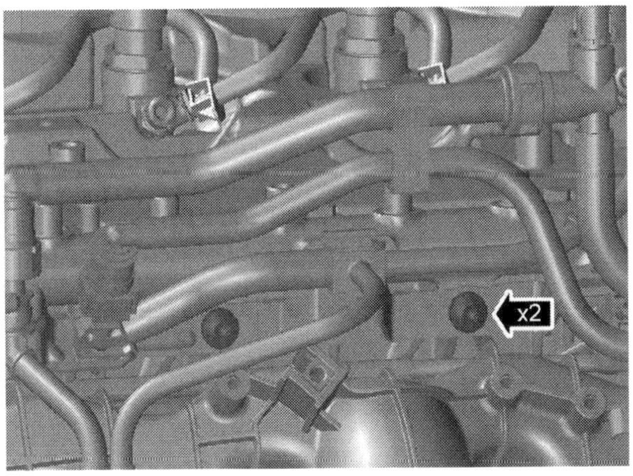

6. Torque Stage 1: 8 Nm, Stage 2: 23 Nm

7. ⚠ CAUTION: Before disconnecting any components, make sure the area is clean and free from foreign material. When disconnected all openings must be sealed.

△ NOTE: Clean the component mating faces.

Torque: 15 Nm

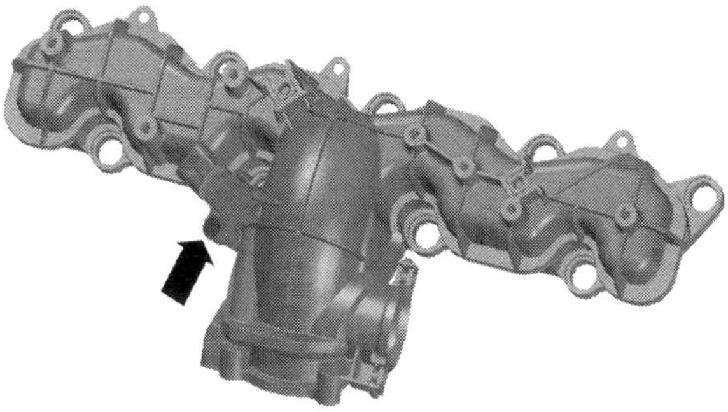

8. *Torque:* 3 Nm

Installation

1. To install, reverse the removal procedure.

Oil Cooler

Removal and Installation

Part(s)

STEP	PART NAME	QUANTITY
Installation Step 1	Renew Part Oil filter and O-ring seal kit	1

Removal

1. Disconnect the battery ground cable.

   Refer to: Startup Battery Disconnect and Connect (414-01 Battery, Mounting and Cables, General Procedures).

2. Remove the oil filter housing.

   Refer to: Oil Filter Housing (303-01 Engine, ID4 2.2L Diesel Removal and Installation).

3. ⚠ CAUTION: Before disconnecting any components, make sure the area is clean and

free from foreign material. When disconnected all openings must be sealed.

Remove the 5 bolts. Remove the oil cooler. Remove and discard the O-ring seals.

Installation

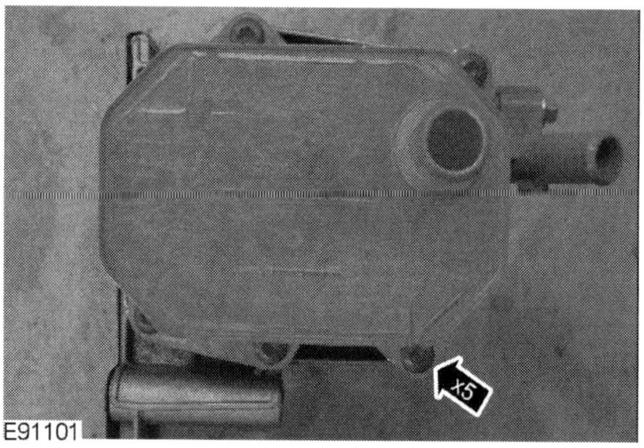

1. NOTES:

△ Remove and discard the blanking caps.

△ Clean the component mating faces.

△ Install a new O-ring seals.

To install, reverse the removal procedure. Tighten the bolts.

Renew Part: Oil filter and O-ring seal kit

Torque: 10 Nm

2. Connect the battery ground cable.

Refer to: Startup Battery Disconnect and Connect (414-01 Battery, Mounting and Cables, General Procedures)

Oil Pan

Removal and Installation

Removal

1. Raise the vehicle on lift.

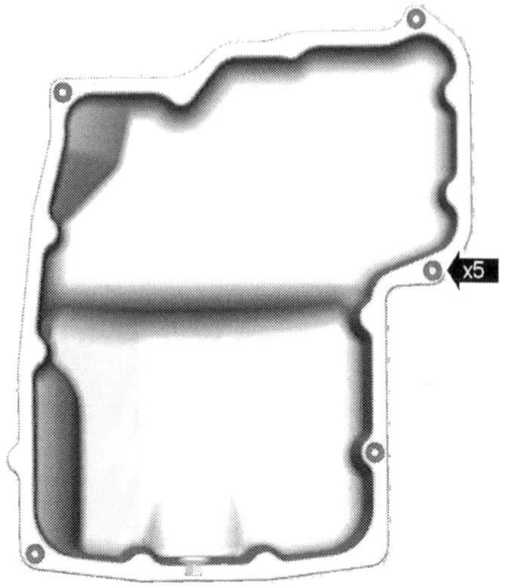

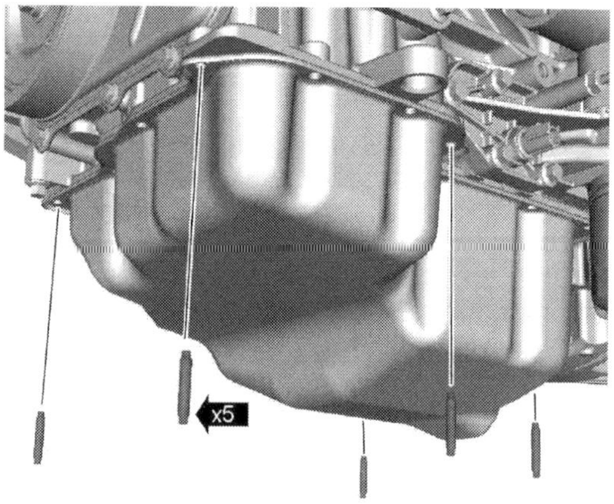

2. Drain the engine oil.

Refer to: Engine Oil Draining and Filling (303-01 Engine - ID4 2.2L Diesel, General Procedures).

3. Remove the 5 oil pan retaining nuts.
4. Remove the 5 oil pan retaining studs.

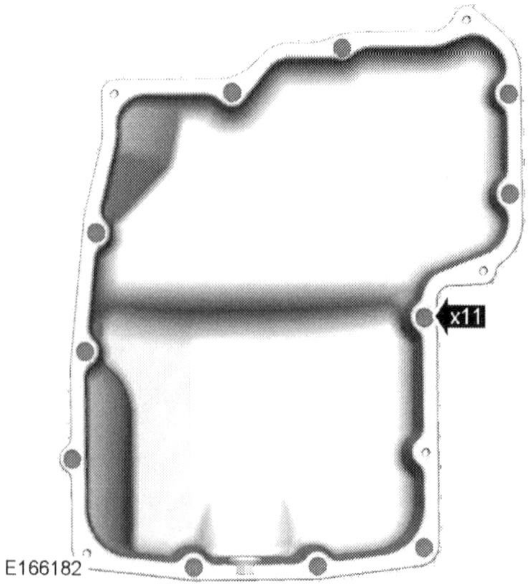

5. Remove the 11 oil pan retaining bolts.

6. CAUTION: Damage to the oil pan may occur during the removal steps, therefore the oil pan must be discarded.

Using a suitable tool release the oil pan in the area indicated.

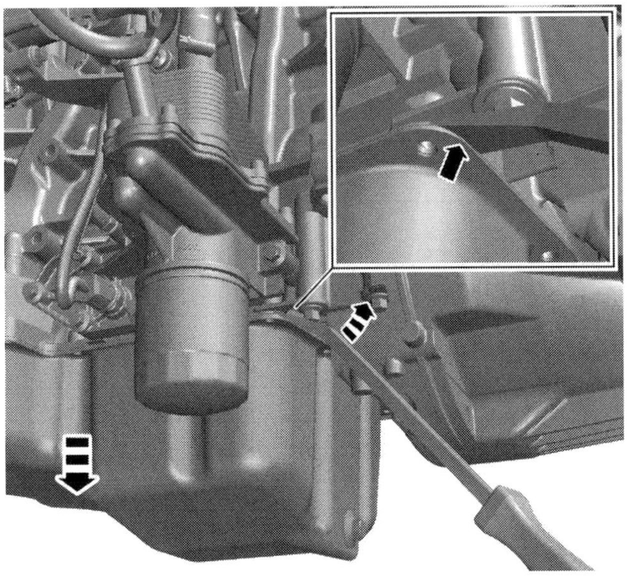

Installation

1. CAUTION: Avoid damage to the oil pan mating face of the cylinder block. Failure to follow this instruction may cause damage to the vehicle.

Make sure that all traces of the old sealant are removed from the mating faces.

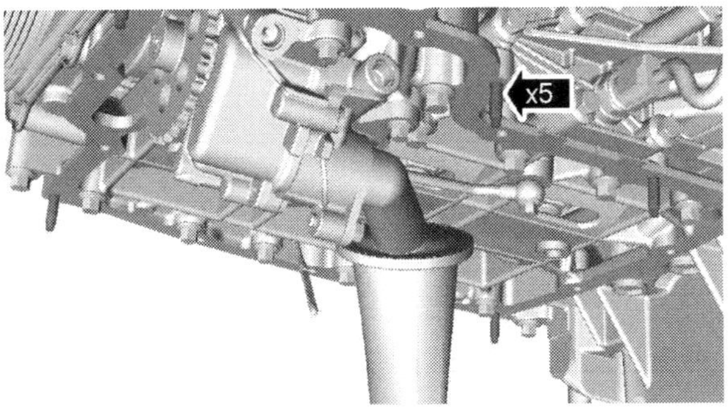

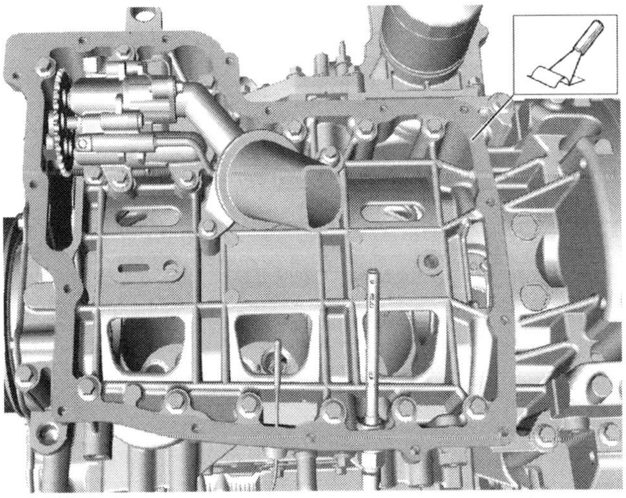

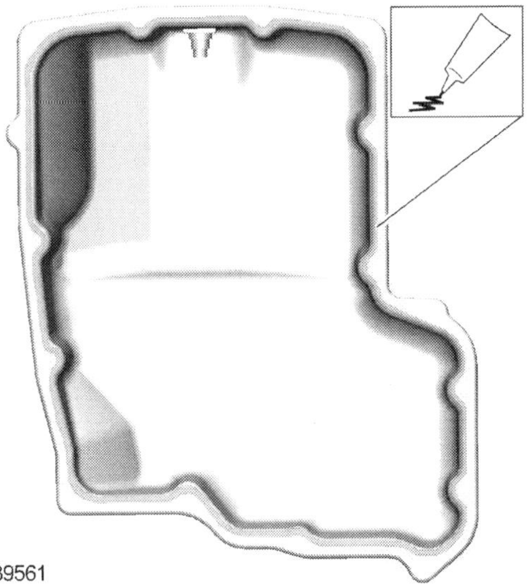

2. Install the 5 oil pan retaining studs.

3. CAUTION: Make sure that a new oil pan is installed.

NOTE: Clean the cylinder block mating faces.

Apply a 3mm bead of sealer WSE-M4G323-A4 to the oil pan.

4. NOTE: Tighten the bolts in the sequence shown.

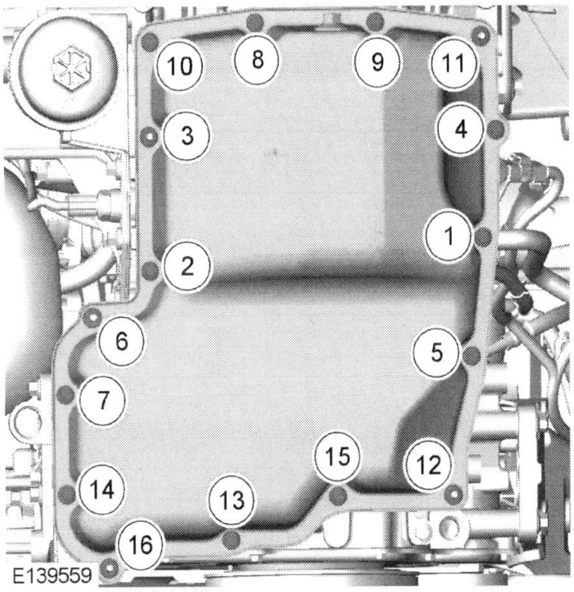

Install the oil pan.

1. Tighten the oil pan bolts.
2. Stage 1: Torque to 7 Nm
3. Stage 2: Torque to 14 Nm

5. Lower the vehicle.
6. Check and top-up the engine oil.

Refer to: Engine Oil Draining and Filling (303-01 Engine - ID4 2.2L Diesel, General Procedures).

Oil Pump

Removal and Installation

Part(s)

STEP	PART NAME	QUANTITY
Installation Step 5	Renew Part Oil pump inlet pipe	1

Removal

1. **Refer to:** Startup Battery Disconnect and Connect (414-01 Battery, Mounting and Cables, General Procedures)

2. WARNING: Do not work on or under a vehicle supported only by a jack. Always support the vehicle on safety stands.

   Raise the vehicle on lift.

3. Remove the engine oil pan.

   Refer to: Oil Pan (303-01 Engine, ID4 2.2L Diesel Removal and Installation)

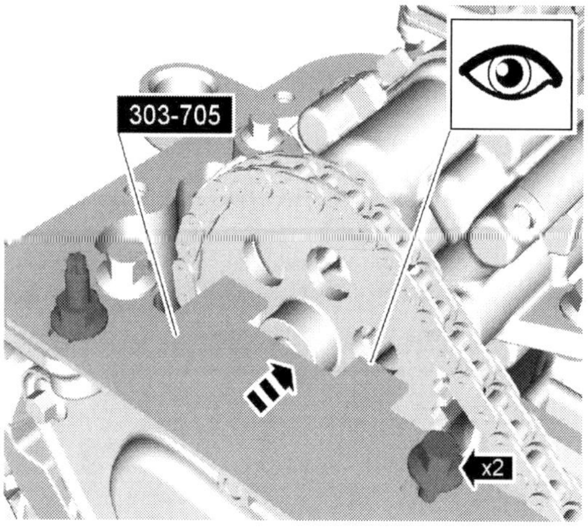

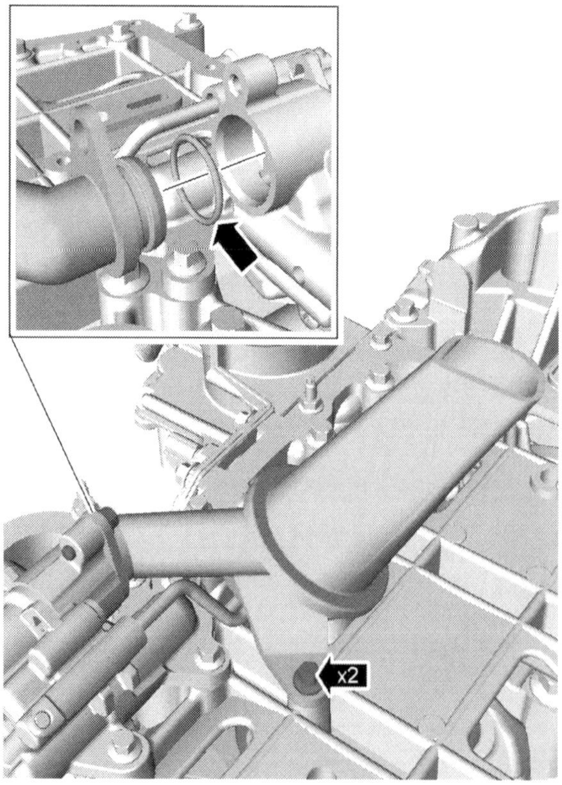

4. NOTE: Install the special tool so that the tool face aligns with the oil pump sprocket.

   Install the special tool to the engine block.

5. Remove and discard the O-ring seal.

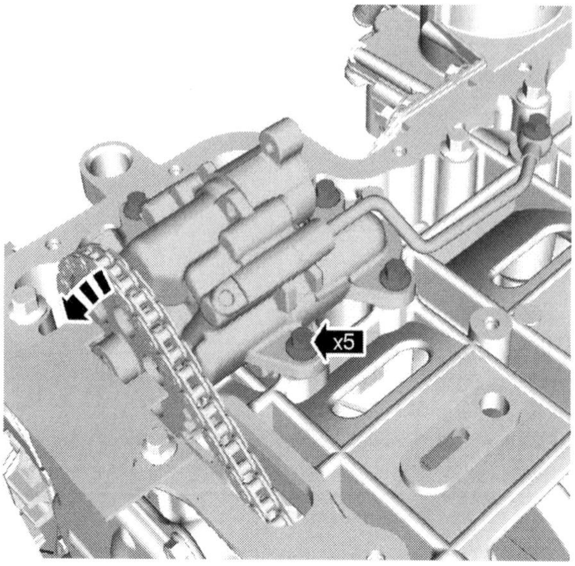

6. Remove the oil pump.

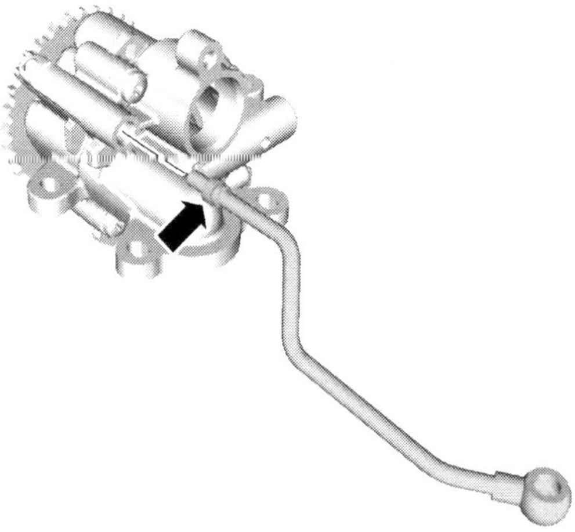

7. NOTE: Do not disassemble further if the component is removed for access only.

   Remove the oil pump outlet pipe.

Installation

1. Install the oil pump outlet pipe.

2. NOTE: Do not tighten the bolts at this stage.

   Install the oil pump so the sprocket touches the special tool.

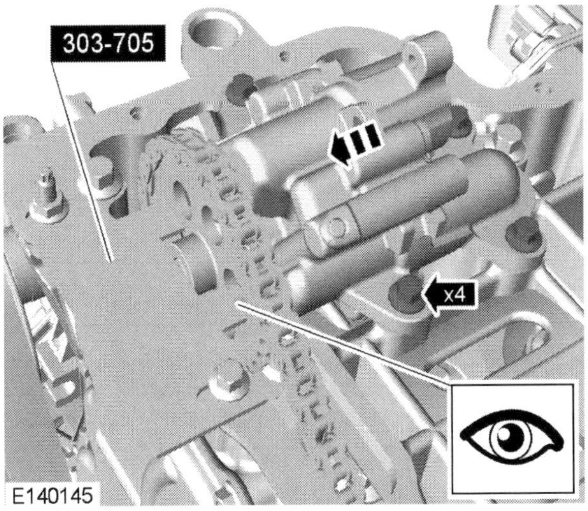

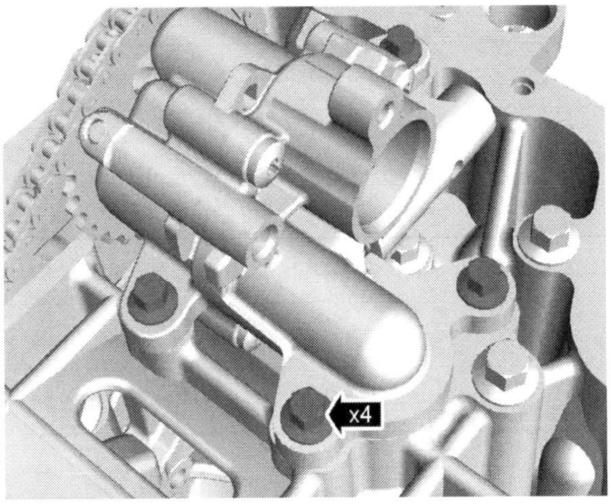

3. *Torque:* 10 Nm
4. Remove the special tool.
5. Install the new O-ring seal.

   Renew Part: Oil pump inlet pipe

   Torque: 10 Nm
6. *Torque:* 10 Nm
7. Install the engine oil pan.

Refer to: Oil Pan (303-01 Engine, ID4 2.2L Diesel Removal and Installation)

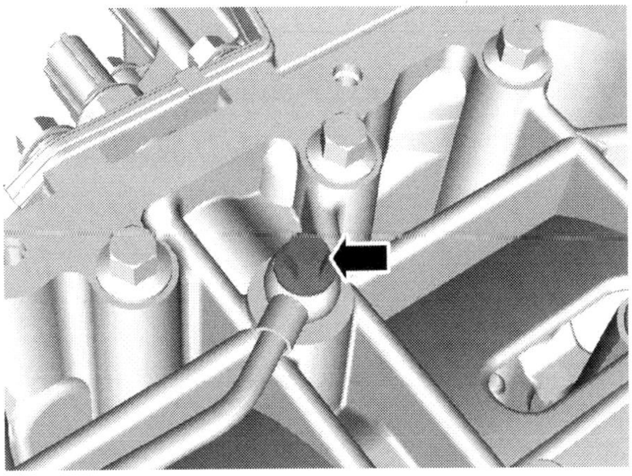

8. **Refer to:** Startup Battery Disconnect and Connect (414-01 Battery, Mounting and Cables, General Procedures)

Timing Chain

Removal and Installation

Special Tool(s)

 E61755	Locking Tool, Fuel Pump Sprocket 303-1151
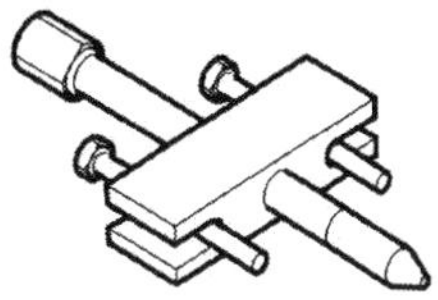 303-1149 E54428	Remover Crankshaft Damper Pulley 303-1149
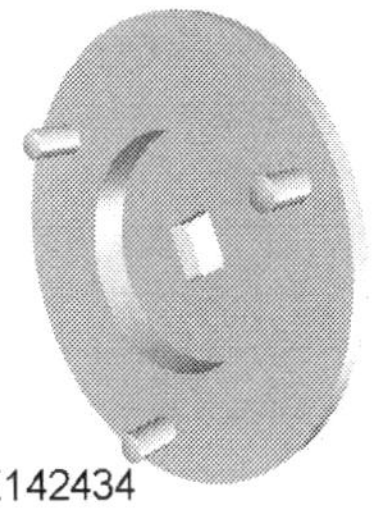 E142434	Remover Crankshaft Damper Oil Seal JLR 303-1619
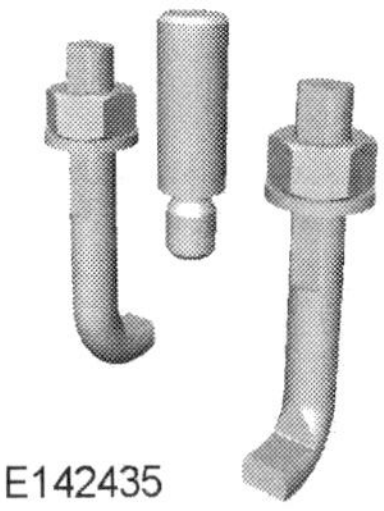 E142435	Locking Tool, Fuel Pump Sprocket JLR 303-249-01
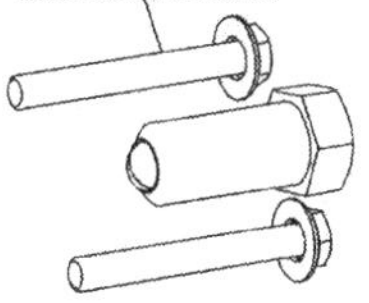 303-1333 E91082	Remover Fuel Pump 303-1333

Removal

1. CAUTION: Do not work on or under a vehicle supported only by a jack. Always support the vehicle on safety stands.

   Raise and support the vehicle.
2. **Refer to:** Timing Cover (303-01 Engine, ID4 2.2L Diesel Removal and Installation)
3. **Refer to:** Timing Cover (303-01 Engine, ID4 2.2L Diesel Removal and Installation)
4. Loosen but do not remove the high pressure fuel pump sprocket bolt.

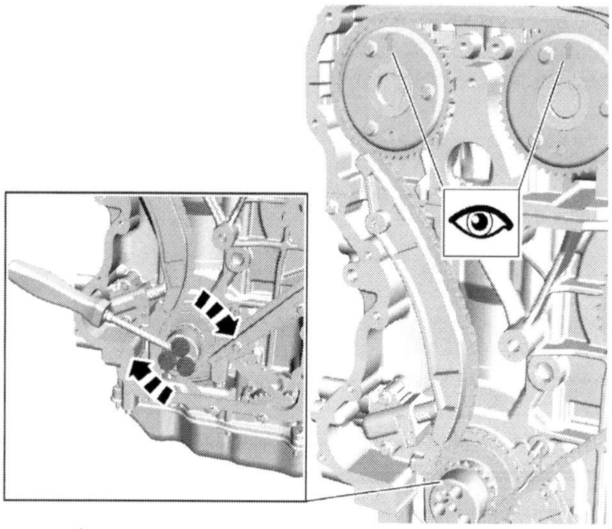

5. CAUTION: Before removing any of the engine timing components rotate the engine so the camshaft sprockets are positioned as shown. Failure to do this may result in damage to the engine.

Position the camshafts.

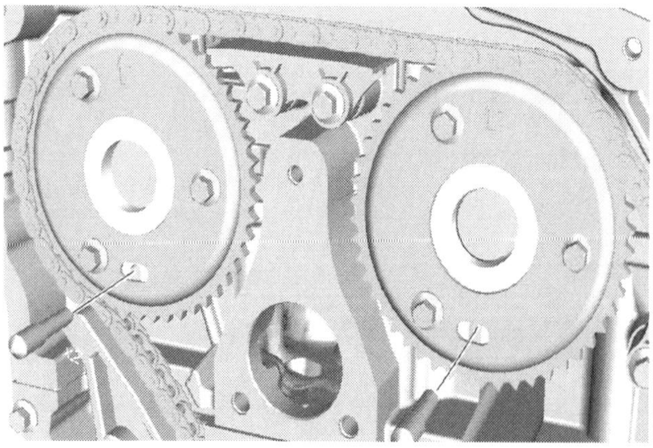

6. Rotate the engine clockwise until the camshaft sprockets are positioned as shown, then install a 6mm drill bit into each sprocket.

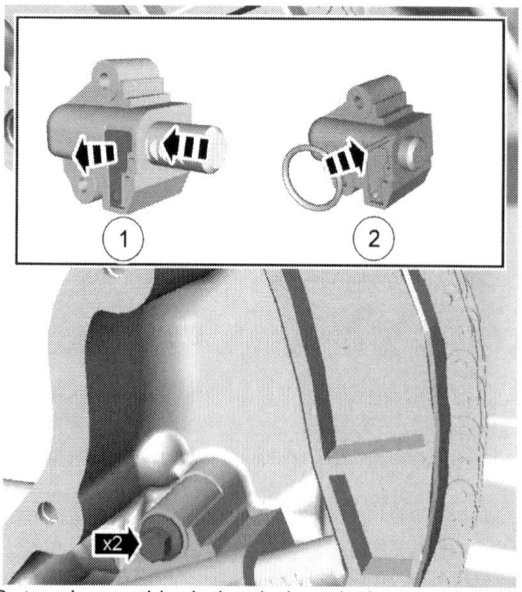

7. Detension and lock the timing chain tensioner, then ......remove.

8. Remove the upper timing chain guide.

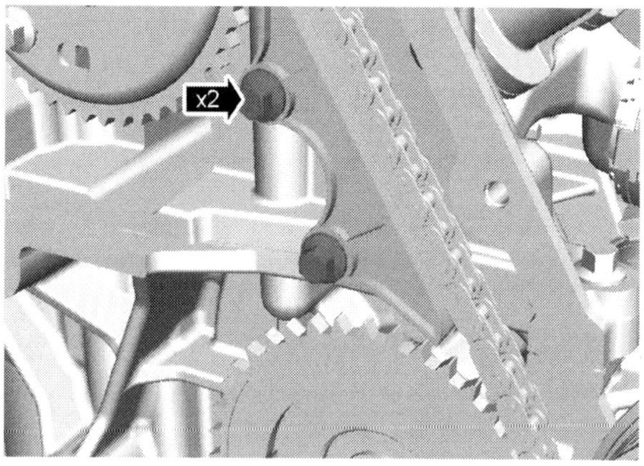

9. Remove the RH timing chain guide.

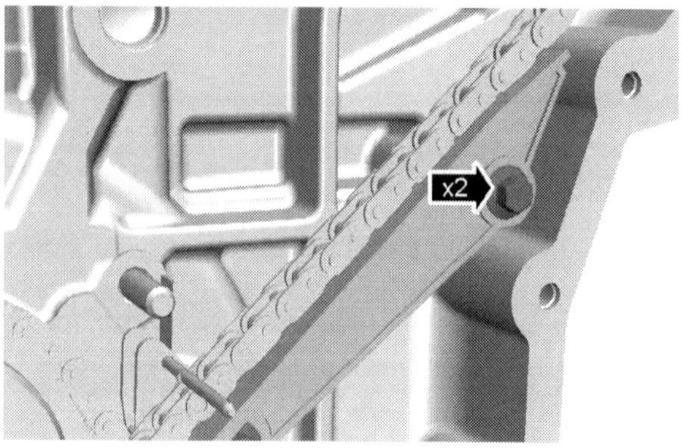

10. Remove the lower timing chain guide.

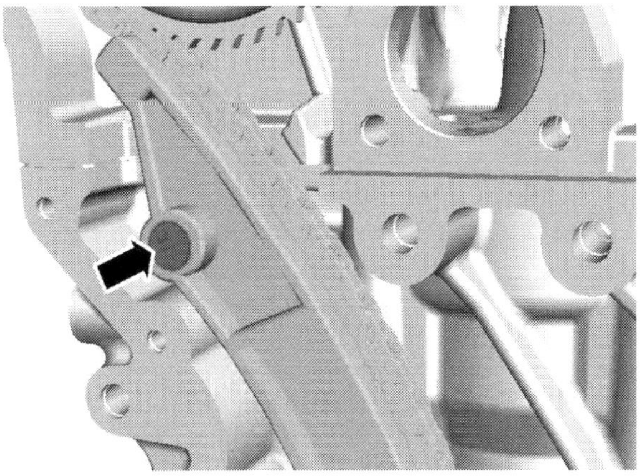

11. Remove the LH timing chain guide.

12. CAUTION: Make sure the timing pins are not removed from the camshaft sprockets.

Remove the timing chain and camshaft sprockets.

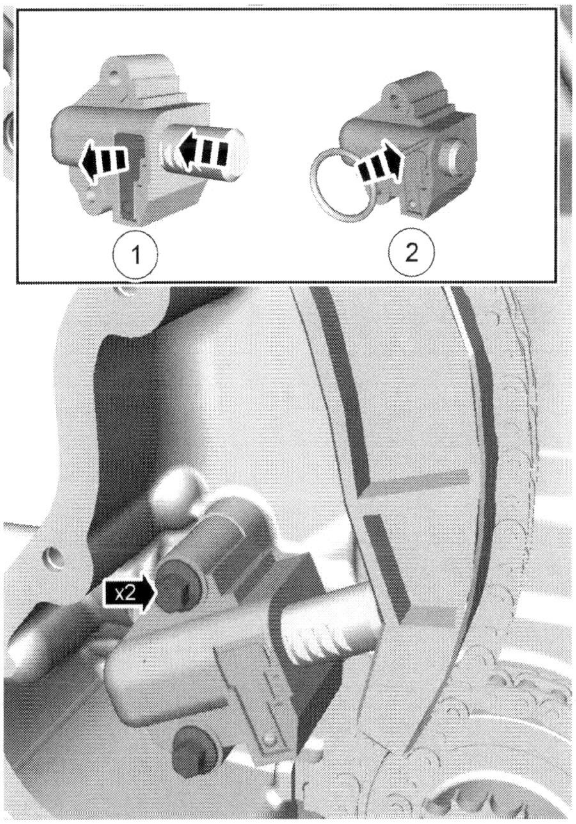

13. Remove the bolt securing the high pressure fuel pump sprocket.

14. Install the special tools.

15. Install the special tools.

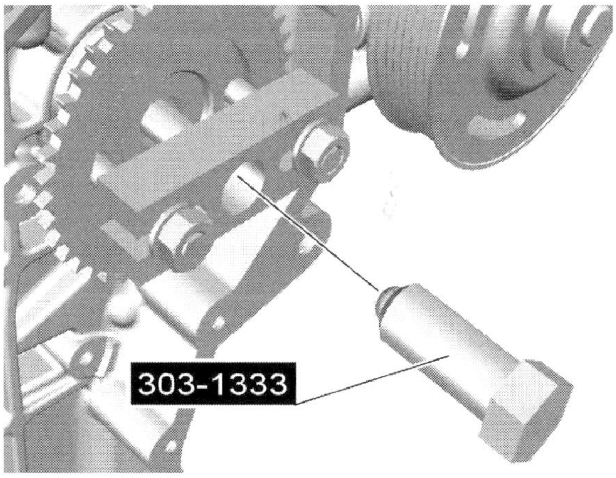

16. Install the special tool.

17. Rotate special tool 303-1333 until the sprocket is released from the high pressure pump.

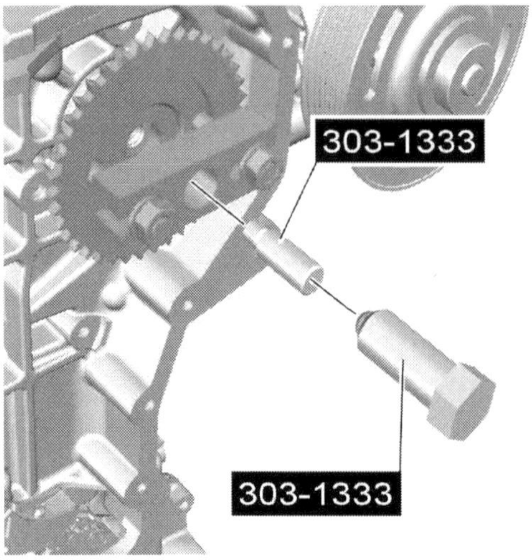

18. Remove the special tools and fuel pump sprocket.

19. Remove the special tools from the fuel pump sprocket.

Installation

1. NOTES:

△ Do not tighten the bolt at this stage.

△ The fuel pump sprocket bolt has a LH thread.

Install the high pressure fuel pump sprocket.

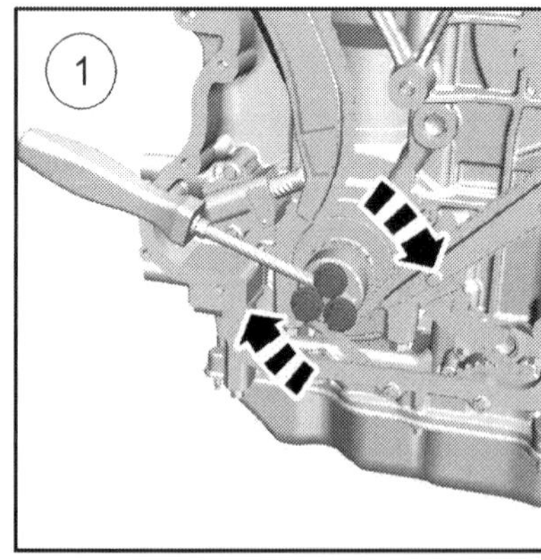

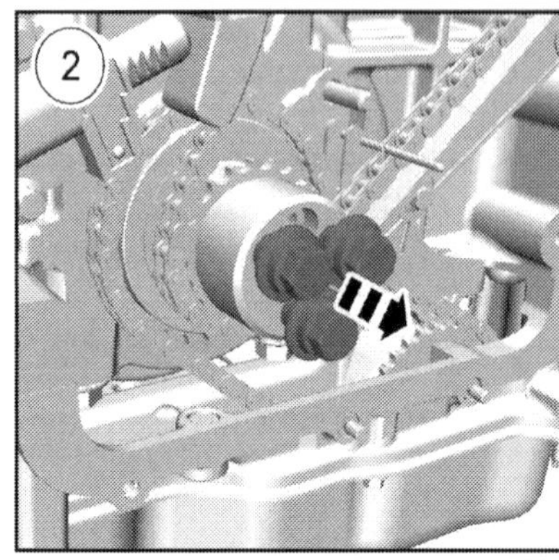

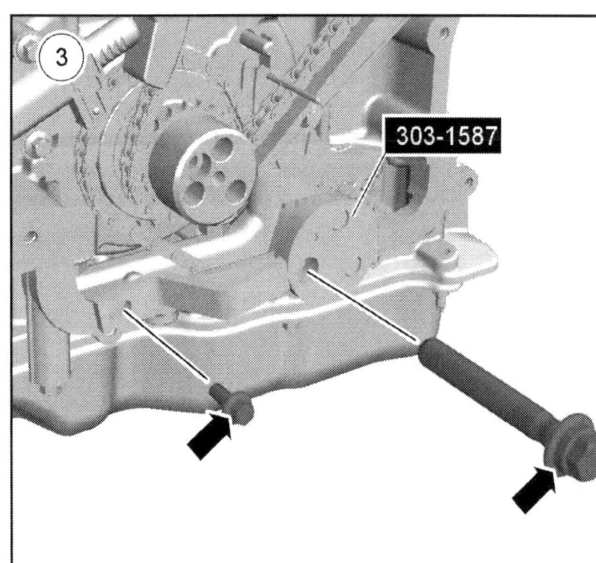

2. CAUTIONS:

⚠ CAUTION: Use a suitable tool to rotate the engine.

⚠ CAUTION: Make sure special tool 303-1587 is not used as a reaction tool. Failure to follow this instruction may result in damage to the engine.

△ NOTE: Do not tighten the bolts at this stage.

Using a suitable tool rotate the engine and align special tool 303-1587 to the crankshaft.

3. ⚠ WARNING: WARNING: Make sure the crankshaft is not rotated with special tool 303-1587 removed.

⚠ CAUTION: Make sure special tool 303-1587 is not used as a reaction tool. Failure to follow this instruction may result in damage to the engine.

Remove the crankshaft timing tool.

4. CAUTIONS:

   ⚠ CAUTION: Make sure the timing pins are not removed from the camshaft sprockets.

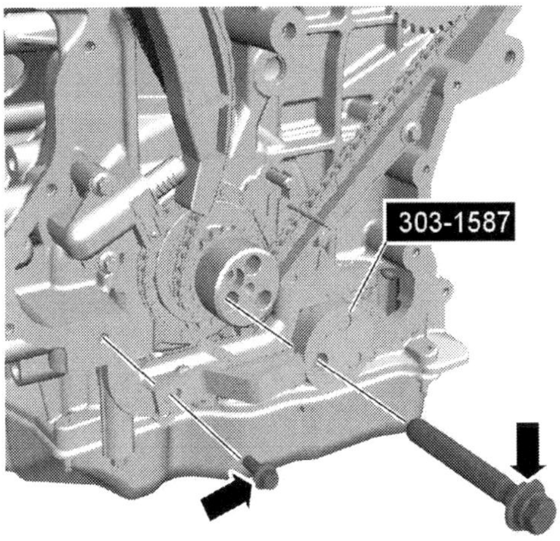

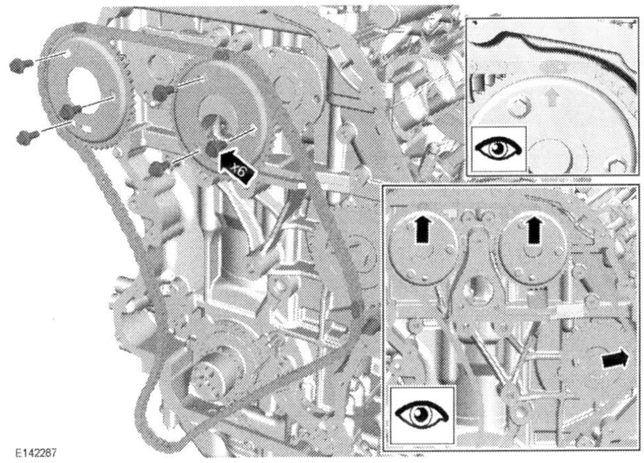

⚠ CAUTION: Make sure the coloured chain links on the timing chain are positioned as shown.

NOTE: Do not tighten the bolts at this stage. Install and align the timing chain.

5. ⚠ CAUTION: Make sure special tool 303-1587 is not used as a reaction tool. Failure to follow this instruction may result in damage to the engine.

Install the crankshaft timing tool and torque the bolts to 10 Nm.

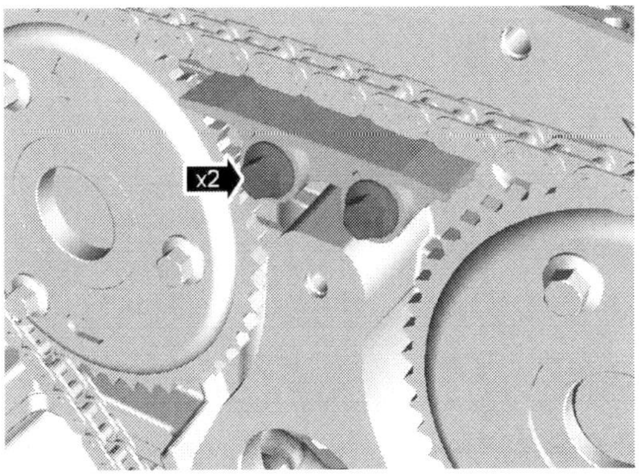

6. Install the upper timing chain guide and torque the bolts to 15 Nm.

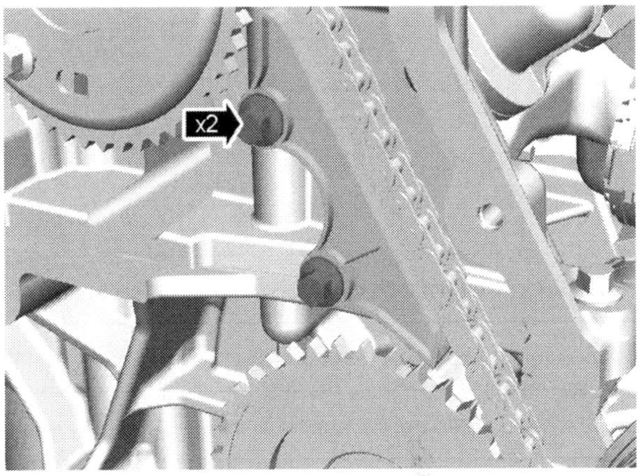

7. Install the RH timing chain guide and torque the bolts to 15 Nm.

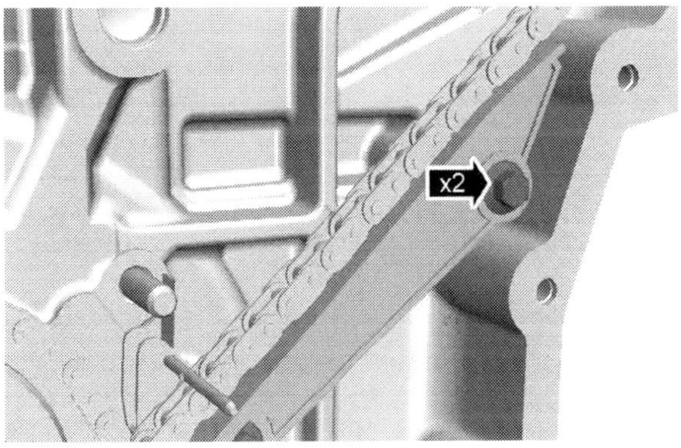

8. Install the lower timing chain guide and torque the bolts to 15 Nm.

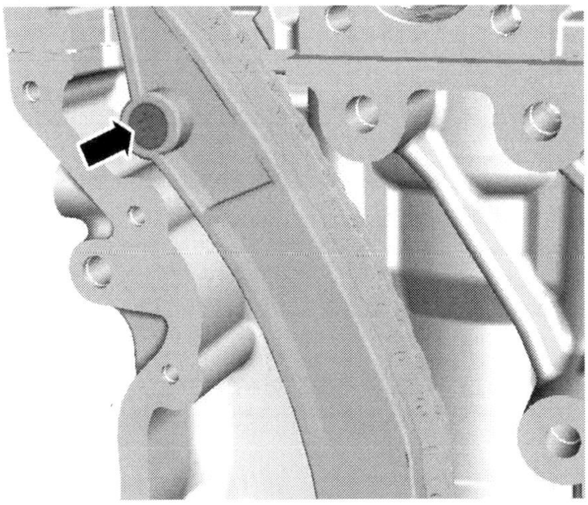

9. Install the LH chain guide and torque the bolt to 31 Nm.

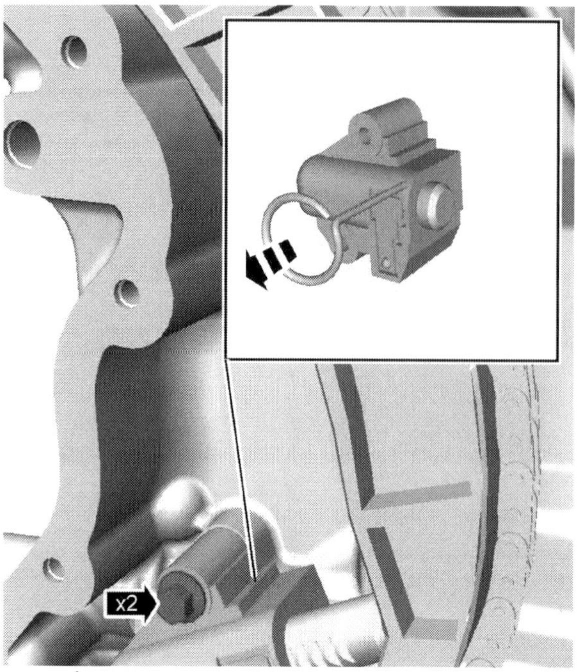

10. WARNING: A new timing chain tensioner must be installed.

    Install the timing chain tensioner and torque the bolts to 15 Nm. Remove the pin to deploy the tensioner.

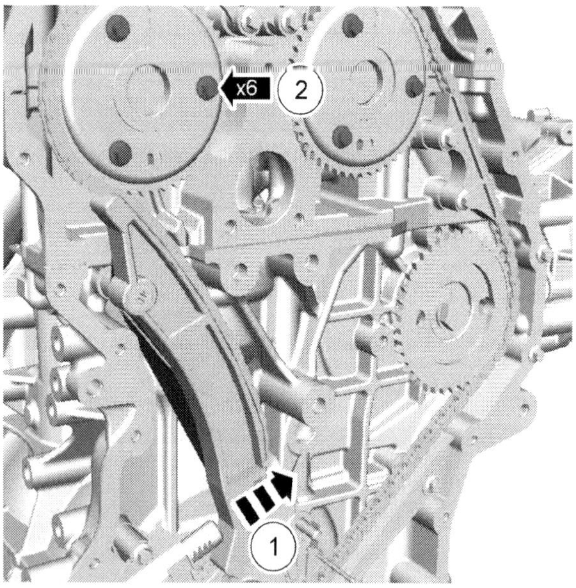

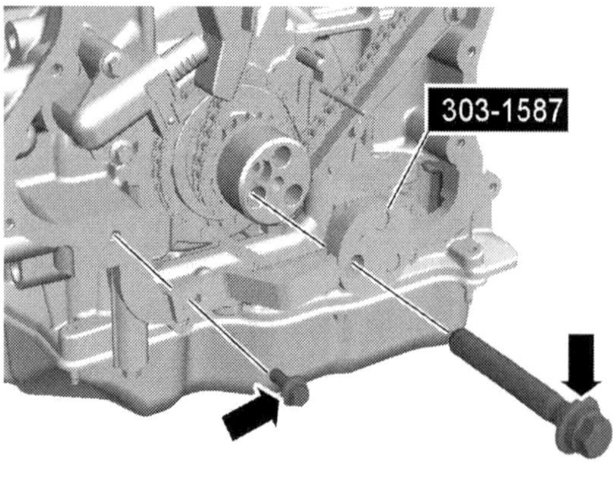

11. Apply tension to the chain guide then torque the camshaft sprocket bolts to 35 Nm.

12. CAUTION: Make sure special tool 303-1587 is not used as a reaction tool. Failure to follow this instruction may result in damage to the engine.

Remove the crankshaft timing tool.

13. Remove the drill bits from the camshaft sprockets.

14. CAUTION: Use a suitable tool to rotate the engine.

    Rotate the crankshaft 2 revolutions.

15. Check the camshaft sprockets are aligned correctly as shown.

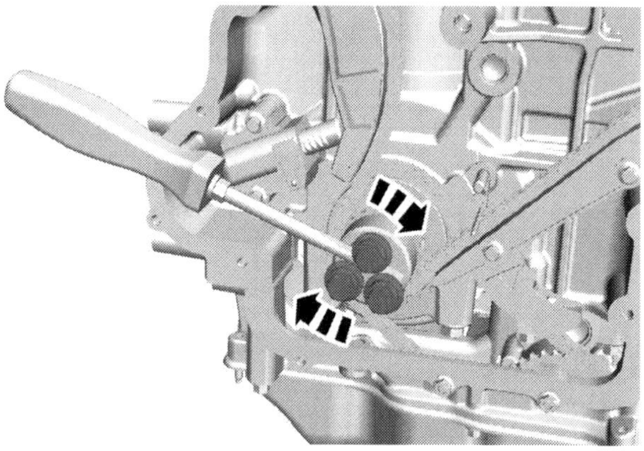

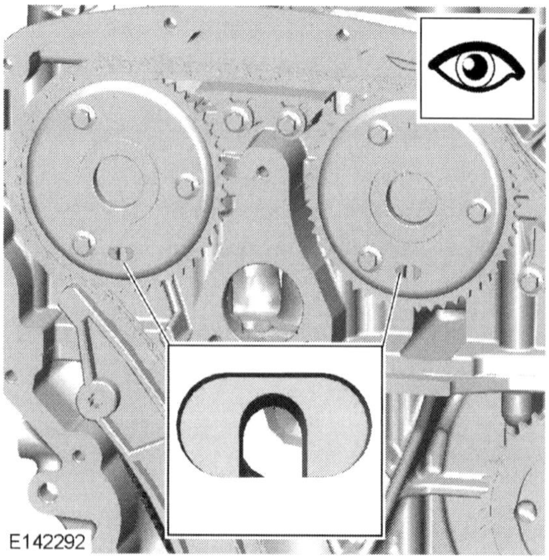

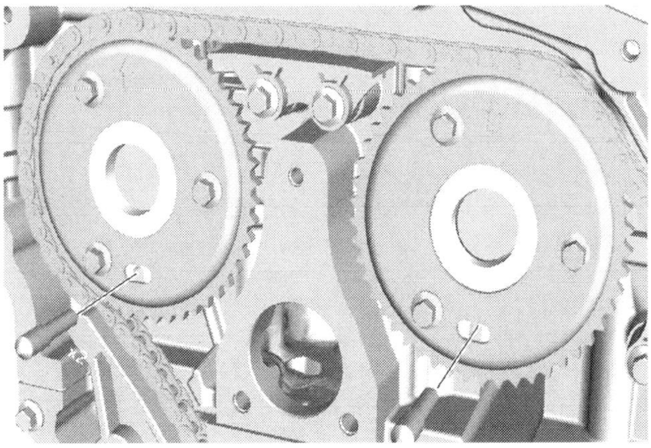

16. Check the timing is correct by installing and removing the drill bits in the camshaft sprockets.

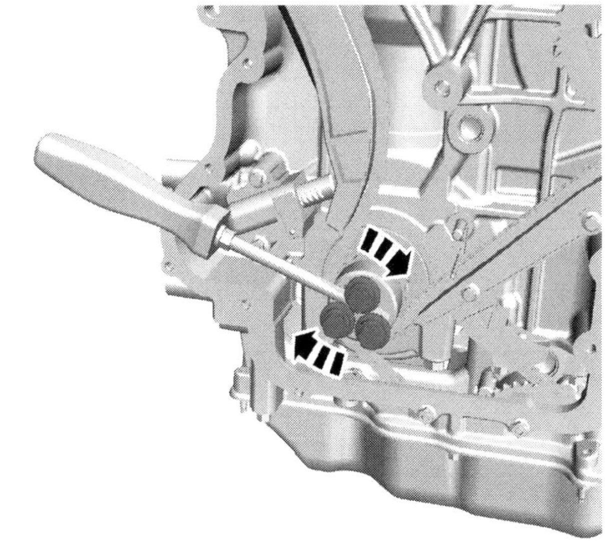

17. CAUTION: Use a suitable tool to rotate the engine.

Check the timing is correct by installing and removing the crankshaft timing tool. Torque the crankshaft locking tool bolts to 10 Nm.

18. **Refer to:** Timing Cover (303-01 Engine, ID4 2.2L Diesel Removal and Installation)

19. **Refer to:** Timing Cover (303-01 Engine, ID4 2.2L Diesel Removal and Installation)

Valve Cover

Removal and Installation

Removal

WARNINGS:

Do not smoke or carry lighted tobacco or open flame of any type when working on or near any fuel related components. Highly flammable vapors are always present and may ignite. Failure to follow these instructions may result in personal injury.

The spilling of fuel is unavoidable during this operation. Make sure that all necessary precautions are taken to prevent fire and explosion.

Do not carry or operate cellular phones when working on or near any fuel related components. Highly flammable vapors are always present and may ignite. Failure to follow these instructions may result in personal injury.

If fuel contacts the eyes, flush the eyes with cold water or eyewash solution and seek immediate medical attention.

Wash hands thoroughly after fuel handling, as prolonged contact may cause irritation. Should irritation develop, seek medical attention.

Do not carry out any repairs to the fuel system with the engine running. The fuel pressure within the system can be as high as 2000 bar (29,008 lb-sq-in). Failure to follow this instruction may result in personal injury.

CAUTIONS:

Diesel fuel injection equipment is manufactured to very precise tolerances and fine clearances. It is therefore essential that absolute cleanliness is observed when working with these components. Always install new blanking plugs to any open orifices or lines. Failure to follow this instruction may result in foreign matter ingress to the fuel injection system.

Before disconnecting any components, make sure the area is clean and free from foreign material. When disconnected all openings must be sealed.

Do not disconnect the fuel injector electrical connectors with the engine running. Failure to follow this instruction may result in damage to the engine.

1. **Refer to:** Startup Battery Disconnect and Connect (414-01 Battery, Mounting and Cables, General Procedures)

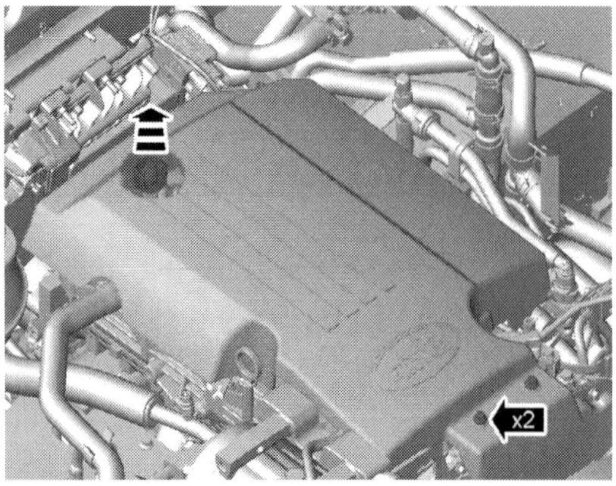

2.

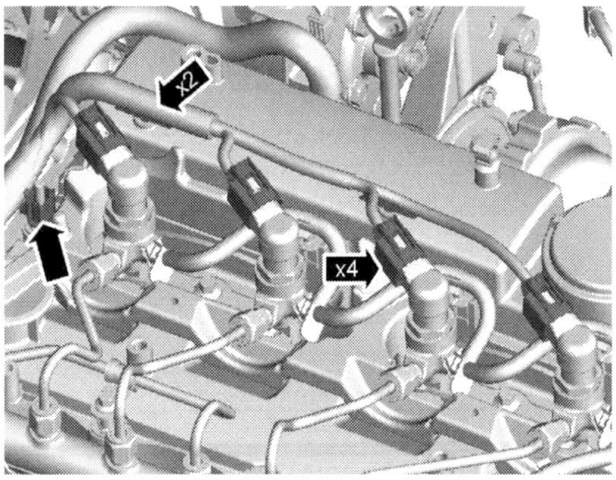

3.

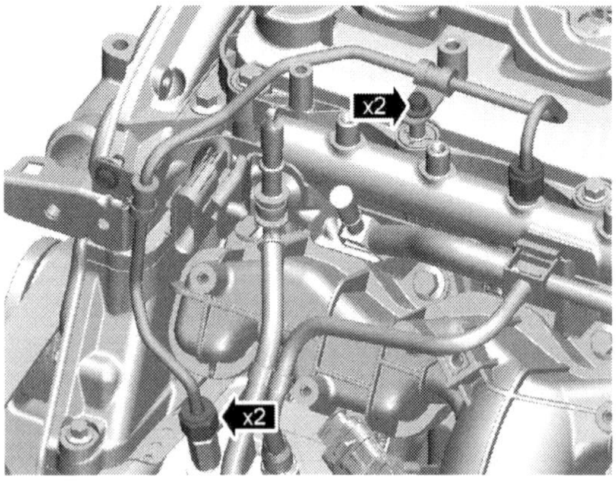

4. WARNING: The spilling of fuel is unavoidable during this operation. Make sure that all necessary precautions are taken to prevent fire and explosion.

CAUTIONS:

Before disconnecting any part of the fuel system, it is imperative that all dust, dirt and debris is removed from around components to prevent ingress of foreign matter into fuel system.

Before disconnecting any components, make sure the area is clean and free from foreign material. When disconnected all openings must be sealed.

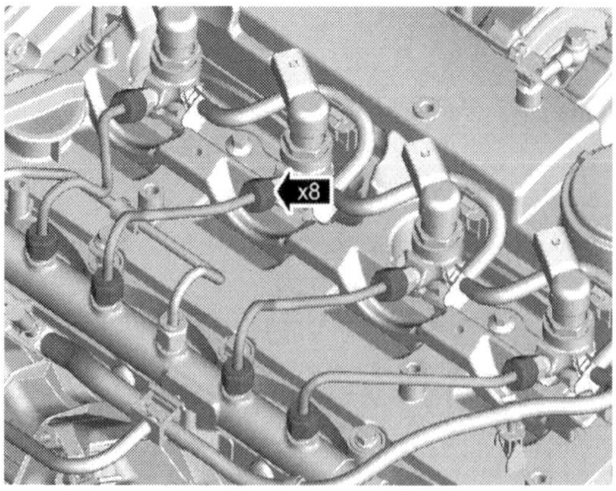

5. WARNING: The spilling of fuel is unavoidable during this operation. Make sure that all necessary precautions are taken to prevent fire and explosion.

CAUTIONS:

Before disconnecting any part of the fuel system, it is imperative that all dust, dirt and debris is removed from around components to prevent ingress of foreign matter into fuel system.

Before disconnecting any components, make sure the area is clean and free from foreign material. When disconnected all openings must be sealed.

6. WARNING: The spilling of fuel is unavoidable during this operation. Make sure that all necessary precautions are taken to prevent fire and explosion.

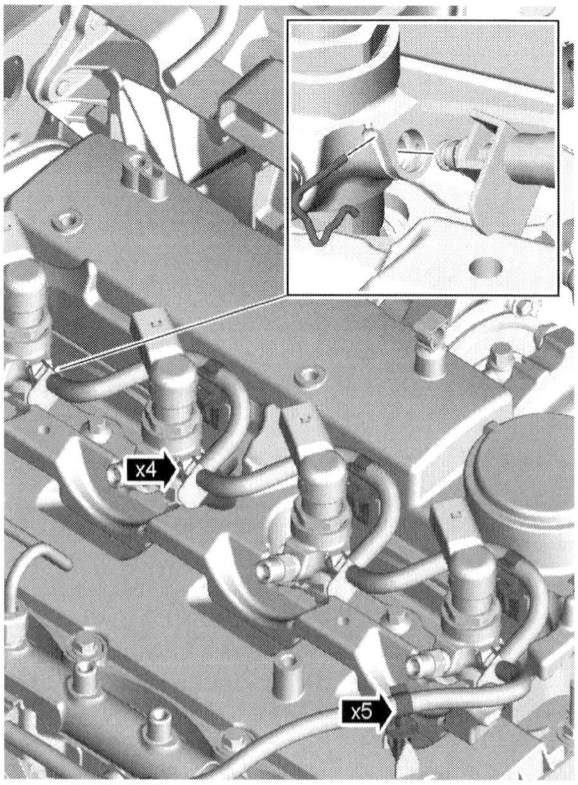

7. CAUTIONS:

⚠ Before disconnecting any components, make sure the area is clean and free from foreign material. When disconnected all openings must be sealed.

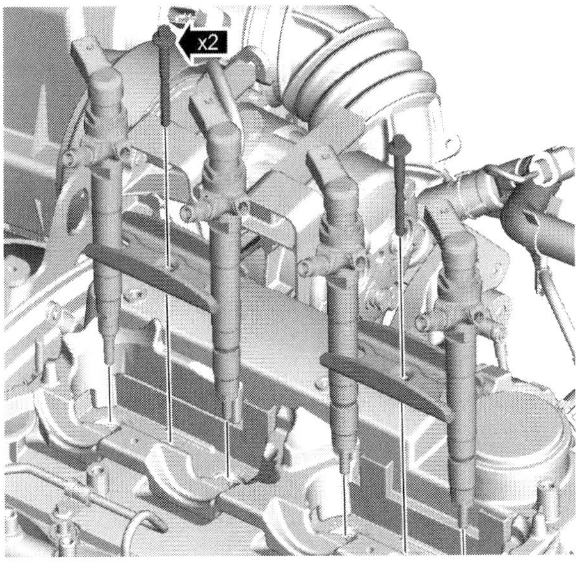

⚠ Before disconnecting any part of the fuel system, it is imperative that all dust, dirt and debris is removed from around components to prevent ingress of foreign matter into fuel system.

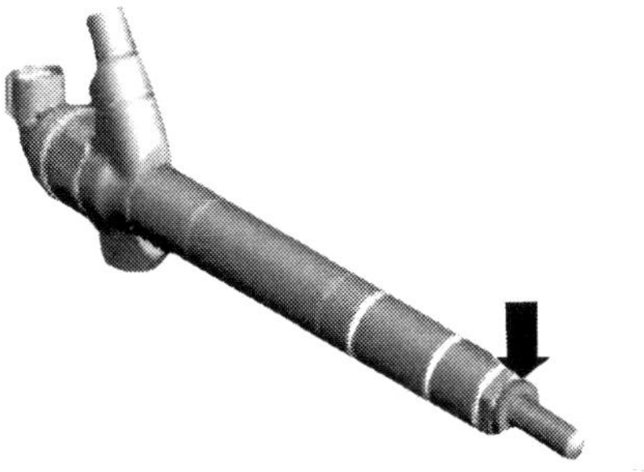

8. ⚠ NOTE: Remove the fuel injector sealing washers.

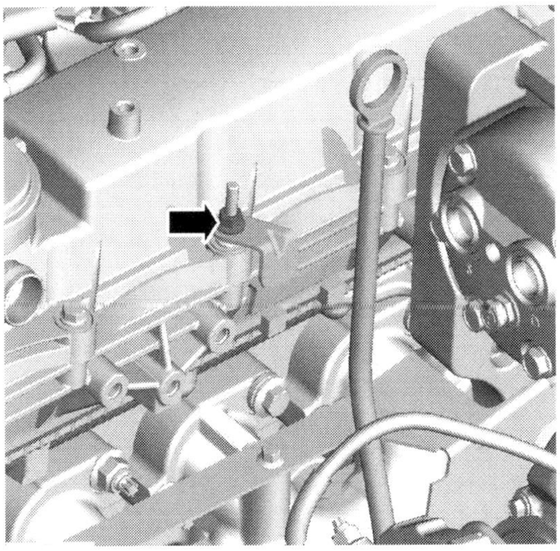

9.

10.

Installation

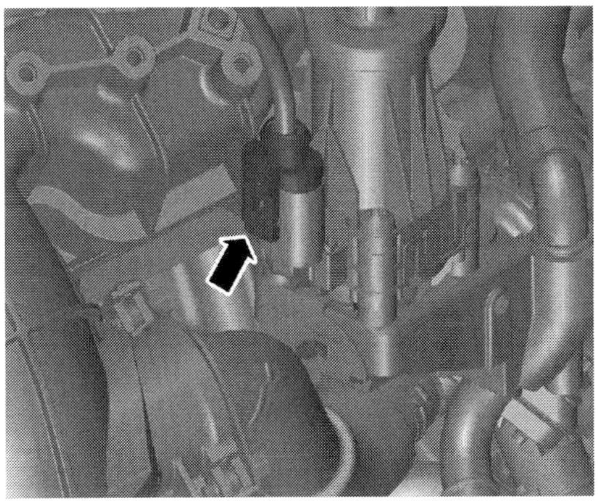

1. Torque: 4 Nm Torque: 11 Nm

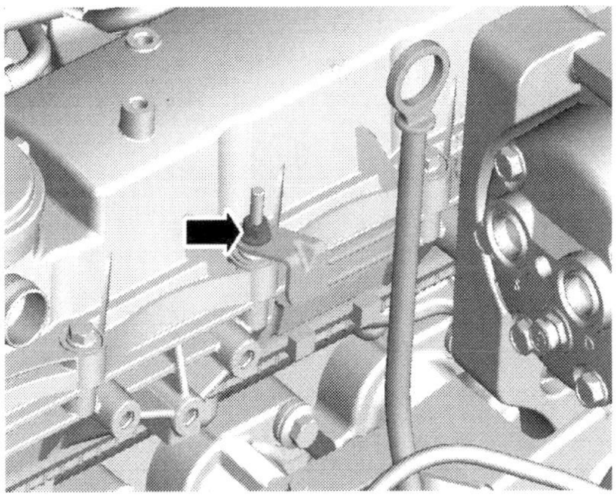

2. Torque: 23 Nm

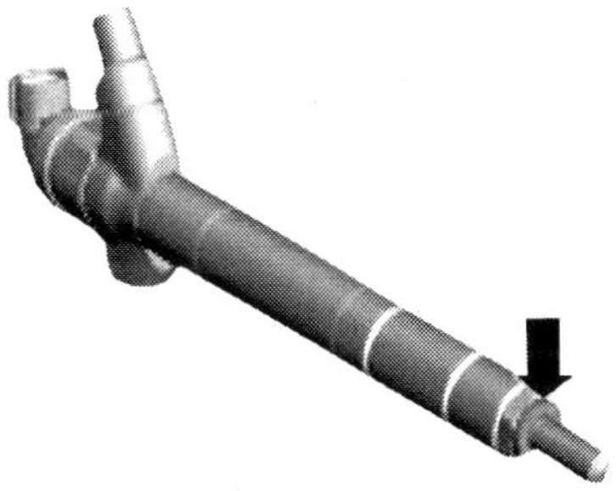

3. ⚠ NOTE: Install new fuel injector sealing washers.

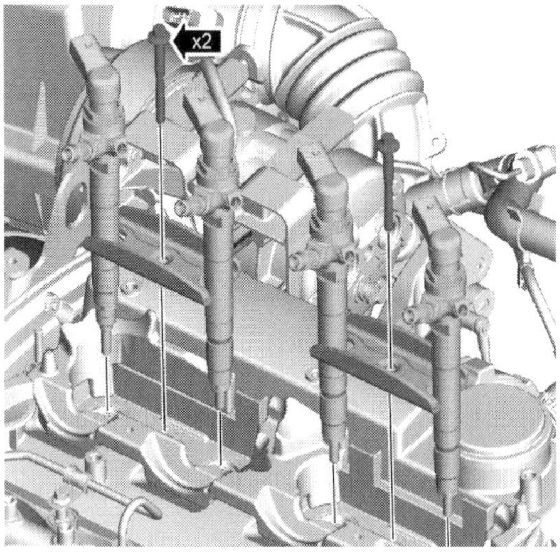

4. CAUTIONS:

Diesel fuel injection equipment is manufactured to very precise tolerances and fine clearances. It is therefore essential that absolute cleanliness is observed when working with these components. Always install new blanking plugs to any open orifices or lines. Failure to follow this instruction may result in foreign matter ingress to the fuel injection system.

Make sure blanking plugs are removed before component is installed

Torque: 6 Nm Torque: 360°

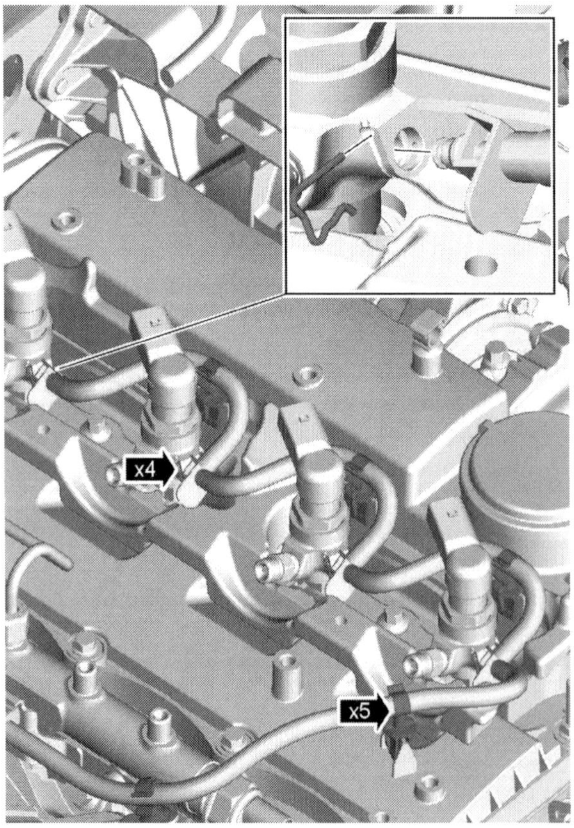

5. CAUTION: Make sure blanking plugs are removed before component is installed.

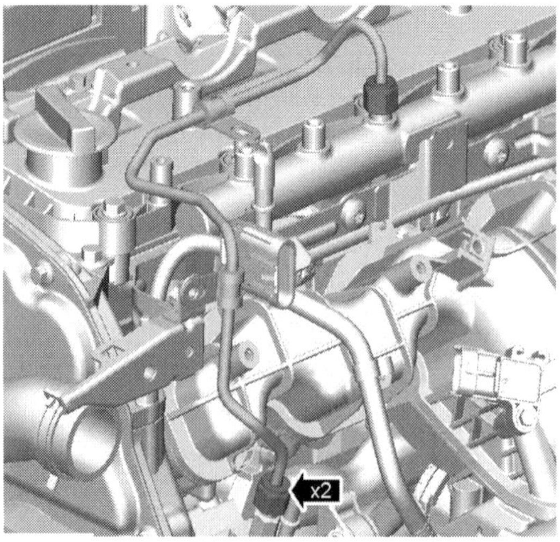

6. CAUTION: Make sure blanking plugs are removed before component is installed.

Torque: 15 Nm

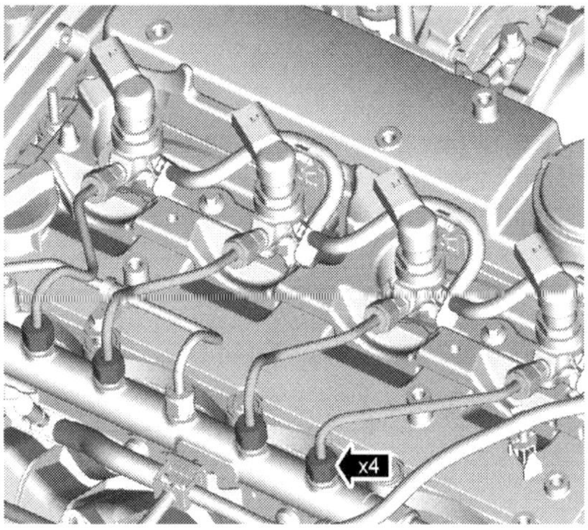

7. CAUTION: Make sure blanking plugs are removed before component is installed.

Torque: 15 Nm

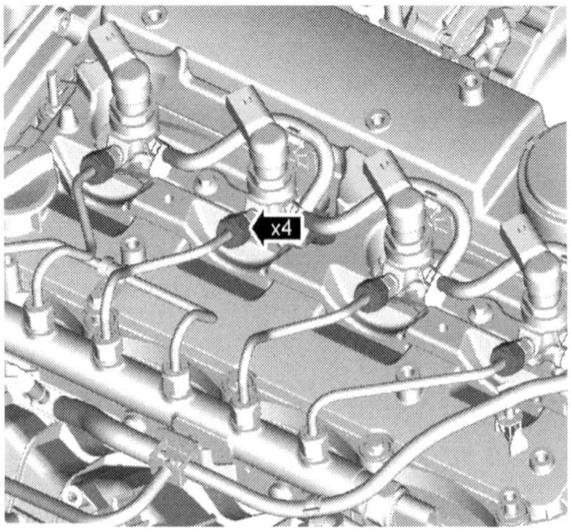

8. CAUTION: Make sure blanking plugs are removed before component is installed.

Torque: 15 Nm

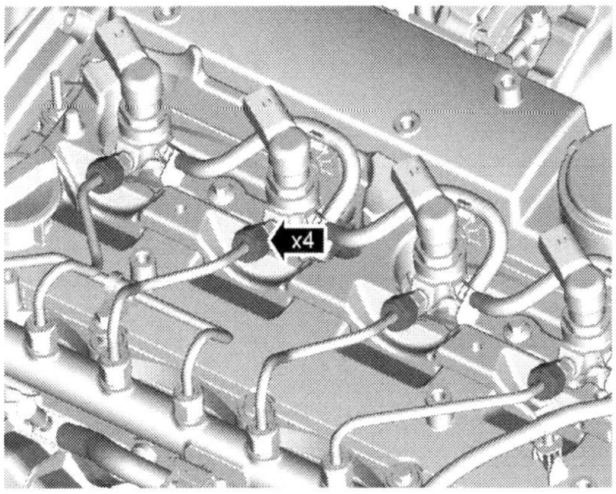

9. *Torque:* 30 Nm

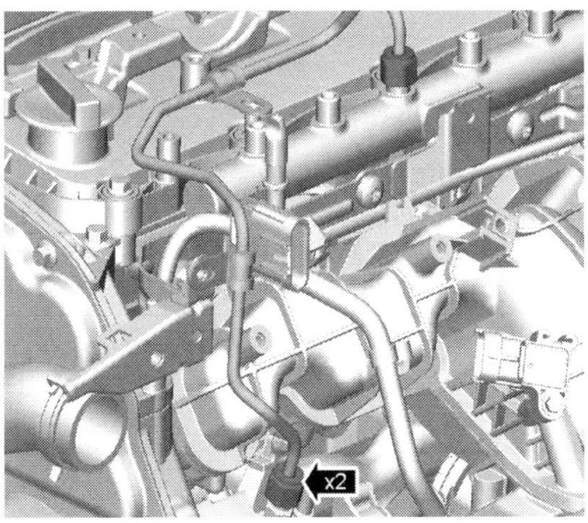

10. *Torque:* 30 Nm

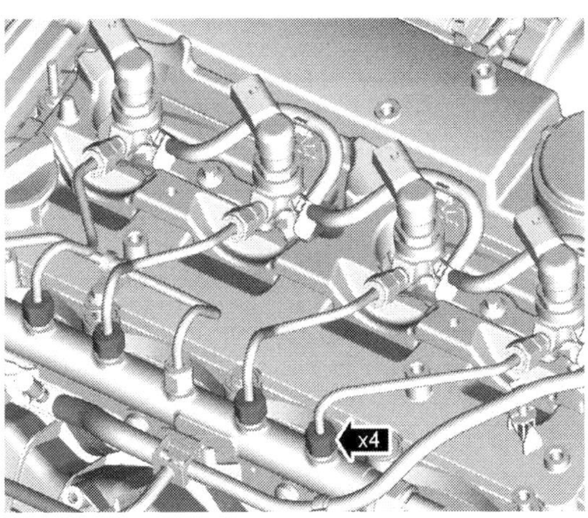

11. *Torque:* 30 Nm

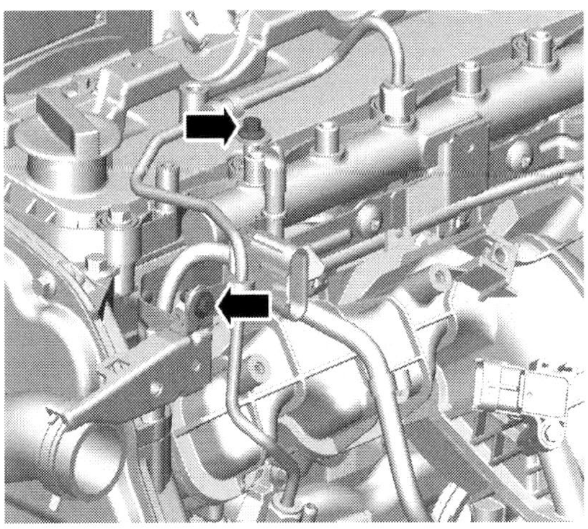

12. *Torque:* : 10 Nm

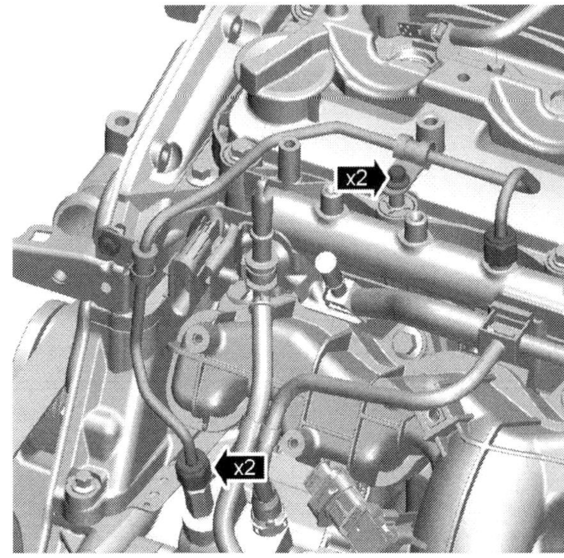

13.

14. **Refer to:** Startup Battery Disconnect and Connect (414-01 Battery, Mounting and Cables, General Procedures).

15. △ NOTE: Check fuel pipes for leaks after starting engine.

16. *Torque:* 10 Nm

Flywheel

Removal and Installation

Removal

1. **Refer to:** Startup Battery Disconnect and Connect (414-01 Battery, Mounting and Cables, General Procedures)

2. WARNING: Do not work on or under a vehicle supported only by a jack. Always support the vehicle on safety stands.

Raise and support the vehicle.

3. **Refer to:** Clutch Disc and Pressure Plate (308-01 Clutch, Vehicles With: MT82 6-Speed Manual Transmission Removal and Installation)

Installation

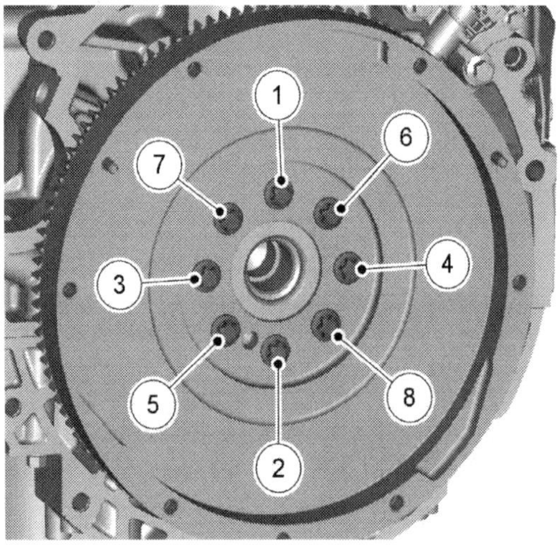

1. NOTES:

Clean the component mating faces.

New bolts must be installed.

Tighten the bolts in the sequence shown.

Install the flywheel.

1. Stage 1: Tighten bolts 1 to 8 to 25 Nm (18 lb.ft).
2. Stage 2: Tighten bolts 1 to 8 to 45 Nm (33 lb.ft).
3. Stage 3: Tighten bolts 1 to 8 a further 45 degrees.

2. **Refer to:** Clutch Disc and Pressure Plate (308-01 Clutch, Vehicles With: MT82 6-Speed Manual Transmission Removal and Installation).

3. **Refer to:** Startup Battery Disconnect and Connect (414-01 Battery, Mounting and Cables, General Procedures).

Oil Filter Housing

Removal and Installation

Part(s)

STEP	PART NAME	QUANTITY
Removal Step 7	Renew Part Oil cooler seal	1

Removal

1. Disconnect the battery ground cable.

   Refer to: Startup Battery Disconnect and Connect (414-01 Battery, Mounting and Cables, General Procedures.)

2. WARNING: Do not work on or under a vehicle supported only by a jack. Always support the vehicle on safety stands.

   Raise and support the vehicle.

3. Drain the cooling system.

   Refer to: Cooling System Draining, Filling and Bleeding (303-03 Engine Cooling, ID4 2.2L Diesel General Procedures).

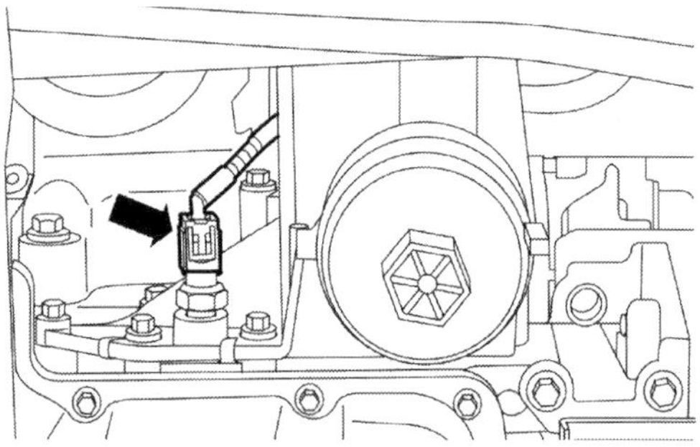

4. Disconnect the engine oil pressure (EOP) sensor electrical connector.

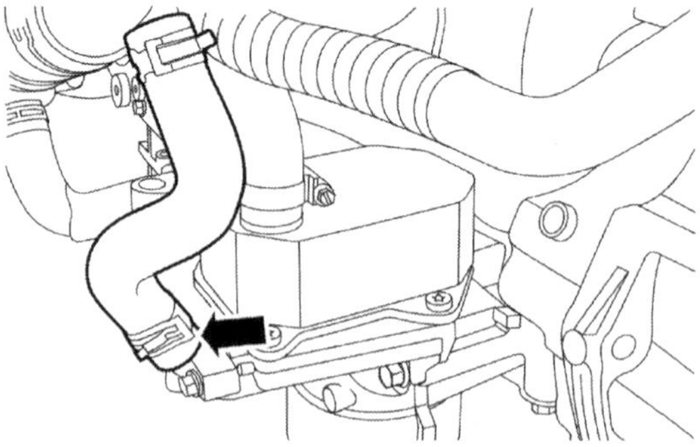

5. Release the clip. Disconnect the coolant hose from the engine oil cooler.

6. Release the clip. Disconnect the coolant hose.

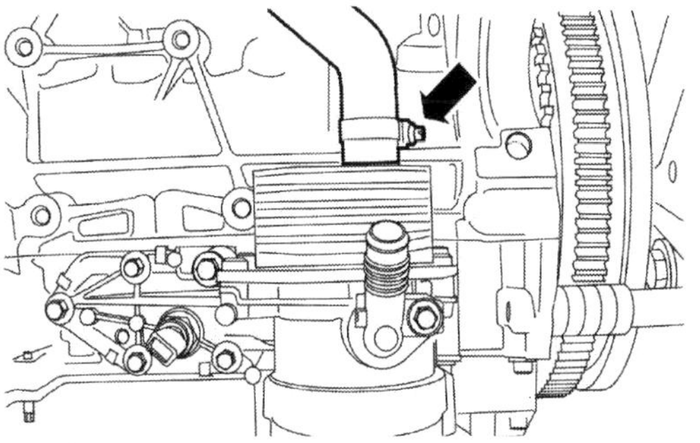

7. CAUTION: Before disconnecting any components, make sure the area is clean and free from foreign material. When disconnected all openings must be sealed.

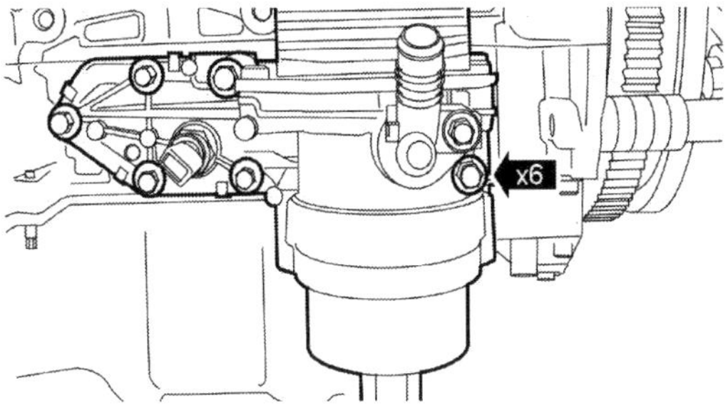

Remove the oil filter housing assembly. Remove the 6 bolts. Remove and discard the seal.

Renew Part: Oil cooler seal

Installation

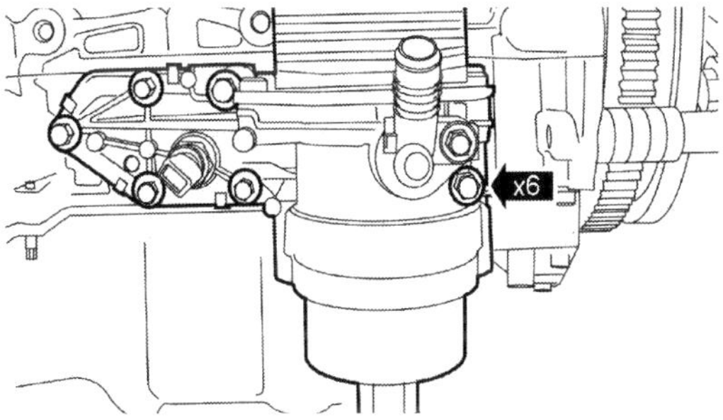

1. NOTES:

Remove and discard the blanking caps.

Clean the component mating faces.

Install a new seal.

To install, reverse the removal procedure. Tighten the bolts.

Torque: 23 Nm

2. Connect the battery ground cable.

Refer to: Startup Battery Disconnect and Connect (414-01 Battery, Mounting and Cables, General Procedures).

Timing Cover

Removal and Installation

Special Tool(s)

 303683	303-682 Aligner, Engine Front Cover

Part(s)

STEP	PART NAME	QUANTITY
Removal Step 6	Renew Part Water outlet connection	1
Installation Step 1	Renew Part Timing chain cover gasket	1
Installation Step 2	Renew Part Timing chain cover	1

Removal

1. **Refer to:** Startup Battery Disconnect and Connect (414-01 Battery, Mounting and Cables, General Procedures).

2. WARNING: Do not work on or under a vehicle supported only by a jack. Always support the vehicle on safety stands.

Raise and support the vehicle.

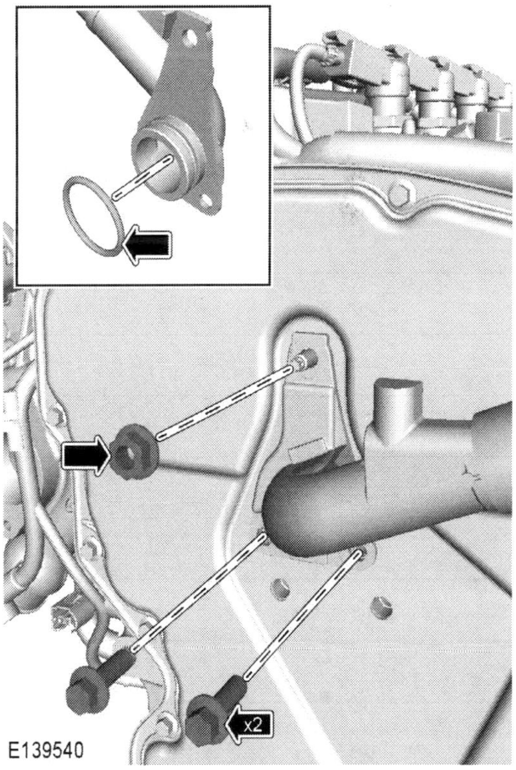

3. **Refer to:** Cooling System Draining, Filling and Bleeding (303-03 Engine Cooling, ID4 2.2L Diesel General Procedures).

4. **Refer to:** Accessory Drive Component Bracket (303-05 Accessory Drive, ID4 2.2L Diesel Removal and Installation).

5. **Refer to:** Crankshaft Front Seal (303-01 Engine, ID4 2.2L Diesel Removal and Installation)

6. ⚠ NOTE: Remove and discard the O-ring seal.

Renew Part: Water outlet connection

Torque: 23 Nm

7. *Torque:* 23 Nm

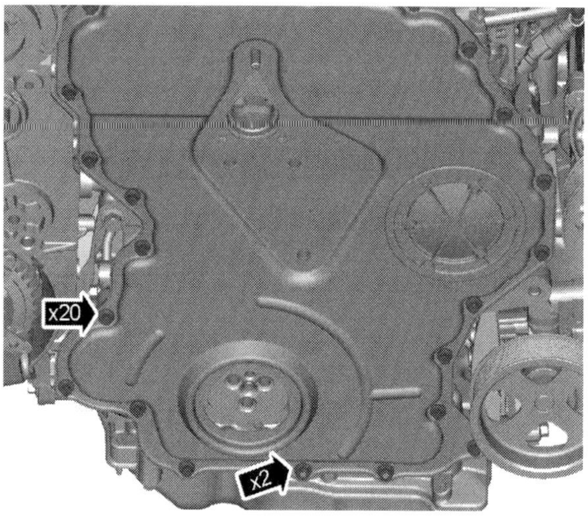

8. ⚠ CAUTION: Make sure that the mating faces are clean and free from foreign material.

9. Remove and discard the gasket.

Installation

1. Install a new gasket.

Renew Part: Timing chain cover gasket

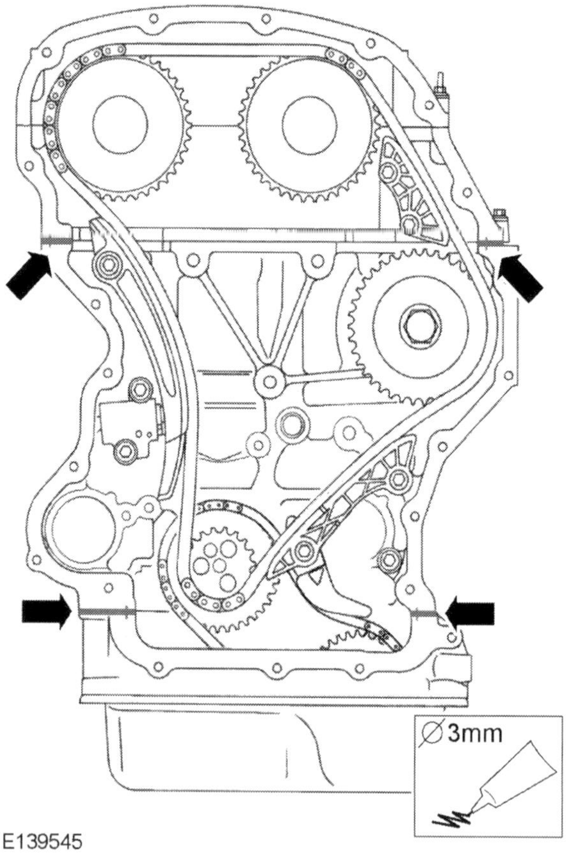

E139545

2. CAUTIONS:

⚠ A new engine timing cover must be installed. Failure to follow this instruction may result in damage to the vehicle.

⚠ Install the engine timing cover within five minutes of applying the sealer. Failure to follow this instruction may result in damage to the vehicle.

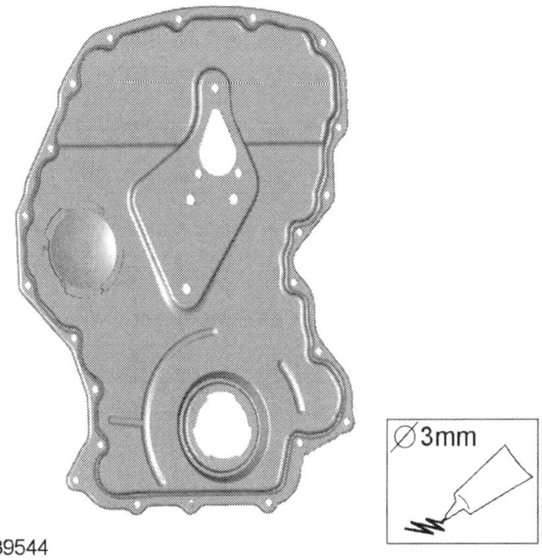

E139544

⚠ Make sure the mating faces are clean, before the sealant is applied.

Renew Part: Timing chain cover

3. ⚠ CAUTION: Make sure the mating faces are clean, before the sealant is applied.

4. ⚠ WARNING: Make sure that the engine timing cover does not come into contact with the engine until correct position is obtained.

Special Tool(s): 303-682

Torque:

Stage 1: 5 Nm

Stage 2: 10 Nm

Stage 3: 10 Nm

5. To install, reverse to removal procedure.

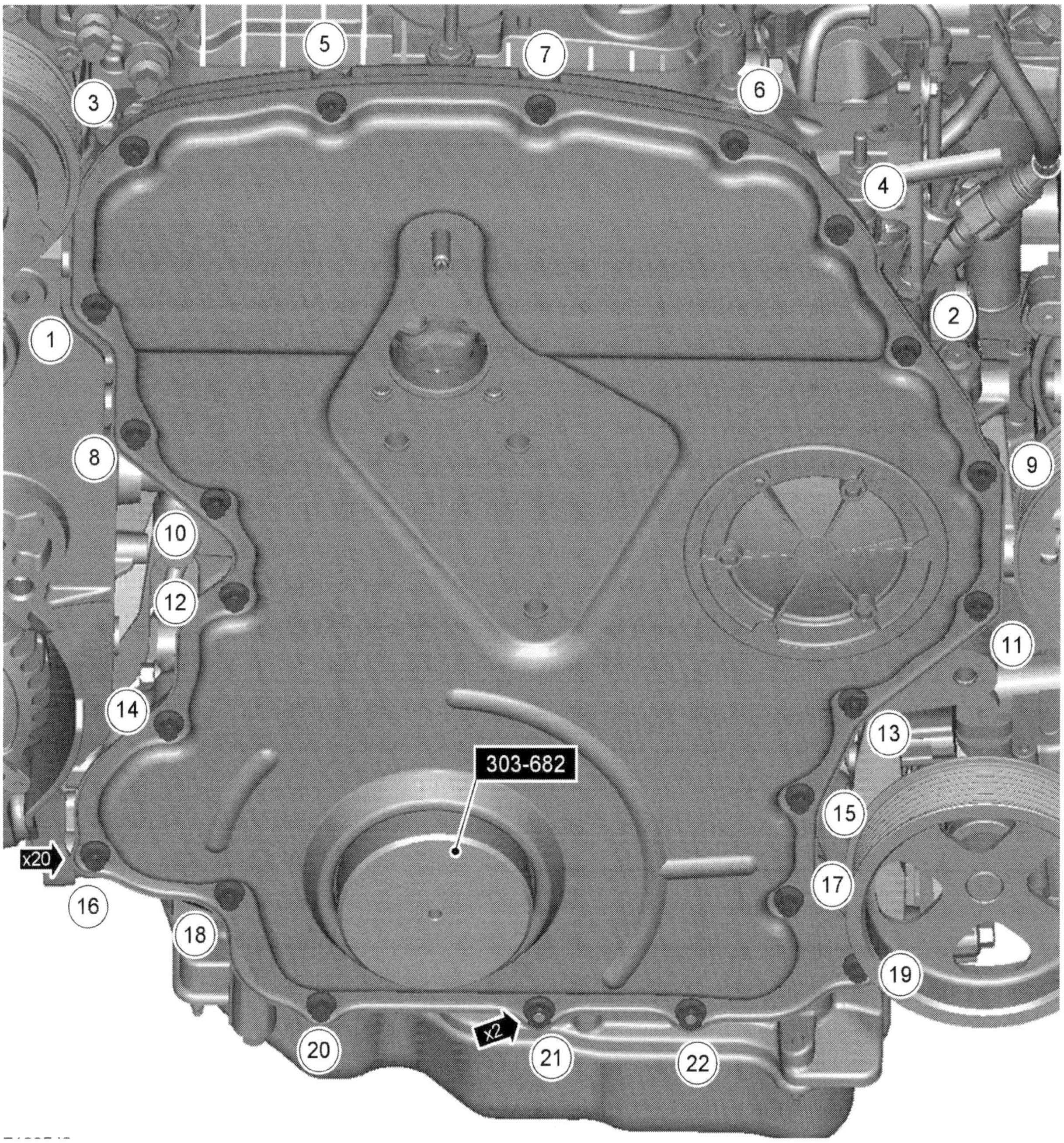

303-03 Engine Cooling - ID4 2.4L and 2.2L Diesel

Specifications

Item	Specification
Anti-freeze	WAS-M97B44-D
Anti-freeze concentration - Will provide frost protection to -40°C (-40°F)	50%
Specific gravity of coolant at 20°C (68°F), to protect against frost down to -40°C (-40°F)	1.068

Item	Capacity
Cooling System	8.2 Liters (17.3 pints) (8.6 quarts)
Radiator	2.3 Liters (4 pints) (2.4 quarts)
Coolant Expansion Tank (To Cold Level)	1.2 Liters (2.1 pints) (1.3 quarts)

Item	Description
Cooling system type	Pressurized, thermostatically controlled with coolant expansion tank
Radiator	Cross flow
Coolant expansion tank	Remote with cold level fill mark
Pressure cap rating	100 kPa (1 bar) (15.9 lbf/in²)
Thermostat: Starts to open	82° C (180°F)
Thermostat: Fully open	96° C (205°F)
Cooling fan	Viscous
Cooling fan control	Bi-metallic
Cooling fan diameter	475 mm (18.70 inches)
Cooling fan direction of rotation	Clockwise when viewed from front of engine
Coolant pump	Centrifugal flow impeller, belt driven from the crankshaft

Torque Specifications

Description	Nm	lb-ft
Coolant bleed screw(s)	3	2
Coolant pump bolts	23	17
Coolant pump outlet manifold	10	7
Coolant pump hose clip	3	2
Viscous coupling bolts	10	7
Expansion tank bolt	10	7
Charge air cooler to radiator bolts	20	15
Charge air cooler inlet and outlet hose clips	3	2
Charge air cooler bracket bolts	9	7
Power steering reservoir nuts	4	3

OVERVIEW

The cooling system employed is the bypass type, which allows coolant to circulate around the engine and the heater circuit while the thermostat is closed. The primary function of the cooling system is to maintain the engine within an optimum temperature range under changing ambient and engine operating conditions. Secondary functions are to provide:

- Heating for the passenger compartment.

Refer to: Climate Control System (412-00, Description and Operation).

- Cooling for:
- The EGR valve and EGR cooler.

Refer to: Engine Emission Control (303-08 Engine Emission Control, ID4 2.2L Diesel Description and Operation)

- The engine oil cooler.

Refer to: Engine (303-01 Engine, ID4 2.2L Diesel Description and Operation)

- The fuel cooler.

Refer to: Fuel Tank and Lines - ID4 2.2L Diesel (310-01 Fuel Tank and Lines, ID4 2.2L Diesel Description and Operation)

The primary components of the engine cooling system are:

- A coolant pump
- An inlet manifold and an outlet connector
- A thermostat
- A radiator
- An expansion tank
- A cooling fan
- Connecting hoses and pipes

Description and Operation

An ECT sensor for the engine management system is installed in the radiator upper hose.

Refer to: Electronic Engine Controls (303-14 Electronic Engine Controls, ID4 2.2L Diesel Description and Operation)

The coolant is circulated by a centrifugal type pump mounted on the front LH (left-hand) side of the engine and driven by the accessory drive belt. An outlet hose supplies coolant from the pump to a distribution manifold on the LH side of the cylinder block. The coolant flows through the cylinder block and cylinder head, to an outlet connector on the front of the cylinder head, then into the radiator upper hose.

The thermostat, installed between the radiator upper and lower hoses, regulates the coolant flow through the radiator to provide a rapid engine warm-up and then maintain a stable engine temperature. The outlet from the thermostat is directed back to the inlet side of the coolant pump.

COMPONENT LOCATION

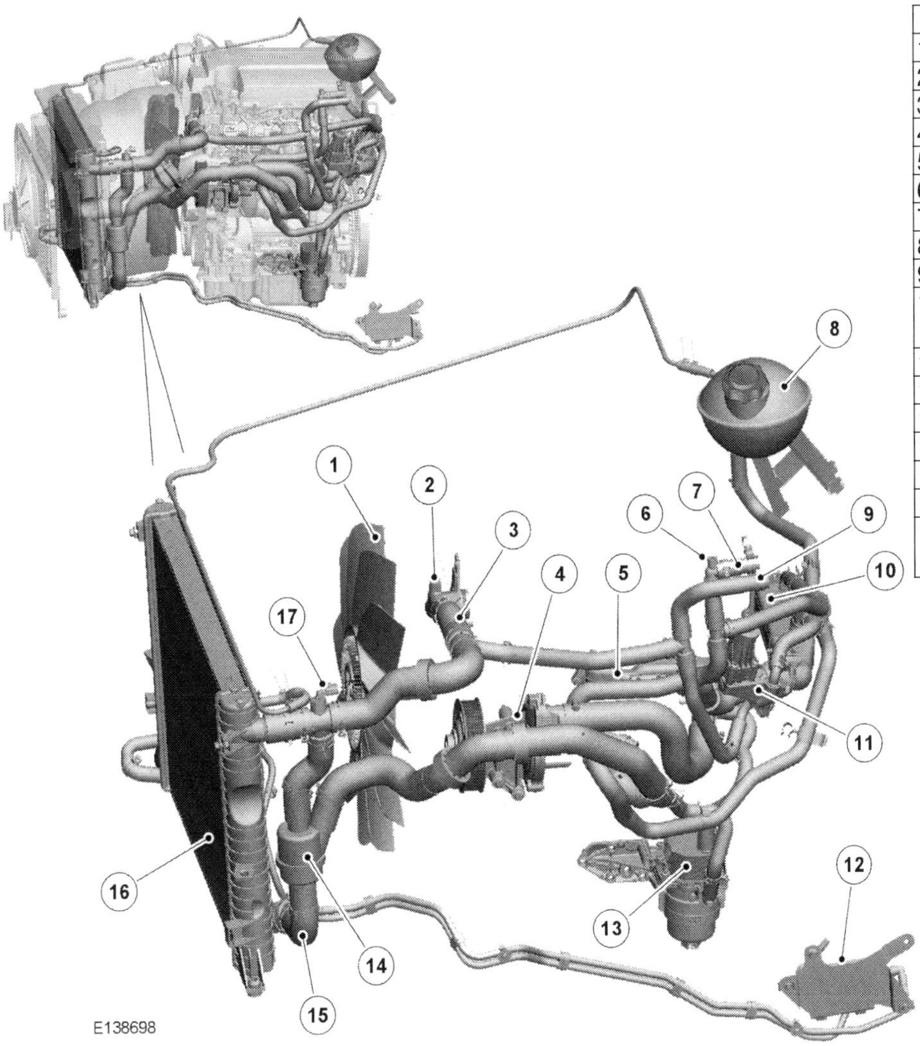

Item	Description
1	Cooling fan
2	Outlet connector
3	Radiator upper hose
4	Coolant pump
5	Distribution manifold
6	Bleed screw
7	Heater return connection
8	Expansion tank
9	Heater feed connection
10	EGR (exhaust gas recirculation) cooler
11	EGR valve
12	Fuel cooler
13	Engine oil cooler
14	Thermostat
15	Radiator lower hose
16	Radiator
17	ECT (engine coolant temperature) sensor

Cooling System Schematic

The radiator is a cross flow type with an aluminum matrix and plastic end tanks. The end tanks have brackets which allow for the attachment of the fan cowl assembly, charge air cooler and, if fitted, A/C (air conditioning) condenser. The bottom of the radiator is located in rubber bushes supported by brackets on the chassis longitudinals. The top of the radiator is located in rubber bushes secured by brackets fitted to the bonnet locking platform.

A connection on the distribution manifold supplies coolant to the engine oil cooler. The return from the engine oil cooler is connected to the hose between the thermostat and the coolant pump inlet.

The radiator upper hose has a connection for the feed to the heater matrix, EGR valve and EGR cooler. The return is connected to an inlet on the coolant pump. The heater matrix outlet hose incorporates a bleed screw to bleed air when filling the cooling system.

The feed for the fuel cooler comes from a connection on the RH (right-hand) end tank of the radiator.

The return from the fuel cooler is connected to the radiator lower hose.

The expansion tank is fitted to the RH suspension turret in the engine compartment. The expansion tank allows for expansion of the coolant when the engine is hot, and retraction as the engine cools down, through a hose connected to an inlet on the coolant pump. A hose connected between the upper left corner of the radiator and the expansion tank ensures there is a constant flow through the expansion tank while the engine is running, to separate out any air in the system. The expansion tank filler cap provides a service fill and top-up point, and pressurizes the system during operation.

The cooling fan generates additional airflow through the radiator matrix, particularly when the vehicle is stationary. The cooling fan is an 11-bladed viscous fan, which is attached to a pulley on the engine front cover and driven by the accessory drive belt. A two-piece shroud provides an air duct between the cooling pack and the fan, and also forms a protective cover.

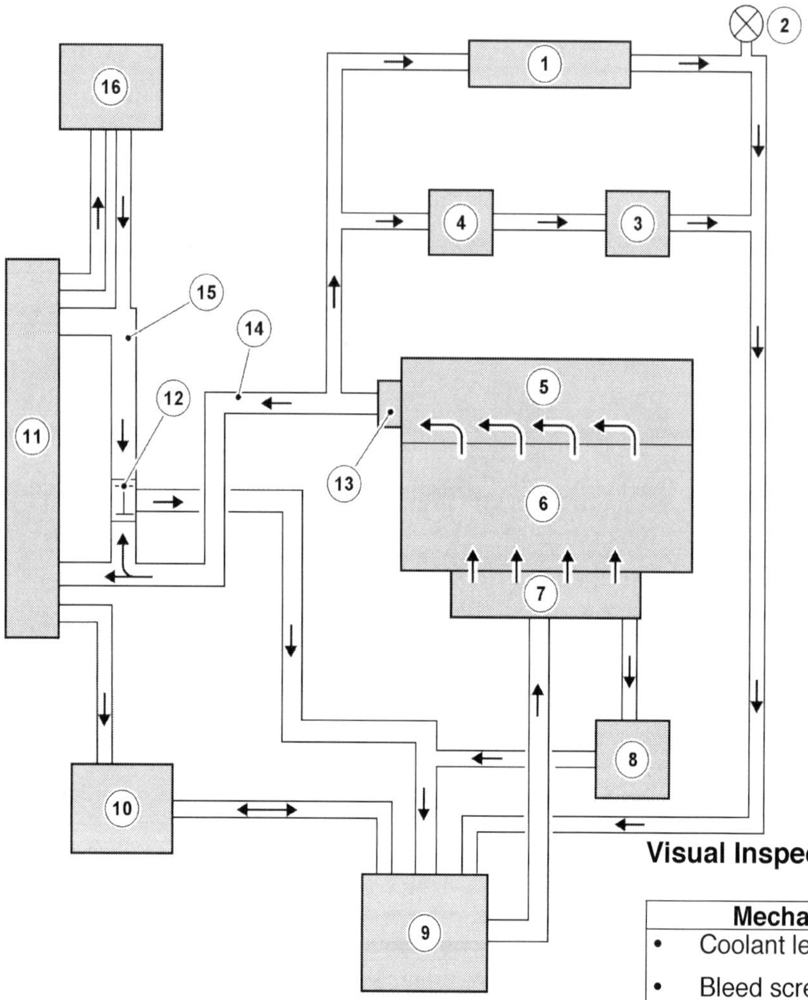

Item	Description
1	Heater matrix
2	Bleed screw
3	EGR cooler
4	EGR valve
5	Cylinder head
6	Cylinder block
7	Distribution manifold
8	Engine oil cooler
9	Coolant pump
10	Expansion tank
11	Radiator
12	Thermostat
13	Outlet connector
14	Radiator upper hose
15	Radiator lower hose
16	Fuel cooler

Engine Cooling

Diagnosis and Testing

Principles of Operation

For a detailed description of the Engine Cooling system, refer to the relevant Description and Operation section in the workshop manual.

Refer to: Engine Cooling (303-03 Engine Cooling, ID4 2.2L Diesel Description and Operation)

Inspection and Verification

⚠ WARNING: Since injury such as scalding could be caused by escaping steam or coolant, do not remove the filler cap from the coolant expansion tank while the system is hot.

⚠ CAUTION: Diagnosis by substitution from a donor vehicle is **NOT** acceptable. Substitution of control modules does not guarantee confirmation of a fault, and may also cause additional faults in the vehicle being tested and/or the donor vehicle.

Visual Inspection

Mechanical	Electrical
• Coolant leaks	• Fuses
• Bleed screw	• Wiring harnesses and connectors
• Radiator	• Engine coolant temperature (ECT) sensor
• Radiator upper hose	• Engine oil temperature (EOT) sensor
• Radiator lower hose	
• Coolant expansion tank	
• Coolant expansion tank cap	
• Heater core	
• Heater feed connection	
• Heater return connection	
• Primary drive belt	
• Coolant pump	
• Viscous fan	
• Exhaust gas recirculation cooler	
• Exhaust gas recirculation valve	
• Cylinder head	
• Cylinder block	
• Engine oil cooler	
• Thermostat	
• Fuel cooler	

NOTES:

⚠ If the control module or a component is suspect and the vehicle remains under manufacturer warranty, refer to the Warranty Policy and Procedures manual (section B1.2), or determine if any prior approval programme is in operation, prior to the installation of a new module/component.

⚠ When performing voltage or resistance tests, always use a digital multimeter accurate to three decimal places, and with an up-to-date calibration certificate. When testing resistance always take the resistance of the digital multimeter leads into account.

⚠ Check and rectify basic faults before beginning diagnostic routines involving pinpoint tests.

1. Visually inspect for obvious signs of damage and system integrity
2. If an obvious cause for an observed or reported concern is found, correct the cause (if possible) before proceeding to the next step
3. If the cause is not visually evident, verify the symptom and refer to the Symptom Chart, alternatively check for Diagnostic Trouble Codes (DTCs) and refer to the DTC Index

DTC Index

For a list of Diagnostic Trouble Codes (DTCs) that could be logged on this vehicle, please refer to Section 100-00.

Refer to: Diagnostic Trouble Code Index - DTC: Engine Control Module (PCM) (100-00 General Information, Description and Operation)

Symptom Chart

Symptom	Possible Causes	Action
Coolant loss	• Bleed screw • Coolant expansion tank cap • Hoses • Hose connections • Radiator • Coolant pump • Heater core • Gaskets • Engine casting cracks • Engine block core plugs	• Check the coolant level and condition. Perform a cooling system pressure test and check for leaks
Overheating	• Low/contaminated coolant • Thermostat stuck closed • Viscous fan - Damaged blades • Restricted air flow over the radiator • Restricted coolant flow through the system • Cylinder head gasket failure • Non approved aftermarket equipment installed	• Check the coolant level and condition. Perform a cooling system pressure test and check for leaks. Check the operation of the thermostat. Check the operation of the viscous fan. Check for obstructions to the air flow through the radiator
Engine not reaching normal temperature	• Thermostat • Viscous fan • Non approved aftermarket equipment installed	• Check the operation of the thermostat. Check the operation of the viscous fan

Cooling System Draining, Filling and Bleeding

General Procedures

⚠ WARNING: Since injury such as scalding could be caused by escaping steam or coolant, do not remove the filler cap from the coolant expansion tank while the system is hot.

CAUTIONS:

⚠ The engine cooling system must be maintained with the correct concentration and type of anti-freeze solution to prevent corrosion and frost damage. Failure to follow this instruction may result in damage to the engine.

⚠ Engine coolant will damage the paint finished surfaces. If spilt, immediately remove the coolant and clean the area with water.

1. Set the heater controls to Hot.
2. ⚠ WARNING: Since injury such as scalding could be caused by escaping steam or coolant, do not remove the filler cap from the coolant expansion tank while the system is hot.

Remove the coolant expansion tank cap.

3. ⚠ WARNING: Since injury such as scalding could be caused by escaping steam or coolant, do not remove the cooling system bleed screw while the system is hot.

Loosen the cooling system bleed screw.

4. ⚠ WARNING: Do not work on or under a vehicle supported only by a jack. Always support the vehicle on safety stands.

Raise and support the vehicle.

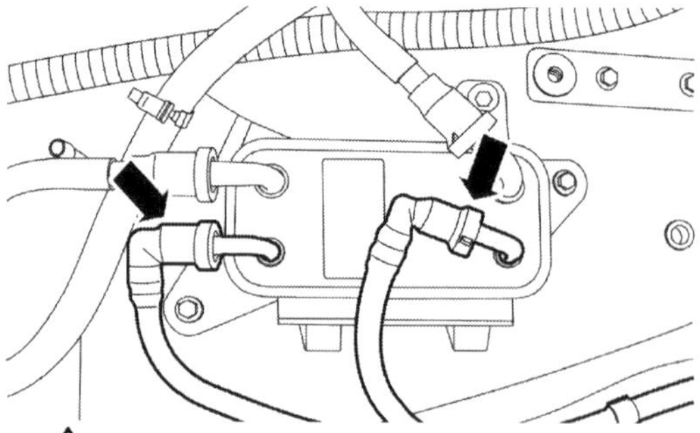

5. ⚠ WARNING: Eye protection must be worn.

△ NOTE: Position a container to collect the coolant.

Disconnect the 2 fuel cooler coolant pipes.

- Allow the coolant to drain.

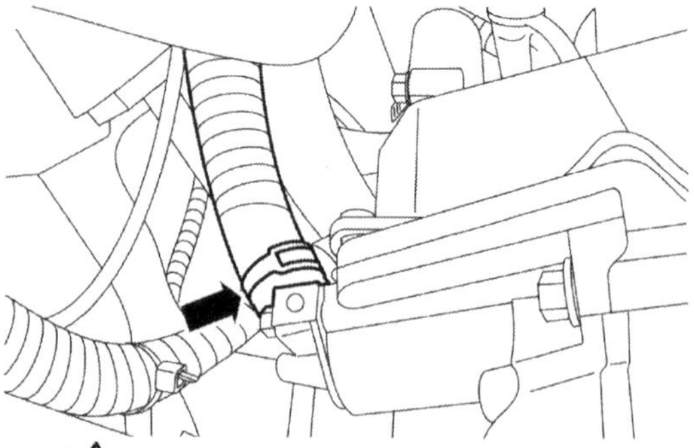

6. ⚠ WARNING: Eye protection must be worn.

Disconnect the oil thermostat coolant hose.

- Allow the coolant to drain.

7. Remove the container.

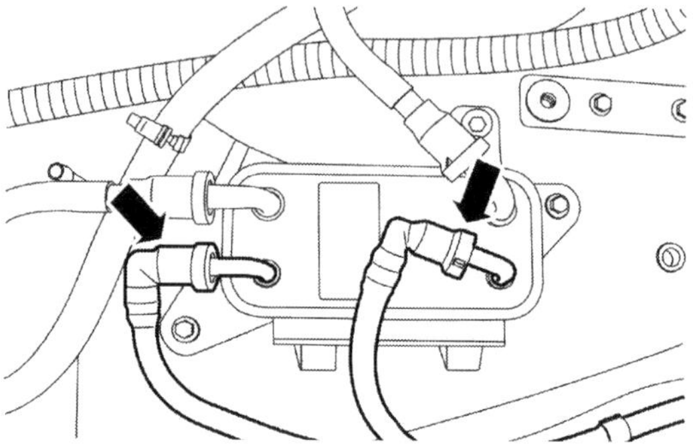

8. Connect the 2 fuel cooler coolant pipes.

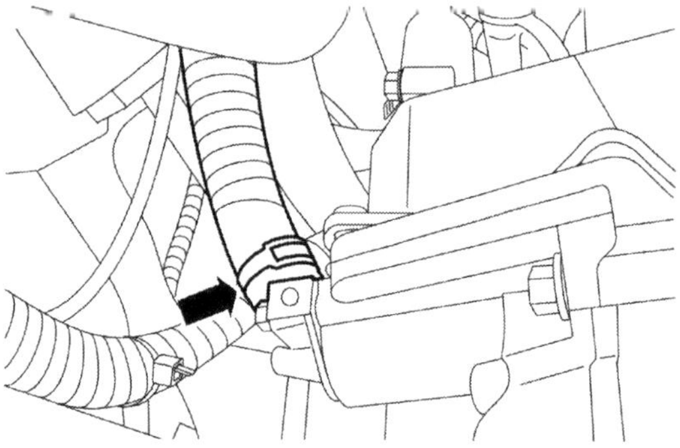

9. Connect the oil thermostat coolant hose.
10. CAUTIONS:

⚠ Engine coolant will damage the paint finished surfaces. If spilt, immediately remove the coolant and clean the area with water.

⚠ Anti-freeze concentration must be maintained at 50%.

Fill the cooling system expansion tank until the coolant level becomes static with in the "COLD FILL RANGE".

11. Tighten the cooling system bleed screw.
12. Install the coolant expansion tank cap.
13. CAUTIONS:

⚠ Make sure the coolant level remains above the "COLD FILL RANGE" lower level mark.

⚠ Observe the engine temperature gauge. If the engine starts to over-heat switch off immediately and allow to cool. Failure to follow this instruction may cause damage to the vehicle.

Start and run the engine.

- Hold the engine speed at 3,000 RPM for one minute.
- Return the engine to idle for five minutes.
- Hold the engine speed at 3,000 RPM for one minute.
- Run the engine until the thermostat opens.

14. ⚠ WARNING: Release the cooling system pressure by slowly turning the coolant expansion tank cap a quarter of a turn. Cover the expansion tank cap with a thick cloth to prevent the possibility of scalding. Failure to follow this instruction may result in personal injury.

△ NOTE: When the coolant bleed is complete and prior to installing the expansion tank cap, top-up the expansion tank to 30mm above the maximum level.

Switch the engine off.

- Remove the coolant expansion cap, allow the level to settle and top-up so that the level is 30mm above the upper level.
- Allow the engine to cool.

15. Clean any spilt coolant from the vehicle.
16. Check and top-up the coolant if required.

Cooling System Draining and Vacuum Filling

General Procedures

⚠ WARNING: To avoid having scalding hot coolant or steam blowing out of the cooling system, use extreme care when removing the coolant pressure cap from a hot cooling system. Wait until the engine has cooled, then wrap a thick cloth around the coolant pressure cap and turn it slowly until the pressure begins to release. Step back while the pressure is released from the system. When certain all the pressure has been released (still with a cloth) turn and remove the coolant pressure cap from the coolant expansion tank. Failure to follow these instructions may result in personal injury.

CAUTIONS:

⚠ The engine cooling system must be maintained with the correct concentration and type of anti-freeze solution to prevent corrosion and frost damage. Failure to follow this instruction may result in damage to the vehicle.

⚠ Engine coolant will damage paint finished surfaces. If spilled, immediately remove coolant and clean area with water.

1. Set the heater controls to maximum HOT.
2. Remove the hood.

Refer to: Hood (501-02, Removal and Installation).

3. ⚠ WARNING: Since injury such as scalding could be caused by escaping steam or coolant, do not remove the filler cap from the coolant expansion tank while the system is hot.

   Remove coolant expansion tank filler cap.

4. Loosen the cooling system bleed screw.
5. ⚠ WARNING: Do not work on or under a vehicle supported only by a jack. Always support the vehicle on safety stands.

   Raise and support the vehicle.

6. ⚠ WARNING: Eye protection must be worn.

   Position container to collect coolant.

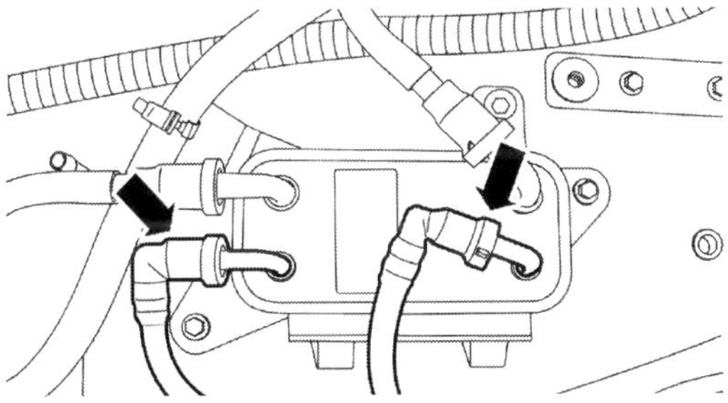

7. Disconnect the 2 fuel cooler coolant pipes.
   - Allow the coolant to drain.

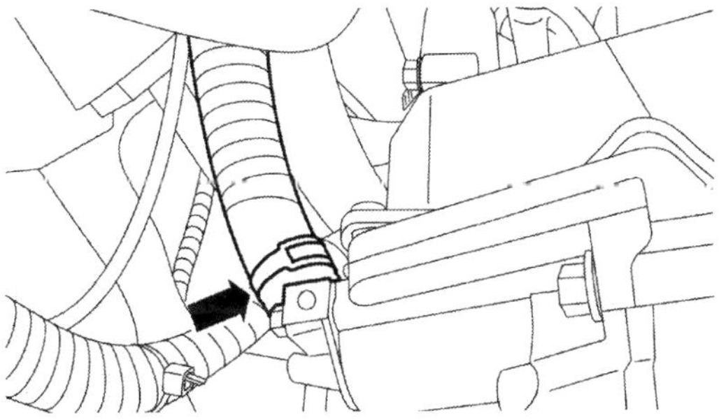

8. Disconnect the coolant hose from the engine oil cooler.
   - Release the clip.
   - Allow the coolant to drain.

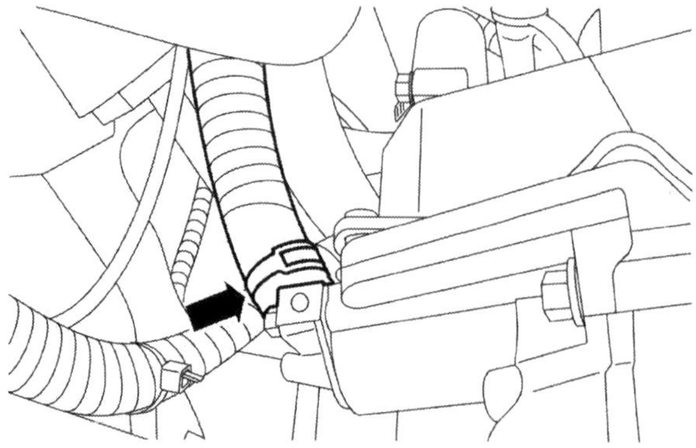

9. Connect coolant hose to engine oil cooler.
    - Install the clip.
10. Connect the 2 fuel cooler coolant pipes.
    - Remove the container.

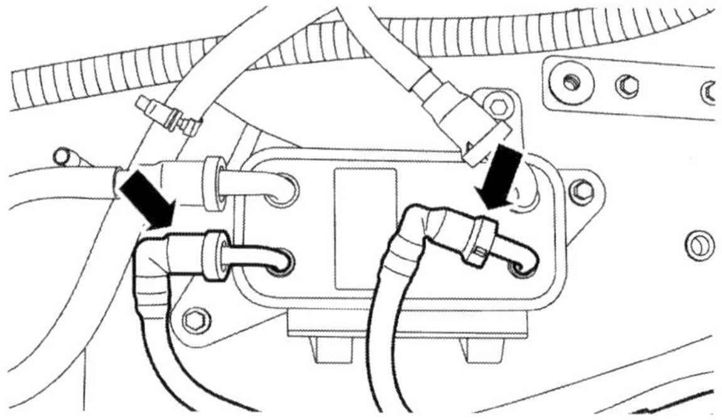

11. Tighten the cooling system bleed screw.
12. ⚠ CAUTION: The engine cooling system must be maintained with the correct concentration and type of anti-freeze solution to prevent corrosion and frost damage. Failure to follow this instruction may result in damage to the vehicle.

Prepare a sufficient amount of coolant to the required concentration and transfer to a container.

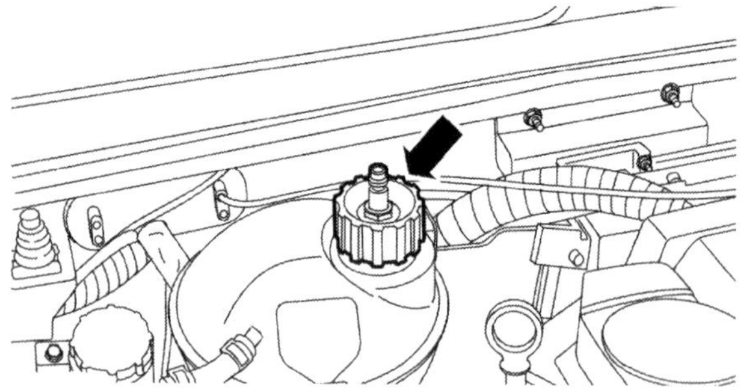

13. Install the cooling system vacuum refill adaptor to the expansion tank.

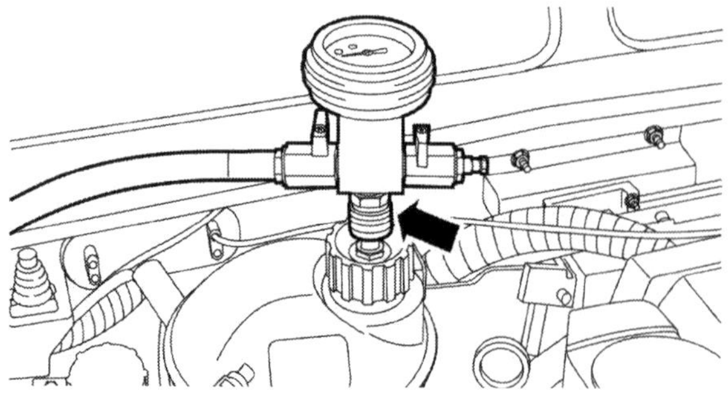

14. Install the vacuum filler gauge to the cooling system vacuum refill adaptor.
15. Install the venturi tube assembly to the vacuum filler gauge.

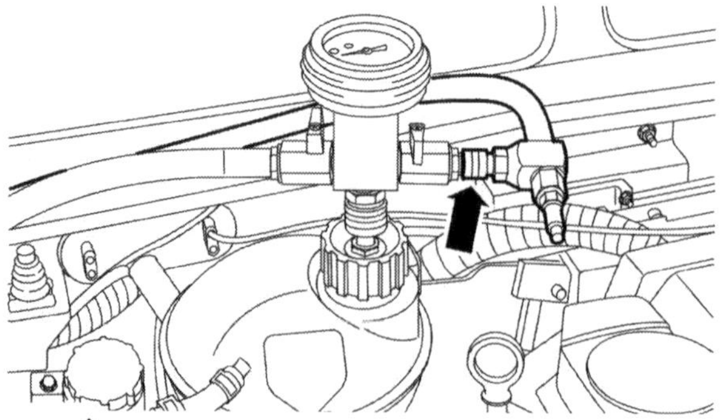

16. △ NOTE: Make sure both valves on gauge assembly are in the closed position.

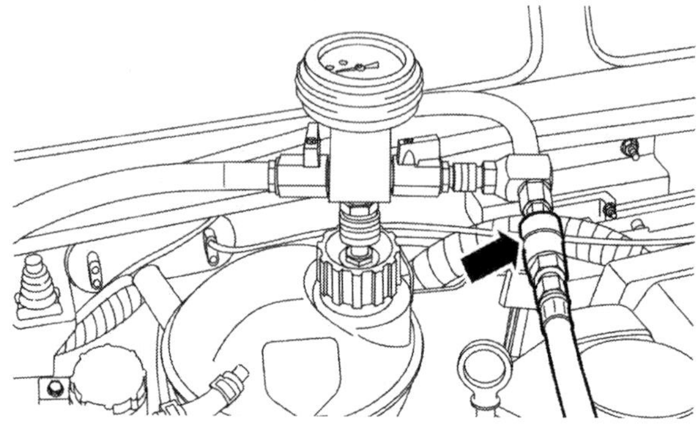

Connect a regulated compressed air supply to the venturi tube assembly.

17. △ NOTE: Make sure air cannot enter the hose.

    Position the coolant pick-up pipe into a container of clean coolant.
18. Position the evacuated air hose into a container.

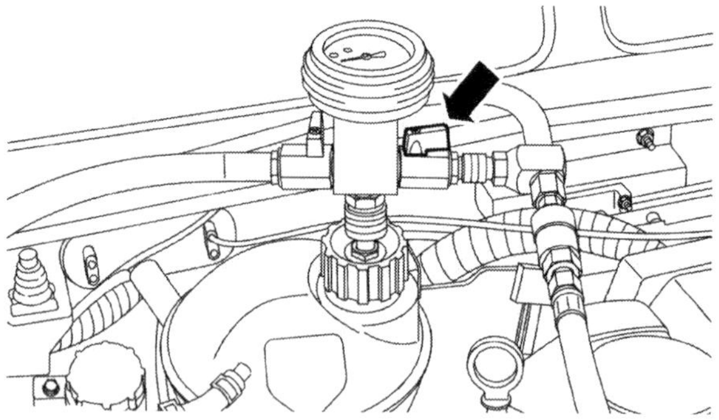

19. NOTES:

△ Make sure the coolant supply valve is in the closed position on the vacuum filler gauge assembly.

△ The coolant vacuum fill tool needs an air pressure of 6 to 8 bar (87 to 116 psi) to operate correctly.

△ Small diameter or long airlines may restrict airflow to the coolant vacuum fill tool.

Open air supply valve.

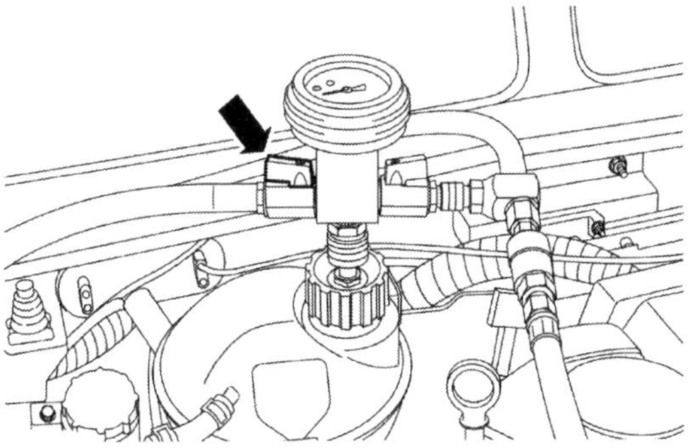

20. Open coolant supply valve for 2 seconds to prime coolant supply hose.

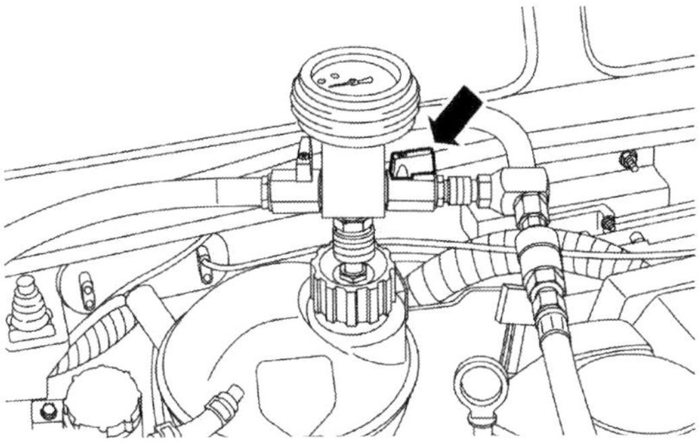

21. Apply air pressure progressively until the arrow on the vacuum filler gauge reaches the green segment.

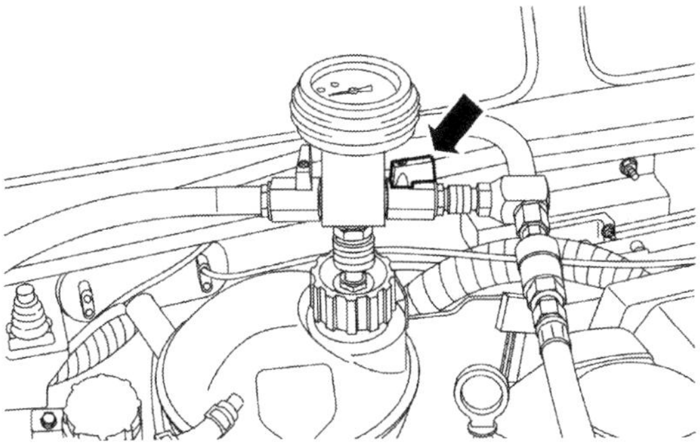

22. Close air supply valve and disconnect air supply.

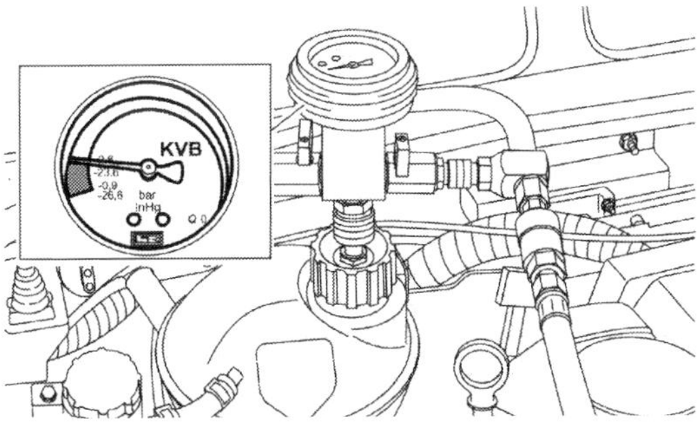

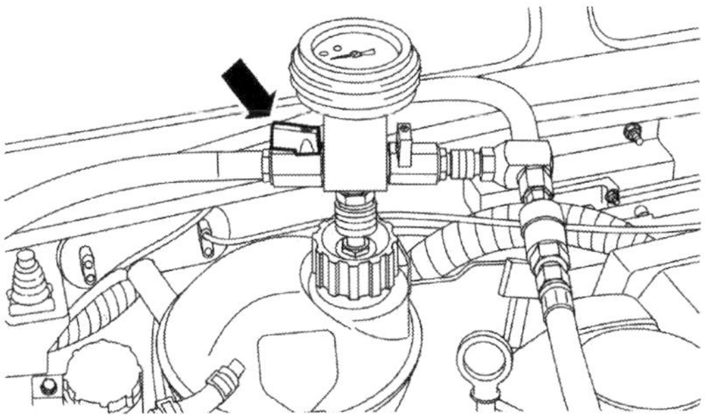

23. Allow 1 minute to check vacuum is held.
24. Open coolant supply valve and allow coolant to be drawn into system.
25. When expansion tank is full and coolant movement has ceased close coolant supply valve.
26. Remove vacuum refill adaptor and gauge from expansion tank.
27. Install the coolant expansion tank filler cap.
28. Install the hood.

Refer to: Hood (501-02, Removal and Installation).

29. CAUTIONS:

⚠ Make sure the coolant level remains above the "COLD FILL RANGE" lower level mark.

⚠ Observe the engine temperature gauge. If the engine starts to over-heat switch off immediately and allow to cool. Failure to follow this instruction may cause damage to the vehicle.

Start and run engine.

- Hold the engine speed at 3,000 RPM for one minute.
- Return the engine to idle for five minutes.
- Hold the engine speed at 3,000 RPM for one minute.
- Run the engine until normal operating temperature is reached.

30. Switch off the engine and allow to cool.
31. ⚠ WARNING: Since injury such as scalding could be caused by escaping steam or coolant, do not remove the filler cap from the coolant expansion tank while the system is hot.

Check and top-up the coolant level if required.

Cooling System Partial Draining and Vacuum Filling

General Procedures

Special Tool(s)

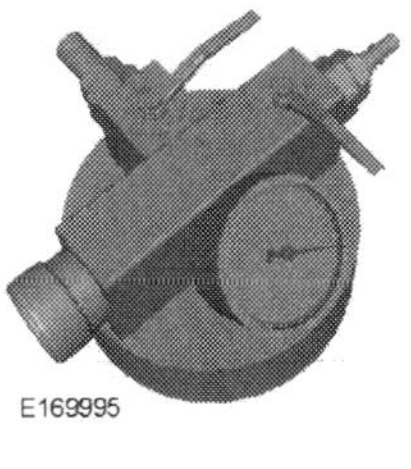	HU-919 Coolant System Vacuum Refill Kit

Draining

⚠ WARNING: Since injury such as scalding could be caused by escaping steam or coolant, do not remove the filler cap from the coolant expansion tank while the system is hot.

CAUTIONS:

The engine cooling system must be maintained with the correct concentration and type of anti-freeze solution to prevent corrosion and frost damage. Failure to follow this instruction may result in damage to the engine.

Engine coolant will damage the paint finished surfaces. If spilt, immediately remove the coolant and clean the area with water.

1. WARNING: Do not work on or under a vehicle supported only by a jack. Always support the vehicle on safety stands.

   Raise and support the vehicle.

2. WARNING: Since injury such as scalding could be caused by escaping steam or coolant, do not remove the filler cap from the coolant expansion tank while the system is hot.

Remove the coolant expansion tank cap.

3. Position a container to collect the fluid.

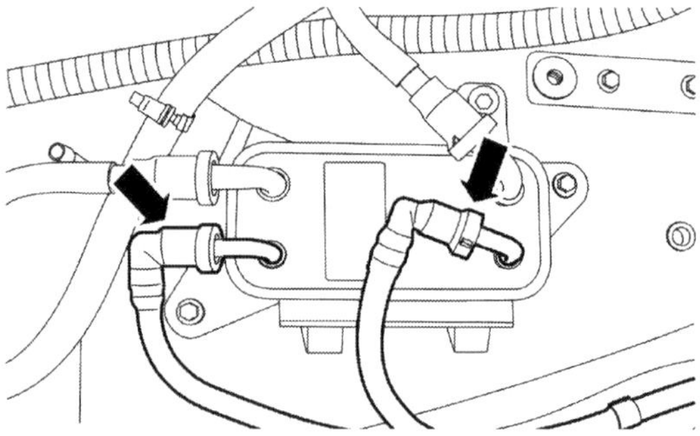

4. WARNING: Eye protection must be worn.

NOTE: Collect the coolant in a clean container and reuse.

   1. Disconnect the fuel cooler coolant pipes.
   2. Allow the coolant to drain.

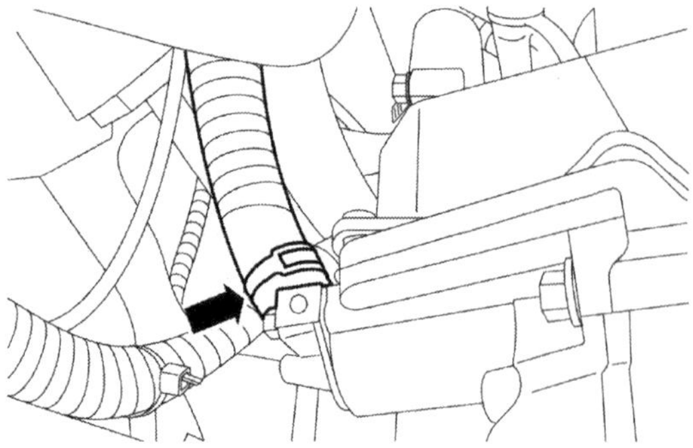

5.

   1. NOTE: Collect the coolant in a clean container and reuse.

   Disconnect the oil thermostat coolant hose.

   2. Allow the coolant to drain.

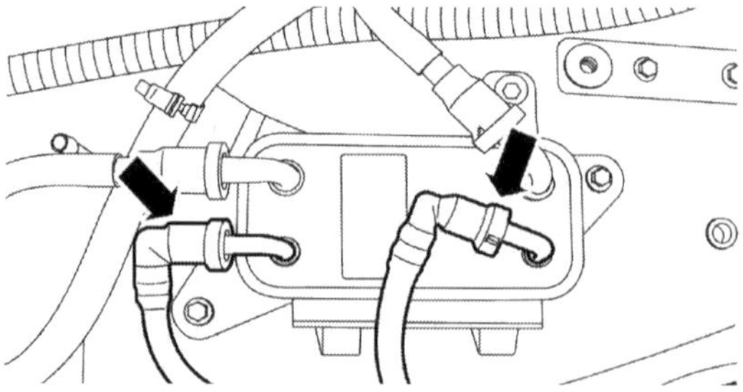

6. Connect the fuel cooler coolant pipes.

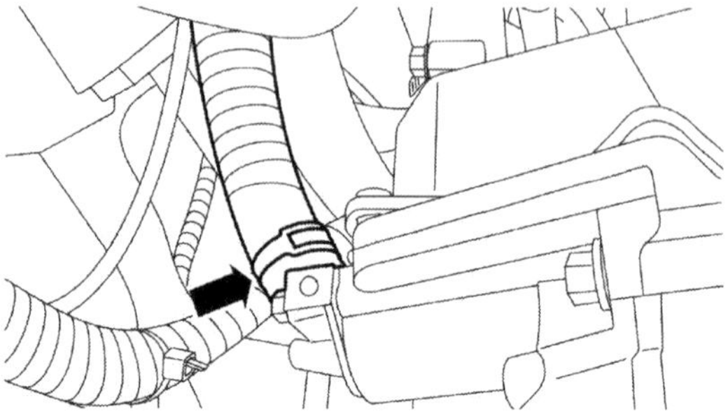

7. Connect the oil thermostat coolant hose.

Filling

WARNING: Since injury such as scalding could be caused by escaping steam or coolant, do not remove the filler cap from the coolant expansion tank while the system is hot.

CAUTIONS:

The engine cooling system must be maintained with the correct concentration and type of anti-freeze solution to prevent corrosion and frost damage. Failure to follow this instruction may result in damage to the engine.

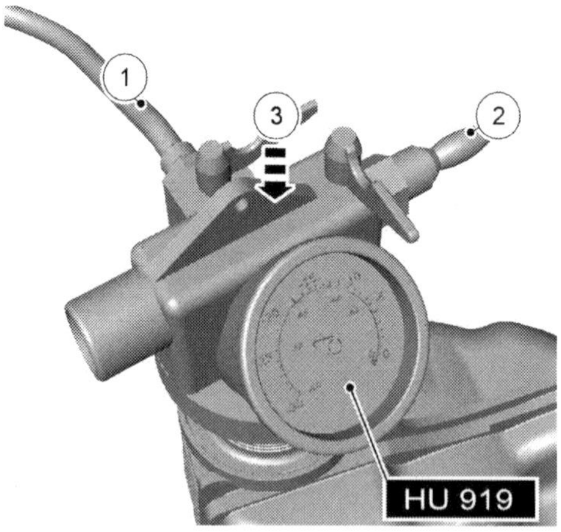

Engine coolant will damage the paint finished surfaces. If spilt, immediately remove the coolant and clean the area with water.

1. Prepare a sufficient amount of coolant to the specified concentration.

2. NOTES:

⚠ Make sure the coolant supply valve is in the closed position on the special tool.

⚠ The special tool needs an air pressure of 6 to 8 bar (87 to 116 psi) to operate correctly

⚠ Small diameter or long airlines may restrict airflow to the coolant vacuum fill tool.

Special Tool(s): HU-919

- Position the hose from the vacuum refill adaptor into a container of clean coolant.
- Connect a regulated compressed air supply to the special tool.
- Move the special tool to the expansion tank.

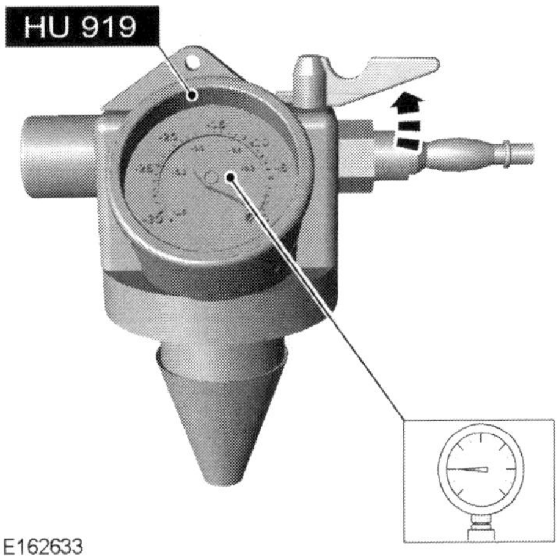

E162633

3. NOTES:

⚠ Make sure the coolant supply valve is in the closed position on the special tool.

⚠ The special tool needs an air pressure of 6 to 8 bar (87 to 116 psi) to operate correctly

⚠ Small diameter or long airlines may restrict airflow to the coolant vacuum fill tool.

Open the air supply valve until -0.8 Bar (-12 psi) is shown on the gauge.

Special Tool(s): HU-919

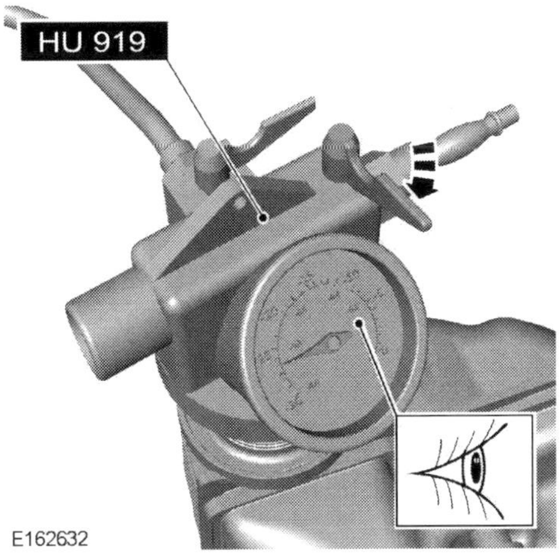

E162632

4.

- *Special Tool(s):* HU-919
- Close the air supply valve.
- Allow 1 minute to check the vacuum is held.

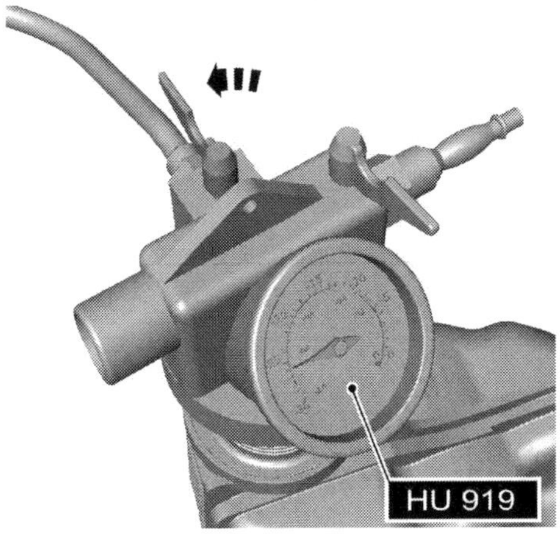

5. NOTES:

⚠ The coolant is to be reused.

⚠ Close the coolant supply valve when the coolant expansion tank MAX mark is reached or coolant movement has stopped.

Open the coolant supply valve and allow the coolant to be drawn into the system.

Special Tool(s): HU-919

6. Remove the special tool.
7. Connect exhaust extraction hoses to the tail pipes.
8. Start and run the engine.
9. Install the coolant expansion tank cap.
10. Hold the engine speed at 2000 revolutions per minute (RPM) until warm air is expelled from the heater.
11. Switch the engine off and allow to cool.
12. Clean any spilt or excess coolant from the vehicle.
13. ⚠ WARNING: Since injury such as scalding could be caused by escaping steam or coolant, allow the vehicle cooling system to cool prior to carrying out this procedure.

Check and top-up the coolant if required.

Coolant Expansion Tank

Removal and Installation

Removal

1. WARNINGS:

⚠ Release the cooling system pressure by slowly turning the coolant expansion tank cap a quarter of a turn. Cover the expansion tank cap with a thick cloth to prevent the possibility of scalding. Failure to follow this instruction may result in personal injury.

⚠ Since injury such as scalding could be caused by escaping steam or coolant, do not remove the filler cap from the coolant expansion tank while the system is hot.

⚠ CAUTION: Engine coolant will damage the paint finished surfaces. If spilt, immediately remove the coolant and clean the area with water.

Release the cooling system pressure.

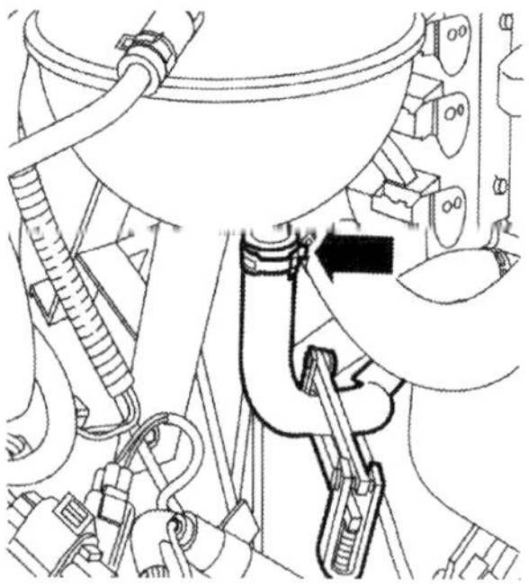

2. Disconnect the coolant hose.
   - Clamp the hose to minimize coolant loss.
   - Position a container to collect the fluid spillage.
   - Release the clip.

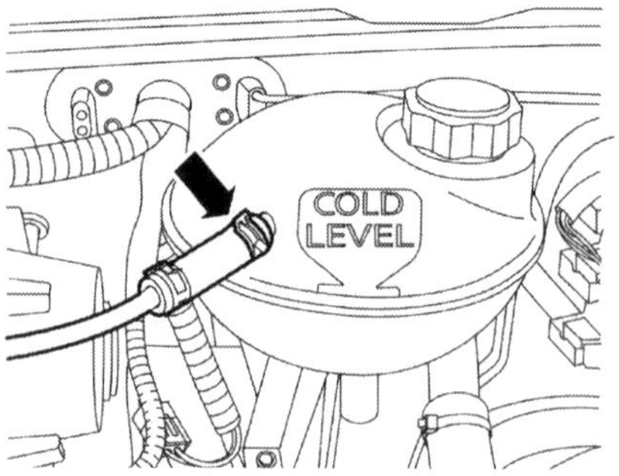

3. Disconnect the radiator bleed hose.
   - Release the clip.
4. Remove the coolant expansion tank.
   - Remove the bolt.

E90029

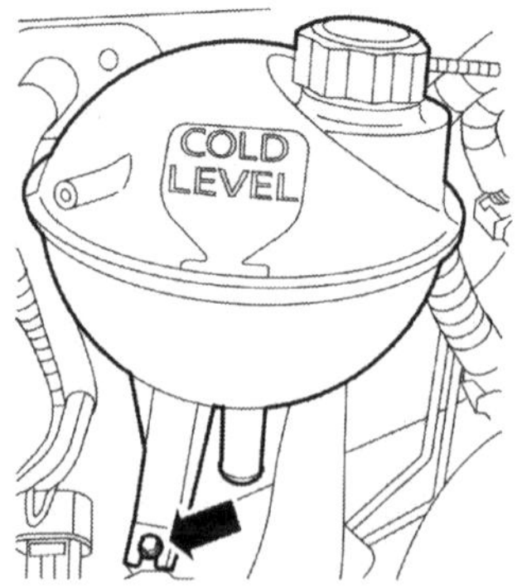

Installation

1. Install the coolant expansion tank.
   - Tighten the bolt to 10 Nm (7 lb.ft).
2. Connect the radiator bleed hose.
   - Secure with the clip.
3. Connect the coolant hose.
   - Secure with the clip.
   - Remove the hose clamp.
4. Check and top up the coolant.
   - Remove the container.

Cooling Fan

Removal and Installation

Special Tool(s)

303-1142 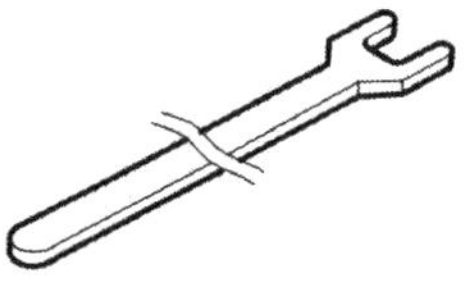 E46076	303-1142 Cooling fan spanner
303-1334  E91076	303-1334 Cooling fan retaining tool

Removal

1. Disconnect the battery ground cable.

Refer to: Battery Disconnect and Connect (414-01, General Procedures).

2. Remove the cooling fan upper shroud.
   - Release the 6 clips.

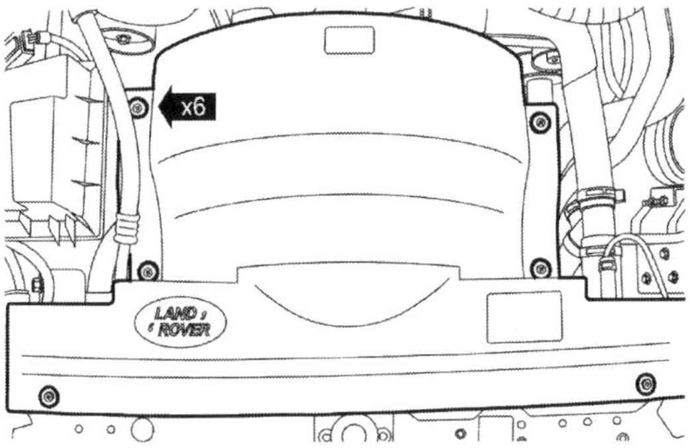

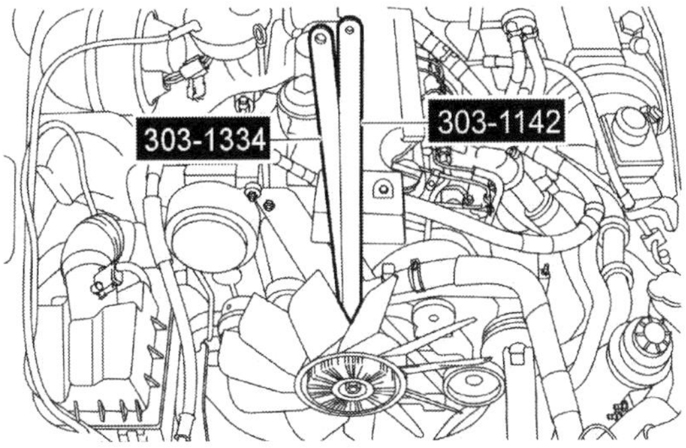

3. Using the special tools, remove the cooling fan.

4. NOTE: Do not disassemble further if the component is removed for access only.

Remove the viscous coupling.

- Remove the 4 bolts.

Installation

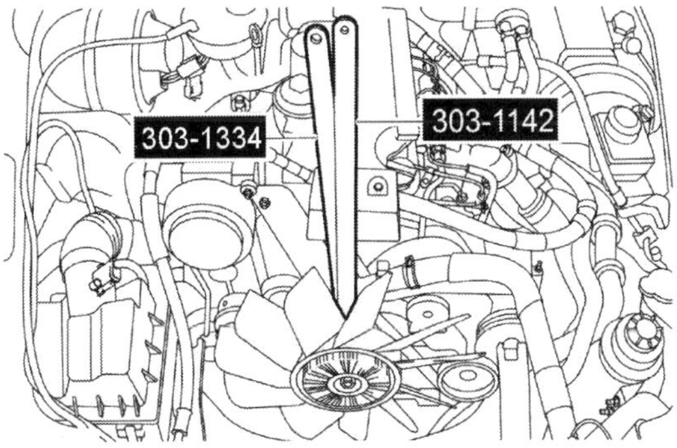

1. Install the viscous coupling.
   - Clean the component mating faces.
   - Tighten the bolts to 10 Nm (7 lb.ft).
2. Using the special tools, install the cooling fan.

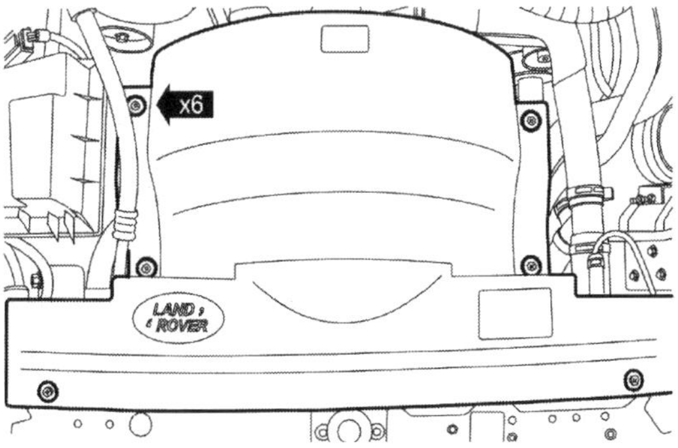

3. Install the cooling fan upper shroud.
   - Secure the 6 clips.
4. Connect the battery ground cable.

Refer to: Battery Disconnect and Connect (414-01, General Procedures).

Cooling Fan Shroud

Removal and Installation

Removal

1. Disconnect the battery ground cable.

   Refer to: Startup Battery Disconnect and Connect (414-01 Battery, Mounting and Cables, General Procedures)

2. Drain the cooling system.

   Refer to: Cooling System Draining, Filling and Bleeding (303-03 Engine Cooling, ID4 2.2L Diesel General Procedures)

3. Remove the cooling fan.

   Refer to: Cooling Fan (303-03 Engine Cooling, ID4 2.2L Diesel Removal and Installation)

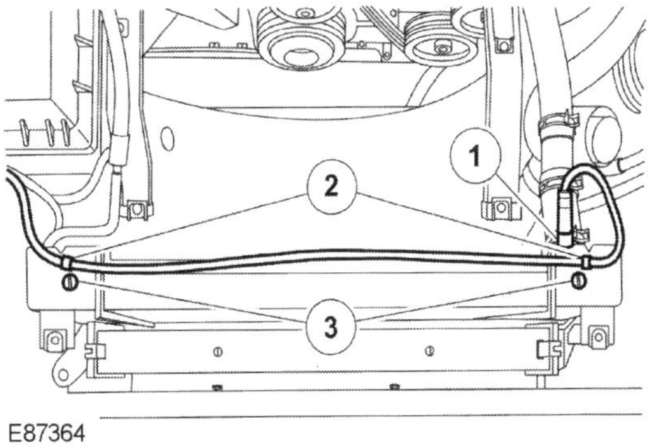

E87364

4. CAUTION: Before disconnecting any components, make sure the area is clean and free from foreign material. When disconnected all openings must be sealed.

Release the cooling fan shroud.

1. Disconnect the coolant expansion tank hose.
2. Release the coolant expansion tank hose from the cooling fan shroud.
3. Release the cooling fan shroud remove the 2 screws.

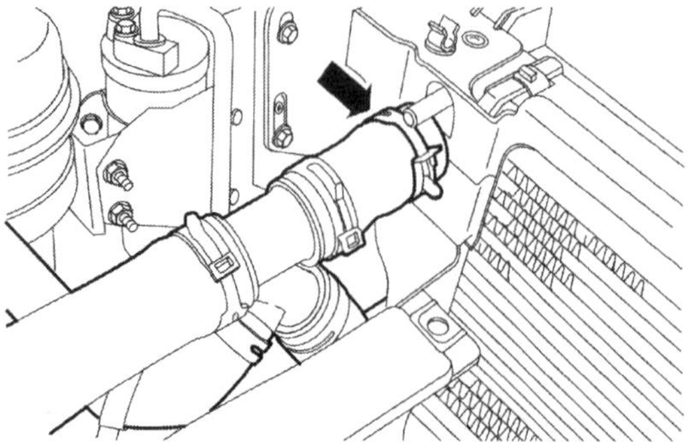

5. CAUTION: Before disconnecting any components, make sure the area is clean and free from foreign material. When disconnected all openings must be sealed.

Disconnect the radiator top hose.

6. Release the thermostat from the cooling fan shroud.

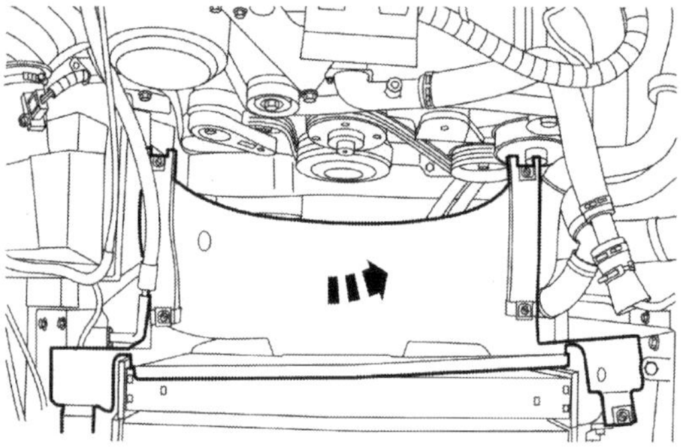

7. Remove the cooling fan shroud.

Installation

1. To install, reverse the removal procedure.
2. Connect the battery ground cable.

Refer to: Battery Connect (414-01 Battery, Mounting and Cables, General Procedures).

Coolant Pump

Removal and Installation

Removal

NOTE: Removal steps in this procedure may contain installation details.

1. **Refer to:** Battery Disconnect and Connect (414-01, General Procedures).
2. **Refer to:** Cooling System Draining, Filling and Bleeding (303-03 Engine Cooling, ID4 2.2L Diesel General Procedures)
3. **Refer to:** Accessory Drive Belt (303-05 Accessory Drive, ID4 2.2L Diesel Removal and Installation)
4. WARNING: Do not work on or under a vehicle supported only by a jack. Always support the vehicle on safety stands.

Raise the vehicle on a twin-post lift.

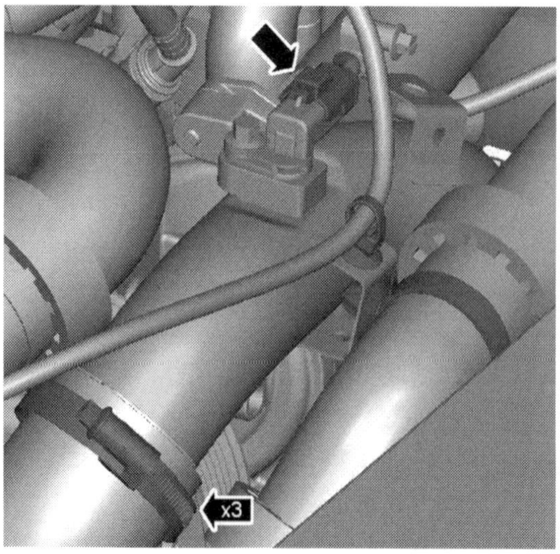

5. *Torque:* 3 Nm

6. CAUTION: Always fit clean plugs to open connections to prevent contamination.

Bolt torque: 25 Nm, Nut torque: 25 Nm, Hose clip torque: 3 Nm

7. *Torque:* 3 Nm

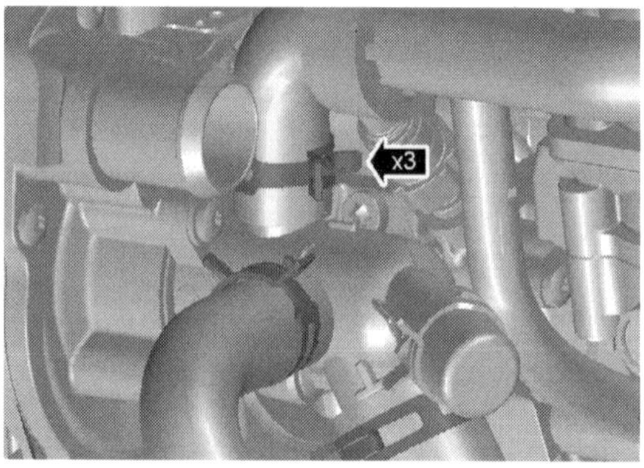

8.

9. *Torque*: 23 Nm

Installation

1. To install, reverse the removal procedure.

Radiator

Removal and Installation

Removal

1. Disconnect the battery ground cable.

Refer to: Startup Battery Disconnect and Connect (414-01 Battery, Mounting and Cables, General Procedures)

2. Remove the cooling fan shroud.

Refer to: Cooling Fan Shroud (303-03 Engine Cooling, ID4 2.2L Diesel Removal and Installation)

3. Remove the hood latch panel.

Refer to: Hood Latch Panel (501-27 Front End Sheet Metal Repairs, Removal and Installation)

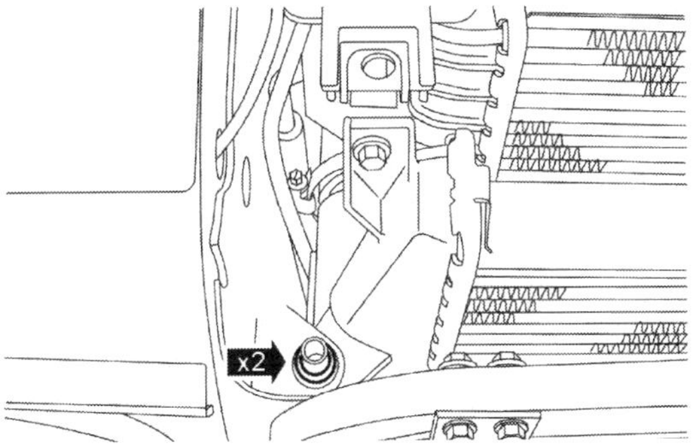

4. NOTE: RH shown, LH similar.

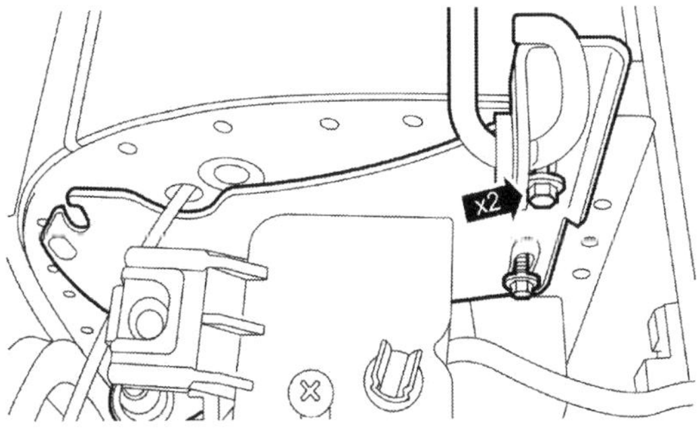

Remove the 2 O-rings.

5. NOTE: RH shown, LH similar.

Remove the 2 charge air cooler brackets. Remove the 4 bolts.

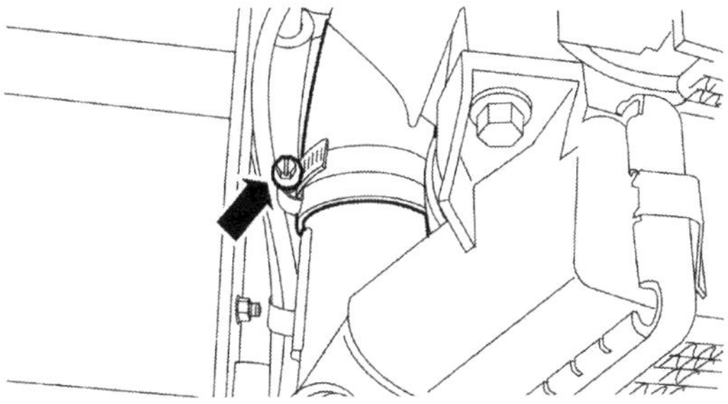

6. CAUTION: Before disconnecting any components, make sure the area is clean and free from foreign material. When disconnected all openings must be sealed.

NOTE: RH shown, LH similar.

Disconnect the charge air cooler inlet hose and charge air cooler outlet hose.

7. CAUTION: Before disconnecting any components, make sure the area is clean and free from foreign material. When disconnected all openings must be sealed.

Disconnect the fuel cooler to radiator hose.

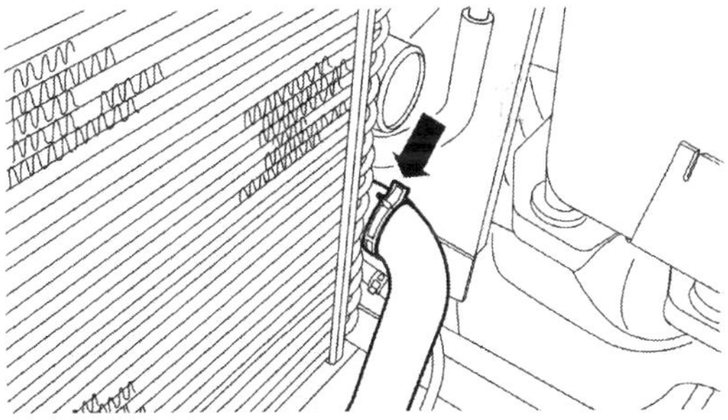

8. CAUTION: Before disconnecting any components, make sure the area is clean and free from foreign material. When disconnected all openings must be sealed.

Disconnect the radiator bottom hose.

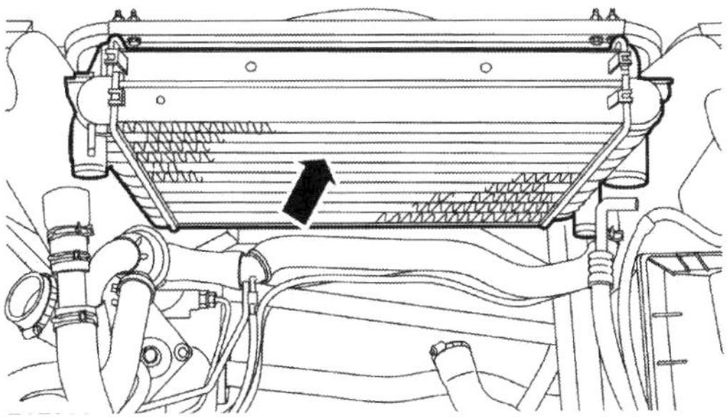

9. CAUTION: Make sure that the radiator and charge air cooler are not damaged when removed.

Remove the radiator and charge air cooler assembly.

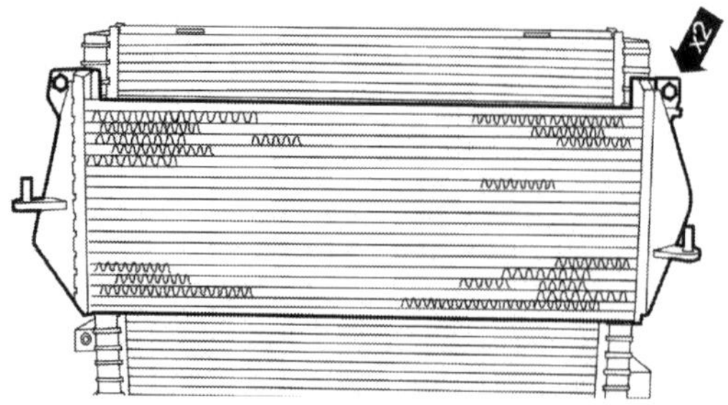

10. Remove the radiator.

Installation

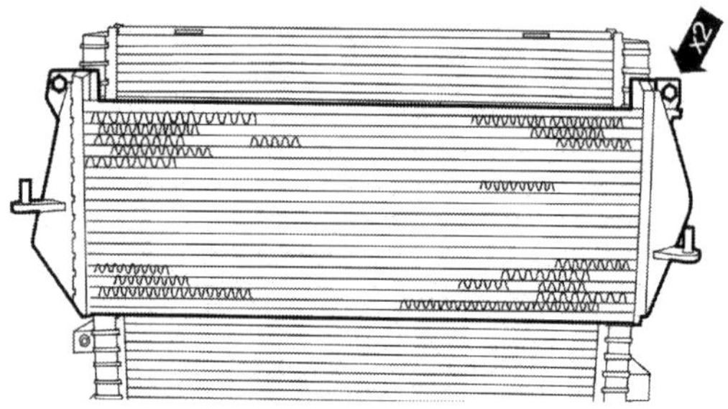

1. To install, reverse the removal procedure. Tighten the bolts.

   Torque: 20 Nm

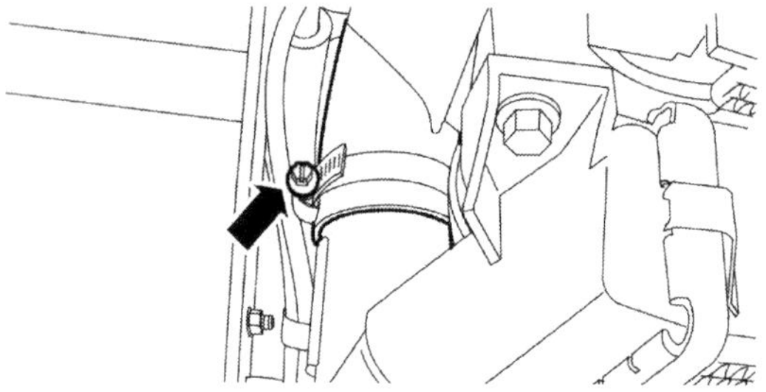

2. NOTES:

△ Remove and discard the blanking caps.

△ RH shown, LH similar.

   Torque: 3 Nm

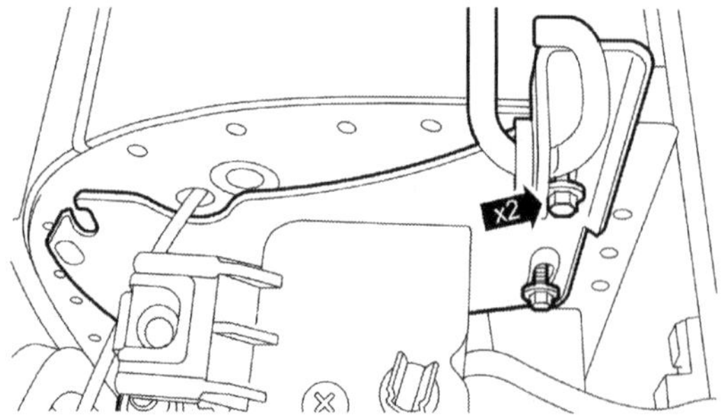

3. *Torque:* 9 Nm

4. Connect the battery ground cable.

Refer to: Startup Battery Disconnect and Connect (414-01 Battery, Mounting and Cables, General Procedures)

Thermostat

Removal and Installation

Removal

1. Disconnect the battery ground cable.

Refer to: Startup Battery Disconnect and Connect (414-01 Battery, Mounting and Cables, General Procedures)

2. Drain the cooling system.

Refer to: Cooling System Draining, Filling and Bleeding (303-03 Engine Cooling; ID4 2.2L Diesel General Procedures)

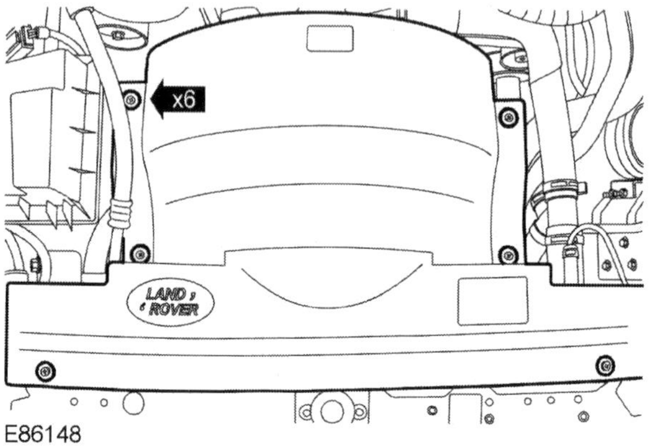

E86148

3. Remove the cooling fan upper shroud.

   - Release the 6 clips.

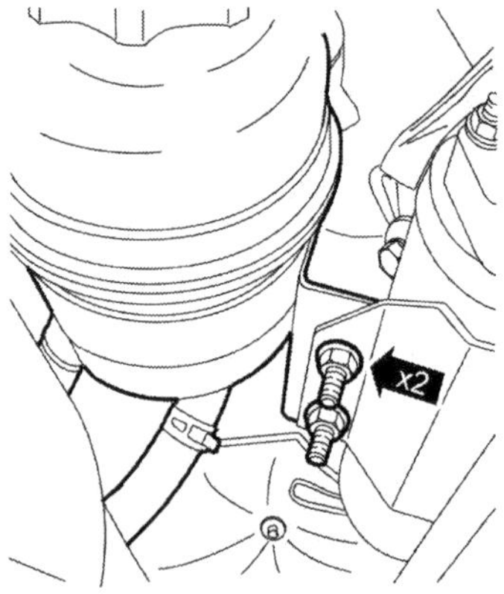

E87600

4. Release the power steering fluid reservoir.

   - Remove the 2 nuts.

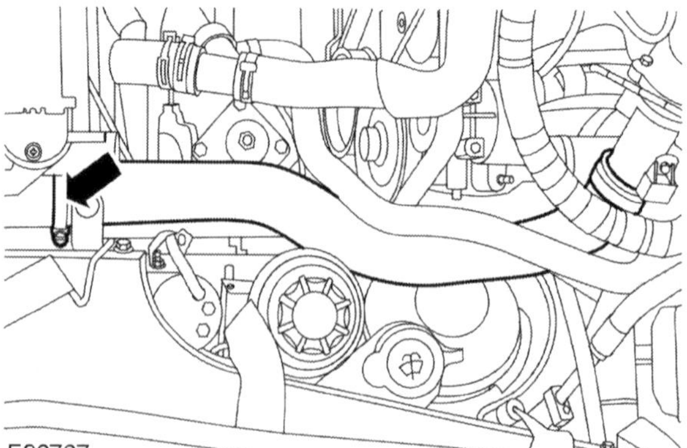

E90767

5. Disconnect the charge air cooler outlet hose.

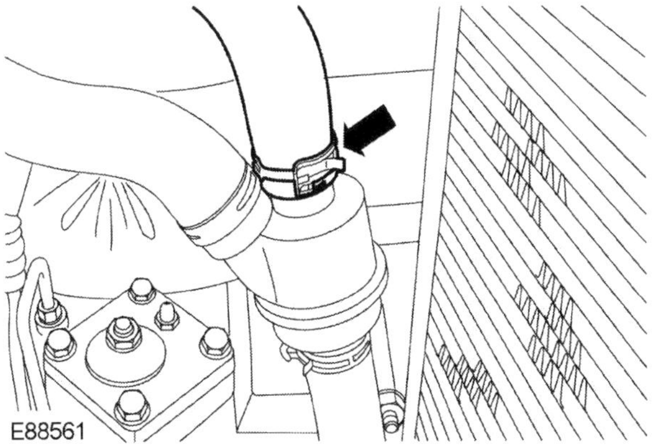

6. △ NOTE: Cooling fan shroud removed for clarity.

Disconnect the thermostat to radiator top hose.

- Release the clip.

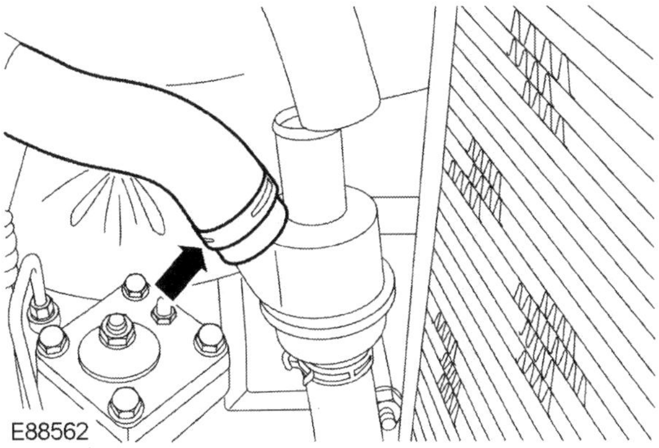

7. △ NOTE: Cooling fan shroud removed for clarity.

Disconnect the thermostat to coolant pump inlet hose.

- Release the clip.

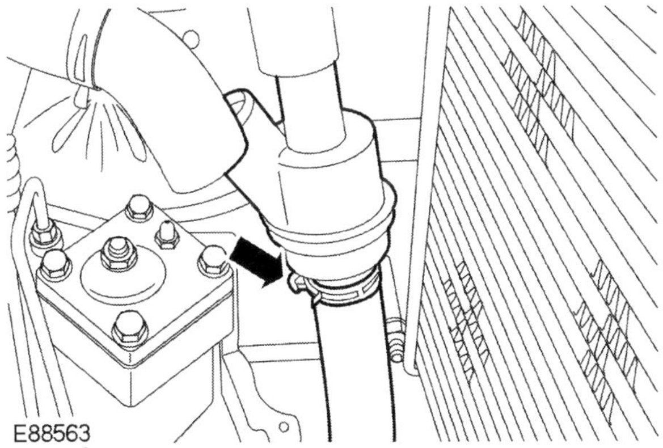

8. △ NOTE: Cooling fan shroud removed for clarity.

Remove the thermostat.

Installation

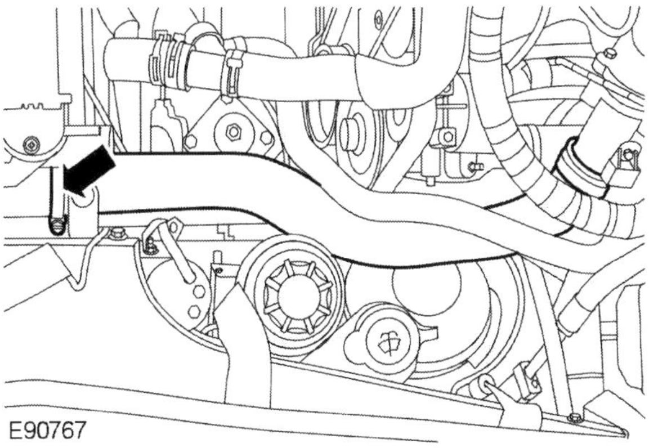

1. To install, reverse the removal procedure.
   - Tighten to 3 Nm (2 lb.ft).

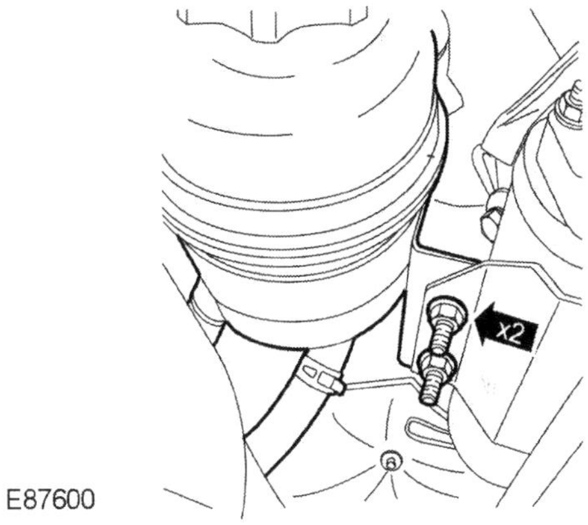

2. Tighten to 4 Nm (3 lb.ft).
3. Connect the battery ground cable.

Refer to: Startup Battery Disconnect and Connect (414-01 Battery, Mounting and Cables, General Procedures)

Specifications

Torque Specifications

Description	Nm	lb-ft
Fuel injector clamp bolt	**A**	—
High-pressure fuel supply line unions		
Stage 1	5	4
Stage 2	35	26
Fuel pressure relief valve	**A**	—
Fuel pump high-pressure fuel supply line bracket bolts	10	7
Fuel pressure relief valve spigot	20	15
Fuel rail to cylinder head		
Stage 1	8	16
Stage 2	23	17
Fuel pump bolts	23	17
Fuel pump drive gear bolts	33	24
Fuel pump drive gear nut	64	47

A = Refer to the procedure for the correct torque sequence

Fuel Charging and Controls

Description and Operation

COMPONENT LOCATION

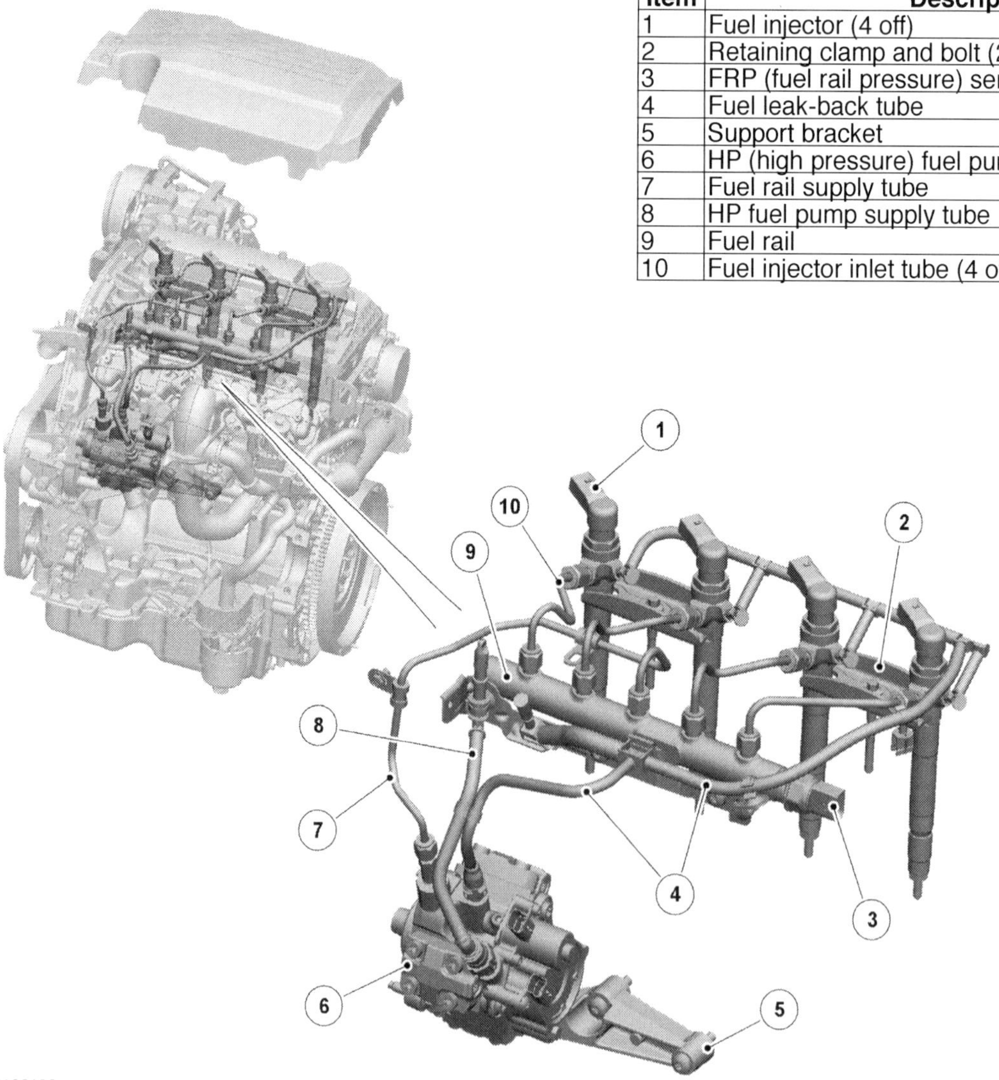

E139186

OVERVIEW

The engine is equipped with a HP common rail fuel injection system. The system contains the following components:

- HP fuel pump
- Fuel rail
- Four fuel injectors
- HP fuel pipes.

The HP fuel pump pressurizes LP (low pressure) fuel received from the fuel tank and delivers it to the fuel rail, which serves all four fuel injectors. Pressure is controlled at the optimum level for smooth operation, up to a maximum pressure of 1800 bar (26100 lbf/in ^{2}). Excess (leak-back) fuel is directed through a return line to the fuel tank.

The required fuel injection pressure is generated independently of engine speed and fuel injection events. The fuel injection timing and volume are calculated by the ECM (engine control module), which then energizes the appropriate fuel injector.

Item	Description
1	Fuel injector (4 off)
2	Retaining clamp and bolt (2 off)
3	FRP (fuel rail pressure) sensor
4	Fuel leak-back tube
5	Support bracket
6	HP (high pressure) fuel pump
7	Fuel rail supply tube
8	HP fuel pump supply tube
9	Fuel rail
10	Fuel injector inlet tube (4 off)

The common rail fuel injection system has the following features:

- High fuel injection pressures for greater atomisation of fuel (increasing performance and lowering emissions)
- Variable injection to optimize combustion in all engine operating conditions
- Low tolerances and high precision throughout the life of the system.

HIGH PRESSURE FUEL PUMP

The HP fuel pump is attached to the front left side of the cylinder block. A support bracket is attached to the rear of the pump. Drive for the pump is provided by the crankshaft via the primary chain and a sprocket on the pump drive shaft.

The HP fuel pump is a three element, radial plunger type pump, with the pumping elements spaced 120 degrees apart and operated by a cam on the drive shaft. The output from the pumping elements is connected to a fuel ring circuit within the pump, and provides a constant pressurized fuel supply to the fuel rail.

A controlled amount of fuel is used to cool and lubricate the pump. This fuel is then bled from the pump into the fuel leak-back tube and the return line to the fuel tank.

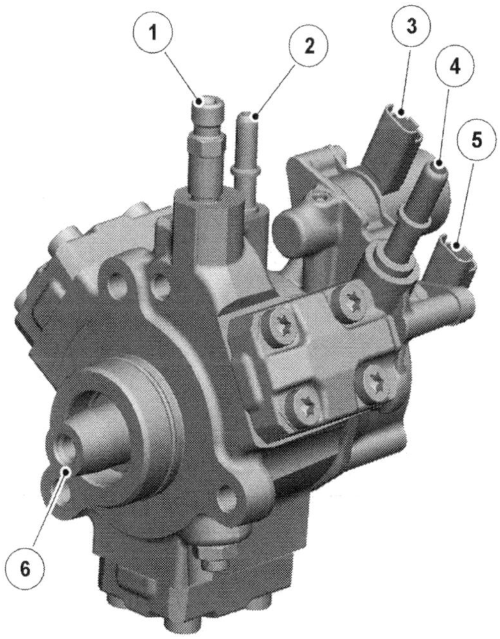

Item	Description
1	HP fuel outlet connection
2	Fuel leak-back connection
3	Electrical connector – VCV (volume control valve)
4	LP fuel inlet connection
5	Electrical connector – fuel temperature sensor
6	Drive shaft

A VCV in the HP fuel pump, controlled by the ECM, regulates the supply to the pumping elements depending on the fuel pressure in the rail. This ensures the delivery from the pump matches the requirements of the engine. When the engine is at idle speed, pump output pressure is in the 220 - 250 bar (3190 - 3625 lbf/in ^{2}) range; at typical driving speeds, pump output pressure is in the 1300 - 1600 bar (18850 - 23200 lbf/in ^{2}) range.

A fuel temperature sensor in the HP fuel pump allows the ECM to monitor the temperature of the fuel delivered from the fuel tank. The ECM uses the temperature input to make fine adjustments to fuel injection quantity and to prevent overheating of the fuel system. If the fuel temperature becomes excessive, the ECM increases the leak-back flow to the fuel tank in order to cool the fuel. If this occurs the driver may notice a loss of performance.

FUEL RAIL

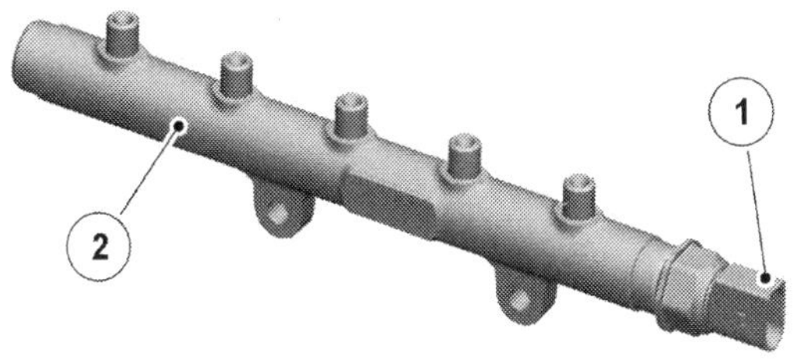

Item	Description
1	FRP sensor
2	Fuel rail

The fuel rail stores HP fuel and minimizes pressure fluctuations in the system.

Pressure fluctuations are induced by the pumping elements of the fuel pump and the opening and closing of the solenoid valves on the fuel injectors. The fuel rail is designed so that it has sufficient volume to absorb pressure fluctuations, but a small enough volume to allow the rapid increase in pressure required for a quick engine start.

Fuel from the HP fuel pump passes through a supply tube to the fuel rail. From the fuel rail, the fuel is then sent to the individual fuel injectors via the four injector inlet tubes, which are all the same length. When fuel is taken from the fuel rail for an injection process, the pressure in the fuel rail remains almost constant.

The FRP sensor is located in the end of the fuel rail. The sensor measures the pressure of the fuel in the fuel rail. This input is then used by the ECM to control the amount of fuel delivered to the fuel rail.

If the FRP sensor fails, the engine will operate at reduced power and a DTC (diagnostic trouble code) will be triggered. The sensor is not serviceable separately, and comes as part of a new fuel rail.

FUEL INJECTORS

The four fuel injectors are located in the cylinder head, one between the four valves in each cylinder. Each injector is sealed into the cylinder head with a copper washer and has connections for HP fuel inlet and leak-back tubes. An electrical connector on the fuel injector connects to the ECM via the engine harness. The fuel injectors are operated directly by the ECM for fuel metering (start of injection and quantity of fuel injected).

Each fuel injector has a solenoid valve, which, when energized, allows a ball valve to lift off its seat. This allows pressurized fuel to lift a needle valve in the injector nozzle and spray a finely atomized jet of fuel into the cylinder. Fuel that spills past the ball valve is directed into the leak-back tube.

Each injector solenoid is controlled separately by the ECM, which provides an earth path to open the injector nozzle at the correct time and for a calculated period to provide a metered injection of fuel into the cylinder. The ECM uses signals from various sensors and a programmed fueling strategy to ensure that the precise amount of fuel is injected at the correct timing for maximum fuel efficiency and minimum emissions.

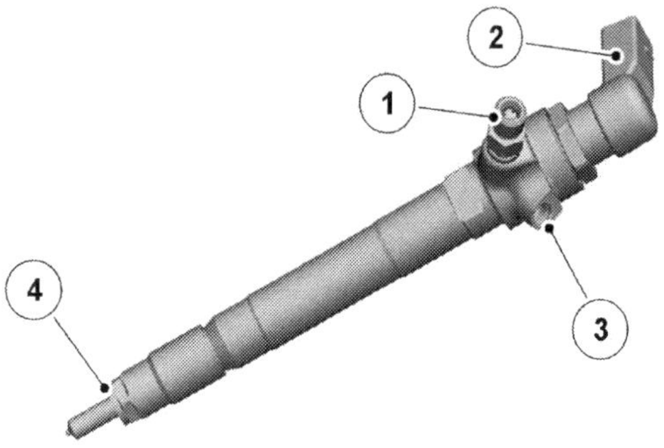

Item	Description
1	HP fuel inlet connection
2	Electrical connector
3	Fuel leak-back connection
4	Copper washer/seal

Fuel Injector Solenoid Valve

At the beginning of an injection process, the solenoid valve is actuated with a higher pick-up current so that it opens quickly. After a short period of time, the pick-up current is reduced to a low holding current.

Inside the fuel injector there are various restrictions with extremely small diameters, which have specific manufacturing tolerances. These manufacturing tolerances are given as part of the fuel injector identification number, which is located on the electrical connector. To ensure optimum fuel metering, the ECM must be informed of a change of injector through the input of the new identification number using Land Rover approved diagnostic equipment.

The identification number of each fuel injector must also be input if the ECM is replaced or if new ECM software is installed. If the correct identification numbers are not entered, the following faults may occur:

- Emissions compliance limits exceeded
- Increased black smoke formation
- Irregular idling
- Increased combustion noise.

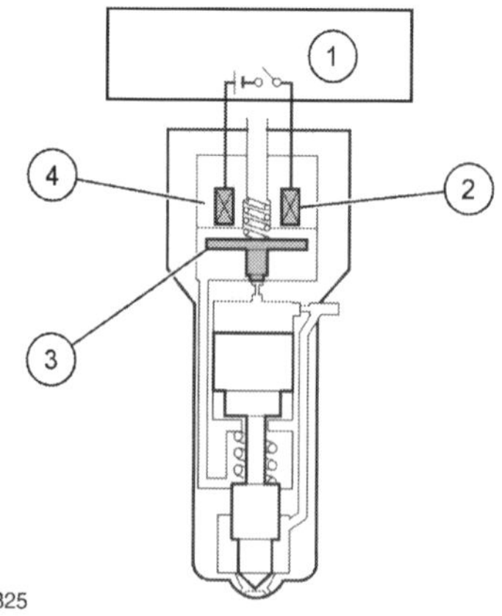

Item	Description
1	-
2	Coil
3	Solenoid armature
4	Solenoid valve

The ECM detects injector faults based on the power consumption of the solenoid valves. In the event of a fuel injector failure, any of the following symptoms may be observed:

- Engine misfire
- Idle faults
- Reduced engine performance
- Reduced fuel economy
- Difficult cold start
- Difficult hot start
- Increased smoke emissions.

Fuel Charging and Controls

Diagnosis and Testing

Principles of Operation

For a detailed description of the Fuel Charging and Controls, refer to the relevant Description and Operation section in the workshop manual.

Refer to: Fuel Charging and Controls (303-04 Fuel Charging and Controls, ID4 2.2L Diesel Description and Operation)

Inspection and Verification

⚠ WARNING: Make sure that all suitable safety precautions are observed when carrying out any work on the fuel system. Failure to observe this warning may result in personal injury.

CAUTIONS:

⚠ Diagnosis by substitution from a donor vehicle is **NOT** acceptable. Substitution of control modules does not guarantee confirmation of a fault, and may also cause additional faults in the vehicle being tested and/or the donor vehicle.

⚠ Make sure that absolute cleanliness is observed when working with these components. Always install blanking plugs to any open orifices or lines.Failure to follow this instruction may result in damage to the vehicle.

NOTES:

⚠ If the control module or a component is suspect and the vehicle remains under manufacturer warranty, refer to the Warranty Policy and Procedures manual (section B1.2), or determine if any prior approval programme is in operation, prior to the installation of a new module/component.

⚠ When performing voltage or resistance tests, always use a digital multimeter accurate to three decimal places, and with an up-to-date calibration certificate. When testing resistance always take the resistance of the digital multimeter leads into account.

⚠ Check and rectify basic faults before beginning diagnostic routines involving pinpoint tests.

1. Visually inspect for obvious signs of damage and system integrity
2. If an obvious cause for an observed or reported concern is found, correct the cause (if possible) before proceeding to the next step
3. If the cause is not visually evident, verify the symptom and refer to the Symptom Chart, alternatively check for Diagnostic Trouble Codes (DTCs) and refer to the DTC Index

Visual Inspection

Mechanical	Electrical
• Low/contaminated fuel • Fuel supply/return pipes • Fuel tank and filler pipe • Fuel leak(s) • Fuel filler cap • Fuel filter • Push connect fittings • Fuel pressure limiting valve • Fuel rail • Fuel injection pump • Exhaust gas recirculation system	• Fuses • Wiring harnesses and connectors • Engine control module (ECM) • Inertia switch • Fuel rail pressure sensor • Fuel temperature sensor • Fuel injector(s) • Exhaust gas recirculation system

DTC Index

For a list of Diagnostic Trouble Codes (DTCs) that could be logged on this vehicle, please refer to Section 100-00.

Refer to: Diagnostic Trouble Code Index - DTC: Engine Control Module (PCM) (100-00 General Information, Description and Operation)

Symptom Chart

Symptom	Possible Causes	Action
Engine cranks, but does not start	• Engine control module relay • Inertia switch tripped • Low/contaminated fuel • Intake air system leaking/restricted • Low pressure fuel system fault • Blocked fuel filter • Engine management system fault	• Check the operation of the engine control module relay • Reset the inertia switch • Check the fuel level and condition. Draw off approximately 1 ltr (2.11 pints) of fuel and allow to stand for 1 minute. Check to make sure there is no separation of the fuel indicating water or other liquid in the fuel • Check the intake air system for leaks/restrictions • Check the low pressure fuel system for leaks/damage • Check the fuel filter • Using the manufacturer approved diagnostic system, check the engine control module for related DTCs and refer to the relevant DTC index

Difficult to start	• Glow plug system fault • Low/contaminated fuel • Intake air system leaking/restricted • Low pressure fuel system fault • Blocked fuel filter • Engine management system fault	• Refer to the electrical circuit diagrams and check the glow plug circuit for short circuit to ground, short circuit to power, open circuit, high resistance • Check the fuel level and condition. Draw off approximately 1 ltr (2.11 pints) of fuel and allow to stand for 1 minute. Check to make sure there is no separation of the fuel indicating water or other liquid in the fuel • Check the intake air system for leaks/restrictions • Check the low pressure fuel system for leaks/damage • Check the fuel filter • Using the manufacturer approved diagnostic system, check the engine control module for related DTCs and refer to the relevant DTC index
Rough idle	• Intake air system leaking/restricted • Low/contaminated fuel • Low pressure fuel system fault • Blocked fuel filter • Engine management system fault	• Check the intake air system for leaks/restrictions • Check the fuel level and condition. Draw off approximately 1 ltr (2.11 pints) of fuel and allow to stand for 1 minute. Check to make sure there is no separation of the fuel indicating water or other liquid in the fuel • Check the low pressure fuel system for leaks/damage • Check the fuel filter • Using the manufacturer approved diagnostic system, check the engine control module for related DTCs and refer to the relevant DTC index
Lack of power when accelerating	• Intake air system leaking/restricted • Restricted exhaust system • Low pressure fuel system fault • Charge air cooler leaking • Charge air cooler hose leaking • Engine management system fault	• Check the intake air system for leaks/restrictions • Check the exhaust system for blockages/restrictions • Check the low pressure fuel system for leaks/damage • Check the charge air cooler for leaks • Check the charge air cooler hoses for leaks • Using the manufacturer approved diagnostic system, check the engine control module for related DTCs and refer to the relevant DTC index
Engine stops/stalls	• Intake air system leaking/restricted • Low/contaminated fuel • Low pressure fuel system fault • High pressure fuel system leak • Engine management system fault	• Check the intake air system for leaks/restrictions • Check the fuel level and condition. Draw off approximately 1 ltr (2.11 pints) of fuel and allow to stand for 1 minute. Check to make sure there is no separation of the fuel indicating water or other liquid in the fuel • Check the low pressure fuel system for leaks/damage • Check the high pressure fuel system for leaks • Using the manufacturer approved diagnostic system, check the engine control module for related DTCs and refer to the relevant DTC index
Engine judders	• Intake air system leaking/restricted • Low/contaminated fuel • Low pressure fuel system fault • High pressure fuel system leak • Engine management system fault	• Check the intake air system for leaks/restrictions • Check the fuel level and condition. Draw off approximately 1 ltr (2.11 pints) of fuel and allow to stand for 1 minute. Check to make sure there is no separation of the fuel indicating water or other liquid in the fuel • Check the low pressure fuel system for leaks/damage • Check the high pressure fuel system for leaks • Using the manufacturer approved diagnostic system, check the engine control module for related DTCs and refer to the relevant DTC index
Excessive fuel consumption	• Low pressure fuel system fault • High pressure fuel system leak • Engine management system fault	• Check the low pressure fuel system for leaks/damage • Check the high pressure fuel system for leaks • Using the manufacturer approved diagnostic system, check the engine control module for related DTCs and refer to the relevant DTC index

Fuel Injection Component Cleaning

General Procedures

General Equipment

Pneumatic vacuum gun

1. WARNINGS:

⚠ Do not carry out any repairs to the fuel system with the engine running. The fuel pressure within the system can be as high as 1700 bar (24,656 lb-sq-in). Failure to follow this instruction may result in personal injury.

⚠ Do not smoke or carry lighted tobacco or open flame of any type when working on or near any fuel related components. Highly flammable vapors are always present and may ignite. Failure to follow these instructions may result in personal injury.

⚠ If fuel contacts the eyes, flush the eyes with cold water or eyewash solution and seek immediate medical attention.

⚠ Place the vehicle in a well ventilated, quarantined area and arrange No Smoking/ Petrol Fumes signs about the vehicle.

⚠ Wait at least 30 seconds after the engine stops before commencing any repair to the high-pressure fuel injection system. Failure to follow this instruction may result in personal injury.

⚠ Wash hands thoroughly after fuel handling, as prolonged contact may cause irritation. Should irritation develop, seek medical attention.

⚠ Do not carry or operate cellular phones when working on or near any fuel related components. Highly flammable vapors are always present and may ignite. Failure to follow these instructions may result in personal injury.

CAUTIONS:

⚠ Before using the cleaning fluid, protect all electrical components and connectors with lint-free non-flocking material.

⚠ Make sure that all parts removed from the vehicle are placed on the lint-free non-flocking material.

⚠ Make sure that any protective clothing worn is clean and made from lint-free non-flocking material.

⚠ Make sure that clean non-plated tools are used. Clean tools using a new brush that will not lose its bristles and fresh cleaning fluid, prior to starting work on the vehicle.

⚠ Use a steel topped workbench and cover it with clean, lint-free non-flocking material.

⚠ Make sure the workshop area in which the vehicle is being worked on is as clean and as dust free as possible. Foreign matter from work on clutches, brakes or from machining or welding operations can contaminate the fuel system and may result in later malfunction.

Using a new brush that will not lose its bristles, brush cleaning fluid onto the components being removed and onto the surrounding area.

2. Using a pneumatic vacuum gun, remove all traces of cleaning fluid and foreign material.
3. Dispose of any used cleaning fluid and the brush after completing the repair.

Fuel Injector

Removal and Installation

Removal

WARNINGS:

⚠ WARNING: The spilling of fuel is unavoidable during this operation. Make sure that all necessary precautions are taken to prevent fire and explosion.

⚠ WARNING: Do not carry or operate cellular phones when working on or near any fuel related components. Highly flammable vapors are always present and may ignite. Failure to follow these instructions may result in personal injury.

⚠ WARNING: Do not smoke or carry lighted tobacco or open flame of any type when working on or near any fuel related components. Highly flammable vapors are always present and may ignite. Failure to follow these instructions may result in personal injury.

⚠ WARNING: If fuel contacts the eyes, flush the eyes with cold water or eyewash solution and seek immediate medical attention.

⚠ WARNING: Wash hands thoroughly after fuel handling, as prolonged contact may cause irritation. Should irritation develop, seek medical attention.

WARNING: Do not carry out any repairs to the fuel system with the engine running. The fuel pressure within the system can be as high as 2000 bar (29,008 lb-sq-in). Failure to follow this instruction may result in personal injury.

CAUTIONS:

CAUTION: Diesel fuel injection equipment is manufactured to very precise tolerances and fine clearances. It is therefore essential that absolute cleanliness is observed when working with these components. Before disconnecting any components, make sure the area is clean and free from foreign material. When disconnected all openings must be sealed. Failure to follow this instruction may result in foreign matter ingress to the fuel injection system.

CAUTION: Always carry out the cleaning process before carrying out any repairs to the fuel injection system components. Failure to follow this instruction may result in foreign matter ingress to the fuel injection system.

CAUTION: Do not disconnect the fuel injector electrical connectors with the engine running. Failure to follow this instruction may result in damage to the engine.

NOTE: Removal steps in this procedure may contain installation details.

1. **Refer to:** Startup Battery Disconnect and Connect (414-01 Battery, Mounting and Cables, General Procedures)
2. **Refer to:** Engine Cover (501-05 Interior Trim and Ornamentation, Removal and Installation)
3. **Refer to:** Fuel Injection Component Cleaning (303-04 Fuel Charging and Controls, ID4 2.2L Diesel General Procedures)

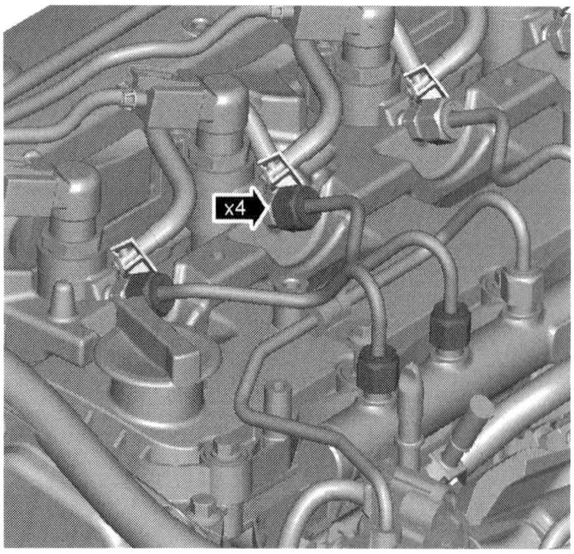

4. CAUTIONS:

CAUTION: Do not allow the unions to hit the olive ends of the high-pressure fuel supply line as this may damage the ends of the line and allow foreign matter to enter the fuel injection system.

Before disconnecting any components, make sure the area is clean and free from foreign material. When disconnected all openings must be sealed.

Torque stage 1: 15 Nm, Stage 2: 30 Nm

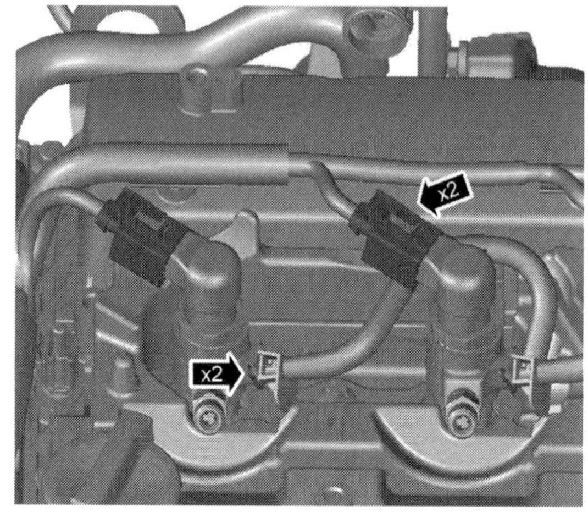

5.

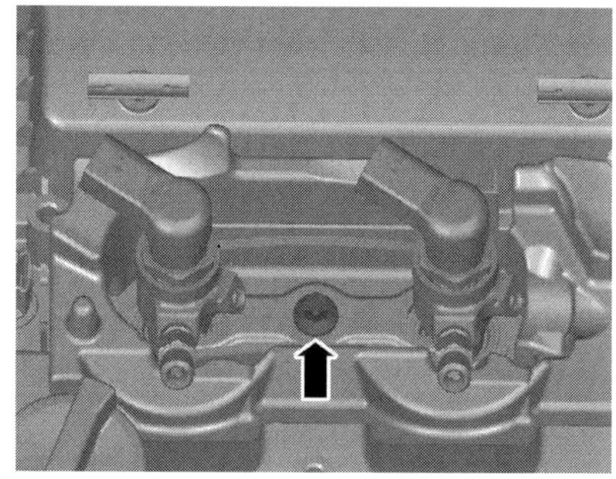

6. CAUTIONS:

CAUTION: Make sure that both fuel injectors are removed. Failure to follow this instruction may result in damage to the vehicle.

Before disconnecting any components, make sure the area is clean and free from foreign material. When disconnected all openings must be sealed.

NOTES:

Note the position of the fuel injector clamp.

Re use bolt only once

Torque stage 1: 6 Nm, Stage 2: 360°

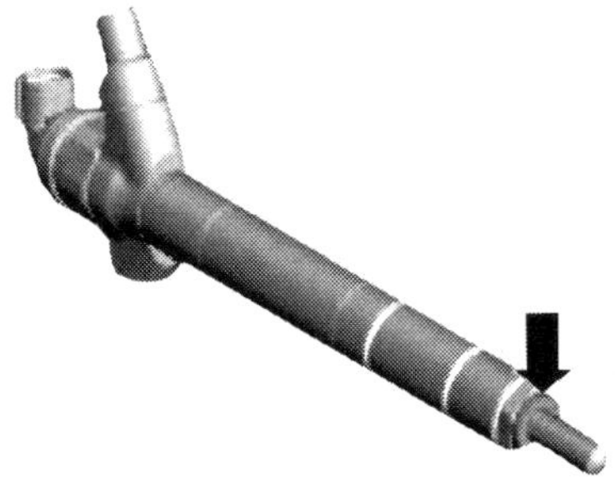

7. ⚠ NOTE: Install a new fuel injector sealing washer.

Installation

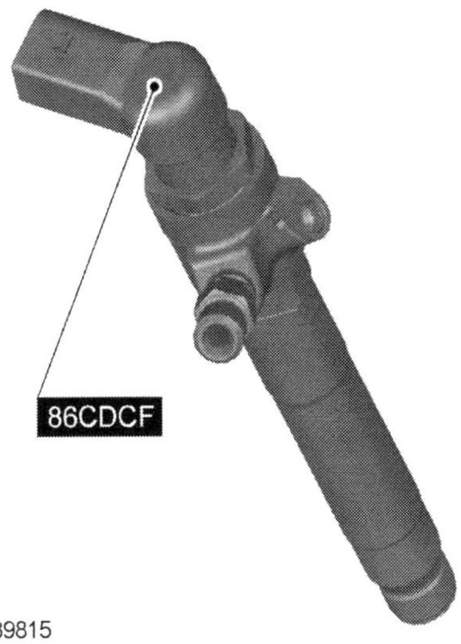

E139815

1. Record the identification code from the new fuel injector to upload into the approved diagnostic system during configuration (graphic shows an example of the identification code only).

2. ⚠ NOTE: Check fuel pipes for leaks after starting engine.

To install, reverse the removal procedure.

Fuel Injection Pump

Removal and Installation

Special Tool(s)

 E61755	Locking Tool, Fuel Pump Sprocket 303-1151
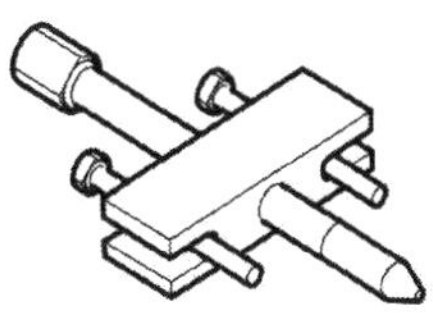 E54428	Remover Crankshaft Damper Pulley 303-1149
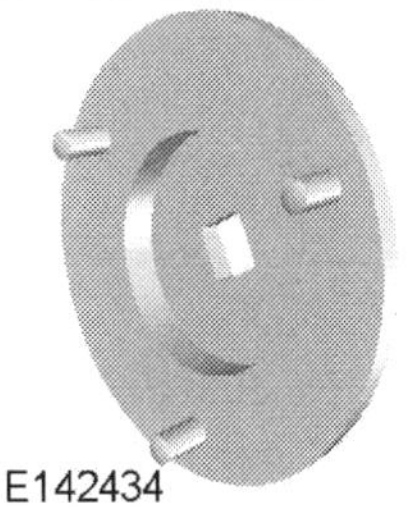 E142434	Remover Crankshaft Damper Oil Seal JLR 303-1619
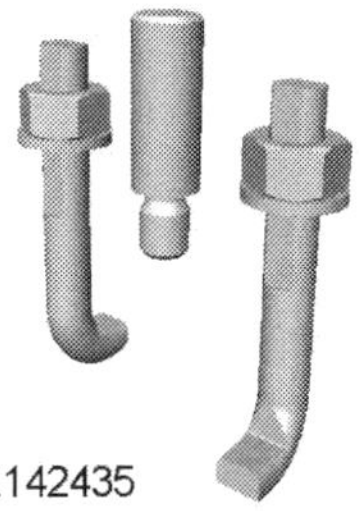 E142435	Locking Tool, Fuel Pump Sprocket JLR 303-249-01
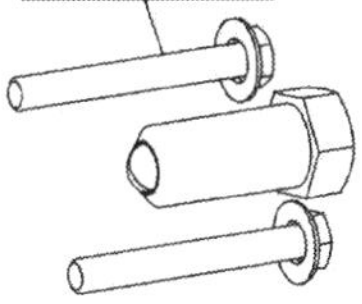 E91082	Remover Fuel Pump 303-1333

Removal

WARNINGS:

⚠ The spilling of fuel is unavoidable during this operation. Make sure that all necessary precautions are taken to prevent fire and explosion.

⚠ Do not carry or operate cellular phones when working on or near any fuel related components. Highly flammable vapors are always present and may ignite. Failure to follow these instructions may result in personal injury.

⚠ Do not smoke or carry lighted tobacco or open flame of any type when working on or near any fuel related components. Highly flammable vapors are always present and may ignite. Failure to follow these instructions may result in personal injury.

⚠ If fuel contacts the eyes, flush the eyes with cold water or eyewash solution and seek immediate medical attention.

⚠ Wash hands thoroughly after fuel handling, as prolonged contact may cause irritation. Should irritation develop, seek medical attention.

⚠ Do not carry out any repairs to the fuel system with the engine running. The fuel pressure within the system can be as high as 2000 bar (29,008 lb-sq-in). Failure to follow this instruction may result in personal injury.

CAUTION: Diesel fuel injection equipment is manufactured to very precise tolerances and fine clearances. It is therefore essential that absolute cleanliness is observed when working with these components. Always install new blanking plugs to any open orifices or lines. Failure to follow this instruction may result in foreign matter ingress to the fuel injection system.

1. **Refer to:** Accessory Drive Component Bracket (303-05 Accessory Drive, ID4 2.2L Diesel Removal and Installation)
2. **Refer to:** Intake Manifold (303-01 Engine, ID4 2.2L Diesel Removal and Installation)

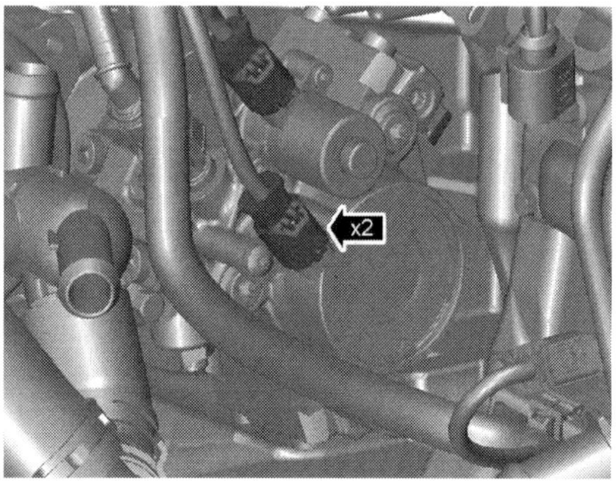

3.

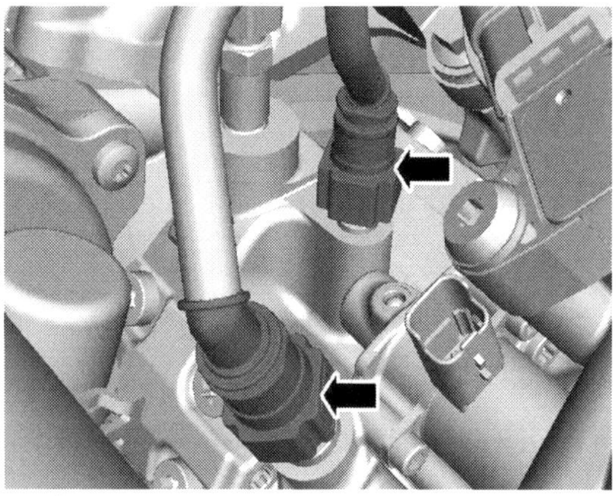

4. WARNING: The spilling of fuel is unavoidable during this operation. Ensure that all necessary precautions are taken to prevent fire and explosion.

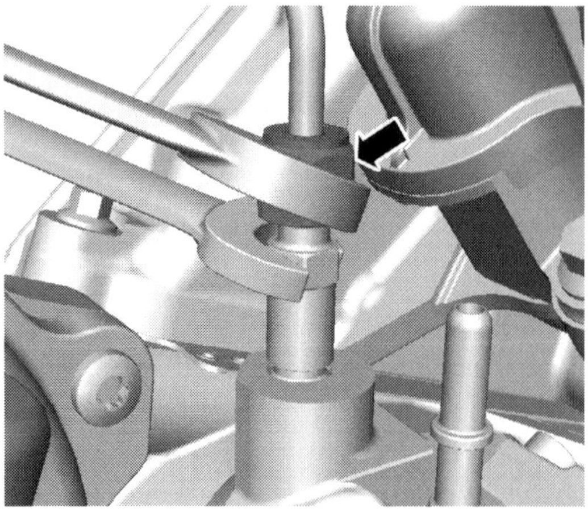

CAUTION: Before disconnecting any components, make sure the area is clean and free from foreign material. When disconnected all openings must be sealed.

5. WARNING: The spilling of fuel is unavoidable during this operation. Ensure that all necessary precautions are taken to prevent fire and explosion.

CAUTIONS:

Make sure the fuel pump connection is supported and does not rotate during the removal of the high pressure fuel pipe, as shown in the illustration.

Before disconnecting any components, make sure the area is clean and free from foreign material. When disconnected all openings must be sealed.

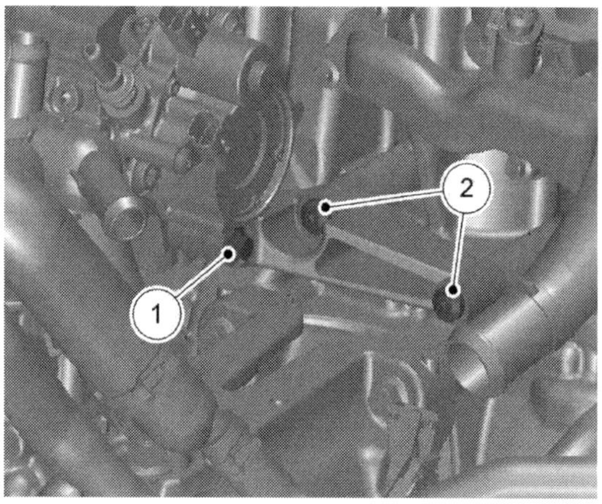

6.

7.

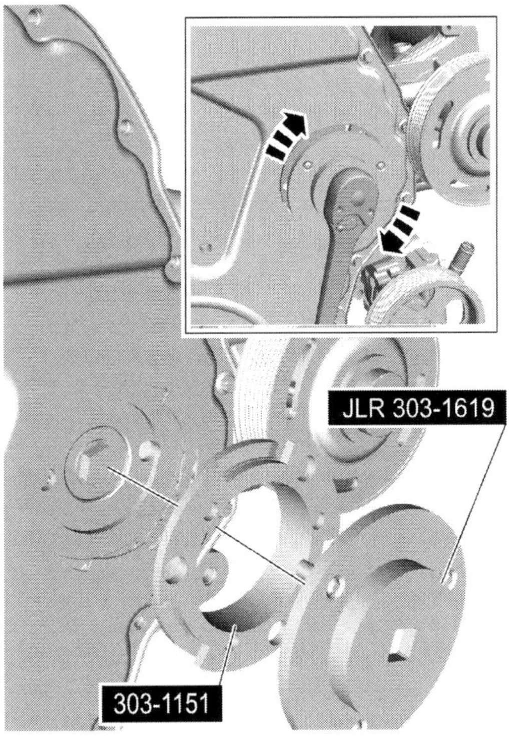

8. Install the special tool.

9. CAUTION: The fuel pump sprocket bolt has a LH thread.

10. Install the special tool.

11. Install the special tool.

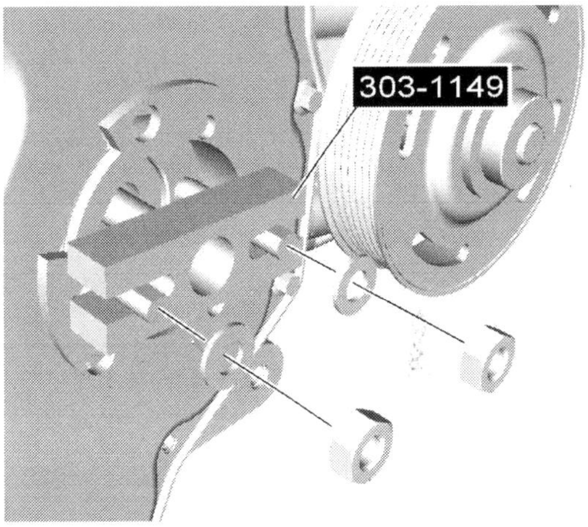

12. Install the special tool.

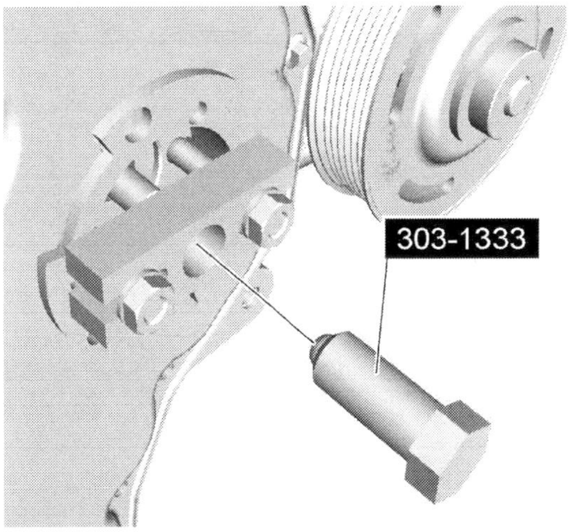

13. Install the special tool.

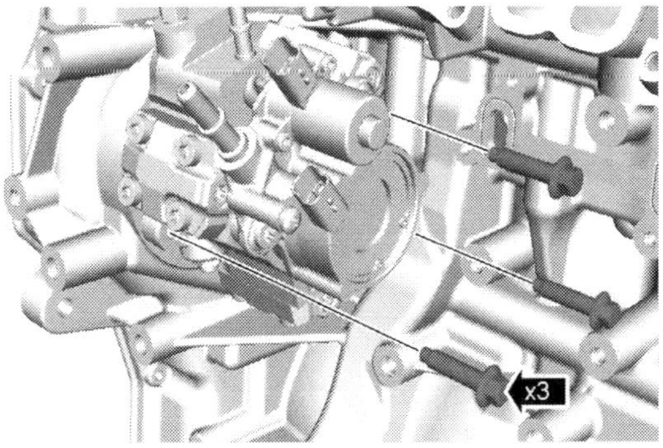

14. Install the special tool.

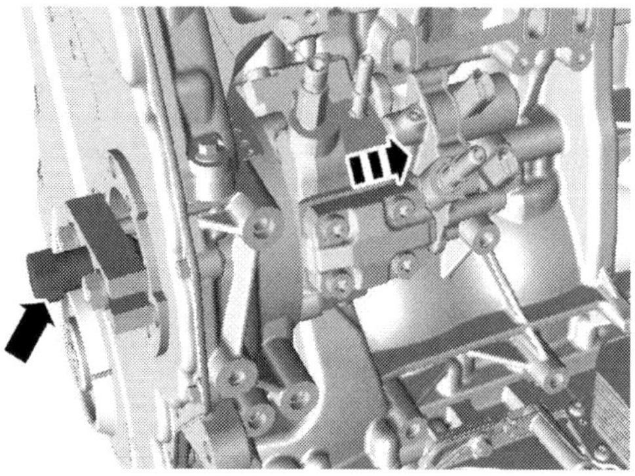

15. ⚠ WARNING: Once the fuel pump sprocket has been released from the taper stop rotating special tool 303-1333-1. Failure to stop at this point may result in damage to the fuel pump.

Rotate special tool 303-1333-1 until the High pressure fuel pump is released from the engine.

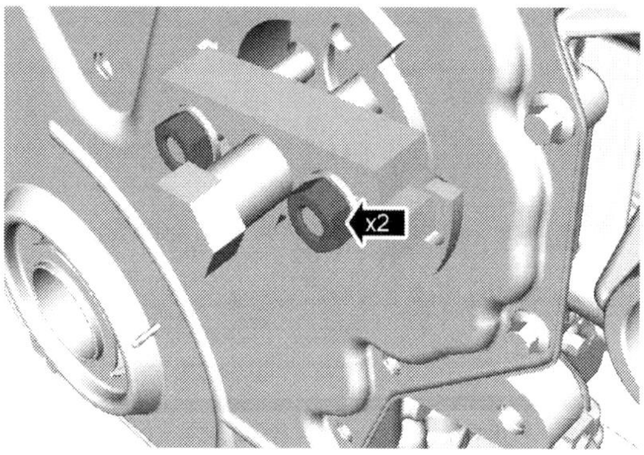

16. Torque both nuts: 8 Nm

17. Remove the special tool.

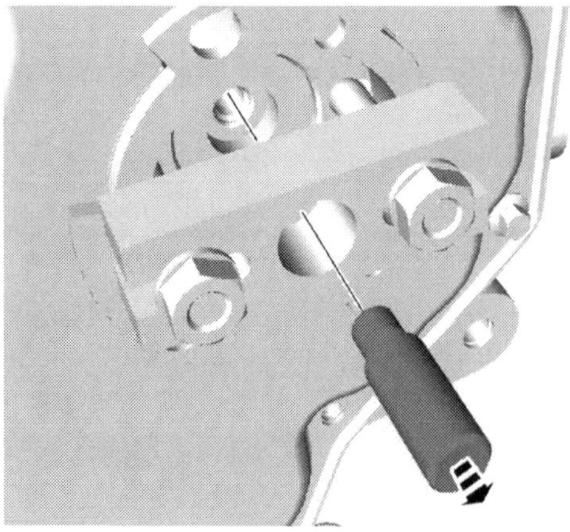

18. Remove the special tool.

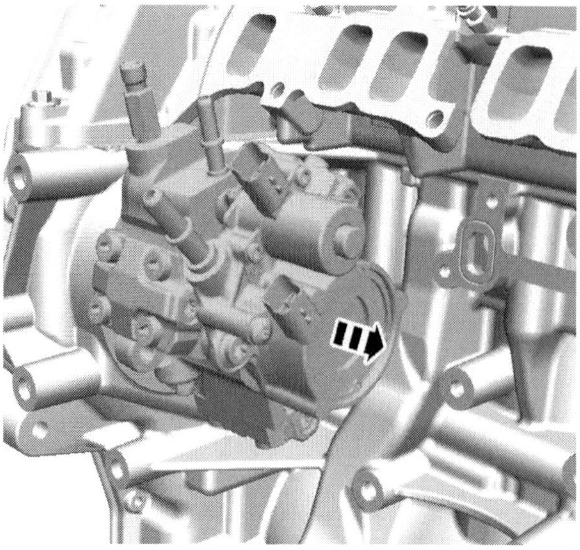

19.

Installation

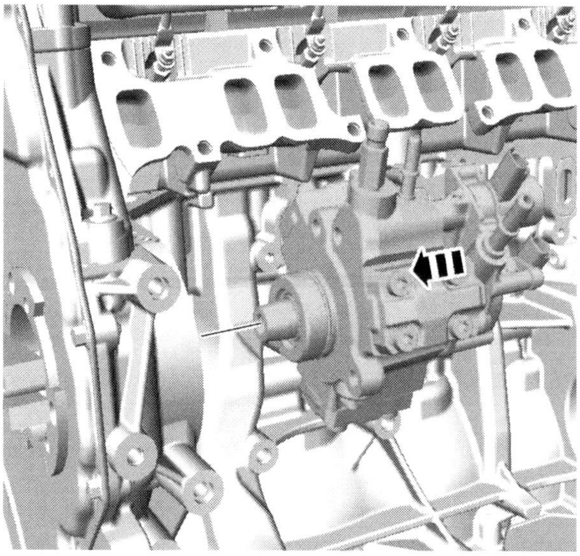

1.

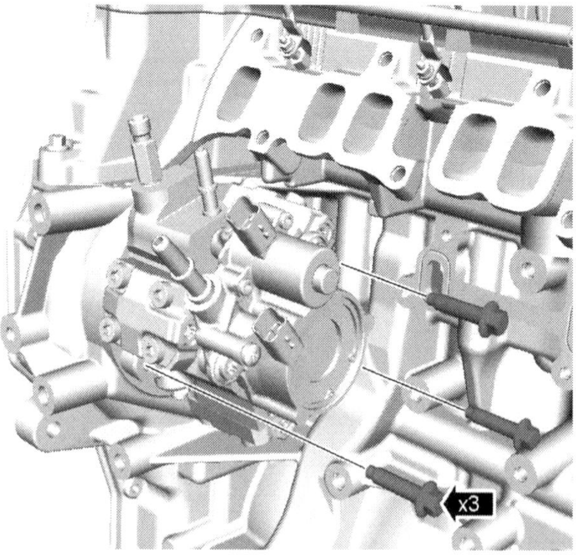

2. Torque the bolt: 23 Nm

3. Remove the special tools.

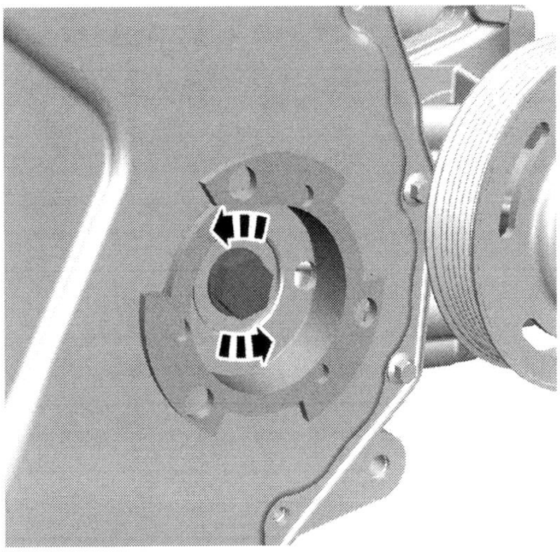

4. ⚠ CAUTION: The fuel pump sprocket bolt has a LH thread.

Torque the bolt: 64 Nm

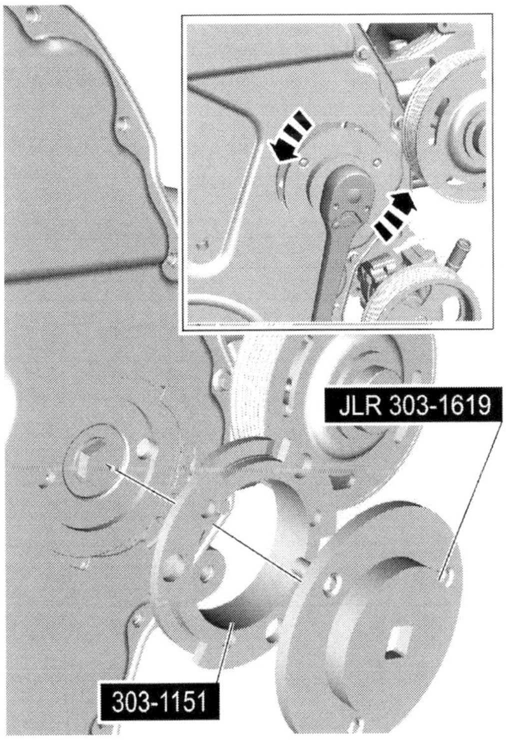

5. Remove the special tools.

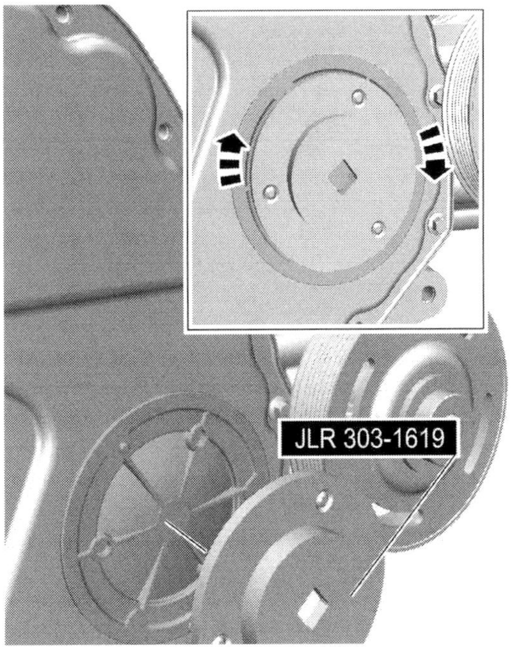

6.

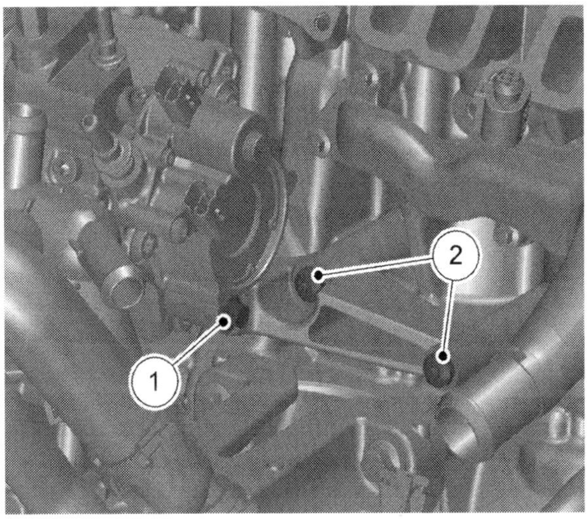

7. Torque: 23 Nm

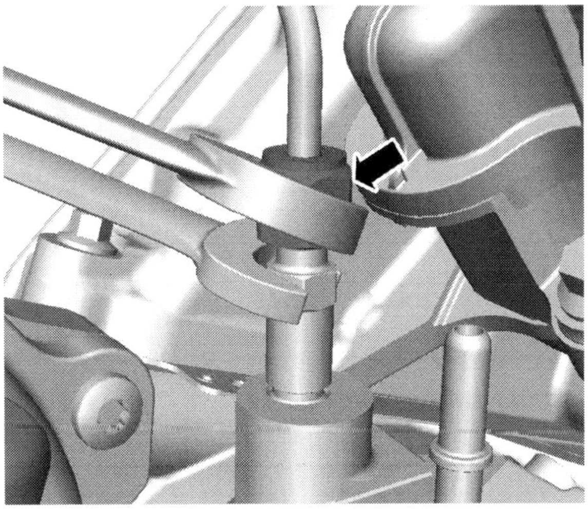

8. ⚠ CAUTION: Make sure the fuel pump connection is supported and does not rotate during the installation of the high pressure fuel pipe, as shown in the illustration.

Torque: 35 Nm

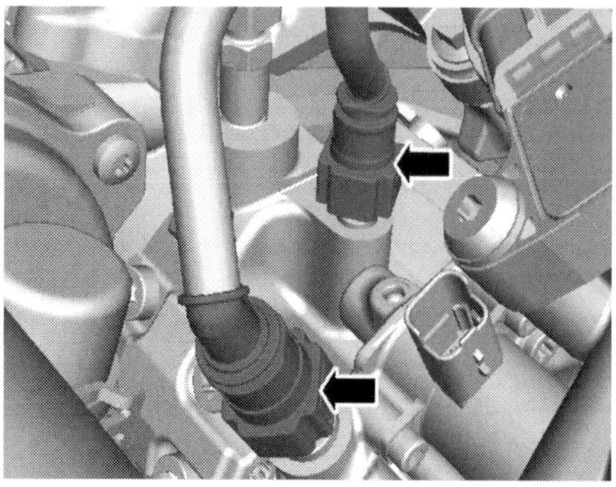

9.

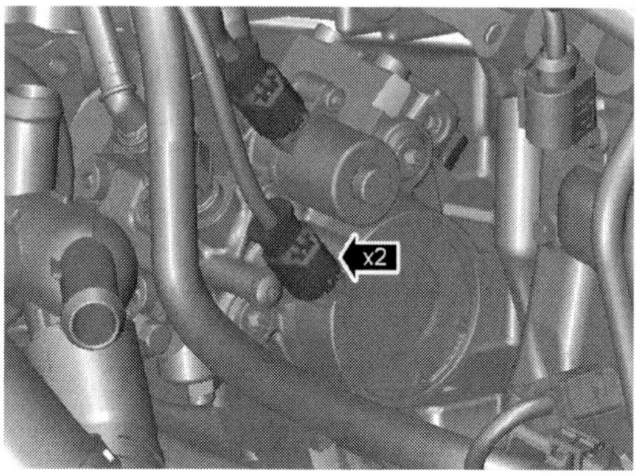

10.

11. Refer to: Intake Manifold (303-01 Engine, ID4 2.2L Diesel Removal and Installation)

12. Refer to: Accessory Drive Component Bracket (303-05 Accessory Drive, ID4 2.2L Diesel Removal and Installation)

13. NOTE: Check fuel pipes for leaks after starting engine.

Fuel Metering Valve

Removal and Installation

Removal

NOTE: Some variation in the illustrations may occur, but the essential information is always correct.

1. WARNINGS:

The spilling of fuel is unavoidable during this operation. Make sure that all necessary precautions are taken to prevent fire and explosion.

Do not carry or operate cellular phones when working on or near any fuel related components. Highly flammable vapors are always present and may ignite. Failure to follow these instructions may result in personal injury.

Do not smoke or carry lighted tobacco or open flame of any type when working on or near any fuel related components. Highly flammable vapors are always present and may ignite. Failure to follow these instructions may result in personal injury.

If fuel contacts the eyes, flush the eyes with cold water or eyewash solution and seek immediate medical attention.

Wash hands thoroughly after fuel handling, as prolonged contact may cause irritation. Should irritation develop, seek medical attention.

Do not carry out any repairs to the fuel system with the engine running. The fuel pressure within the system can be as high as 2000 bar (29,008 lb-sq-in). Failure to follow this instruction may result in personal injury.

CAUTION: Diesel fuel injection equipment is manufactured to very precise tolerances and fine clearances. It is therefore essential that absolute cleanliness is observed when working with these components. Always install new blanking plugs to any open orifices or lines. Failure to follow this instruction may result in foreign matter ingress to the fuel injection system.

2. Open the front door.
3. Remove the front seat cushion assembly.
4. Remove the battery cover.
5. Disconnect the battery ground cable.
6. Open the bonnet.
7. **Refer to:** Intake Manifold (303-01 Engine, ID4 2.2L Diesel Removal and Installation)

Remove the intake manifold

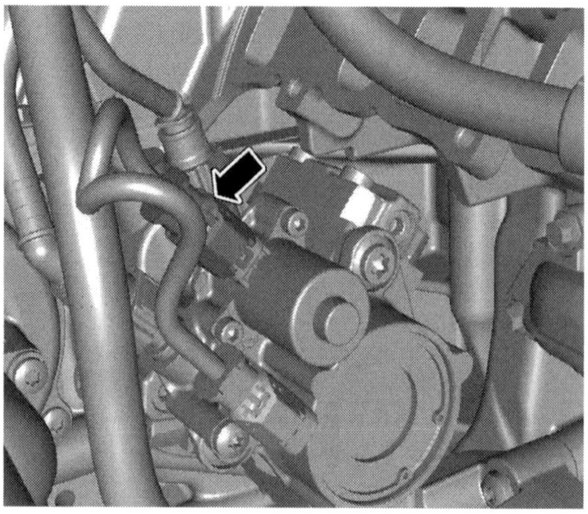

8. Disconnect the volume control valve (VCV) electrical connector.

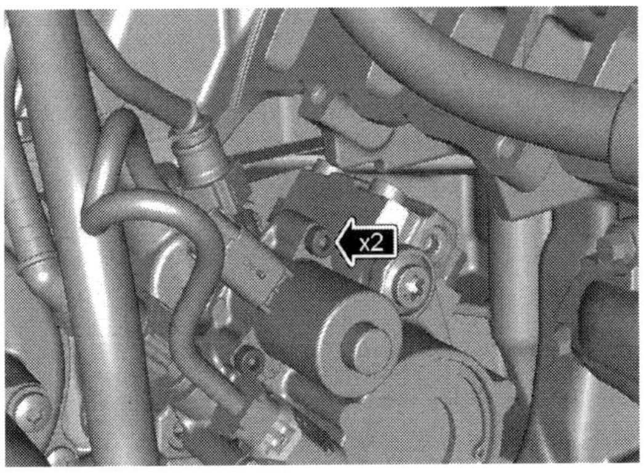

9. Remove the 2 volume control valve securing bolts.

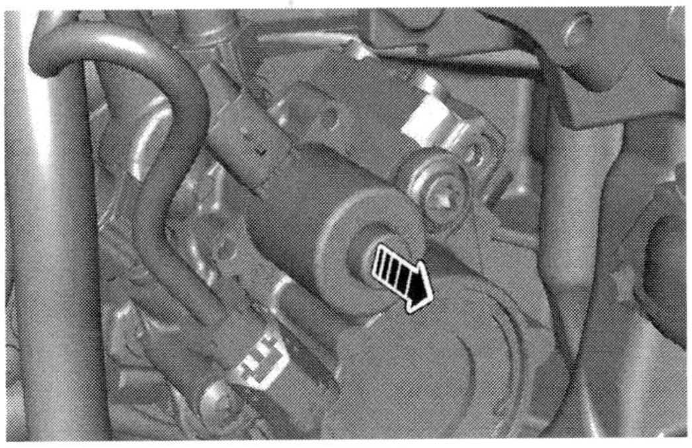

10. Remove the volume control valve.
11. Using the pneumatic vacuum gun, vacuum foreign material from the high-pressure fuel rail.

Installation

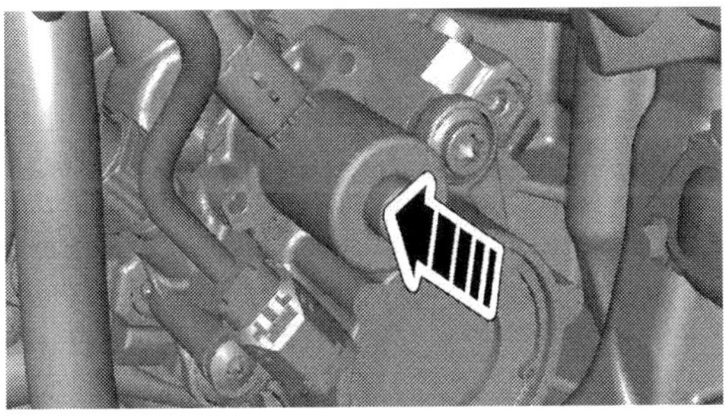

1. Install the volume control valve with new seal and gasket.

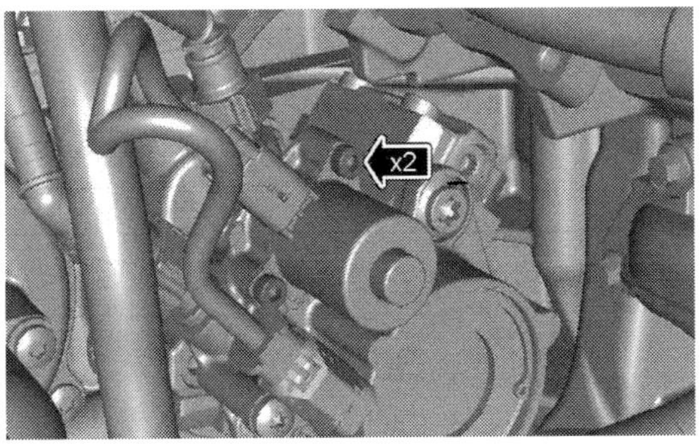

2. Install the 2 volume control valve securing bolts. Stage 1 tighten to 5 Nm (4 lb.ft). Stage 2 tighten to 8 Nm (6 lb.ft).
3. Connect volume control valve electrical connector.
4. **Refer to:** Intake Manifold (303-01 Engine, ID4 2.2L Diesel Removal and Installation)

Remove the intake manifold

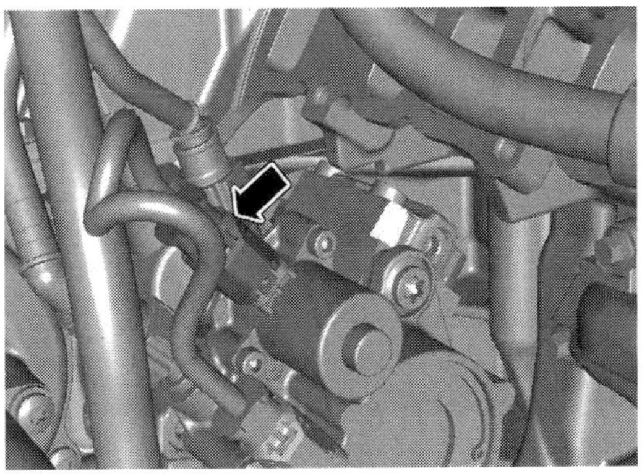

5. Close the bonnet.
6. Connect the battery ground cable.
7. Install the battery cover.
8. Install the front seat cushion assembly.
9. Close the front door.
10. Using the Land Rover approved diagnostic system carry out the pump learn procedure found in the set-up and configuration section.

Fuel Rail

Removal and Installation

Removal

WARNINGS:

WARNING: The spilling of fuel is unavoidable during this operation. Make sure that all necessary precautions are taken to prevent fire and explosion.

WARNING: Do not carry or operate cellular phones when working on or near any fuel related components. Highly flammable vapors are always present and may ignite. Failure to follow these instructions may result in personal injury.

WARNING: Do not smoke or carry lighted tobacco or open flame of any type when working on or near any fuel related components. Highly flammable vapors are always present and may ignite. Failure to follow these instructions may result in personal injury.

WARNING: If fuel contacts the eyes, flush the eyes with cold water or eyewash solution and seek immediate medical attention.

WARNING: Wash hands thoroughly after fuel handling, as prolonged contact may cause irritation. Should irritation develop, seek medical attention.

WARNING: Do not carry out any repairs to the

fuel system with the engine running.

The fuel pressure within the system can be as high as 2000 bar (29,008 lb-sq-in). Failure to follow this instruction may result in personal injury.

CAUTIONS:

CAUTION: Diesel fuel injection equipment is manufactured to very precise tolerances and fine clearances. It is therefore essential that absolute cleanliness is observed when working with these components. Before disconnecting any components, make sure the area is clean and free from foreign material. When disconnected all openings must be sealed. Failure to follow this instruction may result in foreign matter ingress to the fuel injection system.

CAUTION: Always carry out the cleaning process before carrying out any repairs to the fuel injection system components. Failure to follow this instruction may result in foreign matter ingress to the fuel injection system.

CAUTION: Do not disconnect the fuel injector electrical connectors with the engine running. Failure to follow this instruction may result in damage to the engine.

CAUTION: Make sure the workshop area in which the vehicle is being worked on is as clean and as dust free as possible. Foreign matter from work on clutches, brakes or from machining or welding operations can contaminate the fuel system and may result in later malfunction.

NOTE: Removal steps in this procedure may contain installation details.

1. **Refer to:** Startup Battery Disconnect and Connect (414-01 Battery, Mounting and Cables, General Procedures)
2. **Refer to:** Engine Cover (501-05 Interior Trim and Ornamentation, Removal and Installation)
3. **Refer to:** Fuel Injection Component Cleaning (303-04 Fuel Charging and Controls, ID4 2.2L Diesel General Procedures)

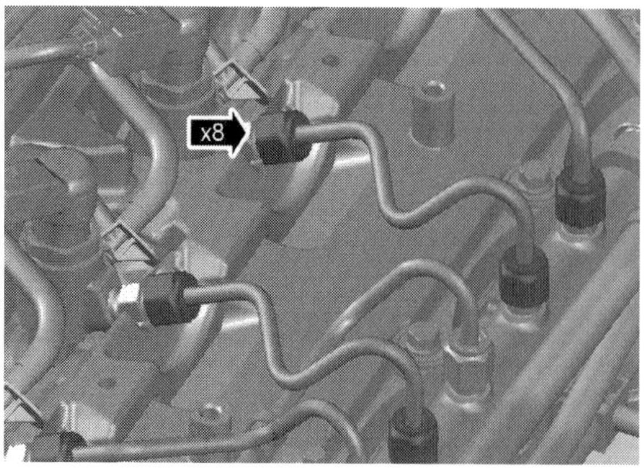

4. CAUTIONS:

CAUTION: Do not allow the unions to hit the olive ends of the high-pressure fuel supply line as this may damage the ends of the line and allow foreign matter to enter the fuel injection system.

Before disconnecting any components, make sure the area is clean and free from foreign material. When disconnected all openings must be sealed.

Union torque stage 1: 15 Nm, Stage 2 torque: 30 Nm

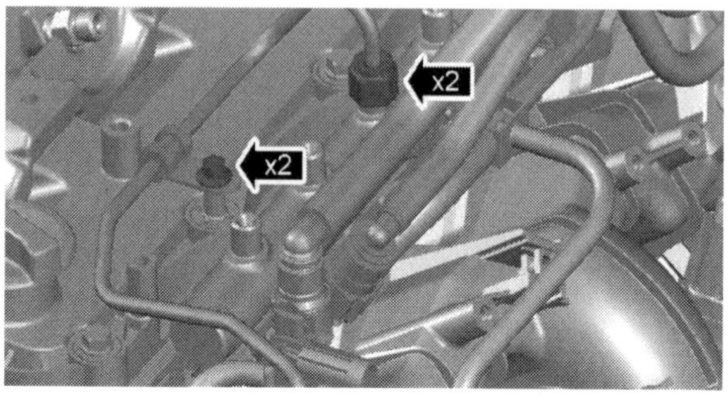

5. CAUTIONS:

CAUTION: Do not allow the unions to hit the olive ends of the high-pressure fuel supply line as this may damage the ends of the line and allow foreign matter to enter the fuel injection system.

Before disconnecting any components, make sure the area is clean and free from foreign material. When disconnected all openings must be sealed.

Union torque stage 1: 15 Nm, Stage 2: 30 Nm, Bracket bolts torque: 10 Nm

6.

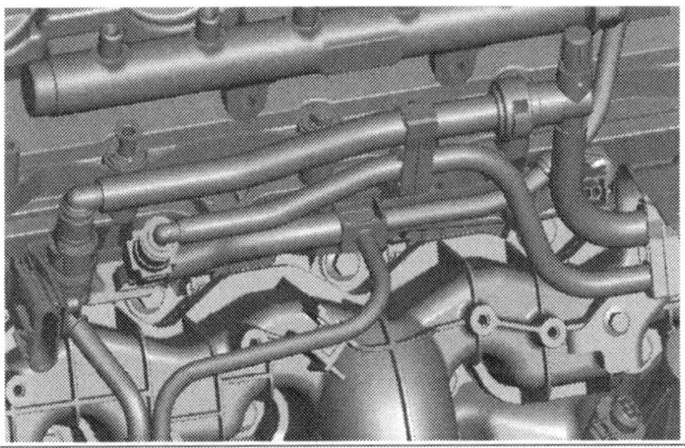

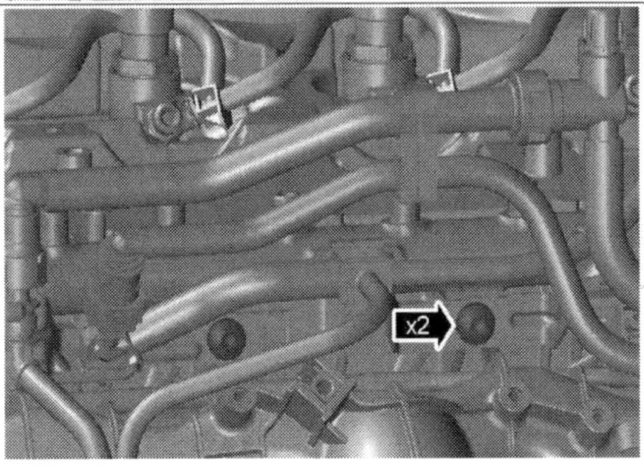

7. Torque stage 1: 8 Nm, Stage 2: 23 Nm

Installation

1. NOTE: Check fuel pipes for leaks after starting engine.

To install, reverse the removal procedure.

Throttle Body

Removal and Installation

Removal

NOTE: Removal steps in this procedure may contain installation details.

1. **Refer to:** Startup Battery Disconnect and Connect (414-01 Battery, Mounting and Cables, General Procedures)
2. **Refer to:** Cooling System Draining, Filling and Bleeding (303-03 Engine Cooling, ID4 2.2L Diesel General Procedures)

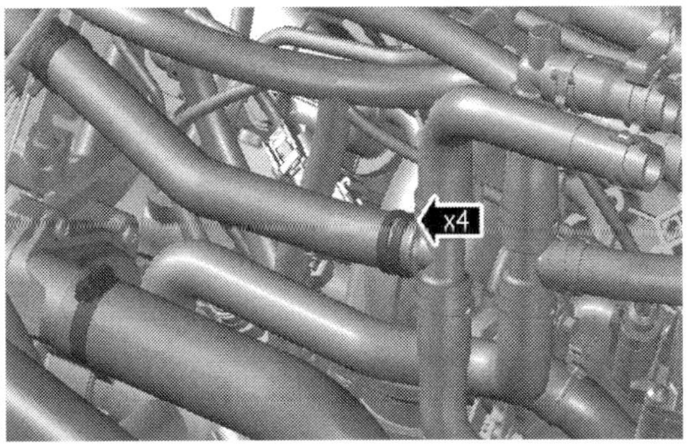

3. CAUTION: Before disconnecting any components, make sure the area is clean and free from foreign material. When disconnected all openings must be sealed.

4.

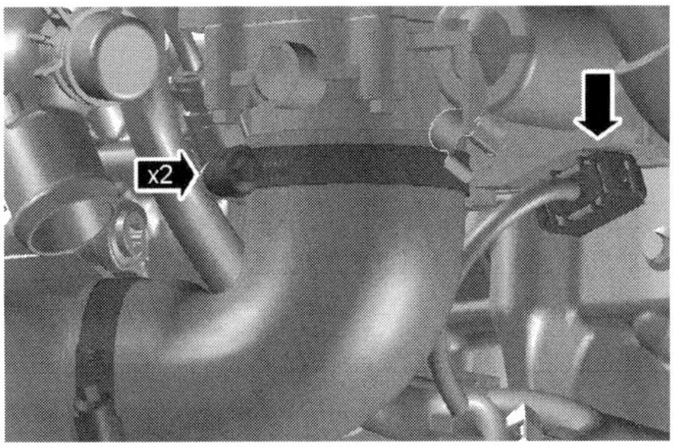

5. Torque: 3 Nm

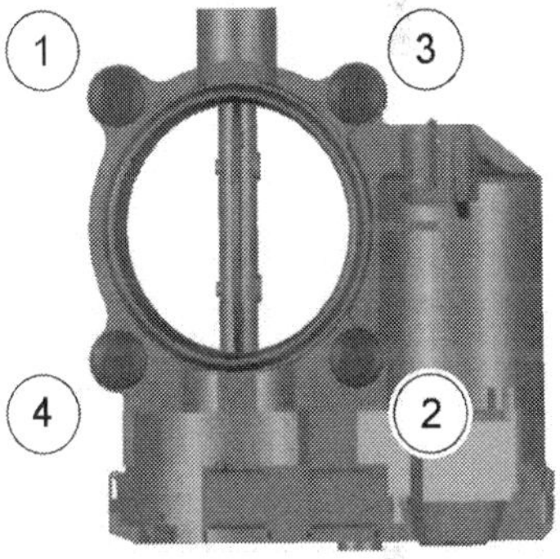

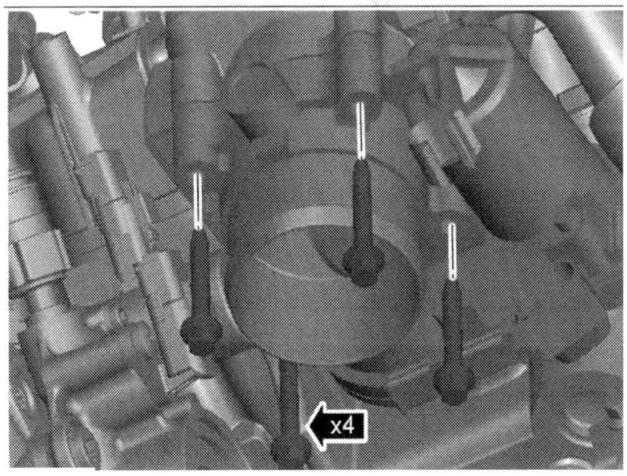

6. NOTES:

Fit new gasket.

Tighten the bolts in the indicated sequence.

Torque Stage 1: 4 Nm, Stage 2: 7 Nm

Installation

1. To install, reverse the removal procedure.

Turbocharger

Specifications

Description	Nm	lb-ft
Turbocharger inlet duct bolts	23	17
Turbocharger nuts	23	17
Turbocharger oil return tube	10	7
Turbocharger oil supply pipe	35	26
Catalytic converter nuts	45	33
Catalytic converter lower mounting bracket bolts	30	22
Front muffler to catalytic converter nuts	30	22
Turbocharger mounting bracket	20	15
Turbocharger heat shield	10	7
Turbocharger to charge air cooler clip	5	4

Description and Operation

COMPONENT LOCATION

OVERVIEW

The variable vane turbocharger, fitted to the exhaust manifold, makes it possible to vary the exhaust gas flow of the turbine, dependent on engine operation. This improves the power transfer to the turbine wheel and compressor, particularly at low engine speeds, thus increasing the boost pressure. The guide vanes are opened progressively as the engine speed increases so that the power transfer always remains in balance with the required turbocharger speed and the required boost pressure level. Variable vanes facilitate better use of the exhaust gas energy so as to further improve the efficiency of the turbocharger, and thus of the engine, compared to the more conventional wastegate control.

Advantages:

- High torque at both high and low engine speeds
- Continuous and optimum adjustment for all engine speeds
- No wastegate valve required, exhaust energy is better utilized, less back-pressure in conjunction with same compressor work
- Low thermal and mechanical load improves engine power output
- Lower emissions
- Optimized fuel consumption over the entire engine speed range.

The position of the guide vanes is adjusted by a DC (direct current) rotary actuator via a rod and actuating lever. Operation of the actuator is controlled by the ECM (engine control module) . A feedback potentiometer in the actuator provides the ECM with a guide vane position signal for closed loop control.

The turbocharger is designed for fail safe operation. If a fault occurs regarding the control of the unit, the vanes default to the fully open position so as to produce minimum boost. The ECM detects any malfunctions in the actuator motor and generates a DTC (diagnostic trouble code) .

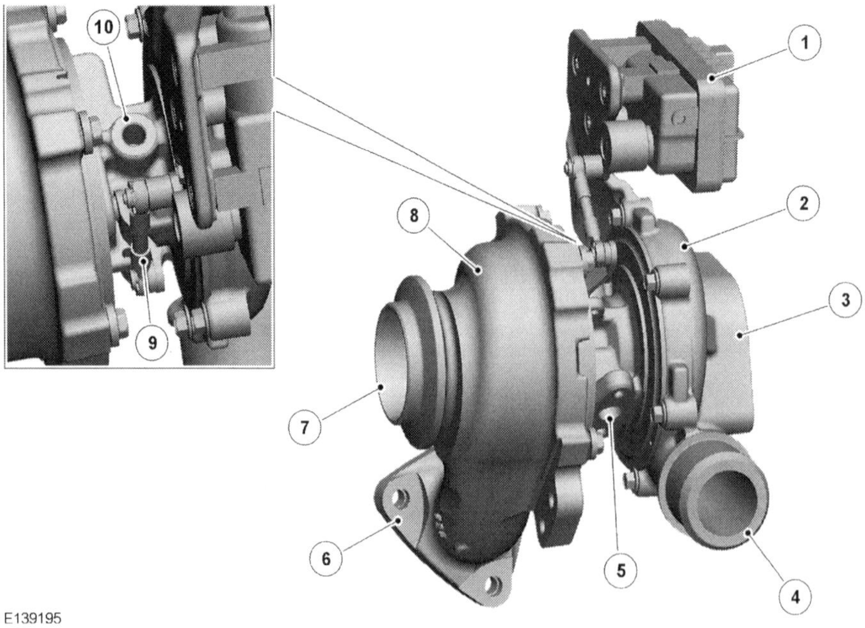

E139195

Item	Description
1	Actuator motor
2	Compressor housing
3	Intake air connection
4	Charge air cooler connection
5	Oil return connection
6	Exhaust manifold connection
7	Catalytic converter/ catalytic converter and DPF (diesel particulate filter) connection
8	Turbine housing
9	Rod and actuating lever
10	Oil feed connection

PRINCIPLES OF OPERATION

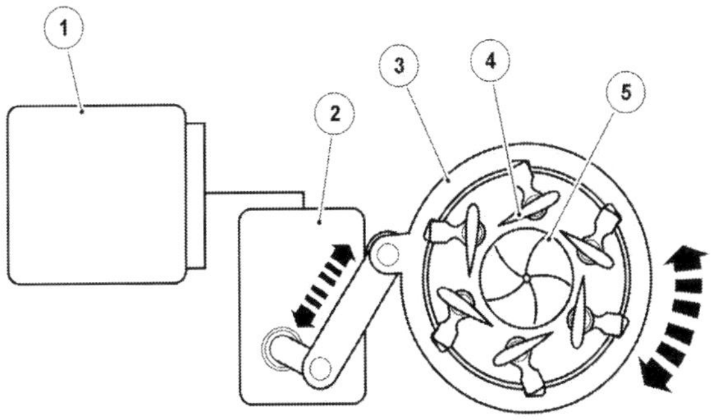

Item	Description
A	Low engine speed
B	Moderate engine speed
C	Maximum engine speed
1	-
2	Actuator motor
3	Adjusting ring
4	Vanes
5	Turbine

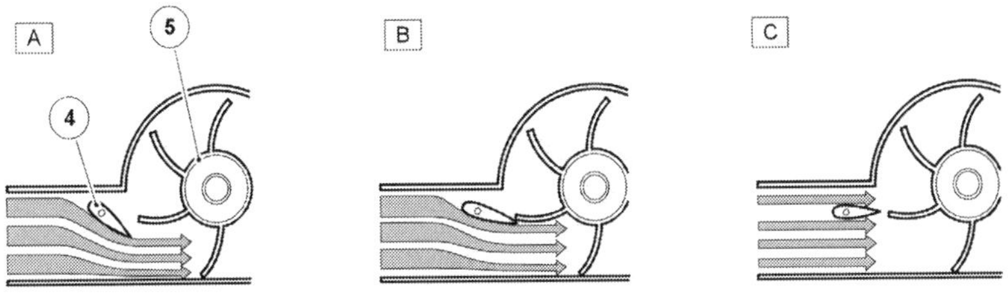

A - Low Engine Speed

At low engine speeds the volume of exhaust gas is low so the vanes are moved towards the closed position to reduce the turbine inlet area. This reduction causes an increase in the gas velocity into the turbine wheel thereby increasing wheel speed and boost.

B - Moderate Engine Speed

As the engine speed increases and the volume of exhaust gas increases the vanes are moved towards the open position to increase the turbine inlet area and maintain the gas velocity.

C - Maximum Engine Speed

At maximum engine speed the vanes are almost fully open maintaining the gas velocity into the turbine wheel.

Barometric Pressure Sensor

When the vehicle is driven at high altitudes the ambient pressure reduces causing the compressor wheel to do less work for the same boost pressure. To prevent the turbine wheel from over-speeding under these conditions, a barometric pressure sensor, located in the ECM, protects the turbocharger by opening the vanes further to reduce the turbine wheel speed. This is known as the altitude margin of the turbocharger.

Turbocharger Lubrication

The rapid acceleration and deceleration response demands of the turbocharger rely greatly on a steady flow of clean oil. The oil supplied from the engine's lubrication system provides lubrication to the turbocharger's spindle and bearings, while also acting as a coolant for the turbocharger center housing.

To maintain the life expectancy of the turbocharger, it is essential that the oil has a free-flow through the turbocharger and unrestricted return to the oil pan.

It is therefore imperative that the engine oil is replenished at regular service intervals with the recommended quality and quantity of oil.

Diagnosis and Testing

Principles of Operation

For a detailed description of the Turbocharger, refer to the relevant Description and Operation section in the workshop manual.

Refer to: Turbocharger (303-04 Fuel Charging and Controls - Turbocharger, ID4 2.2L Diesel Description and Operation)

Inspection and Verification

WARNINGS:

The following tests may involve working in close proximity to hot components. Make sure adequate protection is used. Failure to follow this instruction may result in personal injury.

The turbocharger can continue to rotate after the engine has stopped. Do not attempt to check the turbocharger until one minute has elapsed since the engine was switched off. Failure to follow this instruction may result in personal injury.

CAUTION: Diagnosis by substitution from a donor vehicle is **NOT** acceptable. Substitution of control modules does not guarantee confirmation of a fault, and may also cause additional faults in the vehicle being tested and/or the donor vehicle.

NOTES:

If the control module or a component is suspect and the vehicle remains under manufacturer warranty, refer to the Warranty Policy and Procedures manual (section B1.2), or determine if any prior approval programme is in operation, prior to the installation of a new module/component.

⚠ When performing voltage or resistance tests, always use a digital multimeter accurate to three decimal places, and with an up-to-date calibration certificate. When testing resistance always take the resistance of the digital multimeter leads into account.

⚠ Check and rectify basic faults before beginning diagnostic routines involving pinpoint tests.

1. Visually inspect for obvious signs of damage and system integrity
2. If an obvious cause for an observed or reported concern is found, correct the cause (if possible) before proceeding to the next step
3. If the cause is not visually evident, verify the symptom and refer to the Symptom Chart, alternatively check for Diagnostic Trouble Codes (DTCs) and refer to the DTC Index

Visual Inspection

Mechanical	Electrical
• Intake air system • Variable geometry turbine turbocharger • Variable geometry turbine vane actuator • Charge air cooler • Charge air cooler hoses	• Fuses • Wiring harnesses and connectors • Engine Control Module (ECM) • Variable geometry turbine vane actuator

Symptom Chart

Symptom	Possible Causes	Action
Poor performance (off-boost)	• Intake air system leaking/restricted • Low/contaminated fuel • Engine management system fault	• Check the intake air system for leaks/restrictions. Using the manufacturer approved diagnostic system, perform routine - Turbo, EGR and Air Path Dynamic Test - and perform the relevant corrective actions • Check the fuel level and condition. Draw off approximately 1 ltr (2.11 pints) of fuel and allow to stand for 1 minute. Check to make sure there is no separation of the fuel indicating water or other liquid in the fuel • Using the manufacturer approved diagnostic system, check the engine control module for related DTCs and refer to the relevant DTC index
No boost	• Intake air system leaking/restricted • Charge air cooler leaking • Charge air cooler hose leaking • Variable geometry turbine turbocharger fault • Engine management system fault	• Check the intake air system for leaks/restrictions. Using the manufacturer approved diagnostic system, perform routine - Turbo, EGR and Air Path Dynamic Test - and perform the relevant corrective actions • Check the charge air cooler for leaks • Check the charge air cooler hoses for leaks • Check the mechanical condition of the variable geometry turbine turbocharger. Using the manufacturer approved diagnostic system, check the engine control module for related DTCs and refer to the relevant DTC index. Perform routine - Turbo, EGR and Air Path Dynamic Test - and perform the relevant corrective actions • Using the manufacturer approved diagnostic system, check the engine control module for related DTCs and refer to the relevant DTC index

DTC Index

For a list of Diagnostic Trouble Codes (DTCs) that could be logged on this vehicle, please refer to Section 100-00.

Refer to: Diagnostic Trouble Code Index - DTC: Engine Control Module (PCM) (100-00 General Information, Description and Operation)

Turbocharger - Removal and Installation

Part(s)

STEP	PART NAME	QUANTITY
Installation Step 3	Renew Part Exhaust pipe to turbocharger gasket	1
Installation Step 7	Renew Part Turbocharger oil supply tube to turbocharger sealing washers	2
Installation Step 8	Renew Part Turbocharger oil return pipe	1

Removal

1. **Refer to:** Startup Battery Disconnect and Connect (414-01 Battery, Mounting and Cables, General Procedures)

2. **Refer to:** Generator (414-02 Generator and Regulator, ID4 2.2L Diesel Removal and Installation)

3. ⚠ WARNING: Do not work on or under a vehicle supported only by a jack. Always support the vehicle on safety stands.

Raise and support the vehicle.

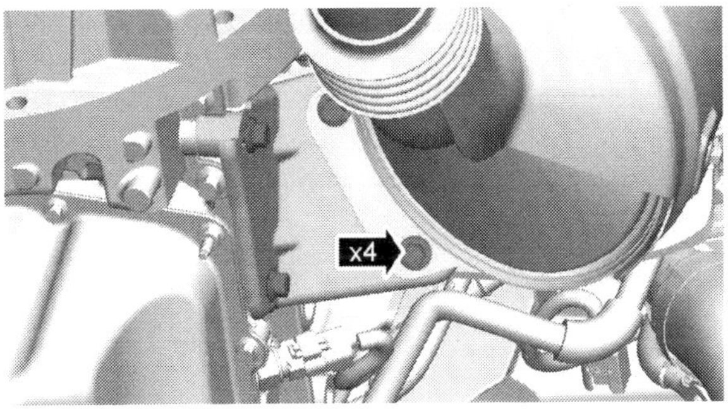

4.

5. With assistance, carefully remove the hood.

6. **Refer to:** Engine Cover (501-05 Interior Trim and Ornamentation, Removal and Installation)

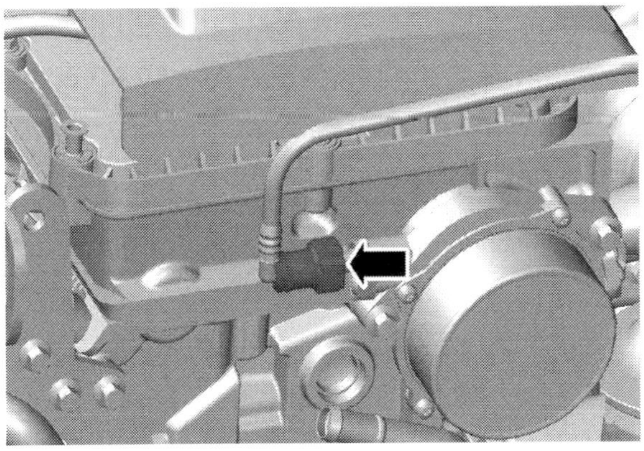

7.

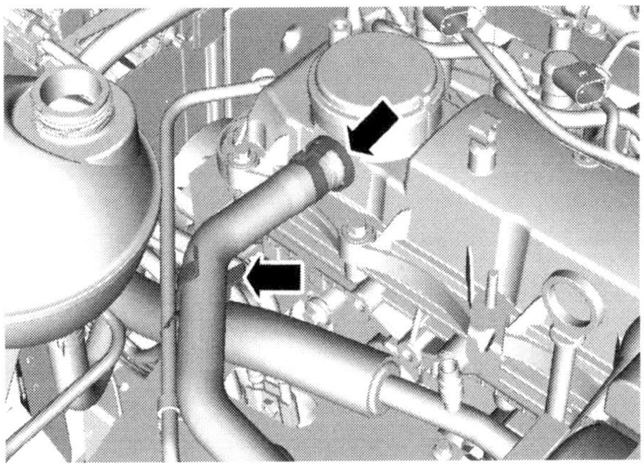

8. △ NOTE: Right-hand drive vehicles.

9.

10.

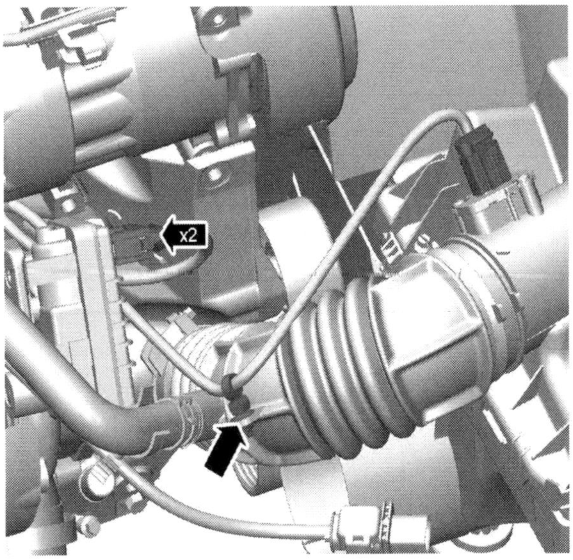

11.

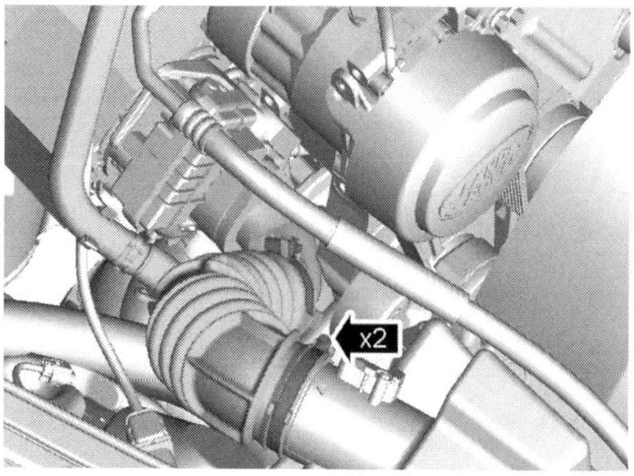

12.

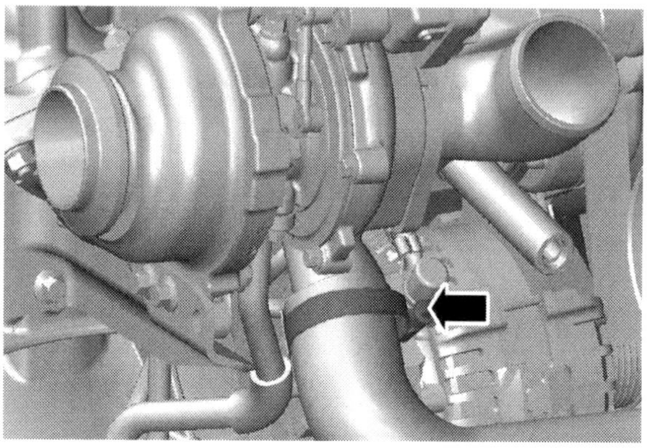

13.

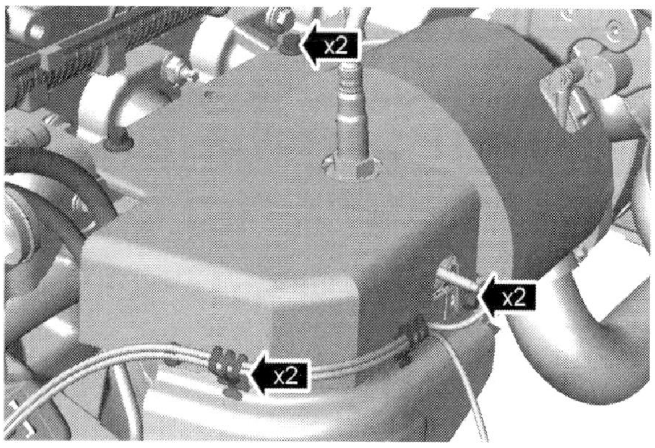

14. △ NOTE: Vehicles with diesel particulate filter (DPF).

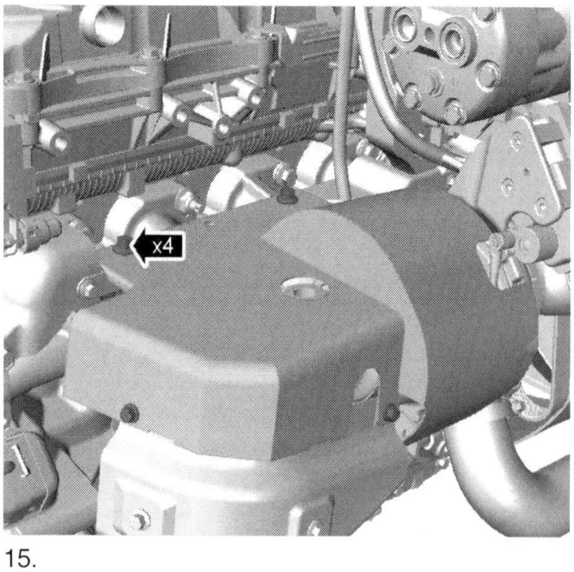

15.

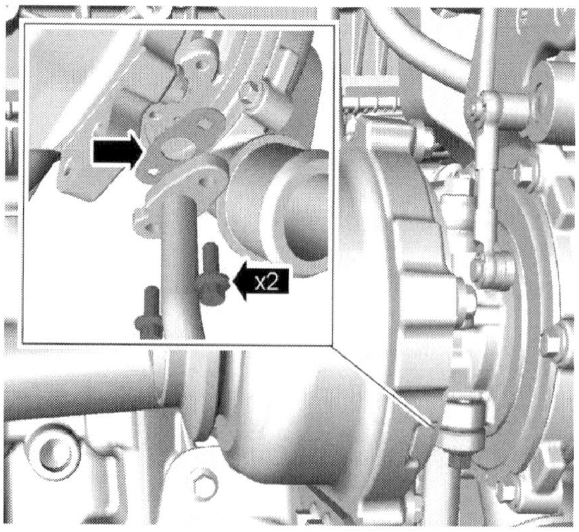

16.

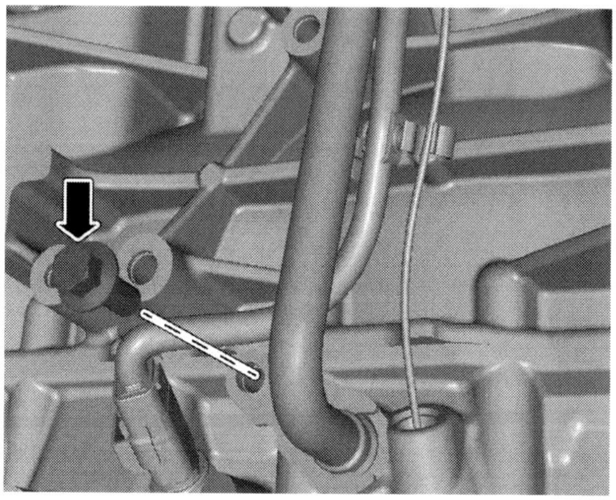

17. Remove and discard the turbocharger oil return pipe.

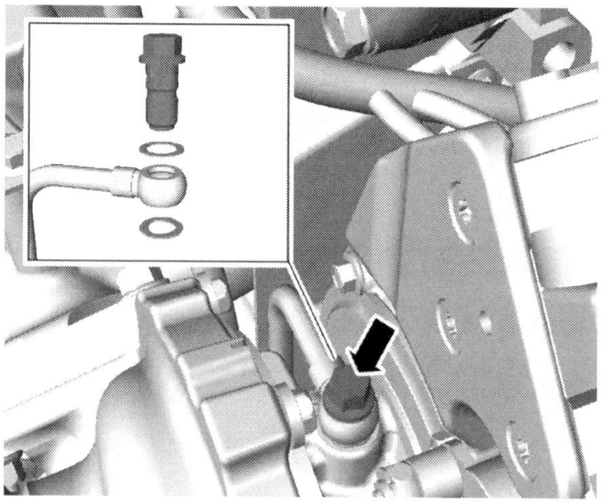

18. Remove and discard the washers.

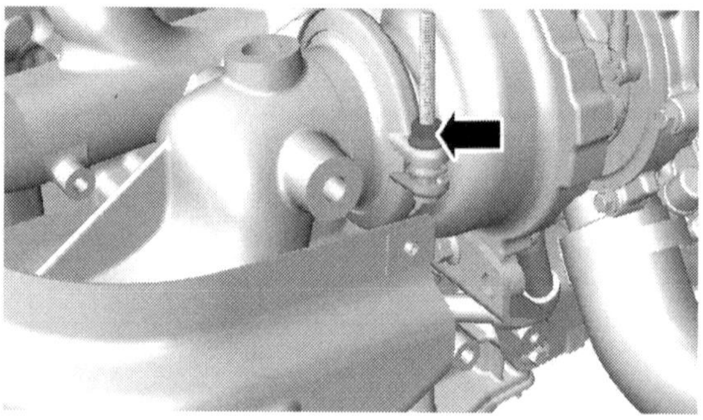

19.

20. NOTE: Remove and discard the gasket.

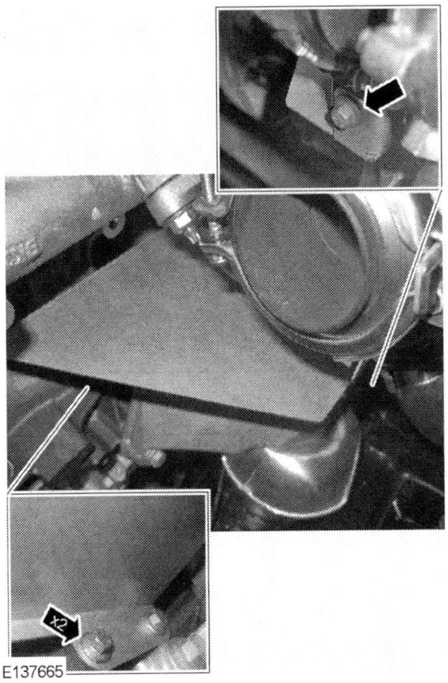

21.

22.

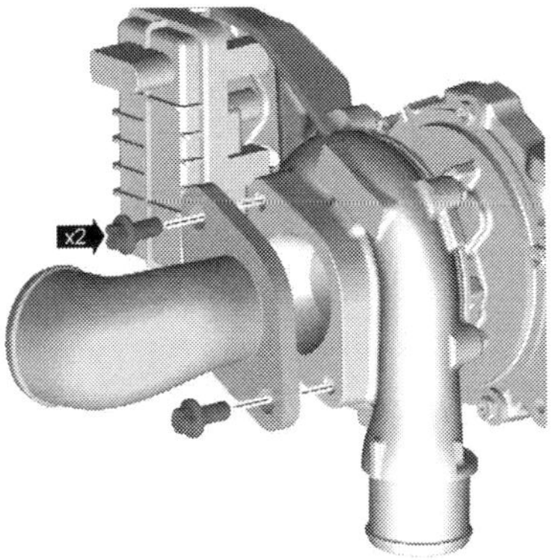

23. NOTE: Do not disassemble further if the component is removed for access only.

Installation

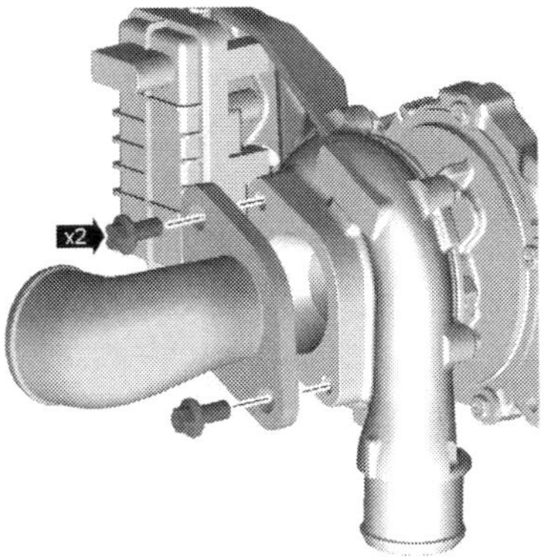

1. CAUTION: Make sure blanking plugs are removed before component is installed.

   NOTE: If there are indications of damage to the O ring seal, a new seal should be installed.

   Torque: 25 Nm

2. *Torque:* 23 Nm

3. NOTE: Install a new gasket.

   Renew Part: Exhaust pipe to turbocharger gasket

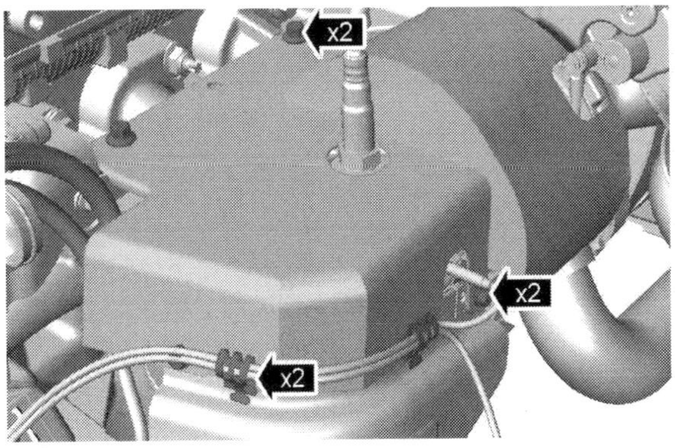

4. NOTE: Vehicles with diesel particulate filter (DPF). *Torque:* 25 Nm

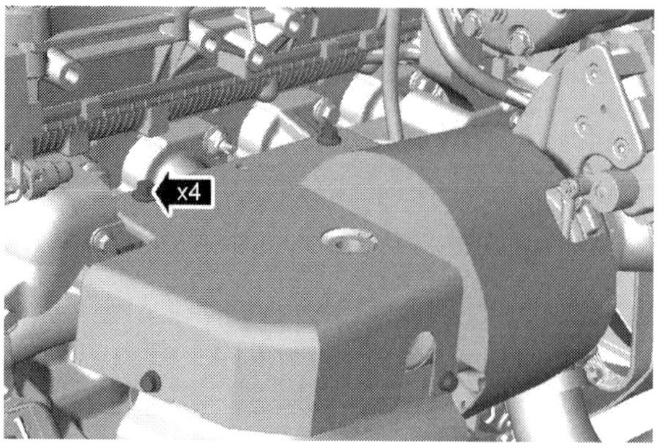

5. *Torque:* 25 Nm

6. *Torque:* 10 Nm

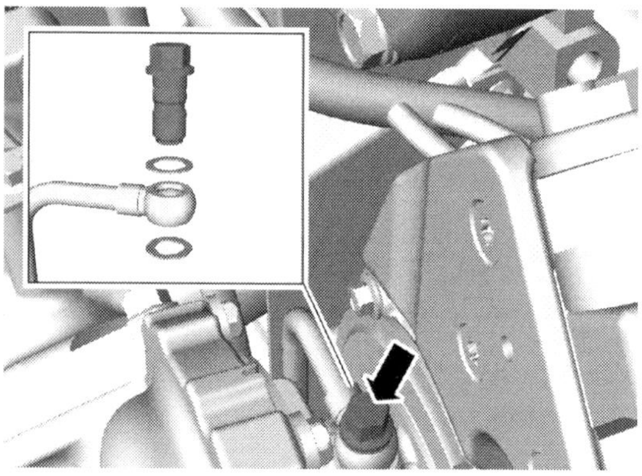

7. CAUTION: Make sure that the oil feed pipe doe not touch the turbo housing when secured.

   Install new washers.

   Renew Part: Turbocharger oil supply tube to turbocharger sealing washers

Torque: 20 Nm

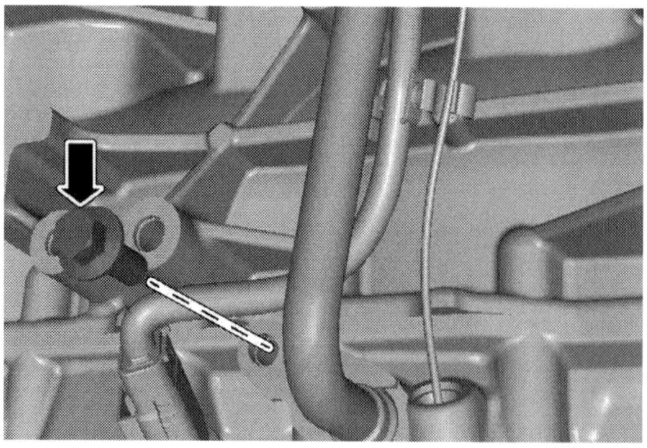

8. Install a new turbocharger oil return pipe.

   Renew Part: Turbocharger oil return pipe

   Torque: 23 Nm

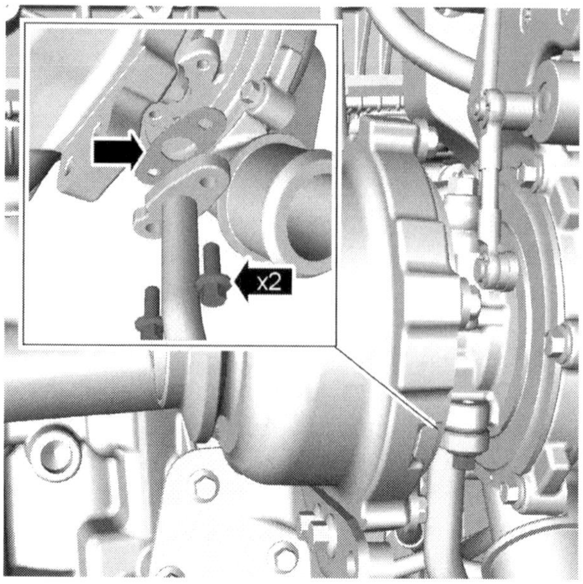

9. CAUTION: Make sure the gasket is installed correctly.

Torque: 10 Nm

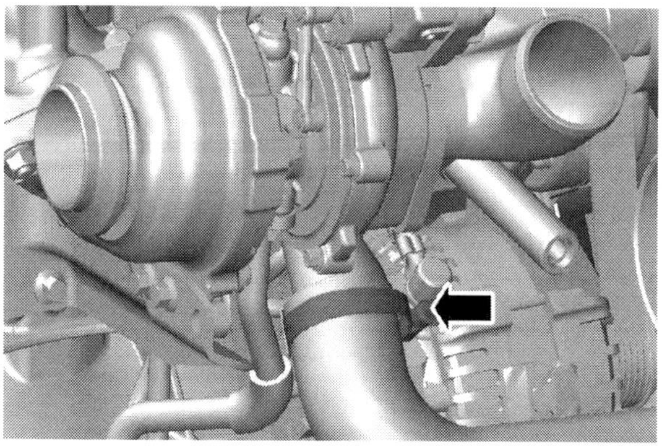

10. *Torque:* 5 Nm

11. *Torque:* 3 Nm

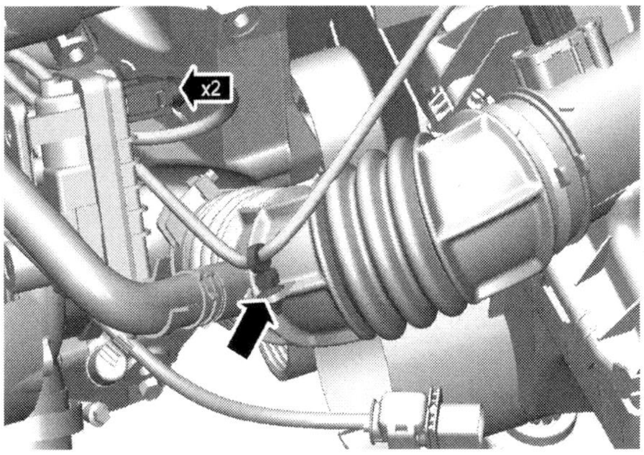

12.

13.

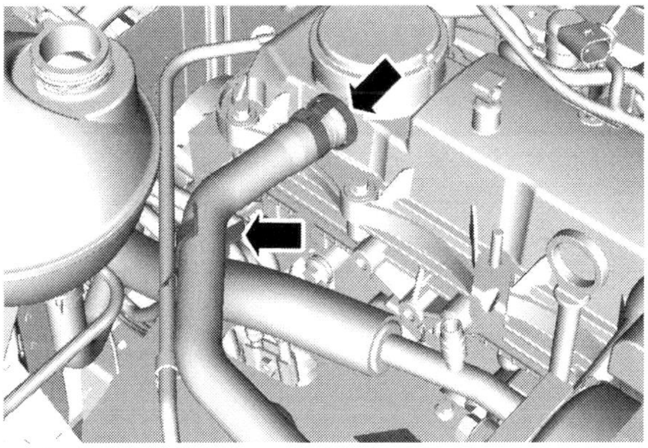

14. NOTE: Right-hand drive vehicles.

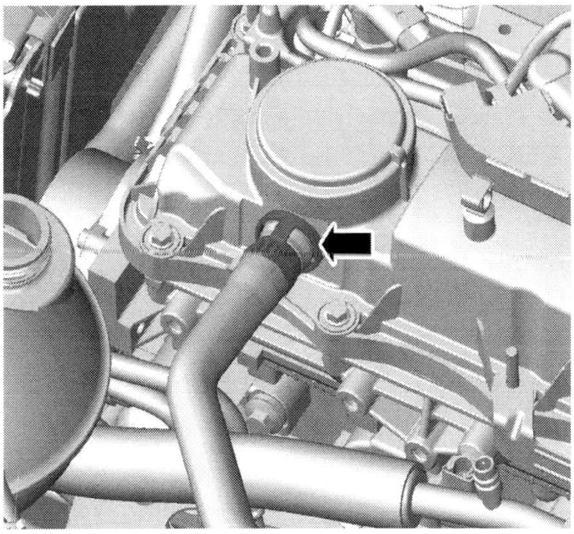

15.

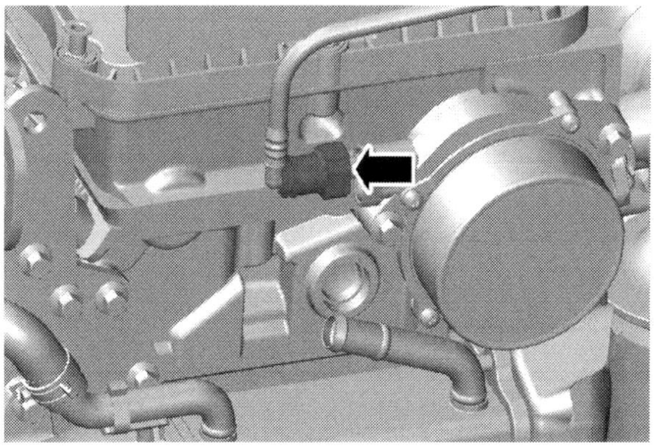

16.

17. **Refer to:** Engine Cover (501-05 Interior Trim and Ornamentation, Removal and Installation)
18. With assistance, carefully install the hood.

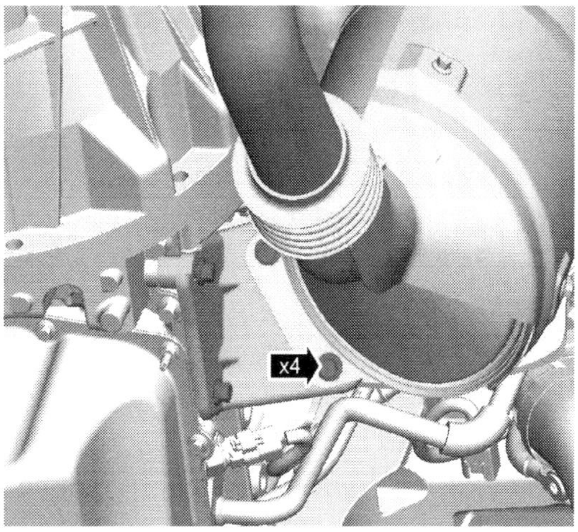

19. *Torque:* 25 Nm
20. **Refer to:** Generator (414-02 Generator and Regulator, ID4 2.2L Diesel Removal and Installation)
21. **Refer to:** Startup Battery Disconnect and Connect (414-01 Battery, Mounting and Cables, General Procedures)

Actuator Rod

Removal and Installation

Removal

NOTE: Removal steps in this procedure may contain installation details.

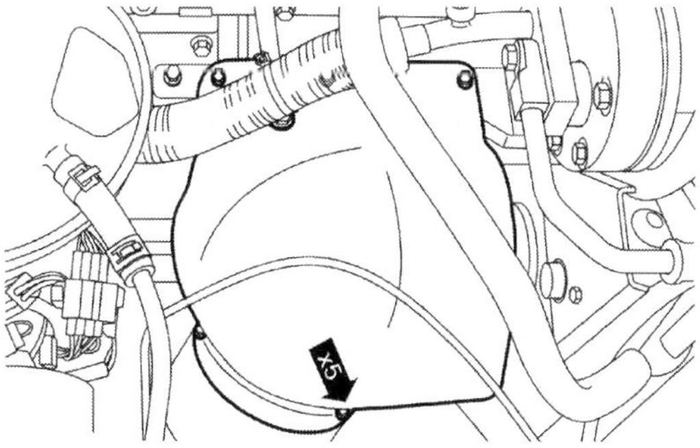

1. Remove the turbocharger heat shield.

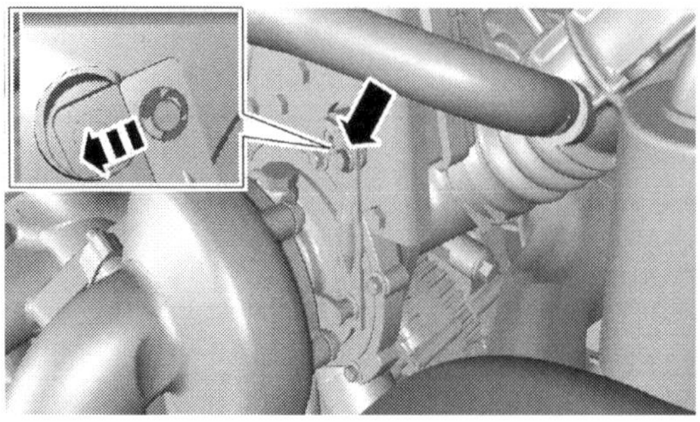

2. CAUTIONS:

⚠ Do not apply sufficient force to deform the turbocharger variable vane actuator crank components.

⚠ Do not apply rotational force to the turbocharger variable vane actuator crank during removal, as the worm gear can be damaged.

Disconnect the connecting rod from the actuator.

- It may be necessary to place a suitable tool between the crank shoulder and the actuator rod, and lever from side to side.

3. NOTE: This may be difficult on the first few cycles due to internal carbon build up.

Using finger pressure, cycle the turbocharger variable vane actuator mechanism to make sure movement is sufficient.

- Cycle 10-20 times as necessary to clear movement.

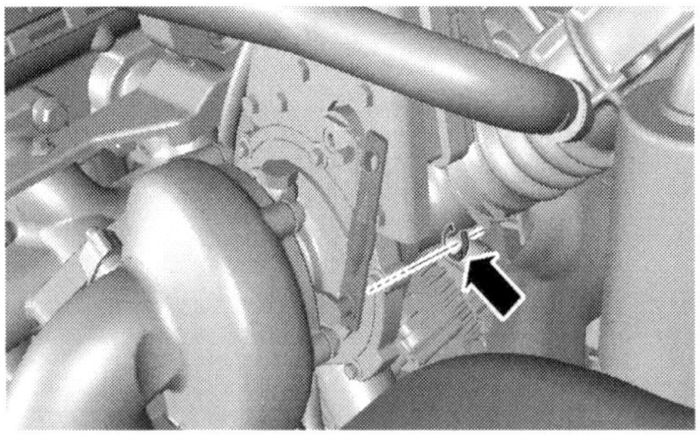

4. ⚠ CAUTION: Do not use excessive force, or lever against the crank arm, as this may bend the crank arm, or damage the turbo mechanism.

Remove the turbocharger actuator connecting rod.

- Use penetrating lubricant spray as required.

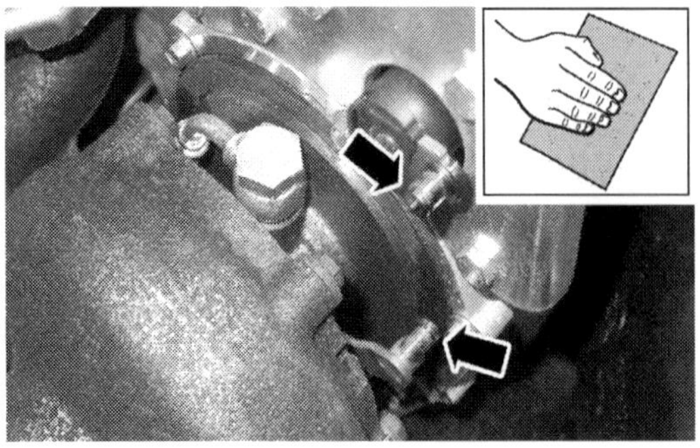

5. NOTE: It may be necessary to use a very fine wet and dry paper to polish and remove corrosion.

Clean the actuator and turbocharger pin surfaces.

- Apply penetrating oil to both turbo and actuator pins.
- Clean the pin surfaces.

Installation

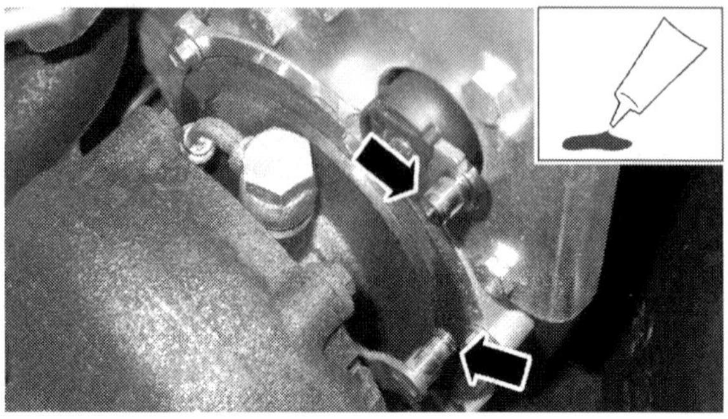

1. Apply specified high temperature lubricant (Land rover Part No. LR029042) to both pins, making sure they are completely covered.

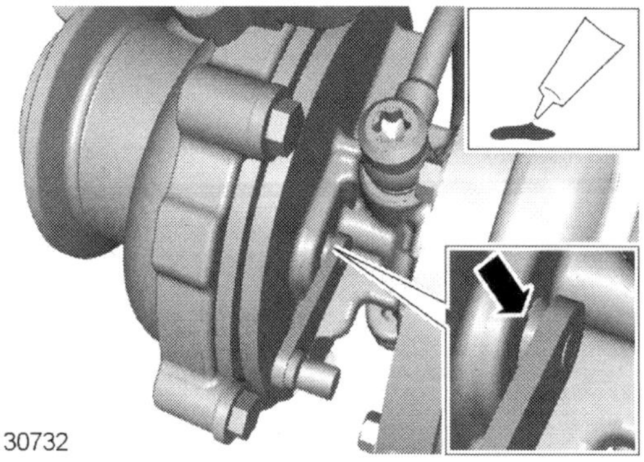

E130732

2. Apply specified high temperature lubricant (Land rover Part No. LR029042) to the internal bush interface with the turbocharger.

3. Install the new turbocharger actuator rod.

- Install new retaining clips.
- Rotate the new retaining clips in the grooves with a suitable tool to check for correct installation.
- Apply grease over the actuator pins, and the actuator rod connections to protect.

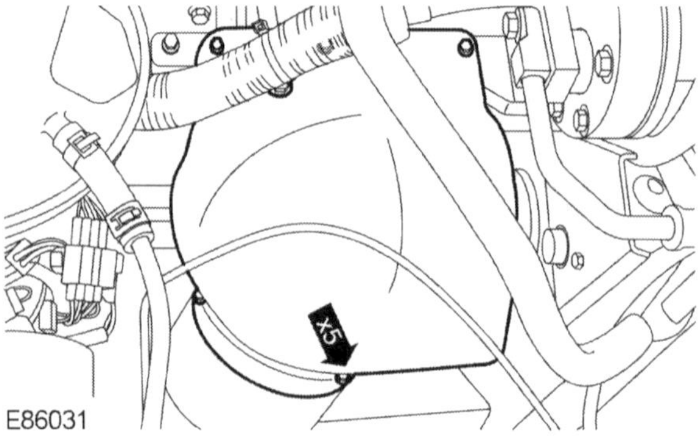

E86031

4. Tighten to 10 Nm (7 lb.ft).

Accessory Drive

Specifications

Torque Specifications

Item	Nm	lb-ft
Accessory drive component bracket bolts	48	35
Engine lifting bracket bolts	10	7
Accessory drive belt tensioner bolt	25	18
Accessory drive belt idler pulley bolt	48	35

Description and Operation

COMPONENT LOCATION

Item	Description
A	Vehicles with A/C (air conditioning)
B	Vehicles without A/C
1	Idler pulley
2	Idler pulley
3	Coolant pump
4	Accessory drive belt
5	Power steering pump
6	Idler pulley
7	Engine crankshaft pulley
8	Engine cooling fan pulley
9	Automatic tensioner
10	Generator
11	Idler pulley
12	A/C compressor

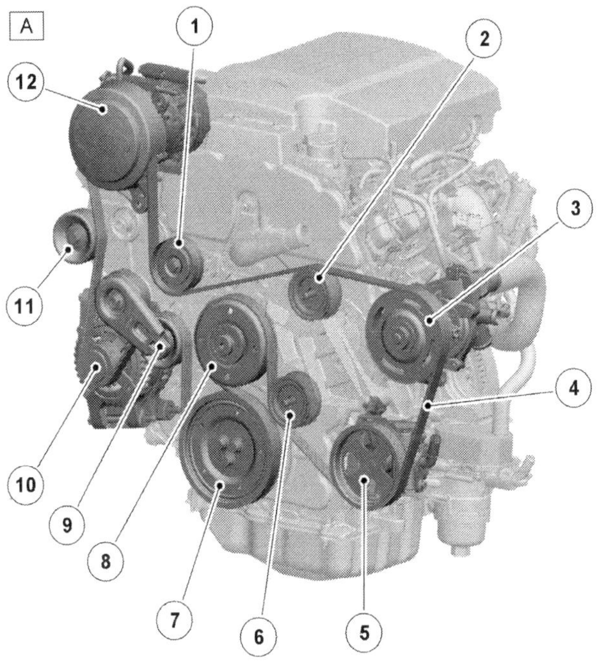

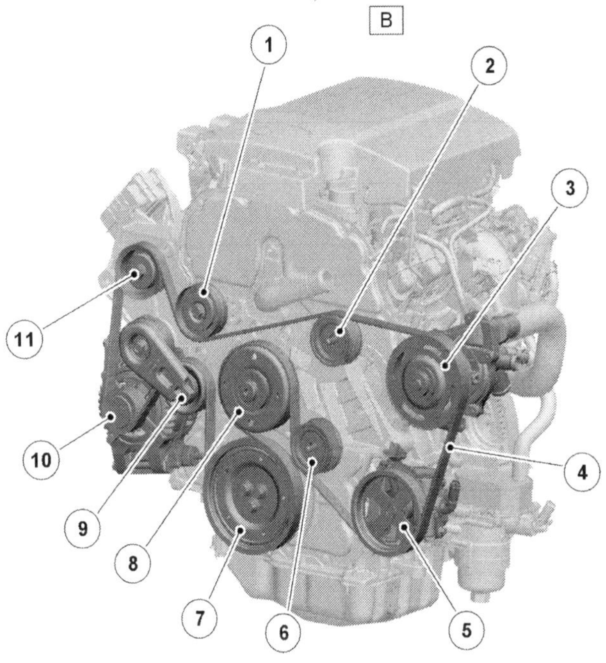

OVERVIEW

The accessory components, which comprise the coolant pump, power steering pump, engine cooling fan, generator and A/C compressor (where fitted), are driven by the engine crankshaft pulley via the accessory drive belt.

The belt is a maintenance free poly-V type belt, automatically pre-loaded by the automatic tensioner and routed over idler pulleys in order to maintain sufficient adhesion about the accessory component pulleys. This ensures a slip-free drive of the accessory components.

Diagnosis and Testing

Principles of Operation

For a detailed description of the Accessory Drive, refer to the relevant Description and Operation section in the workshop manual.

Refer to: Accessory Drive (303-05 Accessory Drive, ID4 2.2L Diesel Description and Operation)

Inspection and Verification

CAUTIONS:

⚠ Diagnosis by substitution from a donor vehicle is **NOT** acceptable. Substitution of control modules does not guarantee confirmation of a fault, and may also cause additional faults in the vehicle being tested and/or the donor vehicle.

⚠ Check and rectify basic faults before beginning diagnostic routines involving pinpoint tests.

1. Visually inspect for obvious signs of damage and system integrity
2. If an obvious cause for an observed or reported concern is found, correct the cause (if possible) before proceeding to the next step
3. If the cause is not visually evident, verify the symptom and refer to the Symptom Chart, alternatively check for Diagnostic Trouble Codes (DTCs) and refer to the DTC Index

⚠ If the engine is run without the primary drive belt installed (for example, to eliminate driven components), DTCs may be set which must be cleared before the vehicle is returned to the owner. The engine must not be run for more than 3 minutes without the primary drive belt installed. Failure to follow this instruction may result in damage to the vehicle.

NOTES:

⚠ If the control module or a component is suspect and the vehicle remains under manufacturer warranty, refer to the Warranty Policy and Procedures manual (section B1.2), or determine if any prior approval programme is in operation, prior to the installation of a new module/ component.

⚠ When performing voltage or resistance tests, always use a digital multimeter accurate to three decimal places, and with an up-to-date calibration certificate. When testing resistance always take the resistance of the digital multimeter leads into account.

Visual Inspection

Mechanical
• Primary drive belt - Cracking/damage/ contamination • Primary drive belt idler pulley(s) • Primary drive belt tensioner • Generator • Engine coolant pump • Power steering pump • Brake vacuum pump • Air conditioning (A/C) compressor • Engine cooling fan • Torsional vibration damper

Symptom Chart

Symptom	Possible Causes	Action
Noise	• Debris on pulley(s) • Primary drive belt condition • Primary drive belt tension • Pulley(s) misaligned • Driven component(s) resisting rotation	⚠ CAUTION: Do not use cleaning chemicals when removing debris from a pulley as the drive belt will become contaminated. • Remove debris from pulley(s) • Check the condition of the primary drive belt • Using the manufacturer approved diagnostic system, perform routine - Noise, vibration and harshness sensor diagnostic test - Accessory drive belt • Check the operation of the primary drive belt tensioner • Check the pulley alignment • Check the driven components for excessive resistance to rotation
Primary drive belt does not hold tension	• Primary drive belt condition • Primary drive belt tensioner fault	• Check the condition of the primary drive belt • Check the operation of the primary drive belt tensioner

DTC Index

For a list of Diagnostic Trouble Codes (DTCs) that could be logged on this vehicle, please refer to Section 100-00.
Refer to: Diagnostic Trouble Code Index - DTC: Engine Control Module (PCM) (100-00 General Information, Description and Operation)

Accessory Drive Belt

Removal and Installation

Removal

1. **Refer to:** Battery Disconnect and Connect (414-01, General Procedures).
2. **Refer to:** Cooling Fan (303-03 Engine Cooling, ID4 2.2L Diesel Removal and Installation)

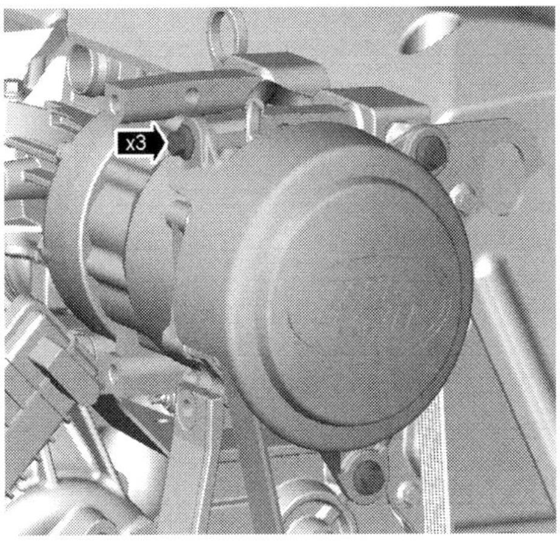

3. Torque: 10 Nm

4.

Installation

1. To install, reverse to removal procedure.

Accessory Drive Belt Pulley

Removal and Installation

Removal

1. **Refer to:** Battery Disconnect and Connect (414-01, General Procedures).
2. **Refer to:** Cooling Fan (303-03 Engine Cooling, ID4 2.2L Diesel Removal and Installation)

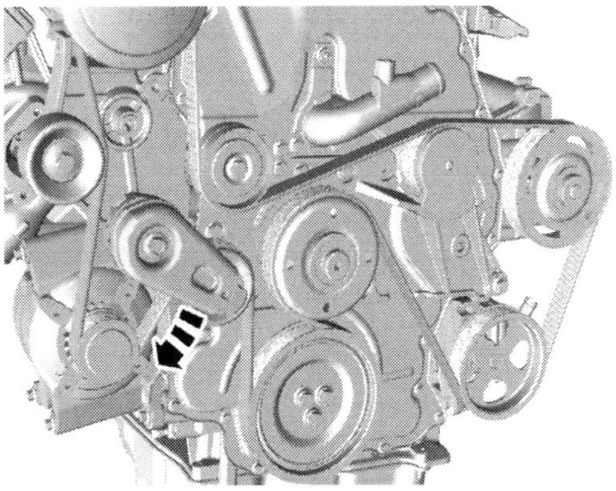

3. Release the accessory drive belt.

4. Remove the centre idler pulley.

Installation

1. Torque: 48 Nm

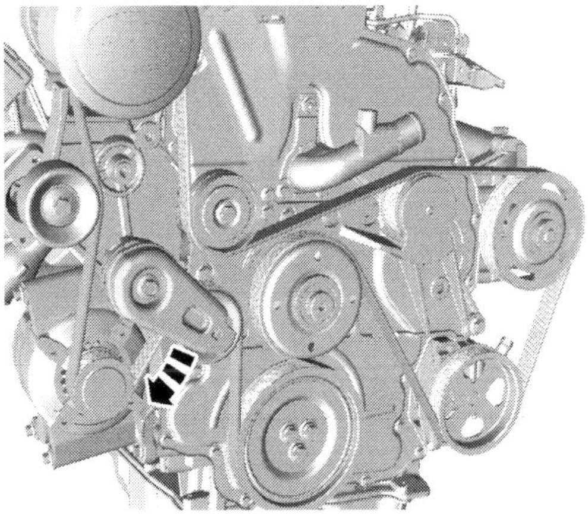

2. Install the accessory drive belt.
3. **Refer to:** Cooling Fan (303-03 Engine Cooling, ID4 2.2L Diesel Removal and Installation)
4. **Refer to:** Battery Disconnect and Connect (414-01, General Procedures).

Accessory Drive Belt Tensioner

Removal and Installation

Removal

1. **Refer to:** Battery Disconnect and Connect (414-01, General Procedures).
2. **Refer to:** Cooling Fan (303-03 Engine Cooling, ID4 2.2L Diesel Removal and Installation)

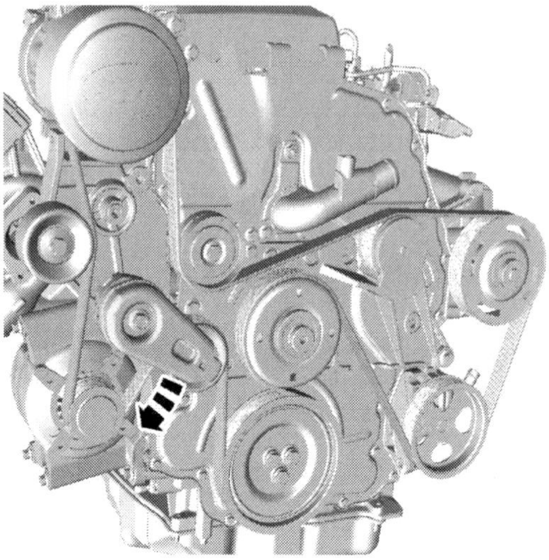

3.

4. Torque: 25 Nm

Installation

1. To install, reverse the removal procedure.

Accessory Drive Component Bracket

Removal and Installation

Removal

1. **Refer to:** Battery Disconnect and Connect (414-01, General Procedures).
2. **Refer to:** Engine Cover (501-05, Removal and Installation).
3. **Refer to:** Accessory Drive Belt (303-05 Accessory Drive, ID4 2.2L Diesel Removal and Installation)
4. WARNING: Do not work on or under a vehicle supported only by a jack. Always support the vehicle on safety stands.

Raise and support the vehicle.

5. Torque: 47 Nm

6. Torque: 48 Nm

7.

8.

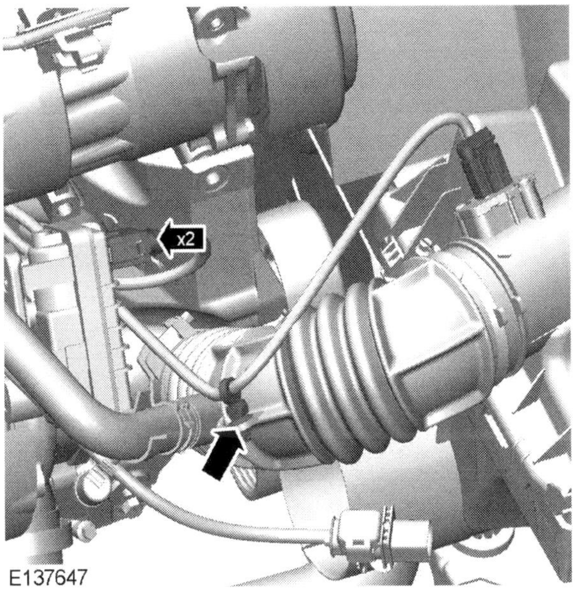

9.

10.

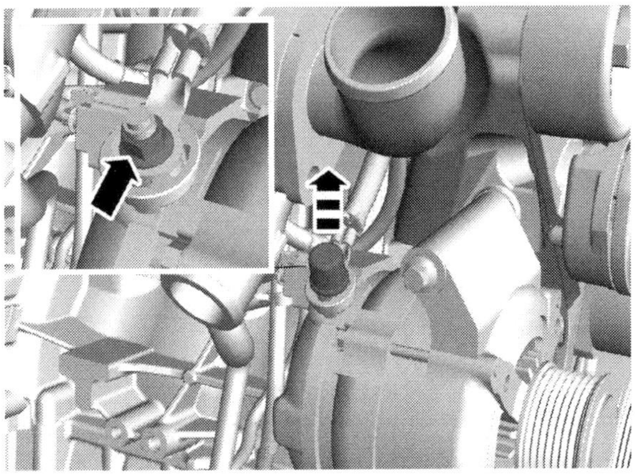

11. Torque: 10 Nm

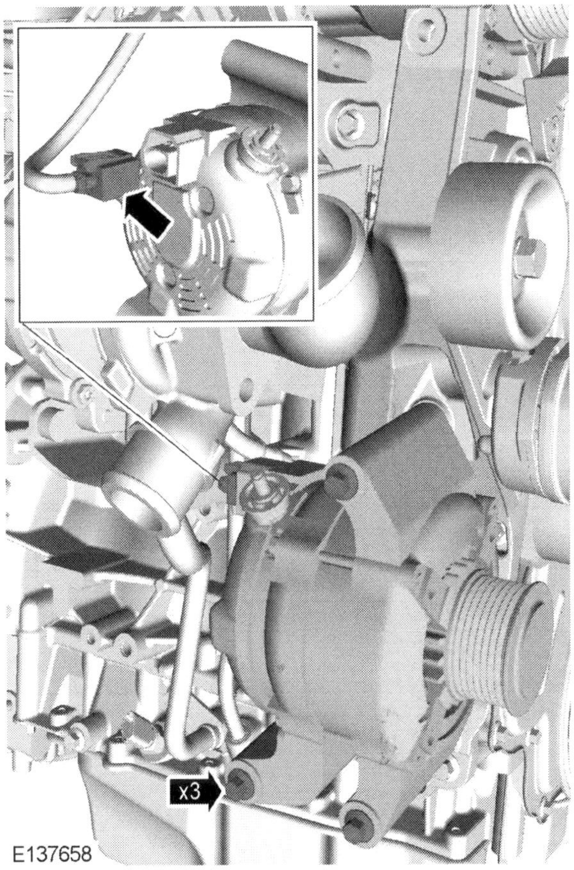

12. Torque: 48 Nm

13. Torque: 47 Nm

Installation

1. To install, reverse to removal procedure.

Specifications

Description	Nm	lb-ft
Starter motor bolts	35	26
Starter motor battery positive cable nut	12	9
Starter motor solenoid cable nut	8	6
Fuel line support bracket nut	8	6

Diagnosis and Testing

Principles of Operation

For a detailed description of the Starting system, refer to the relevant Description and Operation section in the workshop manual.

Refer to: Starting System (303-06 Starting System, ID4 2.2L Diesel Description and Operation)

Inspection and Verification

CAUTION: Diagnosis by substitution from a donor vehicle is **NOT** acceptable. Substitution of control modules does not guarantee confirmation of a fault, and may also cause additional faults in the vehicle being tested and/or the donor vehicle.

NOTES:

If the control module or a component is suspect and the vehicle remains under manufacturer warranty, refer to the Warranty Policy and Procedures manual (section B1.2), or determine if any prior approval programme is in operation, prior to the installation of a new module/component.

When performing voltage or resistance tests, always use a digital multimeter accurate to three decimal places, and with an up-to-date calibration certificate. When testing resistance always take the resistance of the digital multimeter leads into account.

Check and rectify basic faults before beginning diagnostic routines involving pinpoint tests.

1. Visually inspect for obvious signs of damage and system integrity
2. If an obvious cause for an observed or reported concern is found, correct the cause (if possible) before proceeding to the next step
3. If the cause is not visually evident, verify the symptom and refer to the Symptom Chart, alternatively check for Diagnostic Trouble Codes (DTCs) and refer to the DTC Index

Visual Inspection

Mechanical	Electrical
• Starter motor • Engine (rotates freely)	• Fuses • Wiring harnesses and connectors • Engine control module (ECM) • Battery • Starter relay • Ignition switch • Generator

Symptom Chart

Symptom	Possible Causes	Action
The engine does not crank - Starter motor**does not** operate	• Immobiliser armed (immobiliser LED illuminated) • Battery/charging system fault • Starter solenoid circuit short circuit to ground, open circuit, high resistance • Starter motor circuit open circuit, high resistance • Engine management system fault • Engine seized	NOTE: The immobiliser will arm passively 30 seconds after the ignition is set to off. • Operate the unlock switch on the alarm remote to disarm the immobiliser • Refer to the relevant section of the workshop manual and test the battery and charging system • Refer to the electrical circuit diagrams and check the starter solenoid circuit for short circuit to ground, open circuit, high resistance • Refer to the electrical circuit diagrams and check the sterter motor circuit for open circuit, high resistance • Using the manufacturer approved diagnostic system, check the engine control module for related DTCs and refer to the relevant DTC index • Check that the engine rotates freely

The engine does not crank - Starter motor**does** operate	• Starter motor installation • Starter motor internal failure • Flywheel ring gear damaged	• Check the installation of the starter motor • Install a new starter motor as necessary • Install a new flywheel ring gear as necessary
Engine cranks too slowly	• Battery/charging system fault • Starter motor circuit high resistance • Starter motor internal failure • Engine oil grade/condition	• Refer to the relevant section of the workshop manual and test the battery and charging system • Refer to the electrical circuit diagrams and check the sterter motor circuit for high resistance • Install a new starter motor as necessary • Check the engine oil grade/condition
Engine cranks too quickly	• Engine resistance to rotation too low	• Check the mechanical condition of the engine (timing chain, cylinder compressions, etc)
Excessive starter motor noise	• Starter motor installation • Starter motor internal failure • Flywheel ring gear damaged	• Check the installation of the starter motor • Install a new starter motor as necessary • Install a new flywheel ring gear as necessary

DTC Index

For a list of Diagnostic Trouble Codes (DTCs) that could be logged on this vehicle, please refer to Section 100-00.
Refer to: Diagnostic Trouble Code Index - DTC: Engine Control Module (PCM) (100-00 General Information, Description and Operation)

Starter Motor

Removal and Installation

Removal

1. Disconnect the battery ground cable.

Refer to: Battery Disconnect (414-01, General Procedures).

2. ⚠ WARNING: Do not work on or under a vehicle supported only by a jack. Always support the vehicle on safety stands.

Raise and support the vehicle.

Refer to: Lifting (100-02 Jacking and Lifting, Description and Operation)

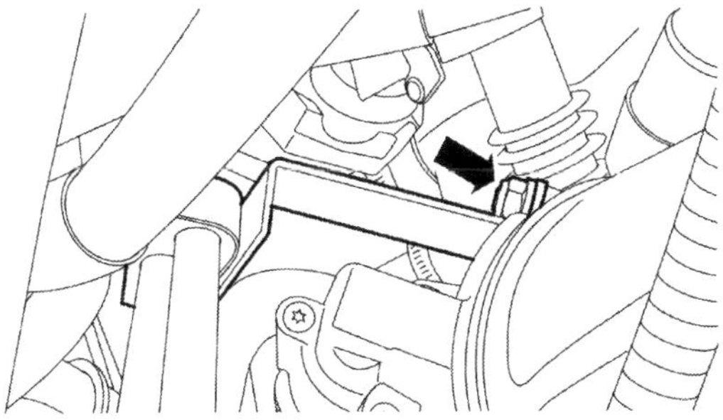

3. Remove the fuel line support bracket.
   - Remove the nut.

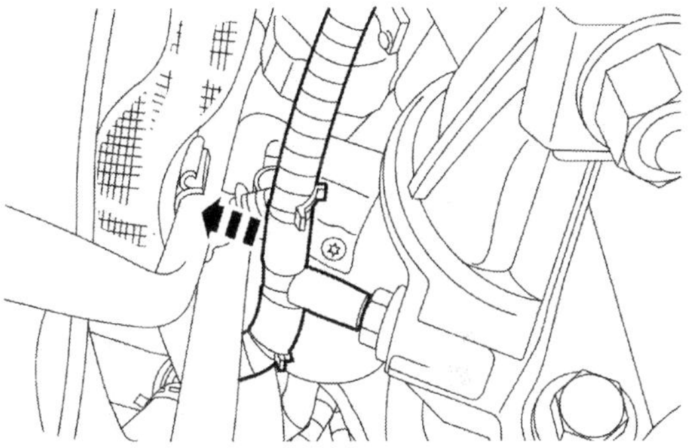

4. Release the starter motor wiring harness clip.

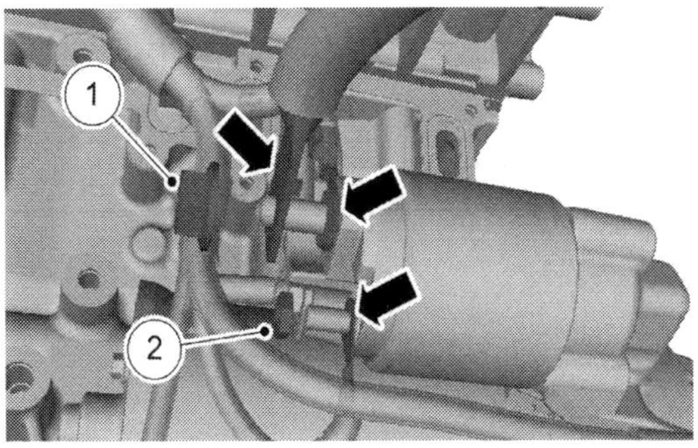

5. Release the 2 battery positive cables and the switch lead from the starter motor solenoid.
   1. Remove the battery positive cable nut.
   2. Release the starter motor switch lead nut.

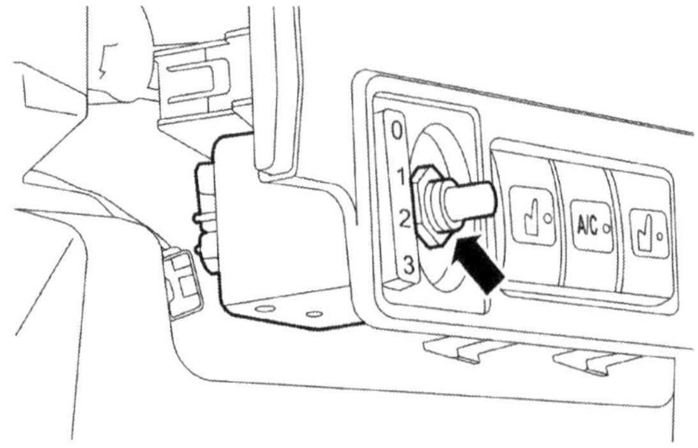

6. Remove the starter motor lower bolt.

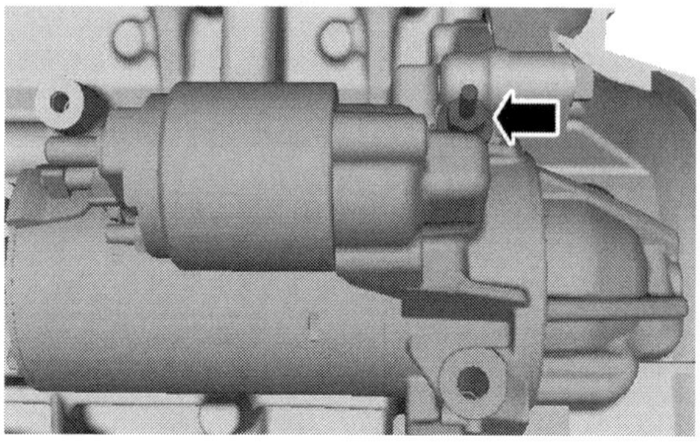

7. Remove the starter motor.

Installation

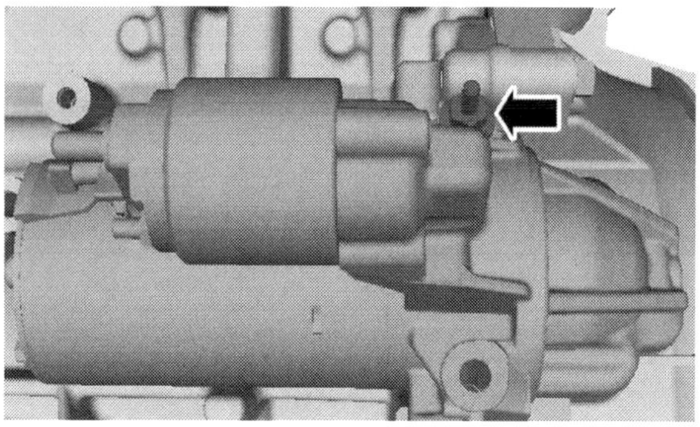

1. Install the starter motor.
   - Tighten to 35 Nm (26 lb.ft).

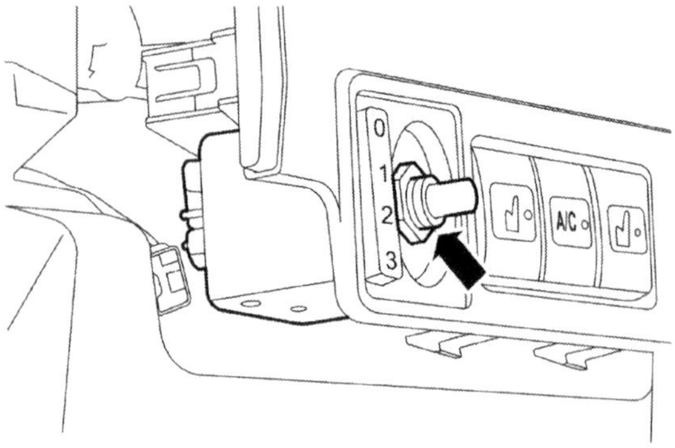

2. Tighten to 35 Nm (26 lb.ft).

3. Secure the 2 battery positive cables and the switch lead to the starter motor solenoid.
   1. Tighten to 12 Nm (9 lb.ft).
   2. Tighten to 8 Nm (6 lb.ft).

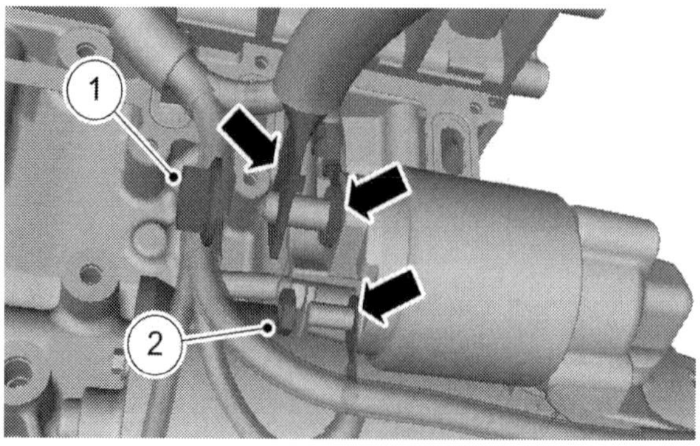

4. Secure the starter motor wiring harness clip.

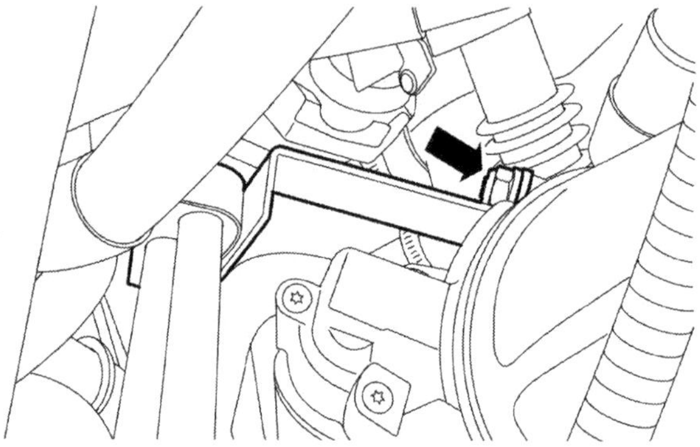

5. Install the fuel line support bracket.
   - Tighten to 8 Nm (6 lb.ft).

6. Connect the battery ground cable.

Refer to: Battery Connect (414-01, General Procedures).

Glow Plug System

Specifications

Item	Nm	lb-ft
Glow plug wiring harness nuts	2	1
Glow plugs	13	10

Description and Operation

COMPONENT LOCATION

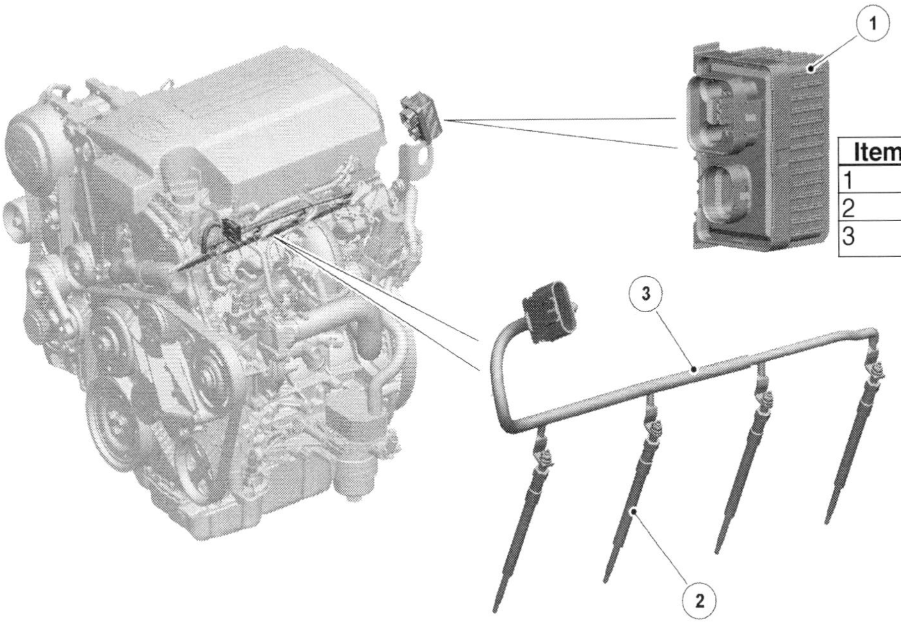

Item	Description
1	Glow plug control module
2	Glow plug (4 off)
3	Glow plug wire assembly

OVERVIEW

A glow plug is installed between the inlet valves of each cylinder, to:

- Preheat the combustion chambers to aid engine starting
- Provide additional heat in the combustion chambers while the engine warms up, to reduce emissions and engine noise.

A glow plug control module, located on the ECM (engine control module) bracket on the engine bulkhead, powers each glow plug individually. Each glow plug is grounded through its fixing in the cylinder head. Operation of the glow plug control module is controlled by the ECM, which also controls the illumination of the glow plug warning indicator in the instrument cluster.

The glow plugs are a low voltage metallic type that operate at a maximum tip temperature of 1050 °C (1922 °F), achievable within 2 seconds.

OPERATION

Power for the glow plugs is supplied to the glow plug control module from a fusible link in the BJB (battery junction box) . Operating power for the glow plug control module is supplied from the power relay in the BJB while the ignition is on.

The power level applied to the glow plugs by the glow plug control module is contained in an input signal from the ECM. The glow plug control module incorporates diagnostic functions for the individual glow plugs and reports the diagnostic status to the ECM.

The ECM uses the signal from the ECT (engine coolant temperature) sensor to determine the preheat time. The lower the temperature, the longer the preheat time. There is a maximum preheat time of 2.1 seconds at -20 °C (-4 °F) or lower. At temperatures above 20 °C (68 °F) there is no preheat phase. The ECM signals the instrument cluster on the high speed CAN (controller area network) bus to illuminate the glow plug warning indicator for the duration of the preheat phase.

Once the engine has started, if the engine temperature is less than 70 °C (158 °F) the glow plug system enters an after-glow phase. The after-glow phase lasts for a maximum of 500 seconds, or until the engine temperature reaches 70 °C (158 °F). The tip temperature of the glow plugs is reduced as the engine warms up. The ECM uses a glow plug control map of engine speed against load to determine the optimum tip temperature during the after-glow phase. At high engine loads the after-glow phase is interrupted to protect the glow plugs and improve their durability.

Glow Plug System

Diagnosis and Testing

Principles of Operation

For a detailed description of the Glow Plug system, refer to the relevant Description and Operation section in the workshop manual.

Refer to: Glow Plug System (303-07 Glow Plug System, Description and Operation)

CONTROL DIAGRAM

NOTE: A = Hardwired; D = High speed CAN bus.

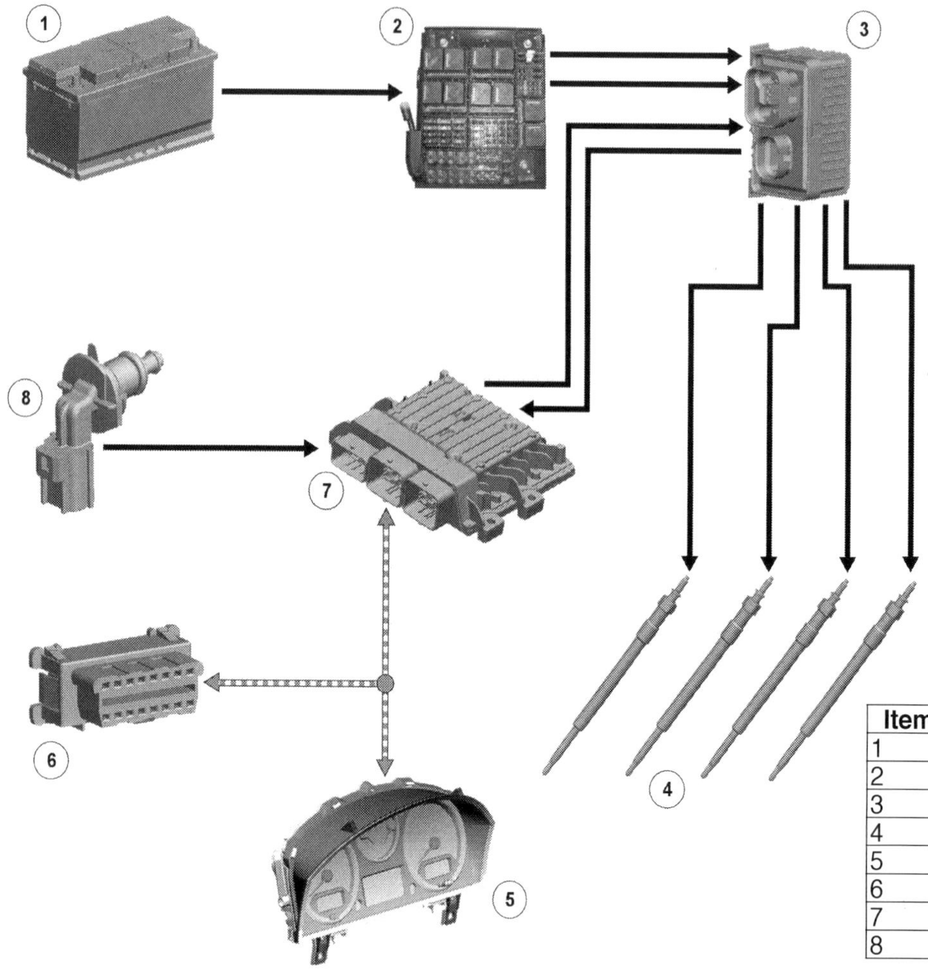

Item	Description
1	Battery
2	-
3	Glow plug control module
4	Glow plug (4 off)
5	Instrument cluster
6	Diagnostic socket
7	-
8	ECT sensor

Inspection and Verification

CAUTION: Diagnosis by substitution from a donor vehicle is **NOT** acceptable. Substitution of control modules does not guarantee confirmation of a fault, and may also cause additional faults in the vehicle being tested and/or the donor vehicle.

NOTES:

If the control module or a component is suspect and the vehicle remains under manufacturer warranty, refer to the Warranty Policy and Procedures manual (section B1.2), or determine if any prior approval programme is in operation, prior to the installation of a new module/ component.

When performing voltage or resistance tests, always use a digital multimeter accurate to three decimal places, and with an up-to-date calibration certificate. When testing resistance always take the resistance of the digital multimeter leads into account.

Check and rectify basic faults before beginning diagnostic routines involving pinpoint tests.

1. Visually inspect for obvious signs of damage and system integrity
2. If an obvious cause for an observed or reported concern is found, correct the cause (if possible) before proceeding to the next step

Visual Inspection

Electrical
• Fuses • Wiring harnesses and connectors • Engine control module (ECM) • Glow plug control module • Glow plugs

3. If the cause is not visually evident, verify the symptom and refer to the Symptom Chart, alternatively check for Diagnostic Trouble Codes (DTCs) and refer to the DTC Index

Symptom Chart

Symptom	Possible Causes	Action
Poor starting in cold ambient temperature	• Glow plugs inoperative/inefficie nt	• Using the manufacturer approved diagnostic system, check the engine control module for related DTCs and refer to the relevant DTC index • Refer to the electrical circuit diagrams and check the glow plug control module power and ground circuits for open circuit, high resistance • Refer to the electrical circuit diagrams and check the glow plugs circuit for open circuit, high resistance
Smoke from exhaust after starting the engine from cold	• After-glow phase inoperative	NOTES: ⚠ After-glow is designed to function at engine temperatures below 50°C (122°F), below 2,500 rpm and for a maximum of 30 seconds after engine starting. ⚠ The glow plug warning indicator is not illuminated during the after-glow phase. ⚠ Exhaust smoke will not be evident from diesel particulate filter equipped vehicles. • Using the manufacturer approved diagnostic system, check the engine control module for related DTCs and refer to the relevant DTC index • Refer to the electrical circuit diagrams and check the glow plug control module power and ground circuits for open circuit, high resistance • Refer to the electrical circuit diagrams and check the glow plugs circuit for open circuit, high resistance
Excessive engine noise, vibration or harshness after starting the engine from cold	• After-glow phase inoperative	NOTES: ⚠ After-glow is designed to function at engine temperatures below 50°C (122°F), below 2,500 rpm and for a maximum of 30 seconds after engine starting. ⚠ The glow plug warning indicator is not illuminated during the after-glow phase. • Using the manufacturer approved diagnostic system, check the engine control module for related DTCs and refer to the relevant DTC index • Refer to the electrical circuit diagrams and check the glow plug control module power and ground circuits for open circuit, high resistance • Refer to the electrical circuit diagrams and check the glow plugs circuit for open circuit, high resistance

DTC Index

For a list of Diagnostic Trouble Codes (DTCs) that could be logged on this vehicle, please refer to Section 100-00.

Refer to: Diagnostic Trouble Code Index - DTC: Engine Control Module (PCM) (100-00 General Information, Description and Operation)

Glow Plug

Removal and Installation

Removal

1. Disconnect the battery ground cable.

Refer to: Battery Disconnect (414-01, General Procedures).

2. Remove the intake manifold.

Refer to: Intake Manifold (303-01, In-vehicle Repair).

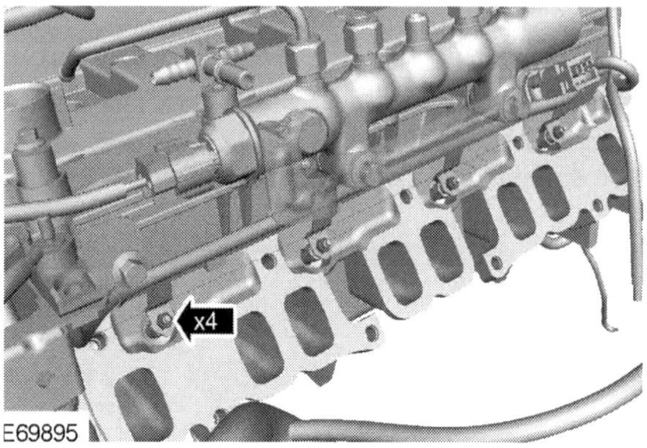

E69895

3. Release the glow plug wiring harness.

    - Remove the 4 nuts.

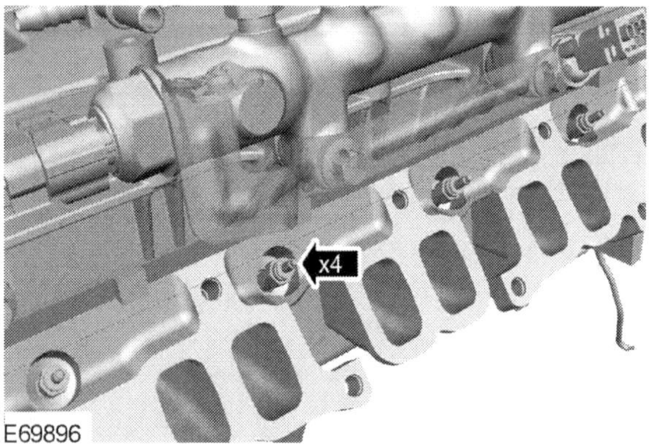

E69896

4. Remove the glow plugs.

Installation

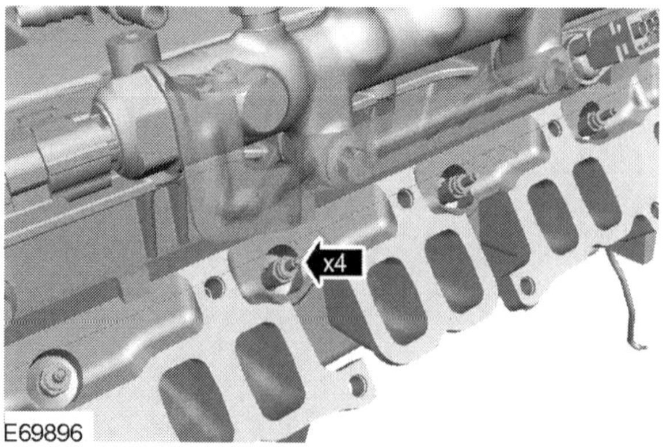

E69896

1. To install, reverse the removal procedure.

- Tighten to 13 Nm (10 lb.ft).

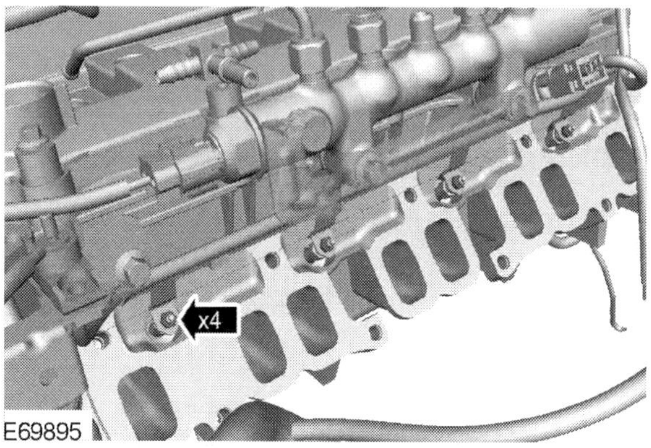

E69895

2. Tighten to 2 Nm (1 lb.ft).

Specifications

Description	Nm	lb-ft
Exhaust gas recirculation (EGR) valve to EGR valve outlet tube nuts	10	7
EGR valve to EGR cooler bolts	23	17
Exhaust manifold to EGR cooler bolts	23	17
EGR valve outlet tube to intake manifold	23	17
DPF Clamp to Turbocharger	10	7
Front muffler to DPF nuts	25	18
Pressure pipe union-Pre DPF	25	18
Pressure pipe union-Post DPF	25	18

Description and Operation

COMPONENT LOCATION - EXHAUST GAS RECIRCULATION

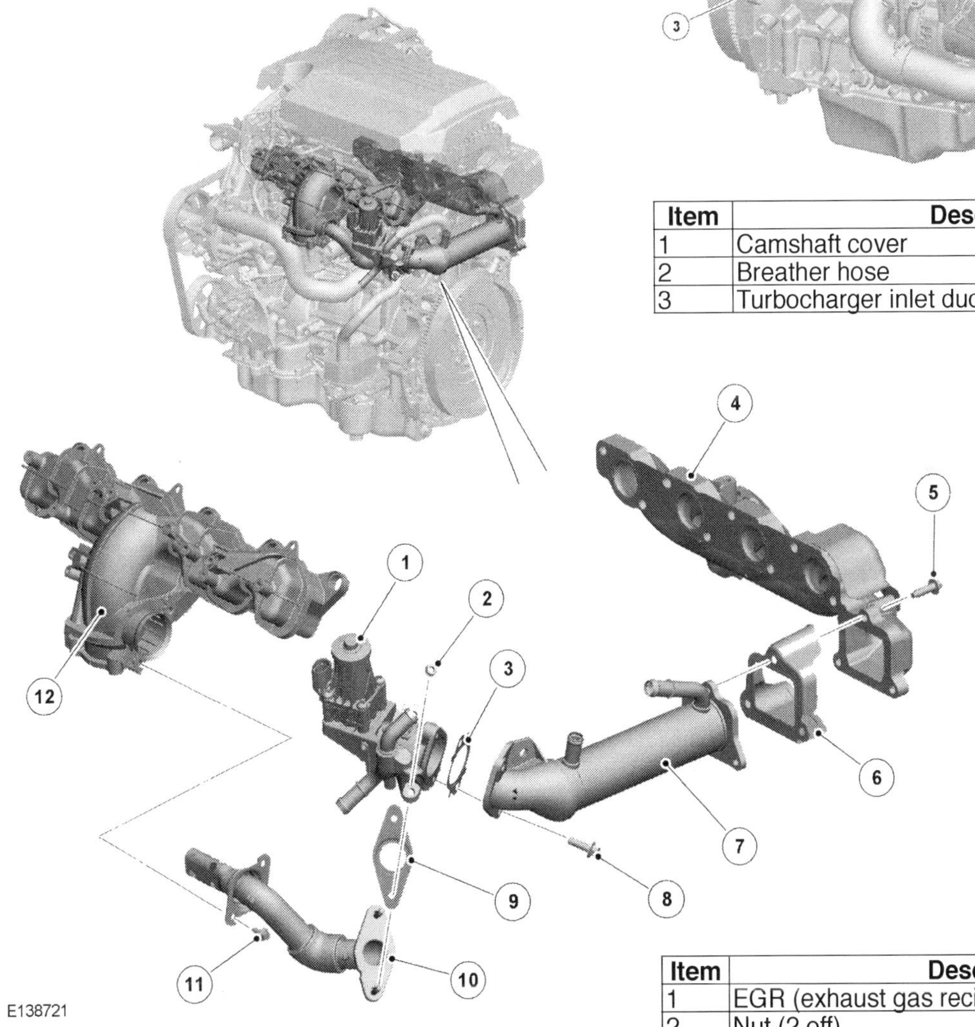

Item	Description
1	EGR (exhaust gas recirculation) valve
2	Nut (2 off)
3	Gasket
4	Exhaust manifold
5	Bolt (3 off)
6	Gasket
7	EGR cooler
8	Bolt (2 off)
9	Gasket
10	EGR outlet tube
11	Bolt (2 off)
12	Intake manifold

OVERVIEW

Engine emission control reduces the level of harmful emissions released to atmosphere from the engine. Engine emission control consists of an EGR system and a PCV (positive crankcase ventilation) system.

The EGR system regulates a supply of exhaust gas into the air charge entering the engine, which lowers the

COMPONENT LOCATION - POSITIVE CRANKCASE VENTILATION

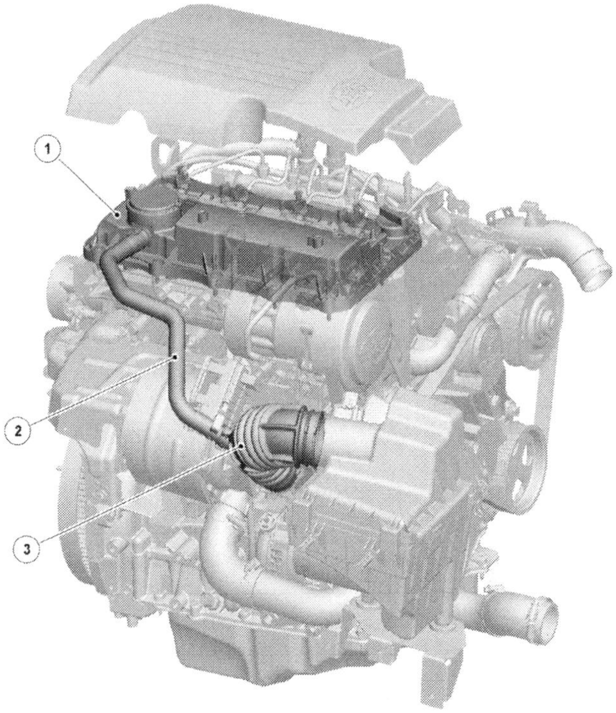

Item	Description
1	Camshaft cover
2	Breather hose
3	Turbocharger inlet duct

combustion temperature and minimizes the formation of NOX (oxides of nitrogen).

The PCV system draws piston blow-by gases into the engine air intake to be added to the air charge. The resultant depression in the engine sump, front cover and cylinder head cover reduces the load on the joint seals in those areas.

EXHAUST GAS RECIRCULATION

The EGR system comprises:

- An EGR valve.
- An EGR cooler.
- An EGR outlet tube.

The EGR cooler is attached to the rear of the exhaust manifold. Coolant from the engine cooling system circulates through a matrix in the EGR cooler to reduce the temperature of the exhaust gas. Attached to the outlet side of the cooler is the EGR valve. The valve is motor driven, under the control of the ECM (engine control module) , to allow varying amounts of exhaust gas into the intake manifold depending on the engine operation. At engine switch off the valve opens and closes several times to clear any deposits which may have accumulated during running.

The EGR outlet tube transports the exhaust gas from the EGR valve to the intake manifold, where it is blended with the incoming charge air stream by a mixer tube.

The ECM monitors the EGR system function and stores fault codes in the event of failure. The EGR valve can also be activated for testing using Land Rover approved diagnostic equipment.

POSITIVE CRANKCASE VENTILATION

The PCV system consists of an oil separator, a PCV valve and a breather hose. The oil separator and the PCV valve are integrated into the RH (right-hand) side of the camshaft cover. The breather hose connects the camshaft cover to the turbocharger inlet duct.

When the engine is running, the depression in the turbocharger inlet duct draws in gases from the engine sump through the oil separator, PCV valve and breather hose. Any oil in the gases is removed by the oil separator and drains back into the oil pan through the drain channels in the cylinder head and cylinder block. The PCV valve prevents reverse flow into the camshaft cover when there is minimal depression in the turbocharger inlet duct.

Camshaft Cover

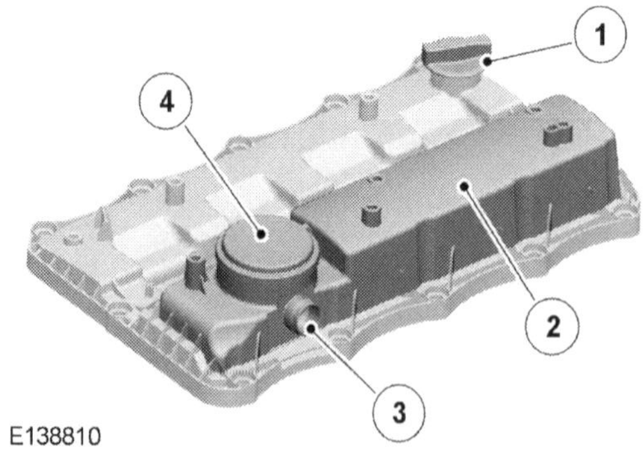

Item	Description
1	Engine oil filler cap
2	Oil separator
3	Ventilation outlet
4	-

Engine Emission Control

Diagnosis and Testing

Principles of Operation

For a detailed description of the Engine Emission Control system, refer to the relevant Description and Operation section in the workshop manual.

Refer to: Engine Emission Control (303-08 Engine Emission Control, ID4 2.2L Diesel Description and Operation)

Inspection and Verification

⚠ CAUTION: Diagnosis by substitution from a donor vehicle is **NOT** acceptable. Substitution of control modules does not guarantee confirmation of a fault, and may also cause additional faults in the vehicle being tested and/or the donor vehicle.

NOTES:

△ If the control module or a component is suspect and the vehicle remains under manufacturer warranty, refer to the Warranty Policy and Procedures manual (section B1.2), or determine if any prior approval programme is in operation, prior to the installation of a new module/component.

△ When performing voltage or resistance tests, always use a digital multimeter accurate to three decimal places, and with an up-to-date calibration certificate. When testing resistance always take the resistance of the digital multimeter leads into account.

△ Check and rectify basic faults before beginning diagnostic routines involving pinpoint tests.

2. Visually inspect for obvious signs of damage and system integrity

Visual Inspection

2. If an obvious cause for an observed or reported concern is found, correct the cause (if possible) before proceeding to the next step
3. If the cause is not visually evident, verify the symptom and refer to the Symptom Chart, alternatively check for Diagnostic Trouble Codes (DTCs) and refer to the DTC Index

Mechanical	Electrical
• Engine breather hoses • Oil separator • Exhaust gas recirculation valve • Exhaust gas recirculation pipes (check for cracks, leaks at joints or gaskets) • Exhaust gas recirculation cooler	• Fuses • Wiring harnesses and connectors • Exhaust gas recirculation valve • Engine control module (ECM)

Symptom Chart

Symptom	Possible Causes	Action
Poor starting Poor/erratic idle Lack of power when accelerating	• Exhaust gas recirculation valve stuck open	• Using the manufacturer approved diagnostic system, check the engine control module for related DTCs and refer to the relevant DTC index. Perform routine - Turbo, EGR and Air Path Dynamic Test - and perform the relevant corrective actions • Using the manufacturer approved diagnostic system, perform routine - Inline diagnostic unit 2 diagnostic test - Exhaust gas recirculation valve.
Engine stops/stalls	• Exhaust gas recirculation valve stuck open • Breather system disconnected/restricted/blocked	• Using the manufacturer approved diagnostic system, check the engine control module for related DTCs and refer to the relevant DTC index. Perform routine - Turbo, EGR and Air Path Dynamic Test - and perform the relevant corrective actions • Using the manufacturer approved diagnostic system, perform routine - Inline diagnostic unit 2 diagnostic test - Exhaust gas recirculation valve. • Check the engine breather system. Check the oil separator
Excessive fuel consumption Excessive black smoke Excessive emissions	• Exhaust gas recirculation valve stuck open/closed • Breather system disconnected/restricted/blocked	NOTE: Exhaust smoke will not be evident from diesel particulate filter equipped vehicles. • Using the manufacturer approved diagnostic system, check the engine control module for related DTCs and refer to the relevant DTC index. Perform routine - Turbo, EGR and Air Path Dynamic Test - and perform the relevant corrective actions • Using the manufacturer approved diagnostic system, perform routine - Inline diagnostic unit 2 diagnostic test - Exhaust gas recirculation valve. • Check the engine breather system. Check the oil separator
Excessive blow-by	• Breather system restricted/blocked	• Check the engine breather system. Check the oil separator
Engine oil leaks	• Breather system disconnected/restricted/blocked	• Check the engine breather system. Check the oil separator

DTC Index

For a list of Diagnostic Trouble Codes (DTCs) that could be logged on this vehicle, please refer to Section 100-00.
Refer to: Diagnostic Trouble Code Index - DTC: Engine Control Module (PCM) (100-00 General Information, Description and Operation)

Exhaust Gas Recirculation Cooler

Removal and Installation

Removal

NOTE: Removal steps in this procedure may contain installation details.

1. **Refer to:** Battery Disconnect and Connect (414-01, General Procedures).
2. WARNING: Do not work on or under a vehicle supported only by a jack. Always support the vehicle on safety stands.

   Raise and support the vehicle.
3. **Refer to:** Cooling System Draining, Filling and Bleeding (303-03 Engine Cooling, ID4 2.2L Diesel General Procedures)

4.

5. Torque: 23 Nm

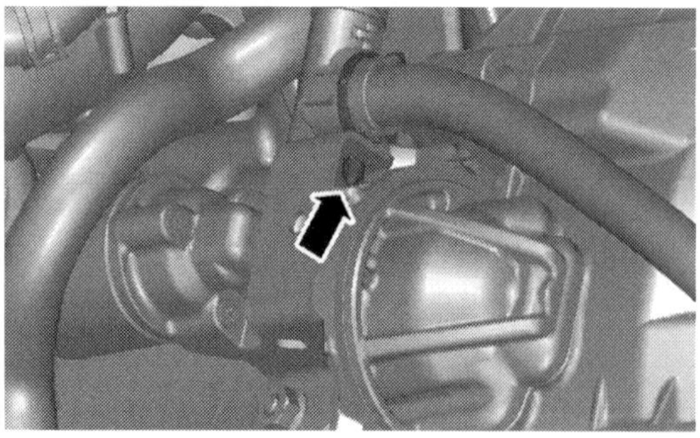

6.

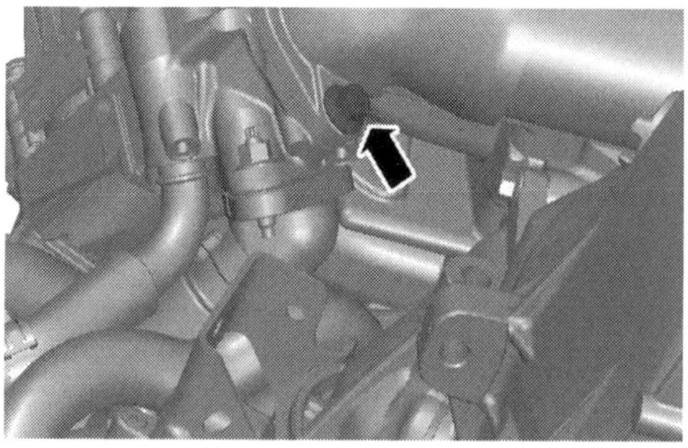

7. Torque: 23 Nm

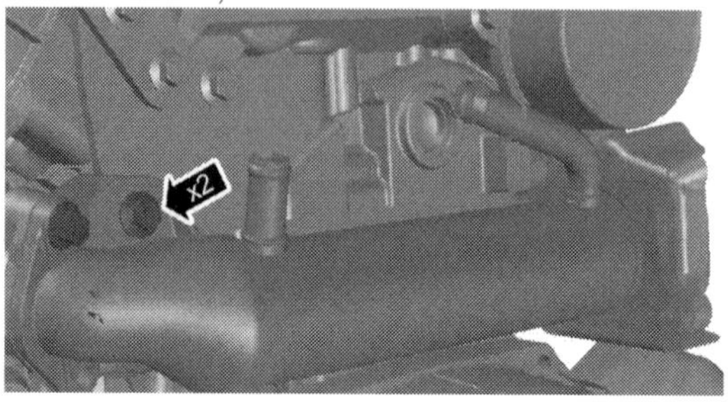

8. NOTES:

   Clean the component mating faces.

   Fit new gasket.

   Torque: 23 Nm

Installation

1. To install, reverse the removal procedure.

Exhaust Gas Recirculation Valve

Removal and Installation

Removal

NOTE: Removal steps in this procedure may contain installation details.

1. **Refer to:** Engine Cover (501-05, Removal and Installation).
2. **Refer to:** Battery Disconnect and Connect (414-01, General Procedures).
3. **Refer to:** Cooling System Draining, Filling and Bleeding (303-03 Engine Cooling, ID4 2.2L Diesel General Procedures)

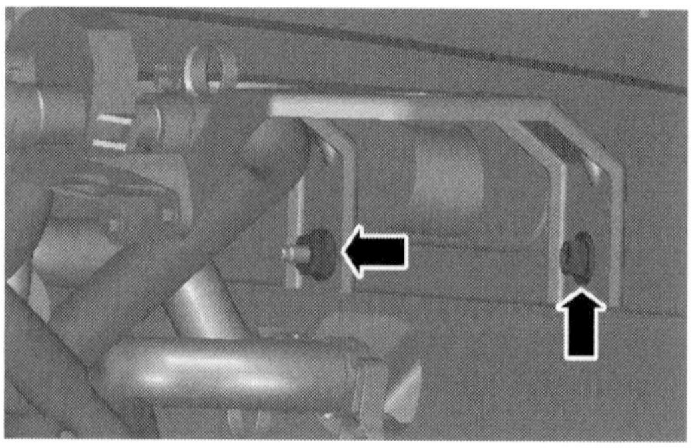

4. Torque: Bolt 22 Nm Torque: Nut 6 Nm

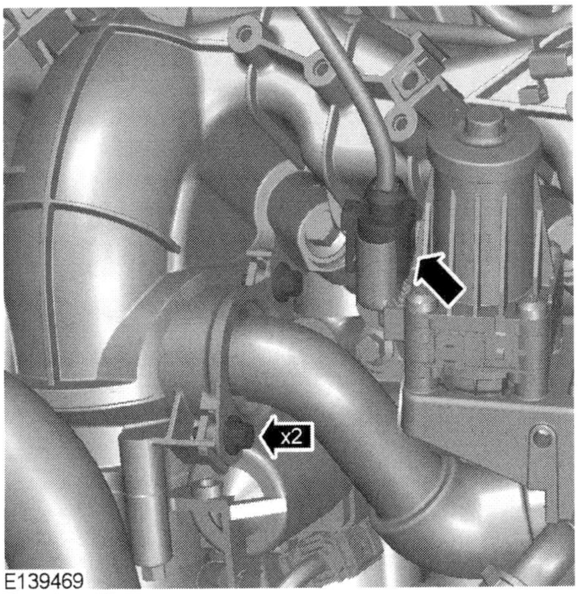

5. Torque: 23 Nm

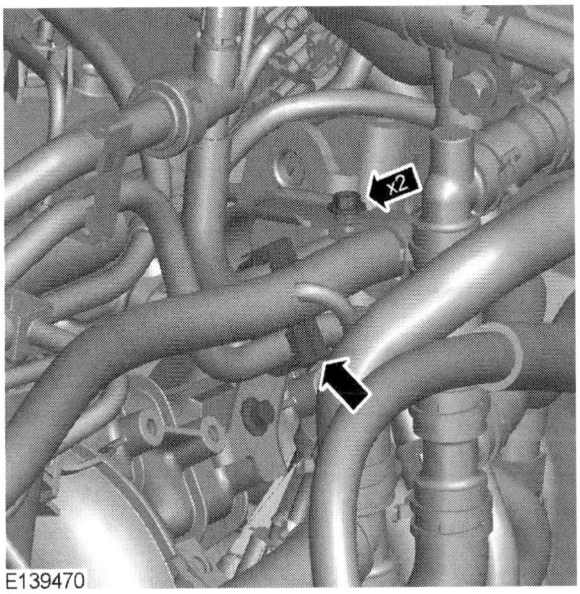

6. Torque: 10 Nm

7.

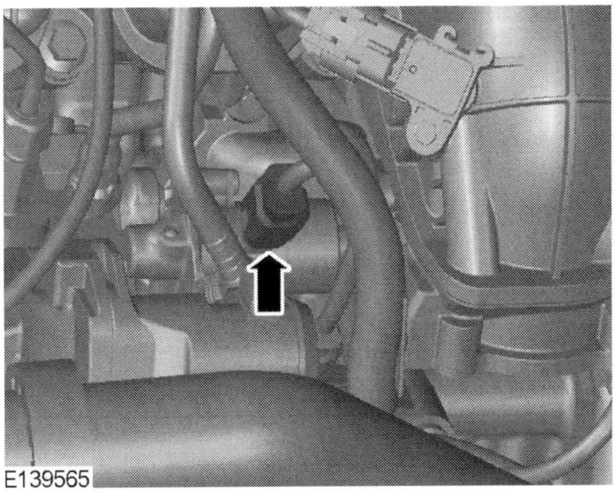

8. Torque: 23 Nm

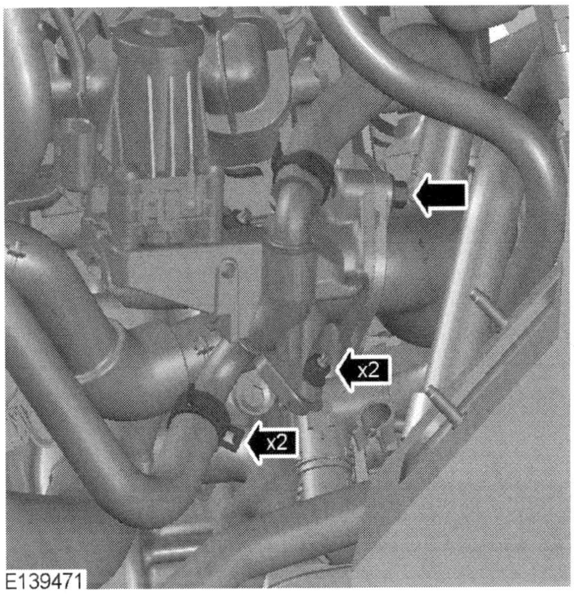

9. Torque: Bolt 23 Nm Torque: Nuts 10 Nm

Installation

1. ⚠ CAUTION: Make sure that the mating faces are clean and free from foreign material.

△ NOTE: Install a new gasket

2. To install, reverse the removal procedure.

Specifications

Torque Specifications

Description	Nm	lb-ft
Air cleaner bolts	13	10
A/C compressor low-pressure and high-pressure refrigerant lines bolt	30	22
Charge air cooler to radiator bolts	20	15
Charge air cooler inlet and outlet hose clips	5	4
Charge air cooler bracket bolts	9	7
Air cleaner outlet pipe clips	3	2

Description and Operation

COMPONENT LOCATION

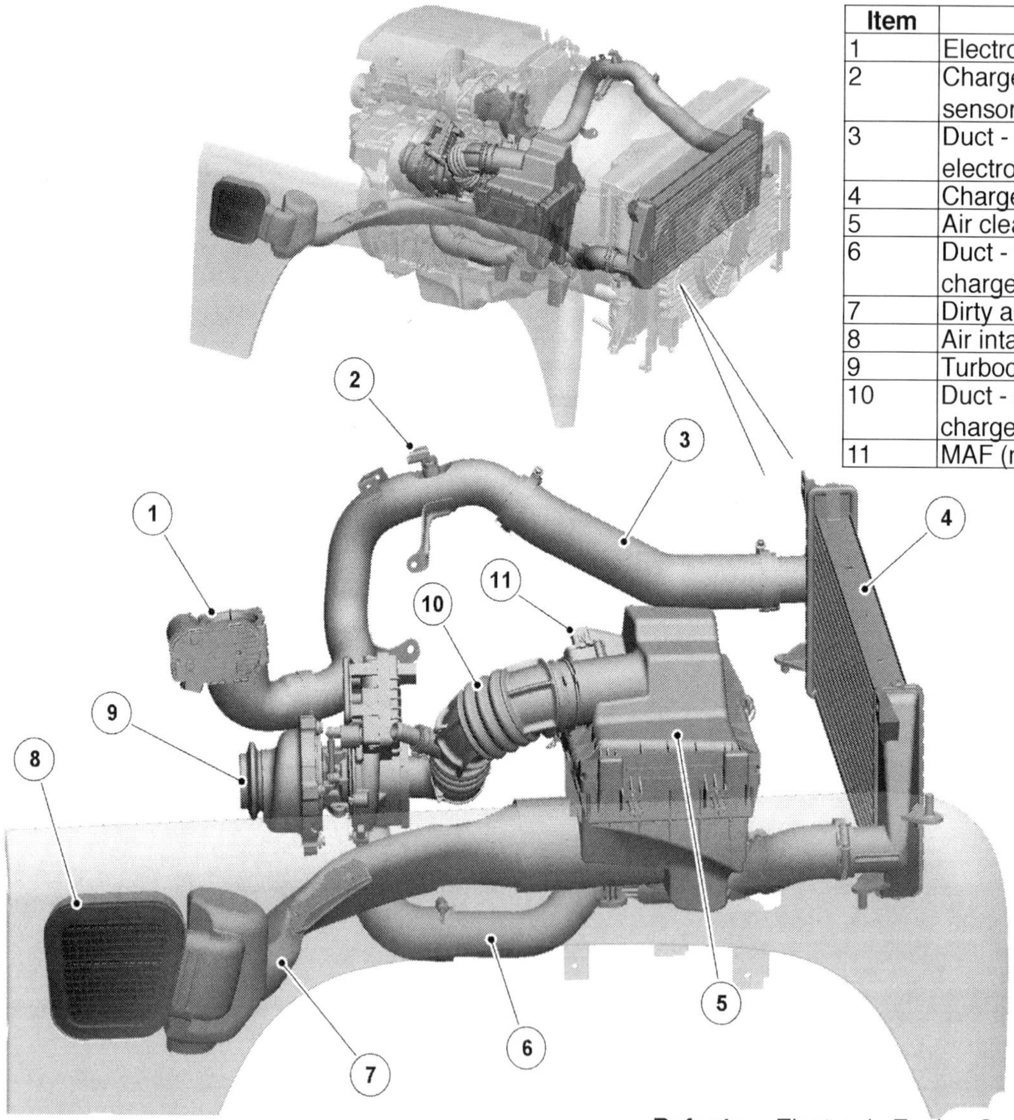

Item	Description
1	Electronic throttle
2	Charge air temperature sensor
3	Duct - charge air cooler to electronic throttle
4	Charge air cooler
5	Air cleaner
6	Duct - turbocharger to charge air cooler
7	Dirty air duct
8	Air intake
9	Turbocharger
10	Duct - air cleaner to turbo-charger
11	MAF (mass air flow) sensor

OVERVIEW

Ambient air is drawn in through the intake in the RH (right-hand) fender, then through the dirty air duct to the air cleaner. From the air cleaner, ducts direct the air through the turbocharger and the charge air cooler to the electronic throttle on the intake manifold.

A charge air temperature sensor is installed in the duct between the charge air cooler and the electronic throttle.

Refer to: Electronic Engine Controls (303-14 Electronic Engine Controls, ID4 2.2L Diesel Description and Operation)

AIR CLEANER

The air cleaner, which contains a serviceable paper filter element, is located on the RH front inner fender. Four rubber mounts attach the air cleaner to brackets on the inner fender.

A MAF sensor is installed in the outlet port of the air cleaner.

Refer to: Electronic Engine Controls (303-14 Electronic Engine Controls, ID4 2.2L Diesel Description and Operation)

CHARGE AIR COOLER - INTERCOOLER

The charge air cooler is located in front of the radiator and is secured to the radiator by two bolts at the top and two location points at the bottom.

Intake Air Distribution and Filtering

Diagnosis and Testing

Principles of Operation

For a detailed description of the Intake Air Distribution and Filtering system, refer to the relevant Description and Operation section in the workshop manual.

Refer to: Intake Air Distribution and Filtering (303-12 Intake Air Distribution and Filtering, ID4 2.2L Diesel Description and Operation)

Inspection and Verification

CAUTION: Diagnosis by substitution from a donor vehicle is **NOT** acceptable. Substitution of control modules does not guarantee confirmation of a fault, and may also cause additional faults in the vehicle being tested and/or the donor vehicle.

NOTES:

If the control module or a component is suspect and the vehicle remains under manufacturer warranty, refer to the Warranty Policy and Procedures manual (section B1.2), or determine if any prior approval programme is in operation, prior to the installation of a new module/component.

When performing voltage or resistance tests, always use a digital multimeter accurate to three decimal places, and with an up-to-date calibration certificate. When testing resistance always take the resistance of the digital multimeter leads into account.

Check and rectify basic faults before beginning diagnostic routines involving pinpoint tests.

1. Visually inspect for obvious signs of damage and system integrity
2. If an obvious cause for an observed or reported concern is found, correct the cause (if possible) before proceeding to the next step
2. If the cause is not visually evident, verify the symptom and refer to the Symptom Chart, alternatively check for Diagnostic Trouble Codes (DTCs) and refer to the DTC Index

Visual Inspection

Mechanical	Electrical
• Hoses and ducts condition/installation • Air cleaner filter element condition/ installation • Vacuum hoses condition/installation • Variable geometry turbine turbocharger • Charge air cooler • Charge air cooler hoses	• Fuses • Wiring harnesses and connectors • Mass air flow sensor • Intake air temperature sensor • Manifold absolute pressure and temperature sensor

Symptom Chart

Symptom	Possible Causes	Action
Vehicle does not start/hard starting	• Intake air system leaking/restricted	• Check the intake air system for blockages or restriction. Rectify as necessary
Poor performance	• Intake air system leaking/restricted • Charge air cooler leaking • Charge air cooler hose leaking • Variable geometry turbine turbocharger fault • Exhaust gas recirculation valve stuck open • Engine management system fault • Exhaust system restricted	• Check the intake air system for leaks/restrictions. Using the manufacturer approved diagnostic system, perform routine - Turbo, EGR and Air Path Dynamic Test - and perform the relevant corrective actions • Check the charge air cooler for leaks • Check the charge air cooler hoses for leaks • Check the mechanical condition of the variable geometry turbine turbocharger. Using the manufacturer approved diagnostic system, check the engine control module for related DTCs and refer to the relevant DTC index. Perform routine - Turbo, EGR and Air Path Dynamic Test - and perform the relevant corrective actions • Using the manufacturer approved diagnostic system, check the engine control module for related DTCs and refer to the relevant DTC index. Perform routine - Turbo, EGR and Air Path Dynamic Test - and perform the relevant corrective actions • Using the manufacturer approved diagnostic system, check the engine control module for related DTCs and refer to the relevant DTC index • Check the exhaust system for evidence of damage or restriction

329

Excessive intake noise	• Intake air system leaking • Charge air cooler leaking • Charge air cooler hose leaking	• Check the intake air system for leaks. Using the manufacturer approved diagnostic system, perform routine - Turbo, EGR and Air Path Dynamic Test - and perform the relevant corrective actions • Check the charge air cooler for leaks • Check the charge air cooler hoses for leaks

DTC Index

For a list of Diagnostic Trouble Codes (DTCs) that could be logged on this vehicle, please refer to Section 100-00. **Refer to:** Diagnostic Trouble Code Index - DTC: Engine Control Module (PCM) (100-00 General Information, Description and Operation)

Charge Air Cooler - Intercooler - Intake Pipe

Removal and Installation

Removal

⚠ NOTE: **Some vehicles may be equipped with a three piece charge air cooler intake pipe assembly. If the charge air cooler intake pipe requires replacement a single piece assembly must be installed.**

1. Disconnect the battery ground cable.

   Refer to: Battery Disconnect and Connect (414-01, General Procedures).

2. With the aid of another technician, remove the hood.

   Refer to: Hood (501-02, Removal and Installation).

3. Remove the cooling fan.

Refer to: Cooling Fan (303-03 Engine Cooling, ID4 2.2L Diesel Removal and Installation)

4. Release the air cleaner outlet pipe.
   - Release the 2 clips.
   - Disconnect the electrical connector.

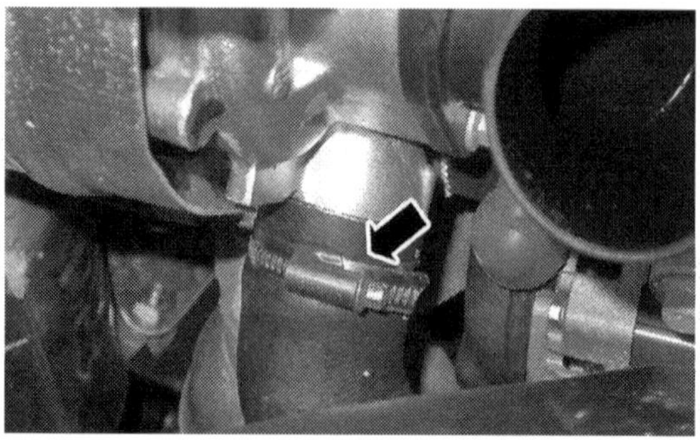

5. Release the charge air cooler intake pipe from the turbocharger.
   - Release the clip.

6. Disconnect the charge air cooler intake pipe from the charge air cooler.
   - Release the clip.

7. ⚠ NOTE: **Vehicles equipped with three piece charge air cooler intake pipe assembly.**

   Remove and discard the charge air cooler hose from the turbocharger end of the pipe.
   - Reposition the hose for access.
   - Release the clip.

8. NOTE: **Vehicles equipped with three piece charge air cooler intake pipe assembly.**

Remove and discard the charge air cooler hose from the charge air cooler end of the pipe.

- Reposition the hose for access.
- Release the clip.

9. CAUTION: Make sure that the air conditioning pipe is not damaged during this procedure.

NOTES:

Three piece charge air cooler intake pipe assembly shown, one piece assembly similar.

All vehicles.

Remove and discard the charge air cooler intake pipe.

- Carefully re-position the fan shroud for access.

Installation

1. CAUTIONS:

Make sure that no other components are damaged during the removal of the peg.

Make sure that no sharp edge is left on the wheel arch liner after the peg has been removed.

NOTE: Components shown removed for clarity.

If equipped, using a suitable tool, carefully remove the peg from the fender splash shield.

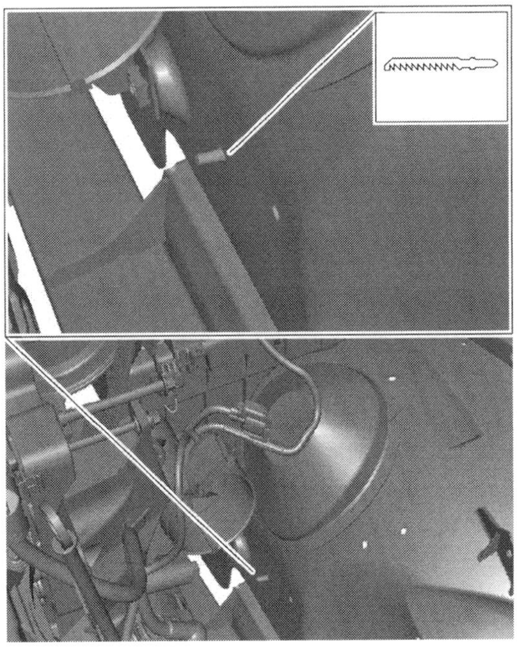

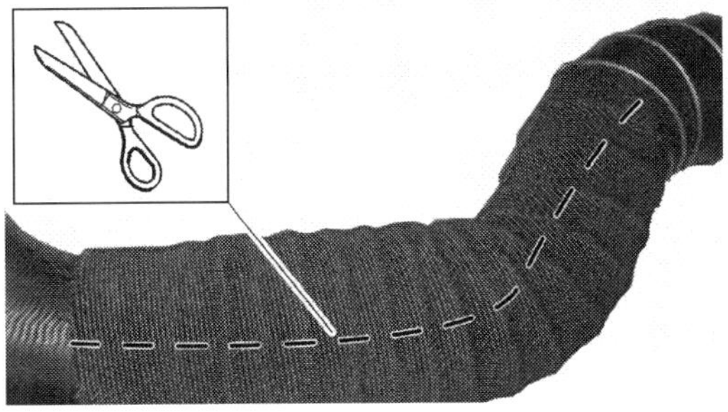

2. CAUTIONS:

Make sure that the pipe is not damaged during the removal of the protective sleeve.

Do not remove the metal coil from the pipe.

If the new charge air cooler intake pipe is supplied with a protective fabric sleeve, use a suitable tool to remove the fabric sleeve.

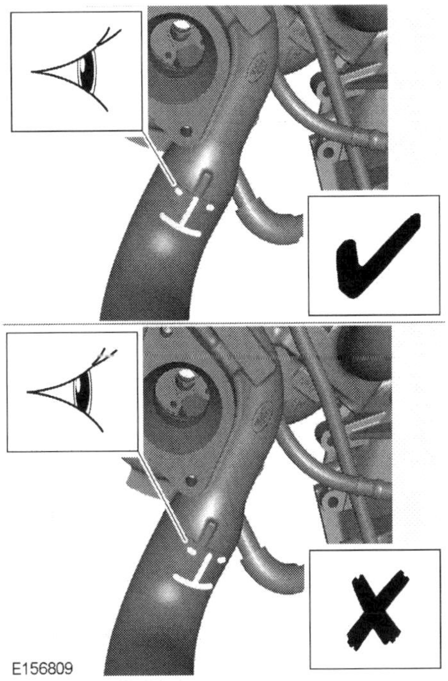

3. CAUTION: Some old charge air cooler intake pipes may be supplied with 2 white dots. The white dots do not show a tolerance range for the alignment of the pipe. Make sure that the centre line is correctly aligned with the timing mark on the component.

Prior to installation, note the timing marks on the new charge air cooler intake pipe.

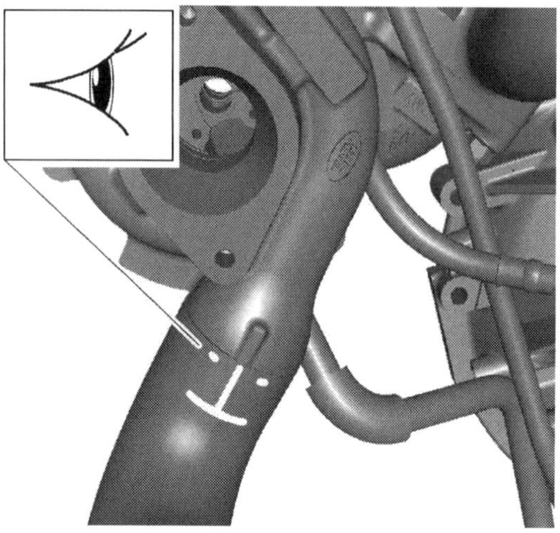

4. NOTES:

Using a suitable brake cleaner, make sure any oil residue is cleaned off the outlet stub of the turbocharger and charge air cooler prior to installation of the new charge air cooler intake pipe.

Hose clip shown removed for clarity.

When installing the charge air cooler intake pipe to the turbocharger make sure that the alignment marks are correctly aligned and that the hose is fully located on to the turbocharger as illustrated.

5. CAUTION: **Make sure that the clip is positioned perpendicular to the pipe when tightened.**

Install the charge air cooler intake pipe to the turbocharger.

- Make sure that the hose clip is tightened to 5 Nm.

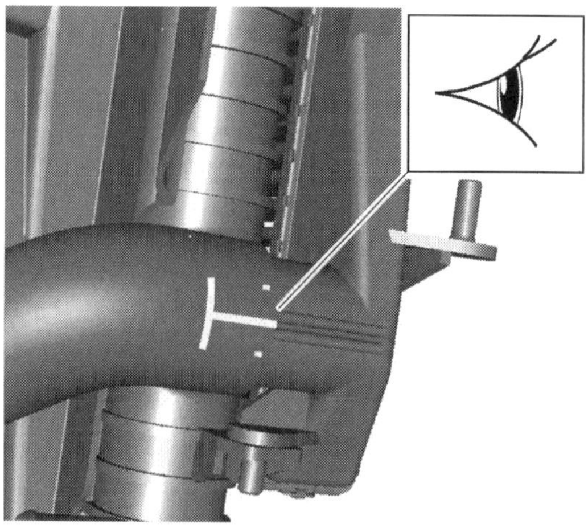

6. NOTE: Hose clip shown removed for clarity.

When installing the charge air cooler intake pipe to the charge air cooler make sure that the alignment marks are correctly aligned and that the hose is fully located on to the charge air cooler as illustrated.

7. CAUTIONS:

Make sure that the clip is positioned perpendicular to the pipe when tightened.

Make sure that the charge air cooler intake pipe is not positioned against any other components.

Install the charge air cooler intake pipe to the charge air cooler.

- Make sure that the hose clip is tightened to 5 Nm.

8. Secure the air cleaner outlet pipe.

- Secure the 2 clips.
- Connect the electrical connector.

9. Install the cooling fan.

   Refer to: Cooling Fan (303-03 Engine Cooling, ID4 2.2L Diesel Removal and Installation)

10. With the aid of another technician, install the hood.

    Refer to: Hood (501-02, Removal and Installation).

11. Connect the battery ground cable.

Refer to: Battery Disconnect and Connect (414-01, General Procedures).

Air Cleaner

Removal and Installation

1. Disconnect the battery ground cable.

Refer to: Startup Battery Disconnect and Connect (414-01 Battery, Mounting and Cables, General Procedures)

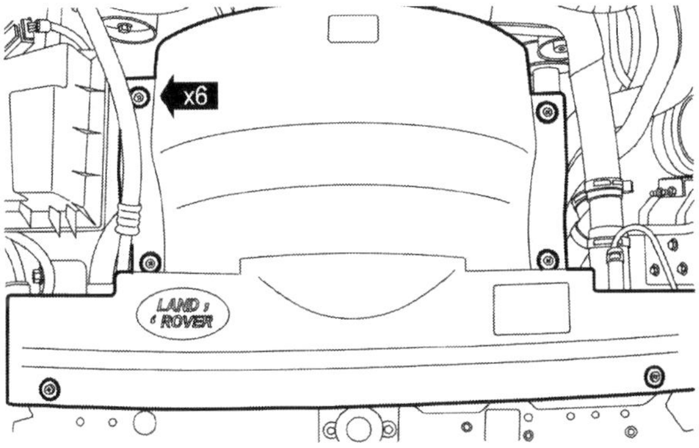

2. Remove the cooling fan upper shroud.
   - Release the 6 clips.
3. Using the special tools, remove the cooling fan.

4. Remove the air cleaner element.

Refer to: Air Cleaner Element (303-12 Intake Air Distribution and Filtering, ID4 2.2L Diesel Removal and Installation)

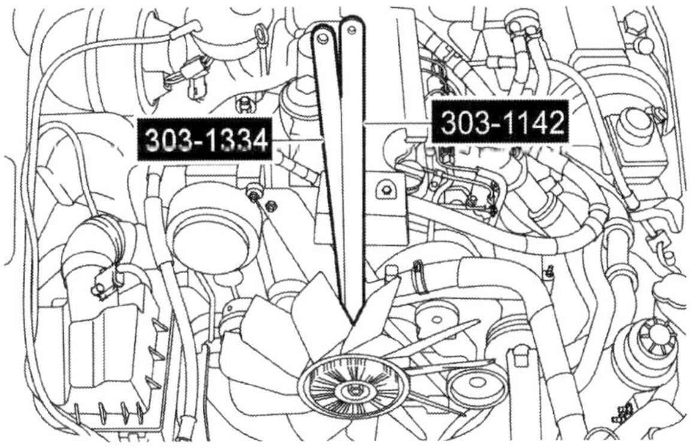

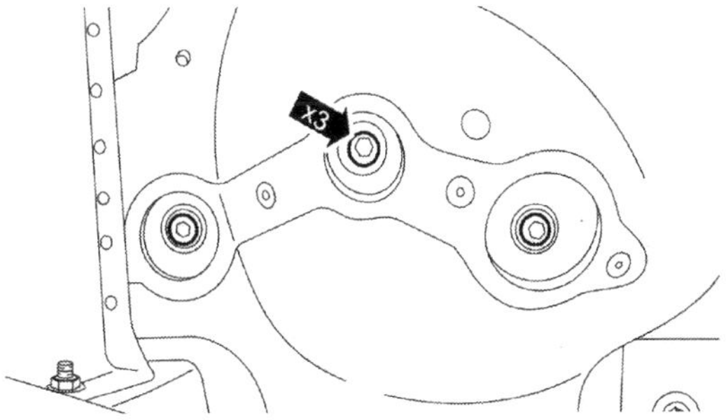

5. Release the air cleaner.
   - Loosen the 3 bolts.

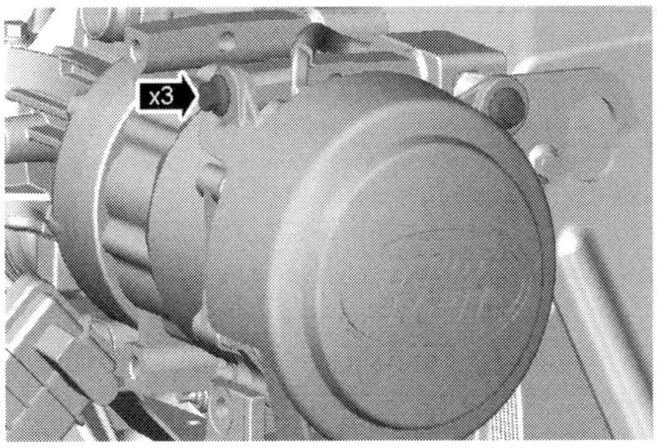

6. Remove the air conditioning (A/C) compressor accessory drive belt cover.

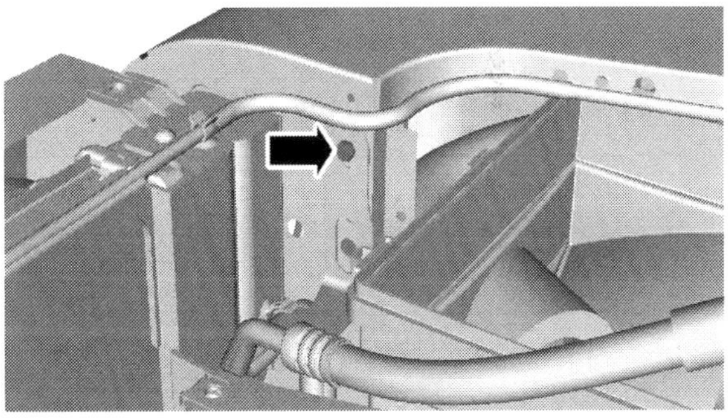

7. Release the A/C line.

8. Remove the air cleaner.
   1. Disconnect the air cleaner intake pipe.
   2. Remove the air cleaner.

Installation

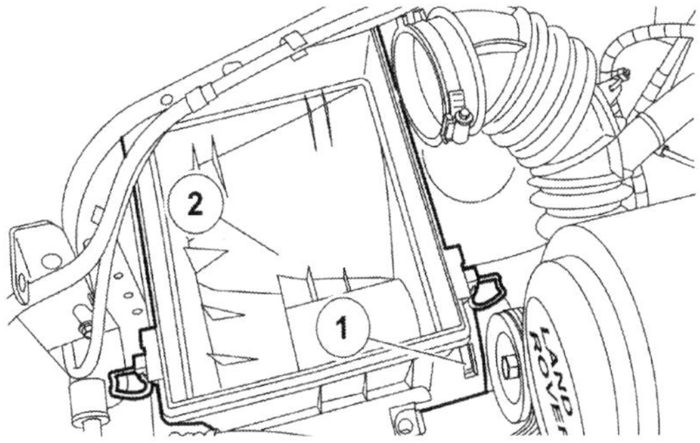

1. Install the air cleaner.
    1. Connect the air cleaner intake pipe.
    2. Install the air cleaner.

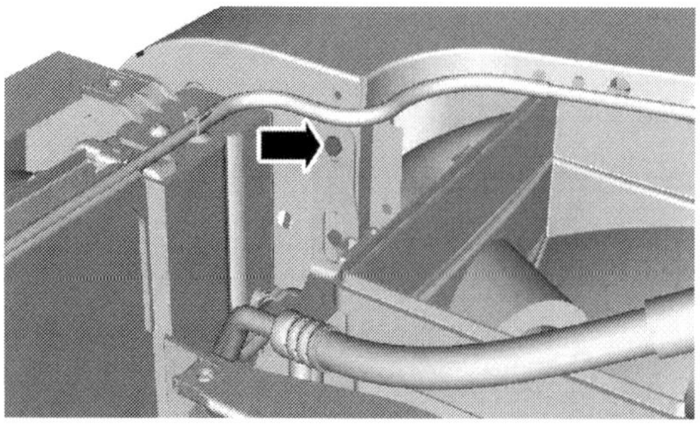

2. Secure the A/C line bracket.
    1. Tighten to 10 Nm (7 lb.ft).

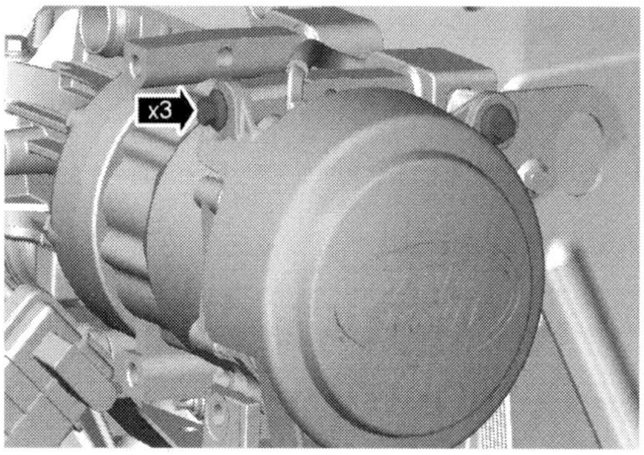

3. Install the A/C compressor accessory drive belt cover.
    1. Tighten to 10 Nm (7 lb.ft).

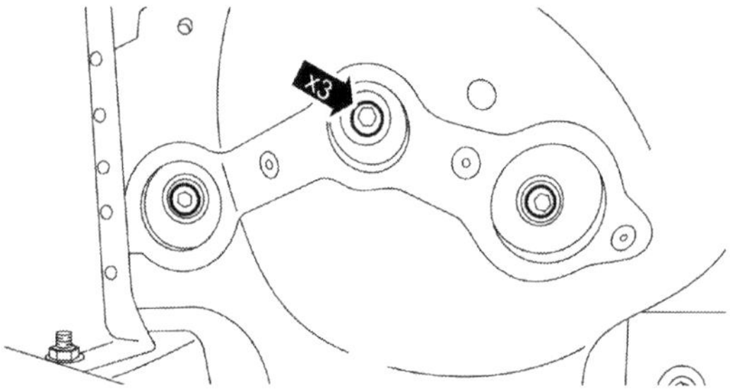

4. Secure the air cleaner.
    - Tighten to 13 Nm (10 lb.ft).
5. Install the air cleaner element.

    Refer to: Air Cleaner Element (303-12 Intake Air Distribution and Filtering, ID4 2.2L Diesel Removal and Installation)

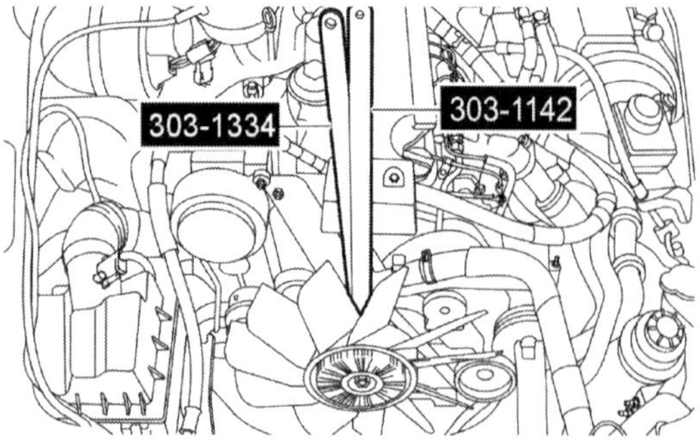

6. 2. Using the special tools, install the cooling fan.

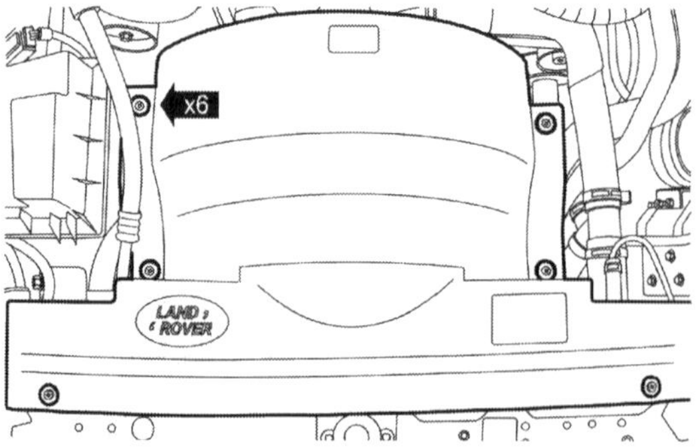

7. Install the cooling fan upper shroud
    - Secure the 6 clips.
8. Connect the battery ground cable.

Refer to: Battery Connect (414-01, General Procedures).

Air Cleaner Element

Removal and Installation

Removal

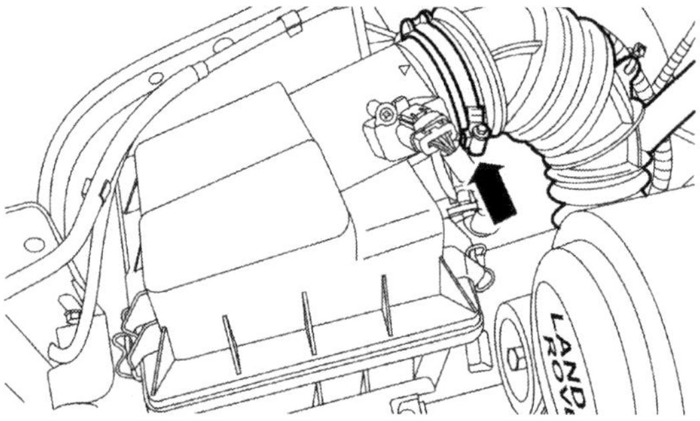

1. Release the air cleaner outlet pipe.
    - Release the clip.

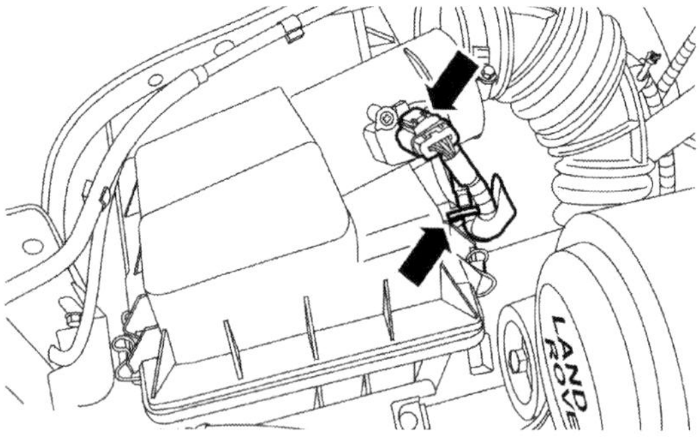

2. Disconnect the mass air flow (MAF) sensor electrical connector.
    - Release the wiring harness.

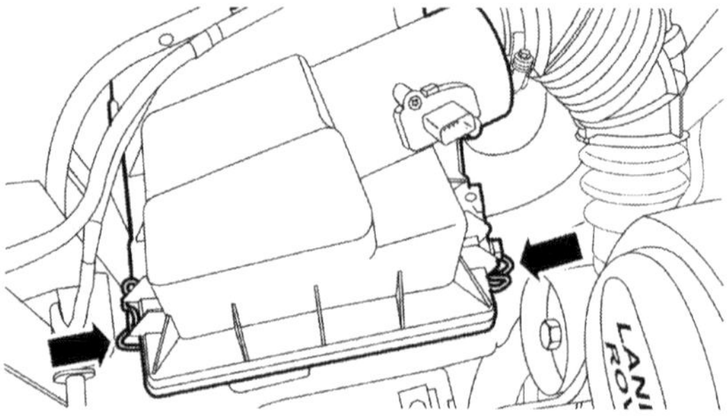

3. Remove the air cleaner housing cover.

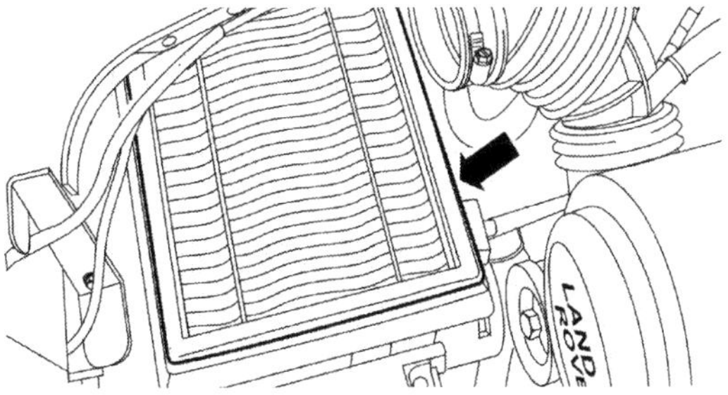

4. Remove the air cleaner element.

Installation

1. CAUTION: Make sure that the air cleaner housing cover rear fixing tangs are correctly located.

NOTE: Clean the base of the air cleaner.

To install, reverse the removal procedure.

Air Cleaner Outlet Pipe

Removal and Installation

Removal

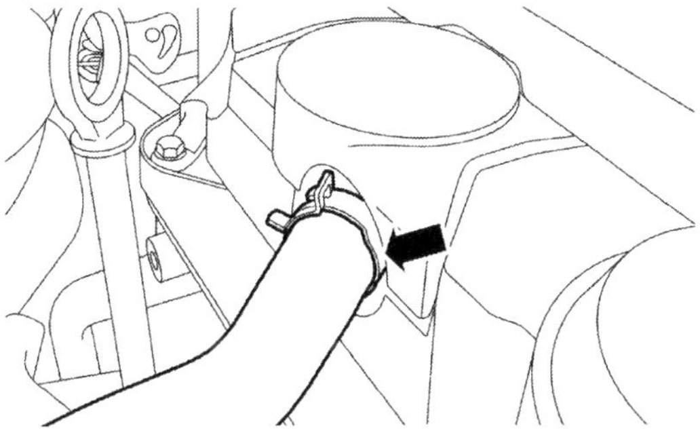

1. Release the positive crankcase ventilation (PCV) hose from the valve cover.
   - Release the clip.

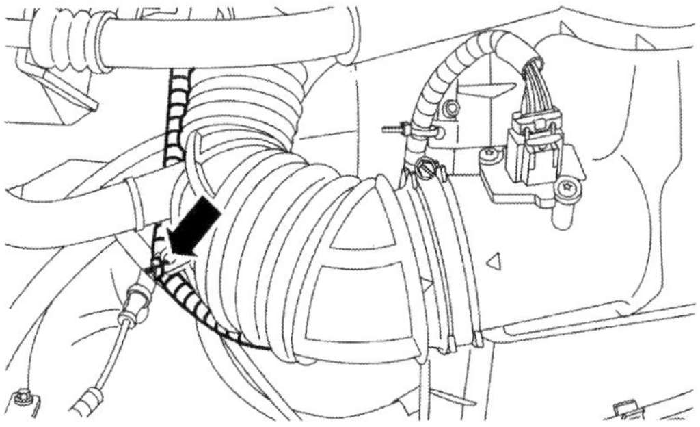

2. Release the mass air flow (MAF) sensor wiring harness.

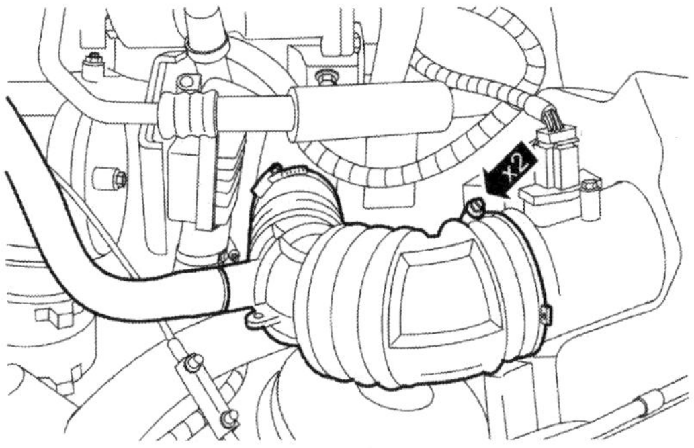

3. Remove the air cleaner outlet pipe.
   - Release the 2 clips.

Installation

1. To install, reverse the removal procedure.

Charge Air Cooler - Intercooler

Removal and Installation

Removal

1. Disconnect the battery ground cable.

Refer to: Startup Battery Disconnect and Connect (414-01 Battery, Mounting and Cables, General Procedures)

2. Remove the radiator grille.

Refer to: Radiator Grille (501-08 Exterior Trim and Ornamentation, Removal and Installation)

3. Release the 6 clips. Remove the cooling fan upper shroud.
4. Remove the hood latch panel.

Refer to: Hood Latch Panel (501-27 Front End Sheet Metal Repairs, Removal and Installation)

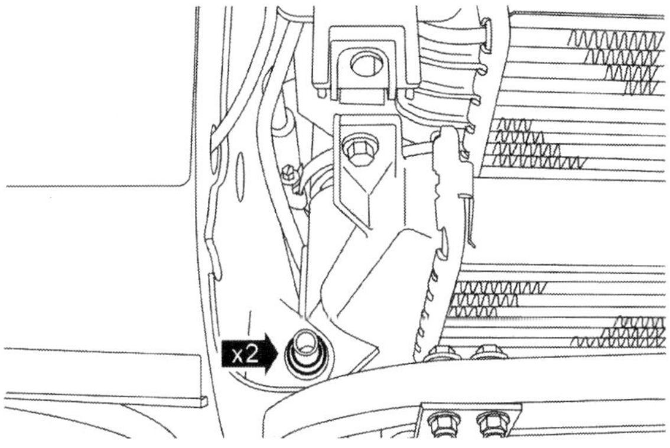

5. NOTE: RH shown, LH similar.

Remove the 2 O-rings.

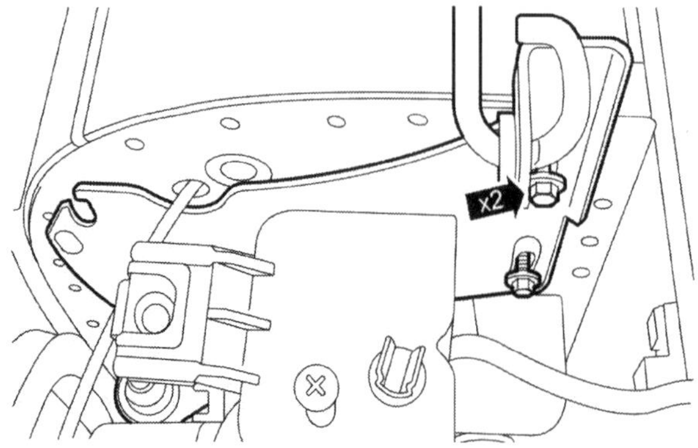

6. NOTE: RH shown, LH similar.

Remove the 2 charge air cooler brackets. Remove the 4 bolts.

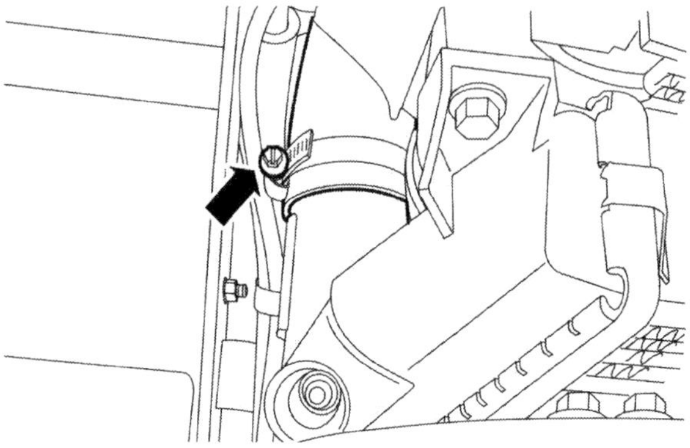

7. CAUTION: Before disconnecting any components, make sure the area is clean and free from foreign material. When disconnected all openings must be sealed.

NOTE: RH shown, LH similar.

Disconnect the charge air cooler inlet hose and charge air cooler outlet hose.

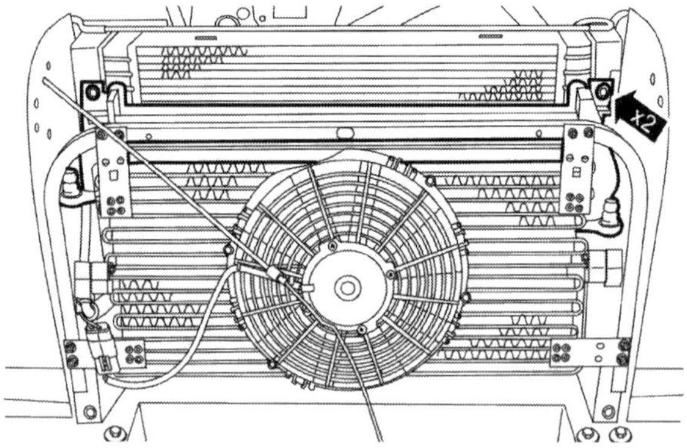

8. Remove the 2 bolts. Remove the charge air cooler.

Installation

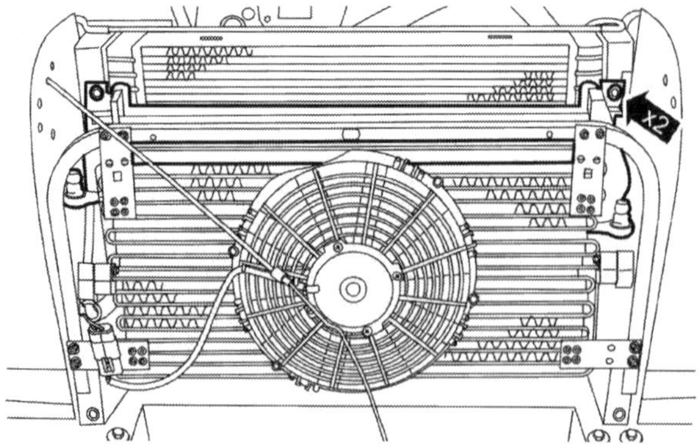

1. To install, reverse the removal procedure. Tighten the bolts.

   Torque: 20 Nm

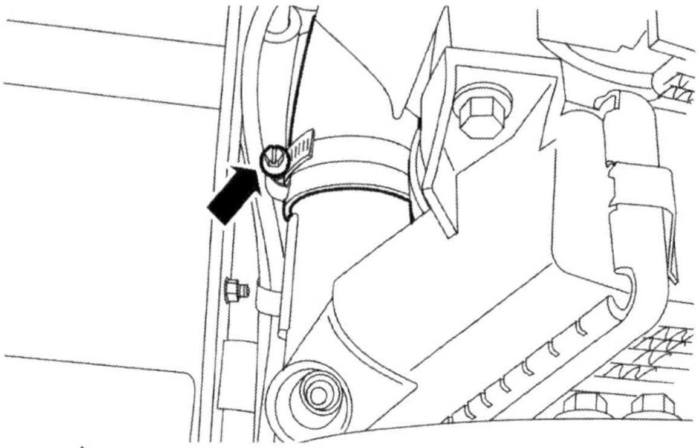

2. NOTE: Remove and discard the blanking caps.

   Tighten the clamp.

   Torque: 5 Nm

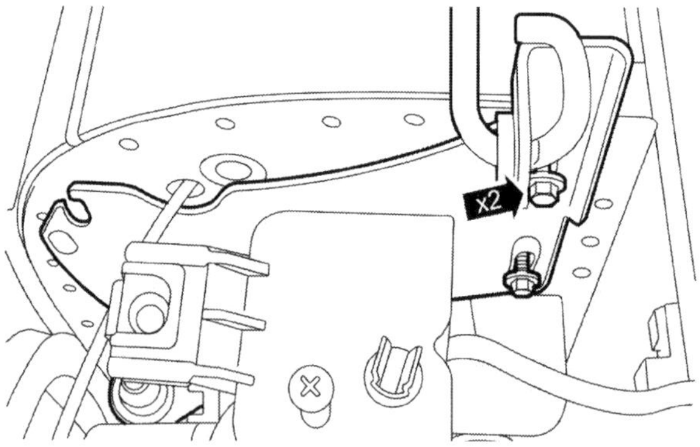

3. Tighten the bolts.

   Torque: 9 Nm

4. Connect the battery ground cable.

Refer to: Startup Battery Disconnect and Connect (414-01 Battery, Mounting and Cables, General Procedures)

Specifications

Torque Specifications

Item	Nm	lb-ft
Crankshaft position (CKP) sensor	9	6
Camshaft position (CMP) sensor	10	7
Manifold absolute pressure and temperature (MAPT) sensor	5	3
Engine oil pressure (EOP) sensor	15	11
Engine control module (ECM)	10	7
Accelerator pedal position sensor	25	18
Engine coolant temperature	N/A	N/A
Engine oil level sensor	27	20
Heated oxygen sensor	48	35
Mass air flow sensor	N/A	N/A
Post catalytic converter temperature sensor	35	26
Pre catalytic converter temperature sensor	35	26
Post DPF exhaust gas temperature sensor	35	26

OVERVIEW

The engine management system is controlled by an ECM and is able to monitor, adapt and precisely control the fuel injection and charge air. The ECM uses multiple sensor inputs and precision control of actuators to achieve optimum performance during all driving conditions.

The ECM processes information from the following input sources:

- APP sensor
- Brake pedal switches
- CMP sensor
- Charge air temperature sensor
- Clutch pedal position switch
- CKP sensor
- ECT sensor
- Electronic throttle position sensor
- HO2S (vehicles fitted with DPF (diesel particulate filter) only)

Description and Operation

COMPONENT LOCATION - SHEET 1 OF 2

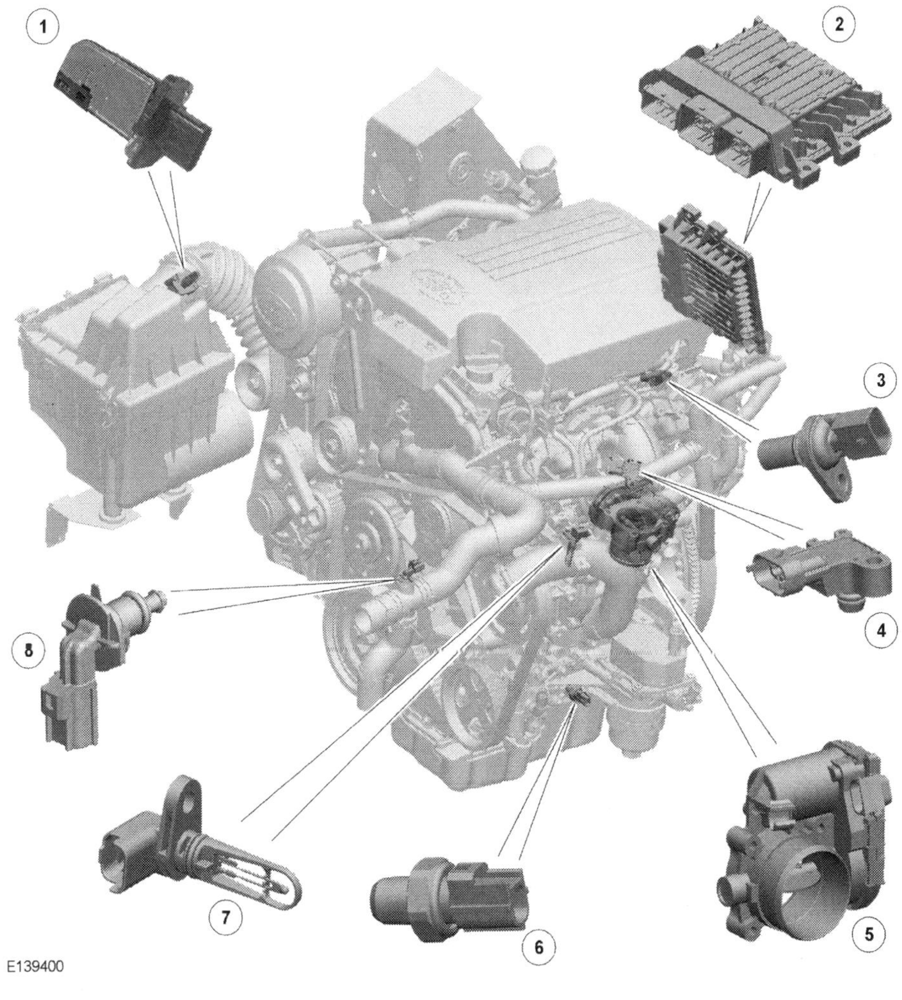

Item	Description
1	MAFT (mass air flow and temperature) sensor
2	-
3	CMP (camshaft position) sensor
4	MAP (manifold absolute pressure) sensor
5	Electronic throttle
6	Oil pressure switch
7	Charge air temperature sensor
8	ECT (engine coolant temperature) sensor

- MAFT sensor
- MAP sensor
- Oil level and temperature sensor (oil temperature input only)
- Oil pressure switch
- ABS (anti-lock brake system) module.

Refer to: Brake System (206-00 Brake System - General Information, Description and Operation)

- A/C (air conditioning) pressure sensor.

Refer to: Climate Control System (412-00, Description and Operation).

- A/C thermostatic switch.

Refer to: Climate Control System (412-00, Description and Operation).

- Anti-theft system module.

Refer to: Anti-Theft - Active (419-01A, Description and Operation).

- Differential pressure sensor.

Refer to: Exhaust System - ID4 2.2L Diesel (309-00 Exhaust System, ID4 2.2L Diesel Description and Operation)

- Exhaust gas temperature sensors.

Refer to: Exhaust System - ID4 2.2L Diesel (309-00 Exhaust System, ID4 2.2L Diesel Description and Operation)

- Fuel rail pressure sensor.

Refer to: Fuel Charging and Controls (303-04 Fuel Charging and Controls, ID4 2.2L Diesel Description and Operation)

- Fuel temperature sensor.

Refer to: Fuel Charging and Controls (303-04 Fuel Charging and Controls, ID4 2.2L Diesel Description and Operation)

The ECM outputs controlling signals to the following actuators:

- Electronic throttle motor
- HO2S heater circuit
- A/C compressor clutch relay.

Refer to: Climate Control System (412-00, Description and Operation).

- A/C cooling fan relay.

Refer to: Climate Control System (412-00, Description and Operation).

COMPONENT LOCATION - SHEET 2 OF 2

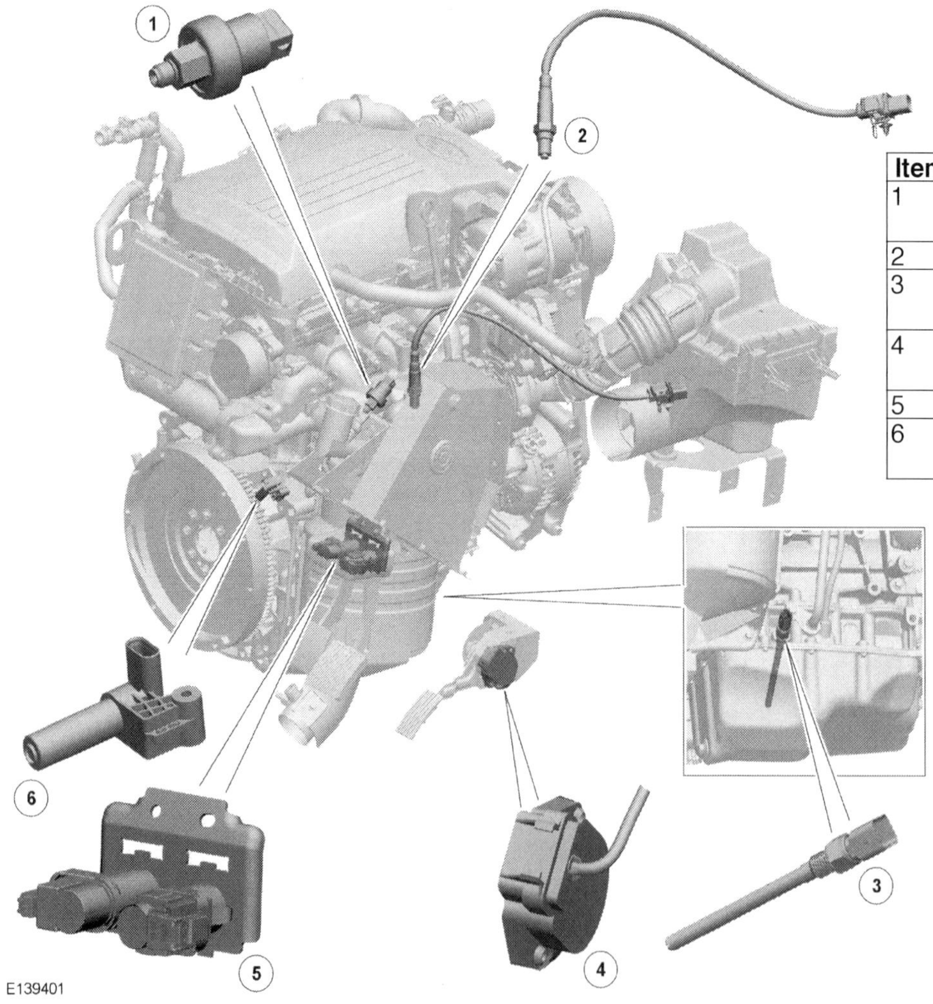

Item	Description
1	Clutch pedal position switch
2	-
3	Oil level and temperature sensor
4	APP (accelerator pedal position) sensor
5	Brake pedal switches
6	CKP (crankshaft position) sensor

- EGR (exhaust gas recirculation) valve.

Refer to: Engine Emission Control (303-08 Engine Emission Control, ID4 2.2L Diesel Description and Operation)

- Fuel injectors.

Refer to: Fuel Charging and Controls (303-04 Fuel Charging and Controls, ID4 2.2L Diesel Description and Operation)

- Fuel pump relay.

Refer to: Fuel Tank and Lines - ID4 2.2L Diesel (310-01 Fuel Tank and Lines, ID4 2.2L Diesel Description and Operation)

- Fuel volume control valve.

Refer to: Fuel Charging and Controls (303-04 Fuel Charging and Controls, ID4 2.2L Diesel Description and Operation)

- Generator.

Refer to: Generator - ID4 2.2L Diesel (414-02, Description and Operation).

- Glow plug control module.

Refer to: Glow Plug System (303-07 Glow Plug System, ID4 2.2L Diesel Description and Operation)

- Instrument cluster.

Refer to: Instrument Cluster (413-01, Description and Operation).

- Starter relay.

Refer to: Starting System (303-06 Starting System, ID4 2.2L Diesel Description and Operation)

- Turbocharger vane actuator.

Refer to: Turbocharger (303-04 Fuel Charging and Controls - Turbocharger, ID4 2.2L Diesel Description and Operation)

ENGINE CONTROL MODULE

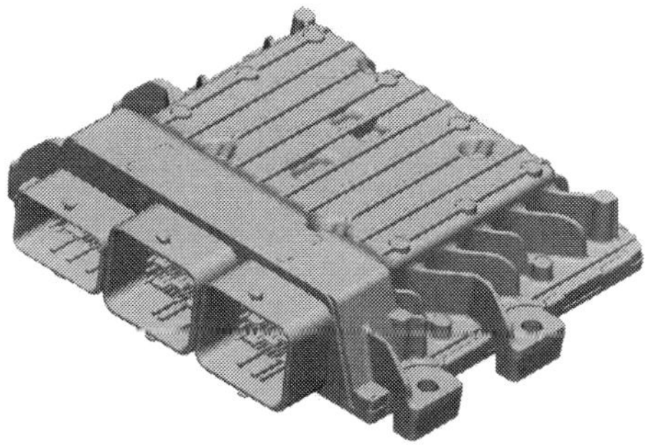

The ECM is installed in the engine compartment, on a bracket attached to the engine bulkhead. Three connectors provide the interface between the ECM and the vehicle and engine harnesses.

The ECM contains data processors and memory microchips. The output signals to the actuators are in the form of ground paths provided by driver circuits within the ECM. The ECM driver circuits produce heat during normal operation and dissipate this heat via the casing. Some sensors receive a regulated voltage supplied by the ECM. This avoids incorrect signals caused by voltage drop during cranking.

The ECM performs self diagnostic routines and stores fault codes in its memory. These fault codes and diagnostics can be accessed using Land Rover approved diagnostic equipment. If the ECM is to be replaced, the new ECM is supplied 'blank' and must be configured to the vehicle using Land Rover approved diagnostic equipment. A 'flash' EEPROM (electrically erasable programmable read only memory) allows the ECM to be configured with market specific or new tune information up to 14 times. If a fifteenth update is required, the ECM must be replaced. The current engine tune data can be accessed and read using Land Rover approved diagnostic equipment.

When a new ECM is fitted, it must also be synchronized to the anti-theft system module using Land Rover approved diagnostic equipment. An ECM cannot be transferred between vehicles.

The ECM is connected to the engine sensors, which allows it to monitor the engine operating conditions. The ECM processes these signals and decides the actions necessary to maintain optimum engine performance in terms of driveability, fuel efficiency and exhaust emissions. The memory of the ECM is programmed with instructions on how to control the engine. This is known as the strategy. The memory also contains data in the form of maps which the ECM uses as a basis for fueling and emission control. By comparing the information from the sensors to the data in the maps, the ECM is able to calculate the various output requirements. The ECM contains an adaptive strategy, which updates the system when components vary due to production tolerances or ageing.

ACCELERATOR PEDAL POSITION SENSOR

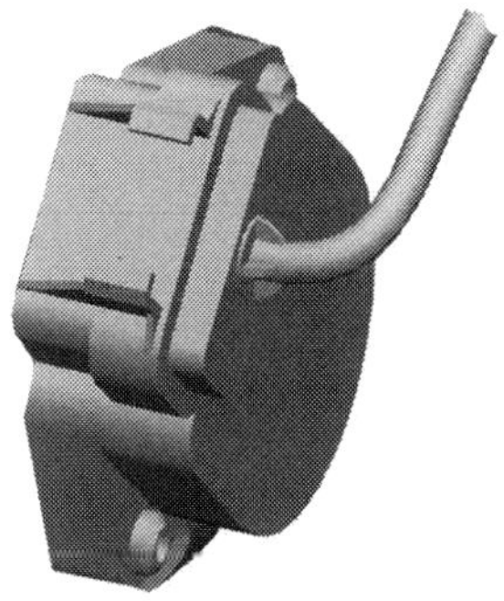

E139403

The APP sensor is a dual rotary potentiometer located on the right side of the accelerator pedal bracket and operated by the accelerator pedal spindle. The potentiometers have independent 5V supply and return connections with the ECM, and a common ground connection.

The ECM uses the two inputs from the APP sensor to determine the position, rate of movement and direction of movement of the accelerator pedal. The ECM then uses this data, along with information from other sensors, to achieve the optimum engine response.

If one of the inputs from the APP sensor fails, or there is a fault with the ground or supply connections, the ECM limits engine response to that produced by the first 28% of accelerator pedal travel.

If both inputs fail, there will be no response to accelerator pedal operation. The engine will idle at a raised value while the brake pedal is released, and at the normal value when the brake pedal is pressed. This allows a restricted limp-home operation.

BRAKE PEDAL SWITCHES

The brake pedal switches are located under a protective cover on the pedal box and operated by the brake pedal. The switches are plunger operated two-pole switches that supply pedal status signals to the ECM. The ECM compares the signals from the two switches to ensure the correct pedal status.

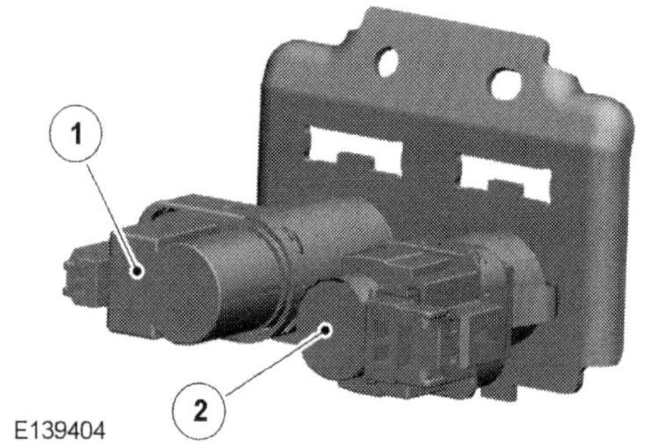

Item	Description
1	Brake diagnostic switch
2	Stop lamp switch

While the brake pedal is released:

- The stoplamp switch contacts are open
- The brake diagnostic switch contacts are closed, connecting a ground to the ECM.

When the brake pedal is pressed:

- The stoplamp switch contacts close and connect an ignition feed from the CJB (central junction box) to the ECM
- The brake diagnostic switch contacts open and disconnect the ground from the ECM.

The ECM uses the brake pedal switch signals to check the plausibility of the APP sensor inputs.

In the event of a brake pedal switch failure and a single APP sensor track fault, there will be no response to accelerator pedal operation. The engine will idle at a raised value while the brake pedal is released, and at the normal value when the brake pedal is pressed. This allows a restricted limp-home operation.

CAMSHAFT POSITION SENSOR

The CMP sensor is installed in the left side of the cylinder head, above the intake ports for cylinder number 4. The sensor tip is located close to a reluctor on the camshaft.

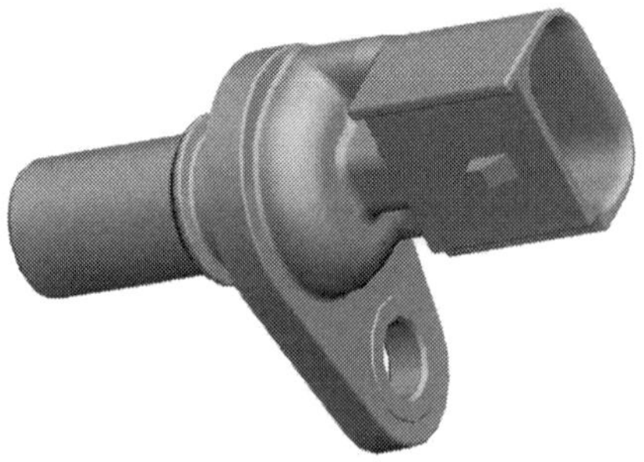

The sensor is a Hall effect sensor, which is used by the ECM at engine start-up to synchronize the CKP sensor signal to identify number one cylinder, to ensure the correct injector timing. Once the ECM has established the injector timing, the CMP sensor signal is no longer used.

The CMP sensor receives a 5V supply from the ECM. Two further connections to the ECM provide ground and signal outputs.

If a fault occurs with the CMP sensor when the engine is running, the engine will continue to run. Depending on the fault, the ECM may also restrict engine performance and signal the instrument cluster, on the high speed CAN (controller area network) bus, to illuminate the MIL (malfunction indicator lamp) . Once the engine is switched off, the engine will crank, but will not restart while the fault is present.

CHARGE AIR TEMPERATURE SENSOR

The charge air temperature sensor is installed in the air duct between the charge air cooler and the electronic throttle.

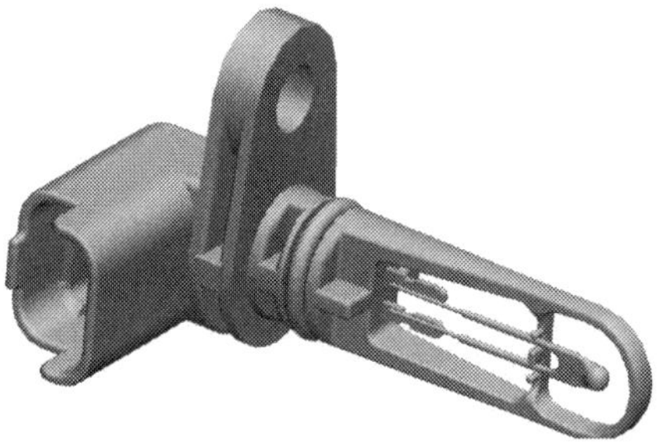

The charge air temperature sensor is a NTC (negative temperature coefficient) thermistor that receives a 5V reference voltage from the ECM. The ECM uses the input from the sensor for fueling calculations.

If the charge air temperature sensor fails, the ECM uses a default charge air temperature of 20 °C (68 °F). In the event of a charge air temperature sensor failure, over fueling may occur, resulting in black smoke from the exhaust.

CLUTCH PEDAL POSITION SWITCH

The clutch pedal position switch is located on the clutch master cylinder.

The clutch pedal position switch is a pressure

transducer type. When the clutch is pressed the switch disconnects a ground connection to the ECM, which reduces engine torque.

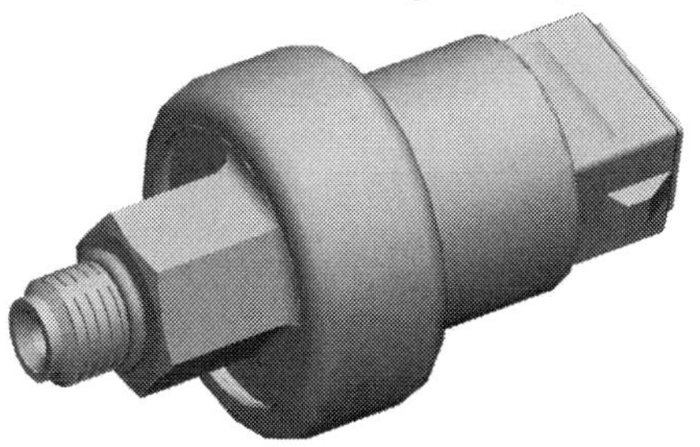

In the event of a clutch pedal switch failure, there may be some loss of driveability in the form of poor acceleration and reduced power.

CRANKSHAFT POSITION SENSOR

The CKP sensor is attached to a bracket on the rear of the cylinder block, in the joint between the cylinder block and the transmission clutch housing. The sensor tip is aligned with a magnetic trigger wheel attached to the crankshaft. The sensor produces a square wave signal, the frequency of which is proportional to engine speed.

The CKP sensor is a Hall effect sensor that receives a 5V supply from the ECM and produces a square wave signal, the frequency of which is proportional to engine speed. The trigger wheel has two missing teeth to provide a reference point for the angular position of the crankshaft.

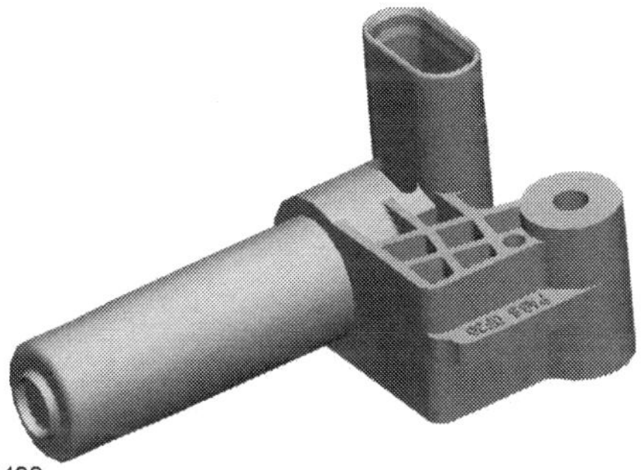

E139408

The ECM uses the signal from the CKP sensor for the following functions:

- Synchronization
- Determining fuel injection timing
- Producing an engine speed signal which is broadcast on the high speed CAN bus for use by other systems.

If there is a fault with the CKP sensor, the following symptoms may be observed:

- Engine will not start
- Rough idle, EGR disabled, poor acceleration and lack of power.

ELECTRONIC THROTTLE

The electronic throttle is installed between the duct from the charge air cooler and the intake manifold, and is used by the ECM for EGR to control emissions, to make engine shutdown smoother, and to provide a second means of stopping the engine if other items fail.

The throttle plate is operated by an electric DC (direct current) motor, attached to the throttle body, which is controlled by the ECM. A position sensor in the DC motor supplies a feedback signal of throttle plate position to the ECM.

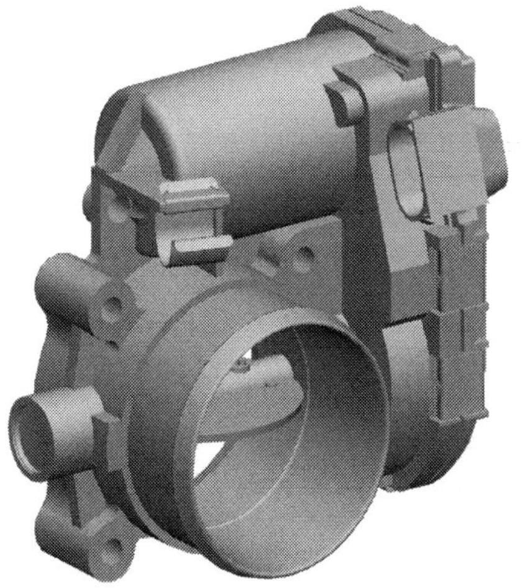

In there is a fault with the DC motor or the position sensor, the ECM will:

- Signal the instrument cluster, on the high speed CAN bus, to illuminate the MIL
- Disable EGR and DPF regeneration (where fitted), which results in increased emissions
- Reduce engine power output.

ENGINE COOLANT TEMPERATURE SENSOR

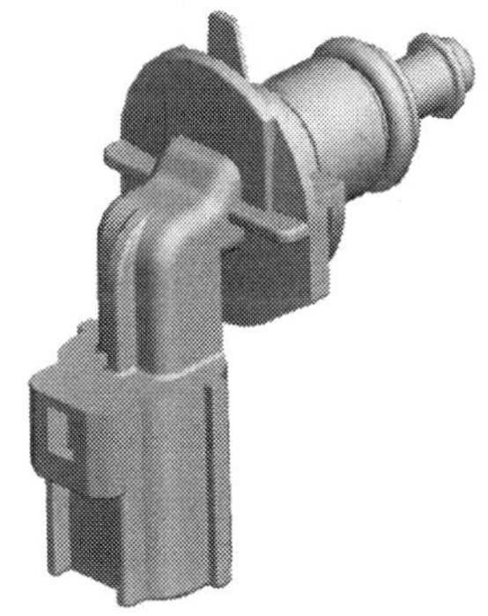

The ECT sensor is installed in the radiator upper hose, in the tee connection with the thermostat hose.

The ECT sensor is a NTC thermistor that receives a 5V reference voltage from the ECM. The ECM uses the temperature information for the following functions:

- Fueling calculations
- Limiting engine operation if engine coolant temperature becomes too high
- Glow plug activation time.

The ECM also transmits coolant temperature information on the high speed CAN bus to the instrument cluster, for temperature gauge operation.

If the ECT sensor fails, the following symptoms may be observed:

- Difficult cold start
- Difficult hot start
- Engine performance compromised
- Temperature gauge inoperative or inaccurate reading.

The ECM may also illuminate the MIL, depending on the fault.

In the event of ECT sensor signal failure, the ECM applies a default value of 88 °C (190 °F) coolant temperature for fueling purposes.

HEATED OXYGEN SENSOR (ONLY FITTED TO VEHICLES WITH DPF)

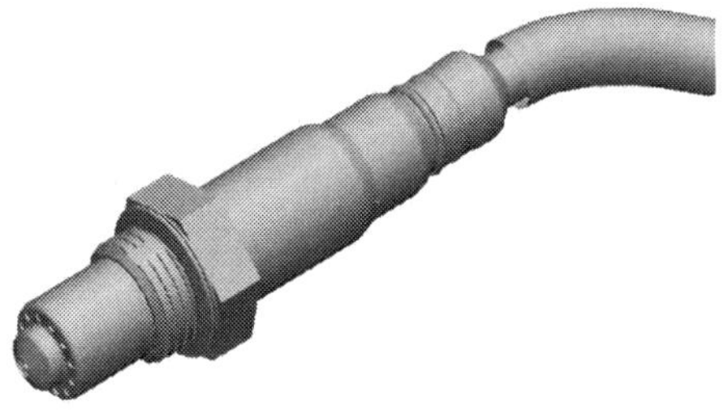

The HO2S is installed in the inlet of the catalytic converter and DPF.

The HO2S allows the ECM to measure the oxygen content of the exhaust gases, for closed loop control of the fuel: air mixture.

The heater element of the HO2S is controlled by a PWM (pulse width modulation) signal from the ECM. The heater element is operated immediately after each engine start and during low load conditions when the temperature of the exhaust gases is insufficient to maintain the required sensor temperature. The PWM duty cycle is carefully controlled to prevent thermal shock to cold sensors. A non-functioning heater delays the sensor's readiness for closed loop control and increases emissions.

If there is a fault with the HO2S (heater or sensor circuits) the ECM illuminates the MIL and defaults to open loop fueling.

MANIFOLD ABSOLUTE PRESSURE SENSOR

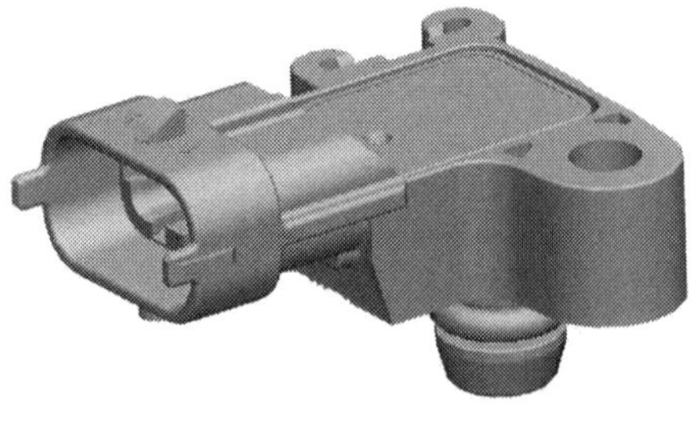

The MAP sensor is installed on the intake manifold. The sensor provides a voltage signal to the ECM relative to the intake manifold pressure. The MAP sensor has a three pin connector that provides a 5V reference supply from the ECM, a signal output to the ECM and a ground.

The MAP sensor uses a diaphragm transducer to measure pressure. The ECM uses the sensor signal for the following functions:

- Maintain manifold boost pressure
- Reduce exhaust smoke emissions when driving at high altitude
- Control of the EGR system.

If the MAP sensor fails, the ECM uses a default pressure of 1013 mbar (14.7 lbf/in^2) absolute and may illuminate the MIL. In the event of a MAP sensor failure, the following symptoms may be observed:

- Engine speed restricted
- Rough running, engine hesitates.
- MIL illuminated.

MASS AIR FLOW AND TEMPERATURE SENSOR

The MAFT sensor is located in the outlet of the air cleaner.

The MAFT sensor receives a 12V supply from the power relay in the BJB (battery junction box) and a ground connection via the ECM. Two further connections to the ECM supply the MAF (mass air flow) and the temperature signals.

The temperature sensor is a NTC thermistor. The ECM uses the temperature signal to correct the fueling map for intake air temperature.

The MAF sensor works on the hot film principle. The sensor output is a digital signal proportional to the mass of the incoming air. The ECM uses this data, in conjunction with signals from other sensors, and information from stored fueling maps, to determine the precise fuel quantity to be injected into the cylinders. The signal is also used as a feedback signal for the EGR system.

The ECM checks the calculated air mass against the engine speed. If the calculated air mass is not plausible, the ECM uses a default air mass figure which is derived from the average engine speed compared to a stored characteristic map. The air mass value will be corrected using values for boost pressure, atmospheric pressure and ambient air temperature.

In the event of an air sensor signal failure, any of the following symptoms may be observed:

- Difficult starting
- Engine stalls after starting
- Delayed engine response
- Emission control inoperative
- Idle speed control inoperative
- Reduced engine performance
- MIL illuminated.

OIL LEVEL AND TEMPERATURE SENSOR

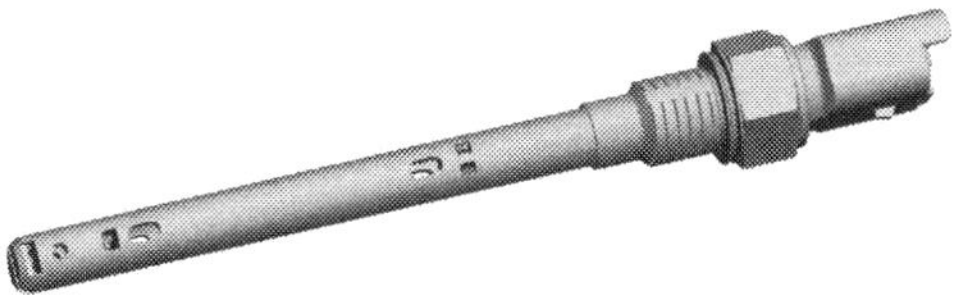

The oil level and temperature sensor is installed in the cylinder block skirt stiffener, mid-way along the RH (right-hand) side of the engine.

The oil level signal is not used, as a mechanical dip stick is fitted.

The ECM uses the oil temperature signal to control the engine oil temperature. In extreme operation the ECM limits engine torque to prevent the oil temperature exceeding 140 °C (284 °F).

OIL PRESSURE SWITCH

The oil pressure switch is located in the oil filter and cooler assembly.

When oil pressure is present, the oil pressure switch makes a ground connection through the switch body to the ECM. The switch operates at a pressure of 0.15 to 0.41 bar (2.2 to 5.9 lbf/in^2). The ECM transmits the status of the oil pressure switch to the instrument cluster, on the high speed CAN bus, for operation of the low oil pressure warning indicator.

If the oil pressure switch fails, the low oil pressure warning indicator is permanently illuminated.

Electronic Engine Controls

Diagnosis and Testing

Principles of Operation

For a detailed description of the Electronic Engine Controls, refer to the relevant Description and Operation section in the workshop manual.

Refer to: Electronic Engine Controls (303-14 Electronic Engine Controls, ID4 2.2L Diesel Description and Operation)

Inspection and Verification

⚠ CAUTION: Diagnosis by substitution from a donor vehicle is **NOT** acceptable. Substitution of control modules does not guarantee confirmation of a fault, and may also cause additional faults in the vehicle being tested and/or the donor vehicle.

NOTES:

△ If the control module or a component is suspect and the vehicle remains under manufacturer warranty, refer to the Warranty Policy and Procedures manual (section B1.2), or determine if any prior approval programme is in operation, prior to the installation of a new module/component.

Visual Inspection

Mechanical	Electrical
• Engine oil level	• Fuses
• Cooling system coolant level	• Wiring harnesses and connectors
• Low/contaminated fuel	• Engine control module (ECM)
• Fuel leaks	• Crankshaft position sensor
• High pressure fuel pump	• Camshaft position sensor
• Primary drive belt	• Engine coolant temperature sensor
	• Fuel temperature sensor
	• Charge air temperature sensor
	• Mass airflow and temperature sensor
	• Fuel rail pressure sensor
	• Manifold absolute pressure sensor
	• Brake pedal switch
	• Clutch pedal switch
	• Fuel volume control valve
	• Fuel injectors
	• CAN bus

CONTROL DIAGRAM - SHEET 1 OF 2

NOTE: A = Hardwired.

Item	Description
1	Battery
2	-
3	Inertia switch
4	Ignition switch
5	-
6	CMP sensor
7	MAFT sensor
8	Charge air temperature sensor
9	MAP sensor
10	ECT sensor
11	Oil pressure switch
12	Oil level and temperature switch
13	FRP (fuel rail pressure) sensor
14	Differential pressure sensor
15	Temperature sensor - pre catalytic converter
16	Temperature sensor - post catalytic converter
17	Temperature sensor - post DPF
18	Anti-theft system module
19	APP sensor
20	Brake switches
21	Clutch switch
22	-
23	A/C thermostatic switch
24	A/C pressure switch

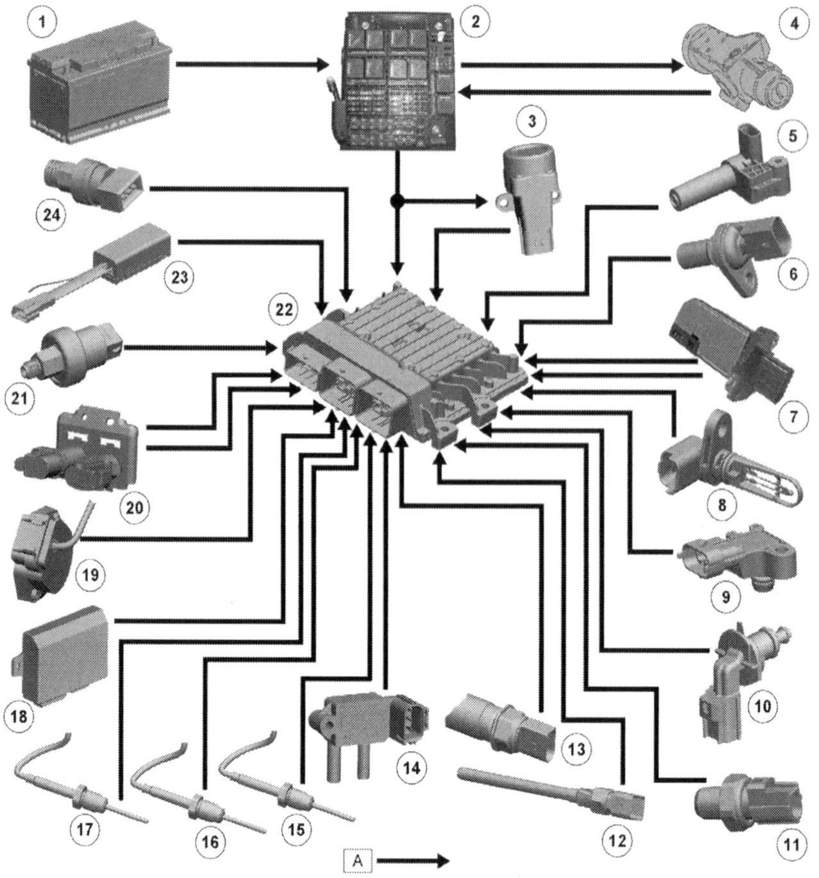

CONTROL DIAGRAM - SHEET 2 OF 2

NOTE: A = Hardwired; D = High speed CAN bus; O = LIN (local interconnect network) bus.

Item	Description
1	-
2	-
3	To A/C compressor clutch relay
4	To A/C cooling fan relay
5	To starter relay
6	To fuel pump relay
7	To main and power relays
8	Instrument cluster
9	ABS module
10	Generator
11	Fuel injector (4 off)
12	Glow plug module
13	Glow plug (4 off)
14	Turbocharger
15	EGR valve
16	High pressure fuel pump
17	Electronic throttle

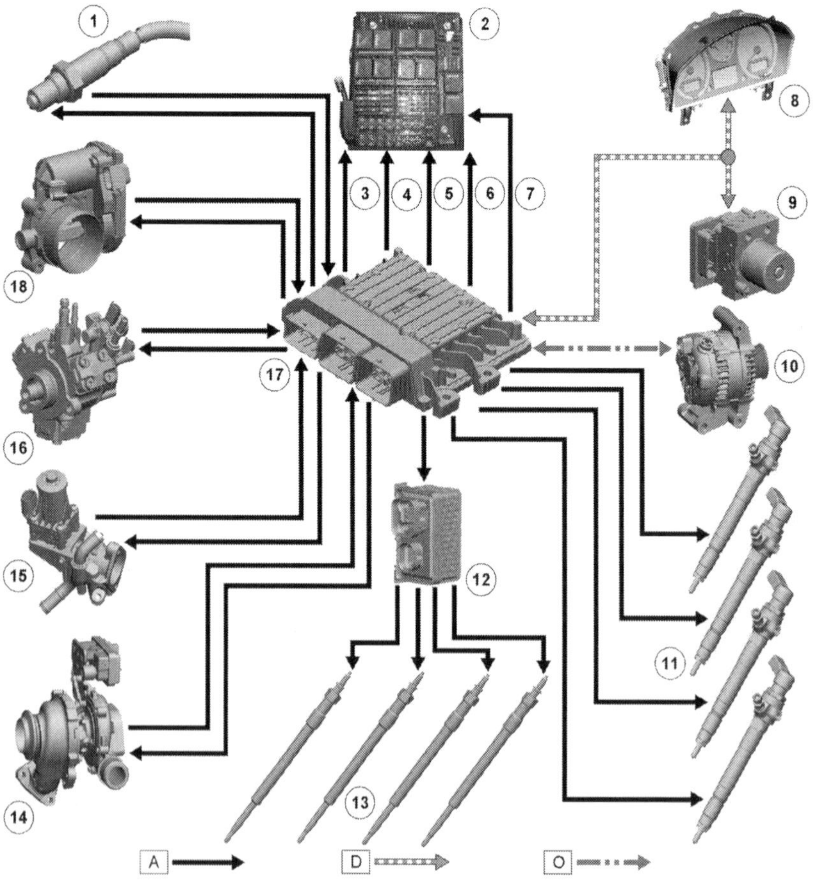

⚠ When performing voltage or resistance tests, always use a digital multimeter accurate to three decimal places, and with an up-to-date calibration certificate. When testing resistance always take the resistance of the digital multimeter leads into account.

⚠ Check and rectify basic faults before beginning diagnostic routines involving pinpoint tests.

1. Verify area of concern.
2. Visually inspect for obvious signs of damage and system integrity.
3. If an obvious cause for an observed or reported concern is found, correct the cause (if possible) before proceeding to the next step.
4. If the cause is not visually evident, verify the symptom and refer to the Symptom Chart, alternatively check for Diagnostic Trouble Codes (DTCs) and refer to the DTC Index.

Symptoms Chart

Symptom	Possible Causes	Action
Engine cranks, but does not start	• Engine management system fault • Low/contaminated fuel • Air intake system fault • Low pressure fuel system fault • High pressure fuel system fault • Camshaft position sensor • Crankshaft position sensor • Exhaust gas recirculation system fault	• Using the manufacturer approved diagnostic system, check the engine control module for related DTCs and refer to the relevant DTC index • Check the fuel level and condition. Draw off approximately 1 ltr (2.11 pints) of fuel and allow to stand for 1 minute. Check to make sure there is no separation of the fuel indicating water or other liquid in the fuel • Check the air intake system for leaks/restrictions • Check the low pressure fuel system for leaks/damage. Check the fuel pump module operation. Check the fuel filter • Check the high pressure fuel system for leaks/damage. Check the fuel volume control valve. Check the high pressure fuel pump. Check the injectors • Check the camshaft position sensor and circuits • Check the crankshaft position sensor and circuits • Check the exhaust gas recirculation system
Difficult to start	• Engine management system fault • Glow plug system fault (in cold ambient temperature) • Low/contaminated fuel • Air intake system fault • Low pressure fuel system fault • High pressure fuel system fault • Exhaust gas recirculation system fault	• Using the manufacturer approved diagnostic system, check the engine control module for related DTCs and refer to the relevant DTC index • Check the glow plug system • Check the fuel level and condition. Draw off approximately 1 ltr (2.11 pints) of fuel and allow to stand for 1 minute. Check to make sure there is no separation of the fuel indicating water or other liquid in the fuel • Check the air intake system for leaks/restrictions • Check the low pressure fuel system for leaks/damage. Check the fuel pump module operation. Check the fuel filter • Check the high pressure fuel system for leaks/damage. Check the fuel volume control valve. Check the high pressure fuel pump. Check the injectors • Check the exhaust gas recirculation system
Rough idle	• Engine management system fault • Low/contaminated fuel • Air intake system fault • Low pressure fuel system fault • High pressure fuel system fault • Exhaust gas recirculation system fault	• Using the manufacturer approved diagnostic system, check the engine control module for related DTCs and refer to the relevant DTC index • Check the fuel level and condition. Draw off approximately 1 ltr (2.11 pints) of fuel and allow to stand for 1 minute. Check to make sure there is no separation of the fuel indicating water or other liquid in the fuel • Check the air intake system for leaks/restrictions • Check the low pressure fuel system for leaks/damage. Check the fuel pump module operation. Check the fuel filter • Check the high pressure fuel system for leaks/damage. Check the fuel volume control valve. Check the high pressure fuel pump. Check the injectors • Check the exhaust gas recirculation system

Lack of power when accelerating	• Engine management system fault • Air intake system fault • Exhaust system / diesel particulate filter restricted • Low pressure fuel system fault • Exhaust gas recirculation system fault • Variable geometry turbocharger fault	• Using the manufacturer approved diagnostic system, check the engine control module for related DTCs and refer to the relevant DTC index • Check the air intake system for leaks/restrictions • Check the exhaust system / diesel particulate filter for restrictions • Check the low pressure fuel system for leaks/damage. Check the fuel pump module operation. Check the fuel filter • Check the variable geometry turbocharger • Check the exhaust gas recirculation system
Engine stops/stalls	• Engine management system fault • Low/contaminated fuel • Air intake system fault • Low pressure fuel system fault • High pressure fuel system fault • Exhaust gas recirculation system fault	• Using the manufacturer approved diagnostic system, check the engine control module for related DTCs and refer to the relevant DTC index • Check the fuel level and condition. Draw off approximately 1 ltr (2.11 pints) of fuel and allow to stand for 1 minute. Check to make sure there is no separation of the fuel indicating water or other liquid in the fuel • Check the air intake system for leaks/restrictions • Check the low pressure fuel system for leaks/damage. Check the fuel pump module operation. Check the fuel filter • Check the high pressure fuel system for leaks/damage. Check the fuel volume control valve. Check the high pressure fuel pump. Check the injectors • Check the exhaust gas recirculation system
Engine judders	• Engine management system fault • Low/contaminated fuel • Air intake system fault • Low pressure fuel system fault • High pressure fuel system fault	• Using the manufacturer approved diagnostic system, check the engine control module for related DTCs and refer to the relevant DTC index • Check the fuel level and condition. Draw off approximately 1 ltr (2.11 pints) of fuel and allow to stand for 1 minute. Check to make sure there is no separation of the fuel indicating water or other liquid in the fuel • Check the air intake system for leaks/restrictions • Check the low pressure fuel system for leaks/damage. Check the fuel pump module operation. Check the fuel filter • Check the high pressure fuel system for leaks/damage. Check the fuel volume control valve. Check the high pressure fuel pump. Check the injectors
Excessive fuel consumption	• Engine management system fault • Low pressure fuel system fault • High pressure fuel system fault • Exhaust gas recirculation system fault	• Using the manufacturer approved diagnostic system, check the engine control module for related DTCs and refer to the relevant DTC index • Check the low pressure fuel system for leaks/damage. Check the fuel pump module operation. Check the fuel filter • Check the high pressure fuel system for leaks/damage. Check the fuel volume control valve. Check the high pressure fuel pump. Check the injectors • Check the exhaust gas recirculation system

DTC Index

For a list of Diagnostic Trouble Codes (DTCs) that could be logged on this vehicle, please refer to Section 100-00.

Refer to: Diagnostic Trouble Code Index - DTC: Engine Control Module (PCM) (100-00 General Information, Description and Operation)

Accelerator Pedal Position Sensor

Removal and Installation

Removal

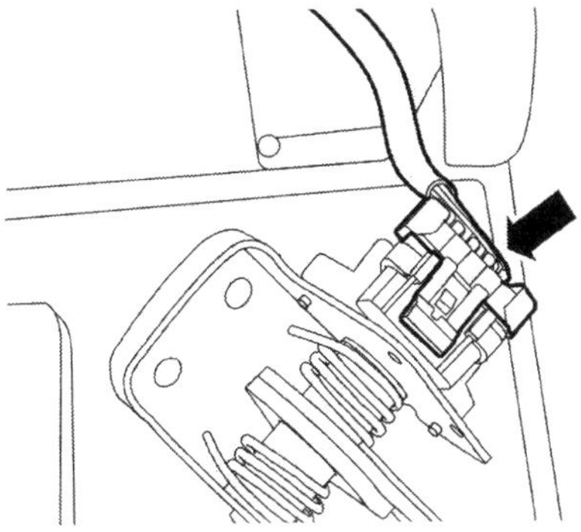

NOTE: Removal steps in this procedure may contain installation details.

Torque: 25 Nm

Installation

1. To install, reverse the removal procedure.

Camshaft Position Sensor

Removal and Installation

Part(s)

STEP	PART NAME	QUANTITY
Installation Step 1	Renew Part Camshaft position sensor	1

Removal

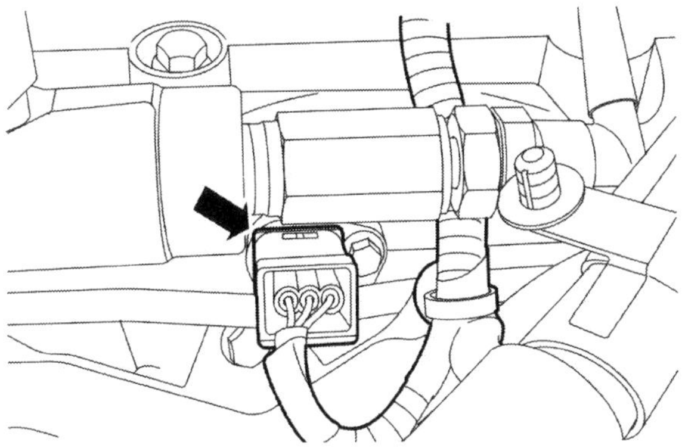

1. Disconnect the camshaft position (CMP) sensor electrical connector.

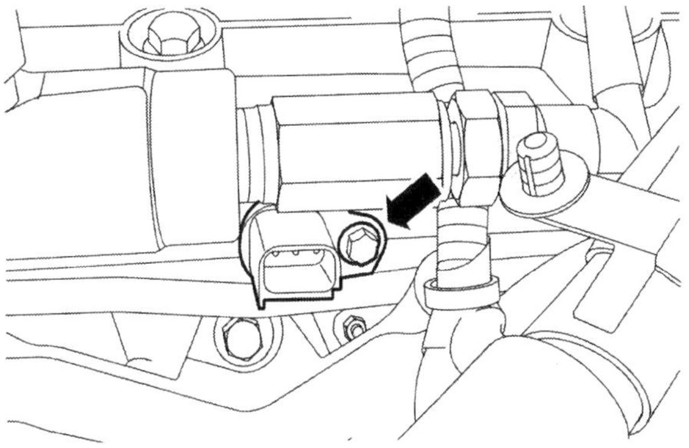

2. Remove the bolt. Remove the CMP sensor. Remove and discard the O-ring seal.

Installation

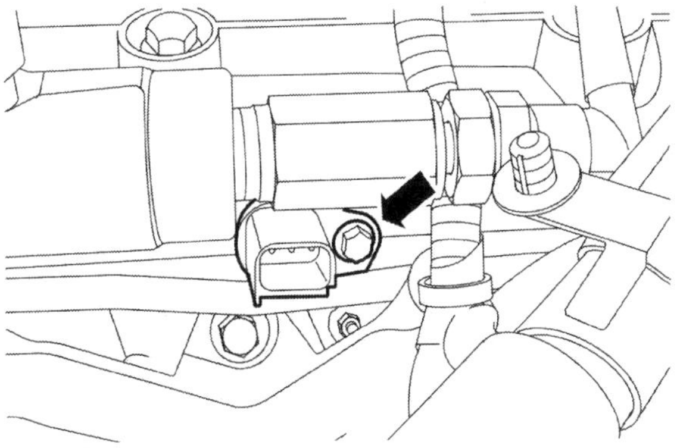

1. NOTE: Install a new O-ring seal.

To install, reverse the removal procedure. Tighten the bolt.

Renew Part: Camshaft position sensor

Torque: 9 Nm

Crankshaft Position Sensor

Removal and Installation

Removal

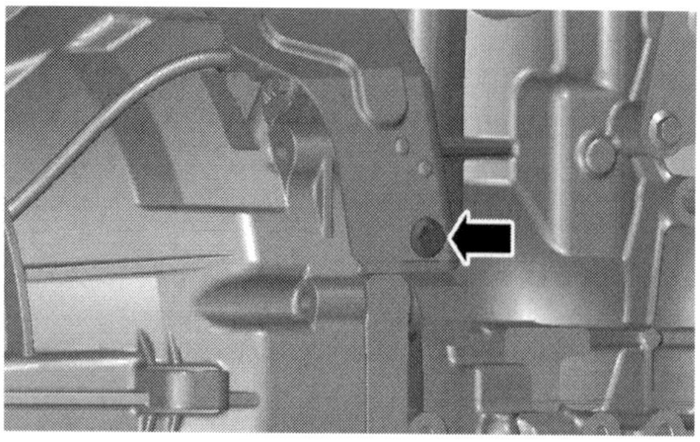

1. Remove the heat shield.

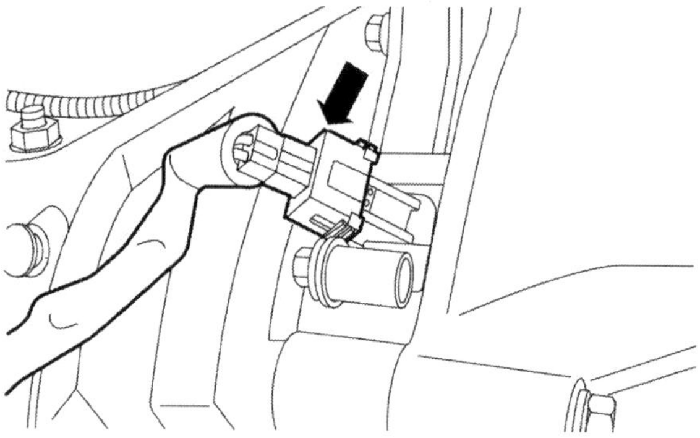

2. Disconnect the crankshaft position (CKP) sensor electrical connector.

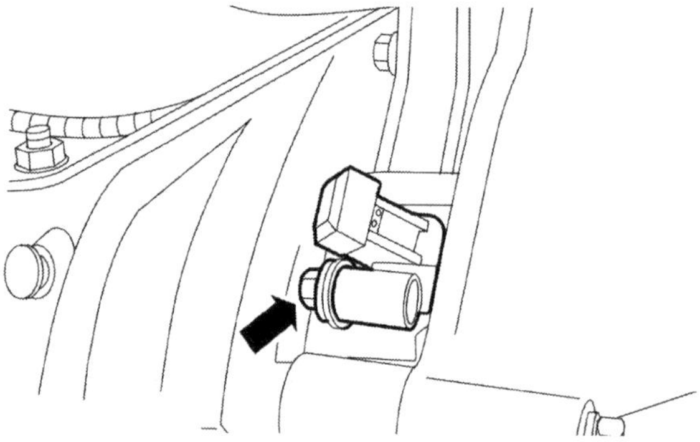

3. Remove and discard the CKP sensor.
   - Remove the bolt.

Installation

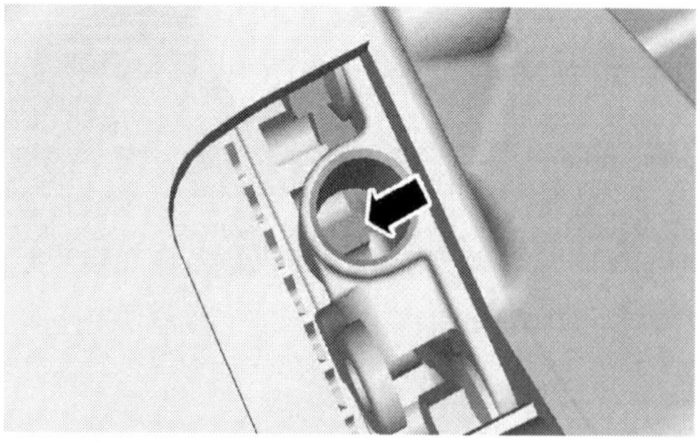

1. NOTE: Only turn the engine in the normal direction of rotation.

Turn the engine until a flywheel trigger tooth is visible through the CKP sensor housing.

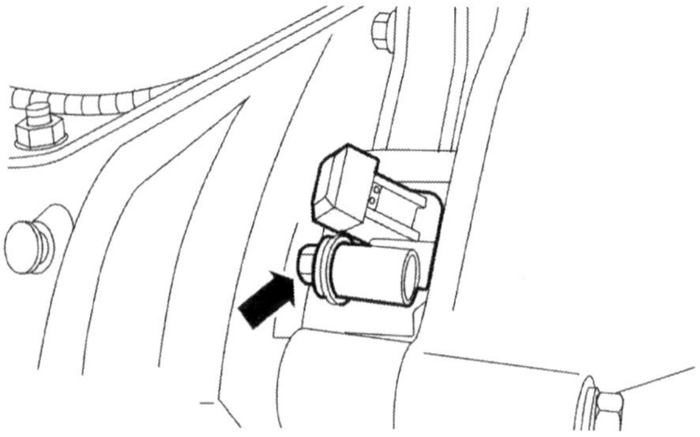

2. CAUTION: The CKP sensor tip must rest on a flywheel trigger tooth. Incorrect installation may result in the CKP sensor being damaged.

NOTE: Make sure that the CKP sensor housing is clean and free from foreign material.

Install the CKP sensor.

- Tighten to 9 Nm.

3. Connect the CKP sensor electrical connector.
4. Install the heat shield.
   - Tighten to 24 Nm.

Cylinder Head Temperature Sensor

Removal and Installation

Special Tool(s)

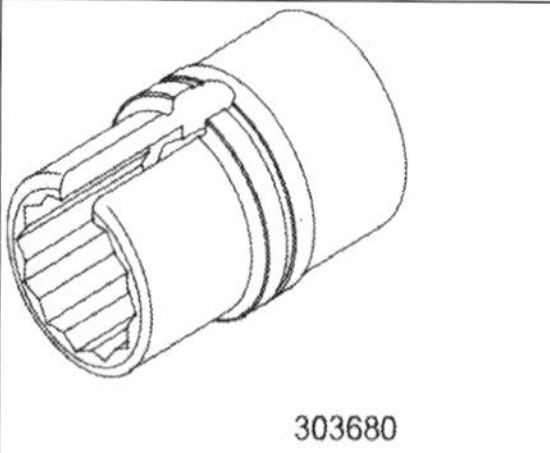	Socket, CHT Sensor 303-680

Removal

1. Remove the engine cover.

   Refer to: Engine Cover (501-05, Removal and Installation).

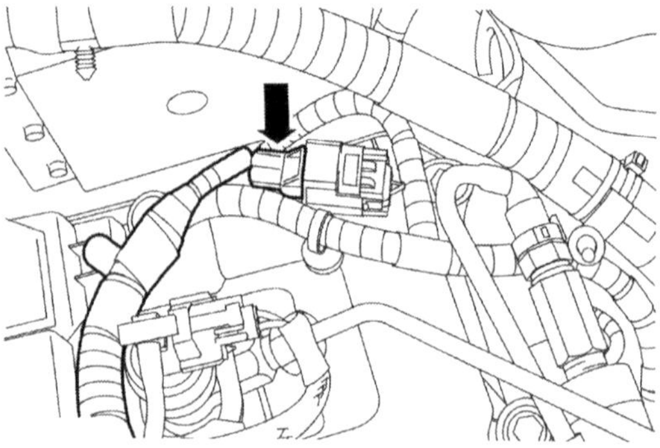

2. Disconnect the cylinder head temperature (CHT) sensor.

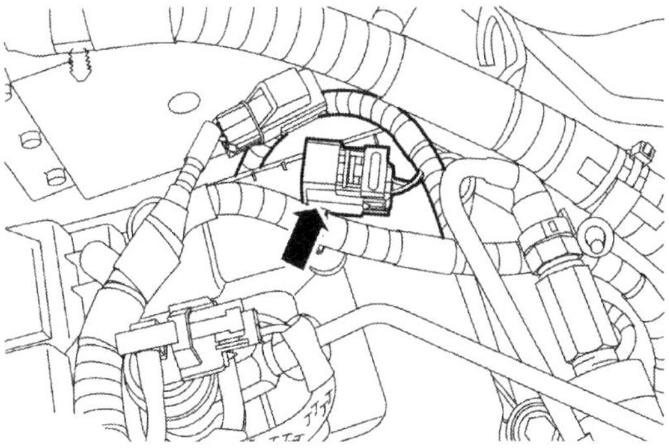

3. Release the CHT sensor wiring harness.

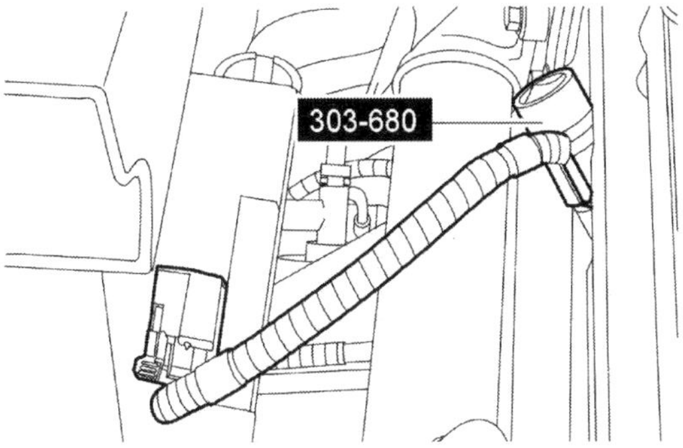

4. Using the special tool, remove the CHT sensor.

Installation

1. To install, reverse the removal procedure.
   - Tighten to 10 Nm (7 lb.ft).

Engine Control Module

Removal and Installation

Removal

1. CAUTION: Make sure that the fuel pump replacement and pilot correction learn are carried out following the download.

NOTE: When a new engine control module (ECM) is installed, the Land Rover approved diagnostic system must be connected to the vehicle and the ECM renewal procedure must be followed. This will allow the vehicle configuration to be uploaded into the new ECM.

2. **Refer to:** Battery Disconnect and Connect (414-01, General Procedures).

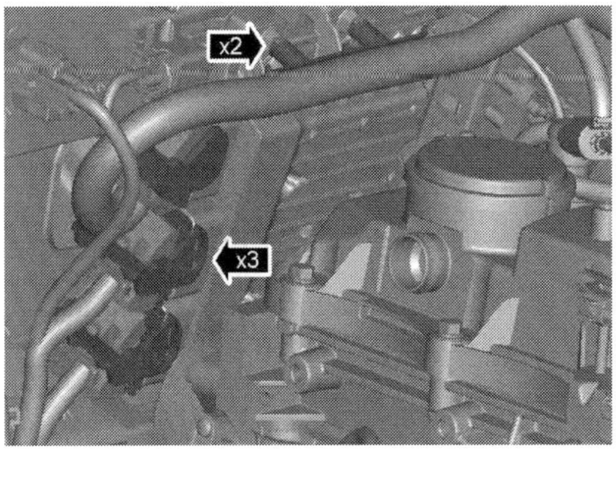

3.

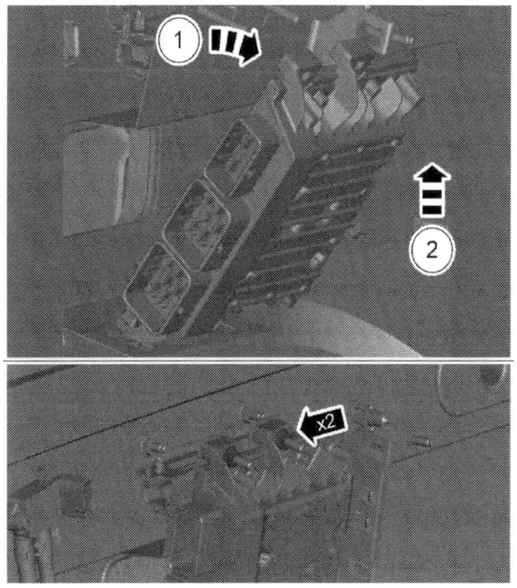

4. Torque: 10 Nm

Installation

1. To install, reverse the removal procedure.

Engine Coolant Temperature Sensor

Removal and Installation

Removal

NOTE: Removal steps in this procedure may contain installation details.

1. **Refer to:** Battery Disconnect and Connect (414-01, General Procedures).

2.

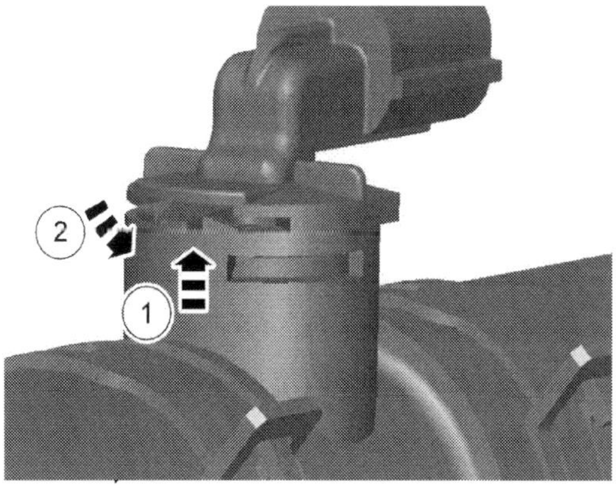

3.

Installation

1. To install, reverse the removal procedure.

Engine Oil Pressure Sensor

Removal and Installation

Removal

1. Raise and support the vehicle.

   Refer to: Lifting (100-02 Jacking and Lifting, Description and Operation)

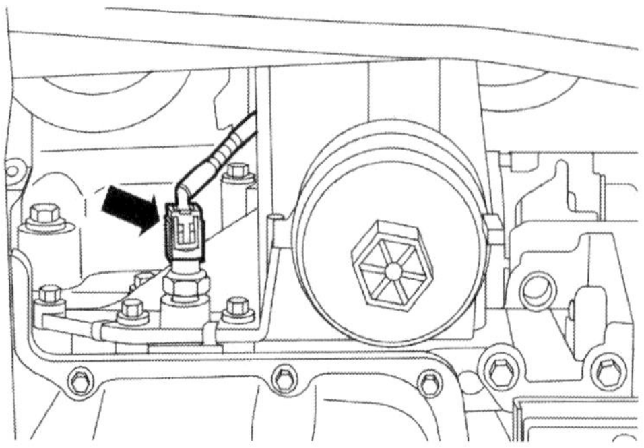

2. Disconnect the engine oil pressure (EOP) sensor electrical connector.

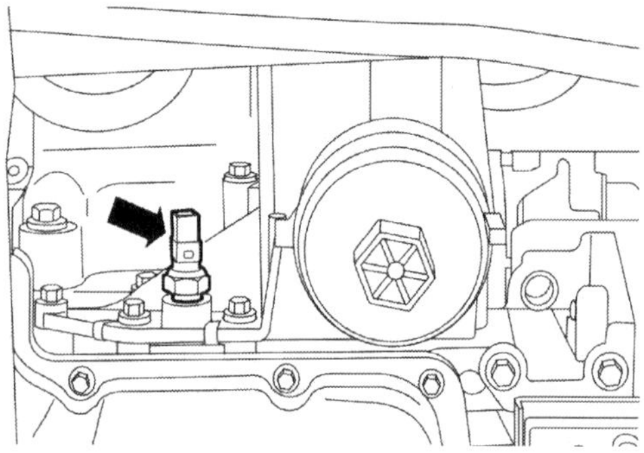

3. Remove the EOP sensor.

Installation

1. Install the EOP sensor.
   - Tighten to 15 Nm (11 lb.ft).
2. Connect the EOP sensor electrical connector.

Engine Oil Level Sensor

Removal and Installation

Removal

NOTE: Removal steps in this procedure may contain installation details.

1. **Refer to:** Startup Battery Disconnect and Connect (414-01 Battery, Mounting and Cables, General Procedures)
2. WARNING: Do not work on or under a vehicle supported only by a jack. Always support the vehicle on safety stands.

Raise and support the vehicle.

3. **Refer to:** Engine Oil Draining and Filling (303-01 Engine - ID4 2.2L Diesel, General Procedures).

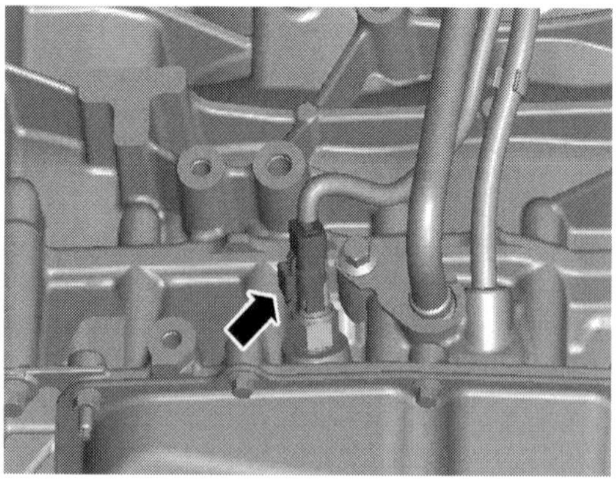

4.

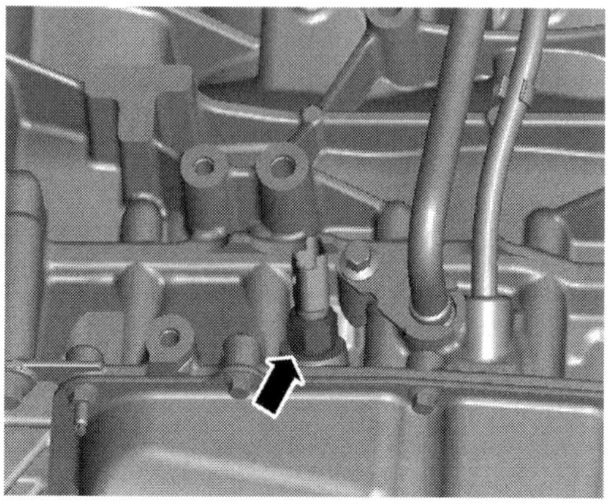

5. Torque: 27 Nm

Installation

1. To install, reverse the removal procedure.

Fuel Rail Pressure Sensor

Removal and Installation

Removal

WARNINGS:

WARNING: The spilling of fuel is unavoidable during this operation. Make sure that all necessary precautions are taken to prevent fire and explosion.

WARNING: Do not carry or operate cellular phones when working on or near any fuel related components. Highly flammable vapors are always present and may ignite. Failure to follow these instructions may result in personal injury.

WARNING: Do not smoke or carry lighted tobacco or open flame of any type when working on or near any fuel related components. Highly flammable vapors are always present and may ignite. Failure to follow these instructions may result in personal injury.

WARNING: If fuel contacts the eyes, flush the eyes with cold water or eyewash solution and seek immediate medical attention.

WARNING: Wash hands thoroughly after fuel handling, as prolonged contact may cause irritation. Should irritation develop, seek medical attention.

WARNING: Do not carry out any repairs to the fuel system with the engine running. The fuel pressure within the system can be as high as 2000 bar (29,008 lb-sq-in). Failure to follow this instruction may result in personal injury.

CAUTIONS:

CAUTION: Diesel fuel injection equipment is manufactured to very precise tolerances and fine clearances. It is therefore essential that absolute cleanliness is observed when working with these components. Always install new blanking plugs to any open orifices or lines. Failure to follow this instruction may result in foreign matter ingress to the fuel injection system.

CAUTION: Always carry out the cleaning process before carrying out any repairs to the fuel injection system components. Failure to follow this instruction may result in foreign matter ingress to the fuel injection system.

CAUTION: Do not disconnect the fuel injector electrical connectors with the engine running. Failure to follow this instruction may result in damage to the engine.

CAUTION: Make sure the workshop area in which the vehicle is being worked on is as clean and as dust free as possible. Foreign matter from work on clutches, brakes or from machining or welding operations can contaminate the fuel system and may result in later malfunction.

NOTE: Removal steps in this procedure may contain installation details.

1. **Refer to:** Fuel Injection Component Cleaning (303-04 Fuel Charging and Controls, General Procedures)

2.

3. CAUTION: Before disconnecting any components, make sure the area is clean and free from foreign material. When disconnected all openings must be sealed.

Installation

1. To install, reverse the removal procedure.

Heated Oxygen Sensor

Removal and Installation

Special Tool(s)

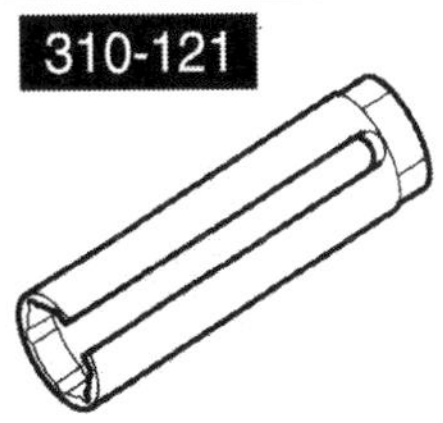 E53465	Wrench, H02S 310-121

Removal

NOTE: Removal steps in this procedure may contain installation details.

1. **Refer to:** Battery Disconnect and Connect (414-01, General Procedures).

2.

3. Torque: 48 Nm Tool: 310-121 (Wrench, H02S)

Installation

1. To install, reverse the removal procedure.

Manifold Absolute Pressure and Temperature Sensor

Removal and Installation

Part(s)

STEP	PART NAME	QUANTITY
Installation Step 1	Renew Part Exhaust manifold sensor	1

Removal

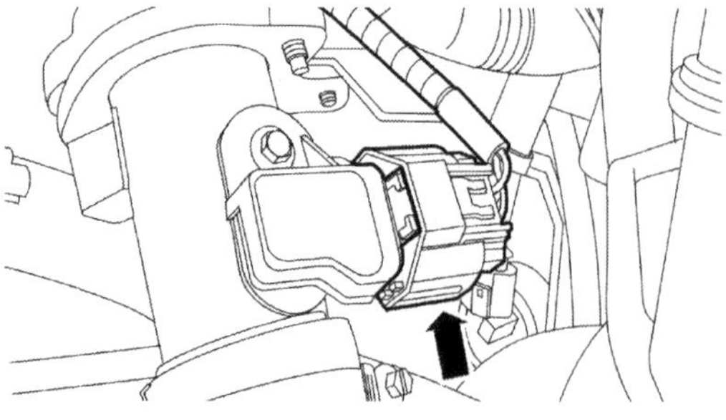

1. Disconnect the manifold absolute pressure and temperature (MAPT) sensor electrical connector.

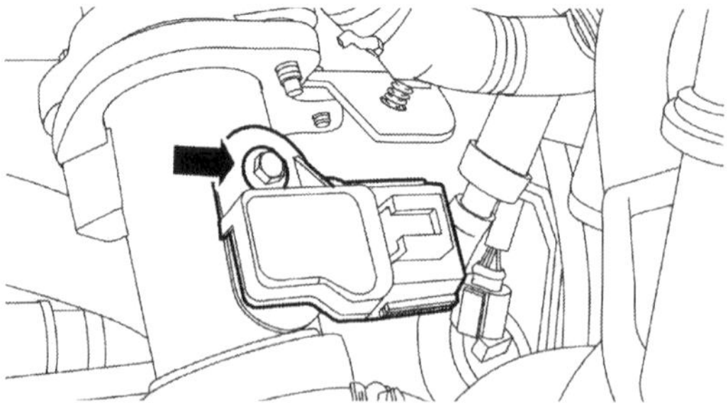

2. CAUTION: Before disconnecting any components, make sure the area is clean and free from foreign material. When disconnected all openings must be sealed.

Remove the bolt. Remove the MAPT sensor. Remove and discard the O-ring seal.

Installation

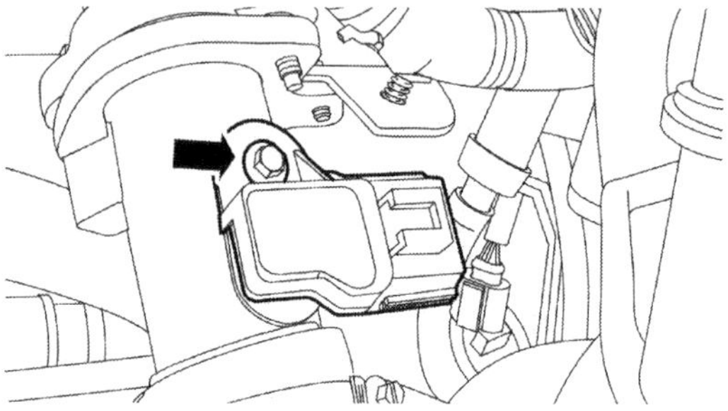

1. NOTES:

Remove and discard the blanking caps.

Install a new O-ring seal.

To install, reverse the removal procedure. Tighten the bolt.

Renew Part: Exhaust manifold sensor

Torque: 3 Nm

Mass Air Flow Sensor

Removal and Installation

Part(s)

STEP	PART NAME	QUANTITY
Installation Step 1	Renew Part Mass air flow sensor O-ring seal	1

Removal

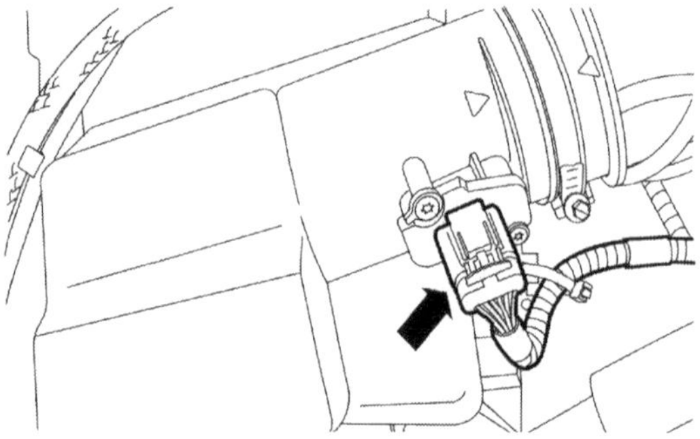

1. Disconnect the mass air flow (MAF) sensor electrical connector.

2. Remove the 2 screws. Remove the MAF sensor. Remove and discard the O-ring seal.

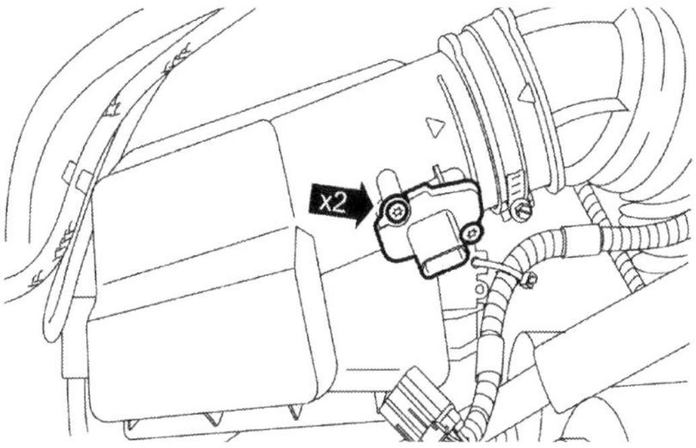

Installation

1. NOTE: Install a new O-ring seal.

To install, reverse the removal procedure.

Renew Part: Mass air flow sensor O-ring seal

Post Catalytic Converter Temperature Sensor

Removal and Installation

Removal

NOTE: Removal steps in this procedure may contain installation details.

1. WARNING: Do not work on or under a vehicle supported only by a jack. Always support the vehicle on safety stands.

Raise and support the vehicle.

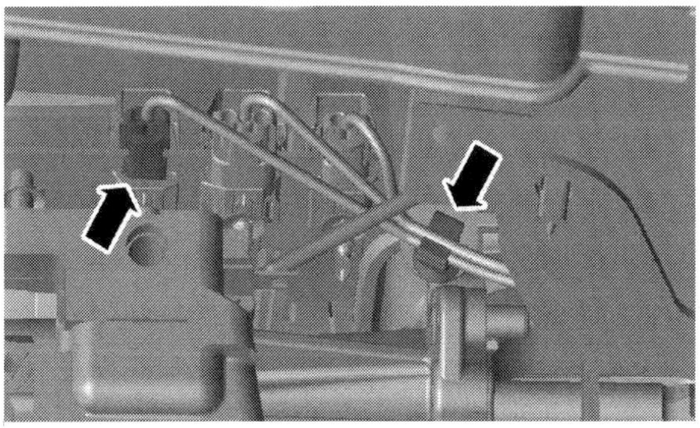

2.

3.

4. CAUTION: If accidentally dropped or knocked install a new sensor

NOTE: If the original sensor is to be installed, apply lubricant meeting specification ESE-M12A4-A to the thread of the sensor.

Torque: 35 Nm

Installation

1. To install, reverse the removal procedure.

Pre Catalytic Converter Temperature Sensor

Removal and Installation

Removal

NOTE: Removal steps in this procedure may contain installation details.

1. WARNING: Do not work on or under a vehicle supported only by a jack. Always support the vehicle on safety stands.

Raise and support the vehicle.

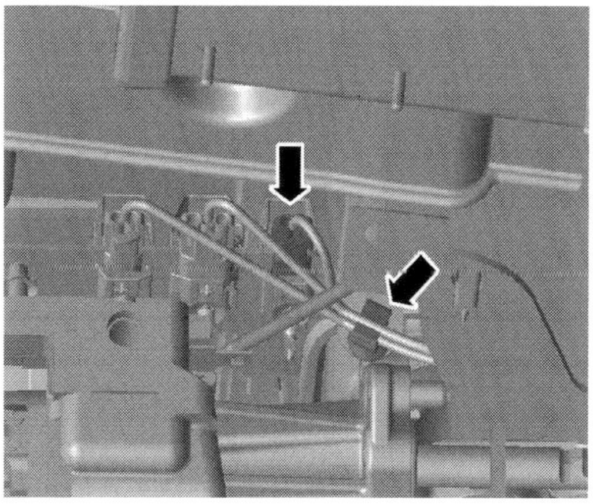

2.

3.

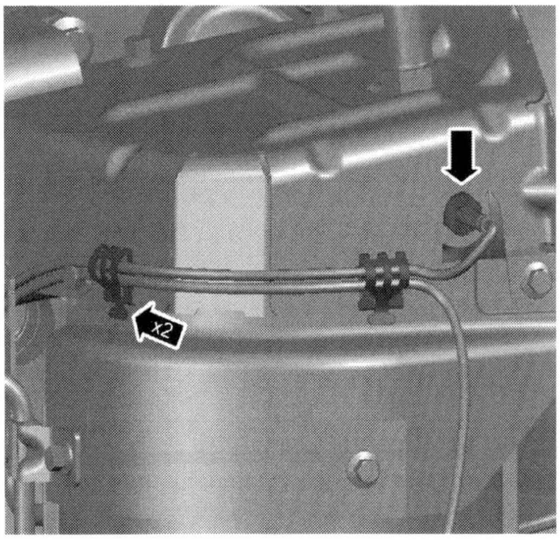

4. CAUTION: If accidentally dropped or knocked install a new sensor.

NOTE: If the original sensor is to be installed, apply lubricant meeting specification ESE-M12A4-A to the thread of the sensor.

Torque: 35 Nm

Installation

1. To install, reverse the removal procedure.

Post DPF Exhaust Gas Temperature Sensor

Removal and Installation

Removal

NOTE: Removal steps in this procedure may contain installation details.

1. WARNING: Do not work on or under a vehicle supported only by a jack. Always support the vehicle on safety stands.

Raise and support the vehicle.

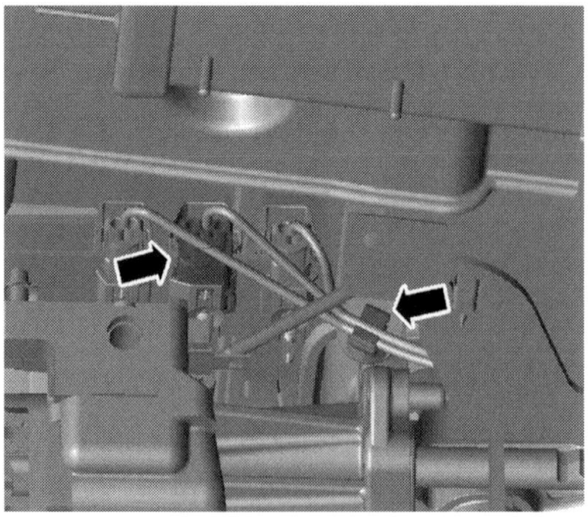

2.

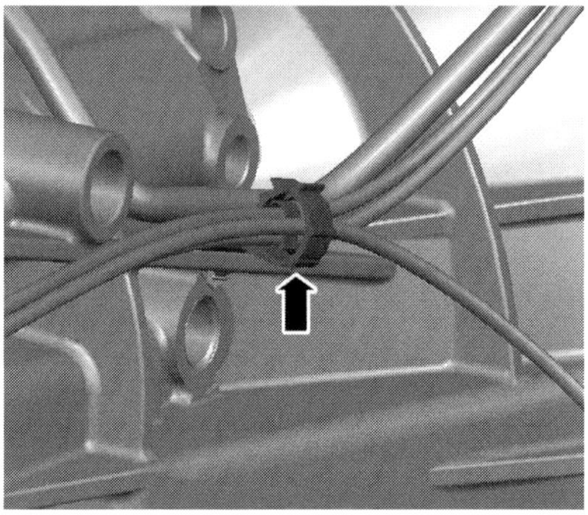

3.

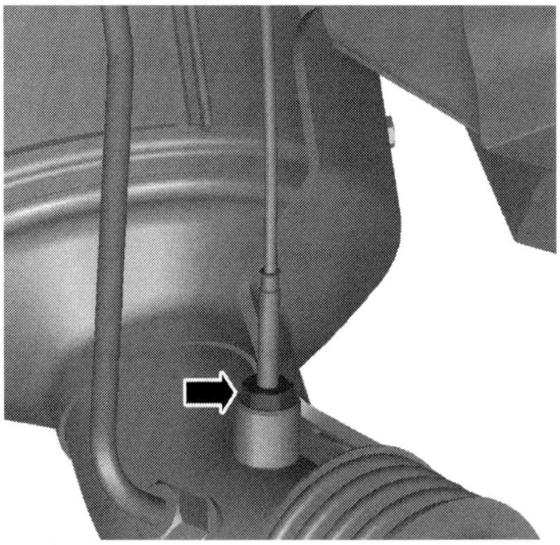

4. CAUTION: If accidentally dropped or knocked install a new sensor.

NOTE: If the original sensor is to be installed, apply lubricant meeting specification ESE-M12A4-A to the thread of the sensor.

Torque: 35 Nm

Installation

1. To install, reverse the removal procedure.

Clutch System Bleeding

General Procedures

WARNING: Brake fluid contains polyglycolethers and polyglycols. Avoid contact with the eyes. Wash hands thoroughly after handling. If brake fluid contacts the eyes, flush the eyes for 15 minutes with cold water. Get medical attention immediately if irritation persists. If taken internally, drink water and induce vomiting. Get medical attention immediately. Failure to follow these instructions may result in personal injury.

CAUTIONS:

Make sure the brake fluid in the clutch system reservoir is maintained between the minimum and maximum levels throughout the bleed procedure using new brake fluid.

If brake fluid comes into contact with the paintwork, it should be washed down immediately with cold water.

1. NOTES:

   Make sure of absolute cleanliness when filling brake fluid.

   Do not re-use brake fluid.

   As necessary, fill the clutch system reservoir with brake fluid.

2. WARNING: Do not work on or under a vehicle supported only by a jack. Always support the vehicle on safety stands.

   Raise and support the vehicle.

3. Remove the bleed nipple cover.

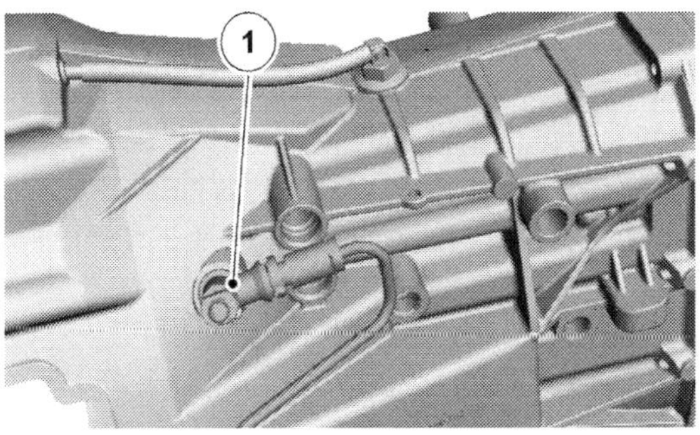

4. Connect a clear bleed tube to the bleed nipple and submerge the free end into a clear container of brake fluid.
5. Bleed the clutch system.
   - Loosen the bleed nipple by 2 turns.
   - Apply steady pressure to the clutch pedal, depressing it through its full stroke.
   - Hold the pedal down and tighten the bleed nipple.
   - Return the pedal to return its normal position.
6. Repeat the procedure until a flow of clean, air-free brake fluid is purged into the container.
7. NOTE: Do not re-use brake fluid.

   Check and top up the clutch system reservoir.
8. Remove the stands and lower the vehicle.

Clutch Pedal Freeplay Adjustment

General Procedures

1. Remove 6 screws and detach pedal box top cover and gasket.
2. Slacken both locknuts on master cylinder push-rod.
3. Check distance from lower edge of clutch pedal to floor. Correct measurement is 140 mm (5.50 in) without floor mat.
4. Adjust pedal stop as necessary to obtain correct setting.
5. Adjust master cylinder push rod to obtain approximately 1.5 mm (1/16 in) free play between push-rod and master cylinder piston.
6. Tighten push rod locknuts.
7. Check operation of clutch pedal and ensure that there is a minimum of 6 mm (0.25 in) of free play before pressure is felt. If necessary, readjust master cylinder push-rod.
8. Fit top cover and gasket to pedal bracket.
9. On RH drive vehicles, refit air cleaner.

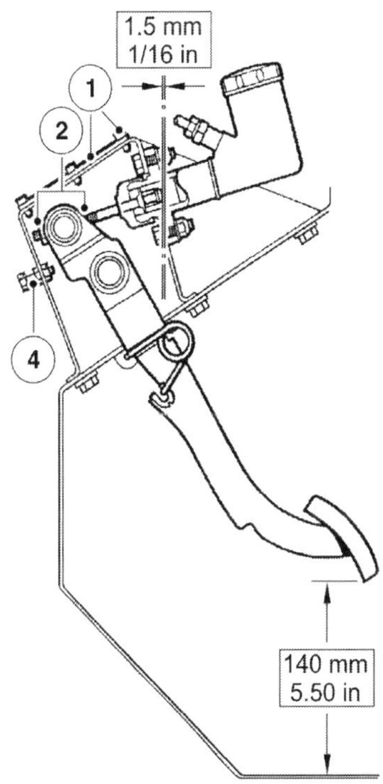

Specifications

Description	Nm	lb-ft
Clutch pressure plate bolts (new bolts must be installed)	29	21

Description and Operation

OVERVIEW

The clutch system is based on the established principle of a single driven plate and diaphragm spring clutch cover assembly hydraulically actuated from the clutch pedal. Depressing the clutch pedal transfers hydraulic fluid through the master cylinder, pipe work, and concentric slave cylinder ultimately actuating the clutch fingers to release the clutch and thus disengage drive from the crankshaft. When your foot is off the pedal, the spring pushes the pressure plate against the clutch disc, which in turn presses against the flywheel; this locks the engine to the transmission input shaft, causing them to rotate at the same speed.

The clutch system is of conventional design comprising the following major components:

- Clutch master cylinder and pressure pipes
- Concentric slave cylinder outlet assembly and peak torque limiter
- Vibration damper (Left hand drive vehicles only)
- Concentric slave cylinder
- Clutch cover assembly
- Clutch driven plate
- Flywheel

CLUTCH MASTER CYLINDER

The clutch master cylinder is attached directly to the pedal box assembly, located in the driver's footwell.

The cylinder contains a piston assembly, with a push rod connected to the clutch pedal and spring. When the clutch pedal is depressed, it pushes on the piston, via a linkage. Pressure builds in the cylinder and lines as the clutch pedal is depressed further.

The cylinder has 2 hydraulic connections:

- A low pressure feed pipe (providing fluid supply from the brake fluid reservoir)
- A high pressure pipe

The pedal travel is constrained by an 'up-stop' contained within the master cylinder and a 'down-stop' contained within the pedal box.

Item	Part Number	Description
1	-	Clutch master cylinder

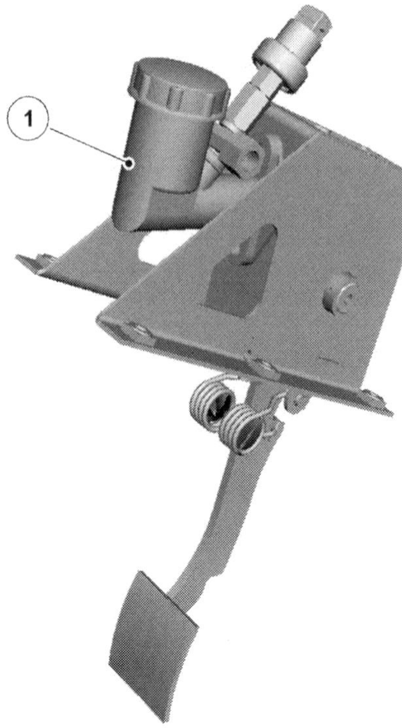

CONCENTRIC SLAVE CYLINDER OUTLET ASSEMBLY

⚠ NOTE: Right hand drive vehicle shown.

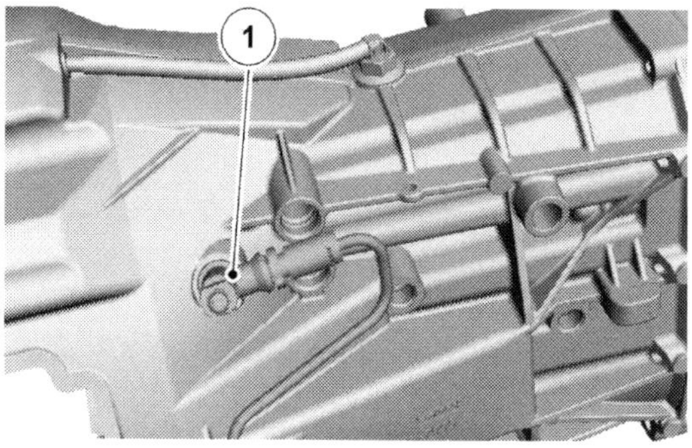

Item	Description
1	Slave cylinder outlet assembly and peak torque limiter

The concentric slave cylinder outlet assembly connects the external pipes with the release system contained within the clutch housing. A securing bracket locates the assembly in the correct orientation and a seal is provided between the assembly and the clutch housing.

Contained within the slave cylinder outlet assembly is a peak torque limiter. This component is designed to restrict the hydraulic fluid flow during the clutch pedal up-stroke. Under normal pedal actuation this restriction can not be detected, but in the event of an unintentional pedal release (e.g. wet shoe slipping off the clutch pedal) the peak torque limiter limits the fluid return rate and protects the transmission and driveline form excessive shock loads, which might cause damage.

On left hand drive vehicles, the hydraulic pipework contains an anti-vibration damper plugged into the peak torque limiter. This is used to reduce pedal roar/ vibrations during clutch operation.

CONCENTRIC SLAVE CYLINDER

The concentric slave cylinder assembly contains the release bearing and the hydraulic slave cylinder. The assembly is attached to the front end of the transmission via 3 bolts. These bolts are asymmetrically positioned to ensure correct angular location of the slave cylinder, which is also spigot-mounted for positional fit. In its free condition the slave cylinder is fully extended, but it positions itself automatically as the clutch housing is fitted to the engine. The assembly requires no setting or adjustment.

CLUTCH COVER ASSEMBLY

The clutch cover assembly comprises a pressure plate, cover and diaphragm and is mounted on and rotates with the flywheel.

The pressure plate is machined to provide a smooth surface for the drive plate to engage on. Lugs on the outer diameter of the pressure plate connect it via leaf springs to the cover. The leaf springs have leaves, which assist in pulling the pressure plate away from the drive plate when the clutch pedal is depressed.

The cover houses all pressure plate components. Shouldered rivets support the diaphragm inside the cover. The rivets heads are chamfered to allow the diaphragm to pivot when pressure is applied to it by the release bearing. Holes in the cover locate on dowels on the flywheel and further holes provide for the attachment of the cover to the flywheel. Larger holes in the cover provide ventilation for the drive plate and pressure plate and flywheel contact surfaces.

The diaphragm comprises a cast ring with fingers. The diaphragm is attached to the cover with shouldered rivets. The inner head of each rivet is chamfered to allow the diaphragm to pivot when the clutch is depressed or released. When pressure is applied to the fingers of the diaphragm by the release bearing, the diaphragm pivots on the rivets and moves away from the pressure plate, releasing the force applied to the pressure plate and allowing the drive plate to slip between the pressure plate and the flywheel.

CLUTCH DRIVEN PLATE

The clutch driven plate is sandwiched between the flywheel and the pressure plate of the clutch cover assembly. The clutch driven plate has a splined hub, which engages with the splines on the primary shaft from the transmission. The splined hub is located in an inner plate, which contains 3 compression pre-damper springs. The inner plate is retained by the springs, which can compress in both directions to cushion engine vibration at idle speed. The inner plate is located on 4 larger compression springs, which are located in a central plate. The hub is sandwiched between the central plate and the friction damper. The friction damper comprises friction washers located between the hub and the central plate. The friction washers reduce transmission noises and vibrations due to engine cyclic excitation.

FLYWHEEL

The single mass flywheel is bolted to the flange of the engine crankshaft. A dowel ensures that the flywheel is correctly located. A ring gear is located on the outside diameter of the flywheel and is seated against a flange. The ring gear is an interference fit on the flywheel and is a serviceable item, which can be replaced if damaged or worn.

The operating face of the flywheel is machined to provide a smooth surface for the clutch driven plate to engage on.

Clutch

Diagnosis and Testing

Principles of Operation

For a detailed description of the clutch, refer to the relevant Description and Operation section in the workshop manual.

Refer to: Clutch (308-01 Clutch, Vehicles With: MT82 6-Speed Manual Transmission Description and Operation)

Inspection and Verification

⚠ CAUTION: Diagnosis by substitution from a donor vehicle is NOT acceptable. Substitution of control modules does not guarantee confirmation of a fault, and may also cause additional faults in the vehicle being tested and/or the donor vehicle.

⚠ NOTE: Check and rectify basic faults before beginning diagnostic routines involving pinpoint tests.

1. Visually inspect for obvious signs of damage and system integrity

Visual Inspection

Mechanical
• Clutch fluid leaks

2. If an obvious cause for an observed or reported concern is found, correct the cause (if possible) before proceeding to the next step

3. If the cause is not visually evident, verify the symptom and refer to the Symptom Chart, alternatively check for Diagnostic Trouble Codes (DTCs) and refer to the DTC Index

Symptom Charts

DTC Index

For a list of Diagnostic Trouble Codes (DTCs) that could be logged on this vehicle, please refer to Section 100-00.

Symptom	Possible Causes	Action
Loss of drive / clutch bite point very high	• Clutch friction plate worn	• Refer to the relevant section of the workshop manual and install a new clutch assembly
Noise from clutch when engaging/disengaging	• Clutch friction plate failure	NOTE: Refer to Technical Service Bulletin LTB00166 for details of affected VIN range • Refer to the relevant section of the workshop manual and install a new clutch assembly
Clutch chatters or rattles at idle	• Clutch friction plate thrust washer worn or broken • Clutch friction plate cracked • Clutch friction plate spring(s) broken	NOTE: Refer to Technical Service Bulletin LTB00477 for details of affected VIN range • Refer to the relevant section of the workshop manual and install a new clutch assembly
Harsh knock from clutch at engine start or stop	• Clutch friction plate thrust washer worn or broken • Clutch friction plate cracked • Clutch friction plate spring(s) broken	NOTE: Refer to Technical Service Bulletin LTB00477 for details of affected VIN range • Refer to the relevant section of the workshop manual and install a new clutch assembly
Difficulty selecting gears - clutch not disengaging completely when pedal fully depressed	• Air in clutch hydraulic system / clutch fluid leak • Clutch pedal height adjusted incorrectly • Clutch friction plate failure	• Refer to the relevant section of the workshop manual and check the clutch fluid level. Check the clutch hydraulic system for leaks. Repair as necessary. Bleed the clutch hydraulic system • Refer to the relevant section of the workshop manual and check the clutch pedal height adjustment • Refer to Technical Service Bulletin LTB00166 and install a new clutch assembly as necessary
Clutch pedal bite point is too low	• Clutch not disengaging fully when pedal depressed • Clutch pedal height adjusted incorrectly	• Refer to the relevant section of the workshop manual and check the clutch fluid level. Check the clutch hydraulic system for leaks. Repair as necessary. Bleed the clutch hydraulic system • Refer to the relevant section of the workshop manual and check the clutch pedal height adjustment
Clutch pedal excessive effort required	• Clutch pedal operation not meeting customer expectation	NOTE: Refer to Technical Service Bulletin LD308-002 for details of affected VIN range • Refer to Technical Service Bulletin LD308-002 and install a clutch pedal assistance spring

MT82 6-Speed Manual Transmission - Clutch Disc and Pressure Plate

Removal and Installation

Special Tool(s)

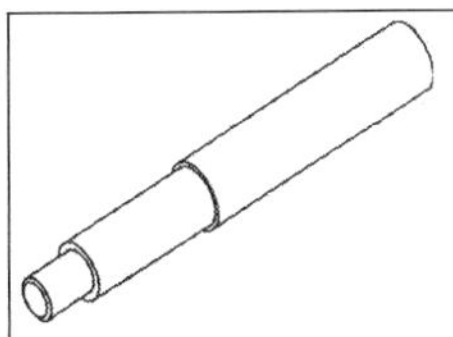 308419	308-419 Clutch Alignment Tool

Part(s)

STEP	PART NAME	QUANTITY
Installation Step 1	Renew Part Clutch cover assembly bolt(s)	6

Removal

1. WARNING: Do not work on or under a vehicle supported only by a jack. Always support the vehicle on safety stands.

Raise and support the vehicle.

2. Disconnect the battery ground cable.

Refer to: Startup Battery Disconnect and Connect (414-01 Battery, Mounting and Cables, General Procedures)

3. Remove the transmission.

Refer to: (308-03 Manual Transmission/Transaxle, Vehicles With: MT82 6-Speed Manual Transmission Specifications)

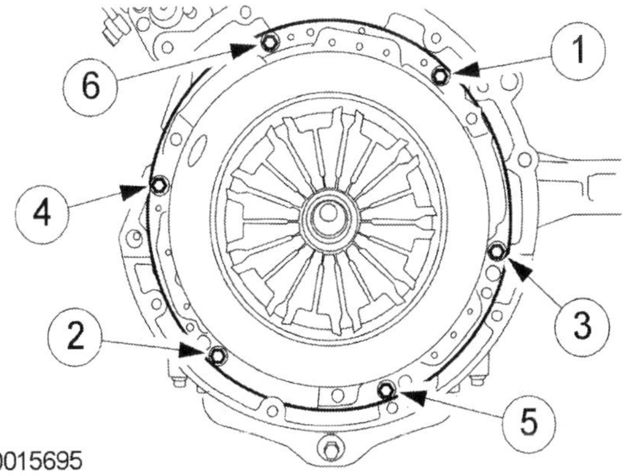

ELE0015695

4. CAUTION: Loosen the clutch pressure plate bolts by 2 turns at a time in the sequence shown. Failure to follow this instruction may result in damage to the vehicle.

Remove the clutch disc and the clutch pressure plate. Remove and discard the 6 bolts.

Installation

1. CAUTIONS:

Make sure new bolts are installed. Failure to follow this instruction may result in damage to the vehicle.

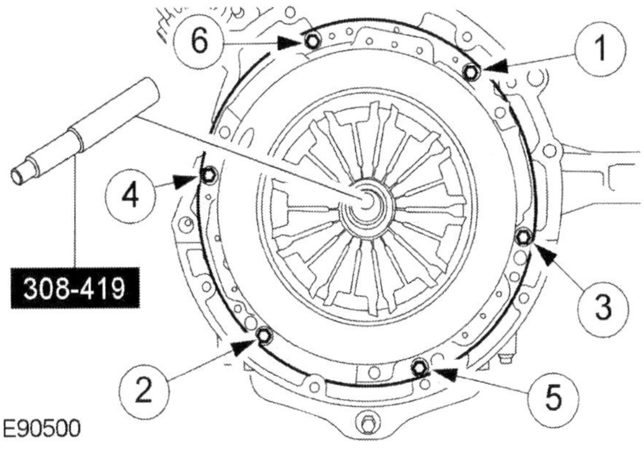

E90500

Apply grease of the correct specification to the pilot bearing. Failure to follow this instruction may result in damage to the vehicle.

Tighten the bolts in the sequence shown. Failure to follow this instruction may result in damage to the vehicle.

NOTE: Clean the component mating faces.

Using the special tool, install the clutch disc and the clutch pressure plate. Tighten the bolts.

Special Tool(s): 308-419

Renew Part: Clutch cover assembly bolt(s)

Torque: 29 Nm

2. Remove the special tool.

3. Install the transmission.

Refer to: (308-03 Manual Transmission/Transaxle, Vehicles With: MT82 6-Speed Manual Transmission Specifications)

4. Connect the battery ground cable.

Refer to: Startup Battery Disconnect and Connect (414-01 Battery, Mounting and Cables, General Procedures)

Pilot Bearing

Removal and Installation

Special Tool(s)

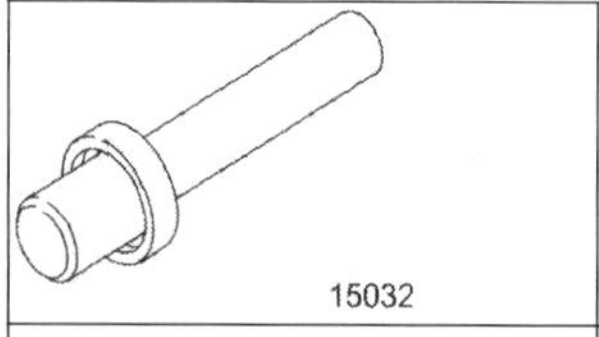	Bearing installer 205-081
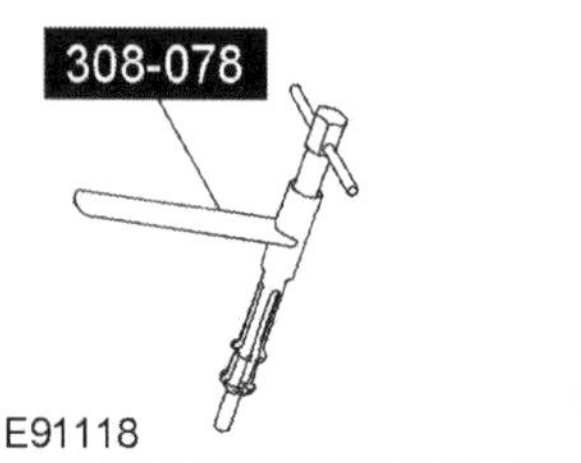	Bearing remover 303-078

Removal

1. WARNING: Do not work on or under a vehicle supported only by a jack. Always support the vehicle on safety stands.

Raise and support the vehicle.

2. Disconnect the battery ground cable.

Refer to: Battery Disconnect (414-01, General Procedures).

3. Remove the clutch disc and pressure plate.

   Refer to: Clutch Disc and Pressure Plate (308-01 Clutch, Vehicles With: MT82 6-Speed Manual Transmission Removal and Installation)

4. Using a suitable tool, break through the pilot bearing.

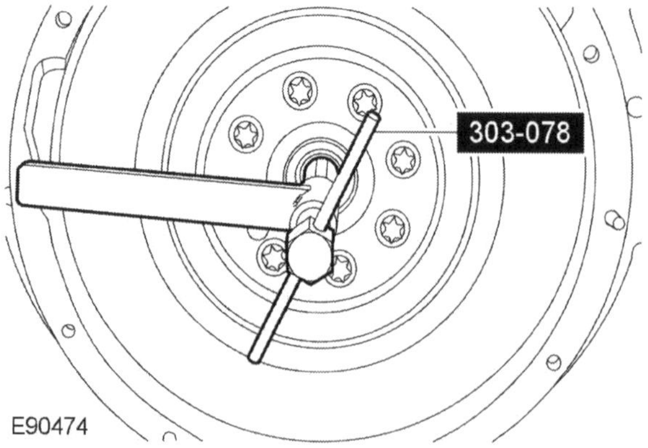

5. Using the special tool, remove the pilot bearing.

Installation

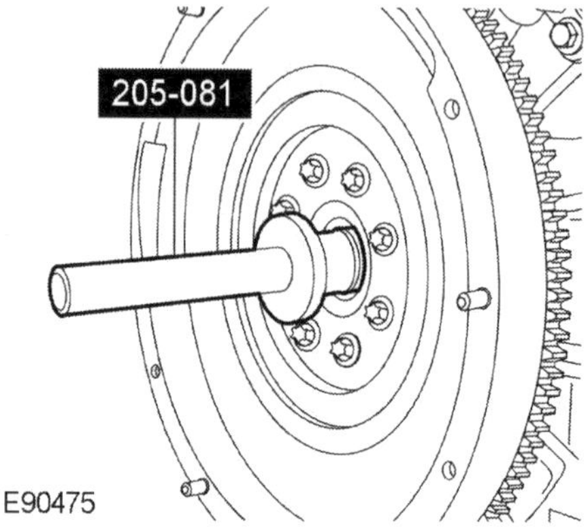

1. Using the special tool, install the pilot bearing.

2. Install the clutch disc and pressure plate.

   Refer to: Clutch Disc and Pressure Plate (308-01 Clutch, Vehicles With: MT82 6-Speed Manual Transmission Removal and Installation)

3. Connect the battery ground cable.

   Refer to: Battery Connect (414-01, General Procedures).

Clutch Controls: MT82 6-Speed Manual Transmission

Specifications

Description	Nm	lb-ft
Clutch slave cylinder bolts	10	7
Clutch master cylinder to pedal box bolts	23	17
Clutch pedal box to bulkhead bolts	23	17
Clutch master cylinder push rod nuts	23	17

Description and Operation

OVERVIEW

Refer to: Clutch (308-01 Clutch, Vehicles With: MT82 6-Speed Manual Transmission Description and Operation)

Clutch Master Cylinder - LHD

Removal and Installation

Part(s)

STEP	PART NAME	QUANTITY
Installation Step 1	Renew Part Clutch master cylinder gasket	1
Installation Step 2	Renew Part Clutch master cylinder lock nut	1
Installation Step 3	Renew Part Clutch pedal sensor washer(s)	1
Installation Step 20	Renew Part Clutch pedal box cover gasket	1

Removal

NOTES:

△ Some variation in the illustrations may occur, but the essential information is always correct.

△ Land Rover approved diagnostic equipment must be used to operate the Antilock Brake System (ABS) solenoid valves and the return pump to ensure correct bleeding of the Hydraulic Control Unit (HCU) and all the vehicle brake circuits.

1. **All Vehicles.** Remove the 2 stop lamp switches.

Refer to: Stoplamp Switch (417-01 Exterior Lighting, Removal and Installation)

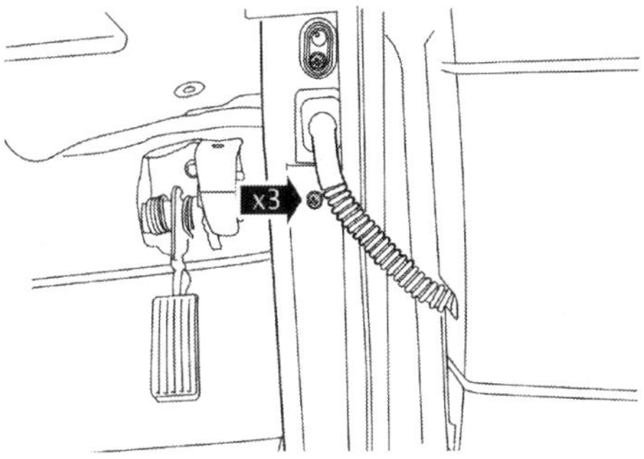

2. Remove 3 screws securing closing panel to lower 'A' pillar.

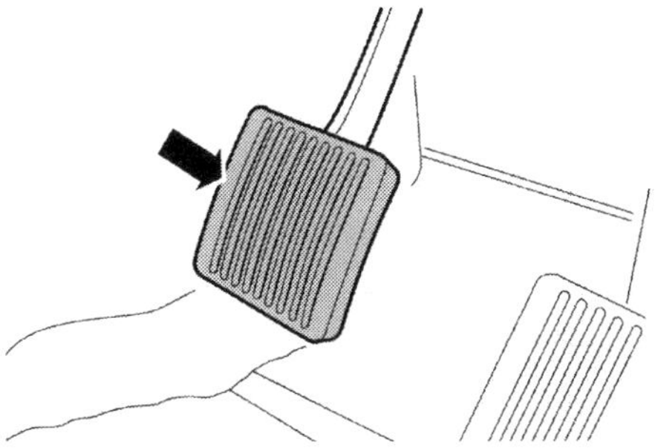

3. Remove clutch pedal rubber.

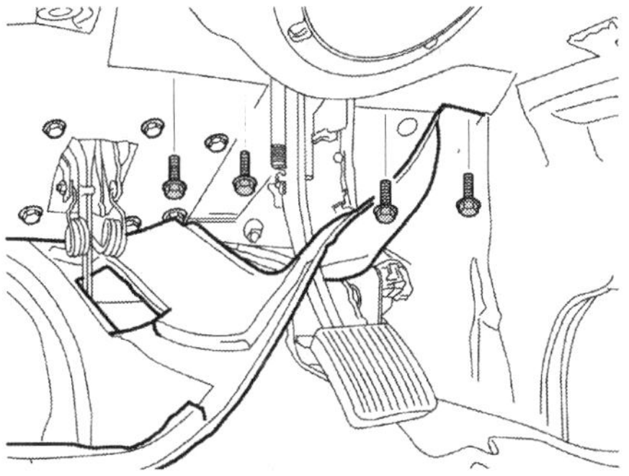

4. Remove 12 bolts securing clutch and brake pedal brackets to body. Release closing panel sufficient to gain access to clutch bracket bolts.

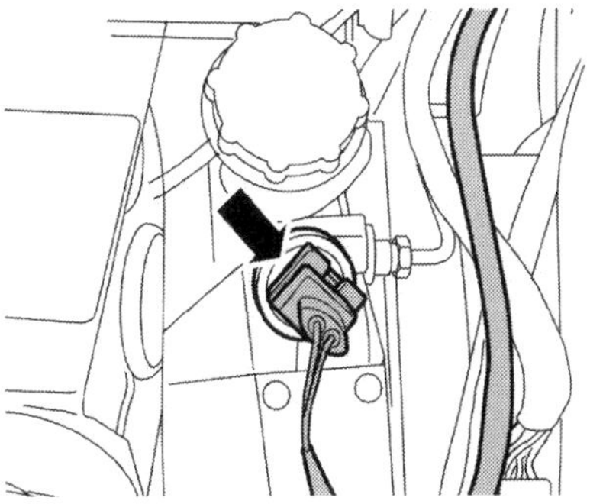

5. Disconnect multiplug clutch pedal sensor.

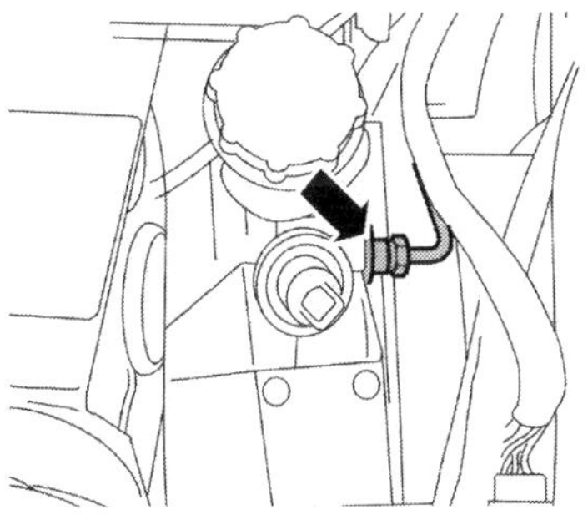

6.

WARNING: A small amount of fluid loss is unavoidable, position an absorbant cloth or container to collect it.

CAUTIONS:

Before disconnecting any components, make sure the area is clean and free from foreign material. When disconnected all openings must be sealed.

Brake fluid will damage paint finished surfaces. If spilled, immediately remove fluid and clean area with water.

Loosen pipe union and release clutch master cylinder fluid pipe.

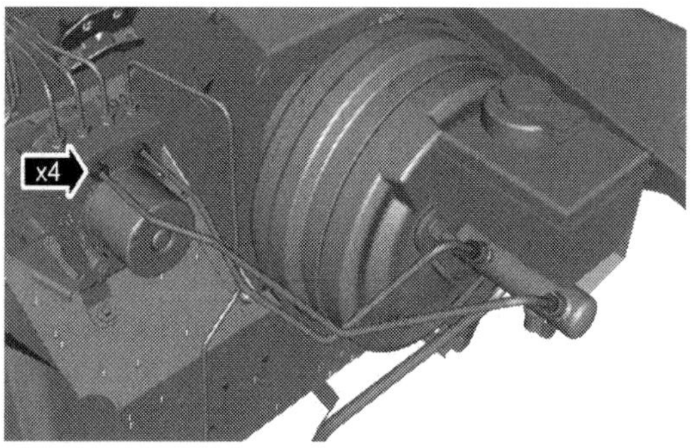

7. WARNING: A small amount of fluid loss is unavoidable, position an absorbant cloth or container to collect it.

CAUTIONS:

Before disconnecting any components, make sure the area is clean and free from foreign material. When disconnected all openings must be sealed.

Brake fluid will damage paint finished surfaces. If spilled, immediately remove fluid and clean area with water.

Vehicles With Anti-lock Brake System (ABS). Remove the 2 brake pipes from the Brake Master Cylinder to the Hydraulic Control Unit (HCU).

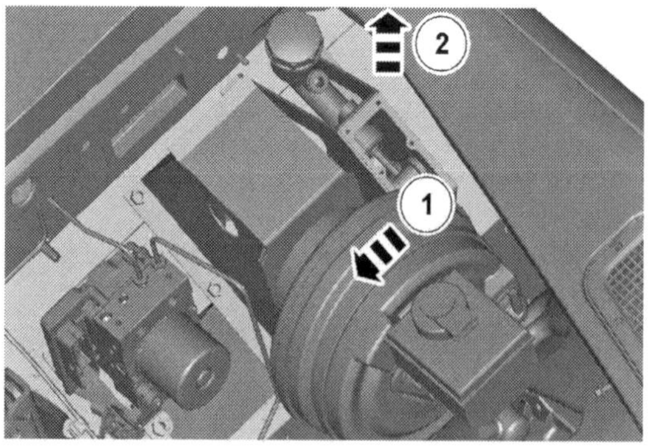

8. CAUTION: Take extra care not to damage any surrounding components.

Remove the clutch pedal box and clutch master cylinder. Carefully reposition the brake pedal box and master cylinder to aid removal.

9. **Vehicles With Brake Pressure Control Valve (BPCV).** Remove the 2 brake pipes from the Brake Master Cylinder to the BPCV. Release the front brake pipe from the Brake Master Cylinder.

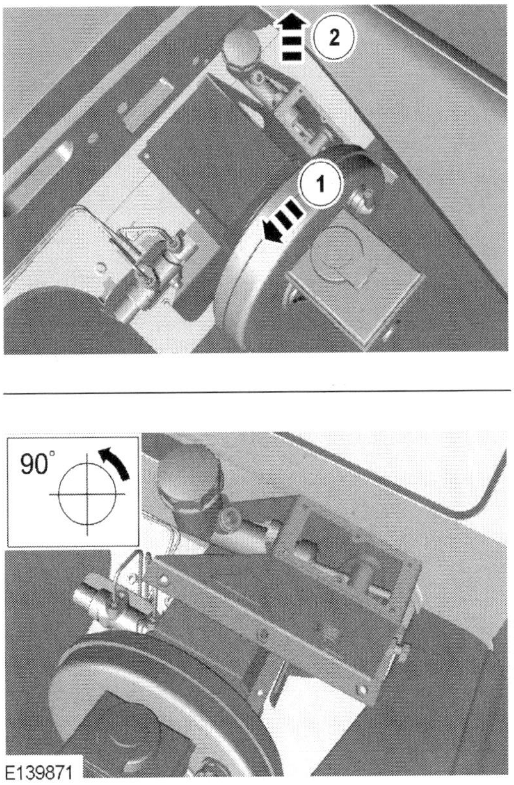

10. ⚠ CAUTION: Take extra care not to damage any surrounding components.

Remove the clutch pedal box and clutch master cylinder. Carefully reposition the brake pedal box and master cylinder to aid removal.

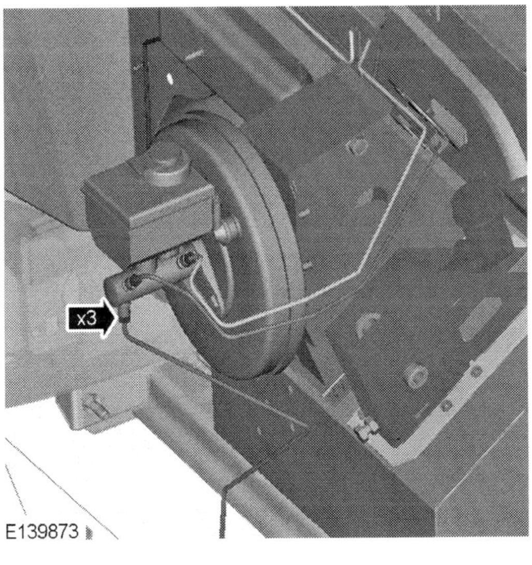

11. **Vehicles With-Out ABS Or BPCV.** Release the 3 brake pipes from the Brake Master Cylinder.

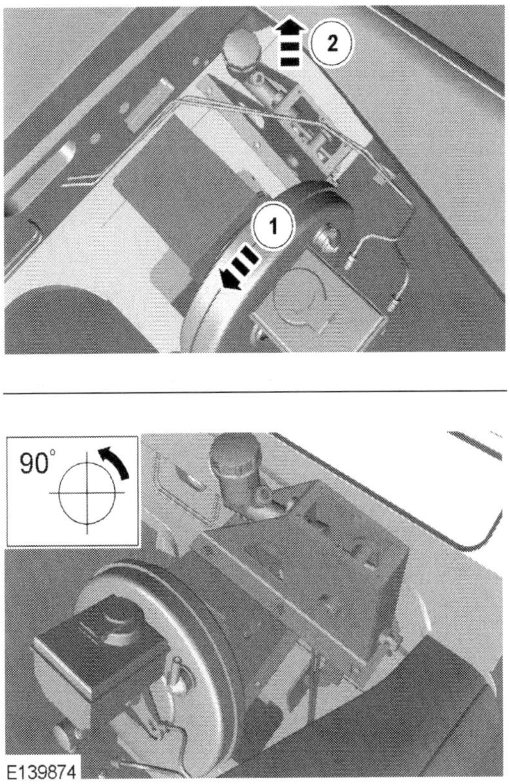

12. ⚠ CAUTION: Take extra care not to damage any surrounding components.

Remove the clutch pedal box and clutch master cylinder. Carefully reposition the brake pedal box and master cylinder to aid removal. Carefully reposition the brake pipes to aid removal.

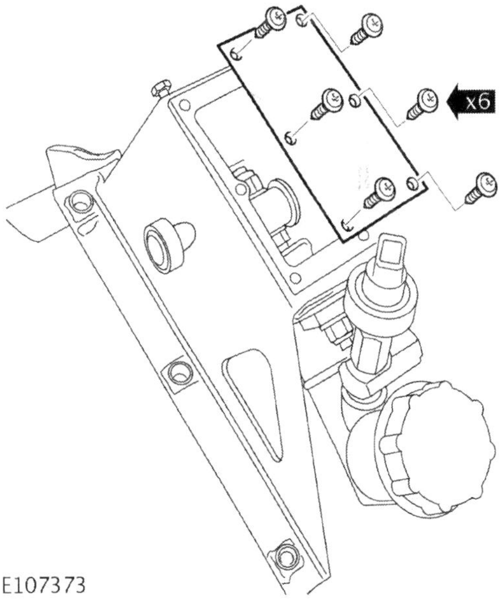

13. **All Vehicles.** Remove the 6 screws, remove pedal box top cover and discard gasket.

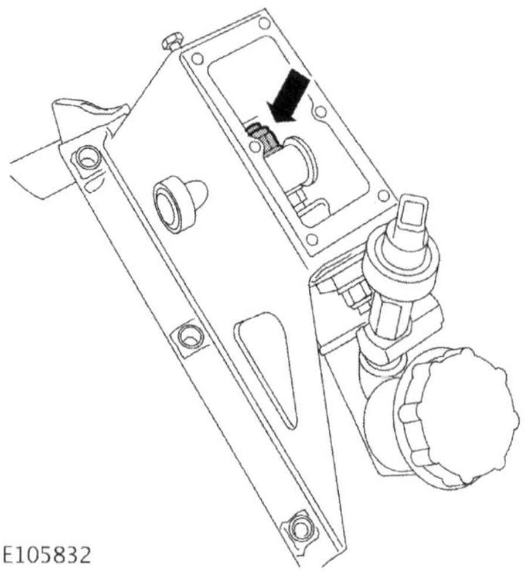

E105832

14. Remove nut securing push rod to clutch pedal trunnion. Move the clutch pedal to the fully down position.

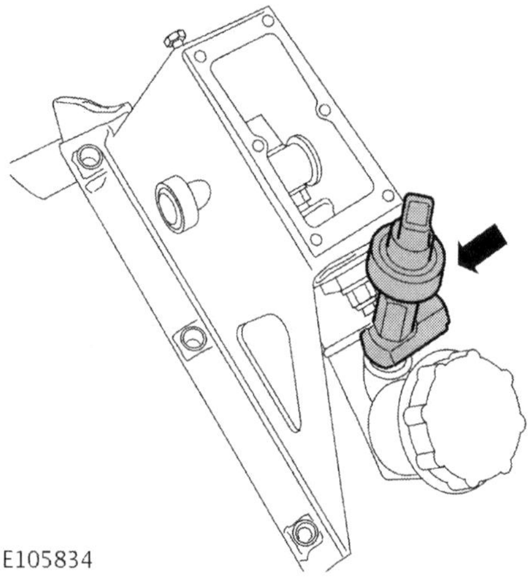

E105834

15. Remove clutch pedal sensor and pipe union from master cylinder and discard sealing washers. Remove 2 nuts and bolts, and remove clutch master cylinder from pedal bracket.

Installation

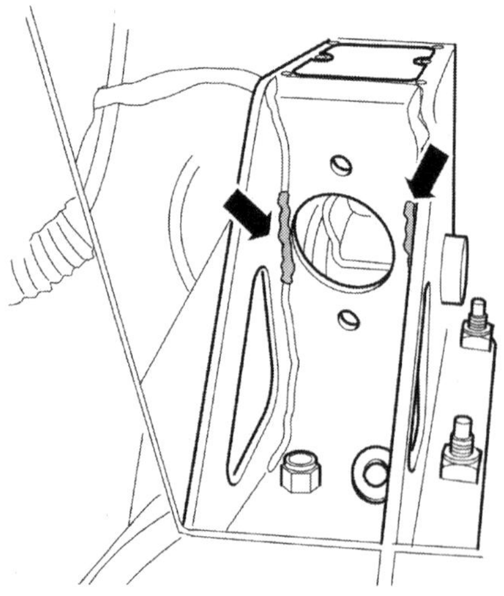

1. **All Vehicles.** Check seal on pedal box at points indicated and seal with RTV if necessary.

Renew Part: Clutch master cylinder gasket

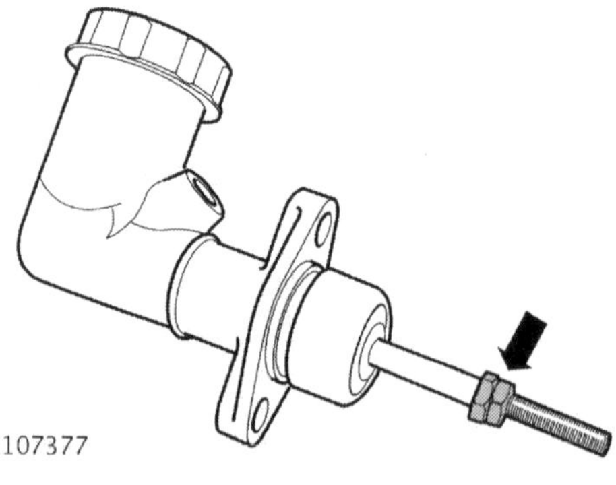

E107377

2. Install new lock nuts to master cylinder.

Renew Part: Clutch master cylinder lock nut

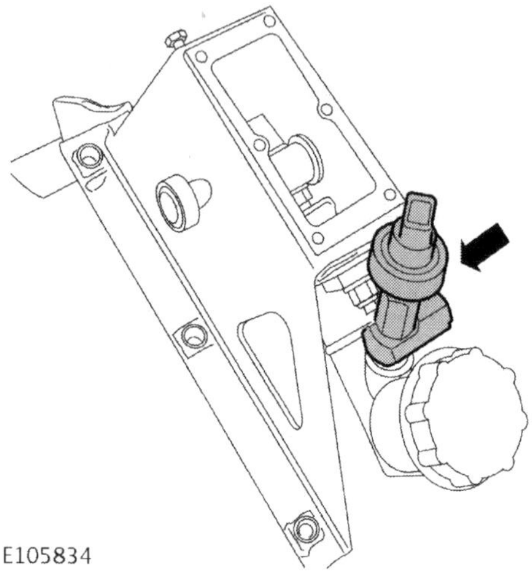

E105834

3. Clean sensor and pipe union. Install new gasket. Position master cylinder to pedal bracket and align push rod to pedal trunnion. Install the clutch master cylinder to the clutch pedal bracket and tighten the nuts and bolts. Install clutch pedal sensor and pipe union to master cylinder using new sealing washers and tighten.

Renew Part: Clutch pedal sensor washer(s)

Bracket bolts 23 Nm

Pipe union nuts 14 Nm

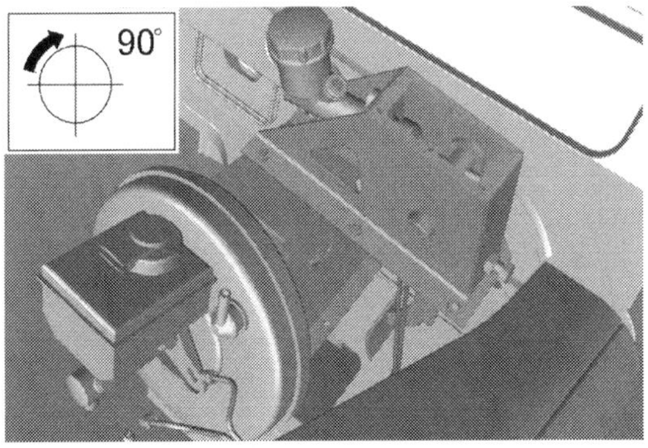

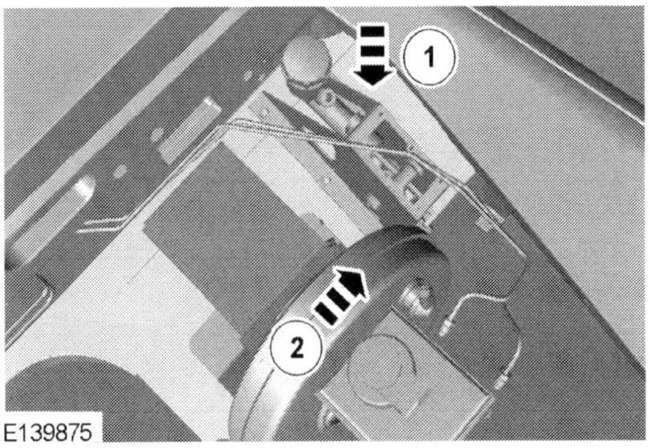

4. CAUTION: Take extra care not to damage any surrounding components.

Vehicles With-Out ABS Or BPCV. Install the clutch pedal box and clutch master cylinder. Carefully reposition the brake pedal box and master cylinder to aid installation. Carefully reposition the brake pipes to aid installation.

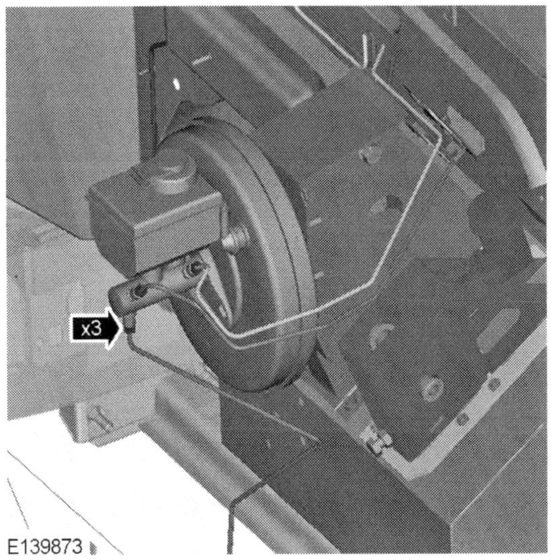

5. Secure the 3 brake pipes to the Brake Master Cylinder. Tighten the 3 brake pipe unions.

Torque: 15 Nm

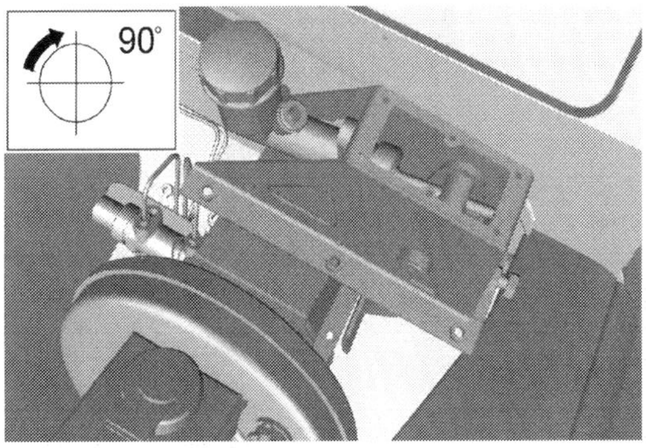

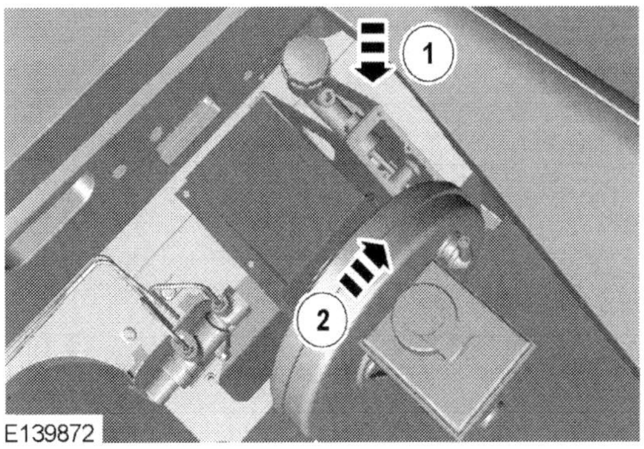

6. CAUTION: Take extra care not to damage any surrounding components.

Vehicles With BPCV. Install the clutch pedal box and clutch master cylinder. Carefully reposition the brake pedal box and master cylinder to aid installation.

7. Install the 2 brake pipes from the Brake Master Cylinder to the BPCV. Tighten the 4 brake pipe unions. Secure the front brake pipe to the Brake Master Cylinder and tighten.

Torque: 15 Nm

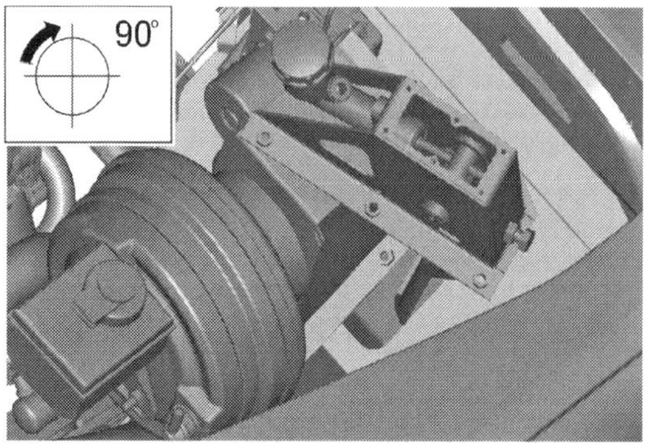

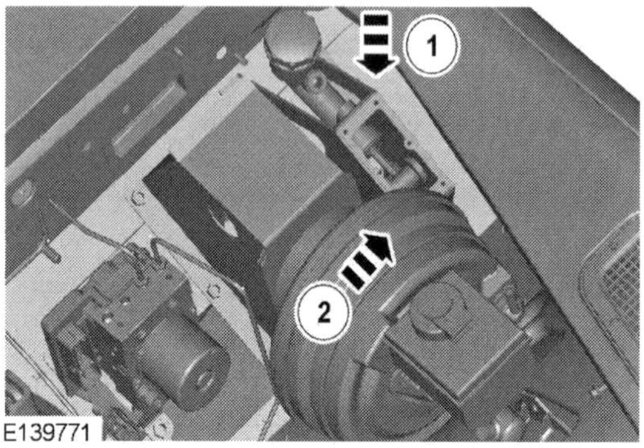

8. CAUTION: Take extra care not to damage any surrounding components.

Vehicles With ABS. Install the clutch pedal box and clutch master cylinder. Carefully reposition the brake pedal box and master cylinder to aid removal.

9. Install the brake pipes from the Brake Master Cylinder to the Hydraulic Control Unit (HCU). Tighten the 4 brake pipe unions.

Torque: 15 Nm

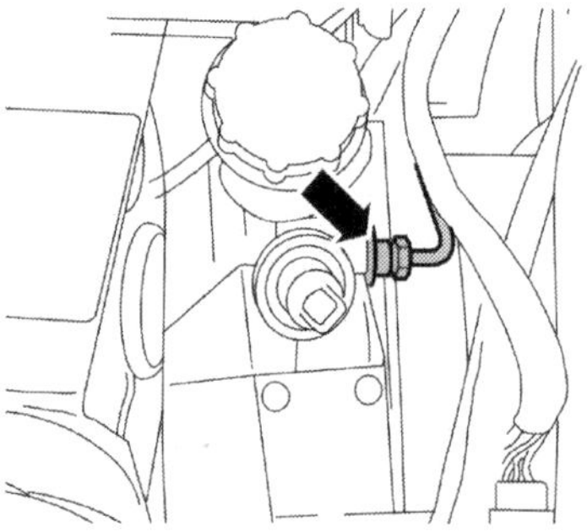

10. **All Vehicles.** Install clutch master cylinder fluid pipe. Clean clutch pipe union. Align pipe and tighten union.

Torque: 15 Nm

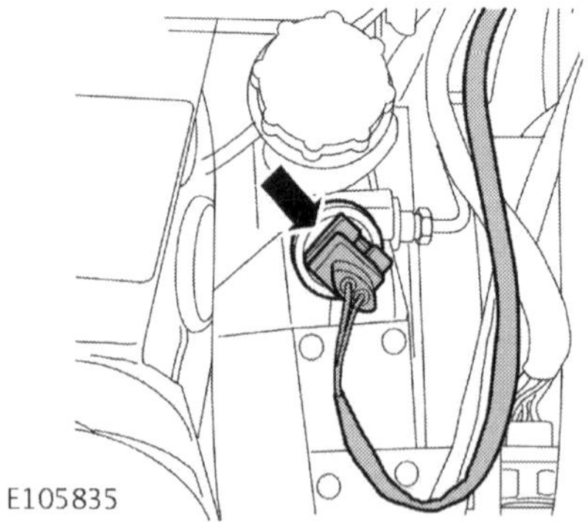

11. Connect multiplug to clutch pedal sensor.

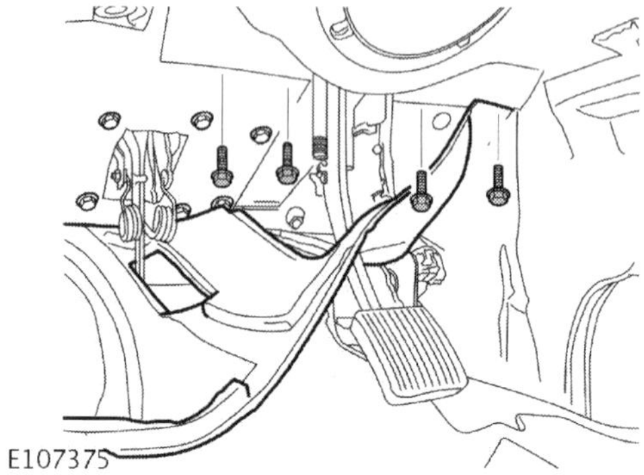

12. Install the 12 bolts securing clutch pedal bracket to body. Tighten the bolts.

Torque: 23 Nm

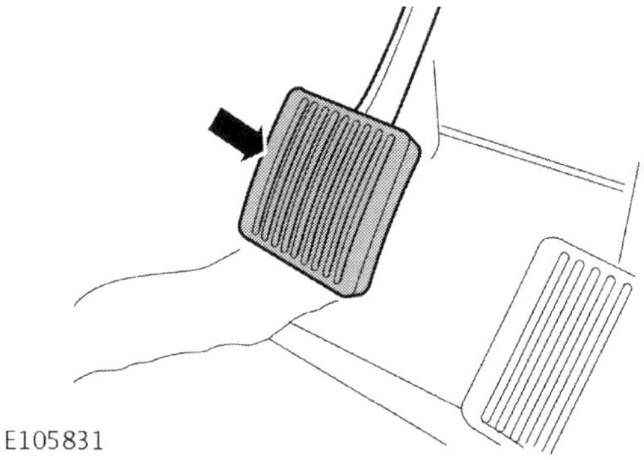

13. Install clutch pedal rubber.

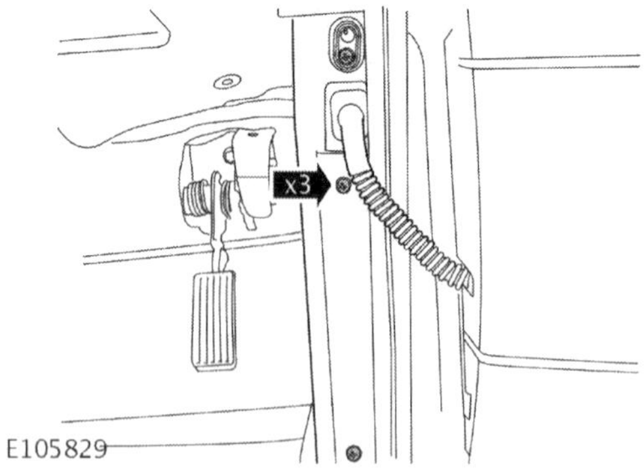

14. Position closing panel and tighten screws to lower 'A' pillar.

15. Install the 2 stoplamp switches.

Refer to: Stoplamp Switch (417-01 Exterior Lighting, Removal and Installation)

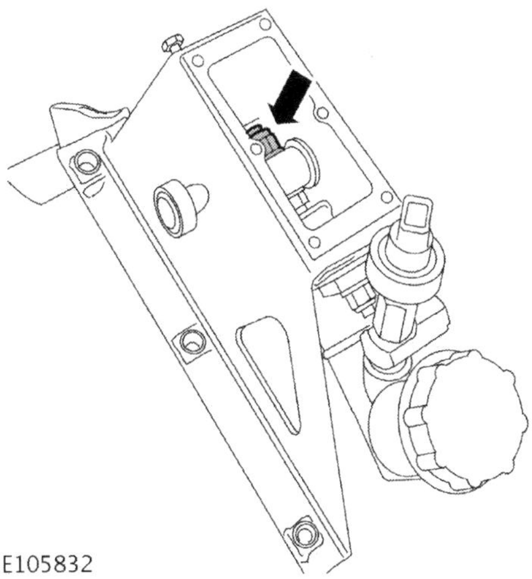

E105832

16. Install and lightly tighten nut securing push rod to clutch pedal trunnion.

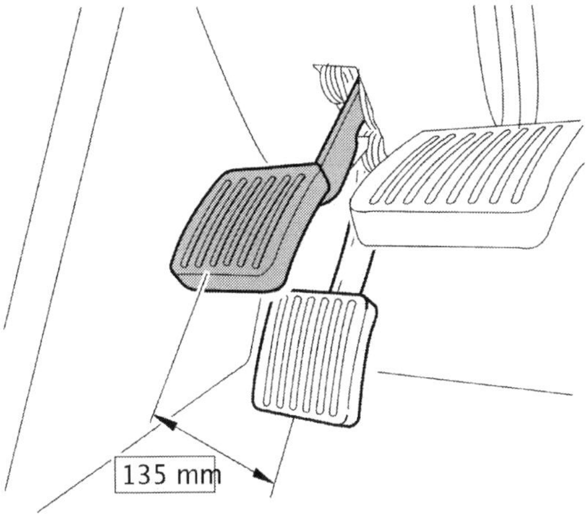

17. Check distance from lower edge of clutch pedal to down position. Correct measurement is 135 mm (5.31 in).

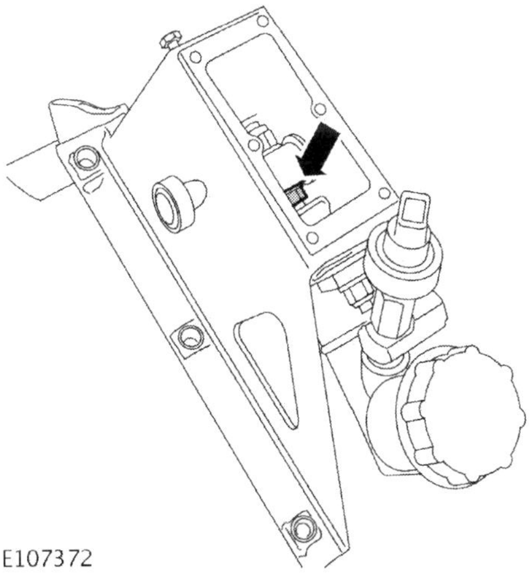

E107372

18. Adjust master cylinder push rod lock nuts until 135mm (5.31 in) is achieved. Tighten lock nuts. Move the clutch pedal to the fully down position. Tighten nylon nut securing push rod to clutch pedal trunnion.

Torque: 23 Nm

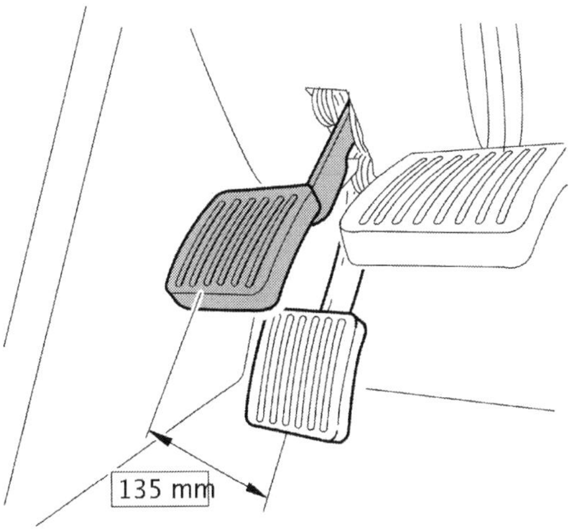

19. Check pedal travel is to correct measurement.

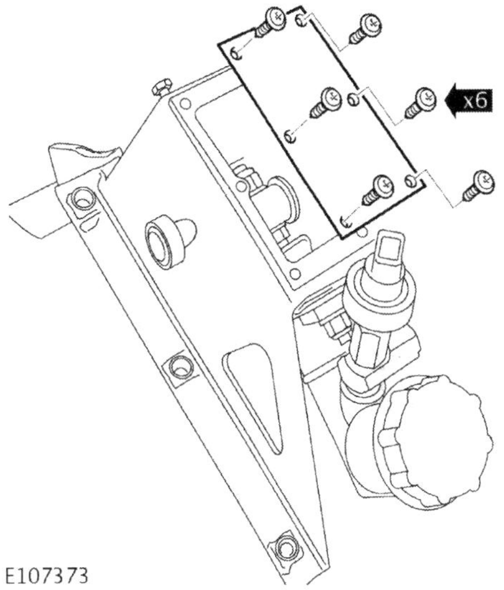

E107373

20. Install a new gasket to cover plate and secure with the 6 screws. Tighten the screw.

Renew Part: Clutch pedal box cover gasket

Torque: 3 Nm

21. Bleed the clutch hydraulic system.

Refer to: Clutch System Bleeding (308-00 Manual Transmission/Transaxle and Clutch - General Information, Vehicles With: MT82 6-Speed Manual Transmission General Procedures)

22. Using the Land Rover approved diagnostic equipment, bleed the complete brake system.

Refer to: Brake System Bleeding (206-00 Brake System - General Information, General Procedures)

Clutch Master Cylinder - RHD

Removal and Installation

Part(s)

STEP	PART NAME	QUANTITY
Installation Step 2	Renew Part Clutch master cylinder lock nut	1
Installation Step 3	Renew Part Clutch pedal sensor washer(s)	1
Installation Step 3	Renew Part Clutch master cylinder gasket	1
Installation Step 23	Renew Part Clutch pedal box cover gasket	1

Removal

NOTE: Some variation in the illustrations may occur, but the essential information is always correct.

1. **All Vehicles.** Remove the 2 stop lamp switches.

Refer to: Stoplamp Switch (417-01 Exterior Lighting, Removal and Installation)

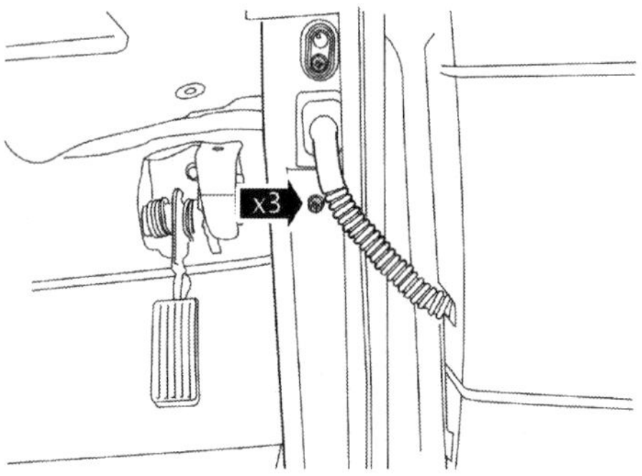

2. Remove 3 screws securing closing panel to lower 'A' pillar.

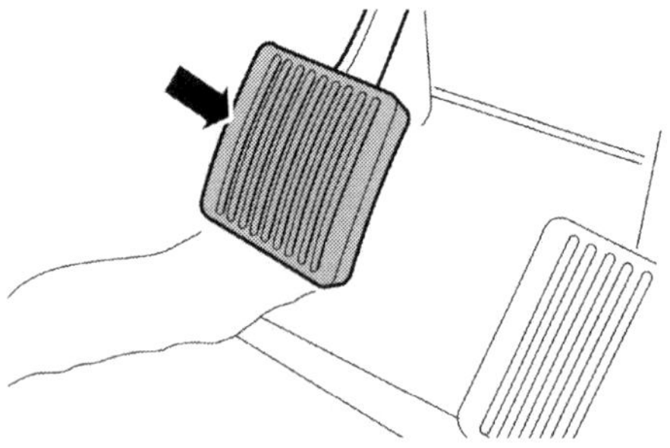

3. Remove clutch pedal rubber.

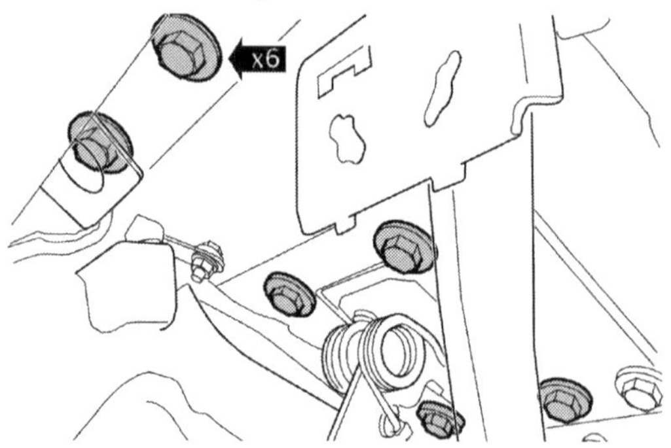

4. Remove 6 bolts securing clutch pedal bracket to body. Release closing panel sufficient to gain access to clutch bracket bolts.

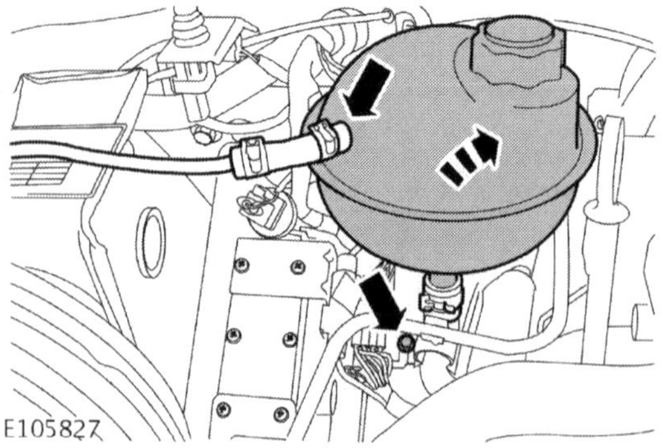

5. WARNING: A small amount of fluid loss in unavoidable, position an absorbant cloth or container to collect it.

Disconnect the expansion tank return hose and plug connections. Remove bolt and tie aside expansion tank.

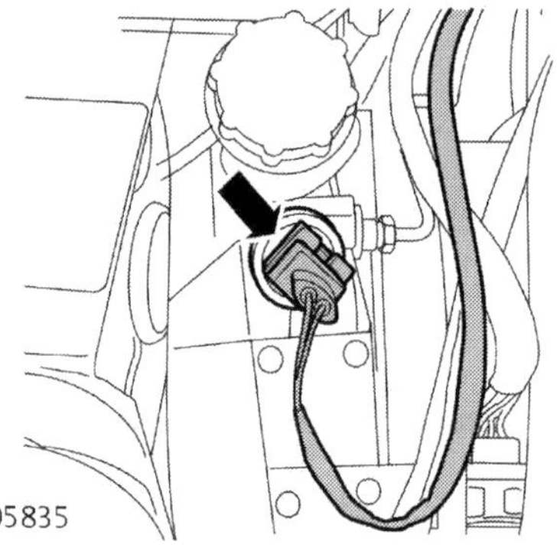

6. Disconnect multiplug clutch pedal sensor.

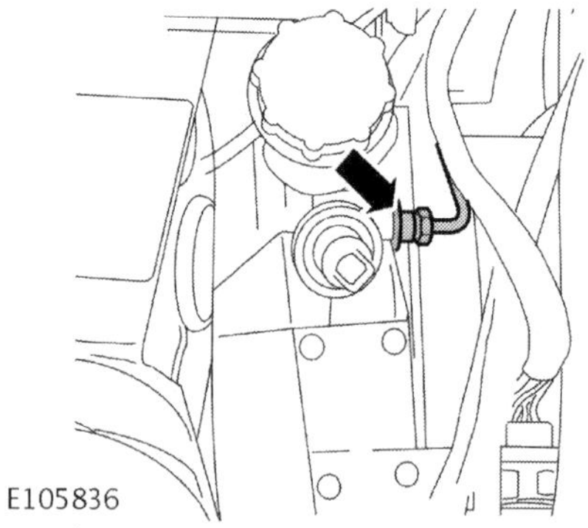

7. WARNING: A small amount of fluid loss is unavoidable, position an absorbant cloth or container to collect it.

CAUTIONS:

Make sure that all openings are sealed. Use new blanking caps.

Brake fluid will damage paint finished surfaces. If spilled, immediately remove fluid and clean area with water.

Loosen pipe union and release clutch master cylinder fluid pipe.

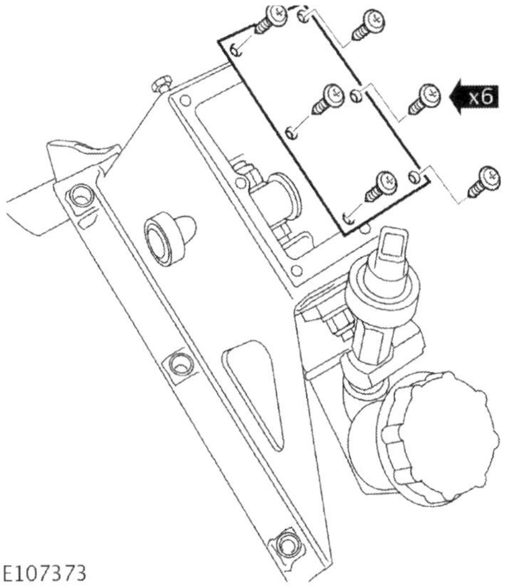

E107373

8. Remove the 6 screws, remove pedal box top cover and discard gasket.

9. **Vehicles With Anti-lock Brake System (ABS).** Remove the brake pipe retaining clip and bracket.

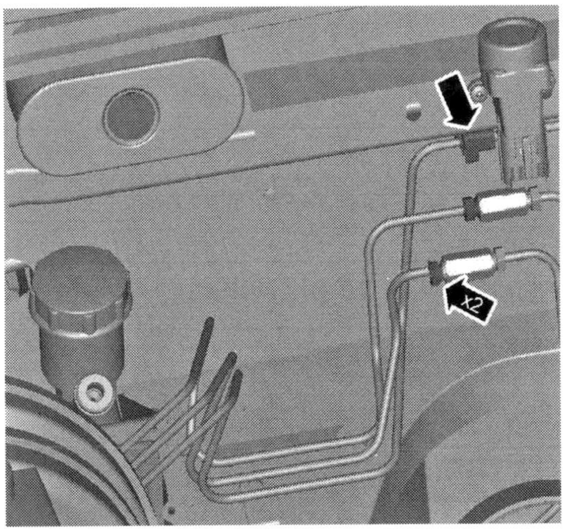

10. ⚠ WARNING: A small amount of fluid loss is unavoidable, position an absorbant cloth or container to collect it.

Release the 3 brake pipes. Loosen the 2 brake pipe unions at intermediate joints and release the pipes. Release brake pipe from the panel clip.

11. ⚠ WARNING: A small amount of fluid loss is unavoidable, position an absorbant cloth or container to collect it.

Release the 3 brake pipes at the Hydraulic Control Unit (HCU). Loosen the 3 brake pipe unions at HCU and release the pipes. Release brake pipes from the securing clip.

12. ⚠ CAUTION: Take extra care not to damage any surrounding components.

Remove the clutch pedal box and clutch master cylinder. Carefully reposition the brake pipes to aid removal.

13. ⚠ WARNING: A small amount of fluid loss is unavoidable, position an absorbant cloth or container to collect it.

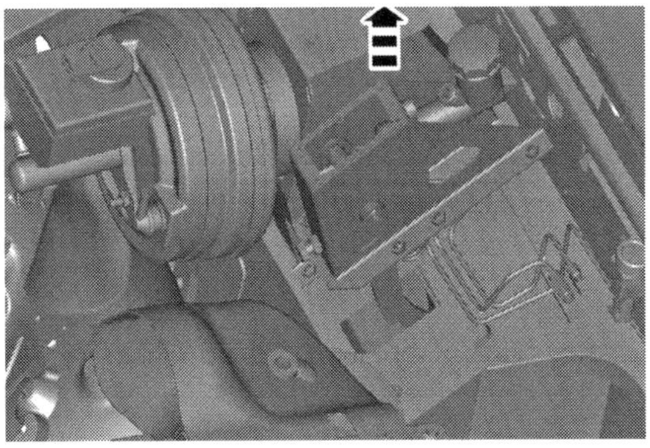

CAUTIONS:

⚠ Make sure that all openings are sealed. Use new blanking caps.

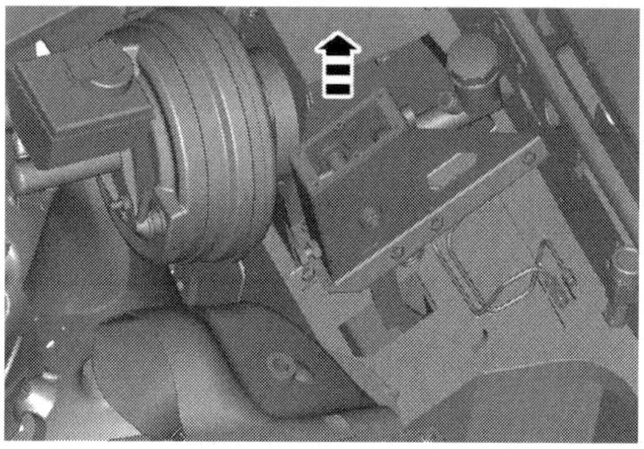

E139882

Brake fluid will damage paint finished surfaces. If spilled, immediately remove fluid and clean area with water.

Vehicles With Brake Pressure Control valve (BPCV). Release the brake pipes from the Brake Master Cylinder and the BPCV. Loosen the brake pipe union at intermediate joint and release the pipes. Release brake pipe from the panel clip. Remove the brake pipe retaining clip and bracket.

14. CAUTION: Take extra care not to damage any surrounding components.

Remove the clutch pedal box and clutch master cylinder. Carefully reposition the brake pipes to aid removal.

15. WARNING: A small amount of fluid loss is unavoidable, position an absorbant cloth or container to collect it.

CAUTIONS:

Make sure that all openings are sealed. Use new blanking caps.

E139880

⚠ Brake fluid will damage paint finished surfaces. If spilled, immediately remove fluid and clean area with water.

Vehicles With-Out ABS Or BPCV. Release the 2 brake pipes from the Brake Master Cylinder Loosen the brake pipe union at intermediate joint and remove the pipe. Release brake pipe from the panel clip. Remove the brake pipe retaining clip and bracket.

16. ⚠ CAUTION: Take extra care not to damage any surrounding components.

Remove the clutch pedal box and clutch master cylinder. Carefully reposition the brake pipe to aid removal.

17. **All Vehicles.** Move the clutch pedal to the fully down position. Remove nut securing push rod to clutch pedal trunnion.

18. Remove clutch pedal sensor and pipe union from master cylinder and discard sealing washers. Remove 2 nuts and bolts, and remove clutch master cylinder from pedal bracket.

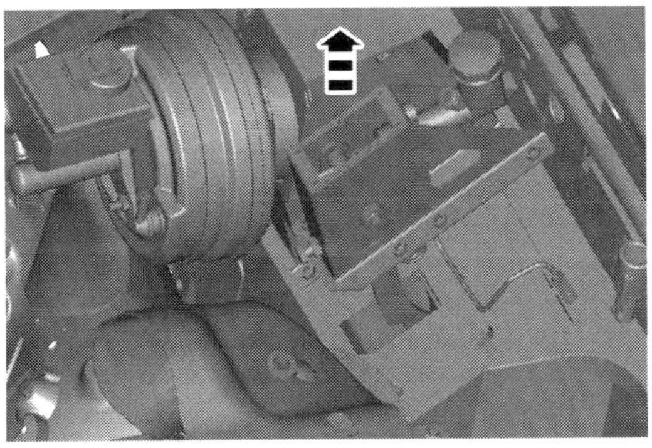

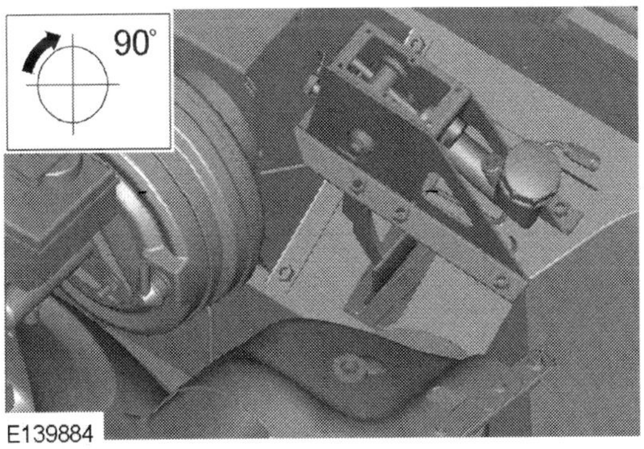

E139884

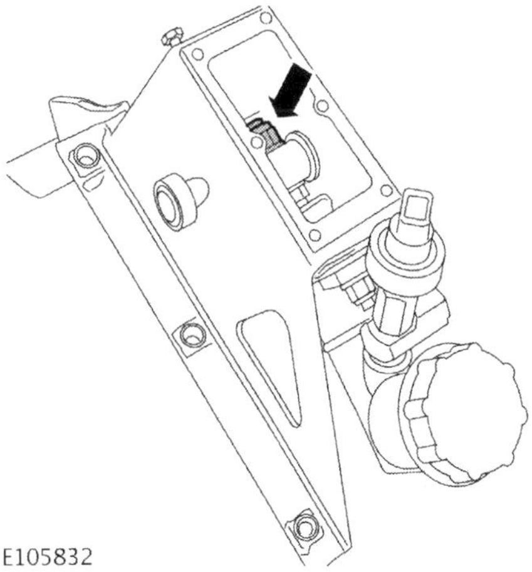

E105832

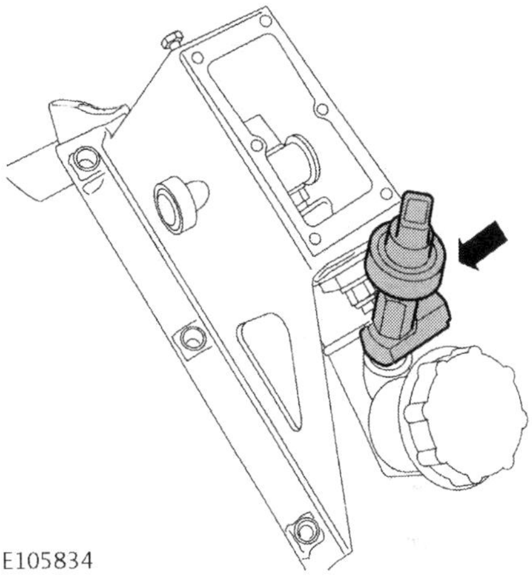

E105834

Installation

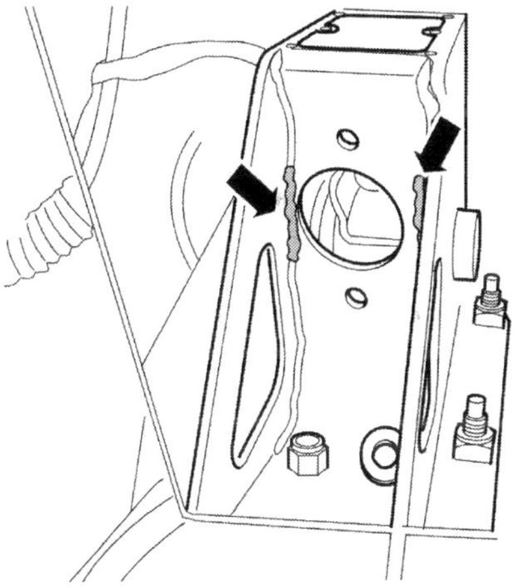

1. **All Vehicles.** Check seal on pedal box at points indicated and seal with RTV if necessary.

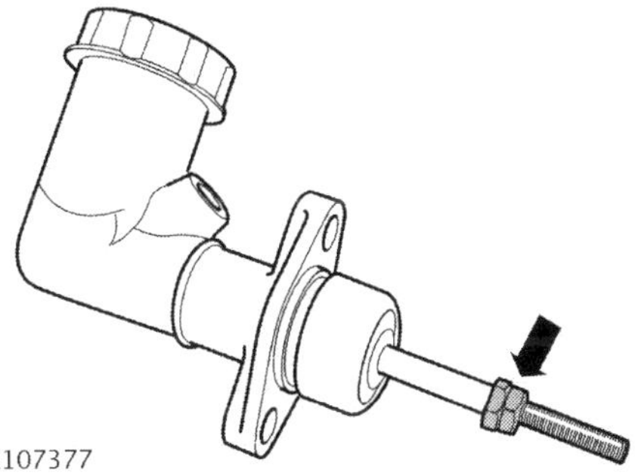

E107377

2. Install new lock nuts to master cylinder.

Renew Part: Clutch master cylinder lock nut

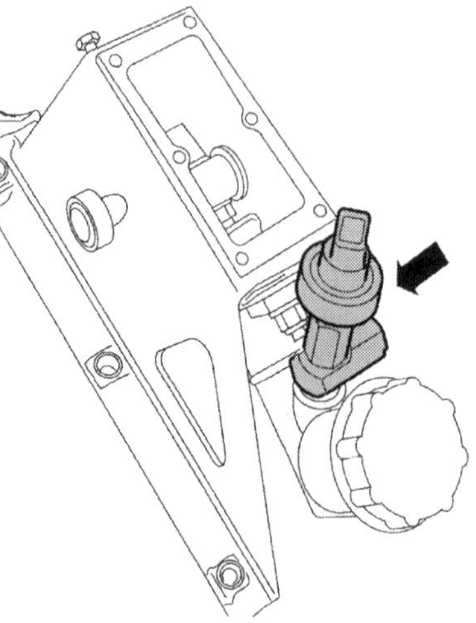

E105834

3. Clean sensor and pipe union. Install new gasket. Position master cylinder to pedal bracket and align push rod to pedal trunnion. Install the clutch master cylinder to the clutch pedal bracket and tighten the nuts and bolts. Install clutch pedal sensor and pipe to master cylinder using new sealing washers and tighten.

Renew Part: Clutch pedal sensor washer(s)

Renew Part: Clutch master cylinder gasket

Clutch pedal bracket bolts and nuts 23 Nm

Clutch pedal sensor 15 Nm

4. ⚠ CAUTION: Take extra care not to damage any surrounding components.

Vehicles With-Out ABS Or BPCV. Install the clutch pedal box and clutch master cylinder. Carefully reposition the brake pipe to aid installation.

5. ⚠ CAUTION: Take extra care not to damage any surrounding components.

Vehicles With BPCV. Install the clutch pedal box and clutch master cylinder. Carefully reposition the brake pipes to aid installation.

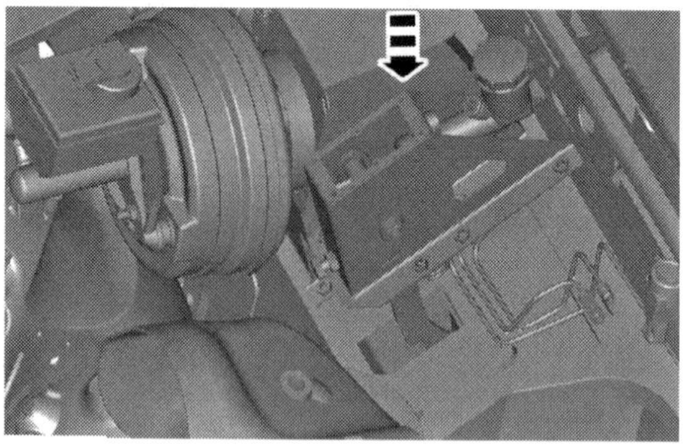

6. ⚠ CAUTION: Take extra care not to damage any surrounding components

Vehicles With ABS. Install the clutch pedal box and clutch master cylinder. Carefully reposition the brake pipes to aid installation.

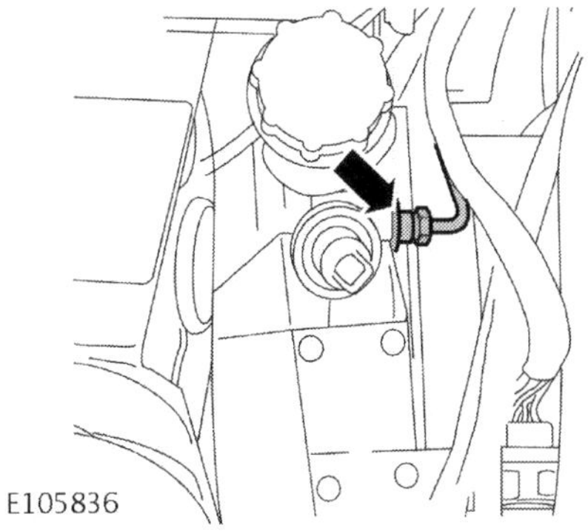

7. **All Vehicles.** Install clutch master cylinder fluid pipe. Clean clutch pipe union. Align pipe and tighten union.

   Torque: 15 Nm

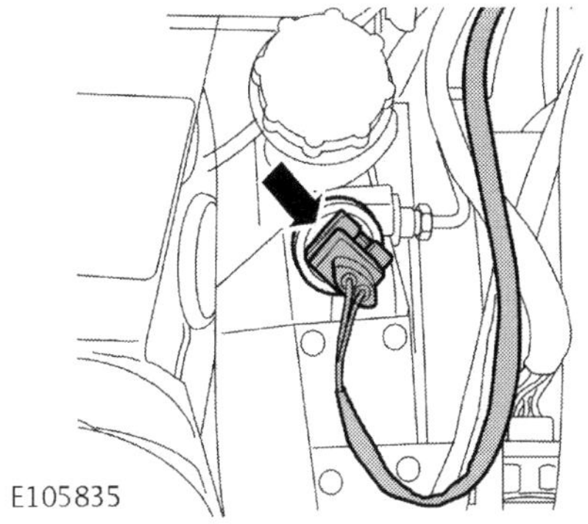

8. Connect multiplug to clutch pedal sensor.

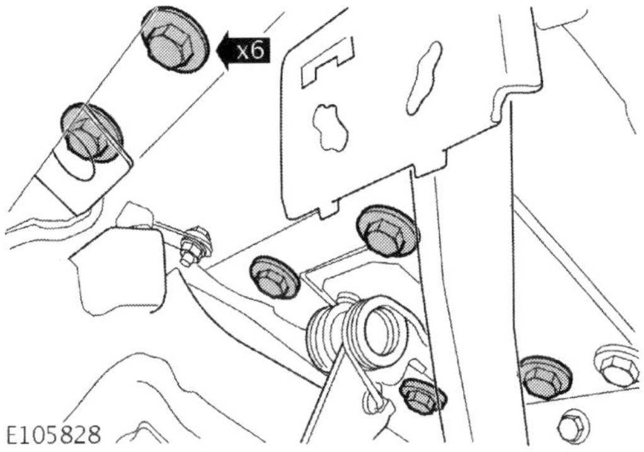

9. Install the 6 bolts securing clutch pedal bracket to body. Tighten the bolts.

   Torque: 23 Nm

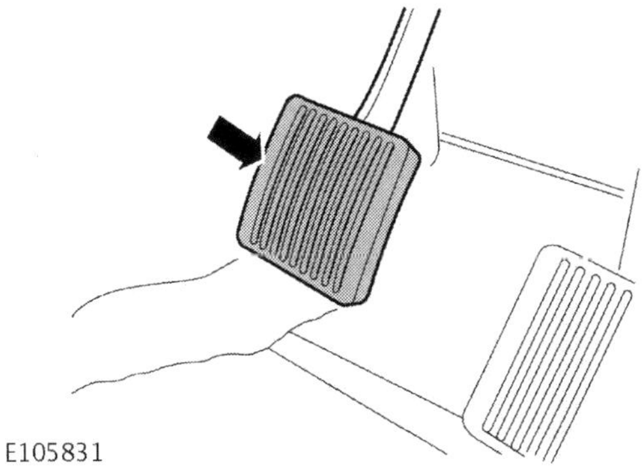

10. Install clutch pedal rubber.

11. Position closing panel and tighten screws to lower 'A' pillar.

12. Install the 2 stoplamp switches.

Refer to: Stoplamp Switch (417-01 Exterior Lighting, Removal and Installation)

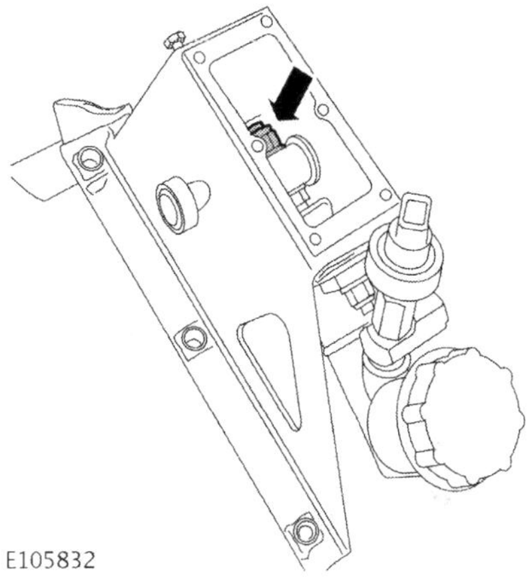

13. Install and lightly tighten nut securing push rod to clutch pedal trunnion.

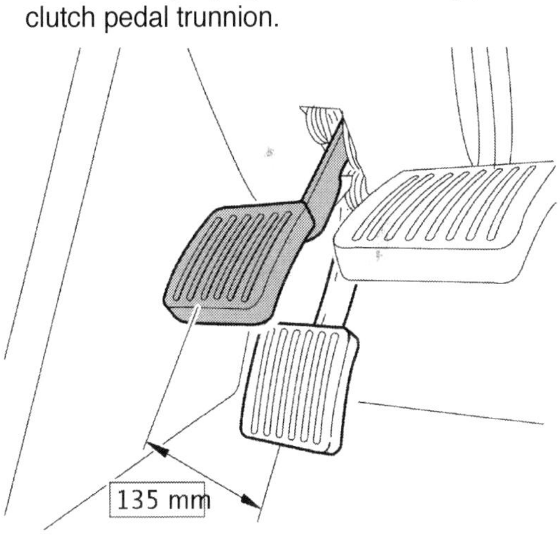

14. Check distance from lower edge of clutch pedal to down position. Correct measurement is 135 mm (5.31 in).

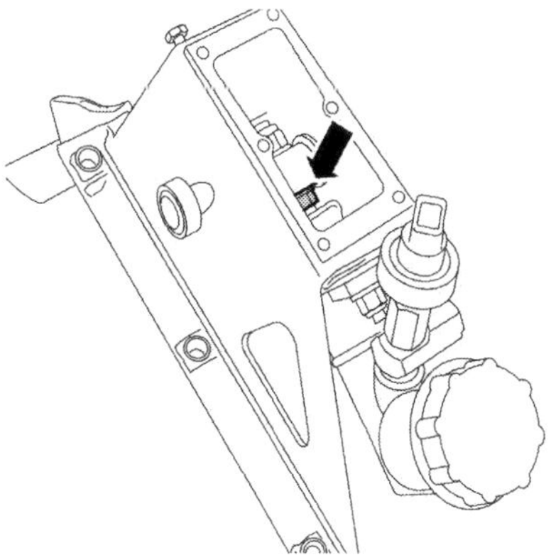

15. Adjust master cylinder push rod lock nuts until 135mm (5.31 in) is achieved. Tighten lock nuts. Move the clutch pedal to the fully down position. Tighten nylon nut securing push rod to clutch pedal trunnion.

    Locknuts 23 Nm

    Nylon 23 Nm

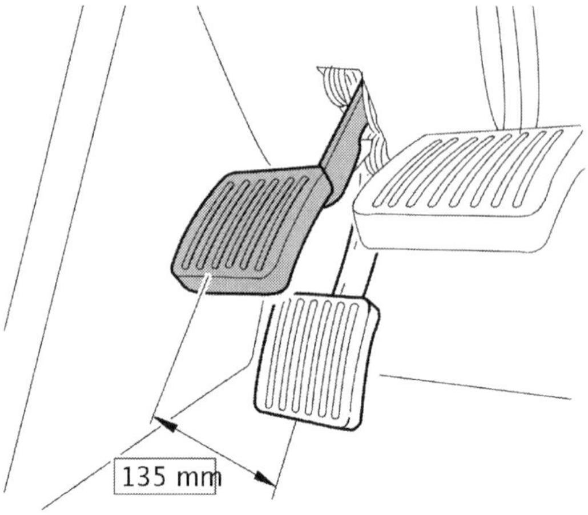

16. Check pedal travel is to correct measurement.

17. Bleed the clutch hydraulic system.

Refer to: Clutch System Bleeding (308-00 Manual Transmission/Transaxle and Clutch - General Information, Vehicles With: MT82 6-Speed Manual Transmission General Procedures)

18. **Vehicles With-Out ABS Or BPCV.** Install the 2 brake pipes to the Brake Master Cylinder and tighten the 2 brake pipe unions. Install the brake pipe union at intermediate joint and tighten. Secure brake pipe to the panel clip. Install the brake pipe retaining clip and bracket.

    Brake pipes 15 Nm

    Brake pipe at intermediate joint 15 Nm

19. **Vehicles With BPCV.** Install the brake pipes at the Brake Master Cylinder and the BPCV and tighten the 2 brake pipe unions. Install the brake pipe union at intermediate joint and tighten. Secure brake pipe to the panel clip. Install the brake pipe retaining clip and bracket.

    Brake pipes 15 Nm

    Brake pipes at intermediate joint 15 Nm

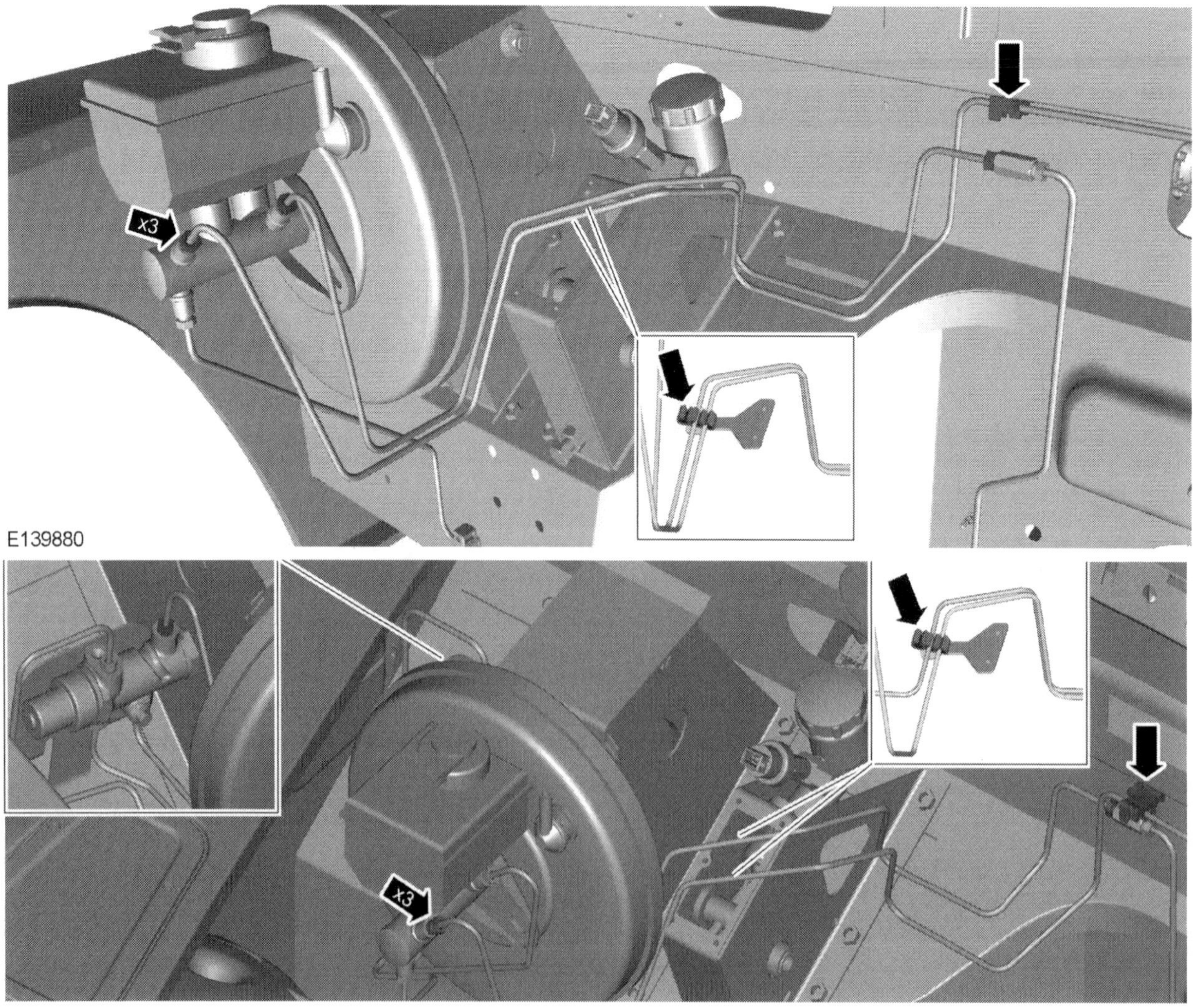

E139880

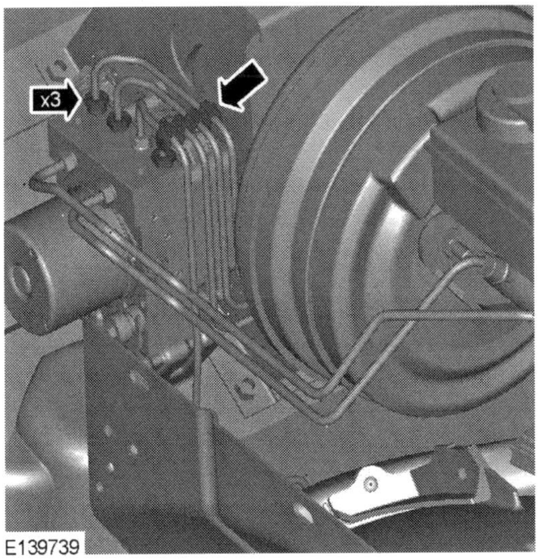

E139739

20. **Vehicles With ABS.** Install the 3 brake pipes at the Hydraulic Control Unit (HCU). Tighten the 3 brake pipe unions. Secure brake pipes to the securing clip.

    Torque: 15 Nm

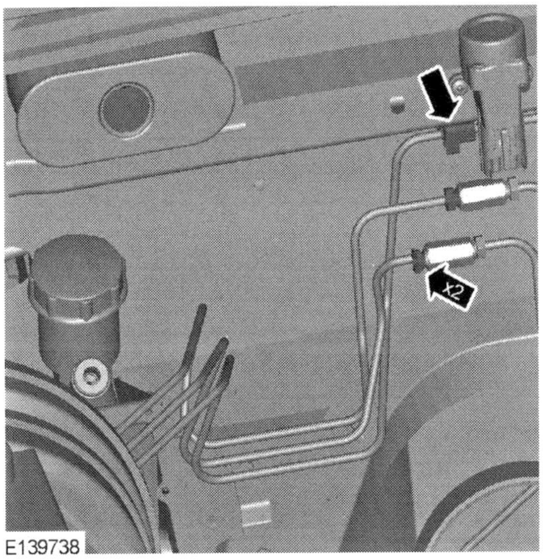

E139738

21. Install the 3 brake pipes. Tighten the 2 brake pipe unions. Secure brake pipe to the panel clip.

    Torque: 15 Nm

22. Install the brake pipe retaining clip bracket.

23. **All Vehicles.** Install a new gasket to cover plate and secure with the 6 screws. Tighten the screws.

Renew Part: Clutch pedal box cover gasket

Torque: 3 Nm

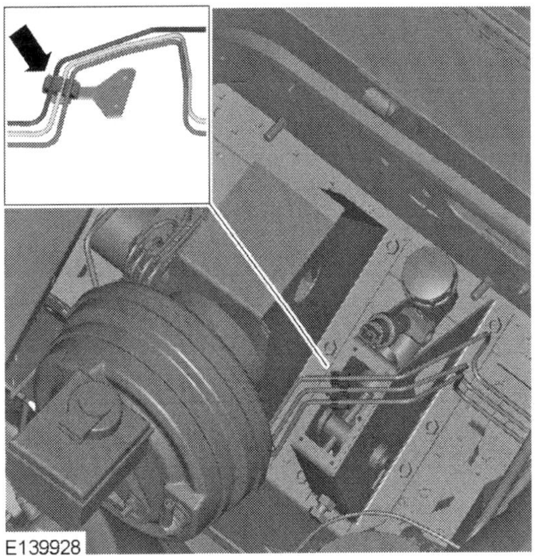

E139928

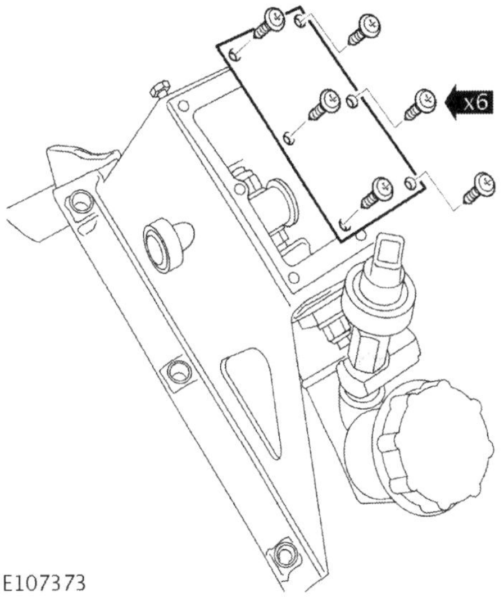

E107373

24. ⚠ NOTE: Land Rover approved diagnostic equipment must be used to operate the Antilock Brake System (ABS) solenoid valves and the return pump to ensure correct bleeding of the Hydraulic Control Unit (HCU) and all the vehicle brake circuits.

Using the Land Rover approved diagnostic equipment, bleed the complete brake system.

Refer to: Brake System Bleeding (206-00 Brake System - General Information, General Procedures)

25. Install the coolant expansion tank. Tighten the bolt. Connect the coolant expansion tank return hose. Check, and if necessary top up the coolant fluid level.

    Torque: 10 Nm

Clutch Slave Cylinder

Removal and Installation

Removal

1. ⚠ WARNING: Do not work on or under a vehicle supported only by a jack. Always support the vehicle on safety stands.

   Raise and support the vehicle.

2. Disconnect the battery ground cable.

Refer to: Battery Disconnect (414-01, General Procedures).

3. Remove the transmission.

Refer to: Transmission (308-03 Manual Transmission/ Transaxle, Vehicles With: MT82 6-Speed Manual Transmission Removal)

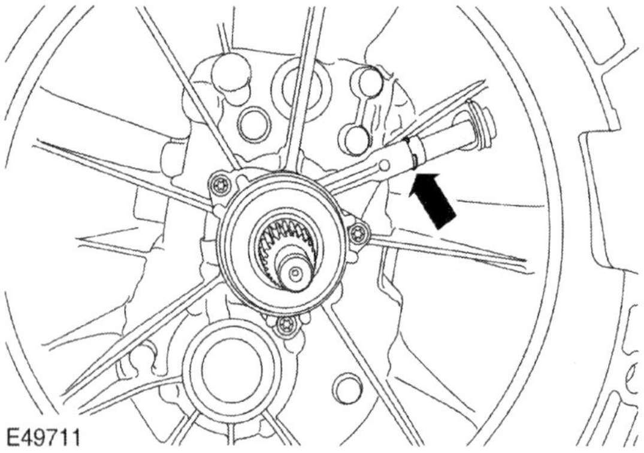

4. Disconnect the fluid pipe.
   - Position an absorbent cloth to collect any fluid spillage.
   - Remove the clip.

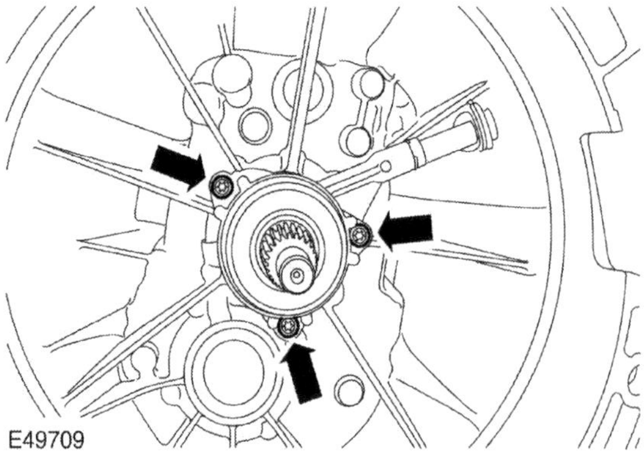

5. Remove the clutch slave cylinder.
   - Remove the 3 bolts.

Installation

1. Install the clutch slave cylinder.
   - Tighten the bolts to 10 Nm (7 lb.ft).
2. Connect the fluid pipe.
   - Install the clip.
3. Install the transmission.

Refer to: Transmission (308-03 Manual Transmission/ Transaxle, Vehicles With: MT82 6-Speed Manual Transmission Installation)

4. Connect the battery ground cable.

Refer to: Battery Connect (414-01, General Procedures).

Manual Transmission/Transaxle Vehicles MT82 6-Speed Manual Transmission

Specifications

Lubricants, Fluids, Sealers and Adhesives

Item	Specification
Manual transmission fluid	WSD-M2C200-C
Super DOT 4 brake fluid	ESD-M6C57-A
Sealant	WSS-M2G348-A10
Grease	ESD-M1C220-A

Capacities

Capacities	Litres
MT82 transmission - Initial fill	2.4
MT82 transmission - Service fill	2.2

Description	Nm	lb-ft
Gearshift yoke securing nut	12	9
Clutch slave cylinder bolts	11	8
Reversing lamp switch	20	15
Transmission extension housing bolts	25	18
Transfer case bolts	45	33
Transmission bolts	40	30
RH transmission mount nut*	30	22
RH transmission mount bolts+	63	46
LH transmission mount nut*	30	22
LH transmission mount bolts+	63	46
Starter motor bolts	35	26
Fuel line support bracket nut	9	7
Clutch slave cylinder hose from the mounting bracket nut	25	18
Transmission wiring harness bracket nut and bolt	47	35
Transfer case breather pipe	15	11
RH earth cables to the transfer case bolt	9	7
LH earth cable to the transfer case nut	45	33
Rear driveshaft to the transfer case nut	50	37
Tail pipe to the intermediate pipe nut	25	18

*If re-using nuts on vehicles built prior to VIN732932, tighten to 48 Nm. If replacing nuts, tighten to 30 Nm.

+If re-using bolts on vehicles built prior to VIN732932, tighten to 85 Nm. If replacing bolts, tighten to 63 Nm.

COMPONENT LOCATION

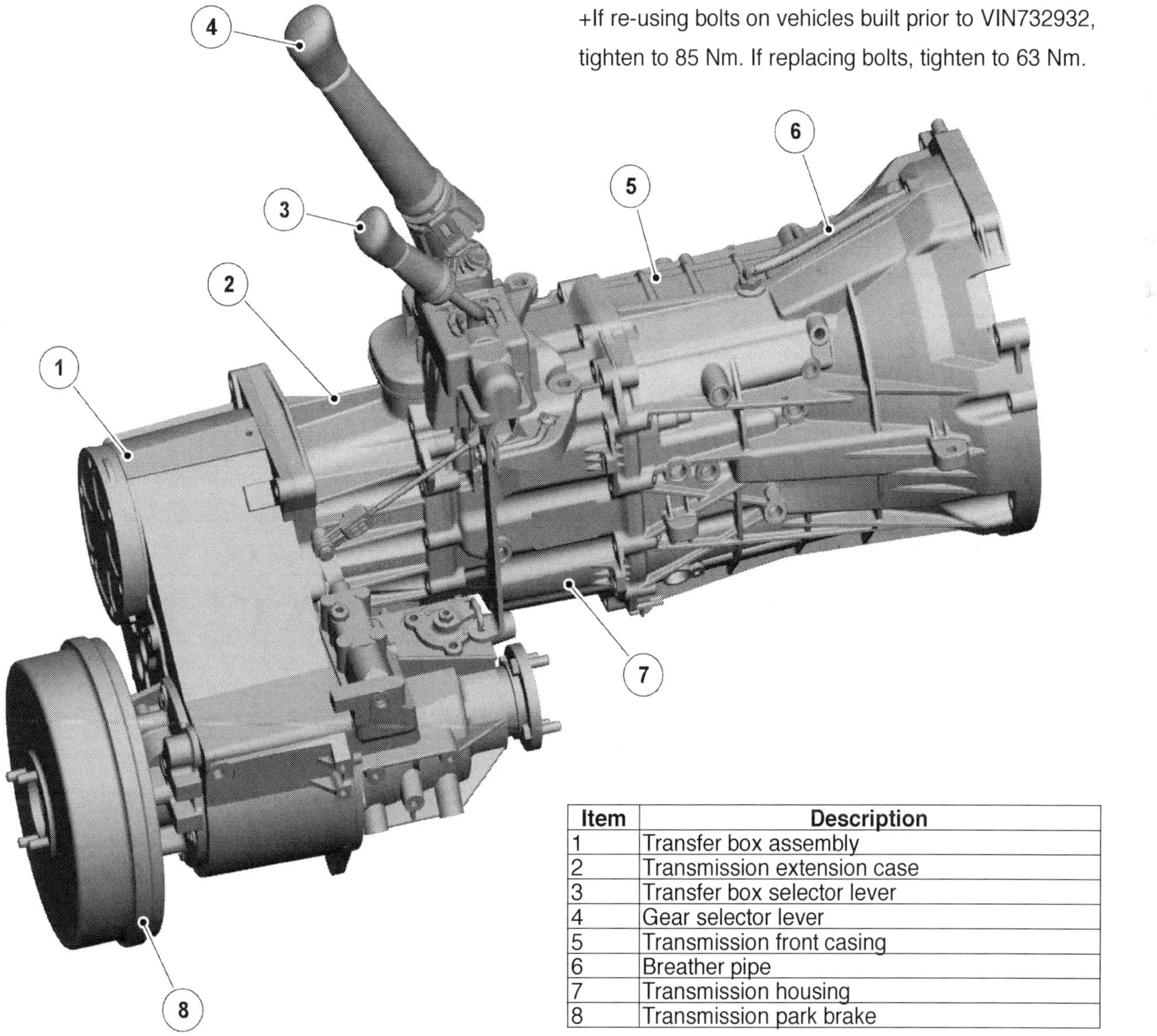

Item	Description
1	Transfer box assembly
2	Transmission extension case
3	Transfer box selector lever
4	Gear selector lever
5	Transmission front casing
6	Breather pipe
7	Transmission housing
8	Transmission park brake

Description and Operation

OVERVIEW

The MT82 manual transmission has 6 forward gears and a reverse. It is mounted longitudinally and has a maximum torque capacity of 360 Nm. The aluminium die-cast transmission housing is bolted to the transfer box via an aluminium die-cast extension case.

The 6th gear ratio has been selected as an overdrive for economy and comfort at higher vehicle speeds. Optimum gear steps ensure highly fuel-efficient utilisation of the engine torque. This 6-speed transmission provides a wide ratio spread supporting both economy and driveability (e.g. low speed maneuvering/trailer towing).

The name MT82 is derived from the distance between the 2 shafts in the transmission:

- **M** stands for manual
- **T** stands for transmission
- **82** is the distance between the 2 shafts in millimeters (mm)

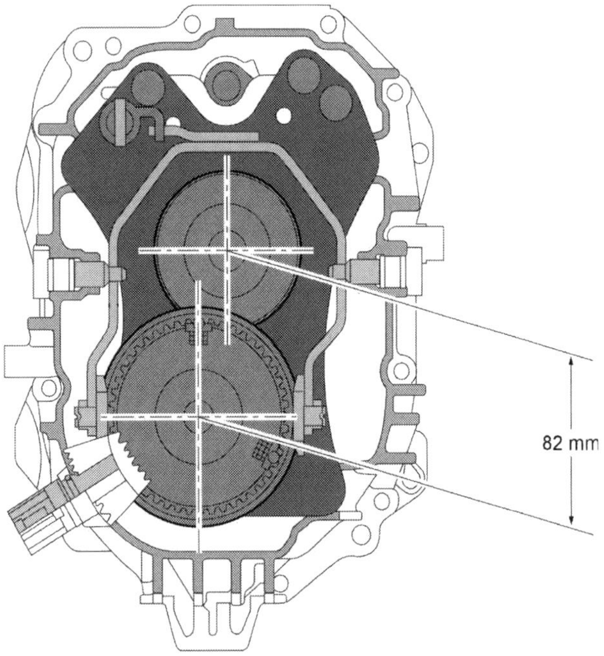

The transmission is a full for life unit and no level check is required at service unless a leak is present.

Technical Data

Input Torque	Ratios							Dry Weight	Oil fill	Oil Specification
	1st	2nd	3rd	4th	5th	6th	Rev			
360 Nm	5.441	2.840	1.721	1.223	1.00	0.742	4.935	50.8 kg	2.4L	WSD-M2C200-C

The input and output shafts are directly connected in 5th gear. This produces a gear ratio of 1:1.

Model Plate Label

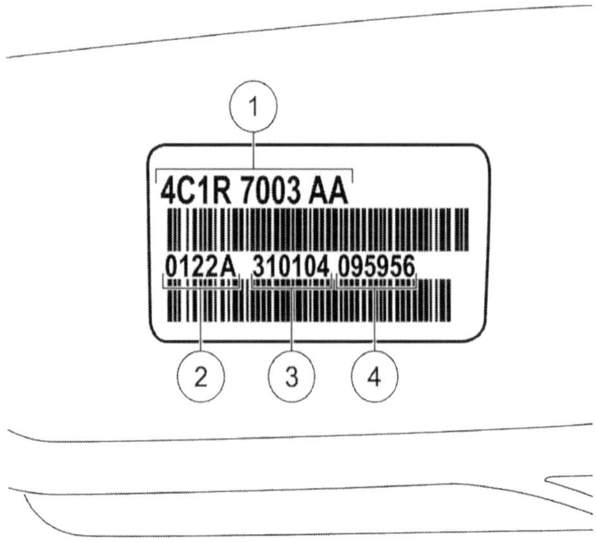

Item	Description
1	Replacement part number
2	Place of manufacture (Halewood)
3	Vehicle build date
4	Build time

The model plate is located on the Right Hand (RH) side of the transmission, near the driveshaft drive flange.

It is only used to identify the transmission. All spare parts orders are still made using the Vehicle Identification number (VIN).

INPUT SHAFT

The input shaft is rotationally mounted in the output shaft on the pilot bearing (3). In order to absorb the axial forces, the input shaft ball bearing (7) is additionally secured.

All the components on the input shaft can be serviced separately.

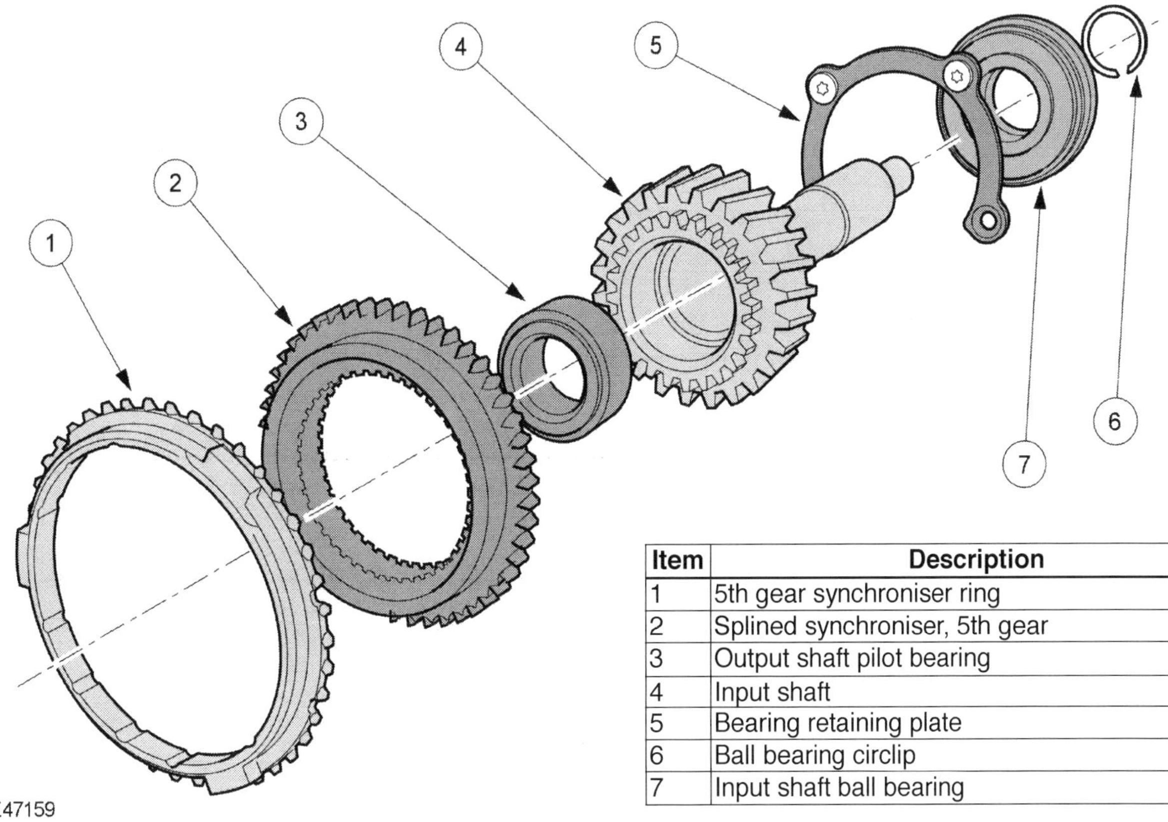

E47159

Item	Description
1	5th gear synchroniser ring
2	Splined synchroniser, 5th gear
3	Output shaft pilot bearing
4	Input shaft
5	Bearing retaining plate
6	Ball bearing circlip
7	Input shaft ball bearing

LAYSHAFT

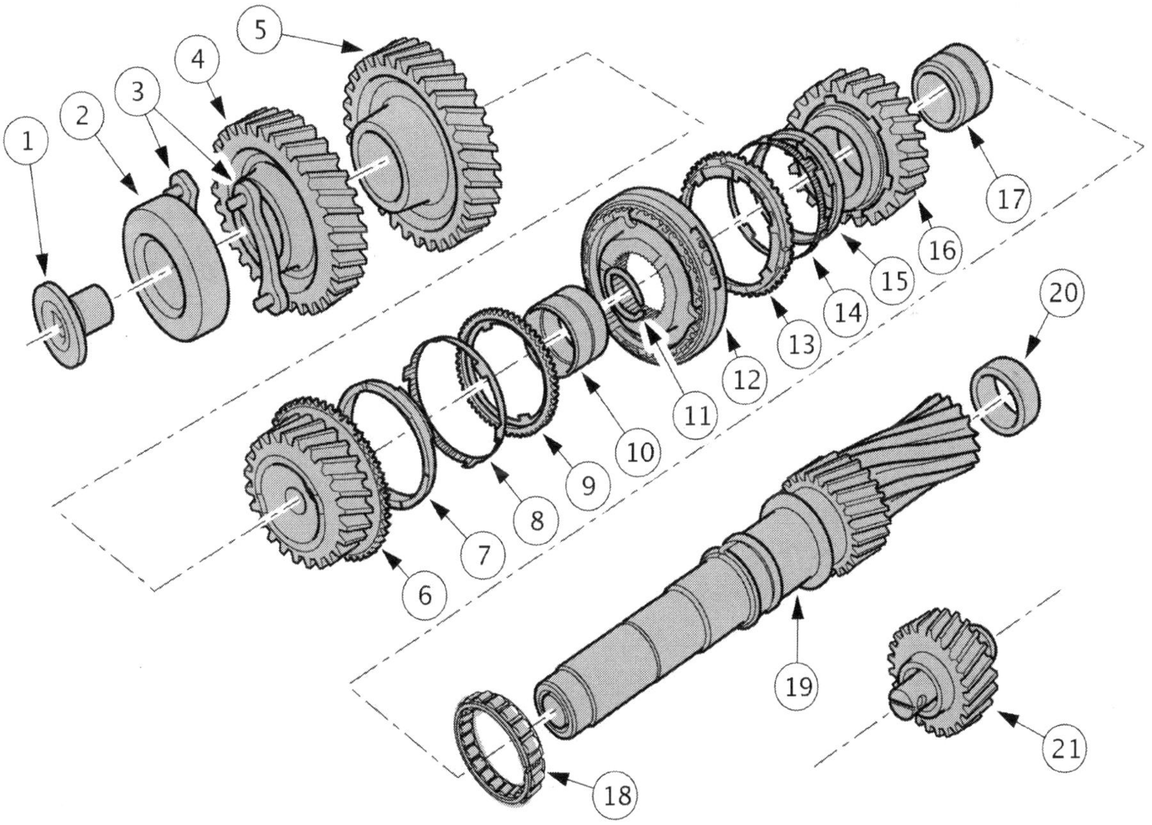

Item	Description
1	Retaining bolt
2	Ball bearing, layshaft
3	Retaining plate - bearing
4	Input pinion, layshaft
5	Gear - 6th gear
6	Gear wheel - 3rd gear
7	3rd gear synchroniser cone
8	Inner synchroniser ring
9	Outer synchroniser ring - 3rd gear
10	Needle bearing
11	Snap ring
12	3rd/4th gear synchroniser assembly
13	Outer synchroniser ring - 4th gear
14	Inner synchroniser ring
15	4th gear synchroniser cone
16	Gear wheel - 4th gear
17	Needle bearing
18	Centre bearing, layshaft
19	Layshaft
20	Roller bearing, layshaft
21	Reverse gear idler

The layshaft transfers the torque from the input shaft onto the output shaft. Gear wheels and gears and the 3rd/4th gear synchroniser assembly are located on the shaft. First, 2nd and reverse gears are an integral part of the shaft.

The layshaft gearwheels and gears can be replaced individually. Because of improved manufacturing tolerances, it is no longer necessary to change the gears and gear wheels in pairs.

The layshaft is a solid shaft. In order to prevent the shaft from moving axially, it is additionally secured with a retaining bolt (1) and a bearing retaining plate (3).

The rotational direction of the output shaft is reversed by the use of the reverse gear idler (21).

OUTPUT SHAFT

The output shaft transfers torque through the output flange, to an extension shaft connected to the transfer box. 1st, 2nd, 6th and reverse gear wheels are located on the output shaft. 3rd and 4th gears are an integral part of the output shaft.

In a similar way to the input shaft, there is a splined synchroniser (4) pushed on the 6th gear, gear wheel. This makes it possible to transfer the torque in 6th gear.

The output shaft gearwheels and gears can be replaced individually.

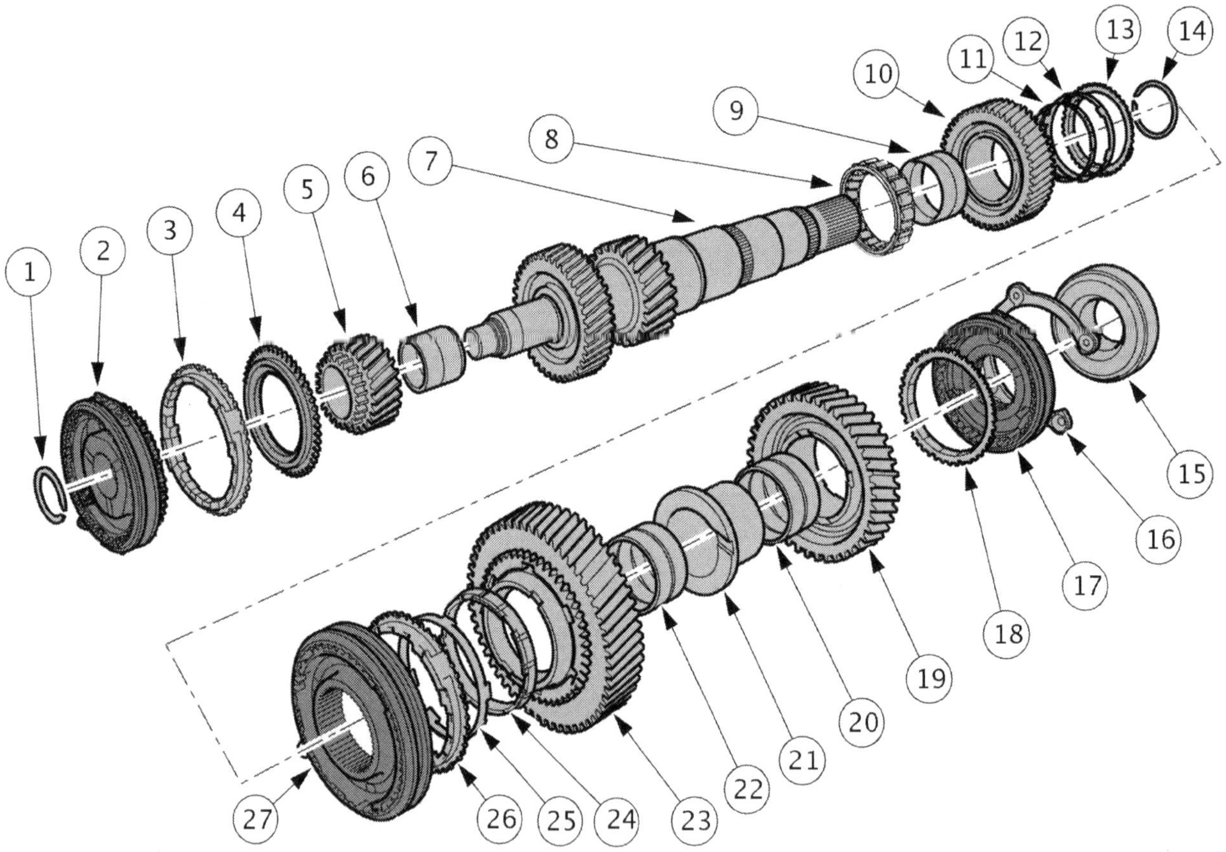

E47148

Item	Description
1	Snap ring
2	5th/6th gear synchroniser assembly
3	6th gear synchroniser ring
4	Splined synchroniser, 6th gear
5	Gear wheel - 6th gear
6	Needle bearing
7	Output shaft
8	Centre bearing - output shaft
9	Needle bearing
10	Gear wheel - 2nd gear
11	2nd gear synchroniser cone
12	Inner synchroniser ring
13	Outer synchroniser ring - 2nd gear
14	Snap ring
15	Ball bearing, output shaft
16	Retaining plate - bearing
17	Reverse gear synchroniser assembly
18	Reverse gear synchroniser ring
19	Gear wheel - reverse gear
20	Needle bearing
21	Inner race - needle bearing
22	Needle bearing
23	Gear wheel - 1st gear
24	1st gear synchroniser cone
25	Inner synchroniser ring - 1st gear
26	Outer synchroniser ring - 1st gear
27	1st/2nd gear synchroniser assembly

TRIPLE SYNCHRONISER ASSEMBLY

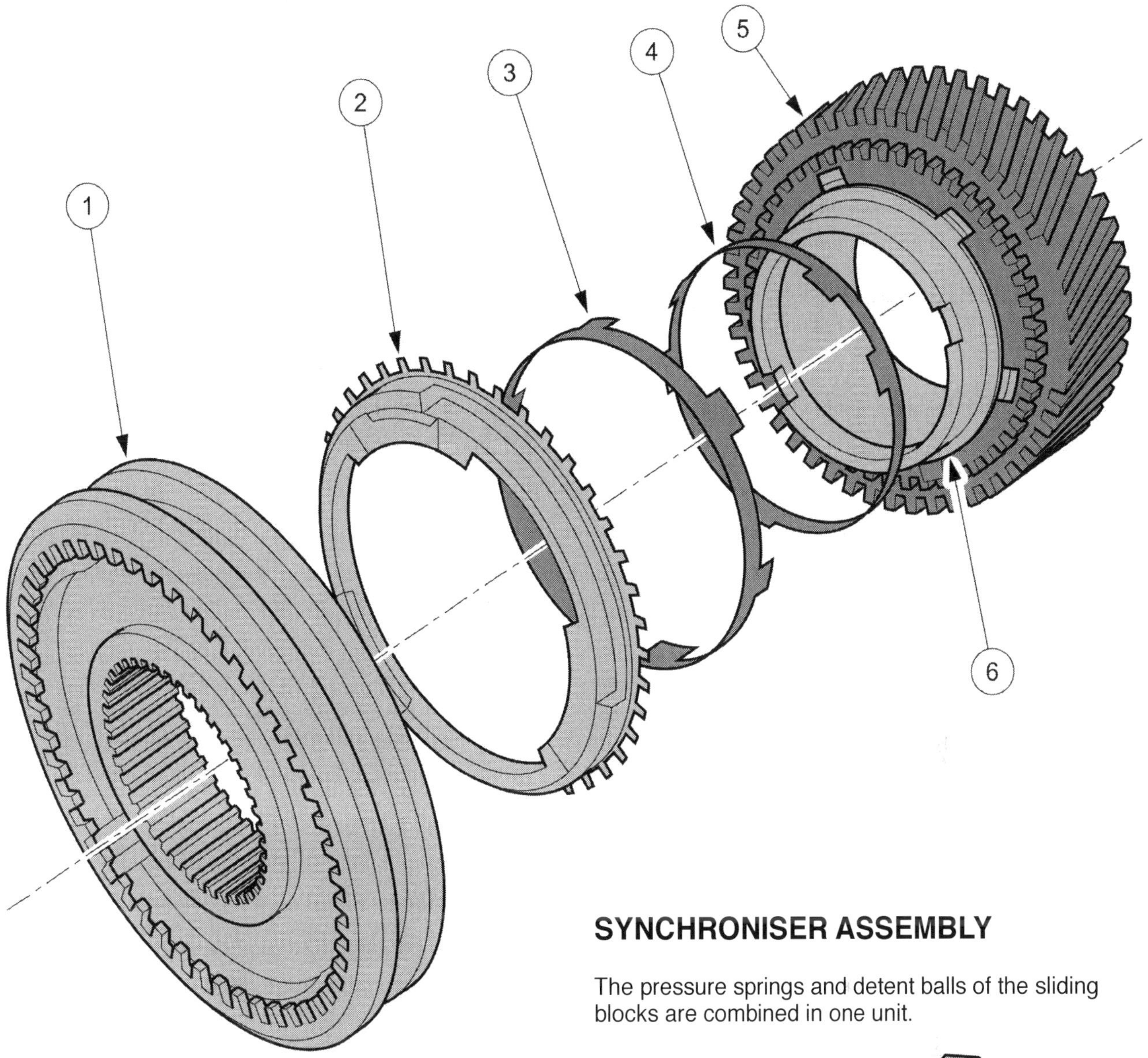

Item	Description
1	1st/2nd gear synchroniser assembly
2	Outer synchroniser ring
3	Inner synchroniser ring
4	Synchroniser cone
5	Gear wheel
6	Conical surface

The Synchronisation assembly consists of 3 friction surfaces. The total friction surface of triple synchronisation is considerably increased by the additional conical surface (6). This leads to a reduction in the force needed to change into 1st or 2nd gear.

As the conical surface is part of the gear wheel, there is no need for an additional synchroniser ring.

SYNCHRONISER ASSEMBLY

The pressure springs and detent balls of the sliding blocks are combined in one unit.

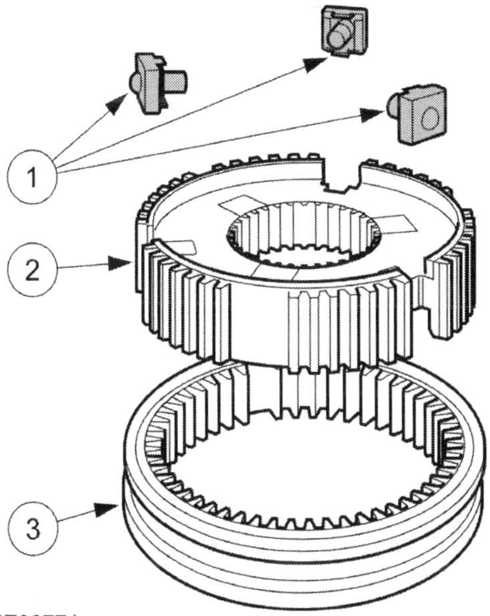

TIE38774

Item	Description
1	Sliding block assembly
2	Synchroniser hub
3	Sliding collar

REVERSE GEAR IDLER

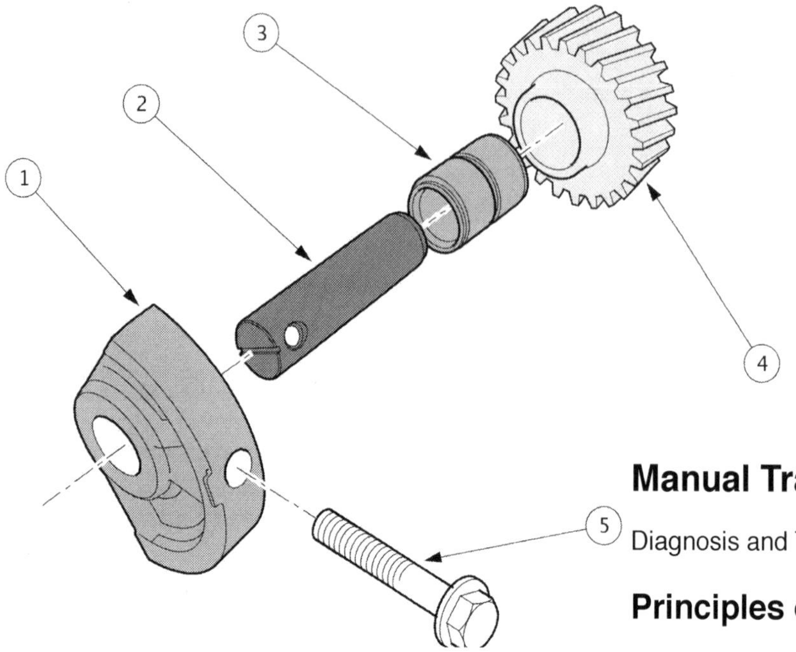

Item	Description
1	Mounting
2	Reverse Gear Idler Shaft
3	Needle Bearing
4	Reverse Gear Idler
5	Retaining Bolt - Reverse Gear Mounting

The reverse gear idler allows the direction of rotation of the output shaft to be reversed. The reverse gear idler turns on a needle bearing, which runs on the reverse gear idler shaft. The shaft is retained by the mounting (1) and a locating bore in the transmission housing.

In order to absorb the radial forces, the reverse gear idler runs on an additional mounting.

If the reverse gear idler can be replaced as an individual unit.

Manual Transmission

Diagnosis and Testing

Principles of Operation

For a detailed description of the manual transmission, refer to the relevant Description and Operation section in the workshop manual.

Refer to: Manual Transmission (308-03 Manual Transmission/Transaxle, Vehicles With: MT82 6-Speed Manual Transmission Description and Operation)

Inspection and Verification

⚠ CAUTION: Diagnosis by substitution from a donor vehicle is NOT acceptable. Substitution of control modules does not guarantee confirmation of a fault, and may also cause additional faults in the vehicle being tested and/or the donor vehicle.

△ NOTE: Check and rectify basic faults before beginning diagnostic routines involving pinpoint tests.

1. Visually inspect for obvious signs of damage and system integrity

Visual Inspection

Mechanical
• Gear selector lever • Manual transmission

2. If an obvious cause for an observed or reported concern is found, correct the cause (if possible) before proceeding to the next step

3. If the cause is not visually evident, verify the symptom and refer to the Symptom Chart, alternatively check for Diagnostic Trouble Codes (DTCs) and refer to the DTC Index.

Symptom Charts

Symptom	Possible Causes	Action
Noisy gear change - all gears	• Gear selector lever binding	• Refer to the relevant section of the workshop manual and install a selector lever bush kit (LR052893)
Noisy gear change into 3rd gear when cold (no fault when warm)	• 3rd/4th gear synchronizer hub worn	• Refer to the relevant section of the workshop manual and install a new 3rd/4th gear synchronizer hub
Noisy gear change into 5th/6th gear (may be worse when cold)	• 5th/6th gear synchronizer hub worn	• Refer to the relevant section of the workshop manual and install a new 5th/6th gear synchronizer hub
Clonk, grinding or loss of drive	• Manual transmission extension shaft and coupling splines worn due to lack of lubrication	• Refer to the relevant section of the workshop manual and install a new manual transmission extension shaft and coupling. Apply anti-seize grease to entire surface if splines
Selection noise when changing from 2nd to 3rd gear	• Transmission wind up due to repeated gear shifting at low engine speed (insufficient time for release of wind up) • Driving on road with the differential lock engaged	• No repair necessary • Disengage the differential lock
Oil leak from the manual transmission selector rod detent cap(s)	• Selector rod detent cap(s) leaking	• Refer to the relevant section of the workshop manual and install new detent caps as necessary. Check the manual transmission fluid level
Oil leak from manual transmission output shaft seal	• Output shaft seal leaking	• Refer to the relevant section of the workshop manual and install a new output shaft seal. Check the manual transmission fluid level
Difficulty selecting gears - stiff gear change	• Gear selector lever binding	• Refer to the relevant section of the workshop manual and install a selector lever bush kit (LR052893)
Stuck in gear (no fault with clutch)	• Mainshaft rail deflection during a high side load during gear selection	NOTE: Refer to Technical Service Bulletins LTB00311 and LTB00373 for details of affected VIN ranges • Refer to Technical Service Bulletins LTB00311 and LTB00373 and install a new transmission selector lever kit
Difficulty selecting gears (particularly reverse) after installing a new manual transmission or a new transmission selector lever kit	• Gearshift yoke pin incorrectly installed • Plastic cup missing from the selector lever	NOTE: Refer to Technical Service Bulletins LTB00311 and LTB00373 for details of affected VIN ranges • Refer to Technical Service Bulletin LTB00311 or LTB00373 and check the gearshift yoke pin for correct orientation • Install the selector lever plastic cup

DTC Index

For a list of Diagnostic Trouble Codes (DTCs) that could be logged on this vehicle, please refer to Section 100-00.

Transmission Draining and Filling

General Procedures

1. WARNING: Do not work on or under a vehicle supported only by a jack. Always support the vehicle on safety stands.

Raise and support the vehicle.

2. NOTE: The oil should be drained when the transmission is warm and the vehicle is standing on a level surface.

Position a container to collect the oil.

3. Clean the area surrounding the transmission drain plug.

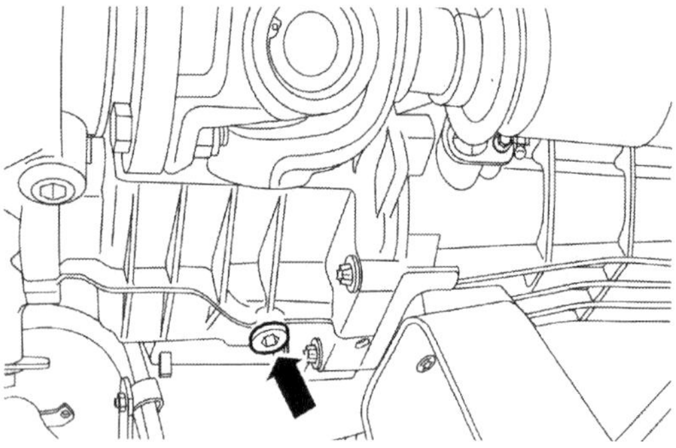

4. Remove the transmission drain plug.
    - Allow the oil to drain.
5. Clean transmission drain plug.
6. Install the transmission drain plug and clean any oil residue from the surrounding area.
    - Tighten to 50 Nm (37 lb.ft).
7. Clean the area surrounding the transmission filler plug.

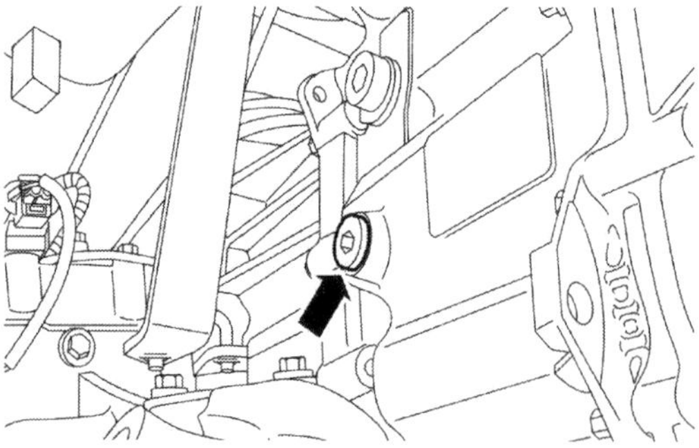

8. Remove the transmission filler plug.
9. Fill the transmission with the correct grade of oil.

Refer to: (308-03 Manual Transmission/Transaxle, Vehicles With: MT82 6-Speed Manual Transmission Specifications)

10. Install the transmission filler plug and clean any oil residue from the surrounding area.
    - Tighten to 35 Nm (26 lb.ft).
11. Remove the container.
12. Lower the vehicle.

Gearshift Control Shaft Seal

In-vehicle Repair

Removal

1. Disconnect the battery ground cable.

Refer to: Battery Disconnect (414-01 Battery, Mounting and Cables, General Procedures).

2. WARNING: Do not work on or under a vehicle supported only by a jack. Always support the vehicle on safety stands.

Raise and support the vehicle.

3. Remove the transfer case.

Refer to: Transfer Case (308-07 Transfer Case, Vehicles With: MT82 6-Speed Manual Transmission Removal)

4. Remove the transmission extension housing.
    - Remove the 10 bolts.

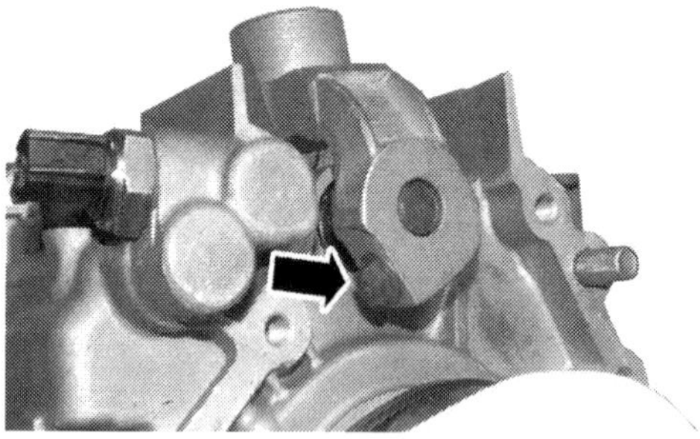

5. Remove the gearshift yoke nut.

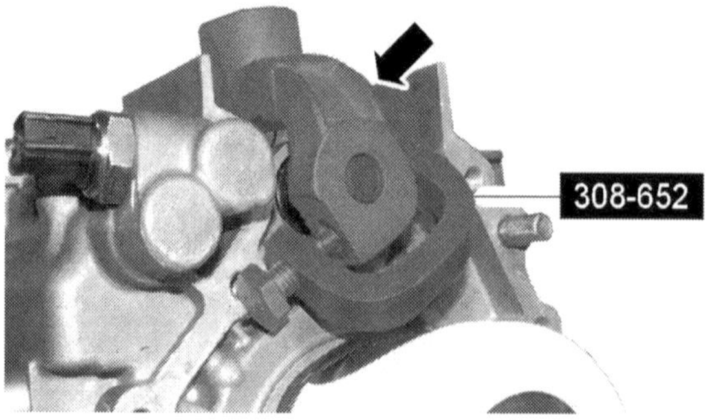

6. Using the special tool, remove the gearshift yoke.
    - Remove the pin.

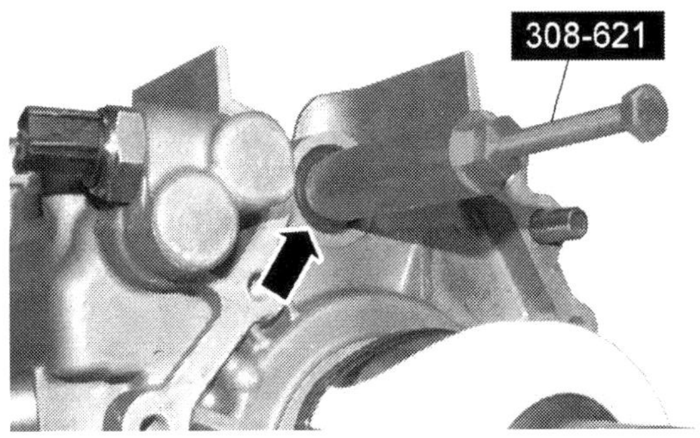

7. Using the special tool, remove the gearshift control shaft seal.

Installation

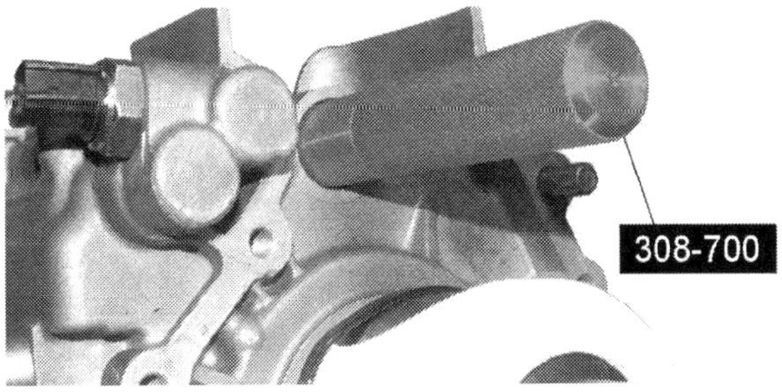

1. Using the special tool, install the gearshift control shaft seal.

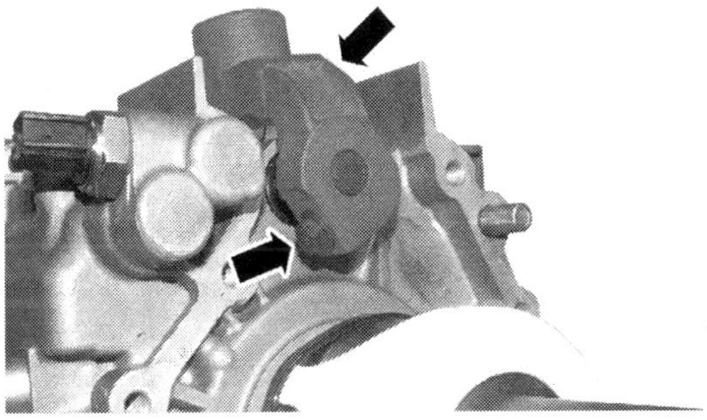

2. Install the gearshift selector yoke.
   - Tighten to 12 Nm (9 lb.ft).

3. Install the transmission extension housing.
   - Tighten to 25 Nm (18 lb.ft).

4. Install the transfer case.

Refer to: Transfer Case (308-07 Transfer Case, Vehicles With: MT82 6-Speed Manual Transmission Installation)

5. Connect the battery ground cable.

Refer to: Battery Connect (414-01 Battery, Mounting and Cables, General Procedures).

Input Shaft Seal

In-vehicle Repair

Special Tool(s)

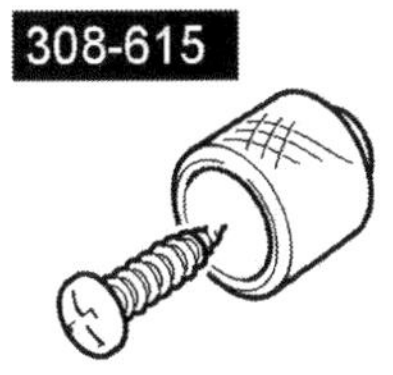 E86871	308-615 Remover, Seals
 E193517	100-012-01 Slide Hammer Adapter
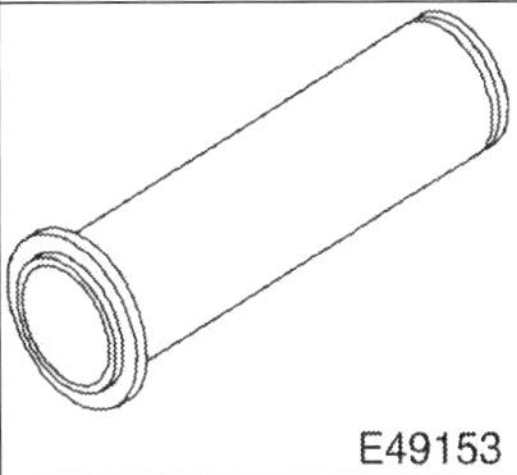 E49153	308-605 Installer, Input Shaft Seal

Removal

1. Disconnect the battery ground cable.

Refer to: Startup Battery Disconnect and Connect (414-01 Battery, Mounting and Cables, General Procedures)

2. ⚠ WARNING: Do not work on or under a vehicle supported only by a jack. Always support the vehicle on safety stands.

Raise and support the vehicle.

3. Drain the transmission.

Refer to: Transmission Draining and Filling (308-03 Manual Transmission/Transaxle, Vehicles With: MT82 6-Speed Manual Transmission General Procedures)

4. Remove the clutch slave cylinder.

Refer to: Clutch Slave Cylinder (308-02 Clutch Controls, Vehicles With: MT82 6-Speed Manual Transmission Removal and Installation)

5. Install the special tool into the seal.

   Special Tool(s): 308-615

6. Using the special tools, remove and discard the input shaft seal.

   Special Tool(s): 100-012-01 , 308-615

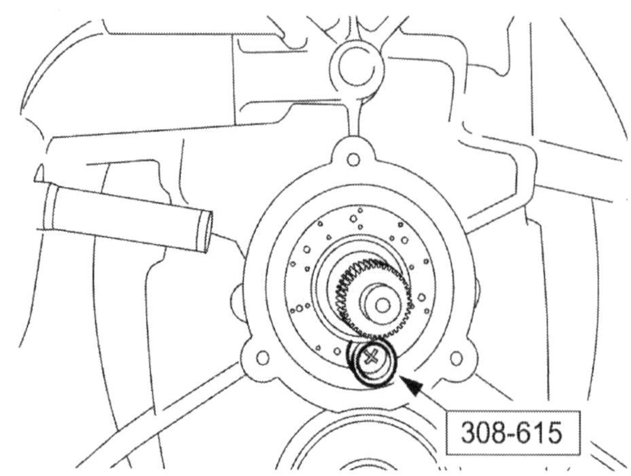

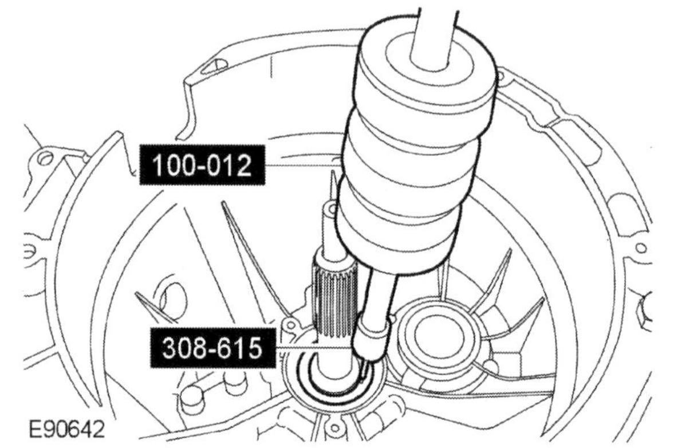

Installation

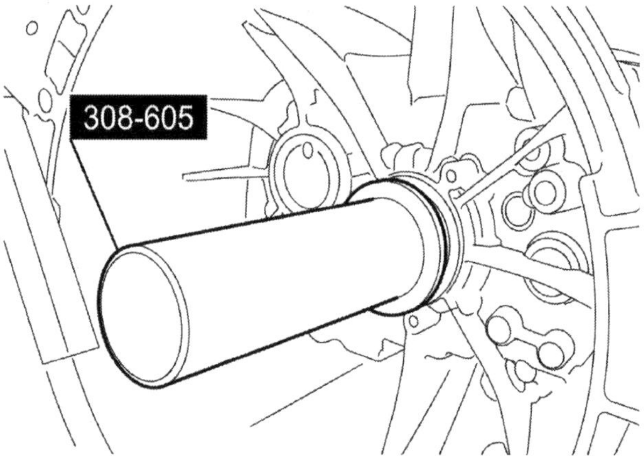

1. Using the special tool, install the new input shaft seal.

   Special Tool(s): 308-605

2. Install the clutch slave cylinder.

Refer to: Clutch Slave Cylinder (308-02 Clutch Controls, Vehicles With: MT82 6-Speed Manual Transmission Removal and Installation)

3. Fill the transmission.

Refer to: Transmission Draining and Filling (308-03 Manual Transmission/Transaxle, Vehicles With: MT82 6-Speed Manual Transmission General Procedures)

Output Shaft Seal

In-vehicle Repair

Special Tool(s)

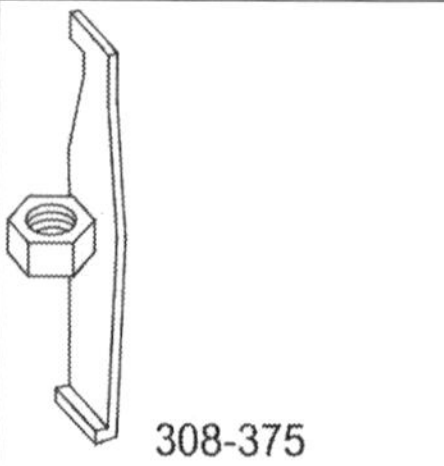 308-375	308-375 Remover, Input and Output Seal
 E193517	100-012-01 Slide Hammer Adapter
 E54135	100-012 Slide Hammer
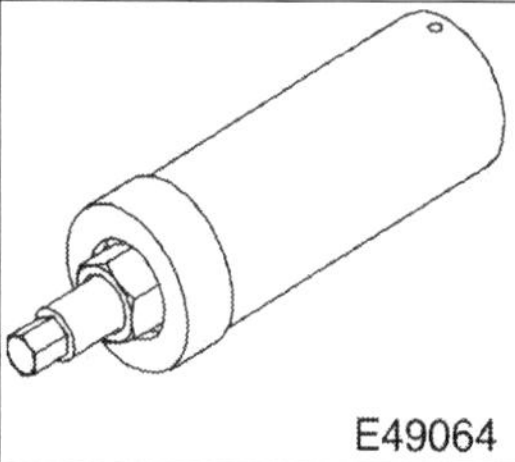 E49064	308-604 Installer, Output Drive Flange Seal

Removal

1. Disconnect the battery ground cable.

Refer to: Startup Battery Disconnect and Connect (414-01 Battery, Mounting and Cables, General Procedures)

2. Remove the gearshift lever.

Refer to: Selector Lever (308-06 Manual Transmission/Transaxle External Controls, Vehicles With: MT82 6-Speed Manual Transmission Removal and Installation)

3. ⚠ WARNING: Do not work on or under a vehicle supported only by a jack. Always support the vehicle on safety stands.

Raise and support the vehicle.

4. Drain the transmission.

Refer to: Transmission Draining and Filling (308-03 Manual Transmission/Transaxle, Vehicles With: MT82 6-Speed Manual Transmission General Procedures)

5. Remove the transfer case.

Refer to: Transfer Case (308-07 Transfer Case, Vehicles With: MT82 6-Speed Manual Transmission Removal)

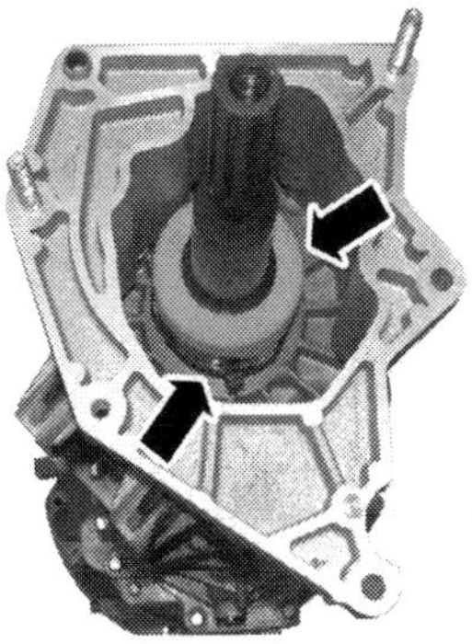

E91522

6. Remove the transmission extension shaft cover. Remove and discard the tie strap.

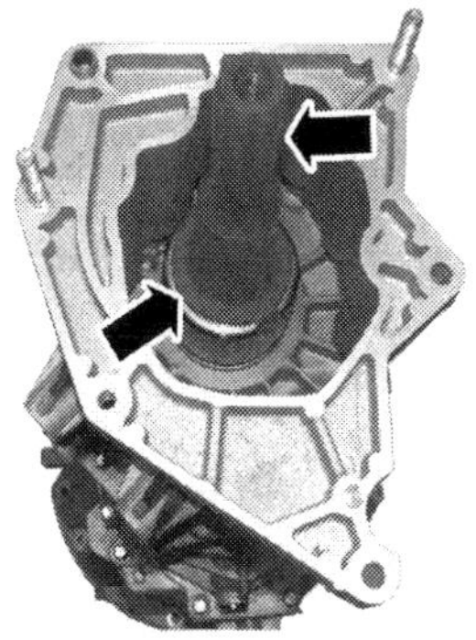

E91523

7. Using a suitable tool, remove the transmission extension shaft. Remove the seal.

E91524

8. Remove the transmission extension housing. Remove the 10 bolts.

9. Using the special tool 308-701, lock the output shaft flange. Remove the bolt.

10. Remove the special tool.

11. Using a suitable puller and special tool, remove the output shaft flange.

12. Using the special tools, remove the output shaft seal.

    Special Tool(s): 308-375 , 100-012-01 , 100-012

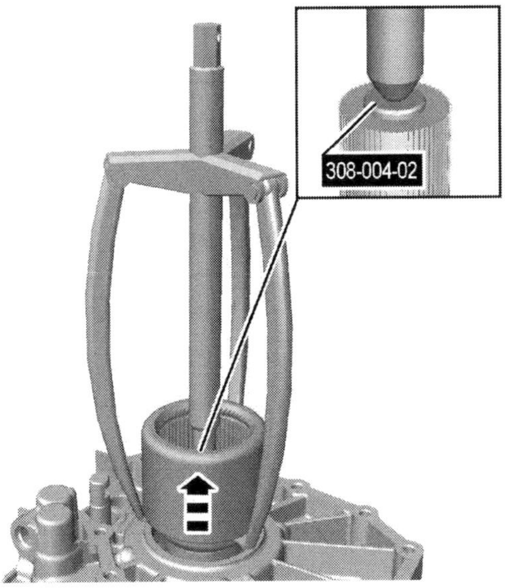

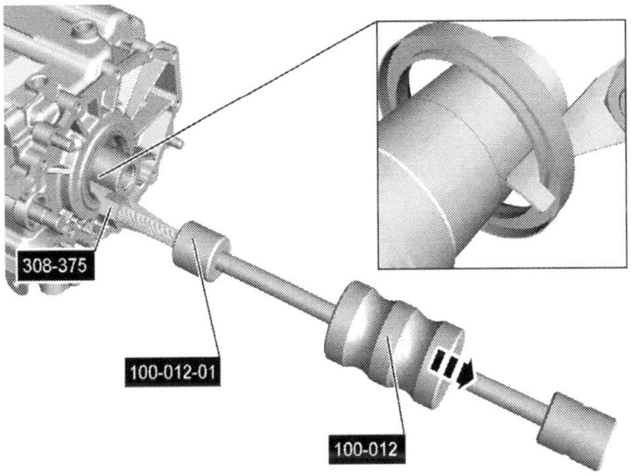

Installation

1. Using the special tool, install the output shaft seal.

   Special Tool(s): 308-604

2. △ NOTE: Note the position of the output flange, as shown in the illustration.

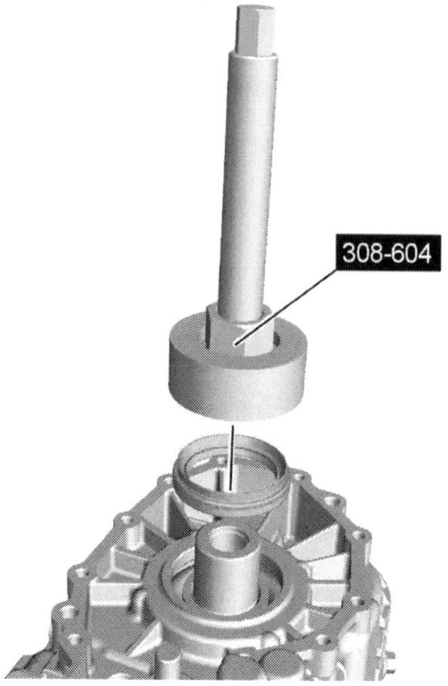

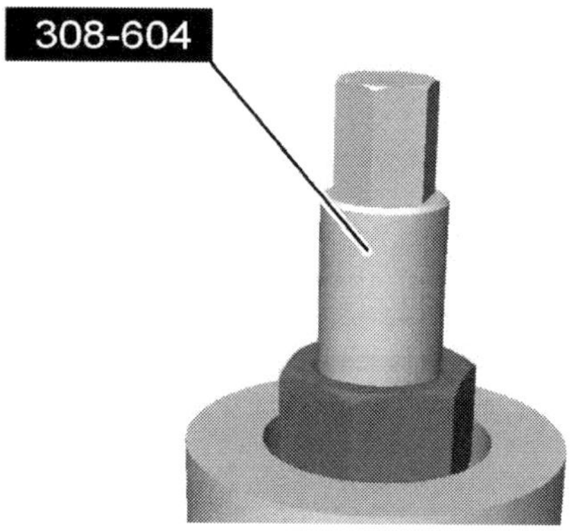

3. NOTE: Apply anti-seize grease to the splines prior to installation.

Using the special tool, install the transmission output flange bolt. Tighten the bolt, then remove the bolt. Apply thread locking compound grease. Tighten the bolt again.

Material: Thread Locking MS (WSK-M2G349-A7 /)

Torque:

Stage 1: 210 Nm

Stage 2: 180 Nm

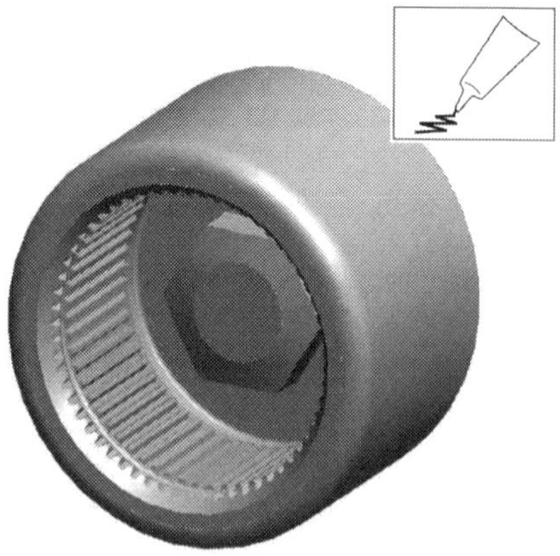

4. NOTE: Make sure that the component mating face is clean for debris and foreign material.

Apply between the circlip grove and base of the flange splines prior to installation of the transmission extension shaft.

Material: Weicon Grease (IYX500050 /)

5. NOTE: Make sure that the component mating face is clean for debris and foreign material.

Apply the entire length of the splines prior to installation of the transfer case.

Material: Weicon Grease (IYX500050 /)

6. Install the transmission extension shaft. Install the seal.

7. ⚠ NOTE: Make sure that the tie strap joint is between the fingers of the extension shaft seal cover and that it has been cut flush.

Install the transmission extension shaft cover. Install the cover. Install the tie strap.

8. Install the transmission extension housing. Tighten the bolts.

   Torque: 25 Nm

9. Install the transfer case.

Refer to: Transfer Case (308-07 Transfer Case, Vehicles With: MT82 6-Speed Manual Transmission Installation)

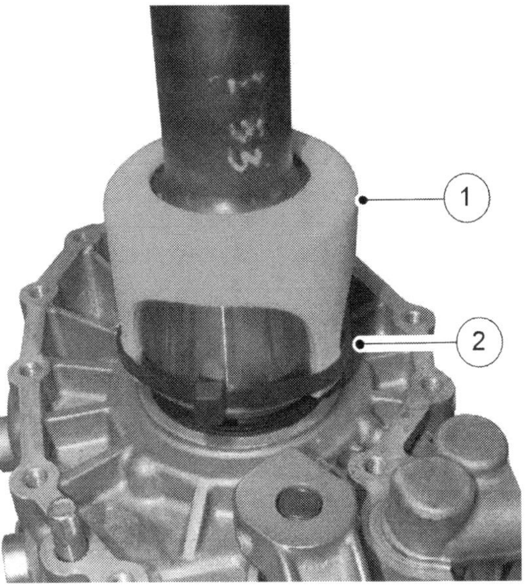

10. Fill the transmission.

Refer to: Transmission Draining and Filling (308-03 Manual Transmission/Transaxle, Vehicles With: MT82 6-Speed Manual Transmission General Procedures)

11. Install the gearshift lever.

Refer to: Selector Lever (308-06 Manual Transmission/Transaxle External Controls, Vehicles With: MT82 6-Speed Manual Transmission Removal and Installation)

12. Connect the battery ground cable.

Refer to: Startup Battery Disconnect and Connect (414-01 Battery, Mounting and Cables, General Procedures)

Selector Shaft Detents

Removal and Installation

Special Tool(s)

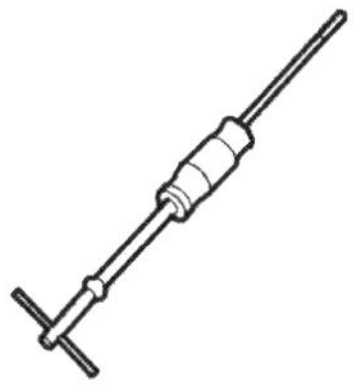 100012	Slide Hammer 100-012
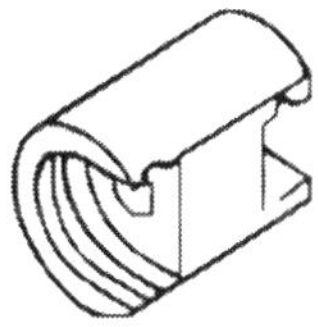	Remover, Detent 308-657
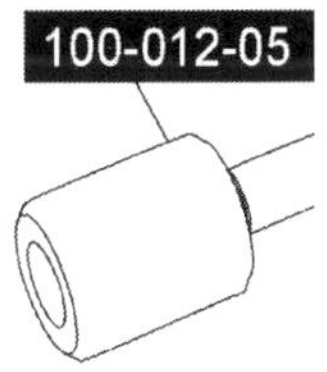 E91077	Adapter, Slide Hammer 100-012-05

General Equipment

Hammer	
Soft Drift	

Removal

1. Remove the transmission.

   Refer to: Transmission (308-03 Manual Transmission/Transaxle, Vehicles With: MT82 6-Speed Manual Transmission Removal)

2. ⚠ CAUTION: Make sure area surrounding component is clean.

Assemble the special tools.

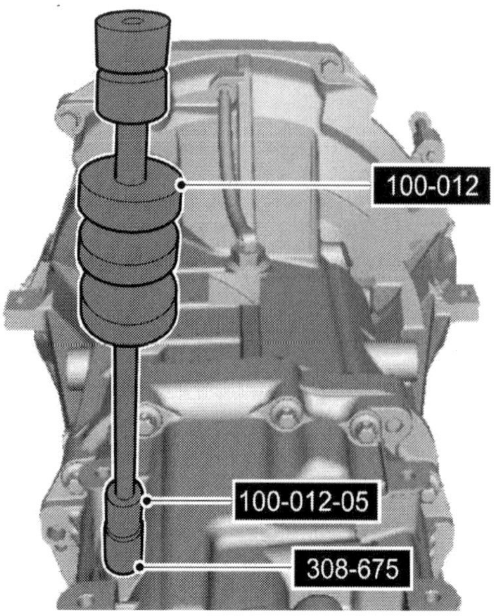

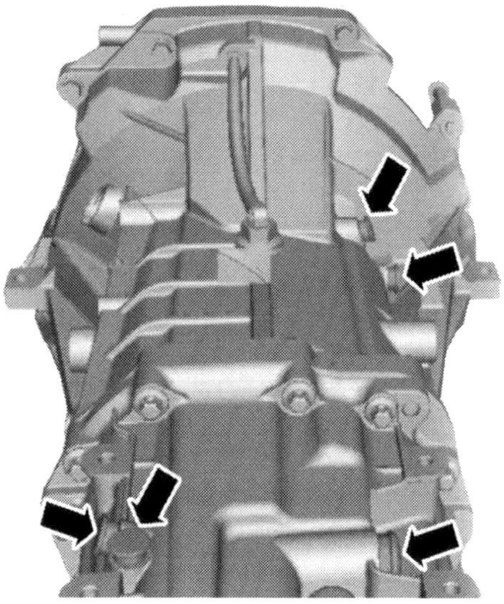

3. CAUTION: Make sure area surrounding component is clean.

Using the special tools, remove the selector rod detents.

Installation

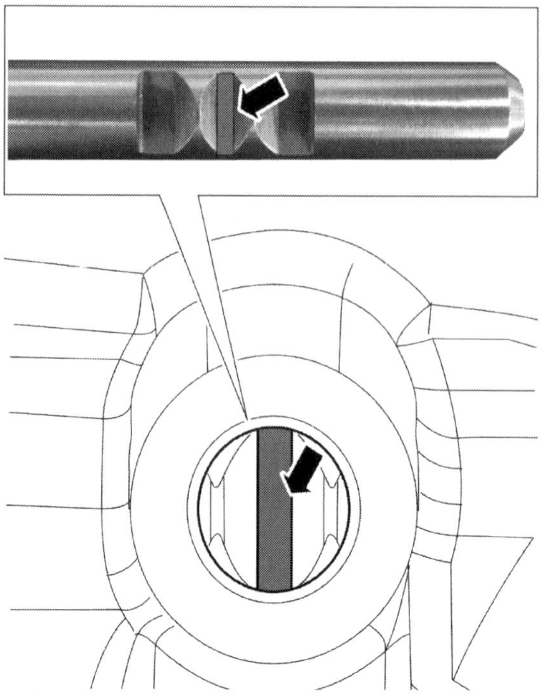

1. NOTE: If transmission is stuck in gear, carefully manipulate selector mechanism using a suitable flat bladed screwdriver.

Make sure all gear selector rods are aligned as shown and the gear selector linkage is free to move.

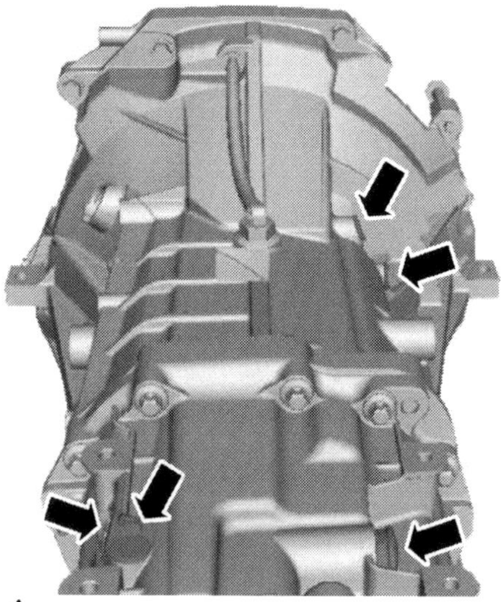

2. CAUTION: Make sure area surrounding component is clean.

   NOTE: New selector rod detents must be seated flush with transmission casing as necessary.

   Using a suitable hammer and soft drift, install the selector rod detents.

3. Install the transmission.

   Refer to: Transmission (308-03 Manual Transmission/Transaxle, Vehicles With: MT82 6-Speed Manual Transmission Installation)

1st-2nd and Reverse Gear Selector Shaft Detents

Removal and Installation

Special Tool(s)

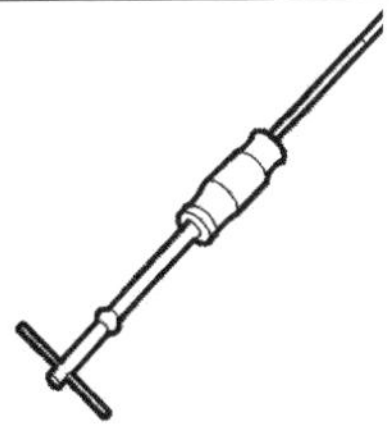 100012	Slide-Hammer 100-012
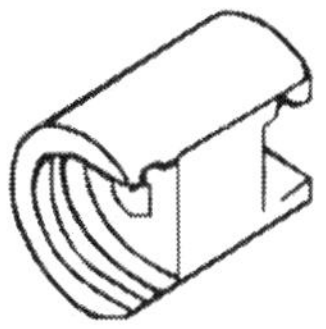	Remover, Detent 308-657
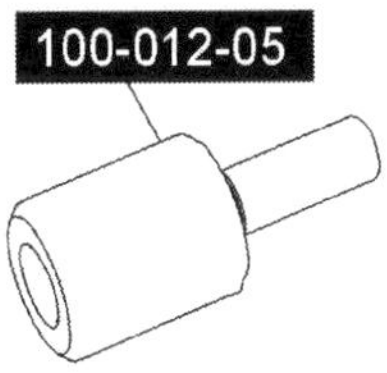 E91077	Adapter, Slide Hammer 100-012-05
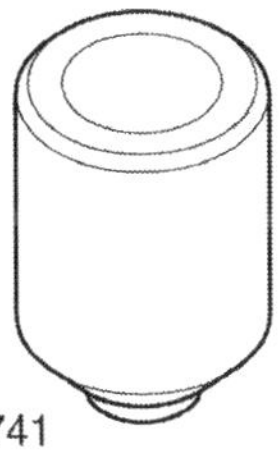 E52741	Remover Seals 308-615

Removal

1. Remove the gearshift lever.

   Refer to: Gearshift Lever (308-06, Removal and Installation).

2. △ NOTE: **Selector shaft detent identification**

Letter **A** shows the old selector shaft detent.
Letter **B** shows the later selector shaft detent.

3. △ NOTE: **Selector shaft detent identification**

Letter **A** shows the old selector shaft detent.
Letter **B** shows the later selector shaft detent.

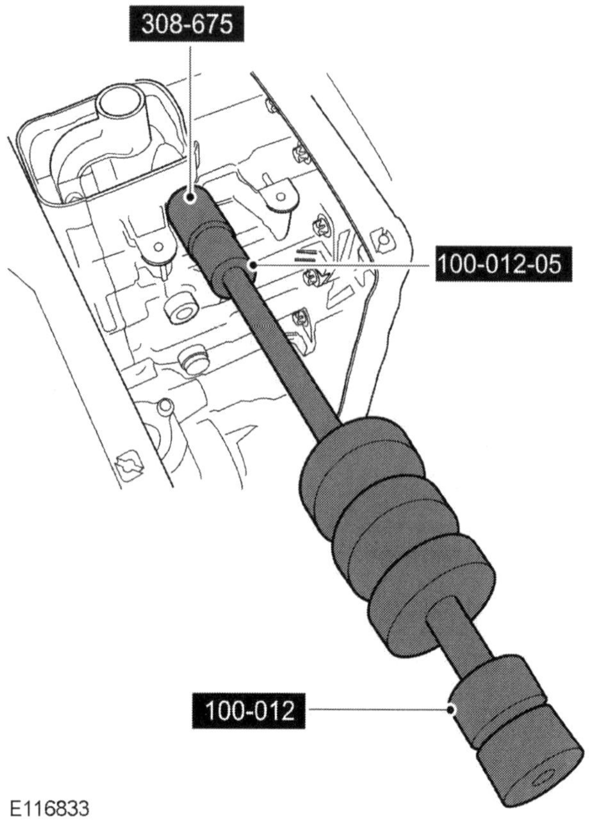

E116833

4. ⚠ CAUTION: Make sure area surrounding component is clean.

Using the special tools, remove the selector shaft detent.

5. ⚠ CAUTION: Make sure area surrounding component is clean.

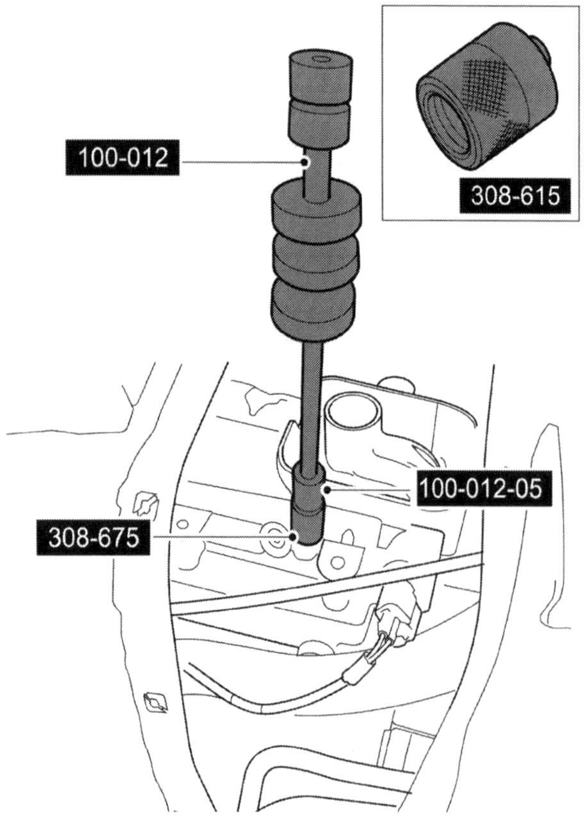

NOTES:

△ If the selector shaft detents are the later type as indicated in the illustration, the detent removal process will require the centre of the detent to be drilled out using a 5.5 mm drill to a depth of approximately 3mm to pierce the detent casing. An M6 bolt should then be inserted through tool 308-615 and into the detent, and using the appropriate special tool, remove the detent.

△ The early type detents indicated in the illustration do not requiring drilling. Use the special tool shown for detent removal.

Using special tools, remove the selector shaft detent.

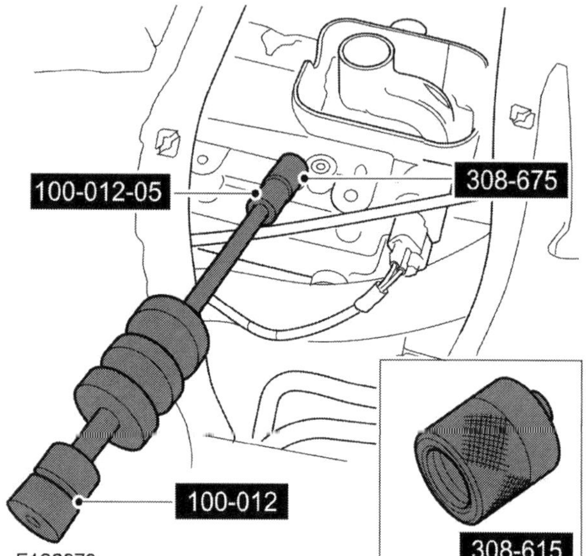

E130876

6. ⚠ CAUTION: Make sure area surrounding component is clean.

Using the appropriate special tools, remove the selector shaft detent.

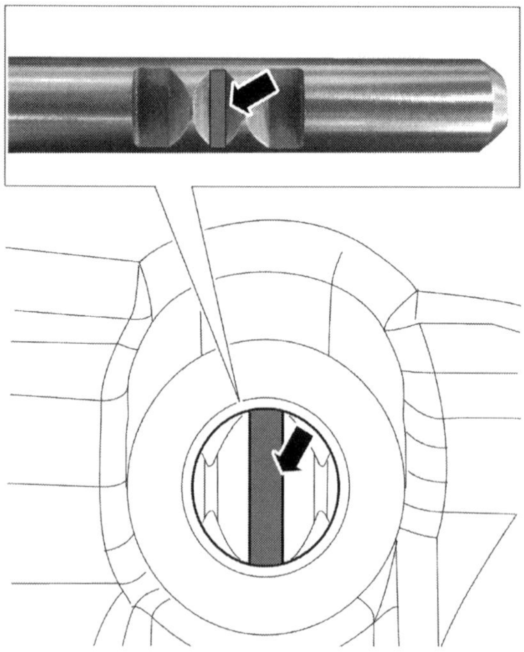

7. With the 1st, 2nd and reverse gear selector rail detents removed, the shafts must be centralized in the neutral position as indicated, viewed through the detent bores in the casting.
   - The shafts can be centralized by carefully moving the selector to the central position using a suitable flat blade screwdriver.
   - With the shafts centralized, check that 3rd and 4th gears and full cross gate movement can be achieved.

Installation

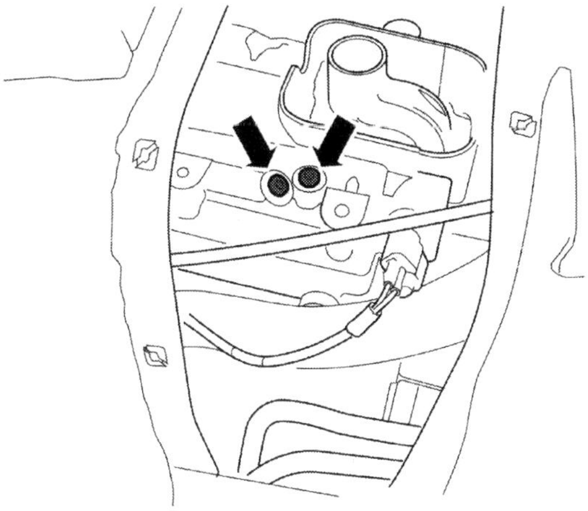

1. ⚠ CAUTION: New selector shaft detents must be seated flush with transmission casing.

Using suitable tools, install selector shaft detents.

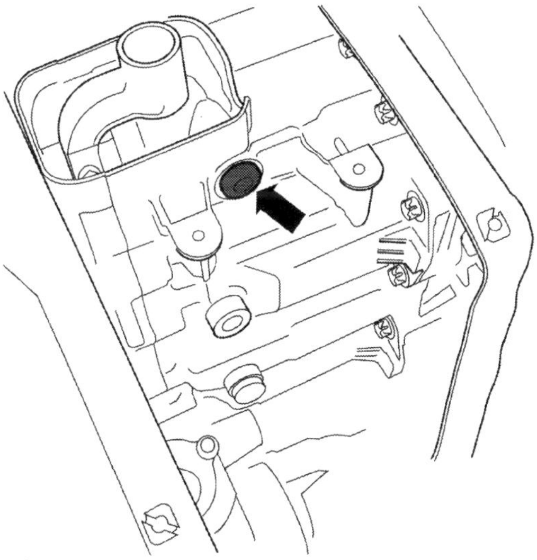

2. ⚠ CAUTION: Make sure component is fully seated against the casing.

Using suitable tools, install selector shaft detent.

3. Install the gearshift lever.

Refer to: Gearshift Lever (308-06, Removal and Installation).£ Using special tools, remove the selector shaft detent.

Manual Transmission/Transaxle - Vehicles With: MT82 6-Speed

Removal

Special Tool(s)

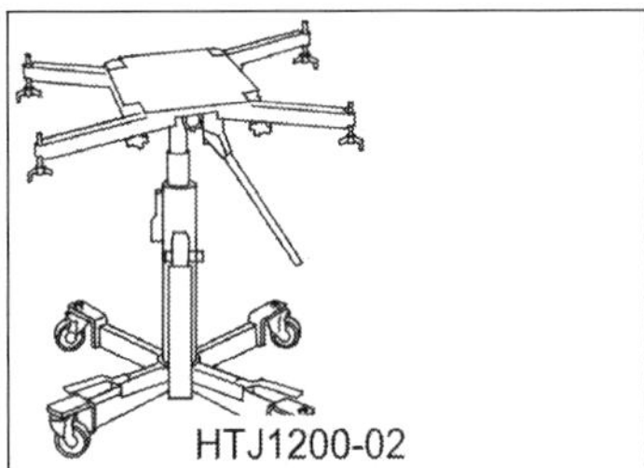 HTJ1200-02	Powertrain Assembly Jack HTJ1200-02
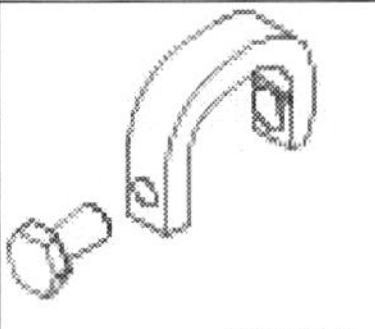 E77040	Remover, Slider 308-652
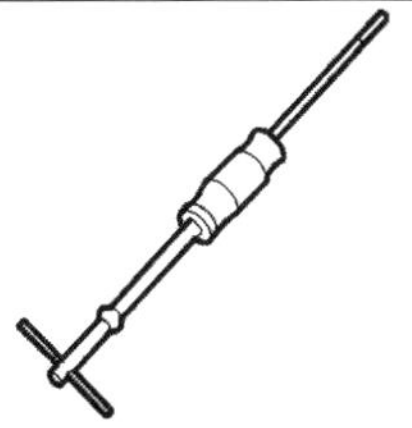 100012	Slide Hammer 100-012
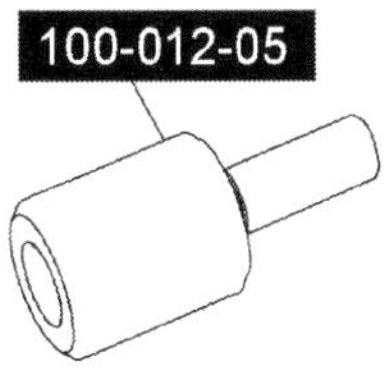 E91077	Adapter Slide Ham-mer 100-012-05

Removal

1. Open the front door.
2. Remove the front seat cushion assembly.

Refer to: Front Driveshaft (205-01 Driveshaft, Removal and Installation)

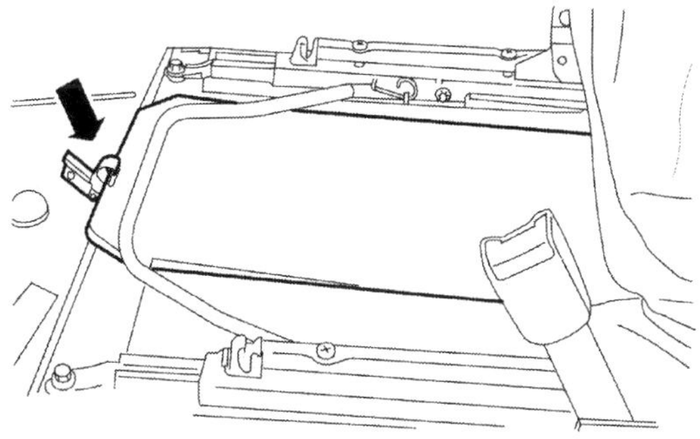

3. Remove the battery cover.
   - Release the clip.
4. Disconnect the battery ground cable.

Refer to: Front Driveshaft (205-01 Driveshaft, Removal and Installation)

5. Remove centre floor console.

Refer to: Front Driveshaft (205-01 Driveshaft, Removal and Installation)

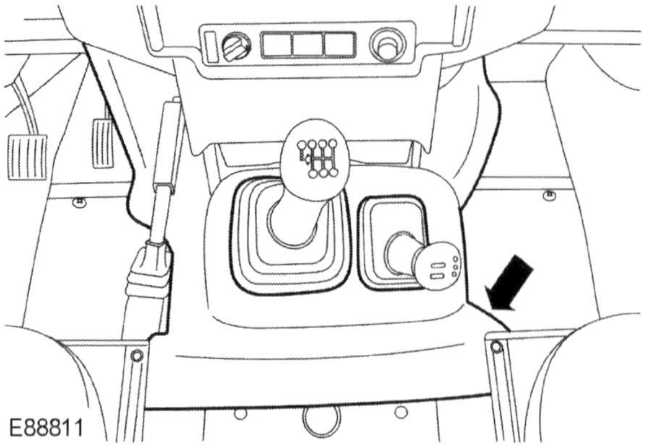
E88811

6. Remove the transmission cover panel carpet.

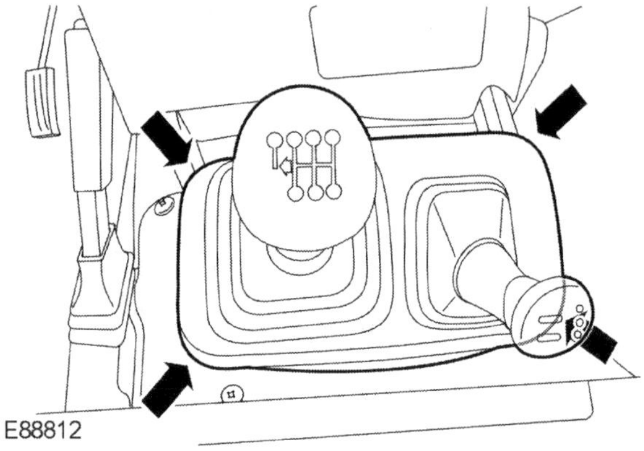
E88812

7. Release the gear selector lever gaiter.

- Release the gear selector lever gaitor from the transmission tunnel.
- Release the gear selector lever gaiter from the high-low selector lever.

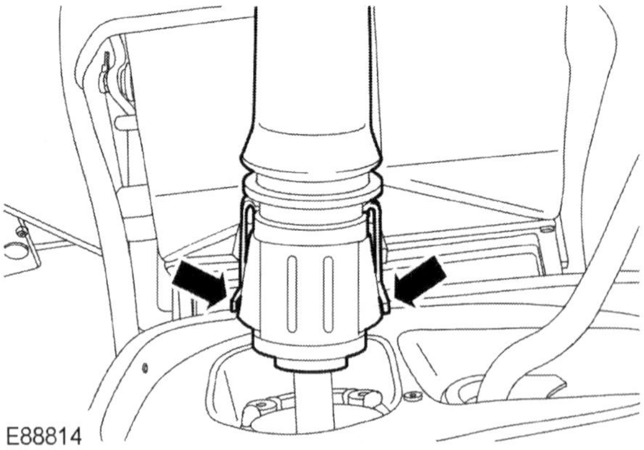
E88814

8. WARNING: The gearshift lever knob will be released suddenly, keep face clear during removal. Failure to follow this instruction may result in personal injury.

Remove the upper gear change lever.

9. Remove the gear selector lever gaiter.

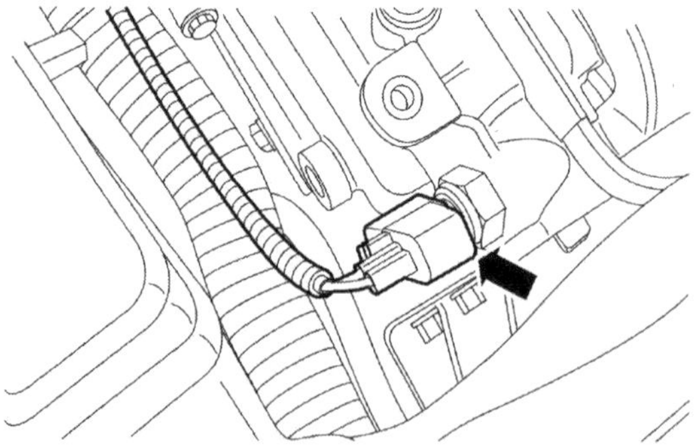

10. Disconnect the reversing lamp switch electrical connector.

11. Open bonnet for access.

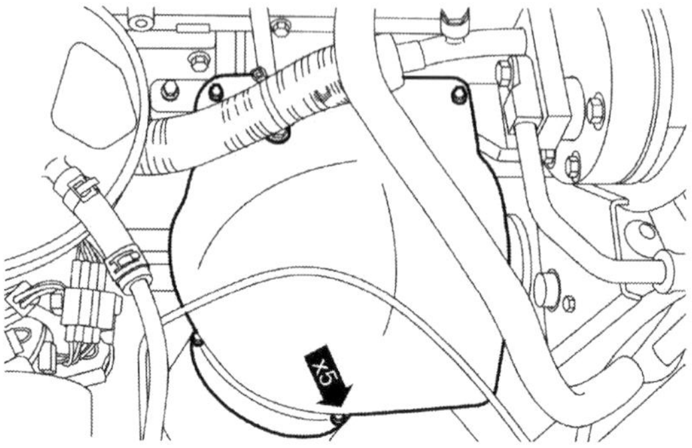

12. Remove the turbocharger heat shield.

- Remove the 5 bolts.

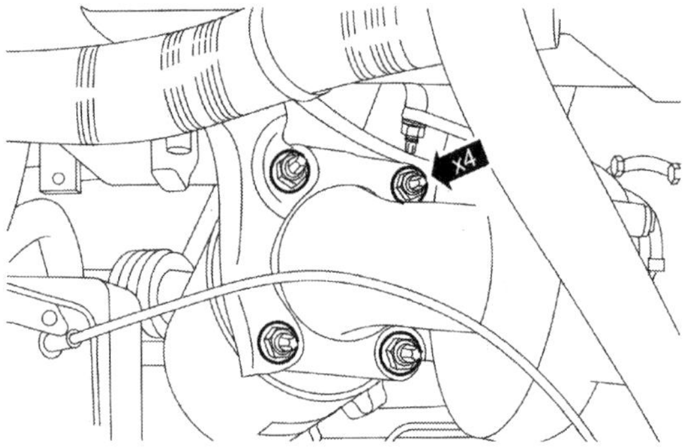

13. Remove the 4 catalytic converter securing studs.

14. WARNING: Do not work on or under a vehicle supported only by a jack. Always support the vehicle on safety stands.

Raise and support the vehicle.

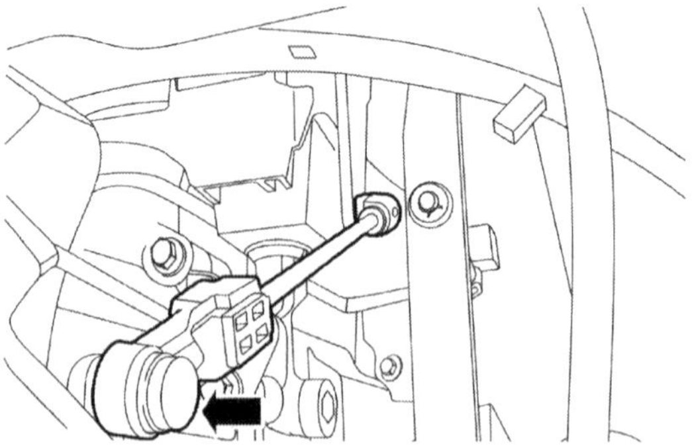

15. Release the high-low selector rod ball joint from the transfer gearbox.

16. Lower the vehicle.

17. Open the front door.

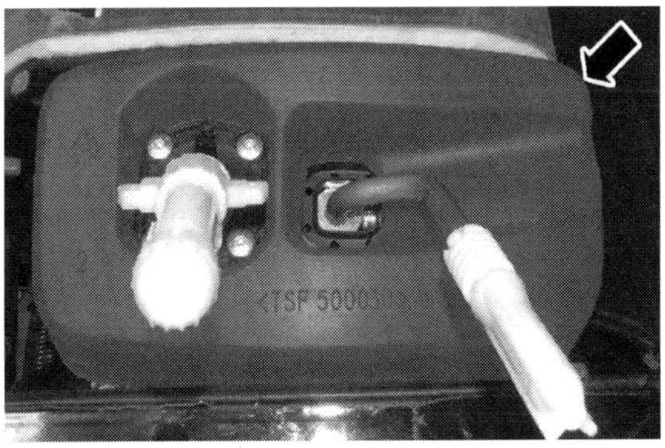

18. Remove the foam pad from around the selector levers.

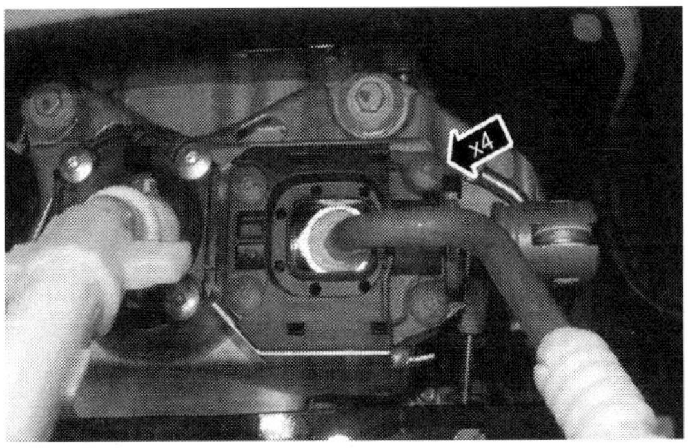

19. Remove the 4 high-low selector lever securing bolts.

20. Remove the high-low selector lever.

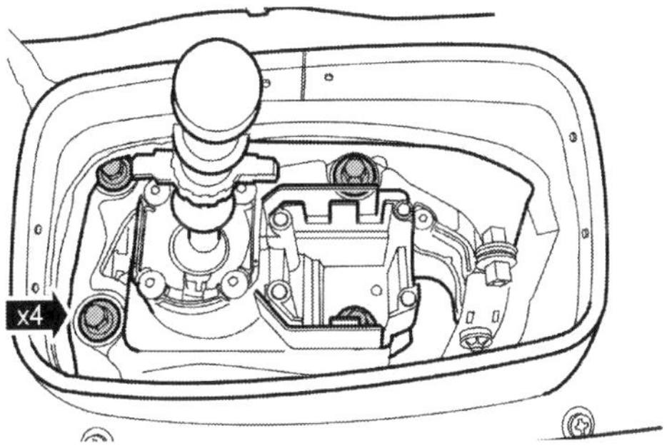

21. Remove the 4 gear selector housing securing bolts.

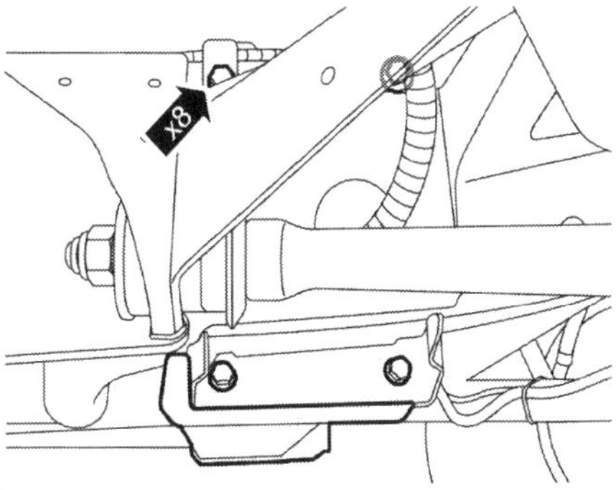

22. NOTE: Left-hand side shown, right-hand side similar.

    Remove the chassis cross member.

    - Remove the 8 bolts.

23. Remove the front driveshaft.

    Refer to: Front Driveshaft (205-01 Driveshaft, Removal and Installation)

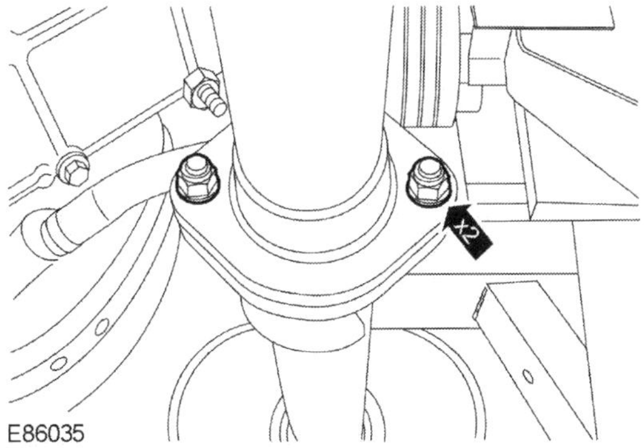

24. Remove the 2 catalytic converter to intermediate pipe securing nuts.

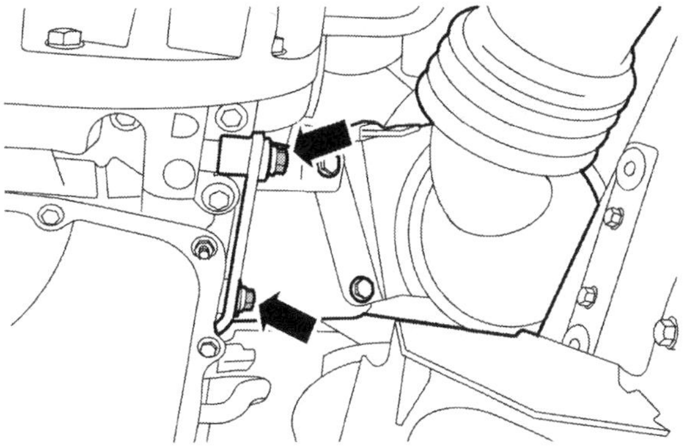

25. Remove the 2 catalytic converter mounting bracket securing bolts.

26. Remove the catalytic converter.

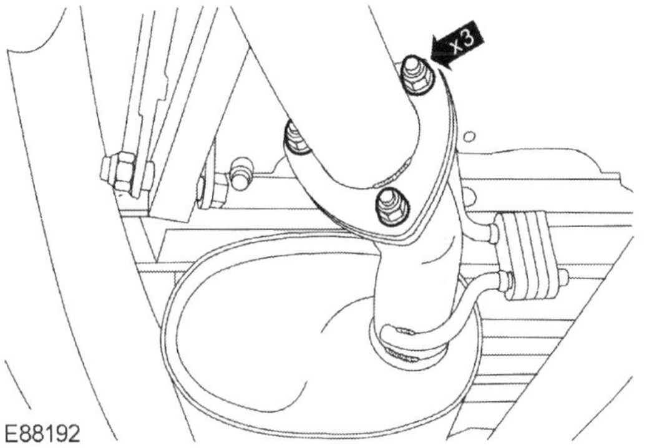

27. Remove the tailpipe and muffler, to intermediate pipe securing nuts.

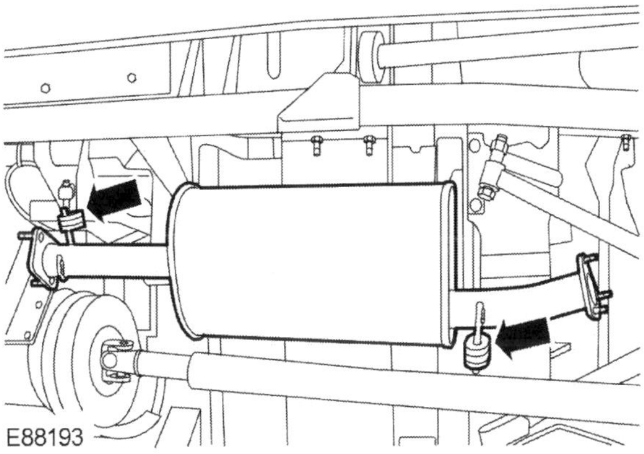

28. Remove the intermediate pipe and muffler.

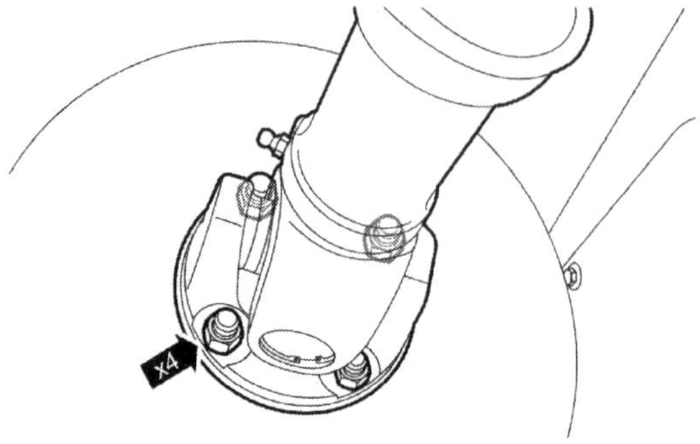

29. Displace and reposition the rear driveshaft.
    - Mark the position of the driveshaft in relation to the drive pinion flange.
    - Remove the rear driveshaft to transfer gearbox securing nuts.

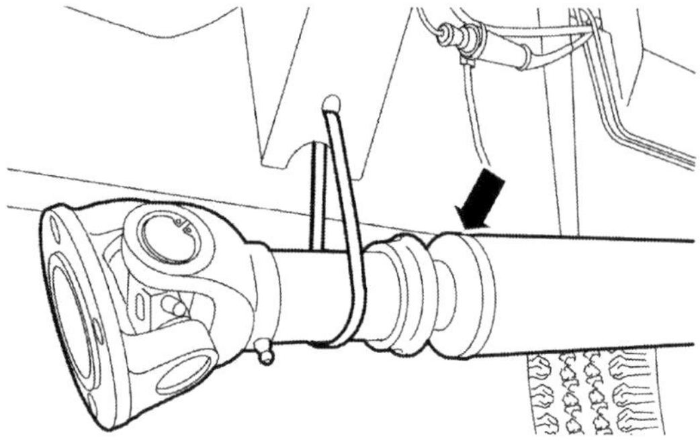

30. Using a suitable tie strap, secure the rear driveshaft to the chassis.

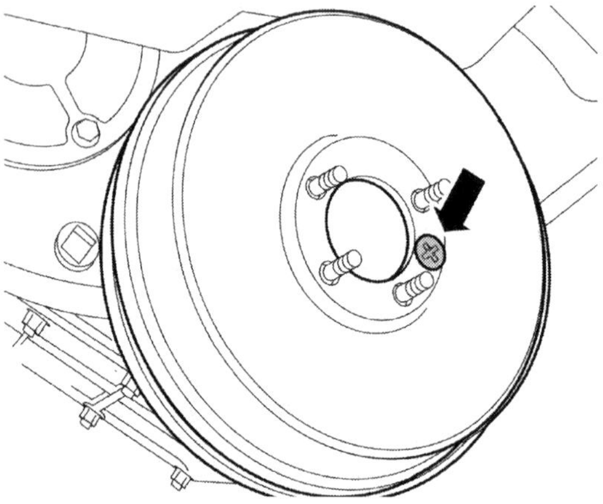

31. Remove the parking brake drum.
    - Remove the securing screw.

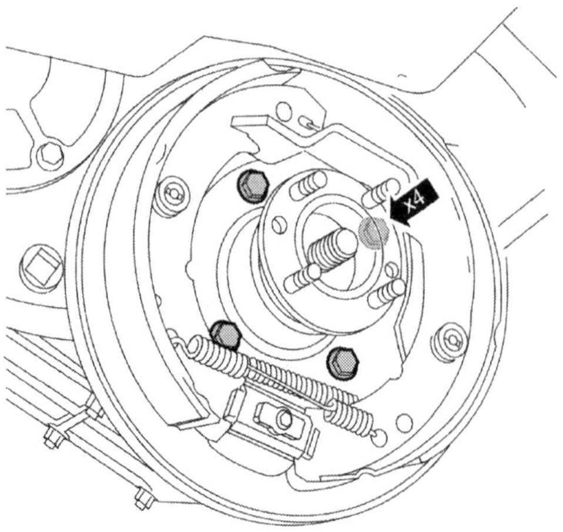

32. Remove the parking brake assembly and tie aside.

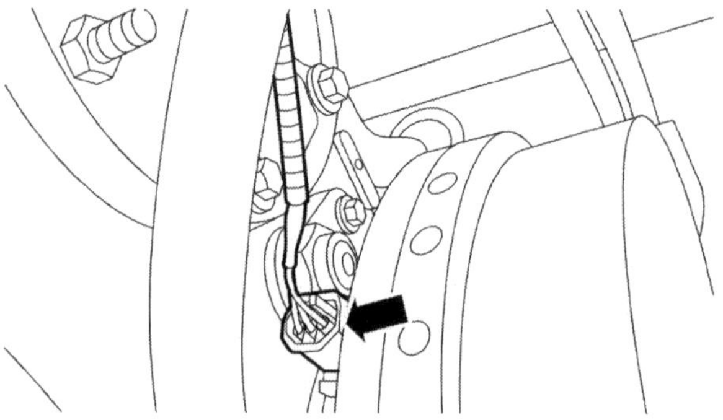

33. Disconnect the electronic speedometer drive electrical connector from the transfer gearbox.

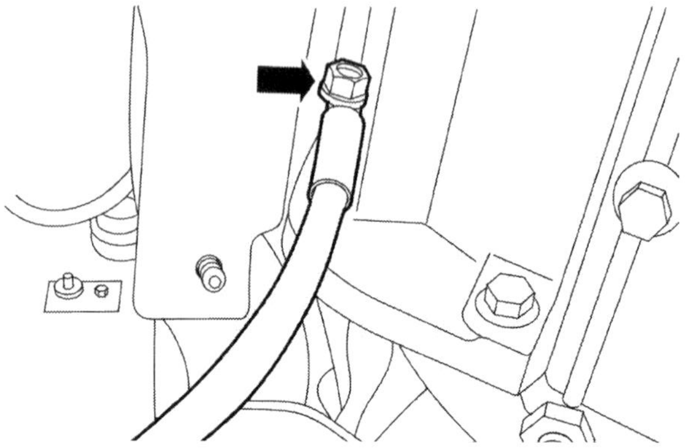

34. Remove the nut and disconnect the LH earth cable from the transfer gearbox.

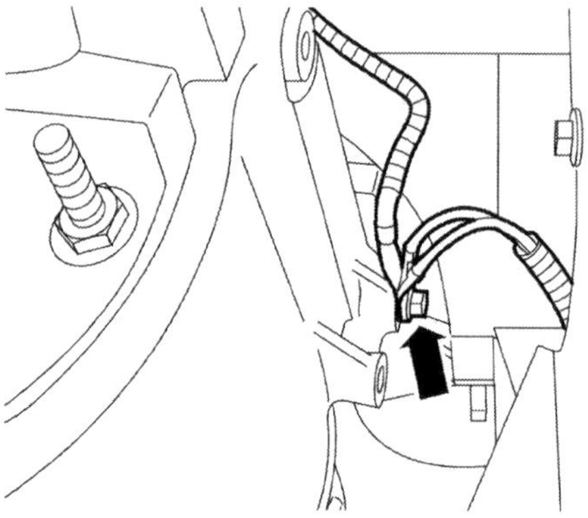

35. Remove the bolt securing the RH earth cables from the transfer gearbox.

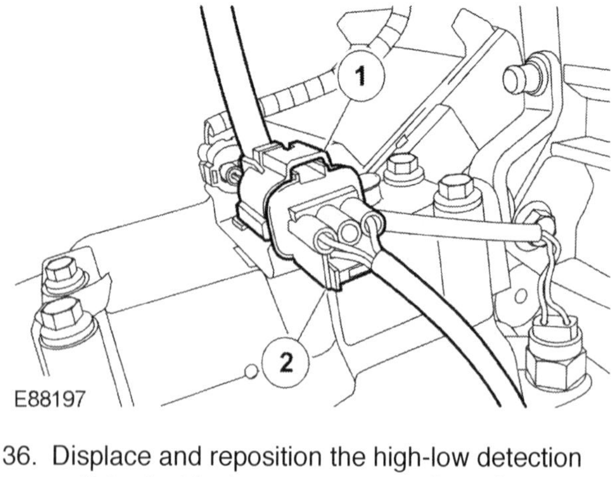

36. Displace and reposition the high-low detection switch electrical connector away from the transfer gearbox.
    - Disconnect the high-low detection switch electrical connector.

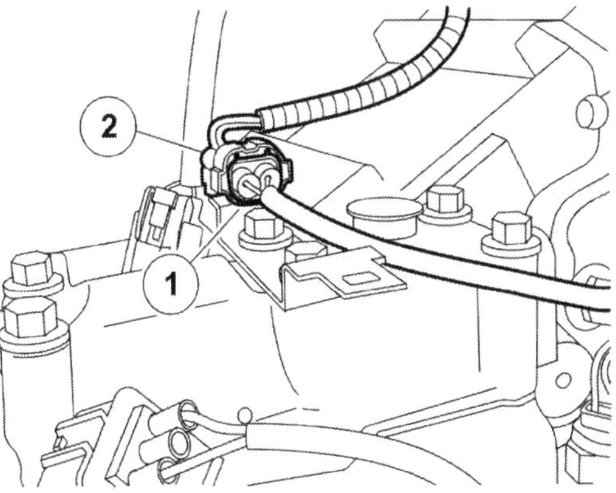

E88198

37. Disconnect the differential lock detection switch electrical connector.

    - Release the electrical connector from the bracket.

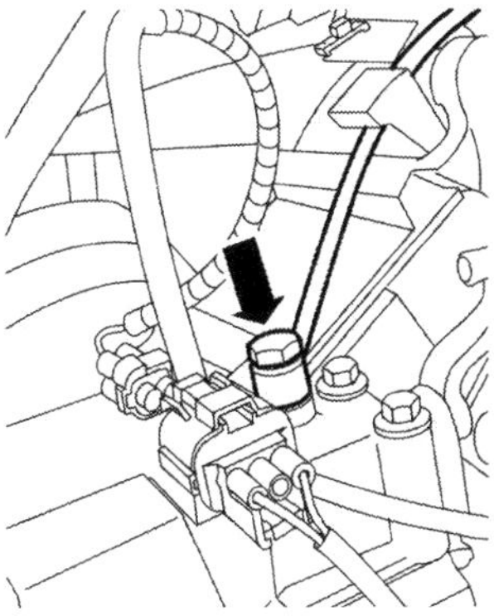

38. Release the breather pipe from transfer gearbox.

    - Remove the securing bolt.

    - Remove and discard the 2 sealing washers.

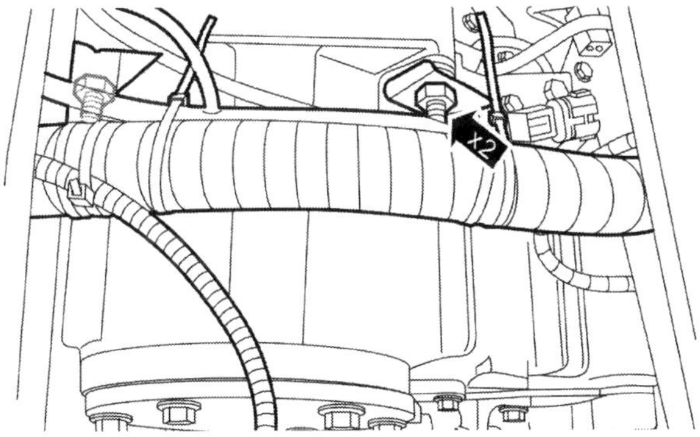

39. Remove the transmission harness retaining bracket securing nut and bolt and reposition the harness.

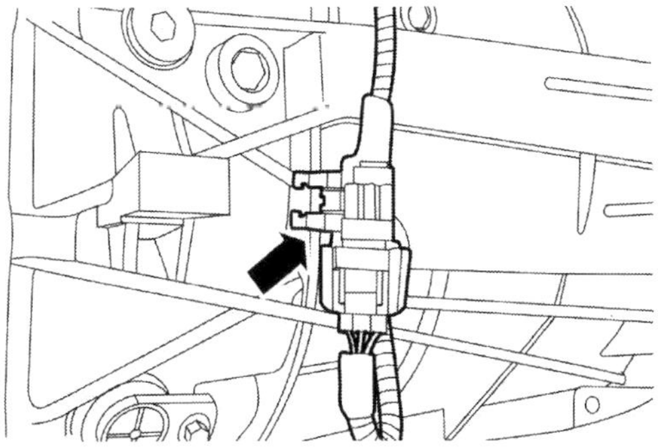

40. Release the transmission wiring harness connector and bracket from the RH side of the transmission.

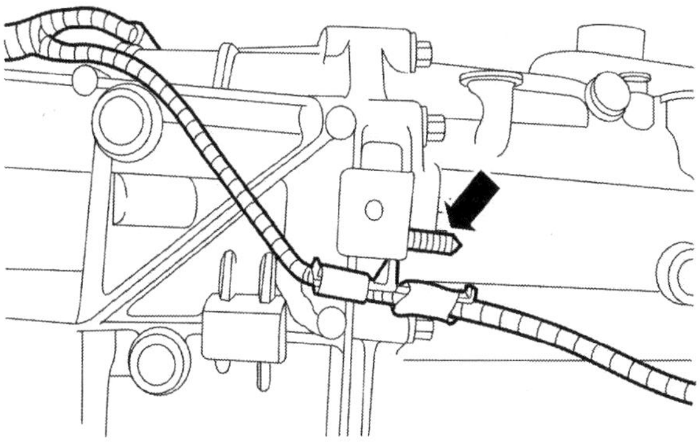

41. Release the transmission wiring harness securing clip from the LH side of the transmission.

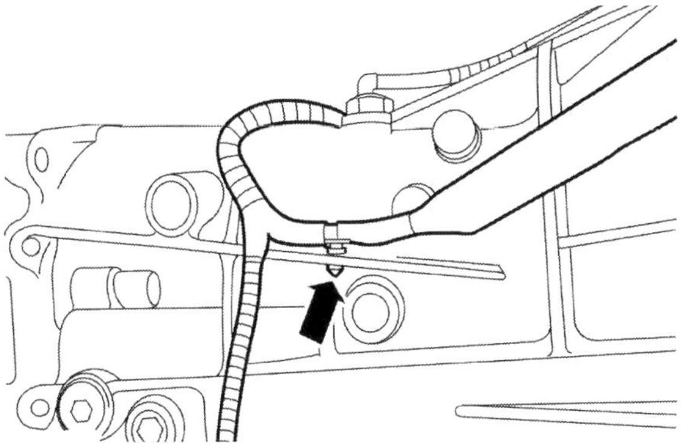

42. Release the transmission wiring harness securing clip from the RH side of the transmission.

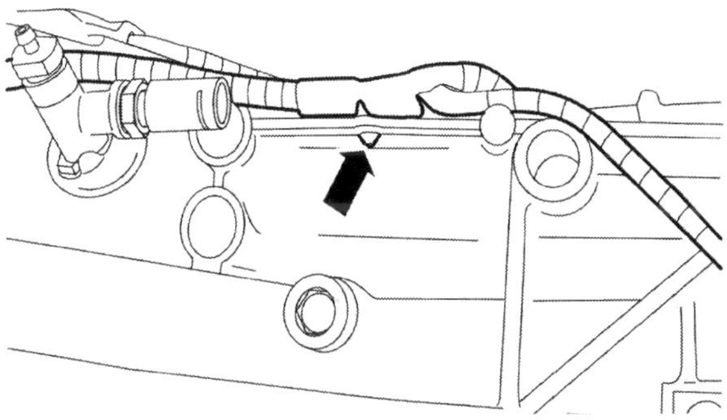

43. Release the transmission wiring harness securing clip from the top of the transmission.

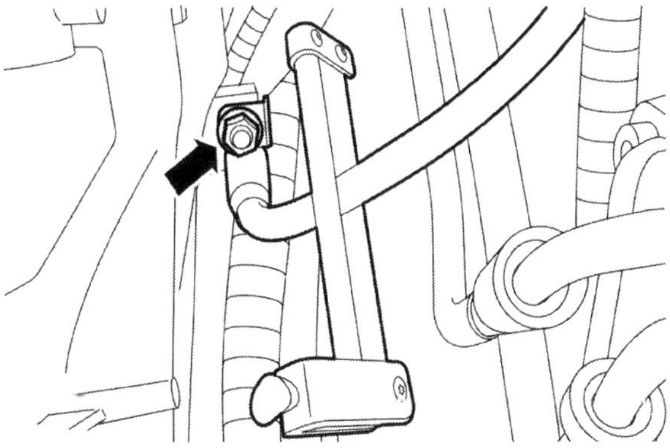

44. Remove the clutch slave cylinder fluid hose mounting bracket securing nut and bolt.

    - Install a suitable pipe clamp to the clutch slave cylinder fluid hose.

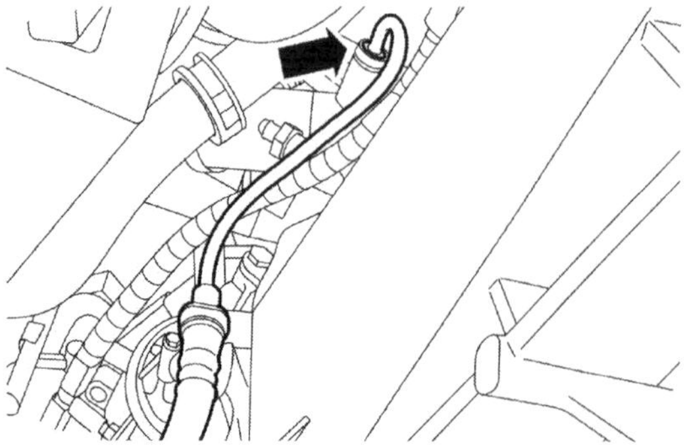

45. ⚠ CAUTION: Make sure that all openings are sealed. Use new blanking caps.

Disconnect the clutch slave cylinder fluid hose from the clutch slave cylinder.

- Remove the clutch slave cylinder line clip.
- Remove and discard the O-ring seal.

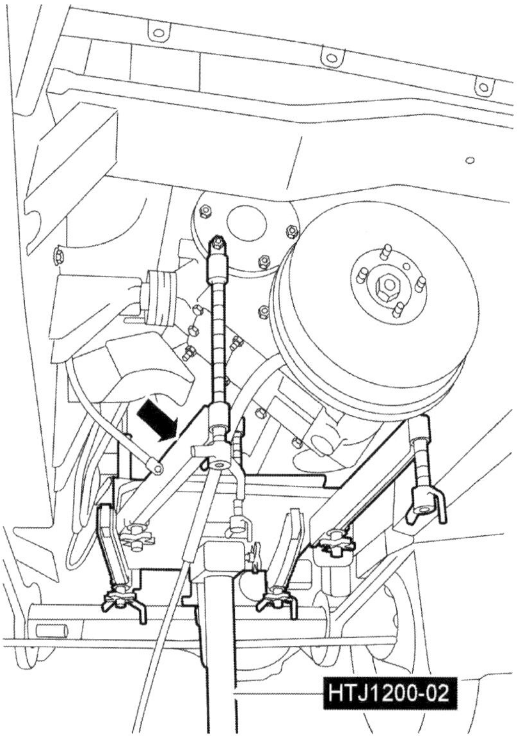

46. Install the special tool HTJ1200-02 to the transmission and transfer gearbox.

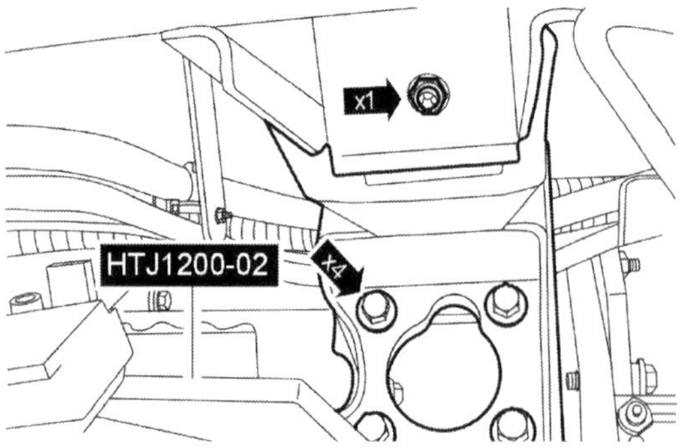

47. Remove the LH transmission mount and mounting bracket.

- Remove the LH transmission mount and mounting bracket securing nuts and bolts.

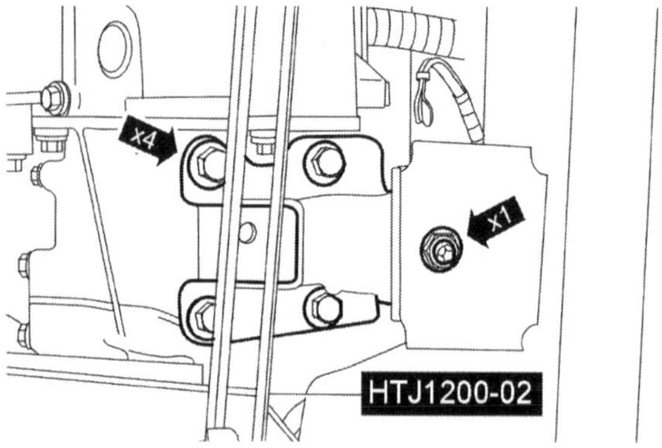

48. Remove the RH transfer gearbox mounting bracket.

- Remove the RH transfer gearbox mount and mounting bracket securing nuts and bolts.

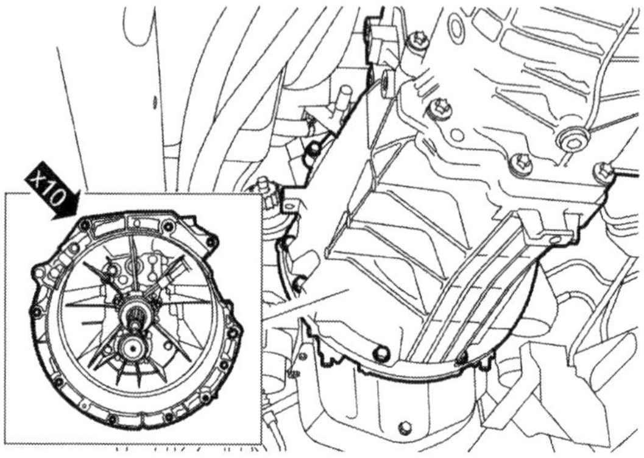

49. Remove the bell housing bolts.

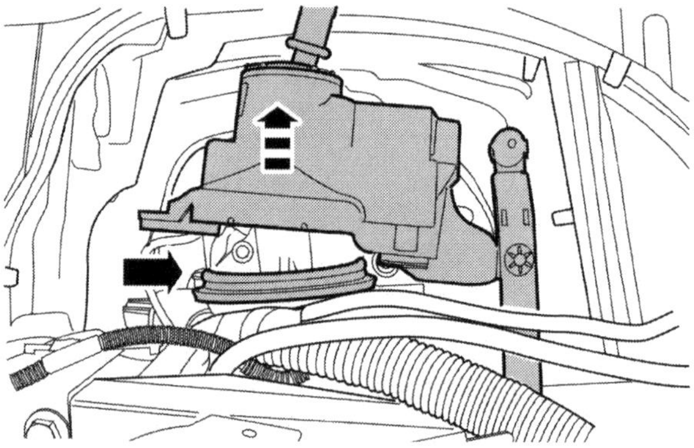

50. Remove the gear selector housing.

- Reposition the transmission to allow access to the gear selector housing.
- Remove from below the vehicle.
- Remove the rubber seal.

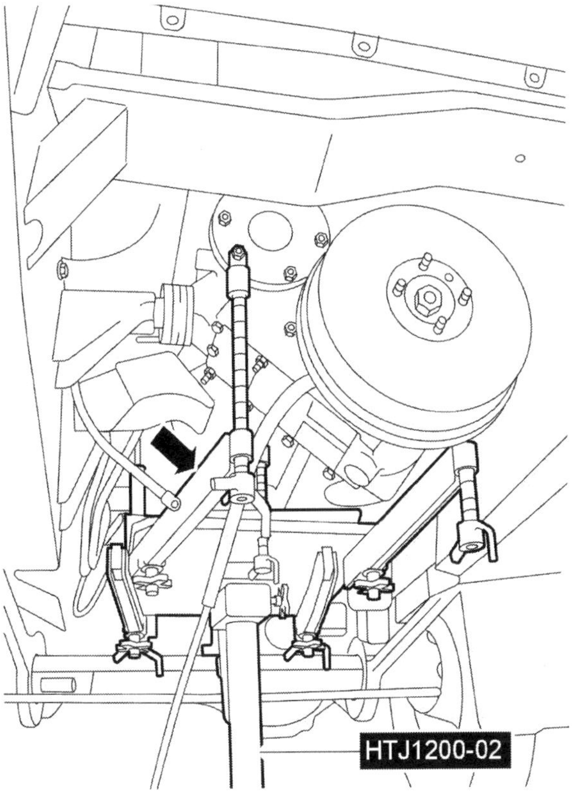

51. CAUTION: Disengage the transmission from the clutch by 30mm before lowering HTJ1200-02 to protect clutch from damage.

Remove the transmission and transfer gearbox from the vehicle.

52. NOTE: Do not disassemble further if the component is removed for access only.

Remove the transmission from the special tool.

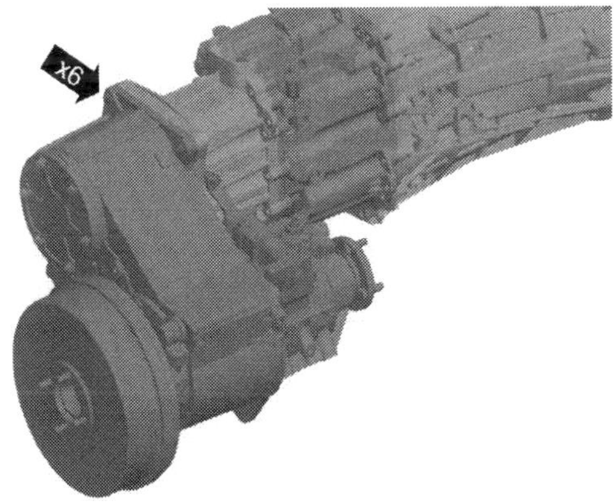

E90815

53. With assistance, remove the transfer case.

- Remove the 4 bolts.
- Remove the 2 nuts.

54. Remove the transmission extension housing.

- Remove the 10 bolts.

55. Remove the transmission extension shaft cover.

- Remove the tie strap.
- Remove the cover.

56. Remove the seal.

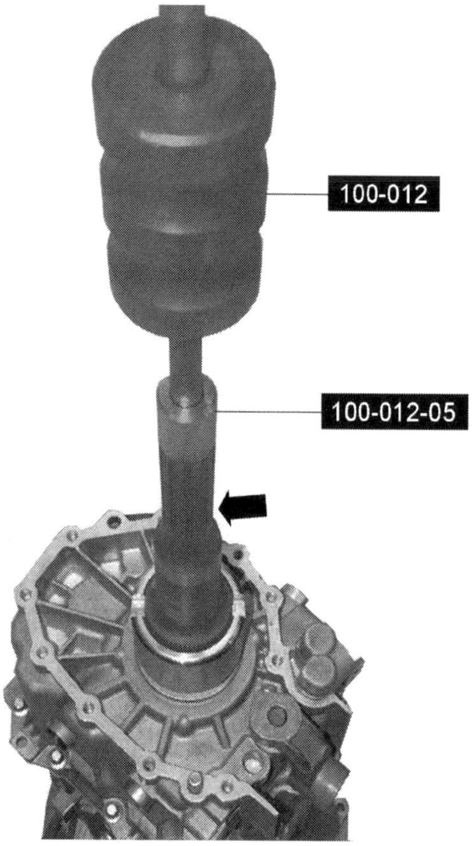

57. Using the special tools, remove the transmission extension shaft.

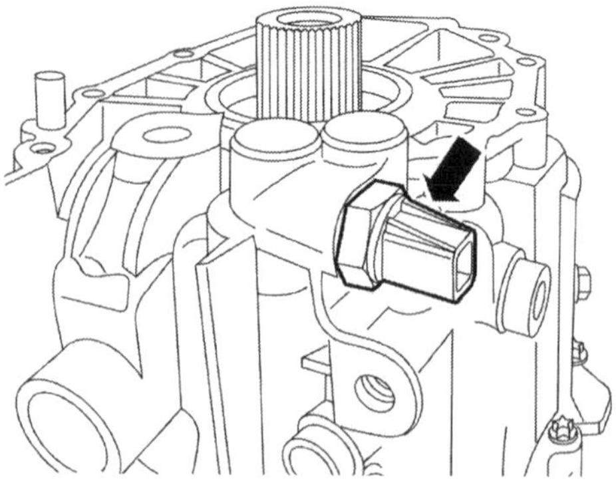

58. Remove the reversing lamp switch.

59. Remove the clutch slave cylinder.

- Remove the connecting pipe securing clip.
- Remove the connecting pipe.
- Remove the 3 bolts.

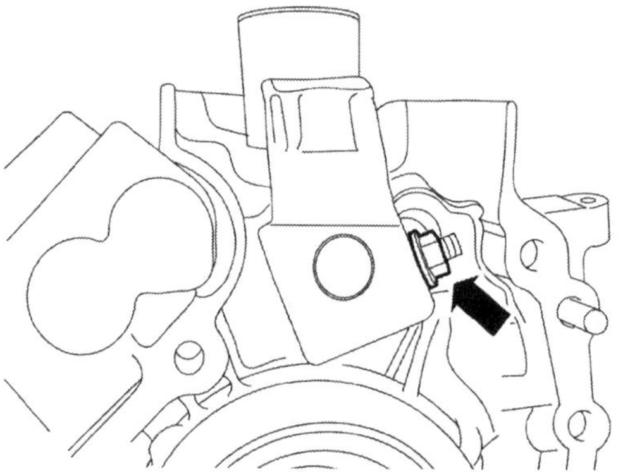

60. Remove the gearshift yoke securing nut.

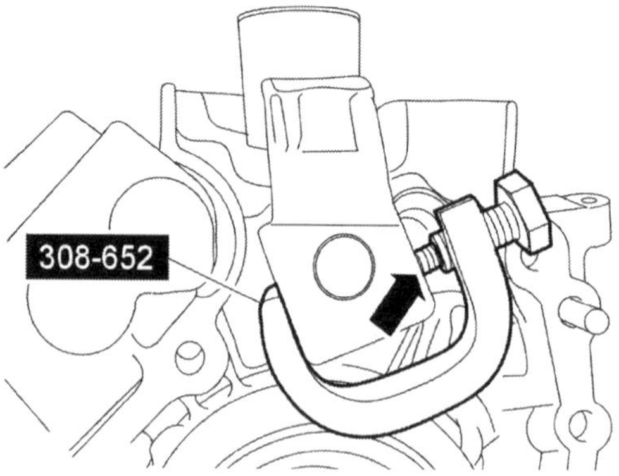

61. Using the special tool, remove the gearshift yoke.

- Remove the pin.

Transmission

Assembly

Special Tool(s)

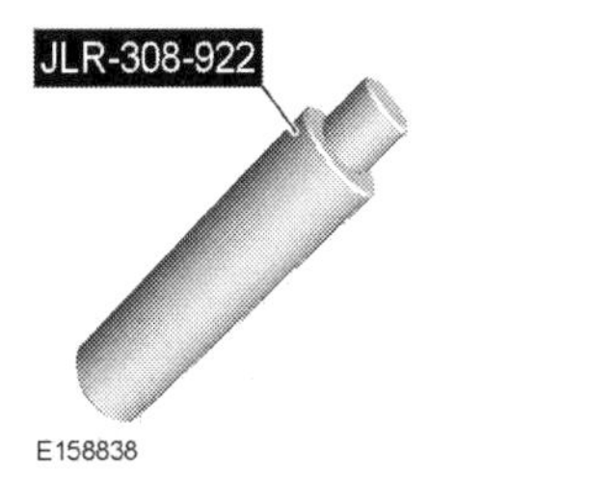	JLR-308-922 Aligner, Countershaft
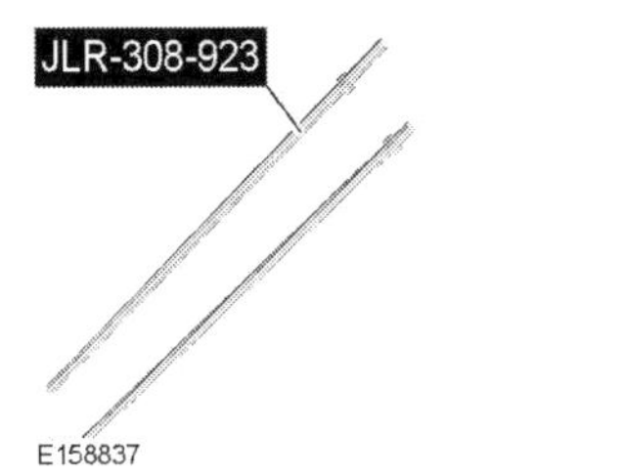	JLR-308-923 Threaded rods, Transmission Bearing Heater
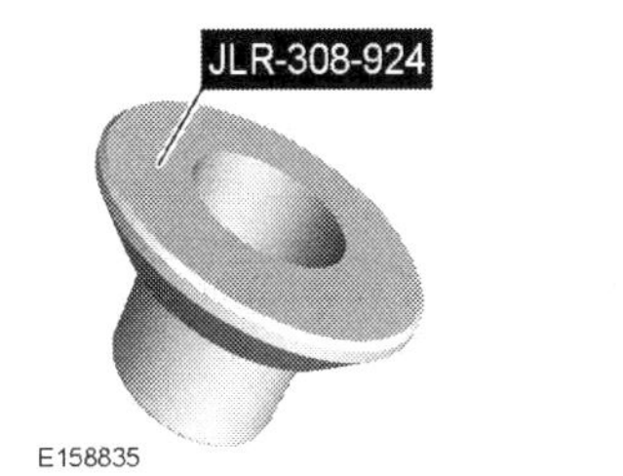	JLR-308-924 Transmission Bearing Heater
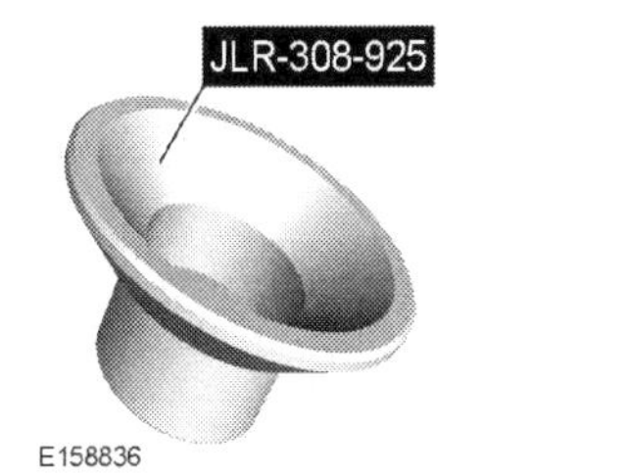	JLR-308-925 Transmission Bearing Heater
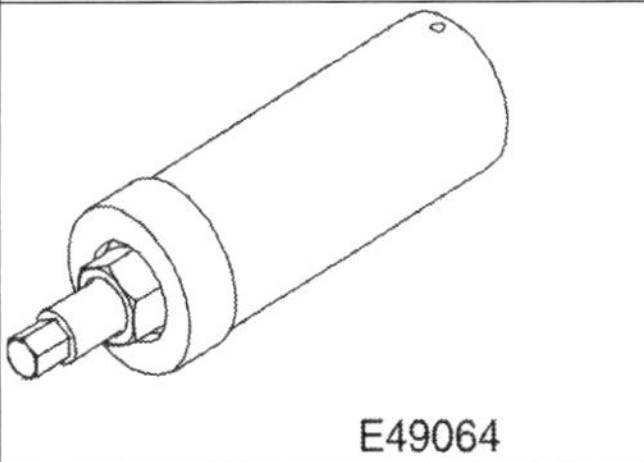	308-604 Installer, Output Drive Flange Seal
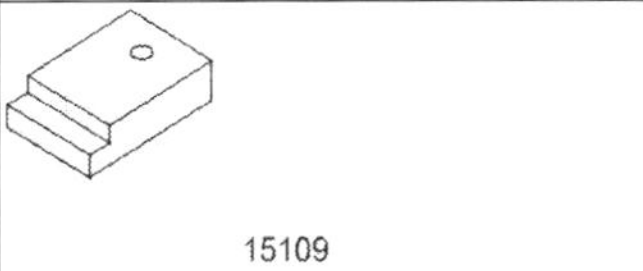	205-332 Protector, Axle Shaft
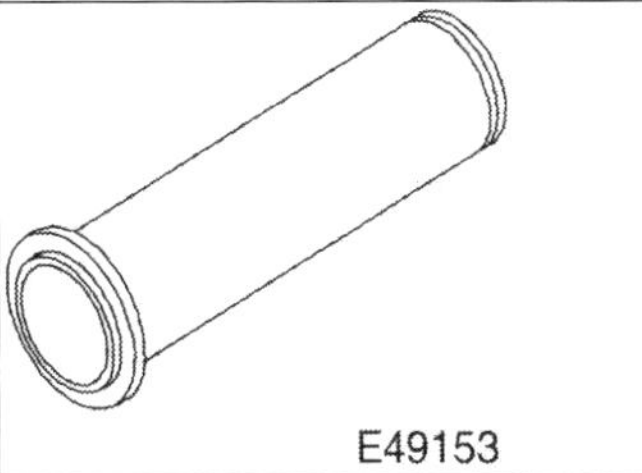	308-605 Installer, Input Shaft Seal
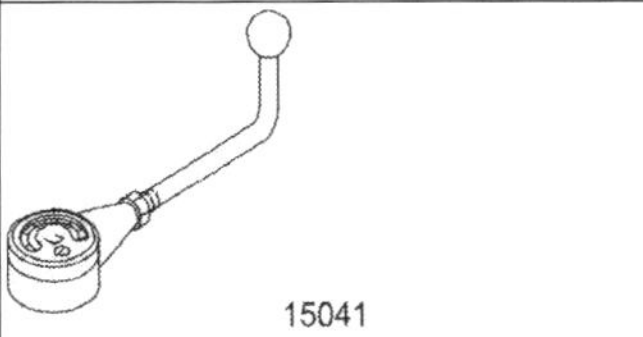	205-067 Preload Gauge
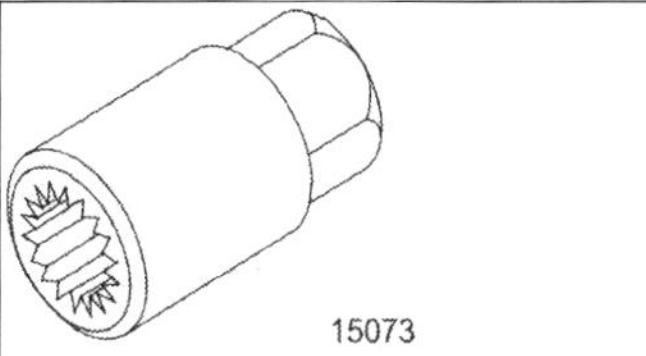	205-175 Socket, Drive Pinion Nut

Part(s)

STEP	PART NAME	QUANTITY
Step 32	Renew Part Circlip	1
Step 47	Renew Part Selector rod pin(s)	1
Step 61	Renew Part Circlip	1
Step 62	Renew Part Transmission output shaft flange seal	1
Step 66	Renew Part Input shaft seal	1
Step 66	Renew Part Transmission countershaft seal	1

1. △ NOTE: All synchronizer rings and needle bearings should be lubricated with clean transmission fluid prior to assembly.

Material: Transmission Oil 75W-90 BO (WSD-M2C200-C /)

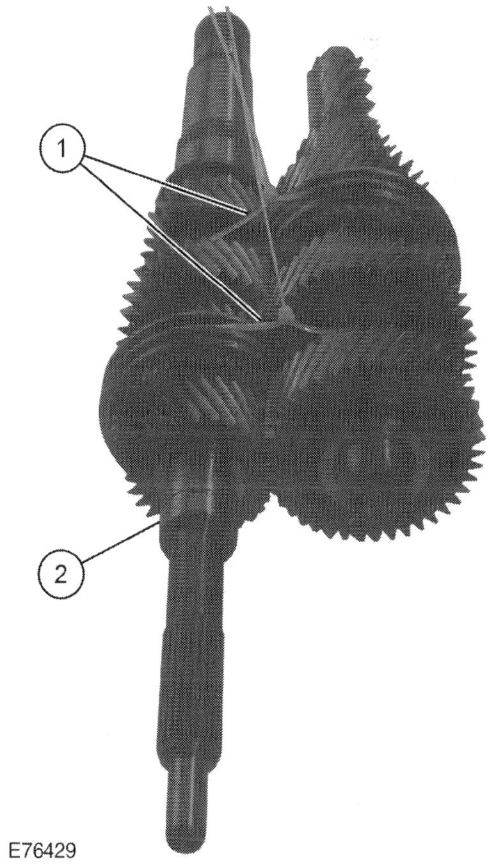

2. △ NOTE: Make sure tie straps are removed after installation.

Secure the shaft assemblies together. Use suitable tie straps at the positions shown. Make sure the input shaft does not become detached.

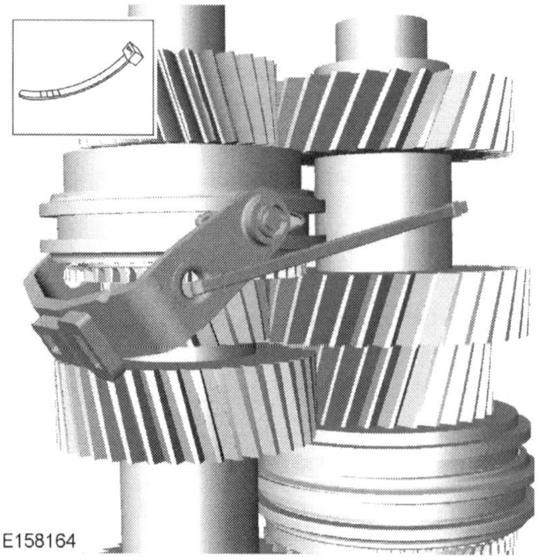

E158164

3. NOTE: Make sure tie straps are removed after installation.

Use a suitable tie strap at the positions shown. Secure the selector fork to the shaft assemblies.

4. WARNING: Care should be taken when using the hot air gun. Failure to follow this instruction may result in personal injury.

NOTES:

Heat up the special tools with a hot air gun to around 100 °C, and then install the shaft assemblies within 5 seconds of the special tool removal.

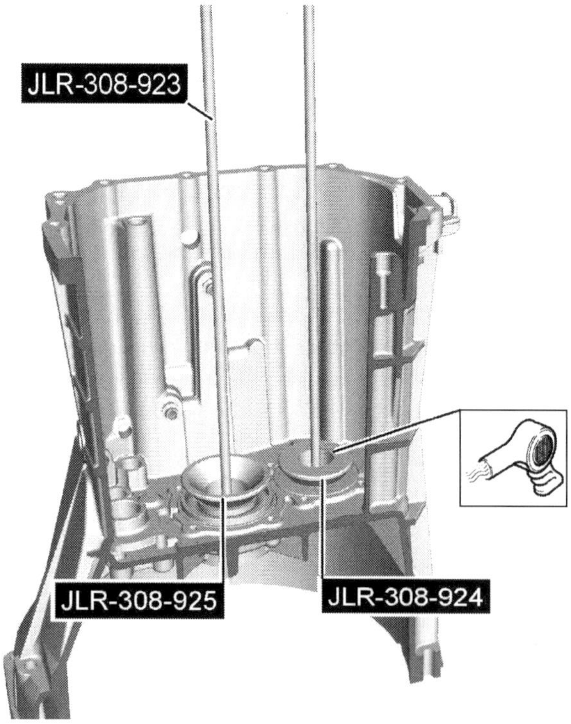

Allow 3 minutes before rotating the shaft assemblies.

Make sure tie straps are removed after installation.

Using the special tools, apply heat to the bearing centre races before installing the shafts. Heat the input shaft bearing for 1 minute. Heat the countershaft bearing for 1 minute. Heat the input shaft bearing for a further 1 minute. Heat the countershaft bearing for a further 1 minute. Heat the input shaft bearing for a further 1 minute.

Special Tool(s): JLR-308-922 , JLR-308-923 , JLR-308-924 , JLR-308-925

5. Remove the special tool.

Special Tool(s): JLR-308-922

E154315

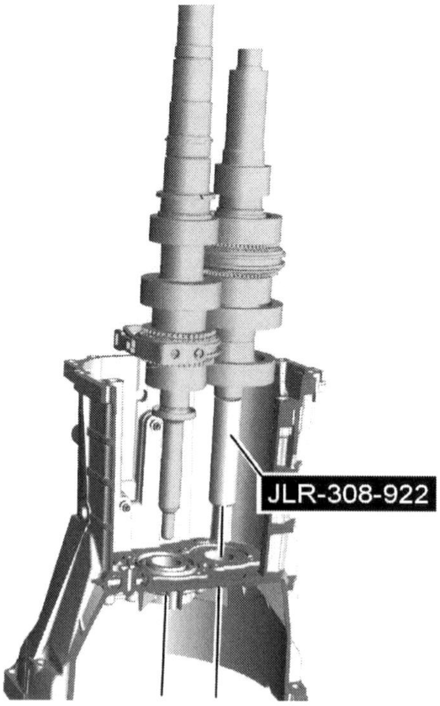

6. Note the position of the countershaft.

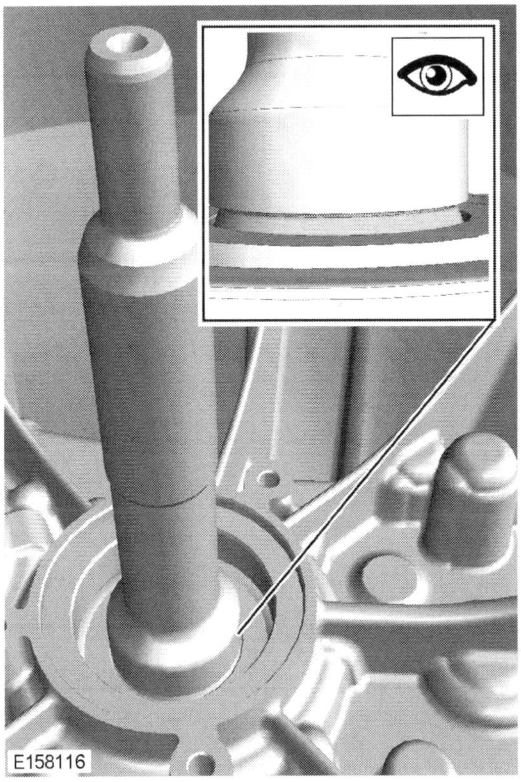

7. Note the position of the input shaft.

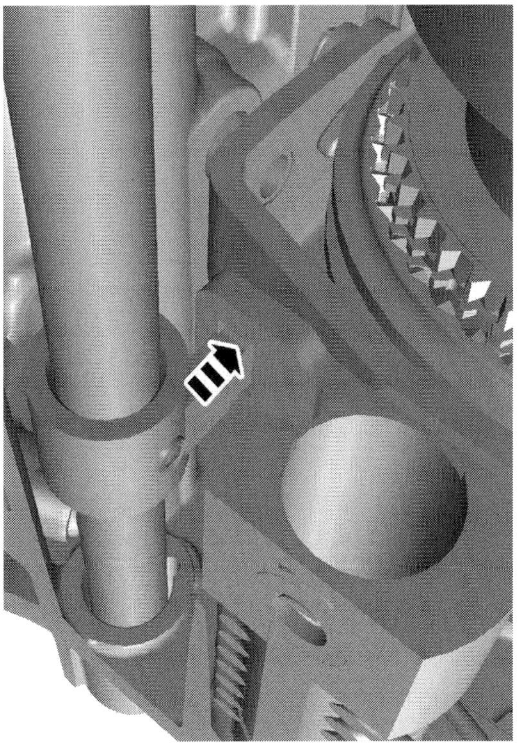

8. Locate the 5th/6th selector rail into the selector fork.

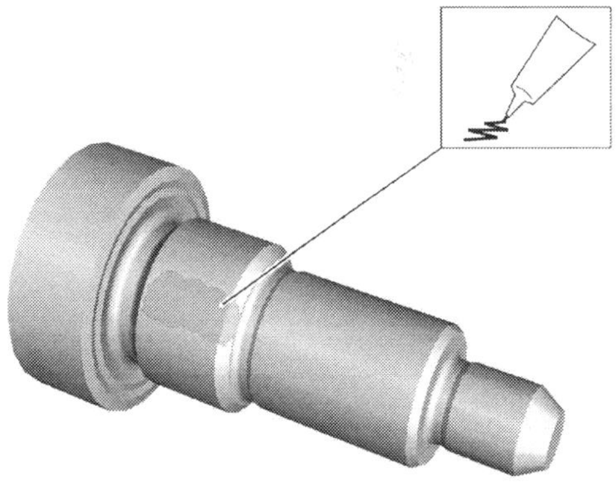

9. Apply Jaguar Land Rover approved thread lock to both sides of the fixing.

Material: Thread Locking MS (WSK-M2G349-A7 /)

10.

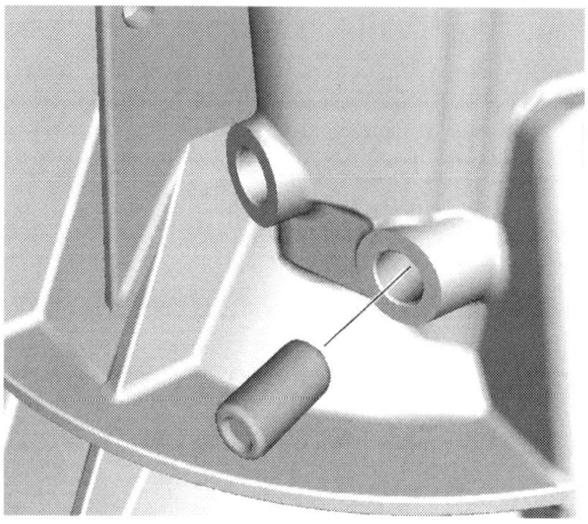

11. △ NOTE: Make sure the detent is installed flush with the transmission casing.

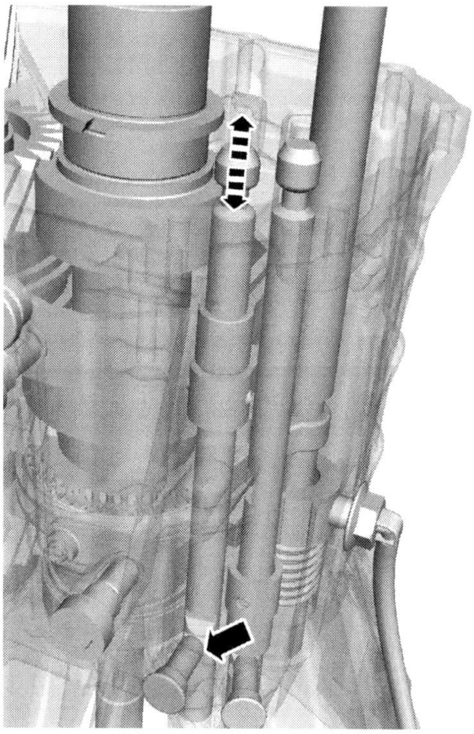

12. △ NOTE: The selector shaft must be raised slightly to allow the detent to be installed.

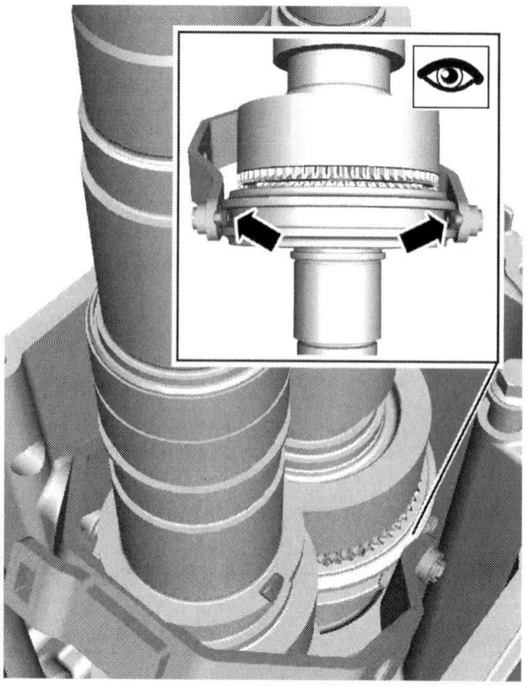

13.

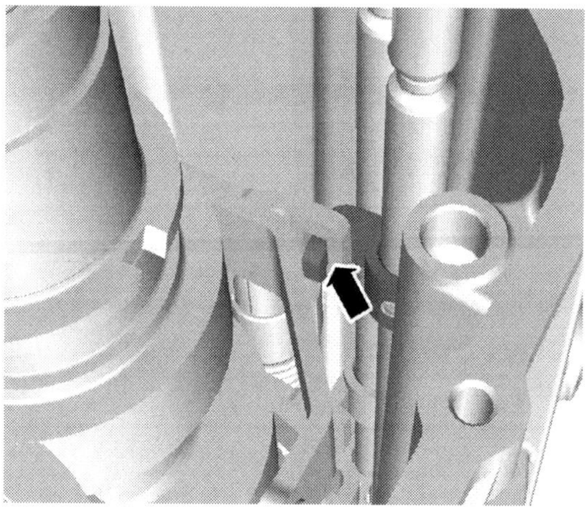

14.

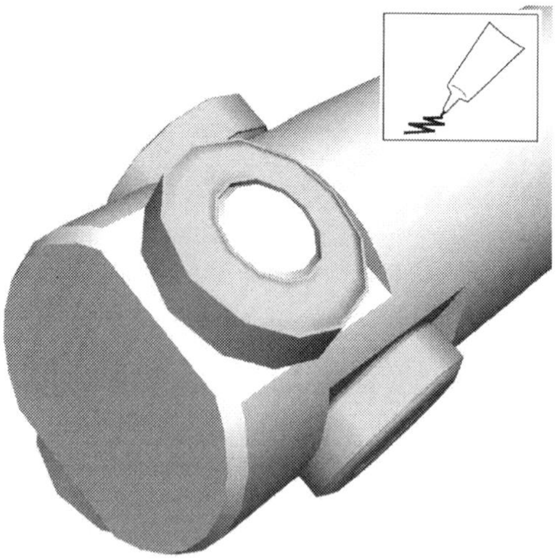

15. △ NOTE: Make sure oil is cleaned off before applying grease.

Lubricate the bearings as indicated with grease, to prevent the bearings falling off the selector during installation.

Material: Grease K-HT (ESD-M1C220-A /)

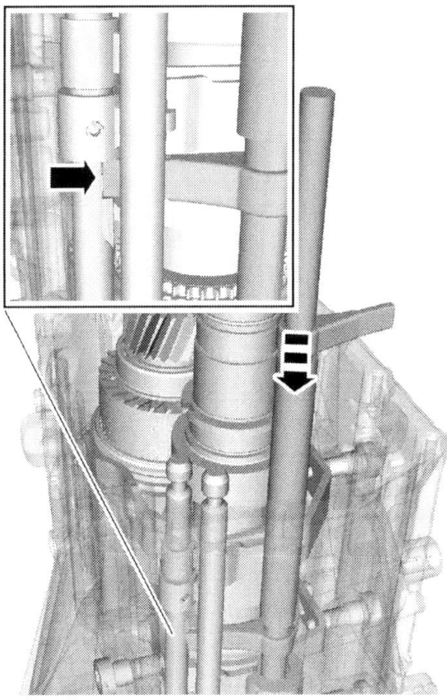

16. Install the main selector.

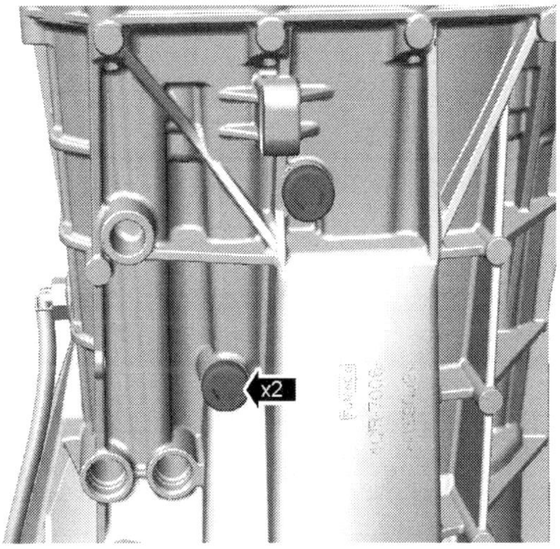

17. Install the 3rd/4th and 5th/6th gear selector bolts. Tighten the bolts.

    Torque: 35 Nm

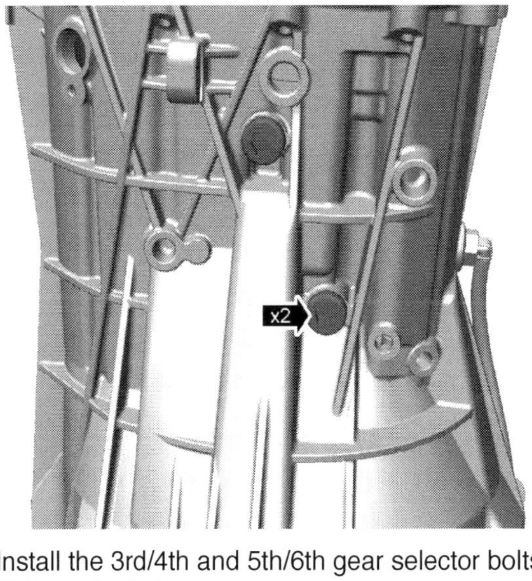

18. Install the 3rd/4th and 5th/6th gear selector bolts. Tighten the bolts.

    Torque: 35 Nm

19. CAUTION: Remove and discard the tie straps.

Install the centre bearings.

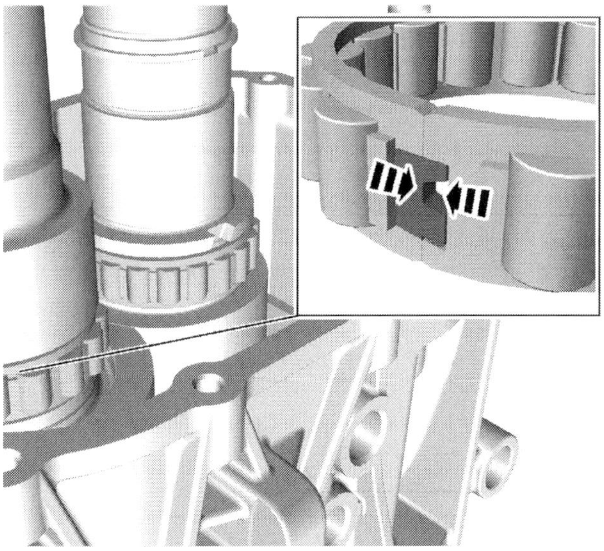

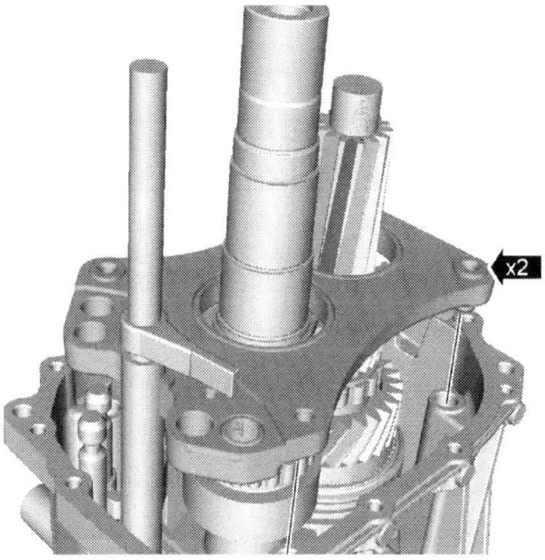

20. Install the centre bearing mounting plate.

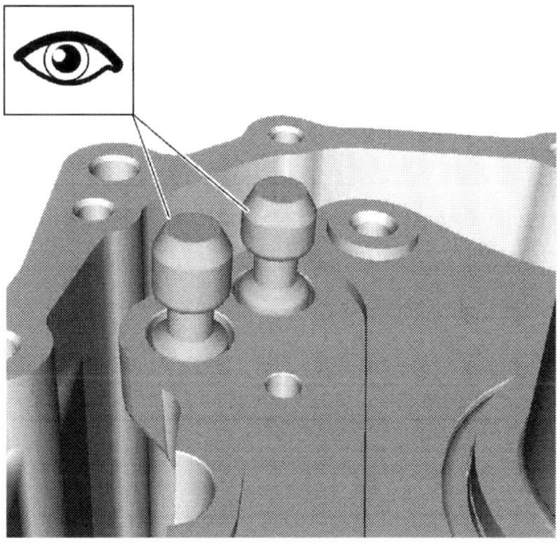

21. Position the rails as indicated.

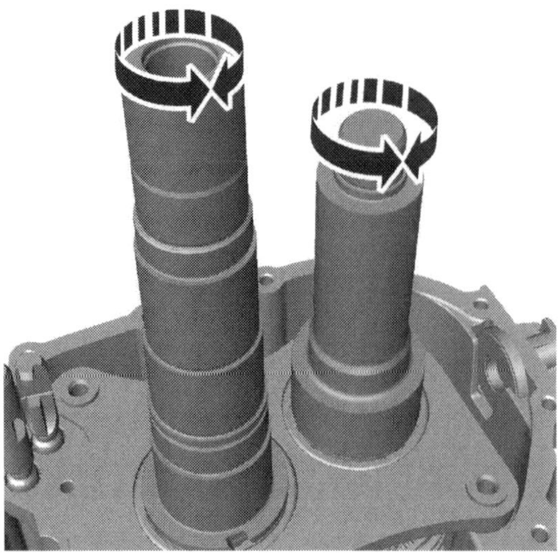

22. Make sure that both shafts are free to rotate.

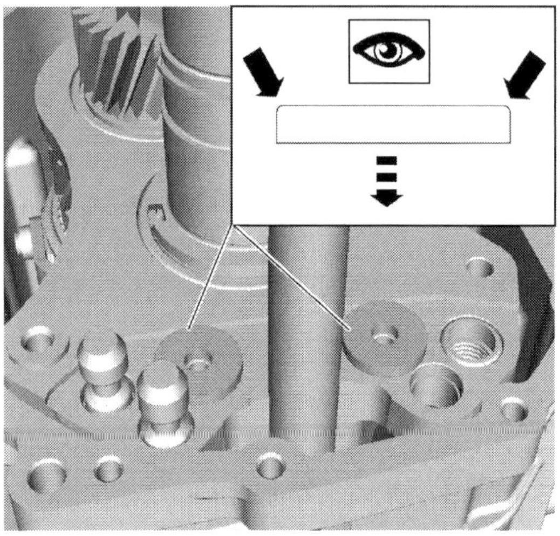

23. ⚠ CAUTION: Note the correct orientation of the washers during the installation.

Install the two blocking plate spacers.

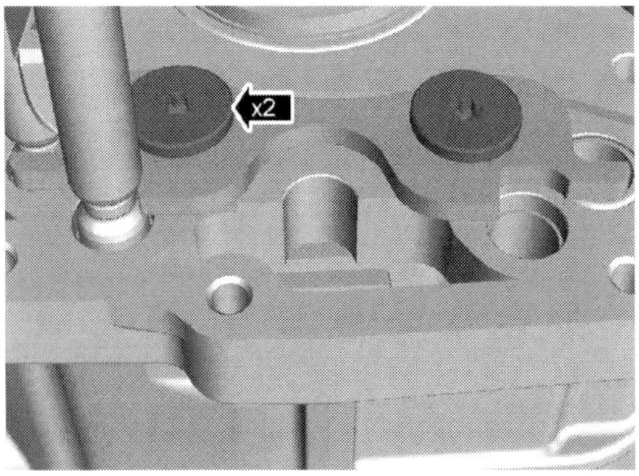

24. ⚠ CAUTION: Make sure the selector shaft locking plate moves freely.

Install the selector shaft locking plate. Tighten the bolts.

Torque: 25 Nm

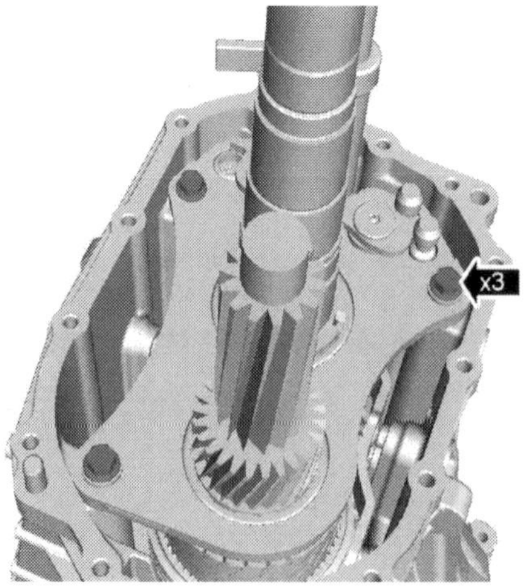

25. △ NOTE: Apply adhesive to the bolts before installing.

Install the centre bearing mounting plate bolts. Tighten the bolts.

Material: Thread Locking MS (WSK-M2G349-A7 /)

Torque: 24 Nm

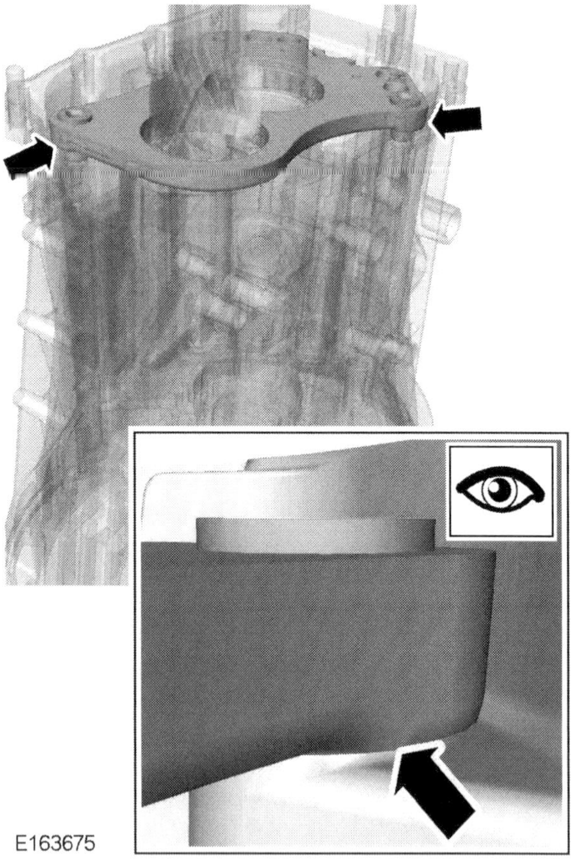

E163675

26. ⚠ CAUTION: Check the centre bearing mounting plate is seated correctly on the transmission casing. If the plate does not seat correctly, remove both gearshafts and begin the assembly process again.

27. ⚠ CAUTION: Make sure that the needle bearings do not fall out during the installation.

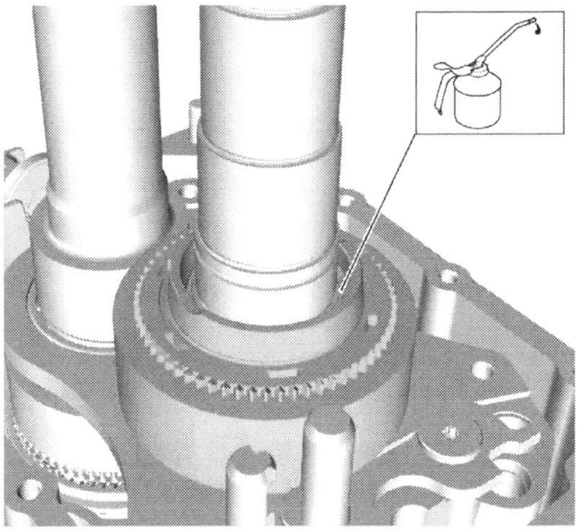

28. Install the 2nd gear and bearing onto the output shaft.

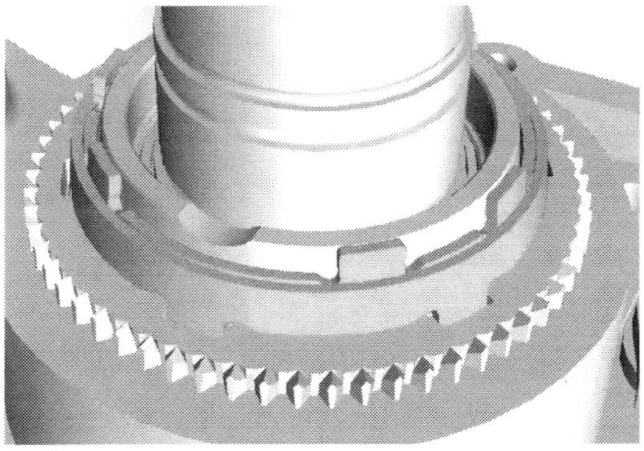

29.

30.

31. CAUTIONS:

⚠ Make sure the synchronizer assembly is correctly aligned to the splines prior to install.

⚠ Make sure the synchronizer assembly is the correct orientation and aligned to the splines prior to install.

⚠ Make sure the output shaft can rotate freely while holding the countershaft.

Using the special tools, install the 1st/2nd gear synchronizer assembly onto the output shaft.

Special Tool(s): 308-604

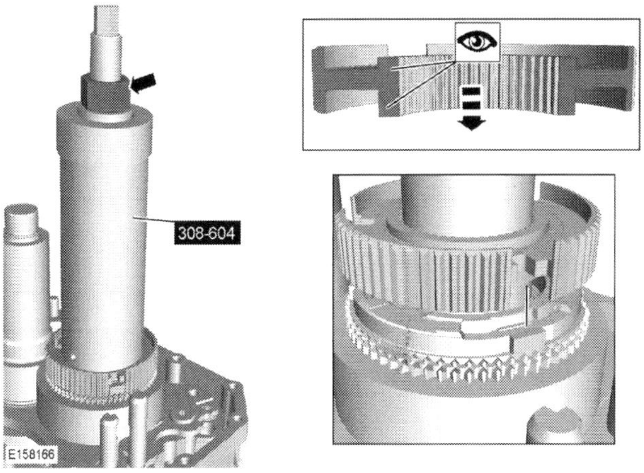

32. ⚠ CAUTION: Make sure the gear rotates freely before installing the snap ring.

Install the 1st/2nd gear synchronizer snap ring. Install a new snap ring.

Renew Part: Circlip

33. ⚠ CAUTION: Make sure the outer sleeve locators are positioned as indicated.

Install the outer sleeve of the synchronizer.

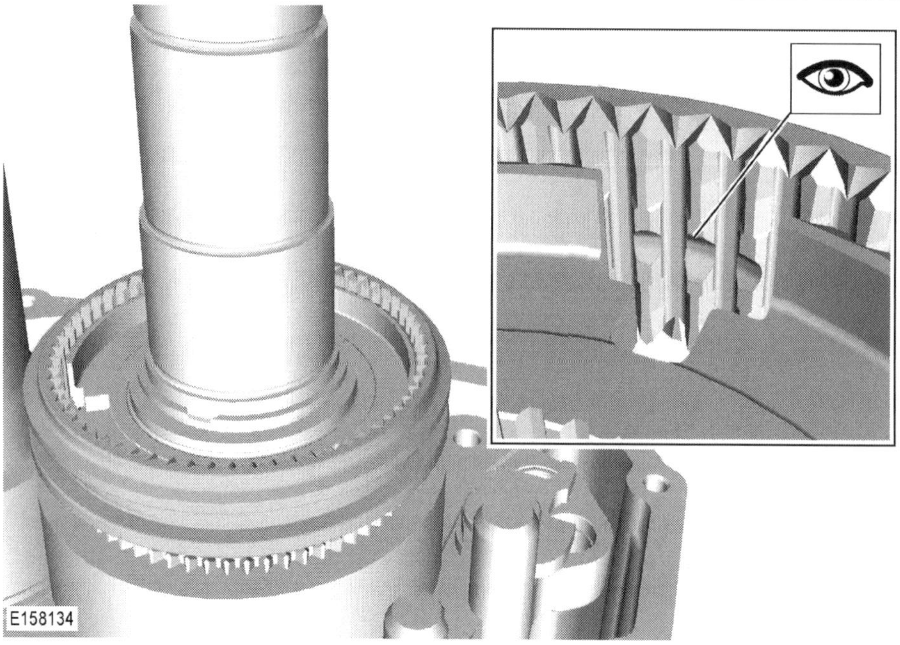

34. Install the spring locators.

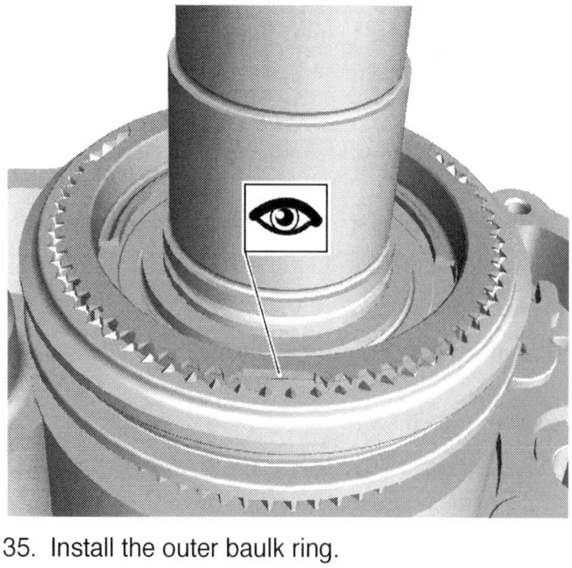

35. Install the outer baulk ring.

36. ⚠ CAUTION: Make sure the baulk rings are seated correctly as indicated.

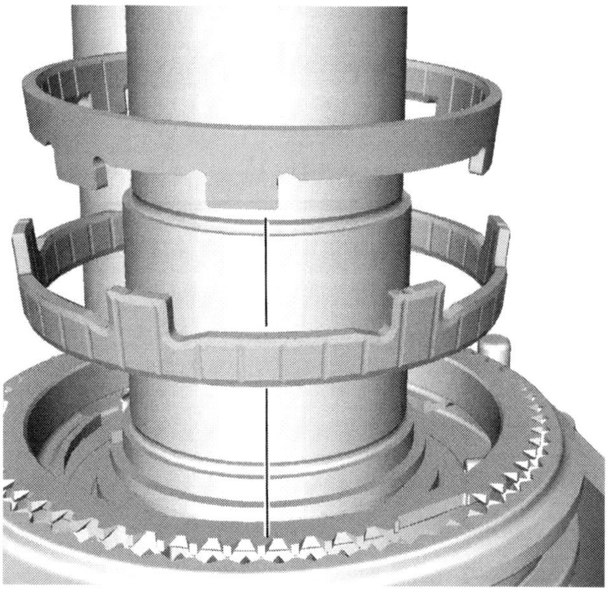

37. Install the 1st gear needle roller bearing.

38. CAUTIONS:

Gently lower the 1st speed gear, rotate the shaft to aid the installation.

Gently lower the 1st gear, rotate the shaft to aid the installation.

Make sure the gear rotates freely.

Install the 1st gear.

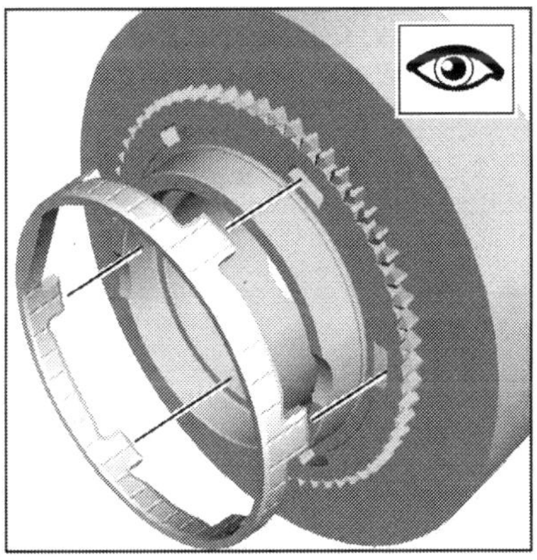

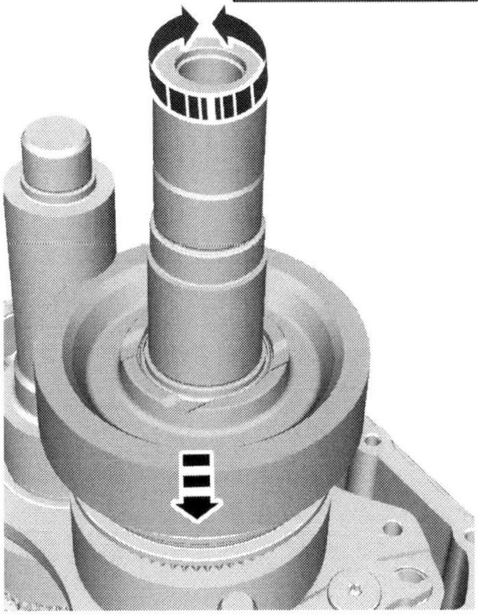

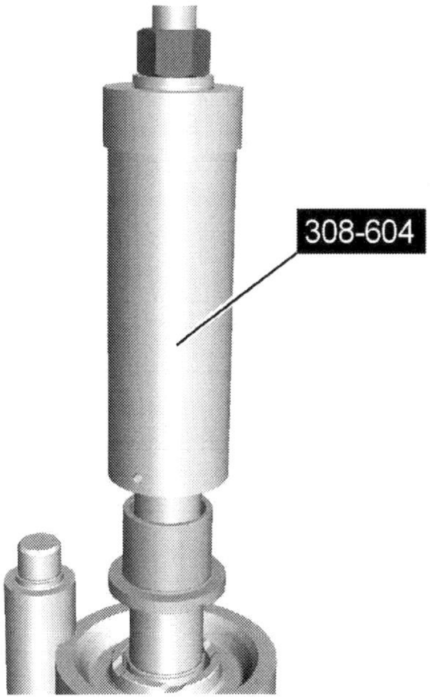

39. CAUTION: Make sure that the shaft rotates freely.

Install the reverse gear inner bearing race.

Special Tool(s): 308-604

40. Install the reverse gear needle roller bearing.

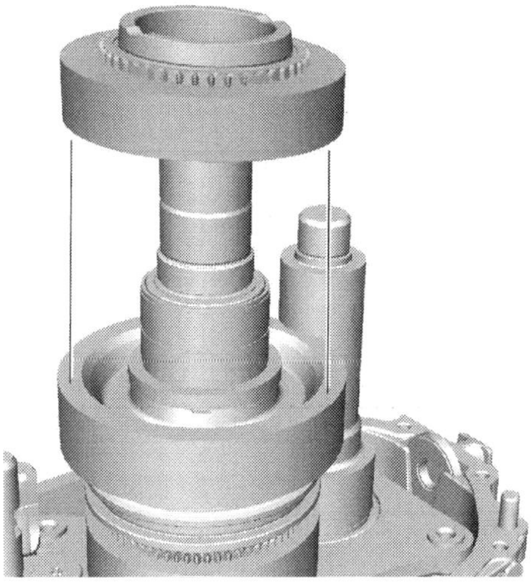

41. CAUTION: Make sure that the shaft rotates freely.

Install the reverse gear onto the output shaft.

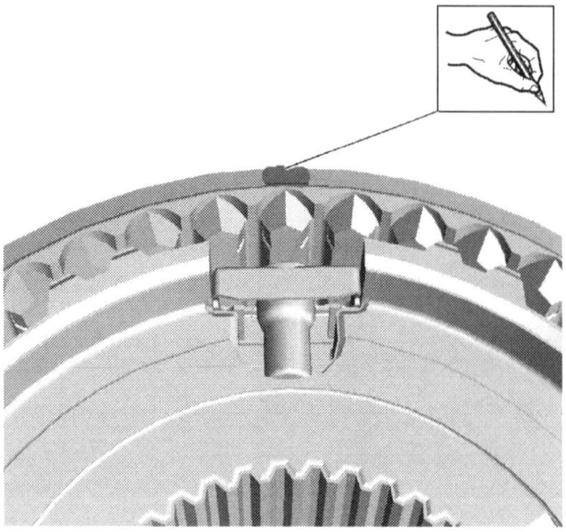

42. Mark the position of one spring locator.

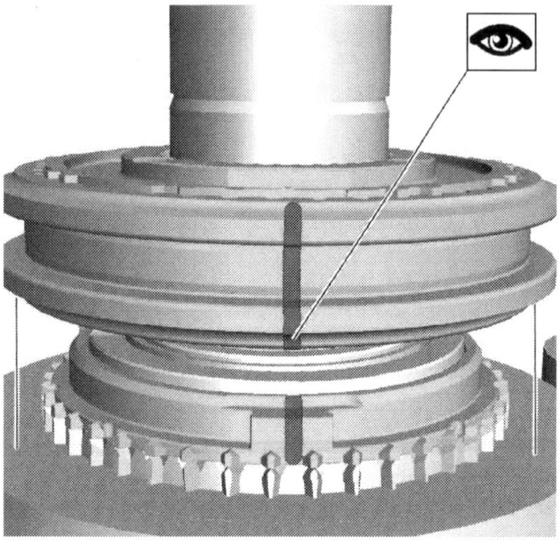

43. Position the reverse gear synchroniser as indicated.

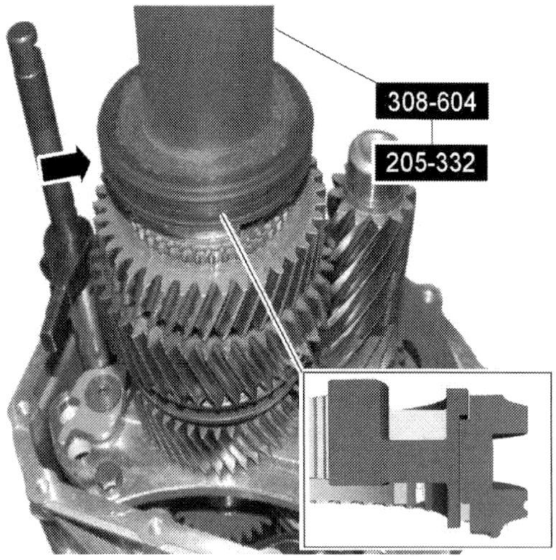

44. ⚠ CAUTION: Make sure the reverse gear rotates freely after installing the gear synchronizer.

Using the special tool, install the reverse gear synchroniser onto the output shaft.

Special Tool(s): 308-604 , 205-332

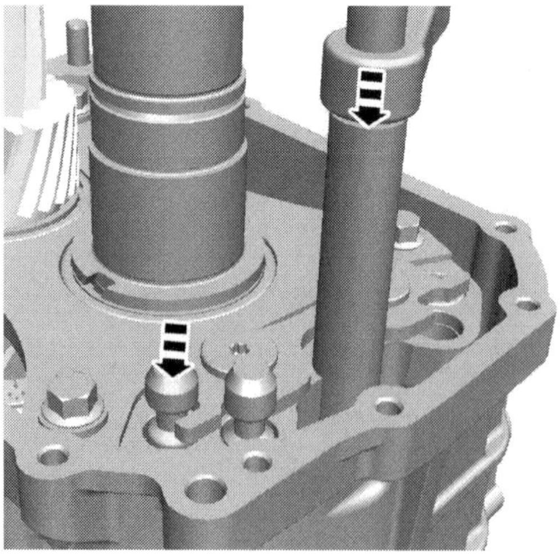

45. Engage 4th gear.

(1)

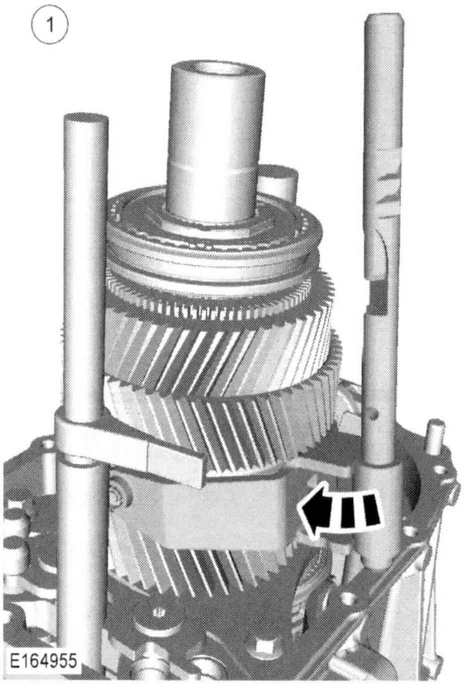

(2)

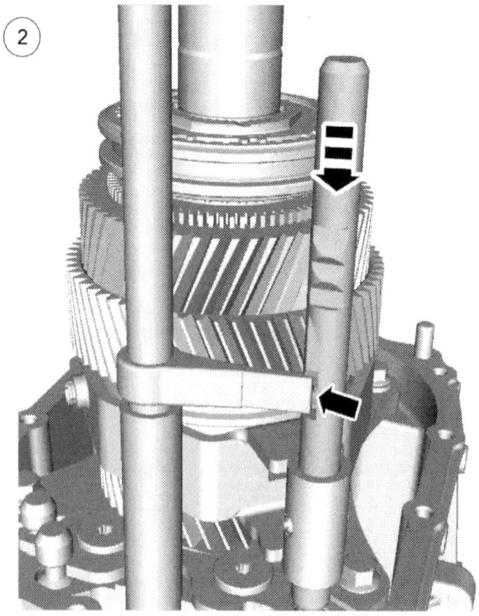

46. CAUTIONS:

⚠ Make sure that there is even protrusion of the pin from each side of the selector.

⚠ Note the orientation of the selector fork.

Install the 1st/2nd gear selector.

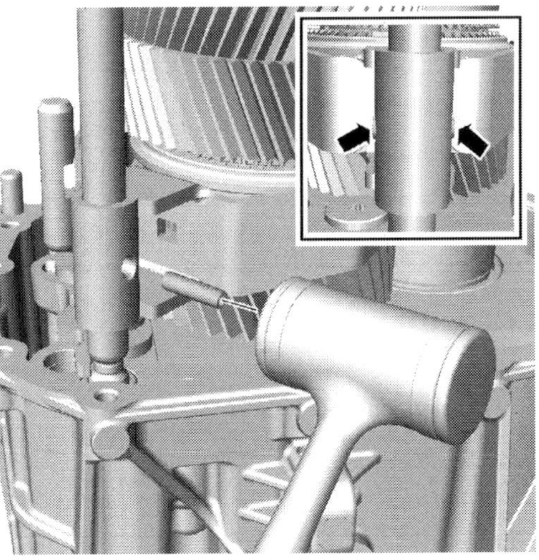

47. ⚠ CAUTION: Make sure the selector rail is supported during the roll pin installation.

Install the roll pin. Install a new pin.

Renew Part: Selector rod pin(s)

48. Install the reverse gear selector fork.

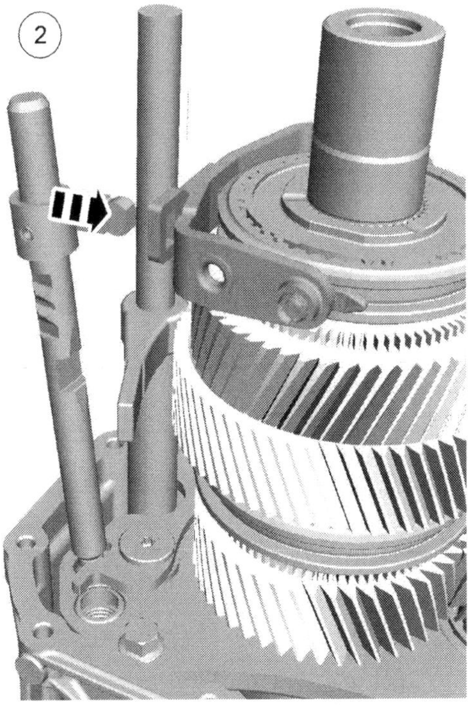

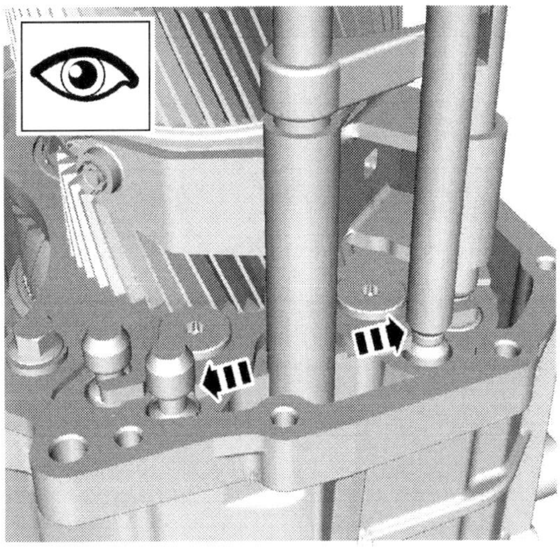

49. ⚠ CAUTION: Make sure the selector shaft locking plate moves freely.

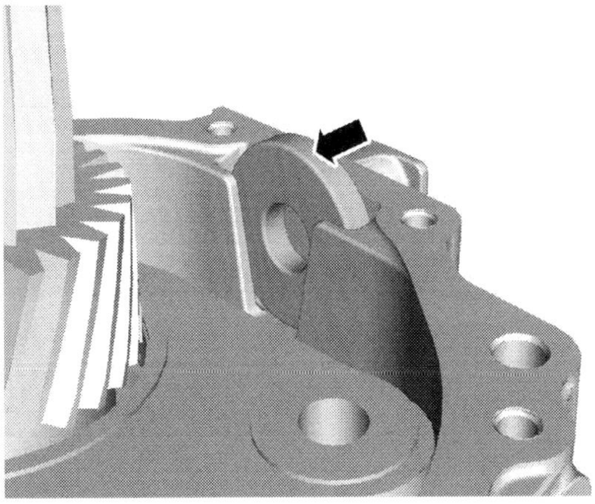

50. Install the magnet.

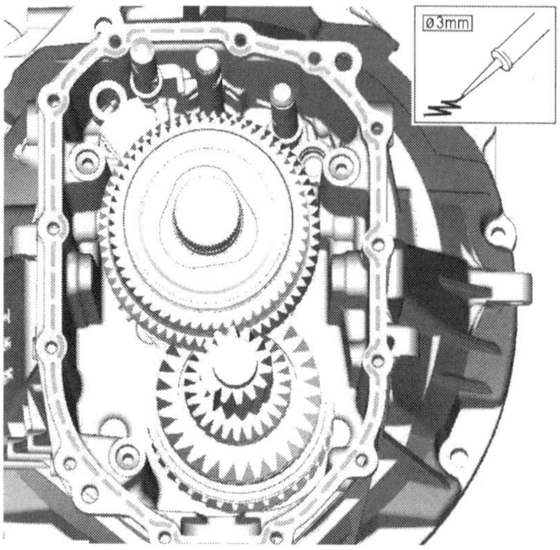

51. CAUTION: Make sure all mating faces are clean and dry before applying sealant.

NOTE: Make sure that there are no gaps when applying the sealant.

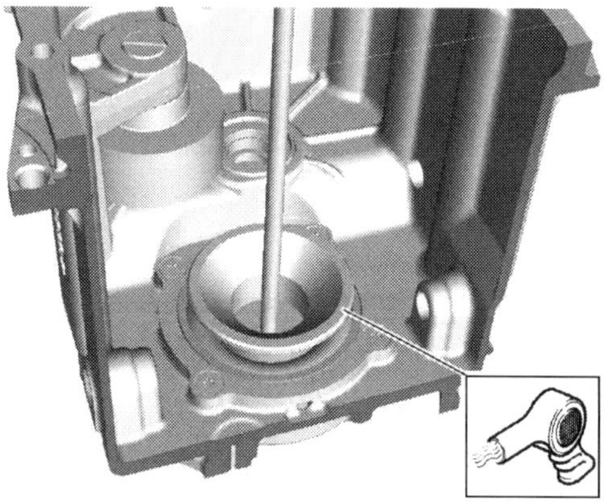

Apply sealant to the mating surfaces.

Material: Flange Sealant - Anaerobic LR (WSK-M2G348-A5 /)

52. WARNING: Care should be taken when using the hot air gun. Failure to follow this instruction may result in personal injury.

NOTE: Heat up the special tool for 4 minutes with a hot air gun to around 100 °C.

Using the special tool, apply heat to the bearing inner race.

53. NOTE: Rotate the output shaft clock-wise to aid the installation of the rear transmission housing.

Install the transmission housing. Apply sealant to the mating surfaces.

Special Tool(s): 308-604

Material: Flange Sealant - Anaerobic LR (WSK-M2G348-A5 /)

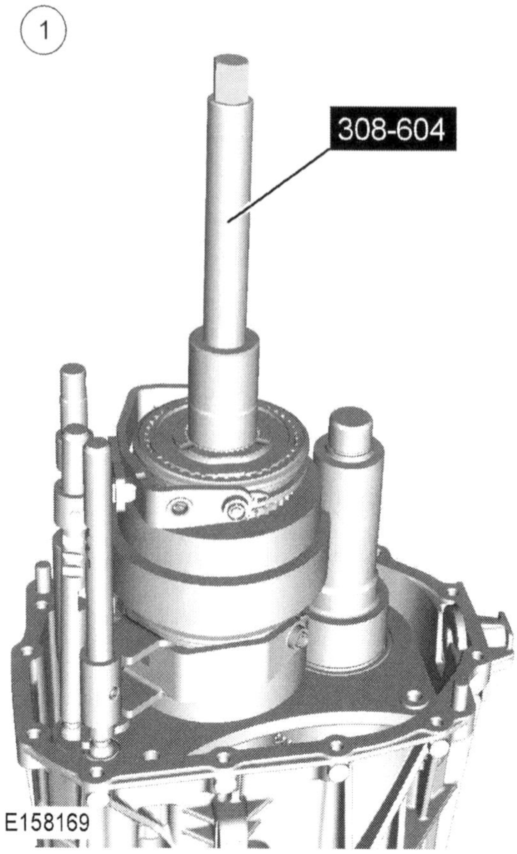

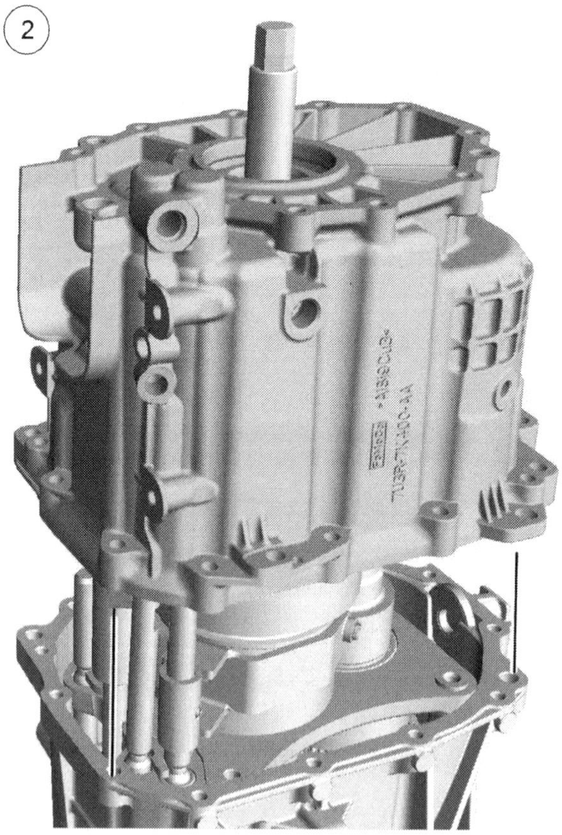

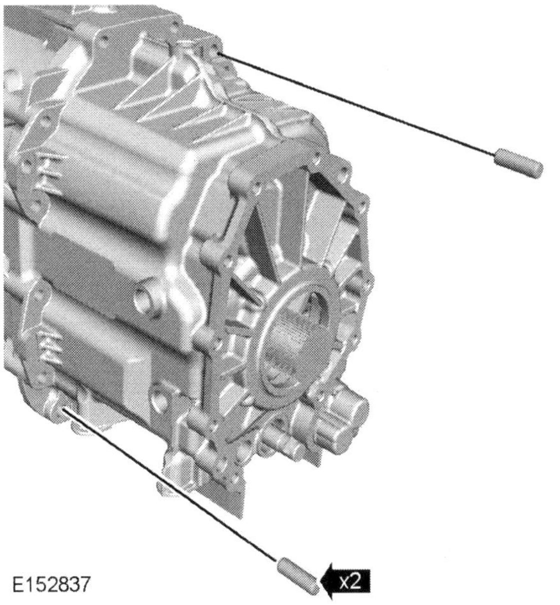

54. Install the locating dowels.

55. Tighten the transmission housing bolts, working diagonally. Tighten the bolts.

Torque: 24 Nm

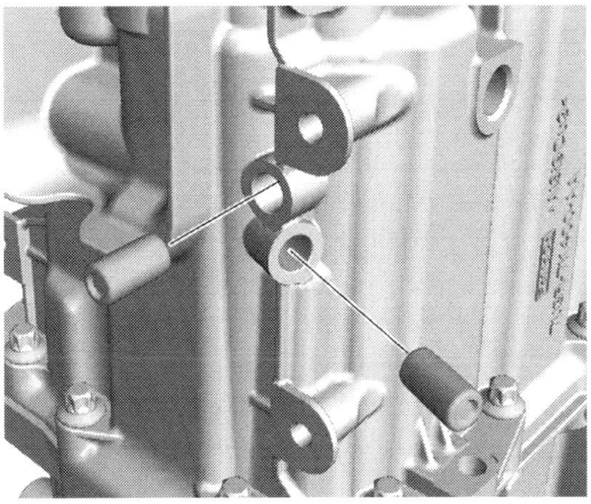

56. Install the two indents.

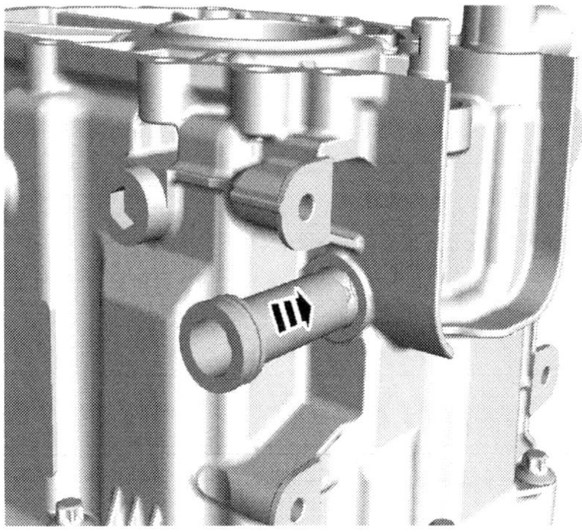

57. Install the indent plunger.

58. Install the reversing lamp switch.

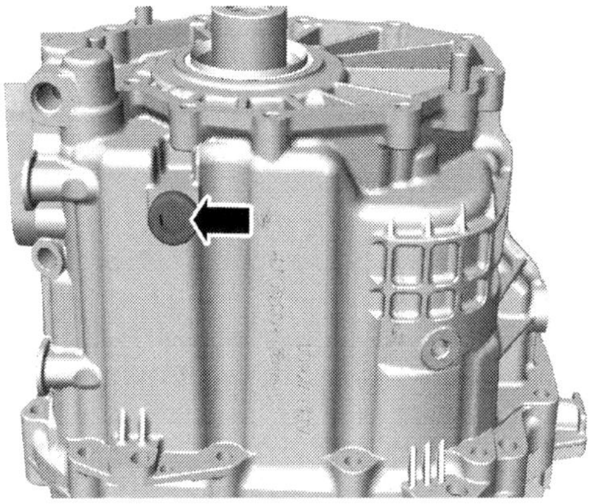

59. Install the reverse gear selector fork mounting bolts. Tighten the bolt.

Torque: 37 Nm

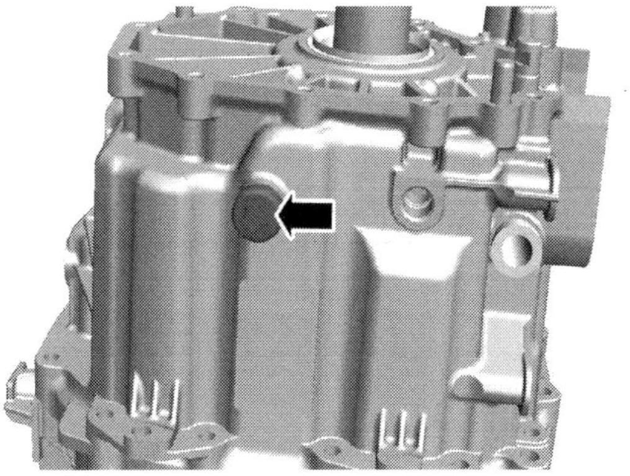

60. Install the reverse gear selector fork mounting bolts. Tighten the bolt.

Torque: 37 Nm

61. Install the input shaft snap ring. Install a new snap ring.

Renew Part: Circlip

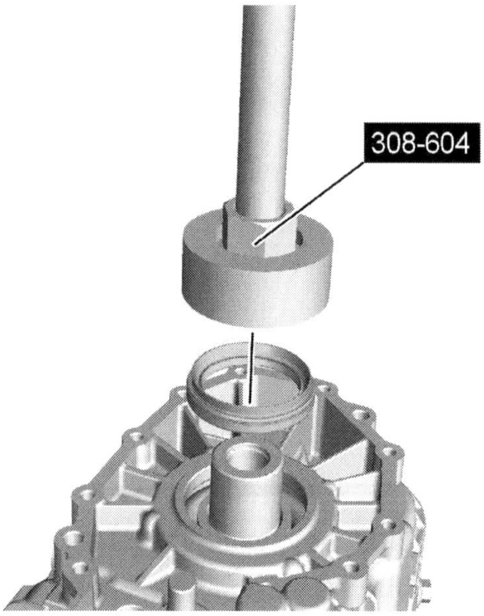

62. Install the transmission output shaft seal.

Special Tool(s): 308-604

Renew Part: Transmission output shaft flange seal

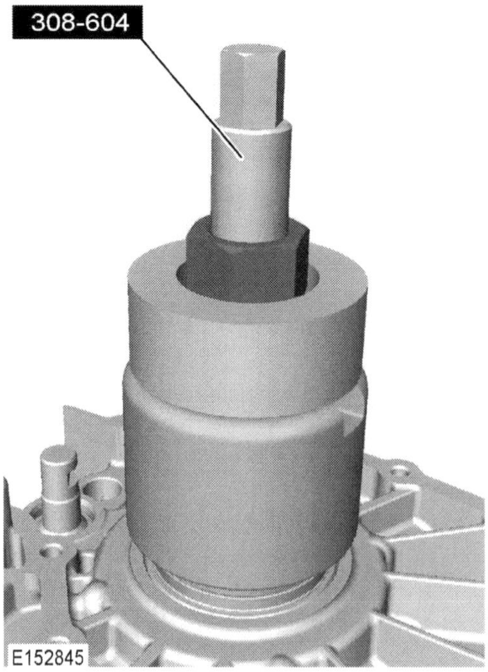

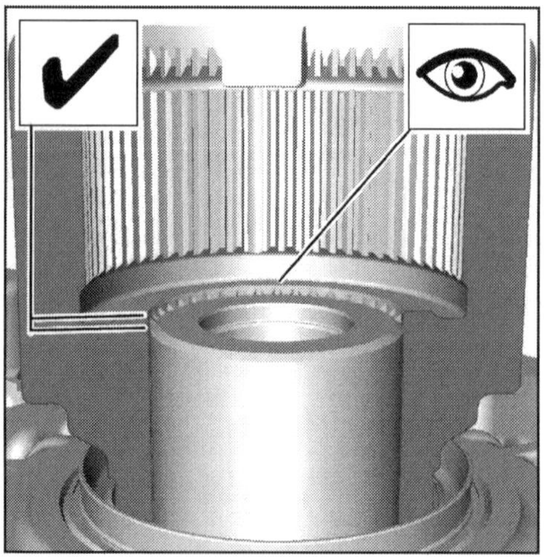

Install the transmission output flange.

63. ⚠ CAUTION: Apply anti seize grease to the splines, failure to follow this step could result in damage to the vehicle.

△ NOTE: Note the position of the output flange.

Install the transmission output flange. *Special Tool(s):* 308-604

Material: Castrol Spheerol (LCX6002 /)

64. △ NOTE: Apply anti-seize grease to the splines prior to installation.

Using the special tool, install the transmission output flange bolt. Tighten the bolt, then remove the bolt. Apply thread locking compound adhesive. Tighten the bolt again.

Material: Thread Locking MS (WSK-M2G349-A7 /)

Torque:

Stage 1: 210 Nm

Stage 2: 180 Nm

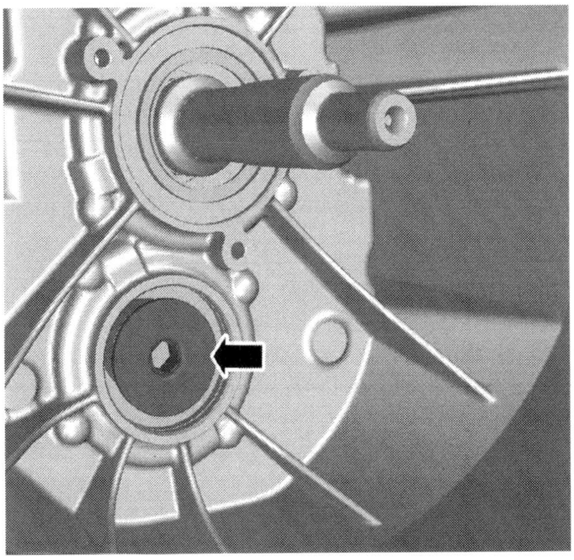

65. ⚠ CAUTION: Inspect the countershaft securing bolt for damage before installation.

NOTES:

⚠ Apply adhesive to the threads.

⚠ Make sure that a gear is selected.

Install the countershaft securing bolt and tighten.

Material: Thread Locking MS (WSK-M2G349-A7 /)

Torque: 90 Nm

66. Using the special tool, install the transmission input shaft seal and countershaft seal.

Special Tool(s): 308-605

Renew Part: Input shaft seal

Renew Part: Transmission countershaft seal

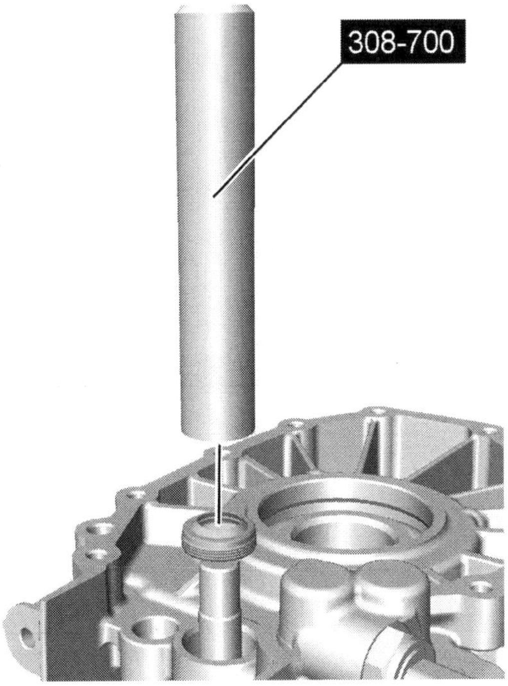

67. Using the special tool, install the main selector shaft seal.

68. Move the main selector shaft to neutral.

69. ⚠ NOTE: Make sure the pin is installed in the orientation illustrated.

Install the selector shaft arm and tighten.

Torque: 12 Nm

70. Rotate the transmission shafts and check the availability of all gears.

71. Using the special tools, check the torque in neutral with the output flange blocked.

Special Tool(s): 205-067 , 205-175

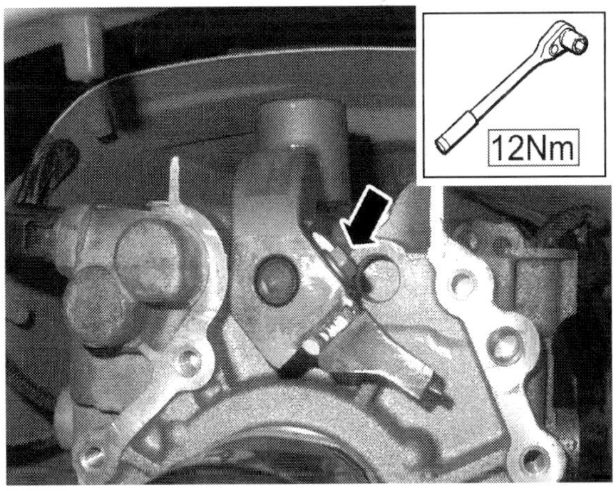

72. Remove the transmission from the stand.

73. ⚠ WARNING: Do not work on or under a vehicle supported only by a jack. Always support the vehicle on safety stands.

Raise and support the vehicle.

74. Install the transmission.

Refer to: Transmission (308-03 Manual Transmission/Transaxle, Vehicles With: MT82 6-Speed Manual Transmission Installation)

75. Fill the transmission.

Refer to: Transmission Draining and Filling (308-03 Manual Transmission/Transaxle, Vehicles With: MT82 6-Speed Manual Transmission General Procedures)

76. Connect the battery ground cable.

Refer to: Startup Battery Disconnect and Connect (414-01 Battery, Mounting and Cables, General Procedures)

Transmission

Removal

1. Open the front door.
2. **Refer to:** Floor Console (501-12 Instrument Panel and Console, Removal and Installation)

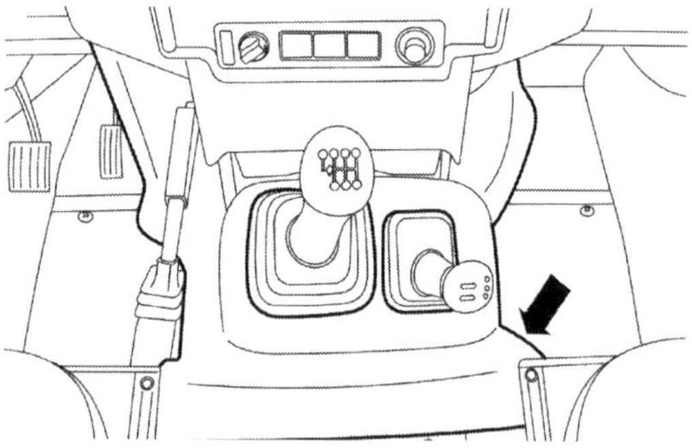

3. Remove the transmission cover panel carpet.

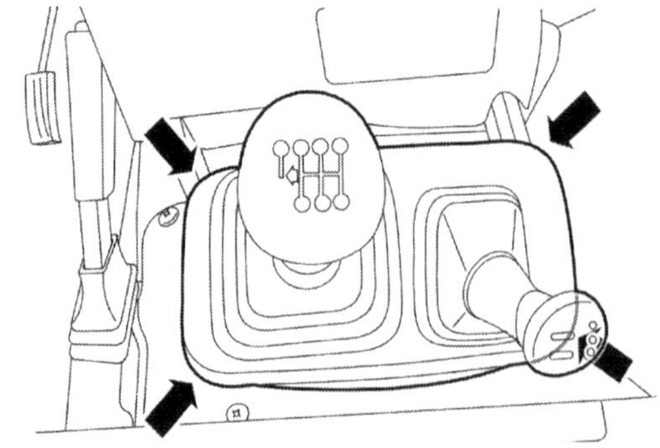

4. Release the gear selector lever gaiter, from the high-low selector lever. Release the gear selector lever gaitor from the transmission tunnel.

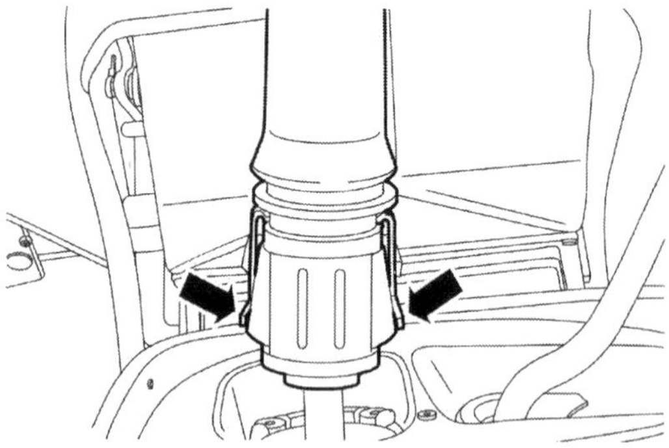

5. ⚠ WARNING: The gearshift lever knob will be released suddenly, keep face clear during removal. Failure to follow this instruction may result in personal injury.

Remove the upper gear change lever.

6. Remove the gear selector lever gaiter.
7. Disconnect the reverse light switch electrical connector.

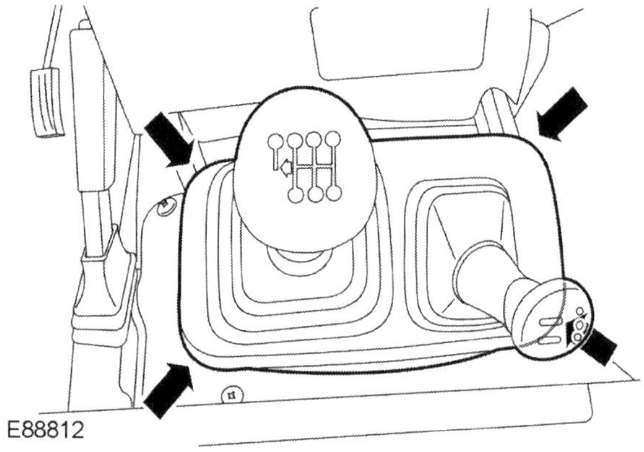

8. Open the bonnet for access.
9. **Refer to:** Diesel Particulate Filter (309-00 Exhaust System, ID4 2.2L Diesel Description and Operation)
10. ⚠ WARNING: Do not work on or under a vehicle supported only by a jack. Always support the vehicle on safety stands.

Raise the vehicle on lift.

11. **Refer to:** Front Muffler (309-00 Exhaust System - ID4 2.2L Diesel, Removal and Installation). ID4 2.2L Diesel

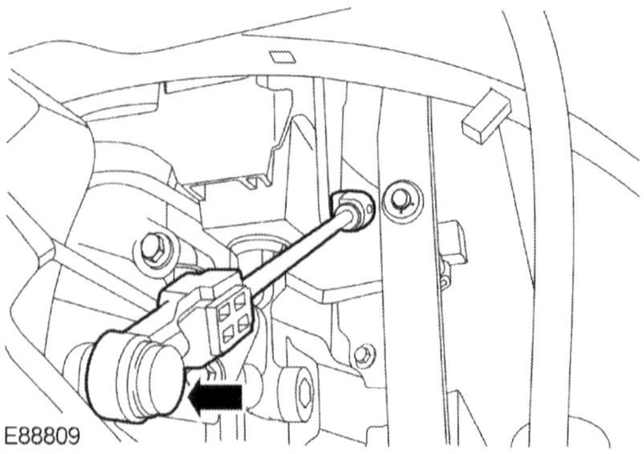

12. Release the high-low selector rod ball joint from the transfer gearbox.

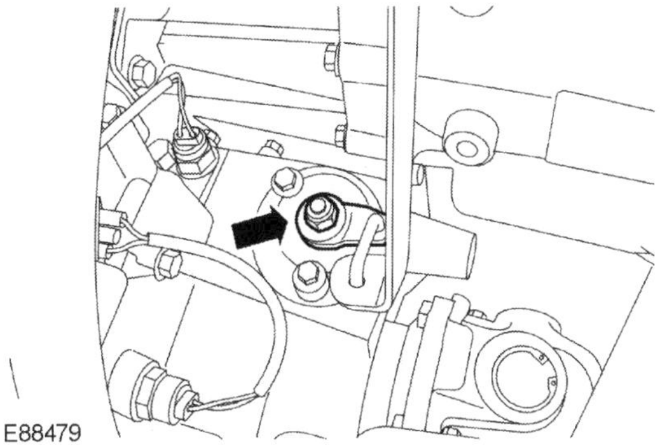

E88479

13. Remove the securing nut at the lower end of pivot arm and disconnect differential lock control operating rod.
14. Lower the vehicle.
15. Open the front door.

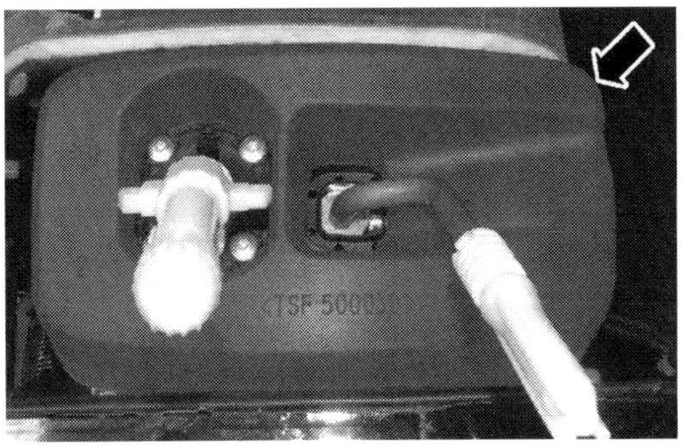

16. Remove the foam pad from around the selector levers.

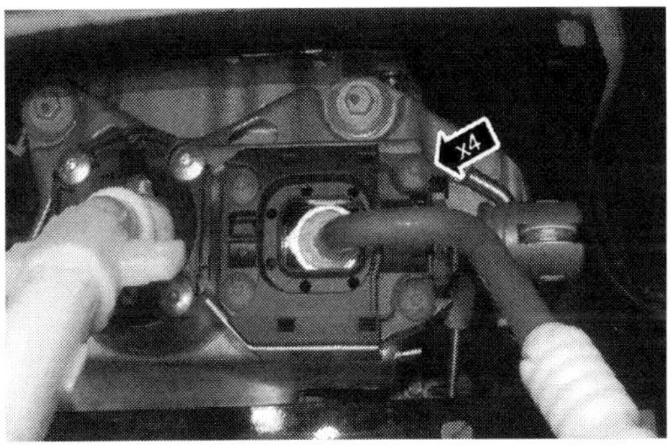

17. Remove the 4 high-low selector lever securing bolts.
18. Remove the high-low selector lever.

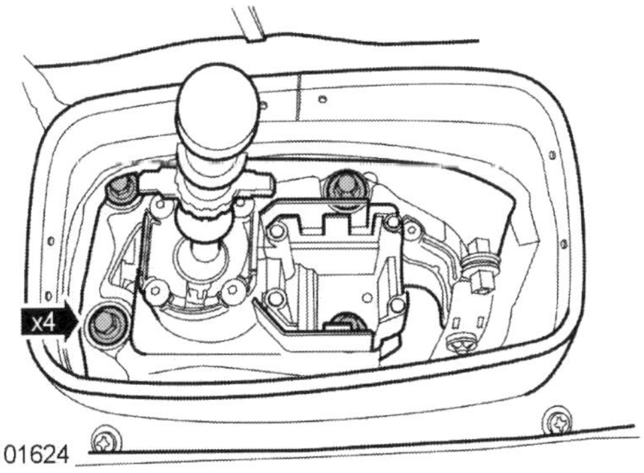

E101624

19. Remove the 4 gear selector housing securing bolts.
20. ⚠ WARNING: Do not work on or under a vehicle supported only by a jack. Always support the vehicle on safety stands.

Raise the vehicle on lift.

21. **Refer to:** Front Driveshaft (205-01 Driveshaft, Removal and Installation)

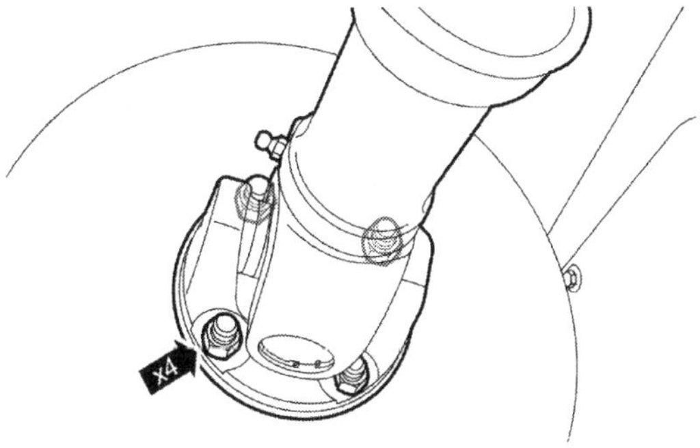

22. Displace and reposition the rear driveshaft. Mark the position of the driveshaft in relation to the drive pinion flange. Remove the rear driveshaft to transfer gearbox securing nuts.

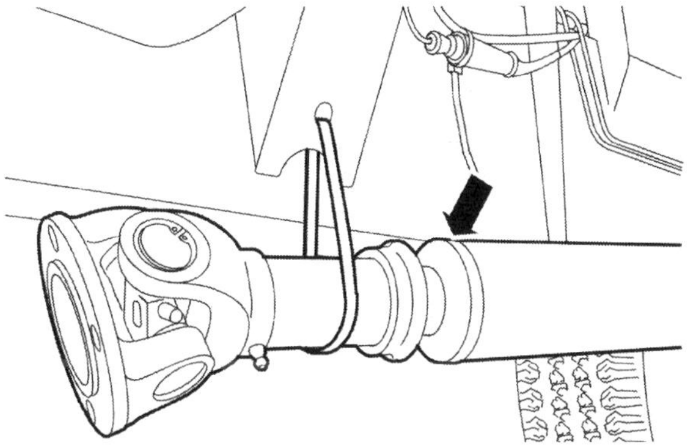

23. Using a suitable tie strap, secure the rear driveshaft to the chassis.

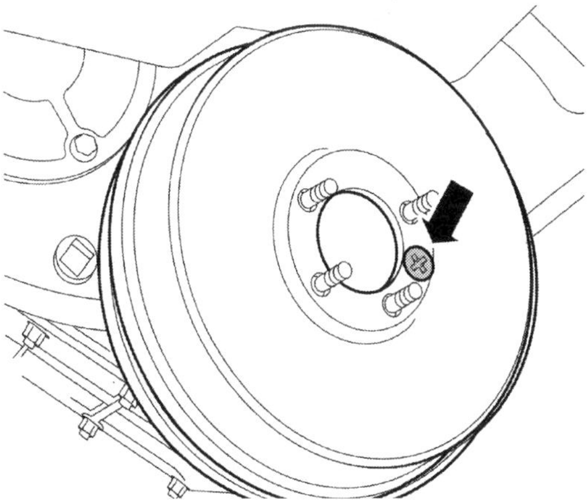

24. Remove the parking brake drum. Remove the securing screw.

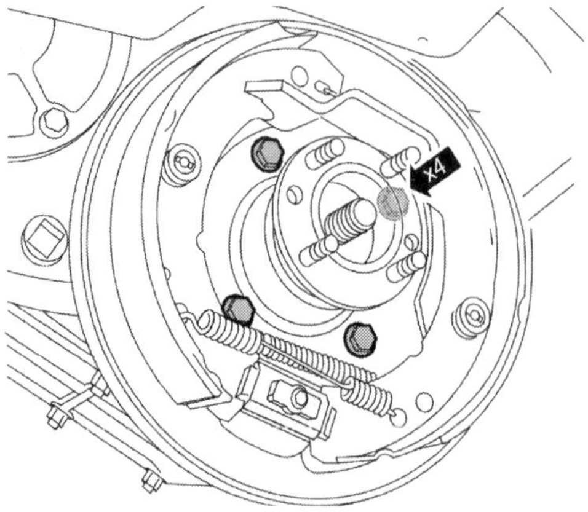

25. Remove the handbrake assembly and tie aside.

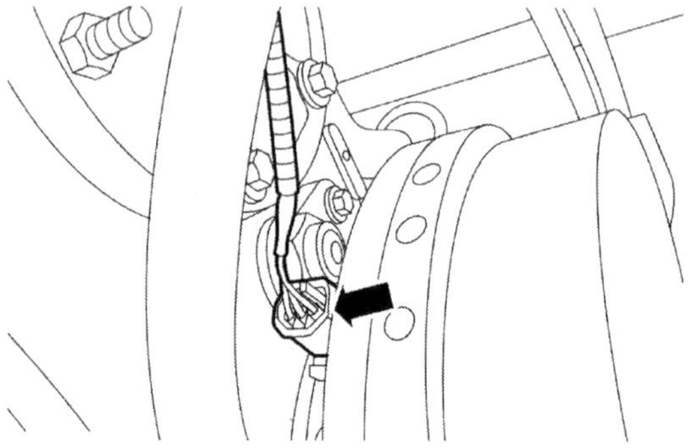

26. Disconnect the electronic speedometer drive electrical connector from the transfer gearbox.

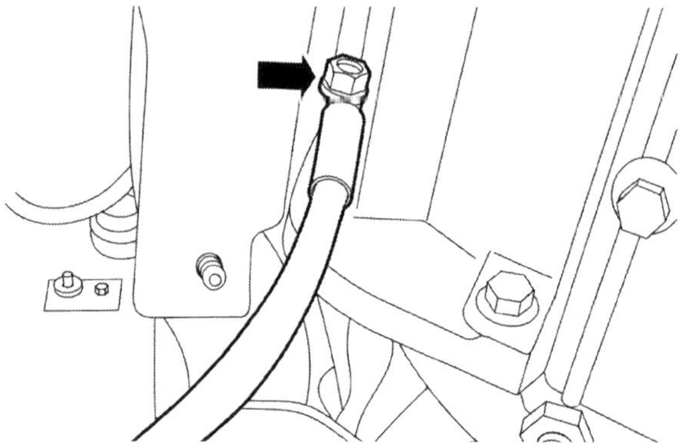

27. Remove the nut and disconnect the LH earth cable from the transfer gearbox.

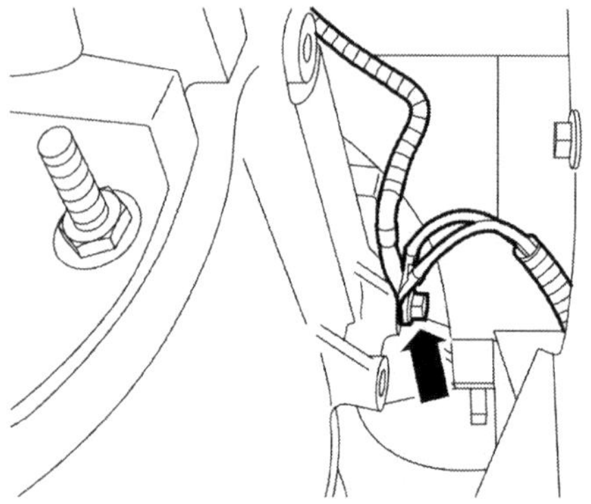

28. Remove the bolt and disconnect the RH earth cables from the transfer gearbox.

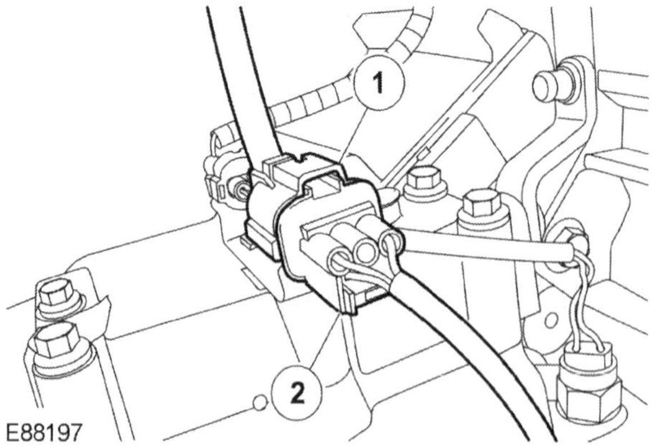

29. Displace and reposition the high-low detection switch electrical connector away from the transfer gearbox. Disconnect the high-low detection switch electrical connector.

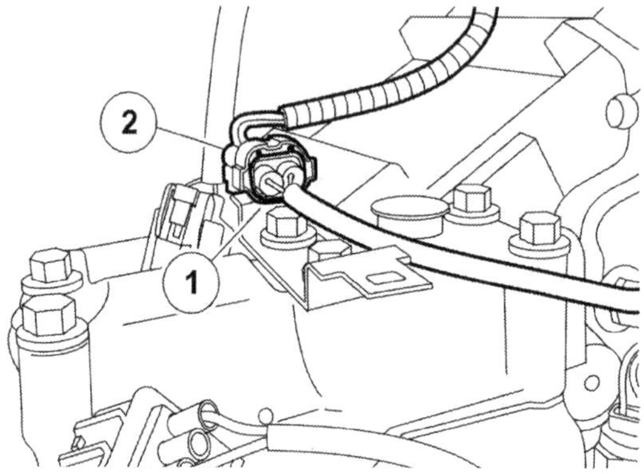

30. Disconnect the differential lock detection switch electrical connector. Release the electrical connector from the bracket.

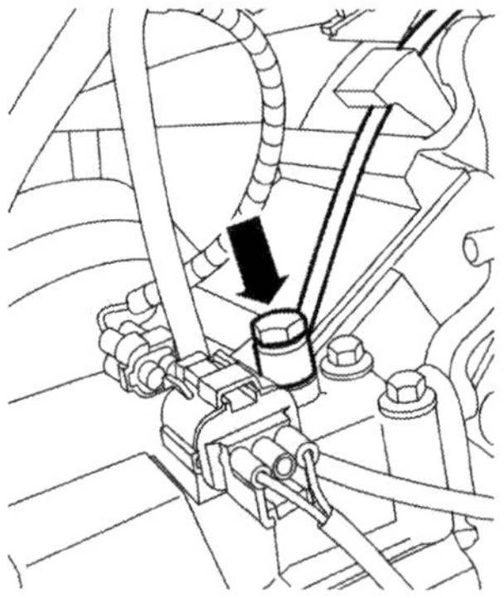

31. Release the breather pipe from the transfer gearbox. Remove the securing bolt. Remove and discard the 2 sealing washers.

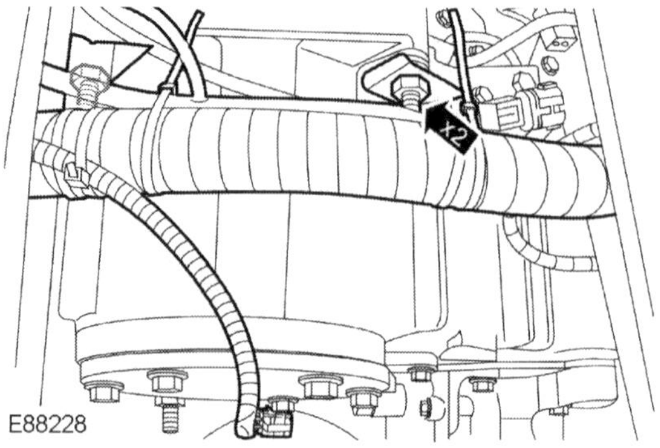

32. Remove the transmission harness retaining bracket securing nut and bolt and reposition the harness.

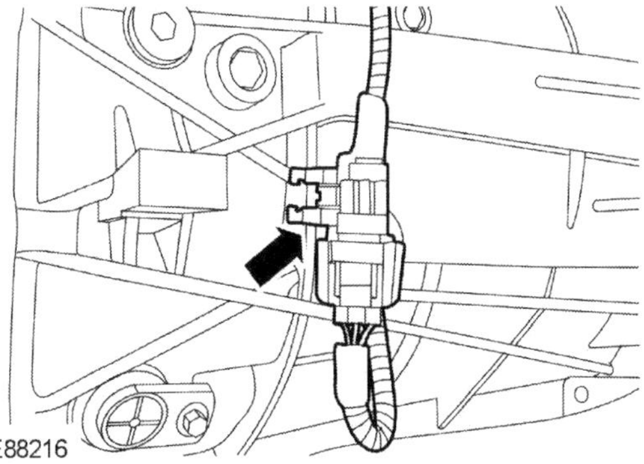
E88216

33. Release the transmission wiring harness connector and bracket from the RH side of the transmission.

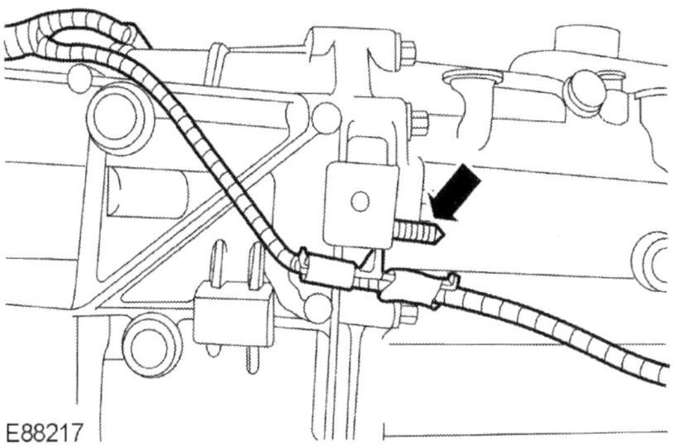
E88217

34. Release the transmission wiring harness securing clip from the LH side of the transmission.

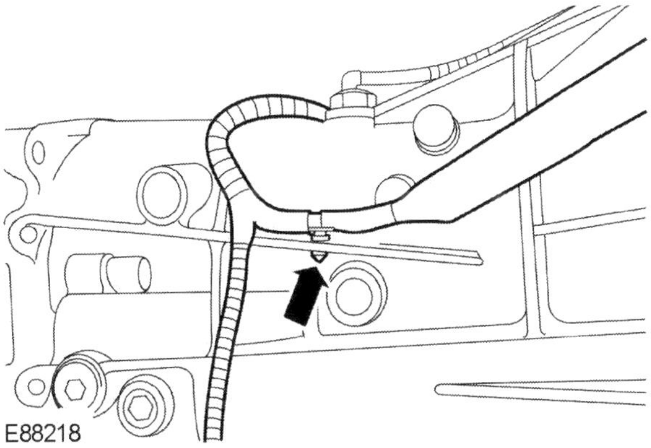
E88218

35. Release the transmission wiring harness securing clip from the RH side of the transmission.

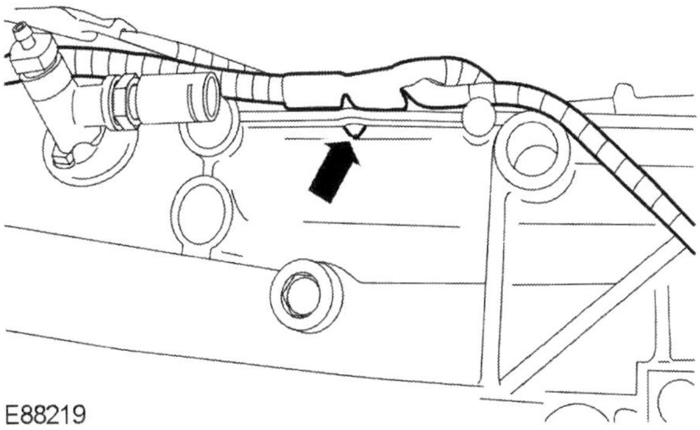
E88219

36. Release the transmission wiring harness securing clip from the top of the transmission.

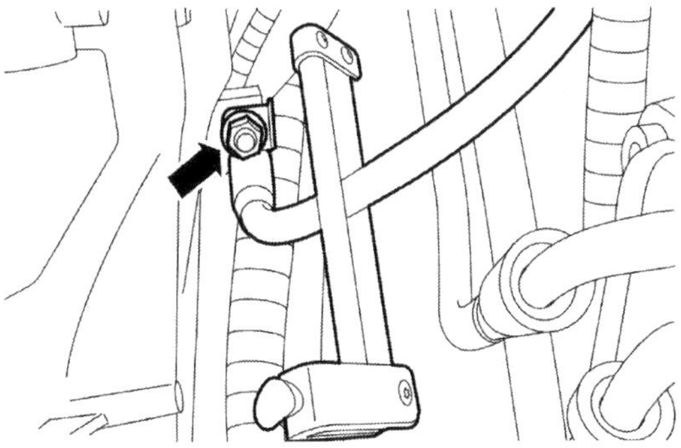

37. Remove the clutch slave cylinder fluid hose mounting bracket securing nut and bolt. Install a suitable pipe clamp to the clutch slave cylinder fluid hose.

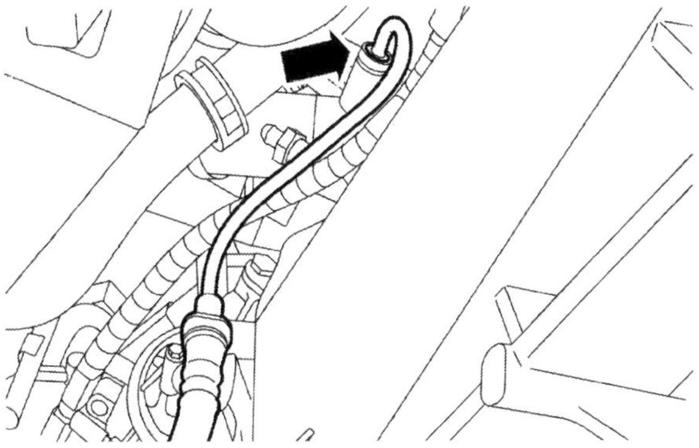

38. Disconnect the clutch slave cylinder fluid hose from the clutch slave cylinder. Remove the clutch slave cylinder line clip. Remove and discard the O-ring seal

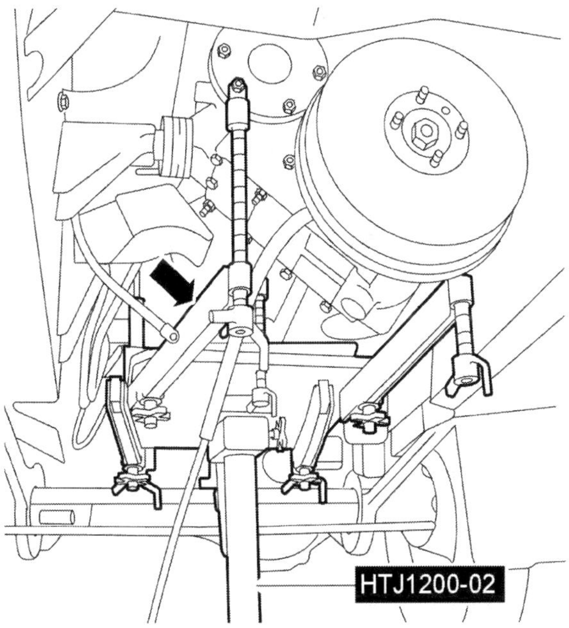

39. Install the special tool HTJ1200-02 to the transmission and transfer gearbox.

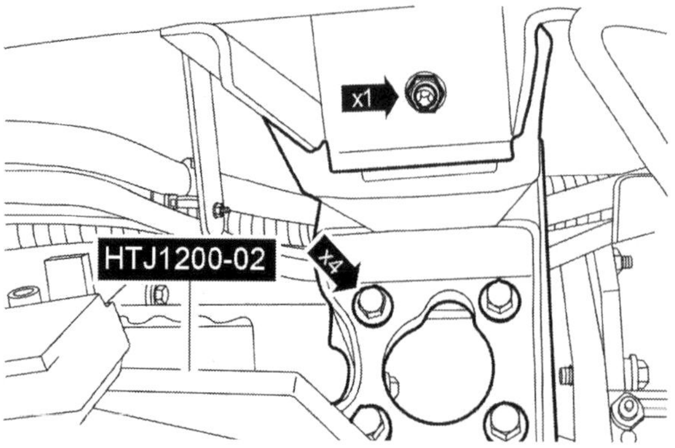

40. Remove the LH transmission mount and mounting bracket. Remove the LH transmission mount and mounting bracket securing nuts and bolts.

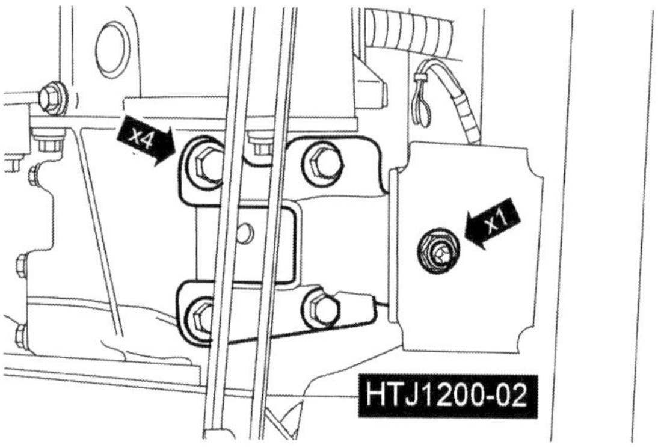

41. Remove the RH transfer gearbox mounting bracket. Remove the RH transfer gearbox mount and mounting bracket securing nuts and bolts.

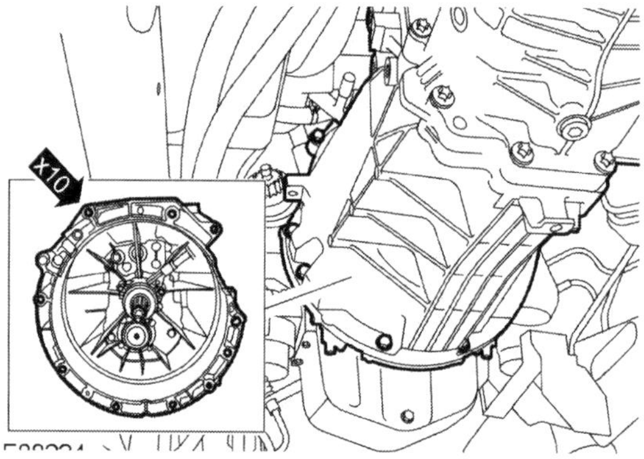

42. Remove the bell housing bolts.

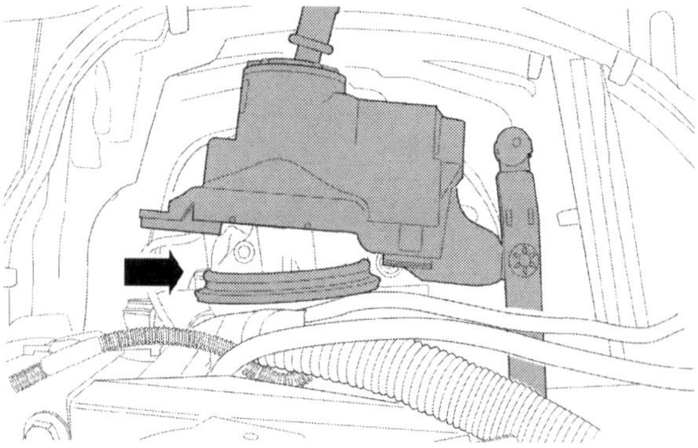

43. Remove the gear selector housing. Reposition the transmission to allow access to the gear selector housing. Remove from below the vehicle. Remove the rubber seal.

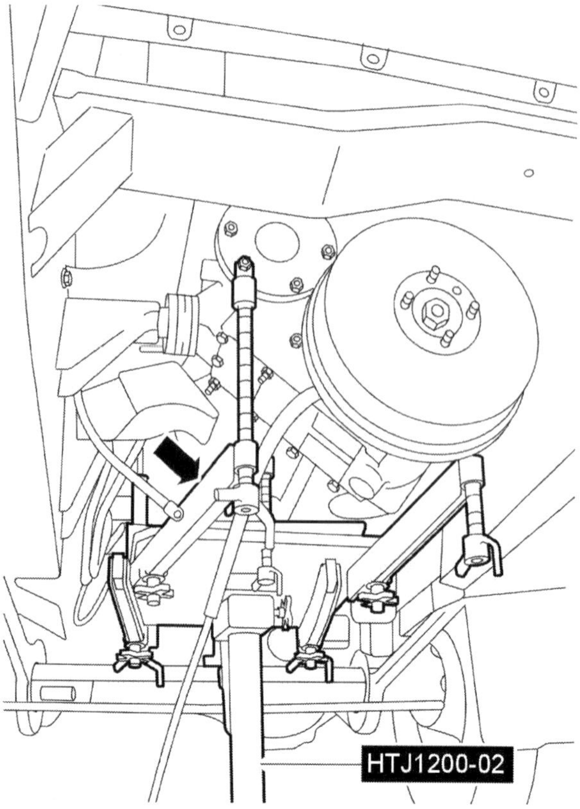

44. WARNING: Disengage the transmission from the clutch by 30mm before lowering HTJ1200-02 to protect clutch from damage.

Remove the transmission and transfer gearbox from the vehicle.

45. NOTE: Do not disassemble further if the component is removed for access only.

Remove the transmission from the special tool.

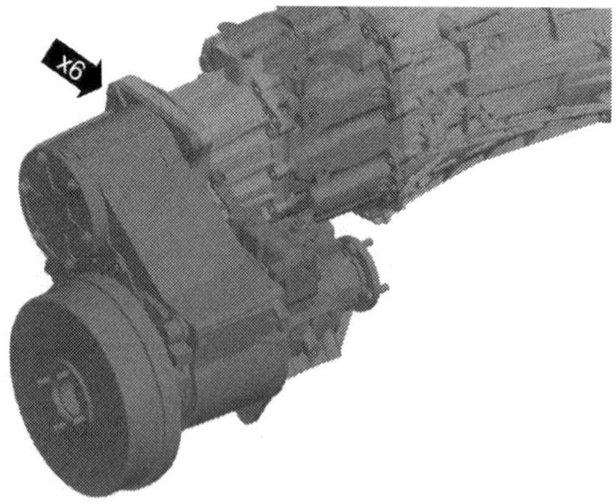

E90815

46. With assistance, remove the transfer case. Remove the four bolts. Remove the two nuts.

47. Remove the 10 bolts. Remove the transmission extension housing.

48. Remove the transmission extension shaft cover. Remove the tie strap. Remove the cover.

49. Remove the seal.

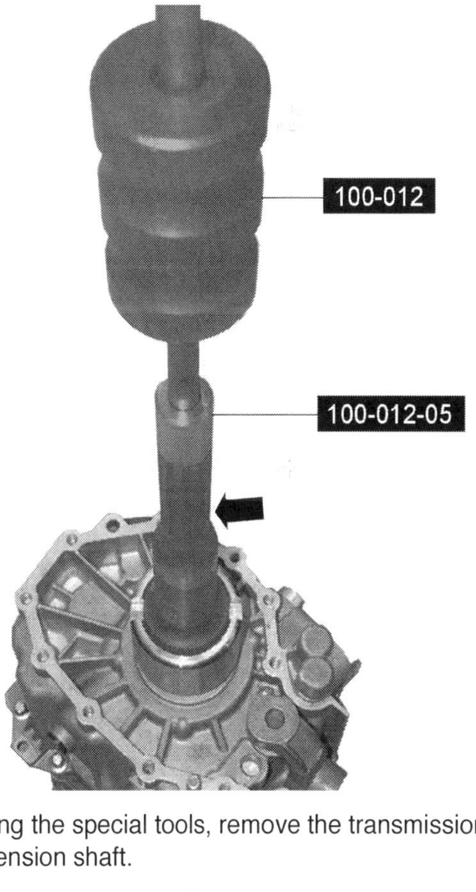

50. Using the special tools, remove the transmission extension shaft.

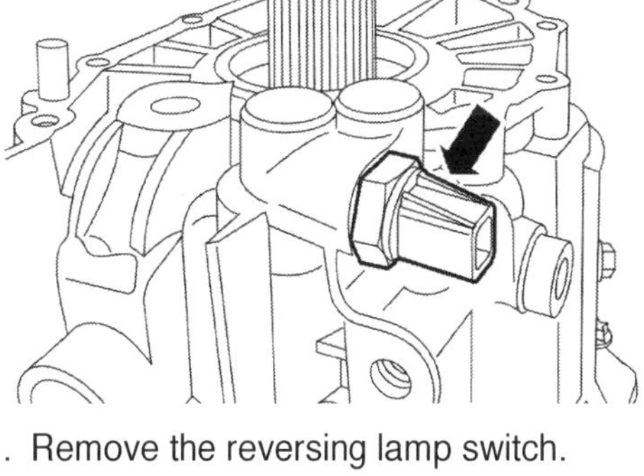

51. Remove the reversing lamp switch.

E75971

52. Remove the clutch slave cylinder. Remove the connecting pipe securing clip. Remove the connecting pipe. Remove the 3 bolts.

E89849

53. Remove the gearshift yoke securing nut.

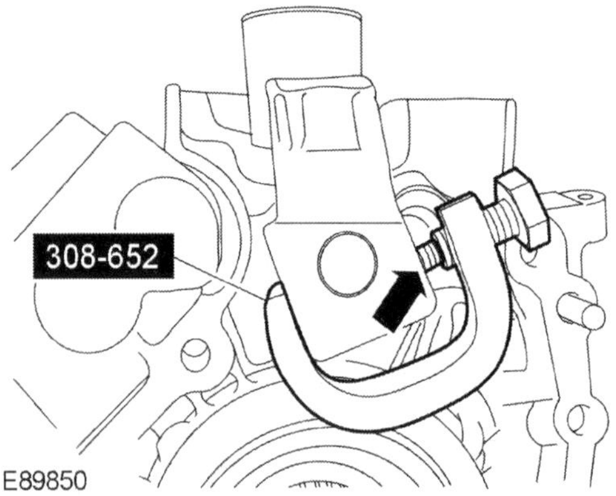

E89850

54. Using the special tool, remove the gearshift yoke. Remove the pin.

Transmission

Installation

Special Tool(s)

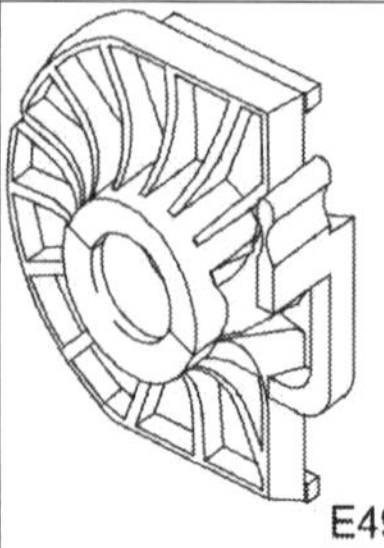 E49085	308-561 Aligner, Gearshift Lever

Part(s)

STEP	PART NAME	QUANTITY
Step 17	Renew Part Clutch slave cylinder fluid hose O-ring seal	1
Step 21	Renew Part Vent tube	1
Step 29	Renew Part Driveshaft to transfer box bolt	4
Step 47	Renew Part Differential lock pivot nut	1

⚠ NOTE: If a new transmission is being fitted then the gearshift lever retaining bolts must be pre installed to the transmission to cut the threads. Failure to follow this instruction will increase the difficulty of fitting the gearshift lever to the transmission.

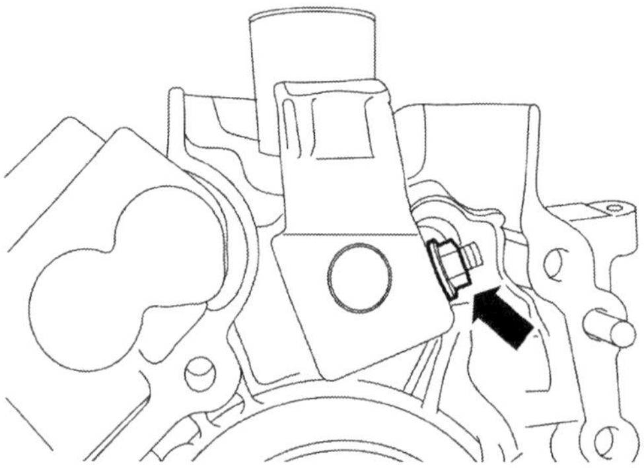

1. ⚠ NOTE: Make sure the pin is fitted from Left to Right when viewed from the rear of the transmission.

Install the gearshift yoke. Install the pin. Tighten the nut.

Torque: 12 Nm

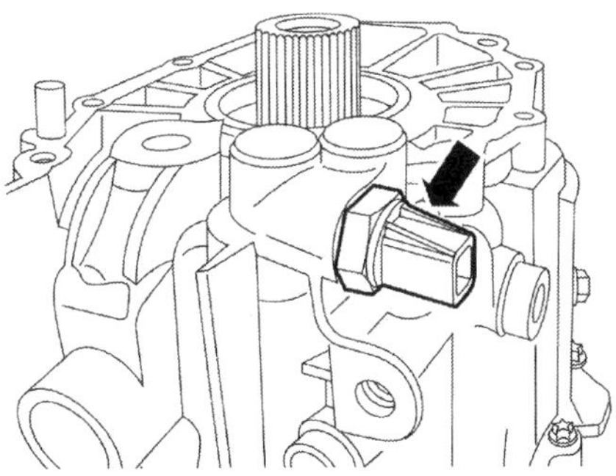

2. Install the clutch slave cylinder. Tighten the bolts.

   Install the connecting pipe. Install the connecting pipe securing clip.

   Torque: 11 Nm

3. Install the reversing lamp switch.

   Torque: 20 Nm

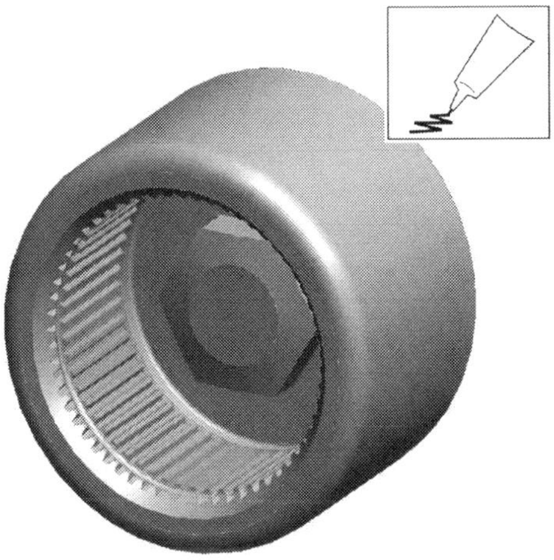

4. △ NOTE: Make sure that the component mating face is clean for debris and foreign material.

Apply suitable grease between the circlip grove and base of the flange splines prior to installation of the transmission extension shaft.

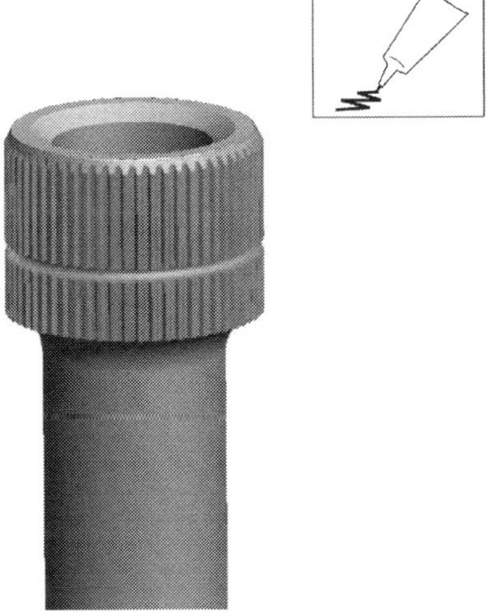

5. △ NOTE: Make sure that the component mating face is clean for debris and foreign material.

Apply suitable grease the entire length of the splines prior to installation of the transfer case.

6. Install the seal.

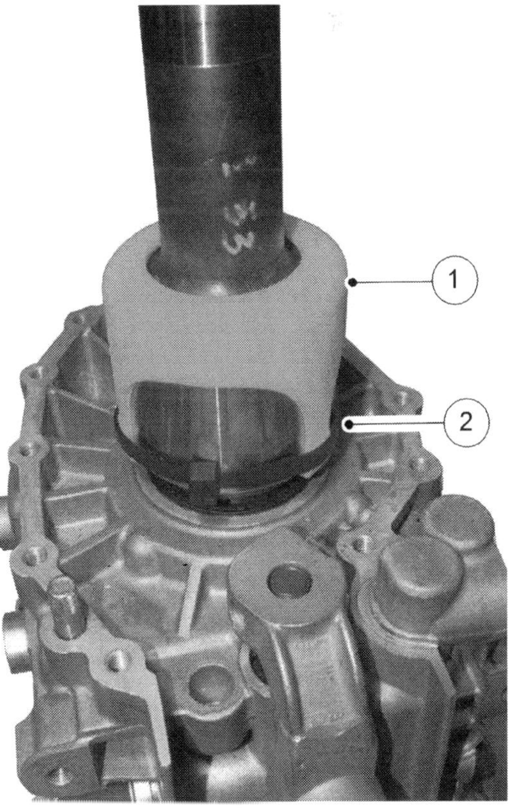

7. NOTES:

△ Apply anti seize grease to the splines.

△ Make sure that the tie strap joint is between the fingers of the extension shaft seal cover and that it has been cut flush.

Install the transmission extension shaft. Install the cover. Install the tie strap.

8. Install the transmission extension housing. Tighten the bolts.

   Torque: 25 Nm

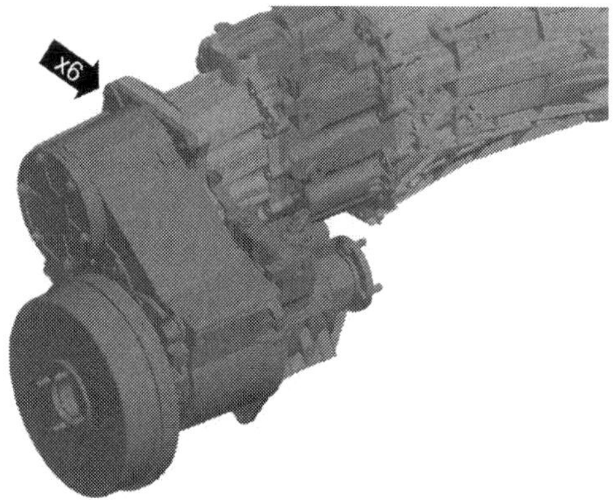

E90815

9. NOTE: Apply sealant STC 50552 to the bolt threads.

With assistance, install the transfer case. Tighten the bolts and nuts.

Bolts 45 Nm

Nuts 45 Nm

10. NOTE: This step must be carried out if a new transmission is being fitted.

Install the gear selector housing securing bolts to cut a new thread. Remove the bolts.

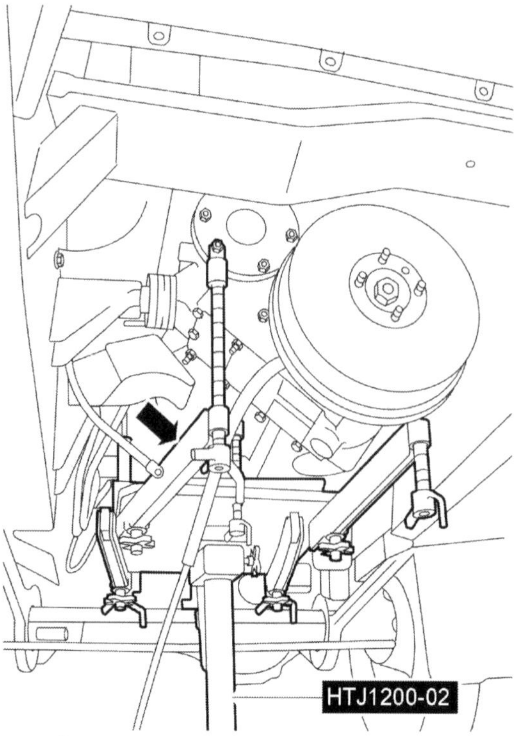

HTJ1200-02

11. CAUTION: Care must be taken when locating the transmission to the clutch.

Install the transmission and transfer gearbox.

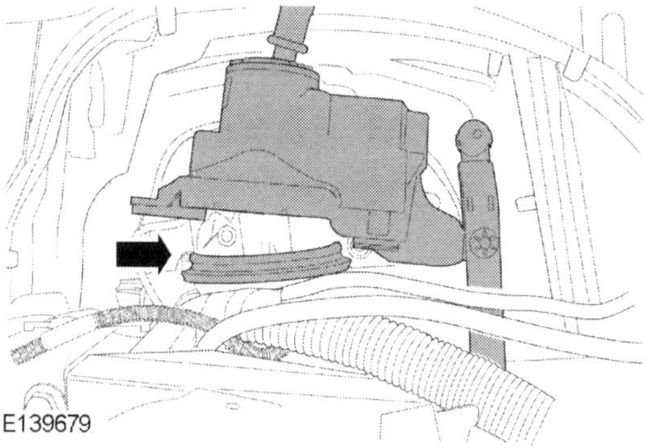

E139679

12. NOTE: Make sure the rubber gasket is fitted to the gearshift lever housing before installation.

Install the gear selector housing.

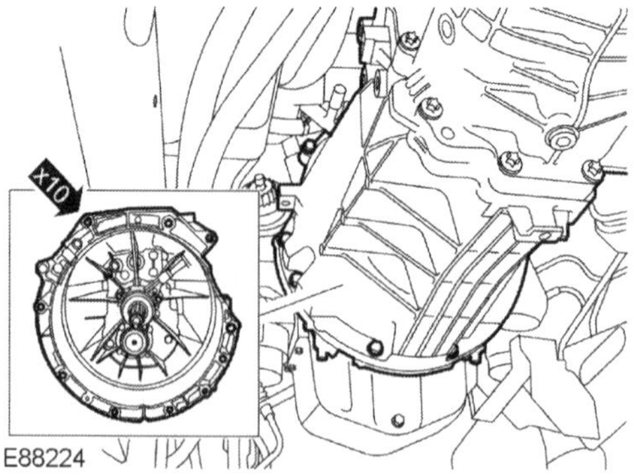

E88224

13. Install the bell housing bolts. Tighten the bolts.

    Torque: 40 Nm

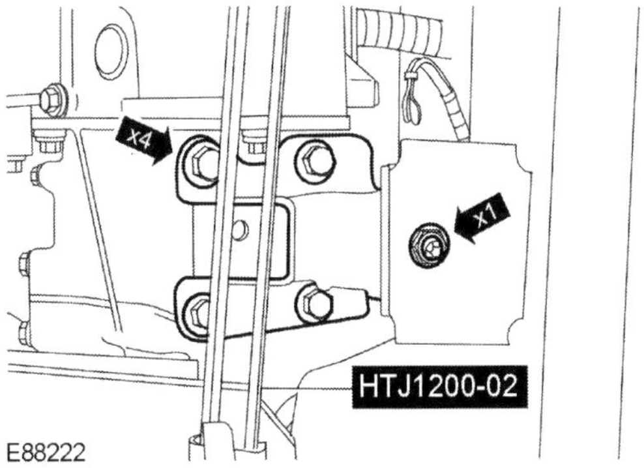

14. Install the RH transfer gearbox mounting bracket. Tighten the nut and bolts.

    Nut 48 Nm

    Bolts 85 Nm

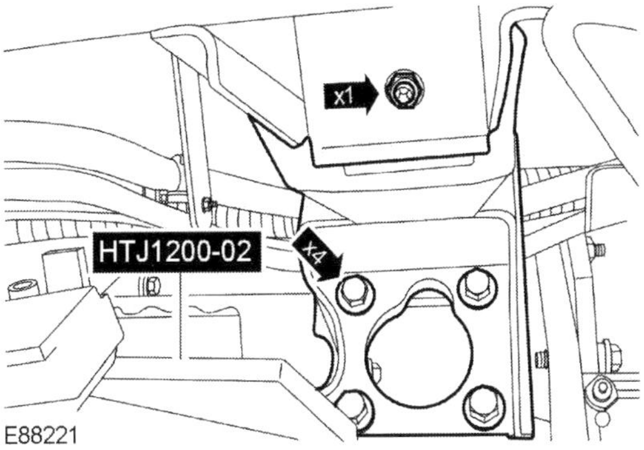

15. Install the LH transmission mount and mounting bracket. Tighten the nut and bolts.

    Nut 48 Nm

    Bolts 85 Nm

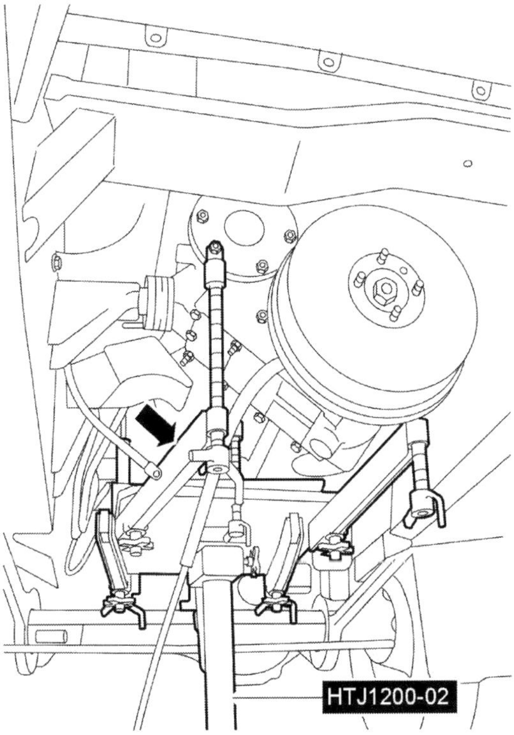

16. Remove the special tool HTJ1200-02 from the transmission and transfer gearbox.

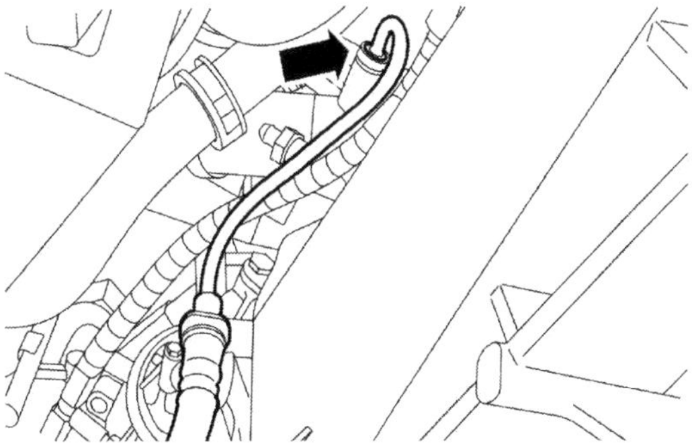

17. Install a new clutch slave cylinder fluid hose O-ring seal. Remove the blanking plugs from the orifices. Connect the clutch slave cylinder fluid hose to the clutch slave cylinder. Install the clutch slave cylinder line clip.

    Renew Part: Clutch slave cylinder fluid hose O-ring seal

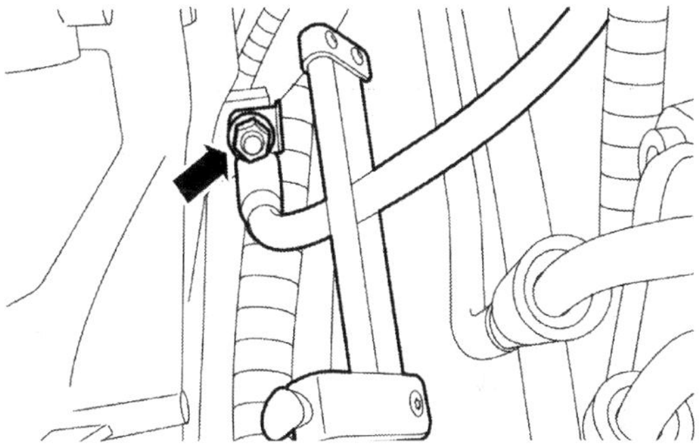

18. Install and fully tighten the clutch slave cylinder fluid hose mounting bracket securing nut and bolt. Remove the pipe clamp from the clutch slave cylinder fluid hose.

    Torque: 25 Nm

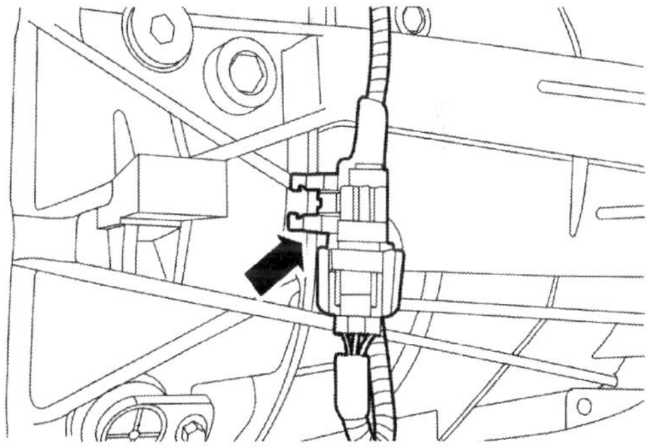

19. Secure the wiring harness to the transmission.

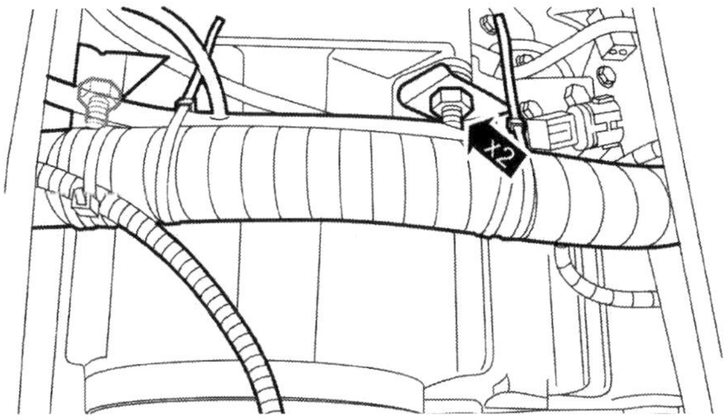

20. Reposition the transmission harness and install the transmission harness retaining bracket securing nut and bolt.

Torque: 47 Nm

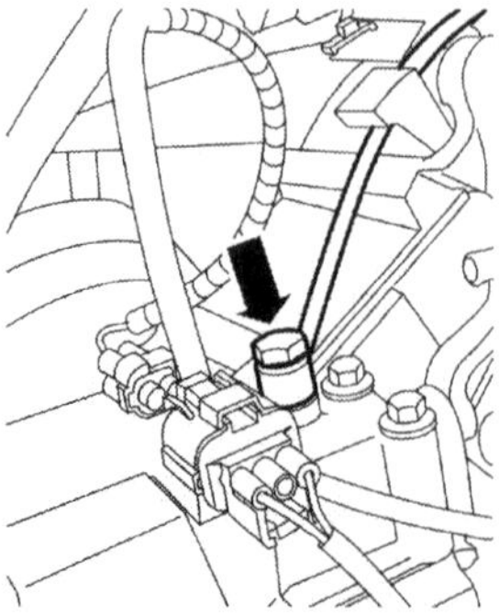

21. Install new sealing washers onto the transfer gearbox breather pipe securing bolt. Install the transfer gearbox breather pipe. Tighten the bolt.

    Renew Part: Vent tube

    Torque: 15 Nm

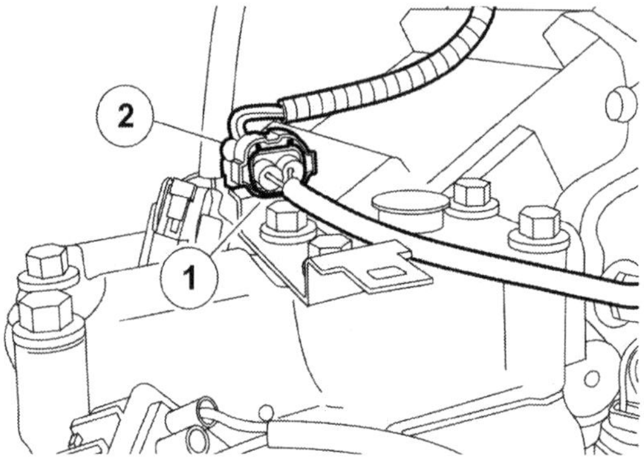

22. Connect the differential lock detection switch electrical connector. Secure the electrical connector to the bracket.

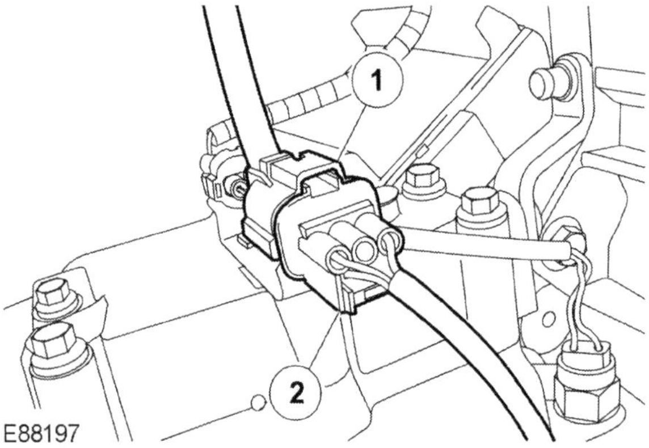

23. Reposition and secure the high-low detection switch electrical connector onto the transfer gearbox. Secure the electrical connector to the bracket.

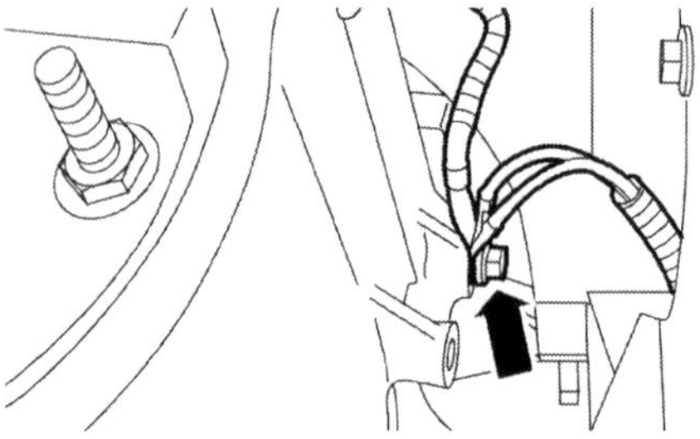

24. Install the RH transfer gearbox earth cables securing bolt. Tighten the bolt.

    Torque: 12 Nm

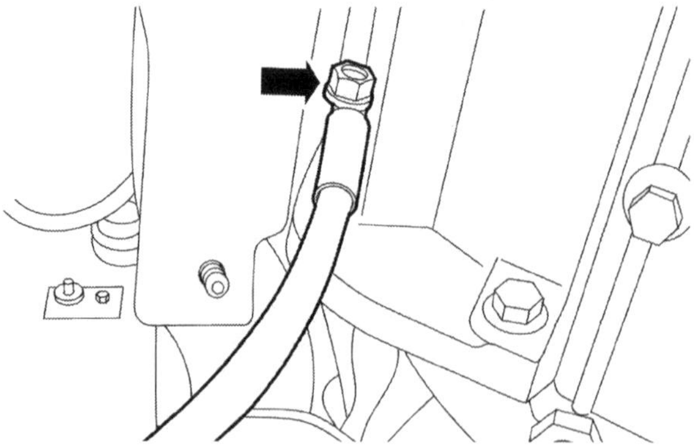

25. Install the LH transfer gearbox earth cable securing nut. Tighten the nut.

    Torque: 45 Nm

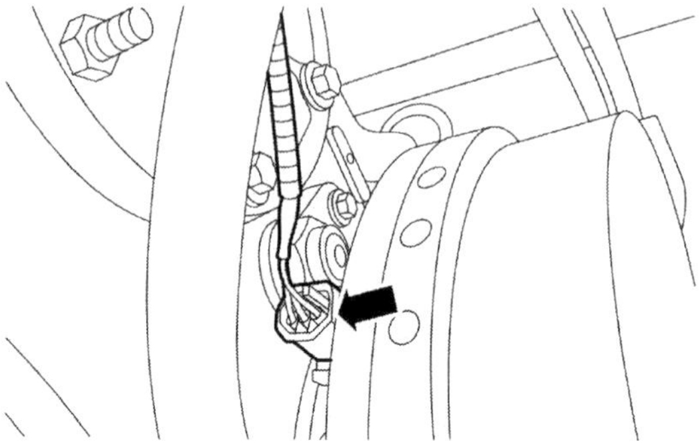

26. Connect the electronic speedometer electrical connector to the transfer gearbox.

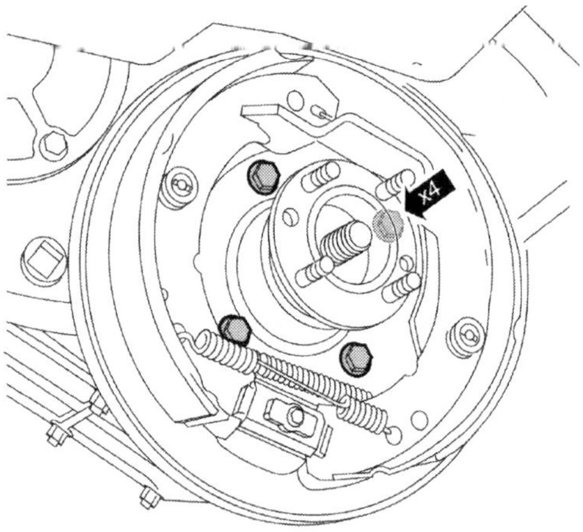

27. Install the parking brake assembly. Install the 4 securing bolts. Tighten the bolts.

    Torque: 73 Nm

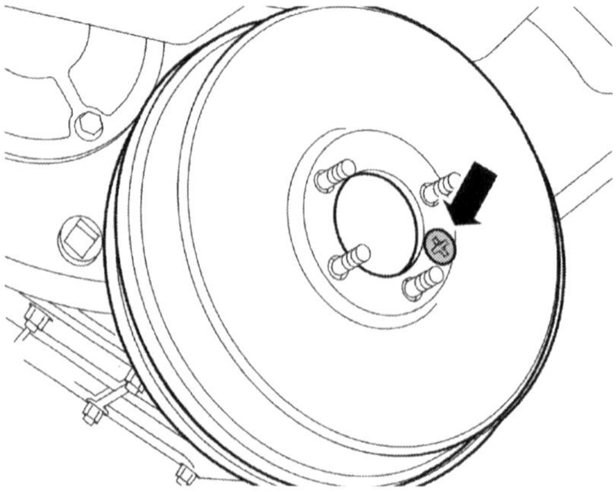

28. Install the parking brake drum. Install the securing screw.

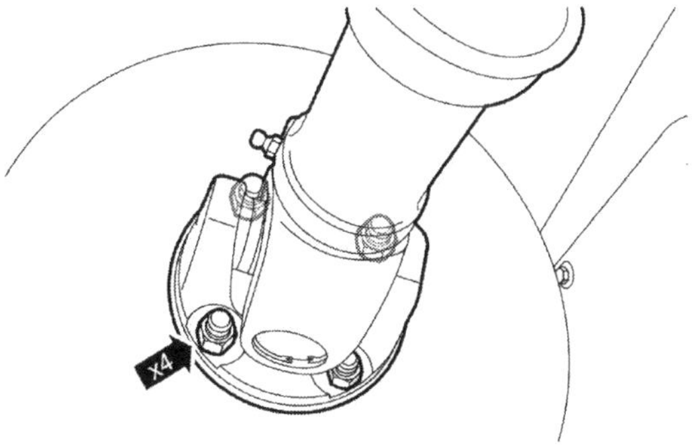

29. △ NOTE: Align to the position noted on removal.

Release the driveshaft from the chassis. Reposition and secure the rear driveshaft to the transfer gearbox. Install the rear driveshaft to transfer gearbox securing bolts. Tighten the bolts.

Renew Part: Driveshaft to transfer box bolt

Torque: 47 Nm

30. **Refer to:** Diesel Particulate Filter (309-00 Exhaust System, ID4 2.2L Diesel Removal and Installation)
31. **Refer to:** Front Muffler (309-00 Exhaust System - ID4 2.2L Diesel, Removal and Installation). ID4 2.2L Diesel
32. **Refer to:** Front Driveshaft (205-01 Driveshaft, Removal and Installation)
33. Lower the vehicle.
34. Close the bonnet.
35. Open the front door.

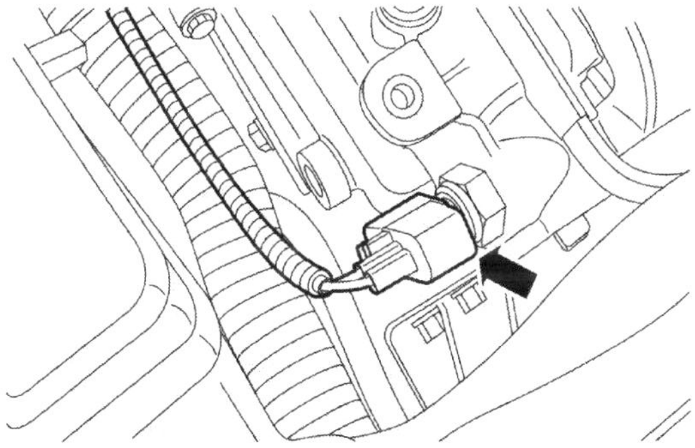

36. Connect the reverse light switch electrical connector.

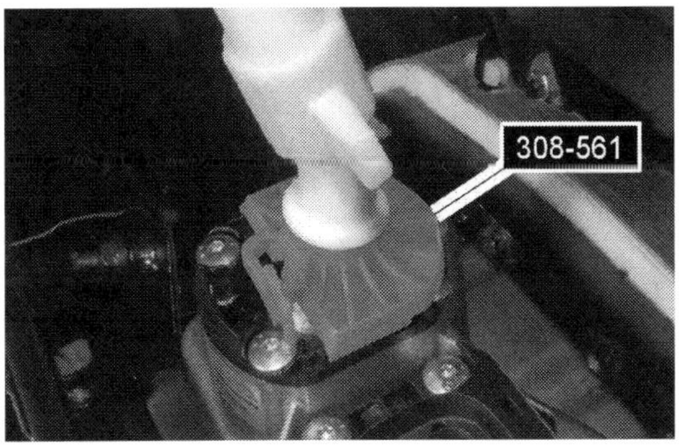

37. NOTES:

△ Make sure that the transmission is in third gear.

△ Make sure the gearshift selector lever ball joint bush and the selector yoke are centralised before installing the special tool. Install

Install the gearshift lever. Install the special tool onto the gear selector lever.

Install the Special Tool(s): 308-561

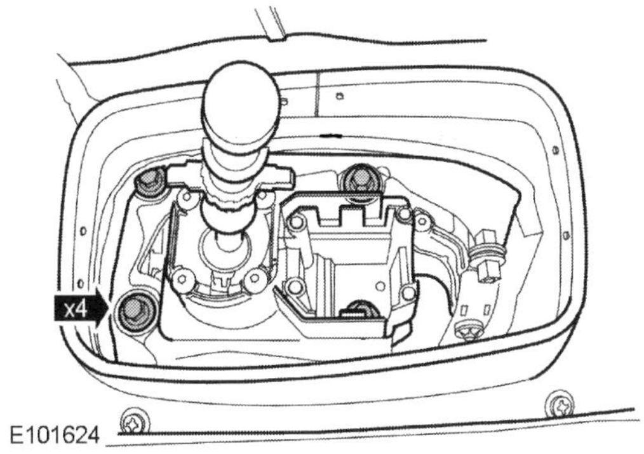

38. Tighten the 4 gear selector housing securing bolts.

Torque: 25 Nm

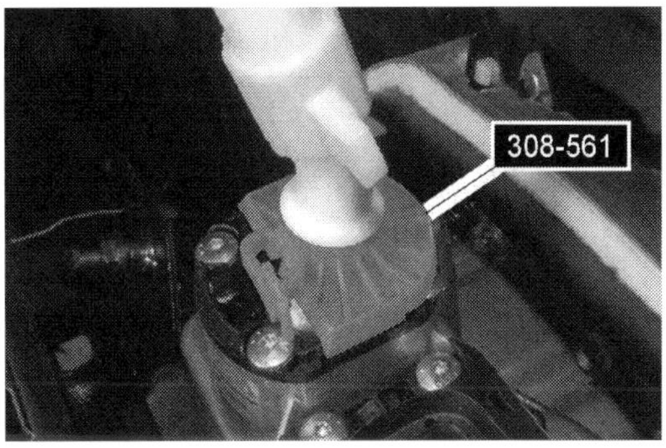

39. Remove the special tool 308-561 from the gear selector lever.

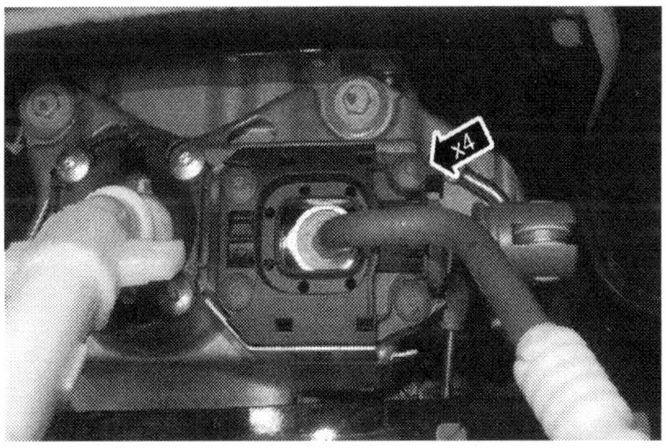

40. Install the high-low selector lever onto the gearbox.

Torque: 7 Nm

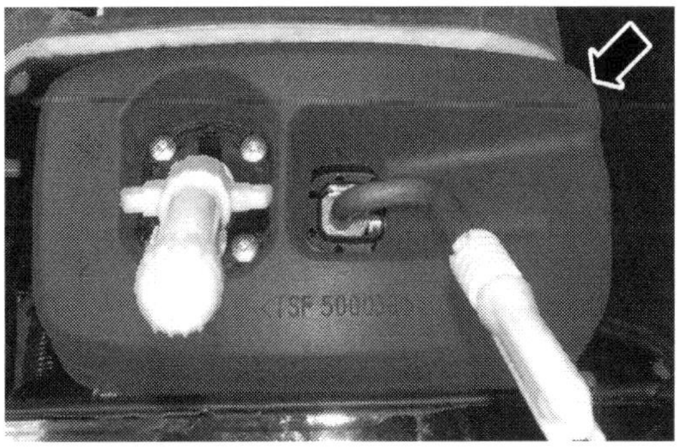

41. Install the foam pad.

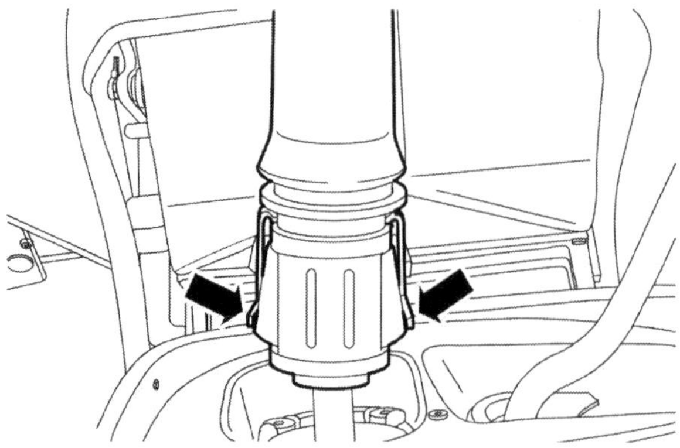

42. Install the upper gear change lever.

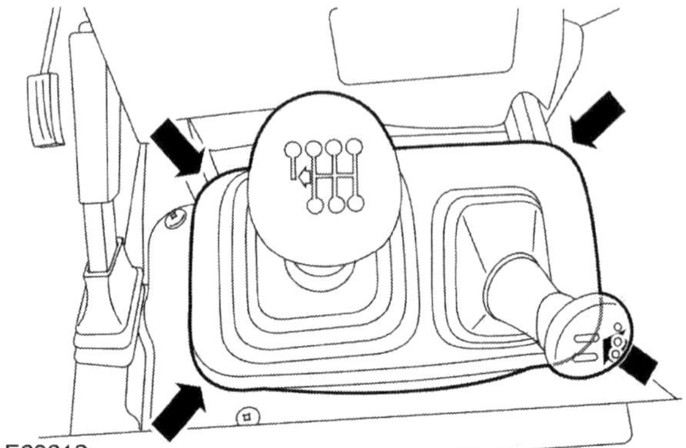

E88812

43. Install the gear selector lever gaiter. Secure the gear selector lever gaitor to the transmission tunnel. Secure the gear selector lever gaitor to the high low selector lever.

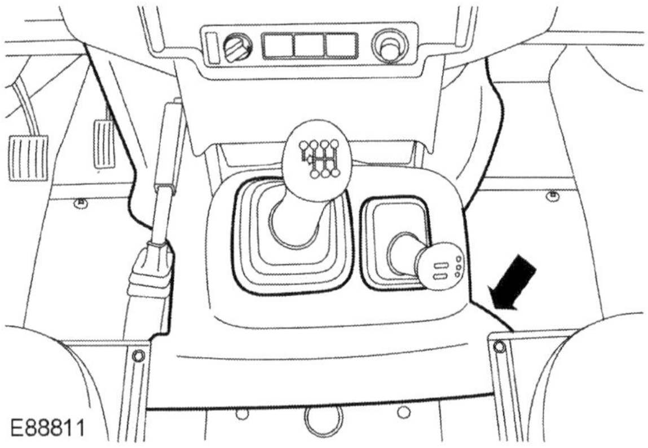

E88811

44. Install the transmission cover panel carpet.

45. **Refer to:** Floor Console (501-12 Instrument Panel and Console, Removal and Installation)

46. Raise the vehicle on lift.

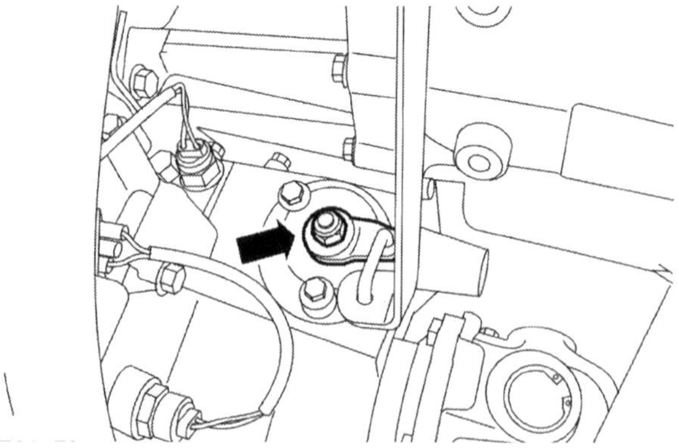

47. Install a new differential lock pivot nut. Tighten the nut.

Renew Part: Differential lock pivot nut

Torque: 25 Nm

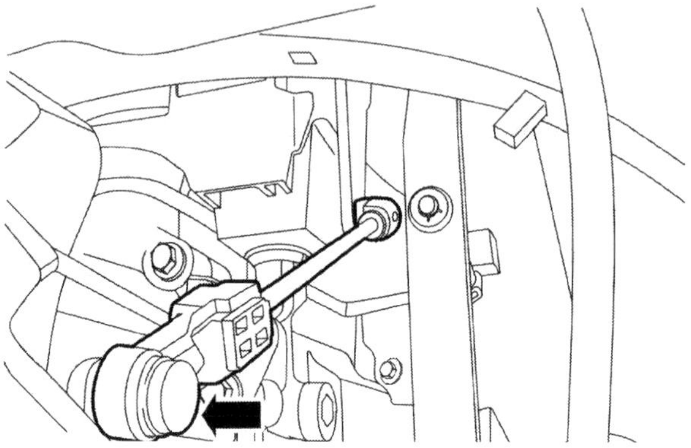

48. ⚠ NOTE: Make sure the rod is fully engaged on the ball joint and not on the foam washer.

Install the high-low selector rod ball joint onto the transfer gearbox.

49. **Refer to:** Clutch System Bleeding (308-00 Manual Transmission/Transaxle and Clutch - General Information, Vehicles With: MT82 6-Speed Manual Transmission General Procedures)

50. Lower the vehicle.

51. Close the front door.

Manual Transmission/Transaxle External Controls - Vehicles MT82 6-Speed Manual Transmission

Specifications

Torque Specifications

Description	Nm	lb-ft
Differential lock pivot nut	25	18
Gearshift lever bolts	25	18
High/Low selector lever screws	7	5

Description and Operation

OVERVIEW

Refer to: Manual Transmission (308-03 Manual Transmission/Transaxle, Vehicles With: MT82 6-Speed Manual Transmission Description and Operation)

Selector Lever

Removal and Installation

Special Tool(s)

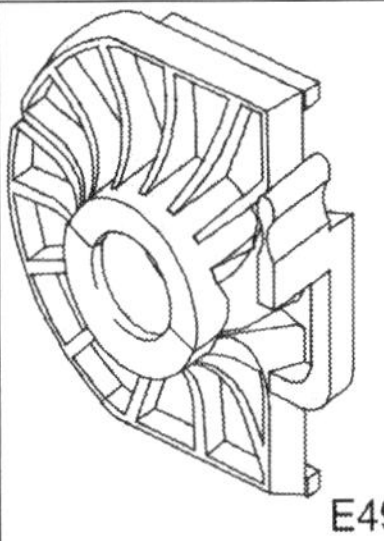 E49085	308-561 Aligner, Gearshift Lever

Part(s)

STEP	PART NAME	QUANTITY
Installation Step 5	Renew Part Differential lock pivot nut	1

Removal

1. Remove the transfer case high/low range linkage.

Refer to: Transfer Case High/Low Range Linkage (308-07 Transfer Case, Vehicles With: MT82 6-Speed Manual Transmission In-vehicle Repair)

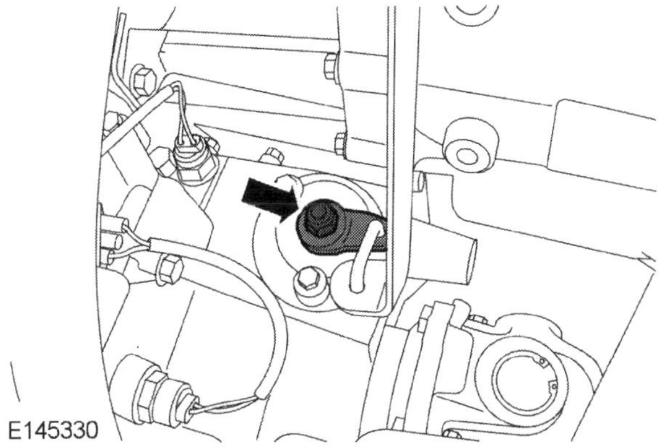

E145330

2. Remove and discard the differential lock pivot nut.

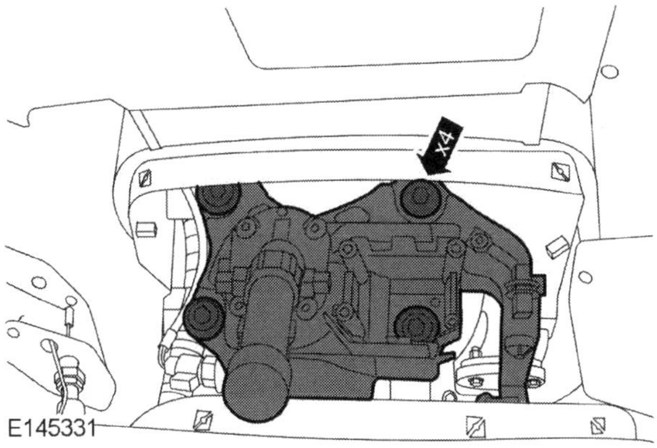

E145331

3. NOTES:

⚠ Do not remove the 4 gearshift lever ball joint securing screws.

⚠ Make sure the rubber seal is not damaged on removal.

Remove the gearshift lever.

Installation

E91087

1. NOTES:

⚠ Make sure the gearshift selector lever ball joint bush and the selector yoke are centralised before installing the special tool.

⚠ Make sure the rubber gasket is fitted to the gearshift lever housing before installation.

⚠ Make sure 3rd gear is engaged on the transmission.

Using the special tool install the gearshift lever.

Special Tool(s): 308-561

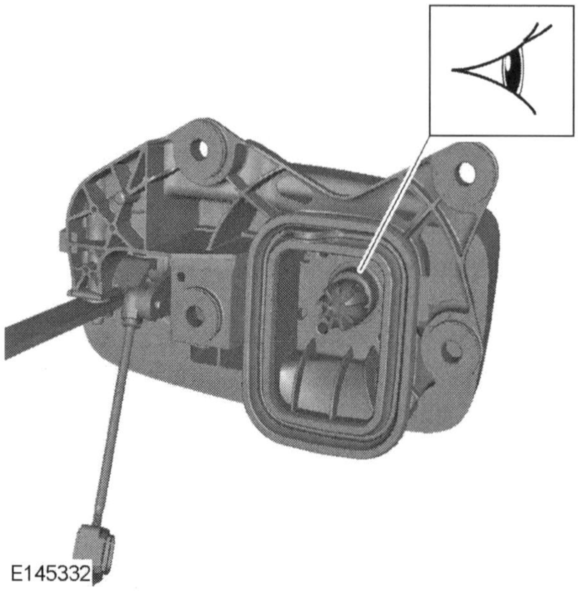
E145332

2. The plastic bush must be connected to the end of the gear lever prior to fitting.

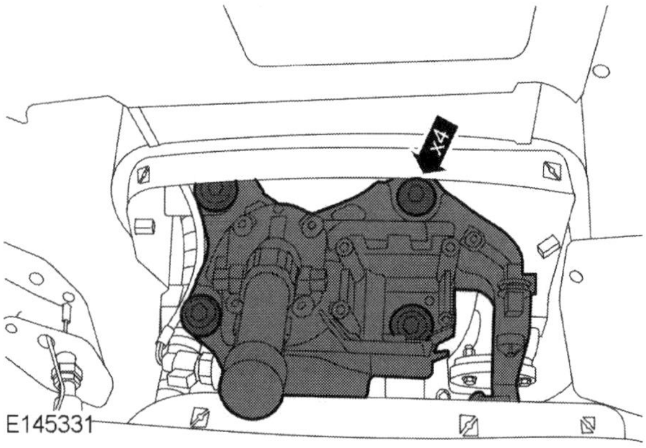

E145331

3. Install the gearshift lever.

Torque: 25 Nm

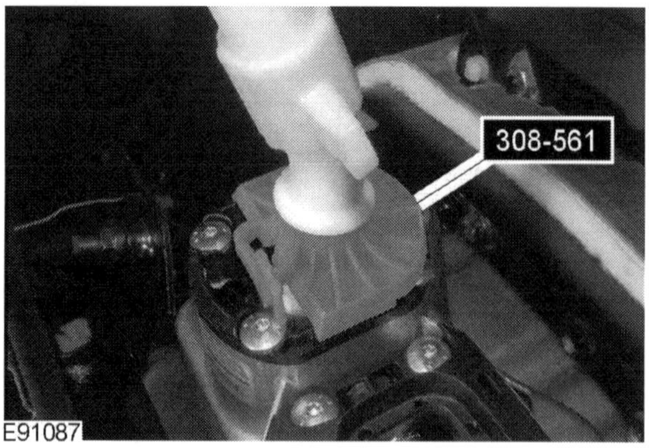

E91087

4. Remove the special tool.

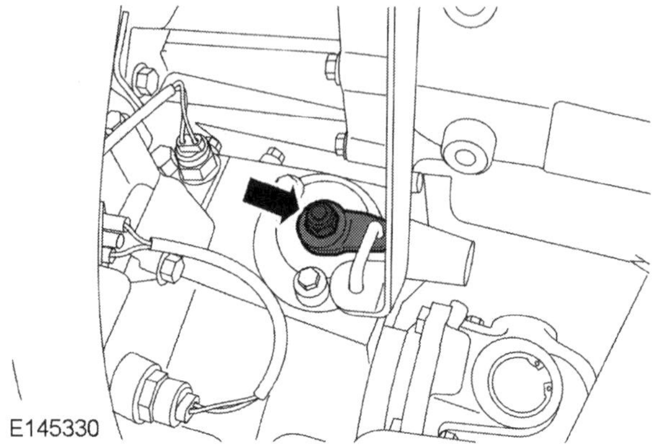
E145330

5. Install a new differential lock pivot nut.

Renew Part: Differential lock pivot nut

Torque: 25 Nm

6. Install the transfer case high/low range linkage.

Refer to: Transfer Case High/Low Range Linkage (308-07 Transfer Case, Vehicles With: MT82 6-Speed Manual Transmission In-vehicle Repair Specifications

Transfer Case - Vehicles MT82 6-Speed Manual Transmission

Specifications

Lubricants

Item	Specification
Recommended oil	SAE 75W/90 or SAE 80W/90, API GL5, MIL-L-2105 (inc B, C and D)
Capacity	2.3 litres, (4.0 pints)

CAUTION: Do not use any lubricant other than that specified

General Specification - LT230Q

Item	Specification
Transfer case type	LT230QRS
Specification	Two speed reduction on main gearbox output, front and rear drive permanently engaged via a lockable differential
Ratios:	
High	1.211:1
Low	3.269:1

Transfer case data

Item	Metric	Imperial
High/low selector finger width	15.90 to 15.95 mm	0.625 to 0.627 in
High/low selector fork finger width	7.37 to 7.47 mm	0.290 to 0.294 in
High/low selector shaft groove width	16.0 to 16.1 mm	0.63 to 0.64 in
High/low selector hub groove width	7.5 to 7.6 mm	0.295 to 0.30 in
Differential lock selector finger width	15.90 to 15.95 mm	0.625 to 0.627 in
Differential lock selector shaft groove width	16.0 to 16.1 mm	0.63 to 0.64 in
Differential lock selector fork finger width	7.92 to 7.9 mm	0.311 to 0.313 in
Differential lock selector spring free length	84.58 mm	3.33 in
Dog clutch selector fork groove width	8.05 to 8.20 mm	0.32 to 0.33 in
Differential front and rear half carrier gears load to turn:		
Used gears	0.45 kg	1.0 lb
New gears	1.72 kg	3.8 lb
Thrust washer thicknesses available	1.05 to 1.45 mm	0.04 to 0.06 in
In increments of:	0.10 mm	0.004 in
Total load to turn - both sun gears fitted:		
Used gears	0.90 kg	2.0 lb
New gears	3.44 kg	7.6 lb
Low range gear to high range hub clearance	0.05 to 0.15 mm	0.002 to 0.006 in
High range gear to high/low hub clearance	0.05 to 0.10 mm	0.002 to 0.004 in
Mainshaft input gear bearing pre-load	0.05 mm	0.002 in
Mainshaft input gear bearing shim thicknesses	3.15 to 4.00 mm	0.12 to 0.16 in
In increments of:	0.05 mm	0.002 in
Differential bearing pre-load	0.05 mm	0.002 in
Differential shim thicknesses available		
In increments of	0.05 mm	0.002 in
Transfer case gear train torque to turn	3.4 Nm max	30 lbf.in max

Torque Specifications

Description	Nm	lb-ft
Drain plug	30	22
Transfer case breather banjo	15	11
Differential carrier bolts	60	44
Differential bearing nut	72	53
Differential lock selector lever nut	25	18
* Front output housing bolts	25	18
* Rear output housing bolts	45	33
*** Output flange nuts	162	120
* Differential lock selector housing bolts	25	18
High/low selector housing bolts	25	18
* Intermediate shaft retaining plate bolt	25	18
*** Intermediate shaft stake nut - Selective length non-collapsible spacer	88	65
* Bottom cover bolts	25	18
* Bearing housing cover plate bolts and stud nut	25	18
Transmission brake backplate bolts	72	54
Interlock solenoid cover bolts - if fitted	10	8
** Neutral warning lamp switch	25	18
Transfer case front drive flange to drive shaft	47	33
Transfer case rear drive flange to drive shaft	50	37
Transfer case to gearbox extension case	47	33

* Apply Loctite 290 to threads   ** Apply Hylomar PL32 to threads   *** New nut must be fitted

Item	Nm	lb-ft
METRIC		
M5	6	5
M6	9	7
M8	25	18
M10	45	33
M12	90	65
M14	105	75
M16	180	130
UNF/UNC		
1/4	9	7
5/16	25	18
3/8	40	30
7/16	80	60
1/2	90	65
5/8	135	100

The table above provided torque values for screws and bolts not previously specified.

Transfer Case

Description and Operation

COMPONENT LOCATION

Item	Description
1	Rear output housing
2	Main casing
3	Transfer box selector lever
4	Breather pipe
5	Front output housing
6	Selector housing

OVERVIEW

The transfer box is a permanent 4-wheel drive, 2-speed ratio reducing transmission, incorporating high and low range outputs with mechanically lockable centre differential (diff-lock). High range has a ratio of 1.211:1 and low range 3.269:1.

High/low range and diff-lock selection are made via a single lever located next to the main gear lever.

The unit is mounted at the rear of the main transmission and transmits drive to the front and rear drive axle differentials via the drive shafts.

Integral with the output shafts is a differential assembly, which compensates for speed differences between the front and rear drive shafts. To improve traction and prevent all the power being transmitted to the drive axle differential offering the least traction, a diff-lock is provided. The diff-lock should only be engaged during severe off-road conditions where traction is poor and should be disengaged as soon as conditions permit, for example, when good traction is available. Selection of differential lock engages, through mechanical linkage, a dog clutch with the front output shaft, this action locks the centre differential and provides a fixed drive, giving equal power to the front and rear drive shafts.

All housings and cover plates are sealed to the main casing by sealant. Output flange rotary seals are protected against mud and water ingress by mud shield fingers.

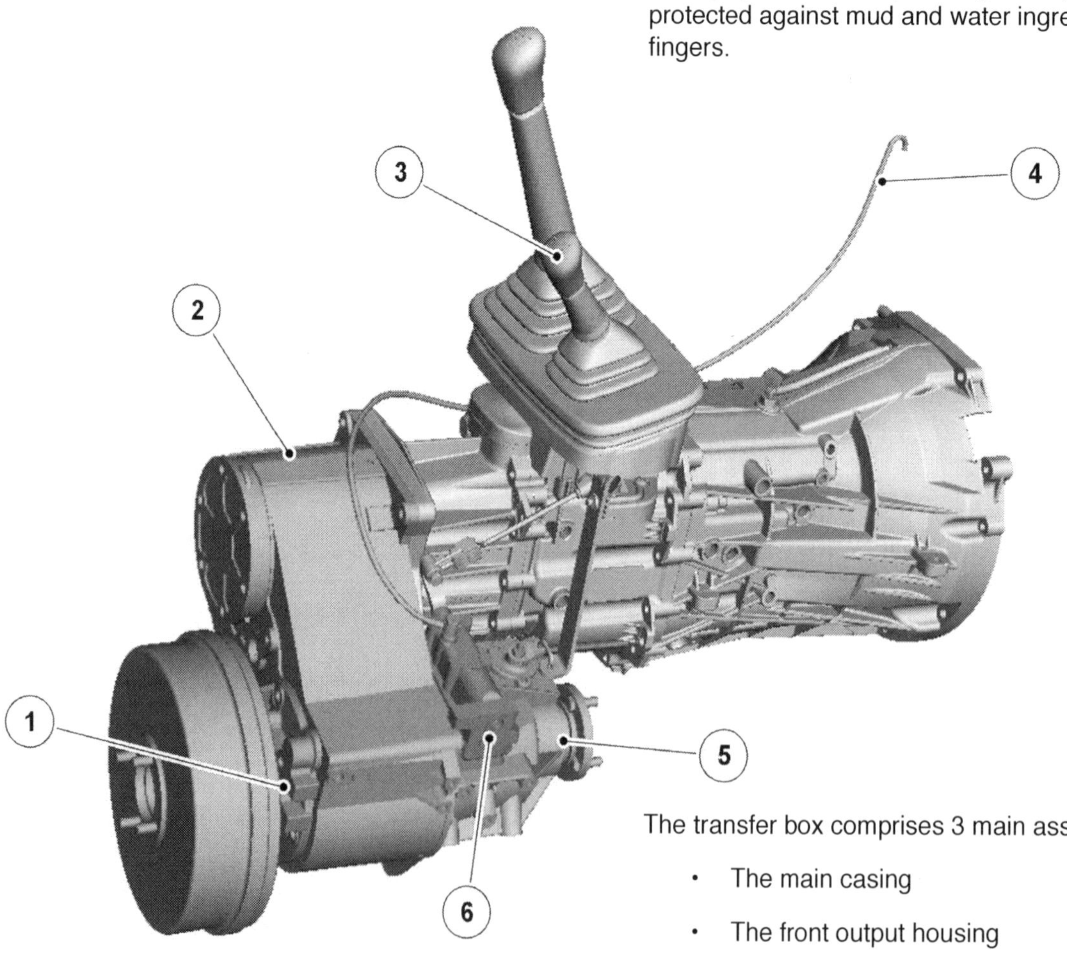

The transfer box comprises 3 main assemblies:

- The main casing
- The front output housing
- The rear output housing

TRANSFER CASE CROSS SECTION

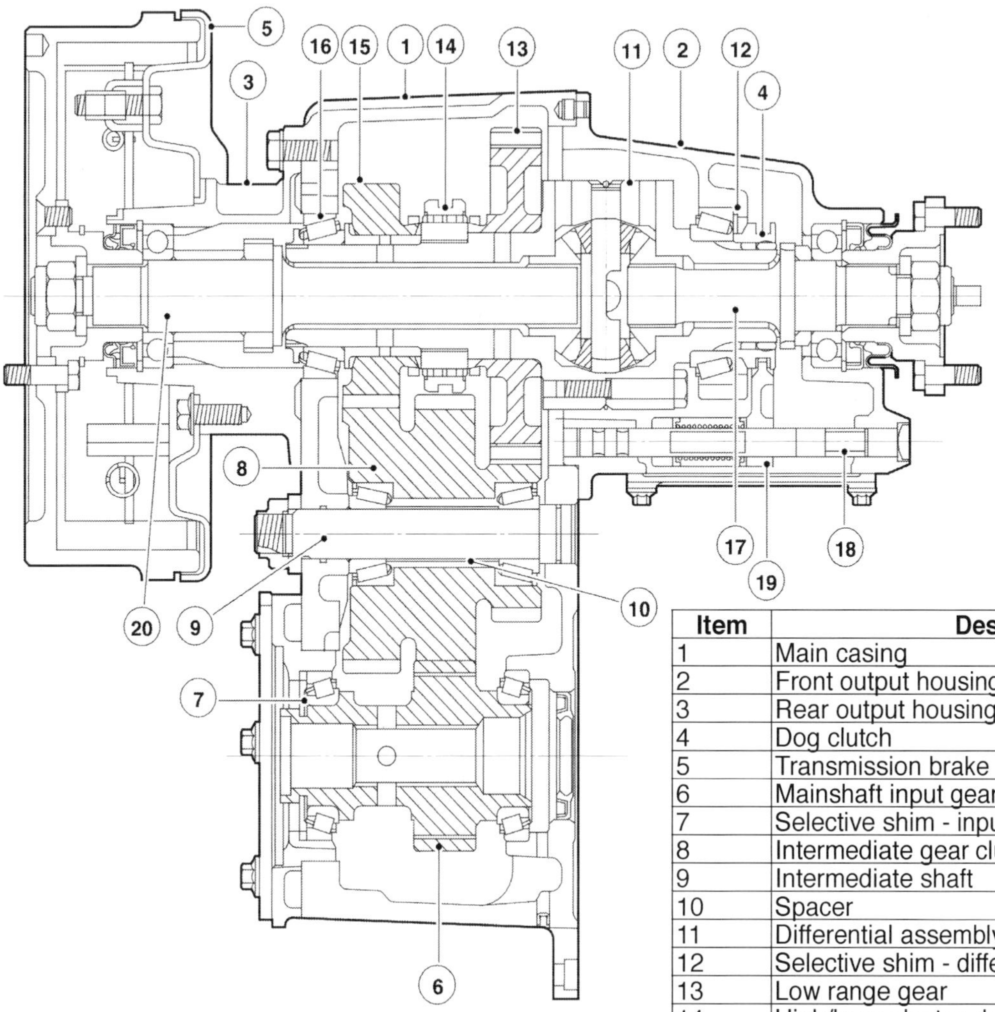

Item	Description
1	Main casing
2	Front output housing
3	Rear output housing
4	Dog clutch
5	Transmission brake
6	Mainshaft input gear
7	Selective shim - input gear bearing pre-load
8	Intermediate gear cluster
9	Intermediate shaft
10	Spacer
11	Differential assembly
12	Selective shim - differential bearing pre-load
13	Low range gear
14	High/low selector sleeve and hub
15	High range gear and bush
16	Differential rear bearing
17	Front output shaft
18	Differential lock selector shaft
19	Selector fork
20	Rear output shaft

THE MAIN CASING

Item	Description
1	Main casing
2	Retaining plate
3	Bolt - retaining plate
4	Stake nut - intermediate shaft
5	Bearings and outer tracks - mainshaft input gear
6	Mainshaft input gear
7	Selective shim
8	Mainshaft input gear bearing housing
9	Cover plate/power take-off cover
10	Bolt - cover plate
11	Oil temperature switch (if fitted)
12	Bearings and outer tracks - intermediate gears
13	Collapsible spacer
14	Intermediate gears
15	Bottom cover plate
16	Bolt - bottom cover plate
17	O-rings - intermediate shaft
18	Intermediate shaft
19	Mainshaft oil seal
20	Locating dowel
21	Detent plug - high/low selector
22	Detent spring - high/low selector
23	Detent ball - high/low selector

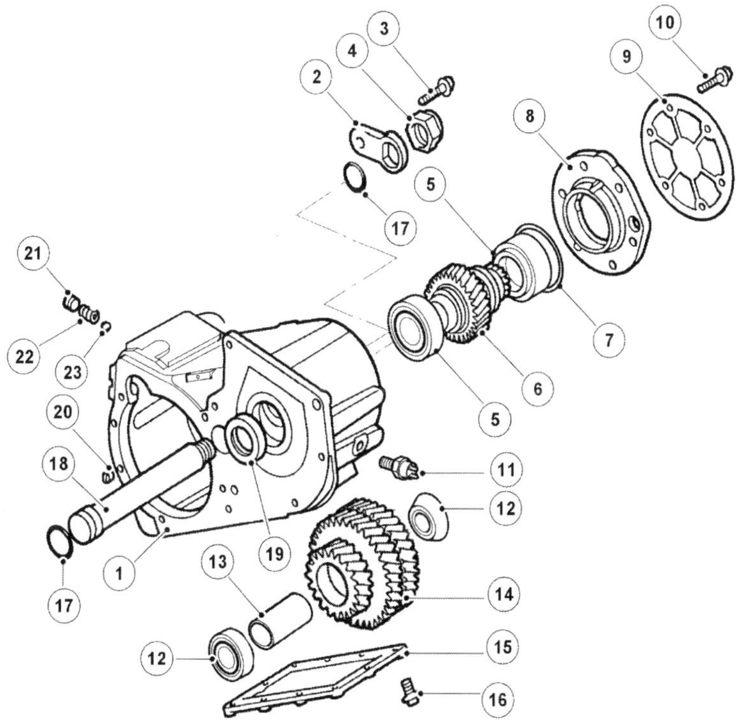

The main casing carries:

- the mainshaft input gear
- the intermediate gears
- the centre differential assembly

The front and rear output housings are bolted to either side of the main casing.

The front and rear output housings are bolted to either side of the main casing.

Mainshaft Input Gear

The transmission output shaft is splined into the mainshaft input gear, which is supported by taper roller bearings.

Input gear bearing pre-load is achieved by the use of a selective shim located in the bearing housing.

Intermediate Gears

The intermediate gear cluster is supported by the taper roller bearings located at each end of the cluster and runs on the intermediate shaft, which in turn, is supported at the front and rear by the main casing.

Intermediate gear bearing pre-load is achieved by means of a collapsible spacer positioned between the bearings, the amount of compression applied to the spacer is by means of a nut on the end of the intermediate shaft.

The bore of the intermediate gear is machined with a shoulder at each end to locate the bearings.

Item	Description
1	Retaining ring
2	Differential carrier - rear half
3	Low range gear
4	High/low hub
5	High/low selector sleeve
6	High/low selector shaft
7	High/low selector fork
8	Setscrew - high/low selector fork
9	High range gear
10	High range gear bush
11	Differential rear bearing
12	Bearing outer track
13	Bearing retaining nut
14	Dished thrust washers
15	Differential planet gears
16	Cross shafts
17	Differential side gears
18	Selective thrust washers
19	Differential carrier - front half
20	Bolt - differential carriers
21	Differential front bearing
22	Bearing outer track
23	Selective shim

The Centre Differential Assembly is supported at the front and rear by taper roller bearings, the front bearing outer track is located in the front output housing and the rear bearing outer track is located in the main casing by the rear output housing. Bearing pre-load is achieved by means of a selective shim located in the front output housing.

The centre differential rear shaft carries the low range gear, high/low selector sleeve and hub, high range gear and bush and the differential rear bearing; these components being secured to the shaft by a special staked nut.

Centre Differential Assembly

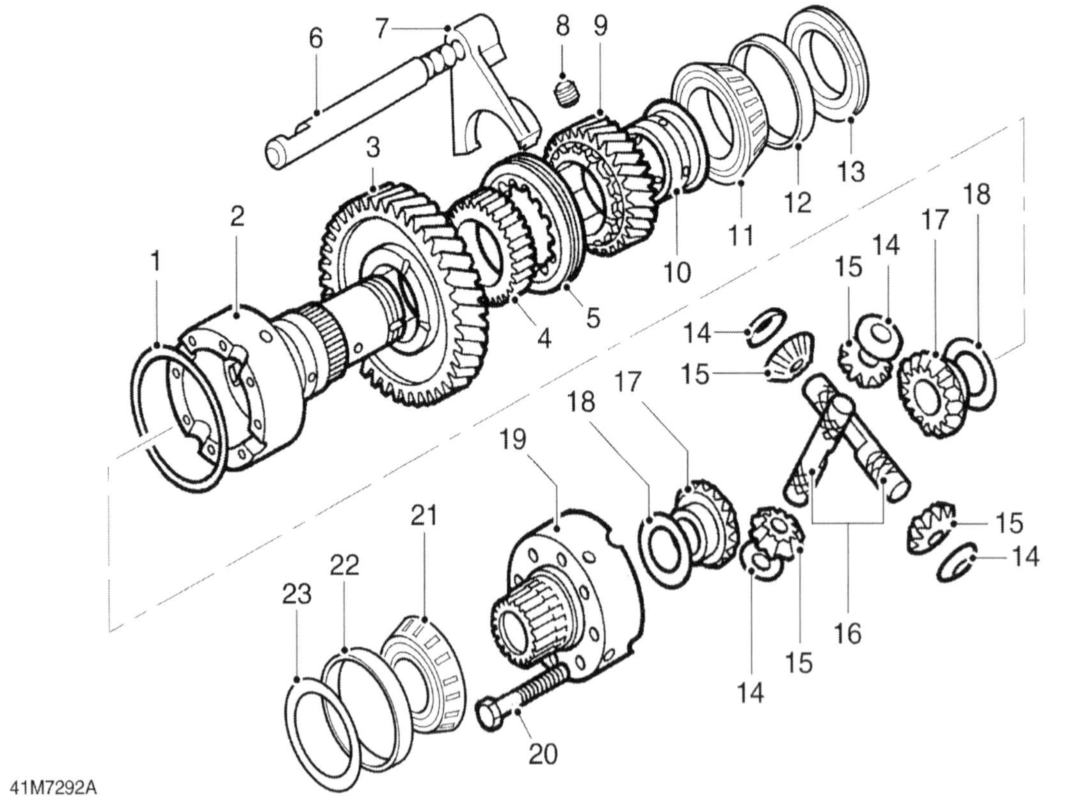

The assembly comprises front and rear half carriers with integral shafts connected to differential side gears and planet gears mounted on cross shafts within the half carriers. Dished, non-selective thrust washers control the engagement of the planet gears with the differential side gears, whilst selective thrust washers are used to control engagement of the differential side gears and 'torque to turn' of the differential. The differential carrier halves are bolted together, a retaining ring providing positive location of the cross shafts.

The high/low selector shaft and fork are located at the side of the differential, movement of the shaft, fork and selector sleeve being controlled by the high/low selector finger. A spring loaded detent ball fitted in the main casing, locates in grooves in the shaft.

The selector fork is fitted with a spring assister and clips to reduce the effort required to move the selector lever.

The Front Output Housing Carries:

- the front output shaft and flange
- the housing and selector
- the differential lock selector shaft and fork

FRONT OUTPUT HOUSING

Item	Description
1	Hollow plug
2	High/low cross shaft housing
3	Bolt - high/low cross shaft housing
4	O-ring
5	High/low cross shaft and lever with cable ball end
6	Dog clutch
7	Front output shaft
8	Differential lock selector shaft
9	Front output housing
10	Spring and clips - differential lock
11	Bolt - Cover plate
12	Cover plate
13	Differential lock selector fork
14	Detent plug and spring - differential lock
15	Detent ball - differential lock
16	Bolt - front output housing
17	Plug
18	Circlip
19	Output shaft flange and mud shield
20	Steel washer
21	Self-locking nut
22	Felt washer
23	Oil seal
24	Output shaft bearing
25	Bearing spacer
26	High/low selector finger
27	Differential lock warning indicator switch
28	Differential lock selector finger and shaft
29	O-rings
30	Differential lock selector housing
31	Selector lever
32	Bolt - housing
33	Washer
34	Self-locking nut

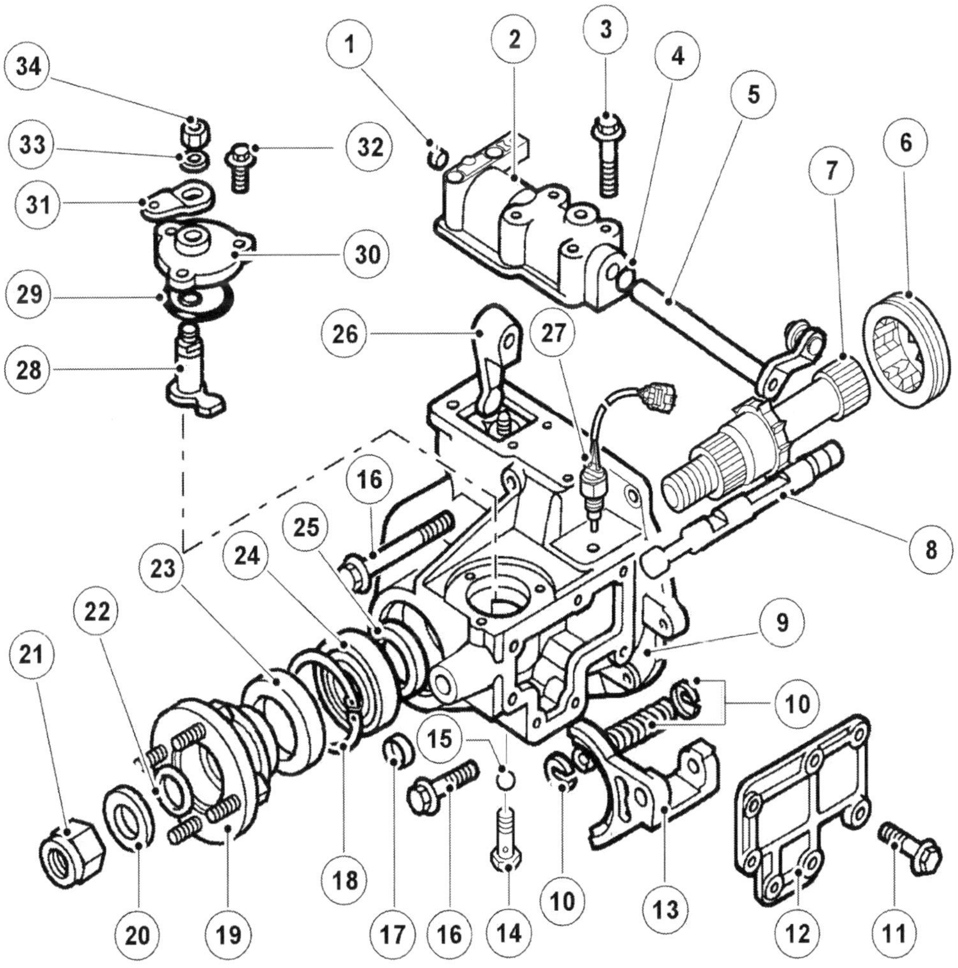

Front Output Shaft and Flange

The front output shaft is supported in the housing by a single bearing and is splined into the differential front sun gear.

Housing and Selector Assembly

A high/low cross shaft is located in a housing bolted to the top of the output housing and is connected to the high/low selector finger, which locates in a slot in the selector shaft.

Differential Lock Selector Shaft and Fork

The differential lock selector housing is bolted to the top of the front output housing, the selector finger passes through the housing, locating in a slot in the differential lock selector shaft. The differential lock selector shaft passes through the selector fork, which is located beneath a plate bolted to the side of the output housing. The selector fork engages the dog clutch sleeve with the differential rear shaft when the splines of the sleeve and differential rear shaft are aligned. A spring loaded detent ball fitted in the output housing locates in grooves in the shaft.

The Rear Output Housing carries the output shaft and flange and the speedometer drive and driven gears. A mechanically operated transmission brake is attached to the housing, the brake drum being attached to the output flange.

The rear output shaft is supported in the rear output housing by a single bearing and is splined into the differential rear shaft. The output shaft also carries the speedometer drive gear, which meshes with the speedometer driven gear located in the rear output housing.

A differential lock warning lamp switch operated by movement of the selector fork and shaft is screwed into the top of the output housing.

LUBRICATION

Lubrication is by splash, oil filler/level and drain plugs being located in the main casing.

Internal pressures caused by thermal expansion and contraction are avoided by the use of a plastic breather pipe venting the interior of the box to atmosphere.
The pipe is attached to the top of the high/low selector housing and is routed in a continuously rising path into the engine compartment where the open end is secured to the engine cylinder block.

REAR OUTPUT HOUSING

Item	Description
1	Rear output shaft
2	Rear output housing
3	Bolt - rear output housing
4	Speedometer drive gear
5	Spacer
6	Output shaft bearing
7	Circlip
8	Oil seal
9	Mud shield
10	Output shaft flange
11	Felt washer
12	Steel washer
13	Self-locking nut
14	Transmission brake backplate
15	Bolt - transmission brake backplate
16	Transmission brake drum
17	Countersunk screw
18	Speedometer driven gear
19	O-ring
20	Speedometer driven gear housing
21	Seal
22	O-ring
23	Vehicle speed sensor (if fitted)
24	Allen screw (if fitted)

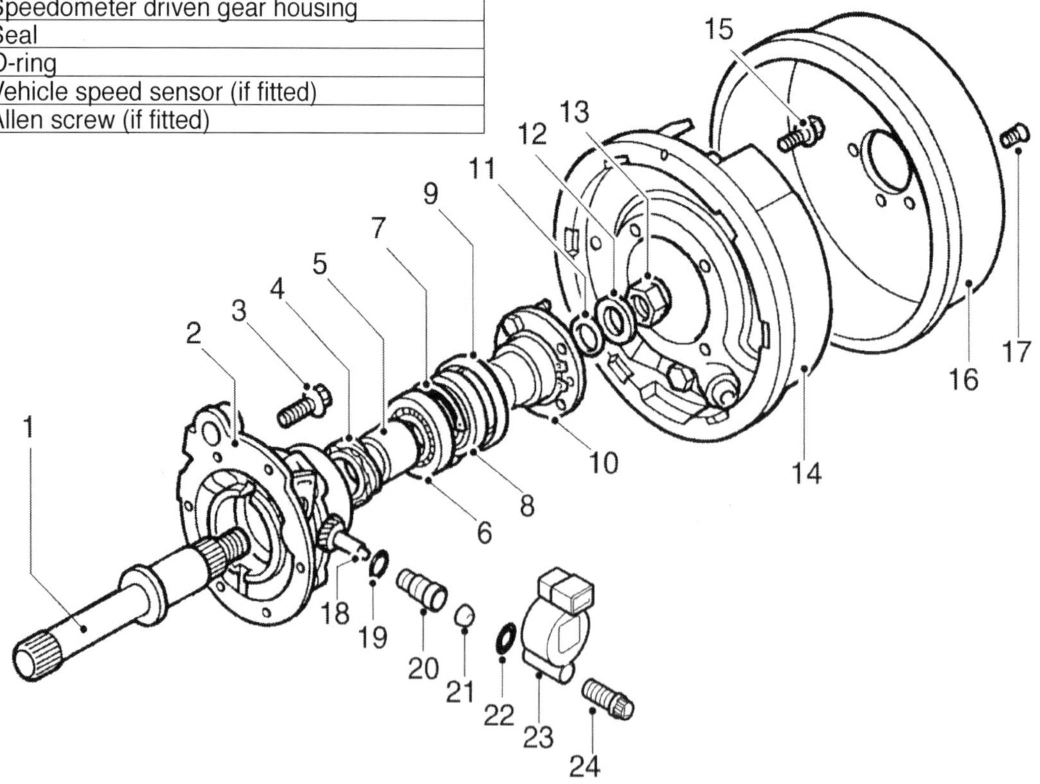

OPERATION

Power Flow - Transfer Box in LOW Range

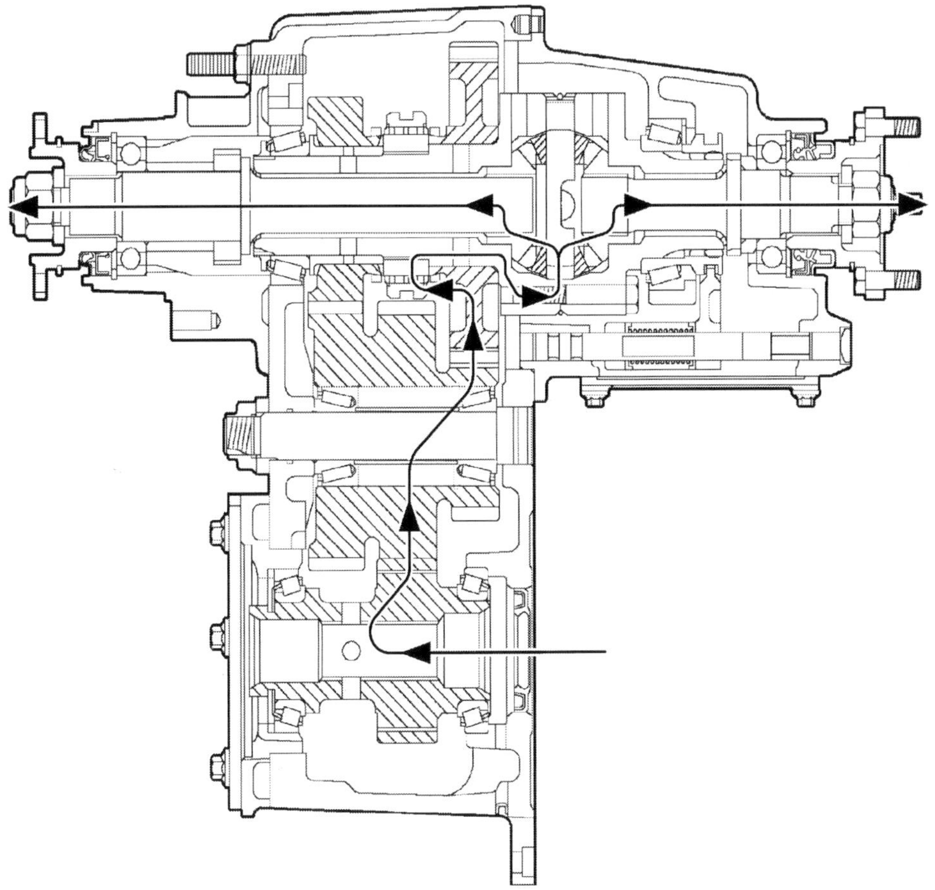

Power Flow - Transfer Box in HIGH Range

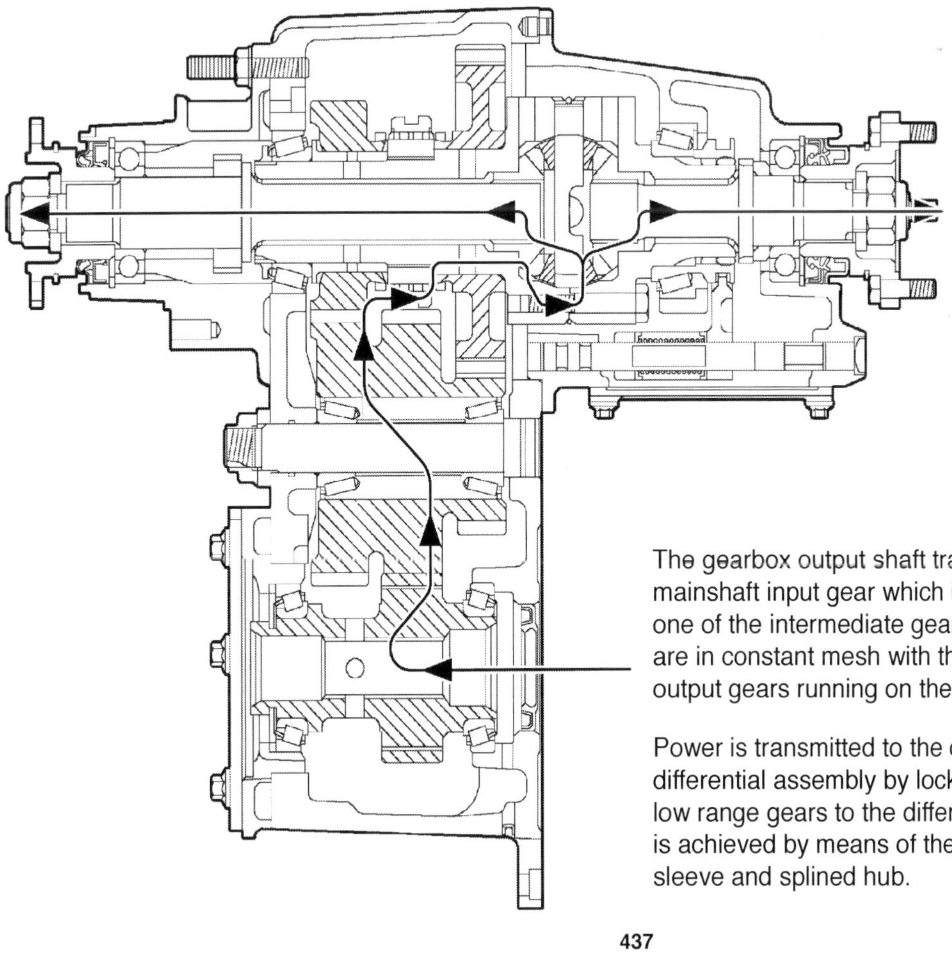

The gearbox output shaft transmits power to the mainshaft input gear which is in constant mesh with one of the intermediate gears. The intermediate gears are in constant mesh with the high and low range output gears running on the differential rear shaft.

Power is transmitted to the output shafts via the differential assembly by locking either the high or low range gears to the differential rear shaft. This is achieved by means of the high/low selector fork, sleeve and splined hub.

Transfer Case

Diagnosis and Testing

Principles of Operation

For a detailed description of the Transfer Case, refer to the relevant Description and Operation section in the workshop manual.

Refer to: Transfer Case (308-07 Transfer Case, Vehicles With: MT82 6-Speed Manual Transmission Description and Operation)

Inspection and Verification

⚠ CAUTION: Diagnosis by substitution from a donor vehicle is **NOT** acceptable. Substitution of control modules does not guarantee confirmation of a fault, and may also cause additional faults in the vehicle being tested and/or the donor vehicle.

NOTES:

△ If a control module or a component is suspect and the vehicle remains under manufacturer warranty, refer to the Warranty Policy and Procedures manual, or determine if any prior approval programme is in operation, prior to the installation of a new module/component.

△ When performing voltage or resistance tests, always use a digital multimeter accurate to three decimal places, and with an up-to-date calibration certificate. When testing resistance always take the resistance of the digital multimeter leads into account.

△ Check and rectify basic faults before beginning diagnostic routines involving pinpoint tests.

Visual Inspection

Mechanical
• Transfer case • Driveshafts • Oil leaks

1. Visually inspect for obvious signs of damage and system integrity
2. If an obvious cause for an observed or reported concern is found, correct the cause (if possible) before proceeding to the next step
3. If the cause is not visually evident, verify the symptom and refer to the Symptom Chart, alternatively check for Diagnostic Trouble Codes (DTCs) and refer to the DTC Index

Symptom Chart

Symptom	Possible Causes	Action
Oil leak from between the manual gearbox and the transfer case	• Manual gearbox output seal or transfer case input seal failure	NOTES: △ After rectifying a manual gearbox oil leak, refer to the relevant section of the workshop manual and check the manual gearbox oil level. Top up as necessary. △ After rectifying a transfer case oil leak, refer to the relevant section of the workshop manual and check the transfer case oil level. Top up as necessary. • Identify the leaking oil type and install a new manual gearbox output seal or transfer case input seal as necessary
Oil leak from the transfer case front output seal	• Transfer case front output seal failure	NOTES: △ After rectifying a transfer case oil leak, refer to the relevant section of the workshop manual and check the transfer case oil level. Top up as necessary. △ When installing a new transfer case front output seal, a new transfer case front output flange should also be installed. • Install a new transfer case front output seal and flange

Oil leak from the transfer case rear output seal	• Transfer case rear output seal failure	NOTE: After rectifying a transfer case oil leak, refer to the relevant section of the workshop manual and check the transfer case oil level. Top up as necessary. • Install a new transfer case rear output seal. Check the parking brake shoes for oil contamination and install new shoes and clean the parking brake drum as necessary
Oil leak from the transfer case cover plate seal	• Transfer case cover plate seal failure	NOTES: After rectifying a transfer case oil leak, refer to the relevant section of the workshop manual and check the transfer case oil level. Top up as necessary. Transfer case cover plate sealant part number: STC4600. Transfer case cover plate fixing thread sealant part number: STC50552. • Remove the cover plate and remove all traces of old sealant. Apply new sealant and re-install the cover plate
Oil leak from the transfer case differential lock switch	• Transfer case differential lock switch sealant failure	NOTES: After rectifying a transfer case oil leak, refer to the relevant section of the workshop manual and check the transfer case oil level. Top up as necessary. Transfer case switch thread sealant part number: STC50552. • Remove the differential lock switch and remove all traces of old sealant. Apply new sealant and re-install the differential lock switch
Oil leak from the transfer case high/low range switch	• Transfer case high/low range switch sealant failure	NOTES: After rectifying a transfer case oil leak, refer to the relevant section of the workshop manual and check the transfer case oil level. Top up as necessary. Transfer case switch thread sealant part number: STC50552. • Remove the high/low range switch and remove all traces of old sealant. Apply new sealant and re-install the high/low range switch
Oil leak from the transfer case detent plug(s)	• Transfer case detent plug(s) sealant failure	NOTES: After rectifying a transfer case oil leak, refer to the relevant section of the workshop manual and check the transfer case oil level. Top up as necessary. Transfer case detent plugs thread sealant part number: STC50552. • Remove the detent plug(s) and remove all traces of old sealant. Apply new sealant and re-install the detent plug(s)

Transfer Case Oil Seal Locations

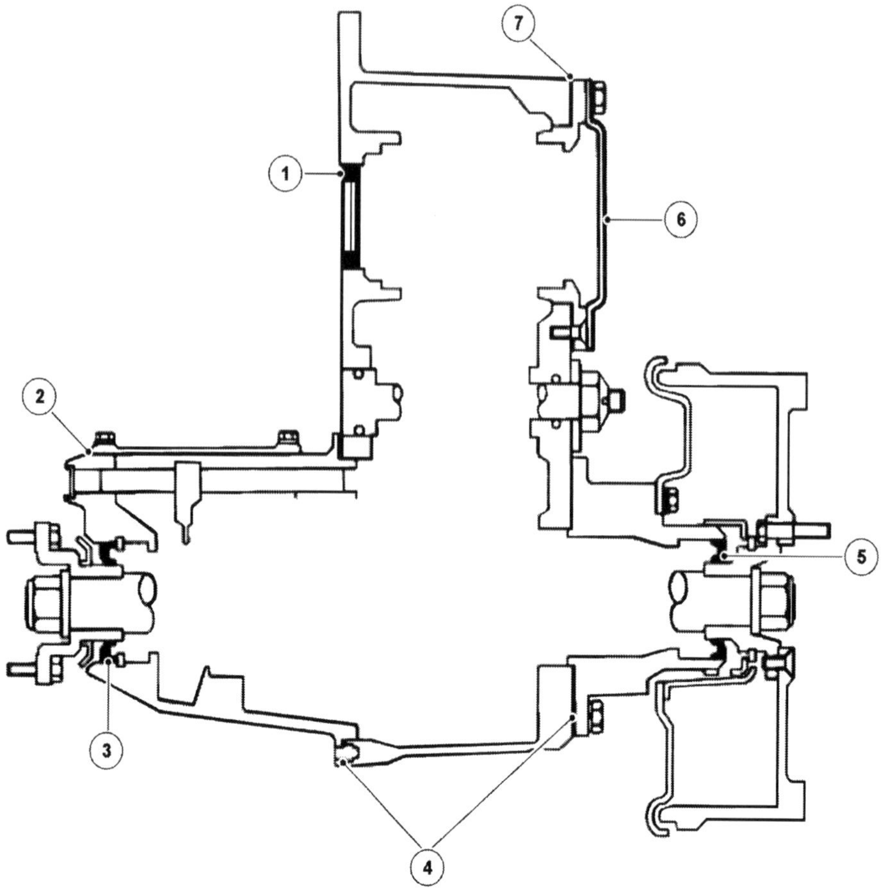

Item	Description
1.	Input seal
2.	Cover plate joints
3.	Front output seal
4.	Front and rear housing joint faces
5.	Rear output seal
6.	Power take off cover plate joint
7.	Bearing plate joint

Transfer Case Draining and Filling

General Procedures

1. WARNING: Do not work on or under a vehicle supported only by a jack. Always support the vehicle on safety stands.

Raise and support the vehicle.

2. Position a container to collect the fluid.

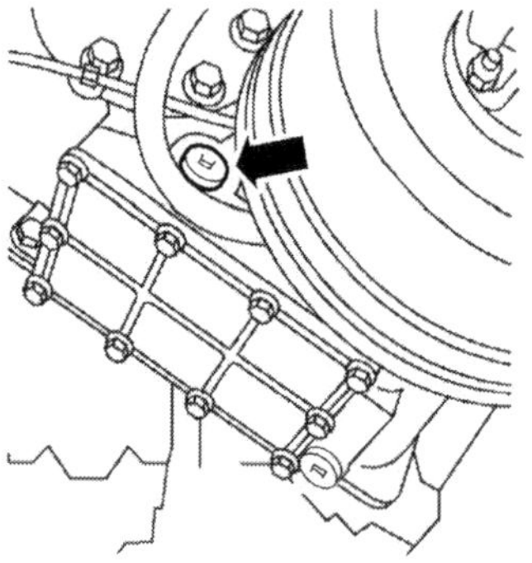

3. NOTE: Clean the area surrounding the transfer case filler/level plug.

Remove the transfer case filler plug to vent the transfer case and assist with draining.

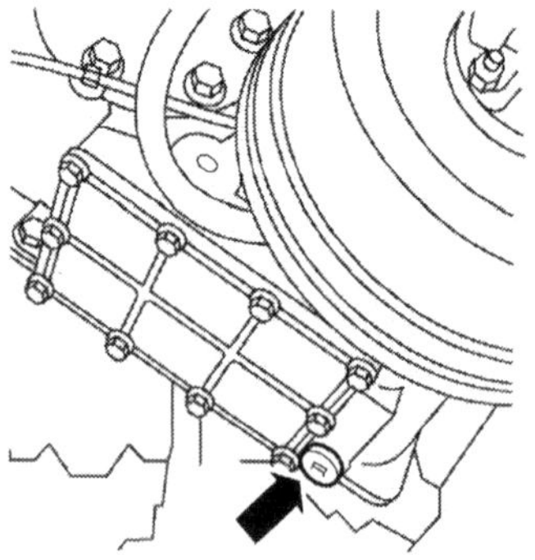

4. NOTE: Clean the area surrounding the transfer case fluid drain plug.

Remove the transfer case fluid drain plug.

5. Allow the fluid to drain.

6. Install the transfer case drain plug and clean any oil residue from the surrounding area.

   • Tighten to 30 Nm (22 lb.ft).

7. Refill the transfer case with the recommended fluid.

Refer to: (308-07 Transfer Case, Vehicles With: MT82 6-Speed Manual Transmission Specifications)

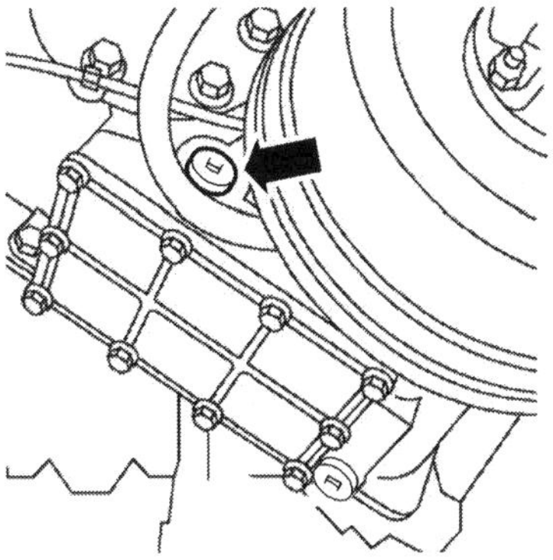

8. Install the transfer case filler plug and clean any oil residue from the surrounding area.
    - Tighten to 30 Nm (22 lb.ft).
9. Remove the container.

Transfer Case High/Low Range Selector Rod Adjustment

General Procedures

Special Tool(s)

308-711 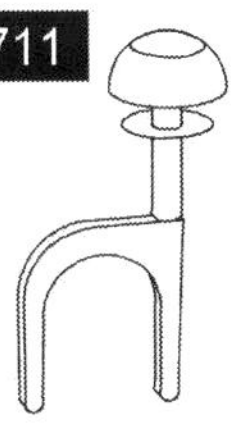 E91269	Locking Tool High/Low Selector 308-711

NOTE: If carrying out this procedure after replacing the transfer case high/low range selector linkage, carry out steps 13 - 19.

1. Remove the floor console.

Refer to: Floor Console (501-12, Removal and Installation).

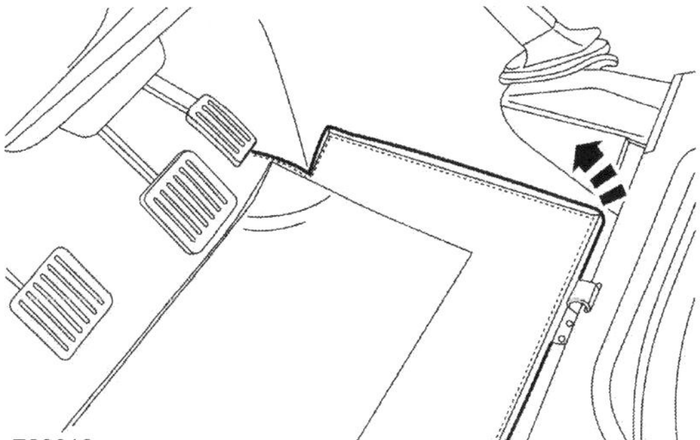

E88810

2. Reposition the LH carpet.

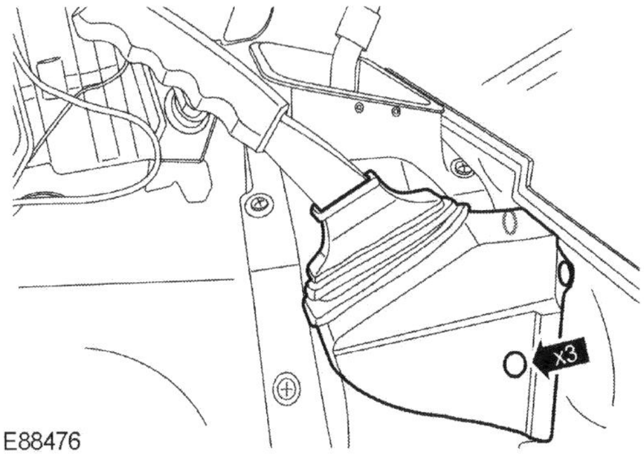

E88476

3. Remove the parking brake lever gaiter.
    - Remove the 3 clips.

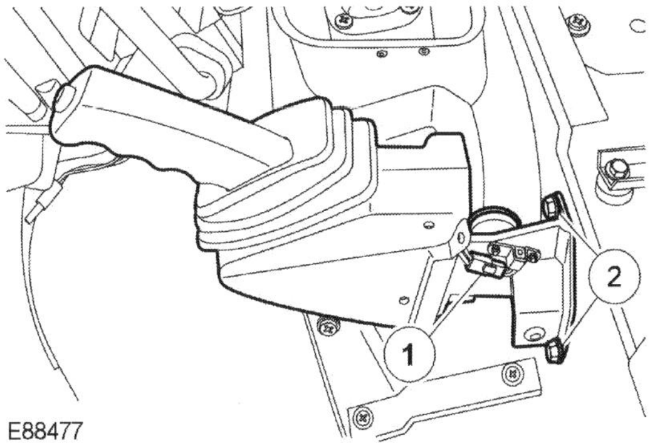

E88477

4. Release the parking brake lever.
    1. Disconnect the electrical connector.
    2. Remove the 2 bolts.

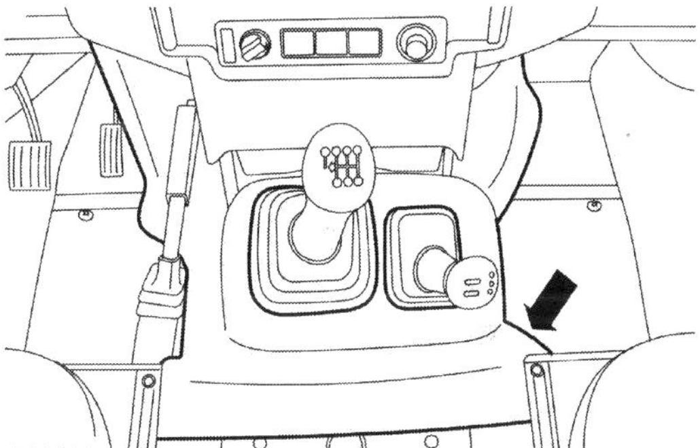

E88811

5. Remove the transmission cover panel floor covering.

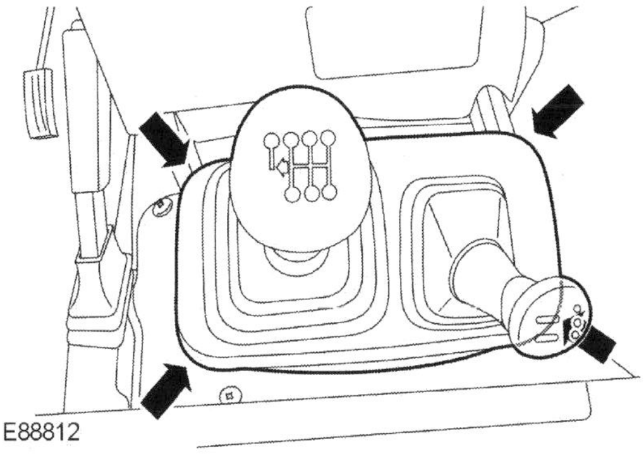

E88812

6. NOTE: Do not detach the gaiter from the selector knobs.

Detach the gaiter from the transmission cover panel.

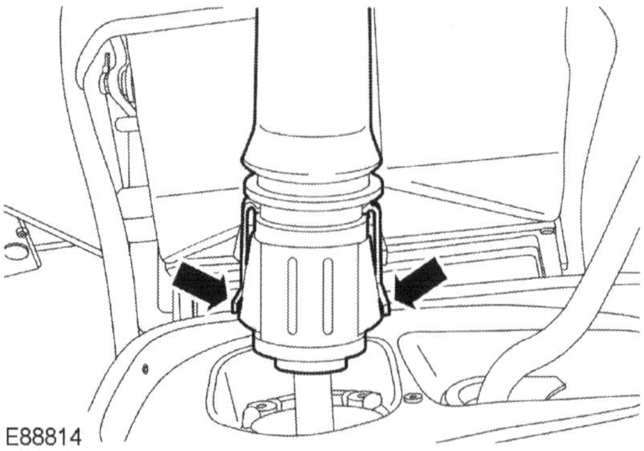

7. WARNING: The gearshift lever knob will be released suddenly, keep face clear during removal. Failure to follow this instruction may result in personal injury.

   Release the gearshift lever knob.

   - Release the 2 clips.

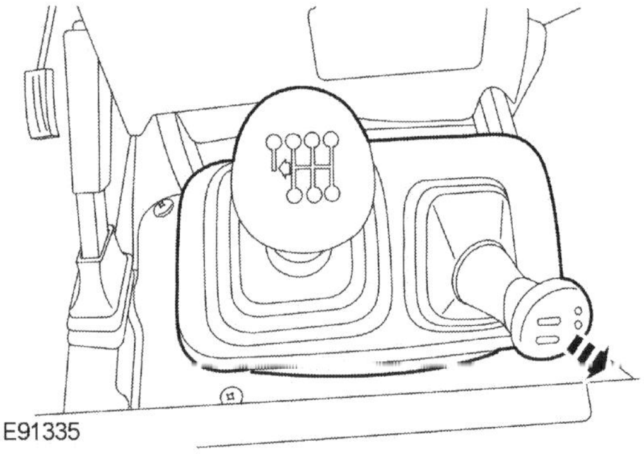

8. WARNING: The high/low range selector lever knob will be released suddenly, keep face clear during removal. Failure to follow this instruction may result in personal injury.

   Remove the gaiter and selector levers.

   - Detach the high/low range selector lever.

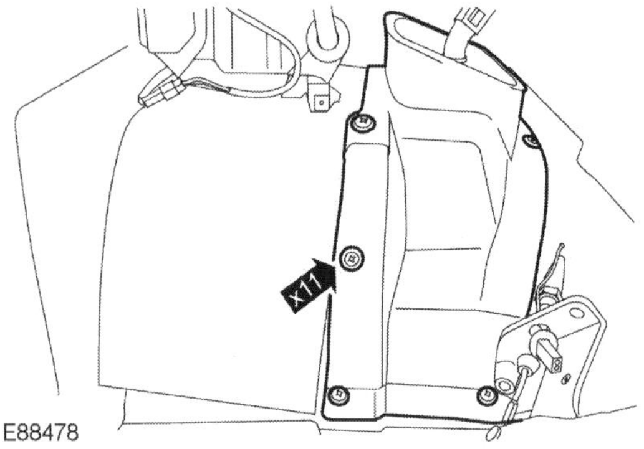

9. Remove the transmission cover panel.

   - Remove the 11 screws.

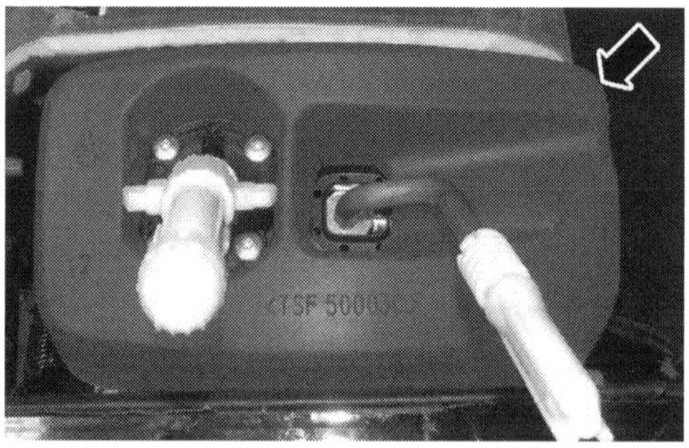

10. NOTE: Position the differential lock to the left, to reduce tension on the insulation.

    Remove the insulation pad.

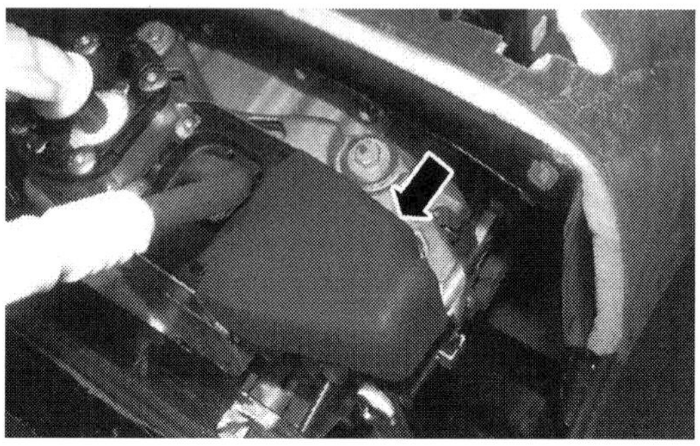

11. Remove the insulation pad support bracket.

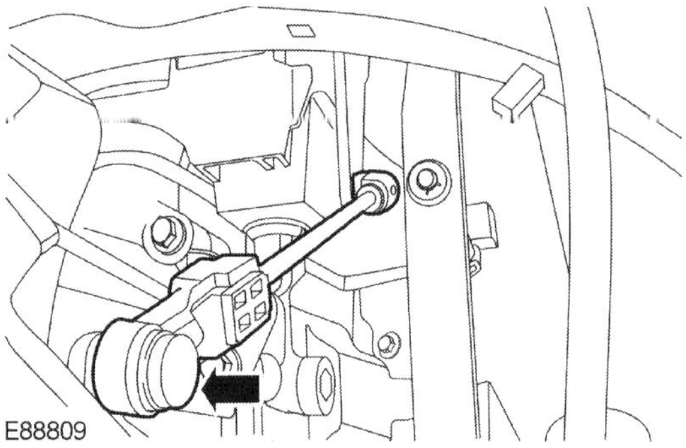

12. Disconnect the transfer case high/low range selector rod.

    - Press the button to release the transfer case high/low range selector rod.

13. Release the adjuster.

    - Release the locking block.

14. WARNING: Do not work on or under a vehicle supported only by a jack. Always support the vehicle on safety stands.

    Raise and support the vehicle.

15. NOTE: It may be necessary to rotate the road wheels to confirm that high range is fully engaged.

Make sure that high range is fully engaged on the transfer case.

16. Using the special tool, set the transfer case high/low range selector lever in high range.
    - Insert the special tool.

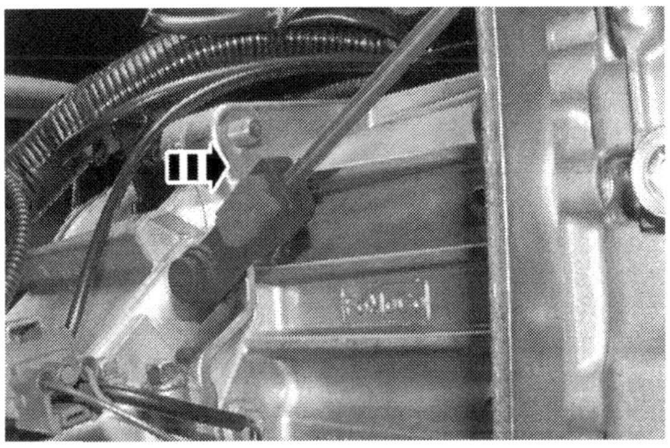

17. NOTE: Make sure the transfer case high/low range selector rod is fully engaged on the ball joint and not the foam washer.

Connect the transfer case high/low range selector rod.

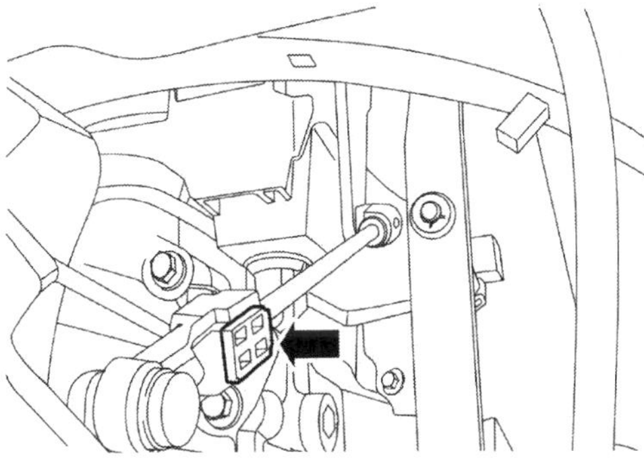

18. Push in the locking block.
    - Fully engage the locking block.

19. Remove the special tool.
20. Install the insulation pad support bracket.
21. NOTE: Position the differential lock to the left, to reduce tension on the insulation.

Install the insulation pad.

22. Install the transmission cover panel.
    - Install the screws.
23. Install the gaiter with the selector levers attached.
    - Install the selector levers.
    - Fully seat the gaiter.
24. Install the transmission cover panel floor covering.

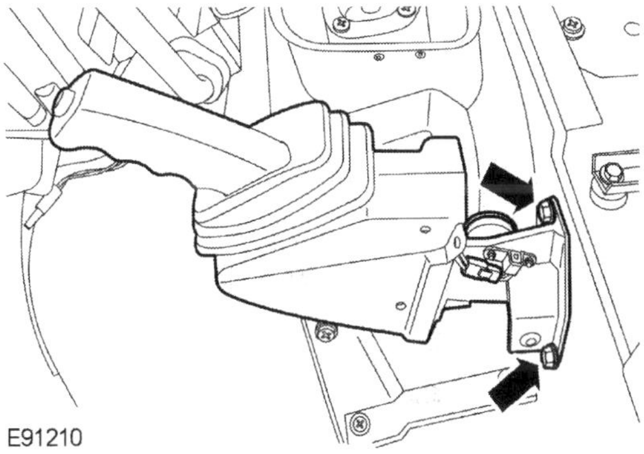

25. Install the parking brake lever.
    - Tighten to 25 Nm (18 lb.ft).
26. Install the parking brake lever gaiter.
    - Install the clips.
27. Reposition the LH carpet.
28. Install the floor console.

Refer to: Floor Console (501-12, Removal and Installation).

Transfer Case Extension Housing

In-vehicle Repair

Removal

1. Drain transfer case oil.

Refer to: Transfer Case Draining and Filling (308-07 Transfer Case, Vehicles With: MT82 6-Speed Manual Transmission General Procedures)

2. Remove transfer case rear output shaft seal.

Refer to: Transfer Case Rear Output Shaft Seal (308-07 Transfer Case, Vehicles With: MT82 6-Speed Manual Transmission In-vehicle Repair)

3. Mark extension housing to main casing for assembly purposes.

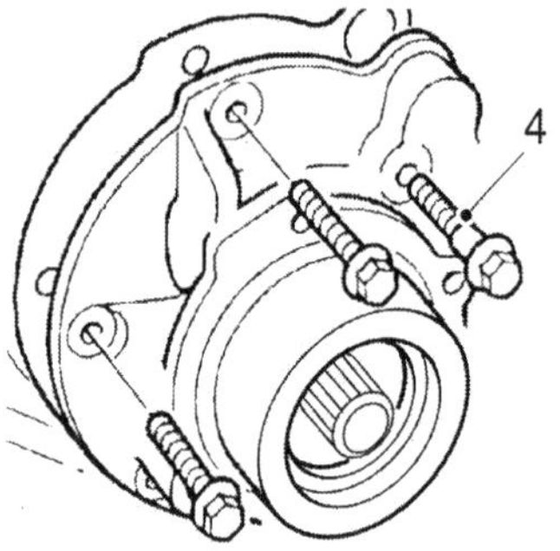

M417563

4. Note position of shoulder bolt, remove 6 bolts securing extension housing to main casing.

Installation

1. Clean extension housing, mating face on main casing and sealant from bolt threads.
2. Apply sealant, Part No. STC 4600 to mating flange on output shaft housing.
3. Fit extension housing to main casing ensuring that splines of output shaft are engaged in differential.
4. Apply sealant, Part No. STC 50552 to bolt threads.
5. Fit bolts and tighten by diagonal selection to 45 Nm (33 lbf.ft).
6. Fit transfer case rear output shaft seal.

Refer to: Transfer Case Rear Output Shaft Seal (308-07 Transfer Case, Vehicles With: MT82 6-Speed Manual Transmission In-vehicle Repair)

7. Fill transfer case with oil.

Refer to: Transfer Case Draining and Filling (308-07 Transfer Case, Vehicles With: MT82 6-Speed Manual Transmission General Procedures)

Transfer Case Front Output Shaft Seal

In-vehicle Repair

Special Tool(s)

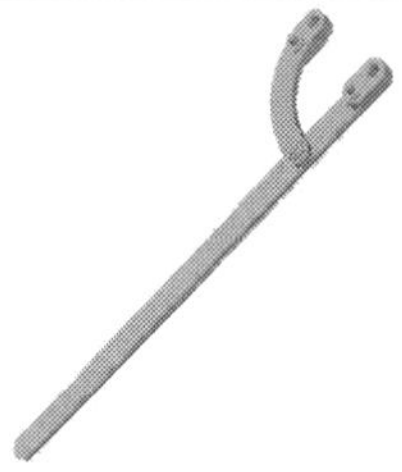 E54574	205-053 Retainer, Differential Pinion Flange
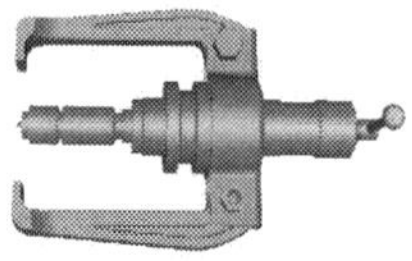 E120295	100-005A Remover, Puller
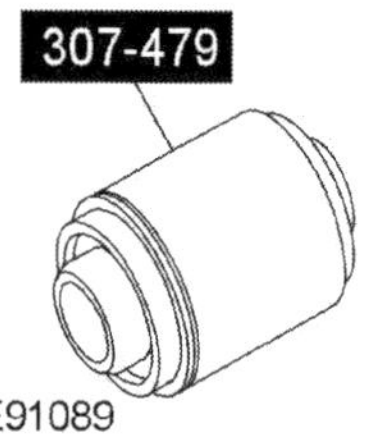 E91089	307-479 Installer, Transfer Case Oil Seal

Part(s)

STEP	PART NAME	QUANTITY
Installation Step 3	Renew Part Transfer case front output shaft flange	1
Installation Step 3	Renew Part Transfer case output shaft flange washer	1
Installation Step 4	Renew Part Output shaft flange nut	1
Installation Step 5	Renew Part Driveshaft flange nut(s)	4

Removal

1. ⚠ WARNING: Do not work on or under a vehicle supported only by a jack. Always support the vehicle on safety stands.

Raise and support the vehicle.

2. Drain the transfer case.

Refer to: Transfer Case Draining and Filling (308-07 Transfer Case, Vehicles With: MT82 6-Speed Manual Transmission General Procedures)

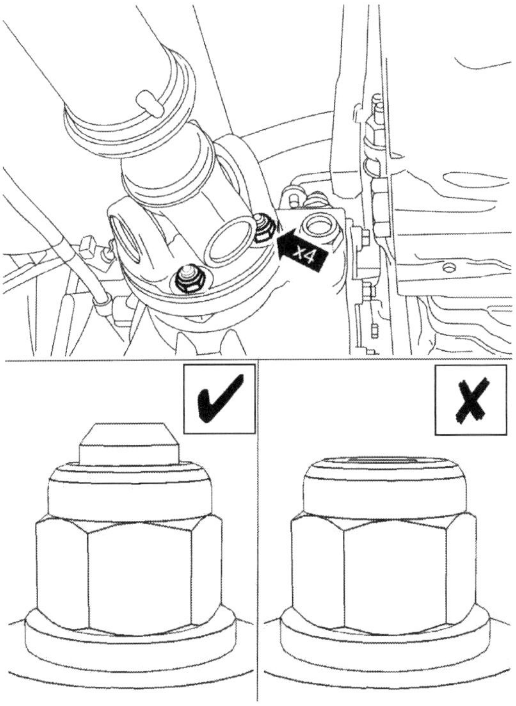

3. Remove and discard the 4 nuts. Release the driveshaft from the transfer case output drive flange.

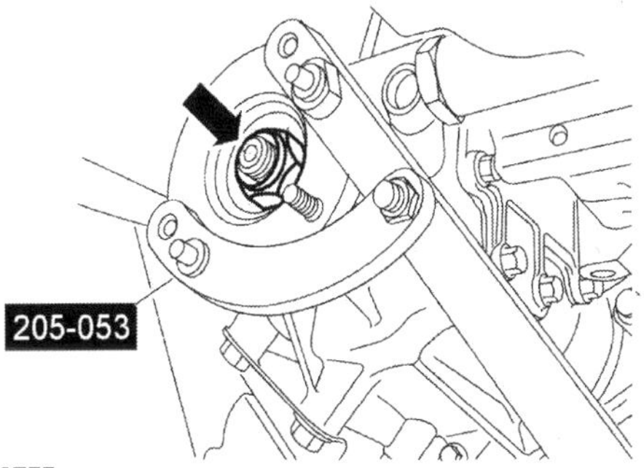

E90755

4. Remove and discard the nut and felt washer. Using the special tool, restrain the transfer case front output flange.

   Special Tool(s): 205-053

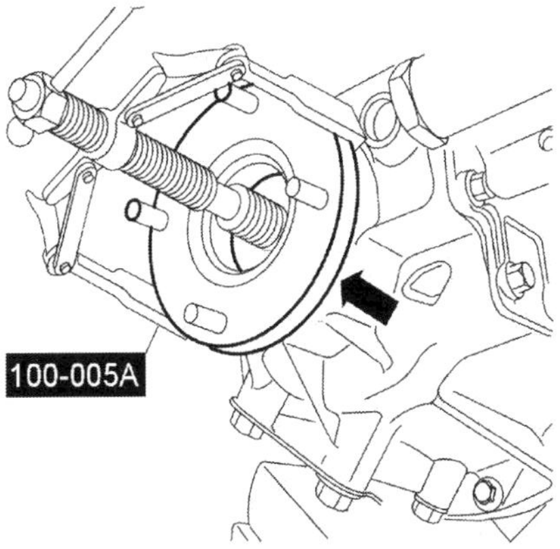

5. Remove and discard the transfer case front output flange. If necessary use the special tool to remove the transfer case front output flange.

   Special Tool(s): 100-005A

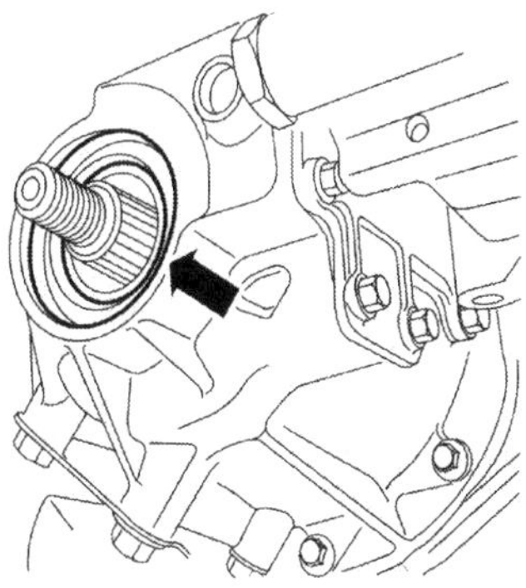

E90757

6. Using a suitable tool, remove and discard the transfer case front output shaft seal.

Installation

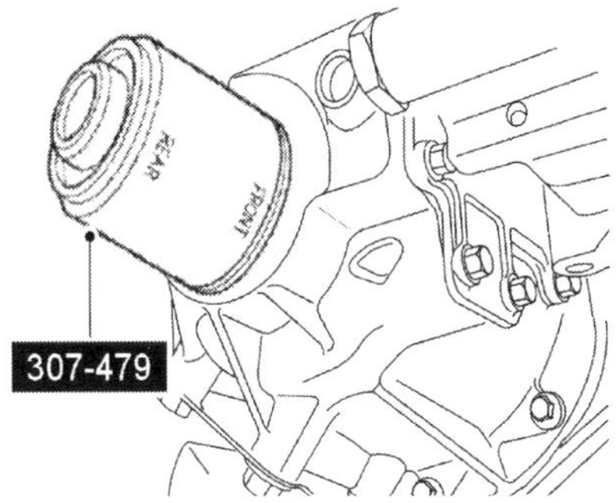

1. NOTES:

⚠ Clean the component mating faces.

⚠ Use end of tool marked 'FRONT' to fit seal.

Using the special tool, install a new transfer case front output shaft seal.

Special Tool(s): 307-479

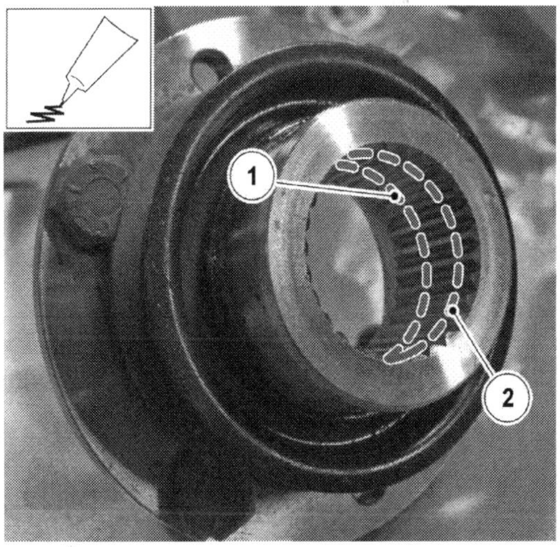

2. ⚠ CAUTION: Make sure that the Loctite 648 is pushed away from the bearing during installation of the drive flange. Failure to follow this instruction may result in damage to the vehicle.

   Hold the drive flange at the stud end and apply Loctite 648 to the splines, as illustrated. Apply a continuous bead along the splines. Apply a second continuous bead 5mm from the end of the flange.

3. NOTES:

Installation of a new felt washer is not required.

Make sure that any excessive Loctite 648 is removed from the drive flange prior to installing the steel washer.

Install the drive flange to the output shaft, making sure that the drive flange is pushed fully home. Install new steel washer

Renew Part: Transfer case front output shaft flange

Renew Part: Transfer case output shaft flange washer

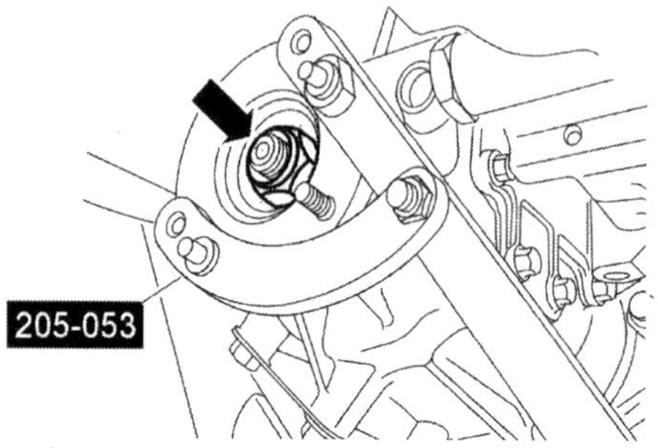

4. CAUTION: Do not use any power tools or air operated equipment to tighten the drive flange nut.

Position LRT-51-003 install and tighten a new drive flange nut.

Special Tool(s): 205-053

Renew Part: Output shaft flange nut

Torque: 165 Nm

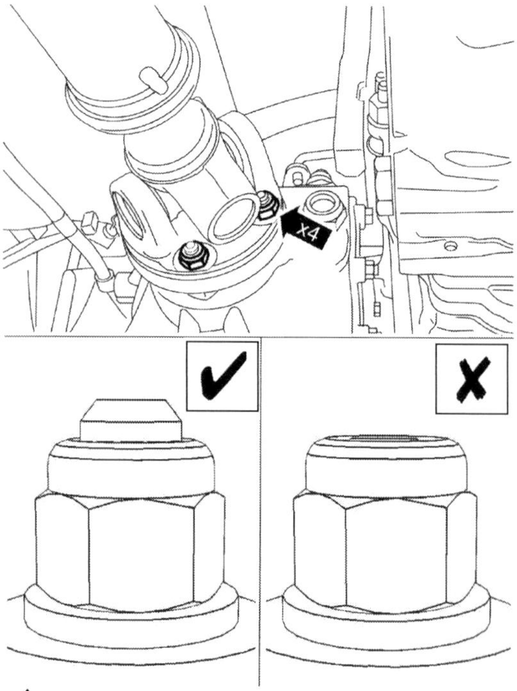

5. NOTE: Install new nuts.

Tighten the nut. Align the position of the driveshaft in relation to the drive pinion flange.

Renew Part: Driveshaft flange nut(s)

Torque: 47 Nm

6. Fill the transfer case.

Refer to: Transfer Case Draining and Filling (308-07 Transfer Case, Vehicles With: MT82 6-Speed Manual Transmission General Procedures)

Transfer Case High/Low Range Linkage

In-vehicle Repair

Removal

1. Remove the floor console.

Refer to: Floor Console (501-12, Removal and Installation).

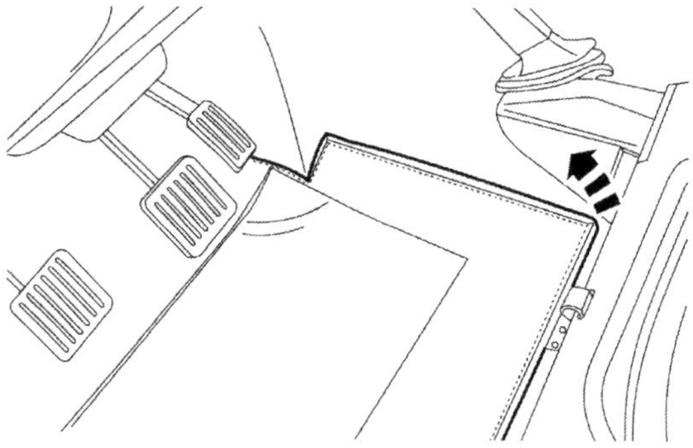

2. Reposition the LH carpet.

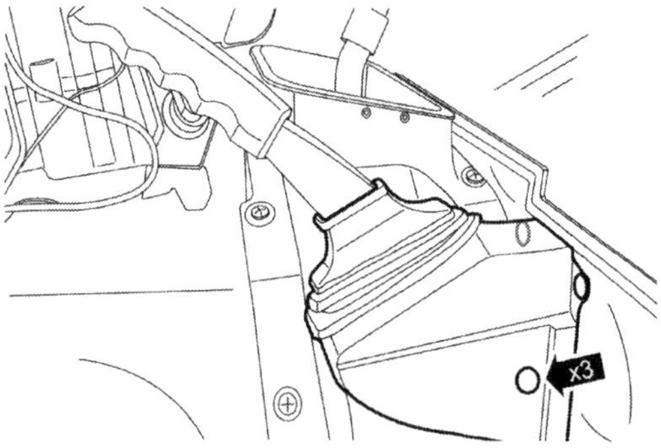

3. Release the parking brake lever gaiter.
   - Remove the 3 clips.

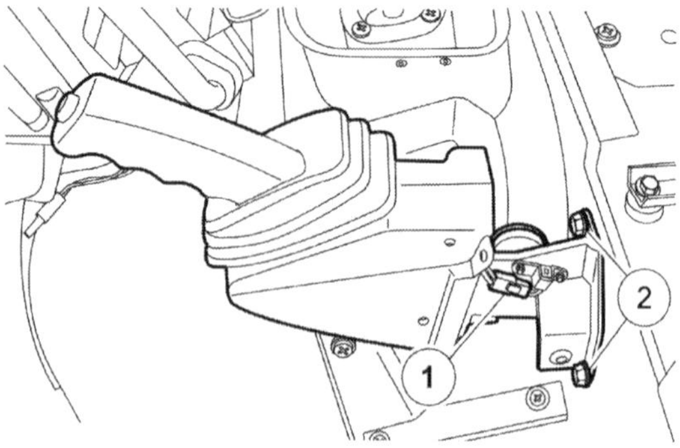

4. Release the parking brake lever.
   1. Disconnect the electrical connector.
   2. Remove the 2 bolts.

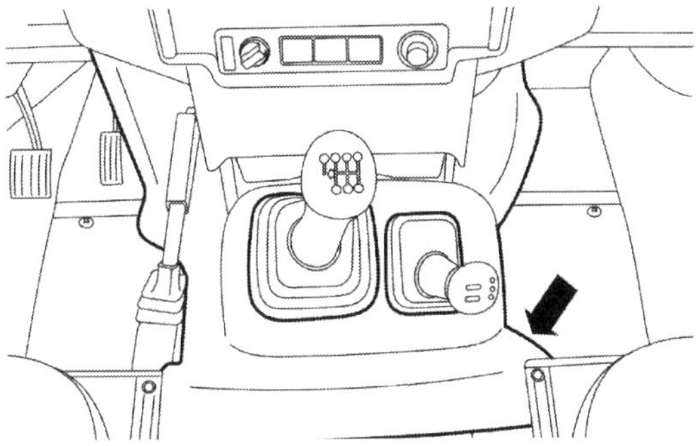

5. Remove the transmission cover panel floor covering.

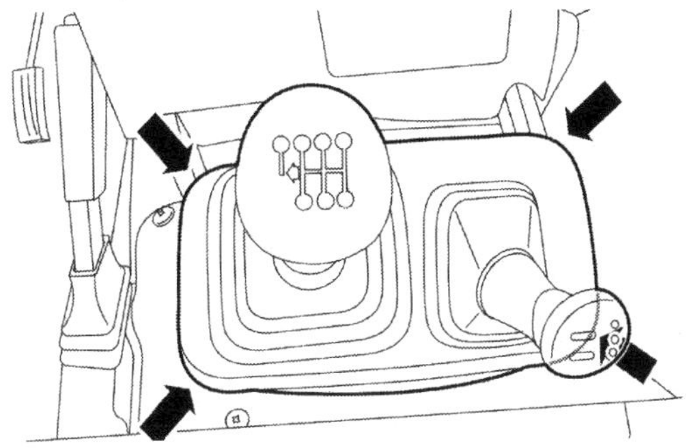

6. NOTE: Do not detach the gaiter from the selector knobs.

Detach the gaiter from the transmission cover panel.

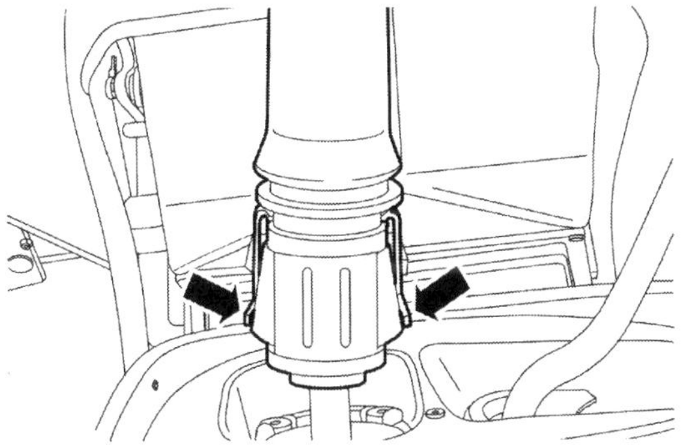

7. WARNING: The gearshift lever knob will be released suddenly, keep face clear during removal. Failure to follow this instruction may result in personal injury.

Release the gearshift lever knob.

- Release the 2 clips.

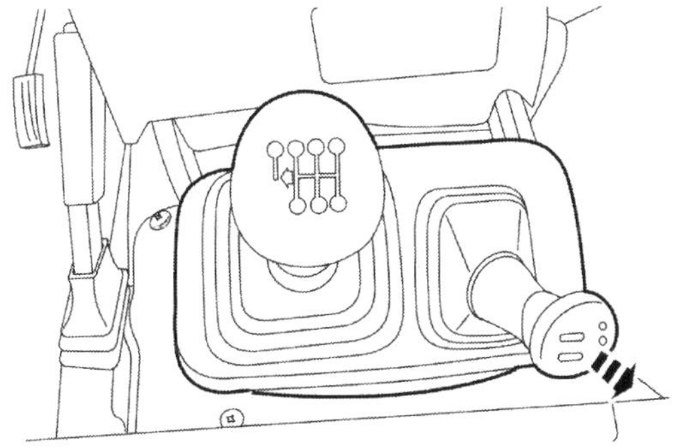

8. WARNING: The high/low range lever knob will be released suddenly, keep face clear during removal. Failure to follow this instruction may result in personal injury.

Remove the gaiter and selector levers.

- Detach the high/low range selector lever.

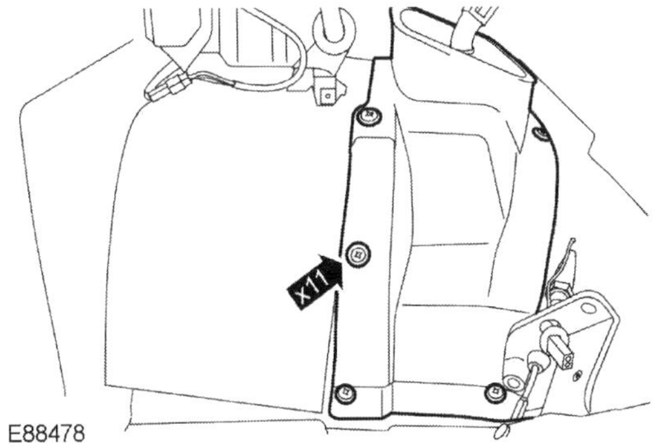

9. Remove the transmission cover panel.

- Remove the 11 screws.

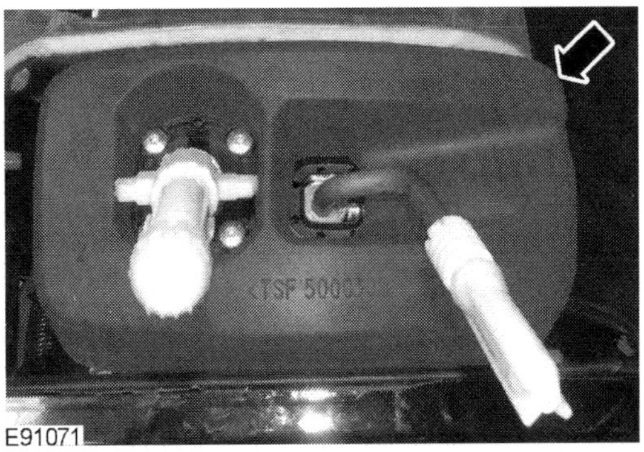

10. NOTE: To reduce tension on the insulation, position the differential lock lever to the left.

Remove the insulation pad.

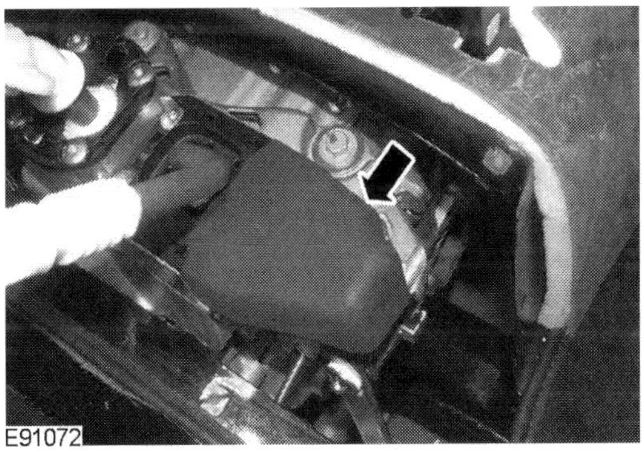

11. Remove the insulation pad support bracket.

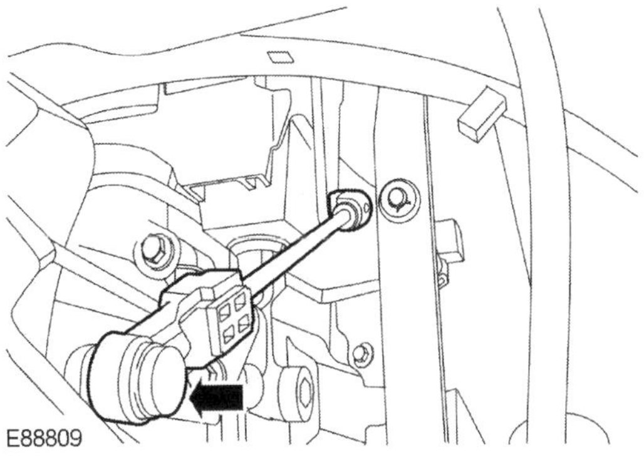

12. Disconnect the high/low range selector rod.

- Press the button to release the fitting before disconnecting the rod.

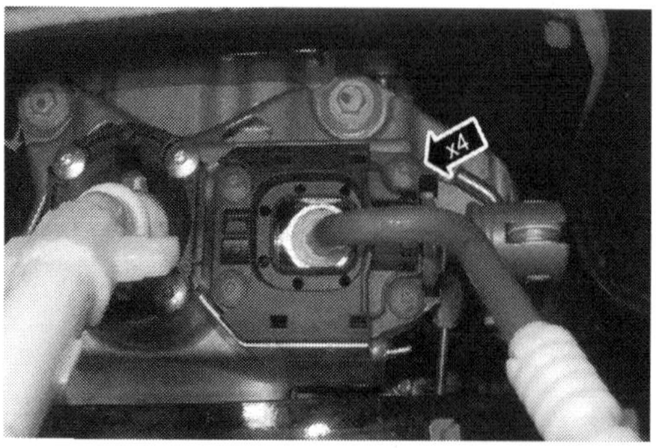

13. Remove the high/low range linkage.
    - Remove the 4 screws.

Installation

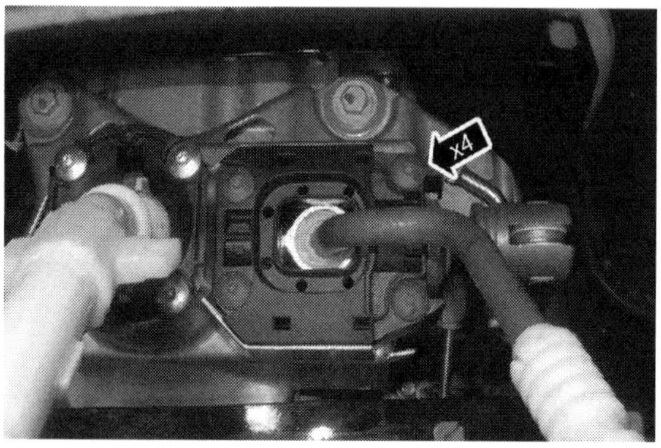

1. Install the high/low range linkage.
    - Tighten to 7 Nm (5 lb.ft).

⚠ NOTE: If adjustment of the transfer case high/low range selector linkage has changed, carry out the adjustment procedure.

Refer to: Transfer Case High/Low Range Selector Rod Adjustment (308-07 Transfer Case, Vehicles With: MT82 6-Speed Manual Transmission General Procedures)

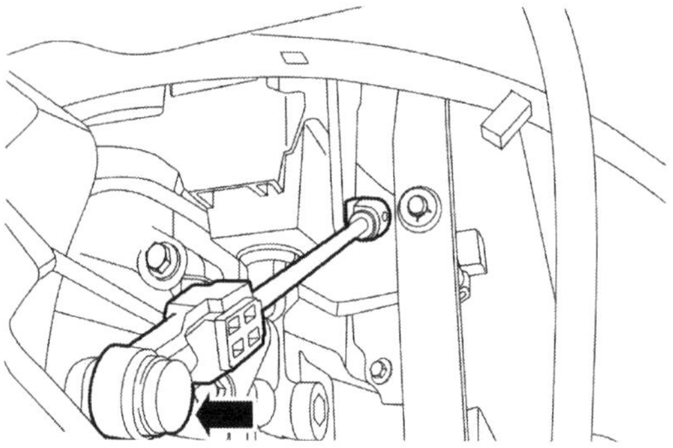

2. ⚠ NOTE: Make sure the rod is fully engaged on the ball joint and not on the foam washer.

Connect the high/low range selector rod.

3. Install the insulation pad support bracket.

4. ⚠ NOTE: Position the differential lock to the left, to reduce tension on the insulation.

Install the insulation pad.

5. Install the transmission cover panel.
    - Install the screws.
6. Install the gaiter with the selector levers attached.
    - Install the selector levers.
    - Fully seat the gaiter.
7. Install the transmission cover panel floor covering.

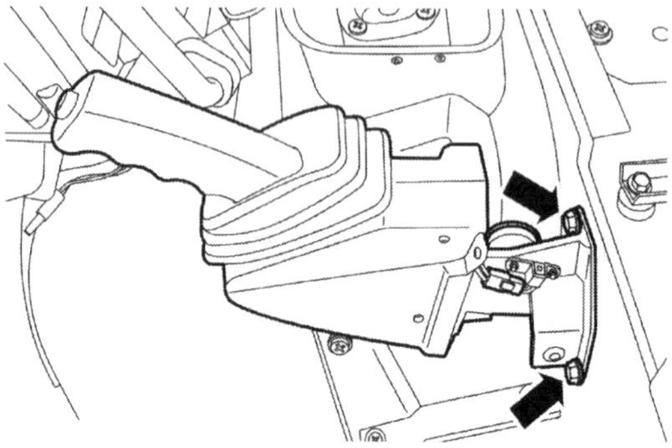

8. Install the parking brake lever.
    - Tighten to 25 Nm (18 lb.ft)
9. Install the parking brake lever gaiter.
    - Install the clips.
10. Reposition the LH carpet.
11. Install the floor console.

Refer to: Floor Console (501-12, Removal and Installation).

Transfer Case Input Shaft Seal

In-vehicle Repair

Special Tool(s)

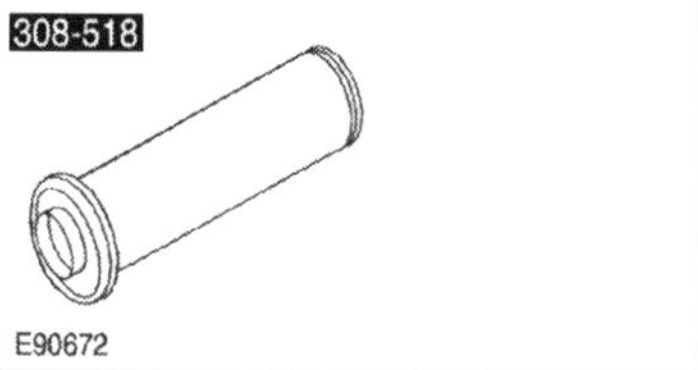	308-518 Installer, Mainshaft Oil Seal

Removal

1. Disconnect the battery ground cable.

Refer to: Startup Battery Disconnect and Connect (414-01 Battery, Mounting and Cables, General Procedures)

2. ⚠ WARNING: Do not work on or under a vehicle supported only by a jack. Always support the vehicle on safety stands.

Raise and support the vehicle.

3. Remove transfer case.

Refer to: Transfer Case (308-07 Transfer Case, Vehicles With: MT82 6-Speed Manual Transmission Removal)

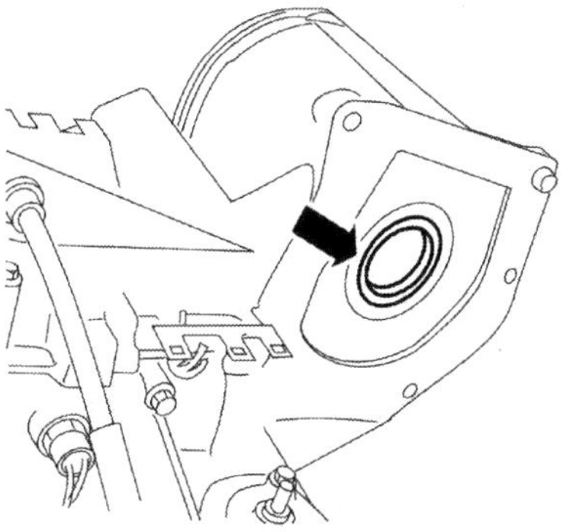

E90673

4. Remove and discard the transfer case input shaft seal.

Installation

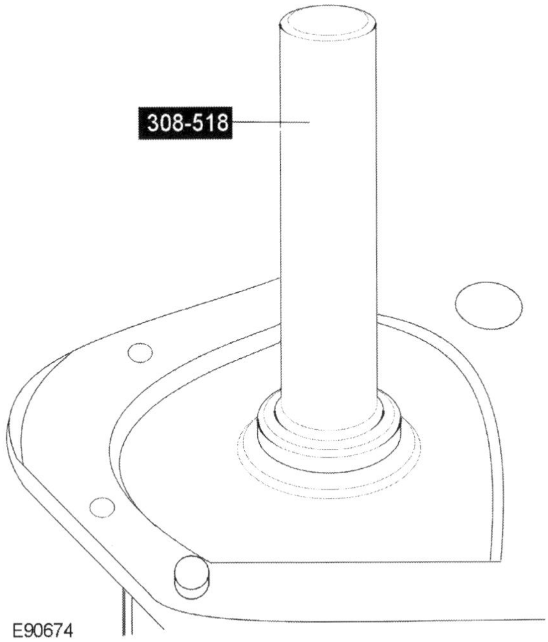

E90674

1. Using the special tool, install a new transfer case input shaft seal.

   Special Tool(s): 308-518

2. Install the transfer case.

Refer to: Transfer Case (308-07 Transfer Case, Vehicles With: MT82 6-Speed Manual Transmission Installation)

3. Connect the battery ground cable.

Refer to: Startup Battery Disconnect and Connect (414-01 Battery, Mounting and Cables, General Procedures)

Transfer Case Lower Cover

In-vehicle Repair

Removal

1. Raise vehicle on ramp.

2. Drain transfer case oil.

Refer to: Transfer Case Draining and Filling (308-07 Transfer Case, Vehicles With: MT82 6-Speed Manual Transmission General Procedures)

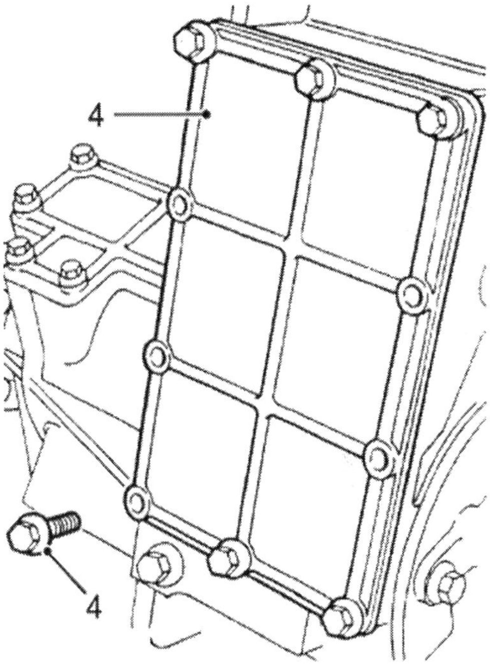

3. Remove 10 bolts securing lower cover to transfer case and remove cover.

Installation

1. Clean lower cover and mating face on transfer case.

2. Clean bolt threads.

3. Apply sealant, Part No. STC 50552 to bolt threads.

4. Apply sealant, Part No. STC 4600 to lower cover sealing face.

5. Position lower cover, fit bolts and tighten to 25 Nm (18 lb.ft).

6. Fill transfer case with oil.

Refer to: Transfer Case Draining and Filling (308-07 Transfer Case, Vehicles With: MT82 6-Speed Manual Transmission General Procedures)

7. Lower vehicle on ramp.

Transfer Case Rear Cover

In-vehicle Repair

Removal

1. Raise vehicle on 4 post ramp.
2. Mark rear cover to bearing housing for assembly purposes.

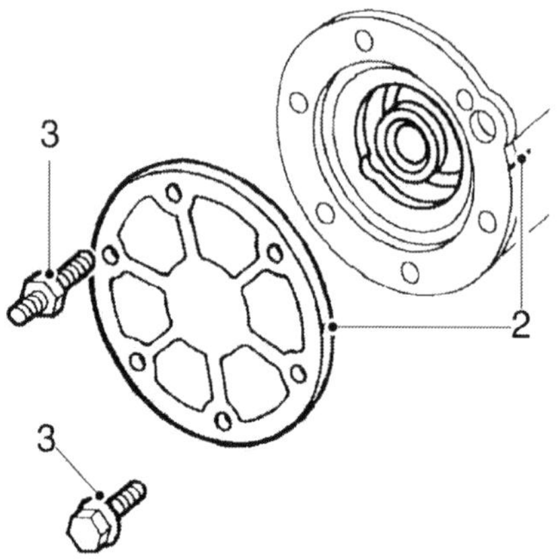

3. Noting fitted position of stud bolt and harness clip bracket, remove 5 bolts and stud bolt securing rear cover.
4. Collect harness clip bracket.
5. Remove rear cover.

Installation

1. Clean rear cover and mating face.
2. Apply sealant, Part No. STC 4600 to rear cover mating face.
3. Align reference marks, fit rear cover.
4. Clean rear cover bolts and apply sealant, Part No. STC 50552 to threads of bolts and stud nut.
5. Position clip bracket, fit stud nut and bolts, tighten by diagonal selection to 25 Nm (18 lb.ft).
6. Check/top-up transfer case oil.

Refer to: Transfer Case Draining and Filling (308-07 Transfer Case, Vehicles With: MT82 6-Speed Manual Transmission General Procedures)

7. Lower vehicle.

Transfer Case Rear Output Shaft Seal

In-vehicle Repair

Part(s)

STEP	PART NAME	QUANTITY
Installation Step 3	Renew Part Transfer case rear output shaft flange oil seal	1
Installation Step 5	Renew Part Transfer case rear output shaft flange	1
Installation Step 5	Renew Part Transfer case output shaft flange washer	1
Installation Step 6	Renew Part Driveshaft flange nut(s)	1

Removal

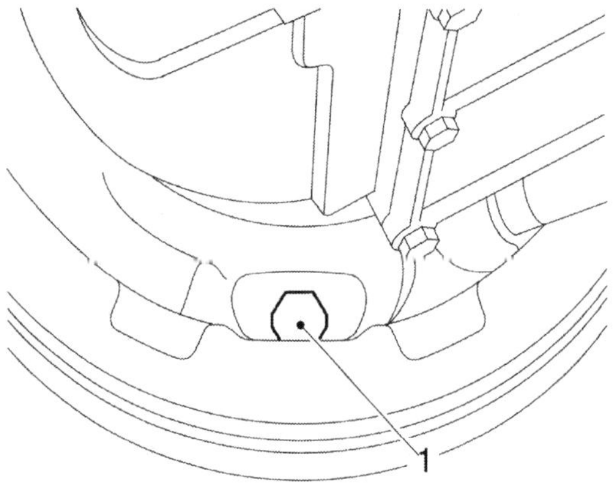

1. Loosen brake shoe adjuster bolt.
2. Remove rear drive shaft.

Refer to: Rear Driveshaft (205-01 Driveshaft, Removal and Installation)

3. Remove screw securing brake drum.
4. Remove brake drum.
5. Restrain transfer case drive flange using LRT-51-003, remove and discard drive flange nut, remove and discard steel and felt washers.
6. Using LRT-99-500 if necessary, remove and discard drive flange from transfer case.
7. Remove and discard rear output shaft oil seal from transfer case.

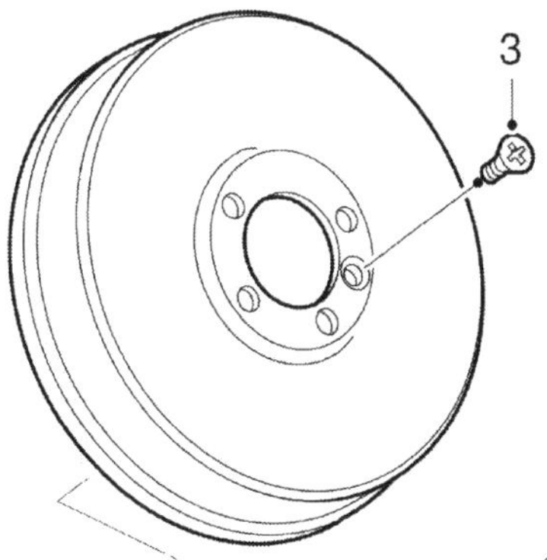

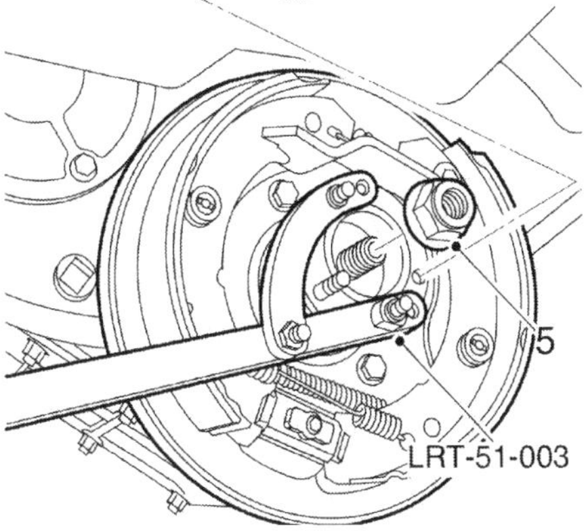

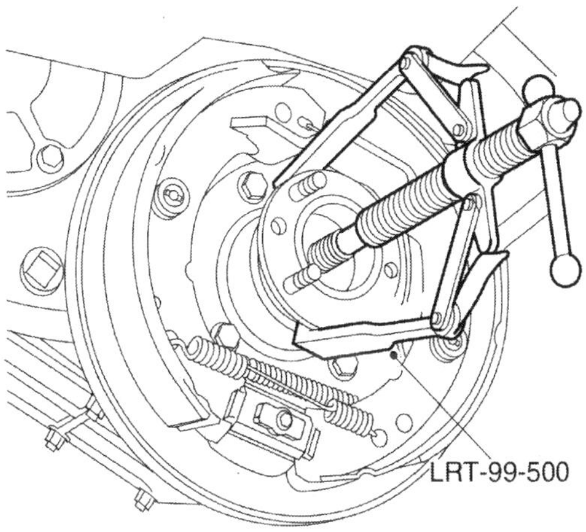

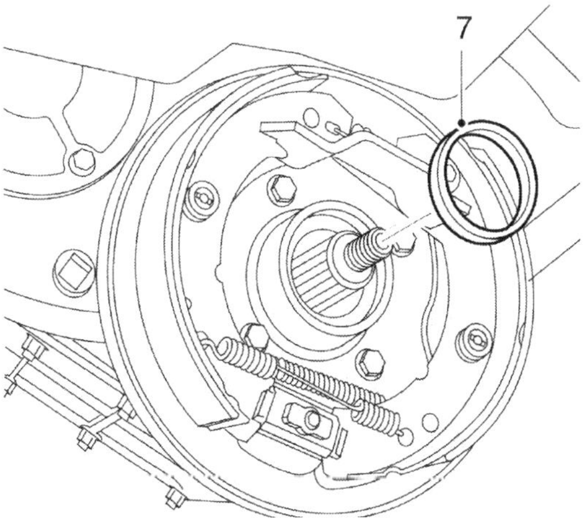

Installation

1. Clean drive flange splines, bearing and washer abutment faces to remove any oil or debris using solvent degreasing cleaner Loctite 7063.

2. Clean the seal recess, output shaft splines, threads and mating faces to remove any oil or debris using solvent degreasing cleaner Loctite 7063.

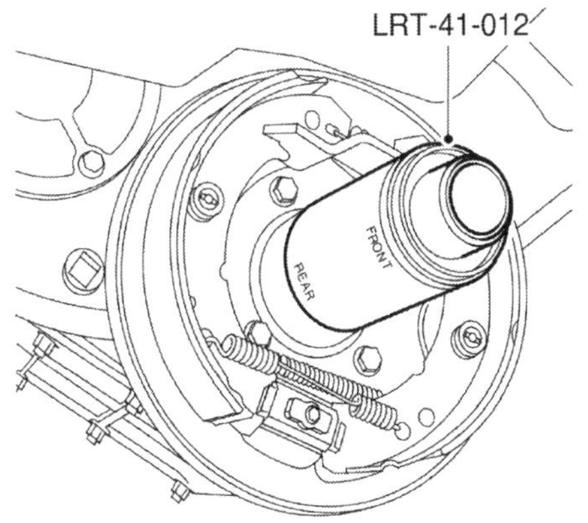

3. ⚠ CAUTION: Oil seal must be fitted dry.

⚠ NOTE: Use end of tool marked 'REAR' to install seal.

Install new seal using LRT-41-012.

Renew Part: Transfer case rear output shaft flange oil seal

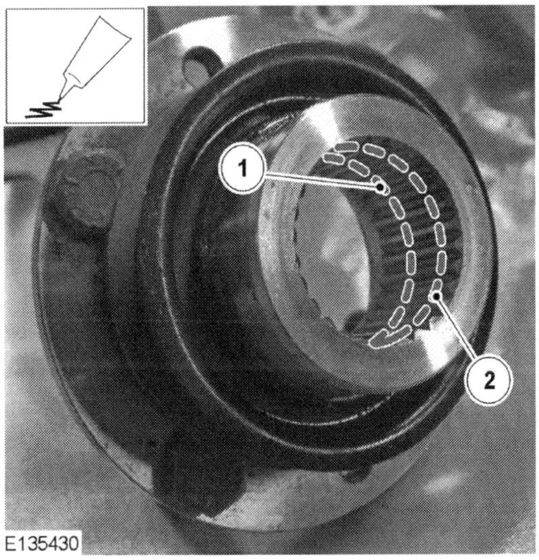

4. ⚠ CAUTION: Make sure that the Loctite 648 is pushed away from the bearing during installation of the new drive flange. Failure to follow this instruction may result in damage to the vehicle.

Hold the drive flange at the stud end and apply Loctite 648 to the splines, as illustrated. Apply a continuous bead along the splines. Apply a second continuous bead 5mm from the end of the flange.

5. NOTES:

⚠ **Installation of a new felt washer is not required.**

⚠ Make sure that any excessive Loctite 648 is removed from the drive flange prior to installing the steel washer.

Install the new drive flange to the output shaft, making sure that the drive flange is pushed fully home. Install new steel washer.

Renew Part: Transfer case rear output shaft flange

Renew Part: Transfer case output shaft flange washer

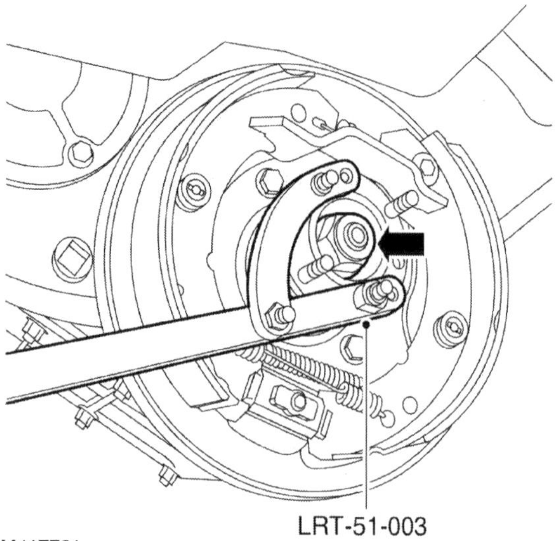

6. CAUTION: Do not use any power tools or air operated equipment to tighten the drive flange nut.

Position LRT-51-003 install and tighten a new drive flange nut.

Renew Part: Driveshaft flange nut(s)

Torque: 165 Nm

7. Position brake drum and tighten screw.

8. Tighten adjuster bolt.

Torque:

Stage 1: 25 Nm

Stage 2: Loosen: 540 °

9. Check that brake drum is free to rotate.

10. Install drive shaft.

Refer to: Rear Driveshaft (205-01 Driveshaft, Removal and Installation)

11. Top-up transfer case oil.

Refer to: Transfer Case Draining and Filling (308-07 Transfer Case, Vehicles With: MT82 6-Speed Manual Transmission General Procedures)

Transfer Case

Removal

Special Tool(s)

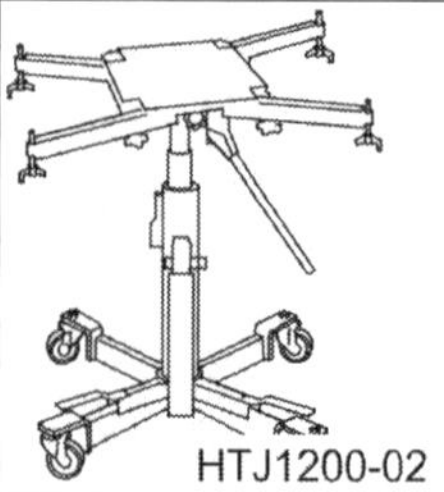	Power train Assembly Jack HTJ1200-02
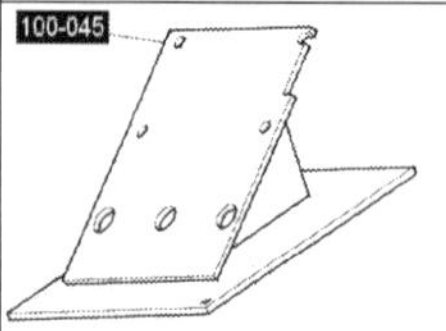	Transfer Case Support 100-045

Removal

1. WARNING: Do not work on or under a vehicle supported only by a jack. Always support the vehicle on safety stands.

Raise and support the vehicle.

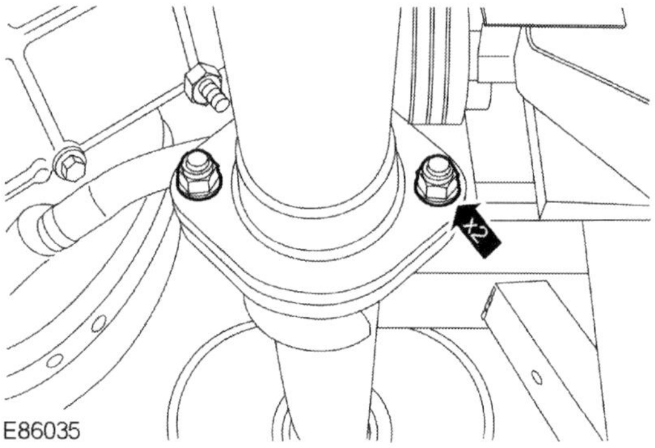

2. Release the intermediate pipe from the catalytic converter.
   - Remove and discard the 2 nuts.

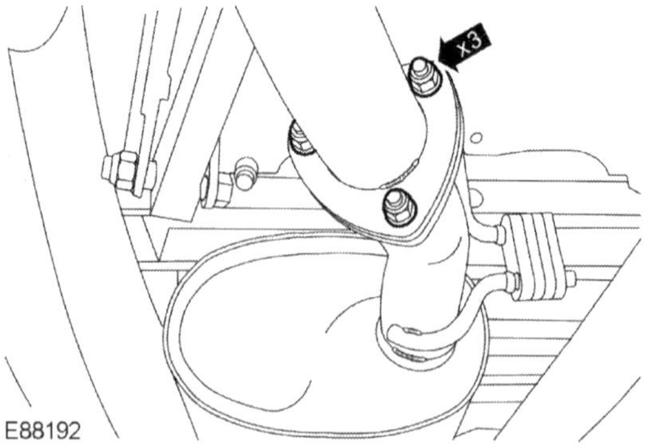

3. Release the tail pipe from the intermediate pipe.
   - Remove and discard the 3 nuts.
   - NOTE: If equipped.

Remove and discard the gasket.

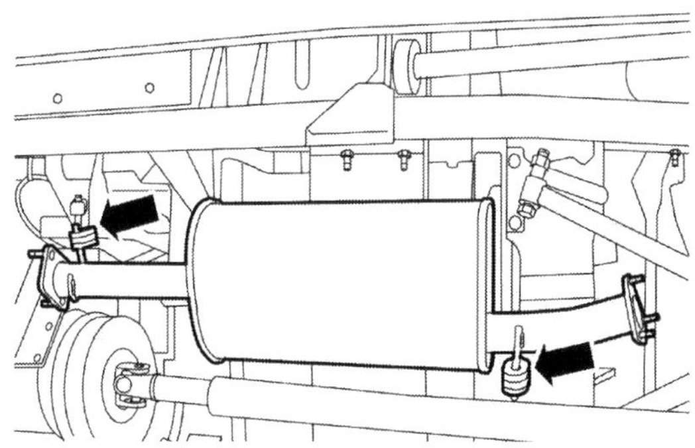

4. Remove the intermediate pipe and muffler.

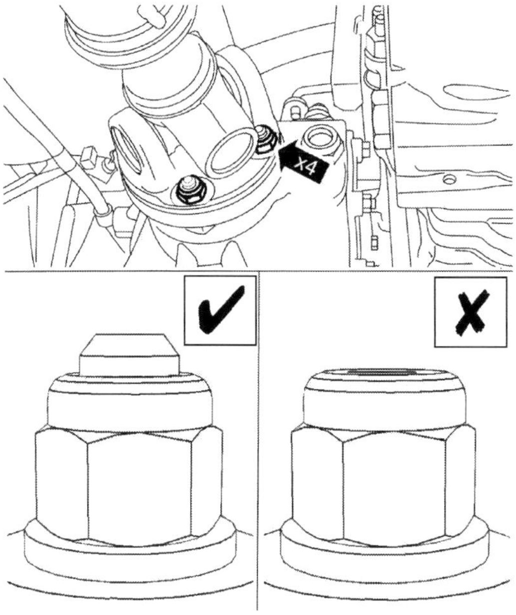

5. NOTE: Mark the front driveshaft to transfer case drive flange.

Remove and discard the 4 nuts.

- Using a suitable tie strap, secure the driveshaft to one side.

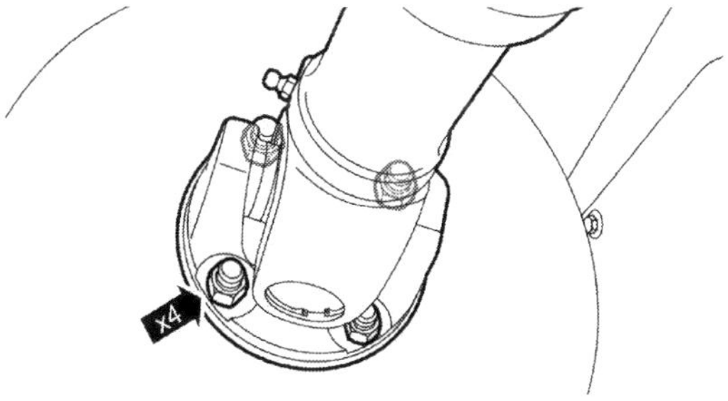

6. Remove and discard the 4 nuts from the rear driveshaft to transfer case.

- Mark the position of the driveshaft in relation to the drive pinion flange.

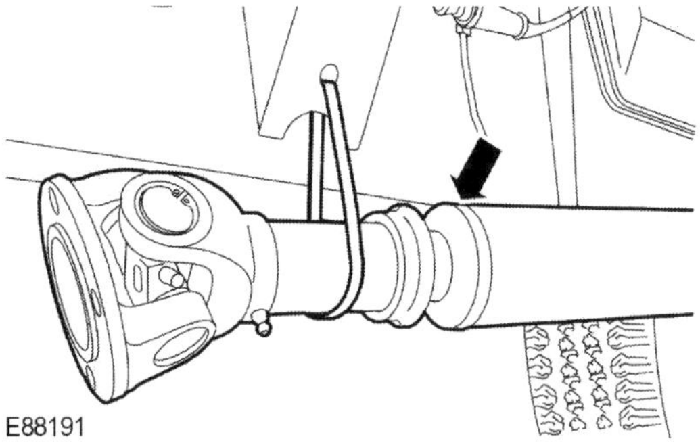

E88191

7. Using a suitable tie strap, secure the rear driveshaft to the chassis.

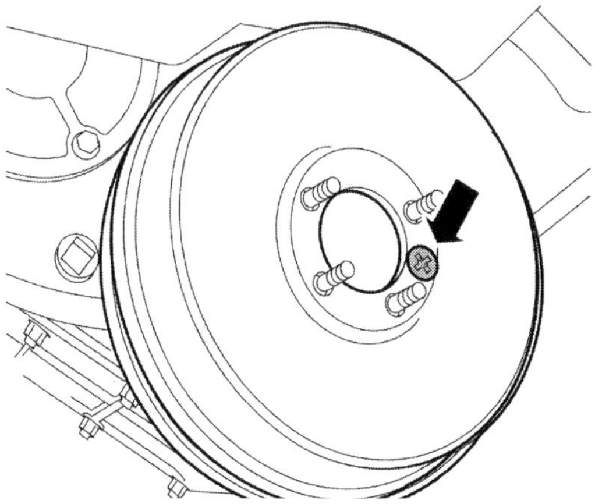

8. Remove the parking brake drum. Remove the securing screw.

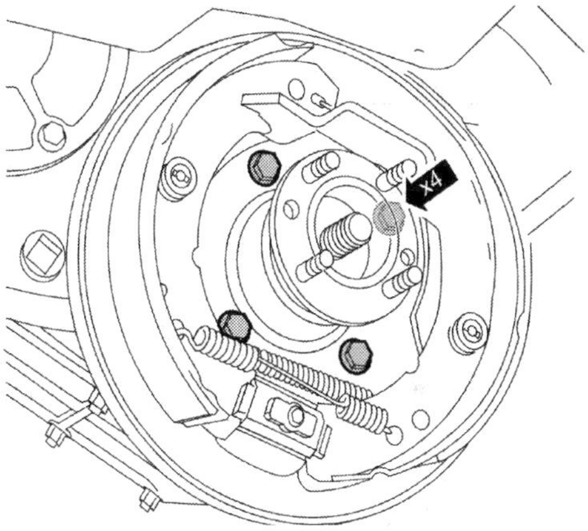

9. Remove the handbrake assembly and tie aside.

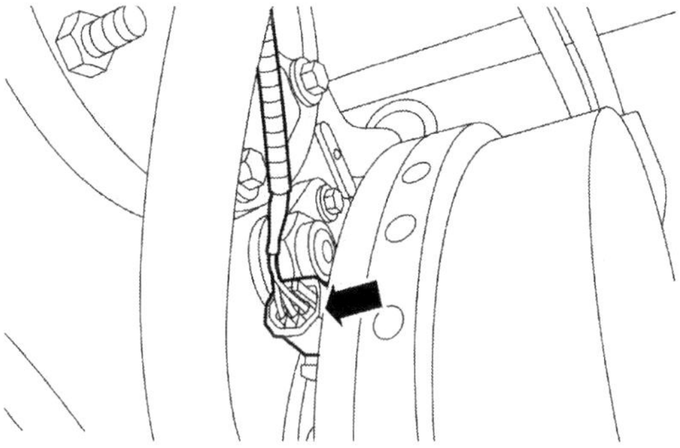

10. Disconnect the electronic speedometer electrical connector from the transfer case.

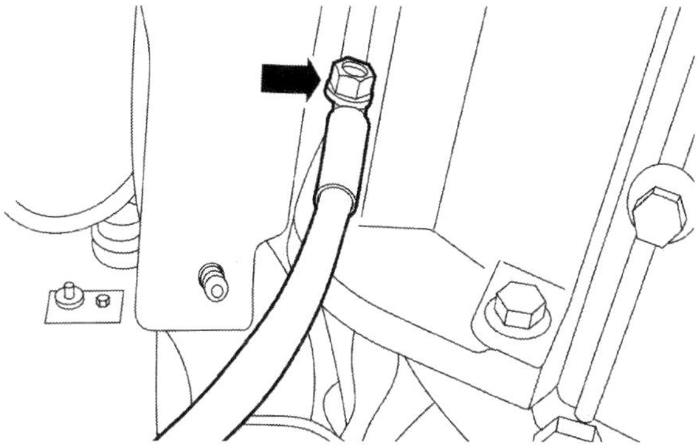

11. Disconnect the LH earth cable from the transfer case.

- Remove the nut.

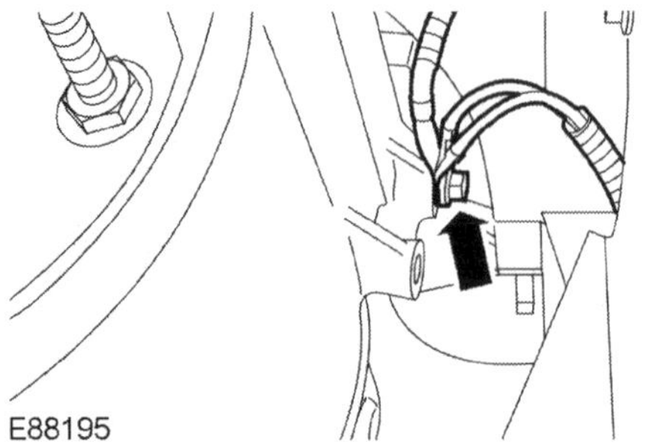

12. Disconnect the RH earth cables from the transfer case.
    - Remove the bolt.

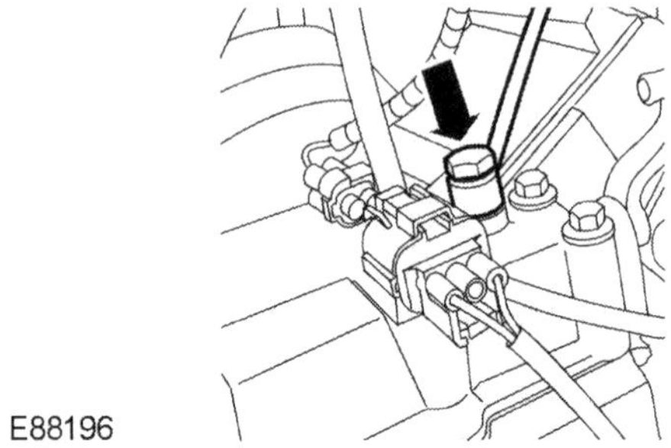

13. CAUTION: Before disconnecting any components, make sure the area is clean and free from foreign material. When disconnected all openings must be sealed.

    Remove the transfer case breather pipe bolt.

    - Remove and discard the sealing washers.

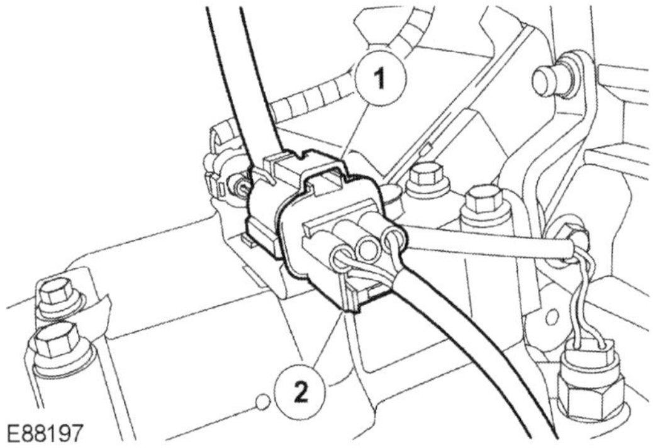

14. Disconnect the transmission range selector warning lamp electrical connector.
    1. Disconnect the electrical connector.
    2. Release the electrical connector from the bracket.

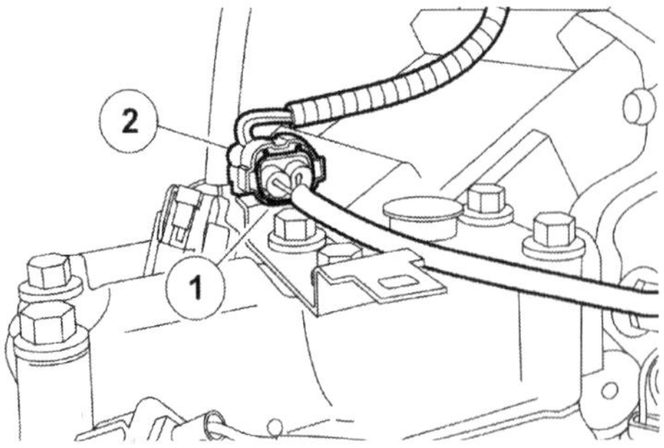

15. Disconnect the differential lock detection switch electrical connector.
    1. Disconnect the electrical connector.
    2. Release the electrical connector from the bracket.

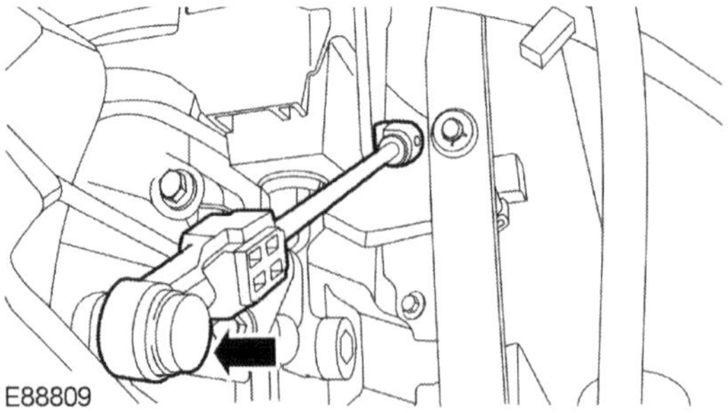

16. Disconnect the transmission range selector rod.
    - Press the button to release the fitting before disconnecting the rod.

17. Remove the retaining clip at the lower end of the pivot arm.

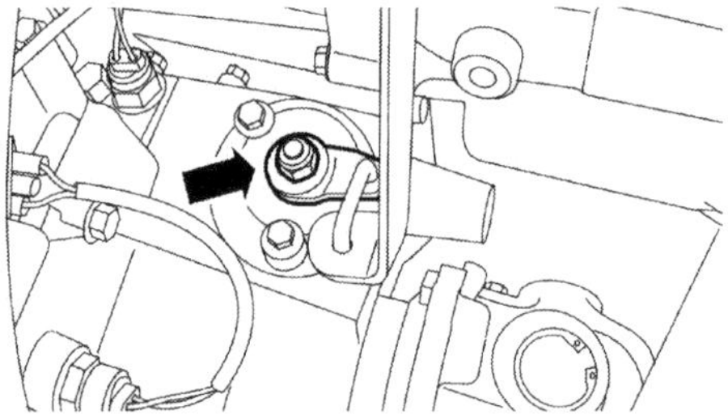

18. Remove the differential lock control operating rod.
    - Remove the nut.

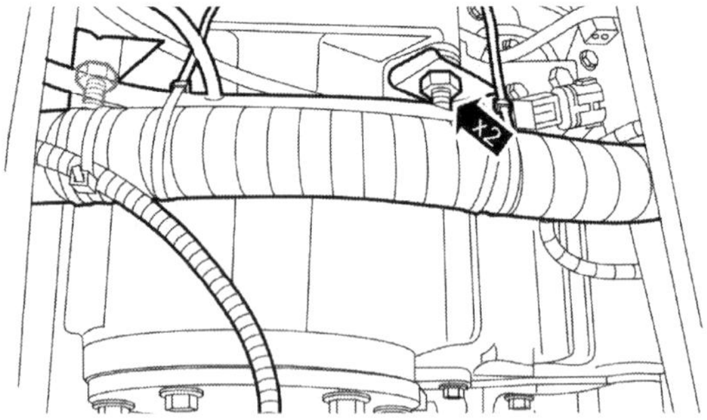

19. Release the transmission wiring harness.
    - Remove the nut and bolt.
20. Drain the transfer case.

    Refer to: Transfer Case Draining and Filling (308-07 Transfer Case, Vehicles With: MT82 6-Speed Manual Transmission General Procedures)

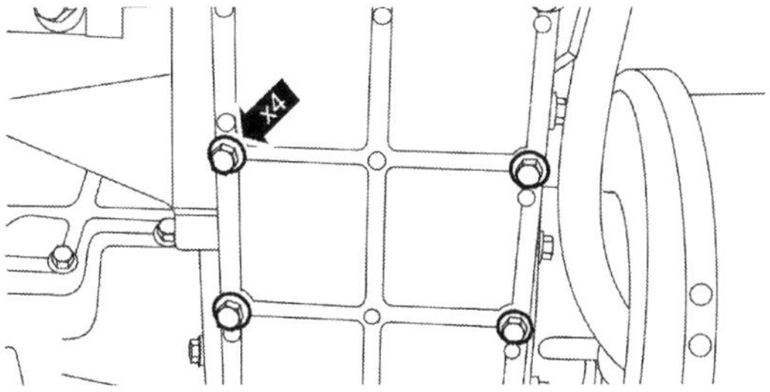

21. Remove the 4 bolts.

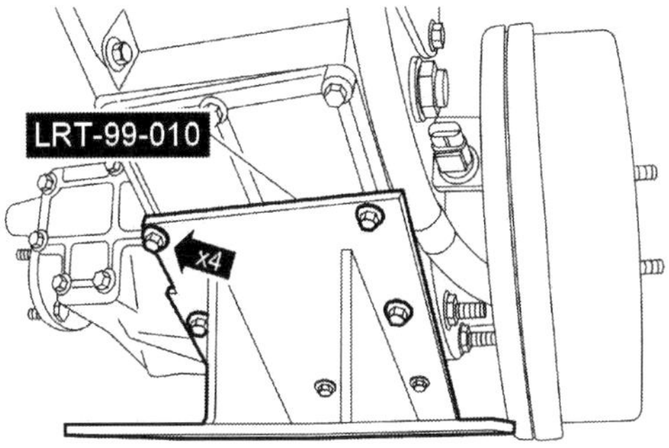

22. Install the special tool.

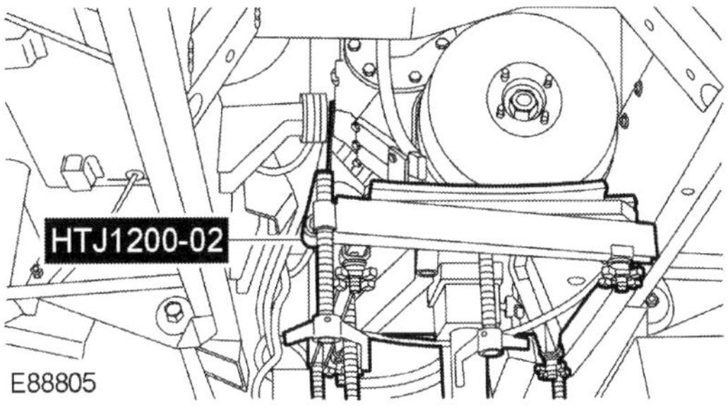

E88805

23. Install the special tool.

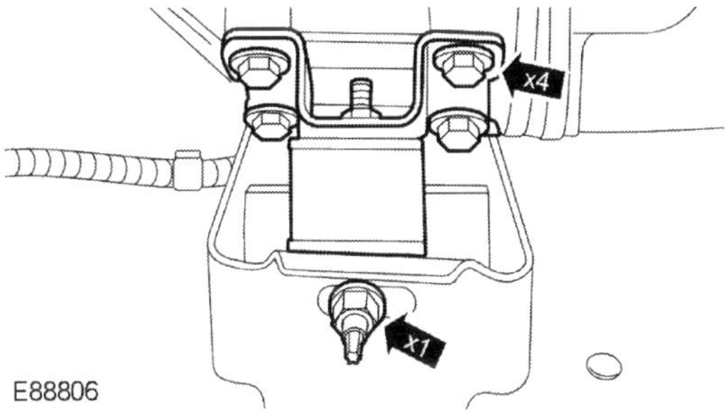

E88806

24. NOTE: Raise the transmission jack.
Remove the right-hand transfer case mount.
  - Remove the 4 bolts.
  - Remove the nut.

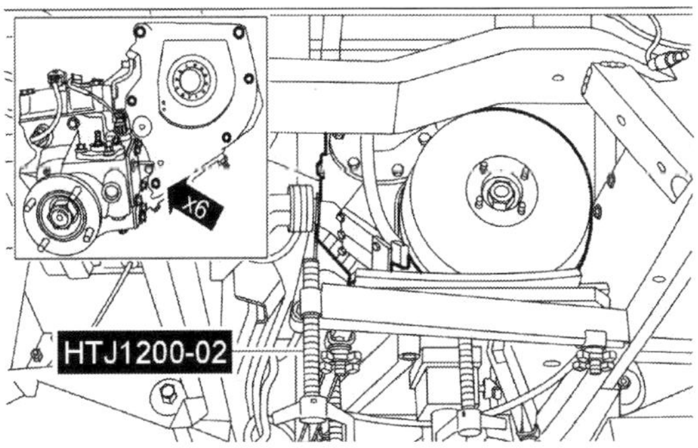

25. With assistance, remove the transfer case.
  - Remove the 4 bolts.
  - Remove the 2 nuts.

Transfer Case

Installation

Special Tool(s)

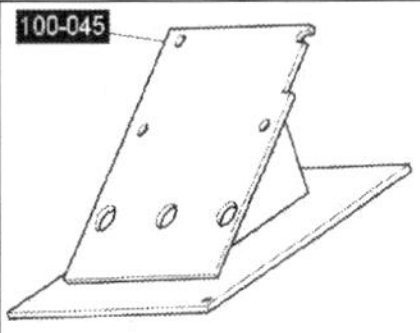 E90632	100-045 Support Bracket - Transfer Case

Part(s)

STEP	PART NAME	QUANTITY
Step 8	Renew Part Retaining clip	1
Step 12	Renew Part Breather hose O-ring seal	1
Step 18	Renew Part Driveshaft flange nut(s)	4
Step 19	Renew Part Driveshaft flange nut(s)	4
Step 21	Renew Part Intermediate exhaust pipe gasket	1
Step 21	Renew Part Intermediate exhaust pipe to rear pipe nut(s)	3
Step 22	Renew Part Intermediate exhaust pipe to catalytic converter nut(s)	2

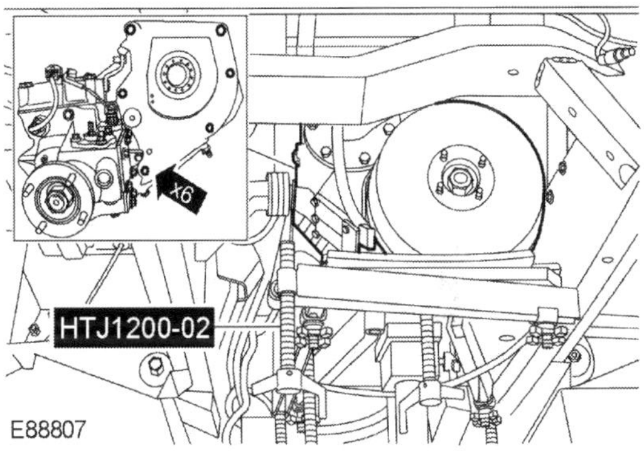

E88807

1. With assistance, install the transfer case. Apply sealant to the bolt threads.

   General Equipment: Transmission jack

   Bolts 45 Nm

   Nuts 45 Nm

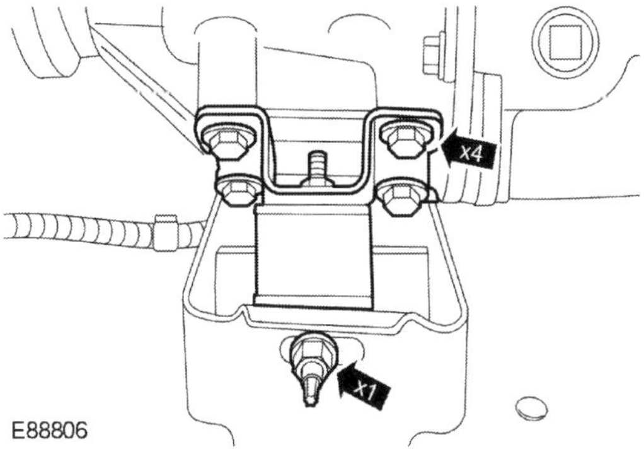

E88806

2. ⚠ NOTE: Lower the transmission jack.

   Install the right-hand transfer case mount.

   Bolts 85 Nm

   Nut 48 Nm

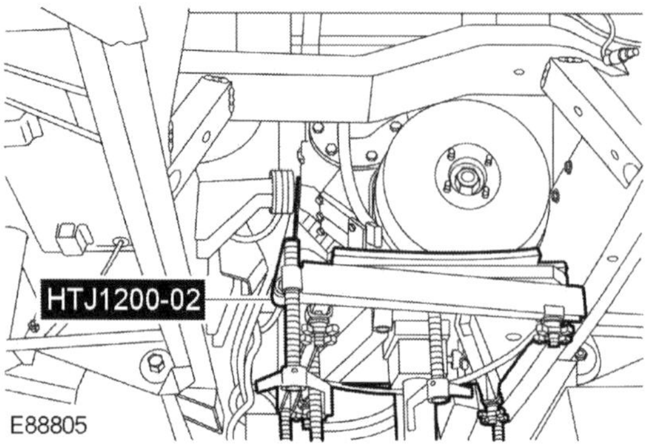

3. Remove the special tool.

   General Equipment: Transmission jack

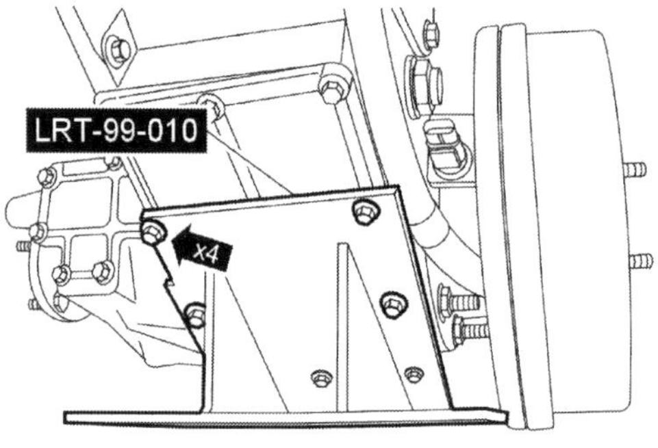

E88804

4. Remove the special tool.

   Special Tool(s): 100-045

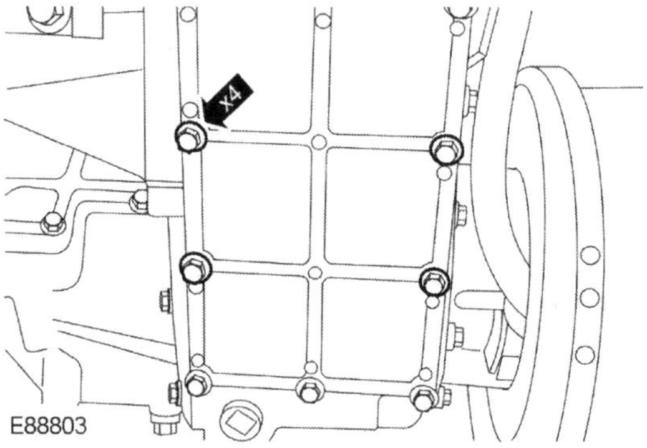

5. Apply sealant to the bolt threads. Install the bolts.

   Torque: 25 Nm

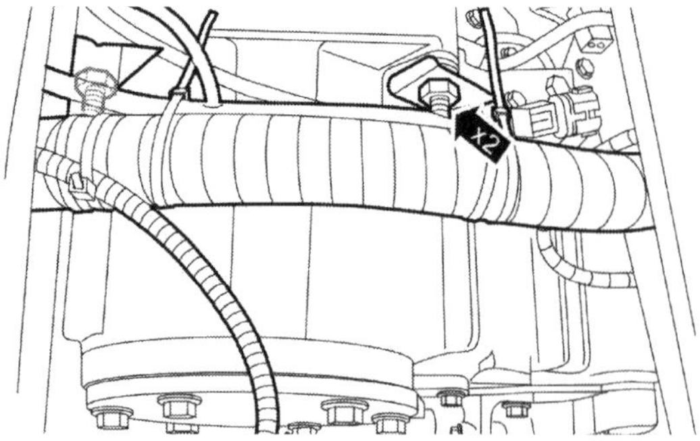

6. Secure the transmission wiring harness.

   Torque: 45 Nm

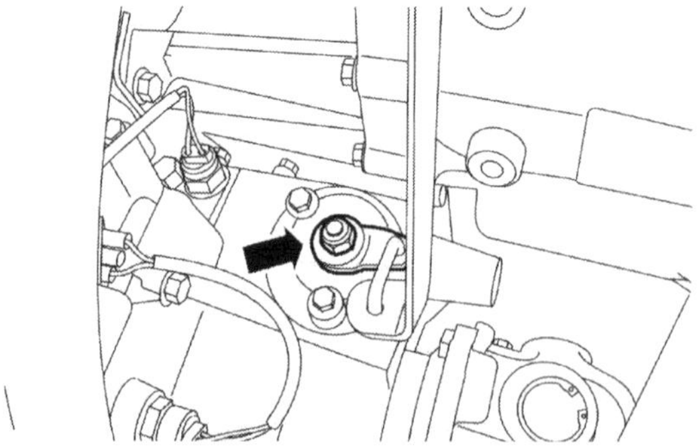

7. Install the differential lock control operating rod.

   Torque: 25 Nm

8. Install the new retaining clip at the lower end of the pivot arm.

   Renew Part: Retaining clip

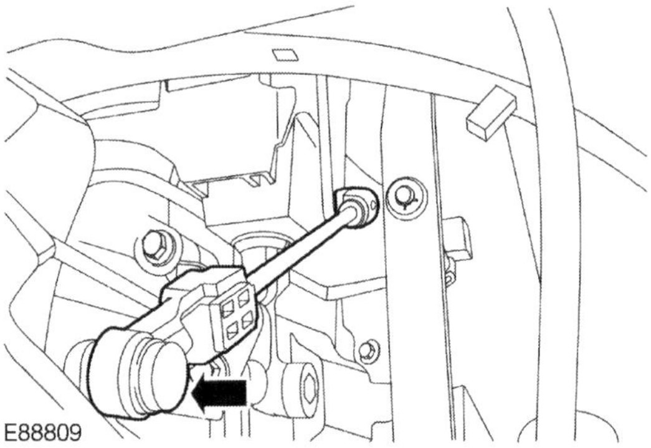

9. ⚠ NOTE: Make sure the rod is fully engaged on the ball joint.

   Connect the high/low range selector rod.

10. Connect the differential lock detection switch electrical connector. Connect the electrical connector. Secure the electrical connector to the bracket.

11. Connect the differential lock warning lamp electrical connector. Connect the electrical connector. Secure the electrical connector to the bracket.

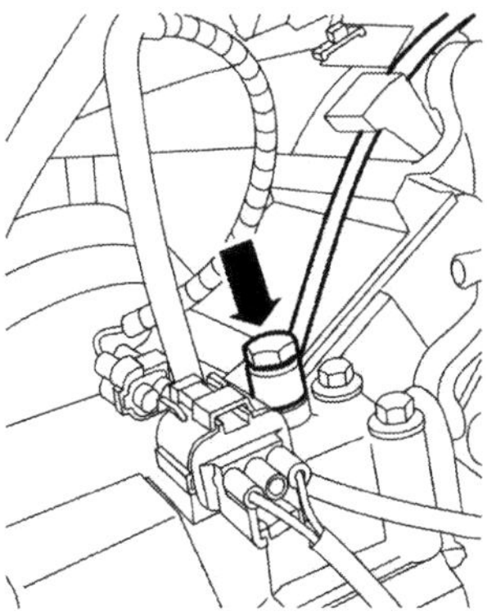

12. ⚠ NOTE: Remove and discard the blanking caps.

    Install new sealing washers. Install the transfer case breather pipe bolt.

    Renew Part: Breather hose O-ring seal

    Torque: 15 Nm

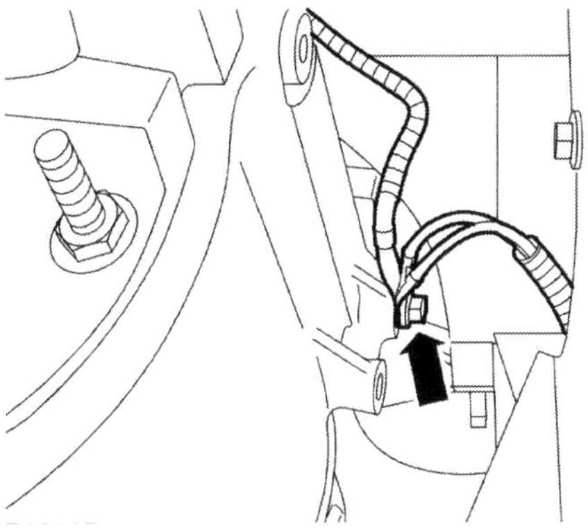

13. *Torque:* 12 Nm

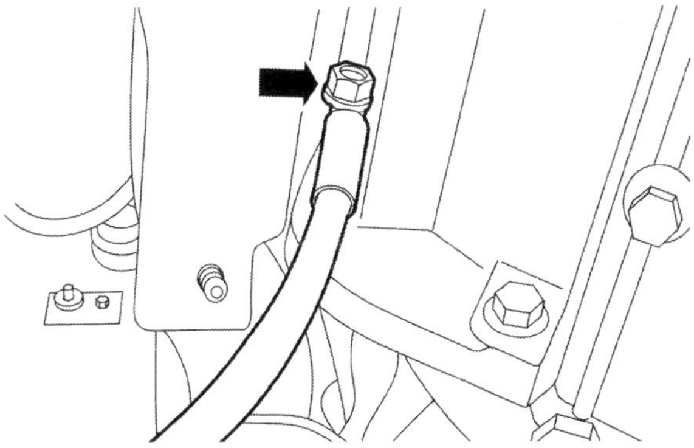

14. *Torque:* 45 Nm

15. Connect the electronic speedometer electrical connector to the transfer case.

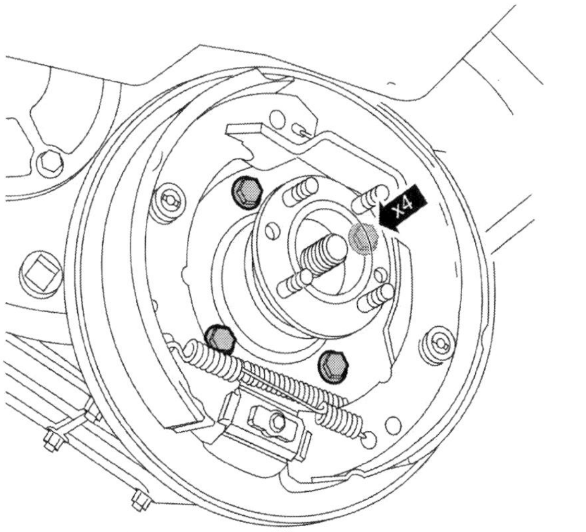

16. Install the parking brake assembly. Install the 4 securing bolts. Tighten the bolts.

    Torque: 73 Nm

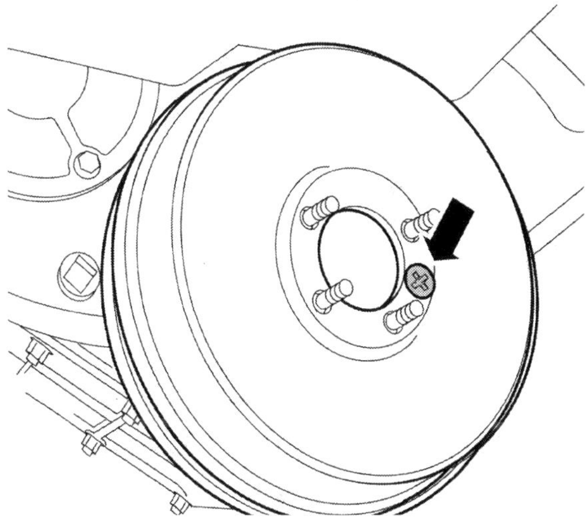

17. Install the parking brake drum. Install the securing screw.

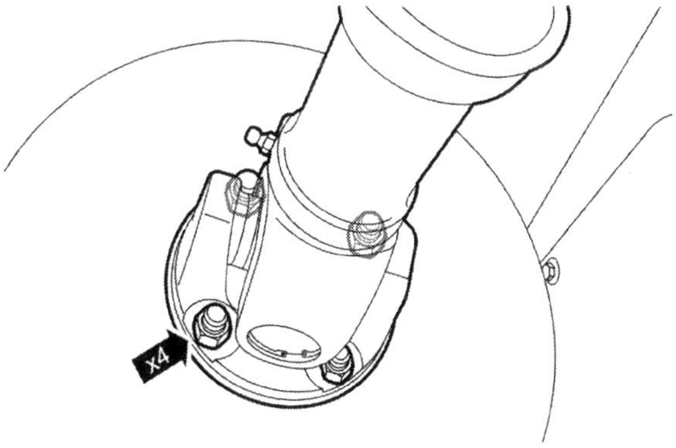

18. NOTES:

⚠ Remove and discard the tie strap.

⚠ Install new nuts.

Align the position of the driveshaft in relation to the drive pinion flange.

Renew Part: Driveshaft flange nut(s)

Torque: 45 Nm

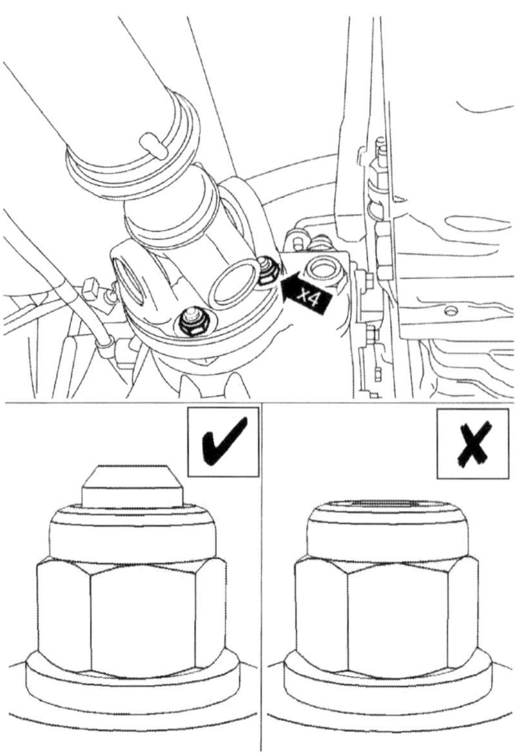

19. NOTES:

△ Install new nuts.

△ Clean the driveshaft drive flanges and mating faces.

Align the position of the driveshaft in relation to the drive pinion flange.

Renew Part: Driveshaft flange nut(s)

Torque: 47 Nm

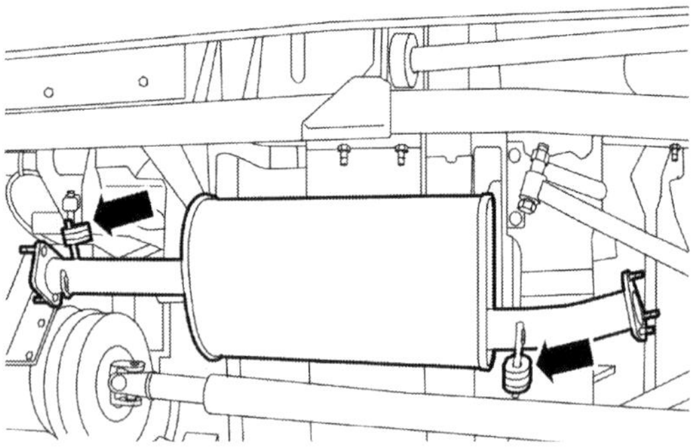

20. Install the intermediate pipe and muffler.

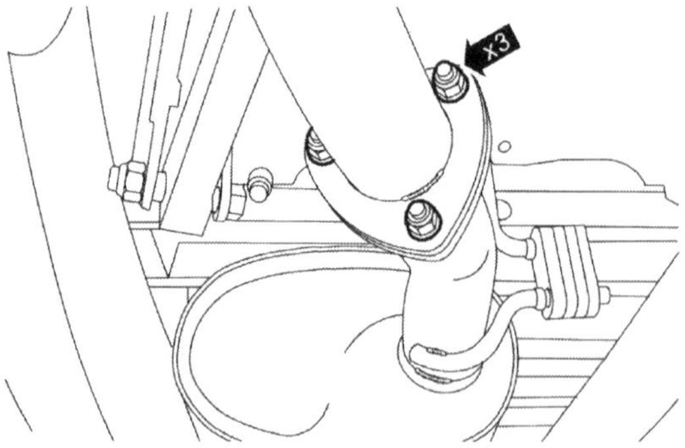

21. Install a new gasket. Tighten the new nuts. Secure the tail pipe to the intermediate pipe.

Renew Part: Intermediate exhaust pipe gasket

Renew Part: Intermediate exhaust pipe to rear pipe nut(s)

Torque: 25 Nm

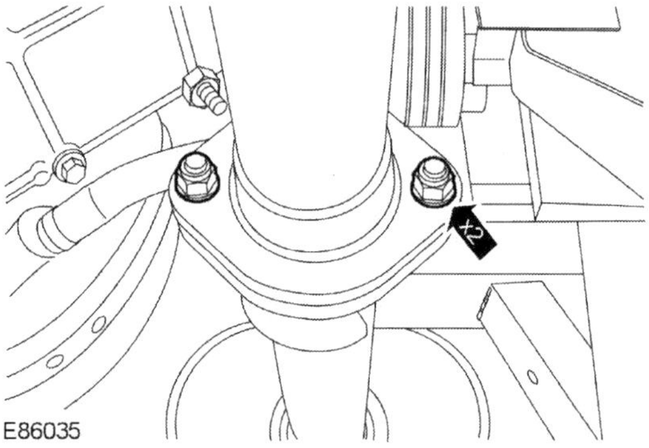

22. Secure the intermediate pipe to the catalytic converter. Loosely install the new nuts.

Renew Part: Intermediate exhaust pipe to catalytic converter nut(s)

Torque: 25 Nm

23. Fill the transfer case.

Refer to: Transfer Case Draining and Filling (308-07 Transfer Case, Vehicles With: MT82 6-Speed Manual Transmission General Procedures)

24. Lower the vehicle.

Differential Lock Indicator Switch Adjustment

General Procedures

NOTE: This procedure is only necessary for switches fitted with a locknut in place of the threaded spacer.

1. Move differential lock selector fork to differential locked position.
2. Connect a 12V test lamp and battery to differential lock indicator switch.
3. Screw switch in until test lamp is illuminated then screw the switch in a further 1/2 turn; tighten locknut.
4. Disengage differential lock, check that test lamp is extinguished.
5. Remove test lamp.

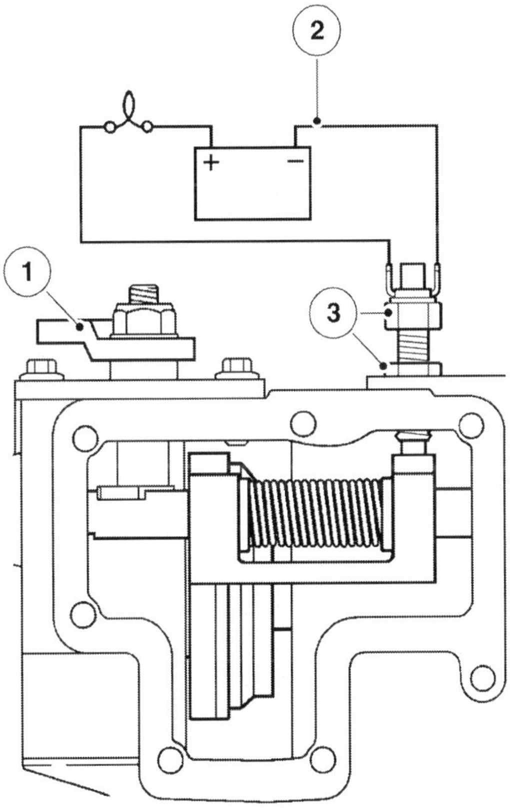

41M7177

Transfer Case Low Range Indicator Switch

Removal and Installation

Part(s)

STEP	PART NAME	QUANTITY
Removal Step 3	Renew Part Transfer case low range indicator switch	1

Removal

1. WARNING: Do not work on or under a vehicle supported only by a jack. Always support the vehicle on safety stands. Failure to follow this instruction may result in personal injury.

   Raise and support the vehicle.

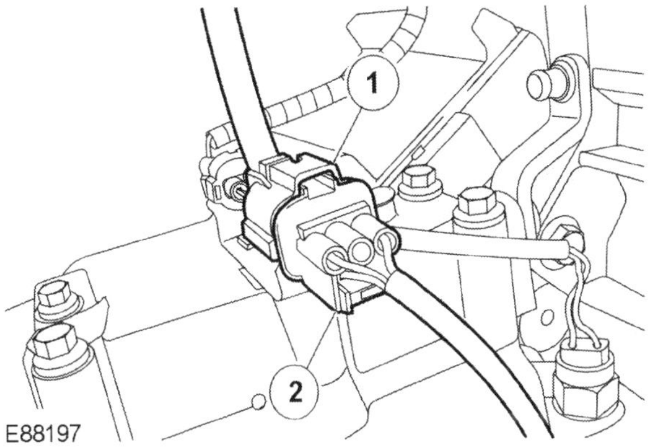

E88197

2. Disconnect the transfer case low range indicator switch electrical connector. Disconnect the electrical connector. Release the electrical connector from the bracket.

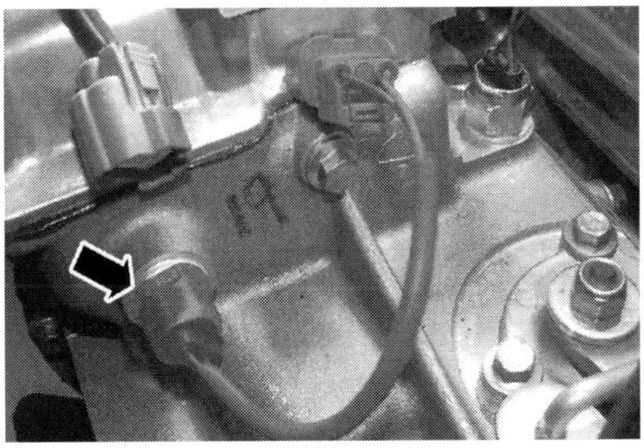

3. Remove the transfer case low range indicator switch. Remove and discard the sealing washer.

Renew Part: Transfer case low range indicator switch

Installation

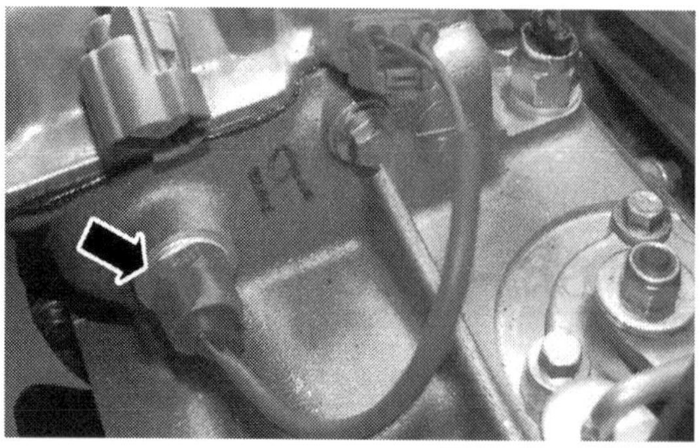

1. To install, reverse the removal procedure. Install a new sealing washer. Tighten.

   Torque: 24 Nm

Differential Lock Indicator Switch

Removal and Installation

Removal

1. ⚠ WARNING: Do not work on or under a vehicle supported only by a jack. Always support the vehicle on safety stands. Failure to follow this instruction may result in personal injury.

Raise and support the vehicle.

2. Disconnect the differential lock indicator switch electrical connector.

3. Remove the differential lock indicator switch. Remove and discard the sealing washer.

Installation

1. To install, reverse the removal procedure. Install a new sealing washer.

   Torque: 24 Nm

Exhaust System - ID4 2.2L Diesel

Important Exhaust System Information

The exhaust systems of the 2.4-litre and 2.2-litre engines are essentially the same. However, the 2.2-litre engines were all prepared to meet the EU5 emissions specification and the exhaust system contains additional components to that end.

The pages that follow relate to the 2.2-litre engine with the EU5 specification. Note that this emissions specification is a legal requirement in most countries and that omitting or tampering with emissions-control components may be subject to criminal penalties.

Specifications

Lubricants, Fluids, Sealers and Adhesives

	Specifications
*Heated oxygen sensor (H02S) removal	WD40 or suitable alternative
+Heated oxygen sensor (H02S) threads	Apply suitable high temperature anti-seize compound to threads of sensor

*** Apply to area around sensor threads prior to removal**

+ Apply anti-seize lubricant to threads of sensor prior to installation

Torque Specifications

Description - Vehicles with DPF	Nm	lb-ft
Diesel particulate filter (DPF) to turbo clamp nut	10	7
DPF heatshield	10	7
DPF pressure pipe unions	25	18
DPF temperature sensors	25	18
+DPF heated oxygen sensor (HO2S)	47	35
DPF lower bracket bolts	25	18
Front muffler to tail pipe nuts	25	18
Front muffler to diesel particulate filter nuts	25	18
Chassis crossmember nuts and bolts	80	59

+ Apply anti-seize lubricant to threads of sensor prior to installation

Description - Vehicles without DPF	Nm	lb-ft
Catalytic Converter to turbo clamp nut	10	7
Catalytic Converter heat shield	10	7
Catalytic Converter lower bracket bolts	25	18
Catalytic Converter top bung	48	35
Catalytic Converter side bung	25	18
Front muffler to tail pipe nuts	25	18
Front muffler to diesel particulate filter nuts	25	18
Chassis crossmember nuts and bolts	80	59

COMPONENT LOCATION - 90 VARIANT

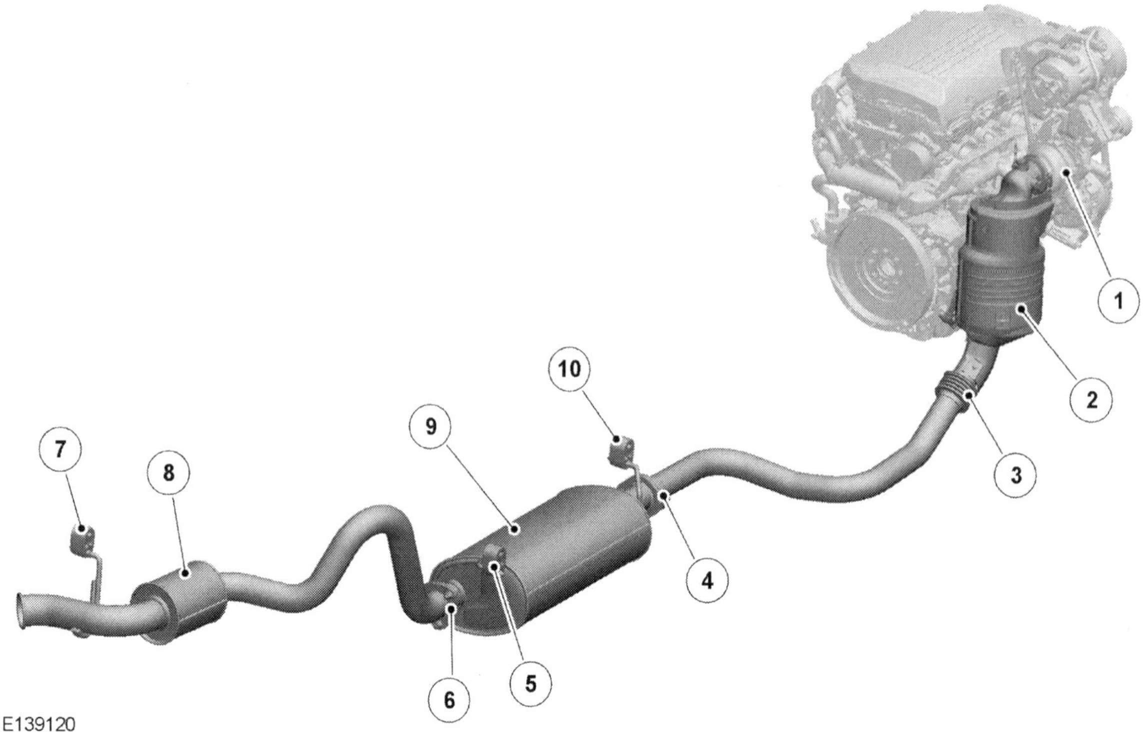

Item	Description
1	Turbocharger
2	Catalytic convertor/catalytic converter and DPF (diesel particulate filter)
3	Flexible joint
4	Front joint
5	Center hanger bar and mounting rubber
6	Rear joint
7	Rear hanger bar and mounting rubber
8	Rear muffler
9	Center muffler
10	Front hanger bar and mounting rubber

Item	Description
1	Turbocharger
2	Catalytic convertor/catalytic converter and DPF
3	Flexible joint
4	Front joint
5	Center hanger bar and mounting rubber
6	Rear joint
7	Rear hanger bar and mounting rubber
8	Rear muffler
9	Center muffler
10	Front hanger bar and mounting rubber

COMPONENT LOCATION - 110 VARIANT

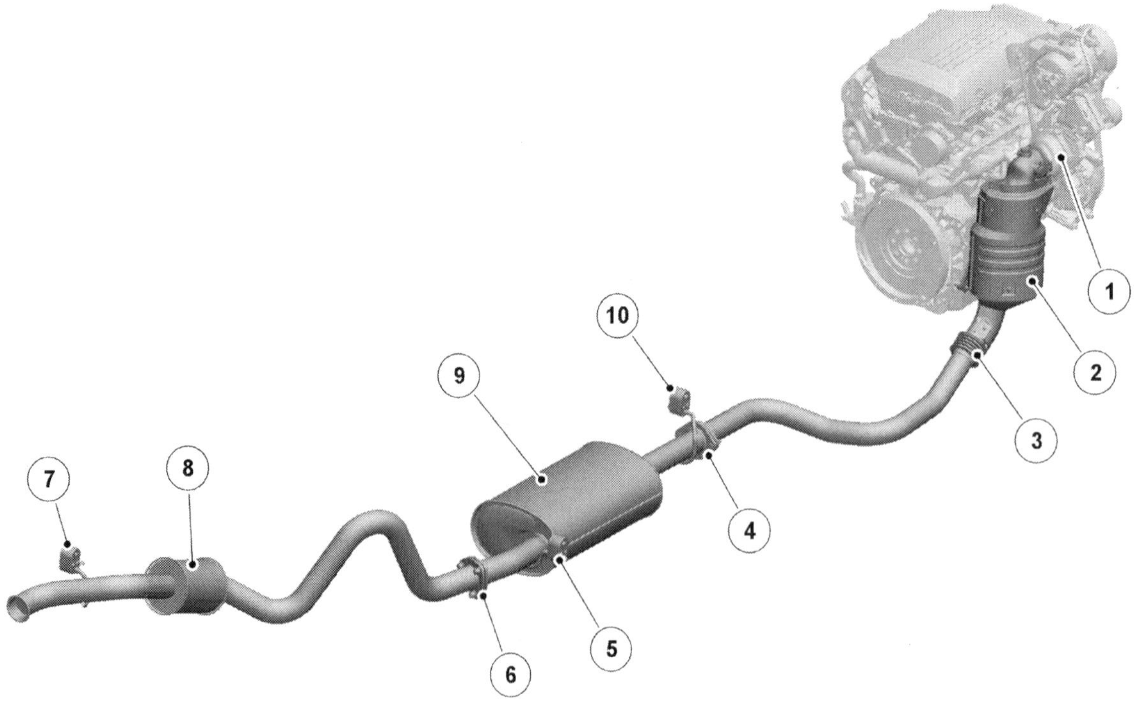

COMPONENT LOCATION - 130 VARIANT

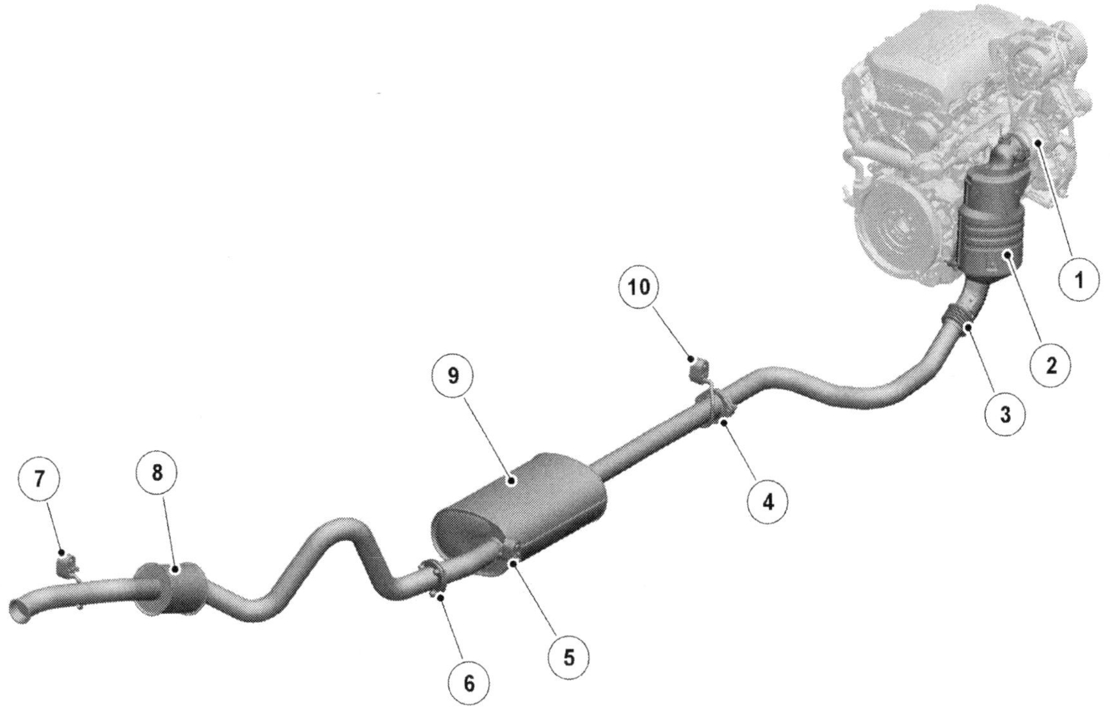

E139122

Item	Description
1	Turbocharger
2	Catalytic convertor/catalytic converter and DPF
3	Flexible joint
4	Front joint
5	Center hanger bar and mounting rubber
6	Rear joint
7	Rear hanger bar and mounting rubber
8	Rear muffler
9	Center muffler
10	Front hanger bar and mounting rubber

OVERVIEW

The exhaust system is fabricated from stainless steel and is supplied as three separate assemblies;

- A front section incorporating a catalytic converter (vehicles to EU4 emission standards) or a combined catalytic converter and DPF (vehicles to EU5 emission standards).
- A center section incorporating a center muffler.
- A rear section incorporating a rear muffler.

The system is attached to the underside of the body with three mounting rubbers, which are located on mild steel hanger bars that are welded to the system. The mounting rubbers locate on corresponding hangers on the underside of the vehicle body.

FRONT SECTION

The front section is connected to the turbocharger by mating flanges on the turbocharger and the catalytic converter/catalytic converter and DPF, which are secured together using a gasket and a U-clamp. Two M08 bolts attach a bracket on the bottom of the catalytic converter/ DPF to the skirt stiffener of the cylinder block.

The body of the catalytic converter/catalytic converter and DPF is covered by an insulation panel. M06 bolts attach upper and lower heat shields to the catalytic converter/catalytic converter and DPF.

On EU5 vehicles, the catalytic converter and DPF incorporates bosses for the attachment of two pressure pipes, and the installation of a HO2S (heated oxygen sensor) and three temperature sensors.

The outlet pipe from the catalytic converter/catalytic converter and DPF incorporates a flexible joint to absorb vibration. The rear of the outlet pipe locates in the inlet pipe of the center section. A loose flange on the outlet pipe locates on two studs in a flange on the inlet pipe of the center section. Two M10 nuts compress and secure the joint.

CENTER SECTION

The center muffler is a 3-pass absorption construction with a capacity of 21.33 liters (1302 in^3). The muffler contains baffles, perforated tubes and E-glass fiber packs which reduce noise as the exhaust gases pass through the muffler. A hanger bar is welded to the muffler inlet pipe, and provides for the location of a mounting rubber. Another hanger bar is welded to the rear of the muffler on 90 variants and to the outlet pipe on 110 and 130 variants.

The outlet pipe of the center section is secured to the rear section using a flange to flange joint. The flange on the outlet pipe has three studs which locate in the rear section and are secured with M10 nuts.

REAR SECTION

The rear section uses a straight through muffler with a volume of 0.73 liters (44.55 in^3). A flange on the inlet pipe connects to the center section. The outlet pipe has a mounting bar welded to the left hand side, which locates in a mounting rubber.

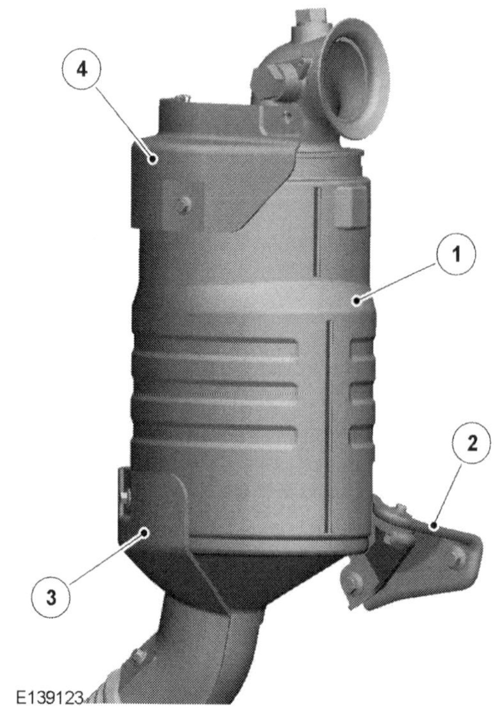

Item	Description
1	Insulation panel
2	Support bracket
3	Lower heat shield
4	Upper heat shield

CATALYTIC CONVERTER

The catalytic converter, which has a capacity of 1.3 liters, is fitted in the front section of the exhaust system. The catalytic converter assembly is common to vehicles with or without the DPF, however, the catalyst coating specification varies depending on the market.

The catalytic converter reduces the carbon monoxide and hydrocarbons content of the exhaust gases. In the catalytic converter the exhaust gases are passed through honeycombed ceramic elements coated with a special surface treatment called a 'washcoat'.

The washcoat increases the surface area of the ceramic elements by a factor of approximately 7000.

On top of the washcoat is a coating containing platinum, which is the active constituent for converting harmful emissions into inert by-products. The platinum adds oxygen to the carbon monoxide and the hydrocarbons in the exhaust gases, to convert them into carbon dioxide and water respectively

Diesel Particulate Filter

Description and Operation

COMPONENT LOCATION - DIESEL PARTICULATE FILTER

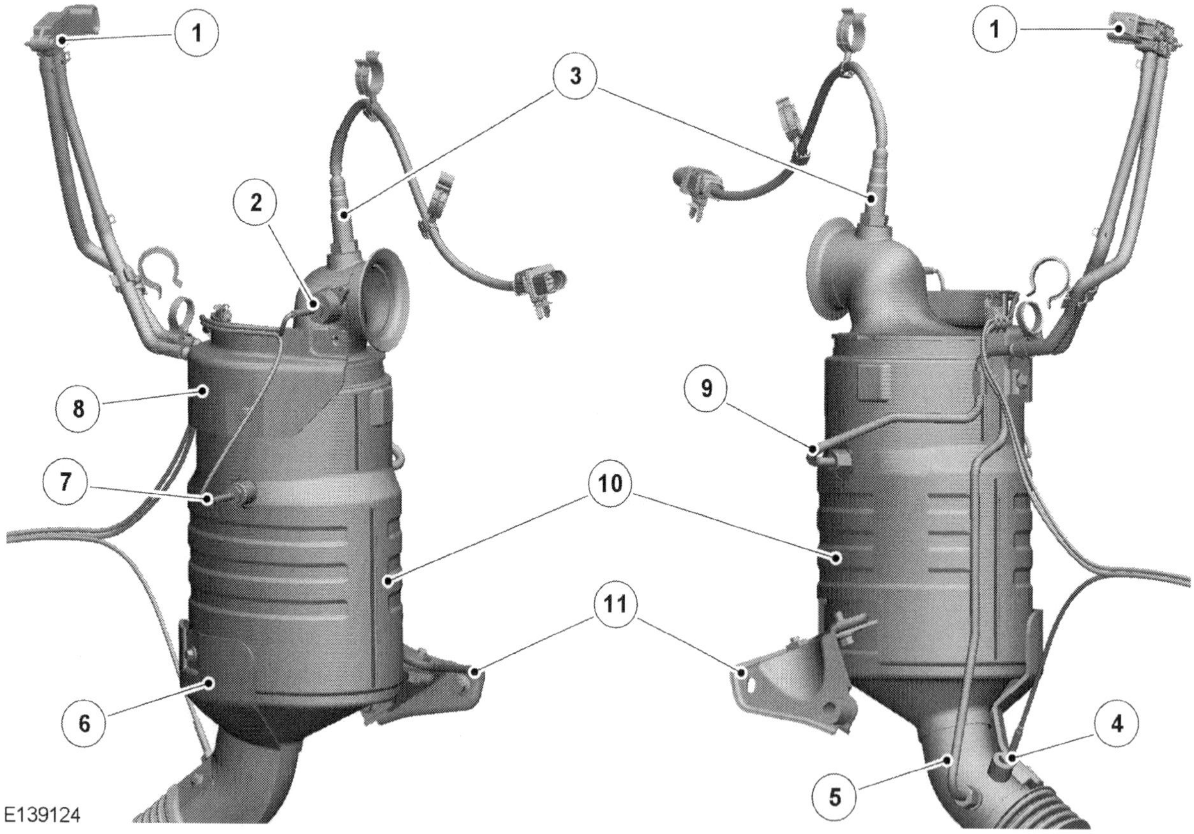

Item	Description
1	Differential pressure sensor
2	Temperature sensor - pre catalytic converter
3	-
4	Temperature sensor - post DPF (diesel particulate filter)
5	Low pressure pipe - post DPF
6	Lower heat shield
7	Temperature sensor - post catalytic converter
8	Upper heat shield
9	High pressure pipe - pre DPF
10	Insulation panel
11	Support bracket

DIFFERENTIAL PRESSURE SENSOR

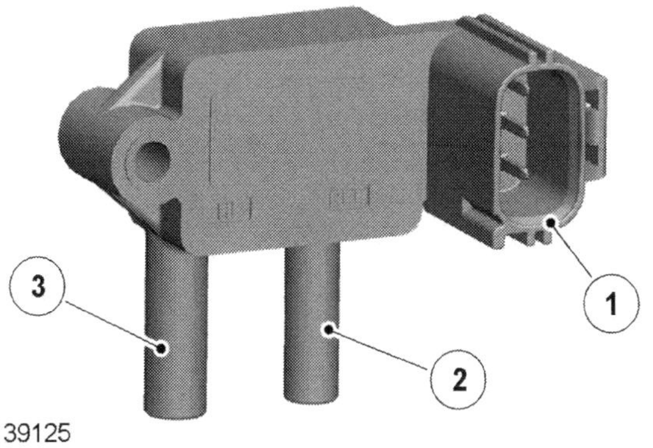

E139125

Item	Description
1	Electrical connector
2	Low pressure - Post DPF connection
3	High pressure - Pre DPF connection

Diesel Particulate Filter

Description and Operation

System Operation

DIESEL PARTICULATE FILTER (DPF)

Two processes are used to regenerate the Diesel Particulate Filter (DPF); passive and active.

Passive Regeneration

Passive regeneration requires no special engine management intervention and occurs during normal engine operation. The passive regeneration involves a slow conversion of the particulate matter deposited in the DPF into carbon dioxide. This process occurs when the DPF temperature exceeds 250°C (482°F) and is a continuous process when the vehicle is being driven at higher engine loads and speeds.

During passive regeneration, only a portion of the particulate matter is converted into carbon dioxide. This is because the chemical reaction, which utilises nitrogen dioxide, is slower than the rate of engine production of particulate matter and is effective from 250°C (482°F).

Above 580°C the conversion efficiency of the particulate into carbon dioxide rapidly increases. These temperatures are generally only be achieved using the active regeneration process.

Active Regeneration

Active regeneration starts when the particulate loading of the DPF reaches a threshold as monitored or determined by the DPF control software. The threshold calculation is based on driving style, distance travelled and back pressure signals from the differential pressure sensor.

Active regeneration generally occurs every 250 miles (400 km) although this is dependant on how the vehicle is driven. For example, if the vehicle is driven at low loads in urban traffic regularly, active regeneration will occur more often. This is due to the rapid build-up of particulate in the DPF than if the vehicle is driven at high speeds when passive regeneration will have occurred.

The DPF software incorporates a mileage trigger which is used as back-up for active regeneration. If active regeneration has not been initiated by a back pressure signal from the differential pressure sensor, regeneration is requested based on distance travelled.

Active regeneration of the DPF is commenced when the temperature of the DPF is increased to the combustion temperature of the particles. The DPF temperature is raised by increasing the exhaust gas temperature. This is achieved by introducing post-injection of fuel after the pilot and main fuel injections have occurred.

It is determined by the DPF software monitoring the signals from the two DPF temperature sensors to establish the temperature of the DPF. Depending on the DPF temperature, the DPF software requests the Engine Control Module (ECM) to perform either one or two post-injections of fuel:

- The first post-injection of fuel retards combustion inside the cylinder which increases the temperature of the exhaust gas.
- The second post-injection of fuel is injected late in the power stroke cycle. The fuel partly combusts in the cylinder, but some unburnt fuel also passes into the exhaust where it creates an exothermic event within the catalytic converter, further increasing the temperature of the DPF.

The active regeneration process takes up to 20 minutes to complete. The first phase increases the DPF temperature to 500°C (932°F). The second phase further increases the DPF temperature to 600°C (1112°F) which is the optimum temperature for particle combustion. This temperature is then maintained for 15-20 minutes to ensure complete oxidation of the particles within the DPF. The oxidation process converts the carbon particles to carbon dioxide.

The active regeneration temperature of the DPF is closely monitored by the DPF software to maintain a target temperature of 600°C (1112°F) at the DPF inlet. The temperature control ensures that the temperatures do not exceed the operational limits of the turbocharger and the catalytic converter. The turbocharger inlet temperature must not exceed 830°C (1526°F) and the catalytic converter brick temperature must not exceed 800°C (1472°F) and the exit temperature must remain below 875°C (1382°F).

During the active regeneration process the following ECM controlled events occur:

- The turbocharger is maintained in the fully open position. This minimizes heat transmission from the exhaust gas to the turbocharger and reduces the rate of exhaust gas flow allowing optimum heating of the DPF. If the driver demands an increase in engine torque, the turbocharger will respond by closing the vanes as necessary.

- The throttle is closed as this assists in increasing the exhaust gas temperature and reduces the rate of exhaust gas flow which has the effect of reducing the time for the DPF to reach the optimum temperature.
- The Exhaust Gas Recirculation (EGR) valve is closed. The use of EGR decreases the exhaust gas temperature and therefore prevents the optimum DPF temperature being achieved.

If, due to vehicle usage and/or driving style, the active regeneration process cannot take place or is unable to regenerate the DPF, the dealer can force regenerate the DPF. This is achieved by either driving the vehicle until the engine is at its normal operating temperature and then driving for a further 20 minutes at speeds of not less than 30 mph (48 km/h).

DPF Control

The DPF requires constant monitoring to ensure that it is operating at its optimum efficiency and does not become blocked. The ECM contains DPF software which controls the monitoring and operation of the DPF system and also monitors other vehicle data to determine regeneration periods and service intervals.

The DPF software can be divided into three separate control software modules; a DPF supervisor module, a DPF fuel management module and a DPF air management module.

These three modules are controlled by a fourth software module known as the DPF co-ordinator module. The co-ordinator module manages the operation of the other modules when an active regeneration is requested.

DPF Fuel Management Module

The DPF fuel management module controls the following functions:

- Timing and quantity of the four split injections per stroke (pilot, main and two post injections).
- Injection pressure and the transition between the three different calibration levels of injection.

The above functions are dependant on the condition of the catalytic converter and the DPF.

The controlled injection determines the required injection level in addition to measuring the activity of the catalytic converter and the DPF. The fuel management calculates the quantity and timing for the four split injections, for each of the three calibration levels for injection pressure, and also manages the transition between the levels.

The two post injections are required to separate the functionality of increasing in-cylinder gas temperatures and the production of hydrocarbons. The first post injection is used to generate the higher in-cylinder gas temperature while simultaneously retaining the same engine torque output produced during normal (non-regeneration) engine operation. The second post injection is used to generate hydrocarbons by allowing unburnt fuel into the catalytic converter without producing increased engine torque.

DPF Air Management Module

The DPF air management module controls the following functions:

- EGR control
- Turbocharger boost pressure control
- Intake air temperature and pressure control.

During active regeneration, the EGR operation is disabled -except for overrun conditions - and the closed-loop activation of the turbocharger boost controller is calculated. The air management module controls the air in the intake manifold to a predetermined level of pressure and temperature. This control is required to achieve the correct in-cylinder conditions for stable and robust combustion of the post injected fuel.

The module controls the intake air temperature by actuating the EGR throttle and by adjustment of the turbocharger boost pressure control.

DPF Co-ordinator Module

The DPF co-ordinator module reacts to a regeneration request from the supervisor module by initiating and co-ordinating the following DPF regeneration requests:

- EGR cut-off - except for overrun condition
- Turbocharger boost pressure control
- Engine load increase
- Control of air pressure and temperature in the intake manifold
- Fuel injection control.

When the supervisor module issues a regeneration request, the co-ordinator module requests EGR cut-off and a regeneration specific turbocharger boost pressure control. It then waits for a feedback signal from the EGR system confirming that the EGR valve is closed.

When the EGR valve is closed, the co-ordinator module initiates requests to increase engine load by controlling the intake air temperature and pressure.

Once confirmation is received that intake conditions are controlled or a calibration time has expired, the co-ordinator module then changes to a state awaiting an accelerator pedal release manoeuvre from the driver. If this occurs or a calibration time has expired, the co-ordinator module generates a request to control fuel injections to increase exhaust gas temperature.

DIFFERENTIAL PRESSURE SENSOR

As the amount of particulate trapped by the DPF increases, the pressure at the inlet side of the DPF increases in comparison to the DPF outlet. The DPF software uses this comparison, in conjunction with other data, to calculate the accumulated amount of trapped particulate.

By measuring the pressure difference between the DPF inlet and outlet and the DPF temperature, the DPF software can determine if the DPF is becoming blocked and requires regeneration.

Component Description

DIESEL PARTICULATE FILTER (DPF)

The DPF system reduces diesel particulate emissions to negligible levels to meet current European stage 5 emission standards.

The particulate emissions are the black fumes emitted from the diesel engine under certain load conditions. The emissions are a complex mixture of solid and liquid components with the majority of the particulate being carbon microspheres on which hydrocarbons from the engine's fuel and lubricant condense.

The DPF system comprises the following components:

- Diesel Particulate Filter (DPF)
- DPF control software incorporated in the ECM
- Differential pressure sensor.

The DPF is located in the exhaust system, downstream of the catalytic converter. Its function is to trap particulate matter in the exhaust gases leaving the engine. A major feature of the DPF is its ability for regeneration. Regeneration is the burning of particulate trapped by the filter to prevent obstruction to the free flow of exhaust gasses. The regeneration process takes place at calculated intervals and is not noticeable by the driver of the vehicle.

Regeneration is most important, since an overfilled filter can damage the engine through excessive exhaust back pressure and can itself be damaged or destroyed. The material trapped in the filter is in the most part carbon particles with some absorbed hydrocarbons.

The DPF uses a filter technology based on a filter with a catalytic coating. The DPF is made from silicon carbide housed in a steel container and has excellent thermal shock resistance and thermal conductivity properties. The DPF is designed for the engine's operating requirements to maintain the optimum back pressure requirements.

The porous surface of the filter consists of thousands of small parallel channels positioned in the longitudinal direction of the exhaust system. Adjacent channels in the filter are alternately plugged at the end. This design forces the exhaust gasses to flow through the porous filter walls, which act as the filter medium. Particulate matter which are too big to pass through the porous surface are collected and stored in the channels.

The collected particulate matter, if not removed, can create an obstruction to exhaust gas flow. The stored particles are removed by a regeneration process which incinerates the particles.

Diesel Particulate Filter (DPF) Temperature Sensors

The DPF temperature sensors are used to provide exhaust gas temperature feedback to the Engine Control Module (ECM) in order to control engine conditions and to effectively monitor and control emissions.

The DPF temperature sensor's information is sent to the ECM using a 0 to 5V signal (lower voltage being higher temperature and higher voltage being lower temperature).

The information is used, in conjunction with other data, to estimate the amount of accumulated particulates and to control the DPF temperature.

Instrument Cluster (IC) Indications

For drivers who make regular short journeys at low speeds, it may not be possible to efficiently regenerate the DPF. In this case, the DPF software will detect a blockage of the DPF from signals from the differential pressure sensor and will alert the driver as follows:

The driver will be alerted to this condition by a message 'EXHAUST FILTER NEARLY FULL'. See 'HANDBOOK'. As detailed in the Owners Handbook, the driver should drive the vehicle until the engine is at its normal operating temperature and then drive for a further 20 minutes at speeds of not less than 30 mph (48 km/h). Successful regeneration of the DPF is indicated to the driver by the 'EXHAUST FILTER NEARLY FULL' message no longer being displayed. If the DPF software detects that the DPF is still blocked, the message will continue to be displayed or an additional message 'EXHAUST FILTER FULL VISIT DEALER' will be displayed. The driver should take the vehicle to an authorized dealer to have the DPF force regenerated using an approved diagnostic system.

Diesel Particulate Filter Side Effects

The following section details some side effects caused by the active regeneration process.

Engine Oil Dilution

Engine oil dilution can occur due to small amounts of fuel entering the engine crankcase during the post-injection phases. This has made it necessary to introduce a calculation based on driving style to reduce oil service intervals if necessary. The driver is alerted to the oil service by a message in the instrument cluster.

The DPF software monitors the driving style and the frequency of the active regeneration and duration. Using this information a calculation can be made on the engine oil dilution. When the DPF software calculates the engine oil dilution has reached a predetermined threshold (fuel being 7% of engine oil volume) a service message is displayed in the IC.

Depending on driving style, some vehicles may require an oil service before the designated interval. If a service message is displayed, the vehicle will be required have a full service and the service interval counter will be reset.

Fuel consumption

During the active regeneration process of the DPF, there will be an increase in fuel consumption.

However, because active regeneration occurs infrequently, the overall effect on fuel consumption is approximately 2%. The additional fuel used during the active regeneration process is accounted for in the instantaneous and average fuel consumption displays in the instrument cluster.

DIFFERENTIAL PRESSURE SENSOR

The differential pressure sensor is used by the DPF software to monitor the condition of the DPF. Two pipe connections on the sensor are connected by pipes to the inlet and outlet ends of the DPF. The pipes allow the sensor to measure the inlet and outlet pressures of the DPF.

Aftermarket DPF cleaning fluids

Recent years have seen the introduction of 'DPF cleaning fluids' to (non JLR approved) aftermarket sales. These products claim to reduce the temperature that the soot reaction takes place. It should be stressed that, during the vehicle development activity, every effort is made to generate DPF regeneration temperatures whilst maintaining safe levels for all other vehicle components. Unauthorized use of the aftermarket fluids produces a significant risk to soot burn rates and DPF peak temperatures real world driving conditions. These fluids are not authorised for JLR use.

Exhaust System

Diagnosis and Testing

Principles of Operation

For a detailed description of the Exhaust System, refer to the relevant Description and Operation section in the workshop manual.

Refer to: Exhaust System - ID4 2.2L Diesel (309-00 Exhaust System, ID4 2.2L Diesel Description and Operation)

Inspection and Verification

⚠ CAUTION: Diagnosis by substitution from a donor vehicle is **NOT** acceptable. Substitution of control modules does not guarantee confirmation of a fault, and may also cause additional faults in the vehicle being tested and/or the donor vehicle.

NOTES:

△ If a control module or a component is suspect and the vehicle remains under manufacturer warranty, refer to the Warranty Policy and Procedures manual, or determine if any prior approval programme is in operation, prior to the installation of a new module/component.

△ When performing voltage or resistance tests, always use a digital multimeter accurate to three decimal places, and with an up-to-date calibration certificate. When testing resistance always take the resistance of the digital multimeter leads into account.

△ Check and rectify basic faults before beginning diagnostic routines involving pinpoint tests.

Visual Inspection

Mechanical
• Exhaust system • Catalytic converter • Silencers • Exhaust leaks • Hanger brackets • Mounting rubbers • Heatshields

1. Visually inspect for obvious signs of damage and system integrity
2. If an obvious cause for an observed or reported concern is found, correct the cause (if possible) before proceeding to the next step
3. If the cause is not visually evident, verify the symptom and refer to the Symptom Chart, alternatively check for Diagnostic Trouble Codes (DTCs) and refer to the DTC Index

DTC Index

For a list of Diagnostic Trouble Codes (DTCs) that could be logged on this vehicle, please refer to Section 100-00.

Refer to: Diagnostic Trouble Code Index - DTC: Engine Control Module (PCM) (100-00 General Information, Description and Operation)

Symptom Chart

Symptom	Possible Causes	Action
Exhaust leaking	• Exhaust system leaking	• Check the integrity of the exhaust system. Rectify as necessary
Exhaust excessively noisy	• Exhaust system leaking • Exhaust system fouling the body, transmission, etc	• Check the integrity of the exhaust system. Rectify as necessary • Check the exhaust system for foul conditions
Engine performance reduced	• Exhaust system restricted/blocked • Catalytic converter restricted/blocked • Engine system fault	• Check the exhaust system for restrictions and blockages. Rectify as necessary • Check the catalytic converter for restrictions and blockages. Rectify as necessary • Using the Jaguar Land Rover approved diagnostic equipment, check the powertrain control module for related DTCs and refer to the relevant DTC index

Diesel Particulate Filter

Diagnosis and Testing

Principle of Operation

For a detailed description of the Diesel Particulate Filter, refer to the relevant Description and Operation section in the workshop manual.

Refer to: Exhaust System - ID4 2.2L Diesel (309-00 Exhaust System, ID4 2.2L Diesel Description and Operation)

Refer to: Diesel Particulate Filter (309-00 Exhaust System, ID4 2.2L Diesel Description and Operation)

Refer to: Diesel Particulate Filter (309-00 Exhaust System, ID4 2.2L Diesel Description and Operation)

Inspection and Verification

⚠ CAUTION: Diagnosis by substitution from a donor vehicle is **NOT** acceptable. Substitution of control modules does not guarantee confirmation of a fault, and may also cause additional faults in the vehicle being tested and/or the donor vehicle.

NOTES:

△ If a control module or a component is suspect and the vehicle remains under manufacturer warranty, refer to the Warranty Policy and Procedures manual, or determine if any prior approval programme is in operation, prior to the installation of a new module/component.

△ When performing voltage or resistance tests, always use a digital multimeter accurate to three decimal places, and with an up-to-date calibration certificate. When testing resistance always take the resistance of the digital multimeter leads into account.

△ Check and rectify basic faults before beginning diagnostic routines involving pinpoint tests.

Visual Inspection

1. Visually inspect for obvious signs of damage and system integrity
2. If an obvious cause for an observed or reported concern is found, correct the cause (if possible) before proceeding to the next step
3. If the cause is not visually evident, verify the symptom and refer to the Symptom Chart, alternatively check for Diagnostic Trouble Codes (DTCs) and refer to the DTC Index

Visual Inspection

Mechanical	Electrical
• Diesel particulate filter • Differential pressure sensor hoses • Differential pressure sensor metal pipes and unions • Differential pressure sensor mounting position integrity • Metal fatigue • Exhaust system joints • Exhaust system mountings • Clearance around components • Turbocharger • Intake air pipes and joints • Exhaust gas recirculation valve	• Fuses • Wiring harnesses and connectors • Engine control module • Differential pressure sensor • Exhaust gas temperature sensors • Heated oxygen sensor

Symptom Chart

Symptom	Possible Causes	Action
Amber diesel particulate filter warning indicator illuminated	• Customer driving routine does not allow the diesel particulate filter to regenerate	NOTE: If DTC P2459-65 is set or the amber diesel particulate filter warning indicator is illuminated, there is no fault with the vehicle. • Drive the vehicle as directed in the Owner's Handbook to perform a diesel particulate filter regeneration. The warning indicator will be extinguished and the system self healed
Red diesel particulate filter warning indicator illuminated	• Engine system fault • Diesel particulate filter full • Air intake system fault • Restricted exhaust system • Leaking exhaust system • Low fuel pressure • Exhaust gas recirculation valve fault • Turbocharger fault	• Using the manufacturer approved diagnostic system, check the engine control module for related DTCs and refer to the relevant DTC index • Using the manufacturer approved diagnostic system, perform routine - Diesel Particulate Filter Dynamic Regeneration • Check the air intake system • Check for a blocked catalytic converter or muffler, install new components as necessary • Check the fuel pressure • Refer to the relevant section of the workshop manual and test the exhaust gas recirculation valve • Refer to the relevant section of the workshop manual and test the turbocharger
Lack of power	• Engine system fault • Air intake system fault • Restricted exhaust system • Low fuel pressure • Exhaust gas recirculation valve fault • Turbocharger fault	• Using the manufacturer approved diagnostic system, check the engine control module for related DTCs and refer to the relevant DTC index • Check the air intake system • Check for a blocked catalytic converter or muffler, install new components as necessary • Check the fuel pressure • Refer to the relevant section of the workshop manual and test the exhaust gas recirculation valve • Refer to the relevant section of the workshop manual and test the turbocharger

DTC Index

For a list of Diagnostic Trouble Codes (DTCs) that could be logged on this vehicle, please refer to Section 100-00.

Refer to: Diagnostic Trouble Code Index - DTC: Engine Control Module (PCM) (100-00 General Information, Description and Operation)

Diesel Particulate Filter (DPF) Regeneration

If the amber diesel particulate filter warning indicator is illuminated, perform the following procedure:

WARNING: The regeneration procedure produces high temperatures in the diesel particulate filter. Heat can be felt radiating from beneath the vehicle; this is normal and not a cause for concern. However, the vehicle should not be parked over combustible material, particularly during dry weather. The heat generated could be sufficient to start a fire when in close proximity to combustible material such as long dry grass, paper, etc.

CAUTION: At all times during this procedure you should observe all relevant speed limits, laws, and regulations.

NOTES:

The ideal speed and conditions for regeneration are 100 km/h (62 mph) to 120 km/h (75 mph), in high range. Keeping a constant speed enables the diesel particulate filter to regenerate more efficiently.

When driving off-road during the regeneration process, greater accelerator pedal use may be required.

1. Drive the vehicle until the engine reaches normal operating temperature. The engine should not be left idling to achieve working temperature
2. Drive the vehicle for a further twenty minutes, keeping the vehicle at a constant speed between 75 km/h (45 mph) and 113 km/h (70 mph)
3. If regeneration is successful the warning indicator will extinguish; continue driving for a further 10 minutes to ensure that the diesel particulate filter is completely clean
4. If the warning indicator remains illuminated, repeat the procedure

Diesel Particulate Filter

Removal and Installation

Part(s)

STEP	PART NAME	QUANTITY
Installation Step 3	Renew Part Diesel Particulate Filter (DPF) to exhaust gasket	1
Installation Step 20	Renew Part Driveshaft flange nut(s)	4

Removal

1. **Refer to:** Startup Battery Disconnect and Connect (414-01 Battery, Mounting and Cables, General Procedures)

2. WARNING: Do not work on or under a vehicle supported only by a jack. Always support the vehicle on safety stands.

Raise and support the vehicle.

3. CAUTION: Discard the nuts.

Remove the nuts.

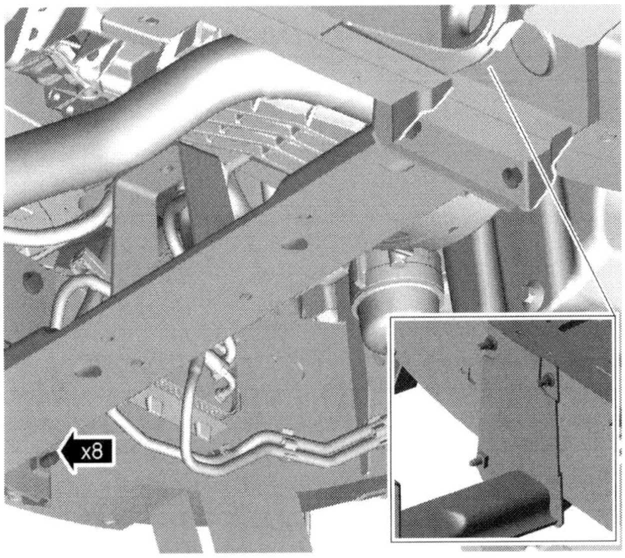

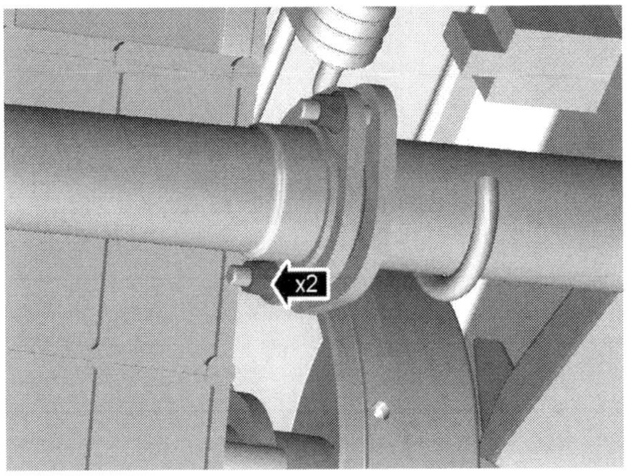

4.

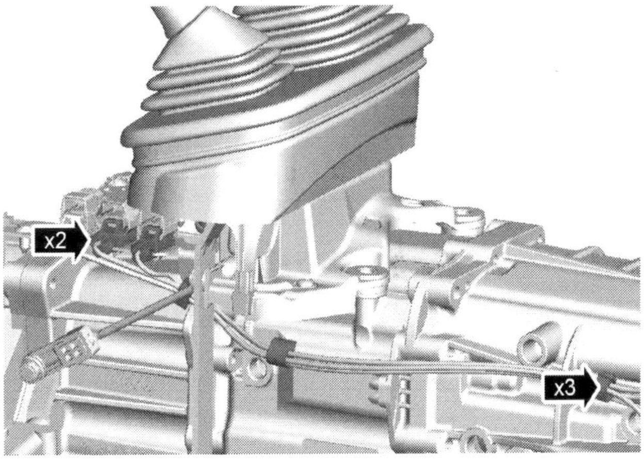

5.

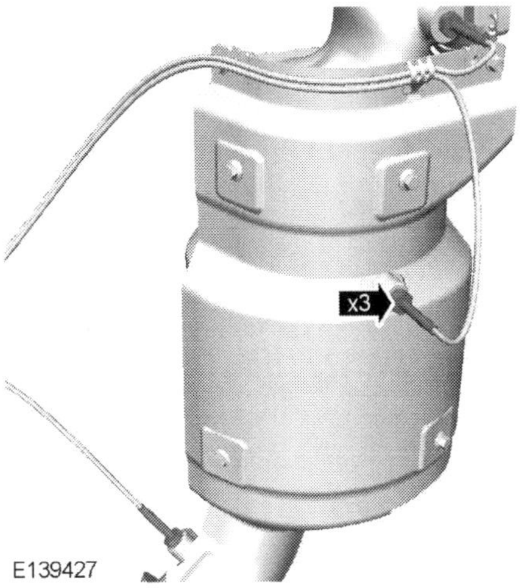

6.

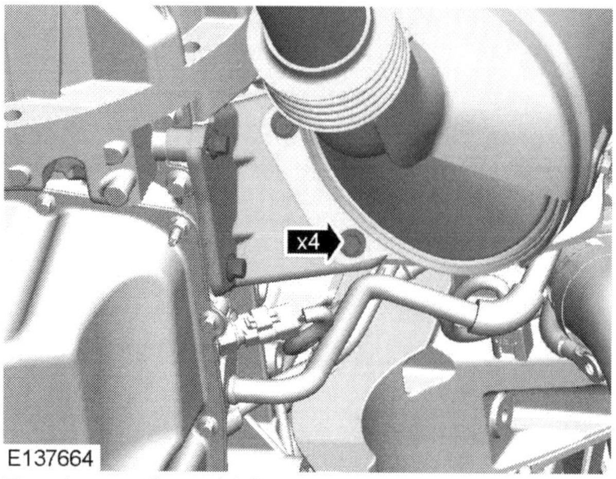

7. Lower the vehicle.

8. **Refer to:** Engine Cover (501-05 Interior Trim and Ornamentation, Removal and Installation)

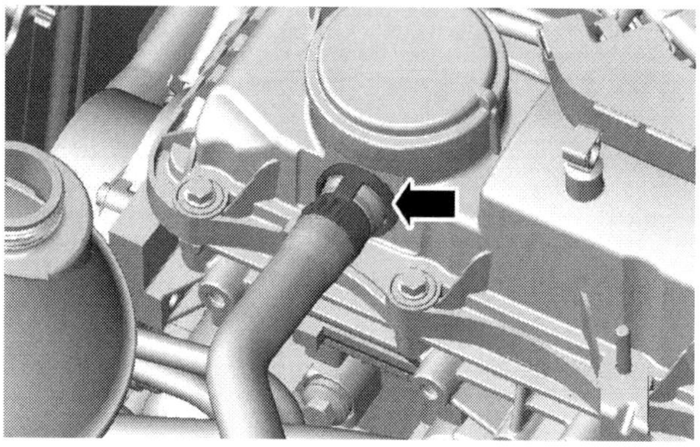

9.

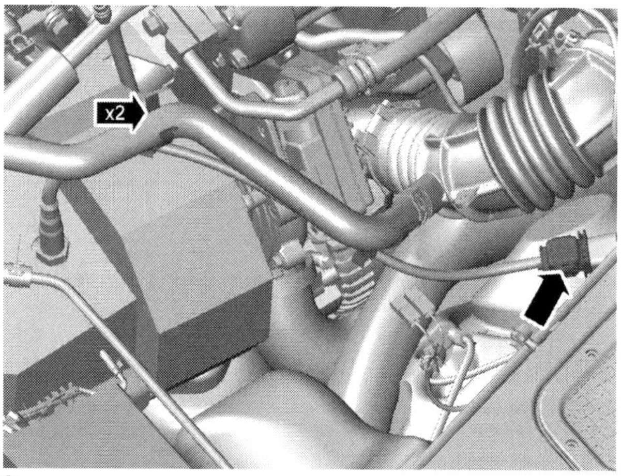

10.

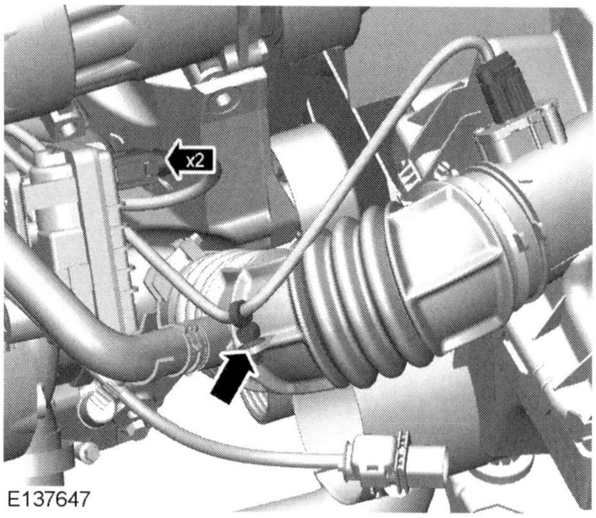

11.

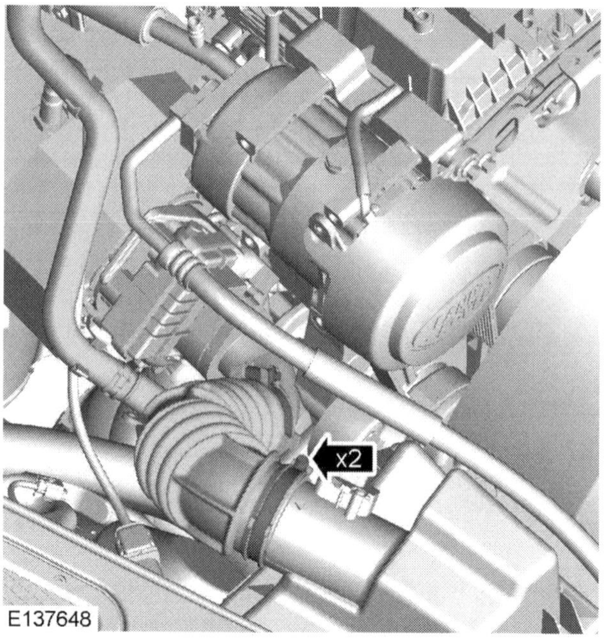

12.

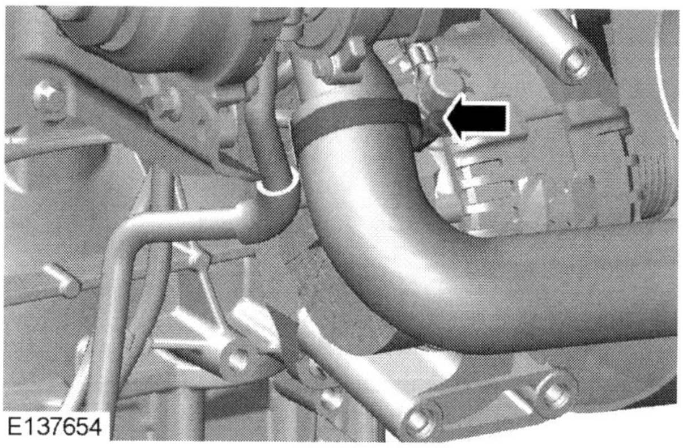

13.

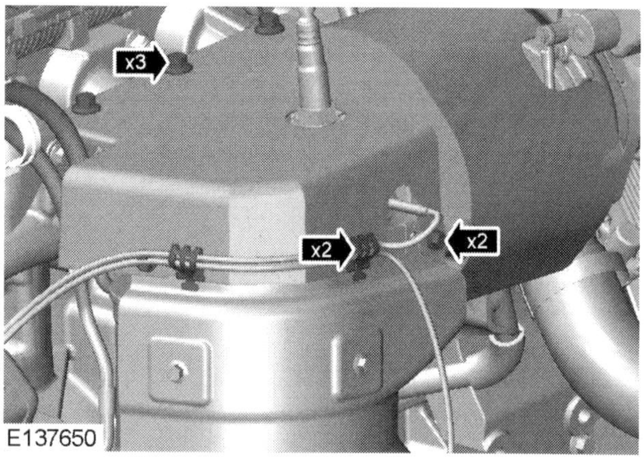

14.

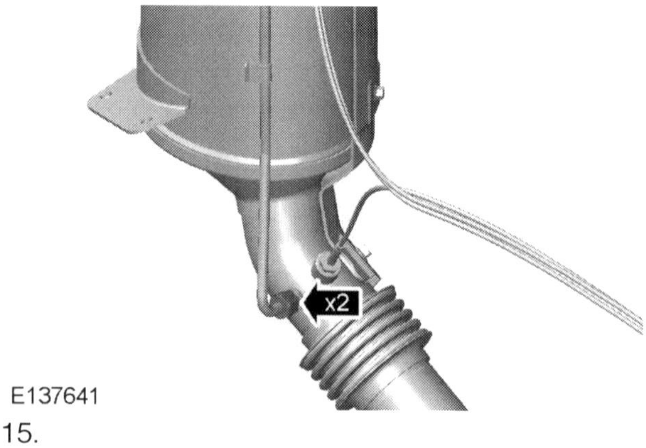

15.

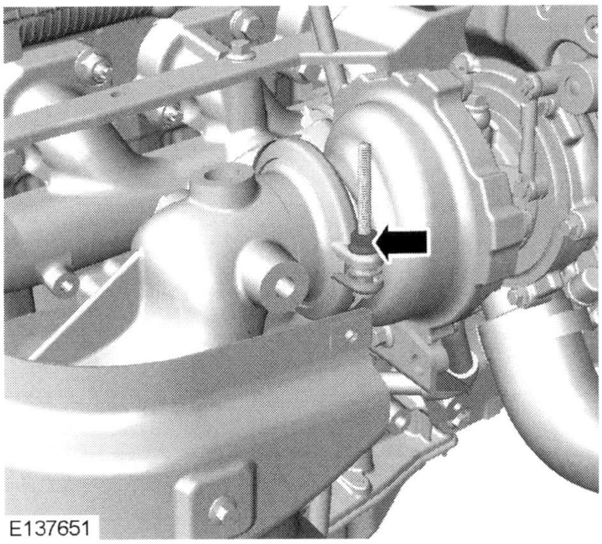

16.

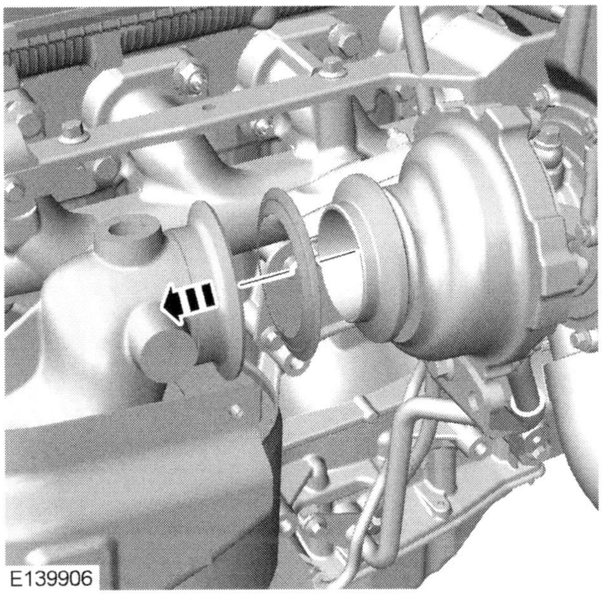

17. Discard the gasket.

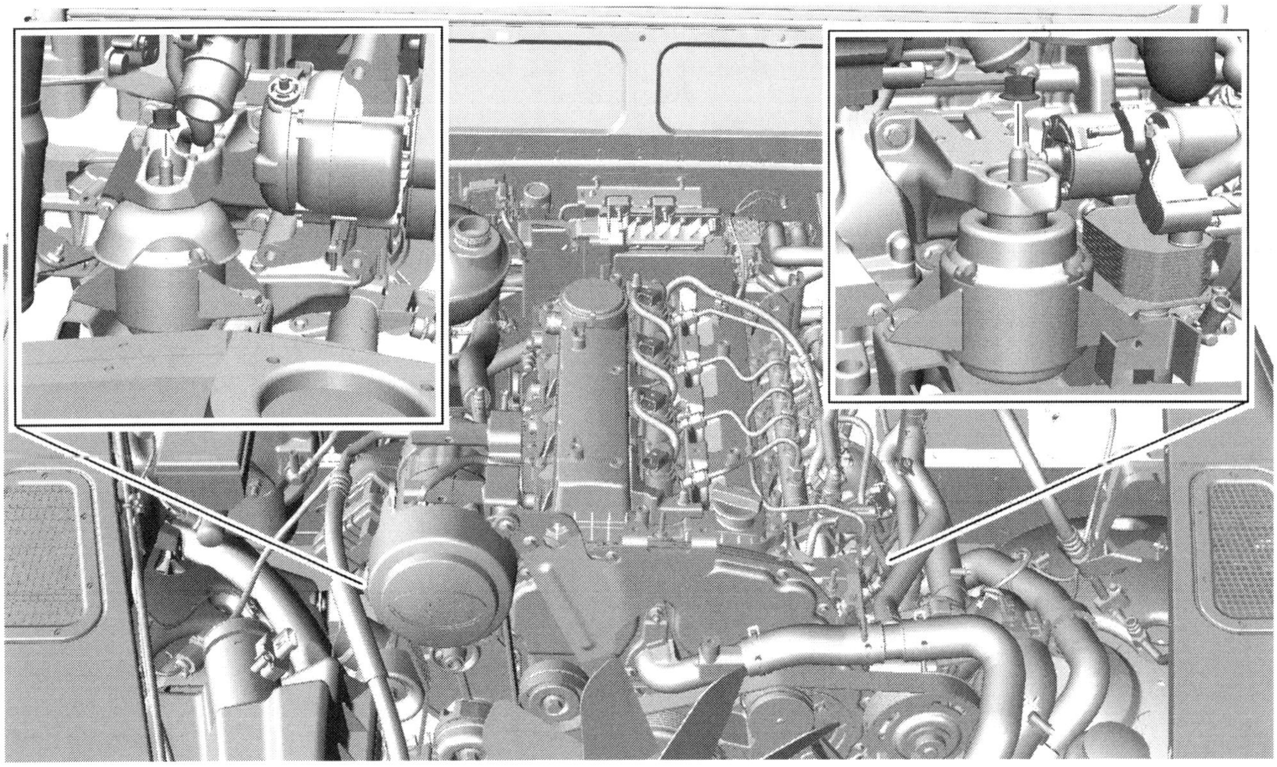

18.

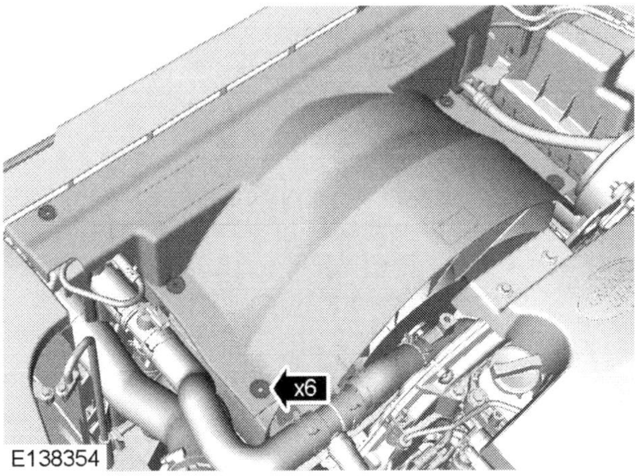

19.

20. ⚠ CAUTION: Raise the engine assembly using suitable lifting equipment. Make sure the attached and surrounding components are not damaged during this step.

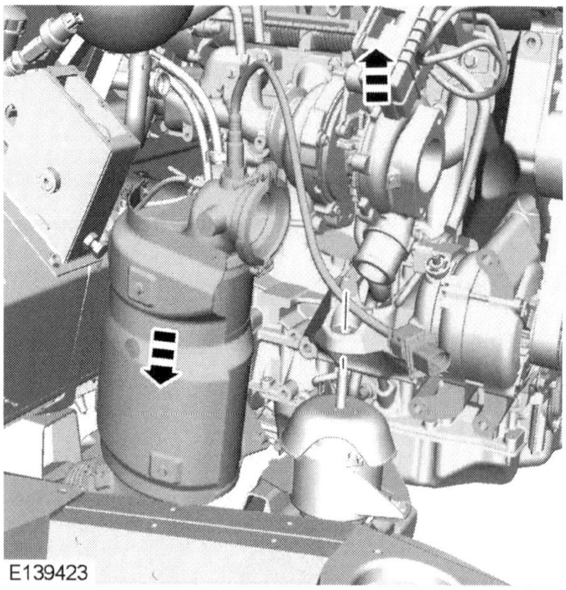

21.

Installation

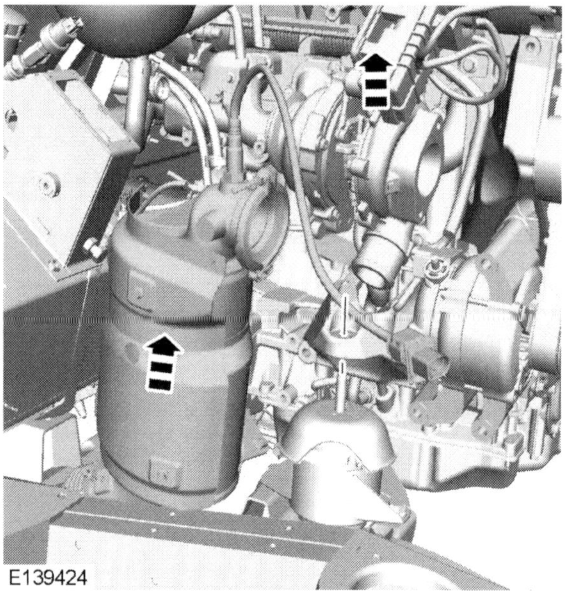

1.

2.

3. Install a new gasket.

Renew Part: Diesel Particulate Filter (DPF) to exhaust gasket

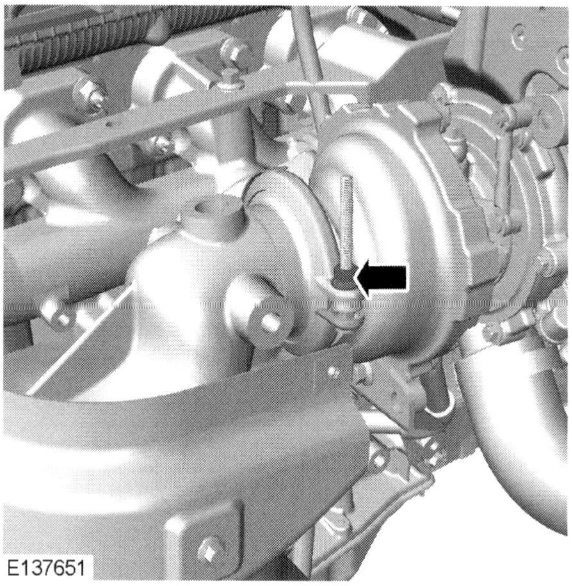

4. *Torque:* 10 Nm

5. *Torque:* 80 Nm

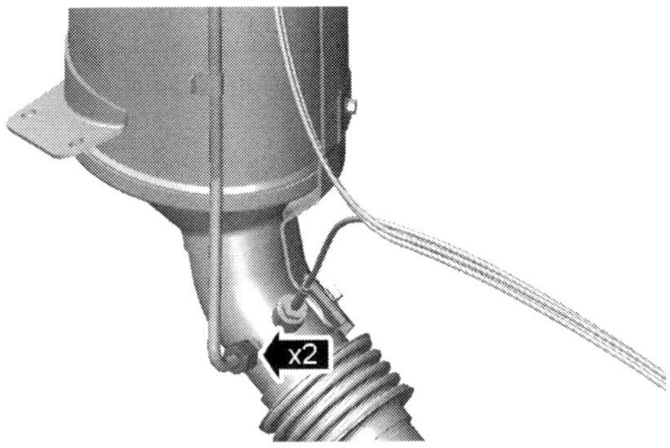

6. *Torque:* 25 Nm

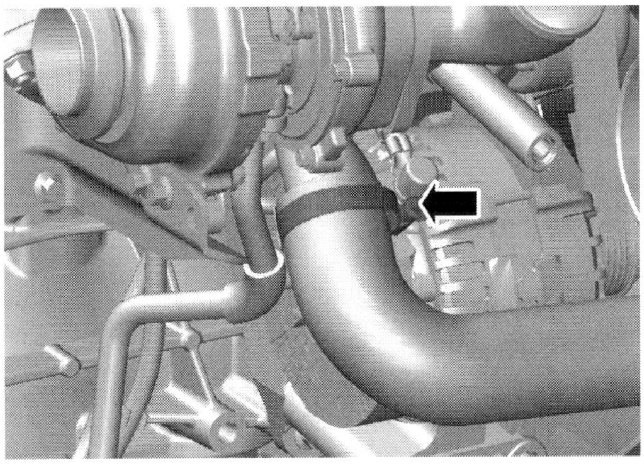

7. *Torque:* 3 Nm

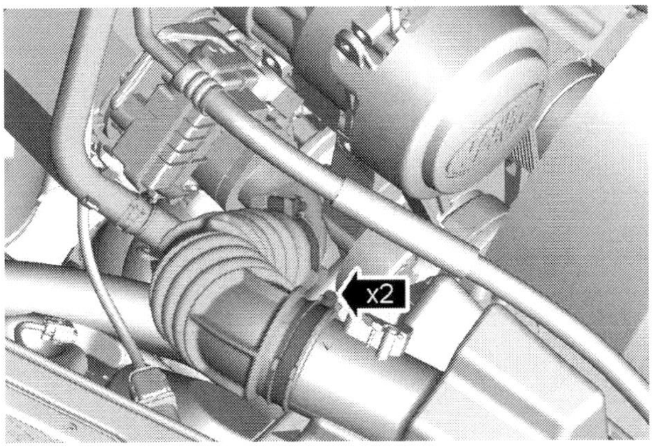

8. *Torque:* 3 Nm

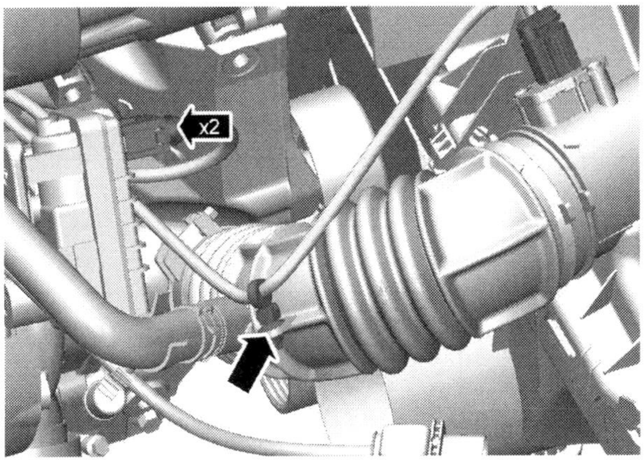

9.

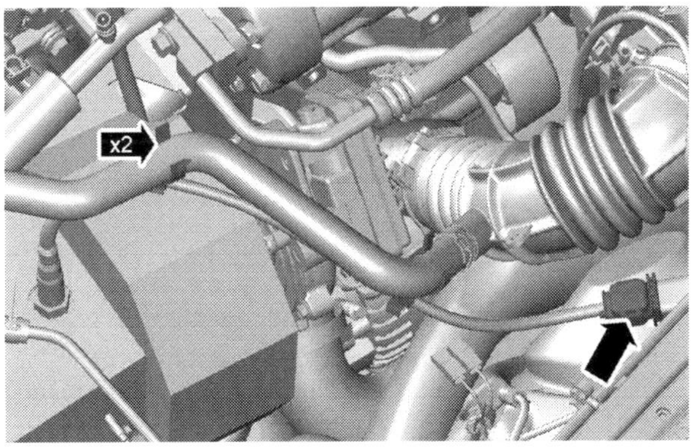

10.

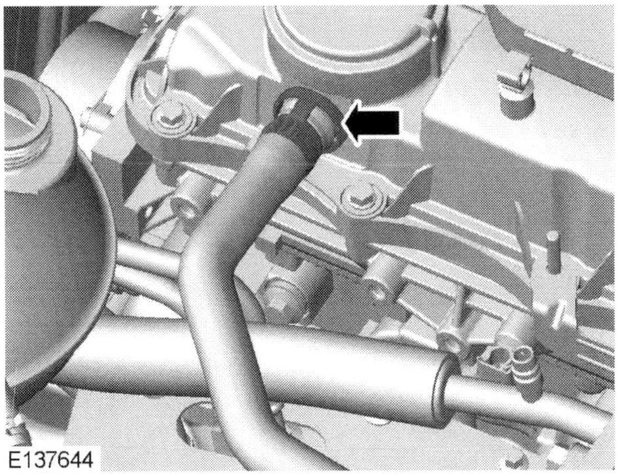

11.

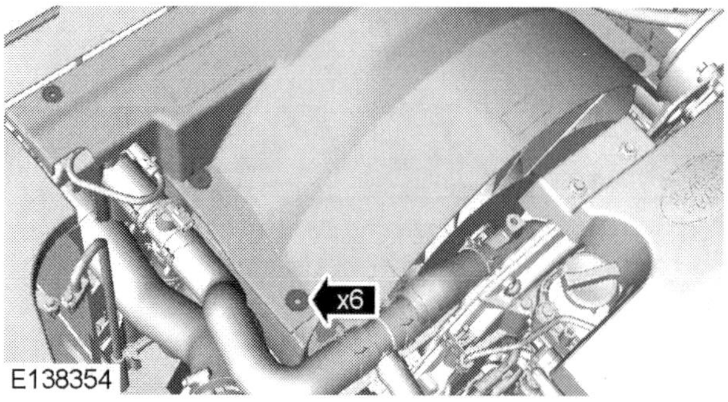

12.

13. **Refer to:** Engine Cover (501-05 Interior Trim and Ornamentation, Removal and Installation)

14. Raise the vehicle on lift.

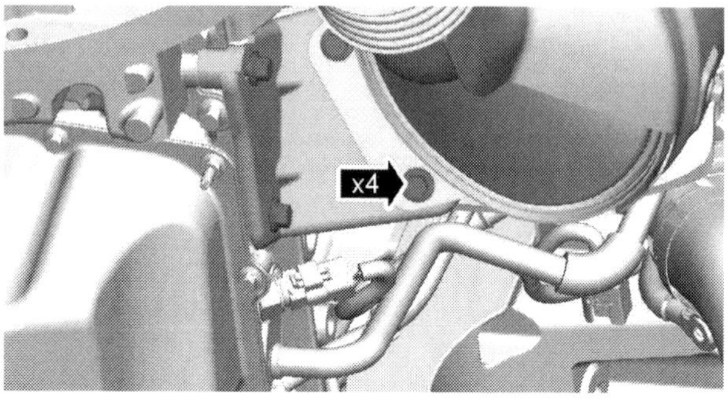

15. *Torque:* 25 Nm

16. *Torque:* 25 Nm

17.

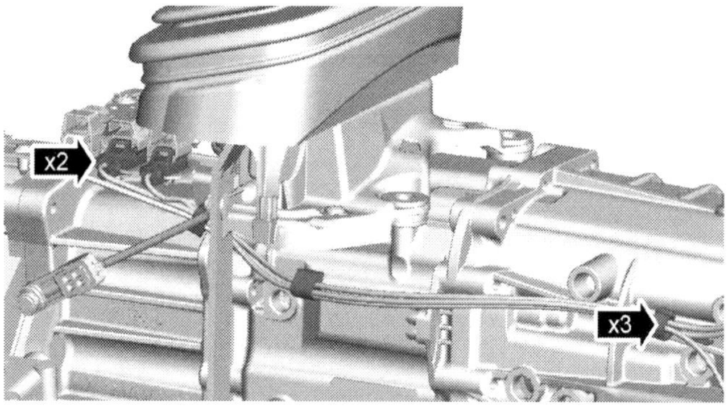

18. *Torque:* 80 Nm

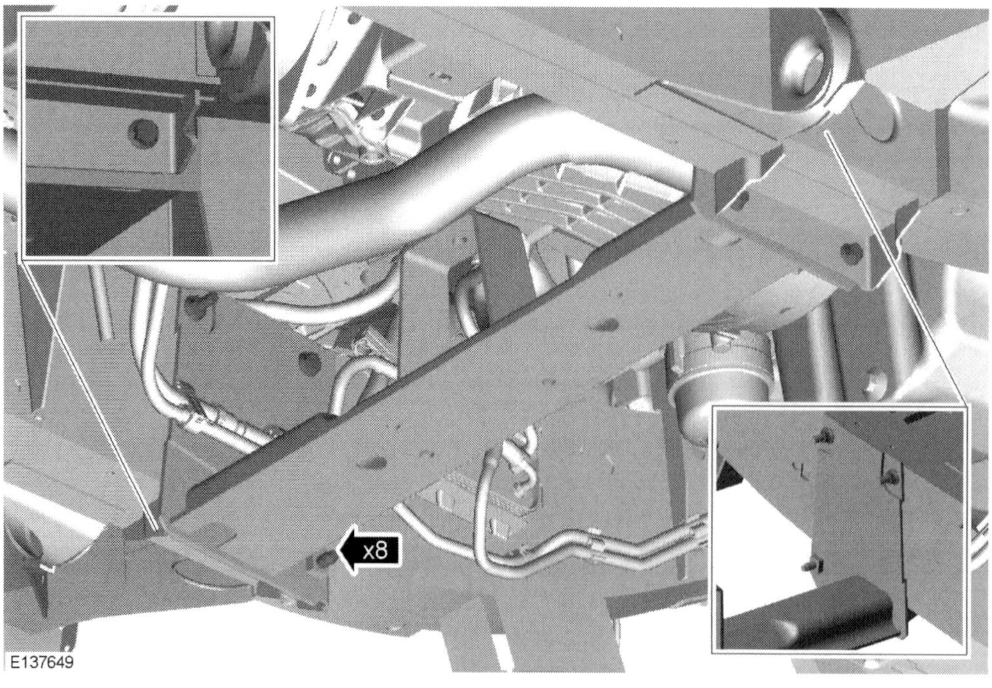

19. *Torque:* 25 Nm

20. ⚠ CAUTION: Install new nuts.

Renew Part: Driveshaft flange nut(s)

Torque: 47 Nm

21. **Refer to:** Startup Battery Disconnect and Connect (414-01 Battery, Mounting and Cables, General Procedures)

22. If a new unit is installed, configure using the approved diagnostic tool.

Muffler and Tailpipe

Removal and Installation

Part(s)

STEP	PART NAME	QUANTITY
Installation Step 2	Renew Part Intermediate exhaust pipe gasket	1
Installation Step 2	Renew Part Intermediate exhaust pipe to rear pipe nut(s)	3

Removal

1. WARNING: Do not work on or under a vehicle supported only by a jack. Always support the vehicle on safety stands.

Raise and support the vehicle.

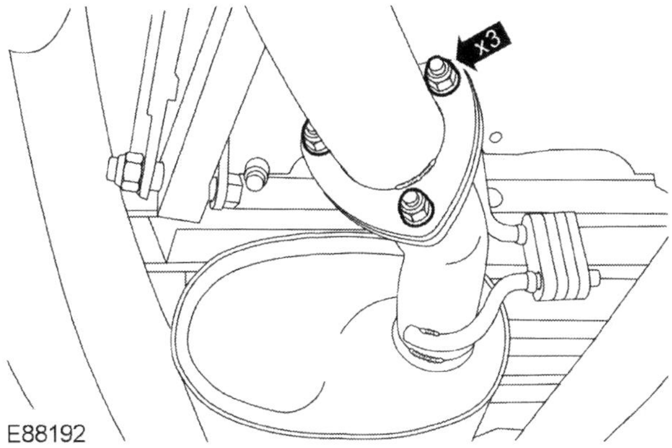

2. Release the front muffler from the tail pipe. Remove and discard the 3 nuts. Remove and discard the gasket.

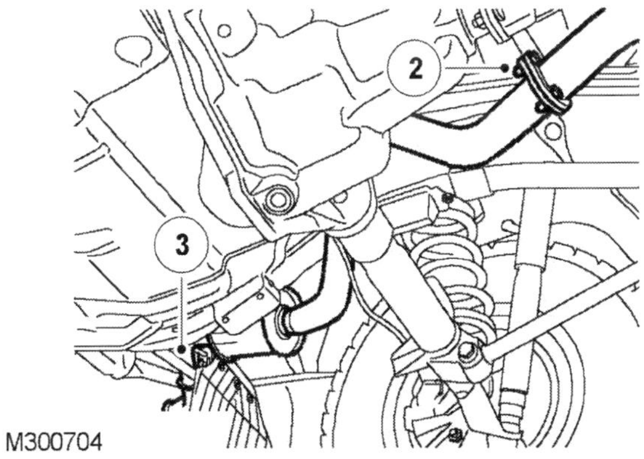

3. Remove the rear muffler and tail pipe. Release the rear muffler from the rubber mounting.

Installation

1. Install the rear muffler and tail pipe. Secure the rear muffler to the rubber mounting.

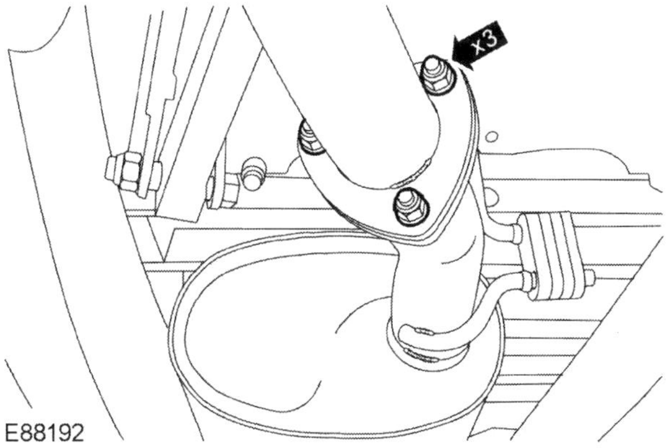

2. NOTES:

Install a new gasket.

Install new nuts.

Secure the front muffler to the tail pipe. Tighten.

Renew Part: Intermediate exhaust pipe gasket

Renew Part: Intermediate exhaust pipe to rear pipe nut(s)

Torque: 25 Nm

Catalytic Converter

Removal and Installation

Removal

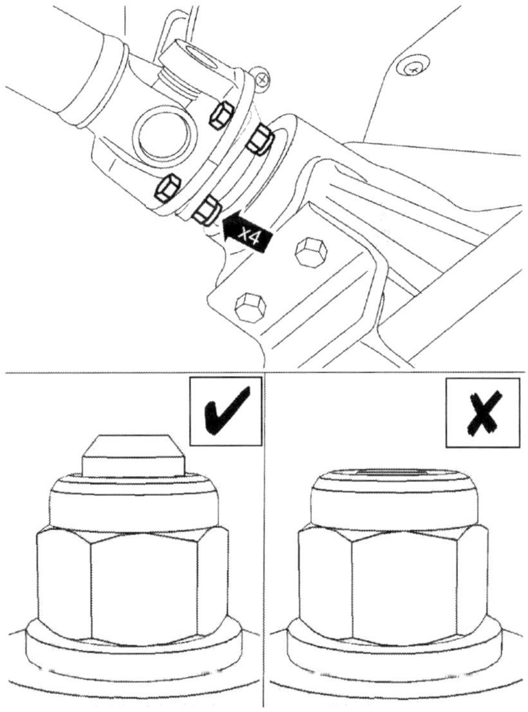

1. **Refer to:** Battery Disconnect and Connect (414-01 Battery, Mounting and Cables, General Procedures).

2. WARNING: Do not work on or under a vehicle supported only by a jack. Always support the vehicle on safety stands.

Raise and support the vehicle.

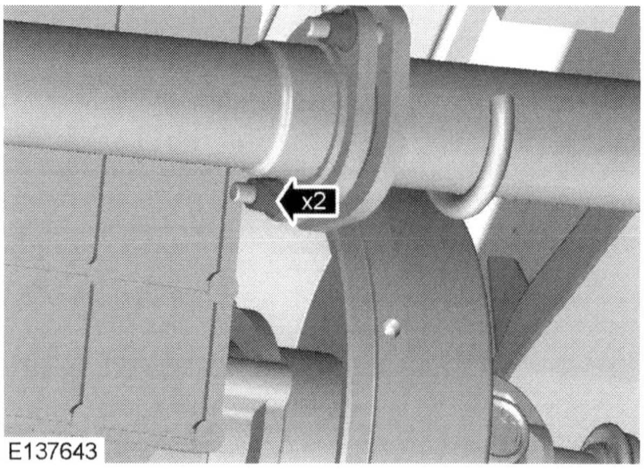

3.

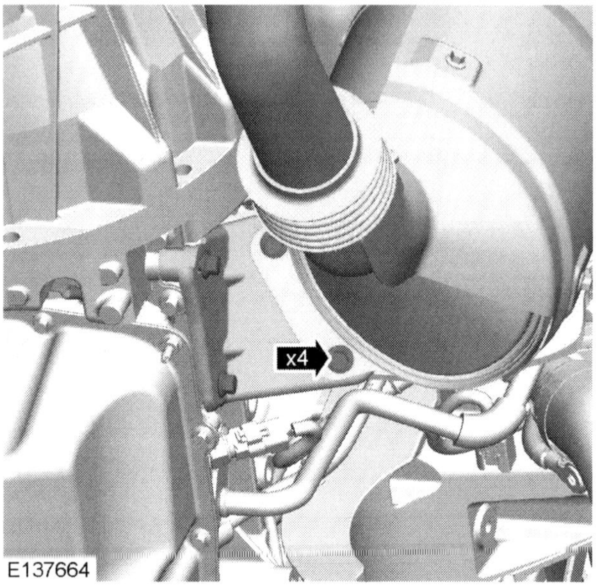

4. Lower the vehicle.

5. **Refer to:** Engine Cover (501-05 Interior Trim and Ornamentation, Removal and Installation).

6.

7.

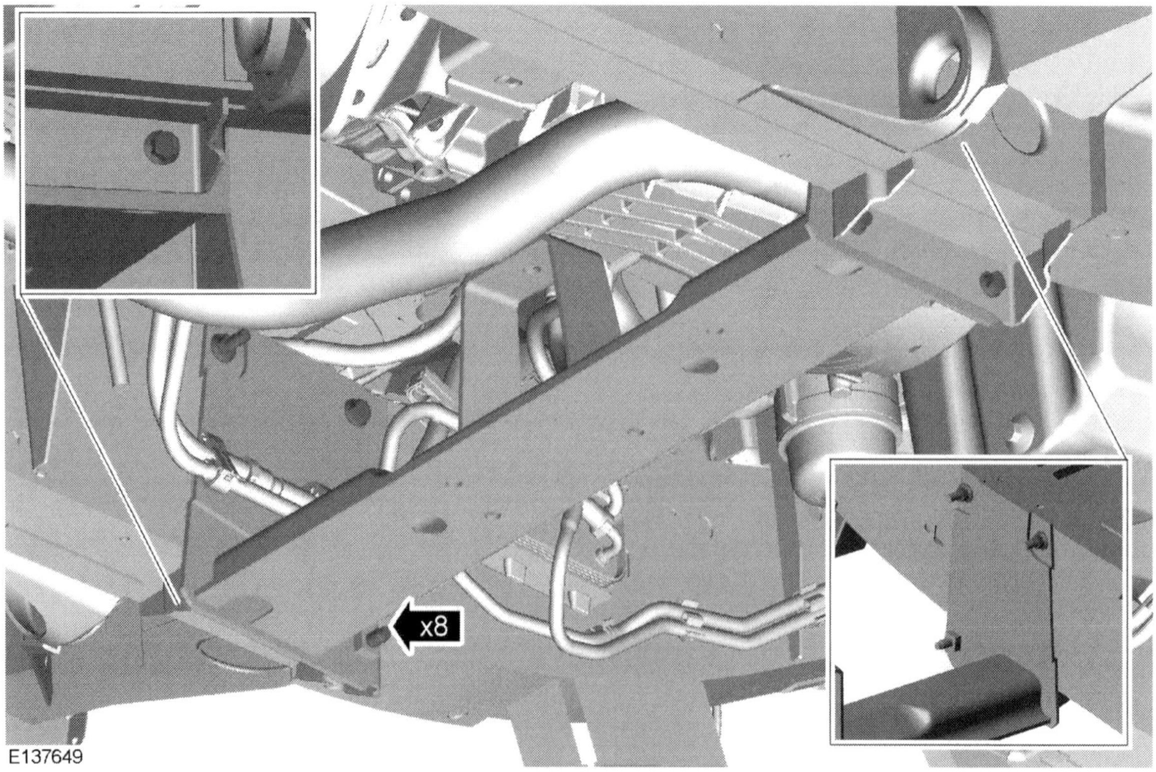

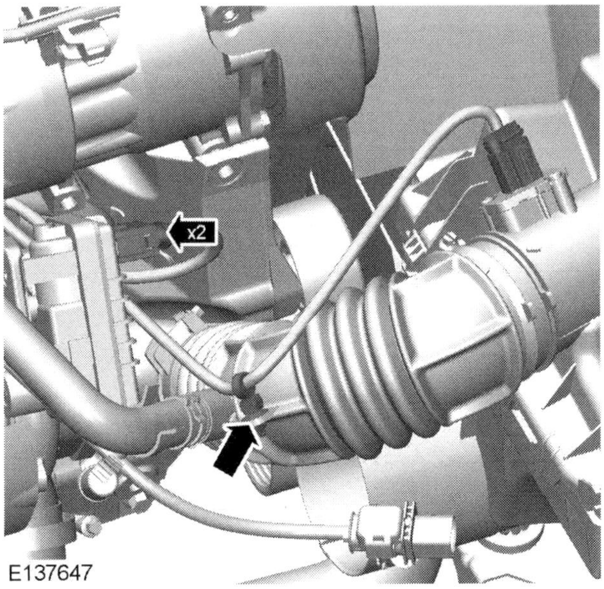

8.

9.

10.

11.

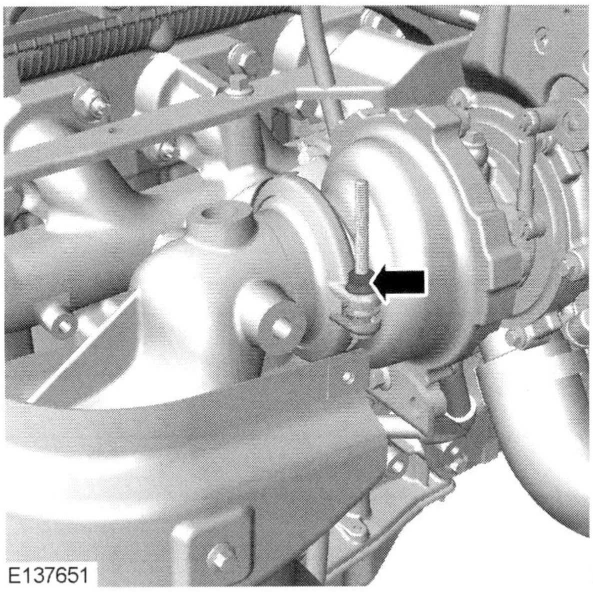

12.

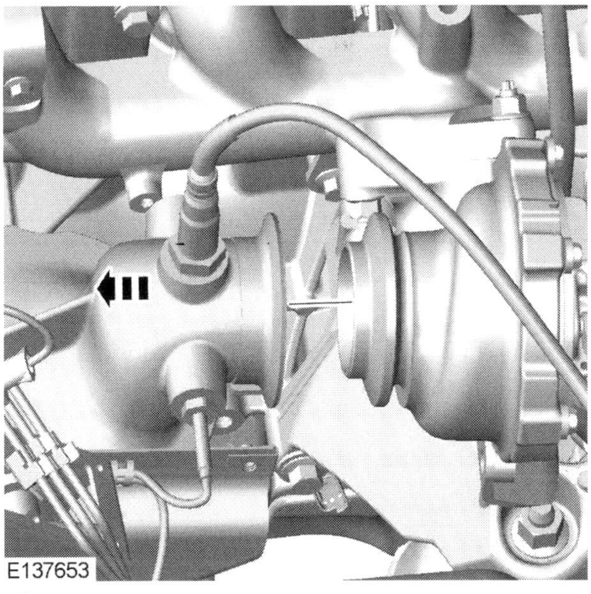

13.

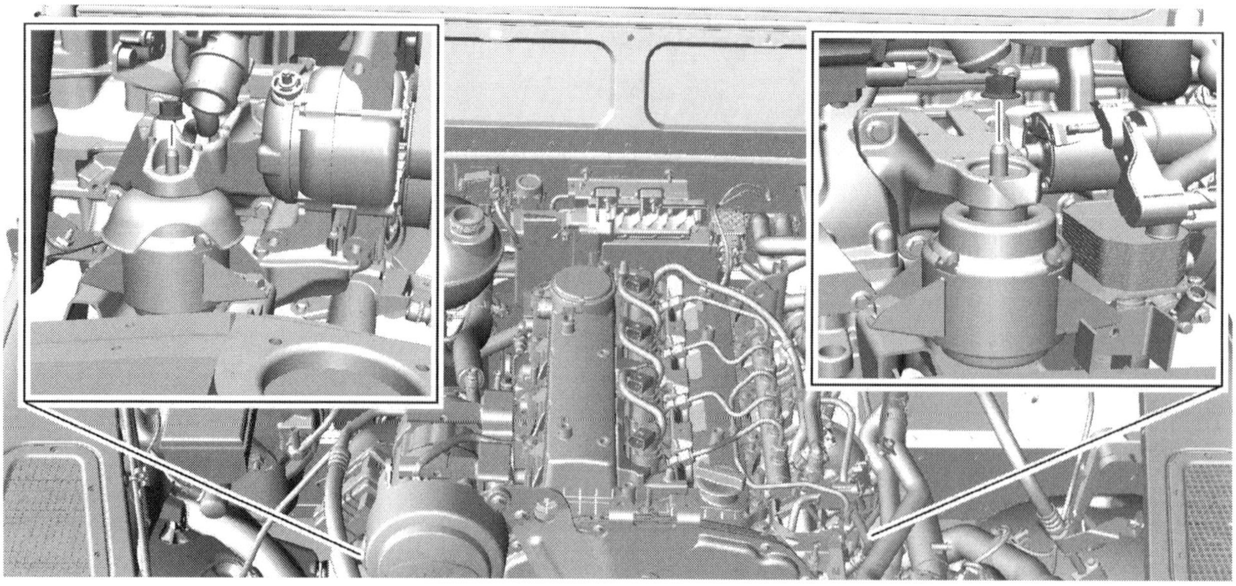

14.

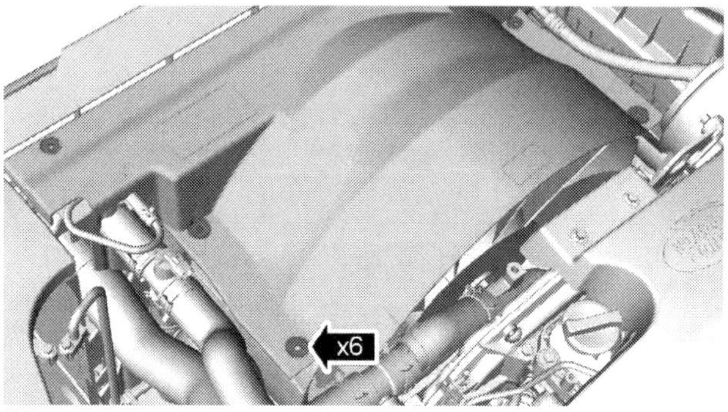

15.

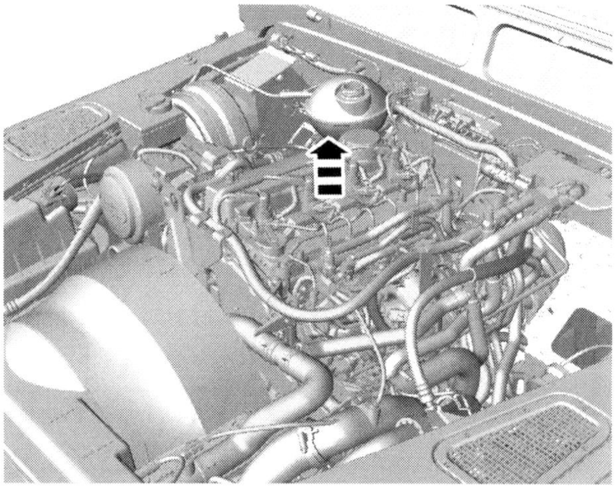

16. CAUTION: Raise the engine assembly using suitable lifting equipment. Make sure the attached and surrounding components are not damaged during this step.

17.

Installation

1.

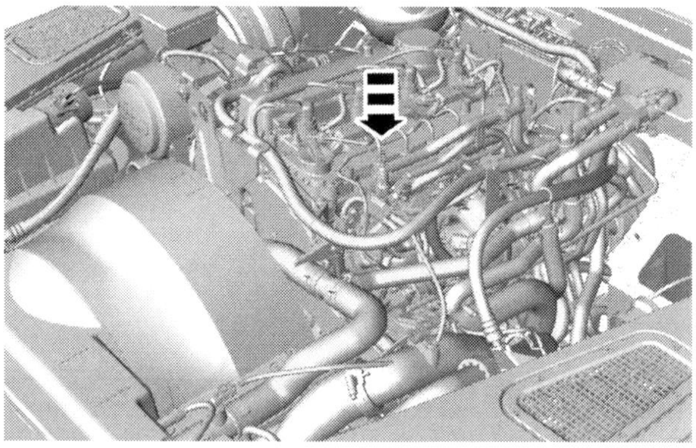

2.

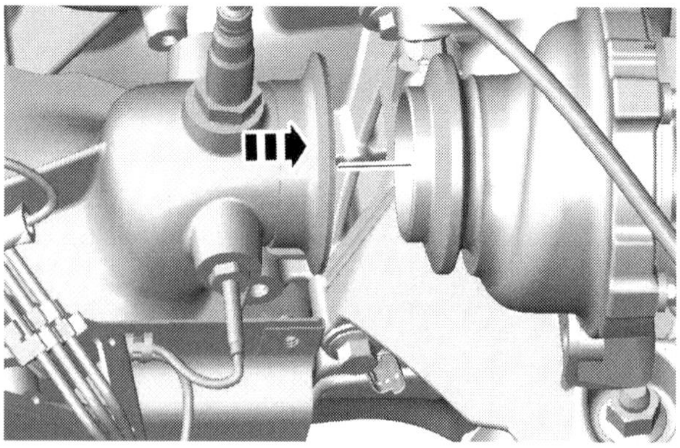

3.

4. *Torque*: 10 Nm

5. *Torque*: 80 Nm

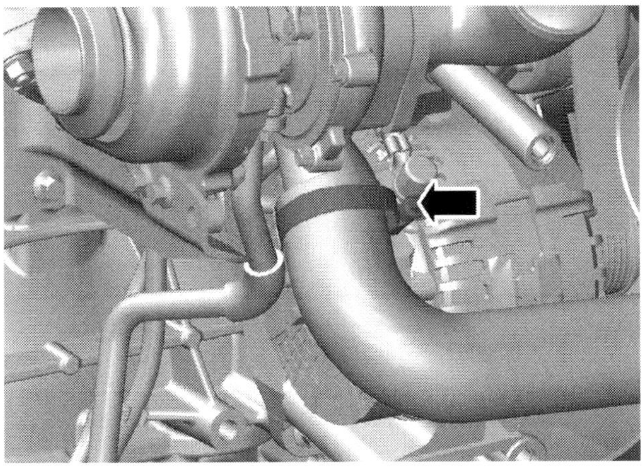

6. *Torque* 3 Nm

7. *Torque* 10 Nm

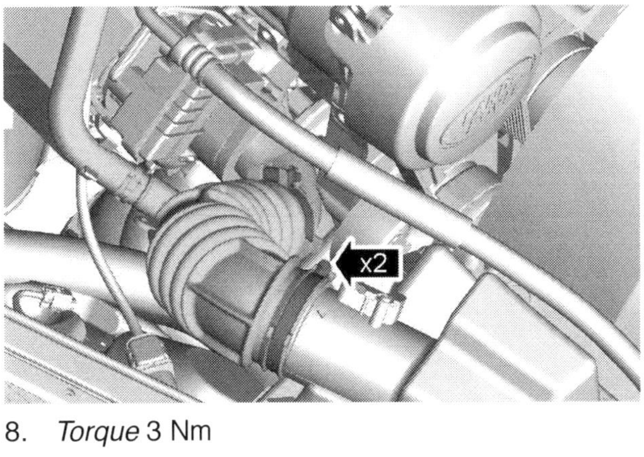

8. *Torque* 3 Nm

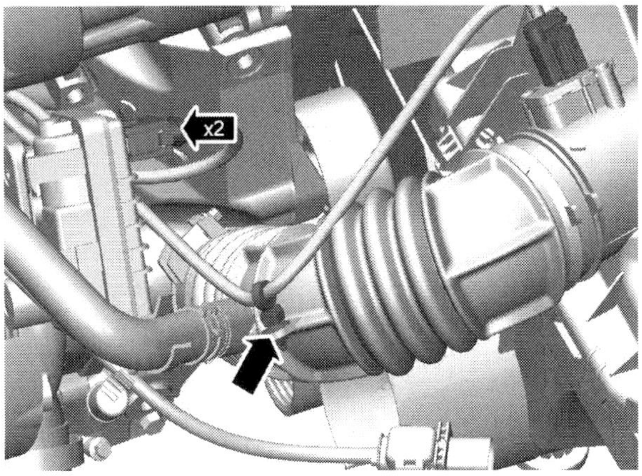

9.

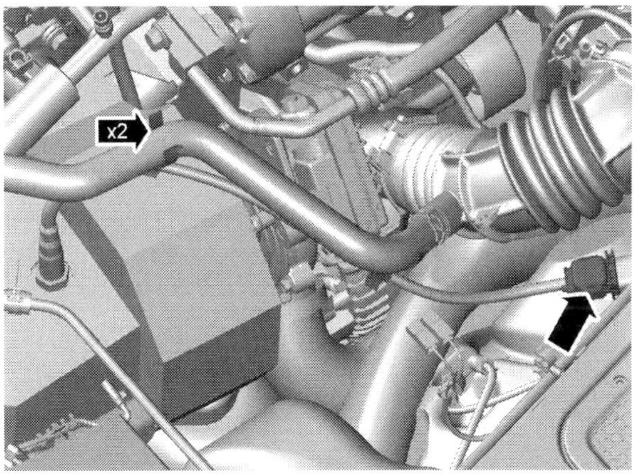

10.

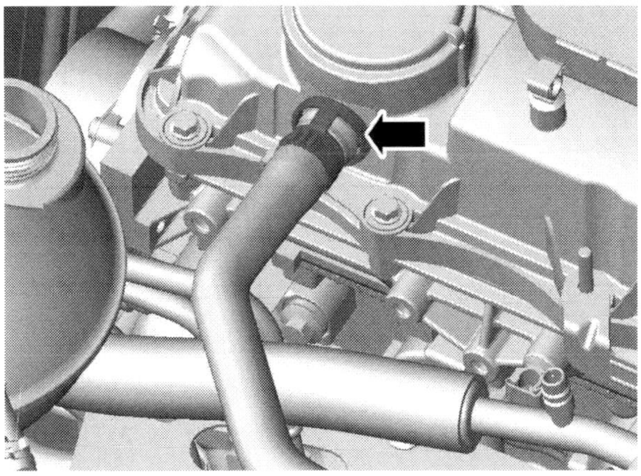

11.

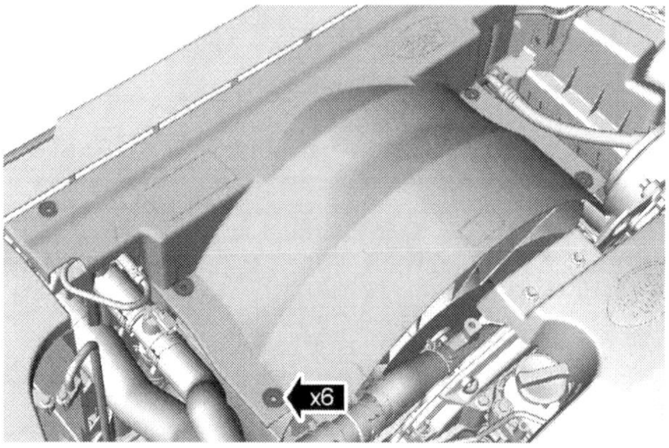

12.

13. **Refer to:** Engine Cover (501-05 Interior Trim and Ornamentation, Removal and Installation).

14. Raise the vehicle on lift.

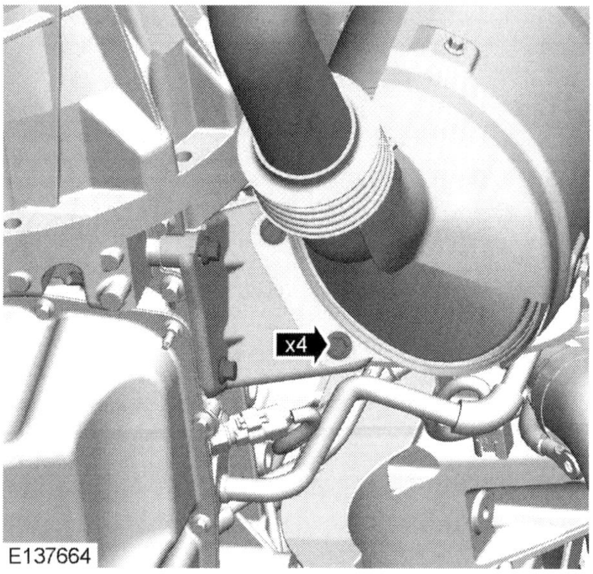

15. *Torque*: 25 Nm

16. *Torque*: 80 Nm

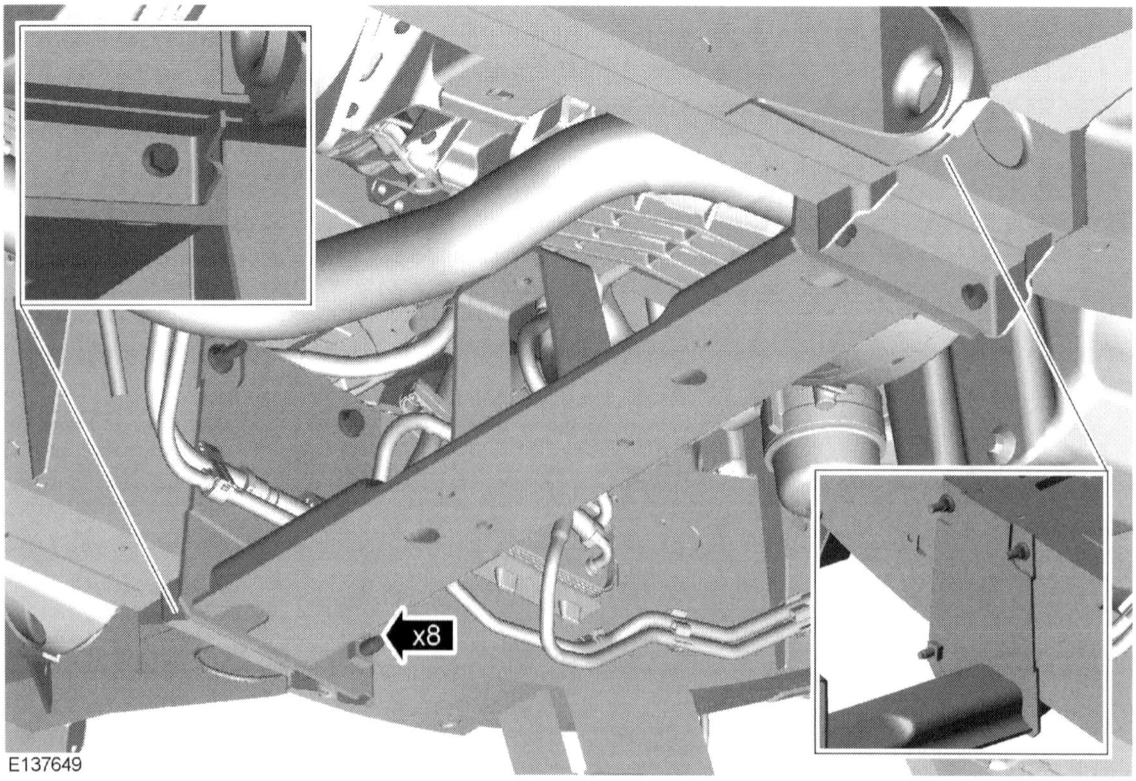

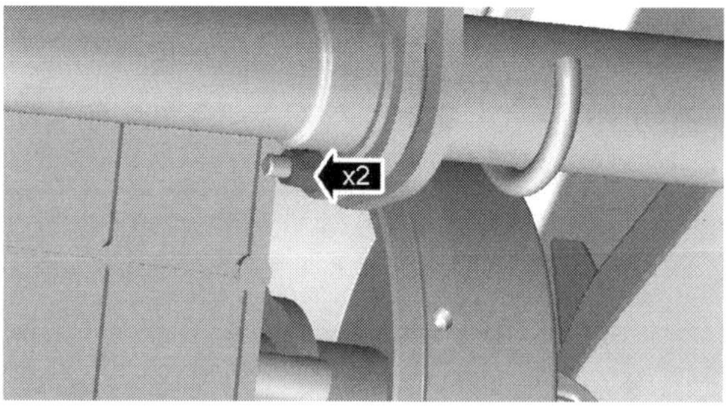

17. *Torque*: 25 Nm

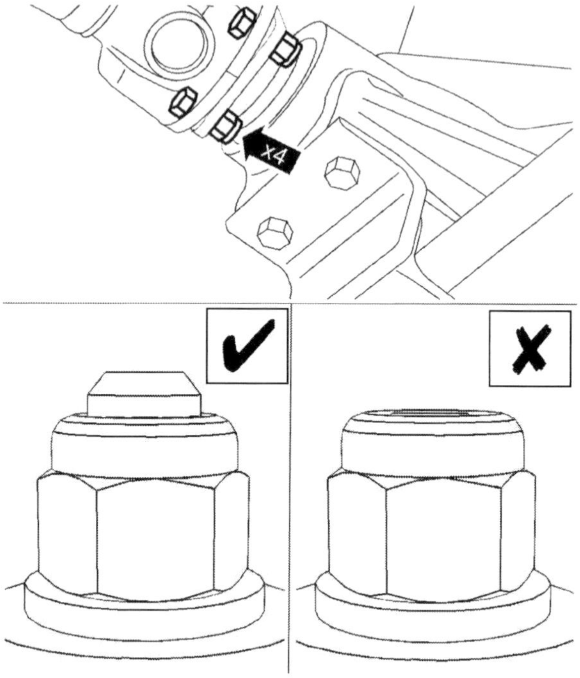

18. *Torque*: 47 Nm

19. **Refer to:** Battery Disconnect and Connect (414-01 Battery, Mounting and Cables, General Procedures).

General Information - Diesel Filter Water Drain Off

General Procedures

WARNINGS:

⚠ The spilling of fuel is unavoidable during this operation. Ensure that all necessary precautions are taken to prevent fire and explosion.

⚠ Do not carry or operate cellular phones when working on or near any fuel related components. Highly flammable vapors are always present and may ignite. Failure to follow these instructions may result in personal injury.

⚠ Do not smoke or carry lighted tobacco or open flame of any type when working on or near any fuel related components. Highly flammable vapors are always present and may ignite. Failure to follow these instructions may result in personal injury.

⚠ If fuel contacts the eyes, flush the eyes with cold water or eyewash solution and seek immediate medical attention.

⚠ Wash hands thoroughly after fuel handling, as prolonged contact may cause irritation. Should irritation develop, seek medical attention.

⚠ CAUTION: Diesel fuel injection equipment is manufactured to very precise tolerances and fine clearances. It is therefore essential that absolute cleanliness is observed when working with these components. Always install blanking plugs to any open orifices or lines. Failure to follow this instruction may result in foreign matter ingress to the fuel injection system.

E91102

1. Drain the fuel filter element.
    - Attach a suitable drain tube to the fuel filter water drain off port.
    - Loosen the water drain off port one complete turn and allow the fuel to drain into a container.
    - Tighten the water drain off port.
    - Remove the drain tube.
    - Remove the container.
2. Bleed the fuel system.

Refer to: Low-Pressure Fuel System Bleeding (310-00 Fuel System - General Information, General Procedures)

Fuel Tank Draining

General Procedures

1. Remove battery cover.
2. Disconnect both leads from battery, earth lead first.

⚠ WARNING: Fuel vapour is highly flammable and in contained spaces is also explosive and toxic. Always have a fire extinguisher containing FOAM, CO2, GAS OR POWDER close at hand when handling or draining fuel.

3. Using a fuel recovery appliance, drain the fuel from the tank into a sealed container. Follow the manufacturers instructions for the connection and safe use of the appliance.
4. Remove filler cap and insert hose into filler neck.
5. Connect bowser earth line. Drain fuel.
6. Connect battery leads.
7. Fit battery cover.

Low-Pressure Fuel System Bleeding

General Procedures

Special Tool(s)

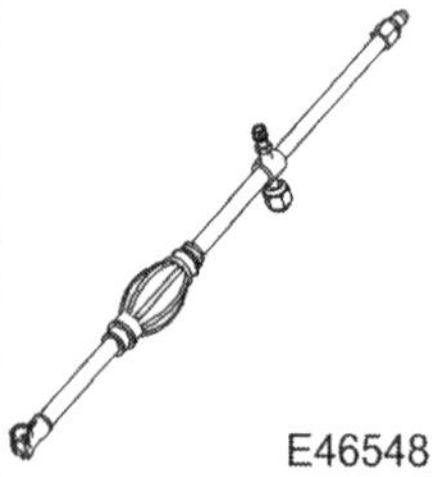 E46548	Hand pressure pump 310-110

WARNINGS:

After carrying out repairs, the fuel system must be checked visually for leaks. Failure to follow this instruction may result in personal injury.

Do not carry or operate cellular phones when working on or near any fuel related components. Highly flammable vapors are always present and may ignite. Failure to follow these instructions may result in personal injury.

Do not smoke or carry lighted tobacco or open flame of any type when working on or near any fuel related components. Highly flammable vapors are always present and may ignite. Failure to follow these instructions may result in personal injury.

If fuel contacts the eyes, flush the eyes with cold water or eyewash solution and seek immediate medical attention.

This procedure involves fuel handling. Be prepared for fuel spillage at all times and always observe fuel handling precautions. Failure to follow these instructions may result in personal injury.

Wash hands thoroughly after fuel handling, as prolonged contact may cause irritation. Should irritation develop, seek medical attention.

CAUTIONS:

Diesel fuel injection equipment is manufactured to very precise tolerances and fine clearances. It is therefore essential that absolute cleanliness is observed when working with these components. Always install blanking plugs to any open orifices or lines. Failure to follow this instruction may result in foreign matter ingress to the fuel injection system.

Make sure the workshop area in which the vehicle is being worked on is as clean and as dust free as possible. Foreign matter from work on clutches, brakes or from machining or welding operations can contaminate the fuel system and may result in later malfunction.

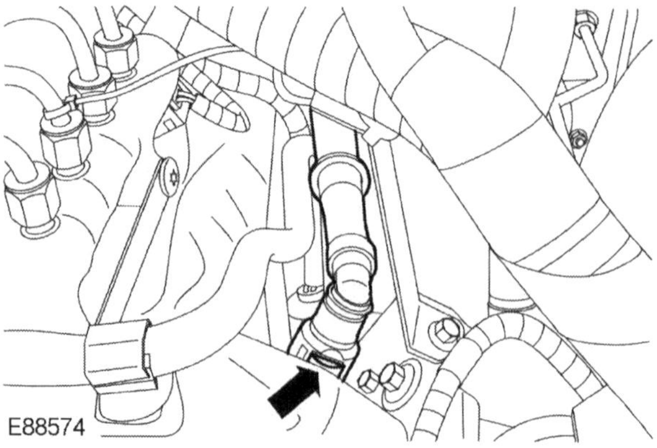
E88574

1. Disconnect the fuel injection pump supply line.

Refer to: Quick Release Coupling (310-00 Fuel System - General Information, General Procedures)

- Position an absorbent cloth to collect any fluid spillage.

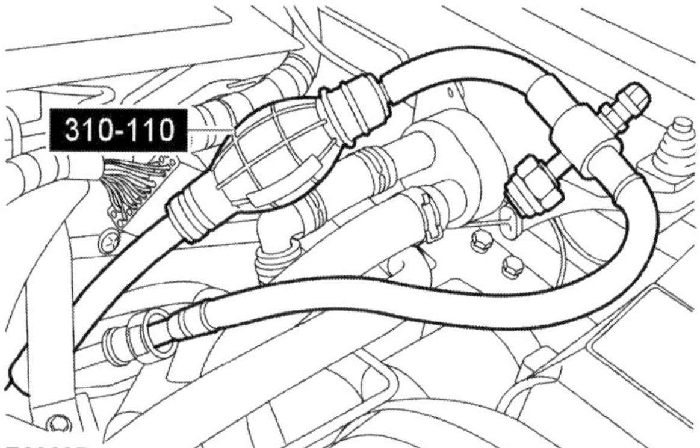

E88225

2. Install the special tool.

- Operate the hand pressure pump special tool until fuel starts to flow through the clear plastic pipe and the special tool becomes firm.

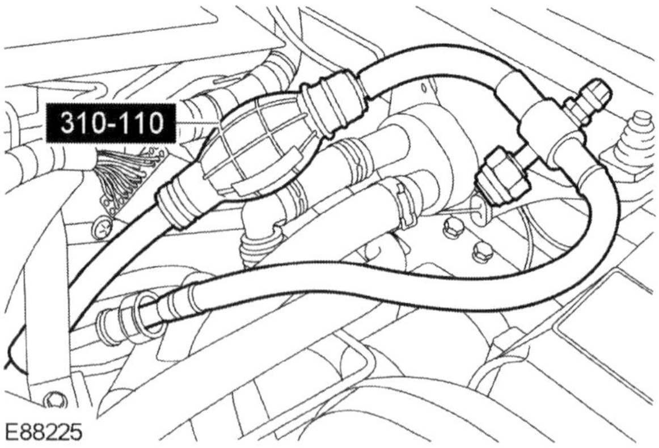

E88225

3. Remove the special tool.
4. Connect the fuel injection pump supply line.

Refer to: Quick Release Coupling (310-00 Fuel System - General Information, General Procedures)

- Remove the absorbent cloth.

Quick Release Coupling

General Procedures

WARNINGS:

After carrying out repairs, the fuel system must be checked visually for leaks. Failure to follow this instruction may result in personal injury.

Do not carry or operate cellular phones when working on or near any fuel related components. Highly flammable vapors are always present and may ignite. Failure to follow these instructions may result in personal injury.

Do not smoke or carry lighted tobacco or open flame of any type when working on or near any fuel related components. Highly flammable vapors are always present and may ignite. Failure to follow these instructions may result in personal injury.

If fuel contacts the eyes, flush the eyes with cold water or eyewash solution and seek immediate medical attention.

This procedure involves fuel handling. Be prepared for fuel spillage at all times and always observe fuel handling precautions. Failure to follow these instructions may result in personal injury.

Wash hands thoroughly after fuel handling, as prolonged contact may cause irritation. Should irritation develop, seek medical attention.

CAUTIONS:

Diesel fuel injection equipment is manufactured to very precise tolerances and fine clearances. It is therefore essential that absolute cleanliness is observed when working with these components. Always install blanking plugs to any open orifices or lines. Failure to follow this instruction may result in foreign matter ingress to the fuel injection system.

Make sure the workshop area in which the vehicle is being worked on is as clean and as dust free as possible. Foreign matter from work on clutches, brakes or from machining or welding operations can contaminate the fuel system and may result in later malfunction.

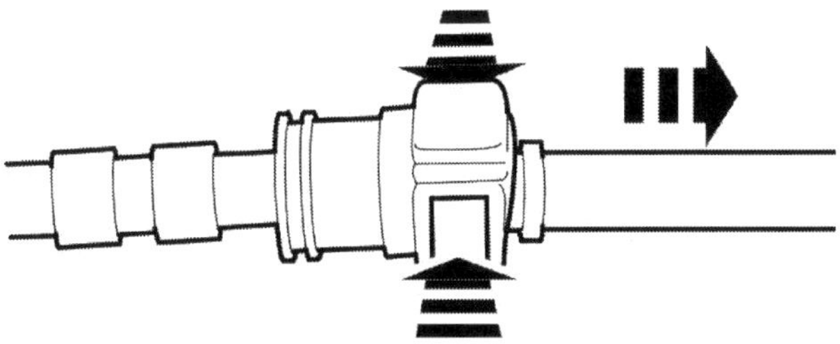

1. Disconnect the fuel line quick release coupling.
   - Push the fuel line quick release buttons and pull the 2 halves of the fuel line to disconnect.

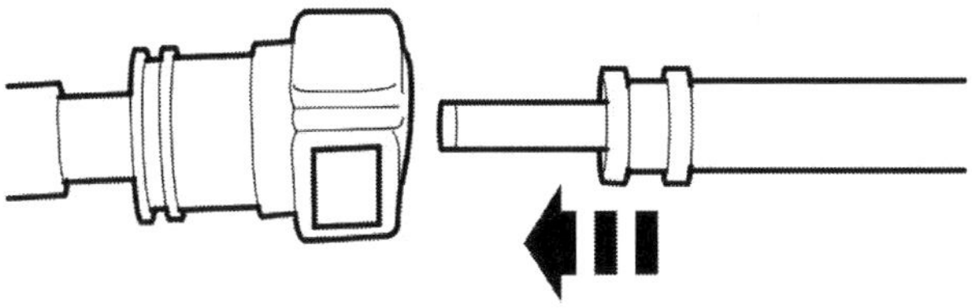

2. CAUTION: Make sure the fuel line quick release coupling clicks into place when connecting the fuel line. To make sure that the quick release coupling is fully seated pull on the 2 halves of the fuel line.

Connect the fuel line quick release coupling.

Fuel Tank and Lines - ID 2.2L and 2.4L Diesel

Specifications

Description	Nm	lb-ft	Psi
Fuel level sender locking ring	35	26	NA
Fuel tank support bracket bolts	45	33	NA
Fuel tank support bracket nuts	25	18	NA
In tank fuel pump supply pressure	NA	NA	6 - 10.5

Description	bar	psi
Maximum fuel pressure	0.9	13

Description and Operation

OVERVIEW

The fuel tank and lines supply fuel to the fuel charging and controls system on the engine.

The fuel tank and lines consists of:

- The fuel filler pipe and hose
- The fuel tank
- The pump and sender assembly
- The fuel filter
- The fuel cooler
- Connecting fuel lines
- An inertia switch

The supply connecting line supplies fuel from the pump and sender assembly in the fuel tank to the HP (high pressure) fuel pump on the engine. The return connecting line returns excess fuel from the HP fuel pump and the fuel injectors to the filter via the fuel cooler. To prevent filter waxing at low temperatures, a thermostatic diverter in the filter routes a proportion of the warm return fuel into the filter and back into the supply line to the engine; the remainder of the fuel is routed back to the fuel tank. At higher temperatures the all of the return fuel is routed back to the fuel tank.

COMPONENT LOCATION

△ NOTE: 110 variant installation shown, installation on 90 and 130 variants similar.

- The pump and sender assembly
- The fuel filter
- The fuel cooler
- Connecting fuel lines
- An inertia switch

Item	Description
1	Inertia switch
2	Fuel tank
3	Fuel filter
4	Fuel cooler

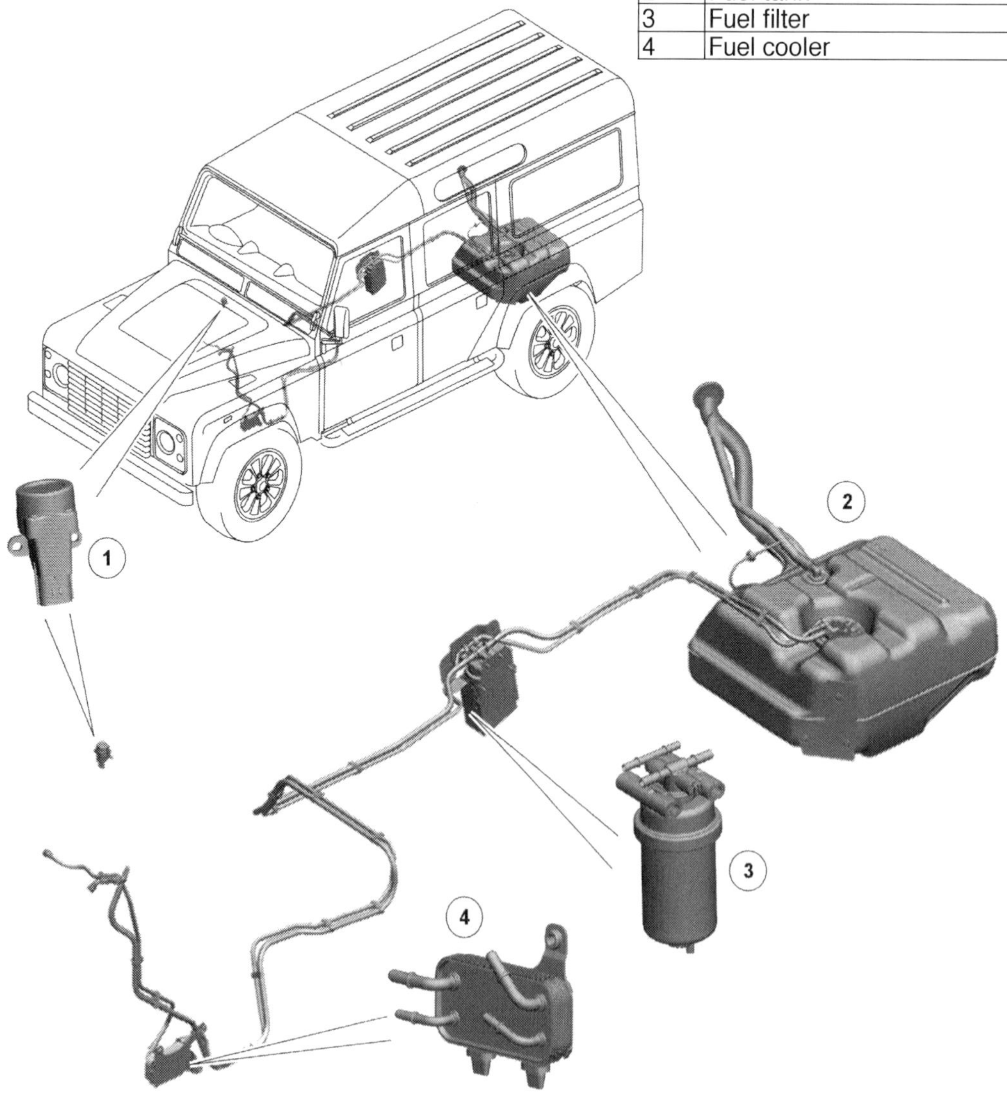

FUEL FILLER AND CAP

The fuel filler is located in the right hand rear quarter panel. The filler is closed by a lockable cap which screws into the filler neck. The cap has a ratchet mechanism to prevent over tightening, and seals against the filler neck to prevent the escape of fuel vapor. The filler cap has a valve which relieves pressure to atmosphere at approximately 0.12 to 0.13 bar (1.8 to 2.0 lbf/in^2) and opens in the opposite direction at approximately 0.04 bar (0.7 lbf/in^2) vacuum.

A HDPE (high density polyethylene) molded filler tube connects the filler to the tank via a flexible hose.

FUEL TANK

The fuel tank is located at the rear underside of the vehicle between the chassis longitudinals.

The cradle is attached to the chassis with six screws. When the cradle is attached to the chassis, the tank is positively secured via foam pads which bear against the central chassis cross beam. A protective cover is fitted to the front right hand corner of the tank and provides additional protection.

The fuel tank is manufactured from HDPE. The tank is a sealed unit with the only internal access being via the pump module flange aperture on the top of the tank.

A heat shield is attached to the front right corner of the fuel tank with three scrivets to protect the tank from heat generated by the exhaust system.

Fuel Tank Capacities

Variant	Capacity (Useable Fuel)
90	55 liters
110/130	70 liters

FUEL TANK BREATHER SYSTEM

The filler tube incorporates a tank vent which allows air and fuel vapor displaced from the tank when filling to vent to atmosphere via the filler neck.

A breather spout within the tank controls the tank 'full' height. When fuel covers the spout it prevents fuel vapor and air from escaping from the tank. This causes the fuel to 'back-up' in the filler tube and shuts off the filler gun. The position of the spout ensures that when the filler gun shuts off, a vapor space of approximately 10% of the tank's total capacity remains.

The breather pipe includes a two-way valve which allows for over-pressure relief one way and allows for air to enter the tank as the system operates in depression.

PUMP AND SENDER ASSEMBLY

The pump and sender assembly is installed in the top of the fuel tank and secured with a gasket, adapter and locking ring. Two fuel connections and an electrical connector are located in the mounting flange for attachment of the fuel lines and the vehicle wiring respectively.

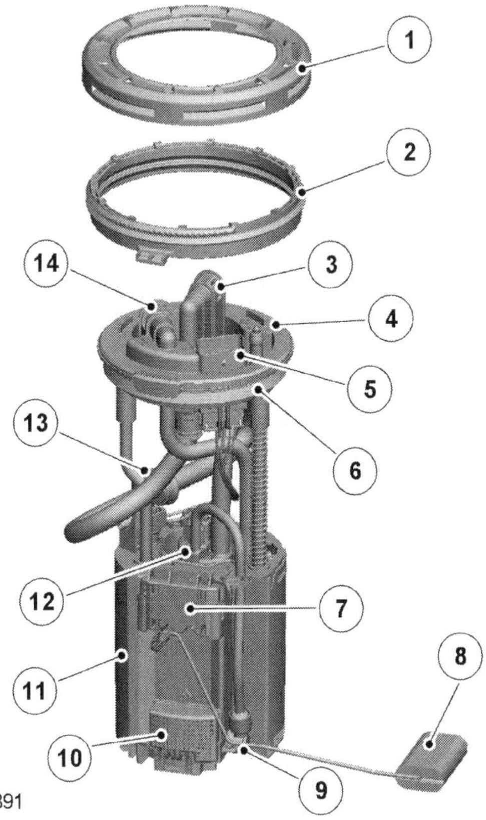

E138891

Item	Description
1	Locking ring
2	Adapter
3	Outlet connection
4	Mounting flange
5	Electrical connector
6	Gasket
7	Sender unit potentiometer
8	Sender unit float
9	Jet pump
10	Screen
11	Base
12	Fuel pump
13	Pressure relief valve
14	Inlet connection

The fuel pump is installed in the base of the assembly, which contains a swirl pot to provide a constant reservoir of fuel for the fuel pump. A jet pump, powered by a feed from the fuel pump, induces fuel into the swirl pot. A coarse mesh screen over the swirl pot inlet prevents the ingress of larger pieces of debris. The primary outlet from the fuel pump is connected to the fuel outlet connector in the mounting flange via a pressure relief valve. The pressure relief valve limits the pressure of the fuel supplied to the engine and releases excess fuel back into the tank. A pipe attached to the fuel inlet connector directs return fuel from the engine into the swirl pot.

The pump is an electric vane pump. Electrical power for the fuel pump is supplied from the fuel pump relay in the BJB (battery junction box) , which is controlled by the ECM (engine control module) . The ECM energizes the fuel pump relay while the ignition switch is in the ignition and crank positions, provided it also receives an ignition on signal from the inertia switch. If an accident occurs that activates the inertia switch, the ignition on signal from it is discontinued and the ECM then de-energizes the fuel pump relay to stop the pump.

The sender unit comprises a potentiometer operated by a float. The float rises and falls with the fuel level in the tank and moves the potentiometer accordingly.

A voltage is supplied to the potentiometer from the instrument pack. The output voltage from the potentiometer varies according to the fuel level and is translated by the instrument pack into a fuel gauge reading.

A warning lamp in the instrument cluster illuminates if the fuel level is at or below 12 liters (3.2 US gallons).

FUEL FILTER

E86472

The fuel filter removes particulate matter from the fuel and also separates water which accumulates at the bottom of the filter.

The fuel filter is positioned on the chassis to the right of the fuel tank. The fuel feed and return to and from the engine passes through the filter. Connections are made using quick fit connectors.

A steel rear cover is attached to the chassis longitudinal by four M8 screws.

The filter is screwed to the back plate by two M8 bolts.

A steel protection cover is fixed onto the back cover using a ¼ turn fastener.

The filter has an internal air bleed feature which allows air into the fuel supply to the engine in small, manageable amounts.

The thermostatic diverter valve is fully closed at 45 degrees Celsius and sends fuel directly to the tank. When the diverter is open fuel is re-circulated through the filter to the engine.

The fuel filter has a replaceable twist-on canister filter element which is sealed to the filter body with rubber seals. The lower part of the canister has a screw-in cap for water draining.

FUEL COOLER

The fuel cooler uses engine coolant, from the radiator, to cool fuel returning to the tank from the HP fuel pump.

The cooler is fixed to a bracket which is attached to the inside of the left hand chassis rail. The bracket has two slots which accept two rubber covered aluminum location pegs on the cooler. An M8 bolt secures the cooler to the bracket.

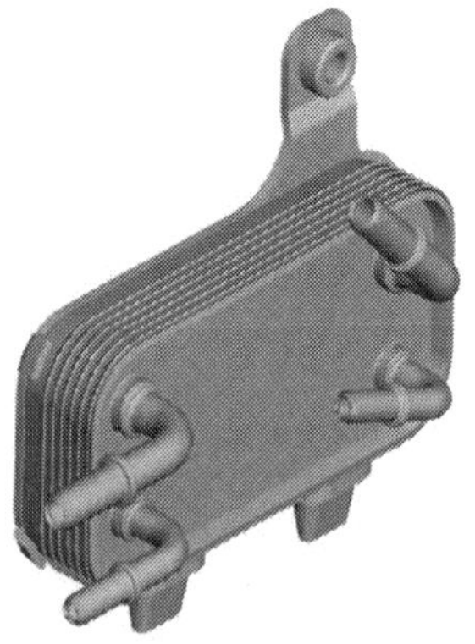

E86473

The cooler has four quick fit connections. Fuel inlet and outlet and coolant inlet feed and return.

BLEED VALVE

The bleed valve is located in the fuel feed line near to the HP fuel pump. It facilitates the priming of the fuel system and removes air during initial and service primes. This protects the HP fuel pump from air locking.

FUEL LINES

Three fuel line assemblies connect the feed and return flow from tank to the filter, filter to the cooler and cooler to the engine.

The fuel lines are constructed from conductive 1 mm thick nylon with 2 mm thick santoprene fire resistant, anti-abrasive coating throughout. The fuel feed line has a larger diameter than the return.

INERTIA SWITCH

The inertia switch, when activated, causes the ECM to disable engine fuel injection and the fuel tank pump

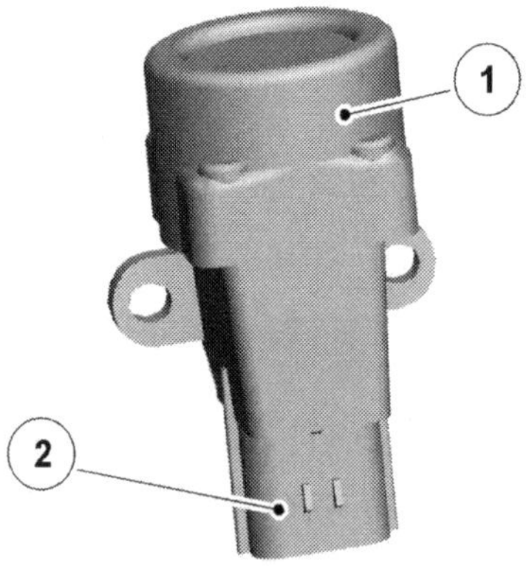

E138892

Item	Description
1	Rubber cap
2	Electrical connector

The inertia switch is installed in the engine compartment, on the engine bulkhead. When the ignition switch is in the ignition on and crank positions, battery voltage from the BJB is supplied to the inertia switch. While the inertia switch is in the de-activated state, the voltage passes through the normally closed contacts of the switch to the ECM. If an impact of a given force occurs, the inertia switch contacts open and disconnect the voltage signal to the ECM.

After it has been activated, the inertia switch can be reset by pressing down on the rubber cap.

LOW FUEL INDICATION AND RUN DRY STRATEGY

The run-dry strategy is used to maintain the systems fuel prime at fuel run out. It ensures the minimum amount of fuel is always left in the swirl pot.

The instrument cluster activates the yellow low fuel warning light (next to the fuel gauge) with 17% of fuel remaining in the tank. The fuel gauge will indicate empty with 12% of fuel left in the tank.

With 6 liters left in the tank the run-dry strategy will be invoked. An engine mis-fire will be induced for approximately 1 mile after which the engine will be shut down. The engine can be re-started in mis-fire mode and will continue to run for a further mile until the engine shuts down again. This can be repeated until the fuel suction port in the tank is uncovered and causes engine fuel starvation and loss of prime. Re-starts after run-dry shut down are not recommended.

CONTROL DIAGRAM

NOTE: A = Hardwired.

Item	Description
1	Battery
2	-
3	Ignition switch
4	Inertia switch
5	-
6	Pump and sender assembly
7	Instrument cluster

Fuel Cooler

Removal and Installation

Removal

WARNINGS:

After carrying out repairs, the fuel system must be checked visually for leaks. Failure to follow this instruction may result in personal injury.

Do not carry or operate cellular phones when working on or near any fuel related components. Highly flammable vapors are always present and may ignite. Failure to follow these instructions may result in personal injury.

Do not smoke or carry lighted tobacco or open flame of any type when working on or near any fuel related components. Highly flammable vapors are always present and may ignite. Failure to follow these instructions may result in personal injury.

If fuel contacts the eyes, flush the eyes with cold water or eyewash solution and seek immediate medical attention.

This procedure involves fuel handling. Be prepared for fuel spillage at all times and always observe fuel handling precautions. Failure to

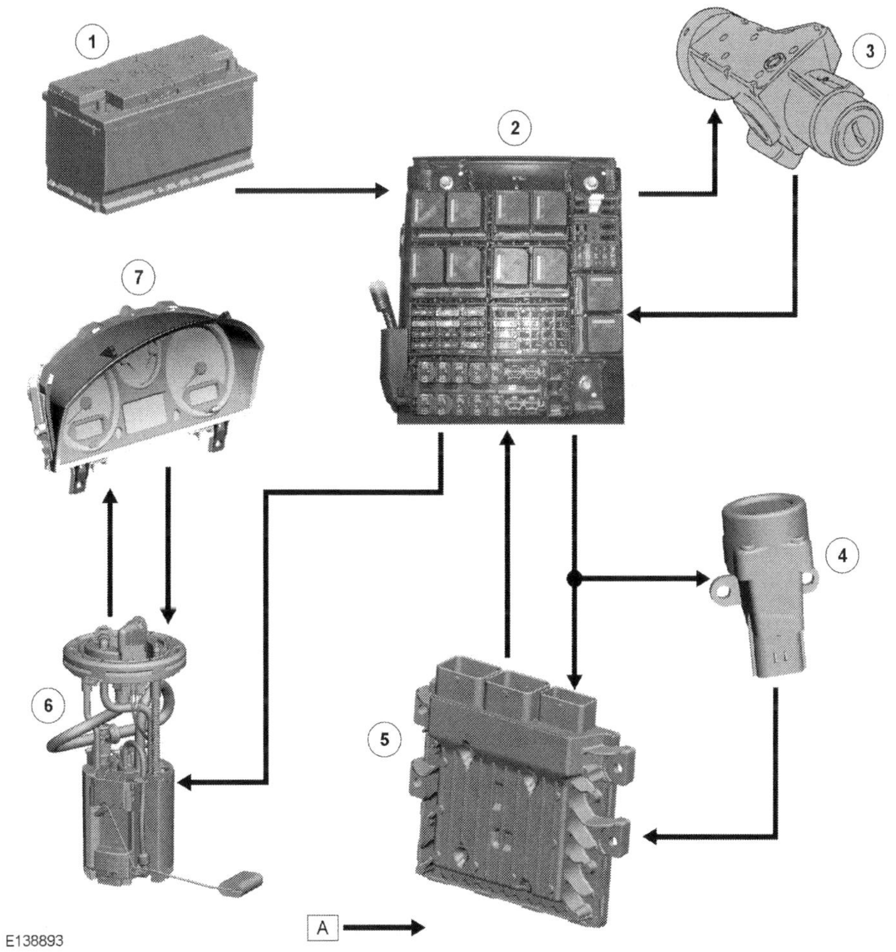

follow these instructions may result in personal injury.

Wash hands thoroughly after fuel handling, as prolonged contact may cause irritation. Should irritation develop, seek medical attention.

CAUTIONS:

Diesel fuel injection equipment is manufactured to very precise tolerances and fine clearances. It is therefore essential that absolute cleanliness is observed when working with these components. Always install blanking plugs to any open orifices or lines. Failure to follow this instruction may result in foreign matter ingress to the fuel injection system.

Make sure the workshop area in which the vehicle is being worked on is as clean and as dust free as possible. Foreign matter from work on clutches, brakes or from machining or welding operations can contaminate the fuel system and may result in later malfunction.

1. WARNING: Do not work on or under a vehicle supported only by a jack. Always support the vehicle on safety stands.

   Raise and support the vehicle.

2. Drain the cooling system.

Refer to: Cooling System Draining, Filling and Bleeding (303-03 Engine Cooling, ID4 2.2L Diesel General Procedures)

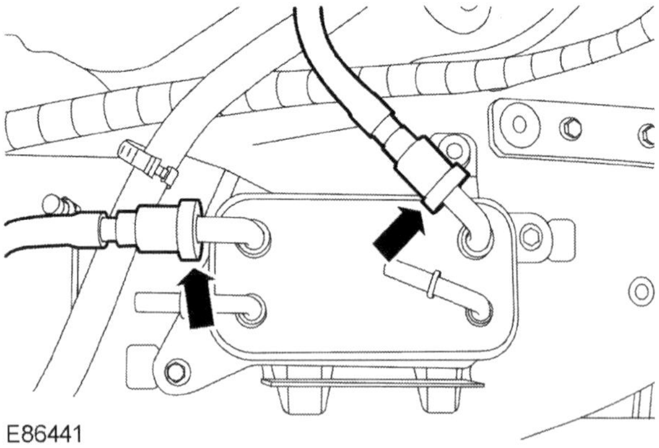

E86441

3. CAUTIONS:

Before disconnecting any components, make sure the area is clean and free from foreign material. When disconnected all openings must be sealed.

Fluid loss is unavoidable, use a suitable container to collect any fluid loss.

Disconnect the 2 fuel lines.

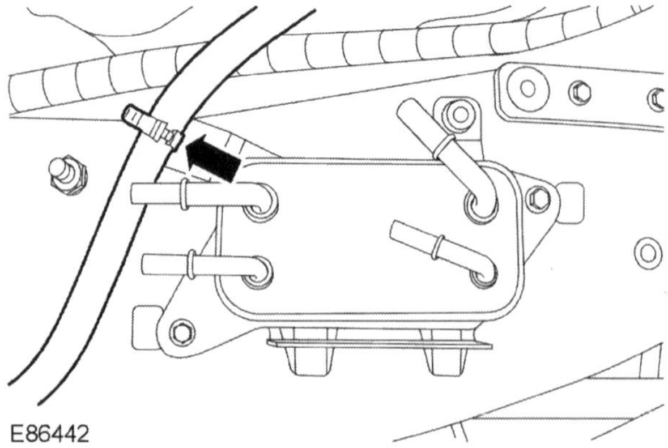

E86442

4. Release the fuel return line from the clip.
5. Remove the fuel cooler. Remove the 2 bolts.

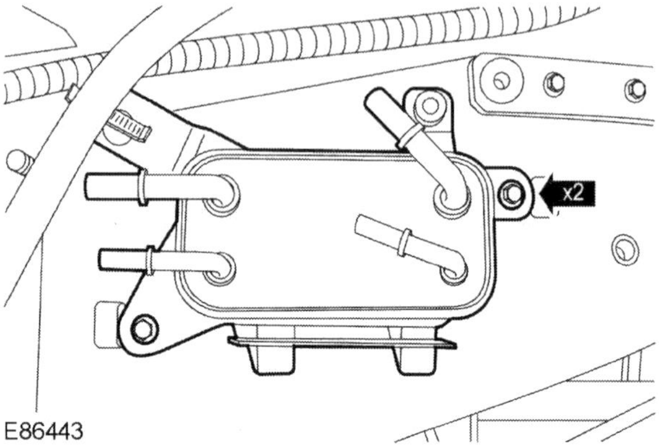

E86443

Installation

1. Install the fuel cooler. Tighten the bolts.

   Torque: 23 Nm

2. Secure the fuel return line in the clip.
3. NOTE: Remove and discard the blanking caps.

   Connect the fuel lines.

4. Bleed the fuel system.

Refer to: Low-Pressure Fuel System Bleeding (310-00 Fuel System - General Information, General Procedures)

5. Fill and bleed the cooling system.

Refer to: Cooling System Draining, Filling and Bleeding (303-03 Engine Cooling, ID4 2.2L Diesel General Procedures)

Fuel Filter

Removal and Installation

Removal

WARNINGS:

After carrying out repairs, the fuel system must be checked visually for leaks. Failure to follow this instruction may result in personal injury.

Do not carry or operate cellular phones when working on or near any fuel related components. Highly flammable vapors are always present and may ignite. Failure to follow these instructions may result in personal injury.

Do not smoke or carry lighted tobacco or open flame of any type when working on or near any fuel related components. Highly flammable vapors are always present and may ignite. Failure to follow these instructions may result in personal injury.

If fuel contacts the eyes, flush the eyes with cold water or eyewash solution and seek immediate medical attention.

This procedure involves fuel handling. Be prepared for fuel spillage at all times and always observe fuel handling precautions. Failure to follow these instructions may result in personal injury.

Wash hands thoroughly after fuel handling, as prolonged contact may cause irritation. Should irritation develop, seek medical attention.

CAUTIONS:

Diesel fuel injection equipment is manufactured to very precise tolerances and fine clearances. It is therefore essential that absolute cleanliness is observed when working with these components. Always install blanking plugs to any open orifices or lines. Failure to follow this instruction may result in foreign matter ingress to the fuel injection system.

Make sure the workshop area in which the vehicle is being worked on is as clean and as dust free as possible. Foreign matter from work on clutches, brakes or from machining or welding operations can contaminate the fuel system and may result in later malfunction.

1. WARNING: Do not work on or under a vehicle supported only by a jack. Always support the vehicle on safety stands.

Raise and support the vehicle.

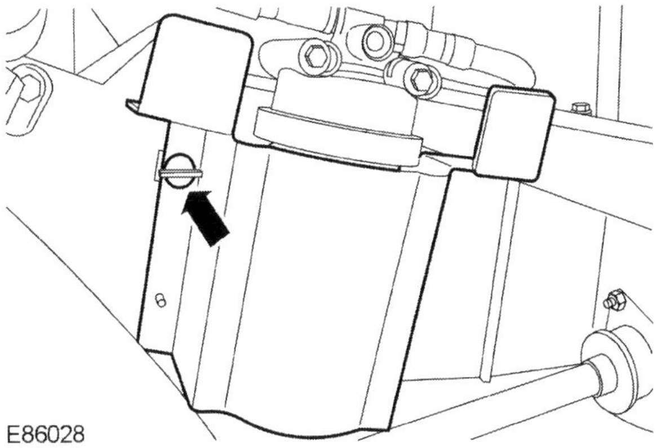

E86028

2. Remove the fuel filter cover.
   - Loosen the clip.

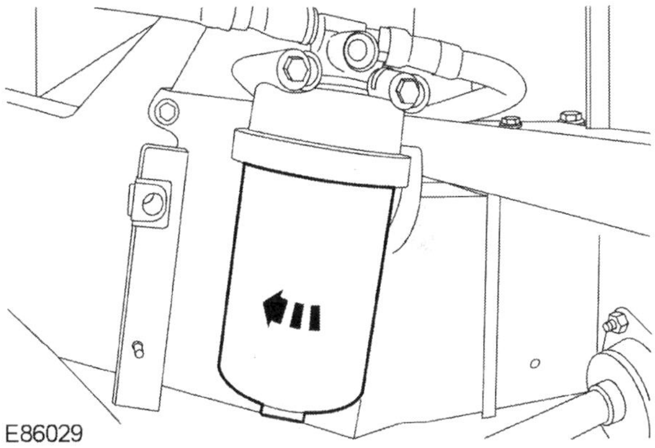

E86029

3. CAUTION: Make sure that the area around the component is clean and free of foreign material.

   Remove the fuel filter.

   - Position a container to collect the fluid spillage.

Installation

1. CAUTION: Clean the component mating faces.

   Install the fuel filter.

   - Remove the container.
2. Install the fuel filter cover.
   - Tighten the clip.
3. Bleed the fuel system.

Refer to: Low-Pressure Fuel System Bleeding (310-00 Fuel System - General Information, General Procedures)

Fuel Level Sensor

Removal and Installation

Special Tool(s)

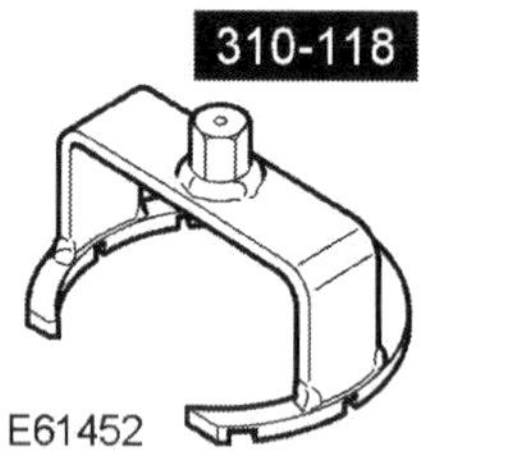 E61452	310-118 Fuel Tank Sender Wrench

Part(s)

STEP	PART NAME	QUANTITY
Installation Step 1	Renew Part Fuel pump O-ring seal	1

Removal

WARNINGS:

After carrying out repairs, the fuel system must be checked visually for leaks. Failure to follow this instruction may result in personal injury.

Do not carry or operate cellular phones when working on or near any fuel related components. Highly flammable vapors are always present and may ignite. Failure to follow these instructions may result in personal injury.

Do not smoke or carry lighted tobacco or open flame of any type when working on or near any fuel related components. Highly flammable vapors are always present and may ignite. Failure to follow these instructions may result in personal injury.

If fuel contacts the eyes, flush the eyes with cold water or eyewash solution and seek immediate medical attention.

This procedure involves fuel handling. Be prepared for fuel spillage at all times and always observe fuel handling precautions. Failure to follow these instructions may result in personal injury.

Wash hands thoroughly after fuel handling, as prolonged contact may cause irritation. Should irritation develop, seek medical attention.

CAUTIONS:

Diesel fuel injection equipment is manufactured to very precise tolerances and fine clearances. It is therefore essential that absolute cleanliness is observed when working with these components. Always install blanking plugs to any open orifices or lines. Failure to follow this instruction may result in foreign matter ingress to the fuel injection system.

Make sure the workshop area in which the vehicle is being worked on is as clean and as dust free as possible. Foreign matter from work on clutches, brakes or from machining or welding operations can contaminate the fuel system and may result in later malfunction.

1. WARNING: Do not work on or under a vehicle supported only by a jack. Always support the vehicle on safety stands.

Raise and support the vehicle.

2. Disconnect the battery ground cable.

Refer to: Startup Battery Disconnect and Connect (414-01 Battery, Mounting and Cables, General Procedures)

3. Remove the fuel tank.

Refer to: Fuel Tank (310-01 Fuel Tank and Lines, ID4 2.2L Diesel Removal and Installation)

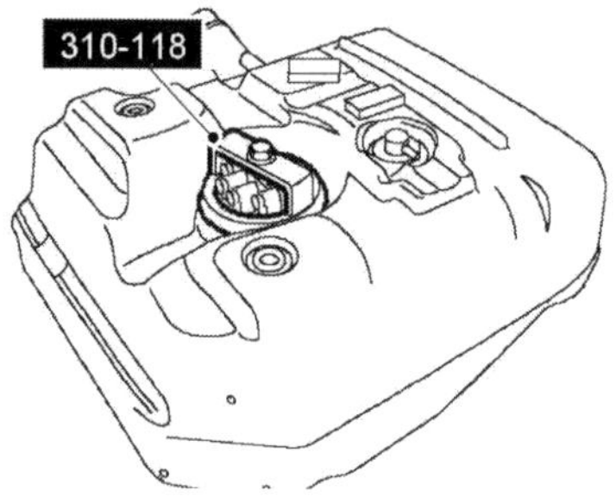

4. Remove the fuel level sender. Using the special tool, remove the locking ring. Remove and discard the seal.

   Special Tool(s): 310-118

Installation

1. NOTE: Clean the component mating faces. Install the fuel level sender. Install a new seal. Using the special tool, tighten the locking ring.

   Special Tool(s): 310-118

   Renew Part: Fuel pump O-ring seal

   Torque: 35 Nm

2. Install the fuel tank.

Refer to: Fuel Tank (310-01 Fuel Tank and Lines, ID4 2.2L Diesel Removal and Installation)

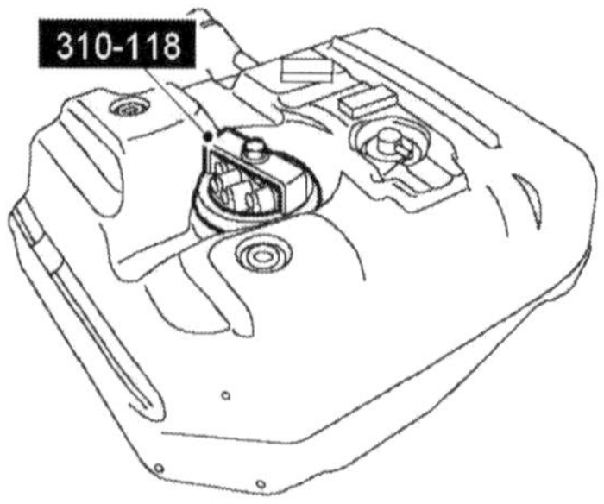

3. Connect the battery ground cable.

Refer to: Startup Battery Disconnect and Connect (414-01 Battery, Mounting and Cables, General Procedures)

Fuel Tank

Removal and Installation

Special Tool(s)

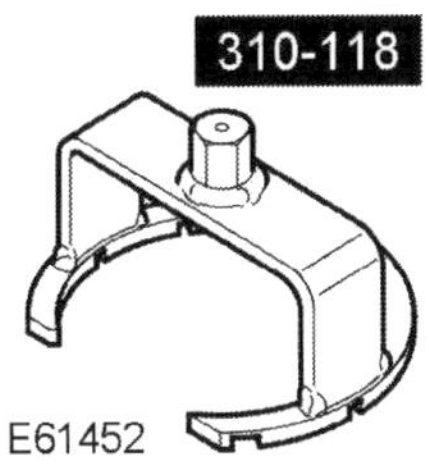 E61452	310-118 Fuel Tank Sender Wrench

Part(s)

STEP	PART NAME	QUANTITY
Installation Step 1	Renew Part Fuel pump O-ring seal	1

Removal

WARNINGS:

After carrying out repairs, the fuel system must be checked visually for leaks. Failure to follow this instruction may result in personal injury.

Do not carry or operate cellular phones when working on or near any fuel related components. Highly flammable vapors are always present and may ignite. Failure to follow these instructions may result in personal injury.

Do not smoke or carry lighted tobacco or open flame of any type when working on or near any fuel related components. Highly flammable vapors are always present and may ignite. Failure to follow these instructions may result in personal injury.

If fuel contacts the eyes, flush the eyes with cold water or eyewash solution and seek immediate medical attention.

This procedure involves fuel handling. Be prepared for fuel spillage at all times and always observe fuel handling precautions. Failure to follow these instructions may result in personal injury.

Wash hands thoroughly after fuel handling, as prolonged contact may cause irritation. Should irritation develop, seek medical attention.

CAUTIONS:

Diesel fuel injection equipment is manufactured to very precise tolerances and fine clearances. It is therefore essential that absolute cleanliness is observed when working with these components. Always install blanking plugs to any open orifices or lines. Failure to follow this instruction may result in foreign matter ingress to the fuel injection system.

Make sure the workshop area in which the vehicle is being worked on is as clean and as dust free as possible. Foreign matter from work on clutches, brakes or from machining or welding operations can contaminate the fuel system and may result in later malfunction.

1. WARNING: Do not work on or under a vehicle supported only by a jack. Always support the vehicle on safety stands.

Raise and support the vehicle.

2. Disconnect the battery ground cable.

Refer to: Startup Battery Disconnect and Connect (414-01 Battery, Mounting and Cables, General Procedures)

3. WARNING: The spilling of fuel is unavoidable during this operation. Make sure that all necessary precautions are taken to prevent fire and explosion.

   Drain the fuel tank.

Refer to: Fuel Tank Draining (310-00 Fuel System - General Information, General Procedures)

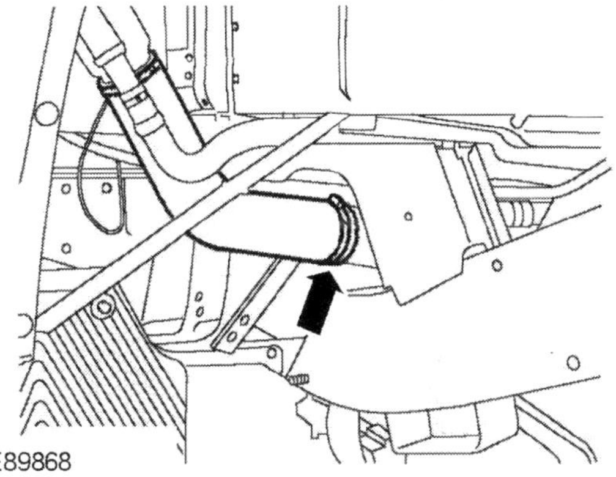

E89868

4. CAUTION: Before disconnecting any components, make sure the area is clean and free from foreign material. When disconnected all openings must be sealed.

Disconnect the fuel tank filler pipe. Loosen the clip.

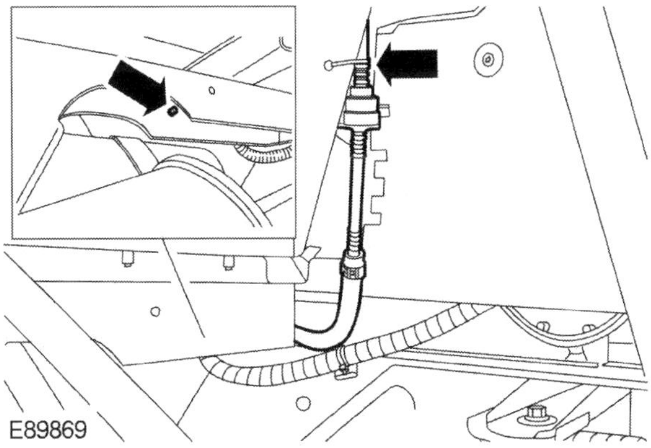

E89869

5. Release the fuel tank vent line. Release from the clip. Cut the cable tie.

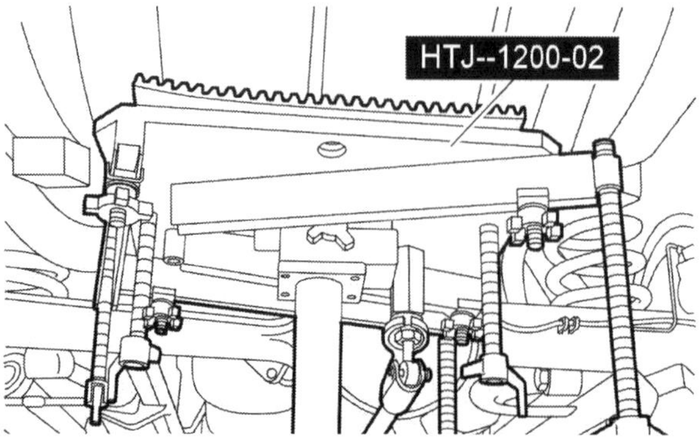

6. Using the special tool, support the fuel tank.

   General Equipment: Transmission jack

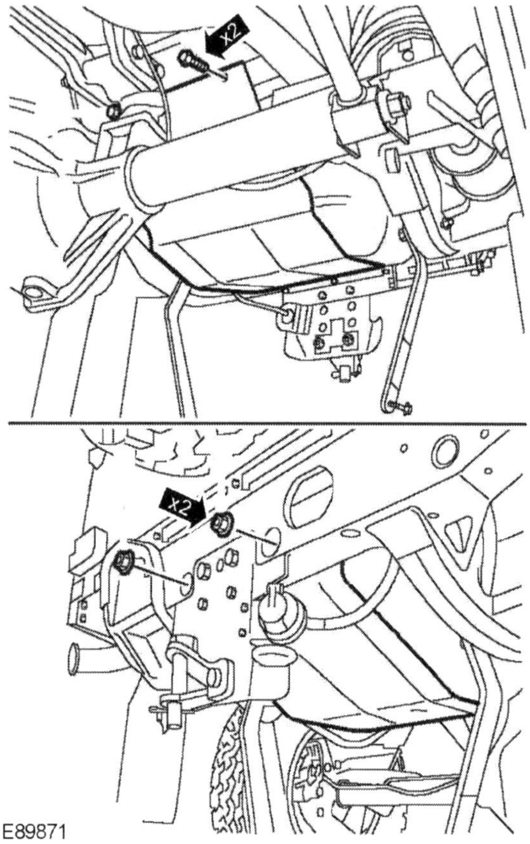

E89871

7. Lower the fuel tank. Remove the 2 bolts. Remove the 2 nuts.

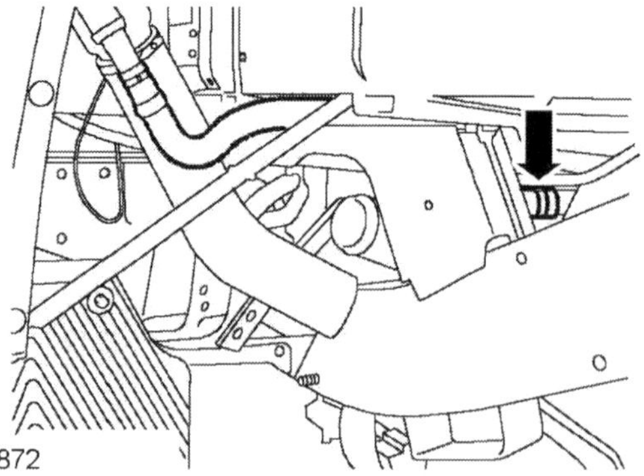

E89872

8. CAUTION: Before disconnecting any components, make sure the area is clean and free from foreign material. When disconnected all openings must be sealed.

   Disconnect the fuel filler pipe breather hose. Release the clip.

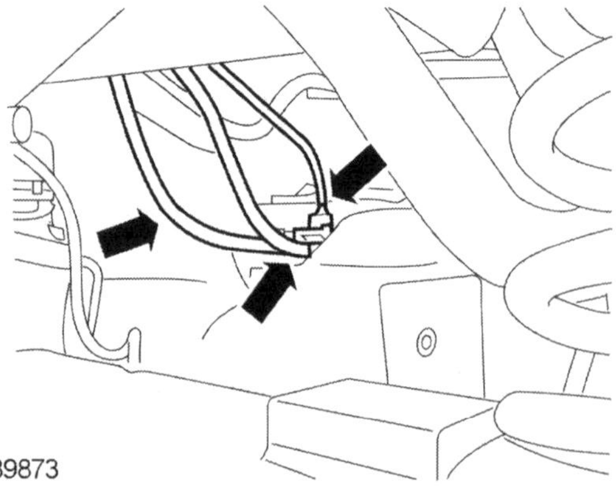

E89873

9. CAUTION: Before disconnecting any components, make sure the area is clean and free from foreign material. When disconnected all openings must be sealed.

   NOTE: Note the fitted position of fuel lines.

   Remove the fuel tank and the fuel tank support bracket. Disconnect the 2 fuel lines. Disconnect the electrical connector.

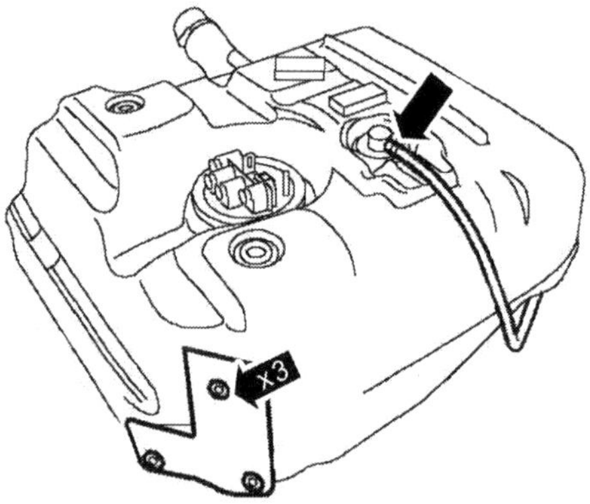

E89874

10. NOTE: Do not disassemble further if the component is removed for access only.

Remove the fuel tank heat shield. Remove the 3 clips.

11. Remove the fuel tank vent line. Release the clip.

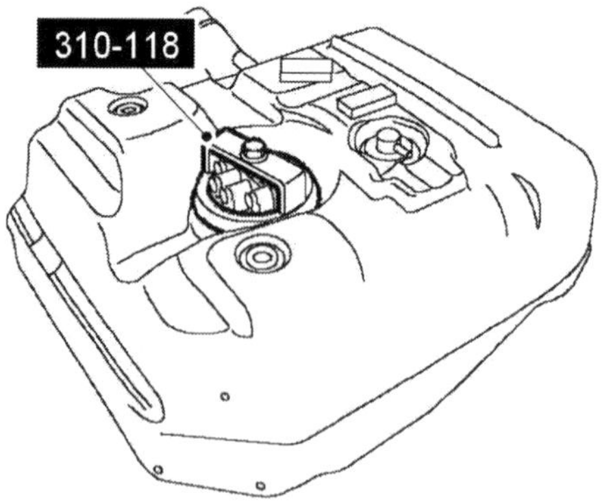

E89875

12. Remove the fuel level sender. Using the special tool, remove the locking ring. Remove and discard the seal.

    Special Tool(s): 310-118

Installation

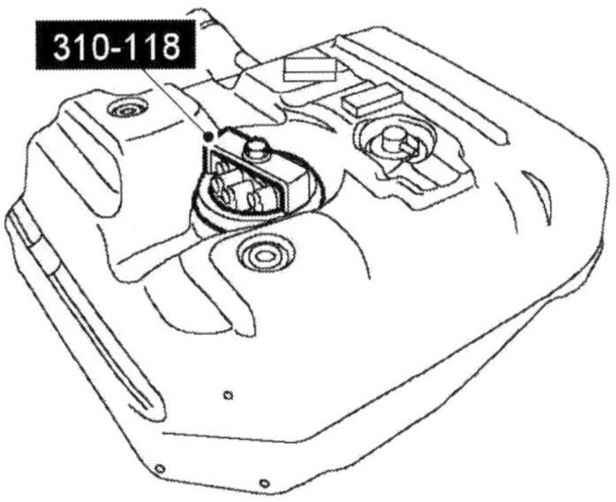

E89875

1. NOTE: Clean the component mating faces.
   Install the fuel level sender. Install a new seal. Using the special tool, tighten the locking ring.

   Special Tool(s): 310-118

   Renew Part: Fuel pump O-ring seal

   Torque: 35 Nm
2. Install the fuel tank vent line. Secure with the clip.
3. Install the fuel tank heat shield. Install the 3 clips.
4. NOTE: Remove and discard the blanking caps.
   Using the special tool, raise the fuel tank and the fuel tank support bracket. Connect the 2 fuel lines. Connect the electrical connector.
5. Connect the fuel filler pipe breather hose. Secure with the clip.
6. Using the special tool, install the fuel tank and the fuel tank support bracket. Tighten the bolts. Tighten the nuts.

   Bolts 45 Nm

   Nuts 25 Nm
7. Secure the fuel tank vent line. Secure in the clip. Secure with a suitable cable tie.
8. Connect the fuel tank filler pipe. Tighten the clip.
9. WARNING: The spilling of fuel is unavoidable during this operation. Make sure that all necessary precautions are taken to prevent fire and explosion.
   Refill the fuel tank.
10. Connect the battery ground cable.

Refer to: Startup Battery Disconnect and Connect (414-01 Battery, Mounting and Cables, General Procedures)

Fuel Tank Filler Pipe

Removal and Installation

Removal

1. Drain fuel tank.

Refer to: Fuel Tank Draining (310-00 Fuel System - General Information, General Procedures)

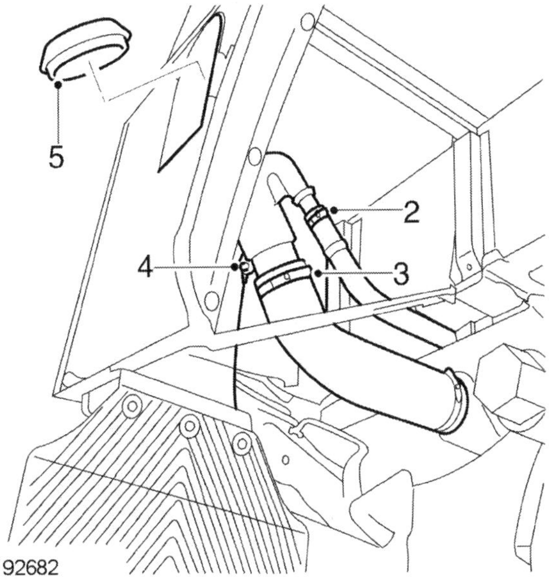

M192682

2. Loosen clip securing breather hose to fuel filler neck and release hose.
3. Loosen clip securing fuel filler hose to neck and release hose.
4. Remove screw and release earth lead from filler neck.
5. Remove grommet securing filler neck to body.
6. Remove filler neck from body.

Installation

1. Fit filler neck to body.
2. Coat rubber grommet with soap solution.
3. Fit rubber grommet securing filler neck to body.
4. Connect earth lead and tighten screw.
5. Connect filler hose to neck and tighten clip.
6. Fit breather hose to filler neck and secure with clip.

General Information

Specifications

Lubricant

Item	Specification
Compressor oil	SP-10
Compressor	4.4 FL. OZ

System Refrigerant Specification/Capacity

Item	Specification
Refrigerant type	R134A
Total system capacity	600g +/- 25g

Description and Operation

COMPONENT LOCATION

△ NOTE: RHD (Right-hand drive) vehicle shown, LHD (left-hand drive) similar.

Item	Description
1	Air inlet duct
2	Air intake and blower assembly
3	Heater assembly and controls
4	A/C (air conditioning) thermostatic switch
5	Water valve and bracket assembly
6	A/C refrigerant pressure switch
7	A/C evaporator
8	A/C thermostatic expansion valve
9	A/C receiver drier
10	A/C condenser assembly
11	A/C compressor

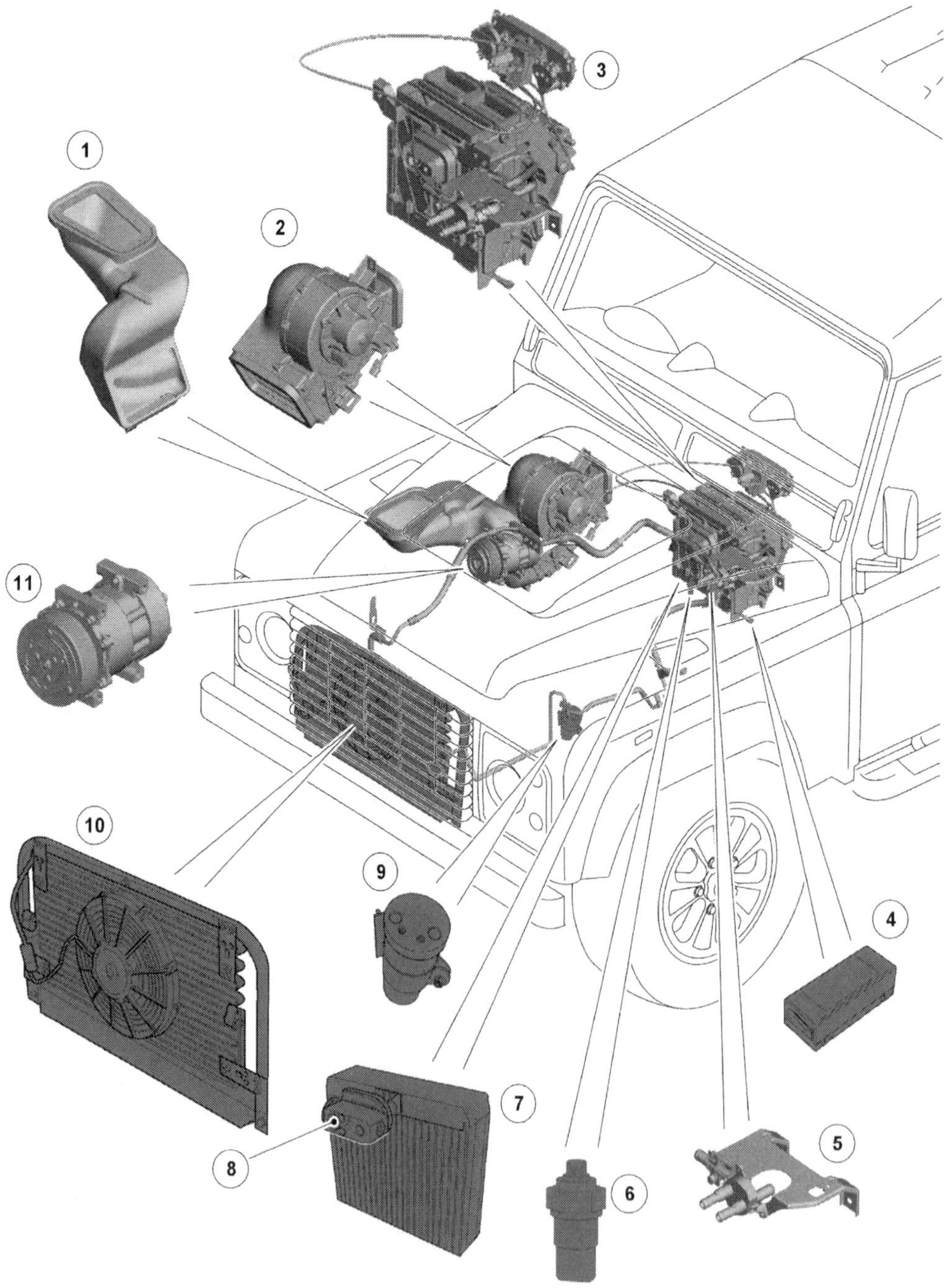

E83737

OVERVIEW

The climate control system features heating, and where fitted, A/C operations. Heating selections are made using the LH (left-hand) rotary control mounted on the center console. The rotary control is connected via a Bowden cable to a water valve in the engine cooling system. The water valve allows a varying amount of engine coolant to flow into the heater matrix. The amount of coolant entering the heater matrix determines the heat transferred to the inlet air and thus the amount of heat transferred into the cabin.

Blower motor speed is controlled using a 4 position sliding switch. The switch allows 3 blower motor speeds to be selected by controlling the ground path for the blower motor in conjunction with a resistor pack. The blower motor and resistor pack are both located in the engine compartment.

Air distribution into the cabin is selected using the RH (right-hand) rotary control mounted on the center console. Air intake into the cabin is controlled by a 2 position sliding switch which is also located on the center console.

Refer to: Air Distribution and Filtering (412-01 Air Distribution and Filtering, Description and Operation)

The A/C system is switched on and off using the switch located in the center console. The system will only operate when the engine is running. If A/C is requested when blower motor speed is set to 0, blower motor speed 1 will be engaged automatically.

The ECM (engine control module) monitors the engine running status and suspends A/C operation if necessary by inhibiting the compressor clutch relay.

The heated seats, heated windshield, and heated rear window systems are not part of the climate control system.

Refer to: Seats (501-10, Description and Operation).

Refer to: Glass, Frames and Mechanisms (501-11, Description and Operation).

AIR INLET DUCT

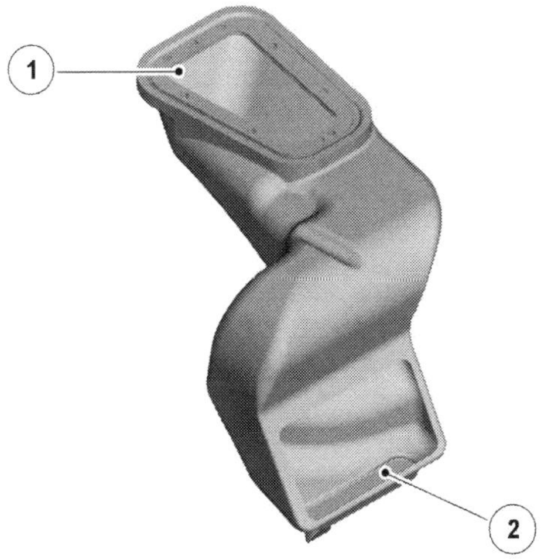

E83738

Item	Description
1	Air inlet
2	Air outlet to blower assembly

The air inlet duct is mounted in the passenger side front fender. Air is drawn into the air inlet duct through a vent on the upper surface of the fender. The air inlet duct directs air into the air intake and blower assembly. The air intake and blower assembly features a Bowden cable operated flap which allows the user to switch between fresh or recirculated air entering the cabin.

AIR INTAKE AND BLOWER ASSEMBLY

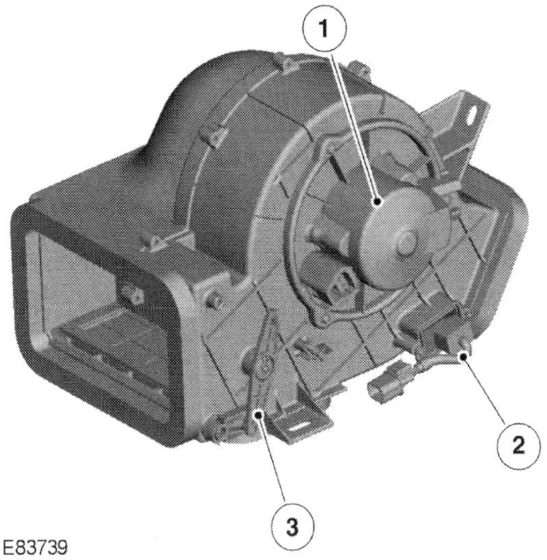

E83739

Item	Description
1	Blower motor
2	Resistor pack
3	Fresh/recirculated air flap

The air intake and blower assembly is mounted on the passenger side of the engine compartment. The blower comprises a centrifugal fan powered by an electric motor. The motor is able to operate at 3 speeds, which are controlled by a resistor pack mounted on the inboard side of the blower motor casing. Air leaving the air intake and blower assembly is directed into the rear of the heater assembly.

HEATER ASSEMBLY

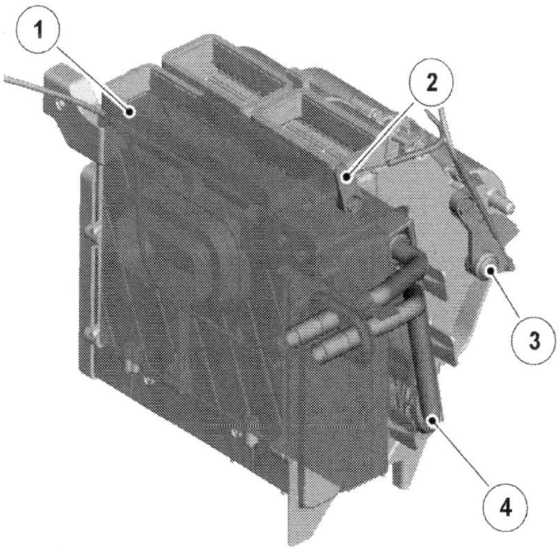

E83740

Item	Description
1	Diffuser plate
2	Windshield/face level air distribution door cam
3	Footwell air distribution door cam
4	Heater matrix

The heater assembly is mounted behind the instrument panel on the vehicle center line. The heater assembly contains an aluminum heater matrix and a diffuser plate to deliver the required air temperature into the cabin. If the vehicle is fitted with A/C, the A/C evaporator replaces the diffuser plate in the heater assembly.

Air distribution into the cabin is controlled by 2 air distribution doors. The doors are operated via a Bowden cable from the RH rotary control on the center console.

WATER VALVE

The water valve is mounted at the rear of the engine compartment. A Bowden cable acts on a lever on the rear face of the water valve to vary the flow of hot engine coolant into the heater matrix. When the water valve is fully closed, no engine coolant is allowed to flow into the heater matrix. In this instance, engine coolant will flow into the water valve through inlet (3) and immediately out through the return (4).

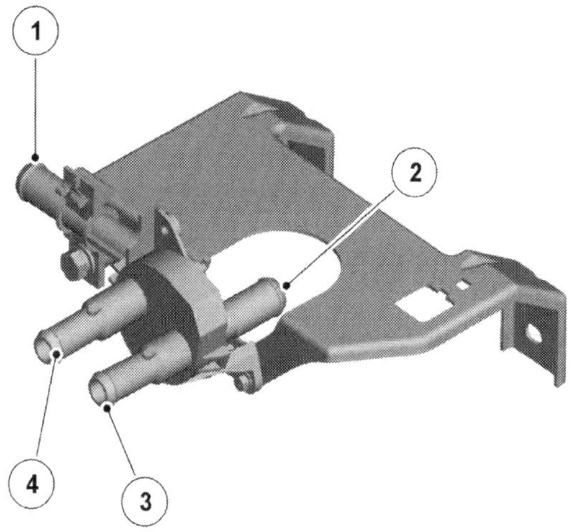

E83741

Item	Description
1	Return from heater matrix
2	Supply to heater matrix
3	Supply from engine cooling system
4	Return to engine cooling system

AIR CONDITIONING COMPRESSOR

The A/C compressor circulates refrigerant around the system by compressing low pressure, low temperature vapor from the evaporator and discharging the resultant high pressure, high temperature vapor to the condenser.

The A/C compressor is a constant displacement unit which is driven by the engine accessory drive belt. To protect the system from excessive pressure, a pressure relief valve is installed in the outlet side of the A/C compressor. The pressure relief valve vents excess pressure into the engine compartment.

Compressor clutch engagement is controlled by the ECM. The ECM receives system pressure inputs from the refrigerant pressure switch and system temperature inputs from the thermostatic switch.

The ECM will de-energize the A/C compressor clutch relay on receipt of a pressure reading above or below the system limits, or a temperature reading that may cause the evaporator to freeze. For more information, refer to the '**Air Conditioning Refrigerant Pressure Switch**' and '**Thermostatic Switch**' sections.

AIR CONDITIONING CONDENSER

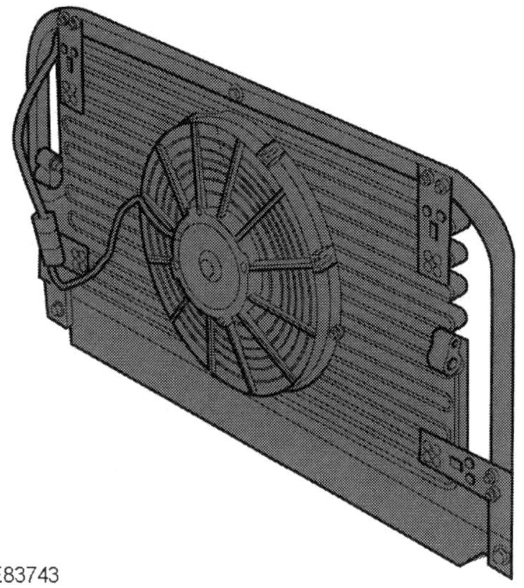

E83743

The condenser transfers heat from the refrigerant to the surrounding air to convert the high pressure vapor from the compressor into a liquid. The condenser is installed immediately in front of the radiator and is held in place by 3 brackets; 1 at the top, 2 at the bottom.

The RH end of the condenser provides a connection to the high pressure line from the A/C compressor. The LH end of the condenser provides a connection to the low pressure line to the receiver drier.

AIR CONDITIONING RECEIVER DRIER

The receiver drier is mounted in the LH front corner of the engine compartment and removes solid impurities and moisture from the refrigerant. It also provides a reservoir for liquid refrigerant to accommodate changes of heat load at the evaporator.

Refrigerant entering the receiver drier passes through a filter and a desiccant pack, then collects in the base of the unit before flowing through the outlet pipe to the evaporator.

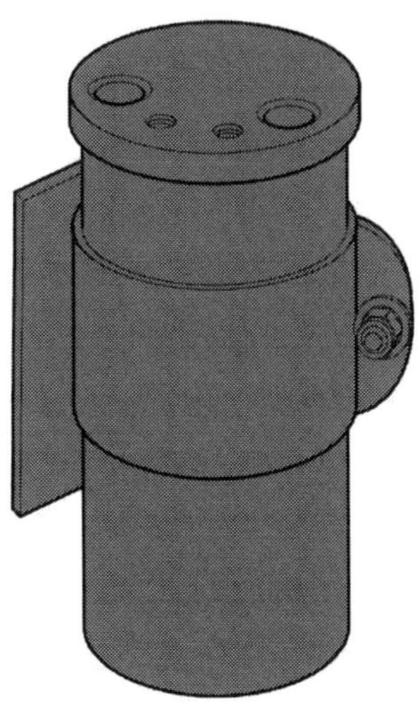

E83744

AIR CONDITIONING EVAPORATOR

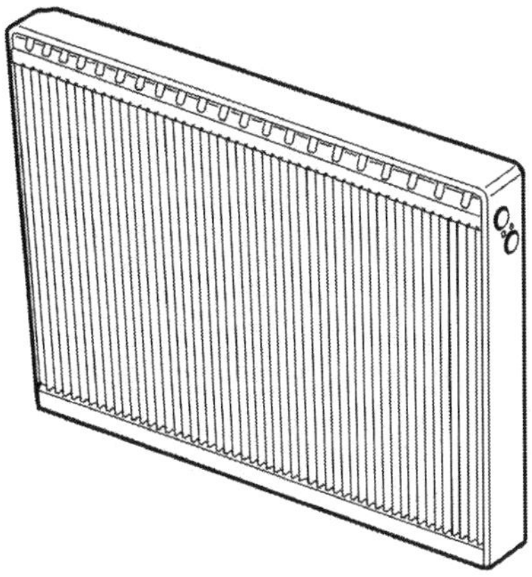

E46923

The evaporator is installed in the heater assembly. Liquid refrigerant enters the evaporator and expands, absorbing large quantities of heat as it changes state to a gas.

Most of the moisture in the air passing through the evaporator condenses into water, which drains out of the vehicle by passing through a drain tube to the underside of the vehicle.

AIR CONDITIONING THERMOSTATIC EXPANSION VALVE

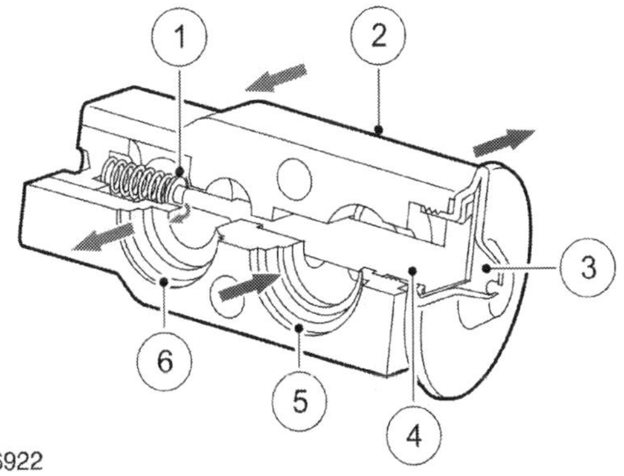

E46922

Item	Description
1	Metering valve
2	Housing
3	Diaphragm
4	Temperature sensor
5	Outlet passage from evaporator
6	Inlet passage to evaporator

The thermostatic expansion valve meters the flow of refrigerant into the evaporator, to match the refrigerant flow with the heat load of the air passing through the evaporator.

The thermostatic expansion valve is a block type valve located behind the heater assembly, and attached to the inlet and outlet ports of the evaporator. The thermostatic expansion valve consists of an aluminum housing containing inlet and outlet passages. A ball and spring metering valve is installed in the inlet passage and a temperature sensor is installed in the outlet passage. The temperature sensor consists of a temperature sensitive tube connected to a diaphragm. The bottom end of the temperature sensitive tube acts on the ball of the metering valve. Pressure on top of the diaphragm is controlled by the evaporator outlet temperature conducted through the temperature sensitive tube. The bottom of the diaphragm senses evaporator outlet pressure.

Liquid refrigerant flows through the metering valve into the evaporator. The restriction across the metering valve reduces the pressure and temperature of the refrigerant. The restriction also changes the liquid stream of refrigerant into a fine spray, to improve the evaporation process. As the refrigerant passes through the evaporator, it absorbs heat from the air flowing through the evaporator. The increase in temperature causes the refrigerant to vaporize and increase in pressure.

The temperature and pressure of the refrigerant leaving the evaporator acts on the diaphragm and temperature sensitive tube, which regulate the metering valve opening and so control the volume of refrigerant flowing through the evaporator. The warmer the air flowing through the evaporator, the more heat available to evaporate refrigerant and thus the greater volume of refrigerant allowed through the metering valve.

AIR CONDITIONING REFRIGERANT PRESSURE SWITCH

The refrigerant pressure switch is mounted monitors the high pressure side of the A/C system. The switch monitors 3 pressure ranges; high, medium and low. The pressure of refrigerant inside the system is provided to the ECM, which controls operation of the A/C compressor accordingly.

The pressure ranges monitored by the switch are as follows:

Pressure Range	Upper Limit (bar)	Lower Limit (bar)
High	32	6
Medium	15.5	12.5
Low	2.3	2

If the pressure in the system falls below 2 bar, or rises above 32 bar, the ECM will suspend A/C compressor operation by de-energizing the main relay, which subsequently de-energizes the A/C compressor clutch relay.

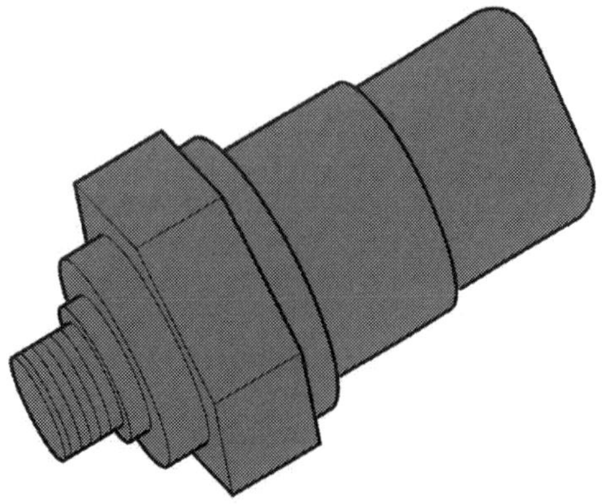

E83745

AIR CONDITIONING THERMOSTATIC SWITCH

The thermostatic switch is mounted on the underside of the heater assembly and measures the temperature of the air coming off the A/C evaporator. This temperature value is fed back to the ECM which controls operation of the A/C compressor accordingly.

E83746

The ECM will suspend A/C compressor operation if the temperature of the air coming off the evaporator drops to 1°C (34 °F). This prevents the evaporator from freezing. The ECM will re-instate A/C compressor operation when the temperature of the air coming off the evaporator rises to 3.5°C (38°F).

CONTROL DIAGRAM

△ NOTE: **A** = Hardwired

Item	Description
1	Ignition switch
2	-
3	Battery
4	Main relay
5	-
6	Cooling fan relay
7	Cooling fan
8	A/C compressor clutch relay
9	A/C compressor clutch
10	A/C refrigerant pressure switch
11	A/C switch
12	A/C thermostatic switch
13	Blower motor relay
14	Blower motor switch
15	Blower motor resistor pack
16	Blower motor
17	-

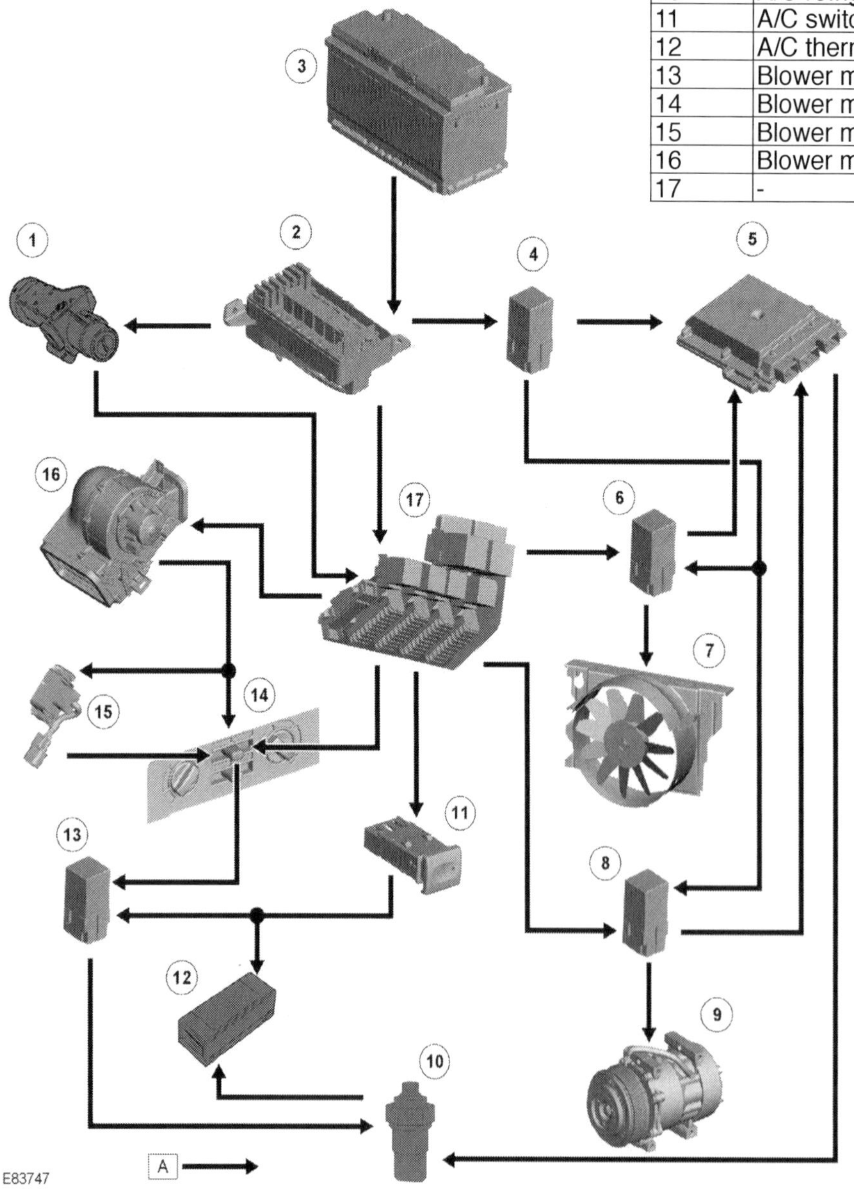

E83747

PRINCIPLES OF OPERATION

Heating

Heating control is achieved by varying the amount of engine coolant fed into the heater matrix. Coolant is bled from the engine cooling system and fed into the heater matrix through the water valve. The water valve is operated via a Bowden cable from the rotary heater control and varies the amount of engine coolant entering the heater matrix in response to heating demand.

If partial heating is requested, a proportion of engine coolant is returned back to the engine cooling system. If no heating is requested, the water valve returns all engine coolant back to the engine cooling system, by-passing the heater matrix.

Heat is transferred to the inlet air as it passes through the heater matrix. The heated air is then delivered into the cabin through a series of ducts, vents, and registers.

Refer to: Air Distribution and Filtering (412-01 Air Distribution and Filtering, Description and Operation)

Air Conditioning

The A/C system transfers heat from the cabin to the outside atmosphere to provide dehumidified cool air. The A/C system is a sealed closed loop system, filled with a charge weight of R134a refrigerant as the heat transfer medium. Oil is added to the refrigerant to lubricate the internal components of the A/C compressor.

Hot, high pressure gas from the compressor flows into the condenser, which allows heat to dissipate causing the gas to condense into a liquid form. The flow of liquid refrigerant into the evaporator is controlled by the thermostatic expansion valve. When in the evaporator the liquid refrigerant expands and absorbs heat from its surroundings, cooling the fins and plates of the A/C evaporator. As inlet air passes across the A/C evaporator surface it is cooled before entering the cabin through a series of ducts, vents, and registers.

Refer to: Air Distribution and Filtering (412-01 Air Distribution and Filtering, Description and Operation)

General Specification

Item	Description
Compressor:	
Make	Sandan
Type	SD7H15

Climate Control System

Diagnosis and Testing

Principles of Operation

For a detailed description of the Climate Control System, refer to the relevant Description and Operation section in the workshop manual.

Refer to: Climate Control System (412-00 Climate Control System - General Information, Description and Operation)

Refer to: Climate Control System (412-00 Climate Control System - General Information, Description and Operation)

Inspection and Verification

WARNING: Servicing must be carried out by personnel familiar with both vehicle system and the charging and testing equipment. All operations must be carried out in a well ventilated area away from open flame and heat sources.

CAUTION: Diagnosis by substitution from a donor vehicle is **NOT** acceptable. Substitution of control modules does not guarantee confirmation of a fault, and may also cause additional faults in the vehicle being tested and/or the donor vehicle.

NOTE: Check and rectify basic faults before beginning diagnostic routines involving pinpoint tests.

1. Visually inspect for obvious signs of damage and system integrity
2. If an obvious cause for an observed or reported concern is found, correct the cause (if possible) before proceeding to the next step
3. If the cause is not visually evident, verify the symptom and refer to the Symptom Chart, alternatively check for Diagnostic Trouble Codes (DTCs) and refer to the DTC Index

Visual Inspection

Mechanical	Electrical
• Coolant level	• Fuse(s)
• Hose(s)	• Wiring harness
• Coolant pump	• Electrical connectors
• Control flap(s)	• Blower
• Duct(s)	• Air conditioning compressor
• Vent(s)	• Electric cooling fan
• Cabin air filter	• HVAC control module (HVAC)
• Drive belt	• Refrigerant pressure sensor
• Air conditioning compressor	
• Thermostatic expansion valve	
• Evaporator	
• Receiver drier	
• Air conditioning condenser	
• Refrigerant pipes	
• Auxiliary drive belt	
• Fuel fired booster heater	
• Fuel fired booster heater fuel pump	
• Fuel fired booster heater fuel pipes	

General Precautions

The refrigerant used in the air conditioning system is HFC (Hydrofluorocarbon) R134a.

WARNINGS:

⚠ R134a is a hazardous liquid and when handled incorrectly can cause serious injury. Suitable protective clothing must be worn when carrying out servicing operations on the air conditioning system.

⚠ R134a is odourless and colourless. Do not handle or discharge in an enclosed area, or in any area where the vapour or liquid can come in contact with naked flame or hot metal. R134a is not flammable, but can form a highly toxic gas.

⚠ Do not smoke or weld in areas where R134a is in use. Inhalation of concentrations of the vapour can cause dizziness, disorientation. uncoordination, narcosis, nausea or vomiting.

⚠ Do not allow fluids other than R134a or compressor lubricant to enter the air conditioning system. Spontaneous combustion may occur.

⚠ R134a splashed on any part of the body will cause immediate freezing of that area. Also refrigerant cylinders and replenishment trolleys when discharging will freeze skin to them if contact is made.

⚠ The refrigerant used in an air conditioning system must be reclaimed in accordance with the recommendations given with a Refrigerant Recovery Recycling Recharging Station.

△ NOTE: Suitable protective clothing comprises: Wrap around safety glasses or helmet, heatproof gloves, rubber apron or waterproof overalls and rubber boots.

Routine servicing, apart from visual checks, is not necessary. The visual inspections are as follows:

Symptom Chart

Symptom	Possible Causes	Action
Air conditioning performance poor or inoperative	• Refrigerant undercharged • Refrigerant overcharged • Thermostatic expansion valve faulty • Receiver drier restricted • Water in refrigerant	• GO to Pinpoint Test A
Air conditioning operates briefly and then switches off	• Electric cooling fan inoperative • Air conditioning condenser airflow obstructed	• Test the operation of the electric cooling fan • Check the air conditioning condenser for external obstructions

Condenser

With a water hose or air line, clean the fins of the condenser to remove flies, leaves, etc. Check the pipe connections for signs of oil leakage.

Compressor

Check pipe connections for signs of oil leakage. Check flexible hoses for swelling. Examine the compressor belt for tightness and condition.

Sight glass

Examine the sight glass for bubbles with the system operating. Check connections for leakage.

Evaporator

Examine the refrigeration connections at the unit.

Remedial Actions

WARNINGS:

⚠ Do not allow a refrigerant container to be heated by a direct flame or to be placed near any heating appliance. A refrigerant container must not be heated above 50° C.

⚠ Do not leave a container of refrigerant without its cap fitted. Do not transport a container of refrigerant that is unrestrained, especially in the boot of a car.

△ NOTE: Due to its low evaporating temperature of -30° C, R134a should be handled with care.

- If liquid R134a strikes the eye, do not rub it. Gently run large quantities of eyewash over the eye to raise the temperature. If eyewash is not available cool, clean water may be used. Cover eye with clean pad and seek immediate medical attention.
- If liquid R134a is splashed on the skin run large quantities of water over the area as soon as possible to raise the temperature. Carry out the same actions if skin comes into contact with discharging cylinders. Wrap affected parts in blankets or similar material and seek immediate medical attention.
- If suspected of being overcome by inhalation of R134a vapour seek fresh air. If unconscious remove to fresh air. Apply artificial respiration and/or oxygen and seek immediate medical attention.

Servicing Precautions

Care must be taken when handling refrigeration system components. Units must not be lifted by their hoses, pipes or capillary lines. Hoses and lines must not be subjected to any twist or stress. Ensure that hoses are positioned in their correct run before fully tightening the couplings, and ensure that all clips and supports are used. Torque wrenches of the correct type must be used when tightening refrigerant connections to the stated value. An additional spanner must be used to hold the union to prevent twisting of the pipe.

Before connecting any hose or pipe ensure that refrigerant oil is applied to the seat of the new '0' ring but not to the threads.

Check the oil trap for the amount of oil lost.

All protective plugs on components must be left in place until immediately prior to connection.

The receiver/drier contains desiccant which absorbs moisture. It must be positively sealed at all times.

⚠ CAUTION: Whenever the refrigerant system is opened, the receiver/drier must be renewed immediately before evacuating and recharging the system.

Use alcohol and a clean cloth to clean dirty connections. Ensure that all new parts fitted are marked for use with R134a.

Refrigerant oil

Use the approved refrigerant lubricating oil - Nippon Denso ND-OIL 8.

⚠ CAUTION: Do not use any other type of refrigerant oil.

Refrigerant oil easily absorbs water and must not be stored for long periods. Do not pour unused oil back into the container. When renewing system components, add the following quantities of refrigerant oil:

Component	Fluid quantity
Condenser	40ml
Evaporator	80ml
Pipe or hose	20ml
Receiver/drier	20ml
Total quantity of refrigerant oil in system	140ml

A new compressor is sealed and pressurised with Nitrogen gas, slowly release the sealing cap, gas pressure should be heard to release as the seal is broken.

△ NOTE: A new compressor should always have its sealing caps in place and these must not be removed until immediately prior to fitting

A new compressor is supplied with an oil fill of 140ml. A calculated quantity of oil must be drained from a new compressor before fitting.

To calculate the quantity of oil to be drained:

- Remove sealing plugs from the OLD compressor.
- Invert compressor and gravity drain oil into measuring cylinder. Rotating the compressor clutch plate will assist complete draining.
- Note the quantity of oil drained (Yml).
- Calculate the quantity (Qml) of oil to be drained from the NEW compressor using the following formula: Xml - (Yml + 20ml) = Qml

Rapid refrigerant discharge

When the air conditioning system is involved in accident damage and the circuit is punctured, the refrigerant is discharged rapidly. The rapid discharge of refrigerant will also result in the loss of most of the oil from the system. The compressor must be removed and all the remaining oil in the compressor drained and refilled as follows:

- Gravity drain all the oil, assist by rotating the clutch plate (not the pulley).
- Refill the compressor with 90ml of new refrigerant oil.
- Plug the inlet and outlet ports.

Servicing Equipment

The following equipment is required for full servicing of the air conditioning system.

Recovery, recycling and charging station Leak detector Thermometer +20° C to -60° C Safety goggles and gloves.

Precautions when handling refrigerant lines

WARNING: Wear eye and hand protection when disconnecting components containing refrigerant. Plug all exposed connections immediately.

- When disconnecting any hose or pipe connection the system must be discharged of all pressure. Proceed cautiously, regardless of gauge readings. Open connections slowly, keeping hands and face well clear, so that no injury occurs if there is liquid in the line. If pressure is noticed, allow it to bleed off slowly.
- Lines, flexible end connections and components must be capped immediately they are opened to prevent the entrance of moisture and dirt.
- Any dirt or grease on fittings must be wiped off with a clean alcohol dampened cloth. Do not use chlorinated solvents such as trichloroethylene. If dirt, grease or moisture cannot be removed from inside the hoses, they must be replaced with new hoses.
- All replacement components and flexible end connections must be sealed, and only opened immediately prior to making the connection.
- Ensure the components are at room temperature before uncapping, to prevent condensation of moisture from the air that enters.
- Components must not remain uncapped for longer than 15 minutes. In the event of delay, the caps must be fitted.
- Receiver/driers must never be left uncapped as they contain Silica Gel crystals which will absorb moisture from the atmosphere. A receiver/ drier left uncapped must not be used, fit a new unit.
- The compressor shaft must not be rotated until the system is entirely assembled and contains a charge of refrigerant.
- A new compressor contains an initial charge of refrigerant oil. The compressor also contains a holding charge of gas when received which should be retained by leaving the seals in place until the pipes are re-connected.
- The receiver/drier should be the last component connected to the system to ensure optimum dehydration and maximum moisture protection of the system.
- All precautions must be taken to prevent damage to fittings and connections. Slight damage could cause a leak with the high pressures used in the system.
- Always use two wrenches of the correct size, one on each fitting when releasing and tightening refrigeration unions.
- Joints and 'O' rings should be coated with refrigeration oil to aid correct seating. Fittings which are not lubricated with refrigerant oil are almost certain to leak.
- All lines must be free of kinks. The efficiency of the system is reduced by a single kink or restriction.
- Flexible hoses should not be bent to a radius less than 90mm radius.
- Flexible hoses should not be within 100mm of the exhaust manifold.
- Completed assemblies must be checked for refrigeration lines touching metal panels. Any direct contact of lines and panels transmits noise and must be eliminated.

Performance Test

WARNING: R134A is hazardous. - Carry out this test with hood and doors or windows open, air conditioning switched on, temperature control set to cold and blower at maximum speed. Set the air supply control to supply fresh air.

1. Close low pressure valve on refrigerant station.
2. Close high pressure valve on refrigerant station.
3. Connect a Refrigerant Station to the high and low pressure servicing connections.
4. Insert dry bulb thermometer into cold air outlet and position dry and wet bulb thermometer close to outside air inlet. Do not spill water from the wet thermometer (psychrometer).
5. Start engine and run it at 1500 rev/min for 10 minutes with air conditioning switched on.
6. Read both pressure gauges and thermometers. Check readings against table below with humidity between 60% and 80%.
7. Switch off air conditioning, stop engine, disconnect test equipment.

Performance range

Intake temperature	Outlet temperature	Low pressure	High pressure
68°F	41 - 50°F	20 - 23 lbf/in²	116 - 190 lbf/in²
20°C	5 - 10°C	1.4 - 1.6 bar	8.0 - 13 bar
75°F	45 - 60°F	21 - 28 lbf/in²	130 - 200 lbf/in²
25°C	7 - 15°C	1.5 - 2.0 bar	9.0 - 14 bar
85°F	46 - 68°F	26 - 40 lbf/in²	190 - 230 lbf/in²
30°C	8 - 20°C	1.8 - 2.8 bar	13 - 16 bar
95°F	52 - 72°F		
35°C	11 - 22°C	2.5 - 3.5 bar	16 - 19 bar

Ambient temperature		Compound gauge readings		High pressure gauge readings	
°C	°F	bar	lbf/in²	bar	lbf/in²
16	60	1,3-1,4	15-20	6,9-10,3	100-150
27	80	1,4-1,7	20-25	9,7-13,1	140-190
38	100	1,7-2,1	25-30	12,4-15,5	180-225
43	110	2,1-2,4	30-35	14,8-17,2	215-250

Pinpoint Tests

PINPOINT TEST A : PRELIMINARY TESTS	
TEST CONDITIONS	**DETAILS/RESULTS/ACTIONS**
A1: PRELIMINARY TEST 1	
NOTES: △ This test is performed with the engine **not** running.	
△ Normal pressure for a correctly charged and switched off system is approximately 4.5 bar on both gauges (system equalised).	
	1. Close the valves on the air conditioning station
	2. Connect the air conditioning station to the vehicle charging ports
	3. Check the pressure values
	Is a pressure registered on both gauges? **Yes** GO to Pinpoint Test B **No** GO to Pinpoint Test D

PINPOINT TEST B : FUNCTIONALITY TESTS	
TEST CONDITIONS	**DETAILS/RESULTS/ACTIONS**
B1: FUNCTIONALITY TEST 1	
NOTES: △ Normal pressures for a correctly charged and working system are 1.0 bar to 2.0 bar (low) and 11.0 bar to 15.0 bar (high).	
△ Normal temperature (measured at the center air vent) for a correctly charged and working system is -7°C to -2°C when ambient temperature is 20°C.	
	1. Close the valves on the air conditioning station
	2. Connect the air conditioning station to the vehicle charging ports
	3. Open all doors and the tailgate
	4. Start the engine
	5. Set the temperature to the lowest setting (all zones)
	6. Set the fan speed to maximum
	7. Set the recirculate switch to off
	8. Insert a temperature probe into the centre air vent
	9. Raise engine speed to 1500rpm and maintain this speed for 5 minutes
	10. Check the temperature value
	11. Check the pressure values
	Are the pressure readings stable and in the green 'normal' region of the gauge? **Yes** Air conditioning system operating normally **No** Air conditioning system fault present. GO to Pinpoint Test C

PINPOINT TEST C : GAUGE TESTS

TEST CONDITIONS	DETAILS/RESULTS/ACTIONS
C1: GAUGE TEST 1	
NOTE: This test is performed with the engine running and the air conditioning set to on.	
	1. Check the pressure values
	Did the gauges register a change in pressure when the air conditioning was switched on? **Yes** GO to 2 **No** Using the manufacturer approved diagnostic system, check the HVAC control module (HVAC) for related DTCs and refer to the relevant DTC index
C2: GAUGE TEST 2	
NOTE: This test is performed with the engine running and the air conditioning set to on.	
	1. Check the pressure values
	Are the pressure gauge readings fluctuating? **Yes** Moisture present in the air conditioning system. Recover the refrigerant. Install a new receiver drier. Refer to the relevant section of the workshop manual and evacuate and recharge the air conditioning system. GO to Pinpoint Test B **No** GO to 3
C3: GAUGE TEST 3	
NOTES: This test is performed with the engine running and the air conditioning set to on.	
Normal pressures for a correctly charged and working system are 1.0 bar to 2.0 bar (low) and 11.0 bar to 15.0 bar (high).	
	1. Check the pressure values
	Are the pressure gauge readings too low? **Yes** GO to 4 **No** GO to 6
C4: GAUGE TEST 4	
NOTE: This test is performed with the engine **not** running.	
	1. Stop the engine
	2. Using the manufacturer approved refrigerant leak detector, check for a refrigerant leak

	Was a refrigerant leak detected? **Yes** Refer to the relevant section of the workshop manual and recover the refrigerant. Repair as necessary. Evacuate and recharge the air conditioning system. GO to Pinpoint Test B **No** GO to 5

C5: GAUGE TEST 5

NOTE: This test is performed with the engine **not** running.

	1. Refer to the relevant section of the workshop manual and recover the refrigerant
	Was the weight of the recovered refrigerant less than specified for the air conditioning system? **Yes** Refer to the relevant section of the workshop manual and evacuate and recharge the air conditioning system. GO to Pinpoint Test B **No** Install a new receiver drier. Refer to the relevant section of the workshop manual and evacuate and recharge the air conditioning system. GO to Pinpoint Test B

C6: GAUGE TEST 6

NOTES:

This test is performed with the engine running and the air conditioning set to on.

Normal pressures for a correctly charged and working system are 1.0 bar to 2.0 bar (low) and 11.0 bar to 15.0 bar (high).

	1. Check the pressure values
	Are the pressure gauge readings too high? **Yes** GO to 7 **No** Test inconclusive. GO to Pinpoint Test B

C7: GAUGE TEST 7

NOTE: This test is performed with the engine **not** running.

	1. Stop the engine
	2. Refer to the relevant section of the workshop manual and recover the refrigerant
	Was the weight of the recovered refrigerant more than specified for the air conditioning system? **Yes** Refer to the relevant section of the workshop manual and evacuate and recharge the air conditioning system. GO to Pinpoint Test B **No** Install a new thermal expansion valve. Refer to the relevant section of the workshop manual and evacuate and recharge the air conditioning system. GO to Pinpoint Test B

PINPOINT TEST D : NITROGEN LEAK TESTS	
TEST CONDITIONS	**DETAILS/RESULTS/ACTIONS**
D1: NITROGEN LEAK TEST	
CAUTION: When charging the system with nitrogen, the pressure should be regulated to 7.0 bar.	
NOTE: This test is performed with the engine **not** running.	
	1. Charge the air conditioning system with nitrogen
	2. Isolate the nitrogen supply
	3. Monitor the pressure gauge and check for leaks
	Has the source of the leak been identified? **Yes** Rectify the leak as necessary. Install a new receiver drier. Refer to the relevant section of the workshop manual and evacuate and recharge the air conditioning system. GO to Pinpoint Test B **No** Refer to the relevant section of the workshop manual and evacuate and recharge the air conditioning system. GO to Pinpoint Test B

DTC Index

For a list of Diagnostic Trouble Codes (DTCs) that could be logged on this vehicle, please refer to Section 100-00.

Air Conditioning System Recovery, Evacuation and Charging

General Procedures

WARNINGS:

The air conditioning system is charged with a high pressure, potentially toxic refrigerant. Repairs or servicing must only be carried out by an operator familiar with both the vehicle system and the charging and testing equipment. All operations must be carried out in a well-ventilated area away from open flame and heat sources. Always wear safety goggles and gloves when opening refrigerant connections.

Wear eye and hand safety protection. Open connections slowly in case liquid or pressure is present. Allow to bleed off slowly.

CAUTION: Overcharging air conditioning system will cause excessive head pressure.

1. Remove dust caps from high and low pressure connectors.
2. Connect high and low pressure hoses to appropriate connections.
3. Open valves on connectors.
4. Turn valves on refrigerant station to correct positions.
5. Turn Process switch to correct position.
6. Turn Main switch to ‘ON’.
7. Allow refrigerant station to recover refrigerant from system.
8. Close valves on refrigerant station.
9. Turn Main switch to ‘OFF’.
10. Close valves on connectors.
11. Disconnect high and low pressure hoses from connectors.
12. Fit dust caps to connectors.
13. Open tap at rear of station to drain refrigerant oil recovered from system.
14. Measure and record quantity of refrigerant oil recovered from system.
15. Close tap at rear of station.

16. WARNING: Refrigerant must always be recycled before reuse, to ensure that the purity of the refrigerant is high enough for safe use in the air conditioning system. Recycling should always be carried out with equipment which is design certified by Underwriter Laboratory Inc. for compliance with SAE-J1991. Other equipment may not recycle refrigerant to the required level of purity. A R134a Refrigerant Recovery Recycling Recharging Station must not be used with any other type of refrigerant. Refrigerant R134a from domestic and commercial sources must not be used in motor vehicle air conditioning systems.

Measure the quantity of oil discharged from the system. Add an equal amount of new refrigerant oil to compressor before evacuation sequence.

17. CAUTIONS:

When a major repair has been carried out, a leak test should be carried out using inert gas.

Whenever the refrigerant system is opened, the receiver/drier must be renewed immediately before evacuating and recharging the system.

Renew the receiver/drier.

18. Remove dust caps from high and low pressure connectors.
19. Connect high and low pressure hoses to appropriate connections.
20. Open valves on connectors.
21. Turn valves on refrigerant station to correct positions.
22. Turn Process switch to correct position.
23. Turn Main switch to 'ON'.
24. Allow refrigerant station to evacuate system.

CAUTION: The system must be Evacuated immediately before recharging commences. Delay between Evacuation and Recharging is not permitted.

25. If the vacuum reading is below 700 mm Hg, 28 in Hg after 15 minutes, suspect a leak in the system. Partially recharge the system and check for leaks using an electronic leak tester. Check suction lines first, then run the compressor for 5 minutes and then check the high pressure lines.

26. WARNINGS:

A R134a Refrigerant Recovery Recycling Recharging station must not be used with any other type of refrigerant.

R134a refrigerant from domestic and commercial sources must not be used in motor vehicle air conditioning systems.

CAUTION: When a major repair has been carried out, a leak test should be carried out using inert gas.

Close valves on refrigerant station.

27. Close valve on oil charger.
28. Disconnect yellow hose from refrigerant station.
29. Remove lid from oil charger.
30. Pour correct quantity of refrigerant oil into oil charger.
31. Fit lid to oil charger.
32. Connect yellow hose to refrigerant station.
33. Open valve on oil charger.
34. Move pointer on refrigerant gauge to mark position of refrigerant drop.
35. Slowly open correct valve on refrigerant station and allow vacuum to pull refrigerant into system.
36. Close valve on refrigerant station when correct amount of refrigerant has been drawn into air conditioning system.
37. If the full charge is not accepted by the system, start the engine and run it at 1,500 rev/min for a minimum of 2 minutes. Switch on the air conditioning system, open the vehicle windows, set the temperature control to cold and the blower switch to maximum.
38. Consult Refrigerant station instruction manual for correct procedure to complete the charge.
39. Turn Main switch to 'OFF'.
40. Close valves on connectors.
41. Disconnect high and low pressure hoses from connectors.
42. Fit dust caps to connectors.
43. Carry out performance test on air conditioning system.

Refrigerant System Tests

General Procedures

1. Place the vehicle in a ventilated, shaded area free from excessive draught, with the doors and windows open.
2. Check that the surface of the condenser is not restricted with dirt, leaves, flies, etc. Do not neglect to check the surface between the condenser and the radiator. Clean as necessary.
3. Switch on the ignition and the air conditioner air flow control. Check that the blower is operating efficiently at low, medium and high speeds. Switch off the blower and the ignition.
4. Check that the evaporator drain tube is open and clear.
5. Check the tension of the compressor driving belt, and adjust if necessary.
6. Inspect all connections for the presence of refrigerant oil. If oil is evident, check for leaks, and repair as necessary.

NOTE: The compressor oil is soluble in Refrigerant R134a and is deposited when the refrigerant evaporates from a leak.

7. Start the engine.
8. Set the temperature controls to cold and switch the air conditioner blower control on and off several times, checking that the magnetic clutch on the compressor engages and releases each time.
9. With the temperature control at maximum cooling and the blower control at high speed, warm up the engine and fast idle at 1000 rev/min.
10. Repeat at 1800 rev/min.
11. Gradually increase the engine speed to the high range and check the sight glass at intervals.
12. Check for frosting on the service valves.
13. Check the high pressure hoses and connections by hand for varying temperature. Low temperature indicates a restriction or blockage at that point.
14. Switch off the air conditioning blower and stop the engine.
15. If the air conditioning equipment is still not satisfactory, carry out a pressure test as previously described in this section

Electronic Leak Detection

General Procedures

CAUTION: When a major repair has been carried out, a leak test should be carried out using an inert gas (see below).

1. Place the vehicle in a well ventilated area but free from draughts, as leakage from the system could be dissipated without detection.
2. Follow the instructions issued by the manufacturer of the particular leak detector being used.
3. Commence searching for leaks by passing the detector probe around all joints and components, refrigerant gas is heavier than air.
4. Insert the probe into an air outlet of the evaporator or into the evaporator drain tube. Switch the air conditioning blower on and off at intervals of ten seconds. Any leaking refrigerant will be gathered in by the blower and detected.
5. Insert the probe between the magnetic clutch and compressor to check the shaft seal for leaks.
6. Check all service valve connections, valve plate, head and base plate joints and back seal plate.
7. Check the condenser for leaks at the pipe unions.
8. If any leaks are found, the system must be discharged before rectification.
9. Rectify any leaks and recheck for leaks during evacuation prior to charging.
10. **Leak test using inert gas**
    1. Connect gas line to recharging station.
    2. Pressurize system to 3 bar (45 lbf/in^2).
    3. Carry out leak test as above.

Air Conditioning Compressor Commissioning

General Procedures

Activation

CAUTION: Failure to follow this instruction may result in damage to the component.

1. Set the ignition to the on position, make sure the Air Conditioning (A/C) is in the off position.
2. Start the engine and allow to run for a minimum of 5 minutes.
3. Set the heater controls to 22°C, with the fan speed set to 75%.
4. Switch on the A/C system.
5. Open all air vents in the instrument panel.
6. Run the A/C system for a minimum of 5 minutes, while the engine is still at idle speed.
7. Once this is achieved the compressor is stabilized, with the oil being distributed evenly throughout the system.

Description and Operation

COMPONENT LOCATION

NOTE: LHD (left-hand drive) vehicle shown, RHD (right-hand drive) similar.

Item	Description
1	LH (left-hand) side window vent
2	Windshield LH vent
3	LH face level register
4	LH face level vent
5	RH (right-hand) face level vent
6	RH face level register
7	Windshield RH vent
8	RH side window vent
9	Recirculated air inlet
10	RH footwell vent
11	LH footwell vent

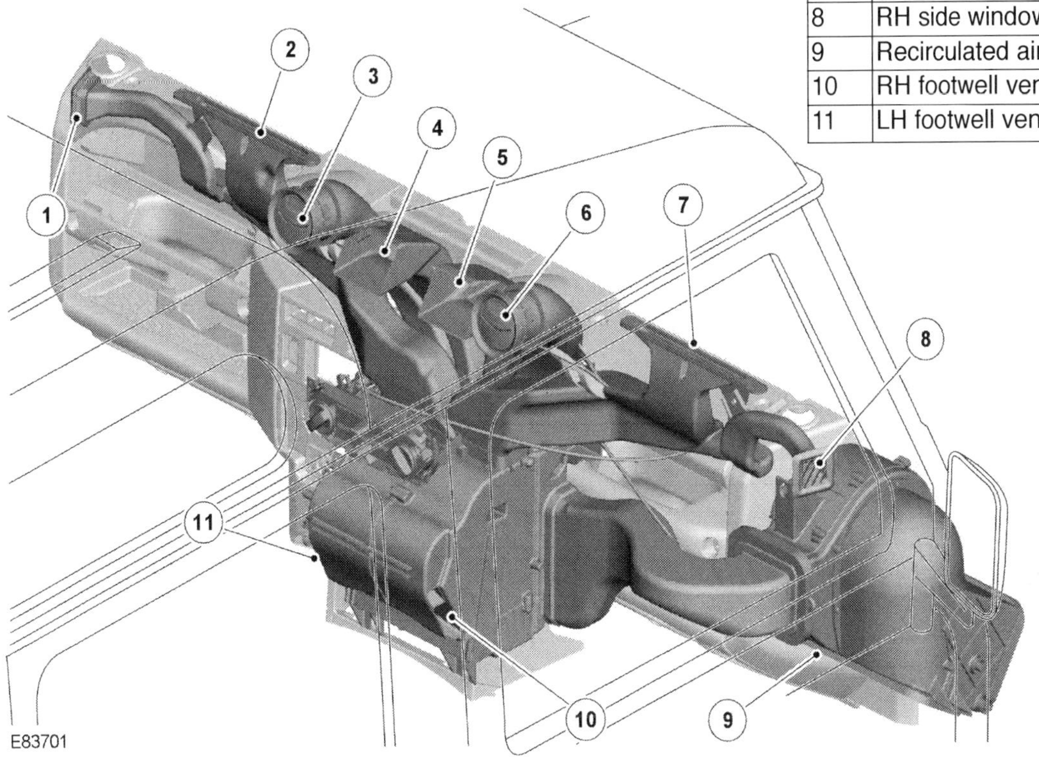

OVERVIEW

Air intake into the cabin is controlled using a 2 position sliding switch located on the center console. Either fresh or recirculated air can be selected. Selections are transmitted to the air intake and blower assembly using a Bowden cable. The air intake and blower assembly is located in the engine compartment.

Fresh air enters the system through the air inlet duct mounted in the passenger side front fender. Inlet air travels through the air intake and blower assembly to the heater assembly where it is distributed into the front of the cabin via a series of ducts, registers and vents.

NOTE: The vehicle is not fitted with a pollen filter.

When selected, recirculated air enters the system through an inlet located in the lower surface of the instrument panel, above the passenger foot well.

Air distribution into the cabin is selected using the RH rotary control mounted on the center console. A Bowden cable transmits selections to 2 air distribution doors using cams mounted on the heater assembly.

Refer to: Climate Control System (412-00 Climate Control System - General Information, Description and Operation)

Air Inlet Duct

Removal and Installation

Part(s)

STEP	PART NAME	QUANTITY
Removal Step 4	Renew Part Air conditioning pipe O-ring seal(s)	1

Removal

NOTES:

Removal steps in this procedure may contain installation details.

Some variation in the illustrations may occur, but the essential information is always correct.

1. **Refer to:** Startup Battery Disconnect and Connect (414-01 Battery, Mounting and Cables, General Procedures)

2. **Refer to:** Air Conditioning System Recovery, Evacuation and Charging (412-00 Climate Control System - General Information, General Procedures)

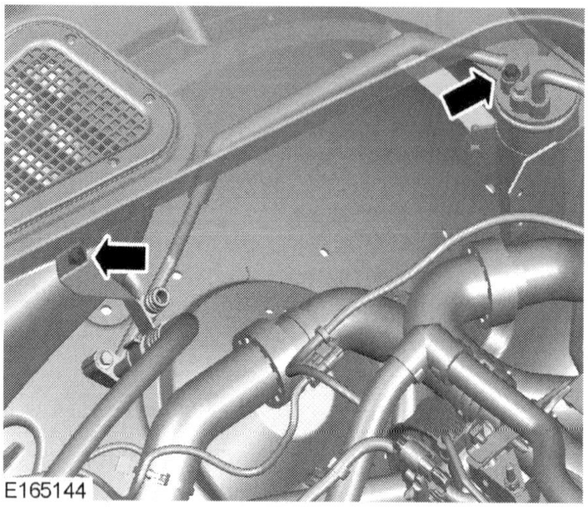

3. ⚠ CAUTION: Before disconnecting any components, make sure the area is clean and free from foreign material. When disconnected all openings must be sealed.

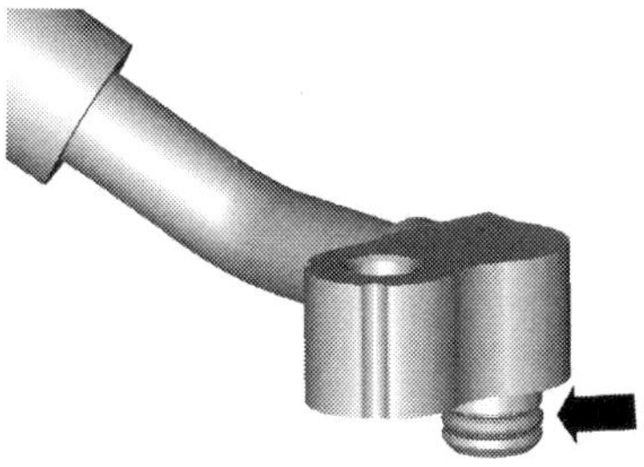

4. NOTES:

⚠ Remove and discard the O-ring seal.

⚠ Install a new O-ring seal.

Renew Part: Air conditioning pipe O-ring seal(s)

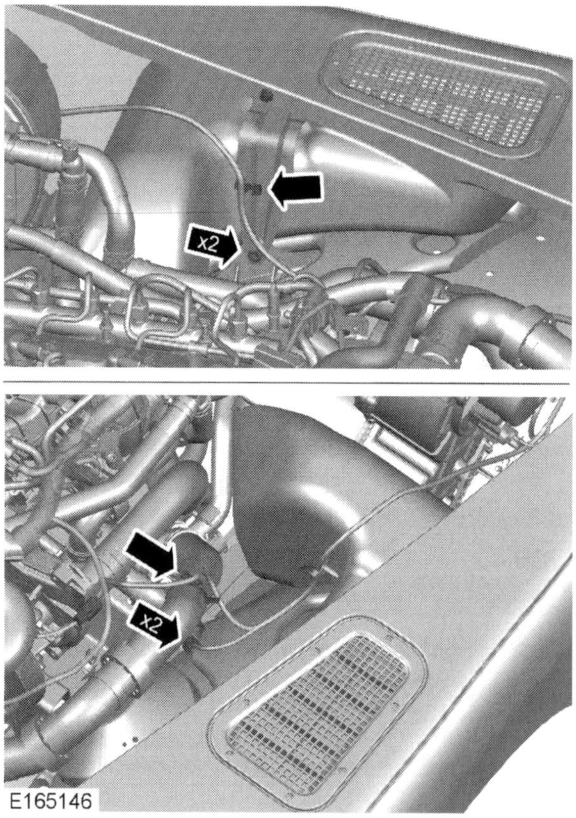

5.

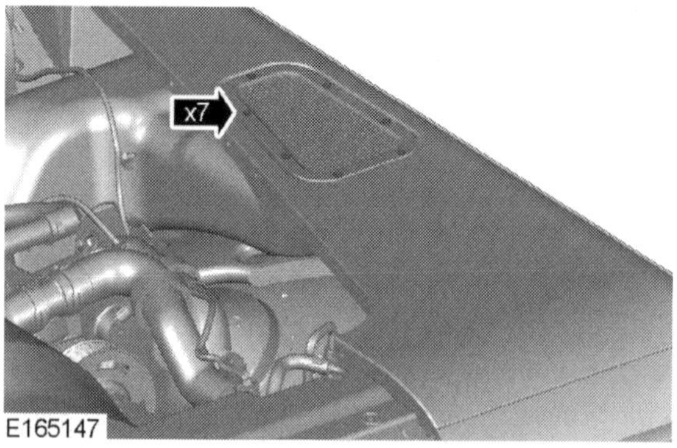

6. Remove the passenger side fresh air intake grille.

7. Remove the 2 air inlet duct upper securing screws.

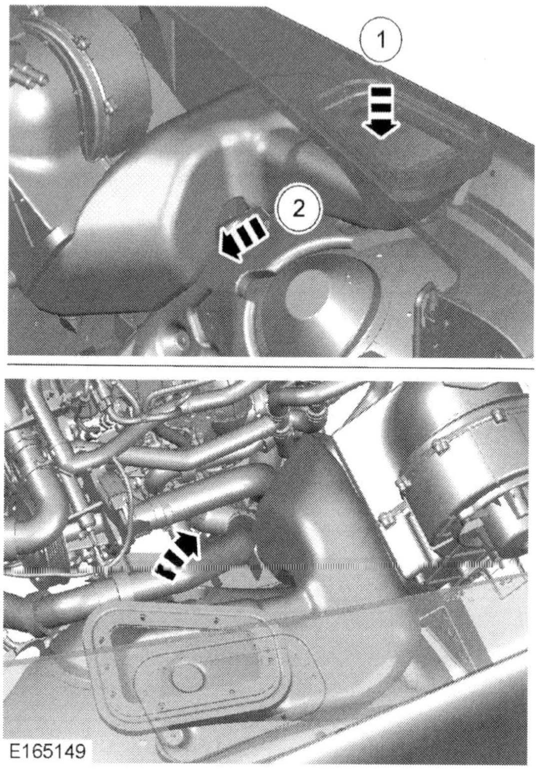

8.

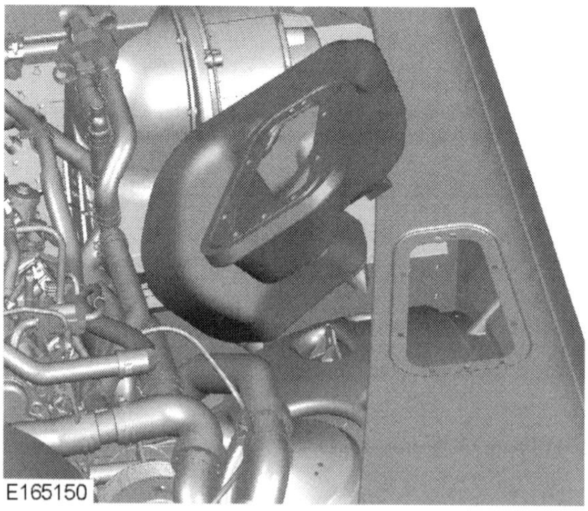

9.

Installation

⚠ NOTE: To prevent water ingress, make sure the air inlet duct is correctly aligned with the blower motor seal.

1. To install, reverse the removal procedure.

Heating and Ventilation -

Specifications

Lubricants, Fluids and Sealers

Item	Specification
Sealant — Blower motor to blower motor housing	WSE-M4G323-A4

Torque specifications

Description	Nm	lb-ft
Blower	15	11
Heater core and evaporator core housing	15	11
Heater control valve nut	6	4
Heater control valve bolt	22	16

Blower - LHD AWD

Removal and Installation

Removal

NOTES:

⚠ Left-hand drive (LHD) vehicle shown, right-hand drive (RHD) similar.

⚠ Removal steps in this procedure may contain installation details.

1. **Refer to:** Air Inlet Duct (412-01 Air Distribution and Filtering, Removal and Installation)

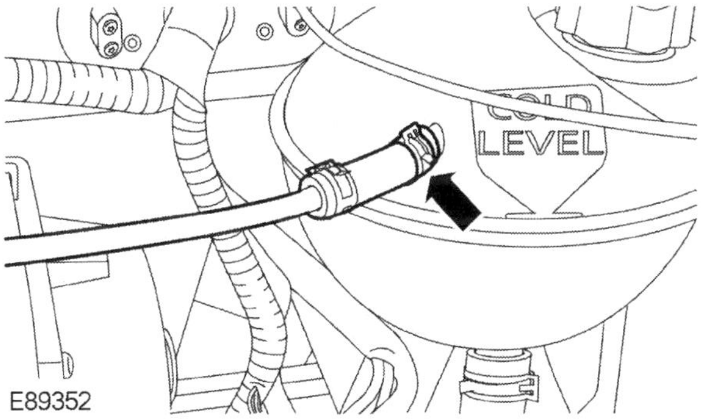

E89352

2. ⚠ NOTE: Left-hand drive (LHD) vehicles only.

Disconnect the coolant expansion tank overflow pipe from the expansion tank.

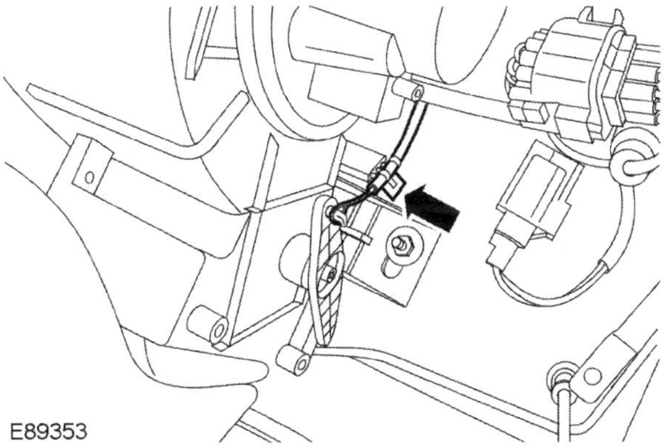

E89353

3. Disconnect the recirculation blend door control cable.

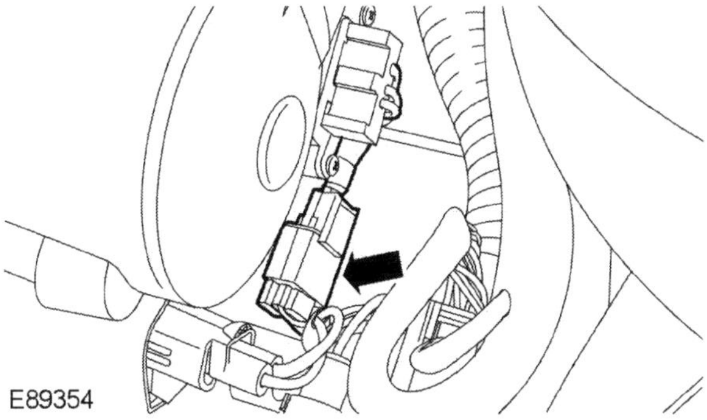

E89354

4. Disconnect the blower motor resistor electrical connector.

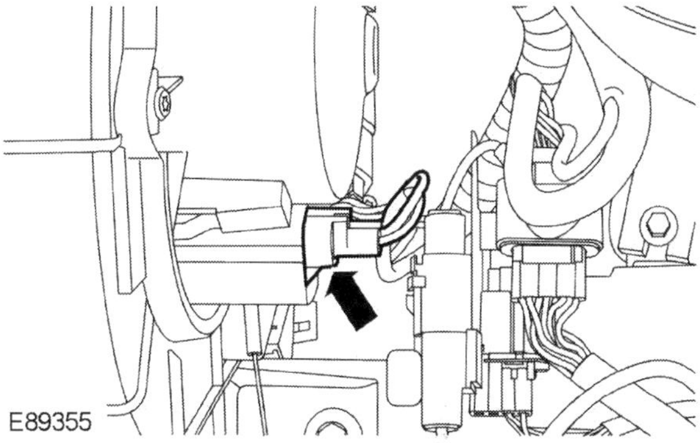

E89355

5. Disconnect the blower motor electrical connector.

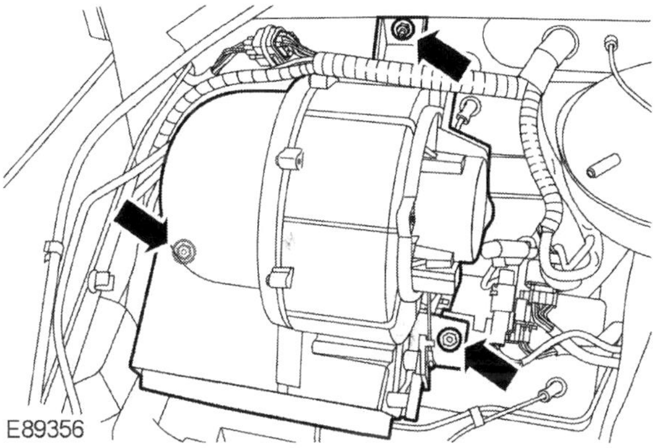

E89356

6. Remove the blower. Tighten to 15 Nm

Installation

1. To install, reverse the removal procedure.

Blower - RHD AWD, RHD

Removal and Installation

Removal

NOTES:

⚠ Removal steps in this procedure may contain installation details.

⚠ Some variation in the illustrations may occur, but the essential information is always correct.

1. **Refer to:** Air Inlet Duct (412-01 Air Distribution and Filtering, Removal and Installation)

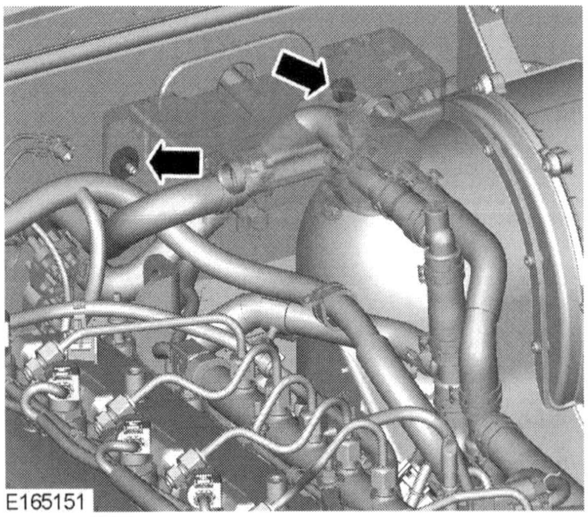

2.

3.

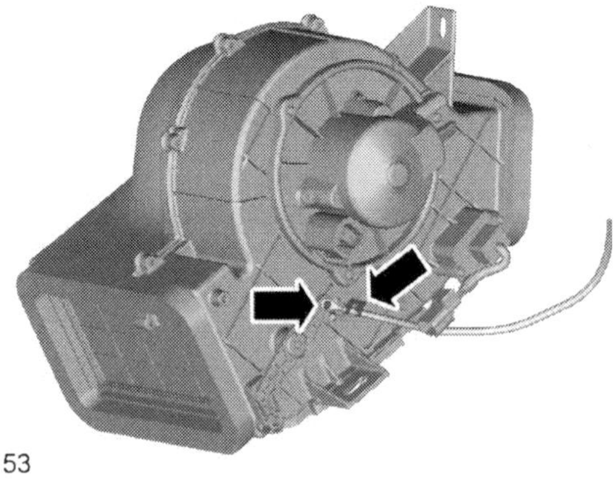

E165153

4. Tighten to 15 Nm

Installation

1. To install, reverse the removal procedure.

Blower Motor - LHD AWD

Removal and Installation

Material: Silicone Sealant (WSE-M4G323-A4)

Removal

△ NOTE: LHD shown, RHD similar.

1. Release the coolant expansion tank to radiator pipe.

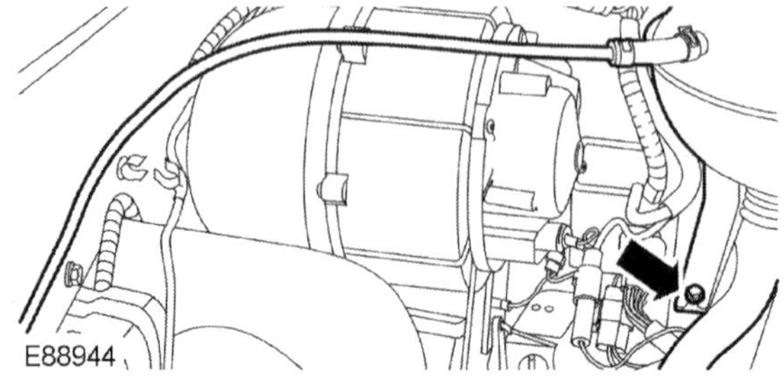

2. Release the coolant expansion tank.
   - Remove the bolt.

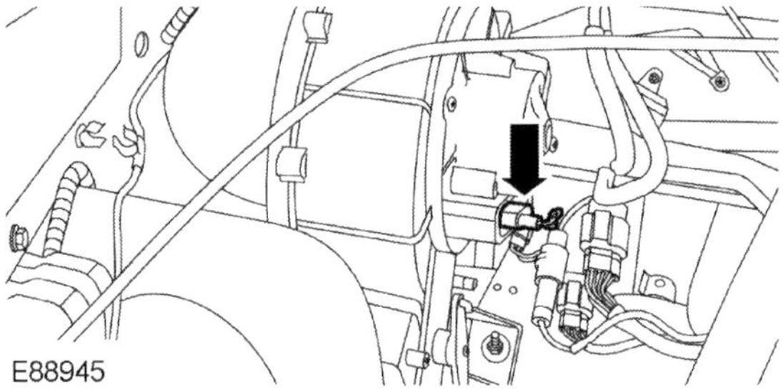

3. Disconnect the blower motor electrical connector.

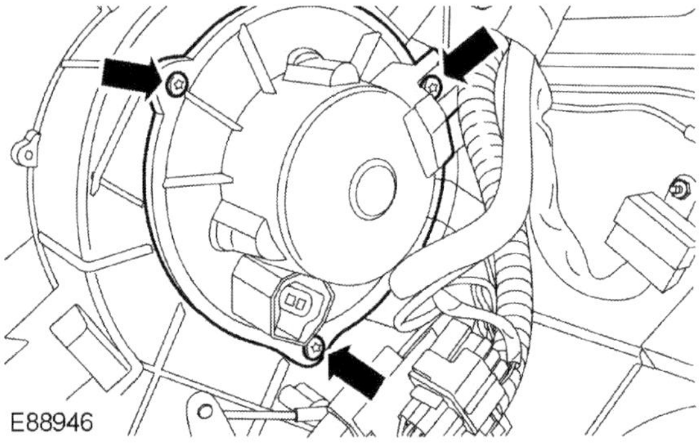

4. Remove the blower motor.
   - Remove the 3 screws.

Installation

1. To install, reverse the removal procedure.
   - Apply a bead of sealant to the mating faces of the blower motor and blower motor housing.

Material: sealant (WSE-M4G323-A4 / Silicone Sealant)

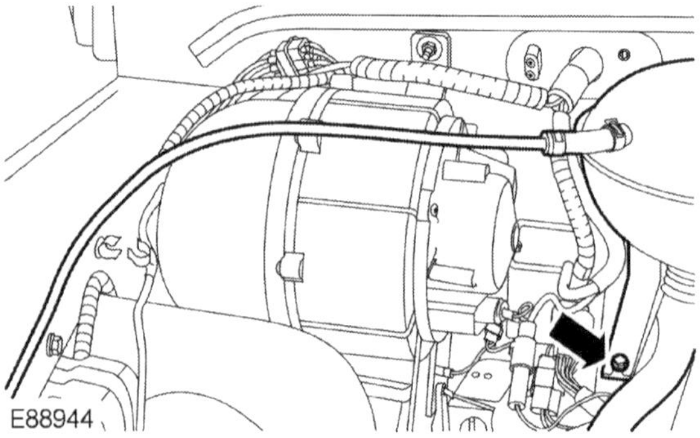

2. Tighten to 10 Nm (7 lb.ft).

Blower Motor - RHD AWD

Removal and Installation

Material: Silicone Sealant (WSE-M4G323-A4)

Removal

NOTES:

△ Removal steps in this procedure may contain installation details.

△ Some variation in the illustrations may occur, but the essential information is always correct.

1. **Refer to:** Blower - RHD AWD, RHD (412-02 Heating and Ventilation, Removal and Installation)

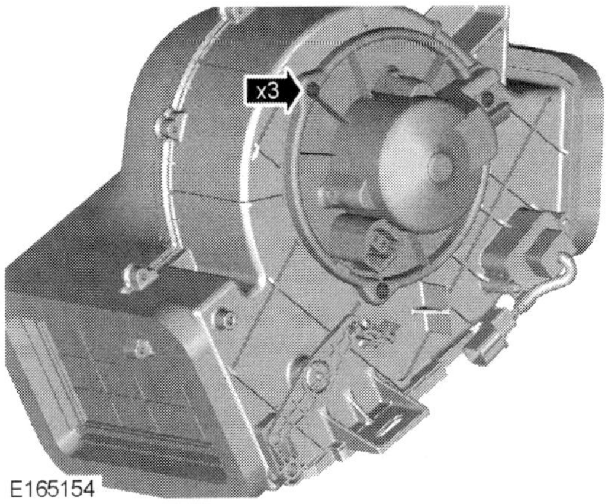

2.

Installation

1. To install, reverse the removal procedure.
   - Apply a bead of sealant to the mating faces of the blower motor and blower motor housing.

Material: sealant (WSE-M4G323-A4 / Silicone Sealant)

Heater Control Valve

Removal and Installation

Removal

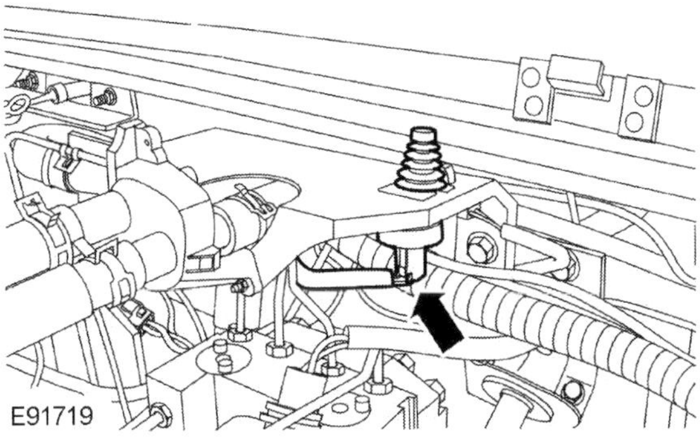

1. Remove the bonnet switch.
   - Disconnect the bonnet switch electrical connector.

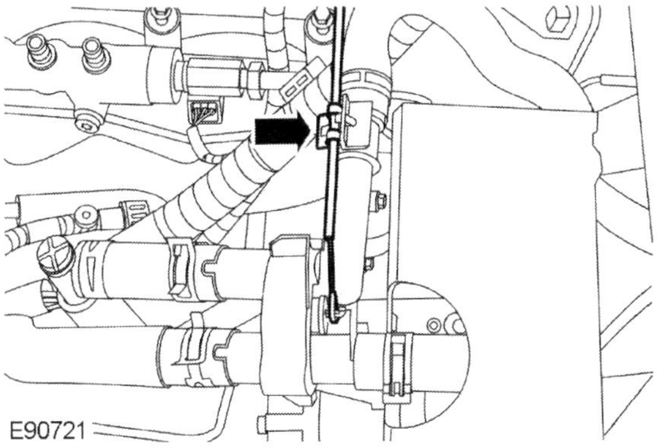

2. △ NOTE: Note the fitted position of the cable prior to removal.

   Disconnect the heater control valve control cable.
   - Release the clip.

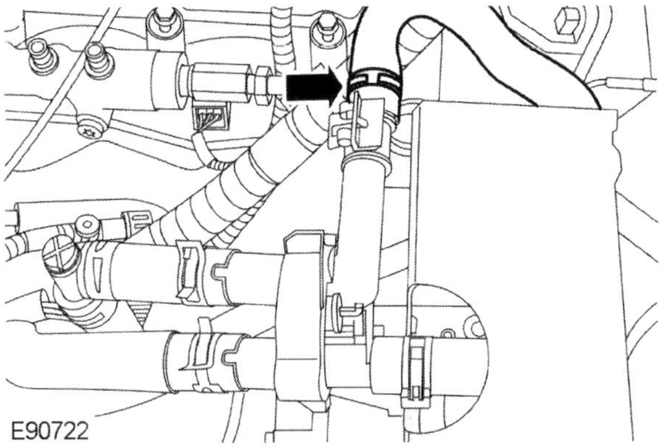

3. Disconnect the heater core feed hose.
   - Using a suitable pipe clamp, clamp the hose.
   - Release the clip.

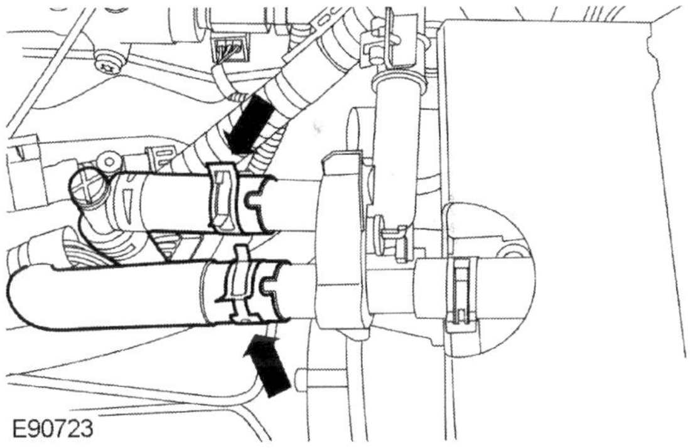

4. Disconnect the heater control valve feed and return hoses.
   - Using a suitable pipe clamp, clamp the hoses.
   - Release the 2 clips.

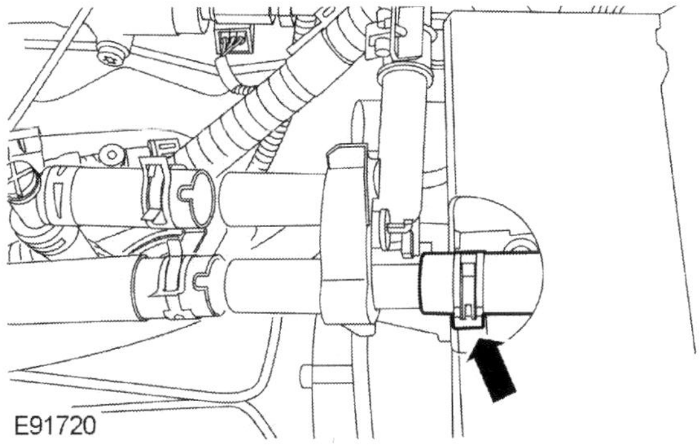

5. Disconnect the heater core return hose.
   - Release the clip.

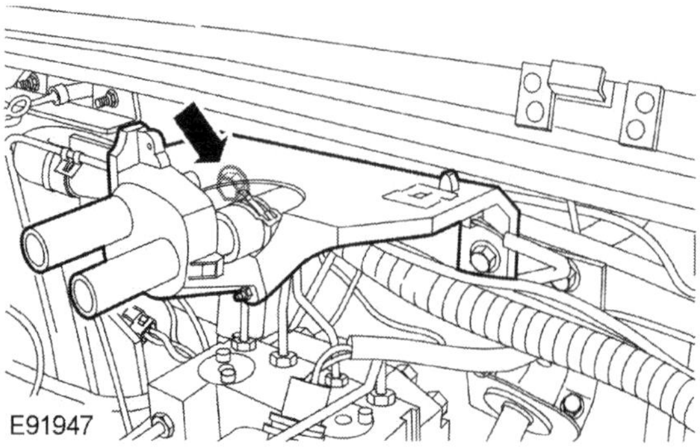

E91947

6. Release the heater control valve.
    - Remove the nut.

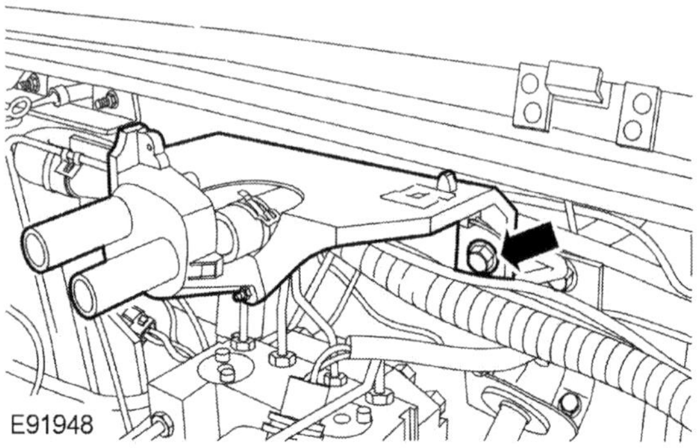

E91948

7. Remove the heater control valve.
    - Remove the bolt.

Installation

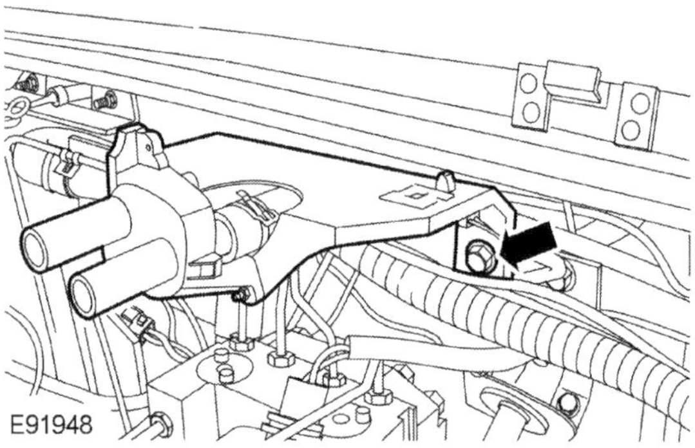

E91948

1. To install, reverse the removal procedure.
    - Tighten to 22 Nm (16 lb.ft).

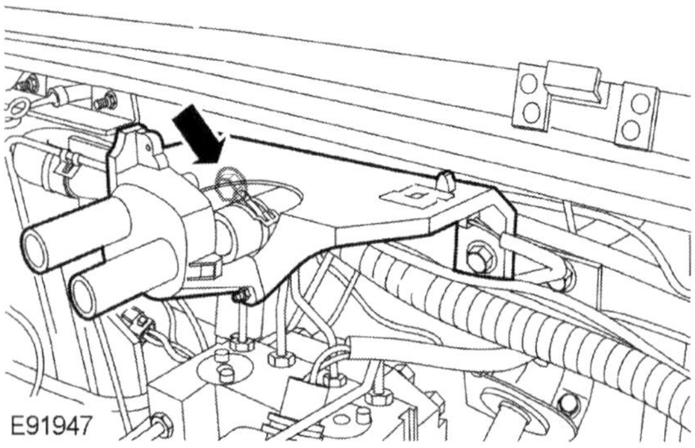

E91947

2. Tighten to 6 Nm (4 lb.ft).

Heater Core

Removal and Installation

Removal

1. Disconnect the battery ground cable.

    Refer to: Startup Battery Disconnect and Connect (414-01 Battery, Mounting and Cables, General Procedures)

2. Remove the heater core housing.

    Refer to: Heater Core Housing (412-02 Heating and Ventilation, Removal and Installation)

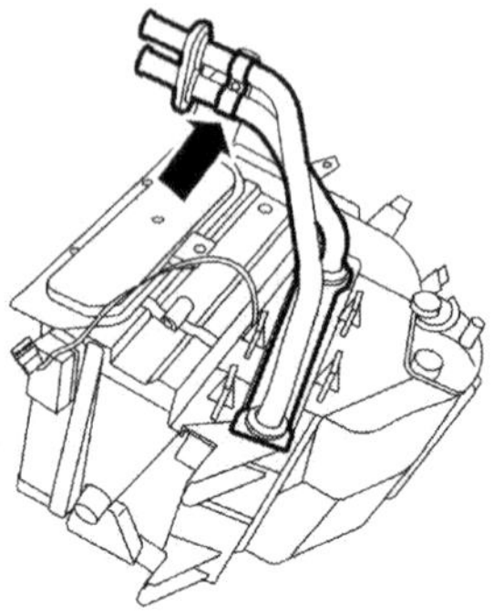

E88665

3. CAUTION: Make sure no damage occurs to the foam seal on the heater core. Failure to follow this caution may result in damage to the component.

    Remove the heater core.
    - Remove the screw.
    - Remove the pipe clamp.

4. NOTE: Do not disassemble further if the component is removed for access only.

Remove the grommet from the heater core pipes.

Installation

1. To install, reverse the removal procedure.
2. Connect the battery ground cable.

Refer to: Battery Connect (414-01 Battery, Mounting and Cables, General Procedures).

Heater Core and Evaporator Housing

Removal and Installation

Part(s)

STEP	PART NAME	QUANTITY
Removal Step 4	Renew Part Thermostatic expansion valve O-ring seal kit	1
Removal Step 7	Renew Part Retaining clip(s)	5

Removal

1. Disconnect the battery ground cable.

    Refer to: Startup Battery Disconnect and Connect (414-01 Battery, Mounting and Cables, General Procedures)

2. Recover the air conditioning (A/C) refrigerant.

   Refer to: Air Conditioning System Recovery, Evacuation and Charging (412-00 Climate Control System - General Information, General Procedures)

3. Remove the engine control module (ECM).

   Refer to: Engine Control Module (303-14 Electronic Engine Controls, ID4 2.2L Diesel Removal and Installation)

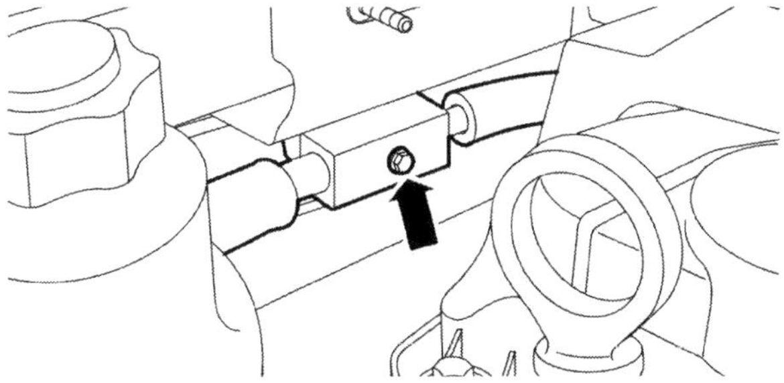

4. CAUTION: Before disconnecting any components, make sure the area is clean and free from foreign material. When disconnected all openings must be sealed.
   Release the thermostatic expansion valve manifold and tube assembly. Remove the bolt. Remove and discard the O-ring seals.

Renew Part: Thermostatic expansion valve O-ring seal kit

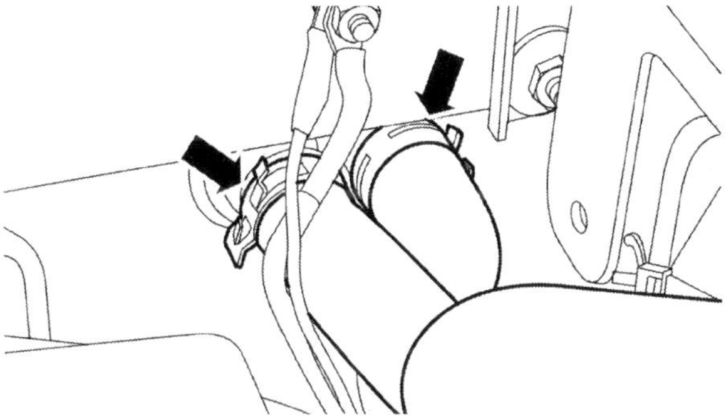

5. Disconnect the heater core hoses. Using a suitable pipe clamp, clamp the heater hoses. Release the 2 clips.

6. Remove the instrument panel.

Refer to: Instrument Panel (501-12 Instrument Panel and Console, Removal and Installation)

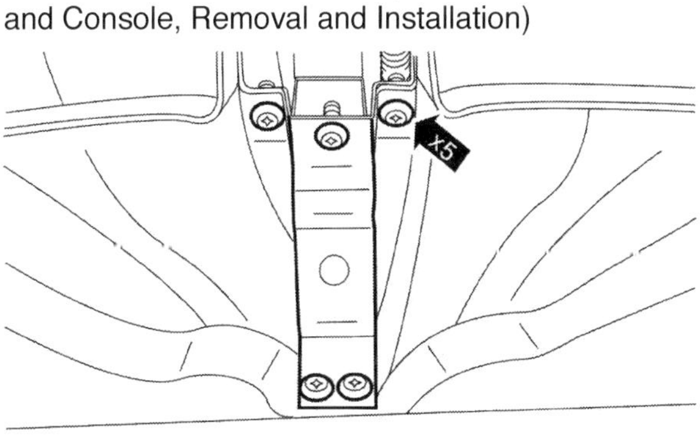

7. NOTE: Make sure the clips are discarded.
   Remove the audio unit support bracket. Remove and discard the 5 clips.

   Renew Part: Retaining clip(s)

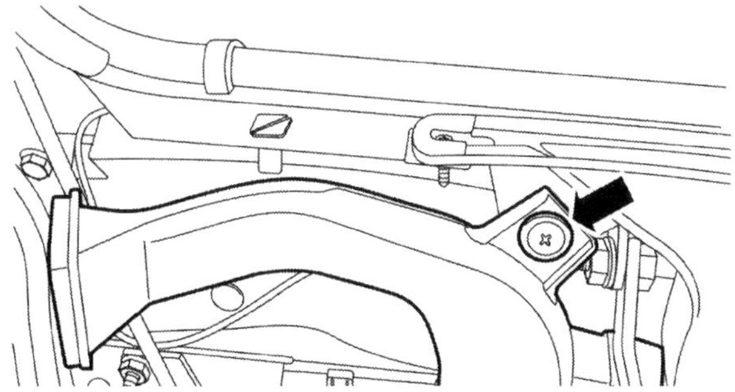

8. Remove the LH demister duct. Remove the clip.

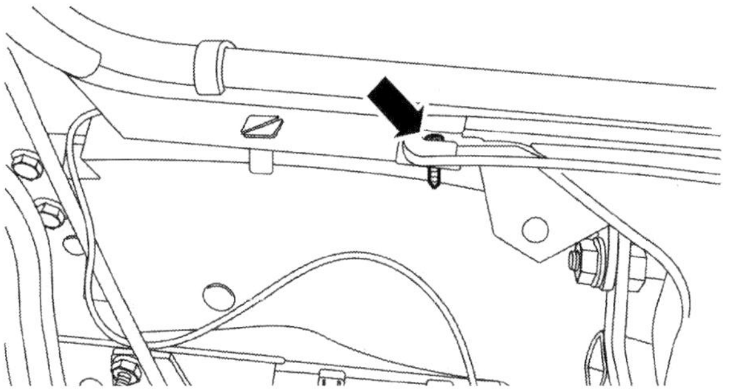

9. Release the LH defrost vent duct. Remove the screw.

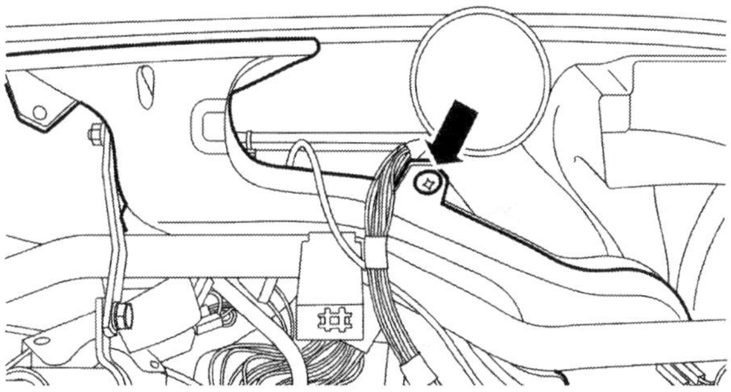

10. Remove the LH defrost vent duct. Remove the clip.

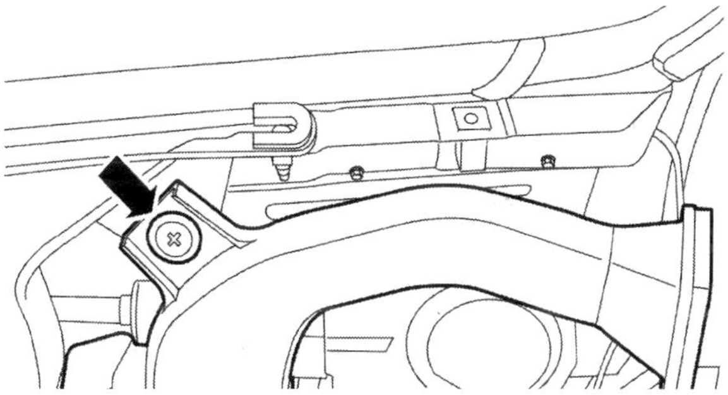

11. Remove the RH demister duct. Remove the clip.

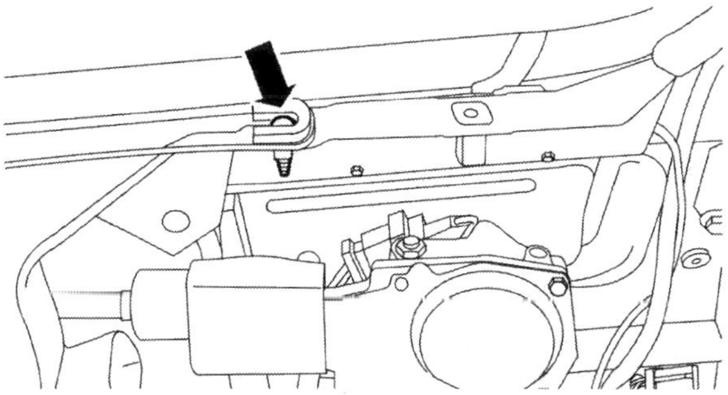

12. Release the RH defrost vent duct. Remove the screw.

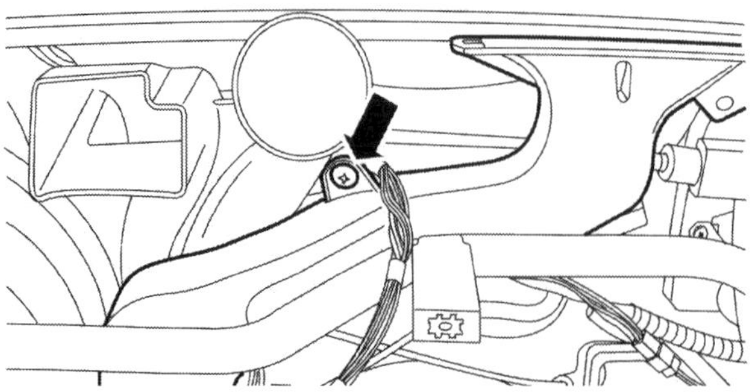

13. Remove the RH vent duct. Remove the clip.

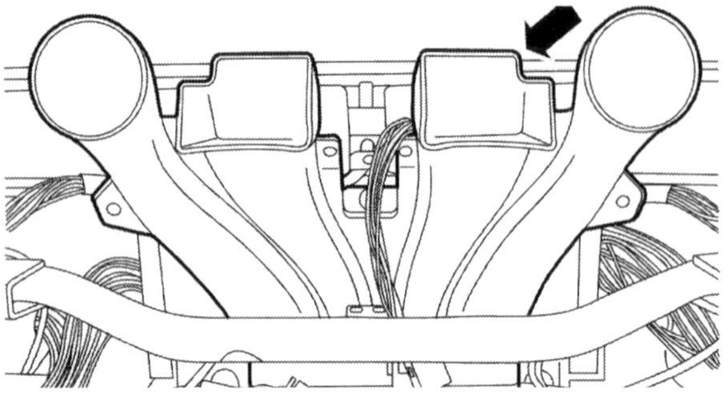

14. Remove the center register duct.

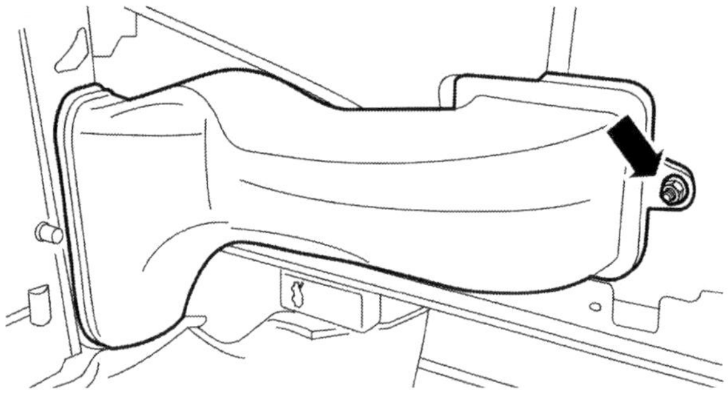

15. Remove the air inlet duct. Remove the nut.

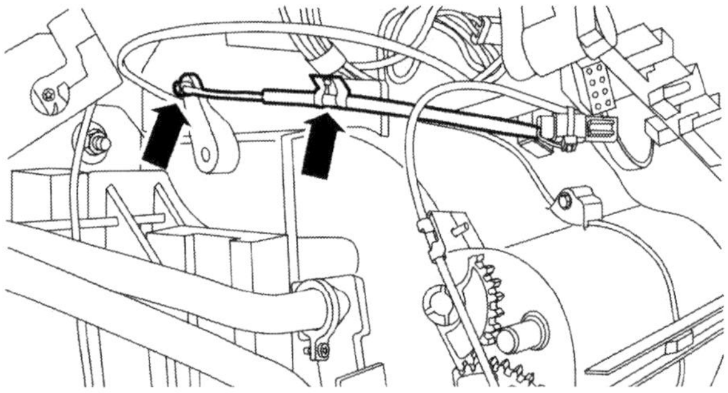

16. Disconnect the defrost duct blend door control cable. Release the clip.

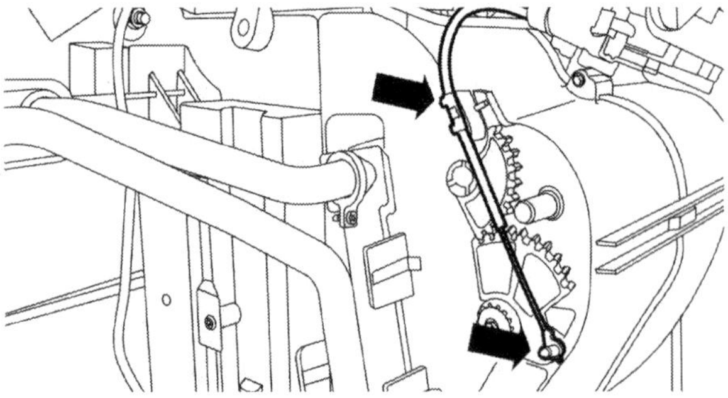

17. Disconnect the footwell duct blend door control cable. Release the clip.

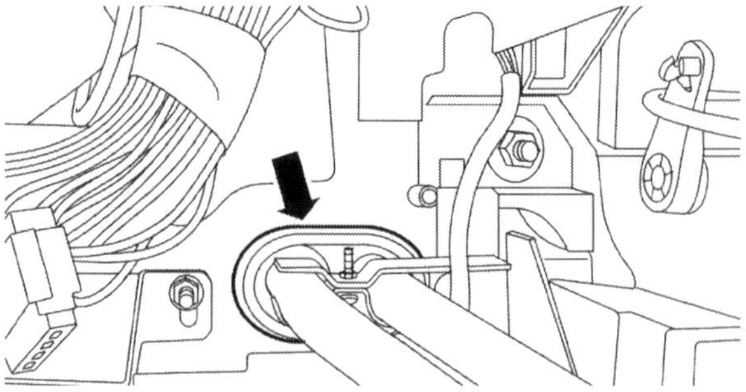

18. Release the grommet from the bulkhead.

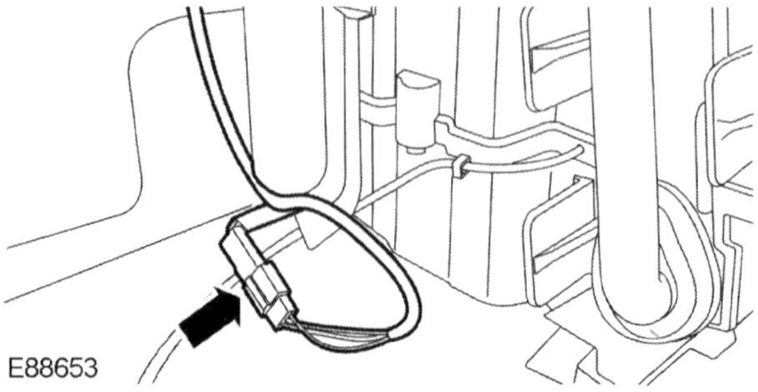

19. Disconnect the de-icing switch electrical connector.

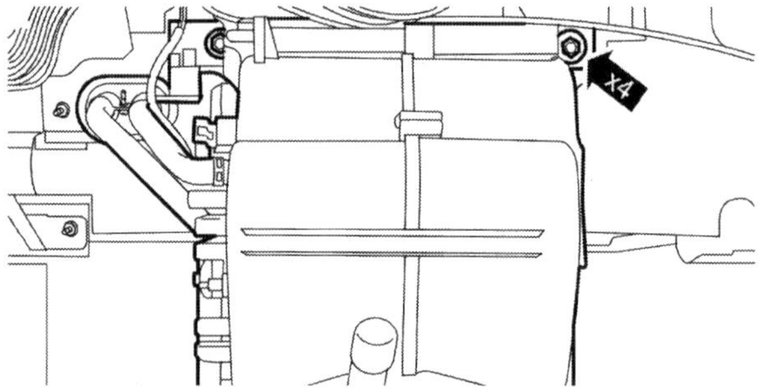

20. Remove the heater core and evaporator core housing. Remove the 4 nuts.

Installation

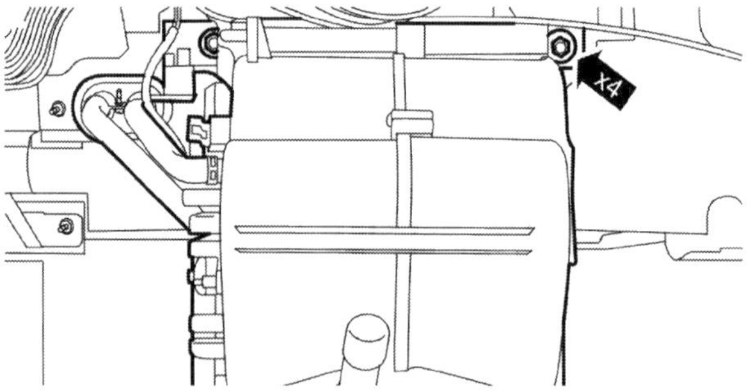

1. To install, reverse the removal procedure. Tighten.

   Torque: 15 Nm

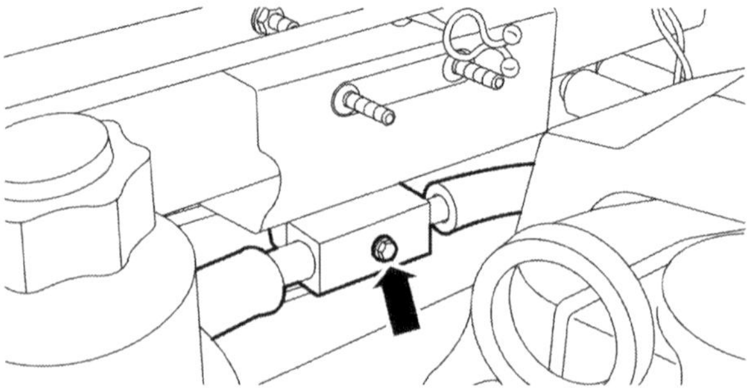

2. Tighten.

   Torque: 8 Nm

3. Connect the battery ground cable.

Refer to: Startup Battery Disconnect & Connect (414-01 Battery, Mounting and Cables, General Procedures)

Specifications

Torque Specifications

Description	Nm	lb-ft
A/C compressor bolts	48	35
A/C compressor low-pressure and high-pressure pipe bolt	30	22
A/C condenser core low-pressure and high-pressure refrigerant lines	8	6
Condenser core and condenser fan assembly bolts	15	11
Condenser fan bolts	2	1
Condenser core nuts	8	6
A/C receiver drier nuts	8	6
A/C receiver drier low-pressure and high-pressure refrigerant line bolts	8	6

Air Conditioning Compressor

Removal and Installation

Removal

NOTES:

⚠ Removal steps in this procedure may contain installation details.

⚠ Some variation in the illustrations may occur, but the essential information is always correct.

1. **Refer to:** Startup Battery Disconnect and Connect (414-01 Battery, Mounting and Cables, General Procedures)

2. **Refer to:** Air Conditioning System Recovery, Evacuation and Charging (412-00 Climate Control System - General Information, General Procedures)

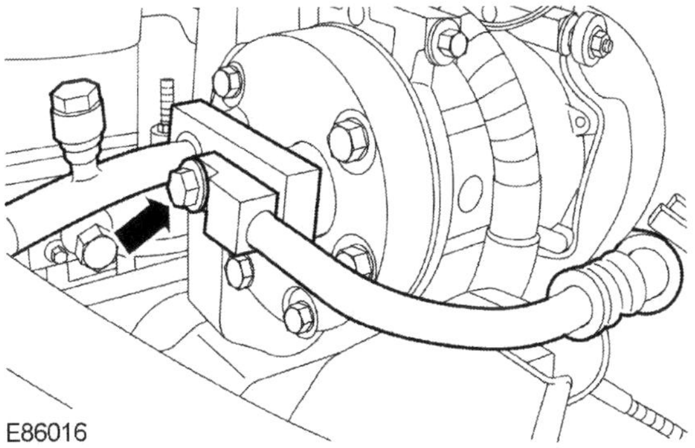

E86016

3. CAUTIONS:

⚠ Before disconnecting any components, make sure the area is clean and free from foreign material. When disconnected all openings must be sealed.

⚠ Lubricate the new seals with clean refrigerant oil.

⚠ NOTE: Install new O-ring seals.

Torque: 30 Nm

4. **Refer to:** Accessory Drive Belt (303-05 Accessory Drive, ID4 2.2L Diesel Removal and Installation)

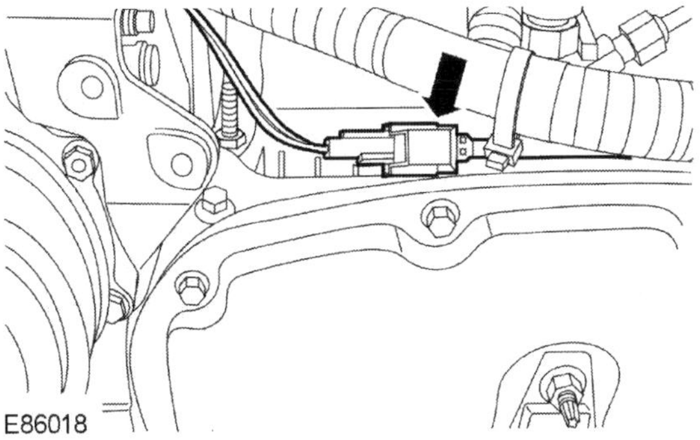

E86018

5.

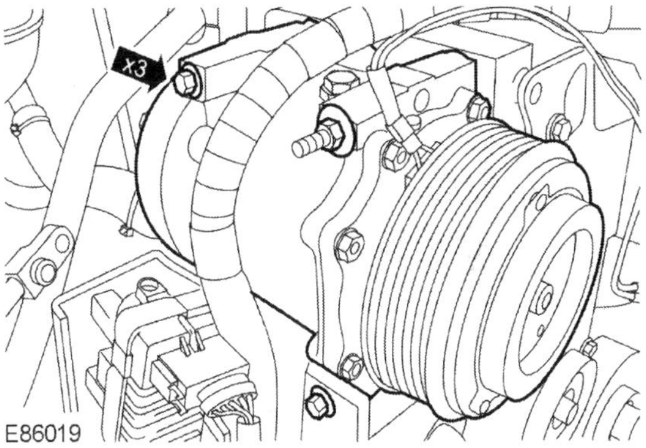

E86019

6. Torque: 23 Nm

Installation

1. ⚠ NOTE: If a new compressor is being installed, the required amount of refrigerant oil must be drained from it.

   To install, reverse the removal procedure.

Air Conditioning Switch

Removal and Installation

Removal

1. Disconnect the battery ground cable.

Refer to: Startup Battery Disconnect and Connect (414-01 Battery, Mounting and Cables, General Procedures)

2. Remove the audio unit.

Refer to: Audio Unit - Vehicles Without: Bluetooth (415-01 Audio Unit, Removal and Installation)

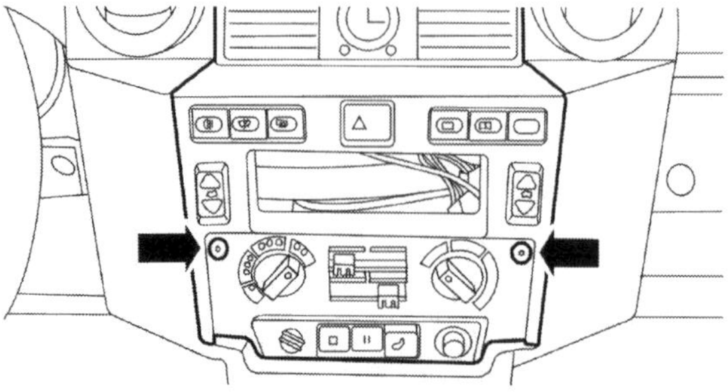

3. Release the instrument panel console.

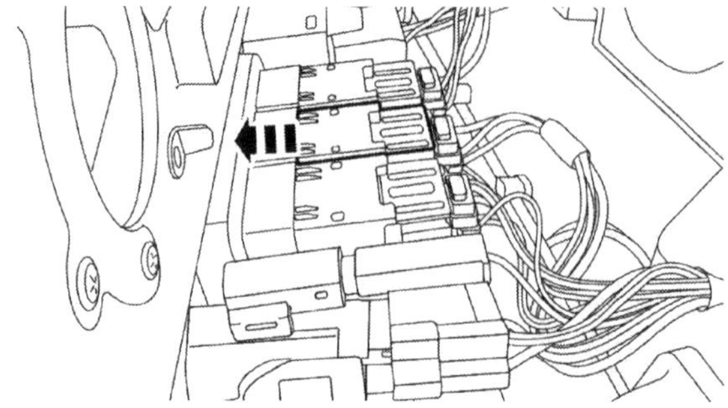

4. Release the air conditioning (A/C) switch.

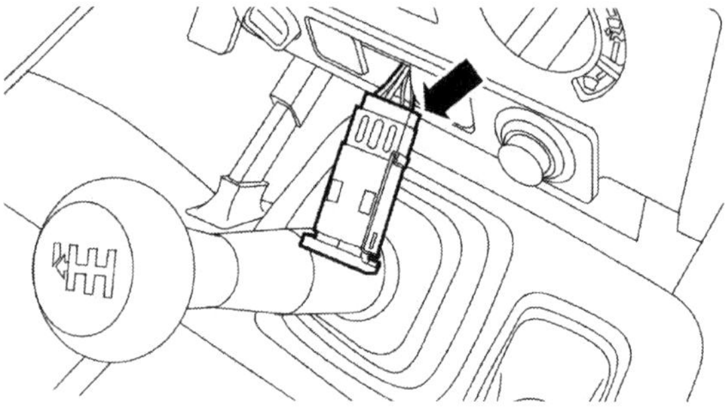

5. Remove the A/C switch.
   - Disconnect the electrical connector.

Installation

1. To install, reverse the removal procedure.
2. Connect the battery ground cable.

   Refer to: Battery Connect (414-01 Battery, Mounting and Cables, General Procedures).

Condenser Core

Removal and Installation

Part(s)

STEP	PART NAME	QUANTITY
Removal Step 5	Renew Part Air conditioning pipes to condenser O-ring seals	2

Removal

1. Disconnect the battery ground cable.

Refer to: Startup Battery Disconnect and Connect (414-01 Battery, Mounting and Cables, General Procedures)

2. Recover the air conditioning (A/C) refrigerant.

   Refer to: Air Conditioning System Recovery, Evacuation and Charging (412-00 Climate Control System - General Information, General Procedures)

3. Remove the radiator grille.

   Refer to: Radiator Grille (501-08 Exterior Trim and Ornamentation, Removal and Installation)

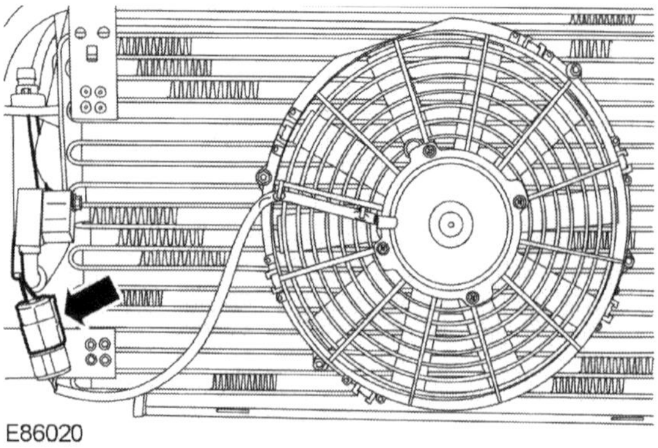

E86020

4. Disconnect the condenser fan electrical connector.

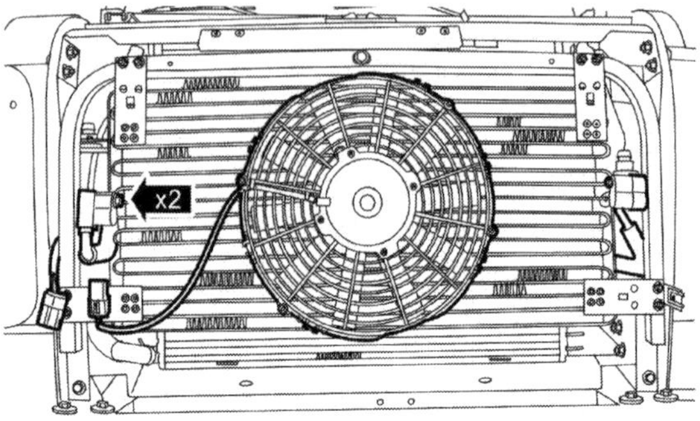

E86024

5. Remove the 2 bolts. Disconnect the A/C condenser core low-pressure and high-pressure refrigerant lines. Remove and discard the O-ring seals.

Renew Part: Air conditioning pipes to condenser O-ring seals

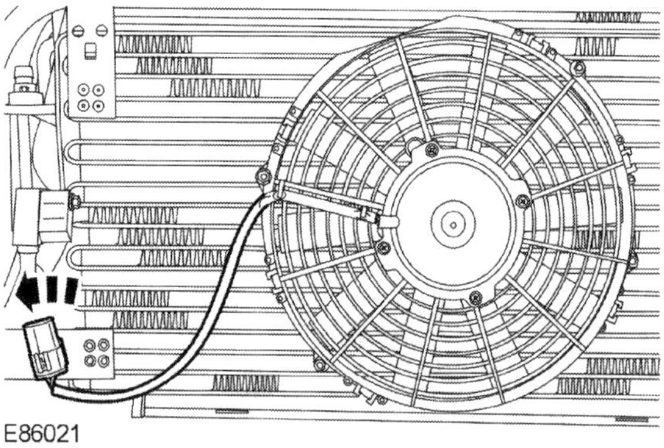

E86021

6. Release the condenser fan electrical connector.

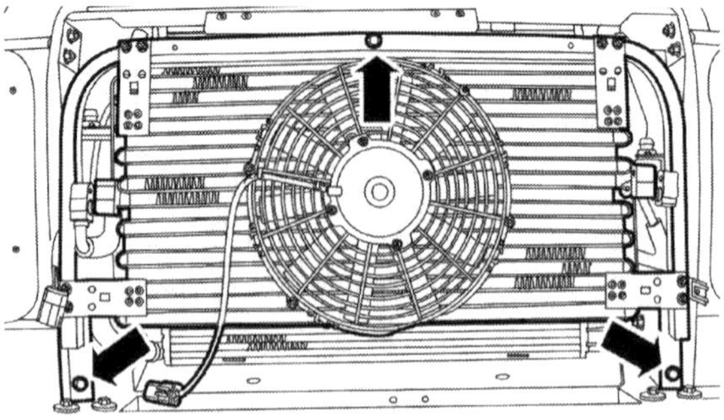

E86025

7. Remove the 3 bolts. Remove the condenser core and condenser fan assembly.

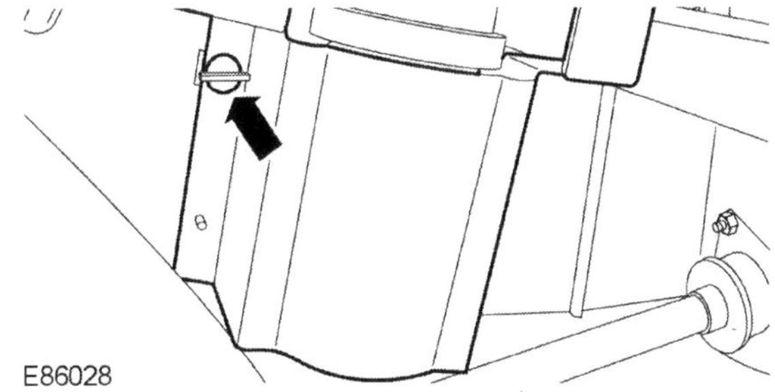

E86028

8. Remove the 4 bolts. Remove the condenser fan.

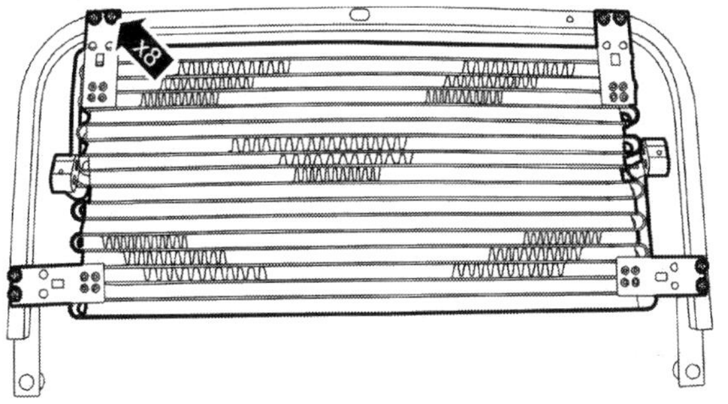

9. Remove the 8 nuts. Remove the condenser core.

Installation

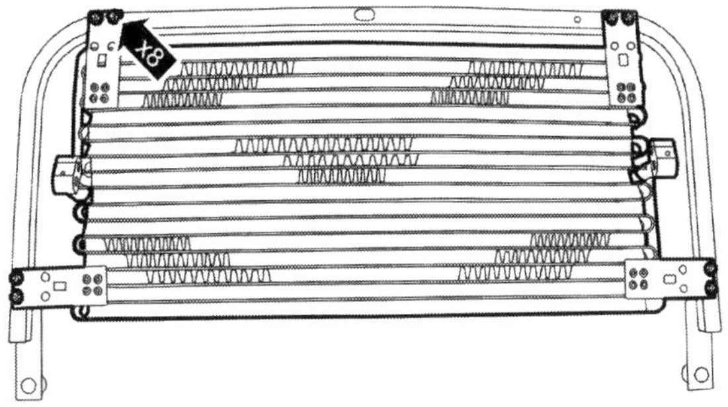

1. To install, reverse the removal procedure.
   Tighten the 8 nuts.

   Torque: 8 Nm

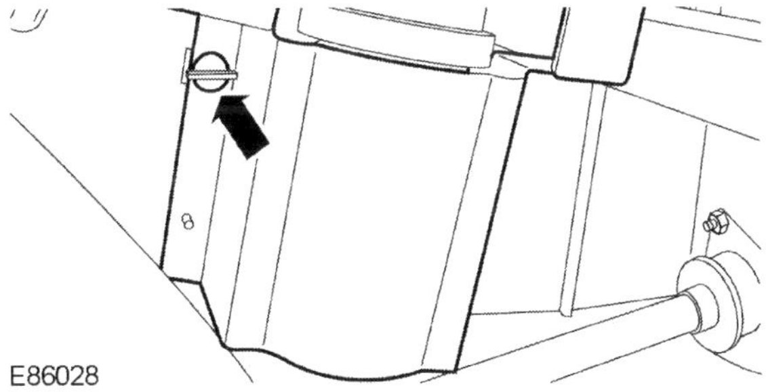

E86028

2. Tighten the 4 bolts.

   Torque: 2 Nm

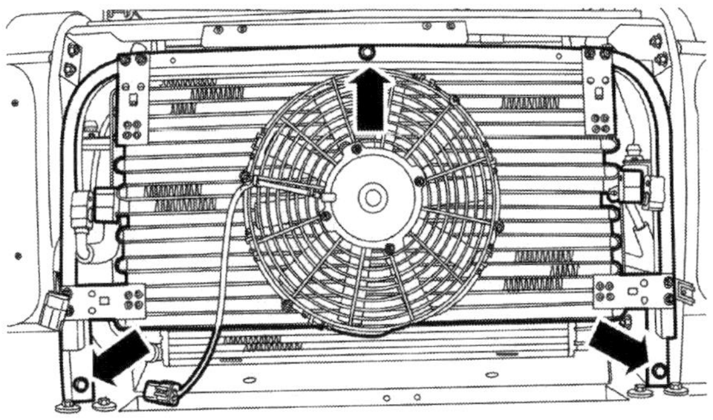

E86025

3. Tighten the 3 bolts.

   Torque: 15 Nm

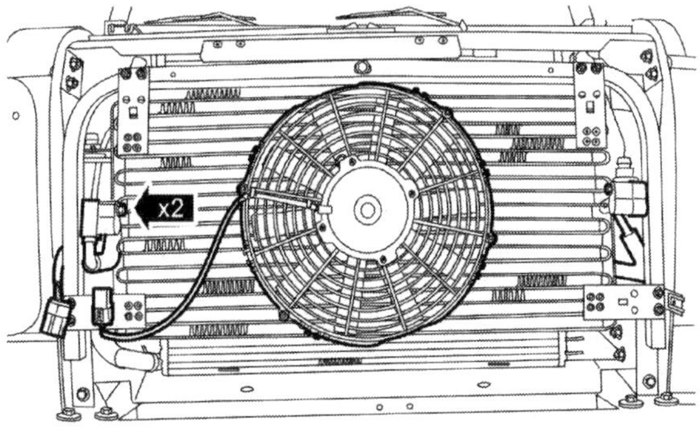

4. Tighten the 2 bolts.

   Torque: 8 Nm

5. Connect the battery ground cable.

   Refer to: Startup Battery Disconnect and Connect (414-01 Battery, Mounting and Cables, General Procedures)

Condenser Fan

Removal and Installation

Removal

1. Disconnect the battery ground cable.

Refer to: Startup Battery Disconnect and Connect (414-01 Battery, Mounting and Cables, General Procedures)

2. Remove the radiator grille.

   Refer to: Radiator Grille (501-08, Removal and Installation).

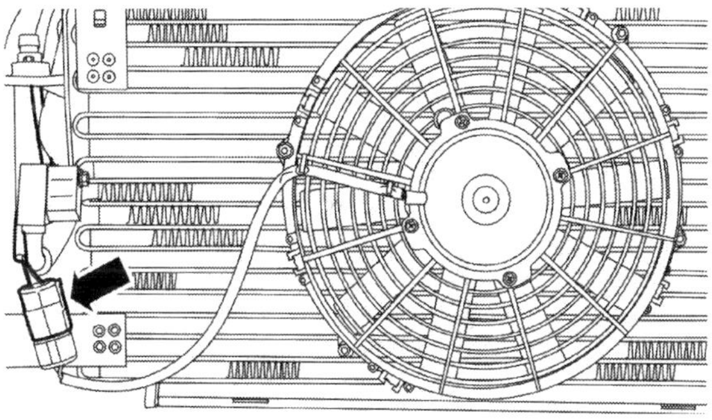

3. Disconnect the condenser fan electrical connector.

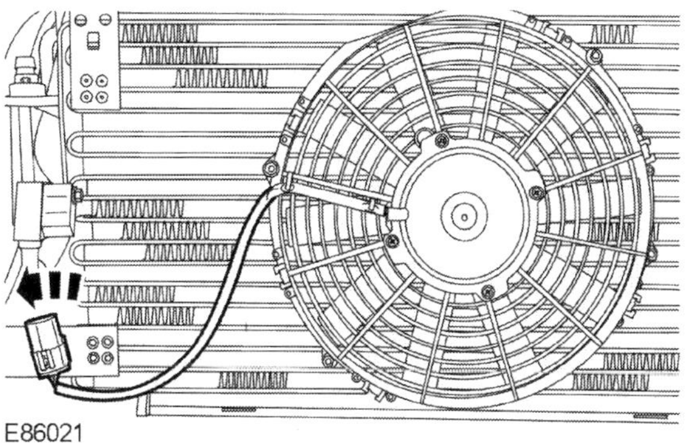

E86021

4. Release the condenser fan electrical connector.

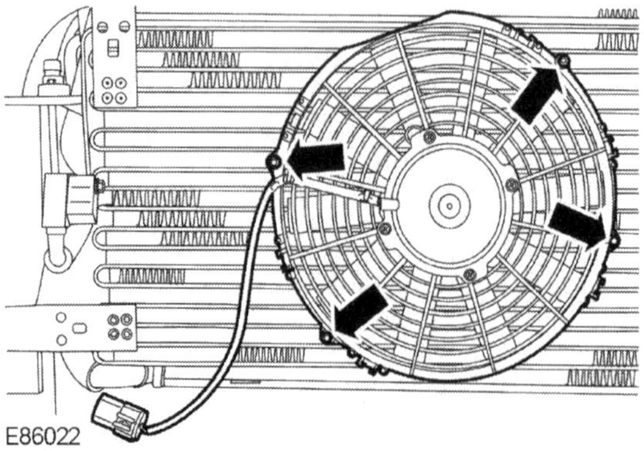

5. Remove the condenser fan.
    - Remove the 4 bolts.

Installation

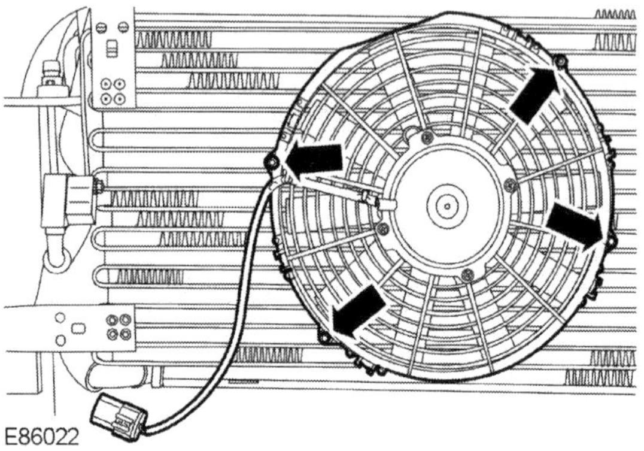

1. To install, reverse the removal procedure.
    - Tighten to 2 Nm (1 lb.ft).
2. Connect the battery ground cable.

    Refer to: Battery Connect (414-01, General Procedures).

De-Icing Switch

Removal and Installation

Removal

1. Disconnect the battery ground cable.

    Refer to: Startup Battery Disconnect and Connect (414-01 Battery, Mounting and Cables, General Procedures)

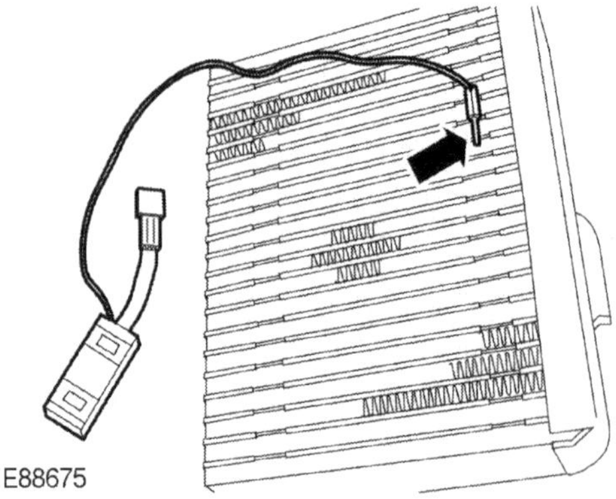

2. Remove the evaporator core.

    Refer to: Evaporator Core (412-03 Air Conditioning, ID4 2.2L Diesel Removal and Installation) 3. Remove the de-icing switch from the evaporator core.

Installation

1. To install, reverse the removal procedure.
2. Connect the battery ground cable.

    Refer to: Startup Battery Disconnect and Connect (414-01 Battery, Mounting and Cables, General Procedures)

Receiver Drier

Removal and Installation

Part(s)

STEP	PART NAME	QUANTITY
Removal Step 3	Renew Part Air conditioning pipes O-ring seals	2

Removal

1. Recover the air conditioning (A/C) refrigerant.

Refer to: Air Conditioning System Recovery, Evacuation and Charging (412-00 Climate Control System - General Information, General Procedures)

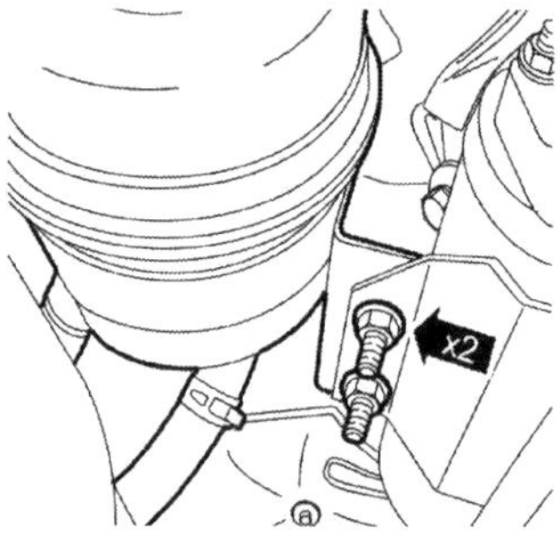

2. Remove the 2 nuts. Release the power steering fluid reservoir.

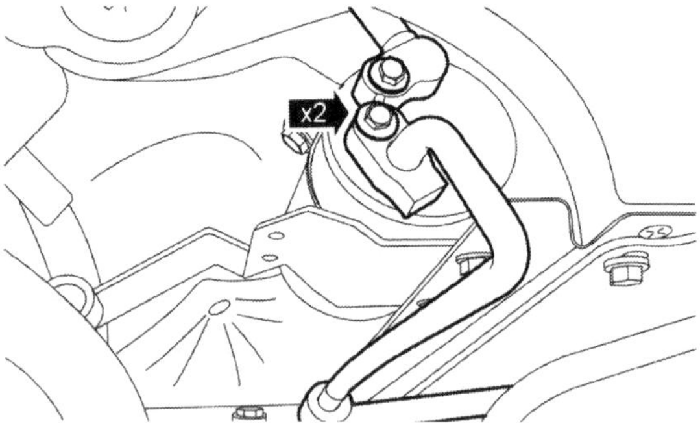

3. CAUTION: Before disconnecting any components, make sure the area is clean and free from foreign material. When disconnected all openings must be sealed.

    Remove the 2 bolts. Disconnect the A/C receiver drier low-pressure and high-pressure refrigerant lines. Remove and discard the O-ring seals.

Renew Part: Air conditioning pipes O-ring seals

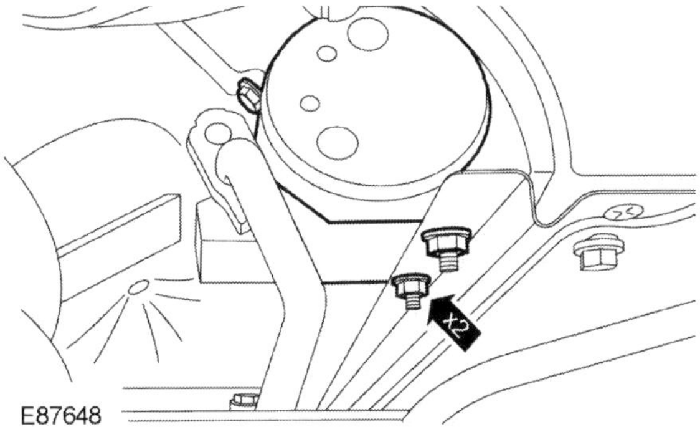

E87648

4. Remove the 2 nuts. Remove the A/C receiver drier.

Installation

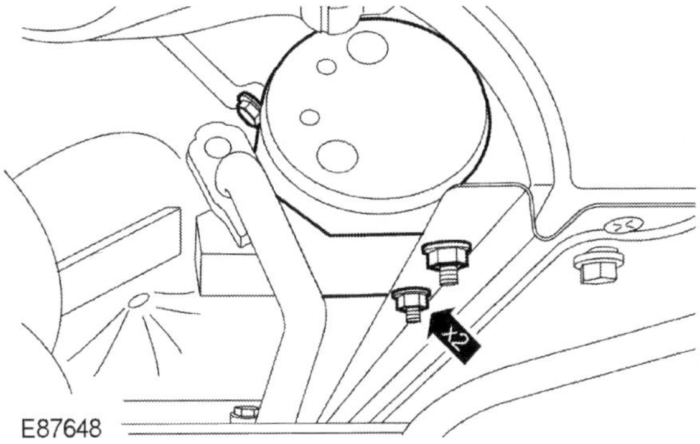

E87648

1. Install the A/C receiver drier. Tighten the nuts.

   Torque: 8 Nm

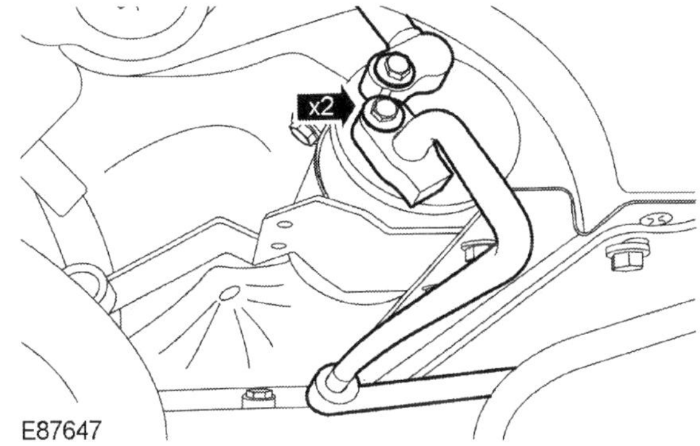

E87647

2. NOTES:

△ Coat the receiver drier O-ring seals in clean refrigerant oil prior to installation.

△ Remove and discard the blanking caps.

Connect the A/C receiver drier low-pressure and high-pressure refrigerant lines. Tighten the bolts.

Torque: 8 Nm

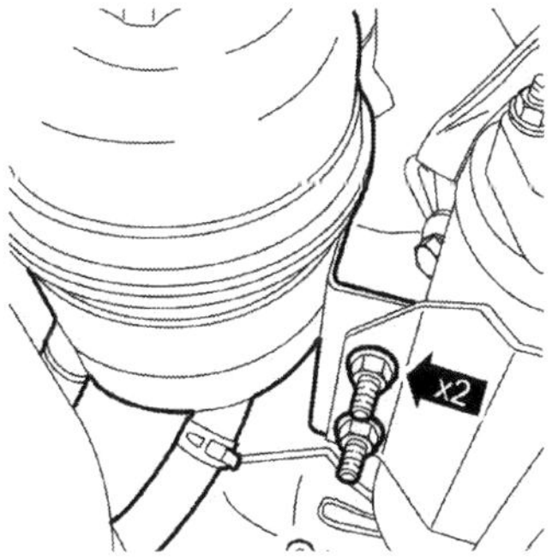

3. Secure the power steering fluid reservoir. Tighten the nuts.

   Torque: 4 Nm

4. Refill the air conditioning (A/C) refrigerant.

   Refer to: Air Conditioning System Recovery, Evacuation and Charging (412-00 Climate Control System - General Information, General Procedures)

Evaporator Core

Removal and Installation

Part(s)

STEP	PART NAME	QUANTITY
Removal Step 9	Renew Part Thermostatic expansion valve O-ring seal kit	1

Removal

1. Disconnect the battery ground cable.

   Refer to: Startup Battery Disconnect and Connect (414-01 Battery, Mounting and Cables, General Procedures)

2. Remove the heater core.

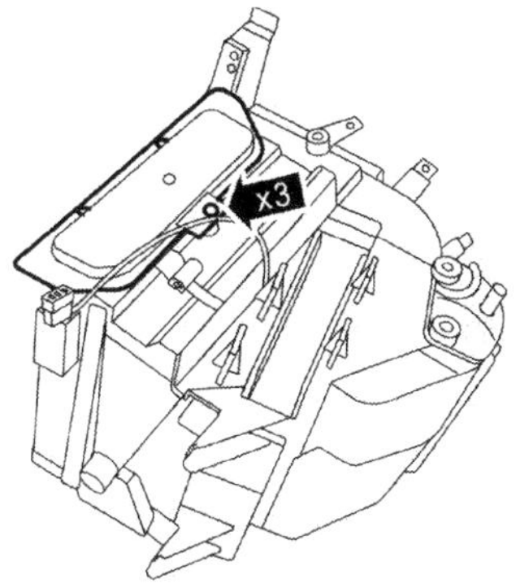

E88666

Refer to: Heater Core (412-02 Heating and Ventilation, Removal and Installation)

3. Remove the 3 screws. Remove the evaporator core end cover.

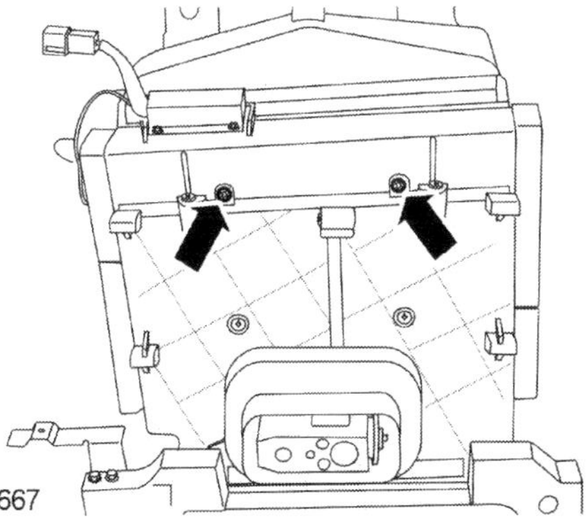

E88667

4. Remove the 2 screws. Release the evaporator core housing lower cover.

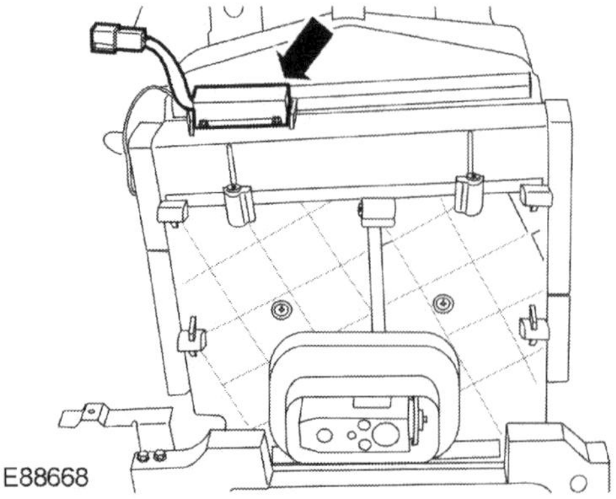

E88668

5. Release the de-icing sensor.

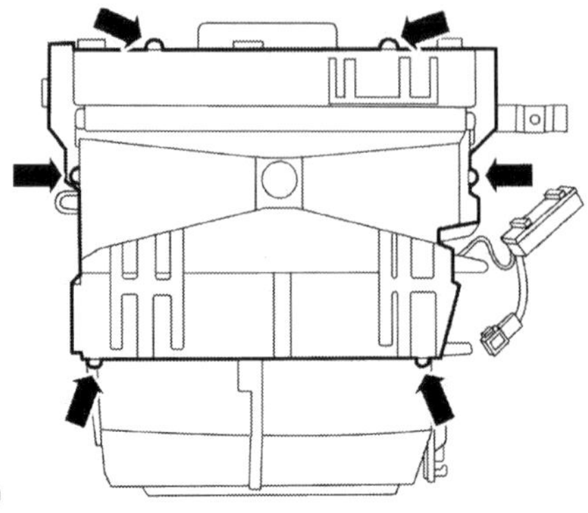

E88669

6. Remove the 6 screws. Remove the evaporator core housing lower cover.

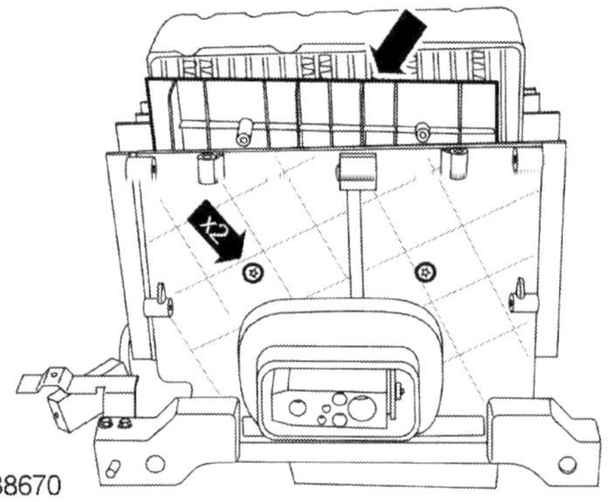

E88670

7. Remove the 2 screws. Remove the air flow deflector.

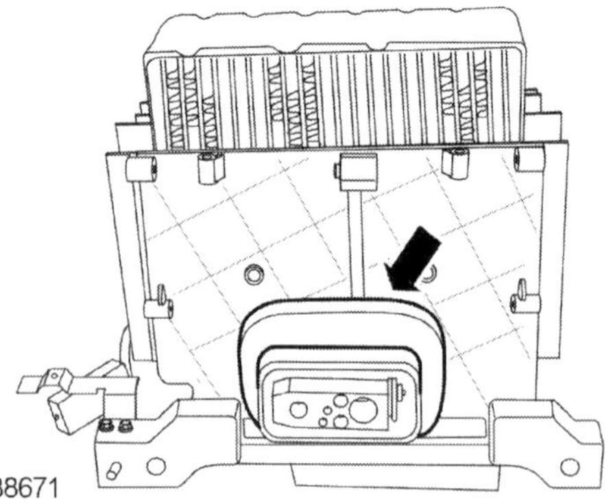

E88671

8. Remove the foam seal.

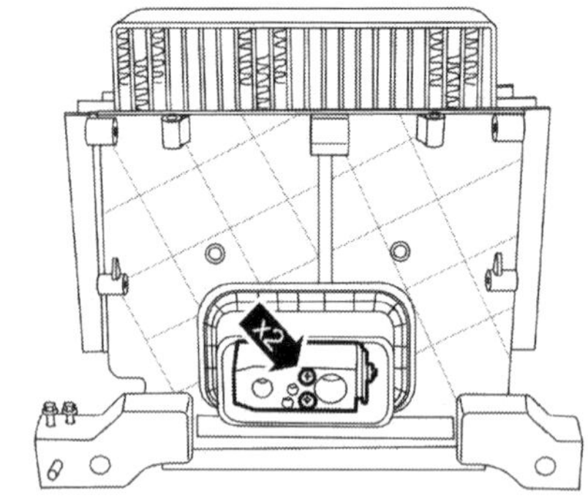

E88672

9. Remove the 2 bolts. Remove and discard the O-ring seals. Remove the thermostatic expansion valve.

Renew Part: Thermostatic expansion valve O-ring seal kit

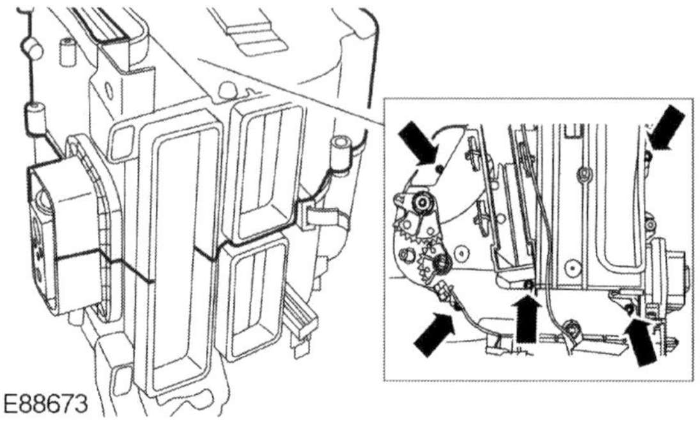

E88673

10. Remove the 5 screws. Separate the evaporator core housing.

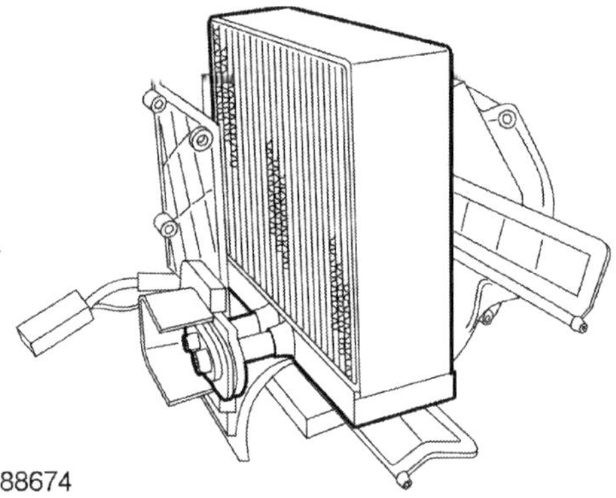

E88674

11. Remove the evaporator core.

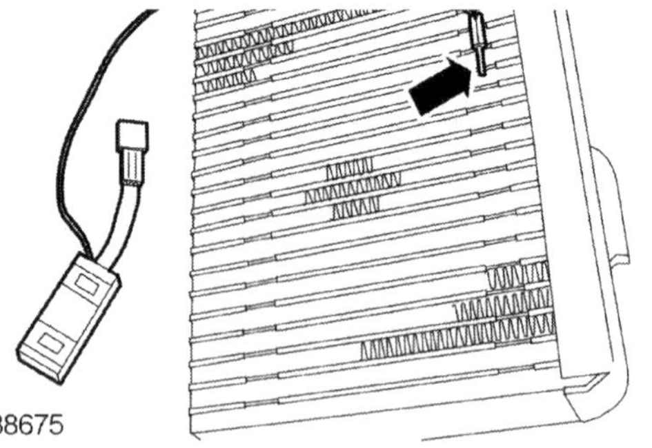

E88675

12. Remove the de-icing switch from the evaporator core.

Installation

1. To install, reverse the removal procedure.

2. Connect the battery ground cable.

Refer to: Startup Battery Disconnect and Connect (414-01 Battery, Mounting and Cables, General Procedures)

Thermostatic Expansion Valve

Removal and Installation

Part(s)

STEP	PART NAME	QUANTITY
Removal Step 4	Renew Part Thermostatic expansion valve O-ring seal kit	1

Removal

1. Disconnect the battery ground cable.

   Refer to: Startup Battery Disconnect and Connect (414-01 Battery, Mounting and Cables, General Procedures)

2. Recover the air conditioning (A/C) refrigerant.

   Refer to: Air Conditioning System Recovery, Evacuation and Charging (412-00 Climate Control System - General Information, General Procedures)

3. Remove the engine control module (ECM).

   Refer to: Engine Control Module (303-14 Electronic Engine Controls, ID4 2.2L Diesel Removal and Installation)

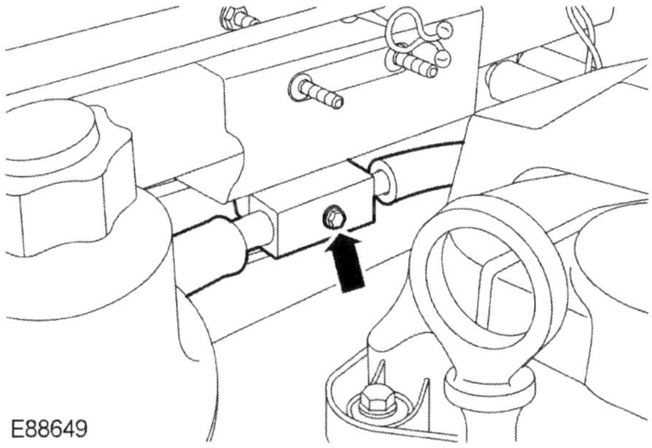

E88649

4. CAUTION: Before disconnecting any components, make sure the area is clean and free from foreign material. When disconnected all openings must be sealed.

   Remove the bolt. Remove and discard the O-ring seal. Release the thermostatic expansion valve manifold and tube assembly.

Renew Part: Thermostatic expansion valve O-ring seal kit

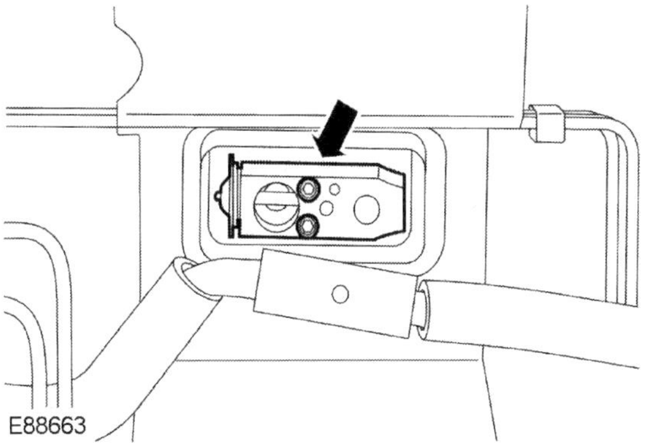

E88663

5. Remove the 2 bolts. Remove and discard the O-ring seals. Remove the thermostatic expansion valve.

Installation

1. To install, reverse the removal procedure.
2. Connect the battery ground cable.

   Refer to: Startup Battery Disconnect and Connect (414-01 Battery, Mounting and Cables, General Procedures)

Climate Control Assembly

Removal and Installation

Removal

1. Remove the instrument panel console.

Refer to: Instrument Panel Console (501-12, Removal and Installation).

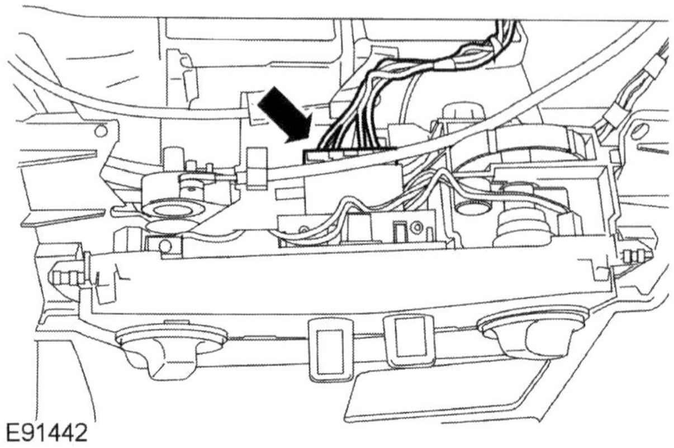

2. Release the climate control assembly.
   - Remove the 4 screws.

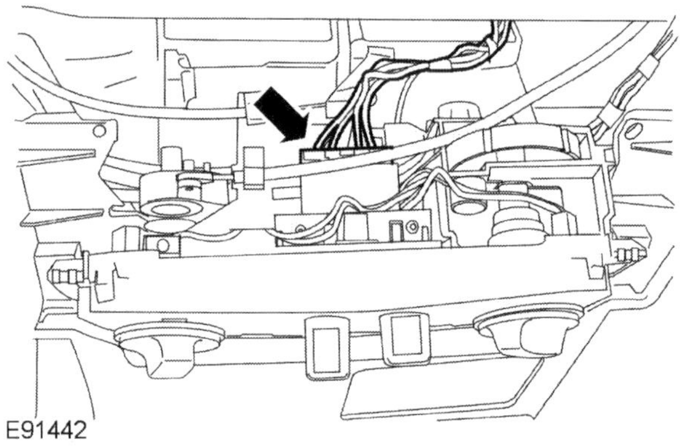

3. Disconnect the blower motor speed control switch electrical connector.

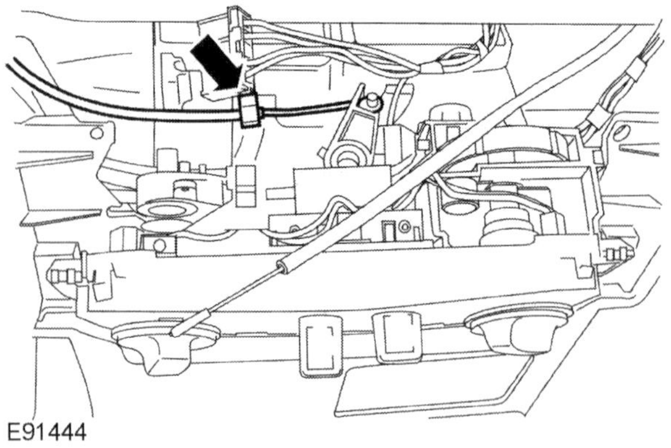

4. Disconnect the recirculation blend door control cable.
   - Release the clip.
5. Disconnect the heater control valve control cable.
   - Release the clip.

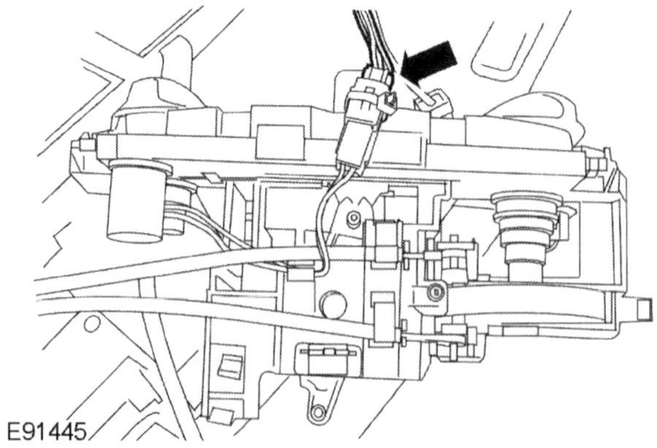

6. Disconnect the climate control assembly illumination electrical connector.

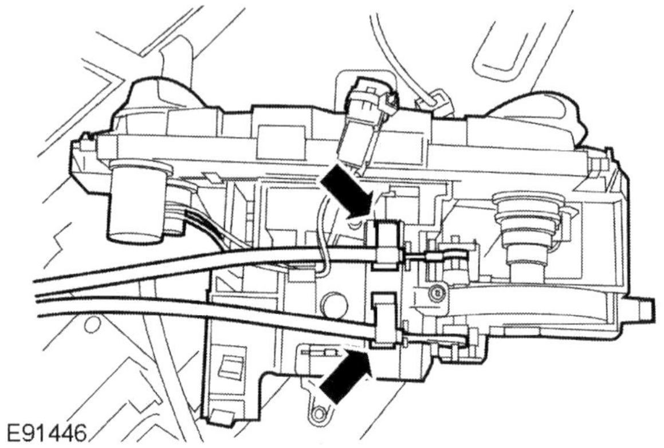

7. Remove the climate control assembly.
   - Release the clips.
   - Disconnect the 2 blend door control cables.

Installation

1. To install, reverse the removal procedure.

Blower Motor Resistor - LHD AWD

Removal and Installation

Removal

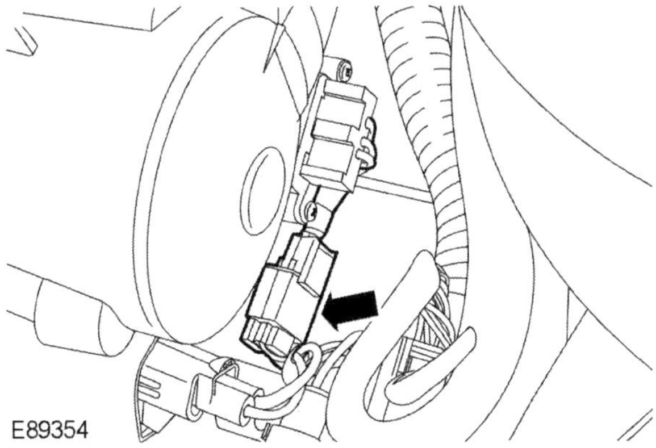

1. Disconnect the electrical connector.

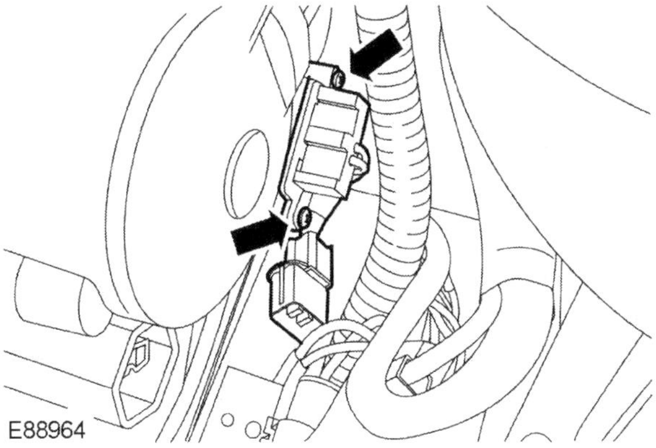

2. Remove the blower motor resistor.
    - Remove the 2 screws.

Installation

1. To install, reverse the removal procedure.

Blower Motor Resistor - RHD AWD

Removal and Installation

Removal

NOTES:

△Removal steps in this procedure may contain installation details.

△ Some variation in the illustrations may occur, but the essential information is always correct.

1. **Refer to:** Blower - RHD AWD, RHD (412-02 Heating and Ventilation, Removal and Installation)

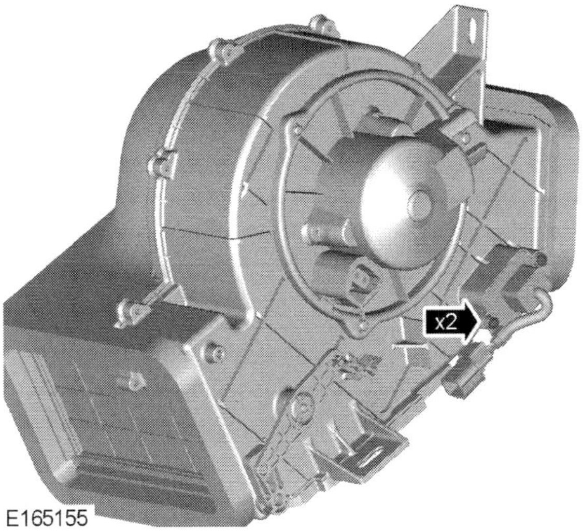

2.

Installation

1. To install, reverse the removal procedure.

Specifications

Torque specifications

Description	Nm	lb.ft
Instrument cluster	2	1
Instrument cluster upper finisher	2	1
Instrument cluster lower finisher	2	1

Description and Operation

COMPONENT LOCATION

Item	Description
1	Tachometer
2	Glow plugs active indicator
3	Seatbelt
4	Malfunction Indicator Lamp (MIL)
5	Left turn signal indicator
6	Engine Coolant Temperature gauge
7	High ECT indicator
8	High beam indicator
9	Hazard warning lamp indicator
10	Fuel level gauge
11	Right turn signal indicator
12	Dynamic Stability Control (DSC) on indicator
13	Dynamic Stability Control (DSC) off indicator
14	Speedometer
15	Differential lock indicator
16	Rear fog lamp indicator
17	Odometer and trip meter display
18	Trip reset button
19	Trailer indicator
20	Low fuel indicator
21	Overspeed warning indicator - market dependent
22	Diesel Particulate Filter (DPF) regeneration indicator
23	Anti-lock Brake System (ABS) indicator
24	Diesel Particulate Filter (DPF) full indicator
25	Brake warning indicator
26	Side lamps on indicator
27	Anti-theft alarm indicator
28	Transfer case low range indicator
29	Ignition/No charge indicator
30	Low oil pressure indicator

OVERVIEW

The Instrument Cluster (IC) is located in the instrument panel, above the steering column. The IC comprises analogue gauges and a number of indicator lamps to display system status.

ANALOGUE GAUGES

The analogue gauges located in the IC are as follows:

- Speedometer
- Tachometer
- Fuel level gauge
- ECT gauge.

Each analogue gauge is driven by an electronic stepper motor. The characteristics of this type of motor produce damping of the pointer needle. All gauges return to their respective zero positions when the ignition is switched off.

INDICATOR LAMPS

Indicator lamps are located in various positions in the IC and can be split into 2 groups; self-controlled and externally controlled.

Self-controlled indicators are dependent on software logic within the IC for activation.

Externally controlled indicators are supplied with current from their respective systems. Engine related externally controlled indicators are illuminated by the IC on receipt of a High Speed (HS) Controller Area Network (CAN) bus message from the Engine Control Module (ECM).

The following table shows the available indicators and indicates if they are subject to an indicator check at ignition on and if they are self or externally controlled.

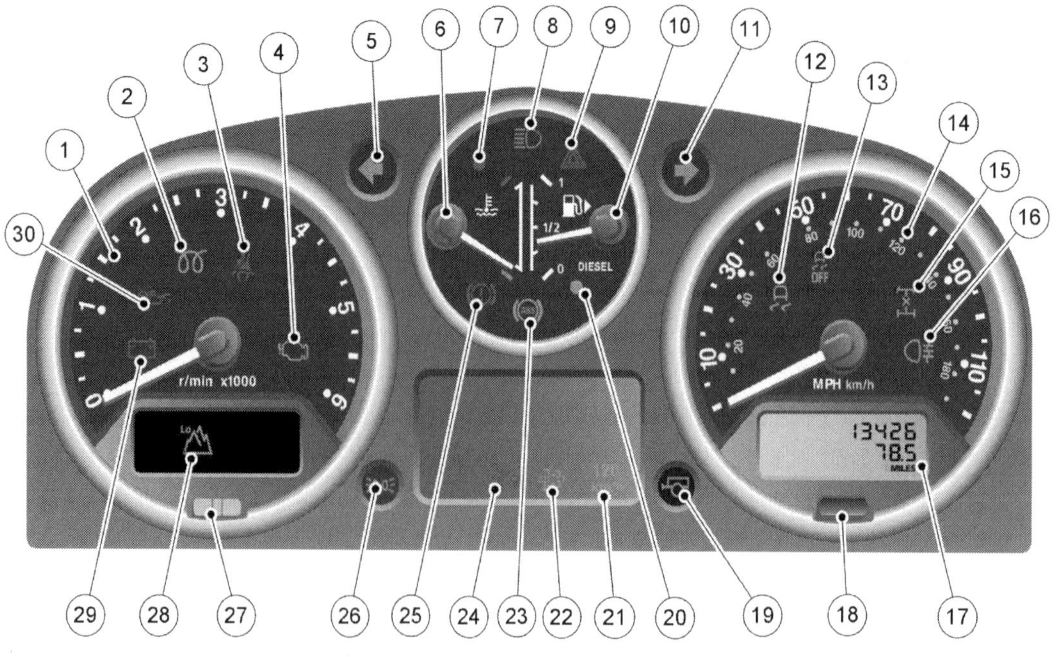

Indicator Lamp	Illumination Color	Bulb Check	Self Controlled (S)/Externally Controlled (E)
Glow plugs active	Amber	No (may illuminate at ignition on to show glow plugs active)	E
seatbelt	Red	No	E
MIL	Amber	* Yes	E
Left turn signal	Green	No	E
High ECT	Red	Yes	S
High beam	Blue	No	E
Hazard warning lamp	Red	No	E
Right turn signal	Green	No	E
DSC on	Amber	No	E
DSC off	Amber	No	E
Traction control	Amber	Yes	E
Differential lock	Amber	No	E
Rear fog lamp	Amber	No	E
Trailer	Green	No	E
Low fuel	Amber	Yes	S
Overspeed warning	Amber	Yes	S
**ABS	Amber	* Yes	E
DPF regeneration	Amber	No	E
DPF full	Red	No	E
Brake warning	Red	Yes	E
Side lamps on	Green	No	E
Anti-theft alarm	Red	No	E
Transfer box low range	Green	No	E
Ignition/No charge	Red	No	E
Low oil pressure	Red	No	E

* Bulb check performed by sub-system module, not IC.

** IC is compatible with both WABCO and Bosch ABS modules.

PRINCIPLES OF OPERATION

Speedometer

The IC receives a hardwired vehicle speed signal from the vehicle speed sensor. The vehicle speed sensor is a Hall effect sensor located on the transfer case. The sensor acts on a reluctor ring located on the transfer case rear output shaft.

Refer to: Transfer Case (308-07 Transfer Case, Vehicles With: MT82 6-Speed Manual Transmission Description and Operation)

Tachometer

The tachometer is driven by an engine speed signal transmitted on the HS CAN bus from the ECM. The signal is derived from the Crankshaft Position (CKP) sensor. The signal is received by the IC microprocessor and the output from the microprocessor drives the tachometer.

Fuel Level Gauge

The IC calculates the amount of fuel in the tank by providing a reference current to the fuel level sensor. The fuel level sensor uses a float operated Magnetic Passive Position Sensor (MAPPS) for measuring fuel tank contents.

The IC measures the returned output from the sensor, which is proportional to the level of the float arm and consequently the amount of fuel in the tank. The IC monitors the signal from the sensor at approximately 20 second intervals. This prevents the gauge needle pointer from continually moving due to the movement of fuel in the tank when cornering or braking.

Engine Coolant Temperature Gauge

The ECT gauge is driven by HS CAN bus messages from the ECM. For normal operating temperatures the gauge needle pointer is positioned centrally in the gauge display. The needle pointer position translates to the following approximate temperatures.

Engine Coolant Temperature °C (°F)	Needle Pointer Position
Ignition off	Park position
40 (104)	Cold (blue segment)
75 - 115 (167 - 239)	Normal (central)
120 (248)	Start of hot (red segment)
125 (257)	End of hot

Glow Plugs Active Indicator

The glow plugs active indicator is illuminated by the IC software on receipt of a HS CAN bus message from the ECM. The indicator illuminates in an amber color when the ignition is turned to position II. The indicator illumination period varies with ECT and if CT is high, will not illuminate.

The indicator is controlled by HS CAN bus messages from the ECM, which equate to the time the glow plugs are energized to pre-heat the combustion chambers. When the glow plug heating time is complete, the indicator is extinguished indicating to the driver that the engine can now be started.

Seatbelt Indicator

The seatbelt indicator is controlled by a hardwired feed from switches located in the front seat seatbelt buckles. The seatbelt indicator is fitted to Gulf specification vehicles only.

CONTROL DIAGRAM - SHEET 1 OF 2

NOTE: **A** = Hardwired; **D** = High Speed (HS) Controller Area Network (CAN) bus.

Item	Description
1	Battery
2	Lighting switch
3	Central Junction Box (CJB)
4	Fuel tank unit
5	Differential lock switch
6	ABS control module
7	Reverse gear switch
8	Hazard warning lamp switch
9	Hazard warning lamp relay
10	IC
11	ECM
12	Ignition switch
13	Battery Junction Box (BJB)

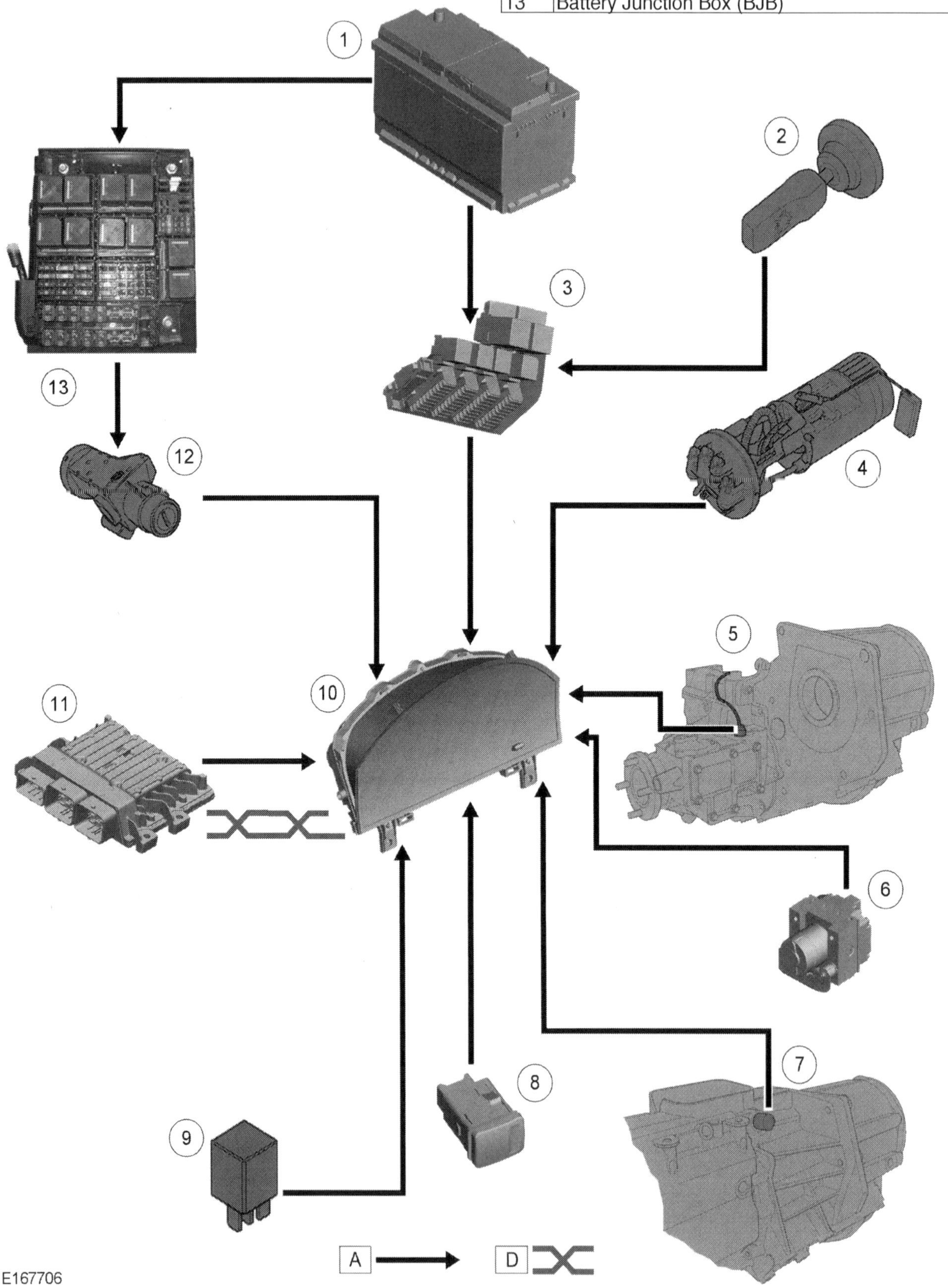

E167706

CONTROL DIAGRAM - SHEET 2 OF 2

NOTE: **A** = Hardwired.

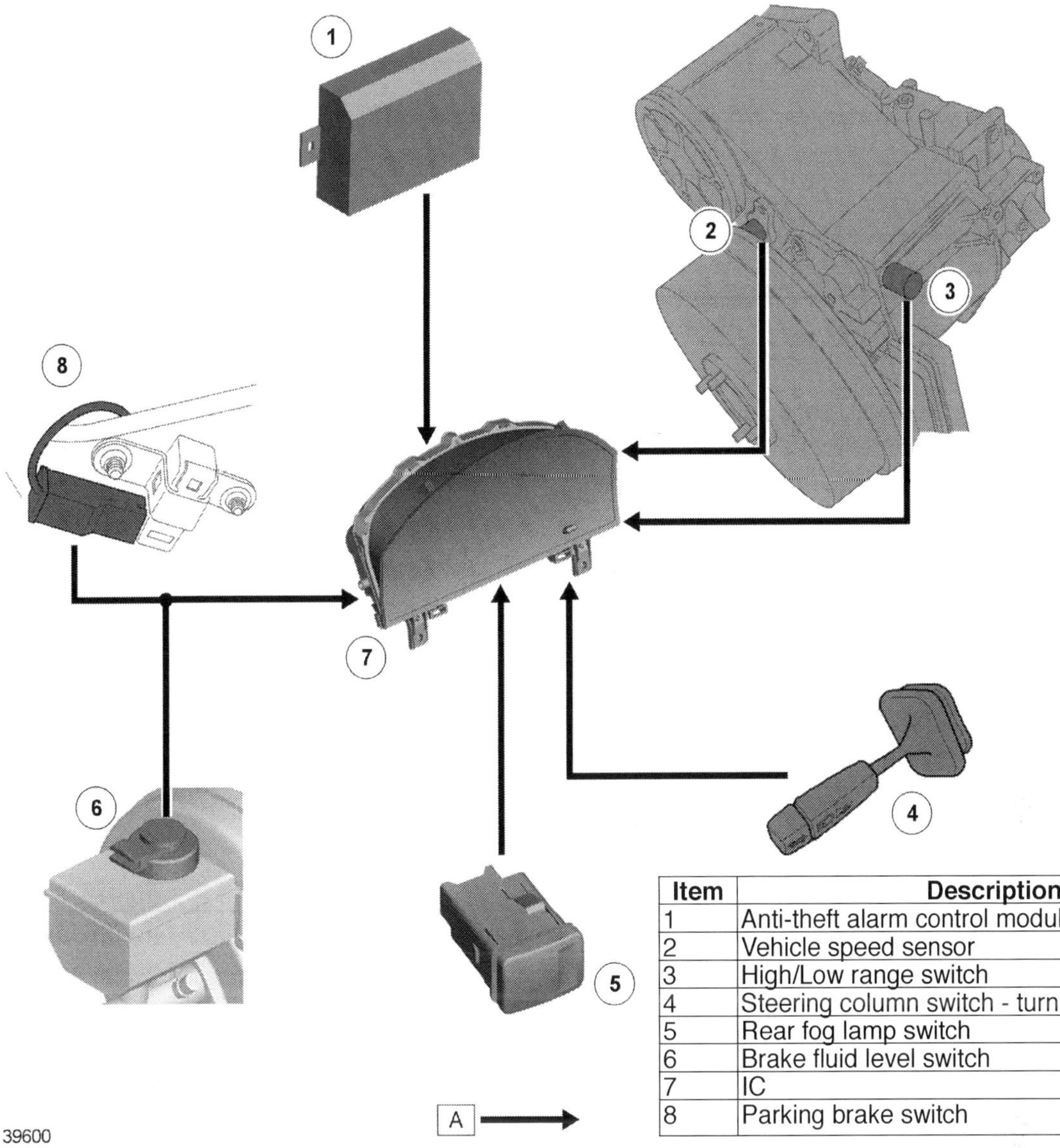

E139600

Item	Description
1	Anti-theft alarm control module
2	Vehicle speed sensor
3	High/Low range switch
4	Steering column switch - turn signal indicators
5	Rear fog lamp switch
6	Brake fluid level switch
7	IC
8	Parking brake switch

Malfunction Indicator Lamp (MIL)

The MIL is controlled by the IC software on receipt of a HS CAN bus message from the ECM. The lamp is illuminated for a bulb check by the ECM when the ignition is moved to position II. The lamp is extinguished when the engine starts.

If the MIL remains illuminated after the engine is started or illuminates while driving, a fault is present and must be investigated at the earliest opportunity. Illumination of the MIL indicates there is an On-Board Diagnostic (OBD) fault which will cause excessive emissions output.

Left and Right Turn Signal Indicators

The turn signal indicators are controlled by the IC software on receipt of hardwired signals from the steering column switch. When the turn signal indicator switch is operated, the IC receives a signal feed from the column switch. The IC software controls the flash rate of the indicator which flashes in a green color.

During normal turn signal indicator operation the indicator flashes slowly, accompanied simultaneously by a sound from the IC sounder. If a fault exists, the IC flashes the indicator at double speed.

High Engine Coolant Temperature Indicator

The high ECT indicator is illuminated on receipt of a HS CAN bus message from the ECM. The indicator illuminates when the ignition is turned to position II for a 3 second bulb check and is extinguished when the engine is started. If the indicator illuminates while driving, a fault in the engine cooling system has become present and the engine must be stopped at the earliest opportunity.

High Beam Indicator

The high beam indicator is controlled by the IC software on receipt of a hardwired signal from the CJB. The signal from the CJB originates from the steering column switch when high beam is selected.

Hazard Warning Lamp indicator

The hazard warning lamp indicator is controlled by the IC software on receipt of a hardwired signal from the hazard warning lamp switch. The hazard warning lamp indicators can operate with the ignition switched off, flashing both the left and right turn signal indicators simultaneously.

DSC on/off Indicators

The DSC on indicator is illuminated by the IC software on receipt of a hardwired signal from the ABS control module. The indicator is illuminated for 3 seconds for a bulb check when the ignition switch is turned to position II. If no fault exists, the indicator is extinguished after the bulb check period.

When traction control is active, the indicator flashes to inform the driver that the system is regulating engine output.

The DSC Off indicator is illuminated when the user presses the DSC off switch in the instrument panel. This behavior is controlled via a HS CAN Bus message from the ABS module. The DSC Off tell tale follows the same prove out strategy as DSC On.

Differential Lock Indicator

The differential lock indicator is illuminated on receipt of a hardwired signal from the differential lock switch. The indicator will illuminate at all times when the differential lock is selected and the ignition switch is in position II.

Rear Fog Lamp Indicator

The rear fog lamp indicator is illuminated on receipt of a hardwired signal from the rear fog lamp switch. The indicator will illuminate at all times when the rear fog lamps are selected on and the ignition switch is in position II.

Trailer Indicator

The trailer indicator is controlled by the IC software on receipt of a hardwired signal from the hazard warning lamp relay. When a trailer is connected, the hazard warning lamp relay energizes and provides a feed to the IC. The feed across the hazard warning lamp relay originates from the steering column switch. The IC software controls the flash rate of the indicator which flashes in a green color. The trailer indicator flashes slowly, accompanied simultaneously by a chime from the IC sounder, at the same rate as the turn signal indicators.

Low Fuel Indicator

The IC calculates the amount of fuel in the tank by providing a reference current to the fuel tank level sensor. If the cluster determines the level of fuel within the tank is at or below 14 liters (3.69 gallons) it will illuminate the low fuel indicator and emit a single chime. For more information, refer to 'Fuel Level Gauge' above.

Overspeed Indicator - market dependent

The amber overspeed indicator is controlled by the IC software. The IC receives a hardwired vehicle speed signal from the vehicle speed sensor on the transfer case. The indicator illuminates if vehicle speed reaches 120 km/h. The indicator also illuminates when the ignition is turned to position II for a 3 second bulb check and is extinguished when the engine is started.

Anti-lock Braking System Indicator

The ABS indicator is controlled by the ABS control module which transmits a hardwired signal to the IC. The indicator is illuminated in an amber color for a 3 second bulb check by the ABS control module when the ignition is turned to position II. If the indicator remains illuminated or illuminates when driving, an ABS fault has occurred and the ABS function will not be available.

The ABS control module will alert the driver that a Diagnostic Trouble Code (DTC) has been stored in its memory during the bulb check process. It will do this by;

- illuminating the indicator for 0.5 seconds
- extinguishing the indicator for 0.5 seconds
- illuminating the indicator for 2 seconds.

If the indicator is illuminated for a sensor fault, the indicator will remain illuminated at the next ignition cycle, even if the fault is rectified. When the vehicle is driven above a speed of 20 km/h (12.5 mph) the indicator will be extinguished. This allows the ABS control module to perform a thorough check of the system and to establish that the output from the replaced sensor is correct.

Diesel Particulate Filter Indicators

The DPF indicators are illuminated on receipt of a HS CAN bus message from the ECM.

If the amber DPF indicator illuminates while driving, the DPF is approaching capacity. If the red DPF indicator illuminates while driving, the DPF is full.

Refer to: Exhaust System - ID4 2.2L Diesel (309-00 Exhaust System, ID4 2.2L Diesel Description and Operation)

Brake Warning Indicator

The brake warning indicator is illuminated for a 3 second bulb check when the ignition is turned to position II. The indicator will also illuminate if the parking brake is on, or the brake fluid falls below a pre-determined level. The IC is hardwired to the parking brake switch and the brake fluid level switch, which are connected in parallel. If either of the conditions above are met a ground path is created, illuminating the indicator.

Side Lights On Indicator

The side lights on indicator is controlled by the lighting switch. When the lighting switch is turned to the side or headlamp position, a hardwired feed is provided to the IC via the CJB. On receipt of the hardwired feed, the IC illuminates the indicator.

Anti-theft Alarm Indicator

Illumination of the anti-theft alarm indicator is controlled directly by the anti-theft alarm control module.

Refer to: Anti-Theft - Active (419-01 Anti-Theft - Active, Description and Operation)

Refer to: Anti-Theft - Passive (419-01 Anti-Theft - Passive, Description and Operation)

Transfer Case Low Range Indicator

The transfer case high/low range switch is hardwired to the IC. When low range is selected, the transfer case provides a feed to IC, which subsequently illuminates the green low range indicator. The low range indicator remains permanently illuminated until high range is selected and the feed from the high/low range switch is removed.

Ignition/No Charge Warning Indicator

The ignition/no charge indicator is controlled by the IC software and illuminated on receipt of a HS CAN bus message from the ECM. The indicator illuminates in a red color when the ignition is turned to position II and is extinguished when the engine is started.

If the indicator remains illuminated after the engine has started or illuminates when driving, the alternator charge output has failed.

Low Oil Pressure Warning Indicator

The low oil pressure indicator is controlled by the IC software and illuminated on receipt of a HS CAN bus message from the ECM. The indicator is illuminated in a red color when the ignition switch is turned to position II. When the engine is started and the oil pressure increases the low oil pressure indicator should extinguish. If the indicator remains illuminated or illuminates when driving the vehicle should be stopped at the earliest opportunity and the engine switched off until the fault is rectified.

Lights on audible warning

The IC receives a 'Driver Door Ajar Status' signal via a hardwired connection from the anti-theft alarm control module. If the Driver door is open, the IC produces an audible chime if the user attempts to exit the vehicle without turning the lights off.

Instrument Cluster Replacement

If a new IC is to be fitted, Land Rover approved diagnostic equipment must be connected to the vehicle and the IC renewal procedure run. This will ensure that vehicle coding data is correctly installed in the new IC. The Land Rover approved diagnostic equipment will also record the current service interval data and restore the settings to the new IC.

Instrument Cluster

Diagnosis and Testing

Principles of Operation

For a detailed description of the Instrument Cluster, refer to the relevant Description and Operation section in the workshop manual.

Refer to: Instrument Cluster (413-01 Instrument Cluster, Diagnosis and Testing)

Inspection and Verification

⚠ CAUTION: Diagnosis by substitution from a donor vehicle is **NOT** acceptable. Substitution of control modules does not guarantee confirmation of a fault, and may also cause additional faults in the vehicle being tested and/or the donor vehicle.

NOTES:

△ If a control module or a component is suspect and the vehicle remains under manufacturer warranty, refer to the Warranty Policy and Procedures manual, or determine if any prior approval programme is in operation, prior to the installation of a new module/component.

△ When performing voltage or resistance tests, always use a digital multimeter accurate to three decimal places, and with an up-to-date calibration certificate. When testing resistance always take the resistance of the digital multimeter leads into account.

△ Check and rectify basic faults before beginning diagnostic routines involving pinpoint tests.

1. Visually inspect for obvious signs of damage and system integrity
2. With the ignition on, check the operation of the audio output from the instrument cluster integrated speakers by operating the turn signal indicators (left and right) and verifying that audible feedback (a ticking sound) is present

Visual Inspection

Electrical
• Fuses
• Wiring harnesses and connectors
• Instrument cluster
• Engine control module
• Anti-lock brake system control module

3. If an obvious cause for an observed or reported concern is found, correct the cause (if possible) before proceeding to the next step
4 If the cause is not visually evident, verify the symptom and refer to the Symptom Chart, alternatively check for Diagnostic Trouble Codes (DTCs) and refer to the DTC Index

DTC Index

For a list of Diagnostic Trouble Codes (DTCs) that could be logged on this vehicle, please refer to Section 100-00.

Refer to: Diagnostic Trouble Code Index - DTC: Instrument Cluster (IPC) (100-00 General Information, Description and Operation)

Instrument Cluster

Removal and Installation

Removal

1. Disconnect the battery ground cable. **Refer to:** Startup Battery Disconnect and Connect (414-01 Battery, Mounting and Cables, General Procedures)

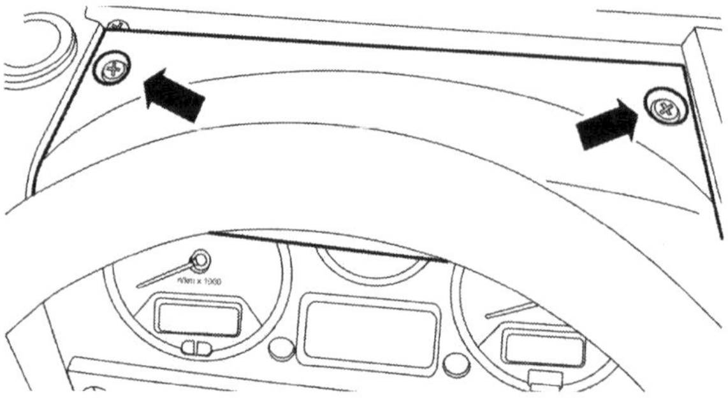

2. Remove the instrument cluster upper finisher.
   - Remove the 2 screws.

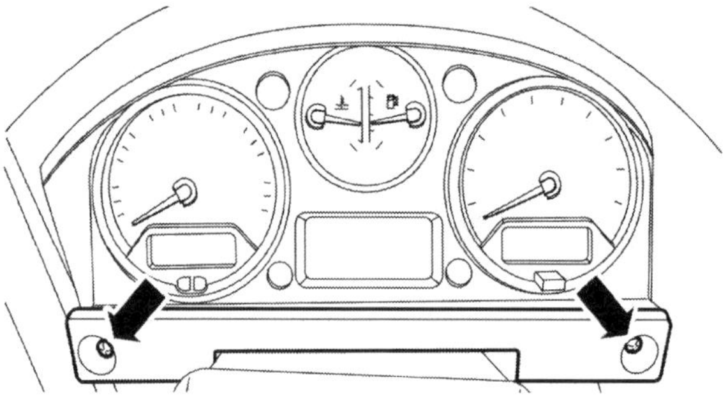

3. Remove the instrument cluster lower finisher.
   - Remove the 2 screws.

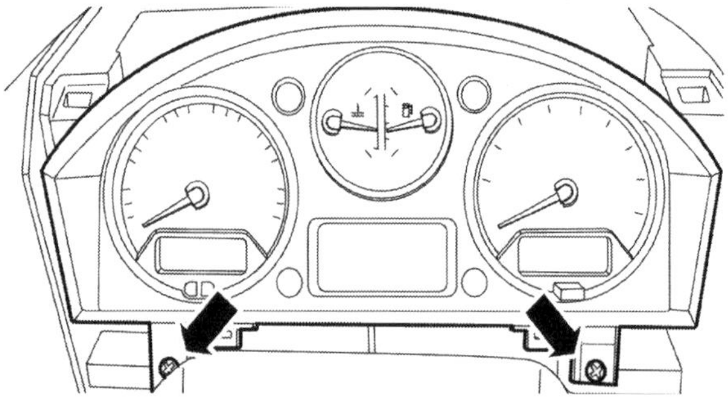

4. Release the instrument cluster.
   - Remove the 2 screws.

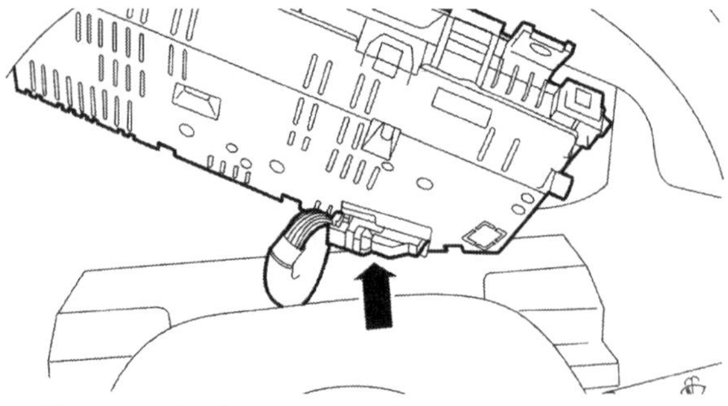

5. Remove the instrument cluster.
   - Disconnect the electrical connector.

Installation

1. Install the instrument cluster.
   - Connect the electrical connector.
2. Secure the instrument cluster.
   - Install the 2 screws.
3. Install the instrument cluster lower finisher.
   - Install the 2 screws.
4. Install the instrument cluster upper finisher.
   - Install the 2 screws.
5. Connect the battery ground cable.

Refer to: Battery Connect (414-01 Battery, Mounting and Cables, General Procedures).

6. Using the approved Land Rover diagnostic equipment, configure the instrument cluster.

Instrument Cluster Lens

Removal and Installation

Removal

⚠ NOTE: Some variation in the illustration may occur, but the essential information is always correct.

1. Fully extend the steering column for access
2. Disconnect the battery ground cable.

Refer to: (414-01 Battery, Mounting and Cables, Specifications)

3. **Refer to:** Instrument Cluster (413-01 Instrument Cluster, Removal and Installation)

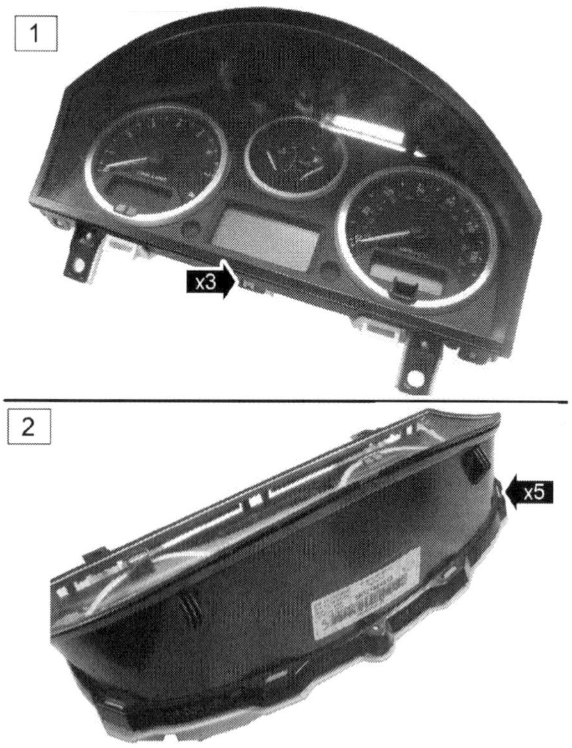

4. Carefully release the 3 clips along the bottom edge and the 5 clips along the top edge.

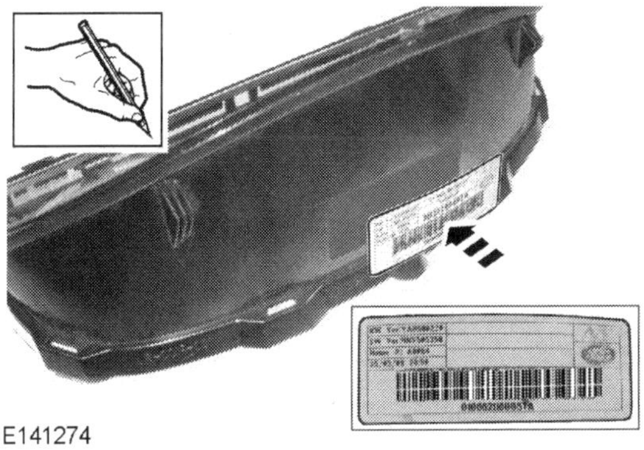

E141274

5. Copy the serial number and build date information from the old label on to the new label.

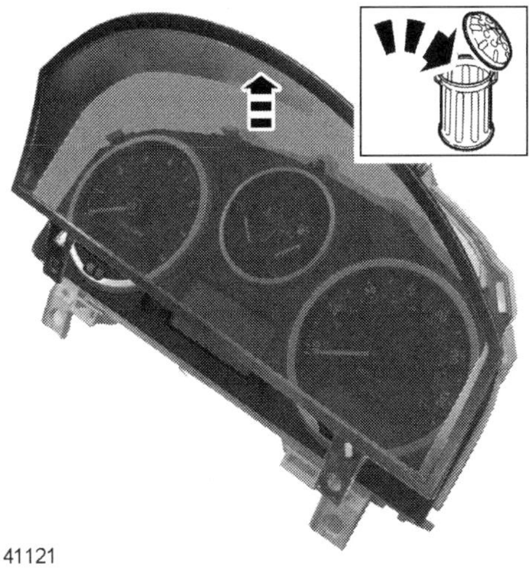

E141121

6. CAUTION: Take care not to damage the dials or buttons when removing the lens.

Installation

NOTE: Take care not to leave finger prints inside new lens when fitting.

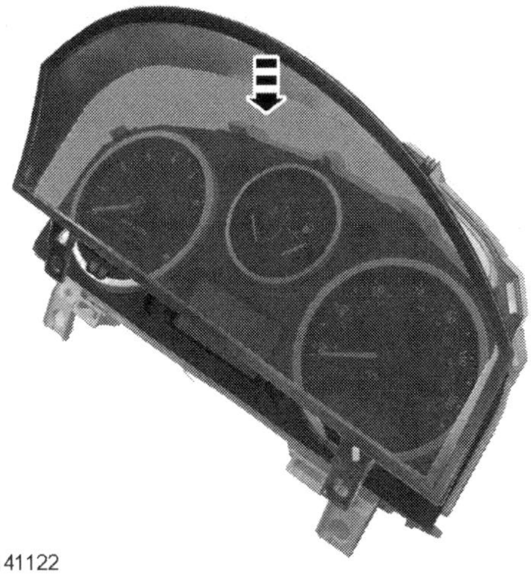

E141122

1.
2. **Refer to:** Instrument Cluster (413-01 Instrument Cluster, Removal and Installation)
3. Connect the battery ground cable

   Refer to: (414-01 Battery, Mounting and Cables, Specifications)
4. Reset the steering column to the original position.

Horn

413-06

Removal and Installation

Removal

1. Remove LH headlamp.

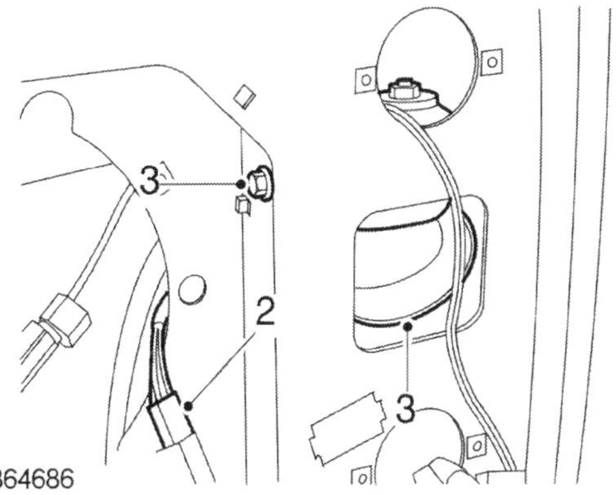

M864686

Refer to: Headlamp Assembly (417-01 Exterior Lighting, Removal and Installation)

2. Disconnect horn multiplug.
3. Remove nut and remove horn from mounting bracket.

Installation

1. Position horn, tighten securing and connect multiplug.
2. Fit LH headlamp.

   Refer to: Headlamp Assembly (417-01 Exterior Lighting, Removal and Installation)

413-07 Clock

Specifications

Torque specifications

Description	Nm	lb-ft
Clock support bracket	1	1

Removal and Installation

Removal

1. Disconnect the battery ground cable.

Refer to: Startup Battery Disconnect and Connect (414-01 Battery, Mounting and Cables, General Procedures)

2. Remove the audio unit.

Refer to: Audio Unit - Vehicles Without: Bluetooth (415-01 Audio Unit, Removal and Installation)

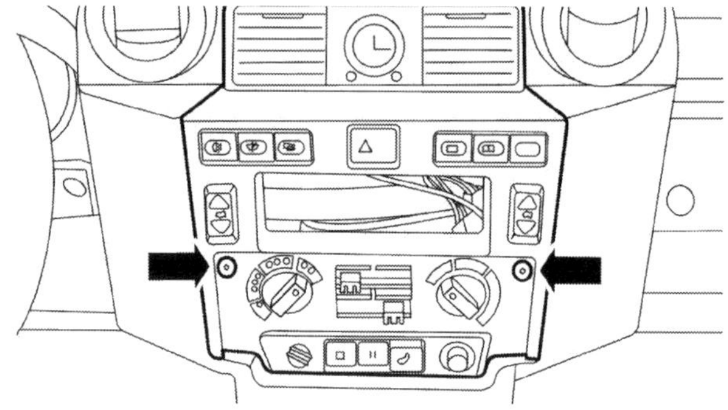

3. Release the instrument panel console.

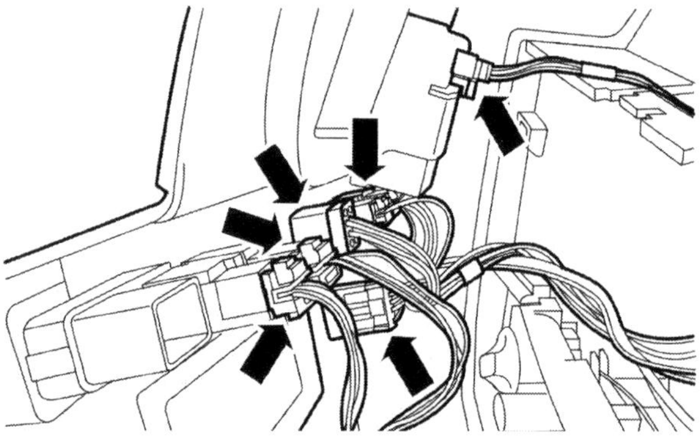

4. Disconnect the 6 electrical connectors.

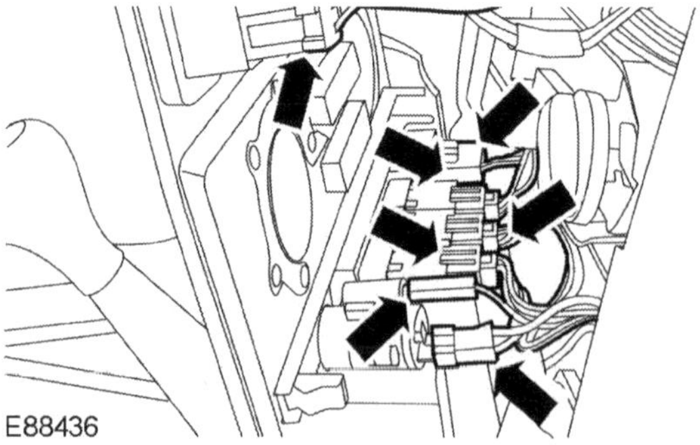

5. Remove the instrument panel console.
   - Disconnect the 7 electrical connectors.

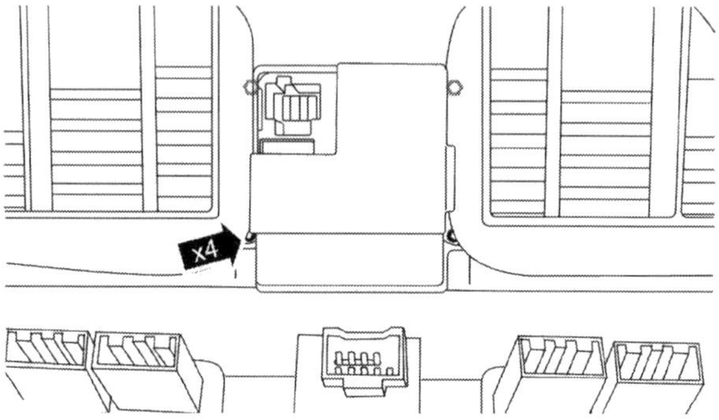

6. Remove the clock support bracket.
   - Remove the 4 screws.

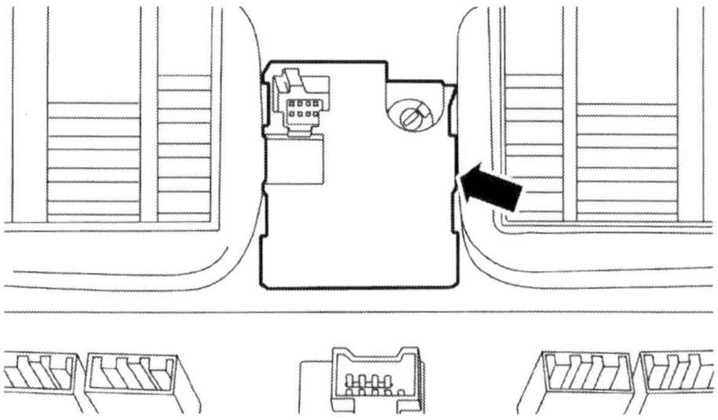

7. Remove the clock.

Installation

1. To install, reverse the removal procedure.
2. Connect the battery ground cable.

413-09 Warning Devices

Service Indicator Reset

General Procedures

Synchronization

△ NOTE: Make sure the Land Rover approved diagnostic equipment is not connected to the vehicle.

1. Set the ignition to the ON position.
2. Open the hood.
3. Open the drivers door.
4. △ NOTE: A service resetting message will **not** appear on the instrument cluster.

Fully depress both the accelerator and brake pedals and hold for 20 seconds.

5. Set the ignition to the OFF position.
6. Set the ignition to the ON position and check the message has been cleared.
7. If the message remains, repeat the process.

Battery Care Requirements

Description and Operation

INTRODUCTION

This document defines the requirements for care and maintenance of batteries, and the standard of battery care at retailers for new vehicles - so has therefore been amended for second users as it goes without saying that all batteries must be handled with care.

This applies to all types of 12 Volt Lead Acid Batteries used in Land Rover vehicles whether they are conventional flooded technology or Absorbed Glass Mat (AGM – also known as Valve Regulated Lead Acid (VRLA)) technology and also applies to both Primary, Secondary and Auxiliary Batteries. AGM batteries offer improved resistance to cycling as seen in stop start applications.

In order to prevent damage to the battery and ensure a satisfactory service life, all processes detailed within this document must be rigorously adhered to.

It is equally important therefore to note the following key points:

- The battery can be discharged by the following mechanisms:
- **Self Discharge**: - A lead acid battery will very slowly discharge itself due to its own internal chemical processes whether it is connected to a vehicle or not.
- **Quiescent Discharge**: - The vehicle electrical systems when connected to the battery will draw charge from the battery.

12 Volt Lead Acid Batteries rely on internal chemical processes to create a voltage and deliver current. These processes and the internal chemical structure of the battery can be damaged if the battery is allowed to discharge over a number of weeks / months, or is left in a discharged state for a lengthy time period.

- **On vehicles with conventional ignition keys**, these must not be left in the ignition lock barrel when the transit relay has been removed, otherwise quiescent current will increase and the battery will discharge more rapidly.
- **For keyless vehicles**, the Smart Key must be stored at least 5m (16 ft) away from the vehicle when the vehicle is parked or stored.
- **AGM Batteries are fully sealed and cannot have the electrolyte level topped up.**

HEALTH AND SAFETY PRECAUTIONS

WARNINGS:

⚠ BATTERY CELLS CONTAIN SULPHURIC ACID AND EXPLOSIVE MIXTURES OF HYDROGEN AND OXYGEN GASES. IT IS THEREFORE ESSENTIAL THAT THE FOLLOWING SAFETY PRECAUTIONS ARE OBSERVED.

⚠ Batteries emit highly explosive hydrogen at all times, particularly during charging. To prevent any potential form of ignition occurring when working in the vicinity of a battery:

- *Do not smoke when working near batteries.*
- *Avoid sparks, short circuits or other sources of ignition in the battery vicinity.*
- *Switch off current before making or breaking electrical connections.*
- *Ensure battery charging area is well ventilated.*
- *Ensure the charger is switched off when: a) connecting to a battery; b) disconnecting from the battery.*
- *Always disconnect the ground cable from the battery terminal first and reconnect it last.*

⚠ Batteries contain poisonous and highly corrosive acid. To prevent personal injury, or damage to clothing or the vehicle, the following working practices should be followed when topping up, checking electrolyte specific gravity, removal, refitting or carrying batteries:

- *Always wear suitable protective clothing (an apron or similar), safety glasses, a face mask and suitable gloves.*
- *If acid is spilled or splashed onto clothing or the body, it must be neutralized immediately and then rinsed with clean water. A solution of baking soda or household ammonia and water may be used as a neutralizer.*
- *In the event of contact with the skin, drench the affected area with water. In the case of contact with the eyes, bathe the affected area with cool clean water for approximately 15 minutes and seek urgent medical attention.*
- *If battery acid is spilled or splashed on any surface of a vehicle, it should be neutralized and rinsed with clean water.*
- *Heat is generated when acid is mixed with water. If it becomes necessary to prepare electrolyte of a desired specific gravity,*

SLOWLY pour the concentrated acid into water (not water into acid), adding small amounts of acid while stirring. Allow the electrolyte to cool if noticeable heat develops. With the exception of lead or lead-lined containers, always use non-metallic receptacles or funnels. Do not store acid in excessively warm locations or in direct sunlight.

⚠ Due to their hazardous contents, the disposal of batteries is strictly controlled. When a battery is scrapped, ensure it is disposed of safely, complying with local environmental regulations. If in doubt, contact your local authority for advice on disposal facilities.

Any actions must be carried out in accordance with the table shown in the **Determining Battery Condition section of this procedure**.

BATTERY MAINTENANCE

- Any battery whether it is in a vehicle or a replacement part must be tested and/or re-charged every 30 days and MUST be re-charged after every 90 day period.
- The battery must always be connected to a battery support unit during any diagnostic sessions, including software updates.

Enhancement to the recommended battery test equipment software, has removed the need to physically remove battery surface charge before completing a Battery Test.

The battery may be tested either on a bench or on the vehicle.

△ NOTE: Make sure the Ignition is OFF and the vehicle is powered down with the modules entering 'sleep mode' before commencing with the battery test.

Battery Test Types

Flooded Battery Testing

In some markets serviceable flooded Batteries are still in use, to test these correctly check the electrolyte level as outlined in the CONFIRMING ELECTROLYTE LEVEL in this procedure before any further testing. These batteries are identified with cell plugs on top of the battery.

All AGM and Flooded Battery Testing

The battery condition must be checked in accordance with the battery test process utilizing a JLR approved tester as outlined in **the Equipment section of this procedure**.

△ NOTE: The Midtronics code from the tester must be recorded on the form.

Any actions must be carried out in accordance with the table shown in the **Determining Battery Condition section of this procedure**. The details must be recorded on the New Vehicle Storage Form which is part of the new vehicle storage document.

⚠ CAUTION: DO NOT connect the tester to any other circuit or chassis point other than the battery negative terminal.

EQUIPMENT

All equipment used must be functionally capable of meeting the compliance requirements. Please refer to the approved equipment document (JLR 000015).

Battery condition can be measured using the appropriate hand-held Midtronics tester as follows:

Battery Type	Battery Tester
	Land Rover
AGM & Flooded	Midtronics EXP1080, GRX-3080, GR8-1180

△ NOTE: All equipment must be calibrated

DETERMINING BATTERY CONDITION

TESTER RESULTS	ACTION
GOOD BATTERY	Return to service.
CHARGE AND RE-TEST	Fully charge battery as per the time to charge table on the EPX-1080. Re-test battery. If same result replace battery.
REPLACE BATTERY OR BAD CELL BATTERY	Repeat the test with the battery disconnected from the vehicle and follow the instructions as per the test result.
REPLACE BATTERY OR BAD CELL BATTERY	For a repeat Bad Cell or Replace Battery test result, **DO NOT ATTEMPT TO RECHARGE, REPLACE THE BATTERY**.
FROZEN BATTERY	Allow the battery to thaw naturally in workshop conditions and re-test.
UNABLE TO DO TEST	Disconnect battery from vehicle and re-test.

BATTERY CHARGING

It is essential that a suitably ventilated defined area exists in each retailer for battery charging.

⚠ CAUTION: It is very important that when charging batteries using a traction charger or other stand-alone chargers that the charger is set for the correct type of battery before charging commences. If the wrong switch is selected the result would be a battery that is not charged fully and / or overheating can occur. Follow the manufacturers operating instructions.

Batteries **MUST BE** tested and if necessary charged every 30 days and charged after 90 days irrespective of any test. It is recommended that retailers always have fully charged batteries ready for use.

⚠ CAUTION: Do not charge AGM batteries with voltages over 14.8 Volts as this will damage the battery.

To bring a discharged but serviceable battery back to a fully charged condition proceed as follows:

- Check and if necessary top-up the battery electrolyte level. (Flooded maintainable batteries only)
- To charge the battery using a JLR approved charger as detailed in the approved equipment document follow the manufacturers operating instructions.

△ NOTE: When using the Midtronics Diagnostic Charger, automatic mode must always be used. After charging and analysis, the charger may display 'Top-Off Charging', Hit STOP To End. Do not stop charging until the current falls to 5A or less, otherwise the battery will not be fully charged.

Following charging, a post charge battery condition test must be carried out in accordance with the table shown in the **Determining Battery Condition section of this procedure**.

△ NOTE: The Midtronics code from the tester must be recorded on the form.

Any actions must be carried out in accordance with the table shown in the **Determining Battery Condition section of this procedure**.

BATTERY REPLACEMENT

If it is determined that a battery requires replacement, always refer to the appropriate section of the workshop manual for instructions on removing and installing the battery from the vehicle.

On in service vehicles fitted with a Battery Monitoring System (BMS), the BMS control module must be reset following the installation of a new battery. The BMS control module reset procedure must be performed using an approved diagnostic system.

CONFIRMING ELECTROLYTE LEVEL

WARNINGS:

⚠ BEFORE CHECKING THE ELECTROLYTE LEVEL AND TOPPING-UP THE BATTERY WITH DISTILLED WATER, REFER TO THE HEALTH AND SAFETY PRECAUTIONS SECTION.

⚠ AGM TECHNOLOGY BATTERIES ARE FULLY SEALED FOR LIFE AND NO ATTEMPT SHOULD BE MADE TO CHECK OR TOP UP THE ELECTROLYTE LEVEL.

NOTES:

△ In hot climates more frequent checks of the battery electrolyte level and condition are required. If necessary, the battery cells can be topped up using distilled water.

△ Only flooded batteries can have the electrolyte level topped up.

On certain flooded battery types the electrolyte level will need to be checked.

1. Make sure the battery is of a type suitable for topping up. These types of batteries will have cell plugs visible on the top face of the battery or a removable access panel to allow access to the cells.
2. On batteries with a clear or opaque case and level marks, check the electrolyte level by visual inspection of the maximum level indicator mark on the battery casing, indicating adequate level above the battery separators.
3. On batteries with black cases, remove the cell plugs or access panel and ensure the electrolyte level is level with the indicator in the cell hole. A flashlight may be required to see the electrolyte level on this type of battery.
4. If the electrolyte level is low, top-up using **distilled water only**.
5. Refit the battery cell plugs.
6. Charge the battery for 1 hour using a recommended charger, if a GR8-1180 or GRX-3080 is used select manual charging for an hour.
7. Carry out a battery test using the recommended test equipment in line with **Battery Test & Equipment Section**.

△ NOTE: Maintenance free and Valve Regulated (AGM) batteries are sealed and therefore cannot be topped up.

⚠ CAUTION: **DO NOT** overfill.

Quiescent Drain

Description and Operation

VEHICLE QUIESCENT CURRENT TESTING

On vehicles equipped with a BMS (Battery Monitoring System) , the diagnostic routine for quiescent drain testing in the approved Land Rover diagnostic system should be utilized.

If a startup battery discharges continuously or is left for a prolonged period of time, it is recommended that a quiescent drain test is performed as described below.

The battery drain should be measured using the approved Land Rover diagnostic system or a Digital Multi Meter (DMM). A procedure for quiescent drain measurement using the diagnostic system is available in the Diagnosis and Testing section of the Workshop Manual. The vehicle should be single locked on. The test must take place in line with the quiescent drain values reference table below.

EQUIPMENT

Approved Jaguar Land Rover diagnostic system.

Method Of Measurement

WARNING: Use Jaguar Land Rover approved diagnostic system only.

Only on vehicles with secure tracker feature installed.

Before disconnecting power to the TCU (Telematic Control Unit Module) the vehicle owner must place the TCU into Service Mode. This must be done for example when disconnecting the startup battery to do maintenance repairs. If the disconnection procedure is not done, the system will send a theft alert to the following:

- The user through the InControl Remote mobile phone application
- The Telematics Service Provider (TSP).

Then the system will enter 'Theft Notification' or 'Stolen Vehicle Tracking' mode during which time the charge in the TCU battery would be used. The emergency services may be falsely alerted that the vehicle has been stolen.

For security purposes 'Service Mode' is the sole responsibility of the vehicle owner as it requires a Personal Identification Number (PIN) code that has been defined by the customer.

When the startup battery is disconnected for a period of time, the vehicle owner can define the time it is to be disconnected for. If the time is exceeded, the vehicle owner must re-activate the 'Service Mode'. (To re-activate the 'Service Mode' the vehicle owner is required to use 'InControl Remote' mobile phone application.)

NOTE:

- Use the reset switch on amp clamps before every test performed.
- Measure the current for the testing on the negative battery to chassis ground connection cable.

Using a Digital Multi Meter (DDM).

QUIESCENT CURRENT TESTING

1. Switch ignition to 'on' or select ignition power mode 6 in keyless vehicles and switch to 'off' (power mode 0) (do not crank)
2. Remove key from ignition switch (if equipped)
3. Open and latch all doors, hood and luggage compartment lid or tailgate
4. Lock the vehicle using the remote function on the remote device or smart key. (Single lock only to avoid volumetric alarm arming)
5. Record the amperage readings after the shutdown period referenced in the Topix on line resource for details
6. Record the final reading on the battery report form.

The preferred method of testing following an excessive current consumption figure is to use a current probe around individual junction box leads to the various suspected circuits to identify a potential cause. This is in preference to the old method of removing fuses for the following reasons:

1. The drain may be caused by a control module remaining active and preventing the quiescent drain from reducing to normal levels
2. The drain may be caused by a relay winding that is activated. Pulling the fuse can allow this to 'reset' and the drain will be lost and go un-diagnosed.

Land Rover Quiescent Drain Values

Brand	Model	Shut down period (Minutes)	Typical battery drain values
Land Rover	All	30	<30 mA (average over 10 Minutes)

The total current drain will be higher if certain approved accessories are installed (for example: tracker, trailer module.)

Charging System

Diagnosis and Testing

Principles of Operation

For a detailed description of the Charging System, refer to the relevant Description and Operation section in the workshop manual.

Refer to: Generator - ID4 2.2L Diesel (414-02 Generator and Regulator, ID4 2.2L Diesel Description and Operation)

Inspection and Verification

⚠ CAUTION: Diagnosis by substitution from a donor vehicle is **NOT** acceptable. Substitution of control modules does not guarantee confirmation of a fault, and may also cause additional faults in the vehicle being tested and/or the donor vehicle.

NOTES:

△ If the control module or a component is suspect determine if any prior approval programme is in operation, prior to the installation of a new module/component.

△ When performing voltage or resistance tests, always use a digital multimeter accurate to three decimal places, and with an up-to-date calibration certificate. When testing resistance always take the resistance of the digital multimeter leads into account.

△ Check and rectify basic faults before beginning diagnostic routines involving pinpoint tests.

△ Check the vehicle battery condition and state of charge before condemning any of the charging system components. For additional information, refer to the battery care manual.

1. Visually inspect for obvious signs of damage and system integrity
2. If an obvious cause for an observed or reported concern is found, correct the cause (if possible) before proceeding to the next step
3. If the cause is not visually evident, verify the symptom and refer to the Symptom Chart, alternatively check for Diagnostic Trouble Codes (DTCs) and refer to the DTC Index

Visual Inspection

Mechanical	Electrical
• Generator • Generator pulley • Primary drive belt • Primary drive belt tensioner	• Fuses • Wiring harnesses and connectors • Mega-fuse • Instrument Cluster (IC) • Engine Control Module (ECM) • Battery • Generator • Engine/generator ground connection

Symptom Chart

Symptom	Possible Causes	Action
Charge warning indicator does not illuminate (ignition on, engine not running)	• Flat/unserviceable battery • Instrument cluster fault	• Refer to the relevant section of the workshop manual and test the battery • Using the Land Rover Approved Diagnostic Equipment, check the Instrument Cluster (IC) for related DTCs and refer to the relevant DTC index
Charge warning indicator illuminates and the battery discharges (engine running)	• Primary drive belt fault • Engine system fault • Generator pulley slipping on shaft • Excessive electrical load • Ignition switch fault (stuck at crank position after engine started) • Generator internal failure • Battery cable open circuit, high resistance	• Check the primary drive belt condition and tension • Using the Land Rover Approved Diagnostic Equipment, check the engine control module for related DTCs and refer to the relevant DTC index • Check the installation of the generator pulley • Check that electrical loads are switched off (glow plugs, non-genuine accessories and equipment) • Check the operation of the ignition switch • GO to Pinpoint Test A • Refer to the electrical circuit diagrams and check the battery power and ground circuits for open circuit, high resistance
Charge warning indicator illuminates intermittently (engine running)	• Primary drive belt fault • Engine system fault • Generator pulley slipping on shaft • Generator internal failure	⚠ NOTE: The use of a boost charger, power pack or jump leads may cause the warning indicator to illuminate temporarily. • Check the primary drive belt condition and tension • Using the Land Rover Approved Diagnostic Equipment, check the engine control module for related DTCs and refer to the relevant DTC index • Check the installation of the generator pulley • GO to Pinpoint Test A
Charge warning indicator does not illuminate and the battery discharges	• Flat/unserviceable battery • Excessive electrical load • Engine system fault • Generator internal failure	• Refer to the relevant section of the workshop manual and test the battery • Check that electrical loads are switched off (glow plugs, non-genuine accessories and equipment) • Using the Land Rover Approved Diagnostic Equipment, check the engine control module for related DTCs and refer to the relevant DTC index • GO to Pinpoint Test A
Noise (mechanical)	• Primary drive belt fault • Generator pulley slipping on shaft • Generator internal failure	• Check the primary drive belt condition and tension • Check the installation of the generator pulley • Disconnect the primary drive belt and check that the generator rotates freely

Pinpoint Tests

PINPOINT TEST A : BATTERY TESTS	
TEST CONDITIONS	**DETAILS/RESULTS/ACTIONS**
A1: BATTERY TEST 1	
NOTES: △ The battery must be fully charged and any battery defects rectified before continuing with generator diagnosis.	
△Any quiescent drain faults must be rectified before continuing with generator diagnosis.	
	1. Using a Midtronics hand held tester or the Midtronics GR-1 diagnostic charger, carry out the "Midtronics battery test" as detailed in the battery care manual. **Refer to:** Battery Care Requirements (414-00 Battery and Charging System - General Information, Description and Operation)
	2. Record battery diagnostic result on the provided form
	Does the battery pass the "Midtronics battery test"? **Yes** GO to Pinpoint Test B **No** Rectify any battery defects before continuing with generator diagnosis GO to Pinpoint Test B

PINPOINT TEST B : GENERATOR TESTS	
TEST CONDITIONS	**DETAILS/RESULTS/ACTIONS**
B1: GENERATOR TEST 1	
△ NOTE: The heated rear window will automatically switch off after 8 minutes.	
	1. Set the ignition to on (engine off)
	2. Set all electrical loads to off (blower, radio, interior lights, etc)
	3. Start the engine
	4. Set the heated rear window to on
	5. Set a multimeter to measure voltage
	6. Connect the multimeter to the battery positive and negative terminals
	7. Wait 90 seconds
	8. Record the voltage value (V1)
	Is the voltage value (V1) **16.00 volts or greater** at idle? **Yes** Install a new generator **No** GO to 2
B2: GENERATOR TEST 2	
	1. Using voltage value (V1)
	Is the voltage value (V1) **13 volts or greater** at idle (but less than 16.00 volts)? **Yes** GO to 3 **No** GO to Pinpoint Test C
B3: GENERATOR TEST 3	
	1. Check the state if the battery/charging warning indicator in the instrument cluster
	Is the battery/charging warning indicator illuminated? **Yes** GO to Pinpoint Test C **No** The generator output is within the expected range. **Do not install a new generator.** If a customer concern is still evident, contact dealer technical support

PINPOINT TEST C : LIN BUS CIRCUIT TESTS	
TEST CONDITIONS	DETAILS/RESULTS/ACTIONS
C1: LIN BUS CIRCUIT TEST 1	
NOTE: The battery/charging warning indicator is driven by the powertrain control module via the CAN bus	
	1. Refer to the electrical circuit diagrams and check the generator LIN bus circuit for short circuit to ground, short circuit to power, open circuit, high resistance
	Are any generator LIN bus circuit faults evident? **Yes** Repair the generator LIN bus circuit as necessary GO to Pinpoint Test B **No** GO to Pinpoint Test D

PINPOINT TEST D : BATTERY/GENERATOR CIRCUIT TESTS	
TEST CONDITIONS	DETAILS/RESULTS/ACTIONS
D1: BATTERY/GENERATOR CIRCUIT TEST 1	
NOTE: The heated rear window will automatically switch off after 8 minutes.	
	1. Set the ignition to on (engine off)
	2. Set all electrical loads to off (blower, radio, interior lights, etc)
	3. Start the engine
	4. Set the heated rear window to on
	5. Set a multimeter to measure voltage
	6. Connect the multimeter to the generator body and the battery negative terminal
	7. Record the voltage value (V2)
	Is the voltage value (V2) **less than 0.30 volts**? **Yes** GO to 2 **No** Stop the engine. Refer to the electrical circuit diagrams and check the generator/engine ground circuit for high resistance. Check the installation of the generator. Rectify as necessary and retest. GO to Pinpoint Test B
D2: BATTERY/GENERATOR CIRCUIT TEST 2	
	1. Connect the multimeter to the battery positive terminal and the generator positive terminal
	2. Record the voltage value (V3)
	Is the voltage value (V3) **less than 0.30 volts**? **Yes** Install a new generator **No** Stop the engine. Refer to the electrical circuit diagrams and check the generator power circuit for high resistance. Rectify as necessary and retest. GO to Pinpoint Test B

DTC Index

For a list of Diagnostic Trouble Codes (DTCs) that could be logged on this vehicle, please refer to Section 100-00.

Refer to: Diagnostic Trouble Code Index - DTC: Engine Control Module (PCM) (100-00 General Information, Description and Operation)

Refer to: Diagnostic Trouble Code Index - DTC: Instrument Cluster (IPC) (100-00 General Information, Description and Operation)

Specifications

Torque specifications

Description	Nm	lb-ft
Battery terminal nuts	5	4
Battery clamp nuts	9	7
Battery ground cable to transmission stud	30	22

Diagnosis and Testing

Principles of Operation

For a detailed description of the battery system and operation, refer to the relevant Description and Operation section of the workshop manual

Refer to: Battery Diagnostics - 12 Volt Midtronics EXP-1080 JLR Hand-Held Battery Diagnostic Tool (414-01 Battery, Mounting and Cables, Diagnosis and Testing)

Inspection and Verification

⚠ CAUTION: Diagnosis by substitution from a donor vehicle is **NOT** acceptable. Substitution of control modules does not guarantee confirmation of a fault and may also cause additional faults in the vehicle being checked and/or the donor vehicle.

NOTES:

△ Generic scan tools may not read the codes listed, or may read only five digit codes. Match the five digits from the scan tool to the first five digits of the seven digit code listed to identify the fault (the last two digits give additional information read by the manufacturer-approved diagnostic system).

△ When performing voltage or resistance tests, always use a digital multimeter that has the resolution ability to view 3 decimal places. For example, on the 2 volts range can measure 1mV or 2 K Ohm range can measure 1 Ohm. When testing resistance always take the resistance of the digital multimeter leads into account.

△ Check and rectify basic faults before beginning diagnostic routines involving pinpoint tests.

△ If DTCs are recorded and, after performing the pinpoint tests, a fault is not present, an intermittent concern may be the cause. Always check for loose connections and corroded terminals.

1. Visually inspect for obvious signs of mechanical or electrical damage.
2. If an obvious cause for an observed or reported concern is found, correct the cause (if possible) before proceeding to the next step.
3. If the cause is not visually evident, check for Diagnostic Trouble Codes (DTCs) and refer to the DTC Index.
4. Check the Jaguar Land Rover (JLR) claims submission system for open campaigns. Refer to the corresponding bulletins and SSMs which may be valid for the specific customer complaint and carry out the recommendations as required.

Visual Inspection

Mechanical	Electrical
• Generator • Drive belt • Drive belt tensioner • Generator pulley • Check the security of the generator fixings	• Generator • Battery • Battery connections • Starter motor • Harnesses and connectors • Fuses • Charge warning indicator function

Symptom Chart

Symptom	Possible Causes	Action
• Battery power to vehicle interrupted	• If an obvious cause for an observed or reported concern is found, correct the cause (if possible) before proceeding to the next step • If the cause is not visually evident, check for Diagnostic Trouble Codes (DTCs) and refer to the DTC Index • Check the Jaguar Land Rover (JLR) claims submission system for open campaigns. Refer to the corresponding bulletins and SSMs which may be valid for the specific customer complaint and carry out the recommendations as required • High resistance between battery terminals and clamps	• Refer to the Battery Diagnostics instructions for the Midtronics GRX-3080 / EXP-1080 / GR8-1180 Diagnostic Battery Chargers in this Section

Further Battery Diagnostics

For further guidance in diagnosing battery issues, refer to the battery diagnostics instructions for the Midtronics GRX-3080 / EXP-1080 / GR8-1180 Diagnostic Battery Chargers in this section

DTC Index

For a list of Diagnostic Trouble Codes (DTCs) that could be logged on this vehicle, please refer to Section 100-00.

Battery Disconnect and Connect

Battery Removal

Removal and Installation

1. Remove LH front seat cushion.

Refer to: Front Seat Cushion (501-10, Removal and Installation).

2. Release retaining clip and remove battery access cover.

3. ⚠ WARNING: During battery removal or before carrying out any repairs or maintenance to electrical components always disconnect negative lead first. If positive lead is disconnected with negative lead in place, accidental contact of a wrench to any grounded metal part could cause a severe spark, possibly resulting in personal injury. Upon installation of battery always connect positive lead first.

   Disconnect both battery leads, negative first.

4. Slacken securing nuts and move battery clamp, with 'J' bolts aside.

5. Remove battery.

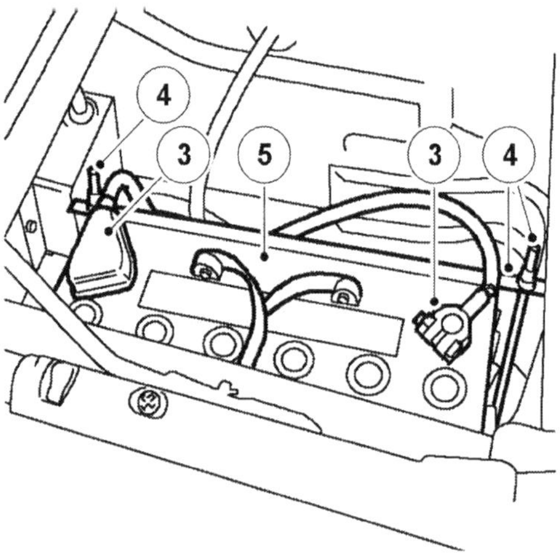

Installation

1. ⚠ NOTE: Coat battery clamps and terminals with petroleum jelly before refitting.

Position battery and secure with clamp. Ensure J bolts are hooked correctly under retaining brackets on base of battery box.

2. Reconnect battery leads, positive lead first.

3. Install battery access cover.

4. Install seat cushion.

Startup Battery Disconnect and Connect

General Procedures

WARNINGS:

⚠ Batteries normally produce explosive gases which may cause personal injury, therefore do not allow lighted substances to come near the battery. When charging or working near the battery always shield your face and protect your eyes. Always provide adequate ventilation. Failure to follow these instructions may result in personal injury.

⚠ Batteries contain sulphuric acid, avoid contact with skin, eyes or clothing. Shield your face and protect your eyes when working near the battery to guard against possible splashing of the acid solution. In case of acid contact with the skin or eyes, flush immediately for a minimum of 15 minutes and seek prompt medical attention. If swallowed, call a physician immediately. Failure to follow these instructions may result in personal injury.

⚠ CAUTION: Make sure the engine is not running before disconnecting the battery ground cable. Failure to follow this instruction may result in damage to the vehicles electrical system.

NOTES:

△ This procedure should be used to disconnect the battery while carrying out repairs that refer to the battery being disconnected.

△ Before disconnecting the battery make sure that no data is required from the engine control module (ECM), as battery cable disconnection will erase any fault codes and idle/drive values held in the keep alive memory (KAM).

1. Obtain and record the audio unit keycode and preset radio frequencies.
2. Remove the LH front seat cushion.

Refer to: Front Seat Cushion (501-10 Seating, Removal and Installation).

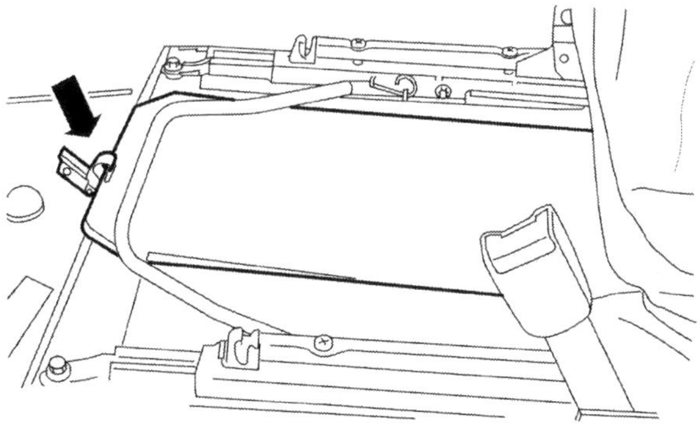

E89791

3. △ NOTE: RH illustration is shown, LH is similar.

Remove the battery cover.

- Release the clip.

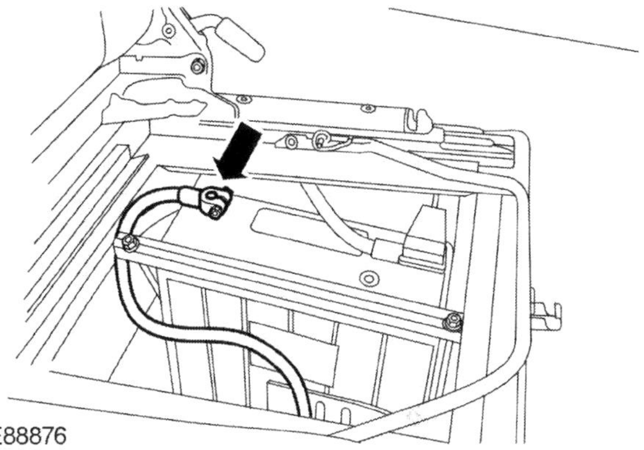

E88876

4. ⚠ CAUTION: Failure to follow the procedure below could result in the alarm sounder self activating when the battery is disconnected.

Disconnect the battery ground cable.

- Loosen the nut.
- Turn the ignition key to the run position.
- Turn the ignition key to the off position.
- Within 17 seconds of turning the ignition key to the off position, release the battery ground cable from the battery.

Generator and Regulator - ID 2.2L and 2.4L Diesel

Specifications

Description	Nm	lb-ft
Generator battery positive cable nut	8	6
Generator bolts	48	35

Description and Operation

COMPONENT LOCATION

Generator output is supplied to the battery, via the starter motor, from a threaded copper post on the generator casing. A single pin electrical connector on the rear of the casing provides a LIN (local interconnect network) bus interface for communication between the regulator and the ECM. The generator mountings provide the electrical ground for the generator.

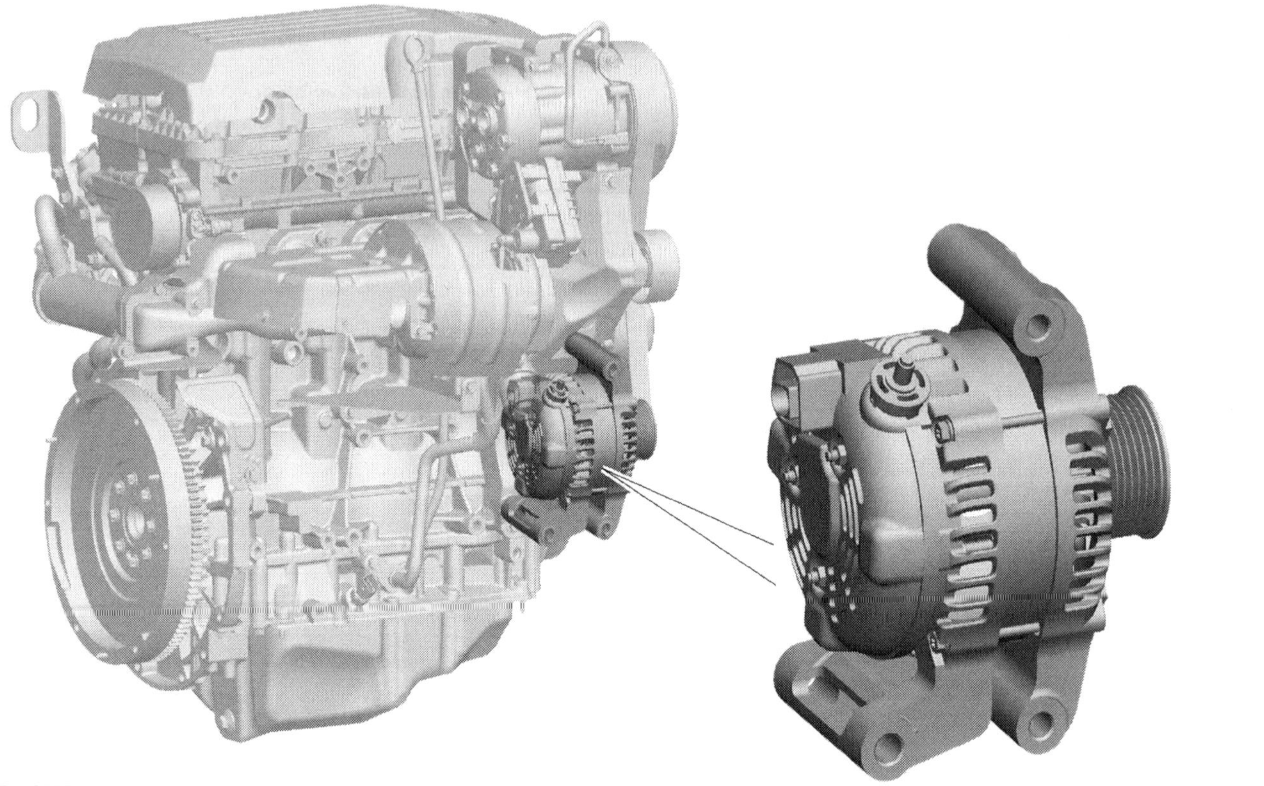

E139192

OVERVIEW

The generator is a self exciting type located at the front RH (right-hand) side of the engine and driven by the accessory drive belt. The generator is rated at 85/150 amps.

The generator pulley incorporates a one-way clutch that reduces NVH (noise, vibration and harshness) and improves the life of the accessory drive belt. The one-way clutch prevents the generator from driving the belt (causing 'belt whip'), due to the high rotational inertia of the generator internal components, during transients in engine speed.

A rectifier and a regulator are incorporated into the generator. The rectifier converts the AC (alternating current) produced in the stator coils of the generator into the DC (direct current) required by the vehicle electrical system. The regulator controls the output voltage from the generator and provides feedback of various parameters, including fault information, to the ECM (engine control module).

Generator output is controlled by the ECM, which sends information regarding charging voltage to the generator on the LIN bus. The ECM calculates the charging voltage using a software based battery temperature model, which uses the intake air temperature, road speed, coolant temperature and engine running timers to estimate the electrolyte temperature. This value is then used to select the appropriate charging voltage based upon the charging characteristic of the battery.

The LIN bus is also used to communicate fault messages from the generator to the ECM. If a fault occurs, a DTC (diagnostic trouble code) is stored in the ECM and, if necessary, after a short delay the ECM sends a high speed CAN (controller area network) bus message to the instrument cluster to illuminate the ignition/no charge warning indicator.

During engine starting, the ignition/no charge warning indicator is illuminated when the ignition is energized, and is extinguished when the engine starts and the ECM detects a generator output voltage.

Generator

Removal and Installation

Removal

1. WARNING: Do not work on or under a vehicle supported only by a jack. Always support the vehicle on safety stands.

Raise and support the vehicle.

2. **Refer to:** Startup Battery Disconnect and Connect (414-01 Battery, Mounting and Cables, General Procedures)

E137656

3.

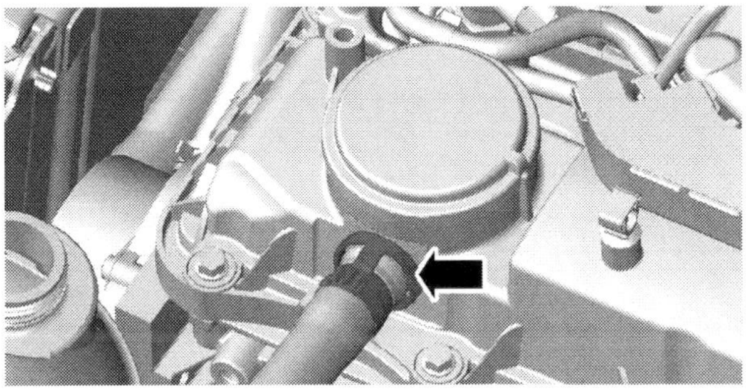

4.

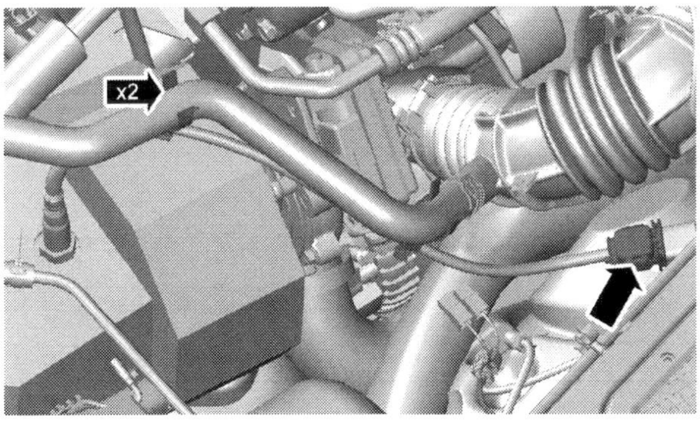

5.

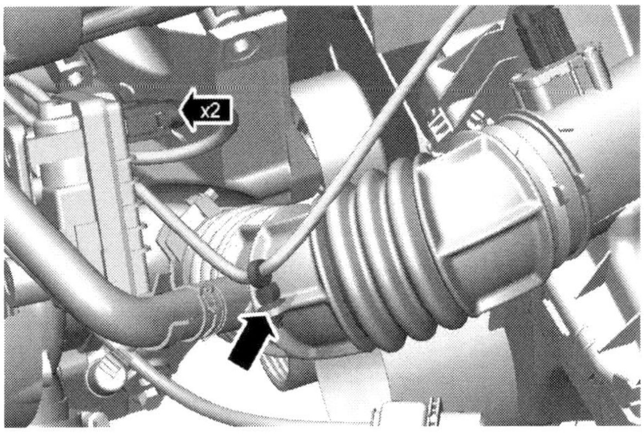

6.

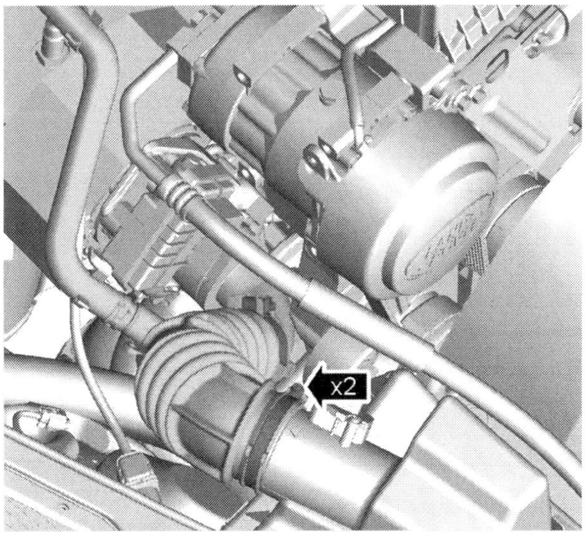

7.

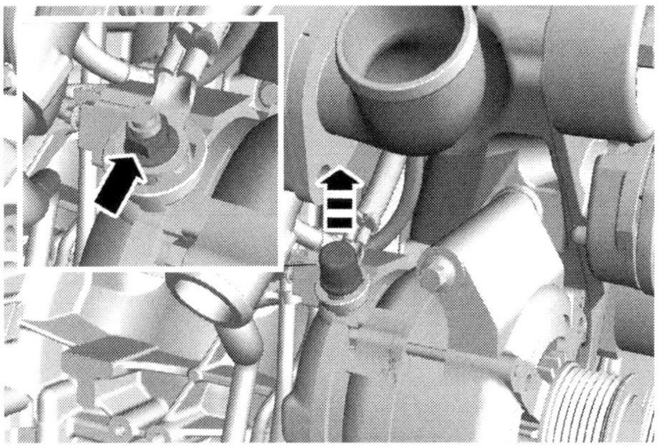

8. Torque: 10 Nm

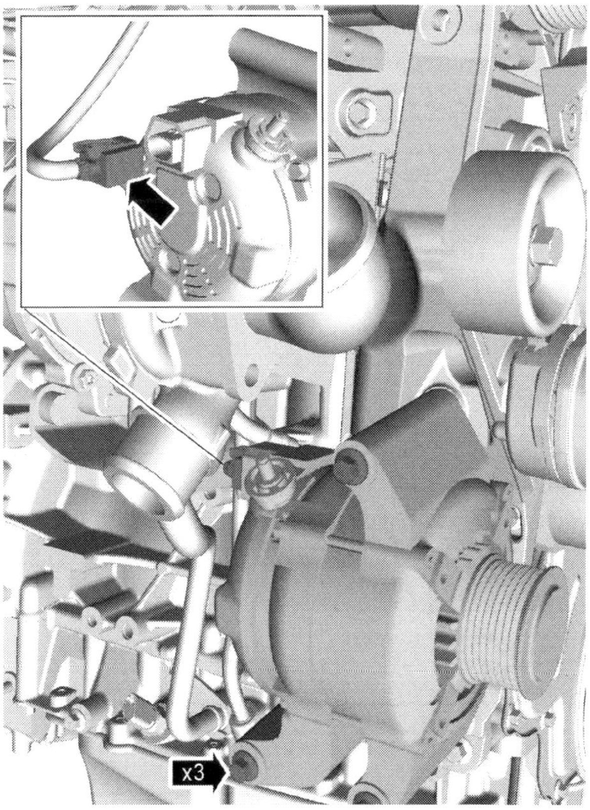

9. Torque: 48 Nm

Installation

1. To install, reverse the removal procedure.

Description and Operation

COMPONENT LOCATION - MODELS FROM 2007MY (NOT SVX (60TH ANNIVERSARY) MODEL)

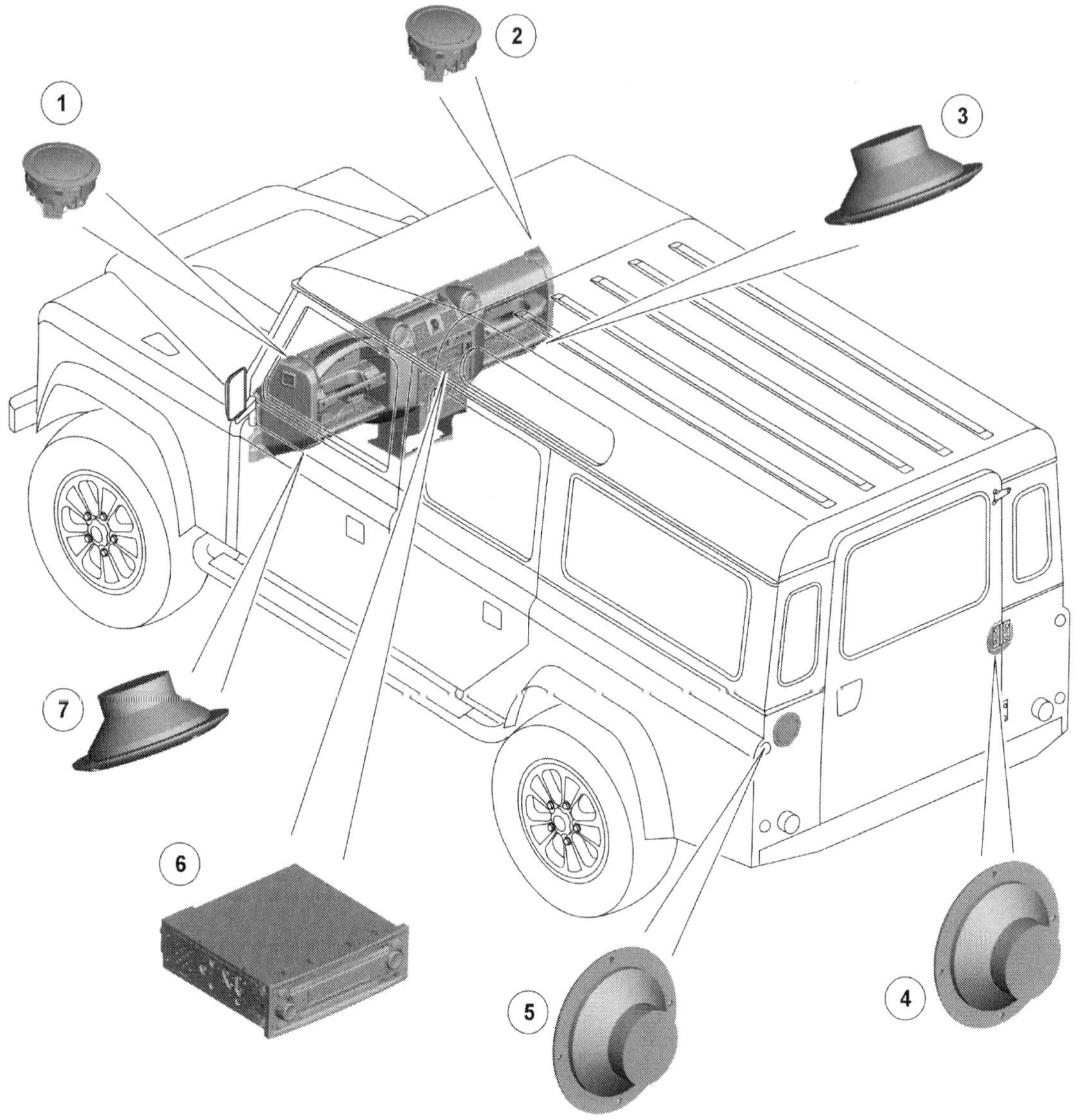

Item	Description
1	Front LH (left-hand) high range speaker
2	Front RH (right-hand) high range speaker
3	Front RH mid range speaker
4	Rear RH mid range speaker
5	Rear LH mid range speaker
6	Audio head unit
7	Front LH mid range speaker

OVERVIEW

The audio system comprises a CD tuner unit which directly drives up to 6 speakers. The audio head unit is located in the center of the instrument panel. The head unit combines the following functions:

- AM/FM radio
- Audio CD player
- MP3 CD player
- RDS traffic information receiver

The audio system has no diagnostic capabilities and is not connected to any of the vehicles other systems.

COMPONENT LOCATION - SVX (60TH ANNIVERSARY) MODEL

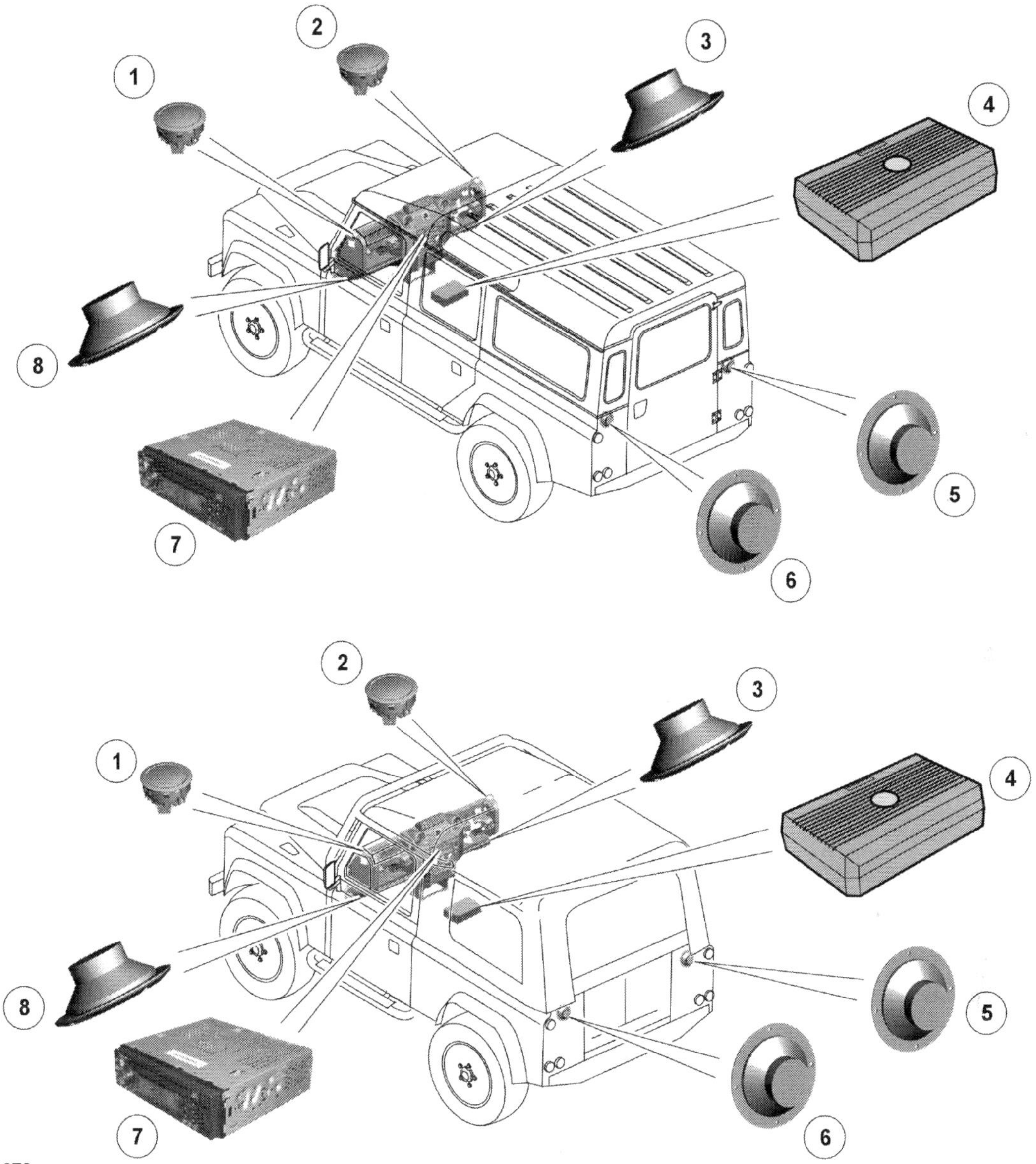

E101273

Item	Description
1	Front LH high range speaker
2	Front RH high range speaker
3	Front RH mid range speaker
4	Sub-woofer
5	Rear RH mid range speaker
6	Rear LH mid range speaker
7	Audio head unit
8	Front LH mid range speaker

SVX (60th Anniversary) Model

The SVX model has an upgraded head unit, speakers and also a sub-woofer. The head unit has the addition of a USB (universal serial bus) port and is also MP3/ WMA compatible.

The 90 model also has two additional speakers located in the rear of the load space.

Clarion® Head Unit DB568RUSB

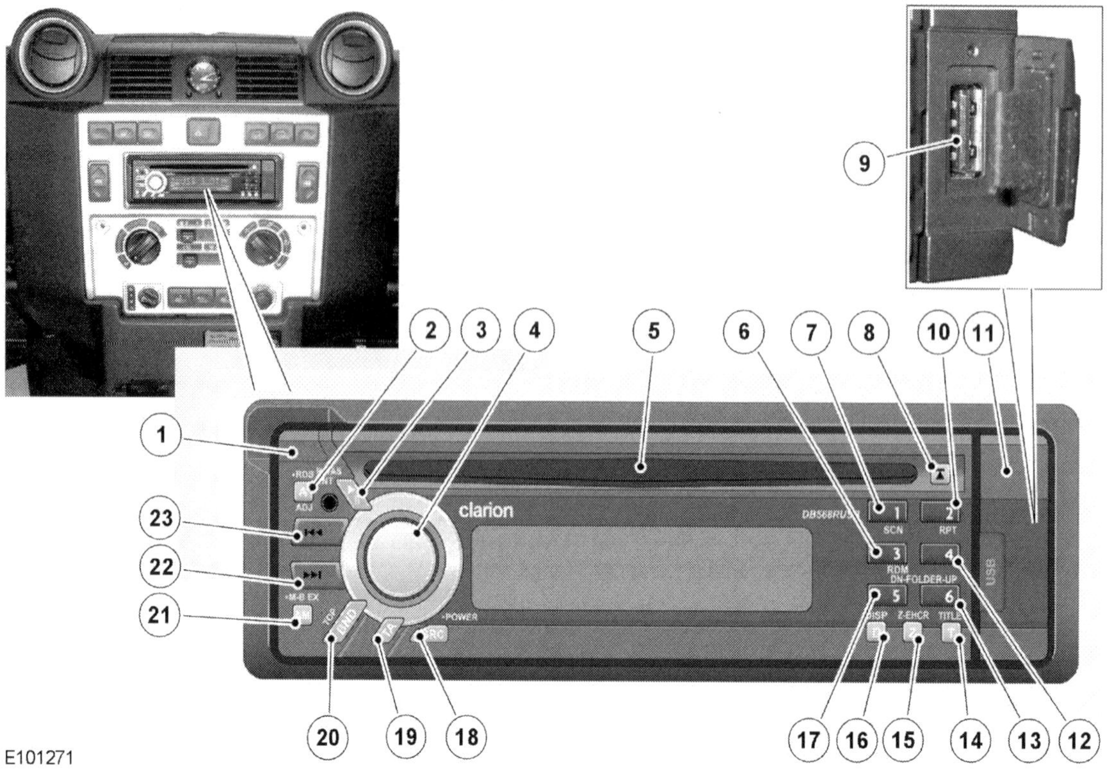

E101271

Item	Description
1	Detachable control panel release button
2	Auto Frequency (AF) button
3	Pre-set scan button/autostore - radio mode or play/pause a track - CD/MP3/WMA/USB mode
4	Volume rotary control and function select
5	CD insertion/eject slot
6	Station pre-set/store 3 button - radio mode or Random play (RDM) button- CD/MP3/WMA/USB mode
7	Station pre-set/store 1 button - radio mode or Scan (SCN) button- CD/MP3/WMA/USB mode
8	CD eject button
9	USB port
10	Station pre-set/store 2 button - radio mode or repeat play button - CD/MP3/WMA/USB mode
11	USB port cover
12	Station pre-set/store 4 button - radio mode
13	Station pre-set/store 6 button or select folder up (UP) button for MP3/WMA/USB folder options and for CD-DA or MP3/WMA selection with a multi-session disc
14	Title display - MP3/WMA/USB mode
15	'Z' sound enhancer button
16	Display change button
17	Station pre-set/store 5 button or select folder down (DN) button for MP3/WMA/USB folder options and for CD-DA or MP3/WMA selection with a multi-session disc
18	Power on/off and mode select button
19	Traffic announcement button
20	Switch band/seek tuning/manual tuning - radio mode or play first track - CD/MP3/WMA/USB mode
21	Audio mode (bass/treble/balance/fade) or Magna bass selection mode
22	Select station - radio mode or select track forwards - CD/USB mode or fast forward - CD/USB mode
23	Select station - radio mode or select track backwards - CD/USB mode or fast reverse - CD/USB mode

The head unit is located in a central position in the instrument panel, above the heater controls. The head unit locates in an ISO standard radio cage which is specific to the Clarion® head unit. Two special tools are required to remove the head unit from the cage once installed.

A link harness is used to connect the Clarion® head unit to the existing In-Car Entertainment (ICE) harness connectors. The link harness plugs into the existing ICE connector on the main harness and has four phono plugs and one connector to provide the connections to the Clarion® head unit. The phono plugs provide the speaker outputs and are colored red and white. Two of the plug wires have a blue tracer for identification. Two phono plugs connect to the top 2 phono sockets in the head unit and the two phono plugs with the blue tracer connect to the center 2 phono sockets on the head unit. The phono sockets on the head unit are also colored red and white to ensure the correct connections are made and the speaker polarity is correct. The connector also plugs into the rear of the head unit. The aerial plug is a male plug which connects into a socket on the rear of the head unit.

If iPod connectivity is installed an additional iPod audio connectivity module and link harness is used. The audio connectivity module is located on the cross car beam to the right of the head unit and is secured with a zip tie. This harness plugs into the existing ICE connector and the head unit link harness connects to the iPod audio connectivity module link harness. The audio connectivity module has a lead for the antenna connection. The lead from the antenna is plugged into the module antenna connection and the antenna lead from the module then connects into the head unit antenna connector.

The Clarion® DB568RUSB head unit features a detachable control panel which can be removed for security. The panel is released by pressing the button at the top LH corner of the unit. The panel will be released and can be removed and stored in the supplied case. The panel can be refitted by locating the RH end of the panel in the slots in the head unit and gently pushing the LH end of the panel towards the unit until the catch engages.

A USB port is located in the RH hand end of the unit and can be accessed by opening a cover on the detachable control panel. The port allows the use of a customer supplied USB memory device to allow access to a large range of music files in both MP3 and WMA formats. The unit can read the files and folder structures and data from the memory device and play the music directly via the USB port. It is not recommended to use this port for iPod connection.

Full operating instructions and unit features are detailed in the owner literature supplied with the vehicle.

An auxiliary input socket is located inside the cubby box. The socket allows for the additional connection of an MP3 player or other storage device via a single pin jack plug.

Specifications

- Power supply voltage: 14.4 Volts (10.8 to 15.6 Volts allowable) - Negative earth
- Current consumption: less than 15 Amperes
- Speaker impedance: 4 Ohms (4 to 8 Ohms allowable)
- Audio maximum power output: 200 Watts (50 Watts X 4 channels)
- Audio power output: 4 X 25 Watts
- USB 1.1 and 2.0 compatible with transfer rate of 12 Mbps maximum
- MP3 and WMA compatible
- FM frequency range: 87.5 to 108 MHz
- MW frequency range: 531 to 1602 kHz
- AM frequency range: 153 to 279 kHz

The speakers installed in the SVX vehicle are upgraded from the standard fitment and are also supplied by Clarion®. These speakers are connected to the existing speaker harnesses by link leads which allow for the different connector specifications required for the Clarion® speakers.

Clarion Sub-Woofer SRV313

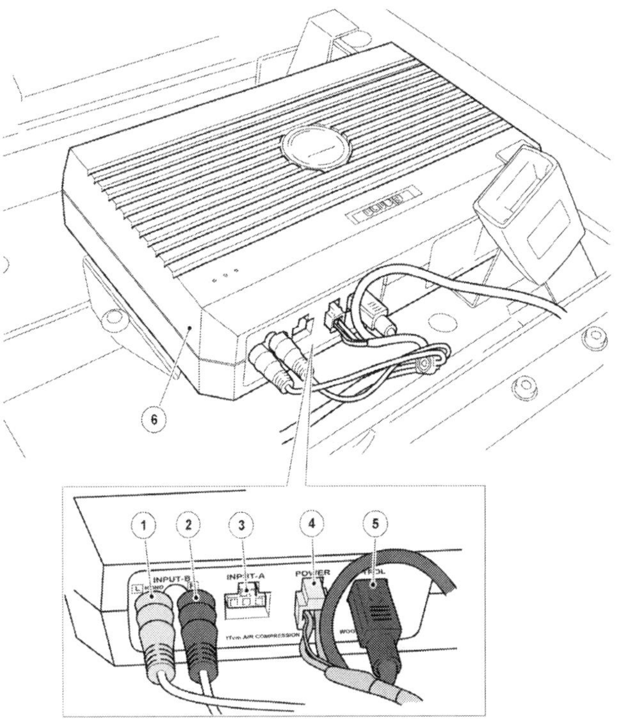

Item	Description
1	LH phono plug speaker input
2	RH phono plug speaker input
3	Not used
4	Power and ground connector
5	Remote control connector
6	Sub-woofer unit

The Clarion® SRV313 sub-woofer is located on the cabin floor below the center cubby box and is secured to the floor with two self-tapping screws.

The sub-woofer is connected to the head unit via a link cable from the head unit. Two phono plugs provide speaker inputs for LH and RH inputs and are colored white and red respectively for identification and connect into the 'Input B' sockets (the 'Input A' socket is not used). A power supply socket provides power and ground for unit power. A 'Remote Control' socket allows for the connection of the remote control unit.

Remote Control and Auxiliary Socket

The remote control unit is located in the front of the cubby box and is accessible by lifting the cubby box lid. The remote control has a rotary control on the LH end which allows for adjustment of the sub-woofer bass output by adjusting the cut-off frequency of the low-pass filter. Four selections are available; 50/75/100 and 120 Hz. The sub-woofer will only output frequencies below the selected Hz level. The rotary control on the RH end of the remote control provides for adjustment of the speaker 'gain' and can be used to adjust the sub-woofer

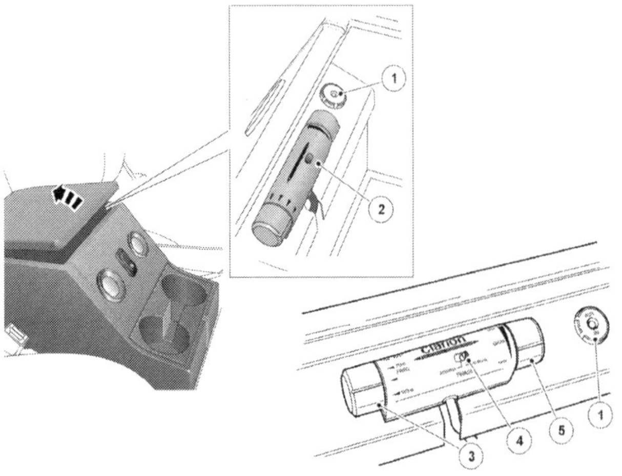

Item	Description
1	Auxiliary socket
2	Remote control
3	Frequency control
4	Phase switch
5	Gain volume control

volume output, independent of the head unit volume control. A switch on the remote control allows for the sub-woofer phase to be switched between normal and reverse. Clarion recommend for this application the setting is the 'normal' phase.

An auxiliary input socket is located adjacent to the remote control. The socket allows for the connection of an additional auxiliary audio input, an MP3 player for example.

Sub-Woofer Specifications

- 120 Watt high power amplifier with low pass filter
- 17 cm counter drive woofer with dualneodymium magnet
- Air compression structure with die-cast aluminum enclosure
- Wired remote control for gain control/low pass filter (selectable 50/75/100/120 Hz) /phase filter
- Dual 4 Ohm voice coils.

CONTROL DIAGRAM - MODELS FROM 2007MY (NOT SVX (60TH ANNIVERSARY) MODEL)

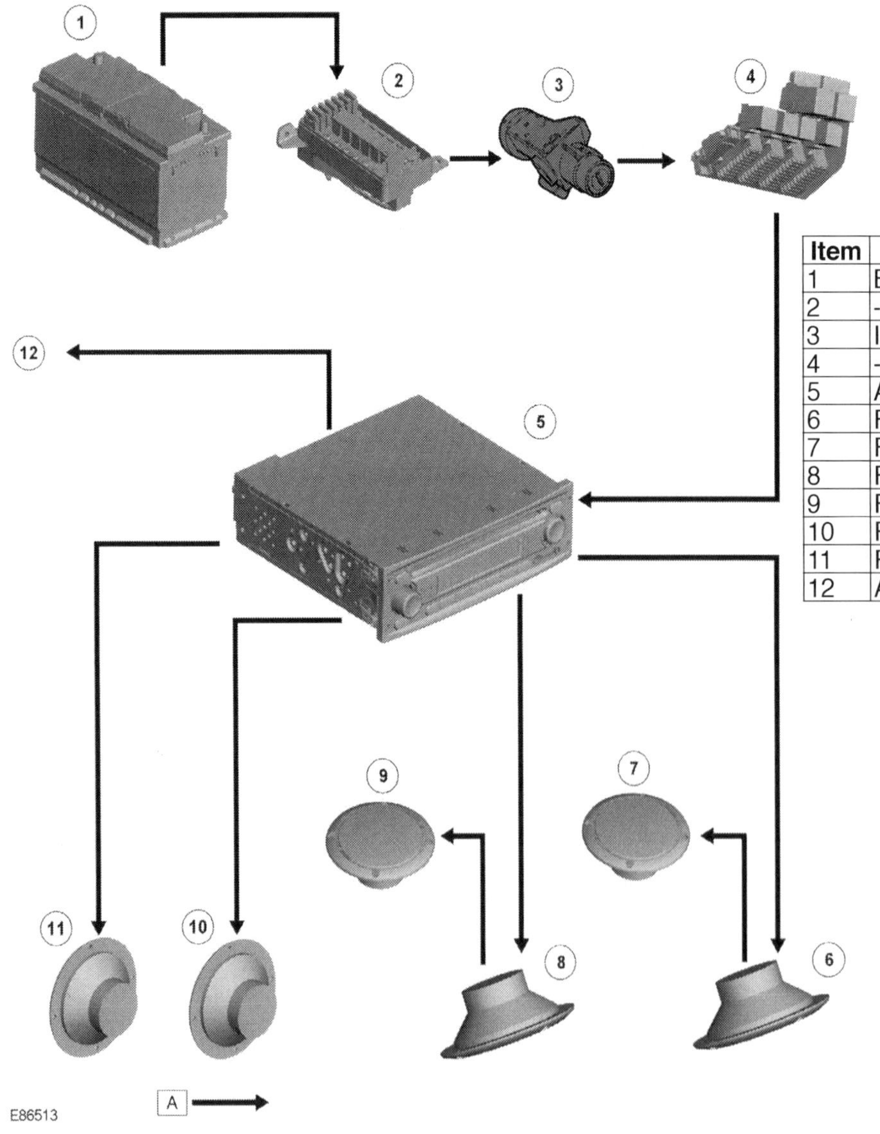

Item	Description
1	Battery
2	-
3	Ignition switch
4	-
5	Audio head unit
6	Front RH mid range speaker
7	Front RH high range speaker
8	Front LH mid range speaker
9	Front LH high range speaker
10	Rear RH mid range speaker
11	Rear LH mid range speaker
12	Antenna

CONTROL DIAGRAM - SVX (60TH ANNIVERSARY) MODEL

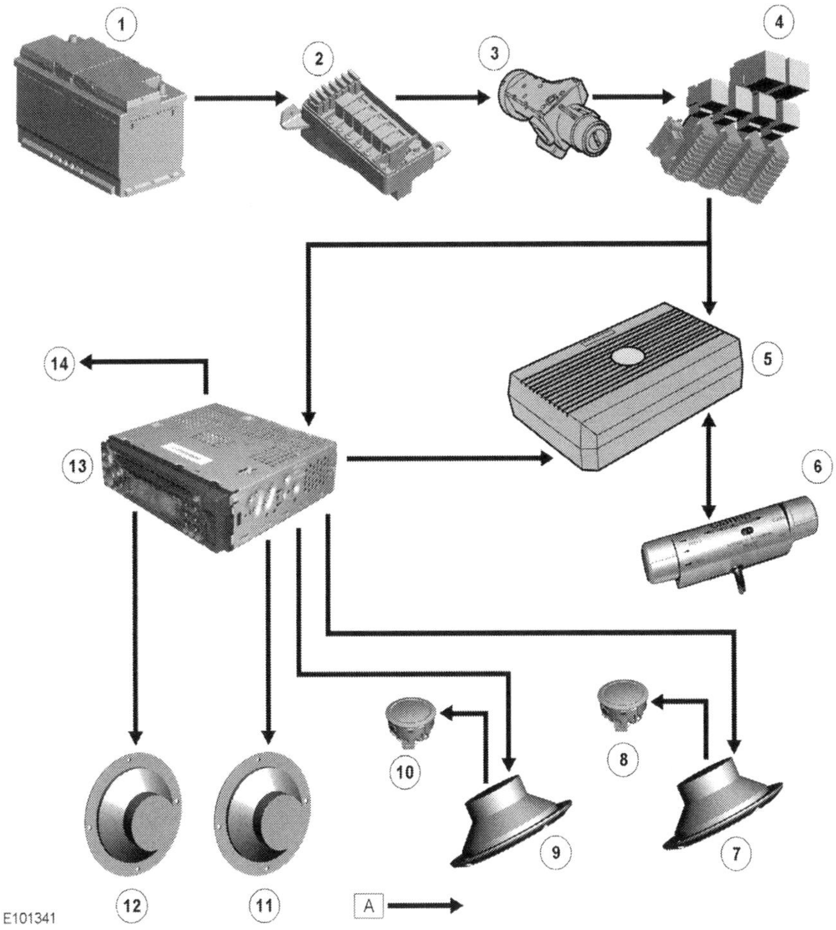

Item	Description
1	Battery
2	-
3	Ignition switch
4	-
5	Sub-woofer
6	Sub-woofer remote control
7	Front RH mid range speaker
8	Front RH high range speaker
9	Front LH mid range speaker
10	Front LH high range speaker
11	Rear RH mid range speaker
12	Rear LH mid range speaker
13	Audio head unit
14	Antenna

Audio Unit - Vehicles With: Bluetooth

Removal and Installation

Removal

NOTE: Some variation in the illustrations may occur, but the essential information is always correct.

1. CAUTION: Do not use excessive force.

Remove the audio unit head to allow access to the bezel.

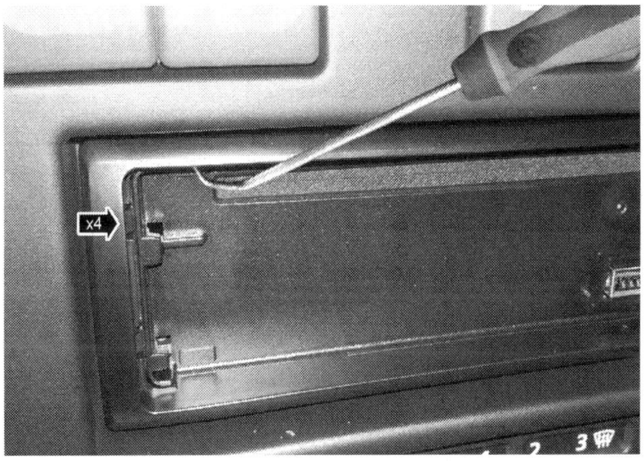

2. CAUTION: Do not use excessive force.
   Using a suitable hook tool, release the audio unit bezel.

   1. Release the locking tangs.

3. NOTE: Operation must be carried out on both sides.

Using a suitable hook tool, release the audio unit.

1. Release the locking tang.
2. Detach the audio unit.

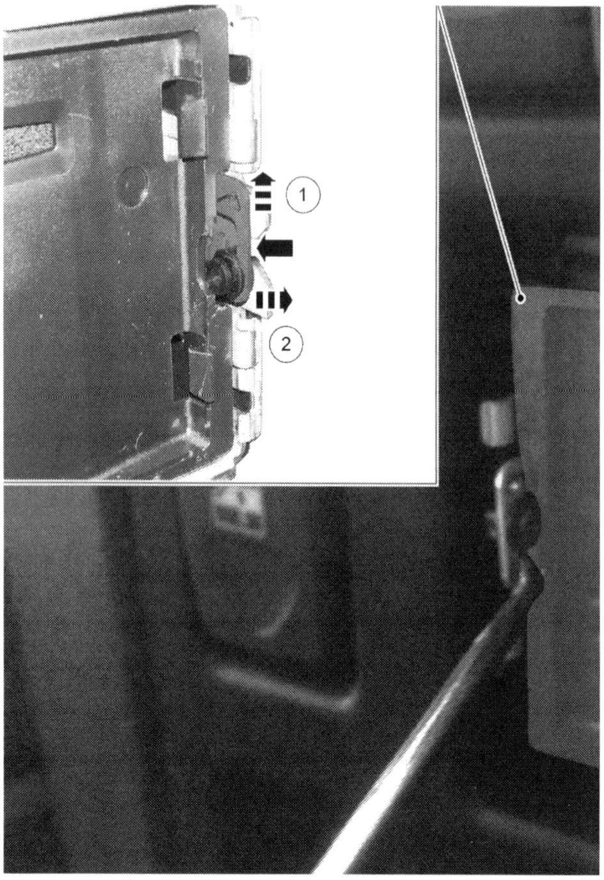

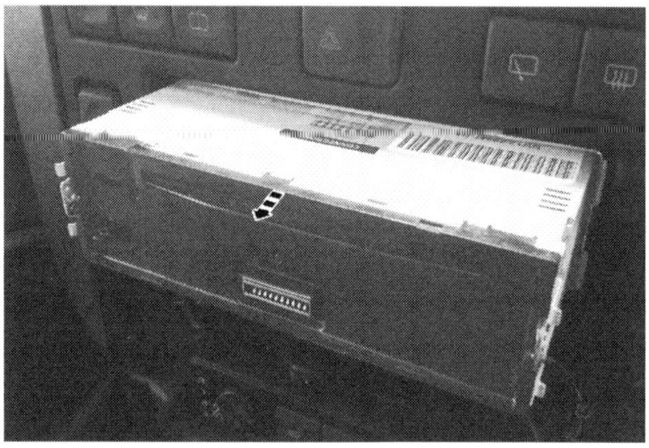

4. Release the audio unit from the fascia.

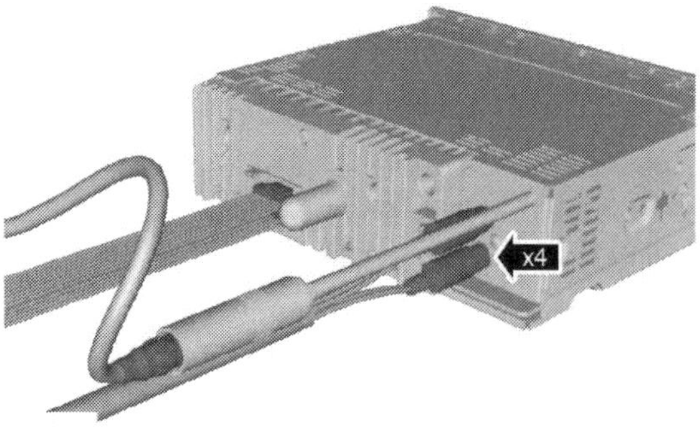

5. Disconnect the electrical connectors and remove the audio unit.

Installation

1. To install, reverse the removal procedure.

Audio Unit - Vehicles Without: Bluetooth

Removal and Installation

Removal

NOTES:

The removal tools used within this procedure can be sourced from the parts catalogue.

Some variation in the illustrations may occur, but the essential information is always correct.

1. Remove the audio unit head to allow access to the bezel.

2. CAUTION: Do not use excessive force.
   Using a suitable hook tool, release the audio unit bezel.
   1. Release the locking tangs.

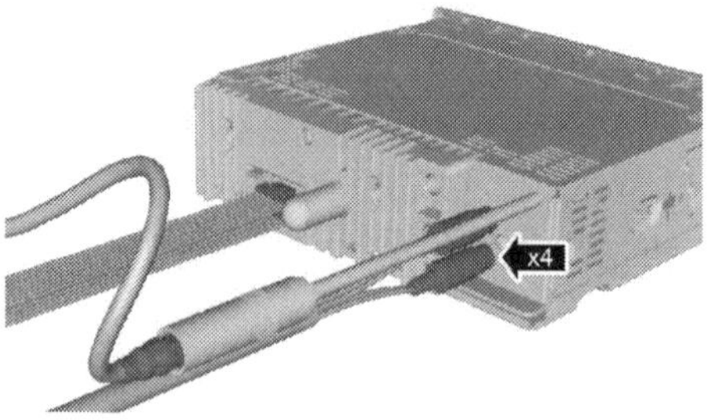

3. Using the audio tools (sourced from the parts catalogue), release the audio unit from the fascia.
4. Disconnect the electrical connectors and remove the audio unit.

Installation

1. To install, reverse the removal procedure.

Navigation System

Description and Operation

COMPONENT LOCATION - SVX (60th ANNIVERSARY) MODEL

Item	Description
1	Personal Travel Assistant navigation unit
2	Cradle
3	Electrical connector
4	External speaker (located behind instrument panel)

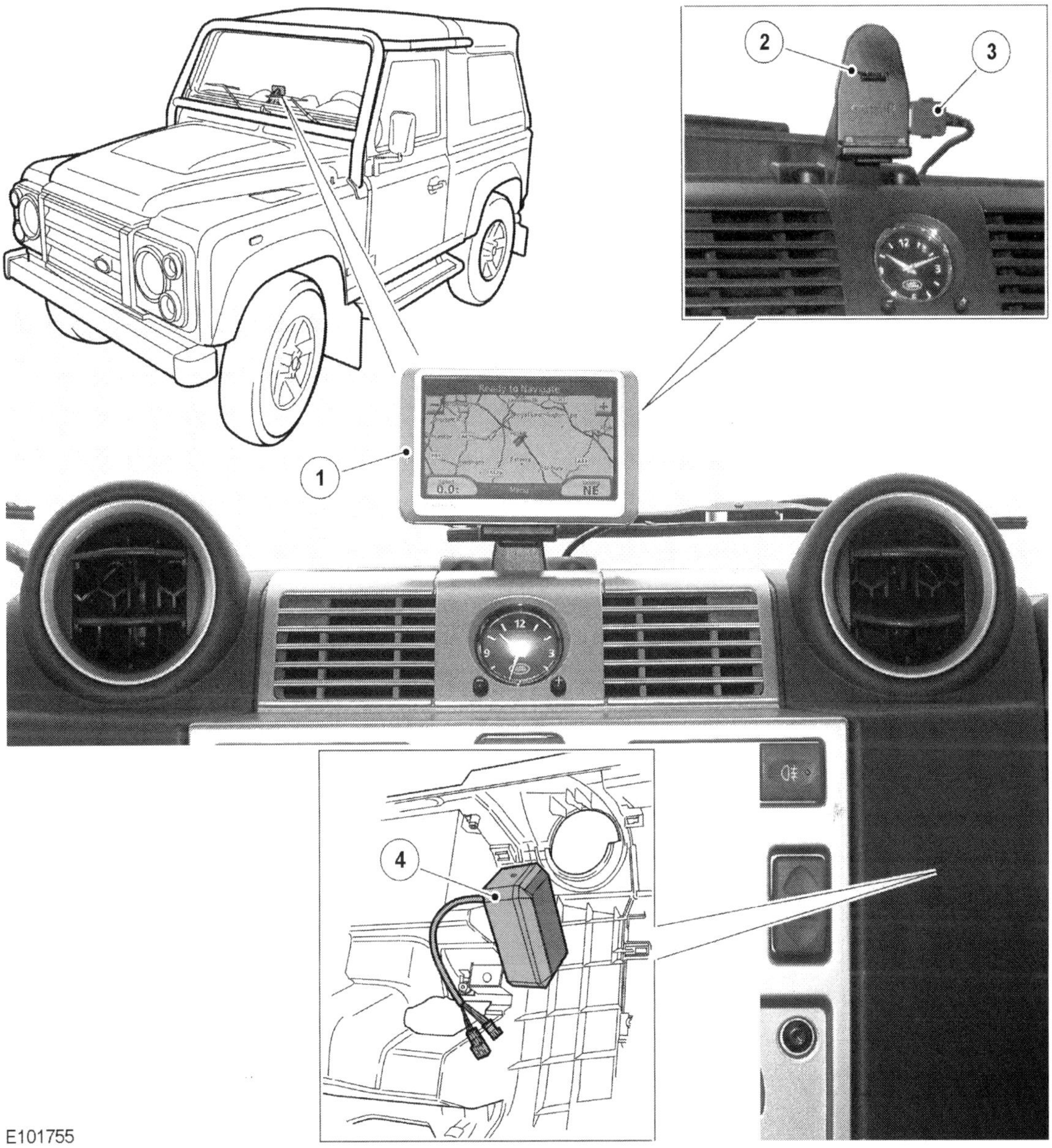

E101755

OVERVIEW

The satellite navigation system is a standard fitment on the SVX model and is available as an accessory fitment on other models. The unit is known as the 'Land Rover Personal Travel Assistant' and its use and operation is covered in a separate instruction book supplied with the vehicle.

The Personal Travel Assistant provides a comprehensive navigation system, which assists the driver by providing a route to a selected destination. The map and turn information which is displayed on the touch screen is supplemented by voice guidance.

The Personal Travel Assistant is also equipped with additional features, which can be accessed via the Tools menu:

- media player
- picture viewer
- calculator
- language guide
- currency converter
- unit converter.

WARNING: In the interests of road safety, only operate, adjust or view the system when it is safe to do so.

The Personal Travel Assistant must be able to locate satellite signals in order to operate. The strength of these signals can vary depending on the location. Near tall buildings or trees, the signal strength may be affected. When a strong signal is being received, the bars at the top left of the main menu are green, when the signal deteriorates, the bars turn red or become clear with a red 'x' appearing on the bars.

DESCRIPTION

On the SVX model, the satellite navigation unit is a 'Garmin™' manufactured unit with Land Rover specific software installed. The unit is self-contained with its own internal battery. The vehicle wiring harness connections provide for connection to the external speaker and the power supply for internal battery charging and powering the unit while it is connected to the vehicle.

The unit uses a remote external speaker for the voice and audio output. The speaker is located below the driver's side center air vent, behind the instrument panel. The unit however, also has its own internal speaker allowing the unit to be used when not in the vehicle.

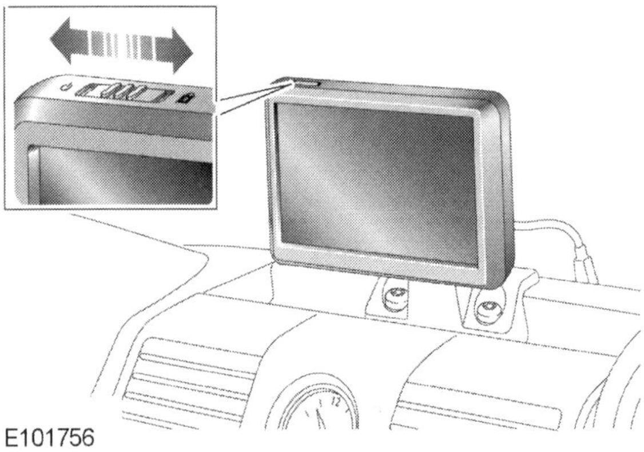

E101756

The unit can be switched on using a spring loaded switch on the top of the unit. To switch the unit on, push the switch to the left. To subsequently switch the unit off, push the switch to the left a second time. The unit can be locked to prevent inadvertent operation by sliding the switch on the top of the unit to the right.

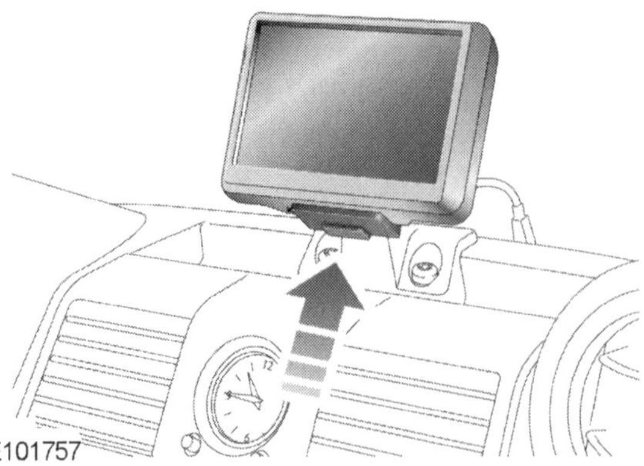

E101757

The satellite navigation unit is mounted on a bracket which is secured to the top of the instrument panel with 2 bolts. The bracket houses a cradle with a connector which allows the unit to be secured into position and also connected electrically to the vehicle electrical system and the navigation speaker. The cradle is mounted on a ball which allows the unit to be adjusted vertically and horizontally so that it can be moved to face the passenger or driver's seat positions. The unit clips into position on the bracket and can be released by pressing a release catch on the bottom of the cradle, tilting the top of the unit forwards and removing the unit from the cradle.

Garmin™ Nuvi 750 Unit Details

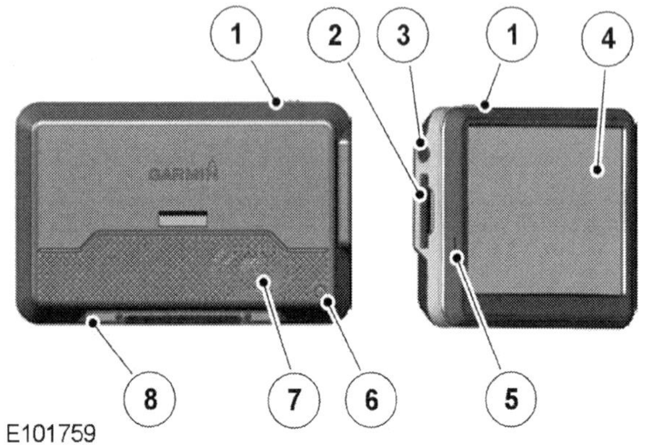

E101759

Item	Description
1	On/Off/Lock switch
2	SD card slot
3	Earphone/audio out jack plug
4	-
5	Microphone (not used)
6	External antenna connector
7	Internal speaker
8	USB (universal serial bus) connector

The Personal Travel Assistant uses a wide-screen, LCD with a white backlight and touch screen functionality. The screen has a 480 X 272 pixel display.

On the LH (left-hand) end of the unit a 3.5 mm jack plug allows for the connection of audio headphones. A slot, below the jack plug, provides for the insertion of an SD memory card. The card can be used to store music and pictures which can be played and displayed on the unit. The unit can play MP3, MP3U and MP3U8 playlist files, display pictures in JPEG format and play audio books in AA format.

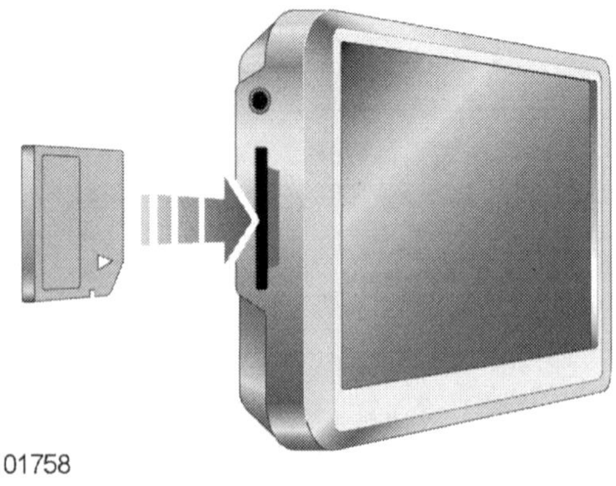

E101758

On the underside of the unit are two plugs. The larger, centrally positioned, plug mates with a similar plug in the cradle and connects the unit to the vehicle electrical system. The smaller plug provides for the connection of a USB cable to allow the unit to be connected to a computer to allow software upgrades to be installed.

The unit contains an integral lithium-ion battery which cannot be replaced.

Trouble Shooting

A battery icon is displayed in the LCD and indicates the status of the internal battery. If the unit is to be discarded, it must be disposed of according to local regulations.

The battery, once fully charged has a battery charge time of between 3 and 7 hours, depending on unit usage.

Problem	Solution
Unable to receive satellite signals	Ensure the unit is not in an enclosed garage or car park and away from tall buildings and trees. Remain stationary for several minutes to allow the unit to establish satellite connection.
The unit will not charge in the vehicle	The unit will only charge if the temperature is between 0 and 45°C (32 and 113°F). At temperatures above or below this range or if the unit is placed in direct sunlight or in a hot environment, the battery will not charge.
How to clear all user data	Hold finger on lower RH (right-hand) corner of touch screen and turn the unit on. Continue to press corner of touch screen until message appears. Touch 'Yes' to clear all user data.
The unit is locked up or frozen	If the touch screen stops functioning, turn the unit off and then on again. If this does not correct the problem, slide the power on/off switch to the left and hold for 8 seconds. The unit will switch off. Switch the unit back on again and it will operate normally.
Touch screen is not responding to touch	Turn the unit off and then back on again. Slide the on/off switch to the right (lock position). Hold a finger on the touch screen for approximately 30 seconds until the calibration screen appears. Follow the on-screen instruction.
The battery gauge seems inaccurate	Allow the unit to fully discharge. Then, fully charge the unit without disrupting the charge cycle.

Antenna

415-02

Antenna

Removal and Installation

Removal

1. Remove the audio unit.

Refer to: Audio Unit - Vehicles Without: Bluetooth (415-01 Audio Unit, Removal and Installation)

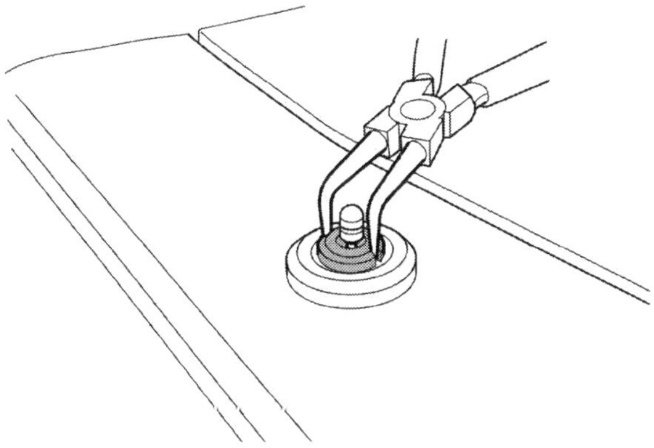

2. Remove the retaining nut from the antenna mast.

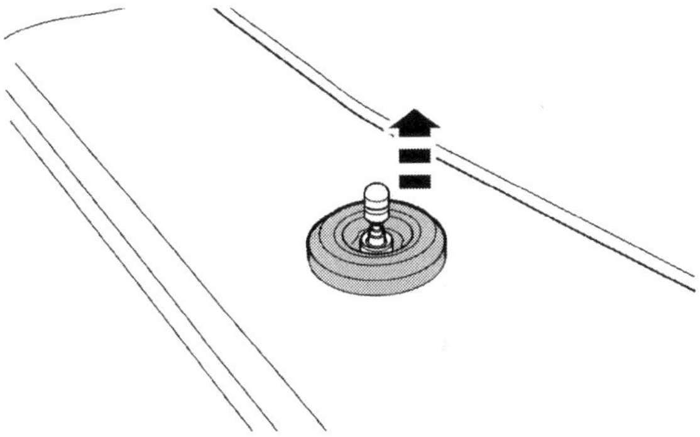

3. Release antenna assembly from the fender.

4. Disconnect the antenna cable.

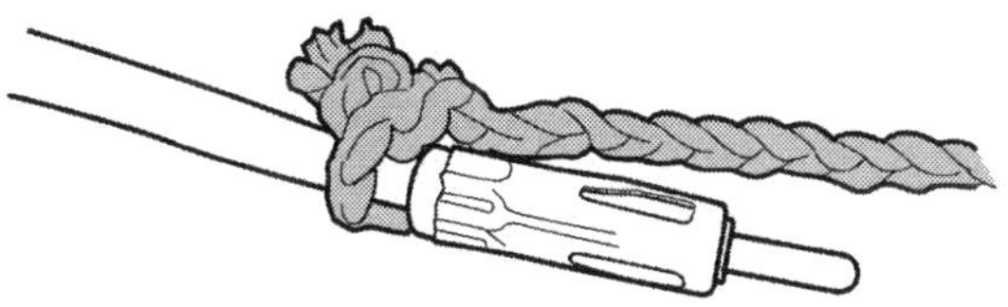

5. Attach a suitable length of string to the antenna cable.

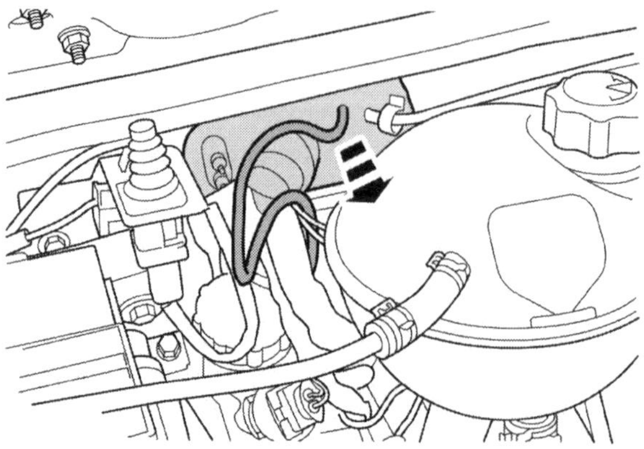

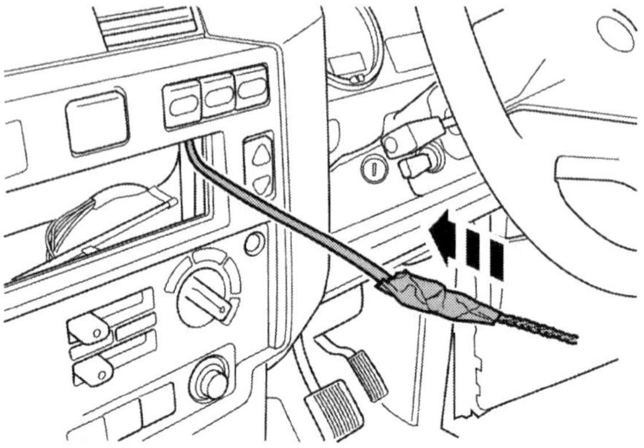

6. Remove the antenna cable.
7. Release the string from the antenna cable.

Installation

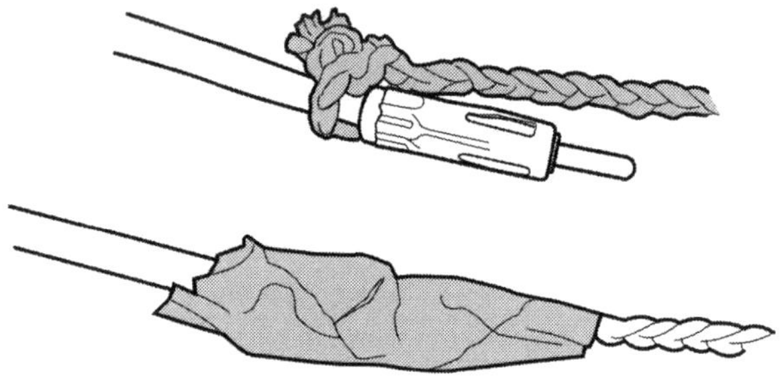

1. NOTE: Using suitable tape, secure the antenna cable and string to aid installation.

   Attach the string to the new Aerial cable.

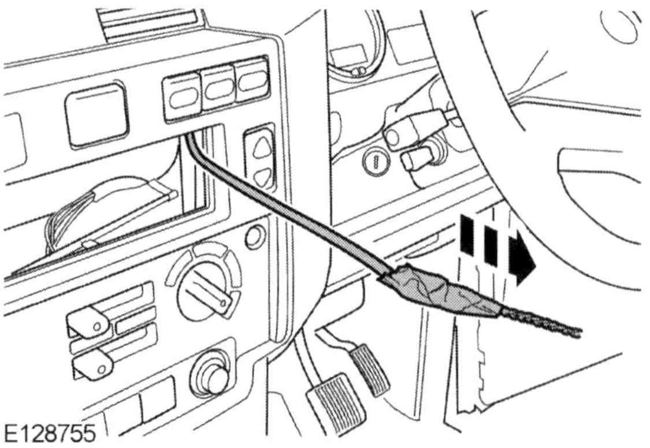

E128755

2. Install the antenna cable.
3. Release the string from the antenna cable.

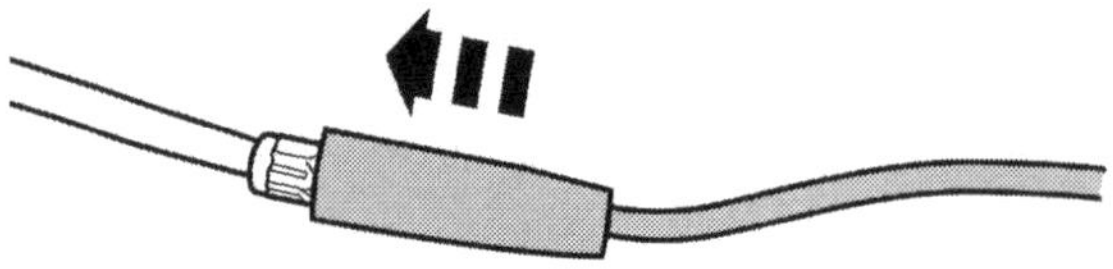

4. Connect the antenna cable.

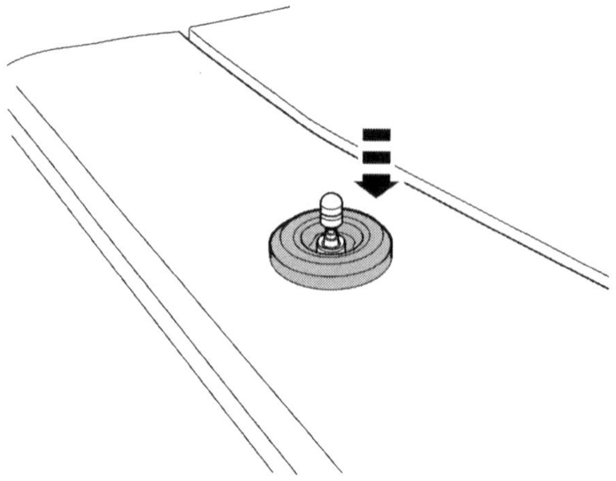

5. Install the antenna assembly in the front fender.

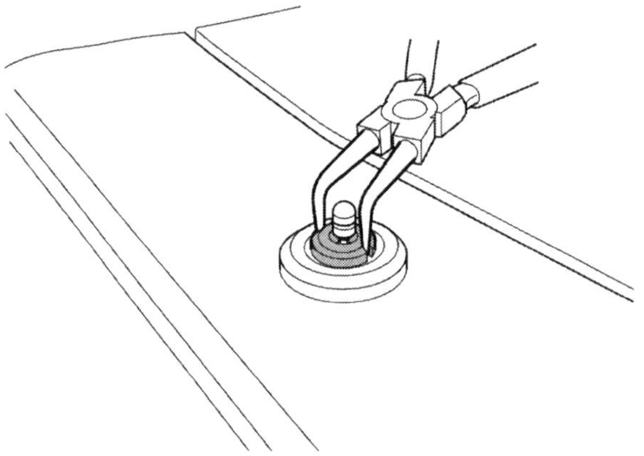

6. Install the antenna assembly retaining nut.
7. Install the audio unit.

Refer to: Audio Unit - Vehicles Without: Bluetooth (415-01 Audio Unit, Removal and Installation)

Speakers

Description and Operation

OVERVIEW

Refer to: Audio System (415-01 Audio Unit,

Description and Operation)

Diagnosis and Testing

Principles of Operation

For a detailed description of the loudspeakers system and operation, refer to the relevant description and operation section of the workshop manual.

Refer to: Audio System (415-01 Audio Unit, Description and Operation)

Inspection and Verification

CAUTION: Diagnosis by substitution from a donor vehicle is **NOT** acceptable. Substitution of control modules does not guarantee confirmation of a fault and may also cause additional faults in the vehicle being checked and/or the donor vehicle.

NOTE: Check and rectify basic faults before beginning diagnostic routines involving pinpoint tests.

1. Visually inspect for obvious signs of damage and system integrity
2. If an obvious cause for an observed or reported concern is found, correct the cause (if possible) before proceeding to the next step
3. If the cause is not visually evident, verify the symptom and refer to the Symptom Chart, alternatively check for Diagnostic Trouble Codes (DTCs) and refer to the DTC Index

Visual Inspection

Mechanical	Electrical
• Check all audio system modules • Compact disc player jammed, not loading • Scratched/dirty compact discs • Speakers • Switch(s) stuck or damaged	• Fuses • Electrical harnesses • Harness connectors • Battery condition, state of charge

Speaker Diagnostics

The symptom chart below should be used when diagnosing speaker faults

Reported Symptom	Symptom Description	Potential Causes (for guidance only)	Recommended Action
• Speaker Buzz	• Periodic high frequency sound (sounds like an insect buzzing)	• Resonance in a trim component, NVH (noise, vibration and harshness) insulating cloth against speaker grill mesh • Lack of retention of interior and exterior trim panels • Mechanical failure of internal speaker component	• 1. Replicate fault and confirm audio source(s) affected: DAB/CD/Bluetooth phone/USB through phone/USB through iPod®/AM/FM radio • 2. If fault is specific to one source, audio source must be investigated further, if issue is only seen on DAB, check DABM module is fully functioning correctly and software is up to date • 3. Confirm if issue is heard on stereo and/or surround sound settings • 4. Isolate the customer symptom to specific vehicle area using balance and fade audio settings. Check for any loose objects in the area, pens, keys, coins, etc • 6. Check SDD/Topix for any audio related SSM/TSB and software updates. Complete service updates as required. Retest to confirm if issue is resolved • 7. Apply pressure by hand to door/trim panel to confirm if there is a change in symptom sound • 8. If issue is still present, or condition changes by applying pressure to the trim, remove internal trim to gain access to relevant speaker • 9. If symptom condition is unchanged by applying pressure to the trim, examine trim to find source of audio issue and repair as appropriate using Squeaks and Rattle kit LTB00389. If audio issue is still present then move to step 10 • 10. Check around front face of speaker for any loose items that may be touching the speaker, debris, loose fixings, etc • 11. If a gasket is present on speaker, check it is correctly
• Speaker Rattle	• Sounds like loose components rattling around in / near the speaker	• Loose trim • Loose items in or around speaker • Loose fixings • Debris inside speaker • Loose harness	
• Speaker Hiss / Static	• Wideband noise (White Noise) / interference (such as experienced from a poorly tuned radio signal)	• Loose connection at speaker • Loose connection at amplifier • Loose connection at audio head unit • Non JLR-approved equipment installed (USB cables) • Damage to harness connected to the speaker • Audio amplifier fault • Audio head unit fault	
• Speaker Crackle	• Electrical crackling noise (such as from a loose electrical connection) • Electrical interference	• Loose connection at speaker • Loose connection at amplifier • Loose connection at audio head unit • Non JLR-approved equipment installed (USB cables) • Damage to harness connected to the speaker • Audio amplifier fault • Audio head unit fault • Internal electrical issue in speaker	

• Speaker Distorted	• No significant extraneous noise, but audio reproduction is not as expected is not clean sounding	• Loose trim • Non JLR-approved equipment installed (USB cables) • Resonance in a trim component, NVH (noise, vibration and harshness) insulating cloth against speaker grill mesh • Lack of retention of interior and exterior trim panels • Mechanical failure of internal speaker component	seated against speaker surface • 12. Check all speaker fixing screws are secure and correctly torqued, refer to workshop manual for correct torque figures • 13. Check harness assembly is not vibrating against speaker unit and that it is clipped and routed correctly • 14. Check harness assembly is not trapped or impeding the speaker unit and rectify as required • 15. Check harness assembly connections are fully inserted and secured to speaker and amplifier • 16. Check for harness damage and repair/replace as required • 17. If audio issue is still present, remove speaker from mountings, retest speaker in hand to see if the fault is still present • 18. Check for debris in speaker and remove. Retest speaker in hand • 19. Check for damage to speaker and replace speaker if damage is present • 20. Connect new speaker to harness, check new speaker in hand to make sure fault has been rectified. If issue has been resolved, reassemble trim with new trim retention clips • 21. Check if audio issue is still present. If issue has been resolved, reassemble trim and retest. If issue is still present, contact JLR Technical Support following the guidelines in the policy and procedures manual

Pinpoint Tests For Suspected Loudspeaker Faults In Harman/Kardon Audio Systems Fitted From 10MY Onwards

PINPOINT TEST A : NO SOUND OUTPUT FROM LOUDSPEAKER(S)	
TEST CONDITIONS	**DETAILS/RESULTS/ACTIONS**
A1: NO SOUND OUTPUT FROM LOUDSPEAKER(S)	
	1. Check loudspeaker operation
	Is the harness connector securely connected to the audio head unit and the loudspeaker unit(s)? **Yes** Proceed to the next step GO to 2 **No** Reconnect wiring harness to audio head unit/loudspeaker unit(s)
A2: NO SOUND OUTPUT FROM LOUDSPEAKER(S)	
	1. Check audio head unit operation
	Is the audio head unit operational? **Yes** Proceed to the next step GO to 3 **No** Check the integrity of the power supply circuits/fuses to the audio head unit and rectify as required
A3: NO SOUND OUTPUT FROM LOUDSPEAKER(S)	
	1. Check which loudspeakers are operational
	Are all loudspeakers working? **Yes** Proceed to the next step GO to 4 **No** Use the fader control to direct audio output to different loudspeaker locations to establish which units are non-operational. Refer to the electrical circuit diagrams and check the circuits between the audio head unit and the affected loudspeaker units for short circuit to ground, open circuit, high resistance. Repair circuit(s) as required. If fault persists, replace non-operational loudspeaker unit(s) as required
A4: NO SOUND OUTPUT FROM LOUDSPEAKER(S)	
	1. Check for signs of water ingress
	Does the loudspeaker drive unit show any signs of water damage? **Yes** Replace the loudspeaker unit(s) as required **No** Proceed to the next step GO to Pinpoint Test B

PINPOINT TEST B : POOR OR WEAK SOUND OUTPUT FROM LOUDSPEAKER(S)	
TEST CONDITIONS	**DETAILS/RESULTS/ACTIONS**
B1: POOR OR WEAK SOUND OUTPUT FROM LOUDSPEAKER(S)	
	1. Check connections to loudspeaker units
	Are the connectors inserted securely into the loudspeaker units? **Yes** Proceed to the next step GO to 2 **No** Make sure all connectors are securely attached to the loudspeaker units
B2: POOR OR WEAK SOUND OUTPUT FROM LOUDSPEAKER(S)	
	1. Check integrity of vehicle power supply fuses
	Are the vehicle/audio head unit power supply fuses functional? **Yes** Proceed to the next step GO to 3 **No** Replace fuse(s) as required
B3: POOR OR WEAK SOUND OUTPUT FROM LOUDSPEAKER(S)	
	1. Check power and ground circuits to the infotainment system components
	Are all the necessary power and ground feeds present? **Yes** Proceed to the next step GO to 4 **No** Refer to the electrical circuit diagrams and check the infotainment power and ground circuits for short circuit to ground, open circuit, high resistance. Repair circuit(s) as required
B4: POOR OR WEAK SOUND OUTPUT FROM LOUDSPEAKER(S)	
	1. Check power supply voltage at power supply connectors
	Is the power supply voltage measured at the power supply connectors between 12 and 14 volts? **Yes** No further action **No** Refer to the electrical circuit diagrams and check the infotainment power supply circuits for short circuit to ground, open circuit, high resistance. Repair circuit(s) as required. Refer to the relevant section of workshop manual and battery care manual. Check battery state of charge and starting/charging system performance and rectify as required

PINPOINT TEST C : LOUDSPEAKER VIBRATING (BUZZING) EXCESSIVELY	
TEST CONDITIONS	**DETAILS/RESULTS/ACTIONS**
C1: LOUDSPEAKER VIBRATING (BUZZING) EXCESSIVELY	
	1. Check for extreme bass/treble settings
	Is the audio system output settings for bass and/or treble set too high? **Yes** Adjust settings to appropriate levels **No** Proceed to the next step GO to 2
C2: LOUDSPEAKER VIBRATING (BUZZING) EXCESSIVELY	
	1. Check loudspeaker unit(s) fixing screws are securely fastened
	Are all loudspeaker fixing screws fully secured to the surrounding trim? **Yes** Proceed to the next step GO to 3 **No** Tighten the fixing screws to the correct torque as directed in the workshop manual
C3: LOUDSPEAKER VIBRATING (BUZZING) EXCESSIVELY	
	1. Check if the wiring harness is resting against the internal surface of the loudspeaker
	Is there any cabling or other parts of the wiring harness resting against the internal surface of the loudspeaker? **Yes** Re-route and secure the wiring harness so that it is not resting against any internal surfaces of the loudspeaker **No** Proceed to the next step GO to 4
C4: LOUDSPEAKER VIBRATING (BUZZING) EXCESSIVELY	
	1. Check security of trim in the vicinity of the loudspeaker units
	Is the trim secure in the vicinity of the loudspeaker units? **Yes** No further action **No** Secure loose trim components to make sure no vibration
C5: AUDIO OUTPUT DISTORTED	
	1. Check for extreme bass/treble settings
	Is the audio system output settings for bass and/or treble set too high? **Yes** Adjust settings to appropriate levels

PINPOINT TEST C : LOUDSPEAKER VIBRATING (BUZZING) EXCESSIVELY	
TEST CONDITIONS	**DETAILS/RESULTS/ACTIONS**
	No No further action

DTC Index

For a list of Diagnostic Trouble Codes (DTCs) that could be set on this vehicle, please refer to Section 100-00.

566

Instrument Panel Speaker

Removal and Installation

Removal

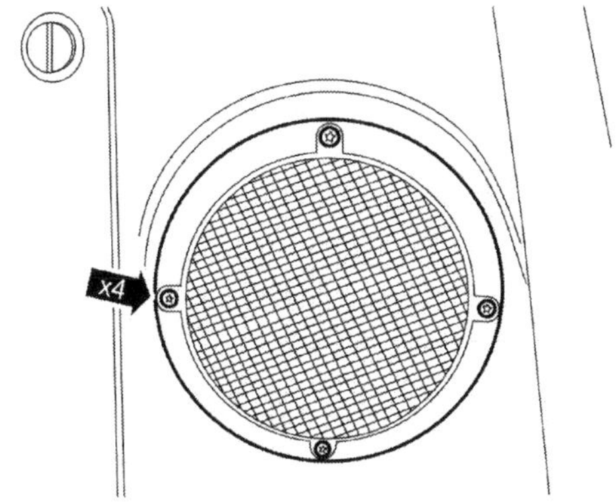

E88719

1. △ NOTE: RH shown, LH similar.

Release the instrument panel speaker.

- Remove the 4 screws.

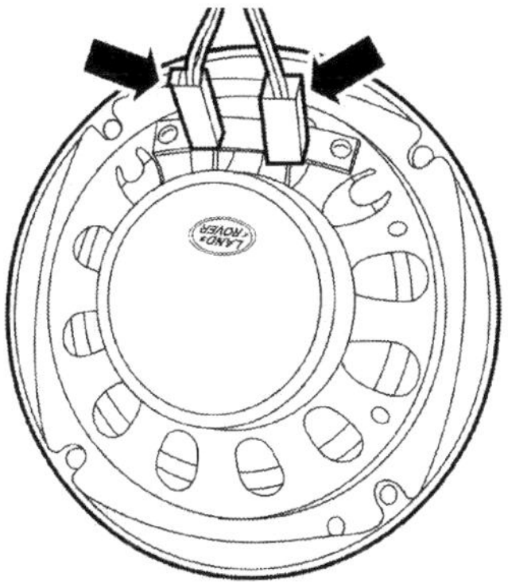

E88720

2. Remove the instrument panel speaker.
   - Disconnect the 2 electrical connectors.

Installation

1. To install, reverse the removal procedure.

D-Pillar Speaker

Removal and Installation

Removal

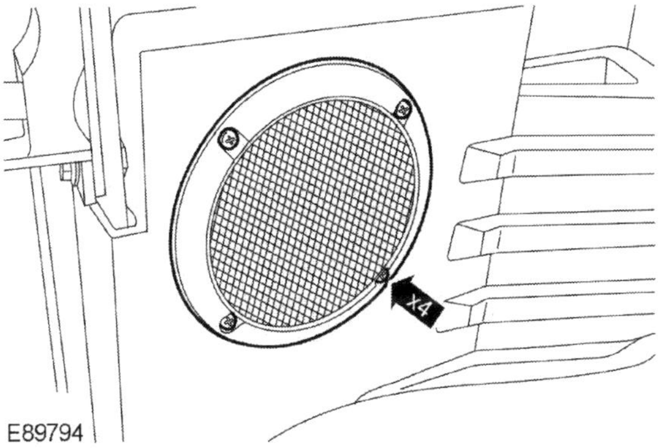

E89794

1. △ NOTE: LH shown, RH similar.

Release the speaker.

- Remove the 4 screws.

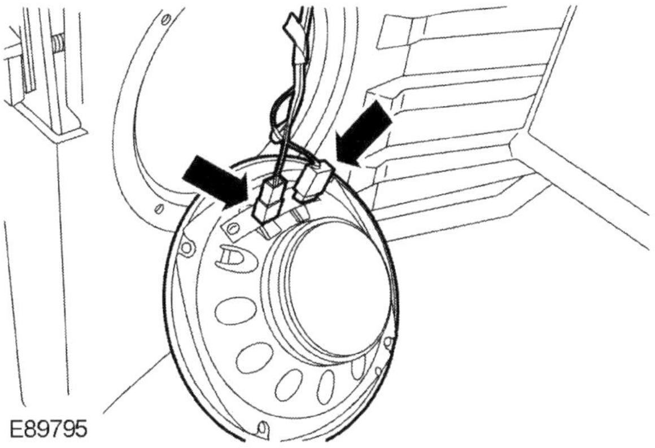

E89795

2. Remove the speaker.
   - Disconnect the 2 electrical connectors.

Installation

1. To install, reverse the removal procedure.

Rear Door Speaker

Removal and Installation

Removal

1. Raise rear seat cushion and fit stowage strap.
2. Carefully prise grille from speaker.
3. Remove 3 screws and withdraw speaker from trim panel.
4. Disconnect 2 electrical leads and remove speaker.

Installation

1. Connect 2 electrical leads, position speaker and secure with screws.

2. Position grille and press firmly to secure.

Specifications

Description and Operation

COMPONENT LOCATION

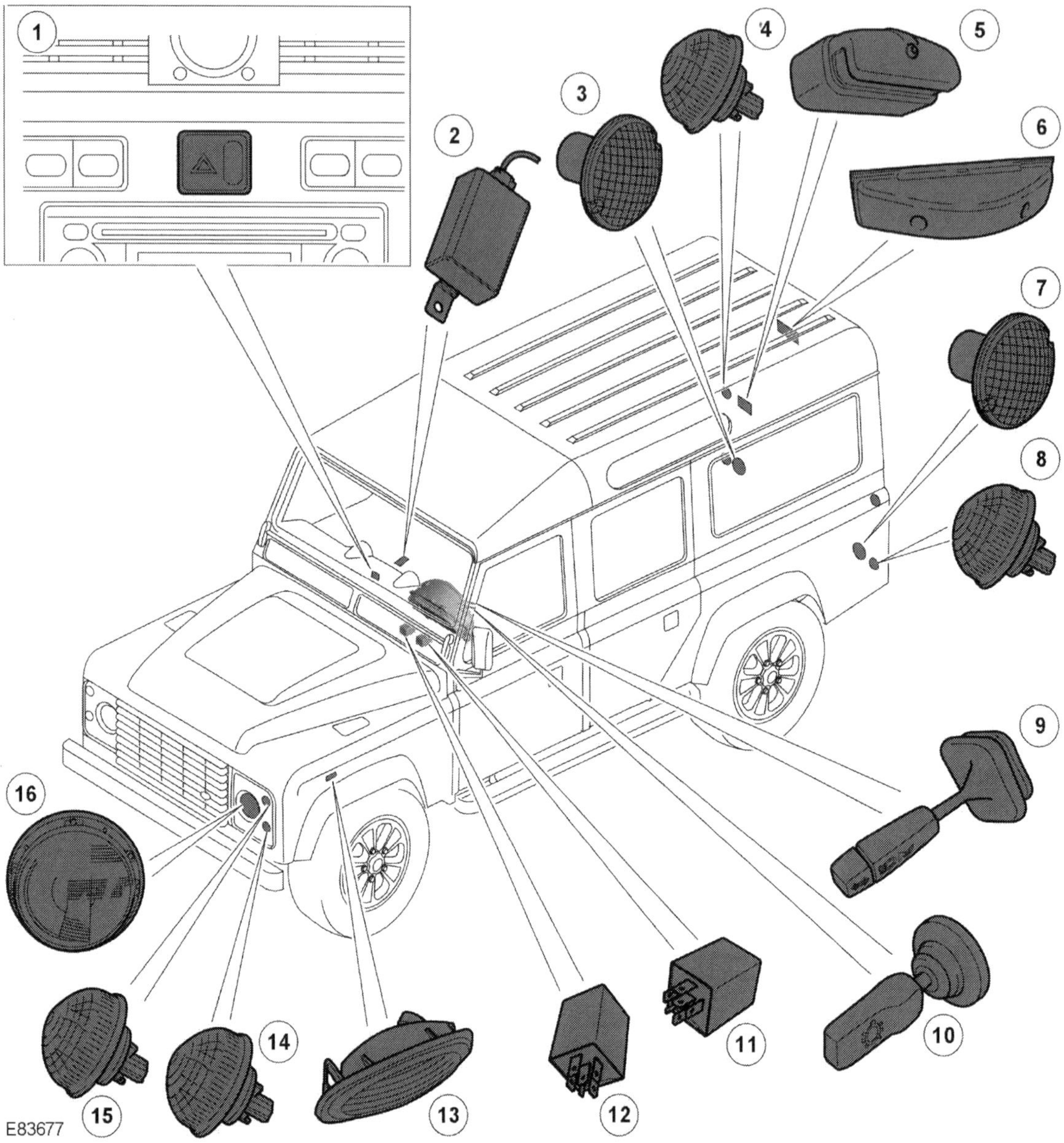

Item	Description
1	Hazard flasher switch
2	Fog lamp control module
3	Rear fog lamp
4	Rear tail/Stop lamp
5	License plate lamp
6	High mounted stop lamp
7	Reverse lamp
8	Rear turn signal indicator lamp
9	Steering column multifunction switch
10	Lighting control switch
11	Headlamp relay
12	Hazard flasher relay
13	Side turn signal indicator lamp
14	Front turn signal indicator lamp
15	Side lamp
16	Headlamp

OVERVIEW

Operation of the exterior lamps is controlled via the lighting control switch and the steering column multifunction switch. The lighting control switch is a 3 way switch mounted on the LH (left-hand) side of the steering column. When pushed forward to the first position the switch will provide a battery feed to the side lamps, tail lamps and number plate lamps irrespective of ignition switch position. When pushed forward to the second position, the lighting control switch will also provide a feed to the headlamp relay.

The steering column multifunction switch is mounted forward of the lighting control switch on the LH side of the steering column and allows the driver to switch between headlamp high beam, low beam and flash operation. The steering column multifunction switch is provided a feed by the headlamp relay, which becomes energized when the ignition switch is turned to position II.

Where fitted, a 4 position rotary headlamp leveling switch is mounted on the instrument panel to allow the driver to raise or lower headlamp alignment according to the load being carried in the vehicle. A motor mounted on the rear of each headlamp adjusts the alignment of the headlamps in response to movements in switch position.

Operation of the turn signal indicator lamps is also controlled using the steering column multifunction switch. The steering column multifunction switch is provided an ignition switch feed from the hazard flasher relay. When the switch is moved to the left or right turn position, the switch contacts close and a feed is provided to the relevant turn signal indicator lamps. A feed is also provided to the instrument cluster to illuminate the turn signal warning indicator.

The hazard flasher switch is located on the instrument panel and operates the LH and RH (right-hand) turn signal indicator lamps simultaneously when pressed. The hazard flasher switch is provided a constant battery feed and will operate the turn signal indicator lamps irrespective of ignition switch position. When pressed, the hazard flasher switch also provides a feed to the instrument cluster to illuminate the hazard flasher warning indicator.

The stop lamps are controlled by the stop lamp switch, which is mounted on top of the brake pedal. When the brake pedal is pressed the switch contacts close allowing an ignition switch feed to power the stop lamps.

The reverse lamp is controlled by the reverse gear switch, which is mounted on the transmission.

CONTROL DIAGRAM

NOTE: **A** = Hardwired

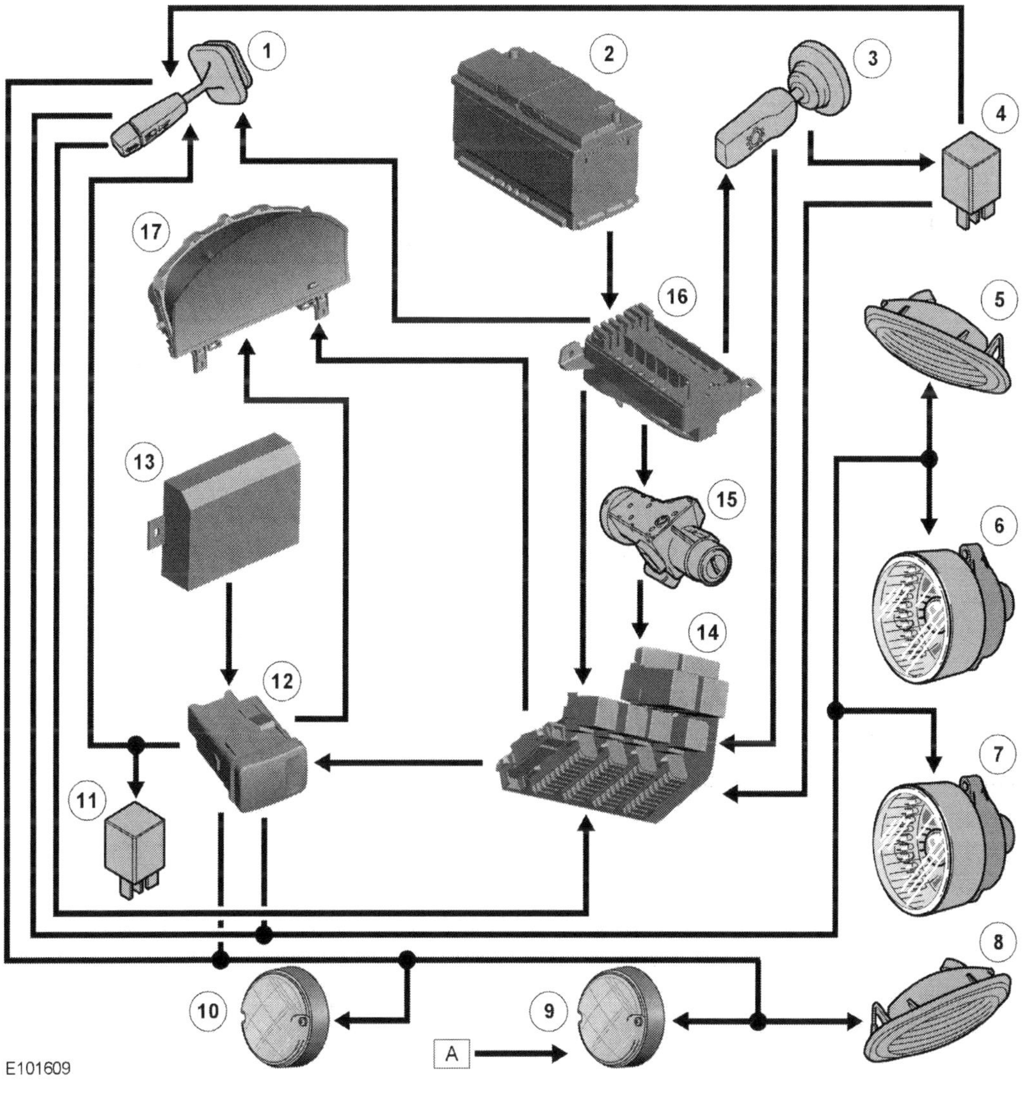

E101609

Item	Description
1	Steering column multifunction switch
2	Battery
3	Lighting control switch
4	Headlamp relay
5	RH side turn signal lamp
6	RH front turn signal lamp
7	LH front turn signal lamp
8	LH side turn signal lamp
9	RH rear turn signal lamp
10	LH rear turn signal lamp
11	Hazard flasher relay
12	Hazard flasher switch
13	Anti-theft system module
14	-
15	Ignition switch
16	-
17	Instrument cluster

When reverse gear is selected the switch contacts close allowing an ignition switch feed to power the reverse lamp.

Operation of the rear fog lamp is controlled by the fog lamp control module, which is located beneath the RH front seat. The control module is provided an ignition switch feed from the energized headlamp relay via the CJB (central junction box) . The control module monitors the condition of the fog lamp switch. When the switch is pressed, the switch contacts close and a ground path is created. When the control module registers the ground path it provides a feed to the rear fog lamp and the instrument cluster.

CONTROL DIAGRAM

⚠ NOTE: **A** = Hardwired

Item	Description
1	LH stop lamp
2	Brake pedal switch
3	RH stop lamp
4	Reverse lamp
5	Reverse gear indicator switch
6	Fog lamp switch
7	Fog lamp control module
8	Fog lamp
9	LH headlamp
10	LH side lamp
11	RH headlamp
12	RH side lamp
13	LH tail lamp
14	License plate lamp
15	RH tail lamp
16	RH headlamp levelling motor (if fitted)
17	-
18	Headlamp levelling switch (if fitted)
19	LH headlamp levelling motor (if fitted)

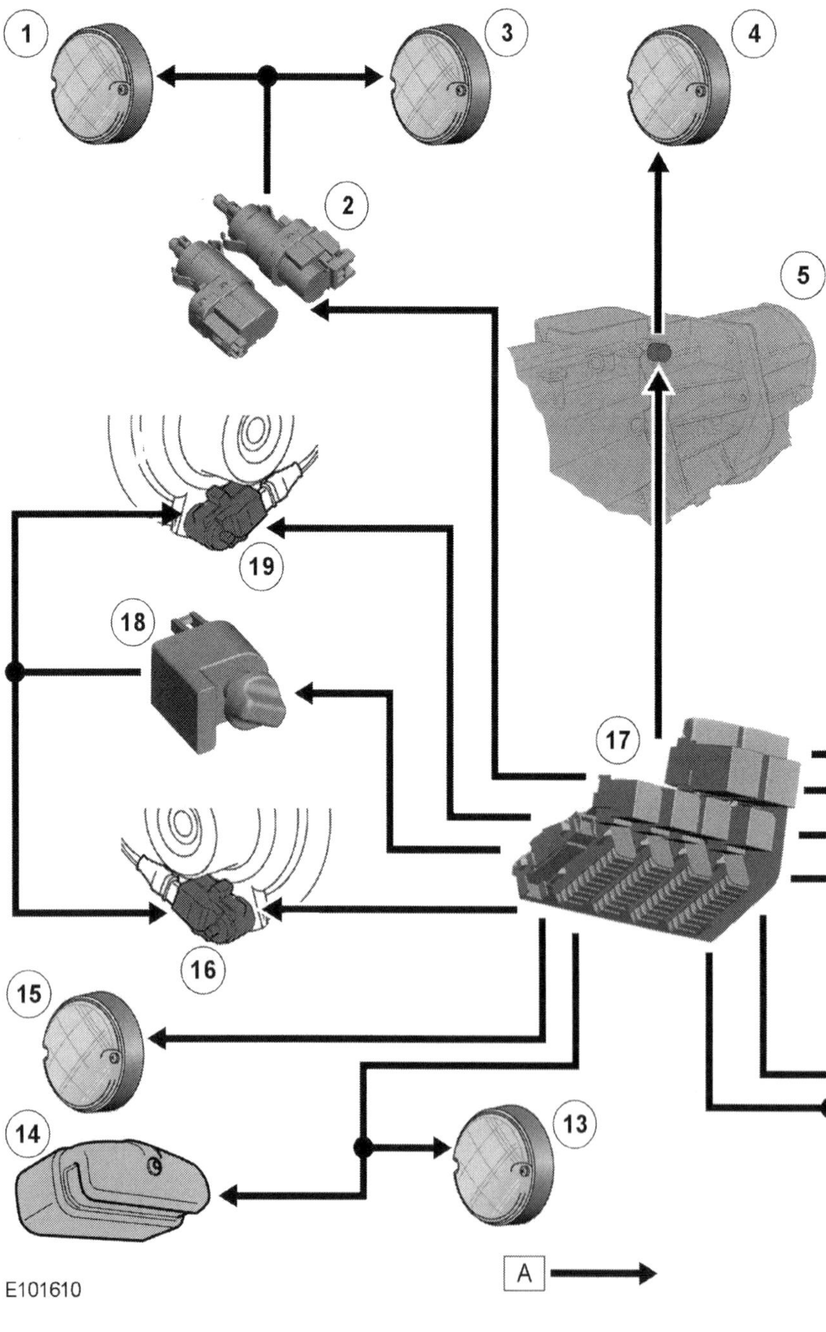

Headlamp Adjustment

General Procedures

1. ⚠ NOTE: Check main beam alignment using beam setting equipment. Should this not be available the beam can be temporarily checked and adjusted as follows.

   Position vehicle, unladen, on level ground with tyres correctly inflated, approximately 4 metres from a wall or screen, marked as illustrated below.

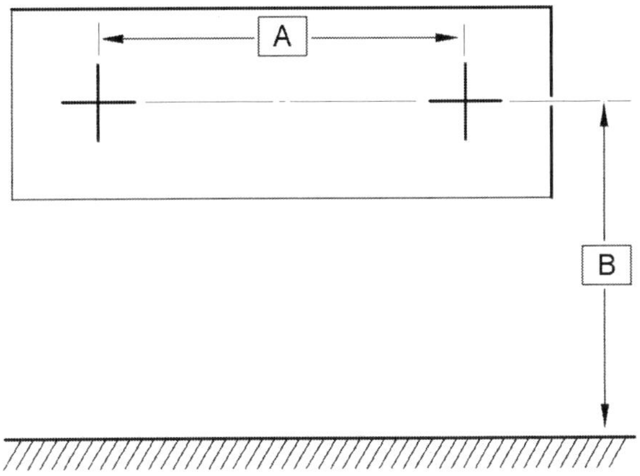

2. The beam centres 'A' are measured horizontally on the vehicle and dimension 'B' vertically from the ground.

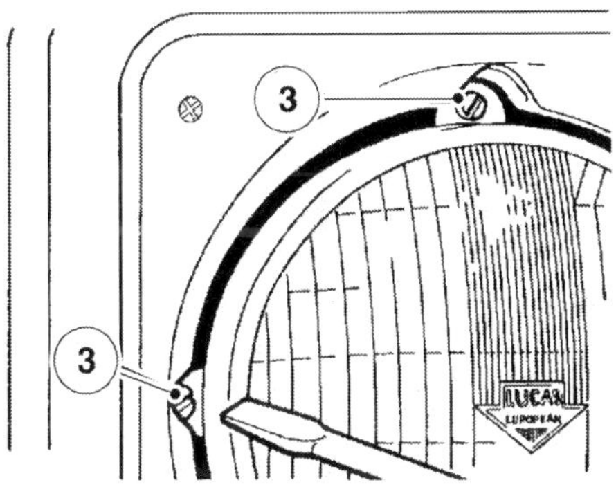

3. Switch on main beam and adjust setting, as necessary, with trimmer screws.

Headlamp Assembly

Removal and Installation

Removal

1. Remove 4 screws and release side and turn signal lamps from fender.
2. Disconnect multiplugs and remove side and turn signal lamps.
3. Remove 2 screws and remove headlamp finisher.
4. Remove 4 screws securing headlamp to fender, disconnect multiplugs and remove headlamp.

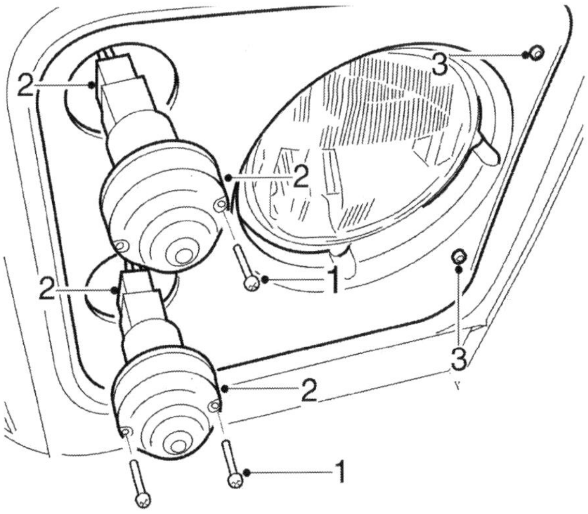

5. Remove headlamp seal.

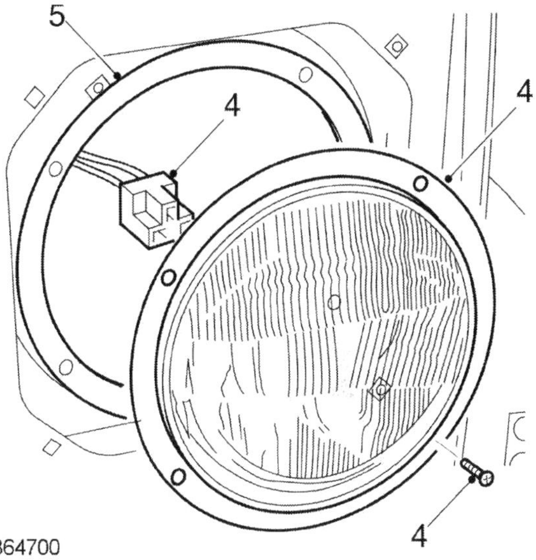

M864700

Installation

1. Fit headlamp seal.
2. Position headlamp, connect multiplugs, fit and tighten headlamp screws.
3. Fit headlamp finisher and secure with screws.
4. Position side and turn signal lamps, connect multiplugs and secure with screws.
5. Reset headlamp alignment.

Refer to: Headlamp Adjustment (417-01 Exterior Lighting, General Procedures)

Headlamp Leveling Motor

Removal and Installation

Removal

1. Remove headlamp assembly.

Refer to: Headlamp Assembly (417-01 Exterior Lighting, Removal and Installation)

2. Disconnect headlamp leveling motor harness plug.

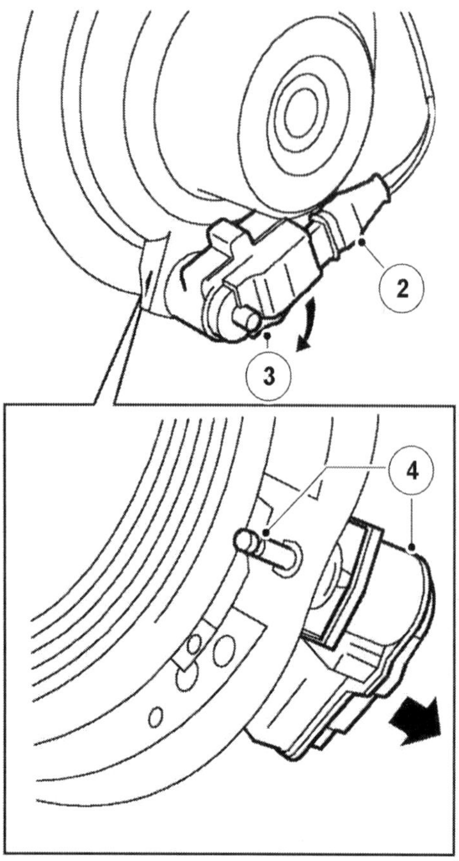

J6447

3. Rotate headlamp leveling motor anti-clockwise to release retaining lugs from mounting bracket on headlamp body.

4. Release headlamp leveling motor spindle from retaining slot on headlamp bezel and withdraw headlamp leveling motor from mounting bracket.

Installation

1. Install headlamp leveling motor into mounting bracket and locate motor spindle into slot in headlamp bezel.

2. Press in headlamp leveling motor and rotate clockwise to engage retaining lugs behind mounting bracket.

3. Connect headlamp leveling motor harness plug.

4. Fit headlamp assembly.

Refer to: Headlamp Assembly (417-01 Exterior Lighting, Removal and Installation)

Headlamp Leveling Switch

Removal and Installation

Removal

1. Disconnect the battery ground cable.

Refer to: Startup Battery Disconnect and Connect (414-01 Battery, Mounting and Cables, General Procedures)

2. Remove the audio unit.

Refer to: Audio Unit - Vehicles Without: Bluetooth (415-01 Audio Unit, Removal and Installation)

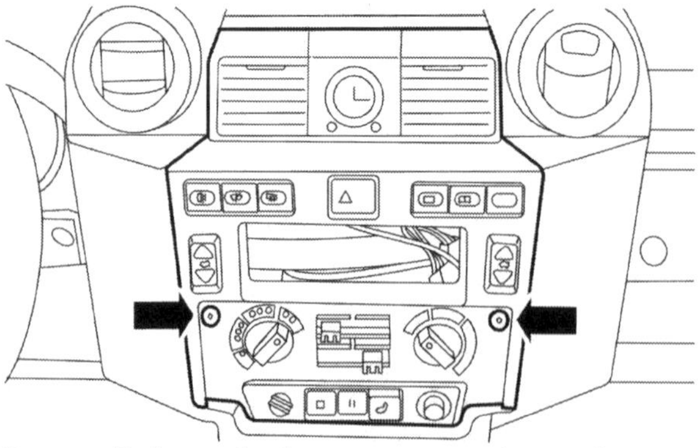

3. Release the instrument panel console.

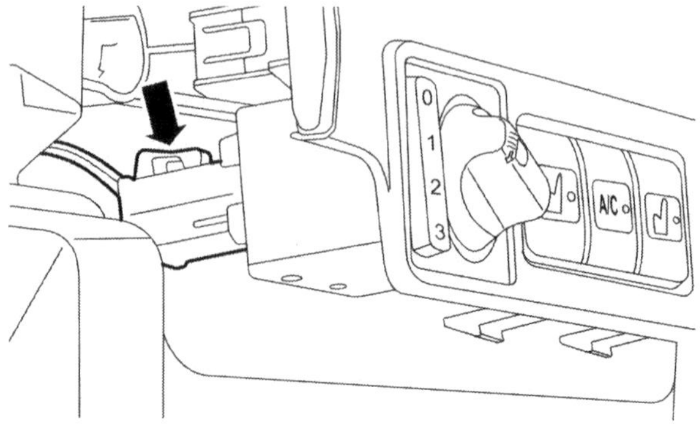

4. Disconnect the electrical connector.

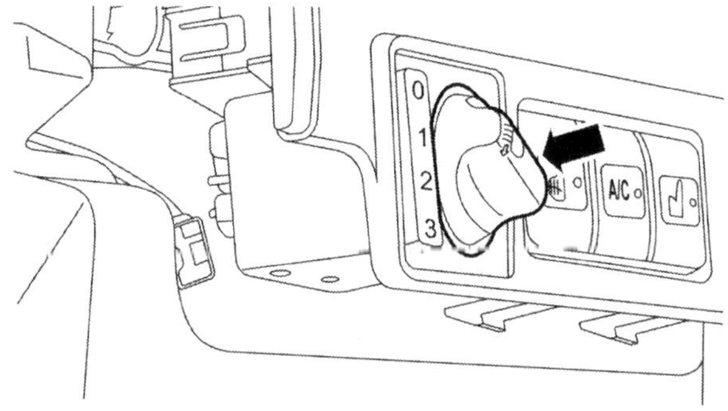

5. Remove the headlamp leveling selector knob.

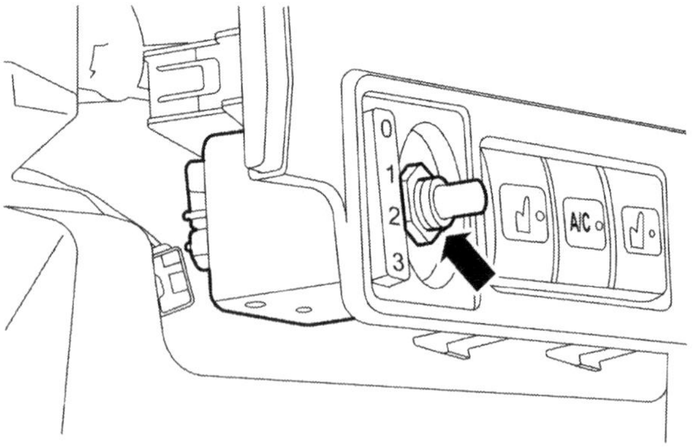

6. Remove the headlamp leveling switch.

   - Remove the nut.

Installation

1. To install, reverse the removal procedure.

2. Connect the battery ground cable.

Refer to: Battery Connect (414-01 Battery, Mounting and Cables, General Procedures).

Headlamp Switch

Removal and Installation

Removal

1. Remove the steering column shrouds.

   Refer to: Steering Column Shrouds (501-05 Interior Trim and Ornamentation, Removal and Installation).

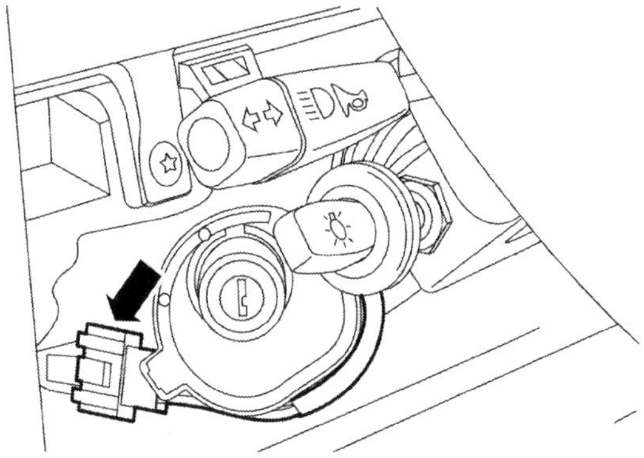

2. Disconnect the electrical connector.

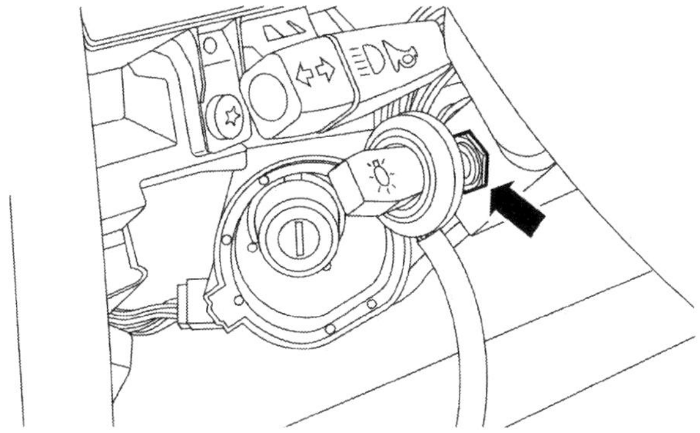

3. Release the nut.

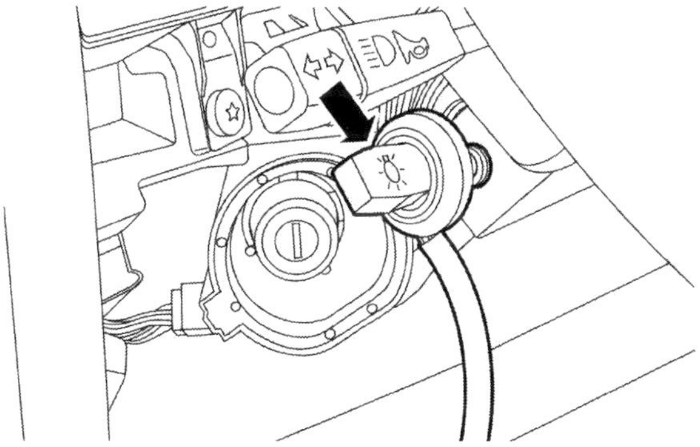

4. Remove the headlamp switch.

Installation

1. To install, reverse the removal procedure.

High Mounted Stoplamp

Removal and Installation

Removal

1. Open rear door.

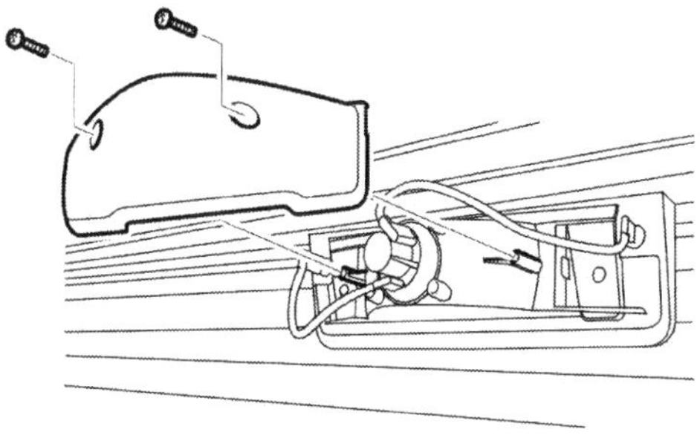

2. Remove 2 screws and cover from the high mounted stoplamp.
3. Remove insulated cover from connector, release catches and disconnect connectors taking care not to damage screen element.
4. Remove nuts securing lamp to rear screen, withdraw high mounted stoplamp, collect seals and washers.
5. Remove rubber backing pad, taking care not to damage screen obscuration band.

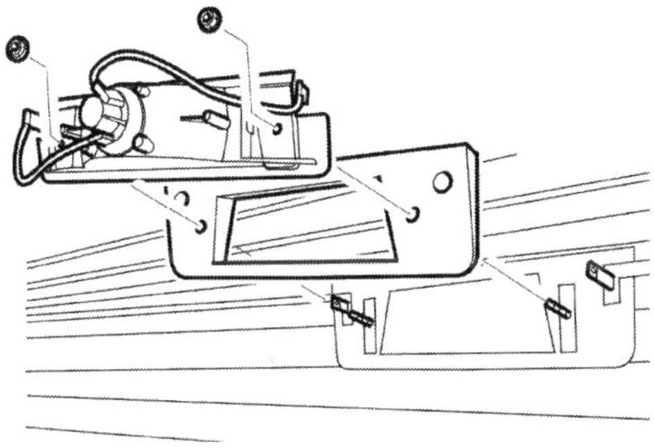

Installation

1. Clean obscuration band and fit new backing pad.
2. Fit high mounted stoplamp, seals and washers. Fit nuts and tighten to 3 Nm (2.2 lbf.ft).
3. Insulate connectors, and connect to rear screen.
4. Fit cover to high mounted stoplamp and secure with screws.

License Plate Lamp

Removal and Installation

Removal

1. Remove single screw and remove lamp cover and lens.
2. Unscrew 2 fixings and remove metal cover to gain access to lamp fixings and harness leads inside vehicle.
3. Disconnect lamp leads from harness.

4. Remove 2 bolts, nuts and washers securing lamp to vehicle body.
5. Remove bulb holder, complete with rubber seal and mounting plinth.

Installation

1. Feed lamp unit leads through vehicle body and secure bulb holder, seal and plinth.
2. Install lamp leads to harness connectors.
3. Install cover to conceal lamp fixings.
4. Install lens and lamp cover.

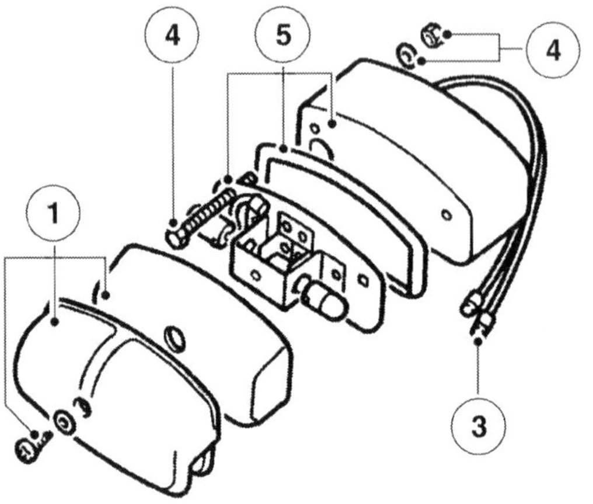

License Plate Lamp Bulb

Removal and Installation

Removal

1. Remove single screw securing lamp cover and lens to bulb holder.
2. Remove bulb/s from holder.

Installation

1. Install new bulb/s.
2. Install lamp lens and cover.

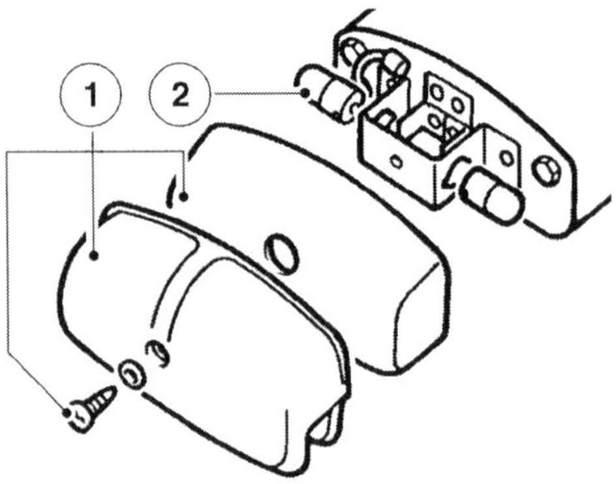

J6379

Rear Fog Lamp Bulb

Removal and Installation

Removal

⚠ NOTE: This procedure covers removal and installation of reversing lamp bulb and rear fog lamp bulb.

1. Remove two screws with sealing washers and remove lens from lamp body.
2. Push and twist bulb to release from holder.
3. Clean interior of lamp and lens.

Installation

1. Fit new bulb.
2. Fit lens to lamp body.

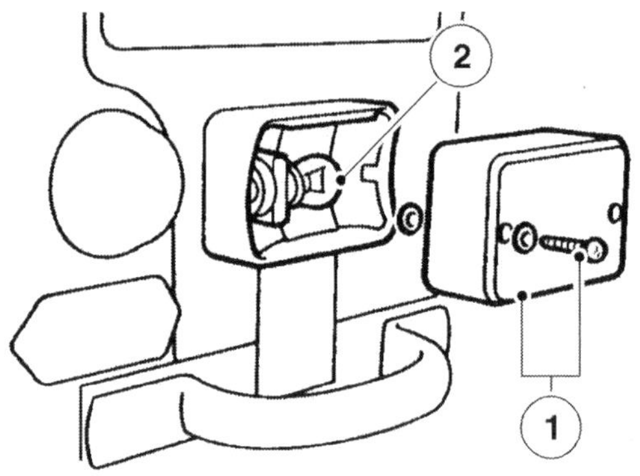

Rear Lamp Assembly

Removal and Installation

Removal

1. Remove 2 screws securing rear lamp to body.
2. Disconnect multiplug and remove lamp.

Installation

1. Position rear lamp and connect multiplug.
2. Fit and tighten lamp securing screws.

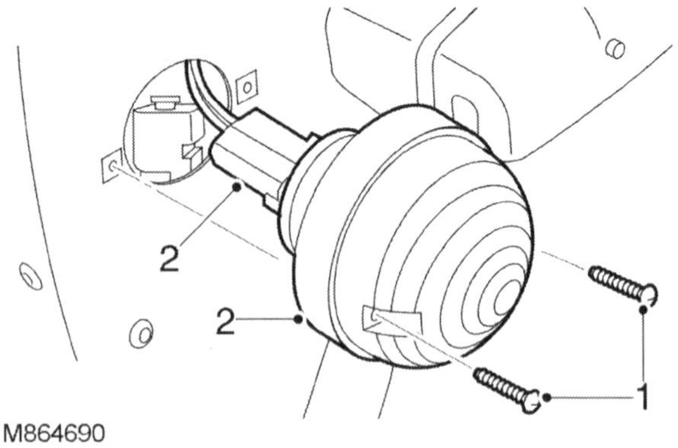

M864690

Reversing Lamp

Removal and Installation

Removal

△ NOTE: This procedure covers removal and installation of the reversing lamp and the rear fog lamp.

1. From underneath the vehicle, lift protective flap and disconnect lamp leads at harness connectors.
2. Unscrew 2 nuts and washers and withdraw lamp unit from vehicle.

Installation

1. Feed lamp leads through grommet and locate lamp studs in vehicle body.
2. Fit retaining nuts and reconnect lamp leads to vehicle harness.

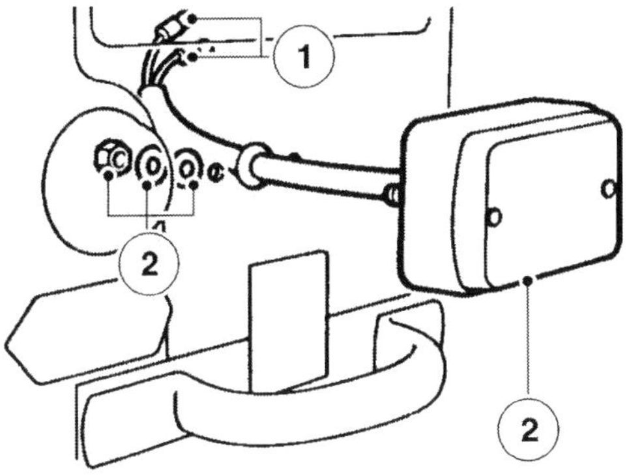

J6382

Reversing Lamp Switch

Removal and Installation

Part(s)

STEP	PART NAME	QUANTITY
Installation Step 1	Renew Part Reversing lamp switch	1

Removal

1. Remove the floor console.

Refer to: Floor Console (501-12 Instrument Panel and Console, Removal and Installation)

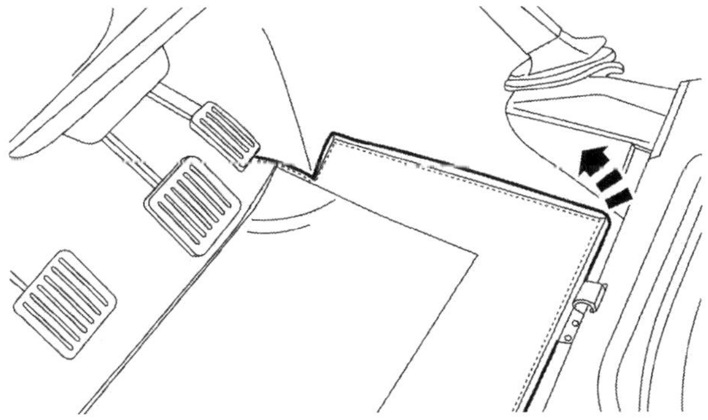

2. Reposition the LH carpet.

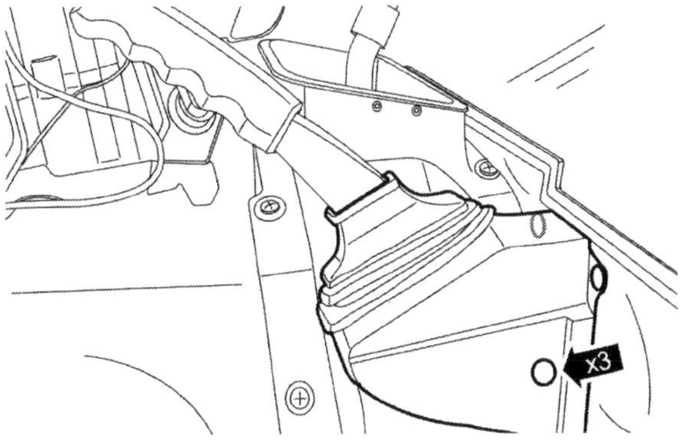

3. Release the parking brake lever gaiter. Remove the 3 clips.

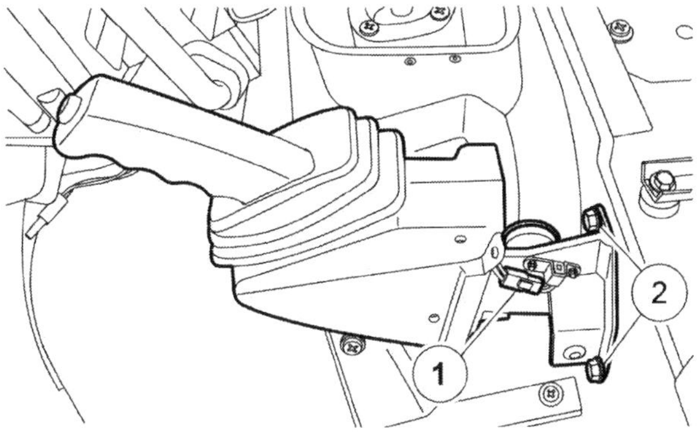

4. Release the parking brake lever. Disconnect the electrical connector. Remove the 2 bolts.

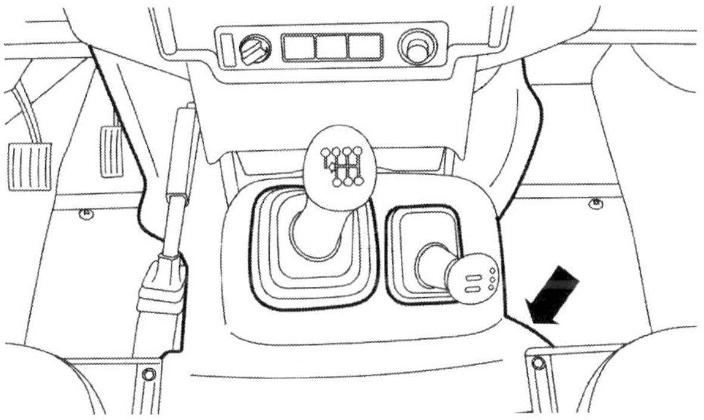

5. Remove the transmission cover panel floor covering.

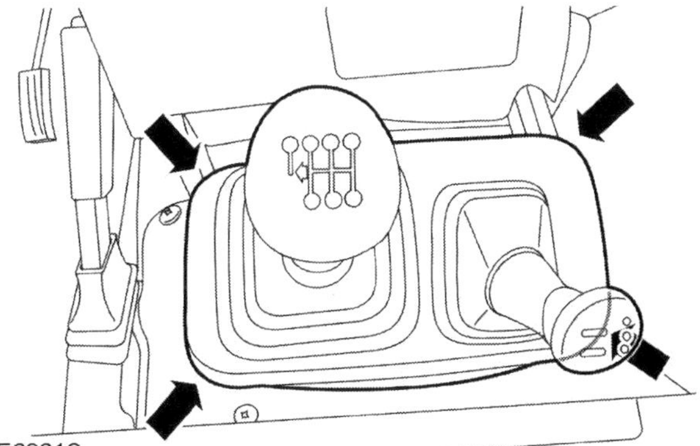
E88812

6. △ NOTE: Do not detach the gaiter from the selector knobs.

Detach the gaiter from the transmission cover panel.

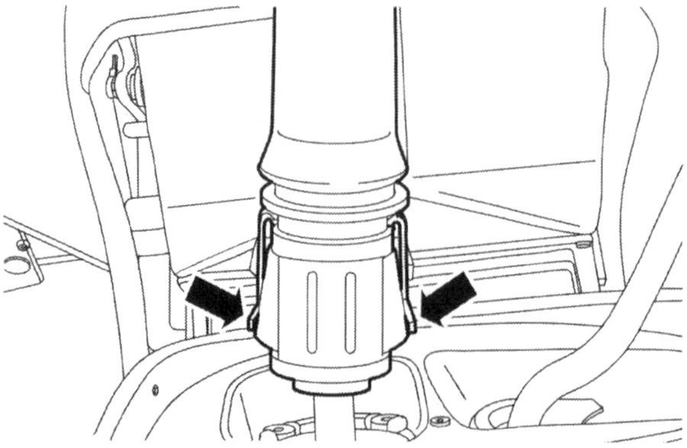

7. WARNING: The gearshift lever knob will be released suddenly, keep face clear during removal. Failure to follow this instruction may result in personal injury.

Release the gearshift lever knob. Release the 2 clips.

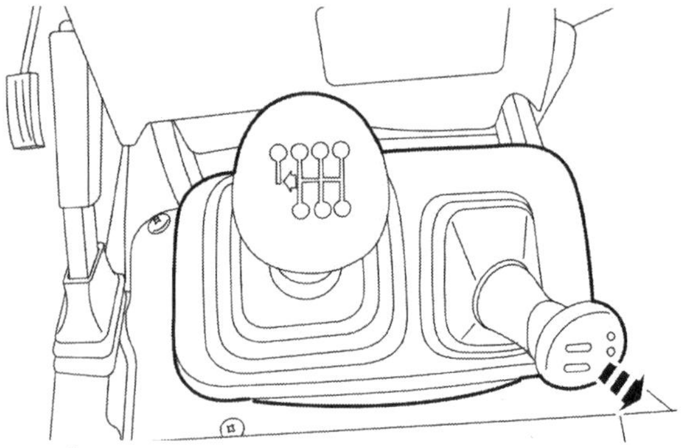

8. WARNING: The high/low range lever knob will be released suddenly, keep face clear during removal. Failure to follow this instruction may result in personal injury.

   Remove the gaiter and selector levers. Detach the high/low range selector lever.

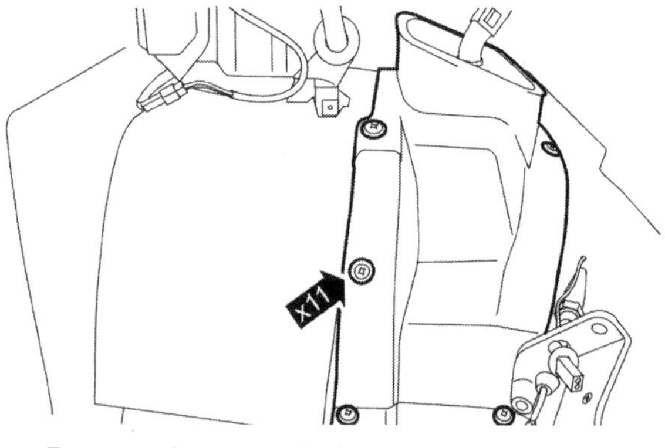

9. Remove the transmission cover panel. Remove the 11 screws.

10. Disconnect reversing lamp switch connector.

11. Remove the reversing lamp switch. Remove and discard the O-ring seal.

Installation

1. Install the reversing lamp switch. Install a new O-ring seal.

Renew Part: Reversing lamp switch

Torque: 20 Nm

2. Connect the reversing lamp switch electrical connector.
3. Install the transmission cover panel. Install the screws.
4. Install the gaiter with the selector levers attached. Install the selector levers. Fully seat the gaiter.
5. Install the transmission cover panel floor covering.
6. Install the parking brake lever.

   Torque: 25 Nm

7. Install the parking brake lever gaiter. Install the clips.

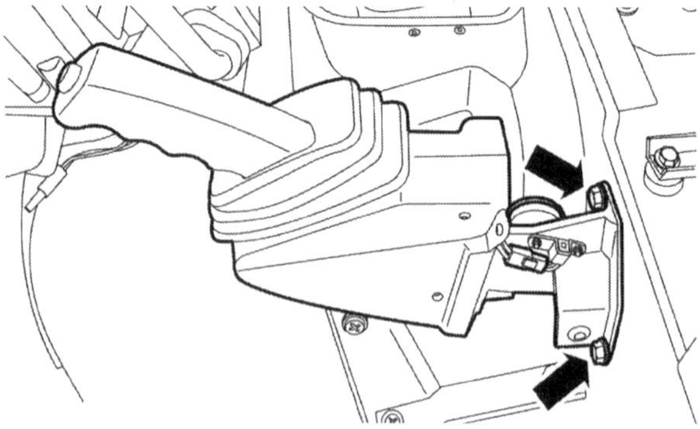

8. Reposition the LH carpet.
9. Install the floor console.

Refer to: Floor Console (501-12 Instrument Panel and Console, Removal and Installation)

Side Marker Lamp

Removal and Installation

Removal

NOTE: This procedure covers removal and installation of the side repeater lamp.

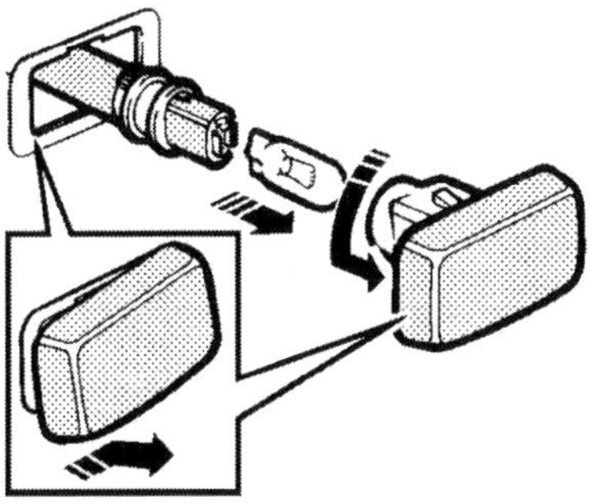

1. Push lens firmly to the right.
2. Lift left edge and withdraw lamp unit from wing.
3. Twist bulb holder and release from lens.
4. Pull bulb from holder.

Installation

1. Fit new bulb, if necessary.
2. Fit bulb holder to lens.
3. Locate lamp unit in wing and push firmly to the left to secure.

Side Repeater Lamp Bulb

Removal and Installation

Removal

1. Remove 2 screws securing side turn signal lamp to front fender.
2. Disconnect multiplug and remove lamp.

Installation

1. Position side turn signal lamp and connect multiplug.
2. Fit and tighten lamp securing screws.

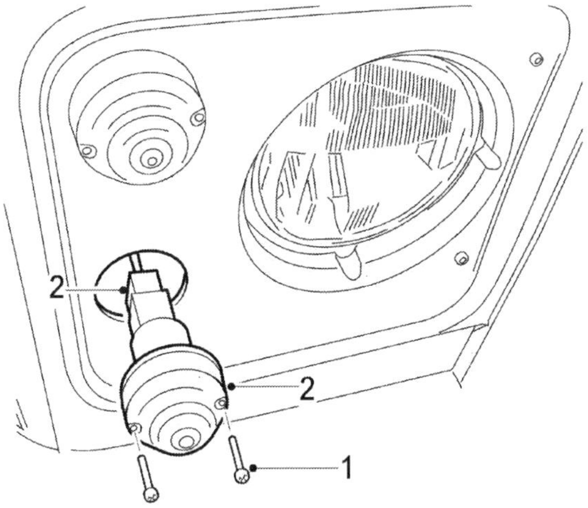

Stoplamp Switch

Removal and Installation

Removal

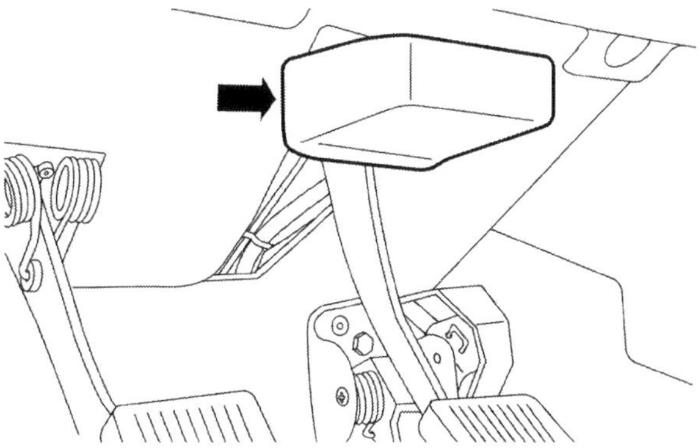

1. Remove the stoplamp switch cover.

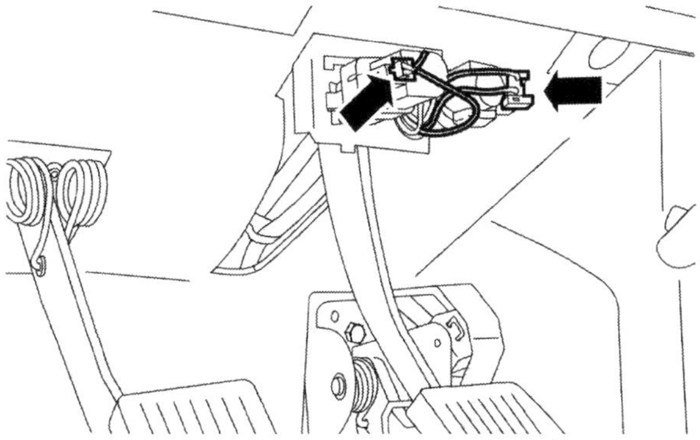

2. Disconnect the electrical connectors.

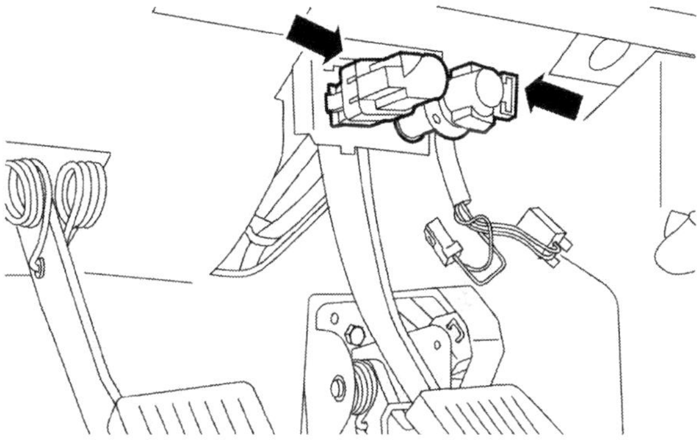

3. Remove the stoplamp switches.

Installation

1. CAUTION: Make sure that the brake pedal is kept in the rest position and is not pressed or moved during the instillation of the stoplamp switch. Failure to follow this instruction may result in damage to the vehicle.

To install, reverse the removal procedure.

Description and Operation

COMPONENT LOCATION

⚠ NOTE: LHD (left-hand drive) vehicle shown, RHD (right-hand drive) similar.

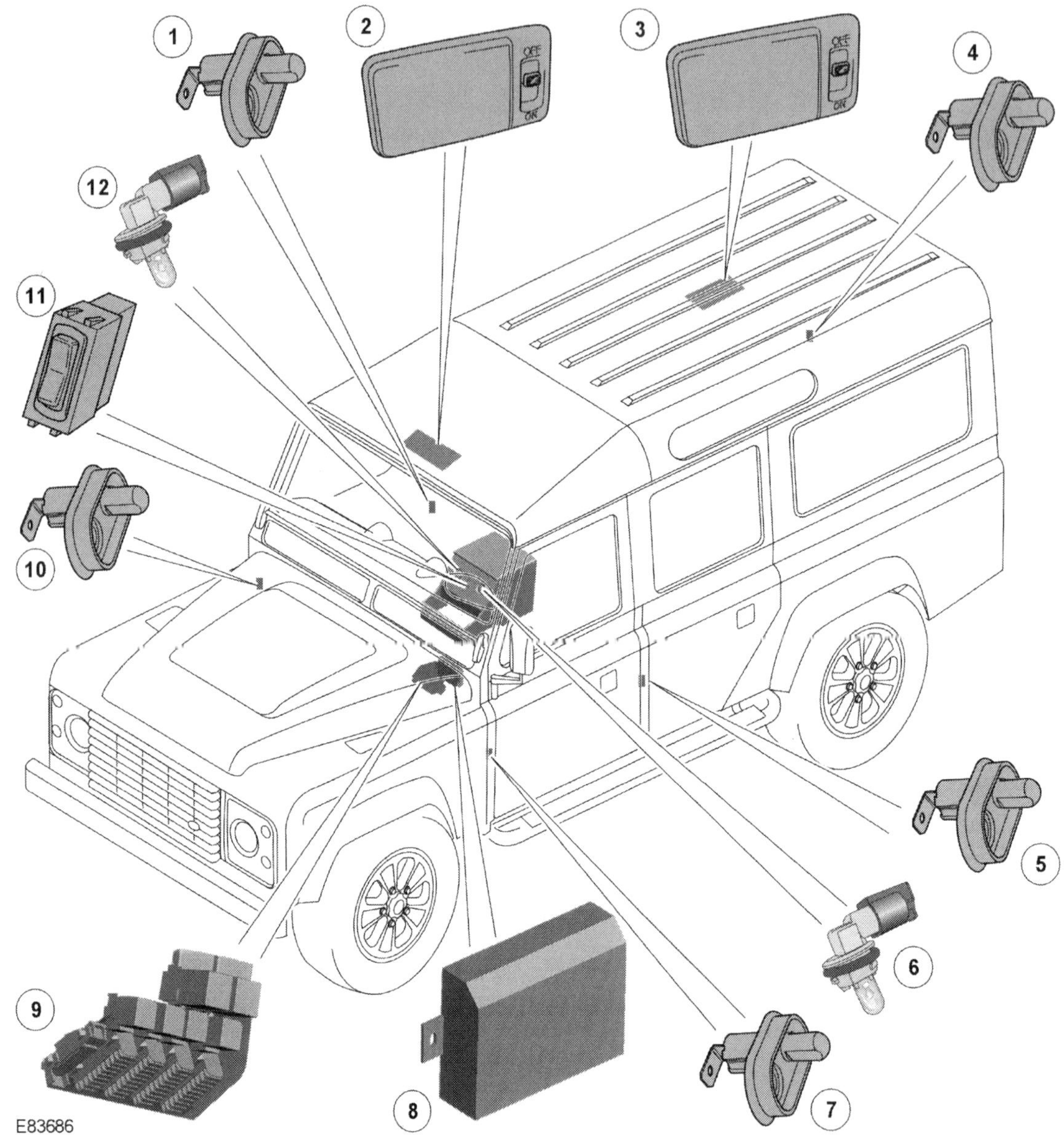

Item	Description
1	RH (right-hand) rear door switch (if fitted)
2	Front interior lamp
3	Rear interior lamp (Not 90 model)
4	Tail door switch (Not 90 model)
5	LH (left-hand) rear door switch (if fitted)
6	Map reading lamp (SVX (60th anniversary) model only)
7	Drivers door switch
8	Anti-theft system module
9	-
10	Passenger door switch
11	Map reading lamps switch (SVX (60 anniversary) model only)
12	Map reading lamp (SVX (60th anniversary) model only)

OVERVIEW

Each interior lamp contains a 3 way switch, allowing always on, always off, or automatic operation. When in the always on position, the interior lamp is provided with a battery voltage feed from the CJB. When in the automatic position, operation of the interior lamps is controlled by the anti-theft system module. The anti-theft system module monitors the condition of the doors, including the tail door, through a series of switches. If any of the doors are opened a ground path is created through the relevant switch. When the anti-theft system module registers a ground path it illuminates the interior lamps. If the door(s) are subsequently closed, the anti-theft system module will extinguish the interior lamps after a period of 15 seconds.

CONTROL DIAGRAM

NOTE: **A** = Hardwired

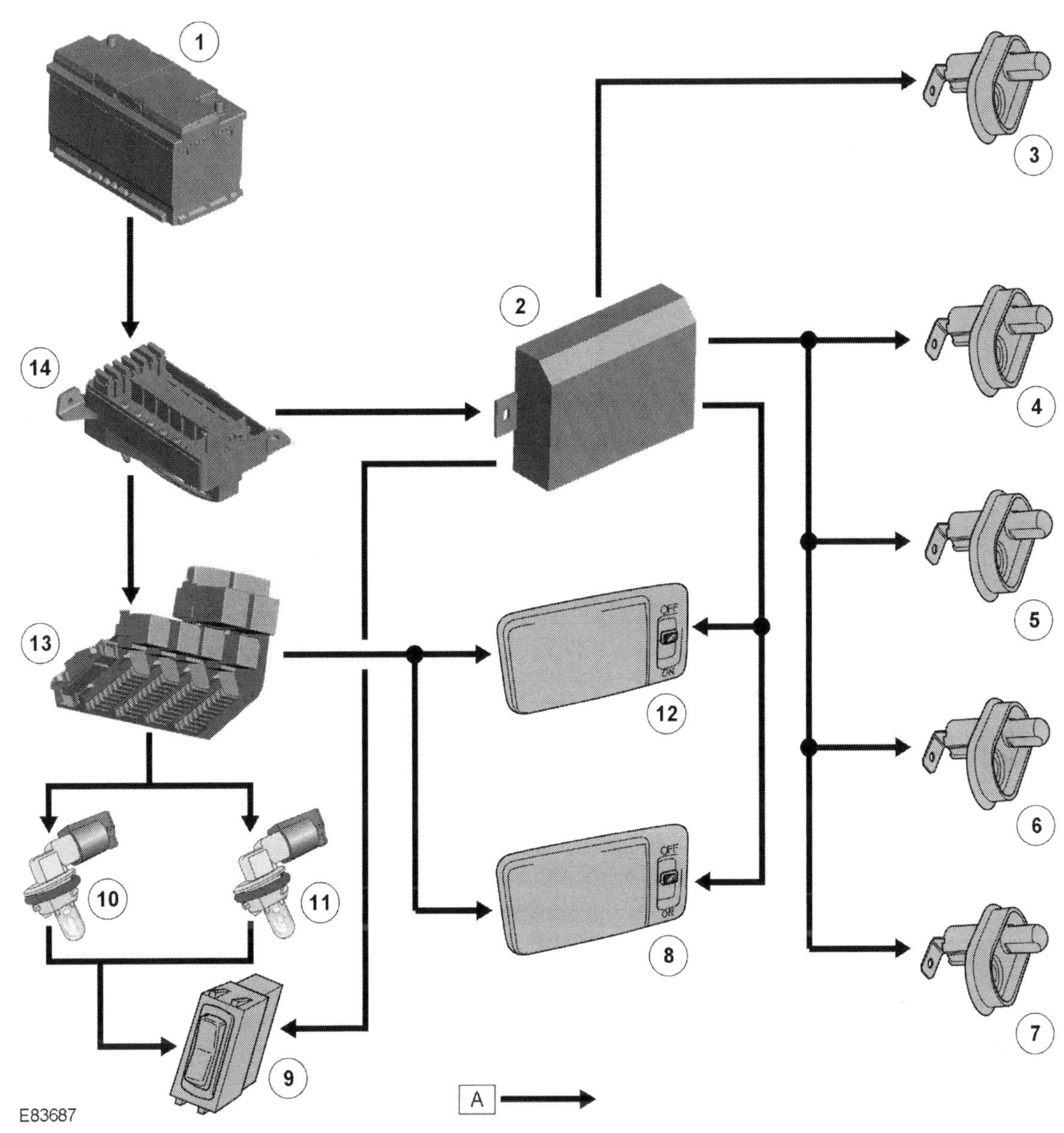

Item	Description
1	Battery
2	Anti-theft system module
3	Drivers door switch
4	Passenger door switch
5	Tail door switch
6	LH rear door switch (if fitted)
7	RH rear door switch (if fitted)
8	Rear interior lamp
9	Map reading lamps switch (SVX (60th anniversary) model only)
10	LH map reading lamp (SVX (60th anniversary) model only)
11	RH map reading lamp (SVX (60th anniversary) model only)
12	Front interior lamp
13	-
14	-

The anti-theft system module will also illuminate the interior lamps if it receives a valid unlock signal from the remote radio frequency (RF) handset.

Refer to: Anti-Theft - Active (419-01 Anti-Theft - Active, Description and Operation)

SVX (60th Anniversary) Model

The SVX model has all the interior lighting features of the standard models but with the addition of 2 map reading lamps.

The map reading lamps are located on the forward face of the cubby box with a control switch positioned between them. The lamps use a W3W bulb. The bulb can be replaced by carefully levering the lamp from the cubby box using a suitable tool and removing the bulb holder from the rear of the lamp.

The map reading lamps are connected directly to the vehicle battery via fuses in the CJB and the BJB (battery junction box) . Therefore the map reading lamps can be operated independent of the ignition switch position. Each lamp receives a battery voltage supply and the lamp illuminates when the switch is operated which completes a ground path.

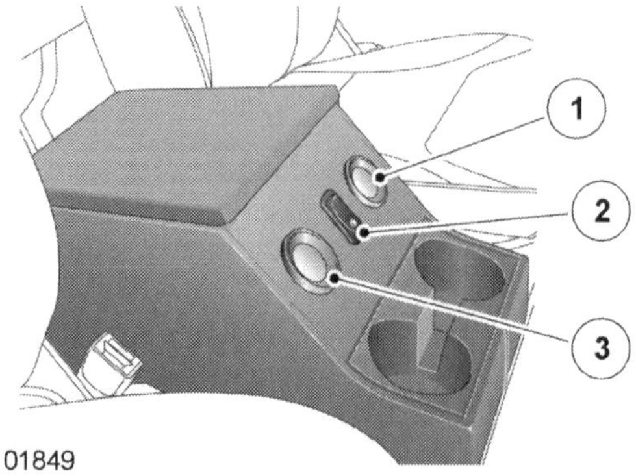

E101849

Item	Description
1	LH map reading lamp
2	Map reading lamps control switch
3	RH map reading lamp

The map reading lamps are connected directly to the vehicle battery via fuses in the CJB and the BJB (battery junction box) . Therefore the map reading lamps can be operated independent of the ignition switch position. Each lamp receives a battery voltage supply and the lamp illuminates when the switch is operated which completes a ground path.

Interior Lamp

Removal and Installation

Removal

1. Prise lens cover from lamp unit.
2. Remove 2 nuts securing lamp unit to headlining and roof mounting bracket.
3. Remove lamp unit from mounting bracket studs and disconnect harness plug.
4. Remove interior lamp.

Installation

1. Reconnect harness plug and locate lamp unit on mounting bracket studs. Ensure lamp switch eyelet is correctly seated on lamp unit base.
2. Secure lamp unit to mounting bracket and fit lamp lens.

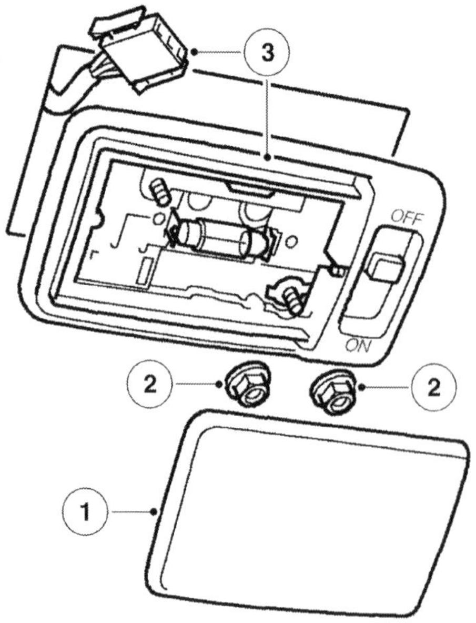

Interior Lamp Bulb

Removal and Installation

Removal

1. Prise lens from lamp unit.
2. Spread bulb holder contacts and release bulb.

Installation

1. Spread bulb holder contacts and fit new bulb.
2. Fit lamp lens.

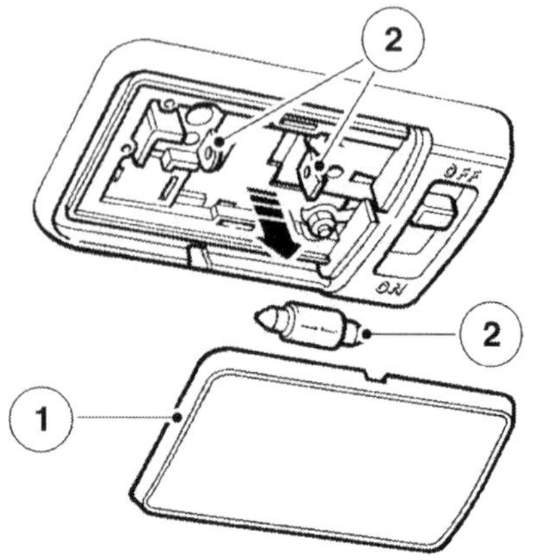

Module Communications Network

Module Communications Network - Communications Network

Description and Operation

OVERVIEW

A high speed CAN (controller area network) bus is used to transfer data between the ECM (engine control module) , instrument cluster and ABS modulator.

The diagnostic socket is also connected to the high speed CAN bus and allows the Land Rover approved diagnostic system to interrogate both the ECM, instrument cluster and ABS modulator software.

The CAN bus is a high-speed broadcast network where control modules automatically transmit information every few microseconds. Information is broadcast down a pair of twisted wires, known as 'CAN high' and 'CAN low'. Information is transmitted on the CAN bus as a voltage difference between the 2 wires.

CONTROL DIAGRAM

NOTE: **D** = High speed CAN bus

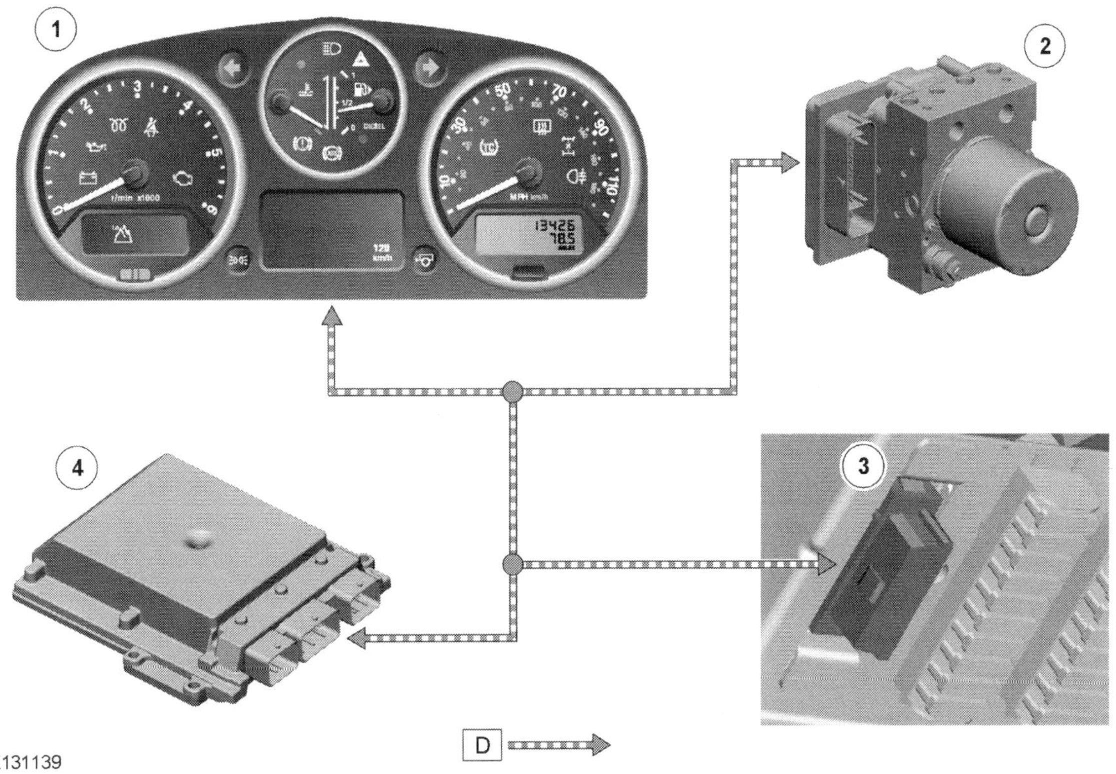

E131139

Item	Description
1	Instrument cluster
2	ABS modulator
3	Diagnostic socket
4	-

Wiring Harness

Description and Operation

Approved probing and repair methods

The purpose of this document is to identify the approved methods to promote an effective and efficient diagnosis and minor repair to the:

- permitted electrical wiring harnesses, connectors and cables
- See Electrical Wiring Harness Repair.
- Media Orientated System Transport (MOST) network harnesses, connectors and fiber optic cables
- See MOST Network Harness Repair.

Replacement Repair Equipment

The repair processes in the following information identifies specific repair equipment needed to complete a repair to the required standard.

Replacement repair equipment can be ordered from the equipment workshop website:

http://jlrequipment.service-solutions.com

Electrical Connector Probing

Only 2 methods of electrical connector probing are allowed.

1. Probing at the rear of unsealed electrical connectors (see illustration E190832).

2. Probing on the conductor crimp of an extracted terminal (see illustration E190928).

- The conductor crimp is the portion crimped to the non-insulated wire.
- This method may be used on sealed or unsealed connectors, but method 1 is preferred for unsealed connectors.

CAUTIONS:

⚠ A suitable sized probe must be used. If the probe is larger than the electrical connector aperture, then damage to the electrical connector will occur.

⚠ The probe must only be inserted into the rear of the electrical connector for a distance sufficient to contact the terminal.

⚠ Take care not to bend or distort any part of the metal terminal wire crimp area with the probe.

⚠ Before extracting any terminals, refer to the **Electrical Connector Terminal Extraction** and **Extraction Tools** sections of this document for more information.

⚠ Probing must **not** be carried out on; either the connector contact area or the portion of the terminal that is crimped to the insulated part of the wire.

⚠ Make sure the terminal is correctly and securely located in the electrical connector after re-installation.

Method 1

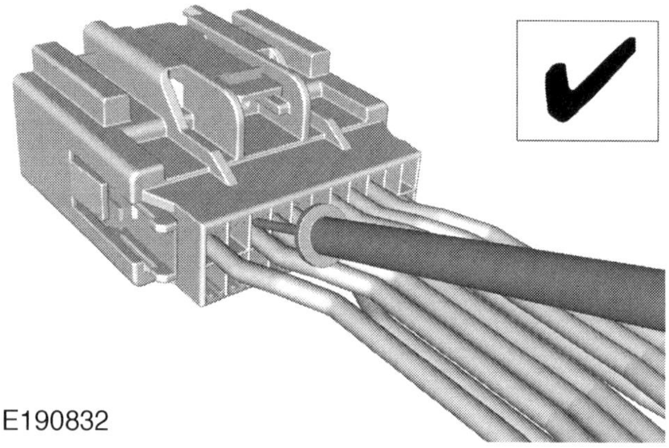

E190832

Method 2

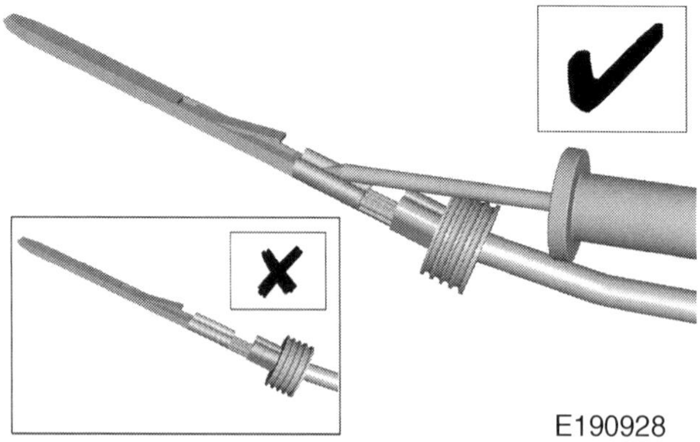

E190928

The following 2 methods of electrical connector probing are not allowed.

1. Probing at the rear of sealed electrical connectors (see illustration method 1).
2. Probing at the electrical connector contact area (see illustration method 2).

Method 1

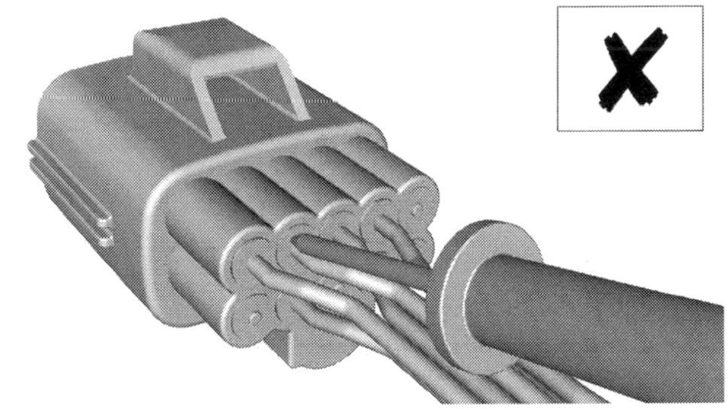

Method 2

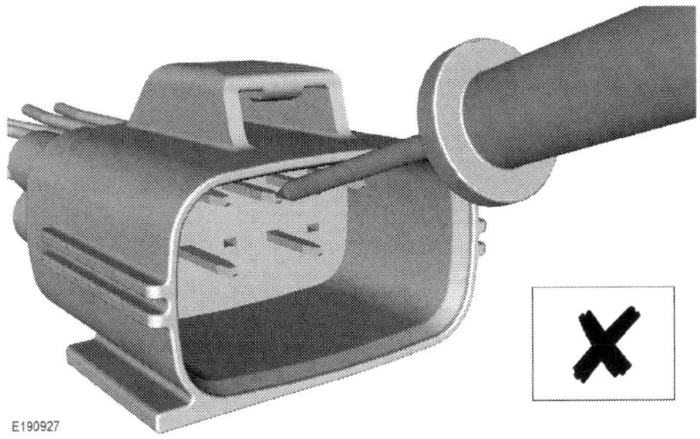

Live Probing on Sealed Connectors

WARNINGS:

This procedure must never be carried out on any of the following. Failure to follow this instruction may result in personal injury;

- Supplemental Restraint System (SRS)
- Pedestrian protection system
- Throttle Control circuits
- Speed Control circuits
- Link lead assembles, which are unique to safety critical circuits such as Anti-lock Brake System (ABS) and thermocouple circuits. An example of this is the ABS wheel speed sensors with molded connectors

This procedure must only be carried on wire with a cross sectional area of 0.5mm squared or less.

CAUTIONS:

The link lead used in this procedure must have a cross sectional area of at least 0.5mm squared.

The battery ground cable must be disconnected before any electrical connectors are disconnected / reconnected.

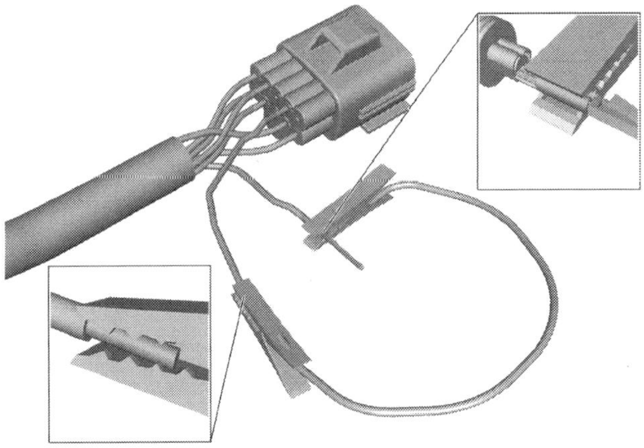

1. Disconnect the battery ground cable.
2. CAUTION: Before extracting any terminals, refer to the **Electrical Connector Terminal Extraction** and **Extraction Tools** sections of this document for more information.

   Extract the terminal to be tested from the electrical connector
3. Insert a substitute wire with the same terminal and cross sectional area of the previously extracted wire, into the vacated position of the electrical connector.
4. CAUTION: The link lead clip must only be attached to the portion of the extracted terminal; that is crimped directly to the section of non-insulated wire.

   Attach one end of the link lead to the original wire terminal.
5. CAUTION: The link lead clip must only be attached to the non-insulated section of the wire.

   Attach the remaining end of the link lead to the substitute wire.
6. Reconnect the electrical connector.
7. Reconnect the battery ground cable and begin the test.

Wiring Harness repair, including Supplementary Restraint System (SRS) Firing Circuit Wiring.

CAUTION: Disconnect the battery before commencing the repair of any wiring harness.

The repair is only appropriate where the damage is localized and is not over the full length of the wiring harness. If the damage is end to end then replacing the whole harness should be considered where appropriate. The damaged section of wiring harness must be removed by cutting either side of the damage at a point where the wiring harness retains its original condition. The number of twists in a twisted pair of wires is important to the functionality of the circuit and as such must be maintained. Note the number of twist over the repair length prior to commencing the repair. The same number of twists must reintroduced after completing the repair. If repairing multiple wires within a harness, stagger the connecting splices and twist the wire between the connecting splice locations to minimise the untwisted length and the effect of increasing the diameter of the harness. On no account must the untwisted length exceed 50 mm each side of the connecting splice. Once each connecting splice repair is completed, it must be covered with black glue lined heat shrink sleeve and shrunk down using the approved heat gun until the glue is visible at both ends. Make sure that connecting splice repairs are not located in any dynamic movement area of a harness.

Once all wiring harness repairs are complete, the wiring harness must be wrapped in insulating tape and returned as close as possible to new vehicle condition. Make sure the wiring harness follows the original routing and all clips and connectors are secured.

Repairs to supplementary restraint system firing circuit wiring.

⚠ WARNING: **Do not use any vehicle with damage to the supplementary restraint system components or wiring harness.**

⚠ CAUTION: **All supplementary restraint system wiring harness repairs must be completed using the approved repair kit only.**

The wiring used for a SRS firing circuit repair must be 0.5 mm² cross sectional area. Both the male and female sides of the damaged connectors must be replaced. Installation of SRS pre-terminated connectors, cut the wiring harness at a suitable point as close as possible to the damaged connector (destroy the removed connector). Identify the correct new pre-terminated connector from the SRS firing circuit wiring harness repair kit and cut to a suitable length. Install to the SRS firing circuit wiring harness using the connecting splices supplied and the repair procedure detailed above, taking into account the staggering of connecting splices and the number of twists. The whole repair should then be covered with yellow heat shrink sleeve supplied to act as SRS firing circuit wiring repair location/ identification. The repair authentication label must be completed by the authorised technician conducting the repair and approved by their supervisor. The label must then be secured around the harness adjacent to or at a suitable point as close as possible to the location of the repair.

Care and neatness are essential requirements in making a perfect repair.

⚠ CAUTION: Under no circumstances should repair be attempted to the following:

- Link lead assembles, which are unique to safety critical circuits such as Anti-lock Brake System (ABS) and thermocouple circuits. An example of this is the ABS wheel speed sensors with moulded connectors.
- Throttle Control circuits.
- Speed Control circuits.
- Screened cables, leads and wiring harness(s).

If any harness(s) with defective electrical connector terminals or cables from the above circuits give cause for concern, new components must be installed.

CAUTIONS:

⚠ Do not attempt to repair or reform a damaged electrical connector terminal. A damaged electrical connector terminal must be replaced using the correct pre-terminated lead.

⚠ A ground point connector with multiple wires to the connector must not be repaired as a complete connector. If a damaged wire is identified, the wire can be repaired individually using the correct pre-terminated lead.

⚠ Do not attempt the repair of damaged battery, hybrid and power cables.

These types of cable generally have a cross sectional area larger than 6mm² and must only be replaced. If the original cable is contained within the harness bundle the original cable must be left in the harness and a replacement cable attached to the harness along the original harness route.

The replacement cable must follow the original cable route to avoid the risk of introducing electrical interference issues. The original cable connections must be cut from the cable at both ends and discarded. The exposed cable ends must be free from sharp edges and strands of wire and must be over taped to prevent injury before being taped back to the main harness.

MOST Network Harness Repair

If a fibre optic cable is damaged, it must not be repaired and must be replaced with a new cable.

Replacement fibre optic cables can only be made using the approved repair equipment and components.

The approved repair kit contains the specially designed fiber optic conductor strippers, which are used to prepare 2.3mm fiber optic cable for the fitment of the brass fiber optic conductor contact. The fiber optic conductor contact crimping pliers must then used be used to crimp the brass contact to the fiber optic conductor core. The approved crimping pliers supply the appropriate pressure to the brass contact to make a secure contact, but not damage the conductor core.

The cut face of the fiber optic core must be protected from damage and contamination at all times.

CAUTION: Fiber optic cables have a maximum bending radius 25mm and must not be kinked or excessively bent.

The performance of fibre optic cables is very dependant upon the quality of the cut surface at connections and to the bending radius of the cables.

MOST Harness Repair Components and Tools - List of Parts

NOTE: Repair components can be ordered via the Jaguar/Land Rover parts ordering system.

Description	Part Number	Quan.
Fiber Optic Conductor Lead	418-676	1
Fiber Optic Conductor Contact	418-677	20
Fiber Optic Conductor Contact Protective Cap	418-678	20
Fiber Optic Conductor Lead Connector - Inner	418-679	10
Fiber Optic Conductor Lead Connector - Outer	418-680	10
MOST Module Protective Cap	418-681	20
Fiber Optic Conductor Lead Connector Protective Cap	418-682	20

Fiber Optic Conductor Stripper

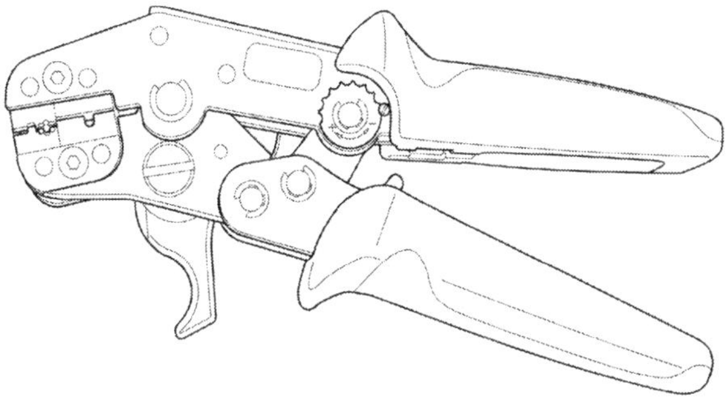

Fiber Optic Conductor Stripper Jaw Positions

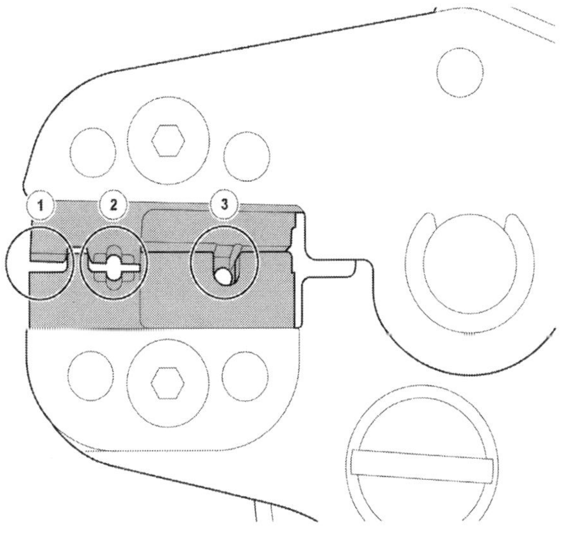

1	Fiber Optic Cable Cutter
2	Fiber Optic Cable Insulation Stripper
3	Fiber Optic Core Cutter

Fiber Optic Core Cutter Locking Screw

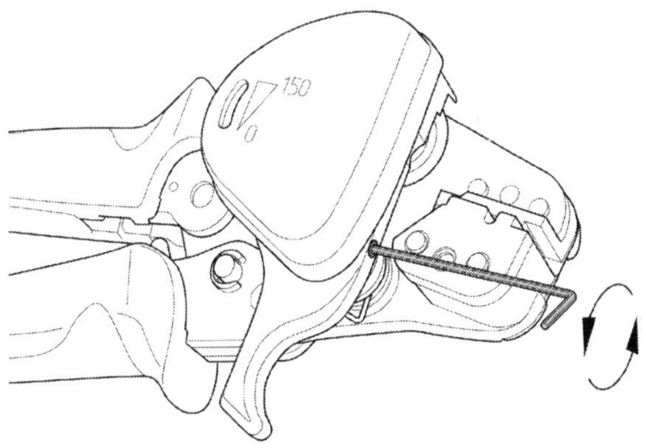

The fiber optic core cutter has a locking screw to protect the cutter wheel when in transit or not in use. A hexagonal key is supplied in the MOST repair kit to release the locking screw.

NOTE: Tighten the transportation locking screw after use.

Fiber Optic Core Cutter Remaining Cut Indicator

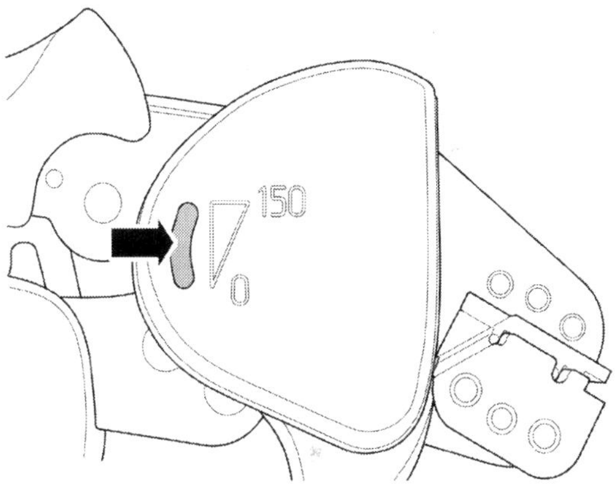

The fiber optic core cutter can be used for approximately 1260 cuts. The indicator line on the remaining cut indicator window only becomes visible when the fiber optic core cutter has 150 cuts or below available.

NOTE: When the fiber optic core cutter has reached the maximum allowed cuts, the cutter will become locked and the fiber optic conductor stripper must then be renewed.

Before using the fiber optic conductor core cutter, make sure it has enough cuts remaining to complete the repair process by viewing the remaining cut indicator.

Fiber Optic Conductor Contact Crimping Pliers

A small amount of effort is required to operate the fiber optic conductor contact crimping pliers and secure a fiber optic conductor contact to the fiber optic conductor core.

The new fiber optic conductor contact is placed into the cramping mechanism in the head of the pliers and the locking arm is repositioned to hold the conductor contact securely in position. The locking arm must locate on to the retaining pin.

The prepared end of the fiber optic core is then inserted into the new conductor contact.

The fiber optic core and conductor contact must be pushed and held against the spring pressure in the cramping mechanism. The grips of the fiber optic conductor contact crimping pliers are then be closed, cramping the conductor contact to the conductor core.

NOTES:

△ Only use the approved fiber optic conductor contact crimping pliers to cramp a new conductor contact.

△ A conductor contact must only be cramped once using the fiber optic conductor contact crimping pliers.

The cramping mechanism inside the head applies the appropriate pressure to the conductor contact at 4 points. This makes a secure contact and does not damage the conductor core.

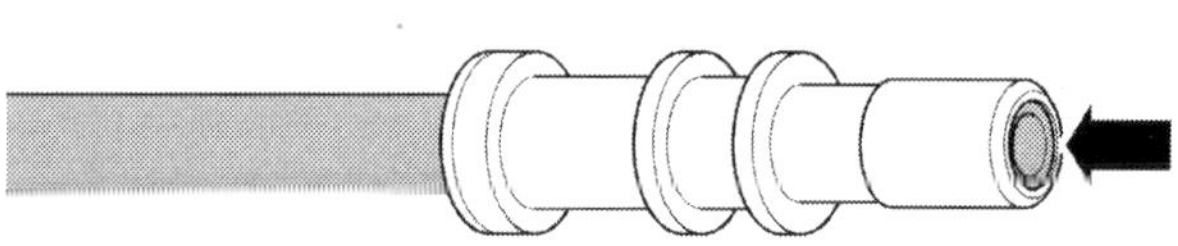

When the new conductor contact has been cramped to the fiber optic core, make sure the fiber optic core end sits 0.01mm to 0.1mm below the height of the conductor contact end.

△ NOTE: Make sure the fiber optic conductor lead contact remains clean and protected at all times. Fit a fiber optic conductor contact protective cap.

MOST Repair Tools

△ NOTE: Replacement repair equipment can be ordered from the equipment workshop website; refer to the Replacement Repair Equipment in the Introduction section.

Description	Part Number	Quantity
MOST Repair Kit	418-673	1
Fiber Optic Conductor Stripper	418-674	1
Fiber Optic Conductor Contact Crimping Pliers	418-675	1
Fiber Optic Conductor Lead Installation Pliers	418-683	2

MOST Harness Repair Procedure

The MOST connector(s) have an anti-backout device which prevents the contact from being released from the connector. The anti-backout device must be released before attempting to remove the terminal from the connector. The anti-backout devices require a special tip to release the device. Please refer to the ERL for the correct tool(s) to use.

The illustration shows an example of a common style of extraction tool being used on a MOST connector(s). Care should be exercised to avoid further damage when removing the terminals from the connector.

CAUTIONS:

⚠ Before releasing the fiber optic cable from the connector housing, mark the IN/OUT assignment of the fiber optic cable.

⚠ To prevent contamination or mechanical damage to the exposed end face of a fiber optic cable, make sure all disconnected MOST connectors and fiber optic cables are fitted with a protective cap.

MOST Connector Terminal Extraction

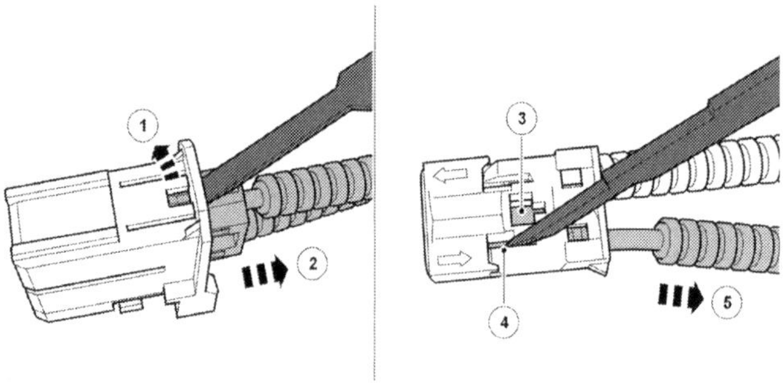

MOST Harness Repair Process

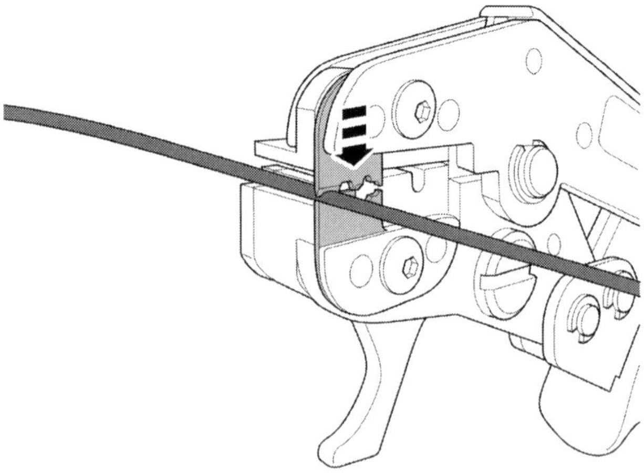

1. Confirm the length of fiber optic conductor lead required to create a new fiber optic cable. Cut the fiber optic conductor lead to the required length using the fiber optic cable cutter.

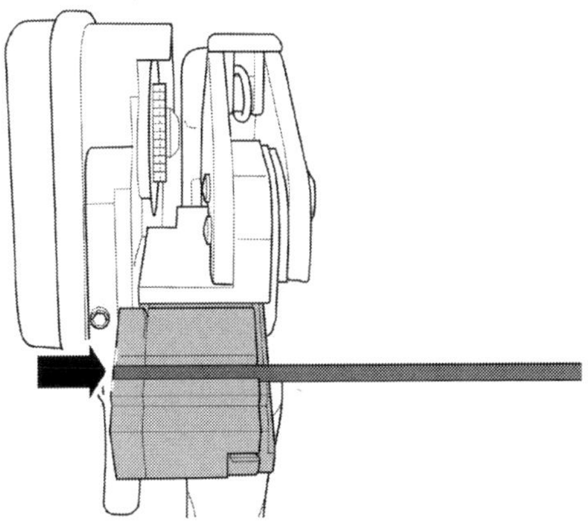

2. Open the fiber optic conductor stripper jaws and insert the fiber optic conductor lead up to the edge of the jaws.

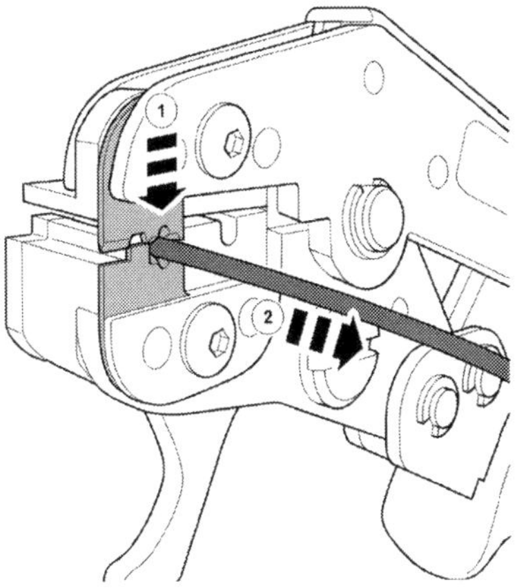

3. Close the fiber optic conductor stripper jaws (1) and carefully pull the fiber optic conductor lead (2) to remove the protective casing.

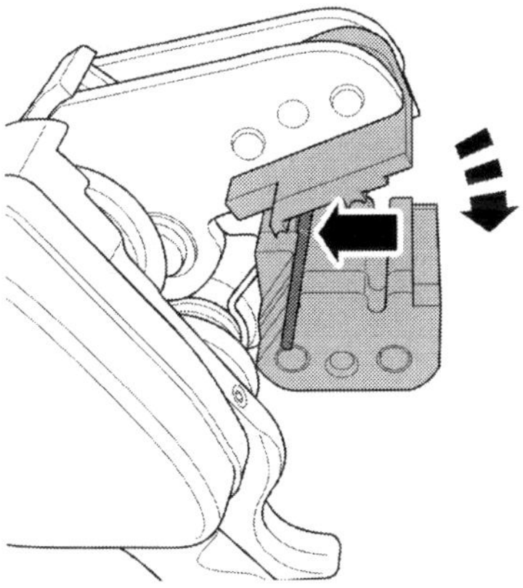

4. △ NOTE: Make sure the protective casing of the fiber optic conductor lead sits against the fiber optic conductor cutter jaw stop.

   Open the fiber optic conductor stripper jaws and insert the fiber optic conductor lead fully into the fiber optic conductor cutter slot. Close the fiber optic conductor stripper jaws.

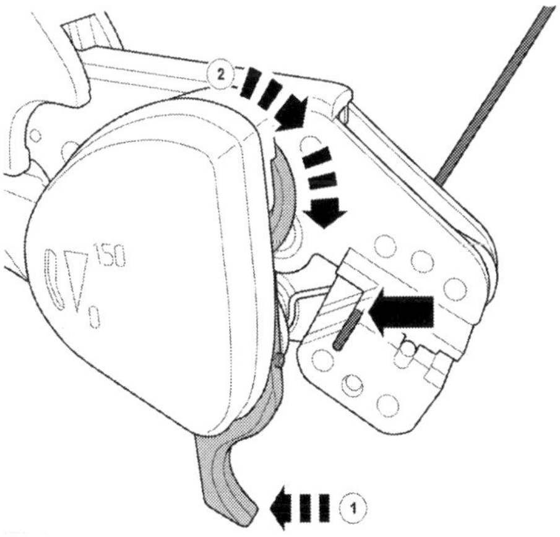

5. Pull the fiber optic conductor core cutter lever (1) to move the cutting wheel (2) and cut the fiber optic conductor core.

6. △ NOTE: The end of the fibre optic core has now been prepared for the fitting of a brass contact.

   Open the fiber optic conductor stripper jaws and remove the fiber optic conductor lead.

7. △ NOTE: Make sure the fiber optic conductor core end remains clean at all times.

   Place fiber optic conductor stripper and fiber optic conductor lead to one side.

8. Open the fiber optic conductor contact pliers and reposition the conductor contact locking arm to the open position.

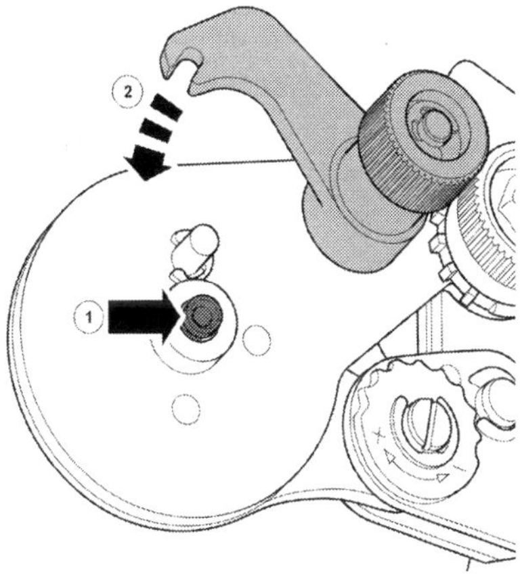

9. △ NOTE: Make sure the locking arm locates on the retaining pin when in the closed position.

   Insert a conductor contact (1) into the fiber optic conductor contact crimping jaws and reposition the conductor contact locking arm to the closed position (2).

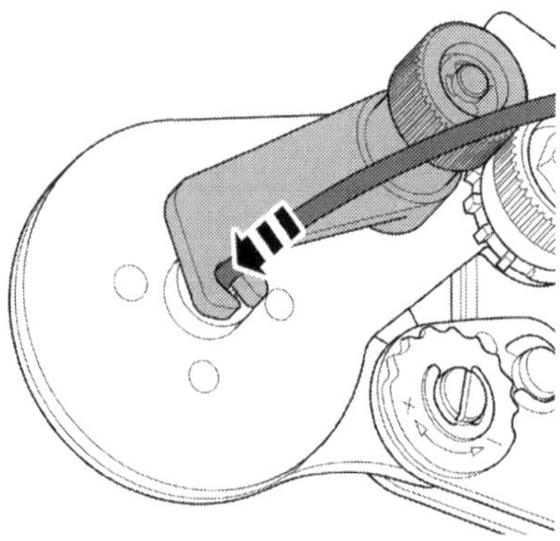

10. Insert the prepared end of the fiber optic conductor lead into the conductor contact.

11. ⚠ CAUTION: Make sure the fiber optic conductor lead is pushed in and held against the spring loaded stop when closing the fiber optic conductor contact pliers. This sets the core to the correct depth in the brass connector. Failure to follow this instruction may result in the fiber optic conductor cable malfunctioning.

    Push the fiber optic conductor lead fully into the conductor contact and close the fiber optic conductor contact pliers.

12. Open the fiber optic conductor contact pliers and reposition the conductor contact locking arm to the open position.

13. Remove the fiber optic conductor cable from fiber optic conductor contact pliers.

14. Place fiber optic conductor contact pliers to one side.

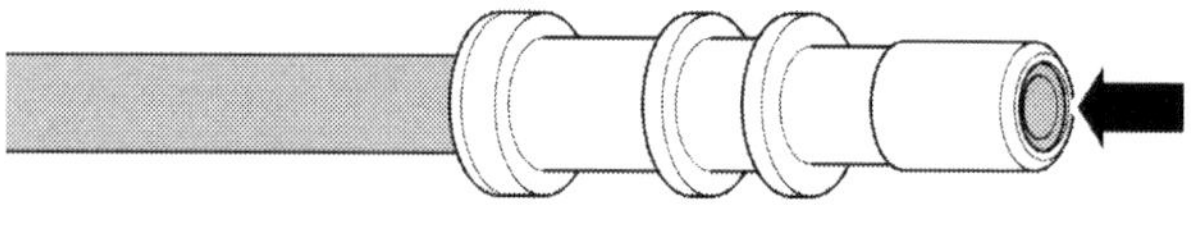

15. ⚠ CAUTION: Make sure the conductor contact has been correctly fitted to the fiber optic conductor core. Failure to follow this instruction may result in the fiber optic cable malfunctioning.

NOTES:

⚠ The fiber optic core end must sit 0.01mm to 0.1mm below the height of the conductor contact end.

⚠ Make sure the fiber optic conductor contact remains clean at all times.

Visually inspect the conductor contact for correct fitment to the fibre optic core.

- Make sure the conductor contact has been visibly crimped at 4 points.
- Pull the conductor contact by hand to make sure it is secure.
- Make sure the end of the fiber optic core sits below the height of the new conductor contact end.
- Fit a fiber optic conductor contact protective cap.

16. Repeat steps 2 to 15 and fit a conductor contact to the opposite end of the fiber optic conductor cable.

17. Measure between the conductor contact ends of the new fibre optic cable. Using a suitable tool cut a length of new protective corrugated tubing to the required length.

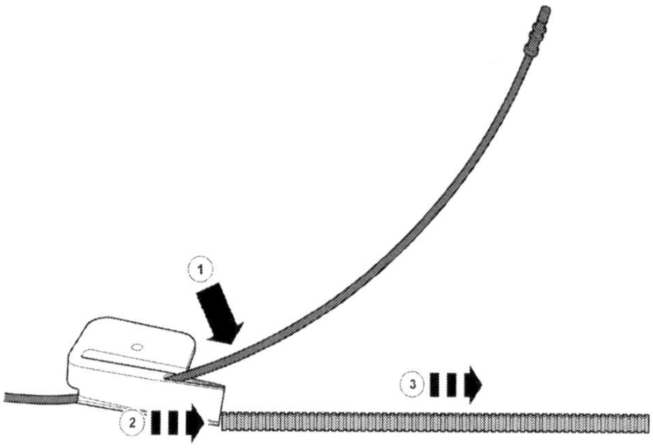

18. Install the fibre optic cable into the corrugated tubing.
    1. Place the fibre optic cable inside the fiber optic conductor cable installation pliers.
    2. Insert the fiber optic conductor cable installation pliers into the corrugated tubing.
    3. Move the fiber optic conductor cable installation pliers down the length of corrugated tubing and install the fibre optic cable.

Air Suspension Overlay Harness Introduction

NOTES:

⚠ This repair applies to L320 / LS, L405 / LG and L494 / LW models only.

⚠ Some variation in the illustrations may occur, but the essential information is always correct.

If after diagnosis the height sensor requires replacement and the area of wiring damage is localized to the height sensor connector, this overlay harness should be used,

1. Disconnect the battery ground cable. For additional information, refer to Workshop manual section 414-00, Specifications - Battery Disconnect/Connect.

2. Disconnect the damaged height sensor electrical connector.

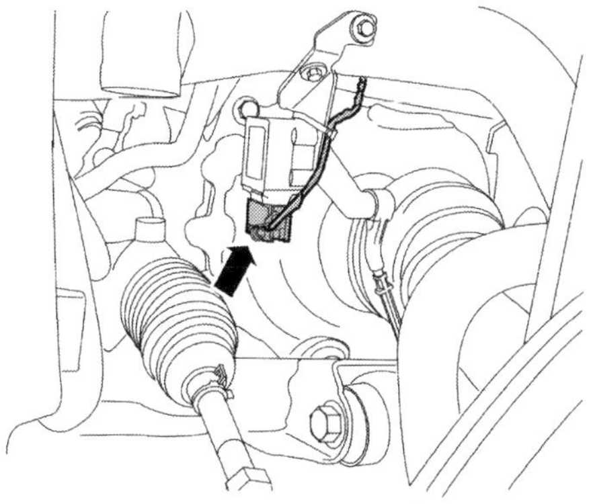

E78443

3. Unclip a sufficient length of the wiring harness to allow easy access during the repair.

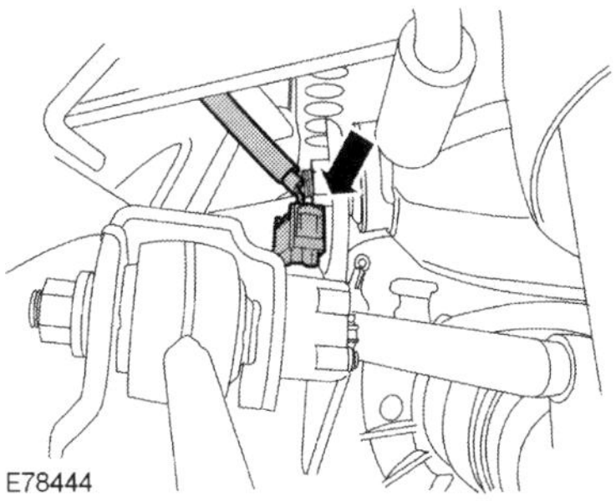

E78444

4. Remove the wiring harness insulation as required.

NOTES:

⚠ It is advisable to cut only one of the wires at a time and to stagger each joint to allow easier insulation.

⚠ The colours of the overlay harness may vary.

5. Cut the wire on the vehicle harness leading to cavity 1 of the height sensor connector in a suitable position and remove 10mm of insulation. For connector location and pin identification, refer to the relevant connector, L320 - C1696/7/8/9 and L405/L494 - C1CD10/11/12/13, in the Connector Details section of the relevant Electrical Library).

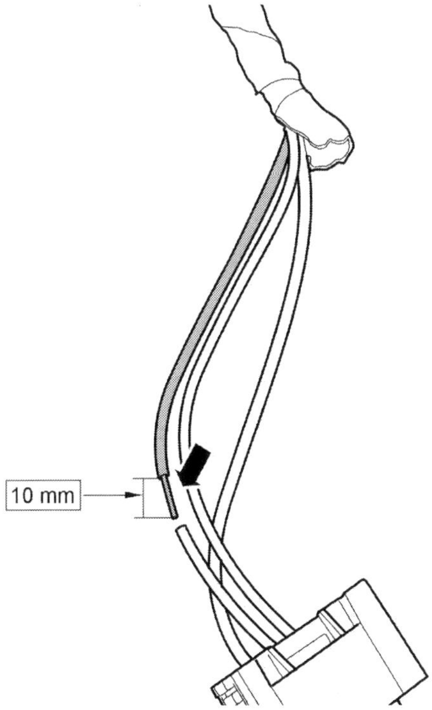

6. Slide over this wire a section of heat shrink sleeving.

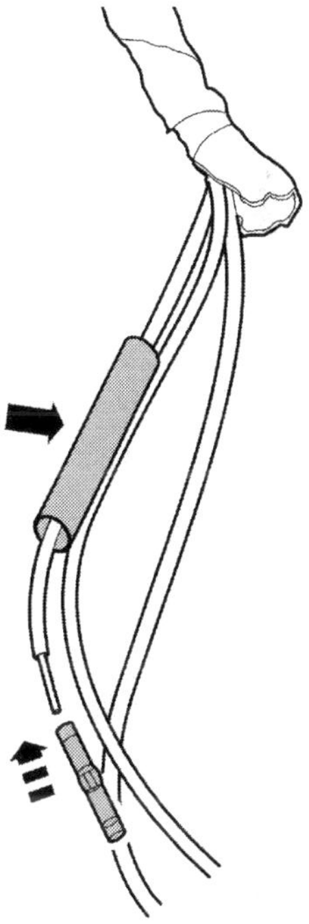

7. Using the inline connector supplied and the correct crimping tool set to the correct jaw size from the harness repair kit, crimp the connector to the wire.

⚠ NOTE: The overlay harness should be cut so that there is no additional length added to the overall length after repair.

8. Select the appropriate wire on the overlay harness that also goes to cavity 1 of the new connector, cut to the correct length and remove 10mm of insulation.

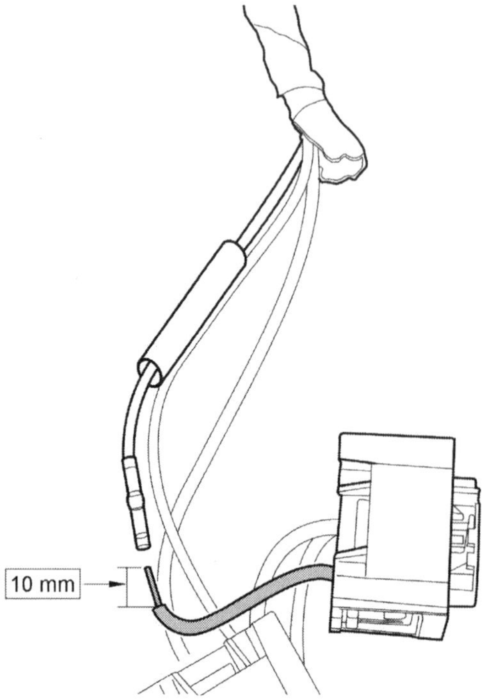

9. Insert the overlay wire into the connector and crimp in place.

CAUTION: Care must be taken when using the heat gun to avoid damage to surrounding areas.

10. Slide the heatshrink over the connector and using a hot air gun carefully apply the heat until the glue appears at both ends.

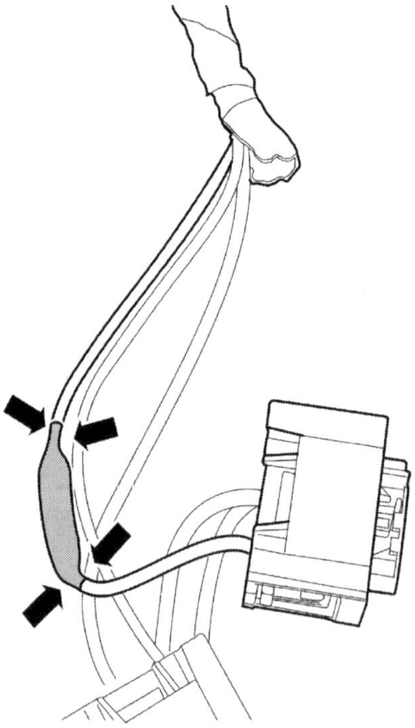

11. Carry out the same process for the wires in cavities 4 and 5 of the connector. For connector location and pin identification, refer to the relevant connector, L320 - C1696/7/8/9 and L405/L494 - C1CD10/11/12/13, in the Connector Details section of the relevant Electrical Library).
12. Discard the damaged connector/section of the harness.
13. Add suitable harness repair tape to the repaired area to within 10mm of the new connector to complete the repair.
14. Correctly route/secure the harness and connect the new connector to the height sensor.
15. Connect the battery ground cable. For additional information, refer to Workshop manual section 414-00, Specifications - Battery Disconnect/ Connect.
16. Clear any Diagnostic Trouble Codes logged in the air suspension control module using Land Rover approved diagnostic equipment and confirm correct operation of the system.

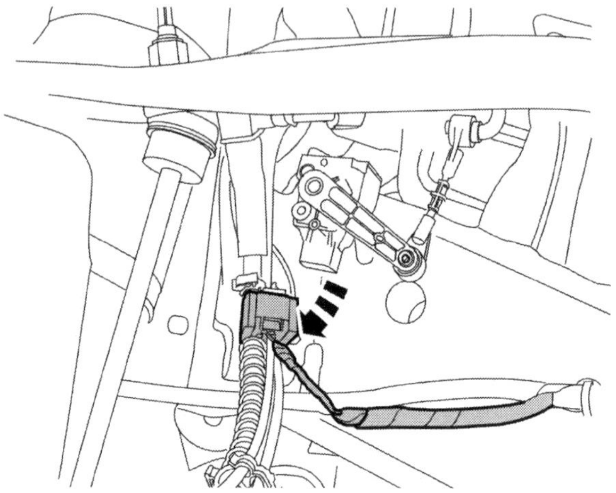

Wiring Harness Overlay

Description and Operation

Overview

Where damage to a vehicle or a wiring harness has occurred; it can be more cost effective to repair the wiring harness with an overlay or a lower level assembly, rather than replacing a full wiring harness. This is especially the case where pre-terminated leads or other wiring harness repair kits are not available.

Description

The overlays and lower level assemblies are available from the parts catalogue and include such items as;

NOTE: Locate the damaged cable / assembly and connectors required. For additional information, refer to: Electrical Wiring Diagrams and Electrical Reference Library.

- SRS repair kits and wiring
- Antenna leads and other communication wiring
- Infotainment and display wiring
- Camera link leads
- Trunk / tailgate links and wiring
- Other special cables

CAUTIONS:

⚠ Any damaged cable / assemblies must not be repaired and must be replaced.

⚠ The replacement overlay must follow the original wiring harness route, to avoid the risk of introducing electrical interference issues.

NOTES:

△ The original cable / assembly must left be in the wiring harness and the overlay taped to the wiring harness along the original route.

△ Where possible, when installing a replacement cable / assembly, cut off the unused ends of the damaged cable / assembly at their exit points from the host harness and discard.

△ The cut ends must be free from sharp edges. Any strands of wire must be taped over, to prevent damaging adjacent wires or cables or any personal injury.

△ The replacement cable / assembly must be attached with coverings to replicate the existing installation of the damaged cable / assembly.

△ If the cut end of the cable / assembly is in an area of the vehicle, which is exposed to moisture, i.e. the engine bay; Cover the cut end with glue-lined heatshrink, squeezed with pliers to seal the end and prevent moisture ingress.

Examples

Some examples of installed wiring harness overlays are illustrated below:

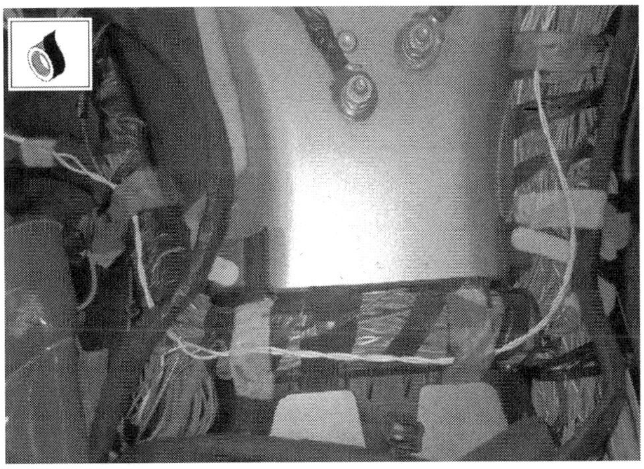

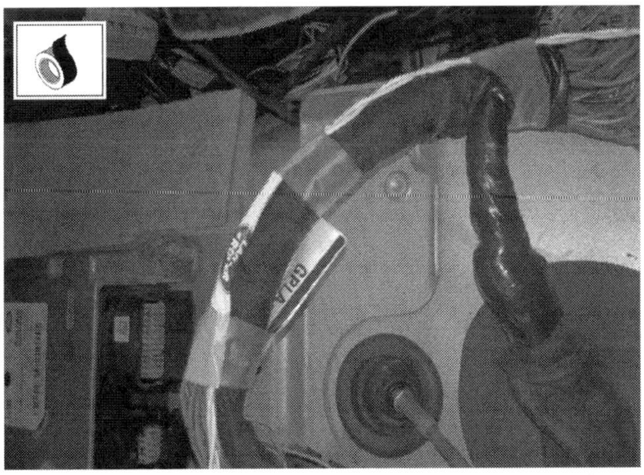

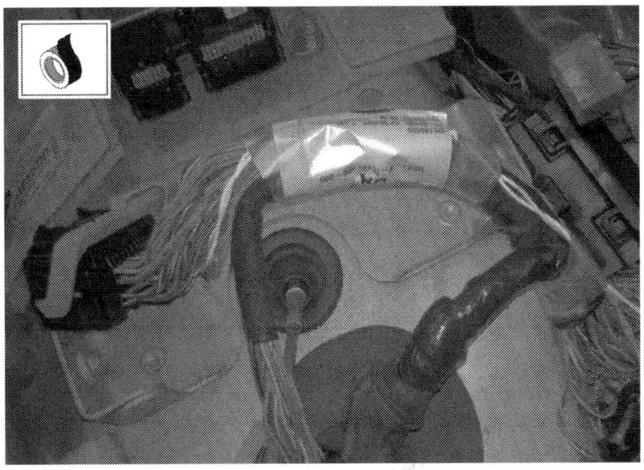

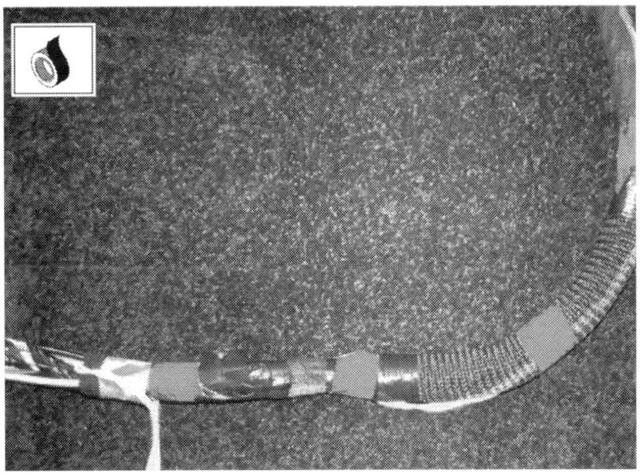

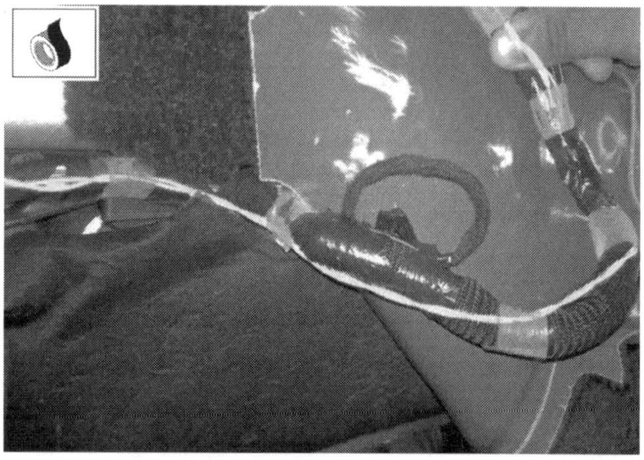

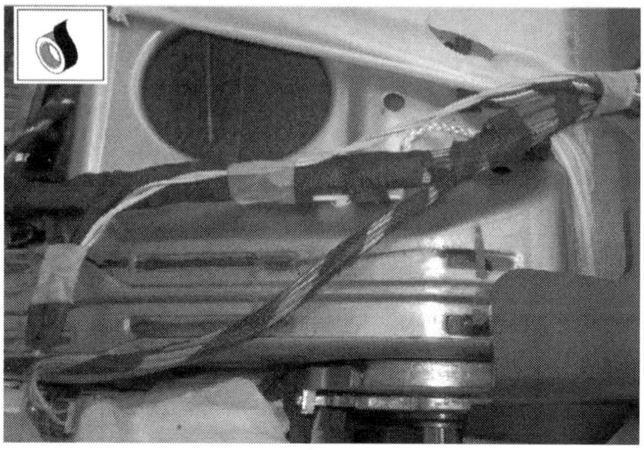

Visible Wiring Harness Overlays

Where wiring harness overlays are visible, with all components in their installed condition, the overlay harness must be fully taped to the original wiring harness, with no gaps in the tape binding. Example illustration below:

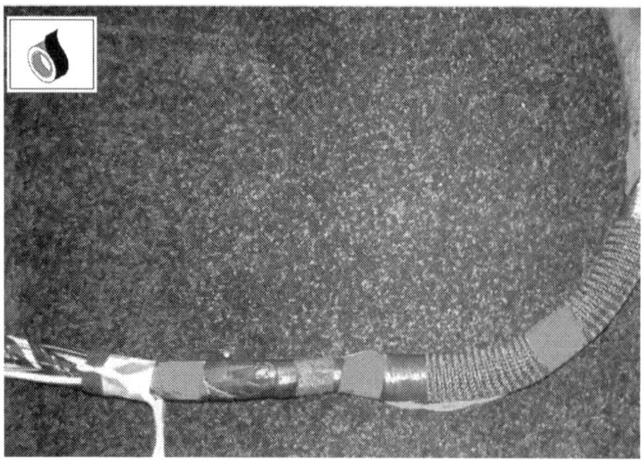

Wiring Harnesses - Engine Wiring Harness

Removal and Installation

Removal

CAUTIONS:

⚠ Make sure that the wiring harness is not twisted or damaged on removal. Failure to follow this instruction may result in damage to the vehicle.

⚠ Take extra care not to damage the wiring harness clips.

△ NOTE: Note the position of the wiring harnesses to aid installation.

1. **Refer to:** Intake Manifold (303-01 Engine, ID4 2.2L Diesel Removal and Installation)
2. ⚠ WARNING: Do not work on or under a vehicle supported only by a jack. Always support the vehicle on safety stands.

Raise and support the vehicle.

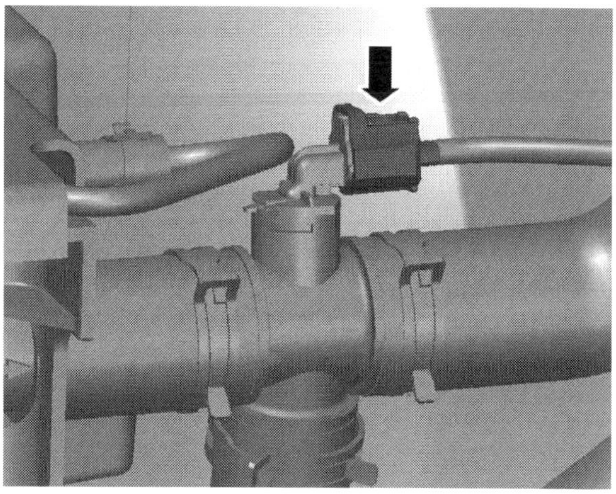

3.

4. Torque: 5 Nm

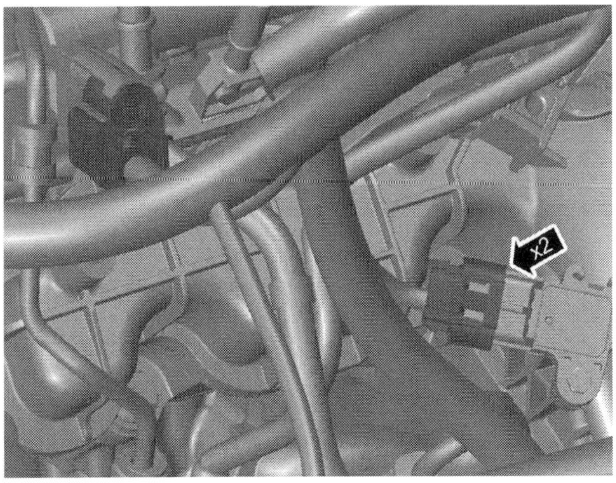

5.

6.

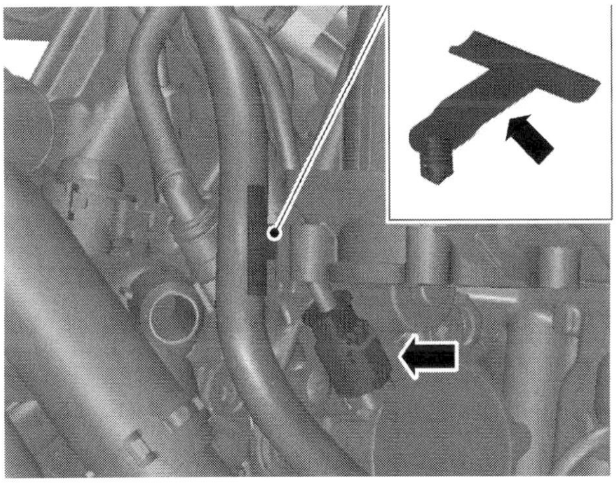

7.

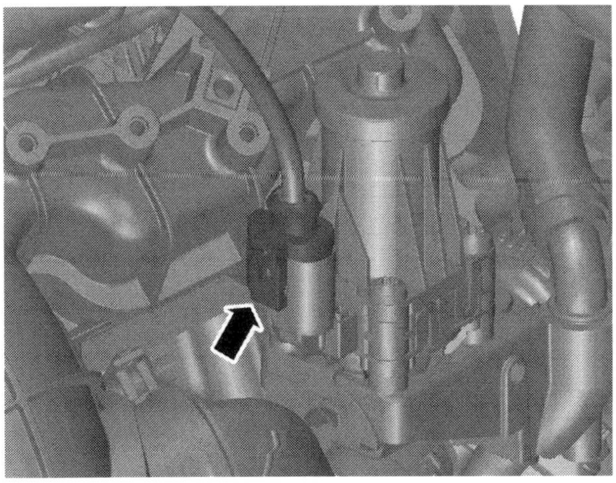

8.

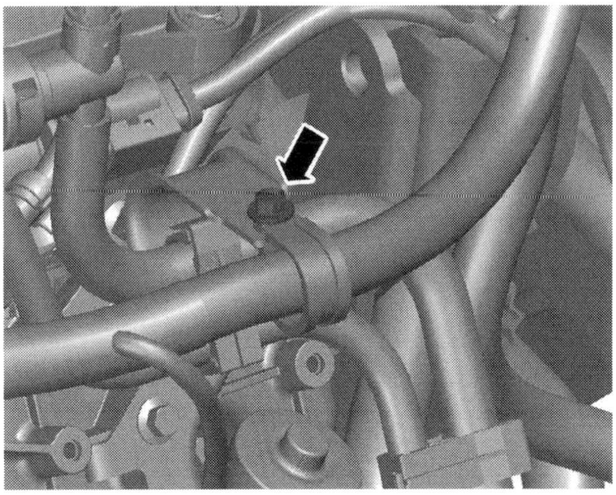

9. Torque: 10 Nm

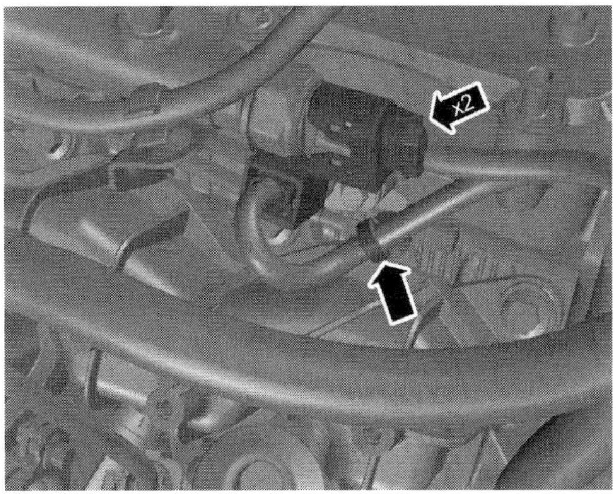

10.

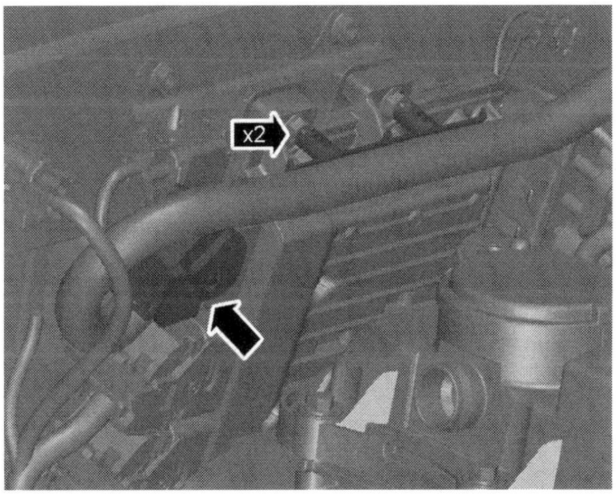

11.

12.

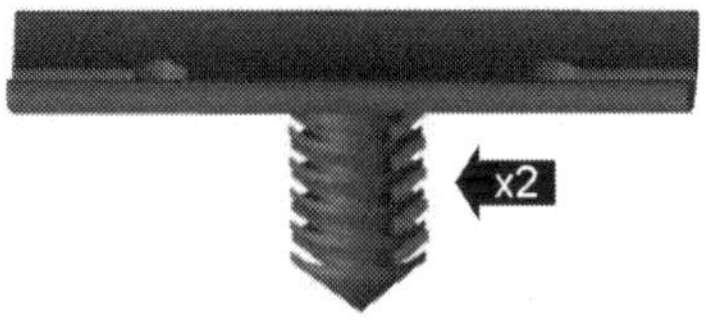

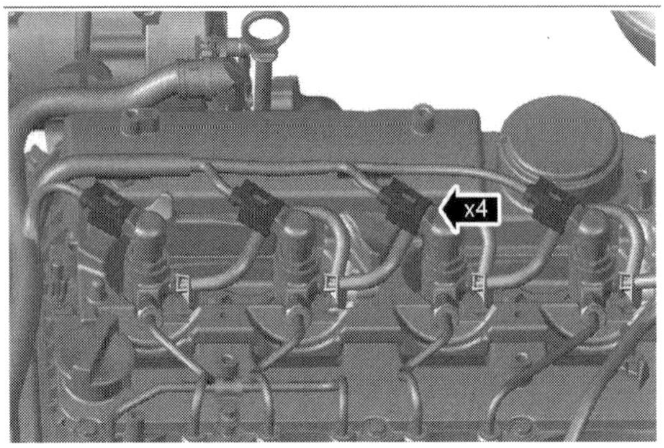

13.

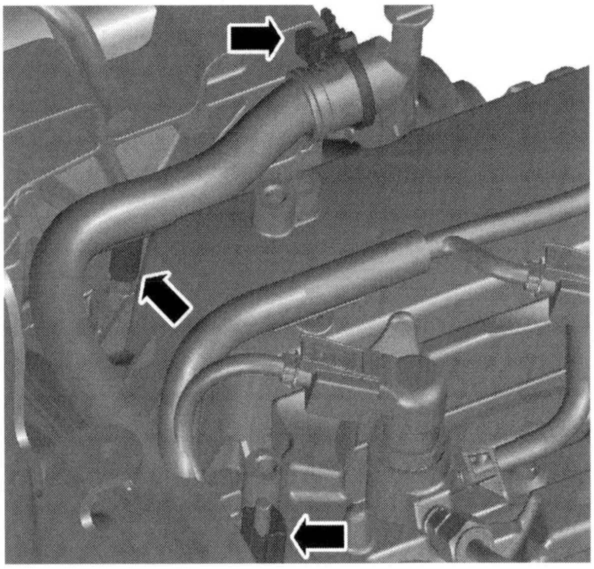

14.

15.

16. Torque: 8 Nm

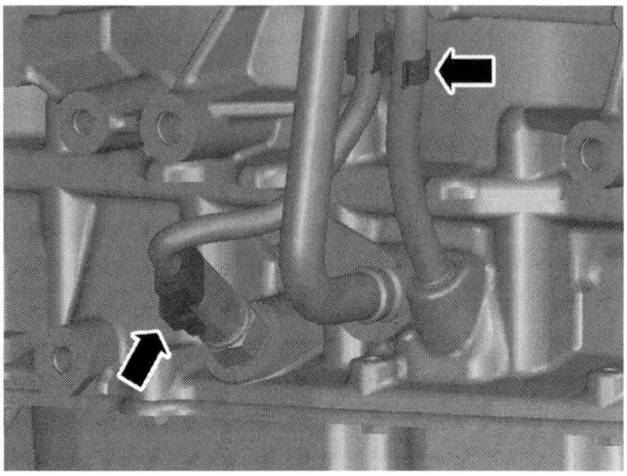

17.

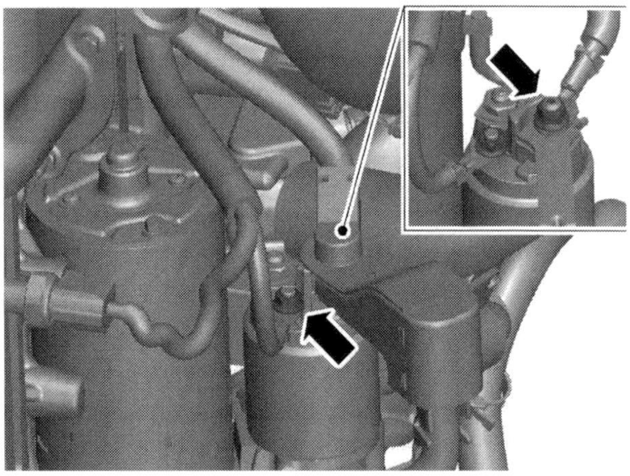

18. Positive cable nut torque: 12 Nm,

Solenoid cable nut torque: 8 Nm

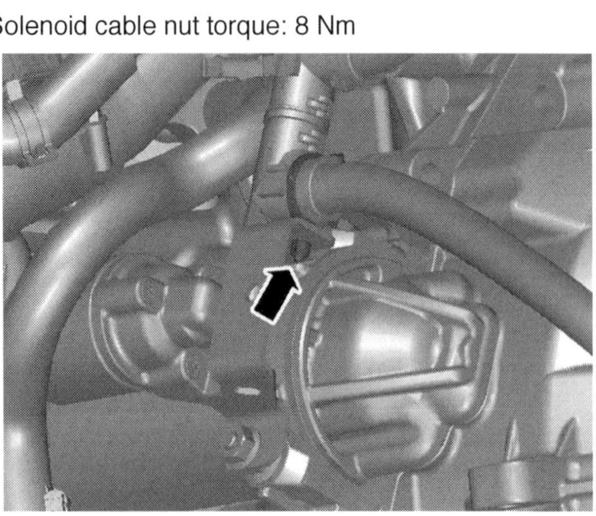

19.

20. Torque: 24 Nm

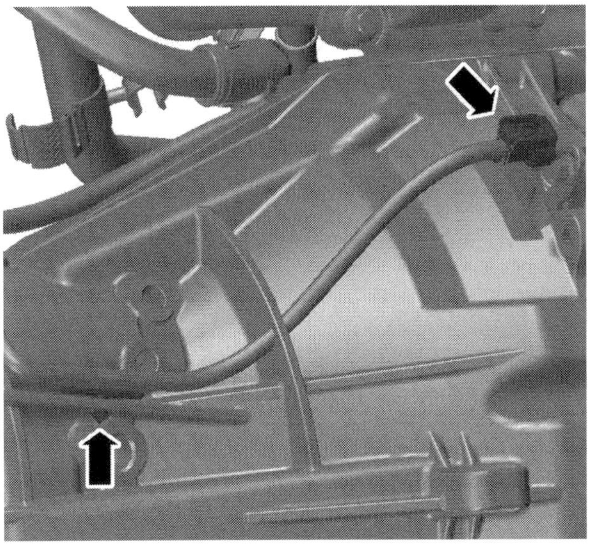

21.

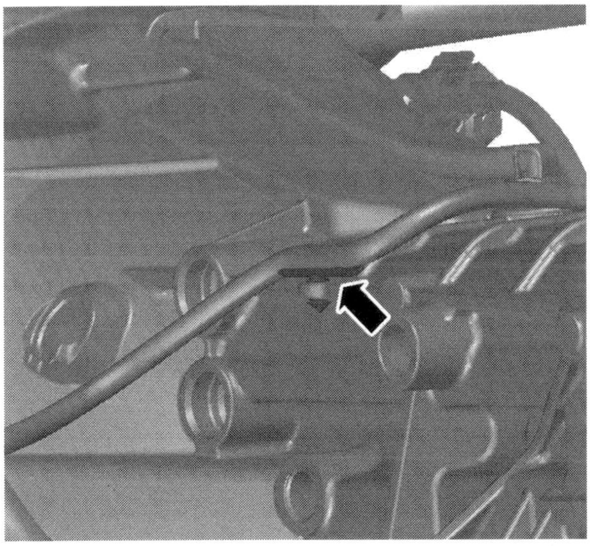

22.

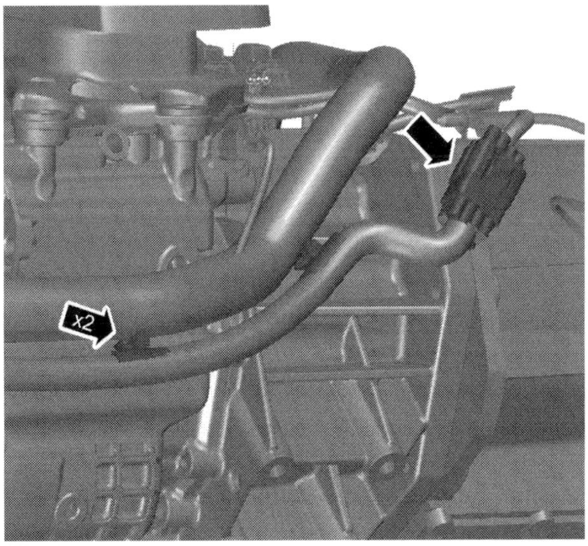

23.

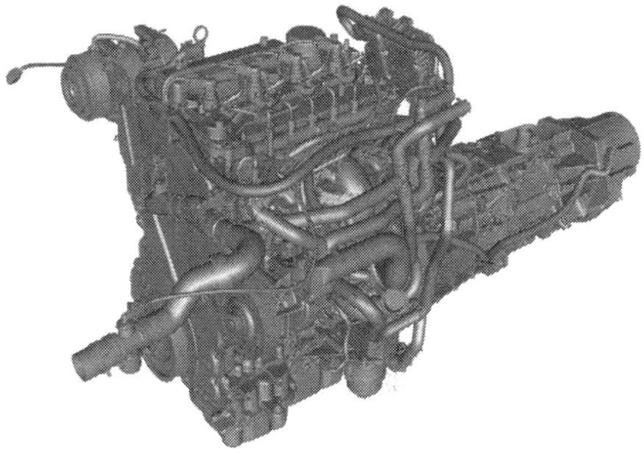

24.

Installation

1. To install, reverse the removal procedure.

419-01A Anti-Theft - Active

Description and Operation

COMPONENT LOCATION

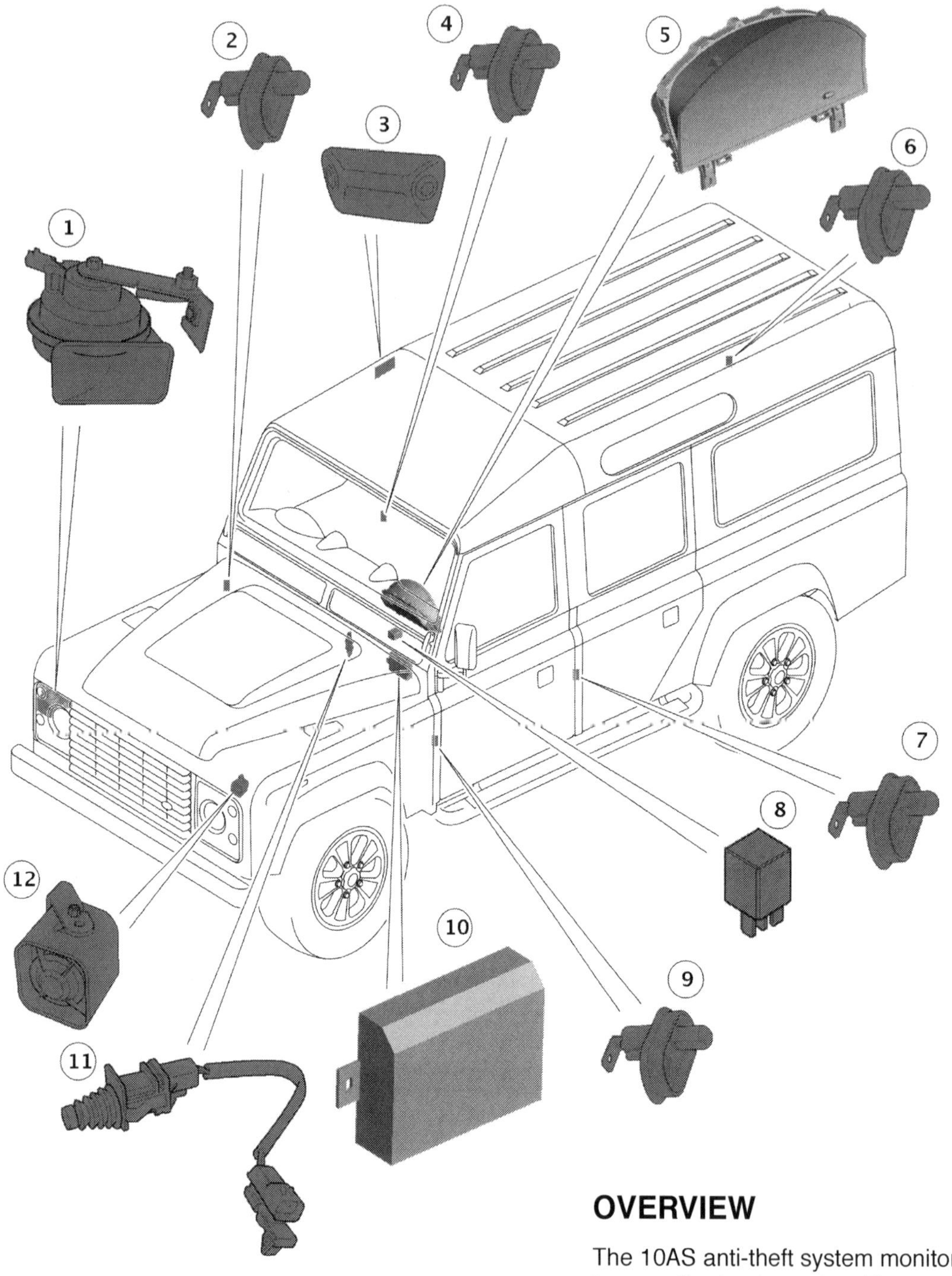

Item	Description
1	Anti-theft alarm horn (if fitted)
2	RH (right-hand) front door switch
3	Intrusion detection module (90/110 station wagon shown)
4	RH rear door switch (if fitted)
5	Instrument cluster
6	Tail door switch
7	LH (left-hand) rear door switch (if fitted)
8	Anti-theft alarm horn relay (if fitted)
9	LH front door switch
10	Anti-theft system module
11	Hood switch
12	Battery backed sounder (if fitted)

OVERVIEW

The 10AS anti-theft system monitors the hinged panels for unauthorized opening. The system also monitors intrusion within the cabin. Depending on market specification, the vehicle may be fitted with either a battery backed sounder or an anti-theft alarm horn.

The controlling software for the active anti-theft system is contained within the anti-theft system module. The anti-theft system module also controls;

- the passive anti-theft system (engine immobilization).

Refer to: Anti-Theft - Passive (419-01 Anti-Theft - Passive, Description and Operation)

- the central locking system.

Refer to: Handles, Locks, Latches and Entry Systems (501-14 Handles, Locks, Latches and Entry Systems, Description and Operation)

- interior lighting.

Refer to: Interior Lighting (417-02 Interior Lighting, Description and Operation)

- the hazard flashers.

Refer to: Exterior Lighting (417-01 Exterior Lighting, Description and Operation)

The active anti-theft system is armed by pressing the lock button on the remote handset or using the vehicle key in the drivers door lock. Using the remote handset will arm both the perimetric and volumetric anti-theft systems. Using the vehicle key will only arm the perimetric anti-theft system.

COMPONENTS

ANTI-THEFT SYSTEM MODULE

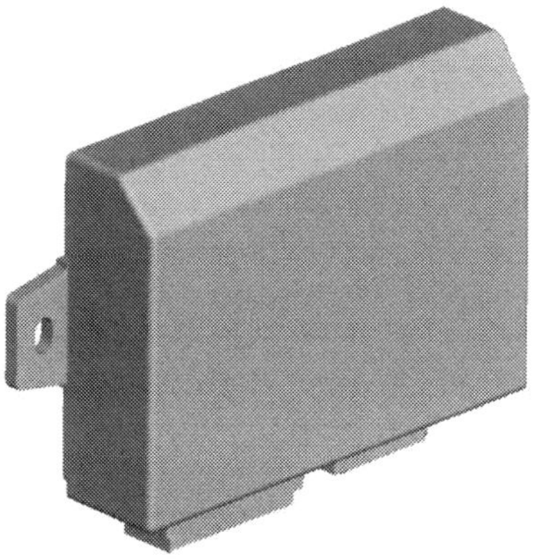

E83589

The anti-theft system module is located behind the instrument cluster and receives a permanent battery supply from the BJB (battery junction box) . The control module incorporates a radio frequency (RF) receiver and antenna to receive signals transmitted from the remote handset.

Depending on market specification, the module will operate at one of two frequencies. This is identified by a label mounted on the module casing. The table shows the frequencies available for each market. The anti-theft system module also incorporates an internal inertia switch.

Frequency	Markets
433 MHz	Europe, Gulf States, South Africa
315 MHz	North America, South East Asia, Japan, Australia

In the event of an impact of sufficient severity to trigger the inertia switch when the ignition switch is in position II, the anti-theft system module will unlock all doors and operate the hazard flashers. The anti-theft system module will remain in this mode of operation for 2 minutes. To de-activate the hazard flashers and restore the anti-theft system module to its normal mode of operation, the ignition switch should be turned to position 0 and back to position II after the 2 minute period has expired.

REMOTE HANDSET

E83590

The remote handset is used to arm and disarm the anti-theft alarm system. The handset comprises a lock and unlock button. Internally, the handset contains a printed circuit board capable of transmitting RF signals, and a battery.

The remote handset RF code comprises 2 parts. The first part is a fixed code that is unique to the handset. The second part is a rolling code which changes in accordance with a pre-determined pattern. The code is received by the anti-theft system module, which is able to store information on up to 4 different handsets.

If the remote handset battery falls below a pre-determined level, the remote handset will transmit an additional RF signal code to the anti-theft system module. The anti-theft system module will then indicate this by:

- Signalling the instrument cluster warning indicator which emits 2 rapid flashes every 0.5 seconds to inform the driver that the remote handset battery needs replacing.
- The remote control will operate only with a second press of the unlock button.
- The turn signal indicators will not flash when the alarm is disarmed.

BATTERY BACKED SOUNDER

Depending on market specification, the vehicle may be fitted with a battery backed sounder. The battery backed sounder is located in the LH front fender, behind the headlamp.

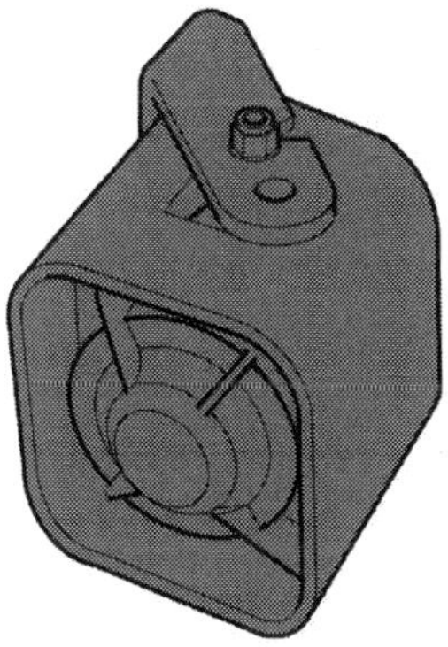

E83591

The battery backed sounder is tamper proof and as the name suggests contains its own power supply. This allows it to operate even if it is disconnected from the vehicle power supply. For added security all wires leading to and from the battery backed sounder are colored black.

If the battery backed sounder is disconnected without first being disarmed it will sound for 4 minutes 30 seconds, if not reset in the interim. The alarm cycle in this period will be made up of a 30 second continuous tone, followed by a 5 second silent interval.

To disarm the battery backed sounder, to allow for its disconnection, follow sequence below:

- Turn the ignition switch to position II then back to position 0,
- Disconnect the vehicle battery within 17 seconds of switching off the ignition,
- Disconnect the battery backed sounder.

If the battery backed sounder is triggered by a battery disconnection it can be reset by turning the starter switch to position II provided the remote control is on the same key ring as the key, or is in close proximity to the ignition switch.

If the battery backed sounder is triggered by the perimetric or volumetric anti-theft systems, it can be reset by either:

- pressing a button on the remote handset, or
- by turning the starter switch to position II provided the remote control is on the same key ring as the key, or is in close proximity to the ignition switch.

ANTI-THEFT ALARM HORN

Vehicles not fitted with a battery backed sounder are fitted with an anti-theft alarm horn. The anti-theft alarm horn is located in the RH front fender, behind the headlamp. Power supply to the anti-theft alarm horn is controlled by the anti-theft system module via a relay located beneath the drivers seat.

Vehicles not fitted with a battery backed sounder are fitted with an anti-theft alarm horn. The anti-theft alarm horn is located in the RH front fender, behind the headlamp.

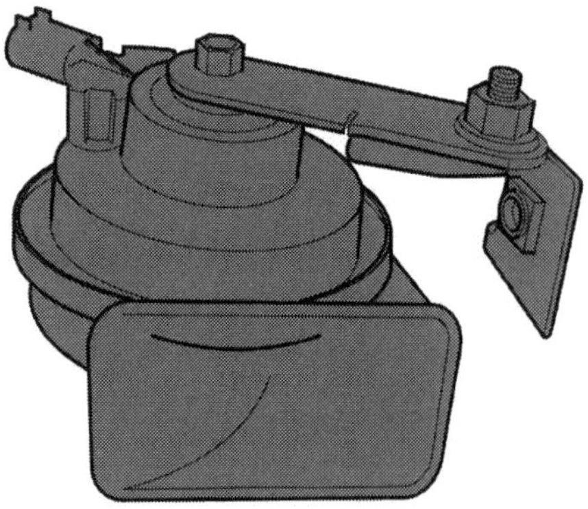

E83592

Power supply to the anti-theft alarm horn is controlled by the anti-theft system module via a relay located beneath the drivers seat.

If the anti-theft system is triggered, the anti-theft alarm horn will emit either a pulsed or continuous tone depending on market legislation.

INTRUSION DETECTION MODULE

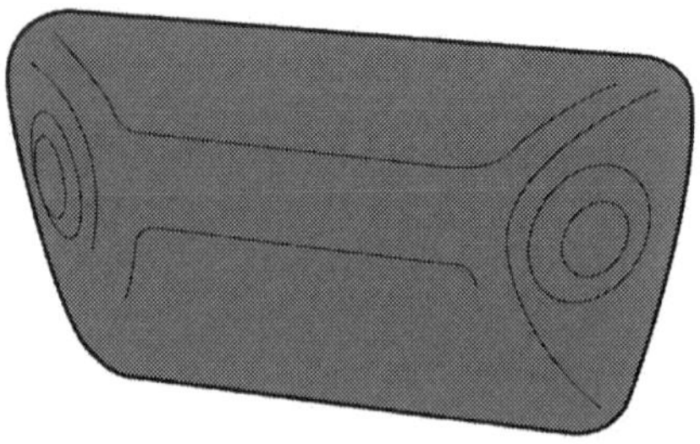

E83593

The intrusion detection module monitors movement within the vehicle cabin. The module is located;

- at the top of the RH B pillar on 90/110 station wagons
- on the front of the headlining, adjacent the interior mirror on 90/110 pick-ups
- on the RH side of the headlining, above the front door on 130 crew cabs.

The intrusion detection module emits a series of ultrasonic sound waves. By measuring the returned sound wave the intrusion detection module can first determine a 'foot print' of the vehicle cabin, and then determine if movement has taken place within this area.

The intrusion detection module is connected directly to the anti-theft system module. If movement is detected in the cabin, the intrusion detection module provides a pulsed voltage signal to the anti-theft module, triggering the alarm.

INSTRUMENT CLUSTER

The instrument cluster contains the anti-theft indicator. The indicator provides a visual anti-theft system status. The anti-theft system is armed, the indicator will flash quickly (8 Hz) for a 10 second period.

During this period, the anti-theft system module will arm the system.

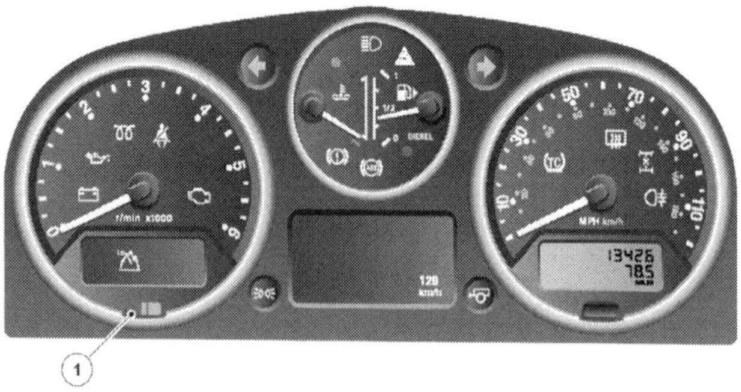

Item	Description
1	Anti-theft indicator

After the 10 second period, the indicator will flash slowly (1 Hz), acting as a deterrent to thieves. Illumination of the anti-theft indicator is controlled by the anti-theft system module.

CONTROL DIAGRAM

CONTROL DIAGRAM - 1 OF 2 - BATTERY BACKED SOUNDER

NOTE: **A** = Hardwired

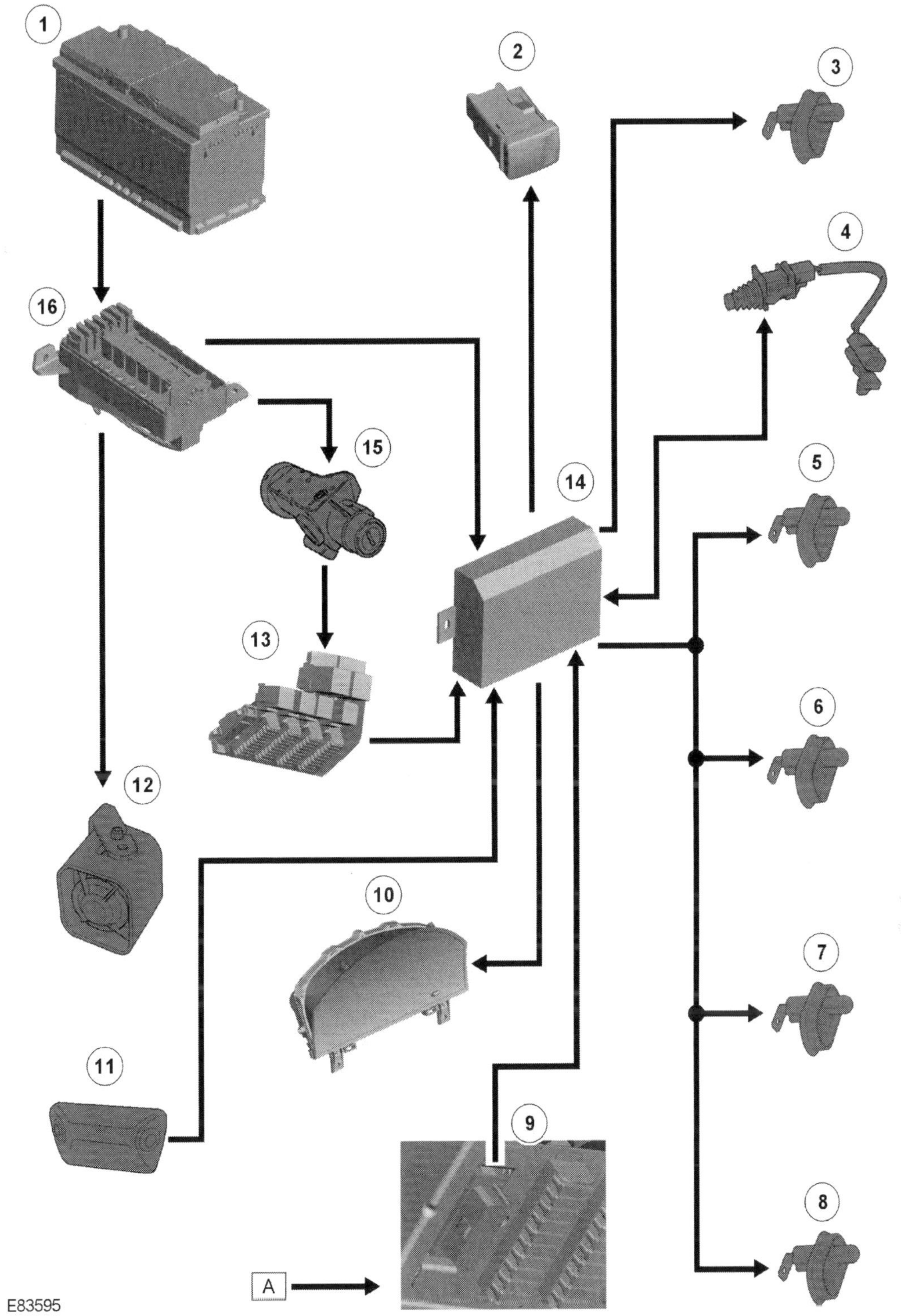

Item	Description
1	Battery
2	Hazard flasher switch
3	Driver's door switch
4	Hood switch
5	Front passenger door switch
6	RH rear door switch (if fitted)
7	LH rear door switch (if fitted)
8	Tail door switch
9	Diagnostic socket

Item	Description
10	Instrument cluster
11	Intrusion detection module
12	Battery backed sounder
13	-
14	Anti-theft system module
15	Ignition switch
16	-

CONTROL DIAGRAM - 2 OF 2 - ANTI-THEFT ALARM HORN

NOTE: **A** = Hardwired

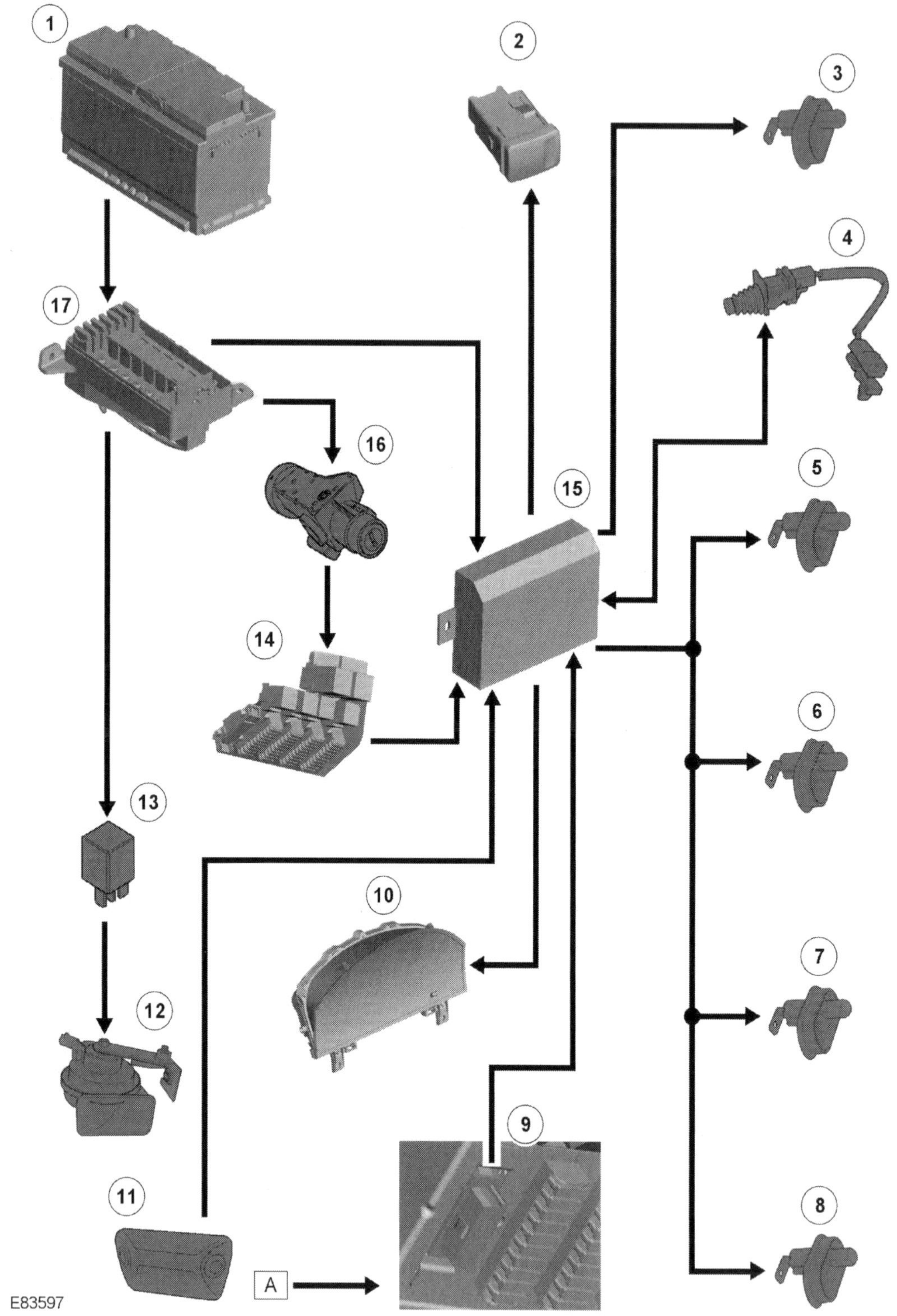

Item	Description
1	Battery
2	Hazard flasher switch
3	Driver's door switch
4	Hood switch
5	Front passenger door switch
6	RH rear door switch (if fitted)
7	LH rear door switch (if fitted)
8	Tail door switch
9	Diagnostic socket
10	Instrument cluster
11	Intrusion detection module
12	Anti-theft alarm horn
13	Anti-theft alarm horn relay
14	-
15	Anti-theft system module
16	Ignition switch
17	-

PRINCIPLES OF OPERATION

On receipt of a valid code from the remote RF handset, the anti-theft system module will arm both the perimetric and volumetric anti-theft systems. The anti-theft system module will also arm the passive immobilization system.

Refer to: Anti-Theft - Passive (419-01 Anti-Theft - Passive, Description and Operation)

Once armed, the anti-theft system can be triggered up to 3 times in any one arming cycle.

Perimetric Protection

Perimetric protection is the monitoring of each hinged panel, namely the doors, tail door and hood, for unauthorized entry. The anti-theft system module monitors the condition of each hinged panel through a series of hardwired connections.

The driver's door switch and the hood switch are connected to the anti-theft system module on individual wires. The remaining hinged panels are wired in parallel and connected to the anti-theft system module on a single wire.

All the switches are open circuit. If a hinged panel is opened, the switch contacts close and a ground path is generated. The switch ground paths are through the vehicle body, except for the hood switch which is provided a ground path via the anti-theft system module. When the anti-theft system module registers a ground path it determines a hinged panel has been opened and triggers the alarm.

Volumetric Protection

Volumetric protection monitors movement within the cabin. When the system is first armed, the anti-theft module suspends volumetric protection for 15 seconds. This allows air in the cabin to settle and prevents nuisance triggers. The anti-theft system module monitors movement in the vehicle cabin using the intrusion detection module. The anti-theft system module will not initiate volumetric protection unless the intrusion detection module detects no disturbances for 15 seconds. If, during the 15 second arming period movement is detected, the 15 second period will re-start from 0.

Volumetric protection can be disabled in one of two ways:

- By locking the vehicle with the key in the driver's door rather than using the remote RF handset.
- By locking the vehicle with the remote RF handset while the driver's door is open. If the driver's door is subsequently shut, only the perimetric system will be armed.

Visual Indications

⚠ NOTE: Visual and audible warnings are dependent on market legislation.

When the anti-theft system is armed the anti-theft system module will flash the hazard flashers 3 times. The hazard flashers are controlled by 2 outputs from the anti-theft system module; one for the LH lamps, one for the RH lamps. To illuminate the hazard flashers, the output from the anti-theft system module is at battery voltage. To extinguish the lamps, the output is driven to ground.

The anti-theft system module will also control operation of the instrument cluster indicator. The module will flash the indicator at a rate of 8 Hz for 10 seconds, followed by a 1 Hz flashing rate.

If the anti-theft system is triggered, the anti-theft system module will flash the hazard flashers for 30 seconds at 0.5 second intervals.

Audible Indications

If the anti-theft system is triggered, the anti-theft system module will activate the sounder. Depending on market legislation, the sounder is either an anti-theft alarm horn or a battery backed sounder. Again, depending on market legislation, the sounder may emit either a continuous or pulsed tone.

The anti-theft system module provides a battery voltage feed to, depending on vehicle specification, the battery backed sounder or the anti-theft alarm horn relay.
To power the sounder, the anti-theft system module removes the feed and provides a ground path.

Description and Operation

COMPONENT LOCATION

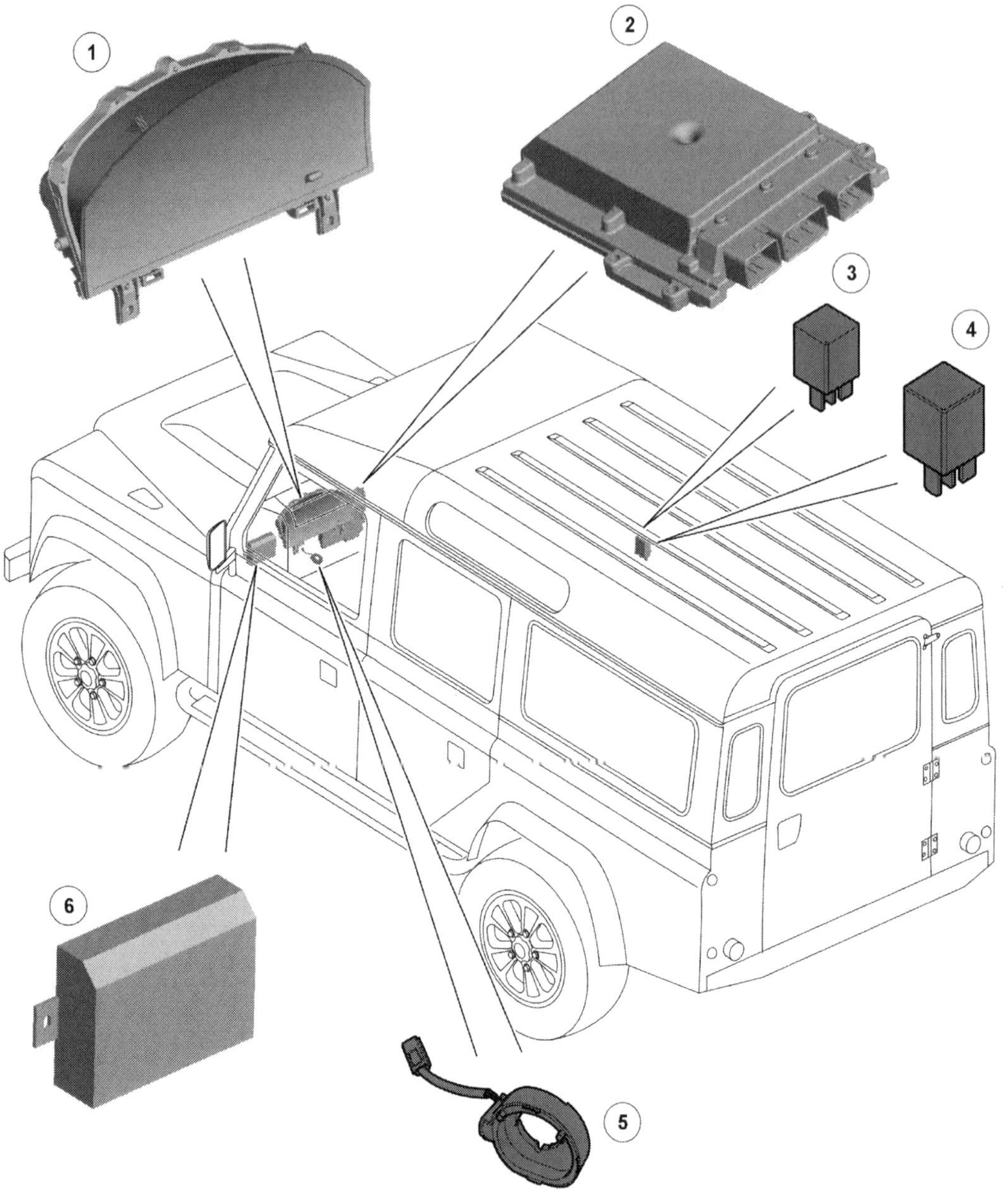

Item	Description
1	Instrument cluster
2	-
3	Starter relay
4	Glow plug relay
5	Transceiver coil
6	Anti-theft system module

OVERVIEW

The 10AS anti-theft system works in conjunction with the ECM to control the passive anti-theft system. The passive anti-theft system immobilizes the engine by inhibiting the starter relay and fuel injectors until a valid code is received from the remote Radio Frequency (RF) handset.

The passive anti-theft system features 2 levels of protection:

- Engine immobilization only
- Engine immobilization plus perimetric and volumetric anti-theft alarm protection.

Refer to: Anti-Theft - Active (419-01 Anti-Theft - Active, Description and Operation)

The anti-theft system module activates the engine immobilization system 30 seconds after the key is removed from the ignition switch and the driver's door is opened. If the key is removed from the ignition switch and the drivers door is not opened, the anti-theft system module will activate the engine immobilization system after a period of 5 minutes.

TRANSCEIVER COIL

The transceiver coil comprises a coil mounted around the ignition barrel. When the ignition switch is turned to position II, the anti-theft system module provides a permanent battery voltage feed and a pulsed voltage feed to the transceiver coil.

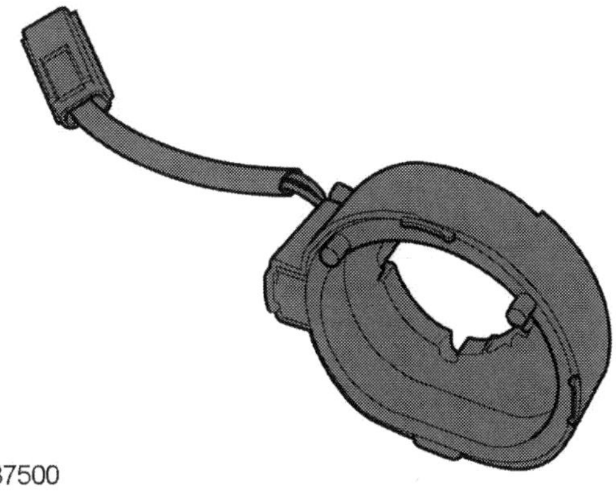

E87500

This creates a magnetic field around the ignition barrel, which collapses and restores in relation to the pulsed power feed from the anti-theft system module. The fluctuating magnetic field activates the remote RF handset which transmits a mobilization signal.

ANTI-THEFT SYSTEM MODULE

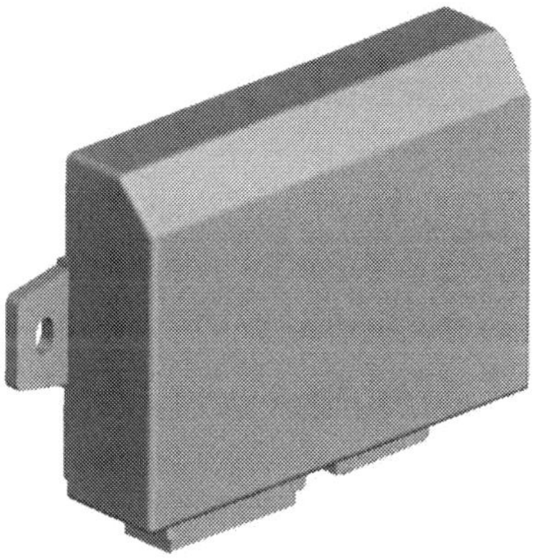

E83589

The anti-theft system module is located behind the instrument cluster and works in conjunction with the ECM to control the passive anti-theft system. The anti-theft system module receives a permanent battery supply and an ignition switch supply from the CJB (central junction box) .

A mobilization code is transmitted to the anti-theft system module by the remote RF handset. The mobilization code is received via the anti-theft system module antenna, which hangs below the module. The anti-theft system module determines if a valid code is received from the remote RF handset by comparing the received code with one stored in its memory.

The anti-theft system module also controls;

- The active anti-theft system.

Refer to: Anti-Theft - Active (419-01 Anti-Theft - Active, Description and Operation)

- The central locking system.

Refer to: Handles, Locks, Latches and Entry Systems (501-14 Handles, Locks, Latches and Entry Systems, Description and Operation)

- Interior lighting.

Refer to: Interior Lighting (417-02 Interior Lighting, Description and Operation)

- The hazard flashers.

Refer to: Exterior Lighting (417-01 Exterior Lighting, Description and Operation)

ENGINE CONTROL MODULE

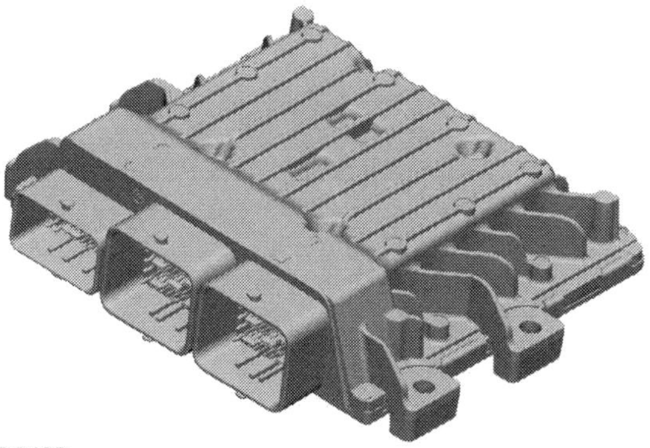

E139402

The ECM is mounted on the engine compartment bulkhead and works in conjunction with the anti-theft system module to control the passive anti-theft system.

PRINCIPLES OF OPERATION

The anti-theft system module receives a permanent battery supply from the CJB. When the vehicle key is placed in the ignition barrel and turned to position II (ignition on), the anti-theft system module also receives an ignition power supply from the CJB.

When the anti-theft system module receives the ignition switch feed it activates the transceiver coil by supplying a battery voltage feed.

The pulsed feed from the anti-theft system module causes the magnetic field created by the transceiver coil to collapse and restore. The fluctuating magnetic field activates the remote RF handset to transmit a mobilization code to the anti-theft system module.

The anti-theft system module receives the mobilization code via its antenna. This code is compared to a value in its memory. If the codes match, the anti-theft system module provides a mobilization signal to the ECM, which will energize the starter relay and the fuel injectors to allow the engine to be started.

Emergency Key Access

If the vehicle is immobilized and the remote RF handset is not available, the emergency key access procedure will allow the engine to be mobilized using the vehicle key. For information on the emergency key access feature, refer to the Owners Handbook.

CONTROL DIAGRAM

NOTE: **A** = Hardwired; **F** = RF Transmission; **AH** = Serial communication line.

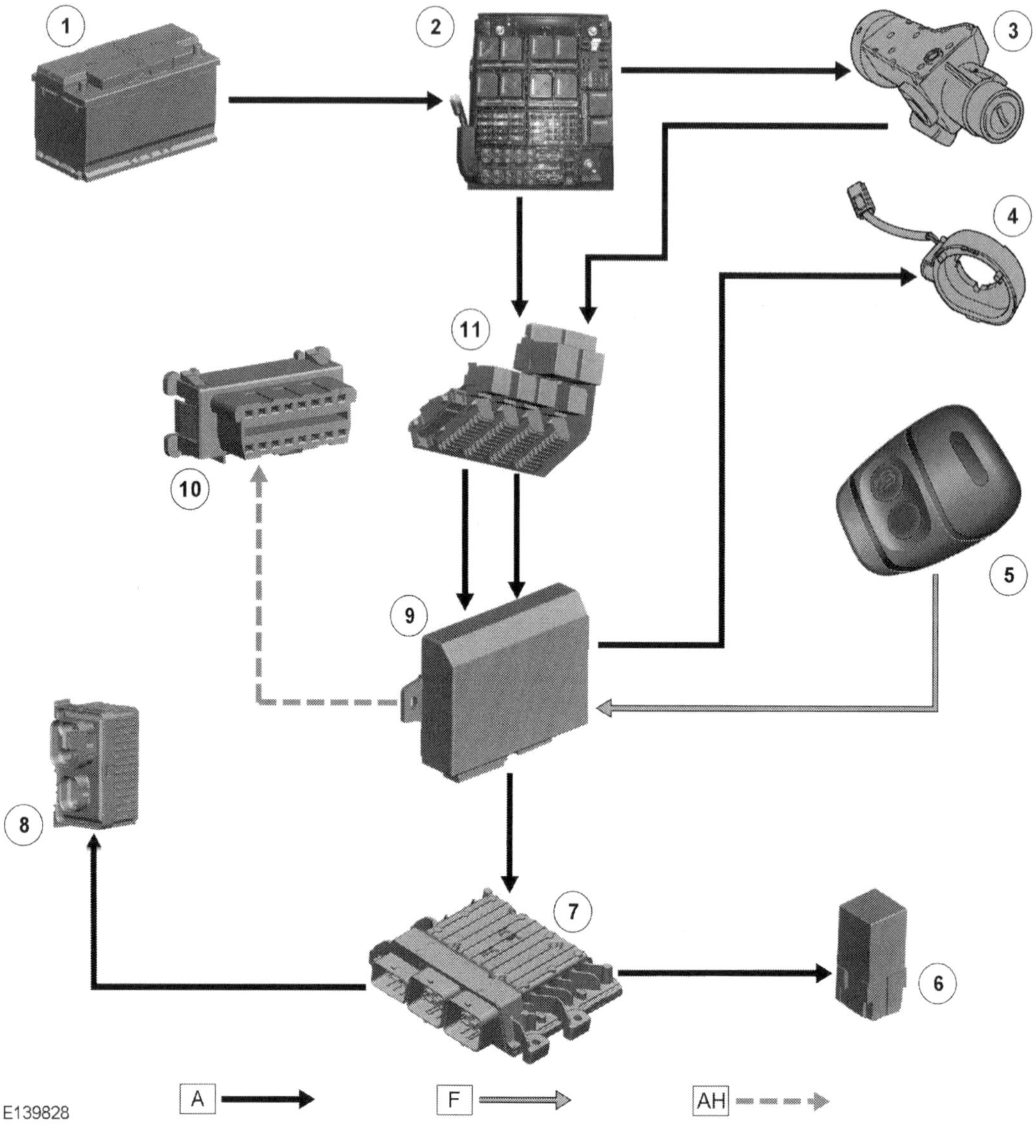

Item	Description
1	Battery
2	-
3	Ignition switch
4	Transceiver coil
5	Remote handset
6	Starter relay
7	-
8	Glow plug control module
9	Anti-theft system module
10	Diagnostic socket
11	-

Passive Anti-Theft System Transceiver

Removal and Installation

Removal - All vehicles

1. Remove the drivers side instrument panel speaker.

Refer to: Instrument Panel Speaker (415-03 Speakers, Removal and Installation)

Left-hand drive vehicles

1. NOTE: Make sure the mating patch lock is attached securely. If not, a new patch lock needs to be installed.

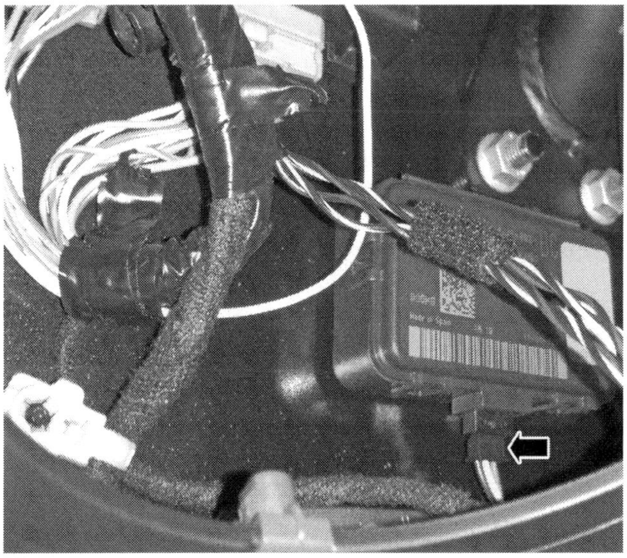

Rotate the component to assist removal. This will help prevent damage to the mating patch lock.

Right-hand drive vehicles

1. NOTE: Make sure the mating patch lock is attached securely. If not, a new patch lock needs to be installed.

   Rotate the component to assist removal. This will help prevent damage to the mating patch lock.

Installation

1. To install, reverse the removal procedure.

Passive Anti-Theft System Module

Removal and Installation

Removal

All vehicles

1. Remove the left front seat cushion.

Refer to: Front Seat Cushion (501-10 Seating, Removal and Installation).

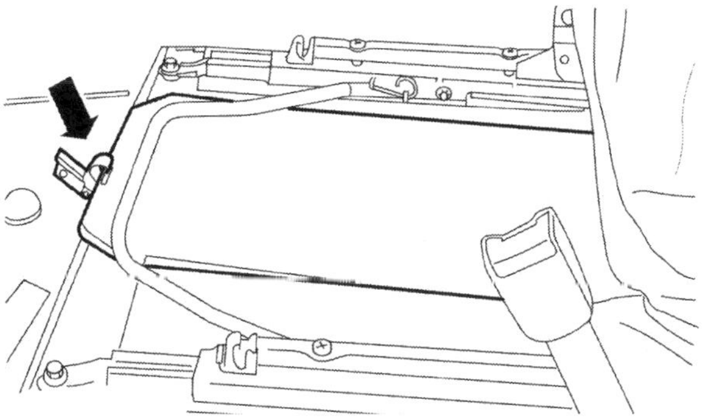

2. Remove the battery access cover.
3. CAUTION: Failure to follow the procedure below could result in the alarm sounder self activating when the battery is disconnected.
   1. Loosen the battery ground cable nut.
   2. Turn the ignition switch to the ON position.
   3. Turn the ignition switch to the OFF position.
   4. Within 17 seconds of turning the ignition switch to the OFF position, release the battery ground cable.

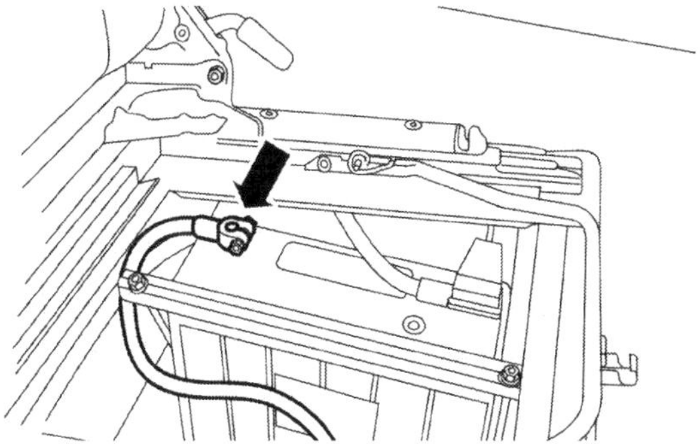

4. Remove the drivers side instrument panel speaker.

Refer to: Instrument Panel Speaker (415-03 Speakers, Removal and Installation)

Left-hand drive vehicles

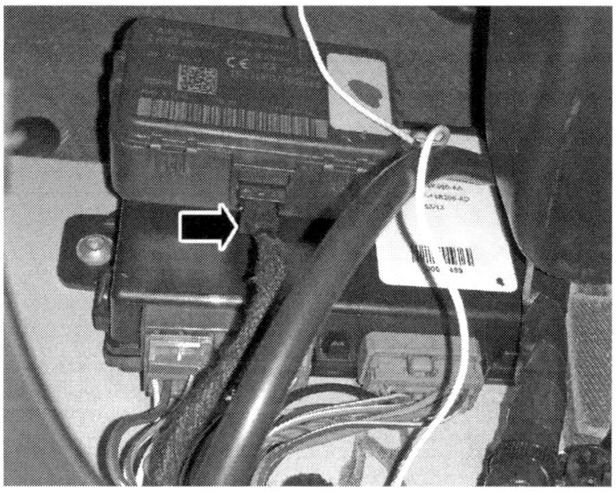

1. NOTE: Install a new locking patch to the PATS module.

   Rotate the component to assist removal. This will help prevent damage to the mating patch lock.

All vehicles

1.

Installation

1. To install, reverse the removal procedure.
2. If a new component is installed, configure using Land Rover approved diagnostic procedure.

Description and Operation

COMPONENT LOCATION

OVERVIEW

The 9.5Ti winch is manufactured by WARN in the U.S.A. The winch features a Gen II Series Wound winch motor.

The winch has 3-stage planetary gearing and full-face contact drum seals, motor and end housing gaskets for extreme duty water resistance. The winch remote control has a thermometric indicator which gives the operator information about motor temperature during the winching operation.

9.5Ti Winch SPECIFICATIONS:

- WARN Part number: 66260 (12V)
- Rated Line Pull: 9,500 lbs. (4310 kgs) single-line
- Intended Use/Application: Vehicle Recovery
- Motor: Gen II Series Wound
- Remote Control: Remote switch, 12 ft (3.7m) lead
- Geartrain: 3-Stage Planetary Gear Ratio: 156:1
- Lubrication: 76 Moly low temperature grease
- Clutch (freespooling): Sliding Ring Gear
- Brake: Automatic Direct Drive Cone
- Wire Rope: 125', 5/16" diameter (38m, 8mm dia.)
- Fairlead: Roller
- Recommended Battery: 650 CCA minimum for winching
- Battery Leads: 2 gauge, 100" (2.54m)
- Finish: High-gloss dark gray powder coat
- Drum Diameter/Length: 2.5"/9.0" (6.4cm/23cm)
- Weight: 89 lbs. (36.3 kgs.).

12V DC PERFORMANCE SPECIFICATIONS:

Line Pull Lbs.(Kgs.)	Line Speed FT./min(M/min.)	Motor Current	Pull by layer Layer/Lbs(Kgs.)
0	62 (18.9)	47 Amps	1/9500 (4313)
2000 (910)	16.3 (4.97)	160 Amps	2/8650 (3927)
4000 (1814)	11.8 (3.60)	230 Amps	3/7920 (3595)
6000 (2720)	9.31 (2.84)	305 Amps	4/7400 (3359)
8000 (3630)	7.59 (2.31)	370 Amps	5/6940 (3150)
9500 (4310)	6.67 (2.03)	425 Amps	-

The above performance figures are based on the first layer of the drum.

System Operation

OPERATION

WARNING: Observe all Warnings and Cautions detailed in the WARN Winch Operator's Guide and the WARN Basic Guide to Winching Techniques before and during winch operation.

The winch motor is powered by the vehicle battery. The motor provides rotational power to the gear mechanism, which in turn rotates the winch drum and winds the cable.

Control Diagram

NOTE: **A** = Hardwired

Vehicle connection

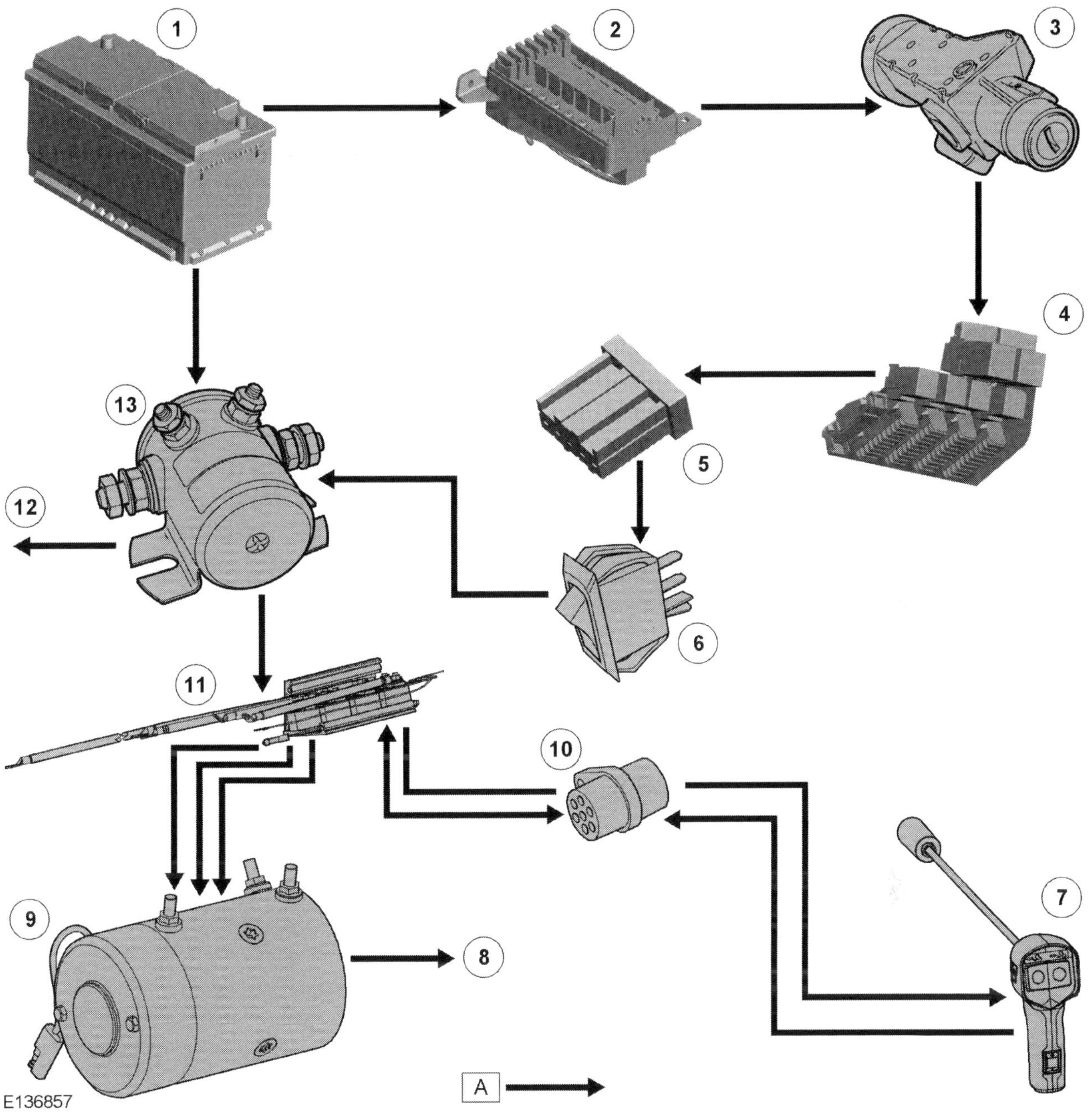

Item	Description
1	Battery
2	Battery Junction Box (BJB)
3	Ignition switch
4	Passenger fusebox
5	Harness connector C0072-1 - Heated rear window switch (ref only)
6	Winch power interrupt solenoid power on/off switch
7	Remote control
8	Ground
9	Winch motor
10	Remote control socket
11	Control box
12	Winch power interrupt solenoid ground
13	Winch power interrupt solenoid

Winch Schematic Electrical Circuit Diagram

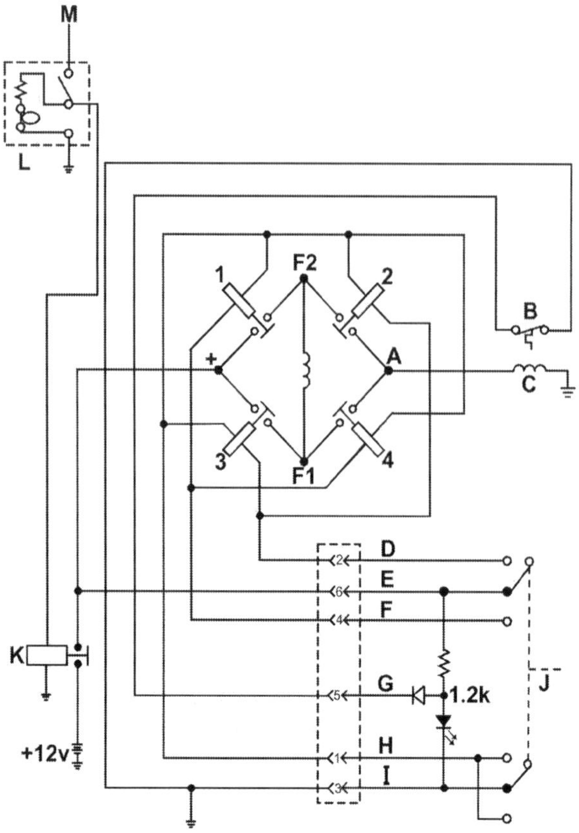

Item	Description
A	Field
B	Thermal protection device - normally closed
C	Armature
D	White wire
E	Red wire
F	Green wire
G	Blue wire
H	Brown wire
I	Black wire
J	Remote control assembly (power in/out)
K	Winch power interrupt solenoid
L	Winch power interrupt solenoid power On/Off switch (mounted on instrument panel)
M	12V Ignition supply from heated rear window switch

A remote control assembly is used to operate the winch. The remote lead plugs into a remote control socket which is connected to the control box and allows the operator to control the winch winding direction while standing at a safe distance from the wire rope.

The remote control includes an LED (light emitting diode) indicator which illuminates when the motor has reached a maximum acceptable temperature. The motor is not disabled and the operator must allow the winch to cool and proceed with winch operation only when the indicator is extinguished.

Winch Motor/Control Box Connections

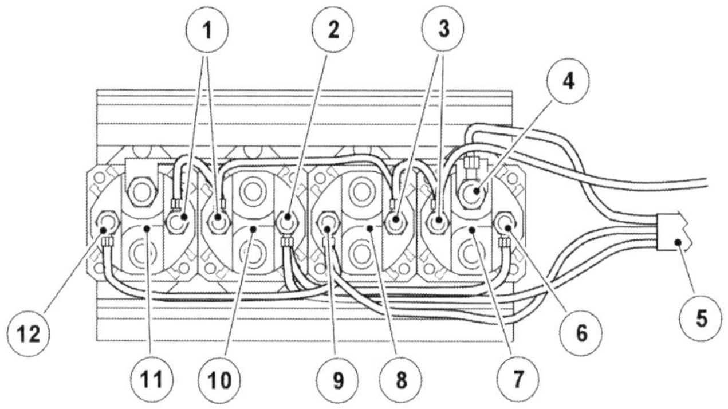

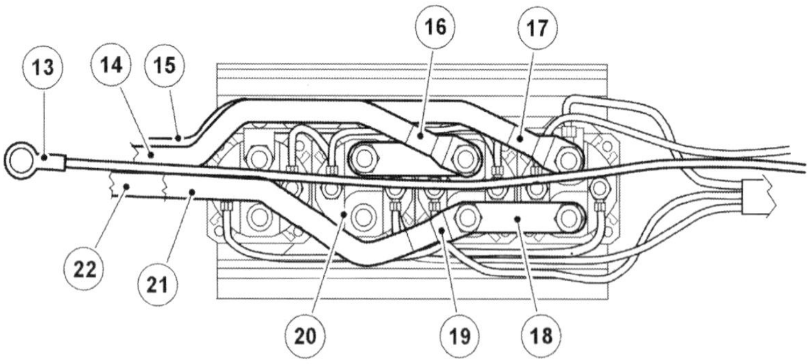

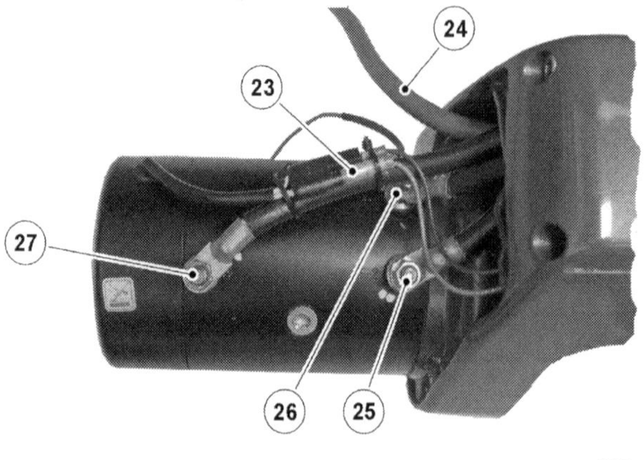

Item	Description
1	Brown wire
2	Green wire
3	Brown wire from socket
4	Red wire from socket
5	Remote control harness
6	Green wire
7	Solenoid No. 1
8	Solenoid No. 2
9	White wire
10	Solenoid No. 3
11	Solenoid No. 4
12	White wire
13	Control ground
14	Armature cable A (to winch motor)
15	Positive (+) cable (to power interrupt solenoid)
16	Armature cable A terminal
17	Battery positive (+) cable terminal
18	Buss bar (3 per assembly)
19	Field cable F2 terminal
20	Field cable F1 terminal
21	Field cable F2 (to winch motor)
22	Field cable F1 (to winch motor)
23	Thermal protection device connector
24	Positive (+) cable (to power interrupt solenoid)
25	Field cable connection F2
26	Field cable connection F1
27	Armature cable connection A

Component Description

DESCRIPTION

Winch

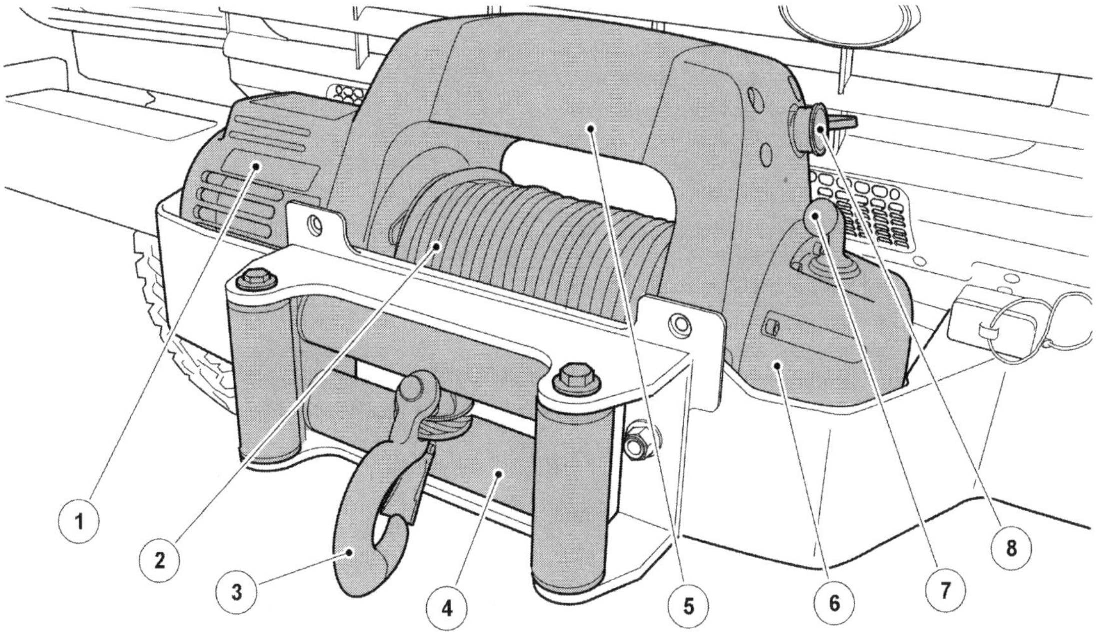

A clutch allows rope to be pulled off the drum without motor operation. Refer to the WARN Basic Guide to Winching Techniques.

The winch is mounted on the front bumper, required for winch fitment. The winch is secured to the bracket with four bolts, washers and nuts.

The winch control box is integrated in the winch housing, above the which drum.

The winch control box manages the power supplies to the winch motor and also receives the input from the remote control for winch operation. A remote control socket is located at the LH (left-hand) end of the winch housing and allows for the connection of the remote control.

The winch will operate with the ignition on, the winch power interrupt solenoid power on/off switch in the 'on' position and the engine not running, but this is not recommended due to excessive battery drain. Therefore it is recommended that the engine is running at all times when the winch is in operation.

Winch/Vehicle Electrical Connections

A blue wire is spliced into harness connector C0072-1, which is the connector for the heated rear window switch. The blue wire is connected to the winch power interrupt solenoid power on/off switch which is located outboard of the center of the instrument panel. From the switch, a white wire is routed from the instrument panel into the battery box where is connects with the winch power interrupt solenoid. This connection provides the ignition on power supply to the winch power interrupt solenoid to energize the solenoid relay when the ignition is on and the winch power interrupt solenoid on/off switch is in the 'on' position.

A red cable is attached from the battery positive (+)

Item	Description
1	Motor
2	Winch drum and wire rope
3	Latching hook
4	Fairlead
5	Control box (hidden)
6	Gear train housing
7	Clutch lever
8	Remote control socket

terminal to the positive terminal on the solenoid. This connection is the main power supply to the winch power interrupt solenoid for the winch motor and control box. The winch power interrupt solenoid is located in the battery box, below the LH front seat. A large red wire from the winch power interrupt solenoid is routed to the winch control box. Power is not supplied to the winch control box until the winch power interrupt solenoid power on/off switch on the instrument panel is switched to the on position, which is confirmed by the switch illuminating.

Winch Power Interrupt Solenoid

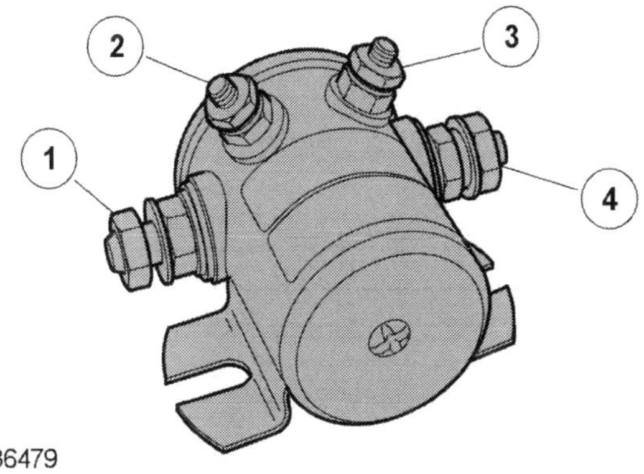

E136479

Item	Description
1	Power to winch motor
2	Ground
3	Solenoid power supply (ignition on)
4	Power supply from vehicle battery positive (+)

Winch Exploded View

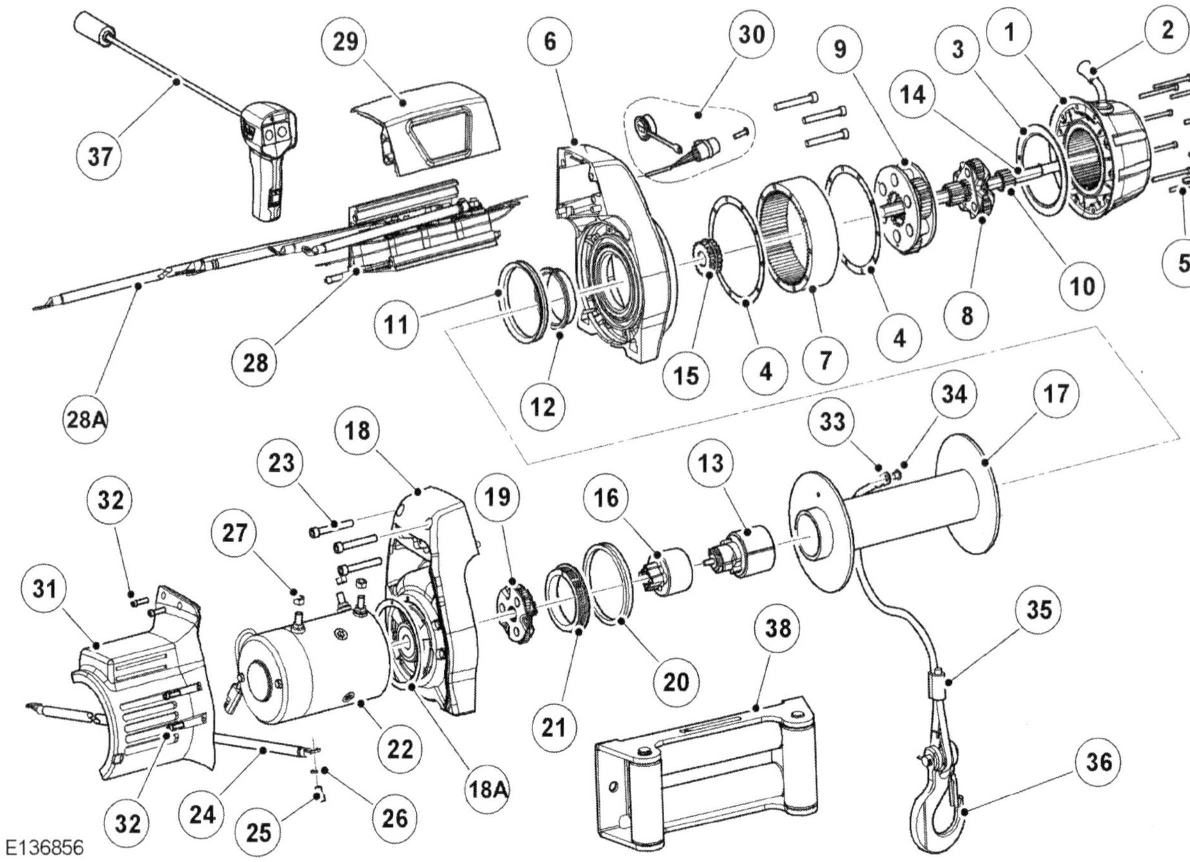

Item	Description
1	End housing assembly
2	Clutch lever
3	Nylon thrust washer
4	Housing gasket
5	Socket head capscrew (10 off)
6	Gear support drum
7	Ring gear
8	Planet carrier assembly (stage 2)
9	Planet carrier assembly (stage 3)
10	Sun gear
11	V-ring seal
12	Drum bushing
13	Brake assembly
14	Drive shaft
15	Drive spline
16	Motor coupler
17	Drum assembly
18	Motor support drum
18A	Motor gasket
19	Planet carrier assembly (stage 1)
20	V-ring seal
21	Drum bushing
22	Motor
23	Socket head capscrew (6 off)
24	Electrical cable (black)
25	Hexagon head capscrew
26	Washer
27	Nut
28	Control pack assembly 9.5Ti
28A	Electrical cable (red)
29	Control pack assembly cover
30	Remote control socket assembly
31	Motor cover
32	Socket head capscrew (4 off)
33	Terminal kit
34	Button head capscrew
35	Wire rope assembly
36	Latching hook
37	Remote control assembly
38	Fairlead roller
39	Seal kit

NOTE: The WARN part numbers for the above components are listed on the WARN web site: www.warn.com

Winch

Diagnosis and Testing

Principle of Operation

For a detailed description of the winch system and operation, refer to the relevant Description and Operation section in the workshop manual.

Refer to: Winch (419-12 Winch, Description and Operation)

Inspection and Verification

WARNING: Observe all Warnings and Cautions detailed in the WARN Winch Operator's Guide before and during winch operation.

CAUTION: Diagnosis by substitution from a donor vehicle is **NOT** acceptable. Substitution of control modules does not guarantee confirmation of a fault, and may also cause additional faults in the vehicle being tested and/or the donor vehicle.

1. Visually inspect for obvious signs of damage and system integrity.

Visual Inspection

Mechanical	Electrical
• Incorrect installation • Winch motor • Gearbox • Wire rope	• Fuses • Battery (650 CCA minimum) • Loose, corroded or damaged electrical connections • Winch power interrupt solenoid • Control box • Remote control

2. If an obvious cause for an observed or reported concern is found, correct the cause (if possible) before proceeding to the next step.

Symptom Chart

Symptom	Possible causes	Action
No winch operation	• Battery/power or ground supply fault • Remote control internal failure • Control box internal failure • Winch motor internal failure	• Refer to the electrical circuit diagrams, check the winch power interrupt solenoid supply fused link ((**Fuse 14** to vin 798535)(**Fuse 15** from vin 798536)) passenger compartment fuse box • For midtronics battery test GO to Pinpoint Test A • For remote control circuit checks GO to Pinpoint Test C • For winch circuit checks GO to Pinpoint Test B
No winch operation (Clicking sound when remote control switch is activated)	• Battery/power or ground supply fault • Control box internal failure	• For midtronics battery test GO to Pinpoint Test A • For winch circuit checks GO to Pinpoint Test B
Winch lacks power, pulls slowly, stalls	• Battery/power or ground supply fault • Winch motor internal failure • Gearbox internal failure	• For midtronics battery test GO to Pinpoint Test A • For winch circuit checks GO to Pinpoint Test B
When the remote control is activated, winch operates in only one direction	• Battery/power or ground supply fault • Remote control internal failure • Control box internal failure	• For remote control circuit checks GO to Pinpoint Test C • For winch circuit checks GO to Pinpoint Test B

Difficulty in spooling rope from drum by hand	• Wire rope incorrectly wound on drum (Rubbing/binding) • Distorted drum flange • Worn drum bushings	• Confirm the rope is spooled correctly onto the drum • Install a new winch as required. **Refer to:** Winch (419-12 Winch, Removal and Installation) Refer to the warranty policy and procedures manual, or determine if any prior approval programme is in operation, prior to the installation of a new module/component
	• Damaged free spool clutch (inside gear train • Corroded clutch ring gear	• Install a new winch gearbox as required. **Refer to:** Winch Gear Assembly (419-12 Winch, Removal and Installation) Refer to the warranty policy and procedures manual, or determine if any prior approval programme is in operation, prior to the installation of a new module/component
Winch does not hold load	• Rope is spooled onto the drum in the wrong direction • Winch drum brake is worn or broken	• Confirm the rope is spooled correctly onto the drum • Confirm the clutch is fully engaged • Install a new winch as required. **Refer to:** Winch (419-12 Winch, Removal and Installation) Refer to the warranty policy and procedures manual, or determine if any prior approval programme is in operation, prior to the installation of a new module/component
Remote control fault	• Winch remote connector terminals damaged • Remote switch internal failure	• For remote control circuit checks GO to Pinpoint Test C
Remote control thermal indicator (LED) illuminated	⚠ NOTE: If the temperature at the motor brushes reaches 205+ C (400+ F) the thermal switch will open, the LED in the remote will illuminate • Thermal trip fault	• Allow the winch motor to cool, check for correct operation • Install a new winch motor as required. **Refer to:** Winch Motor (419-12 Winch, Removal and Installation) Refer to the warranty policy and procedures manual, or determine if any prior approval programme is in operation, prior to the installation of a new module/component
Clutch lever fault	• Cable under load • Clutch or gearbox mechanism internal failure	• Rotate clutch lever on winch. If clutch lever will not rotate, remove tension from the rope and hook by powering out for about 1 second • Attempt to rotate the clutch lever with tension removed. If clutch lever still does not rotate, Install a new winch gearbox as required. **Refer to:** Winch Gear Assembly (419-12 Winch, Removal and Installation) Refer to the warranty policy and procedures manual, or determine if any prior approval programme is in operation, prior to the installation of a new module/component

PINPOINT TEST A : BATTERY TEST	
TEST CONDITIONS	**DETAILS/RESULTS/ACTIONS**
A1: MIDTRONICS BATTERY TEST	
	1. Confirm that the vehicle battery charged, in good condition, and the correct specification (650 CCA minimum)
	2. Using a Midtronics hand held tester or the Midtronics GR-1 diagnostic charger, carry out the "Midtronics battery test" as detailed in the battery care manual. Refer to (Service/ Maintenance information/Battery care manual)
	3. Record battery diagnostic result on the provided form
	Does the battery pass the "Midtronics battery test"? **Yes** Check for correct operation, refer to the symptom chart above if customer concern is still evident **No** Install a new battery as required. Refer to the warranty policy and procedures manual, or determine if any prior approval programme is in operation, prior to the installation of a new module/ component. Check for correct winch operation

PINPOINT TEST B : CIRCUIT CHECKS	
TEST CONDITIONS	**DETAILS/RESULTS/ACTIONS**
B1: WINCH POWER INTERRUPT SOLENOID - CONNECTIONS	
NOTES: △ The winch power interrupt solenoid supplies power to the winch control box	
△ The ignition feed is supplied to the switch to energize the winch power interrupt solenoid	
	1. Refer to the land rover electrical circuit diagrams, check the winch power interrupt solenoid supply fused link (**Fuse 14** to vin 798535)(**Fuse 15** from vin 798536) passenger compartment fuse box
	2. Refer to the winch circuit diagrams **Refer to:** Winch (419-12 Winch, Description and Operation)
	3. Confirm the circuit connections to the winch power interrupt solenoid **(Terminal 1 - power to winch control box, Terminal 2 - solenoid ground, Terminal 3 - solenoid power (From C0072-1 via switch), Terminal 4 - battery supply)** are correctly located, clean and secure
	Are all the connections to the winch power interrupt solenoid, clean and secure? **Yes** GO to 2 **No** Clean and secure the electrical connections. Check for correct winch operation
B2: WINCH POWER INTERRUPT SOLENOID - OPERATION	
	1. With the ignition state on, and the remote control connected
	2. Using a multimeter, check for battery voltage between **terminal 1** (power to winch control box) and **terminal 2** (winch power interrupt solenoid ground)
	Is battery voltage present? (approx. 12 volts) **Yes** GO to 3 **No** Replace the winch power interrupt solenoid as required. **Refer to:** Winch Solenoid (419-12 Winch, Removal and Installation) Refer to the warranty policy and procedures manual, or determine if any prior approval programme is in operation, prior to the installation of a new module/component. Check for correct winch operation

B3: POWER SUPPLY - CONTROL BOX	
⚠ WARNING: Ensure the winch clutch is in free spool mode (Refer to winch operation manual)	
△ NOTE: The winch power interrupt solenoid supplies power to the winch control box	
	1. With the ignition state on, and the remote control connected
	2. Using a multimeter, check for battery voltage between the **red power feed** to the control box and **vehicle ground**
	Is battery voltage present? (approx. 12 volts) **Yes** GO to 4 **No**
	Carry out circuit checks to investigate the loss of supply. Rectify as required. Check for correct winch operation

B4: CONTROL BOX INPUT - REMOTE CONTROL	
⚠ WARNING: Ensure the winch clutch is in free spool mode (Refer to winch operation manual)	
NOTES: △ The winch control box controls the winch motor direction	
△ Power out - The remote control supplies power to remote connector **terminal 2** and grounds **terminal 1**	
△ Power in - The remote control supplies power to remote connector **terminal 4** and grounds **terminal 1**	
	1. Refer to the winch circuit diagrams **Refer to:** Winch (419-12 Winch, Description and Operation)
	2. Using a multimeter, monitor **terminal 2 (White)**, **terminal 4 (Green)** and **terminal 1 (Brown)**
	Do the terminals **2 (Power out) and 4 (Power in)** (Power) and **terminal 1** (Ground) respond correctly when the remote control is used? **Yes** GO to 5 **No** Carry out circuit checks to investigate the loss of supply. Rectify as required. Check for correct winch operation

B5: CONTROL BOX - OUTPUT	
⚠ WARNING: Ensure the winch clutch is in free spool mode (Refer to winch operation manual)	
NOTES: △ When activated the winch control box supplies power to the winch motor	
△ Power out - The control box supplies power to the winch motor terminal **F1** and connects winch motor terminal **F2** to terminal **A**(Armature)	
△ Power in - The control box supplies power to the winch motor terminal **F2** and connects winch motor terminal **F1** to terminal **A**(Armature)	
	1. Refer to the winch circuit diagrams **Refer to:** Winch (419-12 Winch, Description and Operation)
	2. Using a multimeter, monitor terminals **F1**, **F2**

Do the terminals **F1**, and **F2** respond correctly when the remote control is used?

Yes

GO to 6

No

Check the connections to the control box. If all the connections are clean and secure, install a new control box as required.

Refer to: Winch Control Unit (419-12 Winch, Removal and Installation)

Refer to the warranty policy and procedures manual, or determine if any prior approval programme is in operation, prior to the installation of a new module/component. Check for correct winch operation

B6: WINCH MOTOR CONNECTIONS

WARNING: Ensure the winch clutch is in free spool mode (Refer to winch operation manual)

1. Refer to the winch circuit diagrams

Refer to: Winch (419-12 Winch, Description and Operation)

2. Check the connections to the winch motor

Are all the connections clean and secure?

Yes

GO to 7

No

Clean and secure the electrical connections. Check for correct winch operation

B7: WINCH MOTOR TEST

WARNING: Ensure the winch clutch is in free spool mode (Refer to winch operation manual)

1. Disconnect the positive lead from the battery (leave the ground lead attached)
2. Label and disconnect the three cables that run from the control pack to the three posts on the winch motor
3. Stamped next to the three posts on the winch motor will be **A**, **F1** and **F2**. Connect a test cable (Jump leads work well) from **A** to **F1** and supply power from the battery to **F2**. The winch motor should operate
4. Connect a test cable from **A** to **F2** and supply power from the battery to **F1**. The winch motor should operate in the opposite direction

Does the motor operate in both directions?

Yes

Check for correct operation, refer to the symptom chart above if customer concern is still evident

No

Check the connections to the winch motor. If all the connections are clean and secure, install a new winch motor as required.

Refer to: Winch Motor (419-12 Winch, Removal and Installation)

Refer to the warranty policy and procedures manual, or determine if any prior approval programme is in operation, prior to the installation of a new module/component. Check for correct winch operation

PINPOINT TEST C : REMOTE CONTROL	
TEST CONDITIONS	DETAILS/RESULTS/ACTIONS
C1: REMOTE CONTROL - CONTINUITY TEST	
NOTES: △ Visually inspect the winch remote connector terminals for damage, repair or replace as required	
△ The winch remote should be tested for continuity using a multimeter	
	1. With the winch remote **disconnected** check for continuity, using the instructions below
	2. Move the switch to the **power out** position - Check for continuity between terminals **(1) and (3)**, Check for continuity between terminals **(2) and (6)**
	3. Move the switch to the **power in** position - Check for continuity between terminals **(1) and (3)**, Check for continuity between terminals **(4) and (6)**
	Did the winch remote pass the continuity test (Resistance less than 1 ohm)? **Yes** Check for correct operation, refer to the symptom chart above if customer concern is still evident **No** Install a new winch remote as required. Refer to the warranty policy and procedures manual, or determine if any prior approval programme is in operation, prior to the installation of a new module/ component. Check for correct winch operation

Winch

Removal and Installation

Removal

⚠ CAUTION: Note the cable routes and position of the wiring harness clips to aid installation.

1. **Refer to:** Startup Battery Disconnect and Connect (414-01 Battery, Mounting and Cables, General Procedures)

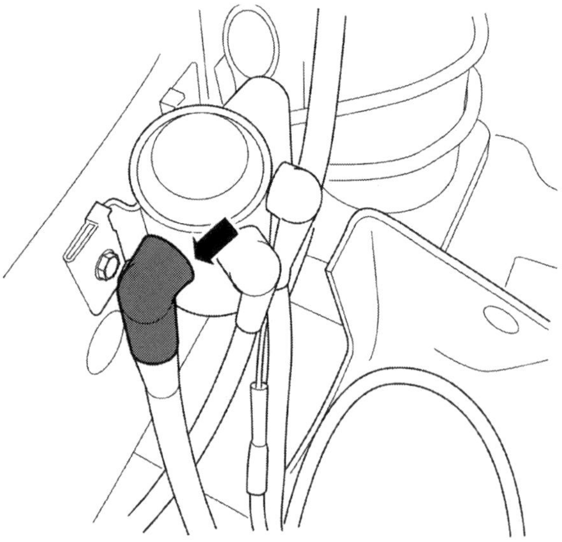

2.

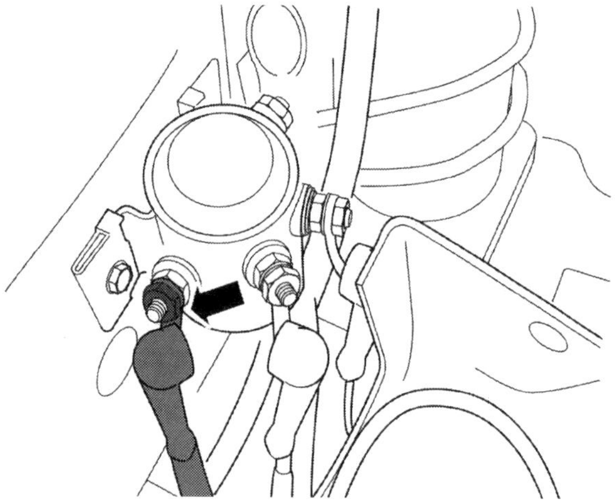

3. ⚠ CAUTION: To prevent damage to components, use an additional wrench when loosening or tightening terminals.

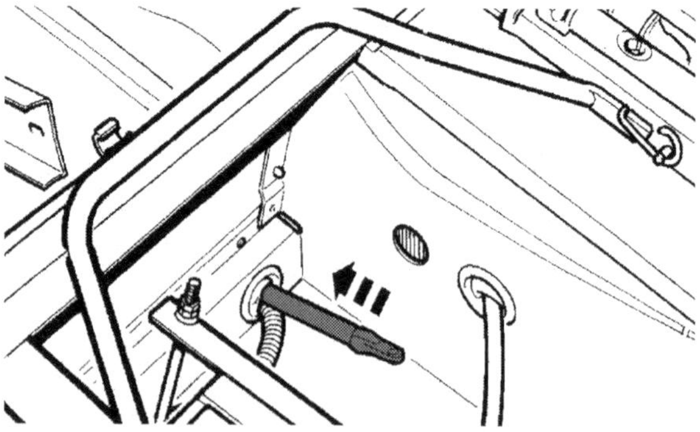

4.

5. **Refer to:** Radiator Grille (501-08, Removal and Installation).

6. ⚠ WARNING: Make sure to support the vehicle with axle stands.

Raise and support the vehicle.

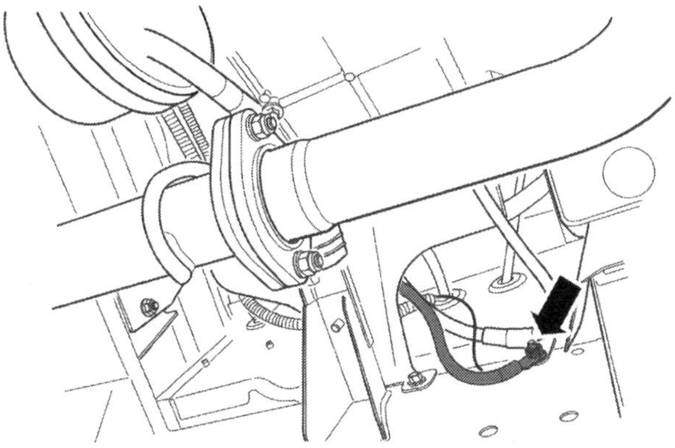

7.

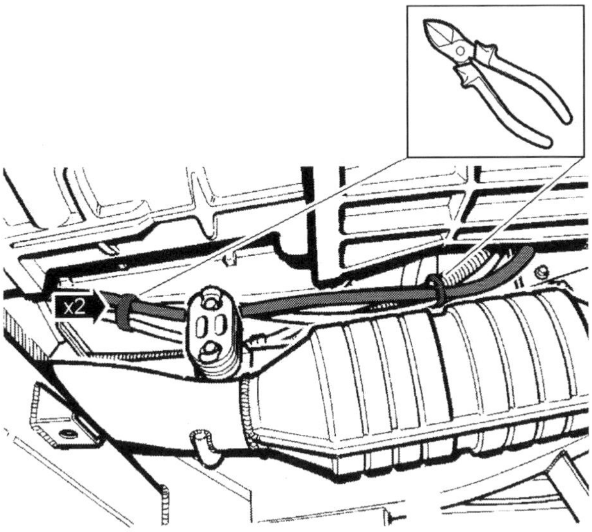

8.

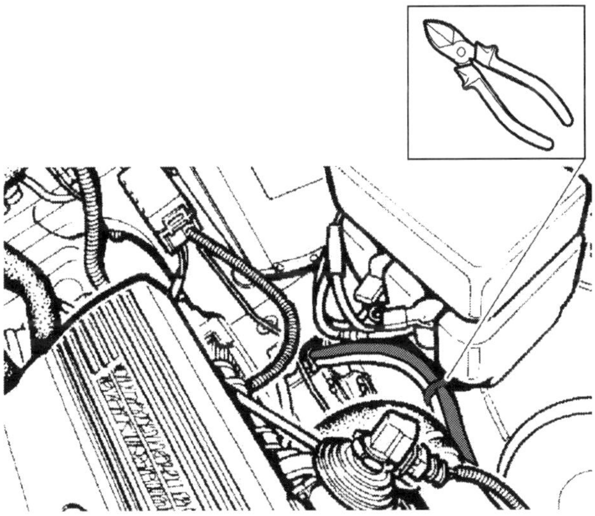

9.

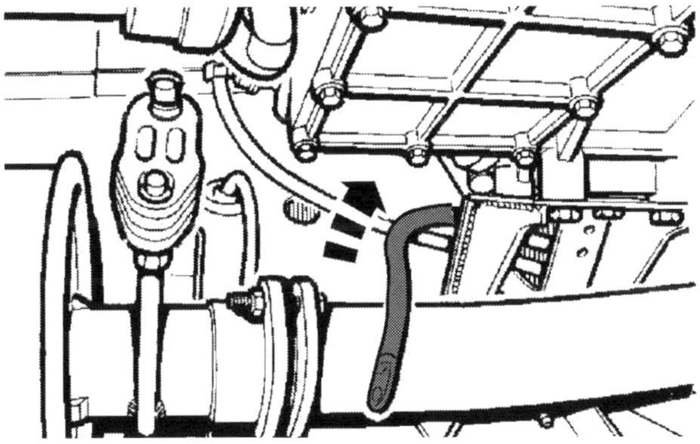

10.

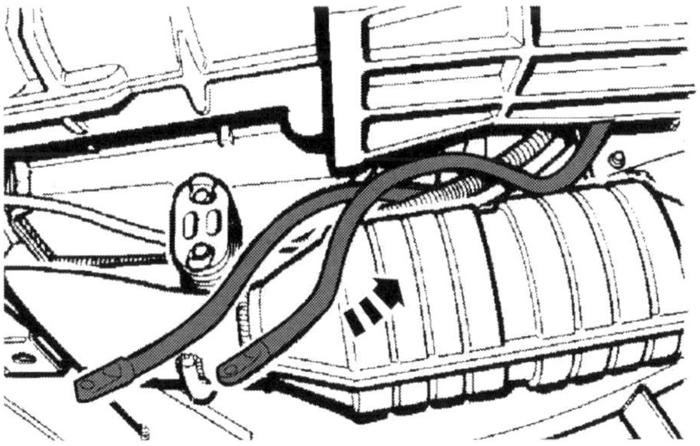

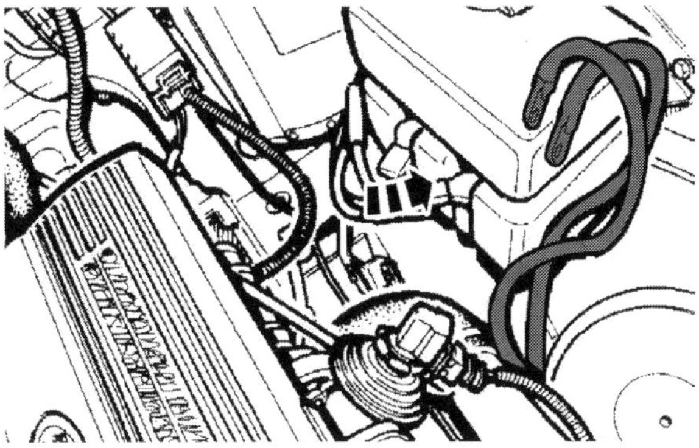

11.

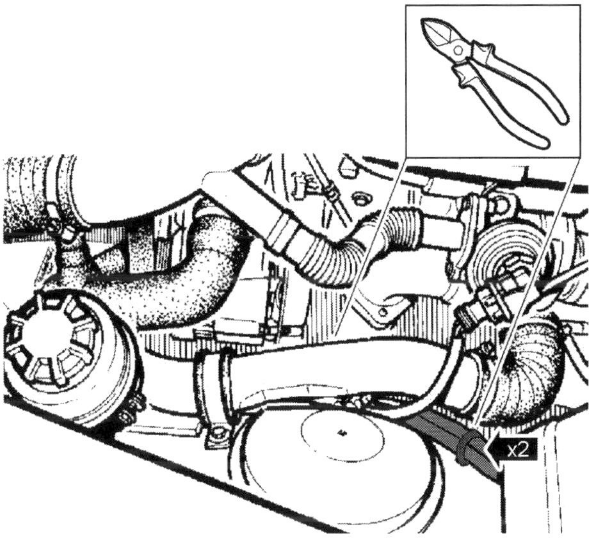

12.

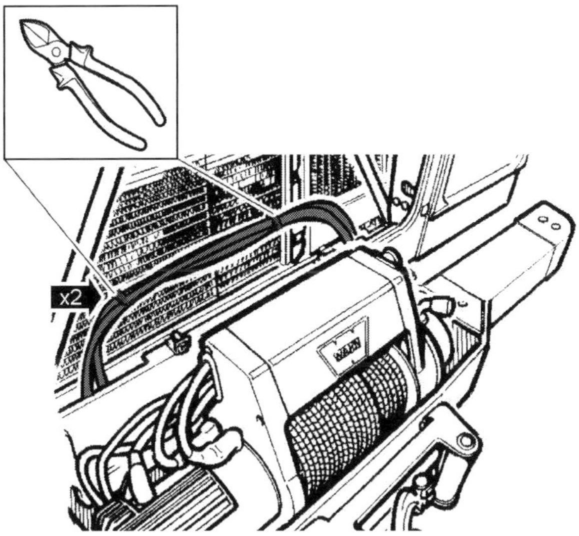

13.

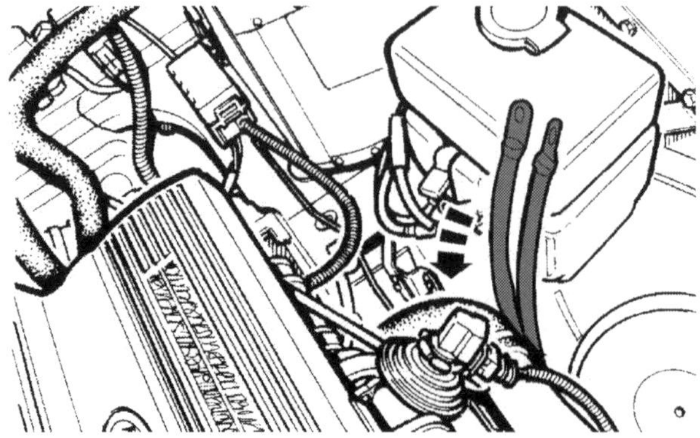

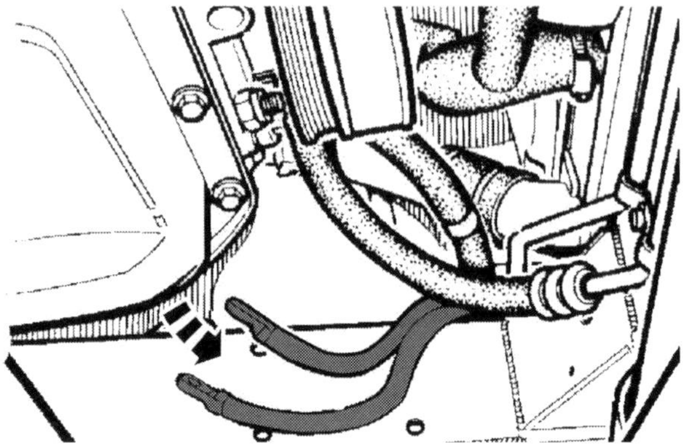

14.

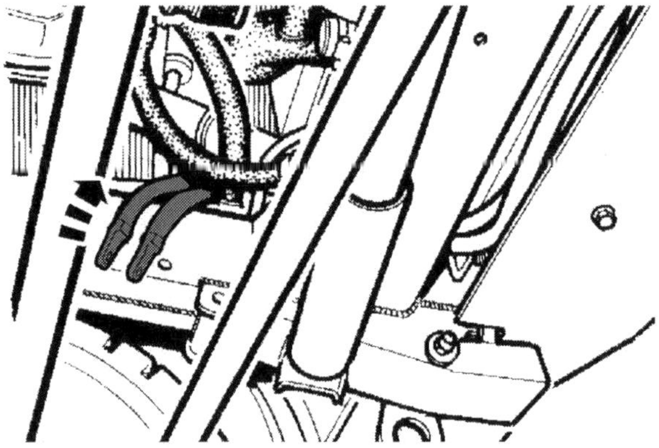

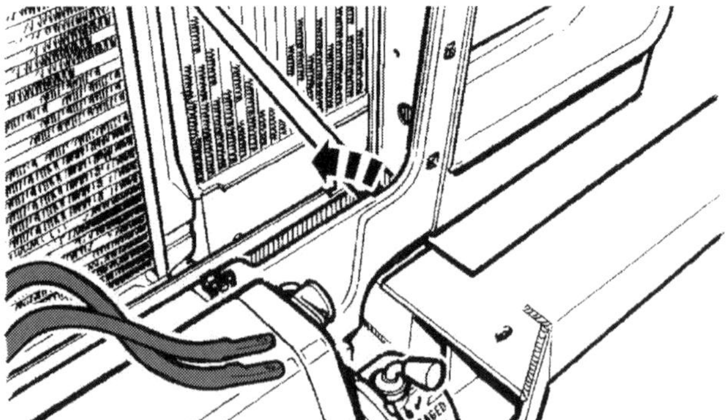

15.

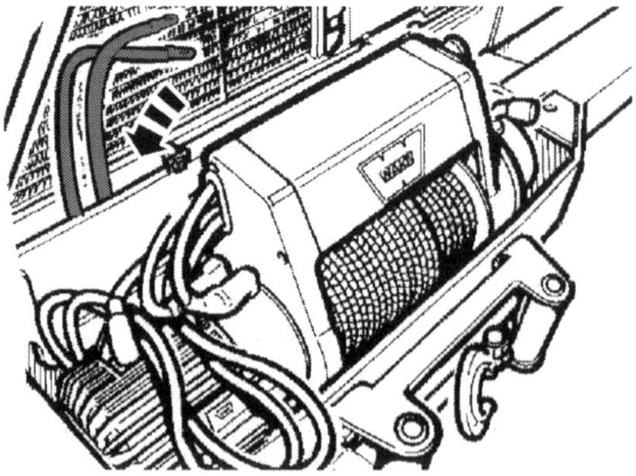

16.

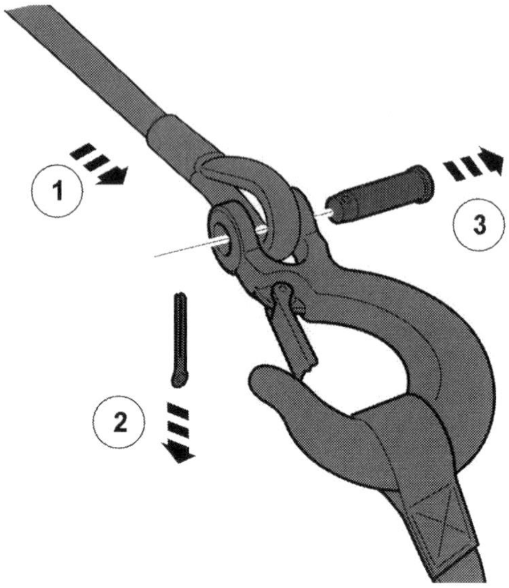

17.

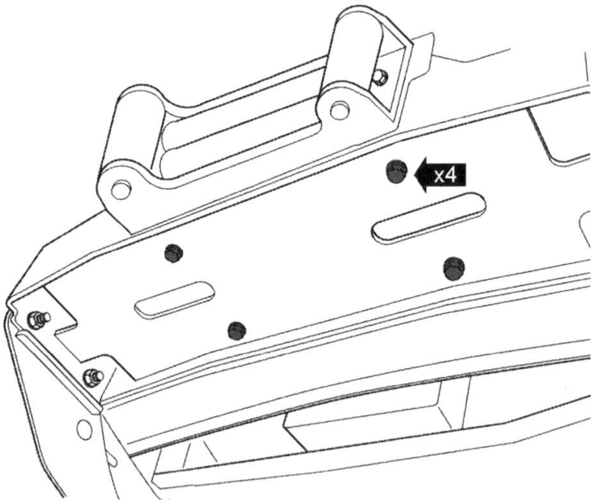

18.

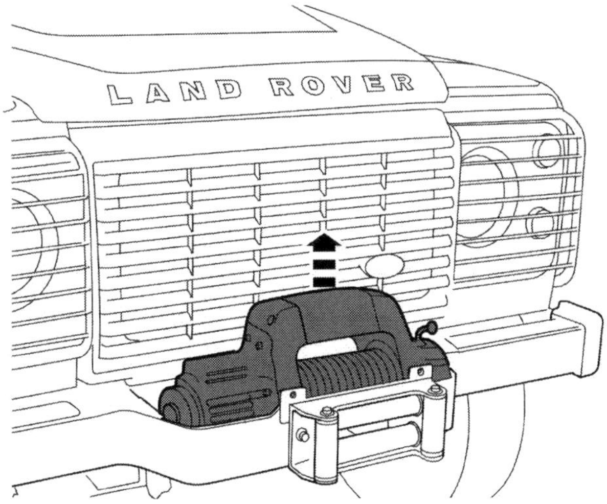

19. ⚠ WARNING: This step requires the aid of another technician.

Installation

1. ⚠ WARNING: This step requires the aid of another technician.

Install the winch.

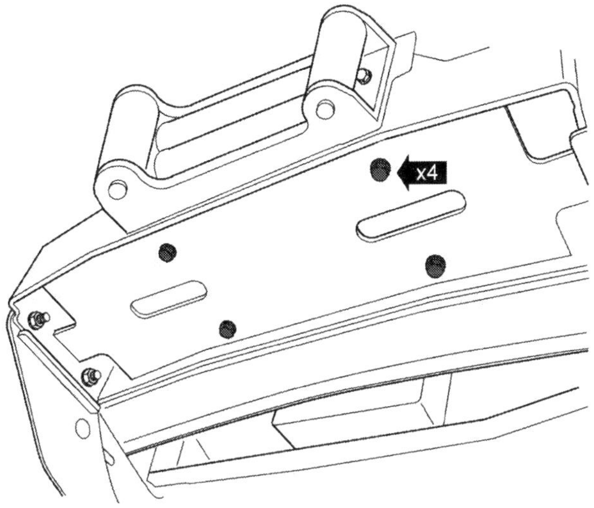

2. Tighten the bolts to 40 Nm.
3. Install the winch hook.
4. CAUTIONS:

⚠ Install the cables following the routes noted on removal.

⚠ Clip in the positions noted on removal.Install the winch power and ground cables.

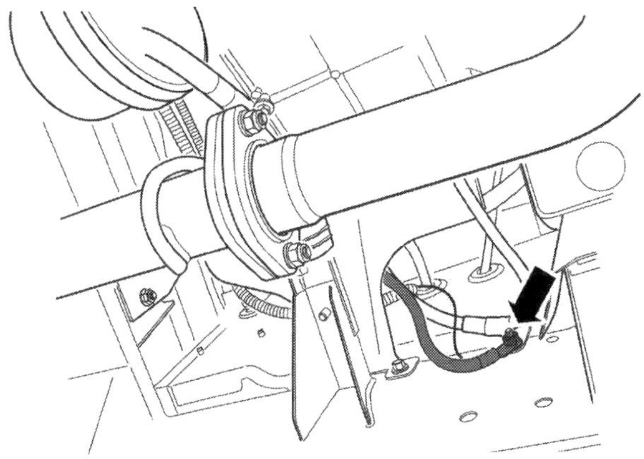

5. Tighten to 25 Nm.
6. Feed the winch power cable into the battery box.

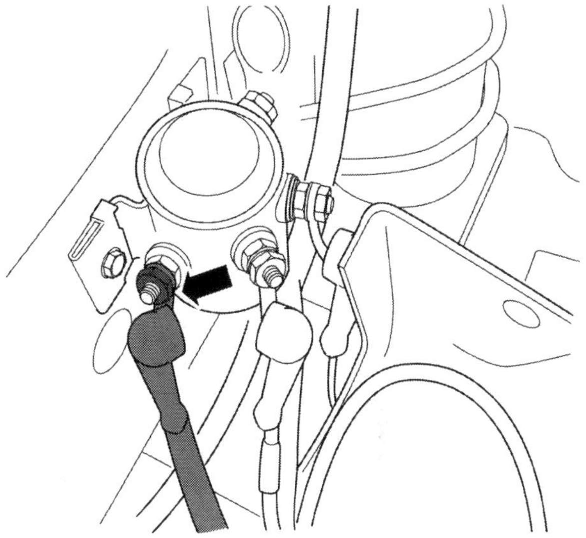

7. ⚠ CAUTION: To prevent damage to components, use an additional wrench when loosening or tightening terminals.

Tighten to 7.3 Nm.

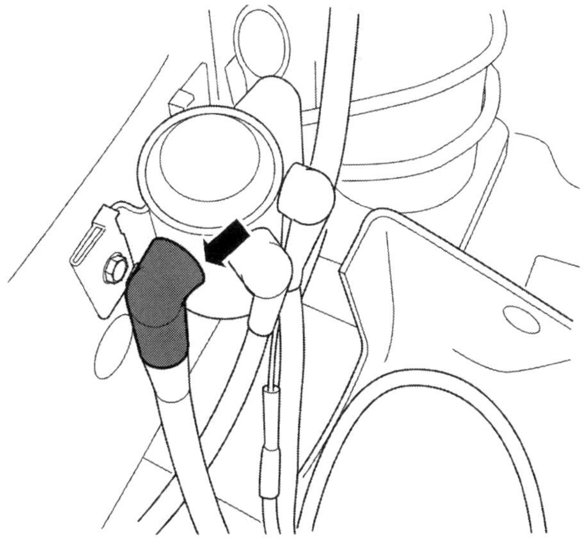

8.
9. **Refer to:** Radiator Grille (501-08, Removal and Installation).
10. **Refer to:** Startup Battery Disconnect and Connect (414-01 Battery, Mounting and Cables, General Procedures)

Licence Plate Panel

Removal and Installation

Removal

△ NOTE: Removal steps in this procedure may contain installation details.

1. If installed: Remove the license plate.
2.

Installation

1. To install, reverse the removal procedure.

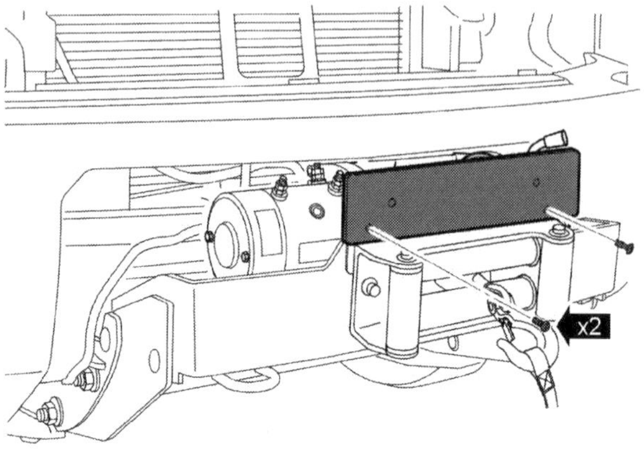

Winch Cable Roller Assembly

Removal and Installation

Removal

△ NOTE: Removal steps in this procedure may contain installation details.

1. **Refer to:** Licence Plate Panel (419-12 Winch, Removal and Installation)

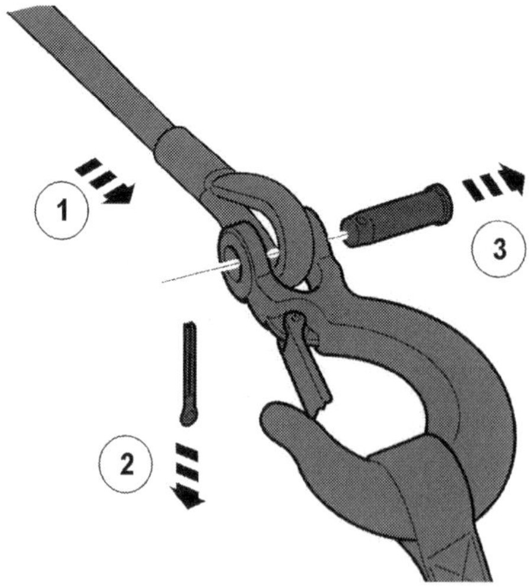

2.

3. Torque: 43 Nm.

Installation

1. To install, reverse the removal procedure.

Winch Control Unit

Removal and Installation

Removal

△ NOTE: Removal steps in this procedure may contain installation details.

1. **Refer to:** Startup Battery Disconnect and Connect (414-01 Battery, Mounting and Cables, General Procedures)
2. **Refer to:** Winch Motor (419-12 Winch, Removal and Installation)

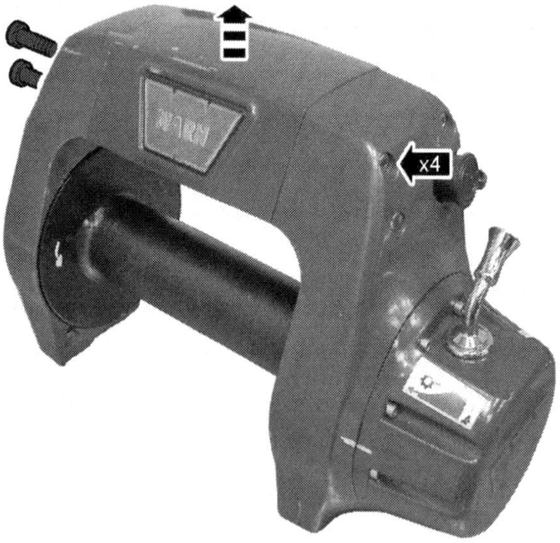

3. Torque: 27 Nm.

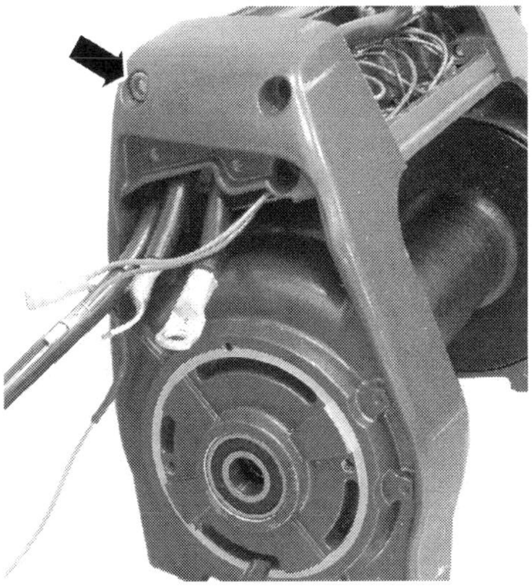

4. CAUTION: Make sure the first stage gear is correctly engaged with the motor coupler.

Torque: 27 Nm.

5.

6. NOTE: Note the orientation of the electrical connectors and wiring harness.

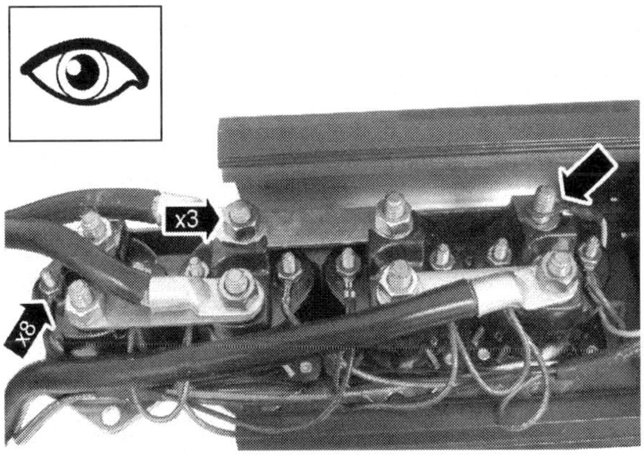

Large terminals 7 Nm

Small terminals 10 Nm

Installation

1. To install, reverse the removal procedure.

Winch Gear Assembly

Removal and Installation

Part(s)

STEP	PART NAME	QUANTITY
Removal Step 4	Renew Part Electric winch	1

Removal

NOTE: Removal steps in this procedure may contain installation details.

1. **Refer to:** Startup Battery Disconnect and Connect (414-01 Battery, Mounting and Cables, General Procedures)
2. **Refer to:** Winch (419-12 Winch, Removal and Installation)

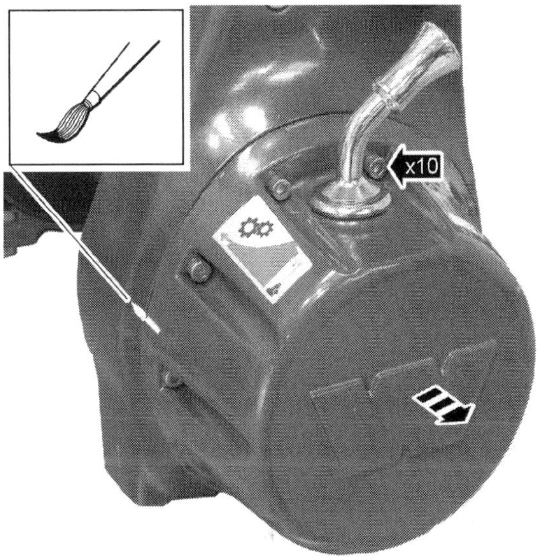

3. NOTE: Mark the components to aid installation.

Torque: 10 Nm.

Torque: 10 Nm.

4. NOTES:

⚠ Remove and discard the gasket.

⚠ Make sure that the mating faces are clean and free of foreign material.

Renew Part: Electric Winch

Installation

1. NOTES:

⚠ Align to the position noted on removal.

⚠ Make sure a new gasket is installed.

To install, reverse the removal procedure.

Winch Motor

Removal and Installation

Removal

⚠ NOTE: Removal steps in this procedure may contain installation details.

1. **Refer to:** Startup Battery Disconnect and Connect (414-01 Battery, Mounting and Cables, General Procedures)
2. **Refer to:** Winch (419-12 Winch, Removal and Installation)

3. Torque: 7 Nm.

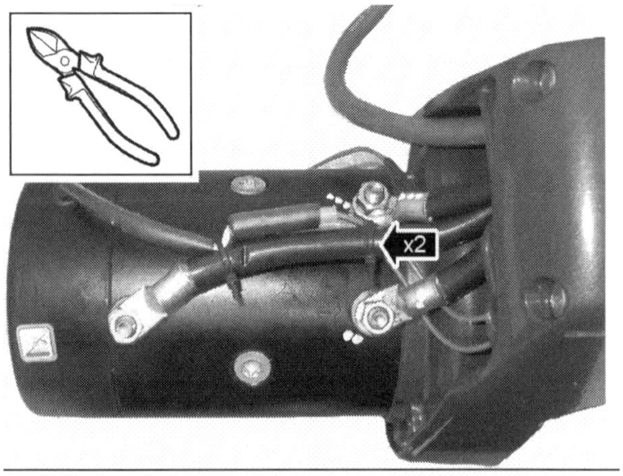

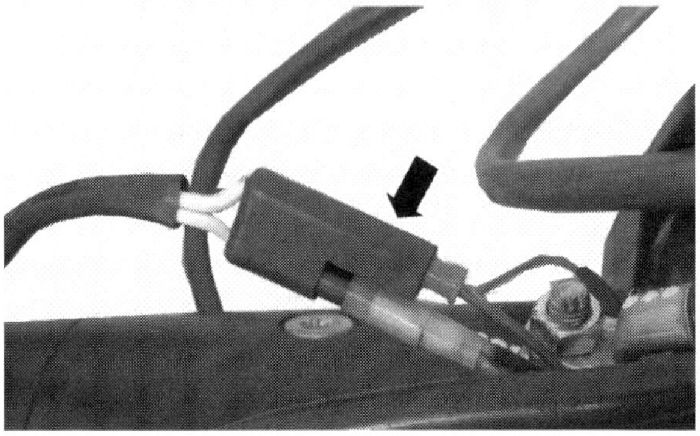

4.

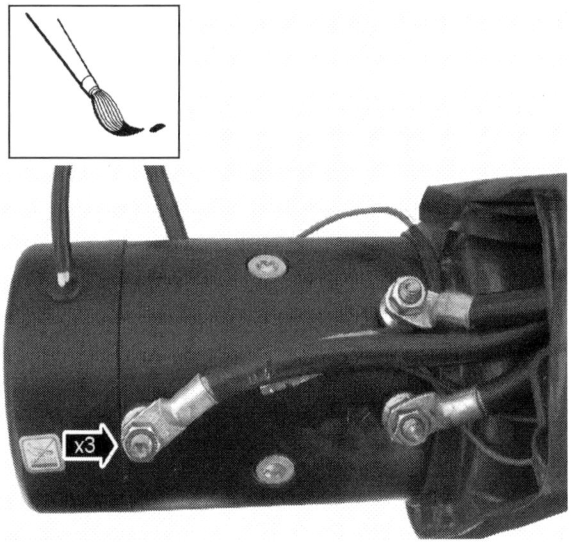

5. CAUTIONS:

⚠ Mark the position of the cables to aid installation.

⚠ To prevent damage to components, use an additional wrench when loosening or tightening terminals.

Torque: 25 Nm.

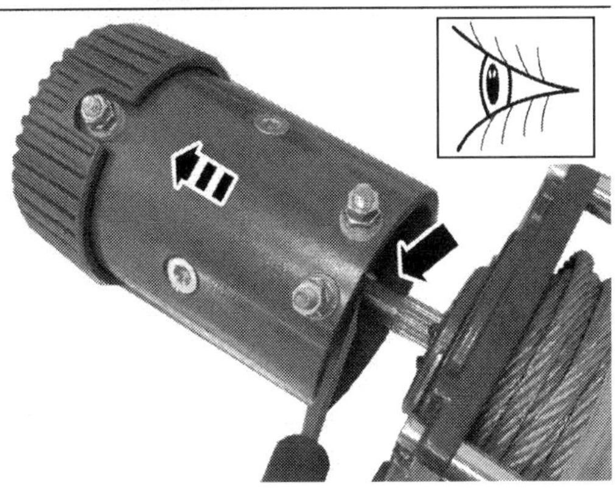

6. CAUTIONS:

⚠ If the armature is dislodged from the motor, the motor brushes will need to be reset to allow reinstallation.

⚠ Make sure that the armature shaft is removed with the motor housing.

⚠ Use a suitable tool if required to move the armature with the motor.

Torque: 8 Nm.

Installation

1. To install, reverse the removal procedure.

Winch Solenoid

Removal and Installation

Removal

⚠ NOTE: Removal steps in this procedure may contain installation details.

1. **Refer to:** Startup Battery Disconnect and Connect (414-01 Battery, Mounting and Cables, General Procedures)

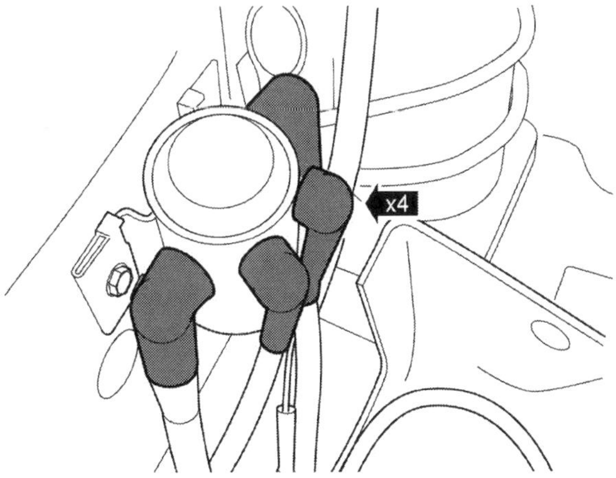

2.

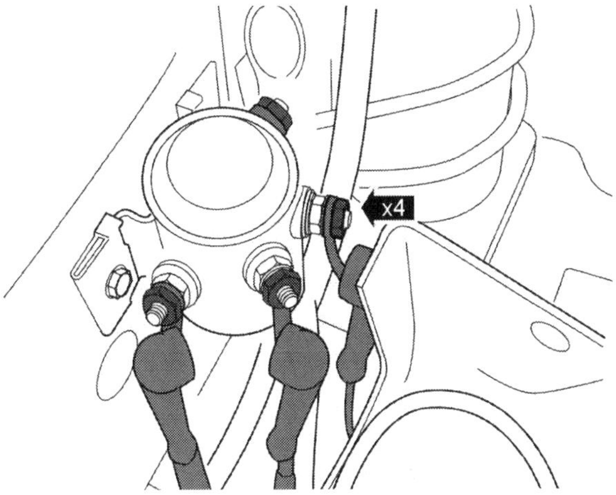

3. CAUTIONS:

⚠ Note the position of the electrical cables prior to disconnection.

⚠ To prevent damage to components, use an additional wrench when loosening or tightening terminals.

Two center terminals 10 Nm

Two outer terminals 7 Nm

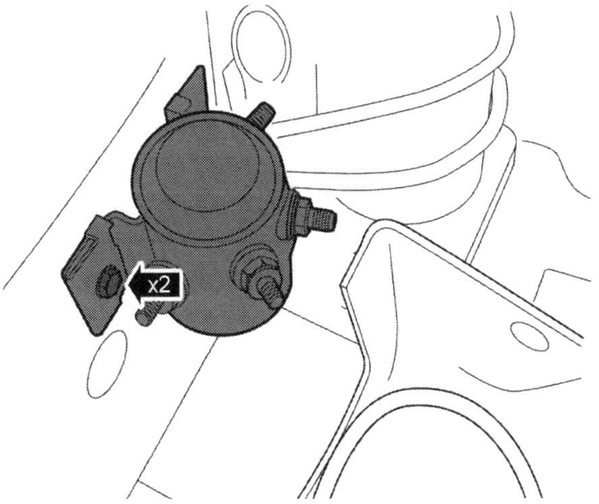

4. Torque: 10 Nm.

Installation

1. To install, reverse the removal procedure.

Front End Body Panels - Fender

Removal and Installation

Removal

1. Remove fender splash shield.

Refer to: Fender Splash Shield (501-02 Front End Body Panels, Removal and Installation)

2. Remove radiator grille.

Refer to: Radiator Grille (501-08 Exterior Trim and Ornamentation, Removal and Installation)

3. Release fixings and remove cooling fan shroud.
4. Remove air cleaner grille from front fender.

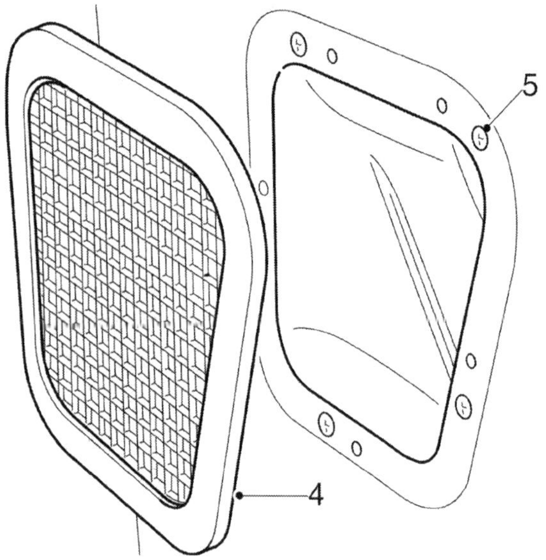

5. Remove 4 screws securing air intake elbow to fender.
6. Remove 7 screws and remove heater air intake grille from fender.
7. Remove 2 bolts securing heater air intake bracket to inner fender. Remove bracket.

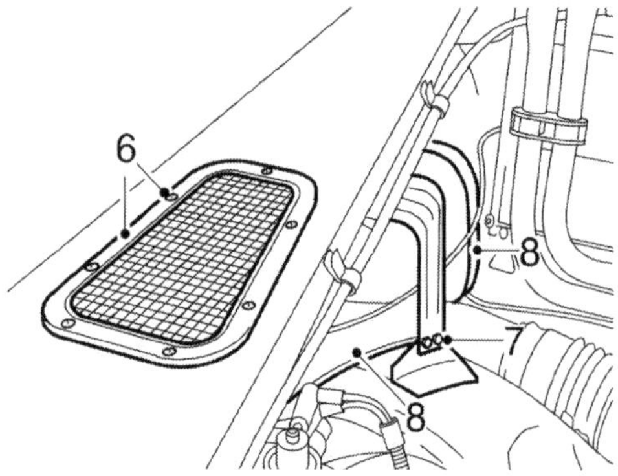

8. Remove heater air intake from fender and collect seal.
9. Remove 2 bolts securing EGR solenoid to inner fender and lay aside.
10. Remove 3 studs securing expansion tank pipe to inner fender.

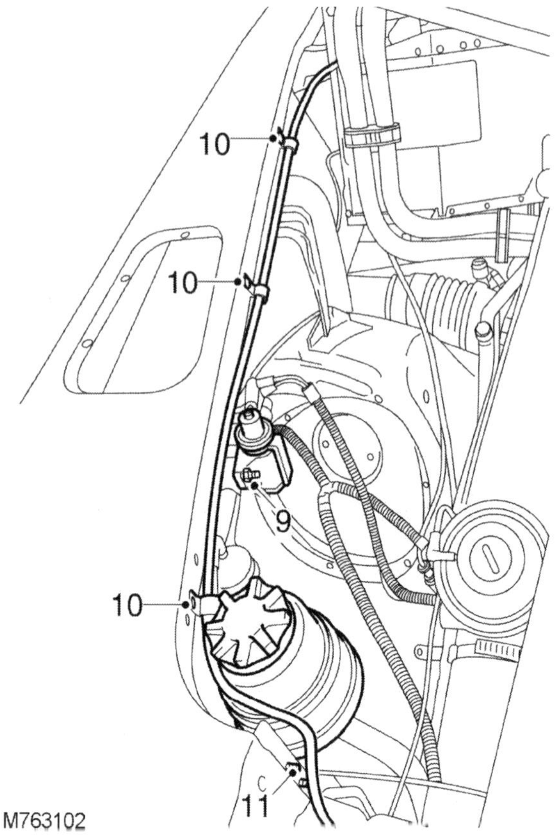

11. Remove 2 bolts securing PAS reservoir to inner fender and lay aside.

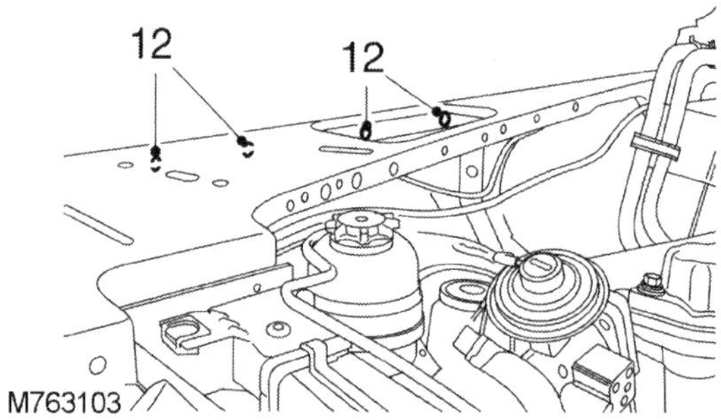

12. Remove 4 bolts securing outer fender to inner fender.

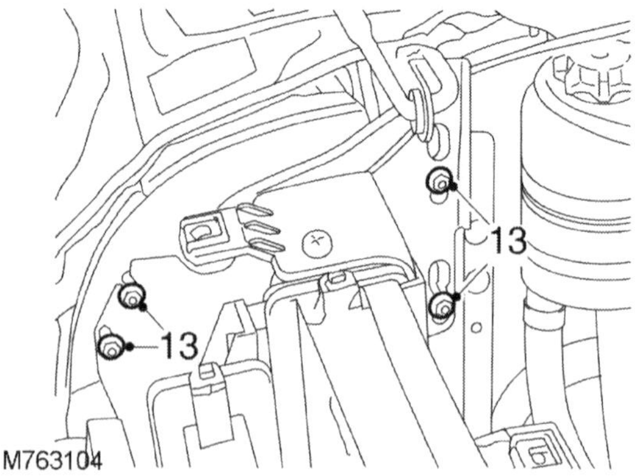

13. Remove 4 bolts securing radiator mounting bracket to fender.
14. Loosen screw and release hood lock inner cable from hood lock.

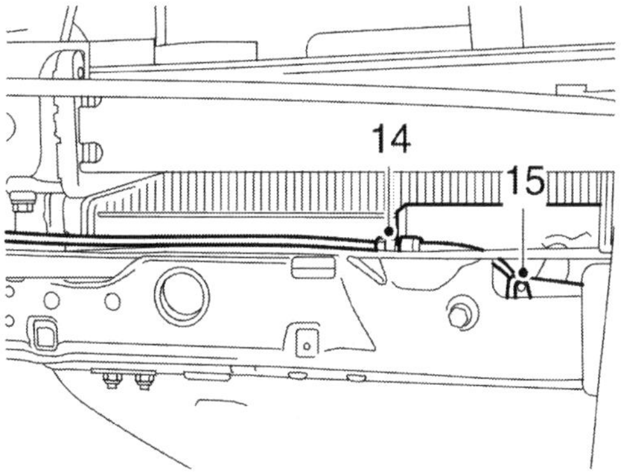

15. Release hood lock outer cable from hood lock.
16. Push centre pins from studs securing wheel arch extension to fender, remove studs and remove wheel arch extension.

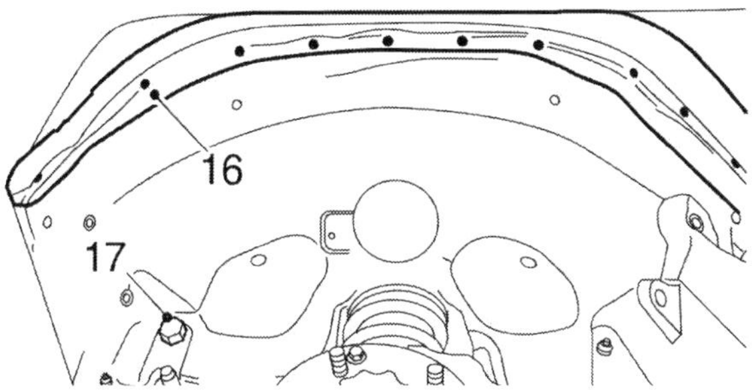

17. Remove bolt securing inner and outer fender to chassis.
18. Remove nut and bolt securing support stay to fender.
19. Remove 4 bolts securing fender to 'A' post.

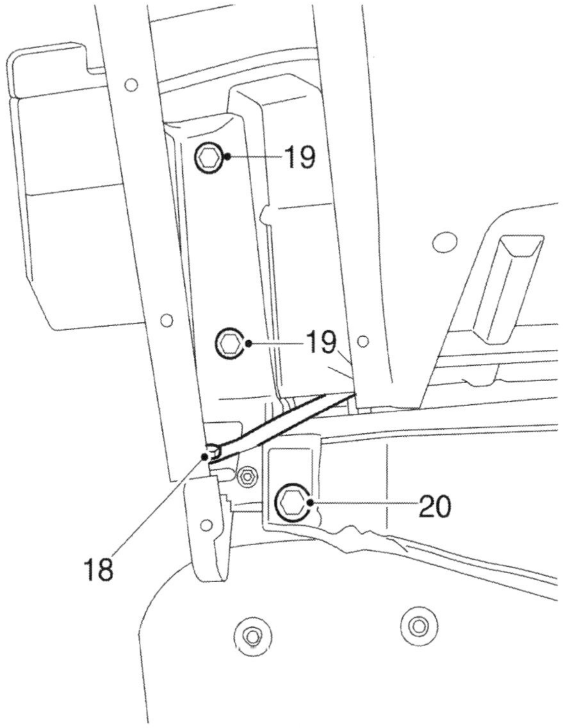

20. Remove bolt securing fender to bulkhead.

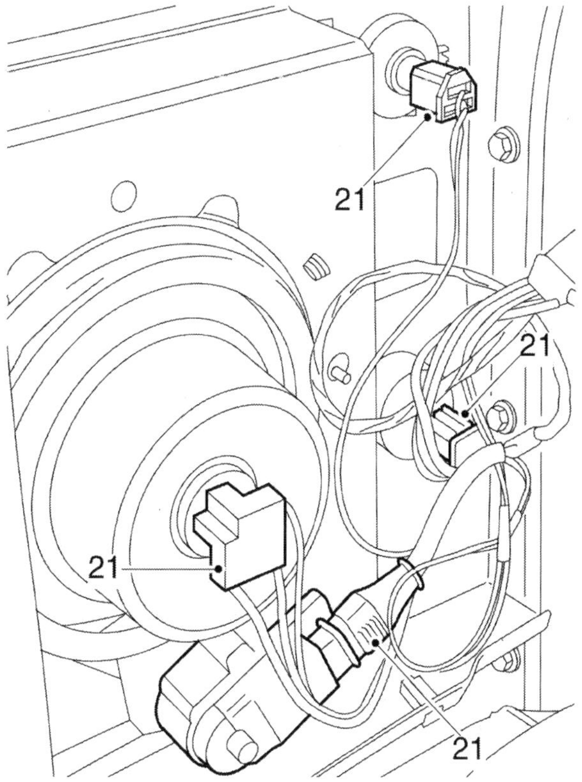

21. Disconnect headlamp, side lamp, repeater and side turn lamp multiplugs.
22. With assistance, remove fender.
23. Remove bolt, remove grille support bracket.
24. Remove 4 screws securing side lamp and side turn lamp.

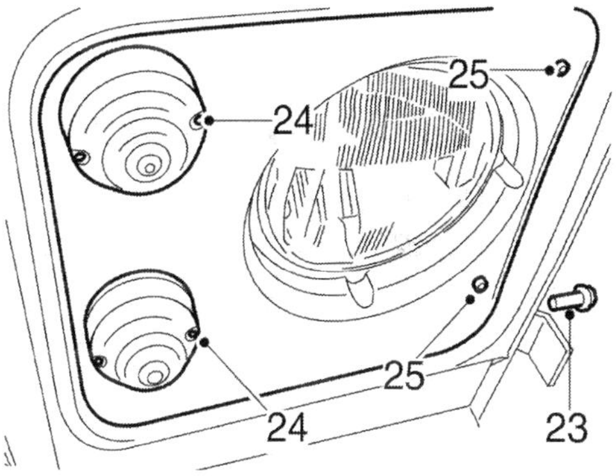

25. Remove 2 screws securing headlamp finisher to fender, remove finisher.

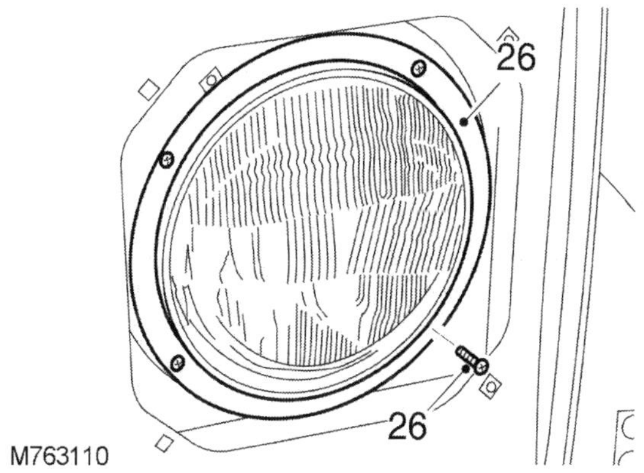

26. Remove 4 screws and remove headlamp from fender, collect sealing ring.

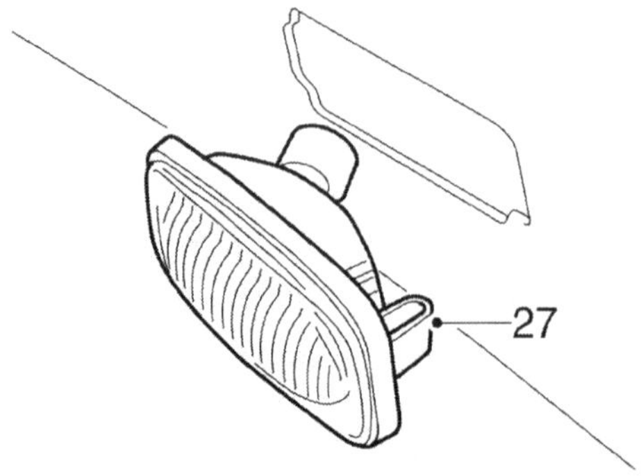

27. Release side repeater lamp from fender.

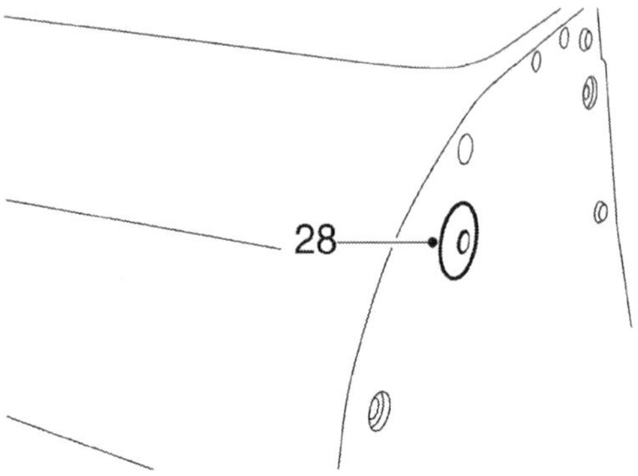

28. Remove hood cable grommet.

Installation

1. Fit hood cable grommet.
2. Fit side repeater lamp to fender.
3. Fit headlamp sealing ring, fit headlamp and tighten screws.
4. Fit headlamp finisher, side lamp and side turn lamps and tighten screws.
5. Fit grille support bracket and tighten bolt.
6. With assistance, position and align fender to body and locate hood cable to fender.
7. Connect headlamp, side lamp, side turn lamp and side repeater lamp multiplugs.
8. Fit bolt securing fender to bulkhead.
9. Fit bolts securing fender to 'A' post.
10. Fit nut and bolt securing support stay to fender.
11. Fit bolt securing inner and outer fender to chassis.
12. Fit bolts securing radiator bracket to fender.
13. Fit bolts securing outer fender to inner fender.
14. Align fender and finally tighten all bolts.
15. Position wheel arch extension to fender, and fit securing studs.
16. Position PAS reservoir to fender and tighten bolts.
17. Position expansion tank pipe and secure with studs.
18. Position EGR solenoid and tighten bolts.
19. Position heater air intake seal and fit air intake.
20. Fit heater air intake bracket and tighten bolts.
21. Fit heater air intake grille and tighten screws.
22. Align air cleaner intake elbow, tighten screws and fit grille.
23. Fit fender splash shield.

Refer to: Fender Splash Shield (501-02 Front End Body Panels, Removal and Installation)

24. Position hood release cable to lock and tighten screw.
25. Fit cooling fan shroud.
26. Fit radiator grille.

Refer to: Radiator Grille (501-08 Exterior Trim and Ornamentation, Removal and Installation)

Fender Splash Shield

Removal and Installation

Removal

1. ⚠ WARNING: Support on safety stands.

Raise front of vehicle.

2. Remove front wheel.
3. Support axle weight with jack.
4. Remove 6 screws securing damper top cover to inner wing, remove cover.
5. Remove nut securing damper to top mounting bracket and collect rubber bush and 2 washers.

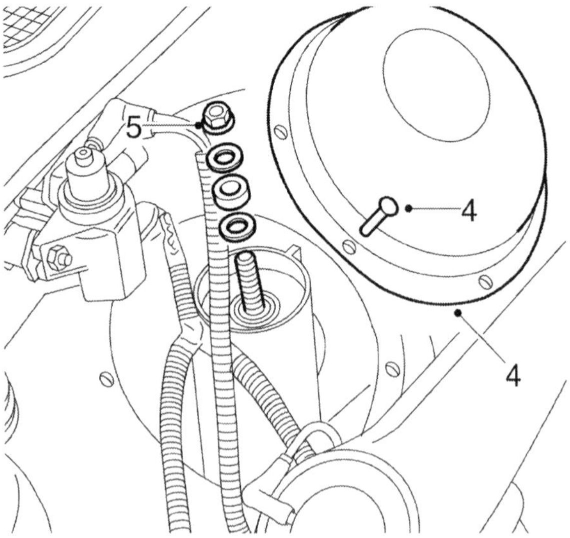

6. Remove 4 nuts securing damper to mounting bracket and collect washers.

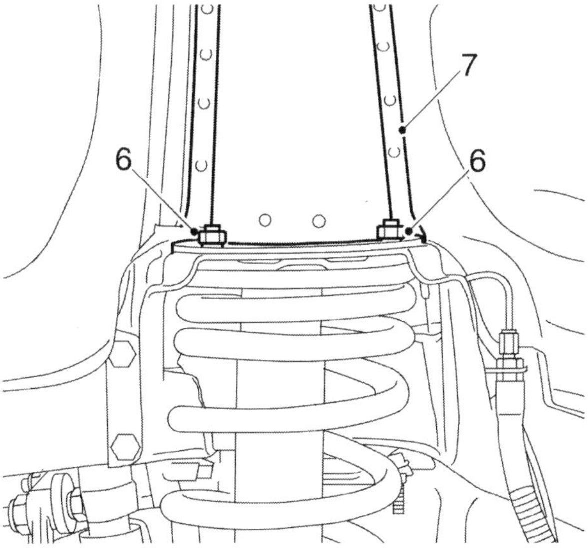

7. Remove damper top cover.

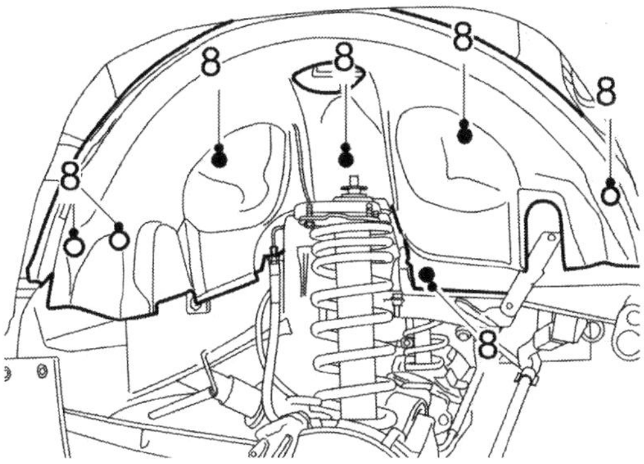

8. Remove 4 screws and 3 studs securing fender splash shield to inner wing. Remove fender splash shield.

Installation

1. Position fender splash shield, secure with screws and studs.
2. Fit damper mounting bracket and tighten nuts.
3. Position damper to mounting bracket, fit rubber bush and washers and tighten nut.
4. Fit damper cover to inner wing and secure with screws.
5. Fit road wheel.
6. Remove stands and lower vehicle.

Hood

Removal and Installation

Removal

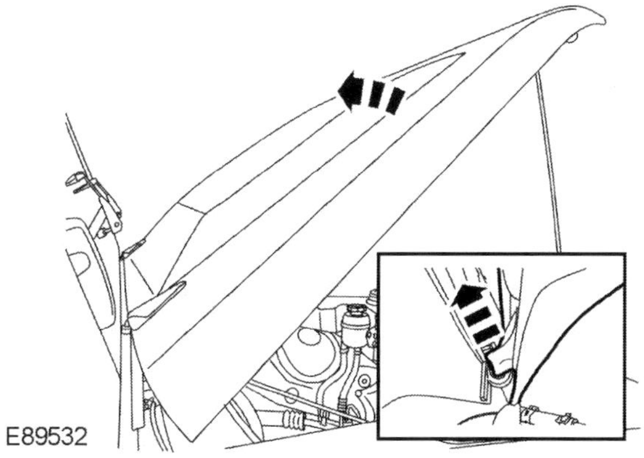

1. Remove the hood.
   - With assistance, remove the hood in an upwards direction.

Installation

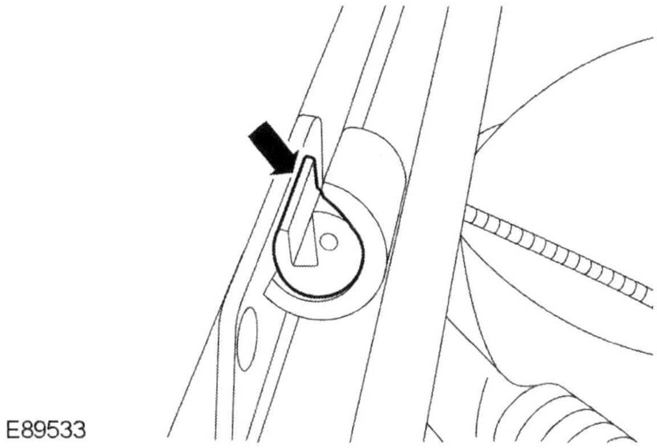

1. To install, reverse the removal procedure.

- Make sure the hood guides are inline with the hinge bracket.

Vehicles Built From: 07/2001

Description and Operation

Front Seat Recline

Front seat recline has been improved by removing the bulkhead situated behind the front seats. This enables the front seats to be reclined further than previous Defender 90's.

This change is only applicable to 90 county station wagon models. Six seat derivatives of this model will provide greater incline of the front seats and increased cabin space.

Soft Top

The soft top is replaced by an improved version manufactured from PVC. The PVC soft top gives the following improvements:

Improvements

- Resistance to shrinkage
- No zips
- Resistant to dust and water ingress
- Weld seams and joints
- Non cotton base construction
- Easy to clean
- Beach bar

A beach bar, currently used on military vehicles only, will be introduced for fitment on all 02MY soft top vehicles. The bar is located behind the front seats and is manufactured from large diameter steel tubing with a diagonal bar which gives additional rigidity.

Tail door

The tail door is now a one piece outer skin pressing located on a reinforced steel, zinc plated frame. This gives improved build quality, water integrity and minimal welding.

A new tail door glass is introduced with a one piece continuous rubber seal. The tail door glass has a revised Heated Rear Window (HRW) element on the glass which removes the need to route wires up the sides of the frame. The HRW element now has integral Lucar connections in the element which provide the earth connections for the new centre high mounted stop lamp.

The inner trim panel is new and accommodates the CDL actuator and linkages.

The new tail door also contains a relay for rear wiper motor operation. Refer to the Wipers section for details.

Because the new tail door uses the current hinges and latch and retains the same geometry as the previous tail door, the new tail door assembly is available as a service item for current vehicles.

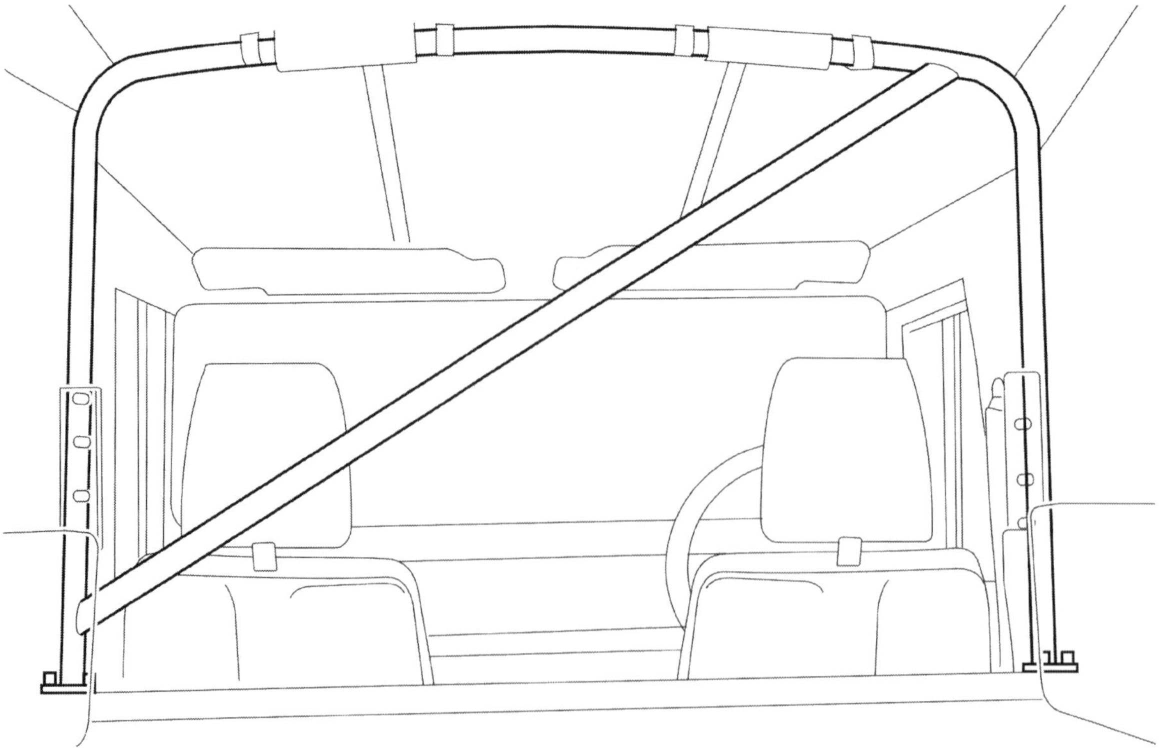

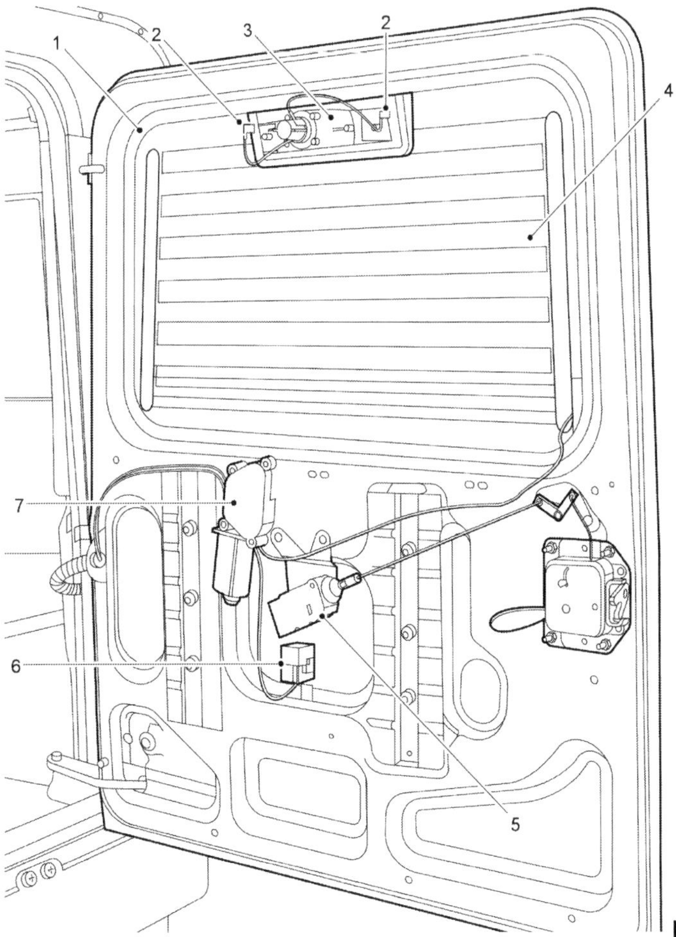

Item	Description
1	Rubber window seal
2	CHMSL earth connection
3	CHMSL
4	Tail door glass and HRW element
5	Tail door CDL actuator
6	Rear wiper relay
7	Wiper motor

Taildoor Striker Adjustment

General Procedures

1. NOTE: Before adjusting the taildoor striker, ensure the spare wheel is fitted to taildoor.

Loosen screws securing taildoor striker.

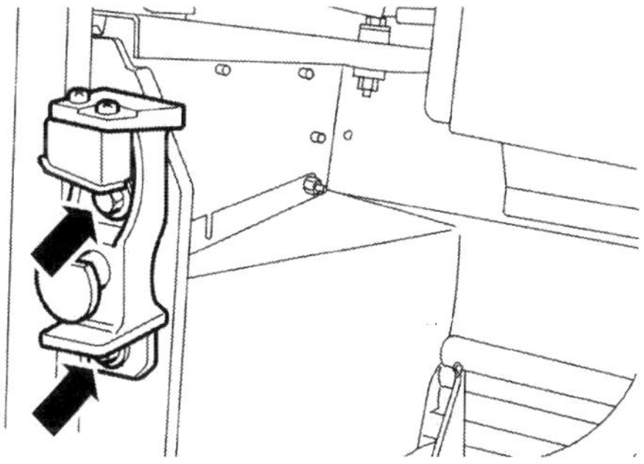

2. Adjust taildoor striker and tighten screws.
3. Close taildoor and check for correct latching with taildoor lock.
4. Carry out further adjustment as necessary. If full adjustment cannot be achieved carry out following.
5. Remove taildoor striker and nut plate.
6. Elongate holes in body metal which is sandwiched between taildoor striker and nut plate.
7. Refit taildoor striker and adjust as necessary.

Door Striker Adjustment

General Procedures

NOTE: This procedure covers adjustment of striker on front and rear doors.

1. Loosen Torx bolts securing striker to B-pillar or C-pillar.

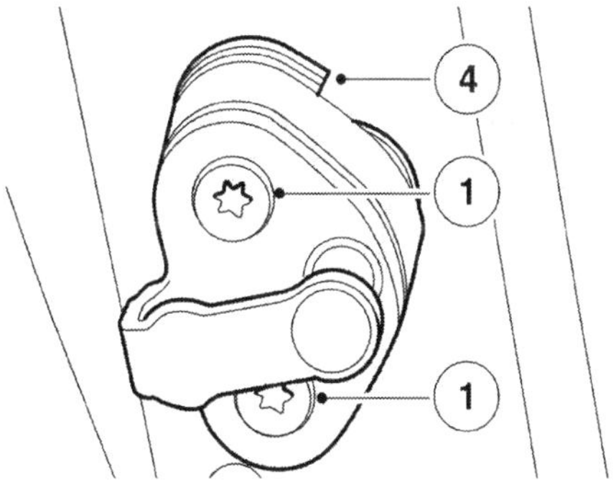

2. Adjust alignment of striker vertically and horizontally, lightly tighten bolts.
3. Close door, check for correct latching with door lock and for centering on striker.
4. Add or remove packing shims as necessary. Fully tighten bolts.
5. Carry out further adjustment as necessary. If full adjustment cannot be achieved carry out following:
6. Remove striker and nut plate.
7. Elongate striker bolt holes in B-pillar or C-pillar to increase movement in direction required.
8. Refit striker and adjust as necessary.

Door

Removal and Installation

Removal

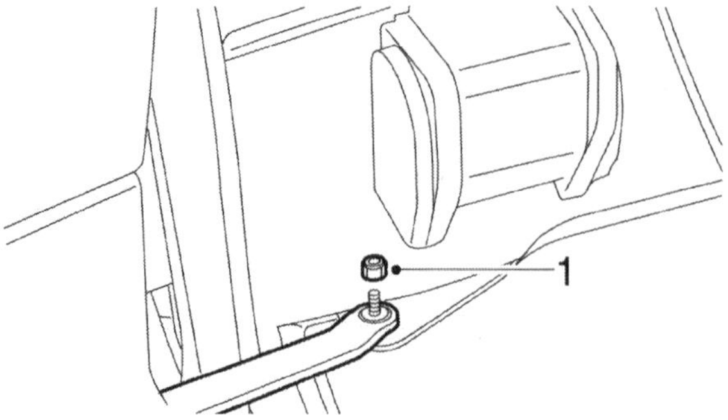

1. Remove nut securing door check strap.

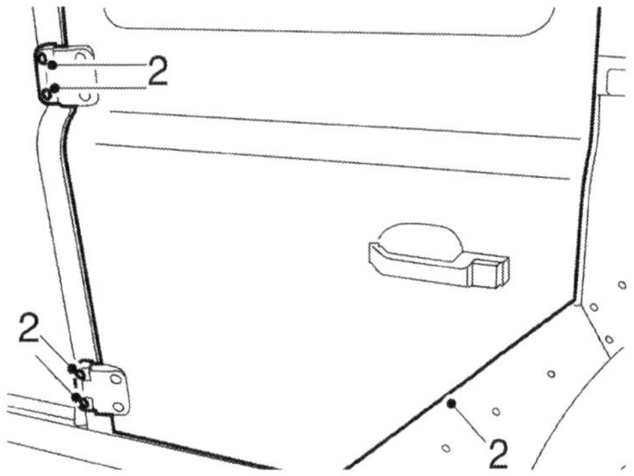

2. With assistance remove 4 bolts securing door to body and remove door.

Installation

1. With assistance position door and fit but do not tighten bolts.
2. Align door to body aperture and tighten bolts. (23 Nm / 17lb-ft)
3. Fit and tighten check strap nut.

Front Door Reinforcement Panel

Removal and Installation

Removal

1. Remove door trim panel and plastic sheet.

Refer to: Front Door Trim Panel (501-05 Interior Trim and Ornamentation, Removal and Installation)

2. Remove plastic sheet.
3. Remove 4 screws securing window regulator to mounting panel.
4. Remove 5 screws securing mounting panel to door frame.

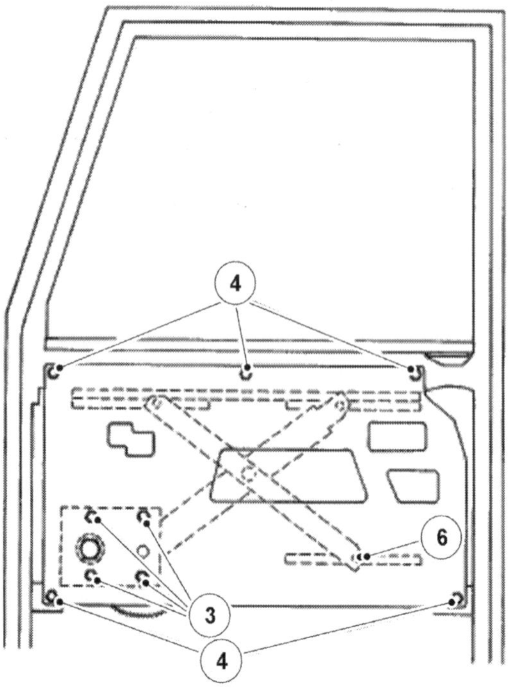

5. Release remote control lever rod from latch mechanism and from plastic clip in mounting panel.

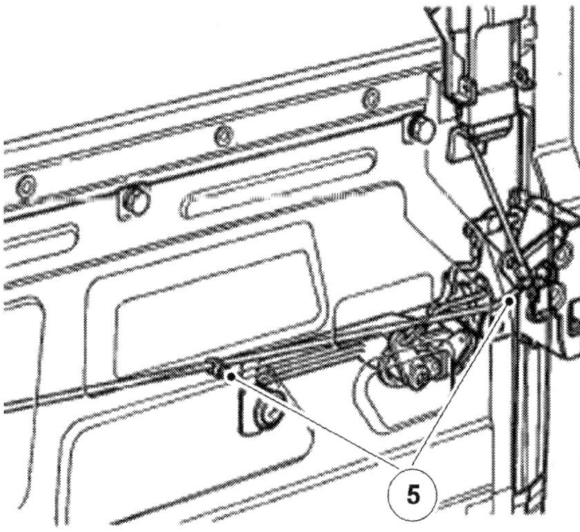

6. Slide window regulator arm from mounting panel channel and remove panel with remote control lever and rod.

Installation

1. Engage window regulator arm in mounting panel channel.
2. Connect remote control rod to latch mechanism and secure with clip.
3. Install mounting panel and retain with 5 screws.
4. Secure window regulator to mounting panel with 4 screws.
5. Raise and lower window to check for free movement.
6. Install plastic sheet, and door trim panel.

Refer to: Front Door Trim Panel (501-05 Interior Trim and Ornamentation, Removal and Installation)

Rear Door Reinforcement Panel

Removal and Installation

Removal

1. Remove rear door trim panel.

Refer to: Rear Door Trim Panel (501-05 Interior Trim and Ornamentation, Removal and Installation)

2. Remove plastic sheet from the door.

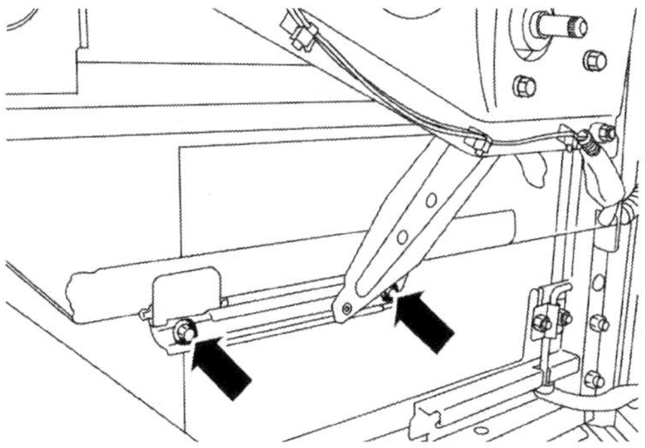

3. Lower the glass and remove 2 bolts securing door glass to regulator.

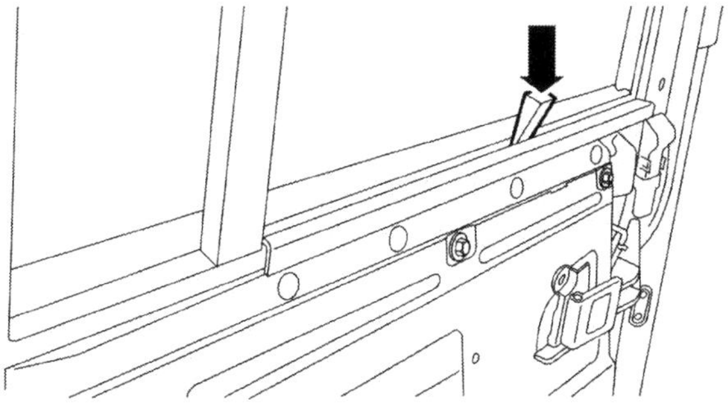

4. Release glass from regulator, raise and wedge in raised position.

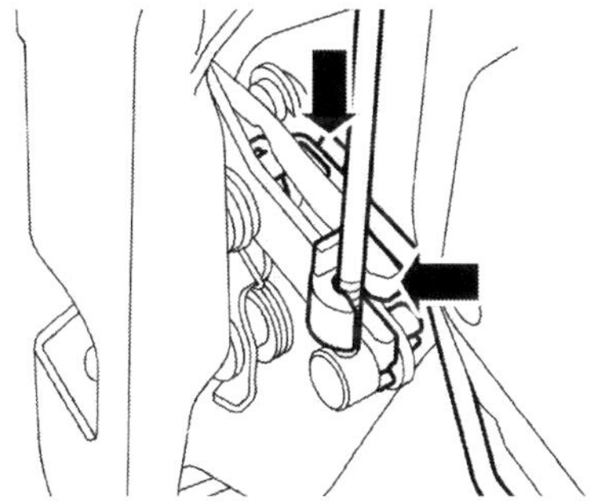

5. Release clip, detach and remove door lock motor link.
6. Release clip and disconnect interior handle link from latch mechanism.

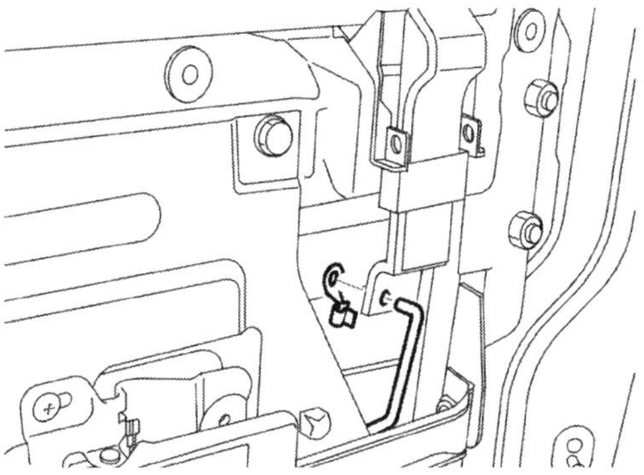

7. Remove clip and disconnect push button link.

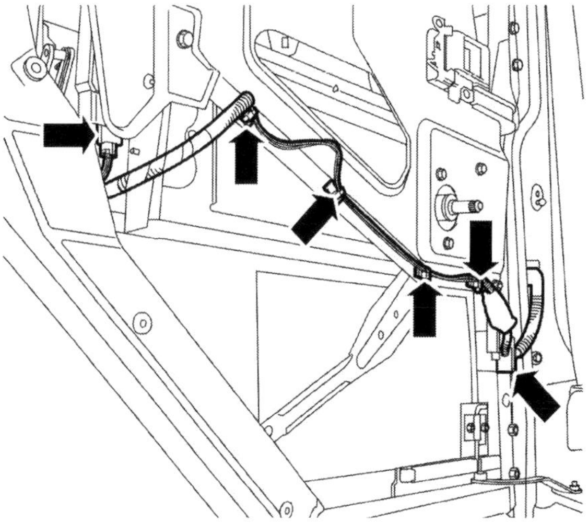

8. Release clips securing harness to door reinforcement panel.
9. Disconnect multiplug from door lock motor.
10. Release harness sleeve and withdraw harness from door.

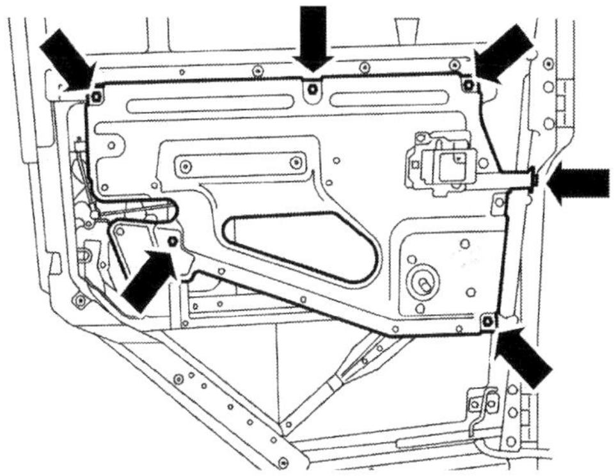

11. Remove 5 bolts and 1 nut securing door reinforcement panel.
12. Remove door reinforcement panel.

Installation

1. Position door reinforcement panel assembly, fit bolts and nut and tighten to 10 Nm (7lbf.ft).
2. Connect multiplug to door lock motor.
3. Secure harness clips and sleeve.
4. Connect interior handle link to latch and secure with clip.
5. Connect link to push button and fit clip.
6. Fit and secure door lock motor link.
7. Remove wedges, lower the glass and locate to regulator. Fit bolts and tighten to 6 Nm (4 lbf.ft).
8. Raise and lower glass to check operation.
9. Fit plastic sheet to the door.
10. Fit rear door trim panel.

Refer to: Rear Door Trim Panel (501-05 Interior Trim and Ornamentation, Removal and Installation)

Taildoor

Removal and Installation

Removal

1. Remove 3 nuts and remove spare wheel.
2. Remove bolt securing check strap to body.
3. Release harness grommet from harness cover, disconnect connectors and multiplug.

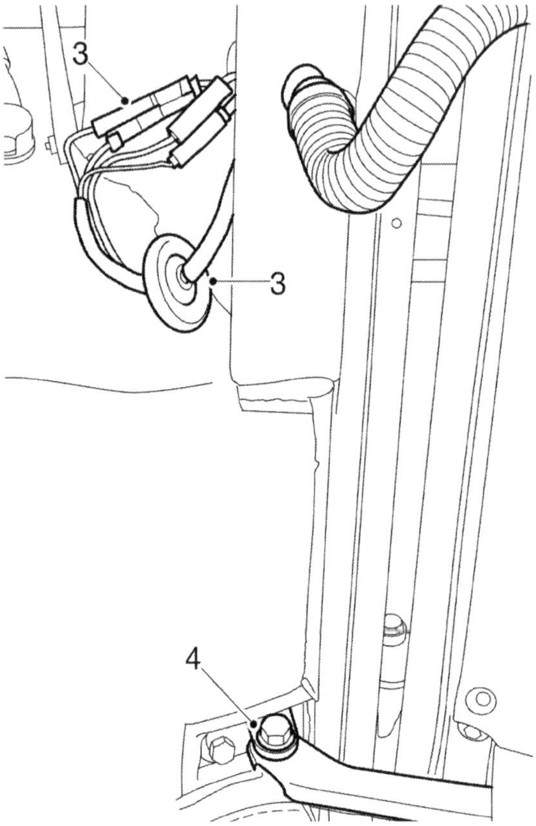

4. Release harness cover and harness from body.
5. With assistance, remove 6 screws securing taildoor hinges to body and remove taildoor.

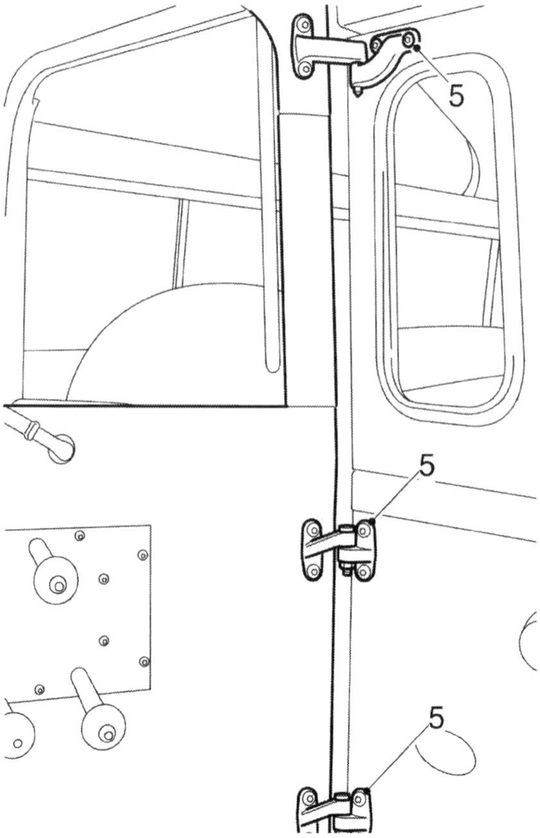

Installation

1. With assistance, position taildoor and fit but do not tighten bolts.
2. Position check strap and tighten bolt.
3. Align taildoor to body and tighten bolts.
4. Position harness through body and secure harness cover.
5. Secure harness connectors and multiplug.
6. Secure harness grommet to body.
7. Position spare wheel to taildoor and tighten nuts.
8. Adjust taildoor striker.

Refer to: Taildoor Striker Adjustment (501-03 Body Closures, General Procedures)

Interior Trim

Description and Operation

Fascia Console

Item	Description
1	Rear fog lamp switch
2	Rear wiper switch
3	Heated front screen switch (if fitted)
4	Hazard warning lamp switch
5	Heated rear window switch
6	Rear washer switch
7	Front fog lamps switch (if fitted)
8	RH front window switch
9	RH heated front seat switch (if fitted)
10	Headlamp levelling control switch
11	Cigar lighter (if fitted)
12	LH heated front seat switch (if fitted)
13	LH front window switch

The console comprises a plastic moulding secured to the fascia with five screws. The new console allows existing switches to be repositioned and has provided the opportunity to commonise the switches with other Land Rover models.

A central cut-out in the console provides for the installation of the radio, where fitted, or a coin tray for markets without a standard radio installation.

Cowl Side Trim Panel

Removal and Installation

Removal

1. Raise the seat cushion and fit stowage strap.
2. Remove the rear quarter trim panel.

Refer to: Rear Quarter Trim Panel (501-05 Interior Trim and Ornamentation, Removal and Installation)

3. Remove the rear door speaker.

Refer to: Rear Door Speaker (415-03 Speakers, Removal and Installation)

4. Peel back carpet and remove 2 screws securing cowl side trim panel to wheel arch.
5. Prise top edge of cowl side trim panel down to release from body channel.
6. Remove the cowl side trim panel.

Installation

1. Position the cowl side trim panel on wheel arch.
2. Press top edge of cowl side trim panel down firmly and locate under body channel.

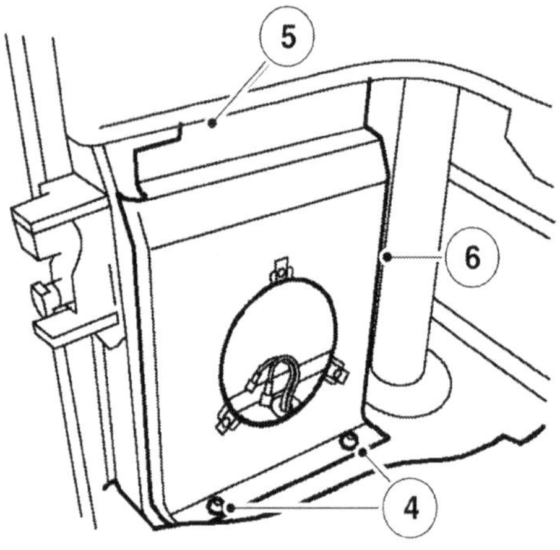

3. Secure cowl side trim panel to wheel arch with screws and reposition carpet.
4. Fit the rear door speaker.

Refer to: Rear Door Speaker (415-03 Speakers, Removal and Installation)

5. Fit the rear quarter trim panel.

Refer to: Rear Quarter Trim Panel (501-05 Interior Trim and Ornamentation, Removal and Installation)

C-Pillar Upper Trim Panel - 90/Station Wagon

Removal and Installation

Removal

1. Remove rear seat squab.

Refer to: Rear Seat Backrest (501-10 Seating, Removal and Installation)

2. Remove rear quarter trim panel.

Refer to: Rear Quarter Trim Panel (501-05 Interior Trim and Ornamentation, Removal and Installation)

3. Prise cap from front seat belt upper guide retaining bolt.
4. Unscrew retaining bolt and remove seat belt from 'B' pillar.
5. Carefully prise trim stud from 'B' pillar.
6. Remove 4 bolts securing side trim panel to body side.
7. Release top edge of side trim panel by striking upwards with the hand to disengage 3 spring clips.

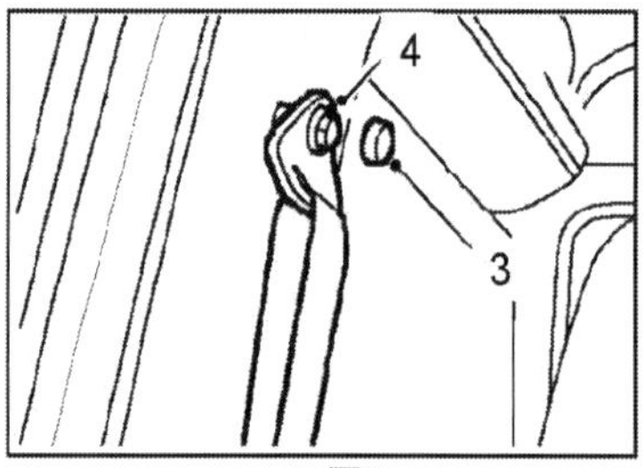

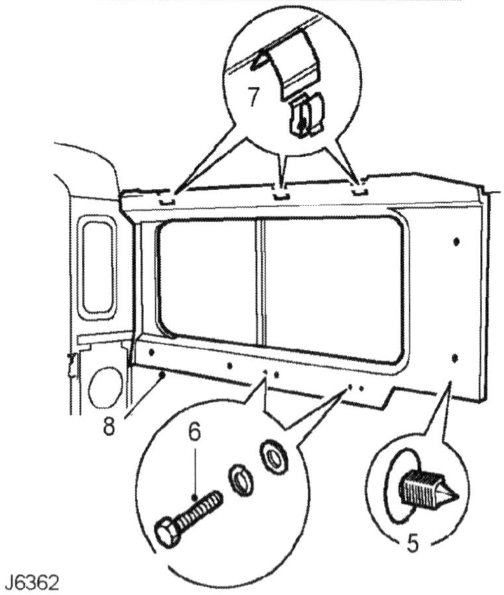

8. Lift trim panel from window surround and remove from vehicle.

Installation

1. Install new spring clips to the 3 brackets on top edge of side trim panel.
2. Align trim panel, ensuring it is correctly positioned around side window.

3. Locate trim panel spring clips on inner cant rail and press firmly down to secure top edge of trim panel.
4. Align bosses of trim panel with body side fixing holes and secure with 4 bolts and washers. Do not fully tighten at this stage.
5. Install trim stud to secure bottom edge of trim panel to 'B' pillar.
6. Secure seat belt guide to 'B' pillar, tighten bolt to 32 Nm (24 lbf.ft) and install cap.
7. Install rear quarter trim panel.

Refer to: Rear Quarter Trim Panel (501-05 Interior Trim and Ornamentation, Removal and Installation)

8. Install rear seat squab.

Refer to: Rear Seat Backrest (501-10 Seating, Removal and Installation)

C-Pillar Upper Trim Panel - 110/ Station Wagon

Removal and Installation

Removal

1. Remove rear bench seat.

Refer to: Third Row Seat (501-10 Seating, Removal and Installation)

2. Remove rear quarter trim panel.

Refer to: Rear Quarter Trim Panel (501-05 Interior Trim and Ornamentation, Removal and Installation)

3. Prise finisher cap from rear seat belt upper guide bracket retaining bolt.
4. Unscrew bolt and remove guide bracket from 'C' pillar.

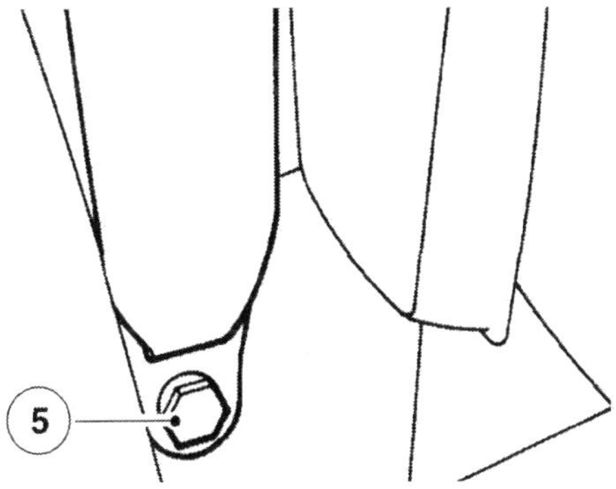

5. Remove bolt securing seat belt to wheel arch mounting.

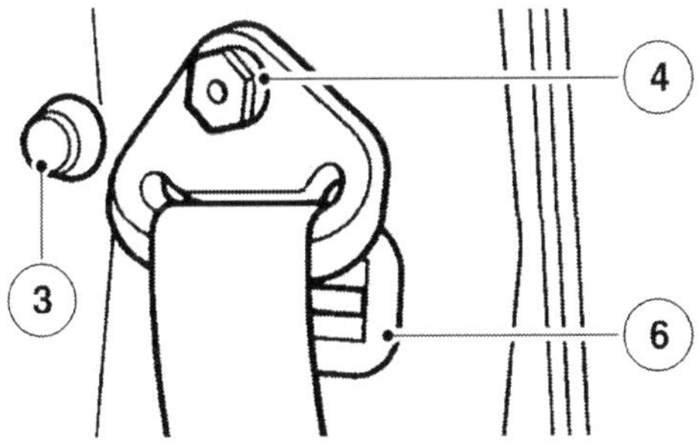

6. Prise trim panel cap fastener from 'C' pillar.
7. Remove seat belt aperture finisher from side trim panel.
8. Remove screw and release seat belt clip from 'C' pillar.
9. Remove fixings securing bottom edge of trim panel to body side.
10. Release top edge of trim panel by striking upwards with the hand to disengage 3 spring clips.

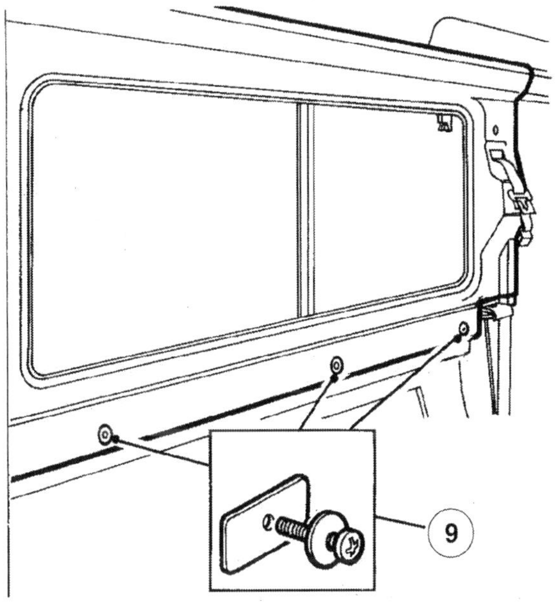

11. Lift trim panel from window surround and feed seat belt through panel aperture.
12. Remove trim panel from vehicle.

Installation

1. Install new spring clips to the 3 brackets on top edge of side trim panel.
2. Feed seat belt through aperture in trim panel and install finisher.
3. Align trim panel, ensuring it is correctly positioned around side window.
4. Locate trim panel spring clips on inner cant rail and press firmly down to secure top edge of trim panel.
5. Align bosses of trim panel with body side fixing holes and secure with bolts or screws. If individual type seats are fitted do not fully tighten fixings at this stage.
6. Install seat belt clip to 'C' pillar.

7. Install trim panel cap fastener to 'C' pillar.
8. Install seat belt to wheel arch. Tighten bolt to 32 Nm (24 lbf.ft).
9. Install seat belt guide bracket to 'C' pillar. Tighten bolt to 32 Nm (24 lbf.ft) and fit finisher cap.
10. Install rear quarter trim panel.

Refer to: Rear Quarter Trim Panel (501-05 Interior Trim and Ornamentation, Removal and Installation)

11. Install rear bench seat.

Refer to: Third Row Seat (501-10 Seating, Removal and Installation)

Front Door Trim Panel

Removal and Installation

Removal

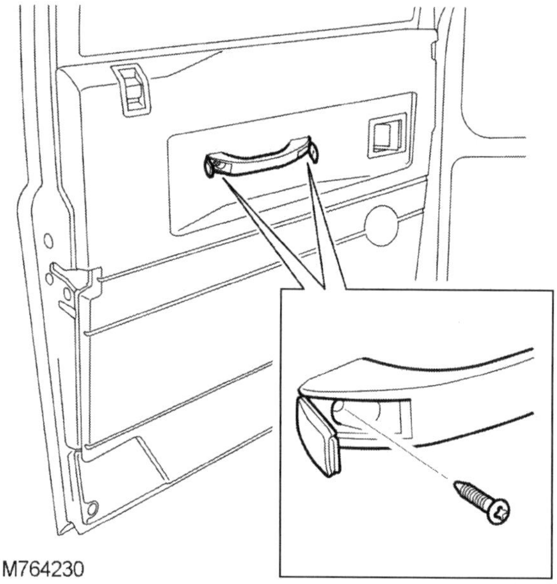

M764230

1. Remove 2 screws securing door pull and remove door pull.

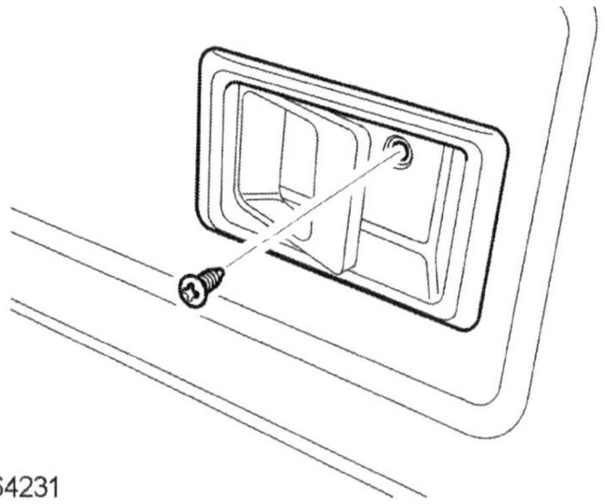

M764231

2. Remove screw securing door release escutcheon and remove escutcheon.
3. Remove sill button guide.

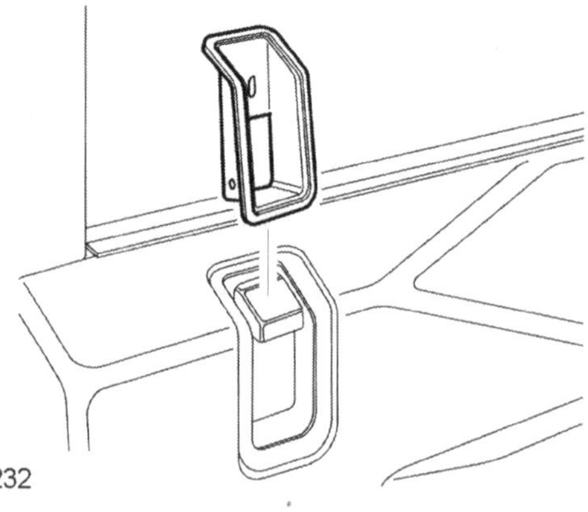

M764232

4. Remove 2 studs securing trim casing.
5. Using a trim removal tool, release 11 fasteners and remove trim casing.
6. △ NOTE: Do not carry out further dismantling if component is removed for access only.

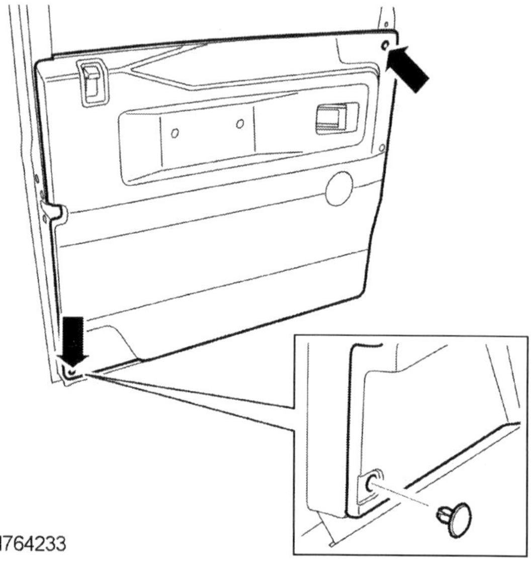

M764233

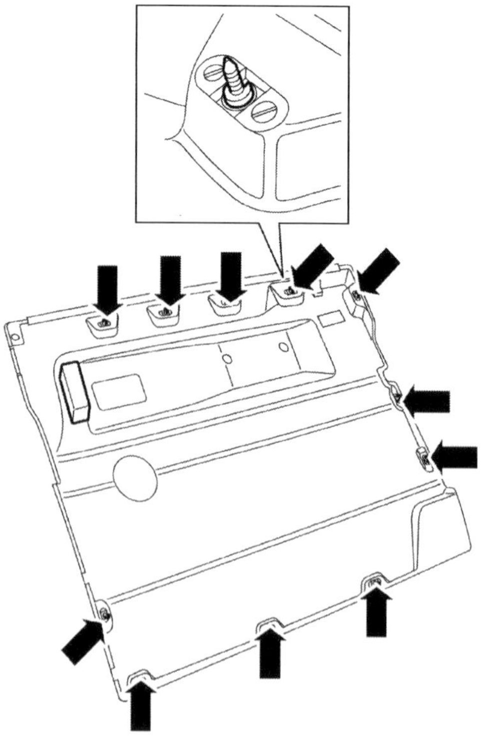

Remove fasteners and anti-vibration pad from trim casing

Installation

1. Install anti-vibration pad and fasteners to trim casing.
2. Position trim casing, align fasteners with holes in door and press firmly into position.
3. Install trim studs to casing.
4. Install sill button guide.
5. Install escutcheon to door handle and secure with screw.
6. Install door pull and secure with screws.

Front Headliner - 90/Station Wagon

Removal and Installation

Removal

1. Remove sun roof headlining finisher.
2. Remove sun visors.

Refer to: Sun Visor (501-05 Interior Trim and Ornamentation, Removal and Installation)

3. Remove interior lamp.

Refer to: Interior Lamp (417-02 Interior Lighting, Removal and Installation)

4. Remove interior mirror and mounting plate.

Refer to: Interior Mirror (501-09 Rear View Mirrors, Removal and Installation)

5. Remove 4 retaining screws and detach both 'A' post trims.
6. Remove caps and unscrew front seat belt fixing bolts from 'B' posts.
7. Carefully prise side trim panel cap fastener from 'B' posts.
8. Carefully prise 2 cap fasteners, from both sides, securing headlining to body at door aperture.
9. Carefully prise out 5 cap fasteners securing front and rear headlinings to roof.
10. From both sides, pull side trim panel inwards enough to release rear corners of front headlining.
11. ⚠ NOTE: Take care not to bend the headlining on removal and refitting.

Lower headlining and remove from vehicle.

Installation

1. With assistance, raise headlining to roof.
2. Carefully pull side trim panel inwards, on both sides, and slide headlining behind trim panel.
3. Position front headlining into recess of rear headlining and secure both to roof mounting brackets with 5 cap fasteners.
4. Secure both sides of front headlining to body at door apertures with cap fasteners.
5. Fit side cap fasteners at 'B' posts.
6. Secure seat belts to 'B' posts and tighten bolts to 32 Nm (24 lbf/ft). Fit caps to bolts.
7. Fit 'A' post trims.
8. Fit interior mirror.

Refer to: Interior Mirror (501-09 Rear View Mirrors, Removal and Installation)

9. Fit interior lamp.

Refer to: Interior Lamp (417-02 Interior Lighting, Removal and Installation)

10. Fit sun visors.

Refer to: Sun Visor (501-05 Interior Trim and Ornamentation, Removal and Installation)

11. Fit sun roof headlining finisher.

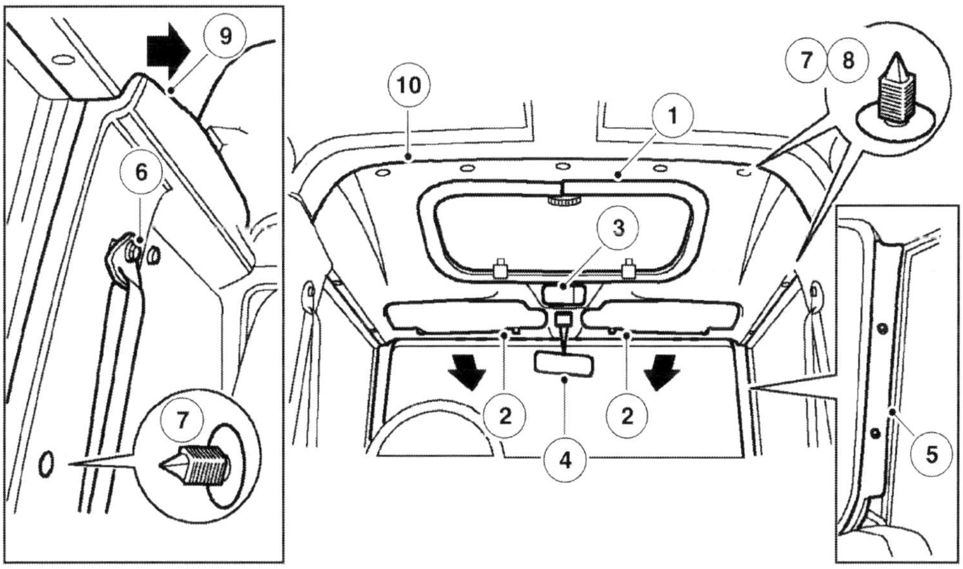

Headliner Rear Trim Panel - 90/110/ Station Wagon

Removal and Installation

Removal

1. Remove rear quarter trim panel.

Refer to: Rear Quarter Trim Panel (501-05 Interior Trim and Ornamentation, Removal and Installation)

2. Remove rear grab handles.

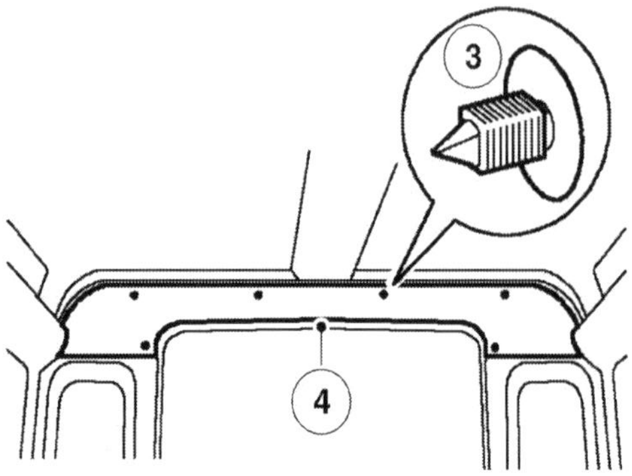

Refer to: Rear Grab Handle - 90/110/Station Wagon (501-05 Interior Trim and Ornamentation, Removal and Installation)

3. Carefully prise out 6 trim studs securing rear end lining to mounting brackets.
4. Remove headliner rear trim panel.

Installation

1. Position rear end lining on mounting brackets and secure with 6 trim studs.
2. Install rear grab handles.

Refer to: Rear Grab Handle - 90/110/Station Wagon (501-05 Interior Trim and Ornamentation, Removal and Installation)

3. Install rear end trim panel.

Refer to: Rear Quarter Trim Panel (501-05 Interior Trim and Ornamentation, Removal and Installation)

Interior Rollover Bar

Removal and Installation

Removal

1. Remove rear seat cushions.

Refer to: Rear Outer Seat Cushion (501-10, Removal and Installation).

2. Remove cowl side trim panels.

Refer to: Cowl Side Trim Panel (501-05 Interior Trim and Ornamentation, Removal and Installation)

3. Remove C-pillar upper trim panel.

Refer to: C-Pillar Upper Trim Panel - 4.0L, 90 (501-05, Removal and Installation).

Refer to: Cowl Side Trim Panel (501-05 Interior Trim and Ornamentation, Removal and Installation)

4. Remove floor console extension.

Refer to: Floor Console Extension (501-12, Removal and Installation).

5. Remove 4 bolts securing centre section of rollover bar and seat belt stalks to floor.
6. Remove 2 torx bolts and nuts from both sides and detach centre section of rollover bar assembly.
7. Remove torx bolt, from both sides, securing rollover bar to body side mounting brackets.
8. Remove 3 bolts and washers, from both sides, securing lower rollover bar brackets to front of wheelarch.
9. Remove 2 torx bolts and washers, from both sides, securing rollover bar upper mountings to roof fixing brackets and carefully lower complete assembly to floor.
10. Unscrew retaining bolts, from both sides, and remove front seat belt inertia reels and rear seat belts from rollover bar mounting.
11. Supporting from both sides, raise and tilt rollover bar rearwards to clear wheel arches.
12. ⚠ NOTE: Assistance may be required to remove rollover bar.

Remove rollover bar from vehicle.

Installation

1. Supporting from both sides, position rollover bar, engage struts in rear corners and front of side panels. Align on floor.
2. Secure front seat belt inertia reels and rear seat belts to rollover bar mountings. Tighten bolts to 32 Nm (24 lbf.ft).
3. Raise rollover bar assembly and secure to roof mounting brackets. Tighten bolts to 45 Nm (33 lbf. ft).
4. Secure rollover bar to wheel arches. Tighten bolts to 45 Nm (33 lbf.ft).
5. Secure rollover bar to body side mounting brackets, tightening bolts to 45 Nm (33 lbf.ft).
6. Secure centre section to main rollover bar, but do not fully tighten fixing bolts at this stage.
7. Secure seat belt stalks and rollover bar floor mounting brackets to chassis. Tighten bolts to 32 Nm (24 lbf ft.).
8. Fully tighten rollover bar centre section retaining bolts to 45 Nm (33 lbf.ft).

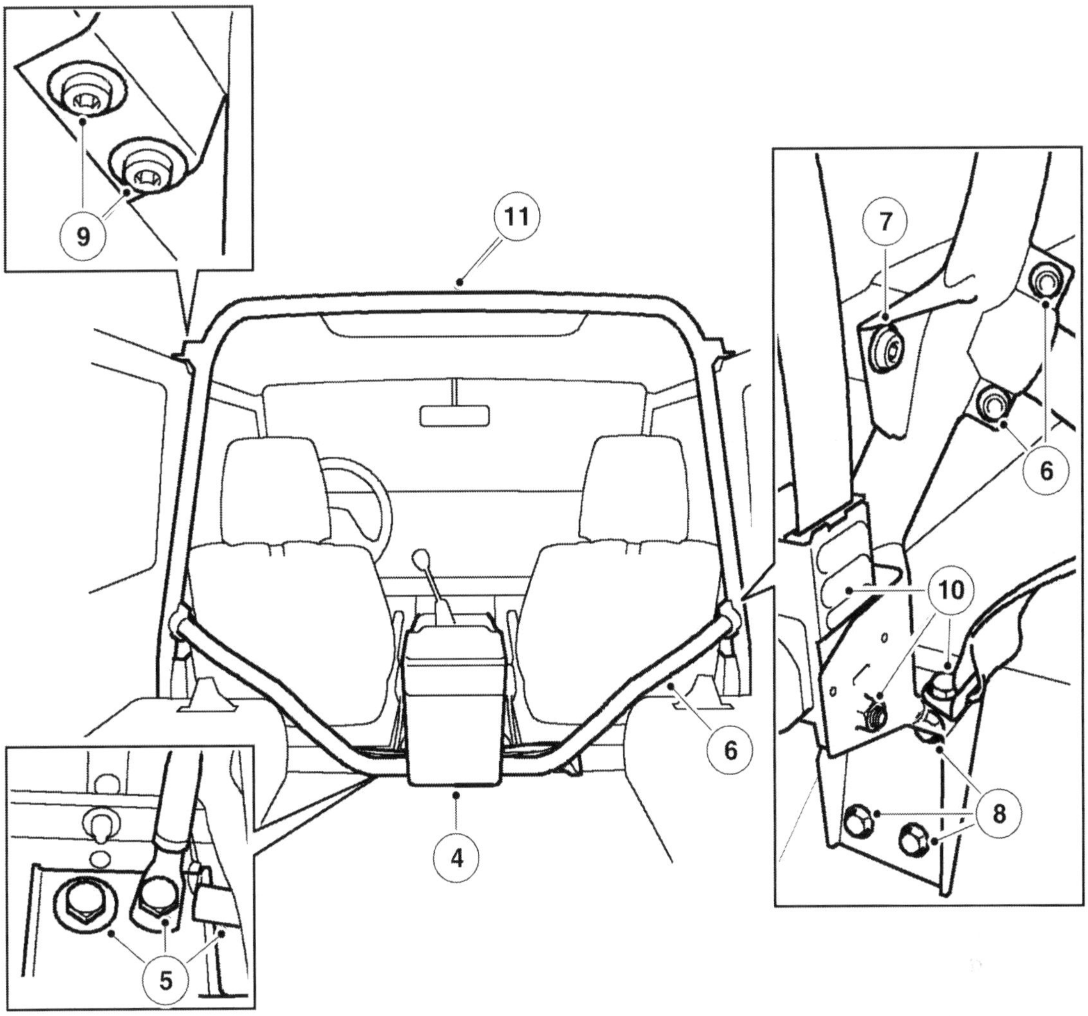

9. Fit floor console extension.

Refer to: Floor Console Extension (501-12, Removal and Installation).

10. Fit C-pillar upper trim panel.

Refer to: C-Pillar Upper Trim Panel - 4.0L, 90 (501-05, Removal and Installation).

Refer to: Cowl Side Trim Panel (501-05 Interior Trim and Ornamentation, Removal and Installation)

11. Fit cowl side trim panels.

Refer to: Cowl Side Trim Panel (501-05 Interior Trim and Ornamentation, Removal and Installation)

12. Fit rear seat cushions.

Refer to: Rear Seat Cushion (501-10 Seating, Removal and Installation)

Rear Door Trim Panel

Removal and Installation

Removal

1. Remove screw securing door handle escutcheon.
2. Remove escutcheon.
3. Remove interior lock button bezel.
4. Remove retaining clip and remove door glass regulator handle.

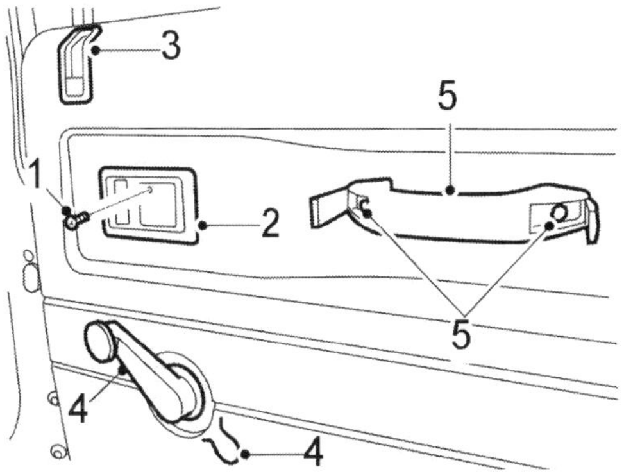

5. Release covers and remove 2 screws securing door pull to door.
6. Carefully release 12 clips securing trim casing to door and remove door casing.

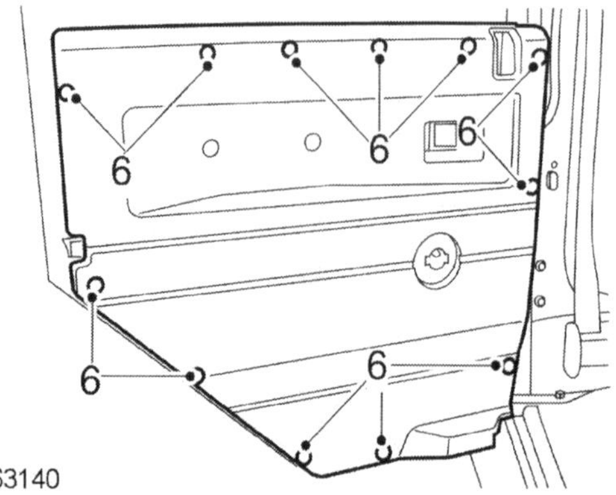

M763140

Installation

1. Install trim casing to door and secure with clips.
2. Install door glass regulator handle.
3. Install escutcheon to door handle and tighten screw.
4. Install door lock button bezel.
5. Install door pull and secure with screws.

Rear Grab Handle - 90/110/Station Wagon

Removal and Installation

Removal

1. Carefully prise top and bottom edges of finisher caps from grab handle and then hinge outwards to gain access to fixing screws.

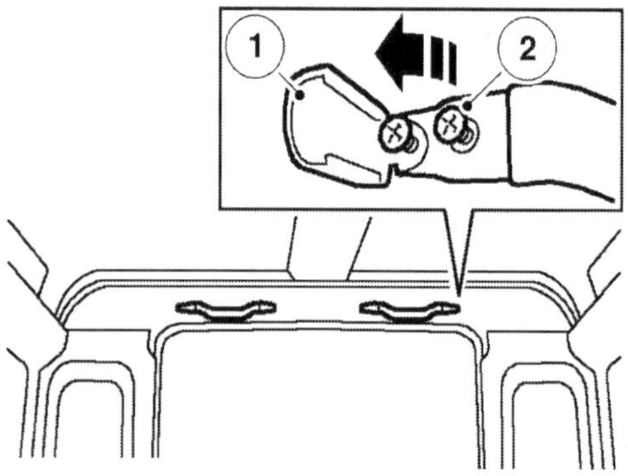

2. Remove 4 fixing screws and detach grab handle from rear end trim panel.

Installation

1. Position grab handle on rear end trim panel and secure with 4 fixing screws.
2. Press finisher caps over screws.

Rear Headliner - 90/Station Wagon

Removal and Installation

Part(s)

STEP	PART NAME	QUANTITY
Installation Step 5	Renew Part Retaining clip(s)	1

Removal

1. Remove C-Pillar upper trim panels.

Refer to: C-Pillar Upper Trim Panel - 90/Station Wagon (501-05 Interior Trim and Ornamentation, Removal and Installation)

2. Remove headliner rear trim panel.

Refer to: Headliner Rear Trim Panel - 90/110/Station Wagon (501-05 Interior Trim and Ornamentation, Removal and Installation)

3. △ NOTE: If required:

Remove rear interior lamp.

Refer to: Interior Lamp (417-02 Interior Lighting, Removal and Installation)

4. Remove the trim clips from the headliner and discard.
5. Release front corners of headliner from cant rail, on both sides.
6. Pull headliner forwards sufficiently to clear rear end lining mounting brackets.
7. Release rear corners of headliner from cant rail, on both sides.
8. △ NOTE: Take care not to bend headliner on removal and refitting.

Remove the rear headliner from roof and remove from vehicle.

Installation

1. With assistance, position rear headliner to roof and locate rear corners in cant rail.
2. Push headliner rearwards and position over rear end lining mounting brackets.
3. Locate front corners of headliner in cant rail.
4. Position front edge of rear headliner over front headliner.
5. Install new trim clips.

Renew Part: Retaining clip(s)

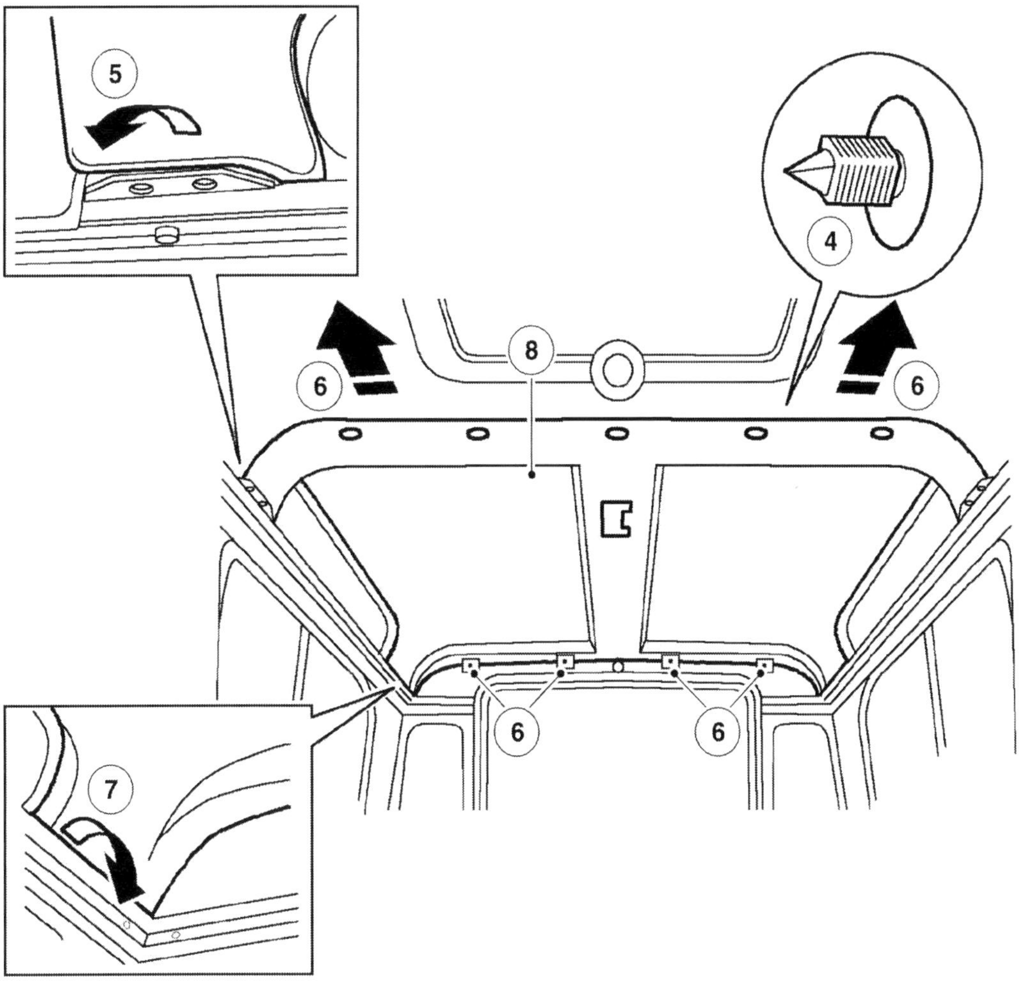

6. Adjust rear headliner to achieve good fit at all corners.

7. △ NOTE: If required:

Install rear interior lamp.

Refer to: Interior Lamp (417-02 Interior Lighting, Removal and Installation)

8. Install headliner rear trim panel.

Refer to: Headliner Rear Trim Panel - 90/110/Station Wagon (501-05 Interior Trim and Ornamentation, Removal and Installation)

9. Install C-pillar upper trim panels.

Refer to: C-Pillar Upper Trim Panel - 90/Station Wagon (501-05 Interior Trim and Ornamentation, Removal and Installation)

Rear Quarter Trim Panel

Removal and Installation

Removal

1. Pull edge of rear quarter trim panel firmly to release 2 spring retaining clips from vertical rail at door aperture.

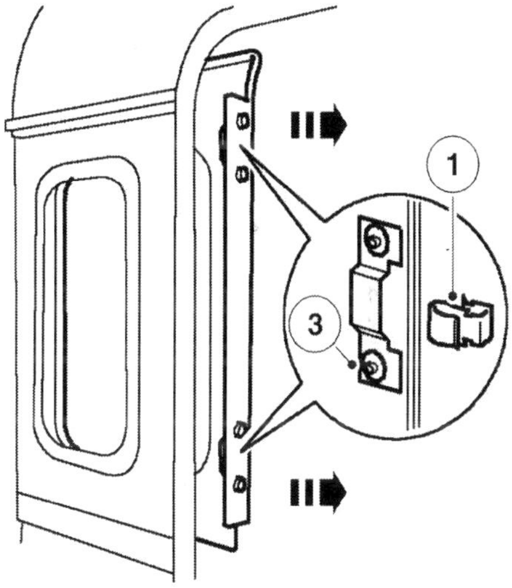

2. Remove rear quarter trim panel.

3. Remove spring clips from the 2 brackets on side of trim panel.

Installation

1. Install the spring clips to the 2 brackets on side of trim panel.

2. Position outboard edge of trim panel in corner to abut quarter trim panel and locate spring clips on vertical rail.

3. Press edge of trim panel firmly to secure in position.

Steering Column Shrouds

Removal and Installation

Removal

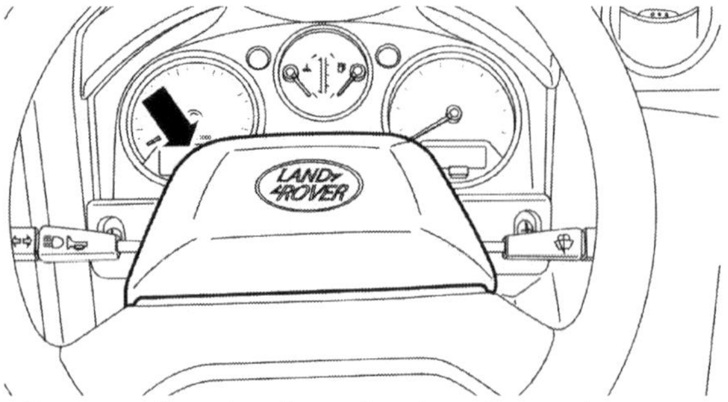

1. Remove the steering wheel center pad.

2. Remove the instrument cluster lower finisher.
   - Remove the 2 screws.

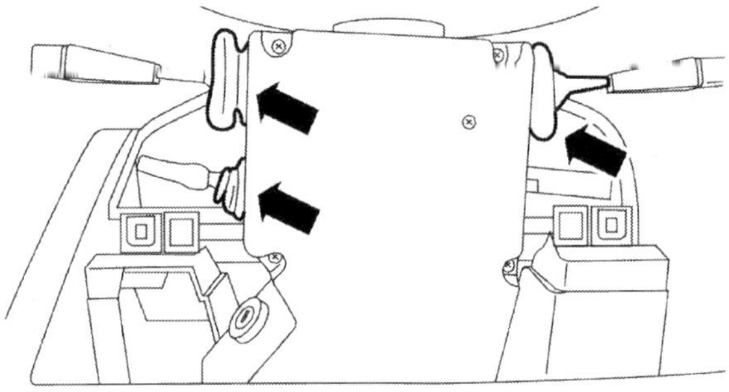

3. Release the 3 grommets from the steering column shrouds.

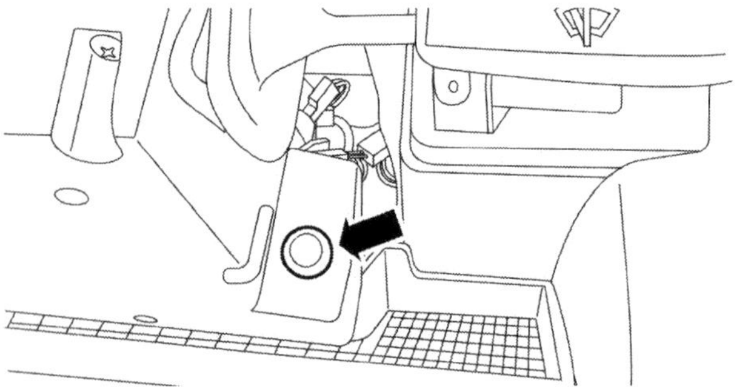

4. Remove the grommet.

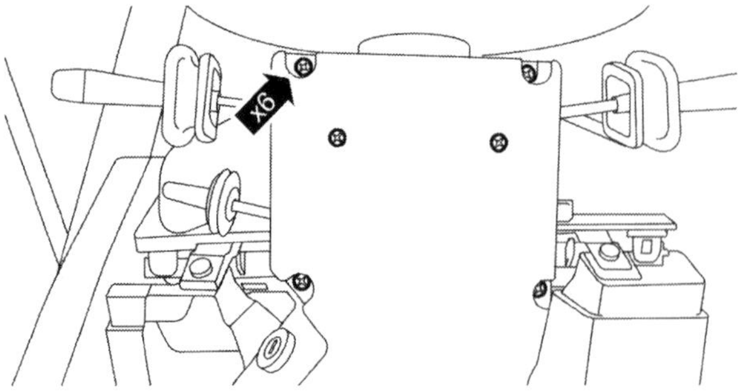

5. Remove the 6 screws.

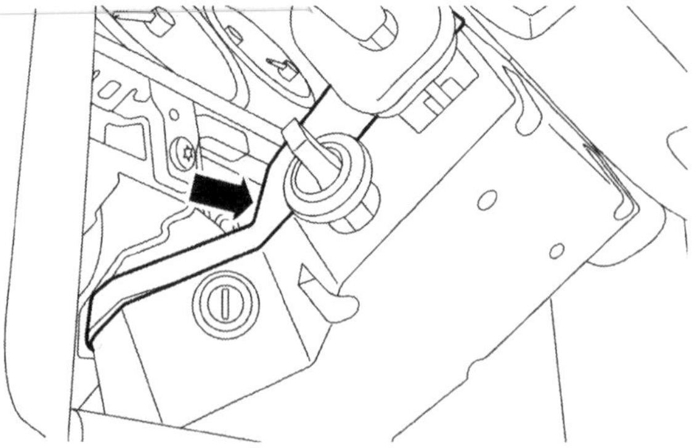

6. Remove the upper steering column shroud.

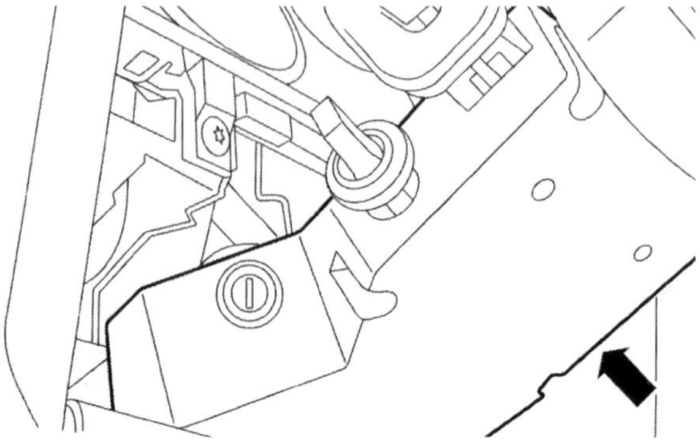

7. Remove the lower steering column shroud.

Installation

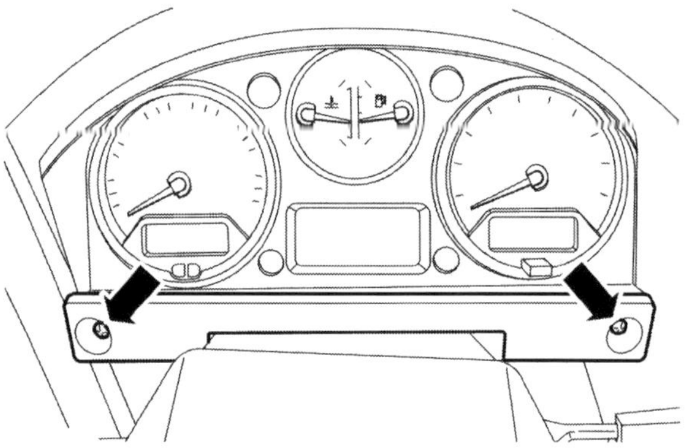

1. To install, reverse the removal procedure.
   - Tighten to 2 Nm (1 lb.ft).

Sun Visor

Removal and Installation

Removal

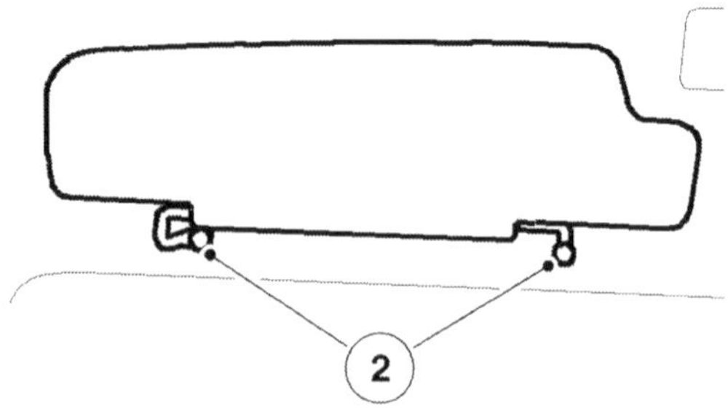

1. Raise sun visor.
2. Remove 2 screws and collect sun visor.

Installation

1. Position sun visor and secure with 2 screws.

Taildoor Trim Panel

Removal and Installation

Removal

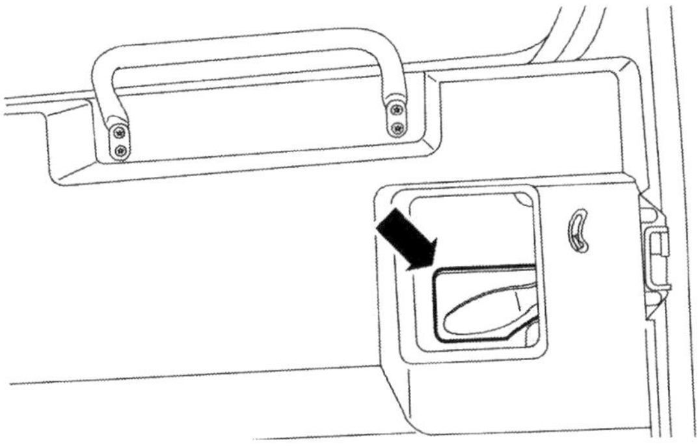

1. Remove the taildoor trim panel trim finisher.

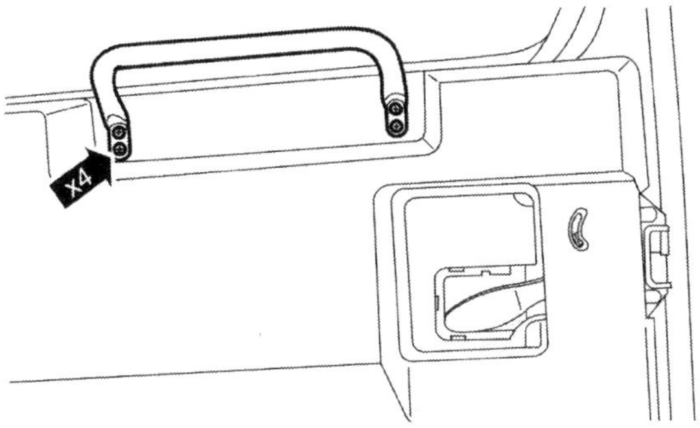

2. Remove the taildoor grab handle.
   - Remove the 4 screws.
3. Remove the taildoor trim panel.
   - Release the 9 clips.
4. △ NOTE: Do not carry out further dismantling if component is removed for access only.

Remove the 9 clips from the taildoor trim panel.

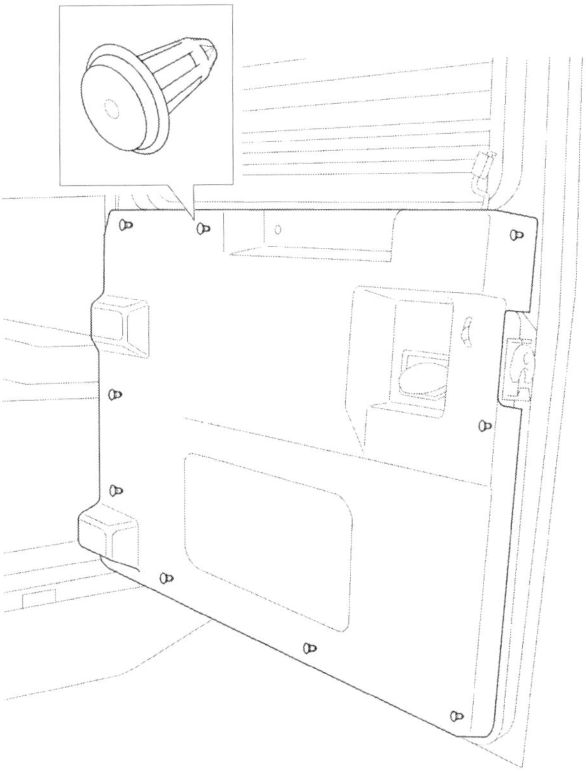

Installation

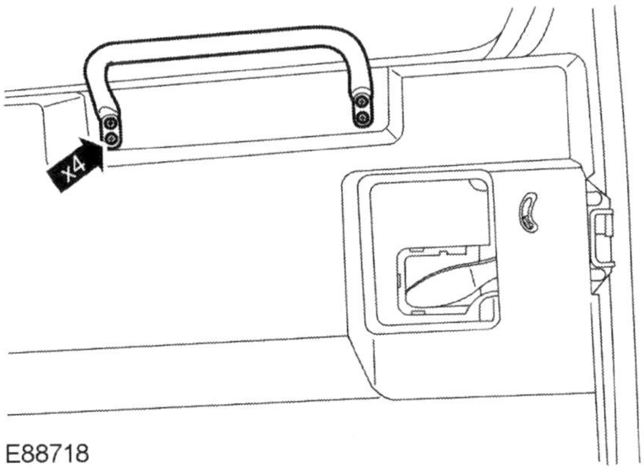

1. To install, reverse the removal procedure.
   - Tighten to 6 Nm (4 lb.ft).

Engine Cover

Removal and Installation

Removal

1. Open the bonnet.

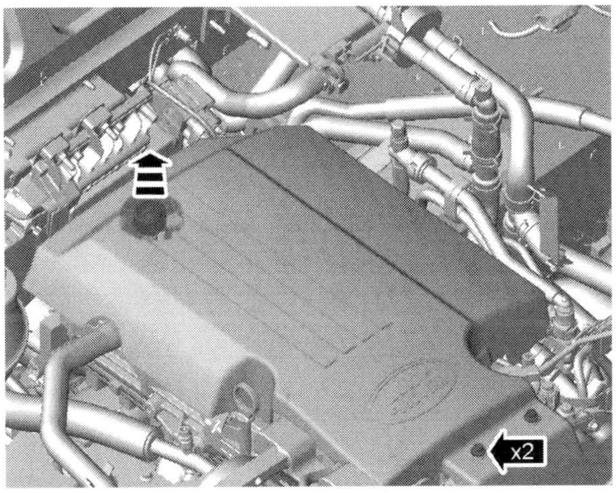

2.

Installation

1. *Torque:* 10 Nm
2. Close the bonnet.

Exterior Rollover Bar

Removal and Installation

Removal

1. Remove Torx bolt, from both sides, securing top of rollover bar to roof mounted bar.
2. With assistance to support rollover bar at windshield, remove 4 Torx bolts, from both sides, securing rollover bar to fender mounting.
3. Remove two Torx bolts, from both sides, securing top bar to roof mounting.
4. Slide top bars rearwards to disengage from main rollover bar assembly.
5. Lift main rollover bar assembly from vehicle.
6. If necessary, remove 2 Torx bolts and nuts, from both sides, and detach lower cross bar from main rollover bar.

Installation

1. Locate lower cross bar into main rollover bar mounting brackets and secure with Torx bolts and nuts tightened to 25 Nm (18 lbf.ft).
2. Ensure rubber gaskets are fitted correctly to rollover bar mounting brackets.
3. With assistance, lift rollover bar assembly and position on fenders.
4. Slide top bars into main rollover bar, position on roof mounting and secure with Torx bolts to 25 Nm (18 lbf ft.).
5. Secure rollover bar to both fender mountings with Torx bolts. Tighten to 25 Nm (18 lbf.ft).
6. Secure top bars to main rollover bar and tighten fixing bolts to 25 Nm (18 lbf.ft).

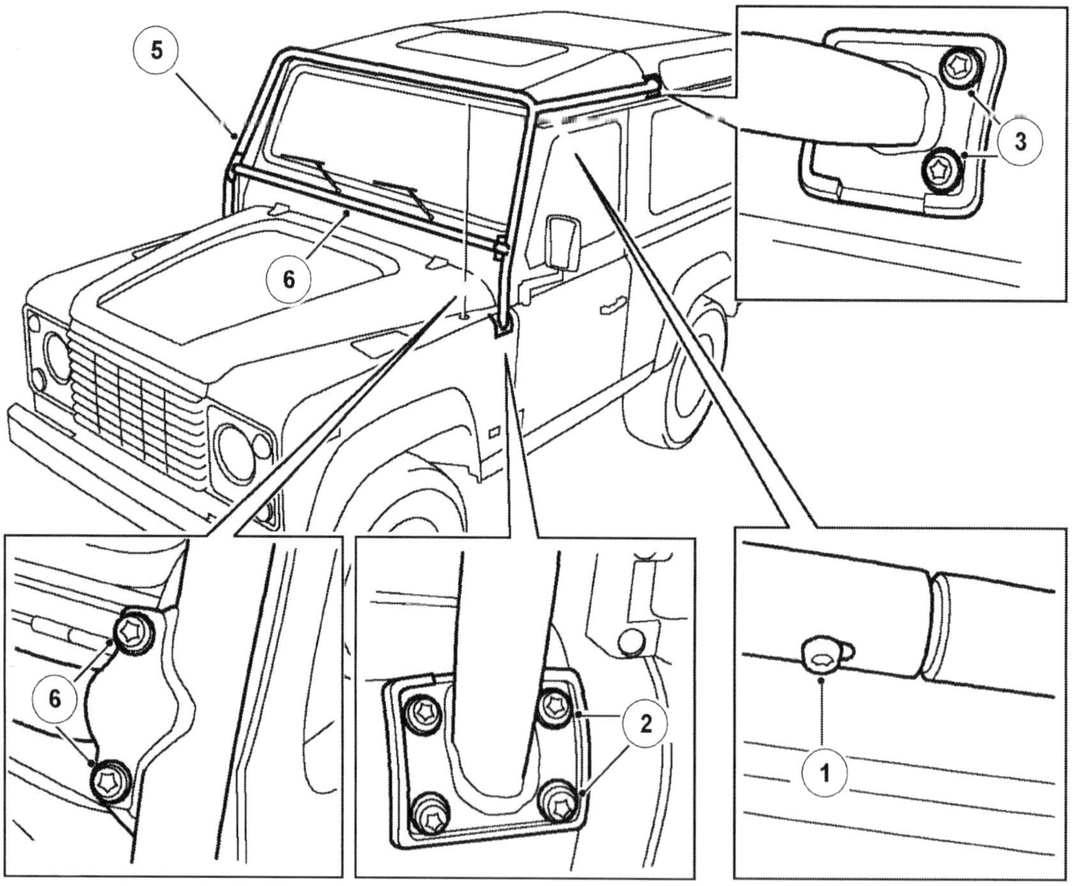

E82155

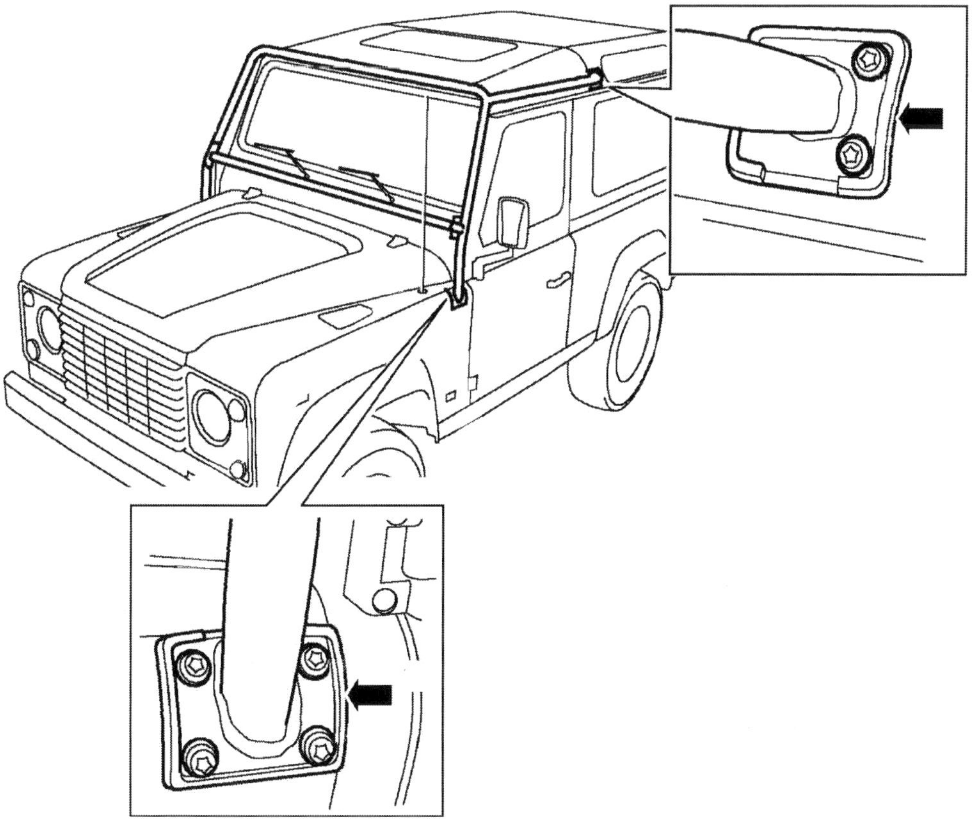

E82105

Radiator Grille

Removal and Installation

Removal

1. Disconnect the battery ground cable.

Refer to: Startup Battery Disconnect and Connect (414-01 Battery, Mounting and Cables, General Procedures)

2. Remove the radiator grille.
   - Remove the 6 screws.

Installation

1. To install, reverse the removal procedure.

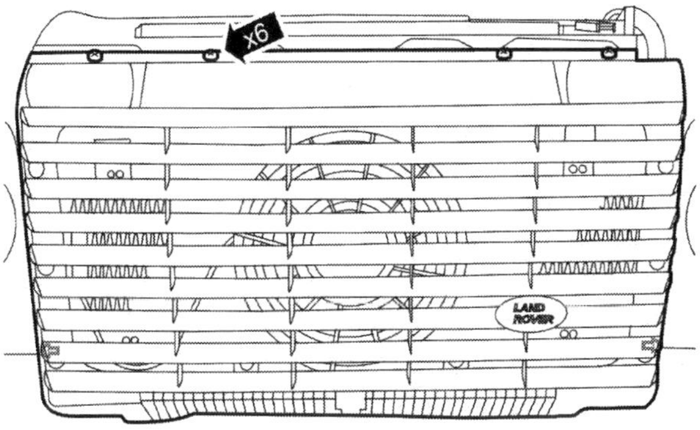

E86147

2. Connect the battery ground cable.

Refer to: Startup Battery Disconnect and Connect (414-01 Battery, Mounting and Cables, General Procedures)

Rear Folding Step

Removal and Installation

Removal

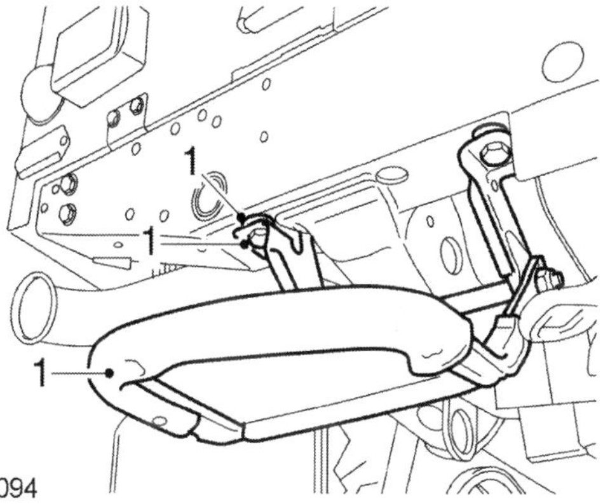

M763094

1. Remove 2 bolts securing step to chassis member, remove step and collect spacers.

Installation

1. Position step, fit spacer washers and tighten bolts.

Towbar

Removal and Installation

Removal

1. Remove 4 bolts securing stabilizer bar to rear undershield.
2. Remove 6 bolts securing rear undershield to chassis.
3. Remove rear undershield.
4. Remove bolt, washer and nut from both sides, securing tow bar lower mounting brackets and towing eyes to chassis.
5. Remove 4 bolts and spring washers securing tow bar centre mounting bracket to chassis.
6. Remove both bolts securing 2 tow bar locating tubes in chassis rear cross member.
7. Remove 2 bolts securing tow bar top mounting brackets to chassis.
8. With assistance, withdraw tow bar assembly from chassis.

Installation

1. With assistance, lift tow bar assembly and locate both tubes in chassis crossmember.
2. Secure top tow bar mounting brackets to chassis but do not fully tighten fixings.
3. Secure tow bar locating tubes to rear of chassis and tighten bolts to 25 Nm (18 lbf.ft).
4. Secure tow bar centre mounting bracket to underside of chassis and tighten bolts to 25 Nm (18 lbf.ft).
5. Position towing eyes to lower mounting brackets and secure to chassis and tighten bolts to 25 Nm (18 lbf.ft).
6. Fully tighten tow bar top mounting bracket bolts to 25 Nm (18 lbf.ft).
7. Fit rear undershield and secure to chassis with bolts.
8. Fit bolts securing rear stabilizer bar and tighten to 30 Nm (22 lbf.ft).

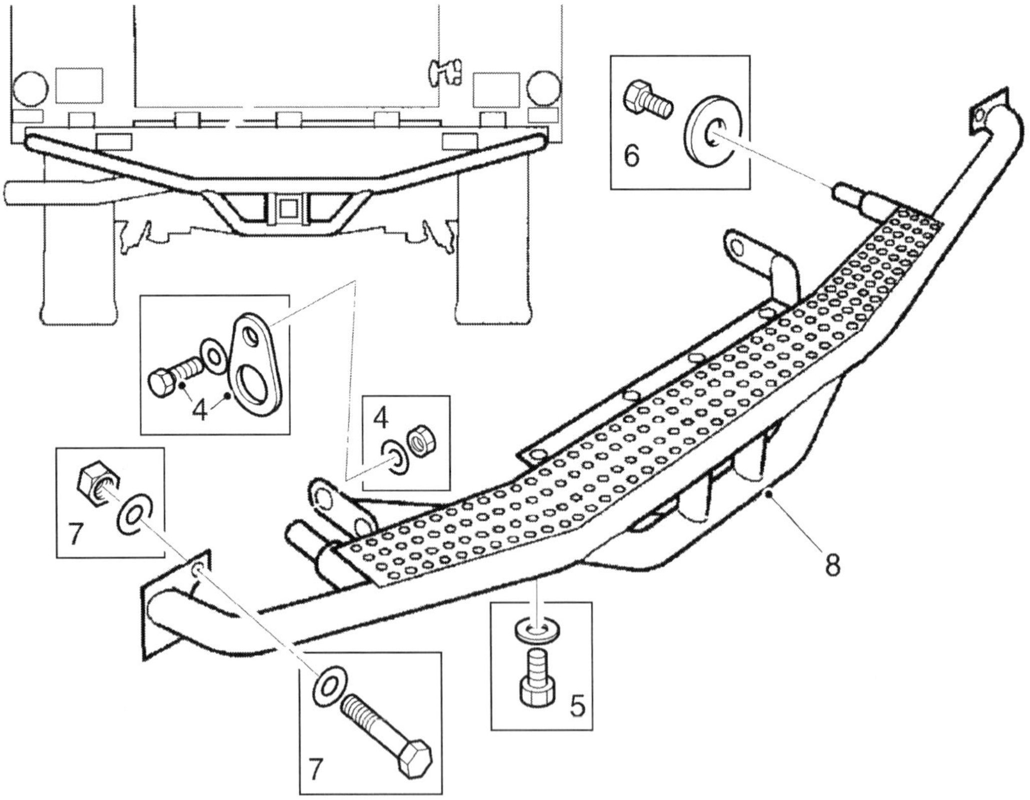

E82156

Rear View Mirrors

501-09

Rear View Mirrors - Interior Mirror

Removal and Installation

Removal

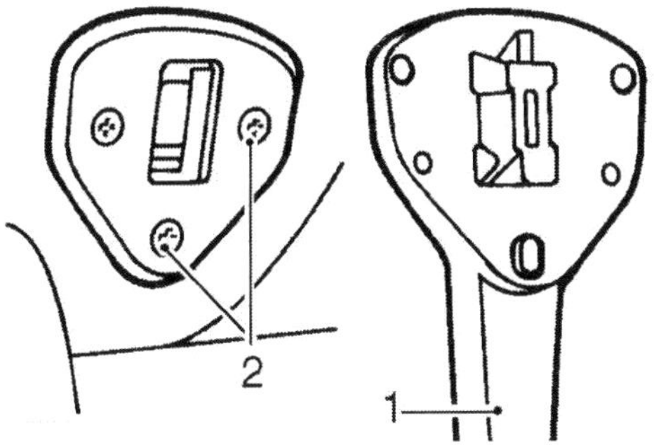

1. Prise interior mirror arm from mounting plate.
2. Remove 3 screws and remove mounting plate from headlining.

Installation

1. Install mounting plate to headlining.
2. Locate lower lug of mirror arm in mounting plate aperture.
3. Press mirror arm firmly to engage the spring clip.

Seating

501-10

Specifications

Torque Specifications

Description	Nm	lb-ft
Rear seat base stop bolt	23	17
Rear seat bolts	23	17
Rear seat nuts	23	17
+ Front seat backrest nuts and bolts	23	17
Front seat bolts	23	17
Third row seat bolts	23	17
Third row seat nuts	23	17
Third row seat nut and bolt	23	17

+ New nuts must be installed

Description and Operation

OVERVIEW

Vehicles from 2007 model year are fitted with an all new seating system. The front seats are manually adjustable for reach and recline. Some vehicles may also feature heated front seats. The heated front seats will only operate when the ignition is switched on.

The second row seats feature a 60/40 split and are able to be folded forward to increase the loadspace in the rear of the vehicle. Defender 110 Station Wagons are fitted with 2 forward facing third row seats. The third row seats can be stowed at the side of the cabin to increase the loadspace in the rear of the vehicle. For more information on seat stowage, refer to the Owners Handbook.

COMPONENT LOCATION

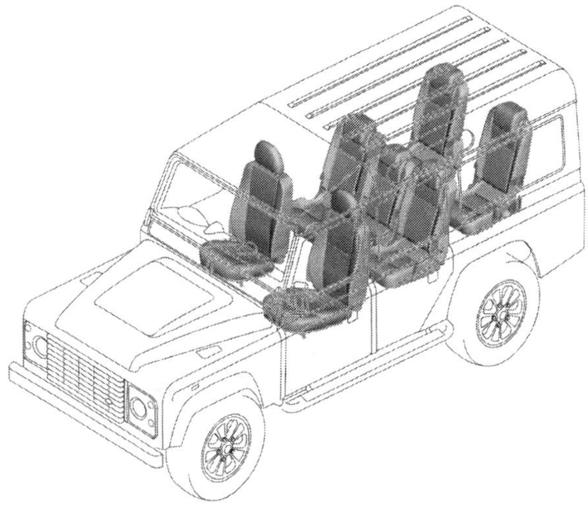

PRINCIPLES OF OPERATION

Heated Front Seats

When the ignition switch is in position II (ignition on) a feed is provided to the window lift relay coil via the CJB (central junction box) . The ground path for the relay coil is controlled by the LH (left-hand) and RH (right-hand) heated seat switches. When either switch is pressed, the relay energizes and provides a battery voltage feed to the appropriate seat heater elements. Both seats contain 2 heating elements; 1 in the cushion, 1 in the backrest, which are wired in series.

The cushion elements in both front seats contain a thermostatically controlled switch. When the cushion element temperature reaches 37 ± 3°C (98 ± 3°F) the thermostat cuts the supply to both the cushion and backrest elements. When the temperature of the cushion element falls below 28 ± 3°C (82 ± 3°F) the thermostat reinstates the supply to both heater elements.

Heater Mats

Diagnosis and Testing

Principles of Operation

For a detailed description of the Heater Mats, refer to the relevant Description and Operation section in the workshop manual.

Refer to: Seats (501-10 Seating, Description and Operation)

Inspection and Verification

⚠ CAUTION: Diagnosis by substitution from a donor vehicle is **NOT** acceptable. Substitution of control modules does not guarantee confirmation of a fault, and may also cause additional faults in the vehicle being tested and/or the donor vehicle.

NOTES:

△ If a control module or a component is at fault and the vehicle remains under manufacturer warranty, refer to the Warranty Policy and Procedures manual, or determine if any prior approval program is in operation, prior to the installation of a new module/component.

△ When performing voltage or resistance tests, always use a digital multimeter accurate to three decimal places, and with an up-to-date calibration certificate. When testing resistance always take the resistance of the digital multimeter leads into account.

△ Check and rectify basic faults before beginning diagnostic routines involving pinpoint tests.

1. Visually inspect for obvious signs of damage and system integrity

Visual Inspection

Mechanical	Electrical
• Seat heater switches	• Fuses • Wiring harnesses and connectors

2. If an obvious cause for an observed or reported concern is found, correct the cause (if possible) before proceeding to the next step
3. If the cause is not visually evident, verify the symptom and refer to the Symptom Chart, alternatively check for Diagnostic Trouble Codes (DTCs) and refer to the DTC Index

Symptom Chart

Symptom	Possible Causes	Action
Seat heating performance poor or inoperative	• Seat heater element circuit short circuit to ground, short circuit to power, open circuit, high resistance	• GO to Pinpoint Test A

Seat Heater Mat Faults - Diagnostic Guidance

- **Seat Heater Sensor Faults** - Each seat has two heating elements: one in the base (seat cushion), and one in the back (squab). There is only one sensor, which is located in the base. This sensor controls both heating elements. If there is a fault with the sensor signal, neither element will operate. If the signal fault is traced to the sensor itself (in the base/seat cushion mat), there is no need to replace the back/squab heater mat.
- **Diagnosis Of Seat Heater Elements Using Resistance Measurements** - The back/squab heater mat consists of a heating element, the base/seat cushion heater mat consists of a heating element with a sensor. Most faults can be confirmed by simple electrical resistance measurements, using the values listed below. It is possible that the sensor could develop a fault causing it to exhibit either higher resistance (resulting in the seat heaters operating at too high a temperature), or lower resistance (resulting in the seat heaters operating at too low a temperature or not at all), but these kinds of faults are uncommon. The maximum resistance of the seat elements are only about 1 ohm each under normal operation, so it is difficult to take an exact reading with a standard multimeter. However, the most likely cause of failure is breaks in a number of the heater element wires, which causes both a significant loss of heating and a corresponding significant increase in resistance. In this situation, both (seat cushion AND squab) elements produce less heat, but contrary to expectation, it is the undamaged element which will show the greatest loss of heating. Given this, suspected seat heater element failures should always be confirmed by electrical resistance tests before replacement. Do not replace components based only on temperature performance.
- **Simultaneous Failure of Both Seat Heater Mats Is Unlikely** - The base/seat cushion and back/squab seat heater elements connect to the HVAC Control Module, but the control module only acts as a splice box and the elements are electrically wired in series (each drawing proportion of battery voltage). The only way for one element to operate while the other is faulty is if one is the elements has a short-circuit fault. Under normal circumstances, where short-circuit faults are present these will be detected by the HVAC Control Module which will

switch off both elements. Similarly, if the sensor (in the base/seat cushion element) is faulty, the HVAC Control Module will not apply voltage to the elements and, if one element is open-circuit, then no voltage can be supplied to the other element. Given this, while it may often be the case that both elements cease to operate, it is extremely unlikely that both heater mats are faulty. Consequently, if a fault is confirmed with one of the seat heater mats, do not also replace the other.

- **Seat Heater Sensor Open Circuit Faults** - Previous investigations have found the mostly likely failure mode to be associated with the seat heater sensor (in the base/seat cushion element) developing an open-circuit fault. This can be measured by disconnecting the 4-way plug from the HVAC Control Module, and (Test 1) testing the circuit into the harness; if an open-circuit fault is detected, then test the circuit at the connector under the seat (Test 2). If the open-circuit fault is present both at Test 1 and Test 2, then assume the seat heater sensor is at fault and replace the base/seat cushion element. If the open circuit fault is detected at Test 1 but NOT at Test 2, then the harness is most likely faulty.

Seat Heater Mat Application Chart

⚠ NOTE: To make sure an accurate resistance reading, calibrated test equipment **must** be used.

DTC Index

For a list of Diagnostic Trouble Codes (DTCs) that could be set on this vehicle, please refer to Section 100-00.

Vehicle / Year	Cushion / Backrest	Heater Mat / NTC Resistor	Left Hand Drive		Right Hand Drive		Minimum Resistance	Maximum Resistance
			Passenger Side Connector / Pin	Driver Side Connector / Pin	Passenger Side Connector / Pin	Driver Side Connector / Pin	Ohms At 20°C ±10°C	Ohms At 20°C ±10°C
Defender (L316)	Cushion	Heater mat	C0237-1 and C0237-2	C0237-1 and C0237-2	C0237-1 and C0237-2	C0237-1 and C0237-2	1.26	1.54
		NTC resistor	-	-	-	-	-	-
	Backrest	Heater mat	-	-	-	-	-	-

Pinpoint Tests

PINPOINT TEST A : CLIMATE SEAT TESTS	
TEST CONDITIONS	DETAILS/RESULTS/ACTIONS
A1: DTC READ	
	1. Using the manufacturer approved diagnostic system, check the climate control system for related DTCs (where possible)
	Are any climate seat DTCs set? **Yes** Perform the relevant corrective actions. Clear the DTCs and retest. If the fault persists, GO to 2 **No** GO to 2
A2: MANUAL CHECK	
⚠ NOTE: On full power the seat should be hot to touch	
	1. If necessary, operate the vehicle air conditioning on full for 10 minutes to reduce the temperature of the vehicle interior
	2. Operate the climate seats on full power
	Do the climate seats operate correctly? **Yes** Using the manufacturer approved diagnostic system, clear the DTCs (where possible) and retest. If the operation of the climate seats is correct no further action is necessary **No** GO to 3

A3: SHORT CIRCUIT TO GROUND	
	1. Refer to the electrical circuit diagrams and the Seat Heater Mat Application Chart (see above) to identify the connector
	2. Disconnect the connector
	3. Refer to the electrical circuit diagrams and check the climate seat element circuit for short circuit to ground
	Is there a short circuit to ground? **Yes** Repair the circuit or install a new climate seat element as necessary. Using the manufacturer approved diagnostic system, clear the DTCs (where possible) and retest **No** GO to 4
A4: CIRCUIT CONTINUITY TEST	
	1. Refer to the electrical circuit diagrams and check the climate seat element circuit for open circuit, high resistance
	Is the climate seat element circuit resistance within specification? **Yes** GO to 5 **No** Repair the circuit or install a new climate seat element as necessary. Using the manufacturer approved diagnostic system, clear the DTCs (where possible) and retest
A5: POWER CONSUMPTION	
△ NOTE: The seat heater power supply cycles on and off dependant on the seat and passenger compartment temperature and may only switch on for 5 seconds in 30 seconds.	
	1. Reconnect the connector
	2. If necessary, operate the vehicle air conditioning on full for 10 minutes to reduce the temperature of the vehicle interior
	3. Connect an inductive ammeter (current clamp) to the climate seat element circuit
	4. Operate the climate seats on full power
	5. Use the Seat Heater Mat Application Chart to calculate a typical current value (using I=V/R, **Current** equals **Volts** divided by **Resistance**). For example: • 12 volts / 0.5 ohms = 24 amps • 12 volts / 1 ohms = 12 amps • 12 volts / 2 ohms = 6 amps
	Does the climate seat circuit draw the correct quantity of current? **Yes** Using the manufacturer approved diagnostic system, clear the DTCs (where possible) and retest. If the operation of the climate seats is correct no further action is necessary **No** GO to 6
A6: RESISTANCE CHECK	
NOTES: △ Make sure the multimeter used is calibrated and a resistance reading of 0 ohms is shown when the test leads are connected together. Alternately, subtract any resistance shown from the result.	
△ The climate seat element circuits should be checked at the seat heater module connector (where possible).	
⚠ Refer to the electrical circuit diagrams and to confirm the total resistance of the circuit the cushion and backrest are connected in series.	

	1. Refer to the electrical circuit diagrams and the Seat Heater Mat Application Chart (see above) to identify the connector and terminals
	2. Disconnect the connector
	3. Using a multimeter, perform resistance measurements of the climate seat element circuit and the NTC resistor circuit (where possible)
	4. Compare the results to the Seat Heater Mat Application Chart
	Are the results within specification at the given ambient temperature? (tolerance ± 0.5 ohms) **Yes** Reconnect the connector. Using the manufacturer approved diagnostic system, clear the DTCs (where possible) and retest. If the fault persists, refer to electrical circuit diagrams and check the power and ground circuits **No** Repair the circuit or install a new climate seat element as necessary. Using the manufacturer approved diagnostic system, clear the DTCs (where possible) and retest

Front Row Seat

Removal and Installation

Removal

1. Remove the front seat cushion.

Refer to: Front Row Seat Cushion (501-10 Seating, Removal and Installation)

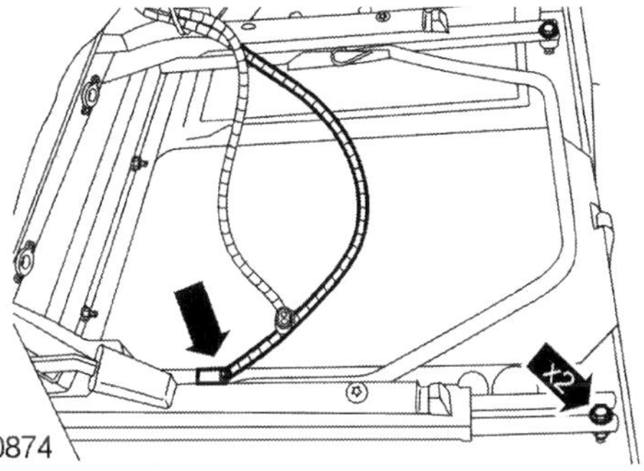

E90874

2. Remove the front 2 bolts from the front seat.
   - Disconnect the front seat heater mat electrical connector.

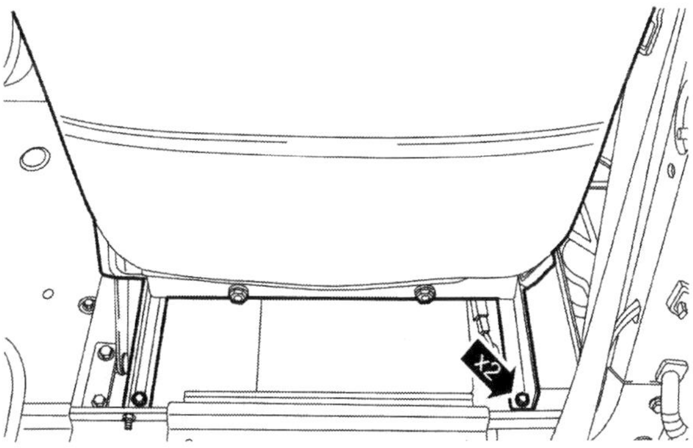

3. Remove the front seat.
   - Slide the front seat forwards.
   - Remove the 2 bolts.
   - Collect the 8 spacer washers.

Installation

1. To install, reverse the removal procedure.
   - Tighten the bolts to 23 Nm (17 lb.ft).

Front Row Seat Backrest

Removal and Installation

Part(s)

STEP	PART NAME	QUANTITY
Installation Step 1	Renew Part Seat base nut	2

Removal

CAUTION: Do not attempt to adjust or dismantle the front seat tilt mechanism, the front seat backrest must be replaced as a complete assembly, failure to follow this instruction may result in damage to the vehicle.

1. Remove the front seat cushion.

Refer to: Front Row Seat Cushion (501-10 Seating, Removal and Installation)

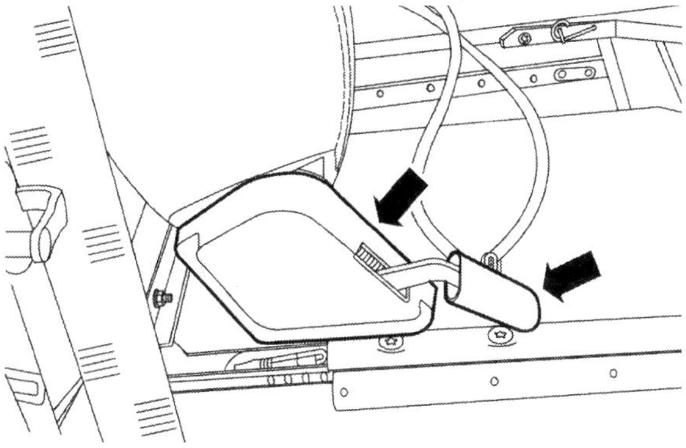

2. Remove the front seat backrest hinge cover. Remove the front seat recliner handle.

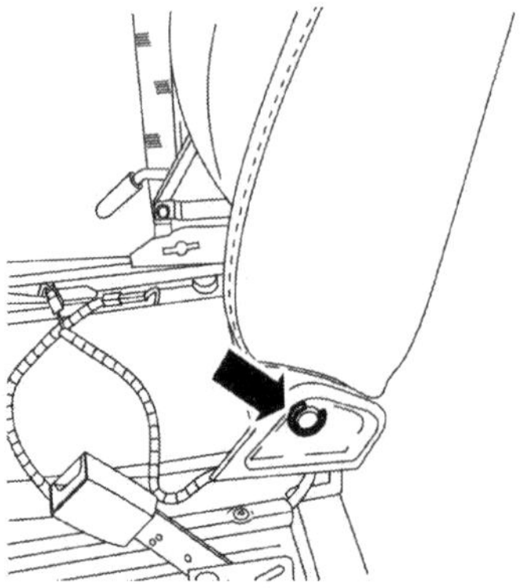

3. Remove the front seat backrest hinge clip.

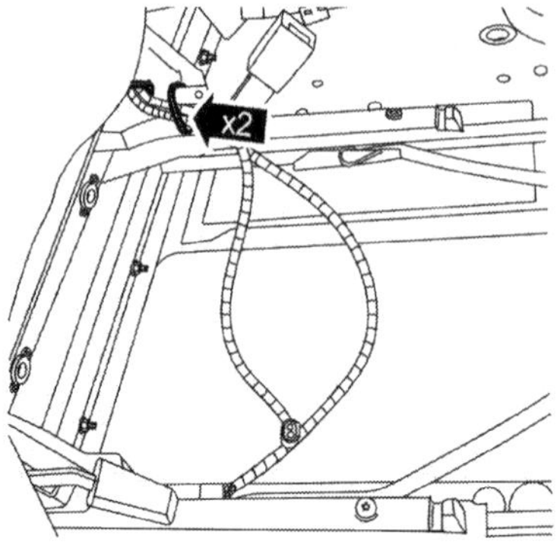

4. Release the front seat heater mat wiring harness. Remove and discard the 2 cable ties.

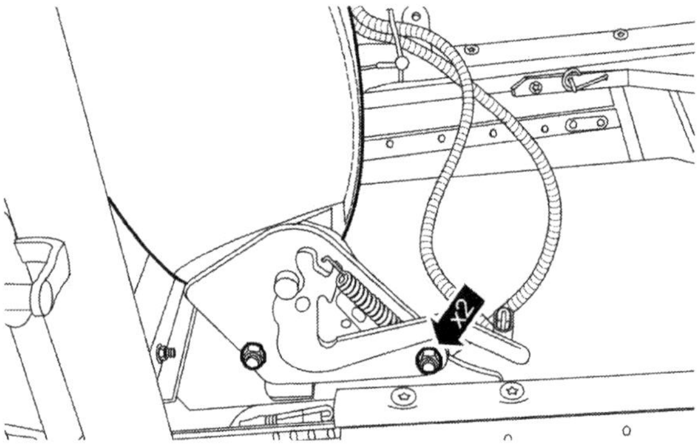

5. △ NOTE: Discard the nuts.

   Remove the front seat backrest. Remove the 2 nuts and bolts.

Installation

1. △ NOTE: New nuts must be installed.

Install the front seat backrest. Tighten the nuts and bolts.

   Renew Part: Seat base nut

   Torque: 23 Nm

2. Secure the front seat heater mat wiring harness. Install new cable ties.
3. Install the front seat backrest hinge clip.
4. Install the front seat backrest hinge cover. Install the front seat recliner handle.

Front Row Seat Backrest Cover

Removal and Installation

Part(s)

STEP	PART NAME	QUANTITY
Removal Step 4	Renew Part Hog rings	11

Removal

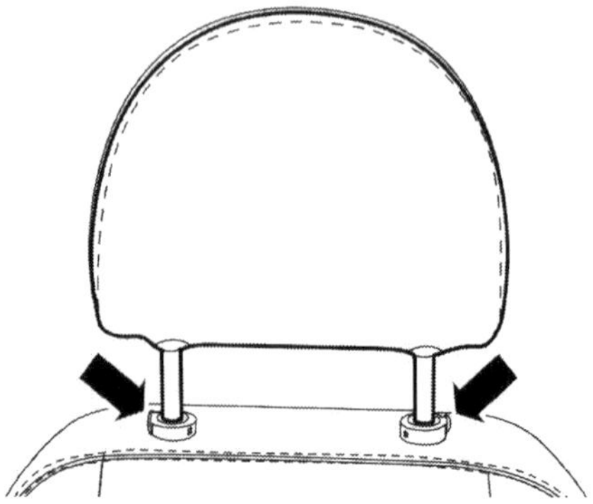

1. Remove the front seat head restraint. Release the 2 clips.

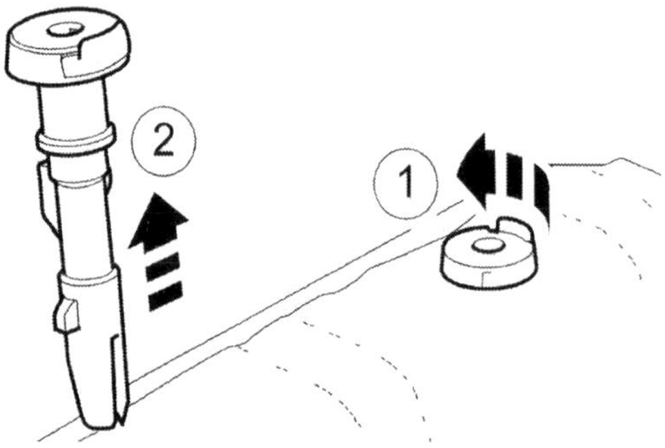

2. Remove the front seat head restraint guides. Turn the front seat head restraint guides 90 degrees counter clockwise.

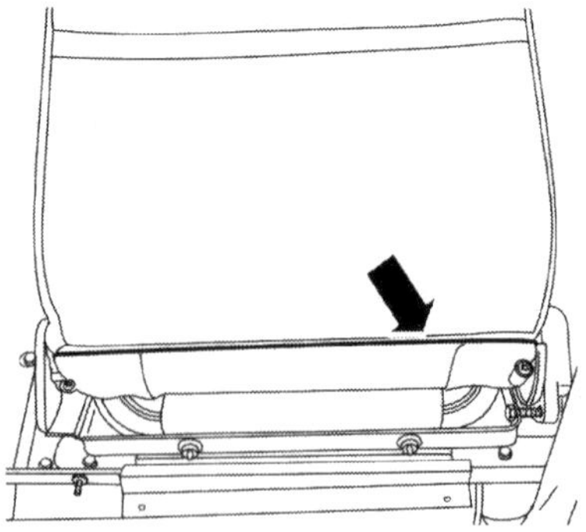

3. Release the front seat cover lower closing strip.

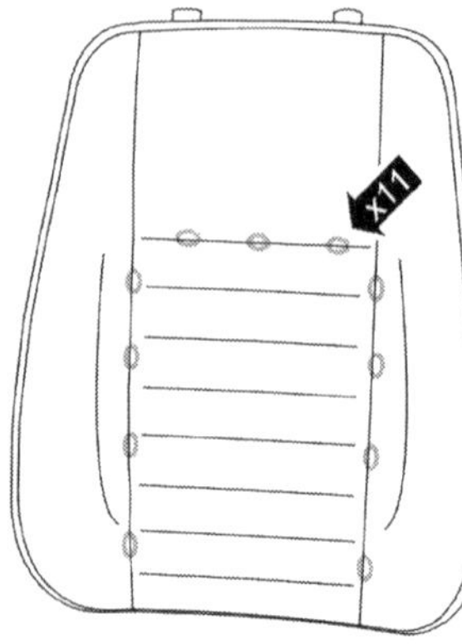

E90856

4. Remove the front seat backrest cover. Remove and discard the 11 hog rings.

   Renew Part: Hog rings

Installation

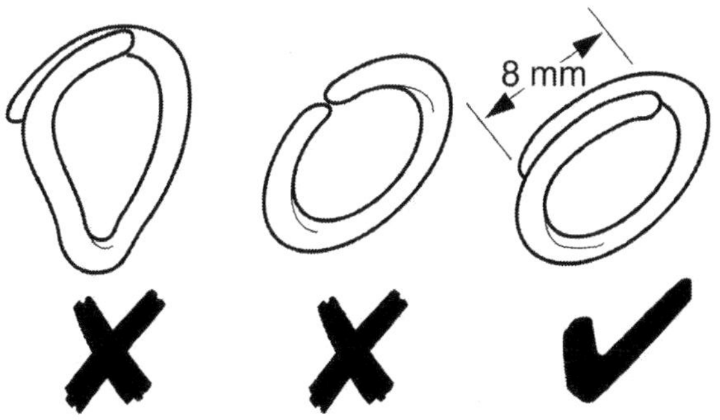

1. CAUTION: Care must be taken when installing the hog rings, failure to follow this instruction may result in damage to the vehicle.

NOTE: Use hog ring pliers to close the hog rings. Do not use any other tool. The hog rings must be closed to overlap as illustrated.

To install, reverse the removal procedure.

Front Seat Backrest Heater Mat

Removal and Installation

Part(s)

STEP	PART NAME	QUANTITY
Installation Step 2	Renew Part Seat back pad	1

Removal

1. Remove the front seat backrest cover.

Refer to: Front Row Seat Backrest Cover (501-10 Seating, Removal and Installation)

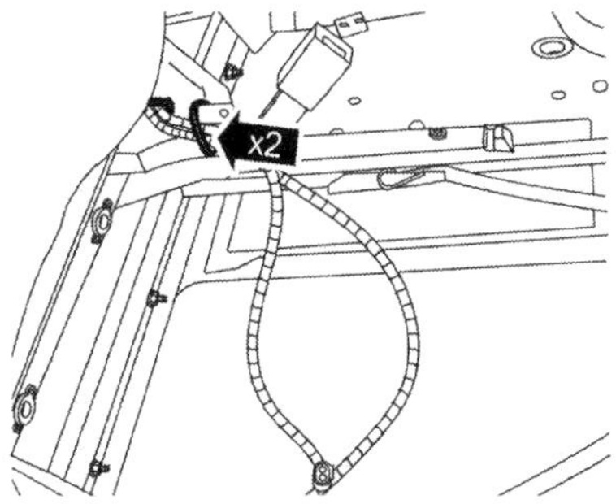

2. Remove and discard the 2 cable ties. Release the front seat heater mat wiring harness.

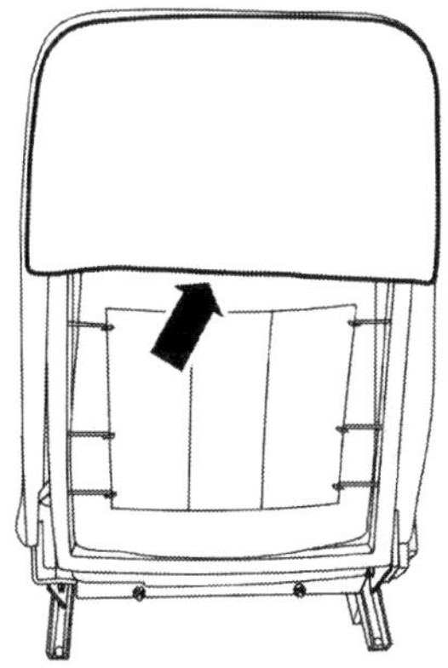

E90850

3. Remove and discard the front seat backrest foam backing.

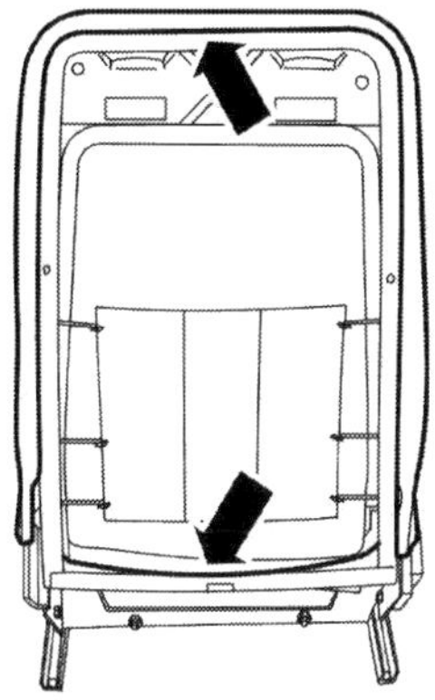

E90851

4. Remove and discard the front seat backrest cushion.

Installation

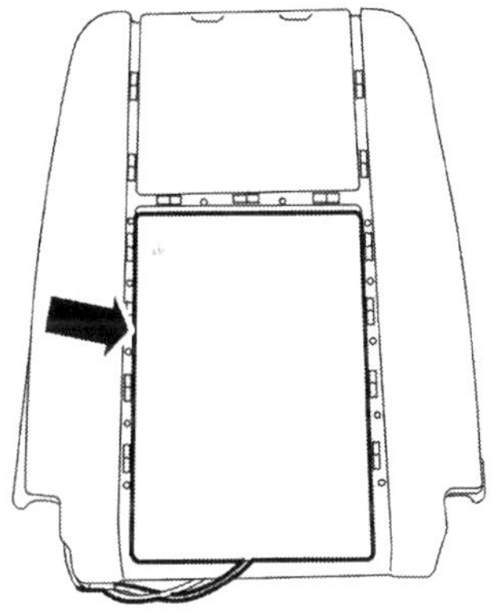

E90852

1. Install the front seat backrest heater mat to the front seat backrest cushion.
2. NOTE: Clean the component mating faces.

   Install the front seat backrest cushion. Secure the front seat backrest cushion using a suitable adhesive.

   Renew Part: Seat back pad
3. Install the front seat backrest foam backing.
4. Secure the front seat heater mat wiring harness. Install new cable ties.
5. Install the front seat backrest cover.

Refer to: Front Row Seat Backrest Cover (501-10 Seating, Removal and Installation)

Front Row Seat Cushion

Removal and Installation

Removal

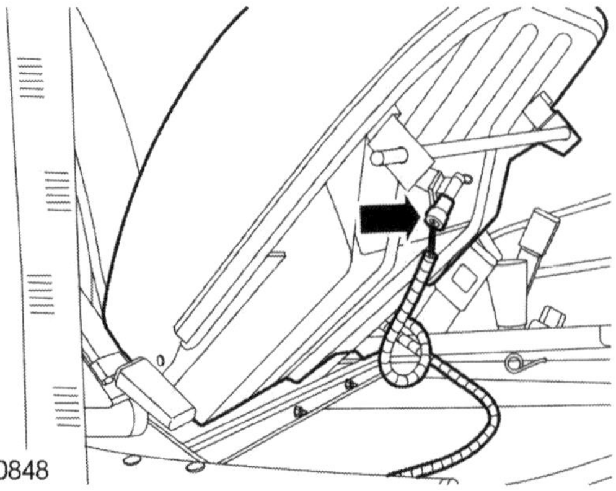

1. Remove the front seat cushion.
    - Lift the front edge of the front seat cushion.
    - Disconnect the front seat cushion heater mat electrical connector.

Installation

1. To install, reverse the removal procedure.

Front Row Seat Cushion Cover

Removal and Installation

Removal

1. NOTE: The front seat cushion cover is part of the front seat cushion and cannot be serviced separately.

   Remove the front seat cushion.

Refer to: Front Row Seat Cushion (501-10 Seating, Removal and Installation)

Installation

1. Install the front seat cushion.

Refer to: Front Row Seat Cushion (501-10 Seating, Removal and Installation)

Front Seat Cushion Heater Mat

Removal and Installation

Removal

1. NOTE: The front seat cushion heater mat is part of the front seat cushion and cannot be serviced separately.

   Remove the front seat cushion.

Refer to: Front Row Seat Cushion (501-10 Seating, Removal and Installation)

Installation

1. Install the front seat cushion.

Refer to: Front Row Seat Cushion (501-10 Seating, Removal and Installation)

Heated Seat Switch

Removal and Installation

Removal

1. Disconnect the battery ground cable.

Refer to: Startup Battery Disconnect and Connect (414-01 Battery, Mounting and Cables, General Procedures)

2. Remove the audio unit.

Refer to: Audio Unit - Vehicles Without: Bluetooth (415-01 Audio Unit, Removal and Installation)

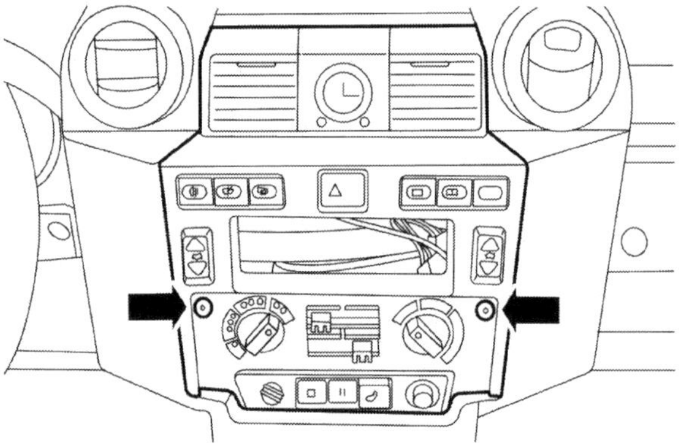

3. Release the instrument panel console.

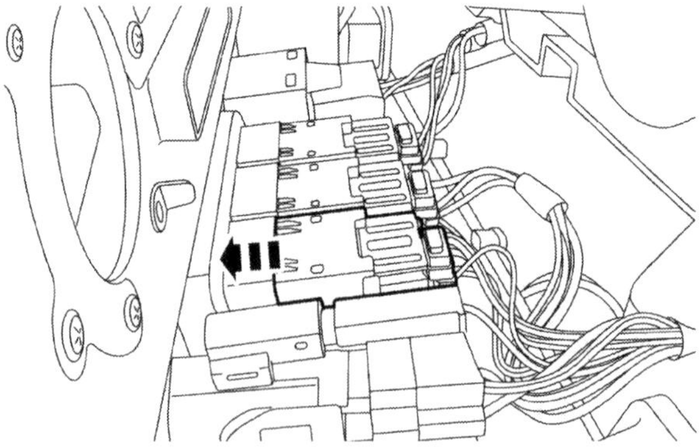

4. Release the heated seat switch.

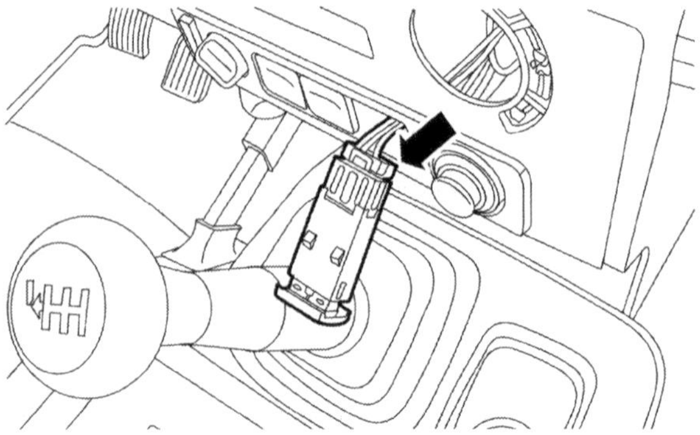

5. Remove the heated seat switch.
   - Disconnect the electrical connector.

Installation

1. To install, reverse the removal procedure.
2. Connect the battery ground cable.

Refer to: Battery Connect (414-01 Battery, Mounting and Cables, General Procedures).

Seating - Rear Seat

Removal and Installation

Removal

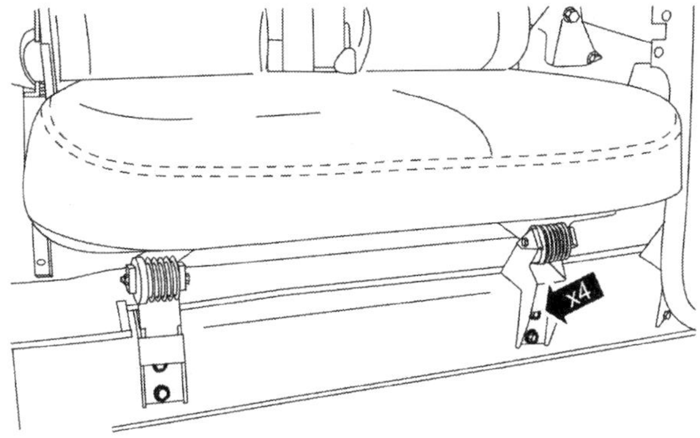

1. Release the front edge of the rear seat.
   - Remove the upper 2 bolts.
   - Loosen the lower 2 bolts.

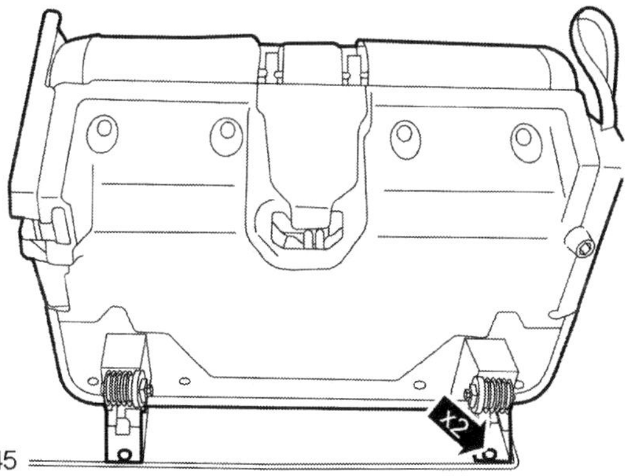

E90845

2. Remove the rear seat.
   - Tilt the rear seat forwards.
   - Remove the 2 nuts.

Installation

1. Install the rear seat.
   - Tighten the nuts to 23 Nm (17 lb.ft).
   - Lock the rear seat back in to position.
2. Secure the front edge of the rear seat.
   - Tighten the bolts to 23 Nm (17 lb.ft).
3. ⚠ CAUTION: Make sure that the seat latch is located correctly to the retaining bar.

△ NOTE: The seat backrest cannot be repositioned to the upright position until the seat has been locked correctly to the retaining bar.

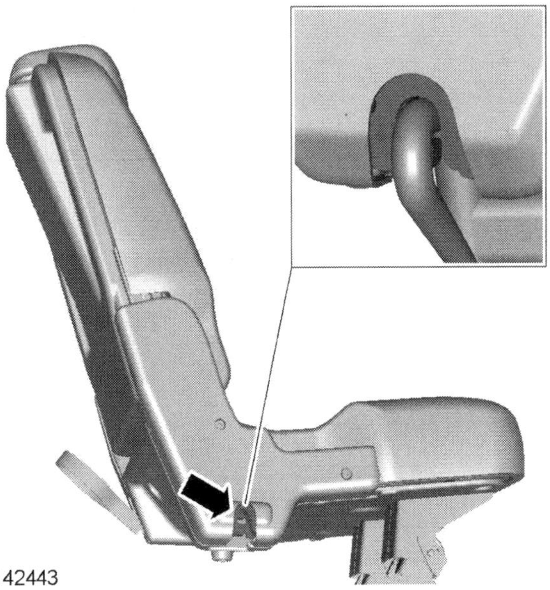

E142443

Rear Seat Backrest

Removal and Installation

Removal

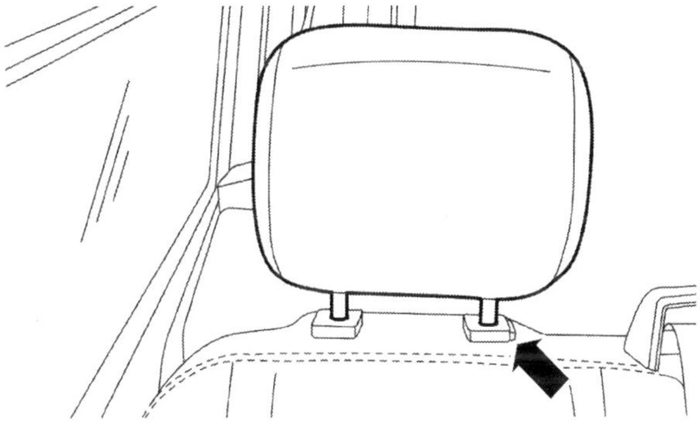

1. Remove the 2 rear seat head restraints.
   - Release the 2 clips.

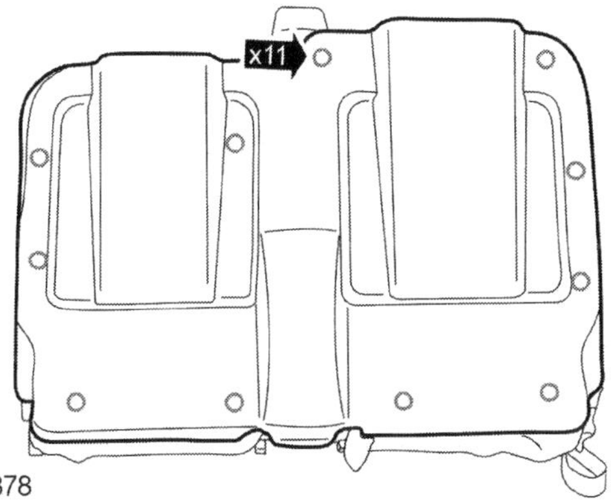

E90878

2. Remove the rear seat backrest trim panel.
   - Release the 11 clips.

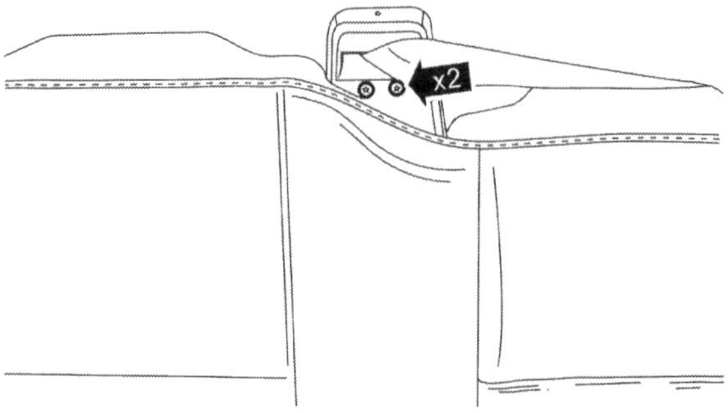

3. Reposition the rear center safety belt guide cover.
   - Remove the 2 screws.

4. Remove the 4 rear seat head restraint guides.
   - Release the 4 clips.

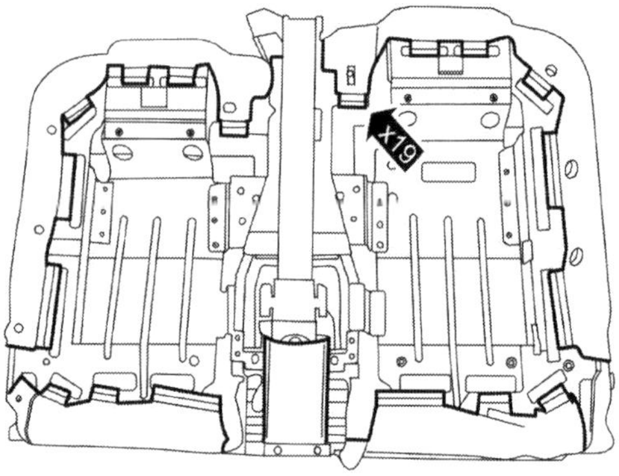

5. Remove the rear seat backrest.
   - Release the 19 clips.

Installation

1. Install the rear seat backrest.
   - Secure the 19 clips.
2. Install the rear seat head restraint guides.
3. Secure the rear center safety belt guide cover.
   - Tighten the screws.
4. Secure the rear seat backrest trim panel.
   - Secure with the clips.
5. Install the rear seat head restraints.

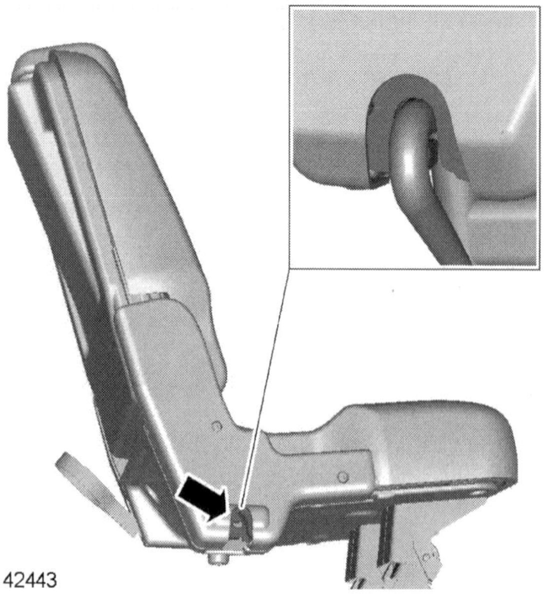

E142443

6. CAUTION: Make sure that the seat latch is located correctly to the retaining bar.

NOTE: The seat backrest cannot be repositioned to the upright position until the seat has been locked correctly to the retaining bar.

Rear Seat Backrest Cover

Removal and Installation

Part(s)

STEP	PART NAME	QUANTITY
Removal Step 2	Renew Part Hog rings	12
Removal Step 3	Renew Part Hog rings	20

Removal

1. Remove the rear seat backrest.

Refer to: Rear Seat Backrest Cover (501-10 Seating, Removal and Installation)

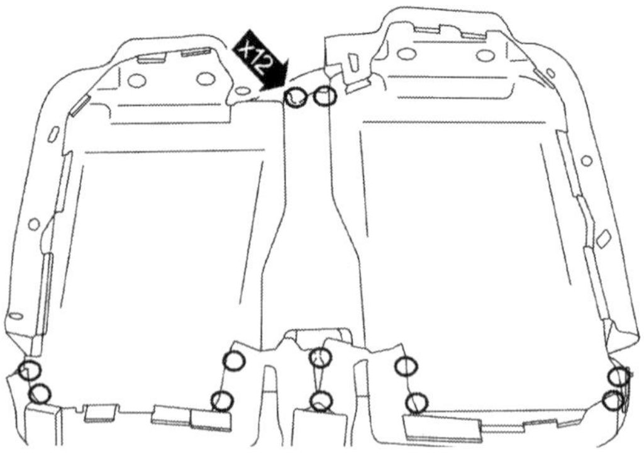

2. Release the rear seat backrest cover. Remove and discard the 12 hog rings.

   Renew Part: Hog rings

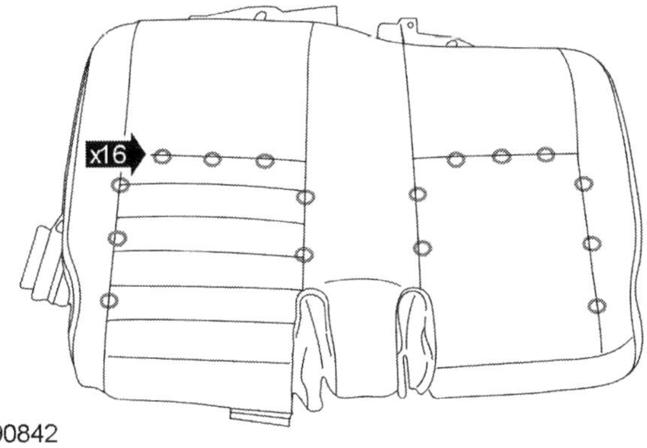

E90842

3. Remove the rear seat back rest cover. Remove and discard the 20 hog rings.

   Renew Part: Hog rings

Installation

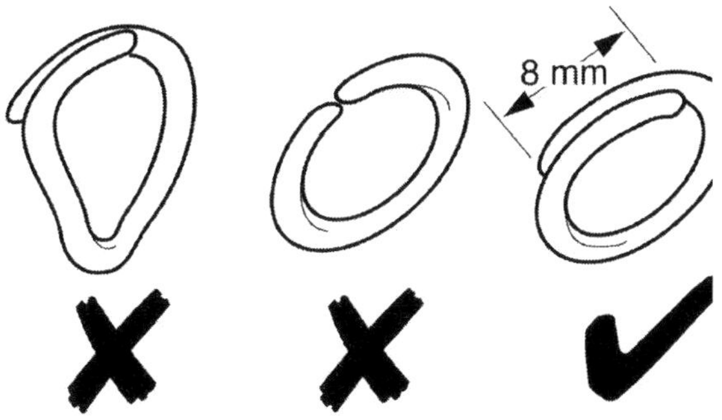

1. NOTE: Use hog ring pliers to close the hog rings. Do not use any other tool. The hog rings must be closed to overlap as illustrated.

   To install, reverse the removal procedure.

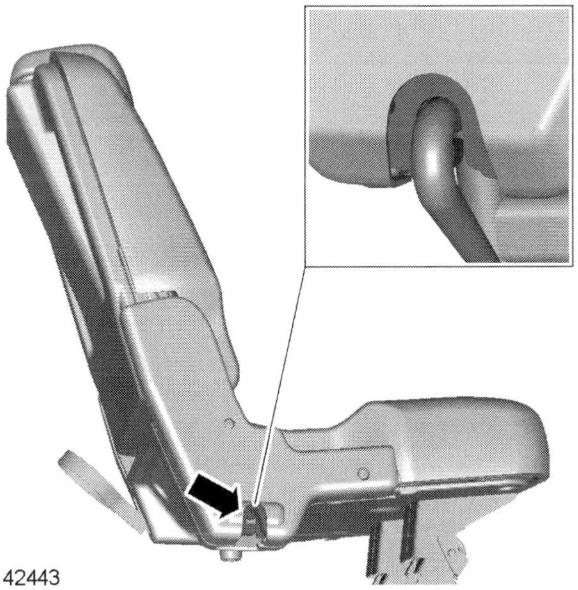

E142443

2. CAUTION: Make sure that the seat latch is located correctly to the retaining bar.

   NOTE: The seat backrest cannot be repositioned to the upright position until the seat has been locked correctly to the retaining bar.

Seating - Rear Seat Cushion

Removal and Installation

Removal

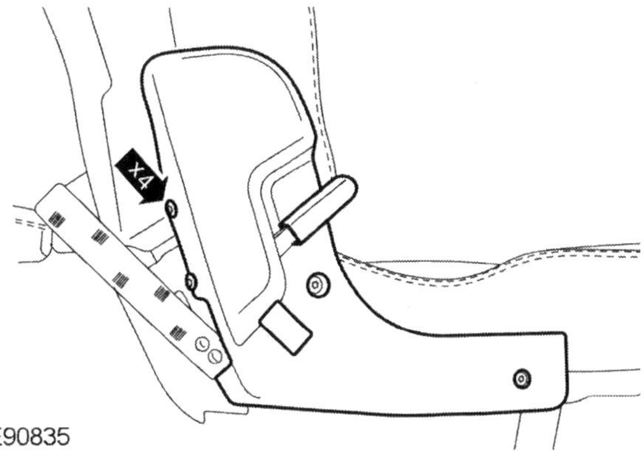

E90835

1. Remove the rear seat outer hinge trim panel.
   - Remove the rear seat recliner handle.
   - Remove the 4 screws.

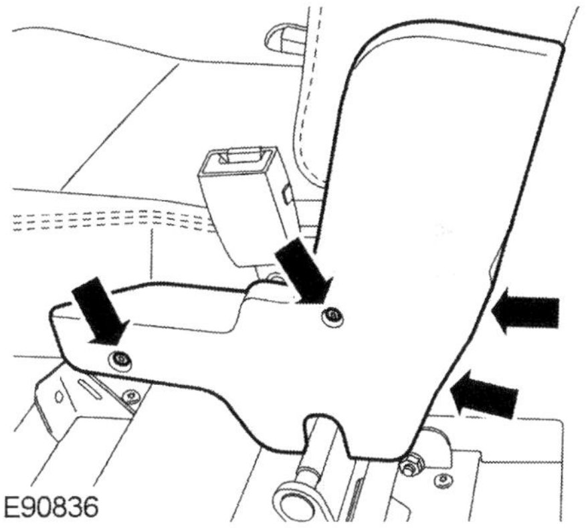

E90836

2. Remove the rear seat inner hinge trim panel.
   - Remove the 4 screws.

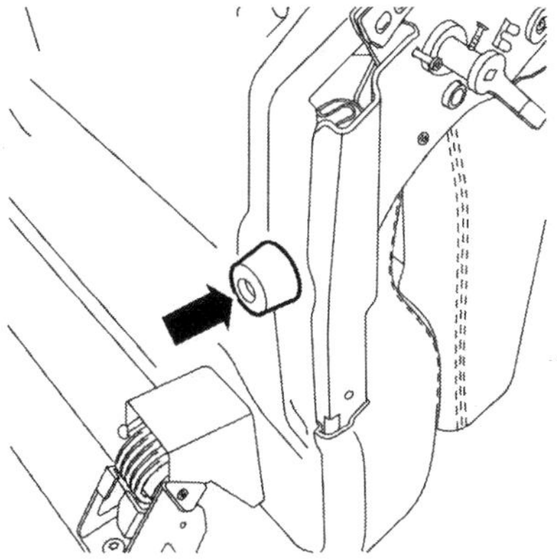

E90837

3. Remove the rear seat base stop.
   - Remove the bolt.
   - Collect the plastic washer.

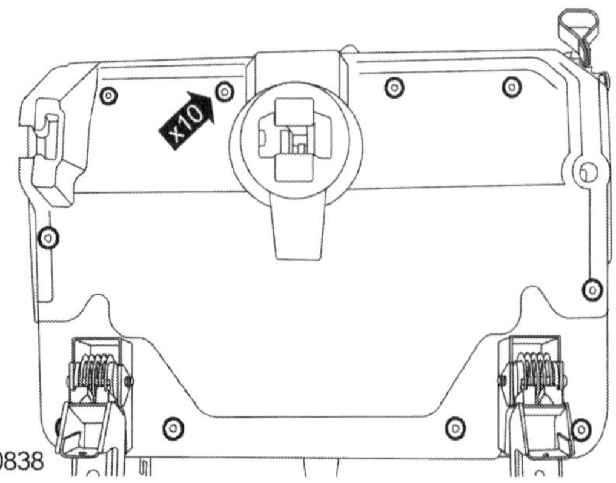

E90838

4. Remove the rear seat cushion trim panel.
   - Remove the 10 clips.

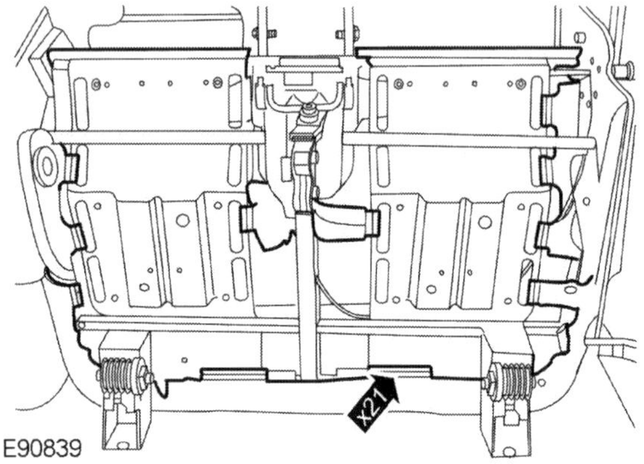

E90839

5. Remove the rear seat cushion.
   - Release the 21 clips.

Installation

1. Install the rear seat cushion.
   - Secure the clips.
2. Install the rear seat cushion trim panel.
   - Install the clips.
3. Install the rear seat base stop.
   - Install the plastic washer.
   - Tighten the bolt to 25 Nm (17 lb.ft).
4. Install the rear seat inner hinge trim panel.
   - Tighten the screws.
5. Install the rear seat outer hinge trim panel.
   - Tighten the screws.
   - Install the rear seat recliner handle.
6. ⚠ CAUTION: Make sure that the seat latch is located correctly to the retaining bar.

△ NOTE: The seat backrest cannot be repositioned to the upright position until the seat has been locked correctly to the retaining bar.

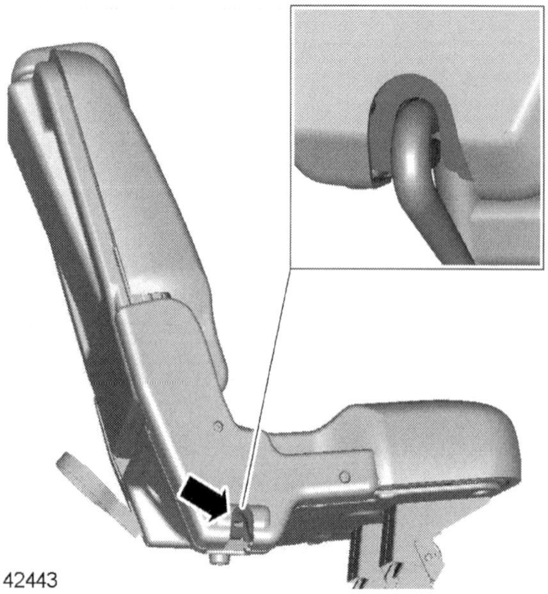
E142443

Rear Seat Cushion Cover

Removal and Installation

Part(s)

STEP	PART NAME	QUANTITY
Removal Step 2	Renew Part Hog rings	12
Removal Step 3	Renew Part Hog rings	20

Removal

1. Remove the rear seat cushion.

Refer to: Rear Seat Cushion Cover (501-10 Seating, Removal and Installation)

E90833

2. Release the under side of the rear seat cushion cover. Remove and discard the 12 hog rings.

   Renew Part: Hog rings

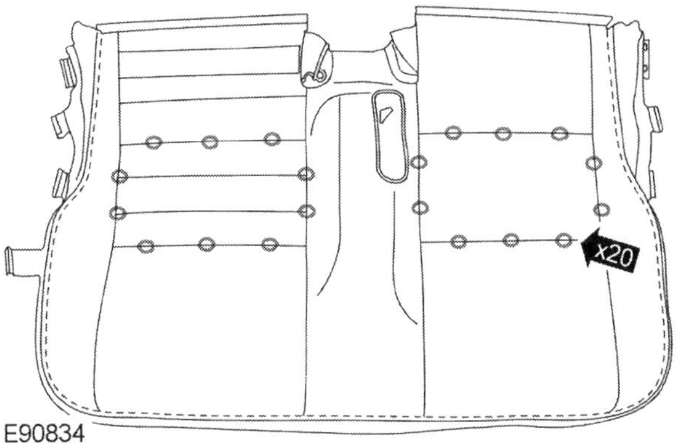

E90834

3. Remove the rear seat cushion cover. Remove and discard the 20 hog rings.

   Renew Part: Hog rings

Installation

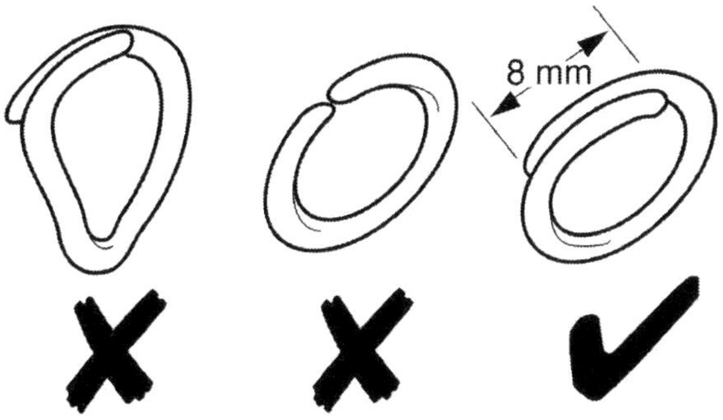

VUJ0005059

1. NOTE: Use hog ring pliers to close the hog rings. Do not use any other tool. The hog rings must be closed to overlap as illustrated.

   To install, reverse the removal procedure.

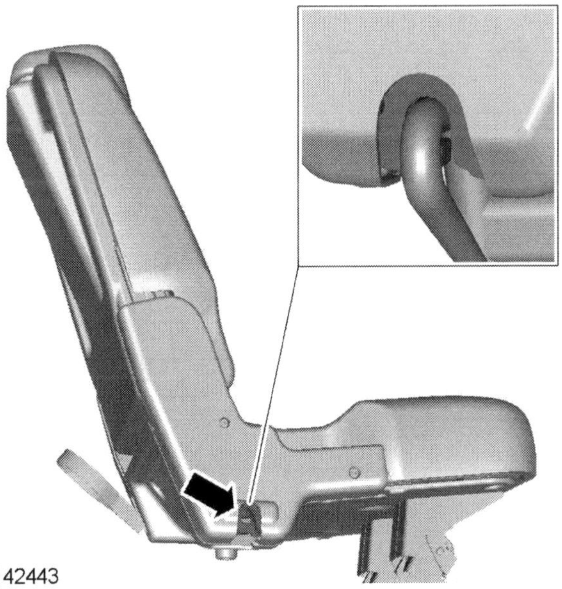

E142443

2. CAUTION: Make sure that the seat latch is located correctly to the retaining bar.

   NOTE: The seat backrest cannot be repositioned to the upright position until the seat has been locked correctly to the retaining bar.

Third Row Seat

Removal and Installation

Removal

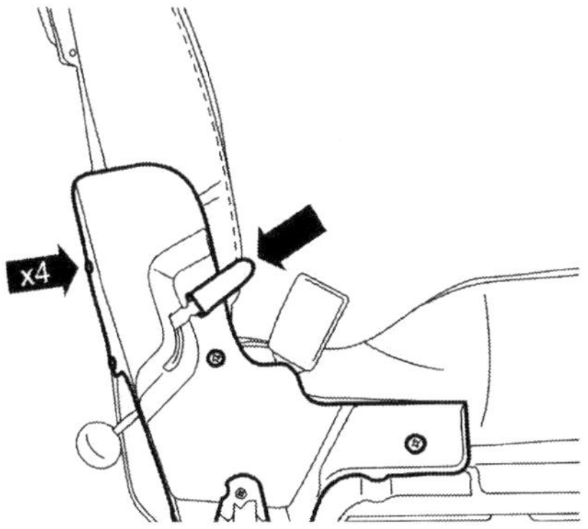

1. Remove the third row seat inner hinge trim panel.
   - Remove the third row seat recliner handle.
   - Remove the 4 screws.

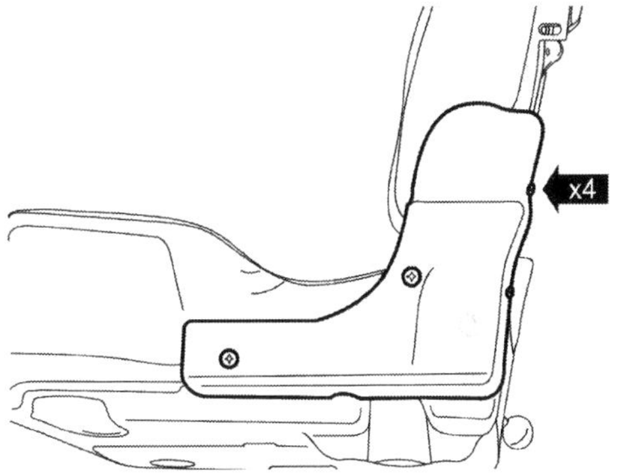

2. Remove the third row seat outer hinge trim panel.
   - Remove the 4 screws.

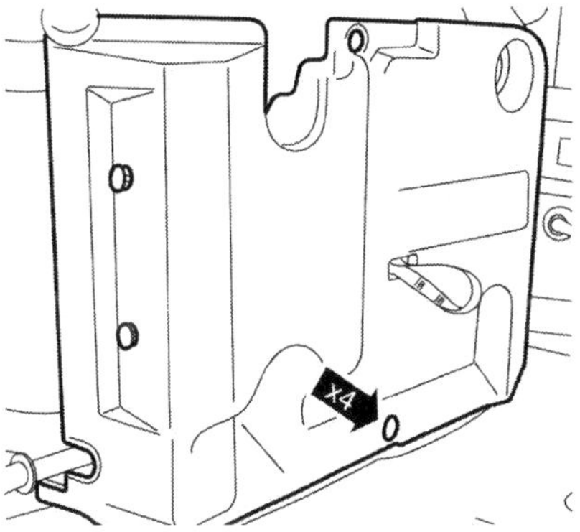

3. Remove the third row seat lower trim panel.
   - Remove the 4 clips.

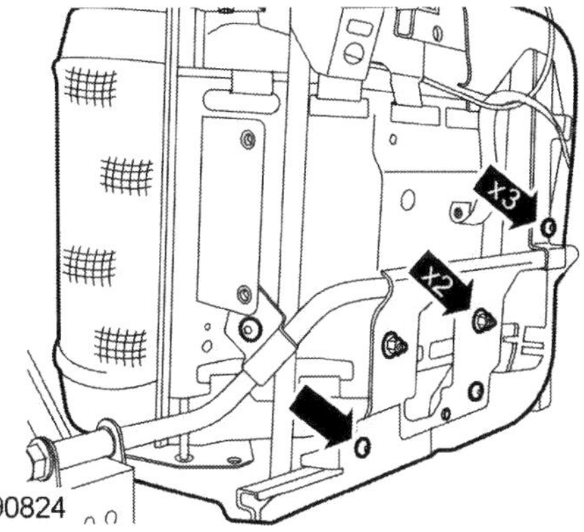

E90824

4. Remove the third row seat.

- Remove the nut and bolt.
- Remove the 2 nuts.
- Remove the 3 bolts.
- Collect the 2 plastic washers.

Installation

1. Install the third row seat.
   - Install the plastic washers.
   - Tighten the nuts to 23 Nm (17 lb.ft).
   - Tighten the bolts to 23 Nm (17 lb.ft).
   - Tighten the nut and bolt to 23 Nm (17 lb.ft).
2. Install the third row seat lower trim panel.
   - Install the clips.
3. Install the third row seat outer hinge trim panel.
   - Tighten the screws.
4. Install the third row seat inner hinge trim panel.
   - Tighten the screws.
   - Install the third row seat recliner handle.

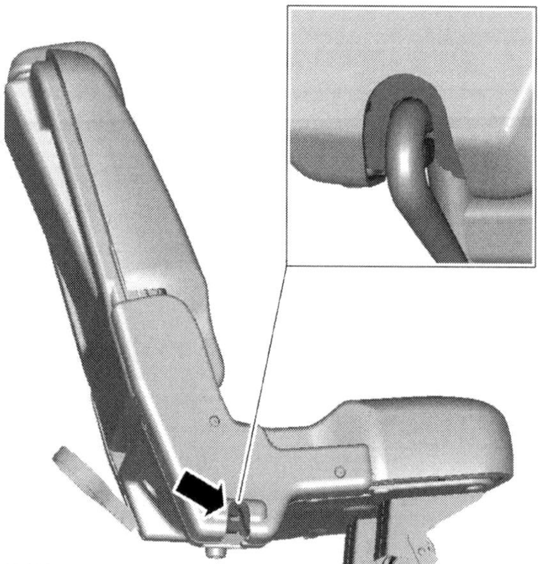

E142443

5. CAUTION: Make sure that the seat latch is located correctly to the retaining bar.

NOTE: The seat backrest cannot be repositioned to the upright position until the seat has been locked correctly to the retaining bar.

Third Row Seat Backrest

Removal and Installation

Removal

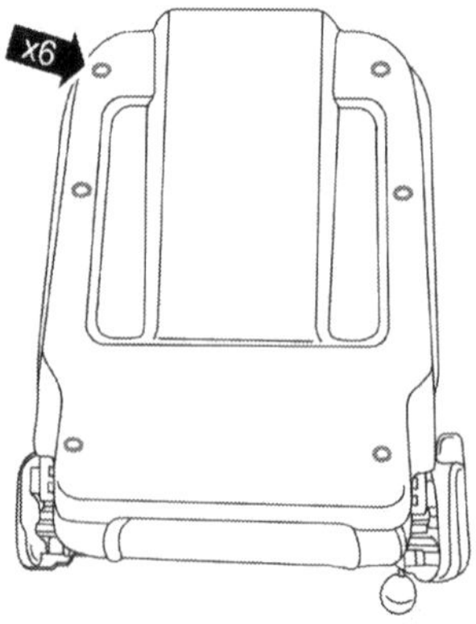

E90829

1. Remove the third row seat backrest trim panel.
   - Release the 6 clips.

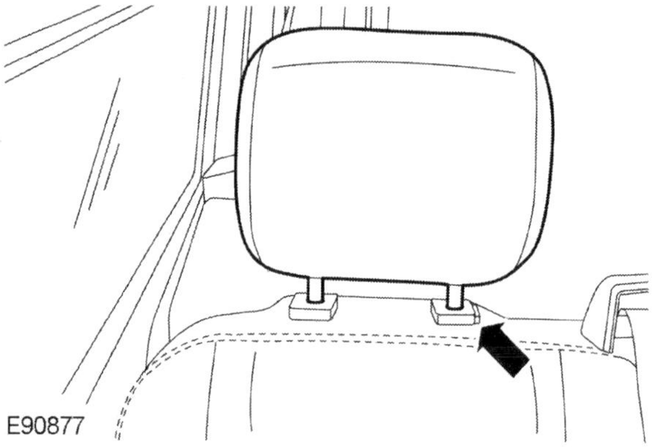

E90877

2. Remove the third row seat head restraint.
   - Release the clip.

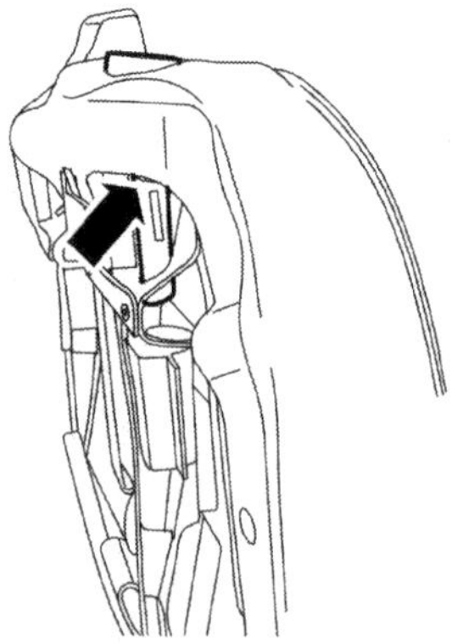

E90880

3. Remove the 2 third row seat head restraint guides.
   - Release the 2 clips.

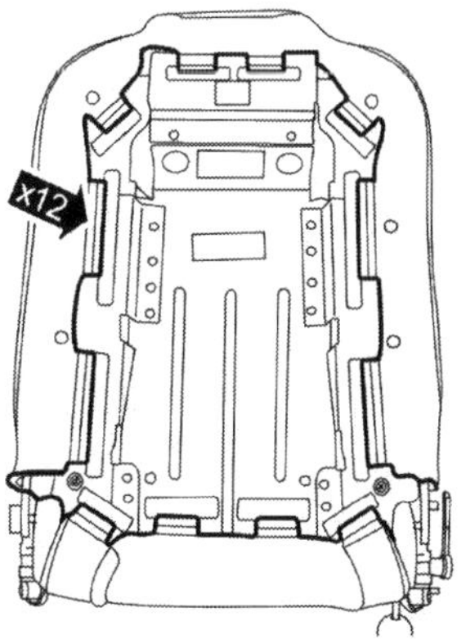

E90830

4. Remove the third row seat backrest.

   - Release the 12 clips.

Installation

1. To install, reverse the removal procedure.

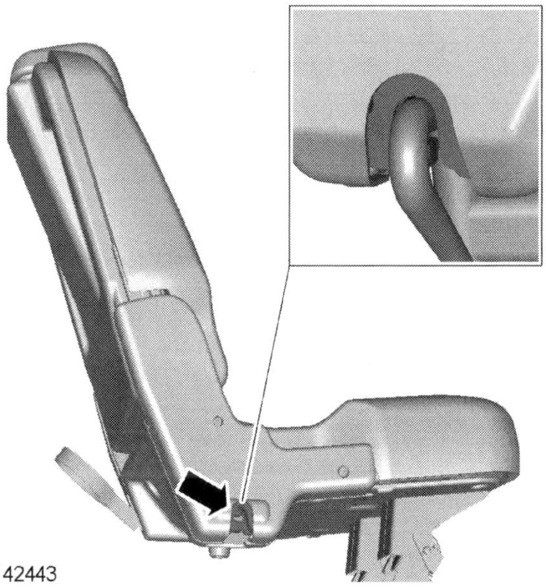

E142443

2. CAUTION: Make sure that the seat latch is located correctly to the retaining bar.

   NOTE: The seat backrest cannot be repositioned to the upright position until the seat has been locked correctly to the retaining bar.

Third Row Seat Backrest Cover

Removal and Installation

Part(s)

STEP	PART NAME	QUANTITY
Removal Step 2	Renew Part Hog rings	4
Removal Step 3	Renew Part Hog rings	9

Removal

1. Remove the third row seat backrest.

Refer to: Third Row Seat Backrest Cover (501-10 Seating, Removal and Installation)

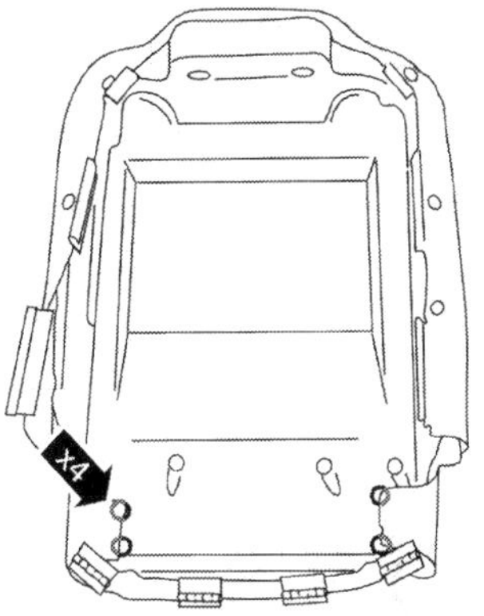

E90831

2. Release the third row seat backrest cover. Remove and discard the 4 hog rings.

   Renew Part: Hog rings

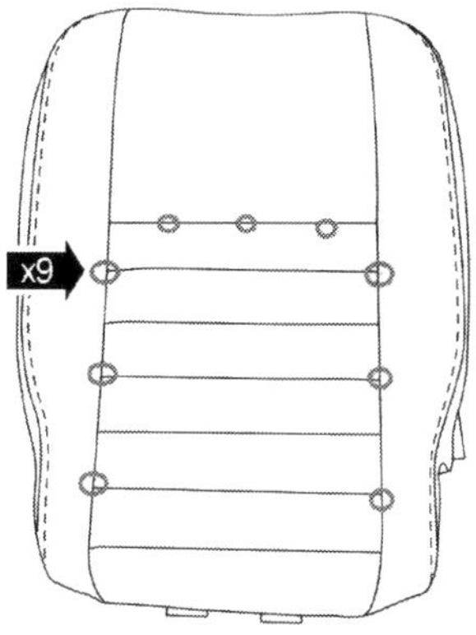

E90832

3. Remove the third row seat backrest cover. Remove and discard the 9 hog rings.

   Renew Part: Hog rings

Installation

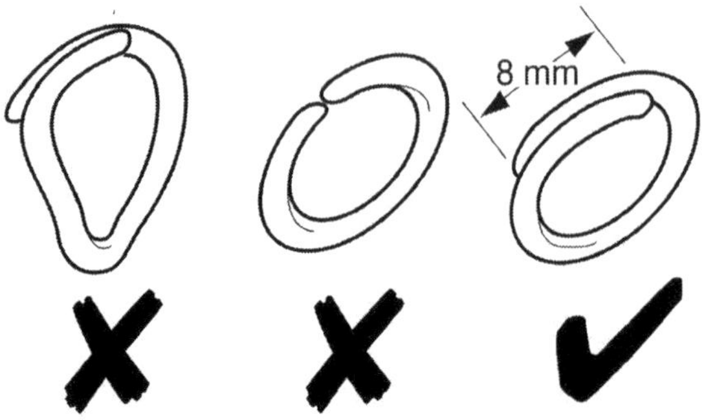

1. NOTE: Use hog ring pliers to close the hog rings. Do not use any other tool. The hog rings must be closed to overlap as illustrated.

   To install, reverse the removal procedure.

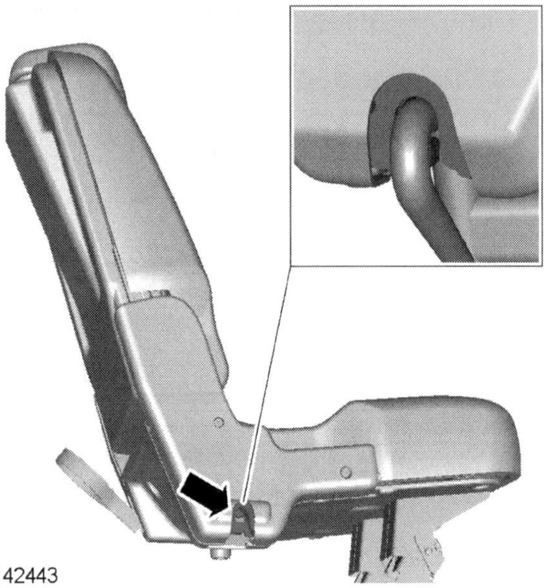

E142443

2. CAUTION: Make sure that the seat latch is located correctly to the retaining bar.

   NOTE: The seat backrest cannot be repositioned to the upright position until the seat has been locked correctly to the retaining bar.

Third Row Seat Cushion

Removal and Installation

Removal

1. Remove the third row seat.

Refer to: Third Row Seat (501-10 Seating, Removal and Installation)

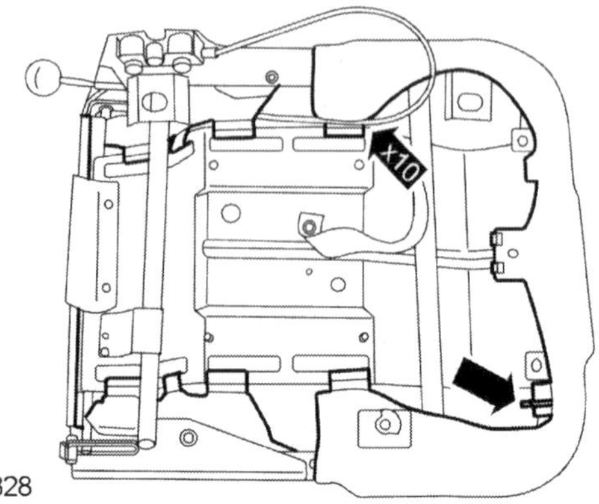

E90828

2. Remove and discard the cable tie. Release the 10 clips. Remove the third row seat cushion.

Installation

1. To install, reverse the removal procedure.

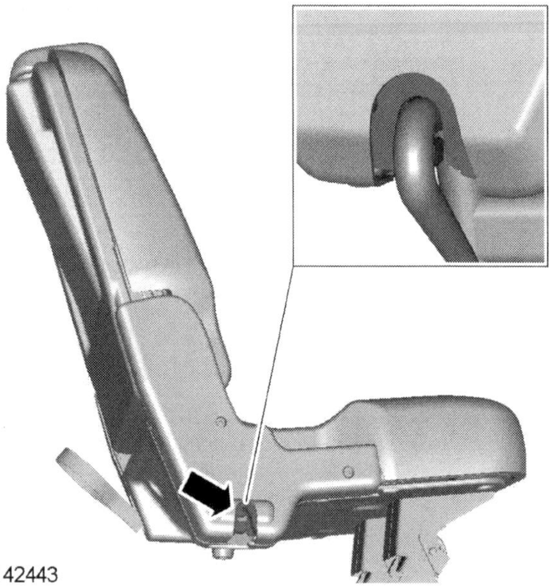

E142443

2. CAUTION: Make sure that the seat latch is located correctly to the retaining bar.

   NOTE: The seat backrest cannot be repositioned to the upright position until the seat has been locked correctly to the retaining bar.

Third Row Seat Cushion Cover

Removal and Installation

Part(s)

STEP	PART NAME	QUANTITY
Removal Step 2	Renew Part Hog rings	4
Removal Step 3	Renew Part Hog rings	10

Removal

1. Remove the third row seat cushion.

Refer to: Third Row Seat Backrest Cover (501-10 Seating, Removal and Installation)

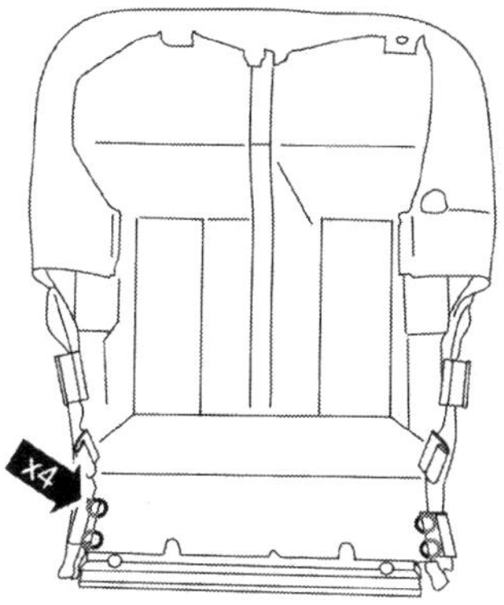

E90825

2. Release the third row seat cushion cover. Remove and discard the 4 hog rings.

   Renew Part: Hog rings

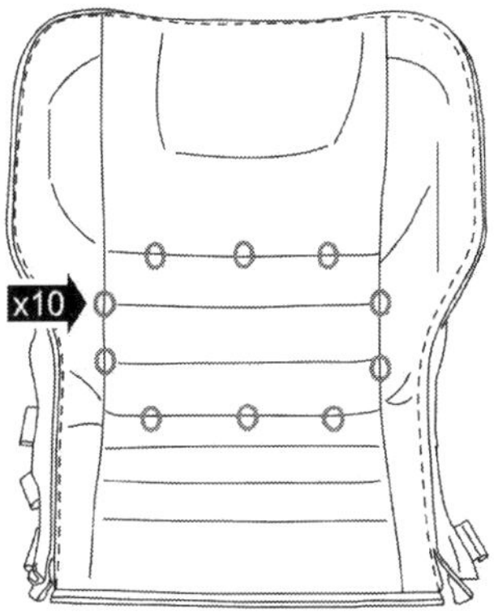

E90826

3. Remove the third row seat cushion cover. Remove and discard the 10 hog rings.

   Renew Part: Hog rings

Installation

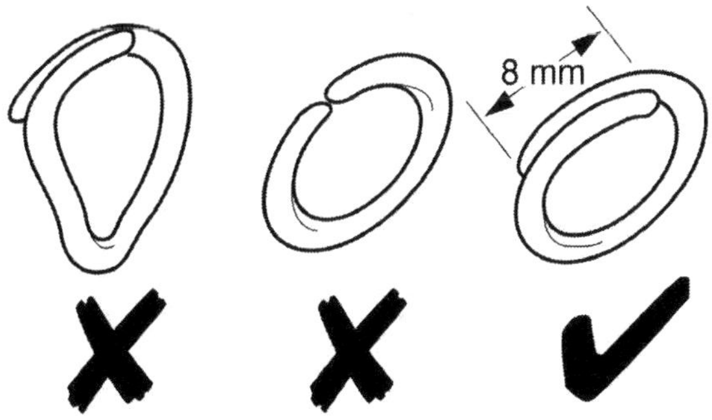

1. NOTE: Use hog ring pliers to close the hog rings. Do not use any other tool. The hog rings must be closed to overlap as illustrated.

   To install, reverse the removal procedure.

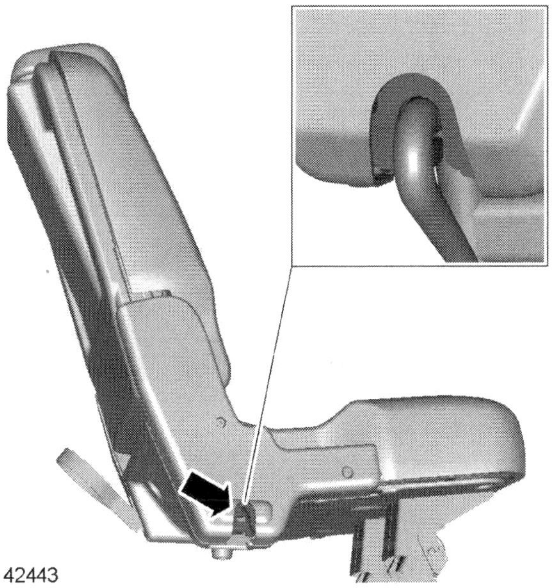

E142443

2. CAUTION: Make sure that the seat latch is located correctly to the retaining bar.

   NOTE: The seat backrest cannot be repositioned to the upright position until the seat has been locked correctly to the retaining bar.

COMPONENT LOCATION

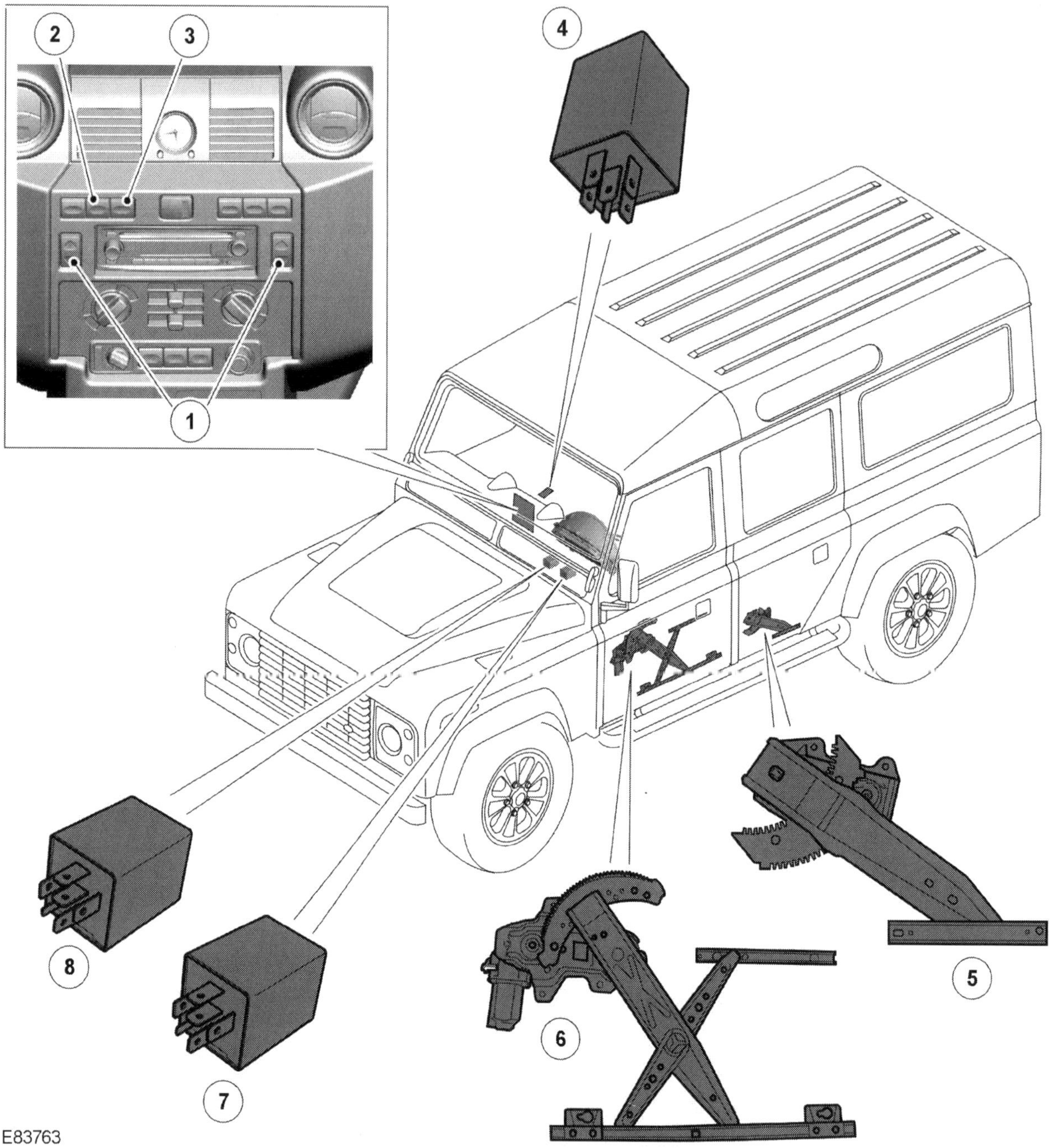

E83763

Item	Description
1	Window lift switches
2	Heated windshield switch
3	Heated rear window switch
4	Heated windshield module
5	Rear door window regulator (if fitted)
6	Front door window regulator and motor assembly
7	Heated rear window relay
8	Window lift relay

OVERVIEW

Window Lift

The front window lift system comprises a motor in each front door and 2 rocker switches mounted in the center console. The window lift system will only operate when the ignition switch is in position II (ignition on) and features anti-trap functionality.

The rear window lift system (if fitted) comprises a handle and manually operated door window regulator in each rear door.

Heated Windshield and Rear Window

The heated windshield contains a single heating element which is powered for up to a maximum of 8 minutes when selected. The system is only operational when the engine is running.

The heated rear window also contains a single heating element and will operate with the ignition switch in position II (ignition on). The heated rear window isn't constrained to an operational time limit.

PRINCIPLES OF OPERATION

Window Lift

The window lift relay coil is energized by an ignition feed from the CJB (central junction box) . The energized relay provides a battery voltage feed to the LH (left-hand) and RH (right-hand) window lift switches.

Each window lift switch has 2 connections with its respective window lift motor; feed and return. These lines switch depending on the direction of travel requested using the switch. In each case, the motor is provided a ground path through its switch.

Each motor has thermal cut-out protection. If the window reaches the top or bottom of its travel, or hits an obstruction, the thermal cut-out will sense the increased load on the motor and cut the power supply to the motor brush contacts. Thermal cut-out time is between 3.5 to 5 seconds, with a re-start time of between 1 to 10 seconds.

Heated Windshield

Operation of the heated windshield is controlled by the heated windshield module. The heated windshield module is located behind the instrument cluster and receives an engine running signal from the ECM (engine control module) . The heated windshield module also monitors the condition of the heated windshield switch. If the switch is pressed, a ground path is created. If both these signals are in place, the heated windshield module provides a feed to the heated windshield element until either;

- the switch is pressed a second time
- 8 minutes elapses
- or the engine is stopped.

Heated Rear Window

The heated rear window switch is located in the center console and is provided an ignition feed from the CJB. When the switch is pressed, a feed is provided to the the heated rear window relay coil. The now energized heated rear window relay provides a battery voltage feed from the CJB to the heated rear window element.

Fixed Window Glass

Diagnosis and Testing

Principles of Operation

For a detailed description of the Glass, Frames and Mechanisms, refer to the relevant Description and Operation section in the workshop manual.

Refer to: Glass, Frames and Mechanisms (501-11 Glass, Frames and Mechanisms, Description and Operation)

Refer to: Glass, Frames and Mechanisms (501-11 Glass, Frames and Mechanisms, Description and Operation)

Inspection and Verification

CAUTION: Diagnosis by substitution from a donor vehicle is **NOT** acceptable. Substitution of control modules does not guarantee confirmation of a fault, and may also cause additional faults in the vehicle being tested and/or the donor vehicle.

NOTES:

If a control module or a component is at fault and the vehicle remains under manufacturer warranty, refer to the Warranty Policy and Procedures manual, or determine if any prior approval program is in operation, prior to the installation of a new module/component.

When performing voltage or resistance tests, always use a digital multimeter that has the resolution ability to view 3 decimal places. For example, on the 2 volts range can measure 1mV or 2 K Ohm range can measure 1 Ohm. When testing resistance always take the resistance of the digital multimeter leads into account.

Check and rectify basic faults before beginning diagnostic routines involving pinpoint tests.

Refer to Section 100-00 General Information for window glass health and safety precautions.

1. Verify the customer concern
2. Visually inspect for obvious signs of damage and system integrity

Visual Inspection

Electrical
• Physical damage to the windshield

3. If an obvious cause for an observed or reported concern is found, correct the cause (if possible) before proceeding to the next step

4. If the cause is not visually evident, verify the symptom and refer to the Symptom Chart, alternatively check for Diagnostic Trouble Codes (DTC (Diagnostic Trouble Code(s)) s) and refer to the DTC Index

Symptom Chart

Symptom	Possible Causes	Action
Scratches	• Debris trapped under a wiper blade • Foreign object damage • Fouling by trim	• GO to Pinpoint Test A
Chips	• Foreign object damage	• GO to Pinpoint Test B
Cracks	• Foreign object damage • Impact damage during assembly	• GO to Pinpoint Test C
Delamination	• Manufacturing defect	• GO to Pinpoint Test D

PANORAMIC ROOF GLASS DAMAGE INSPECTION PROCEDURE

NOTES:

⚠ Photographic evidence of the origin of the glass damage concern must be retained (and available on request) to support both customer conversations and Warranty claims. For further guidance, refer to section A1.20 of the Global Warranty Manual in relation to Glass.

⚠ Damage from an external influence must not be claimed under Warranty. All Warranty claims are subject to audit.

Where customers present vehicles with damage/cracks in the panoramic roof glass, the following guidance is designed to help retailers/authorized repairers determine if cracks in the panoramic roof glass are the result of a manufacturing defect in the glass substrate or, as is more likely, as a result of impact damage, such as a stone chip.

The procedures outlined below should always be completed whenever a vehicle is presented with reports of damage/cracks in the panoramic roof glass. Regardless of if the repair is to be claimed under Warranty or not, the procedures for gathering and recording evidence of the nature of the cause of the damage should be followed.

Preparation

Use a suitable glass cleaner to thoroughly clean the inside and outside of the glass panel. Be careful not to apply excess pressure to the glass panel as this may result in further damage.

Equipment Required

- Paint Marker or China Graph pencil
- Pencil with a sharp point
- Magnifying glass or equivalent

Optimal Viewing Conditions

The inspection should ideally be completed in a well-lit workshop, with the vehicle placed so as to minimize any reflections on the roof glass. To confirm the precise location and root cause of the concern, it may be necessary to survey BOTH the interior and exterior surfaces of the panoramic roof glass panel for signs of cracks/damage.

Damage Inspection Procedure

1. As shown below, use the point of a sharpened pencil to feel for the profile of the crack/breach in the glass (as indicated by a step or rough area on the otherwise smooth surface of the glass). **NOTE:** this should be done on both the interior and exterior surfaces of the roof glass.

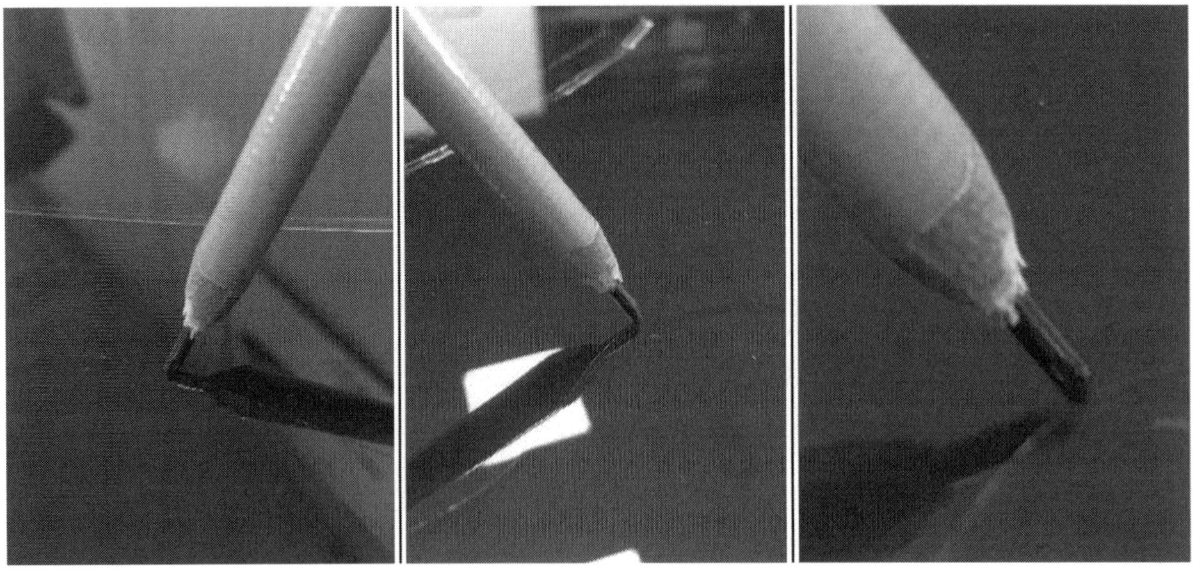

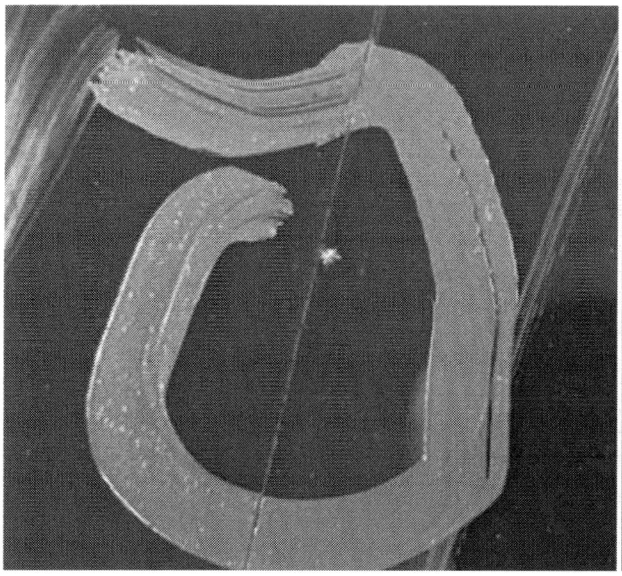

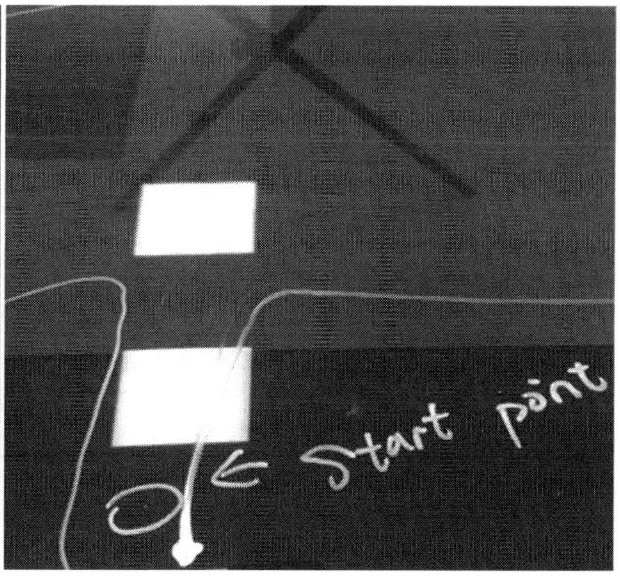

2. Feeling for any step or rough area on the glass, follow the path of the crack with the tip of a sharp pencil. Using a magnifying glass, follow the path of the crack and identify any visual signs of impact damage on the glass surface. If a breach in the glass surface is evident, then the damage to the glass will NOT be covered by the vehicle warranty and the cost of repairs will need to be covered by the customer (or the customer's insurance policy). **NOTE:** not all cracks will necessarily run through the centre of the discernible impact point. See examples of impact points/impact damage shown in the section below.
3. Using a paint marker or chinagraph pencil, circle the origin of the crack/damage identified during the inspection of the glass panel (as shown below). Care should be taken to avoid any further damage when marking-up the glass.
4. Capture the following set of images (see examples below) as supporting evidence that the faulty glass is still installed on the vehicle. Make sure the images are unobstructed, in focus and plainly show the damage to the glass. If necessary carefully remove any damage to the glass.

If necessary. carefully remove any obstructing trims to gain clear access to the damaged portions of the roof glass before capturing the following images:

1. An image showing an overview of the entire glass panel
2. A close-up image of the root cause (impact point) of the crack/damage
3. A close-up image of the glass ID mark/ trademark logo

These images should be retained in case they are requested for Warranty audit purposes or to facilitate discussions with the customer regarding retail or insurance repairs.

Example Of Damage To Roof Glass

Shown below are examples of cracked/damage roof glass. The top row photos show the original impact points that caused the cracks and other damage to the roof glass. **NOTE:** not all cracks will necessarily run through the centre of the discernible impact point. The bottom row shows the magnified images of the impact points.

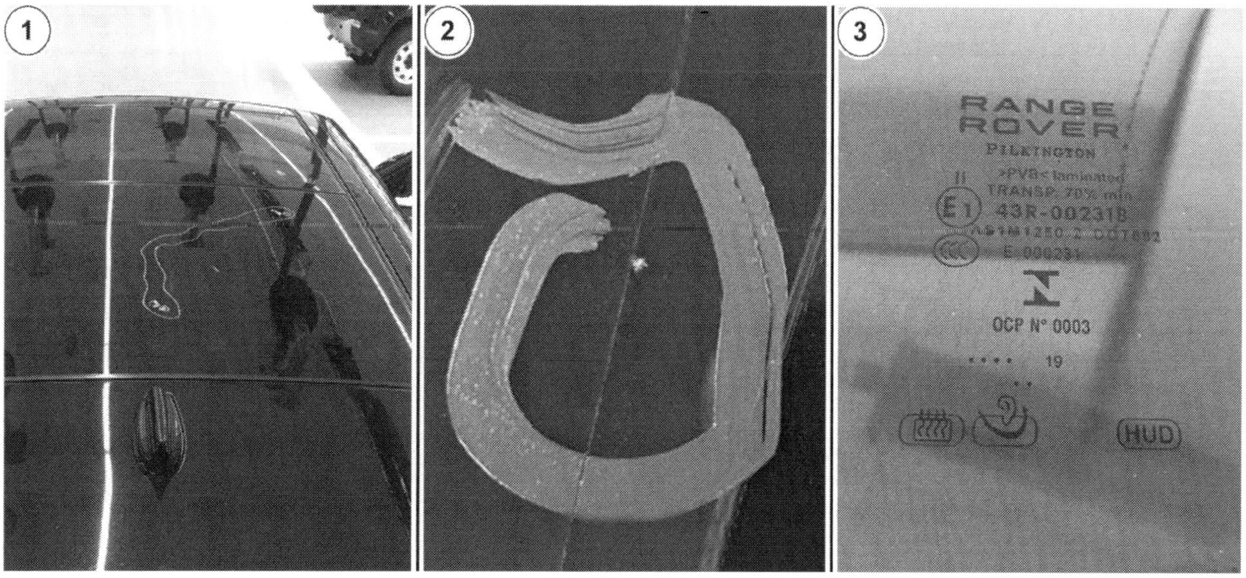

NOTE: All of the examples below show damage to the roof glass that has been caused by some foreign objects that have impacted on the glass. Therefore, none of these instances of damage would be suitable for a Warranty claim. Repairs for these kinds of concerns would need to be undertaken as a retail or insurance repair.

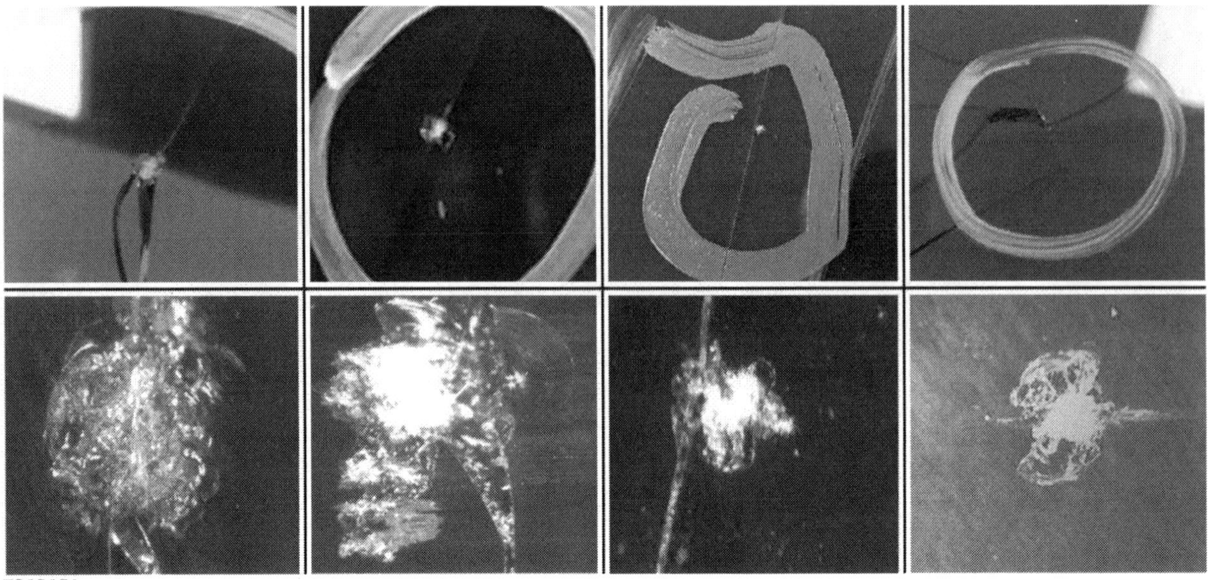

E263651

Pinpoint Tests

PINPOINT TEST A : SCRATCH TESTS	
TEST CONDITIONS	DETAILS/RESULTS/ACTIONS
A1: SCRATCH TEST 1	
NOTE: A scratch will usually be regular in shape, following the line of the object that caused it.	
	1. Probe using the tip of a pencil to identify a groove in the windshield surface.
	Is there a groove? **Yes** Windshield scratched. GO to 2 **No** Defect not valid.
A2: SCRATCH TEST 2	
	1. Check for trim, body panels, or foreign objects that may have caused the scratch.
	Was the scratch caused by a foreign object? **Yes** The damage is not due to a defect or an assembly error. **No** Rectify as appropriate.

PINPOINT TEST B : CHIP TESTS	
TEST CONDITIONS	DETAILS/RESULTS/ACTIONS
B1: CHIP TEST 1	
NOTE: Impact damage may cause a crack to form.	
	1. Assess the damage by probing with the tip of a pencil.

PINPOINT TEST B : CHIP TESTS	
TEST CONDITIONS	**DETAILS/RESULTS/ACTIONS**
	Is the damaged area rough (indicating a breach of the windshield surface)? **Yes** Damage caused by the impact of a foreign object. Not a manufacturing defect. **No** Install a new windshield.

PINPOINT TEST C : CRACK TESTS	
TEST CONDITIONS	**DETAILS/RESULTS/ACTIONS**
C1: CRACK TEST 1	
	1. △ NOTE: A crack will be detectable as a step in the glass. Confirm the presence of a crack by probing with the tip of a pencil.
	Is the windshield cracked? **Yes** Windshield cracked. GO to 2 **No** Windshield not cracked. GO to Pinpoint Test A
C2: CRACK TEST 2	
△ NOTE: Multiple cracks will radiate out from the source.	
	1. Assess the source of the crack by probing with the tip of a pencil.
	Is there evidence of impact damage being the source of the crack? **Yes** GO to Pinpoint Test B **No** Install a new windshield.

PINPOINT TEST D : DELAMINATION TESTS	
TEST CONDITIONS	**DETAILS/RESULTS/ACTIONS**
D1: DELAMINATION TEST 1	
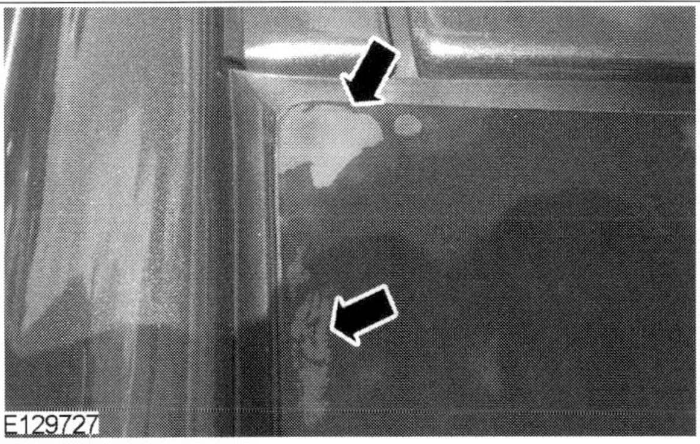	1. Visually assess the windshield for delamination.
	Have the glass laminates separated? **Yes** Install a new windshield. **No** No further action.

Fixed window glass wind noise and water leaks

General

△ NOTE: If wind noise is evident at varying vehicle speeds, or water ingress is evident, perform the following procedure below.

- If wind noise or water leaks occur after bodywork repairs, the cause can be established using the checks described below. A systematic and logical procedure is required to locate wind noise or water leaks. Before beginning extensive checks, a thorough visual inspection must be completed.
- Visual Inspection
    - The following characteristics may indicate existing water leaks:
    - Check the clearance and accurate fit of ancillary components such as the hood, tailgate, doors, and so on.
    - Check for correct fit and possible damage to sealing elements such as blanking plugs, rubber door weather strips, and so on.
    - Check water drain holes for unhindered flow.
- Various tests can be used to provide further information on possible leaks.
    - Water test
    - Washer test
    - Road test
    - Chalk (powder) test

Practical execution of tests and checks

Water test

△ NOTE: Never aim a jet of water directly at a door weather strip or rubber seals.

- Complete the water test with a second person present (in the passenger compartment).
- Use variable washer nozzles (concentrated water jet to fine spray mist).
- Start in the lower section and spray the whole area, working upwards in stages.

Washer test

- Further tests can be completed in the washer system.
- Some leaks originate here, or only occur here.
- The relevant passenger compartment should be checked using a flashlight during the wash procedure.

Road test

- If no leaks are located during the tests above, road tests should be completed on wet roads.
- Road tests under various conditions:
- At various speeds.
- On various road surfaces (asphalt to cobbles).
- With loaded or unloaded vehicle.
- Driving through puddles (splash water).

Chalk test (powder test)

- In this test, the clamping load and the bearing surface of the seal are checked.
- Performing the test:
- Dust the door seal with powder or coat with chalk.
- Coat the bearing surface of the seal with a thin film of petroleum jelly.
- Slowly close the door and open it again.
- Check the width and continuity of the imprint on the door seal.

Other test equipment

- Other equipment such as stethoscopes, UV (ultraviolet) lamps, special mirrors or ultrasound measuring instruments can be used to locate leaks.

Rectifying the leak using recommended tools, auxiliary equipment and materials

- Tools and auxiliary equipment:
- Dry, absorbent cloths
- Variable washer nozzle
- Flashlight, fluorescent tube
- Mirror
- Compressed air
- Seal lip installer
- Wet/dry vacuum cleaner
- Sealing compound compressor
- Remover for interior trim
- Cutter blade or pocket knife
- Wedge (wood or plastic)
- Hot air blower
- Special mirror for concealed leaks
- Air flow checker
- Sealing compound (tape and plastic compound)

- Multi-purpose sticker
- Clinched flange sealer
- Window sealing compound
- Water shield PVC (polyvinyl chloride)
- Double-sided adhesive tape for water shield
- Methylated spirit (available from trade outlets)
- PU (polyurethane) adhesive
- Silicone remover
- Tar remover

Water leaks according to mileage or running time

Increasing mileage has an effect on the problem of leaks in a vehicle. Possible influencing factors are:

- Servicing and maintenance of seals:
    - No maintenance, lack of maintenance or incorrect maintenance
    - Using an incorrect agent
- Damaged seals:
    - As a result of aging, wear or incorrect handling/assembly.
- Heavy soiling of the vehicle:
    - Heavy soiling of a vehicle can seriously impair the function of water drainage channels in particular, and also of rubber seals.
- Age-related factors:
    - Environmental factors
    - UV sunload
    - Extreme climatic conditions
- Corrosion can have a serious impact on bodywork,

Water drainage system (illustration for reference only)

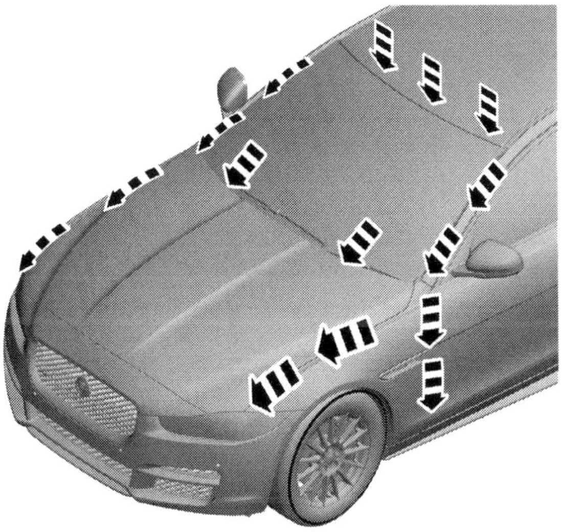

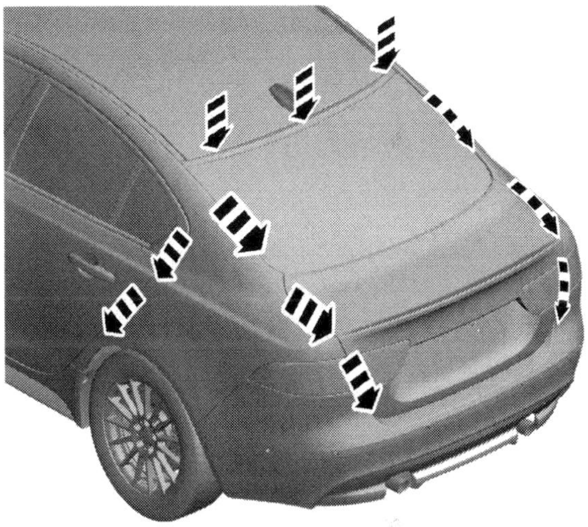

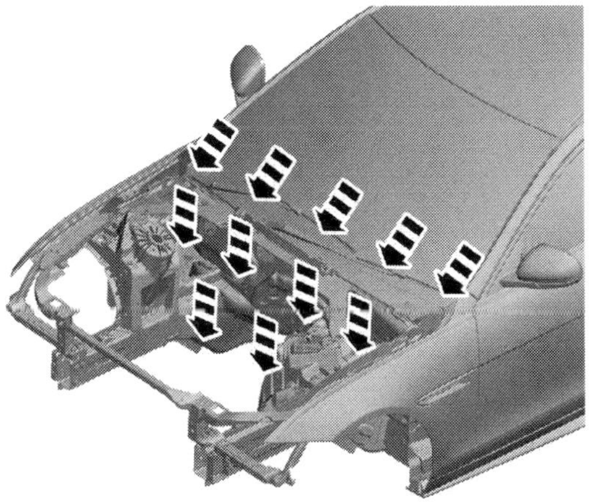

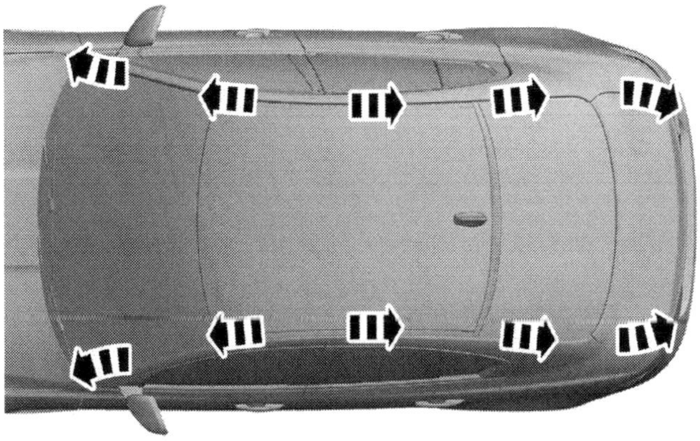

Item	Description
1	Water drainage, front
2	Water drainage, side and rear
3	Engine compartment drainage
4	Roof drainage

in particular as a result of:

- Lightly or heavily rusted seal carriers
- Rusted body seal welds
- Perforation corrosion

Wind noise or water leaks after body repairs

If a vehicle develops wind noise or a water leak after body repairs, the following points must be taken into consideration in particular:

- The correct seating of ancillary components and their seals must be checked.
- The correct alignment of doors/tailgate and tailgate must be checked. The associated seals must not be damaged and must be installed correctly.
- Check that panel seams are correctly sealed.
- The correct seating of rubber grommets must be checked.
- Directly-glazed windows must have correct and complete bonding.

Water drainage system

If a vehicle develops water leaks, then areas into which water is routed or drained should be checked first.

Wind noise or water leaks, diagnosis and corrective action: Front passenger compartment

Windshield

- Diagnosis:
    - Wind noise at varying road speeds and/or ingress of water into A-pillar area or instrument cluster area and rocker panel area.
- Cause:
    - Breaks in adhesive beads.
    - The breaks in adhesive beads can be located from inside by using compressed air. The leak can be identified from outside by the escaping air.
    - The second test method is by means of a water test. The outer trims must be raised carefully using a plastic wedge. The leak should be located from inside by a second assistant.
- Corrective action:
    - Remove the windshield. Remove the existing adhesive bead and reapply using the correct bonding procedure.

Side windows

In the case of side windows, the same problems can arise as for a windshield. The same corrective actions must therefore be used.

Door seal

- Diagnosis:
    - Water ingress in the lower part of the interior door trim or in the rocker panel area.
- Cause:
    - The water shield installed behind the interior door trim exists to drain off water that has entered the door through the drainage holes, either downward or outward. If the water shield seal is damaged or has been installed incorrectly, then water can get into the passenger compartment.
    - In addition to this, the drainage holes can become clogged with leaves, dirt or excess cavity protection agents. Water gathers in the door and ingresses into the passenger compartment.
    - Check water shield for damage or correct fitting.
    - If the water shield needs to be re-bonded, then approved seam sealer should be used.
    - Before the water shield is installed, the drainage holes must be checked for unhindered flow.

Door seals

- Diagnosis:
    - Ingress of water into the rocker panel area
- Cause:
    - Insufficient clamping load between seal and door.
- Corrective action:
    - Check clamping load:
    - The easiest way to check the clamping load of a seal to the respective bearing surface is by means of a paper strip test. This consists of trapping strips of paper at various points between the door and the seal, and fully closing the door. If it is possible to pull out the paper with no great resistance, then the clamping load is too low.
- ⚠ NOTE: When adjusting the clamping load, the profile alignment of the relevant components must always be taken into consideration.

Adjust the clamping load:

- The clamping load is normally adjusted using the striker. When doing so, the edge alignment from the door to the side panel, or from the front door to the rear door must be taken into account.

- Another setting method is to realign the panel flange for the seal mounting. The clamping load is increased by moving the flange towards the door.

- △ NOTE: Do not realign the flange too far in the direction of the door, as this can reduce the bearing surface of the seal to the door.

Check the bearing surface:

  - Apply chalk evenly to the surface of the seal. Evenly coat the bearing surface of the door with petroleum jelly.

  - Close the door fully, the lock must engage. Open the door. The imprint of the chalk (bearing surface) can be identified in the film of petroleum jelly.

  - The bearing surface should be at least 5mm across at all points.

- Other causes:

  - The door seal must completely seal the door where it meets the bodywork.

  - Water can ingress directly or indirectly into the interior of the vehicle if the seal is damaged at any point.

- Corrective action:

  - A damaged or worn door seal must always be installed in full.

  - When installing the seal, the following must be taken into account:

  - Always fit the seal first in the area of the narrow radii (corner points).

  - Next, secure the seal to the flange evenly by tapping lightly with a rubber hammer. The installed seal must not be kinked at any point.

△ NOTE: The prescribed length of a seal must not be shortened.

- Other cause:

  - The door seal is attached to the welded flange all the way round. If this welded flange is uneven or damaged at any point (usually in areas with small radii) then this point could be subject to leaks.

  - A stretched seal carrier can also cause a leak.

  - In both cases, water gets into the vehicle interior under the seal carrier.

- Corrective action:

  - Align the deformed welded flange using a hammer and anvil block, prevent and, if necessary, repair any paint damage.

Roof opening panel - front and rear fixed glass panels

In the case of the roof opening panel - front and rear fixed glass panels, the same problems can arise as for a windshield. The same corrective actions must therefore be used.

Roof opening panel - sliding central glass panel

△ NOTE: In the case of the sliding central glass panel, the rubber seals and the lock actuator or latch mechanism must be checked first of all.

- Diagnosis:

  - Wind noise at varying road speeds and/or ingress of water at the roof opening panel.

- Cause:

  - The roof opening panel frame incorporates drain holes in each front corner allowing water to drain through the A-pillars and each rear corner allowing water to drain through the roof panel channels.

  - Damaged, worn, or incorrectly installed seals on the sliding central glass panel.

  - Incorrect alignment of the sliding central glass panel.

  - Damaged, worn, or incorrectly installed seals on the roof opening panel frame.

  - Water drain channels or water drain holes blocked.

- Corrective action:

  - Check the seals on the sliding central glass panel, install as necessary.

  - Check the correct adjustment of the sliding central glass panel, adjust as necessary, replace any worn or damaged components as necessary.

  - Check the internal seal of the roof opening panel frame, realign or install as necessary.

  - Check the water drain channels and holes for blockages, clear blockages as necessary.

  - Check the drainage system for unhindered flow.

Luggage compartment lid

- Diagnosis:
  - Ingress of water into the luggage area.
- Cause:
  - The leak problems of the luggage compartment lid correspond to those of the doors.
  - In addition to this, the area to be sealed is much bigger. The routing holes for cables and hoses must also be sealed.
  - The rubber grommets for the routing holes must be checked for damage and correct seating (fully unhooked).
  - The grommets around the luggage compartment lid hinges may leak.
- Corrective action:
  - Check the rubber grommets and re-position, reseal or install new, if necessary.

Forced air extraction

- Diagnosis:
  - Ingress of water into side luggage compartment area
- Cause:
  - The forced air extraction for the vehicle interior is located in the quarter panel lower extension.
  - The rubber flap of the forced air extraction must be able to move freely.
- Corrective action:
  - Remove the forced air extraction.
  - Check the seal area between the bodywork and housing, as well as the rubber flap.
  - Install a new seal if necessary.

Rear window

- Diagnosis:
  - Wind noise at varying road speeds and/ or ingress of water into the passenger compartment area.
- Cause:
  - Rear window leaking through breaks in the adhesive bead.
  - Check for leaks in the same way as for a leaking windshield.
- Corrective action:
  - In the case of the rear window, the same problems can arise as for a windshield. The same corrective actions must therefore be used.

DTC Index

For a complete list of all Diagnostic Trouble Codes (DTCs) that could be set on this vehicle, please refer to Section 100-00.

Windshield Water Ingress Diagnostics

Diagnosis and Testing

Principles of Operation

For a detailed description of the Glass, Frames and Mechanisms, refer to the relevant Description and Operation section in the workshop manual.

Inspection and Verification

CAUTION: Diagnosis by substitution from a donor vehicle is **NOT** acceptable. Substitution of control modules does not guarantee confirmation of a fault, and may also cause additional faults in the vehicle being tested and/or the donor vehicle.

NOTES:

If a control module or a component is at fault and the vehicle remains under manufacturer warranty, refer to the Warranty Policy and Procedures manual, or determine if any prior approval program is in operation, prior to the installation of a new module/component.

When performing voltage or resistance tests, always use a digital multimeter that has the resolution ability to view 3 decimal places. For example, on the 2 volts range can measure 1mV or 2 K Ohm range can measure 1 Ohm. When testing resistance always take the resistance of the digital multimeter leads into account.

Check and rectify basic faults before beginning diagnostic routines involving pinpoint tests.

Refer to Section 100-00 General Information for window glass health and safety precautions.

1. Visually inspect for obvious signs of damage and system integrity

Visual Inspection

Mechanical
• Deterioration of window seals

2. If an obvious cause for an observed or reported concern is found, correct the cause (if possible) before proceeding to the next step

3. If the cause is not visually evident, verify the symptom and refer to the Symptom Chart, alternatively check for DTC (Diagnostic Trouble Code(s)) and refer to the DTC Index

Symptom Chart

Symptom	Possible Causes	Action
Ingress of water into A-pillar area or instrument cluster area	• Breaks in seals or adhesive beads	• Identify the area of the leak (see procedures outlined below). Remove the windshield. Remove the existing adhesive bead and reapply using the correct bonding procedure

Windshield Water Ingress

General

NOTES:

If water leaks are noted at the front end of the headliner, this may be as a result of windshield water ingress OR water ingress from the pano roof glass (if installed). To correctly identify the point of water ingress, lower the headliner and look for evidence of water ingress both from the windshield and the pano roof glass BEFORE continuing with the diagnostics described below. For guidance on diagnosis of pano roof glass (if installed) water ingress, see separate section in this manual (REFER to: Section 501-17 Roof Opening Panel - Diagnosis and Testing - Water Ingress Diagnostics).

If water ingress is evident, perform the following procedures below.

If water leaks occur, the cause can be established using the checks described below. A systematic and logical procedure is required to locate water leaks. Before beginning extensive checks, a thorough visual inspection must be completed.

- Visual Inspection
  - The following characteristics may indicate existing water leaks:
  - Check the clearance and accurate fit of ancillary components such as the hood, tailgate, doors, and so on.
  - Check for correct fit and possible damage to sealing elements such as blanking plugs, rubber door weather strips, and so on.
  - Check water drain holes for unhindered flow.

- Various tests can be used to provide further information on possible leaks:
  - The breaks in adhesive beads can be located by using compressed air from inside the vehicle. The leak location can be identified from outside by the escaping air.
  - Water test
  - Washer test
  - Road test

Practical execution of tests and checks

Compressed air test

⚠ NOTE: Complete the compressed air test with a second person present (in the passenger compartment).

- Lower the headliner at the front of the vehicle to give clear access to the upper section of the windshield inside the vehicle.
- The leak may be located behind the exterior A-pillar windshield mouldings (if installed), so these should be removed before commencing testing **Make sure the windshield moulding carrier remains installed** as this may be a source of the water leak and should be in situ for the tests as described below.
- On the outside of the vehicle, use a hand-held spray-bottle to apply a fine mist of soapy water to the areas all around the exterior windshield edge joints (see image A - NOTE: image is NOT vehicle specific).
- From the inside of the vehicle, use an air hose to blow a jet of compressed air around the interior windshield edge joints.
- The leak should be located from outside by a second assistant. Bubbles will form in the affected area and escaping air may be audible
- The leak should be located from outside by a second assistant. Bubbles will form in the affected area and escaping air may be audible (see image B - NOTE: image is NOT vehicle specific).
- Take a photograph of the vehicle showing the location of the water ingress point(s) before proceeding. Make sure that an image is attached to an EPQR (Electronic Product Quality Report) before proceeding with any windshield removals.
- When the leak point(s) have been noted, remove the windshield. NOTE: Depending on the nature of the leak point, it may be possible to reuse the windshield. Consequently, make sure the correct equipment and procedures are utilized during removal of the windshield.
- When the windshield has been removed from the vehicle, it is important to confirm the specific nature of the leak. The image below shows a representation (not to scale) of the materials involved when a windshield is installed. A layer of primer is applied to the windshield glass; a layer of adhesive is applied; finally, the windshield is mounted into the frame. Consequently, there are two possible entry points for water ingress:
- 1) Water ingress between the windshield adhesive layer and the windshield glass (A, below).
- 2) Water ingress between the vehicle body and the windshield adhesive layer (B, below).
- Where possible, note the specific nature of the water ingress (as specified above). Photograph the leak point and attach both the photo and a written description of the location and nature of the water ingress/leak point to the EPQR.
- **If the leak point is identified as being between the vehicle body and the windshield adhesive layer, the existing windshield may be re-used.** Thoroughly remove all traces of adhesive and primer from the existing windshield, clean the windshield and refit.

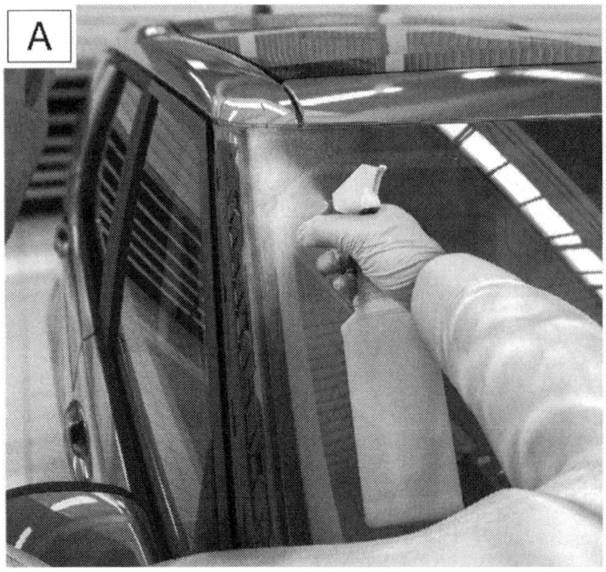

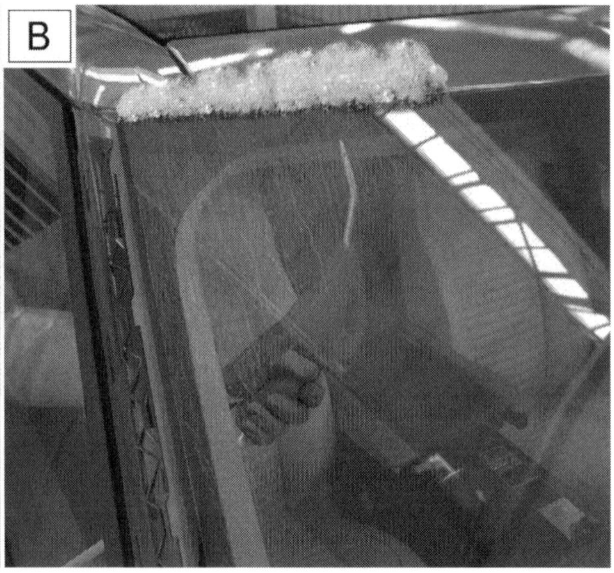

- If the leak point is identified as being between the windshield adhesive layer and the windshield glass, the existing windshield should be replaced and a new windshield installed.

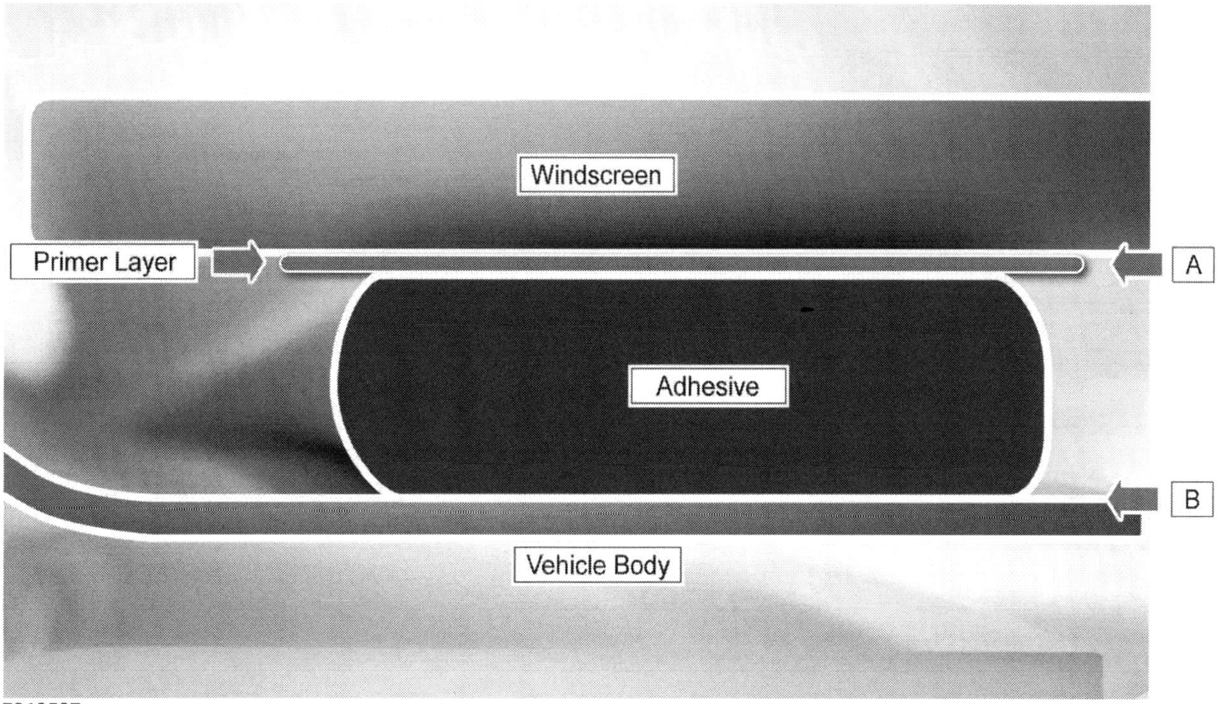

E260527

Water test

NOTES:

⚠ Never aim a jet of water directly at a weather strip or rubber seals.

⚠ DO NOT use a high pressure water hose for water testing.

- Complete the water test with a second person present (in the passenger compartment).
- Use variable washer nozzles (concentrated water jet to fine spray mist).
- Lower the headliner at the front of the vehicle to give clear access to the upper section of the windshield inside the vehicle.
- The outer trims must be raised carefully using a plastic wedge.
- Start in the lower section and spray the whole area, working upwards in stages.
- The leak should be located from inside by a second assistant.

Washer test

- Further tests can be completed in the washer system.
- Some leaks originate here, or only occur here.
- The relevant passenger compartment should be checked using a flashlight during the wash procedure.

Road test

- If no leaks are located during the tests above, road tests should be completed on wet roads.
- Road tests under various conditions:
    - At various speeds.
    - On various road surfaces (asphalt to cobbles).
    - With loaded or unloaded vehicle.
    - Driving through puddles (splash water).

Other test equipment

- Other equipment such as stethoscopes, UV (ultraviolet) lamps, special mirrors or ultrasound measuring instruments can be used to locate leaks.

DTC Index

For a complete list of all Diagnostic Trouble Codes (DTCs) that could be set on this vehicle, please refer to Section 100-00.

Front Door Window

Removal and Installation

Removal

1. Remove front door reinforcement panel.

Refer to: Front Door Reinforcement Panel (501-03 Body Closures, Removal and Installation)

2. Remove window regulator.

Refer to: Front Door Window Regulator and Motor (501-11 Glass, Frames and Mechanisms, Removal and Installation)

3. Push glass up to top of its travel and support with a suitable length of timber.

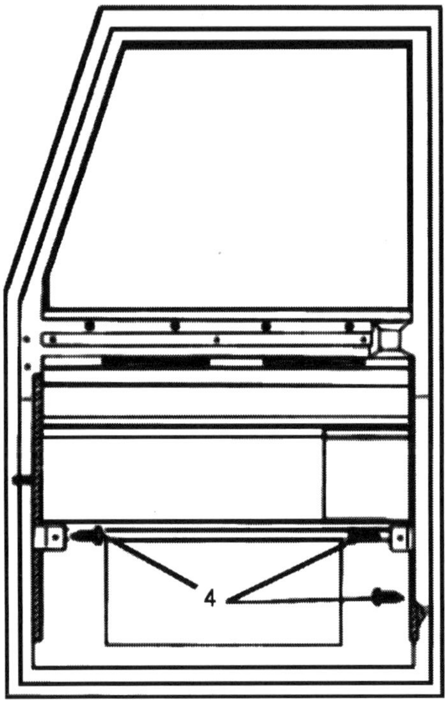

4. Remove 2 self tapping screws securing window glass runner on latch side of door and single screw from hinge side.

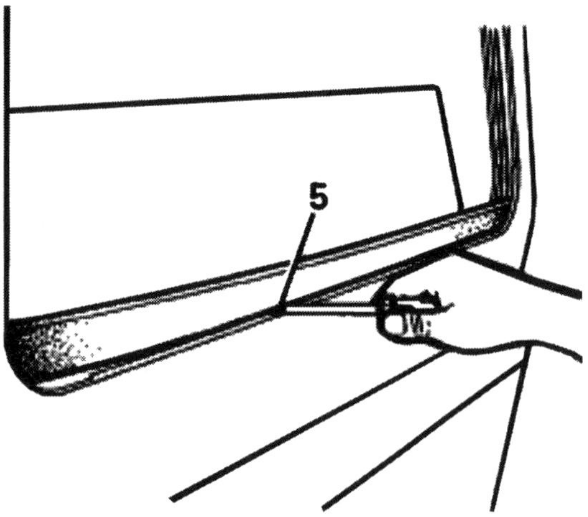

5. Taking care not to damage paint, prise exterior waist weather strip from door.

6. Remove timber support and lower glass to bottom of door.

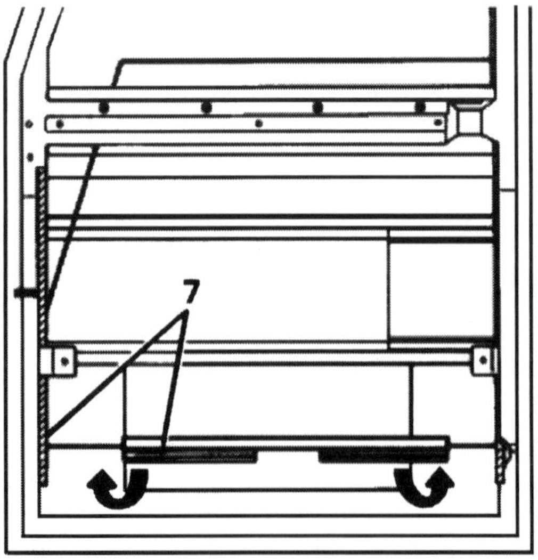

7. Ease runner from glass at hinge side of door, lift glass over bottom edge of door and withdraw.

Installation

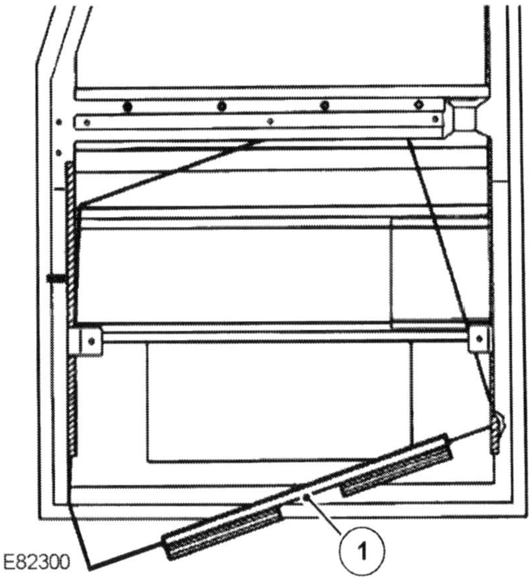

1. Insert glass into runners at an angle as illustrated.

2. Whilst lifting glass, position it squarely in runners, raise to top of travel and insert timber support.

3. Secure hinge side runner with single screw ensuring that packing strip is in position.

4. Locate packing strip and secure opposite runner with 2 screws. Ensure that all 3 screw heads are well below bottom of runners to prevent damage to glass.

5. Locate regulator in window lift channels.

6. Install window regulator.

Refer to: Front Door Window Regulator and Motor (501-11 Glass, Frames and Mechanisms, Removal and Installation)

7. Install front door reinforcement panel.

Refer to: Front Door Reinforcement Panel (501-03 Body Closures, Removal and Installation)

Front Door Window Regulator and Motor

Removal and Installation

Removal

1. Remove door trim panel.

Refer to: Front Door Trim Panel (501-05 Interior Trim and Ornamentation, Removal and Installation)

2. Remove plastic sheet.

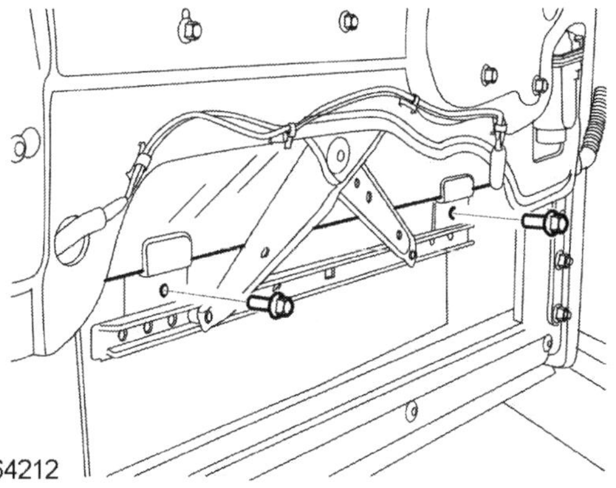

M764212

3. Lower glass and remove 2 bolts securing door glass to regulator.

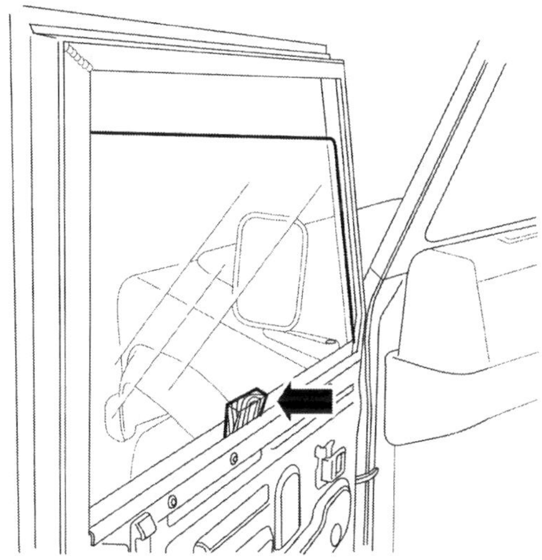

M764213

4. Release glass from regulator, raise and wedge in raised position.

5. Disconnect multiplug from motor.

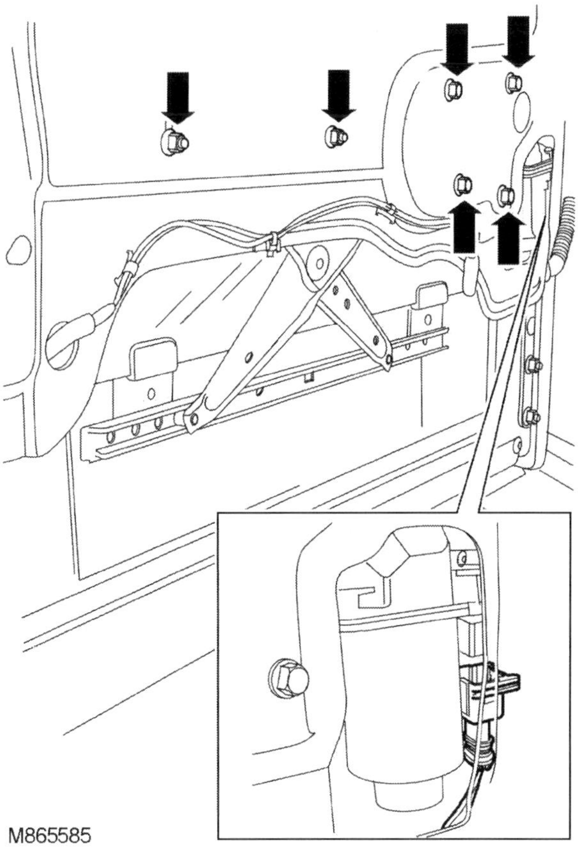

M865585

6. Remove 4 bolts and 2 nuts securing motor and regulator assembly.

7. Release and remove motor and regulator assembly.

Installation

1. Install motor and regulator assembly, install nuts and bolts and tighten to 10 Nm (7 lbf.ft).

2. Connect multiplug to motor.

3. Remove wedges, lower glass and locate to regulator. Install bolts and tighten to 6 Nm (4 lbf.ft).

4. Raise and lower glass to check operation.

5. Install plastic sheet.

6. Install door trim panel.

Refer to: Front Door Trim Panel (501-05 Interior Trim and Ornamentation, Removal and Installation)

Heated Windshield Relay

Removal and Installation

Removal

1. Remove front seat cushion.
2. Release clip and remove compartment cover.

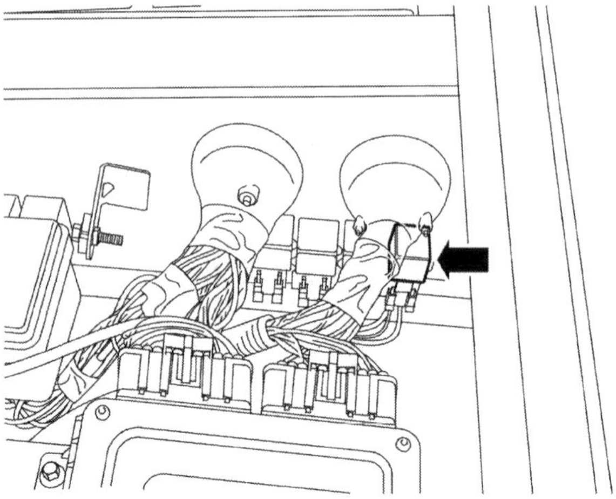

3. Locate and release heated windshield relay from bracket.
4. Remove heated windshield relay.

Installation

1. Fit heated windshield relay.
2. Fit heated windshield relay to mounting bracket.
3. Fit and secure cover.
4. Fit seat cushion.

Rear Door Window Glass

Removal and Installation

Removal

1. Remove rear door reinforcement panel.

Refer to: Rear Door Reinforcement Panel (501-03 Body Closures, Removal and Installation)

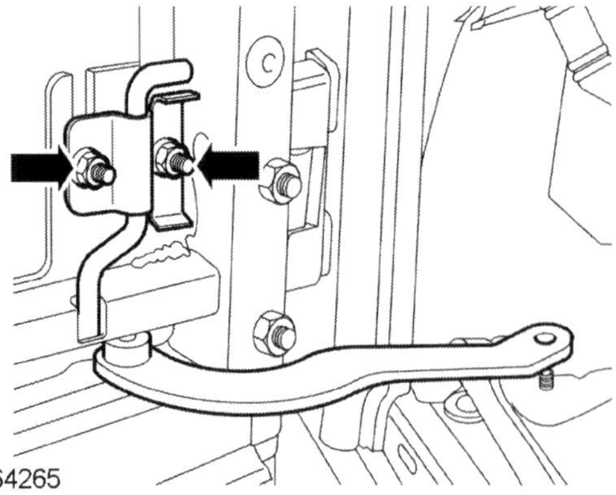

M764265

2. Loosen 2 nuts and remove door check strap torsion bar and plate.
3. Remove check strap.
4. Remove inner and outer waist seal.
5. Remove screw from each glass channel.
6. Lower glass to bottom of door, lift over lower edge and withdraw from door.

Installation

1. Insert glass into channels and carefully push to top of frame.
2. Secure glass channels with screws. Ensure screw

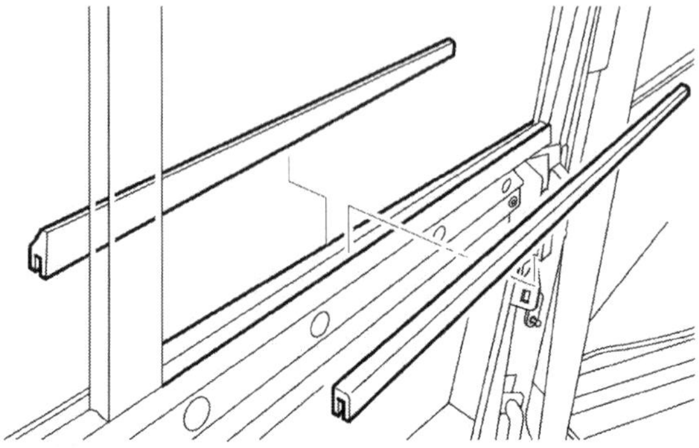

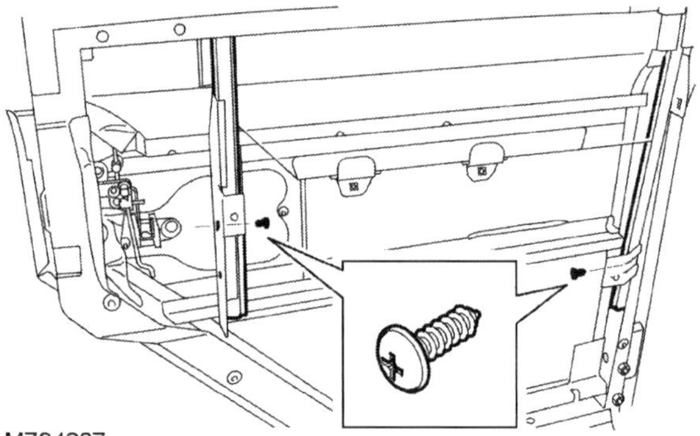

M764267

heads are down below bottom of channels to prevent damage to glass.

3. Install inner and outer waist seal to door.
4. Install check strap.
5. Install check strap torsion bar and tighten nuts to 10 Nm (7 lbf.ft).
6. Install rear door mounting panel.

Refer to: Rear Door Reinforcement Panel (501-03 Body Closures, Removal and Installation)

Rear Door Window Regulator and Motor

Removal and Installation

Removal

1. Remove rear door trim casing.

Refer to: Rear Door Trim Panel (501-05 Interior Trim and Ornamentation, Removal and Installation)

2. Carefully release plastic sheet from rear door.
3. Remove 2 bolts securing door glass to regulator.
4. Raise glass and secure with a suitable wedge.

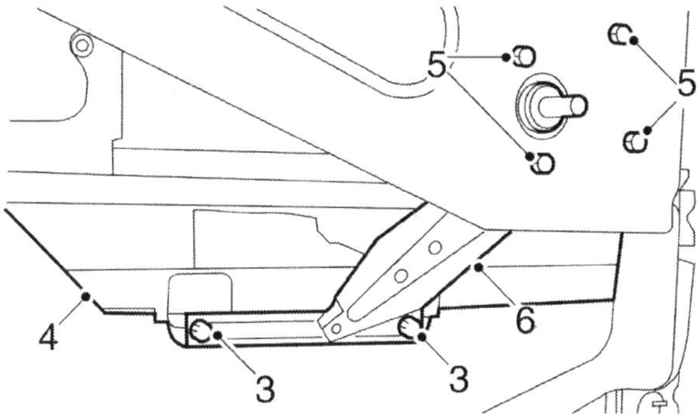

5. Remove 4 bolts securing glass regulator assembly to door.
6. Manoeuvre regulator through access hole at bottom of the door.

Installation

1. Clean mating face of regulator and door glass.
2. Lubricate regulator with grease.
3. Position regulator in door and secure with bolts.
4. Remove wedge holding glass and align glass to regulator.
5. Secure glass to regulator and install and tighten bolts.
6. Install plastic sheet to rear door.
7. Install rear door trim casing.

Refer to: Rear Door Trim Panel (501-05 Interior Trim and Ornamentation, Removal and Installation)

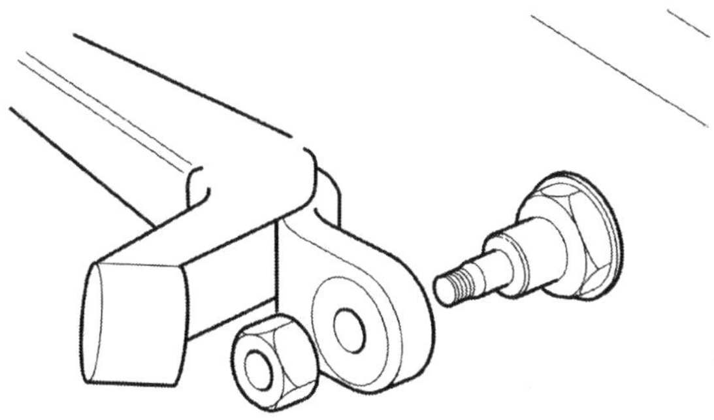

2. Raise nut cover, remove nut and release wiper arm.
3. Open taildoor.
4. Remove high mounted stop lamp.

Refer to: High Mounted Stoplamp (417-01 Exterior Lighting, Removal and Installation)

5. Remove 4 connectors from heated rear window.
6. WARNING: Wear suitable eye protection when removing and refitting glass.

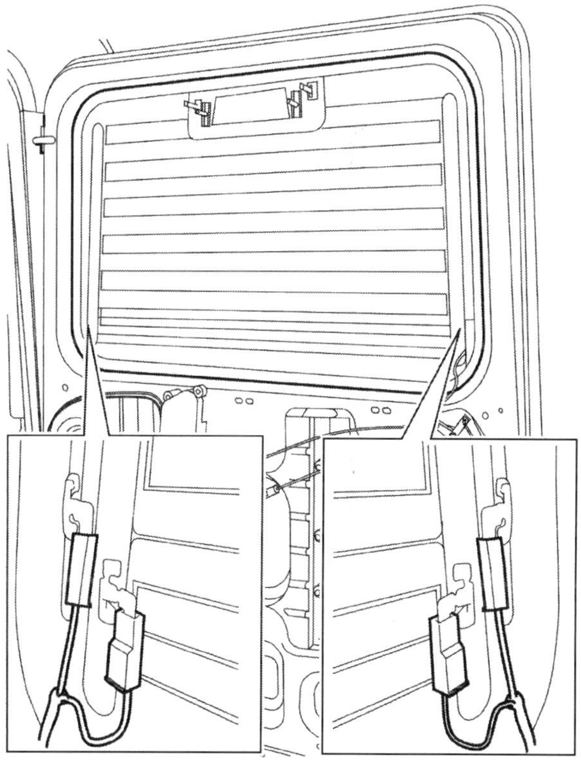

M764253

With assistance, from inside of vehicle apply pressure to glass, relieve seal lip and carefully remove glass.

7. CAUTION: Lay glass on felt covered supports and be careful not to damage the obscuration band. Do not stand on edge as this can cause chips which subsequently develop into cracks.

Remove seal from glass.

Installation

1. Clean glass and install seal. Ensure seal is fully located on glass.
2. Thoroughly clean the taildoor window mounting flange.
3. Install drawstring to seal.
4. Lubricate seal.
5. With assistance, position glass and seal to taildoor window aperture, locate lower edge and Install seal using drawstring.
6. Clean glass and check seal fit.
7. Fit connectors to heated rear window.
8. Install high mounted stop lamp.

Refer to: High Mounted Stoplamp (417-01 Exterior Lighting, Removal and Installation)

9. Install wiper arm to spindle, align blade to window and tighten nut to 18 Nm (13 lbf.ft).
10. Install spare wheel and tighten nuts to 45 Nm (33 lbf.ft).

Window Control Switch

Removal and Installation

Removal

1. Disconnect the battery ground cable.

Refer to: Startup Battery Disconnect and Connect (414-01 Battery, Mounting and Cables, General Procedures)

2. Remove the audio unit.

Refer to: Audio Unit - Vehicles Without: Bluetooth (415-01 Audio Unit, Removal and Installation)

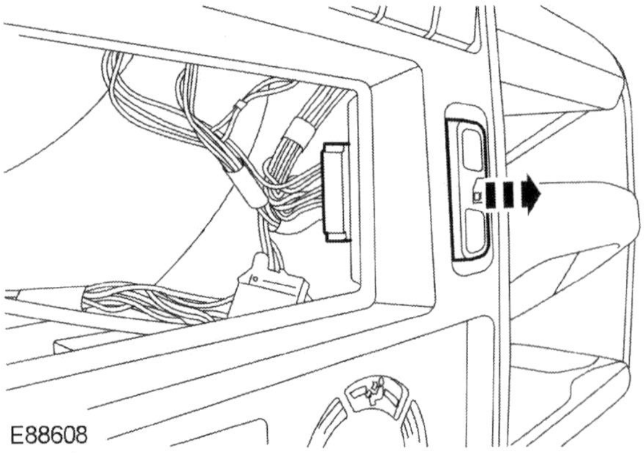

E88608

3. △ NOTE: RH shown, LH similar.

Release the window control switch.

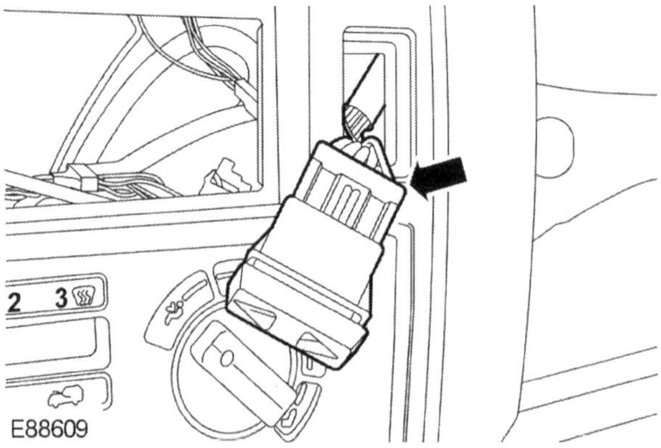

E88609

4. Remove the window control switch.
   - Disconnect the electrical connector.

Installation

1. To install, reverse the removal procedure.
2. Connect the battery ground cable.

Refer to: Battery Connect (414-01 Battery, Mounting and Cables, General Procedures).

Windshield

Removal and Installation

Part(s)

STEP	PART NAME	QUANTITY
Installation Step 1	Renew Part Windshield seal	1

Removal

⚠ WARNING: Wear suitable eye protection when removing and installing glass.

NOTES:

△ RHD shown. LHD is similar.

△ Removal steps in this procedure may contain installation details.

△ Some variation in the illustrations may occur, but the essential information is always correct.

1. Remove the front wiper arms.

Refer to: Front Wiper Pivot Arm (501-16 Wipers and Washers, Removal and Installation)

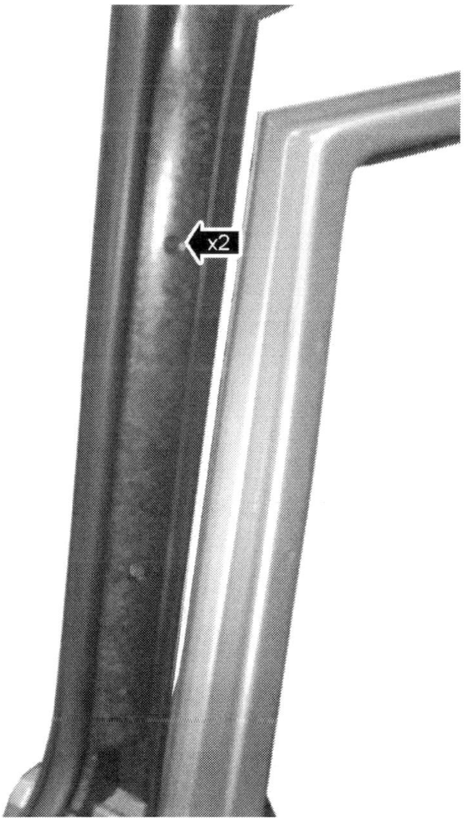

2. △ NOTE: Repeat this step for the other side.

   Remove the A-pillar trim.

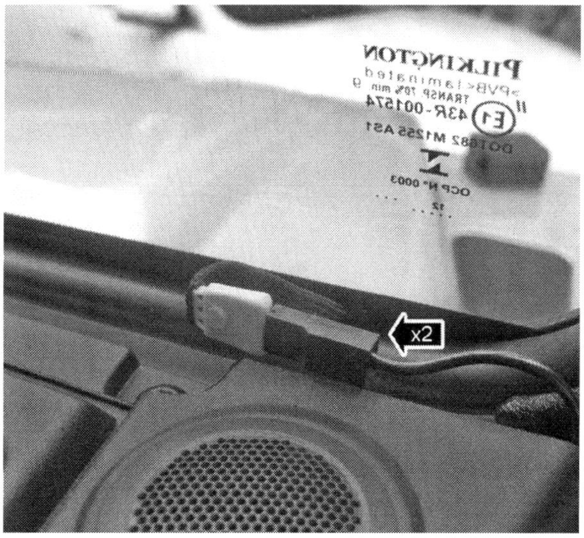

3. △ NOTE: Repeat this step for the other side.

   Disconnect the connector plug for the heated windshield.

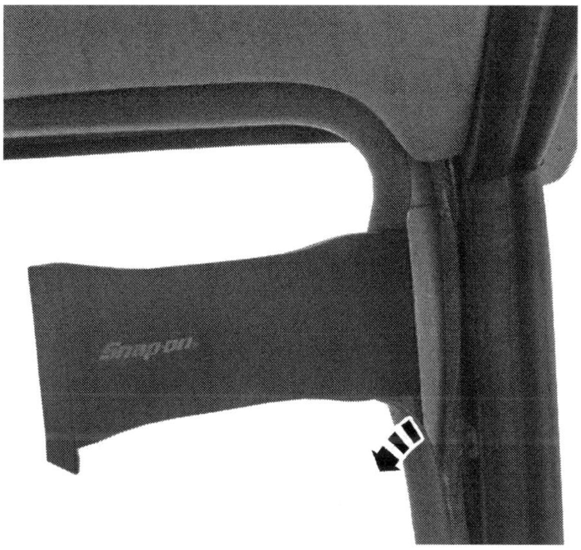

4. Using a suitable trim tool, fold back the lip of the windshield seal (working from the top corner). Using gentle pressure from the interior push the windscreen outwards.

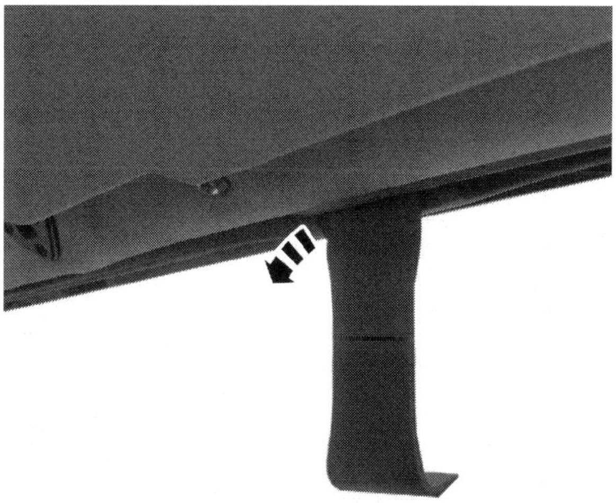

5. Continue along the top of the windscreen fold the seal backwards and easing the windshield outwards.

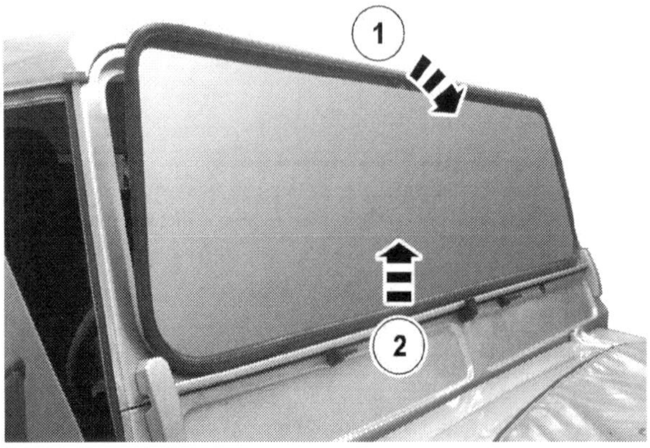

6. Remove the windshield and place on a suitable work surface.

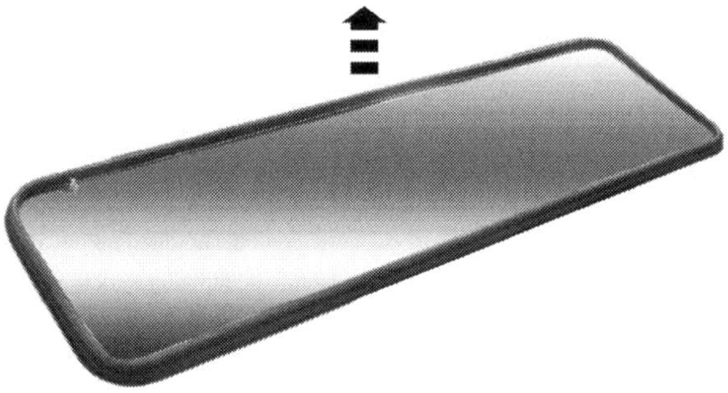

7. Discard the seal.

Installation

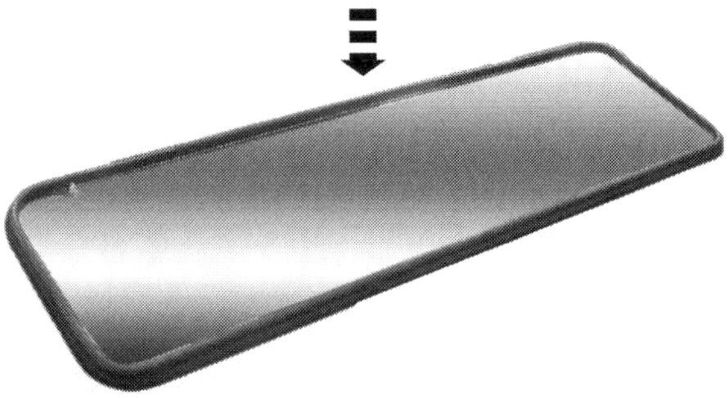

1. Install a new seal onto the windshield.

   Renew Part: Windshield seal

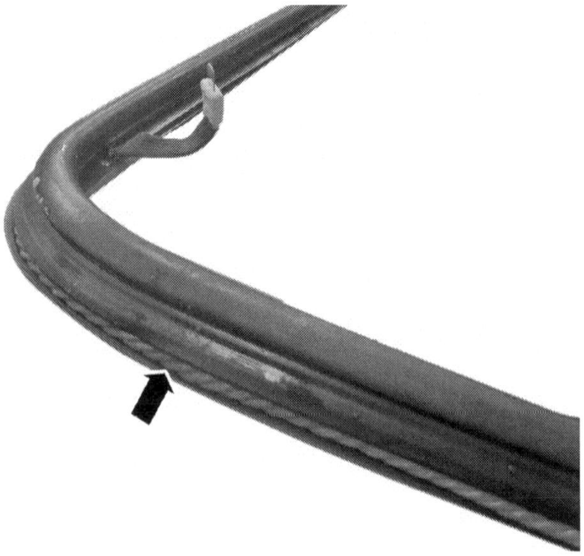

2. Apply a mild soapy solution to the seal. Insert a suitable cord into the seal to aid installation.

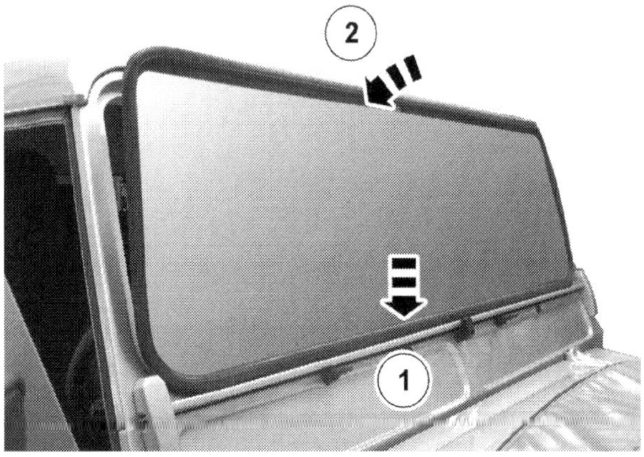

3. Place the windshield into the frame and push downwards to locate the windshield.

4. From the interior of the vehicle pull on the cord to bring the seal to the inside of the frame.

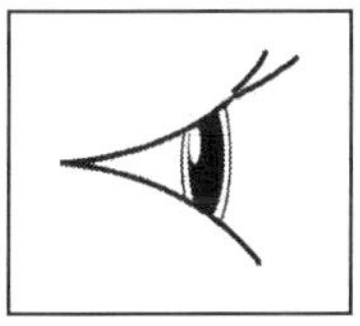

5. Inspect the windshield seal to check for correct fitment.

6. NOTE: Repeat this step for the other side.

   Install the A-pillar trim.

7. Install the front wiper arms.

Refer to: Front Wiper Pivot Arm (501-16 Wipers and Washers, Removal and Installation)

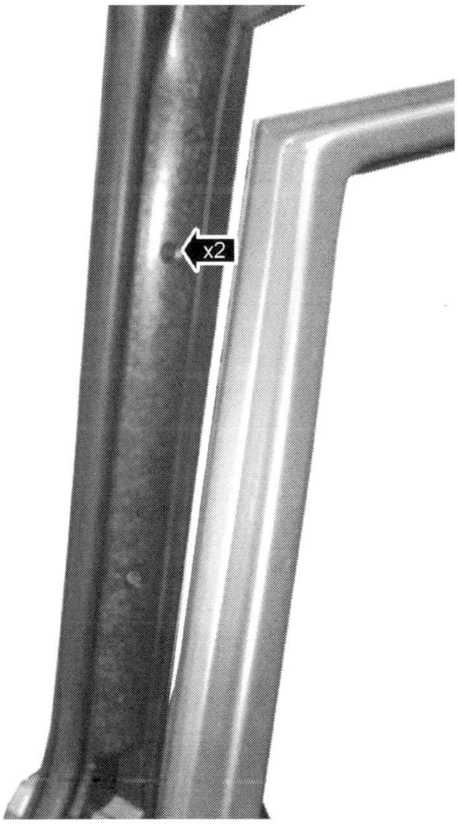

Side Rear Window Glass Assembly

Removal and Installation

Removal

1. **Refer to:** Rear Quarter Trim Panel (501-05 Interior Trim and Ornamentation, Removal and Installation)

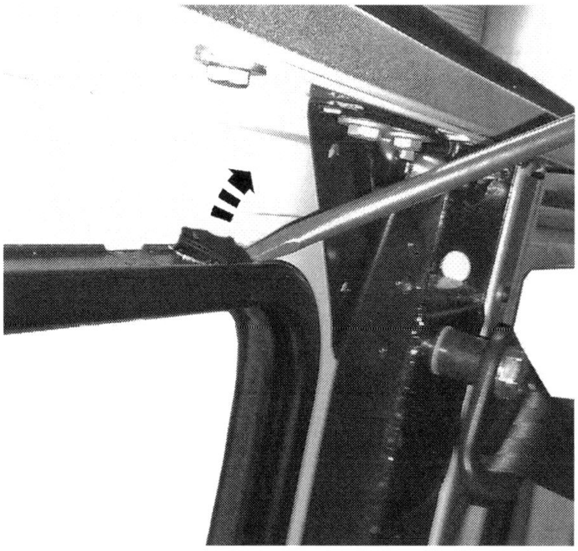

2. Remove all the clips from around the window frame.

E144951

3. Apply protective tape to the body round the window frame.

4. Use a suitable glazing knife to cut the window frame away from the body.

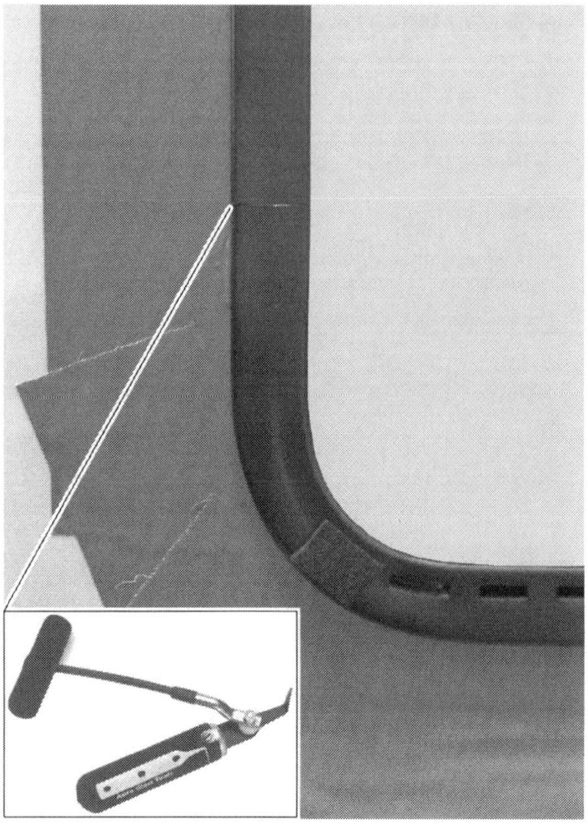

5. With assistance carefully remove the window frame from the body.

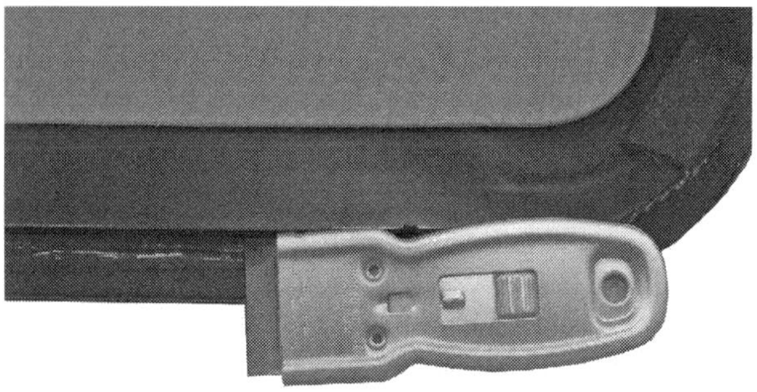

6. Using a suitable tool remove the old adhesive from the window frame and vehicle body.

7. Remove 6 the screws from the window frame as shown.

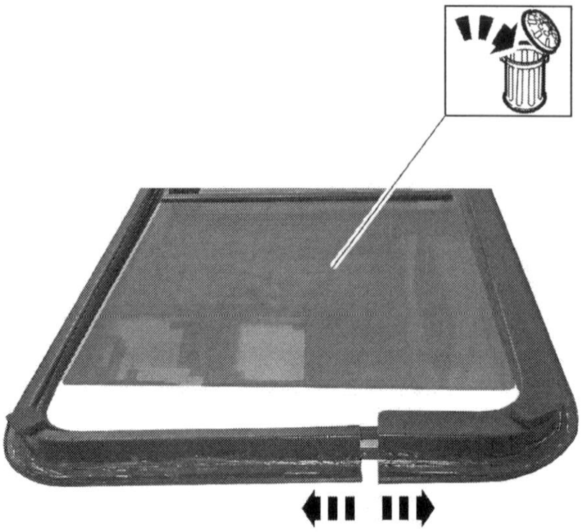

8. Open the two halves of the window frame and remove the glass as shown.

9. Remove the seal.

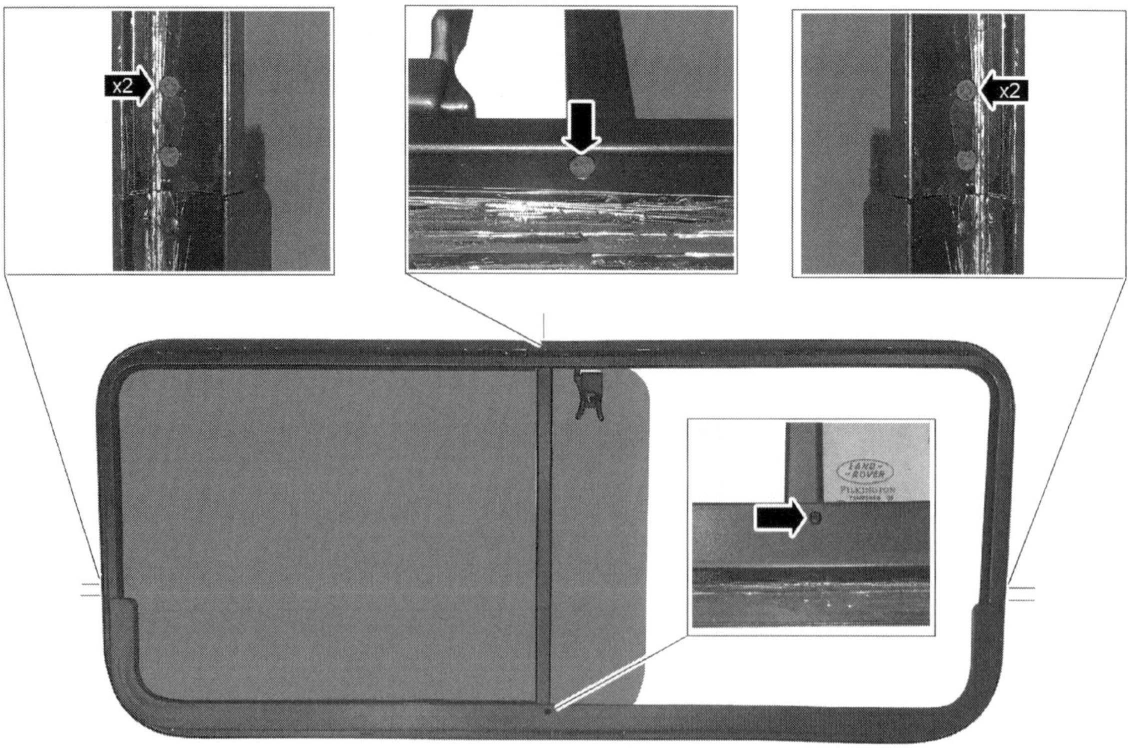

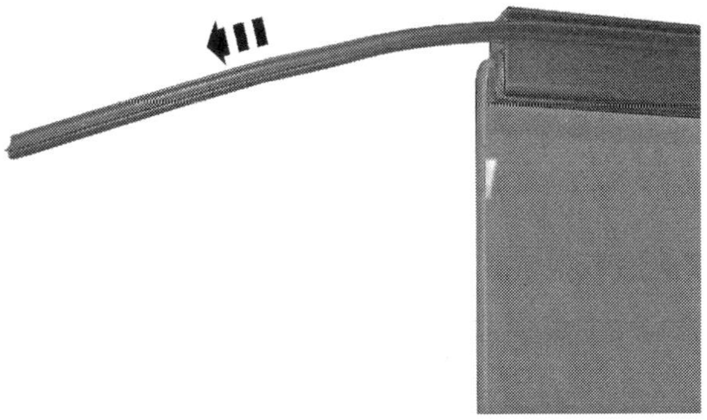

E144941

10. Remove the seal.

Installation

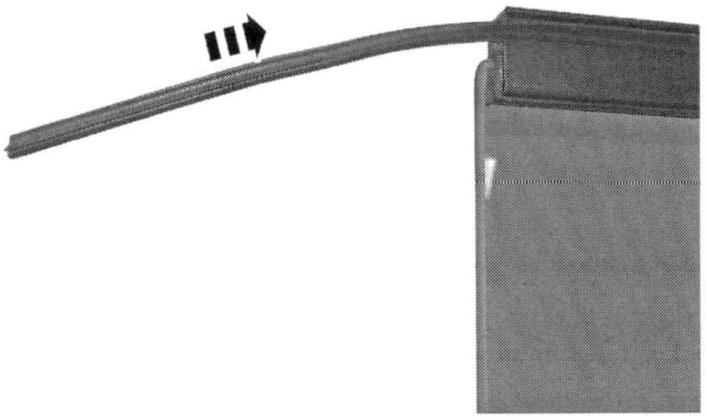

1. Install the seal to the new glass.

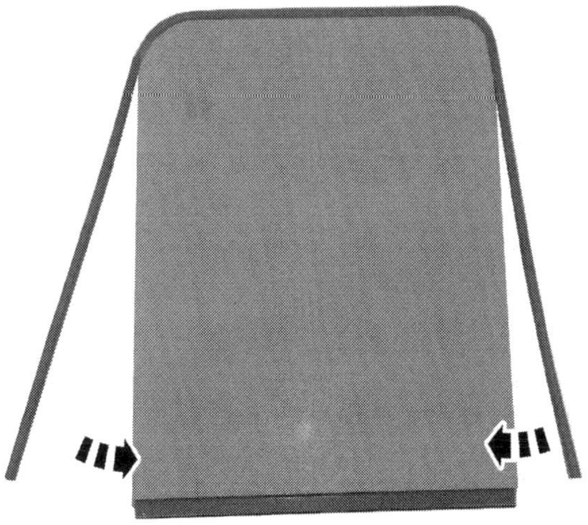

2. Install the seal to the new glass.

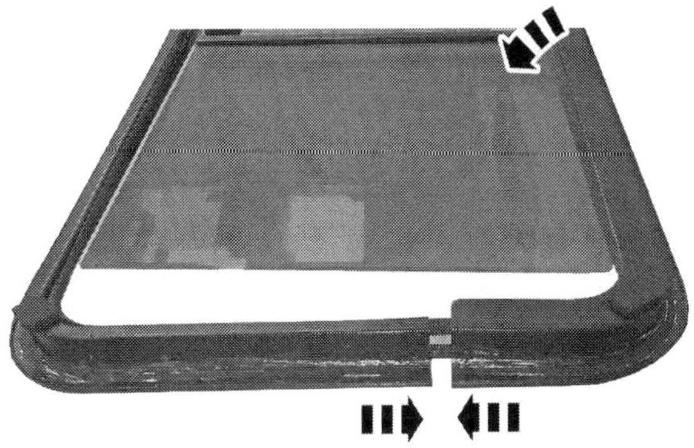

3. Install the new glass into the frame.

4. Install 6 the screws to the window frame as shown.

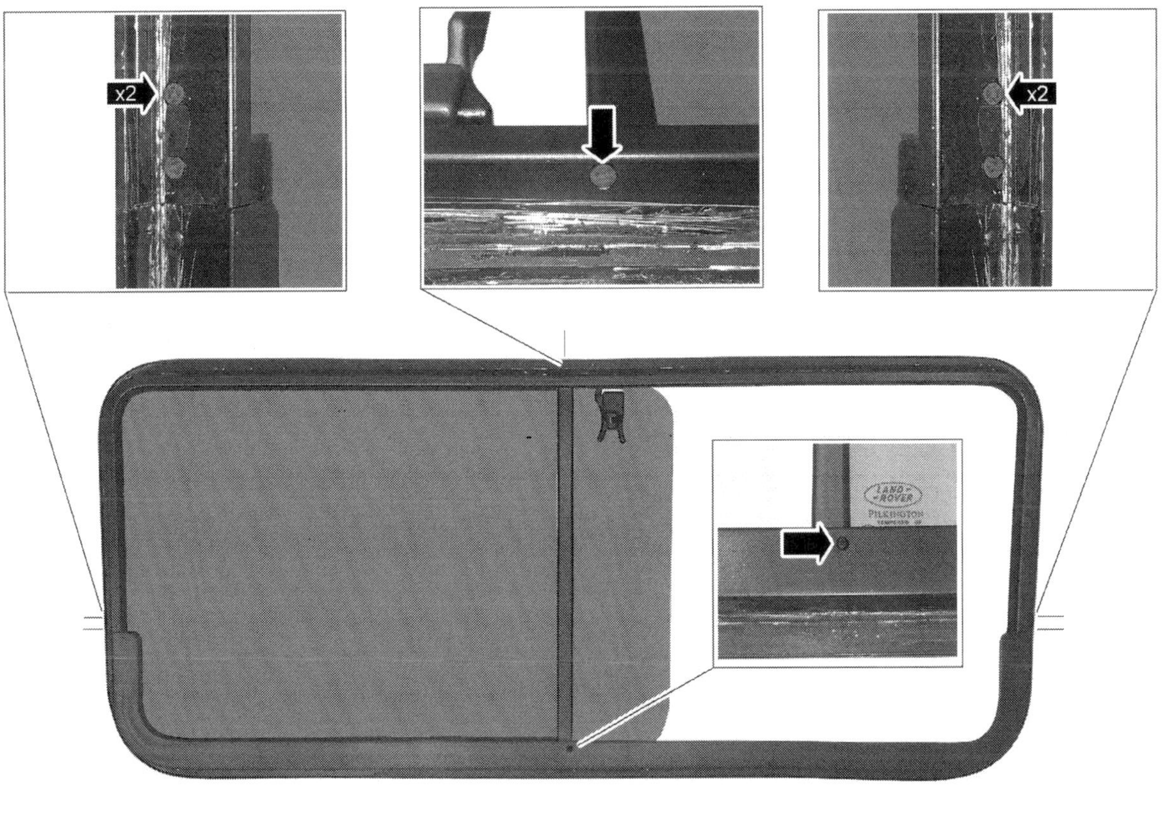

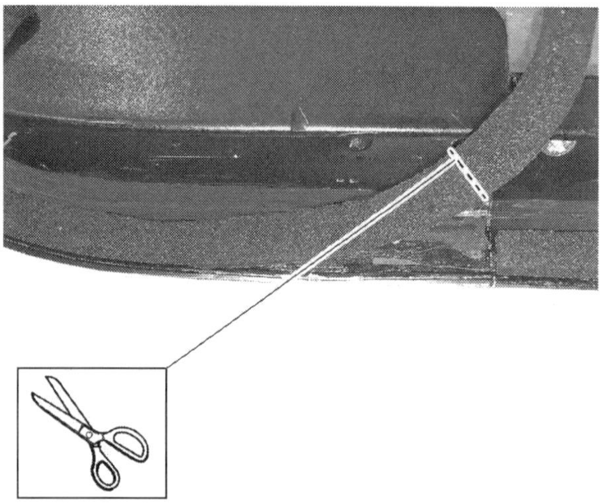

E144945

5. Apply 5mm thick foam tape around the outer edge of the window frame.

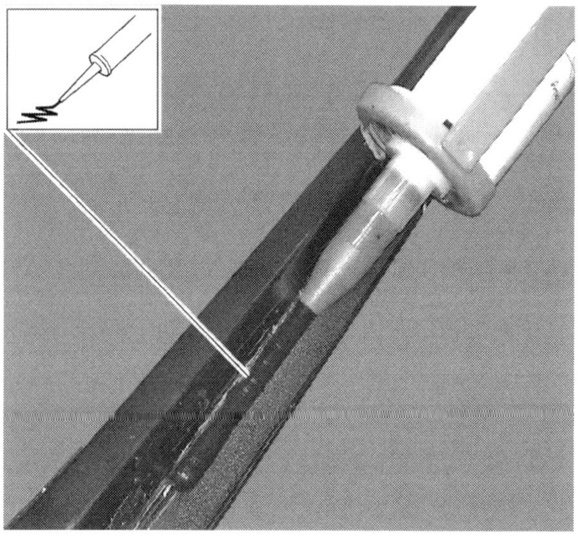

6. Apply a 5mm thick bed of PU adhesive between the foam and inner edge of the window frame in accordance with the instructions included with the PU adhesive kit.

7. With assistance carefully install the window frame into the body.

8. Install the clips evenly spaced around the window frame.

9. Allow adhesive to dry in accordance with the instructions included with the PU adhesive kit before testing for water leaks. Spray water around window and mark any leaks before drying the area and applying additional adhesive.

△ NOTE: If the ambient air temperature falls below 10 degrees C, apply warm air (25 degrees C) continuously for 15 minutes.

10. **Refer to:** Rear Quarter Trim Panel (501-05 Interior Trim and Ornamentation, Removal and Installation)

Specifications

Torque specifications

Description	Nm	lb-ft
Instrument panel Torx 30 screw	2	1
Instrument panel Torx 20 screw	1	1
Instrument panel console	2	1
Grab handle	7	5

Description and Operation

COMPONENT LOCATION

Item	Description
1	Central support
2	Cross car rail
3	Instrument panel

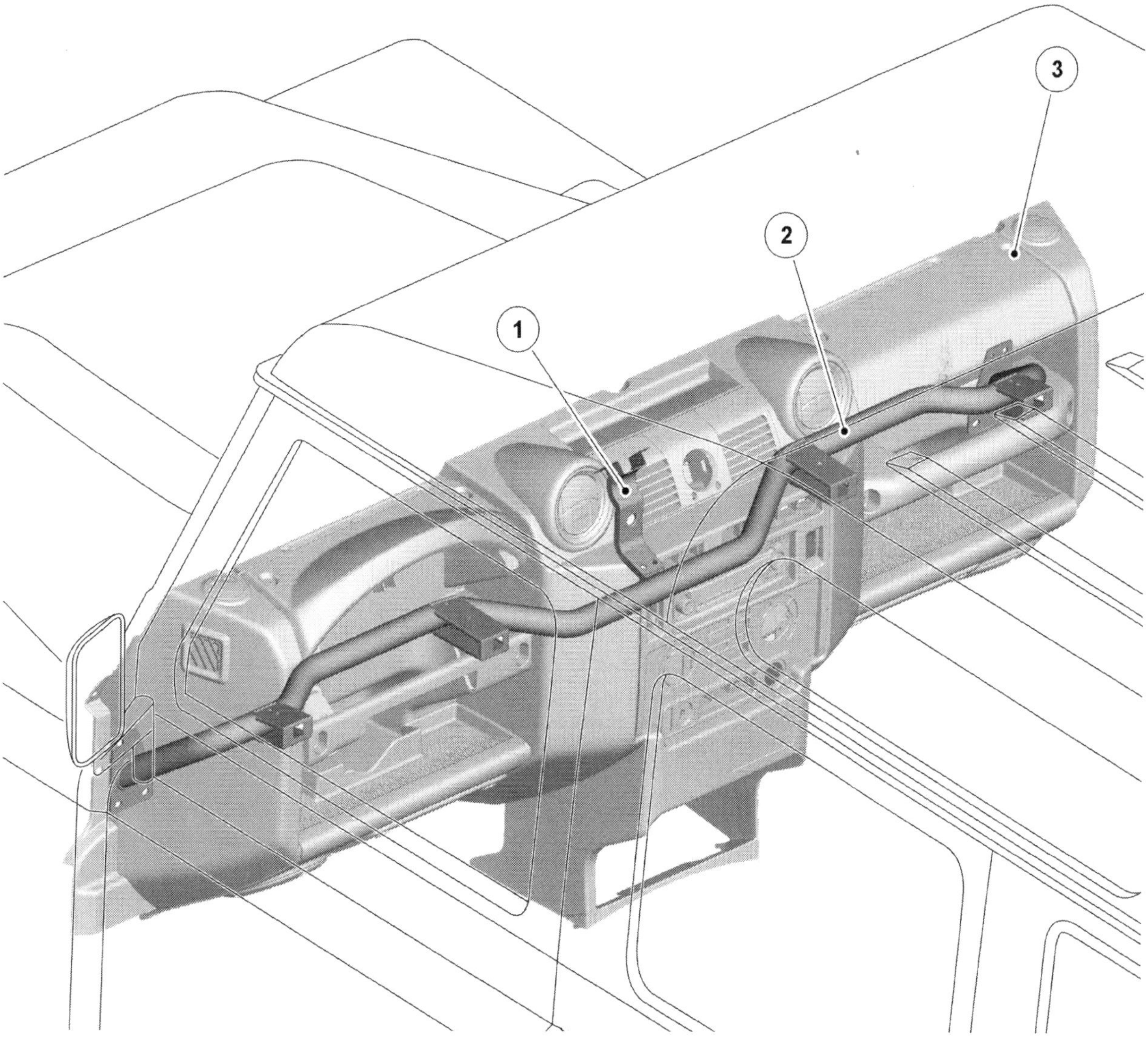

OVERVIEW

Vehicles from 2007 model year feature an all new instrument panel. In addition to providing a location for various system switches and controls, the instrument panel also houses an all new instrument cluster and climate control system.

Refer to: Instrument Cluster (413-01 Instrument Cluster, Description and Operation)

Refer to: Climate Control System (412-00 Climate Control System - General Information, Description and Operation)

Refer to: Air Distribution and Filtering (412-01 Air Distribution and Filtering, Description and Operation)

Mounted behind the instrument panel is a tubular steel cross car rail. Three weld bolts are located on either end of the cabin bulkhead to provide the mounting points for the cross car rail. If the cross car rail is removed, the nuts securing the rail in position must be discarded and replaced with new items.

The cross car rail is non-handed, meaning the same rail can be fitted to LHD (left-hand drive) and RHD (right-hand drive) vehicles.

The cross car rail features 4 mounting brackets. The driver's side brackets provide mounting points for the instrument panel finisher; the passenger side brackets provide mounting points for the instrument panel grab handle. The mounting brackets contain expanding nuts, which must be replaced every time the instrument panel is removed.

A central support bracket is secured to the cross car rail by 2 plastic scrivets. A further plastic scrivet secures the top of the bracket to the cabin bulkhead. The central support bracket provides a mounting point for the audio system head unit and supports the in-board ends of the climate control system air ducts. The air ducts are secured to the central support bracket by a further 2 plastic scrivets. If any of the plastic scrivets are removed, they must be discarded and replaced with new items.

Instrument Panel

Removal and Installation

Removal

1. Disconnect the battery ground cable.

Refer to: Startup Battery Disconnect and Connect (414-01 Battery, Mounting and Cables, General Procedures)

2. Remove the floor console.

Refer to: Floor Console (501-12 Instrument Panel and Console, Removal and Installation)

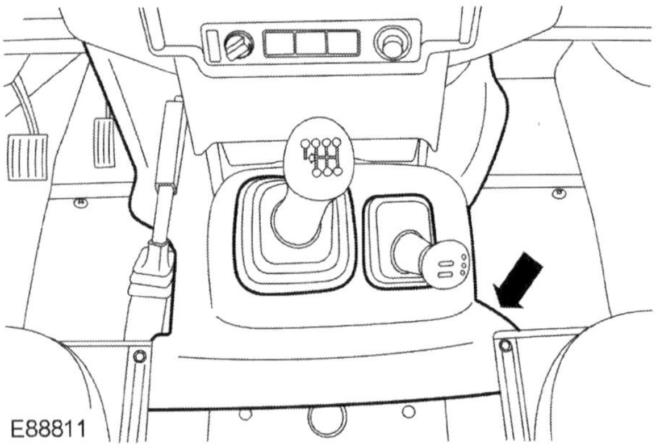

E88811

3. Remove the transmission cover panel floor covering.

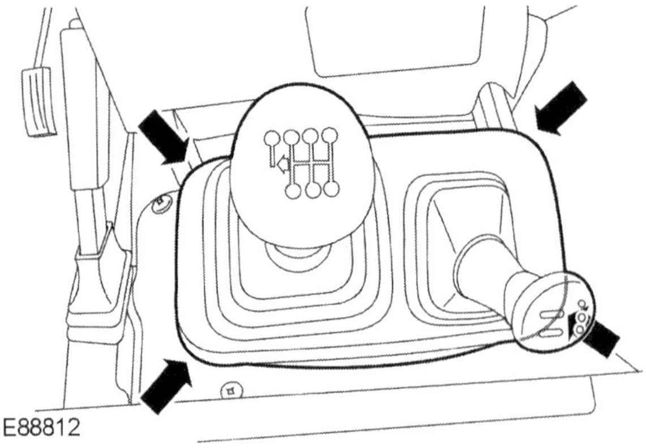

E88812

4. Remove the high/low gear selector lever.

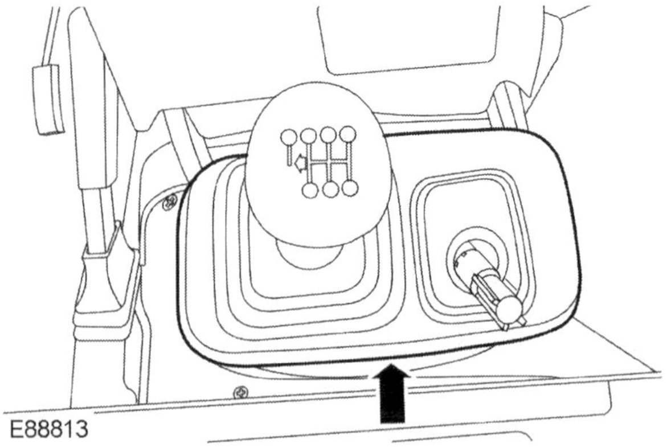

E88813

5. Release the gear shift lever.

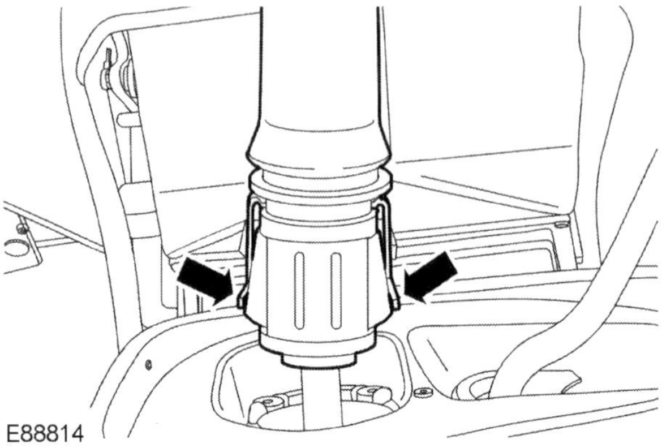

E88814

6. Remove the gear shift lever.

- Release the 2 clips.

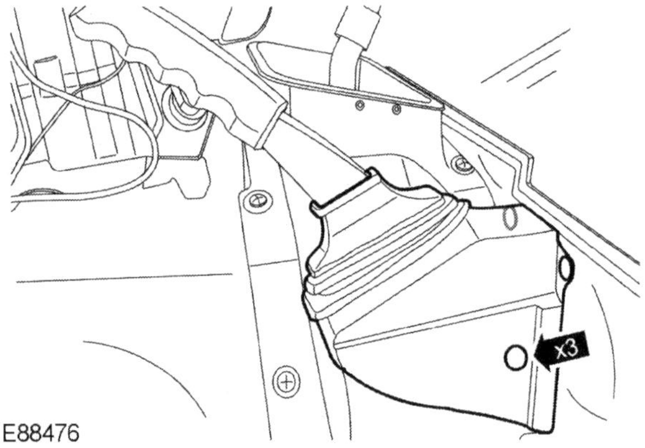

E88476

7. Release the parking brake lever gaiter.

- Remove the 3 clips.

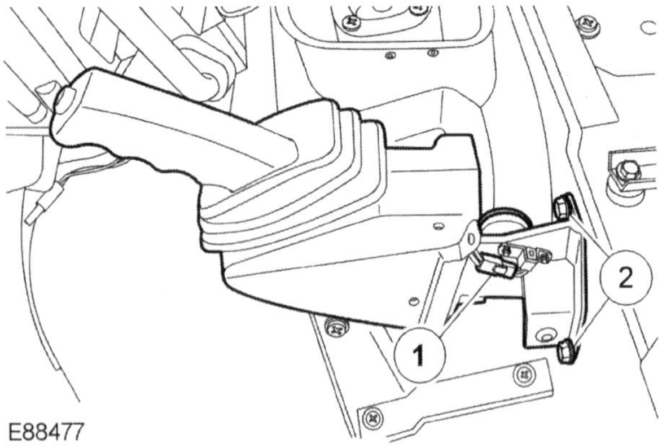

E88477

8. Release the parking brake lever.

   1. Disconnect the electrical connector.
   2. Remove the 2 bolts.

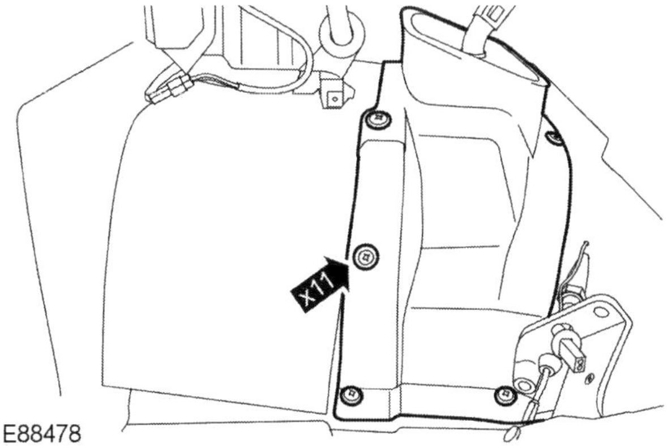

E88478

9. Remove the transmission cover panel.

    - Remove the 11 screws.

10. Remove the steering wheel.

Refer to: Steering Wheel (211-04 Steering Column, Removal and Installation)

11. Remove the instrument cluster.

Refer to: Instrument Cluster (413-01 Instrument Cluster, Removal and Installation)

12. Remove the steering column shrouds.

Refer to: Steering Column Shrouds (501-05 Interior Trim and Ornamentation, Removal and Installation)

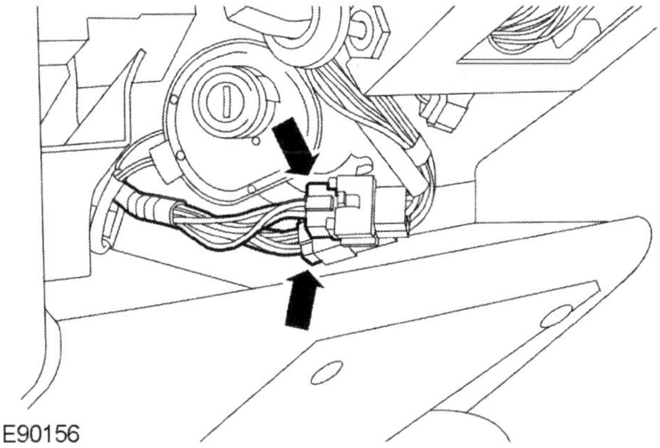

E90156

13. Disconnect the 2 electrical connectors from the multifunction switch.

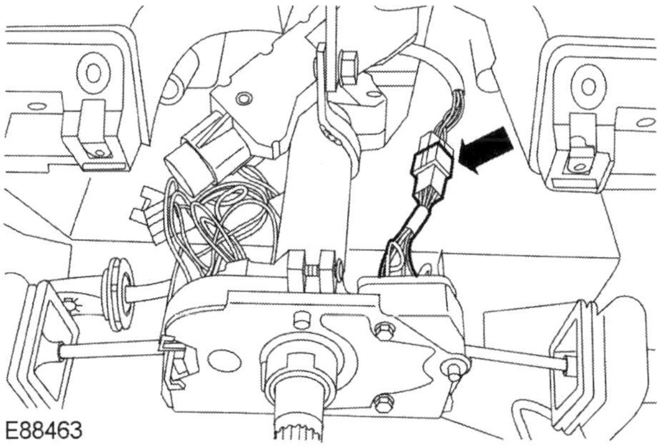

E88463

14. Disconnect the windshield wiper switch electrical connector.

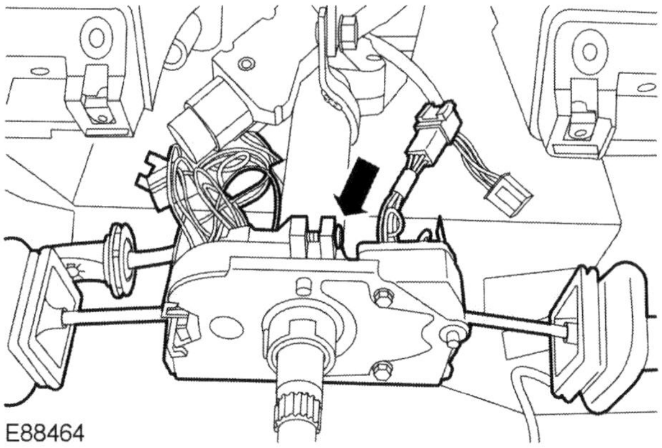

E88464

15. Remove the multifunction switch assembly.

    - Undo but do not remove the multifunction switch clamp screw.

16. Remove the instrument panel console.

Refer to: Instrument Panel Console (501-12 Instrument Panel and Console, Removal and Installation)

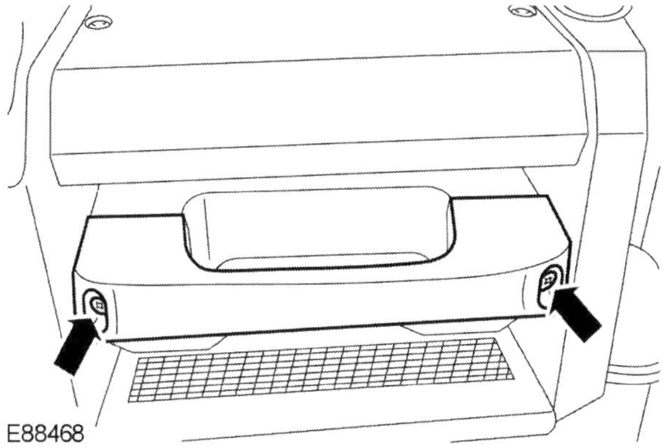

E88468

17. Remove the passenger side grab handle.

    - Remove the 2 screws.

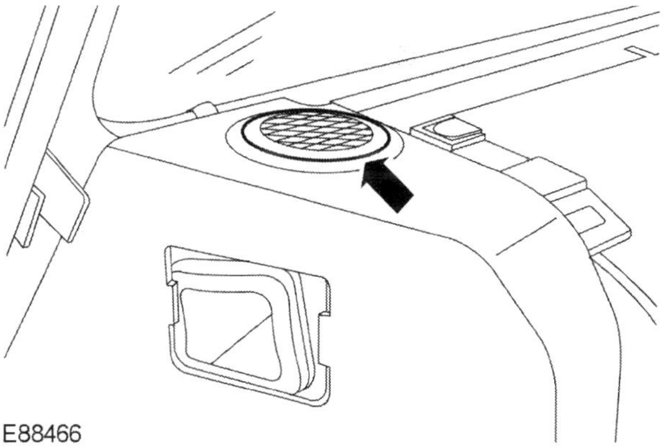

E88466

18. △ NOTE: Left-hand shown, right-hand similar.

Remove both side demister vent finishers.

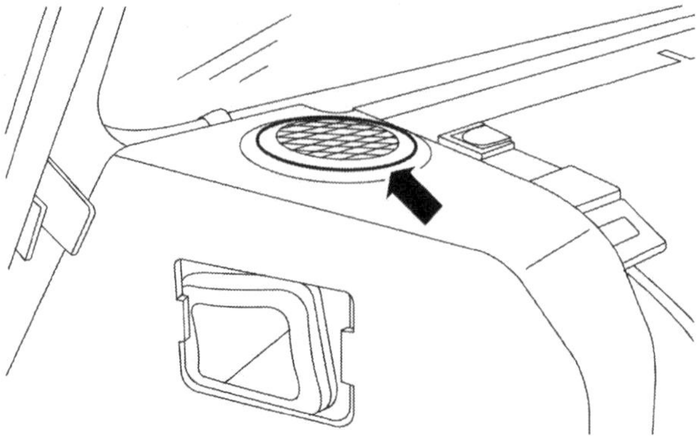

19. △ NOTE: Left-hand shown, right-hand similar.

Release both instrument panel speakers.

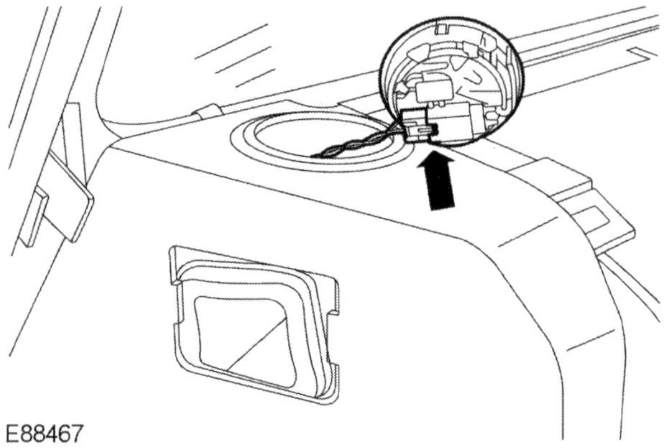

E88467

20. △ NOTE: Left-hand shown, right-hand similar.

Remove both instrument panel speakers.

- Disconnect the electrical connector.

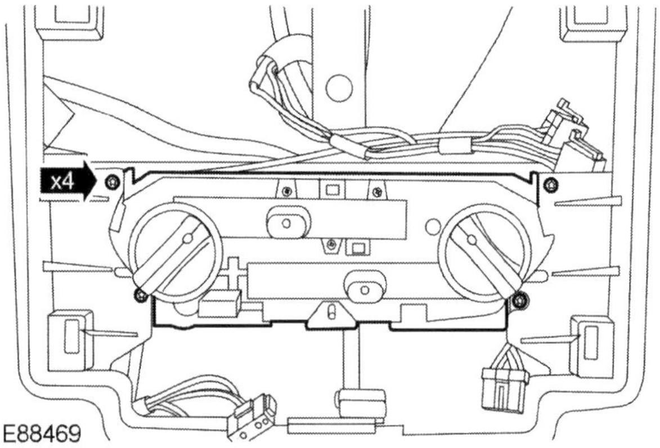

E88469

21. Release the climate control switch panel.

- Remove the 4 screws.

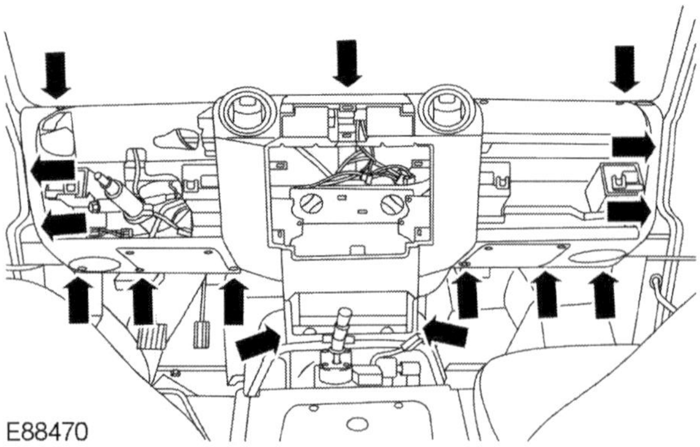

E88470

22. Release the instrument panel.

- Remove the 15 screws.

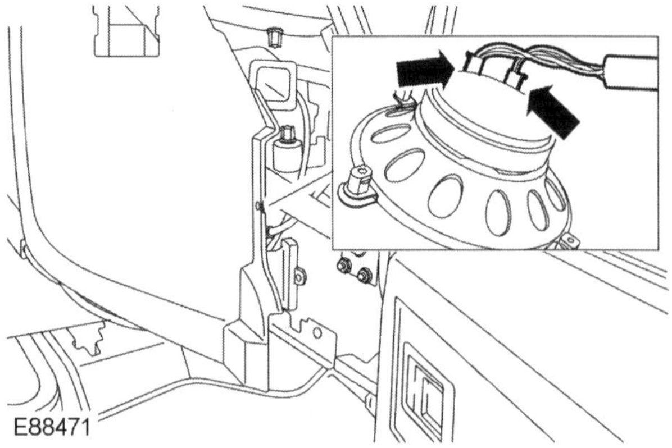

E88471

23. △ NOTE: RH shown, LH similar.

Disconnect the 4 electrical connectors from the instrument panel speakers.

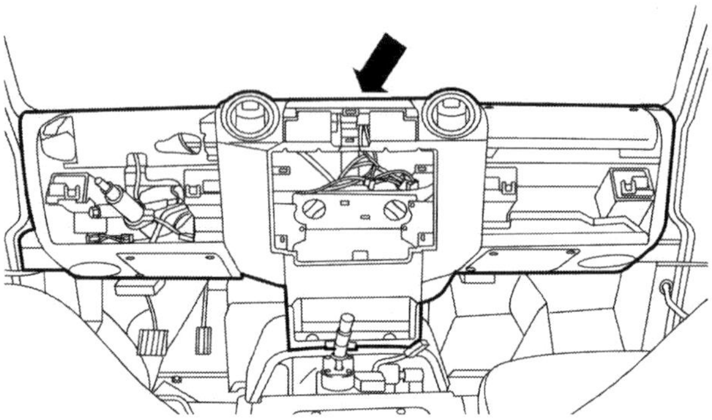

24. With assistance, remove the instrument panel.

Installation

1. With assistance, install the instrument panel.
2. Connect the 4 electrical connectors to the instrument panel speakers.

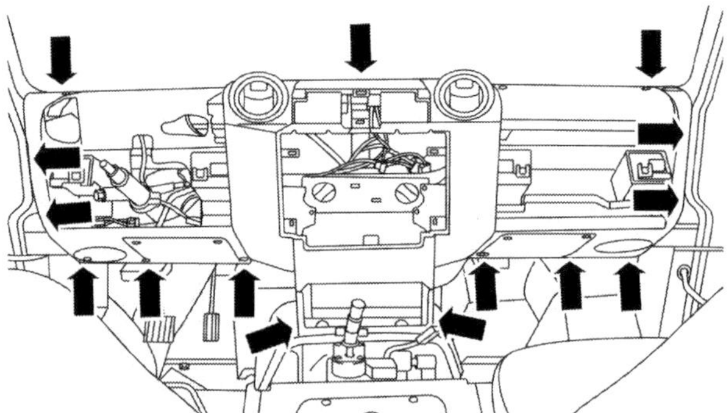

E88470

3. Loosely install the instrument panel screws.

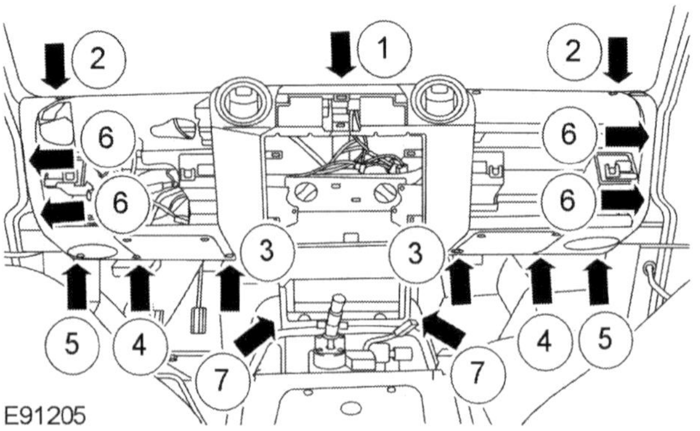

E91205

4. Tighten the screws in the order indicated in the illustration shown.

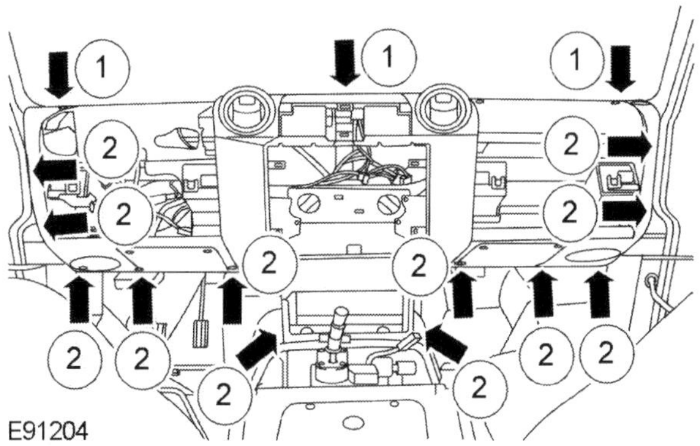

5. Tighten the screws.
   1. Tighten to 2 Nm (1 lb.ft).
   2. Tighten to 1 Nm (1 lb.ft).
6. Secure the climate control switch panel.
   - Install the 4 screws.
7. Install both instrument panel speakers.
   - Connect the electrical connectors.
8. Install both side demister vent finishers.

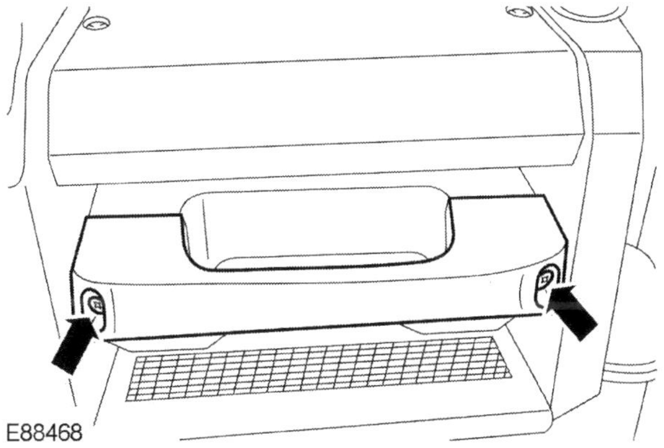

9. Install the passenger side grab handle.
   - Tighten to 7 Nm (5 lb.ft).
10. Install the instrument panel console.

Refer to: Instrument Panel Console (501-12 Instrument Panel and Console, Removal and Installation)

11. Install the multifunction switch assembly.
    - Connect the 3 electrical connectors.
    - Fully tighten the multifunction switch clamp screw.
12. Install the steering column shrouds.

Refer to: Steering Column Shrouds (501-05 Interior Trim and Ornamentation, Removal and Installation)

13. Install the instrument cluster.

Refer to: Instrument Cluster (413-01 Instrument Cluster, Removal and Installation)

14. Install the steering wheel.

Refer to: Steering Wheel (211-04 Steering Column, Removal and Installation)

15. Install the transmission cover panel.
    - Install the 11 screws.

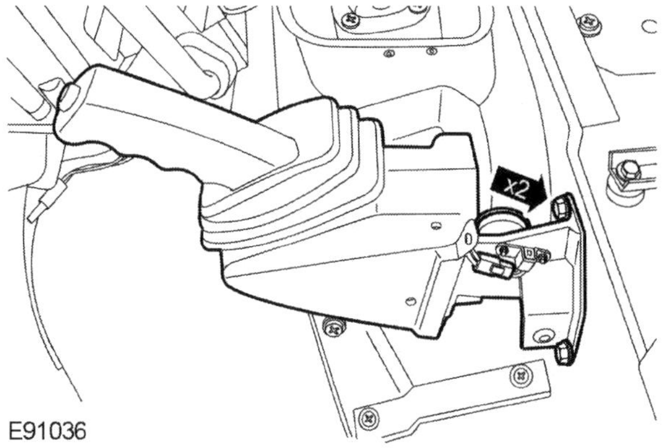

16. Secure the parking brake lever.
    - Tighten to 25 Nm (18 lb.ft).
17. Attach the parking brake lever gaiter.
    - Install the 3 clips.
18. Install the gear shift lever.
19. Attach the gear shift lever.
20. Install the high/low gear selector lever.
21. Install the transmission cover panel floor covering.
22. Install the floor console.

Refer to: Floor Console (501-12 Instrument Panel and Console, Removal and Installation)

23. Connect the battery ground cable.

Refer to: Battery Connect (414-01 Battery, Mounting and Cables, General Procedures).

Instrument Panel Console

Removal and Installation

Removal

1. Disconnect the battery ground cable.

Refer to: Startup Battery Disconnect and Connect (414-01 Battery, Mounting and Cables, General Procedures)

2. Remove the audio unit.

Refer to: Audio Unit - Vehicles Without: Bluetooth (415-01 Audio Unit, Removal and Installation)

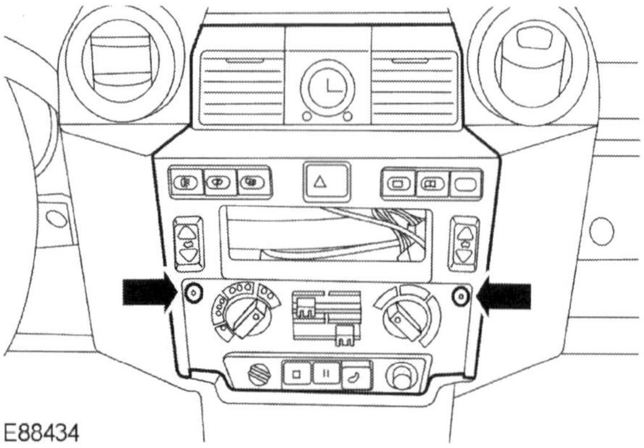

E88434

3. Release the instrument panel console.

   - Remove the 2 screws.

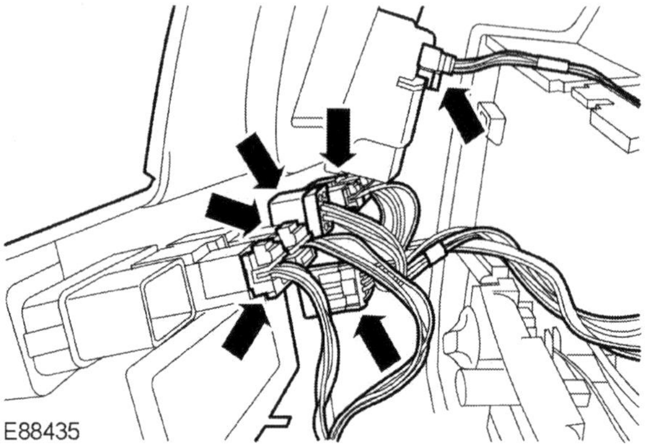

E88435

4. Disconnect the 6 electrical connectors from the instrument panel console.

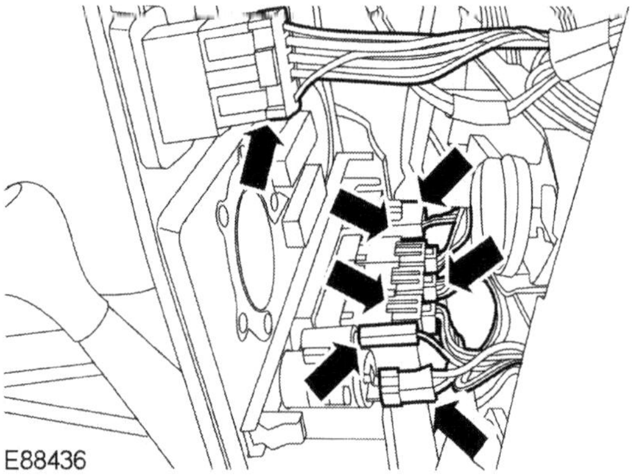

E88436

5. Remove the instrument panel console.

   - Disconnect the 7 electrical connectors.

Installation

1. To install, reverse the removal procedure.
2. Connect the battery ground cable.

Refer to: Battery Connect (414-01 Battery, Mounting and Cables, General Procedures).

In-Vehicle Crossbeam

Removal and Installation

Part(s)

STEP	PART NAME	QUANTITY
Removal Step 3	Renew Part Retaining clip(s)	5
Removal Step 6	Renew Part In-vehicle cross-beam nut(s)	6

Removal

1. Disconnect the battery ground cable.

Refer to: Startup Battery Disconnect and Connect (414-01 Battery, Mounting and Cables, General Procedures)

2. Remove the instrument panel.

Refer to: Instrument Panel (501-12 Instrument Panel and Console, Removal and Installation)

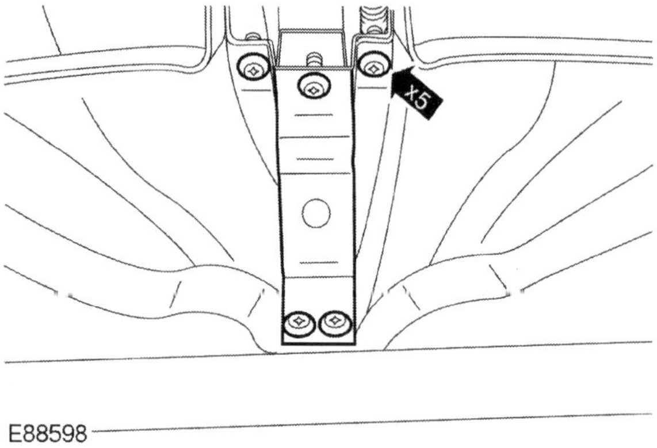

E88598

3. ⚠ NOTE: Make sure the clips are discarded.

Remove and discard the 5 clips. Remove the audio unit support bracket.

Renew Part: Retaining clip(s)

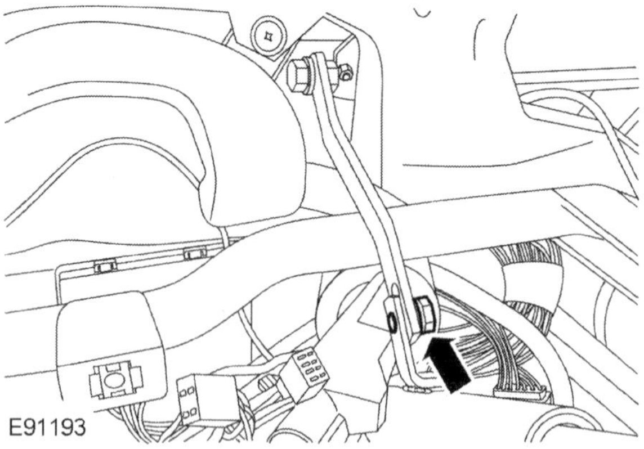

E91193

4. Remove the bolt. Release the steering column bracket.

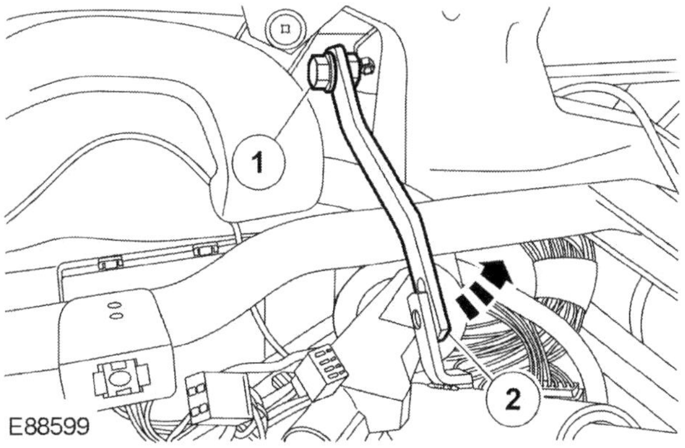

5. Undo but do not remove the bolt. Reposition the bracket. Reposition the steering column bracket.

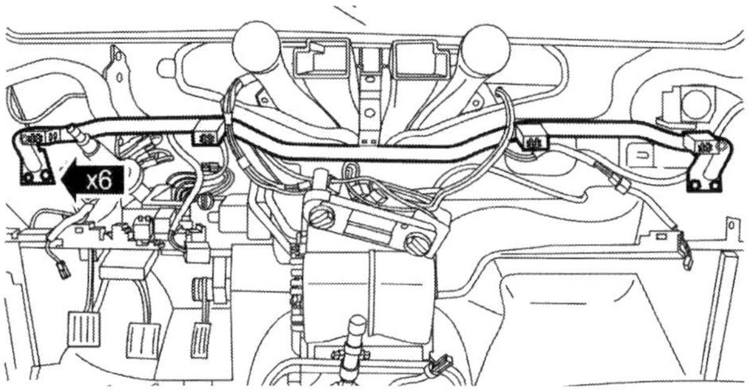

6. Remove and discard the 6 nuts. Remove the in-vehicle crossbeam.

Renew Part: In-vehicle crossbeam nut(s)

Installation

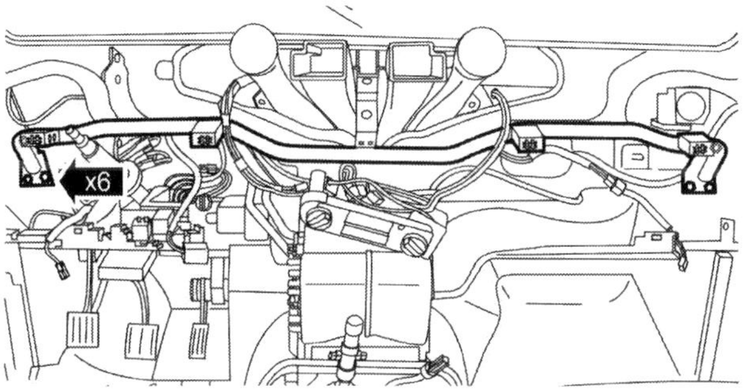

1. To install, reverse the removal procedure. Tighten the nuts.

   Torque: 23 Nm

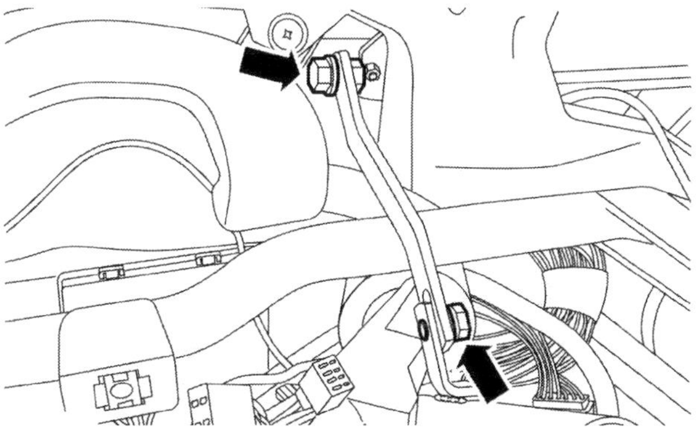

2. Tighten the bolts.

Torque: 22 Nm

3. Connect the battery ground cable.

Refer to: Startup Battery Disconnect and Connect (414-01 Battery, Mounting and Cables, General Procedures)

Floor Console

Removal and Installation

Removal

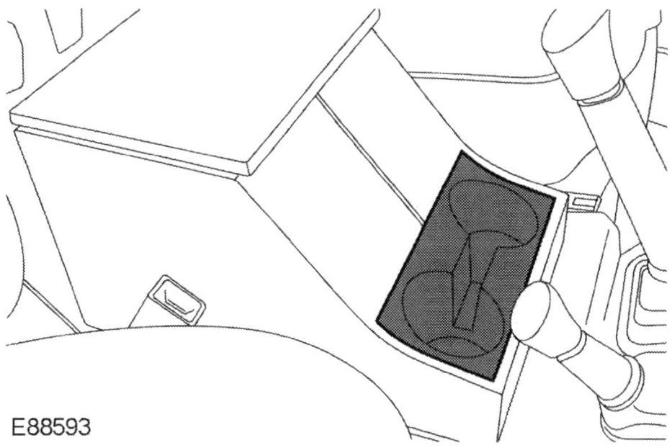

1. Remove the cup holder.

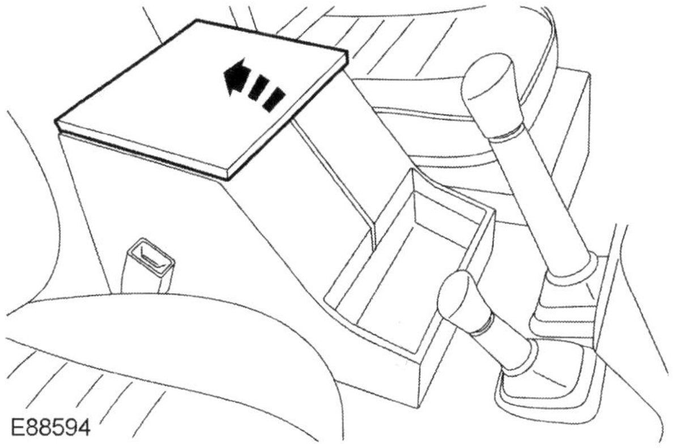

2. Open the floor console lid.

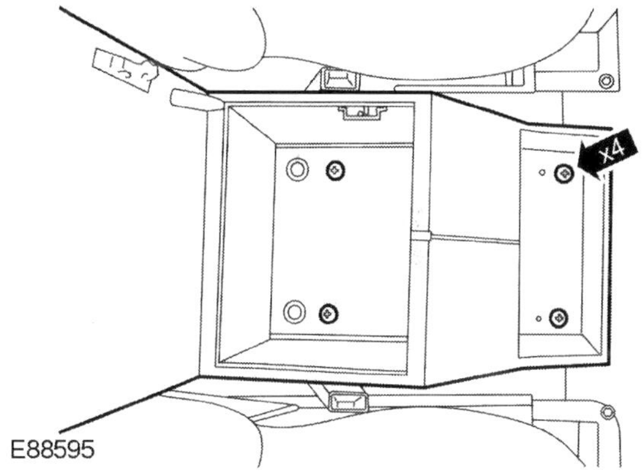

3. Remove the floor console.
   - Remove the 4 screws.

Installation

1. To install, reverse the removal procedure.

Instrument Panel Switches

Removal and Installation

Removal

⚠ NOTE: Rear window wiper switch shown, other instrument panel switches similar.

1. Remove the audio unit.

Refer to: Audio Unit - Vehicles Without: Bluetooth (415-01 Audio Unit, Removal and Installation)

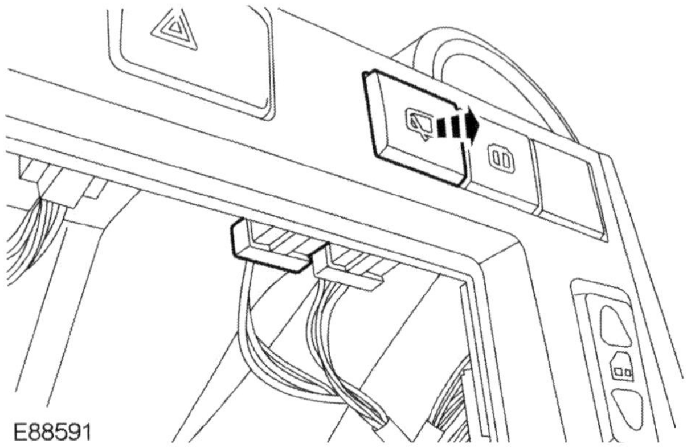

E88591

2. Release the instrument panel switch.

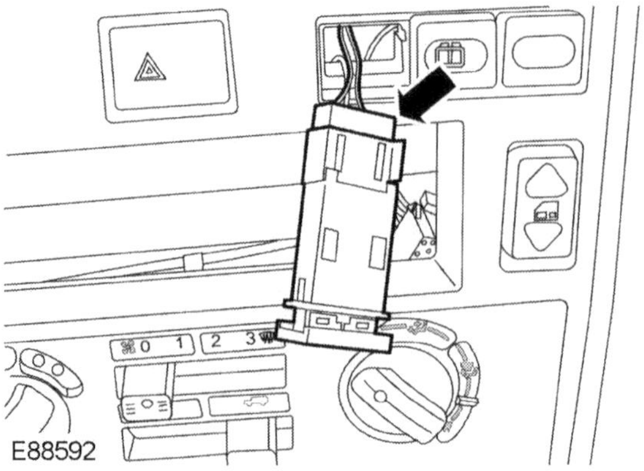

E88592

3. Remove the instrument panel switch.

   • Disconnect the electrical connector.

Installation

1. To install, reverse the removal procedure.

Handles, Locks, Latches and Entry Systems -

Specifications

Torque specifications

Description	Nm	lb.ft
Taildoor latch	10	7

Description and Operation

CENTRAL DOOR LOCKING

Central Door Locking (CDL) operates on all doors on Station Wagon and County Station Wagons. On all other Defender variants, CDL operates on the driver and front and rear passenger doors only. CDL is operated by pressing the appropriate button on the remote handset, using the vehicle key in the drivers door lock or using the drivers door sill button.

The CDL system uses electrically operated door latches, which are controlled by the anti-theft system module located behind the instrument pack. The anti-theft system module receives a permanent power supply from the CJB (central junction box) . Each CDL latch motor is connected by two wires to the anti-theft system module, which alternately supplies power and earth connections to drive each motor to the lock or unlock positions.

If the latch motors are continually operated in a short period of time, the anti-theft system module will suspend all CDL latch motor operation for 15 seconds to prevent the motors from overheating. Before suspension occurs the anti-theft system module ensures that all the doors are left in the unlocked condition.

If the vehicle is locked, and the key is in the ignition switch or the ignition is on and the anti-theft system module inertia switch is tripped, all doors will be unlocked immediately and the hazard warning lamps will operate. To prevent accidental relocking, all latch motors will be inhibited until the anti-theft system module senses that the drivers door is opened with the key removed from the ignition switch.

Locking of the vehicle using the remote handset is inhibited with the key in the ignition switch.

Slam Locking

With the introduction of CDL, slam locking of the drivers door is disabled. If an attempt is made to slam lock the vehicle, the anti-theft system module will unlock all doors.

CDL Using the Drivers Door Sill Button

Each door has a sill button to allow that door to be individually locked from inside the vehicle. The drivers door sill button has additional functionality which allows all doors to be locked from inside the vehicle from this one button.

When the driver door sill lock button is depressed, the CDL system enters a sill locked state, where all doors are locked but immobilisation and perimetric and volumetric sensing is suspended. CDL using the drivers sill lock button can be achieved with or without the key in the ignition switch and without the ignition being on.

If the CDL is in the sill locked state and the lock button on the remote handset is pressed, the volumetric and perimetric functions of the alarm system will become active and active engine immobilisation will be invoked.

CDL Using the Key

The driver and the passenger front doors are fitted with key barrel locks. Using the vehicle key in the passenger door will only lock or unlock that door. Using the key in the drivers door will operate the CDL system and lock or unlock all doors.

⚠ NOTE: The passenger door key barrel is connected to the door latch and has no electrical input to the anti-theft system module. The driver door key barrel is also connected to the door latch and its operation is monitored by the ECU only via the latch motor and the two wires connected to it.

When the vehicle is locked using the key in the drivers door, only CDL and perimetric protection is invoked. Volumetric protection and active immobilisation is not initiated.

CDL Using the Remote Handset

The remote handset has two buttons; lock and unlock. A single press on the lock button will lock all doors and invoke perimetric and volumetric protection and active immobilisation. Remote locking is inhibited if the transponder coil senses that the key is in the ignition switch.

When the vehicle is locked, a single press of the unlock button will unlock all doors and disable perimetric and volumetric protection. Immobilisation is only disabled using the transponder coil and the remote handset or the EKA procedure.

Anti-theft Alarm Indicator

Item	Part Number	Description
1	-	Anti-theft alarm indicator

In addition to the alarm system status indications, the Anti-theft alarm indicator also displays CDL system status.

The system confirms that the drivers door is open by illuminating the indicator for the 10 second confirmation period. If any other door or the hood is open, the indicator remains unlit for the 10 second confirmation period. After the 10 second confirmation period the indicator reverts to the slow flash deterrent mode if the alarm is armed.

Interior Lamp Functionality

The anti-theft system module controls the interior lamp functionality. When the lamps are extinguished, they fade out to the off condition over a 2 second period.

The lamps are turned on by:

- Disarming the alarm using the remote handset
- Opening any door.

When the lamps are on, they will be extinguished by the following actions:

- If all doors are closed, the lamps will be extinguished after a 15 second delay period.
- If any door is left open, the lamps will be extinguished after an 8 minute delay period.
- If one of the above delay periods is active and the ignition is switched on or the alarm is armed, the lamps will be extinguished.

Latch Cable Adjustment

General Procedures

1. Remove door trim panel.

Refer to: Front Door Trim Panel (501-05 Interior Trim and Ornamentation, Removal and Installation)

2. Carefully peel back plastic sheet to expose mechanism.

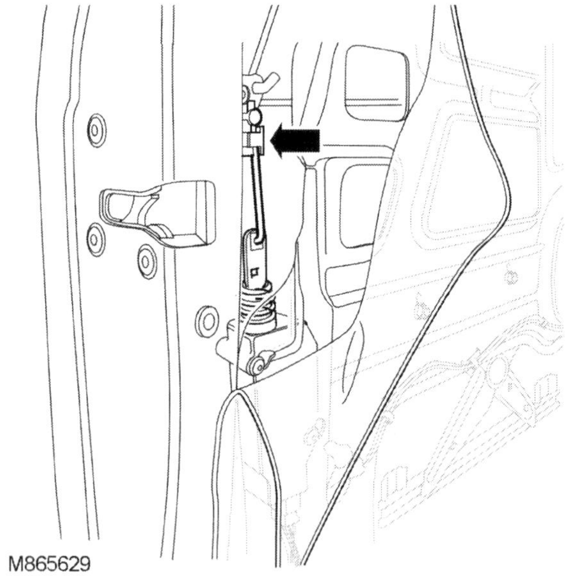

3. Release clip and detach solenoid link from latch.

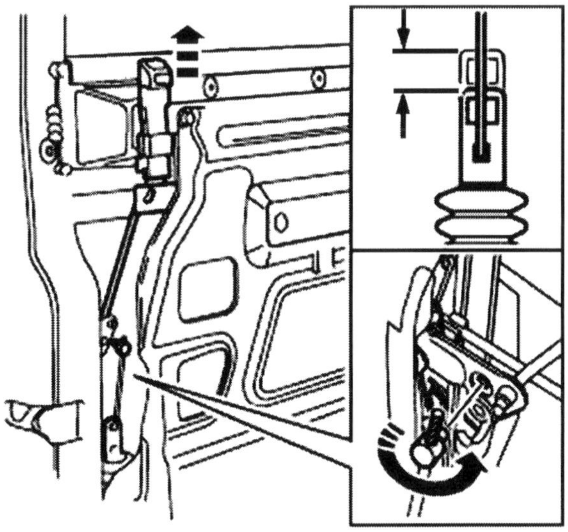

E79255

4. Ensure sill button is fully raised.
5. Fully extend solenoid actuator then lower 3.5 mm ± 1.00 mm.
6. Whilst holding actuator in this position, adjust length of link, until link pin is aligned with hole in latch lever.
7. Connect link and secure with clip.
8. Secure plastic sheet.
9. Install door trim panel.

Refer to: Front Door Trim Panel (501-05 Interior Trim and Ornamentation, Removal and Installation)

Front Door Latch

Removal and Installation

Removal

1. Remove door reinforcing panel and support glass with timber.

Refer to: Front Door Reinforcement Panel (501-03 Body Closures, Removal and Installation)

2. Disconnect control rod from handle operating lever.
3. Disconnect control rod from locking lever on handle.

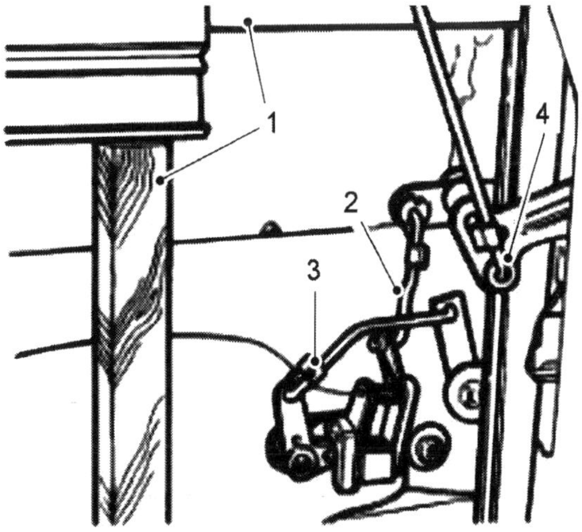

4. Disconnect push button control rod and linkage from latch mechanism.
5. Remove 2 screws and remove handle assembly from door.

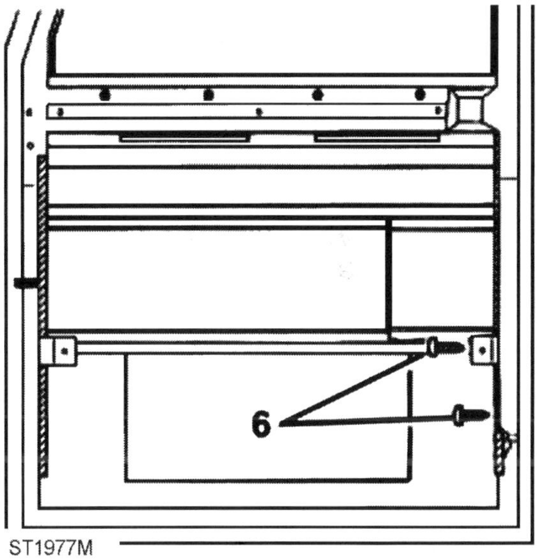

ST1977M

6. Remove 2 self-tapping screws retaining lower end of window glass runner.

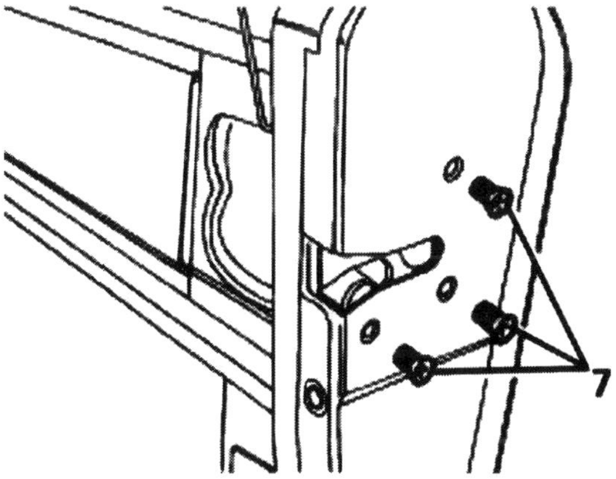

7. Remove 3 screws securing latch assembly to door.
8. Whilst taking care not to damage runner, ease runner away from latch and manoeuvre latch assembly from door.

Installation

1. Carefully ease window runner away, sufficiently to enable latch to be located into position.

2. Secure latch to door with 3 screws.

3. Secure window runner with 2 screws ensuring that packing strip is in position and that screw heads are below bottom of runner to prevent damage to glass.

4. Install handle with 2 screws, ensuring that bezels are in position.

5. Connect control rod to handle operating lever and secure with spring clip.

6. Connect control rod to locking lever and retain with spring clip.

7. Connect push button control rod and linkage to latch lever and secure with spring clip.

8. Install mounting panel.

Refer to: Front Door Reinforcement Panel (501-03 Body Closures, Removal and Installation)

Rear Door Latch

Removal and Installation

Removal

1. Remove door trim panel.

Refer to: Rear Door Trim Panel (501-05 Interior Trim and Ornamentation, Removal and Installation)

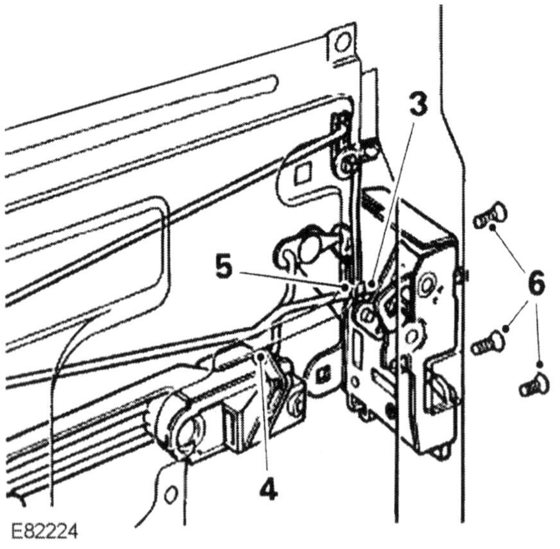

2. Peel-back sufficient of plastic sheet to reveal latch.

3. Release remote control lever rod from latch assembly. 4. Disconnect door outer handle control rod from latch assembly.

5. Disconnect door locking button remote control rod from latch mechanism.

6. Remove 3 retaining screws and withdraw latch assembly from door.

Installation

1. Install latch assembly to door and secure with 3 screws, noting that uppermost screw is longer.

2. Connect remote control levers to latch mechanism reversing instructions 3, 4 and 5.

3. Re-seal plastic sheet and install door trim panel.

Refer to: Rear Door Trim Panel (501-05 Interior Trim and Ornamentation, Removal and Installation)

Door Lock Cylinder

Removal and Installation

Removal

1. Remove door trim panel and plastic sheet.

Refer to: Front Door Trim Panel (501-05 Interior Trim and Ornamentation, Removal and Installation)

2. Remove reinforcement panel.

Refer to: Front Door Reinforcement Panel (501-03 Body Closures, Removal and Installation)

3. Raise and support glass to gain access to latch mechanism.

4. Release spring clip and disconnect rod from lock operating lever.

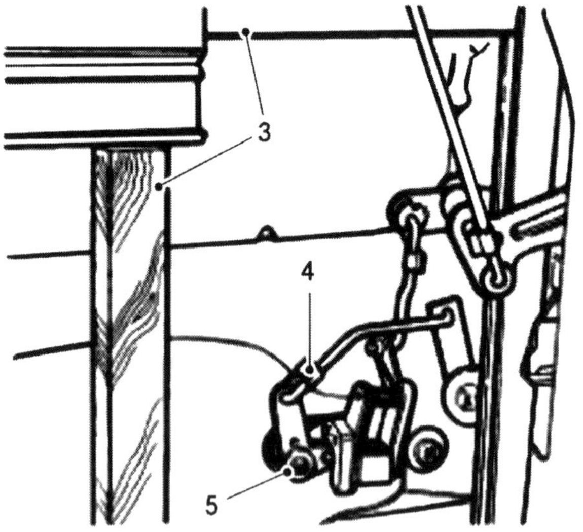

5. Remove single screw and withdraw lock lever assembly.

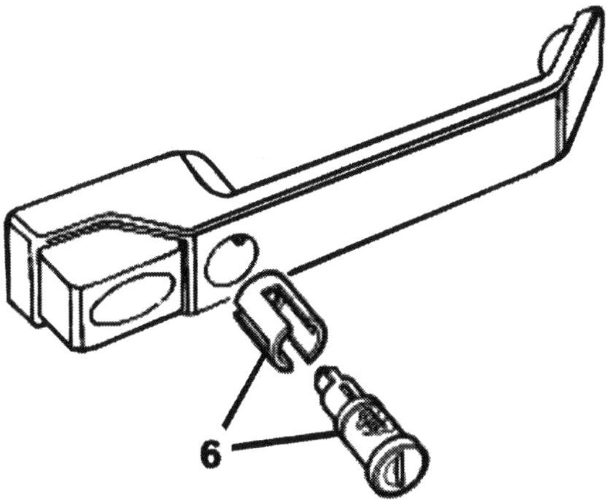

ST1980M

6. Withdraw lock barrel from exterior door handle complete with locking sleeve.

7. To remove barrel from plastic retaining sleeve, depress spring loaded button and withdraw sleeve.

Installation

△ NOTE: If a new barrel is being installed, check that number on barrel coincides with number on accompanying key.

1. Push plastic retaining sleeve over barrel until spring loaded peg locks it into position.
2. Install barrel and plastic sleeve assembly to locking sleeve and insert into exterior handle reversing instruction 6.

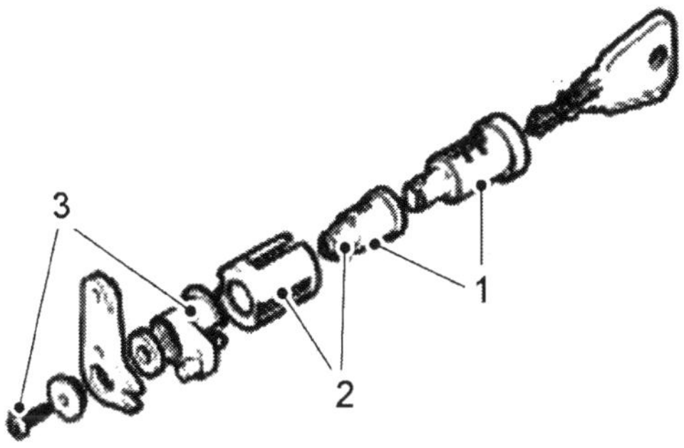

ST1978M

3. Assemble lock lever components as illustrated and from inside of door panel, install them to barrel assembly with single screw.
4. Connect operating rod to lock lever and secure with spring clip, reversing instruction 4.
5. Install front door reinforcement panel.

Refer to: Front Door Reinforcement Panel (501-03 Body Closures, Removal and Installation)

6. Install plastic sheet.
7. Install door trim panel.

Refer to: Front Door Trim Panel (501-05 Interior Trim and Ornamentation, Removal and Installation)

Exterior Front Door Handle

Removal and Installation

Removal

1. Remove door trim panel.

Refer to: Front Door Trim Panel (501-05 Interior Trim and Ornamentation, Removal and Installation)

2. Remove door reinforcement panel and support glass with timber.

Refer to: Front Door Reinforcement Panel (501-03 Body Closures, Removal and Installation)

3. Disconnect operating rod from handle mechanism.
4. Disconnect rod from locking barrel lever.

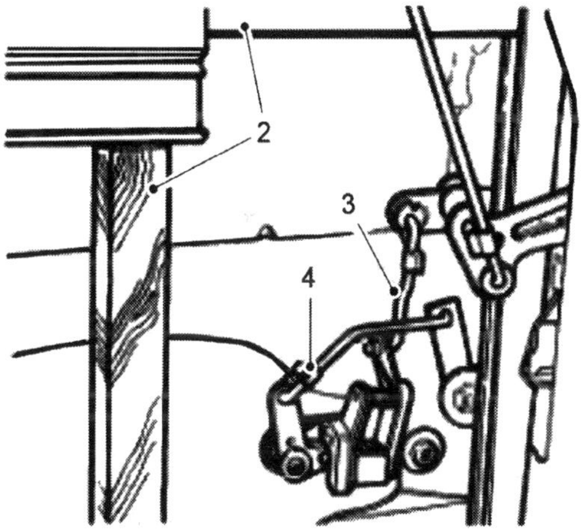

ST1982M

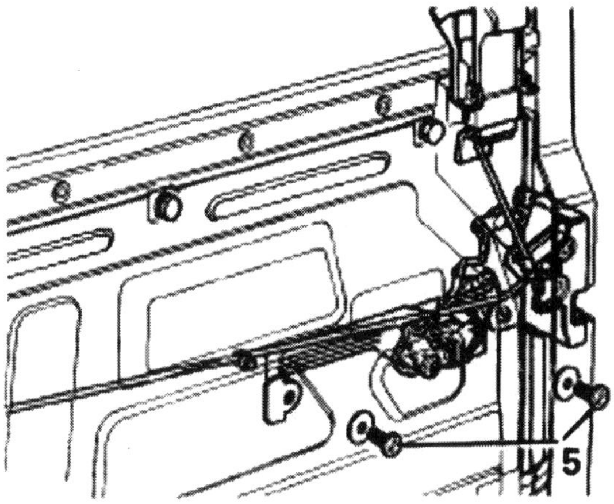

5. Remove 2 screws and withdraw handle assembly.

Installation

1. Install handle to door ensuring that two bezels are in position - flat faces towards door, and secure with 2 screws.
2. Connect rod to handle operating lever and secure with spring clip.
3. Connect rod to locking barrel lever and retain with spring clip.
4. Install door reinforcement panel.

Refer to: Front Door Reinforcement Panel (501-03 Body Closures, Removal and Installation)

Exterior Rear Door Handle

Removal and Installation

Removal

1. Remove door trim panel.

Refer to: Rear Door Trim Panel (501-05 Interior Trim and Ornamentation, Removal and Installation)

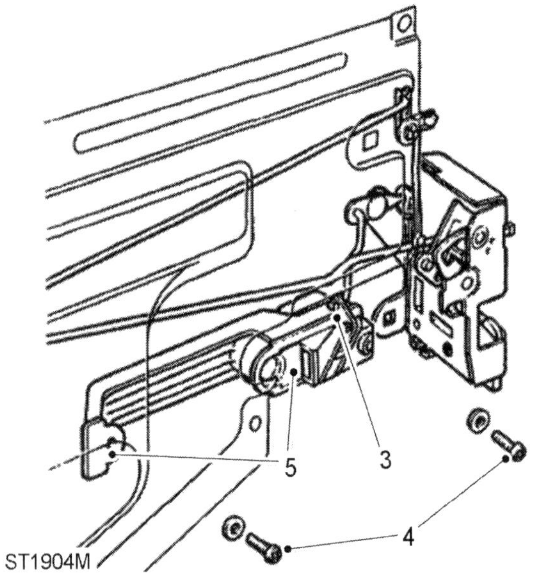

2. Peel back sufficient of plastic sheet to gain access to handle mechanism.
3. Disconnect actuating rod from handle operating lever.
4. Remove 2 screws and withdraw handle and bezels.

Installation

1. Install handle to door ensuring that both bezels are in position - flat faces towards door and secure with 2 screws.
2. Connect actuating rod to handle operating lever and secure with spring clip.
3. Re-seal plastic sheet.
4. Install door trim panel.

Refer to: Rear Door Trim Panel (501-05 Interior Trim and Ornamentation, Removal and Installation)

Interior Rear Door Handle

Removal and Installation

Removal

1. Remove door trim panel.

Refer to: Rear Door Trim Panel (501-05 Interior Trim and Ornamentation, Removal and Installation)

2. Peel-back sufficient of plastic sheet to gain access to remote lever.
3. Remove spring clip and disconnect control rod from push button.
4. Release spring clip and disconnect short locking button control rod from latch mechanism.

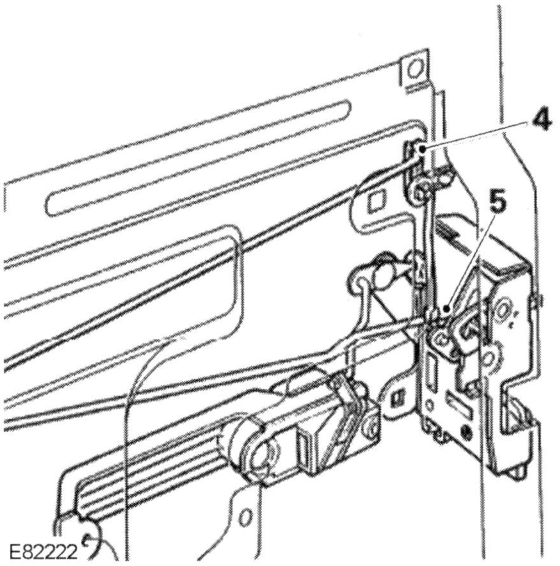

5. Disconnect long remote control rod from latch assembly.
6. Remove 2 screws securing remote control lever to mounting panel.
7. Release control rods from plastic retaining clips located in mounting panel.

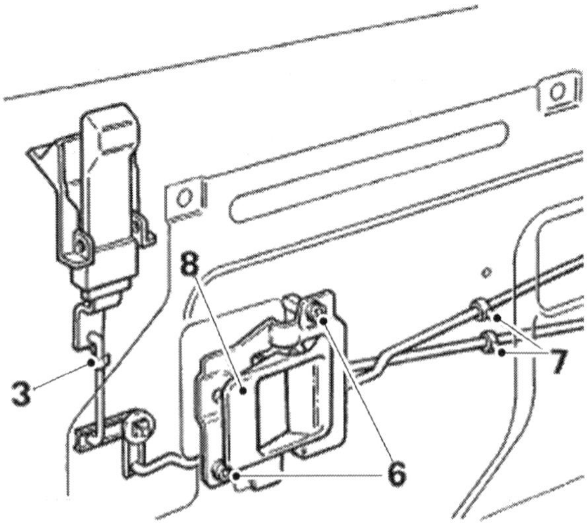

8. Withdraw remote control lever and rods from door.

Installation

1. Install plastic retaining clips to rod assembly into position and secure with 2 screws.
2. Connect control rods to latch assembly and secure with clips.
3. Install plastic retaining rod clips to mounting panel.
4. Connect control rod to push button and secure with clip.
5. Re-seal plastic sheet and install door trim panel.

Refer to: Rear Door Trim Panel (501-05 Interior Trim and Ornamentation, Removal and Installation)

Interior Front Door Handle

Removal and Installation

Removal

1. Remove door trim casing.

Refer to: Front Door Trim Panel (501-05 Interior Trim and Ornamentation, Removal and Installation)

2. Peel back sufficient of plastic sheet to gain access to remote lever.
3. Release spring clip and disconnect control rod from latch mechanism.

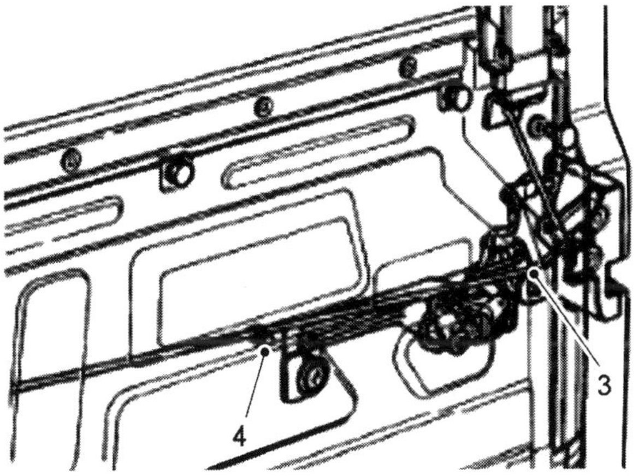

4. Release control rod from plastic clip in mounting panel.

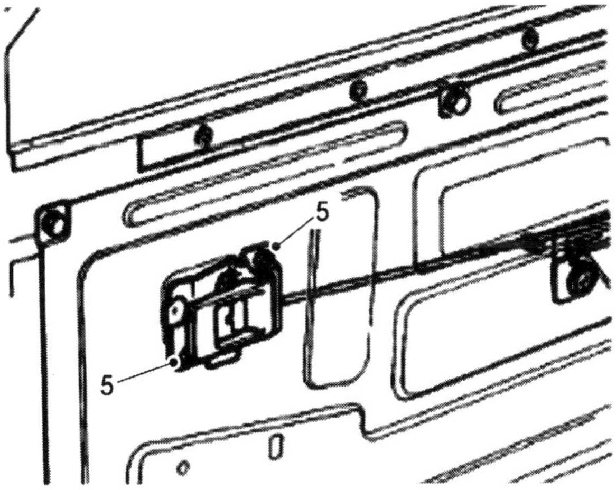

5. Remove 2 screws securing remote control lever to mounting panel and withdraw lever and control rod.

Installation

1. Feed control rod into position and loosely secure lever to mounting panel with 2 screws.
2. Connect control rod to latch mechanism and secure with spring clip.
3. Tighten control lever retaining screws.
4. Secure control rod to plastic clip in mounting panel.
5. Secure plastic sheet.
6. Install door trim casing.

Refer to: Front Door Trim Panel (501-05 Interior Trim and Ornamentation, Removal and Installation)

Hood Latch

Removal and Installation

Removal

1. Open hood.
2. Mark position of guide plate and lock to hood platform.

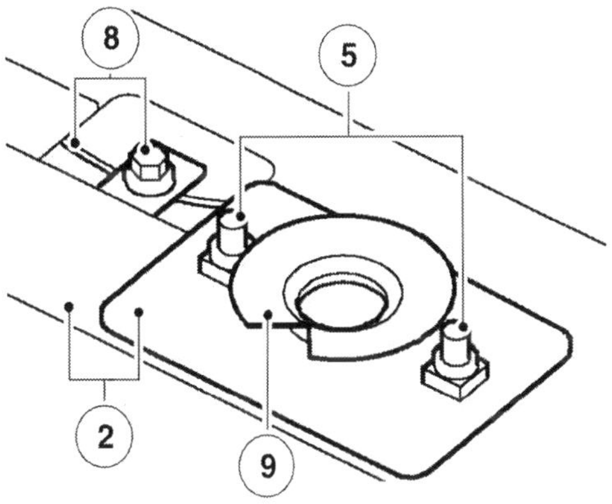

J6371

3. Remove 8 screws securing grille.
4. Remove grille.
5. Remove 2 bolts securing guide plate and lock.
6. Remove guide plate.
7. Release spring securing lock to hood platform.
8. Slacken clamping bolt securing hood release cable.
9. Remove lock.

Installation

1. Fit spring between lock and hood platform.
2. Position lock and guide plate to hood platform and nip up bolts.

3. Position guide plate and lock to position marks and tighten bolts to 10 Nm (7 lbf/ft).

4. Fit cable to lock and tighten clamping bolt.

5. Check operation of release cable and adjust if necessary.

6. Fit grille and tighten securing screws.

7. Close hood.

Hood Latch Release Handle

Removal and Installation

Removal

1. △ NOTE: Hood release cable renew

Open hood.

2. Remove 8 screws securing grille.

3. Remove grille.

4. Slacken clamping bolt securing hood release cable.

5. Remove cable from lock.

6. Release cable from clip on underside of hood platform.

7. Feed cable through valance and collect grommet.

8. Release cable from clip fixed to wheelarch under expansion tank.

9. Loosen clamping nut securing hood release handle to mounting bracket.

10. Remove hood release handle from mounting bracket.

11. Withdraw cable through bulkhead.

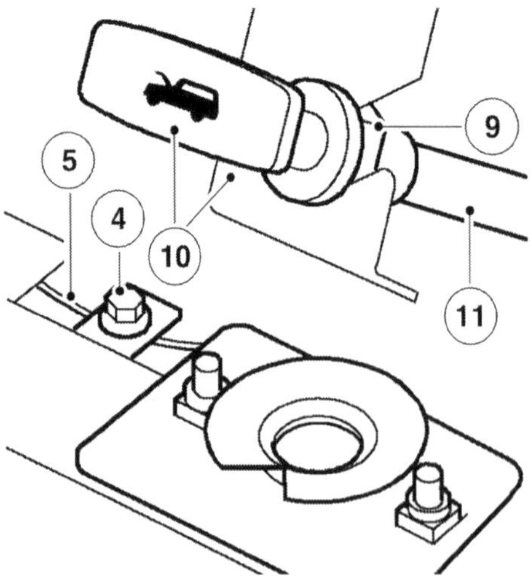

Installation

1. Feed cable through bulkhead and fit hood release handle to mounting bracket. Tighten clamping nut.

2. Route cable behind expansion tank and fit to securing clip.

3. Feed cable through valance.

4. Fit grommet between cable and valance.

5. Fit cable to securing clip on underside of hood platform.

6. Fit cable to lock and tighten clamping bolt.

7. Check operation of release cable and hood lock and adjust if necessary.

8. Fit grille and tighten screws.

9. Close hood.

Front Door Lock Actuator

Removal and Installation

Removal

1. Remove door trim panel.

Refer to: Front Door Trim Panel (501-05 Interior Trim and Ornamentation, Removal and Installation)

2. Carefully peel back plastic sheet to expose mechanism.

3. Release clip, detach and remove lock solenoid link.

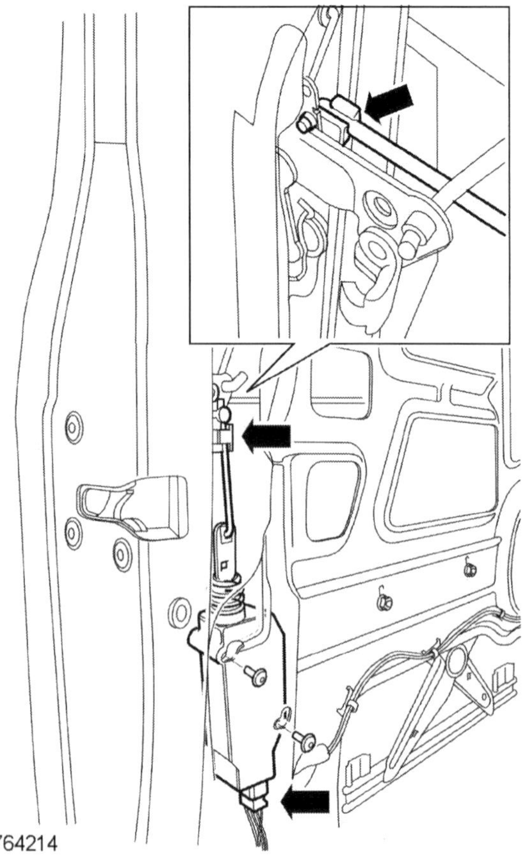

M764214

4. Loosen 2 screws securing solenoid, release solenoid from mounting panel, disconnect multiplug and remove solenoid.

Installation

1. Connect multiplug to solenoid, locate solenoid to mounting panel and tighten screws.
2. Install and secure link.
3. Adjust door latch cable.

Refer to: Latch Cable Adjustment (501-14 Handles, Locks, Latches and Entry Systems, General Procedures)

4. Refit plastic sheet.
5. Fit door trim panel.

Refer to: Front Door Trim Panel (501-05 Interior Trim and Ornamentation, Removal and Installation)

Front Door Push Button Rod and Linkage

Removal and Installation

Removal

1. Remove door trim panel.

Refer to: Front Door Trim Panel (501-05 Interior Trim and Ornamentation, Removal and Installation)

2. Peel back sufficient of plastic sheet to expose mechanism.
3. Release spring clip and disconnect operating rod from latch mechanism.
4. Remove 2 screws and withdraw push button rod and linkage.

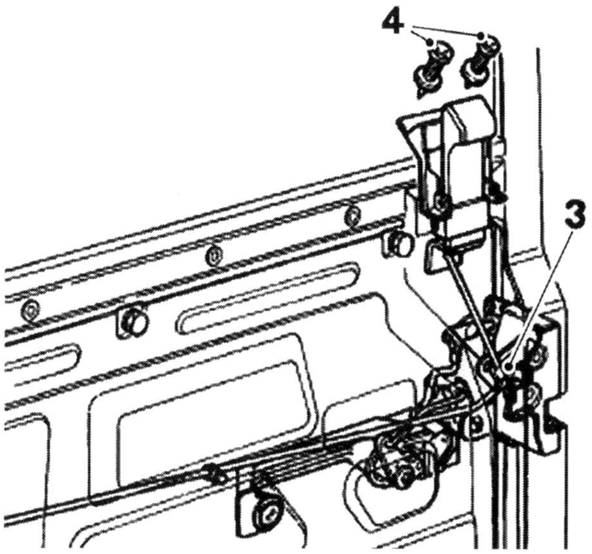

Installation

1. Secure push button rod and linkage to door with 2 screws.
2. Connect operating rod to latch mechanism and secure with spring clip.
3. Re-seal plastic sheet and install door trim panel.

Refer to: Front Door Trim Panel (501-05 Interior Trim and Ornamentation, Removal and Installation)

Rear Door Lock Actuator

Removal and Installation

Removal

1. Remove door trim panel.

Refer to: Rear Door Trim Panel (501-05 Interior Trim and Ornamentation, Removal and Installation)

2. Carefully peel back plastic sheet to expose mechanism
3. Release clip, detach and remove lock solenoid link.

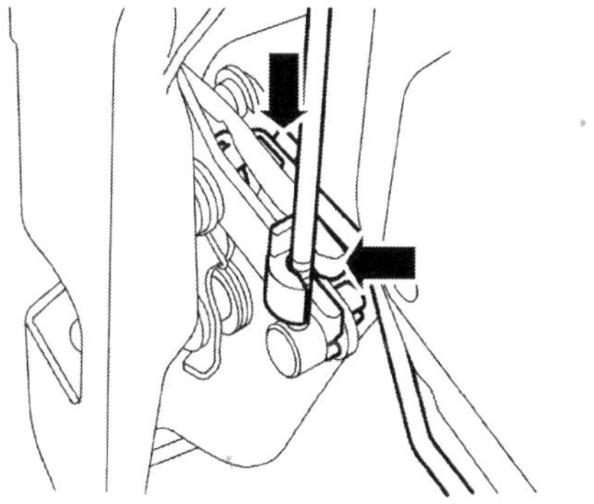

4. Release clip and disconnect interior handle link from latch mechanism.
5. Remove 3 bolts securing mounting panel.

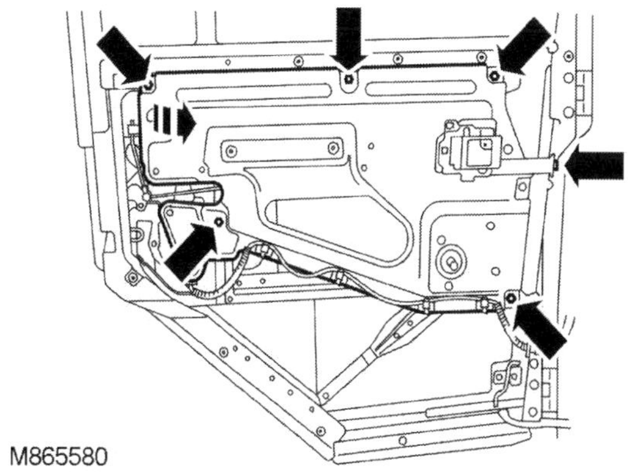

6. Loosen remaining bolts and nut, ease mounting panel away from door and retain in this position for access.
7. Disconnect multiplug from solenoid.

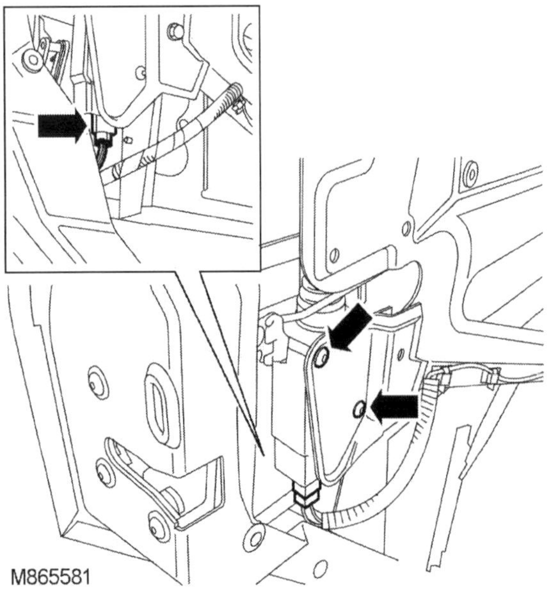

M865581

8. Remove 2 screws securing solenoid.
9. Remove centre pin from bellcrank pivot and release pivot from mounting panel.

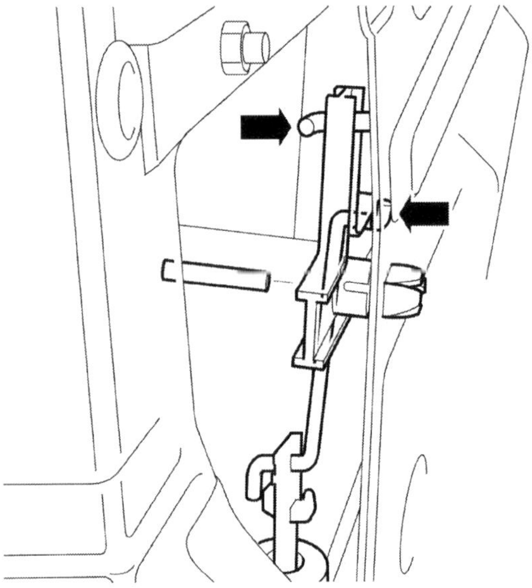

10. Release bellcrank from sill button link and remove from lock solenoid link.
11. Remove door lock solenoid.
12. Remove link from solenoid.

Installation

1. Install link to solenoid, position solenoid, install bellcrank and connect to sill button link.
2. Secure bellcrank to mounting panel and install centre pin.
3. Align solenoid and install and tighten screws.
4. Connect multiplug to solenoid.
5. Install mounting panel bolts and tighten all bolts and nut to 10 Nm (7 lbf.ft).
6. Connect interior handle link to latch and secure with clip.
7. Install and secure solenoid link.
8. Install plastic sheet.
9. Install door trim panel.

Refer to: Rear Door Trim Panel (501-05 Interior Trim and Ornamentation, Removal and Installation)

Rear Door Push Button Rod and Linkage

Removal and Installation

Removal

1. Remove door trim panel.

Refer to: Rear Door Trim Panel (501-05 Interior Trim and Ornamentation, Removal and Installation)

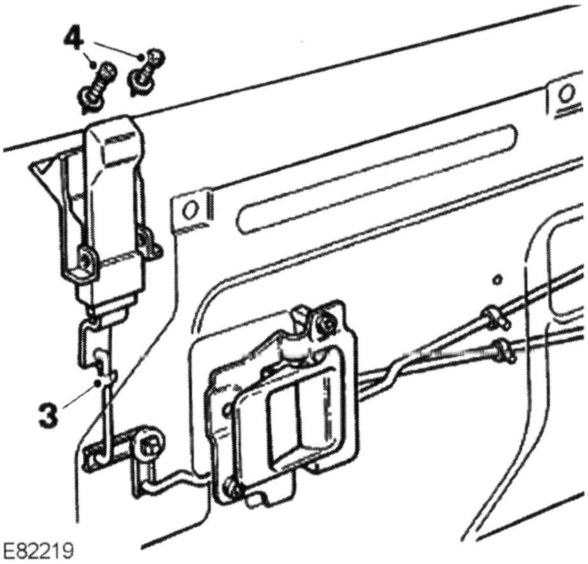

E82219

2. Peel back sufficient of plastic sheet to reveal mechanism.
3. Release spring clip securing button to operating rod and withdraw rod from button.
4. Remove 2 screws securing button to door panel and remove button.

Installation

1. Secure locking button assembly to door panel with 2 screws.
2. Install operating rod to button assembly and secure with spring clip.
3. Re-seal plastic sheet and install door trim panel.

Refer to: Rear Door Trim Panel (501-05 Interior Trim and Ornamentation, Removal and Installation)

Taildoor Latch

Removal and Installation

Removal

1. Remove the taildoor trim panel.

Refer to: Taildoor Trim Panel (501-05 Interior Trim and Ornamentation, Removal and Installation)

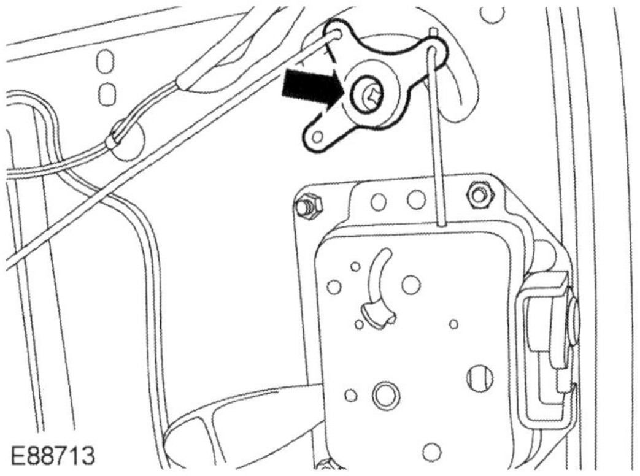

E88713

2. Release the central locking control rod pivot.
   - Remove the screw.

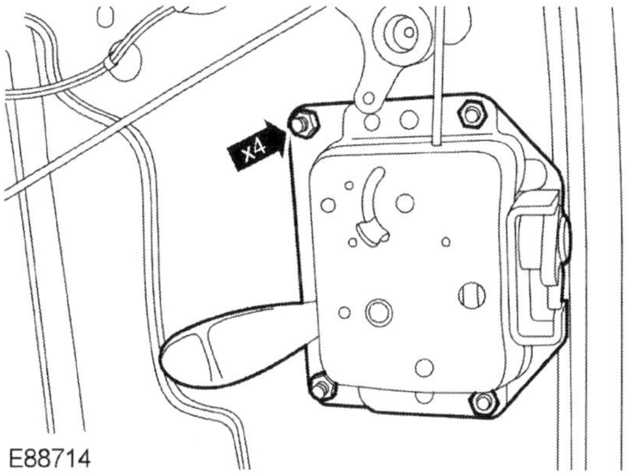

E88714

3. Release the taildoor latch.
   - Remove the 4 nuts.

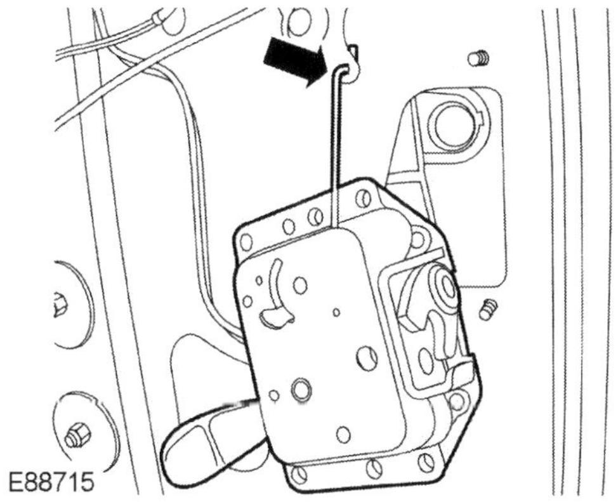

E88715

4. Remove the taildoor latch.
   - Disconnect the taildoor latch from the control rod pivot.

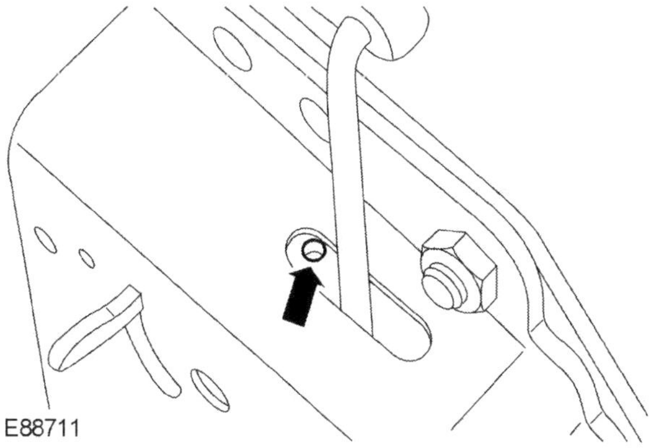

E88711

5. NOTES:

⚠ Do not carry out further dismantling if component is removed for access only.

⚠ Make sure the latch is in the locked position.

Using a suitable tool release the lock cylinder.

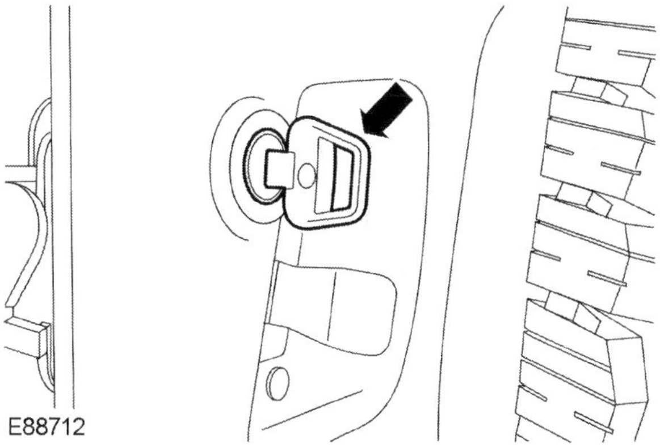

E88712

6. Insert the key and remove the lock cylinder.

Installation

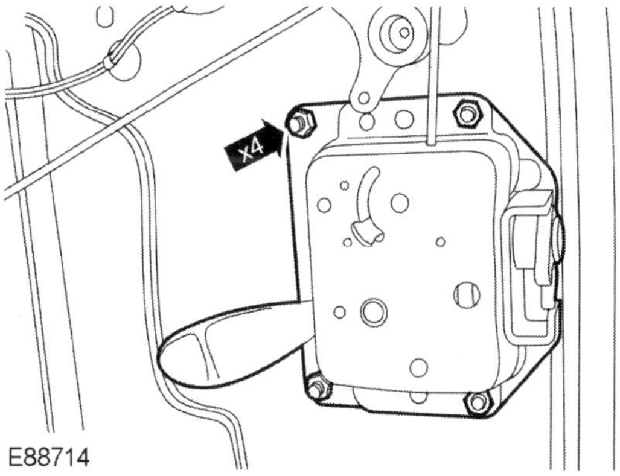

E88714

1. To install, reverse the removal procedure.
   - Tighten to 10 Nm (7 lb.ft).

Taildoor Lock Cylinder

Removal and Installation

Removal

1. Remove the taildoor trim panel.

Refer to: Taildoor Trim Panel (501-05 Interior Trim and Ornamentation, Removal and Installation)

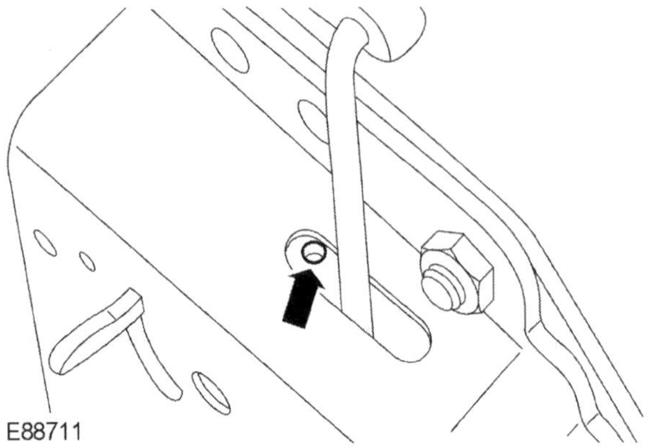

E88711

2. △ NOTE: Make sure the latch is in the locked position.

   Using a suitable tool release the lock cylinder.

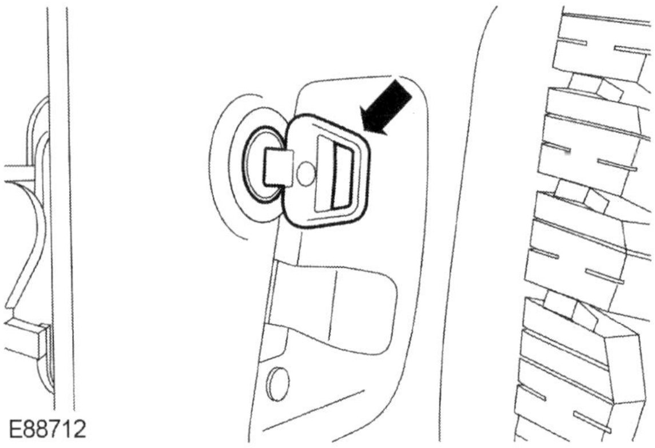

E88712

3. Insert the key and remove the lock cylinder.

Installation

1. To install, reverse the removal procedure.

Taildoor Lock Motor

Removal and Installation

Removal

1. Remove tail door trim panel.

Refer to: Taildoor Trim Panel (501-05 Interior Trim and Ornamentation, Removal and Installation)

2. Disconnect multiplug from taildoor lock motor.
3. Remove bolt and screw securing taildoor lock motor to taildoor.
4. Release clip securing harness to taildoor lock motor mounting bracket.

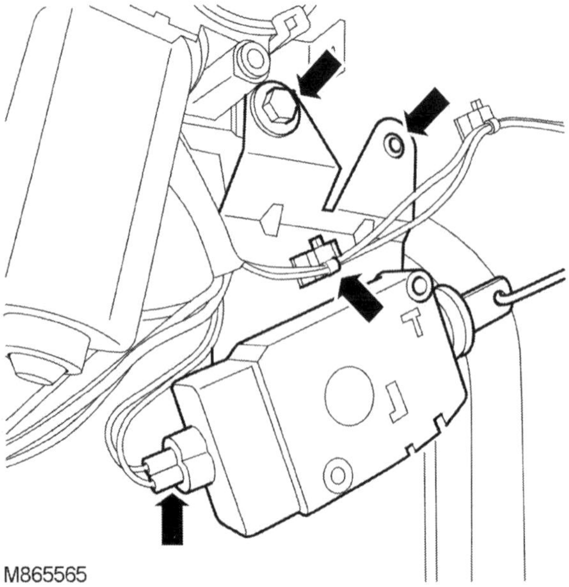

M865565

5. Remove taildoor lock motor assembly from operating rod.

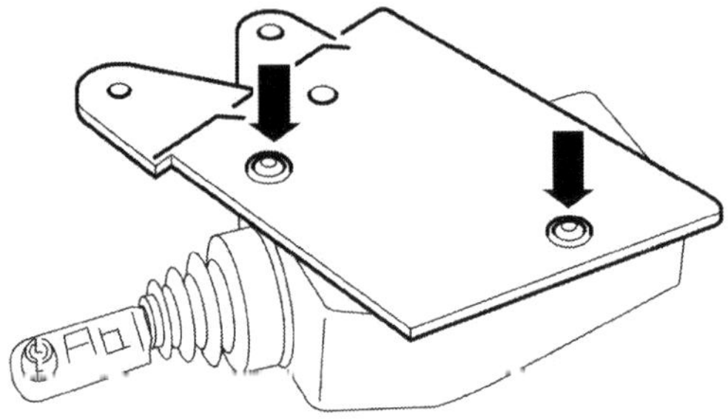

M865566

6. Remove 2 screws securing taildoor lock motor to mounting bracket.

Installation

1. Install taildoor lock motor to mounting bracket and secure with screws.
2. Connect operating rod to taildoor lock motor.
3. Install taildoor lock motor to door, fit screw and tighten bolt to 6 Nm (4 lbf.ft).
4. Position harness to taildoor lock motor mounting bracket and secure with clip.
5. Connect multiplug to taildoor lock motor.
6. Install taildoor trim panel.

Refer to: Taildoor Trim Panel (501-05 Interior Trim and Ornamentation, Removal and Installation)

Description and Operation

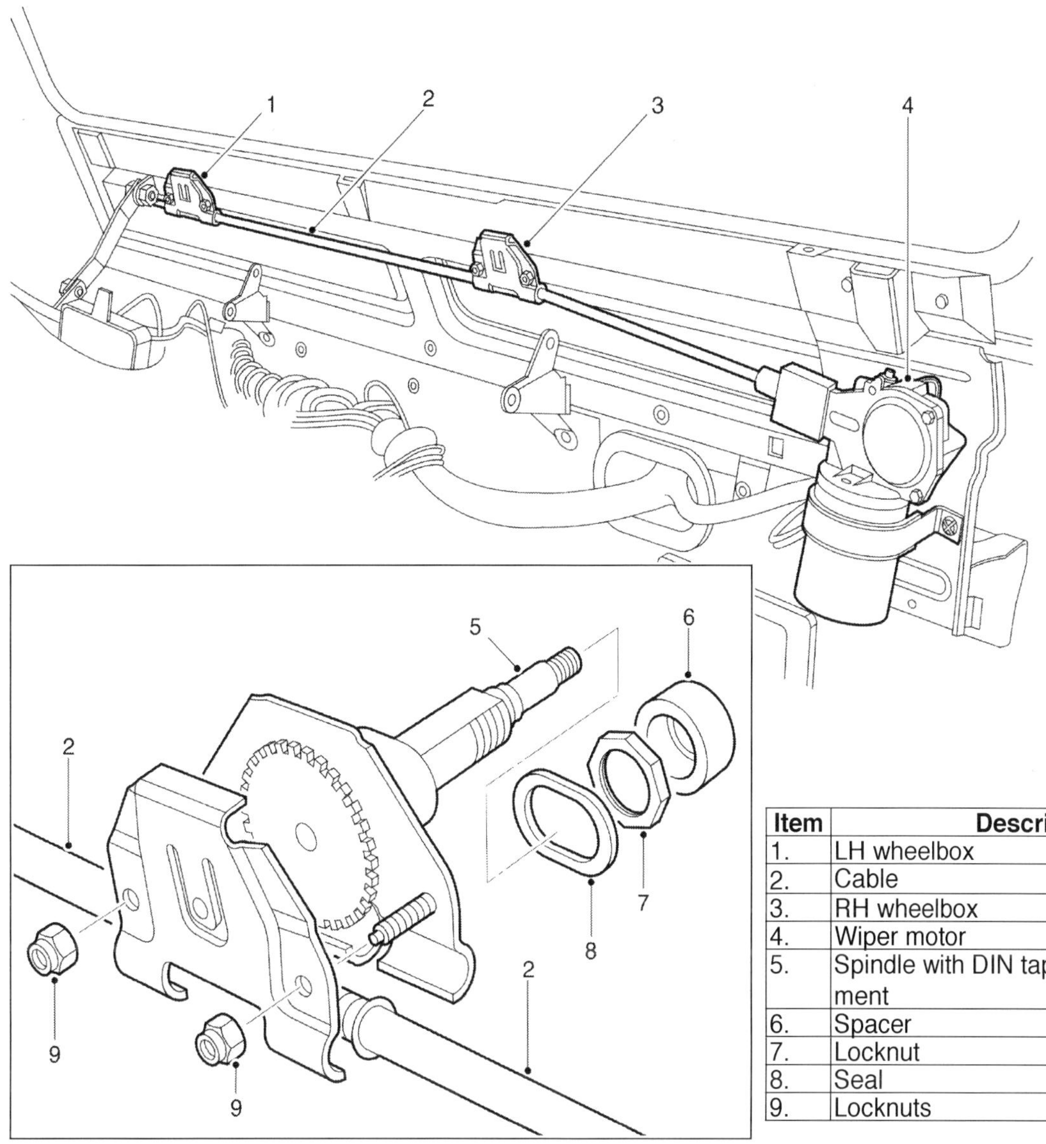

Item	Description
1.	LH wheelbox
2.	Cable
3.	RH wheelbox
4.	Wiper motor
5.	Spindle with DIN taper arm attachment
6.	Spacer
7.	Locknut
8.	Seal
9.	Locknuts

WINDSCREEN WIPERS

Although windscreen wiper functionality remains the same, the mechanical architecture has undergone significant changes, for 2002 Model year. The body structure has undergone some minor changes to house the revised wiper linkage.

A modified wiper motor is introduced which is balanced to reduce operation noise. The brush plate is fitted with capacitors to reduce radio interference and a thermal cut-out is introduced to prevent motor burnout. The motor also contains a new park switch which also reduces operation noise. A common motor is now used on both LHD and RHD variants.

The motor is repositioned on the bulkhead which allows a straighter drive cable run. The straighter run reduces load on the cable allowing a stiffer cable to be used. The stiffer cable and the incorporation of a reduced friction material reduces torsion wind-up of the cable and results in smoother wiper operation and a reduction of blade over travel.

New, larger wheel boxes are fitted in an inverted position from the previous installation. The bulkhead mounting holes are larger with flats for positive location of the wheel box spindle. The wheel boxes now feature a 40 tooth gear ratio in lieu of a 32 tooth gear. This reduces backlash and load and further reduces blade over travel. The wheel box casings are secured with locknuts.

The wheel box spindles now have a splined DIN taper fitting for the new wiper arms which are secured on the spindle with an M8 nut. The taper improves arm retention.

A new grease is introduced which improves wiper operation at very low temperatures.

REAR WIPER

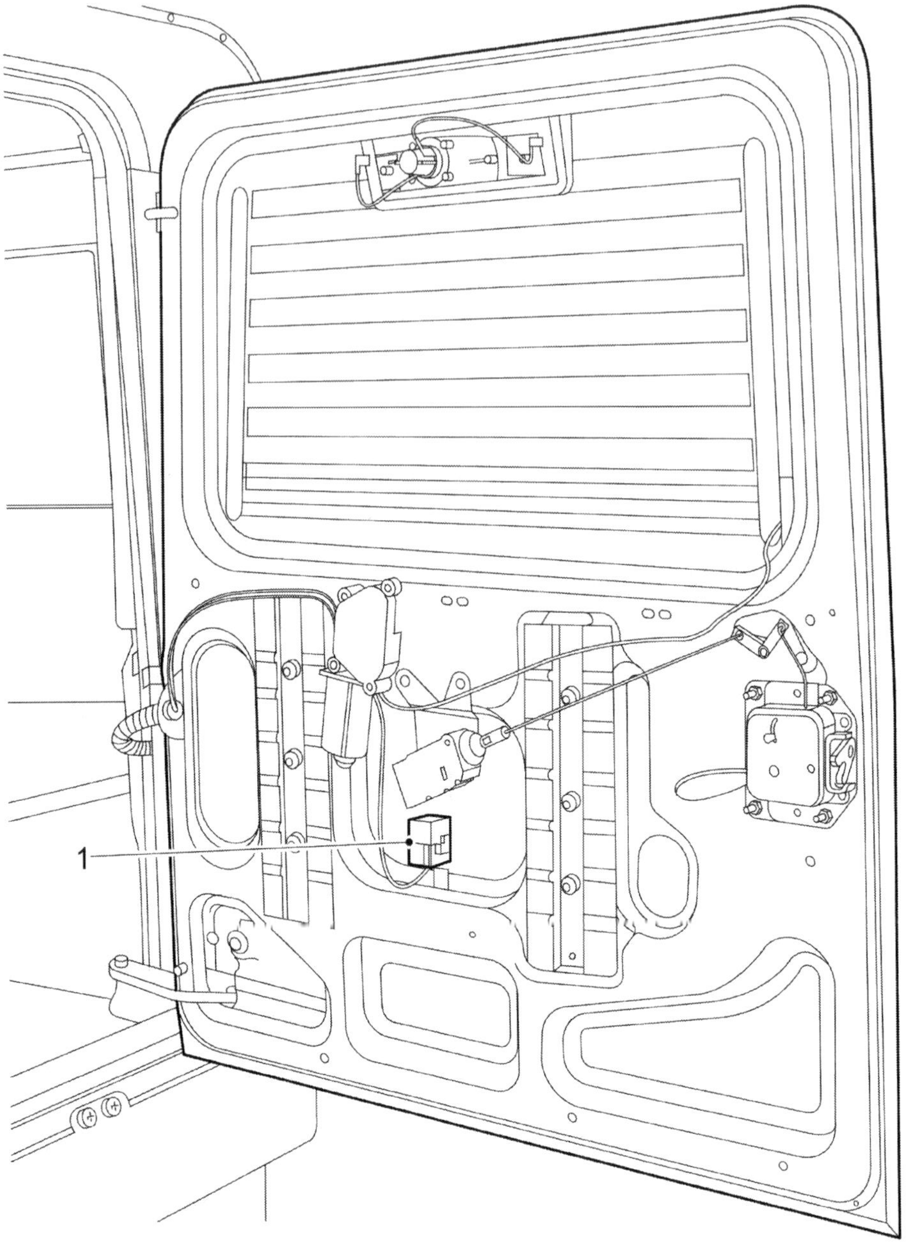

M865569

Item	Part Number	Description
1.	-	Rear wiper relay

The rear wiper system is unchanged except for the addition of a wiper motor relay into the electrical circuit, at 2002 model year.

The new rear wiper relay is located in the tail door, below the wiper motor. The relay coil is energised by a feed supplied from the rear wiper switch. When the relay contacts close, a feed is supplied via the ignition switch (position II) and a fuse in the passenger compartment fuse box, through the relay contacts to the wiper motor.

Front Wiper Pivot Arm

Removal and Installation

Removal

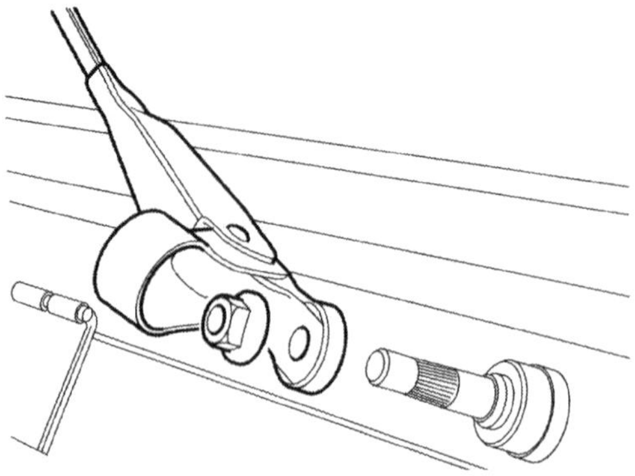

1. Raise nut cover on wiper arm and remove nut.
2. ⚠ NOTE: Do not carry out further dismantling if component is removed for access only.

Remove wiper arm from spindle.

3. Remove wiper blade.

Installation

1. Fit wiper blade.
2. Fit wiper arm to spindle and align blade to windscreen.
3. Tighten nut securing wiper arm to 17 Nm (13 lbf.ft), and fit cover.

Rear Window Washer Jet

Removal and Installation

Removal

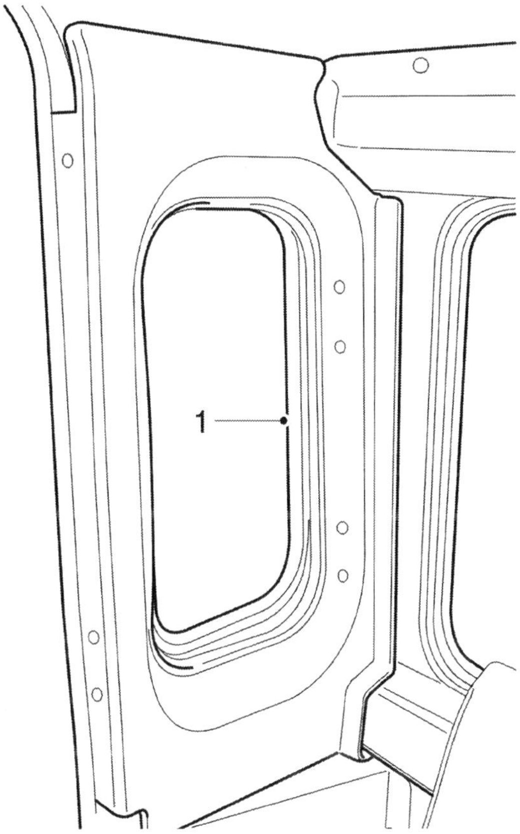

1. Release and remove both rear side window finishers.
2. Remove 8 screws securing both rear grab handles.

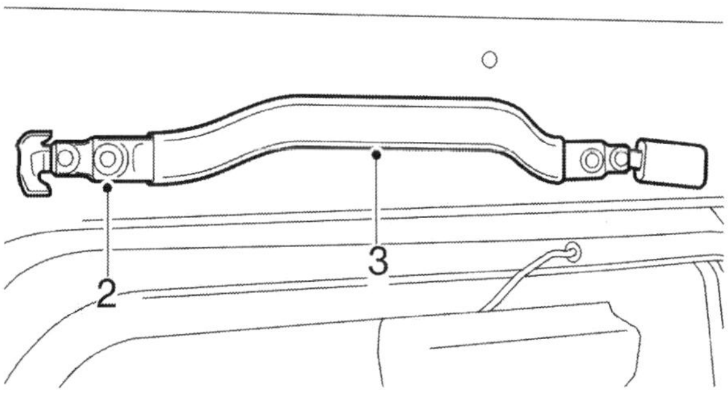

3. Remove both rear grab handles.

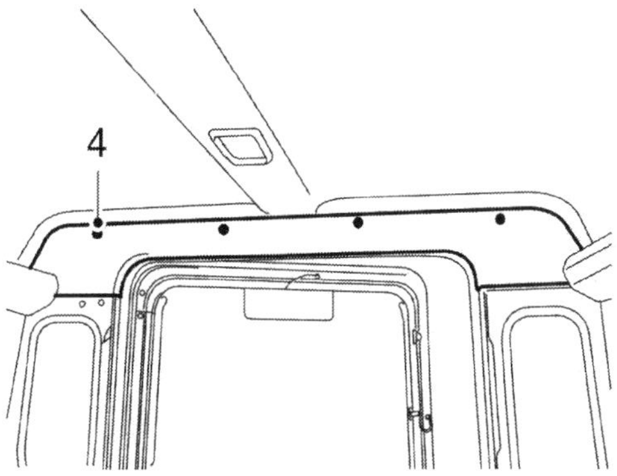

4. Remove 4 trim studs securing upper rear finisher to body.
5. Release and remove upper rear finisher.
6. Position drain tin to collect any fluid spillage.
7. Disconnect washer tube from jet.

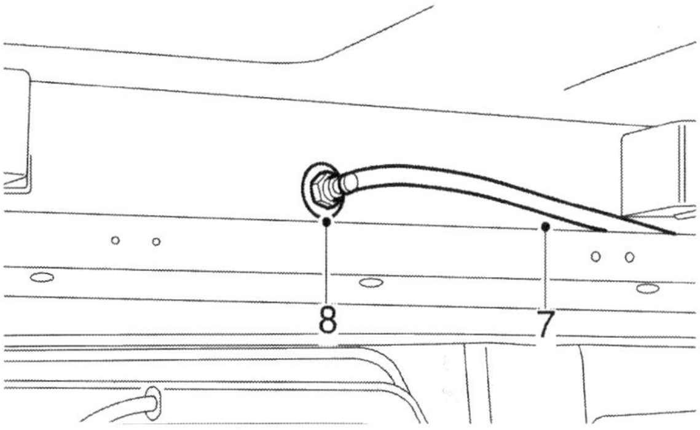

8. Remove nut securing washer jet to body.
9. Remove rear window washer jet.

Installation

1. Fit rear window washer jet to body and secure with nut.
2. Connect washer tube to jet.
3. Remove drain tin.
4. Fit upper rear finisher and secure with trim studs.
5. Fit both rear grab handles and secure with screws.
6. Fit both rear side window finishers.

Rear Window Wiper Motor

Removal and Installation

Removal

1. Remove spare wheel from taildoor.

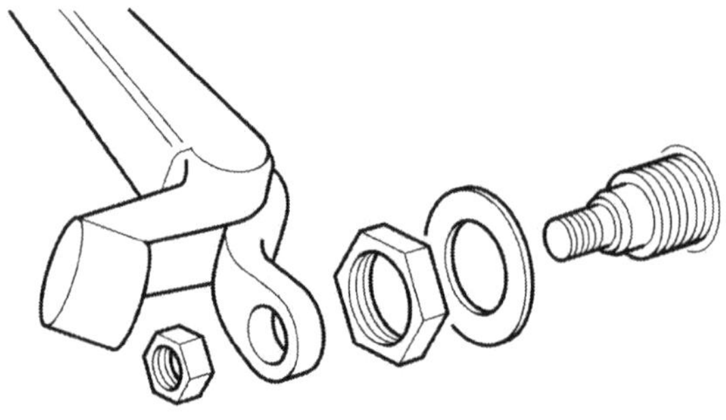

2. Raise nut cover, remove nut and release wiper arm.
3. Remove nut and collect washer securing wiper motor to taildoor.
4. Open taildoor.
5. Remove taildoor trim casing.

Refer to: Taildoor Trim Panel (501-05 Interior Trim and Ornamentation, Removal and Installation)

6. Remove bolt securing door locking solenoid to taildoor.
7. Pivot both assemblies and withdraw wiper motor from taildoor.
8. Disconnect multiplugs from wiper motor.
9. △ NOTE: Do not carry out further dismantling if component is removed for access only.

Remove wiper motor assembly.

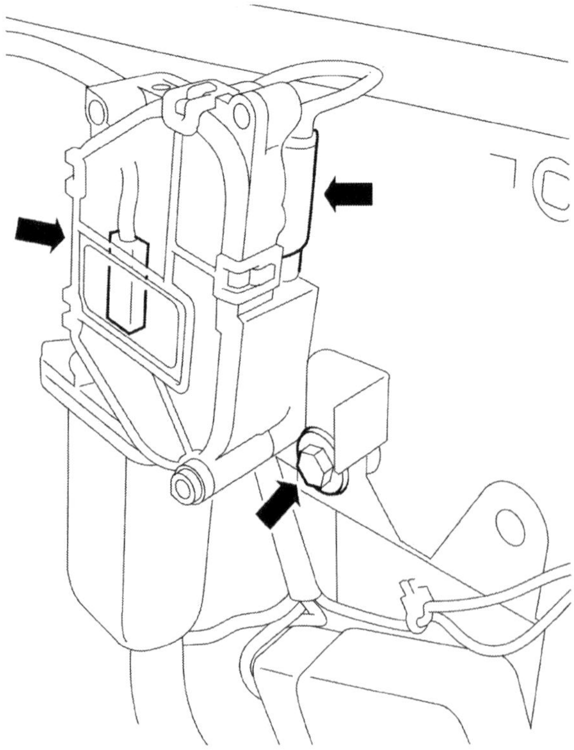

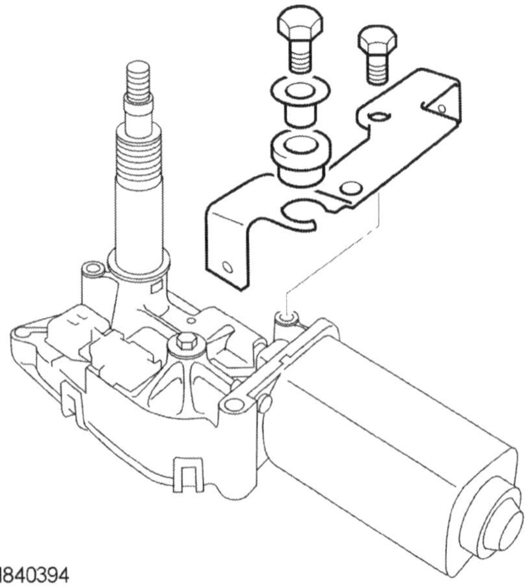

M840394

10. Remove 2 bolts securing wiper motor to mounting bracket, remove bracket.

Installation

1. Fit mounting bracket to motor and tighten bolts to 10 Nm (7 lbf.ft).
2. Fit motor spindle to rear door, connect multiplugs and push motor fully home. Align motor mounting bracket with solenoid mounting bracket, fit bolt and tighten to 10 Nm (7 lbf.ft).
3. Fit taildoor trim casing.

Refer to: Taildoor Trim Panel (501-05 Interior Trim and Ornamentation, Removal and Installation)

4. Fit washer and nut to rear wiper assembly, tighten nut to 3 Nm (2.2 lbf.ft).
5. Fit wiper arm to spindle, align blade to glass and tighten nut to 17 Nm (13 lbf.ft).
6. Close nut cover.
7. Fit spare wheel and tighten nuts to 45 Nm (33 lbf. ft).

Windshield Washer Pump

Removal and Installation

Part(s)

STEP	PART NAME	QUANTITY
Installation Step 1	Renew Part Windshield washer pump seal	1

Removal

1. Remove headlamp.

Refer to: Headlamp Assembly (417-01 Exterior Lighting, Removal and Installation)

2. Position drain tin to collect any fluid spillage.
3. Disconnect multiplug from washer pump.
4. Disconnect hose from washer pump.
5. Release washer pump from reservoir and discard sealing washer.

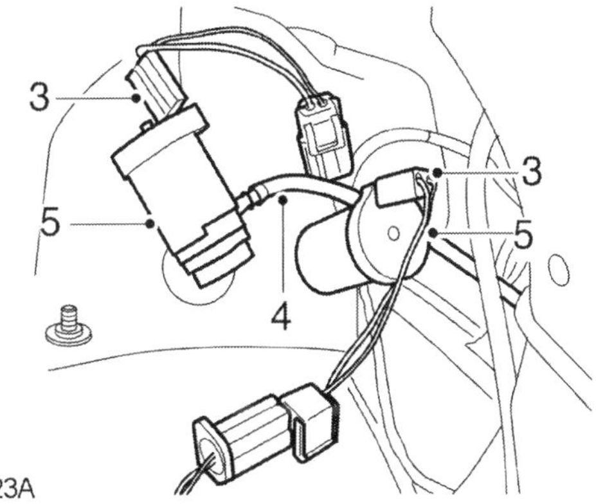

M840323A

Installation

1. Fit new sealing washer and secure pump to reservoir.

Renew Part: Windshield washer pump seal

2. Connect hose and multiplug to washer pump.
3. Fit headlamp.

Refer to: Headlamp Assembly (417-01 Exterior Lighting, Removal and Installation)

4. Fill reservoir with screen cleaning fluid.

Windshield Wiper Motor

Removal and Installation

⚠ NOTE: Some illustrations may show right hand drive. Left hand drive is similar, but the essential information shown is always correct.

1. **Refer to:** Startup Battery Disconnect and Connect (414-01 Battery, Mounting and Cables, General Procedures)
2. **Refer to:** Front Wiper Pivot Arm (501-16 Wipers and Washers, Removal and Installation)

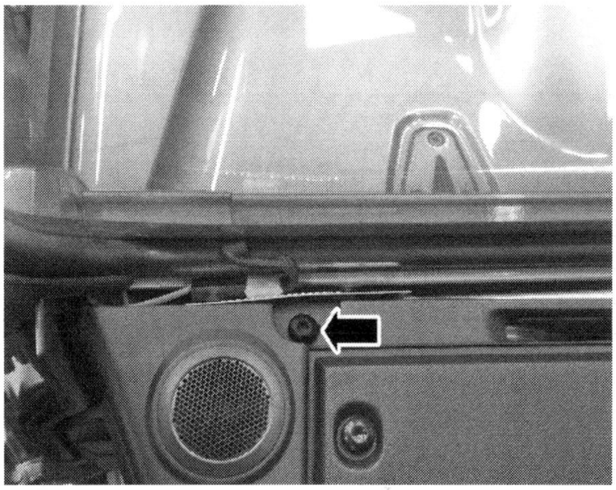

3.

4.

5.

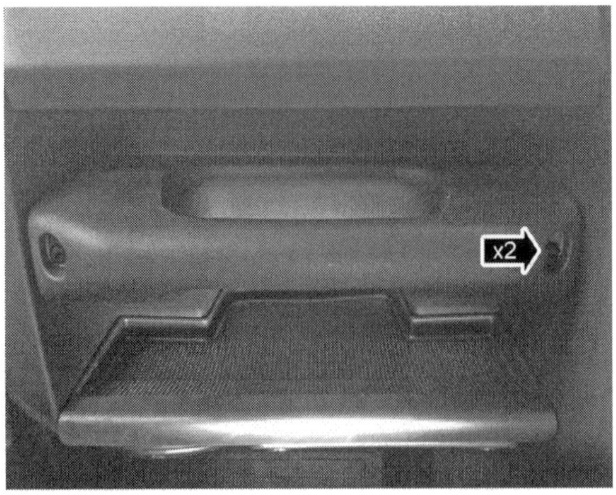

6.

7.

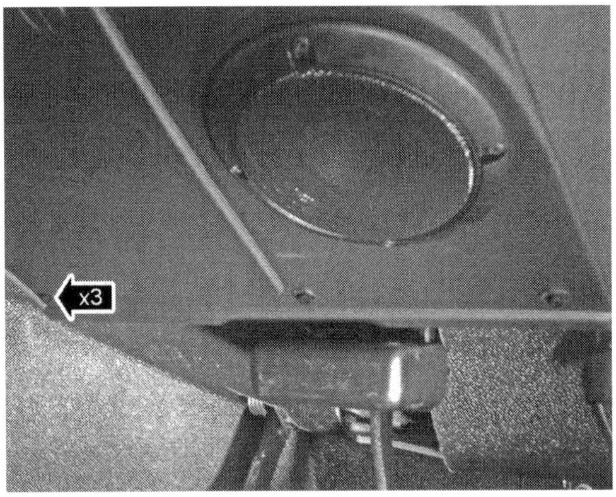

8.

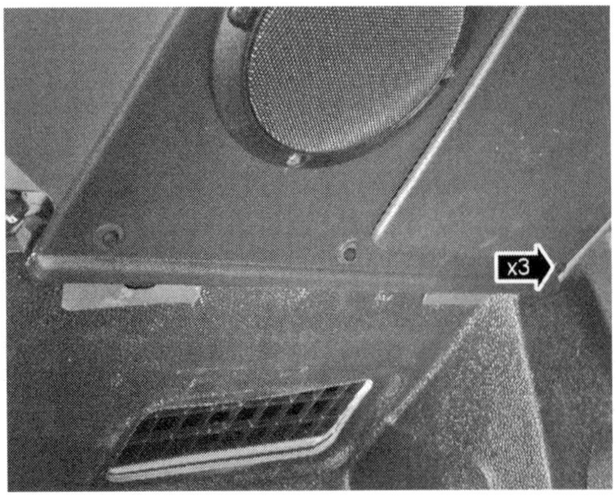

9.

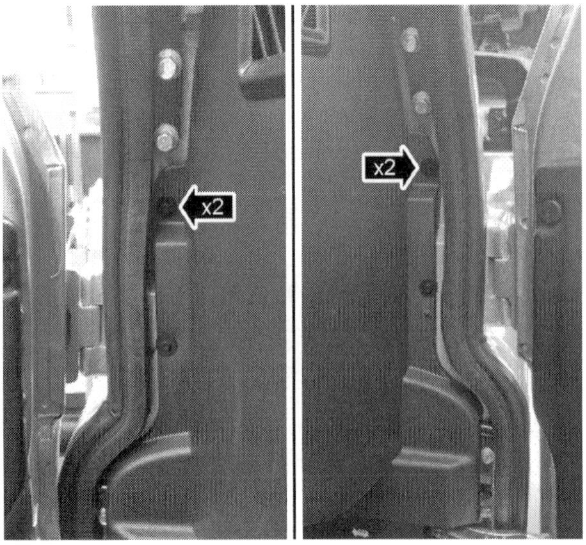

10.

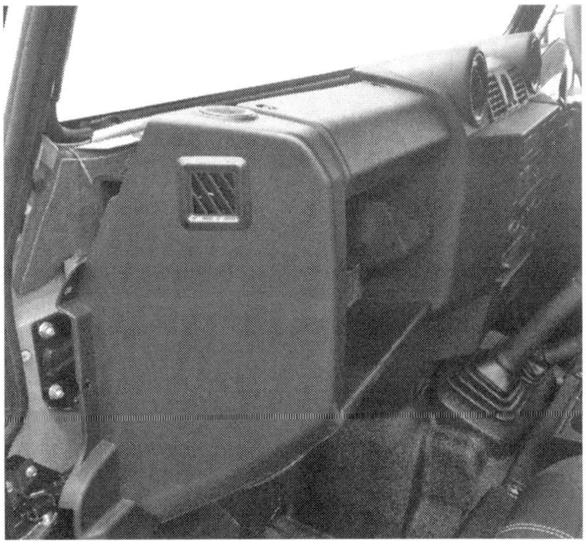

11. CAUTION: Make sure the wiring harness and electrical connectors are not damaged during this operation.

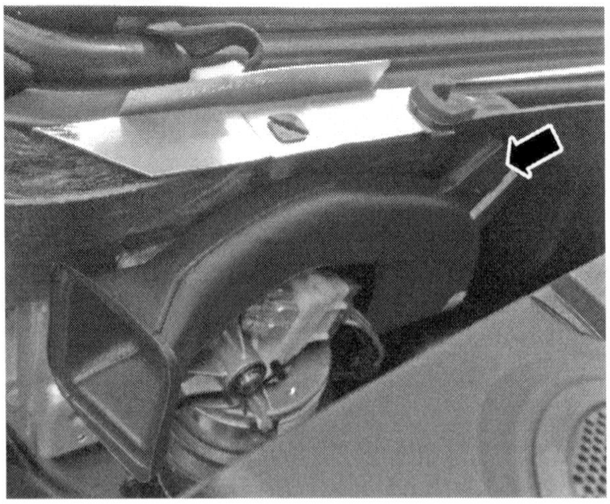

12.

13.

14.

15.

Installation

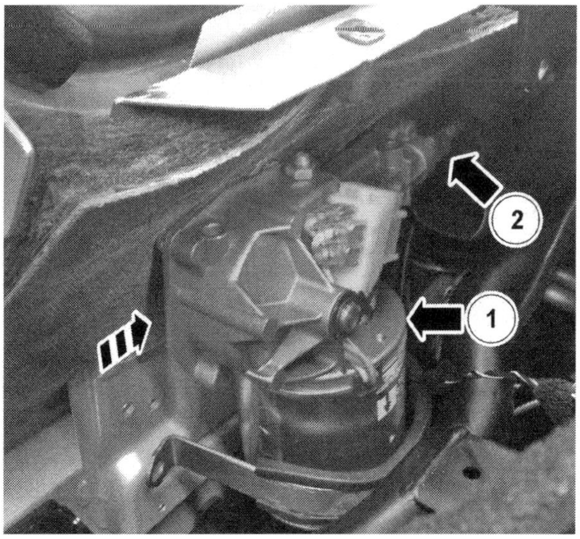

1.

2. Torque 5 Nm

3.

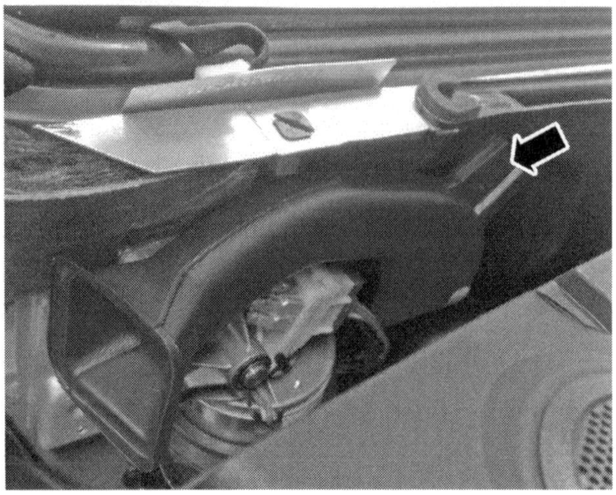

4.

5. CAUTION: Make sure the wiring harness and electrical connectors are not damaged during this operation.

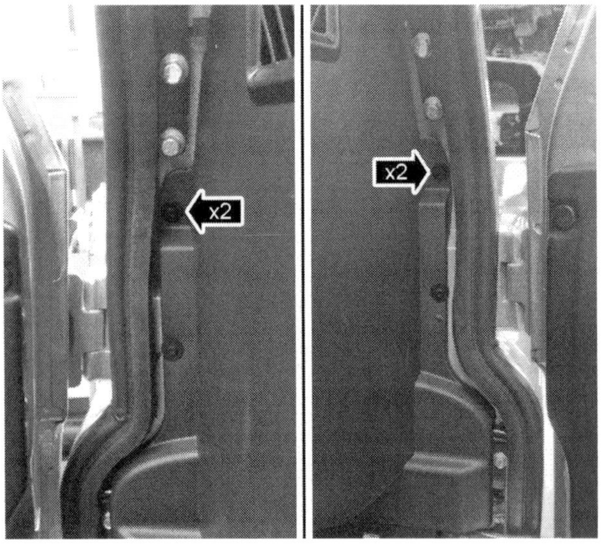

6. Torque 1 Nm

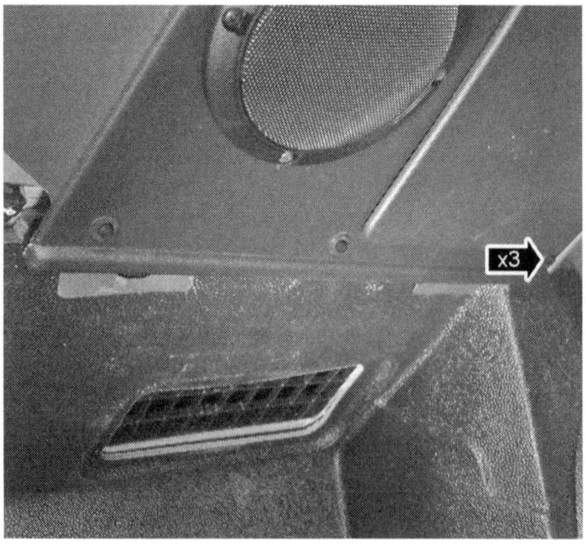

7. Torque 1 Nm

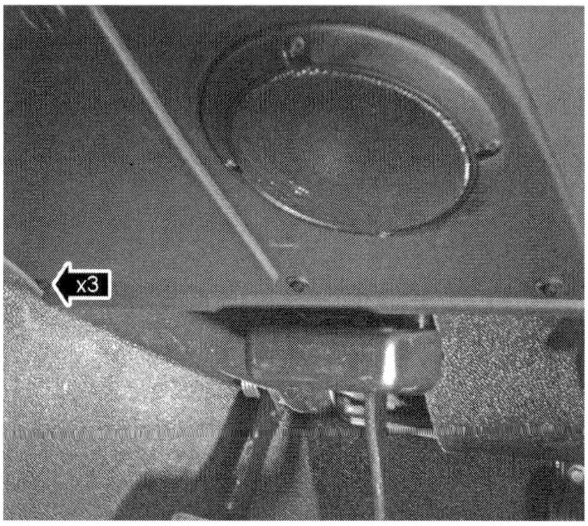

8. Torque 1 Nm

9. Torque 2 Nm

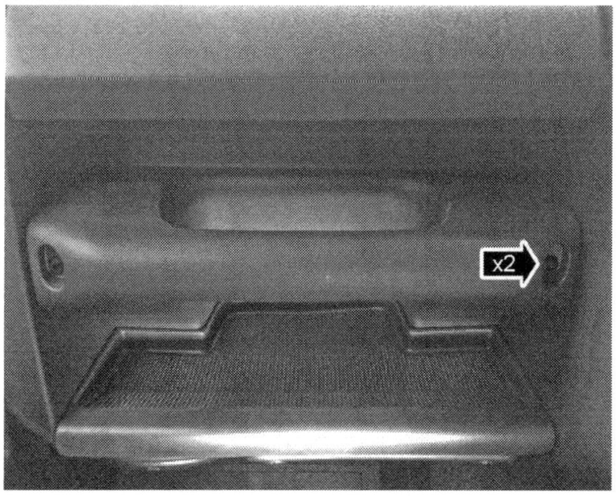

10. Torque 7 Nm

11. Torque 2 Nm

12. Torque 2 Nm

13. Torque 2 Nm

14. **Refer to:** Startup Battery Disconnect and Connect (414-01 Battery, Mounting and Cables, General Procedures)

15. **Refer to:** Front Wiper Pivot Arm (501-16 Wipers and Washers, Removal and Installation)

Windshield Wiper/Washer Switch

Removal and Installation

Removal

1. Remove the steering column shrouds.

Refer to: Steering Column Shrouds (501-05 Interior Trim and Ornamentation, Removal and Installation)

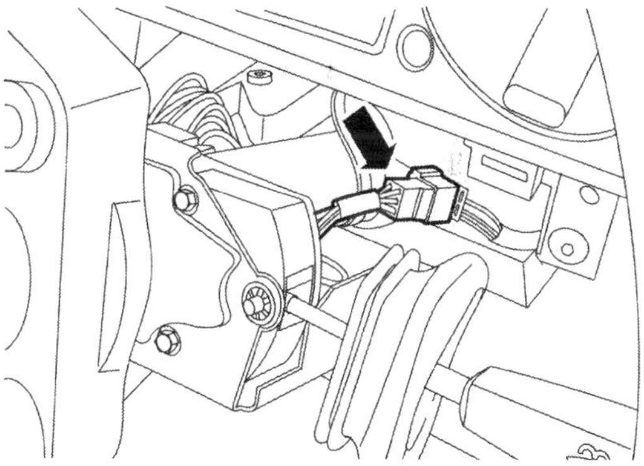

2. Disconnect the electrical connector.

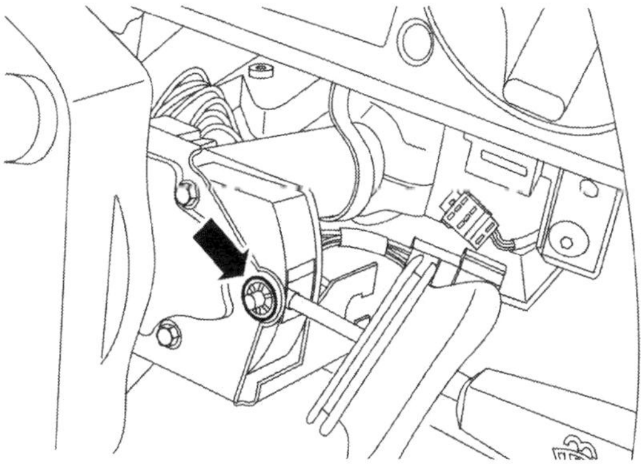

3. Remove and discard the clip.

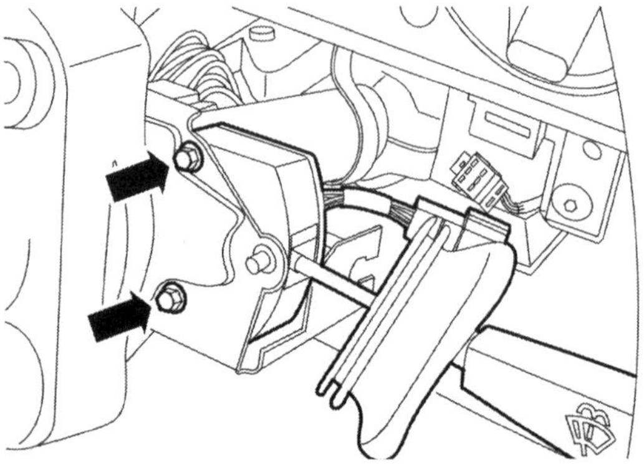

4. Remove the 2 screws. Remove the windshield wiper switch.

Installation

1. To install, reverse the removal procedure.

Wiper Mounting Arm and Pivot Shaft

Removal and Installation

Removal

1. Disconnect the battery.

Refer to: Startup Battery Disconnect and Connect (414-01 Battery, Mounting and Cables, General Procedures)

2. Remove the instrument panel.

Refer to: Instrument Panel (501-12 Instrument Panel and Console, Removal and Installation)

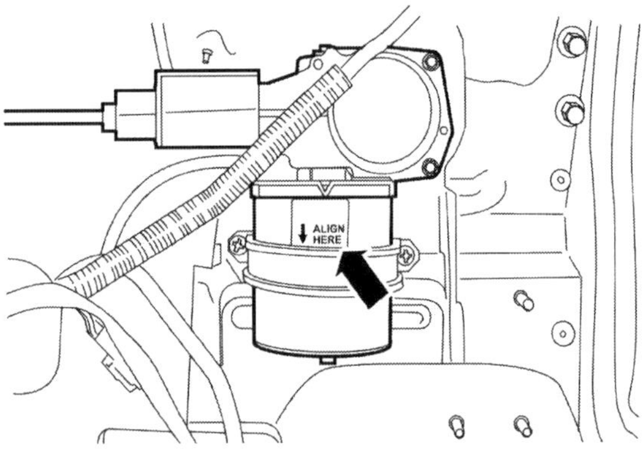

3. Note the fitted position of the component prior to removal.

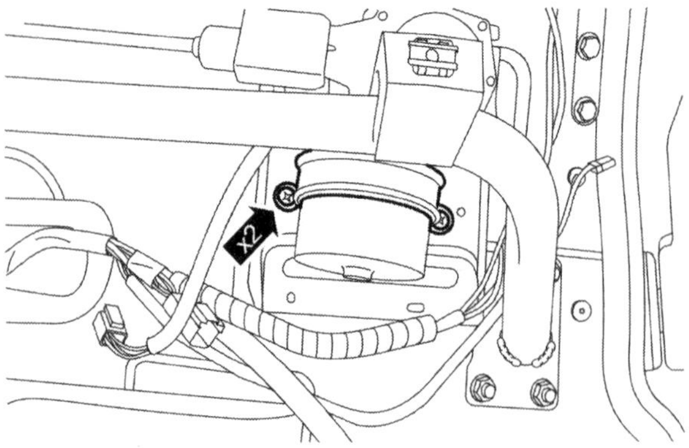

4. Release the windshield wiper motor.

   - Remove the 2 screws.

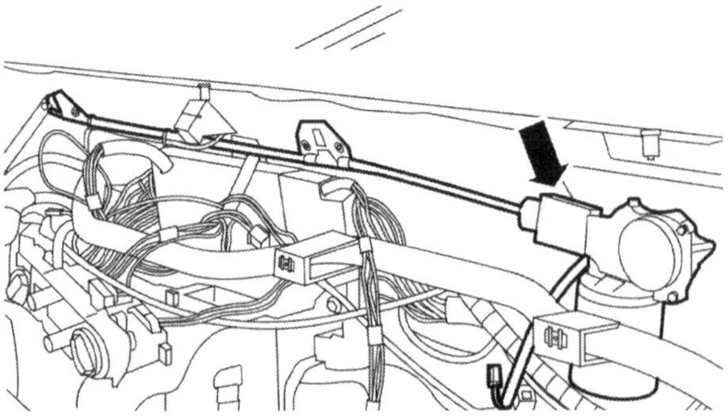

5. Remove the windshield wiper motor.

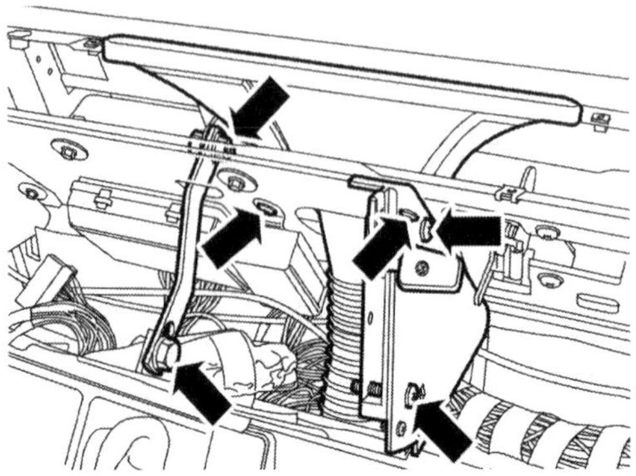

6. Remove 2 screws and move heater fan switch aside.

7. Remove 2 screws securing drivers side demister duct.

8. Disconnect demister tube and remove duct.

9. Remove bolt and nut and bolt securing steering column support bracket and remove bracket.

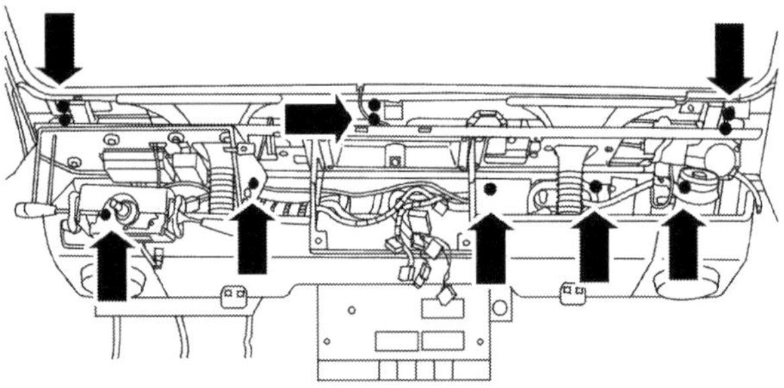

10. Remove 11 screws securing fascia support rail.

11. Remove cable ties securing washer tube to windshield wiper rack tube.

12. Loosen fully 4 nuts securing backplates to wiper mounting arm and pivot shaft and remove windshield wiper rack tubes.

13. Remove seals from wheel box spindles.

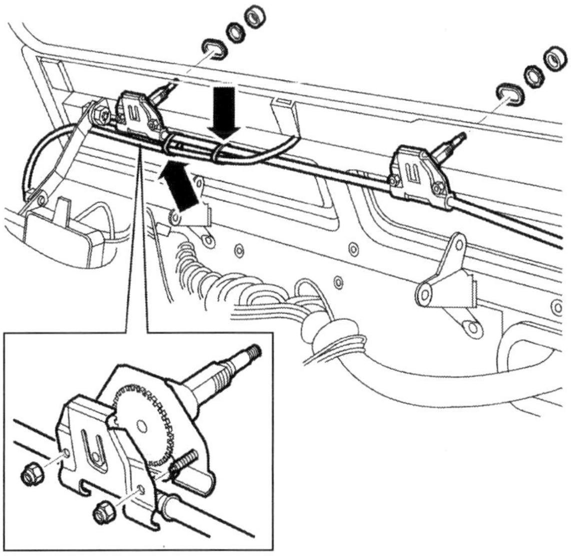

14. Remove 2 nuts securing wiper mounting arm and pivot shaft and remove sealing washers.

15. Carefully pull fascia support rail from bulkhead sufficiently to remove wiper mounting arm and pivot shaft.

Installation

1. Loosen fully but do not remove nuts securing wheel box back plates.

2. Grease wiper mounting arm and pivot shaft.

3. Fit wiper mounting arm and pivot shaft, fit sealing washers and tighten nuts to 5 Nm (3.7 lbf.ft).

4. Fit spindle seals.

5. Fit wiper rack tubes and tighten wiper mounting arm and pivot shaft nuts sufficiently to hold tubes. Tighten wheel box back plate nuts to 6 Nm (4 lbf.ft) after fitting windshield wiper motor.

6. Align washer tube to rack tube and secure with cable ties.

7. Align fascia support rail and secure with screws.

8. Fit steering column support bracket and tighten bolt and nut and bolt to 45 Nm (33 lbf.ft).

9. Fit demister duct to tube, align relay bracket and secure duct with screws.

10. Position heater fan switch and secure with screws.

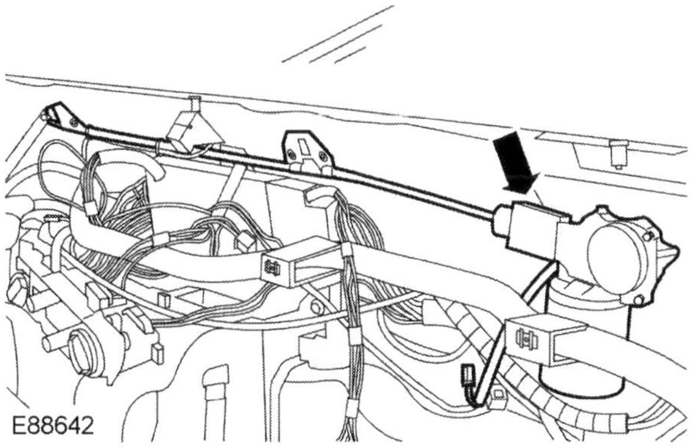

11. Refit the windshield wiper motor.

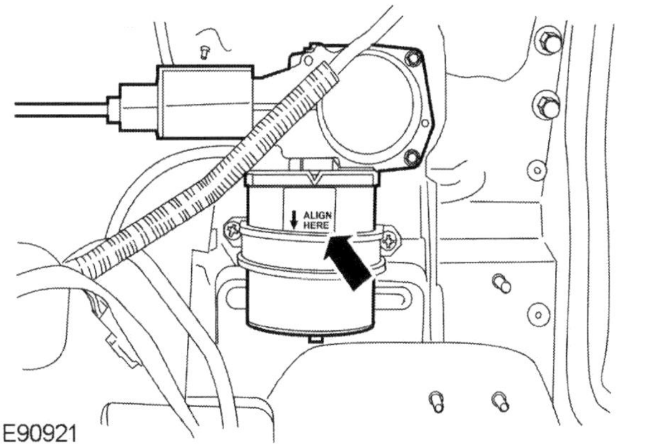

12. Install the windshield wiper motor to its original fitted position.

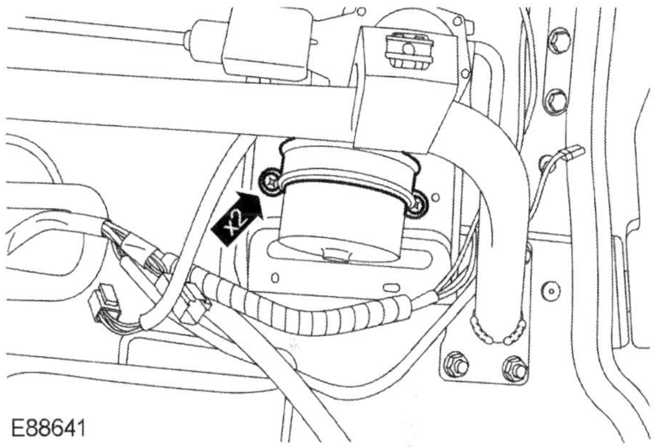

13. Tighten to 5 Nm (4 lb.ft).

14. Refit the dash panel.

Refer to: Instrument Panel (501-12 Instrument Panel and Console, Removal and Installation)

15. Reconnect the battery.

Refer to: Startup Battery Disconnect and Connect (414-01 Battery, Mounting and Cables, General Procedures)

Roof Opening Panel

Removal and Installation

Removal

1. Open roof opening panel fully.
2. Disengage spring lock from operating handle catch.
3. Holding roof opening panel at both sides, release hinges from locating brackets at front of outer frame.
4. Remove roof opening panel.
5. Starting from centre rear, peel headlining finisher from inner frame.
6. Remove 18 screws securing inner frame to outer frame and detach inner frame.
7. △ NOTE: Assistance may be required to remove the sun roof assembly.

Lift outer frame from roof.

Installation

1. Clean roof area around outer frame seating.
2. Position outer frame on roof.
3. Install inner frame to headlining and secure to outer frame with 18 screws but do not fully tighten.
4. Check alignment of inner and outer frames with roof mounting and headlining, adjust as necessary, and fully tighten fixing screws to 10 Nm (7 lbf.ft).
5. Starting from centre rear, Install headlining finisher lip into locating channel of inner frame.
6. Press finisher firmly over inner frame and continue around complete frame, ensuring finisher lies flat on headlining.
7. Fit roof opening panel hinges fully into locating brackets on outer frame and lower the glass panel.
8. Engage operating handle catch with spring lock of outer frame and close roof opening panel.

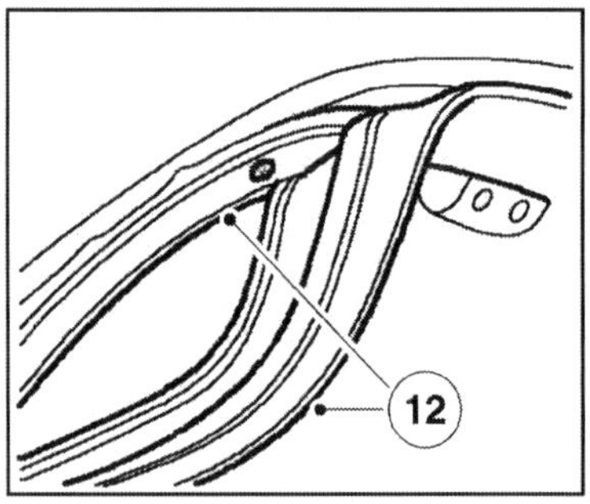

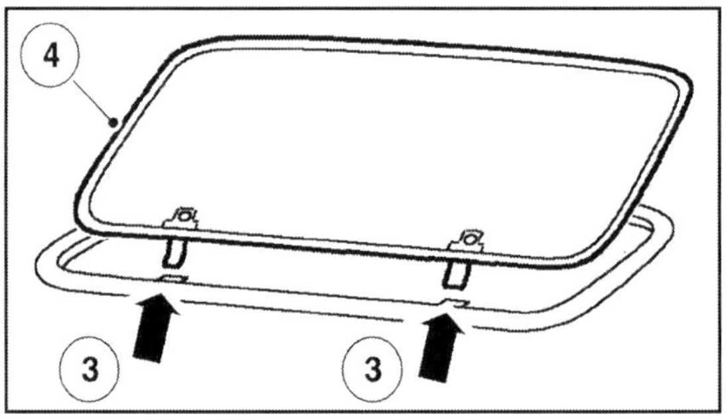

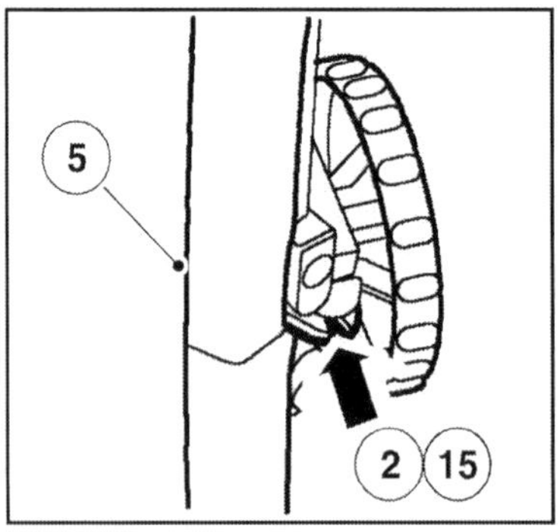

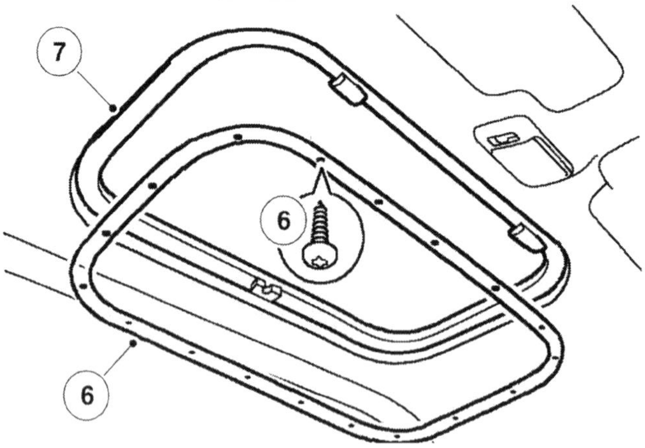

Convertible Top

Description and Operation

SVX (60th ANNIVERSARY) MODEL - CONVERTIBLE TOP

E101948

OVERVIEW

The convertible top is new for the SVX 90 model. The top has removable side windows and rear window which are secured with zips.

DESCRIPTION

The convertible top is manufactured from canvas panels stitched together, to form a taught, waterproof covering over the cab and load area of the vehicle.

It is recommended that fitting and removal of the convertible top is performed by two people.

The rear pivoting support frame is secured with a velcro strap which must removed to allow the frame to be extended to its support position.

The leading edge of the top is located in channels at the top of the windshield. Channels above each door also provide location for the outer edge of the top and once installed are secured in position with further location in an inner channel. Velcro straps near to the seat belt upper anchor point secure the top to the support frame. Three additional velcro straps secure the top onto the door frame.

The convertible top at the rear of the cab area is secured with two flaps which are attached with velcro. From inside the vehicle, the rearmost of the two flaps is wrapped around the support frame. Tension for the top is provided by three straps. The center strap is pulled over the forward frame bar and pulled through the buckle to tension the top. This process is repeated with the two outer straps which provide tension evenly over the top. The rear most flap is wrapped around the support frame, concealing the straps. The forward flap is also wrapped around the support frame in the opposite direction and the velcro attached to the rearmost flap, secures the flaps together.

On the outside of the convertible top, velcro straps are used to secure the top to the support frame where it protrudes from the top.

Convertible Top Attachment - Sheet 1

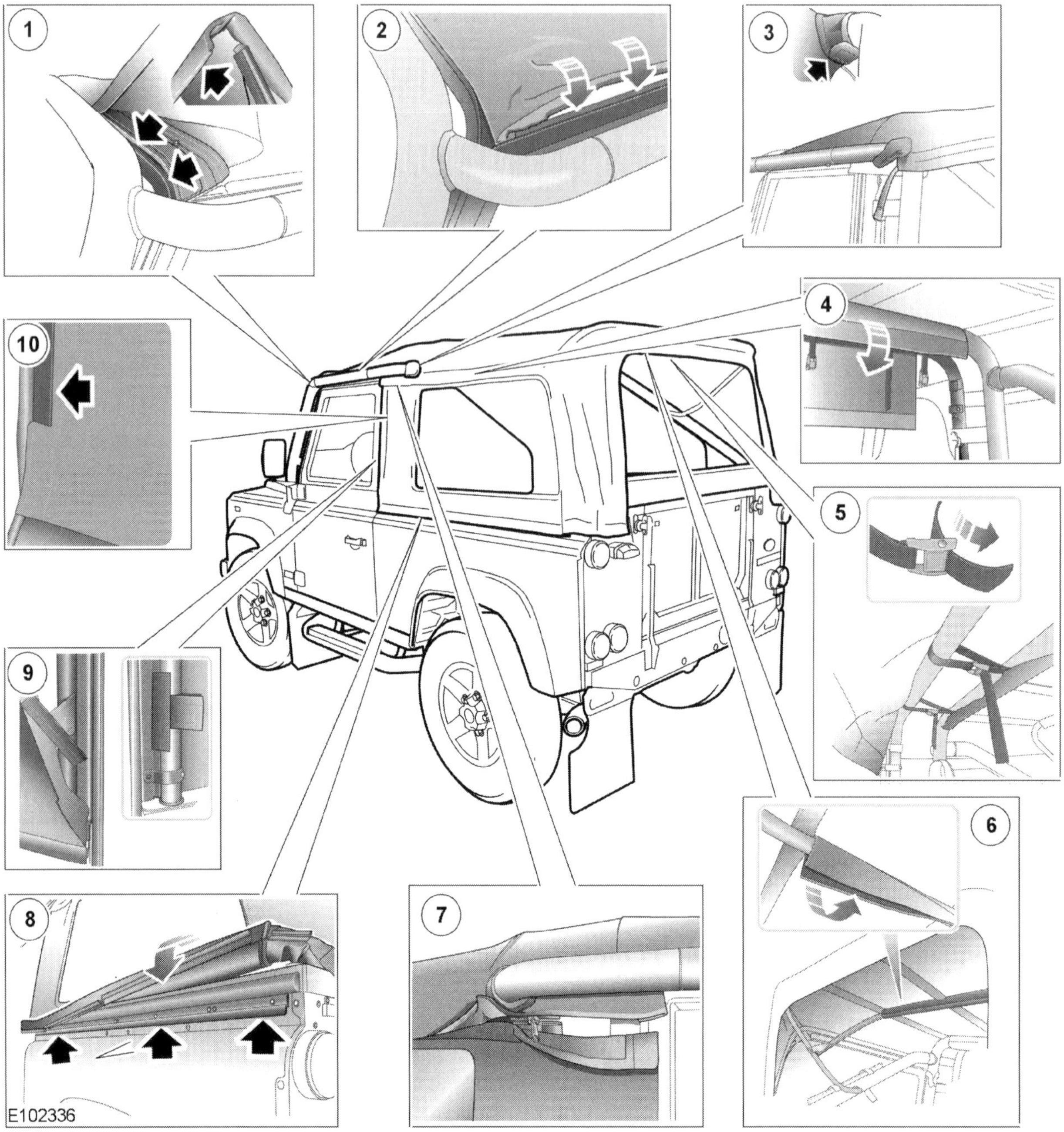

Item	Description
1	Location of leading edge into locating channel above the windscreen
2	Location of outer edges into locating channel above doors
3	Velcro strap attachment to support frame where the support frame protrudes from the hood
4	Rear part of flap extension pull over support frame
5	Three fixing straps located over forward frame bar and tension applied using buckles
6	Velcro fixing at rear of convertible top to hinged frame section
7	Side window zip attachment
8	Attachment to the rail on the body at each side of the load area
9	Forward vertical edge of window panel, elastic strap plate located between frame and vertical body channel
10	Plate positioned across gap between frame and vertical body section

The top is secured at the rear with a flap which locates on the hinged section of the rear support frame and secured with velcro.

The handed side windows are zipped into position in the top. The top is attached to the rail on the body at each side of the load area. The top is pulled down tightly to locate in the rail which applies tension to the top.

An elastic strap with a bridge plate attached, is positioned from the outside of the vehicle, through a gap between the support frame and the vehicle body. The plate is positioned across the support frame and the vehicle body to secure and the elastic strap maintains the tension on the front vertical edge of the side window and ensures they are located tightly into the body channel.

Convertible Top Attachment - Sheet 2

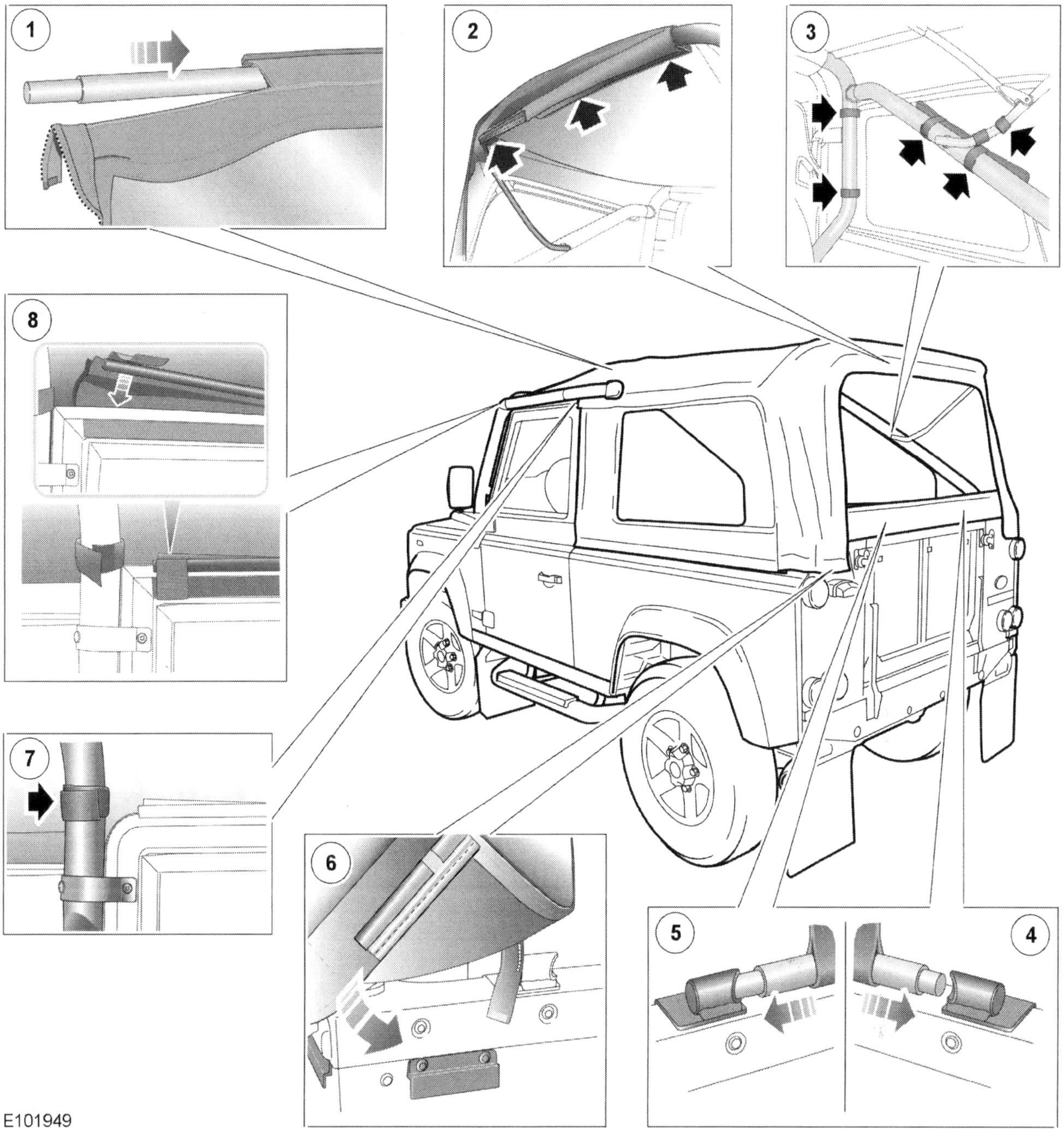

E101949

Item	Description
1	Metal spring loaded rod through the bottom edge stitched section of the window panel
2	Rear section to rear support frame attachment
3	Five velcro fixings on each side of the convertible top to frame and hinged frame
4	Metal spring loaded rod location in RH (right-hand) receiver tube
5	Metal spring loaded rod location in LH (left-hand) receiver tube
6	Rear corner section location in rail on rear body
7	Velcro strap to support frame above safety belt anchor point
8	Location into interior locating channel in rail above door and velcro strap location to door frame

At the rear of the vehicle, each corner of the top is attached to small rails on either side of the tail gate. The top is pulled down tightly which applies tension to the top.

Five velcro straps are located on each side of the convertible top and secure the top to the support frame and the hinged frame.

A metal spring loaded rod locates through a slot in the lower part of the rear window which is zipped in position. The spring loaded rod locates in 2 receiver tubes attached to the top edge of the tail gate.

Removal is a reversal of the fitting process.

Stowage Locker

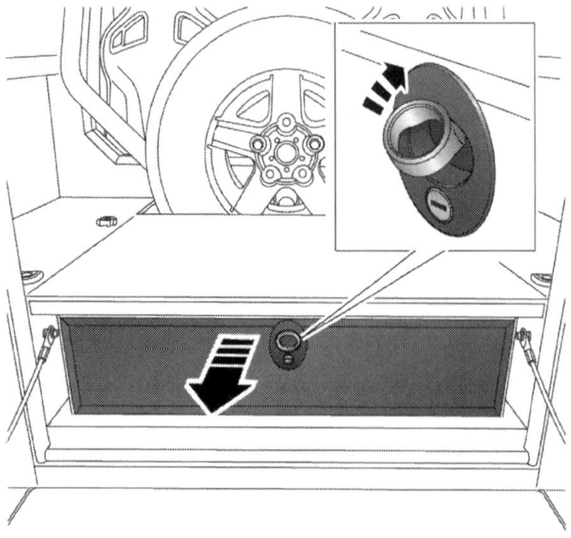

E101950

When the convertible top is removed, it can be carefully folded and stored in the locker in the rear of the load area. The locker has a hinged lid which can be locked for security using the vehicle key.

Ensure the convertible top and window panels are dry before folding and storing in the locker.

Convertible Top Care Points

Do not use an automatic car wash. Doing so will cause damage to the hood.

Some high pressure cleaning systems are sufficiently powerful to damage hood fabrics. Never aim a high pressure water jet at zips or seals.

If the hood is stained, it should be cleaned within 24 hours of the incident occurring, otherwise the stain may become permanent.

Never use spirit, petrol or chlorine based cleaning agents, or wash/wax compounds to clean the hood and windows - wax polishes will become ingrained in the textured finish.

Use a soft brush to remove dust and flaking dirt from the hood (not the windows), then soak with a mild soap and water solution to soften encrusted dirt and remove stains, prior to washing.

Wash using a mild soap and water solution, before rinsing the hood with clean lukewarm or cold water. Rinse until all traces of soap are removed.

Ensure the hood is dry before folding or removing

Improper cleaning and lack of care may cause damage to the hood and window panels, resulting in water penetration.

Bumpers

501-19

Removal and Installation

Removal

1. Remove 6 screws and remove front grille.

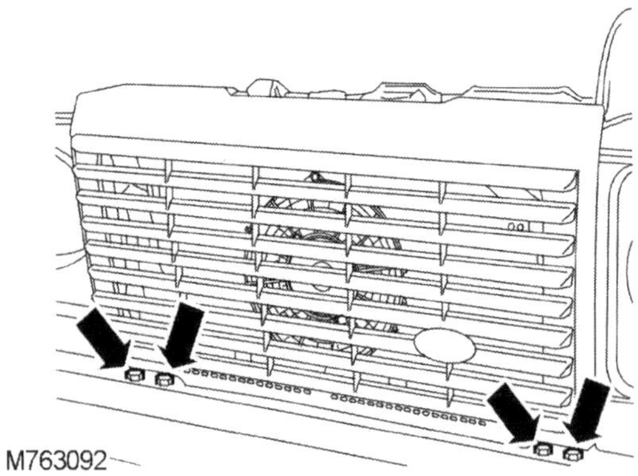

2. Remove 4 through bolts securing front bumper to chassis longitudinals.
3. Remove front bumper.

Installation

1. Install front bumper to chassis longitudinals and secure with bolts.
2. Install front grille and secure with screws.

Specifications

Torque Specifications

Description	Nm	lb-ft
Front safety belt retractor bolt	31	23
Front safety belt upper anchor bolt	31	17
Front safety belt lower anchor bolts	10	7
+ Front safety belt buckle M10 bolt	22	16
Front safety belt buckle M8 bolt	22	16
Rear safety belt retractor bolt	31	17
Rear safety belt upper anchor bolt	31	17
Rear safety belt lower anchor bolt	31	17
Rear center safety belt retractor bolt	31	17
Rear safety belt buckle bolt	31	17
Third row safety belt retractor bolt	31	17
Third row safety belt upper anchor bolt	31	17
Third row safety belt lower anchor bolt	31	17
Third row safety belt buckle bolt	31	17

Description and Operation

COMPONENT LOCATION

OVERVIEW

A three point safety belt is installed at each seat position. All the safety belts have Emergency Locking Retractors (ELR).

ELR retractors incorporate a liftshaft locking system with webbing sensor and car sensor activating mechanisms. The webbing sensor activates the locking system if the webbing is subjected to a sharp pull. The car sensor activates the locking system if the vehicle is subjected to sudden deceleration or a severe tilt angle.

FRONT SAFETY BELTS

The retractor of each front safety belt is attached to the related B pillar. The webbing runs from the retractor through an upper mounting, attached to the B pillar, to an anchor point on the front seat.

The buckle for each front safety belt is to the vehicle body inside of the related front seat frame.

SECOND ROW SAFETY BELTS

The retractor of each outboard second row safety belt is attached to the body immediately behind the D pillar. The webbing runs from the retractor, through an upper mounting on the D pillar, to an anchor point at the front of the related wheel arch.

The retractor for the center second row safety belt is installed in the top of the seat back. The webbing runs from the retractor, over the top of the seat, to an anchor point at the base of the seat frame.

The buckles for the second row safety belts are attached to the related seat frame.

THIRD ROW SAFETY BELTS (110 only)

The retractor of each third row safety belt is attached to the E pillar. The webbing runs from the retractor, through a mounting on the E pillar to an anchor point on the floor. The buckles for the third row safety belts are attached to the related seat frame.

BELTMINDER (GULF MARKETS ONLY)

The belt minder function provides warnings to the driver if the appropriate front safety belts are not fastened when driving. The belt minder function is controlled by the instrument cluster

Refer to: Instrument Cluster (413-01 Instrument Cluster, Description and Operation)

Principles of Operation

For a detailed description of the Safety Belt System, refer to the relevant Description and Operation section in the workshop manual.

Refer to: Seatbelt System (501-20 Seatbelt System, Description and Operation)

Inspection and Verification

CAUTION: Diagnosis by substitution from a donor vehicle is **NOT** acceptable. Substitution of control modules does not guarantee confirmation of a fault, and may also cause additional faults in the vehicle being tested and/or the donor vehicle.

NOTES:

The 'car sense system' of the seatbelt is sensitive to both cornering and braking forces and is designed to lock when it detects this. It is therefore possible to lock when parking the vehicle on high kerbs and inclines. The seatbelt retractor will be locked when removed from the vehicle as it is not in the exact orientation that it was installed on the vehicle. This can be mistaken as a jammed seatbelt retractor.

If a control module or a component is at fault or maybe at fault and the vehicle remains under manufacturer warranty, refer to the Warranty Policy and Procedures manual, or determine if any prior approval program is in operation, prior to the installation of a new module/component.

When performing voltage or resistance tests, always use a digital multimeter that has the resolution ability to view 3 decimal places. For example, on the 2 volts range can measure 1mV or 2 K Ohm range can measure 1 Ohm. When testing resistance always take the resistance of the digital multimeter leads into account.

Check and rectify basic faults before beginning diagnostic routines involving pinpoint tests.

1. Visually inspect for obvious signs of damage and system integrity. Additionally inspect the seatbelt webbing and upper D-loop for signs of dirt/ contamination

Visual Inspection

Mechanical
• Check for the installation of non-standard accessories which may affect or obstruct the function of the seatbelt system • Frayed or damaged webbing • Missing or damaged button stop • Buckles/stalks • Webbing contamination

2. If an obvious cause for an observed or reported concern is found, correct the cause (if possible) before proceeding to the next step
3. If the cause is not visually evident, verify the symptom and refer to the Symptom Chart, alternatively check for Diagnostic Trouble Codes (DTCs) and refer to the DTC Index

Symptom Chart

Symptom	Possible Causes	Action
Seatbelt jammed - Webbing tight	• Backlock effect-in action (webbing retracted quickly and came to sudden stop) • Seatbelt retractor not installed correctly • Rear centre seatbelt only. Mini-button (webbing travel limit stop) missing and seat cushion has been moved causing tight fit	CAUTION: Do not insert any objects or tools into the buckle head. NOTE: The 'car sense system' of the seatbelt is designed to lock when oriented at certain angles. When the seatbelt retractor is removed from the vehicle and not oriented at the exact angle that it is installed in the vehicle, on level ground, it will be locked. This can be mistaken as a jammed seatbelt retractor. • GO to Pinpoint Test A • GO to Pinpoint Test F • GO to Pinpoint Test H

Seat cushion/backrest will not fold/jammed	NOTE: Rear centre seatbelt only. • Mini-button (webbing travel limit stop) missing and seat cushion/ backrest has been moved causing excessive tension	• GO to Pinpoint Test H
Seatbelt jammed - Webbing loose	• Seatbelt webbing trapped in seat • Seatbelt retractor webbing guide loose • Twist in webbing • Interference in webbing routing • D-loop not rotating correctly	• GO to Pinpoint Test B • GO to Pinpoint Test C • GO to Pinpoint Test D • GO to Pinpoint Test E • GO to Pinpoint Test G
Seatbelt - Intermittent jamming	• Seatbelt retractor not installed correctly	CAUTION: Do not insert any objects or tools into the buckle head. NOTE: The 'car sense system' of the seatbelt is designed to lock when oriented at certain angles. When the seatbelt retractor is removed from the vehicle and not oriented at the exact angle that it is installed in the vehicle, on level ground, it will be locked. This can be mistaken as a jammed seatbelt retractor. • GO to Pinpoint Test F

Seatbelt - Slow retraction	• Seatbelt retractor webbing guide loose • Twist in seatbelt webbing • Interference in webbing routing • Seatbelt retractor not installed correctly • D-loop not rotating correctly • Foreign object/debris • Seatbelt webbing dirty/ contaminated	• GO to Pinpoint Test C • GO to Pinpoint Test D • GO to Pinpoint Test E • GO to Pinpoint Test F • GO to Pinpoint Test G • GO to Pinpoint Test E • GO to Pinpoint Test J
Seatbelt - Not retracting	• Seatbelt retractor webbing guide loose • Twist in seatbelt webbing • D-loop not rotating correctly • Interference in webbing routing • Foreign object/debris	• GO to Pinpoint Test C • GO to Pinpoint Test D • GO to Pinpoint Test G • GO to Pinpoint Test E • GO to Pinpoint Test E

Seatbelt - Not extracting	• Backlock effect-in action (webbing retracted quickly and came to sudden stop) • Seatbelt retractor not installed correctly • Seatbelt retractor webbing guide loose • Twist in seatbelt webbing • D-loop not rotating correctly • Interference in webbing routing • Foreign object/debris	CAUTION: Do not insert any objects or tools into the buckle head. NOTE: The 'car sense system' of the seatbelt is designed to lock when oriented at certain angles. When the seatbelt retractor is removed from the vehicle and not oriented at the exact angle that it is installed in the vehicle, on level ground, it will be locked. This can be mistaken as a jammed seatbelt retractor. • GO to Pinpoint Test A • GO to Pinpoint Test F • GO to Pinpoint Test C • GO to Pinpoint Test D • GO to Pinpoint Test G • GO to Pinpoint Test E • GO to Pinpoint Test E
Seatbelt - Noisy during operation	• Seatbelt webbing trapped in seat • Interference in webbing routing (rubbing)	• GO to Pinpoint Test B • GO to Pinpoint Test E
Seatbelt buckle - Not latching / jammed	• Foreign object/debris	CAUTION: Do not insert any objects or tools into the buckle head. • GO to Pinpoint Test I

Inertia Reel Seatbelts

The vehicle is equipped with (two row one), (three row two), and (two row three (seven seat versions only)) inertia reel seatbelts

These seatbelts are "**dual sensitive**" which means that they have:

- **Car sense system - A vehicle motion sensor, which locks the seatbelt webbing under braking, cornering, on steep hills and in adverse camber conditions, when parked on a steep incline or driveway or two wheels on a high pavement. When the seatbelt retractor is removed from the vehicle and not oriented at the exact angle that it is installed in the vehicle, on level ground, it will be locked. This can be mistaken as a jammed seatbelt retractor.**
- **Webb sense system - A webbing motion sensor, which locks when the seatbelt webbing is extracted suddenly**

The seatbelts in the following positions are equipped with an automatic locking retractor function:

The **automatic locking retractor function** is a feature to secure a child seat or heavy load to the seat.

Seatbelt Locking Test

With the vehicle stationary and on level ground take firm hold of the seatbelt webbing (on the tongue side of the upper seatbelt anchor) and withdraw sharply, **the retractor should lock**. Preventing further webbing release **(repeat this test 3 times)**. Any seatbelt retractor which fails to lock **must not be used** and a **new seatbelt must be installed**.

Carline	Market	Seat position	Automatic Locking Retractor Installed	From Model Year
Defender (L316)	All	All	No	2007

Diagnostic Guide Inertia Reel Seatbelts

PINPOINT TEST A : BACKLOCK	
TEST CONDITIONS	**DETAILS/RESULTS/ACTIONS**
A1: BACKLOCK	
CAUTION: Do not insert any objects or tools into the buckle head.	
NOTE: The 'car sense system' of the seatbelt is designed to lock when oriented at certain angles. When the seatbelt retractor is removed from the vehicle and not oriented at the exact angle that it is installed in the vehicle, on level ground, it will be locked. This can be mistaken as a jammed seatbelt retractor.	
	1. Visually inspect the condition of the seatbelt
	2. Draw a maximum of 20mm of the webbing from the seatbelt retractor with moderate force. Then release the webbing
	3. Check for correct operation twice
	Does the webbing move freely then retract correctly? **Yes** No further action required **No** For first row seatbelt GO to Pinpoint Test C For second and third row seatbelts GO to Pinpoint Test B

PINPOINT TEST B : WEBBING - TRAPPED IN SEAT	
TEST CONDITIONS	**DETAILS/RESULTS/ACTIONS**
B1: WEBBING - TRAPPED IN SEAT	
	1. Visually inspect the condition of the seatbelt
	2. Lift the seat base or release the seat backrest as required
	3. Free the trapped webbing, allow the webbing to retract
	4. Check for correct operation **twice**
	Does the webbing move freely then retract correctly? **Yes** No further action required **No** GO to Pinpoint Test C

PINPOINT TEST C : SEATBELT RETRACTOR - WEBBING GUIDE LOOSE	
TEST CONDITIONS	**DETAILS/RESULTS/ACTIONS**
C1: SEATBELT RETRACTOR - WEBBING GUIDE LOOSE	
	1. Refer to 501-20 removal and installation section of the workshop manual, remove any trim panels required to expose the D loop (anchor point) and seatbelt retractor
	2. Check the webbing is not trapped or twisted and is centrally located on the seatbelt retractor spindle
	3. Attempt to withdraw the webbing from the seatbelt retractor
	4. To release the automatic locking retractor, manually wind the webbing onto the spindle until the automatic locking retractor deactivates (clicking stops)
	5. Fully extract webbing
	6. Confirm webbing guide location is correct. Confirm the fixing lugs are correctly located in the retractor frame
	7. Allow webbing to retract
	8. Check for correct operation **twice**

	Does the webbing move freely then retract correctly? **Yes** Refer to the 501-20 removal and installation section of the workshop manual, reinstall any trim panels, make sure there are no obstructions and the webbing does not catch or rub. No further action required **No** GO to Pinpoint Test D

PINPOINT TEST D : TWIST IN WEBBING	
TEST CONDITIONS	**DETAILS/RESULTS/ACTIONS**
D1: TWIST IN WEBBING	
	1. Refer to section 501-20 removal and installation section of the workshop manual, remove any trim panels required to expose the D loop (anchor point)
	2. Twist the webbing back the correct way in the loop
	3. Pass the twist through the pillar loop or escutcheon as required
	4. Check for correct operation **twice**
	Does the webbing move freely then retract correctly? **Yes** Refer to the 501-20 removal and installation section of the workshop manual, reinstall any trim panels, make sure there are no obstructions and the webbing does not catch or rub. No further action required **No** GO to Pinpoint Test E

PINPOINT TEST E : INTERFERENCE - WEBBING ROUTING	
TEST CONDITIONS	**DETAILS/RESULTS/ACTIONS**
E1: INTERFERENCE - WEBBING ROUTING	
	1. Refer to the 501-20 removal and installation section of the workshop manual, remove any trim panels required to expose the D loop (anchor point)
	2. Remove obstructions and foreign objects, make sure the webbing does not catch or rub
	3. Confirm the seatbelt does not contact the wiring harness
	4. Check for correct operation **twice**
	Does the webbing move freely then retract correctly? **Yes** Refer to the 501-20 removal and installation section of the workshop manual, reinstall any trim panels, make sure there are no obstructions and the webbing does not catch or rub. No further action required **No** GO to Pinpoint Test F

PINPOINT TEST F : SEATBELT RETRACTOR - INCORRECT INSTALLATION	
TEST CONDITIONS	**DETAILS/RESULTS/ACTIONS**
F1: SEATBELT RETRACTOR - INCORRECT INSTALLATION	
	1. Refer to the 501-20 removal and installation section of the workshop manual, remove any trim panels required to expose the D loop (anchor point) and the seatbelt retractor
	2. Refer to the 501-20 removal and installation section of the workshop manual, correctly reinstall the seatbelt retractor making sure that the locating “T bar” and “anti rotation pins” are correctly located
	3. Check for correct operation **twice**

	Does the webbing move freely then retract correctly? **Yes** Refer to the 501-20 removal and installation section of the workshop manual, reinstall any trim panels, make sure there are no obstructions and the webbing does not catch or rub. No further action required **No** Replace as required. Refer to the warranty policy and procedures manual, or determine if any prior approval program is in operation, prior to the installation of a new module/ component

PINPOINT TEST G : D-LOOP NOT ROTATING CORRECTLY	
TEST CONDITIONS	**DETAILS/RESULTS/ACTIONS**
G1: D-LOOP NOT ROTATING CORRECTLY	
	1. Refer to the 501-20 removal and installation section of the workshop manual, remove any trim panels required to expose the D loop (anchor point) and the seatbelt retractor
	2. Make sure there are no obstructions and the webbing does not catch or rub, the D loop (anchor point) rotates correctly and if installed the confirm the height adjuster operates correctly
	3. Check for correct operation **twice**
	Does the webbing move freely then retract correctly? **Yes** Refer to the 501-20 removal and installation section of the workshop manual, reinstall any trim panels, make sure there are no obstructions and the webbing does not catch or rub. No further action required **No** Replace as required. Refer to the warranty policy and procedures manual, or determine if any prior approval program is in operation, prior to the installation of a new module/ component

PINPOINT TEST H : MINI BUTTON - MISSING/DAMAGED	
TEST CONDITIONS	**DETAILS/RESULTS/ACTIONS**
H1: MINI BUTTON - MISSING/DAMAGED	
△ NOTE: This test applies to the rear centre seatbelt retractor installed in the **seat back.**	
	1. Refer to the 501-20 removal and installation section of the workshop manual, remove the seat cushion and the plastic escutcheon at the top of the seat back (where the webbing exits to expose the lower anchor fixing point of the center seatbelt)
	2. Remove the lower anchorage of the seatbelt
	3. With the seat back correctly latched, allow up to 20mm webbing to retract, then extract the webbing
	Is the mini-button (webbing travel limit stop) correctly installed to the webbing and in good condition? **Yes** Feed the mini-button back through the plastic escutcheon if required. Correctly reinstall the escutcheon to the seat back, extract the webbing then allow to retract, make sure the mini-button comes to rest outside the escutcheon stop **No** Replace as required. Refer to the warranty policy and procedures manual, or determine if any prior approval program is in operation, prior to the installation of a new module/ component

PINPOINT TEST I : SEATBELT BUCKLE - NOT LATCHING/JAMMED	
TEST CONDITIONS	**DETAILS/RESULTS/ACTIONS**
I1: SEATBELT BUCKLE - NOT LATCHING/JAMMED	

PINPOINT TEST I : SEATBELT BUCKLE - NOT LATCHING/JAMMED	
TEST CONDITIONS	**DETAILS/RESULTS/ACTIONS**
⚠ CAUTION: Do not insert any objects or tools into the buckle head.	
	1. Visually inspect the buckle head for evidence of damage. If damaged replace as required
	2. Press the buckle release (red button) and **(Using a torch)** complete visual inspection for any evidence of debris/material or foreign objects in the buckle head
	3. If required remove the buckle from the vehicle. Remove the seat. Remove the buckle from the seat frame
	4. **Do not insert any objects or tools buckle head.** With the buckle removed invert and attempt to shake out any debris
	5. Attempt to latch the tongue in the buckle
	Does the seatbelt buckle operate correctly **Yes** Reinstall any components, no further action required **No** Replace the buckle, Refer to section 501 20

PINPOINT TEST J : SEATBELT SLOW RETRACTION	
TEST CONDITIONS	**DETAILS/RESULTS/ACTIONS**
J1: D-LOOP CONTAMINATION CHECK/CLEAN	
⚠ NOTE: Part shown removed for clarity. Seatbelt D-Loop can be visually inspected without removing the 'B' pillar trim panel.	
	1. Visually inspect the front seatbelt D-Loop for a build up of dirt or contamination 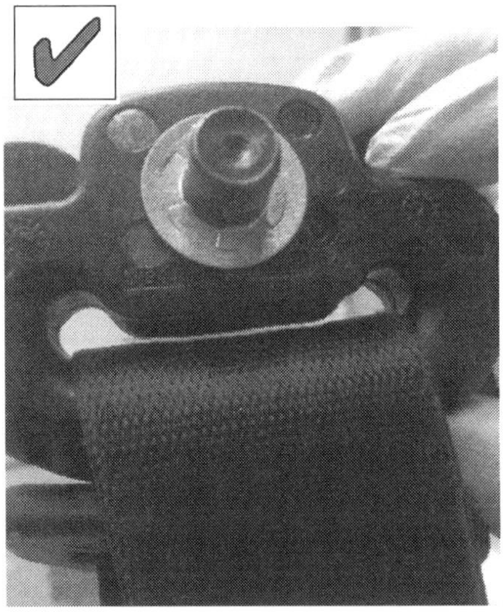 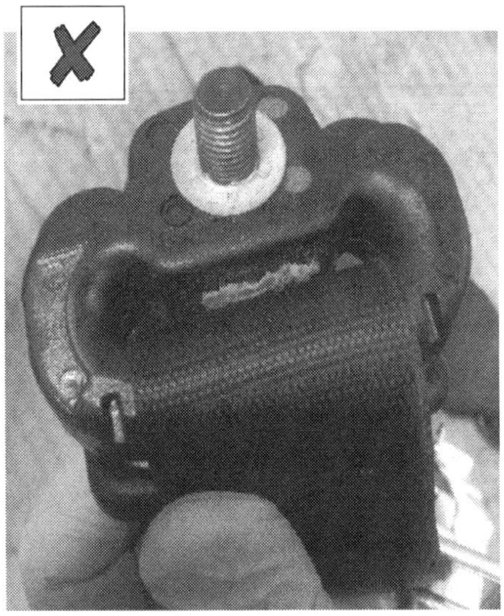
	Does the front seatbelt D-Loop have a build up of dirt or contamination? **Yes** Clean the seatbelt webbing and seatbelt D-loop to remove any dirt or foreign material using warm water CAUTION: Do not use any detergents or allow the water to stain the interior trim or enter the seatbelt retractor system **No** Not the cause of the customers concern, continue diagnostics and follow the symptom chart for the reported failure

DTC Index

For a list of Diagnostic Trouble Codes (DTCs) that could be set on this vehicle, please refer to Section 100-00.

Front Row Safety Belt Buckle

Removal and Installation

Removal

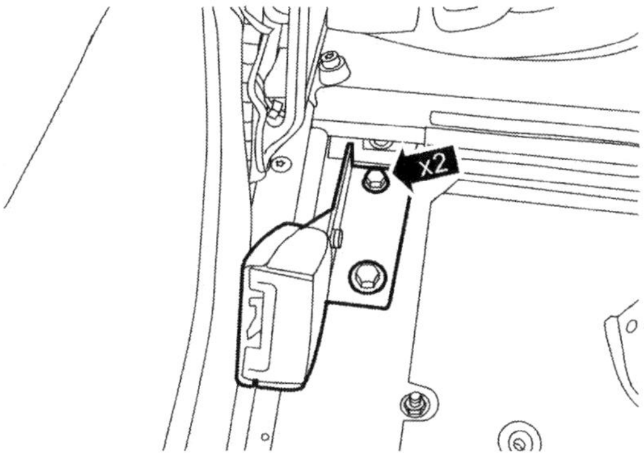

1. Remove the M8 bolt. Remove and discard the M10 bolt. Remove the front safety belt buckle.

Installation

1. Install the front safety belt buckle. Tighten bolts as required.

   M10 bolt 23 Nm

   M8 bolt 23 Nm

Front Safety Belt Retractor

Removal and Installation

Removal

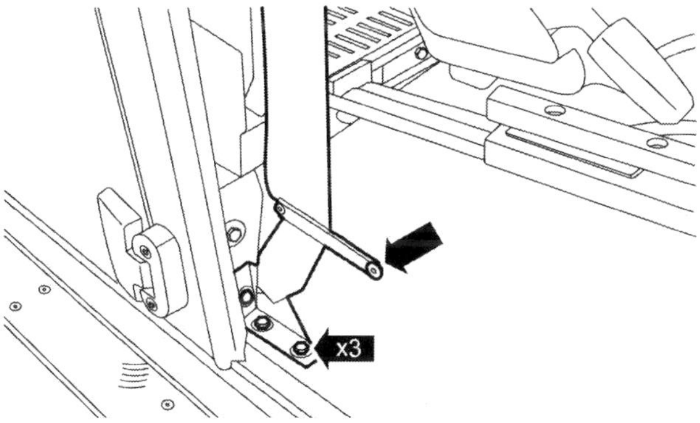

1. Release the front safety belt lower anchor.
   - Remove the 3 bolts.
   - Remove the clip.

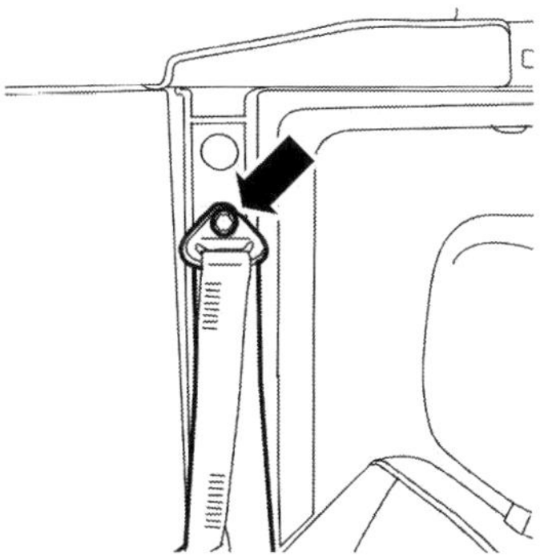

2. Release the front safety belt upper anchor.
   - Remove the bolt cover.
   - Remove the bolt.

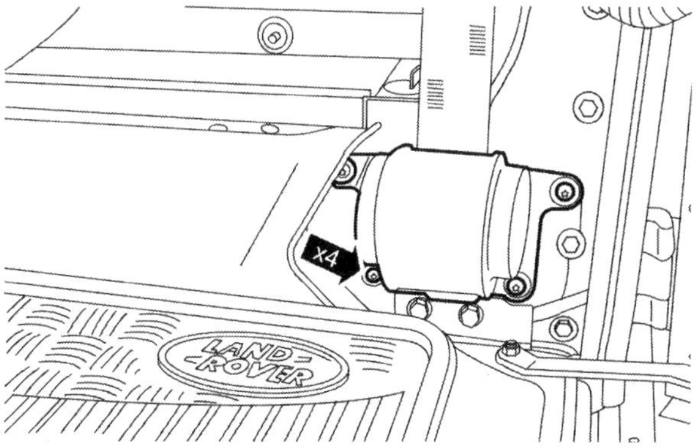

3. Remove the front safety belt retractor trim.
   - Remove the 4 scrivets.

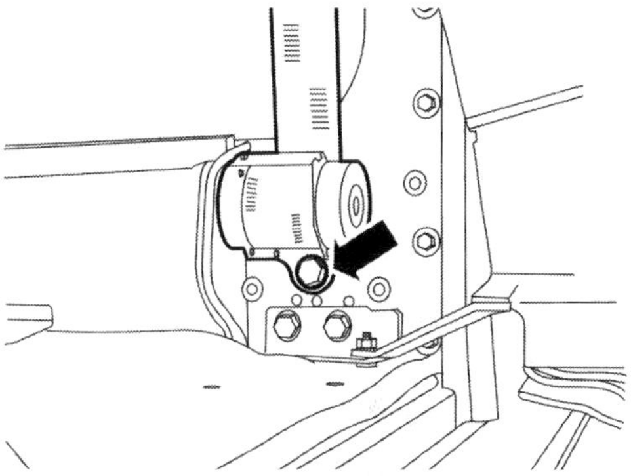

4. Remove the front safety belt retractor.

- Remove the bolt.

Installation

1. Install the front safety belt retractor.
   - Tighten the bolt to 31 Nm (23 lb.ft).
2. Install the front safety belt retractor trim.
   - Install the scrivets.
3. Secure the front safety belt upper anchor.
   - Tighten the bolt to 31 Nm (23 lb.ft).
   - Install the bolt cover.
4. Secure the front safety belt lower anchor.
   - Tighten the bolts to 10 Nm (7 lb.ft).
   - Install the clip.

Rear Center Safety Belt Buckle

Removal and Installation

Removal

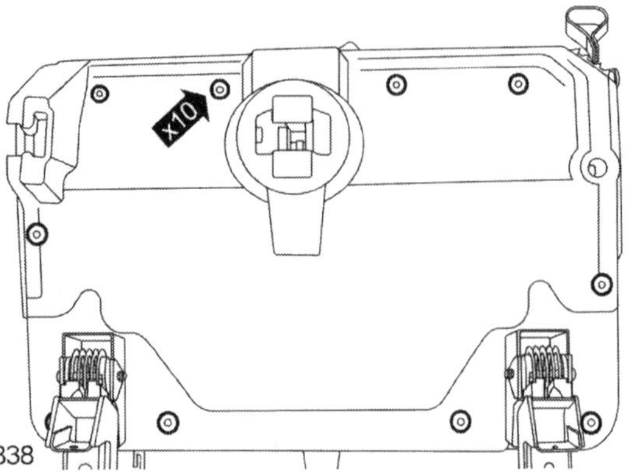

E90838

1. Remove the RH rear seat inner hinge trim panel.
   - Remove the 4 screws.

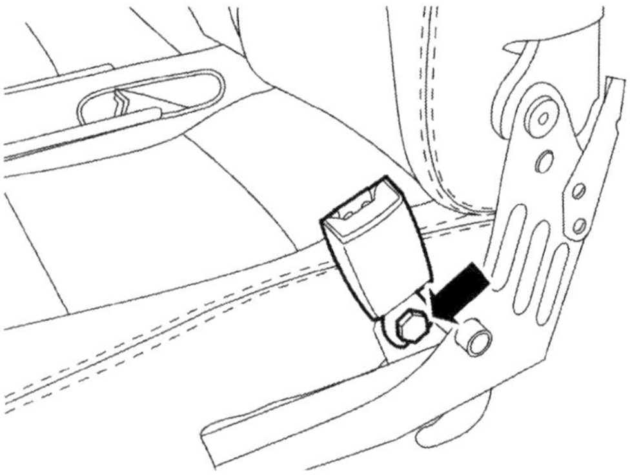

2. Remove the rear center safety belt buckle.
   - Remove the bolt.

Installation

1. To install, reverse the removal procedure.
   - Tighten the bolt to 31 Nm (23 lb.ft)

Rear Center Safety Belt Retractor

Removal and Installation

Removal

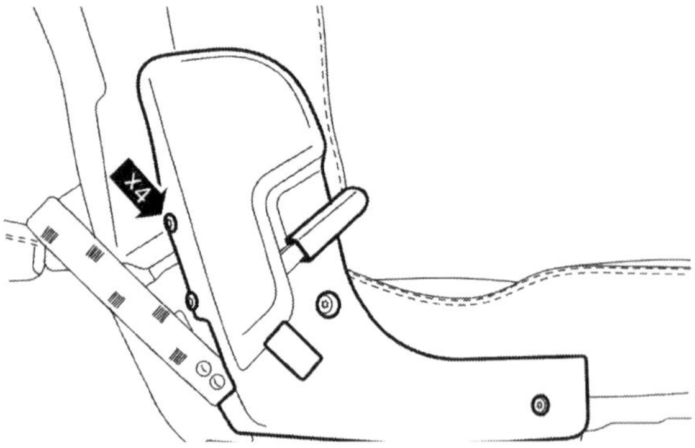

1. Remove the rear seat outer hinge trim panel.
   - Remove the rear seat recliner handle.
   - Remove the 4 screws.

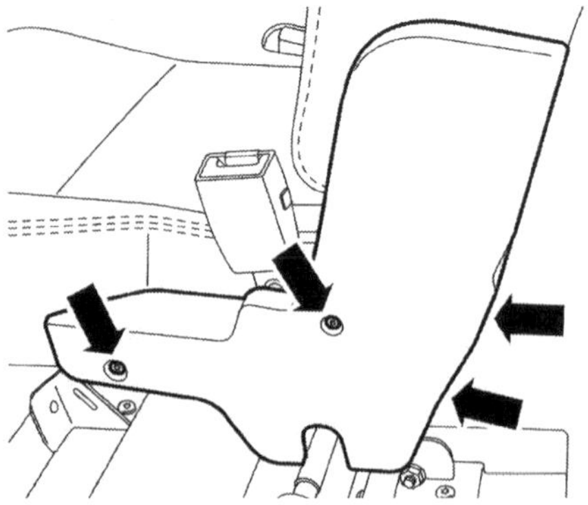

2. Remove the rear seat inner hinge trim panel.
   - Remove the 4 screws.

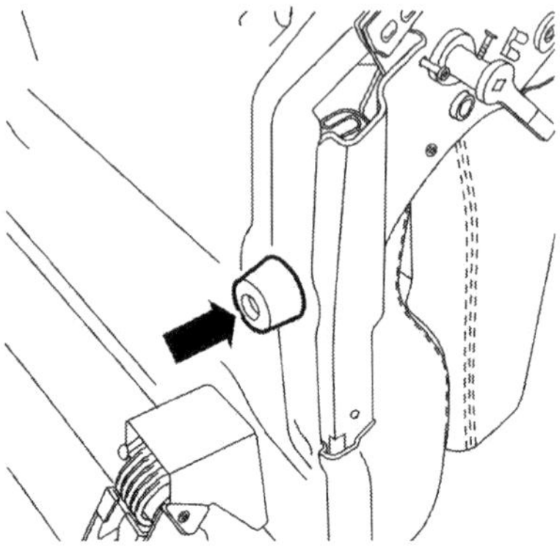

3. Remove the rear seat base stop.
   - Remove the bolt.
   - Collect the plastic washer.

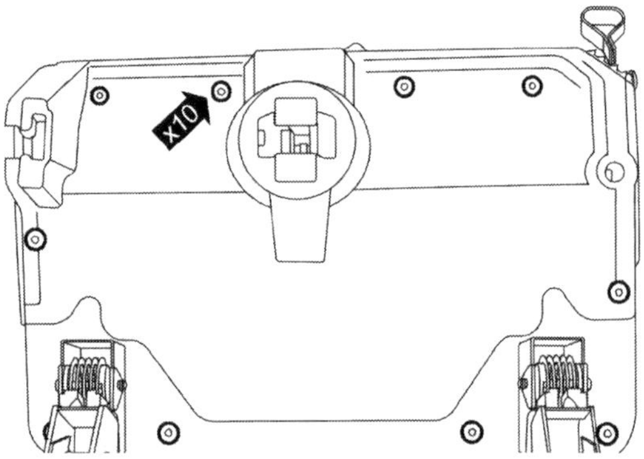

4. Remove the rear seat cushion trim panel.
   - Remove the 10 clips.

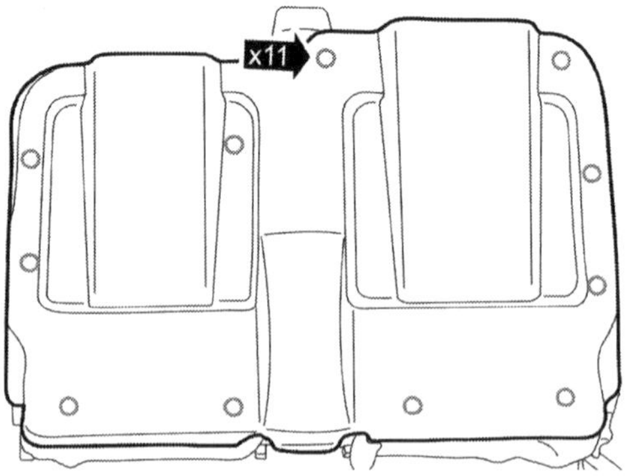

5. Remove the rear seat backrest trim panel.
   - Release the 11 clips.

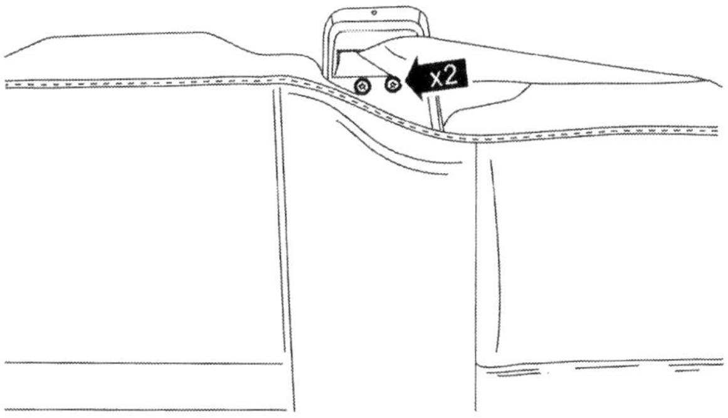

6. Reposition the rear center safety belt guide cover.
   - Remove the 2 screws.

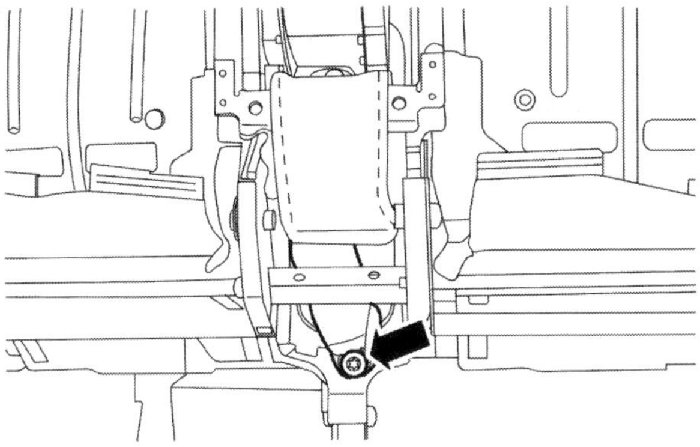

7. Release the rear center safety belt lower anchor.
   - Remove the bolt.

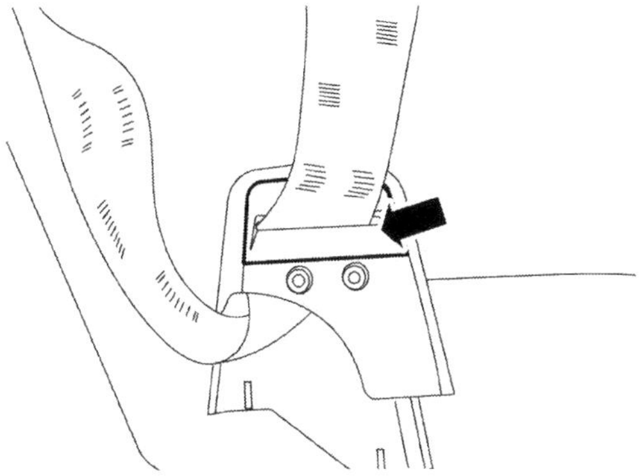

8. Remove the rear center safety belt guide cover.
   - Remove the rear center safety belt guide cover finisher.

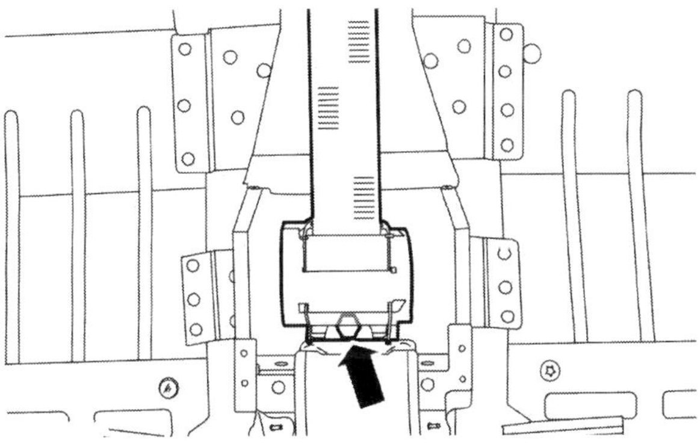

9. Remove the center rear safety belt retractor.
   - Remove the bolt.

Installation

1. Install the center rear safety belt retractor.
   - Tighten the bolt to 31 Nm (23 lb.ft).
2. Install the rear center safety belt guide cover.
   - Install the rear center safety belt guide cover finisher.
3. Secure the rear center safety belt lower anchor.
   - Tighten the bolt to 31 Nm (23 lb.ft).
4. Secure the rear center safety belt guide cover.
   - Tighten the screws.
5. Install the rear seat backrest trim panel.
   - Secure with the clips.
6. Install the rear seat cushion trim panel.
   - Install the clips.
7. Install the rear seat base stop.
   - Install the plastic washer.
   - Tighten the bolt to 23 Nm (17 lb.ft).
8. Install the rear seat inner hinge trim panel.
   - Tighten the screws.
9. Install the rear seat outer hinge trim panel.
   - Tighten the screws.
   - Install the rear seat recliner handle.

Rear Safety Belt Retractor

Removal and Installation

Removal

1. Remove cap from seat belt guide bolt.
2. Remove seat belt upper and lower securing bolts.
3. Remove seat belt guide from D-pillar finisher.
4. Remove 2 studs securing finisher to D-pillar.
5. Release finisher from body upper rail and D-pillar.
6. Pull seat belt through D-pillar finisher.

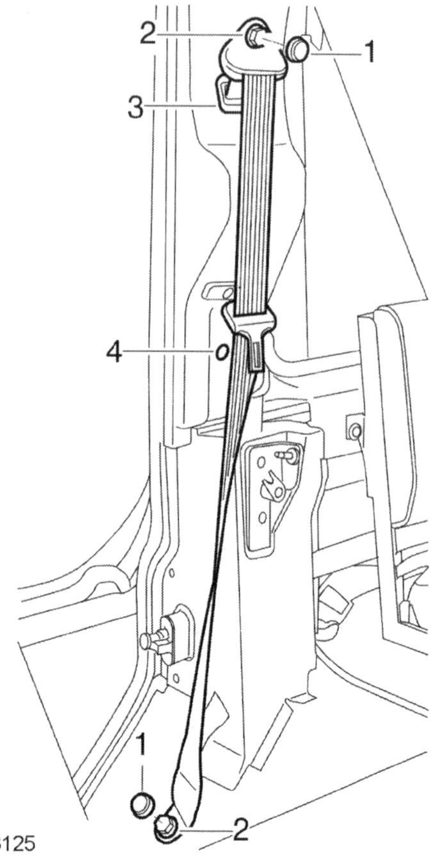

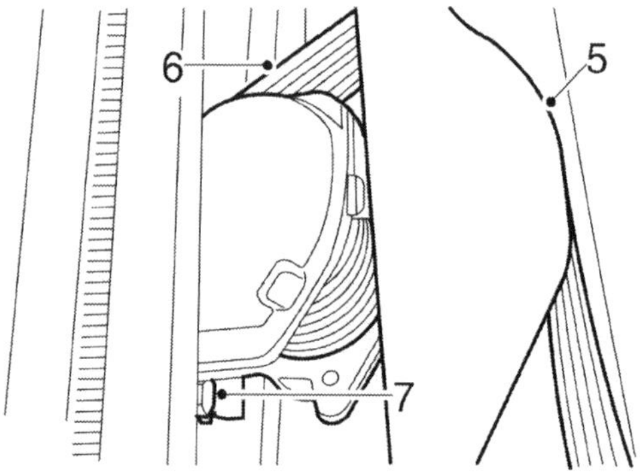

7. Remove bolt securing seat belt to D-pillar and remove belt assembly.

Installation

1. Position seat belt assembly to D-pillar and tighten bolt to 32 Nm (24 lb.ft).
2. Extend seat belt and position through D-pillar post finisher.
3. Install and tighten seat belt upper and lower securing bolts to 32 Nm (24 lb.ft).
4. Secure D-pillar post finisher to body upper rail.
5. Install belt guide to D-pillar post finisher and install finisher retaining studs.
6. Install cap to seat belt bolt.

Rear Left Safety Belt Buckle

Removal and Installation

Removal

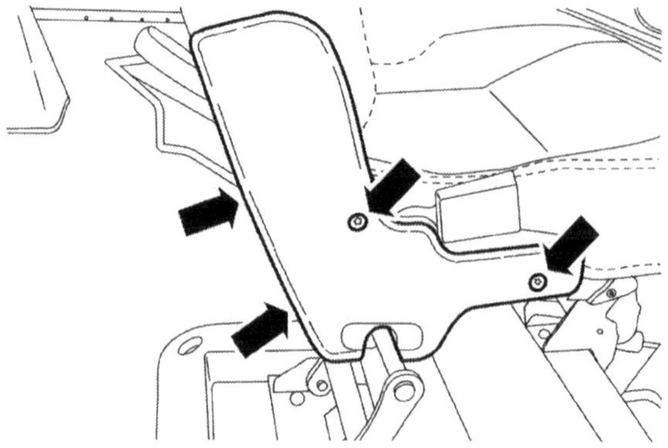

1. Remove the LH rear seat inner hinge trim panel.
   - Remove the 4 screws.

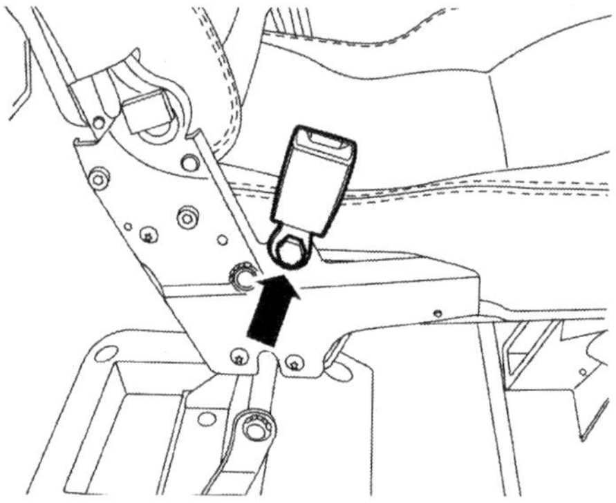

2. Remove the LH rear safety belt buckle.
   - Remove the bolt.

Installation

1. To install, reverse the removal procedure.
   - Tighten the bolt to 31 Nm (23 lb.ft)

Rear Right Safety Belt Buckle

Removal and Installation

Removal

1. Remove the rear seat cushion.

Refer to: Rear Seat Cushion (501-10 Seating, Removal and Installation)

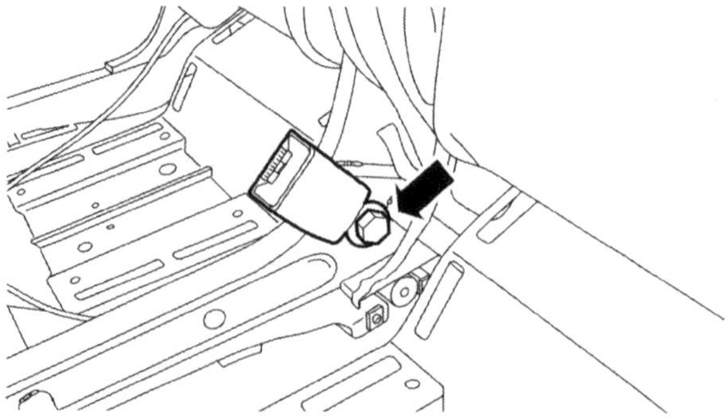

2. Remove the RH rear safety belt buckle.
   - Remove the bolt.

Installation

1. To install, reverse the removal procedure.
   - Tighten the bolt to 31 Nm (23 lb.ft).

Third Row Safety Belt Buckle

Removal and Installation

Removal

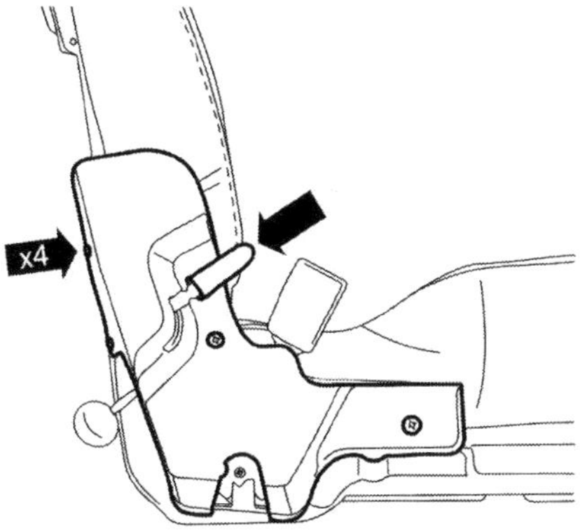

E90821

1. Remove the third row seat inner hinge trim panel.
   - Remove the third row seat recliner handle.
   - Remove the 4 screws.

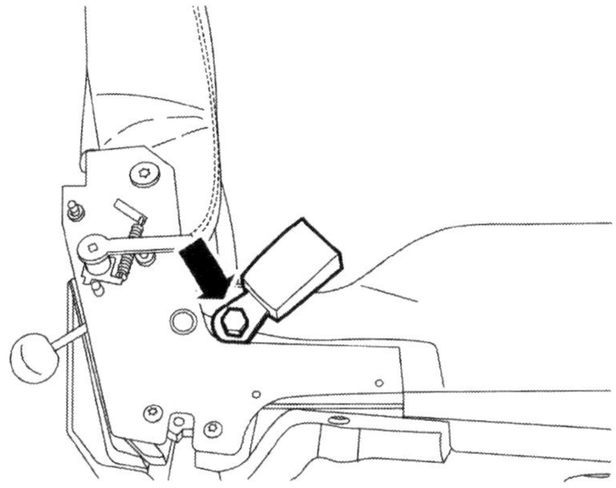

E90910

2. Remove the third row safety belt buckle.
   - Remove the bolt.

Installation

1. To install, reverse the removal procedure.
   - Tighten the bolt to 31 Nm (23 lb.ft).

Third Row Safety Belt Retractor

Removal and Installation

Removal

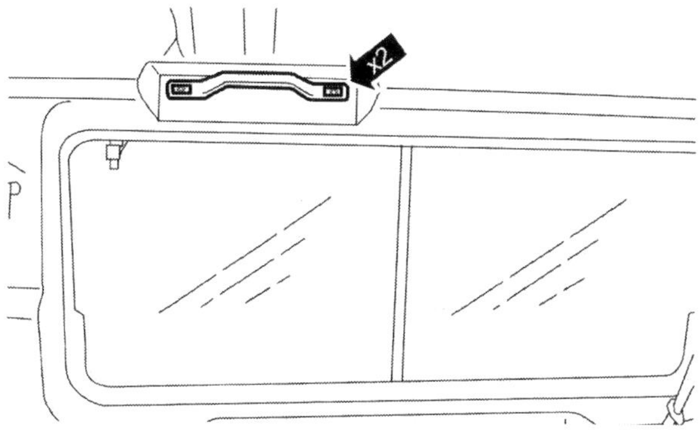

1. Remove the grab handle.
   - Remove the 2 grab handle screw covers.
   - Remove the 2 screws.

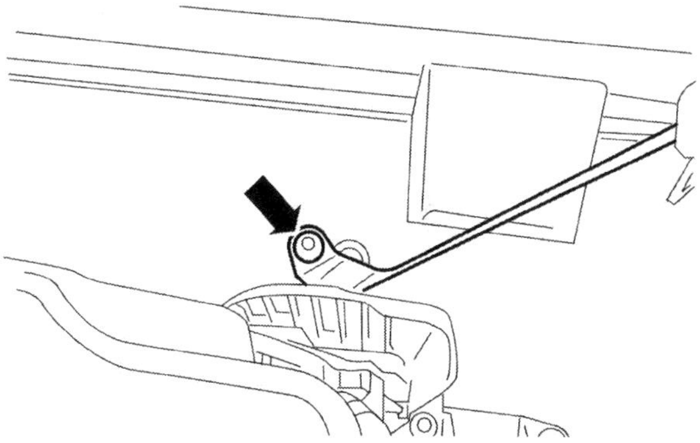

2. Release the third row safety belt lower anchor.
   - Remove the bolt.

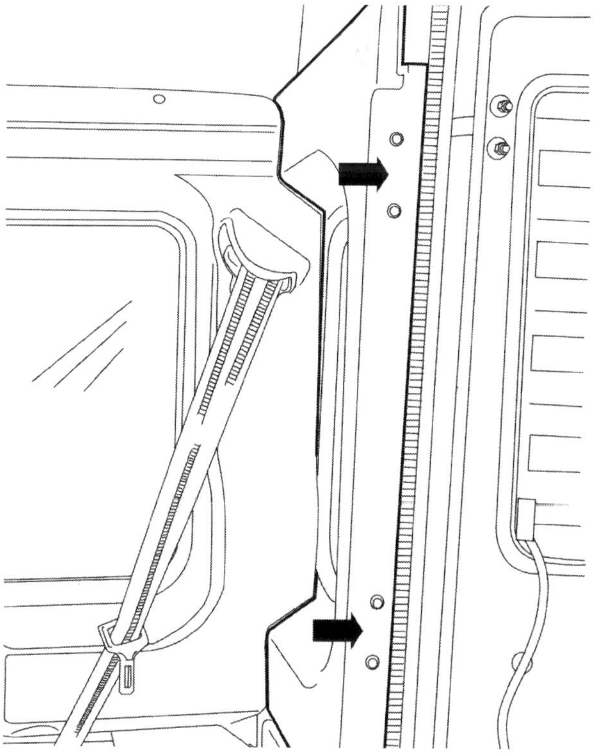

3. Remove the rear quarter trim panel.
   - Release the 2 clips.

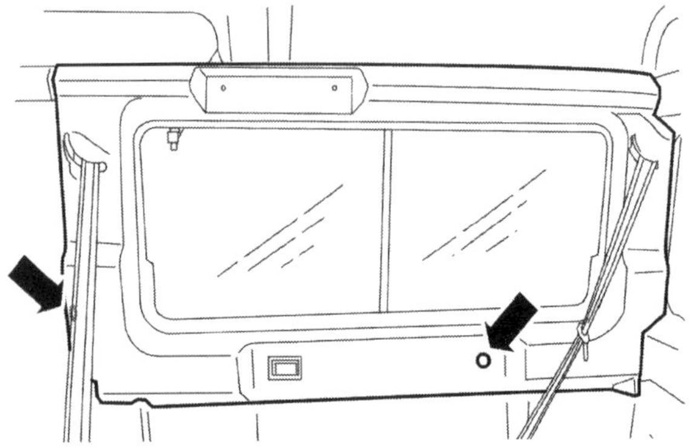

4. Reposition the C-pillar trim panel.
   - Remove the 2 lower clips.
   - Move the C-pillar trim panel upwards to release the 2 upper clips.

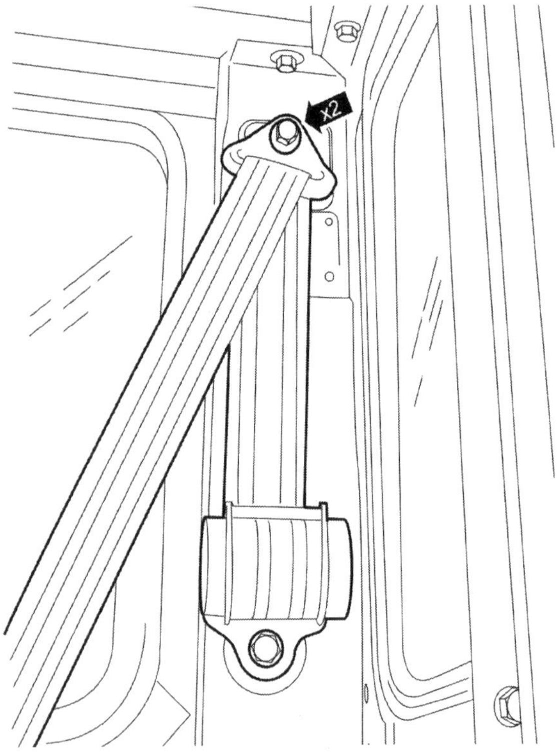

5. Remove the third row safety belt retractor.
   - Remove the 2 bolts.

Installation

1. Install the third row safety belt retractor.
   - Tighten the bolts to 31 Nm (23 lb.ft).
2. Secure the C-pillar trim panel.
   - Install the clips.
3. Install the rear quarter trim panel.
   - Secure with the 2 clips.
4. Secure the third row safety belt lower anchor.
   - Tighten the bolt to 31 Nm (23 lb.ft).
5. Install the grab handle.
   - Tighten the screws.
   - Install the grab handle screw covers.

Front Safety Belt Retractor Mounting Bracket

Removal and Installation

Removal

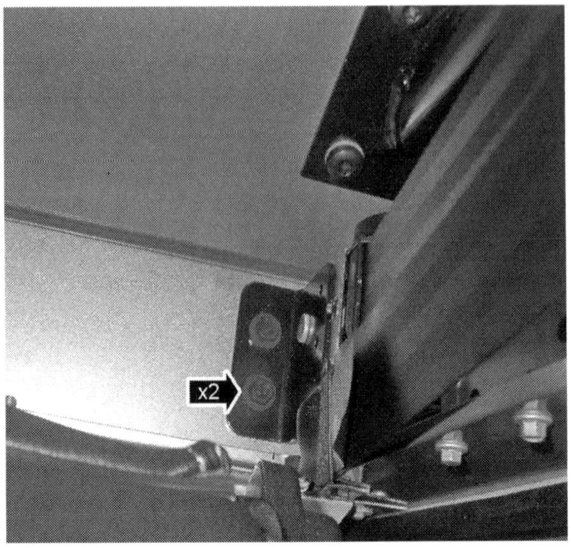

1. Release the front safety belt retractor.
   - Remove the 2 nuts.

2. ⚠ CAUTION: Make sure that the door remains closed until the front safety belt retractor mounting bracket lower bolts are fully tightened.

   Close the front door.

3. ⚠ CAUTION: Make sure that the body of the vehicle is supported to maintain the relative body to chassis alignment during this procedure.

Using a suitable jack, support the body.

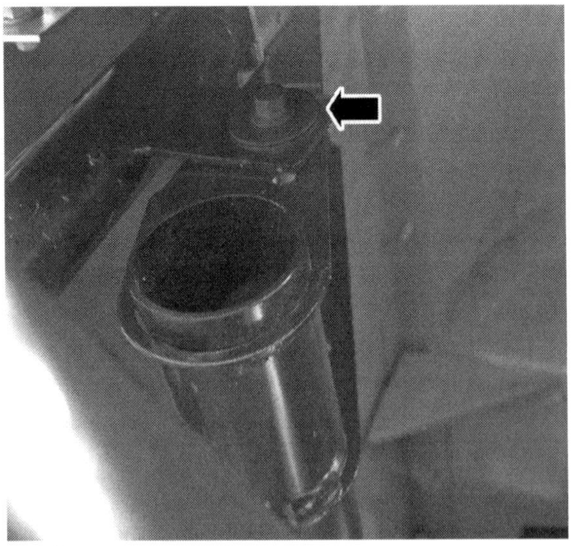

4. ⚠ NOTE: **Vehicles with side steps only.**

Remove the side step rear outer support bracket nut and bolt.

5. NOTE: **Vehicles with side steps only.**

   Remove the side step front inner support bracket bolt.

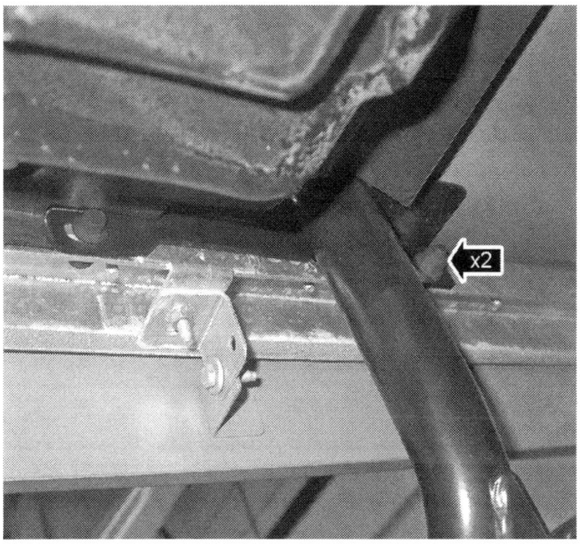

6. NOTES:

Vehicles with side steps only.

This step requires the aid of another technician.

Remove the 2 nuts and bolts from the side step front outer support bracket.

7. CAUTION: Make sure that the front door remains closed.

NOTES:

Up to 4 shims may be fitted to each side of the vehicle.

If a shim(s) is installed, note the orientation of the shim(s) to aid installation.

Remove the 2 nuts and bolts from the front safety belt retractor mounting bracket.

- With assistance remove the side step (if equipped).
- Remove the shim(s) (if equipped).

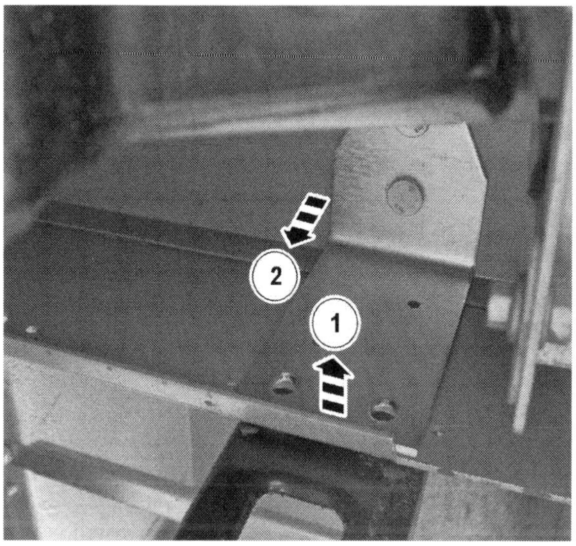

8. CAUTION: Make sure that the front door remains closed.

Remove the front safety belt retractor mounting bracket.

- If required, use a suitable tool to release the bracket.

Installation

1. CAUTION: Make sure that the front door remains closed.

Install the front safety belt retractor mounting bracket.

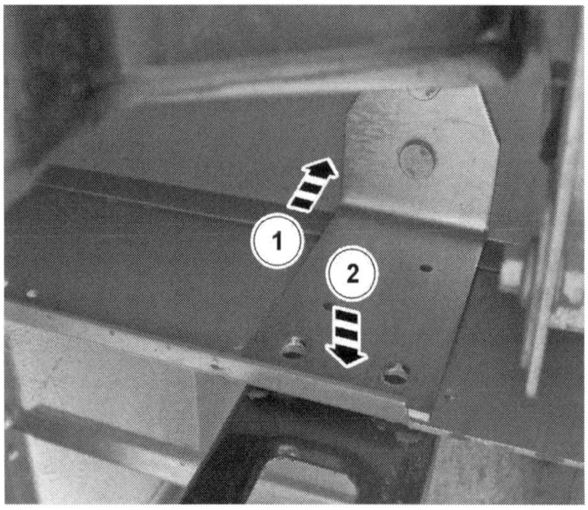

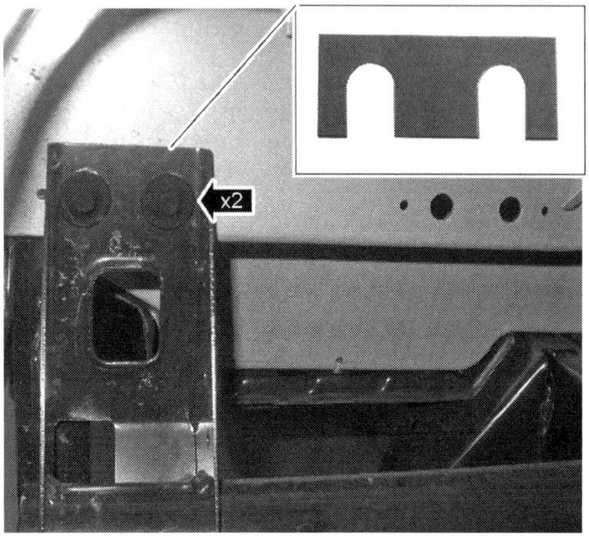

2. CAUTIONS:

⚠ Make sure that the front door remains closed.

⚠ Make sure that the shim is installed to the previously noted position.

⚠ Make sure that the shim(s) is installed to the previously noted position.

⚠ Only tighten the nuts finger-tight at this stage.

Install the 2 nuts and bolts.

- Install the shim(s) (if equipped).

- With assistance install the side step (if equipped).

3. CAUTIONS:

⚠ Make sure that the front door remains closed.

⚠ Only tighten the nuts and bolts finger-tight at this stage.

NOTES:

△ This step requires the aid of another technician.

△ **Vehicles with side steps only.**

Install the 2 nuts and bolts.

4. CAUTIONS:

⚠ Make sure that the front door remains closed.

⚠ Only tighten the bolt finger-tight at this stage.

△ NOTE: **Vehicles with side steps only.**

Install the bolt.

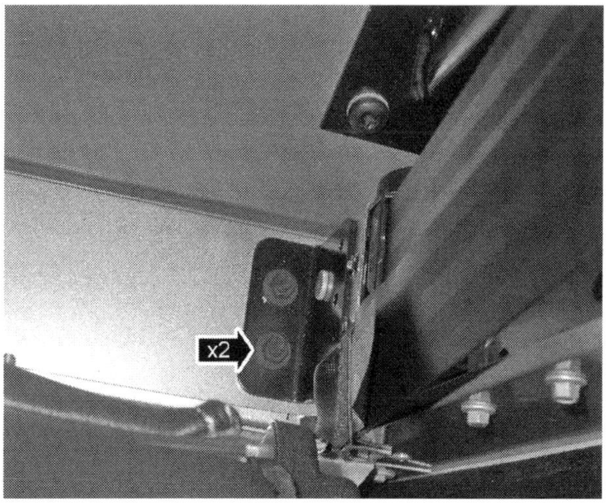

5. CAUTIONS:

Make sure that the front door remains closed.

Only tighten the nut and bolt finger-tight at this stage.

NOTE: **Vehicles with side steps only.**

Install the nut and bolt.

6. CAUTION: Make sure that the front door remains closed.

Tighten the 2 nuts and bolts to 22 Nm.

7. NOTE: **Vehicles with side steps only.**

Tighten the 2 nuts and bolts to 22 Nm.

8. NOTE: **Vehicles with side steps only.**

Tighten the bolt to 22 Nm.

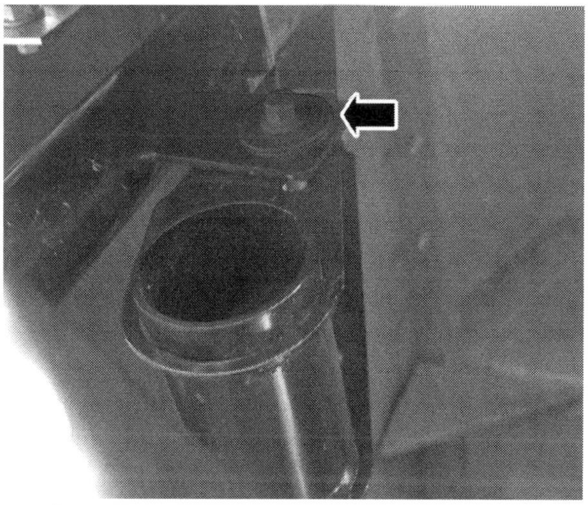

9. NOTE: **Vehicles with side steps only.**

Tighten the nut and bolt to 22 Nm.

10. Lower the body.

11. Open the front door.

12. Secure the front safety belt retractor.

- Tighten the nuts to 13 Nm.

Rear Safety Belt Retractor Mounting Bracket

Removal and Installation

Removal

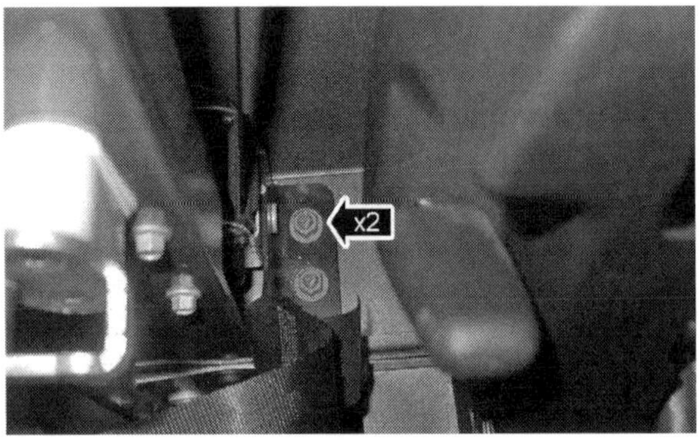

1. Release the rear safety belt retractor.
   - Remove the 2 nuts.

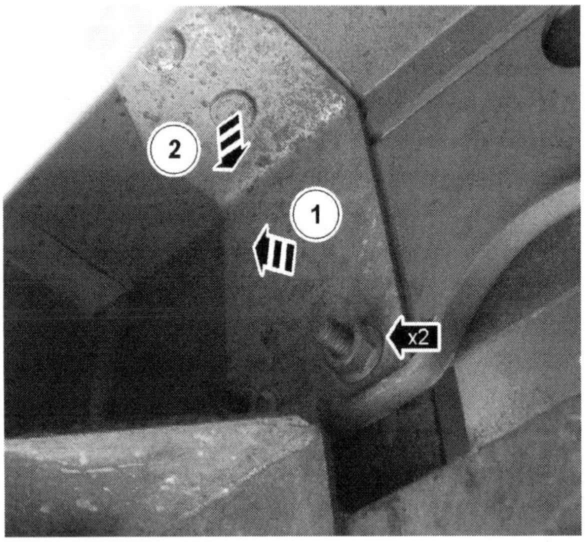

2. Remove the rear safety belt retractor mounting bracket.
   - Remove the 2 nuts and bolts.
   - If required, use a suitable tool to release the bracket.

Installation

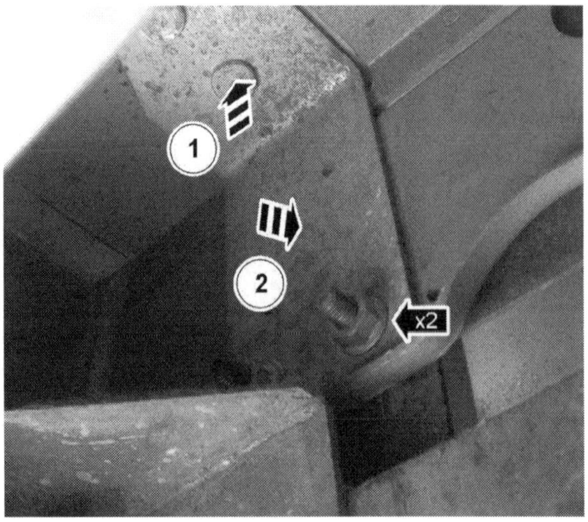

1. Install the rear safety belt retractor mounting bracket.
   - Tighten the 2 nuts and bolts to 22 Nm.

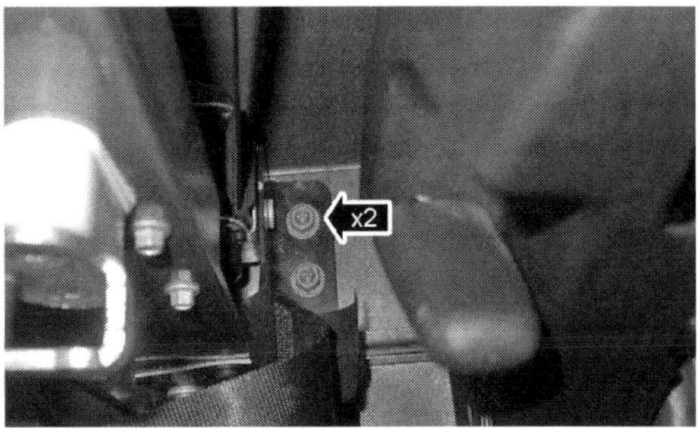

2. Secure the rear safety belt retractor.
   - Tighten the 2 nuts to 13 Nm.

Description and Operation

General Information

Body shells are of riveted, bolted and welded construction and are bolted to the chassis frame.

It is essential that design dimensions and strength are restored in accident rectification. It is important that neither structural weakness nor excessive local stiffness are introduced into the vehicle during body or chassis repair.

Repairs usually involve a combination of operations ranging from straightening procedures to renewal of either individual panels or panel assemblies. The repairer will determine the repair method and this decision will take into account a balance of economics between labour and material costs and the availability of repair facilities in both equipment and skills. It may also involve considerations of vehicles down-time, replacement vehicle availability and repair turn-around time.

It is expected that a repairer will select the best and most economic repair method possible, making use of the facilities available. The instructions given are intended to assist a skilled body repairer by expanding approved procedures for panel replacement with the objective of restoring the vehicle to a safe running condition and effecting a repair which is visually acceptable and which, even to the experienced eye, does not advertise the fact that it has been damaged.

This does not necessarily mean that the repaired vehicle will be identical in all respects with original factory build. Repair facilities cannot always duplicate methods of construction used during production.

It is key to safety and vehicle integrity that only the approved methods, materials and equipment are used. To successfully complete structural repairs to Jaguar Land Rover products, it is necessary for the repairs to follow the correct aftermarket process for adhesive (bonding), riveting and welding to ensure vehicle integrity is maintained after a collision.

Refer to: Adhesives, Fixings and Consumables (501-25 Body Repairs - General Information, Description and Operation)

The panel repairs shown in this section are all based on a 110 Station Wagon. Therefore all illustrations and text relate only to this model. Although certain areas of the vehicle, such as the front end, are relevant to all models.

Operations covered in this Manual do not include reference to testing the vehicle after repair. It is essential that work is inspected and suspension geometry checked after completion and if necessary a road test of the vehicle is carried out, particularly where safety related items are concerned.

Where major units have been disconnected or removed, it is necessary to ensure that fluid levels are checked and topped up when necessary. It is also necessary to ensure that the repaired vehicle is in a roadworthy condition in respect of tyre pressures, lights, washer fluid etc.

Body repairs often involve the removal of mechanical and electrical units as well as associated wiring. Where this is necessary use the relevant section in this manual.

Taking into consideration the differences in body styles, steering and suspension systems as well as engine and suspension layouts, the location of the following components as applicable to a particular vehicle is critical:

- Front suspension upper damper mountings
- Front suspension or sub frame mountings
- Engine mountings on RH and LH chassis longitudinals
- Rear suspension upper damper mountings
- Rear suspension mountings or lower pivots
- Steering rack mountings

Additional points which can be used to check alignment and assembly are:

- Inner holes in cross member - side - main floor
- Holes in valance front assembly
- Body to chassis mounting holes
- Holes in rear floor
- Holes in rear lower panels or extension rear floor
- Fuel tank mountings

Apertures for windscreen, backlight, bonnet and doors can be checked by offering up an undamaged component as a gauge and also by measuring known dimensions.

Refer to: Body and Frame (501-26 Body Repairs - Vehicle Specific Information and Tolerance Checks, Description and Operation)

Straightening

Whenever possible, steel chassis structural members should be cold straightened under tension. Do not attempt to straighten with a single pull, but rework the damaged area using a series of pulls, releasing tension between each stage and using the opportunity to check alignment.

Body jig

Unless damage is limited to cosmetic panels, all repair work to body members must be carried out on a body jig, to ensure that impact damage has not spread into more remote parts of the body structure. Mounting on a jig will also ensure that the straightening and panel replacement procedures do not cause further distortion. If original dimensions cannot be satisfactorily restored by these methods, damaged structural members should be replaced.

⚠ WARNING: Damaged areas should be cut away using a high speed saw, NOT an oxy-acetylene torch.

As a rule, body dimensions are symmetrical about the centre line. A good initial check for distortion is therefore to measure diagonally and to investigate apparent differences in dimensions.

Inspection

Every accident produces individual differences in damage. Each repair is influenced by the extent of the damage and by the facilities and equipment available for its rectification.

Most accident damage can be visually inspected and the approximate extent of the damage assessed. Sometimes deformation will extend beyond the area of direct damage, and the severity of this must be accurately established so that steps may be taken to restore critical body components to their original dimensions.

An initial check of critical dimensions can be carried out by means of drop checks or (preferably) trammels. Gauges are available which will check accurately for body twist. Where repairs necessitate renewal of a critical body component it is recommended that a body jig is used.

Battery care standards

To ensure that the correct maintenance actions are carried out before and during a repair.

Refer to: Battery Care Requirements (414-00 Battery and Charging System - General Information, Description and Operation)

ELECTRONIC CONTROL UNITS (ECUs)

The ECU's fitted to Defender vehicles make it advisable to follow suitable precautions prior to carrying out welding repair operations. All ECU's must be disconnected before any welding operations take place. Harsh conditions of heat and vibration may be generated during these operations which could cause damage to the units.

PAINT PROCEDURES

Replacement Panels

Service panels are supplied with a cathodic primer coating as part of the panel protection, and in compliance with the vehicle's Corrosion Warranty where applicable. DO NOT remove this primer before paint refinishing. In the event of localised surface damage or imperfections, ensure that the minimum of primer is removed during rectification work for effective repair.

Rectify damage by panel beating or straightening. To remove corrosion or paint runs on outer surfaces, abrade primer coat in the affected area as necessary using the following procedure:

- Clean the panel using a solvent wipe.
- Treat exposed areas of metal with an etch phosphate process.
- Re-treat the affected area using either a separate acid-etch primer and two-pack surfacer, or an integrated etch primer/filler.

Bolted Panels

Before fitting bolt-on panels, ensure that all mating and adjacent surfaces on the vehicle and replacement panel are free from damage and distortion. Rectify if necessary as described in this section, and apply preformed strip sealer where specified.

Welded Panels

- Remove primer from the immediate vicinity of new and existing panel flanges, cleaning to bright metal finish.
- On joints to be spot welded, apply weld-through zinc rich primer to joint faces of both flanges. Make spot welds while primer is still wet or according to the manufacturer's instructions.
- Dress accessible weld seams.
- Clean panel using solvent wipe.
- Treat bare metal with an etch phosphate process.
- Re-treat repaired areas.
- It is not satisfactory to use weld-through, zinc rich primers in conjunction with MIG welding.

Sectioned Panels

When replacing part or sectioned panels, the basic procedure is the same as for welded panels described above, with the following variations:

1. Remove primer from both new and existing joint faces, cleaning to a bright metal finish.

2. Where an overlap joint with the existing panel is to be spot welded, apply weld-through, zinc rich primer to both joint faces and spot weld while the primer is still wet or according to the manufacturer's instructions.
3. MIG weld joints where applicable.
4. Clean the panel with a solvent wipe.
5. Treat bare metal areas using an etch phosphate process.
6. Re-prime affected areas as necessary as for rectifying transit damage. See this section.
7. Treat the inner faces of lap or butt joints with a suitable cavity wax. See Sealing and Corrosion Protection.

Clinch Panels (e.g. Door skins etc.)

1. Abrade primer on new and existing panel joint faces, and clean using a solvent wipe.
2. Apply metal-to-metal adhesive where applicable.
3. Where joints are to be spot welded, apply suitable weld-through, zinc rich primer to weld areas.
4. Where joints are to be MIG welded, apply zinc rich primer in adjacent areas but leave the welded area untreated.
5. To retain the panel whilst clinching the flanges, tack spot weld or plug weld as appropriate.
6. Clean the panel with a solvent wipe.
7. Treat bare metal areas with a suitable etch phosphate process.
8. Re-prime affected areas as necessary as for rectifying transit damage. See this section.
9. Replacement doors, bonnets and tailgates must be treated with a suitable seam sealer on clinched seams, following the primer coat.

Paint Refinishing

1. Seal all accessible exterior and interior seams with an approved seam sealer. Certain joints such as sill lower flange seams must be left unsealed.
2. Apply a suitable anti-chip primer where specified.
3. Apply a two-pack paint refinishing system.
4. Repair any damage to underbody sealers either at this stage or before paint operations.

Paint Repairs

Before carrying out paintwork repairs, the vehicle must be thoroughly cleaned using either a steam cleaner or high-pressure washer.

Wash locally repaired areas using a mild water-mixable detergent and wipe them clean with solvent, immediately prior to paint application.

Abrade damaged paintwork where bare metal has been exposed until the metal is clean and extends beyond the area of immediate damage. Treat the bare metal with an etch phosphate to remove all traces of rust and provide a key for new paint coats. Re-treat the affected area using either a separate acid-etch primer and two-pack surfacer or an integrated etch primer/filler, and follow with a two-pack paint system. Those surfaces not receiving paint must be treated with a cavity wax following paint operations.

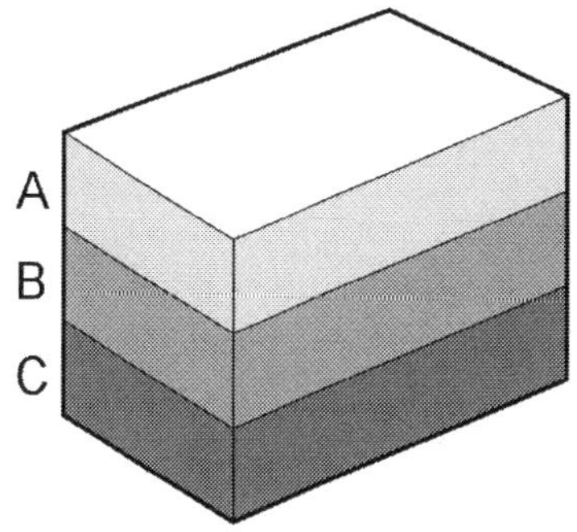

M772108B

A	Two-pack top coat
B	Two-pack primer filler and etch primer
C	Etch phosphate

GENERAL WELDING PRECAUTIONS

The following pages show the procedures to follow when using welding for repairs. No resistance spot welds have been used in any of the repairs.

The aluminium alloy used on all Defender models is a combination of aluminium and magnesium. When converting a MIG welder for use on aluminium it is essential the following components are changed. The materials shown in brackets are the correct materials to use:

When carrying out welding operations the following criteria must be observed:

M772122

Item	Description
A	MIG Plug welds
B	MIG seam weld

Seat Belt Anchorages

Seat belt anchorages are safety critical. When making repairs in these areas it is essential to follow design specifications.

Where possible, the original production assembly should be used, complete with its seat belt anchorages, or the cut line should be so arranged that the original seatbelt anchorage is not disturbed.

⚠ WARNING: Body parts incorporating seat belt anchorages MUST be renewed completely if damaged beyond repair, as the welds in these areas are safety critical and cannot be disturbed.

All welds within 250mm (9.9in.) of seat belt anchorages must be carefully checked for weld quality, including spacing of spot welds. A crack detection process must be carried out in these areas.

PANEL REPLACEMENT PROCEDURE

General

This information is designed to explain the basic panel removal and replacement method. This standard method may vary slightly from one vehicle to another. The main criterion in removal and replacement of body panels is that Land Rover's original standard is maintained as far as possible.

All repair processes and procedures shown relate to Aluminium panels.

Remove Panel

⚠ WARNING: Ensure breathing mask and protective glasses are worn during operations where Aluminium particles are removed.

⚠ CAUTION: Never use a sanding disc that has previously been used on steel, as iron deposits could be left on the surface of the Aluminium.

△ NOTE: In wheel arch areas it may be necessary to soften underbody coating using a hot air gun, prior to exposing spot welds.

1. Expose resistance spot welds. For those spot welds which are not obviously visible, use a rotary drum sander or wire brush fitted to an air drill, or alternatively a hand held wire brush.

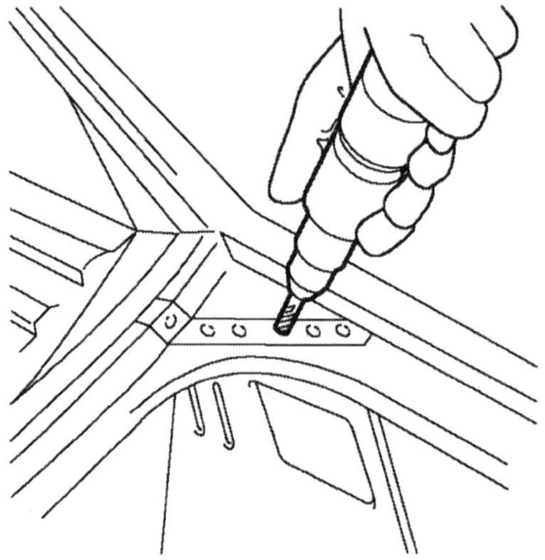

2. Cut out welds using a cobalt drill.

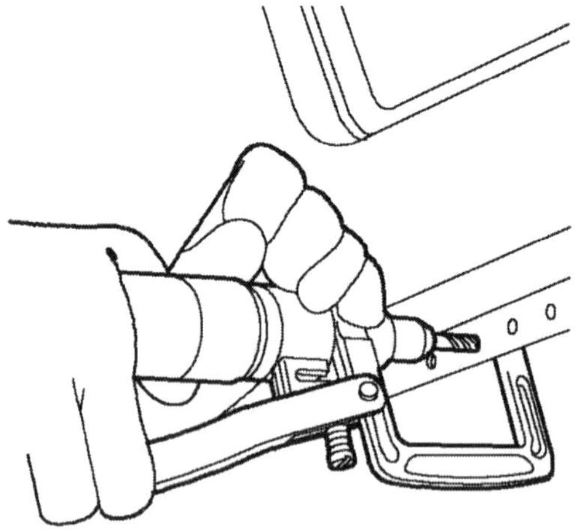

3. A lternatively, use a clamp-tyre spot weld remover.

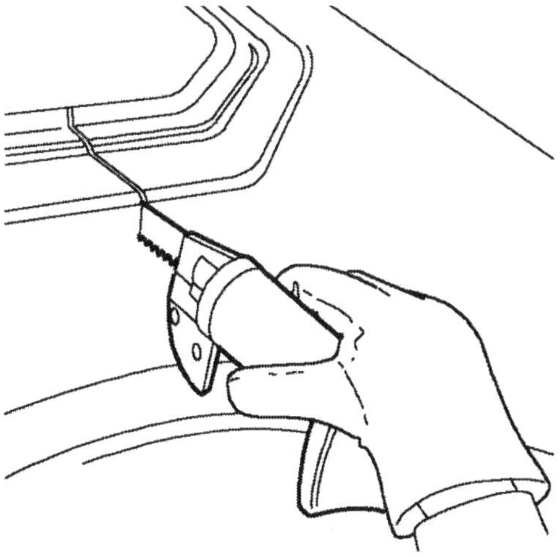

4. Cut away the bulk of the panel as necessary using an air saw.

NOTE: On certain panel joints MIG welds and braze should be removed using a sander where possible, before cutting out the panel bulk.

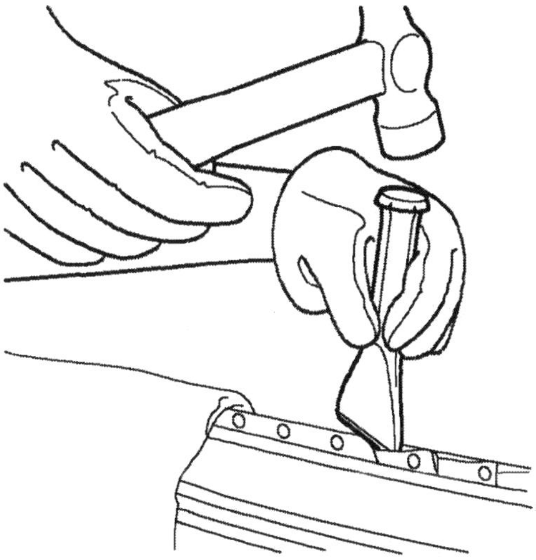

5. Separate the spot welded joints and remove panel remnants using hammer, bolster chisel and pincers.

Prepare old surfaces

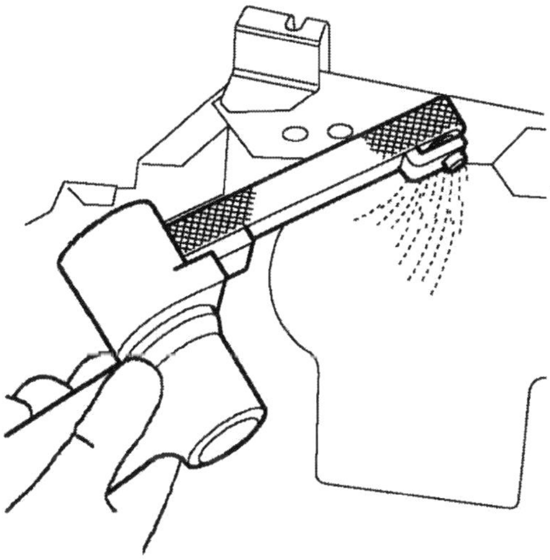

6. Clean all panel joint edges to a bright smooth finish, using a belt type sander.

WARNING: Care must be taken to avoid excessive heat build up which may be caused by this equipment.

NOTE: Prior to sanding, remove remaining sealant using a hot air gun to minimise the risk of toxic fumes caused by generated heat.

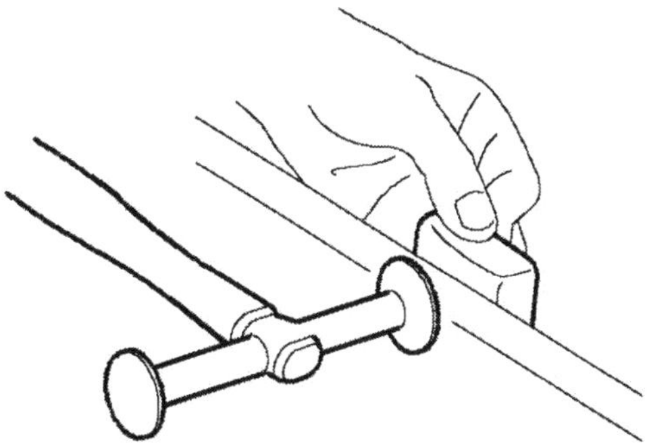

7. Straighten existing joint edges using shaping block and hammer.

Prepare new surfaces

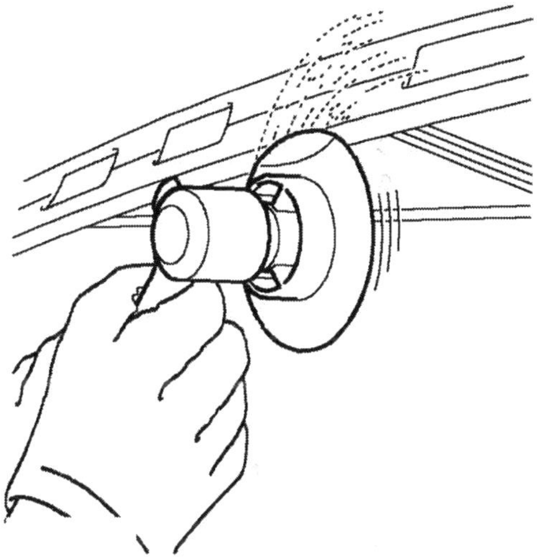

8. Prepare new panel joint edges for welding by sanding to a bright finish,. This must include inner as well as outer face.

9. Drill holes in new panel, in the equivalent spot weld positions.

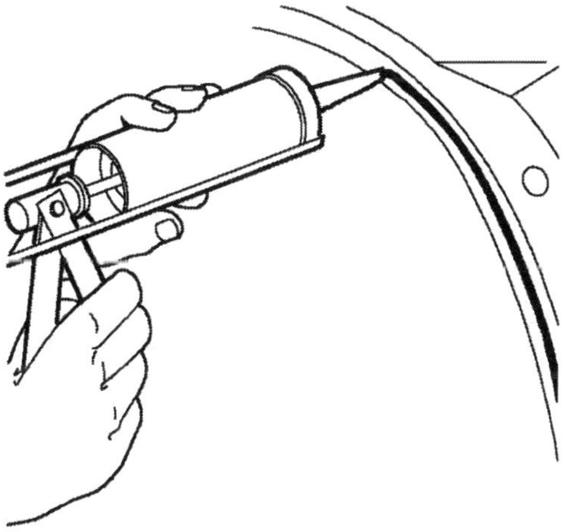

10. Apply adhesive sealant to panel joint surfaces.

CAUTION: Do not use Petroleum Spirit, Alcohol or Paint Thinners to clean mating faces.

11. All mating faces that have sealant applied to them must be cleaned using a suitable solvent. The majority of aluminium sealants have a primer/pre-treatment included with them.

Offer up and align

Offer up new panel and align with associated panels. Clamp into position using welding clamps or Mole grips. Where a joggle or brace joint is being adopted, make a set in the original panel joint edge or insert a brace behind the joint.

NOTE: In cases where access for welding clamps is difficult, it may be necessary to use tack welds.

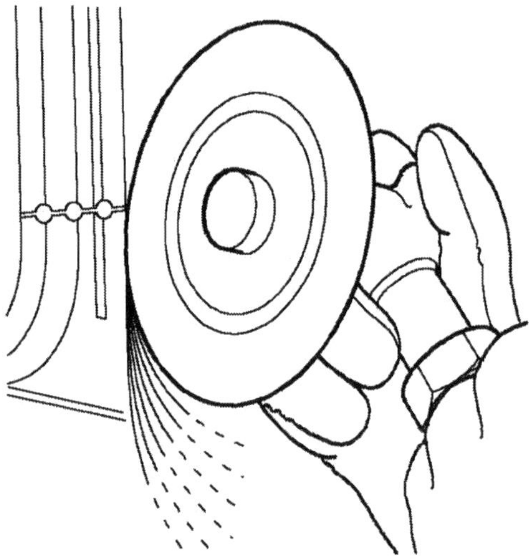

12. Dress MIG tack welds using a sander with 36 grit disc or a belt type sander where access is limited.

Welding

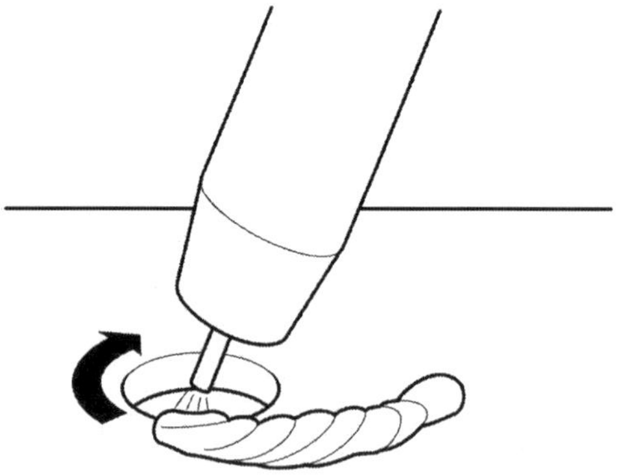

M772123

13. When plug welding, begin weld approximately 15 mm (0.6 in) from hole. This will allow the weld area to be preheated which will enhance weld penetration. It will also allow the operator to see more easily where penetration takes place. Fill hole with weld by moving in a circular direction around the hole.

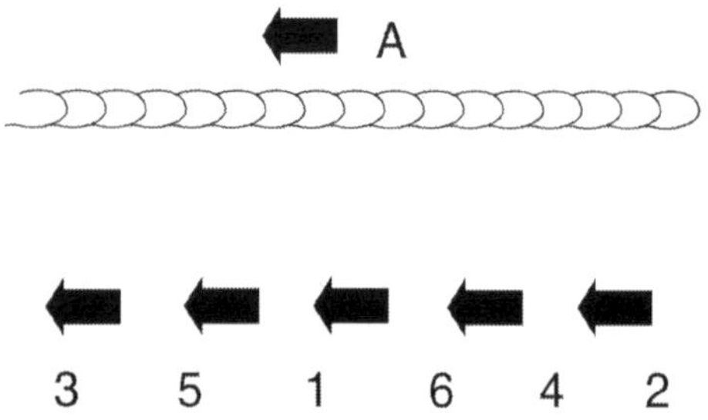

M772124

14. When MIG seam welding, do not carry out lengthy operations. Divide the welds into short operations, this will reduce the chances of overheating the work piece which in turn will avoid distortion and strain.

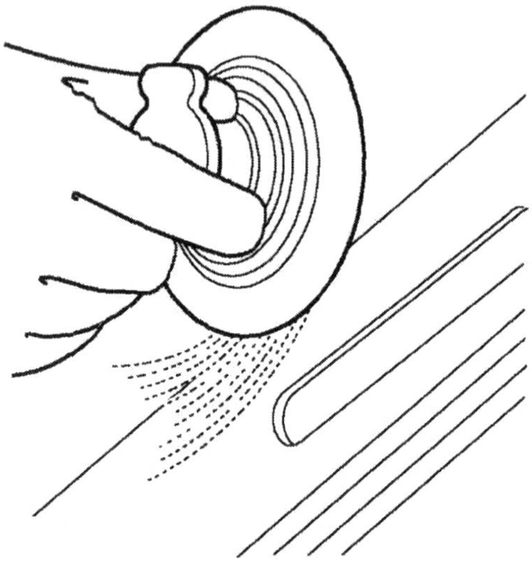

CAUTION: Only use a Stainless Steel wire brush for Aluminium.

15. Dress all welds using a sander with 36 grit disc or a belt type sander and/or wire brush.

Body Trim

The panel repair operations itemise body trim components which must be removed for access during each repair

Because of the unpredictable nature of accident damage, the items listed make no allowance for any difficulties which may be found in removal and only apply to an undamaged vehicle. No allowance is made for any difficulties which may be found during panel removal. Damaged body trim items must be renewed as necessary following body repairs.

SERVICEABLE PANELS

Front end panels

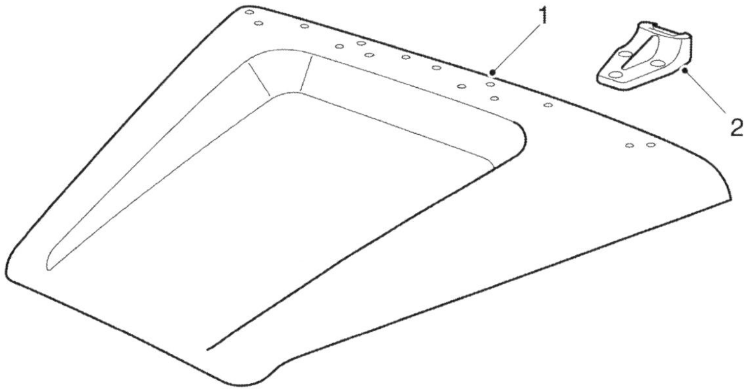

M772097A

Item	Description
1.	Hood
2.	Hinges

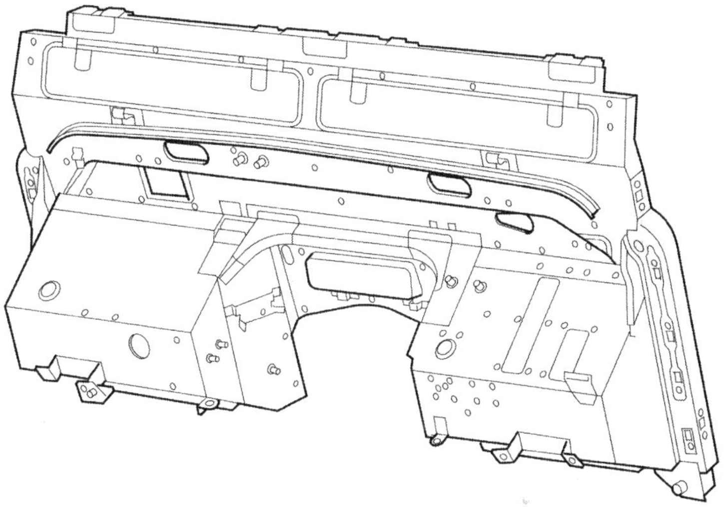

Item	Description
3.	Bulkhead

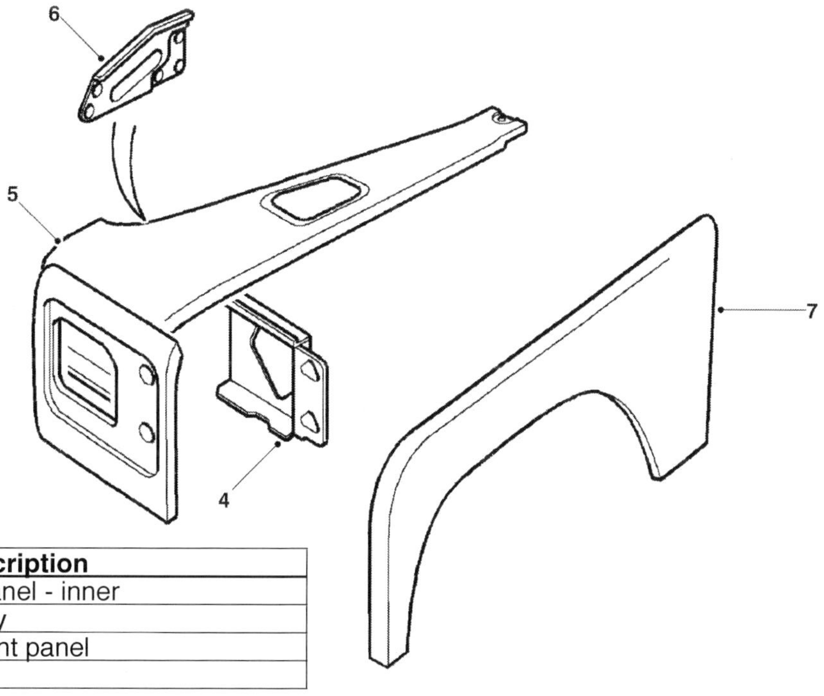

Item	Description
4.	Headlamp mounting panel - inner
5.	Fender upper assembly
6.	Headlamp reinforcement panel
7.	Fender lower

Body side panels

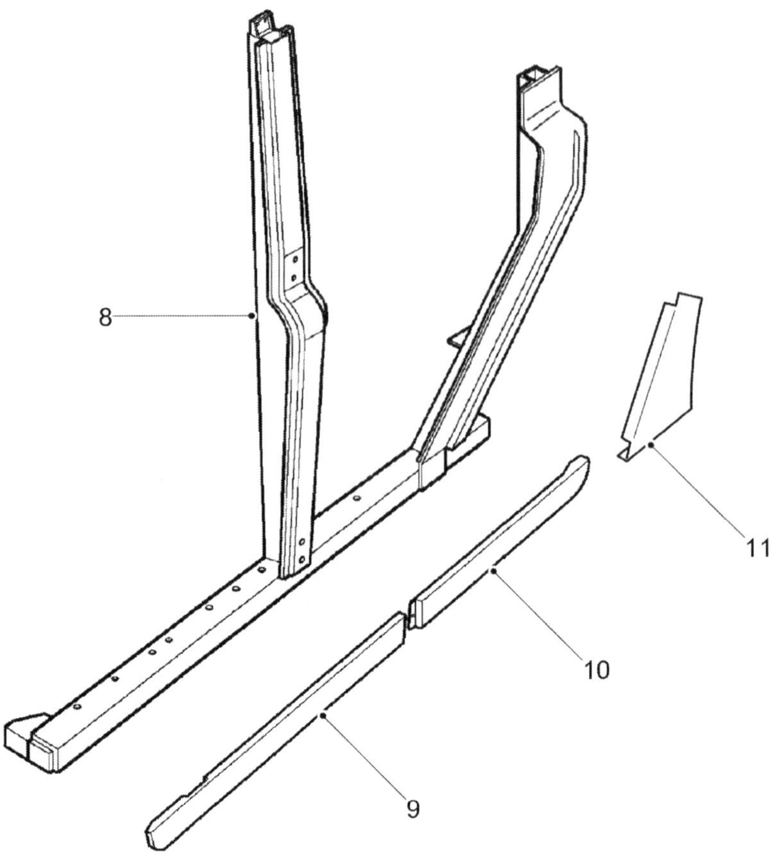

M772110B

Item	Description
8.	'B/C' and 'D' post assembly
9.	Rocker panel - front
10.	Rocker panel - rear
11.	Dogleg - lower panel

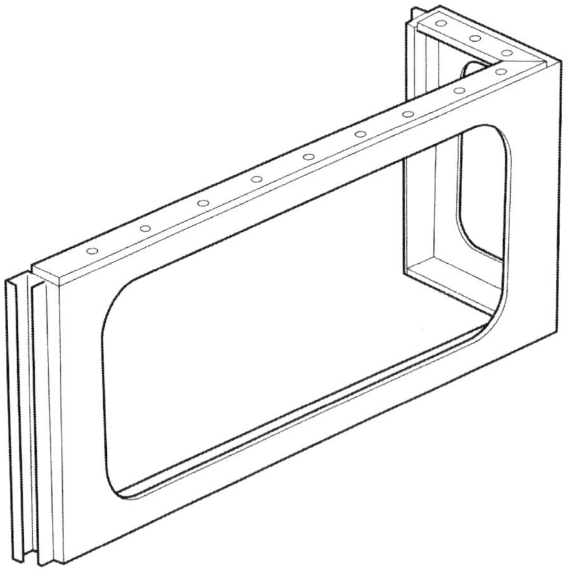

M772104A

Item	Description
12.	Body side rear - upper

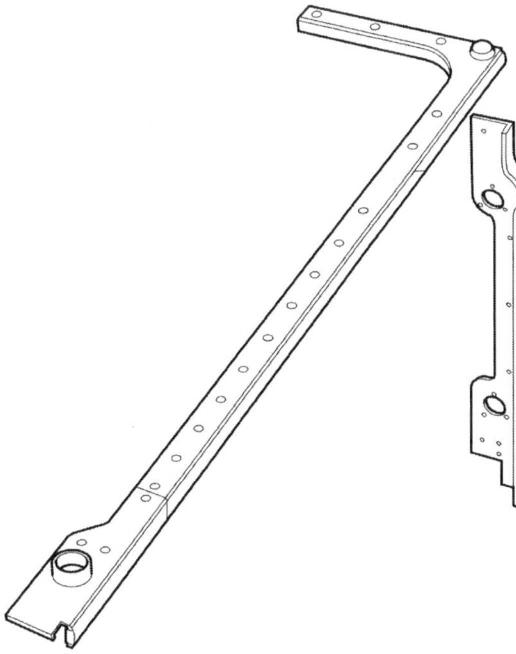

M772100A

Item	Description
13.	Body side cappings

Rear end panels

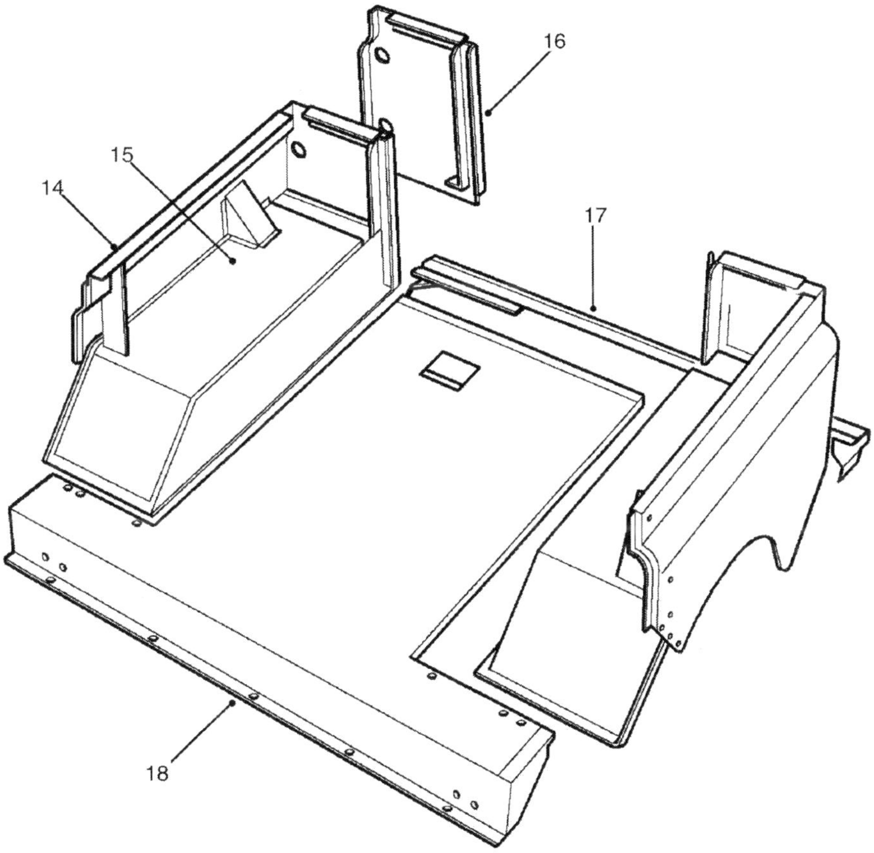

M772094B

Item	Description
14.	Body side panel - lower
15.	Body side panel - lower - assembly
16.	Rear lamp panel
17.	Rear panel
18.	Rear floor

NOTE: From 2002 Model year, the taildoor skin is one panel.

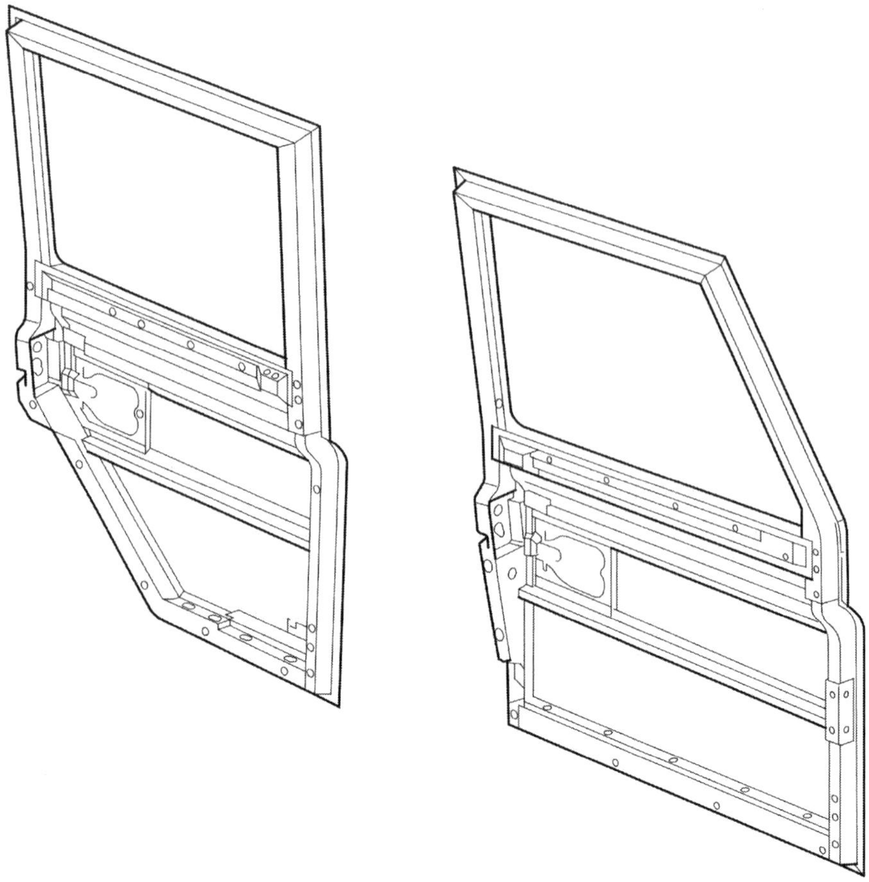

M772077A

Item	Description
19.	Door assemblies

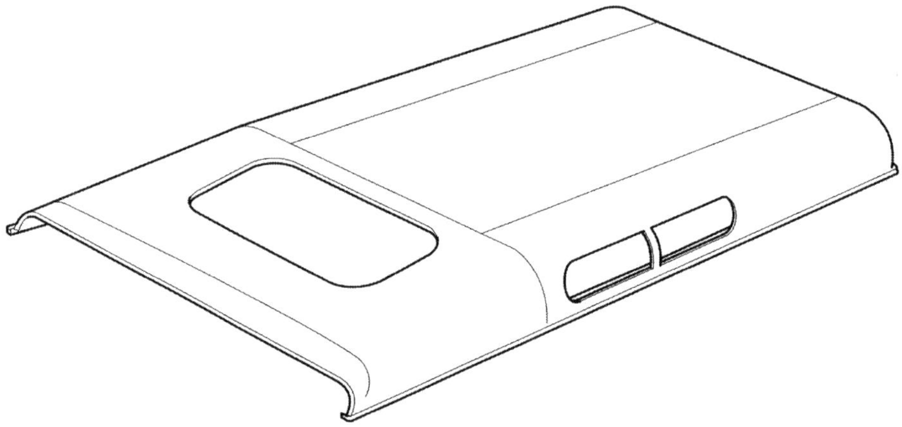

Item	Description
20.	Roof

Adhesives, Fixings and Consumables

Description and Operation

Adhesives, Fixings and Consumables

The information provided within this document is the latest up to date information.

⚠ NOTE: Make sure when referencing existing panel replacement procedures that the data, provided within this document, is also referenced.

Through extensive research, JLR (Jaguar Land Rover) has identified the most appropriate method to rectify aftermarket damage. It is key to safety and vehicle integrity that only the approved methods, materials and equipment are used. To successfully complete structural repairs to JLR products, it is necessary for the repairs to follow the correct aftermarket process for adhesive (bonding), riveting and welding to make sure vehicle integrity is maintained after a collision.

Bonding

This section provides information on the bonded joints used in repair. The following topics are covered:

- Equipment.
- Surface preparation.
- Bonding Application.

It is a pre-requisite that any person undertaking any repairs, which involve panel bonding, has attended JLR Approved Training and has achieved the correct skill level to undertake these processes.

3M 08115

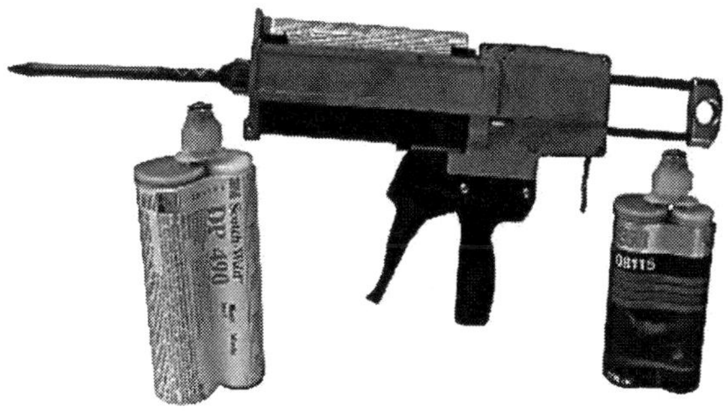

E74991

Equipment

The approved bonding equipment consists of:

- Approved two-pack adhesive applicator.

Surface Preparation

Pre-treating (coupling agent) the panels to be bonded, as described in this section, promotes the improved adhesion of the bonding adhesive.

⚠ CAUTION: Bonding without pre-treatment will reduce the performance of the joint.

- Remove any original adhesive, E-Coat primer (new panels) and paint from the areas to be bonded or welded using a fine bristle disc.
- Wash the surface with soap and water to remove water-soluble contaminants. Follow the soap and water wash with an appropriate VOC compliant product for removal of surface contaminants.
- Straighten all metal, and “dry-fit” the parts.
- Clamp the part in place and check for fit and alignment.
- Remove the panel from the vehicle. All areas to be welded should be coated with JLR approved 'weld-thru' coating. **Adhesive should not be applied to these areas.**

Bonding Application

The approved bonding adhesive is supplied in a two-pack cartridge.

A two-pack cartridge applicator with a disposable mixing nozzle is used for the application of the adhesive.

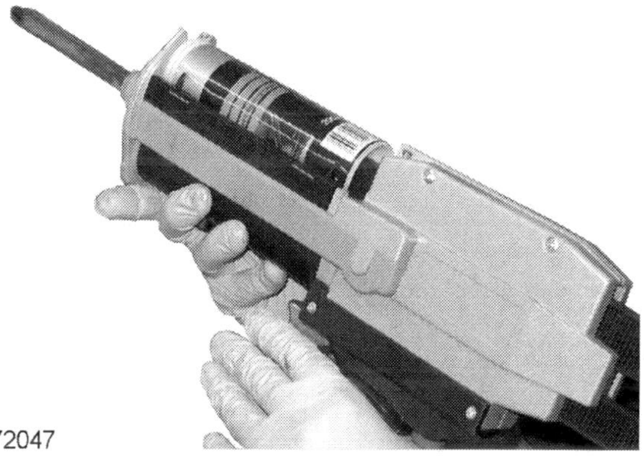

E72047

- Apply an adhesive bead to the prepared surfaces of both pieces to be bonded.
- Using a plastic spreader or acid brush, tool out the adhesive to cover all bare surfaces.
- Apply a 5 mm "zig zag" bead directly to either the replacement panel or vehicle structure.
- Make sure a continuous bead of adhesive surrounds fixing holes.

NOTES:

⚠ On completion, the mixing nozzle must be discarded. Make sure an air tight seal is provided for the cartridges.

⚠ The following technical information and data should be considered representative or typical only.

The following times have been determined with ambient air temperature and substrate temperature @ 70°F (21°C) and are considered typical values.

- **Work Time**-90 Minutes
- **Clamp Time**-4 Hours
- **Cure Time**-24 Hours

NOTE: Panel Bonding Adhesive is a two-part epoxy adhesive which provides extended work-time but can be rapidly cured with heat.

3M Impact Resistant Structural Adhesive - 07333

NOTE: Coupling agent pre-treatment is not required with 07333.

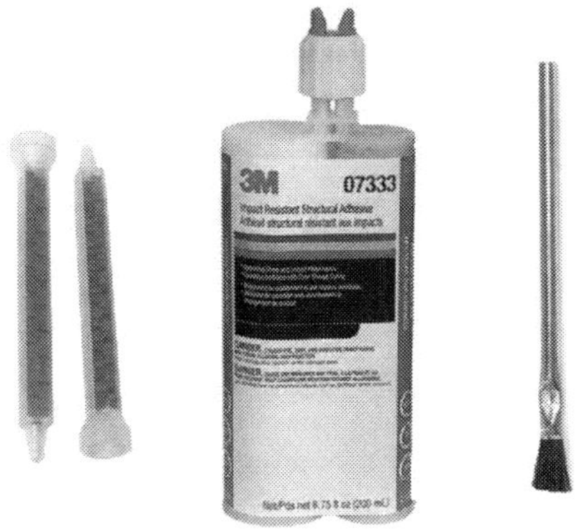

E219632

Surface Preparation

- Remove any original adhesive, or E-Coat from new panels.
- Clean the surface to be treated using approved pre-cleaner/degreaser.

NOTE: No pyrosil required with this adhesive.

Bonding application

NOTE: Set fasteners while the adhesive is still uncured, typically within 2 hours of adhesive application.

- Apply an adhesive bead to the prepared surfaces of both pieces to be bonded.
- Using an acid brush, tool out the adhesive to cover all bare surfaces.
- Apply a 5 mm "zig zag" bead directly to either the replacement panel or vehicle structure.
- Apply a large enough bead to allow the adhesive to fill all voids and squeeze out of the flange seam, indicating that the joint is completely sealed.
- Make sure a continuous bead of adhesive surrounds fixing holes.

Fixings

There are four groups of fixings used, other than conventional nuts and bolts. It is important that the correct procedures are followed for the removal and replacement of these fixings and, where applicable, the correct tools are used.

- Self-Piercing Rivet (SPR)
- Breakstem Fastener (BSF)
- Friction Drilling Screw (FDS)
- Earth Stud

Self-Piercing Rivet (SPR)

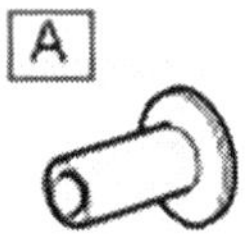

E74390

Self-Piercing Rivets (SPRs) are available in various sizes dependent on the size of the stack of panels to which they are installed. A stack refers to the combined gauge of the panels being riveted.

CAUTION: Operating the Self Piercing Rivet (SPR) installation tool with no SPR or material in position will damage the tool Die.

Removal

Align the plunger and anvil over the fixing and press the trigger. Pressing the trigger will allow the plunger to act on the fixing and remove it from the panel.

Where tool access is impossible, drill out the remnant using a 5.3mm Cryobit drill bit. Remove all debris.

Self-piercing removal tools can be used with an optional, larger C frame for improved access. Installation of the new rivet should be located close to the original rivet location and not in the centre to preserve land for further repairs. The same process is followed for any subsequent replacements. Where access for the tool is restricted and cannot be used, or there is insufficient space on the panel, the fixings should be replaced with Breakstem Fasteners, unless the repair method dictates otherwise.

Installation

NOTES:

Prior to replacing a self-piercing fixing, a test should be performed using identical panel materials. Make sure that all settings are correct and an acceptable joint has been achieved.

Make sure the JLR approved SPRs, die type and load are used during installation.

Load new fixing into the JLR approved tool. Align the tool over the fixing position and press the trigger. As the trigger is pressed, the tool will clamp itself onto the work piece and install the fixing.

Jaguar Land Rover Approved SPR Tools.

Advel ESN50

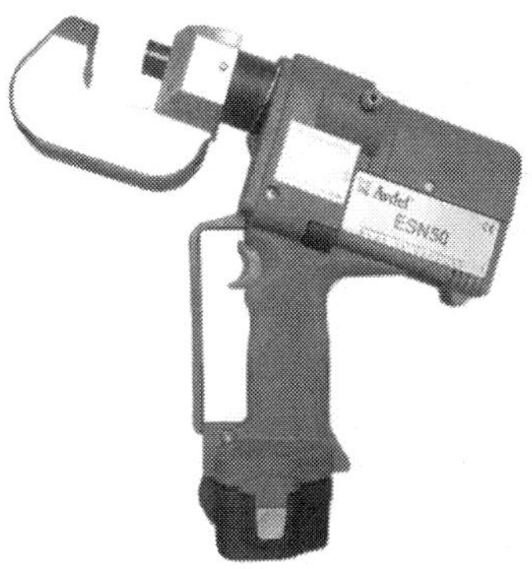

E219633

TKR PNP90 XT2.

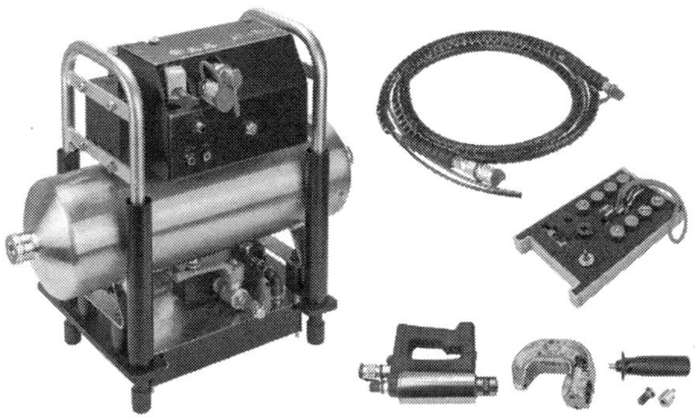

NOTES:

⚠ The TKR PNP90 XT2 **with** the docking station is the JLR approved tool.

⚠ The TKR PNP90 XT2 **without** the docking station is NOT approved.

Pip die

Flat bottom die

Shallow die

Service Repair Fasteners.

NOTES:

⚠ It is important to identify and select the correct fixing by referencing this table prior to installation.

⚠ Make sure the JLR approved Self Piercing Rivets (SPRs), die type and load are used during installation.

⚠ Where original rivet holes cannot be used, replacement rivets should be positioned close to the original rivet location/hole location.

Installation of the new rivet should be located close to the original rivet location and not in the centre to preserve land for further repairs. The same process is followed for any subsequent replacements.

Breakstem Fasteners (BSF)

Breakstem Fasteners are used in repair if the joint/stack cannot be accessed from both sides or require a greater load.

- Breakstem fasteners will cope with joint/stack sizes of 1.5 mm to 12 mm.
- A pre-drilled 6.5 mm hole will be required before installation.

- To identify the correct size and part number of structural breakstem fasteners to be used in repair always refer to TOPIx.

Breakstem Fastener (BSF)

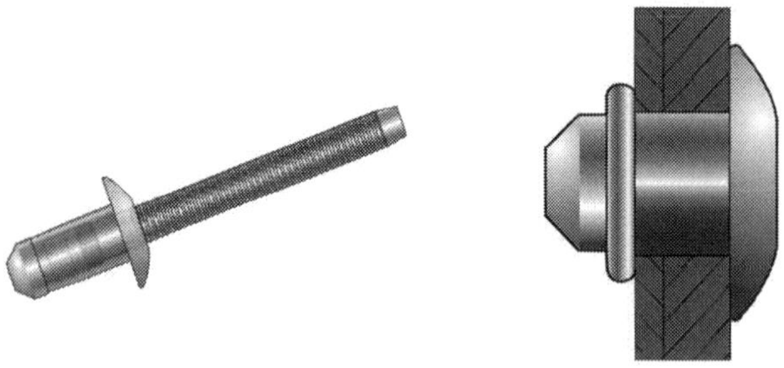

Breakstem Fasteners have the following benefits:

- Exceptional shear and tensile strength
- Large blind side bearing area against the rear sheet
- Interference lock formed by a splined stem
- Spreads the tail bearing/clamp load on the rear sheet
- Provides strong, vibration resistant joints
- No loose stems

When the repair procedure requires the installation of a breakstem fastener the stack size is identified within the repair procedures (millimeters).

Removal

Breakstem Fasteners (BSFs) used in production are removed by removing the centre of the fixing using a 4 mm punch. When the centre has been removed, drill out the remnant using a 6.5 mm Cryobit drill bit. Remove all debris.

△ NOTE: Aftermarket Breakstem Fasteners (BSFs) are removed in the same way as BSFs used in production.

Installation

Make sure a JLR approved installation tool is used during the repair process. Prior to replacement, make sure the fixing hole has been drilled to 6.5mm, using a Cryobit drill bit and all debris has been removed.

Check that you have the correct size Breakstem Fastener (BSF) for replacement and insert it into the JLR approved tool. Use the weight of the tool to apply light pressure as the BSF is inserted into its hole. Press the trigger, which will tighten the BSF in the hole, when a pre-determined pressure is reached the wasted stud will be ejected into the tool.

Genesis G4

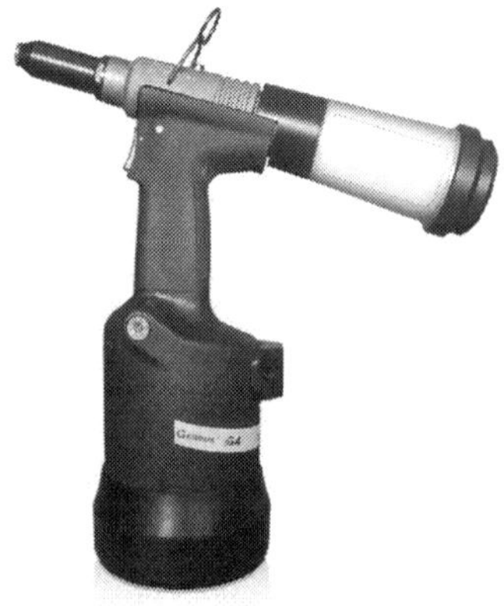

Taurus 4

ProSet XT4

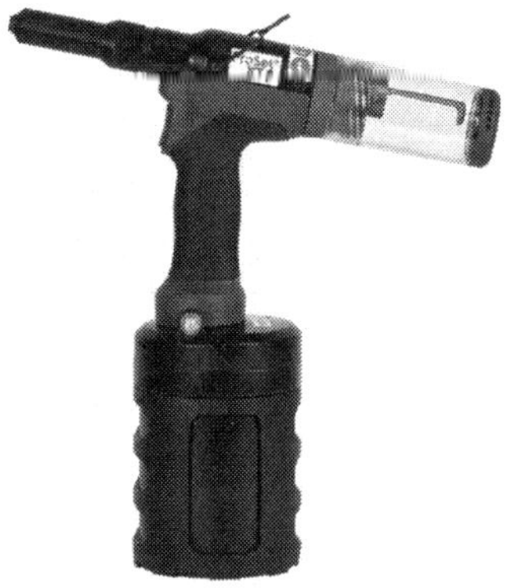

Breakstem Fastener Selection

When the repair procedure requires the installation of a Breakstem Fastener (BSF), the stack code is identified within the repair procedure.

⚠ CAUTION: Always refer to the workshop manual procedure for Breakstem Fastener (BSF) values when completing repairs.

In line with the JLR continuous improvement program, repair procedures will contain the latest fixing data available upon publication.

Documents containing the Breakstem Fastener (BSF) **grip range** will require the BSF selection from the table below.

-	-	-	Self Piercing Rivets	-	-	-
SPR Type No. Jaguar	Jaguar Part Number	SPR Type No. Land Rover	Land Rover Part Number	Rivet Length	Rivet Application Force	Die Description
J1	C2C9705	L1	LRO43550	5 mm	20 kN	Pip
J1A	C2C9705	L1A	LRO43550	5 mm	30 kN	Pip
J1B	C2C9705	L1B	LRO43550	5 mm	30 kN	Flat Bottom
J1C	C2C9705	L1C	LRO43550	5 mm	20 kN	Flat Bottom
J1D	C2C9705	L1D	LRO43550	5 mm	25 kN	Flat Bottom
J1E	C2C9705	L1E	LRO43550	5 mm	25 kN	Pip
J1F	C2C9705	L1F	LRO43550	5 mm	35 kN	Flat Bottom
J1G	C2C9705	L1G	LRO43550	5 mm	35 kN	Pip
J1Hs	C2C9705	L1Hs	LRO43550	5 mm	45 kN	Shallow
J1Is	C2C9705	N/A	N/A	5 mm	40 kN	Shallow
J2	C2C20589	L2	LRO43556	5 mm	30 kN	Pip
J2A	C2C20589	L2A	LRO43556	5 mm	30 kN	Flat Bottom
J2B	C2C20589	L2B	LRO43556	5 mm	25 kN	Flat Bottom
J2C	C2C20589	L2C	LRO43556	5 mm	40 kN	Pip
J2D	C2C20589	L2D	LRO43556	5 mm	35 kN	Flat Bottom
J2E	C2C20589	L2E	LRO43556	5 mm	45 kN	Flat Bottom
J2F	C2C20589	L2F	LRO43556	5 mm	40 kN	Flat Bottom
J2G	C2C20589	L2G	LRO43556	5 mm	45 kN	Pip
J2H	C2C20589	N/A	N/A	5 mm	35 kN	Pip
J3	C2C9706	L3	LRO43551	6 mm	30 kN	Flat Bottom
J3A	C2C9706	L3A	LRO43551	6 mm	25 kN	Flat Bottom
J3B	C2C9706	L3B	LRO43551	6 mm	35 kN	Flat Bottom
J3C	C2C9706	L3C	LRO43551	6 mm	40 kN	Flat Bottom
J3D	C2C9706	L3D	LRO43551	6 mm	35 kN	Pip
J3E	C2C9706	L3E	LRO43551	6 mm	30 kN	Pip
J3F	C2C9706	L3F	LRO43551	6 mm	45 kN	Pip
J3G	C2C9706	L3G	LRO43551	6 mm	40 kN	Pip
J4	C2C9707	L4	LRO43552	6.5 mm	30 kN	Pip
J4A	C2C9707	L4A	LRO43552	6.5 mm	30 kN	Flat Bottom
J4C	C2C9707	L4C	LRO43552	6.5 mm	35 kN	Flat Bottom
J4D	C2C9707	L4D	LRO43552	6.5 mm	35 kN	Pip
J4E	C2C9707	N/A	N/A	6.5 mm	40 kN	Pip
J5	C2C9711	L5	LRO43553	7 mm	50 kN	Flat Bottom
J5B	C2C9711	L5B	LRO43553	7 mm	40 kN	Flat Bottom
J5C	C2C9711	L5C	LRO43553	7 mm	30 kN	Flat Bottom
J5D	C2C9711	L5D	LRO43553	7 mm	30 kN	Pip
J5E	C2C9711	L5E	LRO43553	7 mm	35 kN	Flat Bottom
J5F	C2C9711	L5F	LRO43553	7 mm	45 kN	Flat Bottom
J5G	C2C9711	L5G	LRO43553	7 mm	40 kN	Pip
J5H	C2C9711	L5H	LRO43553	7 mm	45 kN	Pip
J6	C2C20591	L6	LRO43557	8 mm	50 kN	Pip
J6A	C2C20591	L6A	LRO43557	8 mm	40 kN	Pip
J6B	C2C20591	L6B	LRO43557	8 mm	40 kN	Flat Bottom
J6C	C2C20591	L6C	LRO43557	8 mm	50 kN	Flat Bottom
J6Ds	C2C20591	L6Ds	LRO43557	8 mm	45 kN	Shallow
J6E	C2C20591	L6E	LRO43557	8 mm	45 kN	Flat Bottom
J7	C2C9712	L7	LRO43554	8 mm	40 kN	Flat Bottom
J7A	C2C9712	L7A	LRO43554	8 mm	50 kN	Flat Bottom
J7B	C2C9712	L7B	LRO43554	8 mm	45 kN	Flat Bottom
J7C	C2C9712	L7C	LRO43554	8 mm	35 kN	Flat Bottom
J7D	C2C9712	L7D	LRO43554	8 mm	45 kN	Pip
J7E	C2C9712	L7E	LRO43554	8 mm	50 kN	Pip
J8	C2C9713	L8	LRO43555	9 mm	40 kN	Flat Bottom
J8A	C2C9713	L8A	LRO43555	9 mm	45 kN	Flat Bottom
J8B	C2C9713	L8B	LRO43555	9 mm	50 kN	Flat Bottom
J9	C2C9713	L9	LRO43558	9 mm	40 kN	Flat Bottom
J9A	C2C9713	L9A	LRO43558	9 mm	45 kN	Flat Bottom
J9B	C2C9713	L9B	LRO43558	9 mm	34 kN	Flat Bottom
J9C	C2C9713	L9C	LRO43558	9 mm	50 kN	Flat Bottom
J10	C2C20590	L10	LRO99821	7 mm	40 kN	Flat Bottom
J10A	C2C20590	L10A	LRO99821	7 mm	45 kN	Flat Bottom
J10B	C2C20590	L10B	LRO99821	7 mm	35 kN	Flat Bottom
J10C	C2C20590	L10C	LRO99821	7 mm	50 kN	Pip
J10D	C2C20590	L10D	LRO99821	7 mm	45 kN	Pip
J10E	C2C20590	L10E	LRO99821	7 mm	40 kN	Shallow
J10F	C2C20590	L10F	LRO99821	7 mm	50 kN	Flat Bottom
J11	C2P7761	L11	LRO43560	8 mm	35 kN	Flat Bottom
J11A	C2P7761	L11A	LRO43560	8 mm	35 kN	Pip

J11B	C2P7761	L11B	LRO43560	8 mm	40 kN	Pip
J11C	C2P7761	L11C	LRO43560	8 mm	45 kN	Flat Bottom
J11D	C2P7761	L11D	LRO43560	8 mm	40 kN	Flat Bottom
J11E	C2P7761	L11E	LRO43560	8 mm	50 kN	Flat Bottom
J11F	C2P7761	L11F	LRO43560	8 mm	50 kN	Flat Bottom
J12	C2C20592	L12	LRO99822	8.5 mm	50 kN	Flat Bottom
J12A	C2C20592	L12A	LRO99822	8.5 mm	45 kN	Flat Bottom
J12B	C2C20592	L12B	LRO99822	8.5 mm	40 kN	Flat Bottom
J12Cs	C2C20592	L12Cs	LRO99822	8.5 mm	50 kN	Shallow
J12D	C2C20592	L12D	LRO99822	8.5 mm	50 kN	Pip
J13	C2C22612	L13	LRO43559	7 mm	45 kN	Flat Bottom
J13A	C2C22612	L13A	LRO43559	7 mm	50 kN	Flat Bottom
J13B	C2C22612	L13B	LRO43559	7 mm	45 kN	Pip
J13C	C2C22612	L13C	LRO43559	7 mm	40 kN	Flat Bottom
J14	T4N10924	L14	LRO99827	7.5 mm	35 kN	Flat Bottom
J14A	T4N10924	L14A	LRO99827	7.5 mm	50 kN	Flat Bottom
J14B	T4N10924	N/A	N/A	7.5mm	45 kN	Flat Bottom
J14Cs	T4N10924	N/A	N/A	7.5 mm	35 kN	Shallow
J14D	T4N10924	N/A	N/A	7.5 mm	35 kN	Pip
J14Es	T4N10924	N/A	N/A	7.5 mm	40 kN	Shallow
J15	T4N13556	L15	LRO99824	8 mm	40 kN	Pip
J15As	T4N13556	L15As	LRO99824	8 mm	45 kN	Shallow
J15B	T4N13556	L15B	LRO99824	8 mm	45 kN	Flat Bottom
J15C	T4N13556	L15C	LRO99824	8 mm	40 kN	Flat Bottom
J15Ds	T4N13556	L15Ds	LRO99824	8 mm	40 kN	Shallow
J15E	T4N13556	N/A	N/A	8 mm	45 kN	Pip
J15Fs	T4N13556	N/A	N/A	8 mm	50 kN	Shallow
J16	T2H15804	L16	LRO99825	9 mm	45 kN	Flat Bottom
J16As	T2H15804	L16As	LRO99825	9 mm	50 kN	Shallow
J16C	T2H15804	L16C	LRO99825	9 mm	50 kN	Pip
J16D	T2H15804	L16D	LRO99825	9 mm	50 kN	Flat Bottom
J16Es	T2H15804	L16Es	LRO99825	9 mm	40 kN	Shallow
J16Fs	T2H15804	L16Fs	LRO99825	9 mm	45 kN	Shallow
J17	C2C20594	L17	LRO99823	9.5 mm	45 kN	Pip
J17A	C2C20594	L17A	LRO99823	9.5 mm	45 kN	Flat Bottom
J17B	C2C20594	L17B	LRO99823	9.5 mm	50 kN	Flat Bottom
J17C	C2C20594	L17C	LRO99823	9.5 mm	40 kN	Flat Bottom
J18	T4N11943	L18	LRO99823	10 mm	50 kN	Flat Bottom
J18As	T4N11943	N/A	N/A	10 mm	50 kN	Shallow
J18Bs	T4N11943	N/A	N/A	10 mm	60 kN	Shallow
J19	N/A	L19	LRO99830	8.5 mm	35 kN	Flat Bottom
J19A	N/A	L19A	LRO99830	8.5 mm	40 kN	Flat Bottom
J19B	T2H14244	L19B	LRO99830	8.5 mm	45 kN	Flat Bottom
J19C	N/A	L19C	LRO99830	8.5 mm	50 kN	Flat bottom
J19Cs	T2H14244	L19Cs	LRO99830	8.5 mm	50 kN	Shallow
J19Ds	T2H14244	N/A	N/A	8.5 mm	45 kN	Shallow
J19Es	T2H14244	N/A	N/A	8.5 mm	40 kN	Shallow
J20s	T4N10928	L20s	LRO99828	7.5 mm	45 kN	Shallow
J20As	T4N10928	L20As	LRO99828	7.5 mm	50 kN	Shallow
J20Bs	T4N10928	L20Bs	LRO99828	7.5 mm	40 kN	Shallow
J20C	T4N10928	N/A	N/A	7.5 mm	50 kN	Pip
J21s	T4N10929	L21s	LRO99826	9.5 mm	50 kN	Shallow
J22	N/A	L22	LRO99829	9 mm	50 kN	Flat Bottom
J22A	N/A	L22A	LRO99829	9 mm	40 kN	Flat Bottom
J22B	N/A	L22B	LRO99829	9 mm	45 kN	Pip
J22C	T4N10927	L22C	LRO99829	9 mm	45 kN	Flat Bottom
J22Ds	T4N10927	N/A	N/A	9 mm	45 kN	Shallow
J23	T4N10925	L23	LR105719	10.5 mm	50 kN	Flat Bottom
J24S	C2P24963	N/A	N/A		45 kN	Shallow

- **For example:**
- BSF 3.5 = grip range 2.8mm to 4.8mm. **Selection should be towards the center of the grip range.**
- BSF 3.5 would require either BSF B or BSF L (Land Rover) or BSF 2 or BSF 11 (Jaguar).

△ NOTE: Make sure the Breakstem Fastener (BSF) type equivalent is used, if the required selection is unavailable.

Land Rover - Breakstem Fastener Type	Part Number	Jaguar - Breakstem Fastener Type	Part Number	Grip Range	Dimensions	TYPE
BSF A	LR044284	BSF 1	C2S45252	1.5 mm to 3.5 mm	6.5 x 13 mm	Clinch
BSF B	LR044283	BSF 2	C2C36006	2.8 mm to 4.8 mm	6.5 x 15 mm	Clinch
BSF C	LRO44286	BSF 3	C2C36026	4.8 mm to 6.8 mm	6.5 X 17 mm	Clinch
BSF D	LRO43912	BSF 4	C2C36024	6.8 mm to 8.8 mm	6.5 X 19 mm	Clinch
BSF E	LRO44287	BSF 5	T2H15812	7.5 mm to 9.5 mm	6.5 X 20 mm	Clinch
BSF F	LRO44285	BSF 6	C2C36008	8.8 mm to 10.8 mm	6.5 X 18 mm	Clinch
BSF G	LRO44288	BSF 8	T2H15813	3.2 mm to 12.0 mm	6.5 X 27 mm	Countersunk - clinch
BSF H	LRO43932	BSF 7	C2P4773	2.0 mm to 9.5 mm	6.5 X 24 mm	Large head tubular
BSF J	LRO94293	BSF 9	T4N10217	2.0 mm to 9.5 mm	6.5 X 24 mm	Large head tubular
BSF K	N/A	BSF 10	T4K7061	1.5 mm to 3.5 mm	6.5 x 13 mm	Clinch
BSF L	N/A	BSF 11	T4K7059	2.8 mm to 4.8 mm	6.5 x 15 mm	Clinch
BSF M	N/A	BSF 12	T4N9856	4.5 mm to 7.0 mm	6.5 X 17 mm	Clinch
BSF N	N/A	BSF 13	T4K7060	6.8 mm to 8.8 mm	6.5 X 19 mm	Clinch
BSF O	N/A	BSF 14	T4K7063	8.8 mm to 10.8 mm	6.5 X 18 mm	Clinch

Friction Drilling Screws (FDS)

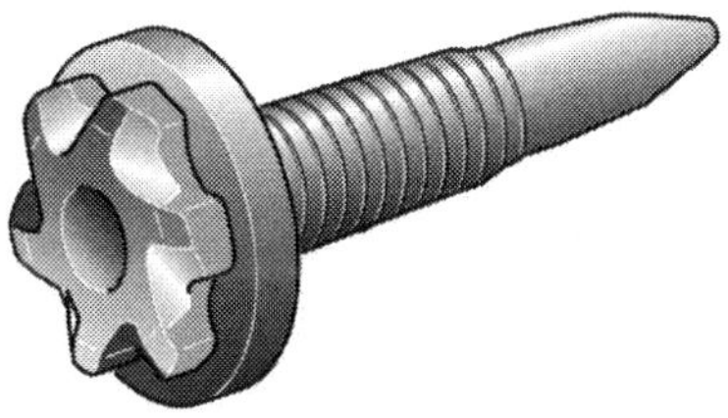

⚠ NOTE: When M5 Friction Drilling Screws (FDS) are removed in service, they must be replaced with a new larger M6 FDS.

This process is also called thermal drilling, flow drilling, form drilling, or friction stir drilling.

This joining method enables high-quality assembly of thin sheet metals made of steel and aluminum. The fixing is pressed against a target material with both high rotational speed and high pressure. This process produces a high local production of heat, which softens the object. The excess material forms a sleeve, which makes it possible to create very strong threaded joint in thin material.

- The main benefits of FDS is one-sided accessibility for application and screw removability.
- The screw joint is able to transfer high pull-out forces and high shearing strength.

Removal

- Screw out the M5 FDS using a battery operated tool or hand tools.
- Clean the area and remove any debris using industrial vacuum cleaner.

Installation

- Position the new M6 FDS in the original position.
- Screw in the FDS until the required torque is reached.

⚠ NOTE: Install the Friction Drilling Screw (FDS) screw using 2-4 Nm of torque until flush with the panel, then apply additional 4 Nm of torque (8-10 Nm total). Overall torque should not exceed 11 Nm.

Description	Jaguar Land Rover Service No.
M6 Friction Drilling Screw (FDS)	T4K7064

Torx Screw and Rivnut

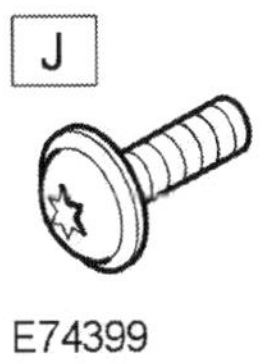

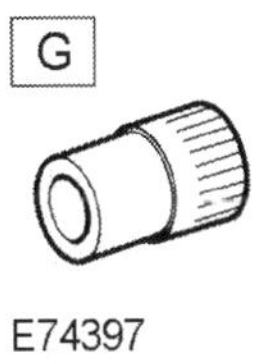

A Torx Screw and Rivnut are used where the installation of a Hemlok would give a poor cosmetic appearance. There are also occasions where the Torx Screw and Rivnut replaces a Breakstem Fastener (BSF), where there is no access for the Genesis G4. The Rivnut acts as the thread.

Description	Jaguar Service Number
Rivnut (G)	C2C 10348
Torx Screw (J)	C2G 1964

Removal

The removal of the Torx Screw is completed using a T30 Torx Driver. The Rivnut is not an original fixing.

Installation

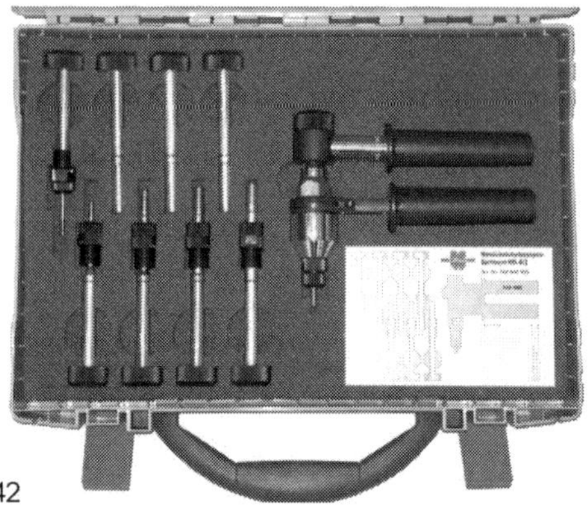

E72042

The Torx Screw is installed using a T30 Torx Driver. The Rivnut is installed using the Wurth HES412 Rivet Nut Thread Setter, (Part No: 964948900).

Earth Stud

Secure and reliable earthing is imperative for trouble free function of electronic components and management systems. A solid permanent joint is needed for long term stability under load, even with high short term amperage values and corresponding heat generation.

The earth bond comprises two components:

- A conical dowel having a male thread.
- A cylindrical flanged bush.

A tensile force is applied to the dowel using a hand hydraulic tool with a pre-determined pulling force. As the dowel is pulled through the bush, the bush remains fixed in the structure and expands within the hole making electrical contact. The tool completes its cycle resulting in a permanent electrical connection.

Removal/Preparation

Damaged or broken studs:

- Locate the broken or cross-threaded stud.
- Dress the broken stud to a safe and acceptable repair standard.
- Complete drilling in close proximity to the broken /cross-threaded stud.

Installation

- Measure and mark stud location.
- Using a stepped cutter, drill a hole in the mounting plate to a predetermined diameter depending on the material and the stack thickness.
- Screw the earth bond stud into the nose of the applicator tool. The flange of the stud should be flush with the panel surface.
- Place the new stud into the drilled hole. Apply pressure to the trigger of the applicator tool until audible click is heard. This will indicate the applicator tool has reached the end of its cycle.

Glenair earth bonding system is the approved equipment to repair the earth studs on JLR vehicles.

Earth Bond Selection

Glenair Part Number	Thread Size	Hole Size	Plate/Material Thickness	Torque Value	Tool Setting
80958	M6	11.5 mm to 11.7 mm	1.5mm to 4mm	8 Nm	PMT6/ PMTC6
80960	M6	10.5 mm to 10.7 mm	> 4mm	8 Nm	PMT6/ PMTC6
80976	M6	11.5 mm to 11.7 mm	1.5mm to 4 mm	8 Nm	PMTC6
80977	M6	10.5 mm to 10.7 mm	> 4mm	8 Nm	PMTC6
81931	M8	15 mm to 15.2 mm	1.5mm to 4 mm	15 Nm	PMT8
81930	M8	13.5 mm to 13.7 mm	> 4mm	15 Nm	PMT8
80959	M10	19 mm to 19.2 mm	2 mm to 4 mm	34 Nm	PMT 10
80961	M10	15 mm to 15.2 mm	> 4mm	34 Nm	PMT 10

Tool Setting Selection

Part Number	Pulling Force	Weight	Length	Optional Test Gauge
PMT6	10 KN	1.28 kg	185 mm	80928
PMTC6	10 KN	1.28 kg	185 mm	80928

Part Number	Pulling Force	Weight	Length	Optional Test Gauge
PMT8	18 KN	1.28 kg	185 mm	80928
PMT10	25 KN	1.28 kg	185 mm	80928

Approved Service Materials For Repair

Brand	Usage	Description	Product Number	Alternative Product
3M	**Sealing**	-	-	-
3M	Sealing	Polyurethane Cartridge Extruded Seam Sealer (gray)	08684	08362
3M	Sealing	Polyurethane Cartridge Extruded Seam Sealer (white)	08689	08368
3M	Sealing	Polyurethane Cartridge Extruded Seam Sealer (black)	08694	08367
3M	Sealing	Polyurethane Cartridge Extruded Seam Sealer (beige)	50740	08365
3M	Sealing	Polyurethane Sachet 310ml (gray)	08782	08361
3M	Sealing	Polyurethane Sachet 310ml (white)	08787	08360
3M	Sealing	Polyurethane Sachet 310ml (black)	08789	N/A
3M	Sealing	Polyurethane Sachet 600ml (gray)	08783	N/A
3M	Sealing	Polyurethane Sachet 600ml (white)	08788	N/A
3M	Sealing	Polyurethane Sachet 600ml (black)	08793	N/A
3M	Sealing	Sprayable Seam Sealer 2K Polyurethane 150ml (gray)	08823	N/A
3M	Sealing	Sprayable Seam Sealer 2K Polyurethane 250ml (gray)	08800	N/A
3M	Sealing	Sprayable Seam Sealer (gray)	08851	08374
3M	Sealing	Super Seam Sealer Can (gray)	08537	08656
3M	Sealing	Super Seam Sealer Brush	08540	N/A
3M	Sealing	Sprayable Seam Sealer Cartridge (gray)	08374	08374
3M	Body Caulk-ing	Body Caulking	08578	08568
3M	Sealing	Caulkable Sealer (white)	08855	08369
-	**Adhesive/ Bonding**	-	-	-
3M	Adhesive/ Bonding	Manual Applicator 50ml	08190	N/A
3M	Adhesive/ Bonding	Impact Resistant Structural Adhesive	07333	N/A
3M	Adhesive/ Bonding	Panel Bonding Adhesive	08115	N/A
3M	Adhesive/ Bonding	Manual Applicator and Nozzle	08117	N/A
-	**Tape and Film**	-	-	-
3M	Tape and Film	Acrylic Tape PT1100 Double Sided (6mm x 40m)	80318	06387
3M	Tape and Film	Acrylic Tape PT1100 Double Sided (9mm x 20m)	80319	N/A
3M	Tape and Film	Acrylic Tape PT1100 Double Sided (12mm x 20m)	80320	06388
3M	Tape and Film	Acrylic Tape PT1100 Double Sided (19mm x 20m)	80322	06389
3M	Tape and Film	Acrylic Tape PT1100 Double Sided (25mm x 20m)	80323	N/A
3M	Tape and Film	Polyolefin Adhesion Promotor	05917	05907
3M	Tape and Film	Abrasion Resistance Film	8210	84906
3M	Tape and Film	Abrasion Resistance Film	8219	84901
-	**Cavity Wax**	-	-	-
3M	Cavity Wax	Body Shultz High Coverage	08964	N/A

Brand	Usage	Description	Product Number
Teroson	-	-	-
-	**Sealing**	-	-
Teroson	Sealing	Terostat 9100 - 1K PUR - Adhesive sealant - white	1896978
Teroson	Sealing	Terostat 9100 - 1K PUR - Adhesive sealant - gray	1896976
Teroson	Sealing	Terostat 9200 - 1K PUR - Adhesive sealant - black	1896974
Teroson	Sealing	Terostat 9320SF - Sprayable seam sealant - gray	1357960
Teroson	Sealing	Terostat 9320SF - Sprayable seam sealant - black	1358054
Teroson	Sealing	Terostat 9320SF - Sprayable seam sealant - ochre	1358059
Teroson	Sealing	Terolan special sealant brushable 1.4 kg tin	799671
Teroson	Sealing	Terostat RB 7 Round profile plastic sealing band	142216
-	-	-	-
Teroson	Sealing	Terostat 9120 SF - white 310 ml	1358213
Teroson	Sealing	Terostat 9120 SF - black 310 ml	1358214
Teroson	Sealing	Terostat 9220 Power adhesive sealant - black 310 ml	2026592
-	**Sealing Tape and Primer**	-	-
Teroson	Sealing	Terotape seam sealing tape 10mm x 16mm	451115
Teroson	Sealing	Terokal primer aerosol 150 ml	267078
-	**Cavity Wax**	-	-
Teroson	Cavity Wax	Terotex 400 1L can	784176
Teroson	Cavity Wax	Terotex 400 10L tin	784149
Teroson	Cavity Wax	Terotex 400 60L barrel	784190
Teroson	Cavity Wax	Teroson 215 cavity spray 500ml aerosol	794224
-	**Underbody Wax**	-	-
Teroson	Underbody Wax	Terotex wax black 1L can	794862
-	**Underbody Coating**	-	-
Teroson	Underbody Coating	Terotex record 2000 underbody coating 1L can - gray	767197
Teroson	Underbody Coating	Terotex record 2000 underbody coating 1L can - black	767198
Teroson	Underbody Coating	Terotex super 3000 aqua anti chip compound 1L can - black	1335491
Teroson	Underbody Coating	Terotex super 3000 aqua anti chip compound 1L can - light	882415
-	**Corrosion Protection**	-	-
Teroson	Corrosion Protection	Zinc spray 400 ml	333170
-	**Anti Corrosive Agent**	-	-
Teroson	Corrosion Protection	Terotex 350 1L can	793958
Teroson	Corrosion Protection	Terotex 350 10L can	793960
Teroson	Corrosion Protection	Terotex 350 60L barrel	793957
-	**Sound Deadening**	-	-
Teroson	Sound Deadening	Terodem SP 100 alu - 50 x 25	150012
Teroson	Sound Deadening	Terodem SP 300 - 50 x 50	150055
Teroson	Sound Deadening	Terodem SP 300 - 100 x 50	150054
-	**Glazing Adhesive**	-	-
-	Glazing Adhesive	Teroson PU 8597 HMLC	1467800

Body Repair Procedure Graphic Key

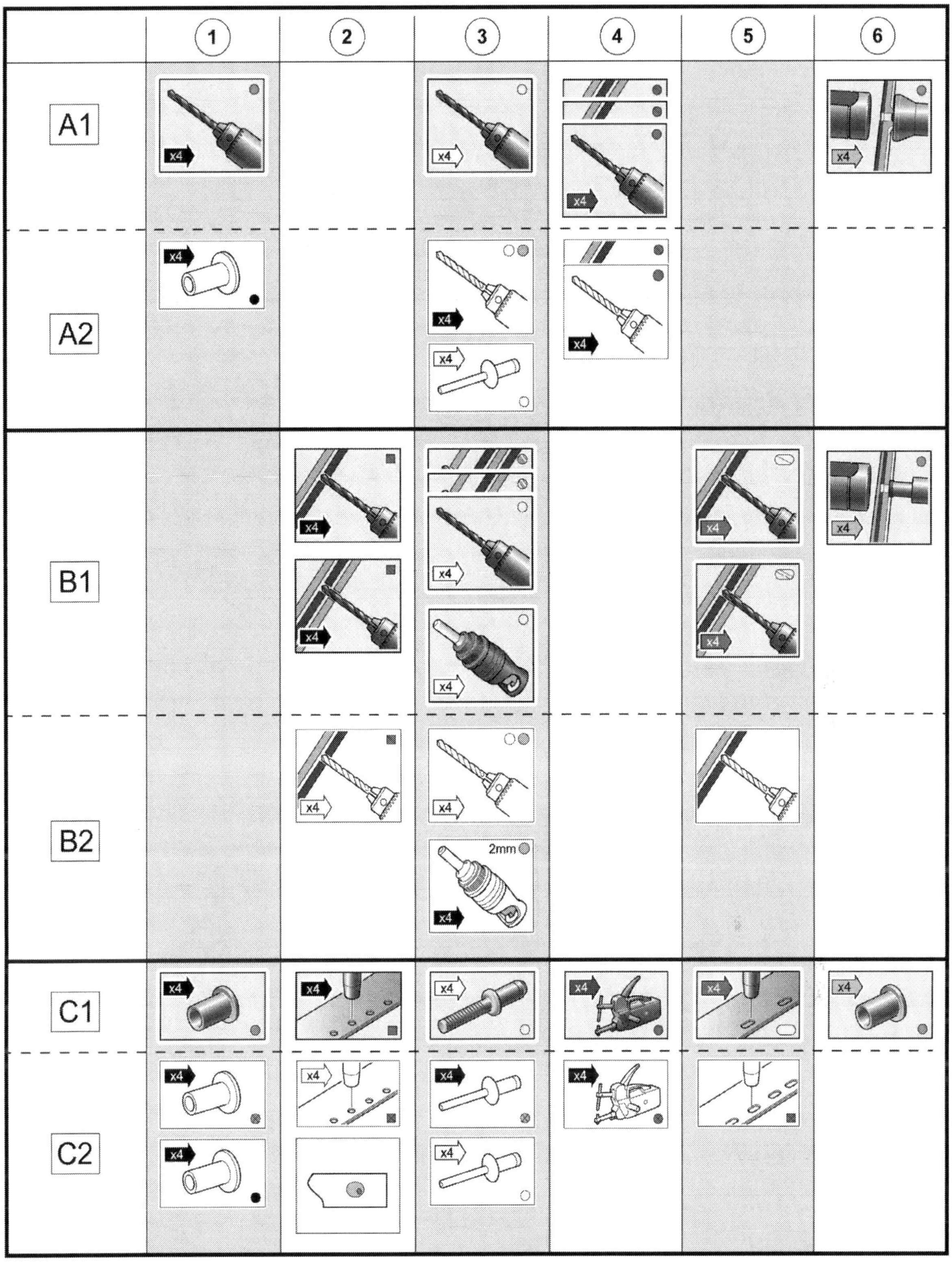

A1/1	Remove Self Piercing Rivet (SPR
A1/3	Remove Breakstem Fastener (BSF)
A1/4	Remove Spot Weld (depths)
A1/6	Planishing
A2/1	Remove Self Piercing Rivet (SPR)
A2/3	Remove Breakstem Fastener (BSF)
A2/4	Remove Spot Weld (depths)
B1/2	Drill for Metal Active Gas (MAG) Plug Weld (depths)
B1/3	Countersink and drill for Breaksterm Fastener (BSF) (depths)
B1/5	Drill for MIG (Metal Inert Gas) Braze Slot Weld (depths)
B1/6	Hole Calibration for Flow Form Rivet (FFR)

B2/2	Drill for Metal Active Gas (MAG) Plug Weld
B2/3	Countersink and drill for Breaksterm Fastener (BSF)
B2/5	Drill for MIG braze slot weld
C1/1	Install Self Piercing Rivet (SPR)
C1/2	Install Metal Active Gas (MAG) Plug Weld
C1/3	Install Breaksterm Fastener (BSF)
C1/4	Install Spot Weld
C1/5	Install MIG Braze Slot weld
C1/6	Install Flow Form Rivet (FFR)
C2/1	Install Self Piercing Rivet (SPR)
C2/2	Install Metal Active Gas (MAG) Plug Weld
C2/3	Install Breaksterm Fastener (BSF)
C2/4	Install Spot Weld
C2/5	Install Slot Weld

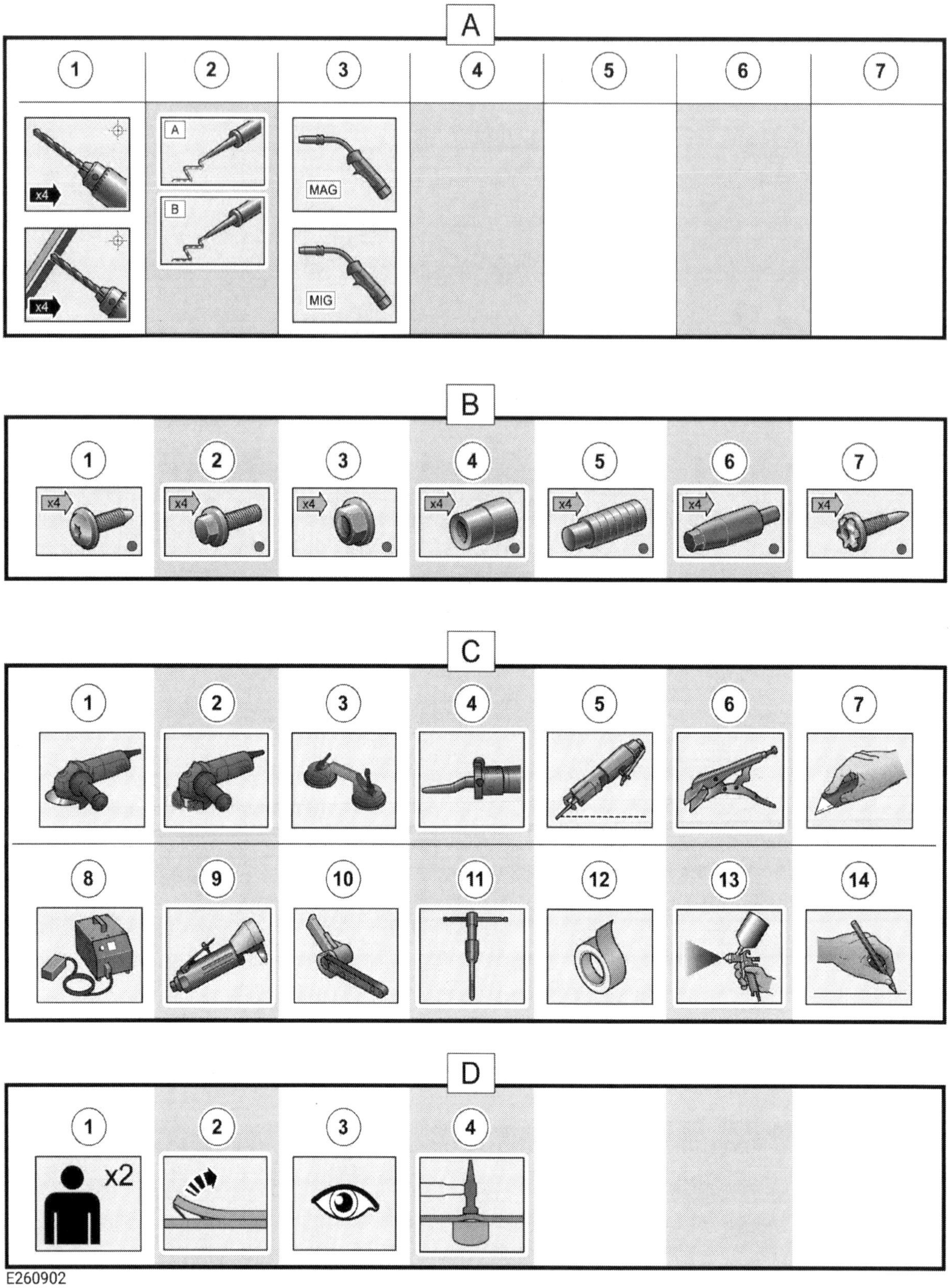

E260902

A1	Drill for Breakstem Fastener (BSF) access
A2	Apply Bonding material/Sealant
A3	MIG/Metal Active Gas (MAG) Weld the Butt Joint
B1	Remove/Install Torque Screw
B2	Remove/Install Bolt
B3	Remove/Install Nut
B4	Remove/Install Rivnut
B5	Install Stud
B6	Install Roof Alignment Pin
B7	Remove/Install Friction Drilling Screw (FDS)
C1	Grinder
C2	Rotary Wire Brush
C3	Suction Clamp
C4	Air Blade

C5	Air Saw
C6	Clamp
C7	Knife
C8	Induction Heater
C9	Circular Saw
C10	Belt Sander
C11	Tap and Die
C12	Protective Tape
C13	Paint Application
C14	Measure and Mark
D1	Assistance Required
D2	Build an Expansion Space
D3	Observation
D4	Hammer and Dolly

Body Closure / Noise Vibration and Harshness Sealing

General Procedures

Repair

⚠ NOTE: If the door is not supplied with the Noise Vibration and Harshness (NVH) pad and seam sealant bead then follow this procedure.

3-door and 5-Door

1. Install a 240 mm x 145 mm Noise, Vibration and Harshness (NVH) pad as indicated.

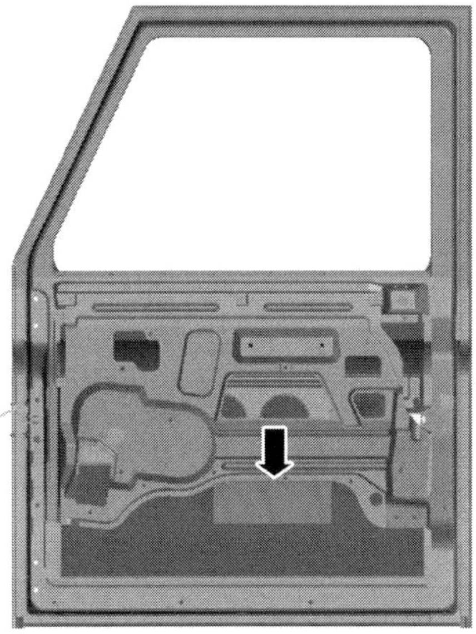

2. Apply the seam sealant to the door as indicated.

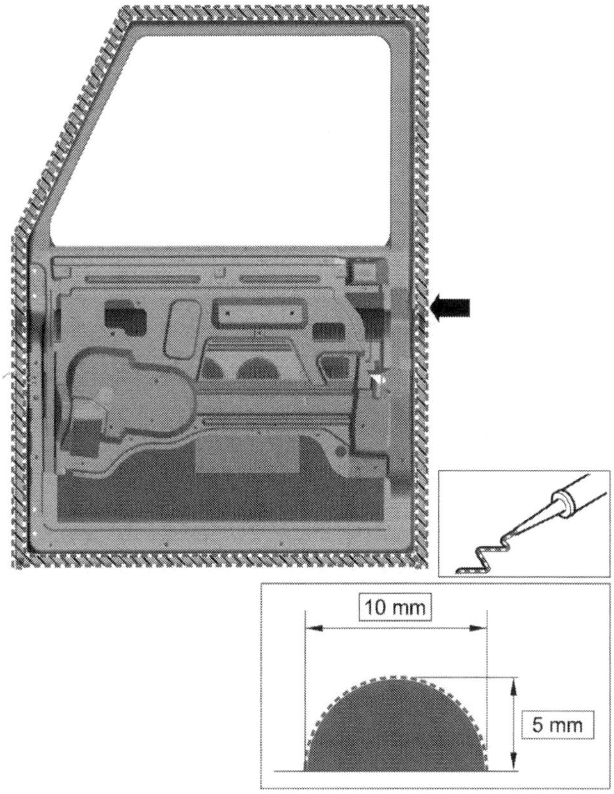

3. Install a 240 mm x 145 mm Noise, Vibration and Harshness (NVH) pad as indicated.

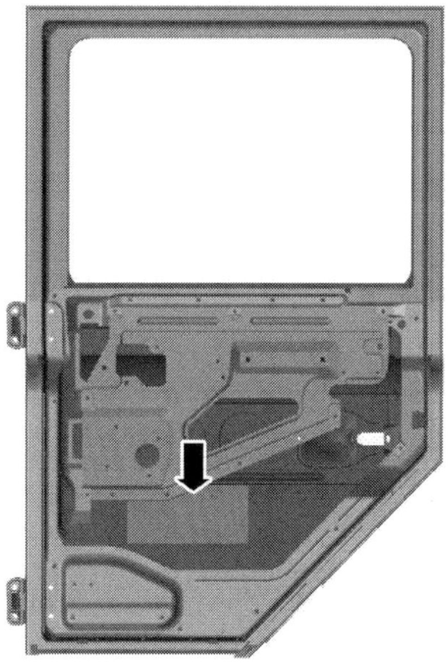

4. Apply the seam sealant to the door as indicated.

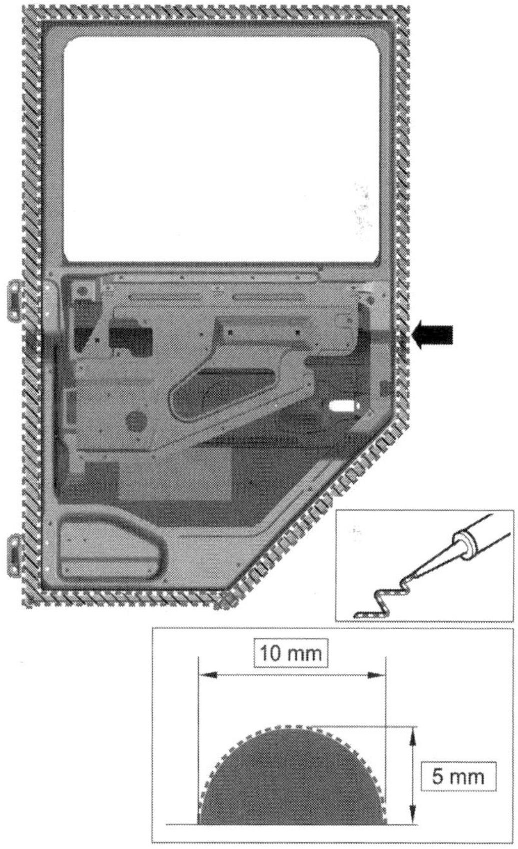

Description and Operation

APPROVED MATERIALS

Joint types

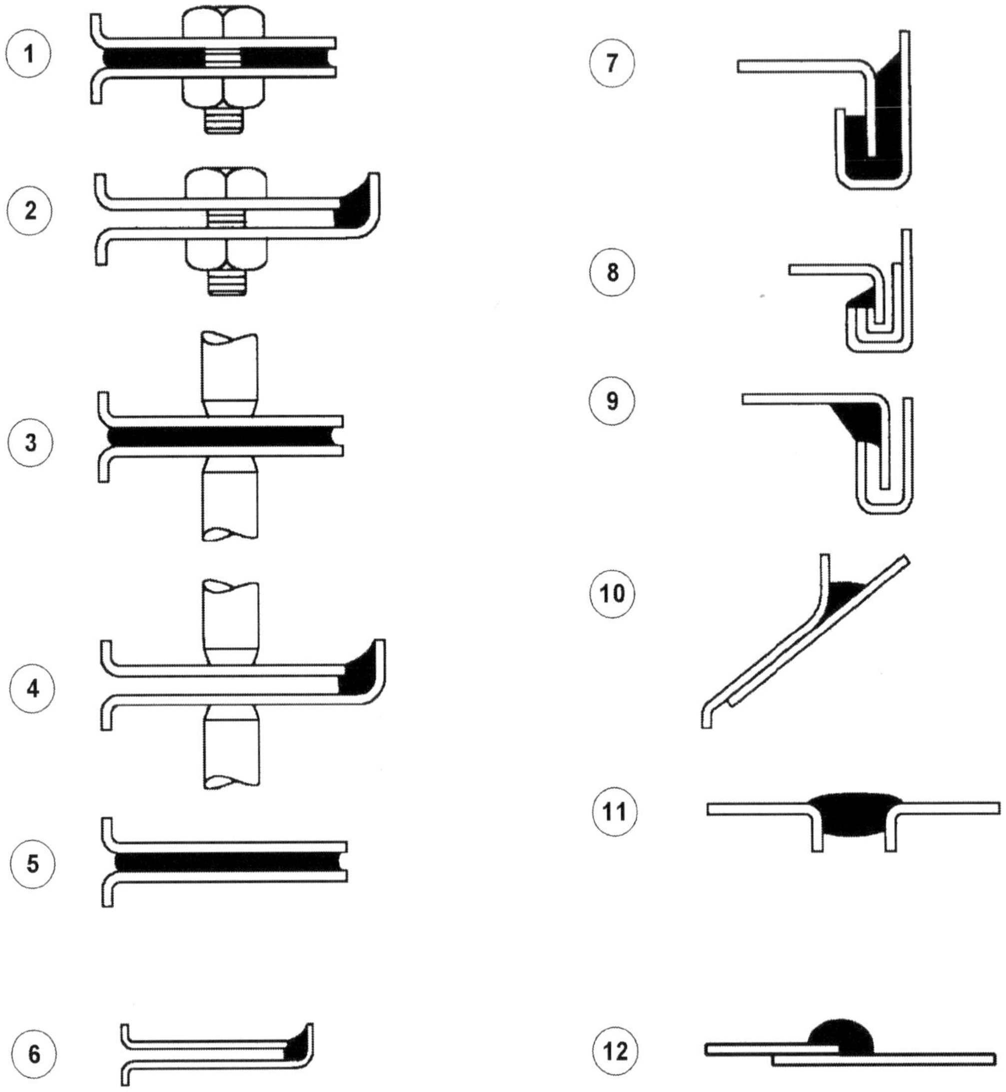

Item	Description
1.	Between bolted panels
2.	Between bolted panel edges
3.	Between spot welded panels
4.	Between spot welded panel edges
5.	Between bonded panels
6.	Between bonded panel edges
7.	Clinch joints (type a)
8.	Clinch joints (type b)
9.	Clinch joints (type c)
10.	Gaps between panels (type a)
11.	Gaps between panels (type b)
12.	Lap joint

The materials listed in this section are those approved to be used in conjunction with the repairs shown in this manual.

If any of these items are not available check with the supplier/manufacturer of these products for a superseded equivalent.

APPLICATION EQUIPMENT

Heavy Duty Manual Gun

Cavity wax application equipment and techniques

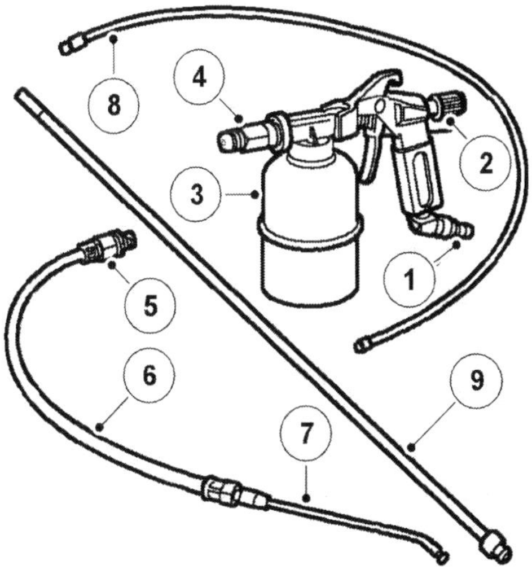

Item	Description
1.	Air inlet
2.	Flow control (spray pattern adjustment)
3.	Pressure cup (1 litre [1.7 pt] capacity). Maximum pressure 140 psi (9.7 bar, 9.8 kg/cm²).
4.	Gun connector
5.	Lance nipple connection
6.	Flexible lance
7.	Rigid directional hook wand (forward cone spray pattern)
8.	Flexible nylon 1100mm (43.3 in.) lance with 360° spray pattern
9.	Rigid 1100mm (43.3 in.) lance with 360° spray pattern

When re-treating wax-injected areas which have been disturbed during repairs, it is necessary to use a compressed air spray gun with integral pressure cup and a selection of interchangeable lances.

The following points must be observed during use, according to the attachments fitted:

- Use the rigid or flexible lance attachments with 360° spray dispersal when treating enclosed areas, to ensure maximum coverage.
- Where openings are restricted, use the hook nozzle to provide a more directional spray

1100 mm (43.3 in.) Rigid Lance: The nozzle on the rigid lance produces a 360° circular spray pattern combined with a forward-directed spray. Although wax is distributed to all box section surfaces in a single stroke, effective and complete coverage is best achieved in long, straight structures and box section cavities by spraying on both outbound and return strokes of the lance.

The rigid lance also provides the positional accuracy required in shaped sections, by allowing visual assessment.

⚠ CAUTION: Do not force the lance into access holes when using this attachment.

1100 mm (43.3 in.) Flexible Nylon Lance: This lance is similar in pattern to the rigid version, but provides the additional penetration needed for curved sections or in places where access is difficult. Its main limitation is a lack of positional accuracy inside box sections.

Carry out spraying on the outward stroke of the lance. Withdraw the lance slowly to ensure sufficient coverage. **DO NOT withdraw the lance too quickly.**

Keep the nylon tube of the lance away from the edges of the access hole to eliminate abrasion and extend the life of the tube. Take care to ensure that spraying ceases just before the nozzle emerges from the access hole. To assist this process, apply RED paint to the final 30mm (1.2 in.) of the nozzle.

Hook Nozzle on Flexible Lance: The rigid hook produces a highly atomised, forward-directed, fully conical spray pattern having long range and good dispersion characteristics. This combination has good directional capabilities for the treatment of short, narrow sections and may also be used for direct spraying of inner wheel arches etc.

Position the flat area at the end of the lance at 180° to the nozzle spray direction. This will help to guide the spray more accurately when it is concealed in a box section or access hole.

△ NOTE: Keep all wax injection/application equipment clean. Use white spirit for this purpose immediately after wax injection operations.

For general spraying move the nozzle in an arc from side to side, to ensure full coverage.

Precautions during Body Repairs and Handling

Take care when handling the vehicle in the workshop. PVC underbody sealers, seam sealers, underbody wax and body panels may be damaged if the vehicle is carelessly lifted.

Always follow the correct lifting, jacking and towing procedures as shown in GENERAL INFORMATION DATA, Information section, paying particular attention to the following points:

Steam Cleaning and Dewaxing

Due to the high temperatures generated by steam cleaning equipment, there is a risk that certain trim items could be damaged and some adhesives and corrosion prevention materials softened or liquified.

Adjust the equipment so that the nozzle temperature does not exceed 90°C (194°F). Take care not to allow the steam jet to dwell on one area, and keep the nozzle at least 300mm (11.8 in.) from panel surfaces.

Do NOT remove wax or lacquer from underbody or underbonnet areas during repairs. Should it be necessary to steam clean these areas, apply a new coating of wax or underbody protection as soon as possible.

CORROSION PROTECTION

The following information details the materials that are applied during manufacture for corrosion protection.

Factory Treatments

The Defender is treated with the following anti-corrosion materials in production:

- A PVC based underbody sealer material which is sprayed onto the underfloor, wheel arches and undersill areas.
- An application of cavity wax which is sprayed into enclosed cavities and box sections.
- A final coating of underbody wax to cover the complete underfloor including components but excluding brake discs, exhaust system and propeller shafts.
- A coat of protective lacquer applied to the whole body.
- A coat of protective wax applied to the engine bay and wheel arch areas.

NOTE: Do not apply wax to engine bay of Td5 models.

In addition to the above measures, all steel parts are zinc-coated both sides.

Underbody Sealer

Underfloor areas and outer sill panels are treated with a Plastisol PVC underbody sealer. This material is not suitable for re-treatment.

When repairing areas of underbody sealer, strip the factory-applied material back to a suitable break point, ensuring that a clean metal surface is exposed and that the edge of the existing material adheres soundly to the panel.

CAUTION: Ensure that suspension units, wheels, tyres, power unit, driveshafts, exhaust and brakes (including all mounting points) are shielded prior to application of fresh underbody sealer

NOTE: Application of new underbody sealer must be carried out between primer and surfacer paint operations. Areas where seam sealer is used should be re-treated as necessary before application of underbody sealer.

Blanking plugs and grommets in the floor pan (except those used for wax injection) MUST be fitted before underbody sealer application. Heat-fusible plugs which have been disturbed should either be refitted with the aid of a hot air blower or replaced with rubber grommets.

Cavity Wax

Cavity wax is applied to certain box sections The information given on the following pages is intended as a guide and shows the areas to be treated with cavity wax, as well as the access holes used during manufacture.

Underbody Wax

A coat of underbody wax is applied to the entire underbody inboard of the sill vertical flanges, and covers all moving and flexible components EXCEPT for wheels and tyres, brakes and exhaust. The wax is applied over paints and underbody sealers.

CAUTION: Old underbody wax must be completely removed from a zone extending at least 200 mm (7.9 in) beyond the area where new underbody sealer is to be applied.

The underbody wax must be reinstated following all repairs affecting floor panels.

Engine Bay Wax

Reinstate protective engine bay wax disturbed during repairs using the approved material.

Wheel Arch Wax

Reinstate protective wheel arch wax disturbed during repairs using the approved material.

Stone Chip Resistant Paint/Primer

Re-treat all areas protected with factory-applied anti-chip primer with suitable approved material in repair.

Inspections during Maintenance Servicing

It is a requirement of the Land Rover Corrosion Warranty that the vehicle body is checked for corrosion by an authorised Land Rover dealer at least once a year, to ensure that the factory-applied protection remains effective.

Service Job Sheets include the following operations to check bodywork for corrosion:

NOTES:

Wash the vehicle and ensure that it is free from deposits prior to inspection. It is part of the owner's responsibility to ensure that the vehicle is kept free of accumulations of mud which could accelerate the onset of corrosion. The Dealer MUST wash the vehicle prior to inspection of bodywork if the customer has offered it in a dirty condition, and pay special attention to areas where access is difficult.

The checks described above are intended to be visual only. It is not intended that the operator should remove trim panels, finishers, rubbing strips or sound deadening materials when checking the vehicle for corrosion and paint damage.

With the vehicle on a lift, and using an inspection or spot lamp, visually check for the following:

NOTE: The presence of small blisters in PVC underbody sealer is acceptable, providing they do not expose bare metal.

Special attention must be paid to signs of damage caused to panels or corrosion material by incorrect jack positioning.

It is essential to follow the correct jacking and lifting procedures. See GENERAL INFORMATION DATA, Information section.

With the vehicle lowered, visually check for evidence of damage and corrosion on all painted areas, in particular the following:

Where bodywork damage or evidence of corrosion is found during inspection, rectify this as soon as is practicable, both to minimise the extent of the damage and to ensure the long term effectiveness of the factory-applied corrosion protection treatment. Where the cost of rectification work is the owners responsibility, the Dealer must advise the owner and endorse the relevant documentation accordingly.

Where corrosion has become evident and is emanating from beneath a removable component (e.g. trim panel, window glass, seat etc.), remove the component as required to permit effective rectification.

Underbody Protection Repairs

When body repairs are carried out, always ensure that full sealing and corrosion protection treatments are restored. This applies both to the damaged area, and also to areas where protection has been indirectly impaired as a result of accident damage or repair operations.

Prior to straightening out or panel beating, remove all corrosion protection material in the damaged area. This applies in particular to panels coated with wax, PVC underbody sealer, sound deadening pads etc.

WARNING: DO NOT use oxy-acetylene gas equipment to remove corrosion prevention materials. Large amounts of fumes and gases are liberated by these materials when they burn.

Equipment for the removal of tough anti-corrosion sealers offers varying degrees of speed and effectiveness. The compressed air-operated scraper (NOT an air chisel) offers a relatively quiet mechanical method of removal using an extremely rapid reciprocating action. During use, direct the operating end of the tool along the work surface.

The most common method is by the use of a hot air blower with integral scraper.

CAUTION: High temperatures can be generated with this equipment which may cause fumes. Always exercise care in its use.

Another tool, and one of the most efficient methods, is the rapid-cutting hot knife. This tool uses a wide blade and is quick and versatile, able to be used easily in profiled sections where access is otherwise awkward.

Use the following procedure when repairing underbody coatings:

Underbody Wax

NOTE: Where repairs include the application of finish paint coats in the areas requiring underbody wax, paint operations must be carried out BEFORE wax application.

After refitting mechanical components, including hoses, pipes and small fixtures, mask off the brake discs and apply a coat of approved underbody wax.

Underbonnet Wax

Where repairs have involved replacement of engine bay panels, treat the entire engine compartment including all components, clips and small fixtures with an approved underbonnet lacquer or wax.

Proprietary Anti-Corrosion Treatments

The application of proprietary anti-corrosion treatments in addition to the factory-applied treatment could invalidate the Corrosion Warranty and should be discouraged. This does not apply to approved, compatible, preservative waxes which may be applied on top of existing coatings.

Fitting Approved Accessories

When fitting accessories it is important that the vehicle's corrosion protection is not affected, either by breaking the protective coating or by introducing a moisture trap.

DO NOT screw self-tapping screws directly into the body panel but fit plastic inserts first. Protect the edges of holes drilled into panels, chassis members and other body parts with a suitable zinc rich or acid etch primer, followed by a protective wax coating brushed onto the surrounding area.

DO NOT affix unpainted metal surfaces of any accessory directly to the vehicle bodywork unless they are suitably protected. Where metal faces are bolted together always interpose a suitable interface material such as weldable zinc rich primer, extruded strip or zinc tape.

CAVITY WAX

Box sections treated with cavity wax are shown in this section. Repairs affecting these areas must include re-treatment with an approved cavity wax, using the access points illustrated. In addition, all interior surfaces which have been disturbed during repairs must be wax injected whether they have been treated in production or not. This includes all box members, cavities, door interiors etc. It is permissible to drill extra holes for access where necessary, provided these are not positioned in load-bearing members. Ensure that such holes are treated with a suitable zinc rich primer, brushed with wax and then sealed with a rubber grommet.

Prior to wax injection, ensure that the cavity to be treated is free from any contamination or foreign matter. Where necessary, clear out any debris using a compressed air supply.

Carry out wax injection after final paint operations.

During application, ensure that the wax covers all flange and seam areas and that it is applied to all repaired areas of both new and existing panels.

△ NOTE: Apply cavity wax AFTER the final paint process and BEFORE refitting of any trim components.

It should also be noted that new panel assemblies and body shells are supplied without wax injection treatment which must be carried out after repairs.

Effective cavity wax protection is vital. Always observe the following points:

- Complete all finish paint operations before wax application.
- Clean body panel areas and blow clean cavities if necessary, before treatment.
- Maintain a temperature of 18°C (64°F) during application and drying.
- Check the spray pattern of injection equipment.
- Mask off all areas not to be wax coated and which could be contaminated by wax overspray.
- Remove body fixings, such as seat belt retractors, if contamination is at all likely.
- Move door glasses to fully closed position before treating door interiors.
- Treat body areas normally covered by trim before refitting items.
- Check that body and door drain holes are clear after the protective wax has dried.
- Keep all equipment clean, especially wax injection nozzles.

The following Illustrations show the treatment areas and Injection holes for Cavity Wax application.

All areas symmetrically opposite to those shown are also treated.

Bulkhead assembly

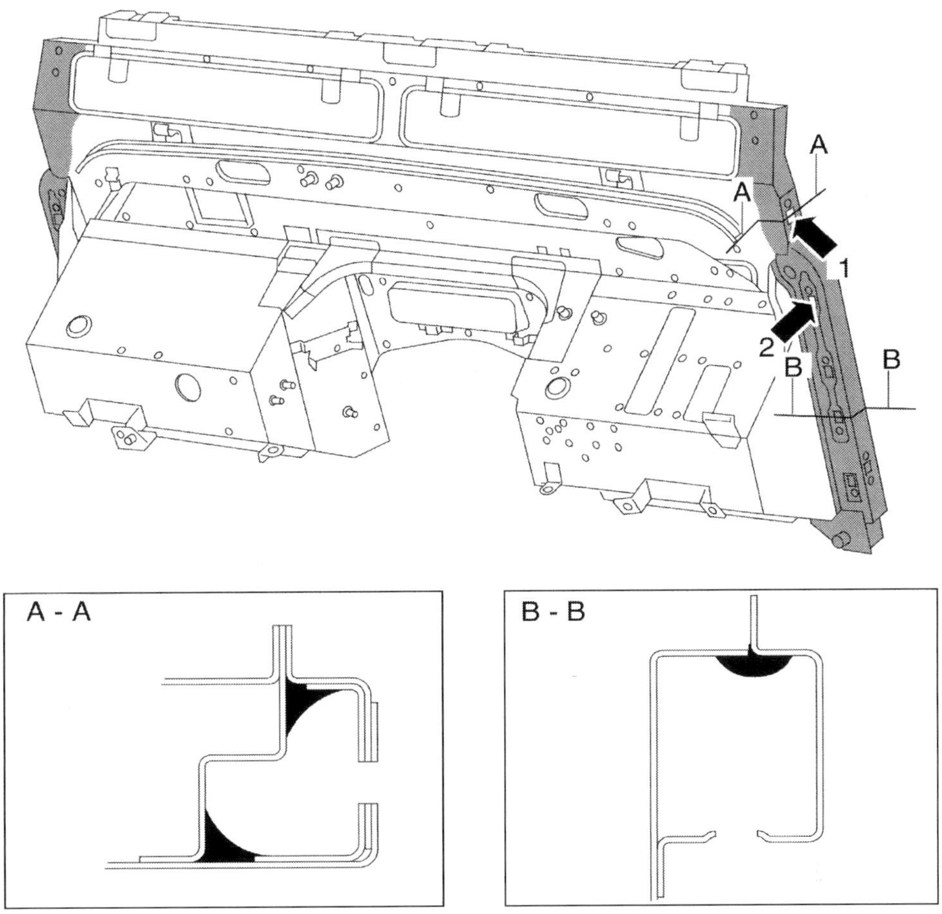

M772072A

Section A-A and B-B show application areas of cavity wax. Arrows 1 and 2 show application holes.

'B/C' post assembly

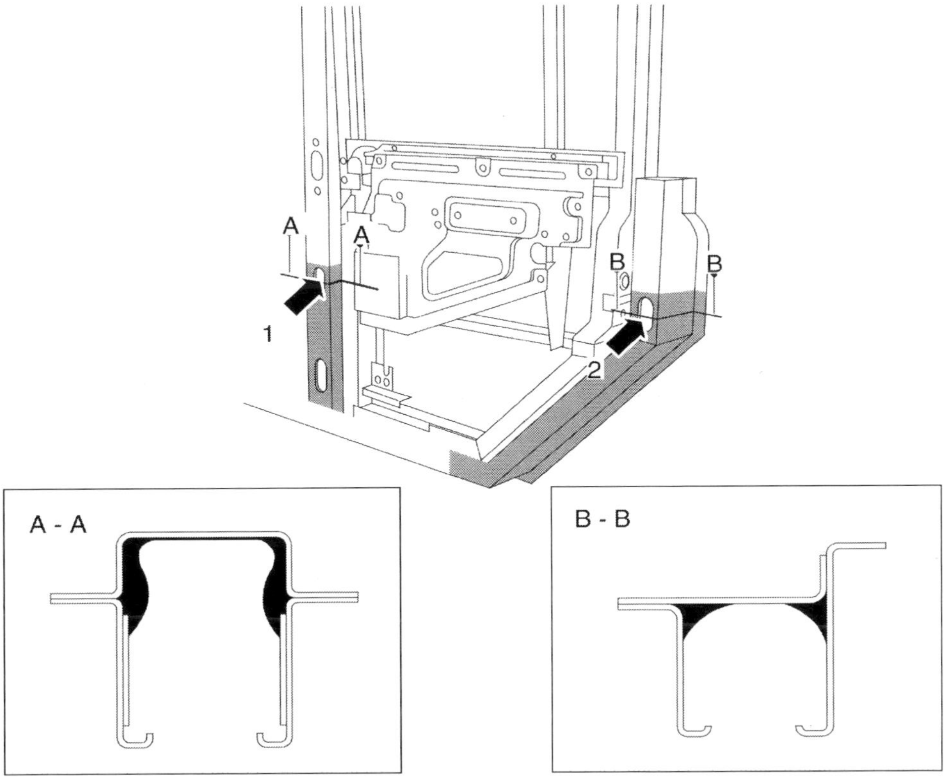

Section A-A shows application area of cavity wax for the 'B/C' post. Arrow 1 shows application hole.

Section B-B shows application area of cavity wax for the 'D' post. Arrow 2 shows application hole.

Front door

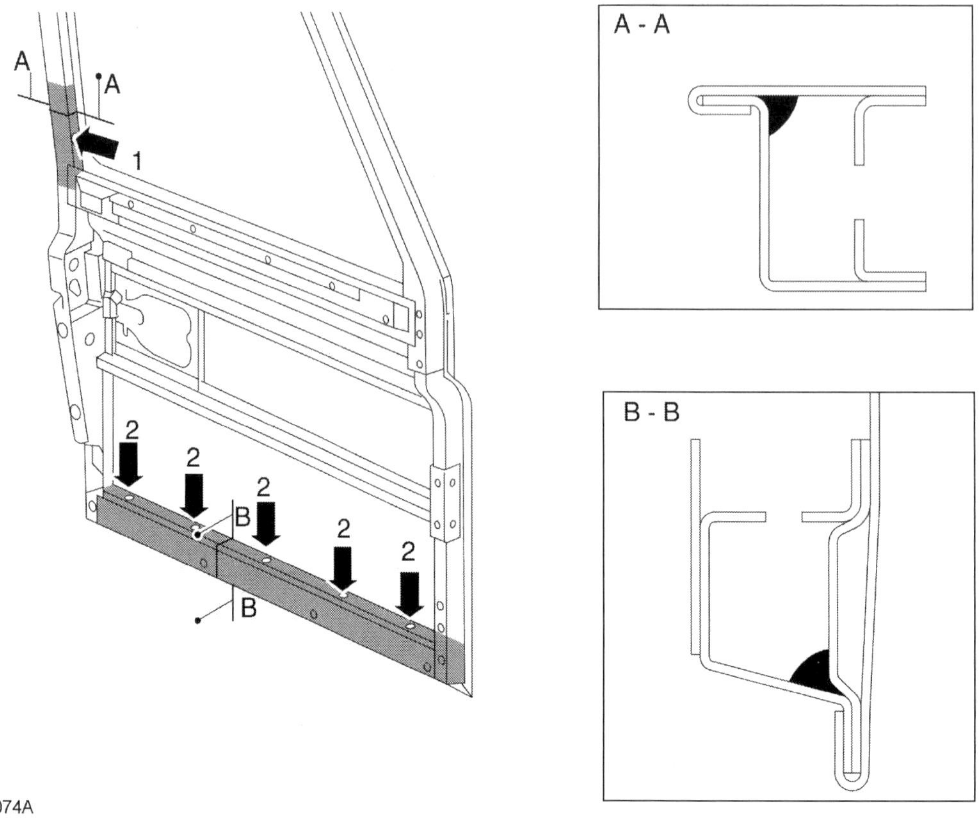

M772074A

Section A-A and B-B show application areas of cavity wax for the front door. Arrows 1 and 2 show the application holes.

Rear door

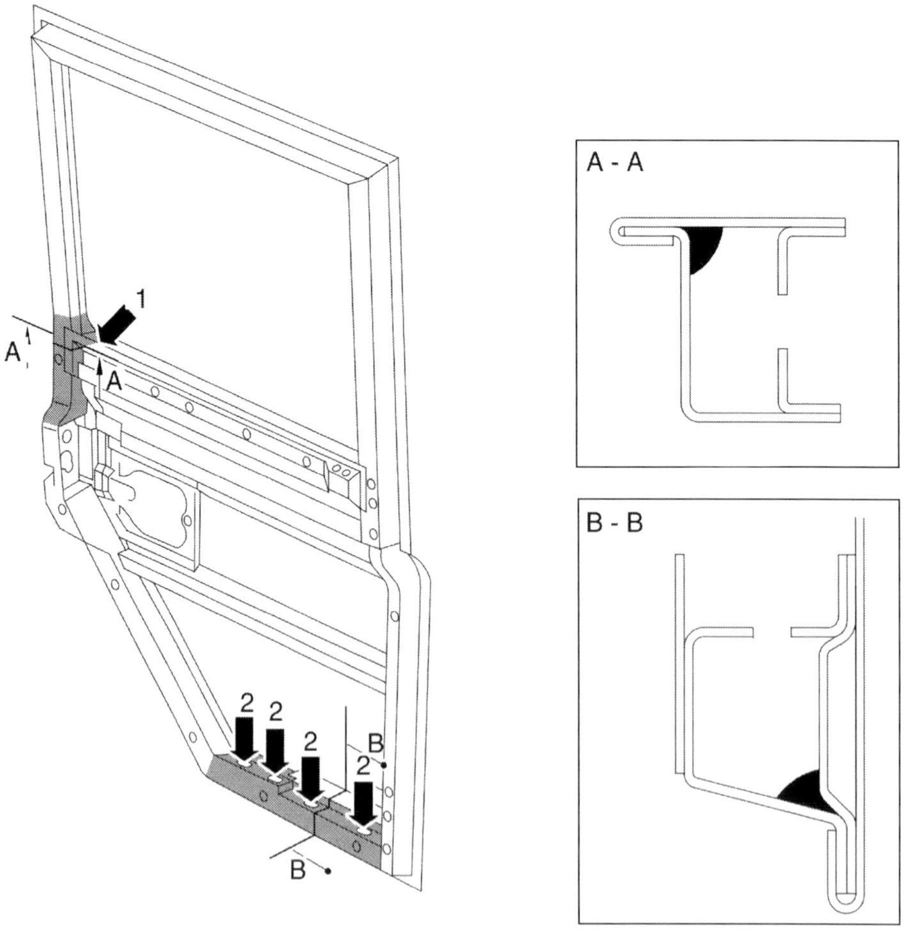

M772075A

SEALANTS AND ADHESIVES

Structural Adhesive

Metal-to-metal adhesive is applied to critical joint areas during factory assembly. The material used is a high-temperature, heat cured, nitrile phenolic which serves both to bond two metal surfaces and also to seal the joint against ingress of dust, water, petrol and fumes. This material is not suited for service use, and should be substituted in repair using a suitable medium strength adhesive.

When separating a joint treated with metal-to-metal adhesive, to avoid distortion it is recommended that the joint be gently heated until the bond weakens sufficiently to permit panel separation.

NOTE: DO NOT carry out MIG welding on a joint area which has been treated with metal-to-metal adhesive until all traces of adhesive have been removed.

The following Illustrations show the treatment areas for Structural Adhesive. All areas symmetrically opposite to those shown are also treated.

Body side capping

Section A-A and B-B show application areas of cavity wax for the rear door. Arrows 1 and 2 show the application holes.

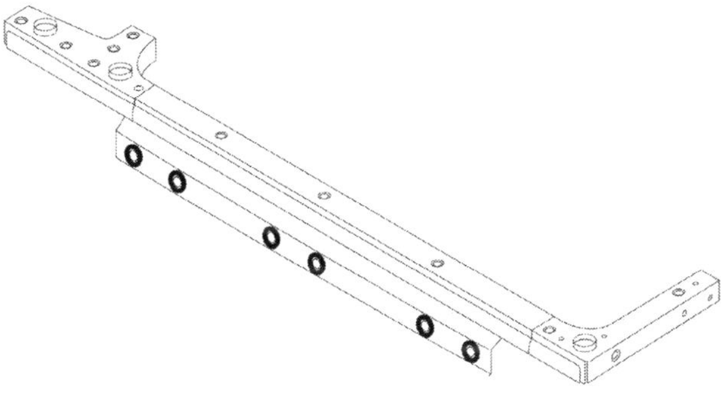

Structural adhesive applied around body side lower fixing holes

Seam sealers

A heat cured, PVC Plastisol sealer is applied to joint areas during factory assembly. This material is not suitable for service use.

Carry out seam sealing after the application of primer and before the surfacer and final paint coats. Ensure that surfaces are first cleaned of all grease and oil. Apply the sealer material to the joint as a bead, either by hand or using an applicator gun. Where necessary, brush sealer well into the joint and wipe smooth using a cloth soaked with solvent such as Shell SBP3. This will ensure an acceptable cosmetic finish.

Apply sealer to ALL accessible joints following repair work. Be aware that damage to a vehicle can often result in deflection to those areas of the body which are remote from the impact. The sealers in these areas can therefore be disturbed by subsequent straightening and repair operations. Check joints in the vicinity of the area undergoing repair for evidence of cracked sealer, clean them out as required and re-treat them with fresh sealer using the following procedure:

Where joints are inaccessible following the reassembly or fitment of components, ensure that a paste-type sealer is applied to such joints. Certain seams also become inaccessible after the completion of panel repairs. In such instances the paint process should be carried out and sealers applied before final assembly.

Provided access is adequate, apply the sealer to both sides of the repair joint. Where access is limited to one side only (e.g. box sections), inject the affected box member with cavity wax.

The following Illustrations show the treatment areas for Seam Sealing.

All areas symmetrically opposite to those shown are also treated.

Seam sealing on bulkhead assembly - front view

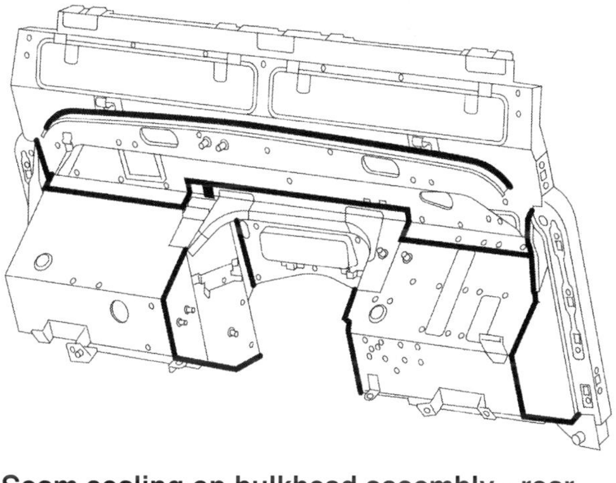

Seam sealing on bulkhead assembly - rear view

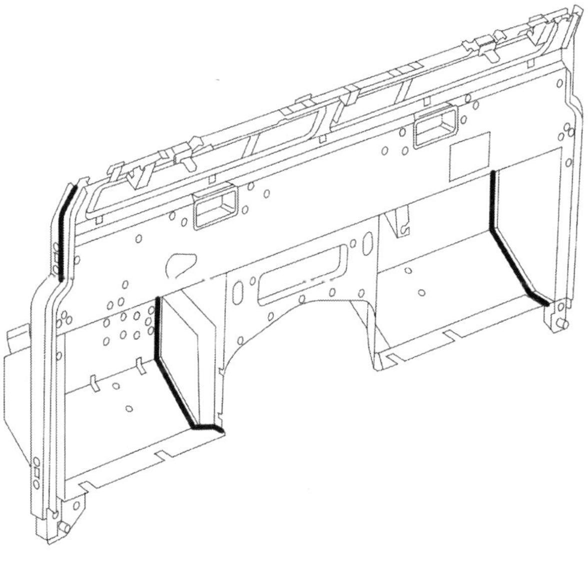

Seam sealer on body side assembly

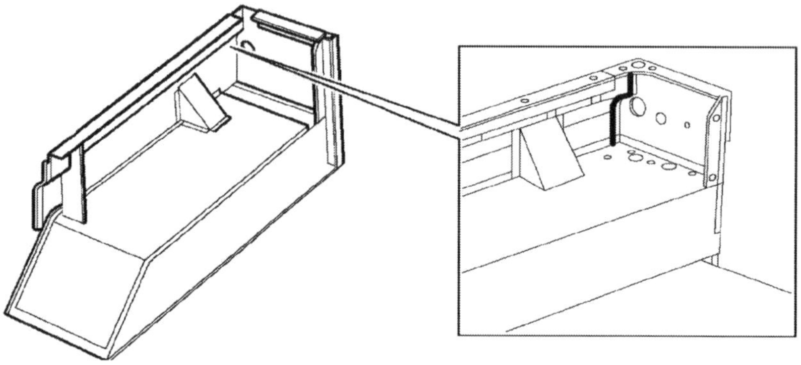

Seam sealer on front door

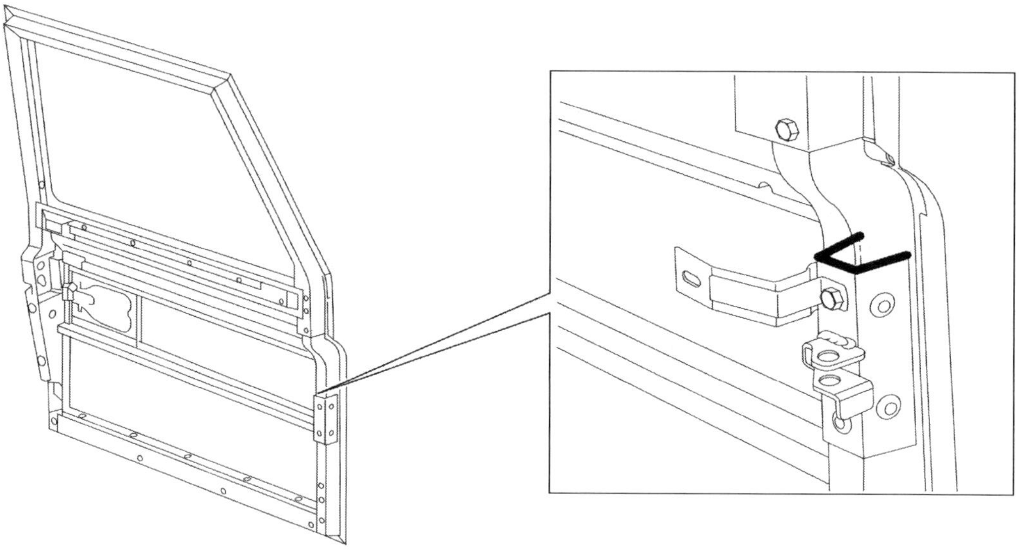

NOTE: Seam sealer to be wiped after application for cosmetic purposes.

Seam sealer on rear end assembly

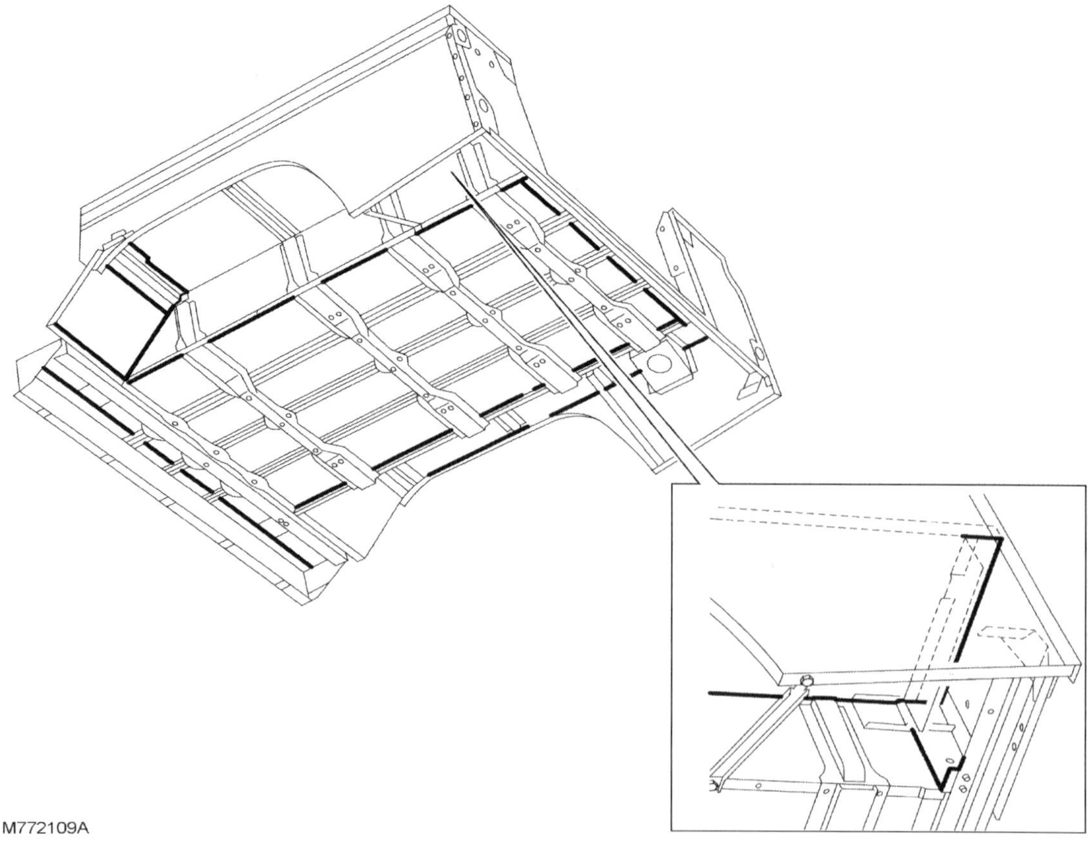

M772109A

CAUTION: Do not block drain holes when applying seam sealer.

Seam sealer on roof assembly

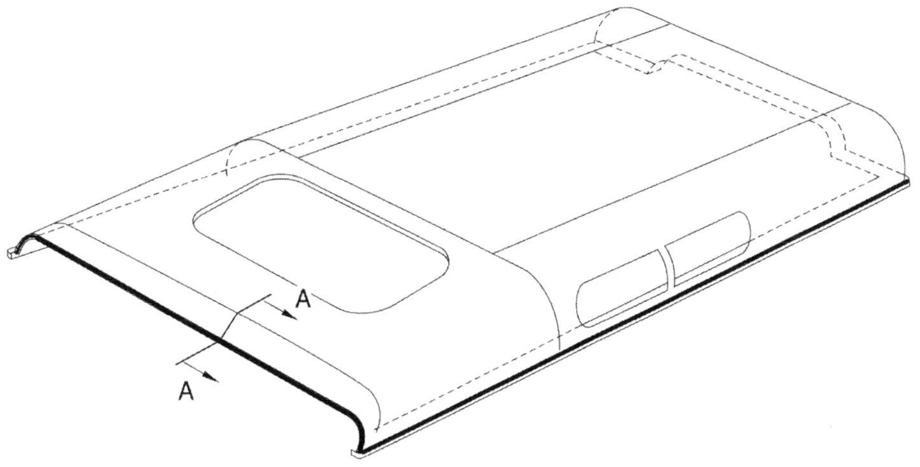

Putty application areas

The following Illustrations show the treatment areas for Putty application. All areas symmetrically opposite to those shown are also treated.

Putty location on bulkhead assembly

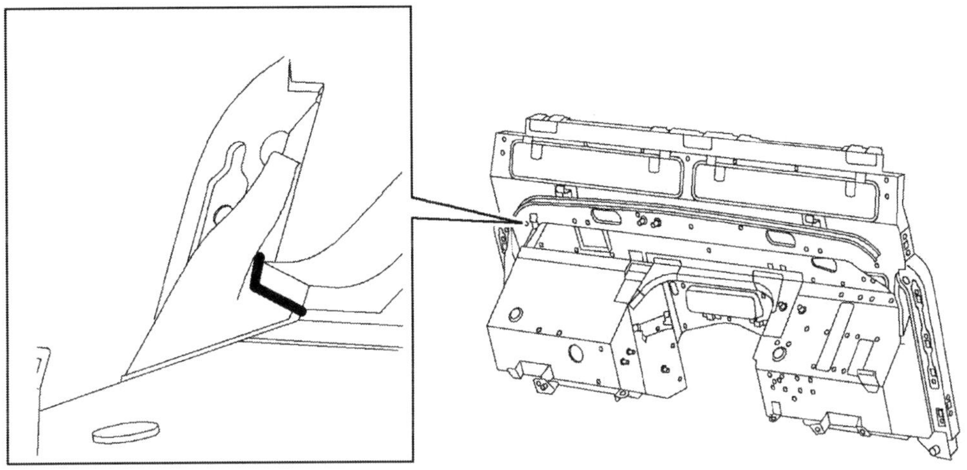

Putty location on rear end

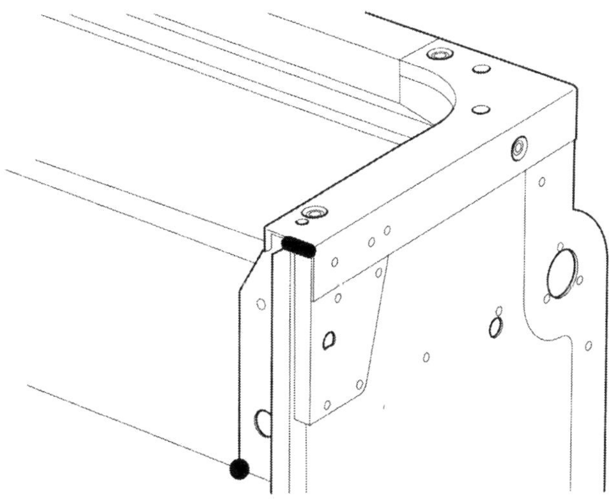

Putty location on underside of vehicle

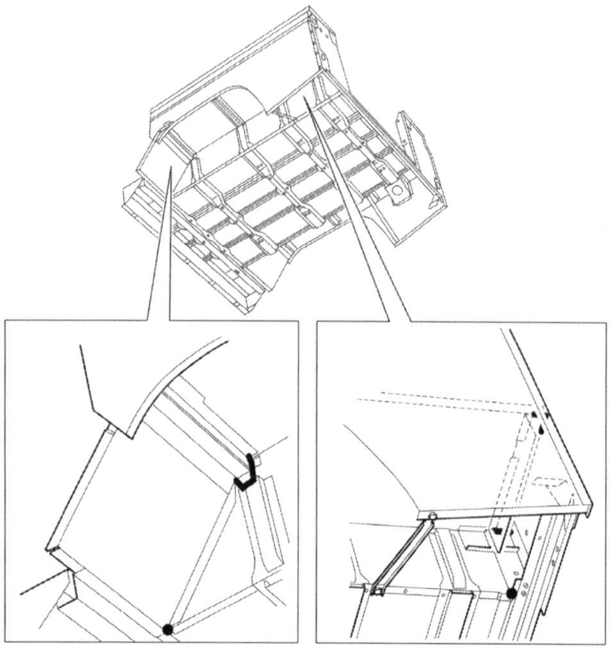

Putty location on front door

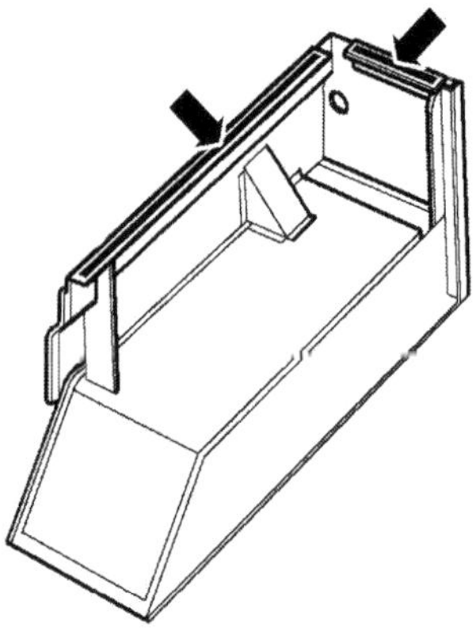

Foam/rubber seal application areas

The following Illustrations show the location of foam/ rubber seals. All areas All areas symmetrically opposite to those shown are also treated.

Body side rear

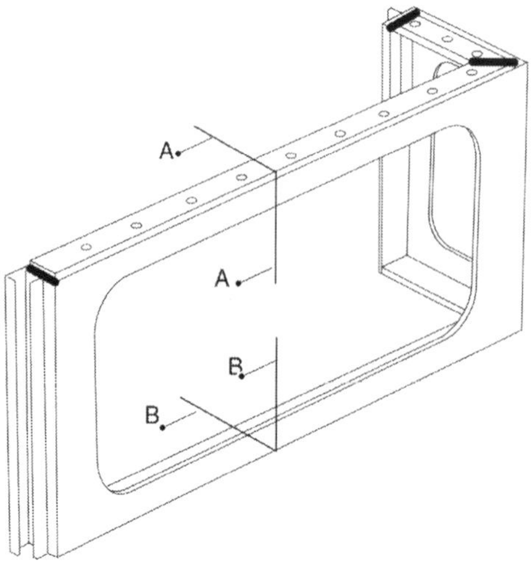

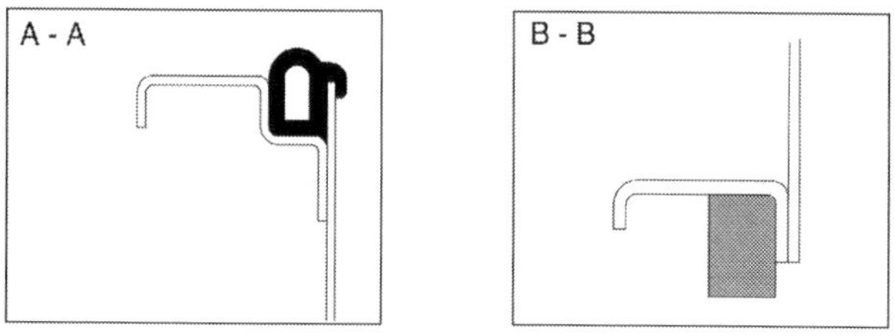

Section A-A shows a rubber seal in position on the body side rear upper assembly.

CAUTION: Ensure rubber seal is correctly seated into channel.

Section B-B shows a foam seal located on the bottom edge of the body side rear upper assembly. It is fixed to the panel using double sided tape.

Body side lower

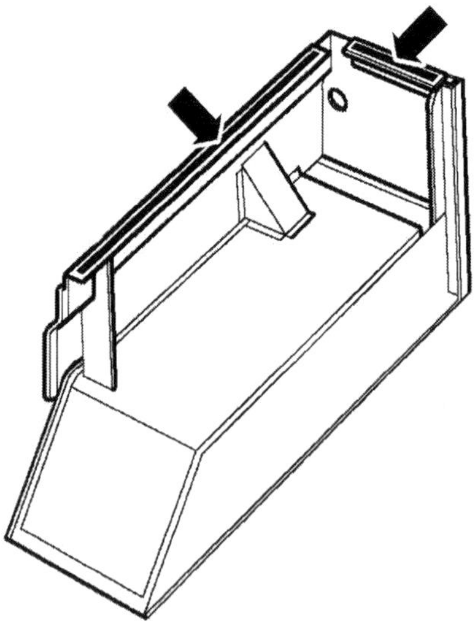

CAUTION: Ensure seals are fitted before refitting assembly.

The arrows indicate the position of two foam seals.

Diagnosis and Testing

Inspection and Verification

1. Visually inspect for obvious signs of mechanical or electrical damage.
2. If an obvious cause for an observed or reported concern is found, correct the cause (if possible) before proceeding to the next step.
3. If the cause is not visually evident, verify the symptom and refer to the Symptom Chart.

Water Leak Diagnostic Procedure

Introduction

This procedure has been developed to aid diagnosis and rectification of water ingress on Defender vehicles with the aim of providing a **right first time fix**.

The procedure provides suggested points of water ingress and guidelines for a recommended fix.

Prerequisites

It is assumed the technician working on the vehicle will be at least **level 3 trained** and normal vehicle service protection equipment will be used where appropriate i.e. seat covers, wing covers etc.

Water Leak Diagnosis

There are certain basic tools required for effective water leak detection the following are a few recommendations.

Basic tools
Hosepipe
Water supply
Pressure Washer
Watering Can
Torch
Mirror (telescopic type)
Air supply
Boning tool (Nylon shaped block for trim removal)

There are several adaptations of tools that can be used, for example a watering can rose attached to a hosepipe to create a spray, or a new sealant tube nozzle attached to a hosepipe can be very effective to direct water into awkward corners, there are also several ready made hosepipe nozzles available from DIY stores which can be switched through several different water patterns, and finally not forgetting a normal car wash. With the exception of a car wash initial diagnosis is more accurate if carried out by **two people**, one inside and one outside the vehicle, the person outside can direct the water onto the areas where the leak is suspected to be entering, and the person inside can inspect with a 12 volt hand lamp to confirm the entry point.

It is worth bearing in mind that the location that the water appears in the interior of the vehicle, may not be the leak source, **for example water lying in the passenger footwell could be entering on the drivers side and running across, behind the fascia**. In order to find the water entry point, trim or components may have to be removed.

Staining

Often when water has been entering over a period of time, the water entry point can be located visually by following the stains or tracks left by the leak.

Sealing Water Leaks

There are different substances that can be used to seal water leaks, putty type sealant and wet/paste sealant. Examples of these are bostic (dum-dum) “303 glasticon” and “betafill 10210” which is a white paste and “terostat 33” which is clear.

NOTE: Do not use silicon based sealers as these will cause problems if any subsequent paint operations are required.

Careful consideration needs be given as to the substance used to seal a water leak, for example an external seam on a white vehicle would require white or clear sealant, dum-dum is best used in concealed places to fill larger gaps. The sealant should be applied in a manner that it does not look unsightly when finished, if used in a box section or under a carpet, applying and smoothing over should be OK, but on the external panels or visible areas the sealant should be «wiped» into the gap and any excess removed with a suitable spirit that will not harm the vehicles paintwork

Once the water entry point has been confirmed, the suspect area must be sufficiently dried prior to the application of any sealing compound the use of a compressed air will assist

Symptom Chart

Water Ingress Paths And Recommended Repair Procedure

⚠ NOTE: Carry out a water leak test and visual inspection of any suspect or wet areas once the water entry point has been determined refer to the symptom chart below.

Symptom - Water Collecting	Possible Water Entry Points	Action - To Be Carried Out Once The Water Entry Point Has Been Confirmed
In footwell (entry point - front area)	• Heater cable grommet • Wiper grommet • Windscreen seal • Footwell seams • Floor seams • Hood release grommet • Pedal box grommet • Blower box • Air intake duct to blower join • Air intake • Clutch/brake pedal box • Windscreen surround frame	• For heater cable grommet GO to Pinpoint Test A • For wiper grommet GO to Pinpoint Test B • For windscreen seal GO to Pinpoint Test C • For footwell seams GO to Pinpoint Test D • For floor seams GO to Pinpoint Test E • For hood release grommet GO to Pinpoint Test F • For pedal box grommet GO to Pinpoint Test G • For blower box GO to Pinpoint Test H • Air intake duct to blower join GO to Pinpoint Test H • For air intake GO to Pinpoint Test I • For clutch/brake pedal box GO to Pinpoint Test J • For windscreen surround frame GO to Pinpoint Test L
In footwell (entry point - front door area)	• Door seal • Water shedder • Door latch/door edge • Rear tub capping • Upper to lower body joint (waist side) • Roof to "A" post joint	• For door seal GO to Pinpoint Test K • For water shedder GO to Pinpoint Test L • For door latch/door edge GO to Pinpoint Test M • For rear tub capping GO to Pinpoint Test N • For upper to lower body joint (waist side) GO to Pinpoint Test O • For roof to "A" post joint GO to Pinpoint Test P
In footwell (entry point - rear door area)	• Door seal • Water shedder • Door latch/door edge • Rear tub capping • Upper to lower body joint (waist side) • Roof to upper body side joint	• For door seal GO to Pinpoint Test K • For water shedder GO to Pinpoint Test L • For door latch/door edge GO to Pinpoint Test M • For rear tub capping GO to Pinpoint Test N • For upper to lower body joint (waist side) GO to Pinpoint Test O • For roof to upper body side joint GO to Pinpoint Test Q

In footwell **(entry point - A post area)**	• Door seal • Windscreen hinge • Main harness grommet • Air-con pipe grommet	• For door seal GO to Pinpoint Test K • For windscreen hinge GO to Pinpoint Test R • For main harness grommet GO to Pinpoint Test S • For air-con pipe grommet GO to Pinpoint Test T
In rear luggage compartment/seating area **(entry point - lower body side/rear)**	• Rivets in body side • Tail lights/rear end capping	• For rivets in body side GO to Pinpoint Test U • For tail lights/rear end capping GO to Pinpoint Test V
In rear luggage compartment/seating area **(entry point - upper body side/rear)**	• Corners of upper body quarter panels • Upper to lower body joint (waist rear)	• Corners of upper body quarter panels GO to Pinpoint Test W • For waist seal rear (waist rear) GO to Pinpoint Test X
In rear luggage compartment/seating area **(entry point - tail door)**	• Rear end door • Upper to lower body joint • Roof to upper body joint	• Rear end door GO to Pinpoint Test Y • For upper to lower body joint GO to Pinpoint Test O • For roof to upper body joint GO to Pinpoint Test Q
In rear luggage compartment/seating area **(entry point - roof)**	• Roof seam • Alpine light's (if installed) • Front corner seams • Drain channel crack	• For roof seam GO to Pinpoint Test Z • For alpine light's (if installed) GO to Pinpoint Test AA • For front corner seams GO to Pinpoint Test AB • For drain channel crack GO to Pinpoint Test AC

PINPOINT TEST A : HEATER CABLE GROMMET	
TEST CONDITIONS	DETAILS/RESULTS/ACTIONS
A1: HEATER CABLE GROMMET	
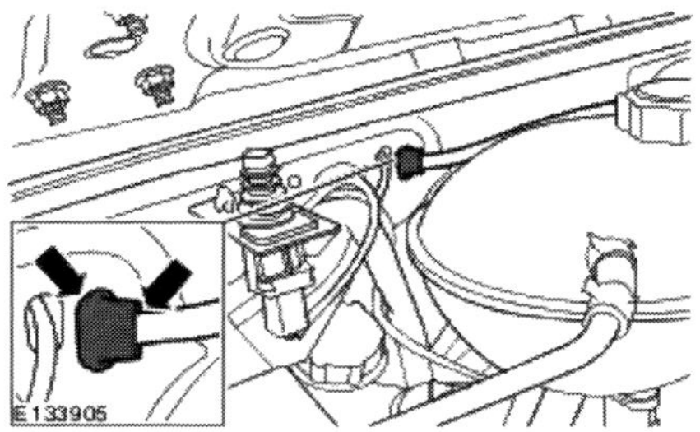	1. Apply sealant around the outer edge of the grommet and onto the grommet to cable joint
	2. Materials required: sealant (clear/body coloured), gloves, spirit wipe, cloth
	3. Remove excess sealant from the body area
	4. Allow the sealant to cure, and retest suspect area for water entry
	Is water still entering the vehicle? **Yes** Carry out a water leak test and visual inspection of any suspect or wet areas, once the water entry point has been determined refer to the "Water Ingress Paths And Recommended Repair Procedure" above **No** Reinstall any trim/panels or component that have been displaced

PINPOINT TEST B : WINDSCREEN WIPER GROMMET	
TEST CONDITIONS	DETAILS/RESULTS/ACTIONS
B1: WINDSCREEN WIPER GROMMET	

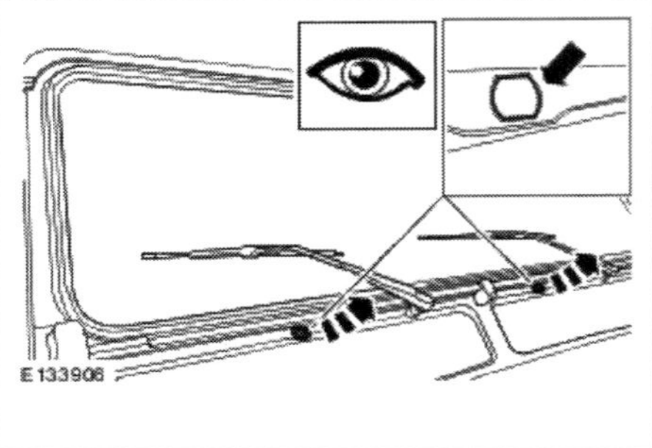	1. Remove the grommets, check the grommet and hole apertures for abnormalities/distortion. Apply sealant to the outer edge of the grommet and reinstall
	2. Remove excess sealant from the body area
	3. Materials required: sealant (clear or black), gloves, spirit wipe, cloth
	4. Allow the sealant to cure, and retest suspect area for water entry
	Is water still entering the vehicle? **Yes** Carry out a water leak test and visual inspection of any suspect or wet areas, once the water entry point has been determined refer to the "Water Ingress Paths And Recommended Repair Procedure" above **No** Reinstall any trim/panels or component that have been displaced

PINPOINT TEST C : WINDSCREEN SEAL

TEST CONDITIONS	DETAILS/RESULTS/ACTIONS
C1: WINDSCREEN SEAL	
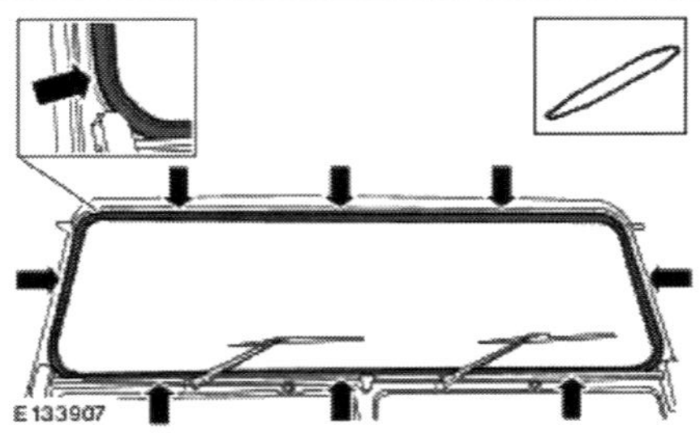	1. Using a "boning" tool, lift the edge of windscreen seal and insert the sealant gun nozzle, apply a bead of sealant between the seal and outer frame
	2. Remove excess sealant from the body area
	3. Materials required: sealant (clear/body coloured), gloves, spirit wipe, cloth, "Boning" tool
	4. Allow the sealant to cure, and retest suspect area for water entry
	Is water still entering the vehicle? **Yes** Carry out a water leak test and visual inspection of any suspect or wet areas, once the water entry point has been determined refer to the "Water Ingress Paths And Recommended Repair Procedure" above **No** Reinstall any trim/panels or component that have been displaced

PINPOINT TEST D : FOOTWELL SEAMS

TEST CONDITIONS	DETAILS/RESULTS/ACTIONS
D1: FOOTWELL SEAMS	
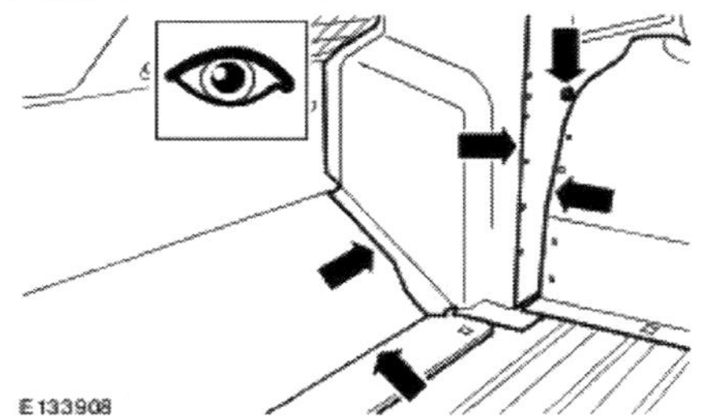	1. △ NOTE: A 12 volt hand lamp shone underneath the vehicle and viewed form the interior may assist in tracing leak paths Lift or remove floor mats/carpets. With the hood open carefully direct water from the top, down the scuttle panel and under the vehicle. There are several seams within the footwell, water entry may be from any or all of these, also visually check and seal, the grommets and apertures in the front of the tunnel area as required
	2. Run sealant along the seams and wipe into joint with a gloved hand. Grommets can be removed and sealant applied around the outer edge before reinstalling
	3. Materials required: sealant (clear/body coloured), gloves, spirit wipe, cloth
	4. Allow the sealant to cure, and retest suspect area for water entry
	Is water still entering the vehicle? **Yes** Carry out a water leak test and visual inspection of any suspect or wet areas, once the water entry point has been determined refer to the "Water Ingress Paths And Recommended Repair Procedure" above **No** Reinstall any trim/panels or component that have been displaced

PINPOINT TEST E : FLOOR SEAMS

TEST CONDITIONS	DETAILS/RESULTS/ACTIONS
E1: FLOOR SEAMS	
NOTE: The floor plate is secured using screws and a foam construction gasket	
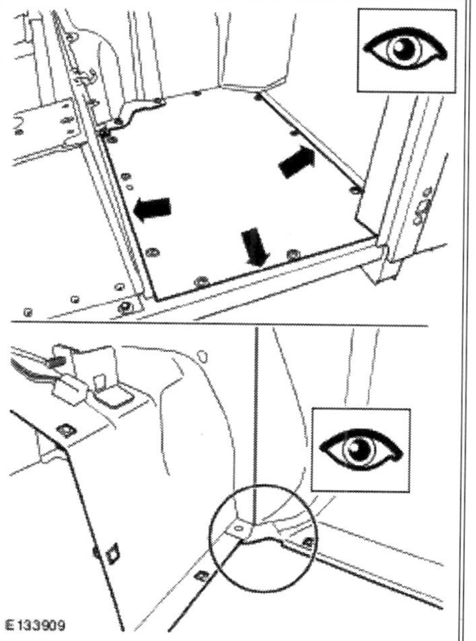	1. Lift or remove floor mats/carpets. With the hood open carefully direct water from the top, down the scuttle panel and under the vehicle
	2. **Pay particular attention to the corners of the floor panel** as there is an overlap condition where multiple panels meet, apply sealant and wipe it into the overlapping panel edges and corners this should prevent further water entry. Sealant can be added to the gasket of the removable floor plate as required
	3. Materials required: sealant (clear/body coloured), gloves, spirit wipe, cloth
	4. Allow the sealant to cure, and retest suspect area for water entry
	Is water still entering the vehicle? **Yes** Carry out a water leak test and visual inspection of any suspect or wet areas, once the water entry point has been determined refer to the "Water Ingress Paths And Recommended Repair Procedure" above **No** Reinstall any trim/panels or component that have been displaced

PINPOINT TEST F : HOOD RELEASE GROMMET

TEST CONDITIONS	DETAILS/RESULTS/ACTIONS
F1: HOOD RELEASE GROMMET	
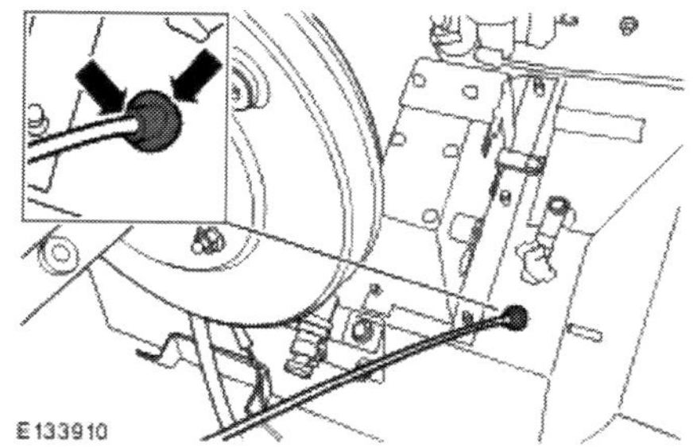	1. Apply sealant around the outer edge of the grommet and onto the grommet to cable joint
	2. Remove excess sealant from the body area
	3. Materials required: sealant (clear/body coloured), gloves, spirit wipe, cloth
	4. Allow the sealant to cure, and retest suspect area for water entry
	Is water still entering the vehicle? **Yes** Carry out a water leak test and visual inspection of any suspect or wet areas, once the water entry point has been determined refer to the "Water Ingress Paths And Recommended Repair Procedure" above **No** Reinstall any trim/panels or component that have been displaced

PINPOINT TEST G : PEDAL BOX GROMMET	
TEST CONDITIONS	DETAILS/RESULTS/ACTIONS
G1: PEDAL BOX GROMMET	
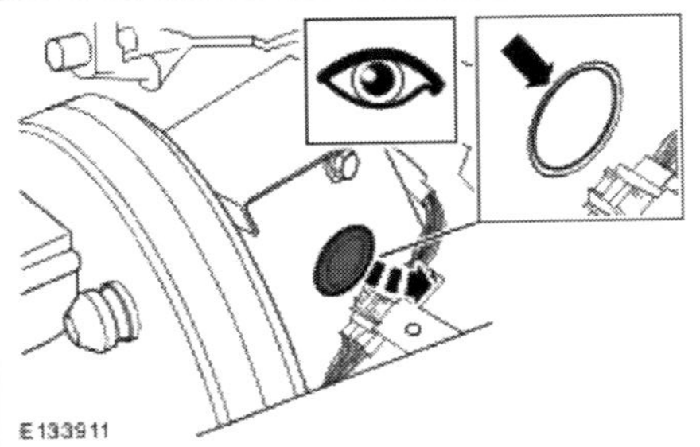 E133911	1. Remove the grommet and check both the grommet and hole aperture for abnormalities/distortion, apply sealant to the outer edge of the grommet and reinstall
	2. Remove excess sealant from the body area
	3. Materials required: sealant (clear/body coloured), gloves, spirit wipe, cloth
	4. Allow the sealant to cure, and retest suspect area for water entry
	Is water still entering the vehicle? **Yes** Carry out a water leak test and visual inspection of any suspect or wet areas, once the water entry point has been determined refer to the "Water Ingress Paths And Recommended Repair Procedure" above **No** Reinstall any trim/panels or component that have been displaced

PINPOINT TEST H : AIR INTAKE DUCT TO BLOWER	
TEST CONDITIONS	DETAILS/RESULTS/ACTIONS
H1: AIR INTAKE DUCT TO BLOWER	
NOTE: The wing install process is to lower the wing into place along the face of the scuttle panel, this can displace the foam seal attached to the blower box	
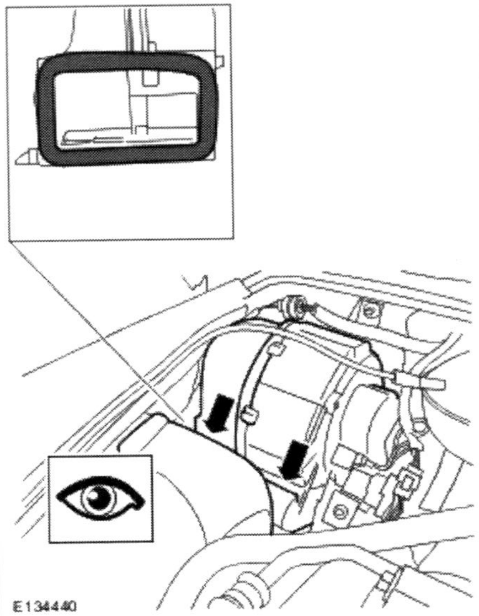 E134440	1. Check the condition and location of the foam gasket joint between the air intake duct, and the blower box installed to the scuttle panel
	2. If possible apply sealant to the aperture with the wing installed
	3. If it is not possible to seal the aperture with the wing installed, remove the wing to complete the repair
	4. Materials required: sealant (clear/body coloured), gloves, spirit wipe, cloth
	5. Allow the sealant to cure, and retest suspect area for water entry
	Is water still entering the vehicle? **Yes** Carry out a water leak test and visual inspection of any suspect or wet areas, once the water entry point has been determined refer to the "Water Ingress Paths And Recommended Repair Procedure" above **No** Reinstall any trim/panels or component that have been displaced

PINPOINT TEST I : BLOWER BOX

TEST CONDITIONS	DETAILS/RESULTS/ACTIONS
I1: BLOWER BOX	
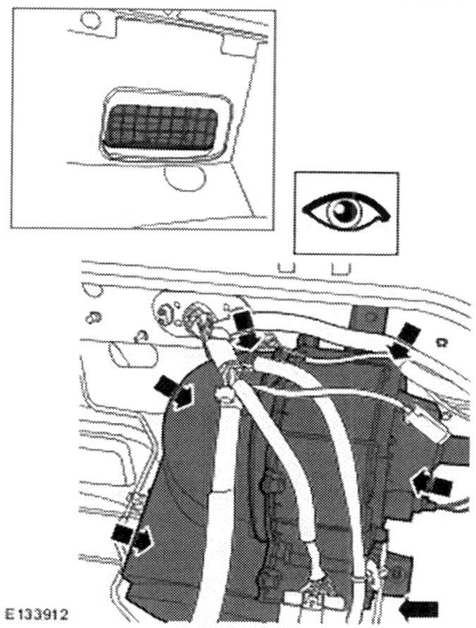	1. Direct water around the blower box, if a leak is found and it is possible to identify the particular area of the leak then sealant can be applied to the seam where the blower box sits on the bulkhead and wiped in if possible
	2. If it is not possible to highlight the actual area of the ingress the remove the blower box and check the integrity of the foam gasket, replace gasket as required, adding a fine bead of sealant to the gasket before carefully reinstalling to ensure a good seal
	3. On some 07 model year vehicles there may be water actually dripping through the blower box grille in the passenger footwell, this is caused by a problem with the sealing of the two halves of the blower box itself, and replacement of the blower box is recommended
	4. Materials required: sealant (clear/body coloured), gloves, spirit wipe, cloth, replacement gasket, blower box

	5. Allow the sealant to cure, and retest suspect area for water entry
	Is water still entering the vehicle? **Yes** Carry out a water leak test and visual inspection of any suspect or wet areas, once the water entry point has been determined refer to the "Water Ingress Paths And Recommended Repair Procedure" above **No** Reinstall any trim/panels or component that have been displaced

PINPOINT TEST J : AIR INTAKE

TEST CONDITIONS	DETAILS/RESULTS/ACTIONS
J1: AIR INTAKE	
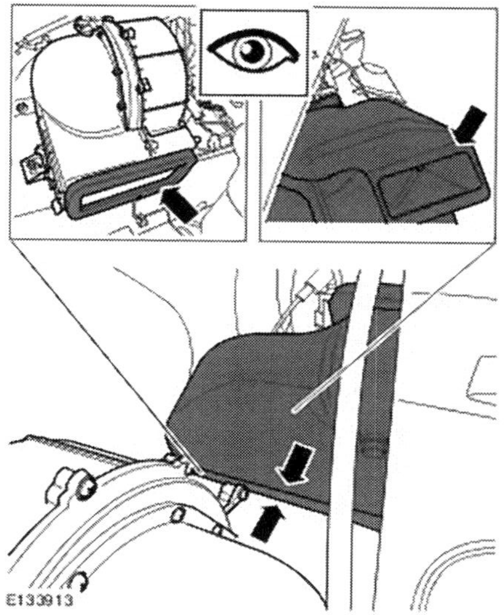	1. Ensure that the interfaces shown in Fig 1 and Fig 2 are correctly made, no ripples or damage. Fig 3 indicates the 2 faces when installed
	2. For repairs to a minor water ingress issue, it may be possible to apply a bead of sealant to affected area and smooth in
	3. For removal and installation of the wing air duct (fig 1). Care must be taken to ensure a good mating interface between the duct and foam gasket on the blower

	4. Materials required: sealant (clear/body coloured), gloves, spirit wipe, cloth
	5. Allow the sealant to cure, and retest suspect area for water entry
	Is water still entering the vehicle? **Yes** Carry out a water leak test and visual inspection of any suspect or wet areas, once the water entry point has been determined refer to the "Water Ingress Paths And Recommended Repair Procedure" above **No** Reinstall any trim/panels or component that have been displaced

PINPOINT TEST K : CLUTCH/BRAKE PEDAL BOX	
TEST CONDITIONS	DETAILS/RESULTS/ACTIONS
K1: CLUTCH/BRAKE PEDAL BOX	
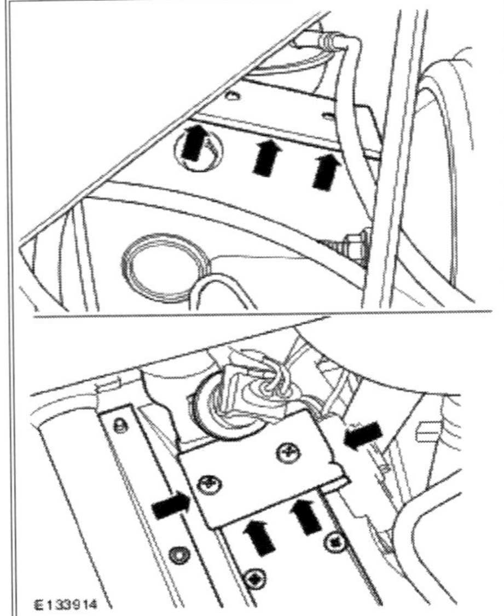 E133914	1. Apply sealant around the pedal box seams, wipe sealant into seams
	2. Remove excess sealant from the body area

	3. Materials required: sealant (clear or black), gloves, spirit wipe, cloth
	4. Allow the sealant to cure, and retest suspect area for water entry
	Is water still entering the vehicle? **Yes** Carry out a water leak test and visual inspection of any suspect or wet areas, once the water entry point has been determined refer to the "Water Ingress Paths And Recommended Repair Procedure" above **No** Reinstall any trim/panels or component that have been displaced

PINPOINT TEST L : WINDSCREEN SURROUND FRAME	
TEST CONDITIONS	DETAILS/RESULTS/ACTIONS
L1: WINDSCREEN SURROUND FRAME	
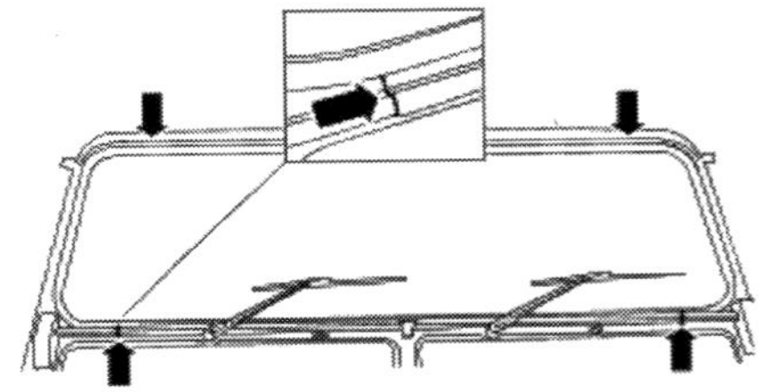 E134933	1. Visually inspect the joint condition between the side casting and the centre extrusion for pinholes and cracks at the top and bottom on both sides
	2. Apply sealant to the joint if pinholes and cracks
	3. Materials required: sealant (clear/body coloured), gloves, spirit wipe, cloth
	4. Remove excess sealant from the body area
	5. Allow the sealant to cure, and retest suspect area for water entry
	Is water still entering the vehicle? **Yes** Carry out a water leak test and visual inspection of any suspect or wet areas, once the water entry point has been determined refer to the "Water Ingress Paths And Recommended Repair Procedure" above **No** Reinstall any trim/panels or component that have been displaced

M1: DOOR APERTURE SEAL (PATCH'S)	
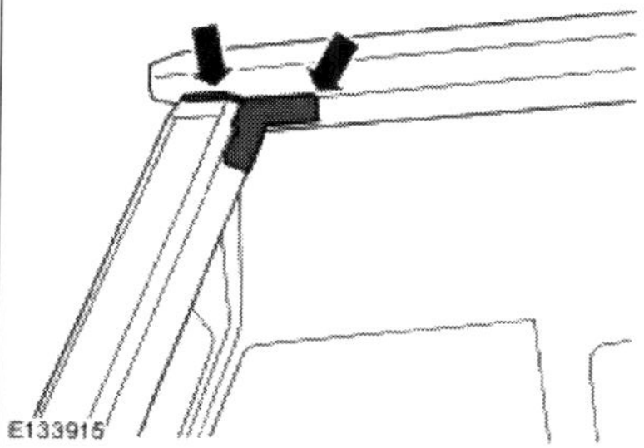 E133915	1. Install the "L" shaped door joint patch ensuring the apex of the patch is aligned with the apex of the roof/A post joint. Once installed the patch should be free from creases/bubbles. Apply a thin bead of sealant approx 30 - 50mm long between roof and A post flange. Smooth sealant into flange area
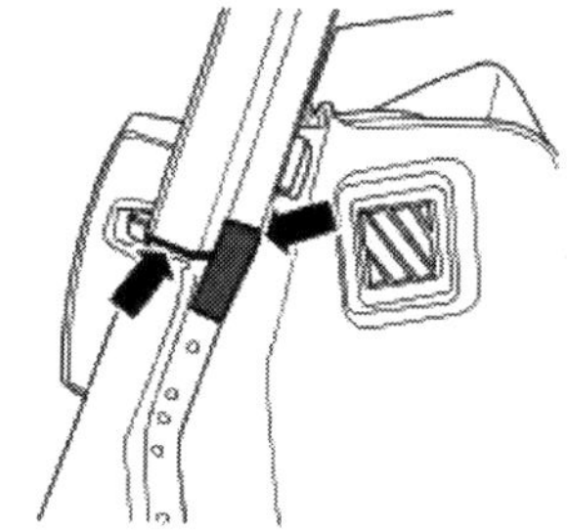 E133916	2. Install the door joint patch ensuring the patch is aligned with the flange. Patch should be free from creases/bubbles. Apply a thin bead of sealant approx 20 - 40mm long across the windscreen casting and scuttle panel joint flange. Apply sealant to the scuttle panel to windscreen casting joint to ensure a water tight seal. Smooth sealant into flange area
 E133917	3. Attach the door joint patch ensuring the patch is aligned with the flange. The patch should be free from creases/bubbles. Apply a thin bead of sealant approx 20 - 40mm long across the body side and lower body flange. Smooth sealant into flange area. Ensure there is no gap between the waist seal and the newly applied sealant
	4. Remove excess sealant from the body area
	5. Materials required: sealant (clear/body coloured) door joint patch(s)
	6. Allow the sealant to cure, and retest suspect area for water entry

	Is water still entering the vehicle? **Yes** Carry out a water leak test and visual inspection of any suspect or wet areas, once the water entry point has been determined refer to the "Water Ingress Paths And Recommended Repair Procedure" above **No** Reinstall any trim/panels or component that have been displaced

PINPOINT TEST N : WATER SHEDDER

TEST CONDITIONS	DETAILS/RESULTS/ACTIONS
N1: WATER SHEDDER	
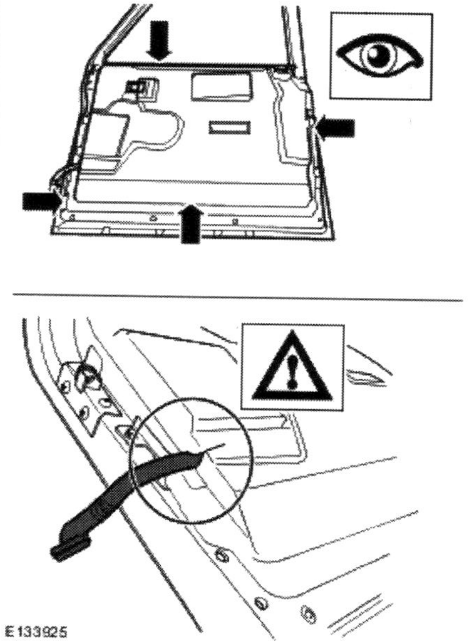 E133925	1. NOTE: Care Point – check that the drainage holes in the bottom of the door are not blocked with sealant/dirt or debris (if vehicle has been off road) Visual check all around edge of shedder, looking for areas of poor adhesion of the butyl sealant between the shedder and the door panel. Due to the complexity of the convoluted tube joint (circled) this is an area for careful inspection, as it a typical leak path. Repairs can be made by rolling or pressing the butyl seal to improve adhesion or by additional sealant / dum-dum in the localized area
	2. Check the condition of the flip seal at the bottom of the door if damaged or misaligned may allow water ingress
	3. Materials required: "paint" roller, sealant, dum-dum
	4. Allow the sealant to cure, and retest suspect area for water entry

TEST CONDITIONS	DETAILS/RESULTS/ACTIONS
	Is water still entering the vehicle? **Yes** Carry out a water leak test and visual inspection of any suspect or wet areas, once the water entry point has been determined refer to the "Water Ingress Paths And Recommended Repair Procedure" above **No** Reinstall any trim/panels or component that have been displaced

PINPOINT TEST O : DOOR LATCH/DOOR EDGE

TEST CONDITIONS	DETAILS/RESULTS/ACTIONS
O1: DOOR LATCH/DOOR EDGE	
⚠ NOTE: Water leak path is visible down the edge of the door and into the A post/footwell/sill area of the vehicle	
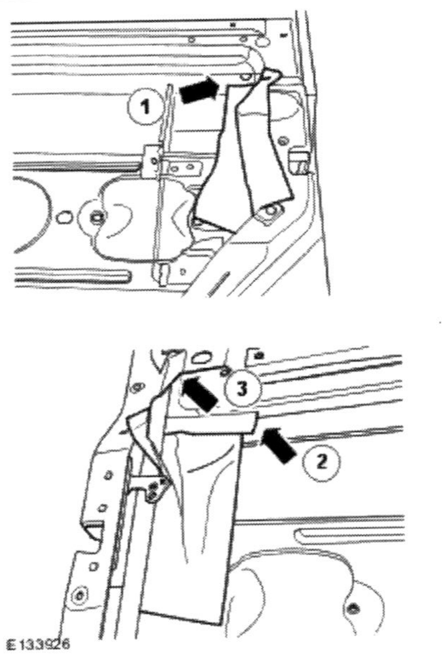	1. ⚠ NOTE: Use of a mirror and torch will assist Remove the door casing. Remove the door shedder to gain access to the latch mechanism. Check along the top of the latch shedder for damage/poor adhesion (Fig 1)
	2. Apply sealant / dum-dum along the top of the latch shedder (fig 2)
	3. Apply sealant to the panel joints above the latch shedder but underneath the glass run channel (Fig 3)

TEST CONDITIONS	DETAILS/RESULTS/ACTIONS
	4. Materials required: sealant, dum-dum, tools (for door casing removal), **door shedder – must be replaced**
	5. Allow the sealant to cure, and retest suspect area for water entry
	Is water still entering the vehicle? **Yes** Carry out a water leak test and visual inspection of any suspect or wet areas, once the water entry point has been determined refer to the "Water Ingress Paths And Recommended Repair Procedure" above **No** Reinstall any trim/panels or component that have been displaced

PINPOINT TEST P : REAR TUB CAPPING

TEST CONDITIONS	DETAILS/RESULTS/ACTIONS
P1: REAR TUB CAPPING	
⚠ NOTE: Whilst it is not necessary to remove the door aperture seal to complete this repair, care must be taken to ensure the repair seals any gaps up to and including the joint seal	
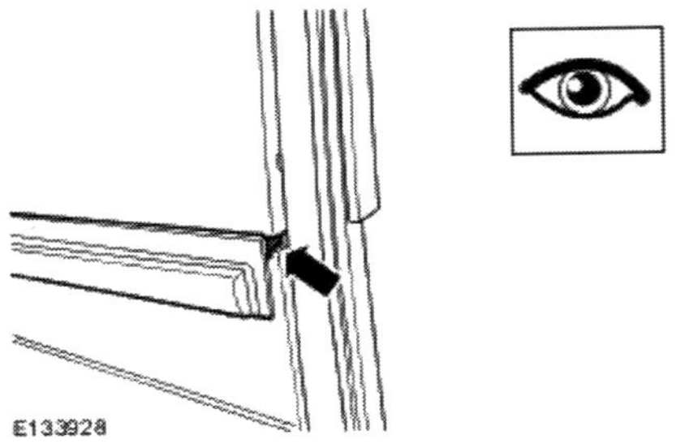	1. Apply small bead of sealant or dum-dum into the capping to body side joint (as arrow indicates)
	2. **Critical visual area = high quality of sealant finish required**
	3. Materials required: sealant, dum-dum
	4. Allow the sealant to cure, and retest suspect area for water entry

	Is water still entering the vehicle? **Yes** Carry out a water leak test and visual inspection of any suspect or wet areas, once the water entry point has been determined refer to the "Water Ingress Paths And Recommended Repair Procedure" above **No** Reinstall any trim/panels or component that have been displaced

PINPOINT TEST Q : UPPER TO LOWER BODY JOINT (WAIST SIDE)

TEST CONDITIONS	DETAILS/RESULTS/ACTIONS
Q1: UPPER TO LOWER BODY JOINT (WAIST SIDE)	
NOTES: The seal between the upper and lower body (waist seal) is made from a type of memory foam, the latest seal is in three sections which lock together, the seal is stuck to the upper body side but is not stuck to the rear tub	
A key area to inspect is the welded joint of the lower body capping. Sometimes the seal can be slightly inboard of the joint leaving a small cavity, by slackening off all the fixings of the body side it is possible with a plastic tool to push the seal back outwards closing the cavity. If it is a small cavity it is possible to seal the cavity with a small amount of sealant wiped into the cavity and any excess cleaned off, if the vehicle is a model that is trimmed inside then the cavity can also be sealed from the inside, a larger amount of sealant may be used if it will be hidden in normal usage by the interior trim	
Should the seal itself require replacement the upper body side can be removed without removing the roof panel itself by removing the upper and lower fixings and slackening the front and rear fixings, the roof panel can then be supported using a temporary support whilst the upper body side is removed to facilitate the seal change	
	1. Care must be taken to ensure that the seal between the roof an upper body side is not compromised, and the seal across the front between the windscreen surround and the roof is still correct after installation
	2. Materials required: sealant (clear or black), gloves, spirit wipe, cloth, tools, seals as required
	3. Allow the sealant to cure, and retest suspect area for water entry

	Is water still entering the vehicle? **Yes** Carry out a water leak test and visual inspection of any suspect or wet areas, once the water entry point has been determined refer to the "Water Ingress Paths And Recommended Repair Procedure" above **No** Reinstall any trim/panels or component that have been displaced

PINPOINT TEST R : ROOF TO "A" POST JOINT

TEST CONDITIONS	DETAILS/RESULTS/ACTIONS
R1: ROOF TO "A" POST JOINT	
NOTE: On some vehicles the gap at the top of the roof is filled with "dum-dum" on later vehicles this was changed to black, repair with original material	
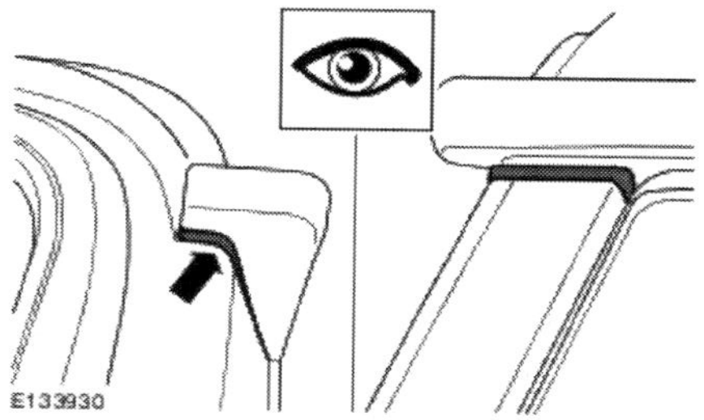	1. Fill any gaps with sealant or dum-dum and wipe off excess
	2. For leaks above this area refer to (front corner seams) GO to Pinpoint Test AC or (Drain Channel Seal) GO to Pinpoint Test AC AD
	3. Materials required: sealant (black or white as appropriate) or dum-dum (black), gloves
	4. Allow the sealant to cure, and retest suspect area for water entry

	Is water still entering the vehicle? **Yes** Carry out a water leak test and visual inspection of any suspect or wet areas, once the water entry point has been determined refer to the "Water Ingress Paths And Recommended Repair Procedure" above **No** Reinstall any trim/panels or component that have been displaced

PINPOINT TEST S : ROOF TO UPPER BODY GASKET

TEST CONDITIONS	DETAILS/RESULTS/ACTIONS
S1: ROOF TO UPPER BODY GASKET	
NOTE: Visually inspect the roof area to ensure that the gasket is installed / seated correctly. Check for lumps / ripples or trapped conditions, as these are the probable cause of the water ingress	
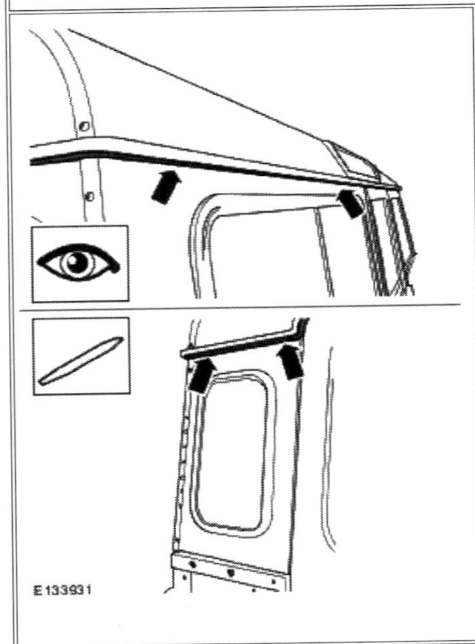	1. Undo/slacken the fixing bolts on the inside of the roof (as required lower the headlining to access the fixings on some derivatives). Working outside of the vehicle, using a non metallic tool (boning tool) prize out the gasket to ensure correct fit. Re-tighten the roof
	2. It is not usually necessary to use sealant in this repair however if the gasket appears to require additional sealing medium, a small run of sealant is acceptable. Care should be taken not to distort the gasket when applying the sealant
	3. Materials required: boning tool, tools, sealant

	4. Allow the sealant to cure, and retest suspect area for water entry
	Is water still entering the vehicle? **Yes** Carry out a water leak test and visual inspection of any suspect or wet areas, once the water entry point has been determined refer to the "Water Ingress Paths And Recommended Repair Procedure" above **No** Reinstall any trim/panels or component that have been displaced

PINPOINT TEST T : WINDSCREEN HINGE

TEST CONDITIONS	DETAILS/RESULTS/ACTIONS
T1: WINDSCREEN HINGE	
NOTES: For pre-2009MY vehicles or if there was a significant leak around this area, it may be pertinent to apply sealant along the scuttle panel to windscreen casting join prior to reinstalling the hinge	
Critical visual area = high quality of sealant finish required	
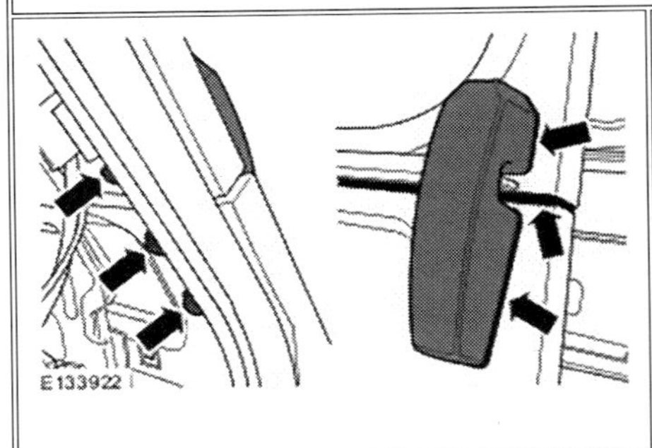	1. From inside the vehicle undo the 3 fixings that hold the windscreen hinge in place. Remove the hinge from the vehicle. Remove the existing gasket and any residue of adhesives or sealant. Re-fit the hinge gasket. Apply a small amount of sealant around each fixing hole
	2. Materials required: sealant, windscreen hinge gasket
	3. Allow the sealant to cure, and retest suspect area for water entry

	Is water still entering the vehicle? **Yes** Carry out a water leak test and visual inspection of any suspect or wet areas, once the water entry point has been determined refer to the "Water Ingress Paths And Recommended Repair Procedure" above **No** Reinstall any trim/panels or component that have been displaced

PINPOINT TEST U : MAIN HARNESS GROMMET

TEST CONDITIONS	DETAILS/RESULTS/ACTIONS
U1: MAIN HARNESS GROMMET	
NOTES: ⚠ Visually inspect the grommet looking for areas of poor seating	
⚠ Inspect the "U" channel to skuttle panel for distortion between spot welds allowing water to seep onto the grommet	
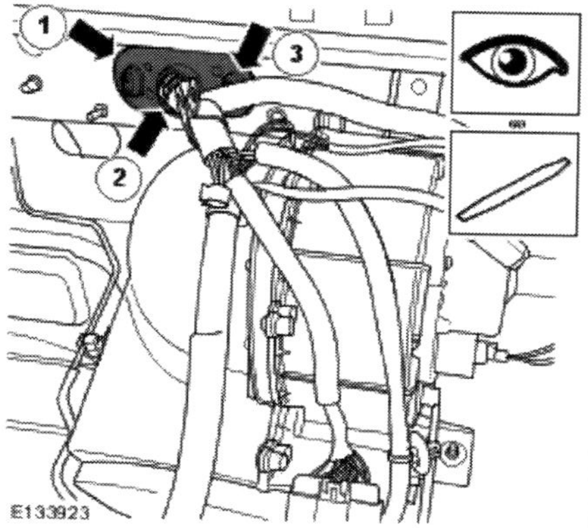	1. Use a boning tool to re-seat the grommet. Additional sealant should be applied around the outer edges of the grommet and the body panel
	2. On occasions it maybe necessary to seal the main harness collar in addition to outer grommet. This will be evident during re-test
	3. Materials required: sealant, boning tool
	4. Allow the sealant to cure, and retest suspect area for water entry

	Is water still entering the vehicle? **Yes** Carry out a water leak test and visual inspection of any suspect or wet areas, once the water entry point has been determined refer to the "Water Ingress Paths And Recommended Repair Procedure" above **No** Reinstall any trim/panels or component that have been displaced

PINPOINT TEST V : AIR CONDITIONING PIPE GROMMET

TEST CONDITIONS	DETAILS/RESULTS/ACTIONS
V1: AIR CONDITIONING PIPE GROMMET	
⚠ NOTE: Visual inspect around grommet looking for areas of poor seating	
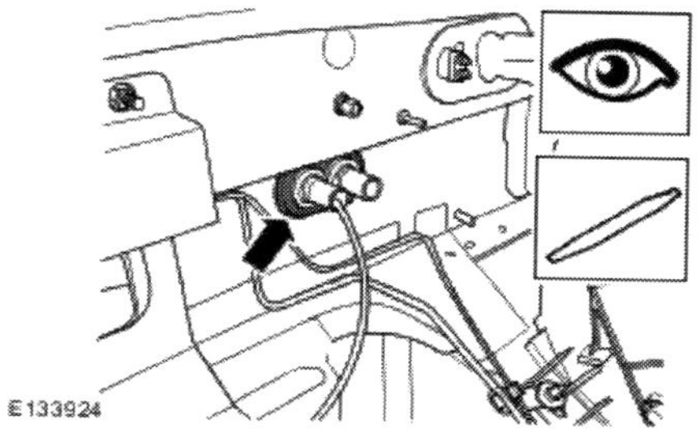	1. Use a boning tool to re-seat the grommet. Additional sealant should be applied around the outer edges of the grommet and the body panel
	2. Materials required: sealant, boning tool
	3. Allow the sealant to cure, and retest suspect area for water entry
	Is water still entering the vehicle? **Yes** Carry out a water leak test and visual inspection of any suspect or wet areas, once the water entry point has been determined refer to the "Water Ingress Paths And Recommended Repair Procedure" above **No** Reinstall any trim/panels or component that have been displaced

PINPOINT TEST W : RIVETS IN BODY SIDE

TEST CONDITIONS	DETAILS/RESULTS/ACTIONS
W1: RIVETS IN BODY SIDE	
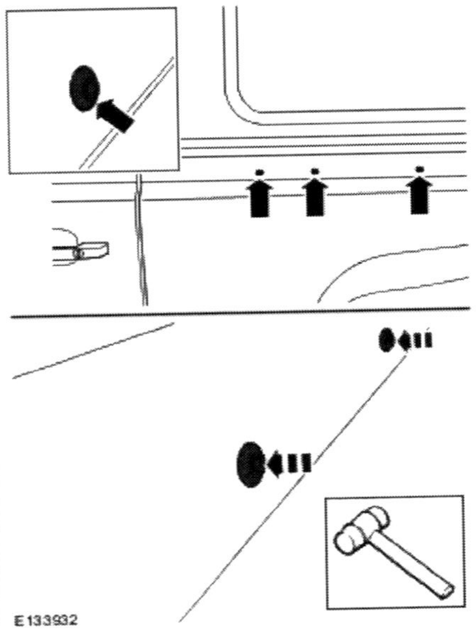 E133932	1. Apply sealant around the rivet and wipe excess away with spirit wipe, the rivet can also be tightened by tapping lightly with a mallet and a nylon block
	2. If leak is still present the rivet should be drilled and replaced, and then painted with a touch up stick
	3. Materials required: sealant (clear/body coloured), gloves, spirit wipe, cloth, hammer, nylon block, rivet, rivet gun, touch up stick
	4. Allow the sealant to cure, and retest suspect area for water entry
	Is water still entering the vehicle? **Yes** Carry out a water leak test and visual inspection of any suspect or wet areas, once the water entry point has been determined refer to the "Water Ingress Paths And Recommended Repair Procedure" above **No** Reinstall any trim/panels or component that have been displaced

PINPOINT TEST X : TAIL LIGHTS/REAR END CAPPING

TEST CONDITIONS	DETAILS/RESULTS/ACTIONS
X1: TAIL LIGHTS/REAR END CAPPING	
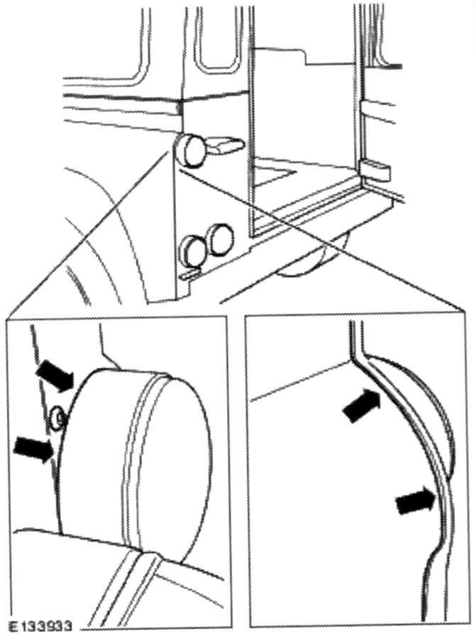 E133933	1. To seal the rear seam, squeeze sealant into the seam then smooth into the gap, wipe sealant into the joint then clean away an excess sealant with a spirit dampened cloth
	2. To seal the rear lamp remove rear lamp and apply sealant around circumference of lamp base, re fit and wipe away any excess sealant
	3. Materials required: sealant (clear/body coloured), gloves, spirit wipe cloth
	4. Allow the sealant to cure, and retest suspect area for water entry
	Is water still entering the vehicle? **Yes** Carry out a water leak test and visual inspection of any suspect or wet areas, once the water entry point has been determined refer to the "Water Ingress Paths And Recommended Repair Procedure" above **No** Reinstall any trim/panels or component that have been displaced

Should the seal itself require replacement the upper body side can be removed without removing the roof panel itself by removing the upper and lower fixings and slackening the front and rear fixings, the roof panel can then be supported using a temporary support whilst the upper body side is removed to facilitate the seal change		
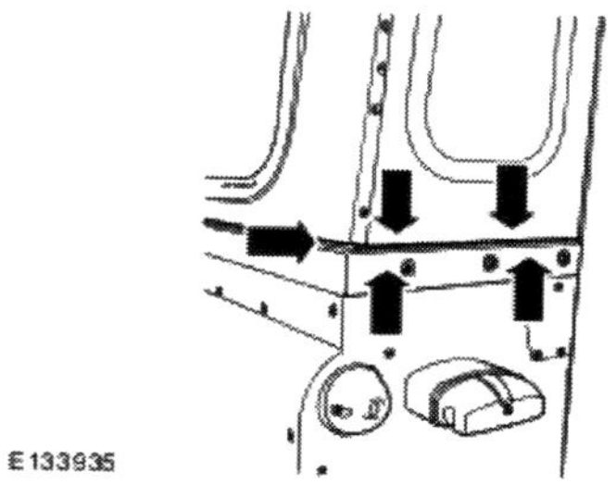	1. Care must be taken to ensure that the seal between the roof an upper body side is not compromised, and the seal across the front between the windscreen surround and the roof is still correctly located	
	2. Materials required: sealant (clear or black), gloves, spirit wipe cloth, tools for removal as required, seals as required	
	3. Allow the sealant to cure, and retest suspect area for water entry	
	Is water still entering the vehicle? **Yes** Carry out a water leak test and visual inspection of any suspect or wet areas, once the water entry point has been determined refer to the "Water Ingress Paths And Recommended Repair Procedure" above **No** Reinstall any trim/panels or component that have been displaced	

E133935

PINPOINT TEST Y CORNERS OF UPPER BODY QUARTER PANELS

TEST CONDITIONS	DETAILS/RESULTS/ACTIONS
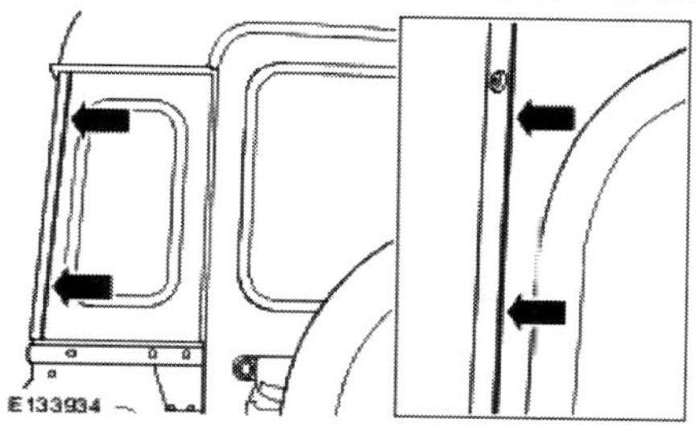 E133934	1. Squeeze sealant into seam then smooth into gap, wipe away an excess sealant
	2. Materials required: sealant (clear/body coloured), gloves, spirit wipe, cloth
	3. Allow the sealant to cure, and retest suspect area for water entry
	Is water still entering the vehicle? **Yes** Carry out a water leak test and visual inspection of any suspect or wet areas, once the water entry point has been determined refer to the "Water Ingress Paths And Recommended Repair Procedure" above **No** Reinstall any trim/panels or component that have been displaced

PINPOINT TEST Z UPPER TO LOWER BODY JOINT (WAIST REAR)

TEST CONDITIONS	DETAILS/RESULTS/ACTIONS
Z1: UPPER TO LOWER BODY JOINT (WAIST REAR)	

NOTES:

The seal between the upper and lower body (waist seal) is made from a type of memory foam, the latest seal is in three sections which lock together, the seal is stuck to the upper body side but is not stuck to the rear tub

Sometimes the seal can be slightly inboard of the joint leaving a small cavity, by slackening off all the fixings of the body side it is possible with a plastic tool to push the seal back outwards closing the cavity. If it is a small cavity it is possible to seal the cavity with a small amount of sealant wiped into the cavity and any excess cleaned off, if the vehicle is a model that is trimmed inside then the cavity can also be sealed from the inside, a larger amount of sealant may be used if it will be hidden in normal usage by the interior trim

PINPOINT TEST AA : REAR END DOOR

TEST CONDITIONS	DETAILS/RESULTS/ACTIONS
AA1: REAR END DOOR	
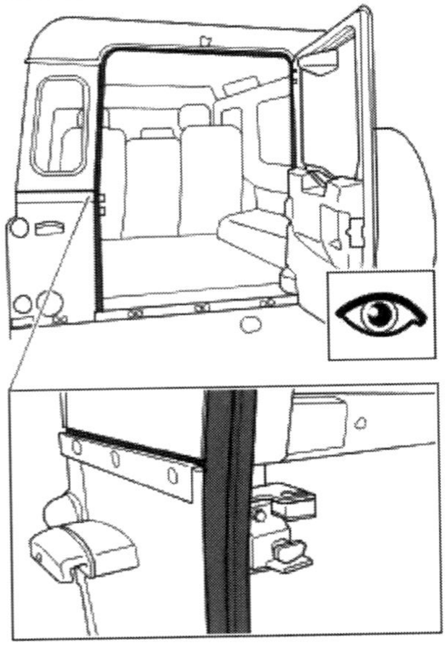 E134439	1. NOTE: Due to the location of the spare wheel, the rear end door may become miss-aligned to the body due to fixing torque relaxation. Check/set the torque of the fixing to 22 newton metres (plus or minus 3 newton metres) Check the profile and set/alignment of rear end door, adjust as required
	2. Check the aperture seal bubble section for splits due to door misalignment
	3. Check condition, location and security of aperture seal replace as required
	4. At the waist join, check for patches and sealant quality (See waist seal instructions above)
	5. Materials required: Aperture seal (replace as required), sealant (clear/body coloured), gloves, spirit wipe cloth, touch up stick
	6. Allow the sealant to cure, and retest suspect area for water entry
	Is water still entering the vehicle? **Yes** Carry out a water leak test and visual inspection of any suspect or wet areas, once the water entry point has been determined refer to the "Water Ingress Paths And Recommended Repair Procedure" above **No** Reinstall any trim/panels or component that have been displaced

PINPOINT TEST AB : ROOF SEAM

TEST CONDITIONS	DETAILS/RESULTS/ACTIONS
AB1: ROOF SEAM	
NOTE: High visibility area excess sealant must be wiped away leaving no residue or evidence of repair	
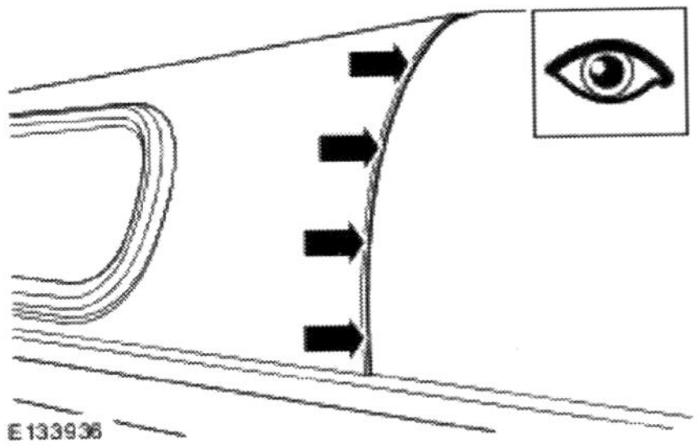 E133936	1. Visual inspect the entire roof seam looking for irregularities in the flange edge condition. Check for gaps between the panels and panel joint conditions. Apply sealant as required
	2. Materials required: sealant (body coloured preferred or clear), cloth, spirit wipe, gloves
	3. Allow the sealant to cure, and retest suspect area for water entry
	Is water still entering the vehicle? **Yes** Carry out a water leak test and visual inspection of any suspect or wet areas, once the water entry point has been determined refer to the "Water Ingress Paths And Recommended Repair Procedure" above **No** Reinstall any trim/panels or component that have been displaced

PINPOINT TEST AB : ROOF SEAM

TEST CONDITIONS	DETAILS/RESULTS/ACTIONS
AB1: ROOF SEAM	
NOTE: High visibility area excess sealant must be wiped away leaving no residue or evidence of repair	
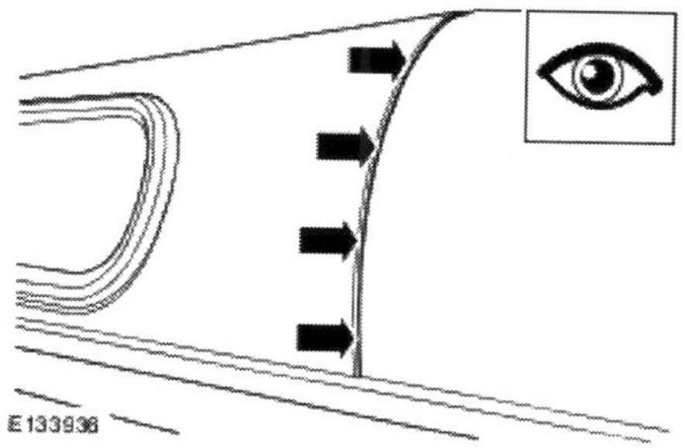	1. Visual inspect the entire roof seam looking for irregularities in the flange edge condition. Check for gaps between the panels and panel joint conditions. Apply sealant as required
	2. Materials required: sealant (body coloured preferred or clear), cloth, spirit wipe, gloves
	3. Allow the sealant to cure, and retest suspect area for water entry
	Is water still entering the vehicle? **Yes** Carry out a water leak test and visual inspection of any suspect or wet areas, once the water entry point has been determined refer to the "Water Ingress Paths And Recommended Repair Procedure" above **No** Reinstall any trim/panels or component that have been displaced

PINPOINT TEST AC : ALPINE LIGHTS

TEST CONDITIONS	DETAILS/RESULTS/ACTIONS
AC1: ALPINE LIGHTS	
NOTE: High visibility area excess sealant must be wiped away leaving no residue or evidence of repair	
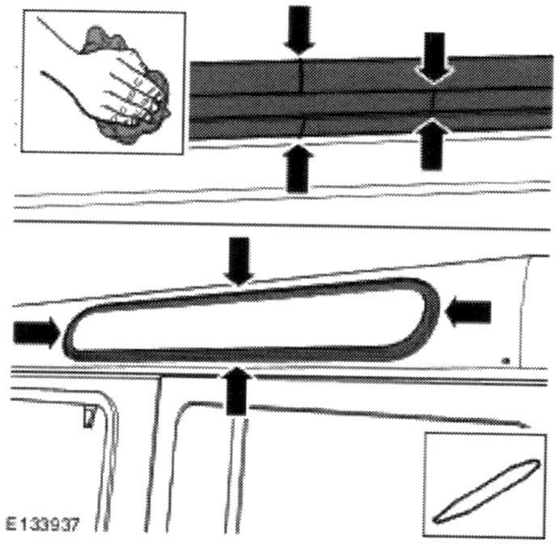	1. **Gasket joint leaks** - Using black sealant, seal between the joint edges of the gasket
	2. **Gasket to roof leaks** - Due to the design of the gasket seal this is a very rare leak path. The leak is probably caused by deformation to the gasket which should be evident by visual check. The gasket should be re-seated and re-tested prior to completing this repair. Using the boning tool carefully lift the outer edge of the gasket seal. Working around the alpine light, lift and apply sealant between the gasket and the roof panel. Care **must** be taken to ensure that the roof is not damaged or scratched whilst sealing the alpine light
	3. For leaks through the centre (lace section) apply sealant between the joint and wipe away an excess sealant
	4. Materials required: sealant (clear/black), boning tool, cloths, spirit wipe, gloves

	5. Allow the sealant to cure, and retest suspect area for water entry
	Is water still entering the vehicle? **Yes** Carry out a water leak test and visual inspection of any suspect or wet areas, once the water entry point has been determined refer to the "Water Ingress Paths And Recommended Repair Procedure" above **No** Reinstall any trim/panels or component that have been displaced

PINPOINT TEST AD : FRONT CORNER SEAMS

TEST CONDITIONS	DETAILS/RESULTS/ACTIONS
AD1: FRONT CORNER SEAMS	
NOTES: Arrow indicates the usual area for front roof cracks (wisp cracks)	
Care point – it may be necessary to touch up the paint work using a paint touch up stick	
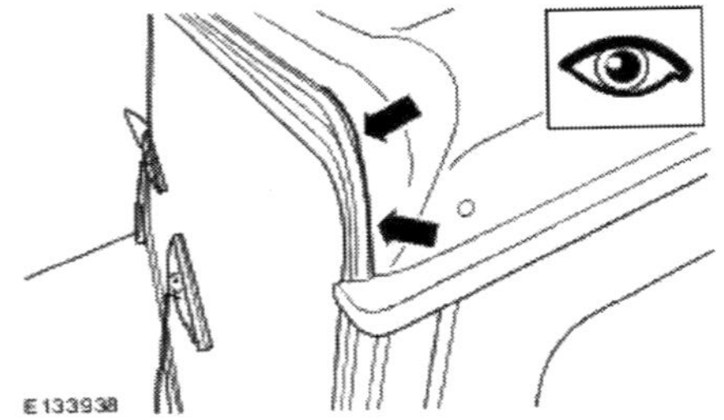 E133938	1. Apply sealant into the cracked area then remove excess sealant
	2. Materials required: sealant (clear/body coloured) cloth, spirit wipe, gloves, paint
	3. Allow the sealant to cure, and retest suspect area for water entry
	Is water still entering the vehicle? **Yes** Carry out a water leak test and visual inspection of any suspect or wet areas, once the water entry point has been determined refer to the "Water Ingress Paths And Recommended Repair Procedure" above **No** Reinstall any trim/panels or component that have been displaced

PINPOINT TEST AE : DRAIN CHANNEL SEAL

TEST CONDITIONS	DETAILS/RESULTS/ACTIONS
AE1: DRAIN CHANNEL SEAL	
NOTE: Before any repairs take place ensure that the issue is not caused by an incorrectly installed or non approved roof rack	
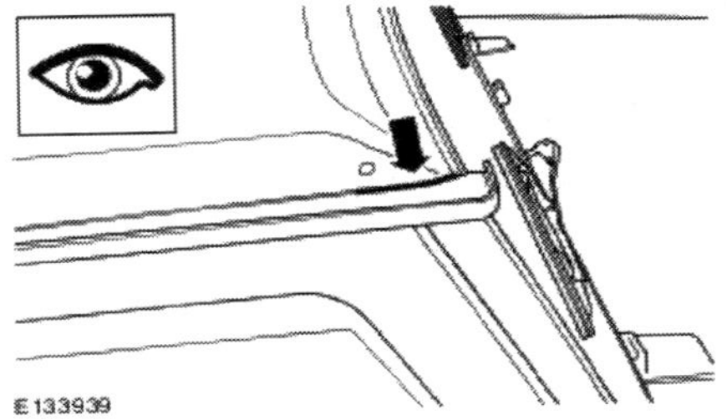 E133939	1. If the crack is minor it can be repaired by applying clear/body coloured sealant and then wiping it into the crack with a gloved hand. If the crack is larger the vehicle may require a body shop repair which will require the body sealant scraping out of the channel, replacing with new body sealant and repaint
	2. Materials required: sealant (clear/body coloured), cloth, spirit wipe, gloves
	3. Allow the sealant to cure, and retest suspect area for water entry
	Is water still entering the vehicle? **Yes** Carry out a water leak test and visual inspection of any suspect or wet areas, once the water entry point has been determined refer to the "Water Ingress Paths And Recommended Repair Procedure" above **No** Reinstall any trim/panels or component that have been displaced

Vehicle Specific Information and Tolerance Checks - Body and Frame

Description and Operation

CHASSIS AND BODY

Chassis

The chassis on all Defender variants are of the box section, ladder type construction, manufactured from 2 mm (14 swg) steel and treated with zinc phosphate, cathodic electro coated, followed by waxing in the rear cross member.

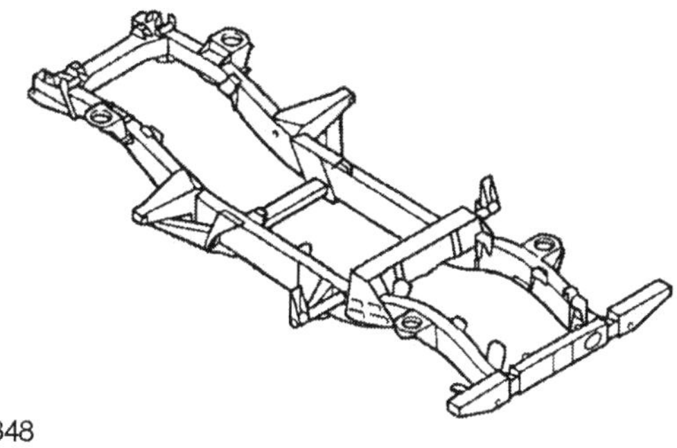

J6348

Outriggers and angled brackets welded to the chassis support suspension and axle components and are also used as body mounting points. A detachable box section cross member, located between the two chassis longitudinals is fitted to facilitate main gearbox and transfer box assembly removal.

Should chassis damage occur, a comprehensive range of components are available, including body support outriggers, cross members and radius arm mounting brackets. ALWAYS fit genuine parts that are fully guaranteed and to original equipment specification, fitted with Land Rover's BS 5135 welding standard.

Body

M764325

All body panels, with the exception of the fascia bulkhead, are manufactured from aluminium alloy. Galvanized steel is used for the front wheel arches to give optimum protection. Most panels are also treated with zinc phosphate and cathodic electro coated with polyester surfacer, and are bolted to the welded chassis.

CHASSIS FRAME ALIGNMENT

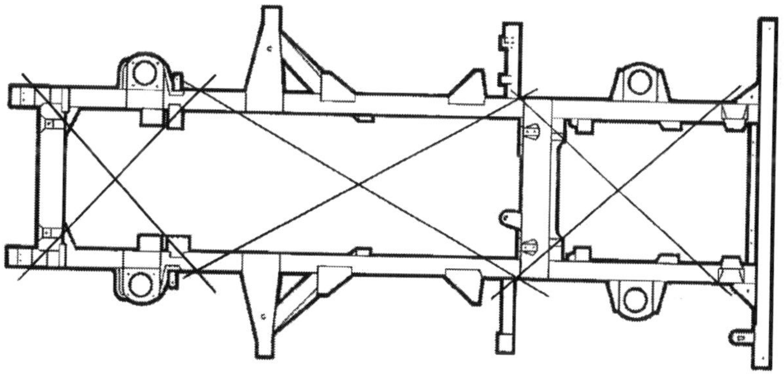

LAND ROVER 90

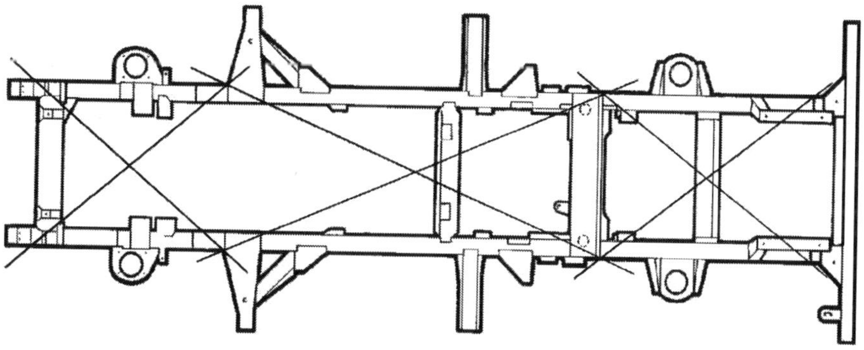

LAND ROVER 110

With the vehicle assembled, a check for chassis squareness can be made as follows:

- **1.** Place the vehicle on a level floor.
- **2.** Mark measuring points at approximately the locations shown in LR4412M ensuring that the marks are exactly opposite on each side of the chassis frame.
- **3.** Hold a plumb line against each of the measuring points in turn and mark the floor directly beneath the plumb-bob.
- **4.** Move the vehicle and measure diagonally between the marks made on the floor, if the chassis is square the diagonals between the related measuring points should agree within 9,50 mm.
- **5.** Chassis frame dimensional checks can be made, with the vehicle upper structure removed, referring to the applicable illustration and associated key.

90 Chassis

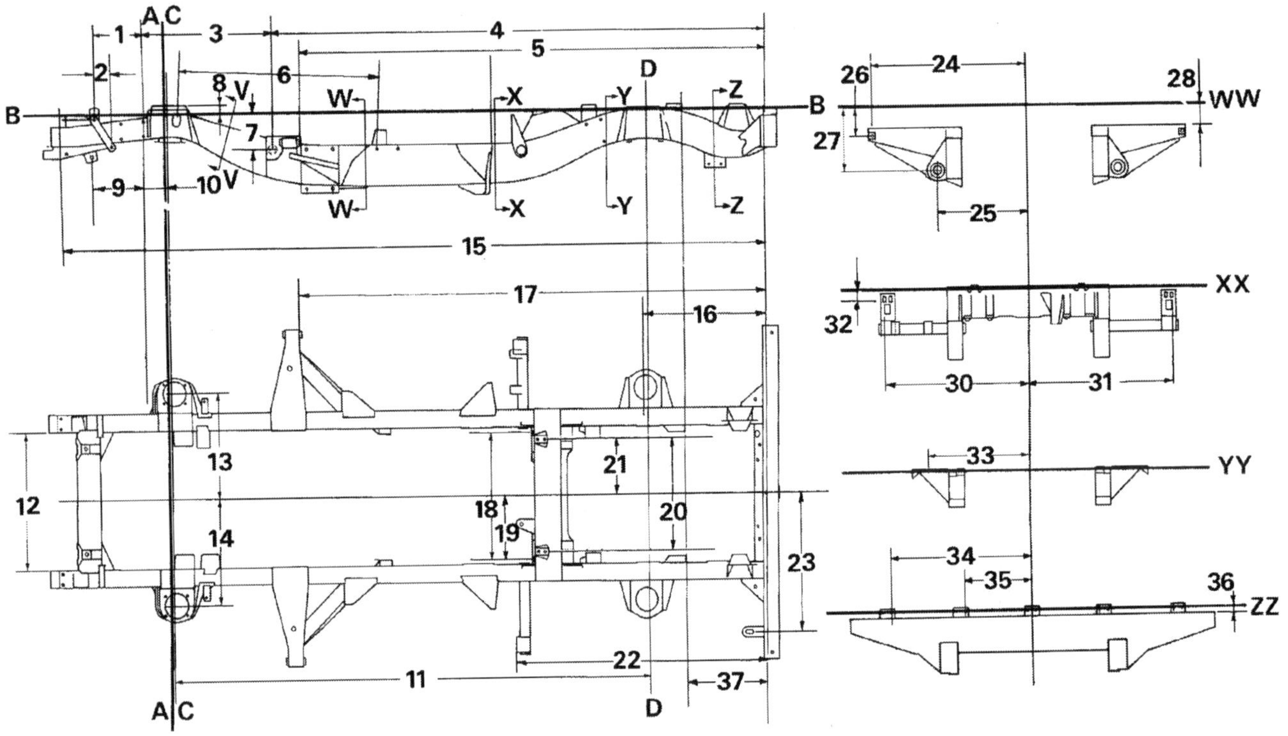

Chassis alignment dimensions

No./Letter	Dimension
A	Front datum
B	Chassis Datum
C	Front axle centre line
D	Rear axle centre line
1.	239,0 - 236,5 mm
2.	82,0 - 79,5 mm
3.	633 mm
4.	2420,6 - 2418,6 mm
5.	2306,4 - 2305,4 mm
6.	981,2 - 978,7 mm
7.	182,7 mm
8.	41,5 - 37,0 mm
9.	252 - 250 mm
10.	110 mm
11.	2360 mm - Wheelbase
12.	636 - 634 mm
13.	488 - 482 mm
14.	488 - 482 mm
15.	3431,1 - 3426,1 mm
16.	588,3 - 586,3 mm
17.	2313,8 - 2311,8 mm
18.	590,5 mm
19.	295,25 mm
20.	519,30 - 517,30 mm
21.	259,80 - 258,50 mm
22.	1242,6 - 1240,6 mm
23.	642,5 - 639,5 mm
24.	750,9 mm
25.	439,5 - 436,5 mm
26.	136,5 mm
27.	299,5 - 295,5 mm
28.	103 - 100 mm
29.	131,5 - 126,5 mm
30.	705,5 - 704,5 mm
31.	705,5 - 704,5 mm
32.	42,2 - 40,2 mm
33.	491 - 486 mm
34.	594,2 - 593,4 mm
35.	283,0 - 282,2 mm
36.	32,25 - 31,25 mm
37.	397 - 395 mm

Engine mounting dimensions - section V-V

⚠ NOTE: The engine mounting bracket dimensions are applicable to all models

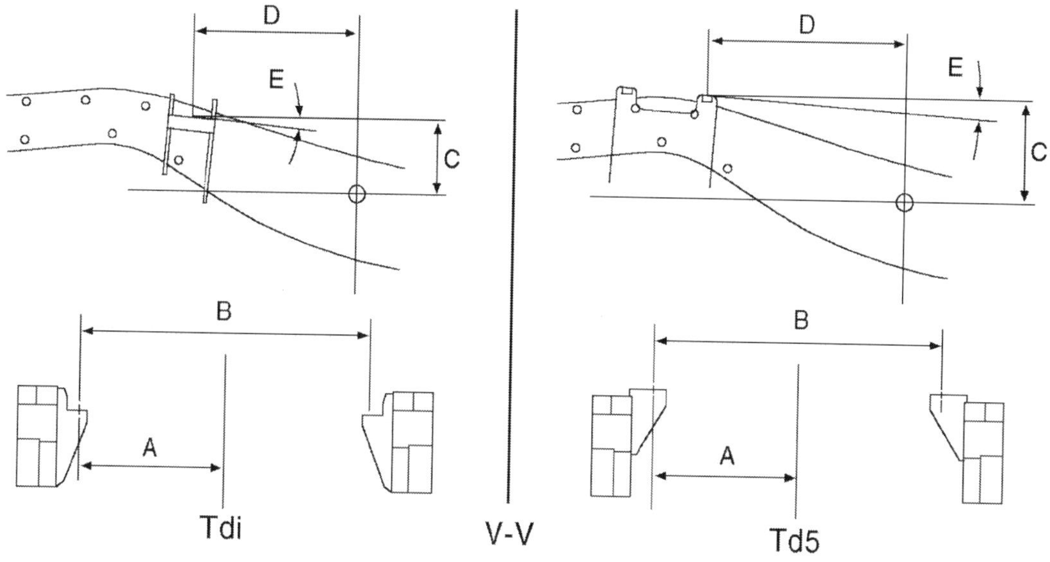

Tdi dimensions

Dimension	value
A	276.5 mm
B	553 mm
C	127.9 mm
D	317.4 mm
E	4 °

Td5 dimensions

Dimension	value
A	273.2 mm
B	546.5 mm
C	187.8 mm
D	347.3 mm
E	4 °

110 Chassis

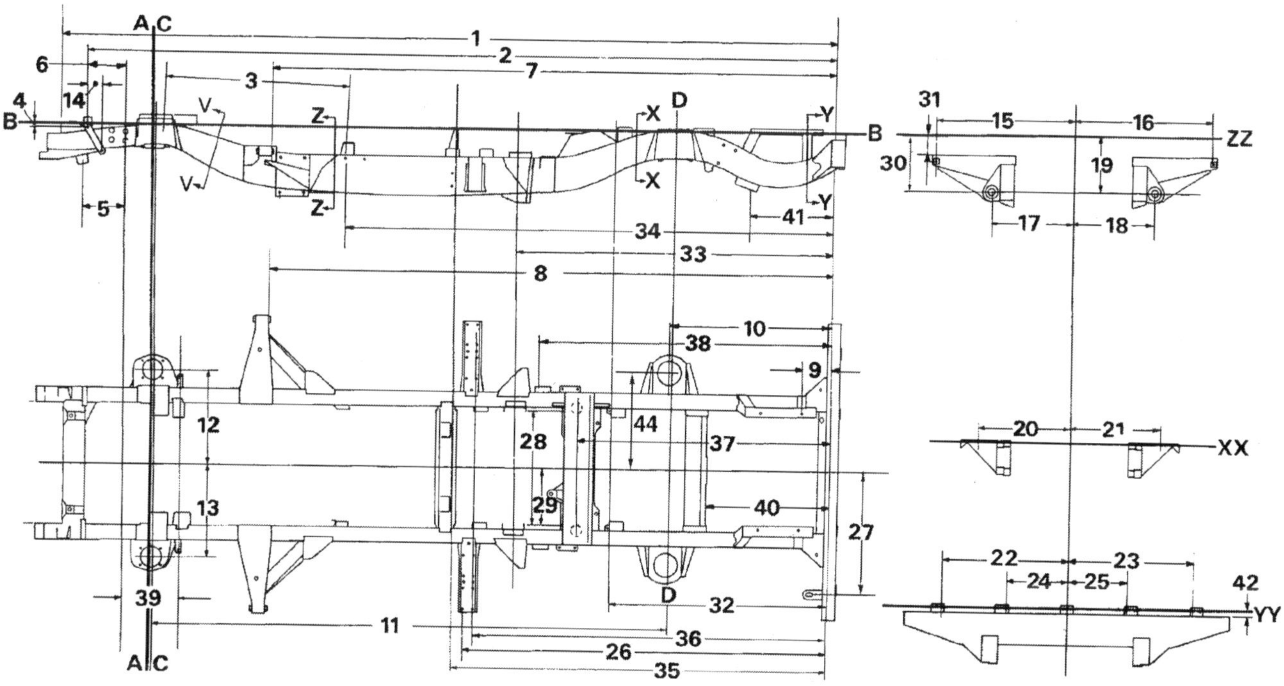

Chassis alignment dimensions

Section V-V is through the engine mountings. Dimensional information can be found in the 90 engine mounting dimensions.

No./Letter	Dimension
A	Front datum
B	Chassis Datum
C	Front axle centre line
D	Rear axle centre line
1.	4148 - 4143 mm
2.	4009,5 - 4005 mm
3.	978,7 - 981,2 mm

4.	22 - 20 mm
5.	252 - 250 mm
6.	239 - 236,5 mm
7.	3023,3 - 3022,3 mm
8.	3030,7 - 3028,7 mm
9.	155 - 153 mm
10.	871,2 - 869,2 mm
11.	2794 mm - Wheelbase
12.	488 - 482 mm
13.	488 - 482 mm
14.	82 - 79,5 mm
15.	750,9 mm
16.	750,9 mm
17.	440,5 - 435,5 mm
18.	440,5 - 435,5 mm
19.	299,5 - 295,5 mm
20.	500 - 495 mm
21.	500 - 495 mm
22.	594,2 - 593,4 mm
23.	594,2 - 593,4 mm

24.	283 - 282,2 mm
25.	283 - 282,2 mm
26.	1970 - 1968 mm
27.	642,9 - 639,5 mm
28.	750,9 mm
29.	290,5 mm
30.	295,5 mm
31.	299,5 - 295,5 mm
32.	103 - 100 mm
33.	1177,5 - 1175,5 mm
34.	1692,5 - 1689,5 mm
35.	2610 - 2606 mm
36.	2040,5 - 2037,5 mm
37.	1912,5 - 1909,5 mm
38.	1359 - 1357 mm
39.	1573 - 1571 mm
40.	270 - 268 mm
41.	665,5 - 663,5 mm
42.	440 - 438 mm
43.	32,25 - 31,25 mm

130 Chassis

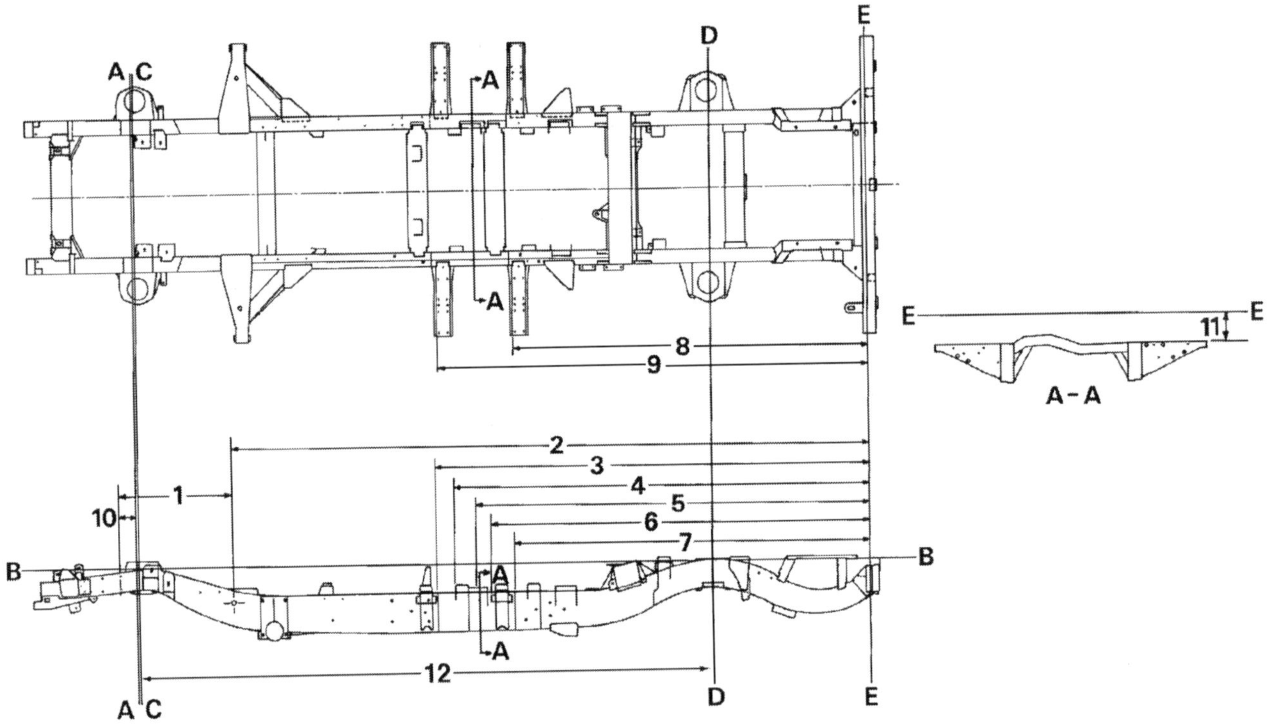

Chassis alignment dimensions

NOTE: The above dimensions are for the Land Rover 130 chassis frame. For additional measurements, refer to the Land Rover 110 chassis frame drawing and alignment dimensions.

No./Letter	Dimension
A	Front datum
B	Chassis Datum
C	Front axle centre line
D	Rear axle centre line
E.	Chassis Datum, section A-A
1.	663,0 mm reference only
2.	3569,3 - 3567,3 mm
3.	2421,8 - 2419,8 mm

4.	2317,5 - 2314,5 mm
5.	2188,3 - 2185,3 mm
6.	2119,5 - 2117,3 mm
7.	1990 - 1988 mm
8.	1970 - 1968 mm
9.	2401,8 - 2399,8 mm
10.	110,0 mm reference
11.	149,7 - 146,7 mm reference dimension
12.	3225,8 mm wheelbase

GAP AND PROFILE INFORMATION

The following information is to be used as a guide to assist the technician in refitting exterior body panels and trim items, to achieve a correctly aligned and cosmetically acceptable vehicle.

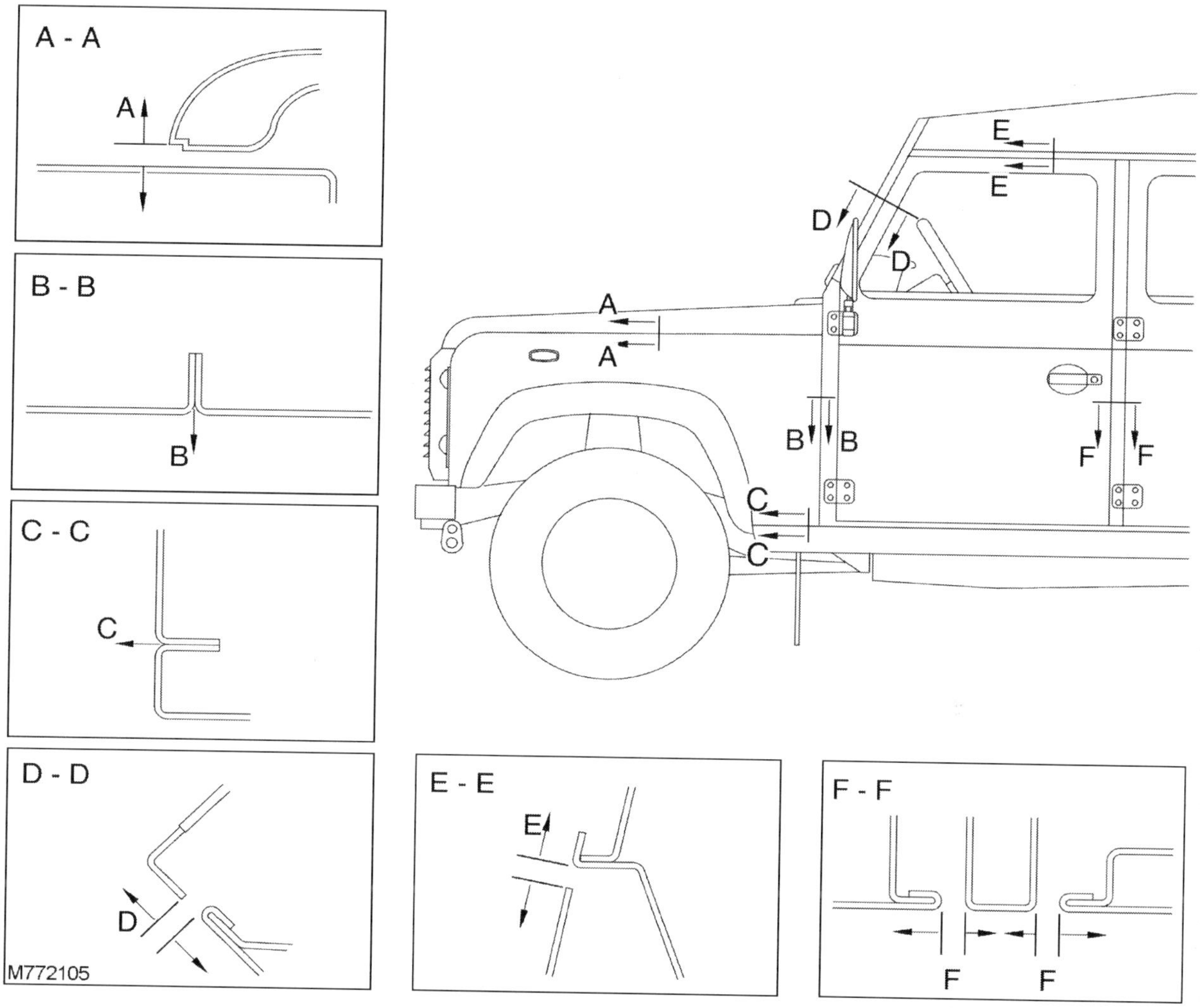

- Section A-A, Bonnet to wing gap A = 3 - 8 mm. To be parallel within 2 mm
- Section B-B, Wing to 'A' post lower, gap B = 0. Profile = ± 1 mm
- Section C-C, Wing to sill, gap C = 0. Profile = ± 1 mm
- Section D-D, Screen aperture to door frame, gap D = 5 - 9 mm
- Section E-E, Door frame to roof, gap E = 7 - 11 mm
- Section F-F, Front and rear door to 'B'/'C' post, gap F = 7 - 11 mm

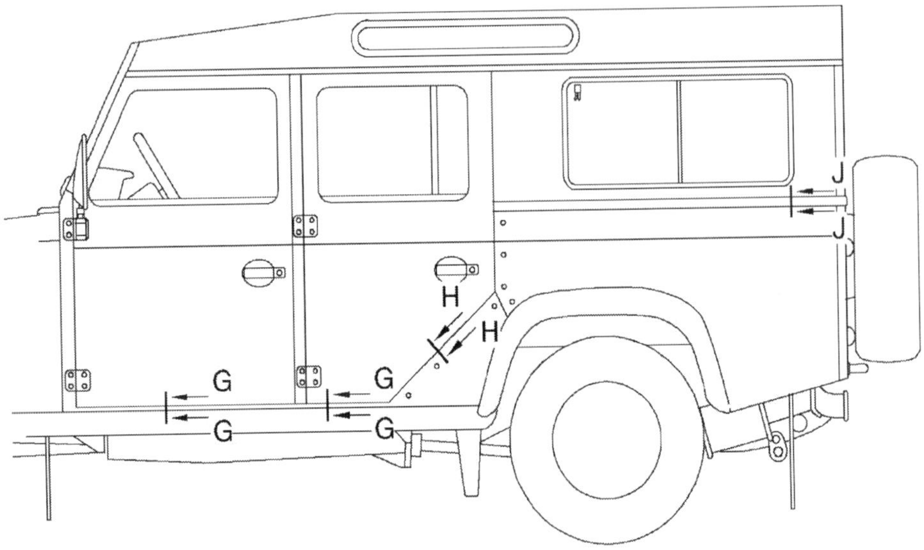

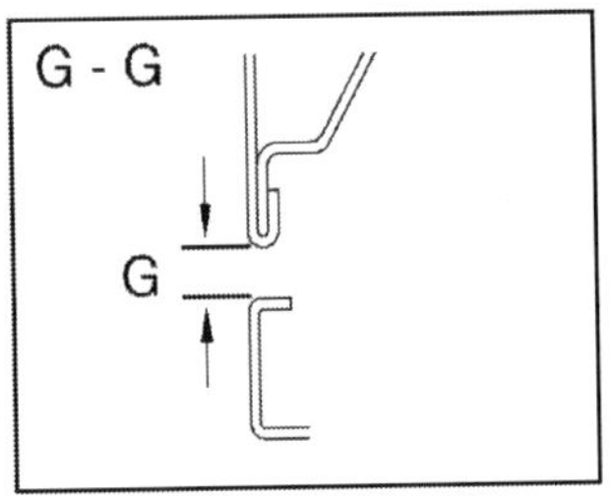

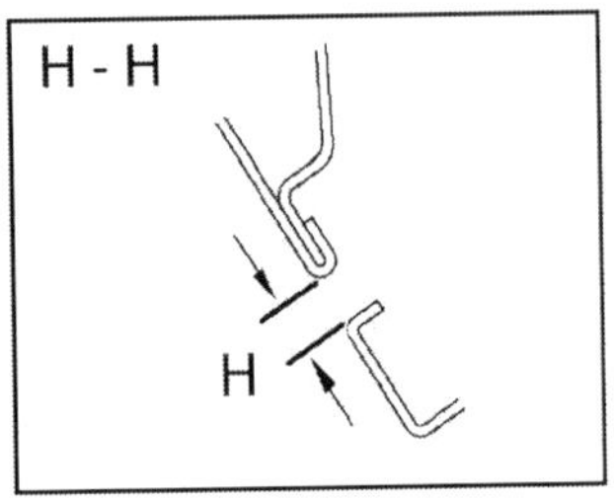

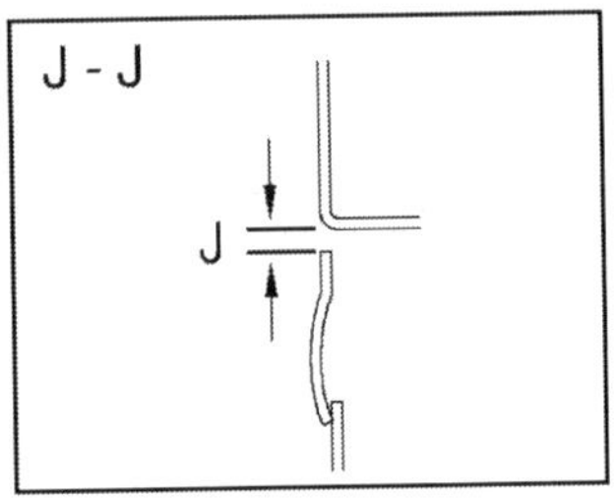

- Section G-G, Door to sill gap G = 7 - 11 mm. Door profile to sill = 0 to = + 2 mm
- Section H-H, Rear door to body gap H = 7 - 11 mm. Door profile to body = 0 to = + 3 mm
- Section J-J, Body side lower to body, gap J = 0 - 4 mm. To be parallel within 2 mm

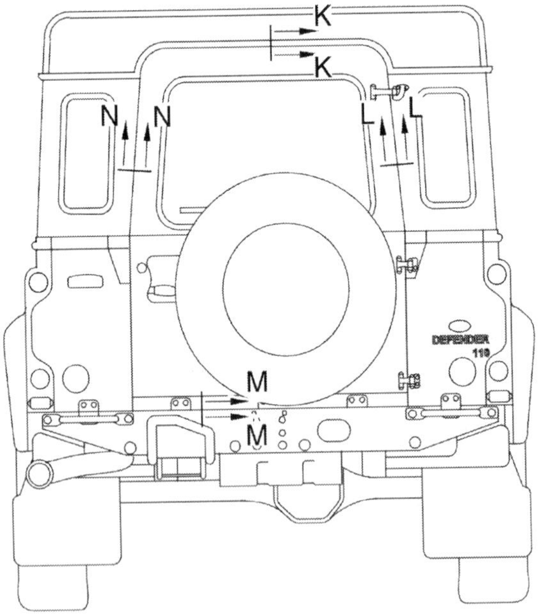

- Section K-K, Roof to taildoor, gap K = 7 - 9 mm. Profile = ± 1 mm
- Section L-L, N-N, Body side to taildoor, gap L = 56 - 7 mm. Profile = ± 1 mm
- Section M-M, Body rear to taildoor, gap M - 7 - 9 mm

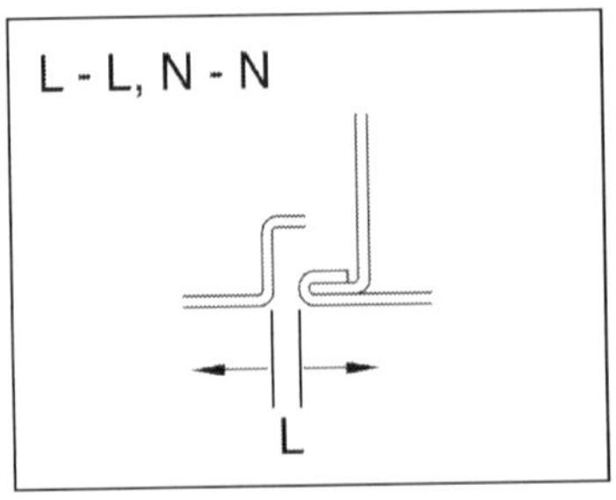

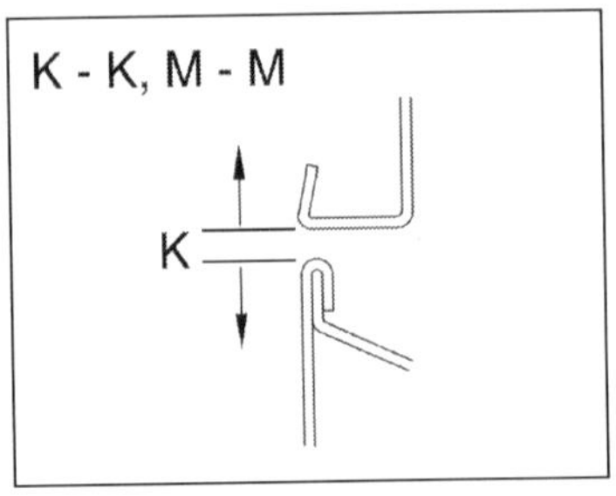

Front Fender Side Panel

Removal and Installation

Removal

1. Disconnect both battery leads, negative lead first.
2. Remove wheelhouse eyebrow.
3. Remove wheelhouse liner.
4. RH side: Remove aerial.
5. Remove air intake hose.
6. Remove bolts securing front fender side panel.

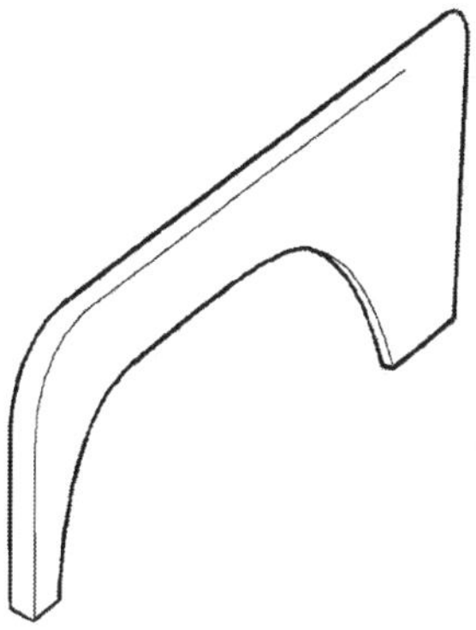

M772098A

Installation

1. Prepare and clean panel joint faces.
2. Reverse removal procedure.

Front Left Fender Top Panel

Removal and Installation

Removal

1. Disconnect both battery leads, negative lead first.
2. Remove headlamp assembly.

Refer to: Headlamp Assembly (417-01 Exterior Lighting, Removal and Installation)

3. Remove headlamp mounting panel.
4. Remove radiator.
    1. Tdi -

    Refer to: Radiator (303-03 Engine Cooling, ID4 2.2L Diesel Removal and Installation)

    2. Td5 -

    Refer to: Radiator (303-03 Engine Cooling, ID4 2.2L Diesel Removal and Installation)

    3. 4.0 -

    Refer to: Radiator (303-03 Engine Cooling, ID4 2.2L Diesel Removal and Installation)

    4. 2.4L Diesel -

    Refer to: Radiator (303-03 Engine Cooling, ID4 2.2L Diesel Removal and Installation)

5. Recover refrigerant from A/C system.

Refer to: Air Conditioning System Recovery, Evacuation and Charging (412-00 Climate Control System - General Information, General Procedures)

6. Remove receiver drier.
    1. Tdi -

    Refer to: Radiator (303-03 Engine Cooling, ID4 2.2L Diesel Removal and Installation)

    2. Td5 -

    Refer to: Radiator (303-03 Engine Cooling, ID4 2.2L Diesel Removal and Installation)

    3. 4.0 -

    Refer to: Radiator (303-03 Engine Cooling, ID4 2.2L Diesel Removal and Installation)

    4. 2.4L Diesel -

    Refer to: Radiator (303-03 Engine Cooling, ID4 2.2L Diesel Removal and Installation)

7. Remove relevant A/C pipes.

8. Remove relevant heater intake pipes.

9. Remove screen washer reservoir.

10. Remove wheelhouse liner.

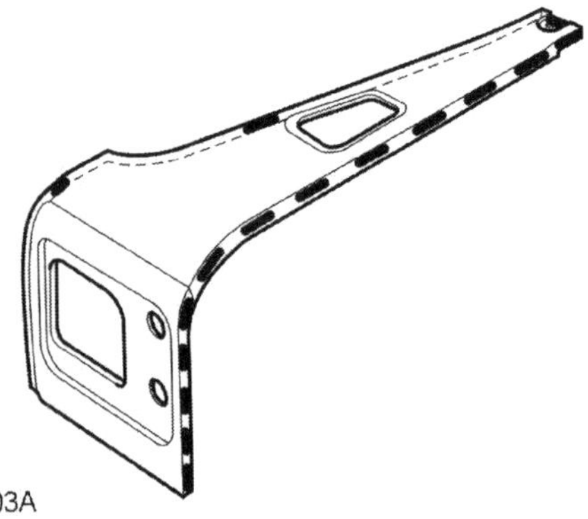

M772103A

11. Remove bolts and release welds securing panel, remove panel.

Installation

1. Prepare and clean panel joint faces.

2. Reverse removal procedure.

3. Adjust headlamps.

   Refer to: Headlamp Adjustment (417-01 Exterior Lighting, General Procedures)

Front Right Fender Top Panel

Removal and Installation

Removal

1. Disconnect both battery leads, negative lead first.

2. Remove headlamp assembly.

   Refer to: Radiator (303-03 Engine Cooling, ID4 2.2L Diesel Removal and Installation)

3. Remove headlamp mounting panel.

4. Remove radiator.

   1. Tdi -

   Refer to: Radiator (303-03 Engine Cooling, ID4 2.2L Diesel Removal and Installation)

   2. Td5 -

   Refer to: Radiator (303-03 Engine Cooling, ID4 2.2L Diesel Removal and Installation)

   3. 4.0 -

   Refer to: Radiator (303-03 Engine Cooling, ID4 2.2L Diesel Removal and Installation)

   4. 2.4L Diesel -

   Refer to: Radiator (303-03 Engine Cooling, ID4 2.2L Diesel Removal and Installation)

5. Recover refrigerant from A/C system.

   Refer to: Radiator (303-03 Engine Cooling, ID4 2.2L Diesel Removal and Installation)

6. Remove relevant A/C pipes.

7. Position PAS reservoir aside.

8. Remove wheelhouse liner.

9. Remove aerial.

10. Remove vacuum solenoid valve.

11. Release clips securing coolant reservoir.

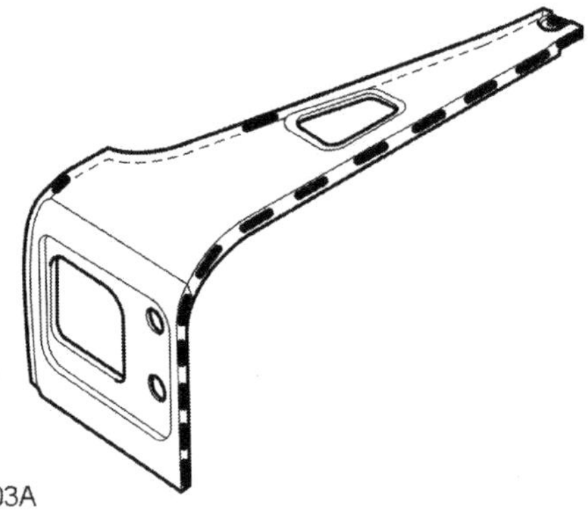

M772103A

12. ⚠ NOTE: LH fender shown.

    Remove bolts and release welds securing panel, remove panel.

Installation

1. Prepare and clean panel joint faces.

2. Reverse removal procedure.

3. Adjust headlamps.

Refer to: Radiator (303-03 Engine Cooling, ID4 2.2L Diesel Removal and Installation)

Front Wheelhouse

Removal and Installation

Removal

1. Remove the front wheel speed sensor and shield, if fitted.

Refer to: Front Wheel Speed Sensor (206-09, Removal and Installation).

2. Remove the front fender top panel.

Refer to: Front Left Fender Top Panel (501-27 Front End Sheet Metal Repairs, Removal and Installation)

or

Refer to: Front Right Fender Top Panel (501-27 Front End Sheet Metal Repairs, Removal and Installation)

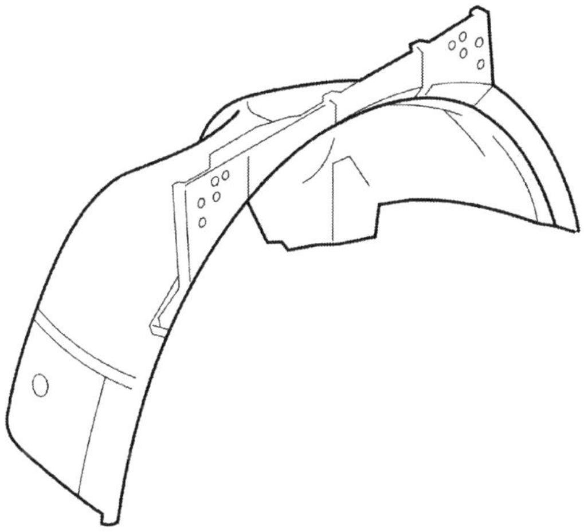

3. Collect the wheelhouse.

Installation

1. Fit the wheelhouse.

2. Fit the front fender top panel.

Refer to: Front Left Fender Top Panel (501-27 Front End Sheet Metal Repairs, Removal and Installation)

or

Refer to: Front Right Fender Top Panel (501-27 Front End Sheet Metal Repairs, Removal and Installation)

3. Fit the front wheel speed sensor and shield, if fitted.

Refer to: Front Wheel Speed Sensor (206-09, Removal and Installation).

Hood Latch Panel

Removal and Installation

Removal

1. Disconnect the battery ground cable.

Refer to: Startup Battery Disconnect and Connect (414-01 Battery, Mounting and Cables, General Procedures)

2. Remove the radiator grille.

Refer to: Radiator Grille (501-08 Exterior Trim and Ornamentation, Removal and Installation)

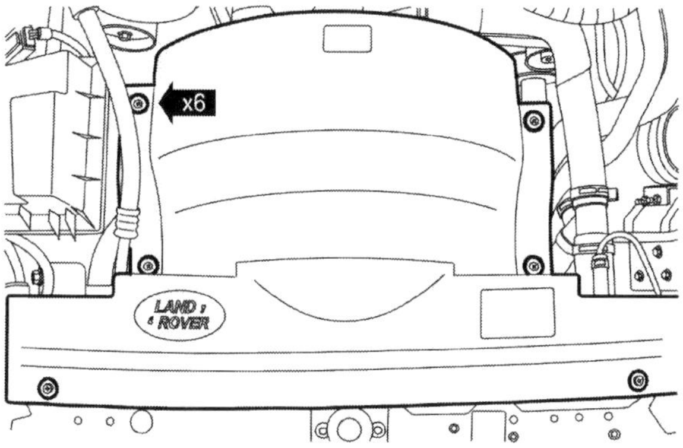

3. Remove the cooling fan upper shroud.

- Release the 6 clips.

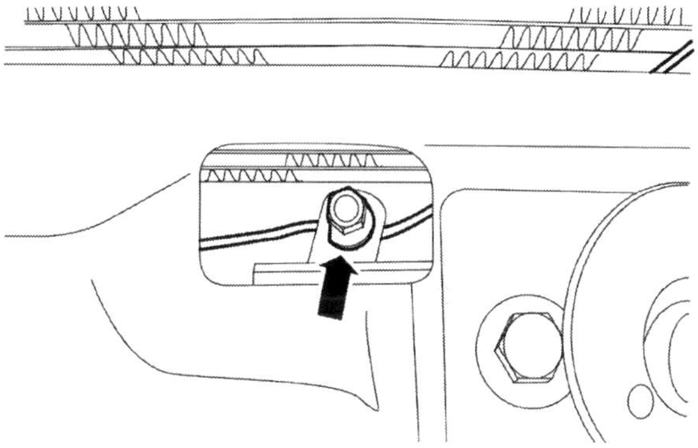

4. Loosen the hood release cable.

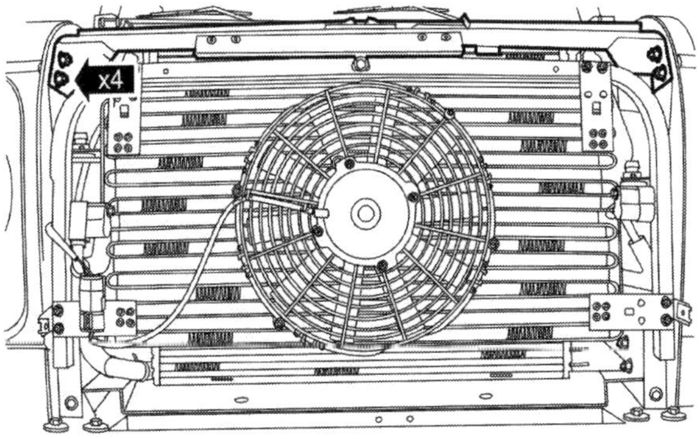

5. Remove the hood latch panel.

- Remove the 4 nuts and studs.

Installation

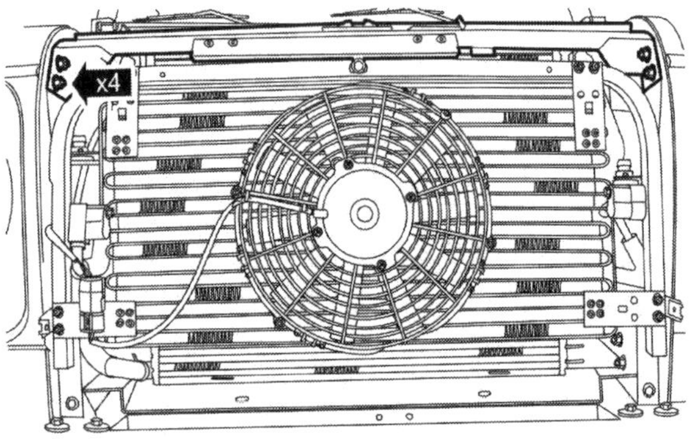

1. To install, reverse the removal procedure.
   - Tighten the nuts to 23 Nm (17 lb.ft).
2. Connect the battery ground cable.

   Refer to: Battery Connect (414-01 Battery, Mounting and Cables, General Procedures).

Windshield Frame

Removal and Installation

Removal

1. Disconnect both battery leads, negative lead first.
2. Raise side of vehicle.
3. Remove front and centre headlining sections.
4. Remove both A-pillar trims.
5. Remove front door aperture seals.
6. Release upper front corner of rear door aperture seal.
7. Remove both wiper pivot arms.

Refer to: Front Wiper Pivot Arm (501-16 Wipers and Washers, Removal and Installation)

8. Remove windshield.
9. Remove all roof fixings, back to the B-pillar.
10. Release sealant securing front edge of roof.

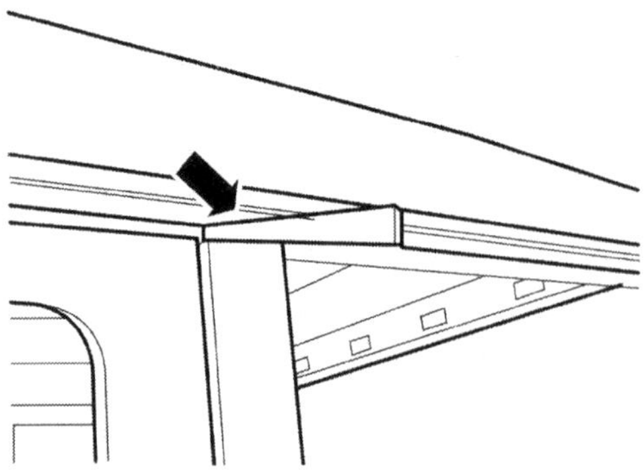

11. Insert wedges into top of both B-pillars. This will allow access for removal of screen aperture.

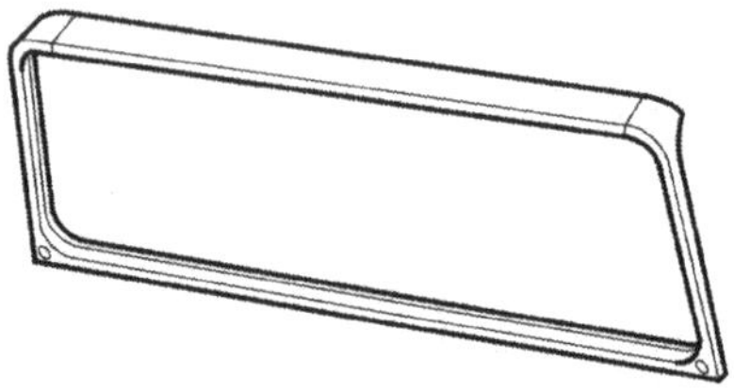

12. With assistance remove windshield frame.

Installation

1. Prepare and clean panel joint faces.
2. Visually check the windshield frame for a casting line in the lower corners. If a casting line is visible follow the next step.
3. Using a suitable tool carefully remove the casting line from the windshield frame as indicated.

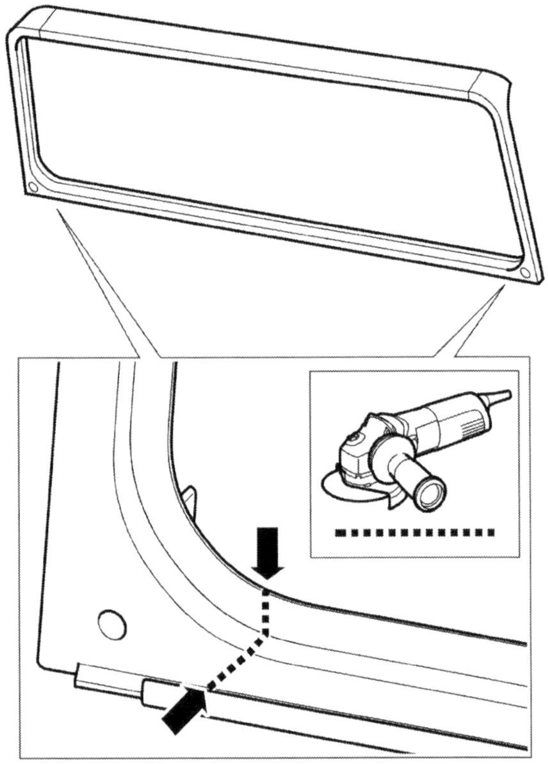

4. ⚠ NOTE: When refitting, ensure windshield frame is correctly fitted onto location pins in bulkhead.

Reverse removal procedure. Fit new foam seal to top edge of windshield frame.

5. ⚠ NOTE: White roof vehicles use a white sealant. Coloured roof vehicles use a black sealant.

When applying sealant to top edge of windshield frame, the correct colour sealant must be used, see note below.

Roof Sheet Metal Repairs

501-28

Roof Panel

Removal and Installation

Removal

1. Disconnect both battery leads, negative lead first.
2. Remove rear bench seats.
3. Remove front, centre and rear sections of headlining.
4. Remove all bolts securing roof.
5. Carefully cut sealant securing front edge of roof to windshield aperture.

Installation

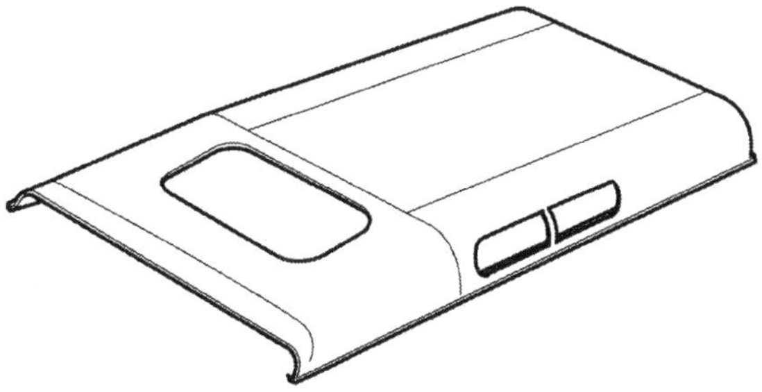

M772115

1. Prepare and clean panel joint faces.
2. Reverse removal procedure.

Side Panel Sheet Metal Repairs

501-29

Rear Rocker Panel

Removal and Installation

Removal

⚠ NOTE: When replacing the rear rocker panel, the dogleg lower panel must also be replaced.

1. Disconnect both battery leads, negative lead first.
2. Remove door aperture seal.
3. Remove rear wheelhouse eyebrow.
4. Remove rivets and bolts securing rocker panel and dogleg panel to the body.

Installation

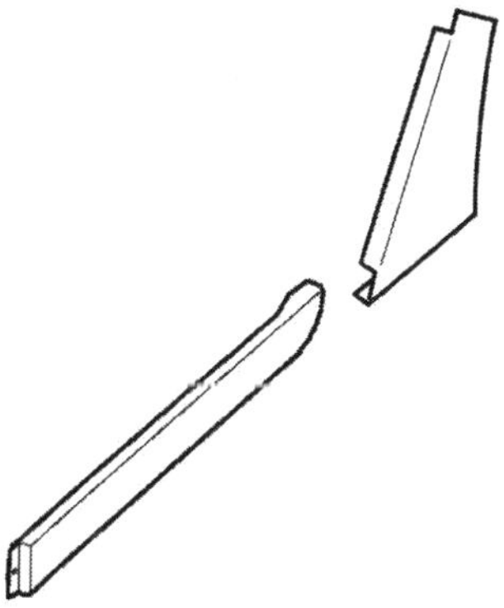

M772112A

1. Prepare and clean panel joint faces.
2. Reverse removal procedure.

Front Rocker Panel

Removal and Installation

Removal

1. Remove bolts securing front rocker panel to body, discard seal.

Installation

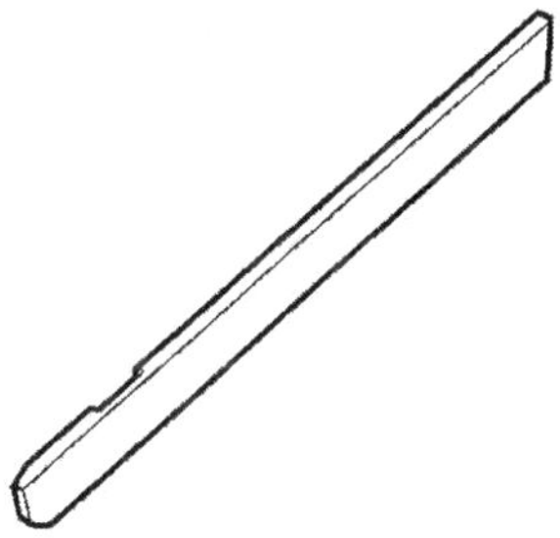

1. Prepare and clean panel joint faces. Refit new seal.
2. Reverse removal procedure.

B-Pillar and C-Pillar Assembly

Removal and Installation

Removal

1. Disconnect both battery leads, negative lead first.
2. Remove rear door and striker.
3. Remove front safety belt retractor.

Refer to: Front Safety Belt Retractor (501-20, Removal and Installation).

Refer to: Front Safety Belt Retractor (501-20 Seatbelt System, Removal and Installation)

4. Remove front and rear carpets.
5. Remove front seat.
6. Remove rear seat.
7. Remove front, centre and rear sections of headlining.
8. △ NOTE: Assembly is held on with bolts.

Remove front and rear rocker panels.

Refer to: Front Rocker Panel (501-29 Side Panel Sheet Metal Repairs, Removal and Installation)

Refer to: Rear Rocker Panel (501-29 Side Panel Sheet Metal Repairs, Removal and Installation)

Installation

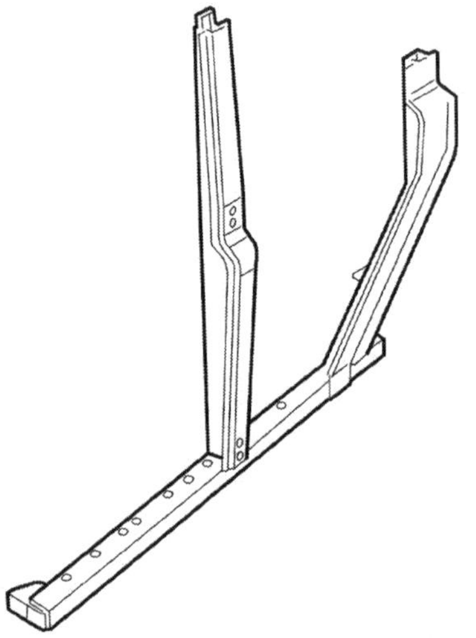

M772113

1. Prepare and clean panel joint faces.
2. Reverse removal procedure.

501-30 Rear End Sheet Metal Repairs

Rear Lamp Mounting Panel

Removal and Installation

Removal

△ NOTE: In this procedure the rear lamp mounting panel is installed in conjunction with:

- Taildoor
- Rear lamp assembly
- Licence plate lamp
- Reversing lamp
- Rear quarter upper panel
- Rear quarter lower panel capping

E161038

1. The rear lamp mounting panel is serviced as indicated.
2. Before commencing this procedure make sure that you are aware of all Health and Safety requirements.

Refer to: Health and Safety Precautions (100-00 General Information, Description and Operation)

3. For further information on the methods, tools and fixings used in this procedure refer to the body repairs - general information section.
4. △ NOTE: Right hand only.

Refer to: Taildoor (501-03 Body Closures, Removal and Installation)

5. △ NOTE: Left hand only.

Refer to: Taildoor Striker Adjustment (501-03 Body Closures, General Procedures)

6. **Refer to:** Rear Lamp Assembly (417-01 Exterior Lighting, Removal and Installation)
7. **Refer to:** License Plate Lamp (417-01 Exterior Lighting, Removal and Installation)

8. **Refer to:** Reversing Lamp (417-01 Exterior Lighting, Removal and Installation)

9. **Refer to:** Rear Quarter Upper Panel (501-30 Rear End Sheet Metal Repairs, Removal and Installation)

10. **Refer to:** Rear Quarter Lower Panel (501-30 Rear End Sheet Metal Repairs, Removal and Installation)

11. Drill out the spot welds and remove the bolts as indicated.

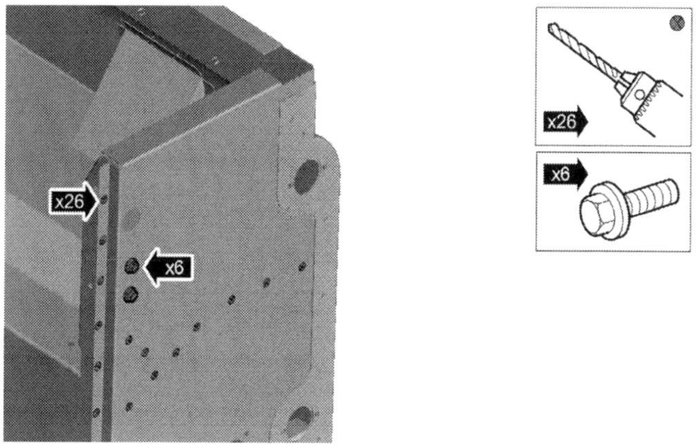

12. Separate the joints and remove the rear lamp mounting panel.

Installation

CAUTION: Where structural adhesive is to be applied between surfaces it is essential compression is applied to the joints during the curing period.

NOTE: Any joints sealed as part of vehicle manufacture must be sealed as part of the repair.

1. Using a fine bristle disc clean and prepare the panel surfaces.

2. Dress the flanges where necessary.

3. Offer up the new rear lamp mounting panel, clamp into position. Check alignment, if correct, proceed to next step, if not rectify and recheck before proceeding.

4. Using a combination of the old rear lamp mounting panel and the new rear lamp mounting panel mark and drill holes in the new rear lamp mounting panel where the metal inert gas (MIG) plug welds are to be installed as indicated.

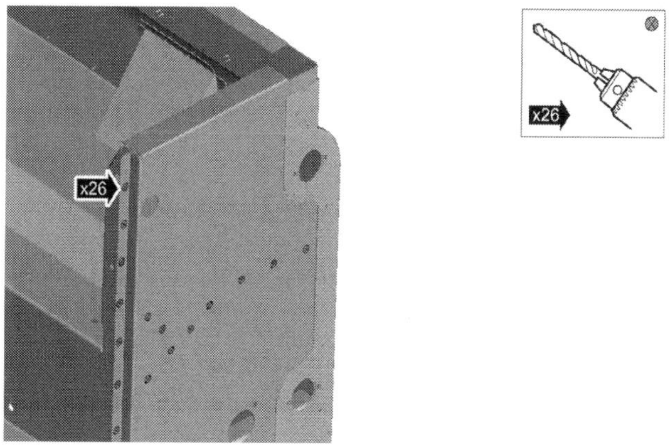

5. Deburr the drilled holes

6. Apply a Land Rover recommended weld through coating at points where the MIG plug welds are to be installed.

7. Offer up the new rear lamp mounting panel, clamp into position. Check alignment, if correct, proceed to next step, if not rectify and recheck before proceeding.

8. MIG plug weld the rear lamp mounting panel, install the bolts and tighten to 25 Nm as indicated.

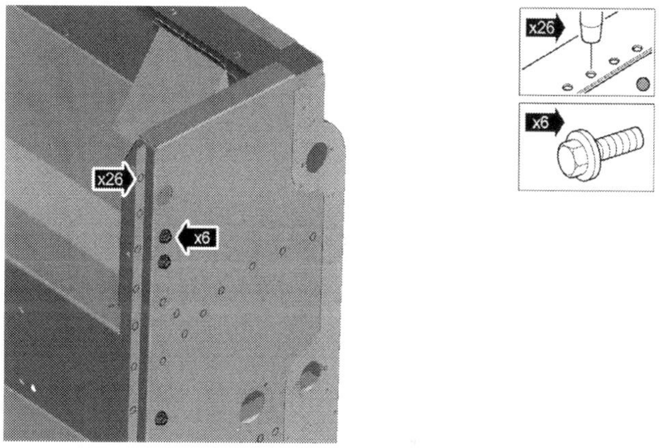

9. Apply a bead of Land Rover recommended seam sealer to the panel joints as indicated.

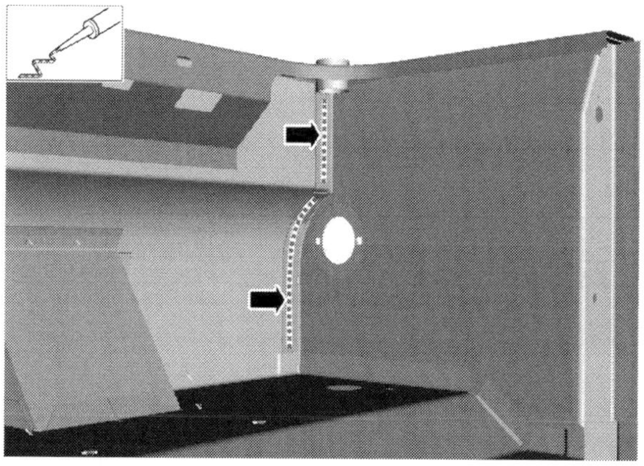

10. Apply a bead of Land Rover recommended seam sealer to the panel joints as indicated.

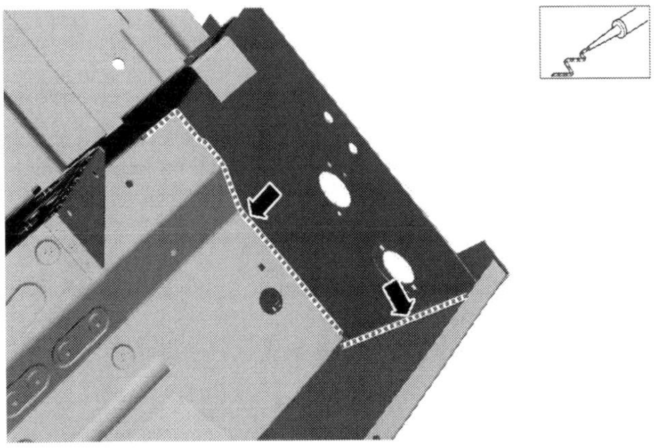

11. Apply the Land Rover recommended sealing putty to the panel joints as indicated.

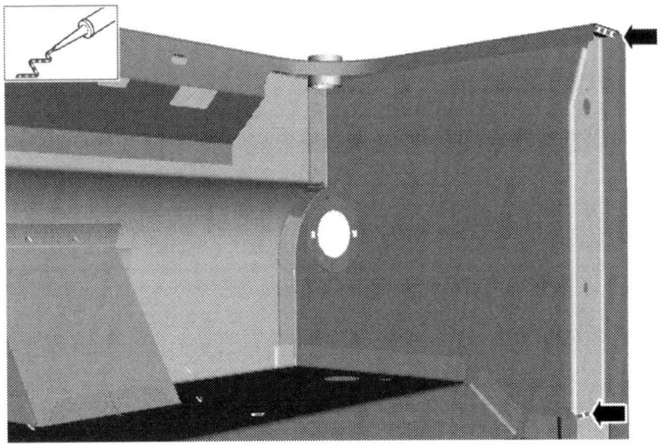

12. Make sure that any open or exposed panel joints are correctly sealed.

13. Make sure corrosion protection is applied to all areas affected by repair.

14. The installation of associated panels and components is the reversal of the removal procedure.

Rear Quarter Upper Panel

Removal and Installation

Removal

1. Release upper edge of front door aperture seal.
2. Release upper edge of rear door aperture seal.
3. Remove fixings from front edge of roof, and along side being installed.
4. Release sealant securing front edge of roof.
5. Insert wedges into top of both B-pillar and C-pillar. This will allow access for removal of body side.
6. RH side: Remove tail door upper hinge.
7. Remove side window.
8. Remove rear sixth light.
9. Remove all bolts securing rear quarter upper panel.

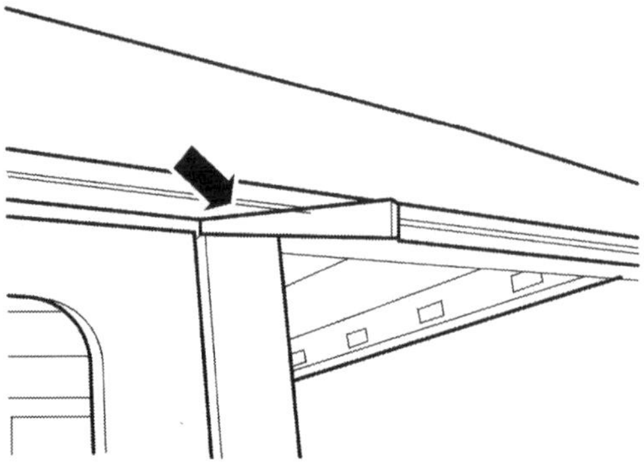

Installation

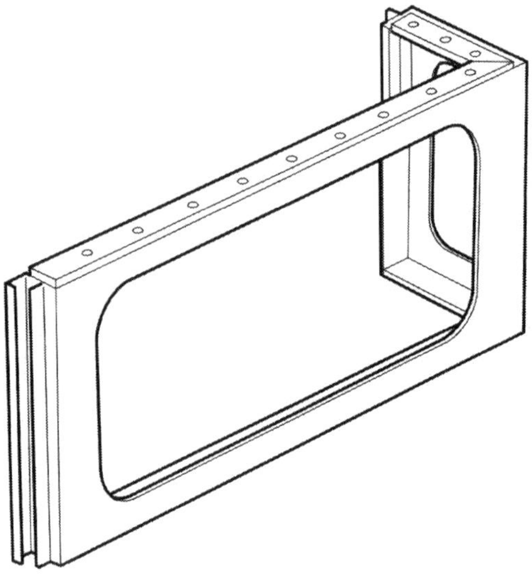

1. Prepare and clean panel joint faces.
2. ⚠ NOTE: When installing, make sure rear quarter upper panel is correctly fitted onto location pins.

To install, reverse removal procedure.

Rear Quarter Lower Panel

Removal and Installation

Removal

⚠ NOTE: In this procedure the bodyside capping and the corner capping are also changed.

1. Disconnect both battery leads, negative lead first.
2. Remove rear quarter upper panel.

Refer to: Rear Quarter Upper Panel (501-30 Rear End Sheet Metal Repairs, Removal and Installation)

3. Remove all rear lamps and reflectors.
4. LH side: Remove licence plate and lamp.
5. LH side: Remove taildoor striker.
6. RH side: Remove fuel filler and grommet.

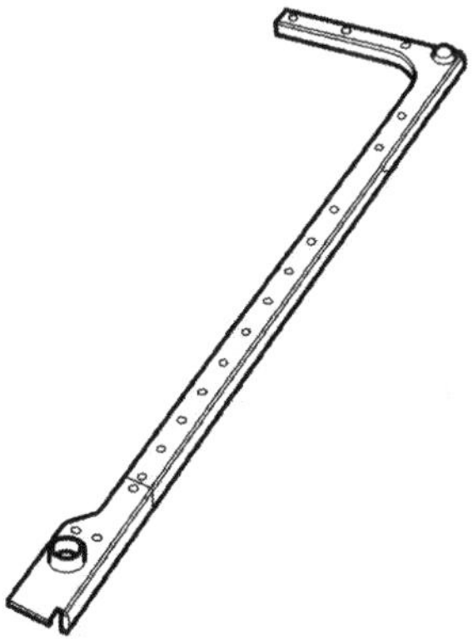

M772101A

7. Remove all bolts and rivets securing rear quarter lower panel.

Installation

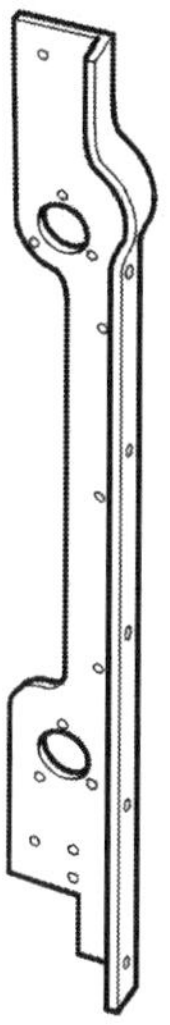

M772102A

1. Prepare and clean panel joint faces.

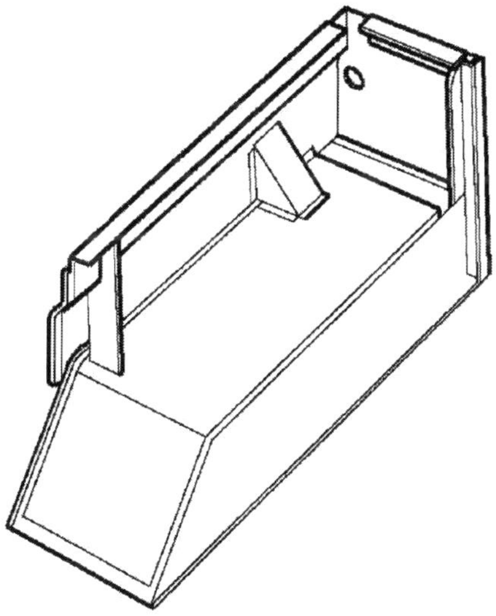

2. ⚠ NOTE: When refitting, ensure rear quarter lower panel is correctly fitted onto location pins.

Reverse removal procedure.

3. Make sure that the rear quarter lower panel brace bolts are installed from the inside as indicated.

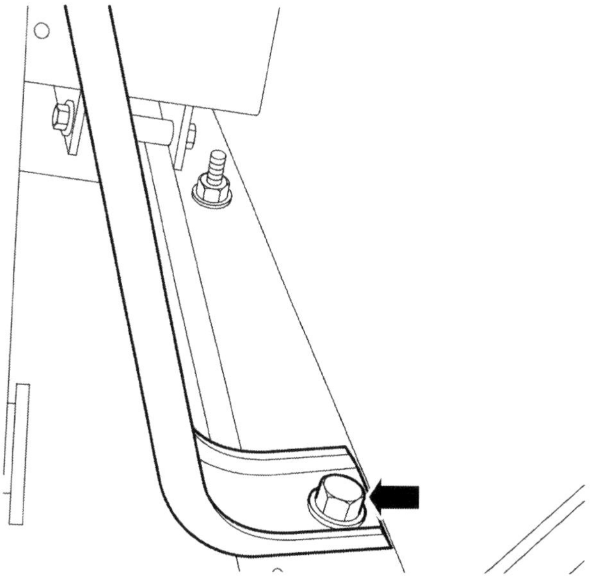

Rear Crossmember Panel

Removal and Installation

Removal

1. Disconnect both battery leads, negative lead first.
2. Remove both rear lamp panels.

Refer to: Rear Lamp Mounting Panel (501-30 Rear End Sheet Metal Repairs, Removal and Installation)

Installation

M772125

1. Prepare and clean panel joint faces.
2. Reverse removal procedure.

Detailed Contents

General Information

100-00 - General Information **7**
How To Use This Manual 7
General Service Information 7
Special Tools 7
Standard Workshop Practices 9
Battery (general) 9
Towing and Transporting a Vehicle 10
Screw Threads 11
Bolt and nut ID 12
Ball and Roller Bearing 13
Pipes and Hoses (general) 14
Electrical precautions (general 16
Oil Seals (general) 17
A/C System Precautions (general) 17
Vehicle Weights 19
Health and Safety Precautions 22
Adhesives and Sealants 22
Chemical handling (general aspects of) 23
Lubricants handling (general aspects of) 27
Special Tool Glossary 31
About This Manual 31
TAS Symbols and Graphics 32
Diagnostic Trouble Code Index - DTC:
Engine Control Module (ECM) 36
Diagnostic Trouble Code Index - DTC:
Instrument Cluster (IPC) 69
Diagnostic Trouble Code Index:
Anti-lock Brake System (ABS) 70

100-01 - Identification Codes **79**
Identification Codes - Locations 79

100-02 - Jacking and Lifting **82**
Jacking 82
Vehicle Recovery 83
Jump Starting 83
Lifting 84

Chassis

204-00 - General Information - Suspension **85**
Specifications 85
Suspension 85
Road Spring data 85
Wheel Alignment specification 86
Torque Values 86
Diagnosis and Testing 87
Principles of Operation 87
General Procedures 89
Front Toe Adjustment 89
Steering Wheel Vibration/Shimmy While Driving 90
Wheel and Tire Assembly Balancing 92

204-01 - Front Suspension **93**
Description and Operation 93
Removal and Installation 93
Front Stabilizer Bar 93
Front Stabilizer Bar Link 94
Panhard Rod 95
Spring 95
Front Wheel Bearing and Wheel Hub 95
Wheel Studs 97
Front Shock Absorber 97
Bump Stop 98
Radius Arm 98

204-02 - Rear Suspension **99**
Description and Operation 99
Rear Axle Suspension 99
Removal and Installation 99
Wheel Bearing and Wheel Hub 99
Rear Suspension Specifications 101
Removal and Installation 101
Spring 101
Wheel Studs 102
Rear Stabilizer Bar 102
Rear Shock Absorber 102
Rear Stabilizer Bar 102
Upper Arm 103
Lower Arm 104
Axle Bump Stop 104

204-4 - Wheels and Tires **105**
Specifications 105
Description and Operation 105
Tire codes 105
Tire Inspection 106
Wheel Balancing and Changing 107
Removal and Installation 110
Wheels and Tires 110

205-01 - Driveshaft **111**
Description and Operation 111
Diagnosis and Testing 111
Principles of Operation 111
Inspection 111
Removal and Installation 112
Front Driveshaft 112
Rear Driveshaft 113
Universal Joint 114

205-02 - Rear Drive Axle and Differential **115**
Specifications 115
Description and Operation 115
Diagnosis and Testing 115
Rear Axle Hub - 90 116
Rear Axle Hub - 110 116
In-vehicle repair 117
Differential Case - 90/110/130 117
Drive Pinion Seal - 110/130 118
Drive Pinion Seal - 90 119
Stub Shaft Pilot Bearing and Seal 120
90 & 110/130 120
Axle Assembly 121
Disassembly and Assembly 121
Axle - 90/110 124
Axle - 110/130 126

205-03 - Front Drive Axle/Differential **131**
Specifications 131
Description and Operation 131
Diagnosis and Testing 132
Front Drive Axle 132
In-vehicle Repair 134
Drive Pinion Seal - 90 134
Axle Assembly 135
Differential Carrier 136
Disassembly and Assembly 139
Front Stub Axle, Constant Velocity Joint and Swivel Pin Housing 139

206-00 - Brake System **143**
General Information 143
Specifications 143
Brake Hydraulic Fluid 143
Description and Operation 144
Component Location/Description 144
Anti-Lock Control 144
Brake Pedal Switch 144
Dynamic Stability Control Switch 145
Wheel Speed Sensors 145
Yaw Rate and Lateral Acceleration Sensor 145
Steering Angle Sensor Module (SASM) 145
Differential Lock Unit 146
Instrument Cluster (IC) Warning Indicator 146
Brake Warning Indicator (Red) 146
ABS Warning Indicator (Amber) 146
Dynamic Stability Control (DSC) Off Indicator 147
Hydraulic Control Unit (HCU) 147
Normal Braking/EBD Mode 147
ABS Banking Mode 147
Active Braking Mode 148
ABS Control Module 148
System Operation 148
Anti-lock Brake System (ABS) 148
Electronic Brake Force Distribution (EBD) 146
Dynamic Stability Control (DSC) 148
Control Diagram 149
Anti-Lock Control 149
Master Cylinder 149
Base Braking System 150
Master Cylinder 150
Hydraulic System - 90 Models 151
Hydraulic System - 110/130 Models 152
Brake Components 152
Front Brake Caliper Assembly 152
Rear Brake Caliper Assembly 153
Brake System Bleeding 153
Parking Brake 154

206-03 - Front Disc Brake **155**
Description and Operation 155
Removal and Installation 155
Front Disc Brake 155
Front Disc Brake Pads 155
Front Brake Caliper 156

206-04 - Rear Disc Brake **157**
Description and Operation 157
Removal and Installation 157
Rear Brake Disc 157
Rear Brake Pads - 90 158
Rear Brake Caliper - 90 159
Rear Brake Caliper - 110/130 160
Rear Brake Pads - 110/130 161

205-05 - Parking Brake and Actuation **162**
Description and Operation 162
Parking Brake 162
General Procedures 162
Parking Brake Shoe and Lining Adjustment 162
Removal and Installation 162
Parking Brake Cable 162
Shoes 163
Parking Brake Switch 164

206-06 - Hydraulic Brake Actuation **165**
Removal and Installation 165
Brake Pedal and Bracket 165
Brake Pressure Control Valve 166
Disassembly and Assembly 167
Brake Master Cylinder 167
Removal and Installation 168
Brake Master Cylinder 168

207-07 - Power Brake Actuation **170**
Specifications 170
Removal and Installation 170
Brake Booster 170
Brake Booster Non-Return Valve 171
Brake Vacuum Pump 171

206-09A - Anti-Lock Control - Traction Control **172**
Description and Operation 172
Anti-Lock Control - Traction Control 172
Diagnosis and Testing 172
Principles of Operation 172
Removal and Installation 172
Hydraulic Control Unit - LHD 172
Front Wheel Speed Sensor 174
Rear Left Wheel Speed Sensor 175
Rear Right Wheel Speed Sensor 176
Hydraulic Control Unit - RHD 177

206-09B - Anti-Lock Control - Stability Assist **178**
Description and Operation 178
Anti-Lock Control - Stability Assist 178
Removal and Installation 178
Yaw Rate Sensor 178

211-00 - Steering System **179**
General Information 179
Specifications 179
Steering System 179
Diagnosis and Testing 179
Steering System 179
Power Steering Pressure Test 182

General Procedures 185
Power Steering System Filling and Bleeding 185
Power Steering System Flushing 185
Steering Gear Adjustment 187
Steering Gear Centralization 188
Steering Lock Stop Adjustment 188

211-02 - Power Steering 189
Specifications 189
Description and Operation 189
Removal and Installation 189
Power Steering Fluid Reservoir 191
Power Steering Pump 191
Steering Angle Sensor 193
Disassembly and Assembly
Steering Gear 194

211-03 - Steering Linkage 199
Removal and Installation 199
Tie Rod End 199
Steering Linkage Damper 200
Sector Shaft Arm Drag Link 200
Steering Gear Drop Arm 201

211-04 - Steering Column 202
Removal and Installation 202
Steering Column 202
Steering Column Lower Shaft 204
Steering Wheel 205

211-05 - Steering Column Switches 206
Description and Operation 206
Overview 206
Left Hand Column Switch 206
Right Hand Column Switch 206
Ignition Switch 207
Removal and Installation 207
Right Steering Column Multifunction Switch 207
Left Steering Column Multifunction Switch 207
Ignition Switch 208

Powertrain

Engine System:
2.2L 2007-2011 and 2.4L 2012-2016 Diesel 209
Important Powertrain information 209
General Procedures 209
Bearing Inspection 209
Camshaft Bearing Journal Clearance 209
Camshaft Bearing Journal Diameter 210
Camshaft End Play 210
Camshaft Lobe Lift 210
Camshaft Surface Inspection 210
Connecting Rod Cleaning 211
Connecting Rod Large End Bore 211
Cylinder Bore Out-of-Round 211
Cylinder Block Distortion 211
Exhaust Manifold Cleaning and Inspection 212
Piston Inspection 212
Piston Pin Diameter 212
Piston Pin to Bore Diameter 212
Piston Ring End Gap 212
Piston Ring-to-Groove Clearance 212
Valve Spring Free Length 213
Valve Stem Diameter 213
Leakage Test Using Smoke Test Equipment 213
Cylinder Head Distortion 214
Cylinder Compression Test 215
Thread repair 216

303-01 - Engine 2.2L 2007-2011 and 2.4L 2012-2016 Diesel 217
Specifications 217
Lubricants, Fluids and Sealers 217
Description and Operation 219
External Views 219
Engine 219
Engine ID Stamp 220
ID Label 220
Cylinder Block 220
Piston Cooling Jets and Oil Cooling 221
Crankshaft 221
Crankshaft Rear Oil Seal 222
Crankshaft Front Oil Seal 222
Connecting Rods and Pistons 222
Piston Range 222
Cylinder Head 223
Camshafts 223
Camshaft Carrier 223
Rocker Arms 223
Lash Adjuster 224
Cylinder Head Gasket 225
Skirt Stiffener 225
Oil Pan 226
Camshaft Drive 226
Drive Gear 226
Front Cover 227
Intake Manifold 227
Exhaust Manifold 227
Camshaft Cover and Engine Cover 228
Lubrication System 228
Oil Pick-up 228
Oil Pump 228
Oil Filter and Cooler Assembly 228
Lubrication System Operation 228
Diagnosis and Testing 229
Principles of Operation 229
Inspection and Verification 229
Oil Pressure Check 229
Engine Oil Draining and Filling 230
Removal and Installation 231
Engine 231
Camshafts 240
Crankshaft Pulley 242
Crankshaft Front Seal 243
Crankshaft Rear Seal 244
Cylinder Head 245
Left Engine Mount 247
Right Engine Mount 248
Exhaust Manifold 249
Intake Manifold 250
Oil Cooler 251
Oil Pan 252

Oil Pump 254
Timing Chain 255
Valve Cover 261
Flywheel 266
Oil Filter Housing 266
Timing Cover 267

303-03 - Engine Cooling - ID4 2.2L Diesel 270
Specifications 270
Overview 270
Component Location 271
Cooling System Schematic 271
Diagnosis and Testing 272
Principles of Operation 272
Inspection and Verification 272
General Procedures 274
Cooling System Draining, Filling and Bleeding 274
Cooling System Draining and Vacuum Filling 275
Cooling System Partial Draining and Vacuum Filling 277
Removal and Installation 280
Coolant Expansion Tank 280
Cooling Fan 280
Cooling Fan Shroud 281
Coolant Pump 282
Radiator 283
Thermostat 284

303-04A - Fuel Charging and Controls
Specifications 286
Description and Operation 286
Component Location 286
High Pressure Pump 287
Fuel Rail 287
Fuel Injectors 287
Diagnosis and Testing 288
Fuel Charging and Controls 288
Principles of Operation 288
Inspection and Verification 288
General Procedures 291
Fuel Injection Component Cleaning 291
Removal and Installation 291
Fuel Injector 291
Fuel Injection Pump 293
Fuel Metering Valve 298
Fuel Rail 299
Throttle Body 301

303-04B - Fuel Charging and Controls - Turbocharger - ID4 2.2L Diesel 302
Specifications 302
Description and Operation 302
Component Location 302
Turbocharger 302
Principles of Operation 303
Barometric Pressure Sensor 303
Turbocharger Lubrication 303
Diagnosis and Testing 303
Removal and Installation 305
Turbocharger 305
Actuator Rod 309

303-05 - Accessory Drive - ID4 2.2L Diesel 311
Specifications 311
Torque Specifications 311
Description and Operation 311
Component Location 311
Diagnosis and Testing 311
Principles of Operation 311
Inspection and Verification 311
Removal and Installation 313
Accessory Drive Belt 313
Accessory Drive Belt Pulley 313
Accessory Drive Belt Tensioner 314
Accessory Drive Component Bracket 314

303-06 - Starting System - 2.2L Diesel & 2.2L Diesel 316
Specifications 316
Description and Operation 316
Principles of Operation 316
Inspection and Verification 316
Removal and Installation 316
Starter Motor 317

303-07 - Glow Plug System - 2.2L Diesel & 2.2L Diesel 319
Specifications 319
Description and Operation 319
Component Location 319
Glow Plug System 319
Diagnosis and Testing 319
Principles of Operation 319
Glow Plug System 319
Inspection and Verification 320
Removal and Installation 322

303-08 - Engine Emission Control - 2.2L Diesel & 2.4L Diesel 323
Specifications 323
Description and Operation 323
Component Location 323
Exhaust Gas Recirculation 323
Positive Crankcase Ventilation 323
Diagnosis and Testing 324
Principles of Operation 324
Inspection and Verification 324
Engine Emission Control 324
Removal and Installation 326
Exhaust Gas Recirculation Cooler 326
Exhaust Gas Recirculation Valve 326

303-12 - Intake Air Distribution and Filtering 328
Specifications 328
Torque Specifications 328
Description and Operation 328
Component Location 328
Air Cleaner 328
Charge Air Cooler - Intercooler 328
Diagnosis and Testing 329
Principles of Operation 329
Inspection and Verification 329

Removal and Installation 330
Charge Air Cooler - Intercooler - Intake Pipe 330
Air Cleaner 333
Air Cleaner Element 334
Air Cleaner Outlet Pipe 335
Charge Air Cooler 335

303-14 - Electronic Engine Controls - 2.2L Diesel & 2.4 Diesel 337
Specifications 337
Torque Specifications 337
Description and Operation 337
Component Location 337
Engine Control Module 339
Accelerator Pedal Position Sensor 339
Brake Pedal Switches 340
Charge Air Temperature Sensor 340
Crankshaft Position Sensor 341
Electronic Throttle 341
Engine Coolant Temperature Sensor 341
Heated Oxygen Sensor (only fitted to Vehicles with DPF) 342
Mass Air Flow and Temperature Sensor 342
Oil Level and Temperature Sensor 343
Oil Pressure Switch 343
Electronic Engine Controls 344
Diagnosis and Testing 344
Principles of Operation 344
Inspection and Verification 344
Removal and Installation 347
Accelerator Pedal Position Sensor 347
Camshaft Position Sensor 347
Crankshaft Position Sensor 348
Cylinder Head Temperature Sensor 349
Engine Control Module 349
Engine Coolant Temperature Sensor 349
Engine Oil Pressure Sensor 350
Engine Oil Level Sensor 350
Fuel Rail Pressure Sensor 350
Heated Oxygen Sensor 351
Manifold Absolute Pressure and Temperature Sensor 352
Mass Air Flow Sensor 352
Post Catalytic Converter Temperature Sensor 353
Pre Catalytic Converter Temperature Sensor 353
Post DPF Exhaust Gas Temperature Sensor 354

308-00 - Manual Transmission/Transaxle and Clutch MT82 6-Speed Manual Transmission 355
General Procedures 355
MT82 6-Speed Manual Transmission 355
Clutch System Bleeding 355
Clutch Pedal Freeplay Adjustment 355

308-01 - Clutch: MT82 6-Speed Manual Transmission 356
Specifications 356
Description and Operation 356
Clutch Master Cylinder 356
Concentric Slave Cylinder Outlet Assembly 356
Concentric Slave Cylinder 357
Clutch Cover Assembly 357
Clutch Driven Plate 357
Flywheel 357
Diagnosis and Testing 357
Principles of Operation 357
Inspection and Verification 357
Clutch 357
Removal and Installation
Clutch Disc and Pressure Plate 359
Pilot Bearing 360

308-02 - Clutch Controls - Vehicles With: MT82 6-Speed Manual Transmission 361
Specifications 361
Description and Operation 361
Removal and Installation 361
Clutch Master Cylinder - LHD 361
Clutch Master Cylinder - RHD 368
Clutch Slave Cylinder 376

308-03 - Manual Transmission/Transaxle MT82 6-Speed Manual Transmission 377
Specifications 377
Lubricants, Fluids, Sealers and Adhesives 377
Component Location 377
Description and Operation 378
Model Plate Label 378
Layshaft 379
Output Shaft 380
Triple Synchroniser Assembly 381
Synchroniser Assembly 381
Reverse Gear Idler 382
Manual Transmission 382
Diagnosis and Testing 382
Inspection and Verification 382
General Procedures 384
Transmission Draining and Filling 384
In-vehicle Repair 384
Gearshift Control Shaft Seal 384
Input Shaft Seal 385
Output Shaft Seal 386
Removal and Installation 387
Selector Shaft Detents 389
1st-2nd and Reverse Gear Selector Shaft Detents 391
Manual Transmission/Tranaxle MT82 6-Speed 393
Removal 393
Assembly 401
Transmission 401
Removal 416
Transmission 416
Installation 422
Transmission 422

308-06 Manual Transmission/Transaxle External Controls - Vehicles MT82 6-Speed Manual Transmission 429
Specifications 429
Torque Specifications 429
Description and Operation 429
Removal and Installation 429
Selector Lever 429

308-07A - Transfer Case - Vehicles MT82 6-Speed Manual Transmission **431**
Specifications 431
Lubricants 431
General Specifications - LT230Q 431
Transfer Case 431
Torque Specifications 431
Description and Operation 432
Component Location 432
Transfer Case 432
Transfer Case Cross Section 433
Main Casing 433
Main Shaft Input Gears 434
Intermediate Gears 434
Centre Differential Assembly 434
Front Output Housing 435
Front Output Shaft and Flange 436
Housing and Selector Assembly 436
Rear Output Housing 436
Lubrication 436
Operation 437
Power Flow - Transfer Box in Low Range 437
Power Flow - Transfer Box in High Range 437
Diagnosis and Testing 438
Principles of Operation 438
Visual Inspection 438
Transfer Case Oil Seal Locations 440
Transfer Case Draining and Filling 440
General Procedures 441
Transfer Case High/Low Range Selector Rod Adjustment 441
In-vehicle Repair 444
Transfer Case Extension Housing 444
Transfer Case Front Output Shaft Seal 444
Transfer Case High/Low Range Linkage 446
Transfer Case Input Shaft Seal 448
Transfer Case Lower Cover 449
Transfer Case Rear Cover 450
Transfer Case Rear Output Shaft Seal 450
Removal 452
Transfer Case 452
Installation 455
Transfer Case 455

308-07B - Four-Wheel Drive Systems - Vehicles With MT82 6-Speed Manual Transmission **459**
General Procedures 459
Differential Lock Indicator Switch Adjustment 459
Removal and Installation 459
Transfer Case Low Range Indicator Switch 459
Differential Lock Indicator Switch 460

309-00 - Exhaust System - ID4 2.2L Diesel **461**
Specifications 461
Lubricants, Fluids, Sealers and Adhesives 461
Torque Specifications 461
Description and Operation 462
Component Location 463
90 Variant 463
110 Variant 463
130 Variant 463
Front Section 463
Centre Section 463
Rear Section 463
Catalytic Converter 464
Diesel Particulate Filter 464
Differential Pressure Sensor 465
Description and Operation 465
Diesel Particulate Filter (DPF) 465
Passive Regeneration 465
Active Regeneration 465
DPF Control 466
DPF Fuel Management Module 466
DPF Air Management Module 466
DPF Co-Ordinator Module 466
Differential Pressure Sensor 466
Component Description 467
Diesel Particulate Filter (DPF) 467
Temperature Sensors 467
Instrument Cluster (IC) Indications 467
Diesel Particulate Filter Side Effects 467
Engine Oil Dilution 467
Fuel Consumption 467
Differential Pressure Sensor 468
DPF Cleaning Fluids 468
Exhaust System - Diagnosis and Testing 468
Principles of Operation 468
Inspection and Verification 468
Diesel Particulate Filter - Diagnosis and Testing 469
Principles of Operation 469
Inspection and Verification 469
Diesel Particulate Filter Regeneration 470
Removal and Installation 471
Diesel Particulate Filter 471
Muffler and Tailpipe 477
Catalytic Converter 477

310-00 - Fuel System **483**
General Procedures 483
Diesel Filter Water Drain Off 483
Fuel Tank Draining 483
Low-Pressure Fuel System Bleeding 484
Quick Release Coupling 485

310-01 - Fuel Tank and Lines - ID4 2.2L Diesel **486**
Specifications 486
Description and Operation 486
Component Location 486
Fuel Filler Cap 487
Fuel Tank 487
Fuel Tank Breather System 487
Fuel Filter 488
Fuel Cooler 488
Bleed Valve 488
Fuel Lines 488
Inertia Switch 488
Low Fuel Indication and Run Dry Strategy 489
Control Diagram 489
Removal and Installation 489
Fuel Cooler 489
Fuel Filter 491
Fuel Level Sensor 492
Fuel Tank 493
Fuel Tank Filler Pipe 495

Electrical

412-00 - Climate Control System **496**
General Information 496
Specifications 496
Lubricant 496
System Refrigerant Specification/Capacity 496
Component Location 496
Air Inlet Duct 497
Air Intake and Blower Assembly 497
Heater Assembly 497
Water Valve 498
Air Conditioning Compressor 498
Air Conditioning Condenser 498
Air Conditioning Receiver Drier 498
Air Conditioning Evaporator 499
Air Conditioning Thermostatic Expansion Valve 499
Air Conditioning Refrigerant Pressure Switch 499
Air Conditioning Thermostatic Switch 500
Control Diagram 500
Principles of Operation 501
Heating 501
Air Conditioning 501
Description and Operation 501
Principles of Operation 501
Inspection and Verification 501
Climate Control System 501
General Precautions 502
Remedial Actions 503
Servicing Precautions 503
Refrigerant Oil 503
Rapid Refrigerant Discharge 504
Servicing Equipment 504
Precautions when Handling Refrigerant Lines 504
Performance Test 504
General Procedures
Air Conditioning System Recovery, Evacuation and Charging 508
Refrigerant System Tests 510
Electronic Leak Detection 510
Air Conditioning Compressor Commissioning 510

412-01 - Air Distribution and Filtering **511**
Description and Operation 511
Component Location 511
Air Inlet Duct 511
Removal and Installation 511
Air Inlet Duct 511

412-02 - Heating and Ventilation **513**
Specifications 513
Lubricants, Fluids and Sealers 513
Torque Specifications 513
Removal and Installation 513
Blower - LHD AWD 513
Blower - RHD AWD, RHD 513
Blower Motor - LHD AWD 514
Blower Motor - RHD AWD 515
Heater Control Valve 515
Heater Core 516
Heater Core and Evaporator Housing 516

412-03 - Air Conditioning - ID4 2.2L Diesel **519**
Specifications 519
Torque Specifications 519
Removal and Installation 519
Air Conditioning Compressor 519
Air Conditioning Switch 519
Condenser Core 520
Condenser Fan 521
De-Icing Switch 522
Receiver Drier 522
Evaporator core 523
Thermostatic Expansion valve 525

412-04 - Control Components **526**
Removal and Installation 526
Climate Control Assembly 526
Blower Motor Resistor - LHD AWD 527
Blower Motor Resistor - RHD AWD 527

413-01- Instrument Cluster **528**
Specifications 528
Torque Specifications 528
Component Location 528
Analogue Gauges 528
Indicator Lamps 528
Principles of Operation 529
Speedometer 529
Tachometer 529
Fuel Level Gauge 529
Engine Temperature Gauge 529
Glow Plugs Active Indicator 529
Seatbelt Indicator 529
Control Diagram 530
Control Diagram 531
Malfunction Indicator Lamp (MIL) 531
Left and Right Turn Signal Indicators 531
High Engine Coolant Temperature Indicator 531
High Beam Indicator 531
Hazard Warning Lamp Indicator 531
DSC on/off Indicators 532
Differential Lock Indicator 532
Rear Fog Lamp Indicator 532
Trailer Indicator 532
Low Fuel Indicator 532
Overspeed Indicator - market dependant 532
Anti-Lock Braking Indicator 532
Diesel Particulate Filter Indicator 532
Brake Warning Indicator 532
Side Lights On 532
Anti-theft Alarm Indicator 532
Transfer Case Low Range Indicator 532
Ignition/No Charge Warning Indicator 533
Low Oil Pressure Warning Indicator 533
Lights on Audible Warning 533
Instrument Cluster Replacement 533

Diagnosis and Testing 533
Principles of Operation 533
Inspection and Verification 533
Visual Inspection 533
Description and Operation 533
Instrument Cluster 533
Removal and Installation 534
Instrument Cluster 534
Instrument Cluster Lens 534
413-06 -Horn **535**
Removal and Installation 535
Horn 535

413-07 - Clock **536**
Specifications 536
Torque Specifications 536
Removal and Installation 536
Clock 536

413-09 - Warning Devices **536**
General Procedures 536
Service Indicator Reset 536

414-00 - Battery and Charging System 537
Description and Operation 537
Battery Care Requirements 537
Health and Safety Precautions 537
Battery Maintenance 537
Battery test Types 538
Flooded Battery Testing 538
Equipment 538
Determining Battery Condition 538
Battery Charging 539
Battery Replacement 539
Confirming Electrolyte Level 539
Description and Operation 540
Quiescent Drain 540
Current Testing 540
Equipment 540
Diagnosis and Testing 541
Principles of Operation 541
Inspection and Verification 541
Charging System 541

414-01 Battery, Mounting and Cables **545**
Specifications 545
Torque Specifications 545
Diagnosis and Testing 545
Principles of Operation 545
Inspection and Verification 545
Removal and Installation 546
Battery 546
General Procedures 547
Removal and Installation 547
Startup Battery

414-02 - Generator and Regulator
ID4 2.2L Diesel **548**
Specifications 548
Description and Operation 548
Component Location 548
Removal and Installation 549
Generator 549

415-01A - Audio Unit **550**
Description and Operation 550
Component Location Models from 2007 MY
(Not SVX (60th Anniversary) Model) 550
SVX (60th Anniversary) Model 551
Clarion Head Unit DB568RUSB 551
Specifications 551
Clarion Sub-Woofer SRV313 553
Control Diagram Models from 2007 MY
(Not SVX (60th Anniversary) Model) 554
SVX (60th Anniversary) Model 555
Removal and Installation 555
Audio Unit - Vehicles With: Bluetooth 555
Audio Unit - Vehicles Without: Bluetooth 556

415-01B - Information and Entertainment System **557**
Description and Operation 557
Navigation System 557
SVX (60th Anniversary) Model 557
Garmin Nuvi Unit Details

415-02 - Antenna **559**
Removal and Installation 559
Antenna 559

415-03 - Speakers **561**
Description and Operation 561
Diagnosis and Testing 561
Speakers 561
Removal and Installation 567
Instrument Panel Speaker 567
D-Pillar Speaker 567
Rear Door Speaker 567

417-01 - Exterior Lighting **568**
Specification 568
Description and Operation 568
Exterior Lighting 568
Control Diagram 568
Control Diagram 570
General Procedures 571
Headlamp Adjustment 571
Removal and Installation 571
Headlamp Assembly 571
Headlamp Levelling Motor 571
Headlamp Levelling Switch 572
Headlamp Switch 573
High Mounted Stoplamp 573
License Plate Lamp 573
License Plate Lamp Bulb 574
Rear Fog Lamp Bulb 574
Rear Lamp Assembly 574
Reversing Lamp 575

Reversing Lamp Switch 575
Side Marker Lamp 576
Side Repeater Lamp Bulb 576
Stoplamp Switch 576

417-02 - Interior Lighting 578
Description and Operation 578
Component Location 578
Control Diagram 579
Interior Lighting 580
Removal and Installation 580
Interior Lamp 580
Interior Lamp Bulb 580

418-00 - Module Communications Network 581
Description and Operation 581
Communications Network 581

418-02 - Wiring Harnesses 582
Description and Operation 582
Wiring Harness 582
Approved probing and repair methods 582
Replacement Equipment 582
Electrical Connector Probing 582
Supplementary Restraint System (SRS) 582
MOST Network Harness Repair 584
MOST Harness Repair Components and Tools
List of Parts 586
Repair Tools 586
MOST Harness Repair Procedure 586
MOST Harness Repair Process 586
Air Suspension Overlay Harness Introduction 588
Description and Operation 590
Wiring Harness Overlay 590
Removal and Installation 592
Engine Wiring Harness 592

419-01A - Anti-Theft - Active 596
Description and Operation 596
Component Location 596
Anti-Theft System Module 597
Remote Handset 597
Battery Backed Sounder 597
Anti-Theft Alarm Horn 598
Intrusion Detection Module 598
Instrument Cluster 598
Control Diagram 599
Battery Backed Sounder 599
Anti-Theft Alarm Horn 600
Principles of Operation 601
Perimetric Protection 601
Volumetric Protection 601
Visual Indications 601
Audible Indications 601

419-01B - Anti-Theft - Passive 602
Description and Operation 602
Component Location 602
Anti-Theft - Passive 602
Transceiver Coil 602
Anti-Theft System Module 603
Engine Module Control 603
Principles of Operation 603
Emergency Key Access 603
Control Diagram 604
Removal and Installation 604
Passive Anti-Theft System Transceiver 604
All Vehicles 604
Left-hand Drive Vehicles 604
Right-hand Drive Vehicles 605
Passive Anti-Theft System Module 605
Passive Anti-Theft System Transceiver 605
Passive Anti-Theft System Module 605

419-12 - Winch 606
Description and Operation 606
Component Location 606
9.5Ti Winch Specifications 606
Systems Operation 606
Control Diagram 607
Winch Schematic Electrical Circuit Diagram 608
Winch Motor/Control Box Connections 608
Component Description 609
Winch vehicle Electrical Connections 609
Winch Power Interrupt Solenoid 609
Exploded View 610
Diagnosis and Testing 611
Principles Operation 611
Visual Inspection 611
Removal and Installation 616
Winch 616
Licence Plate Panel 620
Winch Cable Roller Assembly 620
Winch Control Unit 620
Winch Gear Assembly 621
Winch Motor 622
Winch Solenoid 623

Body and Paint

501-02 - Front End Body Panels 624
Removal and Installation 624
Fender 624
Fender Splash Shield 626
Hood 627

501-03 Body Closures 628
Body Closures - Vehicles Built From: 07/2001 628
Description and Operation 628
Front Seat Recline 628
Soft Top 628
Tail Door 628
General Procedures 629
Taildoor Striker Adjustment 629
Door Striker Adjustment 629
Removal and Installation 630
Door 630
Front Door Reinforcement Panel 630
Rear Door Reinforcement Panel 630
Taildoor 630

501-05 - Interior Trim and Ornamentation 633
Description and Operation 633
Interior Trim 633
Fascia Console 633
Removal and Installation 634
Cowl Side Trim Panel 634
C-Pillar Upper Trim Panel - 90/Station Wagon 634
C-Pillar Upper Trim Panel - 110/Station Wagon 635
Front Door Trim Panel 636
Front Headliner - 90/Station Wagon 637
Headliner Rear Trim Panel - 90/110/Station Wagon 638
Interior Rollover Bar 638
Rear Door Trim Panel 639
Rear Grab Handle - 90/110/Station Wagon 640
Rear Headliner - 90/Station Wagon 640
Rear Quarter Trim Panel 641
Steering Column Shrouds 642
Sun Visor 642
Taildoor Trim Panel 643
Engine Cover 643

501-08 - Exterior Trim and Ornamentation 644
Removal and Installation 644
Exterior Rollover Bar 644
Radiator Grille 645
Rear Folding Step 645
Towbar 645

501-09 - Rear View Mirrors 647
Removal and Installation 647
Interior Mirror 647

501-10 - Seating 647
Specifications 647
Torque Specifications 647
Component Location 647
Description and Operation 647
Principles of Operation 647
Heated Front Seats 647
Diagnosis and Testing 648
Principles of Operation 648
Inspection and Verification 648
Visual Inspection 648
Heater Mats 648
Diagnostic Guidance 648
Seat Heater Mat Faults 648
Application Chart 649
Removal and Installation 651
Front Row Seat 651
Front Row Seat Backrest 651
Front Row Seat Backrest Cover 652
Front Seat Backrest Heater Mat 653
Front Row Seat Cushion 654
Front Row Seat Cushion Cover 654
Front Seat Cushion Heater Mat 654
Heated Seat Switch 654
Rear Seat 655
Rear Seat Backrest 655
Rear Seat Backrest Cover 656
Rear Seat Cushion 657
Rear Seat Cushion Cover 658
Third Row Seat 659
Third Row Seat Backrest 660
Third Row Seat Backrest Cover 661
Third Row Seat Cushion 662
Third Row Seat Cushion Cover 663

501-11 - Glass, Frames and Mechanisms 664
Description and Operation 664
Component Location 664
Window Lift 664
Heated Windshield and Rear Window 665
Principles of Operation 665
Window Lift 665
Heated Windshield 665
Rear Window 665
Diagnosis and Testing 665
Principles of Operation 665
Inspection and Verification 665
Visual Inspection 665
Fixed Window Glass 665
Panoramic Roof Glass Damage Procedure 666
Fixed Window Glass Wind Noise and Water Leaks 670
Practical Execution of tests and Checks 670
Rectifying Leaks 670
Water Leaks according to Mileage 671
Water Drainage System (diagram) 671
Wind Noise or Water leaks After Repair 672
Water Drainage System 672
Windshield 672
Side Windows 672
Door Seal 672
Roof Opening Panel 673
Luggage Compartment Lid 674
Forced Air Extraction 674
Rear Window 674
Diagnosis and Testing 675
Principles of Operation 675
Inspection and Verification 675
Windshield Water Ingress Diagnostics 675
Removal and Installation 678
Front Door Window 678
Front Door Window Regulator and Motor 679
Heated Windshield Relay 680
Rear Door Window Glass 680
Rear Door Window Regulator and Motor 681
Window Control Switch 682
Windshield 683
Side Rear Window Glass Assembly 685

501-12- Instrument Panel and Console 689
Specifications 689
Torque Specifications 689
Description and Operation 689
Component Location 689
Removal and Installation 690
Instrument Panel 690
Instrument Panel Console 693
In-Vehicle Crossbeam 694
Floor Console 695
Instrument Panel Switches 696

501-14 - Handles, Locks, Latches and Entry Systems **697**
Specifications 697
Torque Specifications 697
Description and Operation 697
Central Door Locking 697
Slam Locking 697
CDL Using the Key 697
CDL Using the Remote Handset 697
Anti-Theft Alarm Indicator 698
Interior Lamp Functionality 698
General Procedures 698
Latch Cable Adjustment 698
Removal and Installation 699
Front Door Latch 699
Rear Door Latch 700
Door Lock Cylinder 700
Exterior Front Door Handle 701
Exterior Rear Door Handle 702
Interior Rear Door Handle 702
Interior Front Door Handle 703
Hood Latch 703
Hood Latch Release Handle 704
Front Door Lock Actuator 704
Front Door Push Button Rod and Linkage 705
Rear Door Lock Actuator 705
Rear Door Push Button Rod and Linkage 706
Taildoor Latch 707
Taildoor Lock Cylinder 708
Taildoor Lock Motor 708

501-16 - Wipers and Washers **709**
Description and Operation 709
Wipers and Washers 709
Windscreen Wipers 709
Rear Wiper 710
Removal and Installation 710
Front Wiper Pivot Arm 710
Rear Window Washer Jet 711
Rear Window Wiper Motor 712
Windshield Washer Pump 713
Windshield Wiper Motor 713
Windshield Wiper/Washer Switch 717
Wiper Mounting Arm and Pivot Shaft 718

501-17 -Roof Opening Panel **721**
Removal and Installation 721
Roof Opening Panel 721

501-18 - Convertible Top **722**
Description and Operation 722
Convertible Top 722
Stowage Locker 724
Convertible Care Points 724

501-19 - Bumpers **724**
Removal and Installation 724
Front Bumper 724

501-20 - Seatbelt System **725**
Specifications 725
Torque Specifications 725
Description and Operation 725
Component Location 725
Seatbelt System 725
Front Safety Belt 725
Second Row Safety Belt 725
Third Row Safety Belt 725
Beltminder 725
Diagnosis and Testing 726
Principles of Operation 726
Inspection and Verification 726
Seatbelt System 726
Inertia Reel Seatbelts 728
Seatbelt Locking Test 728
Removal and Installation 733
Front Row Safety Belt Buckle 733
Front Safety Belt Retractor 733
Rear Center Safety Belt Buckle 734
Rear Center Safety Belt Retractor 734
Rear Safety Belt Retractor 735
Rear Left Safety Belt Buckle 736
Rear Right Safety Belt Buckle 736
Third Row Safety Belt Buckle 737
Third Row Safety Belt Retractor 737
Front Safety Belt Retractor Mounting Bracket 738
Rear Safety Belt Retractor Mounting Bracket 742

501-25A - Body Repairs - General Information **743**
Description and Operation 743
Straightening 743
Body Jig 744
Inspection 744
Electronic Control Units (ECU) 744
Paint Procedures 744
Replacement Panels 744
Bolted Panels 744
Welded Panels 744
Sectioned Panels 744
Clinch Panels 745
Paint Refinishing 745
Paint Repairs 745
General Welding 745
Seatbelt Anchorages 746
Panel Replacement Procedure 746
Remove Panel 746
Prepare Old Surfaces 747
Prepare New Surfaces 747
Offer Up and Alignment 748
Welding 748
Body Trim 748
Serviceable Panels 749
Front End Panel 749
Body Side Panels 750
Rear End Panels 751

Body Repairs 753
Adhesives, Fixings and Consumables 753
Body Repair Procedure Graphic Key 763
General Procedures 765
Body Closure / Noise Vibration and Harshness Sealing 765

501-25B - Body Repairs - Corrosion Protection 766
Description and Operation 766
Application Equipment 766
Corrosion Protection and Treatments 768
Cavity Wax 770
Bulkhead Assembly 771
Front Door 772
Rear Door 772
Sealants and Adhesives 773

501-25C - Body Repairs - Water Leaks 780
Diagnosis and Testing 780
Inspection and Verification 780
Water Leak Procedure 777

501-26 - Body Repairs - Vehicle Specific Information and Tolerance Checks 798
Description and Operation 798
Chassis and Body 798
Body and Frame Alignment 798
90 Chassis 796
110 Chassis 797
130 Chassis 798
Gap Profiles 799

501-27 - Front End Sheet Metal Repairs 801
Removal and Installation 801
Front Fender Side Panel 801
Front Left Fender Top Panel 801
Front Right Fender Top Panel 802
Front Wheelhouse 803
Hood Latch Panel 803
Windshield Frame 804

501-28 - Roof Sheet Metal Repairs 805
Removal and Installation 805
Roof Panel 805

501-29 - Side Panel Sheet Metal Repairs 805
Removal and Installation 805
Rear Rocker Panel 805
Front Rocker Panel 805
B-Pillar and C-Pillar Assembly 806

501-30 - Rear End Sheet Metal Repairs 806
Removal and Installation 806
Rear Lamp Mounting Panel 806
Rear Quarter Upper Panel 808
Rear Quarter Lower Panel 808
Rear Crossmember Panel 809

Made in United States
Orlando, FL
04 November 2026

90219735R00450